WIN the complete library of RSMeans Online data!

Register your book below to receive your FREE quarterly updates, PLUS you will be entered into a quarterly drawing to win the COMPLETE RSMeans Online library of 2015 data!

Be sure to keep up-to-date in 2015!

Fill out the card below for RSMeans' free quarterly updates, as well as a chance to win the complete RSMeans Online library. Please provide your name, address, and email below and return this card by mail, or to register online:

http://info.thegordiangroup.com/RSMeans.html

Name _____

email_____ Title_____

Company _____

Street _____

City/Town _____ State/Prov. _____ Zip/Postal Code _____

RSMeans

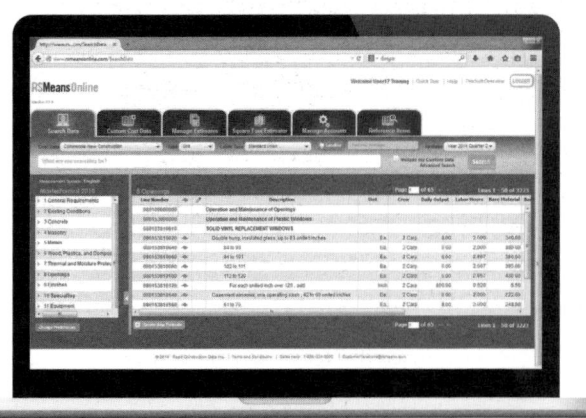

WIN the complete library of RSMeans Online 2015 data!

See other side for details.

RSMeans Mechanical Cost Data

2015 38th annual edition

Melville J. Mossman, PE, Senior Editor

Engineering Director
Bob Mewis, CCC (*1, 14, 41*)

Contributing Editors
Christopher Babbitt
Adrian C. Charest, PE
(*26, 27, 28, 48*)
Gary W. Christensen
Cheryl Elsmore
David Fiske
Robert Fortier, PE
(*2, 31, 32, 33, 34, 35, 44, 46*)
Robert J. Kuchta (*8*)

Numbers in italics are the divisional responsibilities for each editor. Please contact the designated editor directly with any questions.

Thomas Lane (*6, 7*)
Robert C. McNichols
Genevieve Medeiros
Melville J. Mossman, PE (*21, 22, 23*)
Jeannene D. Murphy
Marilyn Phelan, AIA (*9, 10, 11, 12*)
Stephen C. Plotner (*3, 5*)
Siobhan Soucie (*4, 13*)

Vice President Data & Engineering
Chris Anderson

Cover Design
Sara Rutan

Product Manager
Andrea Sillah

Production Manager
Debbie Panarelli

Production
Jill Goodman
Jonathan Forgit
Sara Rutan
Mary Lou Geary

Technical Support
Gary L. Hoitt
Kathryn S. Rodriguez

RSMeans
Construction Publishers & Consultants
700 Longwater Drive
Norwell, MA 02061
United States of America
1-877-795-7394
www.rsmeans.com

Copyright 2014 by RSMeans
All rights reserved.
Cover photo ©
xtrekx/iStock/Thinkstock

Printed in the United States of America
ISSN 1524-3702
ISBN 978-1-940238-62-3

0025 $203.95 per copy (in United States)
Price is subject to change without prior notice

Related RSMeans Products and Services

The engineers at RSMeans suggest the following products and services as companion information resources to *RSMeans Mechanical Cost Data:*

Construction Cost Data Books

Assemblies Cost Data 2015
Square Foot Costs 2015

Reference Books

Estimating Building Costs
RSMeans Estimating Handbook
Green Building: Project Planning & Estimating
How to Estimate with RSMeans Data
Plan Reading & Material Takeoff
Project Scheduling & Management for Construction
Universal Design Ideas for Style, Comfort & Safety

Seminars and In-House Training

Unit Price Estimating
RSMeans Online® Training
Practical Project Management for Construction Professionals
Scheduling with MSProject for Construction Professionals
Mechanical & Electrical Estimating

RSMeans Online Bookstore

Visit RSMeans at **www.rsmeans.com** for a one-stop portal for the most reliable and current resources available on the market. Here you can learn about our more than 20 annual cost data books and eBooks, also available on CD and online versions, as well as our libary of reference books and RSMeans seminars.

RSMeans Electronic Data

Get the information found in RSMeans cost books electronically at **RSMeansOnline.com**.

RSMeans Business Solutions

Cost engineers and analysts conduct facility life cycle and benchmark research studies and predictive cost modeling, as well as offer consultation services for conceptual estimating and real property management. Research studies are designed with the application of proprietary cost and project data from RSMeans' extensive North American databases. Analysts offer building product manufacturers qualitative and quantitative market research, as well as time/motion studies for new products, and Web-based dashboards for market opportunity analysis. Clients are from both the public and private sectors, including federal, state, and municipal agencies; corporations; institutions; construction management firms; hospitals; and associations.

Construction Costs for Software Applications

More than 25 unit price and assemblies cost databases are available through a number of leading estimating and facilities management software partners. For more information see the "Other RSMeans Products and Services" pages at the back of this publication.

RSMeans data is also available to federal, state, and local government agencies as multi-year, multi-seat licenses.

For information on our current partners, visit:
www.rsmeans.com/partners or call 877.795.7394.

Table of Contents

Foreword

Our Mission

Since 1942, RSMeans has been actively engaged in construction cost publishing and consulting throughout North America.

Today, more than 70 years after RSMeans began, our primary objective remains the same: to provide you, the construction and facilities professional, with the most current and comprehensive construction cost data possible.

Whether you are a contractor, owner, architect, engineer, facilities manager, or anyone else who needs a reliable construction cost estimate, you'll find this publication to be a highly useful and necessary tool.

With the constant flow of new construction methods and materials today, it's difficult to find the time to look at and evaluate all the different construction cost possibilities. In addition, because labor and material costs keep changing, last year's cost information is not a reliable basis for today's estimate or budget.

That's why so many construction professionals turn to RSMeans. We keep track of the costs for you, along with a wide range of other key information, from city cost indexes . . . to productivity rates . . . to crew composition . . . to contractor's overhead and profit rates.

RSMeans performs these functions by collecting data from all facets of the industry and organizing it in a format that is instantly accessible to you. From the preliminary budget to the detailed unit price estimate, you'll find the data in this book useful for all phases of construction cost determination.

The Staff, the Organization, and Our Services

When you purchase one of RSMeans' publications, you are, in effect, hiring the services of a full-time staff of construction and engineering professionals.

Our thoroughly experienced and highly qualified staff works daily at collecting, analyzing, and disseminating comprehensive cost information for your needs. These staff members have years of practical construction experience and engineering training prior to joining the firm. As a result, you can count on them not only for accurate cost figures, but also for additional background reference information that will help you create a realistic estimate.

The RSMeans organization is always prepared to help you solve construction problems through its variety of data solutions, including online, CD, and print book formats, as well as cost estimating expertise available via our business solutions, training, and seminars.

Besides a full array of construction cost estimating books, RSMeans also publishes a number of other reference works for the construction industry. Subjects include construction estimating and project and business management, special topics such as green building, and a library of facility management references.

In addition, you can access all of our construction cost data electronically in convenient CD format or on the web. Visit **RSMeansOnline.com** for more information on our 24/7 online cost data.

What's more, you can increase your knowledge and improve your construction estimating and management performance with an RSMeans construction seminar or in-house training program. These two-day seminar programs offer unparalleled opportunities for everyone in your organization to become updated on a wide variety of construction-related issues.

RSMeans is also a worldwide provider of construction cost management and analysis services for commercial and government owners.

In short, RSMeans can provide you with the tools and expertise for constructing accurate and dependable construction estimates and budgets in a variety of ways.

Robert Snow Means Established a Tradition of Quality That Continues Today

Robert Snow Means spent years building RSMeans, making certain he always delivered a quality product.

Today, at RSMeans, we do more than talk about the quality of our data and the usefulness of our books. We stand behind all of our data, from historical cost indexes to construction materials and techniques to current costs.

If you have any questions about our products or services, please call us toll-free at 1-877-795-7394. Our customer service representatives will be happy to assist you. You can also visit our website at **www.rsmeans.com**

How the Book Is Built: An Overview

The Construction Specifications Institute (CSI) and Construction Specifications Canada (CSC) have produced the 2014 edition of MasterFormat®, a system of titles and numbers used extensively to organize construction information.

All unit price data in the RSMeans cost data books is now arranged in the 50-division MasterFormat® 2014 system.

A Powerful Construction Tool

You have in your hands one of the most powerful construction tools available today. A successful project is built on the foundation of an accurate and dependable estimate. This book will enable you to construct just such an estimate.

For the casual user the book is designed to be:

- quickly and easily understood so you can get right to your estimate.
- filled with valuable information so you can understand the necessary factors that go into the cost estimate.

For the regular user, the book is designed to be:

- a handy desk reference that can be quickly referred to for key costs.
- a comprehensive, fully reliable source of current construction costs and productivity rates so you'll be prepared to estimate any project.
- a source book for preliminary project cost, product selections, and alternate materials and methods.

To meet all of these requirements, we have organized the book into the following clearly defined sections.

Quick Start

See our "Quick Start" instructions on the following page to get started right away.

Estimating with RSMeans Unit Price Cost Data

Please refer to these steps for guidance on completing an estimate using RSMeans unit price cost data.

How to Use the Book: The Details

This section contains an in-depth explanation of how the book is arranged . . . and how you can use it to determine a reliable construction cost estimate. It includes information about how we develop our cost figures and how to completely prepare your estimate.

Unit Price Section

All cost data has been divided into the 50 divisions according to the MasterFormat system of classification and numbering. For a listing of these divisions and an outline of their subdivisions, see the Unit Price Section Table of Contents. *Estimating tips are included at the beginning of each division.*

Assemblies Section

The cost data in this section has been organized in an "Assemblies" format. These assemblies are the functional elements of a building and are arranged according to the 7 elements of the UNIFORMAT II classification system. For a complete explanation of a typical "Assemblies" page, see "How RSMeans Assemblies Data Works."

Reference Section

This section includes information on Equipment Rental Costs, Crew Listings, Historical Cost Indexes, City Cost Indexes, Location Factors, Reference Tables, Change Orders, Square Foot Costs, and a listing of Abbreviations.

Equipment Rental Costs: This section contains the average costs to rent and operate hundreds of pieces of construction equipment.

Crew Listings: This section lists all of the crews referenced in the book. For the purposes of this book, a crew is composed of more than one trade classification and/or the addition of power equipment to any trade classification. Power equipment is included in the cost of the crew. Costs are shown both with bare labor rates and with the installing contractor's overhead and profit added. For each, the total crew cost per eight-hour day and the composite cost per labor-hour are listed.

Historical Cost Indexes: These indexes provide you with data to adjust construction costs over time.

City Cost Indexes: All costs in this book are U.S. national averages. Costs vary because of the regional economy. You can adjust costs by CSI Division to over 700 locations throughout the U.S. and Canada by using the data in this section.

Location Factors: You can adjust total project costs to over 900 locations throughout the U.S. and Canada by using the data in this section.

Reference Tables: At the beginning of selected major classifications in the Unit Price section are reference numbers shown in a shaded box. These numbers refer you to related information in the Reference Section. In this section, you'll find reference tables, explanations, estimating information that support how we develop the unit price data, technical data, and estimating procedures.

Change Orders: This section includes information on the factors that influence the pricing of change orders.

Square Foot Costs: This section contains costs for 59 different building types that allow you to make a rough estimate for the overall cost of a project or its major components.

Abbreviations: A listing of abbreviations used throughout this book, along with the terms they represent, is included in this section.

Index

A comprehensive listing of all terms and subjects in this book will help you quickly find what you need when you are not sure where it occurs in MasterFormat.

The Scope of This Book

This book is designed to be as comprehensive and as easy to use as possible. To that end we have made certain assumptions and limited its scope in two key ways:

1. We have established material prices based on a national average.
2. We have computed labor costs based on a 30-city national average of union wage rates.

For a more detailed explanation of how the cost data is developed, see "How To Use the Book: The Details."

Project Size/Type

The material prices in RSMeans cost data books are "contractor's prices." They are the prices that contractors can expect to pay at the lumberyards, suppliers'/distributors' warehouses, etc. Small orders of speciality items would be higher than the costs shown, while very large orders, such as truckload lots, would be less. The variation would depend on the size, timing, and negotiating power of the contractor. The labor costs are primarily for new construction or major renovation rather than repairs or minor alterations. *With reasonable exercise of judgment, the figures can be used for any building work.*

v

Absolute Essentials for a Quick Start

If you feel you are ready to use this book and don't think you will need the detailed instructions that begin on the following page, this Absolute Essentials for a Quick Start page is for you. These steps will allow you to get started estimating in a matter of minutes.

1 Scope
Think through the project that you will be estimating, and identify the many individual work tasks that will need to be covered in your estimate.

2 Quantify
Determine the number of units that will be required for each work task that you identified.

3 Pricing
Locate individual Unit Price line items that match the work tasks you identified. The Unit Price Section Table of Contents that begins on page 1 and the Index in the back of the book will help you find these line items.

4 Multiply
Multiply the Total Incl O&P cost for a Unit Price line item in the book by your quantity for that item. The price you calculate will be an estimate for a completed item of work performed by a subcontractor. Keep adding line items in this manner to build your estimate.

5 Project Overhead
Include project overhead items in your estimate. These items are needed to make the job run and are typically, but not always, provided by the General Contractor. They can be found in Division 1. An alternate method of estimating project overhead costs is to apply a percentage of the total project cost.

Include rented tools not included in crews, waste, rubbish handling, and cleanup.

6 Estimate Summary
Include General Contractor's markup on subcontractors, General Contractor's office overhead and profit, and sales tax on materials and equipment.

Adjust your estimate to the project's location by using the City Cost Indexes or Location Factors found in the Reference Section.

Editors' Note: We urge you to spend time reading and understanding the supporting material in the front of this book. An accurate estimate requires experience, knowledge, and careful calculation. The more you know about how we at RSMeans developed the data, the more accurate your estimate will be. In addition, it is important to take into consideration the reference material in the back of the book such as Equipment Listings, Crew Listings, City Cost Indexes, Location Factors, and Reference Tables.

Estimating with RSMeans Unit Price Cost Data

Following these steps will allow you to complete an accurate estimate using RSMeans Unit Price cost data.

Scope Out the Project

Think through the project and identify those CSI Divisions needed in your estimate.

Identify the individual work tasks that will need to be covered in your estimate.

The Unit Price data in this book has been divided into 50 Divisions according to the CSI MasterFormat® 2014—their titles are listed on the back cover of your book.

The Unit Price Section Table of Contents on page 1 may also be helpful when scoping out your project.

Experienced estimators find it helpful to begin with Division 2 and continue through completion. Division 1 can be estimated after the full project scope is known.

2 Quantify

- Determine the number of units required for each work task that you identified.
- Experienced estimators include an allowance for waste in their quantities. (Waste is not included in RSMeans Unit Price line items unless so stated.)

3 Price the Quantities

- Use the Unit Price Table of Contents, and the Index, to locate individual Unit Price line items for your estimate.
- Reference Numbers indicated within a Unit Price section refer to additional information that you may find useful.
- The crew indicates who is performing the work for that task. Crew codes are expanded in the Crew Listings in the Reference Section to include all trades and equipment that comprise the crew.
- The Daily Output is the amount of work the crew is expected to do in one day.
- The Labor-Hours value is the amount of time it will take for the crew to install one unit of work.
- The abbreviated Unit designation indicates the unit of measure upon which the crew, productivity, and prices are based.
- Bare Costs are shown for materials, labor, and equipment needed to complete the Unit Price line item. Bare costs do not include waste, project overhead, payroll insurance, payroll taxes, main office overhead, or profit.
- The Total Incl O&P cost is the billing rate or invoice amount of the installing contractor or subcontractor who performs the work for the Unit Price line item.

4 Multiply

- Multiply the total number of units needed for your project by the Total Incl O&P cost for each Unit Price line item.
- Be careful that your take off unit of measure matches the unit of measure in the Unit column.
- The price you calculate is an estimate for a completed item of work.
- Keep scoping individual tasks, determining the number of units required for those tasks, matching each task with individual Unit Price line items in the book, and multiply quantities by Total Incl O&P costs.
- An estimate completed in this manner is priced as if a subcontractor, or set of subcontractors, is performing the work. The estimate does not yet include Project Overhead or Estimate Summary components such as General Contractor markups on subcontracted work, General Contractor office overhead and profit, contingency, and location factor.

5 Project Overhead

- Include project overhead items from Division 1 – General Requirements.
- These items are needed to make the job run. They are typically, but not always, provided by the General Contractor. Items include, but are not limited to, field personnel, insurance, performance bond, permits, testing, temporary utilities, field office and storage facilities, temporary scaffolding and platforms, equipment mobilization and demobilization, temporary roads and sidewalks, winter protection, temporary barricades and fencing, temporary security, temporary signs, field engineering and layout, final cleaning and commissioning.
- Each item should be quantified, and matched to individual Unit Price line items in Division 1, and then priced and added to your estimate.
- An alternate method of estimating project overhead costs is to apply a percentage of the total project cost, usually 5% to 15% with an average of 10% (see General Conditions, page ix).
- Include other project related expenses in your estimate such as:
  - Rented equipment not itemized in the Crew Listings
  - Rubbish handling throughout the project (see 02 41 19.19)

6 Estimate Summary

- Include sales tax as required by laws of your state or county.
- Include the General Contractor's markup on self-performed work, usually 5% to 15% with an average of 10%.
- Include the General Contractor's markup on subcontracted work, usually 5% to 15% with an average of 10%.
- Include General Contractor's main office overhead and profit:
  - RSMeans gives general guidelines on the General Contractor's main office overhead (see section 01 31 13.60 and Reference Number R013113-50).
  - RSMeans gives no guidance on the General Contractor's profit.

- Markups will depend on the size of the General Contractor's operations, his projected annual revenue, the level of risk he is taking on, and on the level of competition in the local area and for this project in particular.

- Include a contingency, usually 3% to 5%, if appropriate.
- Adjust your estimate to the project's location by using the City Cost Indexes or the Location Factors in the Reference Section:
  - Look at the rules on the pages for "How to Use the City Cost Indexes" to see how to apply the Indexes for your location.
  - When the proper Index or Factor has been identified for the project's location, convert it to a multiplier by dividing it by 100, and then multiply that multiplier by your estimated total cost. The original estimated total cost will now be adjusted up or down from the national average to a total that is appropriate for your location.

Editors' Notes:
We urge you to spend time reading and understanding the supporting material in the front of this book. An accurate estimate requires experience, knowledge, and careful calculation. The more you know about how we at RSMeans developed the data, the more accurate your estimate will be. In addition, it is important to take into consideration the reference material in the back of the book such as Equipment Listings, Crew Listings, City Cost Indexes, Location Factors, and Reference Tables.

How to Use the Book: The Details

What's Behind the Numbers? The Development of Cost Data

The staff at RSMeans continually monitors developments in the construction industry in order to ensure reliable, thorough, and up-to-date cost information. While overall construction costs may vary relative to general economic conditions, price fluctuations within the industry are dependent upon many factors. Individual price variations may, in fact, be opposite to overall economic trends. Therefore, costs are constantly tracked and complete updates are published yearly. Also, new items are frequently added in response to changes in materials and methods.

Costs—$ (U.S.)

All costs represent U.S. national averages and are given in U.S. dollars. The RSMeans City Cost Indexes can be used to adjust costs to a particular location. The City Cost Indexes for Canada can be used to adjust U.S. national averages to local costs in Canadian dollars. No exchange rate conversion is necessary.

[G] The processes or products identified by the green symbol in our publications have been determined to be environmentally responsible and/or resource-efficient solely by the RSMeans engineering staff. The inclusion of the green symbol does not represent compliance with any specific industry association or standard.

Material Costs

The RSMeans staff contacts manufacturers, dealers, distributors, and contractors all across the U.S. and Canada to determine national average material costs. If you have access to current material costs for your specific location, you may wish to make adjustments to reflect differences from the national average. Included within material costs are fasteners for a normal installation. RSMeans engineers use manufacturers' recommendations, written specifications, and/or standard construction practice for size and spacing of fasteners. Adjustments to material costs may be required for your specific application or location. The manufacturer's warranty is assumed. Extended warranties are not included in the material costs. Material costs do not include sales tax.

Labor Costs

Labor costs are based on the average of wage rates from 30 major U.S. cities. Rates are determined from labor union agreements or prevailing wages for construction trades for the current year. Rates, along with overhead and profit markups, are listed on the inside back cover of this book.

- If wage rates in your area vary from those used in this book, or if rate increases are expected within a given year, labor costs should be adjusted accordingly.

Labor costs reflect productivity based on actual working conditions. In addition to actual installation, these figures include time spent during a normal weekday on tasks such as, material receiving and handling, mobilization at site, site movement, breaks, and cleanup.

Productivity data is developed over an extended period so as not to be influenced by abnormal variations and reflects a typical average.

Equipment Costs

Equipment costs include not only rental, but also operating costs for equipment under normal use. The operating costs include parts and labor for routine servicing such as repair and replacement of pumps, filters, and worn lines. Normal operating expendables, such as fuel, lubricants, tires, and electricity (where applicable), are also included. Extraordinary operating expendables with highly variable wear patterns, such as diamond bits and blades, are excluded. These costs are included under materials. Equipment rental rates are obtained from industry sources throughout North America—contractors, suppliers, dealers, manufacturers, and distributors.

Rental rates can also be treated as reimbursement costs for contractor-owned equipment. Owned equipment costs include depreciation, loan payments, interest, taxes, insurance, storage and major repairs.

Equipment costs do not include operators' wages; nor do they include the cost to move equipment to a job site (mobilization) or from a job site (demobilization).

Equipment Cost/Day—The cost of power equipment required for each crew is included in the Crew Listings in the Reference Section (small tools that are considered as essential everyday tools are not listed out separately). The Crew Listings itemize specialized tools and heavy equipment along with labor trades. The daily cost of itemized equipment included in a crew is based on dividing the weekly bare rental rate by 5 (number of working days per week) and then adding the hourly operating cost times 8 (the number of hours per day). This Equipment Cost/Day is shown in the last column of the Equipment Rental Cost pages in the Reference Section.

Mobilization/Demobilization—The cost to move construction equipment from an equipment yard or rental company to the job site and back again is not included in equipment costs. Mobilization (to the site) and demobilization (from the site) costs can be found in the Unit Price Section. If a piece of equipment is already at the job site, it is not appropriate to utilize mob./demob. costs again in an estimate.

Overhead and Profit

Total Cost including O&P for the *Installing Contractor* is shown in the last column on the Unit Price and/or the Assemblies pages of this book. This figure is the sum of the bare material cost plus 10% for profit, the bare labor cost plus total overhead and profit, and the bare equipment cost plus 10% for profit. Details for the calculation of Overhead and Profit on labor are shown on the inside back cover and in the Reference Section of this book. (See "How RSMeans Unit Price Data Works" for an example of this calculation.)

General Conditions

Cost data in this book is presented in two ways: Bare Costs and Total Cost including O&P (Overhead and Profit). General Conditions, or General Requirements, of the contract should also be added to the Total Cost including O&P when applicable. Costs for General Conditions are listed in Division 1 of the Unit Price Section and the Reference

Section of this book. General Conditions for the *Installing Contractor* may range from 0% to 10% of the Total Cost including O&P. For the *General* or *Prime Contractor*, costs for General Conditions may range from 5% to 15% of the Total Cost including O&P, with a figure of 10% as the most typical allowance. If applicable, the Assemblies and Models sections of this book use costs that include the installing contractor's overhead and profit (O&P).

Factors Affecting Costs

Costs can vary depending upon a number of variables. Here's how we have handled the main factors affecting costs.

Quality—The prices for materials and the workmanship upon which productivity is based represent sound construction work. They are also in line with U.S. government specifications.

Overtime—We have made no allowance for overtime. If you anticipate premium time or work beyond normal working hours, be sure to make an appropriate adjustment to your labor costs.

Productivity—The productivity, daily output, and labor-hour figures for each line item are based on working an eight-hour day in daylight hours in moderate temperatures. For work that extends beyond normal work hours or is performed under adverse conditions, productivity may decrease. (See "How RSMeans Unit Price Data Works" for more on productivity.)

Size of Project—The size, scope of work, and type of construction project will have a significant impact on cost. Economies of scale can reduce costs for large projects. Unit costs can often run higher for small projects.

Location—Material prices in this book are for metropolitan areas. However, in dense urban areas, traffic and site storage limitations may increase costs. Beyond a 20-mile radius of large cities, extra trucking or transportation charges may also increase the material costs slightly. On the other hand, lower wage rates may be in effect. Be sure to consider both of these factors when preparing an estimate, particularly if the job site is located in a central city or remote rural location. In addition, highly specialized subcontract items may require travel and per-diem expenses for mechanics.

Other Factors——
- season of year
- contractor management
- weather conditions
- local union restrictions
- building code requirements
- availability of:
  □ adequate energy
  □ skilled labor
  □ building materials
- owner's special requirements/restrictions
- safety requirements
- environmental considerations

Unpredictable Factors—General business conditions influence "in-place" costs of all items. Substitute materials and construction methods may have to be employed. These may affect the installed cost and/or life cycle costs. Such factors may be difficult to evaluate and cannot necessarily be predicted on the basis of the job's location in a particular section of the country. Thus, where these factors apply, you may find significant but unavoidable cost variations for which you will have to apply a measure of judgment to your estimate.

Rounding of Costs

In general, all unit prices in excess of $5.00 have been rounded to make them easier to use and still maintain adequate precision of the results. The rounding rules we have chosen are in the following table.

Prices from . . .	Rounded to the nearest . . .
$.01 to $5.00	$.01
$5.01 to $20.00	$.05
$20.01 to $100.00	$.50
$100.01 to $300.00	$1.00
$300.01 to $1,000.00	$5.00
$1,000.01 to $10,000.00	$25.00
$10,000.01 to $50,000.00	$100.00
$50,000.01 and above	$500.00

How Subcontracted Items Affect Costs

A considerable portion of all large construction jobs is usually subcontracted. In fact, the percentage done by subcontractors is constantly increasing and may run over 90%. Since the workers employed by these companies do nothing else but install their particular product, they soon become expert in that line. The result is, installation by these firms is accomplished so efficiently that the total in-place cost, even adding the general contractor's overhead and profit, is no more, and often less, than if the principal contractor had handled the installation himself/herself. Companies that deal with construction specialties are anxious to have their product perform well and, consequently, the installation will be the best possible.

Contingencies

The allowance for contingencies generally provides for unforeseen construction difficulties. On alterations or repair jobs, 20% is not too much. If drawings are final and only field contingencies are being considered, 2% or 3% is probably sufficient, and often nothing need be added. Contractually, changes in plans will be covered by extras. The contractor should consider inflationary price trends and possible material shortages during the course of the job. These escalation factors are dependent upon both economic conditions and the anticipated time between the estimate and actual construction. If drawings are not complete or approved, or a budget cost is wanted, it is wise to add 5% to 10%. Contingencies, then, are a matter of judgment. Additional allowances for contingencies are shown in Division 1.

Important Estimating Considerations

The productivity or daily output of each craftsman or crew assumes a well-managed job where tradesmen with the proper tools and equipment, along with the appropriate construction materials, are present. Included are daily mobilization and cleanup time, break time and plan layout time. Unless otherwise indicated, time for material movement on site (for items that can be transported by hand) of up to 200' into the building and to the first or second floor is also included. If material has to be transported by other means, over greater distances, or to higher floors, an additional allowance should be considered by the estimator.

While horizontal movement is typically a sole function of distances, vertical transport introduces other variables that can significantly impact productivity. In an occupied building, the use of elevators (assuming access, size, and required protective measures are acceptable) must be understood at the time of the estimate. For new construction, hoist wait and cycle times can easily be 15 minutes and may result in scheduled access extending beyond the normal work day. Finally, all vertical transport will impose strict weight limits likely to

preclude the use of any motorized material handling.

The productivity, or daily output, also assumes installation that meets manufacturer/designer/ standard specifications. A time allowance for quality control checks, minor adjustments and any task required to ensure the proper function or operation is also included. For items that require connections to services, time is included for positioning, leveling, securing the unit and for making all the necessary connections (and start up where applicable) ensuring a complete installation. Estimating of the services themselves (electrical, plumbing, water, steam, hydraulics, dust collection, etc.) is separate.

In some cases, the estimator must consider the use of a crane and an appropriate crew for the installation of large or heavy items. For those situations where a crane is not included in the assigned crew and as part of the line item cost, equipment rental costs, mobilization and demobilization costs, and operator and support personnel costs must be considered.

Labor-Hours

The labor-hours expressed in this publication are derived by dividing the total daily labor hours for the crew by the daily output. Based on average installation time and the assumptions listed above, the labor-hours include: direct labor, indirect labor, and nonproductive time. A typical day for a craftsman might include, but not be limited to:

- Direct Work
  - □ Measuring and layout
  - □ Preparing materials
  - □ Actual installation
  - □ Quality assurance/quality control
- Indirect Work
  - □ Reading plans or specifications
  - □ Preparing space
  - □ Receiving materials
  - □ Material movement
  - □ Giving or receiving instruction
  - □ Miscellaneous

- Non-Work
  - □ Chatting
  - □ Personal issues
  - □ Breaks
  - □ Interruptions (i.e., sickness, weather, material or equipment shortages, etc.)

If any of the items for a typical day do not apply to the particular work or project situation, the estimator should make any necessary adjustments.

Final Checklist

Estimating can be a straightforward process provided you remember the basics. Here's a checklist of some of the steps you should remember to complete before finalizing your estimate.

Did you remember to . . .

- factor in the City Cost Index for your locale?
- take into consideration which items have been marked up and by how much?
- mark up the entire estimate sufficiently for your purposes?
- read the background information on techniques and technical matters that could impact your project time span and cost?
- include all components of your project in the final estimate?
- double check your figures for accuracy?
- call RSMeans if you have any questions about your estimate or the data you've found in our publications? Remember, RSMeans stands behind its publications. If you have any questions about your estimate . . . about the costs you've used from our books . . . or even about the technical aspects of the job that may affect your estimate, feel free to call the RSMeans editors at 1-877-784-5289.

Free Quarterly Updates

Stay up-to-date throughout 2015 with RSMeans' free cost data updates four times a year. Sign up online to make sure you have access to the newest data. Every quarter we provide the city cost adjustment factors for hundreds of cities and key materials. Register at: **http://info.thegordiangroup.com/RSMeans.html.**

Unit Price Section

Table of Contents

How RSMeans Unit Price Data Works

All RSMeans unit price data is organized in the same way.

It is important to understand the structure, so that you can find information easily and use it correctly.

RSMeans **Line Numbers** consist of 12 characters, which identify a unique location in the database for each task. The first 6 or 8 digits conform to the Construction Specifications Institute MasterFormat® 2014. The remainder of the digits are a further breakdown by RSMeans in order to arrange items in understandable groups of similar tasks. Line numbers are consistent across all RSMeans publications, so a line number in any RSMeans product will always refer to the same unit of work.

RSMeans engineers have created **reference** information to assist you in your estimate. If there is information that applies to a section, it will be indicated at the start of the section. In this case, R033105-10 provides information on the proportionate quantities of formwork, reinforcing, and concrete used in cast-in-place concrete items such as footings, slabs, beams, and columns. The Reference Section is located in the back of the book on the pages with a gray edge.

RSMeans **Descriptions** are shown in a hierarchical structure to make them readable. In order to read a complete description, read up through the indents to the top of the section. Include everything that is above and to the left that is not contradicted by information below. For instance, the complete description for line 03 30 53.40 3550 is "Concrete in place, including forms (4 uses), Grade 60 rebar, concrete (Portland cement Type 1), placement and finishing unless otherwise indicated; Equipment pad (3000 psi), 4' x 4' x 6" thick."

03 30 Cast-In-Place Concrete

03 30 53 – Miscellaneous Cast-In-Place Concrete

03 30 53.40	Concrete In Place	
0010	**CONCRETE IN PLACE**	R033105-10
0020	Including forms (4 uses), Grade 60 rebar, concrete (Portland cement	R033105-20
0050	Type I), placement and finishing unless otherwise indicated	R033105-50
0300	Beams (3500 psi), 5 kip per L.F., 10' span	R033105-65
0350	25' span	R033105-70
0500	Chimney foundations (5000 psi), over 5 C.Y.	R033105-85
0510	(3500 psi), under 5 C.Y.	
0700	Columns, square (4000 psi), 12" x 12", less than 2% reinforcing	
3450	Over 10,000 S.F.	
3500	Add per floor for 3 to 6 stories high	
3520	For 7 to 20 stories high	
3540	Equipment pad (3000 psi), 3' x 3' x 6" thick	
3550	4' x 4' x 6" thick	
3560	5' x 5' x 8" thick	
3570	6' x 6' x 8" thick	

When using **RSMeans data**, it is important to read through an entire section to ensure that you use the data that most closely matches your work. Note that sometimes there is additional information shown in the section that may improve your price. There are frequently lines that further describe, add to, or adjust data for specific situations.

The data published in RSMeans print books represents a "national average" cost. This data should be modified to the project location using the **City Cost Indexes** or **Location Factors** tables found in the Reference Section (see pages 656–704). Use the location factors to adjust estimate totals if the project covers multiple trades. Use the city cost indexes (CCI) for single trade projects or projects where a more detailed analysis is required. All figures in the two tables are derived from the same research. The last row of data in the CCI, the weighted average, is the same as the numbers reported for each location in the location factor table.

Crews include labor or labor and equipment necessary to accomplish each task. In this case, Crew C-14H is used. RSMeans selects a crew to represent the workers and equipment that are typically used for that task. In this case, Crew C-14H consists of one carpenter foreman (outside), two carpenters, one rodman, one laborer, one cement finisher, and one gas engine vibrator. Details of all crews can be found in the reference section.

Crews

Crew No.	Bare Costs		Incl. Subs O&P		Cost Per Labor-Hour	
Crew C-14H	Hr.	Daily	Hr.	Daily	Bare Costs	Incl. O&P
1 Carpenter Foreman (outside)	$48.95	$391.60	$75.35	$602.80	$46.33	$71.08
2 Carpenters	46.95	751.20	72.25	1156.00		
1 Rodman (reinf.)	52.55	420.40	82.10	656.80		
1 Laborer	37.60	300.80	57.85	462.80		
1 Cement Finisher	45.00	360.00	66.65	533.20		
1 Gas Engine Vibrator		31.20		34.32	.65	.71
48 L.H., Daily Totals		$2255.20		$3445.92	$46.98	$71.79

The **Daily Output** is the amount of work that the crew can do in a normal 8-hour workday, including mobilization, layout, movement of materials, and cleanup. In this case, crew C-14H can install thirty 4' x 4' x 6" thick concrete pads in a day. Daily output is variable, based on many factors, including the size of the job, location, and environmental conditions. RSMeans data represents work done in daylight (or adequate lighting) and temperate conditions.

Bare Costs are the costs of materials, labor, and equipment that the installing contractor pays. They represent the cost, in U.S. dollars, for one unit of work. They do not include any markups for profit or labor burden.

Crew	Daily Output	Labor-Hours	Unit	Material	2015 Bare Costs Labor	Equipment	Total	Total Incl O&P
C-14A	15.62	12.804	C.Y.	320	605	48	973	1,325
"	18.55	10.782		335	510	40.50	885.50	1,200
C-14C	32.22	3.476		150	157	.97	307.97	405
"	23.71	4.724		177	213	1.32	391.32	525
C-14A	11.96	16.722		365	790	63	1,218	1,700
	2200	.025		.94	1.07	.33	2.34	3.02
	31800	.002			.07	.02	.09	.13
	21200	.003			.11	.03	.14	.21
C-14H	45	1.067	Ea.	47	49.50	.69	97.19	128
	30	1.600		69.50	74	1.04	144.54	191
	18	2.667		122	124	1.73	247.73	325
	14	3.429		164	159	2.23	325.23	425

The **Total Incl O&P column** is the total cost, including overhead and profit, that the installing contractor will charge the customer. This represents the cost of materials plus 10% profit, the cost of labor plus labor burden and 10% profit, and the cost of equipment plus 10% profit. It does not include the general contractor's overhead and profit. Note: See the inside back cover for details of how RSMeans calculates labor burden.

The **Total column** represents the total bare cost for the installing contractor, in U.S. dollars. In this case, the sum of $69.50 for material + $74.00 for labor + $1.04 for equipment is $144.54.

The figure in the **Labor Hours** column is the amount of labor required to perform one unit of work—in this case the amount of labor required to construct one 4' x 4' equipment pad. This figure is calculated by dividing the number of hours of labor in the crew by the daily output (48 labor hours divided by 30 pads = 1.6 hours of labor per pad). Multiply 1.600 times 60 to see the value in minutes: 60 x 1.6 = 96 minutes. Note: the labor hour figure is not dependent on the crew size. A change in crew size will result in a corresponding change in daily output, but the labor hours per unit of work will not change.

All RSMeans unit cost data includes the typical **Unit of Measure** used for estimating that item. For concrete-in-place the typical unit is cubic yards (C.Y.) or each (Ea.). For installing broadloom carpet it is square yard, and for gypsum board it is square foot. The estimator needs to take special care that the unit in the data matches the unit in the take-off. Unit conversions may be found in the Reference Section.

How RSMeans Unit Price Data Works (Continued)

Sample Estimate

This sample demonstrates the elements of an estimate, including a tally of the RSMeans data lines, and a summary of the markups on a contractor's work to arrive at a total cost to the owner. The RSMeans Location Factor is added at the bottom of the estimate to adjust the cost of the work to a specific location.

Work Performed: The body of the estimate shows the RSMeans data selected, including line number, a brief description of each item, its take-off unit and quantity, and the bare costs of materials, labor, and equipment. This estimate also includes a column titled "SubContract." This data is taken from the RSMeans column "Total Incl O&P," and represents the total that a subcontractor would charge a general contractor for the work, including the sub's markup for overhead and profit.

Division 1, General Requirements: This is the first division numerically, but the last division estimated. Division 1 includes project-wide needs provided by the general contractor. These requirements vary by project, but may include temporary facilities and utilities, security, testing, project cleanup, etc. For small projects a percentage can be used, typically between 5% and 15% of project cost. For large projects the costs may be itemized and priced individually.

Bonds: Bond costs should be added to the estimate. The figures here represent a typical performance bond, ensuring the owner that if the general contractor does not complete the obligations in the construction contract the bonding company will pay the cost for completion of the work.

Location Adjustment: RSMeans published data is based on national average costs. If necessary, adjust the total cost of the project using a location factor from the "Location Factor" table or the "City Cost Index" table. Use location factors if the work is general, covering multiple trades. If the work is by a single trade (e.g., masonry) use the more specific data found in the "City Cost Indexes."

This estimate is based on an interactive spreadsheet. A copy of this spreadsheet is located on the RSMeans website at **http://www.reedconstructiondata.com/rsmeans/extras/546011.** You are free to download it and adjust it to your methodology.

Project Name: Pre-Engineered Steel Building			Architect: As Shown	
Location:	**Anywhere, USA**			
Line Number	**Description**	**Qty**	**Unit**	**Material**
03 30 53.40 3940	Strip footing, 12" x 24", reinforced	34	C.Y.	$4,726.00
03 30 53.40 3950	Strip footing, 12" x 36", reinforced	15	C.Y.	$1,995.00
03 11 13.65 3000	Concrete slab edge forms	500	L.F.	$175.00
03 22 11.10 0200	Welded wire fabric reinforcing	150	C.S.F.	$2,580.00
03 31 13.35 0300	Ready mix concrete, 4000 psi for slab on grade	278	C.Y.	$29,746.00
03 31 13.70 4300	Place, strike off & consolidate concrete slab	278	C.Y.	$0.00
03 35 13.30 0250	Machine float & trowel concrete slab	15,000	S.F.	$0.00
03 15 16.20 0140	Cut control joints in concrete slab	950	L.F.	$57.00
03 39 23.13 0300	Sprayed concrete curing membrane	150	C.S.F.	$1,657.50
Division 03	**Subtotal**			**$40,936.50**
08 36 13.10 2650	Manual 10' x 10' steel sectional overhead door	8	Ea.	$8,800.00
08 36 13.10 2860	Insulation and steel back panel for OH door	800	S.F.	$3,800.00
Division 08	**Subtotal**			**$12,600.00**
13 34 19.50 1100	Pre-Engineered Steel Building, 100' x 150' x 24'	15,000	SF Flr.	$0.00
13 34 19.50 6050	Framing for PESB door opening, 3' x 7'	4	Opng.	$0.00
13 34 19.50 6100	Framing for PESB door opening, 10' x 10'	8	Opng.	$0.00
13 34 19.50 6200	Framing for PESB window opening, 4' x 3'	6	Opng.	$0.00
13 34 19.50 5750	PESB door, 3' x 7', single leaf	4	Opng.	$2,340.00
13 34 19.50 7750	PESB sliding window, 4' x 3' with screen	6	Opng.	$2,280.00
13 34 19.50 6550	PESB gutter, eave type, 26 ga., painted	300	L.F.	$2,085.00
13 34 19.50 8650	PESB roof vent, 12" wide x 10' long	15	Ea.	$540.00
13 34 19.50 6900	PESB insulation, vinyl faced, 4" thick	27,400	S.F.	$11,508.00
Division 13	**Subtotal**			**$18,753.00**
			Subtotal	$72,289.50
Division 01	**General Requirements @ 7%**			5,060.27
			Estimate Subtotal	$77,349.77
			Sales Tax @ 5%	3,867.49
			Subtotal A	81,217.25
			GC O & P	8,121.73
			Subtotal B	89,338.98
			Contingency @ 5%	
			Subtotal C	
			Bond @ $12/1000 +10% O&P	
			Subtotal D	
			Location Adjustment Factor	
			Grand Total	

4

This example shows the cost to construct a pre-engineered steel building. The foundation, doors, windows, and insulation will be installed by the general contractor. A subcontractor will install the structural steel, roofing, and siding.

Labor	Equipment	SubContract	Estimate Total	
		01/01/15	STD	
$3,570.00	$22.10	$0.00		
$1,260.00	$7.80	$0.00		
$1,190.00	$0.00	$0.00		
$4,050.00	$0.00	$0.00		
$0.00	$0.00	$0.00		
$4,753.80	$158.46	$0.00		
$9,000.00	$450.00	$0.00		
$380.00	$85.50	$0.00		
$952.50	$0.00	$0.00		
25,156.30	$723.86	$0.00	$66,816.66	Division 03
$3,320.00	$0.00	$0.00		
$0.00	$0.00	$0.00		
3,320.00	$0.00	$0.00	$15,920.00	Division 08
$0.00	$0.00	$352,500.00		
$0.00	$0.00	$2,280.00		
$0.00	$0.00	$9,200.00		
$0.00	$0.00	$3,330.00		
$672.00	$0.00	$0.00		
$576.00	$67.20	$0.00		
$789.00	$0.00	$0.00		
$3,165.00	$0.00	$0.00		
$9,042.00	$0.00	$0.00		
14,244.00	$67.20	$367,310.00	$400,374.20	Division 13
$42,720.30	$791.06	$367,310.00	$483,110.86	Subtotal
2,990.42	55.37	25,711.70		Gen. Requirements
$45,710.72	$846.43	$393,021.70	$483,110.86	Estimate Subtotal
	42.32	9,825.54		Sales tax
45,710.72	888.76	402,847.24		Subtotal
24,912.34	88.88	40,284.72		GC O&P
70,623.06	977.63	443,131.97	$604,071.64	Subtotal
			30,203.58	Contingency
			$634,275.22	Subtotal
			8,372.43	Bond
			$642,647.66	Subtotal
102.30			14,780.90	Location Adjustment
			$657,428.55	Grand Total

Sales Tax: If the work is subject to state or local sales taxes, the amount must be added to the estimate. Sales tax may be added to material costs, equipment costs, and subcontracted work. In this case, sales tax was added in all three categories. It was assumed that approximately half the subcontracted work would be material cost, so the tax was applied to 50% of the subcontract total.

GC O&P: This entry represents the general contractor's markup on material, labor, equipment, and subcontractor costs. RSMeans' standard markup on materials, equipment, and subcontracted work is 10%. In this estimate, the markup on the labor performed by the GC's workers uses "Skilled Workers Average" shown in Column F on the table "Installing Contractor's Overhead & Profit," which can be found on the inside-back cover of the book.

Contingency: A factor for contingency may be added to any estimate to represent the cost of unknowns that may occur between the time that the estimate is performed and the time the project is constructed. The amount of the allowance will depend on the stage of design at which the estimate is done, and the contractor's assessment of the risk involved. Refer to section 01 21 16.50 for contigency allowances.

Estimating Tips

01 20 00 Price and Payment Procedures

- Allowances that should be added to estimates to cover contingencies and job conditions that are not included in the national average material and labor costs are shown in section 01 21.
- When estimating historic preservation projects (depending on the condition of the existing structure and the owner's requirements), a 15%–20% contingency or allowance is recommended, regardless of the stage of the drawings.

01 30 00 Administrative Requirements

- Before determining a final cost estimate, it is a good practice to review all the items listed in Subdivisions 01 31 and 01 32 to make final adjustments for items that may need customizing to specific job conditions.
- Requirements for initial and periodic submittals can represent a significant cost to the General Requirements of a job. Thoroughly check the submittal specifications when estimating a project to determine any costs that should be included.

01 40 00 Quality Requirements

- All projects will require some degree of Quality Control. This cost is not included in the unit cost of construction listed in each division. Depending upon the terms of the contract, the various costs of inspection and testing can be the responsibility of either the owner or the contractor. Be sure to include the required costs in your estimate.

01 50 00 Temporary Facilities and Controls

- Barricades, access roads, safety nets, scaffolding, security, and many more requirements for the execution of a safe project are elements of direct cost. These costs can easily be overlooked when preparing an estimate. When looking through the major classifications of this subdivision, determine which items apply to each division in your estimate.
- Construction Equipment Rental Costs can be found in the Reference Section in section 01 54 33. Operators' wages are not included in equipment rental costs.
- Equipment mobilization and demobilization costs are not included in equipment rental costs and must be considered separately in section 01 54 36.50.

- The cost of small tools provided by the installing contractor for his workers is covered in the "Overhead" column on the "Installing Contractor's Overhead and Profit" table that lists labor trades, base rates and markups and, therefore, is included in the "Total Incl. O&P" cost of any Unit Price line item.

01 70 00 Execution and Closeout Requirements

- When preparing an estimate, thoroughly read the specifications to determine the requirements for Contract Closeout. Final cleaning, record documentation, operation and maintenance data, warranties and bonds, and spare parts and maintenance materials can all be elements of cost for the completion of a contract. Do not overlook these in your estimate.

Reference Numbers

Reference numbers are shown in shaded boxes at the beginning of some major classifications. These numbers refer to related items in the Reference Section. The reference information may be an estimating procedure, an alternate pricing method, or technical information.

Note: Not all subdivisions listed here necessarily appear in this publication. ■

01 11 Summary of Work

01 11 31 – Professional Consultants

01 11 31.10 Architectural Fees		Crew	Daily Output	Labor-Hours	Unit	Material	2015 Bare Costs Labor	Equipment	Total	Total Incl O&P
0010	**ARCHITECTURAL FEES**									
2000	For "Greening" of building	R011110-10 [G]			Project				3%	3%

01 11 31.30 Engineering Fees		Crew	Daily Output	Labor-Hours	Unit	Material	Labor	Equipment	Total	Total Incl O&P
0010	**ENGINEERING FEES**	R011110-30								
0020	Educational planning consultant, minimum				Project				.50%	.50%
0100	Maximum				"				2.50%	2.50%
0200	Electrical, minimum				Contrct				4.10%	4.10%
0300	Maximum								10.10%	10.10%
0400	Elevator & conveying systems, minimum								2.50%	2.50%
0500	Maximum								5%	5%
0600	Food service & kitchen equipment, minimum								8%	8%
0700	Maximum								12%	12%
1000	Mechanical (plumbing & HVAC), minimum								4.10%	4.10%
1100	Maximum								10.10%	10.10%

01 21 Allowances

01 21 16 – Contingency Allowances

01 21 16.50 Contingencies

		Crew	Daily Output	Labor-Hours	Unit	Material	Labor	Equipment	Total	Total Incl O&P
0010	**CONTINGENCIES**, Add to estimate									
0020	Conceptual stage				Project				20%	20%
0050	Schematic stage								15%	15%
0100	Preliminary working drawing stage (Design Dev.)								10%	10%
0150	Final working drawing stage								3%	3%

01 21 53 – Factors Allowance

01 21 53.50 Factors

		Crew	Daily Output	Labor-Hours	Unit	Material	Labor	Equipment	Total	Total Incl O&P
0010	**FACTORS** Cost adjustments	R012153-10								
0100	Add to construction costs for particular job requirements									
0500	Cut & patch to match existing construction, add, minimum				Costs	2%	3%			
0550	Maximum					5%	9%			
0800	Dust protection, add, minimum					1%	2%			
0850	Maximum					4%	11%			
1100	Equipment usage curtailment, add, minimum					1%	1%			
1150	Maximum					3%	10%			
1400	Material handling & storage limitation, add, minimum					1%	1%			
1450	Maximum					6%	7%			
1700	Protection of existing work, add, minimum					2%	2%			
1750	Maximum					5%	7%			
2000	Shift work requirements, add, minimum						5%			
2050	Maximum						30%			
2300	Temporary shoring and bracing, add, minimum					2%	5%			
2350	Maximum					5%	12%			
2400	Work inside prisons and high security areas, add, minimum						30%			
2450	Maximum						50%			

01 21 55 – Job Conditions Allowance

01 21 55.50 Job Conditions

		Crew	Daily Output	Labor-Hours	Unit	Material	Labor	Equipment	Total	Total Incl O&P
0010	**JOB CONDITIONS** Modifications to applicable									
0020	cost summaries									
0100	Economic conditions, favorable, deduct				Project				2%	2%
0200	Unfavorable, add								5%	5%
0300	Hoisting conditions, favorable, deduct								2%	2%

01 21 Allowances

01 21 55 – Job Conditions Allowance

01 21 55.50 Job Conditions

		Crew	Daily Output	Labor-Hours	Unit	Material	2015 Bare Costs Labor	Equipment	Total	Total Incl O&P
0400	Unfavorable, add				Project				5%	5%
0700	Labor availability, surplus, deduct								1%	1%
0800	Shortage, add								10%	10%
0900	Material storage area, available, deduct								1%	1%
1000	Not available, add								2%	2%
1100	Subcontractor availability, surplus, deduct								5%	5%
1200	Shortage, add								12%	12%
1300	Work space, available, deduct								2%	2%
1400	Not available, add								5%	5%

01 21 57 – Overtime Allowance

01 21 57.50 Overtime

		Crew	Daily Output	Labor-Hours	Unit	Material	2015 Bare Costs Labor	Equipment	Total	Total Incl O&P
0010	**OVERTIME** for early completion of projects or where R012909-90									
0020	labor shortages exist, add to usual labor, up to				Costs		100%			

01 21 61 – Cost Indexes

01 21 61.10 Construction Cost Index

		Crew	Daily Output	Labor-Hours	Unit	Material	2015 Bare Costs Labor	Equipment	Total	Total Incl O&P
0010	**CONSTRUCTION COST INDEX** (Reference) over 930 zip code locations in									
0020	the U.S. and Canada, total bldg. cost, min. (Fayetteville, AR)				%				75.40%	75.40%
0050	Average								100%	100%
0100	Maximum (New York, NY)								131.80%	131.80%

01 21 61.20 Historical Cost Indexes

		Crew	Daily Output	Labor-Hours	Unit	Material	2015 Bare Costs Labor	Equipment	Total	Total Incl O&P
0010	**HISTORICAL COST INDEXES** (See Reference Section)									

01 21 61.30 Labor Index

		Crew	Daily Output	Labor-Hours	Unit	Material	2015 Bare Costs Labor	Equipment	Total	Total Incl O&P
0010	**LABOR INDEX** (Reference) For over 930 zip code locations in									
0020	the U.S. and Canada, minimum (Beaufort, SC)				%		44.90%			
0050	Average						100%			
0100	Maximum (New York, NY)						168.90%			

01 21 61.50 Material Index

		Crew	Daily Output	Labor-Hours	Unit	Material	2015 Bare Costs Labor	Equipment	Total	Total Incl O&P
0010	**MATERIAL INDEX** (Reference) For over 930 zip code locations in									
0020	the U.S. and Canada, minimum (Oil City, PA)				%	92.40%				
0040	Average					100%				
0060	Maximum (Ketchikan, AK)					132%				

01 21 63 – Taxes

01 21 63.10 Taxes

		Crew	Daily Output	Labor-Hours	Unit	Material	2015 Bare Costs Labor	Equipment	Total	Total Incl O&P
0010	**TAXES** R012909-80									
0020	Sales tax, State, average				%	5.04%				
0050	Maximum R012909-85					7.50%				
0200	Social Security, on first $117,000 of wages						7.65%			
0300	Unemployment, combined Federal and State, minimum						.60%			
0350	Average						7.80%			
0400	Maximum						12.87%			

For customer support on your Mechanical Cost Data, call 877.795.7394.

9

01 31 Project Management and Coordination

01 31 13 – Project Coordination

01 31 13.20 Field Personnel

		Crew	Daily Output	Labor-Hours	Unit	Material	2015 Bare Costs Labor	2015 Bare Costs Equipment	Total	Total Incl O&P
0010	**FIELD PERSONNEL**									
0020	Clerk, average				Week		450		450	695
0100	Field engineer, minimum						1,075		1,075	1,650
0120	Average						1,400		1,400	2,150
0140	Maximum						1,575		1,575	2,450
0160	General purpose laborer, average						1,500		1,500	2,325
0180	Project manager, minimum						1,975		1,975	3,075
0200	Average						2,275		2,275	3,525
0220	Maximum						2,600		2,600	4,025
0240	Superintendent, minimum						1,925		1,925	2,975
0260	Average						2,125		2,125	3,275
0280	Maximum						2,400		2,400	3,725
0290	Timekeeper, average						1,225		1,225	1,900

01 31 13.30 Insurance

			Crew	Daily Output	Labor-Hours	Unit	Material	Labor	Equipment	Total	Total Incl O&P
0010	**INSURANCE**	R013113-40									
0020	Builders risk, standard, minimum					Job				.24%	.24%
0050	Maximum	R013113-50								.64%	.64%
0200	All-risk type, minimum									.25%	.25%
0250	Maximum	R013113-60								.62%	.62%
0400	Contractor's equipment floater, minimum					Value				.50%	.50%
0450	Maximum					"				1.50%	1.50%
0600	Public liability, average					Job				2.02%	2.02%
0800	Workers' compensation & employer's liability, average										
0850	by trade, carpentry, general					Payroll		14.93%			
1000	Electrical							5.76%			
1150	Insulation							11.66%			
1450	Plumbing							6.98%			
1550	Sheet metal work (HVAC)							8.82%			

01 31 13.50 General Contractor's Mark-Up

		Crew	Daily Output	Labor-Hours	Unit	Material	Labor	Equipment	Total	Total Incl O&P
0010	**GENERAL CONTRACTOR'S MARK-UP** on Change Orders									
0200	Extra work, by subcontractors, add				%				10%	10%
0250	By General Contractor, add								15%	15%
0400	Omitted work, by subcontractors, deduct all but								5%	5%
0450	By General Contractor, deduct all but								7.50%	7.50%
0600	Overtime work, by subcontractors, add								15%	15%
0650	By General Contractor, add								10%	10%

01 31 13.60 Installing Contractor's Main Office Overhead

		Crew	Daily Output	Labor-Hours	Unit	Material	Labor	Equipment	Total	Total Incl O&P
0010	**INSTALLING CONTRACTOR'S MAIN OFFICE OVERHEAD**									
0020	As percent of direct costs, minimum				%				5%	
0050	Average								13%	
0100	Maximum								30%	

01 31 13.80 Overhead and Profit

		Crew	Daily Output	Labor-Hours	Unit	Material	Labor	Equipment	Total	Total Incl O&P
0010	**OVERHEAD & PROFIT** Allowance to add to items in this									
0020	book that do not include Subs O&P, average				%				25%	
0100	Allowance to add to items in this book that									
0110	do include Subs O&P, minimum				%				5%	5%
0150	Average								10%	10%
0200	Maximum								15%	15%
0300	Typical, by size of project, under $100,000								30%	
0350	$500,000 project								25%	
0400	$2,000,000 project								20%	
0450	Over $10,000,000 project								15%	

01 31 Project Management and Coordination

01 31 13 – Project Coordination

01 31 13.90 Performance Bond

		Crew	Daily Output	Labor-Hours	Unit	Material	2015 Bare Costs Labor	Equipment	Total	Total Incl O&P
0010	**PERFORMANCE BOND**	R013113-80								
0020	For buildings, minimum				Job				.60%	.60%
0100	Maximum				"				2.50%	2.50%

01 32 Construction Progress Documentation

01 32 33 – Photographic Documentation

01 32 33.50 Photographs

		Crew	Daily Output	Labor-Hours	Unit	Material	2015 Bare Costs Labor	Equipment	Total	Total Incl O&P
0010	**PHOTOGRAPHS**									
0020	8" x 10", 4 shots, 2 prints ea., std. mounting				Set	475			475	525
0100	Hinged linen mounts					530			530	580
0200	8" x 10", 4 shots, 2 prints each, in color					425			425	470
0300	For I.D. slugs, add to all above					5.30			5.30	5.85
1500	Time lapse equipment, camera and projector, buy				Ea.	2,650			2,650	2,925
1550	Rent per month				"	1,600			1,600	1,750
1700	Cameraman and film, including processing, B.&W.				Day	1,225			1,225	1,350
1720	Color				"	1,425			1,425	1,550

01 41 Regulatory Requirements

01 41 26 – Permit Requirements

01 41 26.50 Permits

		Crew	Daily Output	Labor-Hours	Unit	Material	2015 Bare Costs Labor	Equipment	Total	Total Incl O&P
0010	**PERMITS**									
0020	Rule of thumb, most cities, minimum				Job				.50%	.50%
0100	Maximum				"				2%	2%

01 45 Quality Control

01 45 23 – Testing and Inspecting Services

01 45 23.50 Testing

		Crew	Daily Output	Labor-Hours	Unit	Material	2015 Bare Costs Labor	Equipment	Total	Total Incl O&P
0010	**TESTING** and Inspecting Services									
5820	Non-destructive metal testing, dye penetrant				Day				310	340
5840	Magnetic particle								310	340
5860	Radiography								450	495
5880	Ultrasonic								310	340
9000	Thermographic testing, for bldg envelope heat loss, average 2,000 S.F.	G			Ea.				500	500

01 51 Temporary Utilities

01 51 13 – Temporary Electricity

01 51 13.80 Temporary Utilities

		Crew	Daily Output	Labor-Hours	Unit	Material	2015 Bare Costs Labor	Equipment	Total	Total Incl O&P
0010	**TEMPORARY UTILITIES**									
0350	Lighting, lamps, wiring, outlets, 40,000 SF building, 8 strings	1 Elec	34	.235	CSF Flr	4.91	12.85		17.76	24.50
0360	16 strings	"	17	.471		9.80	25.50		35.30	49.50
0400	Power for temp lighting only, 6.6 KWH, per month								.92	1.01
0430	11.8 KWH, per month								1.65	1.82
0450	23.6 KWH, per month								3.30	3.63
0600	Power for job duration incl. elevator, etc., minimum								47	51.50
0650	Maximum								110	121

11

01 51 Temporary Utilities

01 51 13 – Temporary Electricity

01 51 13.80 Temporary Utilities	Crew	Daily Output	Labor-Hours	Unit	Material	2015 Bare Costs Labor	Equipment	Total	Total Incl O&P
1000 Toilet, portable, see Equip. Rental 01 54 33 in Reference Section									

01 52 Construction Facilities

01 52 13 – Field Offices and Sheds

01 52 13.20 Office and Storage Space

		Crew	Daily Output	Labor-Hours	Unit	Material	2015 Bare Costs Labor	Equipment	Total	Total Incl O&P
0010	OFFICE AND STORAGE SPACE									
0020	Office trailer, furnished, no hookups, 20' x 8', buy	2 Skwk	1	16	Ea.	10,400	780		11,180	12,600
0250	Rent per month					188			188	206
0300	32' x 8', buy	2 Skwk	.70	22.857		15,500	1,100		16,600	18,800
0350	Rent per month					239			239	262
0400	50' x 10', buy	2 Skwk	.60	26.667		24,100	1,300		25,400	28,500
0450	Rent per month					340			340	375
0500	50' x 12', buy	2 Skwk	.50	32		29,400	1,550		30,950	34,700
0550	Rent per month					410			410	455
0700	For air conditioning, rent per month, add					48.50			48.50	53.50
0800	For delivery, add per mile				Mile	11			11	12.10
0890	Delivery each way				Ea.	200			200	220
0900	Bunk house trailer, 8' x 40' duplex dorm with kitchen, no hookups, buy	2 Carp	1	16		37,300	750		38,050	42,200
0910	9 man with kitchen and bath, no hookups, buy		1	16		38,000	750		38,750	43,000
0920	18 man sleeper with bath, no hookups, buy		1	16		49,000	750		49,750	55,000
1000	Portable buildings, prefab, on skids, economy, 8' x 8'		265	.060	S.F.	25	2.83		27.83	32
1100	Deluxe, 8' x 12'		150	.107	"	20	5		25	29.50
1200	Storage boxes, 20' x 8', buy	2 Skwk	1.80	8.889	Ea.	2,775	430		3,205	3,725
1250	Rent per month					82.50			82.50	91
1300	40' x 8', buy	2 Skwk	1.40	11.429		3,850	555		4,405	5,075
1350	Rent per month					101			101	111

01 52 13.40 Field Office Expense

		Crew	Daily Output	Labor-Hours	Unit	Material	2015 Bare Costs Labor	Equipment	Total	Total Incl O&P
0010	FIELD OFFICE EXPENSE									
0100	Office equipment rental average				Month	200			200	220
0120	Office supplies, average				"	80			80	88
0125	Office trailer rental, see Section 01 52 13.20									
0140	Telephone bill; avg. bill/month incl. long dist.				Month	85			85	93.50
0160	Lights & HVAC				"	160			160	176

01 54 Construction Aids

01 54 16 – Temporary Hoists

01 54 16.50 Weekly Forklift Crew

		Crew	Daily Output	Labor-Hours	Unit	Material	2015 Bare Costs Labor	Equipment	Total	Total Incl O&P
0010	WEEKLY FORKLIFT CREW									
0100	All-terrain forklift, 45' lift, 35' reach, 9000 lb. capacity	A-3P	.20	40	Week		1,950	2,625	4,575	5,850

01 54 19 – Temporary Cranes

01 54 19.50 Daily Crane Crews

		Crew	Daily Output	Labor-Hours	Unit	Material	2015 Bare Costs Labor	Equipment	Total	Total Incl O&P
0010	DAILY CRANE CREWS for small jobs, portal to portal									
0100	12-ton truck-mounted hydraulic crane	A-3H	1	8	Day		415	860	1,275	1,575
0200	25-ton	A-3I	1	8			415	990	1,405	1,725
0300	40-ton	A-3J	1	8			415	1,225	1,640	1,975
0400	55-ton	A-3K	1	16			775	1,625	2,400	2,975
0500	80-ton	A-3L	1	16			775	2,350	3,125	3,750
0600	100-ton	A-3M	1	16			775	2,325	3,100	3,750

01 54 Construction Aids

01 54 19 – Temporary Cranes

01 54 19.50 Daily Crane Crews

		Crew	Daily Output	Labor-Hours	Unit	Material	2015 Bare Costs Labor	Equipment	Total	Total Incl O&P
0900	If crane is needed on a Saturday, Sunday or Holiday									
0910	At time-and-a-half, add				Day		50%			
0920	At double time, add				"		100%			

01 54 19.60 Monthly Tower Crane Crew

		Crew	Daily Output	Labor-Hours	Unit	Material	2015 Bare Costs Labor	Equipment	Total	Total Incl O&P
0010	**MONTHLY TOWER CRANE CREW**, excludes concrete footing									
0100	Static tower crane, 130' high, 106' jib, 6200 lb. capacity	A-3N	.05	176	Month		9,100	24,800	33,900	41,100

01 54 23 – Temporary Scaffolding and Platforms

01 54 23.70 Scaffolding

		Crew	Daily Output	Labor-Hours	Unit	Material	2015 Bare Costs Labor	Equipment	Total	Total Incl O&P
0010	**SCAFFOLDING** R015423-10									
0015	Steel tube, regular, no plank, labor only to erect & dismantle									
0090	Building exterior, wall face, 1 to 5 stories, 6'-4" x 5' frames	3 Carp	8	3	C.S.F.		141		141	217
0200	6 to 12 stories	4 Carp	8	4			188		188	289
0301	13 to 20 stories	5 Clab	8	5			188		188	289
0460	Building interior, wall face area, up to 16' high	3 Carp	12	2			94		94	145
0560	16' to 40' high		10	2.400			113		113	173
0800	Building interior floor area, up to 30' high		150	.160	C.C.F.		7.50		7.50	11.55
0900	Over 30' high	4 Carp	160	.200	"		9.40		9.40	14.45
0906	Complete system for face of walls, no plank, material only rent/mo				C.S.F.	35.50			35.50	39.50
0908	Interior spaces, no plank, material only rent/mo				C.C.F.	4.33			4.33	4.76
0910	Steel tubular, heavy duty shoring, buy									
0920	Frames 5' high 2' wide				Ea.	81.50			81.50	90
0925	5' high 4' wide					93			93	102
0930	6' high 2' wide					93.50			93.50	103
0935	6' high 4' wide					109			109	120
0940	Accessories									
0945	Cross braces				Ea.	15.50			15.50	17.05
0950	U-head, 8" x 8"					19.10			19.10	21
0955	J-head, 4" x 8"					13.90			13.90	15.30
0960	Base plate, 8" x 8"					15.50			15.50	17.05
0965	Leveling jack					33.50			33.50	36.50
1000	Steel tubular, regular, buy									
1100	Frames 3' high 5' wide				Ea.	75			75	82.50
1150	5' high 5' wide					86.50			86.50	95
1200	6'-4" high 5' wide					118			118	129
1350	7'-6" high 6' wide					158			158	173
1500	Accessories cross braces					15.50			15.50	17.05
1550	Guardrail post					16.40			16.40	18.05
1600	Guardrail 7' section					8.70			8.70	9.60
1650	Screw jacks & plates					21.50			21.50	24
1700	Sidearm brackets					30.50			30.50	33.50
1750	8" casters					31			31	34
1800	Plank 2" x 10" x 16'-0"					57.50			57.50	63.50
1900	Stairway section					286			286	315
1910	Stairway starter bar					32			32	35
1920	Stairway inside handrail					58.50			58.50	64
1930	Stairway outside handrail					86.50			86.50	95
1940	Walk-thru frame guardrail					41.50			41.50	46
2000	Steel tubular, regular, rent/mo.									
2100	Frames 3' high 5' wide				Ea.	5			5	5.50
2150	5' high 5' wide					5			5	5.50
2200	6'-4" high 5' wide					5.50			5.50	6.05
2250	7'-6" high 6' wide					10			10	11

For customer support on your Mechanical Cost Data, call 877.795.7394.

13

01 54 23.70 Scaffolding		Crew	Daily Output	Labor-Hours	Unit	Material	2015 Bare Costs Labor	Equipment	Total	Total Incl O&P
2500	Accessories, cross braces				Ea.	1			1	1.10
2550	Guardrail post					1			1	1.10
2600	Guardrail 7' section					1			1	1.10
2650	Screw jacks & plates					2			2	2.20
2700	Sidearm brackets					2			2	2.20
2750	8" casters					8			8	8.80
2800	Outrigger for rolling tower					3			3	3.30
2850	Plank 2" x 10" x 16'-0"					10			10	11
2900	Stairway section					35			35	38.50
2940	Walk-thru frame guardrail					2.50			2.50	2.75
3000	Steel tubular, heavy duty shoring, rent/mo.									
3250	5' high 2' & 4' wide				Ea.	8.50			8.50	9.35
3300	6' high 2' & 4' wide					8.50			8.50	9.35
3500	Accessories, cross braces					1			1	1.10
3600	U - head, 8" x 8"					2.50			2.50	2.75
3650	J - head, 4" x 8"					2.50			2.50	2.75
3700	Base plate, 8" x 8"					1			1	1.10
3750	Leveling jack					2.50			2.50	2.75
5700	Planks, 2" x 10" x 16'-0", labor only to erect & remove to 50' H	3 Carp	72	.333			15.65		15.65	24
5800	Over 50' high	4 Carp	80	.400			18.80		18.80	29
6000	Heavy duty shoring for elevated slab forms to 8'-2" high, floor area									
6100	Labor only to erect & dismantle	4 Carp	16	2	C.S.F.		94		94	145
6110	Materials only, rent.mo				"	44.50			44.50	49
6500	To 14'-8" high									
6600	Labor only to erect & dismantle	4 Carp	10	3.200	C.S.F.		150		150	231
6610	Materials only, rent/mo				"	65			65	71.50

01 54 23.75 Scaffolding Specialties

		Crew	Daily Output	Labor-Hours	Unit	Material	2015 Bare Costs Labor	Equipment	Total	Total Incl O&P
0010	**SCAFFOLDING SPECIALTIES**									
1200	Sidewalk bridge, heavy duty steel posts & beams, including									
1210	parapet protection & waterproofing (material cost is rent/month)									
1220	8' to 10' wide, 2 posts	3 Carp	15	1.600	L.F.	45	75		120	166
1230	3 posts	"	10	2.400	"	69.50	113		182.50	250
1500	Sidewalk bridge using tubular steel scaffold frames including									
1510	planking (material cost is rent/month)	3 Carp	45	.533	L.F.	8.55	25		33.55	48
1600	For 2 uses per month, deduct from all above					50%				
1700	For 1 use every 2 months, add to all above					100%				
1900	Catwalks, 20" wide, no guardrails, 7' span, buy				Ea.	166			166	183
2000	10' span, buy					233			233	256
3720	Putlog, standard, 8' span, with hangers, buy					162			162	178
3730	Rent per month					20			20	22
3750	12' span, buy					203			203	223
3755	Rent per month					25			25	27.50
3760	Trussed type, 16' span, buy					365			365	400
3770	Rent per month					30			30	33
3790	22' span, buy					420			420	465
3795	Rent per month					40			40	44
3800	Rolling ladders with handrails, 30" wide, buy, 2 step					267			267	294
4000	7 step					760			760	835
4050	10 step					1,075			1,075	1,175
4100	Rolling towers, buy, 5' wide, 7' long, 10' high					1,200			1,200	1,325
4200	For 5' high added sections, to buy, add					204			204	224
4300	Complete incl. wheels, railings, outriggers,									

01 54 Construction Aids

01 54 23 – Temporary Scaffolding and Platforms

01 54 23.75 Scaffolding Specialties	Crew	Daily Output	Labor-Hours	Unit	Material	2015 Bare Costs Labor	Equipment	Total	Total Incl O&P
4350	21' high, to buy				Ea.	2,025			2,025	2,225
4400	Rent/month = 5% of purchase cost				"		200		200	220

01 54 36 – Equipment Mobilization

01 54 36.50 Mobilization

		Crew	Daily Output	Labor-Hours	Unit	Material	Labor	Equipment	Total	Total Incl O&P
0010	**MOBILIZATION** (Use line item again for demobilization) R015436-50									
0015	Up to 25 mi. haul dist. (50 mi. RT for mob/demob crew)									
1200	Small equipment, placed in rear of, or towed by pickup truck	A-3A	4	2	Ea.		97	39	136	191
1300	Equipment hauled on 3-ton capacity towed trailer	A-3Q	2.67	3			146	67	213	295
1400	20-ton capacity	B-34U	2	8			355	237	592	795
1500	40-ton capacity	B-34N	2	8			365	375	740	965
1600	50-ton capacity	B-34V	1	24			1,125	1,050	2,175	2,850
1700	Crane, truck-mounted, up to 75 ton (driver only)	1 Eqhv	4	2			103		103	157
1800	Over 75 ton (with chase vehicle)	A-3E	2.50	6.400			294	62.50	356.50	515
2400	Crane, large lattice boom, requiring assembly	B-34W	.50	144			6,275	7,500	13,775	17,800
2500	For each additional 5 miles haul distance, add						10%	10%		
3000	For large pieces of equipment, allow for assembly/knockdown									
3100	For mob/demob of micro-tunneling equip, see Section 33 05 23.19									

01 55 Vehicular Access and Parking

01 55 23 – Temporary Roads

01 55 23.50 Roads and Sidewalks

		Crew	Daily Output	Labor-Hours	Unit	Material	Labor	Equipment	Total	Total Incl O&P
0010	**ROADS AND SIDEWALKS** Temporary									
0050	Roads, gravel fill, no surfacing, 4" gravel depth	B-14	715	.067	S.Y.	4.04	2.67	.51	7.22	9.10
0100	8" gravel depth	"	615	.078	"	8.10	3.10	.59	11.79	14.30
1000	Ramp, 3/4" plywood on 2" x 6" joists, 16" O.C.	2 Carp	300	.053	S.F.	1.58	2.50		4.08	5.60
1100	On 2" x 10" joists, 16" O.C.	"	275	.058	"	2.24	2.73		4.97	6.65

01 56 Temporary Barriers and Enclosures

01 56 13 – Temporary Air Barriers

01 56 13.60 Tarpaulins

		Crew	Daily Output	Labor-Hours	Unit	Material	Labor	Equipment	Total	Total Incl O&P
0010	**TARPAULINS**									
0020	Cotton duck, 10 oz. to 13.13 oz. per S.Y., 6'x8'				S.F.	.83			.83	.91
0050	30'x30'					.59			.59	.65
0100	Polyvinyl coated nylon, 14 oz. to 18 oz., minimum					1.36			1.36	1.50
0150	Maximum*					1.36			1.36	1.50
0200	Reinforced polyethylene 3 mils thick, white					.04			.04	.04
0300	4 mils thick, white, clear or black					.09			.09	.10
0400	5.5 mils thick, clear					.18			.18	.20
0500	White, fire retardant					.44			.44	.48
0600	12 mils, oil resistant, fire retardant					.39			.39	.43
0700	8.5 mils, black					.57			.57	.63
0710	Woven polyethylene, 6 mils thick					.18			.18	.20
0730	Polyester reinforced w/integral fastening system 11 mils thick					.21			.21	.23
0740	Mylar polyester, non-reinforced, 7 mils thick					1.17			1.17	1.29

01 56 13.90 Winter Protection

		Crew	Daily Output	Labor-Hours	Unit	Material	Labor	Equipment	Total	Total Incl O&P
0010	**WINTER PROTECTION**									
0100	Framing to close openings	2 Clab	500	.032	S.F.	.46	1.20		1.66	2.35
0200	Tarpaulins hung over scaffolding, 8 uses, not incl. scaffolding		1500	.011		.24	.40		.64	.88

01 56 Temporary Barriers and Enclosures

01 56 13 – Temporary Air Barriers

01 56 13.90 Winter Protection	Crew	Daily Output	Labor-Hours	Unit	Material	2015 Bare Costs Labor	Equipment	Total	Total Incl O&P
0250	Tarpaulin polyester reinf. w/integral fastening system 11 mils thick	2 Clab	1600	.010	S.F.	.21	.38		.59	.81
0300	Prefab fiberglass panels, steel frame, 8 uses	↓	1200	.013	↓	2.40	.50		2.90	3.41

01 56 16 – Temporary Dust Barriers

01 56 16.10 Dust Barriers, Temporary

		Crew	Daily Output	Labor-Hours	Unit	Material	2015 Bare Costs Labor	Equipment	Total	Total Incl O&P
0010	**DUST BARRIERS, TEMPORARY**									
0020	Spring loaded telescoping pole & head, to 12', erect and dismantle	1 Clab	240	.033	Ea.		1.25		1.25	1.93
0025	Cost per day (based upon 250 days)				Day	.26			.26	.29
0030	To 21', erect and dismantle	1 Clab	240	.033	Ea.		1.25		1.25	1.93
0035	Cost per day (based upon 250 days)				Day	.44			.44	.48
0040	Accessories, caution tape reel, erect and dismantle	1 Clab	480	.017	Ea.		.63		.63	.96
0045	Cost per day (based upon 250 days)				Day	.36			.36	.40
0060	Foam rail and connector, erect and dismantle	1 Clab	240	.033	Ea.		1.25		1.25	1.93
0065	Cost per day (based upon 250 days)				Day	.10			.10	.11
0070	Caution tape	1 Clab	384	.021	C.L.F.	2.70	.78		3.48	4.18
0080	Zipper, standard duty		60	.133	Ea.	8	5		13	16.50
0090	Heavy duty		48	.167	"	9.50	6.25		15.75	20
0100	Polyethylene sheet, 4 mil		37	.216	Sq.	2.90	8.15		11.05	15.70
0110	6 mil	↓	37	.216	"	4.02	8.15		12.17	16.90
1000	Dust partition, 6 mil polyethylene, 1" x 3" frame	2 Carp	2000	.008	S.F.	.30	.38		.68	.90
1080	2" x 4" frame	"	2000	.008	"	.35	.38		.73	.97

01 56 23 – Temporary Barricades

01 56 23.10 Barricades

		Crew	Daily Output	Labor-Hours	Unit	Material	2015 Bare Costs Labor	Equipment	Total	Total Incl O&P
0010	**BARRICADES**									
0020	5' high, 3 rail @ 2" x 8", fixed	2 Carp	20	.800	L.F.	6	37.50		43.50	64.50
0150	Movable		30	.533		5	25		30	44
1000	Guardrail, wooden, 3' high, 1" x 6", on 2" x 4" posts		200	.080		1.27	3.76		5.03	7.20
1100	2" x 6", on 4" x 4" posts	↓	165	.097		2.46	4.55		7.01	9.70
1200	Portable metal with base pads, buy					15.55			15.55	17.10
1250	Typical installation, assume 10 reuses	2 Carp	600	.027	↓	2.55	1.25		3.80	4.74
1300	Barricade tape, polyethylene, 7 mil, 3" wide x 500' long roll				Ea.	25			25	27.50
3000	Detour signs, set up and remove									
3010	Reflective aluminum, MUTCD, 24" x 24", post mounted	1 Clab	20	.400	Ea.	2.46	15.05		17.51	25.50
4000	Roof edge portable barrier stands and warning flags, 50 uses	1 Rohe	9100	.001	L.F.	.07	.03		.10	.12
4010	100 uses	"	9100	.001	"	.03	.03		.06	.09

01 56 26 – Temporary Fencing

01 56 26.50 Temporary Fencing

		Crew	Daily Output	Labor-Hours	Unit	Material	2015 Bare Costs Labor	Equipment	Total	Total Incl O&P
0010	**TEMPORARY FENCING**									
0020	Chain link, 11 ga., 4' high	2 Clab	400	.040	L.F.	2.95	1.50		4.45	5.55
0100	6' high		300	.053		2.95	2.01		4.96	6.35
0200	Rented chain link, 6' high, to 1000' (up to 12 mo.)		400	.040		4.29	1.50		5.79	7.05
0250	Over 1000' (up to 12 mo.)	↓	300	.053		4.29	2.01		6.30	7.80
0350	Plywood, painted, 2" x 4" frame, 4' high	A-4	135	.178		6.20	7.95		14.15	19
0400	4" x 4" frame, 8' high	"	110	.218		12	9.75		21.75	28
0500	Wire mesh on 4" x 4" posts, 4' high	2 Carp	100	.160		9.85	7.50		17.35	22.50
0550	8' high	"	80	.200	↓	14.90	9.40		24.30	31

01 56 29 – Temporary Protective Walkways

01 56 29.50 Protection

		Crew	Daily Output	Labor-Hours	Unit	Material	2015 Bare Costs Labor	Equipment	Total	Total Incl O&P
0010	**PROTECTION**									
0020	Stair tread, 2" x 12" planks, 1 use	1 Carp	75	.107	Tread	5.10	5		10.10	13.30
0100	Exterior plywood, 1/2" thick, 1 use	↓	65	.123	↓	1.97	5.80		7.77	11.05

01 56 Temporary Barriers and Enclosures

01 56 29 – Temporary Protective Walkways

01 56 29.50 Protection		Crew	Daily Output	Labor-Hours	Unit	Material	2015 Bare Costs Labor	2015 Bare Costs Equipment	Total	Total Incl O&P
0200	3/4" thick, 1 use	1 Carp	60	.133	Tread	2.78	6.25		9.03	12.70
2200	Sidewalks, 2" x 12" planks, 2 uses		350	.023	S.F.	.85	1.07		1.92	2.59
2300	Exterior plywood, 2 uses, 1/2" thick		750	.011		.33	.50		.83	1.13
2400	5/8" thick		650	.012		.39	.58		.97	1.32
2500	3/4" thick		600	.013		.46	.63		1.09	1.47

01 58 Project Identification

01 58 13 – Temporary Project Signage

01 58 13.50 Signs

01 58 13.50 Signs		Crew	Daily Output	Labor-Hours	Unit	Material	2015 Bare Costs Labor	2015 Bare Costs Equipment	Total	Total Incl O&P
0010	**SIGNS**									
0020	High intensity reflectorized, no posts, buy				Ea.	25			25	27.50

01 74 Cleaning and Waste Management

01 74 13 – Progress Cleaning

01 74 13.20 Cleaning Up

01 74 13.20 Cleaning Up		Crew	Daily Output	Labor-Hours	Unit	Material	2015 Bare Costs Labor	2015 Bare Costs Equipment	Total	Total Incl O&P
0010	**CLEANING UP**									
0020	After job completion, allow, minimum				Job				.30%	.30%
0040	Maximum				"				1%	1%
0050	Cleanup of floor area, continuous, per day, during const.	A-5	24	.750	M.S.F.	2.18	28.50	2.56	33.24	48.50
0100	Final by GC at end of job	"	11.50	1.565	"	2.31	59	5.35	66.66	99.50

01 91 Commissioning

01 91 13 – General Commissioning Requirements

01 91 13.50 Building Commissioning

01 91 13.50 Building Commissioning		Crew	Daily Output	Labor-Hours	Unit	Material	2015 Bare Costs Labor	2015 Bare Costs Equipment	Total	Total Incl O&P
0010	**BUILDING COMMISSIONING**									
0100	Basic building commissioning, minimum				%				.25%	.25%
0150	Maximum								.50%	.50%
0200	Enhanced building commissioning, minimum								.50%	.50%
0250	Maximum								1%	1%

01 93 Facility Maintenance

01 93 13 – Facility Maintenance Procedures

01 93 13.15 Mechanical Facilities Maintenance

01 93 13.15 Mechanical Facilities Maintenance		Crew	Daily Output	Labor-Hours	Unit	Material	2015 Bare Costs Labor	2015 Bare Costs Equipment	Total	Total Incl O&P
0010	**MECHANICAL FACILITIES MAINTENANCE**									
0100	Air conditioning system maintenance									
0130	Belt, replace	1 Stpi	15	.533	Ea.		32		32	48
0170	Fan, clean		16	.500			30		30	45
0180	Filter, remove, clean, replace		12	.667			40		40	60
0190	Flexible coupling alignment, inspect		40	.200			11.95		11.95	18.05
0200	Gas leak locate and repair		4	2			120		120	180
0250	Pump packing gland, remove and replace		11	.727			43.50		43.50	65.50
0270	Tighten		32	.250			14.95		14.95	22.50
0290	Pump disassemble and assemble		4	2			120		120	180
0300	Air pressure regulator, disassemble, clean, assemble	1 Skwk	4	2			97.50		97.50	150
0310	Repair or replace part	"	6	1.333			65		65	100
0320	Purging system	1 Stpi	16	.500			30		30	45

17

01 93 13 – Facility Maintenance Procedures

01 93 13.15 Mechanical Facilities Maintenance		Crew	Daily Output	Labor-Hours	Unit	Material	2015 Bare Costs Labor	Equipment	Total	Total Incl O&P
0400	Compressor, air, remove or install fan wheel	1 Skwk	20	.400	Ea.		19.45		19.45	30
0410	Disassemble or assemble 2 cylinder, 2 stage		4	2			97.50		97.50	150
0420	4 cylinder, 4 stage		1	8			390		390	600
0430	Repair or replace part		2	4			195		195	300
0700	Demolition, for mech. demolition see Section 23 05 05.10 or 22 05 05.10									
0800	Ductwork, clean									
0810	Rectangular									
0820	6" [G]	1 Shee	187.50	.043	L.F.		2.39		2.39	3.65
0830	8" [G]		140.63	.057			3.18		3.18	4.86
0840	10" [G]		112.50	.071			3.98		3.98	6.10
0850	12" [G]		93.75	.085			4.77		4.77	7.30
0860	14" [G]		80.36	.100			5.55		5.55	8.50
0870	16" [G]		70.31	.114			6.35		6.35	9.75
0900	Round									
0910	4" [G]	1 Shee	358.10	.022	L.F.		1.25		1.25	1.91
0920	6" [G]		238.73	.034			1.87		1.87	2.87
0930	8" [G]		179.05	.045			2.50		2.50	3.82
0940	10" [G]		143.24	.056			3.12		3.12	4.78
0950	12" [G]		119.37	.067			3.75		3.75	5.75
0960	16" [G]		89.52	.089			5		5	7.65
1000	Expansion joint, not screwed, install or remove	1 Stpi	3	2.667	Ea.		159		159	241
1010	Repack	"	6	1.333	"		79.50		79.50	120
1200	Fire protection equipment									
1220	Fire hydrant, replace	Q-1	3	5.333	Ea.		282		282	425
1230	Service, lubricate, inspect, flush, clean	1 Plum	7	1.143			67		67	101
1240	Test	"	11	.727			42.50		42.50	64.50
1310	Inspect valves, pressure, nozzle,	1 Spri	4	2	System		112		112	170
1500	Instrumentation preventive maintenance, rule of thumb									
1510	Annual cost per instrument for calibration & trouble shooting									
1520	Includes flow meters, analytical transmitters, flow control									
1530	chemical feed systems, recorders and distrib. control sys.				Ea.				325	325
1800	Plumbing fixtures, for installation see Section 22 41 00									
1801	Plumbing fixtures									
1840	Open drain with snake	1 Plum	13	.615	Ea.		36		36	54.50
1850	Open drain with toilet auger		16	.500			29.50		29.50	44.50
1860	Grease trap, clean		2	4			235		235	355
1870	Plaster trap, clean		6	1.333			78.50		78.50	118
1881	Clean commode	1 Clab	20	.400		.37	15.05		15.42	23.50
1882	Clean commode seat		48	.167		.18	6.25		6.43	9.85
1883	Clean single sink		28	.286		.24	10.75		10.99	16.80
1884	Clean double sink		20	.400		.37	15.05		15.42	23.50
1885	Clean lavatory		28	.286		.24	10.75		10.99	16.80
1886	Clean bathtub		12	.667		.73	25		25.73	39.50
1887	Clean fiberglass shower		12	.667		.73	25		25.73	39.50
1888	Clean fiberglass tub/shower		8	1		.92	37.50		38.42	59
1889	Clean faucet set		32	.250		.15	9.40		9.55	14.60
1891	Clean shower head		80	.100		.02	3.76		3.78	5.80
1893	Clean water heater		16	.500		1.28	18.80		20.08	30.50
1900	Relief valve, test and adjust	1 Stpi	20	.400			24		24	36
1910	Clean pump, heater, or motor for whirlpool	1 Clab	16	.500		.18	18.80		18.98	29
1920	Clean thermal cover for whirlpool	"	16	.500		1.83	18.80		20.63	31
2000	Repair or replace-steam trap	1 Stpi	8	1			60		60	90
2020	Y-type or bell strainer		6	1.333			79.50		79.50	120

01 93 13.15 Mechanical Facilities Maintenance		Crew	Daily Output	Labor-Hours	Unit	Material	2015 Bare Costs Labor	Equipment	Total	Total Incl O&P
2040	Water trap or vacuum breaker, screwed joints	1 Stpi	13	.615	Ea.		37		37	55.50
2100	Steam specialties, clean									
2120	Air separator with automatic trap, 1" fittings	1 Stpi	12	.667	Ea.		40		40	60
2130	Bucket trap, 2" pipe		7	1.143			68.50		68.50	103
2150	Drip leg, 2" fitting		45	.178			10.60		10.60	16.05
2200	Thermodynamic trap, 1" fittings		50	.160			9.55		9.55	14.45
2210	Thermostatic		65	.123			7.35		7.35	11.10
2240	Screen and seat in y-type strainer, plug type.		25	.320			19.10		19.10	29
2242	Screen and seat in y-type strainer, flange type.		12	.667			40		40	60
2500	Valve, replace broken handwheel		24	.333			19.90		19.90	30
3000	Valve, overhaul, regulator, relief, flushometer, mixing									
3040	Cold water, gas	1 Stpi	5	1.600	Ea.		95.50		95.50	144
3050	Hot water, steam		3	2.667			159		159	241
3080	Globe, gate, check up to 4" cold water, gas		10	.800			48		48	72
3090	Hot water, steam		5	1.600			95.50		95.50	144
3100	Over 4" ID hot or cold line		1.40	5.714			340		340	515
3120	Remove and replace, gate, globe or check up to 4"	Q-5	6	2.667			143		143	217
3130	Over 4"	"	2	8			430		430	650
3150	Repack up to 4"	1 Stpi	13	.615			37		37	55.50
3160	Over 4"	"	4	2			120		120	180

For customer support on your Mechanical Cost Data, call 877.795.7394.

19

Division Notes

	CREW	DAILY OUTPUT	LABOR-HOURS	UNIT	BARE COSTS				TOTAL INCL O&P
					MAT.	LABOR	EQUIP.	TOTAL	

Estimating Tips

02 30 00 Subsurface Investigation

In preparing estimates on structures involving earthwork or foundations, all information concerning soil characteristics should be obtained. Look particularly for hazardous waste, evidence of prior dumping of debris, and previous stream beds.

02 40 00 Demolition and Structure Moving

The costs shown for selective demolition do not include rubbish handling or disposal. These items should be estimated separately using RSMeans data or other sources.

- Historic preservation often requires that the contractor remove materials from the existing structure, rehab them, and replace them. The estimator must be aware of any related measures and precautions that must be taken when doing selective demolition and cutting and patching. Requirements may include special handling and storage, as well as security.

- In addition to Subdivision 02 41 00, you can find selective demolition items in each division. Example: Roofing demolition is in Division 7.

02 40 00 Building Deconstruction

This section provides costs for the careful dismantling and recycling of most of low-rise building materials.

02 50 00 Containment of Hazardous Waste

This section addresses on-site hazardous waste disposal costs.

02 80 00 Hazardous Material Disposal/ Remediation

This subdivision includes information on hazardous waste handling, asbestos remediation, lead remediation, and mold remediation. See reference R028213-20 and R028319-60 for further guidance in using these unit price lines.

02 90 00 Monitoring Chemical Sampling, Testing Analysis

This section provides costs for on-site sampling and testing hazardous waste.

Reference Numbers

Reference numbers are shown in shaded boxes at the beginning of some major classifications. These numbers refer to related items in the Reference Section. The reference information may be an estimating procedure, an alternate pricing method, or technical information.

Note: Not all subdivisions listed here necessarily appear in this publication. ■

02 41 13 – Selective Site Demolition

02 41 13.17 Demolish, Remove Pavement and Curb		Crew	Daily Output	Labor-Hours	Unit	Material	2015 Bare Costs Labor	Equipment	Total	Total Incl O&P
0010	**DEMOLISH, REMOVE PAVEMENT AND CURB** R024119-10									
5010	Pavement removal, bituminous roads, up to 3" thick	B-38	690	.058	S.Y.		2.48	1.85	4.33	5.80
5050	4" to 6" thick		420	.095			4.08	3.04	7.12	9.60
5100	Bituminous driveways		640	.063			2.68	1.99	4.67	6.30
5200	Concrete to 6" thick, hydraulic hammer, mesh reinforced		255	.157			6.70	5	11.70	15.75
5300	Rod reinforced		200	.200			8.55	6.40	14.95	20
5400	Concrete, 7" to 24" thick, plain		33	1.212	C.Y.		52	38.50	90.50	122
5500	Reinforced		24	1.667	"		71.50	53	124.50	168
5600	With hand held air equipment, bituminous, to 6" thick	B-39	1900	.025	S.F.		1	.12	1.12	1.67
5700	Concrete to 6" thick, no reinforcing		1600	.030			1.19	.15	1.34	1.99
5800	Mesh reinforced		1400	.034			1.36	.17	1.53	2.27
5900	Rod reinforced		765	.063			2.50	.30	2.80	4.16

02 41 13.23 Utility Line Removal

		Crew	Daily Output	Labor-Hours	Unit	Material	Labor	Equipment	Total	Total Incl O&P
0010	**UTILITY LINE REMOVAL**									
0015	No hauling, abandon catch basin or manhole	B-6	7	3.429	Ea.		142	52	194	274
0020	Remove existing catch basin or manhole, masonry		4	6			248	91	339	480
0030	Catch basin or manhole frames and covers, stored		13	1.846			76	28	104	148
0040	Remove and reset		7	3.429			142	52	194	274
0900	Hydrants, fire, remove only	B-21A	5	8			375	94.50	469.50	675
0950	Remove and reset	"	2	20			940	237	1,177	1,675

02 41 13.30 Minor Site Demolition

		Crew	Daily Output	Labor-Hours	Unit	Material	Labor	Equipment	Total	Total Incl O&P
0010	**MINOR SITE DEMOLITION** R024119-10									
4000	Sidewalk removal, bituminous, 2" thick	B-6	350	.069	S.Y.		2.83	1.04	3.87	5.45
4010	2-1/2" thick		325	.074			3.05	1.12	4.17	5.90
4050	Brick, set in mortar		185	.130			5.35	1.97	7.32	10.35
4100	Concrete, plain, 4"		160	.150			6.20	2.28	8.48	11.95
4110	Plain, 5"		140	.171			7.05	2.60	9.65	13.70
4120	Plain, 6"		120	.200			8.25	3.04	11.29	16
4200	Mesh reinforced, concrete, 4"		150	.160			6.60	2.43	9.03	12.75
4210	5" thick		131	.183			7.55	2.78	10.33	14.60
4220	6" thick		112	.214			8.85	3.25	12.10	17.15

02 41 19 – Selective Demolition

02 41 19.13 Selective Building Demolition

		Crew	Daily Output	Labor-Hours	Unit	Material	Labor	Equipment	Total	Total Incl O&P
0010	**SELECTIVE BUILDING DEMOLITION**									
0020	Costs related to selective demolition of specific building components									
0025	are included under Common Work Results (XX 05)									
0030	in the component's appropriate division.									

02 41 19.19 Selective Demolition

		Crew	Daily Output	Labor-Hours	Unit	Material	Labor	Equipment	Total	Total Incl O&P
0010	**SELECTIVE DEMOLITION,** Rubbish Handling R024119-10									
0020	The following are to be added to the demolition prices									
0600	Dumpster, weekly rental, 1 dump/week, 6 C.Y. capacity (2 Tons)				Week	415			415	455
0700	10 C.Y. capacity (3 Tons)					480			480	530
0725	20 C.Y. capacity (5 Tons) R024119-20					565			565	625
0800	30 C.Y. capacity (7 Tons)					730			730	800
0840	40 C.Y. capacity (10 Tons)					775			775	850
2000	Load, haul, dump and return, 0 – 50' haul, hand carried	2 Clab	24	.667	C.Y.		25		25	38.50
2005	Wheeled		37	.432			16.25		16.25	25
2040	0 – 100' haul, hand carried		16.50	.970			36.50		36.50	56
2045	Wheeled		25	.640			24		24	37
2080	Haul and return, add per each extra 100' haul, hand carried		35.50	.451			16.95		16.95	26
2085	Wheeled		54	.296			11.15		11.15	17.15

02 41 Demolition

02 41 19 – Selective Demolition

02 41 19.19 Selective Demolition

		Crew	Daily Output	Labor-Hours	Unit	Material	2015 Bare Costs Labor	Equipment	Total	Total Incl O&P
2120	For travel in elevators, up to 10 floors, add	2 Clab	140	.114	C.Y.		4.30		4.30	6.60
2130	0 – 50' haul, incl. up to 5 riser stair, hand carried		23	.696			26		26	40
2135	Wheeled		35	.457			17.20		17.20	26.50
2140	6 – 10 riser stairs, hand carried		22	.727			27.50		27.50	42
2145	Wheeled		34	.471			17.70		17.70	27
2150	11 – 20 riser stairs, hand carried		20	.800			30		30	46.50
2155	Wheeled		31	.516			19.40		19.40	30
2160	21 – 40 riser stairs, hand carried		16	1			37.50		37.50	58
2165	Wheeled		24	.667			25		25	38.50
2170	0 – 100' haul, incl. 5 riser stair, hand carried		15	1.067			40		40	61.50
2175	Wheeled		23	.696			26		26	40
2180	6 – 10 riser stair, hand carried		14	1.143			43		43	66
2185	Wheeled		21	.762			28.50		28.50	44
2190	11 – 20 riser stair, hand carried		12	1.333			50		50	77
2195	Wheeled		18	.889			33.50		33.50	51.50
2200	21 – 40 riser stair, hand carried		8	2			75		75	116
2205	Wheeled		12	1.333			50		50	77
2210	Haul and return, add per each extra 100' haul, hand carried		35.50	.451			16.95		16.95	26
2215	Wheeled		54	.296			11.15		11.15	17.15
2220	For each additional flight of stairs, up to 5 risers, add		550	.029	Flight		1.09		1.09	1.68
2225	6 – 10 risers, add		275	.058			2.19		2.19	3.37
2230	11 – 20 risers, add		138	.116			4.36		4.36	6.70
2235	21 – 40 risers, add		69	.232			8.70		8.70	13.40
3000	Loading & trucking, including 2 mile haul, chute loaded	B-16	45	.711	C.Y.		27.50	15.35	42.85	59
3040	Hand loading truck, 50' haul	"	48	.667			26	14.40	40.40	55.50
3080	Machine loading truck	B-17	120	.267			10.90	6.45	17.35	24
5000	Haul, per mile, up to 8 C.Y. truck	B-34B	1165	.007			.28	.59	.87	1.07
5100	Over 8 C.Y. truck	"	1550	.005			.21	.45	.66	.80

02 41 19.20 Selective Demolition, Dump Charges

		Crew	Daily Output	Labor-Hours	Unit	Material	2015 Bare Costs Labor	Equipment	Total	Total Incl O&P
0010	**SELECTIVE DEMOLITION, DUMP CHARGES** R024119-10									
0020	Dump charges, typical urban city, tipping fees only									
0100	Building construction materials				Ton	74			74	81
0200	Trees, brush, lumber					63			63	69.50
0300	Rubbish only					63			63	69.50
0500	Reclamation station, usual charge					74			74	81

02 41 19.27 Selective Demolition, Torch Cutting

		Crew	Daily Output	Labor-Hours	Unit	Material	2015 Bare Costs Labor	Equipment	Total	Total Incl O&P
0010	**SELECTIVE DEMOLITION, TORCH CUTTING** R024119-10									
0020	Steel, 1" thick plate	E-25	333	.024	L.F.	.84	1.31	.03	2.18	3.24
0040	1" diameter bar	"	600	.013	Ea.	.14	.73	.02	.89	1.44
1000	Oxygen lance cutting, reinforced concrete walls									
1040	12" to 16" thick walls	1 Clab	10	.800	L.F.		30		30	46.50
1080	24" thick walls	"	6	1.333	"		50		50	77

For customer support on your Mechanical Cost Data, call 877.795.7394.

23

02 65 10 – Underground Tank and Contaminated Soil Removal

02 65 10.30 Removal of Underground Storage Tanks		Crew	Daily Output	Labor-Hours	Unit	Material	2015 Bare Costs Labor	2015 Bare Costs Equipment	Total	Total Incl O&P
0010	**REMOVAL OF UNDERGROUND STORAGE TANKS** R026510-20									
0011	Petroleum storage tanks, non-leaking									
0100	Excavate & load onto trailer									
0110	3000 gal. to 5000 gal. tank [G]	B-14	4	12	Ea.		475	91	566	830
0120	6000 gal. to 8000 gal. tank [G]	B-3A	3	13.333			535	345	880	1,200
0130	9000 gal. to 12000 gal. tank [G]	"	2	20	↓		805	515	1,320	1,800
0190	Known leaking tank, add				%				100%	100%
0200	Remove sludge, water and remaining product from tank bottom									
0201	of tank with vacuum truck									
0300	3000 gal. to 5000 gal. tank [G]	A-13	5	1.600	Ea.		78	153	231	286
0310	6000 gal. to 8000 gal. tank [G]		4	2			97	191	288	360
0320	9000 gal. to 12000 gal. tank [G]	↓	3	2.667			130	254	384	475
0390	Dispose of sludge off-site, average				Gal.				6.25	6.80
0400	Insert inert solid CO_2 "dry ice" into tank									
0401	For cleaning/transporting tanks (1.5 lb./100 gal. cap) [G]	1 Clab	500	.016	Lb.	1.17	.60		1.77	2.22
1020	Haul tank to certified salvage dump, 100 miles round trip									
1023	3000 gal. to 5000 gal. tank				Ea.				760	830
1026	6000 gal. to 8000 gal. tank								880	960
1029	9,000 gal. to 12,000 gal. tank				↓				1,050	1,150
1100	Disposal of contaminated soil to landfill									
1110	Minimum				C.Y.				145	160
1111	Maximum				"				400	440
1120	Disposal of contaminated soil to									
1121	bituminous concrete batch plant									
1130	Minimum				C.Y.				80	88
1131	Maximum				"				115	125
2010	Decontamination of soil on site incl poly tarp on top/bottom									
2011	Soil containment berm, and chemical treatment									
2020	Minimum [G]	B-11C	100	.160	C.Y.	7.80	7.05	3.64	18.49	23.50
2021	Maximum [G]	"	100	.160		10.10	7.05	3.64	20.79	26
2050	Disposal of decontaminated soil, minimum				↓				135	150
2055	Maximum								400	440

02 81 20 – Hazardous Waste Handling

02 81 20.10 Hazardous Waste Cleanup/Pickup/Disposal

		Crew	Daily Output	Labor-Hours	Unit	Material	Labor	Equipment	Total	Total Incl O&P
0010	**HAZARDOUS WASTE CLEANUP/PICKUP/DISPOSAL**									
0100	For contractor rental equipment, i.e., Dozer,									
0110	Front end loader, Dump truck, etc., see 01 54 33 Reference Section									
1000	Solid pickup									
1100	55 gal. drums				Ea.				240	265
1120	Bulk material, minimum				Ton				190	210
1130	Maximum				"				595	655
1200	Transportation to disposal site									
1220	Truckload = 80 drums or 25 C.Y. or 18 tons									
1260	Minimum				Mile				3.95	4.45
1270	Maximum				"				7.25	7.35
3000	Liquid pickup, vacuum truck, stainless steel tank									
3100	Minimum charge, 4 hours									
3110	1 compartment, 2200 gallon				Hr.				140	155
3120	2 compartment, 5000 gallon				"				200	225

02 81 Transportation and Disposal of Hazardous Materials

02 81 20 – Hazardous Waste Handling

02 81 20.10 Hazardous Waste Cleanup/Pickup/Disposal	Crew	Daily Output	Labor-Hours	Unit	Material	2015 Bare Costs Labor	Equipment	Total	Total Incl O&P
3400	Transportation in 6900 gallon bulk truck				Mile				7.95	8.75
3410	In teflon lined truck				"				10.20	11.25
5000	Heavy sludge or dry vacuumable material				Hr.				140	160
6000	Dumpsite disposal charge, minimum				Ton				140	155
6020	Maximum				"				415	455

02 82 Asbestos Remediation

02 82 13 – Asbestos Abatement

02 82 13.39 Asbestos Remediation Plans and Methods

		Crew	Daily Output	Labor-Hours	Unit	Material	Labor	Equipment	Total	Total Incl O&P
0010	**ASBESTOS REMEDIATION PLANS AND METHODS**									
0100	Building Survey-Commercial Building				Ea.				2,200	2,400
0200	Asbestos Abatement Remediation Plan				"				1,350	1,475

02 82 13.41 Asbestos Abatement Equipment

		Crew	Daily Output	Labor-Hours	Unit	Material	Labor	Equipment	Total	Total Incl O&P
0010	**ASBESTOS ABATEMENT EQUIPMENT** R028213-20									
0011	Equipment and supplies, buy									
0200	Air filtration device, 2000 CFM				Ea.	960			960	1,050
0250	Large volume air sampling pump, minimum					345			345	380
0260	Maximum					335			335	365
0300	Airless sprayer unit, 2 gun					4,500			4,500	4,950
0350	Light stand, 500 watt					49			49	54
0400	Personal respirators									
0410	Negative pressure, 1/2 face, dual operation, min.				Ea.	26.50			26.50	29
0420	Maximum					29.50			29.50	32.50
0450	P.A.P.R., full face, minimum					124			124	137
0460	Maximum					165			165	182
0470	Supplied air, full face, incl. air line, minimum					168			168	185
0480	Maximum					405			405	445
0500	Personnel sampling pump					226			226	249
1500	Power panel, 20 unit, incl. GFI					600			600	660
1600	Shower unit, including pump and filters					1,275			1,275	1,425
1700	Supplied air system (type C)					3,425			3,425	3,750
1750	Vacuum cleaner, HEPA, 16 gal., stainless steel, wet/dry					435			435	475
1760	55 gallon					560			560	615
1800	Vacuum loader, 9 – 18 ton/hr.					94,500			94,500	104,000
1900	Water atomizer unit, including 55 gal. drum					281			281	310
2000	Worker protection, whole body, foot, head cover & gloves, plastic					8.65			8.65	9.55
2500	Respirator, single use					25			25	27.50
2550	Cartridge for respirator					5.75			5.75	6.30
2570	Glove bag, 7 mil, 50" x 64"					9.20			9.20	10.10
2580	10 mil, 44" x 60"					5.75			5.75	6.35
2590	6 mil, 44" x 60"					5.75			5.75	6.35
3000	HEPA vacuum for work area, minimum					305			305	335
3050	Maximum					815			815	900
6000	Disposable polyethylene bags, 6 mil, 3 C.F.					.82			.82	.90
6300	Disposable fiber drums, 3 C.F.					18.20			18.20	20
6400	Pressure sensitive caution labels, 3" x 5"					3.47			3.47	3.82
6450	11" x 17"					7.35			7.35	8.05
6500	Negative air machine, 1800 CFM					845			845	930

For customer support on your Mechanical Cost Data, call 877.795.7394.

25

02 82 13 – Asbestos Abatement

02 82 13.42 Preparation of Asbestos Containment Area

		Crew	Daily Output	Labor-Hours	Unit	Material	2015 Bare Costs Labor	Equipment	Total	Total Incl O&P
0010	**PREPARATION OF ASBESTOS CONTAINMENT AREA** R028213-20									
0100	Pre-cleaning, HEPA vacuum and wet wipe, flat surfaces	A-9	12000	.005	S.F.	.02	.28		.30	.45
0200	Protect carpeted area, 2 layers 6 mil poly on 3/4" plywood	"	1000	.064		2.04	3.35		5.39	7.45
0300	Separation barrier, 2" x 4" @ 16", 1/2" plywood ea. side, 8' high	2 Carp	400	.040		3.32	1.88		5.20	6.55
0310	12' high		320	.050		3.32	2.35		5.67	7.25
0320	16' high		200	.080		2.30	3.76		6.06	8.35
0400	Personnel decontam. chamber, 2" x 4" @ 16", 3/4" ply ea. side		280	.057		4.34	2.68		7.02	8.90
0450	Waste decontam. chamber, 2" x 4" studs @ 16", 3/4" ply ea. side		360	.044		4.34	2.09		6.43	8
0500	Cover surfaces with polyethylene sheeting									
0501	Including glue and tape									
0550	Floors, each layer, 6 mil	A-9	8000	.008	S.F.	.04	.42		.46	.70
0551	4 mil		9000	.007		.03	.37		.40	.61
0560	Walls, each layer, 6 mil		6000	.011		.04	.56		.60	.92
0561	4 mil		7000	.009		.03	.48		.51	.78
0570	For heights above 12', add						20%			
0575	For heights above 20', add						30%			
0580	For fire retardant poly, add					100%				
0590	For large open areas, deduct					10%	20%			
0600	Seal floor penetrations with foam firestop to 36 sq. in.	2 Carp	200	.080	Ea.	7.75	3.76		11.51	14.30
0610	36 sq. in. to 72 sq. in.		125	.128		15.50	6		21.50	26.50
0615	72 sq. in. to 144 sq. in.		80	.200		31	9.40		40.40	48.50
0620	Wall penetrations, to 36 square inches		180	.089		7.75	4.17		11.92	14.90
0630	36 sq. in. to 72 sq. in.		100	.160		15.50	7.50		23	28.50
0640	72 sq. in. to 144 sq. in.		60	.267		31	12.50		43.50	53.50
0800	Caulk seams with latex	1 Carp	230	.035	L.F.	.17	1.63		1.80	2.70
0900	Set up neg. air machine, 1-2k CFM/25 M.C.F. volume	1 Asbe	4.30	1.860	Ea.		97.50		97.50	152
0950	Set up and remove portable shower unit	2 Asbe	4	4	"		209		209	325

02 82 13.43 Bulk Asbestos Removal

		Crew	Daily Output	Labor-Hours	Unit	Material	2015 Bare Costs Labor	Equipment	Total	Total Incl O&P
0010	**BULK ASBESTOS REMOVAL**									
0020	Includes disposable tools and 2 suits and 1 respirator filter/day/worker									
0200	Boiler insulation	A-9	480	.133	S.F.	.45	7		7.45	11.40
0210	With metal lath, add				%				50%	50%
0300	Boiler breeching or flue insulation	A-9	520	.123	S.F.	.36	6.45		6.81	10.45
0310	For active boiler, add				%				100%	100%
0400	Duct or AHU insulation	A-10B	440	.073	S.F.	.21	3.82		4.03	6.20
0500	Duct vibration isolation joints, up to 24 sq. in. duct	A-9	56	1.143	Ea.	3.30	60		63.30	97
0520	25 sq. in. to 48 sq. in. duct		48	1.333		3.85	70		73.85	113
0530	49 sq. in. to 76 sq. in. duct		40	1.600		4.62	84		88.62	136
0600	Pipe insulation, air cell type, up to 4" diameter pipe		900	.071	L.F.	.21	3.73		3.94	6.05
0610	4" to 8" diameter pipe		800	.080		.23	4.19		4.42	6.80
0620	10" to 12" diameter pipe		700	.091		.26	4.79		5.05	7.75
0630	14" to 16" diameter pipe		550	.116		.34	6.10		6.44	9.85
0650	Over 16" diameter pipe		650	.098	S.F.	.28	5.15		5.43	8.35
0700	With glove bag up to 3" diameter pipe		200	.320	L.F.	9.15	16.75		25.90	36
1000	Pipe fitting insulation up to 4" diameter pipe		320	.200	Ea.	.58	10.50		11.08	16.95
1100	6" to 8" diameter pipe		304	.211		.61	11.05		11.66	17.85
1110	10" to 12" diameter pipe		192	.333		.96	17.45		18.41	28
1120	14" to 16" diameter pipe		128	.500		1.44	26		27.44	42.50
1130	Over 16" diameter pipe		176	.364	S.F.	1.05	19.05		20.10	30.50
1200	With glove bag, up to 8" diameter pipe		75	.853	L.F.	6.25	44.50		50.75	76.50
2000	Scrape foam fireproofing from flat surface		2400	.027	S.F.	.08	1.40		1.48	2.26
2100	Irregular surfaces		1200	.053		.15	2.80		2.95	4.52

02 82 Asbestos Remediation

02 82 13 – Asbestos Abatement

02 82 13.43 Bulk Asbestos Removal		Crew	Daily Output	Labor-Hours	Unit	Material	2015 Bare Costs Labor	Equipment	Total	Total Incl O&P
3000	Remove cementitious material from flat surface	A-9	1800	.036	S.F.	.10	1.86		1.96	3.01
3100	Irregular surface		1000	.064		.13	3.35		3.48	5.35
6000	Remove contaminated soil from crawl space by hand		400	.160	C.F.	.46	8.40		8.86	13.55
6100	With large production vacuum loader	A-12	700	.091	"	.26	4.79	1.09	6.14	8.95
7000	Radiator backing, not including radiator removal	A-9	1200	.053	S.F.	.15	2.80		2.95	4.52
9000	For type B (supplied air) respirator equipment, add				%		10%		10%	10%

02 82 13.44 Demolition In Asbestos Contaminated Area

		Crew	Daily Output	Labor-Hours	Unit	Material	2015 Bare Costs Labor	Equipment	Total	Total Incl O&P
0010	**DEMOLITION IN ASBESTOS CONTAMINATED AREA**									
0200	Ceiling, including suspension system, plaster and lath	A-9	2100	.030	S.F.	.09	1.60		1.69	2.59
0210	Finished plaster, leaving wire lath		585	.109		.32	5.75		6.07	9.30
0220	Suspended acoustical tile		3500	.018		.05	.96		1.01	1.55
0230	Concealed tile grid system		3000	.021		.06	1.12		1.18	1.81
0240	Metal pan grid system		1500	.043		.12	2.24		2.36	3.62
0250	Gypsum board		2500	.026		.07	1.34		1.41	2.17
0260	Lighting fixtures up to 2' x 4'		72	.889	Ea.	2.56	46.50		49.06	75.50
0400	Partitions, non load bearing									
0410	Plaster, lath, and studs	A-9	690	.093	S.F.	.88	4.86		5.74	8.50
0450	Gypsum board and studs	"	1390	.046	"	.13	2.41		2.54	3.91
9000	For type B (supplied air) respirator equipment, add				%		10%		10%	10%

02 82 13.45 OSHA Testing

		Crew	Daily Output	Labor-Hours	Unit	Material	2015 Bare Costs Labor	Equipment	Total	Total Incl O&P
0010	**OSHA TESTING**									
0100	Certified technician, minimum				Day				200	220
0110	Maximum								300	330
0120	Industrial hygienist, minimum								250	250
0130	Maximum								400	440
0200	Asbestos sampling and PCM analysis, NIOSH 7400, minimum	1 Asbe	8	1	Ea.	2.96	52.50		55.46	85
0210	Maximum		4	2		3.26	105		108.26	167
1000	Cleaned area samples		8	1		2.81	52.50		55.31	84.50
1100	PCM air sample analysis, NIOSH 7400, minimum		8	1		31.50	52.50		84	117
1110	Maximum		4	2		3.42	105		108.42	167
1200	TEM air sample analysis, NIOSH 7402, minimum								80	106
1210	Maximum								360	450

02 82 13.46 Decontamination of Asbestos Containment Area

		Crew	Daily Output	Labor-Hours	Unit	Material	2015 Bare Costs Labor	Equipment	Total	Total Incl O&P
0010	**DECONTAMINATION OF ASBESTOS CONTAINMENT AREA**									
0100	Spray exposed substrate with surfactant (bridging)									
0200	Flat surfaces	A-9	6000	.011	S.F.	.36	.56		.92	1.27
0250	Irregular surfaces		4000	.016	"	.31	.84		1.15	1.65
0300	Pipes, beams, and columns		2000	.032	L.F.	.56	1.68		2.24	3.23
1000	Spray encapsulate polyethylene sheeting		8000	.008	S.F.	.34	.42		.76	1.02
1100	Roll down polyethylene sheeting		8000	.008	"		.42		.42	.65
1500	Bag polyethylene sheeting		400	.160	Ea.	.81	8.40		9.21	13.95
2000	Fine clean exposed substrate, with nylon brush		2400	.027	S.F.		1.40		1.40	2.18
2500	Wet wipe substrate		4800	.013			.70		.70	1.09
2600	Vacuum surfaces, fine brush		6400	.010			.52		.52	.82
3000	Structural demolition									
3100	Wood stud walls	A-9	2800	.023	S.F.		1.20		1.20	1.87
3500	Window manifolds, not incl. window replacement		4200	.015			.80		.80	1.24
3600	Plywood carpet protection		2000	.032			1.68		1.68	2.61
4000	Remove custom decontamination facility	A-10A	8	3	Ea.	15.40	158		173.40	262
4100	Remove portable decontamination facility	3 Asbe	12	2	"	13.05	105		118.05	177
5000	HEPA vacuum, shampoo carpeting	A-9	4800	.013	S.F.	.07	.70		.77	1.17
9000	Final cleaning of protected surfaces	A-10A	8000	.003	"		.16		.16	.25

27

02 82 13.47 Asbestos Waste Pkg., Handling, and Disp.

		Crew	Daily Output	Labor-Hours	Unit	Material	2015 Bare Costs Labor	2015 Bare Costs Equipment	Total	Total Incl O&P
0010	**ASBESTOS WASTE PACKAGING, HANDLING, AND DISPOSAL**									
0100	Collect and bag bulk material, 3 C.F. bags, by hand	A-9	400	.160	Ea.	.82	8.40		9.22	13.95
0200	Large production vacuum loader	A-12	880	.073		.86	3.81	.87	5.54	7.85
1000	Double bag and decontaminate	A-9	960	.067		.82	3.49		4.31	6.35
2000	Containerize bagged material in drums, per 3 C.F. drum	"	800	.080		18.20	4.19		22.39	26.50
3000	Cart bags 50' to dumpster	2 Asbe	400	.040			2.09		2.09	3.26
5000	Disposal charges, not including haul, minimum				C.Y.				61	67
5020	Maximum				"				355	395
9000	For type B (supplied air) respirator equipment, add				%				10%	10%

02 82 13.48 Asbestos Encapsulation With Sealants

		Crew	Daily Output	Labor-Hours	Unit	Material	2015 Bare Costs Labor	2015 Bare Costs Equipment	Total	Total Incl O&P
0010	**ASBESTOS ENCAPSULATION WITH SEALANTS**									
0100	Ceilings and walls, minimum	A-9	21000	.003	S.F.	.31	.16		.47	.59
0110	Maximum		10600	.006	"	.45	.32		.77	.99
0300	Pipes to 12" diameter including minor repairs, minimum		800	.080	L.F.	.43	4.19		4.62	7
0310	Maximum		400	.160	"	1.12	8.40		9.52	14.30

02 85 Mold Remediation
02 85 16 – Mold Remediation Preparation and Containment

02 85 16.50 Preparation of Mold Containment Area

		Crew	Daily Output	Labor-Hours	Unit	Material	2015 Bare Costs Labor	2015 Bare Costs Equipment	Total	Total Incl O&P
0010	**PREPARATION OF MOLD CONTAINMENT AREA** R028213-20									
0100	Pre-cleaning, HEPA vacuum and wet wipe, flat surfaces	A-9	12000	.005	S.F.	.02	.28		.30	.45
0300	Separation barrier, 2" x 4" @ 16", 1/2" plywood ea. side, 8' high	2 Carp	400	.040		3.32	1.88		5.20	6.55
0310	12' high		320	.050		3.32	2.35		5.67	7.25
0320	16' high		200	.080		2.30	3.76		6.06	8.35
0400	Personnel decontam. chamber, 2" x 4" @ 16", 3/4" ply ea. side		280	.057		4.34	2.68		7.02	8.90
0450	Waste decontam. chamber, 2" x 4" studs @ 16", 3/4" ply each side		360	.044		4.34	2.09		6.43	8
0500	Cover surfaces with polyethylene sheeting									
0501	Including glue and tape									
0550	Floors, each layer, 6 mil	A-9	8000	.008	S.F.	.04	.42		.46	.70
0551	4 mil		9000	.007		.03	.37		.40	.61
0560	Walls, each layer, 6 mil		6000	.011		.04	.56		.60	.92
0561	4 mil		7000	.009		.03	.48		.51	.78
0570	For heights above 12', add						20%			
0575	For heights above 20', add						30%			
0580	For fire retardant poly, add					100%				
0590	For large open areas, deduct					10%	20%			
0600	Seal floor penetrations with foam firestop to 36 sq. in.	2 Carp	200	.080	Ea.	7.75	3.76		11.51	14.30
0610	36 sq. in. to 72 sq. in.		125	.128		15.50	6		21.50	26.50
0615	72 sq. in. to 144 sq. in.		80	.200		31	9.40		40.40	48.50
0620	Wall penetrations, to 36 square inches		180	.089		7.75	4.17		11.92	14.90
0630	36 sq. in. to 72 sq. in.		100	.160		15.50	7.50		23	28.50
0640	72 sq. in. to 144 sq. in.		60	.267		31	12.50		43.50	53.50
0800	Caulk seams with latex caulk	1 Carp	230	.035	L.F.	.17	1.63		1.80	2.70
0900	Set up neg. air machine, 1-2k CFM/25 M.C.F. volume	1 Asbe	4.30	1.860	Ea.		97.50		97.50	152

02 85 33 – Removal and Disposal of Materials with Mold

02 85 33.50 Demolition in Mold Contaminated Area

		Crew	Daily Output	Labor-Hours	Unit	Material	2015 Bare Costs Labor	2015 Bare Costs Equipment	Total	Total Incl O&P
0010	**DEMOLITION IN MOLD CONTAMINATED AREA**									
0200	Ceiling, including suspension system, plaster and lath	A-9	2100	.030	S.F.	.09	1.60		1.69	2.59
0210	Finished plaster, leaving wire lath		585	.109		.32	5.75		6.07	9.30
0220	Suspended acoustical tile		3500	.018		.05	.96		1.01	1.55

02 85 Mold Remediation

02 85 33 – Removal and Disposal of Materials with Mold

02 85 33.50 Demolition in Mold Contaminated Area	Crew	Daily Output	Labor-Hours	Unit	Material	2015 Bare Costs Labor	Equipment	Total	Total Incl O&P
0230	Concealed tile grid system	A-9	3000	.021	S.F.	.06	1.12		1.18	1.81
0240	Metal pan grid system		1500	.043		.12	2.24		2.36	3.62
0250	Gypsum board		2500	.026		.07	1.34		1.41	2.17
0255	Plywood		2500	.026		.07	1.34		1.41	2.17
0260	Lighting fixtures up to 2' x 4'		72	.889	Ea.	2.56	46.50		49.06	75.50
0400	Partitions, non load bearing									
0410	Plaster, lath, and studs	A-9	690	.093	S.F.	.88	4.86		5.74	8.50
0450	Gypsum board and studs		1390	.046		.13	2.41		2.54	3.91
0465	Carpet & pad		1390	.046		.13	2.41		2.54	3.91
0600	Pipe insulation, air cell type, up to 4" diameter pipe		900	.071	L.F.	.21	3.73		3.94	6.05
0610	4" to 8" diameter pipe		800	.080		.23	4.19		4.42	6.80
0620	10" to 12" diameter pipe		700	.091		.26	4.79		5.05	7.75
0630	14" to 16" diameter pipe		550	.116		.34	6.10		6.44	9.85
0650	Over 16" diameter pipe		650	.098	S.F.	.28	5.15		5.43	8.35
9000	For type B (supplied air) respirator equipment, add				%				10%	10%

For customer support on your Mechanical Cost Data, call 877.795.7394.

For customer support on your Mechanical Cost Data, call 877.795.7394.

29

Division Notes

	CREW	DAILY OUTPUT	LABOR-HOURS	UNIT	BARE COSTS				TOTAL INCL O&P
					MAT.	LABOR	EQUIP.	TOTAL	

Estimating Tips
General
- Carefully check all the plans and specifications. Concrete often appears on drawings other than structural drawings, including mechanical and electrical drawings for equipment pads. The cost of cutting and patching is often difficult to estimate. See Subdivision 03 81 for Concrete Cutting, Subdivision 02 41 19.16 for Cutout Demolition, Subdivision 03 05 05.10 for Concrete Demolition, and Subdivision 02 41 19.19 for Rubbish Handling (handling, loading and hauling of debris).
- Always obtain concrete prices from suppliers near the job site. A volume discount can often be negotiated, depending upon competition in the area. Remember to add for waste, particularly for slabs and footings on grade.

03 10 00 Concrete Forming and Accessories
- A primary cost for concrete construction is forming. Most jobs today are constructed with prefabricated forms. The selection of the forms best suited for the job and the total square feet of forms required for efficient concrete forming and placing are key elements in estimating concrete construction. Enough forms must be available for erection to make efficient use of the concrete placing equipment and crew.
- Concrete accessories for forming and placing depend upon the systems used. Study the plans and specifications to ensure that all special accessory requirements have been included in the cost estimate, such as anchor bolts, inserts, and hangers.
- Included within costs for forms-in-place are all necessary bracing and shoring.

03 20 00 Concrete Reinforcing
- Ascertain that the reinforcing steel supplier has included all accessories, cutting, bending, and an allowance for lapping, splicing, and waste. A good rule of thumb is 10% for lapping, splicing, and waste. Also, 10% waste should be allowed for welded wire fabric.
- The unit price items in the subdivisions for Reinforcing In Place, Glass Fiber Reinforcing, and Welded Wire Fabric include the labor to install accessories such as beam and slab bolsters, high chairs, and bar ties and tie wire. The material cost for these accessories is not included; they may be obtained from the Accessories Division.

03 30 00 Cast-In-Place Concrete
- When estimating structural concrete, pay particular attention to requirements for concrete additives, curing methods, and surface treatments. Special consideration for climate, hot or cold, must be included in your estimate. Be sure to include requirements for concrete placing equipment, and concrete finishing.
- For accurate concrete estimating, the estimator must consider each of the following major components individually: forms, reinforcing steel, ready-mix concrete, placement of the concrete, and finishing of the top surface. For faster estimating, Subdivision 03 30 53.40 for Concrete-In-Place can be used; here, various items of concrete work are presented that include the costs of all five major components (unless specifically stated otherwise).

03 40 00 Precast Concrete
03 50 00 Cast Decks and Underlayment
- The cost of hauling precast concrete structural members is often an important factor. For this reason, it is important to get a quote from the nearest supplier. It may become economically feasible to set up precasting beds on the site if the hauling costs are prohibitive.

Reference Numbers
Reference numbers are shown in shaded boxes at the beginning of some major classifications. These numbers refer to related items in the Reference Section. The reference information may be an estimating procedure, an alternate pricing method, or technical information.

Note: Not all subdivisions listed here necessarily appear in this publication. ■

03 01 Maintenance of Concrete

03 01 30 – Maintenance of Cast-In-Place Concrete

03 01 30.62 Concrete Patching	Crew	Daily Output	Labor-Hours	Unit	Material	2015 Bare Costs Labor	Equipment	Total	Total Incl O&P
0010 **CONCRETE PATCHING**									
0100 Floors, 1/4" thick, small areas, regular grout	1 Cefi	170	.047	S.F.	1.46	2.12		3.58	4.74
0150 Epoxy grout	"	100	.080	"	8.05	3.60		11.65	14.25
2000 Walls, including chipping, cleaning and epoxy grout									
2100 1/4" deep	1 Cefi	65	.123	S.F.	7.65	5.55		13.20	16.65
2150 1/2" deep	↓	50	.160		15.35	7.20		22.55	27.50
2200 3/4" deep	↓	40	.200	↓	23	9		32	39

03 11 Concrete Forming

03 11 13 – Structural Cast-In-Place Concrete Forming

03 11 13.40 Forms In Place, Equipment Foundations

	Crew	Daily Output	Labor-Hours	Unit	Material	2015 Bare Costs Labor	Equipment	Total	Total Incl O&P
0010 **FORMS IN PLACE, EQUIPMENT FOUNDATIONS**									
0020 1 use	C-2	160	.300	SFCA	3.45	13.70		17.15	25
0050 2 use		190	.253		1.90	11.55		13.45	19.90
0100 3 use		200	.240		1.38	11		12.38	18.40
0150 4 use	↓	205	.234	↓	1.12	10.70		11.82	17.75

03 11 13.45 Forms In Place, Footings

	Crew	Daily Output	Labor-Hours	Unit	Material	2015 Bare Costs Labor	Equipment	Total	Total Incl O&P
0010 **FORMS IN PLACE, FOOTINGS**									
0020 Continuous wall, plywood, 1 use	C-1	375	.085	SFCA	6.80	3.81		10.61	13.35
0050 2 use		440	.073		3.74	3.24		6.98	9.10
0100 3 use		470	.068		2.72	3.04		5.76	7.65
0150 4 use		485	.066		2.22	2.94		5.16	6.95
5000 Spread footings, job-built lumber, 1 use		305	.105		2.28	4.68		6.96	9.70
5050 2 use		371	.086		1.27	3.85		5.12	7.30
5100 3 use		401	.080		.91	3.56		4.47	6.50
5150 4 use	↓	414	.077	↓	.74	3.45		4.19	6.10

03 11 13.65 Forms In Place, Slab On Grade

	Crew	Daily Output	Labor-Hours	Unit	Material	2015 Bare Costs Labor	Equipment	Total	Total Incl O&P
0010 **FORMS IN PLACE, SLAB ON GRADE**									
3000 Edge forms, wood, 4 use, on grade, to 6" high	C-1	600	.053	L.F.	.35	2.38		2.73	4.04
6000 Trench forms in floor, wood, 1 use		160	.200	SFCA	1.94	8.90		10.84	15.90
6050 2 use		175	.183		1.06	8.15		9.21	13.70
6100 3 use		180	.178		.77	7.95		8.72	13.05
6150 4 use	↓	185	.173	↓	.63	7.70		8.33	12.55

03 15 Concrete Accessories

03 15 05 – Concrete Forming Accessories

03 15 05.75 Sleeves and Chases

	Crew	Daily Output	Labor-Hours	Unit	Material	2015 Bare Costs Labor	Equipment	Total	Total Incl O&P
0010 **SLEEVES AND CHASES**									
0100 Plastic, 1 use, 12" long, 2" diameter	1 Carp	100	.080	Ea.	2.15	3.76		5.91	8.15
0150 4" diameter		90	.089		6.05	4.17		10.22	13.05
0200 6" diameter		75	.107		10.65	5		15.65	19.40
0250 12" diameter		60	.133		31.50	6.25		37.75	44
5000 Sheet metal, 2" diameter [G]		100	.080		1.36	3.76		5.12	7.30
5100 4" diameter [G]		90	.089		1.70	4.17		5.87	8.25
5150 6" diameter [G]		75	.107		1.82	5		6.82	9.70
5200 12" diameter [G]		60	.133		3.63	6.25		9.88	13.65
6000 Steel pipe, 2" diameter [G]		100	.080		4.55	3.76		8.31	10.80
6100 4" diameter [G]		90	.089		16.30	4.17		20.47	24.50
6150 6" diameter [G]	↓	75	.107		36.50	5		41.50	47.50

03 15 Concrete Accessories

03 15 05 – Concrete Forming Accessories

03 15 05.75 Sleeves and Chases		Crew	Daily Output	Labor-Hours	Unit	Material	2015 Bare Costs Labor	Equipment	Total	Total Incl O&P
6200	12" diameter	G 1 Carp	60	.133	Ea.	86.50	6.25		92.75	105

03 15 16 – Concrete Construction Joints

03 15 16.20 Control Joints, Saw Cut

		Crew	Daily Output	Labor-Hours	Unit	Material	Labor	Equipment	Total	Total Incl O&P
0010	**CONTROL JOINTS, SAW CUT**									
0100	Sawcut control joints in green concrete									
0120	1" depth	C-27	2000	.008	L.F.	.04	.36	.08	.48	.66
0140	1-1/2" depth		1800	.009		.06	.40	.09	.55	.75
0160	2" depth		1600	.010		.08	.45	.10	.63	.88
0180	Sawcut joint reservoir in cured concrete									
0182	3/8" wide x 3/4" deep, with single saw blade	C-27	1000	.016	L.F.	.06	.72	.17	.95	1.31
0184	1/2" wide x 1" deep, with double saw blades		900	.018		.11	.80	.19	1.10	1.52
0186	3/4" wide x 1-1/2" deep, with double saw blades		800	.020		.23	.90	.21	1.34	1.82
0190	Water blast joint to wash away laitance, 2 passes	C-29	2500	.003			.12	.03	.15	.22
0200	Air blast joint to blow out debris and air dry, 2 passes	C-28	2000	.004			.18	.01	.19	.28
0300	For backer rod, see Section 07 91 23.10									
0342	For joint sealant, see Section 07 92 13.20									

03 15 19 – Cast-In Concrete Anchors

03 15 19.10 Anchor Bolts

		Crew	Daily Output	Labor-Hours	Unit	Material	Labor	Equipment	Total	Total Incl O&P
0010	**ANCHOR BOLTS**									
0015	Made from recycled materials									
0025	Single bolts installed in fresh concrete, no templates									
0030	Hooked w/nut and washer, 1/2" diameter, 8" long	G 1 Carp	132	.061	Ea.	1.33	2.85		4.18	5.85
0040	12" long	G	131	.061		1.48	2.87		4.35	6.05
0050	5/8" diameter, 8" long	G	129	.062		2.98	2.91		5.89	7.75
0060	12" long	G	127	.063		3.67	2.96		6.63	8.60
0070	3/4" diameter, 8" long	G	127	.063		3.67	2.96		6.63	8.60
0080	12" long	G	125	.064		4.59	3		7.59	9.65
0090	2-bolt pattern, including job-built 2-hole template, per set									
0100	J-type, incl. hex nut & washer, 1/2" diameter x 6" long	G 1 Carp	21	.381	Set	5.20	17.90		23.10	33.50
0110	12" long	G	21	.381		5.80	17.90		23.70	34
0120	18" long	G	21	.381		6.70	17.90		24.60	35
0130	3/4" diameter x 8" long	G	20	.400		10.20	18.80		29	40
0140	12" long	G	20	.400		12	18.80		30.80	42
0150	18" long	G	20	.400		14.75	18.80		33.55	45.50
0160	1" diameter x 12" long	G	19	.421		20.50	19.75		40.25	53
0170	18" long	G	19	.421		24	19.75		43.75	57
0180	24" long	G	19	.421		28.50	19.75		48.25	62
0190	36" long	G	18	.444		38	21		59	74
0200	1-1/2" diameter x 18" long	G	17	.471		37.50	22		59.50	75.50
0210	24" long	G	16	.500		44	23.50		67.50	84.50
0300	L-type, incl. hex nut & washer, 3/4" diameter x 12" long	G	20	.400		12.90	18.80		31.70	43
0310	18" long	G	20	.400		15.75	18.80		34.55	46.50
0320	24" long	G	20	.400		18.60	18.80		37.40	49.50
0330	30" long	G	20	.400		23	18.80		41.80	54
0340	36" long	G	20	.400		25.50	18.80		44.30	57.50
0350	1" diameter x 12" long	G	19	.421		19.55	19.75		39.30	52
0360	18" long	G	19	.421		23.50	19.75		43.25	56.50
0370	24" long	G	19	.421		28.50	19.75		48.25	62
0380	30" long	G	19	.421		33	19.75		52.75	67
0390	36" long	G	18	.444		37.50	21		58.50	73.50
0400	42" long	G	18	.444		45	21		66	81.50
0410	48" long	G	18	.444		50	21		71	87

For customer support on your Mechanical Cost Data, call 877.795.7394.

33

03 15 19.10 Anchor Bolts		Crew	Daily Output	Labor-Hours	Unit	Material	2015 Bare Costs Labor	Equipment	Total	Total Incl O&P
0420	1-1/4" diameter x 18" long	G	1 Carp	18	.444	Set	30	21		51	65
0430	24" long	G		18	.444		35	21		56	70.50
0440	30" long	G		17	.471		40	22		62	78
0450	36" long	G		17	.471		45	22		67	83.50
0460	42" long	G	2 Carp	32	.500		50.50	23.50		74	91.50
0470	48" long	G		32	.500		57	23.50		80.50	99
0480	54" long	G		31	.516		67	24		91	111
0490	60" long	G		31	.516		73	24		97	118
0500	1-1/2" diameter x 18" long	G		33	.485		51	23		74	91
0510	24" long	G		32	.500		59	23.50		82.50	101
0520	30" long	G		31	.516		66	24		90	110
0530	36" long	G		30	.533		75.50	25		100.50	122
0540	42" long	G		30	.533		85.50	25		110.50	133
0550	48" long	G		29	.552		96	26		122	145
0560	54" long	G		28	.571		116	27		143	170
0570	60" long	G		28	.571		127	27		154	182
0580	1-3/4" diameter x 18" long	G		31	.516		74.50	24		98.50	120
0590	24" long	G		30	.533		87	25		112	134
0600	30" long	G		29	.552		101	26		127	151
0610	36" long	G		28	.571		114	27		141	168
0620	42" long	G		27	.593		128	28		156	184
0630	48" long	G		26	.615		141	29		170	200
0640	54" long	G		26	.615		174	29		203	236
0650	60" long	G		25	.640		187	30		217	252
0660	2" diameter x 24" long	G		27	.593		111	28		139	166
0670	30" long	G		27	.593		125	28		153	181
0680	36" long	G		26	.615		137	29		166	196
0690	42" long	G		25	.640		153	30		183	214
0700	48" long	G		24	.667		175	31.50		206.50	241
0710	54" long	G		23	.696		208	32.50		240.50	280
0720	60" long	G		23	.696		223	32.50		255.50	297
0730	66" long	G		22	.727		239	34		273	315
0740	72" long	G		21	.762		261	36		297	340
1000	4-bolt pattern, including job-built 4-hole template, per set										
1100	J-type, incl. hex nut & washer, 1/2" diameter x 6" long	G	1 Carp	19	.421	Set	7.55	19.75		27.30	39
1110	12" long	G		19	.421		8.75	19.75		28.50	40
1120	18" long	G		18	.444		10.55	21		31.55	43.50
1130	3/4" diameter x 8" long	G		17	.471		17.55	22		39.55	53.50
1140	12" long	G		17	.471		21	22		43	57.50
1150	18" long	G		17	.471		26.50	22		48.50	63.50
1160	1" diameter x 12" long	G		16	.500		38	23.50		61.50	77.50
1170	18" long	G		15	.533		45	25		70	88
1180	24" long	G		15	.533		54	25		79	98
1190	36" long	G		15	.533		73	25		98	119
1200	1-1/2" diameter x 18" long	G		13	.615		72	29		101	124
1210	24" long	G		12	.667		85.50	31.50		117	142
1300	L-type, incl. hex nut & washer, 3/4" diameter x 12" long	G		17	.471		23	22		45	59.50
1310	18" long	G		17	.471		28.50	22		50.50	65.50
1320	24" long	G		17	.471		34.50	22		56.50	72
1330	30" long	G		16	.500		43	23.50		66.50	83
1340	36" long	G		16	.500		48.50	23.50		72	89.50
1350	1" diameter x 12" long	G		16	.500		36	23.50		59.50	76
1360	18" long	G		15	.533		44.50	25		69.50	87.50

03 15 Concrete Accessories

03 15 19 – Cast-In Concrete Anchors

03 15 19.10 Anchor Bolts

			Crew	Daily Output	Labor-Hours	Unit	Material	2015 Bare Costs Labor	Equipment	Total	Total Incl O&P
1370	24" long	G	1 Carp	15	.533	Set	54	25		79	98
1380	30" long	G		15	.533		63.50	25		88.50	109
1390	36" long	G		15	.533		72	25		97	118
1400	42" long	G		14	.571		87	27		114	138
1410	48" long	G		14	.571		97.50	27		124.50	149
1420	1-1/4" diameter x 18" long	G		14	.571		57	27		84	104
1430	24" long	G		14	.571		67	27		94	115
1440	30" long	G		13	.615		77	29		106	129
1450	36" long	G		13	.615		87	29		116	141
1460	42" long	G	2 Carp	25	.640		98	30		128	154
1470	48" long	G		24	.667		112	31.50		143.50	171
1480	54" long	G		23	.696		131	32.50		163.50	195
1490	60" long	G		23	.696		144	32.50		176.50	209
1500	1-1/2" diameter x 18" long	G		25	.640		99	30		129	155
1510	24" long	G		24	.667		115	31.50		146.50	174
1520	30" long	G		23	.696		129	32.50		161.50	193
1530	36" long	G		22	.727		148	34		182	216
1540	42" long	G		22	.727		169	34		203	238
1550	48" long	G		21	.762		189	36		225	263
1560	54" long	G		20	.800		230	37.50		267.50	310
1570	60" long	G		20	.800		251	37.50		288.50	335
1580	1-3/4" diameter x 18" long	G		22	.727		146	34		180	214
1590	24" long	G		21	.762		171	36		207	243
1600	30" long	G		21	.762		199	36		235	273
1610	36" long	G		20	.800		226	37.50		263.50	305
1620	42" long	G		19	.842		254	39.50		293.50	340
1630	48" long	G		18	.889		278	41.50		319.50	370
1640	54" long	G		18	.889		345	41.50		386.50	445
1650	60" long	G		17	.941		370	44		414	480
1660	2" diameter x 24" long	G		19	.842		220	39.50		259.50	305
1670	30" long	G		18	.889		247	41.50		288.50	335
1680	36" long	G		18	.889		272	41.50		313.50	365
1690	42" long	G		17	.941		305	44		349	405
1700	48" long	G		16	1		345	47		392	455
1710	54" long	G		15	1.067		415	50		465	530
1720	60" long	G		15	1.067		445	50		495	565
1730	66" long	G		14	1.143		475	53.50		528.50	605
1740	72" long	G		14	1.143		520	53.50		573.50	655
1990	For galvanized, add					Ea.	75%				

03 15 19.45 Machinery Anchors

			Crew	Daily Output	Labor-Hours	Unit	Material	2015 Bare Costs Labor	Equipment	Total	Total Incl O&P
0010	**MACHINERY ANCHORS**, heavy duty, incl. sleeve, floating base nut,										
0020	lower stud & coupling nut, fiber plug, connecting stud, washer & nut.										
0030	For flush mounted embedment in poured concrete heavy equip. pads.										
0200	Stud & bolt, 1/2" diameter	G	E-16	40	.400	Ea.	72.50	21.50	3.65	97.65	122
0300	5/8" diameter	G		35	.457		80.50	24.50	4.17	109.17	136
0500	3/4" diameter	G		30	.533		93	28.50	4.86	126.36	157
0600	7/8" diameter	G		25	.640		101	34.50	5.85	141.35	177
0800	1" diameter	G		20	.800		117	43	7.30	167.30	212
0900	1-1/4" diameter	G		15	1.067		141	57	9.75	207.75	265

For customer support on your Mechanical Cost Data, call 877.795.7394.

35

03 21 Reinforcement Bars

03 21 11 – Plain Steel Reinforcement Bars

03 21 11.60 Reinforcing In Place		Crew	Daily Output	Labor-Hours	Unit	Material	2015 Bare Costs Labor	Equipment	Total	Total Incl O&P
0010	**REINFORCING IN PLACE**, 50-60 ton lots, A615 Grade 60									
0020	Includes labor, but not material cost, to install accessories									
0030	Made from recycled materials									
0502	Footings, #4 to #7	G 4 Rodm	4200	.008	Lb.	.48	.40		.88	1.16
0552	#8 to #18	G	7200	.004		.48	.23		.71	.89
0602	Slab on grade, #3 to #7	G	4200	.008		.48	.40		.88	1.16
0900	For other than 50 – 60 ton lots									
1000	Under 10 ton job, #3 to #7, add					25%	10%			
1010	#8 to #18, add					20%	10%			
1050	10 – 50 ton job, #3 to #7, add					10%				
1060	#8 to #18, add					5%				
1100	60 – 100 ton job, #3 to #7, deduct					5%				
1110	#8 to #18, deduct					10%				
1150	Over 100 ton job, #3 to #7, deduct					10%				
1160	#8 to #18, deduct					15%				

03 22 Fabric and Grid Reinforcing

03 22 11 – Plain Welded Wire Fabric Reinforcing

03 22 11.10 Plain Welded Wire Fabric

		Crew	Daily Output	Labor-Hours	Unit	Material	Labor	Equipment	Total	Total Incl O&P
0010	**PLAIN WELDED WIRE FABRIC** ASTM A185									
0020	Includes labor, but not material cost, to install accessories									
0030	Made from recycled materials									
0050	Sheets									
0100	6 x 6 - W1.4 x W1.4 (10 x 10) 21 lb. per C.S.F.	G 2 Rodm	35	.457	C.S.F.	14.50	24		38.50	53.50

03 22 13 – Galvanized Welded Wire Fabric Reinforcing

03 22 13.10 Galvanized Welded Wire Fabric

		Crew	Daily Output	Labor-Hours	Unit	Material	Labor	Equipment	Total	Total Incl O&P
0010	**GALVANIZED WELDED WIRE FABRIC**									
0100	Add to plain welded wire pricing for galvanized welded wire				Lb.	.23			.23	.25

03 22 16 – Epoxy-Coated Welded Wire Fabric Reinforcing

03 22 16.10 Epoxy-Coated Welded Wire Fabric

		Crew	Daily Output	Labor-Hours	Unit	Material	Labor	Equipment	Total	Total Incl O&P
0010	**EPOXY-COATED WELDED WIRE FABRIC**									
0100	Add to plain welded wire pricing for epoxy-coated welded wire				Lb.	.21			.21	.23

03 30 Cast-In-Place Concrete

03 30 53 – Miscellaneous Cast-In-Place Concrete

03 30 53.40 Concrete In Place

		Crew	Daily Output	Labor-Hours	Unit	Material	Labor	Equipment	Total	Total Incl O&P
0010	**CONCRETE IN PLACE**									
0020	Including forms (4 uses), Grade 60 rebar, concrete (Portland cement									
0050	Type I), placement and finishing unless otherwise indicated									
0500	Chimney foundations (5000 psi), over 5 C.Y.	C-14C	32.22	3.476	C.Y.	150	157	.97	307.97	405
0510	(3500 psi), under 5 C.Y.	"	23.71	4.724	"	177	213	1.32	391.32	525
3540	Equipment pad (3000 psi), 3' x 3' x 6" thick	C-14H	45	1.067	Ea.	47	49.50	.69	97.19	128
3550	4' x 4' x 6" thick		30	1.600		69.50	74	1.04	144.54	191
3560	5' x 5' x 8" thick		18	2.667		122	124	1.73	247.73	325
3570	6' x 6' x 8" thick		14	3.429		164	159	2.23	325.23	425
3580	8' x 8' x 10" thick		8	6		350	278	3.90	631.90	815
3590	10' x 10' x 12" thick		5	9.600		595	445	6.25	1,046.25	1,350
3800	Footings (3000 psi), spread under 1 C.Y.	C-14C	28	4	C.Y.	166	180	1.12	347.12	460

03 30 Cast-In-Place Concrete

03 30 53 – Miscellaneous Cast-In-Place Concrete

03 30 53.40 Concrete In Place		Crew	Daily Output	Labor-Hours	Unit	Material	2015 Bare Costs Labor	Equipment	Total	Total Incl O&P
3825	1 C.Y. to 5 C.Y.	C-14C	43	2.605	C.Y.	201	117	.73	318.73	405
3850	Over 5 C.Y.	↓	75	1.493		185	67.50	.42	252.92	310
3900	Footings, strip (3000 psi), 18" x 9", unreinforced	C-14L	40	2.400		125	105	.79	230.79	300
3920	18" x 9", reinforced	C-14C	35	3.200		148	144	.90	292.90	385
3925	20" x 10", unreinforced	C-14L	45	2.133		122	93.50	.70	216.20	278
3930	20" x 10", reinforced	C-14C	40	2.800		140	126	.78	266.78	350
3935	24" x 12", unreinforced	C-14L	55	1.745		120	76.50	.58	197.08	250
3940	24" x 12", reinforced	C-14C	48	2.333		139	105	.65	244.65	315
3945	36" x 12", unreinforced	C-14L	70	1.371		116	60	.45	176.45	219
3950	36" x 12", reinforced	C-14C	60	1.867		133	84	.52	217.52	276
4000	Foundation mat (3000 psi), under 10 C.Y.	↓	38.67	2.896		204	131	.81	335.81	425
4050	Over 20 C.Y.	↓	56.40	1.986		178	89.50	.56	268.06	335
4650	Slab on grade (3500 psi), not including finish, 4" thick	C-14E	60.75	1.449		124	67.50	.51	192.01	242
4700	6" thick	"	92	.957	↓	119	44.50	.33	163.83	200
4701	Thickened slab edge (3500 psi), for slab on grade poured									
4702	monolithically with slab; depth is in addition to slab thickness;									
4703	formed vertical outside edge, earthen bottom and inside slope									
4705	8" deep x 8" wide bottom, unreinforced	C-14L	2190	.044	L.F.	3.47	1.92	.01	5.40	6.80
4710	8" x 8", reinforced	C-14C	1670	.067		5.75	3.02	.02	8.79	10.95
4715	12" deep x 12" wide bottom, unreinforced	C-14L	1800	.053		7.05	2.34	.02	9.41	11.35
4720	12" x 12", reinforced	C-14C	1310	.086		11.20	3.85	.02	15.07	18.30
4725	16" deep x 16" wide bottom, unreinforced	C-14L	1440	.067		11.85	2.92	.02	14.79	17.50
4730	16" x 16", reinforced	C-14C	1120	.100		16.80	4.51	.03	21.34	25.50
4735	20" deep x 20" wide bottom, unreinforced	C-14L	1150	.083		17.90	3.66	.03	21.59	25.50
4740	20" x 20", reinforced	C-14C	920	.122		24	5.50	.03	29.53	35
4745	24" deep x 24" wide bottom, unreinforced	C-14L	930	.103		25	4.53	.03	29.56	35
4750	24" x 24", reinforced	C-14C	740	.151	↓	33.50	6.80	.04	40.34	47
4751	Slab on grade (3500 psi), incl. troweled finish, not incl. forms									
4760	or reinforcing, over 10,000 S.F., 4" thick	C-14F	3425	.021	S.F.	1.35	.90	.01	2.26	2.85
4820	6" thick		3350	.021		1.98	.92	.01	2.91	3.57
4840	8" thick		3184	.023		2.71	.97	.01	3.69	4.44
4900	12" thick		2734	.026		4.06	1.13	.01	5.20	6.15
4950	15" thick	↓	2505	.029	↓	5.10	1.23	.01	6.34	7.45

03 31 Structural Concrete

03 31 13 – Heavyweight Structural Concrete

03 31 13.35 Heavyweight Concrete, Ready Mix

		Crew	Daily Output	Labor-Hours	Unit	Material	2015 Bare Costs Labor	Equipment	Total	Total Incl O&P
0010	**HEAVYWEIGHT CONCRETE, READY MIX**, delivered									
0012	Includes local aggregate, sand, Portland cement (Type I) and water									
0015	Excludes all additives and treatments									
0020	2000 psi				C.Y.	97			97	107
0100	2500 psi					99.50			99.50	109
0150	3000 psi					102			102	112
0200	3500 psi					104			104	115
0300	4000 psi					107			107	118
1000	For high early strength (Portland cement Type III), add					10%				
1300	For winter concrete (hot water), add					4.05			4.05	4.46
1410	For mid-range water reducer, add					3.31			3.31	3.64
1420	For high-range water reducer/superplasticizer, add					5.65			5.65	6.20
1430	For retarder, add					3.23			3.23	3.55
1440	For non-Chloride accelerator, add					5.50			5.50	6.05

03 31 13 – Heavyweight Structural Concrete

03 31 13.35 Heavyweight Concrete, Ready Mix

		Crew	Daily Output	Labor-Hours	Unit	Material	2015 Bare Costs Labor	2015 Bare Costs Equipment	Total	Total Incl O&P
1450	For Chloride accelerator, per 1%, add				C.Y.	2.90			2.90	3.19
1460	For fiber reinforcing, synthetic (1 lb./C.Y.), add					6.65			6.65	7.30
1500	For Saturday delivery, add				↓	10.80			10.80	11.90
1510	For truck holding/waiting time past 1st hour per load, add				Hr.	94			94	103
1520	For short load (less than 4 C.Y.), add per load				Ea.	60.50			60.50	66.50
2000	For all lightweight aggregate, add				C.Y.	45%				

03 31 13.70 Placing Concrete

		Crew	Daily Output	Labor-Hours	Unit	Material	2015 Bare Costs Labor	2015 Bare Costs Equipment	Total	Total Incl O&P
0010	**PLACING CONCRETE**									
0020	Includes labor and equipment to place, level (strike off) and consolidate									
1900	Footings, continuous, shallow, direct chute	C-6	120	.400	C.Y.		15.65	.52	16.17	24.50
1950	Pumped	C-20	150	.427			17.25	5.20	22.45	32.50
2000	With crane and bucket	C-7	90	.800			32.50	13.50	46	65
2100	Footings, continuous, deep, direct chute	C-6	140	.343			13.45	.45	13.90	21
2150	Pumped	C-20	160	.400			16.15	4.89	21.04	30
2200	With crane and bucket	C-7	110	.655			27	11.05	38.05	53
2400	Footings, spread, under 1 C.Y., direct chute	C-6	55	.873			34	1.13	35.13	53.50
2450	Pumped	C-20	65	.985			40	12.05	52.05	74.50
2500	With crane and bucket	C-7	45	1.600			65.50	27	92.50	130
2600	Over 5 C.Y., direct chute	C-6	120	.400			15.65	.52	16.17	24.50
2650	Pumped	C-20	150	.427			17.25	5.20	22.45	32.50
2700	With crane and bucket	C-7	100	.720			29.50	12.15	41.65	58.50
2900	Foundation mats, over 20 C.Y., direct chute	C-6	350	.137			5.35	.18	5.53	8.40
2950	Pumped	C-20	400	.160			6.45	1.96	8.41	12
3000	With crane and bucket	C-7	300	.240	↓		9.80	4.05	13.85	19.45

03 35 Concrete Finishing

03 35 13 – High-Tolerance Concrete Floor Finishing

03 35 13.30 Finishing Floors, High Tolerance

		Crew	Daily Output	Labor-Hours	Unit	Material	2015 Bare Costs Labor	2015 Bare Costs Equipment	Total	Total Incl O&P
0010	**FINISHING FLOORS, HIGH TOLERANCE**									
0012	Finishing of fresh concrete flatwork requires that concrete									
0013	first be placed, struck off & consolidated									
0015	Basic finishing for various unspecified flatwork									
0100	Bull float only	C-10	4000	.006	S.F.		.26		.26	.38
0125	Bull float & manual float	↓	2000	.012			.51		.51	.76
0150	Bull float, manual float, & broom finish, w/edging & joints		1850	.013	↓		.55		.55	.83
0200	Bull float, manual float & manual steel trowel	↓	1265	.019	↓		.81		.81	1.21
0210	For specified Random Access Floors in ACI Classes 1, 2, 3 and 4 to achieve									
0215	Composite Overall Floor Flatness and Levelness values up to FF35/FL25									
0250	Bull float, machine float & machine trowel (walk-behind)	C-10C	1715	.014	S.F.		.60	.03	.63	.92
0300	Power screed, bull float, machine float & trowel (walk-behind)	C-10D	2400	.010			.43	.05	.48	.69
0350	Power screed, bull float, machine float & trowel (ride-on)	C-10E	4000	.006	↓		.26	.06	.32	.45
0352	For specified Random Access Floors in ACI Classes 5, 6, 7 and 8 to achieve									
0354	Composite Overall Floor Flatness and Levelness values up to FF50/FL50									
0356	Add for two-dimensional restraightening after power float	C-10	6000	.004	S.F.		.17		.17	.25
0358	For specified Random or Defined Access Floors in ACI Class 9 to achieve									
0360	Composite Overall Floor Flatness and Levelness values up to FF100/FL100									
0362	Add for two-dimensional restraightening after bull float & power float	C-10	3000	.008	S.F.		.34		.34	.51
0364	For specified Superflat Defined Access Floors in ACI Class 9 to achieve									
0366	Minimum Floor Flatness and Levelness values of FF100/FL100									
0368	Add for 2-dim'l restraightening after bull float, power float, power trowel	C-10	2000	.012	S.F.		.51		.51	.76

03 54 Cast Underlayment

03 54 16 – Hydraulic Cement Underlayment

03 54 16.50 Cement Underlayment

03 54 16.50 Cement Underlayment		Crew	Daily Output	Labor-Hours	Unit	Material	2015 Bare Costs Labor	Equipment	Total	Total Incl O&P
0010	**CEMENT UNDERLAYMENT**									
2510	Underlayment, P.C based, self-leveling, 4100 psi, pumped, 1/4" thick	C-8	20000	.003	S.F.	1.55	.12	.04	1.71	1.92
2520	1/2" thick		19000	.003		3.09	.12	.04	3.25	3.63
2530	3/4" thick		18000	.003		4.64	.13	.04	4.81	5.35
2540	1" thick		17000	.003		6.20	.14	.04	6.38	7.05
2550	1-1/2" thick		15000	.004		9.30	.16	.05	9.51	10.50
2560	Hand placed, 1/2" thick	C-18	450	.020		3.09	.76	.13	3.98	4.71
2610	Topping, P.C based, self-leveling, 6100 psi, pumped, 1/4" thick	C-8	20000	.003		2.26	.12	.04	2.42	2.71
2620	1/2" thick		19000	.003		4.52	.12	.04	4.68	5.20
2630	3/4" thick		18000	.003		6.80	.13	.04	6.97	7.70
2660	1" thick		17000	.003		9.05	.14	.04	9.23	10.20
2670	1-1/2" thick		15000	.004		13.55	.16	.05	13.76	15.20
2680	Hand placed, 1/2" thick	C-18	450	.020		4.52	.76	.13	5.41	6.30

03 63 Epoxy Grouting

03 63 05 – Grouting of Dowels and Fasteners

03 63 05.10 Epoxy Only

03 63 05.10 Epoxy Only		Crew	Daily Output	Labor-Hours	Unit	Material	2015 Bare Costs Labor	Equipment	Total	Total Incl O&P
0010	**EPOXY ONLY**									
1500	Chemical anchoring, epoxy cartridge, excludes layout, drilling, fastener									
1530	For fastener 3/4" diam. x 6" embedment	2 Skwk	72	.222	Ea.	5.65	10.80		16.45	23
1535	1" diam. x 8" embedment		66	.242		8.45	11.80		20.25	27.50
1540	1-1/4" diam. x 10" embedment		60	.267		16.90	12.95		29.85	38.50
1545	1-3/4" diam. x 12" embedment		54	.296		28	14.40		42.40	53.50
1550	14" embedment		48	.333		34	16.20		50.20	62
1555	2" diam. x 12" embedment		42	.381		45	18.55		63.55	78
1560	18" embedment		32	.500		56.50	24.50		81	99.50

03 82 Concrete Boring

03 82 13 – Concrete Core Drilling

03 82 13.10 Core Drilling

03 82 13.10 Core Drilling		Crew	Daily Output	Labor-Hours	Unit	Material	2015 Bare Costs Labor	Equipment	Total	Total Incl O&P
0010	**CORE DRILLING**									
0015	Includes bit cost, layout and set-up time									
0020	Reinforced concrete slab, up to 6" thick									
0100	1" diameter core	B-89A	17	.941	Ea.	.18	40.50	6.80	47.48	70
0150	For each additional inch of slab thickness in same hole, add		1440	.011		.03	.48	.08	.59	.86
0200	2" diameter core		16.50	.970		.28	42	7	49.28	72.50
0250	For each additional inch of slab thickness in same hole, add		1080	.015		.05	.64	.11	.80	1.15
0300	3" diameter core		16	1		.38	43	7.25	50.63	75
0350	For each additional inch of slab thickness in same hole, add		720	.022		.06	.96	.16	1.18	1.73
0500	4" diameter core		15	1.067		.49	46	7.70	54.19	80
0550	For each additional inch of slab thickness in same hole, add		480	.033		.08	1.44	.24	1.76	2.58
0700	6" diameter core		14	1.143		.76	49.50	8.25	58.51	86
0750	For each additional inch of slab thickness in same hole, add		360	.044		.13	1.92	.32	2.37	3.45
0900	8" diameter core		13	1.231		1.05	53	8.90	62.95	93
0950	For each additional inch of slab thickness in same hole, add		288	.056		.17	2.40	.40	2.97	4.32
1100	10" diameter core		12	1.333		1.48	57.50	9.65	68.63	101
1150	For each additional inch of slab thickness in same hole, add		240	.067		.25	2.88	.48	3.61	5.25
1300	12" diameter core		11	1.455		1.78	62.50	10.50	74.78	110
1350	For each additional inch of slab thickness in same hole, add		206	.078		.30	3.35	.56	4.21	6.10

39

03 82 13 – Concrete Core Drilling

03 82 13.10 Core Drilling		Crew	Daily Output	Labor-Hours	Unit	Material	2015 Bare Costs			Total Incl O&P
							Labor	Equipment	Total	
1500	14" diameter core	B-89A	10	1.600	Ea.	2.08	69	11.55	82.63	121
1550	For each additional inch of slab thickness in same hole, add		180	.089		.35	3.83	.64	4.82	7
1700	18" diameter core		9	1.778		2.82	76.50	12.85	92.17	135
1750	For each additional inch of slab thickness in same hole, add		144	.111		.47	4.79	.80	6.06	8.80
1754	24" diameter core		8	2		4.02	86.50	14.45	104.97	153
1756	For each additional inch of slab thickness in same hole, add		120	.133		.67	5.75	.96	7.38	10.65
1760	For horizontal holes, add to above						20%	20%		
1770	Prestressed hollow core plank, 8" thick									
1780	1" diameter core	B-89A	17.50	.914	Ea.	.24	39.50	6.60	46.34	68.50
1790	For each additional inch of plank thickness in same hole, add		3840	.004		.03	.18	.03	.24	.34
1794	2" diameter core		17.25	.928		.38	40	6.70	47.08	69.50
1796	For each additional inch of plank thickness in same hole, add		2880	.006		.05	.24	.04	.33	.46
1800	3" diameter core		17	.941		.51	40.50	6.80	47.81	70.50
1810	For each additional inch of plank thickness in same hole, add		1920	.008		.06	.36	.06	.48	.69
1820	4" diameter core		16.50	.970		.66	42	7	49.66	73
1830	For each additional inch of plank thickness in same hole, add		1280	.013		.08	.54	.09	.71	1.02
1840	6" diameter core		15.50	1.032		1.01	44.50	7.45	52.96	78
1850	For each additional inch of plank thickness in same hole, add		960	.017		.13	.72	.12	.97	1.38
1860	8" diameter core		15	1.067		1.40	46	7.70	55.10	81
1870	For each additional inch of plank thickness in same hole, add		768	.021		.17	.90	.15	1.22	1.75
1880	10" diameter core		14	1.143		1.98	49.50	8.25	59.73	87.50
1890	For each additional inch of plank thickness in same hole, add		640	.025		.25	1.08	.18	1.51	2.13
1900	12" diameter core		13.50	1.185		2.37	51	8.55	61.92	91
1910	For each additional inch of plank thickness in same hole, add		548	.029		.30	1.26	.21	1.77	2.50

03 82 16 – Concrete Drilling

03 82 16.10 Concrete Impact Drilling		Crew	Daily Output	Labor-Hours	Unit	Material	2015 Bare Costs			Total Incl O&P
							Labor	Equipment	Total	
0010	**CONCRETE IMPACT DRILLING**									
0020	Includes bit cost, layout and set-up time, no anchors									
0050	Up to 4" deep in concrete/brick floors/walls									
0100	Holes, 1/4" diameter	1 Carp	75	.107	Ea.	.07	5		5.07	7.75
0150	For each additional inch of depth in same hole, add		430	.019		.02	.87		.89	1.36
0200	3/8" diameter		63	.127		.06	5.95		6.01	9.20
0250	For each additional inch of depth in same hole, add		340	.024		.01	1.10		1.11	1.72
0300	1/2" diameter		50	.160		.06	7.50		7.56	11.60
0350	For each additional inch of depth in same hole, add		250	.032		.01	1.50		1.51	2.33
0400	5/8" diameter		48	.167		.09	7.85		7.94	12.15
0450	For each additional inch of depth in same hole, add		240	.033		.02	1.56		1.58	2.43
0500	3/4" diameter		45	.178		.12	8.35		8.47	13
0550	For each additional inch of depth in same hole, add		220	.036		.03	1.71		1.74	2.66
0600	7/8" diameter		43	.186		.16	8.75		8.91	13.60
0650	For each additional inch of depth in same hole, add		210	.038		.04	1.79		1.83	2.79
0700	1" diameter		40	.200		.17	9.40		9.57	14.65
0750	For each additional inch of depth in same hole, add		190	.042		.04	1.98		2.02	3.09
0800	1-1/4" diameter		38	.211		.26	9.90		10.16	15.50
0850	For each additional inch of depth in same hole, add		180	.044		.06	2.09		2.15	3.28
0900	1-1/2" diameter		35	.229		.39	10.75		11.14	16.95
0950	For each additional inch of depth in same hole, add		165	.048		.10	2.28		2.38	3.61
1000	For ceiling installations, add						40%			

40

For customer support on your Mechanical Cost Data, call 877.795.7394.

Estimating Tips
04 05 00 Common Work Results for Masonry

- The terms mortar and grout are often used interchangeably, and incorrectly. Mortar is used to bed masonry units, seal the entry of air and moisture, provide architectural appearance, and allow for size variations in the units. Grout is used primarily in reinforced masonry construction and is used to bond the masonry to the reinforcing steel. Common mortar types are M(2500 psi), S(1800 psi), N(750 psi), and O(350 psi), and conform to ASTM C270. Grout is either fine or coarse and conforms to ASTM C476, and in-place strengths generally exceed 2500 psi. Mortar and grout are different components of masonry construction and are placed by entirely different methods. An estimator should be aware of their unique uses and costs.

- Mortar is included in all assembled masonry line items. The mortar cost, part of the assembled masonry material cost, includes all ingredients, all labor, and all equipment required. Please see reference number R040513-10.

- Waste, specifically the loss/droppings of mortar and the breakage of brick and block, is included in all masonry assemblies in this division. A factor of 25% is added for mortar and 3% for brick and concrete masonry units.

- Scaffolding or staging is not included in any of the Division 4 costs. Refer to Subdivision 01 54 23 for scaffolding and staging costs.

04 20 00 Unit Masonry

- The most common types of unit masonry are brick and concrete masonry. The major classifications of brick are building brick (ASTM C62), facing brick (ASTM C216), glazed brick, fire brick, and pavers. Many varieties of texture and appearance can exist within these classifications, and the estimator would be wise to check local custom and availability within the project area. For repair and remodeling jobs, matching the existing brick may be the most important criteria.

- Brick and concrete block are priced by the piece and then converted into a price per square foot of wall. Openings less than two square feet are generally ignored by the estimator because any savings in units used is offset by the cutting and trimming required.

- It is often difficult and expensive to find and purchase small lots of historic brick. Costs can vary widely. Many design issues affect costs, selection of mortar mix, and repairs or replacement of masonry materials. Cleaning techniques must be reflected in the estimate.

- All masonry walls, whether interior or exterior, require bracing. The cost of bracing walls during construction should be included by the estimator, and this bracing must remain in place until permanent bracing is complete. Permanent bracing of masonry walls is accomplished by masonry itself, in the form of pilasters or abutting wall corners, or by anchoring the walls to the structural frame. Accessories in the form of anchors, anchor slots, and ties are used, but their supply and installation can be by different trades. For instance, anchor slots on spandrel beams and columns are supplied and welded in place by the steel fabricator, but the ties from the slots into the masonry are installed by the bricklayer. Regardless of the installation method, the estimator must be certain that these accessories are accounted for in pricing.

Reference Numbers

Reference numbers are shown in shaded boxes at the beginning of some major classifications. These numbers refer to related items in the Reference Section. The reference information may be an estimating procedure, an alternate pricing method, or technical information.

Note: Not all subdivisions listed here necessarily appear in this publication. ■

04 05 23 – Masonry Accessories

04 05 23.19 Masonry Cavity Drainage, Weepholes, and Vents	Crew	Daily Output	Labor-Hours	Unit	Material	2015 Bare Costs Labor	Equipment	Total	Total Incl O&P
0010 **MASONRY CAVITY DRAINAGE, WEEPHOLES, AND VENTS**									
0020 Extruded aluminum, 4" deep, 2-3/8" x 8-1/8"	1 Bric	30	.267	Ea.	34	12.30		46.30	56.50
0050 5" x 8-1/8"		25	.320		45	14.75		59.75	72
0100 2-1/4" x 25"		25	.320		78	14.75		92.75	108
0150 5" x 16-1/2"		22	.364		62.50	16.80		79.30	94
0175 5" x 24"		22	.364		84	16.80		100.80	118
0200 6" x 16-1/2"		22	.364		90.50	16.80		107.30	125
0250 7-3/4" x 16-1/2"		20	.400		77	18.45		95.45	113
0400 For baked enamel finish, add					35%				
0500 For cast aluminum, painted, add					60%				
1000 Stainless steel ventilators, 6" x 6"	1 Bric	25	.320		210	14.75		224.75	255
1050 8" x 8"		24	.333		232	15.40		247.40	280
1100 12" x 12"		23	.348		266	16.05		282.05	320
1150 12" x 6"		24	.333		254	15.40		269.40	305
1200 Foundation block vent, galv., 1-1/4" thk, 8" high, 16" long, no damper		30	.267		16.20	12.30		28.50	36.50
1250 For damper, add					3.11			3.11	3.42

04 21 Clay Unit Masonry

04 21 13 – Brick Masonry

04 21 13.15 Chimney

	Crew	Daily Output	Labor-Hours	Unit	Material	Labor	Equipment	Total	Total Incl O&P
0010 **CHIMNEY**, excludes foundation, scaffolding, grout and reinforcing									
0100 Brick, 16" x 16", 8" flue	D-1	18.20	.879	V.L.F.	24	37		61	83
0150 16" x 20" with one 8" x 12" flue		16	1		37.50	42		79.50	106
0200 16" x 24" with two 8" x 8" flues		14	1.143		55	48		103	134
0250 20" x 20" with one 12" x 12" flue		13.70	1.168		45.50	49		94.50	125
0300 20" x 24" with two 8" x 12" flues		12	1.333		62.50	56		118.50	154
0350 20" x 32" with two 12" x 12" flues		10	1.600		80	67.50		147.50	191

Estimating Tips

05 05 00 Common Work Results for Metals

- Nuts, bolts, washers, connection angles, and plates can add a significant amount to both the tonnage of a structural steel job and the estimated cost. As a rule of thumb, add 10% to the total weight to account for these accessories.

- Type 2 steel construction, commonly referred to as "simple construction," consists generally of field-bolted connections with lateral bracing supplied by other elements of the building, such as masonry walls or x-bracing. The estimator should be aware, however, that shop connections may be accomplished by welding or bolting. The method may be particular to the fabrication shop and may have an impact on the estimated cost.

05 10 00 Structural Steel

- Steel items can be obtained from two sources: a fabrication shop or a metals service center. Fabrication shops can fabricate items under more controlled conditions than can crews in the field. They are also more efficient and can produce items more economically. Metal service centers serve as a source of long mill shapes to both fabrication shops and contractors.

- Most line items in this structural steel subdivision, and most items in 05 50 00 Metal Fabrications, are indicated as being shop fabricated. The bare material cost for these shop fabricated items is the "Invoice Cost" from the shop and includes the mill base price of steel plus mill extras, transportation to the shop, shop drawings and detailing where warranted, shop fabrication and handling, sandblasting and a shop coat of primer paint, all necessary structural bolts, and delivery to the job site. The bare labor cost and bare equipment cost for these shop fabricated items is for field installation or erection.

- Line items in Subdivision 05 12 23.40 Lightweight Framing, and other items scattered in Division 5, are indicated as being field fabricated. The bare material cost for these field fabricated items is the "Invoice Cost" from the metals service center and includes the mill base price of steel plus mill extras, transportation to the metals service center, material handling, and delivery of long lengths of mill shapes to the job site. Material costs for structural bolts and welding rods should be added to the estimate. The bare labor cost and bare equipment cost for these items is for both field fabrication and field installation or erection, and include time for cutting, welding and drilling in the fabricated metal items. Drilling into concrete and fasteners to fasten field fabricated items to other work are not included and should be added to the estimate.

05 20 00 Steel Joist Framing

- In any given project the total weight of open web steel joists is determined by the loads to be supported and the design. However, economies can be realized in minimizing the amount of labor used to place the joists. This is done by maximizing the joist spacing, and therefore minimizing the number of joists required to be installed on the job. Certain spacings and locations may be required by the design, but in other cases maximizing the spacing and keeping it as uniform as possible will keep the costs down.

05 30 00 Steel Decking

- The takeoff and estimating of metal deck involves more than simply the area of the floor or roof and the type of deck specified or shown on the drawings. Many different sizes and types of openings may exist. Small openings for individual pipes or conduits may be drilled after the floor/roof is installed, but larger openings may require special deck lengths as well as reinforcing or structural support. The estimator should determine who will be supplying this reinforcing. Additionally, some deck terminations are part of the deck package, such as screed angles and pour stops, and others will be part of the steel contract, such as angles attached to structural members and cast-in-place angles and plates. The estimator must ensure that all pieces are accounted for in the complete estimate.

05 50 00 Metal Fabrications

- The most economical steel stairs are those that use common materials, standard details, and most importantly, a uniform and relatively simple method of field assembly. Commonly available A36 channels and plates are very good choices for the main stringers of the stairs, as are angles and tees for the carrier members. Risers and treads are usually made by specialty shops, and it is most economical to use a typical detail in as many places as possible. The stairs should be pre-assembled and shipped directly to the site. The field connections should be simple and straightforward to be accomplished efficiently, and with minimum equipment and labor.

Reference Numbers

Reference numbers are shown in shaded boxes at the beginning of some major classifications. These numbers refer to related items in the Reference Section. The reference information may be an estimating procedure, an alternate pricing method, or technical information.

Note: Not all subdivisions listed here necessarily appear in this publication. ■

05 05 19.10 Chemical Anchors

		Crew	Daily Output	Labor-Hours	Unit	Material	2015 Bare Costs Labor	2015 Bare Costs Equipment	Total	Total Incl O&P
0010	**CHEMICAL ANCHORS**									
0020	Includes layout & drilling									
1430	Chemical anchor, w/rod & epoxy cartridge, 3/4" diam. x 9-1/2" long	B-89A	27	.593	Ea.	9.80	25.50	4.28	39.58	55
1435	1" diameter x 11-3/4" long		24	.667		17.55	29	4.82	51.37	69
1440	1-1/4" diameter x 14" long		21	.762		36.50	33	5.50	75	96.50
1445	1-3/4" diameter x 15" long		20	.800		65	34.50	5.80	105.30	131
1450	18" long		17	.941		78	40.50	6.80	125.30	156
1455	2" diameter x 18" long		16	1		103	43	7.25	153.25	187
1460	24" long		15	1.067		134	46	7.70	187.70	227

05 05 19.20 Expansion Anchors

		Crew	Daily Output	Labor-Hours	Unit	Material	2015 Bare Costs Labor	2015 Bare Costs Equipment	Total	Total Incl O&P
0010	**EXPANSION ANCHORS**									
0100	Anchors for concrete, brick or stone, no layout and drilling									
0200	Expansion shields, zinc, 1/4" diameter, 1-5/16" long, single	G	1 Carp	90	.089	Ea.	.43	4.17		4.60	6.85
0300	1-3/8" long, double	G		85	.094		.54	4.42		4.96	7.40
0400	3/8" diameter, 1-1/2" long, single	G		85	.094		.64	4.42		5.06	7.50
0500	2" long, double	G		80	.100		1.18	4.70		5.88	8.55
0600	1/2" diameter, 2-1/16" long, single	G		80	.100		1.18	4.70		5.88	8.55
0700	2-1/2" long, double	G		75	.107		1.91	5		6.91	9.80
0800	5/8" diameter, 2-5/8" long, single	G		75	.107		2.05	5		7.05	9.95
0900	2-3/4" long, double	G		70	.114		2.70	5.35		8.05	11.20
1000	3/4" diameter, 2-3/4" long, single	G		70	.114		3.09	5.35		8.44	11.65
1100	3-15/16" long, double	G		65	.123		5	5.80		10.80	14.40
2100	Hollow wall anchors for gypsum wall board, plaster or tile									
2300	1/8" diameter, short	G	1 Carp	160	.050	Ea.	.23	2.35		2.58	3.86
2400	Long	G		150	.053		.23	2.50		2.73	4.10
2500	3/16" diameter, short	G		150	.053		.40	2.50		2.90	4.29
2600	Long	G		140	.057		.59	2.68		3.27	4.78
2700	1/4" diameter, short	G		140	.057		.59	2.68		3.27	4.78
2800	Long	G		130	.062		.65	2.89		3.54	5.15
3000	Toggle bolts, bright steel, 1/8" diameter, 2" long	G		85	.094		.19	4.42		4.61	7
3100	4" long	G		80	.100		.25	4.70		4.95	7.55
3200	3/16" diameter, 3" long	G		80	.100		.26	4.70		4.96	7.55
3300	6" long	G		75	.107		.36	5		5.36	8.10
3400	1/4" diameter, 3" long	G		75	.107		.34	5		5.34	8.05
3500	6" long	G		70	.114		.51	5.35		5.86	8.80
3600	3/8" diameter, 3" long	G		70	.114		.84	5.35		6.19	9.15
3700	6" long	G		60	.133		1.36	6.25		7.61	11.15
3800	1/2" diameter, 4" long	G		60	.133		1.88	6.25		8.13	11.70
3900	6" long	G		50	.160		2.45	7.50		9.95	14.25
5000	Screw anchors for concrete, masonry,									
5100	stone & tile, no layout or drilling included									
5700	Lag screw shields, 1/4" diameter, short	G	1 Carp	90	.089	Ea.	.34	4.17		4.51	6.75
5800	Long	G		85	.094		.38	4.42		4.80	7.20
5900	3/8" diameter, short	G		85	.094		.67	4.42		5.09	7.55
6000	Long	G		80	.100		.87	4.70		5.57	8.20
6100	1/2" diameter, short	G		80	.100		.95	4.70		5.65	8.30
6200	Long	G		75	.107		1.22	5		6.22	9.05
6300	5/8" diameter, short	G		70	.114		1.45	5.35		6.80	9.85
6400	Long	G		65	.123		1.86	5.80		7.66	10.95
6600	Lead, #6 & #8, 3/4" long	G		260	.031		.18	1.44		1.62	2.42
6700	#10 - #14, 1-1/2" long	G		200	.040		.37	1.88		2.25	3.30
6800	#16 & #18, 1-1/2" long	G		160	.050		.41	2.35		2.76	4.06

05 05 19 – Post-Installed Concrete Anchors

05 05 19.20 Expansion Anchors

		Crew	Daily Output	Labor-Hours	Unit	Material	2015 Bare Costs Labor	Equipment	Total	Total Incl O&P
6900	Plastic, #6 & #8, 3/4" long	1 Carp	260	.031	Ea.	.04	1.44		1.48	2.26
7000	#8 & #10, 7/8" long		240	.033		.04	1.56		1.60	2.45
7100	#10 & #12, 1" long		220	.036		.05	1.71		1.76	2.69
7200	#14 & #16, 1-1/2" long		160	.050		.07	2.35		2.42	3.69
8000	Wedge anchors, not including layout or drilling									
8050	Carbon steel, 1/4" diameter, 1-3/4" long [G]	1 Carp	150	.053	Ea.	.41	2.50		2.91	4.30
8100	3-1/4" long [G]		140	.057		.54	2.68		3.22	4.72
8150	3/8" diameter, 2-1/4" long [G]		145	.055		.50	2.59		3.09	4.54
8200	5" long [G]		140	.057		.88	2.68		3.56	5.10
8250	1/2" diameter, 2-3/4" long [G]		140	.057		.99	2.68		3.67	5.20
8300	7" long [G]		125	.064		1.70	3		4.70	6.50
8350	5/8" diameter, 3-1/2" long [G]		130	.062		1.86	2.89		4.75	6.50
8400	8-1/2" long [G]		115	.070		3.95	3.27		7.22	9.40
8450	3/4" diameter, 4-1/4" long [G]		115	.070		2.90	3.27		6.17	8.25
8500	10" long [G]		95	.084		6.60	3.95		10.55	13.35
8550	1" diameter, 6" long [G]		100	.080		9.30	3.76		13.06	16.05
8575	9" long [G]		85	.094		12.10	4.42		16.52	20
8600	12" long [G]		75	.107		13.10	5		18.10	22
8650	1-1/4" diameter, 9" long [G]		70	.114		24.50	5.35		29.85	35.50
8700	12" long [G]		60	.133		31.50	6.25		37.75	44
8750	For type 303 stainless steel, add					350%				
8800	For type 316 stainless steel, add					450%				
8950	Self-drilling concrete screw, hex washer head, 3/16" diam. x 1-3/4" long [G]	1 Carp	300	.027	Ea.	.20	1.25		1.45	2.15
8960	2-1/4" long [G]		250	.032		.24	1.50		1.74	2.57
8970	Phillips flat head, 3/16" diam. x 1-3/4" long [G]		300	.027		.21	1.25		1.46	2.16
8980	2-1/4" long [G]		250	.032		.23	1.50		1.73	2.56

05 05 21 – Fastening Methods for Metal

05 05 21.15 Drilling Steel

		Crew	Daily Output	Labor-Hours	Unit	Material	2015 Bare Costs Labor	Equipment	Total	Total Incl O&P
0010	**DRILLING STEEL**									
1910	Drilling & layout for steel, up to 1/4" deep, no anchor									
1920	Holes, 1/4" diameter	1 Sswk	112	.071	Ea.	.09	3.76		3.85	6.65
1925	For each additional 1/4" depth, add		336	.024		.09	1.25		1.34	2.28
1930	3/8" diameter		104	.077		.09	4.05		4.14	7.15
1935	For each additional 1/4" depth, add		312	.026		.09	1.35		1.44	2.44
1940	1/2" diameter		96	.083		.10	4.39		4.49	7.70
1945	For each additional 1/4" depth, add		288	.028		.10	1.46		1.56	2.65
1950	5/8" diameter		88	.091		.14	4.79		4.93	8.45
1955	For each additional 1/4" depth, add		264	.030		.14	1.60		1.74	2.93
1960	3/4" diameter		80	.100		.18	5.25		5.43	9.35
1965	For each additional 1/4" depth, add		240	.033		.18	1.75		1.93	3.25
1970	7/8" diameter		72	.111		.23	5.85		6.08	10.40
1975	For each additional 1/4" depth, add		216	.037		.23	1.95		2.18	3.65
1980	1" diameter		64	.125		.24	6.60		6.84	11.70
1985	For each additional 1/4" depth, add		192	.042		.24	2.19		2.43	4.07
1990	For drilling up, add						40%			

05 05 21.90 Welding Steel

		Crew	Daily Output	Labor-Hours	Unit	Material	2015 Bare Costs Labor	Equipment	Total	Total Incl O&P
0010	**WELDING STEEL**, Structural									
0020	Field welding, 1/8" E6011, cost per welder, no operating engineer	E-14	8	1	Hr.	4.33	54.50	18.25	77.08	120
0200	With 1/2 operating engineer	E-13	8	1.500		4.33	79	18.25	101.58	157
0300	With 1 operating engineer	E-12	8	2		4.33	103	18.25	125.58	194
0500	With no operating engineer, 2# weld rod per ton	E-14	8	1	Ton	4.33	54.50	18.25	77.08	120
0600	8# E6011 per ton	"	2	4		17.30	219	73	309.30	480

For customer support on your Mechanical Cost Data, call 877.795.7394.

45

05 05 21 – Fastening Methods for Metal

05 05 21.90 Welding Steel		Crew	Daily Output	Labor-Hours	Unit	Material	2015 Bare Costs Labor	2015 Bare Costs Equipment	Total	Total Incl O&P
0800	With one operating engineer per welder, 2# E6011 per ton	E-12	8	2	Ton	4.33	103	18.25	125.58	194
0900	8# E6011 per ton	"	2	8	↓	17.30	415	73	505.30	775
1200	Continuous fillet, down welding									
1300	Single pass, 1/8" thick, 0.1#/L.F.	E-14	150	.053	L.F.	.22	2.91	.97	4.10	6.35
1400	3/16" thick, 0.2#/L.F.		75	.107		.43	5.85	1.94	8.22	12.75
1500	1/4" thick, 0.3#/L.F.		50	.160		.65	8.75	2.92	12.32	19.10
1610	5/16" thick, 0.4#/L.F.		38	.211		.87	11.50	3.84	16.21	25
1800	3 passes, 3/8" thick, 0.5#/L.F.		30	.267		1.08	14.55	4.86	20.49	32
2010	4 passes, 1/2" thick, 0.7#/L.F.		22	.364		1.52	19.85	6.65	28.02	43.50
2200	5 to 6 passes, 3/4" thick, 1.3#/L.F.		12	.667		2.81	36.50	12.15	51.46	80
2400	8 to 11 passes, 1" thick, 2.4#/L.F.	↓	6	1.333		5.20	73	24.50	102.70	159
2600	For vertical joint welding, add						20%			
2700	Overhead joint welding, add						300%			
2900	For semi-automatic welding, obstructed joints, deduct						5%			
3000	Exposed joints, deduct				↓		15%			
4000	Cleaning and welding plates, bars, or rods									
4010	to existing beams, columns, or trusses	E-14	12	.667	L.F.	1.08	36.50	12.15	49.73	78

05 05 23 – Metal Fastenings

05 05 23.30 Lag Screws

0010	**LAG SCREWS**									
0020	Steel, 1/4" diameter, 2" long	G	1 Carp	200	.040	Ea.	.09	1.88		1.97	2.99
0100	3/8" diameter, 3" long	G		150	.053		.30	2.50		2.80	4.18
0200	1/2" diameter, 3" long	G		130	.062		.64	2.89		3.53	5.15
0300	5/8" diameter, 3" long	G	↓	120	.067	↓	1.15	3.13		4.28	6.10

05 05 23.35 Machine Screws

0010	**MACHINE SCREWS**									
0020	Steel, round head, #8 x 1" long	G	1 Carp	4.80	1.667	C	3.93	78.50		82.43	124
0110	#8 x 2" long	G		2.40	3.333		6.20	157		163.20	248
0200	#10 x 1" long	G		4	2		5.20	94		99.20	151
0300	#10 x 2" long	G		2	4	↓	8.65	188		196.65	299

05 05 23.50 Powder Actuated Tools and Fasteners

0010	**POWDER ACTUATED TOOLS & FASTENERS**									
0020	Stud driver, .22 caliber, single shot					Ea.	148			148	163
0100	.27 caliber, semi automatic, strip					"	440			440	485
0300	Powder load, single shot, .22 cal, power level 2, brown					C	5.35			5.35	5.90
0400	Strip, .27 cal, power level 4, red						7.70			7.70	8.50
0600	Drive pin, .300 x 3/4" long	G	1 Carp	4.80	1.667	↓	4.11	78.50		82.61	125
0700	.300 x 3" long with washer	G	"	4	2		12.65	94		106.65	159

05 05 23.55 Rivets

0010	**RIVETS**									
0100	Aluminum rivet & mandrel, 1/2" grip length x 1/8" diameter	G	1 Carp	4.80	1.667	C	7.65	78.50		86.15	128
0200	3/16" diameter	G		4	2		11.40	94		105.40	158
0300	Aluminum rivet, steel mandrel, 1/8" diameter	G		4.80	1.667		10.25	78.50		88.75	131
0400	3/16" diameter	G		4	2		16.45	94		110.45	163
0500	Copper rivet, steel mandrel, 1/8" diameter	G		4.80	1.667		9.50	78.50		88	130
1200	Steel rivet and mandrel, 1/8" diameter	G		4.80	1.667		7.40	78.50		85.90	128
1300	3/16" diameter	G	↓	4	2	↓	12	94		106	158
1400	Hand riveting tool, standard					Ea.	70.50			70.50	77.50
1500	Deluxe						365			365	400
1600	Power riveting tool, standard						500			500	555
1700	Deluxe					↓	1,650			1,650	1,825

05 05 Common Work Results for Metals

05 05 23 – Metal Fastenings

05 05 23.70 Structural Blind Bolts		Crew	Daily Output	Labor-Hours	Unit	Material	2015 Bare Costs Labor	Equipment	Total	Total Incl O&P
0010	**STRUCTURAL BLIND BOLTS**	G	1 Sswk	240	.033	Ea.	1.24	1.75		2.99	4.41
0100	1/4" diameter x 1/4" grip	G		216	.037		1.33	1.95		3.28	4.85
0150	1/2" grip	G		232	.034		1.75	1.82		3.57	5.10
0200	3/8" diameter x 1/2" grip	G		208	.038		1.84	2.02		3.86	5.55
0250	3/4" grip	G		224	.036		3.99	1.88		5.87	7.65
0300	1/2" diameter x 1/2" grip	G		200	.040		5.60	2.11		7.71	9.80
0350	3/4" grip	G		216	.037		8.25	1.95		10.20	12.50
0400	5/8" diameter x 3/4" grip	G		192	.042		9.50	2.19		11.69	14.25
0450	1" grip	G									

05 12 Structural Steel Framing

05 12 23 – Structural Steel for Buildings

05 12 23.40 Lightweight Framing

		Crew	Daily Output	Labor-Hours	Unit	Material	2015 Bare Costs Labor	Equipment	Total	Total Incl O&P
0010	**LIGHTWEIGHT FRAMING**									
0015	Made from recycled materials									
0400	Angle framing, field fabricated, 4" and larger	G	E-3	440	.055	Lb.	.77	2.91	.33	4.01	6.25
0450	Less than 4" angles	G		265	.091		.80	4.83	.55	6.18	9.90
0600	Channel framing, field fabricated, 8" and larger	G		500	.048		.80	2.56	.29	3.65	5.65
0650	Less than 8" channels	G		335	.072		.80	3.82	.44	5.06	8
1000	Continuous slotted channel framing system, shop fab, simple framing	G	2 Sswk	2400	.007		4.11	.35		4.46	5.15
1200	Complex framing	G	"	1600	.010		4.64	.53		5.17	6
1250	Plate & bar stock for reinforcing beams and trusses	G					1.46	.		1.46	1.60
1300	Cross bracing, rods, shop fabricated, 3/4" diameter	G	E-3	700	.034		1.59	1.83	.21	3.63	5.15
1310	7/8" diameter	G		850	.028		1.59	1.51	.17	3.27	4.56
1320	1" diameter	G		1000	.024		1.59	1.28	.15	3.02	4.13
1330	Angle, 5" x 5" x 3/8"	G		2800	.009		1.59	.46	.05	2.10	2.60
1350	Hanging lintels, shop fabricated	G		850	.028		1.59	1.51	.17	3.27	4.56
1380	Roof frames, shop fabricated, 3'-0" square, 5' span	G	E-2	4200	.013		1.59	.69	.36	2.64	3.31
1400	Tie rod, not upset, 1-1/2" to 4" diameter, with turnbuckle	G	2 Sswk	800	.020		1.72	1.05		2.77	3.72
1420	No turnbuckle	G		700	.023		1.66	1.20		2.86	3.91
1500	Upset, 1-3/4" to 4" diameter, with turnbuckle	G		800	.020		1.72	1.05		2.77	3.72
1520	No turnbuckle	G		700	.023		1.66	1.20		2.86	3.91

05 12 23.60 Pipe Support Framing

		Crew	Daily Output	Labor-Hours	Unit	Material	2015 Bare Costs Labor	Equipment	Total	Total Incl O&P
0010	**PIPE SUPPORT FRAMING**									
0020	Under 10#/L.F., shop fabricated	G	E-4	3900	.008	Lb.	1.78	.44	.04	2.26	2.75
0200	10.1 to 15#/L.F.	G		4300	.007		1.75	.40	.03	2.18	2.65
0400	15.1 to 20#/L.F.	G		4800	.007		1.72	.35	.03	2.10	2.54
0600	Over 20#/L.F.	G		5400	.006		1.70	.32	.03	2.05	2.45

47

For customer support on your Mechanical Cost Data, call 877.795.7394.

Division Notes

	CREW	DAILY OUTPUT	LABOR-HOURS	UNIT	BARE COSTS				TOTAL INCL O&P
					MAT.	LABOR	EQUIP.	TOTAL	

Estimating Tips

06 05 00 Common Work Results for Wood, Plastics, and Composites

- Common to any wood-framed structure are the accessory connector items such as screws, nails, adhesives, hangers, connector plates, straps, angles, and hold-downs. For typical wood-framed buildings, such as residential projects, the aggregate total for these items can be significant, especially in areas where seismic loading is a concern. For floor and wall framing, the material cost is based on 10 to 25 lbs. per MBF. Hold-downs, hangers, and other connectors should be taken off by the piece.

  Included with material costs are fasteners for a normal installation. RSMeans engineers use manufacturer's recommendations, written specifications, and/or standard construction practice for size and spacing of fasteners. Prices for various fasteners are shown for informational purposes only. Adjustments should be made if unusual fastening conditions exist.

06 10 00 Carpentry

- Lumber is a traded commodity and therefore sensitive to supply and demand in the marketplace. Even in "budgetary" estimating of wood-framed projects, it is advisable to call local suppliers for the latest market pricing.

- Common quantity units for wood-framed projects are "thousand board feet" (MBF). A board foot is a volume of wood, 1" x 1' x 1', or 144 cubic inches. Board-foot quantities are generally calculated using nominal material dimensions— dressed sizes are ignored. Board foot per lineal foot of any stick of lumber can be calculated by dividing the nominal cross-sectional area by 12. As an example, 2,000 lineal feet of 2 x 12 equates to 4 MBF by dividing the nominal area, 2 x 12, by 12, which equals 2, and multiplying by 2,000 to give 4,000 board feet. This simple rule applies to all nominal dimensioned lumber.

- Waste is an issue of concern at the quantity takeoff for any area of construction. Framing lumber is sold in even foot lengths, i.e., 10', 12', 14', 16' and, depending on spans, wall heights, and the grade of lumber, waste is inevitable. A rule of thumb for lumber waste is 5%–10% depending on material quality and the complexity of the framing.

- Wood in various forms and shapes is used in many projects, even where the main structural framing is steel, concrete, or masonry. Plywood as a back-up partition material and 2x boards used as blocking and cant strips around roof edges are two common examples. The estimator should ensure that the costs of all wood materials are included in the final estimate.

06 20 00 Finish Carpentry

- It is necessary to consider the grade of workmanship when estimating labor costs for erecting millwork and interior finish. In practice, there are three grades: premium, custom, and economy. The RSMeans daily output for base and case moldings is in the range of 200 to 250 L.F. per carpenter per day. This is appropriate for most average custom-grade projects. For premium projects, an adjustment to productivity of 25%–50% should be made, depending on the complexity of the job.

Reference Numbers

Reference numbers are shown in shaded boxes at the beginning of some major classifications. These numbers refer to related items in the Reference Section. The reference information may be an estimating procedure, an alternate pricing method, or technical information.

Note: Not all subdivisions listed here necessarily appear in this publication. ■

06 16 36 – Wood Panel Product Sheathing

06 16 36.10 Sheathing	Crew	Daily Output	Labor-Hours	Unit	Material	2015 Bare Costs Labor	Equipment	Total	Total Incl O&P
0010 **SHEATHING**									
0012 Plywood on roofs, CDX									
0030 5/16" thick	2 Carp	1600	.010	S.F.	.56	.47		1.03	1.33
0035 Pneumatic nailed		1952	.008		.56	.39		.95	1.20
0050 3/8" thick		1525	.010		.60	.49		1.09	1.42
0055 Pneumatic nailed		1860	.009		.60	.40		1	1.28
0100 1/2" thick		1400	.011		.66	.54		1.20	1.55
0105 Pneumatic nailed		1708	.009		.66	.44		1.10	1.40
0200 5/8" thick		1300	.012		.78	.58		1.36	1.75
0205 Pneumatic nailed		1586	.010		.78	.47		1.25	1.59
0300 3/4" thick		1200	.013		.93	.63		1.56	1.98
0305 Pneumatic nailed		1464	.011		.93	.51		1.44	1.81
0500 Plywood on walls, with exterior CDX, 3/8" thick		1200	.013		.60	.63		1.23	1.62
0505 Pneumatic nailed		1488	.011		.60	.50		1.10	1.44
0600 1/2" thick		1125	.014		.66	.67		1.33	1.75
0605 Pneumatic nailed		1395	.011		.66	.54		1.20	1.55
0700 5/8" thick		1050	.015		.78	.72		1.50	1.96
0705 Pneumatic nailed		1302	.012		.78	.58		1.36	1.75
0800 3/4" thick		975	.016		.93	.77		1.70	2.21
0805 Pneumatic nailed		1209	.013		.93	.62		1.55	1.98

Estimating Tips

07 10 00 Dampproofing and Waterproofing

■ Be sure of the job specifications before pricing this subdivision. The difference in cost between waterproofing and dampproofing can be great. Waterproofing will hold back standing water. Dampproofing prevents the transmission of water vapor. Also included in this section are vapor retarding membranes.

07 20 00 Thermal Protection

■ Insulation and fireproofing products are measured by area, thickness, volume or R-value. Specifications may give only what the specific R-value should be in a certain situation. The estimator may need to choose the type of insulation to meet that R-value.

07 30 00 Steep Slope Roofing
07 40 00 Roofing and Siding Panels

■ Many roofing and siding products are bought and sold by the square. One square is equal to an area that measures 100 square feet.

This simple change in unit of measure could create a large error if the estimator is not observant. Accessories necessary for a complete installation must be figured into any calculations for both material and labor.

07 50 00 Membrane Roofing
07 60 00 Flashing and Sheet Metal
07 70 00 Roofing and Wall Specialties and Accessories

■ The items in these subdivisions compose a roofing system. No one component completes the installation, and all must be estimated. Built-up or single-ply membrane roofing systems are made up of many products and installation trades. Wood blocking at roof perimeters or penetrations, parapet coverings, reglets, roof drains, gutters, downspouts, sheet metal flashing, skylights, smoke vents, and roof hatches all need to be considered along with the roofing material. Several different installation trades will need to work together on the roofing system. Inherent difficulties in the scheduling and coordination of various trades must be accounted for when estimating labor costs.

07 90 00 Joint Protection

■ To complete the weather-tight shell, the sealants and caulkings must be estimated. Where different materials meet—at expansion joints, at flashing penetrations, and at hundreds of other locations throughout a construction project—they provide another line of defense against water penetration. Often, an entire system is based on the proper location and placement of caulking or sealants. The detailed drawings that are included as part of a set of architectural plans show typical locations for these materials. When caulking or sealants are shown at typical locations, this means the estimator must include them for all the locations where this detail is applicable. Be careful to keep different types of sealants separate, and remember to consider backer rods and primers if necessary.

Reference Numbers

Reference numbers are shown in shaded boxes at the beginning of some major classifications. These numbers refer to related items in the Reference Section. The reference information may be an estimating procedure, an alternate pricing method, or technical information.

Note: Not all subdivisions listed here necessarily appear in this publication. ■

Division 7 – Thermal & Moisture Protection

Estimate with precision with RSMeans Online...
Quick, intuitive and easy-to-use, RSMeans Online gives you instant access to hundreds of thousands of material, labor, and equipment costs from RSMeans' comprehensive database, delivering the information you need to build competitive estimates every time.

Start your free trial today at **www.rsmeansonline.com**

RSMeansOnline

07 65 Flexible Flashing

07 65 10 – Sheet Metal Flashing

07 65 10.10 Sheet Metal Flashing and Counter Flashing	Crew	Daily Output	Labor-Hours	Unit	Material	2015 Bare Costs Labor	Equipment	Total	Total Incl O&P
0010 **SHEET METAL FLASHING AND COUNTER FLASHING**									
0011 Including up to 4 bends									
0020 Aluminum, mill finish, .013" thick	1 Rofc	145	.055	S.F.	.79	2.21		3	4.65
0030 .016" thick		145	.055		.91	2.21		3.12	4.78
0060 .019" thick		145	.055		1.15	2.21		3.36	5.05
0100 .032" thick		145	.055		1.35	2.21		3.56	5.25
0200 .040" thick		145	.055		2.31	2.21		4.52	6.30
0300 .050" thick		145	.055		2.43	2.21		4.64	6.45
0400 Painted finish, add					.29			.29	.32
1600 Copper, 16 oz., sheets, under 1000 lb.	1 Rofc	115	.070		7.65	2.79		10.44	13.20
1700 Over 4000 lb.		155	.052		7.65	2.07		9.72	12
1900 20 oz. sheets, under 1000 lb.		110	.073		9.40	2.92		12.32	15.35
2000 Over 4000 lb.		145	.055		8.95	2.21		11.16	13.60
2200 24 oz. sheets, under 1000 lb.		105	.076		13.30	3.06		16.36	19.85
2300 Over 4000 lb.		135	.059		12.65	2.38		15.03	17.95
2500 32 oz. sheets, under 1000 lb.		100	.080		17.80	3.21		21.01	25
2600 Over 4000 lb.		130	.062		16.90	2.47		19.37	23
5800 Lead, 2.5 lb. per S.F., up to 12" wide		135	.059		4.99	2.38		7.37	9.55
5900 Over 12" wide		135	.059		3.99	2.38		6.37	8.45
8900 Stainless steel sheets, 32 ga.,		155	.052		3	2.07		5.07	6.85
9000 28 ga.		155	.052		3.99	2.07		6.06	7.90
9100 26 ga.		155	.052		4.25	2.07		6.32	8.20
9200 24 ga.		155	.052		4.75	2.07		6.82	8.80
9400 Terne coated stainless steel, .015" thick, 28 ga		155	.052		7.35	2.07		9.42	11.65
9500 .018" thick, 26 ga		155	.052		8.30	2.07		10.37	12.65
9600 Zinc and copper alloy (brass), .020" thick		155	.052		9.30	2.07		11.37	13.80
9700 .027" thick		155	.052		11	2.07		13.07	15.65
9800 .032" thick		155	.052		14	2.07		16.07	18.95
9900 .040" thick		155	.052		18.50	2.07		20.57	24

07 65 13 – Laminated Sheet Flashing

07 65 13.10 Laminated Sheet Flashing	Crew	Daily Output	Labor-Hours	Unit	Material	2015 Bare Costs Labor	Equipment	Total	Total Incl O&P
0010 **LAMINATED SHEET FLASHING**, Including up to 4 bends									
4300 Copper-clad stainless steel, .015" thick, under 500 lb.	1 Rofc	115	.070	S.F.	6.40	2.79		9.19	11.80
4400 Over 2000 lb.		155	.052		6.10	2.07		8.17	10.25
4600 .018" thick, under 500 lb.		100	.080		7.20	3.21		10.41	13.40
4700 Over 2000 lb.		145	.055		6.80	2.21		9.01	11.30
8550 Shower pan, 3 ply copper and fabric, 3 oz.		155	.052		3.50	2.07		5.57	7.40
8600 7 oz.		155	.052		4.20	2.07		6.27	8.15
9300 Stainless steel, paperbacked 2 sides, .005" thick		330	.024		3.62	.97		4.59	5.65

07 65 19 – Plastic Sheet Flashing

07 65 19.10 Plastic Sheet Flashing and Counter Flashing	Crew	Daily Output	Labor-Hours	Unit	Material	2015 Bare Costs Labor	Equipment	Total	Total Incl O&P
0010 **PLASTIC SHEET FLASHING AND COUNTER FLASHING**									
7300 Polyvinyl chloride, black, 10 mil	1 Rofc	285	.028	S.F.	.21	1.13		1.34	2.15
7400 20 mil		285	.028		.28	1.13		1.41	2.23
7600 30 mil		285	.028		.35	1.13		1.48	2.31
7700 60 mil		285	.028		.85	1.13		1.98	2.86
7900 Black or white for exposed roofs, 60 mil		285	.028		1.80	1.13		2.93	3.90
8060 PVC tape, 5" x 45 mils, for joint covers, 100 L.F./roll				Ea.	170			170	187
8850 Polyvinyl chloride, 30 mil	1 Rofc	160	.050	S.F.	1.30	2.01		3.31	4.85

07 71 Roof Specialties

07 71 23 – Manufactured Gutters and Downspouts

07 71 23.10 Downspouts	Crew	Daily Output	Labor-Hours	Unit	Material	2015 Bare Costs Labor	Equipment	Total	Total Incl O&P
0010 **DOWNSPOUTS**									
0020 Aluminum, embossed, .020" thick, 2" x 3"	1 Shee	190	.042	L.F.	.98	2.36		3.34	4.68
0100 Enameled		190	.042		1.37	2.36		3.73	5.10
4800 Steel, galvanized, round, corrugated, 2" or 3" diameter, 28 ga.		190	.042		2.08	2.36		4.44	5.90
4900 4" diameter, 28 ga.		145	.055		2.52	3.09		5.61	7.50
5100 5" diameter, 26 ga.		130	.062		4	3.44		7.44	9.65
5400 6" diameter, 28 ga.		105	.076		4.20	4.26		8.46	11.10
5500 26 ga.		105	.076		4	4.26		8.26	10.90

07 72 Roof Accessories

07 72 26 – Ridge Vents

07 72 26.10 Ridge Vents and Accessories

	Crew	Daily Output	Labor-Hours	Unit	Material	2015 Bare Costs Labor	Equipment	Total	Total Incl O&P
0010 **RIDGE VENTS AND ACCESSORIES**									
2300 Ridge vent strip, mill finish	1 Shee	155	.052	L.F.	3.85	2.89		6.74	8.65

07 72 53 – Snow Guards

07 72 53.10 Snow Guard Options

	Crew	Daily Output	Labor-Hours	Unit	Material	2015 Bare Costs Labor	Equipment	Total	Total Incl O&P
0010 **SNOW GUARD OPTIONS**									
0100 Slate & asphalt shingle roofs, fastened with nails	1 Rofc	160	.050	Ea.	12.55	2.01		14.56	17.20
0200 Standing seam metal roofs, fastened with set screws		48	.167		16.95	6.70		23.65	30
0300 Surface mount for metal roofs, fastened with solder		48	.167		6.05	6.70		12.75	18.05
0400 Double rail pipe type, including pipe		130	.062	L.F.	32	2.47		34.47	39

07 72 73 – Pitch Pockets

07 72 73.10 Pitch Pockets, Variable Sizes

	Crew	Daily Output	Labor-Hours	Unit	Material	2015 Bare Costs Labor	Equipment	Total	Total Incl O&P
0010 **PITCH POCKETS, VARIABLE SIZES**									
0100 Adjustable, 4" to 7", welded corners, 4" deep	1 Rofc	48	.167	Ea.	16.25	6.70		22.95	29.50
0200 Side extenders, 6"	"	240	.033	"	2.50	1.34		3.84	5.05

07 76 Roof Pavers

07 76 16 – Roof Decking Pavers

07 76 16.10 Roof Pavers and Supports

	Crew	Daily Output	Labor-Hours	Unit	Material	2015 Bare Costs Labor	Equipment	Total	Total Incl O&P
0010 **ROOF PAVERS AND SUPPORTS**									
1000 Roof decking pavers, concrete blocks, 2" thick, natural	1 Clab	115	.070	S.F.	3.37	2.62		5.99	7.75
1100 Colors		115	.070	"	3.73	2.62		6.35	8.10
1200 Support pedestal, bottom cap		960	.008	Ea.	3	.31		3.31	3.78
1300 Top cap		960	.008		4.80	.31		5.11	5.80
1400 Leveling shims, 1/16"		1920	.004		1.20	.16		1.36	1.56
1500 1/8"		1920	.004		1.20	.16		1.36	1.56
1600 Buffer pad		960	.008		2.50	.31		2.81	3.23
1700 PVC legs (4" SDR 35)		2880	.003	Inch	.12	.10		.22	.29
2000 Alternate pricing method, system in place		101	.079	S.F.	7	2.98		9.98	12.30

For customer support on your Mechanical Cost Data, call 877.795.7394.

53

07 84 13.10 Firestopping		Crew	Daily Output	Labor-Hours	Unit	Material	2015 Bare Costs Labor	Equipment	Total	Total Incl O&P
0010	**FIRESTOPPING**	R078413-30								
0100	Metallic piping, non insulated									
0110	Through walls, 2" diameter	1 Carp	16	.500	Ea.	17.20	23.50		40.70	55
0120	4" diameter		14	.571		26	27		53	70.50
0130	6" diameter		12	.667		35.50	31.50		67	87
0140	12" diameter		10	.800		62.50	37.50		100	127
0150	Through floors, 2" diameter		32	.250		9.80	11.75		21.55	29
0160	4" diameter		28	.286		14.20	13.40		27.60	36
0170	6" diameter		24	.333		18.50	15.65		34.15	44.50
0180	12" diameter		20	.400		31.50	18.80		50.30	63.50
0190	Metallic piping, insulated									
0200	Through walls, 2" diameter	1 Carp	16	.500	Ea.	24	23.50		47.50	62.50
0210	4" diameter		14	.571		33	27		60	77.50
0220	6" diameter		12	.667		42	31.50		73.50	94
0230	12" diameter		10	.800		68.50	37.50		106	134
0240	Through floors, 2" diameter		32	.250		16.60	11.75		28.35	36.50
0250	4" diameter		28	.286		21	13.40		34.40	43.50
0260	6" diameter		24	.333		25.50	15.65		41.15	52
0270	12" diameter		20	.400		31.50	18.80		50.30	63.50
0280	Non metallic piping, non insulated									
0290	Through walls, 2" diameter	1 Carp	12	.667	Ea.	68.50	31.50		100	124
0300	4" diameter		10	.800		86	37.50		123.50	153
0310	6" diameter		8	1		120	47		167	205
0330	Through floors, 2" diameter		16	.500		54	23.50		77.50	95.50
0340	4" diameter		6	1.333		66.50	62.50		129	170
0350	6" diameter		6	1.333		80.50	62.50		143	185
0370	Ductwork, insulated & non insulated, round									
0380	Through walls, 6" diameter	1 Carp	12	.667	Ea.	35	31.50		66.50	86.50
0390	12" diameter		10	.800		69.50	37.50		107	135
0400	18" diameter		8	1		113	47		160	198
0410	Through floors, 6" diameter		16	.500		18.60	23.50		42.10	56.50
0420	12" diameter		14	.571		34	27		61	79
0430	18" diameter		12	.667		59	31.50		90.50	113
0440	Ductwork, insulated & non insulated, rectangular									
0450	With stiffener/closure angle, through walls, 6" x 12"	1 Carp	8	1	Ea.	28	47		75	104
0460	12" x 24"		6	1.333		37.50	62.50		100	138
0470	24" x 48"		4	2		107	94		201	262
0480	With stiffener/closure angle, through floors, 6" x 12"		10	.800		15.40	37.50		52.90	75
0490	12" x 24"		8	1		28	47		75	103
0500	24" x 48"		6	1.333		54	62.50		116.50	156
0510	Multi trade openings									
0520	Through walls, 6" x 12"	1 Carp	2	4	Ea.	59	188		247	355
0530	12" x 24"	"	1	8		238	375		613	840
0540	24" x 48"	2 Carp	1	16		950	750		1,700	2,200
0550	48" x 96"	"	.75	21.333		3,700	1,000		4,700	5,625
0560	Through floors, 6" x 12"	1 Carp	2	4		39	188		227	330
0570	12" x 24"	"	1	8		157	375		532	755
0580	24" x 48"	2 Carp	.75	21.333		625	1,000		1,625	2,250
0590	48" x 96"	"	.50	32		2,450	1,500		3,950	4,975
0600	Structural penetrations, through walls									
0610	Steel beams, W8 x 10	1 Carp	8	1	Ea.	37.50	47		84.50	114
0620	W12 x 14		6	1.333		59	62.50		121.50	162
0630	W21 x 44		5	1.600		118	75		193	246

For customer support on your Mechanical Cost Data, call 877.795.7394.

07 84 Firestopping

07 84 13 – Penetration Firestopping

07 84 13.10 Firestopping		Crew	Daily Output	Labor-Hours	Unit	Material	2015 Bare Costs Labor	Equipment	Total	Total Incl O&P
0640	W36 x 135	1 Carp	3	2.667	Ea.	287	125		412	510
0650	Bar joists, 18" deep		6	1.333		54.50	62.50		117	156
0660	24" deep		6	1.333		67.50	62.50		130	171
0670	36" deep		5	1.600		101	75		176	228
0680	48" deep		4	2		118	94		212	275
0690	Construction joints, floor slab at exterior wall									
0700	Precast, brick, block or drywall exterior									
0710	2" wide joint	1 Carp	125	.064	L.F.	8.50	3		11.50	13.95
0720	4" wide joint	"	75	.107	"	17	5		22	26.50
0730	Metal panel, glass or curtain wall exterior									
0740	2" wide joint	1 Carp	40	.200	L.F.	20	9.40		29.40	36.50
0750	4" wide joint	"	25	.320	"	28	15		43	53.50
0760	Floor slab to drywall partition									
0770	Flat joint	1 Carp	100	.080	L.F.	8.40	3.76		12.16	15.05
0780	Fluted joint		50	.160		17.40	7.50		24.90	30.50
0790	Etched fluted joint		75	.107		11.10	5		16.10	19.90
0800	Floor slab to concrete/masonry partition									
0810	Flat joint	1 Carp	75	.107	L.F.	18.75	5		23.75	28
0820	Fluted joint	"	50	.160	"	22.50	7.50		30	36
0830	Concrete/CMU wall joints									
0840	1" wide	1 Carp	100	.080	L.F.	10.25	3.76		14.01	17.10
0850	2" wide		75	.107		18.75	5		23.75	28
0860	4" wide		50	.160		38	7.50		45.50	53
0870	Concrete/CMU floor joints									
0880	1" wide	1 Carp	200	.040	L.F.	5.15	1.88		7.03	8.55
0890	2" wide		150	.053		9.40	2.50		11.90	14.20
0900	4" wide		100	.080		17.90	3.76		21.66	25.50

07 91 Preformed Joint Seals

07 91 13 – Compression Seals

07 91 13.10 Compression Seals

		Crew	Daily Output	Labor-Hours	Unit	Material	2015 Bare Costs Labor	Equipment	Total	Total Incl O&P
0010	**COMPRESSION SEALS**									
4900	O-ring type cord, 1/4"	1 Bric	472	.017	L.F.	.37	.78		1.15	1.60
4910	1/2"		440	.018		.95	.84		1.79	2.33
4920	3/4"		424	.019		1.80	.87		2.67	3.31
4930	1"		408	.020		3.40	.91		4.31	5.10
4940	1-1/4"		384	.021		6.30	.96		7.26	8.40
4950	1-1/2"		368	.022		7.95	1		8.95	10.30
4960	1-3/4"		352	.023		13.25	1.05		14.30	16.20
4970	2"		344	.023		18.50	1.07		19.57	22

07 91 16 – Joint Gaskets

07 91 16.10 Joint Gaskets

		Crew	Daily Output	Labor-Hours	Unit	Material	2015 Bare Costs Labor	Equipment	Total	Total Incl O&P
0010	**JOINT GASKETS**									
4400	Joint gaskets, neoprene, closed cell w/adh, 1/8" x 3/8"	1 Bric	240	.033	L.F.	.28	1.54		1.82	2.66
4500	1/4" x 3/4"		215	.037		.59	1.72		2.31	3.27
4700	1/2" x 1"		200	.040		1.30	1.85		3.15	4.25
4800	3/4" x 1-1/2"		165	.048		1.45	2.24		3.69	5

For customer support on your Mechanical Cost Data, call 877.795.7394.

55

07 91 Preformed Joint Seals

07 91 23 – Backer Rods

07 91 23.10 Backer Rods		Crew	Daily Output	Labor-Hours	Unit	Material	2015 Bare Costs Labor	Equipment	Total	Total Incl O&P
0010	**BACKER RODS**									
0030	Backer rod, polyethylene, 1/4" diameter	1 Bric	4.60	1.739	C.L.F.	2.13	80.50		82.63	125
0050	1/2" diameter		4.60	1.739		3.40	80.50		83.90	127
0070	3/4" diameter		4.60	1.739		5.55	80.50		86.05	129
0090	1" diameter		4.60	1.739		8.85	80.50		89.35	133

07 91 26 – Joint Fillers

07 91 26.10 Joint Fillers

		Crew	Daily Output	Labor-Hours	Unit	Material	Labor	Equipment	Total	Total Incl O&P
0010	**JOINT FILLERS**									
4360	Butyl rubber filler, 1/4" x 1/4"	1 Bric	290	.028	L.F.	.22	1.27		1.49	2.19
4365	1/2" x 1/2"		250	.032		.89	1.48		2.37	3.23
4370	1/2" x 3/4"		210	.038		1.34	1.76		3.10	4.15
4375	3/4" x 3/4"		230	.035		2.01	1.61		3.62	4.66
4380	1" x 1"		180	.044		2.68	2.05		4.73	6.05
4390	For coloring, add					12%				
4980	Polyethylene joint backing, 1/4" x 2"	1 Bric	2.08	3.846	C.L.F.	12	178		190	284
4990	1/4" x 6"		1.28	6.250	"	28	288		316	470
5600	Silicone, room temp vulcanizing foam seal, 1/4" x 1/2"		1312	.006	L.F.	.33	.28		.61	.80
5610	1/2" x 1/2"		656	.012		.67	.56		1.23	1.59
5620	1/2" x 3/4"		442	.018		1	.84		1.84	2.38
5630	3/4" x 3/4"		328	.024		1.50	1.13		2.63	3.37
5640	1/8" x 1"		1312	.006		.33	.28		.61	.80
5650	1/8" x 3"		442	.018		1	.84		1.84	2.38
5670	1/4" x 3"		295	.027		2	1.25		3.25	4.11
5680	1/4" x 6"		148	.054		3.99	2.49		6.48	8.20
5690	1/2" x 6"		82	.098		8	4.50		12.50	15.65
5700	1/2" x 9"		52.50	.152		12	7.05		19.05	24
5710	1/2" x 12"		33	.242		15.95	11.20		27.15	34.50

07 92 Joint Sealants

07 92 13 – Elastomeric Joint Sealants

07 92 13.20 Caulking and Sealant Options

		Crew	Daily Output	Labor-Hours	Unit	Material	Labor	Equipment	Total	Total Incl O&P
0010	**CAULKING AND SEALANT OPTIONS**									
0050	Latex acrylic based, bulk				Gal.	26.50			26.50	29.50
0055	Bulk in place 1/4" x 1/4" bead	1 Bric	300	.027	L.F.	.08	1.23		1.31	1.97
0060	1/4" x 3/8"		294	.027		.14	1.26		1.40	2.07
0065	1/4" x 1/2"		288	.028		.19	1.28		1.47	2.17
0075	3/8" x 3/8"		284	.028		.21	1.30		1.51	2.21
0080	3/8" x 1/2"		280	.029		.28	1.32		1.60	2.32
0085	3/8" x 5/8"		276	.029		.35	1.34		1.69	2.43
0095	3/8" x 3/4"		272	.029		.42	1.36		1.78	2.53
0100	1/2" x 1/2"		275	.029		.37	1.34		1.71	2.46
0105	1/2" x 5/8"		269	.030		.47	1.37		1.84	2.61
0110	1/2" x 3/4"		263	.030		.56	1.40		1.96	2.76
0115	1/2" x 7/8"		256	.031		.65	1.44		2.09	2.92
0120	1/2" x 1"		250	.032		.75	1.48		2.23	3.07
0125	3/4" x 3/4"		244	.033		.84	1.51		2.35	3.24
0130	3/4" x 1"		225	.036		1.12	1.64		2.76	3.74
0135	1" x 1"		200	.040		1.50	1.85		3.35	4.47
0190	Cartridges				Gal.	32			32	35.50
0200	11 fl. oz. cartridge				Ea.	2.76			2.76	3.04

For customer support on your Mechanical Cost Data, call 877.795.7394.

07 92 Joint Sealants

07 92 13 – Elastomeric Joint Sealants

07 92 13.20 Caulking and Sealant Options	Crew	Daily Output	Labor-Hours	Unit	Material	2015 Bare Costs Labor	Equipment	Total	Total Incl O&P
0500	1/4" x 1/2"	1 Bric	288	.028	L.F.	.23	1.28		1.51	2.21
0600	1/2" x 1/2"		275	.029		.45	1.34		1.79	2.55
0800	3/4" x 3/4"		244	.033		1.01	1.51		2.52	3.43
0900	3/4" x 1"		225	.036		1.35	1.64		2.99	4
1000	1" x 1"		200	.040		1.69	1.85		3.54	4.68
1400	Butyl based, bulk				Gal.	35			35	38.50
1500	Cartridges				"	38.50			38.50	42
1700	1/4" x 1/2", 154 L.F./gal.	1 Bric	288	.028	L.F.	.23	1.28		1.51	2.21
1800	1/2" x 1/2", 77 L.F./gal.	"	275	.029	"	.45	1.34		1.79	2.55
2300	Polysulfide compounds, 1 component, bulk				Gal.	72			72	79
2600	1 or 2 component, in place, 1/4" x 1/4", 308 L.F./gal.	1 Bric	300	.027	L.F.	.23	1.23		1.46	2.14
2700	1/2" x 1/4", 154 L.F./gal.		288	.028		.47	1.28		1.75	2.47
2900	3/4" x 3/8", 68 L.F./gal.		272	.029		1.06	1.36		2.42	3.23
3000	1" x 1/2", 38 L.F./gal.		250	.032		1.89	1.48		3.37	4.33
3200	Polyurethane, 1 or 2 component				Gal.	49			49	54
3300	Cartridges				"	64			64	70.50
3500	Bulk, in place, 1/4" x 1/4"	1 Bric	300	.027	L.F.	.16	1.23		1.39	2.06
3655	1/2" x 1/4"		288	.028		.32	1.28		1.60	2.31
3800	3/4" x 3/8"		272	.029		.72	1.36		2.08	2.87
3900	1" x 1/2"		250	.032		1.28	1.48		2.76	3.66
4100	Silicone rubber, bulk				Gal.	49			49	54
4200	Cartridges				"	51			51	56

07 92 16 – Rigid Joint Sealants

07 92 16.10 Rigid Joint Sealants

		Crew	Daily Output	Labor-Hours	Unit	Material	Labor	Equipment	Total	Total Incl O&P
0010	**RIGID JOINT SEALANTS**									
5800	Tapes, sealant, PVC foam adhesive, 1/16" x 1/4"				C.L.F.	10			10	11
5900	1/16" x 1/2"					10			10	11
5950	1/16" x 1"					16			16	17.60
6000	1/8" x 1/2"					9.50			9.50	10.45

07 92 19 – Acoustical Joint Sealants

07 92 19.10 Acoustical Sealant

		Crew	Daily Output	Labor-Hours	Unit	Material	Labor	Equipment	Total	Total Incl O&P
0010	**ACOUSTICAL SEALANT**									
0020	Acoustical sealant, elastomeric, cartridges				Ea.	8.50			8.50	9.35
0025	In place, 1/4" x 1/4"	1 Bric	300	.027	L.F.	.35	1.23		1.58	2.26
0030	1/4" x 1/2"		288	.028		.69	1.28		1.97	2.72
0035	1/2" x 1/2"		275	.029		1.39	1.34		2.73	3.57
0040	1/2" x 3/4"		263	.030		2.08	1.40		3.48	4.43
0045	3/4" x 3/4"		244	.033		3.12	1.51		4.63	5.75
0050	1" x 1"		200	.040		5.55	1.85		7.40	8.90

Division Notes

		CREW	DAILY OUTPUT	LABOR-HOURS	UNIT	BARE COSTS				TOTAL INCL O&P
						MAT.	LABOR	EQUIP.	TOTAL	

Estimating Tips

08 10 00 Doors and Frames

All exterior doors should be addressed for their energy conservation (insulation and seals).

- Most metal doors and frames look alike, but there may be significant differences among them. When estimating these items, be sure to choose the line item that most closely compares to the specification or door schedule requirements regarding:
  - □ type of metal
  - □ metal gauge
  - □ door core material
  - □ fire rating
  - □ finish

- Wood and plastic doors vary considerably in price. The primary determinant is the veneer material. Lauan, birch, and oak are the most common veneers. Other variables include the following:
  - □ hollow or solid core
  - □ fire rating
  - □ flush or raised panel
  - □ finish

- Door pricing includes bore for cylindrical lockset and mortise for hinges.

08 30 00 Specialty Doors and Frames

- There are many varieties of special doors, and they are usually priced per each. Add frames, hardware, or operators required for a complete installation.

08 40 00 Entrances, Storefronts, and Curtain Walls

- Glazed curtain walls consist of the metal tube framing and the glazing material. The cost data in this subdivision is presented for the metal tube framing alone or the composite wall. If your estimate requires a detailed takeoff of the framing, be sure to add the glazing cost and any tints.

08 50 00 Windows

- Most metal windows are delivered preglazed. However, some metal windows are priced without glass. Refer to 08 80 00 Glazing for glass pricing. The grade C indicates commercial grade windows, usually ASTM C-35.

- All wood windows and vinyl are priced preglazed. The glazing is insulating glass. Add the cost of screens and grills if required, and not already included.

08 70 00 Hardware

- Hardware costs add considerably to the cost of a door. The most efficient method to determine the hardware requirements for a project is to review the door and hardware schedule together. One type of door may have different hardware, depending on the door usage.

- Door hinges are priced by the pair, with most doors requiring 1-1/2 pairs per door. The hinge prices do not include installation labor, because it is included in door installation.

Hinges are classified according to the frequency of use, base material, and finish.

08 80 00 Glazing

- Different openings require different types of glass. The most common types are:
  - □ float
  - □ tempered
  - □ insulating
  - □ impact-resistant
  - □ ballistic-resistant

- Most exterior windows are glazed with insulating glass. Entrance doors and window walls, where the glass is less than 18" from the floor, are generally glazed with tempered glass. Interior windows and some residential windows are glazed with float glass.

- Coastal communities require the use of impact-resistant glass, dependant on wind speed.

- The insulation or 'u' value is a strong consideration, along with solar heat gain, to determine total energy efficiency.

Reference Numbers

Reference numbers are shown in shaded boxes at the beginning of some major classifications. These numbers refer to related items in the Reference Section. The reference information may be an estimating procedure, an alternate pricing method, or technical information.

Note: Not all subdivisions listed here necessarily appear in this publication. ∎

08 34 13 – Cold Storage Doors

08 34 13.10 Doors for Cold Area Storage	Crew	Daily Output	Labor-Hours	Unit	Material	2015 Bare Costs Labor	2015 Bare Costs Equipment	Total	Total Incl O&P
0010 **DOORS FOR COLD AREA STORAGE**									
0020 Single, 20 ga. galvanized steel									
0300 Horizontal sliding, 5' x 7', manual operation, 3.5" thick	2 Carp	2	8	Ea.	3,250	375		3,625	4,150
0400 4" thick		2	8		3,250	375		3,625	4,150
0500 6" thick		2	8		3,425	375		3,800	4,350
0800 5' x 7', power operation, 2" thick		1.90	8.421		5,600	395		5,995	6,750
0900 4" thick		1.90	8.421		5,700	395		6,095	6,875
1000 6" thick		1.90	8.421		6,500	395		6,895	7,750
1300 9' x 10', manual operation, 2" insulation		1.70	9.412		4,550	440		4,990	5,675
1400 4" insulation		1.70	9.412		4,675	440		5,115	5,825
1500 6" insulation		1.70	9.412		5,575	440		6,015	6,825
1800 Power operation, 2" insulation		1.60	10		7,675	470		8,145	9,175
1900 4" insulation		1.60	10		7,900	470		8,370	9,400
2000 6" insulation		1.70	9.412		8,850	440		9,290	10,400
2300 For stainless steel face, add					25%				
3000 Hinged, lightweight, 3' x 7'-0", 2" thick	2 Carp	2	8	Ea.	1,425	375		1,800	2,125
3050 4" thick		1.90	8.421		1,775	395		2,170	2,550
3300 Polymer doors, 3' x 7'-0"		1.90	8.421		1,350	395		1,745	2,075
3350 6" thick		1.40	11.429		2,350	535		2,885	3,400
3600 Stainless steel, 3' x 7'-0", 4" thick		1.90	8.421		1,725	395		2,120	2,500
3650 6" thick		1.40	11.429		2,950	535		3,485	4,050
3900 Painted, 3' x 7'-0", 4" thick		1.90	8.421		1,275	395		1,670	2,000
3950 6" thick		1.40	11.429		2,350	535		2,885	3,400
5000 Bi-parting, electric operated									
5010 6' x 8' opening, galv. faces, 4" thick for cooler	2 Carp	.80	20	Opng.	7,225	940		8,165	9,375
5050 For freezer, 4" thick		.80	20		7,950	940		8,890	10,200
5300 For door buck framing and door protection, add		2.50	6.400		640	300		940	1,175
6000 Galvanized batten door, galvanized hinges, 4' x 7'		2	8		1,825	375		2,200	2,600
6050 6' x 8'		1.80	8.889		2,475	415		2,890	3,375
6500 Fire door, 3 hr., 6' x 8', single slide		.80	20		8,275	940		9,215	10,600
6550 Double, bi-parting		.70	22.857		12,200	1,075		13,275	15,100

08 91 16 – Operable Louvers

08 91 16.10 Movable Blade Louvers

	Crew	Daily Output	Labor-Hours	Unit	Material	2015 Bare Costs Labor	2015 Bare Costs Equipment	Total	Total Incl O&P
0010 **MOVABLE BLADE LOUVERS**									
0100 PVC, commercial grade, 12" x 12"	1 Shee	20	.400	Ea.	82	22.50		104.50	124
0110 16" x 16"		14	.571		96	32		128	155
0120 18" x 18"		14	.571		99	32		131	158
0130 24" x 24"		14	.571		140	32		172	203
0140 30" x 30"		12	.667		168	37.50		205.50	242
0150 36" x 36"		12	.667		197	37.50		234.50	274
0300 Stainless steel, commercial grade, 12" x 12"		20	.400		219	22.50		241.50	275
0310 16" x 16"		14	.571		265	32		297	340
0320 18" x 18"		14	.571		282	32		314	360
0330 20" x 20"		14	.571		325	32		357	405
0340 24" x 24"		14	.571		385	32		417	475
0350 30" x 30"		12	.667		430	37.50		467.50	530
0360 36" x 36"		12	.667		520	37.50		557.50	630

08 91 Louvers

08 91 19 – Fixed Louvers

08 91 19.10 Aluminum Louvers

	Crew	Daily Output	Labor-Hours	Unit	Material	2015 Bare Costs Labor	2015 Bare Costs Equipment	Total	Total Incl O&P
0010 **ALUMINUM LOUVERS**									
0020 Aluminum with screen, residential, 8" x 8"	1 Carp	38	.211	Ea.	20	9.90		29.90	37
0100 12" x 12"		38	.211		16	9.90		25.90	33
0200 12" x 18"		35	.229		20	10.75		30.75	38.50
0250 14" x 24"		30	.267		29	12.50		41.50	51.50
0300 18" x 24"		27	.296		32	13.90		45.90	56.50
0500 24" x 30"		24	.333		60	15.65		75.65	90
0700 Triangle, adjustable, small		20	.400		55	18.80		73.80	89.50
0800 Large		15	.533		76	25		101	122
1200 Extruded aluminum, see Section 23 37 15.40									
2100 Midget, aluminum, 3/4" deep, 1" diameter	1 Carp	85	.094	Ea.	.75	4.42		5.17	7.65
2150 3" diameter		60	.133		2.47	6.25		8.72	12.35
2200 4" diameter		50	.160		4.95	7.50		12.45	17
2250 6" diameter		30	.267		4.10	12.50		16.60	24
3000 PVC, commercial grade, 12" x 12"	1 Shee	20	.400		97	22.50		119.50	141
3010 12" x 18"		20	.400		109	22.50		131.50	154
3020 12" x 24"		20	.400		146	22.50		168.50	195
3030 14" x 24"		20	.400		152	22.50		174.50	201
3100 Aluminum, commercial grade, 12" x 12"		20	.400		171	22.50		193.50	222
3110 24" x 24"		16	.500		249	28		277	315
3120 24" x 36"		16	.500		300	28		328	375
3130 36" x 24"		16	.500		310	28		338	385
3140 36" x 36"		14	.571		390	32		422	480
3150 36" x 48"		10	.800		400	45		445	510
3160 48" x 36"		10	.800		390	45		435	500
3170 48" x 48"		10	.800		435	45		480	545
3180 60" x 48"		8	1		485	56		541	620
3190 60" x 60"		8	1		600	56		656	745

08 91 19.20 Steel Louvers

	Crew	Daily Output	Labor-Hours	Unit	Material	2015 Bare Costs Labor	2015 Bare Costs Equipment	Total	Total Incl O&P
0010 **STEEL LOUVERS**									
3300 Galvanized Steel, fixed blades, commercial grade, 18" x 18"	1 Shee	20	.400	Ea.	197	22.50		219.50	251
3310 24" x 24"		20	.400		232	22.50		254.50	289
3320 24" x 36"		16	.500		300	28		328	375
3330 36" x 24"		16	.500		298	28		326	375
3340 36" x 36"		14	.571		380	32		412	470
3350 36" x 48"		10	.800		385	45		430	495
3360 48" x 36"		10	.800		385	45		430	495
3370 48" x 48"		10	.800		425	45		470	535
3380 60" x 48"		10	.800		490	45		535	610
3390 60" x 60"		10	.800		590	45		635	715

For customer support on your Mechanical Cost Data, call 877.795.7394.

61

08 95 13.10 Wall Louvers	Crew	Daily Output	Labor-Hours	Unit	Material	2015 Bare Costs Labor	Equipment	Total	Total Incl O&P
0010 **WALL LOUVERS**									
2330 Soffit vent, continuous, 3" wide, aluminum, mill finish	1 Carp	200	.040	L.F.	.67	1.88		2.55	3.63
2340 Baked enamel finish		200	.040	"	5.50	1.88		7.38	8.95
2400 Under eaves vent, aluminum, mill finish, 16" x 4"		48	.167	Ea.	1.90	7.85		9.75	14.15
2500 · 16" x 8"		48	.167	"	2.18	7.85		10.03	14.45

Estimating Tips
General

- Room Finish Schedule: A complete set of plans should contain a room finish schedule. If one is not available, it would be well worth the time and effort to obtain one.

09 20 00 Plaster and Gypsum Board

- Lath is estimated by the square yard plus a 5% allowance for waste. Furring, channels, and accessories are measured by the linear foot. An extra foot should be allowed for each accessory miter or stop.

- Plaster is also estimated by the square yard. Deductions for openings vary by preference, from zero deduction to 50% of all openings over 2 feet in width. The estimator should allow one extra square foot for each linear foot of horizontal interior or exterior angle located below the ceiling level. Also, double the areas of small radius work.

- Drywall accessories, studs, track, and acoustical caulking are all measured by the linear foot. Drywall taping is figured by the square foot. Gypsum wallboard is estimated by the square foot. No material deductions should be made for door or window openings under 32 S.F.

09 60 00 Flooring

- Tile and terrazzo areas are taken off on a square foot basis. Trim and base materials are measured by the linear foot. Accent tiles are listed per each. Two basic methods of installation are used. Mud set is approximately 30% more expensive than thin set. In terrazzo work, be sure to include the linear footage of embedded decorative strips, grounds, machine rubbing, and power cleanup.

- Wood flooring is available in strip, parquet, or block configuration. The latter two types are set in adhesives with quantities estimated by the square foot. The laying pattern will influence labor costs and material waste. In addition to the material and labor for laying wood floors, the estimator must make allowances for sanding and finishing these areas, unless the flooring is prefinished.

- Sheet flooring is measured by the square yard. Roll widths vary, so consideration should be given to use the most economical width, as waste must be figured into the total quantity. Consider also the installation methods available, direct glue down or stretched.

09 70 00 Wall Finishes

- Wall coverings are estimated by the square foot. The area to be covered is measured, length by height of wall above baseboards, to calculate the square footage of each wall. This figure is divided by the number of square feet in the single roll which is being used. Deduct, in full, the areas of openings such as doors and windows. Where a pattern match is required allow 25%–30% waste.

09 80 00 Acoustic Treatment

- Acoustical systems fall into several categories. The takeoff of these materials should be by the square foot of area with a 5% allowance for waste. Do not forget about scaffolding, if applicable, when estimating these systems.

09 90 00 Painting and Coating

- A major portion of the work in painting involves surface preparation. Be sure to include cleaning, sanding, filling, and masking costs in the estimate.

- Protection of adjacent surfaces is not included in painting costs. When considering the method of paint application, an important factor is the amount of protection and masking required. These must be estimated separately and may be the determining factor in choosing the method of application.

Reference Numbers

Reference numbers are shown in shaded boxes at the beginning of some major classifications. These numbers refer to related items in the Reference Section. The reference information may be an estimating procedure, an alternate pricing method, or technical information.

Note: Not all subdivisions listed here necessarily appear in this publication. ∎

09 22 Supports for Plaster and Gypsum Board

09 22 03 – Fastening Methods for Finishes

09 22 03.20 Drilling Plaster/Drywall	Crew	Daily Output	Labor-Hours	Unit	Material	2015 Bare Costs Labor	Equipment	Total	Total Incl O&P
0010 **DRILLING PLASTER/DRYWALL**									
1100 Drilling & layout for drywall/plaster walls, up to 1" deep, no anchor									
1200 Holes, 1/4" diameter	1 Carp	150	.053	Ea.	.01	2.50		2.51	3.86
1300 3/8" diameter		140	.057		.01	2.68		2.69	4.14
1400 1/2" diameter		130	.062		.01	2.89		2.90	4.46
1500 3/4" diameter		120	.067		.01	3.13		3.14	4.84
1600 1" diameter		110	.073		.02	3.41		3.43	5.25
1700 1-1/4" diameter		100	.080		.03	3.76		3.79	5.85
1800 1-1/2" diameter		90	.089		.05	4.17		4.22	6.45
1900 For ceiling installations, add						40%			

09 69 Access Flooring

09 69 13 – Rigid-Grid Access Flooring

09 69 13.10 Access Floors	Crew	Daily Output	Labor-Hours	Unit	Material	2015 Bare Costs Labor	Equipment	Total	Total Incl O&P
0010 **ACCESS FLOORS**									
0015 Access floor package including panel, pedestal, stringers & laminate cover									
0100 Computer room, greater than 6,000 S.F.	4 Carp	750	.043	S.F.	7.50	2		9.50	11.35
0110 Less than 6,000 S.F.	2 Carp	375	.043		8.15	2		10.15	12.10
0120 Office, greater than 6,000 S.F.	4 Carp	1050	.030		5.35	1.43		6.78	8.10
0250 Panels, particle board or steel, 1250# load, no covering, under 6,000 S.F.	2 Carp	600	.027		3.75	1.25		5	6.05
0300 Over 6,000 S.F.		640	.025		3.21	1.17		4.38	5.35
0400 Aluminum, 24" panels		500	.032		33	1.50		34.50	38.50
0600 For carpet covering, add					8.75			8.75	9.65
0700 For vinyl floor covering, add					9.05			9.05	9.95
0900 For high pressure laminate covering, add					7.50			7.50	8.25
0910 For snap on stringer system, add	2 Carp	1000	.016		1.56	.75		2.31	2.88
0950 Office applications, steel or concrete panels,									
0960 no covering, over 6,000 S.F.	2 Carp	960	.017	S.F.	10.55	.78		11.33	12.80
1000 Machine cutouts after initial installation	1 Carp	50	.160	Ea.	20	7.50		27.50	33.50
1050 Pedestals, 6" to 12"	2 Carp	85	.188		8.40	8.85		17.25	23
1100 Air conditioning grilles, 4" x 12"	1 Carp	17	.471		68.50	22		90.50	109
1150 4" x 18"	"	14	.571		93.50	27		120.50	145
1200 Approach ramps, steel	2 Carp	60	.267	S.F.	24.50	12.50		37	46.50
1300 Aluminum	"	40	.400	"	34	18.80		52.80	66.50
1500 Handrail, 2 rail, aluminum	1 Carp	15	.533	L.F.	109	25		134	158

09 91 Painting

09 91 23 – Interior Painting

09 91 23.52 Miscellaneous, Interior	Crew	Daily Output	Labor-Hours	Unit	Material	2015 Bare Costs Labor	Equipment	Total	Total Incl O&P
0010 **MISCELLANEOUS, INTERIOR** R099100-10									
3800 Grilles, per side, oil base, primer coat, brushwork	1 Pord	520	.015	S.F.	.14	.62		.76	1.09
3850 Spray		1140	.007		.15	.28		.43	.59
3880 Paint 1 coat, brushwork		520	.015		.22	.62		.84	1.18
3900 Spray		1140	.007		.24	.28		.52	.70
3920 Paint 2 coats, brushwork		325	.025		.42	.99		1.41	1.97
3940 Spray		650	.012		.48	.50		.98	1.28
3950 Prime & paint 1 coat		325	.025		.36	.99		1.35	1.89
3960 Prime & paint 2 coats		270	.030		.35	1.20		1.55	2.19
4500 Louvers, one side, primer, brushwork		524	.015		.09	.62		.71	1.03

09 91 23.52 Miscellaneous, Interior		Crew	Daily Output	Labor-Hours	Unit	Material	2015 Bare Costs Labor	Equipment	Total	Total Incl O&P
4520	Paint one coat, brushwork	1 Pord	520	.015	S.F.	.10	.62		.72	1.04
4530	Spray		1140	.007		.11	.28		.39	.55
4540	Paint two coats, brushwork		325	.025		.18	.99		1.17	1.70
4550	Spray		650	.012		.20	.50		.70	.97
4560	Paint three coats, brushwork		270	.030		.27	1.20		1.47	2.10
4570	Spray		500	.016		.30	.65		.95	1.30
5000	Pipe, 1" - 4" diameter, primer or sealer coat, oil base, brushwork	2 Pord	1250	.013	L.F.	.10	.52		.62	.89
5100	Spray		2165	.007		.09	.30		.39	.55
5200	Paint 1 coat, brushwork		1250	.013		.11	.52		.63	.90
5300	Spray		2165	.007		.10	.30		.40	.56
5350	Paint 2 coats, brushwork		775	.021		.19	.83		1.02	1.47
5400	Spray		1240	.013		.22	.52		.74	1.02
5420	Paint 3 coats, brushwork		775	.021		.29	.83		1.12	1.58
5450	5" - 8" diameter, primer or sealer coat, brushwork		620	.026		.20	1.04		1.24	1.79
5500	Spray		1085	.015		.32	.60		.92	1.25
5550	Paint 1 coat, brushwork		620	.026		.29	1.04		1.33	1.89
5600	Spray		1085	.015		.33	.60		.93	1.26
5650	Paint 2 coats, brushwork		385	.042		.39	1.68		2.07	2.96
5700	Spray		620	.026		.43	1.04		1.47	2.04
5720	Paint 3 coats, brushwork		385	.042		.58	1.68		2.26	3.16
5750	9" - 12" diameter, primer or sealer coat, brushwork		415	.039		.29	1.56		1.85	2.66
5800	Spray		725	.022		.35	.89		1.24	1.73
5850	Paint 1 coat, brushwork		415	.039		.30	1.56		1.86	2.67
6000	Spray		725	.022		.33	.89		1.22	1.70
6200	Paint 2 coats, brushwork		260	.062		.58	2.48		3.06	4.38
6250	Spray		415	.039		.64	1.56		2.20	3.05
6270	Paint 3 coats, brushwork		260	.062		.86	2.48		3.34	4.69
6300	13" - 16" diameter, primer or sealer coat, brushwork		310	.052		.39	2.08		2.47	3.57
6350	Spray		540	.030		.44	1.20		1.64	2.28
6400	Paint 1 coat, brushwork		310	.052		.40	2.08		2.48	3.58
6450	Spray		540	.030		.44	1.20		1.64	2.29
6500	Paint 2 coats, brushwork		195	.082		.78	3.31		4.09	5.85
6550	Spray		310	.052		.86	2.08		2.94	4.09
6600	Radiators, per side, primer, brushwork	1 Pord	520	.015	S.F.	.09	.62		.71	1.04
6620	Paint, one coat		520	.015		.09	.62		.71	1.04
6640	Two coats		340	.024		.18	.95		1.13	1.63
6660	Three coats		283	.028		.27	1.14		1.41	2.02

For customer support on your Mechanical Cost Data, call 877.795.7394.

65

Division Notes

	CREW	DAILY OUTPUT	LABOR-HOURS	UNIT	BARE COSTS				TOTAL INCL O&P
					MAT.	LABOR	EQUIP.	TOTAL	

Estimating Tips
General

- The items in this division are usually priced per square foot or each.

- Many items in Division 10 require some type of support system or special anchors that are not usually furnished with the item. The required anchors must be added to the estimate in the appropriate division.

- Some items in Division 10, such as lockers, may require assembly before installation. Verify the amount of assembly required. Assembly can often exceed installation time.

10 20 00 Interior Specialties

- Support angles and blocking are not included in the installation of toilet compartments, shower/dressing compartments, or cubicles. Appropriate line items from Divisions 5 or 6 may need to be added to support the installations.

- Toilet partitions are priced by the stall. A stall consists of a side wall, pilaster, and door with hardware. Toilet tissue holders and grab bars are extra.

- The required acoustical rating of a folding partition can have a significant impact on costs. Verify the sound transmission coefficient rating of the panel priced to the specification requirements.

- Grab bar installation does not include supplemental blocking or backing to support the required load. When grab bars are installed at an existing facility, provisions must be made to attach the grab bars to solid structure.

Reference Numbers

Reference numbers are shown in shaded boxes at the beginning of some major classifications. These numbers refer to related items in the Reference Section. The reference information may be an estimating procedure, an alternate pricing method, or technical information.

Note: Not all subdivisions listed here necessarily appear in this publication. ∎

10 31 Manufactured Fireplaces

10 31 13 – Manufactured Fireplace Chimneys

10 31 13.10 Fireplace Chimneys	Crew	Daily Output	Labor-Hours	Unit	Material	2015 Bare Costs Labor	Equipment	Total	Total Incl O&P
0010 **FIREPLACE CHIMNEYS**									
0500 Chimney dbl. wall, all stainless, over 8'-6", 7" diam., add to fireplace	1 Carp	33	.242	V.L.F.	83.50	11.40		94.90	109
0600 10" diameter, add to fireplace		32	.250		132	11.75		143.75	163
0700 12" diameter, add to fireplace		31	.258		158	12.10		170.10	193
0800 14" diameter, add to fireplace		30	.267	↓	227	12.50		239.50	269
1000 Simulated brick chimney top, 4' high, 16" x 16"		10	.800	Ea.	425	37.50		462.50	530
1100 24" x 24"	↓	7	1.143	"	540	53.50		593.50	680

10 31 23 – Prefabricated Fireplaces

10 31 23.10 Fireplace, Prefabricated	Crew	Daily Output	Labor-Hours	Unit	Material	2015 Bare Costs Labor	Equipment	Total	Total Incl O&P
0010 **FIREPLACE, PREFABRICATED**, free standing or wall hung									
0100 With hood & screen, painted	1 Carp	1.30	6.154	Ea.	1,375	289		1,664	1,975
0150 Average		1	8		1,625	375		2,000	2,350
0200 Stainless steel		.90	8.889	↓	3,050	415		3,465	4,025
1500 Simulated logs, gas fired, 40,000 BTU, 2' long, manual safety pilot		7	1.143	Set	425	53.50		478.50	550
1600 Adjustable flame remote pilot		6	1.333		1,100	62.50		1,162.50	1,325
1700 Electric, 1,500 BTU, 1'-6" long, incandescent flame		7	1.143		197	53.50		250.50	299
1800 1,500 BTU, LED flame	↓	6	1.333	↓	300	62.50		362.50	430

10 35 Stoves

10 35 13 – Heating Stoves

10 35 13.10 Woodburning Stoves	Crew	Daily Output	Labor-Hours	Unit	Material	2015 Bare Costs Labor	Equipment	Total	Total Incl O&P
0010 **WOODBURNING STOVES**									
0015 Cast iron, °1500sf	2 Carp	1.30	12.308	Ea.	1,175	580		1,755	2,200
0020 1500-2000sf		1	16		2,025	750		2,775	3,375
0030 2,000 sf	↓	.80	20		2,775	940		3,715	4,500
0050 For gas log lighter, add				↓	45			45	49.50

Estimating Tips
General

- The items in this division are usually priced per square foot or each. Many of these items are purchased by the owner for installation by the contractor. Check the specifications for responsibilities and include time for receiving, storage, installation, and mechanical and electrical hookups in the appropriate divisions.

- Many items in Division 11 require some type of support system that is not usually furnished with the item. Examples of these systems include blocking for the attachment of casework and support angles for ceiling-hung projection screens. The required blocking or supports must be added to the estimate in the appropriate division.

- Some items in Division 11 may require assembly or electrical hookups. Verify the amount of assembly required or the need for a hard electrical connection and add the appropriate costs.

Reference Numbers

Reference numbers are shown in shaded boxes at the beginning of some major classifications. These numbers refer to related items in the Reference Section. The reference information may be an estimating procedure, an alternate pricing method, or technical information.

Note: Not all subdivisions listed here necessarily appear in this publication. ■

11 11 13 – Compressed-Air Vehicle Service Equipment

11 11 13.10 Compressed Air Equipment	Crew	Daily Output	Labor-Hours	Unit	Material	2015 Bare Costs Labor	Equipment	Total	Total Incl O&P
0010 **COMPRESSED AIR EQUIPMENT**									
0030 Compressors, electric, 1-1/2 H.P., standard controls	L-4	1.50	16	Ea.	455	710		1,165	1,60
0550 Dual controls		1.50	16		810	710		1,520	2,00
0600 5 H.P., 115/230 volt, standard controls		1	24		2,575	1,050		3,625	4,47
0650 Dual controls	↓	1	24	↓	3,400	1,050		4,450	5,37

11 11 19 – Vehicle Lubrication Equipment

11 11 19.10 Lubrication Equipment

	Crew	Daily Output	Labor-Hours	Unit	Material	Labor	Equipment	Total	Total Incl O&P
0010 **LUBRICATION EQUIPMENT**									
3000 Lube equipment, 3 reel type, with pumps, not including piping	L-4	.50	48	Set	8,800	2,125		10,925	13,00
3700 Pump lubrication, pneumatic, not incl. air compressor									
3710 Oil/gear lube	Q-1	9.60	1.667	Ea.	860	88		948	1,07
3720 Grease	"	9.60	1.667	"	1,075	88		1,163	1,30

11 11 33 – Vehicle Spray Painting Equipment

11 11 33.10 Spray Painting Equipment

	Crew	Daily Output	Labor-Hours	Unit	Material	Labor	Equipment	Total	Total Incl O&P
0010 **SPRAY PAINTING EQUIPMENT**									
4000 Spray painting booth, 26' long, complete	L-4	.40	60	Ea.	16,400	2,650		19,050	22,10

11 21 73 – Commercial Laundry and Dry Cleaning Equipment

11 21 73.16 Drying and Conditioning Equipment

	Crew	Daily Output	Labor-Hours	Unit	Material	Labor	Equipment	Total	Total Incl O&P
0010 **DRYING AND CONDITIONING EQUIPMENT**									
0100 Dryers, Not including rough-in									
1500 Industrial, 30 lb. capacity	1 Plum	2	4	Ea.	3,175	235		3,410	3,82
1600 50 lb. capacity	"	1.70	4.706	"	3,400	276		3,676	4,17

11 21 73.26 Commercial Washers and Extractors

	Crew	Daily Output	Labor-Hours	Unit	Material	Labor	Equipment	Total	Total Incl O&P
0010 **COMMERCIAL WASHERS AND EXTRACTORS**, not including rough-in									
6000 Combination washer/extractor, 20 lb. capacity	L-6	1.50	8	Ea.	5,950	460		6,410	7,22
6100 30 lb. capacity		.80	15		9,275	860		10,135	11,50
6200 50 lb. capacity		.68	17.647		10,900	1,000		11,900	13,50
6300 75 lb. capacity		.30	40		20,400	2,300		22,700	26,00
6350 125 lb. capacity	↓	.16	75	↓	27,700	4,300		32,000	36,90

11 21 73.33 Coin-Operated Laundry Equipment

	Crew	Daily Output	Labor-Hours	Unit	Material	Labor	Equipment	Total	Total Incl O&P
0010 **COIN-OPERATED LAUNDRY EQUIPMENT**									
0990 Dryer, gas fired									
1000 Commercial, 30 lb. capacity, coin operated, single	1 Plum	3	2.667	Ea.	3,300	157		3,457	3,85
1100 Double stacked	"	2	4	"	7,275	235		7,510	8,35
5290 Clothes washer									
5300 Commercial, coin operated, average	1 Plum	3	2.667	Ea.	1,250	157		1,407	1,60

11 21 83 – Photo Processing Equipment

11 21 83.13 Darkroom Equipment

	Crew	Daily Output	Labor-Hours	Unit	Material	Labor	Equipment	Total	Total Incl O&P
0010 **DARKROOM EQUIPMENT**									
0020 Developing sink, 5" deep, 24" x 48"	Q-1	2	8	Ea.	4,150	425		4,575	5,225
0050 48" x 52"		1.70	9.412		4,425	495		4,920	5,625
0200 10" deep, 24" x 48"		1.70	9.412		1,550	495		2,045	2,450
0250 24" x 108"		1.50	10.667		3,650	565		4,215	4,875
3500 Washers, round, minimum sheet 11" x 14"		2	8		3,275	425		3,700	4,250
3550 Maximum sheet 20" x 24"		1	16		3,600	845		4,445	5,250
3800 Square, minimum sheet 20" x 24"		1	16		3,150	845		3,995	4,750
3900 Maximum sheet 50" x 56"	↓	.80	20	↓	4,900	1,050		5,950	7,000

For customer support on your Mechanical Cost Data, call 877.795.7394.

11 21 Retail and Service Equipment

11 21 83 – Photo Processing Equipment

11 21 83.13 Darkroom Equipment	Crew	Daily Output	Labor-Hours	Unit	Material	2015 Bare Costs Labor	Equipment	Total	Total Incl O&P
4500 Combination tank sink, tray sink, washers, with									
4510 Dry side tables, average	Q-1	.45	35.556	Ea.	10,800	1,875		12,675	14,600

11 30 Residential Equipment

11 30 13 – Residential Appliances

11 30 13.15 Cooking Equipment

11 30 13.15 Cooking Equipment	Crew	Daily Output	Labor-Hours	Unit	Material	2015 Bare Costs Labor	Equipment	Total	Total Incl O&P
0010 COOKING EQUIPMENT									
0020 Cooking range, 30" free standing, 1 oven, minimum	2 Clab	10	1.600	Ea.	440	60		500	580
0050 Maximum		4	4		2,025	150		2,175	2,450
0150 2 oven, minimum		10	1.600		1,225	60		1,285	1,425
0200 Maximum	↓	10	1.600	↓	2,550	60		2,610	2,900

11 30 13.16 Refrigeration Equipment

11 30 13.16 Refrigeration Equipment	Crew	Daily Output	Labor-Hours	Unit	Material	Labor	Equipment	Total	Total Incl O&P
0010 REFRIGERATION EQUIPMENT									
5200 Icemaker, automatic, 20 lb. per day	1 Plum	7	1.143	Ea.	895	67		962	1,075
5350 51 lb. per day	"	2	4	"	1,575	235		1,810	2,100

11 30 13.17 Kitchen Cleaning Equipment

11 30 13.17 Kitchen Cleaning Equipment	Crew	Daily Output	Labor-Hours	Unit	Material	Labor	Equipment	Total	Total Incl O&P
0010 KITCHEN CLEANING EQUIPMENT									
2750 Dishwasher, built-in, 2 cycles, minimum	L-1	4	4	Ea.	238	227		465	600
2800 Maximum		2	8		435	455		890	1,150
2950 4 or more cycles, minimum		4	4		375	227		602	755
2960 Average		4	4		500	227		727	890
3000 Maximum	↓	2	8	↓	1,100	455		1,555	1,875

11 30 13.18 Waste Disposal Equipment

11 30 13.18 Waste Disposal Equipment	Crew	Daily Output	Labor-Hours	Unit	Material	Labor	Equipment	Total	Total Incl O&P
0010 WASTE DISPOSAL EQUIPMENT									
3300 Garbage disposal, sink type, minimum	L-1	10	1.600	Ea.	87.50	90.50		178	232
3350 Maximum	"	10	1.600	"	220	90.50		310.50	380

11 30 13.19 Kitchen Ventilation Equipment

11 30 13.19 Kitchen Ventilation Equipment	Crew	Daily Output	Labor-Hours	Unit	Material	Labor	Equipment	Total	Total Incl O&P
0010 KITCHEN VENTILATION EQUIPMENT									
4150 Hood for range, 2 speed, vented, 30" wide, minimum	L-3	5	3.200	Ea.	70.50	164		234.50	330
4200 Maximum		3	5.333		780	273		1,053	1,275
4300 42" wide, minimum		5	3.200		168	164		332	435
4330 Custom		5	3.200		1,650	164		1,814	2,075
4350 Maximum	↓	3	5.333		2,025	273		2,298	2,650
4500 For ventless hood, 2 speed, add					18.65			18.65	20.50
4650 For vented 1 speed, deduct from maximum				↓	50			50	55

11 30 13.24 Washers

11 30 13.24 Washers	Crew	Daily Output	Labor-Hours	Unit	Material	Labor	Equipment	Total	Total Incl O&P
0010 WASHERS									
5000 Residential, 4 cycle, average	1 Plum	3	2.667	Ea.	875	157		1,032	1,200
6650 Washing machine, automatic, minimum		3	2.667		485	157		642	765
6700 Maximum	↓	1	8	↓	1,525	470		1,995	2,375

11 30 13.25 Dryers

11 30 13.25 Dryers	Crew	Daily Output	Labor-Hours	Unit	Material	Labor	Equipment	Total	Total Incl O&P
0010 DRYERS									
0500 Gas fired residential, 16 lb. capacity, average	1 Plum	3	2.667	Ea.	675	157		832	975
7450 Vent kits for dryers	1 Carp	10	.800	"	39	37.50		76.50	101

11 30 15 – Miscellaneous Residential Appliances

11 30 15.13 Sump Pumps

11 30 15.13 Sump Pumps	Crew	Daily Output	Labor-Hours	Unit	Material	Labor	Equipment	Total	Total Incl O&P
0010 SUMP PUMPS									
6400 Cellar drainer, pedestal, 1/3 H.P., molded PVC base	1 Plum	3	2.667	Ea.	135	157		292	385
6450 Solid brass	"	2	4	"	289	235		524	675

11 30 Residential Equipment

11 30 15 – Miscellaneous Residential Appliances

11 30 15.13 Sump Pumps	Crew	Daily Output	Labor-Hours	Unit	Material	2015 Bare Costs Labor	2015 Bare Costs Equipment	Total	Total Incl O&P
6460	Sump pump, see also Section 22 14 29.16								

11 30 15.23 Water Heaters

		Crew	Daily Output	Labor-Hours	Unit	Material	Labor	Equipment	Total	Total Incl O&P
0010	**WATER HEATERS**									
6900	Electric, glass lined, 30 gallon, minimum	L-1	5	3.200	Ea.	430	181		611	745
6950	Maximum		3	5.333		595	300		895	1,100
7100	80 gallon, minimum		2	8		1,225	455		1,680	2,025
7150	Maximum	▼	1	16		1,700	905		2,605	3,225
7180	Gas, glass lined, 30 gallon, minimum	2 Plum	5	3.200		805	188		993	1,175
7220	Maximum		3	5.333		1,125	315		1,440	1,700
7260	50 gallon, minimum		2.50	6.400		845	375		1,220	1,500
7300	Maximum	▼	1.50	10.667	▼	1,175	625		1,800	2,225

11 30 15.43 Air Quality

		Crew	Daily Output	Labor-Hours	Unit	Material	Labor	Equipment	Total	Total Incl O&P
0010	**AIR QUALITY**									
2450	Dehumidifier, portable, automatic, 15 pint	1 Elec	4	2	Ea.	152	109		261	330
2550	40 pint		3.75	2.133		209	117		326	405
3550	Heater, electric, built-in, 1250 watt, ceiling type, minimum		4	2		107	109		216	282
3600	Maximum		3	2.667		175	146		321	410
3700	Wall type, minimum		4	2		172	109		281	355
3750	Maximum		3	2.667		185	146		331	420
3900	1500 watt wall type, with blower		4	2		172	109		281	355
3950	3000 watt	▼	3	2.667		350	146		496	605
4850	Humidifier, portable, 8 gallons per day					133			133	146
5000	15 gallons per day				▼	211			211	232

11 32 Unit Kitchens

11 32 13 – Metal Unit Kitchens

11 32 13.10 Commercial Unit Kitchens

		Crew	Daily Output	Labor-Hours	Unit	Material	Labor	Equipment	Total	Total Incl O&P
0010	**COMMERCIAL UNIT KITCHENS**									
1500	Combination range, refrigerator and sink, 30" wide, minimum	L-1	2	8	Ea.	1,100	455		1,555	1,900
1550	Maximum		1	16		1,475	905		2,380	3,000
1570	60" wide, average		1.40	11.429		1,525	650		2,175	2,675
1590	72" wide, average		1.20	13.333		1,625	755		2,380	2,900
1600	Office model, 48" wide		2	8		2,025	455		2,480	2,925
1620	Refrigerator and sink only	▼	2.40	6.667	▼	2,525	380		2,905	3,350
1640	Combination range, refrigerator, sink, microwave									
1660	Oven and ice maker	L-1	.80	20	Ea.	4,550	1,125		5,675	6,700

11 41 Foodservice Storage Equipment

11 41 13 – Refrigerated Food Storage Cases

11 41 13.10 Refrigerated Food Cases

		Crew	Daily Output	Labor-Hours	Unit	Material	Labor	Equipment	Total	Total Incl O&P
0010	**REFRIGERATED FOOD CASES**									
0030	Dairy, multi-deck, 12' long	Q-5	3	5.333	Ea.	11,300	287		11,587	12,800
0100	For rear sliding doors, add					1,800			1,800	2,000
0200	Delicatessen case, service deli, 12' long, single deck	Q-5	3.90	4.103		7,750	221		7,971	8,850
0300	Multi-deck, 18 S.F. shelf display		3	5.333		7,150	287		7,437	8,300
0400	Freezer, self-contained, chest-type, 30 C.F.		3.90	4.103		8,525	221		8,746	9,725
0500	Glass door, upright, 78 C.F.		3.30	4.848		10,200	261		10,461	11,600
0600	Frozen food, chest type, 12' long		3.30	4.848		8,225	261		8,486	9,450
0700	Glass door, reach-in, 5 door	▼	3	5.333		14,100	287		14,387	15,900

11 41 Foodservice Storage Equipment

11 41 13 – Refrigerated Food Storage Cases

11 41 13.10 Refrigerated Food Cases

		Crew	Daily Output	Labor-Hours	Unit	Material	2015 Bare Costs Labor	Equipment	Total	Total Incl O&P
0800	Island case, 12' long, single deck	Q-5	3.30	4.848	Ea.	7,275	261		7,536	8,425
0900	Multi-deck		3	5.333		8,500	287		8,787	9,775
1000	Meat case, 12' long, single deck		3.30	4.848		7,275	261		7,536	8,425
1050	Multi-deck		3.10	5.161		10,400	278		10,678	11,800
1100	Produce, 12' long, single deck		3.30	4.848		6,875	261		7,136	7,950
1200	Multi-deck		3.10	5.161		8,725	278		9,003	10,000

11 41 13.20 Refrigerated Food Storage Equipment

		Crew	Daily Output	Labor-Hours	Unit	Material	2015 Bare Costs Labor	Equipment	Total	Total Incl O&P
0010	**REFRIGERATED FOOD STORAGE EQUIPMENT**									
2350	Cooler, reach-in, beverage, 6' long	Q-1	6	2.667	Ea.	3,600	141		3,741	4,175
4300	Freezers, reach-in, 44 C.F.		4	4		4,425	211		4,636	5,175
4500	68 C.F.		3	5.333		4,825	282		5,107	5,725

11 44 Food Cooking Equipment

11 44 13 – Commercial Ranges

11 44 13.10 Cooking Equipment

		Crew	Daily Output	Labor-Hours	Unit	Material	2015 Bare Costs Labor	Equipment	Total	Total Incl O&P
0010	**COOKING EQUIPMENT**									
0020	Bake oven, gas, one section	Q-1	8	2	Ea.	5,450	106		5,556	6,150
0300	Two sections		7	2.286		9,075	121		9,196	10,200
0600	Three sections		6	2.667		11,200	141		11,341	12,500
0900	Electric convection, single deck	L-7	4	7		6,225	320		6,545	7,325
6350	Kettle, w/steam jacket, tilting, w/positive lock, SS, 20 gallons		7	4		8,225	182		8,407	9,325
6600	60 gallons		6	4.667		11,000	212		11,212	12,400

11 46 Food Dispensing Equipment

11 46 83 – Ice Machines

11 46 83.10 Commercial Ice Equipment

		Crew	Daily Output	Labor-Hours	Unit	Material	2015 Bare Costs Labor	Equipment	Total	Total Incl O&P
0010	**COMMERCIAL ICE EQUIPMENT**									
5800	Ice cube maker, 50 pounds per day	Q-1	6	2.667	Ea.	1,600	141		1,741	2,000
6050	500 pounds per day	"	4	4	"	2,775	211		2,986	3,375

11 48 Foodservice Cleaning and Disposal Equipment

11 48 13 – Commercial Dishwashers

11 48 13.10 Dishwashers

			Crew	Daily Output	Labor-Hours	Unit	Material	2015 Bare Costs Labor	Equipment	Total	Total Incl O&P
0010	**DISHWASHERS**										
2700	Dishwasher, commercial, rack type										
2720	10 to 12 racks per hour		Q-1	3.20	5	Ea.	3,525	264		3,789	4,275
2730	Energy star rated, 35 to 40 racks/hour	G		1.30	12.308		4,700	650		5,350	6,150
2740	50 to 60 racks/hour	G		1.30	12.308		10,200	650		10,850	12,200
2800	Automatic, 190 to 230 racks per hour		L-6	.35	34.286		13,900	1,975		15,875	18,300
2820	235 to 275 racks per hour			.25	48		31,500	2,750		34,250	38,800
2840	8,750 to 12,500 dishes per hour			.10	120		55,500	6,875		62,375	71,500

For customer support on your Mechanical Cost Data, call 877.795.7394.

73

11 53 Laboratory Equipment

11 53 13 – Laboratory Fume Hoods

11 53 13.13 Recirculating Laboratory Fume Hoods

11 53 13.13 Recirculating Laboratory Fume Hoods	Crew	Daily Output	Labor-Hours	Unit	Material	2015 Bare Costs Labor	Equipment	Total	Total Incl O&P
0010 **RECIRCULATING LABORATORY FUME HOODS**									
0600 Fume hood, with countertop & base, not including HVAC									
0610 Simple, minimum	2 Carp	5.40	2.963	L.F.	500	139		639	765
0620 Complex, including fixtures		2.40	6.667		805	315		1,120	1,375
0630 Special, maximum	↓	1.70	9.412	↓	840	440		1,280	1,600
0670 Service fixtures, average				Ea.	240			240	264
0680 For sink assembly with hot and cold water, add	1 Plum	1.40	5.714	"	735	335		1,070	1,325

11 53 13.23 Exhaust Hoods

	Crew	Daily Output	Labor-Hours	Unit	Material	Labor	Equipment	Total	Total Incl O&P
0010 **EXHAUST HOODS**									
0650 Ductwork, minimum	2 Shee	1	16	Hood	4,075	895		4,970	5,850
0660 Maximum	"	.50	32	"	6,275	1,800		8,075	9,625

11 53 19 – Laboratory Sterilizers

11 53 19.13 Sterilizers

	Crew	Daily Output	Labor-Hours	Unit	Material	Labor	Equipment	Total	Total Incl O&P
0010 **STERILIZERS**									
0700 Glassware washer, undercounter, minimum	L-1	1.80	8.889	Ea.	6,325	505		6,830	7,700
0710 Maximum	"	1	16		13,300	905		14,205	16,000
1850 Utensil washer-sanitizer	1 Plum	2	4	↓	11,300	235		11,535	12,800

11 53 33 – Emergency Safety Appliances

11 53 33.13 Emergency Equipment

	Crew	Daily Output	Labor-Hours	Unit	Material	Labor	Equipment	Total	Total Incl O&P
0010 **EMERGENCY EQUIPMENT**									
1400 Safety equipment, eye wash, hand held				Ea.	410			410	450
1450 Deluge shower				"	770			770	850

11 53 43 – Service Fittings and Accessories

11 53 43.13 Fittings

	Crew	Daily Output	Labor-Hours	Unit	Material	Labor	Equipment	Total	Total Incl O&P
0010 **FITTINGS**									
1600 Sink, one piece plastic, flask wash, hose, free standing	1 Plum	1.60	5	Ea.	1,950	294		2,244	2,600
1610 Epoxy resin sink, 25" x 16" x 10"	"	2	4	"	221	235		456	600
8000 Alternate pricing method: as percent of lab furniture									
8050 Installation, not incl. plumbing & duct work				% Furn.				22%	22%
8100 Plumbing, final connections, simple system								10%	10%
8110 Moderately complex system								15%	15%
8120 Complex system								20%	20%
8150 Electrical, simple system								10%	10%
8160 Moderately complex system								20%	20%
8170 Complex system								35%	35%

11 71 Medical Sterilizing Equipment

11 71 10 – Medical Sterilizers & Distillers

11 71 10.10 Sterilizers and Distillers

	Crew	Daily Output	Labor-Hours	Unit	Material	Labor	Equipment	Total	Total Incl O&P
0010 **STERILIZERS AND DISTILLERS**									
0700 Distiller, water, steam heated, 50 gal. capacity	1 Plum	1.40	5.714	Ea.	19,200	335		19,535	21,700
5600 Sterilizers, floor loading, 26" x 62" x 42", single door, steam					122,000			122,000	134,500
5650 Double door, steam					206,000			206,000	227,000
5800 General purpose, 20" x 20" x 38", single door					13,100			13,100	14,400
6000 Portable, counter top, steam, minimum					3,550			3,550	3,900
6020 Maximum					4,325			4,325	4,750
6050 Portable, counter top, gas, 17" x 15" x 32-1/2"					39,800			39,800	43,700
6150 Manual washer/sterilizer, 16" x 16" x 26"	1 Plum	2	4	↓	54,500	235		54,735	60,500
6200 Steam generators, electric 10 kW to 180 kW, freestanding									

11 71 Medical Sterilizing Equipment

11 71 10 – Medical Sterilizers & Distillers

11 71 10.10 Sterilizers and Distillers		Crew	Daily Output	Labor-Hours	Unit	Material	2015 Bare Costs Labor	Equipment	Total	Total Incl O&P
6250	Minimum	1 Elec	3	2.667	Ea.	8,775	146		8,921	9,875
6300	Maximum	"	.70	11.429		29,200	625		29,825	33,000
8200	Bed pan washer-sanitizer	1 Plum	2	4	↓	7,500	235		7,735	8,600

11 73 Patient Care Equipment

11 73 10 – Patient Treatment Equipment

11 73 10.10 Treatment Equipment

		Crew	Daily Output	Labor-Hours	Unit	Material	Labor	Equipment	Total	Total Incl O&P
0010	**TREATMENT EQUIPMENT**									
1800	Heat therapy unit, humidified, 26" x 78" x 28"				Ea.	3,600			3,600	3,975
8400	Whirlpool bath, mobile, sst, 18" x 24" x 60"					4,900			4,900	5,400
8450	Fixed, incl. mixing valves	1 Plum	2	4	↓	9,700	235		9,935	11,100

11 74 Dental Equipment

11 74 10 – Dental Office Equipment

11 74 10.10 Diagnostic and Treatment Equipment

		Crew	Daily Output	Labor-Hours	Unit	Material	Labor	Equipment	Total	Total Incl O&P
0010	**DIAGNOSTIC AND TREATMENT EQUIPMENT**									
0020	Central suction system, minimum	1 Plum	1.20	6.667	Ea.	1,575	390		1,965	2,325
0100	Maximum	"	.90	8.889		4,450	520		4,970	5,700
0600	Chair, electric or hydraulic, minimum	1 Skwk	.50	16		2,250	780		3,030	3,675
0700	Maximum		.25	32		7,825	1,550		9,375	11,000
2000	Light, ceiling mounted, minimum		8	1		1,175	48.50		1,223.50	1,350
2100	Maximum	↓	8	1		2,025	48.50		2,073.50	2,300
2200	Unit light, minimum	2 Skwk	5.33	3.002		735	146		881	1,025
2210	Maximum		5.33	3.002		1,575	146		1,721	1,950
2220	Track light, minimum		3.20	5		1,575	243		1,818	2,100
2230	Maximum	↓	3.20	5		2,650	243		2,893	3,300
2300	Sterilizers, steam portable, minimum					1,300			1,300	1,425
2350	Maximum					10,500			10,500	11,600
2600	Steam, institutional					3,275			3,275	3,600
2650	Dry heat, electric, portable, 3 trays				↓	1,225			1,225	1,350

11 78 Mortuary Equipment

11 78 13 – Mortuary Refrigerators

11 78 13.10 Mortuary and Autopsy Equipment

		Crew	Daily Output	Labor-Hours	Unit	Material	Labor	Equipment	Total	Total Incl O&P
0010	**MORTUARY AND AUTOPSY EQUIPMENT**									
0015	Autopsy table, standard	1 Plum	1	8	Ea.	9,750	470		10,220	11,400
0020	Deluxe	"	.60	13.333		15,100	785		15,885	17,800
3200	Mortuary refrigerator, end operated, 2 capacity					12,300			12,300	13,500
3300	6 capacity				↓	22,200			22,200	24,400

11 78 16 – Crematorium Equipment

11 78 16.10 Crematory

		Crew	Daily Output	Labor-Hours	Unit	Material	Labor	Equipment	Total	Total Incl O&P
0010	**CREMATORY**									
1500	Crematory, not including building, 1 place	Q-3	.20	160	Ea.	72,500	8,950		81,450	93,000
1750	2 place	"	.10	320	"	103,500	17,900		121,400	141,000

11 81 Facility Maintenance Equipment

11 81 19 – Vacuum Cleaning Systems

11 81 19.10 Vacuum Cleaning		Crew	Daily Output	Labor-Hours	Unit	Material	2015 Bare Costs Labor	Equipment	Total	Total Incl O&P
0010	**VACUUM CLEANING**									
0020	Central, 3 inlet, residential	1 Skwk	.90	8.889	Total	1,075	430		1,505	1,850
0200	Commercial		.70	11.429		1,225	555		1,780	2,200
0400	5 inlet system, residential		.50	16		1,500	780		2,280	2,850
0600	7 inlet system, commercial		.40	20		1,700	975		2,675	3,375
0800	9 inlet system, residential		.30	26.667		3,750	1,300		5,050	6,125
4010	Rule of thumb: First 1200 S.F., installed								1,425	1,575
4020	For each additional S.F., add				S.F.				.26	.26

11 82 Facility Solid Waste Handling Equipment

11 82 19 – Packaged Incinerators

11 82 19.10 Packaged Gas Fired Incinerators

0010	**PACKAGED GAS FIRED INCINERATORS**									
4400	Incinerator, gas, not incl. chimney, elec. or pipe, 50#/hr., minimum	Q-3	.80	40	Ea.	38,500	2,225		40,725	45,700
4420	Maximum		.70	45.714		40,500	2,550		43,050	48,400
4440	200 lb. per hr., minimum (batch type)		.60	53.333		68,000	2,975		70,975	79,000
4460	Maximum (with feeder)		.50	64		76,000	3,575		79,575	89,000
4480	400 lb. per hr., minimum (batch type)		.30	106		79,000	5,950		84,950	96,000
4500	Maximum (with feeder)		.25	128		101,000	7,150		108,150	122,500
4520	800 lb. per hr., with feeder, minimum		.20	160		121,500	8,950		130,450	147,000
4540	Maximum		.17	188		182,000	10,500		192,500	216,500
4560	1,200 lb. per hr., with feeder, minimum		.15	213		154,000	11,900		165,900	187,000
4580	Maximum		.11	290		200,000	16,300		216,300	244,500
4600	2,000 lb. per hr., with feeder, minimum		.10	320		405,000	17,900		422,900	472,500
4620	Maximum		.05	640		607,000	35,800		642,800	722,000
4700	For heat recovery system, add, minimum		.25	128		81,000	7,150		88,150	100,000
4710	Add, maximum		.11	290		253,000	16,300		269,300	303,000
4720	For automatic ash conveyer, add		.50	64		33,700	3,575		37,275	42,500
4750	Large municipal incinerators, incl. stack, minimum		.25	128	Ton/day	20,500	7,150		27,650	33,300
4850	Maximum		.10	320	"	54,500	17,900		72,400	87,000

11 82 26 – Facility Waste Compactors

11 82 26.10 Compactors

0010	**COMPACTORS**									
0020	Compactors, 115 volt, 250#/hr., chute fed	L-4	1	24	Ea.	12,100	1,050		13,150	15,000
0100	Hand fed		2.40	10		15,100	445		15,545	17,300
0300	Multi-bag, 230 volt, 600#/hr., chute fed		1	24		15,100	1,050		16,150	18,300
0400	Hand fed		1	24		15,300	1,050		16,350	18,500
0500	Containerized, hand fed, 2 to 6 C.Y. containers, 250#/hr.		1	24		15,100	1,050		16,150	18,300
0550	For chute fed, add per floor		1	24		1,400	1,050		2,450	3,175
1000	Heavy duty industrial compactor, 0.5 C.Y. capacity		1	24		10,100	1,050		11,150	12,800
1050	1.0 C.Y. capacity		1	24		15,200	1,050		16,250	18,400
1100	3.0 C.Y. capacity		.50	48		25,900	2,125		28,025	31,800
1150	5.0 C.Y. capacity		.50	48		32,600	2,125		34,725	39,100
1200	Combination shredder/compactor (5,000 lb./hr.)		.50	48		63,500	2,125		65,625	73,000
1400	For handling hazardous waste materials, 55 gallon drum packer, std.					19,700			19,700	21,700
1410	55 gallon drum packer w/HEPA filter					24,600			24,600	27,100
1420	55 gallon drum packer w/charcoal & HEPA filter					32,800			32,800	36,100
1430	All of the above made explosion proof, add					1,450			1,450	1,575
5800	Shredder, industrial, minimum					24,000			24,000	26,400
5850	Maximum					128,500			128,500	141,500

11 82 Facility Solid Waste Handling Equipment

11 82 26 – Facility Waste Compactors

11 82 26.10 Compactors	Crew	Daily Output	Labor-Hours	Unit	Material	2015 Bare Costs Labor	Equipment	Total	Total Incl O&P
5900	Baler, industrial, minimum				Ea.	9,625			9,625	10,600
5950	Maximum				↓	560,500			560,500	616,500

Division Notes

	CREW	DAILY OUTPUT	LABOR-HOURS	UNIT	BARE COSTS				TOTAL INCL O&P
					MAT.	LABOR	EQUIP.	TOTAL	

Estimating Tips
General

- The items and systems in this division are usually estimated, purchased, supplied, and installed as a unit by one or more subcontractors. The estimator must ensure that all parties are operating from the same set of specifications and assumptions, and that all necessary items are estimated and will be provided. Many times the complex items and systems are covered, but the more common ones, such as excavation or a crane, are overlooked for the very reason that everyone assumes nobody could miss them. The estimator should be the central focus and be able to ensure that all systems are complete.

- Another area where problems can develop in this division is at the interface between systems. The estimator must ensure, for instance, that anchor bolts, nuts, and washers are estimated and included for the air-supported structures and pre-engineered buildings to be bolted to their foundations. Utility supply is a common area where essential items or pieces of equipment can be missed or overlooked, because each subcontractor may feel it is another's responsibility. The estimator should also be aware of certain items which may be supplied as part of a package but installed by others, and ensure that the installing contractor's estimate includes the cost of installation. Conversely, the estimator must also ensure that items are not costed by two different subcontractors, resulting in an inflated overall estimate.

13 30 00 Special Structures

- The foundations and floor slab, as well as rough mechanical and electrical, should be estimated, as this work is required for the assembly and erection of the structure. Generally, as noted in the book, the pre-engineered building comes as a shell. Pricing is based on the size and structural design parameters stated in the reference section. Additional features, such as windows and doors with their related structural framing, must also be included by the estimator. Here again, the estimator must have a clear understanding of the scope of each portion of the work and all the necessary interfaces.

Reference Numbers

Reference numbers are shown in shaded boxes at the beginning of some major classifications. These numbers refer to related items in the Reference Section. The reference information may be an estimating procedure, an alternate pricing method, or technical information.

Note: Not all subdivisions listed here necessarily appear in this publication. ∎

13 11 13 – Below-Grade Swimming Pools

13 11 13.50 Swimming Pools	Crew	Daily Output	Labor-Hours	Unit	Material	2015 Bare Costs Labor	Equipment	Total	Total Incl O&P
0010 **SWIMMING POOLS** Residential in-ground, vinyl lined, concrete									
0020 Swimming pools,resi in-ground,vyl lined,conc sides,W/ equip ,sand bot	B-52	300	.187	SF Surf	23.50	8.15	1.98	33.63	40
0100 Metal or polystyrene sides	B-14	410	.117		19.55	4.66	.89	25.10	29.50
0200 Add for vermiculite bottom					1.49			1.49	1.64
0500 Gunite bottom and sides, white plaster finish									
0600 12' x 30' pool	B-52	145	.386	SF Surf	43.50	16.85	4.10	64.45	78.50
0720 16' x 32' pool		155	.361		39	15.75	3.83	58.58	71
0750 20' x 40' pool		250	.224		35	9.75	2.38	47.13	56
0810 Concrete bottom and sides, tile finish									
0820 12' x 30' pool	B-52	80	.700	SF Surf	44	30.50	7.45	81.95	103
0830 16' x 32' pool		95	.589		36.50	25.50	6.25	68.25	86.50
0840 20' x 40' pool		130	.431		29	18.80	4.57	52.37	66
1100 Motel, gunite with plaster finish, incl. medium									
1150 capacity filtration & chlorination	B-52	115	.487	SF Surf	53.50	21	5.15	79.65	97
1200 Municipal, gunite with plaster finish, incl. high									
1250 capacity filtration & chlorination	B-52	100	.560	SF Surf	69.50	24.50	5.95	99.95	120
1350 Add for formed gutters				L.F.	102			102	112
1360 Add for stainless steel gutters				"	300			300	330
1700 Filtration and deck equipment only, as % of total				Total				20%	20%
1800 Deck equipment, rule of thumb, 20' x 40' pool				SF Pool				1.18	1.30
1900 5000 S.F. pool				"				1.73	1.90
3000 Painting pools, preparation + 3 coats, 20' x 40' pool, epoxy	2 Pord	.33	48.485	Total	1,775	1,950		3,725	4,900
3100 Rubber base paint, 18 gallons	"	.33	48.485		1,225	1,950		3,175	4,300
3500 42' x 82' pool, 75 gallons, epoxy paint	3 Pord	.14	171		7,500	6,925		14,425	18,700
3600 Rubber base paint	"	.14	171		5,075	6,925		12,000	16,000

13 11 46 – Swimming Pool Accessories

13 11 46.50 Swimming Pool Equipment

	Crew	Daily Output	Labor-Hours	Unit	Material	2015 Bare Costs Labor	Equipment	Total	Total Incl O&P
0010 **SWIMMING POOL EQUIPMENT**									
0020 Diving stand, stainless steel, 3 meter	2 Carp	.40	40	Ea.	15,100	1,875		16,975	19,500
0300 1 meter	"	2.70	5.926		9,175	278		9,453	10,500
2100 Lights, underwater, 12 volt, with transformer, 300 watt	1 Elec	1	8		330	440		770	1,025
2200 110 volt, 500 watt, standard		1	8		294	440		734	980
2400 Low water cutoff type		1	8		300	440		740	985
2800 Heaters, see Section 23 52 28.10									

13 18 Ice Rinks

13 18 13 – Ice Rink Floor Systems

13 18 13.50 Ice Skating

	Crew	Daily Output	Labor-Hours	Unit	Material	2015 Bare Costs Labor	Equipment	Total	Total Incl O&P
0010 **ICE SKATING** Equipment incl. refrigeration, plumbing & cooling									
0020 coils & concrete slab, 85' x 200' rink									
0300 55° system, 5 mos., 100 ton				Total	575,000			575,000	632,500
0700 90° system, 12 mos., 135 ton				"	650,000			650,000	715,000
1200 Subsoil heating system (recycled from compressor), 85' x 200'	Q-7	.27	118	Ea.	40,000	6,750		46,750	54,000
1300 Subsoil insulation, 2 lb. polystyrene with vapor barrier, 85' x 200'	2 Carp	.14	114	"	30,000	5,375		35,375	41,300

13 18 16 – Ice Rink Dasher Boards

13 18 16.50 Ice Rink Dasher Boards

	Crew	Daily Output	Labor-Hours	Unit	Material	2015 Bare Costs Labor	Equipment	Total	Total Incl O&P
0010 **ICE RINK DASHER BOARDS**									
1000 Dasher boards, 1/2" H.D. polyethylene faced steel frame, 3' acrylic									
1020 screen at sides, 5' acrylic ends, 85' x 200'	F-5	.06	533	Ea.	135,000	25,300		160,300	187,500
1100 Fiberglass & aluminum construction, same sides and ends	"	.06	533	"	155,000	25,300		180,300	209,500

80

For customer support on your Mechanical Cost Data, call 877.795.7394.

13 21 13 – Clean Rooms

13 21 13.50 Clean Room Components

		Daily Output	Labor-Hours	Unit	Material	2015 Bare Costs Labor	Equipment	Total	Total Incl O&P
		Crew							
0010	**CLEAN ROOM COMPONENTS**								
1100	Clean room, soft wall, 12' x 12', Class 100	1 Carp	.18	44.444	Ea.	18,600	2,075		20,675	23,700
1110	Class 1,000		.18	44.444		15,400	2,075		17,475	20,200
1120	Class 10,000		.21	38.095		13,000	1,800		14,800	17,100
1130	Class 100,000		.21	38.095		12,000	1,800		13,800	16,000
2800	Ceiling grid support, slotted channel struts 4'-0" O.C., ea. way				S.F.				5.90	6.50
3000	Ceiling panel, vinyl coated foil on mineral substrate									
3020	Sealed, non-perforated				S.F.				1.27	1.40
4000	Ceiling panel seal, silicone sealant, 150 L.F./gal.	1 Carp	150	.053	L.F.	.34	2.50		2.84	4.22
4100	Two sided adhesive tape	"	240	.033	"	.12	1.56		1.68	2.54
4200	Clips, one per panel				Ea.	.99			.99	1.09
6000	HEPA filter, 2' x 4', 99.97% eff., 3" dp beveled frame (silicone seal)					470			470	515
6040	6" deep skirted frame (channel seal)					440			440	485
6100	99.99% efficient, 3" deep beveled frame (silicone seal)					525			525	580
6140	6" deep skirted frame (channel seal)					455			455	500
6200	99.999% efficient, 3" deep beveled frame (silicone seal)					605			605	665
6240	6" deep skirted frame (channel seal)					485			485	530
7000	Wall panel systems, including channel strut framing									
7020	Polyester coated aluminum, particle board				S.F.				18.20	20
7100	Porcelain coated aluminum, particle board								32	35
7400	Wall panel support, slotted channel struts, to 12' high								16.35	18

13 21 26 – Cold Storage Rooms

13 21 26.50 Refrigeration

		Crew	Daily Output	Labor-Hours	Unit	Material	2015 Bare Costs Labor	Equipment	Total	Total Incl O&P
0010	**REFRIGERATION**									
0020	Curbs, 12" high, 4" thick, concrete	2 Carp	58	.276	L.F.	5.15	12.95		18.10	25.50
1000	Doors, see Section 08 34 13.10									
2400	Finishes, 2 coat portland cement plaster, 1/2" thick	1 Plas	48	.167	S.F.	1.41	7.15		8.56	12.30
2500	For galvanized reinforcing mesh, add	1 Lath	335	.024		.94	1.03		1.97	2.54
2700	3/16" thick latex cement	1 Plas	88	.091		2.40	3.90		6.30	8.50
2900	For glass cloth reinforced ceilings, add	"	450	.018		.57	.76		1.33	1.78
3100	Fiberglass panels, 1/8" thick	1 Carp	149.45	.054		3.18	2.51		5.69	7.35
3200	Polystyrene, plastic finish ceiling, 1" thick		274	.029		2.94	1.37		4.31	5.35
3400	2" thick		274	.029		3.36	1.37		4.73	5.80
3500	4" thick		219	.037		3.71	1.72		5.43	6.70
3800	Floors, concrete, 4" thick	1 Cefi	93	.086		1.32	3.87		5.19	7.20
3900	6" thick	"	85	.094		2.07	4.24		6.31	8.50
4000	Insulation, 1" to 6" thick, cork				B.F.	1.35			1.35	1.49
4100	Urethane					.52			.52	.57
4300	Polystyrene, regular					.53			.53	.58
4400	Bead board					.25			.25	.28
4600	Installation of above, add per layer	2 Carp	657.60	.024	S.F.	.45	1.14		1.59	2.26
4700	Wall and ceiling juncture		298.90	.054	L.F.	2.19	2.51		4.70	6.30
4900	Partitions, galvanized sandwich panels, 4" thick, stock		219.20	.073	S.F.	9.20	3.43		12.63	15.35
5000	Aluminum or fiberglass		219.20	.073	"	10.05	3.43		13.48	16.35
5200	Prefab walk-in, 7'-6" high, aluminum, incl. refrigeration, door & floor									
5210	not incl. partitions, 6' x 6'	2 Carp	54.80	.292	SF Flr.	164	13.70		177.70	202
5500	10' x 10'		82.20	.195		132	9.15		141.15	159
5700	12' x 14'		109.60	.146		119	6.85		125.85	142
5800	12' x 20'		109.60	.146		103	6.85		109.85	125
6100	For 8'-6" high, add					5%				
6300	Rule of thumb for complete units, w/o doors & refrigeration, cooler	2 Carp	146	.110		149	5.15		154.15	172
6400	Freezer		109.60	.146		176	6.85		182.85	205

13 21 Controlled Environment Rooms

13 21 26 – Cold Storage Rooms

13 21 26.50 Refrigeration

13 21 26.50 Refrigeration	Crew	Daily Output	Labor-Hours	Unit	Material	2015 Bare Costs Labor	Equipment	Total	Total Incl O&P
6600	Shelving, plated or galvanized, steel wire type	2 Carp	360	.044	SF Hor.	13.05	2.09		15.14	17.55
6700	Slat shelf type	↓	375	.043		16.10	2		18.10	21
6900	For stainless steel shelving, add				↓	300%				
7000	Vapor barrier, on wood walls	2 Carp	1644	.010	S.F.	.20	.46		.66	.92
7200	On masonry walls	"	1315	.012	"	.49	.57		1.06	1.42
7500	For air curtain doors, see Section 23 34 33.10									

13 24 Special Activity Rooms

13 24 16 – Saunas

13 24 16.50 Saunas and Heaters

13 24 16.50	SAUNAS AND HEATERS	Crew	Daily Output	Labor-Hours	Unit	Material	2015 Bare Costs Labor	Equipment	Total	Total Incl O&P
0010	**SAUNAS AND HEATERS**									
0020	Prefabricated, incl. heater & controls, 7' high, 6' x 4', C/C	L-7	2.20	12.727	Ea.	5,175	580		5,755	6,575
0050	6' x 4', C/P		2	14		4,700	635		5,335	6,125
0400	6' x 5', C/C		2	14		5,875	635		6,510	7,450
0450	6' x 5', C/P		2	14		5,325	635		5,960	6,825
0600	6' x 6', C/C		1.80	15.556		6,225	705		6,930	7,925
0650	6' x 6', C/P		1.80	15.556		5,675	705		6,380	7,300
0800	6' x 9', C/C		1.60	17.500		7,975	795		8,770	10,000
0850	6' x 9', C/P		1.60	17.500		7,225	795		8,020	9,150
1000	8' x 12', C/C		1.10	25.455		11,700	1,150		12,850	14,700
1050	8' x 12', C/P		1.10	25.455		10,500	1,150		11,650	13,400
1200	8' x 8', C/C		1.40	20		9,175	910		10,085	11,500
1250	8' x 8', C/P		1.40	20		8,450	910		9,360	10,700
1400	8' x 10', C/C		1.20	23.333		10,200	1,050		11,250	12,800
1450	8' x 10', C/P		1.20	23.333		9,250	1,050		10,300	11,800
1600	10' x 12', C/C		1	28		12,200	1,275		13,475	15,500
1650	10' x 12', C/P	↓	1	28		11,000	1,275		12,275	14,100
2500	Heaters only (incl. above), wall mounted, to 200 C.F.					685			685	755
2750	To 300 C.F.					930			930	1,025
3000	Floor standing, to 720 C.F., 10,000 watts, w/controls	1 Elec	3	2.667		2,950	146		3,096	3,475
3250	To 1,000 C.F., 16,000 watts	"	3	2.667	↓	3,825	146		3,971	4,425

13 24 26 – Steam Baths

13 24 26.50 Steam Baths and Components

13 24 26.50	STEAM BATHS AND COMPONENTS	Crew	Daily Output	Labor-Hours	Unit	Material	2015 Bare Costs Labor	Equipment	Total	Total Incl O&P
0010	**STEAM BATHS AND COMPONENTS**									
0020	Heater, timer & head, single, to 140 C.F.	1 Plum	1.20	6.667	Ea.	2,200	390		2,590	3,000
0500	To 300 C.F.		1.10	7.273		2,425	425		2,850	3,325
1000	Commercial size, with blow-down assembly, to 800 C.F.		.90	8.889		5,725	520		6,245	7,100
1500	To 2500 C.F.	↓	.80	10		7,650	585		8,235	9,300
2000	Multiple, motels, apts., 2 baths, w/blow-down assm., 500 C.F.	Q-1	1.30	12.308		6,325	650		6,975	7,950
2500	4 baths	"	.70	22.857		10,200	1,200		11,400	13,000
2700	Conversion unit for residential tub, including door				↓	3,550			3,550	3,925

13 34 Fabricated Engineered Structures

13 34 23 – Fabricated Structures

13 34 23.10 Comfort Stations	Crew	Daily Output	Labor-Hours	Unit	Material	2015 Bare Costs Labor	Equipment	Total	Total Incl O&P
0010 **COMFORT STATIONS** Prefab., stock, w/doors, windows & fixt.									
0100 Not incl. interior finish or electrical									
0300 Mobile, on steel frame, 2 unit				S.F.	163			163	180
0350 7 unit					271			271	298
0400 Permanent, including concrete slab, 2 unit	B-12J	50	.320		249	14.30	17.65	280.95	315
0500 6 unit	"	43	.372		191	16.60	20.50	228.10	259
0600 Alternate pricing method, mobile, 2 fixture				Fixture	5,650			5,650	6,225
0650 7 fixture					9,750			9,750	10,700
0700 Permanent, 2 unit	B-12J	.70	22.857		20,500	1,025	1,250	22,775	25,500
0750 6 unit	"	.50	32		17,400	1,425	1,775	20,600	23,300

13 34 23.16 Fabricated Control Booths

	Crew	Daily Output	Labor-Hours	Unit	Material	2015 Bare Costs Labor	Equipment	Total	Total Incl O&P
0010 **FABRICATED CONTROL BOOTHS**									
0100 Guard House, prefab conc. w/bullet resistant doors & windows, roof & wiring									
0110 8' x 8', Level III	L-10	1	24	Ea.	44,200	1,275	655	46,130	51,500
0120 8' x 8', Level IV	"	1	24	"	50,500	1,275	655	52,430	58,500

13 47 Facility Protection

13 47 13 – Cathodic Protection

13 47 13.16 Cathodic Prot. for Underground Storage Tanks

	Crew	Daily Output	Labor-Hours	Unit	Material	2015 Bare Costs Labor	Equipment	Total	Total Incl O&P
0010 **CATHODIC PROTECTION FOR UNDERGROUND STORAGE TANKS**									
1000 Anodes, magnesium type, 9 #	R-15	18.50	2.595	Ea.	41.50	140	16.30	197.80	273
1010 17 #		13	3.692		71.50	199	23	293.50	400
1020 32 #		10	4.800		124	258	30	412	555
1030 48 #		7.20	6.667		164	360	42	566	765
1100 Graphite type w/epoxy cap, 3" x 60" (32 #)	R-22	8.40	4.438		128	206		334	455
1110 4" x 80" (68 #)		6	6.213		241	289		530	705
1120 6" x 72" (80 #)		5.20	7.169		1,500	335		1,835	2,150
1130 6" x 36" (45 #)		9.60	3.883		760	181		941	1,100
2000 Rectifiers, silicon type, air cooled, 28 V/10 A	R-19	3.50	5.714		2,300	315		2,615	3,025
2010 20 V/20 A		3.50	5.714		2,375	315		2,690	3,075
2100 Oil immersed, 28 V/10 A		3	6.667		3,175	365		3,540	4,025
2110 20 V/20 A		3	6.667		3,175	365		3,540	4,050
3000 Anode backfill, coke breeze	R-22	3850	.010	Lb.	.22	.45		.67	.92
4000 Cable, HMWPE, No. 8		2.40	15.533	M.L.F.	500	720		1,220	1,650
4010 No. 6		2.40	15.533		745	720		1,465	1,925
4020 No. 4		2.40	15.533		1,125	720		1,845	2,350
4030 No. 2		2.40	15.533		1,750	720		2,470	3,025
4040 No. 1		2.20	16.945		2,400	790		3,190	3,850
4050 No. 1/0		2.20	16.945		2,975	790		3,765	4,475
4060 No. 2/0		2.20	16.945		4,625	790		5,415	6,300
4070 No. 4/0		2	18.640		6,000	865		6,865	7,925
5000 Test station, 7 terminal box, flush curb type w/lockable cover	R-19	12	1.667	Ea.	71.50	91.50		163	216
5010 Reference cell, 2" dia PVC conduit, cplg., plug, set flush	"	4.80	4.167	"	151	228		379	505

83

Division Notes

	CREW	DAILY OUTPUT	LABOR-HOURS	UNIT	BARE COSTS				TOTAL INCL O&P
					MAT.	LABOR	EQUIP.	TOTAL	

Estimating Tips
General

- Many products in Division 14 will require some type of support or blocking for installation not included with the item itself. Examples are supports for conveyors or tube systems, attachment points for lifts, and footings for hoists or cranes. Add these supports in the appropriate division.

14 10 00 Dumbwaiters
14 20 00 Elevators

- Dumbwaiters and elevators are estimated and purchased in a method similar to buying a car. The manufacturer has a base unit with standard features. Added to this base unit price will be whatever options the owner or specifications require. Increased load capacity, additional vertical travel, additional stops, higher speed, and cab finish options are items to be considered. When developing an estimate for dumbwaiters and elevators, remember that some items needed by the installers may have to be included as part of the general contract.

Examples are:

- □ shaftway
- □ rail support brackets
- □ machine room
- □ electrical supply
- □ sill angles
- □ electrical connections
- □ pits
- □ roof penthouses
- □ pit ladders

Check the job specifications and drawings before pricing.

- Installation of elevators and handicapped lifts in historic structures can require significant additional costs. The associated structural requirements may involve cutting into and repairing finishes, moldings, flooring, etc. The estimator must account for these special conditions.

14 30 00 Escalators and Moving Walks

- Escalators and moving walks are specialty items installed by specialty contractors. There are numerous options associated with these items. For specific options, contact a manufacturer or contractor. In a method similar to estimating dumbwaiters and elevators, you should verify the extent of general contract work and add items as necessary.

14 40 00 Lifts
14 90 00 Other Conveying Equipment

- Products such as correspondence lifts, chutes, and pneumatic tube systems, as well as other items specified in this subdivision, may require trained installers. The general contractor might not have any choice as to who will perform the installation, or when it will be performed. Long lead times are often required for these products, making early decisions in scheduling necessary.

Reference Numbers

Reference numbers are shown in shaded boxes at the beginning of some major classifications. These numbers refer to related items in the Reference Section. The reference information may be an estimating procedure, an alternate pricing method, or technical information.

Note: Not all subdivisions listed here necessarily appear in this publication. ■

14 91 Facility Chutes

14 91 33 – Laundry and Linen Chutes

14 91 33.10 Chutes

14 91 33.10 Chutes	Crew	Daily Output	Labor-Hours	Unit	Material	2015 Bare Costs Labor	Equipment	Total	Total Incl O&P
0011	**CHUTES**, linen, trash or refuse									
0050	Aluminized steel, 16 ga., 18" diameter	2 Shee	3.50	4.571	Floor	1,700	256		1,956	2,27
0100	24" diameter		3.20	5		1,775	280		2,055	2,37
0200	30" diameter		3	5.333		2,125	298		2,423	2,77
0300	36" diameter		2.80	5.714		2,625	320		2,945	3,40
0400	Galvanized steel, 16 ga., 18" diameter		3.50	4.571		1,000	256		1,256	1,50
0500	24" diameter		3.20	5		1,125	280		1,405	1,67
0600	30" diameter		3	5.333		1,275	298		1,573	1,85
0700	36" diameter		2.80	5.714		1,500	320		1,820	2,15
0800	Stainless steel, 18" diameter		3.50	4.571		3,000	256		3,256	3,70
0900	24" diameter		3.20	5		3,150	280		3,430	3,87
1000	30" diameter		3	5.333		3,750	298		4,048	4,57
1005	36" diameter		2.80	5.714		3,950	320		4,270	4,82
1200	Linen chute bottom collector, aluminized steel		4	4	Ea.	1,400	224		1,624	1,90
1300	Stainless steel		4	4		1,800	224		2,024	2,32
1500	Refuse, bottom hopper, aluminized steel, 18" diameter		3	5.333		1,025	298		1,323	1,57
1600	24" diameter		3	5.333		1,250	298		1,548	1,82
1800	36" diameter		3	5.333		2,500	298		2,798	3,20

14 91 82 – Trash Chutes

14 91 82.10 Trash Chutes and Accessories

		Crew	Daily Output	Labor-Hours	Unit	Material	2015 Bare Costs Labor	Equipment	Total	Total Incl O&P
0010	**TRASH CHUTES AND ACCESSORIES**									
2900	Package chutes, spiral type, minimum	2 Shee	4.50	3.556	Floor	2,425	199		2,624	2,975
3000	Maximum	"	1.50	10.667	"	6,325	595		6,920	7,875

14 92 Pneumatic Tube Systems

14 92 10 – Conventional, Automatic and Computer Controlled Pneumatic Tube Systems

14 92 10.10 Pneumatic Tube Systems

		Crew	Daily Output	Labor-Hours	Unit	Material	2015 Bare Costs Labor	Equipment	Total	Total Incl O&P
0010	**PNEUMATIC TUBE SYSTEMS**									
0020	100' long, single tube, 2 stations, stock									
0100	3" diameter	2 Stpi	.12	133	Total	3,375	7,975		11,350	15,700
0300	4" diameter	"	.09	177	"	4,275	10,600		14,875	20,700
0400	Twin tube, two stations or more, conventional system									
0600	2-1/2" round	2 Stpi	62.50	.256	L.F.	38	15.30		53.30	65
0700	3" round		46	.348		38	21		59	73.50
0900	4" round		49.60	.323		48	19.25		67.25	82
1000	4" x 7" oval		37.60	.426		89	25.50		114.50	137
1050	Add for blower		2	8	System	5,200	480		5,680	6,450
1110	Plus for each round station, add		7.50	2.133	Ea.	1,350	127		1,477	1,700
1150	Plus for each oval station, add		7.50	2.133	"	1,350	127		1,477	1,700
1200	Alternate pricing method: base cost, economy model		.75	21.333	Total	5,750	1,275		7,025	8,250
1300	Custom model		.25	64	"	11,500	3,825		15,325	18,500
1500	Plus total system length, add, for economy model		93.40	.171	L.F.	8.25	10.25		18.50	24.50
1600	For custom model		37.60	.426	"	25	25.50		50.50	66
1800	Completely automatic system, 4" round, 15 to 50 stations		.29	55.172	Station	20,100	3,300		23,400	27,100
2200	51 to 144 stations		.32	50		15,600	3,000		18,600	21,700
2400	6" round or 4" x 7" oval, 15 to 50 stations		.24	66.667		25,200	3,975		29,175	33,700
2800	51 to 144 stations		.23	69.565		21,100	4,150		25,250	29,500

For customer support on your Mechanical Cost Data, call 877.795.7394.

Estimating Tips
22 10 00 Plumbing Piping and Pumps

This subdivision is primarily basic pipe and related materials. The pipe may be used by any of the mechanical disciplines, i.e., plumbing, fire protection, heating, and air conditioning.

Note: CPVC plastic piping approved for fire protection is located in 21 11 13.

- The labor adjustment factors listed in Subdivision 22 01 02.20 apply throughout Divisions 21, 22, and 23. CAUTION: the correct percentage may vary for the same items. For example, the percentage add for the basic pipe installation should be based on the maximum height that the craftsman must install for that particular section. If the pipe is to be located 14' above the floor but it is suspended on threaded rod from beams, the bottom flange of which is 18' high (4' rods), then the height is actually 18' and the add is 20%. The pipe coverer, however, does not have to go above the 14', and so the add should be 10%.

- Most pipe is priced first as straight pipe with a joint (coupling, weld, etc.) every 10' and a hanger usually every 10'. There are exceptions with hanger spacing such as for cast iron pipe (5') and plastic pipe (3 per 10'). Following each type of pipe there are several lines listing sizes and the amount to be subtracted to delete couplings and hangers. This is for pipe that is to be buried or supported together on trapeze hangers. The reason that the couplings are deleted is that these runs are usually long, and frequently longer lengths of pipe are used. By deleting the couplings, the estimator is expected to look up and add back the correct reduced number of couplings.

- When preparing an estimate, it may be necessary to approximate the fittings. Fittings usually run between 25% and 50% of the cost of the pipe. The lower percentage is for simpler runs, and the higher number is for complex areas, such as mechanical rooms.

- For historic restoration projects, the systems must be as invisible as possible, and pathways must be sought for pipes, conduit, and ductwork. While installations in accessible spaces (such as basements and attics) are relatively straightforward to estimate, labor costs may be more difficult to determine when delivery systems must be concealed.

22 40 00 Plumbing Fixtures

- Plumbing fixture costs usually require two lines: the fixture itself and its "rough-in, supply, and waste."

- In the Assemblies Section (Plumbing D2010) for the desired fixture, the System Components Group at the center of the page shows the fixture on the first line. The rest of the list (fittings, pipe, tubing, etc.) will total up to what we refer to in the Unit Price section as "Rough-in, supply, waste, and vent." Note that for most fixtures we allow a nominal 5' of tubing to reach from the fixture to a main or riser.

- Remember that gas- and oil-fired units need venting.

Reference Numbers

Reference numbers are shown in shaded boxes at the beginning of some major classifications. These numbers refer to related items in the Reference Section. The reference information may be an estimating procedure, an alternate pricing method, or technical information.

Note: Not all subdivisions listed here necessarily appear in this publication. ■

22 01 Operation and Maintenance of Plumbing

22 01 02 – Labor Adjustments

22 01 02.10 Boilers, General		Crew	Daily Output	Labor-Hours	Unit	Material	2015 Bare Costs Labor	Equipment	Total	Total Incl O&P
0010	**BOILERS, GENERAL**, Prices do not include flue piping, elec. wiring,									
0020	gas or oil piping, boiler base, pad, or tankless unless noted									
0100	Boiler H.P.: 10 KW = 34 lb./steam/hr. = 33,475 BTU/hr.									
0150	To convert SFR to BTU rating: Hot water, 150 x SFR;									
0160	Forced hot water, 180 x SFR; steam, 240 x SFR									

22 01 02.20 Labor Adjustment Factors

		Crew	Daily Output	Labor-Hours	Unit	Material	2015 Bare Costs Labor	Equipment	Total	Total Incl O&P
0010	**LABOR ADJUSTMENT FACTORS**, (For Div. 21, 22 and 23) R220102-20									
0100	Labor factors, The below are reasonable suggestions, however									
0110	each project must be evaluated for its own peculiarities, and									
0120	the adjustments be increased or decreased depending on the									
0130	severity of the special conditions.									
1000	Add to labor for elevated installation (Above floor level)									
1080	10' to 14.5' high R221113-70						10%			
1100	15' to 19.5' high						20%			
1120	20' to 24.5' high						25%			
1140	25' to 29.5' high						35%			
1160	30' to 34.5' high						40%			
1180	35' to 39.5' high						50%			
1200	40' and higher						55%			
2000	Add to labor for crawl space									
2100	3' high						40%			
2140	4' high						30%			
3000	Add to labor for multi-story building									
3100	Add per floor for floors 3 thru 19						2%			
3140	Add per floor for floors 20 and up						4%			
4000	Add to labor for working in existing occupied buildings									
4100	Hospital						35%			
4140	Office building						25%			
4180	School						20%			
4220	Factory or warehouse						15%			
4260	Multi dwelling						15%			
5000	Add to labor, miscellaneous									
5100	Cramped shaft						35%			
5140	Congested area						15%			
5180	Excessive heat or cold						30%			
9000	Labor factors, The above are reasonable suggestions, however									
9010	each project should be evaluated for its own peculiarities.									
9100	Other factors to be considered are:									
9140	Movement of material and equipment through finished areas									
9180	Equipment room									
9220	Attic space									
9260	No service road									
9300	Poor unloading/storage area									
9340	Congested site area/heavy traffic									

For customer support on your Mechanical Cost Data, call 877.795.7394.

22 05 05 – Selective Demolition for Plumbing

22 05 05.10 Plumbing Demolition		Crew	Daily Output	Labor-Hours	Unit	Material	2015 Bare Costs		Total	Total Incl O&P
							Labor	Equipment		
0010	**PLUMBING DEMOLITION** R220105-10									
0400	Air compressor, up thru 2 H.P.	Q-1	10	1.600	Ea.		84.50		84.50	128
0410	3 H.P. thru 7-1/2 H.P. R024119-10		5.60	2.857			151		151	228
0420	10 H.P. thru 15 H.P.		1.40	11.429			605		605	910
0430	20 H.P. thru 30 H.P.	Q-2	1.30	18.462			1,000		1,000	1,525
0500	Backflow preventer, up thru 2" diameter	1 Plum	17	.471			27.50		27.50	41.50
0510	2-1/2" thru 3" diameter	Q-1	10	1.600			84.50		84.50	128
0520	4" thru 6" diameter	"	5	3.200			169		169	255
0530	8" thru 10" diameter	Q-2	3	8			440		440	660
1900	Piping fittings, single connection, up thru 1-1/2" diameter	1 Plum	30	.267			15.65		15.65	23.50
1910	2" thru 4" diameter		14	.571			33.50		33.50	50.50
1980	Pipe hanger/support removal		80	.100			5.85		5.85	8.85
2000	Piping, metal, up thru 1-1/2" diameter		200	.040	L.F.		2.35		2.35	3.55
2050	2" thru 3-1/2" diameter		150	.053			3.13		3.13	4.73
2100	4" thru 6" diameter	2 Plum	100	.160			9.40		9.40	14.20
2150	8" thru 14" diameter	"	60	.267			15.65		15.65	23.50
2153	16" thru 20" diameter	Q-18	70	.343			19.10	.82	19.92	30
2155	24" thru 26" diameter		55	.436			24.50	1.05	25.55	37.50
2156	30" thru 36" diameter		40	.600			33.50	1.44	34.94	52
2160	Plastic pipe with fittings, up thru 1-1/2" diameter	1 Plum	250	.032			1.88		1.88	2.84
2162	2" thru 3" diameter	"	200	.040			2.35		2.35	3.55
2164	4" thru 6" diameter	Q-1	200	.080			4.23		4.23	6.40
2166	8" thru 14" diameter		150	.107			5.65		5.65	8.50
2168	16" diameter		100	.160			8.45		8.45	12.75
2180	Pumps, all fractional horse-power		12	1.333	Ea.		70.50		70.50	106
2184	1 H.P. thru 5 H.P.		6	2.667			141		141	213
2186	7-1/2 H.P. thru 15 H.P.		2.50	6.400			340		340	510
2188	20 H.P. thru 25 H.P.	Q-2	4	6			330		330	495
2190	30 H.P. thru 60 H.P.		.80	30			1,650		1,650	2,475
2192	75 H.P. thru 100 H.P.		.60	40			2,200		2,200	3,300
2194	150 H.P.		.50	48			2,625		2,625	3,975
2230	Temperature maintenance cable	1 Plum	1200	.007	L.F.		.39		.39	.59
3100	Tanks, water heaters and liquid containers									
3110	Up thru 45 gallons	Q-1	22	.727	Ea.		38.50		38.50	58
3120	50 thru 120 gallons		14	1.143			60.50		60.50	91
3130	130 thru 240 gallons		7.60	2.105			111		111	168
3140	250 thru 500 gallons		5.40	2.963			157		157	236
3150	600 thru 1000 gallons	Q-2	1.60	15			820		820	1,250
3160	1100 thru 2000 gallons		.70	34.286			1,875		1,875	2,825
3170	2100 thru 4000 gallons		.50	48			2,625		2,625	3,975
9100	Valve, metal valves or strainers and similar, up thru 1-1/2" diameter	1 Stpi	28	.286			17.05		17.05	26
9110	2" thru 3" diameter	Q-1	11	1.455			77		77	116
9120	4" thru 6" diameter	"	8	2			106		106	160
9130	8" thru 14" diameter	Q-2	8	3			164		164	248
9140	16" thru 20" diameter		2	12			655		655	995
9150	24" diameter		1.20	20			1,100		1,100	1,650
9200	Valve, plastic, up thru 1-1/2" diameter	1 Plum	42	.190			11.20		11.20	16.90
9210	2" thru 3" diameter		15	.533			31.50		31.50	47.50
9220	4" thru 6" diameter		12	.667			39		39	59
9300	Vent flashing and caps		55	.145			8.55		8.55	12.90
9350	Water filter, commercial, 1" thru 1-1/2"	Q-1	2	8			425		425	640
9360	2" thru 2-1/2"	"	1.60	10			530		530	800
9400	Water heaters									

For customer support on your Mechanical Cost Data, call 877.795.7394.

89

22 05 05 – Selective Demolition for Plumbing

22 05 05.10 Plumbing Demolition		Crew	Daily Output	Labor-Hours	Unit	Material	2015 Bare Costs			Total
							Labor	Equipment	Total	Incl O&P
9410	Up thru 245 GPH	Q-1	2.40	6.667	Ea.		350		350	530
9420	250 thru 756 GPH		1.60	10			530		530	800
9430	775 thru 1640 GPH	↓	.80	20			1,050		1,050	1,600
9440	1650 thru 4000 GPH	Q-2	.50	48			2,625		2,625	3,975
9470	Water softener	Q-1	2	8	↓		425		425	640

22 05 23 – General-Duty Valves for Plumbing Piping

22 05 23.10 Valves, Brass

	22 05 23.10 Valves, Brass	Crew	Daily Output	Labor-Hours	Unit	Material	Labor	Equipment	Total	Total Incl O&P
0010	**VALVES, BRASS**									
0032	For motorized valves, see Section 23 09 53.10									
0500	Gas cocks, threaded									
0510	1/4"	1 Plum	26	.308	Ea.	14.70	18.05		32.75	43.5
0520	3/8"		24	.333		14.70	19.55		34.25	45.5
0530	1/2"		24	.333		13.15	19.55		32.70	44
0540	3/4"		22	.364		15.50	21.50		37	49
0550	1"		19	.421		31	24.50		55.50	71.5
0560	1-1/4"		15	.533		44.50	31.50		76	96.5
0570	1-1/2"		13	.615		61	36		97	122
0580	2"	↓	11	.727	↓	103	42.50		145.50	179
0672	For larger sizes use lubricated plug valve, Section 23 05 23.70									

22 05 23.20 Valves, Bronze

	22 05 23.20 Valves, Bronze	Crew	Daily Output	Labor-Hours	Unit	Material	Labor	Equipment	Total	Total Incl O&P
0010	**VALVES, BRONZE** R220523-90									
1020	Angle, 150 lb., rising stem, threaded									
1030	1/8"	1 Plum	24	.333	Ea.	128	19.55		147.55	170
1040	1/4"		24	.333		128	19.55		147.55	170
1050	3/8"		24	.333		130	19.55		149.55	173
1060	1/2"		22	.364		142	21.50		163.50	188
1070	3/4"		20	.400		193	23.50		216.50	249
1080	1"		19	.421		279	24.50		303.50	345
1090	1-1/4"		15	.533		360	31.50		391.50	445
1100	1-1/2"		13	.615		470	36		506	570
1110	2"	↓	11	.727	↓	755	42.50		797.50	895
1300	Ball									
1304	Soldered									
1312	3/8"	1 Plum	21	.381	Ea.	14.95	22.50		37.45	50.5
1316	1/2"		18	.444		14.95	26		40.95	56
1320	3/4"		17	.471		25	27.50		52.50	68.50
1324	1"		15	.533		31	31.50		62.50	82
1328	1-1/4"		13	.615		52	36		88	112
1332	1-1/2"		11	.727		66.50	42.50		109	138
1336	2"		9	.889		83.50	52		135.50	171
1340	2-1/2"		7	1.143		395	67		462	535
1344	3"	↓	5	1.600	↓	455	94		549	640
1350	Single union end									
1358	3/8"	1 Plum	21	.381	Ea.	21.50	22.50		44	57.50
1362	1/2"		18	.444		22.50	26		48.50	64.50
1366	3/4"		17	.471		43	27.50		70.50	89
1370	1"		15	.533		52	31.50		83.50	105
1374	1-1/4"		13	.615		88.50	36		124.50	152
1378	1-1/2"		11	.727		111	42.50		153.50	187
1382	2"	↓	9	.889	↓	173	52		225	269
1398	Threaded, 150 psi									
1400	1/4"	1 Plum	24	.333	Ea.	14.15	19.55		33.70	45

22 05 23 – General-Duty Valves for Plumbing Piping

22 05 23.20 Valves, Bronze		Crew	Daily Output	Labor-Hours	Unit	Material	2015 Bare Costs Labor	Equipment	Total	Total Incl O&P
1430	3/8"	1 Plum	24	.333	Ea.	14.15	19.55		33.70	45
1450	1/2"		22	.364		14.15	21.50		35.65	47.50
1460	3/4"		20	.400		23.50	23.50		47	61
1470	1"		19	.421		33.50	24.50		58	74.50
1480	1-1/4"		15	.533		58.50	31.50		90	112
1490	1-1/2"		13	.615		76.50	36		112.50	139
1500	2"		11	.727		93	42.50		135.50	167
1510	2-1/2"		9	.889		310	52		362	420
1520	3"	▼	8	1	▼	470	58.50		528.50	610
1600	Butterfly, 175 psi, full port, solder or threaded ends									
1610	Stainless steel disc and stem									
1620	1/4"	1 Plum	24	.333	Ea.	17.35	19.55		36.90	48.50
1630	3/8"		24	.333		17.35	19.55		36.90	48.50
1640	1/2"		22	.364		18.65	21.50		40.15	52.50
1650	3/4"		20	.400		29.50	23.50		53	68
1660	1"		19	.421		36	24.50		60.50	77
1670	1-1/4"		15	.533		58	31.50		89.50	112
1680	1-1/2"		13	.615		75	36		111	137
1690	2"	▼	11	.727	▼	94	42.50		136.50	168
1750	Check, swing, class 150, regrinding disc, threaded									
1800	1/8"	1 Plum	24	.333	Ea.	66.50	19.55		86.05	103
1830	1/4"		24	.333		66.50	19.55		86.05	103
1840	3/8"		24	.333		70.50	19.55		90.05	107
1850	1/2"		24	.333		75.50	19.55		95.05	113
1860	3/4"		20	.400		100	23.50		123.50	146
1870	1"		19	.421		144	24.50		168.50	196
1880	1-1/4"		15	.533		208	31.50		239.50	277
1890	1-1/2"		13	.615		242	36		278	320
1900	2"	▼	11	.727		355	42.50		397.50	455
1910	2-1/2"	Q-1	15	1.067		800	56.50		856.50	965
1920	3"	"	13	1.231	▼	1,075	65		1,140	1,275
2000	For 200 lb., add					5%	10%			
2040	For 300 lb., add					15%	15%			
2060	Check swing, 300#, sweat, 3/8" size	1 Plum	24	.333	Ea.	63	19.55		82.55	98.50
2070	1/2"		24	.333		63	19.55		82.55	98.50
2080	3/4"		20	.400		79.50	23.50		103	123
2090	1"		19	.421		114	24.50		138.50	163
2100	1-1/4"		15	.533		159	31.50		190.50	223
2110	1-1/2"		13	.615		194	36		230	268
2120	2"	▼	11	.727		296	42.50		338.50	390
2130	2-1/2"	Q-1	15	1.067		650	56.50		706.50	800
2140	3"	"	13	1.231	▼	870	65		935	1,050
2350	Check, lift, class 150, horizontal composition disc, threaded									
2430	1/4"	1 Plum	24	.333	Ea.	153	19.55		172.55	198
2440	3/8"		24	.333		194	19.55		213.55	243
2450	1/2"		24	.333		169	19.55		188.55	215
2460	3/4"		20	.400		206	23.50		229.50	262
2470	1"		19	.421		292	24.50		316.50	360
2480	1-1/4"		15	.533		395	31.50		426.50	485
2490	1-1/2"		13	.615		470	36		506	570
2500	2"	▼	11	.727	▼	780	42.50		822.50	925
2850	Gate, N.R.S., soldered, 125 psi									
2900	3/8"	1 Plum	24	.333	Ea.	61	19.55		80.55	96.50

22 05 23.20 Valves, Bronze		Crew	Daily Output	Labor-Hours	Unit	Material	2015 Bare Costs Labor	Equipment	Total	Total Incl O&P
2920	1/2"	1 Plum	24	.333	Ea.	54	19.55		73.55	89
2940	3/4"		20	.400		63.50	23.50		87	106
2950	1"		19	.421		76.50	24.50		101	122
2960	1-1/4"		15	.533		126	31.50		157.50	187
2970	1-1/2"		13	.615		142	36		178	212
2980	2"		11	.727		185	42.50		227.50	269
2990	2-1/2"	Q-1	15	1.067		450	56.50		506.50	580
3000	3"	"	13	1.231		560	65		625	715
3350	Threaded, class 150									
3410	1/4"	1 Plum	24	.333	Ea.	75.50	19.55		95.05	113
3420	3/8"		24	.333		75.50	19.55		95.05	113
3430	1/2"		24	.333		71	19.55		90.55	108
3440	3/4"		20	.400		81.50	23.50		105	125
3450	1"		19	.421		102	24.50		126.50	150
3460	1-1/4" size		15	.533		138	31.50		169.50	200
3470	1-1/2"		13	.615		198	36		234	273
3480	2"		11	.727		237	42.50		279.50	325
3490	2-1/2"	Q-1	15	1.067		605	56.50		661.50	750
3500	3" size	"	13	1.231		855	65		920	1,050
3600	Gate, flanged, 150 lb.									
3610	1"	1 Plum	7	1.143	Ea.	1,150	67		1,217	1,350
3620	1-1/2"		6	1.333		1,425	78.50		1,503.50	1,700
3630	2"		5	1.600		2,100	94		2,194	2,450
3634	2-1/2"	Q-1	5	3.200		3,900	169		4,069	4,525
3640	3"	"	4.50	3.556		3,775	188		3,963	4,425
3850	Rising stem, soldered, 300 psi									
3900	3/8"	1 Plum	24	.333	Ea.	141	19.55		160.55	185
3920	1/2"		24	.333		116	19.55		135.55	157
3940	3/4"		20	.400		132	23.50		155.50	181
3950	1"		19	.421		179	24.50		203.50	235
3960	1-1/4"		15	.533		247	31.50		278.50	320
3970	1-1/2"		13	.615		300	36		336	385
3980	2"		11	.727		480	42.50		522.50	595
3990	2-1/2"	Q-1	15	1.067		1,025	56.50		1,081.50	1,200
4000	3"	"	13	1.231		1,575	65		1,640	1,850
4250	Threaded, class 150									
4310	1/4"	1 Plum	24	.333	Ea.	68.50	19.55		88.05	105
4320	3/8"		24	.333		68.50	19.55		88.05	105
4330	1/2"		24	.333		63	19.55		82.55	98.50
4340	3/4"		20	.400		73.50	23.50		97	116
4350	1"		19	.421		98.50	24.50		123	146
4360	1-1/4"		15	.533		134	31.50		165.50	195
4370	1-1/2"		13	.615		169	36		205	240
4380	2"		11	.727		227	42.50		269.50	315
4390	2-1/2"	Q-1	15	1.067		530	56.50		586.50	670
4400	3"	"	13	1.231		740	65		805	910
4500	For 300 psi, threaded, add					100%	15%			
4540	For chain operated type, add					15%				
4850	Globe, class 150, rising stem, threaded									
4920	1/4"	1 Plum	24	.333	Ea.	97	19.55		116.55	136
4940	3/8"		24	.333		95.50	19.55		115.05	135
4950	1/2"		24	.333		95.50	19.55		115.05	135
4960	3/4"		20	.400		99	23.50		122.50	145

22 05 23.20 Valves, Bronze	Crew	Daily Output	Labor-Hours	Unit	Material	2015 Bare Costs Labor	Equipment	Total	Total Incl O&P
4970 1"	1 Plum	19	.421	Ea.	154	24.50		178.50	207
4980 1-1/4"		15	.533		245	31.50		276.50	315
4990 1-1/2"		13	.615		320	36		356	410
5000 2"		11	.727		465	42.50		507.50	580
5010 2-1/2"	Q-1	15	1.067		1,175	56.50		1,231.50	1,350
5020 3"	"	13	1.231		1,675	65		1,740	1,925
5120 For 300 lb. threaded, add					50%	15%			
5130 Globe, 300 lb., sweat, 3/8" size	1 Plum	24	.333	Ea.	121	19.55		140.55	163
5140 1/2"		24	.333		124	19.55		143.55	167
5150 3/4"		20	.400		169	23.50		192.50	221
5160 1"		19	.421		234	24.50		258.50	295
5170 1-1/4"		15	.533		370	31.50		401.50	455
5180 1-1/2"		13	.615		445	36		481	545
5190 2"		11	.727		660	42.50		702.50	790
5200 2-1/2"	Q-1	15	1.067		1,275	56.50		1,331.50	1,500
5210 3"	"	13	1.231		1,650	65		1,715	1,925
5600 Relief, pressure & temperature, self-closing, ASME, threaded									
5640 3/4"	1 Plum	28	.286	Ea.	253	16.75		269.75	305
5650 1"		24	.333		405	19.55		424.55	475
5660 1-1/4"		20	.400		895	23.50		918.50	1,025
5670 1-1/2"		18	.444		1,250	26		1,276	1,425
5680 2"		16	.500		1,350	29.50		1,379.50	1,550
5950 Pressure, poppet type, threaded									
6000 1/2"	1 Plum	30	.267	Ea.	78.50	15.65		94.15	110
6040 3/4"	"	28	.286	"	73.50	16.75		90.25	106
6400 Pressure, water, ASME, threaded									
6440 3/4"	1 Plum	28	.286	Ea.	116	16.75		132.75	153
6450 1"		24	.333		260	19.55		279.55	315
6460 1-1/4"		20	.400		390	23.50		413.50	465
6470 1-1/2"		18	.444		570	26		596	665
6480 2"		16	.500		820	29.50		849.50	950
6490 2-1/2"		15	.533		3,150	31.50		3,181.50	3,525
6900 Reducing, water pressure									
6920 300 psi to 25-75 psi, threaded or sweat									
6940 1/2"	1 Plum	24	.333	Ea.	395	19.55		414.55	465
6950 3/4"		20	.400		405	23.50		428.50	485
6960 1"		19	.421		630	24.50		654.50	730
6970 1-1/4"		15	.533		1,100	31.50		1,131.50	1,250
6980 1-1/2"		13	.615		1,650	36		1,686	1,875
6990 2"		11	.727		2,450	42.50		2,492.50	2,775
7100 For built-in by-pass or 10-35 psi, add					39			39	42.50
7700 High capacity, 250 psi to 25-75 psi, threaded									
7740 1/2"	1 Plum	24	.333	Ea.	540	19.55		559.55	625
7780 3/4"		20	.400		540	23.50		563.50	630
7790 1"		19	.421		775	24.50		799.50	895
7800 1-1/4"		15	.533		1,350	31.50		1,381.50	1,525
7810 1-1/2"		13	.615		1,975	36		2,011	2,225
7820 2"		11	.727		2,875	42.50		2,917.50	3,225
7830 2-1/2"		9	.889		4,100	52		4,152	4,600
7840 3"		8	1		4,925	58.50		4,983.50	5,500
7850 3" flanged (iron body)	Q-1	10	1.600		4,850	84.50		4,934.50	5,450
7860 4" flanged (iron body)	"	8	2		6,300	106		6,406	7,075
7920 For higher pressure, add					25%				

22 05 23.20 Valves, Bronze		Crew	Daily Output	Labor-Hours	Unit	Material	2015 Bare Costs Labor	Equipment	Total	Total Incl O&P
8000	Silent check, bronze trim									
8010	Compact wafer type, for 125 or 150 lb. flanges									
8020	1-1/2"	1 Plum	11	.727	Ea.	335	42.50		377.50	430
8021	2"	"	9	.889		355	52		407	470
8022	2-1/2"	Q-1	9	1.778		370	94		464	550
8023	3"		8	2		415	106		521	615
8024	4"		5	3.200		710	169		879	1,025
8025	5"	Q-2	6	4		940	219		1,159	1,350
8026	6"		5	4.800		1,100	263		1,363	1,625
8027	8"		4.50	5.333		2,250	292		2,542	2,900
8028	10"		4	6		4,325	330		4,655	5,250
8029	12"		3	8		9,800	440		10,240	11,500
8050	For 250 or 300 lb. flanges, thru 6" no change									
8051	For 8" and 10", add				Ea.	40%	10%			
8060	Full flange wafer type, 150 lb.									
8061	1"	1 Plum	14	.571	Ea.	330	33.50		363.50	415
8062	1-1/4"		12	.667		365	39		404	460
8063	1-1/2"		11	.727		415	42.50		457.50	525
8064	2"		9	.889		560	52		612	695
8065	2-1/2"	Q-1	9	1.778		630	94		724	835
8066	3"		8	2		700	106		806	930
8067	4"		5	3.200		1,025	169		1,194	1,375
8068	5"	Q-2	6	4		1,325	219		1,544	1,800
8069	6"	"	5	4.800		1,750	263		2,013	2,325
8080	For 300 lb., add					40%	10%			
8100	Globe type, 150 lb.									
8110	2"	1 Plum	9	.889	Ea.	690	52		742	840
8111	2-1/2"	Q-1	9	1.778		855	94		949	1,075
8112	3"		8	2		1,025	106		1,131	1,275
8113	4"		5	3.200		1,425	169		1,594	1,800
8114	5"	Q-2	6	4		1,725	219		1,944	2,225
8115	6"	"	5	4.800		2,400	263		2,663	3,050
8130	For 300 lb., add					20%	10%			
8140	Screwed end type, 250 lb.									
8141	1/2"	1 Plum	24	.333	Ea.	62.50	19.55		82.05	98
8142	3/4"		20	.400		62.50	23.50		86	104
8143	1"		19	.421		71	24.50		95.50	116
8144	1-1/4"		15	.533		98.50	31.50		130	156
8145	1-1/2"		13	.615		110	36		146	176
8146	2"		11	.727		149	42.50		191.50	229
8350	Tempering, water, sweat connections									
8400	1/2"	1 Plum	24	.333	Ea.	98	19.55		117.55	138
8440	3/4"	"	20	.400	"	126	23.50		149.50	175
8650	Threaded connections									
8700	1/2"	1 Plum	24	.333	Ea.	126	19.55		145.55	169
8740	3/4"		20	.400		770	23.50		793.50	880
8750	1"		19	.421		865	24.50		889.50	995
8760	1-1/4"		15	.533		1,350	31.50		1,381.50	1,525
8770	1-1/2"		13	.615		1,475	36		1,511	1,650
8780	2"		11	.727		2,200	42.50		2,242.50	2,500

22 05 23 – General-Duty Valves for Plumbing Piping

22 05 23.60 Valves, Plastic	Crew	Daily Output	Labor-Hours	Unit	Material	2015 Bare Costs Labor	Equipment	Total	Total Incl O&P
0010 **VALVES, PLASTIC** R220523-90									
1100 Angle, PVC, threaded									
1110 1/4"	1 Plum	26	.308	Ea.	39.50	18.05		57.55	71
1120 1/2"		26	.308		56.50	18.05		74.55	89.50
1130 3/4"		25	.320		67	18.80		85.80	103
1140 1"		23	.348		81.50	20.50		102	121
1150 Ball, PVC, socket or threaded, true union									
1230 1/2"	1 Plum	26	.308	Ea.	40.50	18.05		58.55	72
1240 3/4"		25	.320		40.50	18.80		59.30	73
1250 1"		23	.348		48	20.50		68.50	84
1260 1-1/4"		21	.381		83.50	22.50		106	126
1270 1-1/2"		20	.400		83.50	23.50		107	128
1280 2"		17	.471		110	27.50		137.50	163
1290 2-1/2"	Q-1	26	.615		188	32.50		220.50	256
1300 3"		24	.667		282	35		317	365
1310 4"		20	.800		450	42.50		492.50	560
1360 For PVC, flanged, add					100%	15%			
1450 Double union 1/2"	1 Plum	26	.308		32.50	18.05		50.55	63.50
1460 3/4"		25	.320		42.50	18.80		61.30	75
1470 1"		23	.348		50.50	20.50		71	86.50
1480 1-1/4"		21	.381		69.50	22.50		92	111
1490 1-1/2"		20	.400		84	23.50		107.50	128
1500 2"		17	.471		111	27.50		138.50	164
1650 CPVC, socket or threaded, single union									
1700 1/2"	1 Plum	26	.308	Ea.	58	18.05		76.05	91
1720 3/4"		25	.320		77	18.80		95.80	113
1730 1"		23	.348		87.50	20.50		108	127
1750 1-1/4"		21	.381		140	22.50		162.50	188
1760 1-1/2"		20	.400		140	23.50		163.50	190
1770 2"		17	.471		192	27.50		219.50	254
1780 3"	Q-1	24	.667		760	35		795	890
1840 For CPVC, flanged, add					65%	15%			
1880 For true union, socket or threaded, add					50%	5%			
2050 Polypropylene, threaded									
2100 1/4"	1 Plum	26	.308	Ea.	45	18.05		63.05	77
2120 3/8"		26	.308		45	18.05		63.05	77
2130 1/2"		26	.308		45	18.05		63.05	77
2140 3/4"		25	.320		54.50	18.80		73.30	88.50
2150 1"		23	.348		62.50	20.50		83	99.50
2160 1-1/4"		21	.381		84	22.50		106.50	127
2170 1-1/2"		20	.400		103	23.50		126.50	150
2180 2"		17	.471		138	27.50		165.50	194
2190 3"	Q-1	24	.667		315	35		350	400
2200 4"	"	20	.800		550	42.50		592.50	670
2550 PVC, three way, socket or threaded									
2600 1/2"	1 Plum	26	.308	Ea.	62.50	18.05		80.55	96
2640 3/4"		25	.320		71	18.80		89.80	107
2650 1"		23	.348		76.50	20.50		97	115
2660 1-1/2"		20	.400		181	23.50		204.50	235
2670 2"		17	.471		206	27.50		233.50	269
2680 3"	Q-1	24	.667		430	35		465	530
2740 For flanged, add					60%	15%			

For customer support on your Mechanical Cost Data, call 877.795.7394.

95

22 05 23 – General-Duty Valves for Plumbing Piping

22 05 23.60 Valves, Plastic		Crew	Daily Output	Labor-Hours	Unit	Material	2015 Bare Costs			Total	Total Incl O&P
							Labor	Equipment		Total	
3150	Ball check, PVC, socket or threaded										
3200	1/4"	1 Plum	26	.308	Ea.	45.50	18.05			63.55	78
3220	3/8"		26	.308		45.50	18.05			63.55	78
3240	1/2"		26	.308		45.50	18.05			63.55	78
3250	3/4"		25	.320		51	18.80			69.80	84.50
3260	1"		23	.348		64	20.50			84.50	102
3270	1-1/4"		21	.381		107	22.50			129.50	152
3280	1-1/2"		20	.400		107	23.50			130.50	154
3290	2"		17	.471		146	27.50			173.50	203
3310	3"	Q-1	24	.667		405	35			440	500
3320	4"	"	20	.800		575	42.50			617.50	700
3360	For PVC, flanged, add					50%	15%				
3750	CPVC, socket or threaded										
3800	1/2"	1 Plum	26	.308	Ea.	70.50	18.05			88.55	105
3840	3/4"		25	.320		83.50	18.80			102.30	121
3850	1"		23	.348		99.50	20.50			120	140
3860	1-1/2"		20	.400		171	23.50			194.50	224
3870	2"		17	.471		230	27.50			257.50	295
3880	3"	Q-1	24	.667		620	35			655	735
3920	4"	"	20	.800		840	42.50			882.50	990
3930	For CPVC, flanged, add					40%	15%				
4340	Polypropylene, threaded										
4360	1/2"	1 Plum	26	.308	Ea.	44	18.05			62.05	76
4400	3/4"		25	.320		61	18.80			79.80	95.50
4440	1"		23	.348		65	20.50			85.50	103
4450	1-1/2"		20	.400		125	23.50			148.50	173
4460	2"		17	.471		158	27.50			185.50	216
4500	For polypropylene flanged, add					200%	15%				
4850	Foot valve, PVC, socket or threaded										
4900	1/2"	1 Plum	34	.235	Ea.	73	13.80			86.80	101
4930	3/4"		32	.250		83	14.70			97.70	114
4940	1"		28	.286		107	16.75			123.75	144
4950	1-1/4"		27	.296		206	17.40			223.40	254
4960	1-1/2"		26	.308		206	18.05			224.05	255
4970	2"		24	.333		365	19.55			384.55	430
4980	3"		20	.400		835	23.50			858.50	955
4990	4"		18	.444		1,275	26			1,301	1,475
5000	For flanged, add					25%	10%				
5050	CPVC, socket or threaded										
5060	1/2"	1 Plum	34	.235	Ea.	72	13.80			85.80	101
5070	3/4"		32	.250		83	14.70			97.70	114
5080	1"		28	.286		102	16.75			118.75	139
5090	1-1/4"		27	.296		164	17.40			181.40	207
5100	1-1/2"		26	.308		164	18.05			182.05	208
5110	2"		24	.333		211	19.55			230.55	262
5120	3"		20	.400		430	23.50			453.50	505
5130	4"		18	.444		780	26			806	895
5140	For flanged, add					25%	10%				
5280	Needle valve, PVC, threaded										
5300	1/4"	1 Plum	26	.308	Ea.	55.50	18.05			73.55	88.50
5340	3/8"		26	.308		64	18.05			82.05	98
5360	1/2"		26	.308		64	18.05			82.05	98
5380	For polypropylene, add					10%					

22 05 23 – General-Duty Valves for Plumbing Piping

22 05 23.60 Valves, Plastic		Crew	Daily Output	Labor-Hours	Unit	Material	2015 Bare Costs Labor	Equipment	Total	Total Incl O&P
5800	Y check, PVC, socket or threaded									
5820	1/2"	1 Plum	26	.308	Ea.	55	18.05		73.05	88
5840	3/4"		25	.320		89	18.80		107.80	126
5850	1"		23	.348		96.50	20.50		117	137
5860	1-1/4"		21	.381		151	22.50		173.50	201
5870	1-1/2"		20	.400		165	23.50		188.50	218
5880	2"		17	.471		205	27.50		232.50	268
5890	2-1/2"		15	.533		435	31.50		466.50	525
5900	3"	Q-1	24	.667		435	35		470	530
5910	4"	"	20	.800		710	42.50		752.50	845
5960	For PVC flanged, add					45%	15%			
6350	Y sediment strainer, PVC, socket or threaded									
6400	1/2"	1 Plum	26	.308	Ea.	55	18.05		73.05	88
6440	3/4"		24	.333		58.50	19.55		78.05	93.50
6450	1"		23	.348		69.50	20.50		90	108
6460	1-1/4"		21	.381		117	22.50		139.50	162
6470	1-1/2"		20	.400		117	23.50		140.50	164
6480	2"		17	.471		144	27.50		171.50	200
6490	2-1/2"		15	.533		350	31.50		381.50	435
6500	3"	Q-1	24	.667		350	35		385	440
6510	4"	"	20	.800		585	42.50		627.50	705
6560	For PVC, flanged, add					55%	15%			

22 05 29 – Hangers and Supports for Plumbing Piping and Equipment

22 05 29.10 Hangers & Supp. for Plumb'g/HVAC Pipe/Equip.

		Crew	Daily Output	Labor-Hours	Unit	Material	2015 Bare Costs Labor	Equipment	Total	Total Incl O&P
0010	HANGERS AND SUPPORTS FOR PLUMB'G/HVAC PIPE/EQUIP.									
0011	TYPE numbers per MSS-SP58									
0050	Brackets									
0060	Beam side or wall, malleable iron, TYPE 34									
0070	3/8" threaded rod size	1 Plum	48	.167	Ea.	3.11	9.80		12.91	18.20
0080	1/2" threaded rod size		48	.167		4.39	9.80		14.19	19.65
0090	5/8" threaded rod size		48	.167		8.20	9.80		18	24
0100	3/4" threaded rod size		48	.167		11.30	9.80		21.10	27
0110	7/8" threaded rod size		48	.167		11.70	9.80		21.50	27.50
0120	For concrete installation, add						30%			
0150	Wall, welded steel, medium, TYPE 32									
0160	0 size, 12" wide, 18" deep	1 Plum	34	.235	Ea.	172	13.80		185.80	210
0170	1 size, 18" wide, 24" deep		34	.235		211	13.80		224.80	253
0180	2 size, 24" wide, 30" deep		34	.235		290	13.80		303.80	340
0200	Beam attachment, welded, TYPE 22									
0202	3/8"	Q-15	80	.200	Ea.	7.15	10.55	.72	18.42	24.50
0203	1/2"		76	.211		7.25	11.10	.76	19.11	25.50
0204	5/8"		72	.222		8.45	11.75	.80	21	28
0205	3/4"		68	.235		12.30	12.45	.85	25.60	33
0206	7/8"		64	.250		15.75	13.20	.90	29.85	38.50
0207	1"		56	.286		31.50	15.10	1.03	47.63	58.50
0300	Clamps									
0310	C-clamp, for mounting on steel beam flange, w/locknut, TYPE 23									
0320	3/8" threaded rod size	1 Plum	160	.050	Ea.	2.05	2.94		4.99	6.70
0330	1/2" threaded rod size		160	.050		2.20	2.94		5.14	6.85
0340	5/8" threaded rod size		160	.050		3.70	2.94		6.64	8.50
0350	3/4" threaded rod size		160	.050		4.55	2.94		7.49	9.45
0352	7/8" threaded rod size		140	.057		15	3.35		18.35	21.50

22 05 29 – Hangers and Supports for Plumbing Piping and Equipment

22 05 29.10 Hangers & Supp. for Plumb'g/HVAC Pipe/Equip.	Crew	Daily Output	Labor-Hours	Unit	Material	2015 Bare Costs Labor	Equipment	Total	Total Incl O&P
0400	High temperature to 1050°F, alloy steel								
0410	4" pipe size	Q-1	106	.151	Ea.	26.50	7.95		34.45	41.50
0420	6" pipe size		106	.151		45.50	7.95		53.45	62
0430	8" pipe size		97	.165		50.50	8.70		59.20	68.50
0440	10" pipe size		84	.190		74	10.05		84.05	96.50
0450	12" pipe size		72	.222		83	11.75		94.75	109
0460	14" pipe size		64	.250		209	13.20		222.20	250
0470	16" pipe size		56	.286		230	15.10		245.10	275
0480	Beam clamp, flange type, TYPE 25								
0482	For 3/8" bolt	1 Plum	48	.167	Ea.	3.42	9.80		13.22	18.55
0483	For 1/2" bolt		44	.182		4.39	10.65		15.04	21
0484	For 5/8" bolt		40	.200		4.98	11.75		16.73	23.50
0485	For 3/4" bolt		36	.222		4.98	13.05		18.03	25
0486	For 1" bolt		32	.250		4.98	14.70		19.68	27.50
0500	I-beam, for mounting on bottom flange, strap iron, TYPE 21								
0510	2" flange size	1 Plum	96	.083	Ea.	17.65	4.89		22.54	27
0520	3" flange size		95	.084		21	4.94		25.94	30.50
0530	4" flange size		93	.086		5.75	5.05		10.80	13.95
0540	5" flange size		92	.087		6.30	5.10		11.40	14.65
0550	6" flange size		90	.089		7.15	5.20		12.35	15.75
0560	7" flange size		88	.091		7.95	5.35		13.30	16.80
0570	8" flange size		86	.093		8.55	5.45		14	17.65
0600	One hole, vertical mounting, malleable iron								
0610	1/2" pipe size	1 Plum	160	.050	Ea.	1.08	2.94		4.02	5.60
0620	3/4" pipe size		145	.055		1.17	3.24		4.41	6.20
0630	1" pipe size		136	.059		1.24	3.45		4.69	6.55
0640	1-1/4" pipe size		128	.063		2.19	3.67		5.86	7.95
0650	1-1/2" pipe size		120	.067		2.52	3.91		6.43	8.65
0660	2" pipe size		112	.071		3.64	4.19		7.83	10.35
0670	2-1/2" pipe size		104	.077		7.05	4.52		11.57	14.60
0680	3" pipe size		96	.083		8.55	4.89		13.44	16.80
0690	3-1/2" pipe size		90	.089		10.45	5.20		15.65	19.40
0700	4" pipe size		84	.095		11.90	5.60		17.50	21.50
0750	Riser or extension pipe, carbon steel, TYPE 8								
0756	1/2" pipe size	1 Plum	52	.154	Ea.	2.52	9.05		11.57	16.40
0760	3/4" pipe size		48	.167		2.52	9.80		12.32	17.55
0770	1" pipe size		47	.170		2.64	10		12.64	18
0780	1-1/4" pipe size		46	.174		3.21	10.20		13.41	18.95
0790	1-1/2" pipe size		45	.178		3.42	10.45		13.87	19.50
0800	2" pipe size		43	.186		3.54	10.90		14.44	20.50
0810	2-1/2" pipe size		41	.195		3.68	11.45		15.13	21.50
0820	3" pipe size		40	.200		4	11.75		15.75	22
0830	3-1/2" pipe size		39	.205		4.50	12.05		16.55	23
0840	4" pipe size		38	.211		5.45	12.35		17.80	24.50
0850	5" pipe size		37	.216		6.80	12.70		19.50	26.50
0860	6" pipe size		36	.222		8.35	13.05		21.40	29
0870	8" pipe size		34	.235		14.60	13.80		28.40	37
0880	10" pipe size		32	.250		22.50	14.70		37.20	47
0890	12" pipe size		28	.286		30	16.75		46.75	58.50
0900	For plastic coating 3/4" to 4", add					190%				
0910	For copper plating 3/4" to 4", add					58%				
0950	Two piece, complete, carbon steel, medium weight, TYPE 4								
0960	1/2" pipe size	Q-1	137	.117	Ea.	2.29	6.15		8.44	11.80

22 05 29.10 Hangers & Supp. for Plumb'g/HVAC Pipe/Equip.		Crew	Daily Output	Labor-Hours	Unit	Material	2015 Bare Costs Labor	Equipment	Total	Total Incl O&P
0970	3/4" pipe size	Q-1	134	.119	Ea.	2.29	6.30		8.59	12.05
0980	1" pipe size		132	.121		2.59	6.40		8.99	12.50
0990	1-1/4" pipe size		130	.123		2.80	6.50		9.30	12.90
1000	1-1/2" pipe size		126	.127		3.64	6.70		10.34	14.15
1010	2" pipe size		124	.129		4.48	6.80		11.28	15.25
1020	2-1/2" pipe size		120	.133		4.84	7.05		11.89	15.95
1030	3" pipe size		117	.137		5.35	7.20		12.55	16.80
1040	3-1/2" pipe size		114	.140		10	7.40		17.40	22
1050	4" pipe size		110	.145		7.10	7.70		14.80	19.40
1060	5" pipe size		106	.151		17.40	7.95		25.35	31
1070	6" pipe size		104	.154		19.20	8.15		27.35	33.50
1080	8" pipe size		100	.160		23.50	8.45		31.95	38.50
1090	10" pipe size		96	.167		46	8.80		54.80	64
1100	12" pipe size		89	.180		55.50	9.50		65	76
1110	14" pipe size		82	.195		73.50	10.30		83.80	96.50
1120	16" pipe size	▼	68	.235		95	12.45		107.45	124
1130	For galvanized, add				▼	45%				
1150	Insert, concrete									
1160	Wedge type, carbon steel body, malleable iron nut, galvanized									
1170	1/4" threaded rod size	1 Plum	96	.083	Ea.	9.55	4.89		14.44	17.90
1180	3/8" threaded rod size		96	.083		10.30	4.89		15.19	18.75
1190	1/2" threaded rod size		96	.083		10.65	4.89		15.54	19.15
1200	5/8" threaded rod size		96	.083		11.25	4.89		16.14	19.80
1210	3/4" threaded rod size		96	.083		11.95	4.89		16.84	20.50
1220	7/8" threaded rod size	▼	96	.083	▼	12.35	4.89		17.24	21
1250	Pipe guide sized for insulation									
1260	No. 1, 1" pipe size, 1" thick insulation	1 Stpi	26	.308	Ea.	124	18.40		142.40	164
1270	No. 2, 1-1/4"-2" pipe size, 1" thick insulation		23	.348		151	21		172	198
1280	No. 3, 1-1/4"-2" pipe size, 1-1/2" thick insulation		21	.381		151	23		174	201
1290	No. 4, 2-1/2"-3-1/2" pipe size, 1-1/2" thick insulation		18	.444		151	26.50		177.50	206
1300	No. 5, 4"-5" pipe size, 1-1/2" thick insulation	▼	16	.500		171	30		201	234
1310	No. 6, 5"-6" pipe size, 2" thick insulation	Q-5	21	.762		199	41		240	281
1320	No. 7, 8" pipe size, 2" thick insulation		16	1		251	54		305	355
1330	No. 8, 10" pipe size, 2" thick insulation	▼	12	1.333		370	71.50		441.50	520
1340	No. 9, 12" pipe size, 2" thick insulation	Q-6	17	1.412		370	78.50		448.50	530
1350	No. 10, 12"-14" pipe size, 2-1/2" thick insulation		16	1.500		435	83.50		518.50	605
1360	No. 11, 16" pipe size, 2-1/2" thick insulation		10.50	2.286		435	127		562	670
1370	No. 12, 16"-18" pipe size, 3" thick insulation		9	2.667		640	149		789	930
1380	No. 13, 20" pipe size, 3" thick insulation		7.50	3.200		640	178		818	975
1390	No. 14, 24" pipe size, 3" thick insulation	▼	7	3.429	▼	845	191		1,036	1,225
1400	Bands									
1410	Adjustable band, carbon steel, for non-insulated pipe, TYPE 7									
1420	1/2" pipe size	Q-1	142	.113	Ea.	.30	5.95		6.25	9.35
1430	3/4" pipe size		140	.114		.30	6.05		6.35	9.45
1440	1" pipe size		137	.117		.30	6.15		6.45	9.65
1450	1-1/4" pipe size		134	.119		.33	6.30		6.63	9.90
1460	1-1/2" pipe size		131	.122		.33	6.45		6.78	10.10
1470	2" pipe size		129	.124		.33	6.55		6.88	10.25
1480	2-1/2" pipe size		125	.128		.60	6.75		7.35	10.85
1490	3" pipe size		122	.131		.64	6.95		7.59	11.15
1500	3-1/2" pipe size		119	.134		1.04	7.10		8.14	11.90
1510	4" pipe size		114	.140		1.04	7.40		8.44	12.35
1520	5" pipe size	▼	110	.145		1.66	7.70		9.36	13.45

22 05 29.10 Hangers & Supp. for Plumb'g/HVAC Pipe/Equip.	Crew	Daily Output	Labor-Hours	Unit	Material	2015 Bare Costs Labor	2015 Bare Costs Equipment	Total	Total Incl O&P
1530	6" pipe size	Q-1	108	.148	Ea.	1.94	7.85		9.79	13.95
1540	8" pipe size		104	.154		2.93	8.15		11.08	15.45
1550	For copper plated, add					50%				
1560	For galvanized, add					30%				
1570	For plastic coating, add					30%				
1600	Adjusting nut malleable iron, steel band, TYPE 9									
1610	1/2" pipe size, galvanized band	Q-1	137	.117	Ea.	4.88	6.15		11.03	14.65
1620	3/4" pipe size, galvanized band		135	.119		4.88	6.25		11.13	14.80
1630	1" pipe size, galvanized band		132	.121		5.05	6.40		11.45	15.20
1640	1-1/4" pipe size, galvanized band		129	.124		5.05	6.55		11.60	15.45
1650	1-1/2" pipe size, galvanized band		126	.127		5.20	6.70		11.90	15.90
1660	2" pipe size, galvanized band		124	.129		5.45	6.80		12.25	16.30
1670	2-1/2" pipe size, galvanized band		120	.133		8.60	7.05		15.65	20
1680	3" pipe size, galvanized band		117	.137		9	7.20		16.20	21
1690	3-1/2" pipe size, galvanized band		114	.140		35.50	7.40		42.90	50
1700	4" pipe size, cadmium plated band		110	.145		23	7.70		30.70	36.50
1740	For plastic coated band, add					35%				
1750	For completely copper coated, add					45%				
1800	Clevis, adjustable, carbon steel, for non-insulated pipe, TYPE 1									
1810	1/2" pipe size	Q-1	137	.117	Ea.	.93	6.15		7.08	10.30
1820	3/4" pipe size		135	.119		.93	6.25		7.18	10.45
1830	1" pipe size		132	.121		.98	6.40		7.38	10.75
1840	1-1/4" pipe size		129	.124		1.11	6.55		7.66	11.10
1850	1-1/2" pipe size		126	.127		1.22	6.70		7.92	11.50
1860	2" pipe size		124	.129		1.43	6.80		8.23	11.85
1870	2-1/2" pipe size		120	.133		2.39	7.05		9.44	13.30
1880	3" pipe size		117	.137		2.58	7.20		9.78	13.75
1890	3-1/2" pipe size		114	.140		3	7.40		10.40	14.50
1900	4" pipe size		110	.145		3.24	7.70		10.94	15.15
1910	5" pipe size		106	.151		4.48	7.95		12.43	17
1920	6" pipe size		104	.154		5.45	8.15		13.60	18.20
1930	8" pipe size		100	.160		7.95	8.45		16.40	21.50
1940	10" pipe size		96	.167		14.20	8.80		23	29
1950	12" pipe size		89	.180		18.40	9.50		27.90	34.50
1960	14" pipe size		82	.195		31	10.30		41.30	49.50
1970	16" pipe size		68	.235		49.50	12.45		61.95	73.50
1971	18" pipe size		54	.296		59.50	15.65		75.15	89
1972	20" pipe size		38	.421		123	22		145	170
1980	For galvanized, add					66%				
1990	For copper plated 1/2" to 4", add					77%				
2000	For light weight 1/2" to 4", deduct					13%				
2010	Insulated pipe type, 3/4" to 12" pipe, add					180%				
2020	Insulated pipe type, chrome-moly U-strap, add					530%				
2250	Split ring, malleable iron, for non-insulated pipe, TYPE 11									
2260	1/2" pipe size	Q-1	137	.117	Ea.	4.39	6.15		10.54	14.15
2270	3/4" pipe size		135	.119		4.43	6.25		10.68	14.30
2280	1" pipe size		132	.121		4.70	6.40		11.10	14.80
2290	1-1/4" pipe size		129	.124		5.80	6.55		12.35	16.30
2300	1-1/2" pipe size		126	.127		7.05	6.70		13.75	17.95
2310	2" pipe size		124	.129		8	6.80		14.80	19.10
2320	2-1/2" pipe size		120	.133		11.55	7.05		18.60	23.50
2330	3" pipe size		117	.137		14.45	7.20		21.65	27
2340	3-1/2" pipe size		114	.140		15.05	7.40		22.45	28

22 05 29 – Hangers and Supports for Plumbing Piping and Equipment

22 05 29.10 Hangers & Supp. for Plumb'g/HVAC Pipe/Equip.	Crew	Daily Output	Labor-Hours	Unit	Material	2015 Bare Costs Labor	Equipment	Total	Total Incl O&P
2350	4" pipe size	Q-1	110	.145	Ea.	15.05	7.70		22.75	28
2360	5" pipe size		106	.151		20	7.95		27.95	34
2370	6" pipe size		104	.154		44.50	8.15		52.65	61.50
2380	8" pipe size		100	.160		65.50	8.45		73.95	85
2390	For copper plated, add					8%				
2400	Channels, steel, 3/4" x 1-1/2"	1 Plum	80	.100	L.F.	4.10	5.85		9.95	13.35
2404	1-1/2" x 1-1/2"		70	.114		5.40	6.70		12.10	16.10
2408	1-7/8" x 1-1/2"		60	.133		23	7.85		30.85	37.50
2412	3" x 1-1/2"		50	.160		39.50	9.40		48.90	57.50
2416	Hangers, trapeze channel support, 12 ga. 1-1/2" x 1/1/2", 12" wide, steel		8.80	.909	Ea.	29	53.50		82.50	112
2418	18" wide, steel		8.30	.964		31.50	56.50		88	120
2430	Spring nuts, long, 1/4"		120	.067		1.17	3.91		5.08	7.20
2432	3/8"		100	.080		1.23	4.70		5.93	8.45
2434	1/2"		80	.100		1.33	5.85		7.18	10.30
2436	5/8"		80	.100		3.62	5.85		9.47	12.85
2438	3/4"		75	.107		6.15	6.25		12.40	16.20
2440	Spring nuts, short, 1/4"		120	.067		1.16	3.91		5.07	7.20
2442	3/8"		100	.080		1.20	4.70		5.90	8.40
2444	1/2"		80	.100		1.67	5.85		7.52	10.70
2500	Washer, flat steel									
2502	3/8"	1 Plum	240	.033	Ea.	.09	1.96		2.05	3.05
2503	1/2"		220	.036		.19	2.13		2.32	3.43
2504	5/8"		200	.040		.22	2.35		2.57	3.79
2505	3/4"		180	.044		.54	2.61		3.15	4.53
2506	7/8"		160	.050		.66	2.94		3.60	5.15
2507	1"		140	.057		.94	3.35		4.29	6.10
2508	1-1/4"		120	.067		1.17	3.91		5.08	7.20
2520	Nut, steel, hex									
2522	3/8"	1 Plum	200	.040	Ea.	.14	2.35		2.49	3.70
2523	1/2"		180	.044		.35	2.61		2.96	4.33
2524	5/8"		160	.050		.56	2.94		3.50	5.05
2525	3/4"		140	.057		.94	3.35		4.29	6.10
2526	7/8"		120	.067		1.51	3.91		5.42	7.55
2527	1"		100	.080		2.07	4.70		6.77	9.40
2528	1-1/4"		80	.100		5.20	5.85		11.05	14.55
2532	Turnbuckle, TYPE 13									
2534	3/8"	1 Plum	80	.100	Ea.	4.82	5.85		10.67	14.15
2535	1/2"		72	.111		5.45	6.50		11.95	15.85
2536	5/8"		64	.125		9.45	7.35		16.80	21.50
2537	3/4"		56	.143		12.40	8.40		20.80	26.50
2538	7/8"		48	.167		29.50	9.80		39.30	47.50
2539	1"		40	.200		42	11.75		53.75	64
2540	1-1/4"		32	.250		78	14.70		92.70	108
2650	Rods, carbon steel									
2660	Continuous thread									
2670	1/4" thread size	1 Plum	144	.056	L.F.	1.70	3.26		4.96	6.80
2680	3/8" thread size		144	.056		1.81	3.26		5.07	6.90
2690	1/2" thread size		144	.056		2.86	3.26		6.12	8.10
2700	5/8" thread size		144	.056		4.05	3.26		7.31	9.40
2710	3/4" thread size		144	.056		7.15	3.26		10.41	12.80
2720	7/8" thread size		144	.056		8.95	3.26		12.21	14.80
2721	1" thread size	Q-1	160	.100		15.25	5.30		20.55	25
2722	1-1/8" thread size	"	120	.133		22.50	7.05		29.55	35

For customer support on your Mechanical Cost Data, call 877.795.7394.

101

22 05 29.10 Hangers & Supp. for Plumb'g/HVAC Pipe/Equip.	Crew	Daily Output	Labor-Hours	Unit	Material	2015 Bare Costs Labor	2015 Bare Costs Equipment	Total	Total Incl O&P
2725	1/4" thread size, bright finish	1 Plum	144	.056	L.F.	1.24	3.26		4.50	6.30
2726	1/2" thread size, bright finish	"	144	.056	↓	4.23	3.26		7.49	9.60
2730	For galvanized, add					40%				
2820	Rod couplings									
2821	3/8"	1 Plum	60	.133	Ea.	1.43	7.85		9.28	13.35
2822	1/2"		54	.148		1.99	8.70		10.69	15.35
2823	5/8"		48	.167		2.50	9.80		12.30	17.55
2824	3/4"		44	.182		3.51	10.65		14.16	19.95
2825	7/8"		40	.200		5.95	11.75		17.70	24.50
2826	1"		34	.235		7.85	13.80		21.65	29.50
2827	1-1/8"	↓	30	.267	↓	15.55	15.65		31.20	40.50
2860	Pipe hanger assy, adj. clevis, saddle, rod, clamp, insul. allowance									
2864	1/2" pipe size	Q-5	35	.457	Ea.	30	24.50		54.50	70
2866	3/4" pipe size		34.80	.460		31	24.50		55.50	72
2868	1" pipe size		34.60	.462		18.50	25		43.50	58
2869	1-1/4" pipe size		34.30	.466		18.90	25		43.90	59
2870	1-1/2" pipe size		33.90	.472		19.20	25.50		44.70	59.50
2872	2" pipe size		33.30	.480		21.50	26		47.50	62.50
2874	2-1/2" pipe size		32.30	.495		25	26.50		51.50	67.50
2876	3" pipe size		31.20	.513		30	27.50		57.50	74.50
2880	4" pipe size		30.70	.521		32.50	28		60.50	78
2884	6" pipe size		29.80	.537		41.50	29		70.50	89
2888	8" pipe size		28	.571		52.50	30.50		83	104
2892	10" pipe size		25.20	.635		59.50	34		93.50	117
2896	12" pipe size	↓	23.20	.690	↓	88.50	37		125.50	153
2900	Rolls									
2910	Adjustable yoke, carbon steel with CI roll, TYPE 43									
2918	2" pipe size	Q-1	140	.114	Ea.	10.60	6.05		16.65	21
2920	2-1/2" pipe size		137	.117		10.60	6.15		16.75	21
2930	3" pipe size		131	.122		11.25	6.45		17.70	22
2940	3-1/2" pipe size		124	.129		15.80	6.80		22.60	27.50
2950	4" pipe size		117	.137		15.80	7.20		23	28.50
2960	5" pipe size		110	.145		23	7.70		30.70	36.50
2970	6" pipe size		104	.154		30.50	8.15		38.65	46
2980	8" pipe size		96	.167		43.50	8.80		52.30	61.50
2990	10" pipe size		80	.200		53.50	10.55		64.05	75
3000	12" pipe size		68	.235		85.50	12.45		97.95	113
3010	14" pipe size		56	.286		162	15.10		177.10	201
3020	16" pipe size	↓	48	.333	↓	195	17.60		212.60	241
3050	Chair, carbon steel with CI roll									
3060	2" pipe size	1 Plum	68	.118	Ea.	11.80	6.90		18.70	23.50
3070	2-1/2" pipe size		65	.123		12.80	7.20		20	25
3080	3" pipe size		62	.129		13.80	7.55		21.35	26.50
3090	3-1/2" pipe size		60	.133		17.80	7.85		25.65	31.50
3100	4" pipe size		58	.138		16.25	8.10		24.35	30
3110	5" pipe size		56	.143		19.10	8.40		27.50	33.50
3120	6" pipe size		53	.151		25.50	8.85		34.35	41.50
3130	8" pipe size		50	.160		34.50	9.40		43.90	52
3140	10" pipe size		48	.167		47	9.80		56.80	67
3150	12" pipe size		46	.174		72	10.20		82.20	95
3170	Single pipe roll, (see line 2650 for rods), TYPE 41, 1" pipe size	Q-1	137	.117		7.90	6.15		14.05	18
3180	1-1/4" pipe size		131	.122		8.15	6.45		14.60	18.75
3190	1-1/2" pipe size	↓	129	.124	↓	8.45	6.55		15	19.20

22 05 29 – Hangers and Supports for Plumbing Piping and Equipment

22 05 29.10 Hangers & Supp. for Plumb'g/HVAC Pipe/Equip.		Crew	Daily Output	Labor-Hours	Unit	Material	2015 Bare Costs Labor	Equipment	Total	Total Incl O&P
3200	2" pipe size	Q-1	124	.129	Ea.	8.75	6.80		15.55	19.90
3210	2-1/2" pipe size		118	.136		9.10	7.15		16.25	21
3220	3" pipe size		115	.139		9.40	7.35		16.75	21.50
3230	3-1/2" pipe size		113	.142		10.30	7.50		17.80	22.50
3240	4" pipe size		112	.143		10.60	7.55		18.15	23
3250	5" pipe size		110	.145		12.10	7.70		19.80	25
3260	6" pipe size		101	.158		26	8.35		34.35	41
3270	8" pipe size		90	.178		26.50	9.40		35.90	43.50
3280	10" pipe size		80	.200		35	10.55		45.55	54.50
3290	12" pipe size		68	.235		57.50	12.45		69.95	82
3291	14" pipe size		56	.286		87	15.10		102.10	119
3292	16" pipe size		48	.333		112	17.60		129.60	150
3293	18" pipe size		40	.400		132	21		153	177
3294	20" pipe size		35	.457		156	24		180	209
3296	24" pipe size		30	.533		251	28		279	320
3297	30" pipe size		25	.640		475	34		509	570
3298	36" pipe size	▼	20	.800	▼	580	42.50		622.50	705
3300	Saddles (add vertical pipe riser, usually 3" diameter)									
3310	Pipe support, complete, adjust., CI saddle, TYPE 36									
3320	2-1/2" pipe size	1 Plum	96	.083	Ea.	92.50	4.89		97.39	109
3330	3" pipe size		88	.091		94	5.35		99.35	112
3340	3-1/2" pipe size		79	.101		95.50	5.95		101.45	114
3350	4" pipe size		68	.118		139	6.90		145.90	163
3360	5" pipe size		64	.125		142	7.35		149.35	167
3370	6" pipe size		59	.136		142	7.95		149.95	168
3380	8" pipe size		53	.151		147	8.85		155.85	175
3390	10" pipe size		50	.160		171	9.40		180.40	202
3400	12" pipe size	▼	48	.167	▼	181	9.80		190.80	214
3450	For standard pipe support, one piece, CI deduct						34%			
3460	For stanchion support, CI with steel yoke, deduct						60%			
3550	Insulation shield 1" thick, 1/2" pipe size, TYPE 40	1 Asbe	100	.080	Ea.	2.74	4.19		6.93	9.50
3560	3/4" pipe size		100	.080		3.06	4.19		7.25	9.85
3570	1" pipe size		98	.082		3.30	4.27		7.57	10.30
3580	1-1/4" pipe size		98	.082		3.52	4.27		7.79	10.50
3590	1-1/2" pipe size		96	.083		3.45	4.36		7.81	10.60
3600	2" pipe size		96	.083		3.65	4.36		8.01	10.80
3610	2-1/2" pipe size		94	.085		3.77	4.46		8.23	11.10
3620	3" pipe size		94	.085		4.53	4.46		8.99	11.95
3630	2" thick, 3-1/2" pipe size		92	.087		4.84	4.55		9.39	12.40
3640	4" pipe size		92	.087		6.75	4.55		11.30	14.55
3650	5" pipe size		90	.089		9.15	4.65		13.80	17.30
3660	6" pipe size		90	.089		10.45	4.65		15.10	18.70
3670	8" pipe size		88	.091		17	4.76		21.76	26
3680	10" pipe size		88	.091		21.50	4.76		26.26	31
3690	12" pipe size		86	.093		24.50	4.87		29.37	34.50
3700	14" pipe size		86	.093		40.50	4.87		45.37	52
3710	16" pipe size		84	.095		43.50	4.99		48.49	55.50
3720	18" pipe size		84	.095		48	4.99		52.99	61
3730	20" pipe size		82	.098		53	5.10		58.10	66
3732	24" pipe size	▼	80	.100	▼	62	5.25		67.25	76.50
3750	Covering protection saddle, TYPE 39									
3760	1" covering size									
3770	3/4" pipe size	1 Plum	68	.118	Ea.	4.61	6.90		11.51	15.50

22 05 29.10 Hangers & Supp. for Plumb'g/HVAC Pipe/Equip.	Crew	Daily Output	Labor-Hours	Unit	Material	2015 Bare Costs Labor	Equipment	Total	Total Incl O&P
3780	1" pipe size	1 Plum	68	.118	Ea.	4.61	6.90		11.51	15.50
3790	1-1/4" pipe size		68	.118		4.61	6.90		11.51	15.50
3800	1-1/2" pipe size		66	.121		4.92	7.10		12.02	16.15
3810	2" pipe size		66	.121		4.92	7.10		12.02	16.15
3820	2-1/2" pipe size		64	.125		4.92	7.35		12.27	16.50
3830	3" pipe size		64	.125		6.60	7.35		13.95	18.35
3840	3-1/2" pipe size		62	.129		7.30	7.55		14.85	19.50
3850	4" pipe size		62	.129		7.30	7.55		14.85	19.50
3860	5" pipe size		60	.133		7.30	7.85		15.15	19.85
3870	6" pipe size		60	.133		8.75	7.85		16.60	21.50
3900	1-1/2" covering size									
3910	3/4" pipe size	1 Plum	68	.118	Ea.	5.50	6.90		12.40	16.50
3920	1" pipe size		68	.118		5.50	6.90		12.40	16.50
3930	1-1/4" pipe size		68	.118		5.50	6.90		12.40	16.50
3940	1-1/2" pipe size		66	.121		5.50	7.10		12.60	16.80
3950	2" pipe size		66	.121		5.95	7.10		13.05	17.30
3960	2-1/2" pipe size		64	.125		7.20	7.35		14.55	19
3970	3" pipe size		64	.125		7.20	7.35		14.55	19
3980	3-1/2" pipe size		62	.129		7.60	7.55		15.15	19.80
3990	4" pipe size		62	.129		7.60	7.55		15.15	19.80
4000	5" pipe size		60	.133		7.60	7.85		15.45	20
4010	6" pipe size		60	.133		10.25	7.85		18.10	23
4020	8" pipe size		58	.138		12.85	8.10		20.95	26.50
4022	10" pipe size		56	.143		12.85	8.40		21.25	27
4024	12" pipe size		54	.148		31.50	8.70		40.20	47.50
4028	2" covering size									
4029	2-1/2" pipe size	1 Plum	62	.129	Ea.	7.25	7.55		14.80	19.45
4032	3" pipe size		60	.133		8.55	7.85		16.40	21
4033	4" pipe size		58	.138		8.55	8.10		16.65	21.50
4034	6" pipe size		56	.143		12.25	8.40		20.65	26
4035	8" pipe size		54	.148		14.40	8.70		23.10	29
4080	10" pipe size		58	.138		15.35	8.10		23.45	29
4090	12" pipe size		56	.143		35	8.40		43.40	51
4100	14" pipe size		56	.143		35	8.40		43.40	51
4110	16" pipe size		54	.148		45.50	8.70		54.20	63
4120	18" pipe size		54	.148		45.50	8.70		54.20	63
4130	20" pipe size		52	.154		50.50	9.05		59.55	69.50
4150	24" pipe size		50	.160		59.50	9.40		68.90	79
4160	30" pipe size		48	.167		65	9.80		74.80	86.50
4180	36" pipe size		45	.178		75.50	10.45		85.95	99
4186	2-1/2" covering size									
4187	3" pipe size	1 Plum	58	.138	Ea.	8.65	8.10		16.75	22
4188	4" pipe size		56	.143		9.40	8.40		17.80	23
4189	6" pipe size		52	.154		12.80	9.05		21.85	28
4190	8" pipe size		48	.167		15.85	9.80		25.65	32.50
4191	10" pipe size		44	.182		17.40	10.65		28.05	35.50
4192	12" pipe size		40	.200		37.50	11.75		49.25	59
4193	14" pipe size		36	.222		37.50	13.05		50.55	60.50
4194	16" pipe size		32	.250		48	14.70		62.70	74.50
4195	18" pipe size		28	.286		53.50	16.75		70.25	84.50
4200	Sockets									
4210	Rod end, malleable iron, TYPE 16									
4220	1/4" thread size	1 Plum	240	.033	Ea.	1.42	1.96		3.38	4.51

22 05 29.10 Hangers & Supp. for Plumb'g/HVAC Pipe/Equip.	Crew	Daily Output	Labor-Hours	Unit	Material	2015 Bare Costs Labor	Equipment	Total	Total Incl O&P
4230	3/8" thread size	1 Plum	240	.033	Ea.	1.42	1.96		3.38	4.51
4240	1/2" thread size		230	.035		1.87	2.04		3.91	5.15
4250	5/8" thread size		225	.036		3.50	2.09		5.59	7
4260	3/4" thread size		220	.036		4.91	2.13		7.04	8.60
4270	7/8" thread size		210	.038		6.90	2.24		9.14	11
4290	Strap, 1/2" pipe size, TYPE 26	Q-1	142	.113		1.54	5.95		7.49	10.70
4300	3/4" pipe size		140	.114		1.61	6.05		7.66	10.85
4310	1" pipe size		137	.117		2.37	6.15		8.52	11.90
4320	1-1/4" pipe size		134	.119		2.41	6.30		8.71	12.20
4330	1-1/2" pipe size		131	.122		2.57	6.45		9.02	12.60
4340	2" pipe size		129	.124		2.59	6.55		9.14	12.75
4350	2-1/2" pipe size		125	.128		3.97	6.75		10.72	14.55
4360	3" pipe size		122	.131		5.20	6.95		12.15	16.20
4370	3-1/2" pipe size		119	.134		5.40	7.10		12.50	16.70
4380	4" pipe size		114	.140		5.70	7.40		13.10	17.45
4400	U-bolt, carbon steel									
4410	Standard, with nuts, TYPE 42									
4420	1/2" pipe size	1 Plum	160	.050	Ea.	.98	2.94		3.92	5.50
4430	3/4" pipe size		158	.051		1.03	2.97		4	5.60
4450	1" pipe size		152	.053		1.07	3.09		4.16	5.85
4460	1-1/4" pipe size		148	.054		1.35	3.17		4.52	6.30
4470	1-1/2" pipe size		143	.056		1.43	3.28		4.71	6.55
4480	2" pipe size		139	.058		1.55	3.38		4.93	6.80
4490	2-1/2" pipe size		134	.060		2.53	3.50		6.03	8.10
4500	3" pipe size		128	.063		2.86	3.67		6.53	8.70
4510	3-1/2" pipe size		122	.066		3.08	3.85		6.93	9.20
4520	4" pipe size		117	.068		3.12	4.01		7.13	9.50
4530	5" pipe size		114	.070		3.66	4.12		7.78	10.25
4540	6" pipe size		111	.072		6.95	4.23		11.18	14.05
4550	8" pipe size		109	.073		7.60	4.31		11.91	14.85
4560	10" pipe size		107	.075		13.20	4.39		17.59	21
4570	12" pipe size		104	.077		19.15	4.52		23.67	28
4580	For plastic coating on 1/2" thru 6" size, add					150%				
4700	U-hook, carbon steel, requires mounting screws or bolts									
4710	3/4" thru 2" pipe size									
4720	6" long	1 Plum	96	.083	Ea.	.73	4.89		5.62	8.20
4730	8" long		96	.083		1.06	4.89		5.95	8.55
4740	10" long		96	.083		1.64	4.89		6.53	9.20
4750	12" long		96	.083		2.24	4.89		7.13	9.85
4760	For copper plated, add					50%				
7000	Roof supports									
7006	Duct									
7010	Rectangular, open, 12" off roof									
7020	To 18" wide	Q-9	26	.615	Ea.	150	31		181	213
7030	To 24" wide	"	22	.727		175	36.50		211.50	249
7040	To 36" wide	Q-10	30	.800		200	42		242	284
7050	To 48" wide		28	.857		225	45		270	315
7060	To 60" wide		24	1		250	52		302	355
7100	Equipment									
7120	Equipment support	Q-5	20	.800	Ea.	65	43		108	137
7300	Pipe									
7310	Roller type									
7320	Up to 2-1/2" diam. pipe									

For customer support on your Mechanical Cost Data, call 877.795.7394.

105

22 05 29.10 Hangers & Supp. for Plumb'g/HVAC Pipe/Equip.		Crew	Daily Output	Labor-Hours	Unit	Material	2015 Bare Costs Labor	Equipment	Total	Total Incl O&P
7324	3-1/2" off roof	Q-5	24	.667	Ea.	16	36		52	71.5
7326	Up to 10" off roof	"	20	.800	"	21	43		64	88
7340	2-1/2" to 3-1/2" diam. pipe									
7342	Up to 16" off roof	Q-5	18	.889	Ea.	45.50	48		93.50	122
7360	4" to 5" diam. pipe									
7362	Up to 12" off roof	Q-5	16	1	Ea.	65	54		119	153
7400	Strut/channel type									
7410	Up to 2-1/2" diam. pipe									
7424	3-1/2" off roof	Q-5	24	.667	Ea.	14	36		50	69.5
7426	Up to 10" off roof	"	20	.800	"	19	43		62	86
7440	Strut and roller type									
7452	2-1/2" to 3-1/2" diam. pipe									
7454	Up to 16" off roof	Q-5	18	.889	Ea.	65.50	48		113.50	144
7460	Strut and hanger type									
7470	Up to 3" diam. pipe									
7474	Up to 8" off roof	Q-5	19	.842	Ea.	41.50	45.50		87	115
8000	Pipe clamp, plastic, 1/2" CTS	1 Plum	80	.100		.23	5.85		6.08	9.1
8010	3/4" CTS		73	.110		.24	6.45		6.69	9.9
8020	1" CTS		68	.118		.54	6.90		7.44	11.0
8080	Economy clamp, 1/4" CTS		175	.046		.05	2.68		2.73	4.1
8090	3/8" CTS		168	.048		.07	2.80		2.87	4.3
8100	1/2" CTS		160	.050		.07	2.94		3.01	4.5
8110	3/4" CTS		145	.055		.15	3.24		3.39	5.0
8200	Half clamp, 1/2" CTS		80	.100		.08	5.85		5.93	8.9
8210	3/4" CTS		73	.110		.12	6.45		6.57	9.85
8300	Suspension clamp, 1/2" CTS		80	.100		.23	5.85		6.08	9.1
8310	3/4" CTS		73	.110		.24	6.45		6.69	9.9
8320	1" CTS		68	.118		.54	6.90		7.44	11.0
8400	Insulator, 1/2" CTS		80	.100		.37	5.85		6.22	9.25
8410	3/4" CTS		73	.110		.39	6.45		6.84	10.15
8420	1" CTS		68	.118		.40	6.90		7.30	10.90
8500	J hook clamp with nail 1/2" CTS		240	.033		.13	1.96		2.09	3.09
8501	3/4" CTS	↓	240	.033	↓	.13	1.96		2.09	3.09
8800	Wire cable support system									
8810	Cable with hook terminal and locking device									
8830	2 mm, (.079") dia cable, (100 lb. cap.)									
8840	1 m, (3.3') length, with hook	1 Shee	96	.083	Ea.	3.03	4.66		7.69	10.45
8850	2 m, (6.6') length, with hook		84	.095		3.50	5.35		8.85	12
8860	3 m, (9.9') length, with hook	↓	72	.111		3.97	6.20		10.17	13.85
8870	5 m, (16.4') length, with hook	Q-9	60	.267		4.97	13.45		18.42	26
8880	10 m, (32.8') length, with hook	"	30	.533	↓	7.20	27		34.20	49
8900	3mm, (.118") dia cable, (200 lb. cap.)									
8910	1 m, (3.3') length, with hook	1 Shee	96	.083	Ea.	3.88	4.66		8.54	11.35
8920	2 m, (6.6') length, with hook		84	.095		4.36	5.35		9.71	12.95
8930	3 m, (9.9') length, with hook	↓	72	.111		4.81	6.20		11.01	14.80
8940	5 m, (16.4') length, with hook	Q-9	60	.267		5.95	13.45		19.40	27
8950	10 m, (32.8') length, with hook	"	30	.533	↓	8.55	27		35.55	50.50
9000	Cable system accessories									
9010	Anchor bolt, 3/8", with nut	1 Shee	140	.057	Ea.	1.16	3.20		4.36	6.15
9020	Air duct corner protector		160	.050		.52	2.80		3.32	4.85
9030	Air duct support attachment	↓	140	.057	↓	.99	3.20		4.19	6
9040	Flange clip, hammer-on style									
9044	For flange thickness 3/32" - 9/64", 160 lb. cap.	1 Shee	180	.044	Ea.	.29	2.49		2.78	4.12

22 05 29 – Hangers and Supports for Plumbing Piping and Equipment

22 05 29.10 Hangers & Supp. for Plumb'g/HVAC Pipe/Equip.	Crew	Daily Output	Labor-Hours	Unit	Material	2015 Bare Costs Labor	Equipment	Total	Total Incl O&P
9048	For flange thickness 1/8" - 1/4", 200 lb. cap.	1 Shee	160	.050	Ea.	.29	2.80		3.09	4.60
9052	For flange thickness 5/16" - 1/2", 200 lb. cap.		150	.053		.56	2.98		3.54	5.20
9056	For flange thickness 9/16" - 3/4", 200 lb. cap.		140	.057		.75	3.20		3.95	5.70
9060	Wire insulation protection tube		180	.044	L.F.	.34	2.49		2.83	4.17
9070	Wire cutter				Ea.	41.50			41.50	45.50

22 05 33 – Heat Tracing for Plumbing Piping

22 05 33.20 Temperature Maintenance Cable

		Crew	Daily Output	Labor-Hours	Unit	Material	2015 Bare Costs Labor	Equipment	Total	Total Incl O&P
0010	**TEMPERATURE MAINTENANCE CABLE**									
0040	Components									
0080	Heating cable									
0100	208 V									
0150	140°F	Q-1	1060.80	.015	L.F.	9.20	.80		10	11.35
0200	120 V									
0220	125°F	Q-1	1060.80	.015	L.F.	8.10	.80		8.90	10.10
0300	Power kit w/1 end seal	1 Elec	48.80	.164	Ea.	76.50	8.95		85.45	97.50
0310	Splice kit		35.50	.225		90.50	12.35		102.85	118
0320	End seal		160	.050		8.80	2.74		11.54	13.80
0330	Tee kit w/1 end seal		26.80	.299		96	16.35		112.35	130
0340	Powered splice w/2 end seals		20	.400		102	22		124	145
0350	Powered tee kit w/3 end seals		18	.444		131	24.50		155.50	181
0360	Cross kit w/2 end seals		18.60	.430		135	23.50		158.50	185
0500	Recommended thickness of fiberglass insulation									
0510	Pipe size									
0520	1/2" - 1" use 1" insulation									
0530	1-1/4" - 2" use 1-1/2" insulation									
0540	2-1/2" - 6" use 2" insulation									
0560	NOTE: For pipe sizes 1-1/4" and smaller use 1/4" larger diameter									
0570	insulation to allow room for installation over cable.									

22 05 48 – Vibration and Seismic Controls for Plumbing Piping and Equipment

22 05 48.40 Vibration Absorbers

		Crew	Daily Output	Labor-Hours	Unit	Material	2015 Bare Costs Labor	Equipment	Total	Total Incl O&P
0010	**VIBRATION ABSORBERS**									
0100	Hangers, neoprene flex									
0200	10 – 120 lb. capacity				Ea.	26			26	28.50
0220	75 – 550 lb. capacity					37			37	40.50
0240	250 – 1100 lb. capacity					72			72	79
0260	1000 – 4000 lb. capacity					134			134	148
0500	Spring flex, 60 lb. capacity					36			36	39.50
0520	450 lb. capacity					80.50			80.50	88.50
0540	900 lb. capacity					72			72	79
0560	1100 – 1300 lb. capacity					72			72	79
0600	Rubber in shear									
0610	45 – 340 lb., up to 1/2" rod size	1 Stpi	22	.364	Ea.	14.30	21.50		35.80	49
0620	130 – 700 lb., up to 3/4" rod size		20	.400		31	24		55	70
0630	50 – 1000 lb., up to 3/4" rod size		18	.444		36.50	26.50		63	80
1000	Mounts, neoprene, 45 – 380 lb. capacity					17.25			17.25	19
1020	250 – 1100 lb. capacity					49.50			49.50	54
1040	1000 – 4000 lb. capacity					111			111	122
1100	Spring flex, 60 lb. capacity					70.50			70.50	78
1120	165 lb. capacity					72			72	79
1140	260 lb. capacity					74.50			74.50	82
1160	450 lb. capacity					96.50			96.50	106
1180	600 lb. capacity					99.50			99.50	110

22 05 48.40 Vibration Absorbers		Crew	Daily Output	Labor-Hours	Unit	Material	2015 Bare Costs Labor	2015 Bare Costs Equipment	Total	Total Incl O&P
1200	750 lb. capacity				Ea.	110			110	121
1220	900 lb. capacity					108			108	119
1240	1100 lb. capacity					117			117	129
1260	1300 lb. capacity					117			117	129
1280	1500 lb. capacity					176			176	194
1300	1800 lb. capacity					189			189	208
1320	2200 lb. capacity					189			189	208
1340	2600 lb. capacity					197			197	216
1399	Spring type									
1400	2 Piece									
1410	50 – 1000 lb.	1 Stpi	12	.667	Ea.	130	40		170	203
1420	1100 – 1600 lb.	"	12	.667	"	141	40		181	215
1500	Double spring open									
1510	150 – 450 lb.	1 Stpi	24	.333	Ea.	89	19.90		108.90	128
1520	500 – 1000 lb.		24	.333		89	19.90		108.90	128
1530	1100 – 1600 lb.		24	.333		89	19.90		108.90	128
1540	1700 – 2400 lb.		24	.333		97	19.90		116.90	136
1550	2500 – 3400 lb.		24	.333		165	19.90		184.90	212
2000	Pads, cork rib, 18" x 18" x 1", 10-50 psi					179			179	197
2020	18" x 36" x 1", 10-50 psi					350			350	385
2100	Shear flexible pads, 18" x 18" x 3/8", 20-70 psi					78			78	86
2120	18" x 36" x 3/8", 20-70 psi					179			179	197
2150	Laminated neoprene and cork									
2160	1" thick	1 Stpi	16	.500	S.F.	88.50	30		118.50	143
2200	Neoprene elastomer isolation bearing pad									
2230	464 PSI, 5/8" x 19-11/16" x 39-3/8"	1 Stpi	16	.500	Ea.	295	30		325	370
3000	Note overlap in capacities due to deflections									

22 05 53.10 Piping System Identification Labels

		Crew	Daily Output	Labor-Hours	Unit	Material	2015 Bare Costs Labor	2015 Bare Costs Equipment	Total	Total Incl O&P
0010	**PIPING SYSTEM IDENTIFICATION LABELS**									
0100	Indicate contents and flow direction									
0106	Pipe markers									
0110	Plastic snap around									
0114	1/2" pipe	1 Plum	80	.100	Ea.	6.50	5.85		12.35	16
0116	3/4" pipe		80	.100		6.50	5.85		12.35	16
0118	1" pipe		80	.100		6.50	5.85		12.35	16
0120	2" pipe		75	.107		8.10	6.25		14.35	18.35
0122	3" pipe		70	.114		12.40	6.70		19.10	24
0124	4" pipe		60	.133		12.40	7.85		20.25	25.50
0126	6" pipe		60	.133		13.20	7.85		21.05	26.50
0128	8" pipe		56	.143		17.40	8.40		25.80	32
0130	10" pipe		56	.143		17.40	8.40		25.80	32
0200	Over 10" pipe size		50	.160		21	9.40		30.40	37
1110	Self adhesive									
1114	1" pipe	1 Plum	80	.100	Ea.	3.20	5.85		9.05	12.35
1116	2" pipe		75	.107		3.20	6.25		9.45	12.95
1118	3" pipe		70	.114		5.10	6.70		11.80	15.75
1120	4" pipe		60	.133		5	7.85		12.85	17.30
1122	6" pipe		60	.133		5	7.85		12.85	17.30
1124	8" pipe		56	.143		7.20	8.40		15.60	20.50
1126	10" pipe		56	.143		7.20	8.40		15.60	20.50
1200	Over 10" pipe size		50	.160		10.20	9.40		19.60	25.50

108

For customer support on your Mechanical Cost Data, call 877.795.7394.

22 05 Common Work Results for Plumbing

22 05 53 – Identification for Plumbing Piping and Equipment

22 05 53.10 Piping System Identification Labels		Crew	Daily Output	Labor-Hours	Unit	Material	2015 Bare Costs Labor	Equipment	Total	Total Incl O&P
2000	Valve tags									
2010	Numbered plus identifying legend									
2100	Brass, 2" diameter	1 Plum	40	.200	Ea.	3	11.75		14.75	21
2200	Plastic, 1-1/2" diam.	"	40	.200	"	2.70	11.75		14.45	20.50

22 07 Plumbing Insulation

22 07 16 – Plumbing Equipment Insulation

22 07 16.10 Insulation for Plumbing Equipment

		Crew	Daily Output	Labor-Hours	Unit	Material	2015 Bare Costs Labor	Equipment	Total	Total Incl O&P
0010	**INSULATION FOR PLUMBING EQUIPMENT**									
2900	Domestic water heater wrap kit									
2920	1-1/2" with vinyl jacket, 20-60 gal. [G]	1 Plum	8	1	Ea.	18.60	58.50		77.10	109

22 07 19 – Plumbing Piping Insulation

22 07 19.10 Piping Insulation

		Crew	Daily Output	Labor-Hours	Unit	Material	2015 Bare Costs Labor	Equipment	Total	Total Incl O&P
0010	**PIPING INSULATION**									
0100	Rule of thumb, as a percentage of total mechanical costs				Job				10%	10%
0110	Insulation req'd. is based on the surface size/area to be covered									
0230	Insulated protectors, (ADA)									
0235	For exposed piping under sinks or lavatories									
0240	Vinyl coated foam, velcro tabs									
0245	P Trap, 1-1/4" or 1-1/2"	1 Plum	32	.250	Ea.	22.50	14.70		37.20	46.50
0260	Valve and supply cover									
0265	1/2", 3/8", and 7/16" pipe size	1 Plum	32	.250	Ea.	22	14.70		36.70	46
0280	Tailpiece offset (wheelchair)									
0285	1-1/4" pipe size	1 Plum	32	.250	Ea.	13.65	14.70		28.35	37
0600	Pipe covering (price copper tube one size less than IPS)									
1000	Mineral wool									
1010	Preformed, 1200°F, plain									
1014	1" wall									
1016	1/2" iron pipe size [G]	Q-14	230	.070	L.F.	1.65	3.28		4.93	6.90
1018	3/4" iron pipe size [G]		220	.073		1.72	3.43		5.15	7.25
1022	1" iron pipe size [G]		210	.076		1.77	3.59		5.36	7.55
1024	1-1/4" iron pipe size [G]		205	.078		1.82	3.68		5.50	7.75
1026	1-1/2" iron pipe size [G]		205	.078		1.90	3.68		5.58	7.85
1028	2" iron pipe size [G]		200	.080		2.32	3.77		6.09	8.40
1030	2-1/2" iron pipe size [G]		190	.084		2.46	3.97		6.43	8.90
1032	3" iron pipe size [G]		180	.089		2.55	4.19		6.74	9.30
1034	4" iron pipe size [G]		150	.107		3.14	5.05		8.19	11.30
1036	5" iron pipe size [G]		140	.114		3.55	5.40		8.95	12.30
1038	6" iron pipe size [G]		120	.133		3.79	6.30		10.09	13.95
1040	7" iron pipe size [G]		110	.145		4.43	6.85		11.28	15.50
1042	8" iron pipe size [G]		100	.160		6.10	7.55		13.65	18.45
1044	9" iron pipe size [G]		90	.178		6.65	8.40		15.05	20.50
1046	10" iron pipe size [G]		90	.178		7.50	8.40		15.90	21.50
1050	1-1/2" wall									
1052	1/2" iron pipe size [G]	Q-14	225	.071	L.F.	2.82	3.35		6.17	8.30
1054	3/4" iron pipe size [G]		215	.074		2.89	3.51		6.40	8.65
1056	1" iron pipe size [G]		205	.078		2.93	3.68		6.61	8.95
1058	1-1/4" iron pipe size [G]		200	.080		2.99	3.77		6.76	9.15
1060	1-1/2" iron pipe size [G]		200	.080		3.14	3.77		6.91	9.30
1062	2" iron pipe size [G]		190	.084		3.32	3.97		7.29	9.85

For customer support on your Mechanical Cost Data, call 877.795.7394.

109

22 07 19.10 Piping Insulation		Crew	Daily Output	Labor-Hours	Unit	Material	2015 Bare Costs Labor	Equipment	Total	Total Incl O&P
1064	2-1/2" iron pipe size G	Q-14	180	.089	L.F.	3.62	4.19		7.81	10.50
1066	3" iron pipe size G		165	.097		3.77	4.57		8.34	11.25
1068	4" iron pipe size G		140	.114		4.53	5.40		9.93	13.40
1070	5" iron pipe size G		130	.123		4.84	5.80		10.64	14.35
1072	6" iron pipe size G		110	.145		4.99	6.85		11.84	16.15
1074	7" iron pipe size G		100	.160		5.90	7.55		13.45	18.25
1076	8" iron pipe size G		90	.178		6.55	8.40		14.95	20.50
1078	9" iron pipe size G		85	.188		7.55	8.85		16.40	22
1080	10" iron pipe size G		80	.200		8.30	9.45		17.75	24
1082	12" iron pipe size G		75	.213		9.45	10.05		19.50	26
1084	14" iron pipe size G		70	.229		10.80	10.75		21.55	28.50
1086	16" iron pipe size G		65	.246		12.30	11.60		23.90	31.50
1088	18" iron pipe size G		60	.267		13.85	12.55		26.40	35
1090	20" iron pipe size G		55	.291		15.25	13.70		28.95	38.50
1092	22" iron pipe size G		50	.320		17.45	15.10		32.55	42.50
1094	24" iron pipe size G	▼	45	.356	▼	18.65	16.75		35.40	46.50
1100	2" wall									
1102	1/2" iron pipe size G	Q-14	220	.073	L.F.	3.65	3.43		7.08	9.35
1104	3/4" iron pipe size G		210	.076		3.78	3.59		7.37	9.75
1106	1" iron pipe size G		200	.080		4.01	3.77		7.78	10.25
1108	1-1/4" iron pipe size G		190	.084		4.30	3.97		8.27	10.95
1110	1-1/2" iron pipe size G		190	.084		4.48	3.97		8.45	11.15
1112	2" iron pipe size G		180	.089		4.70	4.19		8.89	11.65
1114	2-1/2" iron pipe size G		170	.094		5.30	4.44		9.74	12.75
1116	3" iron pipe size G		160	.100		5.70	4.71		10.41	13.60
1118	4" iron pipe size G		130	.123		6.60	5.80		12.40	16.35
1120	5" iron pipe size G		120	.133		7.35	6.30		13.65	17.90
1122	6" iron pipe size G		100	.160		7.65	7.55		15.20	20
1124	8" iron pipe size G		80	.200		9.25	9.45		18.70	25
1126	10" iron pipe size G		70	.229		11.35	10.75		22.10	29
1128	12" iron pipe size G		65	.246		12.55	11.60		24.15	32
1130	14" iron pipe size G		60	.267		14.05	12.55		26.60	35
1132	16" iron pipe size G		55	.291		15.90	13.70		29.60	39
1134	18" iron pipe size G		50	.320		17.20	15.10		32.30	42.50
1136	20" iron pipe size G		45	.356		19.80	16.75		36.55	48
1138	22" iron pipe size G		45	.356		21.50	16.75		38.25	49.50
1140	24" iron pipe size G	▼	40	.400	▼	22.50	18.85		41.35	54.50
1150	4" wall									
1152	1-1/4" iron pipe size G	Q-14	170	.094	L.F.	11.65	4.44		16.09	19.70
1154	1-1/2" iron pipe size G		165	.097		12.10	4.57		16.67	20.50
1156	2" iron pipe size G		155	.103		12.35	4.87		17.22	21
1158	4" iron pipe size G		105	.152		15.60	7.20		22.80	28.50
1160	6" iron pipe size G		75	.213		18.40	10.05		28.45	35.50
1162	8" iron pipe size G		60	.267		21	12.55		33.55	42.50
1164	10" iron pipe size G		50	.320		25	15.10		40.10	51
1166	12" iron pipe size G		45	.356		27.50	16.75		44.25	56.50
1168	14" iron pipe size G		40	.400		29.50	18.85		48.35	62
1170	16" iron pipe size G		35	.457		33.50	21.50		55	70
1172	18" iron pipe size G		32	.500		36	23.50		59.50	76
1174	20" iron pipe size G		30	.533		38.50	25		63.50	81.50
1176	22" iron pipe size G		28	.571		43.50	27		70.50	89.50
1178	24" iron pipe size G	▼	26	.615	▼	46	29		75	95.50
4280	Cellular glass, closed cell foam, all service jacket, sealant,									

22 07 19 – Plumbing Piping Insulation

22 07 19.10 Piping Insulation		Crew	Daily Output	Labor-Hours	Unit	Material	2015 Bare Costs Labor	Equipment	Total	Total Incl O&P
6950	5" iron pipe size	G Q-14	140	.114	L.F.	2.22	5.40		7.62	10.85
6960	6" iron pipe size	G	120	.133		2.36	6.30		8.66	12.40
6970	7" iron pipe size	G	110	.145		2.81	6.85		9.66	13.75
6980	8" iron pipe size	G	100	.160		3.85	7.55		11.40	16
6990	9" iron pipe size	G	90	.178		4.06	8.40		12.46	17.50
7000	10" iron pipe size	G	90	.178		4.09	8.40		12.49	17.55
7010	12" iron pipe size	G	80	.200		4.45	9.45		13.90	19.60
7020	14" iron pipe size	G	80	.200		5.35	9.45		14.80	20.50
7030	16" iron pipe size	G	70	.229		6.80	10.75		17.55	24
7040	18" iron pipe size	G	70	.229		7.55	10.75		18.30	25
7050	20" iron pipe size	G	60	.267		8.40	12.55		20.95	29
7060	24" iron pipe size	G	60	.267		10.25	12.55		22.80	31
7080	1-1/2" wall, 1/2" iron pipe size	G	230	.070		1.57	3.28		4.85	6.85
7100	3/4" iron pipe size	G	220	.073		1.57	3.43		5	7.10
7110	1" iron pipe size	G	210	.076		1.68	3.59		5.27	7.45
7120	1-1/4" iron pipe size	G	200	.080		1.83	3.77		5.60	7.85
7130	1-1/2" iron pipe size	G	200	.080		1.92	3.77		5.69	7.95
7140	2" iron pipe size	G	190	.084		2.11	3.97		6.08	8.50
7150	2-1/2" iron pipe size	G	180	.089		2.27	4.19		6.46	9
7160	3" iron pipe size	G	170	.094		2.37	4.44		6.81	9.50
7170	3-1/2" iron pipe size	G	160	.100		2.60	4.71		7.31	10.20
7180	4" iron pipe size	G	140	.114		2.70	5.40		8.10	11.35
7190	5" iron pipe size	G	130	.123		3.02	5.80		8.82	12.35
7200	6" iron pipe size	G	110	.145		3.19	6.85		10.04	14.15
7210	7" iron pipe size	G	100	.160		3.54	7.55		11.09	15.65
7220	8" iron pipe size	G	90	.178		4.47	8.40		12.87	17.95
7230	9" iron pipe size	G	85	.188		4.66	8.85		13.51	18.95
7240	10" iron pipe size	G	80	.200		4.82	9.45		14.27	20
7250	12" iron pipe size	G	75	.213		5.45	10.05		15.50	21.50
7260	14" iron pipe size	G	70	.229		6.55	10.75		17.30	24
7270	16" iron pipe size	G	65	.246		8.50	11.60		20.10	27.50
7280	18" iron pipe size	G	60	.267		9.55	12.55		22.10	30
7290	20" iron pipe size	G	55	.291		9.75	13.70		23.45	32.50
7300	24" iron pipe size	G	50	.320		11.95	15.10		27.05	36.50
7320	2" wall, 1/2" iron pipe size	G	220	.073		2.43	3.43		5.86	8
7340	3/4" iron pipe size	G	210	.076		2.51	3.59		6.10	8.35
7350	1" iron pipe size	G	200	.080		2.67	3.77		6.44	8.80
7360	1-1/4" iron pipe size	G	190	.084		2.82	3.97		6.79	9.30
7370	1-1/2" iron pipe size	G	190	.084		2.94	3.97		6.91	9.45
7380	2" iron pipe size	G	180	.089		3.10	4.19		7.29	9.90
7390	2-1/2" iron pipe size	G	170	.094		3.34	4.44		7.78	10.55
7400	3" iron pipe size	G	160	.100		3.54	4.71		8.25	11.25
7410	3-1/2" iron pipe size	G	150	.107		3.84	5.05		8.89	12.05
7420	4" iron pipe size	G	130	.123		4.13	5.80		9.93	13.60
7430	5" iron pipe size	G	120	.133		4.73	6.30		11.03	15
7440	6" iron pipe size	G	100	.160		4.88	7.55		12.43	17.10
7450	7" iron pipe size	G	90	.178		5.55	8.40		13.95	19.15
7460	8" iron pipe size	G	80	.200		5.95	9.45		15.40	21
7470	9" iron pipe size	G	75	.213		6.50	10.05		16.55	23
7480	10" iron pipe size	G	70	.229		7.10	10.75		17.85	24.50
7490	12" iron pipe size	G	65	.246		7.95	11.60		19.55	27
7500	14" iron pipe size	G	60	.267		10.25	12.55		22.80	31
7510	16" iron pipe size	G	55	.291		11.25	13.70		24.95	34

22 07 19.10 Piping Insulation		Crew	Daily Output	Labor-Hours	Unit	Material	2015 Bare Costs Labor	Equipment	Total	Total Incl O&P
7520	18" iron pipe size [G]	Q-14	50	.320	L.F.	12.55	15.10		27.65	37.50
7530	20" iron pipe size [G]		45	.356		14.25	16.75		31	41.50
7540	24" iron pipe size [G]		40	.400		15.30	18.85		34.15	46.50
7560	2-1/2" wall, 1/2" iron pipe size [G]		210	.076		2.88	3.59		6.47	8.75
7562	3/4" iron pipe size [G]		200	.080		3.01	3.77		6.78	9.15
7564	1" iron pipe size [G]		190	.084		3.14	3.97		7.11	9.65
7566	1-1/4" iron pipe size [G]		185	.086		3.25	4.08		7.33	9.95
7568	1-1/2" iron pipe size [G]		180	.089		3.42	4.19		7.61	10.25
7570	2" iron pipe size [G]		170	.094		3.59	4.44		8.03	10.85
7572	2-1/2" iron pipe size [G]		160	.100		4.13	4.71		8.84	11.90
7574	3" iron pipe size [G]		150	.107		4.35	5.05		9.40	12.65
7576	3-1/2" iron pipe size [G]		140	.114		4.73	5.40		10.13	13.60
7578	4" iron pipe size [G]		120	.133		4.99	6.30		11.29	15.30
7580	5" iron pipe size [G]		110	.145		5.95	6.85		12.80	17.20
7582	6" iron pipe size [G]		90	.178		7.15	8.40		15.55	21
7584	7" iron pipe size [G]		80	.200		7.20	9.45		16.65	22.50
7586	8" iron pipe size [G]		70	.229		7.65	10.75		18.40	25
7588	9" iron pipe size [G]		65	.246		8.35	11.60		19.95	27.50
7590	10" iron pipe size [G]		60	.267		9.10	12.55		21.65	29.50
7592	12" iron pipe size [G]		55	.291		11.05	13.70		24.75	33.50
7594	14" iron pipe size [G]		50	.320		12.90	15.10		28	37.50
7596	16" iron pipe size [G]		45	.356		14.80	16.75		31.55	42.50
7598	18" iron pipe size [G]		40	.400		16.05	18.85		34.90	47
7602	24" iron pipe size [G]		30	.533		21	25		46	62.50
7620	3" wall, 1/2" iron pipe size [G]		200	.080		3.69	3.77		7.46	9.90
7622	3/4" iron pipe size [G]		190	.084		3.90	3.97		7.87	10.50
7624	1" iron pipe size [G]		180	.089		4.13	4.19		8.32	11.05
7626	1-1/4" iron pipe size [G]		175	.091		4.21	4.31		8.52	11.35
7628	1-1/2" iron pipe size [G]		170	.094		4.40	4.44		8.84	11.75
7630	2" iron pipe size [G]		160	.100		4.76	4.71		9.47	12.60
7632	2-1/2" iron pipe size [G]		150	.107		4.95	5.05		10	13.30
7634	3" iron pipe size [G]		140	.114		5.25	5.40		10.65	14.20
7636	3-1/2" iron pipe size [G]		130	.123		5.80	5.80		11.60	15.45
7638	4" iron pipe size [G]		110	.145		6.25	6.85		13.10	17.55
7640	5" iron pipe size [G]		100	.160		7.05	7.55		14.60	19.55
7642	6" iron pipe size [G]		80	.200		7.55	9.45		17	23
7644	7" iron pipe size [G]		70	.229		8.65	10.75		19.40	26.50
7646	8" iron pipe size [G]		60	.267		9.40	12.55		21.95	30
7648	9" iron pipe size [G]		55	.291		10.15	13.70		23.85	32.50
7650	10" iron pipe size [G]		50	.320		10.90	15.10		26	35.50
7652	12" iron pipe size [G]		45	.356		13.55	16.75		30.30	41
7654	14" iron pipe size [G]		40	.400		15.65	18.85		34.50	46.50
7656	16" iron pipe size [G]		35	.457		17.75	21.50		39.25	53
7658	18" iron pipe size [G]		32	.500		18.95	23.50		42.45	57.50
7660	20" iron pipe size [G]		30	.533		20.50	25		45.50	61.50
7662	24" iron pipe size [G]		28	.571		26	27		53	71
7664	26" iron pipe size [G]		26	.615		29	29		58	76.50
7666	30" iron pipe size [G]		24	.667		34	31.50		65.50	86.50
7800	For fiberglass with standard canvas jacket, deduct					5%				
7802	For fittings, add 3 L.F. for each fitting									
7804	plus 4 L.F. for each flange of the fitting									
7810	Finishes									
7814	For single layer of felt, add					10%	10%			

116

For customer support on your Mechanical Cost Data, call 877.795.7394.

22 07 19 – Plumbing Piping Insulation

22 07 19.10 Piping Insulation		Crew	Daily Output	Labor-Hours	Unit	Material	2015 Bare Costs Labor	Equipment	Total	Total Incl O&P
7816	For roofing paper, 45 lb. to 55 lb., add					25%	10%			
7820	Polyethylene tubing flexible closed cell foam, UV resistant									
7828	Standard temperature (-90°F to +212°F)									
7830	3/8" wall, 1/8" iron pipe size	G	1 Asbe	130	.062	L.F.	.22	3.22		3.44	5.25
7831	1/4" iron pipe size	G		130	.062		.23	3.22		3.45	5.25
7832	3/8" iron pipe size	G		130	.062		.26	3.22		3.48	5.30
7833	1/2" iron pipe size	G		126	.063		.29	3.32		3.61	5.45
7834	3/4" iron pipe size	G		122	.066		.34	3.43		3.77	5.70
7835	1" iron pipe size	G		120	.067		.39	3.49		3.88	5.90
7836	1-1/4" iron pipe size	G		118	.068		.47	3.55		4.02	6.05
7837	1-1/2" iron pipe size	G		118	.068		.60	3.55		4.15	6.20
7838	2" iron pipe size	G		116	.069		.79	3.61		4.40	6.45
7839	2-1/2" iron pipe size	G		114	.070		1.01	3.67		4.68	6.80
7840	3" iron pipe size	G		112	.071		1.56	3.74		5.30	7.50
7842	1/2" wall, 1/8" iron pipe size	G		120	.067		.33	3.49		3.82	5.80
7843	1/4" iron pipe size	G		120	.067		.35	3.49		3.84	5.85
7844	3/8" iron pipe size	G		120	.067		.39	3.49		3.88	5.90
7845	1/2" iron pipe size	G		118	.068		.43	3.55		3.98	6
7846	3/4" iron pipe size	G		116	.069		.49	3.61		4.10	6.15
7847	1" iron pipe size	G		114	.070		.55	3.67		4.22	6.30
7848	1-1/4" iron pipe size	G		112	.071		.65	3.74		4.39	6.50
7849	1-1/2" iron pipe size	G		110	.073		.77	3.81		4.58	6.80
7850	2" iron pipe size	G		108	.074		1.11	3.88		4.99	7.25
7851	2-1/2" iron pipe size	G		106	.075		1.37	3.95		5.32	7.65
7852	3" iron pipe size	G		104	.077		1.89	4.03		5.92	8.35
7853	3-1/2" iron pipe size	G		102	.078		2.14	4.11		6.25	8.75
7854	4" iron pipe size	G		100	.080		2.44	4.19		6.63	9.20
7855	3/4" wall, 1/8" iron pipe size	G		110	.073		.50	3.81		4.31	6.50
7856	1/4" iron pipe size	G		110	.073		.53	3.81		4.34	6.55
7857	3/8" iron pipe size	G		108	.074		.61	3.88		4.49	6.70
7858	1/2" iron pipe size	G		106	.075		.73	3.95		4.68	6.95
7859	3/4" iron pipe size	G		104	.077		.88	4.03		4.91	7.20
7860	1" iron pipe size	G		102	.078		1.01	4.11		5.12	7.50
7861	1-1/4" iron pipe size	G		100	.080		1.38	4.19		5.57	8
7862	1-1/2" iron pipe size	G		100	.080		1.56	4.19		5.75	8.20
7863	2" iron pipe size	G		98	.082		1.98	4.27		6.25	8.85
7864	2-1/2" iron pipe size	G		96	.083		2.39	4.36		6.75	9.45
7865	3" iron pipe size	G		94	.085		3.15	4.46		7.61	10.40
7866	3-1/2" iron pipe size	G		92	.087		3.51	4.55		8.06	10.95
7867	4" iron pipe size	G		90	.089		3.92	4.65		8.57	11.55
7868	1" wall, 1/4" iron pipe size	G		100	.080		1.06	4.19		5.25	7.65
7869	3/8" iron pipe size	G		98	.082		1.08	4.27		5.35	7.85
7870	1/2" iron pipe size	G		96	.083		1.35	4.36		5.71	8.30
7871	3/4" iron pipe size	G		94	.085		1.57	4.46		6.03	8.70
7872	1" iron pipe size	G		92	.087		1.91	4.55		6.46	9.20
7873	1-1/4" iron pipe size	G		90	.089		2.17	4.65		6.82	9.65
7874	1-1/2" iron pipe size	G		90	.089		2.48	4.65		7.13	10
7875	2" iron pipe size	G		88	.091		3.23	4.76		7.99	10.95
7876	2-1/2" iron pipe size	G		86	.093		4.22	4.87		9.09	12.25
7877	3" iron pipe size	G		84	.095		4.74	4.99		9.73	12.95
7878	Contact cement, quart can	G				Ea.	10.50			10.50	11.55
7879	Rubber tubing, flexible closed cell foam										
7880	3/8" wall, 1/4" iron pipe size	G	1 Asbe	120	.067	L.F.	.33	3.49		3.82	5.80

For customer support on your Mechanical Cost Data, call 877.795.7394.

117

22 07 19 – Plumbing Piping Insulation

22 07 19.10 Piping Insulation		Crew	Daily Output	Labor-Hours	Unit	Material	2015 Bare Costs Labor	2015 Bare Costs Equipment	Total	Total Incl O&P
7900	3/8" iron pipe size	G	1 Asbe	120	.067	L.F.	.36	3.49		3.85	5.85
7910	1/2" iron pipe size	G		115	.070		.41	3.64		4.05	6.10
7920	3/4" iron pipe size	G		115	.070		.46	3.64		4.10	6.15
7930	1" iron pipe size	G		110	.073		.52	3.81		4.33	6.50
7940	1-1/4" iron pipe size	G		110	.073		.59	3.81		4.40	6.60
7950	1-1/2" iron pipe size	G		110	.073		.73	3.81		4.54	6.75
8100	1/2" wall, 1/4" iron pipe size	G		90	.089		.54	4.65		5.19	7.85
8120	3/8" iron pipe size	G		90	.089		.60	4.65		5.25	7.90
8130	1/2" iron pipe size	G		89	.090		.67	4.71		5.38	8.10
8140	3/4" iron pipe size	G		89	.090		.75	4.71		5.46	8.20
8150	1" iron pipe size	G		88	.091		.82	4.76		5.58	8.30
8160	1-1/4" iron pipe size	G		87	.092		.95	4.81		5.76	8.55
8170	1-1/2" iron pipe size	G		87	.092		1.15	4.81		5.96	8.75
8180	2" iron pipe size	G		86	.093		1.47	4.87		6.34	9.20
8190	2-1/2" iron pipe size	G		86	.093		1.85	4.87		6.72	9.65
8200	3" iron pipe size	G		85	.094		2.06	4.93		6.99	9.90
8210	3-1/2" iron pipe size	G		85	.094		2.86	4.93		7.79	10.80
8220	4" iron pipe size	G		80	.100		3.06	5.25		8.31	11.50
8230	5" iron pipe size	G		80	.100		4.17	5.25		9.42	12.75
8240	6" iron pipe size	G		75	.107		4.17	5.60		9.77	13.30
8300	3/4" wall, 1/4" iron pipe size	G		90	.089		.85	4.65		5.50	8.20
8320	3/8" iron pipe size	G		90	.089		.92	4.65		5.57	8.25
8330	1/2" iron pipe size	G		89	.090		1.10	4.71		5.81	8.55
8340	3/4" iron pipe size	G		89	.090		1.35	4.71		6.06	8.85
8350	1" iron pipe size	G		88	.091		1.54	4.76		6.30	9.10
8360	1-1/4" iron pipe size	G		87	.092		2.05	4.81		6.86	9.75
8370	1-1/2" iron pipe size	G		87	.092		2.32	4.81		7.13	10.05
8380	2" iron pipe size	G		86	.093		2.68	4.87		7.55	10.55
8390	2-1/2" iron pipe size	G		86	.093		3.58	4.87		8.45	11.55
8400	3" iron pipe size	G		85	.094		4.08	4.93		9.01	12.15
8410	3-1/2" iron pipe size	G		85	.094		4.81	4.93		9.74	12.95
8420	4" iron pipe size	G		80	.100		5.10	5.25		10.35	13.80
8430	5" iron pipe size	G		80	.100		6.15	5.25		11.40	14.95
8440	6" iron pipe size	G		80	.100		7.45	5.25		12.70	16.35
8444	1" wall, 1/2" iron pipe size	G		86	.093		2.05	4.87		6.92	9.85
8445	3/4" iron pipe size	G		84	.095		2.48	4.99		7.47	10.50
8446	1" iron pipe size	G		84	.095		2.89	4.99		7.88	10.95
8447	1-1/4" iron pipe size	G		82	.098		3.23	5.10		8.33	11.50
8448	1-1/2" iron pipe size	G		82	.098		3.76	5.10		8.86	12.10
8449	2" iron pipe size	G		80	.100		4.95	5.25		10.20	13.60
8450	2-1/2" iron pipe size	G		80	.100		6.45	5.25		11.70	15.25
8456	Rubber insulation tape, 1/8" x 2" x 30'	G				Ea.	12.05			12.05	13.25
8460	Polyolefin tubing, flexible closed cell foam, UV stabilized, work										
8462	temp.-165°F to +210°F, 0 water vapor transmission										
8464	3/8" wall, 1/8" iron pipe size	G	1 Asbe	140	.057	L.F.	.48	2.99		3.47	5.20
8466	1/4" iron pipe size	G		140	.057		.50	2.99		3.49	5.20
8468	3/8" iron pipe size	G		140	.057		.55	2.99		3.54	5.25
8470	1/2" iron pipe size	G		136	.059		.61	3.08		3.69	5.45
8472	3/4" iron pipe size	G		132	.061		.62	3.17		3.79	5.60
8474	1" iron pipe size	G		130	.062		.83	3.22		4.05	5.90
8476	1-1/4" iron pipe size	G		128	.063		1.02	3.27		4.29	6.20
8478	1-1/2" iron pipe size	G		128	.063		1.25	3.27		4.52	6.50
8480	2" iron pipe size	G		126	.063		1.50	3.32		4.82	6.80

22 07 19 – Plumbing Piping Insulation

22 07 19.10 Piping Insulation		Crew	Daily Output	Labor-Hours	Unit	Material	2015 Bare Costs Labor	Equipment	Total	Total Incl O&P
8482	2-1/2" iron pipe size	G	1 Asbe	123	.065	L.F.	2.17	3.40		5.57	7.70
8484	3" iron pipe size	G		121	.066		2.42	3.46		5.88	8.05
8486	4" iron pipe size	G		118	.068		4.43	3.55		7.98	10.40
8500	1/2" wall, 1/8" iron pipe size	G		130	.062		.71	3.22		3.93	5.80
8502	1/4" iron pipe size	G		130	.062		.76	3.22		3.98	5.85
8504	3/8" iron pipe size	G		130	.062		.81	3.22		4.03	5.90
8506	1/2" iron pipe size	G		128	.063		.89	3.27		4.16	6.10
8508	3/4" iron pipe size	G		126	.063		1.02	3.32		4.34	6.25
8510	1" iron pipe size	G		123	.065		1.18	3.40		4.58	6.60
8512	1-1/4" iron pipe size	G		121	.066		1.39	3.46		4.85	6.95
8514	1-1/2" iron pipe size	G		119	.067		1.67	3.52		5.19	7.35
8516	2" iron pipe size	G		117	.068		2.11	3.58		5.69	7.85
8518	2-1/2" iron pipe size	G		114	.070		2.85	3.67		6.52	8.85
8520	3" iron pipe size	G		112	.071		3.71	3.74		7.45	9.90
8522	4" iron pipe size	G		110	.073		5.15	3.81		8.96	11.60
8534	3/4" wall, 1/8" iron pipe size	G		120	.067		1.08	3.49		4.57	6.65
8536	1/4" iron pipe size	G		120	.067		1.14	3.49		4.63	6.70
8538	3/8" iron pipe size	G		117	.068		1.30	3.58		4.88	7
8540	1/2" iron pipe size	G		114	.070		1.45	3.67		5.12	7.30
8542	3/4" iron pipe size	G		112	.071		1.56	3.74		5.30	7.50
8544	1" iron pipe size	G		110	.073		2.20	3.81		6.01	8.35
8546	1-1/4" iron pipe size	G		108	.074		2.94	3.88		6.82	9.30
8548	1-1/2" iron pipe size	G		108	.074		3.43	3.88		7.31	9.80
8550	2" iron pipe size	G		106	.075		4.34	3.95		8.29	10.90
8552	2-1/2" iron pipe size	G		104	.077		5.10	4.03		9.13	11.85
8554	3" iron pipe size	G		102	.078		6.80	4.11		10.91	13.90
8556	4" iron pipe size	G		100	.080		9	4.19		13.19	16.40
8570	1" wall, 1/8" iron pipe size	G		110	.073		1.95	3.81		5.76	8.10
8572	1/4" iron pipe size	G		108	.074		2	3.88		5.88	8.25
8574	3/8" iron pipe size	G		106	.075		2.04	3.95		5.99	8.40
8576	1/2" iron pipe size	G		104	.077		2.16	4.03		6.19	8.65
8578	3/4" iron pipe size	G		102	.078		2.83	4.11		6.94	9.50
8580	1" iron pipe size	G		100	.080		3.27	4.19		7.46	10.10
8582	1-1/4" iron pipe size	G		97	.082		3.72	4.32		8.04	10.80
8584	1-1/2" iron pipe size	G		97	.082		4.25	4.32		8.57	11.40
8586	2" iron pipe size	G		95	.084		6.75	4.41		11.16	14.25
8588	2-1/2" iron pipe size	G		93	.086		7.85	4.50		12.35	15.60
8590	3" iron pipe size	G		91	.088		9.55	4.60		14.15	17.65
8606	Contact adhesive (R-320)	G				Qt.	28			28	30.50
8608	Contact adhesive (R-320)	G				Gal.	113			113	124
8610	NOTE: Preslit/preglued vs unslit, same price										

22 07 19.30 Piping Insulation Protective Jacketing, PVC

0010	**PIPING INSULATION PROTECTIVE JACKETING, PVC**										
0100	PVC, white, 48" lengths cut from roll goods										
0120	20 mil thick										
0140	Size based on OD of insulation										
0150	1-1/2" ID		Q-14	270	.059	L.F.	.20	2.79		2.99	4.57
0152	2" ID			260	.062		.27	2.90		3.17	4.82
0154	2-1/2" ID			250	.064		.32	3.02		3.34	5.05
0156	3" ID			240	.067		.39	3.14		3.53	5.30
0158	3-1/2" ID			230	.070		.44	3.28		3.72	5.60
0160	4" ID			220	.073		.50	3.43		3.93	5.90

For customer support on your Mechanical Cost Data, call 877.795.7394.

119

22 07 19.30 Piping Insulation Protective Jacketing, PVC		Crew	Daily Output	Labor-Hours	Unit	Material	2015 Bare Costs Labor	Equipment	Total	Total Incl O&P
0162	4-1/2" ID	Q-14	210	.076	L.F.	.56	3.59		4.15	6.20
0164	5" ID		200	.080		.61	3.77		4.38	6.50
0166	5-1/2" ID		190	.084		.67	3.97		4.64	6.95
0168	6" ID		180	.089		.73	4.19		4.92	7.30
0170	6-1/2" ID		175	.091		.79	4.31		5.10	7.55
0172	7" ID		170	.094		.84	4.44		5.28	7.80
0174	7-1/2" ID		164	.098		.91	4.60		5.51	8.15
0176	8" ID		161	.099		.96	4.68		5.64	8.35
0178	8-1/2" ID		158	.101		1.02	4.77		5.79	8.55
0180	9" ID		155	.103		1.08	4.87		5.95	8.80
0182	9-1/2" ID		152	.105		1.14	4.96		6.10	8.95
0184	10" ID		149	.107		1.19	5.05		6.24	9.20
0186	10-1/2" ID		146	.110		1.25	5.15		6.40	9.45
0188	11" ID		143	.112		1.31	5.25		6.56	9.65
0190	11-1/2" ID		140	.114		1.37	5.40		6.77	9.90
0192	12" ID		137	.117		1.42	5.50		6.92	10.10
0194	12-1/2" ID		134	.119		1.49	5.65		7.14	10.40
0195	13" ID		132	.121		1.54	5.70		7.24	10.60
0196	13-1/2" ID		132	.121		1.60	5.70		7.30	10.65
0198	14" ID		130	.123		1.66	5.80		7.46	10.90
0200	15" ID		128	.125		1.78	5.90		7.68	11.10
0202	16" ID		126	.127		1.89	6		7.89	11.40
0204	17" ID		124	.129		2.01	6.10		8.11	11.65
0206	18" ID		122	.131		2.13	6.20		8.33	11.95
0208	19" ID		120	.133		2.24	6.30		8.54	12.25
0210	20" ID		118	.136		2.35	6.40		8.75	12.55
0212	21" ID		116	.138		2.47	6.50		8.97	12.80
0214	22" ID		114	.140		2.58	6.60		9.18	13.15
0216	23" ID		112	.143		2.69	6.75		9.44	13.45
0218	24" ID		110	.145		2.81	6.85		9.66	13.75
0220	25" ID		108	.148		2.93	7		9.93	14.05
0222	26" ID		106	.151		3.04	7.10		10.14	14.45
0224	27" ID		104	.154		3.16	7.25		10.41	14.80
0226	28" ID		102	.157		3.27	7.40		10.67	15.10
0228	29" ID		100	.160		3.39	7.55		10.94	15.50
0230	30" ID		98	.163		3.51	7.70		11.21	15.85
0300	For colors, add				Ea.	10%				
1000	30 mil thick									
1010	Size based on OD of insulation									
1020	2" ID	Q-14	260	.062	L.F.	.40	2.90		3.30	4.96
1022	2-1/2" ID		250	.064		.48	3.02		3.50	5.25
1024	3" ID		240	.067		.58	3.14		3.72	5.55
1026	3-1/2" ID		230	.070		.65	3.28		3.93	5.80
1028	4" ID		220	.073		.74	3.43		4.17	6.15
1030	4-1/2" ID		210	.076		.82	3.59		4.41	6.50
1032	5" ID		200	.080		.91	3.77		4.68	6.85
1034	5-1/2" ID		190	.084		.99	3.97		4.96	7.30
1036	6" ID		180	.089		1.08	4.19		5.27	7.70
1038	6-1/2" ID		175	.091		1.16	4.31		5.47	8
1040	7" ID		170	.094		1.25	4.44		5.69	8.30
1042	7-1/2" ID		164	.098		1.34	4.60		5.94	8.60
1044	8" ID		161	.099		1.43	4.68		6.11	8.85
1046	8-1/2" ID		158	.101		1.51	4.77		6.28	9.10

22 07 Plumbing Insulation

22 07 19 – Plumbing Piping Insulation

22 07 19.30 Piping Insulation Protective Jacketing, PVC	Crew	Daily Output	Labor-Hours	Unit	Material	2015 Bare Costs Labor	Equipment	Total	Total Incl O&P
1048	9" ID	Q-14	155	.103	L.F.	1.60	4.87		6.47	9.35
1050	9-1/2" ID		152	.105		1.68	4.96		6.64	9.55
1052	10" ID		149	.107		1.77	5.05		6.82	9.85
1054	10-1/2" ID		146	.110		1.85	5.15		7	10.10
1056	11" ID		143	.112		1.94	5.25		7.19	10.35
1058	11-1/2" ID		140	.114		2.02	5.40		7.42	10.60
1060	12" ID		137	.117		2.11	5.50		7.61	10.85
1062	12-1/2" ID		134	.119		2.20	5.65		7.85	11.15
1063	13" ID		132	.121		2.29	5.70		7.99	11.40
1064	13-1/2" ID		132	.121		2.37	5.70		8.07	11.50
1066	14" ID		130	.123		2.46	5.80		8.26	11.75
1068	15" ID		128	.125		2.63	5.90		8.53	12.05
1070	16" ID		126	.127		2.80	6		8.80	12.40
1072	17" ID		124	.129		2.97	6.10		9.07	12.70
1074	18" ID		122	.131		3.15	6.20		9.35	13.05
1076	19" ID		120	.133		3.32	6.30		9.62	13.45
1078	20" ID		118	.136		3.49	6.40		9.89	13.80
1080	21" ID		116	.138		3.66	6.50		10.16	14.15
1082	22" ID		114	.140		3.83	6.60		10.43	14.50
1084	23" ID		112	.143		4	6.75		10.75	14.90
1086	24" ID		110	.145		4.17	6.85		11.02	15.25
1088	25" ID		108	.148		4.34	7		11.34	15.60
1090	26" ID		106	.151		4.51	7.10		11.61	16.05
1092	27" ID		104	.154		4.68	7.25		11.93	16.45
1094	28" ID		102	.157		4.85	7.40		12.25	16.85
1096	29" ID		100	.160		5	7.55		12.55	17.25
1098	30" ID		98	.163		5.20	7.70		12.90	17.70
1300	For colors, add				Ea.	10%				
2000	PVC, white, fitting covers									
2020	Fiberglass insulation inserts included with sizes 1-3/4" thru 9-3/4"									
2030	Size is based on OD of insulation									
2040	90° Elbow fitting									
2060	1-3/4"	Q-14	135	.119	Ea.	.47	5.60		6.07	9.20
2062	2"		130	.123		.58	5.80		6.38	9.70
2064	2-1/4"		128	.125		.65	5.90		6.55	9.85
2068	2-1/2"		126	.127		.71	6		6.71	10.10
2070	2-3/4"		123	.130		.77	6.15		6.92	10.40
2072	3"		120	.133		.82	6.30		7.12	10.70
2074	3-3/8"		116	.138		.93	6.50		7.43	11.10
2076	3-3/4"		113	.142		1.04	6.65		7.69	11.55
2078	4-1/8"		110	.145		1.31	6.85		8.16	12.10
2080	4-3/4"		105	.152		1.58	7.20		8.78	12.95
2082	5-1/4"		100	.160		1.84	7.55		9.39	13.75
2084	5-3/4"		95	.168		2.43	7.95		10.38	15
2086	6-1/4"		90	.178		4.05	8.40		12.45	17.50
2088	6-3/4"		87	.184		4.27	8.65		12.92	18.20
2090	7-1/4"		85	.188		5.35	8.85		14.20	19.70
2092	7-3/4"		83	.193		5.60	9.10		14.70	20.50
2094	8-3/4"		80	.200		7.20	9.45		16.65	22.50
2096	9-3/4"		77	.208		9.65	9.80		19.45	26
2098	10-7/8"		74	.216		10.75	10.20		20.95	27.50
2100	11-7/8"		71	.225		12.35	10.60		22.95	30
2102	12-7/8"		68	.235		15.55	11.10		26.65	34.50

22 07 19.30 Piping Insulation Protective Jacketing, PVC		Crew	Daily Output	Labor-Hours	Unit	Material	2015 Bare Costs Labor	Equipment	Total	Total Incl O&P
2104	14-1/8"	Q-14	66	.242	Ea.	17.80	11.45		29.25	37.50
2106	15-1/8"		64	.250		19.15	11.80		30.95	39.50
2108	16-1/8"		63	.254		21	11.95		32.95	41.50
2110	17-1/8"		62	.258		23	12.15		35.15	44.50
2112	18-1/8"		61	.262		30	12.35		42.35	52.50
2114	19-1/8"		60	.267		40.50	12.55		53.05	64
2116	20-1/8"		59	.271		52	12.80		64.80	77.50
2200	45° Elbow fitting									
2220	1-3/4" thru 9-3/4" same price as 90° Elbow fitting									
2320	10-7/8"	Q-14	74	.216	Ea.	10.75	10.20		20.95	27.50
2322	11-7/8"		71	.225		11.85	10.60		22.45	29.50
2324	12-7/8"		68	.235		13.25	11.10		24.35	32
2326	14-1/8"		66	.242		15.10	11.45		26.55	34.50
2328	15-1/8"		64	.250		16.25	11.80		28.05	36
2330	16-1/8"		63	.254		18.55	11.95		30.50	39
2332	17-1/8"		62	.258		21	12.15		33.15	42
2334	18-1/8"		61	.262		25.50	12.35		37.85	47.50
2336	19-1/8"		60	.267		35	12.55		47.55	58
2338	20-1/8"		59	.271		39.50	12.80		52.30	63.50
2400	Tee fitting									
2410	1-3/4"	Q-14	96	.167	Ea.	.88	7.85		8.73	13.20
2412	2"		94	.170		1	8		9	13.60
2414	2-1/4"		91	.176		1.09	8.30		9.39	14.10
2416	2-1/2"		88	.182		1.19	8.55		9.74	14.65
2418	2-3/4"		85	.188		1.31	8.85		10.16	15.25
2420	3"		82	.195		1.42	9.20		10.62	•15.85
2422	3-3/8"		79	.203		1.63	9.55		11.18	16.65
2424	3-3/4"		76	.211		1.85	9.90		11.75	17.50
2426	4-1/8"		73	.219		2.19	10.35		12.54	18.50
2428	4-3/4"		70	.229		2.73	10.75		13.48	19.75
2430	5-1/4"		67	.239		3.28	11.25		14.53	21
2432	5-3/4"		63	.254		4.36	11.95		16.31	23.50
2434	6-1/4"		60	.267		5.75	12.55		18.30	26
2436	6-3/4"		59	.271		7.10	12.80		19.90	27.50
2438	7-1/4"		57	.281		11.45	13.25		24.70	33
2440	7-3/4"		54	.296		12.55	13.95		26.50	35.50
2442	8-3/4"		52	.308		15.30	14.50		29.80	39.50
2444	9-3/4"		50	.320		18	15.10		33.10	43.50
2446	10-7/8"		48	.333		18.25	15.70		33.95	44.50
2448	11-7/8"		47	.340		20.50	16.05		36.55	47.50
2450	12-7/8"		46	.348		22.50	16.40		38.90	50
2452	14-1/8"		45	.356		24.50	16.75		41.25	53
2454	15-1/8"		44	.364		26.50	17.15		43.65	55.50
2456	16-1/8"		43	.372		28.50	17.55		46.05	58.50
2458	17-1/8"		42	.381		30.50	17.95		48.45	61.50
2460	18-1/8"		41	.390		33.50	18.40		51.90	65.50
2462	19-1/8"		40	.400		36.50	18.85		55.35	69.50
2464	20-1/8"		39	.410		40	19.35		59.35	74
4000	Mechanical grooved fitting cover, including insert									
4020	90° Elbow fitting									
4030	3/4" & 1"	Q-14	140	.114	Ea.	3.70	5.40		9.10	12.45
4040	1-1/4" & 1-1/2"		135	.119		4.68	5.60		10.28	13.85
4042	2"		130	.123		7.40	5.80		13.20	17.20

For customer support on your Mechanical Cost Data, call 877.795.7394.

22 07 19 – Plumbing Piping Insulation

22 07 19.30 Piping Insulation Protective Jacketing, PVC		Crew	Daily Output	Labor-Hours	Unit	Material	2015 Bare Costs Labor	Equipment	Total	Total Incl O&P
4044	2-1/2"	Q-14	125	.128	Ea.	8.20	6.05		14.25	18.40
4046	3"		120	.133		9.20	6.30		15.50	19.90
4048	3-1/2"		115	.139		10.65	6.55		17.20	22
4050	4"		110	.145		11.85	6.85		18.70	23.50
4052	5"		100	.160		14.75	7.55		22.30	28
4054	6"		90	.178		22	8.40		30.40	37
4056	8"		80	.200		23.50	9.45		32.95	40.50
4058	10"		75	.213		30.50	10.05		40.55	49
4060	12"		68	.235		43.50	11.10		54.60	65.50
4062	14"		65	.246		43.50	11.60		55.10	66
4064	16"		63	.254		59.50	11.95		71.45	84
4066	18"	↓	61	.262	↓	81.50	12.35		93.85	109
4100	45° Elbow fitting									
4120	3/4" & 1"	Q-14	140	.114	Ea.	3.29	5.40		8.69	12
4130	1-1/4" & 1-1/2"		135	.119		4.19	5.60		9.79	13.30
4140	2"		130	.123		6.60	5.80		12.40	16.30
4142	2-1/2"		125	.128		7.30	6.05		13.35	17.45
4144	3"		120	.133		8.20	6.30		14.50	18.80
4146	3-1/2"		115	.139		8.65	6.55		15.20	19.70
4148	4"		110	.145		10.55	6.85		17.40	22.50
4150	5"		100	.160		13.10	7.55		20.65	26
4152	6"		90	.178		19.35	8.40		27.75	34.50
4154	8"		80	.200		21	9.45		30.45	37.50
4156	10"		75	.213		27	10.05		37.05	45
4158	12"		68	.235		39	11.10		50.10	60
4160	14"		65	.246		50.50	11.60		62.10	73.50
4162	16"		63	.254		61	11.95		72.95	86
4164	18"	↓	61	.262	↓	72	12.35		84.35	99
4200	Tee fitting									
4220	3/4" & 1"	Q-14	93	.172	Ea.	4.82	8.10		12.92	17.90
4230	1-1/4" & 1-1/2"		90	.178		6.10	8.40		14.50	19.75
4240	2"		87	.184		9.60	8.65		18.25	24
4242	2-1/2"		84	.190		10.70	9		19.70	26
4244	3"		80	.200		11.95	9.45		21.40	28
4246	3-1/2"		77	.208		12.65	9.80		22.45	29
4248	4"		73	.219		15.45	10.35		25.80	33
4250	5"		67	.239		17.05	11.25		28.30	36.50
4252	6"		60	.267		25	12.55		37.55	47
4254	8"		54	.296		27.50	13.95		41.45	51.50
4256	10"		50	.320		39.50	15.10		54.60	67
4258	12"		46	.348		56.50	16.40		72.90	88
4260	14"		43	.372		68.50	17.55		86.05	103
4262	16"		42	.381		82	17.95		99.95	119
4264	18"	↓	41	.390	↓	90.50	18.40		108.90	128

22 07 19.40 Pipe Insulation Protective Jacketing, Aluminum

		Crew	Daily Output	Labor-Hours	Unit	Material	2015 Bare Costs Labor	Equipment	Total	Total Incl O&P
0010	**PIPE INSULATION PROTECTIVE JACKETING, ALUMINUM**									
0100	Metal roll jacketing									
0120	Aluminum with polykraft moisture barrier									
0140	Smooth, based on OD of insulation, .016" thick									
0180	1/2" ID	Q-14	220	.073	L.F.	.20	3.43		3.63	5.55
0190	3/4" ID		215	.074		.26	3.51		3.77	5.75
0200	1" ID	↓	210	.076	↓	.32	3.59		3.91	5.95

22 07 19.40 Pipe Insulation Protective Jacketing, Aluminum	Crew	Daily Output	Labor-Hours	Unit	Material	2015 Bare Costs Labor	Equipment	Total	Total Incl O&P
0210	1-1/4" ID	Q-14	205	.078	L.F.	.38	3.68		4.06	6.15
0220	1-1/2" ID		202	.079		.44	3.73		4.17	6.30
0230	1-3/4" ID		199	.080		.50	3.79		4.29	6.45
0240	2" ID		195	.082		.56	3.87		4.43	6.60
0250	2-1/4" ID		191	.084		.62	3.95		4.57	6.85
0260	2-1/2" ID		187	.086		.69	4.03		4.72	7.05
0270	2-3/4" ID		184	.087		.75	4.10		4.85	7.25
0280	3" ID		180	.089		.82	4.19		5.01	7.40
0290	3-1/4" ID		176	.091		.88	4.28		5.16	7.60
0300	3-1/2" ID		172	.093		.94	4.38		5.32	7.90
0310	3-3/4" ID		169	.095		1	4.46		5.46	8.05
0320	4" ID		165	.097		1.06	4.57		5.63	8.25
0330	4-1/4" ID		161	.099		1.12	4.68		5.80	8.55
0340	4-1/2" ID		157	.102		1.19	4.80		5.99	8.80
0350	4-3/4" ID		154	.104		1.25	4.90		6.15	9
0360	5" ID		150	.107		1.31	5.05		6.36	9.30
0370	5-1/4" ID		146	.110		1.37	5.15		6.52	9.55
0380	5-1/2" ID		143	.112		1.43	5.25		6.68	9.75
0390	5-3/4" ID		139	.115		1.49	5.45		6.94	10.10
0400	6" ID		135	.119		1.55	5.60		7.15	10.40
0410	6-1/4" ID		133	.120		1.61	5.65		7.26	10.60
0420	6-1/2" ID		131	.122		1.67	5.75		7.42	10.80
0430	7" ID		128	.125		1.79	5.90		7.69	11.10
0440	7-1/4" ID		125	.128		1.85	6.05		7.90	11.45
0450	7-1/2" ID		123	.130		1.91	6.15		8.06	11.65
0460	8" ID		121	.132		2.03	6.25		8.28	11.95
0470	8-1/2" ID		119	.134		2.15	6.35		8.50	12.20
0480	9" ID		116	.138		2.27	6.50		8.77	12.60
0490	9-1/2" ID		114	.140		2.39	6.60		8.99	12.95
0500	10" ID		112	.143		2.51	6.75		9.26	13.25
0510	10-1/2" ID		110	.145		2.63	6.85		9.48	13.55
0520	11" ID		107	.150		2.75	7.05		9.80	14
0530	11-1/2" ID		105	.152		2.88	7.20		10.08	14.35
0540	12" ID		103	.155		3	7.30		10.30	14.70
0550	12-1/2" ID		100	.160		3.12	7.55		10.67	15.20
0560	13" ID		99	.162		3.24	7.60		10.84	15.40
0570	14" ID		98	.163		3.49	7.70		11.19	15.85
0580	15" ID		96	.167		3.73	7.85		11.58	16.35
0590	16" ID		95	.168		3.98	7.95		11.93	16.75
0600	17" ID		93	.172		4.22	8.10		12.32	17.25
0610	18" ID		92	.174		4.38	8.20		12.58	17.55
0620	19" ID		90	.178		4.70	8.40		13.10	18.20
0630	20" ID		89	.180		4.95	8.45		13.40	18.65
0640	21" ID		87	.184		5.20	8.65		13.85	19.20
0650	22" ID		86	.186		5.45	8.75		14.20	19.60
0660	23" ID		84	.190		5.70	9		14.70	20.50
0670	24" ID		83	.193		5.90	9.10		15	20.50
0710	For smooth .020" thick, add					27%	10%			
0720	For smooth .024" thick, add					52%	20%			
0730	For smooth .032" thick, add					104%	33%			
0800	For stucco embossed, add					1%				
0820	For corrugated, add					2.50%				
0900	White aluminum with polysurlyn moisture barrier									

22 07 19 – Plumbing Piping Insulation

22 07 19.40 Pipe Insulation Protective Jacketing, Aluminum	Crew	Daily Output	Labor-Hours	Unit	Material	2015 Bare Costs Labor	Equipment	Total	Total Incl O&P
0910	Smooth, % is an add to polykraft lines of same thickness								
0940	For smooth .016" thick, add				L.F.	35%			
0960	For smooth .024" thick, add				"	22%			
1000	Aluminum fitting covers								
1010	Size is based on OD of insulation								
1020	90° LR elbow, 2 piece								
1100	1-1/2"	Q-14	140	.114	Ea.	3.45	5.40		8.85	12.20
1110	1-3/4"		135	.119		3.45	5.60		9.05	12.50
1120	2"		130	.123		4.34	5.80		10.14	13.80
1130	2-1/4"		128	.125		4.34	5.90		10.24	13.90
1140	2-1/2"		126	.127		4.34	6		10.34	14.05
1150	2-3/4"		123	.130		4.34	6.15		10.49	14.30
1160	3"		120	.133		4.71	6.30		11.01	15
1170	3-1/4"		117	.137		4.71	6.45		11.16	15.25
1180	3-1/2"		115	.139		5.55	6.55		12.10	16.35
1190	3-3/4"		113	.142		5.70	6.65		12.35	16.70
1200	4"		110	.145		6	6.85		12.85	17.25
1210	4-1/4"		108	.148		6.15	7		13.15	17.60
1220	4-1/2"		106	.151		6.15	7.10		13.25	17.85
1230	4-3/4"		104	.154		6.15	7.25		13.40	18.05
1240	5"		102	.157		6.95	7.40		14.35	19.10
1250	5-1/4"		100	.160		8	7.55		15.55	20.50
1260	5-1/2"		97	.165		9.95	7.75		17.70	23
1270	5-3/4"		95	.168		9.95	7.95		17.90	23.50
1280	6"		92	.174		8.50	8.20		16.70	22
1290	6-1/4"		90	.178		8.50	8.40		16.90	22.50
1300	6-1/2"		87	.184		10.25	8.65		18.90	25
1310	7"		85	.188		12.40	8.85		21.25	27.50
1320	7-1/4"		84	.190		12.40	9		21.40	27.50
1330	7-1/2"		83	.193		18.25	9.10		27.35	34
1340	8"		82	.195		13.60	9.20		22.80	29.50
1350	8-1/2"		80	.200		25	9.45		34.45	42
1360	9"		78	.205		25	9.65		34.65	42.50
1370	9-1/2"		77	.208		18.50	9.80		28.30	36
1380	10"		76	.211		18.50	9.90		28.40	36
1390	10-1/2"		75	.213		18.90	10.05		28.95	36.50
1400	11"		74	.216		18.90	10.20		29.10	37
1410	11-1/2"		72	.222		21.50	10.45		31.95	40
1420	12"		71	.225		21.50	10.60		32.10	40
1430	12-1/2"		69	.232		38	10.95		48.95	59
1440	13"		68	.235		38	11.10		49.10	59.50
1450	14"		66	.242		51.50	11.45		62.95	74.50
1460	15"		64	.250		53.50	11.80		65.30	77.50
1470	16"		63	.254		58.50	11.95		70.45	83
2000	45° Elbow, 2 piece								
2010	2-1/2"	Q-14	126	.127	Ea.	3.59	6		9.59	13.25
2020	2-3/4"		123	.130		3.59	6.15		9.74	13.50
2030	3"		120	.133		4.05	6.30		10.35	14.25
2040	3-1/4"		117	.137		4.05	6.45		10.50	14.50
2050	3-1/2"		115	.139		4.62	6.55		11.17	15.30
2060	3-3/4"		113	.142		4.62	6.65		11.27	15.50
2070	4"		110	.145		4.70	6.85		11.55	15.80
2080	4-1/4"		108	.148		5.35	7		12.35	16.75

For customer support on your Mechanical Cost Data, call 877.795.7394.

125

22 07 19.40 Pipe Insulation Protective Jacketing, Aluminum	Crew	Daily Output	Labor-Hours	Unit	Material	2015 Bare Costs Labor	Equipment	Total	Total Incl O&P
2090	4-1/2"	Q-14	106	.151	Ea.	5.35	7.10		12.45	17
2100	4-3/4"		104	.154		5.35	7.25		12.60	17.20
2110	5"		102	.157		6.30	7.40		13.70	18.45
2120	5-1/4"		100	.160		6.30	7.55		13.85	18.70
2130	5-1/2"		97	.165		6.60	7.75		14.35	19.35
2140	6"		92	.174		6.60	8.20		14.80	20
2150	6-1/2"		87	.184		9.10	8.65		17.75	23.50
2160	7"		85	.188		9.10	8.85		17.95	24
2170	7-1/2"		83	.193		9.20	9.10		18.30	24.50
2180	8"		82	.195		9.20	9.20		18.40	24.50
2190	8-1/2"		80	.200		11.45	9.45		20.90	27.50
2200	9"		78	.205		11.45	9.65		21.10	27.50
2210	9-1/2"		77	.208		15.90	9.80		25.70	33
2220	10"		76	.211		15.90	9.90		25.80	33
2230	10-1/2"		75	.213		14.85	10.05		24.90	32
2240	11"		74	.216		14.85	10.20		25.05	32
2250	11-1/2"		72	.222		17.85	10.45		28.30	36
2260	12"		71	.225		17.85	10.60		28.45	36
2270	13"		68	.235		21	11.10		32.10	40.50
2280	14"		66	.242		25.50	11.45		36.95	46.50
2290	15"		64	.250		39.50	11.80		51.30	62
2300	16"		63	.254		43	11.95		54.95	66
2310	17"		62	.258		42	12.15		54.15	65
2320	18"		61	.262		48	12.35		60.35	72
2330	19"		60	.267		58.50	12.55		71.05	84
2340	20"		59	.271		56.50	12.80		69.30	82
2350	21"		58	.276		61	13		74	87
3000	Tee, 4 piece									
3010	2-1/2"	Q-14	88	.182	Ea.	29	8.55		37.55	45.50
3020	2-3/4"		86	.186		29	8.75		37.75	45.50
3030	3"		84	.190		32.50	9		41.50	50
3040	3-1/4"		82	.195		32.50	9.20		41.70	50.50
3050	3-1/2"		80	.200		34.50	9.45		43.95	52
3060	4"		78	.205		34.50	9.65		44.15	53
3070	4-1/4"		76	.211		36.50	9.90		46.40	55.50
3080	4-1/2"		74	.216		36.50	10.20		46.70	56
3090	4-3/4"		72	.222		36.50	10.45		46.95	56.50
3100	5"		70	.229		37.50	10.75		48.25	58.50
3110	5-1/4"		68	.235		37.50	11.10		48.60	59
3120	5-1/2"		66	.242		39.50	11.45		50.95	61.50
3130	6"		64	.250		39.50	11.80		51.30	62
3140	6-1/2"		60	.267		43.50	12.55		56.05	67
3150	7"		58	.276		43.50	13		56.50	67.50
3160	7-1/2"		56	.286		48	13.45		61.45	74
3170	8"		54	.296		48	13.95		61.95	74.50
3180	8-1/2"		52	.308		49.50	14.50		64	77
3190	9"		50	.320		49.50	15.10		64.60	78
3200	9-1/2"		49	.327		36.50	15.40		51.90	64.50
3210	10"		48	.333		36.50	15.70		52.20	65
3220	10-1/2"		47	.340		39	16.05		55.05	68
3230	11"		46	.348		39	16.40		55.40	68.50
3240	11-1/2"		45	.356		41	16.75		57.75	71.50
3250	12"		44	.364		41	17.15		58.15	72

22 07 19 – Plumbing Piping Insulation

22 07 19.40 Pipe Insulation Protective Jacketing, Aluminum		Crew	Daily Output	Labor-Hours	Unit	Material	2015 Bare Costs Labor	Equipment	Total	Total Incl O&P
3260	13"	Q-14	43	.372	Ea.	43	17.55		60.55	75
3270	14"		42	.381		45.50	17.95		63.45	78
3280	15"		41	.390		49	18.40		67.40	82.50
3290	16"		40	.400		50.50	18.85		69.35	85
3300	17"		39	.410		58	19.35		77.35	93.50
3310	18"		38	.421		60	19.85		79.85	97
3320	19"		37	.432		66.50	20.50		87	105
3330	20"		36	.444		68	21		89	107
3340	22"		35	.457		86.50	21.50		108	129
3350	23"		34	.471		89	22		111	132
3360	24"		31	.516		91	24.50		115.50	138

22 07 19.50 Pipe Insulation Protective Jacketing, St. Stl.

		Crew	Daily Output	Labor-Hours	Unit	Material	2015 Bare Costs Labor	Equipment	Total	Total Incl O&P
0010	**PIPE INSULATION PROTECTIVE JACKETING, STAINLESS STEEL**									
0100	Metal roll jacketing									
0120	Type 304 with moisture barrier									
0140	Smooth, based on OD of insulation, .010" thick									
0260	2-1/2" ID	Q-14	250	.064	L.F.	2.12	3.02		5.14	7.05
0270	2-3/4" ID		245	.065		2.31	3.08		5.39	7.35
0280	3" ID		240	.067		2.50	3.14		5.64	7.65
0290	3-1/4" ID		235	.068		2.69	3.21		5.90	7.95
0300	3-1/2" ID		230	.070		2.88	3.28		6.16	8.25
0310	3-3/4" ID		225	.071		3.07	3.35		6.42	8.60
0320	4" ID		220	.073		3.26	3.43		6.69	8.95
0330	4-1/4" ID		215	.074		3.44	3.51		6.95	9.25
0340	4-1/2" ID		210	.076		3.63	3.59		7.22	9.60
0350	5" ID		200	.080		4.01	3.77		7.78	10.25
0360	5-1/2" ID		190	.084		4.39	3.97		8.36	11.05
0370	6" ID		180	.089		4.76	4.19		8.95	11.75
0380	6-1/2" ID		175	.091		5.15	4.31		9.46	12.35
0390	7" ID		170	.094		5.50	4.44		9.94	12.95
0400	7-1/2" ID		164	.098		5.90	4.60		10.50	13.65
0410	8" ID		161	.099		6.25	4.68		10.93	14.20
0420	8-1/2" ID		158	.101		6.65	4.77		11.42	14.75
0430	9" ID		155	.103		7.05	4.87		11.92	15.35
0440	9-1/2" ID		152	.105		7.40	4.96		12.36	15.85
0450	10" ID		149	.107		7.80	5.05		12.85	16.45
0460	10-1/2" ID		146	.110		8.15	5.15		13.30	17.05
0470	11" ID		143	.112		8.55	5.25		13.80	17.60
0480	12" ID		137	.117		9.30	5.50		14.80	18.75
0490	13" ID		132	.121		10.05	5.70		15.75	19.95
0500	14" ID		130	.123		10.80	5.80		16.60	21
0700	For smooth .016" thick, add					45%	33%			
1000	Stainless steel, Type 316, fitting covers									
1010	Size is based on OD of insulation									
1020	90° LR Elbow, 2 piece									
1100	1-1/2"	Q-14	126	.127	Ea.	12.50	6		18.50	23
1110	2-3/4"		123	.130		12.50	6.15		18.65	23.50
1120	3"		120	.133		13.05	6.30		19.35	24
1130	3-1/4"		117	.137		13.05	6.45		19.50	24.50
1140	3-1/2"		115	.139		13.75	6.55		20.30	25.50
1150	3-3/4"		113	.142		14.50	6.65		21.15	26.50
1160	4"		110	.145		15.95	6.85		22.80	28

For customer support on your Mechanical Cost Data, call 877.795.7394.

127

22 07 19.50 Pipe Insulation Protective Jacketing, St. Stl.	Crew	Daily Output	Labor-Hours	Unit	Material	2015 Bare Costs Labor	Equipment	Total	Total Incl O&P
1170	4-1/4"	Q-14	108	.148	Ea.	21	7		28	34.50
1180	4-1/2"		106	.151		21	7.10		28.10	34.50
1190	5"		102	.157		21.50	7.40		28.90	35.50
1200	5-1/2"		97	.165		32.50	7.75		40.25	47.50
1210	6"		92	.174		35.50	8.20		43.70	52
1220	6-1/2"		87	.184		49	8.65		57.65	67
1230	7"		85	.188		49	8.85		57.85	67.50
1240	7-1/2"		83	.193		57	9.10		66.10	76.50
1250	8"		80	.200		57	9.45		66.45	77
1260	8-1/2"		80	.200		59	9.45		68.45	79.50
1270	9"		78	.205		89.50	9.65		99.15	114
1280	9-1/2"		77	.208		88	9.80		97.80	112
1290	10"		76	.211		88	9.90		97.90	112
1300	10-1/2"		75	.213		105	10.05		115.05	132
1310	11"		74	.216		101	10.20		111.20	127
1320	12"		71	.225		114	10.60		124.60	143
1330	13"		68	.235		158	11.10		169.10	191
1340	14"		66	.242		159	11.45		170.45	193
2000	45° Elbow, 2 piece									
2010	2-1/2"	Q-14	126	.127	Ea.	10.40	6		16.40	21
2020	2-3/4"		123	.130		10.40	6.15		16.55	21
2030	3"		120	.133		11.20	6.30		17.50	22
2040	3-1/4"		117	.137		11.20	6.45		17.65	22.50
2050	3-1/2"		115	.139		11.35	6.55		17.90	22.50
2060	3-3/4"		113	.142		11.35	6.65		18	23
2070	4"		110	.145		14.60	6.85		21.45	26.50
2080	4-1/4"		108	.148		20.50	7		27.50	33.50
2090	4-1/2"		106	.151		20.50	7.10		27.60	33.50
2100	4-3/4"		104	.154		20.50	7.25		27.75	34
2110	5"		102	.157		21	7.40		28.40	34.50
2120	5-1/2"		97	.165		21	7.75		28.75	35
2130	6"		92	.174		24.50	8.20		32.70	40
2140	6-1/2"		87	.184		24.50	8.65		33.15	40.50
2150	7"		85	.188		42	8.85		50.85	60
2160	7-1/2"		83	.193		42.50	9.10		51.60	61
2170	8"		82	.195		42.50	9.20		51.70	61.50
2180	8-1/2"		80	.200		50	9.45		59.45	69.50
2190	9"		78	.205		50	9.65		59.65	70
2200	9-1/2"		77	.208		58.50	9.80		68.30	80
2210	10"		76	.211		58.50	9.90		68.40	80
2220	10-1/2"		75	.213		70.50	10.05		80.55	93
2230	11"		74	.216		70.50	10.20		80.70	93.50
2240	12"		71	.225		76.50	10.60		87.10	101
2250	13"		68	.235		90	11.10		101.10	116

22 11 13 – Facility Water Distribution Piping

22 11 13.14 Pipe, Brass

	Crew	Daily Output	Labor-Hours	Unit	Material	2015 Bare Costs Labor	Equipment	Total	Total Incl O&P
0010 **PIPE, BRASS**, Plain end									
0900 Field threaded, coupling & clevis hanger assembly 10' O.C.									
0920 Regular weight									
1120 1/2" diameter	1 Plum	48	.167	L.F.	7.15	9.80		16.95	22.50
1140 3/4" diameter		46	.174		9.30	10.20		19.50	25.50
1160 1" diameter		43	.186		13.30	10.90		24.20	31
1180 1-1/4" diameter	Q-1	72	.222		19.90	11.75		31.65	40
1200 1-1/2" diameter		65	.246		23.50	13		36.50	45.50
1220 2" diameter		53	.302		33	15.95		48.95	60.50
1240 2-1/2" diameter		41	.390		51	20.50		71.50	87.50
1260 3" diameter		31	.516		71.50	27.50		99	120
1300 4" diameter	Q-2	37	.649		126	35.50		161.50	192
1930 To delete coupling & hanger, subtract									
1940 1/2" diam.					14%	46%			
1950 3/4" diam. to 1-1/2" diam.					8%	47%			
1960 2" diam. to 4" diam.					10%	37%			

22 11 13.16 Pipe Fittings, Brass

	Crew	Daily Output	Labor-Hours	Unit	Material	2015 Bare Costs Labor	Equipment	Total	Total Incl O&P
0010 **PIPE FITTINGS, BRASS**, Rough bronze, threaded.									
1000 Standard wt., 90° Elbow									
1040 1/8"	1 Plum	13	.615	Ea.	18	36		54	74.50
1060 1/4"		13	.615		18	36		54	74.50
1080 3/8"		13	.615		18	36		54	74.50
1100 1/2"		12	.667		18	39		57	79
1120 3/4"		11	.727		24	42.50		66.50	91
1140 1"		10	.800		39	47		86	114
1160 1-1/4"	Q-1	17	.941		63.50	49.50		113	145
1180 1-1/2"		16	1		78.50	53		131.50	166
1200 2"		14	1.143		126	60.50		186.50	230
1220 2-1/2"		11	1.455		305	77		382	450
1240 3"		8	2		465	106		571	675
1260 4"	Q-2	11	2.182		945	120		1,065	1,225
1280 5"		8	3		2,625	164		2,789	3,125
1300 6"		7	3.429		3,850	188		4,038	4,525
1500 45° Elbow, 1/8"	1 Plum	13	.615		22	36		58	79
1540 1/4"		13	.615		22	36		58	79
1560 3/8"		13	.615		22	36		58	79
1580 1/2"		12	.667		22	39		61	83.50
1600 3/4"		11	.727		31.50	42.50		74	99
1620 1"		10	.800		53.50	47		100.50	130
1640 1-1/4"	Q-1	17	.941		85	49.50		134.50	169
1660 1-1/2"		16	1		107	53		160	198
1680 2"		14	1.143		173	60.50		233.50	282
1700 2-1/2"		11	1.455		330	77		407	480
1720 3"		8	2		510	106		616	720
1740 4"	Q-2	11	2.182		1,125	120		1,245	1,425
1760 5"		8	3		2,125	164		2,289	2,600
1780 6"		7	3.429		2,950	188		3,138	3,525
2000 Tee, 1/8"	1 Plum	9	.889		21	52		73	103
2040 1/4"		9	.889		21	52		73	103
2060 3/8"		9	.889		21	52		73	103
2080 1/2"		8	1		21	58.50		79.50	112
2100 3/4"		7	1.143		30	67		97	134

For customer support on your Mechanical Cost Data, call 877.795.7394.

129

22 11 13 - Facility Water Distribution Piping

22 11 13.16 Pipe Fittings, Brass

		Crew	Daily Output	Labor-Hours	Unit	Material	2015 Bare Costs Labor	Equipment	Total	Total Incl O&P
2120	1"	1 Plum	6	1.333	Ea.	54.50	78.50		133	178
2140	1-1/4"	Q-1	10	1.600		93.50	84.50		178	231
2160	1-1/2"		9	1.778		105	94		199	258
2180	2"		8	2		175	106		281	350
2200	2-1/2"		7	2.286		415	121		536	640
2220	3"		5	3.200		635	169		804	955
2240	4"	Q-2	7	3.429		1,575	188		1,763	2,000
2260	5"		5	4.800		3,275	263		3,538	4,000
2280	6"		4	6		5,150	330		5,480	6,150
2500	Coupling, 1/8"	1 Plum	26	.308		15.05	18.05		33.10	44
2540	1/4"		22	.364		15.05	21.50		36.55	48.50
2560	3/8"		18	.444		15.05	26		41.05	56
2580	1/2"		15	.533		15.05	31.50		46.55	64
2600	3/4"		14	.571		21	33.50		54.50	74
2620	1"		13	.615		36	36		72	94
2640	1-1/4"	Q-1	22	.727		60	38.50		98.50	125
2660	1-1/2"		20	.800		78.50	42.50		121	150
2680	2"		18	.889		130	47		177	213
2700	2-1/2"		14	1.143		221	60.50		281.50	335
2720	3"		10	1.600		305	84.50		389.50	465
2740	4"	Q-2	12	2		630	110		740	860
2760	5"		10	2.400		1,175	131		1,306	1,500
2780	6"		9	2.667		1,675	146		1,821	2,075
3000	Union, 125 lb.									
3020	1/8"	1 Plum	12	.667	Ea.	51.50	39		90.50	116
3040	1/4"		12	.667		51.50	39		90.50	116
3060	3/8"		12	.667		51.50	39		90.50	116
3080	1/2"		11	.727		51.50	42.50		94	121
3100	3/4"		10	.800		70.50	47		117.50	149
3120	1"		9	.889		106	52		158	196
3140	1-1/4"	Q-1	16	1		154	53		207	249
3160	1-1/2"		15	1.067		183	56.50		239.50	286
3180	2"		13	1.231		247	65		312	370
3200	2-1/2"		10	1.600		675	84.50		759.50	870
3220	3"		7	2.286		1,050	121		1,171	1,325
3240	4"	Q-2	10	2.400		3,250	131		3,381	3,775
3320	For 250 lb. (navy pattern), add					100%				

22 11 13.23 Pipe/Tube, Copper

		Crew	Daily Output	Labor-Hours	Unit	Material	2015 Bare Costs Labor	Equipment	Total	Total Incl O&P
0010	**PIPE/TUBE, COPPER**, Solder joints R221113-50									
0100	Solder									
0120	Solder, lead free, roll				Lb.	9.35			9.35	10.30
1000	Type K tubing, couplings & clevis hanger assemblies 10' O.C.									
1100	1/4" diameter	1 Plum	84	.095	L.F.	3.71	5.60		9.31	12.55
1120	3/8" diameter		82	.098		4.31	5.75		10.06	13.40
1140	1/2" diameter		78	.103		4.78	6		10.78	14.35
1160	5/8" diameter		77	.104		5.90	6.10		12	15.65
1180	3/4" diameter		74	.108		7.80	6.35		14.15	18.15
1200	1" diameter		66	.121		10.35	7.10		17.45	22
1220	1-1/4" diameter		56	.143		12.20	8.40		20.60	26
1240	1-1/2" diameter		50	.160		15.60	9.40		25	31.50
1260	2" diameter		40	.200		23.50	11.75		35.25	43.50
1280	2-1/2" diameter	Q-1	60	.267		35	14.10		49.10	60

22 11 13.23 Pipe/Tube, Copper		Crew	Daily Output	Labor-Hours	Unit	Material	2015 Bare Costs Labor	Equipment	Total	Total Incl O&P
1300	3" diameter	Q-1	54	.296	L.F.	48	15.65		63.65	76
1320	3-1/2" diameter		42	.381		64	20		84	101
1330	4" diameter		38	.421		80	22		102	122
1340	5" diameter		32	.500		152	26.50		178.50	207
1360	6" diameter	Q-2	38	.632		222	34.50		256.50	297
1380	8" diameter	"	34	.706		385	38.50		423.50	485
1390	For other than full hard temper, add					13%				
1440	For silver solder, add						15%			
1800	For medical clean, (oxygen class), add					12%				
1950	To delete cplgs. & hngrs., 1/4"-1" pipe, subtract					27%	60%			
1960	1-1/4"-3" pipe, subtract					14%	52%			
1970	3-1/2"-5" pipe, subtract					10%	60%			
1980	6"-8" pipe, subtract					19%	53%			
2000	Type L tubing, couplings & clevis hanger assemblies 10' O.C.									
2100	1/4" diameter	1 Plum	88	.091	L.F.	2.77	5.35		8.12	11.10
2120	3/8" diameter		84	.095		3.50	5.60		9.10	12.30
2140	1/2" diameter		81	.099		3.70	5.80		9.50	12.80
2160	5/8" diameter		79	.101		5.35	5.95		11.30	14.85
2180	3/4" diameter		76	.105		5.20	6.20		11.40	15.05
2200	1" diameter		68	.118		7.85	6.90		14.75	19.05
2220	1-1/4" diameter		58	.138		10.35	8.10		18.45	23.50
2240	1-1/2" diameter		52	.154		12.95	9.05		22	28
2260	2" diameter		42	.190		18.80	11.20		30	37.50
2280	2-1/2" diameter	Q-1	62	.258		29.50	13.65		43.15	53
2300	3" diameter		56	.286		37	15.10		52.10	63.50
2320	3-1/2" diameter		43	.372		53	19.65		72.65	87.50
2340	4" diameter		39	.410		65.50	21.50		87	105
2360	5" diameter		34	.471		120	25		145	170
2380	6" diameter	Q-2	40	.600		166	33		199	232
2400	8" diameter	"	36	.667		285	36.50		321.50	370
2410	For other than full hard temper, add					21%				
2590	For silver solder, add						15%			
2900	For medical clean, (oxygen class), add					12%				
2940	To delete cplgs. & hngrs., 1/4"-1" pipe, subtract					37%	63%			
2960	1-1/4"-3" pipe, subtract					12%	53%			
2970	3-1/2"-5" pipe, subtract					12%	63%			
2980	6"-8" pipe, subtract					24%	55%			
3000	Type M tubing, couplings & clevis hanger assemblies 10' O.C.									
3100	1/4" diameter	1 Plum	90	.089	L.F.	3	5.20		8.20	11.20
3120	3/8" diameter		87	.092		3.05	5.40		8.45	11.50
3140	1/2" diameter		84	.095		3.05	5.60		8.65	11.80
3160	5/8" diameter		81	.099		4.49	5.80		10.29	13.70
3180	3/4" diameter		78	.103		4.13	6		10.13	13.65
3200	1" diameter		70	.114		6.60	6.70		13.30	17.40
3220	1-1/4" diameter		60	.133		8.80	7.85		16.65	21.50
3240	1-1/2" diameter		54	.148		11.55	8.70		20.25	26
3260	2" diameter		44	.182		17.35	10.65		28	35
3280	2-1/2" diameter	Q-1	64	.250		25.50	13.20		38.70	48
3300	3" diameter		58	.276		34	14.55		48.55	59.50
3320	3-1/2" diameter		45	.356		47.50	18.80		66.30	80.50
3340	4" diameter		40	.400		60	21		81	98
3360	5" diameter		36	.444		120	23.50		143.50	168
3370	6" diameter	Q-2	42	.571		166	31.50		197.50	230

For customer support on your Mechanical Cost Data, call 877.795.7394.

131

22 11 13 – Facility Water Distribution Piping

22 11 13.23 Pipe/Tube, Copper		Crew	Daily Output	Labor-Hours	Unit	Material	2015 Bare Costs Labor	Equipment	Total	Total Incl O&P
3380	8" diameter	Q-2	38	.632	L.F.	285	34.50		319.50	370
3440	For silver solder, add						15%			
3960	To delete cplgs. & hngrs., 1/4"-1" pipe, subtract					35%	65%			
3970	1-1/4"-3" pipe, subtract					19%	56%			
3980	3-1/2"-5" pipe, subtract					13%	65%			
3990	6"-8" pipe, subtract					28%	58%			
4000	Type DWV tubing, couplings & clevis hanger assemblies 10' O.C.									
4100	1-1/4" diameter	1 Plum	60	.133	L.F.	9.15	7.85		17	22
4120	1-1/2" diameter		54	.148		11.15	8.70		19.85	25.50
4140	2" diameter		44	.182		14.60	10.65		25.25	32
4160	3" diameter	Q-1	58	.276		26.50	14.55		41.05	51
4180	4" diameter		40	.400		44	21		65	80
4200	5" diameter		36	.444		107	23.50		130.50	154
4220	6" diameter	Q-2	42	.571		153	31.50		184.50	216
4240	8" diameter	"	38	.632		310	34.50		344.50	395
4730	To delete cplgs. & hngrs., 1-1/4"-2" pipe, subtract					16%	53%			
4740	3"-4" pipe, subtract					13%	60%			
4750	5"-8" pipe, subtract					23%	58%			
5200	ACR tubing, type L, hard temper, cleaned and									
5220	capped, no couplings or hangers									
5240	3/8" OD				L.F.	1.51			1.51	1.66
5250	1/2" OD					2.30			2.30	2.53
5260	5/8" OD					2.81			2.81	3.09
5270	3/4" OD					3.97			3.97	4.37
5280	7/8" OD					4.41			4.41	4.85
5290	1-1/8" OD					6.40			6.40	7.05
5300	1-3/8" OD					8.60			8.60	9.45
5310	1-5/8" OD					11			11	12.10
5320	2-1/8" OD					17.30			17.30	19.05
5330	2-5/8" OD					26			26	28.50
5340	3-1/8" OD					34.50			34.50	38
5350	3-5/8" OD					45			45	49.50
5360	4-1/8" OD					57.50			57.50	63
5380	ACR tubing, type L, hard, cleaned and capped									
5381	No couplings or hangers									
5384	3/8"	1 Stpi	160	.050	L.F.	1.51	2.99		4.50	6.15
5385	1/2"		160	.050		2.30	2.99		5.29	7.05
5386	5/8"		160	.050		2.81	2.99		5.80	7.60
5387	3/4"		130	.062		3.97	3.68		7.65	9.90
5388	7/8"		130	.062		4.41	3.68		8.09	10.40
5389	1-1/8"		115	.070		6.40	4.16		10.56	13.35
5390	1-3/8"		100	.080		8.60	4.78		13.38	16.65
5391	1-5/8"		90	.089		11	5.30		16.30	20
5392	2-1/8"		80	.100		17.30	6		23.30	28
5393	2-5/8"	Q-5	125	.128		26	6.90		32.90	39
5394	3-1/8"		105	.152		34.50	8.20		42.70	50.50
5395	4-1/8"		95	.168		57.50	9.05		66.55	76.50
5800	Refrigeration tubing, dryseal, 50' coils									
5840	1/8" OD				Coil	32			32	35.50
5850	3/16" OD					37.50			37.50	41.50
5860	1/4" OD					43.50			43.50	47.50
5870	5/16" OD					58			58	64
5880	3/8" OD					60			60	66

For customer support on your Mechanical Cost Data, call 877.795.7394.

22 11 13 – Facility Water Distribution Piping

22 11 13.23 Pipe/Tube, Copper

		Crew	Daily Output	Labor-Hours	Unit	Material	2015 Bare Costs Labor	Equipment	Total	Total Incl O&P
5890	1/2" OD				Coil	87			87	96
5900	5/8" OD					112			112	124
5910	3/4" OD					133			133	147
5920	7/8" OD					199			199	219
5930	1-1/8" OD					300			300	330
5940	1-3/8" OD					520			520	575
5950	1-5/8" OD					660			660	730
9400	Sub assemblies used in assembly systems									
9410	Chilled water unit, coil connections per unit under 10 ton	Q-5	.80	20	System	1,175	1,075		2,250	2,925
9420	Chilled water unit, coil connections per unit 10 ton and up		1	16		1,900	860		2,760	3,400
9430	Chilled water dist. piping per ton, less than 61 ton systems		26	.615		21	33		54	73
9440	Chilled water dist. piping per ton, 61 through 120 ton systems	Q-6	31	.774		48.50	43		91.50	119
9450	Chilled water dist. piping/ton, 135 ton systems and up	Q-8	25.40	1.260		64.50	71.50	2.27	138.27	181
9510	Refrigerant piping/ton of cooling for remote condensers	Q-5	2	8		375	430		805	1,075
9520	Refrigerant piping per ton up to 10 ton w/remote condensing unit		2.40	6.667		176	360		536	735
9530	Refrigerant piping per ton, 20 ton w/remote condensing unit		2	8		259	430		689	935
9540	Refrigerant piping per ton, 40 ton w/remote condensing unit		1.90	8.421		330	455		785	1,050
9550	Refrigerant piping per ton, 75-80 ton w/remote condensing unit	Q-6	2.40	10		485	560		1,045	1,375
9560	Refrigerant piping per ton, 100 ton w/remote condensing unit	"	2.20	10.909		610	610		1,220	1,600

22 11 13.25 Pipe/Tube Fittings, Copper

		Crew	Daily Output	Labor-Hours	Unit	Material	2015 Bare Costs Labor	Equipment	Total	Total Incl O&P
0010	**PIPE/TUBE FITTINGS, COPPER**, Wrought unless otherwise noted									
0020	For silver solder, add						15%			
0040	Solder joints, copper x copper									
0070	90° elbow, 1/4"	1 Plum	22	.364	Ea.	3.28	21.50		24.78	35.50
0090	3/8"		22	.364		3.11	21.50		24.61	35.50
0100	1/2"		20	.400		1.10	23.50		24.60	36.50
0110	5/8"		19	.421		2.46	24.50		26.96	40
0120	3/4"		19	.421		2.46	24.50		26.96	40
0130	1"		16	.500		6.05	29.50		35.55	51
0140	1-1/4"		15	.533		9	31.50		40.50	57.50
0150	1-1/2"		13	.615		14.05	36		50.05	70
0160	2"		11	.727		25.50	42.50		68	92.50
0170	2-1/2"	Q-1	13	1.231		51.50	65		116.50	155
0180	3"		11	1.455		64.50	77		141.50	187
0190	3-1/2"		10	1.600		225	84.50		309.50	375
0200	4"		9	1.778		152	94		246	310
0210	5"		6	2.667		645	141		786	925
0220	6"	Q-2	9	2.667		865	146		1,011	1,175
0230	8"	"	8	3		3,475	164		3,639	4,075
0250	45° elbow, 1/4"	1 Plum	22	.364		5.90	21.50		27.40	38.50
0270	3/8"		22	.364		5	21.50		26.50	37.50
0280	1/2"		20	.400		2.01	23.50		25.51	37.50
0290	5/8"		19	.421		9.55	24.50		34.05	48
0300	3/4"		19	.421		3.52	24.50		28.02	41.50
0310	1"		16	.500		8.85	29.50		38.35	54
0320	1-1/4"		15	.533		11.85	31.50		43.35	60.50
0330	1-1/2"		13	.615		14.25	36		50.25	70
0340	2"		11	.727		24	42.50		66.50	90.50
0350	2-1/2"	Q-1	13	1.231		50.50	65		115.50	154
0360	3"		13	1.231		70.50	65		135.50	176
0370	3-1/2"		10	1.600		125	84.50		209.50	266
0380	4"		9	1.778		151	94		245	310

For customer support on your Mechanical Cost Data, call 877.795.7394.

133

22 11 13 – Facility Water Distribution Piping

22 11 13.25 Pipe/Tube Fittings, Copper		Crew	Daily Output	Labor-Hours	Unit	Material	2015 Bare Costs Labor	Equipment	Total	Total Incl O&P
0390	5"	Q-1	6	2.667	Ea.	545	141		686	815
0400	6"	Q-2	9	2.667		860	146		1,006	1,175
0410	8"	"	8	3		3,400	164		3,564	3,975
0450	Tee, 1/4"	1 Plum	14	.571		6.55	33.50		40.05	57.50
0470	3/8"		14	.571		5.25	33.50		38.75	56.50
0480	1/2"		13	.615		1.87	36		37.87	56.50
0490	5/8"		12	.667		12.05	39		51.05	72.50
0500	3/4"		12	.667		4.51	39		43.51	64
0510	1"		10	.800		13.95	47		60.95	86.50
0520	1-1/4"		9	.889		18.85	52		70.85	99.50
0530	1-1/2"		8	1		29	58.50		87.50	121
0540	2"		7	1.143		45.50	67		112.50	151
0550	2-1/2"	Q-1	8	2		91.50	106		197.50	261
0560	3"		7	2.286		131	121		252	325
0570	3-1/2"		6	2.667		380	141		521	635
0580	4"		5	3.200		320	169		489	605
0590	5"		4	4		975	211		1,186	1,400
0600	6"	Q-2	6	4		1,325	219		1,544	1,800
0610	8"	"	5	4.800		5,125	263		5,388	6,050
0612	Tee, reducing on the outlet, 1/4"	1 Plum	15	.533		16.45	31.50		47.95	65.50
0613	3/8"		15	.533		15.40	31.50		46.90	64.50
0614	1/2"		14	.571		10.05	33.50		43.55	61.50
0615	5/8"		13	.615		32	36		68	90
0616	3/4"		12	.667		6.45	39		45.45	66
0617	1"		11	.727		18.50	42.50		61	85
0618	1-1/4"		10	.800		20.50	47		67.50	93.50
0619	1-1/2"		9	.889		21.50	52		73.50	103
0620	2"		8	1		39	58.50		97.50	132
0621	2-1/2"	Q-1	9	1.778		107	94		201	259
0622	3"		8	2		110	106		216	281
0623	4"		6	2.667		203	141		344	435
0624	5"		5	3.200		1,125	169		1,294	1,500
0625	6"	Q-2	7	3.429		1,550	188		1,738	1,975
0626	8"	"	6	4		5,950	219		6,169	6,875
0630	Tee, reducing on the run, 1/4"	1 Plum	15	.533		13.90	31.50		45.40	63
0631	3/8"		15	.533		19.60	31.50		51.10	69
0632	1/2"		14	.571		11.35	33.50		44.85	63
0633	5/8"		13	.615		19.40	36		55.40	76
0634	3/4"		12	.667		15.15	39		54.15	75.50
0635	1"		11	.727		16.65	42.50		59.15	83
0636	1-1/4"		10	.800		26	47		73	100
0637	1-1/2"		9	.889		46	52		98	130
0638	2"		8	1		59	58.50		117.50	154
0639	2-1/2"	Q-1	9	1.778		152	94		246	310
0640	3"		8	2		202	106		308	380
0641	4"		6	2.667		425	141		566	680
0642	5"		5	3.200		1,075	169		1,244	1,425
0643	6"	Q-2	7	3.429		1,625	188		1,813	2,075
0644	8"	"	6	4		5,950	219		6,169	6,875
0650	Coupling, 1/4"	1 Plum	24	.333		.75	19.55		20.30	30.50
0670	3/8"		24	.333		1	19.55		20.55	30.50
0680	1/2"		22	.364		.83	21.50		22.33	33
0690	5/8"		21	.381		2.50	22.50		25	37

22 11 13 – Facility Water Distribution Piping

22 11 13.25 Pipe/Tube Fittings, Copper		Crew	Daily Output	Labor-Hours	Unit	Material	2015 Bare Costs Labor	Equipment	Total	Total Incl O&P
0700	3/4"	1 Plum	21	.381	Ea.	1.67	22.50		24.17	36
0710	1"		18	.444		5.55	26		31.55	45.50
0715	1-1/4"		17	.471		5.85	27.50		33.35	48
0716	1-1/2"		15	.533		7.70	31.50		39.20	56
0718	2"		13	.615		12.90	36		48.90	68.50
0721	2-1/2"	Q-1	15	1.067		27.50	56.50		84	115
0722	3"		13	1.231		38.50	65		103.50	141
0724	3-1/2"		8	2		74	106		180	242
0726	4"		7	2.286		81.50	121		202.50	272
0728	5"		6	2.667		187	141		328	420
0731	6"	Q-2	8	3		310	164		474	590
0732	8"	"	7	3.429		1,000	188		1,188	1,375
0741	Coupling, reducing, concentric									
0743	1/2"	1 Plum	23	.348	Ea.	2.16	20.50		22.66	33.50
0745	3/4"		21.50	.372		4.70	22		26.70	38
0747	1"		19.50	.410		5.95	24		29.95	43
0748	1-1/4"		18	.444		8.30	26		34.30	48.50
0749	1-1/2"		16	.500		10.80	29.50		40.30	56.50
0751	2"		14	.571		23	33.50		56.50	75.50
0752	2-1/2"		13	.615		47	36		83	107
0753	3"	Q-1	14	1.143		53.50	60.50		114	150
0755	4"	"	8	2		113	106		219	284
0757	5"	Q-2	7.50	3.200		585	175		760	905
0759	6"		7	3.429		1,075	188		1,263	1,475
0761	8"		6.50	3.692		2,150	202		2,352	2,650
0771	Cap, sweat									
0773	1/2"	1 Plum	40	.200	Ea.	.80	11.75		12.55	18.65
0775	3/4"		38	.211		1.48	12.35		13.83	20.50
0777	1"		32	.250		3.52	14.70		18.22	26
0778	1-1/4"		29	.276		4.66	16.20		20.86	29.50
0779	1-1/2"		26	.308		6.80	18.05		24.85	35
0781	2"		22	.364		12.30	21.50		33.80	45.50
0791	Flange, sweat									
0793	3"	Q-1	22	.727	Ea.	174	38.50		212.50	249
0795	4"		18	.889		243	47		290	340
0797	5"		12	1.333		455	70.50		525.50	605
0799	6"	Q-2	18	1.333		475	73		548	630
0801	8"	"	16	1.500		1,000	82		1,082	1,225
0850	Unions, 1/4"	1 Plum	21	.381		26	22.50		48.50	63
0870	3/8"		21	.381		26.50	22.50		49	63
0880	1/2"		19	.421		14.05	24.50		38.55	53
0890	5/8"		18	.444		60	26		86	106
0900	3/4"		18	.444		17.60	26		43.60	59
0910	1"		15	.533		30.50	31.50		62	81
0920	1-1/4"		14	.571		50	33.50		83.50	105
0930	1-1/2"		12	.667		65.50	39		104.50	131
0940	2"		10	.800		113	47		160	195
0950	2-1/2"	Q-1	12	1.333		245	70.50		315.50	375
0960	3"	"	10	1.600		635	84.50		719.50	825
0980	Adapter, copper x male IPS, 1/4"	1 Plum	20	.400		10.85	23.50		34.35	47.50
0990	3/8"		20	.400		5.40	23.50		28.90	41.50
1000	1/2"		18	.444		2.29	26		28.29	42
1010	3/4"		17	.471		3.84	27.50		31.34	45.50

For customer support on your Mechanical Cost Data, call 877.795.7394.

135

22 11 Facility Water Distribution

22 11 13 – Facility Water Distribution Piping

22 11 13.25 Pipe/Tube Fittings, Copper		Crew	Daily Output	Labor-Hours	Unit	Material	2015 Bare Costs Labor	Equipment	Total	Total Incl O&P
1020	1"	1 Plum	15	.533	Ea.	9.95	31.50		41.45	58.50
1030	1-1/4"		13	.615		14.55	36		50.55	70.50
1040	1-1/2"		12	.667		16.70	39		55.70	77.50
1050	2"		11	.727		28.50	42.50		71	95.50
1060	2-1/2"	Q-1	10.50	1.524		108	80.50		188.50	241
1070	3"		10	1.600		135	84.50		219.50	276
1080	3-1/2"		9	1.778		161	94		255	320
1090	4"		8	2		182	106		288	360
1200	5", cast		6	2.667		4,600	141		4,741	5,275
1210	6", cast	Q-2	8.50	2.824		5,125	155		5,280	5,850
1214	Adapter, copper x female IPS									
1216	1/2"	1 Plum	18	.444	Ea.	3.64	26		29.64	43.50
1218	3/4"		17	.471		4.99	27.50		32.49	47
1220	1"		15	.533		11.55	31.50		43.05	60
1221	1-1/4"		13	.615		16.70	36		52.70	73
1222	1-1/2"		12	.667		26	39		65	88
1224	2"		11	.727		35.50	42.50		78	104
1250	Cross, 1/2"		10	.800		18	47		65	91
1260	3/4"		9.50	.842		30.50	49.50		80	108
1270	1"		8	1		38	58.50		96.50	131
1280	1-1/4"		7.50	1.067		85.50	62.50		148	189
1290	1-1/2"		6.50	1.231		122	72.50		194.50	243
1300	2"		5.50	1.455		237	85.50		322.50	390
1310	2-1/2"	Q-1	6.50	2.462		530	130		660	780
1320	3"	"	5.50	2.909		620	154		774	915
1500	Tee fitting, mechanically formed, (Type 1, 'branch sizes up to 2 in.')									
1520	1/2" run size, 3/8" to 1/2" branch size	1 Plum	80	.100	Ea.		5.85		5.85	8.85
1530	3/4" run size, 3/8" to 3/4" branch size		60	.133			7.85		7.85	11.80
1540	1" run size, 3/8" to 1" branch size		54	.148			8.70		8.70	13.15
1550	1-1/4" run size, 3/8" to 1-1/4" branch size		48	.167			9.80		9.80	14.80
1560	1-1/2" run size, 3/8" to 1-1/2" branch size		40	.200			11.75		11.75	17.75
1570	2" run size, 3/8" to 2" branch size		35	.229			13.40		13.40	20.50
1580	2-1/2" run size, 1/2" to 2" branch size		32	.250			14.70		14.70	22
1590	3" run size, 1" to 2" branch size		26	.308			18.05		18.05	27.50
1600	4" run size, 1" to 2" branch size		24	.333			19.55		19.55	29.50
1640	Tee fitting, mechanically formed, (Type 2, branches 2-1/2" thru 4")									
1650	2-1/2" run size, 2-1/2" branch size	1 Plum	12.50	.640	Ea.		37.50		37.50	56.50
1660	3" run size, 2-1/2" to 3" branch size		12	.667			39		39	59
1670	3-1/2" run size, 2-1/2" to 3-1/2" branch size		11	.727			42.50		42.50	64.50
1680	4" run size, 2-1/2" to 4" branch size		10.50	.762			44.50		44.50	67.50
1698	5" run size, 2" to 4" branch size		9.50	.842			49.50		49.50	74.50
1700	6" run size, 2" to 4" branch size		8.50	.941			55.50		55.50	83.50
1710	8" run size, 2" to 4" branch size		7	1.143			67		67	101
1800	ACR fittings, OD size									
1802	Tee, straight									
1808	5/8"	1 Stpi	12	.667	Ea.	1.87	40		41.87	62
1810	3/4"		12	.667		12.05	40		52.05	73.50
1812	7/8"		10	.800		4.51	48		52.51	77
1813	1"		10	.800		33	48		81	108
1814	1-1/8"		10	.800		13.95	48		61.95	87.50
1816	1-3/8"		9	.889		18.85	53		71.85	101
1818	1-5/8"		8	1		29	60		89	122
1820	2-1/8"		7	1.143		45.50	68.50		114	153

136

For customer support on your Mechanical Cost Data, call 877.795.7394.

22 11 13 – Facility Water Distribution Piping

22 11 13.25 Pipe/Tube Fittings, Copper		Crew	Daily Output	Labor-Hours	Unit	Material	2015 Bare Costs Labor	Equipment	Total	Total Incl O&P
1822	2-5/8"	Q-5	8	2	Ea.	91.50	108		199.50	263
1824	3-1/8"		7	2.286		131	123		254	330
1826	4-1/8"		5	3.200		320	172		492	610
1830	90° elbow									
1836	5/8"	1 Stpi	19	.421	Ea.	3.67	25		28.67	42
1838	3/4"		19	.421		6.65	25		31.65	45.50
1840	7/8"		16	.500		6.60	30		36.60	52.50
1842	1-1/8"		16	.500		8.85	30		38.85	55
1844	1-3/8"		15	.533		9	32		41	58
1846	1-5/8"		13	.615		14.05	37		51.05	71
1848	2-1/8"		11	.727		25.50	43.50		69	93.50
1850	2-5/8"	Q-5	13	1.231		51.50	66		117.50	157
1852	3-1/8"		11	1.455		64.50	78		142.50	189
1854	4-1/8"		9	1.778		170	95.50		265.50	330
1860	Coupling									
1866	5/8"	1 Stpi	21	.381	Ea.	1.15	23		24.15	36
1868	3/4"		21	.381		2.19	23		25.19	37
1870	7/8"		18	.444		1.67	26.50		28.17	42
1871	1"		18	.444		5.70	26.50		32.20	46.50
1872	1-1/8"		18	.444		3.33	26.50		29.83	43.50
1874	1-3/8"		17	.471		5.85	28		33.85	49
1876	1-5/8"		15	.533		7.70	32		39.70	56.50
1878	2-1/8"		13	.615		12.90	37		49.90	69.50
1880	2-5/8"	Q-5	15	1.067		29	57.50		86.50	119
1882	3-1/8"		13	1.231		39.50	66		105.50	144
1884	4-1/8"		7	2.286		90	123		213	285
2000	DWV, solder joints, copper x copper									
2030	90° Elbow, 1-1/4"	1 Plum	13	.615	Ea.	10.45	36		46.45	66
2050	1-1/2"		12	.667		14.55	39		53.55	75
2070	2"		10	.800		28	47		75	102
2090	3"	Q-1	10	1.600		56.50	84.50		141	191
2100	4"	"	9	1.778		335	94		429	505
2150	45° Elbow, 1-1/4"	1 Plum	13	.615		9.65	36		45.65	65
2170	1-1/2"		12	.667		8	39		47	68
2180	2"		10	.800		18.35	47		65.35	91
2190	3"	Q-1	10	1.600		36.50	84.50		121	169
2200	4"	"	9	1.778		184	94		278	345
2250	Tee, Sanitary, 1-1/4"	1 Plum	9	.889		20.50	52		72.50	102
2270	1-1/2"		8	1		26	58.50		84.50	117
2290	2"		7	1.143		30	67		97	134
2310	3"	Q-1	7	2.286		138	121		259	335
2330	4"	"	6	2.667		330	141		471	575
2400	Coupling, 1-1/4"	1 Plum	14	.571		4.91	33.50		38.41	56
2420	1-1/2"		13	.615		6.10	36		42.10	61
2440	2"		11	.727		8.45	42.50		50.95	74
2460	3"	Q-1	11	1.455		15.40	77		92.40	133
2480	4"	"	10	1.600		49	84.50		133.50	182
2602	Traps, see Section 22 13 16.60									
3500	Compression joint fittings									
3510	As used for plumbing and oil burner work									
3520	Fitting price includes nuts and sleeves									
3540	Sleeve, 1/8"				Ea.	.22			.22	.24
3550	3/16"					.22			.22	.24

For customer support on your Mechanical Cost Data, call 877.795.7394.

137

22 11 13 – Facility Water Distribution Piping

22 11 13.25 Pipe/Tube Fittings, Copper	Crew	Daily Output	Labor-Hours	Unit	Material	2015 Bare Costs Labor	Equipment	Total	Total Incl O&P
3560	1/4"				Ea.	.08			.08	.09
3570	5/16"					.23			.23	.25
3580	3/8"					.46			.46	.51
3600	1/2"					.61			.61	.67
3620	Nut, 1/8"					.32			.32	.35
3630	3/16"					.34			.34	.37
3640	1/4"					.33			.33	.36
3650	5/16"					.49			.49	.54
3660	3/8"					.54			.54	.59
3670	1/2"					.95			.95	1.05
3710	Union, 1/8"	1 Plum	26	.308		2	18.05		20.05	29.50
3720	3/16"		24	.333		1.92	19.55		21.47	31.50
3730	1/4"		24	.333		2.01	19.55		21.56	31.50
3740	5/16"		23	.348		2.58	20.50		23.08	34
3750	3/8"		22	.364		2.60	21.50		24.10	35
3760	1/2"		22	.364		4	21.50		25.50	36.50
3780	5/8"		21	.381		4.43	22.50		26.93	39
3820	Union tee, 1/8"		17	.471		7.60	27.50		35.10	50
3830	3/16"		16	.500		4.42	29.50		33.92	49.50
3840	1/4"		15	.533		4	31.50		35.50	52
3850	5/16"		15	.533		5.65	31.50		37.15	53.50
3860	3/8"		15	.533		5.70	31.50		37.20	54
3870	1/2"		15	.533		9.65	31.50		41.15	58
3910	Union elbow, 1/4"		24	.333		2.47	19.55		22.02	32
3920	5/16"		23	.348		4.34	20.50		24.84	36
3930	3/8"		22	.364		4.19	21.50		25.69	36.50
3940	1/2"		22	.364		6.85	21.50		28.35	39.50
3980	Female connector, 1/8"		26	.308		1.55	18.05		19.60	29
4000	3/16" x 1/8"		24	.333		1.98	19.55		21.53	31.50
4010	1/4" x 1/8"		24	.333		1.75	19.55		21.30	31.50
4020	1/4"		24	.333		2.35	19.55		21.90	32
4030	3/8" x 1/4"		22	.364		2.35	21.50		23.85	34.50
4040	1/2" x 3/8"		22	.364		3.75	21.50		25.25	36
4050	5/8" x 1/2"		21	.381		5.30	22.50		27.80	40
4090	Male connector, 1/8"		26	.308		1.25	18.05		19.30	29
4100	3/16" x 1/8"		24	.333		1.29	19.55		20.84	31
4110	1/4" x 1/8"		24	.333		1.34	19.55		20.89	31
4120	1/4"		24	.333		1.34	19.55		20.89	31
4130	5/16" x 1/8"		23	.348		1.58	20.50		22.08	32.50
4140	5/16" x 1/4"		23	.348		1.79	20.50		22.29	33
4150	3/8" x 1/8"		22	.364		1.87	21.50		23.37	34
4160	3/8" x 1/4"		22	.364		2.15	21.50		23.65	34.50
4170	3/8"		22	.364		2.15	21.50		23.65	34.50
4180	3/8" x 1/2"		22	.364		2.95	21.50		24.45	35.50
4190	1/2" x 3/8"		22	.364		2.86	21.50		24.36	35
4200	1/2"		22	.364		3.21	21.50		24.71	35.50
4210	5/8" x 1/2"		21	.381		3.71	22.50		26.21	38
4240	Male elbow, 1/8"		26	.308		2.58	18.05		20.63	30.50
4250	3/16" x 1/8"		24	.333		2.85	19.55		22.40	32.50
4260	1/4" x 1/8"		24	.333		2.15	19.55		21.70	32
4270	1/4"		24	.333		2.45	19.55		22	32
4280	3/8" x 1/4"		22	.364		3.11	21.50		24.61	35.50
4290	3/8"		22	.364		5.35	21.50		26.85	38

22 11 13 – Facility Water Distribution Piping

22 11 13.25 Pipe/Tube Fittings, Copper	Crew	Daily Output	Labor-Hours	Unit	Material	2015 Bare Costs Labor	Equipment	Total	Total Incl O&P
4300	1/2" x 1/4"	1 Plum	22	.364	Ea.	5.15	21.50		26.65	37.50
4310	1/2" x 3/8"		22	.364		8.10	21.50		29.60	41
4340	Female elbow, 1/8"		26	.308		5.55	18.05		23.60	33.50
4350	1/4" x 1/8"		24	.333		3.15	19.55		22.70	33
4360	1/4"		24	.333		4.13	19.55		23.68	34
4370	3/8" x 1/4"		22	.364		4.07	21.50		25.57	36.50
4380	1/2" x 3/8"		22	.364		6.95	21.50		28.45	39.50
4390	1/2"		22	.364		9.35	21.50		30.85	42.50
4420	Male run tee, 1/4" x 1/8"		15	.533		3.86	31.50		35.36	52
4430	5/16" x 1/8"		15	.533		6.65	31.50		38.15	55
4440	3/8" x 1/4"		15	.533		5.95	31.50		37.45	54
4480	Male branch tee, 1/4" x 1/8"		15	.533		4.47	31.50		35.97	52.50
4490	1/4"		15	.533		4.81	31.50		36.31	53
4500	3/8" x 1/4"		15	.533		5.60	31.50		37.10	53.50
4510	1/2" x 3/8"		15	.533		8.70	31.50		40.20	57
4520	1/2"		15	.533		13.95	31.50		45.45	63
4800	Flare joint fittings								
4810	Refrigeration fittings								
4820	Flare joint nuts and labor not incl. in price. Add 1 nut per jnt.								
4830	90° Elbow, 1/4"				Ea.	2.32			2.32	2.55
4840	3/8"					3.32			3.32	3.65
4850	1/2"					6.20			6.20	6.85
4860	5/8"					9.10			9.10	10
4870	3/4"					13.80			13.80	15.15
5030	Tee, 1/4"					2.99			2.99	3.29
5040	5/16"					3.58			3.58	3.94
5050	3/8"					3.77			3.77	4.15
5060	1/2"					5.85			5.85	6.45
5070	5/8"					8.05			8.05	8.85
5080	3/4"					8.50			8.50	9.35
5140	Union, 3/16"					1.37			1.37	1.51
5150	1/4"					.94			.94	1.03
5160	5/16"					1.70			1.70	1.87
5170	3/8"					1.58			1.58	1.74
5180	1/2"					2.35			2.35	2.59
5190	5/8"					3.78			3.78	4.16
5200	3/4"					10.55			10.55	11.60
5260	Long flare nut, 3/16"	1 Stpi	42	.190		1.74	11.40		13.14	19.10
5270	1/4"		41	.195		.96	11.65		12.61	18.65
5280	5/16"		40	.200		1.74	11.95		13.69	19.95
5290	3/8"		39	.205		1.66	12.25		13.91	20.50
5300	1/2"		38	.211		2.61	12.60		15.21	22
5310	5/8"		37	.216		5.85	12.90		18.75	26
5320	3/4"		34	.235		11.65	14.05		25.70	34
5380	Short flare nut, 3/16"		42	.190		2.88	11.40		14.28	20.50
5390	1/4"		41	.195		.68	11.65		12.33	18.35
5400	5/16"		40	.200		.81	11.95		12.76	18.95
5410	3/8"		39	.205		1.06	12.25		13.31	19.65
5420	1/2"		38	.211		1.58	12.60		14.18	20.50
5430	5/8"		36	.222		2.20	13.30		15.50	22.50
5440	3/4"		34	.235		7.35	14.05		21.40	29
5500	90° Elbow flare by MIPS, 1/4"					2.59			2.59	2.85
5510	3/8"					2.82			2.82	3.10

139

For customer support on your Mechanical Cost Data, call 877.795.7394.

22 11 13 – Facility Water Distribution Piping

22 11 13.25 Pipe/Tube Fittings, Copper		Crew	Daily Output	Labor-Hours	Unit	Material	2015 Bare Costs Labor	Equipment	Total	Total Incl O&P
5520	1/2"				Ea.	4.97			4.97	5.45
5530	5/8"					9.95			9.95	10.95
5540	3/4"					16.65			16.65	18.30
5600	Flare by FIPS, 1/4"					4.73			4.73	5.20
5610	3/8"					5.15			5.15	5.65
5620	1/2"					7.35			7.35	8.10
5670	Flare by sweat, 1/4"					2.64			2.64	2.90
5680	3/8"					3.43			3.43	3.77
5690	1/2"					7			7	7.70
5700	5/8"					22			22	24
5760	Tee flare by IPS, 1/4"					4.73			4.73	5.20
5770	3/8"					6.65			6.65	7.30
5780	1/2"					11.10			11.10	12.25
5790	5/8"					8.25			8.25	9.05
5850	Connector, 1/4"					1.02			1.02	1.12
5860	3/8"					1.53			1.53	1.68
5870	1/2"					2.49			2.49	2.74
5880	5/8"					4.34			4.34	4.77
5890	3/4"					6.50			6.50	7.15
5950	Seal cap, 1/4"					.84			.84	.92
5960	3/8"					1.01			1.01	1.11
5970	1/2"					1.26			1.26	1.39
5980	5/8"					2.62			2.62	2.88
5990	3/4"					7.65			7.65	8.40
6000	Water service fittings									
6010	Flare joints nut and labor are included in the fitting price.									
6020	90° Elbow, C x C, 3/8"	1 Plum	19	.421	Ea.	44.50	24.50		69	86.50
6030	1/2"		18	.444		44.50	26		70.50	88.50
6040	3/4"		16	.500		57	29.50		86.50	107
6050	1"		15	.533		108	31.50		139.50	166
6080	2"		10	.800		390	47		437	500
6090	90° Elbow, C x MPT, 3/8"		19	.421		34	24.50		58.50	75
6100	1/2"		18	.444		34	26		60	77
6110	3/4"		16	.500		39.50	29.50		69	87.50
6120	1"		15	.533		107	31.50		138.50	166
6130	1-1/4"		13	.615		220	36		256	298
6140	1-1/2"		12	.667		220	39		259	300
6150	2"		10	.800		268	47		315	365
6160	90° Elbow, C x FPT, 3/8"		19	.421		36.50	24.50		61	77.50
6170	1/2"		18	.444		36.50	26		62.50	79.50
6180	3/4"		16	.500		45	29.50		74.50	94
6190	1"		15	.533		104	31.50		135.50	162
6200	1-1/4"		13	.615		108	36		144	174
6210	1-1/2"		12	.667		415	39		454	520
6220	2"		10	.800		600	47		647	730
6230	Tee, C x C x C, 3/8"		13	.615		65	36		101	126
6240	1/2"		12	.667		65	39		104	131
6250	3/4"		11	.727		76.50	42.50		119	149
6260	1"		10	.800		135	47		182	220
6330	Tube nut, C x nut seat, 3/8"		40	.200		13.90	11.75		25.65	33
6340	1/2"		38	.211		13.90	12.35		26.25	34
6350	3/4"		34	.235		13.20	13.80		27	35.50
6360	1"		32	.250		24	14.70		38.70	48.50

For customer support on your Mechanical Cost Data, call 877.795.7394.

22 11 13 – Facility Water Distribution Piping

22 11 13.25 Pipe/Tube Fittings, Copper

		Crew	Daily Output	Labor-Hours	Unit	Material	2015 Bare Costs Labor	2015 Bare Costs Equipment	Total	Total Incl O&P
6380	Coupling, C x C, 3/8"	1 Plum	19	.421	Ea.	42	24.50		66.50	84
6390	1/2"		18	.444		42	26		68	86
6400	3/4"		16	.500		50.50	29.50		80	100
6410	1"		15	.533		93.50	31.50		125	151
6420	1-1/4"		12	.667		151	39		190	226
6430	1-1/2"		12	.667		229	39		268	310
6440	2"		10	.800		345	47		392	450
6450	Adapter, C x FPT, 3/8"		19	.421		32	24.50		56.50	72.50
6460	1/2"		18	.444		32	26		58	74.50
6470	3/4"		16	.500		36.50	29.50		66	84.50
6480	1"		15	.533		78	31.50		109.50	133
6490	1-1/4"		13	.615		101	36		137	167
6500	1-1/2"		12	.667		166	39		205	241
6510	2"		10	.800		212	47		259	305
6520	Adapter, C x MPT, 3/8"		19	.421		27	24.50		51.50	67
6530	1/2"		18	.444		27	26		53	69
6540	3/4"		16	.500		37.50	29.50		67	85.50
6550	1"		15	.533		66	31.50		97.50	120
6560	1-1/4"		13	.615		164	36		200	236
6570	1-1/2"		12	.667		164	39		203	240
6580	2"		10	.800		223	47		270	315
6992	Tube connector fittings, See Section 22 11 13.76 for plastic ftng.									
7000	Insert type Brass/copper, 100 psi @ 180°F, CTS									
7010	Adapter MPT 3/8" x 3/8" CTS	1 Plum	29	.276	Ea.	2.93	16.20		19.13	27.50
7020	1/2" x 1/2"		26	.308		2.95	18.05		21	31
7030	3/4" x 1/2"		26	.308		3.77	18.05		21.82	31.50
7040	3/4" x 3/4"		25	.320		4.35	18.80		23.15	33.50
7050	Adapter CTS 1/2" x 1/2" sweat		24	.333		3.96	19.55		23.51	34
7060	3/4" x 3/4" sweat		22	.364		1.39	21.50		22.89	33.50
7070	Coupler center set 3/8" CTS		25	.320		1.44	18.80		20.24	30
7080	1/2" CTS		23	.348		3.96	20.50		24.46	35.50
7090	3/4" CTS		22	.364		1.39	21.50		22.89	33.50
7100	Elbow 90°, copper 3/8"		25	.320		3.16	18.80		21.96	32
7110	1/2" CTS		23	.348		2.02	20.50		22.52	33
7120	3/4" CTS		22	.364		2.46	21.50		23.96	34.50
7130	Tee copper 3/8" CTS		17	.471		3.75	27.50		31.25	45.50
7140	1/2" CTS		15	.533		2.58	31.50		34.08	50.50
7150	3/4" CTS		14	.571		3.96	33.50		37.46	55
7160	3/8" x 3/8" x 1/2"		16	.500		2.98	29.50		32.48	48
7170	1/2" x 3/8" x 1/2"		15	.533		1.93	31.50		33.43	49.50
7180	3/4" x 1/2" x 3/4"		14	.571		2.98	33.50		36.48	54

22 11 13.27 Pipe/Tube, Grooved Joint for Copper

		Crew	Daily Output	Labor-Hours	Unit	Material	2015 Bare Costs Labor	2015 Bare Costs Equipment	Total	Total Incl O&P
0010	**PIPE/TUBE, GROOVED JOINT FOR COPPER**									
4000	Fittings: coupling material required at joints not incl. in fitting price.									
4001	Add 1 selected coupling, material only, per joint for installed price.									
4010	Coupling, rigid style									
4018	2" diameter	1 Plum	50	.160	Ea.	21	9.40		30.40	37.50
4020	2-1/2" diameter	Q-1	80	.200		24	10.55		34.55	42.50
4022	3" diameter		67	.239		27	12.60		39.60	48.50
4024	4" diameter		50	.320		40	16.90		56.90	69.50
4026	5" diameter		40	.400		68	21		89	107
4028	6" diameter	Q-2	50	.480		89.50	26.50		116	138

22 11 13.27 Pipe/Tube, Grooved Joint for Copper

		Crew	Daily Output	Labor-Hours	Unit	Material	2015 Bare Costs Labor	Equipment	Total	Total Incl O&P
4100	Elbow, 90° or 45°									
4108	2" diameter	1 Plum	25	.320	Ea.	37.50	18.80		56.30	69.50
4110	2-1/2" diameter	Q-1	40	.400		40.50	21		61.50	76.50
4112	3" diameter		33	.485		57	25.50		82.50	101
4114	4" diameter		25	.640		130	34		164	194
4116	5" diameter		20	.800		370	42.50		412.50	470
4118	6" diameter	Q-2	25	.960		590	52.50		642.50	730
4200	Tee									
4208	2" diameter	1 Plum	17	.471	Ea.	61.50	27.50		89	109
4210	2-1/2" diameter	Q-1	27	.593		65	31.50		96.50	119
4212	3" diameter		22	.727		97	38.50		135.50	165
4214	4" diameter		17	.941		213	49.50		262.50	310
4216	5" diameter		13	1.231		595	65		660	755
4218	6" diameter	Q-2	17	1.412		735	77.50		812.50	925
4300	Reducer, concentric									
4310	3" x 2-1/2" diameter	Q-1	35	.457	Ea.	54	24		78	95.50
4312	4" x 2-1/2" diameter		32	.500		110	26.50		136.50	161
4314	4" x 3" diameter		29	.552		110	29		139	165
4316	5" x 3" diameter		25	.640		310	34		344	390
4318	5" x 4" diameter		22	.727		310	38.50		348.50	400
4320	6" x 3" diameter	Q-2	28	.857		335	47		382	440
4322	6" x 4" diameter		26	.923		335	50.50		385.50	445
4324	6" x 5" diameter		24	1		335	55		390	455
4350	Flange, w/groove gasket									
4351	ANSI class 125 and 150									
4355	2" diameter	1 Plum	23	.348	Ea.	162	20.50		182.50	209
4356	2-1/2" diameter	Q-1	37	.432		169	23		192	220
4358	3" diameter		31	.516		177	27.50		204.50	235
4360	4" diameter		23	.696		194	37		231	269
4362	5" diameter		19	.842		260	44.50		304.50	355
4364	6" diameter	Q-2	23	1.043		283	57		340	395

22 11 13.29 Pipe, Fittings and Valves, Copper, Pressed-Joint

		Crew	Daily Output	Labor-Hours	Unit	Material	2015 Bare Costs Labor	Equipment	Total	Total Incl O&P
0010	**PIPE, FITTINGS AND VALVES, COPPER, PRESSED-JOINT**									
0040	Pipe/tube includes coupling & clevis type hanger assy's, 10' O.C.									
0120	Type K									
0130	1/2" diameter	1 Plum	78	.103	L.F.	4.98	6		10.98	14.60
0134	3/4" diameter		74	.108		8.05	6.35		14.40	18.45
0138	1" diameter		66	.121		10.65	7.10		17.75	22.50
0142	1-1/4" diameter		56	.143		12.70	8.40		21.10	26.50
0146	1-1/2" diameter		50	.160		16.80	9.40		26.20	32.50
0150	2" diameter		40	.200		24.50	11.75		36.25	45
0154	2-1/2" diameter	Q-1	60	.267		40	14.10		54.10	65.50
0158	3" diameter		54	.296		54	15.65		69.65	83
0162	4" diameter		38	.421		86	22		108	128
0180	To delete cplgs. & hngrs., 1/2" pipe, subtract					19%	48%			
0184	3/4" -2" pipe, subtract					14%	46%			
0186	2-1/2"-4" pipe, subtract					24%	34%			
0220	Type L									
0230	1/2" diameter	1 Plum	81	.099	L.F.	3.90	5.80		9.70	13.05
0234	3/4" diameter		76	.105		5.45	6.20		11.65	15.35
0238	1" diameter		68	.118		8.10	6.90		15	19.40
0242	1-1/4" diameter		58	.138		10.80	8.10		18.90	24

22 11 Facility Water Distribution

22 11 13 – Facility Water Distribution Piping

22 11 13.29 Pipe, Fittings and Valves, Copper, Pressed-Joint	Crew	Daily Output	Labor-Hours	Unit	Material	2015 Bare Costs Labor	Equipment	Total	Total Incl O&P
0246	1-1/2" diameter	1 Plum	52	.154	L.F.	14.15	9.05		23.20	29.50
0250	2" diameter		42	.190		20	11.20		31.20	39
0254	2-1/2" diameter	Q-1	62	.258		35	13.65		48.65	58.50
0258	3" diameter		56	.286		43.50	15.10		58.60	70.50
0262	4" diameter		39	.410		71.50	21.50		93	112
0280	To delete cplgs. & hngrs., 1/2" - 1"pipe, subtract					21%	52%			
0284	1-1/4"-2" pipe, subtract					17%	46%			
0286	2-1/2"-4" pipe, subtract					23%	35%			
0320	Type M									
0330	1/2" diameter	1 Plum	84	.095	L.F.	3.25	5.60		8.85	12.05
0334	3/4" diameter		78	.103		4.38	6		10.38	13.90
0338	1" diameter		70	.114		6.85	6.70		13.55	17.70
0342	1-1/4" diameter		60	.133		9.25	7.85		17.10	22
0346	1-1/2" diameter		54	.148		12.75	8.70		21.45	27
0350	2" diameter		44	.182		18.55	10.65		29.20	36.50
0354	2-1/2" diameter	Q-1	64	.250		31	13.20		44.20	54
0358	3" diameter		58	.276		40.50	14.55		55.05	66.50
0362	4" diameter		40	.400		66	21		87	105
0380	To delete cplgs. & hngrs., 1/2" pipe, subtract					32%	49%			
0384	3/4"-2" pipe, subtract					21%	46%			
0386	2-1/2"-4" pipe, subtract					25%	36%			
1600	Fittings									
1610	Press joints, copper x copper									
1620	Note: Reducing fittings show most expensive size combination.									
1800	90° elbow, 1/2"	1 Plum	36.60	.219	Ea.	3.11	12.85		15.96	23
1810	3/4"		27.50	.291		5.25	17.10		22.35	32
1820	1"		25.90	.309		10.50	18.15		28.65	39
1830	1-1/4"		20.90	.383		21	22.50		43.50	57.50
1840	1-1/2"		18.30	.437		40	25.50		65.50	83
1850	2"		15.70	.510		55.50	30		85.50	106
1860	2-1/2"	Q-1	25.90	.618		161	32.50		193.50	228
1870	3"		22	.727		203	38.50		241.50	281
1880	4"		16.30	.982		251	52		303	355
2000	45° elbow, 1/2"	1 Plum	36.60	.219		3.61	12.85		16.46	23.50
2010	3/4"		27.50	.291		4.44	17.10		21.54	31
2020	1"		25.90	.309		14.20	18.15		32.35	43
2030	1-1/4"		20.90	.383		20.50	22.50		43	56.50
2040	1-1/2"		18.30	.437		33	25.50		58.50	75
2050	2"		15.70	.510		46	30		76	95.50
2060	2-1/2"	Q-1	25.90	.618		107	32.50		139.50	167
2070	3"		22	.727		152	38.50		190.50	226
2080	4"		16.30	.982		212	52		264	310
2200	Tee, 1/2"	1 Plum	27.50	.291		4.74	17.10		21.84	31
2210	3/4"		20.70	.386		8.35	22.50		30.85	43.50
2220	1"		19.40	.412		15	24		39	53
2230	1-1/4"		15.70	.510		24	30		54	71.50
2240	1-1/2"		13.80	.580		49	34		83	106
2250	2"		11.80	.678		61	40		101	127
2260	2-1/2"	Q-1	19.40	.825		199	43.50		242.50	285
2270	3"		16.50	.970		249	51		300	350
2280	4"		12.20	1.311		350	69.50		419.50	490
2400	Tee, reducing on the outlet									
2410	3/4"	1 Plum	20.70	.386	Ea.	7	22.50		29.50	42

22 11 13.29 Pipe, Fittings and Valves, Copper, Pressed-Joint		Crew	Daily Output	Labor-Hours	Unit	Material	2015 Bare Costs Labor	Equipment	Total	Total Incl O&P
2420	1″	1 Plum	19.40	.412	Ea.	17.15	24		41.15	55.50
2430	1-1/4″		15.70	.510		24	30		54	71.50
2440	1-1/2″		13.80	.580		52.50	34		86.50	109
2450	2″		11.80	.678		84	40		124	153
2460	2-1/2″	Q-1	19.40	.825		278	43.50		321.50	370
2470	3″		16.50	.970		305	51		356	415
2480	4″		12.20	1.311		420	69.50		489.50	565
2600	Tee, reducing on the run									
2610	3/4″	1 Plum	20.70	.386	Ea.	13.35	22.50		35.85	49
2620	1″		19.40	.412		26	24		50	65
2630	1-1/4″		15.70	.510		50	30		80	100
2640	1-1/2″		13.80	.580		76.50	34		110.50	136
2650	2″		11.80	.678		84	40		124	153
2660	2-1/2″	Q-1	19.40	.825		284	43.50		327.50	380
2670	3″		16.50	.970		340	51		391	455
2680	4″		12.20	1.311		485	69.50		554.50	640
2800	Coupling, 1/2″	1 Plum	36.60	.219		2.80	12.85		15.65	22.50
2810	3/4″		27.50	.291		4.23	17.10		21.33	30.50
2820	1″		25.90	.309		8.55	18.15		26.70	37
2830	1-1/4″		20.90	.383		10.60	22.50		33.10	45.50
2840	1-1/2″		18.30	.437		19.70	25.50		45.20	60.50
2850	2″		15.70	.510		25	30		55	72.50
2860	2-1/2″	Q-1	25.90	.618		80.50	32.50		113	138
2870	3″		22	.727		102	38.50		140.50	171
2880	4″		16.30	.982		142	52		194	235
3000	Union, 1/2″	1 Plum	36.60	.219		23	12.85		35.85	45
3010	3/4″		27.50	.291		30.50	17.10		47.60	59.50
3020	1″		25.90	.309		48.50	18.15		66.65	81
3030	1-1/4″		20.90	.383		69.50	22.50		92	111
3040	1-1/2″		18.30	.437		94.50	25.50		120	143
3050	2″		15.70	.510		154	30		184	214
3200	Adapter, tube to MPT									
3210	1/2″	1 Plum	15.10	.530	Ea.	3.65	31		34.65	51
3220	3/4″		13.60	.588		6.60	34.50		41.10	59.50
3230	1″		11.60	.690		12.05	40.50		52.55	74.50
3240	1-1/4″		9.90	.808		26.50	47.50		74	101
3250	1-1/2″		9	.889		37	52		89	120
3260	2″		7.90	1.013		71.50	59.50		131	169
3270	2-1/2″	Q-1	13.40	1.194		161	63		224	274
3280	3″		10.60	1.509		203	79.50		282.50	345
3290	4″		7.70	2.078		235	110		345	425
3400	Adapter, tube to FPT									
3410	1/2″	1 Plum	15.10	.530	Ea.	4.40	31		35.40	52
3420	3/4″		13.60	.588		7.15	34.50		41.65	60
3430	1″		11.60	.690		13.15	40.50		53.65	75.50
3440	1-1/4″		9.90	.808		29.50	47.50		77	104
3450	1-1/2″		9	.889		43	52		95	127
3460	2″		7.90	1.013		73.50	59.50		133	171
3470	2-1/2″	Q-1	13.40	1.194		216	63		279	335
3480	3″		10.60	1.509		283	79.50		362.50	430
3490	4″		7.70	2.078		365	110		475	565
3600	Flange									
3620	1″	1 Plum	36.20	.221	Ea.	123	12.95		135.95	155

144

For customer support on your Mechanical Cost Data, call 877.795.7394.

22 11 13 – Facility Water Distribution Piping

22 11 13.29 Pipe, Fittings and Valves, Copper, Pressed-Joint		Crew	Daily Output	Labor-Hours	Unit	Material	2015 Bare Costs Labor	Equipment	Total	Total Incl O&P
3630	1-1/4"	1 Plum	29.30	.273	Ea.	178	16.05		194.05	220
3640	1-1/2"		25.60	.313		197	18.35		215.35	245
3650	2"		22	.364		221	21.50		242.50	275
3660	2-1/2"	Q-1	36.20	.442		247	23.50		270.50	310
3670	3"		30.80	.519		299	27.50		326.50	370
3680	4"		22.80	.702		335	37		372	425
3800	Cap, 1/2"	1 Plum	53.10	.151		5.95	8.85		14.80	19.90
3810	3/4"		39.80	.201		10.05	11.80		21.85	29
3820	1"		37.50	.213		15.50	12.50		28	36
3830	1-1/4"		30.40	.263		17.55	15.45		33	43
3840	1-1/2"		26.60	.301		27	17.65		44.65	56.50
3850	2"		22.80	.351		33	20.50		53.50	67.50
3860	2-1/2"	Q-1	37.50	.427		106	22.50		128.50	150
3870	3"		31.90	.502		134	26.50		160.50	187
3880	4"		23.60	.678		178	36		214	250
4000	Reducer									
4010	3/4"	1 Plum	27.50	.291	Ea.	12.65	17.10		29.75	40
4020	1"		25.90	.309		22	18.15		40.15	52
4030	1-1/4"		20.90	.383		32.50	22.50		55	70
4040	1-1/2"		18.30	.437		39	25.50		64.50	82
4050	2"		15.70	.510		53	30		83	104
4060	2-1/2"	Q-1	25.90	.618		159	32.50		191.50	225
4070	3"		22	.727		206	38.50		244.50	284
4080	4"		16.30	.982		265	52		317	370
4100	Stub out, 1/2"	1 Plum	50	.160		7.85	9.40		17.25	23
4110	3/4"		35	.229		13.90	13.40		27.30	36
4120	1"		30	.267		19.05	15.65		34.70	44.50
6000	Valves									
6200	Ball valve									
6210	1/2"	1 Plum	25.60	.313	Ea.	43.50	18.35		61.85	75
6220	3/4"		19.20	.417		61	24.50		85.50	104
6230	1"		18.10	.442		75	26		101	122
6240	1-1/4"		14.70	.544		122	32		154	183
6250	1-1/2"		12.80	.625		169	36.50		205.50	241
6260	2"		11	.727		269	42.50		311.50	360
6400	Check valve									
6410	1/2"	1 Plum	30	.267	Ea.	32.50	15.65		48.15	59.50
6420	3/4"		22.50	.356		41.50	21		62.50	77.50
6430	1"		21.20	.377		48	22		70	86.50
6440	1-1/4"		17.20	.465		69.50	27.50		97	118
6450	1-1/2"		15	.533		98.50	31.50		130	157
6460	2"		12.90	.620		180	36.50		216.50	253
6600	Butterfly valve, lug type									
6660	2-1/2"	Q-1	9	1.778	Ea.	154	94		248	310
6670	3"		8	2		188	106		294	365
6680	4"		5	3.200		235	169		404	515

For customer support on your Mechanical Cost Data, call 877.795.7394.

145

22 11 13.44 Pipe, Steel		Crew	Daily Output	Labor-Hours	Unit	Material	2015 Bare Costs Labor	Equipment	Total	Total Incl O&P
0010	**PIPE, STEEL**	R221113-50								
0020	All pipe sizes are to Spec. A-53 unless noted otherwise	R221113-70								
0032	Schedule 10, see Line 22 11 13.48 0500									
0050	Schedule 40, threaded, with couplings, and clevis hanger									
0060	assemblies sized for covering, 10' O.C.									
0540	Black, 1/4" diameter	1 Plum	66	.121	L.F.	6.55	7.10		13.65	17.95
0550	3/8" diameter		65	.123		7.30	7.20		14.50	18.95
0560	1/2" diameter		63	.127		3.32	7.45		10.77	14.90
0570	3/4" diameter		61	.131		3.91	7.70		11.61	15.95
0580	1" diameter		53	.151		5.05	8.85		13.90	18.95
0590	1-1/4" diameter	Q-1	89	.180		6.20	9.50		15.70	21
0600	1-1/2" diameter		80	.200		7.10	10.55		17.65	24
0610	2" diameter		64	.250		9	13.20		22.20	30
0620	2-1/2" diameter		50	.320		14	16.90		30.90	41
0630	3" diameter		43	.372		17.80	19.65		37.45	49
0640	3-1/2" diameter		40	.400		24.50	21		45.50	59
0650	4" diameter		36	.444		27	23.50		50.50	65
0809	A-106, gr. A/B, seamless w/cplgs. & clevis hanger assemblies									
0811	1/4" diameter	1 Plum	66	.121	L.F.	8.60	7.10		15.70	20
0812	3/8" diameter		65	.123		8.55	7.20		15.75	20.50
0813	1/2" diameter		63	.127		9.55	7.45		17	22
0814	3/4" diameter		61	.131		10.95	7.70		18.65	23.50
0815	1" diameter		53	.151		12.60	8.85		21.45	27.50
0816	1-1/4" diameter	Q-1	89	.180		15.10	9.50		24.60	31
0817	1-1/2" diameter		80	.200		21.50	10.55		32.05	40
0819	2" diameter		64	.250		19.05	13.20		32.25	41
0821	2-1/2" diameter		50	.320		23	16.90		39.90	.51
0822	3" diameter		43	.372		30	19.65		49.65	62.50
0823	4" diameter		36	.444		46.50	23.50		70	86.50
1220	To delete coupling & hanger, subtract									
1230	1/4" diam. to 3/4" diam.					31%	56%			
1240	1" diam. to 1-1/2" diam.					23%	51%			
1250	2" diam. to 4" diam.					23%	41%			
1280	All pipe sizes are to Spec. A-53 unless noted otherwise									
1281	Schedule 40, threaded, with couplings and clevis hanger									
1282	assemblies sized for covering, 10' O. C.									
1290	Galvanized, 1/4" diameter	1 Plum	66	.121	L.F.	9.10	7.10		16.20	21
1300	3/8" diameter		65	.123		9.95	7.20		17.15	22
1310	1/2" diameter		63	.127		3.69	7.45		11.14	15.30
1320	3/4" diameter		61	.131		4.24	7.70		11.94	16.30
1330	1" diameter		53	.151		5.75	8.85		14.60	19.70
1340	1-1/4" diameter	Q-1	89	.180		7	9.50		16.50	22
1350	1-1/2" diameter		80	.200		8.05	10.55		18.60	25
1360	2" diameter		64	.250		10.35	13.20		23.55	31.50
1370	2-1/2" diameter		50	.320		16.30	16.90		33.20	43.50
1380	3" diameter		43	.372		20.50	19.65		40.15	52.50
1390	3-1/2" diameter		40	.400		26	21		47	61
1400	4" diameter		36	.444		30	23.50		53.50	68.50
1750	To delete coupling & hanger, subtract									
1760	1/4" diam. to 3/4" diam.					31%	56%			
1770	1" diam. to 1-1/2" diam.					23%	51%			
1780	2" diam. to 4" diam.					23%	41%			

22 11 13.44 Pipe, Steel

	Crew	Daily Output	Labor-Hours	Unit	Material	2015 Bare Costs Labor	Equipment	Total	Total Incl O&P
2000	Welded, sch. 40, on yoke & roll hanger assy's, sized for covering, 10' O.C.								
2040	Black, 1" diameter	Q-15	93	.172	L.F.	4.72	9.10	.62	14.44	19.65
2050	1-1/4" diameter		84	.190		5.90	10.05	.69	16.64	22.50
2060	1-1/2" diameter		76	.211		6.45	11.10	.76	18.31	24.50
2070	2" diameter		61	.262		7.85	13.85	.95	22.65	30.50
2080	2-1/2" diameter		47	.340		12.75	18	1.23	31.98	42.50
2090	3" diameter		43	.372		15.45	19.65	1.34	36.44	48
2100	3-1/2" diameter		39	.410		18.55	21.50	1.48	41.53	54.50
2110	4" diameter		37	.432		21.50	23	1.56	46.06	60
2120	5" diameter		32	.500		34.50	26.50	1.81	62.81	80
2130	6" diameter	Q-16	36	.667		43	36.50	1.60	81.10	104
2140	8" diameter		29	.828		68.50	45.50	1.99	115.99	146
2150	10" diameter		24	1		89.50	55	2.40	146.90	183
2160	12" diameter		19	1.263		105	69	3.03	177.03	224
2170	14" diameter, (two rod roll type hanger for 14" diam. and up)		15	1.600		98.50	87.50	3.84	189.84	244
2180	16" diameter, (two rod roll type hanger)		13	1.846		148	101	4.43	253.43	320
2190	18" diameter, (two rod roll type hanger)		11	2.182		141	120	5.25	266.25	340
2200	20" diameter, (two rod roll type hanger)		9	2.667		143	146	6.40	295.40	385
2220	24" diameter, (two rod roll type hanger)		8	3		183	164	7.20	354.20	460
2345	Sch. 40, A-53, gr. A/B, ERW, welded w/hngrs.								
2346	2" diameter	Q-15	61	.262	L.F.	7.70	13.85	.95	22.50	30.50
2347	2-1/2" diameter		47	.340		10.10	18	1.23	29.33	39.50
2348	3" diameter		43	.372		12.60	19.65	1.34	33.59	45
2349	4" diameter		38	.421		17.15	22	1.52	40.67	54
2350	6" diameter	Q-16	37	.649		28	35.50	1.56	65.06	86
2351	8" diameter		29	.828		67.50	45.50	1.99	114.99	145
2352	10" diameter		24	1		88	55	2.40	145.40	182
2560	To delete hanger, subtract								
2570	1" diam. to 1-1/2" diam.					15%	34%		
2580	2" diam. to 3-1/2" diam.					9%	21%		
2590	4" diam. to 12" diam.					5%	12%		
2596	14" diam. to 24" diam.					3%	10%		
2640	Galvanized, 1" diameter	Q-15	93	.172	L.F.	6.45	9.10	.62	16.17	21.50
2650	1-1/4" diameter		84	.190		8.25	10.05	.69	18.99	25
2660	1-1/2" diameter		76	.211		9.20	11.10	.76	21.06	28
2670	2" diameter		61	.262		11.65	13.85	.95	26.45	35
2680	2-1/2" diameter		47	.340		16	18	1.23	35.23	46
2690	3" diameter		43	.372		21	19.65	1.34	41.99	54
2700	3-1/2" diameter		39	.410		25.50	21.50	1.48	48.48	62
2710	4" diameter		37	.432		30	23	1.56	54.56	69
2720	5" diameter		32	.500		56.50	26.50	1.81	84.81	104
2730	6" diameter	Q-16	36	.667		59.50	36.50	1.60	97.60	122
2740	8" diameter		29	.828		66	45.50	1.99	113.49	143
2750	10" diameter		24	1		105	55	2.40	162.40	200
2760	12" diameter		19	1.263		135	69	3.03	207.03	257
3160	To delete hanger, subtract								
3170	1" diam. to 1-1/2" diam.					30%	34%		
3180	2" diam. to 3-1/2" diam.					19%	21%		
3190	4" diam. to 12" diam.					10%	12%		
3250	Flanged, 150 lb. weld neck, on yoke & roll hangers								
3260	sized for covering, 10' O.C.								
3290	Black, 1" diameter	Q-15	70	.229	L.F.	9.90	12.10	.83	22.83	30
3300	1-1/4" diameter		64	.250		11.10	13.20	.90	25.20	33

22 11 13 – Facility Water Distribution Piping

22 11 13.44 Pipe, Steel	Crew	Daily Output	Labor-Hours	Unit	Material	2015 Bare Costs Labor	Equipment	Total	Total Incl O&P
3310	1-1/2" diameter	Q-15	58	.276	L.F.	11.65	14.55	1	27.20	36
3320	2" diameter		45	.356		13.90	18.80	1.28	33.98	45
3330	2-1/2" diameter		36	.444		19.50	23.50	1.60	44.60	59
3340	3" diameter		32	.500		23	26.50	1.81	51.31	67
3350	3-1/2" diameter		29	.552		28	29	1.99	58.99	77
3360	4" diameter		26	.615		31	32.50	2.22	65.72	85.50
3370	5" diameter		21	.762		49.50	40.50	2.75	92.75	119
3380	6" diameter	Q-16	25	.960		59	52.50	2.30	113.80	147
3390	8" diameter		19	1.263		96.50	69	3.03	168.53	214
3400	10" diameter		16	1.500		138	82	3.60	223.60	279
3410	12" diameter		14	1.714		165	94	4.11	263.11	330
3470	For 300 lb. flanges, add					63%				
3480	For 600 lb. flanges, add					310%				
3960	To delete flanges & hanger, subtract									
3970	1" diam. to 2" diam.					76%	65%			
3980	2-1/2" diam. to 4" diam.					62%	59%			
3990	5" diam. to 12" diam.					60%	46%			
4040	Galvanized, 1" diameter	Q-15	70	.229	L.F.	11.65	12.10	.83	24.58	32
4050	1-1/4" diameter		64	.250		13.45	13.20	.90	27.55	35.50
4060	1-1/2" diameter		58	.276		14.40	14.55	1	29.95	39
4070	2" diameter		45	.356		16.90	18.80	1.28	36.98	48.50
4080	2-1/2" diameter		36	.444		23	23.50	1.60	48.10	62.50
4090	3" diameter		32	.500		28	26.50	1.81	56.31	72.50
4100	3-1/2" diameter		29	.552		33.50	29	1.99	64.49	83
4110	4" diameter		26	.615		39.50	32.50	2.22	74.22	95
4120	5" diameter		21	.762		71	40.50	2.75	114.25	143
4130	6" diameter	Q-16	25	.960		76	52.50	2.30	130.80	166
4140	8" diameter		19	1.263		84	69	3.03	156.03	200
4150	10" diameter		16	1.500		139	82	3.60	224.60	281
4160	12" diameter		14	1.714		165	94	4.11	263.11	330
4220	For 300 lb. flanges, add					65%				
4240	For 600 lb. flanges, add					350%				
4660	To delete flanges & hanger, subtract									
4670	1" diam. to 2" diam.					73%	65%			
4680	2-1/2" diam. to 4" diam.					58%	59%			
4690	5" diam. to 12" diam.					43%	46%			
4750	Schedule 80, threaded, with couplings, and clevis hanger assemblies									
4760	sized for covering, 10' O.C.									
4790	Black, 1/4" diameter	1 Plum	54	.148	L.F.	10.95	8.70		19.65	25
4800	3/8" diameter		53	.151		13.90	8.85		22.75	28.50
4810	1/2" diameter		52	.154		4.77	9.05		13.82	18.90
4820	3/4" diameter		50	.160		5.45	9.40		14.85	20
4830	1" diameter		45	.178		7.20	10.45		17.65	23.50
4840	1-1/4" diameter	Q-1	75	.213		9.45	11.25		20.70	27.50
4850	1-1/2" diameter		69	.232		10.65	12.25		22.90	30.50
4860	2" diameter		56	.286		14.10	15.10		29.20	38.50
4870	2-1/2" diameter		44	.364		21	19.20		40.20	52
4880	3" diameter		38	.421		27	22		49	63
4890	3-1/2" diameter		35	.457		33	24		57	72.50
4900	4" diameter		32	.500		38.50	26.50		65	82.50
5061	A-106, gr. A/B seamless with cplgs. & clevis hanger assemblies, 1/4" diam.	1 Plum	63	.127		10.55	7.45		18	23
5062	3/8" diameter		62	.129		12.25	7.55		19.80	25
5063	1/2" diameter		61	.131		10.15	7.70		17.85	23

For customer support on your Mechanical Cost Data, call 877.795.7394.

22 11 13 – Facility Water Distribution Piping

22 11 13.44 Pipe, Steel		Crew	Daily Output	Labor-Hours	Unit	Material	2015 Bare Costs Labor	Equipment	Total	Total Incl O&P
5064	3/4" diameter	1 Plum	57	.140	L.F.	11.75	8.25		20	25.50
5065	1" diameter	↓	51	.157		13.75	9.20		22.95	29
5066	1-1/4" diameter	Q-1	85	.188		18.15	9.95		28.10	35
5067	1-1/2" diameter		77	.208		26.50	11		37.50	45.50
5071	2" diameter		61	.262		32.50	13.85		46.35	56.50
5072	2-1/2" diameter		48	.333		24.50	17.60		42.10	53.50
5073	3" diameter		41	.390		32	20.50		52.50	66
5074	4" diameter	↓	35	.457	↓	46.50	24		70.50	87.50
5430	To delete coupling & hanger, subtract									
5440	1/4" diam. to 1/2" diam.					31%	54%			
5450	3/4" diam. to 1-1/2" diam.					28%	49%			
5460	2" diam. to 4" diam.					21%	40%			
5510	Galvanized, 1/4" diameter	1 Plum	54	.148	L.F.	9.90	8.70		18.60	24
5520	3/8" diameter		53	.151		10.85	8.85		19.70	25.50
5530	1/2" diameter		52	.154		5.30	9.05		14.35	19.50
5540	3/4" diameter		50	.160		6.40	9.40		15.80	21.50
5550	1" diameter	↓	45	.178		8.50	10.45		18.95	25
5560	1-1/4" diameter	Q-1	75	.213		11.35	11.25		22.60	29.50
5570	1-1/2" diameter		69	.232		13.15	12.25		25.40	33
5580	2" diameter		56	.286		17.45	15.10		32.55	42
5590	2-1/2" diameter		44	.364		26	19.20		45.20	57.50
5600	3" diameter		38	.421		34	22		56	70.50
5610	3-1/2" diameter		35	.457		53	24		77	95
5620	4" diameter	↓	32	.500	↓	40	26.50		66.50	84
5930	To delete coupling & hanger, subtract									
5940	1/4" diam. to 1/2" diam.					31%	54%			
5950	3/4" diam. to 1-1/2" diam.					28%	49%			
5960	2" diam. to 4" diam.					21%	40%			
6000	Welded, on yoke & roller hangers									
6010	sized for covering, 10' O.C.									
6040	Black, 1" diameter	Q-15	85	.188	L.F.	7.35	9.95	.68	17.98	24
6050	1-1/4" diameter		79	.203		9.50	10.70	.73	20.93	27.50
6060	1-1/2" diameter		72	.222		10.75	11.75	.80	23.30	30.50
6070	2" diameter		57	.281		13.55	14.85	1.01	29.41	38.50
6080	2-1/2" diameter		44	.364		20	19.20	1.31	40.51	52.50
6090	3" diameter		40	.400		25.50	21	1.44	47.94	61.50
6100	3-1/2" diameter		34	.471		31	25	1.70	57.70	73.50
6110	4" diameter		33	.485		36	25.50	1.75	63.25	80
6120	5" diameter, A-106B	↓	26	.615		51.50	32.50	2.22	86.22	108
6130	6" diameter, A-106B	Q-16	30	.800		70.50	44	1.92	116.42	146
6140	8" diameter, A-106B		25	.960		106	52.50	2.30	160.80	199
6150	10" diameter, A-106B		20	1.200		172	65.50	2.88	240.38	292
6160	12" diameter, A-106B	↓	15	1.600	↓	385	87.50	3.84	476.34	560
6540	To delete hanger, subtract									
6550	1" diam. to 1-1/2" diam.					30%	14%			
6560	2" diam. to 3" diam.					23%	9%			
6570	3-1/2" diam. to 5" diam.					12%	6%			
6580	6" diam. to 12" diam.					10%	4%			
6610	Galvanized, 1" diameter	Q-15	85	.188	L.F.	8.60	9.95	.68	19.23	25
6650	1-1/4" diameter		79	.203		11.45	10.70	.73	22.88	29.50
6660	1-1/2" diameter		72	.222		13.20	11.75	.80	25.75	33
6670	2" diameter		57	.281		16.85	14.85	1.01	32.71	42
6680	2-1/2" diameter	↓	44	.364		25	19.20	1.31	45.51	58

22 11 13.44 Pipe, Steel		Crew	Daily Output	Labor-Hours	Unit	Material	2015 Bare Costs Labor	Equipment	Total	Total Incl O&P
6690	3" diameter	Q-15	40	.400	L.F.	32.50	21	1.44	54.94	69
6700	3-1/2" diameter		34	.471		51.50	25	1.70	78.20	96
6710	4" diameter		33	.485		37.50	25.50	1.75	64.75	82
6720	5" diameter, A-106B		26	.615		54	32.50	2.22	88.72	111
6730	6" diameter, A-106B	Q-16	30	.800		74	44	1.92	119.92	149
6740	8" diameter, A-106B		25	.960		139	52.50	2.30	193.80	235
6750	10" diameter, A-106B		20	1.200		256	65.50	2.88	324.38	385
6760	12" diameter, A-106B		15	1.600		515	87.50	3.84	606.34	700
7150	To delete hanger, subtract									
7160	1" diam. to 1-1/2" diam.					26%	14%			
7170	2" diam. to 3" diam.					16%	9%			
7180	3-1/2" diam. to 5" diam.					10%	6%			
7190	6" diam. to 12" diam.					7%	4%			
7250	Flanged, 300 lb. weld neck, on yoke & roll hangers									
7260	sized for covering, 10' O.C.									
7290	Black, 1" diameter	Q-15	66	.242	L.F.	13.75	12.80	.88	27.43	35.50
7300	1-1/4" diameter		61	.262		15.95	13.85	.95	30.75	39.50
7310	1-1/2" diameter		54	.296		17.20	15.65	1.07	33.92	43.50
7320	2" diameter		42	.381		22	20	1.38	43.38	56
7330	2-1/2" diameter		33	.485		29.50	25.50	1.75	56.75	73
7340	3" diameter		29	.552		35	29	1.99	65.99	84.50
7350	3-1/2" diameter		24	.667		47	35	2.41	84.41	107
7360	4" diameter		23	.696		52	37	2.51	91.51	115
7370	5" diameter		19	.842		75.50	44.50	3.04	123.04	153
7380	6" diameter	Q-16	23	1.043		95	57	2.50	154.50	193
7390	8" diameter		17	1.412		147	77.50	3.39	227.89	283
7400	10" diameter		14	1.714		247	94	4.11	345.11	420
7410	12" diameter		12	2		480	110	4.80	594.80	700
7470	For 600 lb. flanges, add					100%				
7940	To delete flanges & hanger, subtract									
7950	1" diam. to 1-1/2" diam.					75%	66%			
7960	2" diam. to 3" diam.					62%	60%			
7970	3-1/2" diam. to 5" diam.					54%	66%			
7980	6" diam. to 12" diam.					55%	62%			
8040	Galvanized, 1" diameter	Q-15	66	.242	L.F.	15.05	12.80	.88	28.73	37
8050	1-1/4" diameter		61	.262		17.85	13.85	.95	32.65	41.50
8060	1-1/2" diameter		54	.296		19.65	15.65	1.07	36.37	46
8070	2" diameter		42	.381		25	20	1.38	46.38	59.50
8080	2-1/2" diameter		33	.485		34.50	25.50	1.75	61.75	78.50
8090	3" diameter		29	.552		42	29	1.99	72.99	92
8100	3-1/2" diameter		24	.667		67	35	2.41	104.41	130
8110	4" diameter		23	.696		53.50	37	2.51	93.01	117
8120	5" diameter, A-106B		19	.842		78	44.50	3.04	125.54	156
8130	6" diameter, A-106B	Q-16	23	1.043		98	57	2.50	157.50	197
8140	8" diameter, A-106B		17	1.412		180	77.50	3.39	260.89	320
8150	10" diameter, A-106B		14	1.714		330	94	4.11	428.11	510
8160	12" diameter, A-106B		12	2		610	110	4.80	724.80	840
8240	For 600 lb. flanges, add					100%				
8300	To delete flanges & hangers, subtract									
8310	1" diam. to 1-1/2" diam.					72%	66%			
8320	2" diam. to 3" diam.					59%	60%			
8330	3-1/2" diam. to 5" diam.					51%	66%			
8340	6" diam. to 12" diam.					49%	62%			

22 11 13 – Facility Water Distribution Piping

22 11 13.44 Pipe, Steel	Crew	Daily Output	Labor-Hours	Unit	Material	2015 Bare Costs Labor	Equipment	Total	Total Incl O&P
9000 Threading pipe labor, one end, all schedules through 80									
9010 1/4" through 3/4" pipe size	1 Plum	80	.100	Ea.		5.85		5.85	8.85
9020 1" through 2" pipe size		73	.110			6.45		6.45	9.70
9030 2-1/2" pipe size		53	.151			8.85		8.85	13.40
9040 3" pipe size		50	.160			9.40		9.40	14.20
9050 3-1/2" pipe size	Q-1	89	.180			9.50		9.50	14.35
9060 4" pipe size		73	.219			11.60		11.60	17.50
9070 5" pipe size		53	.302			15.95		15.95	24
9080 6" pipe size		46	.348			18.40		18.40	28
9090 8" pipe size		29	.552			29		29	44
9100 10" pipe size		21	.762			40.50		40.50	61
9110 12" pipe size		13	1.231			65		65	98
9120 Cutting pipe labor, one cut									
9124 Shop fabrication, machine cut									
9126 Schedule 40, straight pipe									
9128 2" pipe size or less	1 Stpi	62	.129	Ea.		7.70		7.70	11.65
9130 2-1/2" pipe size		56	.143			8.55		8.55	12.90
9132 3" pipe size		42	.190			11.40		11.40	17.20
9134 4" pipe size		31	.258			15.40		15.40	23.50
9136 5" pipe size		26	.308			18.40		18.40	28
9138 6" pipe size		19	.421			25		25	38
9140 8" pipe size		14	.571			34		34	51.50
9142 10" pipe size		10	.800			48		48	72
9144 12" pipe size		7	1.143			68.50		68.50	103
9146 14" pipe size	Q-5	10.50	1.524			82		82	124
9148 16" pipe size		8.60	1.860			100		100	151
9150 18" pipe size		7	2.286			123		123	186
9152 20" pipe size		5.80	2.759			148		148	224
9154 24" pipe size		4	4			215		215	325
9160 Schedule 80, straight pipe									
9164 2" pipe size or less	1 Stpi	42	.190	Ea.		11.40		11.40	17.20
9166 2-1/2" pipe size		37	.216			12.90		12.90	19.50
9168 3" pipe size		31	.258			15.40		15.40	23.50
9170 4" pipe size		23	.348			21		21	31.50
9172 5" pipe size		18	.444			26.50		26.50	40
9174 6" pipe size		14.60	.548			32.50		32.50	49.50
9176 8" pipe size		10	.800			48		48	72
9178 10" pipe size	Q-5	14	1.143			61.50		61.50	93
9180 12" pipe size	"	10	1.600			86		86	130
9200 Welding labor per joint									
9210 Schedule 40,									
9230 1/2" pipe size	Q-15	32	.500	Ea.		26.50	1.81	28.31	42
9240 3/4" pipe size		27	.593			31.50	2.14	33.64	50
9250 1" pipe size		23	.696			37	2.51	39.51	58.50
9260 1-1/4" pipe size		20	.800			42.50	2.89	45.39	67
9270 1-1/2" pipe size		19	.842			44.50	3.04	47.54	70.50
9280 2" pipe size		16	1			53	3.61	56.61	84
9290 2-1/2" pipe size		13	1.231			65	4.44	69.44	103
9300 3" pipe size		12	1.333			70.50	4.81	75.31	111
9310 4" pipe size		10	1.600			84.50	5.80	90.30	134
9320 5" pipe size		9	1.778			94	6.40	100.40	149
9330 6" pipe size		8	2			106	7.20	113.20	168
9340 8" pipe size		5	3.200			169	11.55	180.55	268

For customer support on your Mechanical Cost Data, call 877.795.7394.

151

22 11 13.44 Pipe, Steel

		Crew	Daily Output	Labor-Hours	Unit	Material	2015 Bare Costs Labor	Equipment	Total	Total Incl O&P
9350	10" pipe size	Q-15	4	4	Ea.		211	14.45	225.45	335
9360	12" pipe size		3	5.333			282	19.25	301.25	445
9370	14" pipe size		2.60	6.154			325	22	347	515
9380	16" pipe size		2.20	7.273			385	26.50	411.50	610
9390	18" pipe size		2	8			425	29	454	670
9400	20" pipe size		1.80	8.889			470	32	502	745
9410	22" pipe size		1.70	9.412			495	34	529	790
9420	24" pipe size		1.50	10.667			565	38.50	603.50	895
9450	Schedule 80,									
9460	1/2" pipe size	Q-15	27	.593	Ea.	31.50	2.14		33.64	50
9470	3/4" pipe size		23	.696		37	2.51		39.51	58.50
9480	1" pipe size		20	.800		42.50	2.89		45.39	67
9490	1-1/4" pipe size		19	.842		44.50	3.04		47.54	70.50
9500	1-1/2" pipe size		18	.889		47	3.21		50.21	74.50
9510	2" pipe size		15	1.067		56.50	3.85		60.35	89
9520	2-1/2" pipe size		12	1.333		70.50	4.81		75.31	111
9530	3" pipe size		11	1.455		77	5.25		82.25	122
9540	4" pipe size		8	2		106	7.20		113.20	168
9550	5" pipe size		6	2.667		141	9.65		150.65	224
9560	6" pipe size		5	3.200		169	11.55		180.55	268
9570	8" pipe size		4	4		211	14.45		225.45	335
9580	10" pipe size		3	5.333		282	19.25		301.25	445
9590	12" pipe size		2	8		425	29		454	670
9600	14" pipe size	Q-16	2.60	9.231		505	22		527	790
9610	16" pipe size		2.30	10.435		570	25		595	895
9620	18" pipe size		2	12		655	29		684	1,025
9630	20" pipe size		1.80	13.333		730	32		762	1,125
9640	22" pipe size		1.60	15		820	36		856	1,300
9650	24" pipe size		1.50	16		875	38.50		913.50	1,375

22 11 13.45 Pipe Fittings, Steel, Threaded

		Crew	Daily Output	Labor-Hours	Unit	Material	2015 Bare Costs Labor	Equipment	Total	Total Incl O&P
0010	**PIPE FITTINGS, STEEL, THREADED**									
0020	Cast Iron									
0040	Standard weight, black									
0060	90° Elbow, straight									
0070	1/4"	1 Plum	16	.500	Ea.	9.70	29.50		39.20	55
0080	3/8"		16	.500		14.05	29.50		43.55	60
0090	1/2"		15	.533		6.15	31.50		37.65	54.50
0100	3/4"		14	.571		6.40	33.50		39.90	57.50
0110	1"		13	.615		7.60	36		43.60	63
0120	1-1/4"	Q-1	22	.727		10.75	38.50		49.25	70
0130	1-1/2"		20	.800		14.90	42.50		57.40	80.50
0140	2"		18	.889		23.50	47		70.50	96.50
0150	2-1/2"		14	1.143		56	60.50		116.50	153
0160	3"		10	1.600		91.50	84.50		176	229
0170	3-1/2"		8	2		248	106		354	435
0180	4"		6	2.667		169	141		310	400
0250	45° Elbow, straight									
0260	1/4"	1 Plum	16	.500	Ea.	11.95	29.50		41.45	57.50
0270	3/8"		16	.500		12.85	29.50		42.35	58.50
0280	1/2"		15	.533		9.35	31.50		40.85	58
0300	3/4"		14	.571		9.50	33.50		43	61
0320	1"		13	.615		11.10	36		47.10	66.50

152

For customer support on your Mechanical Cost Data, call 877.795.7394.

22 11 13 – Facility Water Distribution Piping

22 11 13.45 Pipe Fittings, Steel, Threaded		Crew	Daily Output	Labor-Hours	Unit	Material	2015 Bare Costs Labor	Equipment	Total	Total Incl O&P
0330	1-1/4"	Q-1	22	.727	Ea.	14.95	38.50		53.45	74.50
0340	1-1/2"		20	.800		24.50	42.50		67	91
0350	2"		18	.889		28.50	47		75.50	103
0360	2-1/2"		14	1.143		74	60.50		134.50	173
0370	3"		10	1.600		118	84.50		202.50	257
0380	3-1/2"		8	2		271	106		377	460
0400	4"		6	2.667		244	141		385	480
0500	Tee, straight									
0510	1/4"	1 Plum	10	.800	Ea.	15.20	47		62.20	87.50
0520	3/8"		10	.800		14.75	47		61.75	87.50
0530	1/2"		9	.889		9.60	52		61.60	89.50
0540	3/4"		9	.889		11.20	52		63.20	91.50
0550	1"		8	1		9.95	58.50		68.45	99.50
0560	1-1/4"	Q-1	14	1.143		18.15	60.50		78.65	111
0570	1-1/2"		13	1.231		23.50	65		88.50	124
0580	2"		11	1.455		33	77		110	152
0590	2-1/2"		9	1.778		85.50	94		179.50	236
0600	3"		6	2.667		131	141		272	355
0610	3-1/2"		5	3.200		264	169		433	545
0620	4"		4	4		264	211		475	610
0660	Tee, reducing, run or outlet									
0661	1/2"	1 Plum	9	.889	Ea.	23.50	52		75.50	105
0662	3/4"		9	.889		20.50	52		72.50	102
0663	1"		8	1		23.50	58.50		82	114
0664	1-1/4"	Q-1	14	1.143		32.50	60.50		93	127
0665	1-1/2"		13	1.231		38.50	65		103.50	141
0666	2"		11	1.455		60	77		137	182
0667	2-1/2"		9	1.778		112	94		206	265
0668	3"		6	2.667		169	141		310	400
0669	3-1/2"		5	3.200		400	169		569	695
0670	4"		4	4		445	211		656	810
0674	Reducer, concentric									
0675	3/4"	1 Plum	18	.444	Ea.	17.20	26		43.20	58.50
0676	1"	"	15	.533		11.65	31.50		43.15	60.50
0677	1-1/4"	Q-1	26	.615		34.50	32.50		67	87
0678	1-1/2"		24	.667		54.50	35		89.50	113
0679	2"		21	.762		62	40.50		102.50	130
0680	2-1/2"		18	.889		88.50	47		135.50	168
0681	3"		14	1.143		139	60.50		199.50	244
0682	3-1/2"		12	1.333		237	70.50		307.50	365
0683	4"		10	1.600		258	84.50		342.50	410
0687	Reducer, eccentric									
0688	3/4"	1 Plum	16	.500	Ea.	36	29.50		65.50	84.50
0689	1"	"	14	.571		39	33.50		72.50	93.50
0690	1-1/4"	Q-1	25	.640		62.50	34		96.50	120
0691	1-1/2"		22	.727		82.50	38.50		121	149
0692	2"		20	.800		119	42.50		161.50	195
0693	2-1/2"		16	1		164	53		217	260
0694	3"		12	1.333		258	70.50		328.50	390
0695	3-1/2"		10	1.600		360	84.50		444.50	530
0696	4"		9	1.778		465	94		559	655
0700	Standard weight, galvanized cast iron									
0720	90° Elbow, straight									

22 11 13.45 Pipe Fittings, Steel, Threaded		Crew	Daily Output	Labor-Hours	Unit	Material	2015 Bare Costs Labor	Equipment	Total	Total Incl O&P
0730	1/4"	1 Plum	16	.500	Ea.	15.95	29.50		45.45	62
0740	3/8"		16	.500		15.95	29.50		45.45	62
0750	1/2"		15	.533		18.20	31.50		49.70	67.50
0760	3/4"		14	.571		17.85	33.50		51.35	70
0770	1"		13	.615		20.50	36		56.50	77
0780	1-1/4"	Q-1	22	.727		32	38.50		70.50	93.50
0790	1-1/2"		20	.800		44	42.50		86.50	113
0800	2"		18	.889		65	47		112	143
0810	2-1/2"		14	1.143		134	60.50		194.50	238
0820	3"		10	1.600		203	84.50		287.50	350
0830	3-1/2"		8	2		325	106		431	515
0840	4"		6	2.667		370	141		511	625
0900	45° Elbow, straight									
0910	1/4"	1 Plum	16	.500	Ea.	18.15	29.50		47.65	64.50
0920	3/8"		16	.500		18.15	29.50		47.65	64.50
0930	1/2"		15	.533		18.15	31.50		49.65	67.50
0940	3/4"		14	.571		21	33.50		54.50	73.50
0950	1"		13	.615		25.50	36		61.50	82.50
0960	1-1/4"	Q-1	22	.727		38	38.50		76.50	100
0970	1-1/2"		20	.800		54	42.50		96.50	124
0980	2"		18	.889		76	47		123	155
0990	2-1/2"		14	1.143		122	60.50		182.50	225
1000	3"		10	1.600		238	84.50		322.50	390
1010	3-1/2"		8	2		435	106		541	640
1020	4"		6	2.667		435	141		576	695
1100	Tee, straight									
1110	1/4"	1 Plum	10	.800	Ea.	16.75	47		63.75	89.50
1120	3/8"		10	.800		19.30	47		66.30	92
1130	1/2"		9	.889		18.60	52		70.60	99.50
1140	3/4"		9	.889		25.50	52		77.50	107
1150	1"		8	1		28	58.50		86.50	119
1160	1-1/4"	Q-1	14	1.143		49	60.50		109.50	145
1170	1-1/2"		13	1.231		64.50	65		129.50	169
1180	2"		11	1.455		80.50	77		157.50	205
1190	2-1/2"		9	1.778		164	94		258	320
1200	3"		6	2.667		365	141		506	615
1210	3-1/2"		5	3.200		460	169		629	760
1220	4"		4	4		515	211		726	890
1300	Extra heavy weight, black									
1310	Couplings, steel straight									
1320	1/4"	1 Plum	19	.421	Ea.	3.77	24.50		28.27	41.50
1330	3/8"		19	.421		4.10	24.50		28.60	42
1340	1/2"		19	.421		5.55	24.50		30.05	43.50
1350	3/4"		18	.444		5.90	26		31.90	46
1360	1"		15	.533		7.55	31.50		39.05	56
1370	1-1/4"	Q-1	26	.615		12.10	32.50		44.60	62.50
1380	1-1/2"		24	.667		12.10	35		47.10	66.50
1390	2"		21	.762		18.45	40.50		58.95	81.50
1400	2-1/2"		18	.889		27.50	47		74.50	101
1410	3"		14	1.143		32.50	60.50		93	127
1420	3-1/2"		12	1.333		43.50	70.50		114	154
1430	4"		10	1.600		51.50	84.50		136	185
1510	90° Elbow, straight									

154

For customer support on your Mechanical Cost Data, call 877.795.7394.

22 11 13 – Facility Water Distribution Piping

22 11 13.45 Pipe Fittings, Steel, Threaded	Crew	Daily Output	Labor-Hours	Unit	Material	2015 Bare Costs Labor	Equipment	Total	Total Incl O&P
1520	1/2"	1 Plum	15	.533	Ea.	31.50	31.50		63	82.50
1530	3/4"		14	.571		32	33.50		65.50	86
1540	1"		13	.615		39	36		75	97.50
1550	1-1/4"	Q-1	22	.727		58	38.50		96.50	122
1560	1-1/2"		20	.800		72	42.50		114.50	143
1580	2"		18	.889		89	47		136	169
1590	2-1/2"		14	1.143		216	60.50		276.50	330
1600	3"		10	1.600		285	84.50		369.50	445
1610	4"		6	2.667		585	141		726	855
1650	45° Elbow, straight									
1660	1/2"	1 Plum	15	.533	Ea.	45	31.50		76.50	97
1670	3/4"		14	.571		43.50	33.50		77	98.50
1680	1"		13	.615		52.50	36		88.50	112
1690	1-1/4"	Q-1	22	.727		86	38.50		124.50	153
1700	1-1/2"		20	.800		95	42.50		137.50	168
1710	2"		18	.889		135	47		182	219
1720	2-1/2"		14	1.143		233	60.50		293.50	345
1800	Tee, straight									
1810	1/2"	1 Plum	9	.889	Ea.	50	52		102	134
1820	3/4"		9	.889		49.50	52		101.50	134
1830	1"		8	1		59.50	58.50		118	154
1840	1-1/4"	Q-1	14	1.143		89	60.50		149.50	189
1850	1-1/2"		13	1.231		115	65		180	224
1860	2"		11	1.455		142	77		219	272
1870	2-1/2"		9	1.778		305	94		399	475
1880	3"		6	2.667		415	141		556	675
1890	4"		4	4		810	211		1,021	1,225
4000	Standard weight, black									
4010	Couplings, steel straight, merchants									
4030	1/4"	1 Plum	19	.421	Ea.	1.01	24.50		25.51	38.50
4040	3/8"		19	.421		1.22	24.50		25.72	39
4050	1/2"		19	.421		1.30	24.50		25.80	39
4060	3/4"		18	.444		1.64	26		27.64	41.50
4070	1"		15	.533		2.31	31.50		33.81	50
4080	1-1/4"	Q-1	26	.615		2.95	32.50		35.45	52.50
4090	1-1/2"		24	.667		3.73	35		38.73	57
4100	2"		21	.762		5.35	40.50		45.85	67
4110	2-1/2"		18	.889		16.75	47		63.75	89.50
4120	3"		14	1.143		23.50	60.50		84	117
4130	3-1/2"		12	1.333		41.50	70.50		112	152
4140	4"		10	1.600		41.50	84.50		126	174
4166	Plug, 1/4"	1 Plum	38	.211		2.73	12.35		15.08	21.50
4167	3/8"		38	.211		2.73	12.35		15.08	21.50
4168	1/2"		38	.211		2.73	12.35		15.08	21.50
4169	3/4"		32	.250		8.25	14.70		22.95	31
4170	1"		30	.267		8.80	15.65		24.45	33
4171	1-1/4"	Q-1	52	.308		10.10	16.25		26.35	35.50
4172	1-1/2"		48	.333		14.35	17.60		31.95	42.50
4173	2"		42	.381		18.55	20		38.55	51
4176	2-1/2"		36	.444		27.50	23.50		51	65.50
4180	4"		20	.800		47.50	42.50		90	116
4200	Standard weight, galvanized									
4210	Couplings, steel straight, merchants									

For customer support on your Mechanical Cost Data, call 877.795.7394.

155

22 11 13.45 Pipe Fittings, Steel, Threaded	Crew	Daily Output	Labor-Hours	Unit	Material	2015 Bare Costs Labor	Equipment	Total	Total Incl O&P
4230	1/4"	1 Plum	19	.421	Ea.	1.16	24.50		25.66	39
4240	3/8"		19	.421		1.49	24.50		25.99	39
4250	1/2"		19	.421		1.58	24.50		26.08	39
4260	3/4"		18	.444		1.99	26		27.99	41.50
4270	1"		15	.533		2.78	31.50		34.28	50.50
4280	1-1/4"	Q-1	26	.615		3.55	32.50		36.05	53
4290	1-1/2"		24	.667		4.41	35		39.41	58
4300	2"		21	.762		6.60	40.50		47.10	68.50
4310	2-1/2"		18	.889		20.50	47		67.50	93.50
4320	3"		14	1.143		27.50	60.50		88	121
4330	3-1/2"		12	1.333		48	70.50		118.50	159
4340	4"		10	1.600		48	84.50		132.50	181
4370	Plug, galvanized, square head									
4374	1/2"	1 Plum	38	.211	Ea.	6.60	12.35		18.95	26
4375	3/4"		32	.250		6.60	14.70		21.30	29.50
4376	1"		30	.267		6.60	15.65		22.25	31
4377	1-1/4"	Q-1	52	.308		11	16.25		27.25	36.50
4378	1-1/2"		48	.333		14.90	17.60		32.50	43
4379	2"		42	.381		18.50	20		38.50	51
4380	2-1/2"		36	.444		39	23.50		62.50	78.50
4381	3"		28	.571		50	30		80	101
4382	4"		20	.800		137	42.50		179.50	215
4700	Nipple, black									
4710	1/2" x 4" long	1 Plum	19	.421	Ea.	2.21	24.50		26.71	40
4712	3/4" x 4" long		18	.444		2.67	26		28.67	42.50
4714	1" x 4" long		15	.533		3.70	31.50		35.20	51.50
4716	1-1/4" x 4" long	Q-1	26	.615		4.63	32.50		37.13	54
4718	1-1/2" x 4" long		24	.667		5.45	35		40.45	59
4720	2" x 4" long		21	.762		7.60	40.50		48.10	69.50
4722	2-1/2" x 4" long		18	.889		20.50	47		67.50	94
4724	3" x 4" long		14	1.143		26	60.50		86.50	120
4726	4" x 4" long		10	1.600		35.50	84.50		120	167
4800	Nipple, galvanized									
4810	1/2" x 4" long	1 Plum	19	.421	Ea.	2.71	24.50		27.21	40.50
4812	3/4" x 4" long		18	.444		3.35	26		29.35	43
4814	1" x 4" long		15	.533		4.49	31.50		35.99	52.50
4816	1-1/4" x 4" long	Q-1	26	.615		5.50	32.50		38	55
4818	1-1/2" x 4" long		24	.667		7	35		42	60.50
4820	2" x 4" long		21	.762		8.85	40.50		49.35	71
4822	2-1/2" x 4" long		18	.889		23.50	47		70.50	96.50
4824	3" x 4" long		14	1.143		31	60.50		91.50	125
4826	4" x 4" long		10	1.600		41.50	84.50		126	174
5000	Malleable iron, 150 lb.									
5020	Black									
5040	90° elbow, straight									
5060	1/4"	1 Plum	16	.500	Ea.	4.78	29.50		34.28	50
5070	3/8"		16	.500		4.78	29.50		34.28	50
5080	1/2"		15	.533		3.30	31.50		34.80	51
5090	3/4"		14	.571		3.99	33.50		37.49	55
5100	1"		13	.615		6.95	36		42.95	62
5110	1-1/4"	Q-1	22	.727		11.45	38.50		49.95	70.50
5120	1-1/2"		20	.800		15.05	42.50		57.55	80.50
5130	2"		18	.889		26	47		73	99.50

22 11 13 – Facility Water Distribution Piping

22 11 13.45 Pipe Fittings, Steel, Threaded	Crew	Daily Output	Labor-Hours	Unit	Material	2015 Bare Costs Labor	Equipment	Total	Total Incl O&P
5140	2-1/2"	Q-1	14	1.143	Ea.	58	60.50		118.50	155
5150	3"		10	1.600		84.50	84.50		169	221
5160	3-1/2"		8	2		232	106		338	415
5170	4"		6	2.667		182	141		323	415
5250	45° elbow, straight									
5270	1/4"	1 Plum	16	.500	Ea.	7.20	29.50		36.70	52.50
5280	3/8"		16	.500		7.20	29.50		36.70	52.50
5290	1/2"		15	.533		5.45	31.50		36.95	53.50
5300	3/4"		14	.571		6.75	33.50		40.25	58
5310	1"		13	.615		8.50	36		44.50	64
5320	1-1/4"	Q-1	22	.727		15.05	38.50		53.55	74.50
5330	1-1/2"		20	.800		18.60	42.50		61.10	84.50
5340	2"		18	.889		28	47		75	102
5350	2-1/2"		14	1.143		81.50	60.50		142	181
5360	3"		10	1.600		188	84.50		272.50	335
5370	3-1/2"		8	2		208	106		314	390
5380	4"		6	2.667		208	141		349	440
5450	Tee, straight									
5470	1/4"	1 Plum	10	.800	Ea.	6.95	47		53.95	78.50
5480	3/8"		10	.800		6.95	47		53.95	78.50
5490	1/2"		9	.889		4.42	52		56.42	84
5500	3/4"		9	.889		6.35	52		58.35	86
5510	1"		8	1		10.85	58.50		69.35	100
5520	1-1/4"	Q-1	14	1.143		17.60	60.50		78.10	110
5530	1-1/2"		13	1.231		22	65		87	122
5540	2"		11	1.455		37.50	77		114.50	157
5550	2-1/2"		9	1.778		80.50	94		174.50	231
5560	3"		6	2.667		119	141		260	345
5570	3-1/2"		5	3.200		275	169		444	555
5580	4"		4	4		286	211		497	635
5601	Tee, reducing, on outlet									
5602	1/2"	1 Plum	9	.889	Ea.	9.65	52		61.65	89.50
5603	3/4"		9	.889		10.90	52		62.90	91
5604	1"		8	1		17.45	58.50		75.95	108
5605	1-1/4"	Q-1	14	1.143		30.50	60.50		91	125
5606	1-1/2"		13	1.231		34.50	65		99.50	136
5607	2"		11	1.455		48	77		125	169
5608	2-1/2"		9	1.778		121	94		215	275
5609	3"		6	2.667		166	141		307	395
5610	3-1/2"		5	3.200		390	169		559	685
5611	4"		4	4		330	211		541	685
5650	Coupling									
5670	1/4"	1 Plum	19	.421	Ea.	5.95	24.50		30.45	44
5680	3/8"		19	.421		5.95	24.50		30.45	44
5690	1/2"		19	.421		4.58	24.50		29.08	42.50
5700	3/4"		18	.444		5.35	26		31.35	45.50
5710	1"		15	.533		8	31.50		39.50	56.50
5720	1-1/4"	Q-1	26	.615		10.40	32.50		42.90	60.50
5730	1-1/2"		24	.667		14.05	35		49.05	68.50
5740	2"		21	.762		21	40.50		61.50	84
5750	2-1/2"		18	.889		57.50	47		104.50	134
5760	3"		14	1.143		77.50	60.50		138	177
5770	3-1/2"		12	1.333		156	70.50		226.50	278

For customer support on your Mechanical Cost Data, call 877.795.7394.

157

22 11 13.45 Pipe Fittings, Steel, Threaded		Crew	Daily Output	Labor-Hours	Unit	Material	2015 Bare Costs Labor	Equipment	Total	Total Incl O&P
5780	4"	Q-1	10	1.600	Ea.	156	84.50		240.50	300
5840	Reducer, concentric, 1/4"	1 Plum	19	.421		6.25	24.50		30.75	44.50
5850	3/8"		19	.421		8	24.50		32.50	46.50
5860	1/2"		19	.421		6.30	24.50		30.80	44.50
5870	3/4"		16	.500		7.50	29.50		37	53
5880	1"		15	.533		12.60	31.50		44.10	61.50
5890	1-1/4"	Q-1	26	.615		14.10	32.50		46.60	64.50
5900	1-1/2"		24	.667		20	35		55	75.50
5910	2"		21	.762		29	40.50		69.50	93
5911	2-1/2"		18	.889		65	47		112	143
5912	3"		14	1.143		79	60.50		139.50	178
5913	3-1/2"		12	1.333		253	70.50		323.50	385
5914	4"		10	1.600		190	84.50		274.50	335
5981	Bushing, 1/4"	1 Plum	19	.421		1.15	24.50		25.65	39
5982	3/8"		19	.421		6.10	24.50		30.60	44
5983	1/2"		19	.421		6.10	24.50		30.60	44
5984	3/4"		16	.500		6.15	29.50		35.65	51.50
5985	1"		15	.533		8.70	31.50		40.20	57
5986	1-1/4"	Q-1	26	.615		10.85	32.50		43.35	61
5987	1-1/2"		24	.667		9.35	35		44.35	63.50
5988	2"		21	.762		11.65	40.50		52.15	74
5989	Cap, 1/4"	1 Plum	38	.211		4.79	12.35		17.14	24
5991	3/8"		38	.211		5.85	12.35		18.20	25
5992	1/2"		38	.211		3.27	12.35		15.62	22.50
5993	3/4"		32	.250		4.53	14.70		19.23	27
5994	1"		30	.267		5.50	15.65		21.15	29.50
5995	1-1/4"	Q-1	52	.308		7.15	16.25		23.40	32.50
5996	1-1/2"		48	.333		9.95	17.60		27.55	37.50
5997	2"		42	.381		14.50	20		34.50	46.50
6000	For galvanized elbows, tees, and couplings add					20%				
6058	For galvanized reducers, caps and bushings add					20%				
6100	90° Elbow, galvanized, 150 lb., reducing									
6110	3/4" x 1/2"	1 Plum	15.40	.519	Ea.	9.50	30.50		40	56.50
6112	1" x 3/4"		14	.571		12.20	33.50		45.70	64
6114	1" x 1/2"		14.50	.552		12.95	32.50		45.45	63.50
6116	1-1/4" x 1"	Q-1	24.20	.661		20	35		55	75
6118	1-1/4" x 3/4"		25.40	.630		24.50	33.50		58	77.50
6120	1-1/4" x 1/2"		26.20	.611		26	32.50		58.50	77
6122	1-1/2" x 1-1/4"		21.60	.741		32.50	39		71.50	95
6124	1-1/2" x 1"		23.50	.681		32.50	36		68.50	90.50
6126	1-1/2" x 3/4"		24.60	.650		32.50	34.50		67	88
6128	2" x 1-1/2"		20.50	.780		38	41		79	105
6130	2" x 1-1/4"		21	.762		44	40.50		84.50	109
6132	2" x 1"		22.80	.702		45.50	37		82.50	106
6134	2" x 3/4"		23.90	.669		46.50	35.50		82	105
6136	2-1/2" x 2"		12.30	1.301		130	68.50		198.50	246
6138	2-1/2" x 1-1/2"		12.50	1.280		144	67.50		211.50	261
6140	3" x 2-1/2"		8.60	1.860		235	98.50		333.50	405
6142	3" x 2"		11.80	1.356		204	71.50		275.50	335
6144	4" x 3"		8.20	1.951		535	103		638	740
6160	90° Elbow, black, 150 lb., reducing									
6170	1" x 3/4"	1 Plum	14	.571	Ea.	8.35	33.50		41.85	59.50
6174	1-1/2" x 1"	Q-1	23.50	.681		19.75	36		55.75	76

22 11 13 – Facility Water Distribution Piping

22 11 13.45 Pipe Fittings, Steel, Threaded	Crew	Daily Output	Labor-Hours	Unit	Material	2015 Bare Costs Labor	Equipment	Total	Total Incl O&P
6178	1-1/2" x 3/4"	Q-1	24.60	.650	Ea.	22.50	34.50		57	77
6182	2" x 1-1/2"		20.50	.780		28.50	41		69.50	94
6186	2" x 1"		22.80	.702		32.50	37		69.50	92
6190	2" x 3/4"		23.90	.669		34.50	35.50		70	91.50
6194	2-1/2" x 2"	▼	12.30	1.301	▼	78.50	68.50		147	191
7000	Union, with brass seat									
7010	1/4"	1 Plum	15	.533	Ea.	25.50	31.50		57	75.50
7020	3/8"		15	.533		16.15	31.50		47.65	65.50
7030	1/2"		14	.571		14.60	33.50		48.10	66.50
7040	3/4"		13	.615		16.80	36		52.80	73
7050	1"	▼	12	.667		22	39		61	83
7060	1-1/4"	Q-1	21	.762		31.50	40.50		72	95.50
7070	1-1/2"		19	.842		39	44.50		83.50	110
7080	2"		17	.941		45.50	49.50		95	125
7090	2-1/2"		13	1.231		136	65		201	247
7100	3"	▼	9	1.778	▼	163	94		257	320
7120	Union, galvanized									
7124	1/2"	1 Plum	14	.571	Ea.	18.65	33.50		52.15	71
7125	3/4"		13	.615		22.50	36		58.50	79
7126	1"	▼	12	.667		29	39		68	91
7127	1-1/4"	Q-1	21	.762		42	40.50		82.50	108
7128	1-1/2"		19	.842		51	44.50		95.50	124
7129	2"		17	.941		59	49.50		108.50	140
7130	2-1/2"		13	1.231		207	65		272	325
7131	3"	▼	9	1.778	▼	286	94		380	455
7500	Malleable iron, 300 lb.									
7520	Black									
7540	90° Elbow, straight, 1/4"	1 Plum	16	.500	Ea.	16.75	29.50		46.25	63
7560	3/8"		16	.500		14.90	29.50		44.40	61
7570	1/2"		15	.533		19.30	31.50		50.80	68.50
7580	3/4"		14	.571		22	33.50		55.50	74.50
7590	1"		13	.615		28	36		64	85
7600	1-1/4"	Q-1	22	.727		40.50	38.50		79	103
7610	1-1/2"		20	.800		48	42.50		90.50	117
7620	2"		18	.889		69	47		116	147
7630	2-1/2"		14	1.143		179	60.50		239.50	288
7640	3"		10	1.600		205	84.50		289.50	355
7650	4"	▼	6	2.667		480	141		621	745
7700	45° Elbow, straight, 1/4"	1 Plum	16	.500		25	29.50		54.50	72
7720	3/8"		16	.500		28	29.50		57.50	75
7730	1/2"		15	.533		28	31.50		59.50	78
7740	3/4"		14	.571		31.50	33.50		65	85
7750	1"	▼	13	.615		34	36		70	92
7760	1-1/4"	Q-1	22	.727		54.50	38.50		93	118
7770	1-1/2"		20	.800		71.50	42.50		114	143
7780	2"		18	.889		108	47		155	190
7790	2-1/2"		14	1.143		241	60.50		301.50	355
7800	3"		10	1.600		315	84.50		399.50	480
7810	4"	▼	6	2.667		720	141		861	1,000
7850	Tee, straight, 1/4"	1 Plum	10	.800		21	47		68	94
7870	3/8"		10	.800		22.50	47		69.50	95.50
7880	1/2"		9	.889		28.50	52		80.50	111
7890	3/4"	▼	9	.889	▼	30.50	52		82.50	113

For customer support on your Mechanical Cost Data, call 877.795.7394.

159

22 11 13.45 Pipe Fittings, Steel, Threaded		Crew	Daily Output	Labor-Hours	Unit	Material	2015 Bare Costs Labor	Equipment	Total	Total Incl O&P
7900	1"	1 Plum	8	1	Ea.	37	58.50		95.50	129
7910	1-1/4"	Q-1	14	1.143		53	60.50		113.50	149
7920	1-1/2"		13	1.231		62	65		127	166
7930	2"		11	1.455		91	77		168	216
7940	2-1/2"		9	1.778		222	94		316	385
7950	3"		6	2.667		315	141		456	560
7960	4"		4	4		900	211		1,111	1,300
8050	Couplings, straight, 1/4"	1 Plum	19	.421		15.10	24.50		39.60	54
8070	3/8"		19	.421		15.10	24.50		39.60	54
8080	1/2"		19	.421		16.85	24.50		41.35	56
8090	3/4"		18	.444		19.40	26		45.40	61
8100	1"		15	.533		22	31.50		53.50	72
8110	1-1/4"	Q-1	26	.615		26.50	32.50		59	78
8120	1-1/2"		24	.667		39.50	35		74.50	96.50
8130	2"		21	.762		57	40.50		97.50	124
8140	2-1/2"		18	.889		105	47		152	186
8150	3"		14	1.143		151	60.50		211.50	257
8160	4"		10	1.600		275	84.50		359.50	430
8200	Galvanized									
8220	90° Elbow, straight, 1/4"	1 Plum	16	.500	Ea.	30	29.50		59.50	77.50
8222	3/8"		16	.500		30.50	29.50		60	78
8224	1/2"		15	.533		36.50	31.50		68	87.50
8226	3/4"		14	.571		41	33.50		74.50	95.50
8228	1"		13	.615		52.50	36		88.50	112
8230	1-1/4"	Q-1	22	.727		82.50	38.50		121	149
8232	1-1/2"		20	.800		89.50	42.50		132	163
8234	2"		18	.889		150	47		197	236
8236	2-1/2"		14	1.143		335	60.50		395.50	455
8238	3"		10	1.600		395	84.50		479.50	565
8240	4"		6	2.667		1,175	141		1,316	1,525
8280	45° Elbow, straight									
8282	1/2"	1 Plum	15	.533	Ea.	54	31.50		85.50	107
8284	3/4"		14	.571		62.50	33.50		96	120
8286	1"		13	.615		68.50	36		104.50	130
8288	1-1/4"	Q-1	22	.727		106	38.50		144.50	175
8290	1-1/2"		20	.800		137	42.50		179.50	215
8292	2"		18	.889		190	47		237	280
8310	Tee, straight, 1/4"	1 Plum	10	.800	Ea.	42.50	47		89.50	118
8312	3/8"		10	.800		43.50	47		90.50	119
8314	1/2"		9	.889		54.50	52		106.50	139
8316	3/4"		9	.889		60	52		112	145
8318	1"		8	1		74	58.50		132.50	170
8320	1-1/4"	Q-1	14	1.143		106	60.50		166.50	207
8322	1-1/2"		13	1.231		110	65		175	219
8324	2"		11	1.455		175	77		252	310
8326	2-1/2"		9	1.778		555	94		649	750
8328	3"		6	2.667		630	141		771	910
8330	4"		4	4		1,625	211		1,836	2,100
8380	Couplings, straight, 1/4"	1 Plum	19	.421		19.65	24.50		44.15	59
8382	3/8"		19	.421		28.50	24.50		53	68.50
8384	1/2"		19	.421		29	24.50		53.50	69.50
8386	3/4"		18	.444		33	26		59	75.50
8388	1"		15	.533		43.50	31.50		75	95.50

22 11 13.45 Pipe Fittings, Steel, Threaded		Crew	Daily Output	Labor-Hours	Unit	Material	2015 Bare Costs Labor	Equipment	Total	Total Incl O&P
8390	1-1/4"	Q-1	26	.615	Ea.	56.50	32.50		89	112
8392	1-1/2"		24	.667		80	35		115	141
8394	2"		21	.762		96.50	40.50		137	167
8396	2-1/2"		18	.889		254	47		301	350
8398	3"		14	1.143		305	60.50		365.50	425
8399	4"		10	1.600		465	84.50		549.50	640
8529	Black									
8530	Reducer, concentric, 1/4"	1 Plum	19	.421	Ea.	20	24.50		44.50	59.50
8531	3/8"		19	.421		20	24.50		44.50	59.50
8532	1/2"		17	.471		24	27.50		51.50	68
8533	3/4"		16	.500		34.50	29.50		64	82
8534	1"		15	.533		41	31.50		72.50	92.50
8535	1-1/4"	Q-1	26	.615		61.50	32.50		94	117
8536	1-1/2"		24	.667		63.50	35		98.50	123
8537	2"		21	.762		92	40.50		132.50	162
8550	Cap, 1/4"	1 Plum	38	.211		15.50	12.35		27.85	35.50
8551	3/8"		38	.211		15.50	12.35		27.85	35.50
8552	1/2"		34	.235		15.35	13.80		29.15	38
8553	3/4"		32	.250		19.40	14.70		34.10	43.50
8554	1"		30	.267		25.50	15.65		41.15	51.50
8555	1-1/4"	Q-1	52	.308		28	16.25		44.25	55.50
8556	1-1/2"		48	.333		41	17.60		58.60	72
8557	2"		42	.381		57	20		77	93.50
8570	Plug, 1/4"	1 Plum	38	.211		3.47	12.35		15.82	22.50
8571	3/8"		38	.211		3.64	12.35		15.99	22.50
8572	1/2"		34	.235		3.82	13.80		17.62	25
8573	3/4"		32	.250		4.76	14.70		19.46	27.50
8574	1"		30	.267		7.30	15.65		22.95	31.50
8575	1-1/4"	Q-1	52	.308		15	16.25		31.25	41
8576	1-1/2"		48	.333		17.15	17.60		34.75	45.50
8577	2"		42	.381		27	20		47	60.50
9500	Union with brass seat, 1/4"	1 Plum	15	.533		34	31.50		65.50	85
9530	3/8"		15	.533		34.50	31.50		66	85.50
9540	1/2"		14	.571		28	33.50		61.50	81.50
9550	3/4"		13	.615		31	36		67	88.50
9560	1"		12	.667		40.50	39		79.50	104
9570	1-1/4"	Q-1	21	.762		66	40.50		106.50	134
9580	1-1/2"		19	.842		69	44.50		113.50	143
9590	2"		17	.941		86	49.50		135.50	170
9600	2-1/2"		13	1.231		276	65		341	405
9610	3"		9	1.778		355	94		449	535
9620	4"		5	3.200		1,125	169		1,294	1,500
9630	Union, all iron, 1/4"	1 Plum	15	.533		53	31.50		84.50	106
9650	3/8"		15	.533		53	31.50		84.50	106
9660	1/2"		14	.571		58.50	33.50		92	115
9670	3/4"		13	.615		60	36		96	121
9680	1"		12	.667		58	39		97	123
9690	1-1/4"	Q-1	21	.762		113	40.50		153.50	185
9700	1-1/2"		19	.842		138	44.50		182.50	219
9710	2"		17	.941		174	49.50		223.50	267
9720	2-1/2"		13	1.231		310	65		375	440
9730	3"		9	1.778		585	94		679	785
9750	For galvanized unions, add					15%				

22 11 13 – Facility Water Distribution Piping

22 11 13.45 Pipe Fittings, Steel, Threaded	Crew	Daily Output	Labor-Hours	Unit	Material	2015 Bare Costs Labor	Equipment	Total	Total Incl O&P
9757 Forged steel, 3000 lb.									
9758 Black									
9760 90° Elbow, 1/4"	1 Plum	16	.500	Ea.	20.50	29.50		50	67
9761 3/8"		16	.500		20.50	29.50		50	67
9762 1/2"		15	.533		15.70	31.50		47.20	65
9763 3/4"		14	.571		19.70	33.50		53.20	72
9764 1"		13	.615		29.50	36		65.50	86.50
9765 1-1/4"	Q-1	22	.727		55.50	38.50		94	119
9766 1-1/2"		20	.800		72	42.50		114.50	143
9767 2"		18	.889		87	47		134	167
9780 45° Elbow 1/4"	1 Plum	16	.500		26	29.50		55.50	73
9781 3/8"		16	.500		26	29.50		55.50	73
9782 1/2"		15	.533		25.50	31.50		57	75.50
9783 3/4"		14	.571		29.50	33.50		63	83
9784 1"		13	.615		40.50	36		76.50	99
9785 1-1/4"	Q-1	22	.727		56	38.50		94.50	120
9786 1-1/2"		20	.800		80	42.50		122.50	152
9787 2"		18	.889		110	47		157	193
9800 Tee, 1/4"	1 Plum	10	.800		25	47		72	98.50
9801 3/8"		10	.800		25	47		72	98.50
9802 1/2"		9	.889		22.50	52		74.50	104
9803 3/4"		9	.889		30.50	52		82.50	113
9804 1"		8	1		40	58.50		98.50	133
9805 1-1/4"	Q-1	14	1.143		77.50	60.50		138	176
9806 1-1/2"		13	1.231		90.50	65		155.50	198
9807 2"		11	1.455		116	77		193	244
9820 Reducer, concentric, 1/4"	1 Plum	19	.421		12.55	24.50		37.05	51.50
9821 3/8"		19	.421		13.05	24.50		37.55	52
9822 1/2"		17	.471		13.05	27.50		40.55	56
9823 3/4"		16	.500		15.50	29.50		45	61.50
9824 1"		15	.533		20	31.50		51.50	69.50
9825 1-1/4"	Q-1	26	.615		35.50	32.50		68	88
9826 1-1/2"		24	.667		38.50	35		73.50	95
9827 2"		21	.762		55.50	40.50		96	122
9840 Cap, 1/4"	1 Plum	38	.211		7.30	12.35		19.65	26.50
9841 3/8"		38	.211		7.70	12.35		20.05	27
9842 1/2"		34	.235		7.50	13.80		21.30	29.50
9843 3/4"		32	.250		10.45	14.70		25.15	33.50
9844 1"		30	.267		16	15.65		31.65	41
9845 1-1/4"	Q-1	52	.308		26	16.25		42.25	53
9846 1-1/2"		48	.333		31	17.60		48.60	60.50
9847 2"		42	.381		44.50	20		64.50	79.50
9860 Plug, 1/4"	1 Plum	38	.211		3.47	12.35		15.82	22.50
9861 3/8"		38	.211		3.64	12.35		15.99	22.50
9862 1/2"		34	.235		3.82	13.80		17.62	25
9863 3/4"		32	.250		4.76	14.70		19.46	27.50
9864 1"		30	.267		7.30	15.65		22.95	31.50
9865 1-1/4"	Q-1	52	.308		15	16.25		31.25	41
9866 1-1/2"		48	.333		17.15	17.60		34.75	45.50
9867 2"		42	.381		27	20		47	60.50
9880 Union, bronze seat, 1/4"	1 Plum	15	.533		57	31.50		88.50	110
9881 3/8"		15	.533		57	31.50		88.50	110
9882 1/2"		14	.571		54.50	33.50		88	111

22 11 13.45 Pipe Fittings, Steel, Threaded

		Crew	Daily Output	Labor-Hours	Unit	Material	2015 Bare Costs Labor	2015 Bare Costs Equipment	Total	Total Incl O&P
9883	3/4"	1 Plum	13	.615	Ea.	73	36		109	135
9884	1"	▼	12	.667		82.50	39		121.50	150
9885	1-1/4"	Q-1	21	.762		160	40.50		200.50	237
9886	1-1/2"		19	.842		165	44.50		209.50	248
9887	2"	▼	17	.941		195	49.50		244.50	290
9900	Coupling, 1/4"	1 Plum	19	.421		8	24.50		32.50	46.50
9901	3/8"		19	.421		8	24.50		32.50	46.50
9902	1/2"		17	.471		6.55	27.50		34.05	48.50
9903	3/4"		16	.500		8.55	29.50		38.05	54
9904	1"	▼	15	.533		14.85	31.50		46.35	64
9905	1-1/4"	Q-1	26	.615		25	32.50		57.50	76.50
9906	1-1/2"		24	.667		32	35		67	88
9907	2"	▼	21	.762	▼	40	40.50		80.50	105

22 11 13.47 Pipe Fittings, Steel

		Crew	Daily Output	Labor-Hours	Unit	Material	2015 Bare Costs Labor	2015 Bare Costs Equipment	Total	Total Incl O&P
0010	**PIPE FITTINGS, STEEL**, Flanged, Welded & Special									
0020	Flanged joints, C.I., standard weight, black. One gasket & bolt									
0040	set, mat'l only, required at each joint, not included (see line 0620)									
0060	90° Elbow, straight, 1-1/2" pipe size	Q-1	14	1.143	Ea.	550	60.50		610.50	695
0080	2" pipe size		13	1.231		325	65		390	455
0090	2-1/2" pipe size		12	1.333		350	70.50		420.50	490
0100	3" pipe size		11	1.455		292	77		369	435
0110	4" pipe size		8	2		355	106		461	550
0120	5" pipe size	▼	7	2.286		855	121		976	1,125
0130	6" pipe size	Q-2	9	2.667		565	146		711	840
0140	8" pipe size		8	3		975	164		1,139	1,325
0150	10" pipe size		7	3.429		2,150	188		2,338	2,625
0160	12" pipe size	▼	6	4	▼	4,300	219		4,519	5,075
0171	90° Elbow, reducing									
0172	2-1/2" by 2" pipe size	Q-1	12	1.333	Ea.	1,075	70.50		1,145.50	1,275
0173	3" by 2-1/2" pipe size		11	1.455		990	77		1,067	1,225
0174	4" by 3" pipe size		8	2		800	106		906	1,050
0175	5" by 3" pipe size	▼	7	2.286		1,725	121		1,846	2,075
0176	6" by 4" pipe size	Q-2	9	2.667		1,075	146		1,221	1,400
0177	8" by 6" pipe size		8	3		1,450	164		1,614	1,850
0178	10" by 8" pipe size		7	3.429		2,750	188		2,938	3,300
0179	12" by 10" pipe size	▼	6	4		5,300	219		5,519	6,150
0200	45° Elbow, straight, 1-1/2" pipe size	Q-1	14	1.143		665	60.50		725.50	820
0220	2" pipe size		13	1.231		470	65		535	615
0230	2-1/2" pipe size		12	1.333		500	70.50		570.50	655
0240	3" pipe size		11	1.455		485	77		562	650
0250	4" pipe size		8	2		545	106		651	760
0260	5" pipe size	▼	7	2.286		1,300	121		1,421	1,625
0270	6" pipe size	Q-2	9	2.667		895	146		1,041	1,200
0280	8" pipe size		8	3		1,300	164		1,464	1,700
0290	10" pipe size		7	3.429		2,775	188		2,963	3,325
0300	12" pipe size	▼	6	4	▼	4,225	219		4,444	4,975
0310	Cross, straight									
0311	2-1/2" pipe size	Q-1	6	2.667	Ea.	1,025	141		1,166	1,350
0312	3" pipe size		5	3.200		1,075	169		1,244	1,425
0313	4" pipe size		4	4		1,400	211		1,611	1,875
0314	5" pipe size	▼	3	5.333		2,950	282		3,232	3,675
0315	6" pipe size	Q-2	5	4.800		2,950	263		3,213	3,650

22 11 Facility Water Distribution

22 11 13 – Facility Water Distribution Piping

22 11 13.47 Pipe Fittings, Steel

		Crew	Daily Output	Labor-Hours	Unit	Material	2015 Bare Costs Labor	Equipment	Total	Total Incl O&P
0316	8" pipe size	Q-2	4	6	Ea.	4,750	330		5,080	5,725
0317	10" pipe size		3	8		5,650	440		6,090	6,850
0318	12" pipe size	▼	2	12		8,600	655		9,255	10,500
0350	Tee, straight, 1-1/2" pipe size	Q-1	10	1.600		635	84.50		719.50	825
0370	2" pipe size		9	1.778		355	94		449	530
0380	2-1/2" pipe size		8	2		515	106		621	725
0390	3" pipe size		7	2.286		360	121		481	580
0400	4" pipe size		5	3.200		550	169		719	860
0410	5" pipe size	▼	4	4		1,475	211		1,686	1,950
0420	6" pipe size	Q-2	6	4		800	219		1,019	1,200
0430	8" pipe size		5	4.800		1,375	263		1,638	1,900
0440	10" pipe size		4	6		3,675	330		4,005	4,525
0450	12" pipe size	▼	3	8	▼	5,775	440		6,215	7,000
0459	Tee, reducing on outlet									
0460	2-1/2" by 2" pipe size	Q-1	8	2	Ea.	1,075	106		1,181	1,325
0461	3" by 2-1/2" pipe size		7	2.286		1,075	121		1,196	1,350
0462	4" by 3" pipe size		5	3.200		1,125	169		1,294	1,500
0463	5" by 4" pipe size	▼	4	4		2,475	211		2,686	3,050
0464	6" by 4" pipe size	Q-2	6	4		1,100	219		1,319	1,525
0465	8" by 6" pipe size		5	4.800		1,700	263		1,963	2,275
0466	10" by 8" pipe size		4	6		4,600	330		4,930	5,550
0467	12" by 10" pipe size	▼	3	8		6,725	440		7,165	8,050
0476	Reducer, concentric									
0477	3" by 2-1/2"	Q-1	12	1.333	Ea.	650	70.50		720.50	820
0478	4" by 3"		9	1.778		730	94		824	940
0479	5" by 4"	▼	8	2		1,125	106		1,231	1,400
0480	6" by 4"	Q-2	10	2.400		975	131		1,106	1,275
0481	8" by 6"		9	2.667		1,225	146		1,371	1,575
0482	10" by 8"		8	3		2,225	164		2,389	2,700
0483	12" by 10"	▼	7	3.429	▼	4,375	188		4,563	5,075
0492	Reducer, eccentric									
0493	4" by 3"	Q-1	8	2	Ea.	1,200	106		1,306	1,475
0494	5" by 4"	"	7	2.286		1,825	121		1,946	2,175
0495	6" by 4"	Q-2	9	2.667		1,125	146		1,271	1,450
0496	8" by 6"		8	3		1,450	164		1,614	1,850
0497	10" by 8"		7	3.429		3,600	188		3,788	4,250
0498	12" by 10"	▼	6	4		4,875	219		5,094	5,700
0500	For galvanized elbows and tees, add					100%				
0520	For extra heavy weight elbows and tees, add					140%				
0620	Gasket and bolt set, 150#, 1/2" pipe size	1 Plum	20	.400		2.75	23.50		26.25	38.50
0622	3/4" pipe size		19	.421		2.90	24.50		27.40	40.50
0624	1" pipe size		18	.444		2.96	26		28.96	43
0626	1-1/4" pipe size		17	.471		4.02	27.50		31.52	46
0628	1-1/2" pipe size		15	.533		4.16	31.50		35.66	52
0630	2" pipe size		13	.615		5.95	36		41.95	61
0640	2-1/2" pipe size		12	.667		6.15	39		45.15	66
0650	3" pipe size		11	.727		6.35	42.50		48.85	71.50
0660	3-1/2" pipe size		9	.889		13.05	52		65.05	93.50
0670	4" pipe size		8	1		14.50	58.50		73	104
0680	5" pipe size		7	1.143		18.85	67		85.85	122
0690	6" pipe size		6	1.333		21.50	78.50		100	142
0700	8" pipe size		5	1.600		21.50	94		115.50	166
0710	10" pipe size		4.50	1.778		45	104		149	208

22 11 13 – Facility Water Distribution Piping

22 11 13.47 Pipe Fittings, Steel		Crew	Daily Output	Labor-Hours	Unit	Material	2015 Bare Costs Labor	Equipment	Total	Total Incl O&P
0720	12" pipe size	1 Plum	4.20	1.905	Ea.	41.50	112		153.50	215
0730	14" pipe size		4	2		40.50	117		157.50	222
0740	16" pipe size		3	2.667		46	157		203	287
0750	18" pipe size		2.70	2.963		88.50	174		262.50	360
0760	20" pipe size		2.30	3.478		144	204		348	470
0780	24" pipe size		1.90	4.211		180	247		427	575
0790	26" pipe size		1.60	5		244	294		538	715
0810	30" pipe size		1.40	5.714		475	335		810	1,025
0830	36" pipe size		1.10	7.273		890	425		1,315	1,625
0850	For 300 lb. gasket set, add					40%				
2000	Flanged unions, 125 lb., black, 1/2" pipe size	1 Plum	17	.471	Ea.	82.50	27.50		110	133
2040	3/4" pipe size		17	.471		109	27.50		136.50	162
2050	1" pipe size		16	.500		105	29.50		134.50	161
2060	1-1/4" pipe size	Q-1	28	.571		125	30		155	183
2070	1-1/2" pipe size		27	.593		115	31.50		146.50	175
2080	2" pipe size		26	.615		135	32.50		167.50	198
2090	2-1/2" pipe size		24	.667		183	35		218	255
2100	3" pipe size		22	.727		207	38.50		245.50	286
2110	3-1/2" pipe size		18	.889		350	47		397	455
2120	4" pipe size		16	1		282	53		335	390
2130	5" pipe size		14	1.143		660	60.50		720.50	815
2140	6" pipe size	Q-2	19	1.263		615	69		684	785
2150	8" pipe size	"	16	1.500		1,425	82		1,507	1,700
2200	For galvanized unions, add					150%				
2290	Threaded flange									
2300	Cast iron									
2310	Black, 125 lb., per flange									
2320	1" pipe size	1 Plum	27	.296	Ea.	47	17.40		64.40	78.50
2330	1-1/4" pipe size	Q-1	44	.364		56.50	19.20		75.70	91
2340	1-1/2" pipe size		40	.400		52	21		73	89.50
2350	2" pipe size		36	.444		52	23.50		75.50	93
2360	2-1/2" pipe size		28	.571		61	30		91	113
2370	3" pipe size		20	.800		78.50	42.50		121	151
2380	3-1/2" pipe size		16	1		110	53		163	201
2390	4" pipe size		12	1.333		106	70.50		176.50	222
2400	5" pipe size		10	1.600		149	84.50		233.50	292
2410	6" pipe size	Q-2	14	1.714		169	94		263	330
2420	8" pipe size		12	2		267	110		377	460
2430	10" pipe size		10	2.400		475	131		606	720
2440	12" pipe size		8	3		1,050	164		1,214	1,400
2460	For galvanized flanges, add					95%				
2490	Blind flange									
2492	Cast iron									
2494	Black, 125 lb., per flange									
2496	1" pipe size	1 Plum	27	.296	Ea.	75	17.40		92.40	109
2500	1-1/2" pipe size	Q-1	40	.400		84.50	21		105.50	125
2502	2" pipe size		36	.444		95	23.50		118.50	141
2504	2-1/2" pipe size		28	.571		102	30		132	158
2506	3" pipe size		20	.800		.125	42.50		167.50	202
2508	4" pipe size		12	1.333		161	70.50		231.50	283
2510	5" pipe size		10	1.600		261	84.50		345.50	415
2512	6" pipe size	Q-2	14	1.714		284	94		378	450
2514	8" pipe size		12	2		445	110		555	655

22 11 13 – Facility Water Distribution Piping

22 11 13.47 Pipe Fittings, Steel		Crew	Daily Output	Labor-Hours	Unit	Material	2015 Bare Costs Labor	Equipment	Total	Total Incl O&P
2516	10" pipe size	Q-2	10	2.400	Ea.	660	131		791	925
2518	12" pipe size	▼	8	3	▼	1,250	164		1,414	1,625
2520	For galvanized flanges, add					80%				
2570	Threaded flange									
2580	Forged steel,									
2590	Black 150 lb., per flange									
2600	1/2" pipe size	1 Plum	30	.267	Ea.	26.50	15.65		42.15	52.50
2610	3/4" pipe size		28	.286		26.50	16.75		43.25	54.50
2620	1" pipe size	▼	27	.296		26.50	17.40		43.90	55.50
2630	1-1/4" pipe size	Q-1	44	.364		26.50	19.20		45.70	58
2640	1-1/2" pipe size		40	.400		26.50	21		47.50	61
2650	2" pipe size		36	.444		29.50	23.50		53	68
2660	2-1/2" pipe size		28	.571		36	30		66	85.50
2670	3" pipe size		20	.800		36.50	42.50		79	104
2690	4" pipe size		12	1.333		42.50	70.50		113	153
2700	5" pipe size	▼	10	1.600		66.50	84.50		151	202
2710	6" pipe size	Q-2	14	1.714		73	94		167	223
2720	8" pipe size		12	2		126	110		236	305
2730	10" pipe size	▼	10	2.400	▼	228	131		359	450
2860	Black 300 lb., per flange									
2870	1/2" pipe size	1 Plum	30	.267	Ea.	29	15.65		44.65	55.50
2880	3/4" pipe size		28	.286		29	16.75		45.75	57.50
2890	1" pipe size	▼	27	.296		29	17.40		46.40	58.50
2900	1-1/4" pipe size	Q-1	44	.364		29	19.20		48.20	61
2910	1-1/2" pipe size		40	.400		29	21		50	64
2920	2" pipe size		36	.444		34	23.50		57.50	73
2930	2-1/2" pipe size		28	.571		47.50	30		77.50	97.50
2940	3" pipe size		20	.800		50.50	42.50		93	120
2960	4" pipe size	▼	12	1.333		73	70.50		143.50	186
2970	6" pipe size	Q-2	14	1.714	▼	136	94		230	292
3000	Weld joint, butt, carbon steel, standard weight									
3040	90° elbow, long radius									
3050	1/2" pipe size	Q-15	16	1	Ea.	45	53	3.61	101.61	133
3060	3/4" pipe size		16	1		45	53	3.61	101.61	133
3070	1" pipe size		16	1		21	53	3.61	77.61	107
3080	1-1/4" pipe size		14	1.143		21	60.50	4.13	85.63	119
3090	1-1/2" pipe size		13	1.231		21	65	4.44	90.44	126
3100	2" pipe size		10	1.600		22.50	84.50	5.80	112.80	159
3110	2-1/2" pipe size		8	2		27.50	106	7.20	140.70	198
3120	3" pipe size		7	2.286		29	121	8.25	158.25	223
3130	4" pipe size		5	3.200		48	169	11.55	228.55	320
3136	5" pipe size	▼	4	4		103	211	14.45	328.45	450
3140	6" pipe size	Q-16	5	4.800		106	263	11.50	380.50	525
3150	8" pipe size		3.75	6.400		200	350	15.35	565.35	765
3160	10" pipe size		3	8		400	440	19.20	859.20	1,125
3170	12" pipe size		2.50	9.600		590	525	23	1,138	1,475
3180	14" pipe size		2	12		970	655	29	1,654	2,100
3190	16" pipe size		1.50	16		1,325	875	38.50	2,238.50	2,850
3191	18" pipe size		1.25	19.200		1,525	1,050	46	2,621	3,325
3192	20" pipe size		1.15	20.870		2,450	1,150	50	3,650	4,475
3194	24" pipe size	▼	1.02	23.529	▼	3,450	1,300	56.50	4,806.50	5,800
3200	45° Elbow, long									
3210	1/2" pipe size	Q-15	16	1	Ea.	62.50	53	3.61	119.11	152

22 11 13 – Facility Water Distribution Piping

22 11 13.47 Pipe Fittings, Steel		Crew	Daily Output	Labor-Hours	Unit	Material	2015 Bare Costs Labor	Equipment	Total	Total Incl O&P
3220	3/4" pipe size	Q-15	16	1	Ea.	62.50	53	3.61	119.11	152
3230	1" pipe size		16	1		22	53	3.61	78.61	108
3240	1-1/4" pipe size		14	1.143		22	60.50	4.13	86.63	120
3250	1-1/2" pipe size		13	1.231		22	65	4.44	91.44	127
3260	2" pipe size		10	1.600		22	84.50	5.80	112.30	159
3270	2-1/2" pipe size		8	2		26.50	106	7.20	139.70	197
3280	3" pipe size		7	2.286		27.50	121	8.25	156.75	222
3290	4" pipe size		5	3.200		49.50	169	11.55	230.05	320
3296	5" pipe size		4	4		75	211	14.45	300.45	420
3300	6" pipe size	Q-16	5	4.800		97.50	263	11.50	372	515
3310	8" pipe size		3.75	6.400		163	350	15.35	528.35	725
3320	10" pipe size		3	8		325	440	19.20	784.20	1,025
3330	12" pipe size		2.50	9.600		460	525	23	1,008	1,325
3340	14" pipe size		2	12		615	655	29	1,299	1,700
3341	16" pipe size		1.50	16		1,100	875	38.50	2,013.50	2,600
3342	18" pipe size		1.25	19.200		1,575	1,050	46	2,671	3,375
3343	20" pipe size		1.15	20.870		1,625	1,150	50	2,825	3,550
3345	24" pipe size		1.05	22.857		2,450	1,250	55	3,755	4,650
3346	26" pipe size		.85	28.235		2,700	1,550	68	4,318	5,375
3347	30" pipe size		.45	53.333		2,975	2,925	128	6,028	7,825
3349	36" pipe size		.38	63.158		3,275	3,450	152	6,877	9,025
3350	Tee, straight									
3352	For reducing tees and concentrics see starting line 4600									
3360	1/2" pipe size	Q-15	10	1.600	Ea.	110	84.50	5.80	200.30	255
3370	3/4" pipe size		10	1.600		110	84.50	5.80	200.30	255
3380	1" pipe size		10	1.600		54.50	84.50	5.80	144.80	194
3390	1-1/4" pipe size		9	1.778		68.50	94	6.40	168.90	225
3400	1-1/2" pipe size		8	2		68.50	106	7.20	181.70	243
3410	2" pipe size		6	2.667		54.50	141	9.65	205.15	284
3420	2-1/2" pipe size		5	3.200		75.50	169	11.55	256.05	350
3430	3" pipe size		4	4		84	211	14.45	309.45	430
3440	4" pipe size		3	5.333		118	282	19.25	419.25	575
3446	5" pipe size		2.50	6.400		195	340	23	558	750
3450	6" pipe size	Q-16	3	8		203	440	19.20	662.20	905
3460	8" pipe size		2.50	9.600		355	525	23	903	1,200
3470	10" pipe size		2	12		695	655	29	1,379	1,800
3480	12" pipe size		1.60	15		975	820	36	1,831	2,375
3481	14" pipe size		1.30	18.462		1,700	1,000	44.50	2,744.50	3,450
3482	16" pipe size		1	24		1,900	1,325	57.50	3,282.50	4,150
3483	18" pipe size		.80	30		3,025	1,650	72	4,747	5,875
3484	20" pipe size		.75	32		4,750	1,750	77	6,577	7,950
3486	24" pipe size		.70	34.286		6,150	1,875	82.50	8,107.50	9,675
3487	26" pipe size		.55	43.636		6,775	2,400	105	9,280	11,200
3488	30" pipe size		.30	80		7,425	4,375	192	11,992	15,000
3490	36" pipe size		.25	96		8,200	5,250	230	13,680	17,200
3491	Eccentric reducer, 1-1/2" pipe size	Q-15	14	1.143		63.50	60.50	4.13	128.13	166
3492	2" pipe size		11	1.455		43	77	5.25	125.25	169
3493	2-1/2" pipe size		9	1.778		50	94	6.40	150.40	204
3494	3" pipe size		8	2		59	106	7.20	172.20	233
3495	4" pipe size		6	2.667		75.50	141	9.65	226.15	305
3496	6" pipe size	Q-16	5	4.800		232	263	11.50	506.50	665
3497	8" pipe size		4	6		345	330	14.40	689.40	890
3498	10" pipe size		3	8		435	440	19.20	894.20	1,150

22 11 Facility Water Distribution

22 11 13 – Facility Water Distribution Piping

22 11 13.47 Pipe Fittings, Steel

		Crew	Daily Output	Labor-Hours	Unit	Material	2015 Bare Costs Labor	Equipment	Total	Total Incl O&P
3499	12" pipe size	Q-16	2.50	9.600	Ea.	605	525	23	1,153	1,475
3501	Cap, 1-1/2" pipe size	Q-15	28	.571		24.50	30	2.06	56.56	75
3502	2" pipe size		22	.727		22	38.50	2.63	63.13	85
3503	2-1/2" pipe size		18	.889		30	47	3.21	80.21	107
3504	3" pipe size		16	1		33.50	53	3.61	90.11	120
3505	4" pipe size		12	1.333		45	70.50	4.81	120.31	161
3506	6" pipe size	Q-16	10	2.400		92	131	5.75	228.75	305
3507	8" pipe size		8	3		138	164	7.20	309.20	410
3508	10" pipe size		6	4		212	219	9.60	440.60	575
3509	12" pipe size		5	4.800		283	263	11.50	557.50	720
3511	14" pipe size		4	6		293	330	14.40	637.40	835
3512	16" pipe size		4	6		455	330	14.40	799.40	1,000
3513	18" pipe size		3	8		665	440	19.20	1,124.20	1,425
3517	Weld joint, butt, carbon steel, extra strong									
3519	90° elbow, long									
3520	1/2" pipe size	Q-15	13	1.231	Ea.	55.50	65	4.44	124.94	164
3530	3/4" pipe size		12	1.333		55.50	70.50	4.81	130.81	173
3540	1" pipe size		11	1.455		27	77	5.25	109.25	152
3550	1-1/4" pipe size		10	1.600		27	84.50	5.80	117.30	164
3560	1-1/2" pipe size		9	1.778		27	94	6.40	127.40	179
3570	2" pipe size		8	2		27.50	106	7.20	140.70	198
3580	2-1/2" pipe size		7	2.286		38.50	121	8.25	167.75	234
3590	3" pipe size		6	2.667		49.50	141	9.65	200.15	278
3600	4" pipe size		4	4		81.50	211	14.45	306.95	425
3606	5" pipe size		3.50	4.571		195	242	16.50	453.50	595
3610	6" pipe size	Q-16	4.50	5.333		206	292	12.80	510.80	680
3620	8" pipe size		3.50	6.857		390	375	16.45	781.45	1,025
3630	10" pipe size		2.50	9.600		830	525	23	1,378	1,725
3640	12" pipe size		2.25	10.667		1,025	585	25.50	1,635.50	2,025
3650	45° Elbow, long									
3660	1/2" pipe size	Q-15	13	1.231	Ea.	62	65	4.44	131.44	171
3670	3/4" pipe size		12	1.333		62	70.50	4.81	137.31	179
3680	1" pipe size		11	1.455		29	77	5.25	111.25	153
3690	1-1/4" pipe size		10	1.600		29	84.50	5.80	119.30	166
3700	1-1/2" pipe size		9	1.778		29	94	6.40	129.40	181
3710	2" pipe size		8	2		29	106	7.20	142.20	199
3720	2-1/2" pipe size		7	2.286		64	121	8.25	193.25	262
3730	3" pipe size		6	2.667		37	141	9.65	187.65	265
3740	4" pipe size		4	4		58.50	211	14.45	283.95	400
3746	5" pipe size		3.50	4.571		139	242	16.50	397.50	535
3750	6" pipe size	Q-16	4.50	5.333		160	292	12.80	464.80	630
3760	8" pipe size		3.50	6.857		277	375	16.45	668.45	890
3770	10" pipe size		2.50	9.600		535	525	23	1,083	1,400
3780	12" pipe size		2.25	10.667		790	585	25.50	1,400.50	1,775
3800	Tee, straight									
3810	1/2" pipe size	Q-15	9	1.778	Ea.	136	94	6.40	236.40	298
3820	3/4" pipe size		8.50	1.882		131	99.50	6.80	237.30	300
3830	1" pipe size		8	2		52	106	7.20	165.20	225
3840	1-1/4" pipe size		7	2.286		52	121	8.25	181.25	249
3850	1-1/2" pipe size		6	2.667		52	141	9.65	202.65	281
3860	2" pipe size		5	3.200		59.50	169	11.55	240.05	335
3870	2-1/2" pipe size		4	4		99	211	14.45	324.45	445
3880	3" pipe size		3.50	4.571		124	242	16.50	382.50	520

168

For customer support on your Mechanical Cost Data, call 877.795.7394.

22 11 13 – Facility Water Distribution Piping

22 11 13.47 Pipe Fittings, Steel

		Crew	Daily Output	Labor-Hours	Unit	Material	2015 Bare Costs Labor	Equipment	Total	Total Incl O&P
3890	4" pipe size	Q-15	2.50	6.400	Ea.	149	340	23	512	700
3896	5" pipe size	▼	2.25	7.111		375	375	25.50	775.50	1,000
3900	6" pipe size	Q-16	2.25	10.667		286	585	25.50	896.50	1,225
3910	8" pipe size		2	12		545	655	29	1,229	1,625
3920	10" pipe size		1.75	13.714		845	750	33	1,628	2,100
3930	12" pipe size	▼	1.50	16		1,225	875	38.50	2,138.50	2,725
4000	Eccentric reducer, 1-1/2" pipe size	Q-15	10	1.600		17.30	84.50	5.80	107.60	153
4010	2" pipe size		9	1.778		47.50	94	6.40	147.90	202
4020	2-1/2" pipe size		8	2		71.50	106	7.20	184.70	246
4030	3" pipe size		7	2.286		59	121	8.25	188.25	256
4040	4" pipe size		5	3.200		95.50	169	11.55	276.05	375
4046	5" pipe size	▼	4.70	3.404		266	180	12.30	458.30	580
4050	6" pipe size	Q-16	4.50	5.333		259	292	12.80	563.80	740
4060	8" pipe size		3.50	6.857		390	375	16.45	781.45	1,025
4070	10" pipe size		2.50	9.600		665	525	23	1,213	1,550
4080	12" pipe size		2.25	10.667		845	585	25.50	1,455.50	1,825
4090	14" pipe size		2.10	11.429		1,575	625	27.50	2,227.50	2,700
4100	16" pipe size	▼	1.90	12.632		2,025	690	30.50	2,745.50	3,300
4151	Cap, 1-1/2" pipe size	Q-15	24	.667		24.50	35	2.41	61.91	82.50
4152	2" pipe size		18	.889		22	47	3.21	72.21	98.50
4153	2-1/2" pipe size		16	1		30	53	3.61	86.61	116
4154	3" pipe size		14	1.143		33.50	60.50	4.13	98.13	132
4155	4" pipe size	▼	10	1.600		45	84.50	5.80	135.30	184
4156	6" pipe size	Q-16	9	2.667		92	146	6.40	244.40	330
4157	8" pipe size		7	3.429		138	188	8.25	334.25	445
4158	10" pipe size		5	4.800		212	263	11.50	486.50	640
4159	12" pipe size	▼	4	6	▼	283	330	14.40	627.40	820
4190	Weld fittings, reducing, standard weight									
4200	Welding ring w/spacer pins, 2" pipe size				Ea.	2.40			2.40	2.64
4210	2-1/2" pipe size					2.45			2.45	2.70
4220	3" pipe size					2.50			2.50	2.75
4230	4" pipe size					2.65			2.65	2.92
4236	5" pipe size					3.35			3.35	3.69
4240	6" pipe size					3.60			3.60	3.96
4250	8" pipe size					4.20			4.20	4.62
4260	10" pipe size					4.70			4.70	5.15
4270	12" pipe size					5.40			5.40	5.95
4280	14" pipe size					6.45			6.45	7.10
4290	16" pipe size					7.60			7.60	8.35
4300	18" pipe size					8.60			8.60	9.45
4310	20" pipe size					8.95			8.95	9.85
4330	24" pipe size					11.10			11.10	12.20
4340	26" pipe size					13.70			13.70	15.05
4350	30" pipe size					16.75			16.75	18.45
4370	36" pipe size				▼	20.50			20.50	22.50
4600	Tee, reducing on outlet									
4601	2-1/2" x 2" pipe size	Q-15	5	3.200	Ea.	100	169	11.55	280.55	380
4602	3" x 2-1/2" pipe size		4	4		118	211	14.45	343.45	465
4604	4" x 3" pipe size		3	5.333		124	282	19.25	425.25	585
4605	5" x 4" pipe size	▼	2.50	6.400		268	340	23	631	830
4606	6" x 5" pipe size	Q-16	3	8		370	440	19.20	829.20	1,100
4607	8" x 6" pipe size		2.50	9.600		460	525	23	1,008	1,325
4608	10" x 8" pipe size	▼	2	12	▼	850	655	29	1,534	1,950

For customer support on your Mechanical Cost Data, call 877.795.7394.

169

22 11 13.47 Pipe Fittings, Steel

		Crew	Daily Output	Labor-Hours	Unit	Material	2015 Bare Costs Labor	2015 Bare Costs Equipment	Total	Total Incl O&P
4609	12" x 10" pipe size	Q-16	1.60	15	Ea.	1,250	820	36	2,106	2,675
4610	16" x 12" pipe size		1.50	16		2,000	875	38.50	2,913.50	3,575
4611	14" x 12" pipe size	↓	1.52	15.789	↓	1,900	865	38	2,803	3,425
4618	Reducer, concentric									
4619	2-1/2" by 2" pipe size	Q-15	10	1.600	Ea.	42.50	84.50	5.80	132.80	181
4620	3" by 2-1/2" pipe size		9	1.778		35	94	6.40	135.40	188
4621	3-1/2" by 3" pipe size		8	2		89.50	106	7.20	202.70	266
4622	4" by 2-1/2" pipe size		7	2.286		54.50	121	8.25	183.75	251
4623	5" by 3" pipe size	↓	7	2.286		114	121	8.25	243.25	315
4624	6" by 4" pipe size	Q-16	6	4		94	219	9.60	322.60	445
4625	8" by 6" pipe size		5	4.800		124	263	11.50	398.50	545
4626	10" by 8" pipe size		4	6		240	330	14.40	584.40	775
4627	12" by 10" pipe size	↓	3	8	↓	296	440	19.20	755.20	1,000
4660	Reducer, eccentric									
4662	3" x 2" pipe size	Q-15	8	2	Ea.	59	106	7.20	172.20	233
4664	4" x 3" pipe size		6	2.667		71.50	141	9.65	222.15	305
4666	4" x 2" pipe size	↓	6	2.667		87.50	141	9.65	238.15	320
4670	6" x 4" pipe size	Q-16	5	4.800		148	263	11.50	422.50	570
4672	6" x 3" pipe size		5	4.800		232	263	11.50	506.50	665
4676	8" x 6" pipe size		4	6		188	330	14.40	532.40	720
4678	8" x 4" pipe size		4	6		345	330	14.40	689.40	890
4682	10" x 8" pipe size		3	8		247	440	19.20	706.20	950
4684	10" x 6" pipe size		3	8		435	440	19.20	894.20	1,150
4688	12" x 10" pipe size		2.50	9.600		385	525	23	933	1,250
4690	12" x 8" pipe size		2.50	9.600		605	525	23	1,153	1,475
4691	14" x 12" pipe size		2.20	10.909		675	600	26	1,301	1,675
4693	16" x 14" pipe size		1.80	13.333		1,025	730	32	1,787	2,250
4694	16" x 12" pipe size		2	12		1,375	655	29	2,059	2,525
4696	18" x 16" pipe size	↓	1.60	15	↓	1,100	820	36	1,956	2,525
5000	Weld joint, socket, forged steel, 3000 lb., schedule 40 pipe									
5010	90° elbow, straight									
5020	1/4" pipe size	Q-15	22	.727	Ea.	25	38.50	2.63	66.13	88.50
5030	3/8" pipe size		22	.727		25	38.50	2.63	66.13	88.50
5040	1/2" pipe size		20	.800		13.95	42.50	2.89	59.34	82.50
5050	3/4" pipe size		20	.800		14.55	42.50	2.89	59.94	83
5060	1" pipe size		20	.800		18.70	42.50	2.89	64.09	87.50
5070	1-1/4" pipe size		18	.889		36	47	3.21	86.21	114
5080	1-1/2" pipe size		16	1		41.50	53	3.61	98.11	130
5090	2" pipe size		12	1.333		61.50	70.50	4.81	136.81	179
5100	2-1/2" pipe size		10	1.600		174	84.50	5.80	264.30	325
5110	3" pipe size		8	2		300	106	7.20	413.20	500
5120	4" pipe size	↓	6	2.667	↓	795	141	9.65	945.65	1,100
5130	45° Elbow, straight									
5134	1/4" pipe size	Q-15	22	.727	Ea.	25	38.50	2.63	66.13	88.50
5135	3/8" pipe size		22	.727		25	38.50	2.63	66.13	88.50
5136	1/2" pipe size		20	.800		18.60	42.50	2.89	63.99	87.50
5137	3/4" pipe size		20	.800		21.50	42.50	2.89	66.89	91
5140	1" pipe size		20	.800		28.50	42.50	2.89	73.89	98
5150	1-1/4" pipe size		18	.889		39	47	3.21	89.21	117
5160	1-1/2" pipe size		16	1		47.50	53	3.61	104.11	136
5170	2" pipe size		12	1.333		76.50	70.50	4.81	151.81	195
5180	2-1/2" pipe size		10	1.600		201	84.50	5.80	291.30	355
5190	3" pipe size	↓	8	2	↓	335	106	7.20	448.20	535

22 11 13 – Facility Water Distribution Piping

22 11 13.47 Pipe Fittings, Steel		Crew	Daily Output	Labor-Hours	Unit	Material	2015 Bare Costs Labor	2015 Bare Costs Equipment	Total	Total Incl O&P
5200	4" pipe size	Q-15	6	2.667	Ea.	660	141	9.65	810.65	950
5250	Tee, straight									
5254	1/4" pipe size	Q-15	15	1.067	Ea.	27.50	56.50	3.85	87.85	120
5255	3/8" pipe size		15	1.067		27.50	56.50	3.85	87.85	120
5256	1/2" pipe size		13	1.231		17.30	65	4.44	86.74	122
5257	3/4" pipe size		13	1.231		21	65	4.44	90.44	126
5260	1" pipe size		13	1.231		29	65	4.44	98.44	134
5270	1-1/4" pipe size		12	1.333		44	70.50	4.81	119.31	159
5280	1-1/2" pipe size		11	1.455		58.50	77	5.25	140.75	186
5290	2" pipe size		8	2		84.50	106	7.20	197.70	261
5300	2-1/2" pipe size		6	2.667		243	141	9.65	393.65	490
5310	3" pipe size		5	3.200		580	169	11.55	760.55	910
5320	4" pipe size		4	4		935	211	14.45	1,160.45	1,350
5350	For reducing sizes, add					60%				
5450	Couplings									
5451	1/4" pipe size	Q-15	23	.696	Ea.	15.90	37	2.51	55.41	76
5452	3/8" pipe size		23	.696		15.90	37	2.51	* 55.41	76
5453	1/2" pipe size		21	.762		7.35	40.50	2.75	50.60	72
5454	3/4" pipe size		21	.762		9.50	40.50	2.75	52.75	74.50
5460	1" pipe size		20	.800		10.50	42.50	2.89	55.89	78.50
5470	1-1/4" pipe size		20	.800		18.60	42.50	2.89	63.99	87.50
5480	1-1/2" pipe size		18	.889		20.50	47	3.21	70.71	97
5490	2" pipe size		14	1.143		33	60.50	4.13	97.63	132
5500	2-1/2" pipe size		12	1.333		73.50	70.50	4.81	148.81	192
5510	3" pipe size		9	1.778		165	94	6.40	265.40	330
5520	4" pipe size		7	2.286		248	121	8.25	377.25	465
5570	Union, 1/4" pipe size		21	.762		34	40.50	2.75	77.25	102
5571	3/8" pipe size		21	.762		34	40.50	2.75	77.25	102
5572	1/2" pipe size		19	.842		28.50	44.50	3.04	76.04	102
5573	3/4" pipe size		19	.842		33.50	44.50	3.04	81.04	107
5574	1" pipe size		19	.842		43	44.50	3.04	90.54	118
5575	1-1/4" pipe size		17	.941		69.50	49.50	3.40	122.40	155
5576	1-1/2" pipe size		15	1.067		75	56.50	3.85	135.35	172
5577	2" pipe size		11	1.455		108	77	5.25	190.25	241
5600	Reducer, 1/4" pipe size		23	.696		40.50	37	2.51	80.01	103
5601	3/8" pipe size		23	.696		43	37	2.51	82.51	105
5602	1/2" pipe size		21	.762		26	40.50	2.75	69.25	92.50
5603	3/4" pipe size		21	.762		26	40.50	2.75	69.25	92.50
5604	1" pipe size		21	.762		36.50	40.50	2.75	79.75	104
5605	1-1/4" pipe size		19	.842		46.50	44.50	3.04	94.04	122
5607	1-1/2" pipe size		17	.941		50	49.50	3.40	102.90	134
5608	2" pipe size		13	1.231		55	65	4.44	124.44	163
5612	Cap, 1/4" pipe size		46	.348		15.15	18.40	1.26	34.81	46
5613	3/8" pipe size		46	.348		15.15	18.40	1.26	34.81	46
5614	1/2" pipe size		42	.381		9.25	20	1.38	30.63	42
5615	3/4" pipe size		42	.381		10.80	20	1.38	32.18	44
5616	1" pipe size		42	.381		16.35	20	1.38	37.73	50
5617	1-1/4" pipe size		38	.421		19.30	22	1.52	42.82	56
5618	1-1/2" pipe size		34	.471		27	25	1.70	53.70	69
5619	2" pipe size		26	.615		41.50	32.50	2.22	76.22	97
5630	T-O-L, 1/4" pipe size, nozzle		23	.696		7.10	37	2.51	46.61	66
5631	3/8" pipe size, nozzle		23	.696		7.20	37	2.51	46.71	66
5632	1/2" pipe size, nozzle		22	.727		7.10	38.50	2.63	48.23	68.50

For customer support on your Mechanical Cost Data, call 877.795.7394.

171

22 11 13.47 Pipe Fittings, Steel		Crew	Daily Output	Labor-Hours	Unit	Material	2015 Bare Costs Labor	Equipment	Total	Total Incl O&P
5633	3/4" pipe size, nozzle	Q-15	21	.762	Ea.	8.20	40.50	2.75	51.45	73
5634	1" pipe size, nozzle		20	.800		9.55	42.50	2.89	54.94	77.50
5635	1-1/4" pipe size, nozzle		18	.889		15.10	47	3.21	65.31	91
5636	1-1/2" pipe size, nozzle		16	1		15.10	53	3.61	71.71	101
5637	2" pipe size, nozzle		12	1.333		17.20	70.50	4.81	92.51	130
5638	2-1/2" pipe size, nozzle		10	1.600		57.50	84.50	5.80	147.80	198
5639	4" pipe size, nozzle		6	2.667		131	141	9.65	281.65	370
5640	W-O-L, 1/4" pipe size, nozzle		23	.696		16.80	37	2.51	56.31	76.50
5641	3/8" pipe size, nozzle		23	.696		15.85	37	2.51	55.36	75.50
5642	1/2" pipe size, nozzle		22	.727		14.70	38.50	2.63	55.83	77
5643	3/4" pipe size, nozzle		21	.762		15.50	40.50	2.75	58.75	81
5644	1" pipe size, nozzle		20	.800		16.20	42.50	2.89	61.59	85
5645	1-1/4" pipe size, nozzle		18	.889		19.25	47	3.21	69.46	95.50
5646	1-1/2" pipe size, nozzle		16	1		19.25	53	3.61	75.86	105
5647	2" pipe size, nozzle		12	1.333		19.45	70.50	4.81	94.76	133
5648	2-1/2" pipe size, nozzle		10	1.600		44.50	84.50	5.80	134.80	183
5649	3" pipe size, nozzle		8	2		48.50	106	7.20	161.70	221
5650	4" pipe size, nozzle		6	2.667		61.50	141	9.65	212.15	291
5651	5" pipe size, nozzle		5	3.200		152	169	11.55	332.55	435
5652	6" pipe size, nozzle		4	4		177	211	14.45	402.45	530
5653	8" pipe size, nozzle		3	5.333		340	282	19.25	641.25	820
5654	10" pipe size, nozzle		2.60	6.154		490	325	22	837	1,050
5655	12" pipe size, nozzle		2.20	7.273		930	385	26.50	1,341.50	1,625
5674	S-O-L, 1/4" pipe size, outlet		23	.696		9.45	37	2.51	48.96	68.50
5675	3/8" pipe size, outlet		23	.696		9.45	37	2.51	48.96	68.50
5676	1/2" pipe size, outlet		22	.727		8.80	38.50	2.63	49.93	70.50
5677	3/4" pipe size, outlet		21	.762		8.95	40.50	2.75	52.20	74
5678	1" pipe size, outlet		20	.800		9.90	42.50	2.89	55.29	78
5679	1-1/4" pipe size, outlet		18	.889		16.55	47	3.21	66.76	92.50
5680	1-1/2" pipe size, outlet		16	1		16.55	53	3.61	73.16	102
5681	2" pipe size, outlet		12	1.333		18.80	70.50	4.81	94.11	132
6000	Weld-on flange, forged steel									
6020	Slip-on, 150 lb. flange (welded front and back)									
6050	1/2" pipe size	Q-15	18	.889	Ea.	17.50	47	3.21	67.71	94
6060	3/4" pipe size		18	.889		17.50	47	3.21	67.71	94
6070	1" pipe size		17	.941		17.50	49.50	3.40	70.40	98
6080	1-1/4" pipe size		16	1		17.50	53	3.61	74.11	103
6090	1-1/2" pipe size		15	1.067		17.50	56.50	3.85	77.85	108
6100	2" pipe size		12	1.333		19.45	70.50	4.81	94.76	133
6110	2-1/2" pipe size		10	1.600		25	84.50	5.80	115.30	162
6120	3" pipe size		9	1.778		26.50	94	6.40	126.90	178
6130	3-1/2" pipe size		7	2.286		33	121	8.25	162.25	228
6140	4" pipe size		6	2.667		33	141	9.65	183.65	260
6150	5" pipe size		5	3.200		57.50	169	11.55	238.05	330
6160	6" pipe size	Q-16	6	4		54.50	219	9.60	283.10	400
6170	8" pipe size		5	4.800		82.50	263	11.50	357	500
6180	10" pipe size		4	6		143	330	14.40	487.40	670
6190	12" pipe size		3	8		212	440	19.20	671.20	915
6191	14" pipe size		2.50	9.600		281	525	23	829	1,125
6192	16" pipe size		1.80	13.333		440	730	32	1,202	1,625
6200	300 lb. flange									
6210	1/2" pipe size	Q-15	17	.941	Ea.	23.50	49.50	3.40	76.40	104
6220	3/4" pipe size		17	.941		23.50	49.50	3.40	76.40	104

22 11 13 – Facility Water Distribution Piping

22 11 13.47 Pipe Fittings, Steel	Crew	Daily Output	Labor-Hours	Unit	Material	2015 Bare Costs Labor	Equipment	Total	Total Incl O&P
6230	1" pipe size	Q-15	16	1	Ea.	23.50	53	3.61	80.11	109
6240	1-1/4" pipe size		13	1.231		23.50	65	4.44	92.94	128
6250	1-1/2" pipe size		12	1.333		23.50	70.50	4.81	98.81	137
6260	2" pipe size		11	1.455		31	77	5.25	113.25	156
6270	2-1/2" pipe size		9	1.778		35	94	6.40	135.40	188
6280	3" pipe size		7	2.286		38	121	8.25	167.25	233
6290	4" pipe size		6	2.667		55.50	141	9.65	206.15	285
6300	5" pipe size	▼	4	4		95	211	14.45	320.45	440
6310	6" pipe size	Q-16	5	4.800		95.50	263	11.50	370	515
6320	8" pipe size		4	6		163	330	14.40	507.40	690
6330	10" pipe size		3.40	7.059		278	385	16.95	679.95	910
6340	12" pipe size	▼	2.80	8.571	▼	340	470	20.50	830.50	1,100
6400	Welding neck, 150 lb. flange									
6410	1/2" pipe size	Q-15	40	.400	Ea.	25.50	21	1.44	47.94	61.50
6420	3/4" pipe size		36	.444		25.50	23.50	1.60	50.60	65.50
6430	1" pipe size		32	.500		25.50	26.50	1.81	53.81	70
6440	1-1/4" pipe size		29	.552		25.50	29	1.99	56.49	74
6450	1-1/2" pipe size		26	.615		25.50	32.50	2.22	60.22	79.50
6460	2" pipe size		20	.800		29.50	42.50	2.89	74.89	99
6470	2-1/2" pipe size		16	1		32	53	3.61	88.61	119
6480	3" pipe size		14	1.143		35	60.50	4.13	99.63	134
6500	4" pipe size		10	1.600		42.50	84.50	5.80	132.80	181
6510	5" pipe size	▼	8	2		66.50	106	7.20	179.70	241
6520	6" pipe size	Q-16	10	2.400		64	131	5.75	200.75	275
6530	8" pipe size		7	3.429		113	188	8.25	309.25	415
6540	10" pipe size		6	4		180	219	9.60	408.60	540
6550	12" pipe size		5	4.800		262	263	11.50	536.50	695
6551	14" pipe size		4.50	5.333		380	292	12.80	684.80	870
6552	16" pipe size		3	8		570	440	19.20	1,029.20	1,300
6553	18" pipe size		2.50	9.600		800	525	23	1,348	1,700
6554	20" pipe size		2.30	10.435		975	570	25	1,570	1,975
6556	24" pipe size		2	12		1,300	655	29	1,984	2,475
6557	26" pipe size		1.70	14.118		1,425	775	34	2,234	2,800
6558	30" pipe size		.90	26.667		1,650	1,450	64	3,164	4,075
6559	36" pipe size	▼	.75	32	▼	1,900	1,750	77	3,727	4,800
6560	300 lb. flange									
6570	1/2" pipe size	Q-15	36	.444	Ea.	31	23.50	1.60	56.10	71.50
6580	3/4" pipe size		34	.471		31	25	1.70	57.70	73.50
6590	1" pipe size		30	.533		31	28	1.93	60.93	78.50
6600	1-1/4" pipe size		28	.571		31	30	2.06	63.06	82
6610	1-1/2" pipe size		24	.667		31	35	2.41	68.41	89.50
6620	2" pipe size		18	.889		38.50	47	3.21	88.71	117
6630	2-1/2" pipe size		14	1.143		44	60.50	4.13	108.63	144
6640	3" pipe size		12	1.333		44.50	70.50	4.81	119.81	160
6650	4" pipe size		8	2		73	106	7.20	186.20	248
6660	5" pipe size	▼	7	2.286		111	121	8.25	240.25	315
6670	6" pipe size	Q-16	9	2.667		113	146	6.40	265.40	355
6680	8" pipe size		6	4		195	219	9.60	423.60	555
6690	10" pipe size		5	4.800		355	263	11.50	629.50	800
6700	12" pipe size		4	6		455	330	14.40	799.40	1,000
6710	14" pipe size		3.50	6.857		980	375	16.45	1,371.45	1,650
6720	16" pipe size	▼	2	12	▼	1,150	655	29	1,834	2,300
7740	Plain ends for plain end pipe, mechanically coupled									

22 11 13.47 Pipe Fittings, Steel	Crew	Daily Output	Labor-Hours	Unit	Material	2015 Bare Costs Labor	Equipment	Total	Total Incl O&P
7750 Cplg. & labor required at joints not included, add 1 per									
7760 joint for installed price, see line 9180									
7770 Malleable iron, painted, unless noted otherwise									
7800 90° Elbow 1"				Ea.	112			112	124
7810 1-1/2"					133			133	146
7820 2"					198			198	218
7830 2-1/2"					232			232	255
7840 3"					240			240	264
7860 4"					272			272	300
7870 5" welded steel					315			315	345
7880 6"					395			395	435
7890 8" welded steel					745			745	815
7900 10" welded steel					945			945	1,050
7910 12" welded steel					1,050			1,050	1,150
7970 45° Elbow 1"					67			67	73.50
7980 1-1/2"					100			100	110
7990 2"					210			210	231
8000 2-1/2"					210			210	231
8010 3"					240			240	264
8030 4"					251			251	276
8040 5" welded steel					315			315	345
8050 6"					360			360	395
8060 8"					420			420	460
8070 10" welded steel					435			435	480
8080 12" welded steel					785			785	865
8140 Tee, straight 1"					126			126	138
8150 1-1/2"					162			162	178
8160 2"					162			162	178
8170 2-1/2"					211			211	232
8180 3"					325			325	355
8200 4"					465			465	510
8210 5" welded steel					635			635	695
8220 6"					550			550	605
8230 8" welded steel					795			795	875
8240 10" welded steel					1,250			1,250	1,375
8250 12" welded steel					1,450			1,450	1,600
8340 Segmentally welded steel, painted									
8390 Wye 2"				Ea.	201			201	221
8400 2-1/2"					201			201	221
8410 3"					225			225	247
8430 4"					335			335	370
8440 5"					410			410	450
8450 6"					560			560	620
8460 8"					730			730	800
8470 10"					1,075			1,075	1,175
8480 12"					1,675			1,675	1,825
8540 Wye, lateral 2"					218			218	240
8550 2-1/2"					251			251	276
8560 3"					298			298	330
8580 4"					410			410	450
8590 5"					695			695	765
8600 6"					720			720	790
8610 8"					1,200			1,200	1,325

22 11 13 – Facility Water Distribution Piping

22 11 13.47 Pipe Fittings, Steel		Crew	Daily Output	Labor-Hours	Unit	Material	2015 Bare Costs Labor	Equipment	Total	Total Incl O&P
8620	10"				Ea.	1,200			1,200	1,325
8630	12"					2,300			2,300	2,525
8690	Cross, 2"					212			212	234
8700	2-1/2"					212			212	234
8710	3"					253			253	279
8730	4"					345			345	380
8740	5"					490			490	540
8750	6"					650			650	715
8760	8"					830			830	915
8770	10"					1,225			1,225	1,325
8780	12"					1,750			1,750	1,925
8800	Tees, reducing 2" x 1"					138			138	152
8810	2" x 1-1/2"					138			138	152
8820	3" x 1"					138			138	152
8830	3" x 1-1/2"					138			138	152
8840	3" x 2"					139			139	153
8850	4" x 1"					203			203	223
8860	4" x 1-1/2"					203			203	223
8870	4" x 2"					200			200	220
8880	4" x 2-1/2"					206			206	227
8890	4" x 3"					206			206	227
8900	6" x 2"					300			300	330
8910	6" x 3"					325			325	355
8920	6" x 4"					355			355	390
8930	8" x 2"					365			365	405
8940	8" x 3"					375			375	410
8950	8" x 4"					400			400	440
8960	8" x 5"					430			430	475
8970	8" x 6"					465			465	515
8980	10" x 4"					620			620	680
8990	10" x 6"					620			620	680
9000	10" x 8"					655			655	720
9010	12" x 6"					955			955	1,050
9020	12" x 8"					955			955	1,050
9030	12" x 10"					995			995	1,100
9080	Adapter nipples 3" long									
9090	1"				Ea.	19.40			19.40	21.50
9100	1-1/2"					19.40			19.40	21.50
9110	2"					19.40			19.40	21.50
9120	2-1/2"					22.50			22.50	24.50
9130	3"					27			27	30
9140	4"					44.50			44.50	49
9150	6"					117			117	129
9180	Coupling, mechanical, plain end pipe to plain end pipe or fitting									
9190	1"	Q-1	29	.552	Ea.	68	29		97	119
9200	1-1/2"		28	.571		68	30		98	120
9210	2"		27	.593		68	31.50		99.50	122
9220	2-1/2"		26	.615		68	32.50		100.50	124
9230	3"		25	.640		99	34		133	160
9240	3-1/2"		24	.667		114	35		149	179
9250	4"		22	.727		114	38.50		152.50	184
9260	5"	Q-2	28	.857		163	47		210	250
9270	6"		24	1		199	55		254	300

For customer support on your Mechanical Cost Data, call 877.795.7394.

175

22 11 13.47 Pipe Fittings, Steel

		Crew	Daily Output	Labor-Hours	Unit	Material	2015 Bare Costs Labor	Equipment	Total	Total Incl O&P
9280	8"	Q-2	19	1.263	Ea.	350	69		419	490
9290	10"		16	1.500		455	82		537	625
9300	12"	↓	12	2	↓	570	110		680	795
9310	Outlets for precut holes through pipe wall									
9331	Strapless type, with gasket									
9332	4" to 8" pipe x 1/2"	1 Plum	13	.615	Ea.	74	36		110	136
9333	4" to 8" pipe x 3/4"		13	.615		79	36		115	142
9334	10" pipe and larger x 1/2"		11	.727		74	42.50		116.50	146
9335	10" pipe and larger x 3/4"	↓	11	.727	↓	79	42.50		121.50	152
9341	Thermometer wells with gasket									
9342	4" to 8" pipe, 6" stem	1 Plum	14	.571	Ea.	111	33.50		144.50	174
9343	8" pipe and larger, 6" stem	"	13	.615	"	111	36		147	178
9940	For galvanized fittings for plain end pipe, add					20%				

22 11 13.48 Pipe, Fittings and Valves, Steel, Grooved-Joint

		Crew	Daily Output	Labor-Hours	Unit	Material	2015 Bare Costs Labor	Equipment	Total	Total Incl O&P
0010	**PIPE, FITTINGS AND VALVES, STEEL, GROOVED-JOINT**									
0012	Fittings are ductile iron. Steel fittings noted.									
0020	Pipe includes coupling & clevis type hanger assemblies, 10' O.C.									
0500	Schedule 10, black									
0550	2" diameter	1 Plum	43	.186	L.F.	7.15	10.90		18.05	24.50
0560	2-1/2" diameter	Q-1	61	.262		8.85	13.85		22.70	31
0570	3" diameter		55	.291		10.15	15.35		25.50	34
0580	3-1/2" diameter		53	.302		12.80	15.95		28.75	38
0590	4" diameter		49	.327		13.40	17.25		30.65	41
0600	5" diameter	↓	40	.400		16.15	21		37.15	50
0610	6" diameter	Q-2	46	.522		20.50	28.50		49	65.50
0620	8" diameter	"	41	.585	↓	28	32		60	79.50
0700	To delete couplings & hangers, subtract									
0710	2" diam. to 5" diam.					25%	20%			
0720	6" diam. to 8" diam.					27%	15%			
1000	Schedule 40, black									
1040	3/4" diameter	1 Plum	71	.113	L.F.	5.50	6.60		12.10	16
1050	1" diameter		63	.127		5.30	7.45		12.75	17.05
1060	1-1/4" diameter		58	.138		6.60	8.10		14.70	19.50
1070	1-1/2" diameter		51	.157		7.35	9.20		16.55	22
1080	2" diameter	↓	40	.200		8.70	11.75		20.45	27.50
1090	2-1/2" diameter	Q-1	57	.281		13.65	14.85		28.50	37.50
1100	3" diameter		50	.320		16.75	16.90		33.65	44
1110	4" diameter		45	.356		23.50	18.80		42.30	54.50
1120	5" diameter	↓	37	.432		38	23		61	76.50
1130	6" diameter	Q-2	42	.571		48.50	31.50		80	101
1140	8" diameter		37	.649		78.50	35.50		114	140
1150	10" diameter		31	.774		106	42.50		148.50	181
1160	12" diameter		27	.889		118	48.50		166.50	204
1170	14" diameter		20	1.200		122	65.50		187.50	235
1180	16" diameter		17	1.412		183	77.50		260.50	320
1190	18" diameter		14	1.714		183	94		277	345
1200	20" diameter		12	2		214	110		324	400
1210	24" diameter	↓	10	2.400	↓	241	131		372	465
1740	To delete coupling & hanger, subtract									
1750	3/4" diam. to 2" diam.					65%	27%			
1760	2-1/2" diam. to 5" diam.					41%	18%			
1770	6" diam. to 12" diam.					31%	13%			

22 11 13 – Facility Water Distribution Piping

22 11 13.48 Pipe, Fittings and Valves, Steel, Grooved-Joint		Crew	Daily Output	Labor-Hours	Unit	Material	2015 Bare Costs Labor	Equipment	Total	Total Incl O&P
1780	14" diam. to 24" diam.					35%	10%			
1800	Galvanized									
1840	3/4" diameter	1 Plum	71	.113	L.F.	5.80	6.60		12.40	16.35
1850	1" diameter		63	.127		7	7.45		14.45	18.95
1860	1-1/4" diameter		58	.138		8.95	8.10		17.05	22
1870	1-1/2" diameter		51	.157		10.10	9.20		19.30	25
1880	2" diameter		40	.200		12.50	11.75		24.25	31.50
1890	2-1/2" diameter	Q-1	57	.281		17.60	14.85		32.45	42
1900	3" diameter		50	.320		22.50	16.90		39.40	50
1910	4" diameter		45	.356		32	18.80		50.80	63.50
1920	5" diameter		37	.432		60	23		83	100
1930	6" diameter	Q-2	42	.571		64.50	31.50		96	119
1940	8" diameter		37	.649		76	35.50		111.50	137
1950	10" diameter		31	.774		122	42.50		164.50	198
1960	12" diameter		27	.889		148	48.50		196.50	237
2540	To delete coupling & hanger, subtract									
2550	3/4" diam. to 2" diam.					36%	27%			
2560	2-1/2" diam. to 5" diam.					19%	18%			
2570	6" diam. to 12" diam.					14%	13%			
2600	Schedule 80, black									
2610	3/4" diameter	1 Plum	65	.123	L.F.	6.60	7.20		13.80	18.15
2650	1" diameter		61	.131		8.10	7.70		15.80	20.50
2660	1-1/4" diameter		55	.145		10.40	8.55		18.95	24.50
2670	1-1/2" diameter		49	.163		11.80	9.60		21.40	27.50
2680	2" diameter		38	.211		14.80	12.35		27.15	35
2690	2-1/2" diameter	Q-1	54	.296		21	15.65		36.65	47
2700	3" diameter		48	.333		27	17.60		44.60	56
2710	4" diameter		44	.364		38	19.20		57.20	71
2720	5" diameter		35	.457		55.50	24		79.50	97.50
2730	6" diameter	Q-2	40	.600		74.50	33		107.50	131
2740	8" diameter		35	.686		113	37.50		150.50	181
2750	10" diameter		29	.828		181	45.50		226.50	268
2760	12" diameter		24	1		395	55		450	520
3240	To delete coupling & hanger, subtract									
3250	3/4" diam. to 2" diam.					30%	25%			
3260	2-1/2" diam. to 5" diam.					14%	17%			
3270	6" diam. to 12" diam.					12%	12%			
3300	Galvanized									
3310	3/4" diameter	1 Plum	65	.123	L.F.	7.55	7.20		14.75	19.20
3350	1" diameter		61	.131		9.35	7.70		17.05	22
3360	1-1/4" diameter		55	.145		12.30	8.55		20.85	26.50
3370	1-1/2" diameter		46	.174		14.30	10.20		24.50	31
3380	2" diameter		38	.211		18.60	12.35		30.95	39
3390	2-1/2" diameter	Q-1	54	.296		26.50	15.65		42.15	53
3400	3" diameter		48	.333		34.50	17.60		52.10	64.50
3410	4" diameter		44	.364		40.50	19.20		59.70	73.50
3420	5" diameter		35	.457		57.50	24		81.50	99.50
3430	6" diameter	Q-2	40	.600		78.50	33		111.50	136
3440	8" diameter		35	.686		147	37.50		184.50	218
3450	10" diameter		29	.828		271	45.50		316.50	365
3460	12" diameter		24	1		530	55		585	670
3920	To delete coupling & hanger, subtract									
3930	3/4" diam. to 2" diam.					30%	25%			

For customer support on your Mechanical Cost Data, call 877.795.7394.

177

22 11 13.48 Pipe, Fittings and Valves, Steel, Grooved-Joint	Crew	Daily Output	Labor-Hours	Unit	Material	2015 Bare Costs Labor	Equipment	Total	Total Incl O&P
3940	2-1/2" diam. to 5" diam.					15%	17%			
3950	6" diam. to 12" diam.					11%	12%			
3990	Fittings: coupling material required at joints not incl. in fitting price.									
3994	Add 1 selected coupling, material only, per joint for installed price.									
4000	Elbow, 90° or 45°, painted									
4030	3/4" diameter	1 Plum	50	.160	Ea.	60	9.40		69.40	80
4040	1" diameter		50	.160		32	9.40		41.40	49
4050	1-1/4" diameter		40	.200		32	11.75		43.75	53
4060	1-1/2" diameter		33	.242		32	14.25		46.25	56.50
4070	2" diameter		25	.320		32	18.80		50.80	63.50
4080	2-1/2" diameter	Q-1	40	.400		32	21		53	67
4090	3" diameter		33	.485		56.50	25.50		82	101
4100	4" diameter		25	.640		61.50	34		95.50	119
4110	5" diameter		20	.800		146	42.50		188.50	225
4120	6" diameter	Q-2	25	.960		172	52.50		224.50	269
4130	8" diameter		21	1.143		360	62.50		422.50	490
4140	10" diameter		18	1.333		625	73		698	800
4150	12" diameter		15	1.600		1,000	87.50		1,087.50	1,225
4170	14" diameter		12	2		860	110		970	1,100
4180	16" diameter		11	2.182		1,125	120		1,245	1,400
4190	18" diameter	Q-3	15	2.133		1,425	119		1,544	1,725
4200	20" diameter		13	2.462		1,875	138		2,013	2,250
4210	24" diameter		11	2.909		2,700	163		2,863	3,225
4250	For galvanized elbows, add					26%				
4690	Tee, painted									
4700	3/4" diameter	1 Plum	38	.211	Ea.	64.50	12.35		76.85	89.50
4740	1" diameter		33	.242		50	14.25		64.25	76.50
4750	1-1/4" diameter		27	.296		50	17.40		67.40	81.50
4760	1-1/2" diameter		22	.364		50	21.50		71.50	87
4770	2" diameter		17	.471		50	27.50		77.50	96.50
4780	2-1/2" diameter	Q-1	27	.593		50	31.50		81.50	103
4790	3" diameter		22	.727		68	38.50		106.50	133
4800	4" diameter		17	.941		103	49.50		152.50	189
4810	5" diameter		13	1.231		241	65		306	365
4820	6" diameter	Q-2	17	1.412		278	77.50		355.50	420
4830	8" diameter		14	1.714		610	94		704	810
4840	10" diameter		12	2		1,225	110		1,335	1,525
4850	12" diameter		10	2.400		1,700	131		1,831	2,075
4851	14" diameter		9	2.667		1,150	146		1,296	1,500
4852	16" diameter		8	3		1,300	164		1,464	1,675
4853	18" diameter	Q-3	11	2.909		1,625	163		1,788	2,050
4854	20" diameter		10	3.200		2,325	179		2,504	2,850
4855	24" diameter		8	4		3,550	224		3,774	4,250
4900	For galvanized tees, add					24%				
4906	Couplings, rigid style, painted									
4908	1" diameter	1 Plum	100	.080	Ea.	25	4.70		29.70	34.50
4909	1-1/4" diameter		100	.080		25	4.70		29.70	34.50
4910	1-1/2" diameter		67	.119		25	7		32	38
4912	2" diameter		50	.160		31.50	9.40		40.90	48.50
4914	2-1/2" diameter	Q-1	80	.200		36	10.55		46.55	55.50
4916	3" diameter		67	.239		41.50	12.60		54.10	65
4918	4" diameter		50	.320		58	16.90		74.90	89.50
4920	5" diameter		40	.400		75	21		96	115

For customer support on your Mechanical Cost Data, call 877.795.7394.

22 11 Facility Water Distribution

22 11 13 – Facility Water Distribution Piping

22 11 13.48 Pipe, Fittings and Valves, Steel, Grooved-Joint	Crew	Daily Output	Labor-Hours	Unit	Material	2015 Bare Costs Labor	Equipment	Total	Total Incl O&P
4922	6" diameter	Q-2	50	.480	Ea.	99	26.50		125.50	149
4924	8" diameter		42	.571		156	31.50		187.50	219
4926	10" diameter		35	.686		305	37.50		342.50	390
4928	12" diameter		32	.750		340	41		381	430
4930	14" diameter		24	1		274	55		329	385
4931	16" diameter		20	1.200		355	65.50		420.50	490
4932	18" diameter		18	1.333		410	73		483	565
4933	20" diameter		16	1.500		565	82		647	745
4934	24" diameter	↓	13	1.846	↓	720	101		821	950
4940	Flexible, standard, painted									
4950	3/4" diameter	1 Plum	100	.080	Ea.	17.80	4.70		22.50	26.50
4960	1" diameter		100	.080		17.80	4.70		22.50	26.50
4970	1-1/4" diameter		80	.100		23.50	5.85		29.35	34.50
4980	1-1/2" diameter		67	.119		25.50	7		32.50	38.50
4990	2" diameter	↓	50	.160		27	9.40		36.40	44
5000	2-1/2" diameter	Q-1	80	.200		31.50	10.55		42.05	51
5010	3" diameter		67	.239		35	12.60		47.60	57.50
5020	3-1/2" diameter		57	.281		50	14.85		64.85	77.50
5030	4" diameter		50	.320		50.50	16.90		67.40	81
5040	5" diameter	↓	40	.400		76.50	21		97.50	117
5050	6" diameter	Q-2	50	.480		90.50	26.50		117	139
5070	8" diameter		42	.571		147	31.50		178.50	210
5090	10" diameter		35	.686		240	37.50		277.50	320
5110	12" diameter		32	.750		272	41		313	360
5120	14" diameter		24	1		390	55		445	515
5130	16" diameter		20	1.200		510	65.50		575.50	660
5140	18" diameter		18	1.333		595	73		668	765
5150	20" diameter		16	1.500		940	82		1,022	1,150
5160	24" diameter	↓	13	1.846	↓	1,025	101		1,126	1,275
5176	Lightweight style, painted									
5178	1-1/2" diameter	1 Plum	67	.119	Ea.	22.50	7		29.50	35
5180	2" diameter	"	50	.160		27	9.40		36.40	44
5182	2-1/2" diameter	Q-1	80	.200		31.50	10.55		42.05	51
5184	3" diameter		67	.239		35	12.60		47.60	57.50
5186	3-1/2" diameter		57	.281		43	14.85		57.85	70
5188	4" diameter		50	.320		50.50	16.90		67.40	81
5190	5" diameter	↓	40	.400		76.50	21		97.50	117
5192	6" diameter	Q-2	50	.480		90.50	26.50		117	139
5194	8" diameter		42	.571		147	31.50		178.50	210
5196	10" diameter		35	.686		297	37.50		334.50	380
5198	12" diameter	↓	32	.750		335	41		376	425
5200	For galvanized couplings, add				↓	33%				
5220	Tee, reducing, painted									
5225	2" x 1-1/2" diameter	Q-1	38	.421	Ea.	105	22		127	149
5226	2-1/2" x 2" diameter		28	.571		105	30		135	161
5227	3" x 2-1/2" diameter		23	.696		92.50	37		129.50	158
5228	4" x 3" diameter		18	.889		124	47		171	208
5229	5" x 4" diameter	↓	15	1.067		266	56.50		322.50	375
5230	6" x 4" diameter	Q-2	18	1.333		293	73		366	430
5231	8" x 6" diameter		15	1.600		610	87.50		697.50	800
5232	10" x 8" diameter		13	1.846		770	101		871	1,000
5233	12" x 10" diameter		11	2.182		1,175	120		1,295	1,475
5234	14" x 12" diameter	↓	10	2.400	↓	840	131		971	1,125

For customer support on your Mechanical Cost Data, call 877.795.7394.

179

22 11 13 – Facility Water Distribution Piping

	22 11 13.48 Pipe, Fittings and Valves, Steel, Grooved-Joint	Crew	Daily Output	Labor-Hours	Unit	Material	2015 Bare Costs Labor	Equipment	Total	Total Incl O&P
5235	16" x 12" diameter	Q-2	9	2.667	Ea.	1,050	146		1,196	1,375
5236	18" x 12" diameter	Q-3	12	2.667		1,250	149		1,399	1,600
5237	18" x 16" diameter		11	2.909		1,575	163		1,738	1,975
5238	20" x 16" diameter		10	3.200		2,050	179		2,229	2,550
5239	24" x 20" diameter	▼	9	3.556	▼	3,250	199		3,449	3,875
5240	Reducer, concentric, painted									
5241	2-1/2" x 2" diameter	Q-1	43	.372	Ea.	37.50	19.65		57.15	71
5242	3" x 2-1/2" diameter		35	.457		45	24		69	86
5243	4" x 3" diameter		29	.552		54.50	29		83.50	104
5244	5" x 4" diameter	▼	22	.727		75	38.50		113.50	141
5245	6" x 4" diameter	Q-2	26	.923		87	50.50		137.50	172
5246	8" x 6" diameter		23	1.043		224	57		281	335
5247	10" x 8" diameter		20	1.200		440	65.50		505.50	585
5248	12" x 10" diameter	▼	16	1.500	▼	790	82		872	990
5255	Eccentric, painted									
5256	2-1/2" x 2" diameter	Q-1	42	.381	Ea.	78	20		98	117
5257	3" x 2-1/2" diameter		34	.471		89	25		114	136
5258	4" x 3" diameter		28	.571		109	30		139	165
5259	5" x 4" diameter	▼	21	.762		147	40.50		187.50	223
5260	6" x 4" diameter	Q-2	25	.960		172	52.50		224.50	269
5261	8" x 6" diameter		22	1.091		345	60		405	470
5262	10" x 8" diameter		19	1.263		915	69		984	1,100
5263	12" x 10" diameter	▼	15	1.600	▼	1,250	87.50		1,337.50	1,500
5270	Coupling, reducing, painted									
5272	2" x 1-1/2" diameter	1 Plum	52	.154	Ea.	36	9.05		45.05	53
5274	2-1/2" x 2" diameter	Q-1	82	.195		47	10.30		57.30	67
5276	3" x 2" diameter		69	.232		53	12.25		65.25	77
5278	4" x 2" diameter		52	.308		84	16.25		100.25	117
5280	5" x 4" diameter	▼	42	.381		94.50	20		114.50	135
5282	6" x 4" diameter	Q-2	52	.462		142	25.50		167.50	195
5284	8" x 6" diameter	"	44	.545	▼	213	30		243	279
5290	Outlet coupling, painted									
5294	1-1/2" x 1" pipe size	1 Plum	65	.123	Ea.	47.50	7.20		54.70	63.50
5296	2" x 1" pipe size	"	48	.167		48.50	9.80		58.30	68.50
5298	2-1/2" x 1" pipe size	Q-1	78	.205		75	10.85		85.85	99
5300	2-1/2" x 1-1/4" pipe size		70	.229		84	12.10		96.10	111
5302	3" x 1" pipe size		65	.246		95.50	13		108.50	125
5304	4" x 3/4" pipe size		48	.333		106	17.60		123.60	143
5306	4" x 1-1/2" pipe size	▼	46	.348		150	18.40		168.40	193
5308	6" x 1-1/2" pipe size	Q-2	44	.545	▼	211	30		241	278
5320	Outlet, strap-on T, painted									
5324	Threaded female branch									
5330	2" pipe x 1-1/2" branch size	1 Plum	50	.160	Ea.	34.50	9.40		43.90	52
5334	2-1/2" pipe x 1-1/2" branch size	Q-1	80	.200		42	10.55		52.55	62.50
5338	3" pipe x 1-1/2" branch size		67	.239		44	12.60		56.60	67
5342	3" pipe x 2" branch size		59	.271		50.50	14.35		64.85	77
5346	4" pipe x 2" branch size		50	.320		53	16.90		69.90	84
5350	4" pipe x 2-1/2" branch size		47	.340		55.50	18		73.50	88
5354	4" pipe x 3" branch size		43	.372		59	19.65		78.65	94.50
5358	5" pipe x 3" branch size	▼	40	.400		73	21		94	112
5362	6" pipe x 2" branch size	Q-2	50	.480		66.50	26.50		93	113
5366	6" pipe x 3" branch size		47	.511		76	28		104	126
5370	6" pipe x 4" branch size		44	.545	▼	84	30		114	138

22 11 13.48 Pipe, Fittings and Valves, Steel, Grooved-Joint	Crew	Daily Output	Labor-Hours	Unit	Material	2015 Bare Costs Labor	Equipment	Total	Total Incl O&P
5374	8" pipe x 2-1/2" branch size	Q-2	42	.571	Ea.	116	31.50		147.50	176
5378	8" pipe x 4" branch size		38	.632	↓	126	34.50		160.50	192
5390	Grooved branch									
5394	2" pipe x 1-1/4" branch size	1 Plum	78	.103	Ea.	34.50	6		40.50	47
5398	2" pipe x 1-1/2" branch size	"	50	.160		34.50	9.40		43.90	52
5402	2-1/2" pipe x 1-1/2" branch size	Q-1	80	.200		42	10.55		52.55	62.50
5406	3" pipe x 1-1/2" branch size		67	.239		44	12.60		56.60	67
5410	3" pipe x 2" branch size		59	.271		50.50	14.35		64.85	77
5414	4" pipe x 2" branch size		50	.320		53	16.90		69.90	84
5418	4" pipe x 2-1/2" branch size		47	.340		55.50	18		73.50	88
5422	4" pipe x 3" branch size		43	.372		59	19.65		78.65	94.50
5426	5" pipe x 3" branch size	↓	40	.400		73	21		94	112
5430	6" pipe x 2" branch size	Q-2	50	.480		66.50	26.50		93	113
5434	6" pipe x 3" branch size	Q-1	78	.205		76	10.85		86.85	100
5438	6" pipe x 4" branch size	Q-2	44	.545		84	30		114	138
5442	8" pipe x 2-1/2" branch size		42	.571		116	31.50		147.50	176
5446	8" pipe x 4" branch size	↓	38	.632	↓	126	34.50		160.50	192
5750	Flange, w/groove gasket, black steel									
5754	See Line 22 11 13.47 0620 for gasket & bolt set						·			
5760	ANSI class 125 and 150, painted									
5780	2" pipe size	1 Plum	23	.348	Ea.	106	20.50		126.50	148
5790	2-1/2" pipe size	Q-1	37	.432		132	23		155	180
5800	3" pipe size		31	.516		142	27.50		169.50	197
5820	4" pipe size		23	.696		189	37		226	264
5830	5" pipe size		19	.842		220	44.50		264.50	310
5840	6" pipe size	Q-2	23	1.043		240	57		297	350
5850	8" pipe size		17	1.412		271	77.50		348.50	415
5860	10" pipe size		14	1.714		430	94		524	610
5870	12" pipe size		12	2		560	110		670	780
5880	14" pipe size		10	2.400		1,100	131		1,231	1,400
5890	16" pipe size		9	2.667		1,275	146		1,421	1,625
5900	18" pipe size		6	4		1,575	219		1,794	2,050
5910	20" pipe size		5	4.800		1,875	263		2,138	2,475
5920	24" pipe size	↓	4.50	5.333	↓	2,400	292		2,692	3,100
5940	ANSI class 350, painted									
5946	2" pipe size	1 Plum	23	.348	Ea.	132	20.50		152.50	177
5948	2-1/2" pipe size	Q-1	37	.432		153	23		176	204
5950	3" pipe size		31	.516		209	27.50		236.50	271
5952	4" pipe size		23	.696		278	37		315	360
5954	5" pipe size	↓	19	.842		315	44.50		359.50	415
5956	6" pipe size	Q-2	23	1.043		370	57		427	490
5958	8" pipe size		17	1.412		420	77.50		497.50	580
5960	10" pipe size	↓	14	1.714		670	94		764	880
5962	12" pipe size	1 Plum	12	.667	↓	720	39		759	850
6100	Cross, painted									
6110	2" diameter	1 Plum	12.50	.640	Ea.	88.50	37.50		126	154
6112	2-1/2" diameter	Q-1	20	.800		88.50	42.50		131	161
6114	3" diameter		15.50	1.032		157	54.50		211.50	256
6116	4" diameter	↓	12.50	1.280		261	67.50		328.50	390
6118	6" diameter	Q-2	12.50	1.920		680	105		785	910
6120	8" diameter		10.50	2.286		900	125		1,025	1,175
6122	10" diameter		9	2.667		1,475	146		1,621	1,850
6124	12" diameter	↓	7.50	3.200		2,125	175		2,300	2,625

22 11 13 – Facility Water Distribution Piping

22 11 13.48 Pipe, Fittings and Valves, Steel, Grooved-Joint		Crew	Daily Output	Labor-Hours	Unit	Material	2015 Bare Costs Labor	Equipment	Total	Total Incl O&P
7400	Suction diffuser									
7402	Grooved end inlet x flanged outlet									
7410	3" x 3"	Q-1	50	.320	Ea.	860	16.90		876.90	970
7412	4" x 4"		38	.421		1,175	22		1,197	1,300
7414	5" x 5"		30	.533		1,350	28		1,378	1,550
7416	6" x 6"	Q-2	38	.632		1,725	34.50		1,759.50	1,925
7418	8" x 8"		27	.889		3,200	48.50		3,248.50	3,575
7420	10" x 10"		20	1.200		4,350	65.50		4,415.50	4,875
7422	12" x 12"		16	1.500		7,150	82		7,232	7,975
7424	14" x 14"		15	1.600		7,575	87.50		7,662.50	8,475
7426	16" x 14"		14	1.714		7,800	94		7,894	8,725
7500	Strainer, tee type, painted									
7506	2" pipe size	1 Plum	38	.211	Ea.	575	12.35		587.35	650
7508	2-1/2" pipe size	Q-1	62	.258		600	13.65		613.65	680
7510	3" pipe size		50	.320		675	16.90		691.90	770
7512	4" pipe size		38	.421		765	22		787	875
7514	5" pipe size		30	.533		1,100	28		1,128	1,275
7516	6" pipe size	Q-2	38	.632		1,200	34.50		1,234.50	1,375
7518	8" pipe size		27	.889		1,850	48.50		1,898.50	2,100
7520	10" pipe size		20	1.200		2,700	65.50		2,765.50	3,075
7522	12" pipe size		16	1.500		3,475	82		3,557	3,950
7524	14" pipe size		15	1.600		10,200	87.50		10,287.50	11,400
7526	16" pipe size		14	1.714		12,700	94		12,794	14,100
7570	Expansion joint, max. 3" travel									
7572	2" diameter	1 Plum	38	.211	Ea.	720	12.35		732.35	815
7574	3" diameter	Q-1	50	.320		830	16.90		846.90	940
7576	4" diameter	"	38	.421		1,100	22		1,122	1,225
7578	6" diameter	Q-2	38	.632		1,700	34.50		1,734.50	1,925
7800	Ball valve w/handle, carbon steel trim									
7810	1-1/2" pipe size	1 Plum	50	.160	Ea.	144	9.40		153.40	172
7812	2" pipe size	"	38	.211		159	12.35		171.35	194
7814	2-1/2" pipe size	Q-1	62	.258		345	13.65		358.65	400
7816	3" pipe size		50	.320		560	16.90		576.90	640
7818	4" pipe size		38	.421		870	22		892	990
7820	6" pipe size	Q-2	30	.800		2,650	44		2,694	3,000
7830	With gear operator									
7834	2-1/2" pipe size	Q-1	62	.258	Ea.	700	13.65		713.65	790
7836	3" pipe size		50	.320		995	16.90		1,011.90	1,125
7838	4" pipe size		38	.421		1,275	22		1,297	1,425
7840	6" pipe size	Q-2	30	.800		3,000	44		3,044	3,350
7870	Check valve									
7874	2-1/2" pipe size	Q-1	62	.258	Ea.	274	13.65		287.65	320
7876	3" pipe size		50	.320		325	16.90		341.90	380
7878	4" pipe size		38	.421		340	22		362	410
7880	5" pipe size		30	.533		570	28		598	670
7882	6" pipe size	Q-2	38	.632		675	34.50		709.50	795
7884	8" pipe size		27	.889		920	48.50		968.50	1,075
7886	10" pipe size		20	1.200		2,450	65.50		2,515.50	2,800
7888	12" pipe size		16	1.500		2,925	82		3,007	3,325
7900	Plug valve, balancing, w/lever operator									
7906	3" pipe size	Q-1	50	.320	Ea.	680	16.90		696.90	770
7908	4" pipe size	"	38	.421		835	22		857	955
7909	6" pipe size	Q-2	30	.800		1,275	44		1,319	1,475

22 11 13 — Facility Water Distribution Piping

22 11 13.48 Pipe, Fittings and Valves, Steel, Grooved-Joint	Crew	Daily Output	Labor-Hours	Unit	Material	2015 Bare Costs Labor	Equipment	Total	Total Incl O&P
7916	With gear operator									
7920	3" pipe size	Q-1	50	.320	Ea.	1,275	16.90		1,291.90	1,450
7922	4" pipe size	"	38	.421		1,325	22		1,347	1,500
7924	6" pipe size	Q-2	38	.632		1,800	34.50		1,834.50	2,025
7926	8" pipe size		27	.889		2,775	48.50		2,823.50	3,125
7928	10" pipe size		20	1.200		3,475	65.50		3,540.50	3,925
7930	12" pipe size		16	1.500		5,325	82		5,407	5,975
8000	Butterfly valve, 2 position handle, with standard trim									
8010	1-1/2" pipe size	1 Plum	50	.160	Ea.	257	9.40		266.40	297
8020	2" pipe size	"	38	.211		257	12.35		269.35	300
8030	3" pipe size	Q-1	50	.320		370	16.90		386.90	430
8050	4" pipe size	"	38	.421		405	22		427	480
8070	6" pipe size	Q-2	38	.632		820	34.50		854.50	955
8080	8" pipe size		27	.889		1,100	48.50		1,148.50	1,300
8090	10" pipe size		20	1.200		1,800	65.50		1,865.50	2,100
8200	With stainless steel trim									
8240	1-1/2" pipe size	1 Plum	50	.160	Ea.	325	9.40		334.40	375
8250	2" pipe size	"	38	.211		325	12.35		337.35	380
8270	3" pipe size	Q-1	50	.320		440	16.90		456.90	505
8280	4" pipe size	"	38	.421		475	22		497	560
8300	6" pipe size	Q-2	38	.632		890	34.50		924.50	1,025
8310	8" pipe size		27	.889		2,200	48.50		2,248.50	2,500
8320	10" pipe size		20	1.200		3,675	65.50		3,740.50	4,125
8322	12" pipe size		16	1.500		4,200	82		4,282	4,750
8324	14" pipe size		15	1.600		5,400	87.50		5,487.50	6,050
8326	16" pipe size		14	1.714		7,800	94		7,894	8,725
8328	18" pipe size	Q-3	12	2.667		9,600	149		9,749	10,800
8330	20" pipe size		11	2.909		11,800	163		11,963	13,200
8332	24" pipe size		10	3.200		15,800	179		15,979	17,700
8336	Note: sizes 8" up w/manual gear operator									
9000	Cut one groove, labor									
9010	3/4" pipe size	Q-1	152	.105	Ea.		5.55		5.55	8.40
9020	1" pipe size		140	.114			6.05		6.05	9.10
9030	1-1/4" pipe size		124	.129			6.80		6.80	10.30
9040	1-1/2" pipe size		114	.140			7.40		7.40	11.20
9050	2" pipe size		104	.154			8.15		8.15	12.25
9060	2-1/2" pipe size		96	.167			8.80		8.80	13.30
9070	3" pipe size		88	.182			9.60		9.60	14.50
9080	3-1/2" pipe size		83	.193			10.20		10.20	15.40
9090	4" pipe size		78	.205			10.85		10.85	16.35
9100	5" pipe size		72	.222			11.75		11.75	17.75
9110	6" pipe size		70	.229			12.10		12.10	18.25
9120	8" pipe size		54	.296			15.65		15.65	23.50
9130	10" pipe size		38	.421			22		22	33.50
9140	12" pipe size		30	.533			28		28	42.50
9150	14" pipe size		20	.800			42.50		42.50	64
9160	16" pipe size		19	.842			44.50		44.50	67
9170	18" pipe size		18	.889			47		47	71
9180	20" pipe size		17	.941			49.50		49.50	75
9190	24" pipe size		15	1.067			56.50		56.50	85
9210	Roll one groove									
9220	3/4" pipe size	Q-1	266	.060	Ea.		3.18		3.18	4.80
9230	1" pipe size		228	.070			3.71		3.71	5.60

For customer support on your Mechanical Cost Data, call 877.795.7394.

183

22 11 13.48 Pipe, Fittings and Valves, Steel, Grooved-Joint

		Crew	Daily Output	Labor-Hours	Unit	Material	2015 Bare Costs Labor	Equipment	Total	Total Incl O&P
9240	1-1/4" pipe size	Q-1	200	.080	Ea.		4.23		4.23	6.40
9250	1-1/2" pipe size		178	.090			4.75		4.75	7.15
9260	2" pipe size		116	.138			7.30		7.30	11
9270	2-1/2" pipe size		110	.145			7.70		7.70	11.60
9280	3" pipe size		100	.160			8.45		8.45	12.75
9290	3-1/2" pipe size		94	.170			9		9	13.60
9300	4" pipe size		86	.186			9.85		9.85	14.85
9310	5" pipe size		84	.190			10.05		10.05	15.20
9320	6" pipe size		80	.200			10.55		10.55	15.95
9330	8" pipe size		66	.242			12.80		12.80	19.35
9340	10" pipe size		58	.276			14.55		14.55	22
9350	12" pipe size		46	.348			18.40		18.40	28
9360	14" pipe size		30	.533			28		28	42.50
9370	16" pipe size		28	.571			30		30	45.50
9380	18" pipe size		27	.593			31.50		31.50	47.50
9390	20" pipe size		25	.640			34		34	51
9400	24" pipe size	▼	23	.696	▼		37		37	55.50

22 11 13.60 Tubing, Stainless Steel

		Crew	Daily Output	Labor-Hours	Unit	Material	2015 Bare Costs Labor	Equipment	Total	Total Incl O&P
0010	**TUBING, STAINLESS STEEL**									
5000	Tubing									
5010	Type 304, no joints, no hangers									
5020	.035 wall									
5021	1/4"	1 Plum	160	.050	L.F.	3.12	2.94		6.06	7.85
5022	3/8"		160	.050		3.94	2.94		6.88	8.75
5023	1/2"		160	.050		5.05	2.94		7.99	10.05
5024	5/8"		160	.050		6.20	2.94		9.14	11.30
5025	3/4"		133	.060		7.50	3.53		11.03	13.60
5026	7/8"		133	.060		7.90	3.53		11.43	14.05
5027	1"	▼	114	.070	▼	8.70	4.12		12.82	15.75
5040	.049 wall									
5041	1/4"	1 Plum	160	.050	L.F.	3.67	2.94		6.61	8.45
5042	3/8"		160	.050		5.05	2.94		7.99	10.05
5043	1/2"		160	.050		5.60	2.94		8.54	10.65
5044	5/8"		160	.050		7.60	2.94		10.54	12.80
5045	3/4"		133	.060		8.10	3.53		11.63	14.25
5046	7/8"		133	.060		8.45	3.53		11.98	14.60
5047	1"	▼	114	.070	▼	8.95	4.12		13.07	16.05
5060	.065 wall									
5061	1/4"	1 Plum	160	.050	L.F.	4.11	2.94		7.05	8.95
5062	3/8"		160	.050		5.65	2.94		8.59	10.70
5063	1/2"		160	.050		6.85	2.94		9.79	12
5064	5/8"		160	.050		8.30	2.94		11.24	13.55
5065	3/4"		133	.060		8.85	3.53		12.38	15.05
5066	7/8"		133	.060		10.15	3.53		13.68	16.50
5067	1"	▼	114	.070	▼	9.95	4.12		14.07	17.10
5210	Type 316									
5220	.035 wall									
5221	1/4"	1 Plum	160	.050	L.F.	3.96	2.94		6.90	8.80
5222	3/8"		160	.050		5	2.94		7.94	9.95
5223	1/2"		160	.050		7	2.94		9.94	12.15
5224	5/8"		160	.050		9.10	2.94		12.04	14.45
5225	3/4"	▼	133	.060		10.50	3.53		14.03	16.90

22 11 13 – Facility Water Distribution Piping

22 11 13.60 Tubing, Stainless Steel

		Crew	Daily Output	Labor-Hours	Unit	Material	2015 Bare Costs Labor	Equipment	Total	Total Incl O&P
5226	7/8"	1 Plum	133	.060	L.F.	15	3.53		18.53	22
5227	1"		114	.070		11.90	4.12		16.02	19.30
5240	.049 wall									
5241	1/4"	1 Plum	160	.050	L.F.	5.45	2.94		8.39	10.45
5242	3/8"		160	.050		7.15	2.94		10.09	12.30
5243	1/2"		160	.050		7.85	2.94		10.79	13.10
5244	5/8"		160	.050		10.90	2.94		13.84	16.45
5245	3/4"		133	.060		10.50	3.53		14.03	16.90
5246	7/8"		133	.060		13.50	3.53		17.03	20
5247	1"		114	.070		12.30	4.12		16.42	19.75
5260	.065 wall									
5261	1/4"	1 Plum	160	.050	L.F.	5.20	2.94		8.14	10.15
5262	3/8"		160	.050		8.30	2.94		11.24	13.55
5263	1/2"		160	.050		8.90	2.94		11.84	14.25
5264	5/8"		160	.050		12.40	2.94		15.34	18.10
5265	3/4"		133	.060		13.35	3.53		16.88	20
5266	7/8"		133	.060		15.80	3.53		19.33	23
5267	1"		114	.070		15.05	4.12		19.17	23

22 11 13.61 Tubing Fittings, Stainless Steel

		Crew	Daily Output	Labor-Hours	Unit	Material	2015 Bare Costs Labor	Equipment	Total	Total Incl O&P
0010	**TUBING FITTINGS, STAINLESS STEEL**									
8200	Tube fittings, compression type									
8202	Type 316									
8204	90° elbow									
8206	1/4"	1 Plum	24	.333	Ea.	17.60	19.55		37.15	49
8207	3/8"		22	.364		21.50	21.50		43	56
8208	1/2"		22	.364		35.50	21.50		57	71
8209	5/8"		21	.381		40	22.50		62.50	78
8210	3/4"		21	.381		63	22.50		85.50	103
8211	7/8"		20	.400		96.50	23.50		120	142
8212	1"		20	.400		120	23.50		143.50	168
8220	Union tee									
8222	1/4"	1 Plum	15	.533	Ea.	25	31.50		56.50	75
8224	3/8"		15	.533		32	31.50		63.50	83
8225	1/2"		15	.533		49.50	31.50		81	102
8226	5/8"		14	.571		55.50	33.50		89	112
8227	3/4"		14	.571		74	33.50		107.50	132
8228	7/8"		13	.615		148	36		184	218
8229	1"		13	.615		159	36		195	230
8234	Union									
8236	1/4"	1 Plum	24	.333	Ea.	12.20	19.55		31.75	43
8237	3/8"		22	.364		17.40	21.50		38.90	51
8238	1/2"		22	.364		26	21.50		47.50	60.50
8239	5/8"		21	.381		34	22.50		56.50	71
8240	3/4"		21	.381		42	22.50		64.50	80
8241	7/8"		20	.400		68.50	23.50		92	111
8242	1"		20	.400		72.50	23.50		96	116
8250	Male connector									
8252	1/4" x 1/4"	1 Plum	24	.333	Ea.	7.80	19.55		27.35	38
8253	3/8" x 3/8"		22	.364		12.20	21.50		33.70	45.50
8254	1/2" x 1/2"		22	.364		18.05	21.50		39.55	52
8256	3/4" x 3/4"		21	.381		27.50	22.50		50	64.50
8258	1" x 1"		20	.400		48.50	23.50		72	88.50

22 11 13.64 Pipe, Stainless Steel		Crew	Daily Output	Labor-Hours	Unit	Material	2015 Bare Costs Labor	Equipment	Total	Total Incl O&P
0010	**PIPE, STAINLESS STEEL**									
0020	Welded, with clevis type hanger assemblies, 10' O.C.									
0500	Schedule 5, type 304									
0540	1/2" diameter	Q-15	128	.125	L.F.	8.25	6.60	.45	15.30	19.55
0550	3/4" diameter		116	.138		10.25	7.30	.50	18.05	23
0560	1" diameter		103	.155		11.35	8.20	.56	20.11	25.50
0570	1-1/4" diameter		93	.172		15.45	9.10	.62	25.17	31.50
0580	1-1/2" diameter		85	.188		18.60	9.95	.68	29.23	36.50
0590	2" diameter		69	.232		31	12.25	.84	44.09	53.50
0600	2-1/2" diameter		53	.302		33.50	15.95	1.09	50.54	62
0610	3" diameter		48	.333		40	17.60	1.20	58.80	72
0620	4" diameter		44	.364		51	19.20	1.31	71.51	86.50
0630	5" diameter		36	.444		103	23.50	1.60	128.10	151
0640	6" diameter	Q-16	42	.571		95.50	31.50	1.37	128.37	154
0650	8" diameter		34	.706		146	38.50	1.69	186.19	221
0660	10" diameter		26	.923		202	50.50	2.22	254.72	300
0670	12" diameter		21	1.143		265	62.50	2.74	330.24	390
0700	To delete hangers, subtract									
0710	1/2" diam. to 1-1/2" diam.					8%	19%			
0720	2" diam. to 5" diam.					4%	9%			
0730	6" diam. to 12" diam.					3%	4%			
0750	For small quantities, add				L.F.	10%				
1250	Schedule 5, type 316									
1290	1/2" diameter	Q-15	128	.125	L.F.	10.30	6.60	.45	17.35	22
1300	3/4" diameter		116	.138		14.75	7.30	.50	22.55	28
1310	1" diameter		103	.155		16.45	8.20	.56	25.21	31
1320	1-1/4" diameter		93	.172		20	9.10	.62	29.72	37
1330	1-1/2" diameter		85	.188		27.50	9.95	.68	38.13	46.50
1340	2" diameter		69	.232		42.50	12.25	.84	55.59	66
1350	2-1/2" diameter		53	.302		54.50	15.95	1.09	71.54	85
1360	3" diameter		48	.333		69	17.60	1.20	87.80	104
1370	4" diameter		44	.364		85.50	19.20	1.31	106.01	124
1380	5" diameter		36	.444		163	23.50	1.60	188.10	216
1390	6" diameter	Q-16	42	.571		159	31.50	1.37	191.87	224
1400	8" diameter		34	.706		239	38.50	1.69	279.19	320
1410	10" diameter		26	.923		340	50.50	2.22	392.72	455
1420	12" diameter		21	1.143		440	62.50	2.74	505.24	585
1490	For small quantities, add					10%				
1940	To delete hanger, subtract									
1950	1/2" diam. to 1-1/2" diam.					5%	19%			
1960	2" diam. to 5" diam.					3%	9%			
1970	6" diam. to 12" diam.					2%	4%			
2000	Schedule 10, type 304									
2040	1/4" diameter	Q-15	131	.122	L.F.	5.40	6.45	.44	12.29	16.20
2050	3/8" diameter		128	.125		6.45	6.60	.45	13.50	17.55
2060	1/2" diameter		125	.128		6.75	6.75	.46	13.96	18.15
2070	3/4" diameter		113	.142		7.80	7.50	.51	15.81	20.50
2080	1" diameter		100	.160		11.15	8.45	.58	20.18	25.50
2090	1-1/4" diameter		91	.176		12.70	9.30	.63	22.63	28.50
2100	1-1/2" diameter		83	.193		14.10	10.20	.70	25	31.50
2110	2" diameter		67	.239		17.10	12.60	.86	30.56	39
2120	2-1/2" diameter		51	.314		22	16.55	1.13	39.68	50.50

22 11 13 – Facility Water Distribution Piping

22 11 13.64 Pipe, Stainless Steel		Crew	Daily Output	Labor-Hours	Unit	Material	2015 Bare Costs Labor	Equipment	Total	Total Incl O&P
2130	3" diameter	Q-15	46	.348	L.F.	26.50	18.40	1.26	46.16	58.50
2140	4" diameter		42	.381		32.50	20	1.38	53.88	67.50
2150	5" diameter		35	.457		43	24	1.65	68.65	86
2160	6" diameter	Q-16	40	.600		51	33	1.44	85.44	107
2170	8" diameter		33	.727		88.50	40	1.75	130.25	159
2180	10" diameter		25	.960		122	52.50	2.30	176.80	216
2190	12" diameter		21	1.143		152	62.50	2.74	217.24	265
2250	For small quantities, add					10%				
2650	To delete hanger, subtract									
2660	1/4" diam. to 3/4" diam.					9%	22%			
2670	1" diam. to 2" diam.					4%	15%			
2680	2-1/2" diam. to 5" diam.					3%	8%			
2690	6" diam. to 12" diam.					3%	4%			
2750	Schedule 10, type 316									
2790	1/4" diameter	Q-15	131	.122	L.F.	6.50	6.45	.44	13.39	17.40
2800	3/8" diameter		128	.125		7.60	6.60	.45	14.65	18.80
2810	1/2" diameter		125	.128		8.45	6.75	.46	15.66	20
2820	3/4" diameter		113	.142		9.95	7.50	.51	17.96	23
2830	1" diameter		100	.160		14.10	8.45	.58	23.13	29
2840	1-1/4" diameter		91	.176		17.90	9.30	.63	27.83	34.50
2850	1-1/2" diameter		83	.193		19.75	10.20	.70	30.65	38
2860	2" diameter		67	.239		24	12.60	.86	37.46	46.50
2870	2-1/2" diameter		51	.314		31	16.55	1.13	48.68	60.50
2880	3" diameter		46	.348		36.50	18.40	1.26	56.16	69.50
2890	4" diameter		42	.381		46.50	20	1.38	67.88	83
2900	5" diameter		35	.457		56.50	24	1.65	82.15	100
2910	6" diameter	Q-16	40	.600		68.50	33	1.44	102.94	127
2920	8" diameter		33	.727		130	40	1.75	171.75	205
2930	10" diameter		25	.960		171	52.50	2.30	225.80	270
2940	12" diameter		21	1.143		214	62.50	2.74	279.24	335
2990	For small quantities, add					10%				
3430	To delete hanger, subtract									
3440	1/4" diam. to 3/4" diam.					6%	22%			
3450	1" diam. to 2" diam.					3%	15%			
3460	2-1/2" diam. to 5" diam.					2%	8%			
3470	6" diam. to 12" diam.					2%	4%			
3500	Threaded, couplings and clevis hanger assemblies, 10' O.C.									
3520	Schedule 40, type 304									
3540	1/4" diameter	1 Plum	54	.148	L.F.	10.85	8.70		19.55	25
3550	3/8" diameter		53	.151		11.05	8.85		19.90	25.50
3560	1/2" diameter		52	.154		12.85	9.05		21.90	28
3570	3/4" diameter		51	.157		13.35	9.20		22.55	28.50
3580	1" diameter		45	.178		19.30	10.45		29.75	37
3590	1-1/4" diameter	Q-1	76	.211		27	11.10		38.10	46.50
3600	1-1/2" diameter		69	.232		29.50	12.25		41.75	50.50
3610	2" diameter		57	.281		45	14.85		59.85	72
3620	2-1/2" diameter		44	.364		77	19.20		96.20	114
3630	3" diameter		38	.421		100	22		122	144
3640	4" diameter	Q-2	51	.471		138	26		164	190
3740	For small quantities, add					10%				
4200	To delete couplings & hangers, subtract									
4210	1/4" diam. to 3/4" diam.					15%	56%			
4220	1" diam. to 2" diam.					18%	49%			

22 11 13 – Facility Water Distribution Piping

22 11 13.64 Pipe, Stainless Steel	Crew	Daily Output	Labor-Hours	Unit	Material	2015 Bare Costs Labor	Equipment	Total	Total Incl O&P
4230	2-1/2" diam. to 4" diam.					34%	40%			
4250	Schedule 40, type 316									
4290	1/4" diameter	1 Plum	54	.148	L.F.	11.65	8.70		20.35	26
4300	3/8" diameter		53	.151		12.70	8.85		21.55	27.50
4310	1/2" diameter		52	.154		15.50	9.05		24.55	30.50
4320	3/4" diameter		51	.157		17.95	9.20		27.15	33.50
4330	1" diameter		45	.178		25	10.45		35.45	43.50
4340	1-1/4" diameter	Q-1	76	.211		34	11.10		45.10	54
4350	1-1/2" diameter		69	.232		38.50	12.25		50.75	61
4360	2" diameter		57	.281		53	14.85		67.85	81
4370	2-1/2" diameter		44	.364		90.50	19.20		109.70	129
4380	3" diameter		38	.421		116	22		138	162
4390	4" diameter	Q-2	51	.471		153	26		179	207
4490	For small quantities, add					10%				
4900	To delete couplings & hangers, subtract									
4910	1/4" diam. to 3/4" diam.					12%	56%			
4920	1" diam. to 2" diam.					14%	49%			
4930	2-1/2" diam. to 4" diam.					27%	40%			
5000	Schedule 80, type 304									
5040	1/4" diameter	1 Plum	53	.151	L.F.	18.10	8.85		26.95	33.50
5050	3/8" diameter		52	.154		20.50	9.05		29.55	36.50
5060	1/2" diameter		51	.157		25	9.20		34.20	41.50
5070	3/4" diameter		48	.167		26.50	9.80		36.30	44
5080	1" diameter		43	.186		33	10.90		43.90	53
5090	1-1/4" diameter	Q-1	73	.219		44.50	11.60		56.10	66.50
5100	1-1/2" diameter		67	.239		56	12.60		68.60	81
5110	2" diameter		54	.296		69.50	15.65		85.15	100
5190	For small quantities, add					10%				
5700	To delete couplings & hangers, subtract									
5710	1/4" diam. to 3/4" diam.					10%	53%			
5720	1" diam. to 2" diam.					14%	47%			
5750	Schedule 80, type 316									
5790	1/4" diameter	1 Plum	53	.151	L.F.	21.50	8.85		30.35	37
5800	3/8" diameter		52	.154		26.50	9.05		35.55	42.50
5810	1/2" diameter		51	.157		31	9.20		40.20	48.50
5820	3/4" diameter		48	.167		34	9.80		43.80	52.50
5830	1" diameter		43	.186		44.50	10.90		55.40	65.50
5840	1-1/4" diameter	Q-1	73	.219		64	11.60		75.60	88
5850	1-1/2" diameter		67	.239		68.50	12.60		81.10	94.50
5860	2" diameter		54	.296		86	15.65		101.65	119
5950	For small quantities, add					10%				
7000	To delete couplings & hangers, subtract									
7010	1/4" diam. to 3/4" diam.					9%	53%			
7020	1" diam. to 2" diam.					14%	47%			
8000	Weld joints with clevis type hanger assemblies, 10' O.C.									
8010	Schedule 40, type 304									
8050	1/8" pipe size	Q-15	126	.127	L.F.	8.70	6.70	.46	15.86	20
8060	1/4" pipe size		125	.128		9.60	6.75	.46	16.81	21.50
8070	3/8" pipe size		122	.131		9.60	6.95	.47	17.02	21.50
8080	1/2" pipe size		118	.136		10.95	7.15	.49	18.59	23.50
8090	3/4" pipe size		109	.147		10.80	7.75	.53	19.08	24
8100	1" pipe size		95	.168		15.15	8.90	.61	24.66	31
8110	1-1/4" pipe size		86	.186		20.50	9.85	.67	31.02	38

188

For customer support on your Mechanical Cost Data, call 877.795.7394.

22 11 13 – Facility Water Distribution Piping

22 11 13.64 Pipe, Stainless Steel	Crew	Daily Output	Labor-Hours	Unit	Material	2015 Bare Costs Labor	Equipment	Total	Total Incl O&P
8120	1-1/2" pipe size	Q-15	78	.205	L.F.	22	10.85	.74	33.59	41
8130	2" pipe size		62	.258		32.50	13.65	.93	47.08	57.50
8140	2-1/2" pipe size		49	.327		48.50	17.25	1.18	66.93	80.50
8150	3" pipe size		44	.364		61	19.20	1.31	81.51	98
8160	3-1/2" pipe size		44	.364		72.50	19.20	1.31	93.01	110
8170	4" pipe size		39	.410		83	21.50	1.48	105.98	125
8180	5" pipe size		32	.500		110	26.50	1.81	138.31	163
8190	6" pipe size	Q-16	37	.649		128	35.50	1.56	165.06	196
8200	8" pipe size		29	.828		179	45.50	1.99	226.49	268
8210	10" pipe size		24	1		305	55	2.40	362.40	420
8220	12" pipe size		20	1.200		490	65.50	2.88	558.38	645
8300	Schedule 40, type 316									
8310	1/8" pipe size	Q-15	126	.127	L.F.	9.25	6.70	.46	16.41	21
8320	1/4" pipe size		125	.128		10.15	6.75	.46	17.36	22
8330	3/8" pipe size		122	.131		10.95	6.95	.47	18.37	23
8340	1/2" pipe size		118	.136		13.15	7.15	.49	20.79	26
8350	3/4" pipe size		109	.147		14.85	7.75	.53	23.13	28.50
8360	1" pipe size		95	.168		20	8.90	.61	29.51	36.50
8370	1-1/4" pipe size		86	.186		26	9.85	.67	36.52	44
8380	1-1/2" pipe size		78	.205		29.50	10.85	.74	41.09	49.50
8390	2" pipe size		62	.258		38.50	13.65	.93	53.08	63.50
8400	2-1/2" pipe size		49	.327		56	17.25	1.18	74.43	89
8410	3" pipe size		44	.364		69	19.20	1.31	89.51	106
8420	3-1/2" pipe size		44	.364		80.50	19.20	1.31	101.01	119
8430	4" pipe size		39	.410		87	21.50	1.48	109.98	130
8440	5" pipe size		32	.500		120	26.50	1.81	148.31	174
8450	6" pipe size	Q-16	37	.649		138	35.50	1.56	175.06	207
8460	8" pipe size		29	.828		184	45.50	1.99	231.49	273
8470	10" pipe size		24	1		245	55	2.40	302.40	355
8480	12" pipe size		20	1.200		330	65.50	2.88	398.38	470
8500	Schedule 80, type 304									
8510	1/4" pipe size	Q-15	110	.145	L.F.	16.75	7.70	.53	24.98	30.50
8520	3/8" pipe size		109	.147		19.05	7.75	.53	27.33	33.50
8530	1/2" pipe size		106	.151		23	7.95	.54	31.49	38
8540	3/4" pipe size		96	.167		24	8.80	.60	33.40	40.50
8550	1" pipe size		87	.184		28.50	9.70	.66	38.86	47
8560	1-1/4" pipe size		81	.198		34	10.45	.71	45.16	53.50
8570	1-1/2" pipe size		74	.216		43.50	11.40	.78	55.68	66
8580	2" pipe size		58	.276		53	14.55	1	68.55	81.50
8590	2-1/2" pipe size		46	.348		68	18.40	1.26	87.66	104
8600	3" pipe size		41	.390		78.50	20.50	1.41	100.41	119
8610	4" pipe size		33	.485		96.50	25.50	1.75	123.75	146
8630	6" pipe size	Q-16	30	.800		171	44	1.92	216.92	256
8640	Schedule 80, type 316									
8650	1/4" pipe size	Q-15	110	.145	L.F.	19.95	7.70	.53	28.18	34
8660	3/8" pipe size		109	.147		24.50	7.75	.53	32.78	39.50
8670	1/2" pipe size		106	.151		29	7.95	.54	37.49	44
8680	3/4" pipe size		96	.167		30.50	8.80	.60	39.90	47.50
8690	1" pipe size		87	.184		38.50	9.70	.66	48.86	58
8700	1-1/4" pipe size		81	.198		50.50	10.45	.71	61.66	72
8710	1-1/2" pipe size		74	.216		53	11.40	.78	65.18	76
8720	2" pipe size		58	.276		64	14.55	1	79.55	93.50
8730	2-1/2" pipe size		46	.348		80	18.40	1.26	99.66	117

22 11 13.64 Pipe, Stainless Steel		Crew	Daily Output	Labor-Hours	Unit	Material	2015 Bare Costs Labor	2015 Bare Costs Equipment	Total	Total Incl O&P
8740	3" pipe size	Q-15	41	.390	L.F.	92.50	20.50	1.41	114.41	135
8760	4" pipe size	↓	33	.485		122	25.50	1.75	149.25	174
8770	6" pipe size	Q-16	30	.800	↓	219	44	1.92	264.92	310
9100	Threading pipe labor, sst, one end, schedules 40 & 80									
9110	1/4" through 3/4" pipe size	1 Plum	61.50	.130	Ea.		7.65		7.65	11.55
9120	1" through 2" pipe size		55.90	.143			8.40		8.40	12.70
9130	2-1/2" pipe size		41.50	.193			11.30		11.30	17.10
9140	3" pipe size	↓	38.50	.208			12.20		12.20	18.40
9150	3-1/2" pipe size	Q-1	68.40	.234			12.35		12.35	18.65
9160	4" pipe size		73	.219			11.60		11.60	17.50
9170	5" pipe size		40.70	.393			21		21	31.50
9180	6" pipe size		35.40	.452			24		24	36
9190	8" pipe size		22.30	.717			38		38	57
9200	10" pipe size		16.10	.994			52.50		52.50	79.50
9210	12" pipe size	↓	12.30	1.301	↓		68.50		68.50	104
9250	Welding labor per joint for stainless steel									
9260	Schedule 5 and 10									
9270	1/4" pipe size	Q-15	36	.444	Ea.		23.50	1.60	25.10	37.50
9280	3/8" pipe size		35	.457			24	1.65	25.65	38.50
9290	1/2" pipe size		35	.457			24	1.65	25.65	38.50
9300	3/4" pipe size		28	.571			30	2.06	32.06	48
9310	1" pipe size		25	.640			34	2.31	36.31	53.50
9320	1-1/4" pipe size		22	.727			38.50	2.63	41.13	61
9330	1-1/2" pipe size		21	.762			40.50	2.75	43.25	64
9340	2" pipe size		18	.889			47	3.21	50.21	74.50
9350	2-1/2" pipe size		12	1.333			70.50	4.81	75.31	111
9360	3" pipe size		9.73	1.644			87	5.95	92.95	138
9370	4" pipe size		7.37	2.171			115	7.85	122.85	182
9380	5" pipe size		6.15	2.602			137	9.40	146.40	218
9390	6" pipe size		5.71	2.802			148	10.10	158.10	235
9400	8" pipe size		3.69	4.336			229	15.65	244.65	360
9410	10" pipe size		2.91	5.498			290	19.85	309.85	460
9420	12" pipe size	↓	2.31	6.926	↓		365	25	390	585
9500	Schedule 40									
9510	1/4" pipe size	Q-15	28	.571	Ea.		30	2.06	32.06	48
9520	3/8" pipe size		27	.593			31.50	2.14	33.64	50
9530	1/2" pipe size		25.40	.630			33.50	2.27	35.77	53
9540	3/4" pipe size		22.22	.720			38	2.60	40.60	60.50
9550	1" pipe size		20.25	.790			41.50	2.85	44.35	66
9560	1-1/4" pipe size		18.82	.850			45	3.07	48.07	71.50
9570	1-1/2" pipe size		17.78	.900			47.50	3.25	50.75	75.50
9580	2" pipe size		15.09	1.060			56	3.83	59.83	88.50
9590	2-1/2" pipe size		7.96	2.010			106	7.25	113.25	168
9600	3" pipe size		6.43	2.488			131	9	140	209
9610	4" pipe size		4.88	3.279			173	11.85	184.85	275
9620	5" pipe size		4.26	3.756			198	13.55	211.55	315
9630	6" pipe size		3.77	4.244			224	15.30	239.30	355
9640	8" pipe size		2.44	6.557			345	23.50	368.50	550
9650	10" pipe size		1.92	8.333			440	30	470	700
9660	12" pipe size	↓	1.52	10.526	↓		555	38	593	880
9750	Schedule 80									
9760	1/4" pipe size	Q-15	21.55	.742	Ea.		39	2.68	41.68	62
9770	3/8" pipe size	↓	20.75	.771	↓		40.50	2.78	43.28	64.50

22 11 13 – Facility Water Distribution Piping

22 11 13.64 Pipe, Stainless Steel

		Crew	Daily Output	Labor-Hours	Unit	Material	2015 Bare Costs Labor	Equipment	Total	Total Incl O&P
9780	1/2" pipe size	Q-15	19.54	.819	Ea.		43.50	2.96	46.46	69
9790	3/4" pipe size		17.09	.936			49.50	3.38	52.88	78
9800	1" pipe size		15.58	1.027			54.50	3.71	58.21	86
9810	1-1/4" pipe size		14.48	1.105			58.50	3.99	62.49	92.50
9820	1-1/2" pipe size		13.68	1.170			62	4.22	66.22	98
9830	2" pipe size		11.61	1.378			73	4.98	77.98	115
9840	2-1/2" pipe size		6.12	2.614			138	9.45	147.45	219
9850	3" pipe size		4.94	3.239			171	11.70	182.70	271
9860	4" pipe size		3.75	4.267			225	15.40	240.40	355
9870	5" pipe size		3.27	4.893			259	17.65	276.65	410
9880	6" pipe size		2.90	5.517			291	19.90	310.90	460
9890	8" pipe size		1.87	8.556			450	31	481	720
9900	10" pipe size		1.48	10.811			570	39	609	905
9910	12" pipe size		1.17	13.675			720	49.50	769.50	1,150
9920	Schedule 160, 1/2" pipe size		17	.941			49.50	3.40	52.90	78.50
9930	3/4" pipe size		14.81	1.080			57	3.90	60.90	90.50
9940	1" pipe size		13.50	1.185			62.50	4.28	66.78	99
9950	1-1/4" pipe size		12.55	1.275			67.50	4.60	72.10	107
9960	1-1/2" pipe size		11.85	1.350			71.50	4.87	76.37	113
9970	2" pipe size		10	1.600			84.50	5.80	90.30	134
9980	3" pipe size		4.28	3.738			198	13.50	211.50	315
9990	4" pipe size		3.25	4.923			260	17.75	277.75	415

22 11 13.66 Pipe Fittings, Stainless Steel

		Crew	Daily Output	Labor-Hours	Unit	Material	2015 Bare Costs Labor	Equipment	Total	Total Incl O&P
0010	**PIPE FITTINGS, STAINLESS STEEL**									
0100	Butt weld joint, schedule 5, type 304									
0120	90° Elbow, long									
0140	1/2"	Q-15	17.50	.914	Ea.	19.80	48.50	3.30	71.60	98.50
0150	3/4"		14	1.143		19.80	60.50	4.13	84.43	118
0160	1"		12.50	1.280		21	67.50	4.62	93.12	130
0170	1-1/4"		11	1.455		28.50	77	5.25	110.75	153
0180	1-1/2"		10.50	1.524		24	80.50	5.50	110	155
0190	2"		9	1.778		28.50	94	6.40	128.90	181
0200	2-1/2"		6	2.667		66	141	9.65	216.65	296
0210	3"		4.86	3.292		72.50	174	11.90	258.40	355
0220	3-1/2"		4.27	3.747		182	198	13.55	393.55	515
0230	4"		3.69	4.336		99	229	15.65	343.65	470
0240	5"		3.08	5.195		385	274	18.75	677.75	860
0250	6"	Q-16	4.29	5.594		297	305	13.45	615.45	805
0260	8"		2.76	8.696		625	475	21	1,121	1,425
0270	10"		2.18	11.009		970	605	26.50	1,601.50	2,025
0280	12"		1.73	13.873		1,375	760	33.50	2,168.50	2,700
0320	For schedule 5, type 316, add					30%				
0600	45° Elbow, long									
0620	1/2"	Q-15	17.50	.914	Ea.	19.80	48.50	3.30	71.60	98.50
0630	3/4"		14	1.143		19.80	60.50	4.13	84.43	118
0640	1"		12.50	1.280		21	67.50	4.62	93.12	130
0650	1-1/4"		11	1.455		28.50	77	5.25	110.75	153
0660	1-1/2"		10.50	1.524		24	80.50	5.50	110	155
0670	2"		9	1.778		28.50	94	6.40	128.90	181
0680	2-1/2"		6	2.667		66	141	9.65	216.65	296
0690	3"		4.86	3.292		72.50	174	11.90	258.40	355
0700	3-1/2"		4.27	3.747		182	198	13.55	393.55	515

22 11 13.66 Pipe Fittings, Stainless Steel

		Crew	Daily Output	Labor-Hours	Unit	Material	Labor	Equipment	Total	Total Incl O&P
0710	4"	Q-15	3.69	4.336	Ea.	83.50	229	15.65	328.15	455
0720	5"	↓	3.08	5.195		310	274	18.75	602.75	775
0730	6"	Q-16	4.29	5.594		209	305	13.45	527.45	710
0740	8"		2.76	8.696		440	475	21	936	1,225
0750	10"		2.18	11.009		770	605	26.50	1,401.50	1,775
0760	12"	↓	1.73	13.873		965	760	33.50	1,758.50	2,225
0800	For schedule 5, type 316, add				↓	25%				
1100	Tee, straight									
1130	1/2"	Q-15	11.66	1.372	Ea.	59.50	72.50	4.95	136.95	180
1140	3/4"		9.33	1.715		59.50	90.50	6.20	156.20	209
1150	1"		8.33	1.921		62.50	101	6.95	170.45	230
1160	1-1/4"		7.33	2.183		50.50	115	7.90	173.40	238
1170	1-1/2"		7	2.286		49.50	121	8.25	178.75	246
1180	2"		6	2.667		51.50	141	9.65	202.15	281
1190	2-1/2"		4	4		124	211	14.45	349.45	475
1200	3"		3.24	4.938		198	261	17.85	476.85	635
1210	3-1/2"		2.85	5.614		260	297	20.50	577.50	760
1220	4"		2.46	6.504		141	345	23.50	509.50	700
1230	5"	↓	2	8		445	425	29	899	1,150
1240	6"	Q-16	2.85	8.421		350	460	20	830	1,100
1250	8"		1.84	13.043		750	715	31.50	1,496.50	1,925
1260	10"		1.45	16.552		1,200	905	39.50	2,144.50	2,750
1270	12"	↓	1.15	20.870		1,675	1,150	50	2,875	3,625
1320	For schedule 5, type 316, add				↓	25%				
2000	Butt weld joint, schedule 10, type 304									
2020	90° elbow, long									
2040	1/2"	Q-15	17	.941	Ea.	13.50	49.50	3.40	66.40	93.50
2050	3/4"		14	1.143		13.50	60.50	4.13	78.13	110
2060	1"		12.50	1.280		14.25	67.50	4.62	86.37	123
2070	1-1/4"		11	1.455		19.50	77	5.25	101.75	143
2080	1-1/2"		10.50	1.524		16.50	80.50	5.50	102.50	146
2090	2"		9	1.778		19.50	94	6.40	119.90	171
2100	2-1/2"		6	2.667		45	141	9.65	195.65	273
2110	3"		4.86	3.292		42	174	11.90	227.90	320
2120	3-1/2"		4.27	3.747		124	198	13.55	335.55	450
2130	4"		3.69	4.336		70.50	229	15.65	315.15	440
2140	5"		3.08	5.195		263	274	18.75	555.75	725
2150	6"	Q-16	4.29	5.594		203	305	13.45	521.45	705
2160	8"		2.76	8.696		430	475	21	926	1,225
2170	10"		2.18	11.009		660	605	26.50	1,291.50	1,675
2180	12"	↓	1.73	13.873	↓	940	760	33.50	1,733.50	2,200
2500	45° elbow, long									
2520	1/2"	Q-15	17.50	.914	Ea.	13.50	48.50	3.30	65.30	91.50
2530	3/4"		14	1.143		13.50	60.50	4.13	78.13	110
2540	1"		12.50	1.280		14.25	67.50	4.62	86.37	123
2550	1-1/4"		11	1.455		19.50	77	5.25	101.75	143
2560	1-1/2"		10.50	1.524		16.50	80.50	5.50	102.50	146
2570	2"		9	1.778		19.50	94	6.40	119.90	171
2580	2-1/2"		6	2.667		45	141	9.65	195.65	273
2590	3"		4.86	3.292		34	174	11.90	219.90	315
2600	3-1/2"		4.27	3.747		124	198	13.55	335.55	450
2610	4"		3.69	4.336		57	229	15.65	301.65	425
2620	5"	↓	3.08	5.195	↓	210	274	18.75	502.75	665

22 11 13 – Facility Water Distribution Piping

22 11 13.66 Pipe Fittings, Stainless Steel	Crew	Daily Output	Labor-Hours	Unit	Material	2015 Bare Costs Labor	Equipment	Total	Total Incl O&P
2630	6"	Q-16	4.29	5.594	Ea.	143	305	13.45	461.45	635
2640	8"		2.76	8.696		300	475	21	796	1,075
2650	10"		2.18	11.009		525	605	26.50	1,156.50	1,525
2660	12"		1.73	13.873		655	760	33.50	1,448.50	1,900
2670	Reducer, concentric									
2674	1" x 3/4"	Q-15	13.25	1.208	Ea.	29.50	64	4.36	97.86	133
2676	2" x 1-1/2"	"	9.75	1.641		24	86.50	5.90	116.40	164
2678	6" x 4"	Q-16	4.91	4.888		72	268	11.75	351.75	495
2680	Caps									
2682	1"	Q-15	25	.640	Ea.	30	34	2.31	66.31	86.50
2684	1-1/2"		21	.762		36	40.50	2.75	79.25	104
2685	2"		18	.889		34	47	3.21	84.21	112
2686	4"		7.38	2.168		54	115	7.85	176.85	241
2687	6"	Q-16	8.58	2.797		86.50	153	6.70	246.20	335
3000	Tee, straight									
3030	1/2"	Q-15	11.66	1.372	Ea.	40.50	72.50	4.95	117.95	159
3040	3/4"		9.33	1.715		40.50	90.50	6.20	137.20	188
3050	1"		8.33	1.921		43	101	6.95	150.95	208
3060	1-1/4"		7.33	2.183		49	115	7.90	171.90	236
3070	1-1/2"		7	2.286		49	121	8.25	178.25	245
3080	2"		6	2.667		51	141	9.65	201.65	280
3090	2-1/2"		4	4		35.50	211	14.45	260.95	375
3100	3"		3.24	4.938		67	261	17.85	345.85	490
3110	3-1/2"		2.85	5.614		98.50	297	20.50	416	580
3120	4"		2.46	6.504		96	345	23.50	464.50	650
3130	5"		2	8		305	425	29	759	1,000
3140	6"	Q-16	2.85	8.421		240	460	20	720	980
3150	8"		1.84	13.043		510	715	31.50	1,256.50	1,675
3151	10"		1.45	16.552		825	905	39.50	1,769.50	2,325
3152	12"		1.15	20.870		1,150	1,150	50	2,350	3,025
3154	For schedule 10, type 316, add					25%				
3281	Butt weld joint, schedule 40, type 304									
3284	90° Elbow, long, 1/2"	Q-15	12.70	1.260	Ea.	15.75	66.50	4.55	86.80	123
3288	3/4"		11.10	1.441		15.75	76	5.20	96.95	138
3289	1"		10.13	1.579		16.50	83.50	5.70	105.70	150
3290	1-1/4"		9.40	1.702		22	90	6.15	118.15	167
3300	1-1/2"		8.89	1.800		17.25	95	6.50	118.75	170
3310	2"		7.55	2.119		25	112	7.65	144.65	204
3320	2-1/2"		3.98	4.020		47.50	212	14.50	274	390
3330	3"		3.21	4.984		61	263	18	342	485
3340	3-1/2"		2.83	5.654		225	299	20.50	544.50	720
3350	4"		2.44	6.557		105	345	23.50	473.50	665
3360	5"		2.13	7.512		355	395	27	777	1,025
3370	6"	Q-16	2.83	8.481		310	465	20.50	795.50	1,075
3380	8"		1.83	13.115		615	720	31.50	1,366.50	1,775
3390	10"		1.44	16.667		1,300	915	40	2,255	2,850
3400	12"		1.14	21.053		1,650	1,150	50.50	2,850.50	3,625
3410	For schedule 40, type 316, add					25%				
3460	45° Elbow, long, 1/2"	Q-15	12.70	1.260	Ea.	15.75	66.50	4.55	86.80	123
3470	3/4"		11.10	1.441		15.75	76	5.20	96.95	138
3480	1"		10.13	1.579		16.50	83.50	5.70	105.70	150
3490	1-1/4"		9.40	1.702		22	90	6.15	118.15	167
3500	1-1/2"		8.89	1.800		17.25	95	6.50	118.75	170

For customer support on your Mechanical Cost Data, call 877.795.7394.

193

22 11 13.66 Pipe Fittings, Stainless Steel		Crew	Daily Output	Labor-Hours	Unit	Material	2015 Bare Costs Labor	Equipment	Total	Total Incl O&P
3510	2"	Q-15	7.55	2.119	Ea.	25	112	7.65	144.65	204
3520	2-1/2"		3.98	4.020		47.50	212	14.50	274	390
3530	3"		3.21	4.984		48	263	18	329	475
3540	3-1/2"		2.83	5.654		225	299	20.50	544.50	720
3550	4"		2.44	6.557		75	345	23.50	443.50	635
3560	5"		2.13	7.512		248	395	27	670	900
3570	6"	Q-16	2.83	8.481		218	465	20.50	703.50	960
3580	8"		1.83	13.115		430	720	31.50	1,181.50	1,575
3590	10"		1.44	16.667		910	915	40	1,865	2,425
3600	12"		1.14	21.053		1,150	1,150	50.50	2,350.50	3,075
3610	For schedule 40, type 316, add					25%				
3660	Tee, straight 1/2"	Q-15	8.46	1.891	Ea.	40.50	100	6.85	147.35	203
3670	3/4"		7.40	2.162		40.50	114	7.80	162.30	226
3680	1"		6.74	2.374		43	125	8.55	176.55	245
3690	1-1/4"		6.27	2.552		100	135	9.20	244.20	325
3700	1-1/2"		5.92	2.703		46	143	9.75	198.75	277
3710	2"		5.03	3.181		68.50	168	11.50	248	340
3720	2-1/2"		2.65	6.038		86.50	320	22	428.50	600
3730	3"		2.14	7.477		87	395	27	509	720
3740	3-1/2"		1.88	8.511		164	450	30.50	644.50	895
3750	4"		1.62	9.877		164	520	35.50	719.50	1,000
3760	5"		1.42	11.268		380	595	40.50	1,015.50	1,350
3770	6"	Q-16	1.88	12.766		365	700	30.50	1,095.50	1,475
3780	8"		1.22	19.672		735	1,075	47	1,857	2,475
3790	10"		.96	25		1,425	1,375	60	2,860	3,725
3800	12"		.76	31.579		1,900	1,725	76	3,701	4,800
3810	For schedule 40, type 316, add					25%				
3820	Tee, reducing on outlet, 3/4" x 1/2"	Q-15	7.73	2.070	Ea.	59.50	109	7.45	175.95	239
3822	1" x 1/2"		7.24	2.210		58.50	117	8	183.50	249
3824	1" x 3/4"		6.96	2.299		54	121	8.30	183.30	252
3826	1-1/4" x 1"		6.43	2.488		132	131	9	272	355
3828	1-1/2" x 1/2"		6.58	2.432		79.50	128	8.80	216.30	291
3830	1-1/2" x 3/4"		6.35	2.520		74.50	133	9.10	216.60	293
3832	1-1/2" x 1"		6.18	2.589		57	137	9.35	203.35	280
3834	2" x 1"		5.50	2.909		102	154	10.50	266.50	355
3836	2" x 1-1/2"		5.30	3.019		85.50	159	10.90	255.40	345
3838	2-1/2" x 2"		3.15	5.079		129	268	18.35	415.35	565
3840	3" x 1-1/2"		2.72	5.882		131	310	21	462	640
3842	3" x 2"		2.65	6.038		109	320	22	451	625
3844	4" x 2"		2.10	7.619		272	405	27.50	704.50	940
3846	4" x 3"		1.77	9.040		197	480	32.50	709.50	975
3848	5" x 4"		1.48	10.811		455	570	39	1,064	1,400
3850	6" x 3"	Q-16	2.19	10.959		505	600	26.50	1,131.50	1,500
3852	6" x 4"		2.04	11.765		435	645	28	1,108	1,475
3854	8" x 4"		1.46	16.438		1,025	900	39.50	1,964.50	2,525
3856	10" x 8"		.69	34.783		1,725	1,900	83.50	3,708.50	4,850
3858	12" x 10"		.55	43.636		2,275	2,400	105	4,780	6,250
3950	Reducer, concentric, 3/4" x 1/2"	Q-15	11.85	1.350		30	71.50	4.87	106.37	146
3952	1" x 3/4"		10.60	1.509		31.50	79.50	5.45	116.45	161
3954	1-1/4" x 3/4"		10.19	1.570		77.50	83	5.65	166.15	216
3956	1-1/4" x 1"		9.76	1.639		39	86.50	5.90	131.40	181
3958	1-1/2" x 3/4"		9.88	1.619		65.50	85.50	5.85	156.85	207
3960	1-1/2" x 1"		9.47	1.690		52.50	89.50	6.10	148.10	200

22 11 13 – Facility Water Distribution Piping

22 11 13.66 Pipe Fittings, Stainless Steel		Crew	Daily Output	Labor-Hours	Unit	Material	2015 Bare Costs Labor	Equipment	Total	Total Incl O&P
3962	2" x 1"	Q-15	8.65	1.850	Ea.	30	97.50	6.70	134.20	188
3964	2" x 1-1/2"		8.16	1.961		26.50	104	7.10	137.60	193
3966	2-1/2" x 1"		5.71	2.802		113	148	10.10	271.10	360
3968	2-1/2" x 2"		5.21	3.071		58	162	11.10	231.10	320
3970	3" x 1"		4.88	3.279		78	173	11.85	262.85	360
3972	3" x 1-1/2"		4.72	3.390		40	179	12.25	231.25	325
3974	3" x 2"		4.51	3.548		37.50	187	12.80	237.30	340
3976	4" x 2"		3.69	4.336		53.50	229	15.65	298.15	420
3978	4" x 3"		2.77	5.776		40	305	21	366	525
3980	5" x 3"		2.56	6.250		237	330	22.50	589.50	785
3982	5" x 4"		2.27	7.048		191	370	25.50	586.50	800
3984	6" x 3"	Q-16	3.57	6.723		122	370	16.15	508.15	705
3986	6" x 4"		3.19	7.524		101	410	18.05	529.05	750
3988	8" x 4"		2.44	9.836		315	540	23.50	878.50	1,175
3990	8" x 6"		2.22	10.811		236	590	26	852	1,175
3992	10" x 6"		1.91	12.565		440	690	30	1,160	1,575
3994	10" x 8"		1.61	14.907		365	815	36	1,216	1,675
3995	12" x 6"		1.63	14.724		675	805	35.50	1,515.50	2,000
3996	12" x 8"		1.41	17.021		620	930	41	1,591	2,125
3997	12" x 10"		1.27	18.898		435	1,025	45.50	1,505.50	2,100
4000	Socket weld joint, 3000 lb., type 304									
4100	90° Elbow									
4140	1/4"	Q-15	13.47	1.188	Ea.	44	63	4.29	111.29	148
4150	3/8"		12.97	1.234		57	65	4.45	126.45	166
4160	1/2"		12.21	1.310		62.50	69	4.73	136.23	179
4170	3/4"		10.68	1.498		72	79	5.40	156.40	205
4180	1"		9.74	1.643		108	87	5.95	200.95	257
4190	1-1/4"		9.05	1.768		189	93.50	6.40	288.90	355
4200	1-1/2"		8.55	1.871		229	99	6.75	334.75	410
4210	2"		7.26	2.204		370	116	7.95	493.95	590
4300	45° Elbow									
4340	1/4"	Q-15	13.47	1.188	Ea.	83	63	4.29	150.29	191
4350	3/8"		12.97	1.234		83	65	4.45	152.45	194
4360	1/2"		12.21	1.310		83	69	4.73	156.73	201
4370	3/4"		10.68	1.498		94	79	5.40	178.40	229
4380	1"		9.74	1.643		136	87	5.95	228.95	288
4390	1-1/4"		9.05	1.768		224	93.50	6.40	323.90	395
4400	1-1/2"		8.55	1.871		225	99	6.75	330.75	405
4410	2"		7.26	2.204		410	116	7.95	533.95	635
4500	Tee									
4540	1/4"	Q-15	8.97	1.784	Ea.	58	94	6.45	158.45	213
4550	3/8"		8.64	1.852		69	98	6.70	173.70	231
4560	1/2"		8.13	1.968		85	104	7.10	196.10	258
4570	3/4"		7.12	2.247		98	119	8.10	225.10	296
4580	1"		6.48	2.469		132	130	8.90	270.90	350
4590	1-1/4"		6.03	2.653		234	140	9.60	383.60	480
4600	1-1/2"		5.69	2.812		335	149	10.15	494.15	605
4610	2"		4.83	3.313		510	175	11.95	696.95	835
5000	Socket weld joint, 3000 lb., type 316									
5100	90° Elbow									
5140	1/4"	Q-15	13.47	1.188	Ea.	54	63	4.29	121.29	159
5150	3/8"		12.97	1.234		63	65	4.45	132.45	172
5160	1/2"		12.21	1.310		76	69	4.73	149.73	194

For customer support on your Mechanical Cost Data, call 877.795.7394.

195

22 11 13.66 Pipe Fittings, Stainless Steel		Crew	Daily Output	Labor-Hours	Unit	Material	2015 Bare Costs Labor	Equipment	Total	Total Incl O&P
5170	3/4"	Q-15	10.68	1.498	Ea.	100	79	5.40	184.40	236
5180	1"		9.74	1.643		142	87	5.95	234.95	294
5190	1-1/4"		9.05	1.768		253	93.50	6.40	352.90	425
5200	1-1/2"		8.55	1.871		286	99	6.75	391.75	470
5210	2"		7.26	2.204		485	116	7.95	608.95	720
5300	45° Elbow									
5340	1/4"	Q-15	13.47	1.188	Ea.	104	63	4.29	171.29	214
5350	3/8"		12.97	1.234		104	65	4.45	173.45	217
5360	1/2"		12.21	1.310		104	69	4.73	177.73	224
5370	3/4"		10.68	1.498		121	79	5.40	205.40	259
5380	1"		9.74	1.643		182	87	5.95	274.95	340
5390	1-1/4"		9.05	1.768		258	93.50	6.40	357.90	430
5400	1-1/2"		8.55	1.871		292	99	6.75	397.75	475
5410	2"		7.26	2.204		435	116	7.95	558.95	665
5500	Tee									
5540	1/4"	Q-15	8.97	1.784	Ea.	73	94	6.45	173.45	230
5550	3/8"		8.64	1.852		89.50	98	6.70	194.20	253
5560	1/2"		8.13	1.968		99.50	104	7.10	210.60	274
5570	3/4"		7.12	2.247		123	119	8.10	250.10	325
5580	1"		6.48	2.469		188	130	8.90	326.90	415
5590	1-1/4"		6.03	2.653		299	140	9.60	448.60	555
5600	1-1/2"		5.69	2.812		420	149	10.15	579.15	700
5610	2"		4.83	3.313		670	175	11.95	856.95	1,025
5700	For socket weld joint, 6000 lb., type 304 and 316, add					100%				
6000	Threaded companion flange									
6010	Stainless steel, 150 lb., type 304									
6020	1/2" diam.	1 Plum	30	.267	Ea.	45	15.65		60.65	73
6030	3/4" diam.		28	.286		50	16.75		66.75	80.50
6040	1" diam.		27	.296		55	17.40		72.40	87
6050	1-1/4" diam.	Q-1	44	.364		71	19.20		90.20	107
6060	1-1/2" diam.		40	.400		71	21		92	110
6070	2" diam.		36	.444		93	23.50		116.50	138
6080	2-1/2" diam.		28	.571		130	30		160	189
6090	3" diam.		20	.800		136	42.50		178.50	214
6110	4" diam.		12	1.333		186	70.50		256.50	310
6130	6" diam.	Q-2	14	1.714		335	94		429	505
6140	8" diam.	"	12	2		645	110		755	875
6150	For type 316 add					40%				
6260	Weld flanges, stainless steel, type 304									
6270	Slip on, 150 lb. (welded, front and back)									
6280	1/2" diam.	Q-15	12.70	1.260	Ea.	40	66.50	4.55	111.05	150
6290	3/4" diam.		11.11	1.440		41	76	5.20	122.20	166
6300	1" diam.		10.13	1.579		45	83.50	5.70	134.20	182
6310	1-1/4" diam.		9.41	1.700		61	90	6.15	157.15	210
6320	1-1/2" diam.		8.89	1.800		61	95	6.50	162.50	218
6330	2" diam.		7.55	2.119		79	112	7.65	198.65	264
6340	2-1/2" diam.		3.98	4.020		111	212	14.50	337.50	460
6350	3" diam.		3.21	4.984		119	263	18	400	550
6370	4" diam.		2.44	6.557		162	345	23.50	530.50	730
6390	6" diam.	Q-16	1.89	12.698		247	695	30.50	972.50	1,350
6400	8" diam.	"	1.22	19.672		465	1,075	47	1,587	2,200
6410	For type 316, add					40%				
6530	Weld neck 150 lb.									

196

For customer support on your Mechanical Cost Data, call 877.795.7394.

22 11 Facility Water Distribution

22 11 13 – Facility Water Distribution Piping

22 11 13.66 Pipe Fittings, Stainless Steel		Crew	Daily Output	Labor-Hours	Unit	Material	2015 Bare Costs Labor	Equipment	Total	Total Incl O&P
6540	1/2" diam.	Q-15	25.40	.630	Ea.	20.50	33.50	2.27	56.27	75.50
6550	3/4" diam.		22.22	.720		25	38	2.60	65.60	87.50
6560	1" diam.		20.25	.790		27	41.50	2.85	71.35	95.50
6570	1-1/4" diam.		18.82	.850		43.50	45	3.07	91.57	119
6580	1-1/2" diam.		17.78	.900		38	47.50	3.25	88.75	117
6590	2" diam.		15.09	1.060		40.50	56	3.83	100.33	133
6600	2-1/2" diam.		7.96	2.010		65	106	7.25	178.25	240
6610	3" diam.		6.43	2.488		66.50	131	9	206.50	282
6630	4" diam.		4.88	3.279		101	173	11.85	285.85	385
6640	5" diam.		4.26	3.756		132	198	13.55	343.55	460
6650	6" diam.	Q-16	5.66	4.240		145	232	10.20	387.20	520
6652	8" diam.		3.65	6.575		261	360	15.80	636.80	850
6654	10" diam.		2.88	8.333		365	455	20	840	1,125
6656	12" diam.		2.28	10.526		705	575	25.50	1,305.50	1,675
6670	For type 316 add					23%				
7000	Threaded joint, 150 lb., type 304									
7030	90° elbow									
7040	1/8"	1 Plum	13	.615	Ea.	25.50	36		61.50	82.50
7050	1/4"		13	.615		25.50	36		61.50	82.50
7070	3/8"		13	.615		29.50	36		65.50	87
7080	1/2"		12	.667		26.50	39		65.50	88
7090	3/4"		11	.727		31.50	42.50		74	99
7100	1"		10	.800		43	47		90	119
7110	1-1/4"	Q-1	17	.941		66.50	49.50		116	149
7120	1-1/2"		16	1		77	53		130	165
7130	2"		14	1.143		111	60.50		171.50	214
7140	2-1/2"		11	1.455		271	77		348	415
7150	3"		8	2		395	106		501	595
7160	4"	Q-2	11	2.182		670	120		790	915
7180	45° elbow									
7190	1/8"	1 Plum	13	.615	Ea.	37	36		73	95.50
7200	1/4"		13	.615		37	36		73	95.50
7210	3/8"		13	.615		37.50	36		73.50	95.50
7220	1/2"		12	.667		38	39		77	101
7230	3/4"		11	.727		41	42.50		83.50	110
7240	1"		10	.800		47.50	47		94.50	123
7250	1-1/4"	Q-1	17	.941		66	49.50		115.50	148
7260	1-1/2"		16	1		85.50	53		138.50	174
7270	2"		14	1.143		121	60.50		181.50	224
7280	2-1/2"		11	1.455		380	77		457	535
7290	3"		8	2		560	106		666	775
7300	4"	Q-2	11	2.182		1,000	120		1,120	1,275
7320	Tee, straight									
7330	1/8"	1 Plum	9	.889	Ea.	39	52		91	122
7340	1/4"		9	.889		39	52		91	122
7350	3/8"		9	.889		41.50	52		93.50	125
7360	1/2"		8	1		39.50	58.50		98	132
7370	3/4"		7	1.143		42	67		109	148
7380	1"		6.50	1.231		54	72.50		126.50	169
7390	1-1/4"	Q-1	11	1.455		91	77		168	216
7400	1-1/2"		10	1.600		116	84.50		200.50	256
7410	2"		9	1.778		145	94		239	300
7420	2-1/2"		7	2.286		390	121		511	610

For customer support on your Mechanical Cost Data, call 877.795.7394.

197

22 11 13.66 Pipe Fittings, Stainless Steel		Crew	Daily Output	Labor-Hours	Unit	Material	2015 Bare Costs Labor	Equipment	Total	Total Incl O&P
7430	3"	Q-1	5	3.200	Ea.	590	169		759	905
7440	4"	Q-2	7	3.429	↓	1,475	188		1,663	1,875
7460	Coupling, straight									
7470	1/8"	1 Plum	19	.421	Ea.	10.25	24.50		34.75	49
7480	1/4"		19	.421		12.10	24.50		36.60	51
7490	3/8"		19	.421		14.50	24.50		39	53.50
7500	1/2"		19	.421		19.30	24.50		43.80	58.50
7510	3/4"		18	.444		25.50	26		51.50	68
7520	1"	↓	15	.533		41.50	31.50		73	93
7530	1-1/4"	Q-1	26	.615		66.50	32.50		99	123
7540	1-1/2"		24	.667		74.50	35		109.50	135
7550	2"		21	.762		123	40.50		163.50	197
7560	2-1/2"		18	.889		287	47		334	385
7570	3"	↓	14	1.143		390	60.50		450.50	520
7580	4"	Q-2	16	1.500		550	82		632	730
7600	Reducer, concentric, 1/2"	1 Plum	12	.667		20.50	39		59.50	82
7610	3/4"		11	.727		26.50	42.50		69	94
7612	1"	↓	10	.800		44.50	47		91.50	120
7614	1-1/4"	Q-1	17	.941		94.50	49.50		144	179
7616	1-1/2"		16	1		103	53		156	193
7618	2"		14	1.143		161	60.50		221.50	268
7620	2-1/2"		11	1.455		400	77		477	555
7622	3"	↓	8	2		445	106		551	650
7624	4"	Q-2	11	2.182	↓	735	120		855	990
7710	Union									
7720	1/8"	1 Plum	12	.667	Ea.	47.50	39		86.50	112
7730	1/4"		12	.667		47.50	39		86.50	112
7740	3/8"		12	.667		55.50	39		94.50	120
7750	1/2"		11	.727		69.50	42.50		112	141
7760	3/4"		10	.800		96	47		143	176
7770	1"	↓	9	.889		138	52		190	231
7780	1-1/4"	Q-1	16	1		340	53		393	455
7790	1-1/2"		15	1.067		370	56.50		426.50	490
7800	2"		13	1.231		465	65		530	615
7810	2-1/2"		10	1.600		880	84.50		964.50	1,100
7820	3"	↓	7	2.286		1,175	121		1,296	1,475
7830	4"	Q-2	10	2.400	↓	1,625	131		1,756	1,975
7838	Caps									
7840	1/2"	1 Plum	24	.333	Ea.	12.90	19.55		32.45	43.50
7841	3/4"		22	.364		18.80	21.50		40.30	52.50
7842	1"	↓	20	.400		30	23.50		53.50	68.50
7843	1-1/2"	Q-1	32	.500		76	26.50		102.50	124
7844	2"	"	28	.571		96	30		126	152
7845	4"	Q-2	22	1.091		380	60		440	505
7850	For 150 lb., type 316, add				↓	25%				
8750	Threaded joint, 2000 lb., type 304									
8770	90° Elbow									
8780	1/8"	1 Plum	13	.615	Ea.	32.50	36		68.50	90
8790	1/4"		13	.615		32.50	36		68.50	90
8800	3/8"		13	.615		40	36		76	98.50
8810	1/2"		12	.667		54.50	39		93.50	119
8820	3/4"		11	.727		63	42.50		105.50	134
8830	1"	↓	10	.800		83.50	47		130.50	163

For customer support on your Mechanical Cost Data, call 877.795.7394.

22 11 13 – Facility Water Distribution Piping

22 11 13.66 Pipe Fittings, Stainless Steel	Crew	Daily Output	Labor-Hours	Unit	Material	2015 Bare Costs Labor	Equipment	Total	Total Incl O&P
8840 1-1/4"	Q-1	17	.941	Ea.	136	49.50		185.50	225
8850 1-1/2"		16	1		218	53		271	320
8860 2"		14	1.143		305	60.50		365.50	425
8880 45° Elbow									
8890 1/8"	1 Plum	13	.615	Ea.	67.50	36		103.50	129
8900 1/4"		13	.615		67.50	36		103.50	129
8910 3/8"		13	.615		84.50	36		120.50	148
8920 1/2"		12	.667		84.50	39		123.50	152
8930 3/4"		11	.727		93	42.50		135.50	167
8940 1"		10	.800		111	47		158	193
8950 1-1/4"	Q-1	17	.941		194	49.50		243.50	288
8960 1-1/2"		16	1		260	53		313	365
8970 2"		14	1.143		299	60.50		359.50	420
8990 Tee, straight									
9000 1/8"	1 Plum	9	.889	Ea.	42	52		94	125
9010 1/4"		9	.889		42	52		94	125
9020 3/8"		9	.889		54	52		106	138
9030 1/2"		8	1		69	58.50		127.50	165
9040 3/4"		7	1.143		78.50	67		145.50	187
9050 1"		6.50	1.231		108	72.50		180.50	228
9060 1-1/4"	Q-1	11	1.455		182	77		259	315
9070 1-1/2"		10	1.600		305	84.50		389.50	465
9080 2"		9	1.778		420	94		514	605
9100 For couplings and unions use 3000 lb., type 304									
9120 2000 lb., type 316									
9130 90° Elbow									
9140 1/8"	1 Plum	13	.615	Ea.	40.50	36		76.50	99
9150 1/4"		13	.615		40.50	36		76.50	99
9160 3/8"		13	.615		47	36		83	107
9170 1/2"		12	.667		61	39		100	127
9180 3/4"		11	.727		75	42.50		117.50	147
9190 1"		10	.800		111	47		158	194
9200 1-1/4"	Q-1	17	.941		202	49.50		251.50	297
9210 1-1/2"		16	1		233	53		286	335
9220 2"		14	1.143		395	60.50		455.50	525
9240 45° Elbow									
9250 1/8"	1 Plum	13	.615	Ea.	85.50	36		121.50	149
9260 1/4"		13	.615		85.50	36		121.50	149
9270 3/8"		13	.615		85.50	36		121.50	149
9280 1/2"		12	.667		85.50	39		124.50	153
9300 3/4"		11	.727		96.50	42.50		139	171
9310 1"		10	.800		146	47		193	231
9320 1-1/4"	Q-1	17	.941		208	49.50		257.50	305
9330 1-1/2"		16	1		285	53		338	395
9340 2"		14	1.143		350	60.50		410.50	475
9360 Tee, straight									
9370 1/8"	1 Plum	9	.889	Ea.	51.50	52		103.50	136
9380 1/4"		9	.889		51.50	52		103.50	136
9390 3/8"		9	.889		63	52		115	148
9400 1/2"		8	1		78	58.50		136.50	175
9410 3/4"		7	1.143		96.50	67		163.50	207
9420 1"		6.50	1.231		147	72.50		219.50	271
9430 1-1/4"	Q-1	11	1.455		239	77		316	380

22 11 13.66 Pipe Fittings, Stainless Steel	Crew	Daily Output	Labor-Hours	Unit	Material	2015 Bare Costs Labor	Equipment	Total	Total Incl O&P
9440 1-1/2"	Q-1	10	1.600	Ea.	345	84.50		429.50	505
9450 2"	↓	9	1.778	↓	545	94		639	740
9470 For couplings and unions use 3000 lb., type 316									
9490 3000 lb., type 304									
9510 Coupling									
9520 1/8"	1 Plum	19	.421	Ea.	12.80	24.50		37.30	51.50
9530 1/4"		19	.421		13.90	24.50		38.40	53
9540 3/8"		19	.421		16.20	24.50		40.70	55.50
9550 1/2"		19	.421		19.50	24.50		44	59
9560 3/4"		18	.444		26	26		52	68
9570 1"	↓	15	.533		44	31.50		75.50	96
9580 1-1/4"	Q-1	26	.615		107	32.50		139.50	166
9590 1-1/2"		24	.667		125	35		160	191
9600 2"	↓	21	.762	↓	165	40.50		205.50	242
9620 Union									
9630 1/8"	1 Plum	12	.667	Ea.	98	39		137	167
9640 1/4"		12	.667		98	39		137	167
9650 3/8"		12	.667		105	39		144	174
9660 1/2"		11	.727		105	42.50		147.50	180
9670 3/4"		10	.800		128	47		175	212
9680 1"	↓	9	.889		198	52		250	296
9690 1-1/4"	Q-1	16	1		350	53		403	465
9700 1-1/2"		15	1.067		410	56.50		466.50	540
9710 2"	↓	13	1.231	↓	550	65		615	705
9730 3000 lb., type 316									
9750 Coupling									
9770 1/8"	1 Plum	19	.421	Ea.	14.40	24.50		38.90	53.50
9780 1/4"		19	.421		16.25	24.50		40.75	55.50
9790 3/8"		19	.421		17.40	24.50		41.90	56.50
9800 1/2"		19	.421		24.50	24.50		49	64.50
9810 3/4"		18	.444		36.50	26		62.50	79.50
9820 1"	↓	15	.533		60.50	31.50		92	115
9830 1-1/4"	Q-1	26	.615		136	32.50		168.50	199
9840 1-1/2"		24	.667		159	35		194	228
9850 2"	↓	21	.762	↓	219	40.50		259.50	300
9870 Union									
9880 1/8"	1 Plum	12	.667	Ea.	108	39		147	177
9890 1/4"		12	.667		108	39		147	177
9900 3/8"		12	.667		125	39		164	196
9910 1/2"		11	.727		125	42.50		167.50	202
9920 3/4"		10	.800		163	47		210	250
9930 1"	↓	9	.889		257	52		309	360
9940 1-1/4"	Q-1	16	1		450	53		503	575
9950 1-1/2"		15	1.067		555	56.50		611.50	695
9960 2"	↓	13	1.231		705	65		770	875

22 11 13.74 Pipe, Plastic

	Crew	Daily Output	Labor-Hours	Unit	Material	2015 Bare Costs Labor	Equipment	Total	Total Incl O&P
0010 **PIPE, PLASTIC**									
0020 Fiberglass reinforced, couplings 10' O.C., clevis hanger assy's, 3 per 10'									
0080 General service									
0120 2" diameter	Q-1	59	.271	L.F.	12.55	14.35		26.90	35.50
0140 3" diameter		52	.308		16.10	16.25		32.35	42.50
0150 4" diameter	↓	48	.333	↓	21	17.60		38.60	49.50

200

For customer support on your Mechanical Cost Data, call 877.795.7394.

22 11 13 – Facility Water Distribution Piping

22 11 13.74 Pipe, Plastic		Crew	Daily Output	Labor-Hours	Unit	Material	2015 Bare Costs Labor	Equipment	Total	Total Incl O&P
0160	6" diameter	Q-1	39	.410	L.F.	35.50	21.50		57	72
0170	8" diameter	Q-2	49	.490		55	27		82	101
0180	10" diameter		41	.585		80	32		112	137
0190	12" diameter		36	.667		103	36.50		139.50	168
0600	PVC, high impact/pressure, cplgs. 10' O.C., clevis hanger assy's, 3 per 10'									
1020	Schedule 80									
1070	1/2" diameter	1 Plum	50	.160	L.F.	6.75	9.40		16.15	21.50
1080	3/4" diameter		47	.170		7.80	10		17.80	23.50
1090	1" diameter		43	.186		9.65	10.90		20.55	27
1100	1-1/4" diameter		39	.205		11.85	12.05		23.90	31.50
1110	1-1/2" diameter		34	.235		13.40	13.80		27.20	36
1120	2" diameter	Q-1	55	.291		16.65	15.35		32	41.50
1140	3" diameter		50	.320		31	16.90		47.90	60
1150	4" diameter		46	.348		43.50	18.40		61.90	76
1170	6" diameter		38	.421		94.50	22		116.50	138
1730	To delete coupling & hangers, subtract									
1740	1/2" diam.					62%	80%			
1750	3/4" diam. to 1-1/4" diam.					58%	73%			
1760	1-1/2" diam. to 6" diam.					40%	57%			
1800	PVC, couplings 10' O.C., clevis hanger assemblies, 3 per 10'									
1820	Schedule 40									
1860	1/2" diameter	1 Plum	54	.148	L.F.	4.88	8.70		13.58	18.50
1870	3/4" diameter		51	.157		5.20	9.20		14.40	19.65
1880	1" diameter		46	.174		5.85	10.20		16.05	22
1890	1-1/4" diameter		42	.190		6.55	11.20		17.75	24
1900	1-1/2" diameter		36	.222		6.85	13.05		19.90	27.50
1910	2" diameter	Q-1	59	.271		8.05	14.35		22.40	30.50
1920	2-1/2" diameter		56	.286		10.35	15.10		25.45	34.50
1930	3" diameter		53	.302		12.60	15.95		28.55	38
1940	4" diameter		48	.333		16.10	17.60		33.70	44
1950	5" diameter		43	.372		26.50	19.65		46.15	58.50
1960	6" diameter		39	.410		27.50	21.50		49	62.50
1970	8" diameter	Q-2	48	.500		36.50	27.50		64	81.50
1980	10" diameter		43	.558		75	30.50		105.50	129
1990	12" diameter		42	.571		90	31.50		121.50	147
2000	14" diameter		31	.774		143	42.50		185.50	222
2010	16" diameter		23	1.043		205	57		262	315
2340	To delete coupling & hangers, subtract									
2360	1/2" diam. to 1-1/4" diam.					65%	74%			
2370	1-1/2" diam. to 6" diam.					44%	57%			
2380	8" diam. to 12" diam.					41%	53%			
2390	14" diam. to 16" diam.					48%	45%			
2420	Schedule 80									
2440	1/4" diameter	1 Plum	58	.138	L.F.	4.35	8.10		12.45	17.05
2450	3/8" diameter		55	.145		4.35	8.55		12.90	17.70
2460	1/2" diameter		50	.160		5	9.40		14.40	19.70
2470	3/4" diameter		47	.170		5.35	10		15.35	21
2480	1" diameter		43	.186		6.10	10.90		17	23
2490	1-1/4" diameter		39	.205		6.85	12.05		18.90	25.50
2500	1-1/2" diameter		34	.235		7.30	13.80		21.10	29
2510	2" diameter	Q-1	55	.291		8.20	15.35		23.55	32
2520	2-1/2" diameter		52	.308		9.85	16.25		26.10	35.50
2530	3" diameter		50	.320		13.95	16.90		30.85	41

For customer support on your Mechanical Cost Data, call 877.795.7394.

201

22 11 13.74 Pipe, Plastic		Crew	Daily Output	Labor-Hours	Unit	Material	2015 Bare Costs Labor	Equipment	Total	Total Incl O&P
2540	4" diameter	Q-1	46	.348	L.F.	19.55	18.40		37.95	49.50
2550	5" diameter		42	.381		29.50	20		49.50	62.50
2560	6" diameter		38	.421		35.50	22		57.50	73
2570	8" diameter	Q-2	47	.511		45	28		73	92
2580	10" diameter		42	.571		106	31.50		137.50	164
2590	12" diameter		38	.632		128	34.50		162.50	194
2830	To delete coupling & hangers, subtract									
2840	1/4" diam. to 1/2" diam.					66%	80%			
2850	3/4" diam. to 1-1/4" diam.					61%	73%			
2860	1-1/2" diam. to 6" diam.					41%	57%			
2870	8" diam. to 12" diam.					31%	50%			
2900	Schedule 120									
2910	1/2" diameter	1 Plum	50	.160	L.F.	5.50	9.40		14.90	20.50
2950	3/4" diameter		47	.170		6	10		16	21.50
2960	1" diameter		43	.186		7.05	10.90		17.95	24.50
2970	1-1/4" diameter		39	.205		8.30	12.05		20.35	27.50
2980	1-1/2" diameter		33	.242		9.05	14.25		23.30	31.50
2990	2" diameter	Q-1	54	.296		10.75	15.65		26.40	35.50
3000	2-1/2" diameter		52	.308		15.85	16.25		32.10	42
3010	3" diameter		49	.327		19.10	17.25		36.35	47
3020	4" diameter		45	.356		27.50	18.80		46.30	58.50
3030	6" diameter		37	.432		49.50	23		72.50	89
3240	To delete coupling & hangers, subtract									
3250	1/2" diam. to 1-1/4" diam.					52%	74%			
3260	1-1/2" diam. to 4" diam.					30%	57%			
3270	6" diam.					17%	50%			
3300	PVC, pressure, couplings 10' O.C., clevis hanger assy's, 3 per 10'									
3310	SDR 26, 160 psi									
3350	1-1/4" diameter	1 Plum	42	.190	L.F.	6.20	11.20		17.40	24
3360	1-1/2" diameter	"	36	.222		6.40	13.05		19.45	27
3370	2" diameter	Q-1	59	.271		7.30	14.35		21.65	29.50
3380	2-1/2" diameter		56	.286		9.80	15.10		24.90	34
3390	3" diameter		53	.302		12.30	15.95		28.25	37.50
3400	4" diameter		48	.333		16.50	17.60		34.10	44.50
3420	6" diameter		39	.410		30.50	21.50		52	66
3430	8" diameter	Q-2	48	.500		45	27.50		72.50	91
3660	To delete coupling & clevis hanger assy's, subtract									
3670	1-1/4" diam.					63%	68%			
3680	1-1/2" diam. to 4" diam.					48%	57%			
3690	6" diam. to 8" diam.					60%	54%			
3720	SDR 21, 200 psi, 1/2" diameter	1 Plum	54	.148	L.F.	4.77	8.70		13.47	18.40
3740	3/4" diameter		51	.157		5.10	9.20		14.30	19.50
3750	1" diameter		46	.174		5.60	10.20		15.80	21.50
3760	1-1/4" diameter		42	.190		6.40	11.20		17.60	24
3770	1-1/2" diameter		36	.222		6.75	13.05		19.80	27
3780	2" diameter	Q-1	59	.271		7.65	14.35		22	30
3790	2-1/2" diameter		56	.286		11.90	15.10		27	36
3800	3" diameter		53	.302		13.10	15.95		29.05	38.50
3810	4" diameter		48	.333		20	17.60		37.60	49
3830	6" diameter		39	.410		34	21.50		55.50	69.50
3840	8" diameter	Q-2	48	.500		50	27.50		77.50	96.50
4000	To delete coupling & hangers, subtract									
4010	1/2" diam. to 3/4" diam.					71%	77%			

22 11 13 – Facility Water Distribution Piping

22 11 13.74 Pipe, Plastic	Crew	Daily Output	Labor-Hours	Unit	Material	2015 Bare Costs Labor	2015 Bare Costs Equipment	Total	Total Incl O&P
4020　　1" diam. to 1-1/4" diam.					63%	70%			
4030　　1-1/2" diam. to 6" diam.					44%	57%			
4040　　8" diam.					46%	54%			
4100　DWV type, schedule 40, couplings 10' O.C., clevis hanger assy's, 3 per 10'									
4210　　ABS, schedule 40, foam core type									
4212　　　Plain end black									
4214　　　　1-1/2" diameter	1 Plum	39	.205	L.F.	5.20	12.05		17.25	24
4216　　　　2" diameter	Q-1	62	.258		5.60	13.65		19.25	26.50
4218　　　　3" diameter		56	.286		8	15.10		23.10	32
4220　　　　4" diameter		51	.314		10.20	16.55		26.75	36
4222　　　　6" diameter		42	.381		18	20		38	50.50
4240　　　To delete coupling & hangers, subtract									
4244　　　　1-1/2" diam. to 6" diam.					43%	48%			
4400　　PVC									
4410　　　1-1/4" diameter	1 Plum	42	.190	L.F.	5.35	11.20		16.55	23
4420　　　1-1/2" diameter	"	36	.222		5.30	13.05		18.35	25.50
4460　　　2" diameter	Q-1	59	.271		5.65	14.35		20	27.50
4470　　　3" diameter		53	.302		8.05	15.95		24	33
4480　　　4" diameter		48	.333		10	17.60		27.60	37.50
4490　　　6" diameter		39	.410		16.40	21.50		37.90	50.50
4500　　　8" diameter	Q-2	48	.500		21.50	27.50		49	65.50
4510　　To delete coupling & hangers, subtract									
4520　　　1-1/4" diam. to 1-1/2" diam.					48%	60%			
4530　　　2" diam. to 8" diam.					42%	54%			
4550　　PVC, schedule 40, foam core type									
4552　　　Plain end, white									
4554　　　　1-1/2" diameter	1 Plum	39	.205	L.F.	5	12.05		17.05	23.50
4556　　　　2" diameter	Q-1	62	.258		5.35	13.65		19	26.50
4558　　　　3" diameter		56	.286		7.45	15.10		22.55	31
4560　　　　4" diameter		51	.314		9.10	16.55		25.65	35
4562　　　　6" diameter		42	.381		14.55	20		34.55	46.50
4564　　　　8" diameter	Q-2	51	.471		18.75	26		44.75	59.50
4568　　　　10" diameter		48	.500		23.50	27.50		51	67
4570　　　　12" diameter		46	.522		28	28.50		56.50	74
4580　　To delete coupling & hangers, subtract									
4582　　　1-1/2" dia to 2" dia					58%	54%			
4584　　　3" dia to 12" dia					46%	42%			
4800　PVC, clear pipe, cplgs. 10' O.C., clevis hanger assy's 3 per 10', Sched. 40									
4840　　　1/4" diameter	1 Plum	59	.136	L.F.	4.91	7.95		12.86	17.40
4850　　　3/8" diameter		56	.143		5.25	8.40		13.65	18.40
4860　　　1/2" diameter		54	.148		5.90	8.70		14.60	19.65
4870　　　3/4" diameter		51	.157		6.60	9.20		15.80	21
4880　　　1" diameter		46	.174		8.15	10.20		18.35	24.50
4890　　　1-1/4" diameter		42	.190		9.60	11.20		20.80	27.50
4900　　　1-1/2" diameter		36	.222		10.70	13.05		23.75	31.50
4910　　　2" diameter	Q-1	59	.271		13	14.35		27.35	36
4920　　　2-1/2" diameter		56	.286		19	15.10		34.10	44
4930　　　3" diameter		53	.302		23.50	15.95		39.45	49.50
4940　　　3-1/2" diameter		50	.320		32.50	16.90		49.40	61
4950　　　4" diameter		48	.333		33.50	17.60		51.10	63.50
5250　　To delete coupling & hangers, subtract									
5260　　　1/4" diam. to 3/8" diam.					60%	81%			
5270　　　1/2" diam. to 3/4" diam.					41%	77%			

22 11 13.74 Pipe, Plastic	Crew	Daily Output	Labor-Hours	Unit	Material	2015 Bare Costs Labor	Equipment	Total	Total Incl O&P
5280	1" diam. to 1-1/2" diam.					26%	67%			
5290	2" diam. to 4" diam.					16%	58%			
5300	CPVC, socket joint, couplings 10' O.C., clevis hanger assemblies, 3 per 10'									
5302	Schedule 40									
5304	1/2" diameter	1 Plum	54	.148	L.F.	5.90	8.70		14.60	19.65
5305	3/4" diameter		51	.157		6.75	9.20		15.95	21.50
5306	1" diameter		46	.174		8.10	10.20		18.30	24.50
5307	1-1/4" diameter		42	.190		9.65	11.20		20.85	27.50
5308	1-1/2" diameter		36	.222		10.85	13.05		23.90	31.50
5309	2" diameter	Q-1	59	.271		12.85	14.35		27.20	35.50
5310	2-1/2" diameter		56	.286		19.90	15.10		35	45
5311	3" diameter		53	.302		23.50	15.95		39.45	50
5312	4" diameter		48	.333		31.50	17.60		49.10	61.50
5314	6" diameter		43	.372		57.50	19.65		77.15	92.50
5318	To delete coupling & hangers, subtract									
5319	1/2" diam. to 3/4" diam.					37%	77%			
5320	1" diam. to 1-1/4" diam.					27%	70%			
5321	1-1/2" diam. to 3" diam.					21%	57%			
5322	4" diam. to 6" diam.					16%	57%			
5324	Schedule 80									
5325	1/2" diameter	1 Plum	50	.160	L.F.	6.60	9.40		16	21.50
5326	3/4" diameter		47	.170		7	10		17	23
5327	1" diameter		43	.186		8.70	10.90		19.60	26
5328	1-1/4" diameter		39	.205		10.60	12.05		22.65	30
5329	1-1/2" diameter		34	.235		12.05	13.80		25.85	34.50
5330	2" diameter	Q-1	55	.291		14.70	15.35		30.05	39
5331	2-1/2" diameter		52	.308		22.50	16.25		38.75	49
5332	3" diameter		50	.320		26	16.90		42.90	54
5333	4" diameter		46	.348		35.50	18.40		53.90	67
5334	6" diameter		38	.421		67.50	22		89.50	108
5335	8" diameter	Q-2	47	.511		170	28		198	230
5339	To delete couplings & hangers, subtract									
5340	1/2" diam. to 3/4" diam.					44%	77%			
5341	1" diam. to 1-1/4" diam.					32%	71%			
5342	1-1/2" diam. to 4" diam.					25%	58%			
5343	6" diam. to 8" diam.					20%	53%			
5360	CPVC, threaded, couplings 10' O.C., clevis hanger assemblies, 3 per 10'									
5380	Schedule 40									
5460	1/2" diameter	1 Plum	54	.148	L.F.	6.70	8.70		15.40	20.50
5470	3/4" diameter		51	.157		8.15	9.20		17.35	23
5480	1" diameter		46	.174		9.55	10.20		19.75	26
5490	1-1/4" diameter		42	.190		10.80	11.20		22	29
5500	1-1/2" diameter		36	.222		11.80	13.05		24.85	32.50
5510	2" diameter	Q-1	59	.271		14	14.35		28.35	37
5520	2-1/2" diameter		56	.286		21	15.10		36.10	46
5530	3" diameter		53	.302		25.50	15.95		41.45	52
5540	4" diameter		48	.333		38.50	17.60		56.10	69
5550	6" diameter		43	.372		60.50	19.65		80.15	96.50
5730	To delete coupling & hangers, subtract									
5740	1/2" diam. to 3/4" diam.					37%	77%			
5750	1" diam. to 1-1/4" diam.					27%	70%			
5760	1-1/2" diam. to 3" diam.					21%	57%			
5770	4" diam. to 6" diam.					16%	57%			

22 11 13.74 Pipe, Plastic	Crew	Daily Output	Labor-Hours	Unit	Material	2015 Bare Costs Labor	Equipment	Total	Total Incl O&P
5800	Schedule 80									
5860	1/2" diameter	1 Plum	50	.160	L.F.	7.40	9.40		16.80	22.50
5870	3/4" diameter		47	.170		8.40	10		18.40	24.50
5880	1" diameter		43	.186		10.15	10.90		21.05	27.50
5890	1-1/4" diameter		39	.205		11.70	12.05		23.75	31
5900	1-1/2" diameter		34	.235		13	13.80		26.80	35.50
5910	2" diameter	Q-1	55	.291		15.85	15.35		31.20	40.50
5920	2-1/2" diameter		52	.308		23.50	16.25		39.75	50.50
5930	3" diameter		50	.320		27.50	16.90		44.40	56
5940	4" diameter		46	.348		42.50	18.40		60.90	75
5950	6" diameter		38	.421		71	22		93	112
5960	8" diameter	Q-2	47	.511		169	28		197	229
6060	To delete couplings & hangers, subtract									
6070	1/2" diam. to 3/4" diam.					44%	77%			
6080	1" diam. to 1-1/4" diam.					32%	71%			
6090	1-1/2" diam. to 4" diam.					25%	58%			
6100	6" diam. to 8" diam.					20%	53%			
6240	CTS, 1/2" diameter	1 Plum	54	.148	L.F.	5.50	8.70		14.20	19.20
6250	3/4" diameter		51	.157		6.75	9.20		15.95	21.50
6260	1" diameter		46	.174		10.10	10.20		20.30	26.50
6270	1 1/4"		42	.190		14.75	11.20		25.95	33
6280	1 1/2" diameter		36	.222		18.35	13.05		31.40	39.50
6290	2" diameter	Q-1	59	.271		28.50	14.35		42.85	52.50
6370	To delete coupling & hangers, subtract									
6380	1/2" diam.					51%	79%			
6390	3/4" diam.					40%	76%			
6392	1" thru 2" diam.					72%	68%			
7280	PEX, flexible, no couplings or hangers									
7282	Note: For labor costs add 25% to the couplings and fittings labor total.									
7285	For fittings see section 23 83 16.10 7000									
7300	Non-barrier type, hot/cold tubing rolls									
7310	1/4" diameter x 100'				L.F.	.49			.49	.54
7350	3/8" diameter x 100'					.55			.55	.61
7360	1/2" diameter x 100'					.61			.61	.67
7370	1/2" diameter x 500'					.61			.61	.67
7380	1/2" diameter x 1000'					.61			.61	.67
7400	3/4" diameter x 100'					1.11			1.11	1.22
7410	3/4" diameter x 500'					1.11			1.11	1.22
7420	3/4" diameter x 1000'					1.11			1.11	1.22
7460	1" diameter x 100'					1.90			1.90	2.09
7470	1" diameter x 300'					1.90			1.90	2.09
7480	1" diameter x 500'					1.90			1.90	2.09
7500	1-1/4" diameter x 100'					3.23			3.23	3.55
7510	1-1/4" diameter x 300'					3.23			3.23	3.55
7540	1-1/2" diameter x 100'					4.39			4.39	4.83
7550	1-1/2" diameter x 300'					4.39			4.39	4.83
7596	Most sizes available in red or blue									
7700	Non-barrier type, hot/cold tubing straight lengths									
7710	1/2" diameter x 20'				L.F.	.60			.60	.66
7750	3/4" diameter x 20'					1.10			1.10	1.21
7760	1" diameter x 20'					1.90			1.90	2.09
7770	1-1/4" diameter x 20'					3.23			3.23	3.55
7780	1-1/2" diameter x 20'					4.39			4.39	4.83

22 11 13 – Facility Water Distribution Piping

22 11 13.74 Pipe, Plastic		Crew	Daily Output	Labor-Hours	Unit	Material	2015 Bare Costs Labor	2015 Bare Costs Equipment	Total	Total Incl O&P
7790	2" diameter				L.F.	8.60			8.60	9.45
7796	Most sizes available in red or blue									
9000	Polypropylene pipe									
9002	For fusion weld fittings and accessories see line 22 11 13.76 9400									
9004	Note: sizes 1/2" thru 4" use socket fusion									
9005	Sizes 6" thru 10" use butt fusion									
9010	SDR 7.4, (domestic hot water piping)									
9011	Enhanced to minimize thermal expansion and high temperature life									
9016	13' lengths, size is ID, includes joints 13' O.C. and hangers 3 per 10'									
9020	3/8" diameter	1 Plum	53	.151	L.F.	2.35	8.85		11.20	16
9022	1/2" diameter		52	.154		2.80	9.05		11.85	16.75
9024	3/4" diameter		50	.160		3.31	9.40		12.71	17.85
9026	1" diameter		45	.178		4.34	10.45		14.79	20.50
9028	1-1/4" diameter		40	.200		6.20	11.75		17.95	24.50
9030	1-1/2" diameter		35	.229		8.75	13.40		22.15	30
9032	2" diameter	Q-1	58	.276		12.40	14.55		26.95	35.50
9034	2-1/2" diameter		55	.291		17	15.35		32.35	41.50
9036	3" diameter		52	.308		22.50	16.25		38.75	49
9038	3-1/2" diameter		49	.327		31	17.25		48.25	60.50
9040	4" diameter		46	.348		34	18.40		52.40	65.50
9042	6" diameter		39	.410		46	21.50		67.50	83
9044	8" diameter	Q-2	48	.500		69.50	27.50		97	118
9046	10" diameter	"	43	.558		110	30.50		140.50	167
9050	To delete joint & hangers, subtract									
9052	3/8" diam. to 1" diam.					45%	65%			
9054	1-1/4" diam. to 4" diam.					15%	45%			
9056	6" diam. to 10" diam.					5%	24%			
9060	SDR 11, (domestic cold water piping)									
9062	13' lengths, size is ID, includes joints 13' O.C. and hangers 3 per 10'									
9064	1/2" diameter	1 Plum	57	.140	L.F.	2.48	8.25		10.73	15.20
9066	3/4" diameter		54	.148		2.80	8.70		11.50	16.25
9068	1" diameter		49	.163		3.41	9.60		13.01	18.20
9070	1-1/4" diameter		45	.178		4.83	10.45		15.28	21
9072	1-1/2" diameter		40	.200		6.40	11.75		18.15	25
9074	2" diameter	Q-1	62	.258		9	13.65		22.65	30.50
9076	2-1/2" diameter		59	.271		11.85	14.35		26.20	34.50
9078	3" diameter		56	.286		15.90	15.10		31	40.50
9080	3-1/2" diameter		53	.302		23	15.95		38.95	49.50
9082	4" diameter		50	.320		26	16.90		42.90	54
9084	6" diameter		47	.340		31.50	18		49.50	61.50
9086	8" diameter	Q-2	51	.471		47.50	26		73.50	91.50
9088	10" diameter	"	46	.522		72.50	28.50		101	123
9090	To delete joint & hangers, subtract									
9092	1/2" diam. to 1" diam.					45%	65%			
9094	1-1/4" diam. to 4" diam.					15%	45%			
9096	6" diam. to 10" diam.					5%	24%			

22 11 13.76 Pipe Fittings, Plastic

		Crew	Daily Output	Labor-Hours	Unit	Material	2015 Bare Costs Labor	2015 Bare Costs Equipment	Total	Total Incl O&P
0010	**PIPE FITTINGS, PLASTIC**									
0030	Epoxy resin, fiberglass reinforced, general service									
0100	3"	Q-1	20.80	.769	Ea.	113	40.50		153.50	186
0110	4"		16.50	.970		115	51		166	204
0120	6"		10.10	1.584		225	83.50		308.50	375

22 11 13 – Facility Water Distribution Piping

22 11 13.76 Pipe Fittings, Plastic		Crew	Daily Output	Labor-Hours	Unit	Material	2015 Bare Costs Labor	Equipment	Total	Total Incl O&P
0130	8"	Q-2	9.30	2.581	Ea.	415	141		556	670
0140	10"		8.50	2.824		520	155		675	810
0150	12"		7.60	3.158		750	173		923	1,075
0170	Elbow, 90°, flanged									
0172	2"	Q-1	23	.696	Ea.	197	37		234	273
0173	3"		16	1		226	53		279	330
0174	4"		13	1.231		294	65		359	425
0176	6"		8	2		535	106		641	745
0177	8"	Q-2	9	2.667		965	146		1,111	1,300
0178	10"		7	3.429		1,325	188		1,513	1,725
0179	12"		5	4.800		1,775	263		2,038	2,350
0186	Elbow, 45°, flanged									
0188	2"	Q-1	23	.696	Ea.	197	37		234	273
0189	3"		16	1		226	53		279	330
0190	4"		13	1.231		294	65		359	425
0192	6"		8	2		535	106		641	745
0193	8"	Q-2	9	2.667		915	146		1,061	1,225
0194	10"		7	3.429		1,150	188		1,338	1,550
0195	12"		5	4.800		1,500	263		1,763	2,050
0352	Tee, flanged									
0354	2"	Q-1	17	.941	Ea.	267	49.50		316.50	370
0355	3"		10	1.600		355	84.50		439.50	520
0356	4"		8	2		395	106		501	595
0358	6"		5	3.200		680	169		849	1,000
0359	8"	Q-2	6	4		1,300	219		1,519	1,750
0360	10"		5	4.800		1,875	263		2,138	2,450
0361	12"		4	6		2,575	330		2,905	3,350
0365	Wye, flanged									
0367	2"	Q-1	17	.941	Ea.	530	49.50		579.50	655
0368	3"		10	1.600		725	84.50		809.50	930
0369	4"		8	2		970	106		1,076	1,225
0371	6"		5	3.200		1,375	169		1,544	1,750
0372	8"	Q-2	6	4		2,250	219		2,469	2,800
0373	10"		5	4.800		3,475	263		3,738	4,225
0374	12"		4	6		5,075	330		5,405	6,075
0380	Couplings									
0410	2"	Q-1	33.10	.483	Ea.	22.50	25.50		48	63.50
0420	3"		20.80	.769		24.50	40.50		65	88.50
0430	4"		16.50	.970		33	51		84	114
0440	6"		10.10	1.584		78.50	83.50		162	212
0450	8"	Q-2	9.30	2.581		133	141		274	360
0460	10"		8.50	2.824		195	155		350	450
0470	12"		7.60	3.158		262	173		435	550
0473	High corrosion resistant couplings, add					30%				
0474	Reducer, concentric, flanged									
0475	2" x 1-1/2"	Q-1	30	.533	Ea.	510	28		538	610
0476	3" x 2"		24	.667		560	35		595	670
0477	4" x 3"		19	.842		625	44.50		669.50	755
0479	6" x 4"		15	1.067		625	56.50		681.50	775
0480	8" x 6"	Q-2	16	1.500		1,025	82		1,107	1,250
0481	10" x 8"		13	1.846		955	101		1,056	1,200
0482	12" x 10"		11	2.182		1,475	120		1,595	1,800
0486	Adapter, bell x male or female									

22 11 13.76 Pipe Fittings, Plastic		Crew	Daily Output	Labor-Hours	Unit	Material	2015 Bare Costs Labor	Equipment	Total	Total Incl O&P
0488	2"	Q-1	28	.571	Ea.	25.50	30		55.50	73.50
0489	3"		20	.800		38	42.50		80.50	106
0491	4"		17	.941		49.50	49.50		99	130
0492	6"		12	1.333		107	70.50		177.50	224
0493	8"	Q-2	15	1.600		154	87.50		241.50	300
0494	10"	"	11	2.182		221	120		341	425
0528	Flange									
0532	2"	Q-1	46	.348	Ea.	36.50	18.40		54.90	68
0533	3"		32	.500		46.50	26.50		73	91
0534	4"		26	.615		60.50	32.50		93	116
0536	6"		16	1		114	53		167	206
0537	8"	Q-2	18	1.333		175	73		248	305
0538	10"		14	1.714		248	94		342	415
0539	12"		10	2.400		310	131		441	540
2100	PVC schedule 80, socket joint									
2110	90° elbow, 1/2"	1 Plum	30.30	.264	Ea.	2.48	15.50		17.98	26
2130	3/4"		26	.308		3.17	18.05		21.22	31
2140	1"		22.70	.352		4.95	20.50		25.45	36.50
2150	1-1/4"		20.20	.396		6.85	23.50		30.35	42.50
2160	1-1/2"		18.20	.440		7.30	26		33.30	47
2170	2"	Q-1	33.10	.483		8.85	25.50		34.35	48
2180	3"		20.80	.769		23.50	40.50		64	87
2190	4"		16.50	.970		35.50	51		86.50	117
2200	6"		10.10	1.584		101	83.50		184.50	237
2210	8"	Q-2	9.30	2.581		277	141		418	520
2250	45° elbow, 1/2"	1 Plum	30.30	.264		4.67	15.50		20.17	28.50
2270	3/4"		26	.308		7.15	18.05		25.20	35.50
2280	1"		22.70	.352		10.70	20.50		31.20	43
2290	1-1/4"		20.20	.396		13.60	23.50		37.10	50
2300	1-1/2"		18.20	.440		16.10	26		42.10	56.50
2310	2"	Q-1	33.10	.483		21	25.50		46.50	61.50
2320	3"		20.80	.769		53.50	40.50		94	120
2330	4"		16.50	.970		96	51		147	184
2340	6"		10.10	1.584		121	83.50		204.50	259
2350	8"	Q-2	9.30	2.581		262	141		403	500
2400	Tee, 1/2"	1 Plum	20.20	.396		7	23.50		30.50	42.50
2420	3/4"		17.30	.462		7.35	27		34.35	49
2430	1"		15.20	.526		9.15	31		40.15	56.50
2440	1-1/4"		13.50	.593		25.50	35		60.50	80.50
2450	1-1/2"		12.10	.661		25.50	39		64.50	86.50
2460	2"	Q-1	20	.800		31.50	42.50		74	98.50
2470	3"		13.90	1.151		43	61		104	139
2480	4"		11	1.455		49.50	77		126.50	171
2490	6"		6.70	2.388		169	126		295	375
2500	8"	Q-2	6.20	3.871		395	212		607	750
2510	Flange, socket, 150 lb., 1/2"	1 Plum	55.60	.144		13.60	8.45		22.05	27.50
2514	3/4"		47.60	.168		14.50	9.85		24.35	31
2518	1"		41.70	.192		16.15	11.25		27.40	35
2522	1-1/2"		33.30	.240		16.65	14.10		30.75	40
2526	2"	Q-1	60.60	.264		22.50	13.95		36.45	46
2530	4"		30.30	.528		49	28		77	96
2534	6"		18.50	.865		77	45.50		122.50	154
2538	8"	Q-2	17.10	1.404		138	77		215	267

For customer support on your Mechanical Cost Data, call 877.795.7394.

22 11 13 – Facility Water Distribution Piping

22 11 13.76 Pipe Fittings, Plastic	Crew	Daily Output	Labor-Hours	Unit	Material	2015 Bare Costs Labor	Equipment	Total	Total Incl O&P
2550	Coupling, 1/2"	1 Plum	30.30	.264	Ea.	4.49	15.50		19.99	28.50
2570	3/4"		26	.308		6.05	18.05		24.10	34
2580	1"		22.70	.352		6.25	20.50		26.75	38
2590	1-1/4"		20.20	.396		9.50	23.50		33	45.50
2600	1-1/2"		18.20	.440		10.25	26		36.25	50.50
2610	2"	Q-1	33.10	.483		11	25.50		36.50	50.50
2620	3"		20.80	.769		31	40.50		71.50	95.50
2630	4"		16.50	.970		39	51		90	121
2640	6"		10.10	1.584		84	83.50		167.50	218
2650	8"	Q-2	9.30	2.581		114	141		255	340
2660	10"		8.50	2.824		400	155		555	675
2670	12"		7.60	3.158		450	173		623	755
2700	PVC (white), schedule 40, socket joints									
2760	90° elbow, 1/2"	1 Plum	33.30	.240	Ea.	.48	14.10		14.58	22
2770	3/4"		28.60	.280		.54	16.40		16.94	25.50
2780	1"		25	.320		.97	18.80		19.77	29.50
2790	1-1/4"		22.20	.360		1.71	21		22.71	34
2800	1-1/2"		20	.400		1.85	23.50		25.35	37.50
2810	2"	Q-1	36.40	.440		2.89	23		25.89	38
2820	2-1/2"		26.70	.599		8.80	31.50		40.30	57.50
2830	3"		22.90	.699		10.50	37		47.50	67
2840	4"		18.20	.879		18.80	46.50		65.30	90.50
2850	5"		12.10	1.322		48.50	70		118.50	159
2860	6"		11.10	1.441		60	76		136	181
2870	8"	Q-2	10.30	2.330		154	128		282	365
2980	45° elbow, 1/2"	1 Plum	33.30	.240		.80	14.10		14.90	22.50
2990	3/4"		28.60	.280		1.23	16.40		17.63	26.50
3000	1"		25	.320		1.47	18.80		20.27	30
3010	1-1/4"		22.20	.360		2.03	21		23.03	34
3020	1-1/2"		20	.400		2.60	23.50		26.10	38.50
3030	2"	Q-1	36.40	.440		3.38	23		26.38	38.50
3040	2-1/2"		26.70	.599		8.80	31.50		40.30	57.50
3050	3"		22.90	.699		13.65	37		50.65	70.50
3060	4"		18.20	.879		24.50	46.50		71	97
3070	5"		12.10	1.322		48.50	70		118.50	159
3080	6"		11.10	1.441		60.50	76		136.50	182
3090	8"	Q-2	10.30	2.330		146	128		274	355
3180	Tee, 1/2"	1 Plum	22.20	.360		.60	21		21.60	32.50
3190	3/4"		19	.421		.69	24.50		25.19	38.50
3200	1"		16.70	.479		1.29	28		29.29	44
3210	1-1/4"		14.80	.541		2.01	31.50		33.51	50
3220	1-1/2"		13.30	.602		2.45	35.50		37.95	56
3230	2"	Q-1	24.20	.661		3.56	35		38.56	57
3240	2-1/2"		17.80	.899		11.75	47.50		59.25	84.50
3250	3"		15.20	1.053		15.45	55.50		70.95	101
3260	4"		12.10	1.322		28	70		98	136
3270	5"		8.10	1.975		67.50	104		171.50	232
3280	6"		7.40	2.162		94	114		208	276
3290	8"	Q-2	6.80	3.529		218	193		411	530
3380	Coupling, 1/2"	1 Plum	33.30	.240		.32	14.10		14.42	22
3390	3/4"		28.60	.280		.44	16.40		16.84	25.50
3400	1"		25	.320		.77	18.80		19.57	29.50
3410	1-1/4"		22.20	.360		1.06	21		22.06	33

For customer support on your Mechanical Cost Data, call 877.795.7394.

209

22 11 13.76 Pipe Fittings, Plastic		Crew	Daily Output	Labor-Hours	Unit	Material	2015 Bare Costs Labor	Equipment	Total	Total Incl O&P
3420	1-1/2"	1 Plum	20	.400	Ea.	1.13	23.50		24.63	36.50
3430	2"	Q-1	36.40	.440		1.72	23		24.72	37
3440	2-1/2"		26.70	.599		3.81	31.50		35.31	52
3450	3"		22.90	.699		5.95	37		42.95	62
3460	4"		18.20	.879		8.65	46.50		55.15	79.50
3470	5"		12.10	1.322		15.80	70		85.80	122
3480	6"		11.10	1.441		27.50	76		103.50	145
3490	8"	Q-2	10.30	2.330		51	128		179	249
3600	Cap, schedule 40, PVC socket 1/2"	1 Plum	60.60	.132		.44	7.75		8.19	12.20
3610	3/4"		51.90	.154		.50	9.05		9.55	14.20
3620	1"		45.50	.176		.80	10.30		11.10	16.50
3630	1-1/4"		40.40	.198		1.13	11.60		12.73	18.80
3640	1-1/2"		36.40	.220		1.23	12.90		14.13	21
3650	2"	Q-1	66.10	.242		1.49	12.80		14.29	21
3660	2-1/2"		48.50	.330		4.73	17.45		22.18	31.50
3670	3"		41.60	.385		5.10	20.50		25.60	36
3680	4"		33.10	.483		11.75	25.50		37.25	51.50
3690	6"		20.20	.792		28	42		70	94
3700	8"	Q-2	18.60	1.290		71	70.50		141.50	185
3710	Reducing insert, schedule 40, socket weld									
3712	3/4"	1 Plum	31.50	.254	Ea.	.50	14.90		15.40	23
3713	1"		27.50	.291		.92	17.10		18.02	27
3715	1-1/2"		22	.364		1.31	21.50		22.81	33.50
3716	2"	Q-1	40	.400		2.17	21		23.17	34.50
3717	4"		20	.800		11.45	42.50		53.95	76.50
3718	6"		12.20	1.311		28.50	69.50		98	136
3719	8"	Q-2	11.30	2.124		112	116		228	299
3730	Reducing insert, socket weld x female/male thread									
3732	1/2"	1 Plum	38.30	.209	Ea.	2.28	12.25		14.53	21
3733	3/4"		32.90	.243		1.41	14.25		15.66	23
3734	1"		28.80	.278		1.98	16.30		18.28	26.50
3736	1-1/2"		23	.348		3.56	20.50		24.06	35
3737	2"	Q-1	41.90	.382		3.81	20		23.81	34.50
3738	4"	"	20.90	.766		36.50	40.50		77	101
3742	Male adapter, socket weld x male thread									
3744	1/2"	1 Plum	38.30	.209	Ea.	.44	12.25		12.69	19
3745	3/4"		32.90	.243		.48	14.25		14.73	22
3746	1"		28.80	.278		.87	16.30		17.17	25.50
3748	1-1/2"		23	.348		1.41	20.50		21.91	32.50
3749	2"	Q-1	41.90	.382		1.85	20		21.85	32.50
3750	4"	"	20.90	.766		10.20	40.50		50.70	72
3754	Female adapter, socket weld x female thread									
3756	1/2"	1 Plum	38.30	.209	Ea.	.54	12.25		12.79	19.10
3757	3/4"		32.90	.243		.69	14.25		14.94	22.50
3758	1"		28.80	.278		.80	16.30		17.10	25.50
3760	1-1/2"		23	.348		1.41	20.50		21.91	32.50
3761	2"	Q-1	41.90	.382		1.90	20		21.90	32.50
3762	4"	"	20.90	.766		10.70	40.50		51.20	73
3800	PVC, schedule 80, socket joints									
3810	Reducing insert									
3812	3/4"	1 Plum	28.60	.280	Ea.	1.45	16.40		17.85	26.50
3813	1"		25	.320		4.14	18.80		22.94	33
3815	1-1/2"		20	.400		9	23.50		32.50	45.50

22 11 13 – Facility Water Distribution Piping

22 11 13.76 Pipe Fittings, Plastic		Crew	Daily Output	Labor-Hours	Unit	Material	2015 Bare Costs Labor	Equipment	Total	Total Incl O&P
3816	2"	Q-1	36.40	.440	Ea.	12.60	23		35.60	49
3817	4"		18.20	.879		48	46.50		94.50	123
3818	6"		11.10	1.441		67	76		143	189
3819	8"	Q-2	10.20	2.353		385	129		514	620
3830	Reducing insert, socket weld x female/male thread									
3832	1/2"	1 Plum	34.80	.230	Ea.	9.05	13.50		22.55	30.50
3833	3/4"		29.90	.268		5.55	15.70		21.25	29.50
3834	1"		26.10	.307		8.70	18		26.70	36.50
3836	1-1/2"		20.90	.383		10.95	22.50		33.45	46
3837	2"	Q-1	38	.421		16	22		38	51
3838	4"	"	19	.842		77.50	44.50		122	153
3844	Adapter, male socket x male thread									
3846	1/2"	1 Plum	34.80	.230	Ea.	3.50	13.50		17	24.50
3847	3/4"		29.90	.268		3.86	15.70		19.56	28
3848	1"		26.10	.307		6.65	18		24.65	34.50
3850	1-1/2"		20.90	.383		11.20	22.50		33.70	46.50
3851	2"	Q-1	38	.421		16.20	22		38.20	51.50
3852	4"	"	19	.842		36.50	44.50		81	107
3860	Adapter, female socket x female thread									
3862	1/2"	1 Plum	34.80	.230	Ea.	4.20	13.50		17.70	25
3863	3/4"		29.90	.268		6.25	15.70		21.95	30.50
3864	1"		26.10	.307		9.20	18		27.20	37
3866	1-1/2"		20.90	.383		18.25	22.50		40.75	54
3867	2"	Q-1	38	.421		32	22		54	68.50
3868	4"	"	19	.842		97.50	44.50		142	174
3872	Union, socket joints									
3874	1/2"	1 Plum	25.80	.310	Ea.	9.20	18.20		27.40	37.50
3875	3/4"		22.10	.362		11.70	21.50		33.20	45
3876	1"		19.30	.415		13.35	24.50		37.85	51.50
3878	1-1/2"		15.50	.516		30	30.50		60.50	79
3879	2"	Q-1	28.10	.569		40.50	30		70.50	90
3888	Cap									
3890	1/2"	1 Plum	54.50	.147	Ea.	4.39	8.60		12.99	17.85
3891	3/4"		46.70	.171		4.63	10.05		14.68	20.50
3892	1"		41	.195		8.20	11.45		19.65	26.50
3894	1-1/2"		32.80	.244		9.90	14.30		24.20	32.50
3895	2"	Q-1	59.50	.269		26	14.20		40.20	50
3896	4"		30	.533		79	28		107	130
3897	6"		18.20	.879		196	46.50		242.50	286
3898	8"	Q-2	16.70	1.437		252	78.50		330.50	395
4500	DWV, ABS, non pressure, socket joints									
4540	1/4 Bend, 1-1/4"	1 Plum	20.20	.396	Ea.	2.57	23.50		26.07	38
4560	1-1/2"	"	18.20	.440		1.92	26		27.92	41
4570	2"	Q-1	33.10	.483		3.05	25.50		28.55	42
4580	3"		20.80	.769		7.70	40.50		48.20	70
4590	4"		16.50	.970		15.95	51		66.95	95
4600	6"		10.10	1.584		65	83.50		148.50	198
4650	1/8 Bend, same as 1/4 Bend									
4800	Tee, sanitary									
4820	1-1/4"	1 Plum	13.50	.593	Ea.	3.42	35		38.42	56.50
4830	1-1/2"	"	12.10	.661		2.92	39		41.92	61.50
4840	2"	Q-1	20	.800		4.52	42.50		47.02	69
4850	3"		13.90	1.151		12.40	61		73.40	106

For customer support on your Mechanical Cost Data, call 877.795.7394.

211

22 11 13.76 Pipe Fittings, Plastic		Crew	Daily Output	Labor-Hours	Unit	Material	2015 Bare Costs Labor	Equipment	Total	Total Incl O&P
4860	4"	Q-1	11	1.455	Ea.	22	77		99	140
4862	Tee, sanitary, reducing, 2" x 1-1/2"		22	.727		3.94	38.50		42.44	62.50
4864	3" x 2"		15.30	1.046		9.05	55.50		64.55	93.50
4868	4" x 3"	↓	12.10	1.322	↓	20	70		90	127
4870	Combination Y and 1/8 bend									
4872	1-1/2"	1 Plum	12.10	.661	Ea.	7	39		46	66
4874	2"	Q-1	20	.800		7.85	42.50		50.35	72.50
4876	3"		13.90	1.151		18.50	61		79.50	113
4878	4"		11	1.455		35.50	77		112.50	155
4880	3" x 1-1/2"		15.50	1.032		18.05	54.50		72.55	102
4882	4" x 3"	↓	12.10	1.322		28	70		98	136
4900	Wye, 1-1/4"	1 Plum	13.50	.593		3.94	35		38.94	57
4902	1-1/2"	"	12.10	.661		4.49	39		43.49	63.50
4904	2"	Q-1	20	.800		5.85	42.50		48.35	70.50
4906	3"		13.90	1.151		13.80	61		74.80	107
4908	4"		11	1.455		28	77		105	147
4910	6"		6.70	2.388		81	126		207	280
4918	3" x 1-1/2"		15.50	1.032		11.05	54.50		65.55	94.50
4920	4" x 3"		12.10	1.322		22	70		92	130
4922	6" x 4"	↓	6.90	2.319		66.50	123		189.50	259
4930	Double Wye, 1-1/2"	1 Plum	9.10	.879		13.95	51.50		65.45	93.50
4932	2"	Q-1	16.60	.964		16.65	51		67.65	95.50
4934	3"		10.40	1.538		38	81.50		119.50	165
4936	4"		8.25	1.939		74	102		176	237
4940	2" x 1-1/2"		16.80	.952		14.55	50.50		65.05	92
4942	3" x 2"		10.60	1.509		28	79.50		107.50	151
4944	4" x 3"		8.45	1.893		58.50	100		158.50	216
4946	6" x 4"		7.25	2.207		90	117		207	275
4950	Reducer bushing, 2" x 1-1/2"		36.40	.440		1.56	23		24.56	36.50
4952	3" x 1-1/2"		27.30	.586		6.65	31		37.65	54.50
4954	4" x 2"		18.20	.879		12.70	46.50		59.20	84
4956	6" x 4"	↓	11.10	1.441		34.50	76		110.50	153
4960	Couplings, 1-1/2"	1 Plum	18.20	.440		.95	26		26.95	40
4962	2"	Q-1	33.10	.483		1.28	25.50		26.78	40
4963	3"		20.80	.769		3.58	40.50		44.08	65.50
4964	4"		16.50	.970		6.45	51		57.45	84.50
4966	6"		10.10	1.584		27	83.50		110.50	156
4970	2" x 1-1/2"		33.30	.480		2.75	25.50		28.25	41.50
4972	3" x 1-1/2"		21	.762		7.65	40.50		48.15	69.50
4974	4" x 3"	↓	16.70	.958		12.15	50.50		62.65	90
4978	Closet flange, 4"	1 Plum	32	.250		6.65	14.70		21.35	29.50
4980	4" x 3"	"	34	.235	↓	7.55	13.80		21.35	29.50
5000	DWV, PVC, schedule 40, socket joints									
5040	1/4 bend, 1-1/4"	1 Plum	20.20	.396	Ea.	5.20	23.50		28.70	40.50
5060	1-1/2"	"	18.20	.440		1.48	26		27.48	40.50
5070	2"	Q-1	33.10	.483		2.33	25.50		27.83	41
5080	3"		20.80	.769		6.85	40.50		47.35	69
5090	4"		16.50	.970		13.05	51		64.05	92
5100	6"	↓	10.10	1.584		46	83.50		129.50	177
5105	8"	Q-2	9.30	2.581		65	141		206	286
5106	10"	"	8.50	2.824		290	155		445	555
5110	1/4 bend, long sweep, 1-1/2"	1 Plum	18.20	.440		3.42	26		29.42	43
5112	2"	Q-1	33.10	.483		3.83	25.50		29.33	42.50

22 11 13.76	Pipe Fittings, Plastic	Crew	Daily Output	Labor-Hours	Unit	Material	2015 Bare Costs Labor	Equipment	Total	Total Incl O&P
5114	3"	Q-1	20.80	.769	Ea.	8.75	40.50		49.25	71
5116	4"	↓	16.50	.970		16.65	51		67.65	96
5150	1/8 bend, 1-1/4"	1 Plum	20.20	.396		3.40	23.50		26.90	38.50
5170	1-1/2"	"	18.20	.440		1.44	26		27.44	40.50
5180	2"	Q-1	33.10	.483		2.08	25.50		27.58	41
5190	3"		20.80	.769		6.15	40.50		46.65	68.50
5200	4"		16.50	.970		10.40	51		61.40	89
5210	6"	↓	10.10	1.584		42	83.50		125.50	172
5215	8"	Q-2	9.30	2.581		158	141		299	385
5216	10"		8.50	2.824		206	155		361	460
5217	12"	↓	7.60	3.158		284	173		457	570
5250	Tee, sanitary 1-1/4"	1 Plum	13.50	.593		5.20	35		40.20	58.50
5254	1-1/2"	"	12.10	.661		2.58	39		41.58	61.50
5255	2"	Q-1	20	.800		3.80	42.50		46.30	68
5256	3"		13.90	1.151		10	61		71	103
5257	4"		11	1.455		17.75	77		94.75	136
5259	6"	↓	6.70	2.388		74	126		200	273
5261	8"	Q-2	6.20	3.871	↓	173	212		385	510
5276	Tee, sanitary, reducing									
5281	2" x 1-1/2" x 1-1/2"	Q-1	23	.696	Ea.	3.35	37		40.35	59
5282	2" x 1-1/2" x 2"		22	.727		4.05	38.50		42.55	62.50
5283	2" x 2" x 1-1/2"		22	.727		3.35	38.50		41.85	61.50
5284	3" x 3" x 1-1/2"		15.50	1.032		7.05	54.50		61.55	90.50
5285	3" x 3" x 2"		15.30	1.046		7.45	55.50		62.95	91.50
5286	4" x 4" x 1-1/2"		12.30	1.301		19.10	68.50		87.60	125
5287	4" x 4" x 2"		12.20	1.311		15.85	69.50		85.35	122
5288	4" x 4" x 3"		12.10	1.322		21.50	70		91.50	129
5291	6" x 6" x 4"	↓	6.90	2.319	↓	71.50	123		194.50	264
5294	Tee, double sanitary									
5295	1-1/2"	1 Plum	9.10	.879	Ea.	5.75	51.50		57.25	84.50
5296	2"	Q-1	16.60	.964		7.75	51		58.75	85.50
5297	3"		10.40	1.538		21.50	81.50		103	147
5298	4"	↓	8.25	1.939	↓	35	102		137	194
5303	Wye, reducing									
5304	2" x 1-1/2" x 1-1/2"	Q-1	23	.696	Ea.	6.55	37		43.55	62.50
5305	2" x 2" x 1-1/2"		22	.727		5.70	38.50		44.20	64.50
5306	3" x 3" x 2"		15.30	1.046		18.50	55.50		74	104
5307	4" x 4" x 2"		12.20	1.311		13.65	69.50		83.15	120
5309	4" x 4" x 3"	↓	12.10	1.322		18.40	70		88.40	125
5314	Combination Y & 1/8 bend, 1-1/2"	1 Plum	12.10	.661		6.25	39		45.25	65.50
5315	2"	Q-1	20	.800		6.70	42.50		49.20	71.50
5317	3"		13.90	1.151		16.60	61		77.60	110
5318	4"		11	1.455		33	77		110	152
5319	6"	↓	6.70	2.388		140	126		266	345
5320	8"	Q-2	6.20	3.871		202	212		414	540
5321	10"		5.70	4.211		790	231		1,021	1,225
5322	12"	↓	5.10	4.706	↓	930	258		1,188	1,425
5324	Combination Y & 1/8 bend, reducing									
5325	2" x 2" x 1-1/2"	Q-1	22	.727	Ea.	8.80	38.50		47.30	67.50
5327	3" x 3" x 1-1/2"		15.50	1.032		15.65	54.50		70.15	99.50
5328	3" x 3" x 2"		15.30	1.046		11.35	55.50		66.85	96
5329	4" x 4" x 2"	↓	12.20	1.311		17.85	69.50		87.35	125
5331	Wye, 1-1/4"	1 Plum	13.50	.593	↓	6.65	35		41.65	60

22 11 13.76 Pipe Fittings, Plastic		Crew	Daily Output	Labor-Hours	Unit	Material	2015 Bare Costs Labor	2015 Bare Costs Equipment	Total	Total Incl O&P
5332	1-1/2"	1 Plum	12.10	.661	Ea.	4.87	39		43.87	64
5333	2"	Q-1	20	.800		4.63	42.50		47.13	69
5334	3"		13.90	1.151		12.50	61		73.50	106
5335	4"		11	1.455		22.50	77		99.50	141
5336	6"		6.70	2.388		68.50	126		194.50	267
5337	8"	Q-2	6.20	3.871		124	212		336	455
5338	10"		5.70	4.211		268	231		499	645
5339	12"		5.10	4.706		440	258		698	870
5341	2" x 1-1/2"	Q-1	22	.727		5.70	38.50		44.20	64.50
5342	3" x 1-1/2"		15.50	1.032		8.20	54.50		62.70	91.50
5343	4" x 3"		12.10	1.322		18.40	70		88.40	125
5344	6" x 4"		6.90	2.319		50.50	123		173.50	241
5345	8" x 6"	Q-2	6.40	3.750		93.50	205		298.50	415
5347	Double wye, 1-1/2"	1 Plum	9.10	.879		10.55	51.50		62.05	89.50
5348	2"	Q-1	16.60	.964		12.20	51		63.20	90.50
5349	3"		10.40	1.538		25	81.50		106.50	151
5350	4"		8.25	1.939		50.50	102		152.50	211
5353	Double wye, reducing									
5354	2" x 2" x 1-1/2" x 1-1/2"	Q-1	16.80	.952	Ea.	10.75	50.50		61.25	88
5355	3" x 3" x 2" x 2"		10.60	1.509		18.45	79.50		97.95	141
5356	4" x 4" x 3" x 3"		8.45	1.893		40	100		140	195
5357	6" x 6" x 4" x 4"		7.25	2.207		138	117		255	330
5374	Coupling, 1-1/4"	1 Plum	20.20	.396		3.15	23.50		26.65	38.50
5376	1-1/2"	"	18.20	.440		.72	26		26.72	40
5378	2"	Q-1	33.10	.483		.95	25.50		26.45	39.50
5380	3"		20.80	.769		3.29	40.50		43.79	65
5390	4"		16.50	.970		5.65	51		56.65	83.50
5400	6"		10.10	1.584		18.40	83.50		101.90	146
5402	8"	Q-2	9.30	2.581		31.50	141		172.50	249
5404	2" x 1-1/2"	Q-1	33.30	.480		2.13	25.50		27.63	41
5406	3" x 1-1/2"		21	.762		6.25	40.50		46.75	68
5408	4" x 3"		16.70	.958		10.15	50.50		60.65	87.50
5410	Reducer bushing, 2" x 1-1/4"		36.50	.438		2.82	23		25.82	38
5411	2" x 1-1/2"		36.40	.440		1.22	23		24.22	36.50
5412	3" x 1-1/2"		27.30	.586		5.85	31		36.85	53.50
5413	3" x 2"		27.10	.590		3.06	31		34.06	50.50
5414	4" x 2"		18.20	.879		10.20	46.50		56.70	81
5415	4" x 3"		16.70	.958		5.20	50.50		55.70	82
5416	6" x 4"		11.10	1.441		27	76		103	145
5418	8" x 6"	Q-2	10.20	2.353		46	129		175	246
5425	Closet flange 4"	Q-1	32	.500		7.20	26.50		33.70	48
5426	4" x 3"	"	34	.471		5.80	25		30.80	44
5450	Solvent cement for PVC, industrial grade, per quart				Qt.	27.50			27.50	30.50
5500	CPVC, Schedule 80, threaded joints									
5540	90° Elbow, 1/4"	1 Plum	32	.250	Ea.	12.15	14.70		26.85	35.50
5560	1/2"		30.30	.264		7.05	15.50		22.55	31.50
5570	3/4"		26	.308		10.55	18.05		28.60	39
5580	1"		22.70	.352		14.80	20.50		35.30	47.50
5590	1-1/4"		20.20	.396		28.50	23.50		52	66.50
5600	1-1/2"		18.20	.440		30.50	26		56.50	73
5610	2"	Q-1	33.10	.483		41	25.50		66.50	83.50
5620	2-1/2"		24.20	.661		128	35		163	193
5630	3"		20.80	.769		137	40.50		177.50	212

22 11 13 – Facility Water Distribution Piping

22 11 13.76 Pipe Fittings, Plastic		Crew	Daily Output	Labor-Hours	Unit	Material	2015 Bare Costs Labor	Equipment	Total	Total Incl O&P
5640	4"	Q-1	16.50	.970	Ea.	214	51		265	315
5650	6"	↓	10.10	1.584	↓	252	83.50		335.50	405
5660	45° Elbow same as 90° Elbow									
5700	Tee, 1/4"	1 Plum	22	.364	Ea.	23.50	21.50		45	58
5702	1/2"		20.20	.396		23.50	23.50		47	61
5704	3/4"		17.30	.462		34	27		61	78.50
5706	1"		15.20	.526		36.50	31		67.50	87
5708	1-1/4"		13.50	.593		37	35		72	93
5710	1-1/2"	↓	12.10	.661		38.50	39		77.50	101
5712	2"	Q-1	20	.800		43	42.50		85.50	112
5714	2-1/2"		16.20	.988		211	52		263	310
5716	3"		13.90	1.151		245	61		306	360
5718	4"		11	1.455		580	77		657	750
5720	6"	↓	6.70	2.388		670	126		796	925
5730	Coupling, 1/4"	1 Plum	32	.250		15.45	14.70		30.15	39
5732	1/2"		30.30	.264		12.75	15.50		28.25	37.50
5734	3/4"		26	.308		20.50	18.05		38.55	50
5736	1"		22.70	.352		23.50	20.50		44	57
5738	1-1/4"		20.20	.396		25	23.50		48.50	62
5740	1-1/2"	↓	18.20	.440		26.50	26		52.50	68.50
5742	2"	Q-1	33.10	.483		31.50	25.50		57	73
5744	2-1/2"		24.20	.661		56.50	35		91.50	115
5746	3"		20.80	.769		65.50	40.50		106	134
5748	4"		16.50	.970		133	51		184	225
5750	6"	↓	10.10	1.584		182	83.50		265.50	325
5752	8"	Q-2	9.30	2.581	↓	390	141		531	640
5900	CPVC, Schedule 80, socket joints									
5904	90° Elbow, 1/4"	1 Plum	32	.250	Ea.	11.55	14.70		26.25	34.50
5906	1/2"		30.30	.264		4.53	15.50		20.03	28.50
5908	3/4"		26	.308		5.80	18.05		23.85	34
5910	1"		22.70	.352		9.20	20.50		29.70	41
5912	1-1/4"		20.20	.396		19.85	23.50		43.35	57
5914	1-1/2"	↓	18.20	.440		22	26		48	63.50
5916	2"	Q-1	33.10	.483		26.50	25.50		52	68
5918	2-1/2"		24.20	.661		61.50	35		96.50	121
5920	3"		20.80	.769		69.50	40.50		110	138
5922	4"		16.50	.970		126	51		177	216
5924	6"		10.10	1.584		252	83.50		335.50	405
5926	8"	↓	9.30	1.720		620	91		711	815
5930	45° Elbow, 1/4"	1 Plum	32	.250		17.20	14.70		31.90	41
5932	1/2"		30.30	.264		5.55	15.50		21.05	29.50
5934	3/4"		26	.308		7.95	18.05		26	36.50
5936	1"		22.70	.352		12.75	20.50		33.25	45
5938	1-1/4"		20.20	.396		25	23.50		48.50	62.50
5940	1-1/2"	↓	18.20	.440		25.50	26		51.50	67
5942	2"	Q-1	33.10	.483		29	25.50		54.50	70.50
5944	2-1/2"		24.20	.661		59	35		94	118
5946	3"		20.80	.769		75.50	40.50		116	145
5948	4"		16.50	.970		104	51		155	192
5950	6"		10.10	1.584		305	83.50		388.50	460
5952	8"	↓	9.30	1.720		665	91		756	865
5960	Tee, 1/4"	1 Plum	22	.364		10.60	21.50		32.10	43.50
5962	1/2"	↓	20.20	.396		10.60	23.50		34.10	46.50

For customer support on your Mechanical Cost Data, call 877.795.7394.

215

22 11 13.76 Pipe Fittings, Plastic		Crew	Daily Output	Labor-Hours	Unit	Material	2015 Bare Costs Labor	Equipment	Total	Total Incl O&P
5964	3/4"	1 Plum	17.30	.462	Ea.	10.80	27		37.80	53
5966	1"		15.20	.526		13.25	31		44.25	61
5968	1-1/4"		13.50	.593		28	35		63	83
5970	1-1/2"		12.10	.661		32	39		71	93.50
5972	2"	Q-1	20	.800		36	42.50		78.50	104
5974	2-1/2"		16.20	.988		90.50	52		142.50	179
5976	3"		13.90	1.151		90.50	61		151.50	192
5978	4"		11	1.455		121	77		198	249
5980	6"		6.70	2.388		315	126		441	535
5982	8"	Q-2	6.20	3.871		900	212		1,112	1,300
5990	Coupling, 1/4"	1 Plum	32	.250		12.30	14.70		27	35.50
5992	1/2"		30.30	.264		4.79	15.50		20.29	29
5994	3/4"		26	.308		6.70	18.05		24.75	35
5996	1"		22.70	.352		9	20.50		29.50	41
5998	1-1/4"		20.20	.396		13.50	23.50		37	50
6000	1-1/2"		18.20	.440		17	26		43	57.50
6002	2"	Q-1	33.10	.483		19.75	25.50		45.25	60
6004	2-1/2"		24.20	.661		44	35		79	102
6006	3"		20.80	.769		47.50	40.50		88	114
6008	4"		16.50	.970		62.50	51		113.50	147
6010	6"		10.10	1.584		147	83.50		230.50	288
6012	8"	Q-2	9.30	2.581		400	141		541	655
6200	CTS, 100 psi at 180°F, hot and cold water									
6230	90° Elbow, 1/2"	1 Plum	20	.400	Ea.	.48	23.50		23.98	36
6250	3/4"		19	.421		.80	24.50		25.30	38.50
6251	1"		16	.500		2.19	29.50		31.69	47
6252	1-1/4"		15	.533		4.65	31.50		36.15	52.50
6253	1-1/2"		14	.571		8.40	33.50		41.90	60
6254	2"	Q-1	23	.696		16.10	37		53.10	73
6260	45° Elbow, 1/2"	1 Plum	20	.400		.62	23.50		24.12	36
6280	3/4"		19	.421		1.06	24.50		25.56	38.50
6281	1"		16	.500		2.71	29.50		32.21	47.50
6282	1-1/4"		15	.533		5.75	31.50		37.25	54
6283	1-1/2"		14	.571		8.30	33.50		41.80	59.50
6284	2"	Q-1	23	.696		17.40	37		54.40	74.50
6290	Tee, 1/2"	1 Plum	13	.615		.62	36		36.62	55
6310	3/4"		12	.667		1.17	39		40.17	60.50
6311	1"		11	.727		6.10	42.50		48.60	71.50
6312	1-1/4"		10	.800		9.40	47		56.40	81.50
6313	1-1/2"		10	.800		12.25	47		59.25	84.50
6314	2"	Q-1	17	.941		19.80	49.50		69.30	97
6320	Coupling, 1/2"	1 Plum	22	.364		.40	21.50		21.90	32.50
6340	3/4"		21	.381		.53	22.50		23.03	34.50
6341	1"		18	.444		2.38	26		28.38	42
6342	1-1/4"		17	.471		3.07	27.50		30.57	45
6343	1-1/2"		16	.500		4.34	29.50		33.84	49.50
6344	2"	Q-1	28	.571		8.25	30		38.25	54.50
6360	Solvent cement for CPVC, commercial grade, per quart				Qt.	45			45	49.50
7340	PVC flange, slip-on, Sch 80 std., 1/2"	1 Plum	22	.364	Ea.	13.60	21.50		35.10	47
7350	3/4"		21	.381		14.50	22.50		37	50
7360	1"		18	.444		16.15	26		42.15	57.50
7370	1-1/4"		17	.471		16.65	27.50		44.15	60
7380	1-1/2"		16	.500		17	29.50		46.50	63

22 11 13 – Facility Water Distribution Piping

22 11 13.76 Pipe Fittings, Plastic	Crew	Daily Output	Labor-Hours	Unit	Material	2015 Bare Costs Labor	Equipment	Total	Total Incl O&P
7390	2"	Q-1	26	.615	Ea.	22.50	32.50		55	74
7400	2-1/2"		24	.667		35	35		70	91.50
7410	3"		18	.889		38.50	47		85.50	114
7420	4"		15	1.067		49	56.50		105.50	139
7430	6"		10	1.600		77	84.50		161.50	213
7440	8"	Q-2	11	2.182		138	120		258	330
7550	Union, schedule 40, socket joints, 1/2"	1 Plum	19	.421		5.20	24.50		29.70	43.50
7560	3/4"		18	.444		5.35	26		31.35	45.50
7570	1"		15	.533		5.55	31.50		37.05	53.50
7580	1-1/4"		14	.571		16.65	33.50		50.15	69
7590	1-1/2"		13	.615		18.60	36		54.60	75
7600	2"	Q-1	20	.800		25	42.50		67.50	91.50
7992	Polybutyl/polyethyl pipe, for copper fittings see Line 22 11 13.25 7000									
8000	Compression type, PVC, 160 psi cold water									
8010	Coupling, 3/4" CTS	1 Plum	21	.381	Ea.	3.48	22.50		25.98	38
8020	1" CTS		18	.444		4.32	26		30.32	44.50
8030	1-1/4" CTS		17	.471		6.10	27.50		33.60	48
8040	1-1/2" CTS		16	.500		8.30	29.50		37.80	53.50
8050	2" CTS		15	.533		11.65	31.50		43.15	60.50
8060	Female adapter, 3/4" FPT x 3/4" CTS		23	.348		5.70	20.50		26.20	37.50
8070	3/4" FPT x 1" CTS		21	.381		6.60	22.50		29.10	41.50
8080	1" FPT x 1" CTS		20	.400		6.60	23.50		30.10	43
8090	1-1/4" FPT x 1-1/4" CTS		18	.444		8.65	26		34.65	49
8100	1-1/2" FPT x 1-1/2" CTS		16	.500		9.90	29.50		39.40	55.50
8110	2" FPT x 2" CTS		13	.615		14.35	36		50.35	70.50
8130	Male adapter, 3/4" MPT x 3/4" CTS		23	.348		4.74	20.50		25.24	36
8140	3/4" MPT x 1" CTS		21	.381		5.65	22.50		28.15	40
8150	1" MPT x 1" CTS		20	.400		5.65	23.50		29.15	41.50
8160	1-1/4" MPT x 1-1/4" CTS		18	.444		7.60	26		33.60	48
8170	1-1/2" MPT x 1-1/2" CTS		16	.500		9.15	29.50		38.65	54.50
8180	2" MPT x 2" CTS		13	.615		11.85	36		47.85	67.50
8200	Spigot adapter, 3/4" IPS x 3/4" CTS		23	.348		2.70	20.50		23.20	34
8210	3/4" IPS x 1" CTS		21	.381		3.30	22.50		25.80	37.50
8220	1" IPS x 1" CTS		20	.400		3.30	23.50		26.80	39
8230	1-1/4" IPS x 1-1/4" CTS		18	.444		4.99	26		30.99	45
8240	1-1/2" IPS x 1-1/2" CTS		16	.500		5.25	29.50		34.75	50.50
8250	2" IPS x 2" CTS		13	.615		6.45	36		42.45	61.50
8270	Price includes insert stiffeners									
8280	250 psi is same price as 160 psi									
8300	Insert type, nylon, 160 & 250 psi, cold water									
8310	Clamp ring stainless steel, 3/4" IPS	1 Plum	115	.070	Ea.	2.62	4.08		6.70	9.05
8320	1" IPS		107	.075		2.67	4.39		7.06	9.60
8330	1-1/4" IPS		101	.079		2.69	4.65		7.34	9.95
8340	1-1/2" IPS		95	.084		3.61	4.94		8.55	11.40
8350	2" IPS		85	.094		4.15	5.50		9.65	12.90
8370	Coupling, 3/4" IPS		22	.364		.89	21.50		22.39	33
8380	1" IPS		19	.421		.92	24.50		25.42	38.50
8390	1-1/4" IPS		18	.444		1.37	26		27.37	41
8400	1-1/2" IPS		17	.471		1.62	27.50		29.12	43.50
8410	2" IPS		16	.500		3.60	29.50		33.10	48.50
8430	Elbow, 90°, 3/4" IPS		22	.364		1.77	21.50		23.27	34
8440	1" IPS		19	.421		1.96	24.50		26.46	39.50
8450	1-1/4" IPS		18	.444		2.19	26		28.19	42

For customer support on your Mechanical Cost Data, call 877.795.7394.

217

22 11 13 – Facility Water Distribution Piping

22 11 13.76 Pipe Fittings, Plastic		Crew	Daily Output	Labor-Hours	Unit	Material	2015 Bare Costs Labor	Equipment	Total	Total Incl O&P
8460	1-1/2" IPS	1 Plum	17	.471	Ea.	2.58	27.50		30.08	44.50
8470	2" IPS		16	.500		3.60	29.50		33.10	48.50
8490	Male adapter, 3/4" IPS x 3/4" MPT		25	.320		.89	18.80		19.69	29.50
8500	1" IPS x 1" MPT		21	.381		.92	22.50		23.42	35
8510	1-1/4" IPS x 1-1/4" MPT		20	.400		1.45	23.50		24.95	37
8520	1-1/2" IPS x 1-1/2" MPT		18	.444		1.62	26		27.62	41.50
8530	2" IPS x 2" MPT		15	.533		3.11	31.50		34.61	51
8550	Tee, 3/4" IPS		14	.571		1.71	33.50		35.21	52.50
8560	1" IPS		13	.615		2.23	36		38.23	57
8570	1-1/4" IPS		12	.667		3.48	39		42.48	63
8580	1-1/2" IPS		11	.727		3.95	42.50		46.45	69
8590	2" IPS		10	.800		7.80	47		54.80	79.50
8610	Insert type, PVC, 100 psi @ 180°F, hot & cold water									
8620	Coupler, male, 3/8" CTS x 3/8" MPT	1 Plum	29	.276	Ea.	.72	16.20		16.92	25.50
8630	3/8" CTS x 1/2" MPT		28	.286		.72	16.75		17.47	26.50
8640	1/2" CTS x 1/2" MPT		27	.296		.72	17.40		18.12	27.50
8650	1/2" CTS x 3/4" MPT		26	.308		2.29	18.05		20.34	30
8660	3/4" CTS x 1/2" MPT		25	.320		2.07	18.80		20.87	31
8670	3/4" CTS x 3/4" MPT		25	.320		2.46	18.80		21.26	31
8700	Coupling, 3/8" CTS x 1/2" CTS		25	.320		4.58	18.80		23.38	33.50
8710	1/2" CTS		23	.348		5.50	20.50		26	37
8730	3/4" CTS		22	.364		11.60	21.50		33.10	45
8750	Elbow 90°, 3/8" CTS		25	.320		4	18.80		22.80	33
8760	1/2" CTS		23	.348		4.73	20.50		25.23	36
8770	3/4" CTS		22	.364		7.25	21.50		28.75	40
8800	Rings, crimp, copper, 3/8" CTS		120	.067		.22	3.91		4.13	6.15
8810	1/2" CTS		117	.068		.22	4.01		4.23	6.30
8820	3/4" CTS		115	.070		.29	4.08		4.37	6.45
8850	Reducer tee, bronze, 3/8" x 3/8" x 1/2" CTS		17	.471		4.35	27.50		31.85	46.50
8860	1/2" x 1/2" x 3/4" CTS		15	.533		5.55	31.50		37.05	53.50
8870	3/4" x 1/2" x 1/2" CTS		14	.571		5.55	33.50		39.05	56.50
8890	3/4" x 3/4" x 1/2" CTS		14	.571		5.65	33.50		39.15	57
8900	1" x 1/2" x 1/2" CTS		14	.571		9.35	33.50		42.85	61
8930	Tee, 3/8" CTS		17	.471		3.31	27.50		30.81	45
8940	1/2" CTS		15	.533		2.27	31.50		33.77	50
8950	3/4" CTS		14	.571		3.50	33.50		37	54.50
8960	Copper rings included in fitting price									
9000	Flare type, assembled, acetal, hot & cold water									
9010	Coupling, 1/4" & 3/8" CTS	1 Plum	24	.333	Ea.	3.36	19.55		22.91	33
9020	1/2" CTS		22	.364		3.84	21.50		25.34	36
9030	3/4" CTS		21	.381		5.70	22.50		28.20	40.50
9040	1" CTS		18	.444		7.25	26		33.25	47.50
9050	Elbow 90°, 1/4" CTS		26	.308		3.73	18.05		21.78	31.50
9060	3/8" CTS		24	.333		4	19.55		23.55	34
9070	1/2" CTS		22	.364		4.73	21.50		26.23	37
9080	3/4" CTS		21	.381		7.25	22.50		29.75	42
9090	1" CTS		18	.444		9.15	26		35.15	49.50
9110	Tee, 1/4" CTS		16	.500		4.09	29.50		33.59	49
9114	3/8" CTS		15	.533		4.13	31.50		35.63	52
9120	1/2" CTS		14	.571		5.25	33.50		38.75	56.50
9130	3/4" CTS		13	.615		8.25	36		44.25	63.50
9140	1" CTS		12	.667		11.05	39		50.05	71
9400	Polypropylene, fittings and accessories									

218

For customer support on your Mechanical Cost Data, call 877.795.7394.

22 11 13.76 Pipe Fittings, Plastic	Crew	Daily Output	Labor-Hours	Unit	Material	2015 Bare Costs Labor	Equipment	Total	Total Incl O&P
9404	Fittings fusion welded, sizes are I.D.								
9408	Note: sizes 1/2" thru 4" use socket fusion								
9410	Sizes 6" thru 10" use butt fusion								
9416	Coupling								
9420	3/8"	1 Plum	39	.205	Ea.	.88	12.05		12.93	19.15
9422	1/2"		37.40	.214		1.15	12.55		13.70	20
9424	3/4"		35.40	.226		1.28	13.25		14.53	21.50
9426	1"		29.70	.269		1.69	15.80		17.49	26
9428	1-1/4"		27.60	.290		2.02	17		19.02	27.50
9430	1-1/2"		24.80	.323		4.25	18.95		23.20	33
9432	2"	Q-1	43	.372		8.50	19.65		28.15	39
9434	2-1/2"		35.60	.449		9.50	23.50		33	46.50
9436	3"		30.90	.518		21	27.50		48.50	64.50
9438	3-1/2"		27.80	.576		34	30.50		64.50	83.50
9440	4"		25	.640		44.50	34		78.50	100
9442	Reducing coupling, female to female								
9446	2" to 1-1/2"	Q-1	49	.327	Ea.	12.25	17.25		29.50	39.50
9448	2-1/2" to 2"		41.20	.388		13.45	20.50		33.95	46
9450	3" to 2-1/2"		33.10	.483		18.15	25.50		43.65	58.50
9470	Reducing bushing, female to female								
9472	1/2" to 3/8"	1 Plum	38.20	.209	Ea.	1.15	12.30		13.45	19.80
9474	3/4" to 3/8" or 1/2"		36.50	.219		1.28	12.85		14.13	21
9476	1" to 3/4" or 1/2"		33.10	.242		1.71	14.20		15.91	23.50
9478	1-1/4" to 3/4" or 1"		28.70	.279		2.63	16.35		18.98	27.50
9480	1-1/2" to 1/2" thru 1-1/4"		26.20	.305		4.36	17.90		22.26	32
9482	2" to 1/2" thru 1-1/2"	Q-1	43	.372		8.70	19.65		28.35	39
9484	2-1/2" to 1/2" thru 2"		41.20	.388		9.75	20.50		30.25	41.50
9486	3" to 1-1/2" thru 2-1/2"		33.10	.483		21.50	25.50		47	62.50
9488	3-1/2" to 2" thru 3"		29.20	.548		34.50	29		63.50	81.50
9490	4" to 2-1/2" thru 3-1/2"		26.20	.611		54	32.50		86.50	108
9491	6" to 4" SDR 7.4		16.50	.970		65	51		116	149
9492	6" to 4" SDR 11		16.50	.970		65	51		116	149
9493	8" to 6" SDR 7.4		10.10	1.584		95.50	83.50		179	231
9494	8" to 6" SDR 11		10.10	1.584		87.50	83.50		171	223
9495	10" to 8" SDR 7.4		7.80	2.051		131	108		239	310
9496	10" to 8" SDR 11		7.80	2.051		114	108		222	290
9500	90° Elbow								
9504	3/8"	1 Plum	39	.205	Ea.	.97	12.05		13.02	19.25
9506	1/2"		37.40	.214		1.23	12.55		13.78	20.50
9508	3/4"		35.40	.226		1.58	13.25		14.83	21.50
9510	1"		29.70	.269		2.28	15.80		18.08	26.50
9512	1-1/4"		27.60	.290		3.51	17		20.51	29.50
9514	1-1/2"		24.80	.323		7.55	18.95		26.50	37
9516	2"	Q-1	43	.372		11.60	19.65		31.25	42.50
9518	2-1/2"		35.60	.449		26	23.50		49.50	64.50
9520	3"		30.90	.518		43	27.50		70.50	88.50
9522	3-1/2"		27.80	.576		61	30.50		91.50	113
9524	4"		25	.640		93.50	34		127.50	154
9526	6" SDR 7.4		5.55	2.883		107	152		259	345
9528	6" SDR 11		5.55	2.883		83.50	152		235.50	320
9530	8" SDR 7.4	Q-2	8.10	2.963		415	162		577	705
9532	8" SDR 11		8.10	2.963		355	162		517	635
9534	10" SDR 7.4		7.50	3.200		635	175		810	965

For customer support on your Mechanical Cost Data, call 877.795.7394.

219

22 11 13 – Facility Water Distribution Piping

22 11 13.76 Pipe Fittings, Plastic		Crew	Daily Output	Labor-Hours	Unit	Material	2015 Bare Costs Labor	Equipment	Total	Total Incl O&P
9536	10" SDR 11	Q-2	7.50	3.200	Ea.	575	175		750	900
9551	45° Elbow									
9554	3/8"	1 Plum	39	.205	Ea.	.94	12.05		12.99	19.25
9556	1/2"		37.40	.214		1.23	12.55		13.78	20.50
9558	3/4"		35.40	.226		1.58	13.25		14.83	21.50
9564	1"		29.70	.269		2.28	15.80		18.08	26.50
9566	1-1/4"		27.60	.290		3.51	17		20.51	29.50
9568	1-1/2"		24.80	.323		7.55	18.95		26.50	37
9570	2"	Q-1	43	.372		11.45	19.65		31.10	42
9572	2-1/2"		35.60	.449		25.50	23.50		49	64
9574	3"		30.90	.518		47	27.50		74.50	93
9576	3-1/2"		27.80	.576		67	30.50		97.50	120
9578	4"		25	.640		103	34		137	164
9580	6" SDR 7.4		5.55	2.883		119	152		271	360
9582	6" SDR 11		5.55	2.883		98	152		250	340
9584	8" SDR 7.4	Q-2	8.10	2.963		325	162		487	605
9586	8" SDR 11		8.10	2.963		284	162		446	555
9588	10" SDR 7.4		7.50	3.200		525	175		700	840
9590	10" SDR 11		7.50	3.200		435	175		610	740
9600	Tee									
9604	3/8"	1 Plum	26	.308	Ea.	1.22	18.05		19.27	29
9606	1/2"		24.90	.321		1.66	18.85		20.51	30.50
9608	3/4"		23.70	.338		2.28	19.80		22.08	32.50
9610	1"		19.90	.402		2.90	23.50		26.40	38.50
9612	1-1/4"		18.50	.432		4.43	25.50		29.93	43.50
9614	1-1/2"		16.60	.482		12.65	28.50		41.15	56.50
9616	2"	Q-1	26.80	.597		18.10	31.50		49.60	67.50
9618	2-1/2"		23.80	.672		30.50	35.50		66	87
9620	3"		20.60	.777		50	41		91	117
9622	3-1/2"		18.40	.870		78	46		124	156
9624	4"		16.70	.958		104	50.50		154.50	191
9626	6" SDR 7.4		3.70	4.324		129	228		357	485
9628	6" SDR 11		3.70	4.324		98	228		326	455
9630	8" SDR 7.4	Q-2	5.40	4.444		355	243		598	760
9632	8" SDR 11		5.40	4.444		310	243		553	715
9634	10" SDR 7.4		5	4.800		610	263		873	1,075
9636	10" SDR 11		5	4.800		530	263		793	975
9638	For reducing tee use same tee price									
9660	End cap									
9662	3/8"	1 Plum	78	.103	Ea.	1.22	6		7.22	10.45
9664	1/2"		74.60	.107		1.80	6.30		8.10	11.50
9666	3/4"		71.40	.112		2.28	6.60		8.88	12.45
9668	1"		59.50	.134		2.76	7.90		10.66	14.95
9670	1-1/4"		55.60	.144		4.36	8.45		12.81	17.55
9672	1-1/2"		49.50	.162		6	9.50		15.50	21
9674	2"	Q-1	80	.200		10.05	10.55		20.60	27
9676	2-1/2"		71.40	.224		14.55	11.85		26.40	34
9678	3"		61.70	.259		33	13.70		46.70	56.50
9680	3-1/2"		55.20	.290		39.50	15.30		54.80	66.50
9682	4"		50	.320		60.50	16.90		77.40	92
9684	6" SDR 7.4		26.50	.604		83	32		115	139
9686	6" SDR 11		26.50	.604		83	32		115	139
9688	8" SDR 7.4	Q-2	16.30	1.472		83	80.50		163.50	214

22 11 13 – Facility Water Distribution Piping

22 11 13.76 Pipe Fittings, Plastic		Crew	Daily Output	Labor-Hours	Unit	Material	2015 Bare Costs Labor	2015 Bare Costs Equipment	Total	Total Incl O&P
9690	8" SDR 11	Q-2	16.30	1.472	Ea.	72	80.50		152.50	201
9692	10" SDR 7.4		14.90	1.611		124	88		212	270
9694	10" SDR 11		14.90	1.611		106	88		194	249
9800	Accessories and tools									
9802	Pipe clamps for suspension, not including rod or beam clamp									
9804	3/8"	1 Plum	74	.108	Ea.	2.13	6.35		8.48	11.95
9805	1/2"		70	.114		2.67	6.70		9.37	13.10
9806	3/4"		68	.118		3.07	6.90		9.97	13.85
9807	1"		66	.121		3.35	7.10		10.45	14.45
9808	1-1/4"		64	.125		3.44	7.35		10.79	14.90
9809	1-1/2"		62	.129		3.76	7.55		11.31	15.60
9810	2"	Q-1	110	.145		4.66	7.70		12.36	16.75
9811	2-1/2"		104	.154		5.95	8.15		14.10	18.80
9812	3"		98	.163		6.45	8.65		15.10	20
9813	3-1/2"		92	.174		7.05	9.20		16.25	21.50
9814	4"		86	.186		7.90	9.85		17.75	23.50
9815	6"		70	.229		9.65	12.10		21.75	29
9816	8"	Q-2	100	.240		34.50	13.15		47.65	58
9817	10"	"	94	.255		39.50	14		53.50	64.50
9820	Pipe cutter									
9822	For 3/8" thru 1-1/4"				Ea.	118			118	129
9824	For 1-1/2" thru 4"				"	310			310	340
9826	Note: Pipes may be cut with standard									
9827	iron saw with blades for plastic.									
9982	For plastic hangers see Line 22 05 29.10 8000									
9986	For copper/brass fittings see Line 22 11 13.25 7000									

22 11 13.78 Pipe, High Density Polyethylene Plastic (HDPE)

		Crew	Daily Output	Labor-Hours	Unit	Material	2015 Bare Costs Labor	2015 Bare Costs Equipment	Total	Total Incl O&P
0010	**PIPE, HIGH DENSITY POLYETHYLENE PLASTIC (HDPE)**									
0020	Not incl. hangers, trenching, backfill, hoisting or digging equipment.									
0030	Standard length is 40', add a weld for each joint									
0040	Single wall									
0050	Straight									
0054	1" diameter DR 11				L.F.	.79			.79	.87
0058	1-1/2" diameter DR 11					1			1	1.10
0062	2" diameter DR 11					1.67			1.67	1.84
0066	3" diameter DR 11					2.01			2.01	2.21
0070	3" diameter DR 17					1.61			1.61	1.77
0074	4" diameter DR 11					3.36			3.36	3.70
0078	4" diameter DR 17					3.36			3.36	3.70
0082	6" diameter DR 11					8.35			8.35	9.20
0086	6" diameter DR 17					5.55			5.55	6.10
0090	8" diameter DR 11					13.90			13.90	15.30
0094	8" diameter DR 26					6.50			6.50	7.15
0098	10" diameter DR 11					22			22	24
0102	10" diameter DR 26					10.05			10.05	11.05
0106	12" diameter DR 11					32			32	35
0110	12" diameter DR 26					15.05			15.05	16.60
0114	16" diameter DR 11					48.50			48.50	53.50
0118	16" diameter DR 26					22			22	24
0122	18" diameter DR 11					62			62	68
0126	18" diameter DR 26					28.50			28.50	31.50
0130	20" diameter DR 11					75.50			75.50	83

For customer support on your Mechanical Cost Data, call 877.795.7394.

221

22 11 13.78 Pipe, High Density Polyethylene Plastic (HDPE)		Crew	Daily Output	Labor-Hours	Unit	Material	2015 Bare Costs Labor	Equipment	Total	Total Incl O&P
0134	20" diameter DR 26				L.F.	33.50			33.50	37
0138	22" diameter DR 11					92			92	101
0142	22" diameter DR 26					42			42	46
0146	24" diameter DR 11					109			109	120
0150	24" diameter DR 26					48.50			48.50	53.50
0154	28" diameter DR 17					100			100	111
0158	28" diameter DR 26					67			67	73.50
0162	30" diameter DR 21					94			94	103
0166	30" diameter DR 26					77			77	84.50
0170	36" diameter DR 26					77			77	84.50
0174	42" diameter DR 26					149			149	164
0178	48" diameter DR 26					196			196	216
0182	54" diameter DR 26				▼	246			246	271
0300	90° Elbow									
0304	1" diameter DR 11				Ea.	5.60			5.60	6.15
0308	1-1/2" diameter DR 11					7			7	7.70
0312	2" diameter DR 11					7			7	7.70
0316	3" diameter DR 11					14			14	15.40
0320	3" diameter DR 17					14			14	15.40
0324	4" diameter DR 11					19.60			19.60	21.50
0328	4" diameter DR 17					19.60			19.60	21.50
0332	6" diameter DR 11					45			45	49.50
0336	6" diameter DR 17					45			45	49.50
0340	8" diameter DR 11					111			111	122
0344	8" diameter DR 26					98			98	108
0348	10" diameter DR 11					415			415	455
0352	10" diameter DR 26					380			380	420
0356	12" diameter DR 11					435			435	480
0360	12" diameter DR 26					395			395	435
0364	16" diameter DR 11					515			515	570
0368	16" diameter DR 26					510			510	560
0372	18" diameter DR 11					650			650	715
0376	18" diameter DR 26					610			610	670
0380	20" diameter DR 11					765			765	840
0384	20" diameter DR 26					740			740	810
0388	22" diameter DR 11					790			790	870
0392	22" diameter DR 26					765			765	840
0396	24" diameter DR 11					890			890	980
0400	24" diameter DR 26					865			865	950
0404	28" diameter DR 17					1,075			1,075	1,200
0408	28" diameter DR 26					1,025			1,025	1,125
0412	30" diameter DR 17					1,525			1,525	1,675
0416	30" diameter DR 26					1,400			1,400	1,550
0420	36" diameter DR 26					1,775			1,775	1,950
0424	42" diameter DR 26					2,300			2,300	2,525
0428	48" diameter DR 26					2,675			2,675	2,950
0432	54" diameter DR 26				▼	6,375			6,375	7,000
0500	45° Elbow									
0512	2" diameter DR 11				Ea.	5.60			5.60	6.15
0516	3" diameter DR 11					14			14	15.40
0520	3" diameter DR 17					14			14	15.40
0524	4" diameter DR 11					19.60			19.60	21.50
0528	4" diameter DR 17				▼	19.60			19.60	21.50

For customer support on your Mechanical Cost Data, call 877.795.7394.

22 11 13 – Facility Water Distribution Piping

22 11 13.78 Pipe, High Density Polyethylene Plastic (HDPE)	Crew	Daily Output	Labor-Hours	Unit	Material	2015 Bare Costs Labor	Equipment	Total	Total Incl O&P
0532	6" diameter DR 11				Ea.	45			45	49.50
0536	6" diameter DR 17					45			45	49.50
0540	8" diameter DR 11					111			111	122
0544	8" diameter DR 26					64.50			64.50	71
0548	10" diameter DR 11					415			415	455
0552	10" diameter DR 26					380			380	420
0556	12" diameter DR 11					435			435	480
0560	12" diameter DR 26					395			395	435
0564	16" diameter DR 11					229			229	252
0568	16" diameter DR 26					216			216	238
0572	18" diameter DR 11					243			243	268
0576	18" diameter DR 26					223			223	245
0580	20" diameter DR 11					370			370	405
0584	20" diameter DR 26					350			350	385
0588	22" diameter DR 11					510			510	560
0592	22" diameter DR 26					485			485	530
0596	24" diameter DR 11					625			625	685
0600	24" diameter DR 26					600			600	660
0604	28" diameter DR 17					710			710	780
0608	28" diameter DR 26					690			690	755
0612	30" diameter DR 17					880			880	965
0616	30" diameter DR 26					855			855	940
0620	36" diameter DR 26					1,075			1,075	1,200
0624	42" diameter DR 26					1,400			1,400	1,525
0628	48" diameter DR 26					1,525			1,525	1,675
0632	54" diameter DR 26					2,025			2,025	2,225
0700	Tee									
0704	1" diameter DR 11				Ea.	7.30			7.30	8.05
0708	1-1/2" diameter DR 11					10.25			10.25	11.30
0712	2" diameter DR 11					8.80			8.80	9.65
0716	3" diameter DR 11					16.10			16.10	17.70
0720	3" diameter DR 17					16.10			16.10	17.70
0724	4" diameter DR 11					23.50			23.50	26
0728	4" diameter DR 17					23.50			23.50	26
0732	6" diameter DR 11					58.50			58.50	64.50
0736	6" diameter DR 17					58.50			58.50	64.50
0740	8" diameter DR 11					145			145	159
0744	8" diameter DR 17					145			145	159
0748	10" diameter DR 11					430			430	475
0752	10" diameter DR 17					430			430	475
0756	12" diameter DR 11					575			575	635
0760	12" diameter DR 17					575			575	635
0764	16" diameter DR 11					305			305	335
0768	16" diameter DR 17					251			251	276
0772	18" diameter DR 11					430			430	470
0776	18" diameter DR 17					355			355	390
0780	20" diameter DR 11					525			525	575
0784	20" diameter DR 17					430			430	470
0788	22" diameter DR 11					670			670	740
0792	22" diameter DR 17					530			530	585
0796	24" diameter DR 11					835			835	915
0800	24" diameter DR 17					705			705	775
0804	28" diameter DR 17					1,350			1,350	1,500

For customer support on your Mechanical Cost Data, call 877.795.7394.

223

22 11 13 – Facility Water Distribution Piping

22 11 13.78 Pipe, High Density Polyethylene Plastic (HDPE)	Crew	Daily Output	Labor-Hours	Unit	Material	2015 Bare Costs Labor	Equipment	Total	Total Incl O&P
0812	30" diameter DR 17				Ea.	1,550			1,550	1,725
0820	36" diameter DR 17					2,575			2,575	2,825
0824	42" diameter DR 26					2,875			2,875	3,150
0828	48" diameter DR 26					3,075			3,075	3,375
1000	Flange adptr, w/back-up ring and 1/2 cost of plated bolt set									
1004	1" diameter DR 11				Ea.	28			28	30.50
1008	1-1/2" diameter DR 11					28			28	30.50
1012	2" diameter DR 11					17.55			17.55	19.35
1016	3" diameter DR 11					20.50			20.50	22.50
1020	3" diameter DR 17					20.50			20.50	22.50
1024	4" diameter DR 11					28			28	30.50
1028	4" diameter DR 17					28			28	30.50
1032	6" diameter DR 11					39.50			39.50	43.50
1036	6" diameter DR 17					39.50			39.50	43.50
1040	8" diameter DR 11					57			57	63
1044	8" diameter DR 26					57			57	63
1048	10" diameter DR 11					91			91	100
1052	10" diameter DR 26					91			91	100
1056	12" diameter DR 11					133			133	147
1060	12" diameter DR 26					133			133	147
1064	16" diameter DR 11					286			286	315
1068	16" diameter DR 26					286			286	315
1072	18" diameter DR 11					370			370	410
1076	18" diameter DR 26					370			370	410
1080	20" diameter DR 11					515			515	565
1084	20" diameter DR 26					515			515	565
1088	22" diameter DR 11					560			560	615
1092	22" diameter DR 26					555			555	610
1096	24" diameter DR 17					605			605	665
1100	24" diameter DR 32.5					605			605	665
1104	28" diameter DR 15.5					820			820	900
1108	28" diameter DR 32.5					820			820	900
1112	30" diameter DR 11					950			950	1,050
1116	30" diameter DR 21					950			950	1,050
1120	36" diameter DR 26					1,050			1,050	1,150
1124	42" diameter DR 26					1,175			1,175	1,300
1128	48" diameter DR 26					1,450			1,450	1,600
1132	54" diameter DR 26					1,750			1,750	1,925
1200	Reducer									
1208	2" x 1-1/2" diameter DR 11				Ea.	8.40			8.40	9.25
1212	3" x 2" diameter DR 11					8.40			8.40	9.25
1216	4" x 2" diameter DR 11					9.80			9.80	10.80
1220	4" x 3" diameter DR 11					12.60			12.60	13.85
1224	6" x 4" diameter DR 11					29.50			29.50	32.50
1228	8" x 6" diameter DR 11					45			45	49.50
1232	10" x 8" diameter DR 11					77			77	85
1236	12" x 8" diameter DR 11					126			126	139
1240	12" x 10" diameter DR 11					101			101	111
1244	14" x 12" diameter DR 11					112			112	123
1248	16" x 14" diameter DR 11					143			143	157
1252	18" x 16" diameter DR 11					174			174	191
1256	20" x 18" diameter DR 11					345			345	380
1260	22" x 20" diameter DR 11					425			425	465

For customer support on your Mechanical Cost Data, call 877.795.7394.

22 11 13 – Facility Water Distribution Piping

22 11 13.78 Pipe, High Density Polyethylene Plastic (HDPE)	Crew	Daily Output	Labor-Hours	Unit	Material	2015 Bare Costs Labor	Equipment	Total	Total Incl O&P
1264 24" x 22" diameter DR 11				Ea.	475			475	525
1268 26" x 24" diameter DR 11					560			560	615
1272 28" x 24" diameter DR 11					715			715	785
1276 32" x 28" diameter DR 17					925			925	1,025
1280 36" x 32" diameter DR 17				↓	1,250			1,250	1,375
4000 Welding labor per joint, not including welding machine									
4010 Pipe joint size (cost based on thickest wall for each diam.)									
4030 1" pipe size	4 Skwk	273	.117	Ea.		5.70		5.70	8.80
4040 1-1/2" pipe size		175	.183			8.90		8.90	13.75
4050 2" pipe size		128	.250			12.15		12.15	18.80
4060 3" pipe size		100	.320			15.55		15.55	24
4070 4" pipe size	↓	77	.416			20		20	31
4080 6" pipe size	5 Skwk	63	.635			31		31	47.50
4090 8" pipe size		48	.833			40.50		40.50	62.50
4100 10" pipe size	↓	40	1			48.50		48.50	75
4110 12" pipe size	6 Skwk	41	1.171			57		57	88
4120 16" pipe size		34	1.412			68.50		68.50	106
4130 18" pipe size	↓	32	1.500			73		73	113
4140 20" pipe size	8 Skwk	37	1.730			84		84	130
4150 22" pipe size		35	1.829			89		89	137
4160 24" pipe size		34	1.882			91.50		91.50	141
4170 28" pipe size		33	1.939			94.50		94.50	146
4180 30" pipe size		32	2			97.50		97.50	150
4190 36" pipe size		31	2.065			100		100	155
4200 42" pipe size	↓	30	2.133			104		104	160
4210 48" pipe size	9 Skwk	33	2.182			106		106	164
4220 54" pipe size	"	31	2.323	↓		113		113	175
4300 Note: Cost for set up each time welder is moved.									
4301 Add 50% of a weld cost									
4310 Welder usually remains stationary with pipe moved through it.									
4340 Weld machine, rental per day based on diam. capacity									
4350 1" thru 2" diameter				Ea.			40.50	40.50	44.50
4360 3" thru 4" diameter							46	46	50.50
4370 6" thru 8" diameter							103	103	113
4380 10" thru 12" diameter							178	178	196
4390 16" thru 18" diameter							259	259	285
4400 20" thru 24" diameter							500	500	550
4410 28" thru 32" diameter							545	545	600
4420 36" diameter							570	570	625
4430 42" thru 54" diameter				↓			890	890	980
5000 Dual wall contained pipe									
5040 Straight									
5054 1" DR 11 x 3" DR 11				L.F.	7.10			7.10	7.80
5058 1" DR 11 x 4" DR 11					7.55			7.55	8.30
5062 1-1/2" DR 11 x 4" DR 17					7.75			7.75	8.55
5066 2" DR 11 x 4" DR 17					8.35			8.35	9.20
5070 2" DR 11 x 6" DR 17					12.50			12.50	13.75
5074 3" DR 11 x 6" DR 17					13.95			13.95	15.35
5078 3" DR 11 x 6" DR 26					12			12	13.20
5086 3" DR 17 x 8" DR 17					17.70			17.70	19.50
5090 4" DR 11 x 8" DR 17					20.50			20.50	23
5094 4" DR 17 x 8" DR 26					16.40			16.40	18.05
5098 6" DR 11 x 10" DR 17				↓	32			32	35.50

For customer support on your Mechanical Cost Data, call 877.795.7394.

225

22 11 13.78 Pipe, High Density Polyethylene Plastic (HDPE)	Crew	Daily Output	Labor-Hours	Unit	Material	2015 Bare Costs Labor	Equipment	Total	Total Incl O&P
5102	6" DR 17 x 10" DR 26				L.F.	24			24	26.50
5106	6" DR 26 x 10" DR 26					23			23	25
5110	8" DR 17 x 12" DR 26					34			34	37.50
5114	8" DR 26 x 12" DR 32.5					29			29	32
5118	10" DR 17 x 14" DR 26					47.50			47.50	52.50
5122	10" DR 17 x 16" DR 26					52			52	57.50
5126	10" DR 26 x 16" DR 26					47.50			47.50	52
5130	12" DR 26 x 16" DR 26					54			54	59
5134	12" DR 17 x 18" DR 26					68.50			68.50	75.50
5138	12" DR 26 x 18" DR 26					61.50			61.50	67.50
5142	14" DR 26 x 20" DR 32.5					67.50			67.50	74
5146	16" DR 26 x 22" DR 32.5					102			102	112
5150	18" DR 26 x 24" DR 32.5					93.50			93.50	103
5154	20" DR 32.5 x 28" DR 32.5					121			121	133
5158	22" DR 32.5 x 30" DR 32.5					121			121	133
5162	24" DR 32.5 x 32" DR 32.5					155			155	170
5166	36" DR 32.5 x 42" DR 32.5				▼	261			261	287
5300	Force transfer coupling									
5354	1" DR 11 x 3" DR 11				Ea.	255			255	280
5358	1" DR 11 x 4" DR 17					267			267	294
5362	1-1/2" DR 11 x 4" DR 17					280			280	310
5366	2" DR 11 x 4" DR 17					293			293	320
5370	2" DR 11 x 6" DR 17					425			425	470
5374	3" DR 11 x 6" DR 17					425			425	470
5378	3" DR 11 x 6" DR 26					425			425	470
5382	3" DR 11 x 8" DR 11					450			450	495
5386	3" DR 11 x 8" DR 17					450			450	495
5390	4" DR 11 x 8" DR 17					450			450	495
5394	4" DR 17 x 8" DR 26					395			395	435
5398	6" DR 11 x 10" DR 17					570			570	625
5402	6" DR 17 x 10" DR 26					470			470	515
5406	6" DR 26 x 10" DR 26					470			470	515
5410	8" DR 17 x 12" DR 26					625			625	685
5414	8" DR 26 x 12" DR 32.5					505			505	555
5418	10" DR 17 x 14" DR 26					820			820	900
5422	10" DR 17 x 16" DR 26					820			820	900
5426	10" DR 26 x 16" DR 26					820			820	900
5430	12" DR 26 x 16" DR 26					1,025			1,025	1,125
5434	12" DR 17 x 18" DR 26					1,050			1,050	1,150
5438	12" DR 26 x 18" DR 26					1,050			1,050	1,150
5442	14" DR 26 x 20" DR 32.5					1,100			1,100	1,225
5446	16" DR 26 x 22" DR 32.5					1,150			1,150	1,250
5450	18" DR 26 x 24" DR 32.5					1,250			1,250	1,350
5454	20" DR 32.5 x 28" DR 32.5					1,500			1,500	1,650
5458	22" DR 32.5 x 30" DR 32.5					1,675			1,675	1,825
5462	24" DR 32.5 x 32" DR 32.5					1,800			1,800	1,975
5466	36" DR 32.5 x 42" DR 32.5				▼	3,000			3,000	3,300
5600	90° Elbow									
5654	1" DR 11 x 3" DR 11				Ea.	210			210	231
5658	1" DR 11 x 4" DR 17					201			201	222
5662	1-1/2" DR 11 x 4" DR 17					220			220	242
5666	2" DR 11 x 4" DR 17					233			233	257
5670	2" DR 11 x 6" DR 17					297			297	325

For customer support on your Mechanical Cost Data, call 877.795.7394.

22 11 13.78 Pipe, High Density Polyethylene Plastic (HDPE)	Crew	Daily Output	Labor-Hours	Unit	Material	2015 Bare Costs Labor	Equipment	Total	Total Incl O&P
5674	3" DR 11 x 6" DR 17				Ea.	340			340	375
5678	3" DR 17 x 6" DR 26					271			271	298
5682	3" DR 17 x 8" DR 11					550			550	605
5686	3" DR 17 x 8" DR 17					425			425	470
5690	4" DR 11 x 8" DR 17					500			500	550
5694	4" DR 17 x 8" DR 26					400			400	440
5698	6" DR 11 x 10" DR 17					670			670	735
5702	6" DR 17 x 10" DR 26					510			510	565
5706	6" DR 26 x 10" DR 26					480			480	525
5710	8" DR 17 x 12" DR 26					760			760	835
5714	8" DR 26 x 12" DR 32.5					705			705	775
5718	10" DR 17 x 14" DR 26					1,200			1,200	1,325
5722	10" DR 17 x 16" DR 26					1,150			1,150	1,250
5726	10" DR 26 x 16" DR 26					1,050			1,050	1,175
5730	12" DR 26 x 16" DR 26					1,175			1,175	1,300
5734	12" DR 17 x 18" DR 26					1,400			1,400	1,550
5738	12" DR 26 x 18" DR 26					1,300			1,300	1,425
5742	14" DR 26 x 20" DR 32.5					1,700			1,700	1,875
5746	16" DR 26 x 22" DR 32.5					1,700			1,700	1,875
5750	18" DR 26 x 24" DR 32.5					2,100			2,100	2,300
5754	20" DR 32.5 x 28" DR 32.5					2,425			2,425	2,650
5758	22" DR 32.5 x 30" DR 32.5					3,300			3,300	3,625
5762	24" DR 32.5 x 32" DR 32.5					3,200			3,200	3,525
5766	36" DR 32.5 x 42" DR 32.5					5,200			5,200	5,725
5800	45° Elbow									
5804	1" DR 11 x 3" DR 11				Ea.	126			126	139
5808	1" DR 11 x 4" DR 17					125			125	138
5812	1-1/2" DR 11 x 4" DR 17					134			134	148
5816	2" DR 11 x 4" DR 17					147			147	161
5820	2" DR 11 x 6" DR 17					184			184	203
5824	3" DR 11 x 6" DR 17					204			204	225
5828	3" DR 17 x 6" DR 26					169			169	186
5832	3" DR 17 x 8" DR 11					315			315	345
5836	3" DR 17 x 8" DR 17					250			250	275
5840	4" DR 11 x 8" DR 17					292			292	320
5844	4" DR 17 x 8" DR 26					238			238	262
5848	6" DR 11 x 10" DR 17					385			385	425
5852	6" DR 17 x 10" DR 26					300			300	335
5856	6" DR 26 x 10" DR 26					285			285	315
5860	8" DR 17 x 12" DR 26					460			460	505
5864	8" DR 26 x 12" DR 32.5					430			430	475
5868	10" DR 17 x 14" DR 26					720			720	790
5872	10" DR 17 x 16" DR 26					705			705	775
5876	10" DR 26 x 16" DR 26					660			660	725
5880	12" DR 26 x 16" DR 26					745			745	820
5884	12" DR 17 x 18" DR 26					875			875	965
5888	12" DR 26 x 18" DR 26					820			820	900
5892	14" DR 26 x 20" DR 32.5					1,050			1,050	1,150
5896	16" DR 26 x 22" DR 32.5					1,100			1,100	1,225
5900	18" DR 26 x 24" DR 32.5					1,300			1,300	1,425
5904	20" DR 32.5 x 28" DR 32.5					1,600			1,600	1,750
5908	22" DR 32.5 x 30" DR 32.5					2,025			2,025	2,250
5912	24" DR 32.5 x 32" DR 32.5					2,025			2,025	2,225

22 11 13 – Facility Water Distribution Piping

22 11 13.78 Pipe, High Density Polyethylene Plastic (HDPE)	Crew	Daily Output	Labor-Hours	Unit	Material	2015 Bare Costs Labor	Equipment	Total	Total Incl O&P
5916	36" DR 32.5 x 42" DR 32.5				Ea.	3,125			3,125	3,450
6000	Access port with 4" riser									
6050	1" DR 11 x 4" DR 17				Ea.	275			275	305
6054	1-1/2" DR 11 x 4" DR 17					281			281	310
6058	2" DR 11 x 6" DR 17					340			340	375
6062	3" DR 11 x 6" DR 17					345			345	380
6066	3" DR 17 x 6" DR 26					340			340	370
6070	3" DR 17 x 8" DR 11					385			385	425
6074	3" DR 17 x 8" DR 17					370			370	405
6078	4" DR 11 x 8" DR 17					380			380	420
6082	4" DR 17 x 8" DR 26					365			365	400
6086	6" DR 11 x 10" DR 17					475			475	525
6090	6" DR 17 x 10" DR 26					440			440	485
6094	6" DR 26 x 10" DR 26					430			430	475
6098	8" DR 17 x 12" DR 26					460			460	510
6102	8" DR 26 x 12" DR 32.5					440			440	485
6200	End termination with vent plug									
6204	1" DR 11 x 3" DR 11				Ea.	228			228	250
6208	1" DR 11 x 4" DR 17					244			244	268
6212	1-1/2" DR 11 x 4" DR 17					244			244	268
6216	2" DR 11 x 4" DR 17					265			265	291
6220	2" DR 11 x 6" DR 17					300			300	330
6224	3" DR 11 x 6" DR 17					300			300	330
6228	3" DR 17 x 6" DR 26					300			300	330
6232	3" DR 17 x 8" DR 11					415			415	460
6236	3" DR 17 x 8" DR 17					415			415	460
6240	4" DR 11 x 8" DR 17					415			415	460
6244	4" DR 17 x 8" DR 26					365			365	400
6248	6" DR 11 x 10" DR 17					495			495	545
6252	6" DR 17 x 10" DR 26					405			405	450
6256	6" DR 26 x 10" DR 26					405			405	450
6260	8" DR 17 x 12" DR 26					605			605	665
6264	8" DR 26 x 12" DR 32.5					480			480	530
6268	10" DR 17 x 14" DR 26					625			625	690
6272	10" DR 17 x 16" DR 26					625			625	690
6276	10" DR 26 x 16" DR 26					625			625	690
6280	12" DR 26 x 16" DR 26					625			625	690
6284	12" DR 17 x 18" DR 26					825			825	905
6288	12" DR 26 x 18" DR 26					825			825	905
6292	14" DR 26 x 20" DR 32.5					840			840	925
6296	16" DR 26 x 22" DR 32.5					975			975	1,075
6300	18" DR 26 x 24" DR 32.5					1,000			1,000	1,100
6304	20" DR 32.5 x 28" DR 32.5					1,275			1,275	1,425
6308	22" DR 32.5 x 30" DR 32.5					1,425			1,425	1,550
6312	24" DR 32.5 x 32" DR 32.5					1,575			1,575	1,725
6316	36" DR 32.5 x 42" DR 32.5					2,275			2,275	2,500
6600	Tee									
6604	1" DR 11 x 3" DR 11				Ea.	232			232	255
6608	1" DR 11 x 4" DR 17					288			288	315
6612	1-1/2" DR 11 x 4" DR 17					320			320	350
6616	2" DR 11 x 4" DR 17					350			350	385
6620	2" DR 11 x 6" DR 17					425			425	465
6624	3" DR 11 x 6" DR 17					490			490	540

22 11 13 – Facility Water Distribution Piping

22 11 13.78 Pipe, High Density Polyethylene Plastic (HDPE)	Crew	Daily Output	Labor-Hours	Unit	Material	2015 Bare Costs Labor	Equipment	Total	Total Incl O&P
6628	3" DR 17 x 6" DR 26				Ea.	475			475	525
6632	3" DR 17 x 8" DR 11					575			575	635
6636	3" DR 17 x 8" DR 17					550			550	605
6640	4" DR 11 x 8" DR 17					595			595	655
6644	4" DR 17 x 8" DR 26					575			575	630
6648	6" DR 11 x 10" DR 17					745			745	820
6652	6" DR 17 x 10" DR 26					705			705	780
6656	6" DR 26 x 10" DR 26					700			700	770
6660	8" DR 17 x 12" DR 26					700			700	770
6664	8" DR 26 x 12" DR 32.5					915			915	1,000
6668	10" DR 17 x 14" DR 26					1,050			1,050	1,150
6672	10" DR 17 x 16" DR 26					1,100			1,100	1,225
6676	10" DR 26 x 16" DR 26					1,100			1,100	1,225
6680	12" DR 26 x 16" DR 26					1,200			1,200	1,325
6684	12" DR 17 x 18" DR 26					1,200			1,200	1,325
6688	12" DR 26 x 18" DR 26					1,325			1,325	1,450
6692	14" DR 26 x 20" DR 32.5					1,750			1,750	1,925
6696	16" DR 26 x 22" DR 32.5					2,150			2,150	2,350
6700	18" DR 26 x 24" DR 32.5					2,325			2,325	2,575
6704	20" DR 32.5 x 28" DR 32.5					2,950			2,950	3,250
6708	22" DR 32.5 x 30" DR 32.5					3,500			3,500	3,850
6712	24" DR 32.5 x 32" DR 32.5					4,200			4,200	4,600
6716	36" DR 32.5 x 42" DR 32.5				▼	8,525			8,525	9,400
6800	Wye									
6816	2" DR 11 x 4" DR 17				Ea.	440			440	485
6820	2" DR 11 x 6" DR 17					500			500	550
6824	3" DR 11 x 6" DR 17					505			505	555
6828	3" DR 17 x 6" DR 26					490			490	540
6832	3" DR 17 x 8" DR 11					615			615	675
6836	3" DR 17 x 8" DR 17					575			575	630
6840	4" DR 11 x 8" DR 17					625			625	690
6844	4" DR 17 x 8" DR 26					595			595	655
6848	6" DR 11 x 10" DR 17					775			775	855
6852	6" DR 17 x 10" DR 26					720			720	795
6856	6" DR 26 x 10" DR 26					715			715	790
6860	8" DR 17 x 12" DR 26					1,025			1,025	1,125
6864	8" DR 26 x 12" DR 32.5					995			995	1,100
6868	10" DR 17 x 14" DR 26					1,250			1,250	1,350
6872	10" DR 17 x 16" DR 26					1,325			1,325	1,450
6876	10" DR 26 x 16" DR 26					1,300			1,300	1,425
6880	12" DR 26 x 16" DR 26					1,450			1,450	1,600
6884	12" DR 17 x 18" DR 26					1,625			1,625	1,800
6888	12" DR 26 x 18" DR 26					1,650			1,650	1,825
6892	14" DR 26 x 20" DR 32.5					2,025			2,025	2,225
6896	16" DR 26 x 22" DR 32.5					2,900			2,900	3,200
6900	18" DR 26 x 24" DR 32.5					3,425			3,425	3,750
6904	20" DR 32.5 x 28" DR 32.5					4,300			4,300	4,725
6908	22" DR 32.5 x 30" DR 32.5					4,875			4,875	5,375
6912	24" DR 32.5 x 32" DR 32.5				▼	5,550			5,550	6,125
9000	Welding labor per joint, not including welding machine									
9010	Pipe joint size, outer pipe (cost based on the thickest walls)									
9020	Straight pipe									
9050	3" pipe size	4 Skwk	96	.333	Ea.		16.20		16.20	25

22 11 13 – Facility Water Distribution Piping

22 11 13.78 Pipe, High Density Polyethylene Plastic (HDPE)	Crew	Daily Output	Labor-Hours	Unit	Material	2015 Bare Costs Labor	Equipment	Total	Total Incl O&P
9060	4" pipe size	4 Skwk	77	.416	Ea.		20		20	31
9070	6" pipe size	5 Skwk	60	.667			32.50		32.50	50
9080	8" pipe size	"	40	1			48.50		48.50	75
9090	10" pipe size	6 Skwk	41	1.171			57		57	88
9100	12" pipe size		39	1.231			60		60	92.50
9110	14" pipe size		38	1.263			61.50		61.50	95
9120	16" pipe size		35	1.371			66.50		66.50	103
9130	18" pipe size	8 Skwk	45	1.422			69		69	107
9140	20" pipe size		42	1.524			74		74	115
9150	22" pipe size		40	1.600			78		78	120
9160	24" pipe size		38	1.684			82		82	127
9170	28" pipe size		37	1.730			84		84	130
9180	30" pipe size		36	1.778			86.50		86.50	134
9190	32" pipe size		35	1.829			89		89	137
9200	42" pipe size		32	2			97.50		97.50	150
9300	Note: Cost for set up each time welder is moved.									
9301	Add 100% of weld labor cost									
9310	For handling between fitting welds add 50% of weld labor cost									
9320	Welder usually remains stationary with pipe moved through it									
9360	Weld machine, rental per day based on diam. capacity									
9380	3" thru 4" diameter				Ea.			63.50	63.50	70
9390	6" thru 8" diameter							207	207	228
9400	10" thru 12" diameter							310	310	340
9410	14" thru 18" diameter							420	420	460
9420	20" thru 24" diameter							750	750	825
9430	28" thru 32" diameter							880	880	970
9440	42" thru 58" diameter							910	910	1,000

22 11 19 – Domestic Water Piping Specialties

22 11 19.10 Flexible Connectors

		Crew	Daily Output	Labor-Hours	Unit	Material	2015 Bare Costs Labor	Equipment	Total	Total Incl O&P
0010	**FLEXIBLE CONNECTORS**, Corrugated, 7/8" O.D., 1/2" I.D.									
0050	Gas, seamless brass, steel fittings									
0200	12" long	1 Plum	36	.222	Ea.	17.40	13.05		30.45	39
0220	18" long		36	.222		21.50	13.05		34.55	43.50
0240	24" long		34	.235		25.50	13.80		39.30	49
0260	30" long		34	.235		27.50	13.80		41.30	51.50
0280	36" long		32	.250		30.50	14.70		45.20	55.50
0320	48" long		30	.267		38.50	15.65		54.15	66
0340	60" long		30	.267		46	15.65		61.65	74
0360	72" long		30	.267		53	15.65		68.65	82
2000	Water, copper tubing, dielectric separators									
2100	12" long	1 Plum	36	.222	Ea.	17.35	13.05		30.40	39
2220	15" long		36	.222		19.30	13.05		32.35	40.50
2240	18" long		36	.222		21	13.05		34.05	42.50
2260	24" long		34	.235		26	13.80		39.80	49.50

22 11 19.14 Flexible Metal Hose

		Crew	Daily Output	Labor-Hours	Unit	Material	2015 Bare Costs Labor	Equipment	Total	Total Incl O&P
0010	**FLEXIBLE METAL HOSE**, Connectors, standard lengths									
0100	Bronze braided, bronze ends									
0120	3/8" diameter x 12"	1 Stpi	26	.308	Ea.	21	18.40		39.40	51
0140	1/2" diameter x 12"		24	.333		22	19.90		41.90	54
0160	3/4" diameter x 12"		20	.400		32	24		56	71
0180	1" diameter x 18"		19	.421		38	25		63	80
0200	1-1/2" diameter x 18"		13	.615		52	37		89	113

22 11 19 – Domestic Water Piping Specialties

22 11 19.14 Flexible Metal Hose	Crew	Daily Output	Labor-Hours	Unit	Material	2015 Bare Costs Labor	Equipment	Total	Total Incl O&P
0220	2" diameter x 18"	1 Stpi	11	.727	Ea.	68	43.50		111.50	141
1000	Carbon steel ends									
1020	1/4" diameter x 12"	1 Stpi	28	.286	Ea.	15.50	17.05		32.55	43
1040	3/8" diameter x 12"		26	.308		15.90	18.40		34.30	45.50
1060	1/2" diameter x 12"		24	.333		17.55	19.90		37.45	49.50
1080	1/2" diameter x 24"		24	.333		47.50	19.90		67.40	82
1120	3/4" diameter x 12"		20	.400		28.50	24		52.50	67.50
1140	3/4" diameter x 24"		20	.400		58.50	24		82.50	101
1160	3/4" diameter x 36"		20	.400		69.50	24		93.50	112
1180	1" diameter x 18"		19	.421		35.50	25		60.50	77
1200	1" diameter x 30"		19	.421		67.50	25		92.50	112
1220	1" diameter x 36"		19	.421		87.50	25		112.50	135
1240	1-1/4" diameter x 18"		15	.533		49.50	32		81.50	102
1260	1-1/4" diameter x 36"		15	.533		109	32		141	168
1280	1-1/2" diameter x 18"		13	.615		69.50	37		106.50	132
1300	1-1/2" diameter x 36"		13	.615		118	37		155	186
1320	2" diameter x 24"		11	.727		96.50	43.50		140	172
1340	2" diameter x 36"		11	.727		146	43.50		189.50	227
1360	2-1/2" diameter x 24"		9	.889		221	53		274	325
1380	2-1/2" diameter x 36"		9	.889		560	53		613	695
1400	3" diameter x 24"		7	1.143		305	68.50		373.50	440
1420	3" diameter x 36"	↓	7	1.143	↓	670	68.50		738.50	840
2000	Carbon steel braid, carbon steel solid ends									
2100	1/2" diameter x 12"	1 Stpi	24	.333	Ea.	42.50	19.90		62.40	77
2120	3/4" diameter x 12"		20	.400		64.50	24		88.50	107
2140	1" diameter x 12"		19	.421		92	25		117	139
2160	1-1/4" diameter x 12"		15	.533		28	32		60	79
2180	1-1/2" diameter x 12"	↓	13	.615	↓	30	37		67	88.50
3000	Stainless steel braid, welded on carbon steel ends									
3100	1/2" diameter x 12"	1 Stpi	24	.333	Ea.	59	19.90		78.90	95
3120	3/4" diameter x 12"		20	.400		71	24		95	115
3140	3/4" diameter x 24"		20	.400		82	24		106	126
3160	3/4" diameter x 36"		20	.400		94	24		118	140
3180	1" diameter x 12"		19	.421		84.50	25		109.50	131
3200	1" diameter x 24"		19	.421		102	25		127	150
3220	1" diameter x 36"		19	.421		117	25		142	167
3240	1-1/4" diameter x 12"		15	.533		128	32		160	188
3260	1-1/4" diameter x 24"		15	.533		139	32		171	201
3280	1-1/4" diameter x 36"		15	.533		158	32		190	222
3300	1-1/2" diameter x 12"		13	.615		44	37		81	104
3320	1-1/2" diameter x 24"		13	.615		152	37		189	223
3340	1-1/2" diameter x 36"	↓	13	.615	↓	178	37		215	252
3400	Metal stainless steel braid, over corrugated stainless steel, flanged ends									
3410	150 PSI									
3420	1/2" diameter x 12"	1 Stpi	24	.333	Ea.	120	19.90		139.90	162
3430	1" diameter x 12"		20	.400		180	24		204	234
3440	1-1/2" diameter x 12"		15	.533		204	32		236	272
3450	2-1/2" diameter x 9"		12	.667		108	40		148	179
3460	3" diameter x 9"		9	.889		93	53		146	182
3470	4" diameter x 9"		7	1.143		113	68.50		181.50	227
3480	4" diameter x 30"		5	1.600		560	95.50		655.50	760
3490	4" diameter x 36"		4.80	1.667		580	99.50		679.50	790
3500	6" diameter x 11"	↓	5	1.600	↓	188	95.50		283.50	350

22 11 19 – Domestic Water Piping Specialties

22 11 19.14 Flexible Metal Hose

		Crew	Daily Output	Labor-Hours	Unit	Material	2015 Bare Costs Labor	2015 Bare Costs Equipment	Total	Total Incl O&P
3510	6" diameter x 36"	1 Stpi	3.80	2.105	Ea.	685	126		811	940
3520	8" diameter x 12"		4	2		410	120		530	630
3530	10" diameter x 13"		3	2.667		625	159		784	930
3540	12" diameter x 14"	Q-5	4	4		950	215		1,165	1,375
6000	Molded rubber with helical wire reinforcement									
6010	150 PSI									
6020	1-1/2" diameter x 12"	1 Stpi	15	.533	Ea.	88	32		120	145
6030	2" diameter x 12"		12	.667		240	40		280	325
6040	3" diameter x 12"		8	1		251	60		311	365
6050	4" diameter x 12"		6	1.333		320	79.50		399.50	470
6060	6" diameter x 18"		4	2		475	120		595	705
6070	8" diameter x 24"		3	2.667		665	159		824	975
6080	10" diameter x 24"		2	4		810	239		1,049	1,250
6090	12" diameter x 24"	Q-5	3	5.333		925	287		1,212	1,450
7000	Molded teflon with stainless steel flanges									
7010	150 PSI									
7020	2-1/2" diameter x 3-3/16"	Q-1	7.80	2.051	Ea.	2,325	108		2,433	2,725
7030	3" diameter x 3-5/8"		6.50	2.462		1,625	130		1,755	1,975
7040	4" diameter x 3-5/8"		5	3.200		2,100	169		2,269	2,575
7050	6" diameter x 4"		4.30	3.721		2,950	197		3,147	3,550
7060	8" diameter x 6"		3.80	4.211		4,700	222		4,922	5,500

22 11 19.18 Mixing Valve

		Crew	Daily Output	Labor-Hours	Unit	Material	2015 Bare Costs Labor	2015 Bare Costs Equipment	Total	Total Incl O&P
0010	**MIXING VALVE**, Automatic, water tempering.									
0040	1/2" size	1 Stpi	19	.421	Ea.	555	25		580	650
0050	3/4" size		18	.444		555	26.50		581.50	650
0100	1" size		16	.500		845	30		875	975
0120	1-1/4" size		13	.615		1,150	37		1,187	1,325
0140	1-1/2" size		10	.800		1,400	48		1,448	1,600
0160	2" size		8	1		1,750	60		1,810	2,025
0170	2-1/2" size		6	1.333		1,750	79.50		1,829.50	2,050
0180	3" size		4	2		4,125	120		4,245	4,725
0190	4" size		3	2.667		4,125	159		4,284	4,800

22 11 19.22 Pressure Reducing Valve

		Crew	Daily Output	Labor-Hours	Unit	Material	2015 Bare Costs Labor	2015 Bare Costs Equipment	Total	Total Incl O&P
0010	**PRESSURE REDUCING VALVE**, Steam, pilot operated.									
0100	Threaded, iron body									
0200	1-1/2" size	1 Stpi	8	1	Ea.	1,825	60		1,885	2,100
0220	2" size	"	5	1.600	"	2,100	95.50		2,195.50	2,450
1000	Flanged, iron body, 125 lb. flanges									
1020	2" size	1 Stpi	8	1	Ea.	2,100	60		2,160	2,425
1040	2-1/2" size	"	4	2		2,525	120		2,645	2,975
1060	3" size	Q-5	4.50	3.556		3,075	191		3,266	3,675
1080	4" size	"	3	5.333		4,425	287		4,712	5,300
1500	For 250 lb. flanges, add					5%				

22 11 19.26 Pressure Regulators

		Crew	Daily Output	Labor-Hours	Unit	Material	2015 Bare Costs Labor	2015 Bare Costs Equipment	Total	Total Incl O&P
0010	**PRESSURE REGULATORS**									
0100	Gas appliance regulators									
0106	Main burner and pilot applications									
0108	Rubber seat poppet type									
0109	1/8" pipe size	1 Stpi	24	.333	Ea.	16	19.90		35.90	47.50
0110	1/4" pipe size		24	.333		18.40	19.90		38.30	50
0112	3/8" pipe size		24	.333		19.50	19.90		39.40	51.50
0113	1/2" pipe size		24	.333		21.50	19.90		41.40	54

22 11 19 – Domestic Water Piping Specialties

22 11 19.26 Pressure Regulators		Crew	Daily Output	Labor-Hours	Unit	Material	2015 Bare Costs Labor	Equipment	Total	Total Incl O&P
0114	3/4" pipe size	1 Stpi	20	.400	Ea.	25.50	24		49.50	64
0122	Lever action type									
0123	3/8" pipe size	1 Stpi	24	.333	Ea.	23.50	19.90		43.40	56
0124	1/2" pipe size		24	.333		23.50	19.90		43.40	56
0125	3/4" pipe size		20	.400		46.50	24		70.50	87
0126	1" pipe size	↓	19	.421	↓	46.50	25		71.50	89
0132	Double diaphragm type									
0133	3/8" pipe size	1 Stpi	24	.333	Ea.	35.50	19.90		55.40	69.50
0134	1/2" pipe size		24	.333		49	19.90		68.90	84
0135	3/4" pipe size		20	.400		82	24		106	127
0136	1" pipe size		19	.421		100	25		125	148
0137	1-1/4" pipe size		15	.533		350	32		382	435
0138	1-1/2" pipe size		13	.615		640	37		677	760
0139	2" pipe size	↓	11	.727		640	43.50		683.50	770
0140	2-1/2" pipe size	Q-5	15	1.067		1,250	57.50		1,307.50	1,450
0141	3" pipe size		13	1.231		1,250	66		1,316	1,475
0142	4" pipe size (flanged)	↓	8	2	↓	2,225	108		2,333	2,575
0160	Main burner only									
0162	Straight-thru-flow design									
0163	1/2" pipe size	1 Stpi	24	.333	Ea.	40	19.90		59.90	74
0164	3/4" pipe size		20	.400		54.50	24		78.50	96
0165	1" pipe size		19	.421		77.50	25		102.50	124
0166	1-1/4" pipe size	↓	15	.533	↓	77.50	32		109.50	134
0200	Oil, light, hot water, ordinary steam, threaded									
0220	Bronze body, 1/4" size	1 Stpi	24	.333	Ea.	154	19.90		173.90	199
0230	3/8" size		24	.333		166	19.90		185.90	213
0240	1/2" size		24	.333		206	19.90		225.90	257
0250	3/4" size		20	.400		246	24		270	305
0260	1" size		19	.421		375	25		400	450
0270	1-1/4" size		15	.533		505	32		537	605
0280	1-1/2" size		13	.615		540	37		577	650
0290	2" size		11	.727		985	43.50		1,028.50	1,150
0320	Iron body, 1/4" size		24	.333		120	19.90		139.90	162
0330	3/8" size		24	.333		141	19.90		160.90	185
0340	1/2" size		24	.333		150	19.90		169.90	195
0350	3/4" size		20	.400		182	24		206	236
0360	1" size		19	.421		237	25		262	299
0370	1-1/4" size		15	.533		330	32		362	410
0380	1-1/2" size		13	.615		365	37		402	460
0390	2" size	↓	11	.727	↓	530	43.50		573.50	650
0500	Oil, heavy, viscous fluids, threaded									
0520	Bronze body, 3/8" size	1 Stpi	24	.333	Ea.	264	19.90		283.90	320
0530	1/2" size		24	.333		340	19.90		359.90	405
0540	3/4" size		20	.400		380	24		404	450
0550	1" size		19	.421		470	25		495	560
0560	1-1/4" size		15	.533		665	32		697	785
0570	1-1/2" size		13	.615		760	37		797	895
0600	Iron body, 3/8" size		24	.333		210	19.90		229.90	261
0620	1/2" size		24	.333		255	19.90		274.90	310
0630	3/4" size		20	.400		289	24		313	355
0640	1" size		19	.421		340	25		365	415
0650	1-1/4" size		15	.533		495	32		527	595
0660	1-1/2" size	↓	13	.615		530	37		567	635

For customer support on your Mechanical Cost Data, call 877.795.7394.

233

22 11 19.26 Pressure Regulators		Crew	Daily Output	Labor-Hours	Unit	Material	2015 Bare Costs Labor	Equipment	Total	Total Incl O&P
0800	Process steam, wet or super heated, monel trim, threaded									
0820	Bronze body, 1/4" size	1 Stpi	24	.333	Ea.	600	19.90		619.90	690
0830	3/8" size		24	.333		600	19.90		619.90	690
0840	1/2" size		24	.333		690	19.90		709.90	790
0850	3/4" size		20	.400		810	24		834	930
0860	1" size		19	.421		1,025	25		1,050	1,175
0870	1-1/4" size		15	.533		1,275	32		1,307	1,450
0880	1-1/2" size		13	.615		1,525	37		1,562	1,725
0920	Iron body, max 125 PSIG press out, 1/4" size		24	.333		795	19.90		814.90	905
0930	3/8" size		24	.333		795	19.90		814.90	905
0940	1/2" size		24	.333		990	19.90		1,009.90	1,125
0950	3/4" size		20	.400		1,200	24		1,224	1,350
0960	1" size		19	.421		1,525	25		1,550	1,725
0970	1-1/4" size		15	.533		1,850	32		1,882	2,075
0980	1-1/2" size		13	.615		2,225	37		2,262	2,475
3000	Steam, high capacity, bronze body, stainless steel trim									
3020	Threaded, 1/2" diameter	1 Stpi	24	.333	Ea.	1,975	19.90		1,994.90	2,200
3030	3/4" diameter		24	.333		2,150	19.90		2,169.90	2,375
3040	1" diameter		19	.421		2,400	25		2,425	2,700
3060	1-1/4" diameter		15	.533		2,500	32		2,532	2,800
3080	1-1/2" diameter		13	.615		3,025	37		3,062	3,375
3100	2" diameter		11	.727		3,725	43.50		3,768.50	4,150
3120	2-1/2" diameter	Q-5	12	1.333		4,650	71.50		4,721.50	5,225
3140	3" diameter	"	11	1.455		5,075	78		5,153	5,700
3500	Flanged connection, iron body, 125 lb. W.S.P.									
3520	3" diameter	Q-5	11	1.455	Ea.	5,800	78		5,878	6,525
3540	4" diameter	"	5	3.200	"	7,325	172		7,497	8,300
9002	For water pressure regulators, see Section 22 05 23.20									

22 11 19.30 Pressure and Temperature Safety Plug

		Crew	Daily Output	Labor-Hours	Unit	Material	Labor	Equipment	Total	Total Incl O&P
0010	**PRESSURE & TEMPERATURE SAFETY PLUG**									
1000	3/4" external thread, 3/8" diam. element									
1020	Carbon steel									
1050	7-1/2" insertion	1 Stpi	32	.250	Ea.	45	14.95		59.95	72
1120	304 stainless steel									
1150	7-1/2" insertion	1 Stpi	32	.250	Ea.	50.50	14.95		65.45	78
1220	316 Stainless steel									
1250	7-1/2" insertion	1 Stpi	32	.250	Ea.	69.50	14.95		84.45	99

22 11 19.32 Pressure and Temperature Measurement Plug

		Crew	Daily Output	Labor-Hours	Unit	Material	Labor	Equipment	Total	Total Incl O&P
0010	**PRESSURE & TEMPERATURE MEASUREMENT PLUG**									
0020	A permanent access port for insertion of									
0030	a pressure or temperature measuring probe									
0100	Plug, brass									
0110	1/4" MNPT, 1-1/2" long	1 Stpi	32	.250	Ea.	5.40	14.95		20.35	28.50
0120	3" long		31	.258		10.45	15.40		25.85	35
0140	1/2" MNPT, 1-1/2" long		30	.267		8.65	15.95		24.60	33.50
0150	3" long		29	.276		12.90	16.50		29.40	39
0200	Pressure gauge probe adapter									
0210	1/8" diameter, 1-1/2" probe				Ea.	26			26	28.50
0220	3" probe				"	53			53	58
0300	Temperature gauge, 5" stem									
0310	Analog				Ea.	69.50			69.50	76.50
0330	Digital				"	75.50			75.50	83

234

For customer support on your Mechanical Cost Data, call 877.795.7394.

22 11 Facility Water Distribution

22 11 19 – Domestic Water Piping Specialties

22 11 19.32 Pressure and Temperature Measurement Plug	Crew	Daily Output	Labor-Hours	Unit	Material	2015 Bare Costs Labor	Equipment	Total	Total Incl O&P
0400 Pressure gauge, compound									
0410 1/4" MNPT				Ea.	30.50			30.50	33.50
0500 Pressure and temperature test kit									
0510 Contains 2 thermometers and 2 pressure gauges									
0520 Kit				Ea.	355			355	390

22 11 19.34 Sleeves and Escutcheons

	Crew	Daily Output	Labor-Hours	Unit	Material	2015 Bare Costs Labor	Equipment	Total	Total Incl O&P
0010 **SLEEVES & ESCUTCHEONS**									
0100 Pipe sleeve									
0110 Steel, w/water stop, 12" long, with link seal									
0120 2" diam. for 1/2" carrier pipe	1 Plum	8.40	.952	Ea.	59.50	56		115.50	150
0130 2-1/2" diam. for 3/4" carrier pipe		8	1		67	58.50		125.50	163
0140 2-1/2" diam. for 1" carrier pipe		8	1		63.50	58.50		122	159
0150 3" diam. for 1-1/4" carrier pipe		7.20	1.111		81.50	65		146.50	188
0160 3-1/2" diam. for 1-1/2" carrier pipe		6.80	1.176		82	69		151	194
0170 4" diam. for 2" carrier pipe		6	1.333		88.50	78.50		167	216
0180 4" diam. for 2-1/2" carrier pipe		6	1.333		89.50	78.50		168	217
0190 5" diam. for 3" carrier pipe		5.40	1.481		102	87		189	243
0200 6" diam. for 4" carrier pipe	▼	4.80	1.667		122	98		220	283
0210 10" diam. for 6" carrier pipe	Q-1	8	2		244	106		350	430
0220 12" diam. for 8" carrier pipe		7.20	2.222		325	117		442	535
0230 14" diam. for 10" carrier pipe		6.40	2.500		345	132		477	575
0240 16" diam. for 12" carrier pipe		5.80	2.759		380	146		526	640
0250 18" diam. for 14" carrier pipe		5.20	3.077		705	163		868	1,025
0260 24" diam. for 18" carrier pipe		4	4		1,050	211		1,261	1,475
0270 24" diam. for 20" carrier pipe		4	4		885	211		1,096	1,300
0280 30" diam. for 24" carrier pipe	▼	3.20	5	▼	1,300	264		1,564	1,825
0500 Wall sleeve									
0510 Ductile iron with rubber gasket seal									
0520 3"	1 Plum	8.40	.952	Ea.	700	56		756	855
0530 4"		7.20	1.111		750	65		815	925
0540 6"		6	1.333		920	78.50		998.50	1,150
0550 8"		4	2		1,125	117		1,242	1,425
0560 10"		3	2.667		1,375	157		1,532	1,750
0570 12"	▼	2.40	3.333		1,625	196		1,821	2,100
5000 Escutcheon									
5100 Split ring, pipe									
5110 Chrome plated									
5120 1/2"	1 Plum	160	.050	Ea.	1.09	2.94		4.03	5.65
5130 3/4"		160	.050		1.21	2.94		4.15	5.75
5140 1"		135	.059		1.30	3.48		4.78	6.70
5150 1-1/2"		115	.070		1.83	4.08		5.91	8.15
5160 2"		100	.080		2.03	4.70		6.73	9.35
5170 4"		80	.100		4.50	5.85		10.35	13.80
5180 6"	▼	68	.118	▼	7.50	6.90		14.40	18.70
5400 Shallow flange type									
5410 Chrome plated steel									
5420 1/2" CTS	1 Plum	180	.044	Ea.	.32	2.61		2.93	4.29
5430 3/4" CTS		180	.044		.34	2.61		2.95	4.31
5440 1/2" IPS		180	.044		.36	2.61		2.97	4.34
5450 3/4" IPS		180	.044		.40	2.61		3.01	4.38
5460 1" IPS		175	.046		.46	2.68		3.14	4.56
5470 1-1/2" IPS	▼	170	.047	▼	.76	2.76		3.52	5

22 11 19 – Domestic Water Piping Specialties

22 11 19.34 Sleeves and Escutcheons	Crew	Daily Output	Labor-Hours	Unit	Material	2015 Bare Costs Labor	2015 Bare Costs Equipment	Total	Total Incl O&P
5480	2" IPS	1 Plum	160	.050	Ea.	.91	2.94		3.85	5.45

22 11 19.38 Water Supply Meters

		Crew	Daily Output	Labor-Hours	Unit	Material	Labor	Equipment	Total	Total Incl O&P
0010	**WATER SUPPLY METERS**									
1000	Detector, serves dual systems such as fire and domestic or									
1020	process water, wide range cap., UL and FM approved									
1100	3" mainline x 2" by-pass, 400 GPM	Q-1	3.60	4.444	Ea.	7,450	235		7,685	8,525
1140	4" mainline x 2" by-pass, 700 GPM	"	2.50	6.400		7,450	340		7,790	8,675
1180	6" mainline x 3" by-pass, 1600 GPM	Q-2	2.60	9.231		11,400	505		11,905	13,300
1220	8" mainline x 4" by-pass, 2800 GPM		2.10	11.429		16,900	625		17,525	19,400
1260	10" mainline x 6" by-pass, 4400 GPM		2	12		24,100	655		24,755	27,500
1300	10" x 12" mainlines x 6" by-pass, 5400 GPM		1.70	14.118		32,700	775		33,475	37,200
2000	Domestic/commercial, bronze									
2020	Threaded									
2060	5/8" diameter, to 20 GPM	1 Plum	16	.500	Ea.	50	29.50		79.50	99.50
2080	3/4" diameter, to 30 GPM		14	.571		91	33.50		124.50	151
2100	1" diameter, to 50 GPM		12	.667		138	39		177	211
2300	Threaded/flanged									
2340	1-1/2" diameter, to 100 GPM	1 Plum	8	1	Ea.	340	58.50		398.50	460
2360	2" diameter, to 160 GPM	"	6	1.333	"	460	78.50		538.50	625
2600	Flanged, compound									
2640	3" diameter, 320 GPM	Q-1	3	5.333	Ea.	3,125	282		3,407	3,850
2660	4" diameter, to 500 GPM		1.50	10.667		5,000	565		5,565	6,350
2680	6" diameter, to 1,000 GPM		1	16		7,975	845		8,820	10,000
2700	8" diameter, to 1,800 GPM		.80	20		12,500	1,050		13,550	15,300
7000	Turbine									
7260	Flanged									
7300	2" diameter, to 160 GPM	1 Plum	7	1.143	Ea.	610	67		677	770
7320	3" diameter, to 450 GPM	Q-1	3.60	4.444		1,275	235		1,510	1,750
7340	4" diameter, to 650 GPM	"	2.50	6.400		2,100	340		2,440	2,825
7360	6" diameter, to 1800 GPM	Q-2	2.60	9.231		3,775	505		4,280	4,925
7380	8" diameter, to 2500 GPM		2.10	11.429		5,975	625		6,600	7,525
7400	10" diameter, to 5500 GPM		1.70	14.118		8,075	775		8,850	10,100

22 11 19.42 Backflow Preventers

		Crew	Daily Output	Labor-Hours	Unit	Material	Labor	Equipment	Total	Total Incl O&P
0010	**BACKFLOW PREVENTERS**, Includes valves									
0020	and four test cocks, corrosion resistant, automatic operation									
1000	Double check principle									
1010	Threaded, with ball valves									
1020	3/4" pipe size	1 Plum	16	.500	Ea.	231	29.50		260.50	299
1030	1" pipe size		14	.571		264	33.50		297.50	340
1040	1-1/2" pipe size		10	.800		560	47		607	690
1050	2" pipe size		7	1.143		655	67		722	820
1080	Threaded, with gate valves									
1100	3/4" pipe size	1 Plum	16	.500	Ea.	1,075	29.50		1,104.50	1,250
1120	1" pipe size		14	.571		1,100	33.50		1,133.50	1,250
1140	1-1/2" pipe size		10	.800		1,400	47		1,447	1,625
1160	2" pipe size		7	1.143		1,725	67		1,792	2,000
1200	Flanged, valves are gate									
1210	3" pipe size	Q-1	4.50	3.556	Ea.	2,800	188		2,988	3,350
1220	4" pipe size	"	3	5.333		2,900	282		3,182	3,625
1230	6" pipe size	Q-2	3	8		4,275	440		4,715	5,350
1240	8" pipe size		2	12		7,775	655		8,430	9,550
1250	10" pipe size		1	24		11,500	1,325		12,825	14,700

For customer support on your Mechanical Cost Data, call 877.795.7394.

22 11 19 – Domestic Water Piping Specialties

22 11 19.42 Backflow Preventers	Crew	Daily Output	Labor-Hours	Unit	Material	2015 Bare Costs Labor	Equipment	Total	Total Incl O&P
1300	Flanged, valves are OS&Y								
1370	1" pipe size	1 Plum	5	1.600	Ea.	1,200	94		1,294	1,450
1374	1-1/2" pipe size		5	1.600		1,525	94		1,619	1,825
1378	2" pipe size		4.80	1.667		1,825	98		1,923	2,175
1380	3" pipe size	Q-1	4.50	3.556		2,875	188		3,063	3,425
1400	4" pipe size	"	3	5.333		3,475	282		3,757	4,250
1420	6" pipe size	Q-2	3	8		5,450	440		5,890	6,650
1430	8" pipe size	"	2	12		10,800	655		11,455	12,800
4000	Reduced pressure principle									
4100	Threaded, bronze, valves are ball									
4120	3/4" pipe size	1 Plum	16	.500	Ea.	445	29.50		474.50	535
4140	1" pipe size		14	.571		480	33.50		513.50	575
4150	1-1/4" pipe size		12	.667		845	39		884	990
4160	1-1/2" pipe size		10	.800		960	47		1,007	1,125
4180	2" pipe size		7	1.143		1,075	67		1,142	1,300
5000	Flanged, bronze, valves are OS&Y									
5060	2-1/2" pipe size	Q-1	5	3.200	Ea.	4,025	169		4,194	4,675
5080	3" pipe size		4.50	3.556		4,650	188		4,838	5,375
5100	4" pipe size		3	5.333		5,500	282		5,782	6,475
5120	6" pipe size	Q-2	3	8		8,750	440		9,190	10,300
5200	Flanged, iron, valves are gate									
5210	2-1/2" pipe size	Q-1	5	3.200	Ea.	2,650	169		2,819	3,175
5220	3" pipe size		4.50	3.556		2,775	188		2,963	3,325
5230	4" pipe size		3	5.333		3,750	282		4,032	4,550
5240	6" pipe size	Q-2	3	8		5,275	440		5,715	6,475
5250	8" pipe size		2	12		9,475	655		10,130	11,400
5260	10" pipe size		1	24		13,400	1,325		14,725	16,700
5600	Flanged, iron, valves are OS&Y									
5660	2-1/2" pipe size	Q-1	5	3.200	Ea.	3,025	169		3,194	3,575
5680	3" pipe size		4.50	3.556		3,175	188		3,363	3,775
5700	4" pipe size		3	5.333		3,975	282		4,257	4,800
5720	6" pipe size	Q-2	3	8		5,775	440		6,215	7,000
5740	8" pipe size		2	12		10,100	655		10,755	12,100
5760	10" pipe size		1	24		13,600	1,325		14,925	16,900

22 11 19.50 Vacuum Breakers

		Crew	Daily Output	Labor-Hours	Unit	Material	2015 Bare Costs Labor	Equipment	Total	Total Incl O&P
0010	**VACUUM BREAKERS**									
0013	See also backflow preventers Section 22 11 19.42									
1000	Anti-siphon continuous pressure type									
1010	Max. 150 PSI - 210°F									
1020	Bronze body									
1030	1/2" size	1 Stpi	24	.333	Ea.	187	19.90		206.90	236
1040	3/4" size		20	.400		187	24		211	242
1050	1" size		19	.421		194	25		219	251
1060	1-1/4" size		15	.533		380	32		412	470
1070	1-1/2" size		13	.615		470	37		507	570
1080	2" size		11	.727		485	43.50		528.50	595
1200	Max. 125 PSI with atmospheric vent									
1210	Brass, in-line construction									
1220	1/4" size	1 Stpi	24	.333	Ea.	117	19.90		136.90	159
1230	3/8" size	"	24	.333		117	19.90		136.90	159
1260	For polished chrome finish, add					13%				
2000	Anti-siphon, non-continuous pressure type									

For customer support on your Mechanical Cost Data, call 877.795.7394.

237

22 11 19 – Domestic Water Piping Specialties

22 11 19.50 Vacuum Breakers	Crew	Daily Output	Labor-Hours	Unit	Material	2015 Bare Costs Labor	Equipment	Total	Total Incl O&P
2010 Hot or cold water 125 PSI - 210°F									
2020 Bronze body									
2030 1/4" size	1 Stpi	24	.333	Ea.	64	19.90		83.90	101
2040 3/8" size		24	.333		64	19.90		83.90	101
2050 1/2" size		24	.333		72.50	19.90		92.40	110
2060 3/4" size		20	.400		86.50	24		110.50	131
2070 1" size		19	.421		134	25		159	185
2080 1-1/4" size		15	.533		235	32		267	305
2090 1-1/2" size		13	.615		276	37		313	360
2100 2" size		11	.727		430	43.50		473.50	535
2110 2-1/2" size		8	1		1,225	60		1,285	1,450
2120 3" size	↓	6	1.333	↓	1,625	79.50		1,704.50	1,925
2150 For polished chrome finish, add					50%				
3000 Air gap fitting									
3020 1/2" NPT size	1 Plum	19	.421	Ea.	50	24.50		74.50	92.50
3030 1" NPT size	"	15	.533		57.50	31.50		89	111
3040 2" NPT size	Q-1	21	.762		113	40.50		153.50	185
3050 3" NPT size		14	1.143		223	60.50		283.50	335
3060 4" NPT size	↓	10	1.600	↓	223	84.50		307.50	375

22 11 19.54 Water Hammer Arresters/Shock Absorbers

	Crew	Daily Output	Labor-Hours	Unit	Material	Labor	Equipment	Total	Total Incl O&P
0010 **WATER HAMMER ARRESTERS/SHOCK ABSORBERS**									
0490 Copper									
0500 3/4" male I.P.S. For 1 to 11 fixtures	1 Plum	12	.667	Ea.	28	39		67	90
0600 1" male I.P.S. For 12 to 32 fixtures		8	1		45.50	58.50		104	139
0700 1-1/4" male I.P.S. For 33 to 60 fixtures		8	1		47	58.50		105.50	140
0800 1-1/2" male I.P.S. For 61 to 113 fixtures		8	1		67.50	58.50		126	163
0900 2" male I.P.S. For 114 to 154 fixtures		8	1		98.50	58.50		157	197
1000 2-1/2" male I.P.S. For 155 to 330 fixtures	↓	4	2	↓	305	117		422	510
4000 Bellows type									
4010 3/4" FNPT, to 11 fixture units	1 Plum	10	.800	Ea.	253	47		300	350
4020 1" FNPT, to 32 fixture units		8.80	.909		510	53.50		563.50	640
4030 1" FNPT, to 60 fixture units		8.80	.909		770	53.50		823.50	925
4040 1" FNPT, to 113 fixture units		8.80	.909		1,925	53.50		1,978.50	2,200
4050 1" FNPT, to 154 fixture units		6.60	1.212		2,175	71		2,246	2,500
4060 1-1/2" FNPT, to 300 fixture units	↓	5	1.600	↓	2,675	94		2,769	3,100

22 11 19.64 Hydrants

	Crew	Daily Output	Labor-Hours	Unit	Material	Labor	Equipment	Total	Total Incl O&P
0010 **HYDRANTS**									
0050 Wall type, moderate climate, bronze, encased									
0200 3/4" IPS connection	1 Plum	16	.500	Ea.	745	29.50		774.50	860
0300 1" IPS connection		14	.571		850	33.50		883.50	985
0500 Anti-siphon type, 3/4" connection	↓	16	.500	↓	640	29.50		669.50	750
1000 Non-freeze, bronze, exposed									
1100 3/4" IPS connection, 4" to 9" thick wall	1 Plum	14	.571	Ea.	500	33.50		533.50	600
1120 10" to 14" thick wall		12	.667		545	39		584	660
1140 15" to 19" thick wall		12	.667		605	39		644	725
1160 20" to 24" thick wall	↓	10	.800	↓	640	47		687	775
1200 For 1" IPS connection, add					15%	10%			
1240 For 3/4" adapter type vacuum breaker, add				Ea.	63			63	69.50
1280 For anti-siphon type, add				"	132			132	145
2000 Non-freeze bronze, encased, anti-siphon type									
2100 3/4" IPS connection, 5" to 9" thick wall	1 Plum	14	.571	Ea.	1,275	33.50		1,308.50	1,450
2120 10" to 14" thick wall	↓	12	.667	↓	1,300	39		1,339	1,500

22 11 19 – Domestic Water Piping Specialties

22 11 19.64 Hydrants	Crew	Daily Output	Labor-Hours	Unit	Material	2015 Bare Costs Labor	Equipment	Total	Total Incl O&P
2140	15" to 19" thick wall	1 Plum	12	.667	Ea.	1,375	39		1,414	1,550
2160	20" to 24" thick wall		10	.800		1,425	47		1,472	1,650
2200	For 1" IPS connection, add					10%	10%			
3000	Ground box type, bronze frame, 3/4" IPS connection									
3080	Non-freeze, all bronze, polished face, set flush									
3100	2 feet depth of bury	1 Plum	8	1	Ea.	950	58.50		1,008.50	1,150
3120	3 feet depth of bury		8	1		1,025	58.50		1,083.50	1,225
3140	4 feet depth of bury		8	1		1,075	58.50		1,133.50	1,300
3160	5 feet depth of bury		7	1.143		1,175	67		1,242	1,375
3180	6 feet depth of bury		7	1.143		1,250	67		1,317	1,475
3200	7 feet depth of bury		6	1.333		1,300	78.50		1,378.50	1,575
3220	8 feet depth of bury		5	1.600		1,375	94		1,469	1,675
3240	9 feet depth of bury		4	2		1,450	117		1,567	1,775
3260	10 feet depth of bury		4	2		1,525	117		1,642	1,850
3400	For 1" IPS connection, add					15%	10%			
3450	For 1-1/4" IPS connection, add					325%	14%			
3500	For 1-1/2" connection, add					370%	18%			
3550	For 2" connection, add					445%	24%			
3600	For tapped drain port in box, add					86			86	94.50
4000	Non-freeze, CI body, bronze frame & scoriated cover									
4010	with hose storage									
4100	2 feet depth of bury	1 Plum	7	1.143	Ea.	1,750	67		1,817	2,025
4120	3 feet depth of bury		7	1.143		1,825	67		1,892	2,100
4140	4 feet depth of bury		7	1.143		1,875	67		1,942	2,175
4160	5 feet depth of bury		6.50	1.231		1,925	72.50		1,997.50	2,200
4180	6 feet depth of bury		6	1.333		1,950	78.50		2,028.50	2,275
4200	7 feet depth of bury		5.50	1.455		2,025	85.50		2,110.50	2,350
4220	8 feet depth of bury		5	1.600		2,100	94		2,194	2,450
4240	9 feet depth of bury		4.50	1.778		2,150	104		2,254	2,525
4260	10 feet depth of bury		4	2		2,225	117		2,342	2,625
4280	For 1" IPS connection, add					390			390	425
4300	For tapped drain port in box, add					86			86	94.50
5000	Moderate climate, all bronze, polished face									
5020	and scoriated cover, set flush									
5100	3/4" IPS connection	1 Plum	16	.500	Ea.	655	29.50		684.50	770
5120	1" IPS connection	"	14	.571		810	33.50		843.50	940
5200	For tapped drain port in box, add					86			86	94.50
6000	Ground post type, all non-freeze, all bronze, aluminum casing									
6010	guard, exposed head, 3/4" IPS connection									
6100	2 feet depth of bury	1 Plum	8	1	Ea.	925	58.50		983.50	1,125
6120	3 feet depth of bury		8	1		1,000	58.50		1,058.50	1,200
6140	4 feet depth of bury		8	1		1,075	58.50		1,133.50	1,275
6160	5 feet depth of bury		7	1.143		1,150	67		1,217	1,375
6180	6 feet depth of bury		7	1.143		1,225	67		1,292	1,450
6200	7 feet depth of bury		6	1.333		1,325	78.50		1,403.50	1,575
6220	8 feet depth of bury		5	1.600		1,400	94		1,494	1,700
6240	9 feet depth of bury		4	2		1,500	117		1,617	1,825
6260	10 feet depth of bury		4	2		1,575	117		1,692	1,900
6300	For 1" IPS connection, add					40%	10%			
6350	For 1-1/4" IPS connection, add					140%	14%			
6400	For 1-1/2" IPS connection, add					225%	18%			
6450	For 2" IPS connection, add					315%	24%			

22 11 23.10 General Utility Pumps	Crew	Daily Output	Labor-Hours	Unit	Material	2015 Bare Costs Labor	Equipment	Total	Total Incl O&P
0010 **GENERAL UTILITY PUMPS**									
2000 Single stage									
3000 Double suction,									
3140 50 HP, 5"D. x 6"S.	Q-2	.33	72.727	Ea.	9,175	3,975		13,150	16,100
3180 60 HP, 6"D. x 8"S.	Q-3	.30	106		13,400	5,950		19,350	23,700
3190 75 HP, to 2500 GPM		.28	114		20,100	6,375		26,475	31,800
3220 100 HP, to 3000 GPM		.26	123		25,500	6,875		32,375	38,500
3240 150 HP, to 4000 GPM		.24	133		36,000	7,450		43,450	51,000
4000 Centrifugal, end suction, mounted on base									
4010 Horizontal mounted, with drip proof motor, rated @ 100' head									
4020 Vertical split case, single stage									
4040 100 GPM, 5 HP, 1-1/2" discharge	Q-1	1.70	9.412	Ea.	3,675	495		4,170	4,775
4050 200 GPM, 10 HP, 2" discharge		1.30	12.308		4,000	650		4,650	5,375
4060 250 GPM, 10 HP, 3" discharge		1.28	12.500		4,450	660		5,110	5,900
4070 300 GPM, 15 HP, 2" discharge	Q-2	1.56	15.385		5,225	845		6,070	7,025
4080 500 GPM, 20 HP, 4" discharge		1.44	16.667		6,025	915		6,940	8,000
4090 750 GPM, 30 HP, 4" discharge		1.20	20		6,350	1,100		7,450	8,625
4100 1050 GPM, 40 HP, 5" discharge		1	24		9,175	1,325		10,500	12,100
4110 1500 GPM, 60 HP, 6" discharge		.60	40		13,400	2,200		15,600	18,000
4120 2000 GPM, 75 HP, 6" discharge		.50	48		14,600	2,625		17,225	20,000
4130 3000 GPM, 100 HP, 8" discharge		.40	60		18,900	3,275		22,175	25,800
4200 Horizontal split case, single stage									
4210 100 GPM, 7.5 HP, 1-1/2" discharge	Q-1	1.70	9.412	Ea.	3,725	495		4,220	4,850
4220 250 GPM, 15 HP, 2-1/2" discharge	"	1.30	12.308		7,900	650		8,550	9,650
4230 500 GPM, 20 HP, 4" discharge	Q-2	1.60	15		9,625	820		10,445	11,900
4240 750 GPM, 25 HP, 5" discharge		1.54	15.584		10,100	855		10,955	12,400
4250 1000 GPM, 40 HP, 5" discharge		1.20	20		14,000	1,100		15,100	17,000
4260 1500 GPM, 50 HP, 6" discharge	Q-3	1.42	22.535		16,400	1,250		17,650	20,000
4270 2000 GPM, 75 HP, 8" discharge		1.14	28.070		22,300	1,575		23,875	27,000
4280 3000 GPM, 100 HP, 10" discharge		.96	33.333		25,000	1,875		26,875	30,300
4290 3500 GPM, 150 HP, 10" discharge		.86	37.209		32,900	2,075		34,975	39,400
4300 4000 GPM, 200 HP, 10" discharge		.66	48.485		39,300	2,700		42,000	47,300
4330 Horizontal split case, two stage, 500' head									
4340 100 GPM, 40 HP, 1-1/2" discharge	Q-2	1.70	14.118	Ea.	17,000	775		17,775	19,900
4350 200 GPM, 50 HP, 1-1/2" discharge	"	1.44	16.667		17,800	915		18,715	21,000
4360 300 GPM, 75 HP, 2" discharge	Q-3	1.57	20.382		24,500	1,150		25,650	28,600
4370 400 GPM, 100 HP, 3" discharge		1.14	28.070		28,900	1,575		30,475	34,200
4380 800 GPM, 200 HP, 4" discharge		.86	37.209		38,300	2,075		40,375	45,300
5000 Centrifugal, in-line									
5006 Vertical mount, iron body, 125 lb. flgd, 1800 RPM TEFC mtr									
5010 Single stage									
5012 .5 HP, 1-1/2" suction & discharge	Q-1	3.20	5	Ea.	1,350	264		1,614	1,900
5014 .75 HP, 2" suction & discharge		2.80	5.714		2,600	300		2,900	3,300
5015 1 HP, 3" suction & discharge		2.50	6.400		2,900	340		3,240	3,700
5016 1.5 HP, 3" suction & discharge		2.30	6.957		2,950	370		3,320	3,800
5018 2 HP, 4" suction & discharge		2.10	7.619		3,825	405		4,230	4,800
5020 3 HP, 5" suction & discharge		1.90	8.421		4,475	445		4,920	5,600
5030 5 HP, 6" suction & discharge		1.60	10		5,100	530		5,630	6,400
5040 7.5 HP, 6" suction & discharge		1.30	12.308		5,850	650		6,500	7,425
5050 10 HP, 8" suction & discharge	Q-2	1.80	13.333		7,800	730		8,530	9,675
5060 15 HP, 8" suction & discharge		1.70	14.118		9,125	775		9,900	11,200
5064 20 HP, 8" suction & discharge		1.60	15		10,600	820		11,420	12,900

22 11 23 – Domestic Water Pumps

22 11 23.10 General Utility Pumps

		Crew	Daily Output	Labor-Hours	Unit	Material	2015 Bare Costs Labor	2015 Bare Costs Equipment	Total	Total Incl O&P
5070	30 HP, 8" suction & discharge	Q-2	1.50	16	Ea.	13,500	875		14,375	16,100
5080	40 HP, 8" suction & discharge		1.40	17.143		16,200	940		17,140	19,200
5090	50 HP, 8" suction & discharge		1.30	18.462		16,400	1,000		17,400	19,500
5094	60 HP, 8" suction & discharge		.90	26.667		17,200	1,450		18,650	21,200
5100	75 HP, 8" suction & discharge	Q-3	.60	53.333		19,700	2,975		22,675	26,200
5110	100 HP, 8" suction & discharge	"	.50	64		22,500	3,575		26,075	30,100

22 11 23.11 Miscellaneous Pumps

		Crew	Daily Output	Labor-Hours	Unit	Material	2015 Bare Costs Labor	2015 Bare Costs Equipment	Total	Total Incl O&P
0010	**MISCELLANEOUS PUMPS**									
0020	Water pump, portable, gasoline powered									
0100	170 GPH, 2" discharge	Q-1	11	1.455	Ea.	670	77		747	850
0110	343 GPH, 3" discharge		10.50	1.524		755	80.50		835.50	950
0120	608 GPH, 4" discharge		10	1.600		1,075	84.50		1,159.50	1,325
0500	Pump, propylene body, housing and impeller									
0510	22 GPM, 1/3 HP, 40' HD	1 Plum	5	1.600	Ea.	860	94		954	1,075
0520	33 GPM, 1/2 HP, 40' HD	Q-1	5	3.200		790	169		959	1,125
0530	53 GPM, 3/4 HP, 40' HD	"	4	4		1,275	211		1,486	1,725
0600	Rotary pump, CI									
0614	282 GPH, 3/4 HP, 1" discharge	Q-1	4.50	3.556	Ea.	1,225	188		1,413	1,625
0618	277 GPH, 1 HP, 1" discharge		4	4		1,225	211		1,436	1,675
0624	1100 GPH, 1.5 HP, 1-1/4" discharge		3.60	4.444		3,225	235		3,460	3,900
0628	1900 GPH, 2 HP, 1-1/4" discharge		3.20	5		4,550	264		4,814	5,425
1000	Turbine pump, CI									
1010	50 GPM, 2 HP, 3" discharge	Q-1	.80	20	Ea.	3,675	1,050		4,725	5,625
1020	100 GPM, 3 HP, 4" discharge	Q-2	.96	25		6,050	1,375		7,425	8,725
1030	250 GPM, 15 HP, 6" discharge	"	.94	25.532		7,500	1,400		8,900	10,400
1040	500 GPM, 25 HP, 6" discharge	Q-3	1.22	26.230		7,550	1,475		9,025	10,600
1050	1000 GPM, 50 HP, 8" discharge		1.14	28.070		10,400	1,575		11,975	13,900
1060	2000 GPM, 100 HP, 10" discharge		1	32		12,400	1,800		14,200	16,300
1070	3000 GPM, 150 HP, 10" discharge		.80	40		19,900	2,225		22,125	25,300
1080	4000 GPM, 200 HP, 12" discharge		.70	45.714		21,700	2,550		24,250	27,800
1090	6000 GPM, 300 HP, 14" discharge		.60	53.333		33,200	2,975		36,175	41,000
1100	10,000 GPM, 300 HP, 18" discharge		.58	55.172		38,300	3,075		41,375	46,900
2000	Centrifugal stainless steel pumps									
2100	100 GPM, 100' TDH, 3 HP	Q-1	1.80	8.889	Ea.	10,900	470		11,370	12,700
2130	250 GPM, 100' TDH, 10 HP	"	1.28	12.500		15,300	660		15,960	17,800
2160	500 GPM, 100' TDH, 20 HP	Q-2	1.44	16.667		17,500	915		18,415	20,600
2200	Vertical turbine stainless steel pumps									
2220	100 GPM, 100' TDH, 7.5 HP	Q-2	.95	25.263	Ea.	61,000	1,375		62,375	69,500
2240	250 GPM, 100' TDH, 15 HP		.94	25.532		68,000	1,400		69,400	76,500
2260	500 GPM, 100' TDH, 20 HP		.93	25.806		74,500	1,425		75,925	84,000
2280	750 GPM, 100' TDH, 30 HP	Q-3	1.19	26.891		81,000	1,500		82,500	91,500
2300	1000 GPM, 100' TDH, 40 HP	"	1.16	27.586		87,500	1,550		89,050	98,500
2320	100 GPM, 200' TDH, 15 HP	Q-2	.94	25.532		94,000	1,400		95,400	105,500
2340	250 GPM, 200' TDH, 25 HP	Q-3	1.22	26.230		100,500	1,475		101,975	112,500
2360	500 GPM, 200' TDH, 40 HP		1.16	27.586		107,000	1,550		108,550	120,500
2380	750 GPM, 200' TDH, 60 HP		1.13	28.319		108,000	1,575		109,575	121,000
2400	1000 GPM, 200' TDH, 75 HP		1.12	28.571		111,500	1,600		113,100	125,000

22 11 23.13 Domestic-Water Packaged Booster Pumps

		Crew	Daily Output	Labor-Hours	Unit	Material	2015 Bare Costs Labor	2015 Bare Costs Equipment	Total	Total Incl O&P
0010	**DOMESTIC-WATER PACKAGED BOOSTER PUMPS**									
0200	Pump system, with diaphragm tank, control, press. switch									
0300	1 HP pump	Q-1	1.30	12.308	Ea.	6,500	650		7,150	8,125
0400	1-1/2 HP pump		1.25	12.800		6,550	675		7,225	8,250

For customer support on your Mechanical Cost Data, call 877.795.7394.

241

22 11 Facility Water Distribution

22 11 23 – Domestic Water Pumps

22 11 23.13 Domestic-Water Packaged Booster Pumps	Crew	Daily Output	Labor-Hours	Unit	Material	2015 Bare Costs Labor	Equipment	Total	Total Incl O&P
0420	2 HP pump	Q-1	1.20	13.333	Ea.	6,725	705		7,430	8,475
0440	3 HP pump	↓	1.10	14.545		6,825	770		7,595	8,650
0460	5 HP pump	Q-2	1.50	16		7,550	875		8,425	9,625
0480	7-1/2 HP pump		1.42	16.901		8,400	925		9,325	10,700
0500	10 HP pump	↓	1.34	17.910	↓	8,800	980		9,780	11,200
2000	Pump system, variable speed, base, controls, starter									
2010	Duplex, 100' head									
2020	400 GPM, 7-1/2HP, 4" discharge	Q-2	.70	34.286	Ea.	38,900	1,875		40,775	45,500
2025	Triplex, 100' head									
2030	1000 GPM, 15HP, 6" discharge	Q-2	.50	48	Ea.	56,500	2,625		59,125	66,500
2040	1700 GPM, 30HP, 6" discharge	"	.30	80	"	68,500	4,375		72,875	81,500

22 12 Facility Potable-Water Storage Tanks

22 12 21 – Facility Underground Potable-Water Storage Tanks

22 12 21.13 Fiberglass, Underground Potable-Water Storage Tanks

		Crew	Daily Output	Labor-Hours	Unit	Material	2015 Bare Costs Labor	Equipment	Total	Total Incl O&P
0010	**FIBERGLASS, UNDERGROUND POTABLE-WATER STORAGE TANKS**									
0020	Excludes excavation, backfill, & piping									
0030	Single wall									
2000	600 gallon capacity	B-21B	3.75	10.667	Ea.	3,300	435	174	3,909	4,475
2010	1,000 gallon capacity		3.50	11.429		4,250	465	187	4,902	5,600
2020	2,000 gallon capacity		3.25	12.308		6,200	500	201	6,901	7,800
2030	4,000 gallon capacity		3	13.333		8,575	545	218	9,338	10,500
2040	6,000 gallon capacity		2.65	15.094		9,625	615	247	10,487	11,800
2050	8,000 gallon capacity		2.30	17.391		11,500	710	284	12,494	14,100
2060	10,000 gallon capacity		2	20		13,200	815	325	14,340	16,200
2070	12,000 gallon capacity		1.50	26.667		18,300	1,100	435	19,835	22,300
2080	15,000 gallon capacity		1	40		21,200	1,625	655	23,480	26,500
2090	20,000 gallon capacity		.75	53.333		27,300	2,175	870	30,345	34,300
2100	25,000 gallon capacity		.50	80		40,800	3,275	1,300	45,375	51,500
2110	30,000 gallon capacity		.35	114		49,400	4,675	1,875	55,950	63,500
2120	40,000 gallon capacity	↓	.30	133	↓	71,500	5,450	2,175	79,125	89,500

22 12 23 – Facility Indoor Potable-Water Storage Tanks

22 12 23.13 Facility Steel, Indoor Pot.-Water Storage Tanks

		Crew	Daily Output	Labor-Hours	Unit	Material	2015 Bare Costs Labor	Equipment	Total	Total Incl O&P
0010	**FACILITY STEEL, INDOOR POT.-WATER STORAGE TANKS**									
2000	Galvanized steel, 15 gal., 14" diam., 26" LOA	1 Plum	12	.667	Ea.	1,425	39		1,464	1,625
2060	30 gal., 14" diam. x 49" LOA		11	.727		1,700	42.50		1,742.50	1,925
2080	80 gal., 20" diam. x 64" LOA		9	.889		2,625	52		2,677	2,975
2100	135 gal., 24" diam. x 75" LOA		6	1.333		3,825	78.50		3,903.50	4,325
2120	240 gal., 30" diam. x 86" LOA		4	2		7,475	117		7,592	8,400
2140	300 gal., 36" diam. x 76" LOA	↓	3	2.667		10,600	157		10,757	11,900
2160	400 gal., 36" diam. x 100" LOA	Q-1	4	4		13,100	211		13,311	14,700
2180	500 gal., 36" diam., x 126" LOA	"	3	5.333		16,200	282		16,482	18,200
3000	Glass lined, P.E., 80 gal., 20" diam. x 60" LOA	1 Plum	9	.889		2,800	52		2,852	3,150
3060	140 gal., 24" diam. x 72" LOA		6	1.333		3,825	78.50		3,903.50	4,350
3080	200 gal., 30" diam. x 71" LOA		4	2		5,225	117		5,342	5,900
3100	350 gal., 36" diam. x 86" LOA	↓	3	2.667		6,300	157		6,457	7,150
3120	450 gal., 42" diam. x 78" LOA	Q-1	4	4		8,575	211		8,786	9,750
3140	600 gal., 48" diam. x 81" LOA		3	5.333		10,500	282		10,782	11,900
3160	750 gal., 48" diam. x 105" LOA		3	5.333		11,700	282		11,982	13,300
3180	900 gal., 54" diam. x 95" LOA	↓	2.50	6.400		16,400	340		16,740	18,500

22 12 Facility Potable-Water Storage Tanks

22 12 23 – Facility Indoor Potable-Water Storage Tanks

22 12 23.13 Facility Steel, Indoor Pot.-Water Storage Tanks	Crew	Daily Output	Labor-Hours	Unit	Material	2015 Bare Costs Labor	Equipment	Total	Total Incl O&P
3200	1250 gal., 54" diam. x 129" LOA	Q-1	2	8	Ea.	21,000	425		21,425	23,700
3220	1800 gal., 54" diam. x 181" LOA		1.50	10.667		23,500	565		24,065	26,800
3240	2000 gal., 60" diam. x 165" LOA		1	16		27,900	845		28,745	32,000
3260	3500 gal., 72" diam. x 201" LOA	Q-2	1.50	16		34,100	875		34,975	38,900

22 13 Facility Sanitary Sewerage

22 13 16 – Sanitary Waste and Vent Piping

22 13 16.40 Pipe Fittings, Cast Iron for Drainage

		Crew	Daily Output	Labor-Hours	Unit	Material	2015 Bare Costs Labor	Equipment	Total	Total Incl O&P
0010	**PIPE FITTINGS, CAST IRON FOR DRAINAGE**, Special									
1000	Drip pan elbow, (safety valve discharge elbow)									
1010	Cast iron, threaded inlet									
1014	2-1/2"	Q-1	8	2	Ea.	1,300	106		1,406	1,575
1015	3"		6.40	2.500		1,400	132		1,532	1,750
1017	4"		4.80	3.333		1,900	176		2,076	2,350
1018	Cast iron, flanged inlet									
1019	6"	Q-2	3.60	6.667	Ea.	1,950	365		2,315	2,700
1020	8"	"	2.60	9.231	"	2,050	505		2,555	3,025

22 13 16.60 Traps

		Crew	Daily Output	Labor-Hours	Unit	Material	2015 Bare Costs Labor	Equipment	Total	Total Incl O&P
0010	**TRAPS**									
4700	Copper, drainage, drum trap									
4800	3" x 5" solid, 1-1/2" pipe size	1 Plum	16	.500	Ea.	106	29.50		135.50	162
4840	3" x 6" swivel, 1-1/2" pipe size	"	16	.500	"	160	29.50		189.50	221
5100	P trap, standard pattern									
5200	1-1/4" pipe size	1 Plum	18	.444	Ea.	78.50	26		104.50	126
5240	1-1/2" pipe size		17	.471		72	27.50		99.50	121
5260	2" pipe size		15	.533		111	31.50		142.50	170
5280	3" pipe size		11	.727		281	42.50		323.50	375
5340	With cleanout, swivel joint and slip joint									
5360	1-1/4" pipe size	1 Plum	18	.444	Ea.	99.50	26		125.50	149
5400	1-1/2" pipe size		17	.471		106	27.50		133.50	159
5420	2" pipe size		15	.533		177	31.50		208.50	242
6710	ABS DWV P trap, solvent weld joint									
6720	1-1/2" pipe size	1 Plum	18	.444	Ea.	6.30	26		32.30	46.50
6722	2" pipe size		17	.471		8.30	27.50		35.80	50.50
6724	3" pipe size		15	.533		32.50	31.50		64	83.50
6726	4" pipe size		14	.571		65	33.50		98.50	122
6732	PVC DWV P trap, solvent weld joint									
6733	1-1/2" pipe size	1 Plum	18	.444	Ea.	4.95	26		30.95	45
6734	2" pipe size		17	.471		6.70	27.50		34.20	49
6735	3" pipe size		15	.533		22.50	31.50		54	72.50
6736	4" pipe size		14	.571		51.50	33.50		85	108
6760	PP DWV, dilution trap, 1-1/2" pipe size		16	.500		226	29.50		255.50	294
6770	P trap, 1-1/2" pipe size		17	.471		56.50	27.50		84	104
6780	2" pipe size		16	.500		83	29.50		112.50	136
6790	3" pipe size		14	.571		153	33.50		186.50	219
6800	4" pipe size		13	.615		257	36		293	340
6830	S trap, 1-1/2" pipe size		16	.500		46.50	29.50		76	95.50
6840	2" pipe size		15	.533		69.50	31.50		101	124
6850	Universal trap, 1-1/2" pipe size		14	.571		92	33.50		125.50	152
6860	PVC DWV hub x hub, basin trap, 1-1/4" pipe size		18	.444		40	26		66	83.50
6870	Sink P trap, 1-1/2" pipe size		18	.444		9.65	26		35.65	50

For customer support on your Mechanical Cost Data, call 877.795.7394.

243

22 13 16 – Sanitary Waste and Vent Piping

22 13 16.60 Traps		Crew	Daily Output	Labor-Hours	Unit	Material	2015 Bare Costs Labor	2015 Bare Costs Equipment	Total	Total Incl O&P
6880	Tubular S trap, 1-1/2" pipe size	1 Plum	17	.471	Ea.	23.50	27.50		51	67
6890	PVC sch. 40 DWV, drum trap									
6900	1-1/2" pipe size	1 Plum	16	.500	Ea.	23.50	29.50		53	70.50
6910	P trap, 1-1/2" pipe size		18	.444		5.85	26		31.85	46
6920	2" pipe size		17	.471		7.90	27.50		35.40	50
6930	3" pipe size		15	.533		27	31.50		58.50	77
6940	4" pipe size		14	.571		61	33.50		94.50	118
6950	P trap w/clean out, 1-1/2" pipe size		18	.444		9.65	26		35.65	50
6960	2" pipe size		17	.471		16.40	27.50		43.90	59.50
6970	P trap adjustable, 1-1/2" pipe size		17	.471		7.75	27.50		35.25	50
6980	P trap adj. w/union & cleanout, 1-1/2" pipe size		16	.500		30.50	29.50		60	78

22 13 16.80 Vent Flashing and Caps

		Crew	Daily Output	Labor-Hours	Unit	Material	2015 Bare Costs Labor	2015 Bare Costs Equipment	Total	Total Incl O&P
0010	**VENT FLASHING AND CAPS**									
0120	Vent caps									
0140	Cast iron									
0180	2-1/2" - 3-5/8" pipe	1 Plum	21	.381	Ea.	45	22.50		67.50	83.50
0190	4" - 4-1/8" pipe	"	19	.421	"	55	24.50		79.50	98
0900	Vent flashing									
1000	Aluminum with lead ring									
1020	1-1/4" pipe	1 Plum	20	.400	Ea.	8	23.50		31.50	44.50
1030	1-1/2" pipe		20	.400		8.45	23.50		31.95	45
1040	2" pipe		18	.444		8.60	26		34.60	49
1050	3" pipe		17	.471		9.55	27.50		37.05	52
1060	4" pipe		16	.500		11.50	29.50		41	57
1350	Copper with neoprene ring									
1400	1-1/4" pipe	1 Plum	20	.400	Ea.	61.50	23.50		85	103
1430	1-1/2" pipe		20	.400		61.50	23.50		85	103
1440	2" pipe		18	.444		61.50	26		87.50	107
1450	3" pipe		17	.471		74.50	27.50		102	124
1460	4" pipe		16	.500		74.50	29.50		104	127
2000	Galvanized with neoprene ring									
2020	1-1/4" pipe	1 Plum	20	.400	Ea.	16.95	23.50		40.45	54
2030	1-1/2" pipe		20	.400		16.95	23.50		40.45	54
2040	2" pipe		18	.444		17.55	26		43.55	59
2050	3" pipe		17	.471		18.55	27.50		46.05	62
2060	4" pipe		16	.500		21.50	29.50		51	68
2980	Neoprene, one piece									
3000	1-1/4" pipe	1 Plum	24	.333	Ea.	5.05	19.55		24.60	35
3030	1-1/2" pipe		24	.333		5.10	19.55		24.65	35
3040	2" pipe		23	.348		5.10	20.50		25.60	36.50
3050	3" pipe		21	.381		5.85	22.50		28.35	40.50
3060	4" pipe		20	.400		8.65	23.50		32.15	45
4000	Lead, 4#, 8" skirt, vent through roof									
4100	2" pipe	1 Plum	18	.444	Ea.	41.50	26		67.50	85.50
4110	3" pipe		17	.471		47.50	27.50		75	93.50
4120	4" pipe		16	.500		57	29.50		86.50	107
4130	6" pipe		14	.571		81	33.50		114.50	140

22 13 26 – Sanitary Waste Separators

22 13 26.10 Separators

		Crew	Daily Output	Labor-Hours	Unit	Material	2015 Bare Costs Labor	2015 Bare Costs Equipment	Total	Total Incl O&P
0010	**SEPARATORS**, Entrainment eliminator, steel body, 150 PSIG									
0100	1/4" size	1 Stpi	24	.333	Ea.	259	19.90		278.90	315
0120	1/2" size		24	.333		268	19.90		287.90	325

22 13 26 – Sanitary Waste Separators

22 13 26.10 Separators		Crew	Daily Output	Labor-Hours	Unit	Material	2015 Bare Costs Labor	Equipment	Total	Total Incl O&P
0140	3/4" size	1 Stpi	20	.400	Ea.	280	24		304	345
0160	1" size		19	.421		289	25		314	360
0180	1-1/4" size		15	.533		305	32		337	390
0200	1-1/2" size		13	.615		340	37		377	425
0220	2" size	↓	11	.727		375	43.50		418.50	480
0240	2-1/2" size	Q-5	15	1.067		1,600	57.50		1,657.50	1,850
0260	3" size		13	1.231		1,800	66		1,866	2,100
0280	4" size		10	1.600		2,050	86		2,136	2,375
0300	5" size		6	2.667		2,425	143		2,568	2,900
0320	6" size	↓	3	5.333		2,650	287		2,937	3,325
0340	8" size	Q-6	4.40	5.455		3,350	305		3,655	4,150
0360	10" size	"	4	6		4,750	335		5,085	5,725
1000	For 300 PSIG, add				↓	15%				

22 13 29 – Sanitary Sewerage Pumps

22 13 29.13 Wet-Pit-Mounted, Vertical Sewerage Pumps

		Crew	Daily Output	Labor-Hours	Unit	Material	2015 Bare Costs Labor	Equipment	Total	Total Incl O&P
0010	**WET-PIT-MOUNTED, VERTICAL SEWERAGE PUMPS**									
0020	Controls incl. alarm/disconnect panel w/wire. Excavation not included									
0260	Simplex, 9 GPM at 60 PSIG, 91 gal. tank				Ea.	3,325			3,325	3,675
0300	Unit with manway, 26" I.D., 18" high					3,700			3,700	4,075
0340	26" I.D., 36" high					3,750			3,750	4,125
0380	43" I.D., 4' high					3,975			3,975	4,375
0600	Simplex, 9 GPM at 60 PSIG, 150 gal. tank, indoor					3,525			3,525	3,900
0700	Unit with manway, 26" I.D., 36" high					4,225			4,225	4,650
0740	26" I.D., 4' high					4,450			4,450	4,900
2000	Duplex, 18 GPM at 60 PSIG, 150 gal. tank, indoor					7,050			7,050	7,750
2060	Unit with manway, 43" I.D., 4' high					8,100			8,100	8,925
2400	For core only				↓	1,875			1,875	2,050
3000	Indoor residential type installation									
3020	Simplex, 9 GPM at 60 PSIG, 91 gal. HDPE tank				Ea.	3,350			3,350	3,675

22 13 29.14 Sewage Ejector Pumps

		Crew	Daily Output	Labor-Hours	Unit	Material	2015 Bare Costs Labor	Equipment	Total	Total Incl O&P
0010	**SEWAGE EJECTOR PUMPS**, With operating and level controls									
0100	Simplex system incl. tank, cover, pump 15' head									
0500	37 gal. PE tank, 12 GPM, 1/2 HP, 2" discharge	Q-1	3.20	5	Ea.	480	264		744	925
0510	3" discharge		3.10	5.161		520	273		793	980
0530	87 GPM, .7 HP, 2" discharge		3.20	5		735	264		999	1,200
0540	3" discharge		3.10	5.161		795	273		1,068	1,275
0600	45 gal. coated stl. tank, 12 GPM, 1/2 HP, 2" discharge		3	5.333		855	282		1,137	1,375
0610	3" discharge		2.90	5.517		890	291		1,181	1,425
0630	87 GPM, .7 HP, 2" discharge		3	5.333		1,100	282		1,382	1,625
0640	3" discharge		2.90	5.517		1,150	291		1,441	1,725
0660	134 GPM, 1 HP, 2" discharge		2.80	5.714		1,175	300		1,475	1,750
0680	3" discharge		2.70	5.926		1,250	315		1,565	1,850
0700	70 gal. PE tank, 12 GPM, 1/2 HP, 2" discharge		2.60	6.154		920	325		1,245	1,525
0710	3" discharge		2.40	6.667		980	350		1,330	1,600
0730	87 GPM, 0.7 HP, 2" discharge		2.50	6.400		1,200	340		1,540	1,800
0740	3" discharge		2.30	6.957		1,275	370		1,645	1,950
0760	134 GPM, 1 HP, 2" discharge		2.20	7.273		1,300	385		1,685	2,000
0770	3" discharge		2	8		1,375	425		1,800	2,175
0800	75 gal. coated stl. tank, 12 GPM, 1/2 HP, 2" discharge		2.40	6.667		1,025	350		1,375	1,650
0810	3" discharge		2.20	7.273		1,075	385		1,460	1,750
0830	87 GPM, .7 HP, 2" discharge		2.30	6.957		1,300	370		1,670	1,975
0840	3" discharge		2.10	7.619		1,350	405		1,755	2,100

For customer support on your Mechanical Cost Data, call 877.795.7394.

245

22 13 29.14 Sewage Ejector Pumps	Crew	Daily Output	Labor-Hours	Unit	Material	2015 Bare Costs Labor	2015 Bare Costs Equipment	Total	Total Incl O&P
0860	134 GPM, 1 HP, 2" discharge	Q-1	2	8	Ea.	1,375	425		1,800	2,175
0880	3" discharge	↓	1.80	8.889	↓	1,450	470		1,920	2,300
1040	Duplex system incl. tank, covers, pumps									
1060	110 gal. fiberglass tank, 24 GPM, 1/2 HP, 2" discharge	Q-1	1.60	10	Ea.	1,850	530		2,380	2,825
1080	3" discharge		1.40	11.429		1,950	605		2,555	3,025
1100	174 GPM, .7 HP, 2" discharge		1.50	10.667		2,400	565		2,965	3,475
1120	3" discharge		1.30	12.308		2,475	650		3,125	3,700
1140	268 GPM, 1 HP, 2" discharge		1.20	13.333		2,600	705		3,305	3,925
1160	3" discharge	↓	1	16		2,700	845		3,545	4,225
1260	135 gal. coated stl. tank, 24 GPM, 1/2 HP, 2" discharge	Q-2	1.70	14.118		1,900	775		2,675	3,275
2000	3" discharge		1.60	15		2,025	820		2,845	3,475
2640	174 GPM, .7 HP, 2" discharge		1.60	15		2,500	820		3,320	4,000
2660	3" discharge		1.50	16		2,625	875		3,500	4,225
2700	268 GPM, 1 HP, 2" discharge		1.30	18.462		2,700	1,000		3,700	4,500
3040	3" discharge		1.10	21.818		2,875	1,200		4,075	4,975
3060	275 gal. coated stl. tank, 24 GPM, 1/2 HP, 2" discharge		1.50	16		2,375	875		3,250	3,950
3080	3" discharge		1.40	17.143		2,425	940		3,365	4,075
3100	174 GPM, .7 HP, 2" discharge		1.40	17.143		3,075	940		4,015	4,800
3120	3" discharge		1.30	18.462		3,250	1,000		4,250	5,100
3140	268 GPM, 1 HP, 2" discharge		1.10	21.818		3,375	1,200		4,575	5,500
3160	3" discharge	↓	.90	26.667	↓	3,550	1,450		5,000	6,100
3260	Pump system accessories, add									
3300	Alarm horn and lights, 115 V mercury switch	Q-1	8	2	Ea.	97	106		203	266
3340	Switch, mag. contactor, alarm bell, light, 3 level control		5	3.200		475	169		644	780
3380	Alternator, mercury switch activated	↓	4	4	↓	855	211		1,066	1,250

22 14 29.13 Wet-Pit-Mounted, Vertical Sump Pumps

		Crew	Daily Output	Labor-Hours	Unit	Material	2015 Bare Costs Labor	2015 Bare Costs Equipment	Total	Total Incl O&P
0010	**WET-PIT-MOUNTED, VERTICAL SUMP PUMPS**									
0400	Molded PVC base, 21 GPM at 15' head, 1/3 HP	1 Plum	5	1.600	Ea.	135	94		229	291
0800	Iron base, 21 GPM at 15' head, 1/3 HP		5	1.600		164	94		258	320
1200	Solid brass, 21 GPM at 15' head, 1/3 HP	↓	5	1.600	↓	289	94		383	460
2000	Sump pump, single stage									
2010	25 GPM, 1 HP, 1-1/2" discharge	Q-1	1.80	8.889	Ea.	3,825	470		4,295	4,925
2020	75 GPM, 1-1/2 HP, 2" discharge		1.50	10.667		4,050	565		4,615	5,300
2030	100 GPM, 2 HP, 2-1/2" discharge		1.30	12.308		4,125	650		4,775	5,525
2040	150 GPM, 3 HP, 3" discharge		1.10	14.545		4,125	770		4,895	5,700
2050	200 GPM, 3 HP, 3" discharge	↓	1	16		4,375	845		5,220	6,100
2060	300 GPM, 10 HP, 4" discharge	Q-2	1.20	20		4,725	1,100		5,825	6,850
2070	500 GPM, 15 HP, 5" discharge		1.10	21.818		5,375	1,200		6,575	7,700
2080	800 GPM, 20 HP, 6" discharge		1	24		6,350	1,325		7,675	8,950
2090	1000 GPM, 30 HP, 6" discharge		.85	28.235		6,975	1,550		8,525	10,000
2100	1600 GPM, 50 HP, 8" discharge	↓	.72	33.333		10,900	1,825		12,725	14,800
2110	2000 GPM, 60 HP, 8" discharge	Q-3	.85	37.647		11,100	2,100		13,200	15,400
2202	For general purpose float switch, copper coated float, add	Q-1	5	3.200	↓	108	169		277	375

22 14 29.16 Submersible Sump Pumps

0010	**SUBMERSIBLE SUMP PUMPS**									
1000	Elevator sump pumps, automatic									
1010	Complete systems, pump, oil detector, controls and alarm									
1020	1-1/2"discharge, does not include the sump pit/tank.									

22 14 Facility Storm Drainage

22 14 29 – Sump Pumps

22 14 29.16 Submersible Sump Pumps

22 14 29.16 Submersible Sump Pumps		Crew	Daily Output	Labor-Hours	Unit	Material	2015 Bare Costs Labor	Equipment	Total	Total Incl O&P
1040	1/3 HP, 115 V	1 Plum	4.40	1.818	Ea.	1,875	107		1,982	2,200
1050	1/2 HP, 115 V		4	2		1,925	117		2,042	2,300
1060	1/2 HP, 230 V		4	2		1,975	117		2,092	2,350
1070	3/4 HP, 115 V		3.60	2.222		2,025	130		2,155	2,425
1080	3/4 HP, 230 V		3.60	2.222		2,075	130		2,205	2,475
1100	Sump pump only									
1110	1/3 HP, 115 V	1 Plum	6.40	1.250	Ea.	170	73.50		243.50	298
1120	1/2 HP, 115 V		5.80	1.379		243	81		324	390
1130	1/2 HP, 230 V		5.80	1.379		287	81		368	435
1140	3/4 HP, 115 V		5.40	1.481		325	87		412	485
1150	3/4 HP, 230 V		5.40	1.481		370	87		457	540
1200	Oil detector, control and alarm only									
1210	115 V	1 Plum	8	1	Ea.	1,700	58.50		1,758.50	1,975
1220	230 V	"	8	1	"	1,700	58.50		1,758.50	1,975
7000	Sump pump, automatic									
7100	Plastic, 1-1/4" discharge, 1/4 HP	1 Plum	6.40	1.250	Ea.	138	73.50		211.50	263
7140	1/3 HP		6	1.333		200	78.50		278.50	340
7160	1/2 HP		5.40	1.481		246	87		333	400
7180	1-1/2" discharge, 1/2 HP		5.20	1.538		281	90.50		371.50	445
7500	Cast iron, 1-1/4" discharge, 1/4 HP		6	1.333		194	78.50		272.50	330
7540	1/3 HP		6	1.333		229	78.50		307.50	370
7560	1/2 HP		5	1.600		277	94		371	445

22 14 53 – Rainwater Storage Tanks

22 14 53.13 Fiberglass, Rainwater Storage Tank

22 14 53.13 Fiberglass, Rainwater Storage Tank		Crew	Daily Output	Labor-Hours	Unit	Material	2015 Bare Costs Labor	Equipment	Total	Total Incl O&P
0010	**FIBERGLASS, RAINWATER STORAGE TANK**									
2000	600 gallon	B-21B	3.75	10.667	Ea.	3,300	435	174	3,909	4,475
2010	1,000 gallon		3.50	11.429		4,250	465	187	4,902	5,600
2020	2,000 gallon		3.25	12.308		6,200	500	201	6,901	7,800
2030	4,000 gallon		3	13.333		8,575	545	218	9,338	10,500
2040	6,000 gallon		2.65	15.094		9,625	615	247	10,487	11,800
2050	8,000 gallon		2.30	17.391		11,500	710	284	12,494	14,100
2060	10,000 gallon		2	20		13,200	815	325	14,340	16,200
2070	12,000 gallon		1.50	26.667		18,300	1,100	435	19,835	22,300
2080	15,000 gallon		1	40		21,200	1,625	655	23,480	26,500
2090	20,000 gallon		.75	53.333		27,300	2,175	870	30,345	34,300
2100	25,000 gallon		.50	80		40,800	3,275	1,300	45,375	51,500
2110	30,000 gallon		.35	114		49,400	4,675	1,875	55,950	63,500
2120	40,000 gallon		.30	133		71,500	5,450	2,175	79,125	89,500

22 15 General Service Compressed-Air Systems

22 15 13 – General Service Compressed-Air Piping

22 15 13.10 Compressor Accessories

22 15 13.10 Compressor Accessories		Crew	Daily Output	Labor-Hours	Unit	Material	2015 Bare Costs Labor	Equipment	Total	Total Incl O&P
0010	**COMPRESSOR ACCESSORIES**									
1700	Refrigerated air dryers with ambient air filters									
1710	10 CFM	Q-5	8	2	Ea.	960	108		1,068	1,200
1720	25 CFM		6.60	2.424		1,075	130		1,205	1,375
1730	50 CFM		6.20	2.581		1,950	139		2,089	2,350
1740	75 CFM		5.80	2.759		2,525	148		2,673	3,000
1750	100 CFM		5.60	2.857		2,675	154		2,829	3,175
3460	Air filter, regulator, lubricator combination									

For customer support on your Mechanical Cost Data, call 877.795.7394.

247

22 15 13 – General Service Compressed-Air Piping

22 15 13.10 Compressor Accessories		Crew	Daily Output	Labor-Hours	Unit	Material	2015 Bare Costs Labor	Equipment	Total	Total Incl O&P
3470	Flush mount									
3480	Adjustable range 0-140 psi									
3500	1/8" NPT, 34 SCFM	1 Stpi	17	.471	Ea.	108	28		136	162
3510	1/4" NPT, 61 SCFM		17	.471		179	28		207	240
3520	3/8" NPT, 85 SCFM		16	.500		184	30		214	248
3530	1/2" NPT, 150 SCFM		15	.533		179	32		211	245
3540	3/4" NPT, 171 SCFM		14	.571		237	34		271	315
3550	1" NPT, 150 SCFM		13	.615		370	37		407	460
4000	Couplers, air line, sleeve type									
4010	Female, connection size NPT									
4020	1/4"	1 Stpi	38	.211	Ea.	9.95	12.60		22.55	30
4030	3/8"		36	.222		8.55	13.30		21.85	29.50
4040	1/2"		35	.229		21.50	13.65		35.15	44.50
4050	3/4"		34	.235		21.50	14.05		35.55	45
4100	Male									
4110	1/4"	1 Stpi	38	.211	Ea.	9	12.60		21.60	29
4120	3/8"		36	.222		10.60	13.30		23.90	31.50
4130	1/2"		35	.229		21	13.65		34.65	44
4140	3/4"		34	.235		21.50	14.05		35.55	45
4150	Coupler, combined male and female halves									
4160	1/2"	1 Stpi	17	.471	Ea.	43	28		71	89.50
4170	3/4"	"	15	.533	"	43.50	32		75.50	95.50

22 15 19 – General Service Packaged Air Compressors and Receivers

22 15 19.10 Air Compressors

		Crew	Daily Output	Labor-Hours	Unit	Material	2015 Bare Costs Labor	Equipment	Total	Total Incl O&P
0010	**AIR COMPRESSORS**									
5250	Air, reciprocating air cooled, splash lubricated, tank mounted									
5300	Single stage, 1 phase, 140 psi									
5303	1/2 HP, 30 gal. tank	1 Stpi	3	2.667	Ea.	1,950	159		2,109	2,400
5305	3/4 HP, 30 gal. tank		2.60	3.077		1,950	184		2,134	2,425
5307	1 HP, 30 gal. tank		2.20	3.636		2,450	217		2,667	3,025
5309	2 HP, 30 gal. tank	Q-5	4	4		2,625	215		2,840	3,225
5310	3 HP, 30 gal. tank		3.60	4.444		3,350	239		3,589	4,050
5314	3 HP, 60 gal. tank		3.50	4.571		3,550	246		3,796	4,275
5320	5 HP, 60 gal. tank		3.20	5		3,725	269		3,994	4,500
5330	5 HP, 80 gal. tank		3	5.333		4,100	287		4,387	4,925
5340	7.5 HP, 80 gal. tank		2.60	6.154		5,300	330		5,630	6,325
5600	2 stage pkg., 3 phase									
5650	6 CFM at 125 psi 1-1/2 HP, 60 gal. tank	Q-5	3	5.333	Ea.	3,375	287		3,662	4,125
5670	10.9 CFM at 125 psi, 3 HP, 80 gal. tank		1.50	10.667		3,800	575		4,375	5,050
5680	38.7 CFM at 125 psi, 10 HP, 120 gal. tank		.60	26.667		6,825	1,425		8,250	9,675
5690	105 CFM at 125 psi, 25 HP, 250 gal. tank	Q-6	.60	40		14,100	2,225		16,325	18,900
5800	With single stage pump									
5850	8.3 CFM at 125 psi, 2 HP, 80 gal. tank	Q-6	3.50	6.857	Ea.	3,775	380		4,155	4,725
5860	38.7 CFM at 125 psi, 10 HP, 120 gal. tank	"	.90	26.667	"	6,625	1,475		8,100	9,550
6000	Reciprocating, 2 stage, tank mtd, 3 Ph., Cap. rated @175 PSIG									
6050	Pressure lubricated, hvy. duty, 9.7 CFM, 3 HP, 120 gal. tank	Q-5	1.30	12.308	Ea.	5,775	660		6,435	7,350
6054	5 CFM, 1-1/2 HP, 80 gal. tank		2.80	5.714		3,575	305		3,880	4,425
6056	6.4 CFM, 2 HP, 80 gal. tank		2	8		3,775	430		4,205	4,800
6058	8.1 CFM, 3 HP, 80 gal. tank		1.70	9.412		3,800	505		4,305	4,950
6059	14.8 CFM, 5 HP, 80 gal. tank		1	16		4,100	860		4,960	5,800
6060	16.5 CFM, 5 HP, 120 gal. tank		1	16		5,900	860		6,760	7,775
6063	13 CFM, 6 HP, 80 gal. tank		.90	17.778		5,875	955		6,830	7,925

22 15 General Service Compressed-Air Systems

22 15 19 – General Service Packaged Air Compressors and Receivers

22 15 19.10 Air Compressors

	Crew	Daily Output	Labor-Hours	Unit	Material	2015 Bare Costs Labor	Equipment	Total	Total Incl O&P
6066	19.8 CFM, 7.5 HP, 80 gal. tank	Q-5	.80	20	Ea.	5,875	1,075		6,950	8,100
6070	25.8 CFM, 7-1/2 HP, 120 gal. tank		.80	20		7,975	1,075		9,050	10,400
6078	34.8 CFM, 10 HP, 80 gal. tank		.70	22.857		8,050	1,225		9,275	10,700
6080	34.8 CFM, 10 HP, 120 gal. tank		.60	26.667		8,500	1,425		9,925	11,600
6090	53.7 CFM, 15 HP, 120 gal. tank	Q-6	.80	30		10,300	1,675		11,975	13,800
6100	76.7 CFM, 20 HP, 120 gal. tank		.70	34.286		13,500	1,900		15,400	17,700
6104	76.7 CFM, 20 HP, 240 gal. tank		.68	35.294		14,800	1,975		16,775	19,300
6110	90.1 CFM, 25 HP, 120 gal. tank		.63	38.095		14,100	2,125		16,225	18,700
6120	101 CFM, 30 HP, 120 gal. tank		.57	42.105		15,200	2,350		17,550	20,300
6130	101 CFM, 30 HP, 250 gal. tank		.52	46.154		16,400	2,575		18,975	22,000
6200	Oil-less, 13.6 CFM, 5 HP, 120 gal. tank	Q-5	.88	18.182		16,900	980		17,880	20,100
6210	13.6 CFM, 5 HP, 250 gal. tank		.80	20		18,200	1,075		19,275	21,600
6220	18.2 CFM, 7.5 HP, 120 gal. tank		.73	21.918		16,900	1,175		18,075	20,400
6230	18.2 CFM, 7.5 HP, 250 gal. tank		.67	23.881		18,200	1,275		19,475	22,000
6250	30.5 CFM, 10 HP, 120 gal. tank		.57	28.070		19,900	1,500		21,400	24,200
6260	30.5 CFM, 10 HP, 250 gal. tank		.53	30.189		21,200	1,625		22,825	25,800
6270	41.3 CFM, 15 HP, 120 gal. tank	Q-6	.70	34.286		21,700	1,900		23,600	26,800
6280	41.3 CFM, 15 HP, 250 gal. tank	"	.67	35.821		22,900	2,000		24,900	28,200

22 31 Domestic Water Softeners

22 31 13 – Residential Domestic Water Softeners

22 31 13.10 Residential Water Softeners

		Crew	Daily Output	Labor-Hours	Unit	Material	2015 Bare Costs Labor	Equipment	Total	Total Incl O&P
0010	**RESIDENTIAL WATER SOFTENERS**									
7350	Water softener, automatic, to 30 grains per gallon	2 Plum	5	3.200	Ea.	405	188		593	730
7400	To 100 grains per gallon	"	4	4	"	660	235		895	1,075

22 35 Domestic Water Heat Exchangers

22 35 30 – Water Heating by Steam

22 35 30.10 Water Heating Transfer Package

		Crew	Daily Output	Labor-Hours	Unit	Material	2015 Bare Costs Labor	Equipment	Total	Total Incl O&P
0010	**WATER HEATING TRANSFER PACKAGE**, Complete controls,									
0020	expansion tank, converter, air separator									
1000	Hot water, 180°F enter, 200°F leaving, 15# steam									
1010	One pump system, 28 GPM	Q-6	.75	32	Ea.	20,000	1,775		21,775	24,800
1020	35 GPM		.70	34.286		21,800	1,900		23,700	26,900
1040	55 GPM		.65	36.923		25,800	2,050		27,850	31,400
1060	130 GPM		.55	43.636		32,500	2,425		34,925	39,500
1080	255 GPM		.40	60		42,900	3,350		46,250	52,000
1100	550 GPM		.30	80		58,000	4,450		62,450	70,500
1120	800 GPM		.25	96		69,000	5,350		74,350	84,000
1220	Two pump system, 28 GPM		.70	34.286		27,300	1,900		29,200	32,900
1240	35 GPM		.65	36.923		31,900	2,050		33,950	38,200
1260	55 GPM		.60	40		34,400	2,225		36,625	41,300
1280	130 GPM		.50	48		45,200	2,675		47,875	54,000
1300	255 GPM		.35	68.571		58,000	3,825		61,825	70,000
1320	550 GPM		.25	96		73,000	5,350		78,350	88,000
1340	800 GPM		.20	120		94,000	6,700		100,700	113,500

For customer support on your Mechanical Cost Data, call 877.795.7394.

249

22 41 Residential Plumbing Fixtures

22 41 39 – Residential Faucets, Supplies and Trim

22 41 39.10 Faucets and Fittings	Crew	Daily Output	Labor-Hours	Unit	Material	2015 Bare Costs Labor	Equipment	Total	Total Incl O&P
0010 **FAUCETS AND FITTINGS**									
5000 Sillcock, compact, brass, IPS or copper to hose	1 Plum	24	.333	Ea.	9.70	19.55		29.25	40
6000 Stop and waste valves, bronze									
6100 Angle, solder end 1/2"	1 Plum	24	.333	Ea.	12.05	19.55		31.60	43
6110 3/4"		20	.400		14.15	23.50		37.65	51
6300 Straightway, solder end 3/8"		24	.333		12.35	19.55		31.90	43
6310 1/2"		24	.333		12.35	19.55		31.90	43
6320 3/4"		20	.400		13.30	23.50		36.80	50
6410 Straightway, threaded 1/2"		24	.333		15.95	19.55		35.50	47
6420 3/4"		20	.400		18.30	23.50		41.80	55.50
6430 1"		19	.421		13.40	24.50		37.90	52.50

22 51 Swimming Pool Plumbing Systems

22 51 19 – Swimming Pool Water Treatment Equipment

22 51 19.50 Swimming Pool Filtration Equipment

	Crew	Daily Output	Labor-Hours	Unit	Material	2015 Bare Costs Labor	Equipment	Total	Total Incl O&P
0010 **SWIMMING POOL FILTRATION EQUIPMENT**									
0900 Filter system, sand or diatomite type, incl. pump, 6,000 gal./hr.	2 Plum	1.80	8.889	Total	1,975	520		2,495	2,975
1020 Add for chlorination system, 800 S.F. pool		3	5.333	Ea.	181	315		496	675
1040 5,000 S.F. pool		3	5.333	"	1,850	315		2,165	2,500

22 52 Fountain Plumbing Systems

22 52 16 – Fountain Pumps

22 52 16.10 Fountain Water Pumps

	Crew	Daily Output	Labor-Hours	Unit	Material	2015 Bare Costs Labor	Equipment	Total	Total Incl O&P
0010 **FOUNTAIN WATER PUMPS**									
0100 Pump w/controls									
0200 Single phase, 100' cord, 1/2 H.P. pump	2 Skwk	4.40	3.636	Ea.	1,275	177		1,452	1,675
0300 3/4 H.P. pump		4.30	3.721		2,175	181		2,356	2,675
0400 1 H.P. pump		4.20	3.810		2,350	185		2,535	2,875
0500 1-1/2 H.P. pump		4.10	3.902		2,800	190		2,990	3,375
0600 2 H.P. pump		4	4		3,775	195		3,970	4,475
0700 Three phase, 200' cord, 5 H.P. pump		3.90	4.103		5,175	200		5,375	6,000
0800 7-1/2 H.P. pump		3.80	4.211		9,000	205		9,205	10,200
0900 10 H.P. pump		3.70	4.324		13,300	210		13,510	15,000
1000 15 H.P. pump		3.60	4.444		16,800	216		17,016	18,800
2000 DESIGN NOTE: Use two horsepower per surface acre.									

22 52 33 – Fountain Ancillary

22 52 33.10 Fountain Miscellaneous

	Crew	Daily Output	Labor-Hours	Unit	Material	2015 Bare Costs Labor	Equipment	Total	Total Incl O&P
0010 **FOUNTAIN MISCELLANEOUS**									
1300 Lights w/mounting kits, 200 watt	2 Skwk	18	.889	Ea.	1,050	43		1,093	1,225
1400 300 watt		18	.889		1,275	43		1,318	1,475
1500 500 watt		18	.889		1,425	43		1,468	1,650
1600 Color blender		12	1.333		555	65		620	710

250

For customer support on your Mechanical Cost Data, call 877.795.7394.

22 62 19 – Vacuum Equipment for Laboratory and Healthcare Facilities

22 62 19.70 Healthcare Vacuum Equipment	Crew	Daily Output	Labor-Hours	Unit	Material	2015 Bare Costs Labor	Equipment	Total	Total Incl O&P
0010 HEALTHCARE VACUUM EQUIPMENT									
0300 Dental oral									
0310 Duplex									
0330 165 SCFM with 77 gal. separator	Q-2	1.30	18.462	Ea.	52,000	1,000		53,000	58,500
1100 Vacuum system									
1110 Vacuum outlet alarm panel	1 Plum	3.20	2.500	Ea.	1,100	147		1,247	1,450
2000 Medical, with receiver									
2100 Rotary vane type, lubricated, with controls									
2110 Simplex									
2120 1.5 HP, 80 gal tank	Q-1	8	2	Ea.	6,525	106		6,631	7,325
2130 2 HP, 80 gal tank		7	2.286		6,750	121		6,871	7,600
2140 3 HP, 80 gal tank		6.60	2.424		7,125	128		7,253	8,025
2150 5 HP, 80 gal tank		6	2.667		7,775	141		7,916	8,775
2200 Duplex									
2210 1 HP, 80 gal tank	Q-1	8.40	1.905	Ea.	10,400	101		10,501	11,700
2220 1.5 HP, 80 gal tank		7.80	2.051		10,700	108		10,808	12,000
2230 2 HP, 80 gal tank		6.80	2.353		11,600	124		11,724	12,900
2240 3 HP, 120 gal tank		6	2.667		12,600	141		12,741	14,000
2250 5 HP, 120 gal tank		5.40	2.963		14,300	157		14,457	15,900
2260 7.5 HP, 200 gal tank	Q-2	7	3.429		20,400	188		20,588	22,800
2270 10 HP, 200 gal tank		6.40	3.750		23,600	205		23,805	26,300
2280 15 HP, 200 gal tank		5.80	4.138		42,200	227		42,427	46,800
2290 20 HP, 200 gal tank		5	4.800		48,100	263		48,363	53,500
2300 25 HP, 200 gal tank		4	6		56,000	330		56,330	62,000
2400 Triplex									
2410 7.5 HP, 200 gal tank	Q-2	6.40	3.750	Ea.	32,400	205		32,605	36,000
2420 10 HP, 200 gal tank		5.70	4.211		37,800	231		38,031	41,900
2430 15 HP, 200 gal tank		4.90	4.898		64,000	268		64,268	70,500
2440 20 HP, 200 gal tank		4	6		73,000	330		73,330	80,500
2450 25 HP, 200 gal tank		3.70	6.486		84,000	355		84,355	93,000
2500 Quadruplex									
2510 7.5 HP, 200 gal tank	Q-2	5.30	4.528	Ea.	40,400	248		40,648	44,800
2520 10 HP, 200 gal tank		4.70	5.106		46,600	280		46,880	51,500
2530 15 HP, 200 gal tank		4	6		84,000	330		84,330	93,000
2540 20 HP, 200 gal tank		3.60	6.667		96,000	365		96,365	106,000
2550 25 HP, 200 gal tank		3.20	7.500		111,500	410		111,910	123,000
4000 Liquid ring type , water sealed, with controls									
4200 Duplex									
4210 1.5 HP, 120 gal tank	Q-1	6	2.667	Ea.	20,600	141		20,741	22,900
4220 3 HP, 120 gal tank	"	5.50	2.909		21,500	154		21,654	23,900
4230 4 HP, 120 gal tank	Q-2	6.80	3.529		22,900	193		23,093	25,500
4240 5 HP, 120 gal tank		6.50	3.692		24,200	202		24,402	26,900
4250 7.5 HP, 200 gal tank		6.20	3.871		29,400	212		29,612	32,700
4260 10 HP, 200 gal tank		5.80	4.138		36,700	227		36,927	40,700
4270 15 HP, 200 gal tank		5.10	4.706		44,500	258		44,758	49,400
4280 20 HP, 200 gal tank		4.60	5.217		57,500	286		57,786	64,000
4290 30 HP, 200 gal tank		4	6		69,000	330		69,330	76,000

For customer support on your Mechanical Cost Data, call 877.795.7394.

251

22 63 13 – Gas Piping for Laboratory and Healthcare Facilities

22 63 13.70 Healthcare Gas Piping	Crew	Daily Output	Labor-Hours	Unit	Material	2015 Bare Costs Labor	Equipment	Total	Total Incl O&P
0010 **HEALTHCARE GAS PIPING**									
1000 Nitrogen or oxygen system									
1010 Cylinder manifold									
1020 5 cylinder	1 Plum	.80	10	Ea.	5,950	585		6,535	7,425
1026 10 cylinder	"	.40	20	"	7,175	1,175		8,350	9,650
3000 Outlets and valves									
3010 Recessed, wall mounted									
3012 Single outlet	1 Plum	3.20	2.500	Ea.	71.50	147		218.50	300
3100 Ceiling outlet									
3190 Zone valve with box									
3210 2" valve size	1 Plum	3.20	2.500	Ea.	420	147		567	680
4000 Alarm panel, medical gases and vacuum									
4010 Alarm panel	1 Plum	3.20	2.500	Ea.	1,100	147		1,247	1,450

Estimating Tips

The labor adjustment factors listed in Subdivision 22 01 02.20 also apply to Division 23.

23 10 00 Facility Fuel Systems

- The prices in this subdivision for above- and below-ground storage tanks do not include foundations or hold-down slabs, unless noted. The estimator should refer to Divisions 3 and 31 for foundation system pricing. In addition to the foundations, required tank accessories, such as tank gauges, leak detection devices, and additional manholes and piping, must be added to the tank prices.

23 50 00 Central Heating Equipment

- When estimating the cost of an HVAC system, check to see who is responsible for providing and installing the temperature control system. It is possible to overlook controls, assuming that they would be included in the electrical estimate.
- When looking up a boiler, be careful on specified capacity. Some manufacturers rate their products on output while others use input.

- Include HVAC insulation for pipe, boiler, and duct (wrap and liner).
- Be careful when looking up mechanical items to get the correct pressure rating and connection type (thread, weld, flange).

23 70 00 Central HVAC Equipment

- Combination heating and cooling units are sized by the air conditioning requirements. (See Reference No. R236000-20 for preliminary sizing guide.)
- A ton of air conditioning is nominally 400 CFM.
- Rectangular duct is taken off by the linear foot for each size, but its cost is usually estimated by the pound. Remember that SMACNA standards now base duct on internal pressure.
- Prefabricated duct is estimated and purchased like pipe: straight sections and fittings.
- Note that cranes or other lifting equipment are not included on any lines in

Division 23. For example, if a crane is required to lift a heavy piece of pipe into place high above a gym floor, or to put a rooftop unit on the roof of a four-story building, etc., it must be added. Due to the potential for extreme variation—from nothing additional required to a major crane or helicopter—we feel that including a nominal amount for "lifting contingency" would be useless and detract from the accuracy of the estimate. When using equipment rental cost data from RSMeans, do not forget to include the cost of the operator(s).

Reference Numbers

Reference numbers are shown in shaded boxes at the beginning of some major classifications. These numbers refer to related items in the Reference Section. The reference information may be an estimating procedure, an alternate pricing method, or technical information.

Note: Not all subdivisions listed here necessarily appear in this publication. ∎

*Note: **Trade Service**, in part, has been used as a reference source for some of the material prices used in Division 23.*

23 05 02 – HVAC General

23 05 02.10 Air Conditioning, General	Crew	Daily Output	Labor-Hours	Unit	Material	2015 Bare Costs Labor	Equipment	Total	Total Incl O&P
0010 **AIR CONDITIONING, GENERAL** Prices are for standard efficiencies (SEER 13)									
0020 for upgrade to SEER 14 add					10%				

23 05 05 – Selective Demolition for HVAC

23 05 05.10 HVAC Demolition

		Crew	Daily Output	Labor-Hours	Unit	Material	Labor	Equipment	Total	Total Incl O&P
0010	**HVAC DEMOLITION** R220105-10									
0100	Air conditioner, split unit, 3 ton	Q-5	2	8	Ea.		430		430	650
0150	Package unit, 3 ton R024119-10	Q-6	3	8			445		445	675
0190	Rooftop, self contained, up to 5 ton	1 Plum	1.20	6.667			390		390	590
0250	Air curtain	Q-9	20	.800	L.F.		40.50		40.50	61.50
0254	Air filters, up thru 16,000 CFM		20	.800	Ea.		40.50		40.50	61.50
0256	20,000 thru 60,000 CFM		16	1			50.50		50.50	77
0297	Boiler blowdown	Q-5	8	2			108		108	162
0298	Boilers									
0300	Electric, up thru 148 kW	Q-19	2	12	Ea.		650		650	975
0310	150 thru 518 kW	"	1	24			1,300		1,300	1,950
0320	550 thru 2000 kW	Q-21	.40	80			4,450		4,450	6,700
0330	2070 kW and up	"	.30	106			5,925		5,925	8,925
0340	Gas and/or oil, up thru 150 MBH	Q-7	2.20	14.545			825		825	1,250
0350	160 thru 2000 MBH		.80	40			2,275		2,275	3,425
0360	2100 thru 4500 MBH		.50	64			3,650		3,650	5,500
0370	4600 thru 7000 MBH		.30	106			6,075		6,075	9,175
0380	7100 thru 12,000 MBH		.16	200			11,400		11,400	17,200
0390	12,200 thru 25,000 MBH		.12	266			15,200		15,200	22,900
0400	Central station air handler unit, up thru 15 ton	Q-5	1.60	10			540		540	810
0410	17.5 thru 30 ton	"	.80	20			1,075		1,075	1,625
0430	Computer room unit									
0434	Air cooled split, up thru 10 ton	Q-5	.67	23.881	Ea.		1,275		1,275	1,950
0436	12 thru 23 ton		.53	30.189			1,625		1,625	2,450
0440	Chilled water, up thru 10 ton		1.30	12.308			660		660	1,000
0444	12 thru 23 ton		1	16			860		860	1,300
0450	Glycol system, up thru 10 ton		.53	30.189			1,625		1,625	2,450
0454	12 thru 23 ton		.40	40			2,150		2,150	3,250
0460	Water cooled, not including condenser, up thru 10 ton		.80	20			1,075		1,075	1,625
0464	12 thru 23 ton		.60	26.667			1,425		1,425	2,175
0600	Condenser, up thru 50 ton		1	16			860		860	1,300
0610	51 thru 100 ton	Q-6	.90	26.667			1,475		1,475	2,250
0620	101 thru 1000 ton	"	.70	34.286			1,900		1,900	2,875
0660	Condensing unit, up thru 10 ton	Q-5	1.25	12.800			690		690	1,050
0670	11 thru 50 ton	"	.40	40			2,150		2,150	3,250
0680	60 thru 100 ton	Q-6	.30	80			4,450		4,450	6,725
0700	Cooling tower, up thru 400 ton		.80	30			1,675		1,675	2,525
0710	450 thru 600 ton		.53	45.283			2,525		2,525	3,825
0720	700 thru 1300 ton		.40	60			3,350		3,350	5,050
0780	Dehumidifier, up thru 155 lb./hr.	Q-1	8	2			106		106	160
0790	240 lb./hr. and up	"	2	8			425		425	640
1560	Ductwork									
1570	Metal, steel, sst, fabricated	Q-9	1000	.016	Lb.		.81		.81	1.23
1580	Aluminum, fabricated		485	.033	"		1.66		1.66	2.54
1590	Spiral, prefabricated		400	.040	L.F.		2.01		2.01	3.08
1600	Fiberglass, prefabricated		400	.040			2.01		2.01	3.08
1610	Flex, prefabricated		500	.032			1.61		1.61	2.46
1620	Glass fiber reinforced plastic, prefabricated		280	.057			2.88		2.88	4.40

23 05 05 – Selective Demolition for HVAC

23 05 05.10 HVAC Demolition	Crew	Daily Output	Labor-Hours	Unit	Material	2015 Bare Costs Labor	Equipment	Total	Total Incl O&P
1630	Diffusers, registers or grills, up thru 20" max dimension	1 Shee	50	.160	Ea.		8.95		8.95	13.70
1640	21 thru 36" max dimension		36	.222			12.45		12.45	19
1650	Above 36" max dimension		30	.267			14.90		14.90	23
1700	Evaporator, up thru 12,000 BTUH	Q-5	5.30	3.019			162		162	245
1710	12,500 thru 30,000 BTUH	"	2.70	5.926			320		320	480
1720	31,000 BTUH and up	Q-6	1.50	16			890		890	1,350
1730	Evaporative cooler, up thru 5 H.P.	Q-9	2.70	5.926			298		298	455
1740	10 thru 30 H.P.	"	.67	23.881			1,200		1,200	1,850
1750	Exhaust systems									
1760	Exhaust components	1 Shee	8	1	System		56		56	85.50
1770	Weld fume hoods	"	20	.400	Ea.		22.50		22.50	34
2120	Fans, up thru 1 H.P. or 2000 CFM	Q-9	8	2			101		101	154
2124	1-1/2 thru 10 H.P. or 20,000 CFM		5.30	3.019			152		152	232
2128	15 thru 30 H.P. or above 20,000 CFM		4	4			201		201	310
2150	Fan coil air conditioner, chilled water, up thru 7.5 ton	Q-5	14	1.143			61.50		61.50	93
2154	Direct expansion, up thru 10 ton		8	2			108		108	162
2158	11 thru 30 ton		2	8			430		430	650
2170	Flue shutter damper	Q-9	8	2			101		101	154
2200	Furnace, electric	Q-20	2	10			510		510	780
2300	Gas or oil, under 120 MBH	Q-9	4	4			201		201	310
2340	Over 120 MBH	"	3	5.333			269		269	410
2730	Heating and ventilating unit	Q-5	2.70	5.926			320		320	480
2740	Heater, electric, wall, baseboard and quartz	1 Elec	10	.800			44		44	65.50
2750	Heater, electric, unit, cabinet, fan and convector	"	8	1			54.50		54.50	82
2760	Heat exchanger, shell and tube type	Q-5	1.60	10			540		540	810
2770	Plate type	Q-6	.60	40			2,225		2,225	3,375
2810	Heat pump									
2820	Air source, split, up thru 10 ton	Q-5	.90	17.778	Ea.		955		955	1,450
2830	15 thru 25 ton	Q-6	.80	30			1,675		1,675	2,525
2850	Single package, up thru 12 ton	Q-5	1	16			860		860	1,300
2860	Water source, up thru 15 ton	"	.90	17.778			955		955	1,450
2870	20 thru 50 ton	Q-6	.80	30			1,675		1,675	2,525
2910	Heat recovery package, up thru 20,000 CFM	Q-5	2	8			430		430	650
2920	25,000 CFM and up		1.20	13.333			715		715	1,075
2930	Heat transfer package, up thru 130 GPM		.80	20			1,075		1,075	1,625
2934	255 thru 800 GPM		.42	38.095			2,050		2,050	3,100
2940	Humidifier		10.60	1.509			81		81	123
2961	Hydronic unit heaters, up thru 200 MBH		14	1.143			61.50		61.50	93
2962	Above 200 MBH		8	2			108		108	162
2964	Valance units		32	.500			27		27	40.50
2966	Radiant floor heating									
2967	System valves, controls, manifolds	Q-5	16	1	Ea.		54		54	81
2968	Per room distribution		8	2			108		108	162
2970	Hydronic heating, baseboard radiation		16	1			54		54	81
2976	Convectors and free standing radiators		18	.889			48		48	72
2980	Induced draft fan, up thru 1 H.P.	Q-9	4.60	3.478			175		175	268
2984	1-1/2 H.P. thru 7-1/2 H.P.	"	2.20	7.273			365		365	560
2988	Infrared unit	Q-5	16	1			54		54	81
2992	Louvers	1 Shee	46	.174	S.F.		9.75		9.75	14.85
3000	Mechanical equipment, light items. Unit is weight, not cooling.	Q-5	.90	17.778	Ton		955		955	1,450
3600	Heavy items		1.10	14.545	"		780		780	1,175
3720	Make-up air unit, up thru 6000 CFM		3	5.333	Ea.		287		287	435
3730	6500 thru 30,000 CFM		1.60	10			540		540	810

23 05 05 – Selective Demolition for HVAC

23 05 05.10 HVAC Demolition		Crew	Daily Output	Labor-Hours	Unit	Material	2015 Bare Costs Labor	2015 Bare Costs Equipment	Total	Total Incl O&P
3740	35,000 thru 75,000 CFM	Q-6	1	24	Ea.		1,350		1,350	2,025
3800	Mixing boxes, constant and VAV	Q-9	18	.889			45		45	68.50
4000	Packaged terminal air conditioner, up thru 18,000 BTUH	Q-5	8	2			108		108	162
4010	24,000 thru 48,000 BTUH	"	2.80	5.714			305		305	465
5000	Refrigerant compressor, reciprocating or scroll									
5010	Up thru 5 ton	1 Stpi	6	1.333	Ea.		79.50		79.50	120
5020	5.08 thru 10 ton	Q-5	6	2.667			143		143	217
5030	15 thru 50 ton	"	3	5.333			287		287	435
5040	60 thru 130 ton	Q-6	2.80	8.571			480		480	720
5090	Remove refrigerant from system	1 Stpi	40	.200	Lb.		11.95		11.95	18.05
5100	Rooftop air conditioner, up thru 10 ton	Q-5	1.40	11.429	Ea.		615		615	930
5110	12 thru 40 ton	Q-6	1	24			1,350		1,350	2,025
5120	50 thru 140 ton		.50	48			2,675		2,675	4,050
5130	150 thru 300 ton		.30	80			4,450		4,450	6,725
6000	Self contained single package air conditioner, up thru 10 ton	Q-5	1.60	10			540		540	810
6010	15 thru 60 ton	Q-6	1.20	20			1,125		1,125	1,675
6100	Space heaters, up thru 200 MBH	Q-5	10	1.600			86		86	130
6110	Over 200 MBH		5	3.200			172		172	260
6200	Split ductless, both sections		8	2			108		108	162
6300	Steam condensate meter	1 Stpi	11	.727			43.50		43.50	65.50
6600	Thru-the-wall air conditioner	L-2	8	2			82.50		82.50	127
7000	Vent chimney, prefabricated, up thru 12" diameter	Q-9	94	.170	V.L.F.		8.55		8.55	13.10
7010	14" thru 36" diameter		40	.400			20		20	31
7020	38" thru 48" diameter		32	.500			25		25	38.50
7030	54" thru 60" diameter	Q-10	14	1.714			89.50		89.50	137
7400	Ventilators, up thru 14" neck diameter	Q-9	58	.276	Ea.		13.90		13.90	21
7410	16" thru 50" neck diameter		40	.400			20		20	31
7450	Relief vent, up thru 24" x 96"		22	.727			36.50		36.50	56
7460	48" x 60" thru 96" x 144"		10	1.600			80.50		80.50	123
8000	Water chiller up thru 10 ton	Q-5	2.50	6.400			345		345	520
8010	15 thru 100 ton	Q-6	.48	50			2,800		2,800	4,200
8020	110 thru 500 ton	Q-7	.29	110			6,275		6,275	9,475
8030	600 thru 1000 ton		.23	139			7,925		7,925	12,000
8040	1100 ton and up		.20	160			9,100		9,100	13,700
8400	Window air conditioner	1 Carp	16	.500			23.50		23.50	36

23 05 23 – General-Duty Valves for HVAC Piping

23 05 23.20 Valves, Bronze/Brass

		Crew	Daily Output	Labor-Hours	Unit	Material	Labor	Equipment	Total	Total Incl O&P
0010	**VALVES, BRONZE/BRASS**									
0020	Brass									
1300	Ball combination valves, shut-off and union									
1310	Solder, with strainer, drain and PT ports									
1320	1/2"	1 Stpi	17	.471	Ea.	75	28		103	125
1330	3/4"		16	.500		78	30		108	131
1340	1"		14	.571		123	34		157	187
1350	1-1/4"		12	.667		142	40		182	216
1360	1-1/2"		10	.800		221	48		269	315
1370	2"		8	1		260	60		320	375
1410	Threaded, with strainer, drain and PT ports									
1420	1/2"	1 Stpi	20	.400	Ea.	75	24		99	119
1430	3/4"		18	.444		78	26.50		104.50	126
1440	1"		17	.471		123	28		151	178
1450	1-1/4"		12	.667		142	40		182	216

23 05 23 – General-Duty Valves for HVAC Piping

23 05 23.20 Valves, Bronze/Brass		Crew	Daily Output	Labor-Hours	Unit	Material	2015 Bare Costs Labor	Equipment	Total	Total Incl O&P
1460	1-1/2"	1 Stpi	11	.727	Ea.	221	43.50		264.50	310
1470	2"	↓	9	.889	↓	260	53		313	365

23 05 23.30 Valves, Iron Body

		Crew	Daily Output	Labor-Hours	Unit	Material	2015 Bare Costs Labor	Equipment	Total	Total Incl O&P
0010	**VALVES, IRON BODY** R220523-90									
0022	For grooved joint, see Section 22 11 13.48									
0100	Angle, 125 lb.									
0110	Flanged									
0116	2"	1 Plum	5	1.600	Ea.	1,200	94		1,294	1,475
0118	4"	Q-1	3	5.333		2,000	282		2,282	2,625
0120	6"	Q-2	3	8		4,000	440		4,440	5,025
0122	8"	"	2.50	9.600		5,550	525		6,075	6,900
0560	Butterfly, lug type, pneumatic operator, 2" size	1 Stpi	14	.571		475	34		509	575
0570	3"	Q-1	8	2		510	106		616	725
0580	4"	"	5	3.200		610	169		779	925
0590	6"	Q-2	5	4.800		790	263		1,053	1,250
0600	8"		4.50	5.333		1,050	292		1,342	1,600
0610	10"		4	6		1,300	330		1,630	1,925
0620	12"		3	8		1,925	440		2,365	2,750
0630	14"		2.30	10.435		2,175	570		2,745	3,275
0640	18"		1.50	16		3,700	875		4,575	5,400
0650	20"	↓	1	24		4,725	1,325		6,050	7,175
0790	Butterfly, lug type, gear operated, 2" size, 200 lb. except noted	1 Plum	14	.571		390	33.50		423.50	480
0800	2-1/2"	Q-1	9	1.778		395	94		489	575
0810	3"		8	2		415	106		521	615
0820	4"	↓	5	3.200		470	169		639	770
0830	5"	Q-2	5	4.800		580	263		843	1,025
0840	6"		5	4.800		645	263		908	1,100
0850	8"		4.50	5.333		835	292		1,127	1,350
0860	10"		4	6		1,175	330		1,505	1,775
0870	12"		3	8		1,600	440		2,040	2,425
0880	14", 150 lb.		2.30	10.435		3,075	570		3,645	4,250
0890	16", 150 lb.		1.75	13.714		4,700	750		5,450	6,275
0900	18", 150 lb.		1.50	16		5,900	875		6,775	7,800
0910	20", 150 lb.		1	24		7,875	1,325		9,200	10,600
0930	24", 150 lb.	↓	.75	32	↓	13,400	1,750		15,150	17,400
1020	Butterfly, wafer type, gear actuator, 200 lb.									
1030	2"	1 Plum	14	.571	Ea.	96	33.50		129.50	156
1040	2-1/2"	Q-1	9	1.778		97.50	94		191.50	249
1050	3"		8	2		101	106		207	271
1060	4"	↓	5	3.200		113	169		282	380
1070	5"	Q-2	5	4.800		126	263		389	535
1080	6"		5	4.800		143	263		406	555
1090	8"		4.50	5.333		178	292		470	635
1100	10"		4	6		239	330		569	760
1110	12"	↓	3	8	↓	360	440		800	1,050
1200	Wafer type, lever actuator, 200 lb.									
1220	2"	1 Plum	14	.571	Ea.	156	33.50		189.50	222
1230	2-1/2"	Q-1	9	1.778		161	94		255	320
1240	3"		8	2		190	106		296	370
1250	4"	↓	5	3.200		207	169		376	485
1260	5"	Q-2	5	4.800		320	263		583	750
1270	6"	↓	5	4.800	↓	350	263		613	780

For customer support on your Mechanical Cost Data, call 877.795.7394.

257

23 05 23 – General-Duty Valves for HVAC Piping

23 05 23.30 Valves, Iron Body		Crew	Daily Output	Labor-Hours	Unit	Material	2015 Bare Costs Labor	Equipment	Total	Total Incl O&P
1280	8"	Q-2	4.50	5.333	Ea.	540	292		832	1,025
1290	10"		4	6		755	330		1,085	1,325
1300	12"	↓	3	8	↓	1,250	440		1,690	2,025
1650	Gate, 125 lb., N.R.S.									
2150	Flanged									
2200	2"	1 Plum	5	1.600	Ea.	715	94		809	930
2240	2-1/2"	Q-1	5	3.200		735	169		904	1,075
2260	3"		4.50	3.556		825	188		1,013	1,200
2280	4"	↓	3	5.333		1,175	282		1,457	1,725
2290	5"	Q-2	3.40	7.059		2,025	385		2,410	2,800
2300	6"		3	8		2,025	440		2,465	2,875
2320	8"		2.50	9.600		3,450	525		3,975	4,600
2340	10"		2.20	10.909		6,075	600		6,675	7,575
2360	12"		1.70	14.118		8,325	775		9,100	10,400
2370	14"		1.30	18.462		12,000	1,000		13,000	14,700
2380	16"	↓	1	24	↓	17,100	1,325		18,425	20,800
2420	For 250 lb. flanged, add					200%	10%			
3550	OS&Y, 125 lb., flanged									
3600	2"	1 Plum	5	1.600	Ea.	475	94		569	665
3640	2-1/2"	Q-1	5	3.200		480	169		649	785
3660	3"		4.50	3.556		530	188		718	865
3670	3-1/2"		3	5.333		680	282		962	1,175
3680	4"	↓	3	5.333		770	282		1,052	1,275
3690	5"	Q-2	3.40	7.059		1,250	385		1,635	1,950
3700	6"		3	8		1,250	440		1,690	2,025
3720	8"		2.50	9.600		2,175	525		2,700	3,200
3740	10"		2.20	10.909		4,050	600		4,650	5,350
3760	12"		1.70	14.118		5,500	775		6,275	7,250
3770	14"		1.30	18.462		12,000	1,000		13,000	14,700
3780	16"		1	24		16,900	1,325		18,225	20,600
3790	18"		.80	30		20,700	1,650		22,350	25,300
3800	20"		.60	40		22,900	2,200		25,100	28,500
3830	24"	↓	.50	48	↓	24,900	2,625		27,525	31,400
3900	For 175 lb., flanged, add					200%	10%			
4350	Globe, OS&Y									
4540	Class 125, flanged									
4550	2"	1 Plum	5	1.600	Ea.	935	94		1,029	1,175
4560	2-1/2"	Q-1	5	3.200		940	169		1,109	1,275
4570	3"		4.50	3.556		1,150	188		1,338	1,525
4580	4"	↓	3	5.333		1,650	282		1,932	2,225
4590	5"	Q-2	3.40	7.059		2,975	385		3,360	3,850
4600	6"		3	8		2,975	440		3,415	3,925
4610	8"		2.50	9.600		5,825	525		6,350	7,225
4612	10"		2.20	10.909		9,125	600		9,725	10,900
4614	12"	↓	1.70	14.118	↓	9,900	775		10,675	12,100
5040	Class 250, flanged									
5050	2"	1 Plum	4.50	1.778	Ea.	1,500	104		1,604	1,800
5060	2-1/2"	Q-1	4.50	3.556		1,950	188		2,138	2,425
5070	3"		4	4		2,000	211		2,211	2,550
5080	4"	↓	2.70	5.926		2,950	315		3,265	3,725
5090	5"	Q-2	3	8		5,475	440		5,915	6,675
5100	6"		2.70	8.889		5,275	485		5,760	6,550
5110	8"	↓	2.20	10.909		8,975	600		9,575	10,800

23 05 23 – General-Duty Valves for HVAC Piping

23 05 23.30 Valves, Iron Body		Crew	Daily Output	Labor-Hours	Unit	Material	2015 Bare Costs Labor	Equipment	Total	Total Incl O&P
5120	10"	Q-2	2	12	Ea.	13,400	655		14,055	15,800
5130	12"	↓	1.60	15		19,600	820		20,420	22,800
5240	Valve sprocket rim w/chain, for 2" valve	1 Stpi	30	.267		127	15.95		142.95	164
5250	2-1/2" valve		27	.296		127	17.70		144.70	167
5260	3-1/2" valve		25	.320		127	19.10		146.10	169
5270	6" valve		20	.400		127	24		151	176
5280	8" valve		18	.444		156	26.50		182.50	211
5290	12" valve		16	.500		156	30		186	216
5300	16" valve		12	.667		215	40		255	296
5310	20" valve		10	.800		215	48		263	310
5320	36" valve	↓	8	1	↓	380	60		440	510
5450	Swing check, 125 lb., threaded									
5500	2"	1 Plum	11	.727	Ea.	430	42.50		472.50	540
5540	2-1/2"	Q-1	15	1.067		555	56.50		611.50	695
5550	3"	↓	13	1.231		590	65		655	750
5560	4"	↓	10	1.600	↓	955	84.50		1,039.50	1,175
5950	Flanged									
5994	1"	1 Plum	7	1.143	Ea.	210	67		277	335
5998	1-1/2"		6	1.333		299	78.50		377.50	450
6000	2"	↓	5	1.600		395	94		489	575
6040	2-1/2"	Q-1	5	3.200		380	169		549	675
6050	3"		4.50	3.556		410	188		598	735
6060	4"	↓	3	5.333		605	282		887	1,100
6070	6"	Q-2	3	8		1,025	440		1,465	1,800
6080	8"		2.50	9.600		1,950	525		2,475	2,950
6090	10"		2.20	10.909		3,325	600		3,925	4,550
6100	12"		1.70	14.118		5,175	775		5,950	6,875
6110	18"		1.30	18.462		23,700	1,000		24,700	27,600
6114	24"	↓	.75	32	↓	52,000	1,750		53,750	60,000
6160	For 250 lb. flanged, add					200%	20%			
6600	Silent check, bronze trim									
6610	Compact wafer type, for 125 or 150 lb. flanges									
6630	1-1/2"	1 Plum	11	.727	Ea.	136	42.50		178.50	214
6640	2"	"	9	.889		165	52		217	261
6650	2-1/2"	Q-1	9	1.778		180	94		274	340
6660	3"		8	2		192	106		298	370
6670	4"	↓	5	3.200		245	169		414	525
6680	5"	Q-2	6	4		325	219		544	685
6690	6"		6	4		590	219		809	975
6700	8"		4.50	5.333		730	292		1,022	1,250
6710	10"		4	6		1,300	330		1,630	1,925
6720	12"	↓	3	8	↓	2,425	440		2,865	3,300
6740	For 250 or 300 lb. flanges, thru 6" no change									
6741	For 8" and 10", add				Ea.	11%	10%			
6750	Twin disc									
6752	2"	1 Plum	9	.889	Ea.	300	52		352	410
6754	4"	Q-1	5	3.200		490	169		659	795
6756	6"	Q-2	5	4.800		730	263		993	1,200
6758	8"		4.50	5.333		1,125	292		1,417	1,675
6760	10"		4	6		1,800	330		2,130	2,475
6762	12"		3	8		2,325	440		2,765	3,200
6764	18"		1.50	16		10,000	875		10,875	12,300
6766	24"	↓	.75	32	↓	14,100	1,750		15,850	18,200

For customer support on your Mechanical Cost Data, call 877.795.7394.

259

23 05 23.30 Valves, Iron Body

		Crew	Daily Output	Labor-Hours	Unit	Material	2015 Bare Costs Labor	2015 Bare Costs Equipment	Total	Total Incl O&P
6800	Full flange type, 150 lb.									
6810	1"	1 Plum	14	.571	Ea.	153	33.50		186.50	219
6811	1-1/4"		12	.667		161	39		200	236
6812	1-1/2"		11	.727		187	42.50		229.50	271
6813	2"		9	.889		214	52		266	315
6814	2-1/2"	Q-1	9	1.778		260	94		354	430
6815	3"		8	2		269	106		375	455
6816	4"		5	3.200		335	169		504	620
6817	5"	Q-2	6	4		525	219		744	905
6818	6"		5	4.800		690	263		953	1,150
6819	8"		4.50	5.333		1,175	292		1,467	1,725
6820	10"		4	6		1,475	330		1,805	2,125
6840	For 250 lb., add					30%	10%			
6900	Globe type, 125 lb.									
6910	2"	1 Plum	9	.889	Ea.	300	52		352	410
6911	2-1/2"	Q-1	9	1.778		405	94		499	585
6912	3"		8	2		475	106		581	685
6913	4"		5	3.200		640	169		809	960
6914	5"	Q-2	6	4		890	219		1,109	1,300
6915	6"		5	4.800		1,125	263		1,388	1,625
6916	8"		4.50	5.333		2,050	292		2,342	2,700
6917	10"		4	6		2,575	330		2,905	3,325
6918	12"		3	8		4,275	440		4,715	5,350
6919	14"		2.30	10.435		5,825	570		6,395	7,300
6920	16"		1.75	13.714		6,475	750		7,225	8,225
6921	18"		1.50	16		9,350	875		10,225	11,600
6922	20"		1	24		11,300	1,325		12,625	14,500
6923	24"		.75	32		14,000	1,750		15,750	18,100
6940	For 250 lb., add					40%	10%			
6980	Screwed end type, 125 lb.									
6981	1"	1 Plum	19	.421	Ea.	61.50	24.50		86	105
6982	1-1/4"		15	.533		80	31.50		111.50	136
6983	1-1/2"		13	.615		99	36		135	164
6984	2"		11	.727		134	42.50		176.50	212

23 05 23.50 Valves, Multipurpose

		Crew	Daily Output	Labor-Hours	Unit	Material	2015 Bare Costs Labor	2015 Bare Costs Equipment	Total	Total Incl O&P
0010	**VALVES, MULTIPURPOSE**									
0100	Functions as a shut off, balancing, check & metering valve									
1000	Cast iron body									
1010	Threaded									
1020	1-1/2" size	1 Stpi	11	.727	Ea.	380	43.50		423.50	485
1030	2" size	"	8	1		460	60		520	595
1040	2-1/2" size	Q-5	5	3.200		500	172		672	810
1200	Flanged									
1210	3" size	Q-5	4.20	3.810	Ea.	575	205		780	940
1220	4" size	"	3	5.333		1,175	287		1,462	1,700
1230	5" size	Q-6	3.80	6.316		1,375	350		1,725	2,050
1240	6" size		3	8		1,725	445		2,170	2,550
1250	8" size		2.50	9.600		2,975	535		3,510	4,075
1260	10" size		2.20	10.909		4,250	610		4,860	5,600
1270	12" size		2.10	11.429		7,675	635		8,310	9,375
1280	14" size		2	12		11,100	670		11,770	13,200

23 05 23 – General-Duty Valves for HVAC Piping

23 05 23.70 Valves, Semi-Steel		Crew	Daily Output	Labor-Hours	Unit	Material	2015 Bare Costs Labor	Equipment	Total	Total Incl O&P
0010	**VALVES, SEMI-STEEL** R220523-90									
1020	Lubricated plug valve, threaded, 200 psi									
1030	1/2"	1 Plum	18	.444	Ea.	560	26		586	655
1040	3/4"		16	.500		560	29.50		589.50	660
1050	1"		14	.571		375	33.50		408.50	460
1060	1-1/4"		12	.667		415	39		454	515
1070	1-1/2"		11	.727		440	42.50		482.50	550
1080	2"		8	1		535	58.50		593.50	680
1090	2-1/2"	Q-1	5	3.200		590	169		759	900
1100	3"	"	4.50	3.556		845	188		1,033	1,225
6990	Flanged, 200 psi									
7000	2"	1 Plum	8	1	Ea.	217	58.50		275.50	325
7010	2-1/2"	Q-1	5	3.200		310	169		479	595
7020	3"		4.50	3.556		375	188		563	700
7030	4"		3	5.333		470	282		752	945
7036	5"		2.50	6.400		840	340		1,180	1,425
7040	6"	Q-2	3	8		1,100	440		1,540	1,850
7050	8"		2.50	9.600		1,575	525		2,100	2,525
7060	10"		2.20	10.909		2,450	600		3,050	3,575
7070	12"		1.70	14.118		4,700	775		5,475	6,350

23 05 23.80 Valves, Steel		Crew	Daily Output	Labor-Hours	Unit	Material	2015 Bare Costs Labor	Equipment	Total	Total Incl O&P
0010	**VALVES, STEEL** R220523-90									
0800	Cast									
1350	Check valve, swing type, 150 lb., flanged									
1370	1"	1 Plum	10	.800	Ea.	375	47		422	485
1400	2"	"	8	1		685	58.50		743.50	840
1440	2-1/2"	Q-1	5	3.200		890	169		1,059	1,225
1450	3"		4.50	3.556		805	188		993	1,175
1460	4"		3	5.333		1,225	282		1,507	1,750
1470	6"	Q-2	3	8		1,900	440		2,340	2,725
1480	8"		2.50	9.600		3,300	525		3,825	4,425
1490	10"		2.20	10.909		4,875	600		5,475	6,275
1500	12"		1.70	14.118		7,100	775		7,875	9,000
1510	14"		1.30	18.462		10,600	1,000		11,600	13,200
1520	16"		1	24		11,900	1,325		13,225	15,100
1540	For 300 lb., flanged, add					50%	15%			
1548	For 600 lb., flanged, add					110%	20%			
1571	300 lb., 2"	1 Plum	7.40	1.081		775	63.50		838.50	945
1572	2-1/2"	Q-1	4.20	3.810		1,150	201		1,351	1,575
1573	3"		4	4		1,150	211		1,361	1,600
1574	4"		2.80	5.714		1,575	300		1,875	2,200
1575	6"	Q-2	2.90	8.276		3,075	455		3,530	4,050
1576	8"		2.40	10		4,525	550		5,075	5,800
1577	10"		2.10	11.429		6,800	625		7,425	8,450
1578	12"		1.60	15		9,850	820		10,670	12,100
1579	14"		1.20	20		14,700	1,100		15,800	17,800
1581	16"		.90	26.667		18,700	1,450		20,150	22,800
1950	Gate valve, 150 lb., flanged									
2000	2"	1 Plum	8	1	Ea.	780	58.50		838.50	950
2040	2-1/2"	Q-1	5	3.200		1,100	169		1,269	1,475
2050	3"		4.50	3.556		1,100	188		1,288	1,500
2060	4"		3	5.333		1,350	282		1,632	1,925

For customer support on your Mechanical Cost Data, call 877.795.7394.

261

23 05 23 – General-Duty Valves for HVAC Piping

23 05 23.80 Valves, Steel		Crew	Daily Output	Labor-Hours	Unit	Material	2015 Bare Costs Labor	Equipment	Total	Total Incl O&P
2070	6"	Q-2	3	8	Ea.	2,250	440		2,690	3,125
2080	8"		2.50	9.600		3,425	525		3,950	4,550
2090	10"		2.20	10.909		5,050	600		5,650	6,450
2100	12"		1.70	14.118		6,900	775		7,675	8,750
2110	14"		1.30	18.462		11,300	1,000		12,300	13,900
2120	16"		1	24		14,600	1,325		15,925	18,000
2130	18"		.80	30		19,100	1,650		20,750	23,500
2140	20"		.60	40		22,000	2,200		24,200	27,500
2650	300 lb., flanged									
2700	2"	1 Plum	7.40	1.081	Ea.	1,075	63.50		1,138.50	1,275
2740	2-1/2"	Q-1	4.20	3.810		1,475	201		1,676	1,900
2750	3"		4	4		1,475	211		1,686	1,925
2760	4"		2.80	5.714		2,050	300		2,350	2,700
2770	6"	Q-2	2.90	8.276		3,450	455		3,905	4,475
2780	8"		2.40	10		5,400	550		5,950	6,775
2790	10"		2.10	11.429		7,250	625		7,875	8,925
2800	12"		1.60	15		10,500	820		11,320	12,800
2810	14"		1.20	20		20,900	1,100		22,000	24,700
2820	16"		.90	26.667		26,200	1,450		27,650	31,000
2830	18"		.70	34.286		37,400	1,875		39,275	43,900
2840	20"		.50	48		41,600	2,625		44,225	49,800
3650	Globe valve, 150 lb., flanged									
3700	2"	1 Plum	8	1	Ea.	980	58.50		1,038.50	1,175
3740	2-1/2"	Q-1	5	3.200		1,250	169		1,419	1,625
3750	3"		4.50	3.556		1,250	188		1,438	1,650
3760	4"		3	5.333		1,825	282		2,107	2,425
3770	6"	Q-2	3	8		2,875	440		3,315	3,800
3780	8"		2.50	9.600		5,350	525		5,875	6,675
3790	10"		2.20	10.909		9,825	600		10,425	11,700
3800	12"		1.70	14.118		13,500	775		14,275	16,000
4080	300 lb., flanged									
4100	2"	1 Plum	7.40	1.081	Ea.	1,325	63.50		1,388.50	1,550
4140	2-1/2"	Q-1	4.20	3.810		1,775	201		1,976	2,275
4150	3"		4	4		1,775	211		1,986	2,300
4160	4"		2.80	5.714		2,475	300		2,775	3,175
4170	6"	Q-2	2.90	8.276		4,450	455		4,905	5,575
4180	8"		2.40	10		7,575	550		8,125	9,150
4190	10"		2.10	11.429		14,500	625		15,125	16,900
4200	12"		1.60	15		17,100	820		17,920	20,100
4680	600 lb., flanged									
4700	2"	1 Plum	7	1.143	Ea.	1,825	67		1,892	2,100
4740	2-1/2"	Q-1	4	4		2,925	211		3,136	3,550
4750	3"		3.60	4.444		2,925	235		3,160	3,575
4760	4"		2.50	6.400		4,500	340		4,840	5,450
4770	6"	Q-2	2.60	9.231		9,500	505		10,005	11,300
4780	8"	"	2.10	11.429		15,900	625		16,525	18,400
4800	Silent check, 316 S.S. trim									
4810	Full flange type, 150 lb.									
4811	1"	1 Plum	14	.571	Ea.	535	33.50		568.50	640
4812	1-1/4"		12	.667		535	39		574	650
4813	1-1/2"		11	.727		590	42.50		632.50	710
4814	2"		9	.889		710	52		762	860
4815	2-1/2"	Q-1	9	1.778		855	94		949	1,075

23 05 23 – General-Duty Valves for HVAC Piping

23 05 23.80 Valves, Steel		Crew	Daily Output	Labor-Hours	Unit	Material	2015 Bare Costs Labor	Equipment	Total	Total Incl O&P
4816	3"	Q-1	8	2	Ea.	995	106		1,101	1,250
4817	4"		5	3.200		1,350	169		1,519	1,750
4818	5"	Q-2	6	4		1,700	219		1,919	2,200
4819	6"		5	4.800		2,200	263		2,463	2,825
4820	8"		4.50	5.333		3,175	292		3,467	3,950
4821	10"		4	6		4,475	330		4,805	5,425
4840	For 300 lb., add					20%	10%			
4860	For 600 lb., add					50%	15%			
4900	Globe type, 150 lb.									
4910	2"	1 Plum	9	.889	Ea.	840	52		892	1,000
4911	2-1/2"	Q-1	9	1.778		1,000	94		1,094	1,275
4912	3"		8	2		1,125	106		1,231	1,400
4913	4"		5	3.200		1,500	169		1,669	1,900
4914	5"	Q-2	6	4		1,875	219		2,094	2,400
4915	6"		5	4.800		2,325	263		2,588	2,950
4916	8"		4.50	5.333		3,250	292		3,542	4,025
4917	10"		4	6		5,625	330		5,955	6,700
4918	12"		3	8		6,900	440		7,340	8,250
4919	14"		2.30	10.435		10,500	570		11,070	12,400
4920	16"		1.75	13.714		14,600	750		15,350	17,200
4921	18"		1.50	16		15,400	875		16,275	18,200
4940	For 300 lb., add					30%	10%			
4960	For 600 lb., add					60%	15%			
5150	Forged									
5340	Ball valve, 1500 psi, threaded, 1/4" size	1 Plum	24	.333	Ea.	62	19.55		81.55	97.50
5350	3/8"		24	.333		76	19.55		95.55	113
5360	1/2"		24	.333		76.50	19.55		96.05	114
5370	3/4"		20	.400		102	23.50		125.50	148
5380	1"		19	.421		132	24.50		156.50	183
5390	1-1/4"		15	.533		208	31.50		239.50	276
5400	1-1/2"		13	.615		221	36		257	298
5410	2"		11	.727		305	42.50		347.50	400
5460	Ball valve, 800 lb., socket weld, 1/4" size	Q-15	19	.842		76	44.50	3.04	123.54	154
5470	3/8"		19	.842		76	44.50	3.04	123.54	154
5480	1/2"		19	.842		94	44.50	3.04	141.54	173
5490	3/4"		19	.842		125	44.50	3.04	172.54	207
5500	1"		15	1.067		156	56.50	3.85	216.35	260
5510	1-1/4"		13	1.231		208	65	4.44	277.44	330
5520	1-1/2"		11	1.455		271	77	5.25	353.25	420
5530	2"		8.50	1.882		360	99.50	6.80	466.30	550
5550	Ball valve, 150 lb., flanged									
5560	4"	Q-1	3	5.333	Ea.	1,075	282		1,357	1,600
5570	6"	Q-2	3	8		2,225	440		2,665	3,100
5580	8"	"	2.50	9.600		4,450	525		4,975	5,700
5650	Check valve, class 800, horizontal, socket									
5651	,Socket									
5652	1/4"	Q-15	19	.842	Ea.	94.50	44.50	3.04	142.04	174
5654	3/8"		19	.842		94.50	44.50	3.04	142.04	174
5656	1/2"		19	.842		94.50	44.50	3.04	142.04	174
5658	3/4"		19	.842		101	44.50	3.04	148.54	181
5660	1"		15	1.067		119	56.50	3.85	179.35	220
5662	1-1/4"		13	1.231		233	65	4.44	302.44	360
5664	1-1/2"		11	1.455		233	77	5.25	315.25	380

For customer support on your Mechanical Cost Data, call 877.795.7394.

263

23 05 23.80 Valves, Steel		Crew	Daily Output	Labor-Hours	Unit	Material	2015 Bare Costs Labor	Equipment	Total	Total Incl O&P
5666	2"	Q-15	8.50	1.882	Ea.	325	99.50	6.80	431.30	515
5698	Threaded									
5700	1/4"	1 Plum	24	.333	Ea.	94.50	19.55		114.05	134
5720	3/8"		24	.333		94.50	19.55		114.05	134
5730	1/2"		24	.333		94.50	19.55		114.05	134
5740	3/4"		20	.400		101	23.50		124.50	147
5750	1"		19	.421		119	24.50		143.50	169
5760	1-1/4"		15	.533		233	31.50		264.50	305
5770	1-1/2"		13	.615		233	36		269	310
5780	2"		11	.727		325	42.50		367.50	425
5840	For class 150, flanged, add					100%	15%			
5860	For class 300, flanged, add					120%	20%			
6100	Gate, class 800, OS&Y, socket									
6102	3/8"	Q-15	19	.842	Ea.	65.50	44.50	3.04	113.04	142
6103	1/2"		19	.842		65.50	44.50	3.04	113.04	142
6104	3/4"		19	.842		72	44.50	3.04	119.54	149
6105	1"		15	1.067		87	56.50	3.85	147.35	185
6106	1-1/4"		13	1.231		165	65	4.44	234.44	284
6107	1-1/2"		11	1.455		165	77	5.25	247.25	305
6108	2"		8.50	1.882		185	99.50	6.80	291.30	360
6118	Threaded									
6120	3/8"	1 Plum	24	.333	Ea.	65.50	19.55		85.05	102
6130	1/2"		24	.333		65.50	19.55		85.05	102
6140	3/4"		20	.400		72	23.50		95.50	115
6150	1"		19	.421		87	24.50		111.50	134
6160	1-1/4"		15	.533		165	31.50		196.50	229
6170	1-1/2"		13	.615		165	36		201	236
6180	2"		11	.727		185	42.50		227.50	269
6260	For OS&Y, flanged, add					100%	20%			
6700	Globe, OS&Y, class 800, socket									
6710	1/4"	Q-15	19	.842	Ea.	101	44.50	3.04	148.54	181
6720	3/8"		19	.842		101	44.50	3.04	148.54	181
6730	1/2"		19	.842		101	44.50	3.04	148.54	181
6740	3/4"		19	.842		116	44.50	3.04	163.54	198
6750	1"		15	1.067		151	56.50	3.85	211.35	255
6760	1-1/4"		13	1.231		295	65	4.44	364.44	430
6770	1-1/2"		11	1.455		295	77	5.25	377.25	445
6780	2"		8.50	1.882		380	99.50	6.80	486.30	570
6860	For OS&Y, flanged, add					300%	20%			
6880	Threaded									
6882	1/4"	1 Plum	24	.333	Ea.	101	19.55		120.55	141
6884	3/8"		24	.333		101	19.55		120.55	141
6886	1/2"		24	.333		101	19.55		120.55	141
6888	3/4"		20	.400		116	23.50		139.50	164
6890	1"		19	.421		151	24.50		175.50	204
6892	1-1/4"		15	.533		295	31.50		326.50	375
6894	1-1/2"		13	.615		295	36		331	380
6896	2"		11	.727		380	42.50		422.50	480

23 05 23.90 Valves, Stainless Steel

		Crew	Daily Output	Labor-Hours	Unit	Material	2015 Bare Costs Labor	Equipment	Total	Total Incl O&P
0010	**VALVES, STAINLESS STEEL** R220523-90									
1610	Ball, threaded 1/4"	1 Stpi	24	.333	Ea.	50	19.90		69.90	85
1620	3/8"		24	.333		50	19.90		69.90	85

23 05 23.90 Valves, Stainless Steel		Crew	Daily Output	Labor-Hours	Unit	Material	2015 Bare Costs Labor	Equipment	Total	Total Incl O&P
1630	1/2"	1 Stpi	22	.364	Ea.	50	21.50		71.50	88
1640	3/4"		20	.400		82.50	24		106.50	127
1650	1"		19	.421		101	25		126	149
1660	1-1/4		15	.533		192	32		224	259
1670	1-1/2"		13	.615		200	37		237	276
1680	2"		11	.727		260	43.50		303.50	350
1700	Check, 200 lb., threaded									
1710	1/4"	1 Plum	24	.333	Ea.	143	19.55		162.55	188
1720	1/2"		22	.364		143	21.50		164.50	190
1730	3/4"		20	.400		148	23.50		171.50	198
1750	1"		19	.421		192	24.50		216.50	249
1760	1-1/2"		13	.615		360	36		396	450
1770	2"		11	.727		610	42.50		652.50	735
1800	150 lb., flanged									
1810	2-1/2"	Q-1	5	3.200	Ea.	1,875	169		2,044	2,300
1820	3"		4.50	3.556		1,875	188		2,063	2,325
1830	4"		3	5.333		2,775	282		3,057	3,500
1840	6"	Q-2	3	8		4,925	440		5,365	6,075
1850	8"	"	2.50	9.600		10,100	525		10,625	11,900
2100	Gate, OS&Y, 150 lb., flanged									
2120	1/2"	1 Plum	18	.444	Ea.	595	26		621	695
2140	3/4"		16	.500		575	29.50		604.50	680
2150	1"		14	.571		725	33.50		758.50	850
2160	1-1/2"		11	.727		1,400	42.50		1,442.50	1,625
2170	2"		8	1		1,650	58.50		1,708.50	1,925
2180	2-1/2"	Q-1	5	3.200		2,025	169		2,194	2,475
2190	3"		4.50	3.556		2,025	188		2,213	2,500
2200	4"		3	5.333		2,975	282		3,257	3,700
2205	5"		2.80	5.714		4,800	300		5,100	5,725
2210	6"	Q-2	3	8		5,575	440		6,015	6,775
2220	8"		2.50	9.600		9,675	525		10,200	11,400
2230	10"		2.30	10.435		16,800	570		17,370	19,400
2240	12"		1.90	12.632		22,500	690		23,190	25,900
2260	For 300 lb., flanged, add					120%	15%			
2600	600 lb., flanged									
2620	1/2"	1 Plum	16	.500	Ea.	213	29.50		242.50	279
2640	3/4"		14	.571		230	33.50		263.50	305
2650	1"		12	.667		277	39		316	365
2660	1-1/2"		10	.800		445	47		492	555
2670	2"		7	1.143		610	67		677	770
2680	2-1/2"	Q-1	4	4		7,375	211		7,586	8,450
2690	3"	"	3.60	4.444		7,375	235		7,610	8,475
3100	Globe, OS&Y, 150 lb., flanged									
3120	1/2"	1 Plum	18	.444	Ea.	570	26		596	665
3140	3/4"		16	.500		620	29.50		649.50	725
3150	1"		14	.571		810	33.50		843.50	940
3160	1-1/2"		11	.727		1,275	42.50		1,317.50	1,475
3170	2"		8	1		1,650	58.50		1,708.50	1,925
3180	2-1/2"	Q-1	5	3.200		3,425	169		3,594	4,000
3190	3"		4.50	3.556		3,425	188		3,613	4,025
3200	4"		3	5.333		5,450	282		5,732	6,425
3210	6"	Q-2	3	8		9,200	440		9,640	10,800
5000	Silent check, 316 S.S. body and trim									

23 05 23 – General-Duty Valves for HVAC Piping

23 05 23.90 Valves, Stainless Steel		Crew	Daily Output	Labor-Hours	Unit	Material	2015 Bare Costs Labor	Equipment	Total	Total Incl O&P
5010	Compact wafer type, 300 lb.									
5020	1"	1 Plum	14	.571	Ea.	895	33.50		928.50	1,025
5021	1-1/4"		12	.667		1,000	39		1,039	1,150
5022	1-1/2"		11	.727		1,000	42.50		1,042.50	1,175
5023	2"		9	.889		1,150	52		1,202	1,325
5024	2-1/2"	Q-1	9	1.778		1,575	94		1,669	1,875
5025	3"	"	8	2		1,800	106		1,906	2,150
5100	Full flange wafer type, 150 lb.									
5110	1"	1 Plum	14	.571	Ea.	630	33.50		663.50	740
5111	1-1/4"		12	.667		675	39		714	805
5112	1-1/2"		11	.727		815	42.50		857.50	960
5113	2"		9	.889		1,000	52		1,052	1,175
5114	2-1/2"	Q-1	9	1.778		1,200	94		1,294	1,450
5115	3"		8	2		1,350	106		1,456	1,650
5116	4"		5	3.200		2,025	169		2,194	2,475
5117	5"	Q-2	6	4		2,350	219		2,569	2,900
5118	6"		5	4.800		2,950	263		3,213	3,625
5119	8"		4.50	5.333		4,900	292		5,192	5,850
5120	10"		4	6		6,800	330		7,130	7,975
5200	Globe type, 300 lb.									
5210	2"	1 Plum	9	.889	Ea.	1,125	52		1,177	1,300
5211	2-1/2"	Q-1	9	1.778		1,375	94		1,469	1,675
5212	3"		8	2		1,700	106		1,806	2,025
5213	4"		5	3.200		2,375	169		2,544	2,850
5214	5"	Q-2	6	4		3,300	219		3,519	3,950
5215	6"		5	4.800		3,725	263		3,988	4,500
5216	8"		4.50	5.333		5,275	292		5,567	6,250
5217	10"		4	6		7,475	330		7,805	8,725
5300	Screwed end type, 300 lb.									
5310	1/2"	1 Plum	24	.333	Ea.	79.50	19.55		99.05	117
5311	3/4"		20	.400		94	23.50		117.50	140
5312	1"		19	.421		109	24.50		133.50	158
5313	1-1/4"		15	.533		146	31.50		177.50	208
5314	1-1/2"		13	.615		157	36		193	227
5315	2"		11	.727		194	42.50		236.50	278

23 05 93 – Testing, Adjusting, and Balancing for HVAC

23 05 93.10 Balancing, Air

	23 05 93.10 Balancing, Air	Crew	Daily Output	Labor-Hours	Unit	Material	2015 Bare Costs Labor	Equipment	Total	Total Incl O&P
0010	**BALANCING, AIR** (Subcontractor's quote incl. material and labor) R230500-10									
0900	Heating and ventilating equipment									
1000	Centrifugal fans, utility sets				Ea.				410	410
1100	Heating and ventilating unit								615	615
1200	In-line fan								615	615
1300	Propeller and wall fan								116	116
1400	Roof exhaust fan								274	274
2000	Air conditioning equipment, central station								890	890
2100	Built-up low pressure unit								820	820
2200	Built-up high pressure unit								960	960
2300	Built-up high pressure dual duct								1,500	1,500
2400	Built-up variable volume								1,775	1,775
2500	Multi-zone A.C. and heating unit								615	615
2600	For each zone over one, add								137	137
2700	Package A.C. unit								340	340

23 05 93 – Testing, Adjusting, and Balancing for HVAC

23 05 93.10 Balancing, Air	Crew	Daily Output	Labor-Hours	Unit	Material	2015 Bare Costs Labor	Equipment	Total	Total Incl O&P
2800 Rooftop heating and cooling unit				Ea.				480	480
3000 Supply, return, exhaust, registers & diffusers, avg. height ceiling								82	82
3100 High ceiling								123	123
3200 Floor height								68.50	68.50
3300 Off mixing box								54.50	54.50
3500 Induction unit								89	89
3600 Lab fume hood								410	410
3700 Linear supply								205	205
3800 Linear supply high								239	239
4000 Linear return								68.50	68.50
4100 Light troffers								82	82
4200 Moduline - master								82	82
4300 Moduline - slaves								41	41
4400 Regenerators								545	545
4500 Taps into ceiling plenums								103	103
4600 Variable volume boxes								82	82

23 05 93.20 Balancing, Water

	Crew	Daily Output	Labor-Hours	Unit	Material	Labor	Equipment	Total	Total Incl O&P
0010 **BALANCING, WATER** (Subcontractor's quote incl. material and labor) R230500-10									
0050 Air cooled condenser				Ea.				253	253
0080 Boiler								510	510
0100 Cabinet unit heater								86.50	86.50
0200 Chiller								615	615
0300 Convector								72	72
0400 Converter								360	360
0500 Cooling tower								470	470
0600 Fan coil unit, unit ventilator								130	130
0700 Fin tube and radiant panels								144	144
0800 Main and duct re-heat coils								134	134
0810 Heat exchanger								134	134
0900 Main balancing cocks								108	108
1000 Pumps								320	320
1100 Unit heater								101	101

23 05 93.50 Piping, Testing

	Crew	Daily Output	Labor-Hours	Unit	Material	Labor	Equipment	Total	Total Incl O&P
0010 **PIPING, TESTING**									
0100 Nondestructive testing									
0110 Nondestructive hydraulic pressure test, isolate & 1 hr. hold									
0120 1" - 4" pipe									
0140 0 – 250 L.F.	1 Stpi	1.33	6.015	Ea.		360		360	545
0160 250 – 500 L.F.	"	.80	10			600		600	900
0180 500 – 1000 L.F.	Q-5	1.14	14.035			755		755	1,150
0200 1000 – 2000 L.F.	"	.80	20			1,075		1,075	1,625
0300 6" - 10" pipe									
0320 0 – 250 L.F.	Q-5	1	16	Ea.		860		860	1,300
0340 250 – 500 L.F.		.73	21.918			1,175		1,175	1,775
0360 500 – 1000 L.F.		.53	30.189			1,625		1,625	2,450
0380 1000 – 2000 L.F.		.38	42.105			2,275		2,275	3,425
1000 Pneumatic pressure test, includes soaping joints									
1120 1" - 4" pipe									
1140 0 – 250 L.F.	Q-5	2.67	5.993	Ea.	10.25	320		330.25	495
1160 250 – 500 L.F.		1.33	12.030		20.50	645		665.50	1,000
1180 500 – 1000 L.F.		.80	20		31	1,075		1,106	1,650
1200 1000 – 2000 L.F.		.50	32		41	1,725		1,766	2,650

23 05 93 – Testing, Adjusting, and Balancing for HVAC

23 05 93.50 Piping, Testing		Crew	Daily Output	Labor-Hours	Unit	Material	2015 Bare Costs Labor	Equipment	Total	Total Incl O&P
1300	6" - 10" pipe									
1320	0 – 250 L.F.	Q-5	1.33	12.030	Ea.	10.25	645		655.25	985
1340	250 – 500 L.F.		.67	23.881		20.50	1,275		1,295.50	1,975
1360	500 – 1000 L.F.		.40	40		41	2,150		2,191	3,300
1380	1000 – 2000 L.F.		.25	64		51.50	3,450		3,501.50	5,250
2000	X-Ray of welds									
2110	2" diam.	1 Stpi	8	1	Ea.	13.50	60		73.50	105
2120	3" diam.		8	1		13.50	60		73.50	105
2130	4" diam.		8	1		20.50	60		80.50	113
2140	6" diam.		8	1		20.50	60		80.50	113
2150	8" diam.		6.60	1.212		20.50	72.50		93	132
2160	10" diam.		6	1.333		27	79.50		106.50	150
3000	Liquid penetration of welds									
3110	2" diam.	1 Stpi	14	.571	Ea.	4.50	34		38.50	56.50
3120	3" diam.		13.60	.588		4.50	35		39.50	58
3130	4" diam.		13.40	.597		4.50	35.50		40	59
3140	6" diam.		13.20	.606		4.50	36		40.50	59.50
3150	8" diam.		13	.615		6.75	37		43.75	63
3160	10" diam.		12.80	.625		6.75	37.50		44.25	64

23 07 HVAC Insulation

23 07 13 – Duct Insulation

23 07 13.10 Duct Thermal Insulation

			Crew	Daily Output	Labor-Hours	Unit	Material	2015 Bare Costs Labor	Equipment	Total	Total Incl O&P
0010	**DUCT THERMAL INSULATION**										
0100	Rule of thumb, as a percentage of total mechanical costs					Job				10%	10%
0110	Insulation req'd. is based on the surface size/area to be covered										
3000	Ductwork										
3020	Blanket type, fiberglass, flexible										
3030	Fire rated for grease and hazardous exhaust ducts										
3060	1-1/2" thick		Q-14	300	.053	S.F.	4.54	2.51		7.05	8.90
3090	Fire rated for plenums										
3100	1/2" x24" x 25'		Q-14	7.20	2.222	Roll	167	105		272	345
3110	1/2" x24" x 25'			360	.044	S.F.	3.35	2.09		5.44	6.95
3120	1/2" x 48" x 25'			3.80	4.211	Roll	335	198		533	680
3126	1/2" x 48" x 25'			380	.042	S.F.	3.35	1.98		5.33	6.75
3140	FSK vapor barrier wrap, .75 lb. density										
3160	1" thick	G	Q-14	350	.046	S.F.	.18	2.15		2.33	3.55
3170	1-1/2" thick	G		320	.050		.22	2.36		2.58	3.91
3180	2" thick	G		300	.053		.26	2.51		2.77	4.20
3190	3" thick	G		260	.062		.36	2.90		3.26	4.92
3200	4" thick	G		242	.066		.51	3.12		3.63	5.40
3210	Vinyl jacket, same as FSK										
3280	Unfaced, 1 lb. density										
3310	1" thick	G	Q-14	360	.044	S.F.	.19	2.09		2.28	3.47
3320	1-1/2" thick	G		330	.048		.29	2.28		2.57	3.88
3330	2" thick	G		310	.052		.35	2.43		2.78	4.18
3400	FSK facing, 1 lb. density										
3420	1-1/2" thick	G	Q-14	310	.052	S.F.	.23	2.43		2.66	4.04
3430	2" thick	G	"	300	.053	"	.35	2.51		2.86	4.30
3450	FSK facing, 1.5 lb. density										
3470	1-1/2" thick	G	Q-14	300	.053	S.F.	.36	2.51		2.87	4.31

23 07 HVAC Insulation

23 07 13 – Duct Insulation

23 07 13.10 Duct Thermal Insulation

		Crew	Daily Output	Labor-Hours	Unit	Material	2015 Bare Costs Labor	Equipment	Total	Total Incl O&P
3480	2" thick	**G** Q-14	290	.055	S.F.	.43	2.60		3.03	4.52
3730	Sheet insulation									
3760	Polyethylene foam, closed cell, UV resistant									
3770	Standard temperature (-90°F to +212°F)									
3771	1/4" thick	**G** Q-14	450	.036	S.F.	2.01	1.68		3.69	4.82
3772	3/8" thick	**G**	440	.036		2.48	1.71		4.19	5.40
3773	1/2" thick	**G**	420	.038		3.05	1.80		4.85	6.15
3774	3/4" thick	**G**	400	.040		4.36	1.89		6.25	7.75
3775	1" thick	**G**	380	.042		5.90	1.98		7.88	9.55
3776	1-1/2" thick	**G**	360	.044		9.30	2.09		11.39	13.50
3777	2" thick	**G**	340	.047		12.30	2.22		14.52	17
3778	2-1/2" thick	**G**	320	.050		15.80	2.36		18.16	21
3779	Adhesive (see line 7878)									
3780	Foam, rubber									
3782	1" thick	**G** 1 Stpi	50	.160	S.F.	2.89	9.55		12.44	17.65
3795	Finishes									
3800	Stainless steel woven mesh	Q-14	100	.160	S.F.	.74	7.55		8.29	12.55
3810	For .010" stainless steel, add		160	.100		2.88	4.71		7.59	10.50
3820	18 oz. fiberglass cloth, pasted on		170	.094		.67	4.44		5.11	7.65
3900	8 oz. canvas, pasted on		180	.089		.21	4.19		4.40	6.75
3940	For .016" aluminum jacket, add		200	.080		.92	3.77		4.69	6.85
7000	Board insulation									
7020	Mineral wool, 1200° F									
7022	6 lb. density, plain									
7024	1" thick	**G** Q-14	370	.043	S.F.	.33	2.04		2.37	3.53
7026	1-1/2" thick	**G**	350	.046		.50	2.15		2.65	3.90
7028	2" thick	**G**	330	.048		.67	2.28		2.95	4.30
7030	3" thick	**G**	300	.053		.83	2.51		3.34	4.82
7032	4" thick	**G**	280	.057		1.33	2.69		4.02	5.65
7038	8 lb. density, plain									
7040	1" thick	**G** Q-14	360	.044	S.F.	.40	2.09		2.49	3.70
7042	1-1/2" thick	**G**	340	.047		.61	2.22		2.83	4.12
7044	2" thick	**G**	320	.050		.81	2.36		3.17	4.56
7046	3" thick	**G**	290	.055		1.21	2.60		3.81	5.40
7048	4" thick	**G**	270	.059		1.61	2.79		4.40	6.10
7060	10 lb. density, plain									
7062	1" thick	**G** Q-14	350	.046	S.F.	.55	2.15		2.70	3.96
7064	1-1/2" thick	**G**	330	.048		.83	2.28		3.11	4.47
7066	2" thick	**G**	310	.052		1.11	2.43		3.54	5
7068	3" thick	**G**	280	.057		1.66	2.69		4.35	6
7070	4" thick	**G**	260	.062		2.21	2.90		5.11	6.95
7878	Contact cement, quart can				Ea.	10.50			10.50	11.55

23 07 16 – HVAC Equipment Insulation

23 07 16.10 HVAC Equipment Thermal Insulation

		Crew	Daily Output	Labor-Hours	Unit	Material	2015 Bare Costs Labor	Equipment	Total	Total Incl O&P
0010	**HVAC EQUIPMENT THERMAL INSULATION**									
0100	Rule of thumb, as a percentage of total mechanical costs				Job				10%	10%
0110	Insulation req'd. is based on the surface size/area to be covered									
1000	Boiler, 1-1/2" calcium silicate only	**G** Q-14	110	.145	S.F.	4.64	6.85		11.49	15.75
1020	Plus 2" fiberglass	**G** "	80	.200	"	5.65	9.45		15.10	21
2000	Breeching, 2" calcium silicate									
2020	Rectangular	**G** Q-14	42	.381	S.F.	9.10	17.95		27.05	38
2040	Round	**G** "	38.70	.413	"	9.50	19.50		29	41

For customer support on your Mechanical Cost Data, call 877.795.7394.

269

23 07 HVAC Insulation

23 07 16 – HVAC Equipment Insulation

23 07 16.10 HVAC Equipment Thermal Insulation		Crew	Daily Output	Labor-Hours	Unit	Material	2015 Bare Costs Labor	Equipment	Total	Total Incl O&P
2300	Calcium silicate block, + 200°F to + 1200°F									
2310	On irregular surfaces, valves and fittings									
2340	1" thick	G	Q-14	30	.533	S.F.	4.12	25		29.12	43.50
2360	1-1/2" thick	G		25	.640		4.64	30		34.64	52
2380	2" thick	G		22	.727		6.10	34.50		40.60	60
2400	3" thick	G	↓	18	.889	↓	9.35	42		51.35	75.50
2410	On plane surfaces									
2420	1" thick	G	Q-14	126	.127	S.F.	4.12	6		10.12	13.85
2430	1-1/2" thick	G		120	.133		4.64	6.30		10.94	14.90
2440	2" thick	G		100	.160		6.10	7.55		13.65	18.45
2450	3" thick	G	↓	70	.229	↓	9.35	10.75		20.10	27

23 09 Instrumentation and Control for HVAC

23 09 13 – Instrumentation and Control Devices for HVAC

23 09 13.60 Water Level Controls

		Crew	Daily Output	Labor-Hours	Unit	Material	2015 Bare Costs Labor	Equipment	Total	Total Incl O&P
0010	**WATER LEVEL CONTROLS**									
1000	Electric water feeder	1 Stpi	12	.667	Ea.	293	40		333	380
2000	Feeder cut-off combination									
2100	Steam system up to 5000 sq. ft.	1 Stpi	12	.667	Ea.	625	40		665	750
2200	Steam system above 5000 sq. ft.	↓	12	.667	↓	775	40		815	910
2300	Steam and hot water, high pressure		10	.800		910	48		958	1,075
3000	Low water cut-off for hot water boiler, 50 psi maximum									
3100	1" top & bottom equalizing pipes, manual reset	1 Stpi	14	.571	Ea.	350	34		384	435
3200	1" top & bottom equalizing pipes		14	.571		350	34		384	435
3300	2-1/2" side connection for nipple-to-boiler	↓	14	.571		315	34		349	395
4000	Low water cut-off for low pressure steam with quick hook-up ftgs.									
4100	For installation in gauge glass tappings	1 Stpi	16	.500	Ea.	255	30		285	325
4200	Built-in type, 2-1/2" tap - 3-1/8" insertion		16	.500		210	30		240	276
4300	Built-in type, 2-1/2" tap - 1-3/4" insertion		16	.500		225	30		255	293
4400	Side connection to 2-1/2" tapping		16	.500	↓	240	30		270	310
5000	Pump control, low water cut-off and alarm switch	↓	14	.571	↓	675	34		709	790
9000	Water gauges, complete									
9010	Rough brass, wheel type									
9020	125 PSI at 350°F									
9030	3/8" pipe size	1 Stpi	11	.727	Ea.	62	43.50		105.50	134
9040	1/2" pipe size	"	10	.800	"	62.50	48		110.50	141
9060	200 PSI at 400°F									
9070	3/8" pipe size	1 Stpi	11	.727	Ea.	77.50	43.50		121	151
9080	1/2" pipe size		10	.800		77.50	48		125.50	158
9090	3/4" pipe size	↓	9	.889	↓	99	53		152	189
9130	Rough brass, chain lever type									
9140	250 PSI at 400°F									
9200	Polished brass, wheel type									
9210	200 PSI at 400°F									
9220	3/8" pipe size	1 Stpi	11	.727	Ea.	149	43.50		192.50	230
9230	1/2" pipe size		10	.800		151	48		199	238
9240	3/4" pipe size	↓	9	.889	↓	159	53		212	255
9260	Polished brass, chain lever type									
9270	250 PSI at 400°F									
9280	1/2" pipe size	1 Stpi	10	.800	Ea.	219	48		267	315
9290	3/4" pipe size	"	9	.889	"	237	53		290	340

23 09 13 – Instrumentation and Control Devices for HVAC

23 09 13.60 Water Level Controls		Crew	Daily Output	Labor-Hours	Unit	Material	2015 Bare Costs Labor	Equipment	Total	Total Incl O&P
9400	Bronze, high pressure, ASME									
9410	1/2" pipe size	1 Stpi	10	.800	Ea.	256	48		304	355
9420	3/4" pipe size	"	9	.889	"	289	53		342	400
9460	Chain lever type									
9470	1/2" pipe size	1 Stpi	10	.800	Ea.	350	48		398	455
9480	3/4" pipe size	"	9	.889	"	390	53		443	510
9500	316 stainless steel, high pressure, ASME									
9510	500 PSI at 450°F									
9520	1/2" pipe size	1 Stpi	10	.800	Ea.	835	48		883	990
9530	3/4" pipe size	"	9	.889	"	820	53		873	980

23 09 23 – Direct-Digital Control System for HVAC

23 09 23.10 Control Components/DDC Systems

		Crew	Daily Output	Labor-Hours	Unit	Material	2015 Bare Costs Labor	Equipment	Total	Total Incl O&P
0010	**CONTROL COMPONENTS/DDC SYSTEMS** (Sub's quote incl. M & L)									
0100	Analog inputs									
0110	Sensors (avg. 50' run in 1/2" EMT)									
0120	Duct temperature				Ea.				415	415
0130	Space temperature								665	665
0140	Duct humidity, +/- 3%								695	695
0150	Space humidity, +/- 2%								1,075	1,075
0160	Duct static pressure								565	565
0170	CFM/Transducer								765	765
0172	Water temperature								655	655
0174	Water flow								2,400	2,400
0176	Water pressure differential								975	975
0177	Steam flow								2,400	2,400
0178	Steam pressure								1,025	1,025
0180	KW/Transducer								1,350	1,350
0182	KWH totalization (not incl. elec. meter pulse xmtr.)								625	625
0190	Space static pressure				▼				1,075	1,075
1000	Analog outputs (avg. 50' run in 1/2" EMT)									
1010	P/I Transducer				Ea.				635	635
1020	Analog output, matl. in MUX								305	305
1030	Pneumatic (not incl. control device)								645	645
1040	Electric (not incl. control device)				▼				380	380
2000	Status (Alarms)									
2100	Digital inputs (avg. 50' run in 1/2" EMT)									
2110	Freeze				Ea.				435	435
2120	Fire								395	395
2130	Differential pressure, (air)								595	595
2140	Differential pressure, (water)								975	975
2150	Current sensor								435	435
2160	Duct high temperature thermostat								570	570
2170	Duct smoke detector				▼				705	705
2200	Digital output (avg. 50' run in 1/2" EMT)									
2210	Start/stop				Ea.				340	340
2220	On/off (maintained contact)				"				585	585
3000	Controller MUX panel, incl. function boards									
3100	48 point				Ea.				5,275	5,275
3110	128 point				"				7,225	7,225
3200	DDC controller (avg. 50' run in conduit)									
3210	Mechanical room									
3214	16 point controller (incl. 120 volt/1 phase power supply)				Ea.				3,275	3,275

For customer support on your Mechanical Cost Data, call 877.795.7394.

271

23 09 23 – Direct-Digital Control System for HVAC

23 09 23.10 Control Components/DDC Systems		Crew	Daily Output	Labor-Hours	Unit	Material	2015 Bare Costs Labor	Equipment	Total	Total Incl O&P
3229	32 point controller (incl. 120 volt/1 phase power supply)				Ea.				5,425	5,425
3230	Includes software programming and checkout									
3260	Space									
3266	VAV terminal box (incl. space temp. sensor)				Ea.				840	840
3280	Host computer (avg. 50' run in conduit)									
3281	Package complete with PC, keyboard,									
3282	printer, monitor, basic software				Ea.				3,150	3,150
4000	Front end costs									
4100	Computer (P.C.) with software program				Ea.				6,350	6,350
4200	Color graphics software								3,925	3,925
4300	Color graphics slides								490	490
4350	Additional printer								980	980
4400	Communications trunk cable				L.F.				3.80	3.80
4500	Engineering labor, (not incl. dftg.)				Point				94	94
4600	Calibration labor								120	120
4700	Start-up, checkout labor								120	120
4800	Programming labor, as req'd									
5000	Communications bus (data transmission cable)									
5010	#18 twisted shielded pair in 1/2" EMT conduit				C.L.F.				380	380
8000	Applications software									
8050	Basic maintenance manager software (not incl. data base entry)				Ea.				1,950	1,950
8100	Time program				Point				6.85	6.85
8120	Duty cycle								13.65	13.65
8140	Optimum start/stop								41.50	41.50
8160	Demand limiting								20.50	20.50
8180	Enthalpy program								41.50	41.50
8200	Boiler optimization				Ea.				1,225	1,225
8220	Chiller optimization				"				1,625	1,625
8240	Custom applications									
8260	Cost varies with complexity									

23 09 33 – Electric and Electronic Control System for HVAC

23 09 33.10 Electronic Control Systems

0010	**ELECTRONIC CONTROL SYSTEMS**	R230500-10								
0020	For electronic costs, add to Section 23 09 43.10					Ea.				15%	15%

23 09 43 – Pneumatic Control System for HVAC

23 09 43.10 Pneumatic Control Systems

0010	**PNEUMATIC CONTROL SYSTEMS**		R230500-10								
0011	Including a nominal 50 ft. of tubing. Add control panelboard if req'd.										
0100	Heating and ventilating, split system										
0200	Mixed air control, economizer cycle, panel readout, tubing										
0220	Up to 10 tons	G	Q-19	.68	35.294	Ea.	4,200	1,900		6,100	7,500
0240	For 10 to 20 tons	G		.63	37.915		4,500	2,050		6,550	8,050
0260	For over 20 tons	G		.58	41.096		4,875	2,225		7,100	8,700
0270	Enthalpy cycle, up to 10 tons			.50	48.387		4,650	2,625		7,275	9,050
0280	For 10 to 20 tons			.46	52.174		5,000	2,825		7,825	9,750
0290	For over 20 tons			.42	56.604		5,425	3,050		8,475	10,600
0300	Heating coil, hot water, 3 way valve,										
0320	Freezestat, limit control on discharge, readout		Q-5	.69	23.088	Ea.	3,125	1,250		4,375	5,300
0500	Cooling coil, chilled water, room										
0520	Thermostat, 3 way valve		Q-5	2	8	Ea.	1,400	430		1,830	2,175
0600	Cooling tower, fan cycle, damper control,										
0620	Control system including water readout in/out at panel		Q-19	.67	35.821	Ea.	5,525	1,925		7,450	9,000

23 09 43 – Pneumatic Control System for HVAC

23 09 43.10 Pneumatic Control Systems		Crew	Daily Output	Labor-Hours	Unit	Material	2015 Bare Costs Labor	Equipment	Total	Total Incl O&P
1000	Unit ventilator, day/night operation,									
1100	freezestat, ASHRAE, cycle 2	Q-19	.91	26.374	Ea.	3,050	1,425		4,475	5,525
2000	Compensated hot water from boiler, valve control,									
2100	readout and reset at panel, up to 60 GPM	Q-19	.55	43.956	Ea.	5,725	2,375		8,100	9,875
2120	For 120 GPM		.51	47.059		6,125	2,550		8,675	10,600
2140	For 240 GPM		.49	49.180		6,400	2,650		9,050	11,100
3000	Boiler room combustion air, damper to 5 S.F., controls		1.37	17.582		2,750	950		3,700	4,450
3500	Fan coil, heating and cooling valves, 4 pipe control system		3	8		1,250	435		1,685	2,025
3600	Heat exchanger system controls	▼	.86	27.907	▼	2,675	1,500		4,175	5,225
3900	Multizone control (one per zone), includes thermostat, damper									
3910	motor and reset of discharge temperature	Q-5	.51	31.373	Ea.	2,750	1,675		4,425	5,600
4000	Pneumatic thermostat, including controlling room radiator valve	"	2.43	6.593		830	355		1,185	1,450
4040	Program energy saving optimizer [G]	Q-19	1.21	19.786		6,900	1,075		7,975	9,200
4060	Pump control system	"	3	8		1,275	435		1,710	2,050
4080	Reheat coil control system, not incl. coil	Q-5	2.43	6.593	▼	1,075	355		1,430	1,725
4500	Air supply for pneumatic control system									
4600	Tank mounted duplex compressor, starter, alternator,									
4620	piping, dryer, PRV station and filter									
4630	1/2 HP	Q-19	.68	35.139	Ea.	10,300	1,900		12,200	14,200
4640	3/4 HP		.64	37.383		10,800	2,025		12,825	14,900
4650	1 HP		.61	39.539		11,800	2,150		13,950	16,100
4660	1-1/2 HP		.58	41.739		12,500	2,250		14,750	17,200
4680	3 HP		.55	43.956		17,100	2,375		19,475	22,400
4690	5 HP	▼	.42	57.143	▼	29,800	3,100		32,900	37,500
4800	Main air supply, includes 3/8" copper main and labor	Q-5	1.82	8.791	C.L.F.	345	475		820	1,100
4810	If poly tubing used, deduct								30%	30%
7000	Static pressure control for air handling unit, includes pressure									
7010	sensor, receiver controller, readout and damper motors	Q-19	.64	37.383	Ea.	8,175	2,025		10,200	12,000
7020	If return air fan requires control, add								70%	70%
8600	VAV boxes, incl. thermostat, damper motor, reheat coil & tubing	Q-5	1.46	10.989		1,275	590		1,865	2,300
8610	If no reheat coil, deduct				▼				204	204

23 09 53 – Pneumatic and Electric Control System for HVAC

23 09 53.10 Control Components

		Crew	Daily Output	Labor-Hours	Unit	Material	2015 Bare Costs Labor	Equipment	Total	Total Incl O&P
0010	**CONTROL COMPONENTS**									
0500	Aquastats									
0508	Immersion type									
0510	High/low limit, breaks contact w/temp rise	1 Stpi	4	2	Ea.	305	120		425	515
0514	Sequencing, break 2 switches w/temp rise		4	2		236	120		356	440
0518	Circulating, makes contact w/temp rise	▼	4	2	▼	221	120		341	425
0600	Carbon monoxide detector system									
0606	Panel	1 Stpi	4	2	Ea.	1,075	120		1,195	1,350
0610	Sensor		7.30	1.096		720	65.50		785.50	895
0680	Controller for VAV box, includes actuator	▼	7.30	1.096	▼	278	65.50		343.50	405
0700	Controller, receiver									
0730	Pneumatic, panel mount, single input	1 Plum	8	1	Ea.	515	58.50		573.50	655
0740	With conversion mounting bracket		8	1		515	58.50		573.50	655
0750	Dual input, with control point adjustment	▼	7	1.143		715	67		782	885
0850	Electric, single snap switch	1 Elec	4	2		465	109		574	675
0860	Dual snap switches		3	2.667		625	146		771	910
0870	Humidity controller		8	1		275	54.50		329.50	380
0880	Load limiting controller		8	1		1,025	54.50		1,079.50	1,200
0890	Temperature controller	▼	8	1		570	54.50		624.50	710

For customer support on your Mechanical Cost Data, call 877.795.7394.

273

23 09 53 – Pneumatic and Electric Control System for HVAC

23 09 53.10 Control Components	Crew	Daily Output	Labor-Hours	Unit	Material	2015 Bare Costs Labor	Equipment	Total	Total Incl O&P
0900	Control panel readout									
0910	Panel up to 12 indicators	1 Stpi	1.20	6.667	Ea.	365	400		765	1,000
0914	Panel up to 24 indicators		.86	9.302		450	555		1,005	1,325
0918	Panel up to 48 indicators		.50	16		1,075	955		2,030	2,650
1000	Enthalpy control, boiler water temperature control									
1010	governed by outdoor temperature, with timer	1 Elec	3	2.667	Ea.	420	146		566	680
1300	Energy control/monitor									
1320	BTU computer meter and controller	1 Stpi	1	8	Ea.	1,350	480		1,830	2,225
1600	Flow meters									
1610	Gas	1 Stpi	4	2	Ea.	159	120		279	355
1620	Liquid	"	4	2	"	159	120		279	355
1640	Freezestat									
1644	20' sensing element, adjustable	1 Stpi	3.50	2.286	Ea.	97.50	137		234.50	315
2000	Gauges, pressure or vacuum									
2100	2" diameter dial	1 Stpi	32	.250	Ea.	8.20	14.95		23.15	31.50
2200	2-1/2" diameter dial		32	.250		8.40	14.95		23.35	32
2300	3-1/2" diameter dial		32	.250		10.40	14.95		25.35	34
2400	4-1/2" diameter dial		32	.250		10.70	14.95		25.65	34.50
2700	Flanged iron case, black ring									
2800	3-1/2" diameter dial	1 Stpi	32	.250	Ea.	102	14.95		116.95	135
2900	4-1/2" diameter dial		32	.250		105	14.95		119.95	139
3000	6" diameter dial		32	.250		167	14.95		181.95	206
3010	Steel case, 0 – 300 psi									
3012	2" diam. dial	1 Stpi	16	.500	Ea.	7.60	30		37.60	53.50
3014	4" diam. dial	"	16	.500	"	21	30		51	68
3020	Aluminum case, 0 – 300 psi									
3022	3-1/2" diam. dial	1 Stpi	16	.500	Ea.	10.15	30		40.15	56
3024	4-1/2" diam. dial		16	.500		112	30		142	169
3026	6" diam. dial		16	.500		184	30		214	247
3028	8-1/2" diam. dial		16	.500		315	30		345	390
3030	Brass case, 0 – 300 psi									
3032	2" diam. dial	1 Stpi	16	.500	Ea.	44.50	30		74.50	94
3034	4-1/2" diam. dial	"	16	.500	"	110	30		140	166
3040	Steel case, high pressure, 0 -10,000 psi									
3042	4-1/2" diam. dial	1 Stpi	16	.500	Ea.	291	30		321	365
3044	6-1/2" diam. dial		16	.500		305	30		335	380
3046	8-1/2" diam. dial		16	.500		390	30		420	475
3080	Pressure gauge, differential, magnehelic									
3084	0 – 2" W.C., with air filter kit	1 Stpi	6	1.333	Ea.	105	79.50		184.50	236
3300	For compound pressure-vacuum, add					18%				
3350	Humidistat									
3360	Pneumatic operation									
3361	Room humidistat, direct acting	1 Stpi	12	.667	Ea.	380	40		420	475
3362	Room humidistat, reverse acting		12	.667		380	40		420	475
3363	Room humidity transmitter		17	.471		405	28		433	490
3364	Duct mounted controller		12	.667		440	40		480	545
3365	Duct mounted transmitter		12	.667		400	40		440	500
3366	Humidity indicator, 3-1/2"		28	.286		156	17.05		173.05	197
3390	Electric operated	1 Shee	8	1		70.50	56		126.50	163
3400	Relays									
3430	Pneumatic/electric	1 Plum	16	.500	Ea.	395	29.50		424.50	480
3440	Pneumatic proportioning		8	1		273	58.50		331.50	390
3450	Pneumatic switching		12	.667		179	39		218	256

23 09 53 – Pneumatic and Electric Control System for HVAC

23 09 53.10 Control Components		Crew	Daily Output	Labor-Hours	Unit	Material	2015 Bare Costs Labor	Equipment	Total	Total Incl O&P
3460	Selector, 3 point	1 Plum	6	1.333	Ea.	128	78.50		206.50	258
3470	Pneumatic time delay	↓	8	1	↓	345	58.50		403.50	470
3500	Sensor, air operated									
3520	Humidity	1 Plum	16	.500	Ea.	440	29.50		469.50	530
3540	Pressure	↓	16	.500		67.50	29.50		97	119
3560	Temperature	↓	12	.667	↓	192	39		231	270
3600	Electric operated									
3620	Humidity	1 Elec	8	1	Ea.	157	54.50		211.50	255
3650	Pressure		8	1		248	54.50		302.50	355
3680	Temperature	↓	10	.800	↓	122	44		166	201
3700	Switches									
3710	Minimum position									
3720	Electrical	1 Stpi	4	2	Ea.	21	120		141	203
3730	Pneumatic	"	4	2	"	93.50	120		213.50	283
4000	Thermometers									
4100	Dial type, 3-1/2" diameter, vapor type, union connection	1 Stpi	32	.250	Ea.	335	14.95		349.95	395
4120	Liquid type, union connection		32	.250		650	14.95		664.95	740
4130	Remote reading, 15' capillary		32	.250		251	14.95		265.95	299
4500	Stem type, 6-1/2" case, 2" stem, 1/2" NPT		32	.250		68	14.95		82.95	97
4520	4" stem, 1/2" NPT		32	.250		83	14.95		97.95	114
4600	9" case, 3-1/2" stem, 3/4" NPT		28	.286		115	17.05		132.05	152
4620	6" stem, 3/4" NPT		28	.286		133	17.05		150.05	172
4640	8" stem, 3/4" NPT		28	.286		217	17.05		234.05	265
4660	12" stem, 1" NPT	↓	26	.308	↓	184	18.40		202.40	230
4670	Bi-metal, dial type, steel case brass stem									
4672	2" dial, 4" - 9" stem	1 Stpi	16	.500	Ea.	45	30		75	94.50
4673	2-1/2" dial, 4" - 9" stem		16	.500		45	30		75	94.50
4674	3-1/2" dial, 4" - 9" stem	↓	16	.500	↓	49.50	30		79.50	99.50
4680	Mercury filled, industrial, union connection type									
4682	Angle stem, 7" scale	1 Stpi	16	.500	Ea.	57	30		87	108
4683	9" scale		16	.500		153	30		183	213
4684	12" scale		16	.500		197	30		227	261
4686	Straight stem, 7" scale		16	.500		150	30		180	210
4687	9" scale		16	.500		167	30		197	229
4688	12" scale	↓	16	.500	↓	206	30		236	272
4690	Mercury filled, industrial, separable socket type, with well									
4692	Angle stem, with socket, 7" scale	1 Stpi	16	.500	Ea.	131	30		161	189
4693	9" scale		16	.500		126	30		156	184
4694	12" scale		16	.500		231	30		261	299
4696	Straight stem, with socket, 7" scale		16	.500		91.50	30		121.50	145
4697	9" scale		16	.500		112	30		142	168
4698	12" scale	↓	16	.500	↓	174	30		204	236
5000	Thermostats									
5030	Manual	1 Stpi	8	1	Ea.	51	60		111	146
5040	1 set back, electric, timed G		8	1		37	60		97	131
5050	2 set back, electric, timed G		8	1		186	60		246	295
5100	Locking cover	↓	20	.400		18.80	24		42.80	56.50
5200	24 hour, automatic, clock G	1 Shee	8	1		165	56		221	268
5220	Electric, low voltage, 2 wire	1 Elec	13	.615		49.50	33.50		83	105
5230	3 wire		10	.800		36	44		80	106
5236	Heating/cooling, low voltage, with clock	↓	8	1	↓	259	54.50		313.50	365
5240	Pneumatic									
5250	Single temp., single pressure	1 Stpi	8	1	Ea.	252	60		312	370

23 09 53 – Pneumatic and Electric Control System for HVAC

23 09 53.10 Control Components		Crew	Daily Output	Labor-Hours	Unit	Material	2015 Bare Costs Labor	2015 Bare Costs Equipment	Total	Total Incl O&P
5251	Dual pressure	1 Stpi	8	1	Ea.	355	60		415	480
5252	Dual temp., dual pressure		8	1		325	60		385	450
5253	Reverse acting w/averaging element		8	1		261	60		321	375
5254	Heating-cooling w/deadband		8	1		565	60		625	710
5255	Integral w/piston top valve actuator		8	1		231	60		291	345
5256	Dual temp., dual pressure		8	1		236	60		296	350
5257	Low limit, 8' averaging element		8	1		216	60		276	330
5258	Room single temp. proportional		8	1		90.50	60		150.50	190
5260	Dual temp, direct acting for VAV fan		8	1		248	60		308	365
5262	Capillary tube type, 20', 30°F to 100°F	↓	8	1	↓	249	60		309	365
5300	Transmitter, pneumatic									
5320	Temperature averaging element	Q-1	8	2	Ea.	151	106		257	325
5350	Pressure differential	1 Plum	7	1.143		1,275	67		1,342	1,500
5370	Humidity, duct		8	1		400	58.50		458.50	530
5380	Room		12	.667		405	39		444	505
5390	Temperature, with averaging element	↓	6	1.333	↓	207	78.50		285.50	345
5500	Timer/time clocks									
5510	7 day, 12 hr. battery	1 Stpi	2.60	3.077	Ea.	201	184		385	500
5600	Recorders									
5610	Hydrograph humidity, wall mtd., aluminum case, grad. chart									
5612	8"	1 Stpi	1.60	5	Ea.	815	299		1,114	1,350
5614	10"	"	1.60	5	"	810	299		1,109	1,350
5620	Thermo-hydrograph, wall mtd., grad. chart, 5' SS probe									
5622	8"	1 Stpi	1.60	5	Ea.	785	299		1,084	1,325
5624	10"	"	1.60	5	"	1,250	299		1,549	1,825
5630	Time of operation, wall mtd., 1 pen cap, 24 hr. clock									
5632	6"	1 Stpi	1.60	5	Ea.	420	299		719	910
5634	8"	"	1.60	5	"	660	299		959	1,175
5640	Pressure & vacuum, bourdon tube type									
5642	Flush mount, 1 pen									
5644	4"	1 Stpi	1.60	5	Ea.	450	299		749	945
5645	6"		1.60	5		450	299		749	945
5646	8"		1.60	5		725	299		1,024	1,250
5648	10"	↓	1.60	5	↓	955	299		1,254	1,500
6000	Valves, motorized zone									
6100	Sweat connections, 1/2" C x C	1 Stpi	20	.400	Ea.	168	24		192	220
6110	3/4" C x C		20	.400		168	24		192	221
6120	1" C x C		19	.421		219	25		244	279
6140	1/2" C x C, with end switch, 2 wire		20	.400		166	24		190	218
6150	3/4" C x C, with end switch, 2 wire		20	.400		177	24		201	231
6160	1" C x C, with end switch, 2 wire	↓	19	.421	↓	202	25		227	260
7090	Valves, motor controlled, including actuator									
7100	Electric motor actuated									
7200	Brass, two way, screwed									
7210	1/2" pipe size	L-6	36	.333	Ea.	296	19.10		315.10	355
7220	3/4" pipe size		30	.400		525	23		548	610
7230	1" pipe size		28	.429		605	24.50		629.50	700
7240	1-1/2" pipe size		19	.632		665	36		701	785
7250	2" pipe size	↓	16	.750	↓	975	43		1,018	1,150
7350	Brass, three way, screwed									
7360	1/2" pipe size	L-6	33	.364	Ea.	320	21		341	385
7370	3/4" pipe size		27	.444		375	25.50		400.50	450
7380	1" pipe size	↓	25.50	.471		485	27		512	575

23 09 53 – Pneumatic and Electric Control System for HVAC

23 09 53.10 Control Components		Crew	Daily Output	Labor-Hours	Unit	Material	2015 Bare Costs Labor	Equipment	Total	Total Incl O&P
7384	1-1/4" pipe size	L-6	21	.571	Ea.	620	33		653	730
7390	1-1/2" pipe size		17	.706		705	40.50		745.50	835
7400	2" pipe size	↓	14	.857	↓	945	49		994	1,100
7550	Iron body, two way, flanged									
7560	2-1/2" pipe size	L-6	4	3	Ea.	1,400	172		1,572	1,800
7570	3" pipe size		3	4		1,475	229		1,704	1,975
7580	4" pipe size	↓	2	6	↓	2,550	345		2,895	3,350
7850	Iron body, three way, flanged									
7860	2-1/2" pipe size	L-6	3	4	Ea.	1,425	229		1,654	1,900
7870	3" pipe size		2.50	4.800		1,650	275		1,925	2,225
7880	4" pipe size	↓	2	6	↓	2,050	345		2,395	2,775
8000	Pneumatic, air operated									
8050	Brass, two way, screwed									
8060	1/2" pipe size, class 250	1 Plum	24	.333	Ea.	207	19.55		226.55	258
8070	3/4" pipe size, class 250		20	.400		248	23.50		271.50	310
8080	1" pipe size, class 250		19	.421		290	24.50		314.50	360
8090	1-1/4" pipe size, class 125		15	.533		360	31.50		391.50	445
8100	1-1/2" pipe size, class 125		13	.615		460	36		496	560
8110	2" pipe size, class 125	↓	11	.727	↓	535	42.50		577.50	650
8180	Brass, three way, screwed									
8190	1/2" pipe size, class 250	1 Plum	22	.364	Ea.	228	21.50		249.50	283
8200	3/4" pipe size, class 250		18	.444		282	26		308	350
8210	1" pipe size, class 250		17	.471		330	27.50		357.50	405
8214	1-1/4" pipe size, class 250		14	.571		465	33.50		498.50	560
8220	1-1/2" pipe size, class 125		11	.727		550	42.50		592.50	670
8230	2" pipe size, class 125	↓	9	.889	↓	665	52		717	810
8450	Iron body, two way, flanged									
8510	2-1/2" pipe size, class 125	Q-1	5	3.200	Ea.	1,125	169		1,294	1,500
8520	3" pipe size, class 125		4.50	3.556		1,225	188		1,413	1,625
8530	4" pipe size, class 125	↓	3	5.333		1,575	282		1,857	2,150
8540	5" pipe size, class 125	Q-2	3.40	7.059		3,375	385		3,760	4,300
8550	6" pipe size, class 125	"	3	8	↓	3,925	440		4,365	4,975
8560	Iron body, three way, flanged									
8570	2-1/2" pipe size, class 125	Q-1	4.50	3.556	Ea.	1,225	188		1,413	1,625
8580	3" pipe size, class 125		4	4		1,400	211		1,611	1,875
8590	4" pipe size, class 125	↓	2.50	6.400		2,825	340		3,165	3,600
8600	6" pipe size, class 125	Q-2	3	8	↓	4,300	440		4,740	5,375
9005	Pneumatic system misc. components									
9010	Adding/subtracting repeater	1 Stpi	20	.400	Ea.	198	24		222	253
9020	Adjustable ratio network		16	.500		186	30		216	250
9030	Comparator	↓	20	.400	↓	390	24		414	465
9040	Cumulator									
9041	Air switching	1 Stpi	20	.400	Ea.	179	24		203	233
9042	Averaging		20	.400		167	24		191	219
9043	2:1 ratio		20	.400		375	24		399	450
9044	Sequencing		20	.400		114	24		138	162
9045	Two-position		16	.500		224	30		254	291
9046	Two-position pilot	↓	16	.500	↓	425	30		455	515
9050	Damper actuator									
9051	For smoke control	1 Stpi	8	1	Ea.	340	60		400	465
9052	For ventilation	"	8	1		252	60		312	365
9053	Series duplex pedestal mounted									
9054	For inlet vanes on fans, compressors	1 Stpi	6	1.333	Ea.	1,925	79.50		2,004.50	2,250

23 09 53.10 Control Components	Crew	Daily Output	Labor-Hours	Unit	Material	2015 Bare Costs Labor	Equipment	Total	Total Incl O&P
9060	Enthalpy logic center	1 Stpi	16	.500	Ea.	375	30		405	460
9070	High/low pressure selector		22	.364		72	21.50		93.50	113
9080	Signal limiter		20	.400		114	24		138	162
9081	Transmitter		28	.286		43.50	17.05		60.55	74
9090	Optimal start									
9092	Mass temperature transmitter	1 Stpi	18	.444	Ea.	171	26.50		197.50	228
9100	Step controller with positioner, time delay restrictor									
9110	recycler air valve, switches and cam settings									
9112	With cabinet									
9113	6 points	1 Stpi	4	2	Ea.	1,150	120		1,270	1,450
9114	6 points, 6 manual sequences		2	4		1,000	239		1,239	1,450
9115	8 points		3	2.667		1,250	159		1,409	1,650
9200	Pressure controller and switches									
9210	High static pressure limit	1 Stpi	8	1	Ea.	315	60		375	440
9220	Pressure transmitter		8	1		269	60		329	385
9230	Differential pressure transmitter		8	1		1,050	60		1,110	1,275
9240	Static pressure transmitter		8	1		1,100	60		1,160	1,300
9250	Proportional-only control, single input		6	1.333		515	79.50		594.50	685
9260	Proportional plus integral, single input		6	1.333		770	79.50		849.50	970
9270	Proportional-only, dual input		6	1.333		715	79.50		794.50	905
9280	Proportional plus integral, dual input		6	1.333		940	79.50		1,019.50	1,150
9281	Time delay for above proportional units		16	.500		187	30		217	251
9290	Proportional/2 position controller units		6	1.333		545	79.50		624.50	720
9300	Differential pressure control direct/reverse		6	1.333		480	79.50		559.50	645
9310	Air pressure reducing valve									
9311	1/8" size	1 Stpi	18	.444	Ea.	40	26.50		66.50	84
9315	Precision valve, 1/4" size		17	.471		222	28		250	288
9320	Air flow controller		12	.667		208	40		248	288
9330	Booster relay volume amplifier		18	.444		123	26.50		149.50	175
9340	Series restrictor, straight		60	.133		6.30	7.95		14.25	19
9341	T-fitting		40	.200		9.15	11.95		21.10	28
9342	In-line adjustable		48	.167		40.50	9.95		50.45	60
9350	Diode tee		40	.200		21	11.95		32.95	41
9351	Restrictor tee		40	.200		21	11.95		32.95	41
9360	Pneumatic gradual switch		8	1		355	60		415	480
9361	Selector switch		8	1		128	60		188	230
9370	Electro-pneumatic motor driven servo		10	.800		1,150	48		1,198	1,350
9390	Fan control switch and mounting base		16	.500		127	30		157	185
9400	Circulating pump sequencer		14	.571		1,475	34		1,509	1,675
9410	Pneumatic tubing, fittings and accessories									
9414	Tubing, urethane									
9415	1/8" OD x 1/16" ID	1 Stpi	120	.067	L.F.	.84	3.98		4.82	6.90
9416	1/4" OD x 1/8" ID		115	.070		1.65	4.16		5.81	8.10
9417	5/32" OD x 3/32" ID		110	.073		.94	4.35		5.29	7.60
9420	Coupling, straight									
9422	Barb x barb									
9423	1/4" x 5/32"	1 Stpi	160	.050	Ea.	.93	2.99		3.92	5.55
9424	1/4" x 1/4"		158	.051		.72	3.03		3.75	5.35
9425	3/8" x 3/8"		154	.052		.69	3.10		3.79	5.45
9426	3/8" x 1/4"		150	.053		.65	3.19		3.84	5.55
9427	1/2" x 1/4"		148	.054		6	3.23		9.23	11.50
9428	1/2" x 3/8"		144	.056		1.25	3.32		4.57	6.40
9429	1/2" x 1/2"		140	.057		.82	3.41		4.23	6.05

23 09 53 – Pneumatic and Electric Control System for HVAC

23 09 53.10 Control Components	Crew	Daily Output	Labor-Hours	Unit	Material	2015 Bare Costs Labor	Equipment	Total	Total Incl O&P
9440	Tube x tube								
9441	1/4" x 1/4"	1 Stpi	100	.080	Ea.	2.57	4.78		7.35	10.05
9442	3/8" x 1/4"		96	.083		4.77	4.98		9.75	12.75
9443	3/8" x 3/8"		92	.087		3.19	5.20		8.39	11.35
9444	1/2" x 3/8"		88	.091		5.25	5.45		10.70	14
9445	1/2" x 1/2"		84	.095		5.85	5.70		11.55	15
9450	Elbow coupling								
9452	Barb x barb								
9454	1/4" x 1/4"	1 Stpi	158	.051	Ea.	1.46	3.03		4.49	6.20
9455	3/8" x 3/8"		154	.052		3.91	3.10		7.01	9
9456	1/2" x 1/2"		140	.057		1.73	3.41		5.14	7.05
9460	Tube x tube								
9462	1/2" x 1/2"	1 Stpi	84	.095	Ea.	8.40	5.70		14.10	17.85
9470	Tee coupling								
9472	Barb x barb x barb								
9474	1/4" x 1/4" x 5/32"	1 Stpi	108	.074	Ea.	1.70	4.43		6.13	8.55
9475	1/4" x 1/4" x 1/4"		104	.077		1.15	4.60		5.75	8.20
9476	3/8" x 3/8" x 5/32"		102	.078		3.59	4.69		8.28	11
9477	3/8" x 3/8" x 1/4"		99	.081		1.70	4.83		6.53	9.15
9478	3/8" x 3/8" x 3/8"		98	.082		3.91	4.88		8.79	11.65
9479	1/2" x 1/2" x 1/4"		96	.083		4.65	4.98		9.63	12.60
9480	1/2" x 1/2" x 3/8"		95	.084		3.88	5.05		8.93	11.85
9481	1/2" x 1/2" x 1/2"		92	.087		5.20	5.20		10.40	13.60
9484	Tube x tube x tube								
9485	1/4" x 1/4" x 1/4"	1 Stpi	66	.121	Ea.	5.25	7.25		12.50	16.75
9486	3/8" x 1/4" x 1/4"		64	.125		15.55	7.45		23	28.50
9487	3/8" x 3/8" x 1/4"		61	.131		7.40	7.85		15.25	19.95
9488	3/8" x 3/8" x 3/8"		60	.133		7.75	7.95		15.70	20.50
9489	1/2" x 1/2" x 3/8"		58	.138		16.65	8.25		24.90	31
9490	1/2" x 1/2" x 1/2"		56	.143		11.25	8.55		19.80	25.50
9492	Needle valve								
9494	Tube x tube								
9495	1/4" x 1/4"	1 Stpi	86	.093	Ea.	9.05	5.55		14.60	18.35
9496	3/8" x 3/8"	"	82	.098	"	13.45	5.85		19.30	23.50
9600	Electronic system misc. components								
9610	Electric motor damper actuator	1 Elec	8	1	Ea.	600	54.50		654.50	735
9620	Damper position indicator		10	.800		149	44		193	230
9630	Pneumatic-electronic transducer		12	.667		78	36.50		114.50	140
9700	Modulating step controller								
9701	2-5 steps	1 Elec	5.30	1.509	Ea.	810	82.50		892.50	1,025
9702	6-10 steps		3.20	2.500		221	137		358	450
9703	11-15 steps		2.70	2.963		455	162		617	745
9704	16-20 steps		1.60	5		680	274		954	1,150
9710	Staging network								
9740	Accessory unit regulator								
9750	Accessory power supply	1 Elec	18	.444	Ea.	229	24.50		253.50	288
9760	Step down transformer	"	16	.500	"	228	27.50		255.50	292

23 11 13 – Facility Fuel-Oil Piping

23 11 13.10 Fuel Oil Specialties	Crew	Daily Output	Labor-Hours	Unit	Material	2015 Bare Costs Labor	2015 Bare Costs Equipment	Total	Total Incl O&P
0010 **FUEL OIL SPECIALTIES**									
0020 Foot valve, single poppet, metal to metal construction									
0040 Bevel seat, 1/2" diameter	1 Stpi	20	.400	Ea.	61.50	24		85.50	104
0060 3/4" diameter		18	.444		71	26.50		97.50	119
0080 1" diameter		16	.500		91	30		121	145
0100 1-1/4" diameter		15	.533		118	32		150	178
0120 1-1/2" diameter		13	.615		163	37		200	236
0140 2" diameter		11	.727		169	43.50		212.50	252
0160 Foot valve, double poppet, metal to metal construction									
0164 1" diameter	1 Stpi	15	.533	Ea.	186	32		218	253
0166 1-1/2" diameter	"	12	.667	"	223	40		263	305
0400 Fuel fill box, flush type									
0408 Nonlocking, watertight									
0410 1-1/2" diameter	1 Stpi	12	.667	Ea.	22	40		62	84
0440 2" diameter		10	.800		18.75	48		66.75	92.50
0450 2-1/2" diameter		9	.889		62.50	53		115.50	149
0460 3" diameter		7	1.143		55.50	68.50		124	164
0470 4" diameter		5	1.600		76.50	95.50		172	229
0500 Locking inner cover									
0510 2" diameter	1 Stpi	8	1	Ea.	84	60		144	183
0520 2-1/2" diameter		7	1.143		113	68.50		181.50	227
0530 3" diameter		5	1.600		121	95.50		216.50	277
0540 4" diameter		4	2		155	120		275	350
0600 Fuel system components									
0620 Spill container	1 Stpi	4	2	Ea.	790	120		910	1,050
0640 Fill adapter, 4", straight drop		8	1		69	60		129	166
0680 Fill cap, 4"		30	.267		37	15.95		52.95	65
0700 Extractor fitting, 4" x 1-1/2"		8	1		425	60		485	560
0740 Vapor hose adapter, 4"		8	1		111	60		171	212
0760 Wood gage stick, 10'					14.50			14.50	15.95
1000 Oil filters, 3/8" IPT, 20 gal. per hour	1 Stpi	20	.400		34.50	24		58.50	74
1020 32 gal. per hour		18	.444		46.50	26.50		73	91.50
1040 40 gal. per hour		16	.500		52	30		82	103
1060 50 gal. per hour		14	.571		60	34		94	118
2000 Remote tank gauging system, self contained									
2100 Single tank kit/8 sensors inputs w/printer	1 Stpi	2.50	3.200	Ea.	3,800	191		3,991	4,475
2120 Two tank kit/8 sensors inputs w/printer		2	4		5,125	239		5,364	5,975
3000 Valve, ball check, globe type, 3/8" diameter		24	.333		12.80	19.90		32.70	44
3500 Fusible, 3/8" diameter		24	.333		13.10	19.90		33	44.50
3600 1/2" diameter		24	.333		31.50	19.90		51.40	65
3610 3/4" diameter		20	.400		70	24		94	113
3620 1" diameter		19	.421		205	25		230	264
4000 Nonfusible, 3/8" diameter		24	.333		22.50	19.90		42.40	55
4500 Shutoff, gate type, lever handle, spring-fusible kit									
4520 1/4" diameter	1 Stpi	14	.571	Ea.	34.50	34		68.50	89.50
4540 3/8" diameter		12	.667		33.50	40		73.50	97
4560 1/2" diameter		10	.800		49.50	48		97.50	127
4570 3/4" diameter		8	1		86	60		146	185
4580 Lever handle, requires weight and fusible kit									
4600 1" diameter	1 Stpi	9	.889	Ea.	261	53		314	365
4620 1-1/4" diameter		8	1		320	60		380	440
4640 1-1/2" diameter		7	1.143		300	68.50		368.50	440
4660 2" diameter		6	1.333		545	79.50		624.50	720

23 11 Facility Fuel Piping

23 11 13 – Facility Fuel-Oil Piping

23 11 13.10 Fuel Oil Specialties

		Crew	Daily Output	Labor-Hours	Unit	Material	2015 Bare Costs Labor	Equipment	Total	Total Incl O&P
4680	For fusible link, weight and braided wire, add				Ea.	5%				
5000	Vent alarm, whistling signal					26			26	28.50
5500	Vent protector/breather, 1-1/4" diameter	1 Stpi	32	.250		13.60	14.95		28.55	37.50
5520	1-1/2" diameter		32	.250		15.20	14.95		30.15	39
5540	2" diameter		32	.250		30.50	14.95		45.45	56
5560	3" diameter		28	.286		46.50	17.05		63.55	77
5580	4" diameter		24	.333		56	19.90		75.90	91.50
5600	Dust cap, breather, 2"	↓	40	.200	↓	10.90	11.95		22.85	30
8000	Fuel oil and tank heaters									
8020	Electric, capacity rated at 230 volts									
8040	Immersion element in steel manifold									
8060	96 GPH at 50°F rise	Q-5	6.40	2.500	Ea.	1,175	134		1,309	1,500
8070	128 GPH at 50°F rise		6.20	2.581		1,225	139		1,364	1,550
8080	160 GPH at 50°F rise		5.90	2.712		1,350	146		1,496	1,725
8090	192 GPH at 50°F rise		5.50	2.909		1,475	156		1,631	1,850
8100	240 GPH at 50°F rise		5.10	3.137		1,700	169		1,869	2,125
8110	288 GPH at 50°F rise		4.60	3.478		1,925	187		2,112	2,400
8120	384 GPH at 50°F rise		3.10	5.161		2,250	278		2,528	2,900
8130	480 GPH at 50°F rise		2.30	6.957		2,825	375		3,200	3,675
8140	576 GPH at 50°F rise	↓	2.10	7.619	↓	2,825	410		3,235	3,725
8300	Suction stub, immersion type									
8320	75" long, 750 watts	1 Stpi	14	.571	Ea.	805	34		839	935
8330	99" long, 2000 watts		12	.667		1,125	40		1,165	1,300
8340	123" long, 3000 watts	↓	10	.800	↓	1,225	48		1,273	1,400
8660	Steam, cross flow, rated at 5 PSIG									
8680	42 GPH	Q-5	7	2.286	Ea.	1,850	123		1,973	2,225
8690	73 GPH		6.70	2.388		2,150	128		2,278	2,575
8700	112 GPH		6.20	2.581		2,250	139		2,389	2,675
8710	158 GPH		5.80	2.759		2,450	148		2,598	2,925
8720	187 GPH		4	4		3,350	215		3,565	4,025
8730	270 GPH		3.60	4.444		3,625	239		3,864	4,325
8740	365 GPH	Q-6	4.90	4.898		4,250	273		4,523	5,075
8750	635 GPH		3.70	6.486		5,375	360		5,735	6,450
8760	845 GPH		2.50	9.600		7,850	535		8,385	9,450
8770	1420 GPH		1.60	15		12,800	835		13,635	15,400
8780	2100 GPH	↓	1.10	21.818	↓	17,600	1,225		18,825	21,100

23 11 23 – Facility Natural-Gas Piping

23 11 23.10 Gas Meters

		Crew	Daily Output	Labor-Hours	Unit	Material	2015 Bare Costs Labor	Equipment	Total	Total Incl O&P
0010	**GAS METERS**									
4000	Residential									
4010	Gas meter, residential, 3/4" pipe size	1 Plum	14	.571	Ea.	228	33.50		261.50	300
4020	Gas meter, residential, 1" pipe size		12	.667		183	39		222	260
4030	Gas meter, residential, 1-1/4" pipe size	↓	10	.800	↓	183	47		230	272

23 11 23.20 Gas Piping, Flexible (Csst)

		Crew	Daily Output	Labor-Hours	Unit	Material	2015 Bare Costs Labor	Equipment	Total	Total Incl O&P
0010	**GAS PIPING, FLEXIBLE (CSST)**									
0100	Tubing with lightning protection									
0110	3/8"	1 Stpi	65	.123	L.F.	2.47	7.35		9.82	13.80
0120	1/2"		62	.129		2.77	7.70		10.47	14.70
0130	3/4"		60	.133		3.60	7.95		11.55	16
0140	1"		55	.145		5.15	8.70		13.85	18.75
0150	1-1/4"		50	.160		6.45	9.55		16	21.50
0160	1-1/2"		45	.178		11.30	10.60		21.90	28.50

For customer support on your Mechanical Cost Data, call 877.795.7394.

281

23 11 23.20 Gas Piping, Flexible (Csst)	Crew	Daily Output	Labor-Hours	Unit	Material	2015 Bare Costs Labor	Equipment	Total	Total Incl O&P
0170 2"	1 Stpi	40	.200	L.F.	16.25	11.95		28.20	36
0200 Tubing for underground/underslab burial									
0210 3/8"	1 Stpi	65	.123	L.F.	3.43	7.35		10.78	14.85
0220 1/2"		62	.129		3.91	7.70		11.61	15.95
0230 3/4"		60	.133		4.96	7.95		12.91	17.50
0240 1"		55	.145		6.70	8.70		15.40	20.50
0250 1-1/4"		50	.160		8.95	9.55		18.50	24.50
0260 1-1/2"		45	.178		16.80	10.60		27.40	34.50
0270 2"	↓	40	.200	↓	20	11.95		31.95	40
3000 Fittings									
3010 Straight									
3100 Tube to NPT									
3110 3/8"	1 Stpi	29	.276	Ea.	9.90	16.50		26.40	36
3120 1/2"		27	.296		10.65	17.70		28.35	38
3130 3/4"		25	.320		14.50	19.10		33.60	45
3140 1"		23	.348		22.50	21		43.50	56
3150 1-1/4"		20	.400		50	24		74	91
3160 1-1/2"		17	.471		102	28		130	156
3170 2"	↓	15	.533	↓	174	32		206	240
3200 Coupling									
3210 3/8"	1 Stpi	29	.276	Ea.	17.20	16.50		33.70	44
3220 1/2"		27	.296		19.60	17.70		37.30	48
3230 3/4"		25	.320		27	19.10		46.10	58.50
3240 1"		23	.348		41.50	21		62.50	77.50
3250 1-1/4"		20	.400		92.50	24		116.50	138
3260 1-1/2"		17	.471		194	28		222	256
3270 2"	↓	15	.533	↓	330	32		362	415
3300 Flange fitting									
3310 3/8"	1 Stpi	25	.320	Ea.	15.15	19.10		34.25	45.50
3320 1/2"		22	.364		14.45	21.50		35.95	49
3330 3/4"		19	.421		18.75	25		43.75	58.50
3340 1"		16	.500		26	30		56	73.50
3350 1-1/4"	↓	12	.667	↓	60	40		100	126
3400 90° Flange valve									
3410 3/8"	1 Stpi	25	.320	Ea.	29	19.10		48.10	61
3420 1/2"		22	.364		29.50	21.50		51	65.50
3430 3/4"	↓	19	.421	↓	36	25		61	77.50
4000 Tee									
4120 1/2"	1 Stpi	20.50	.390	Ea.	31	23.50		54.50	69
4130 3/4"		19	.421		41	25		66	83
4140 1"	↓	17.50	.457	↓	74	27.50		101.50	123
5000 Reducing									
5110 Tube to NPT									
5120 3/4" to 1/2" NPT	1 Stpi	26	.308	Ea.	15.60	18.40		34	45
5130 1" to 3/4" NPT	"	24	.333	"	24	19.90		43.90	56.50
5200 Reducing tee									
5210 1/2" x 3/8" x 3/8"	1 Stpi	21	.381	Ea.	40.50	23		63.50	79.50
5220 3/4" x 1/2" x 1/2"		20	.400		38.50	24		62.50	78
5230 1" x 3/4" x 1/2"		18	.444		63.50	26.50		90	110
5240 1-1/4" x 1-1/4" x 1"		15.60	.513		169	30.50		199.50	233
5250 1-1/2" x 1-1/2" x 1-1/4"		13.30	.602		258	36		294	340
5260 2" x 2" x 1-1/2"	↓	11.80	.678		390	40.50		430.50	485
5300 Manifold with four ports and mounting bracket									

For customer support on your Mechanical Cost Data, call 877.795.7394.

23 11 Facility Fuel Piping

23 11 23 – Facility Natural-Gas Piping

23 11 23.20 Gas Piping, Flexible (Csst)		Crew	Daily Output	Labor-Hours	Unit	Material	2015 Bare Costs Labor	2015 Bare Costs Equipment	Total	Total Incl O&P
5302	Labor to mount manifold does not include making pipe									
5304	connections which are included in fitting labor.									
5310	3/4" x 1/2" x 1/2"(4)	1 Stpi	76	.105	Ea.	33.50	6.30		39.80	46
5330	1-1/4" x 1" x 3/4"(4)		72	.111		48	6.65		54.65	63
5350	2" x 1-1/2" x 1"(4)		68	.118		60	7.05		67.05	76.50
5600	Protective striker plate									
5610	Quarter plate, 3" x 2"	1 Stpi	88	.091	Ea.	.77	5.45		6.22	9.05
5620	Half plate, 3" x 7"		82	.098		1.50	5.85		7.35	10.45
5630	Full plate, 3" x 12"		78	.103		3.11	6.15		9.26	12.65

23 12 Facility Fuel Pumps

23 12 13 – Facility Fuel-Oil Pumps

23 12 13.10 Pump and Motor Sets

		Crew	Daily Output	Labor-Hours	Unit	Material	2015 Bare Costs Labor	2015 Bare Costs Equipment	Total	Total Incl O&P
0010	**PUMP AND MOTOR SETS**									
1810	Light fuel and diesel oils									
1820	20 GPH 1/3 HP	Q-5	6	2.667	Ea.	1,050	143		1,193	1,375
1830	27 GPH, 1/3 HP		6	2.667		1,050	143		1,193	1,375
1840	80 GPH, 1/3 HP		5	3.200		1,050	172		1,222	1,400
1850	145 GPH, 1/2 HP		4	4		1,100	215		1,315	1,550
1860	277 GPH, 1 HP		4	4		1,225	215		1,440	1,675
1870	700 GPH, 1-1/2 HP		3	5.333		2,750	287		3,037	3,450
1880	1000 GPH, 2 HP		3	5.333		3,225	287		3,512	3,975
1890	1800 GPH, 5 HP		1.80	8.889		4,550	480		5,030	5,750

23 13 Facility Fuel-Storage Tanks

23 13 13 – Facility Underground Fuel-Oil, Storage Tanks

23 13 13.09 Single-Wall Steel Fuel-Oil Tanks

		Crew	Daily Output	Labor-Hours	Unit	Material	2015 Bare Costs Labor	2015 Bare Costs Equipment	Total	Total Incl O&P
0010	**SINGLE-WALL STEEL FUEL-OIL TANKS**									
5000	Tanks, steel ugnd., sti-p3, not incl. hold-down bars									
5500	Excavation, pad, pumps and piping not included									
5510	Single wall, 500 gallon capacity, 7 ga. shell	Q-5	2.70	5.926	Ea.	2,050	320		2,370	2,725
5520	1,000 gallon capacity, 7 ga. shell	"	2.50	6.400		3,825	345		4,170	4,725
5530	2,000 gallon capacity, 1/4" thick shell	Q-7	4.60	6.957		6,200	395		6,595	7,425
5535	2,500 gallon capacity, 7 ga. shell	Q-5	3	5.333		6,850	287		7,137	7,950
5540	5,000 gallon capacity, 1/4" thick shell	Q-7	3.20	10		11,500	570		12,070	13,600
5560	10,000 gallon capacity, 1/4" thick shell		2	16		12,300	910		13,210	15,000
5580	15,000 gallon capacity, 5/16" thick shell		1.70	18.824		19,000	1,075		20,075	22,500
5600	20,000 gallon capacity, 5/16" thick shell		1.50	21.333		26,700	1,225		27,925	31,100
5610	25,000 gallon capacity, 3/8" thick shell		1.30	24.615		37,900	1,400		39,300	43,800
5620	30,000 gallon capacity, 3/8" thick shell		1.10	29.091		38,400	1,650		40,050	44,700
5630	40,000 gallon capacity, 3/8" thick shell		.90	35.556		42,000	2,025		44,025	49,300
5640	50,000 gallon capacity, 3/8" thick shell		.80	40		46,600	2,275		48,875	55,000

23 13 Facility Fuel-Storage Tanks

23 13 13 – Facility Underground Fuel-Oil, Storage Tanks

23 13 13.13 Dbl-Wall Steel, Undrgrnd Fuel-Oil, Stor. Tanks	Crew	Daily Output	Labor-Hours	Unit	Material	2015 Bare Costs Labor	Equipment	Total	Total Incl O&P
0010 **DOUBLE-WALL STEEL, UNDERGROUND FUEL-OIL, STORAGE TANKS**									
6200 Steel, underground, 360°, double wall, U.L. listed,									
6210 with sti-P3 corrosion protection,									
6220 (dielectric coating, cathodic protection, electrical									
6230 isolation) 30 year warranty,									
6240 not incl. manholes or hold-downs.									
6250 500 gallon capacity	Q-5	2.40	6.667	Ea.	5,575	360		5,935	6,700
6260 1,000 gallon capacity	"	2.25	7.111		6,900	380		7,280	8,150
6270 2,000 gallon capacity	Q-7	4.16	7.692		9,050	440		9,490	10,600
6280 3,000 gallon capacity		3.90	8.205		10,900	465		11,365	12,700
6290 4,000 gallon capacity		3.64	8.791		13,800	500		14,300	15,900
6300 5,000 gallon capacity		2.91	10.997		19,900	625		20,525	22,800
6310 6,000 gallon capacity		2.42	13.223		25,600	750		26,350	29,200
6320 8,000 gallon capacity		2.08	15.385		24,600	875		25,475	28,300
6330 10,000 gallon capacity		1.82	17.582		27,700	1,000		28,700	32,000
6340 12,000 gallon capacity		1.70	18.824		30,600	1,075		31,675	35,300
6350 15,000 gallon capacity		1.33	24.060		31,900	1,375		33,275	37,200
6360 20,000 gallon capacity		1.33	24.060		32,700	1,375		34,075	38,000
6370 25,000 gallon capacity		1.16	27.586		72,500	1,575		74,075	82,500
6380 30,000 gallon capacity		1.03	31.068		87,500	1,775		89,275	98,500
6390 40,000 gallon capacity		.80	40		112,500	2,275		114,775	127,000
6395 50,000 gallon capacity		.73	43.836		136,500	2,500		139,000	154,500
6400 For hold-downs 500-2000 gal., add		16	2	Set	183	114		297	375
6410 For hold-downs 3000-6000 gal., add		12	2.667		370	152		522	640
6420 For hold-downs 8000-12,000 gal., add		11	2.909		445	166		611	740
6430 For hold-downs 15,000 gal., add		9	3.556		640	202		842	1,000
6440 For hold-downs 20,000 gal., add		8	4		735	228		963	1,150
6450 For hold-downs 20,000 gal. plus, add		6	5.333		960	305		1,265	1,500
6500 For manways, add				Ea.	1,800			1,800	2,000
6600 In place with hold-downs									
6652 550 gallon capacity	Q-5	1.84	8.696	Ea.	5,775	470		6,245	7,050

23 13 13.23 Glass-Fiber-Reinfcd-Plastic, Fuel-Oil, Storage

	Crew	Daily Output	Labor-Hours	Unit	Material	Labor	Equipment	Total	Total Incl O&P
0010 **GLASS-FIBER-REINFCD-PLASTIC, UNDERGRND. FUEL-OIL, STORAGE**									
0210 Fiberglass, underground, single wall, U.L. listed, not including									
0220 manway or hold-down strap									
0225 550 gallon capacity	Q-5	2.67	5.993	Ea.	3,200	320		3,520	3,975
0230 1,000 gallon capacity	"	2.46	6.504		4,125	350		4,475	5,050
0240 2,000 gallon capacity	Q-7	4.57	7.002		5,950	400		6,350	7,150
0245 3,000 gallon capacity		3.90	8.205		7,125	465		7,590	8,550
0250 4,000 gallon capacity		3.55	9.014		8,275	515		8,790	9,875
0255 5,000 gallon capacity		3.20	10		9,200	570		9,770	11,000
0260 6,000 gallon capacity		2.67	11.985		9,400	680		10,080	11,300
0270 8,000 gallon capacity		2.29	13.974		11,200	795		11,995	13,600
0280 10,000 gallon capacity		2	16		12,800	910		13,710	15,500
0282 12,000 gallon capacity		1.88	17.021		14,500	970		15,470	17,500
0284 15,000 gallon capacity		1.68	19.048		20,500	1,075		21,575	24,100
0290 20,000 gallon capacity		1.45	22.069		26,300	1,250		27,550	30,900
0300 25,000 gallon capacity		1.28	25		54,000	1,425		55,425	61,500
0320 30,000 gallon capacity		1.14	28.070		77,500	1,600		79,100	87,500
0340 40,000 gallon capacity		.89	35.955		86,000	2,050		88,050	97,500
0360 48,000 gallon capacity		.81	39.506		128,000	2,250		130,250	144,000
0500 For manway, fittings and hold-downs, add					20%	15%			

23 13 13 - Facility Underground Fuel-Oil, Storage Tanks

23 13 13.23 Glass-Fiber-Reinfcd-Plastic, Fuel-Oil, Storage		Crew	Daily Output	Labor-Hours	Unit	Material	2015 Bare Costs Labor	Equipment	Total	Total Incl O&P
0600	For manways, add				Ea.	2,250			2,250	2,475
1000	For helical heating coil, add	Q-5	2.50	6.400	↓	4,775	345		5,120	5,775
1020	Fiberglass, underground, double wall, U.L. listed									
1030	includes manways, not incl. hold-down straps									
1040	600 gallon capacity	Q-5	2.42	6.612	Ea.	7,200	355		7,555	8,450
1050	1,000 gallon capacity	"	2.25	7.111		9,825	380		10,205	11,400
1060	2,500 gallon capacity	Q-7	4.16	7.692		14,800	440		15,240	17,000
1070	3,000 gallon capacity		3.90	8.205		16,000	465		16,465	18,300
1080	4,000 gallon capacity		3.64	8.791		16,200	500		16,700	18,700
1090	6,000 gallon capacity		2.42	13.223		21,100	750		21,850	24,400
1100	8,000 gallon capacity		2.08	15.385		23,500	875		24,375	27,200
1110	10,000 gallon capacity		1.82	17.582		27,300	1,000		28,300	31,500
1120	12,000 gallon capacity		1.70	18.824		34,100	1,075		35,175	39,100
1122	15,000 gallon capacity		1.52	21.053		45,300	1,200		46,500	51,500
1124	20,000 gallon capacity		1.33	24.060		56,000	1,375		57,375	63,500
1126	25,000 gallon capacity		1.16	27.586		76,500	1,575		78,075	87,000
1128	30,000 gallon capacity	↓	1.03	31.068		92,000	1,775		93,775	104,000
1140	For hold-down straps, add					2%	10%			
1150	For hold-downs 500-4000 gal., add	Q-7	16	2	Set	445	114		559	660
1160	For hold-downs 5000-15000 gal., add		8	4		895	228		1,123	1,325
1170	For hold-downs 20,000 gal., add		5.33	6.004		1,350	340		1,690	2,000
1180	For hold-downs 25,000 gal., add		4	8		1,800	455		2,255	2,650
1190	For hold-downs 30,000 gal., add	↓	2.60	12.308	↓	2,675	700		3,375	4,000
2210	Fiberglass, underground, single wall, U.L. listed, including									
2220	hold-down straps, no manways									
2225	550 gallon capacity	Q-5	2	8	Ea.	3,625	430		4,055	4,650
2230	1,000 gallon capacity	"	1.88	8.511		4,575	460		5,035	5,725
2240	2,000 gallon capacity	Q-7	3.55	9.014		6,400	515		6,915	7,825
2250	4,000 gallon capacity		2.90	11.034		8,725	630		9,355	10,600
2260	6,000 gallon capacity		2	16		10,300	910		11,210	12,700
2270	8,000 gallon capacity		1.78	17.978		12,100	1,025		13,125	14,900
2280	10,000 gallon capacity		1.60	20		13,700	1,150		14,850	16,800
2282	12,000 gallon capacity		1.52	21.053		15,400	1,200		16,600	18,800
2284	15,000 gallon capacity		1.39	23.022		21,400	1,300		22,700	25,500
2290	20,000 gallon capacity		1.14	28.070		27,700	1,600		29,300	32,800
2300	25,000 gallon capacity		.96	33.333		56,000	1,900		57,900	64,500
2320	30,000 gallon capacity	↓	.80	40	↓	80,000	2,275		82,275	91,500
3020	Fiberglass, underground, double wall, U.L. listed									
3030	includes manways and hold-down straps									
3040	600 gallon capacity	Q-5	1.86	8.602	Ea.	7,650	465		8,115	9,100
3050	1,000 gallon capacity	"	1.70	9.412		10,300	505		10,805	12,100
3060	2,500 gallon capacity	Q-7	3.29	9.726		15,200	555		15,755	17,500
3070	3,000 gallon capacity		3.13	10.224		16,500	580		17,080	19,000
3080	4,000 gallon capacity		2.93	10.922		16,700	620		17,320	19,300
3090	6,000 gallon capacity		1.86	17.204		22,000	980		22,980	25,700
3100	8,000 gallon capacity		1.65	19.394		24,400	1,100		25,500	28,600
3110	10,000 gallon capacity		1.48	21.622		28,200	1,225		29,425	32,900
3120	12,000 gallon capacity		1.40	22.857		34,900	1,300		36,200	40,400
3122	15,000 gallon capacity		1.28	25		46,200	1,425		47,625	53,000
3124	20,000 gallon capacity		1.06	30.189		57,000	1,725		58,725	65,500
3126	25,000 gallon capacity		.90	35.556		78,500	2,025		80,525	89,500
3128	30,000 gallon capacity	↓	.74	43.243	↓	95,000	2,450		97,450	107,500

23 13 23.13 Vertical, Steel, Abvground Fuel-Oil, Stor. Tanks	Crew	Daily Output	Labor-Hours	Unit	Material	2015 Bare Costs Labor	Equipment	Total	Total Incl O&P
0010 **VERTICAL, STEEL, ABOVEGROUND FUEL-OIL, STORAGE TANKS**									
4000 Fixed roof oil storage tanks, steel, (1 BBL=42 gal. w/foundation 3'D x 1'W)									
4200 5,000 barrels				Ea.				194,000	213,500
4300 24,000 barrels								333,500	367,000
4500 56,000 barrels								729,000	802,000
4600 110,000 barrels								1,060,000	1,166,000
4800 143,000 barrels								1,250,000	1,375,000
4900 225,000 barrels								1,360,000	1,496,000
5100 Floating roof gasoline tanks, steel, 5,000 barrels (w/foundation 3'D x 1'W)								204,000	225,000
5200 25,000 barrels								381,000	419,000
5400 55,000 barrels								839,000	923,000
5500 100,000 barrels								1,253,000	1,379,000
5700 150,000 barrels								1,532,000	1,685,000
5800 225,000 barrels								2,300,000	2,783,000

23 13 23.16 Horizontal, Stl, Abvgrd Fuel-Oil, Storage Tanks

	Crew	Daily Output	Labor-Hours	Unit	Material	Labor	Equipment	Total	Total Incl O&P
0010 **HORIZONTAL, STEEL, ABOVEGROUND FUEL-OIL, STORAGE TANKS**									
3000 Steel, storage, above ground, including cradles, coating,									
3020 fittings, not including foundation, pumps or piping									
3040 Single wall, 275 gallon	Q-5	5	3.200	Ea.	490	172		662	800
3060 550 gallon	"	2.70	5.926		3,750	320		4,070	4,600
3080 1,000 gallon	Q-7	5	6.400		4,025	365		4,390	4,975
3100 1,500 gallon		4.75	6.737		8,775	385		9,160	10,300
3120 2,000 gallon		4.60	6.957		10,600	395		10,995	12,300
3140 5,000 gallon		3.20	10		19,100	570		19,670	21,900
3150 10,000 gallon		2	16		35,100	910		36,010	40,000
3160 15,000 gallon		1.70	18.824		45,100	1,075		46,175	51,000
3170 20,000 gallon		1.45	22.069		58,500	1,250		59,750	66,000
3180 25,000 gallon		1.30	24.615		68,000	1,400		69,400	77,000
3190 30,000 gallon		1.10	29.091		81,500	1,650		83,150	92,500
3320 Double wall, 500 gallon capacity	Q-5	2.40	6.667		2,625	360		2,985	3,425
3330 2000 gallon capacity	Q-7	4.15	7.711		9,975	440		10,415	11,700
3340 4000 gallon capacity		3.60	8.889		17,800	505		18,305	20,400
3350 6000 gallon capacity		2.40	13.333		21,000	760		21,760	24,300
3360 8000 gallon capacity		2	16		27,000	910		27,910	31,100
3370 10000 gallon capacity		1.80	17.778		30,200	1,000		31,200	34,700
3380 15000 gallon capacity		1.50	21.333		45,900	1,225		47,125	52,500
3390 20000 gallon capacity		1.30	24.615		52,500	1,400		53,900	59,500
3400 25000 gallon capacity		1.15	27.826		63,500	1,575		65,075	72,500
3410 30000 gallon capacity		1	32		69,500	1,825		71,325	79,500

23 13 23.26 Horizontal, Conc., Abvgrd Fuel-Oil, Stor. Tanks

	Crew	Daily Output	Labor-Hours	Unit	Material	Labor	Equipment	Total	Total Incl O&P
0010 **HORIZONTAL, CONCRETE, ABOVEGROUND FUEL-OIL, STORAGE TANKS**									
0050 Concrete, storage, above ground, including pad & pump									
0100 500 gallon	F-3	2	20	Ea.	10,000	960	325	11,285	12,800
0200 1,000 gallon	"	2	20		14,000	960	325	15,285	17,200
0300 2,000 gallon	F-4	2	24		18,000	1,150	565	19,715	22,200
0400 4,000 gallon		2	24		23,000	1,150	565	24,715	27,700
0500 8,000 gallon		2	24		36,000	1,150	565	37,715	42,000
0600 12,000 gallon		2	24		48,000	1,150	565	49,715	55,500

23 21 20 – Hydronic HVAC Piping Specialties

23 21 20.10 Air Control	Crew	Daily Output	Labor-Hours	Unit	Material	2015 Bare Costs Labor	Equipment	Total	Total Incl O&P
0010 **AIR CONTROL**									
0030 Air separator, with strainer									
0040 2" diameter	Q-5	6	2.667	Ea.	1,175	143		1,318	1,525
0080 2-1/2" diameter		5	3.200		1,325	172		1,497	1,700
0100 3" diameter		4	4		2,050	215		2,265	2,575
0120 4" diameter		3	5.333		2,950	287		3,237	3,650
0130 5" diameter	Q-6	3.60	6.667		3,750	370		4,120	4,675
0140 6" diameter		3.40	7.059		4,500	395		4,895	5,550
0160 8" diameter		3	8		6,700	445		7,145	8,050
0180 10" diameter		2.20	10.909		10,400	610		11,010	12,300
0200 12" diameter		1.70	14.118		17,300	785		18,085	20,200
0210 14" diameter		1.30	18.462		20,600	1,025		21,625	24,200
0220 16" diameter		1	24		30,700	1,350		32,050	35,800
0230 18" diameter		.80	30		40,400	1,675		42,075	46,900
0240 20" diameter		.60	40		45,100	2,225		47,325	53,000
0300 Without strainer									
0310 2" diameter	Q-5	6	2.667	Ea.	730	143		873	1,025
0320 2-1/2" diameter		5	3.200		875	172		1,047	1,225
0330 3" diameter		4	4		1,225	215		1,440	1,675
0340 4" diameter		3	5.333		1,925	287		2,212	2,550
0350 5" diameter		2.40	6.667		2,600	360		2,960	3,425
0360 6" diameter	Q-6	3.40	7.059		3,000	395		3,395	3,900
0370 8" diameter		3	8		4,125	445		4,570	5,200
0380 10" diameter		2.20	10.909		6,075	610		6,685	7,600
0390 12" diameter		1.70	14.118		9,350	785		10,135	11,500
0400 14" diameter		1.30	18.462		14,000	1,025		15,025	17,000
0410 16" diameter		1	24		19,500	1,350		20,850	23,500
0420 18" diameter		.80	30		25,400	1,675		27,075	30,500
0430 20" diameter		.60	40		30,500	2,225		32,725	37,000
1000 Micro-bubble separator for total air removal, closed loop system									
1010 Requires bladder type tank in system.									
1020 Water (hot or chilled) or glycol system									
1030 Threaded									
1040 3/4" diameter	1 Stpi	20	.400	Ea.	90	24		114	135
1050 1" diameter		19	.421		101	25		126	150
1060 1-1/4" diameter		16	.500		140	30		170	199
1070 1-1/2" diameter		13	.615		182	37		219	257
1080 2" diameter		11	.727		885	43.50		928.50	1,050
1090 2-1/2" diameter	Q-5	15	1.067		985	57.50		1,042.50	1,150
1100 3" diameter		13	1.231		1,400	66		1,466	1,650
1110 4" diameter		10	1.600		1,525	86		1,611	1,800
1230 Flanged									
1250 2" diameter	1 Stpi	8	1	Ea.	1,125	60		1,185	1,350
1260 2-1/2" diameter	Q-5	5	3.200		1,225	172		1,397	1,600
1270 3" diameter		4.50	3.556		1,650	191		1,841	2,125
1280 4" diameter		3	5.333		1,825	287		2,112	2,425
1290 5" diameter	Q-6	3.40	7.059		2,925	395		3,320	3,800
1300 6" diameter	"	3	8		3,550	445		3,995	4,575
1400 Larger sizes available									
1590 With extended tank dirt catcher									
1600 Flanged									
1620 2" diameter	1 Stpi	7.40	1.081	Ea.	1,475	64.50		1,539.50	1,725
1630 2-1/2" diameter	Q-5	4.20	3.810		1,575	205		1,780	2,025

23 21 20 – Hydronic HVAC Piping Specialties

23 21 20.10 Air Control

		Crew	Daily Output	Labor-Hours	Unit	Material	2015 Bare Costs Labor	Equipment	Total	Total Incl O&P
1640	3" diameter	Q-5	4	4	Ea.	2,050	215		2,265	2,575
1650	4" diameter	↓	2.80	5.714		2,225	305		2,530	2,925
1660	5" diameter	Q-6	3.10	7.742		3,275	430		3,705	4,250
1670	6" diameter	"	2.90	8.276	↓	4,025	460		4,485	5,125
2000	Boiler air fitting, separator									
2010	1-1/4"	Q-5	22	.727	Ea.	224	39		263	305
2020	1-1/2"	↓	20	.800		430	43		473	540
2021	2"	↓	18	.889	↓	745	48		793	890
2400	Compression tank air fitting									
2410	For tanks 9" to 24" diameter	1 Stpi	15	.533	Ea.	76.50	32		108.50	132
2420	For tanks 100 gallon or larger	"	12	.667	"	287	40		327	375

23 21 20.14 Air Purging Scoop

		Crew	Daily Output	Labor-Hours	Unit	Material	2015 Bare Costs Labor	Equipment	Total	Total Incl O&P
0010	**AIR PURGING SCOOP**, With tappings.									
0020	For air vent and expansion tank connection									
0100	1" pipe size, threaded	1 Stpi	19	.421	Ea.	24.50	25		49.50	65
0110	1-1/4" pipe size, threaded		15	.533		24.50	32		56.50	75
0120	1-1/2" pipe size, threaded		13	.615		53.50	37		90.50	114
0130	2" pipe size, threaded	↓	11	.727		61	43.50		104.50	133
0140	2-1/2" pipe size, threaded	Q-5	15	1.067		131	57.50		188.50	231
0150	3" pipe size, threaded		13	1.231		171	66		237	288
0160	4" 150 lb. flanges	↓	3	5.333	↓	380	287		667	855

23 21 20.18 Automatic Air Vent

		Crew	Daily Output	Labor-Hours	Unit	Material	2015 Bare Costs Labor	Equipment	Total	Total Incl O&P
0010	**AUTOMATIC AIR VENT**									
0020	Cast iron body, stainless steel internals, float type									
0060	1/2" NPT inlet, 300 psi	1 Stpi	12	.667	Ea.	109	40		149	180
0140	3/4" NPT inlet, 300 psi		12	.667		109	40		149	180
0180	1/2" NPT inlet, 250 psi		10	.800		350	48		398	455
0220	3/4" NPT inlet, 250 psi		10	.800		350	48		398	455
0260	1" NPT inlet, 250 psi	↓	10	.800		510	48		558	630
0340	1-1/2" NPT inlet, 250 psi	Q-5	12	1.333		1,075	71.50		1,146.50	1,300
0380	2" NPT inlet, 250 psi	"	12	1.333	↓	1,075	71.50		1,146.50	1,300
0600	Forged steel body, stainless steel internals, float type									
0640	1/2" NPT inlet, 750 psi	1 Stpi	12	.667	Ea.	1,150	40		1,190	1,325
0680	3/4" NPT inlet, 750 psi		12	.667		1,150	40		1,190	1,325
0760	3/4" NPT inlet, 1000 psi	↓	10	.800		1,725	48		1,773	1,975
0800	1" NPT inlet, 1000 psi	Q-5	12	1.333		1,725	71.50		1,796.50	2,000
0880	1-1/2" NPT inlet, 1000 psi		10	1.600		4,875	86		4,961	5,475
0920	2" NPT inlet, 1000 psi		10	1.600	↓	4,875	86		4,961	5,475
1100	Formed steel body, noncorrosive									
1110	1/8" NPT inlet 150 psi	1 Stpi	32	.250	Ea.	12.45	14.95		27.40	36
1120	1/4" NPT inlet 150 psi		32	.250		42	14.95		56.95	69
1130	3/4" NPT inlet 150 psi	↓	32	.250	↓	42	14.95		56.95	69
1300	Chrome plated brass, automatic/manual, for radiators									
1310	1/8" NPT inlet, nickel plated brass	1 Stpi	32	.250	Ea.	7.25	14.95		22.20	30.50

23 21 20.22 Circuit Sensor

		Crew	Daily Output	Labor-Hours	Unit	Material	2015 Bare Costs Labor	Equipment	Total	Total Incl O&P
0010	**CIRCUIT SENSOR**, Flow meter									
0020	Metering stations									
0040	Wafer orifice insert type									
0060	2-1/2" pipe size	Q-5	12	1.333	Ea.	210	71.50		281.50	340
0100	3" pipe size		11	1.455		238	78		316	380
0140	4" pipe size		8	2		273	108		381	460
0180	5" pipe size	↓	7.30	2.192	↓	345	118		463	560

23 21 20 – Hydronic HVAC Piping Specialties

23 21 20.22 Circuit Sensor		Crew	Daily Output	Labor-Hours	Unit	Material	2015 Bare Costs Labor	Equipment	Total	Total Incl O&P
0220	6" pipe size	Q-5	6.40	2.500	Ea.	415	134		549	660
0260	8" pipe size	Q-6	5.30	4.528		575	253		828	1,000
0280	10" pipe size		4.60	5.217		660	291		951	1,175
0360	12" pipe size		4.20	5.714		1,100	320		1,420	1,675
23 21 20.26 Circuit Setter										
0010	**CIRCUIT SETTER**, Balance valve									
0012	Bronze body, soldered									
0013	1/2" size	1 Stpi	24	.333	Ea.	71	19.90		90.90	109
0014	3/4" size	"	20	.400	"	75	24		99	119
0018	Threaded									
0019	1/2" pipe size	1 Stpi	22	.364	Ea.	74	21.50		95.50	114
0020	3/4" pipe size		20	.400		79	24		103	123
0040	1" pipe size		18	.444		102	26.50		128.50	152
0050	1-1/4" pipe size		15	.533		150	32		182	213
0060	1-1/2" pipe size		12	.667		177	40		217	255
0080	2" pipe size		10	.800		253	48		301	350
0100	2-1/2" pipe size	Q-5	15	1.067		575	57.50		632.50	715
0120	3" pipe size	"	10	1.600		805	86		891	1,025
0130	Cast iron body, flanged									
0136	3" pipe size, flanged	Q-5	4	4	Ea.	805	215		1,020	1,200
0140	4" pipe size	"	3	5.333		1,200	287		1,487	1,750
0200	For differential meter, accurate to 1%, add					930			930	1,025
0300	For differential meter 400 psi/200°F continuous, add					2,300			2,300	2,525
23 21 20.30 Cocks, Drains and Specialties										
0010	**COCKS, DRAINS AND SPECIALTIES**									
1000	Boiler drain									
1010	Pipe thread to hose									
1020	Bronze									
1030	1/2" size	1 Stpi	36	.222	Ea.	9.40	13.30		22.70	30.50
1040	3/4" size	"	34	.235	"	9.85	14.05		23.90	32
1100	Solder to hose									
1110	Bronze									
1120	1/2" size	1 Stpi	46	.174	Ea.	11.90	10.40		22.30	29
1130	3/4" size	"	44	.182	"	12.95	10.85		23.80	30.50
1600	With built-in vacuum breaker									
1610	1/2" I.P. or solder	1 Stpi	36	.222	Ea.	40	13.30		53.30	64
1630	With tamper proof vacuum breaker									
1640	1/2" I.P. or solder	1 Stpi	36	.222	Ea.	45.50	13.30		58.80	70
1650	3/4" I.P. or solder	"	34	.235	"	46	14.05		60.05	71.50
3000	Cocks									
3010	Air, lever or tee handle									
3020	Bronze, single thread									
3030	1/8" size	1 Stpi	52	.154	Ea.	10.85	9.20		20.05	26
3040	1/4" size		46	.174		11.45	10.40		21.85	28.50
3050	3/8" size		40	.200		11.60	11.95		23.55	31
3060	1/2" size		36	.222		13.70	13.30		27	35
3100	Bronze, double thread									
3110	1/8" size	1 Stpi	26	.308	Ea.	13.70	18.40		32.10	43
3120	1/4" size		22	.364		14.20	21.50		35.70	48.50
3130	3/8" size		18	.444		14.80	26.50		41.30	56.50
3140	1/2" size		15	.533		18.15	32		50.15	68
4000	Steam									

For customer support on your Mechanical Cost Data, call 877.795.7394.

289

23 21 20.30 Cocks, Drains and Specialties

		Crew	Daily Output	Labor-Hours	Unit	Material	2015 Bare Costs Labor	Equipment	Total	Total Incl O&P
4010	Bronze									
4020	1/8" size	1 Stpi	26	.308	Ea.	31.50	18.40		49.90	63
4030	1/4" size		22	.364		27.50	21.50		49	63
4040	3/8" size		18	.444		40	26.50		66.50	84
4050	1/2" size		15	.533		42.50	32		74.50	94.50
4060	3/4" size		14	.571		50	34		84	107
4070	1" size		13	.615		65	37		102	127
4080	1-1/4" size		11	.727		100	43.50		143.50	176
4090	1-1/2" size		10	.800		129	48		177	214
4100	2" size		9	.889		213	53		266	315
4300	Bronze, 3 way									
4320	1/4" size	1 Stpi	15	.533	Ea.	47.50	32		79.50	100
4330	3/8" size		12	.667		47.50	40		87.50	112
4340	1/2" size		10	.800		47.50	48		95.50	124
4350	3/4" size		9.50	.842		81	50.50		131.50	166
4360	1" size		8.50	.941		98	56		154	193
4370	1-1/4" size		7.50	1.067		147	63.50		210.50	258
4380	1-1/2" size		6.50	1.231		170	73.50		243.50	298
4390	2" size		6	1.333		252	79.50		331.50	395
4500	Gauge cock, brass									
4510	1/4" FPT	1 Stpi	24	.333	Ea.	11.15	19.90		31.05	42.50
4512	1/4" MPT	"	24	.333	"	13.75	19.90		33.65	45
4600	Pigtail, steam syphon									
4604	1/4"	1 Stpi	24	.333	Ea.	19.70	19.90		39.60	51.50
4650	Snubber valve									
4654	1/4"	1 Stpi	22	.364	Ea.	11.75	21.50		33.25	46
4660	Nipple, black steel									
4664	1/4" x 3"	1 Stpi	37	.216	Ea.	2.83	12.90		15.73	22.50

23 21 20.34 Dielectric Unions

		Crew	Daily Output	Labor-Hours	Unit	Material	2015 Bare Costs Labor	Equipment	Total	Total Incl O&P
0010	**DIELECTRIC UNIONS**, Standard gaskets for water and air									
0020	250 psi maximum pressure									
0280	Female IPT to sweat, straight									
0300	1/2" pipe size	1 Plum	24	.333	Ea.	9	19.55		28.55	39.50
0340	3/4" pipe size		20	.400		6.20	23.50		29.70	42.50
0360	1" pipe size		19	.421		8.45	24.50		32.95	47
0380	1-1/4" pipe size		15	.533		13.10	31.50		44.60	62
0400	1-1/2" pipe size		13	.615		19.70	36		55.70	76
0420	2" pipe size		11	.727		26.50	42.50		69	93.50
0580	Female IPT to brass pipe thread, straight									
0600	1/2" pipe size	1 Plum	24	.333	Ea.	12.10	19.55		31.65	43
0640	3/4" pipe size		20	.400		13.30	23.50		36.80	50
0660	1" pipe size		19	.421		24.50	24.50		49	64.50
0680	1-1/4" pipe size		15	.533		30	31.50		61.50	80.50
0700	1-1/2" pipe size		13	.615		44	36		80	103
0720	2" pipe size		11	.727		86	42.50		128.50	159
0780	Female IPT to female IPT, straight									
0800	1/2" pipe size	1 Plum	24	.333	Ea.	13.85	19.55		33.40	44.50
0840	3/4" pipe size		20	.400		15.60	23.50		39.10	52.50
0860	1" pipe size		19	.421		21	24.50		45.50	60.50
0880	1-1/4" pipe size		15	.533		28	31.50		59.50	78.50
0900	1-1/2" pipe size		13	.615		43	36		79	102
0920	2" pipe size		11	.727		63	42.50		105.50	134

23 21 20.34 Dielectric Unions

		Crew	Daily Output	Labor-Hours	Unit	Material	2015 Bare Costs Labor	Equipment	Total	Total Incl O&P
2000	175 psi maximum pressure									
2180	Female IPT to sweat									
2240	2" pipe size	1 Plum	9	.889	Ea.	142	52		194	235
2260	2-1/2" pipe size	Q-1	15	1.067		155	56.50		211.50	256
2280	3" pipe size		14	1.143		213	60.50		273.50	325
2300	4" pipe size	↓	11	1.455	↓	565	77		642	735
2480	Female IPT to brass pipe									
2500	1-1/2" pipe size	1 Plum	11	.727	Ea.	176	42.50		218.50	259
2540	2" pipe size	"	9	.889		207	52		259	305
2560	2-1/2" pipe size	Q-1	15	1.067		305	56.50		361.50	420
2580	3" pipe size		14	1.143		355	60.50		415.50	480
2600	4" pipe size	↓	11	1.455	↓	635	77		712	815

23 21 20.38 Expansion Couplings

		Crew	Daily Output	Labor-Hours	Unit	Material	2015 Bare Costs Labor	Equipment	Total	Total Incl O&P
0010	**EXPANSION COUPLINGS**, Hydronic									
0100	Copper to copper, sweat									
1000	Baseboard riser fitting, 5" stub by coupling 12" long									
1020	1/2" diameter	1 Stpi	24	.333	Ea.	20.50	19.90		40.40	52.50
1040	3/4" diameter		20	.400		29	24		53	68
1060	1" diameter		19	.421		40.50	25		65.50	82.50
1080	1-1/4" diameter	↓	15	.533	↓	52.50	32		84.50	106
1180	9" Stub by tubing 8" long									
1200	1/2" diameter	1 Stpi	24	.333	Ea.	18.30	19.90		38.20	50
1220	3/4" diameter		20	.400		24.50	24		48.50	62.50
1240	1" diameter		19	.421		33.50	25		58.50	75
1260	1-1/4" diameter	↓	15	.533	↓	47	32		79	100

23 21 20.42 Expansion Joints

		Crew	Daily Output	Labor-Hours	Unit	Material	2015 Bare Costs Labor	Equipment	Total	Total Incl O&P
0010	**EXPANSION JOINTS**									
0100	Bellows type, neoprene cover, flanged spool									
0140	6" face to face, 1-1/4" diameter	1 Stpi	11	.727	Ea.	255	43.50		298.50	345
0160	1-1/2" diameter	"	10.60	.755		255	45		300	350
0180	2" diameter	Q-5	13.30	1.203		258	64.50		322.50	380
0190	2-1/2" diameter		12.40	1.290		267	69.50		336.50	400
0200	3" diameter		11.40	1.404		299	75.50		374.50	445
0210	4" diameter		8.40	1.905		325	102		427	510
0220	5" diameter		7.60	2.105		395	113		508	605
0230	6" diameter		6.80	2.353		405	127		532	640
0240	8" diameter		5.40	2.963		475	159		634	760
0250	10" diameter		5	3.200		650	172		822	975
0260	12" diameter		4.60	3.478		745	187		932	1,100
0480	10" face to face, 2" diameter		13	1.231		370	66		436	510
0500	2-1/2" diameter		12	1.333		390	71.50		461.50	540
0520	3" diameter		11	1.455		400	78		478	560
0540	4" diameter		8	2		455	108		563	660
0560	5" diameter		7	2.286		540	123		663	780
0580	6" diameter		6	2.667		560	143		703	830
0600	8" diameter		5	3.200		670	172		842	995
0620	10" diameter		4.60	3.478		735	187		922	1,100
0640	12" diameter		4	4		910	215		1,125	1,325
0660	14" diameter		3.80	4.211		1,125	226		1,351	1,600
0680	16" diameter		2.90	5.517		1,300	297		1,597	1,875
0700	18" diameter		2.50	6.400		1,475	345		1,820	2,150
0720	20" diameter		2.10	7.619		1,550	410		1,960	2,325

For customer support on your Mechanical Cost Data, call 877.795.7394.

291

23 21 20.42 Expansion Joints

		Crew	Daily Output	Labor-Hours	Unit	Material	2015 Bare Costs Labor	Equipment	Total	Total Incl O&P
0740	24" diameter	Q-5	1.80	8.889	Ea.	1,800	480		2,280	2,700
0760	26" diameter		1.40	11.429		2,025	615		2,640	3,150
0780	30" diameter		1.20	13.333		2,250	715		2,965	3,550
0800	36" diameter		1	16		2,800	860		3,660	4,375
1000	Bellows with internal sleeves and external covers									
1010	Stainless steel, 150 lb.									
1020	With male threads									
1040	3/4" diameter	1 Stpi	20	.400	Ea.	80	24		104	124
1050	1" diameter		19	.421		86	25		111	133
1060	1-1/4" diameter		16	.500		93	30		123	147
1070	1-1/2" diameter		13	.615		104	37		141	170
1080	2" diameter		11	.727		111	43.50		154.50	188
1110	With flanged ends									
1120	1-1/4" diameter	1 Stpi	12	.667	Ea.	103	40		143	173
1130	1-1/2" diameter		11	.727		118	43.50		161.50	195
1140	2" diameter		9	.889		160	53		213	256
1150	3" diameter	Q-5	8	2		223	108		331	410
1160	4" diameter	"	5	3.200		730	172		902	1,075
1170	5" diameter	Q-6	6	4		870	223		1,093	1,300
1180	6" diameter		5	4.800		1,025	268		1,293	1,550
1190	8" diameter		4.50	5.333		1,400	297		1,697	2,000
1200	10" diameter		4	6		1,775	335		2,110	2,450
1210	12" diameter		3.60	6.667		2,750	370		3,120	3,600

23 21 20.46 Expansion Tanks

		Crew	Daily Output	Labor-Hours	Unit	Material	2015 Bare Costs Labor	Equipment	Total	Total Incl O&P
0010	**EXPANSION TANKS**									
1400	Plastic, corrosion resistant, see Plumbing Cost Data									
1507	Underground fuel-oil storage tanks, see Section 23 13 13									
1512	Tank leak detection systems, see Section 28 33 33.50									
2000	Steel, liquid expansion, ASME, painted, 15 gallon capacity	Q-5	17	.941	Ea.	640	50.50		690.50	775
2020	24 gallon capacity		14	1.143		715	61.50		776.50	880
2040	30 gallon capacity		12	1.333		715	71.50		786.50	895
2060	40 gallon capacity		10	1.600		835	86		921	1,050
2080	60 gallon capacity		8	2		1,000	108		1,108	1,250
2100	80 gallon capacity		7	2.286		1,075	123		1,198	1,350
2120	100 gallon capacity		6	2.667		1,450	143		1,593	1,825
2130	120 gallon capacity		5	3.200		1,550	172		1,722	1,950
2140	135 gallon capacity		4.50	3.556		1,625	191		1,816	2,075
2150	175 gallon capacity		4	4		2,550	215		2,765	3,125
2160	220 gallon capacity		3.60	4.444		2,875	239		3,114	3,525
2170	240 gallon capacity		3.30	4.848		3,000	261		3,261	3,700
2180	305 gallon capacity		3	5.333		4,225	287		4,512	5,075
2190	400 gallon capacity		2.80	5.714		5,200	305		5,505	6,200
2360	Galvanized									
2370	15 gallon capacity	Q-5	17	.941	Ea.	1,125	50.50		1,175.50	1,300
2380	24 gallon capacity		14	1.143		1,150	61.50		1,211.50	1,375
2390	30 gallon capacity		12	1.333		1,350	71.50		1,421.50	1,575
2400	40 gallon capacity		10	1.600		1,575	86		1,661	1,850
2410	60 gallon capacity		8	2		1,825	108		1,933	2,150
2420	80 gallon capacity		7	2.286		2,075	123		2,198	2,450
2430	100 gallon capacity		6	2.667		2,700	143		2,843	3,200
2440	120 gallon capacity		5	3.200		2,900	172		3,072	3,450
2450	135 gallon capacity		4.50	3.556		3,050	191		3,241	3,675

23 21 20.46 Expansion Tanks

		Crew	Daily Output	Labor-Hours	Unit	Material	2015 Bare Costs Labor	2015 Bare Costs Equipment	Total	Total Incl O&P
2460	175 gallon capacity	Q-5	4	4	Ea.	4,950	215		5,165	5,775
2470	220 gallon capacity		3.60	4.444		5,675	239		5,914	6,575
2480	240 gallon capacity		3.30	4.848		5,900	261		6,161	6,875
2490	305 gallon capacity		3	5.333		8,600	287		8,887	9,900
2500	400 gallon capacity		2.80	5.714		10,600	305		10,905	12,200
3000	Steel ASME expansion, rubber diaphragm, 19 gal. cap. accept.		12	1.333		2,425	71.50		2,496.50	2,775
3020	31 gallon capacity		8	2		2,700	108		2,808	3,125
3040	61 gallon capacity		6	2.667		3,800	143		3,943	4,400
3060	79 gallon capacity		5	3.200		3,875	172		4,047	4,500
3080	119 gallon capacity		4	4		4,100	215		4,315	4,825
3100	158 gallon capacity		3.80	4.211		5,675	226		5,901	6,600
3120	211 gallon capacity		3.30	4.848		6,575	261		6,836	7,625
3140	317 gallon capacity		2.80	5.714		8,575	305		8,880	9,900
3160	422 gallon capacity		2.60	6.154		12,700	330		13,030	14,500
3180	528 gallon capacity		2.40	6.667		13,900	360		14,260	15,800

23 21 20.50 Float Valves

		Crew	Daily Output	Labor-Hours	Unit	Material	2015 Bare Costs Labor	2015 Bare Costs Equipment	Total	Total Incl O&P
0010	**FLOAT VALVES**									
0020	With ball and bracket									
0030	Single seat, threaded									
0040	Brass body									
0050	1/2"	1 Stpi	11	.727	Ea.	100	43.50		143.50	176
0060	3/4"		9	.889		112	53		165	203
0070	1"		7	1.143		144	68.50		212.50	261
0080	1-1/2"		4.50	1.778		211	106		317	390
0090	2"		3.60	2.222		215	133		348	435
0300	For condensate receivers, CI, in-line mount									
0320	1" inlet	1 Stpi	7	1.143	Ea.	144	68.50		212.50	261
0360	For condensate receiver, CI, external float, flanged tank mount									
0370	3/4" inlet	1 Stpi	5	1.600	Ea.	112	95.50		207.50	267

23 21 20.54 Flow Check Control

		Crew	Daily Output	Labor-Hours	Unit	Material	2015 Bare Costs Labor	2015 Bare Costs Equipment	Total	Total Incl O&P
0010	**FLOW CHECK CONTROL**									
0100	Bronze body, soldered									
0110	3/4" size	1 Stpi	20	.400	Ea.	64	24		88	107
0120	1" size	"	19	.421	"	76.50	25		101.50	122
0200	Cast iron body, threaded									
0210	3/4" size	1 Stpi	20	.400	Ea.	49.50	24		73.50	90
0220	1" size		19	.421		55.50	25		80.50	99
0230	1-1/4" size		15	.533		67.50	32		99.50	122
0240	1-1/2" size		13	.615		103	37		140	169
0250	2" size		11	.727		148	43.50		191.50	229
0300	Flanged inlet, threaded outlet									
0310	2-1/2" size	1 Stpi	8	1	Ea.	310	60		370	430
0320	3" size	Q-5	9	1.778		365	95.50		460.50	550
0330	4" size	"	7	2.286		715	123		838	975

23 21 20.58 Hydronic Heating Control Valves

		Crew	Daily Output	Labor-Hours	Unit	Material	2015 Bare Costs Labor	2015 Bare Costs Equipment	Total	Total Incl O&P
0010	**HYDRONIC HEATING CONTROL VALVES**									
0050	Hot water, nonelectric, thermostatic									
0100	Radiator supply, 1/2" diameter	1 Stpi	24	.333	Ea.	68.50	19.90		88.40	105
0120	3/4" diameter		20	.400		71.50	24		95.50	115
0140	1" diameter		19	.421		88	25		113	135
0160	1-1/4" diameter		15	.533		120	32		152	180
0500	For low pressure steam, add					25%				

For customer support on your Mechanical Cost Data, call 877.795.7394.

293

23 21 20.58 Hydronic Heating Control Valves		Crew	Daily Output	Labor-Hours	Unit	Material	2015 Bare Costs Labor	Equipment	Total	Total Incl O&P
1000	Manual, radiator supply									
1010	1/2" pipe size, angle union	1 Stpi	24	.333	Ea.	52.50	19.90		72.40	88
1020	3/4" pipe size, angle union		20	.400		66	24		90	109
1030	1" pipe size, angle union		19	.421		85.50	25		110.50	132
1100	Radiator, balancing, straight, sweat connections									
1110	1/2" pipe size	1 Stpi	24	.333	Ea.	20	19.90		39.90	52
1120	3/4" pipe size		20	.400		28	24		52	67
1130	1" pipe size		19	.421		45.50	25		70.50	88
1200	Steam, radiator, supply									
1210	1/2" pipe size, angle union	1 Stpi	24	.333	Ea.	49	19.90		68.90	84
1220	3/4" pipe size, angle union		20	.400		55.50	24		79.50	97.50
1230	1" pipe size, angle union		19	.421		64.50	25		89.50	109
1240	1-1/4" pipe size, angle union		15	.533		82.50	32		114.50	139
8000	System balancing and shut-off									
8020	Butterfly, quarter turn, calibrated, threaded or solder									
8040	Bronze, -30°F to +350°F, pressure to 175 psi									
8060	1/2" size	1 Stpi	22	.364	Ea.	18.65	21.50		40.15	53.50
8070	3/4" size		20	.400		29.50	24		53.50	68.50
8080	1" size		19	.421		36	25		61	77.50
8090	1-1/4" size		15	.533		58	32		90	112
8100	1-1/2" size		13	.615		75	37		112	138
8110	2" size		11	.727		94	43.50		137.50	169

23 21 20.62 Liquid Drainers

		Crew	Daily Output	Labor-Hours	Unit	Material	2015 Bare Costs Labor	Equipment	Total	Total Incl O&P
0010	**LIQUID DRAINERS**									
0100	Guided lever type									
0110	Cast iron body, threaded									
0120	1/2" pipe size	1 Stpi	22	.364	Ea.	109	21.50		130.50	153
0130	3/4" pipe size		20	.400		224	24		248	282
0140	1" pipe size		19	.421		460	25		485	545
0150	1-1/2" pipe size		11	.727		1,000	43.50		1,043.50	1,175
0160	2" pipe size		8	1		1,000	60		1,060	1,200
0200	Forged steel body, threaded									
0210	1/2" pipe size	1 Stpi	18	.444	Ea.	1,100	26.50		1,126.50	1,250
0220	3/4" pipe size		16	.500		1,100	30		1,130	1,250
0230	1" pipe size		14	.571		1,650	34		1,684	1,850
0240	1-1/2" pipe size	Q-5	10	1.600		4,775	86		4,861	5,375
0250	2" pipe size	"	8	2		4,775	108		4,883	5,400
0300	Socket weld									
0310	1/2" pipe size	Q-17	16	1	Ea.	1,250	54	3.61	1,307.61	1,450
0320	3/4" pipe size		14	1.143		1,250	61.50	4.13	1,315.63	1,475
0330	1" pipe size		12	1.333		1,825	71.50	4.81	1,901.31	2,125
0340	1-1/2" pipe size		9	1.778		4,950	95.50	6.40	5,051.90	5,600
0350	2" pipe size		7	2.286		4,950	123	8.25	5,081.25	5,650
0400	Flanged									
0410	1/2" pipe size	1 Stpi	12	.667	Ea.	2,600	40		2,640	2,900
0420	3/4" pipe size		10	.800		2,600	48		2,648	2,925
0430	1" pipe size		7	1.143		3,275	68.50		3,343.50	3,700
0440	1-1/2" pipe size	Q-5	9	1.778		6,650	95.50		6,745.50	7,450
0450	2" pipe size	"	7	2.286		6,650	123		6,773	7,475
1000	Fixed lever and snap action type									
1100	Cast iron body, threaded									
1110	1/2" pipe size	1 Stpi	22	.364	Ea.	170	21.50		191.50	220

23 21 20 – Hydronic HVAC Piping Specialties

23 21 20.62 Liquid Drainers		Crew	Daily Output	Labor-Hours	Unit	Material	2015 Bare Costs Labor	Equipment	Total	Total Incl O&P
1120	3/4" pipe size	1 Stpi	20	.400	Ea.	445	24		469	525
1130	1" pipe size	↓	19	.421	↓	445	25		470	530
1200	Forged steel body, threaded									
1210	1/2" pipe size	1 Stpi	18	.444	Ea.	1,175	26.50		1,201.50	1,350
1220	3/4" pipe size		16	.500		1,175	30		1,205	1,350
1230	1" pipe size		14	.571		2,925	34		2,959	3,250
1240	1-1/4" pipe size	↓	12	.667	↓	2,925	40		2,965	3,250
2000	High pressure high leverage type									
2100	Forged steel body, threaded									
2110	1/2" pipe size	1 Stpi	18	.444	Ea.	2,325	26.50		2,351.50	2,600
2120	3/4" pipe size		16	.500		2,325	30		2,355	2,600
2130	1" pipe size		14	.571		3,875	34		3,909	4,325
2140	1-1/4" pipe size		12	.667		3,875	40		3,915	4,325
2150	1-1/2" pipe size	Q-5	10	1.600		5,850	86		5,936	6,550
2160	2" pipe size	"	8	2	↓	5,850	108		5,958	6,575
2200	Socket weld									
2210	1/2" pipe size	Q-17	16	1	Ea.	2,500	54	3.61	2,557.61	2,825
2220	3/4" pipe size		14	1.143		2,500	61.50	4.13	2,565.63	2,850
2230	1" pipe size		12	1.333		4,075	71.50	4.81	4,151.31	4,600
2240	1-1/4" pipe size		11	1.455		4,075	78	5.25	4,158.25	4,600
2250	1-1/2" pipe size		9	1.778		6,025	95.50	6.40	6,126.90	6,775
2260	2" pipe size	↓	7	2.286	↓	6,025	123	8.25	6,156.25	6,825
2300	Flanged									
2310	1/2" pipe size	1 Stpi	12	.667	Ea.	3,350	40		3,390	3,750
2320	3/4" pipe size		10	.800		3,350	48		3,398	3,775
2330	1" pipe size		7	1.143		5,300	68.50		5,368.50	5,925
2340	1-1/4" pipe size		6	1.333		5,300	79.50		5,379.50	5,950
2350	1-1/2" pipe size	Q-5	9	1.778		7,250	95.50		7,345.50	8,125
2360	2" pipe size	"	7	2.286	↓	7,250	123		7,373	8,150
3000	Guided lever dual gravity type									
3100	Cast iron body, threaded									
3110	1/2" pipe size	1 Stpi	22	.364	Ea.	425	21.50		446.50	500
3120	3/4" pipe size		20	.400		425	24		449	500
3130	1" pipe size		19	.421		615	25		640	715
3140	1-1/2" pipe size		11	.727		1,175	43.50		1,218.50	1,375
3150	2" pipe size	↓	8	1	↓	1,175	60		1,235	1,400
3200	Forged steel body, threaded									
3210	1/2" pipe size	1 Stpi	18	.444	Ea.	1,275	26.50		1,301.50	1,450
3220	3/4" pipe size		16	.500		1,275	30		1,305	1,450
3230	1" pipe size	↓	14	.571		1,875	34		1,909	2,125
3240	1-1/2" pipe size	Q-5	10	1.600		5,000	86		5,086	5,625
3250	2" pipe size	"	8	2	↓	5,000	108		5,108	5,650
3300	Socket weld									
3310	1/2" pipe size	Q-17	16	1	Ea.	1,425	54	3.61	1,482.61	1,650
3320	3/4" pipe size		14	1.143		1,425	61.50	4.13	1,490.63	1,675
3330	1" pipe size		12	1.333		2,075	71.50	4.81	2,151.31	2,400
3340	1-1/2" pipe size		9	1.778		5,175	95.50	6.40	5,276.90	5,850
3350	2" pipe size	↓	7	2.286	↓	5,175	123	8.25	5,306.25	5,900
3400	Flanged									
3410	1/2" pipe size	1 Stpi	12	.667	Ea.	2,300	40		2,340	2,600
3420	3/4" pipe size		10	.800		2,300	48		2,348	2,625
3430	1" pipe size	↓	7	1.143		3,025	68.50		3,093.50	3,425
3440	1-1/2" pipe size	Q-5	9	1.778		6,400	95.50		6,495.50	7,200

For customer support on your Mechanical Cost Data, call 877.795.7394.

295

23 21 20 – Hydronic HVAC Piping Specialties

23 21 20.62 Liquid Drainers	Crew	Daily Output	Labor-Hours	Unit	Material	2015 Bare Costs Labor	Equipment	Total	Total Incl O&P
3450	2" pipe size	Q-5	7	2.286	Ea.	6,400	123		6,523	7,225

23 21 20.66 Monoflow Tee Fitting

		Crew	Daily Output	Labor-Hours	Unit	Material	Labor	Equipment	Total	Total Incl O&P
0010	**MONOFLOW TEE FITTING**									
1100	For one pipe hydronic, supply and return									
1110	Copper, soldered									
1120	3/4" x 1/2" size	1 Stpi	13	.615	Ea.	20.50	37		57.50	78
1130	1" x 1/2" size		12	.667		20.50	40		60.50	82.50
1140	1" x 3/4" size		11	.727		20.50	43.50		64	88
1150	1-1/4" x 1/2" size		11	.727		21.50	43.50		65	89
1160	1-1/4" x 3/4" size		10	.800		21.50	48		69.50	95.50
1170	1-1/2" x 3/4" size		10	.800		30.50	48		78.50	106
1180	1-1/2" x 1" size		9	.889		30.50	53		83.50	114
1190	2" x 3/4" size		9	.889		47.50	53		100.50	132
1200	2" x 1" size		8	1		47.50	60		107.50	142

23 21 20.70 Steam Traps

		Crew	Daily Output	Labor-Hours	Unit	Material	Labor	Equipment	Total	Total Incl O&P
0010	**STEAM TRAPS**									
0030	Cast iron body, threaded									
0040	Inverted bucket									
0050	1/2" pipe size	1 Stpi	12	.667	Ea.	157	40		197	232
0070	3/4" pipe size		10	.800		278	48		326	375
0100	1" pipe size		9	.889		420	53		473	545
0120	1-1/4" pipe size		8	1		635	60		695	790
0130	1-1/2" pipe size		7	1.143		800	68.50		868.50	985
0140	2" pipe size		6	1.333		1,250	79.50		1,329.50	1,500
0200	With thermic vent & check valve									
0202	1/2" pipe size	1 Stpi	12	.667	Ea.	185	40		225	264
0204	3/4" pipe size		10	.800		315	48		363	415
0206	1" pipe size		9	.889		455	53		508	580
0208	1-1/4" pipe size		8	1		690	60		750	850
0210	1-1/2" pipe size		7	1.143		765	68.50		833.50	950
0212	2" pipe size		6	1.333		1,175	79.50		1,254.50	1,425
1000	Float & thermostatic, 15 psi									
1010	3/4" pipe size	1 Stpi	16	.500	Ea.	141	30		171	200
1020	1" pipe size		15	.533		169	32		201	234
1030	1-1/4" pipe size		13	.615		187	37		224	262
1040	1-1/2" pipe size		9	.889		298	53		351	410
1060	2" pipe size		6	1.333		590	79.50		669.50	770
1100	For vacuum breaker, add					69			69	76
1110	Differential condensate controller									
1120	With union and manual metering valve									
1130	1/2" pipe size	1 Stpi	24	.333	Ea.	595	19.90		614.90	685
1140	3/4" pipe size		20	.400		690	24		714	795
1150	1" pipe size		19	.421		925	25		950	1,075
1160	1-1/4" pipe size		15	.533		1,250	32		1,282	1,425
1170	1-1/2" pipe size		13	.615		1,500	37		1,537	1,700
1180	2" pipe size		11	.727		1,950	43.50		1,993.50	2,225
1290	Brass body, threaded									
1300	Thermostatic, angle union, 25 psi									
1310	1/2" pipe size	1 Stpi	24	.333	Ea.	57	19.90		76.90	92.50
1320	3/4" pipe size		20	.400		91	24		115	136
1330	1" pipe size		19	.421		139	25		164	191
1990	Carbon steel body, threaded									

296

For customer support on your Mechanical Cost Data, call 877.795.7394.

23 21 Hydronic Piping and Pumps

23 21 20 – Hydronic HVAC Piping Specialties

23 21 20.70 Steam Traps	Crew	Daily Output	Labor-Hours	Unit	Material	2015 Bare Costs Labor	Equipment	Total	Total Incl O&P
2000 Controlled disc, integral strainer & blow down									
2010 Threaded									
2020 3/8" pipe size	1 Stpi	24	.333	Ea.	266	19.90		285.90	320
2030 1/2" pipe size		24	.333		266	19.90		285.90	320
2040 3/4" pipe size		20	.400		355	24		379	430
2050 1" pipe size		19	.421		460	25		485	545
2100 Socket weld									
2110 3/8" pipe size	Q-17	17	.941	Ea.	320	50.50	3.40	373.90	430
2120 1/2" pipe size		16	1		320	54	3.61	377.61	435
2130 3/4" pipe size		14	1.143		415	61.50	4.13	480.63	555
2140 1" pipe size		12	1.333		520	71.50	4.81	596.31	685
2170 Flanged, 600 lb. ASA									
2180 3/8" pipe size	1 Stpi	12	.667	Ea.	715	40		755	845
2190 1/2" pipe size		12	.667		715	40		755	845
2200 3/4" pipe size		10	.800		855	48		903	1,000
2210 1" pipe size		7	1.143		1,025	68.50		1,093.50	1,225
5000 Forged steel body									
5010 Inverted bucket									
5020 Threaded									
5030 1/2" pipe size	1 Stpi	12	.667	Ea.	480	40		520	590
5040 3/4" pipe size		10	.800		835	48		883	990
5050 1" pipe size		9	.889		1,250	53		1,303	1,450
5060 1-1/4" pipe size		8	1		1,825	60		1,885	2,100
5070 1-1/2" pipe size		7	1.143		2,575	68.50		2,643.50	2,925
5080 2" pipe size		6	1.333		4,550	79.50		4,629.50	5,125
5100 Socket weld									
5110 1/2" pipe size	Q-17	16	1	Ea.	580	54	3.61	637.61	725
5120 3/4" pipe size		14	1.143		935	61.50	4.13	1,000.63	1,125
5130 1" pipe size		12	1.333		1,350	71.50	4.81	1,426.31	1,600
5140 1-1/4" pipe size		10	1.600		1,925	86	5.80	2,016.80	2,225
5150 1-1/2" pipe size		9	1.778		2,675	95.50	6.40	2,776.90	3,100
5160 2" pipe size		7	2.286		4,650	123	8.25	4,781.25	5,300
5200 Flanged									
5210 1/2" pipe size	1 Stpi	12	.667	Ea.	895	40		935	1,050
5220 3/4" pipe size		10	.800		1,475	48		1,523	1,675
5230 1" pipe size		7	1.143		2,000	68.50		2,068.50	2,300
5240 1-1/4" pipe size		6	1.333		2,700	79.50		2,779.50	3,075
5250 1-1/2" pipe size	Q-5	9	1.778		3,450	95.50		3,545.50	3,950
5260 2" pipe size	"	7	2.286		5,425	123		5,548	6,150
6000 Stainless steel body									
6010 Inverted bucket									
6020 Threaded									
6030 1/2" pipe size	1 Stpi	12	.667	Ea.	159	40		199	235
6040 3/4" pipe size		10	.800		209	48		257	300
6050 1" pipe size		9	.889		745	53		798	900
6100 Threaded with thermic vent									
6110 1/2" pipe size	1 Stpi	12	.667	Ea.	182	40		222	260
6120 3/4" pipe size		10	.800		233	48		281	330
6130 1" pipe size		9	.889		775	53		828	930
6200 Threaded with check valve									
6210 1/2" pipe size	1 Stpi	12	.667	Ea.	228	40		268	310
6220 3/4" pipe size	"	10	.800	"	310	48		358	410
6300 Socket weld									

For customer support on your Mechanical Cost Data, call 877.795.7394.

297

23 21 20.70 Steam Traps		Crew	Daily Output	Labor-Hours	Unit	Material	2015 Bare Costs Labor	Equipment	Total	Total Incl O&P
6310	1/2" pipe size	Q-17	16	1	Ea.	185	54	3.61	242.61	288
6320	3/4" pipe size	"	14	1.143	"	239	61.50	4.13	304.63	360

23 21 20.74 Strainers, Basket Type

		Crew	Daily Output	Labor-Hours	Unit	Material	2015 Bare Costs Labor	Equipment	Total	Total Incl O&P
0010	**STRAINERS, BASKET TYPE**, Perforated stainless steel basket									
0100	Brass or monel available									
2000	Simplex style									
2300	Bronze body									
2320	Screwed, 3/8" pipe size	1 Stpi	22	.364	Ea.	172	21.50		193.50	222
2340	1/2" pipe size		20	.400		295	24		319	360
2360	3/4" pipe size		17	.471		400	28		428	485
2380	1" pipe size		15	.533		400	32		432	490
2400	1-1/4" pipe size		13	.615		575	37		612	690
2420	1-1/2" pipe size		12	.667		580	40		620	695
2440	2" pipe size		10	.800		855	48		903	1,000
2460	2-1/2" pipe size	Q-5	15	1.067		770	57.50		827.50	935
2480	3" pipe size	"	14	1.143		1,125	61.50		1,186.50	1,325
2600	Flanged, 2" pipe size	1 Stpi	6	1.333		820	79.50		899.50	1,025
2620	2-1/2" pipe size	Q-5	4.50	3.556		1,275	191		1,466	1,725
2640	3" pipe size		3.50	4.571		1,450	246		1,696	1,975
2660	4" pipe size		3	5.333		2,450	287		2,737	3,125
2680	5" pipe size	Q-6	3.40	7.059		3,750	395		4,145	4,725
2700	6" pipe size		3	8		4,775	445		5,220	5,925
2710	8" pipe size		2.50	9.600		7,750	535		8,285	9,325
3600	Iron body									
3700	Screwed, 3/8" pipe size	1 Stpi	22	.364	Ea.	98.50	21.50		120	142
3720	1/2" pipe size		20	.400		102	24		126	148
3740	3/4" pipe size		17	.471		130	28		158	186
3760	1" pipe size		15	.533		133	32		165	194
3780	1-1/4" pipe size		13	.615		173	37		210	246
3800	1-1/2" pipe size		12	.667		191	40		231	270
3820	2" pipe size		10	.800		228	48		276	325
3840	2-1/2" pipe size	Q-5	15	1.067		305	57.50		362.50	420
3860	3" pipe size	"	14	1.143		370	61.50		431.50	500
4000	Flanged, 2" pipe size	1 Stpi	6	1.333		355	79.50		434.50	510
4020	2-1/2" pipe size	Q-5	4.50	3.556		475	191		666	815
4040	3" pipe size		3.50	4.571		500	246		746	920
4060	4" pipe size		3	5.333		760	287		1,047	1,275
4080	5" pipe size	Q-6	3.40	7.059		1,150	395		1,545	1,875
4100	6" pipe size		3	8		1,475	445		1,920	2,300
4120	8" pipe size		2.50	9.600		2,675	535		3,210	3,750
4140	10" pipe size		2.20	10.909		5,875	610		6,485	7,400
6000	Cast steel body									
6400	Screwed, 1" pipe size	1 Stpi	15	.533	Ea.	195	32		227	263
6410	1-1/4" pipe size		13	.615		295	37		332	380
6420	1-1/2" pipe size		12	.667		320	40		360	410
6440	2" pipe size		10	.800		405	48		453	515
6460	2-1/2" pipe size	Q-5	15	1.067		570	57.50		627.50	715
6480	3" pipe size	"	14	1.143		750	61.50		811.50	920
6560	Flanged, 2" pipe size	1 Stpi	6	1.333		685	79.50		764.50	875
6580	2-1/2" pipe size	Q-5	4.50	3.556		1,025	191		1,216	1,450
6600	3" pipe size		3.50	4.571		1,100	246		1,346	1,600
6620	4" pipe size		3	5.333		1,575	287		1,862	2,150

For customer support on your Mechanical Cost Data, call 877.795.7394.

23 21 20.74 Strainers, Basket Type		Crew	Daily Output	Labor-Hours	Unit	Material	2015 Bare Costs Labor	Equipment	Total	Total Incl O&P
6640	6" pipe size	Q-6	3	8	Ea.	2,875	445		3,320	3,825
6660	8" pipe size	"	2.50	9.600	↓	4,625	535		5,160	5,900
7000	Stainless steel body									
7200	Screwed, 1" pipe size	1 Stpi	15	.533	Ea.	283	32		315	360
7210	1-1/4" pipe size		13	.615		440	37		477	540
7220	1-1/2" pipe size		12	.667		440	40		480	545
7240	2" pipe size	↓	10	.800		655	48		703	790
7260	2-1/2" pipe size	Q-5	15	1.067		925	57.50		982.50	1,100
7280	3" pipe size	"	14	1.143		1,275	61.50		1,336.50	1,500
7400	Flanged, 2" pipe size	1 Stpi	6	1.333		1,125	79.50		1,204.50	1,350
7420	2-1/2" pipe size	Q-5	4.50	3.556		2,025	191		2,216	2,550
7440	3" pipe size		3.50	4.571		2,075	246		2,321	2,675
7460	4" pipe size	↓	3	5.333		3,275	287		3,562	4,025
7480	6" pipe size	Q-6	3	8		5,700	445		6,145	6,950
7500	8" pipe size	"	2.50	9.600	↓	8,300	535		8,835	9,925
8100	Duplex style									
8200	Bronze body									
8240	Screwed, 3/4" pipe size	1 Stpi	16	.500	Ea.	1,175	30		1,205	1,325
8260	1" pipe size		14	.571		1,175	34		1,209	1,325
8280	1-1/4" pipe size		12	.667		2,825	40		2,865	3,150
8300	1-1/2" pipe size		11	.727		2,825	43.50		2,868.50	3,175
8320	2" pipe size	↓	9	.889		3,700	53		3,753	4,125
8340	2-1/2" pipe size	Q-5	14	1.143		4,775	61.50		4,836.50	5,350
8420	Flanged, 2" pipe size	1 Stpi	6	1.333		3,975	79.50		4,054.50	4,500
8440	2-1/2" pipe size	Q-5	4.50	3.556		5,475	191		5,666	6,325
8460	3" pipe size		3.50	4.571		5,975	246		6,221	6,950
8480	4" pipe size	↓	3	5.333		8,575	287		8,862	9,850
8500	5" pipe size	Q-6	3.40	7.059	↓	18,600	395		18,995	21,000
8520	6" pipe size	"	3	8	↓	19,100	445		19,545	21,700
8700	Iron body									
8740	Screwed, 3/4" pipe size	1 Stpi	16	.500	Ea.	1,050	30		1,080	1,200
8760	1" pipe size		14	.571		1,050	34		1,084	1,200
8780	1-1/4" pipe size		12	.667		1,175	40		1,215	1,325
8800	1-1/2" pipe size		11	.727		1,175	43.50		1,218.50	1,350
8820	2" pipe size	↓	9	.889		1,975	53		2,028	2,250
8840	2-1/2" pipe size	Q-5	14	1.143		2,175	61.50		2,236.50	2,500
9000	Flanged, 2" pipe size	1 Stpi	6	1.333		2,100	79.50		2,179.50	2,450
9020	2-1/2" pipe size	Q-5	4.50	3.556		2,250	191		2,441	2,775
9040	3" pipe size		3.50	4.571		2,450	246		2,696	3,075
9060	4" pipe size	↓	3	5.333		4,150	287		4,437	4,975
9080	5" pipe size	Q-6	3.40	7.059		9,025	395		9,420	10,500
9100	6" pipe size		3	8		9,025	445		9,470	10,600
9120	8" pipe size		2.50	9.600		16,700	535		17,235	19,200
9140	10" pipe size		2.20	10.909		23,900	610		24,510	27,100
9160	12" pipe size		1.70	14.118		26,200	785		26,985	30,100
9170	14" pipe size		1.40	17.143		31,000	955		31,955	35,600
9180	16" pipe size	↓	1	24	↓	38,200	1,350		39,550	44,000
9300	Cast steel body									
9340	Screwed, 1" pipe size	1 Stpi	14	.571	Ea.	1,350	34		1,384	1,525
9360	1-1/2" pipe size		11	.727		2,175	43.50		2,218.50	2,475
9380	2" pipe size		9	.889		2,875	53		2,928	3,250
9460	Flanged, 2" pipe size	↓	6	1.333		3,225	79.50		3,304.50	3,675
9480	2-1/2" pipe size	Q-5	4.50	3.556	↓	5,200	191		5,391	6,000

23 21 20.74 Strainers, Basket Type

		Crew	Daily Output	Labor-Hours	Unit	Material	2015 Bare Costs Labor	Equipment	Total	Total Incl O&P
9500	3" pipe size	Q-5	3.50	4.571	Ea.	5,650	246		5,896	6,600
9520	4" pipe size		3	5.333		6,975	287		7,262	8,100
9540	6" pipe size	Q-6	3	8		13,100	445		13,545	15,100
9560	8" pipe size	"	2.50	9.600		31,500	535		32,035	35,400
9700	Stainless steel body									
9740	Screwed, 1" pipe size	1 Stpi	14	.571	Ea.	1,900	34		1,934	2,125
9760	1-1/2" pipe size		11	.727		2,850	43.50		2,893.50	3,225
9780	2" pipe size		9	.889		4,000	53		4,053	4,475
9860	Flanged, 2" pipe size		6	1.333		4,350	79.50		4,429.50	4,925
9880	2-1/2" pipe size	Q-5	4.50	3.556		7,300	191		7,491	8,325
9900	3" pipe size		3.50	4.571		7,950	246		8,196	9,125
9920	4" pipe size		3	5.333		10,400	287		10,687	11,800
9940	6" pipe size	Q-6	3	8		15,800	445		16,245	18,100
9960	8" pipe size	"	2.50	9.600		44,200	535		44,735	49,400

23 21 20.76 Strainers, Y Type, Bronze Body

		Crew	Daily Output	Labor-Hours	Unit	Material	2015 Bare Costs Labor	Equipment	Total	Total Incl O&P
0010	**STRAINERS, Y TYPE, BRONZE BODY**									
0050	Screwed, 125 lb., 1/4" pipe size	1 Stpi	24	.333	Ea.	23	19.90		42.90	55.50
0070	3/8" pipe size		24	.333		27.50	19.90		47.40	60.50
0100	1/2" pipe size		20	.400		27.50	24		51.50	66.50
0120	3/4" pipe size		19	.421		34	25		59	75.50
0140	1" pipe size		17	.471		50	28		78	97.50
0150	1-1/4" pipe size		15	.533		81	32		113	137
0160	1-1/2" pipe size		14	.571		108	34		142	171
0180	2" pipe size		13	.615		144	37		181	214
0182	3" pipe size		12	.667		845	40		885	990
0200	300 lb., 2-1/2" pipe size	Q-5	17	.941		510	50.50		560.50	635
0220	3" pipe size		16	1		1,000	54		1,054	1,175
0240	4" pipe size		15	1.067		2,300	57.50		2,357.50	2,600
0500	For 300 lb. rating 1/4" thru 2", add					15%				
1000	Flanged, 150 lb., 1-1/2" pipe size	1 Stpi	11	.727	Ea.	520	43.50		563.50	640
1020	2" pipe size	"	8	1		705	60		765	865
1030	2-1/2" pipe size	Q-5	5	3.200		950	172		1,122	1,300
1040	3" pipe size		4.50	3.556		1,175	191		1,366	1,575
1060	4" pipe size		3	5.333		1,775	287		2,062	2,375
1080	5" pipe size	Q-6	3.40	7.059		1,775	395		2,170	2,550
1100	6" pipe size		3	8		3,400	445		3,845	4,400
1106	8" pipe size		2.60	9.231		3,725	515		4,240	4,875
1500	For 300 lb. rating, add					40%				

23 21 20.78 Strainers, Y Type, Iron Body

		Crew	Daily Output	Labor-Hours	Unit	Material	2015 Bare Costs Labor	Equipment	Total	Total Incl O&P
0010	**STRAINERS, Y TYPE, IRON BODY**									
0050	Screwed, 250 lb., 1/4" pipe size	1 Stpi	20	.400	Ea.	11.05	24		35.05	48
0070	3/8" pipe size		20	.400		11.05	24		35.05	48
0100	1/2" pipe size		20	.400		11.05	24		35.05	48
0120	3/4" pipe size		18	.444		12.95	26.50		39.45	54.50
0140	1" pipe size		16	.500		18.25	30		48.25	65
0150	1-1/4" pipe size		15	.533		24	32		56	74.50
0160	1-1/2" pipe size		12	.667		30	40		70	92.50
0180	2" pipe size		8	1		44.50	60		104.50	139
0200	2-1/2" pipe size	Q-5	12	1.333		268	71.50		339.50	400
0220	3" pipe size		11	1.455		289	78		367	440
0240	4" pipe size		5	3.200		490	172		662	800
0500	For galvanized body, add					50%				

300

For customer support on your Mechanical Cost Data, call 877.795.7394.

23 21 20 – Hydronic HVAC Piping Specialties

23 21 20.78 Strainers, Y Type, Iron Body

		Crew	Daily Output	Labor-Hours	Unit	Material	2015 Bare Costs Labor	Equipment	Total	Total Incl O&P
1000	Flanged, 125 lb., 1-1/2" pipe size	1 Stpi	11	.727	Ea.	107	43.50		150.50	183
1020	2" pipe size	"	8	1		113	60		173	214
1030	2-1/2" pipe size	Q-5	5	3.200		115	172		287	385
1040	3" pipe size		4.50	3.556		186	191		377	495
1060	4" pipe size		3	5.333		305	287		592	770
1080	5" pipe size	Q-6	3.40	7.059		385	395		780	1,025
1100	6" pipe size		3	8		615	445		1,060	1,350
1120	8" pipe size		2.50	9.600		830	535		1,365	1,725
1140	10" pipe size		2	12		1,875	670		2,545	3,050
1160	12" pipe size		1.70	14.118		2,275	785		3,060	3,700
1170	14" pipe size		1.30	18.462		4,175	1,025		5,200	6,150
1180	16" pipe size		1	24		5,925	1,350		7,275	8,525
1500	For 250 lb. rating, add					20%				
2000	For galvanized body, add					50%				
2500	For steel body, add					40%				

23 21 20.80 Suction Diffusers

		Crew	Daily Output	Labor-Hours	Unit	Material	2015 Bare Costs Labor	Equipment	Total	Total Incl O&P
0010	**SUCTION DIFFUSERS**									
0100	Cast iron body with integral straightening vanes, strainer									
1000	Flanged									
1010	2" inlet, 1-1/2" pump side	1 Stpi	6	1.333	Ea.	325	79.50		404.50	475
1020	2" pump side	"	5	1.600		335	95.50		430.50	515
1030	3" inlet, 2" pump side	Q-5	6.50	2.462		405	132		537	645
1040	2-1/2" pump side		5.30	3.019		590	162		752	895
1050	3" pump side		4.50	3.556		600	191		791	950
1060	4" inlet, 3" pump side		3.50	4.571		685	246		931	1,125
1070	4" pump side		3	5.333		795	287		1,082	1,300
1080	5" inlet, 4" pump side	Q-6	3.80	6.316		905	350		1,255	1,525
1090	5" pump side		3.40	7.059		1,075	395		1,470	1,800
1100	6" inlet, 4" pump side		3.30	7.273		940	405		1,345	1,625
1110	5" pump side		3.10	7.742		1,125	430		1,555	1,900
1120	6" pump side		3	8		1,200	445		1,645	1,975
1130	8" inlet, 6" pump side		2.70	8.889		1,325	495		1,820	2,225
1140	8" pump side		2.50	9.600		2,325	535		2,860	3,350
1150	10" inlet, 6" pump side		2.40	10		2,675	560		3,235	3,800
1160	8" pump side		2.30	10.435		2,775	580		3,355	3,925
1170	10" pump side		2.20	10.909		3,875	610		4,485	5,175
1180	12" inlet, 8" pump side		2	12		4,075	670		4,745	5,500
1190	10" pump side		1.80	13.333		4,525	745		5,270	6,100
1200	12" pump side		1.70	14.118		5,100	785		5,885	6,800
1210	14" inlet, 12" pump side		1.50	16		5,350	890		6,240	7,225
1220	14" pump side		1.30	18.462		7,525	1,025		8,550	9,825

23 21 20.84 Thermoflo Indicator

		Crew	Daily Output	Labor-Hours	Unit	Material	2015 Bare Costs Labor	Equipment	Total	Total Incl O&P
0010	**THERMOFLO INDICATOR**, For balancing									
1000	Sweat connections, 1-1/4" pipe size	1 Stpi	12	.667	Ea.	650	40		690	775
1020	1-1/2" pipe size		10	.800		660	48		708	800
1040	2" pipe size		8	1		695	60		755	850
1060	2-1/2" pipe size		7	1.143		1,050	68.50		1,118.50	1,275
2000	Flange connections, 3" pipe size	Q-5	5	3.200		1,275	172		1,447	1,650
2020	4" pipe size		4	4		1,525	215		1,740	2,000
2030	5" pipe size		3.50	4.571		1,900	246		2,146	2,475
2040	6" pipe size		3	5.333		2,025	287		2,312	2,675
2060	8" pipe size		2	8		2,475	430		2,905	3,375

For customer support on your Mechanical Cost Data, call 877.795.7394.

301

23 21 20.88 Venturi Flow

		Crew	Daily Output	Labor-Hours	Unit	Material	2015 Bare Costs Labor	Equipment	Total	Total Incl O&P
0010	**VENTURI FLOW**, Measuring device									
0050	1/2" diameter	1 Stpi	24	.333	Ea.	281	19.90		300.90	340
0100	3/4" diameter		20	.400		257	24		281	320
0120	1" diameter		19	.421		276	25		301	345
0140	1-1/4" diameter		15	.533		340	32		372	425
0160	1-1/2" diameter		13	.615		355	37		392	445
0180	2" diameter		11	.727		365	43.50		408.50	470
0200	2-1/2" diameter	Q-5	16	1		500	54		554	630
0220	3" diameter		14	1.143		515	61.50		576.50	665
0240	4" diameter		11	1.455		775	78		853	970
0260	5" diameter	Q-6	4	6		1,025	335		1,360	1,625
0280	6" diameter		3.50	6.857		1,125	380		1,505	1,825
0300	8" diameter		3	8		1,450	445		1,895	2,275
0320	10" diameter		2	12		3,425	670		4,095	4,775
0330	12" diameter		1.80	13.333		4,950	745		5,695	6,575
0340	14" diameter		1.60	15		5,600	835		6,435	7,450
0350	16" diameter		1.40	17.143		6,250	955		7,205	8,325
0500	For meter, add					2,125			2,125	2,350

23 21 20.94 Weld End Ball Joints

		Crew	Daily Output	Labor-Hours	Unit	Material	2015 Bare Costs Labor	Equipment	Total	Total Incl O&P
0010	**WELD END BALL JOINTS**, Steel									
0050	2-1/2" diameter	Q-17	13	1.231	Ea.	615	66	4.44	685.44	785
0100	3" diameter		12	1.333		720	71.50	4.81	796.31	905
0120	4" diameter		11	1.455		1,175	78	5.25	1,258.25	1,400
0140	5" diameter	Q-18	14	1.714		1,350	95.50	4.11	1,449.61	1,625
0160	6" diameter		12	2		2,000	112	4.80	2,116.80	2,375
0180	8" diameter		9	2.667		3,025	149	6.40	3,180.40	3,550
0200	10" diameter		8	3		4,075	167	7.20	4,249.20	4,725
0220	12" diameter		6	4		6,500	223	9.60	6,732.60	7,500

23 21 20.98 Zone Valves

		Crew	Daily Output	Labor-Hours	Unit	Material	2015 Bare Costs Labor	Equipment	Total	Total Incl O&P
0010	**ZONE VALVES**									
1000	Bronze body									
1010	2 way, 125 psi									
1020	Standard, 65' pump head									
1030	1/2" soldered	1 Stpi	24	.333	Ea.	124	19.90		143.90	166
1040	3/4" soldered		20	.400		124	24		148	172
1050	1" soldered		19	.421		133	25		158	184
1060	1-1/4" soldered		15	.533		155	32		187	219
1200	High head, 125' pump head									
1210	1/2" soldered	1 Stpi	24	.333	Ea.	174	19.90		193.90	221
1220	3/4" soldered		20	.400		180	24		204	234
1230	1" soldered		19	.421		196	25		221	254
1300	Geothermal, 150' pump head									
1310	3/4" soldered	1 Stpi	20	.400	Ea.	207	24		231	263
1320	1" soldered		19	.421		219	25		244	279
1330	3/4" threaded		20	.400		207	24		231	263
3000	3 way, 125 psi									
3200	Bypass, 65' pump head									
3210	1/2" soldered	1 Stpi	24	.333	Ea.	187	19.90		206.90	236
3220	3/4" soldered		20	.400		188	24		212	243
3230	1" soldered		19	.421		213	25		238	272
9000	Transformer, for up to 3 zone valves		20	.400		26	24		50	65

23 21 Hydronic Piping and Pumps

23 21 23 – Hydronic Pumps

23 21 23.13 In-Line Centrifugal Hydronic Pumps	Crew	Daily Output	Labor-Hours	Unit	Material	2015 Bare Costs Labor	Equipment	Total	Total Incl O&P
0010 **IN-LINE CENTRIFUGAL HYDRONIC PUMPS**									
0600 Bronze, sweat connections, 1/40 HP, in line									
0640 3/4" size	Q-1	16	1	Ea.	218	53		271	320
1000 Flange connection, 3/4" to 1-1/2" size									
1040 1/12 HP	Q-1	6	2.667	Ea.	565	141		706	835
1060 1/8 HP		6	2.667		970	141		1,111	1,300
1100 1/3 HP		6	2.667		1,075	141		1,216	1,425
1140 2" size, 1/6 HP		5	3.200		1,400	169		1,569	1,775
1180 2-1/2" size, 1/4 HP		5	3.200		1,775	169		1,944	2,200
1220 3" size, 1/4 HP		4	4		1,850	211		2,061	2,375
1260 1/3 HP		4	4		2,275	211		2,486	2,825
1300 1/2 HP		4	4		2,325	211		2,536	2,875
1340 3/4 HP		4	4		2,550	211		2,761	3,125
1380 1 HP		4	4		4,100	211		4,311	4,825
2000 Cast iron, flange connection									
2040 3/4" to 1-1/2" size, in line, 1/12 HP	Q-1	6	2.667	Ea.	365	141		506	615
2060 1/8 HP		6	2.667		610	141		751	885
2100 1/3 HP		6	2.667		680	141		821	960
2140 2" size, 1/6 HP		5	3.200		745	169		914	1,075
2180 2-1/2" size, 1/4 HP		5	3.200		960	169		1,129	1,300
2220 3" size, 1/4 HP		4	4		975	211		1,186	1,400
2260 1/3 HP		4	4		1,325	211		1,536	1,775
2300 1/2 HP		4	4		1,375	211		1,586	1,825
2340 3/4 HP		4	4		1,575	211		1,786	2,075
2380 1 HP		4	4		2,275	211		2,486	2,825
2600 For nonferrous impeller, add					3%				
3000 High head, bronze impeller									
3030 1-1/2" size 1/2 HP	Q-1	5	3.200	Ea.	1,150	169		1,319	1,525
3040 1-1/2" size 3/4 HP		5	3.200		1,250	169		1,419	1,625
3050 2" size 1 HP		4	4		1,525	211		1,736	2,000
3090 2" size 1-1/2 HP		4	4		1,900	211		2,111	2,400
4000 Close coupled, end suction, bronze impeller									
4040 1-1/2" size, 1-1/2 HP, to 40 GPM	Q-1	3	5.333	Ea.	2,250	282		2,532	2,900
4090 2" size, 2 HP, to 50 GPM		3	5.333		2,675	282		2,957	3,350
4100 2" size, 3 HP, to 90 GPM		2.30	6.957		2,775	370		3,145	3,600
4190 2-1/2" size, 3 HP, to 150 GPM		2	8		3,000	425		3,425	3,950
4300 3" size, 5 HP, to 225 GPM		1.80	8.889		3,425	470		3,895	4,475
4410 3" size, 10 HP, to 350 GPM		1.60	10		5,025	530		5,555	6,325
4420 4" size, 7-1/2 HP, to 350 GPM		1.60	10		5,075	530		5,605	6,375
4520 4" size, 10 HP, to 600 GPM	Q-2	1.70	14.118		5,125	775		5,900	6,800
4530 5" size, 15 HP, to 1000 GPM		1.70	14.118		5,125	775		5,900	6,825
4610 5" size, 20 HP, to 1350 GPM		1.50	16		5,450	875		6,325	7,325
4620 5" size, 25 HP, to 1550 GPM		1.50	16		7,450	875		8,325	9,525
5000 Base mounted, bronze impeller, coupling guard									
5040 1-1/2" size, 1-1/2 HP, to 40 GPM	Q-1	2.30	6.957	Ea.	5,950	370		6,320	7,100
5090 2" size, 2 HP, to 50 GPM		2.30	6.957		6,675	370		7,045	7,900
5100 2" size, 3 HP, to 90 GPM		2	8		7,325	425		7,750	8,725
5190 2-1/2" size, 3 HP, to 150 GPM		1.80	8.889		7,900	470		8,370	9,375
5300 3" size, 5 HP, to 225 GPM		1.60	10		9,050	530		9,580	10,800
5410 4" size, 5 HP, to 350 GPM		1.50	10.667		9,850	565		10,415	11,700
5420 4" size, 7-1/2 HP, to 350 GPM		1.50	10.667		11,100	565		11,665	13,100
5520 5" size, 10 HP, to 600 GPM	Q-2	1.60	15		14,400	820		15,220	17,200

For customer support on your Mechanical Cost Data, call 877.795.7394.

303

23 21 Hydronic Piping and Pumps

23 21 23 – Hydronic Pumps

23 21 23.13 In-Line Centrifugal Hydronic Pumps		Crew	Daily Output	Labor-Hours	Unit	Material	2015 Bare Costs Labor	Equipment	Total	Total Incl O&P
5530	5" size, 15 HP, to 1000 GPM	Q-2	1.60	15	Ea.	15,900	820		16,720	18,800
5610	6" size, 20 HP, to 1350 GPM	↓	1.40	17.143		17,600	940		18,540	20,700
5620	6" size, 25 HP, to 1550 GPM	↓	1.40	17.143	↓	19,700	940		20,640	23,100
5800	The above pump capacities are based on 1800 RPM,									
5810	at a 60 foot head. Increasing the RPM									
5820	or decreasing the head will increase the GPM.									

23 21 29 – Automatic Condensate Pump Units

23 21 29.10 Condensate Removal Pump System

			Crew	Daily Output	Labor-Hours	Unit	Material	Labor	Equipment	Total	Total Incl O&P
0010	**CONDENSATE REMOVAL PUMP SYSTEM**										
0020	Pump with 1 gal. ABS tank										
0100	115 V										
0120	1/50 HP, 200 GPH	G	1 Stpi	12	.667	Ea.	197	40		237	277
0140	1/18 HP, 270 GPH	G	↓	10	.800		210	48		258	305
0160	1/5 HP, 450 GPH	G	↓	8	1	↓	470	60		530	605
0200	230 V										
0260	1/5 HP, 450 GPH	G	1 Stpi	8	1	Ea.	520	60		580	660

23 22 Steam and Condensate Piping and Pumps

23 22 13 – Steam and Condensate Heating Piping

23 22 13.23 Aboveground Steam and Condensate Piping

		Crew	Daily Output	Labor-Hours	Unit	Material	Labor	Equipment	Total	Total Incl O&P
0010	**ABOVEGROUND STEAM AND CONDENSATE HEATING PIPING**									
0020	Condensate meter									
0100	500 lb. per hour	1 Stpi	14	.571	Ea.	3,275	34		3,309	3,650
0140	1500 lb. per hour		7	1.143		4,100	68.50		4,168.50	4,600
0160	3000 lb. per hour	↓	5	1.600		4,525	95.50		4,620.50	5,125
0200	12,000 lb. per hour	Q-5	3.50	4.571	↓	5,400	246		5,646	6,300

23 22 23 – Steam Condensate Pumps

23 22 23.10 Condensate Return System

		Crew	Daily Output	Labor-Hours	Unit	Material	Labor	Equipment	Total	Total Incl O&P
0010	**CONDENSATE RETURN SYSTEM**									
2000	Simplex									
2010	With pump, motor, CI receiver, float switch									
2020	3/4 HP, 15 GPM	Q-1	1.80	8.889	Ea.	6,425	470		6,895	7,750
2100	Duplex									
2110	With 2 pumps and motors, CI receiver, float switch, alternator									
2120	3/4 HP, 15 GPM, 15 gal. CI rcvr.	Q-1	1.40	11.429	Ea.	7,000	605		7,605	8,600
2130	1 HP, 25 GPM		1.20	13.333		8,400	705		9,105	10,300
2140	1-1/2 HP, 45 GPM		1	16		9,750	845		10,595	12,000
2150	1-1/2 HP, 60 GPM	↓	1	16	↓	10,900	845		11,745	13,300

23 23 Refrigerant Piping

23 23 13 – Refrigerant Piping Valves

23 23 13.10 Valves		Crew	Daily Output	Labor-Hours	Unit	Material	2015 Bare Costs Labor	Equipment	Total	Total Incl O&P
0010	**VALVES**									
8100	Check valve, soldered									
8110	5/8"	1 Stpi	36	.222	Ea.	87	13.30		100.30	116
8114	7/8"		26	.308		93	18.40		111.40	130
8118	1-1/8"		18	.444		124	26.50		150.50	176
8122	1-3/8"		14	.571		199	34		233	271
8126	1-5/8"		13	.615		243	37		280	325
8130	2-1/8"		12	.667		284	40		324	370
8134	2-5/8"	Q-5	22	.727		405	39		444	505
8138	3-1/8"	"	20	.800		540	43		583	655
8500	Refrigeration valve, packless, soldered									
8510	1/2"	1 Stpi	38	.211	Ea.	46.50	12.60		59.10	70
8514	5/8"		36	.222		46.50	13.30		59.80	71
8518	7/8"		26	.308		113	18.40		131.40	152
8520	Packed, soldered									
8522	1-1/8"	1 Stpi	18	.444	Ea.	161	26.50		187.50	217
8526	1-3/8"		14	.571		284	34		318	360
8530	1-5/8"		13	.615		315	37		352	400
8534	2-1/8"		12	.667		500	40		540	610
8538	2-5/8"	Q-5	22	.727		700	39		739	825
8542	3-1/8"		20	.800		785	43		828	930
8546	4-1/8"		18	.889		1,375	48		1,423	1,600
8600	Solenoid valve, flange/solder									
8610	1/2"	1 Stpi	38	.211	Ea.	186	12.60		198.60	224
8614	5/8"		36	.222		253	13.30		266.30	298
8618	3/4"		30	.267		297	15.95		312.95	350
8622	7/8"		26	.308		365	18.40		383.40	430
8626	1-1/8"		18	.444		475	26.50		501.50	560
8630	1-3/8"		14	.571		590	34		624	700
8634	1-5/8"		13	.615		825	37		862	960
8638	2-1/8"		12	.667		900	40		940	1,050
8800	Thermostatic valve, flange/solder									
8810	1/2 – 3 ton, 3/8" x 5/8"	1 Stpi	9	.889	Ea.	182	53		235	280
8814	4 – 5 ton, 1/2" x 7/8"		7	1.143		213	68.50		281.50	335
8818	6 – 8 ton, 5/8" x 7/8"		5	1.600		213	95.50		308.50	380
8822	7 – 12 ton, 7/8" x 1-1/8"		4	2		249	120		369	455
8826	15 – 20 ton, 7/8" x 1-3/8"		3.20	2.500		249	149		398	500

23 23 16 – Refrigerant Piping Specialties

23 23 16.10 Refrigerant Piping Component Specialties

		Crew	Daily Output	Labor-Hours	Unit	Material	2015 Bare Costs Labor	Equipment	Total	Total Incl O&P
0010	**REFRIGERANT PIPING COMPONENT SPECIALTIES**									
0600	Accumulator									
0610	3/4"	1 Stpi	8.80	.909	Ea.	70.50	54.50		125	160
0614	7/8"		6.40	1.250		80.50	74.50		155	202
0618	1-1/8"		4.80	1.667		111	99.50		210.50	272
0622	1-3/8"		4	2		145	120		265	340
0626	1-5/8"		3.20	2.500		157	149		306	400
0630	2-1/8"		2.40	3.333		375	199		574	715
0700	Condensate drip/drain pan, 10" x 9" x 2-3/4"		24	.333		143	19.90		162.90	188
1000	Filter dryer									
1010	Replaceable core type, solder									
1020	1/2"	1 Stpi	20	.400	Ea.	96.50	24		120.50	142
1030	5/8"		19	.421		96.50	25		121.50	144

For customer support on your Mechanical Cost Data, call 877.795.7394.

305

23 23 16 – Refrigerant Piping Specialties

23 23 16.10 Refrigerant Piping Component Specialties		Crew	Daily Output	Labor-Hours	Unit	Material	2015 Bare Costs Labor	Equipment	Total	Total Incl O&P
1040	7/8"	1 Stpi	18	.444	Ea.	127	26.50		153.50	179
1050	1-1/8"		15	.533		134	32		166	195
1060	1-3/8"		14	.571		136	34		170	201
1070	1-5/8"		12	.667		133	40		173	206
1080	2-1/8"		10	.800		286	48		334	385
1090	2-5/8"		9	.889		575	53		628	710
1100	3-1/8"		8	1		620	60		680	770
1200	Sealed in-line, solder									
1210	1/4"	1 Stpi	22	.364	Ea.	30.50	21.50		52	66.50
1220	3/8"		21	.381		25.50	23		48.50	62.50
1230	1/2"		20	.400		30.50	24		54.50	69.50
1260	5/8"		19	.421		43.50	25		68.50	85.50
1270	7/8"		18	.444		45	26.50		71.50	89.50
1290	1-1/8"		15	.533		68.50	32		100.50	124
4000	P-Trap, suction line, solder									
4010	5/8"	1 Stpi	19	.421	Ea.	33.50	25		58.50	74.50
4020	3/4"		19	.421		49.50	25		74.50	92.50
4030	7/8"		18	.444		37	26.50		63.50	81
4040	1-1/8"		15	.533		56.50	32		88.50	110
4050	1-3/8"		14	.571		102	34		136	164
4060	1-5/8"		12	.667		214	40		254	296
4070	2-1/8"		10	.800		440	48		488	550
5000	Sightglass									
5010	Moisture and liquid indicator, solder									
5020	1/4"	1 Stpi	22	.364	Ea.	18	21.50		39.50	53
5030	3/8"		21	.381		18.15	23		41.15	54.50
5040	1/2"		20	.400		23	24		47	61
5050	5/8"		19	.421		23.50	25		48.50	64
5060	7/8"		18	.444		35	26.50		61.50	78.50
5070	1-1/8"		15	.533		38	32		70	89.50
5080	1-3/8"		14	.571		66.50	34		100.50	125
5090	1-5/8"		12	.667		75	40		115	143
5100	2-1/8"		10	.800		94.50	48		142.50	176
7400	Vacuum pump set									
7410	Two stage, high vacuum continuous duty	1 Stpi	8	1	Ea.	3,875	60		3,935	4,375

23 23 16.16 Refrigerant Line Sets

		Crew	Daily Output	Labor-Hours	Unit	Material	2015 Bare Costs Labor	Equipment	Total	Total Incl O&P
0010	**REFRIGERANT LINE SETS**									
0100	Copper tube									
0110	1/2" insulation, both tubes									
0120	Combination 1/4" and 1/2" tubes									
0130	10' set	Q-5	42	.381	Ea.	70.50	20.50		91	109
0140	20' set		40	.400		110	21.50		131.50	154
0150	30' set		37	.432		148	23.50		171.50	197
0160	40' set		35	.457		193	24.50		217.50	250
0170	50' set		32	.500		229	27		256	293
0180	100' set		22	.727		520	39		559	635
0300	Combination 1/4" and 3/4" tubes									
0310	10' set	Q-5	40	.400	Ea.	78.50	21.50		100	119
0320	20' set		38	.421		139	22.50		161.50	187
0330	30' set		35	.457		208	24.50		232.50	266
0340	40' set		33	.485		272	26		298	340
0350	50' set		30	.533		345	28.50		373.50	420

23 23 16 – Refrigerant Piping Specialties

23 23 16.16 Refrigerant Line Sets		Crew	Daily Output	Labor-Hours	Unit	Material	2015 Bare Costs Labor	Equipment	Total	Total Incl O&P
0380	100' set	Q-5	20	.800	Ea.	780	43		823	925
0500	Combination 3/8" & 3/4" tubes									
0510	10' set	Q-5	28	.571	Ea.	90.50	30.50		121	146
0520	20' set		36	.444		141	24		165	191
0530	30' set		34	.471		191	25.50		216.50	249
0540	40' set		31	.516		252	28		280	320
0550	50' set		28	.571		296	30.50		326.50	370
0580	100' set		18	.889		875	48		923	1,025
0700	Combination 3/8" & 1-1/8" tubes									
0710	10' set	Q-5	36	.444	Ea.	158	24		182	210
0720	20' set		33	.485		265	26		291	330
0730	30' set		31	.516		355	28		383	430
0740	40' set		28	.571		535	30.50		565.50	630
0750	50' set		26	.615		615	33		648	730
0900	Combination 1/2" & 3/4" tubes									
0910	10' set	Q-5	37	.432	Ea.	95.50	23.50		119	140
0920	20' set		35	.457		169	24.50		193.50	223
0930	30' set		33	.485		255	26		281	320
0940	40' set		30	.533		335	28.50		363.50	415
0950	50' set		27	.593		420	32		452	515
0980	100' set		17	.941		950	50.50		1,000.50	1,125
2100	Combination 1/2" & 1-1/8" tubes									
2110	10' set	Q-5	35	.457	Ea.	163	24.50		187.50	216
2120	20' set		31	.516		279	28		307	345
2130	30' set		29	.552		420	29.50		449.50	505
2140	40' set		25	.640		565	34.50		599.50	670
2150	50' set		14	1.143		625	61.50		686.50	780
2300	For 1" thick insulation add					30%	15%			

23 23 23 – Refrigerants

23 23 23.10 Anti-Freeze

		Crew	Daily Output	Labor-Hours	Unit	Material	Labor	Equipment	Total	Total Incl O&P
0010	**ANTI-FREEZE**, Inhibited									
0900	Ethylene glycol concentrated									
1000	55 gallon drums, small quantities					Gal.	11.95			11.95	13.10
1200	Large quantities						9.40			9.40	10.35
2000	Propylene glycol, for solar heat, small quantities						15.55			15.55	17.10
2100	Large quantities						10.35			10.35	11.35

23 23 23.20 Refrigerant

		Crew	Daily Output	Labor-Hours	Unit	Material	Labor	Equipment	Total	Total Incl O&P
0010	**REFRIGERANT**									
4420	R-22, 30 lb. disposable cylinder					Lb.	35			35	38.50
4428	R-134A, 30 lb. disposable cylinder						13.45			13.45	14.80
4434	R-407C, 30 lb. disposable cylinder						28			28	30.50
4440	R-408A, 25 lb. disposable cylinder						31			31	34
4450	R-410A, 25 lb. disposable cylinder						22.50			22.50	24.50
4470	R-507, 25 lb. disposable cylinder						17.75			17.75	19.55

For customer support on your Mechanical Cost Data, call 877.795.7394.

307

23 31 HVAC Ducts and Casings

23 31 13 – Metal Ducts

23 31 13.13 Rectangular Metal Ducts	Crew	Daily Output	Labor-Hours	Unit	Material	2015 Bare Costs Labor	Equipment	Total	Total Incl O&P
0010 RECTANGULAR METAL DUCTS									
0020 Fabricated rectangular, includes fittings, joints, supports,									
0021 allowance for flexible connections and field sketches.									
0030 Does not include "as-built dwgs." or insulation. R233100-30									
0031 NOTE: Fabrication and installation are combined									
0040 as LABOR cost. Approx. 25% fittings assumed. R233100-40									
0042 Fabrication/Inst. is to commercial quality standards									
0043 (SMACNA or equiv.) for structure, sealing, leak testing, etc. R233100-50									
0050 Add to labor for elevated installation									
0051 of fabricated ductwork									
0052 10' to 15' high						6%			
0053 15' to 20' high						12%			
0054 20' to 25' high						15%			
0055 25' to 30' high						21%			
0056 30' to 35' high						24%			
0057 35' to 40' high						30%			
0058 Over 40' high						33%			
0072 For duct insulation see Line 23 07 13.10 3000									
0100 Aluminum, alloy 3003-H14, under 100 lb.	Q-10	75	.320	Lb.	3.20	16.70		19.90	29
0110 100 to 500 lb.		80	.300		1.88	15.65		17.53	26
0120 500 to 1,000 lb.		95	.253		1.82	13.20		15.02	22
0140 1,000 to 2,000 lb.		120	.200		1.77	10.45		12.22	17.90
0150 2,000 to 5,000 lb.		130	.185		1.77	9.65		11.42	16.70
0160 Over 5,000 lb.		145	.166		1.77	8.65		10.42	15.15
0500 Galvanized steel, under 200 lb.		235	.102		.65	5.35		6	8.85
0520 200 to 500 lb.		245	.098		.64	5.10		5.74	8.50
0540 500 to 1,000 lb.		255	.094		.62	4.91		5.53	8.20
0560 1,000 to 2,000 lb.		265	.091		.61	4.73		5.34	7.90
0570 2,000 to 5,000 lb.		275	.087		.61	4.56		5.17	7.60
0580 Over 5,000 lb.		285	.084		.61	4.40		5.01	7.35
0600 For large quantities special prices available from supplier									
1000 Stainless steel, type 304, under 100 lb.	Q-10	165	.145	Lb.	6.35	7.60		13.95	18.55
1020 100 to 500 lb.		175	.137		4.05	7.15		11.20	15.40
1030 500 to 1,000 lb.		190	.126		2.95	6.60		9.55	13.35
1040 1,000 to 2,000 lb.		200	.120		2.89	6.25		9.14	12.80
1050 2,000 to 5,000 lb.		225	.107		2.39	5.55		7.94	11.15
1060 Over 5,000 lb.		235	.102		1.96	5.35		7.31	10.30
1080 Note: Minimum order cost exceeds per lb. cost for min. wt.									
1100 For medium pressure ductwork, add				Lb.		15%			
1200 For high pressure ductwork, add						40%			
1210 For welded ductwork, add						85%			
1220 For 30% fittings, add						11%			
1224 For 40% fittings, add						34%			
1228 For 50% fittings, add						56%			
1232 For 60% fittings, add						79%			
1236 For 70% fittings, add						101%			
1240 For 80% fittings, add						124%			
1244 For 90% fittings, add						147%			
1248 For 100% fittings, add						169%			
1252 Note: Fittings add includes time for detailing and installation.									
1276 Duct tape, 2" x 60 yd roll				Ea.	6.50			6.50	7.15

23 31 13 – Metal Ducts

23 31 13.16 Round and Flat-Oval Spiral Ducts	Crew	Daily Output	Labor-Hours	Unit	Material	2015 Bare Costs Labor	Equipment	Total	Total Incl O&P
0010 **ROUND AND FLAT-OVAL SPIRAL DUCTS**									
1280 Add to labor for elevated installation									
1282 of prefabricated (purchased) ductwork									
1283 10' to 15' high						10%			
1284 15' to 20' high						20%			
1285 20' to 25' high						25%			
1286 25' to 30' high						35%			
1287 30' to 35' high						40%			
1288 35' to 40' high						50%			
1289 Over 40' high						55%			
5400 Spiral preformed, steel, galv., straight lengths, Max 10" spwg.									
5410 4" diameter, 26 ga.	Q-9	360	.044	L.F.	1.70	2.24		3.94	5.30
5416 5" diameter, 26 ga.		320	.050		1.82	2.52		4.34	5.85
5420 6" diameter, 26 ga.		280	.057		1.82	2.88		4.70	6.40
5425 7" diameter, 26 ga.		240	.067		2.12	3.36		5.48	7.50
5430 8" diameter, 26 ga.		200	.080		2.42	4.03		6.45	8.80
5440 10" diameter, 26 ga.		160	.100		3.02	5.05		8.07	11
5450 12" diameter, 26 ga.		120	.133		3.63	6.70		10.33	14.25
5460 14" diameter, 26 ga.		80	.200		4.23	10.05		14.28	20
5480 16" diameter, 24 ga.		60	.267		5.65	13.45		19.10	27
5490 18" diameter, 24 ga.		50	.320		6.35	16.10		22.45	31.50
5500 20" diameter, 24 ga.	Q-10	65	.369		7.10	19.30		26.40	37.50
5510 22" diameter, 24 ga.		60	.400		7.80	21		28.80	40.50
5520 24" diameter, 24 ga.		55	.436		8.50	23		31.50	44.50
5540 30" diameter, 22 ga.		45	.533		12.40	28		40.40	56
5600 36" diameter, 22 ga.		40	.600		14.90	31.50		46.40	64.50
5800 Connector, 4" diameter	Q-9	100	.160	Ea.	3.20	8.05		11.25	15.80
5810 5" diameter		94	.170		3.20	8.55		11.75	16.60
5820 6" diameter		88	.182		3.20	9.15		12.35	17.50
5840 8" diameter		78	.205		3.60	10.35		13.95	19.75
5860 10" diameter		70	.229		4.20	11.50		15.70	22
5880 12" diameter		50	.320		4.70	16.10		20.80	29.50
5900 14" diameter		44	.364		5.05	18.30		23.35	33.50
5920 16" diameter		40	.400		5.50	20		25.50	37
5930 18" diameter		37	.432		5.85	22		27.85	40
5940 20" diameter		34	.471		6.65	23.50		30.15	43.50
5950 22" diameter		31	.516		7.80	26		33.80	48
5960 24" diameter		28	.571		8.25	29		37.25	53
5980 30" diameter		22	.727		14.70	36.50		51.20	72
6000 36" diameter		18	.889		17.90	45		62.90	88
6300 Elbow, 45°, 4" diameter		60	.267		6.05	13.45		19.50	27
6310 5" diameter		52	.308		6.70	15.50		22.20	31
6320 6" diameter		44	.364		6.70	18.30		25	35.50
6340 8" diameter		28	.571		8.10	29		37.10	53
6360 10" diameter		18	.889		8.50	45		53.50	78
6380 12" diameter		13	1.231		9.20	62		71.20	105
6400 14" diameter		11	1.455		12.75	73		85.75	126
6420 16" diameter		10	1.600		20	80.50		100.50	145
6430 18" diameter		9.60	1.667		50	84		134	183
6440 20" diameter	Q-10	14	1.714		55	89.50		144.50	198
6450 22" diameter		13	1.846		65	96.50		161.50	219
6460 24" diameter		12	2		75	104		179	243

23 31 13.16 Round and Flat-Oval Spiral Ducts		Crew	Daily Output	Labor-Hours	Unit	Material	2015 Bare Costs Labor	Equipment	Total	Total Incl O&P
6480	30" diameter	Q-10	9	2.667	Ea.	108	139		247	330
6500	36" diameter	↓	7	3.429		132	179		311	420
6600	Elbow, 90°, 4" diameter	Q-9	60	.267		4.60	13.45		18.05	25.50
6610	5" diameter		52	.308		4.60	15.50		20.10	28.50
6620	6" diameter		44	.364		4.60	18.30		22.90	33
6625	7" diameter		36	.444		5.15	22.50		27.65	39.50
6630	8" diameter		28	.571		5.65	29		34.65	50
6640	10" diameter		18	.889		7.40	45		52.40	76.50
6650	12" diameter		13	1.231		9.25	62		71.25	105
6660	14" diameter		11	1.455		15.85	73		88.85	129
6670	16" diameter		10	1.600		18	80.50		98.50	143
6676	18" diameter	↓	9.60	1.667		65	84		149	200
6680	20" diameter	Q-10	14	1.714		70	89.50		159.50	214
6684	22" diameter		13	1.846		75	96.50		171.50	230
6690	24" diameter		12	2		80	104		184	248
6700	30" diameter		9	2.667		144	139		283	370
6710	36" diameter	↓	7	3.429	↓	223	179		402	520
6720	End cap									
6722	4"	Q-9	110	.145	Ea.	4.38	7.30		11.68	16
6724	6"		94	.170		4.38	8.55		12.93	17.90
6725	8"		88	.182		5.50	9.15		14.65	20
6726	10"		78	.205		6.85	10.35		17.20	23.50
6727	12"		70	.229		7.45	11.50		18.95	26
6728	14"		50	.320		10.45	16.10		26.55	36
6729	16"		44	.364		11.65	18.30		29.95	41
6730	18"		40	.400		16	20		36	48.50
6731	20"		37	.432		32.50	22		54.50	69.50
6732	22"		34	.471		36.50	23.50		60	76.50
6733	24"		30	.533		41	27		68	86
6800	Reducing coupling, 6" x 4"		46	.348		18.80	17.50		36.30	47.50
6820	8" x 6"		40	.400		18.80	20		38.80	51.50
6840	10" x 8"		32	.500		20.50	25		45.50	61
6860	12" x 10"		24	.667		23.50	33.50		57	77.50
6880	14" x 12"		20	.800		25	40.50		65.50	89
6900	16" x 14"		18	.889		32	45		77	104
6920	18" x 16"	↓	16	1		39	50.50		89.50	120
6940	20" x 18"	Q-10	24	1		45.50	52		97.50	130
6950	22" x 20"		23	1.043		45	54.50		99.50	133
6960	24" x 22"		22	1.091		58.50	57		115.50	152
6980	30" x 28"		18	1.333		112	69.50		181.50	230
7000	36" x 34"	↓	16	1.500		143	78.50		221.50	277
7100	Tee, 90°, 4" diameter	Q-9	40	.400		12.65	20		32.65	45
7110	5" diameter		34.60	.462		19	23.50		42.50	56.50
7120	6" diameter		30	.533		19	27		46	62
7130	8" diameter		19	.842		24.50	42.50		67	92
7140	10" diameter		12	1.333		34	67		101	141
7150	12" diameter		9	1.778		41	89.50		130.50	182
7160	14" diameter		7	2.286		62.50	115		177.50	245
7170	16" diameter		6.70	2.388		72	120		192	264
7176	18" diameter	↓	6	2.667		105	134		239	320
7180	20" diameter	Q-10	8.70	2.759		125	144		269	360
7190	24" diameter		7	3.429		196	179		375	490
7200	30" diameter	↓	6	4	↓	239	209		448	585

23 31 13.16 Round and Flat-Oval Spiral Ducts	Crew	Daily Output	Labor-Hours	Unit	Material	2015 Bare Costs Labor	Equipment	Total	Total Incl O&P
7210	36" diameter	Q-10	5	4.800	Ea.	305	251		556	720
7250	Saddle Tap									
7254	4" dia. on 6" dia.	Q-9	36	.444	Ea.	25	22.50		47.50	61
7256	5" dia on 7" dia		35.80	.447		25	22.50		47.50	61.50
7258	6" dia. on 8" dia.		35.40	.452		22	23		45	59
7260	6" dia. on 10" dia.		35.20	.455		22	23		45	59
7262	6" dia. on 12" dia.		35	.457		22.50	23		45.50	60
7264	7" dia. on 9" dia.		35.20	.455		25.50	23		48.50	63
7266	8" dia. on 10" dia.		35	.457		25.50	23		48.50	63
7268	8" dia. on 12" dia.		34.80	.460		22.50	23		45.50	60.50
7270	8" dia. on 14" dia.		34.50	.464		28.50	23.50		52	67
7272	10" dia. on 12" dia.		34.20	.468		26	23.50		49.50	64.50
7274	10" dia. on 14" dia.		33.80	.473		28.50	24		52.50	68
7276	10" dia. on 16" dia.		33.40	.479		28.50	24		52.50	68.50
7278	10" dia. on 18" dia.		33	.485		28.50	24.50		53	69
7280	12" dia. on 14" dia.		32.60	.491		33	24.50		57.50	74.50
7282	12" dia. on 18" dia.		32.20	.497		36.50	25		61.50	78
7284	14" dia. on 16" dia.		31.60	.506		38	25.50		63.50	81
7286	14" dia. on 20" dia.		31	.516		38.50	26		64.50	82
7288	16" dia. on 20" dia.		30	.533		45	27		72	90.50
7290	18" dia. on 20" dia.		28	.571		52.50	29		81.50	102
7292	20" dia. on 22" dia.		26	.615		69.50	31		100.50	124
7294	22" dia. on 24" dia.	▼	24	.667	▼	74.50	33.50		108	133
7400	Steel, PVC coated both sides, straight lengths									
7410	4" diameter, 26 ga.	Q-9	360	.044	L.F.	4.97	2.24		7.21	8.85
7416	5" diameter, 26 ga.		240	.067		3.60	3.36		6.96	9.10
7420	6" diameter, 26 ga.		280	.057		3.60	2.88		6.48	8.35
7440	8" diameter, 26 ga.		200	.080		4.90	4.03		8.93	11.55
7460	10" diameter, 26 ga.		160	.100		6.10	5.05		11.15	14.40
7480	12" diameter, 26 ga.		120	.133		7.30	6.70		14	18.30
7500	14" diameter, 26 ga.		80	.200		8.50	10.05		18.55	25
7520	16" diameter, 24 ga.		60	.267		9.70	13.45		23.15	31
7540	18" diameter, 24 ga.	▼	45	.356		10.90	17.90		28.80	39.50
7560	20" diameter, 24 ga.	Q-10	65	.369		12.10	19.30		31.40	43
7580	24" diameter, 24 ga.		55	.436		14.50	23		37.50	51
7600	30" diameter, 22 ga.		45	.533		22	28		50	66.50
7620	36" diameter, 22 ga.	▼	40	.600	▼	26	31.50		57.50	77
7890	Connector, 4" diameter	Q-9	100	.160	Ea.	2.60	8.05		10.65	15.15
7896	5" diameter		83	.193		2.90	9.70		12.60	18.05
7900	6" diameter		88	.182		3.30	9.15		12.45	17.65
7920	8" diameter		78	.205		3.90	10.35		14.25	20
7940	10" diameter		70	.229		4.50	11.50		16	22.50
7960	12" diameter		50	.320		5.30	16.10		21.40	30.50
7980	14" diameter		44	.364		5.60	18.30		23.90	34
8000	16" diameter		40	.400		6.10	20		26.10	37.50
8020	18" diameter		37	.432		6.80	22		28.80	41
8040	20" diameter		34	.471		8.90	23.50		32.40	46
8060	24" diameter		28	.571		11	29		40	56
8080	30" diameter		22	.727		14.70	36.50		51.20	72
8100	36" diameter		18	.889		17.90	45		62.90	88
8390	Elbow, 45°, 4" diameter		60	.267		23	13.45		36.45	46
8396	5" diameter		36	.444		23	22.50		45.50	59.50
8400	6" diameter	▼	44	.364	▼	23	18.30		41.30	53.50

For customer support on your Mechanical Cost Data, call 877.795.7394.

311

23 31 HVAC Ducts and Casings

23 31 13 – Metal Ducts

23 31 13.16 Round and Flat-Oval Spiral Ducts

		Crew	Daily Output	Labor-Hours	Unit	Material	2015 Bare Costs Labor	Equipment	Total	Total Incl O&P
8420	8" diameter	Q-9	28	.571	Ea.	23	29		52	69.50
8440	10" diameter		18	.889		23	45		68	94
8460	12" diameter		13	1.231		23	62		85	120
8480	14" diameter		11	1.455		23	73		96	138
8500	16" diameter		10	1.600		23	80.50		103.50	149
8520	18" diameter		9	1.778		25	89.50		114.50	165
8540	20" diameter	Q-10	13	1.846		75.50	96.50		172	231
8560	24" diameter		11	2.182		97.50	114		211.50	281
8580	30" diameter		9	2.667		129	139		268	355
8600	36" diameter		7	3.429		157	179		336	445
8800	Elbow, 90°, 4" diameter	Q-9	60	.267		25	13.45		38.45	48
8804	5" diameter		52	.308		25	15.50		40.50	51
8806	6" diameter		44	.364		25	18.30		43.30	55.50
8808	8" diameter		28	.571		25	29		54	71.50
8810	10" diameter		18	.889		25	45		70	96
8812	12" diameter		13	1.231		25	62		87	122
8814	14" diameter		11	1.455		25	73		98	140
8816	16" diameter		10	1.600		29	80.50		109.50	155
8818	18" diameter		9.60	1.667		35.50	84		119.50	167
8820	20" diameter	Q-10	14	1.714		45	89.50		134.50	187
8822	24" diameter		12	2		104	104		208	274
8824	30" diameter		9	2.667		175	139		314	405
8826	36" diameter		7	3.429		285	179		464	590
9000	Reducing coupling									
9040	6" x 4"	Q-9	46	.348	Ea.	19.50	17.50		37	48.50
9060	8" x 6"		40	.400		19.50	20		39.50	52.50
9080	10" x 8"		32	.500		20.50	25		45.50	61
9100	12" x 10"		24	.667		23.50	33.50		57	77.50
9120	14" x 12"		20	.800		25	40.50		65.50	89
9140	16" x 14"		18	.889		32	45		77	104
9160	18" x 16"		16	1		39	50.50		89.50	120
9180	20" x 18"	Q-10	24	1		45.50	52		97.50	130
9200	24" x 22"		22	1.091		58.50	57		115.50	152
9220	30" x 28"		18	1.333		104	69.50		173.50	220
9240	36" x 34"		16	1.500		143	78.50		221.50	277
9800	Steel, stainless, straight lengths									
9810	3" diameter, 26 Ga.	Q-9	400	.040	L.F.	14.20	2.01		16.21	18.70
9820	4" diameter, 26 Ga.		360	.044		16.45	2.24		18.69	21.50
9830	5" diameter, 26 Ga.		320	.050		17.75	2.52		20.27	23.50
9840	6" diameter, 26 Ga.		280	.057		17.75	2.88		20.63	24
9850	7" diameter, 26 Ga.		240	.067		29.50	3.36		32.86	37.50
9860	8" diameter, 26 Ga.		200	.080		23	4.03		27.03	31.50
9870	9" diameter, 26 Ga.		180	.089		33.50	4.48		37.98	44
9880	10" diameter, 26 Ga.		160	.100		43.50	5.05		48.55	55.50
9890	12" diameter, 26 Ga.		120	.133		61.50	6.70		68.20	78.50
9900	14" diameter, 26 Ga.		80	.200		65	10.05		75.05	87

23 31 13.17 Round Grease Duct

0010	**ROUND GREASE DUCT**									
1280	Add to labor for elevated installation									
1282	of prefabricated (purchased) ductwork									
1283	10' to 15' high						10%			
1284	15' to 20' high						20%			

312

For customer support on your Mechanical Cost Data, call 877.795.7394.

23 31 13 – Metal Ducts

23 31 13.17 Round Grease Duct		Crew	Daily Output	Labor-Hours	Unit	Material	2015 Bare Costs Labor	Equipment	Total	Total Incl O&P
1285	20' to 25' high						25%			
1286	25' to 30' high						35%			
1287	30' to 35' high						40%			
1288	35' to 40' high						50%			
1289	Over 40' high						55%			
4020	Zero clearance, 2 hr. fire rated, UL listed, 3" double wall									
4030	304 interior, aluminized exterior, all sizes are ID									
4032	For 316 interior and aluminized exterior, add					6%				
4034	For 304 interior and 304 exterior, add					21%				
4036	For 316 interior and 316 exterior, add					40%				
4040	Straight									
4052	6"	Q-9	34.20	.468	L.F.	74.50	23.50		98	118
4054	8"		29.64	.540		85.50	27		112.50	136
4056	10"		27.36	.585		95.50	29.50		125	150
4058	12"		25.10	.637		110	32		142	169
4060	14"		23.94	.668		123	33.50		156.50	187
4062	16"		22.80	.702		138	35.50		173.50	206
4064	18"		21.66	.739		157	37		194	229
4066	20"	Q-10	20.52	1.170		178	61		239	290
4068	22"		19.38	1.238		201	64.50		265.50	320
4070	24"		18.24	1.316		227	68.50		295.50	355
4072	26"		17.67	1.358		245	71		316	375
4074	28"		17.10	1.404		261	73.50		334.50	400
4076	30"		15.96	1.504		274	78.50		352.50	420
4078	32"		15.39	1.559		295	81.50		376.50	450
4080	36"		14.25	1.684		330	88		418	495
4100	Adjustable section, 30" long									
4104	6"	Q-9	17.10	.936	Ea.	226	47		273	320
4106	8"		14.82	1.080		282	54.50		336.50	395
4108	10"		13.68	1.170		320	59		379	440
4110	12"		12.54	1.276		360	64		424	495
4112	14"		11.97	1.337		405	67.50		472.50	550
4114	16"		11.40	1.404		455	70.50		525.50	610
4116	18"		10.83	1.477		515	74.50		589.50	680
4118	20"	Q-10	10.26	2.339		585	122		707	830
4120	22"		9.69	2.477		660	129		789	925
4122	24"		9.12	2.632		750	137		887	1,025
4124	26"		8.83	2.718		805	142		947	1,100
4126	28"		8.55	2.807		860	147		1,007	1,175
4128	30"		8.26	2.906		915	152		1,067	1,225
4130	32"		7.98	3.008		970	157		1,127	1,325
4132	36"		6.84	3.509		1,075	183		1,258	1,475
4140	Tee, 90°, grease									
4144	6"	Q-9	13.68	1.170	Ea.	400	59		459	530
4146	8"		12.54	1.276		435	64		499	580
4148	10"		11.97	1.337		490	67.50		557.50	645
4150	12"		11.40	1.404		570	70.50		640.50	735
4152	14"		10.26	1.559		645	78.50		723.50	830
4154	16"		9.12	1.754		715	88.50		803.50	920
4156	18"		7.98	2.005		840	101		941	1,075
4158	20"		9.69	1.651		970	83		1,053	1,200
4160	22"		8.26	1.937		1,075	97.50		1,172.50	1,350
4162	24"		6.84	2.339		1,200	118		1,318	1,500

23 31 13.17 Round Grease Duct		Crew	Daily Output	Labor-Hours	Unit	Material	2015 Bare Costs Labor	Equipment	Total	Total Incl O&P
4164	26"	Q-9	6.55	2.443	Ea.	1,325	123		1,448	1,650
4166	28"	Q-10	6.27	3.828		1,450	200		1,650	1,900
4168	30"		5.98	4.013		1,725	210		1,935	2,225
4170	32"		5.70	4.211		2,025	220		2,245	2,550
4172	36"		5.13	4.678		2,250	244		2,494	2,850
4180	Cleanout Tee Cap									
4184	6"	Q-9	21.09	.759	Ea.	81.50	38		119.50	148
4186	8"		19.38	.826		86	41.50		127.50	158
4188	10"		18.24	.877		92.50	44		136.50	170
4190	12"		17.10	.936		99	47		146	181
4192	14"		15.96	1.003		105	50.50		155.50	192
4194	16"		14.25	1.123		116	56.50		172.50	214
4196	18"		13.68	1.170		122	59		181	224
4198	20"	Q-10	15.39	1.559		135	81.50		216.50	273
4200	22"		12.54	1.914		146	100		246	315
4202	24"		11.97	2.005		160	105		265	335
4204	26"		11.40	2.105		174	110		284	360
4206	28"		10.83	2.216		189	116		305	385
4208	30"		10.26	2.339		206	122		328	415
4210	32"		9.69	2.477		224	129		353	445
4212	36"		8.55	2.807		263	147		410	515
4220	Elbow, 45°									
4224	6"	Q-9	17.10	.936	Ea.	282	47		329	380
4226	8"		14.82	1.080		315	54.50		369.50	435
4228	10"		13.68	1.170		360	59		419	485
4230	12"		12.54	1.276		410	64		474	550
4232	14"		11.97	1.337		460	67.50		527.50	610
4234	16"		11.40	1.404		520	70.50		590.50	685
4236	18"		10.83	1.477		590	74.50		664.50	765
4238	20"	Q-10	10.26	2.339		670	122		792	920
4240	22"		9.69	2.477		745	129		874	1,025
4242	24"		9.12	2.632		855	137		992	1,150
4244	26"		8.83	2.718		905	142		1,047	1,200
4246	28"		8.55	2.807		1,075	147		1,222	1,400
4248	30"		8.26	2.906		1,075	152		1,227	1,400
4250	32"		7.98	3.008		1,300	157		1,457	1,675
4252	36"		6.84	3.509		1,725	183		1,908	2,175
4260	Elbow, 90°									
4264	6"	Q-9	17.10	.936	Ea.	565	47		612	690
4266	8"		14.82	1.080		635	54.50		689.50	785
4268	10"		13.68	1.170		720	59		779	885
4270	12"		12.54	1.276		815	64		879	1,000
4272	14"		11.97	1.337		925	67.50		992.50	1,125
4274	16"		11.40	1.404		1,050	70.50		1,120.50	1,250
4276	18"		10.83	1.477		1,275	74.50		1,349.50	1,525
4278	20"	Q-10	10.26	2.339		1,325	122		1,447	1,650
4280	22"		9.69	2.477		1,500	129		1,629	1,850
4282	24"		9.12	2.632		1,700	137		1,837	2,075
4284	26"		8.83	2.718		1,825	142		1,967	2,225
4286	28"		8.55	2.807		1,925	147		2,072	2,325
4288	30"		8.26	2.906		2,150	152		2,302	2,575
4290	32"		7.98	3.008		2,575	157		2,732	3,100
4292	36"		6.84	3.509		3,475	183		3,658	4,100

23 31 13 – Metal Ducts

23 31 13.17 **Round Grease Duct**		Crew	Daily Output	Labor-Hours	Unit	Material	2015 Bare Costs Labor	Equipment	Total	Total Incl O&P
4300	Support strap									
4304	6"	Q-9	25.50	.627	Ea.	99	31.50		130.50	158
4306	8"		22	.727		116	36.50		152.50	183
4308	10"		20.50	.780		119	39.50		158.50	191
4310	12"		18.80	.851		128	43		171	206
4312	14"		17.90	.894		138	45		183	220
4314	16"		17	.941		139	47.50		186.50	225
4316	18"		16.20	.988		141	49.50		190.50	231
4318	20"	Q-10	15.40	1.558		156	81.50		237.50	296
4320	22"	"	14.50	1.655		173	86.50		259.50	320
4340	Plate Support Assembly									
4344	6"	Q-9	14.82	1.080	Ea.	139	54.50		193.50	235
4346	8"		12.54	1.276		162	64		226	276
4348	10"		11.40	1.404		176	70.50		246.50	300
4350	12"		10.26	1.559		185	78.50		263.50	325
4352	14"		9.69	1.651		219	83		302	370
4354	16"		9.12	1.754		231	88.50		319.50	390
4356	18"		8.55	1.871		243	94		337	410
4358	20"	Q-10	9.12	2.632		255	137		392	490
4360	22"		8.55	2.807		261	147		408	510
4362	24"		7.98	3.008		265	157		422	530
4364	26"		7.69	3.121		298	163		461	580
4366	28"		7.41	3.239		325	169		494	620
4368	30"		7.12	3.371		365	176		541	670
4370	32"		6.84	3.509		400	183		583	720
4372	36"		5.70	4.211		480	220		700	865
4380	Wall Guide Assembly									
4384	6"	Q-9	19.10	.838	Ea.	184	42		226	267
4386	8"		16.50	.970		208	49		257	305
4388	10"		15.30	1.046		220	52.50		272.50	325
4390	12"		14.10	1.135		230	57		287	340
4392	14"		13.40	1.194		249	60		309	365
4394	16"		12.75	1.255		252	63		315	375
4396	18"		12.10	1.322		255	66.50		321.50	385
4398	20"	Q-10	11.50	2.087		285	109		394	480
4400	22"		10.90	2.202		305	115		420	515
4402	24"		10.20	2.353		345	123		468	570
4404	26"		9.90	2.424		380	127		507	615
4406	28"		9.60	2.500		395	131		526	635
4408	30"		9.30	2.581		450	135		585	695
4410	32"		9	2.667		475	139		614	740
4412	36"		7.60	3.158		525	165		690	825
4420	Hood Transition, Flanged or Unflanged									
4424	6"	Q-9	15.60	1.026	Ea.	78	51.50		129.50	165
4426	8"		13.50	1.185		89	59.50		148.50	189
4428	10"		12.50	1.280		99	64.50		163.50	208
4430	12"		11.40	1.404		119	70.50		189.50	239
4432	14"		10.90	1.468		128	74		202	253
4434	16"		10.40	1.538		138	77.50		215.50	269
4436	18"		9.80	1.633		157	82		239	299
4438	20"	Q-10	9.40	2.553		167	133		300	390
4440	22"		8.80	2.727		176	142		318	410
4442	24"		8.30	2.892		197	151		348	450

For customer support on your Mechanical Cost Data, call 877.795.7394.

315

23 31 13.17 Round Grease Duct

		Crew	Daily Output	Labor-Hours	Unit	Material	2015 Bare Costs Labor	Equipment	Total	Total Incl O&P
4444	26"	Q-10	8.10	2.963	Ea.	206	155		361	460
4446	28"		7.80	3.077		217	161		378	485
4448	30"		7.50	3.200		235	167		402	515
4450	32"		7.30	3.288		246	172		418	535
4452	36"	▼	6.20	3.871	▼	274	202		476	610
4460	Fan Adapter									
4464	6"	Q-9	15	1.067	Ea.	219	53.50		272.50	325
4466	8"		13	1.231		249	62		311	370
4468	10"		12	1.333		272	67		339	400
4470	12"		11	1.455		305	73		378	445
4472	14"		10.50	1.524		330	76.50		406.50	480
4474	16"		10	1.600		360	80.50		440.50	520
4476	18"	▼	9.50	1.684		385	85		470	555
4478	20"	Q-10	9	2.667		415	139		554	670
4480	22"		8.50	2.824		445	147		592	715
4482	24"		8	3		470	157		627	755
4484	26"		7.70	3.117		495	163		658	795
4486	28"		7.50	3.200		525	167		692	835
4488	30"		7.20	3.333		555	174		729	875
4490	32"		7	3.429		585	179		764	915
4492	36"	▼	6	4	▼	640	209		849	1,025
4500	Tapered Increaser/Reducer									
4530	6" x 10" dia	Q-9	13.40	1.194	Ea.	282	60		342	400
4540	6" x 20" dia		10.10	1.584		282	80		362	430
4550	8" x 10" dia		13.20	1.212		340	61		401	470
4560	8" x 20" dia		9.90	1.616		340	81.50		421.50	500
4570	10" x 20" dia	▼	10.60	1.509		395	76		471	550
4580	10" x 28" dia	Q-10	8.60	2.791		395	146		541	660
4590	12" x 20" dia		8.80	2.727		455	142		597	720
4600	14" x 28" dia		8.40	2.857		515	149		664	795
4610	16" x 20" dia		8.60	2.791		570	146		716	850
4620	16" x 30" dia		8.30	2.892		570	151		721	855
4630	18" x 24" dia		8.80	2.727		625	142		767	910
4640	18" x 30" dia		8.10	2.963		625	155		780	925
4650	20" x 24" dia		8.30	2.892		680	151		831	980
4660	20" x 32" dia		8	3		680	157		837	990
4670	24" x 26" dia		8	3		800	157		957	1,125
4680	24" x 36" dia		7.30	3.288		800	172		972	1,150
4690	28" x 30" dia		7.90	3.038		915	159		1,074	1,250
4700	28" x 36" dia		7.10	3.380		915	177		1,092	1,275
4710	30" x 36" dia	▼	6.90	3.478	▼	970	182		1,152	1,350
4730	Many intermediate standard sizes of tapered fittings are available									

23 31 13.19 Metal Duct Fittings

		Crew	Daily Output	Labor-Hours	Unit	Material	2015 Bare Costs Labor	Equipment	Total	Total Incl O&P
0010	**METAL DUCT FITTINGS**									
0350	Duct collar, spin-in type									
0352	Without damper									
0354	Round									
0356	4"	1 Shee	32	.250	Ea.	2.61	14		16.61	24.50
0357	5"		30	.267		3.16	14.90		18.06	26.50
0358	6"		28	.286		3.49	16		19.49	28.50
0359	7"		26	.308		3.80	17.20		21	30.50
0360	8"	▼	24	.333		3.80	18.65		22.45	32.50

23 31 HVAC Ducts and Casings

23 31 13 – Metal Ducts

23 31 13.19 Metal Duct Fittings

		Crew	Daily Output	Labor-Hours	Unit	Material	2015 Bare Costs Labor	Equipment	Total	Total Incl O&P
0361	9"	1 Shee	22	.364	Ea.	3.80	20.50		24.30	35
0362	10"		20	.400		4.18	22.50		26.68	38.50
0363	12"		17	.471		5.25	26.50		31.75	46
2000	Fabrics for flexible connections, with metal edge		100	.080	L.F.	3.44	4.48		7.92	10.65
2100	Without metal edge		160	.050	"	2.46	2.80		5.26	7

23 31 16 – Nonmetal Ducts

23 31 16.13 Fibrous-Glass Ducts

		Crew	Daily Output	Labor-Hours	Unit	Material	2015 Bare Costs Labor	Equipment	Total	Total Incl O&P
0010	**FIBROUS-GLASS DUCTS**									
1280	Add to labor for elevated installation									
1282	of prefabricated (purchased) ductwork									
1283	10' to 15' high						10%			
1284	15' to 20' high						20%			
1285	20' to 25' high						25%			
1286	25' to 30' high						35%			
1287	30' to 35' high						40%			
1288	35' to 40' high						50%			
1289	Over 40' high						55%			
3490	Rigid fiberglass duct board, foil reinf. kraft facing									
3500	Rectangular, 1" thick, alum. faced, (FRK), std. weight	Q-10	350	.069	SF Surf	.79	3.58		4.37	6.30

23 31 16.16 Thermoset Fiberglass-Reinforced Plastic Ducts

		Crew	Daily Output	Labor-Hours	Unit	Material	2015 Bare Costs Labor	Equipment	Total	Total Incl O&P
0010	**THERMOSET FIBERGLASS-REINFORCED PLASTIC DUCTS**									
1280	Add to labor for elevated installation									
1282	of prefabricated (purchased) ductwork									
1283	10' to 15' high						10%			
1284	15' to 20' high						20%			
1285	20' to 25' high						25%			
1286	25' to 30' high						35%			
1287	30' to 35' high						40%			
1288	35' to 40' high						50%			
1289	Over 40' high						55%			
3550	Rigid fiberglass reinforced plastic, FM approved									
3552	for acid fume and smoke exhaust system, nonflammable									
3554	Straight, 4" diameter	Q-9	200	.080	L.F.	10.90	4.03		14.93	18.15
3555	6" diameter		146	.110		14.25	5.50		19.75	24
3556	8" diameter		106	.151		17.30	7.60		24.90	30.50
3557	10" diameter		85	.188		21	9.50		30.50	37.50
3558	12" diameter		68	.235		24.50	11.85		36.35	44.50
3559	14" diameter		50	.320		36	16.10		52.10	64
3560	16" diameter		40	.400		40	20		60	75
3561	18" diameter		31	.516		44.50	26		70.50	88.50
3562	20" diameter	Q-10	44	.545		49	28.50		77.50	97.50
3563	22" diameter		40	.600		53	31.50		84.50	107
3564	24" diameter		37	.649		57	34		91	115
3565	26" diameter		34	.706		61.50	37		98.50	125
3566	28" diameter		28	.857		66.50	45		111.50	142
3567	30" diameter		24	1		70.50	52		122.50	158
3568	32" diameter		22.60	1.062		75	55.50		130.50	167
3569	34" diameter		21.60	1.111		79	58		137	176
3570	36" diameter		20.60	1.165		83.50	61		144.50	185
3571	38" diameter		19.80	1.212		129	63.50		192.50	238
3572	42" diameter		18.30	1.311		144	68.50		212.50	263
3573	46" diameter		17	1.412		159	73.50		232.50	288

For customer support on your Mechanical Cost Data, call 877.795.7394.

317

23 31 16.16 Thermoset Fiberglass-Reinforced Plastic Ducts		Crew	Daily Output	Labor-Hours	Unit	Material	2015 Bare Costs Labor	Equipment	Total	Total Incl O&P
3574	50" diameter	Q-10	15.90	1.509	L.F.	173	79		252	310
3575	54" diameter		14.90	1.611		190	84		274	340
3576	60" diameter		13.60	1.765		218	92		310	380
3577	64" diameter		12.80	1.875		320	98		418	500
3578	68" diameter		12.20	1.967		340	103		443	530
3579	72" diameter		11.90	2.017		365	105		470	560
3584	Note: joints are cemented with									
3586	fiberglass resin, included in material cost.									
3590	Elbow, 90°, 4" diameter	Q-9	20.30	.788	Ea.	52.50	39.50		92	119
3591	6" diameter		13.80	1.159		73.50	58.50		132	170
3592	8" diameter		10	1.600		81	80.50		161.50	212
3593	10" diameter		7.80	2.051		106	103		209	275
3594	12" diameter		6.50	2.462		135	124		259	335
3595	14" diameter		5.50	2.909		165	146		311	405
3596	16" diameter		4.80	3.333		196	168		364	475
3597	18" diameter		4.40	3.636		281	183		464	590
3598	20" diameter	Q-10	5.90	4.068		296	212		508	650
3599	22" diameter		5.40	4.444		365	232		597	755
3600	24" diameter		4.90	4.898		420	256		676	850
3601	26" diameter		4.60	5.217		425	272		697	880
3602	28" diameter		4.20	5.714		530	298		828	1,050
3603	30" diameter		3.90	6.154		590	320		910	1,150
3604	32" diameter		3.70	6.486		660	340		1,000	1,250
3605	34" diameter		3.45	6.957		775	365		1,140	1,400
3606	36" diameter		3.20	7.500		890	390		1,280	1,575
3607	38" diameter		3.07	7.818		1,275	410		1,685	2,025
3608	42" diameter		2.70	8.889		1,525	465		1,990	2,375
3609	46" diameter		2.54	9.449		1,775	495		2,270	2,700
3610	50" diameter		2.33	10.300		2,050	540		2,590	3,075
3611	54" diameter		2.14	11.215		2,350	585		2,935	3,500
3612	60" diameter		1.94	12.371		3,125	645		3,770	4,425
3613	64" diameter		1.82	13.187		4,550	690		5,240	6,050
3614	68" diameter		1.71	14.035		5,050	735		5,785	6,700
3615	72" diameter		1.62	14.815		5,600	775		6,375	7,325
3626	Elbow, 45°, 4" diameter	Q-9	22.30	.717		50.50	36		86.50	111
3627	6" diameter		15.20	1.053		70	53		123	158
3628	8" diameter		11	1.455		74.50	73		147.50	194
3629	10" diameter		8.60	1.860		78.50	93.50		172	230
3630	12" diameter		7.20	2.222		86	112		198	266
3631	14" diameter		6.10	2.623		105	132		237	320
3632	16" diameter		5.28	3.030		125	153		278	370
3633	18" diameter		4.84	3.306		147	166		313	415
3634	20" diameter	Q-10	6.50	3.692		170	193		363	480
3635	22" diameter		5.94	4.040		232	211		443	575
3636	24" diameter		5.40	4.444		264	232		496	645
3637	26" diameter		5.06	4.743		299	248		547	710
3638	28" diameter		4.62	5.195		325	271		596	770
3639	30" diameter		4.30	5.581		375	291		666	855
3640	32" diameter		4.07	5.897		415	310		725	925
3641	34" diameter		3.80	6.316		490	330		820	1,050
3642	36" diameter		3.52	6.818		565	355		920	1,175
3643	38" diameter		3.38	7.101		795	370		1,165	1,450
3644	42" diameter		2.97	8.081		950	420		1,370	1,700

23 31 16 – Nonmetal Ducts

23 31 16.16 Thermoset Fiberglass-Reinforced Plastic Ducts	Crew	Daily Output	Labor-Hours	Unit	Material	2015 Bare Costs Labor	Equipment	Total	Total Incl O&P
3645	46" diameter	Q-10	2.80	8.571	Ea.	1,125	450		1,575	1,900
3646	50" diameter		2.56	9.375		1,300	490		1,790	2,175
3647	54" diameter		2.35	10.213		1,475	535		2,010	2,450
3648	60" diameter		2.13	11.268		1,975	590		2,565	3,075
3649	64" diameter		2	12		2,850	625		3,475	4,100
3650	68" diameter		1.88	12.766		3,150	665		3,815	4,500
3651	72" diameter		1.78	13.483		3,500	705		4,205	4,925
3660	Tee, 90°, 4" diameter	Q-9	14.06	1.138		29	57.50		86.50	119
3661	6" diameter		9.52	1.681		44	84.50		128.50	178
3662	8" diameter		6.98	2.292		60	115		175	242
3663	10" diameter		5.45	2.936		78.50	148		226.50	315
3664	12" diameter		4.60	3.478		98	175		273	375
3665	14" diameter		3.86	4.145		143	209		352	475
3666	16" diameter		3.40	4.706		170	237		407	545
3667	18" diameter		3.05	5.246		201	264		465	625
3668	20" diameter	Q-10	4.12	5.825		236	305		541	725
3669	22" diameter		3.78	6.349		269	330		599	800
3670	24" diameter		3.46	6.936		305	360		665	890
3671	26" diameter		3.20	7.500		345	390		735	980
3672	28" diameter		2.96	8.108		385	425		810	1,075
3673	30" diameter		2.75	8.727		430	455		885	1,175
3674	32" diameter		2.54	9.449		475	495		970	1,275
3675	34" diameter		2.40	10		520	520		1,040	1,375
3676	36" diameter		2.27	10.573		570	550		1,120	1,475
3677	38" diameter		2.15	11.163		860	585		1,445	1,825
3678	42" diameter		1.91	12.565		1,025	655		1,680	2,150
3679	46" diameter		1.78	13.483		1,225	705		1,930	2,400
3680	50" diameter		1.63	14.724		1,425	770		2,195	2,725
3681	54" diameter		1.50	16		1,650	835		2,485	3,075
3682	60" diameter		1.35	17.778		2,050	930		2,980	3,675
3683	64" diameter		1.27	18.898		3,075	985		4,060	4,900
3684	68" diameter		1.20	20		3,450	1,050		4,500	5,375
3685	72" diameter		1.13	21.239		3,850	1,100		4,950	5,925
3690	For Y @ 45°, add					22%				
3700	Blast gate, 4" diameter	Q-9	22.30	.717		140	36		176	209
3701	6" diameter		15.20	1.053		158	53		211	255
3702	8" diameter		11	1.455		212	73		285	345
3703	10" diameter		8.60	1.860		243	93.50		336.50	410
3704	12" diameter		7.20	2.222		296	112		408	495
3705	14" diameter		6.10	2.623		330	132		462	560
3706	16" diameter		5.28	3.030		365	153		518	635
3707	18" diameter		4.84	3.306		375	166		541	670
3708	20" diameter	Q-10	6.50	3.692		430	193		623	770
3709	22" diameter		5.94	4.040		465	211		676	830
3710	24" diameter		5.40	4.444		500	232		732	905
3711	26" diameter		5.06	4.743		535	248		783	970
3712	28" diameter		4.62	5.195		575	271		846	1,050
3713	30" diameter		4.30	5.581		610	291		901	1,125
3714	32" diameter		4.07	5.897		670	310		980	1,200
3715	34" diameter		3.80	6.316		730	330		1,060	1,300
3716	36" diameter		3.52	6.818		785	355		1,140	1,400
3717	38" diameter		3.38	7.101		1,050	370		1,420	1,750
3718	42" diameter		2.97	8.081		1,225	420		1,645	1,975

For customer support on your Mechanical Cost Data, call 877.795.7394.

319

23 31 16 – Nonmetal Ducts

23 31 16.16 Thermoset Fiberglass-Reinforced Plastic Ducts		Crew	Daily Output	Labor-Hours	Unit	Material	2015 Bare Costs Labor	Equipment	Total	Total Incl O&P
3719	46" diameter	Q-10	2.80	8.571	Ea.	1,400	450		1,850	2,200
3720	50" diameter		2.56	9.375		1,600	490		2,090	2,500
3721	54" diameter		2.35	10.213		1,800	535		2,335	2,800
3722	60" diameter		2.13	11.268		2,100	590		2,690	3,200
3723	64" diameter		2	12		2,325	625		2,950	3,500
3724	68" diameter		1.88	12.766		2,575	665		3,240	3,875
3725	72" diameter	▽	1.78	13.483		2,825	705		3,530	4,200
3740	Reducers, 4" diameter	Q-9	24.50	.653		35	33		68	89
3741	6" diameter		16.70	.958		52	48		100	131
3742	8" diameter		12	1.333		70.50	67		137.50	181
3743	10" diameter		9.50	1.684		87	85		172	226
3744	12" diameter		7.90	2.025		104	102		206	271
3745	14" diameter		6.70	2.388		121	120		241	320
3746	16" diameter		5.80	2.759		139	139		278	365
3747	18" diameter	▽	5.30	3.019		157	152		309	405
3748	20" diameter	Q-10	7.20	3.333		174	174		348	455
3749	22" diameter		6.50	3.692		280	193		473	605
3750	24" diameter		5.90	4.068		305	212		517	660
3751	26" diameter		5.60	4.286		330	224		554	700
3752	28" diameter		5.10	4.706		360	246		606	770
3753	30" diameter		4.70	5.106		385	267		652	830
3754	32" diameter		4.50	5.333		410	279		689	875
3755	34" diameter		4.20	5.714		435	298		733	935
3756	36" diameter		3.90	6.154		460	320		780	995
3757	38" diameter		3.70	6.486		700	340		1,040	1,300
3758	42" diameter		3.30	7.273		770	380		1,150	1,425
3759	46" diameter		3.10	7.742		845	405		1,250	1,550
3760	50" diameter		2.80	8.571		915	450		1,365	1,675
3761	54" diameter		2.60	9.231		990	480		1,470	1,825
3762	60" diameter		2.34	10.256		1,125	535		1,660	2,050
3763	64" diameter		2.20	10.909		1,725	570		2,295	2,775
3764	68" diameter		2.06	11.650		1,825	610		2,435	2,925
3765	72" diameter	▽	1.96	12.245		1,950	640		2,590	3,125
3780	Flange, per each, 4" diameter	Q-9	40.60	.394		84.50	19.85		104.35	124
3781	6" diameter		27.70	.578		98.50	29		127.50	154
3782	8" diameter		20.50	.780		123	39.50		162.50	196
3783	10" diameter		16.30	.982		148	49.50		197.50	239
3784	12" diameter		13.90	1.151		170	58		228	276
3785	14" diameter		11.30	1.416		196	71.50		267.50	325
3786	16" diameter		9.70	1.649		221	83		304	370
3787	18" diameter	▽	8.30	1.928		245	97		342	415
3788	20" diameter	Q-10	11.30	2.124		269	111		380	465
3789	22" diameter		10.40	2.308		293	121		414	510
3790	24" diameter		9.50	2.526		320	132		452	550
3791	26" diameter		8.80	2.727		345	142		487	600
3792	28" diameter		8.20	2.927		370	153		523	640
3793	30" diameter		7.60	3.158		390	165		555	680
3794	32" diameter		7.20	3.333		415	174		589	720
3795	34" diameter		6.80	3.529		440	184		624	765
3796	36" diameter		6.40	3.750		465	196		661	810
3797	38" diameter		6.10	3.934		490	205		695	855
3798	42" diameter		5.50	4.364		535	228		763	940
3799	46" diameter	▽	5	4.800	▽	585	251		836	1,025

23 31 HVAC Ducts and Casings

23 31 16 – Nonmetal Ducts

23 31 16.16 Thermoset Fiberglass-Reinforced Plastic Ducts

		Crew	Daily Output	Labor-Hours	Unit	Material	2015 Bare Costs Labor	Equipment	Total	Total Incl O&P
3800	50" diameter	Q-10	4.60	5.217	Ea.	635	272		907	1,125
3801	54" diameter		4.20	5.714		685	298		983	1,200
3802	60" diameter		3.80	6.316		760	330		1,090	1,350
3803	64" diameter		3.60	6.667		805	350		1,155	1,425
3804	68" diameter		3.40	7.059		855	370		1,225	1,500
3805	72" diameter	▼	3.20	7.500	▼	905	390		1,295	1,600

23 31 16.19 PVC Ducts

		Crew	Daily Output	Labor-Hours	Unit	Material	2015 Bare Costs Labor	Equipment	Total	Total Incl O&P
0010	**PVC DUCTS**									
4000	Rigid plastic, corrosive fume resistant PVC									
4020	Straight, 6" diameter	Q-9	220	.073	L.F.	11.55	3.66		15.21	18.30
4040	8" diameter		160	.100		16.75	5.05		21.80	26
4060	10" diameter		120	.133		21	6.70		27.70	33.50
4070	12" diameter		100	.160		24.50	8.05		32.55	39.50
4080	14" diameter		70	.229		30.50	11.50		42	51
4090	16" diameter		62	.258		35.50	13		48.50	59
4100	18" diameter	▼	58	.276		48.50	13.90		62.40	74.50
4110	20" diameter	Q-10	75	.320		59.50	16.70		76.20	91
4130	24" diameter	"	55	.436	▼	62	23		85	104
4250	Coupling, 6" diameter	Q-9	88	.182	Ea.	23.50	9.15		32.65	40
4270	8" diameter		78	.205		30	10.35		40.35	49
4290	10" diameter		70	.229		35.50	11.50		47	56.50
4300	12" diameter		55	.291		38.50	14.65		53.15	65
4310	14" diameter		44	.364		50	18.30		68.30	83
4320	16" diameter		40	.400		52.50	20		72.50	89
4330	18" diameter	▼	38	.421		68	21		89	108
4340	20" diameter	Q-10	55	.436		82	23		105	125
4360	24" diameter	"	45	.533		103	28		131	156
4470	Elbow, 90°, 6" diameter	Q-9	44	.364		108	18.30		126.30	146
4490	8" diameter		28	.571		121	29		150	177
4510	10" diameter		18	.889		147	45		192	230
4520	12" diameter		15	1.067		167	53.50		220.50	265
4530	14" diameter		11	1.455		193	73		266	325
4540	16" diameter	▼	10	1.600		251	80.50		331.50	400
4550	18" diameter	Q-10	15	1.600		350	83.50		433.50	510
4560	20" diameter		14	1.714		520	89.50		609.50	705
4580	24" diameter	▼	12	2	▼	640	104		744	860
4750	Elbow 45°, use 90° and deduct					25%				

23 31 16.21 Polypropylene Ducts

		Crew	Daily Output	Labor-Hours	Unit	Material	2015 Bare Costs Labor	Equipment	Total	Total Incl O&P
0010	**POLYPROPYLENE DUCTS**									
0020	Rigid plastic, corrosive fume resistant, flame retardant, PP									
0110	Straight, 4" diameter	Q-9	200	.080	L.F.	7.55	4.03		11.58	14.45
0120	6" diameter		146	.110		10.75	5.50		16.25	20.50
0140	8" diameter		106	.151		13.75	7.60		21.35	27
0160	10" diameter		85	.188		20.50	9.50		30	37
0170	12" diameter		68	.235		39	11.85		50.85	61
0180	14" diameter		50	.320		44	16.10		60.10	73
0190	16" diameter		40	.400		59.50	20		79.50	96.50
0200	18" diameter	▼	31	.516		97	26		123	146
0210	20" diameter	Q-10	44	.545		123	28.50		151.50	179
0220	22" diameter		40	.600		137	31.50		168.50	199
0230	24" diameter		37	.649		186	34		220	257
0250	28" diameter	▼	28	.857	▼	221	45		266	310

For customer support on your Mechanical Cost Data, call 877.795.7394.

321

23 31 16.21 Polypropylene Ducts	Crew	Daily Output	Labor-Hours	Unit	Material	2015 Bare Costs Labor	Equipment	Total	Total Incl O&P
0270	32" diameter	Q-10	24	1	L.F.	249	52		301	355
0400	Coupling									
0410	4" diameter	Q-9	23	.696	Ea.	21	35		56	76.50
0420	6" diameter		16	1		21.50	50.50		72	101
0430	8" diameter		11.40	1.404		31	70.50		101.50	143
0440	10" diameter		9	1.778		41	89.50		130.50	182
0450	12" diameter		8	2		48	101		149	207
0460	14" diameter		6.60	2.424		93.50	122		215.50	290
0470	16" diameter		5.70	2.807		103	141		244	330
0480	18" diameter		5.20	3.077		123	155		278	375
0490	20" diameter	Q-10	6.80	3.529		134	184		318	430
0500	22" diameter		6.20	3.871		340	202		542	685
0510	24" diameter		5.80	4.138		435	216		651	810
0520	28" diameter		5.50	4.364		440	228		668	835
0530	32" diameter		5.20	4.615		490	241		731	910
0800	Elbow, 90°									
0810	4" diameter	Q-9	22	.727	Ea.	35	36.50		71.50	94.50
0820	6" diameter		15.40	1.039		49.50	52.50		102	135
0830	8" diameter		11	1.455		60	73		133	178
0840	10" diameter		8.60	1.860		86.50	93.50		180	238
0850	12" diameter		7.70	2.078		225	105		330	405
0860	14" diameter		6.40	2.500		260	126		386	480
0870	16" diameter		5.40	2.963		273	149		422	530
0880	18" diameter	Q-10	7.50	3.200		705	167		872	1,025
0890	20" diameter		6.60	3.636		775	190		965	1,150
0896	22" diameter		6.10	3.934		1,225	205		1,430	1,675
0900	24" diameter		5.60	4.286		1,100	224		1,324	1,575
0910	28" diameter		5.40	4.444		2,200	232		2,432	2,750
0920	32" diameter		5.20	4.615		2,850	241		3,091	3,525
1000	Elbow, 45°									
1020	4" diameter	Q-9	22	.727	Ea.	23	36.50		59.50	81.50
1030	6" diameter		15.40	1.039		41.50	52.50		94	126
1040	8" diameter		11	1.455		48	73		121	165
1050	10" diameter		8.60	1.860		65	93.50		158.50	215
1060	12" diameter		7.70	2.078		174	105		279	350
1070	14" diameter		6.40	2.500		194	126		320	405
1080	16" diameter		5.40	2.963		226	149		375	475
1090	18" diameter	Q-10	7.50	3.200		445	167		612	745
1100	20" diameter		6.60	3.636		575	190		765	925
1106	22" diameter		6.10	3.934		800	205		1,005	1,200
1110	24" diameter		5.60	4.286		760	224		984	1,175
1120	28" diameter		5.40	4.444		1,475	232		1,707	1,950
1130	32" diameter		5.20	4.615		1,850	241		2,091	2,400
1200	Tee									
1210	4" diameter	Q-9	16.50	.970	Ea.	150	49		199	240
1220	6" diameter		11.60	1.379		183	69.50		252.50	310
1230	8" diameter		8.30	1.928		228	97		325	400
1240	10" diameter		6.50	2.462		315	124		439	535
1250	12" diameter		5.80	2.759		415	139		554	665
1260	14" diameter		4.80	3.333		540	168		708	850
1270	16" diameter		4.05	3.951		600	199		799	960
1280	18" diameter	Q-10	5.60	4.286		790	224		1,014	1,200
1290	20" diameter		4.90	4.898		855	256		1,111	1,325

23 31 HVAC Ducts and Casings

23 31 16 – Nonmetal Ducts

23 31 16.21 Polypropylene Ducts		Crew	Daily Output	Labor-Hours	Unit	Material	2015 Bare Costs Labor	Equipment	Total	Total Incl O&P
1300	22" diameter	Q-10	4.40	5.455	Ea.	1,025	285		1,310	1,550
1310	24" diameter		4.20	5.714		1,300	298		1,598	1,875
1320	28" diameter		4.10	5.854		2,300	305		2,605	3,000
1330	32" diameter		4	6		2,650	315		2,965	3,400

23 33 Air Duct Accessories

23 33 13 – Dampers

23 33 13.13 Volume-Control Dampers

	23 33 13.13	Crew	Daily Output	Labor-Hours	Unit	Material	Labor	Equipment	Total	Total Incl O&P
0010	**VOLUME-CONTROL DAMPERS**									
5990	Multi-blade dampers, opposed blade, 8" x 6"	1 Shee	24	.333	Ea.	21.50	18.65		40.15	52.50
5994	8" x 8"		22	.364		22.50	20.50		43	55.50
5996	10" x 10"		21	.381		26.50	21.50		48	62
6000	12" x 12"		21	.381		29.50	21.50		51	65
6020	12" x 18"		18	.444		39.50	25		64.50	81.50
6030	14" x 10"		20	.400		29	22.50		51.50	65.50
6031	14" x 14"		17	.471		35.50	26.50		62	79
6032	16" x 10"		18	.444		31	25		56	72
6033	16" x 12"		17	.471		35.50	26.50		62	79
6035	16" x 16"		16	.500		44	28		72	91.50
6036	18" x 14"		16	.500		44	28		72	91.50
6037	18" x 16"		15	.533		48	30		78	98.50
6038	18" x 18"		15	.533		52	30		82	103
6040	18" x 24"		12	.667		67	37.50		104.50	131
6060	18" x 28"		10	.800		78	45		123	154
6068	20" x 6"		18	.444		28	25		53	69
6069	20" x 8"		16	.500		32.50	28		60.50	78.50
6070	20" x 16"		14	.571		52	32		84	106
6071	20" x 18"		13	.615		56	34.50		90.50	115
6072	20" x 20"		13	.615		62.50	34.50		97	122
6073	22" x 6"		20	.400		29	22.50		51.50	65.50
6074	22" x 18"		14	.571		62.50	32		94.50	118
6075	22" x 20"		12	.667		67.50	37.50		105	132
6076	24" x 16"		11	.727		61	40.50		101.50	130
6077	22" x 22"		10	.800		72.50	45		117.50	149
6078	24" x 20"		8	1		72.50	56		128.50	166
6080	24" x 24"		8	1		85	56		141	179
6100	24" x 28"		6	1.333		98	74.50		172.50	222
6110	26" x 26"		6	1.333		95	74.50		169.50	219
6120	28" x 28"	Q-9	11	1.455		107	73		180	229
6130	30" x 18"		10	1.600		84	80.50		164.50	216
6132	30" x 24"		7	2.286		107	115		222	293
6133	30" x 30"		6.60	2.424		137	122		259	335
6135	32" x 32"		6.40	2.500		154	126		280	360
6150	34" x 34"		6	2.667		172	134		306	395
6151	36" x 12"		10	1.600		70	80.50		150.50	200
6152	36" x 16"		8	2		96	101		197	259
6156	36" x 32"		6.30	2.540		172	128		300	385
6157	36" x 34"		6.20	2.581		181	130		311	400
6158	36" x 36"		6	2.667		189	134		323	415
6160	44" x 28"		5.80	2.759		194	139		333	425
6180	48" x 36"		5.60	2.857		253	144		397	500

For customer support on your Mechanical Cost Data, call 877.795.7394.

323

23 33 13.13 Volume-Control Dampers		Crew	Daily Output	Labor-Hours	Unit	Material	2015 Bare Costs Labor	Equipment	Total	Total Incl O&P
6200	56" x 36"	Q-9	5.40	2.963	Ea.	310	149		459	570
6220	60" x 36"		5.20	3.077		330	155		485	600
6240	60" x 44"	↓	5	3.200	↓	390	161		551	675
7500	Variable volume modulating motorized damper, incl. elect. mtr.									
7504	8" x 6"	1 Shee	15	.533	Ea.	118	30		148	176
7506	10" x 6"		14	.571		118	32		150	179
7510	10" x 10"		13	.615		122	34.50		156.50	187
7520	12" x 12"		12	.667		127	37.50		164.50	196
7522	12" x 16"		11	.727		127	40.50		167.50	202
7524	16" x 10"		12	.667		125	37.50		162.50	194
7526	16" x 14"		10	.800		130	45		175	212
7528	16" x 18"		9	.889		135	49.50		184.50	224
7540	18" x 12"		10	.800		130	45		175	212
7542	18" x 18"		8	1		138	56		194	237
7544	20" x 14"		8	1		141	56		197	241
7546	20" x 18"		7	1.143		147	64		211	260
7560	24" x 12"		8	1		143	56		199	243
7562	24" x 18"		7	1.143		154	64		218	267
7568	28" x 10"		7	1.143		143	64		207	255
7580	28" x 16"		6	1.333		163	74.50		237.50	293
7590	30" x 14"		5	1.600		151	89.50		240.50	305
7600	30" x 18"		4	2		198	112		310	390
7610	30" x 24"	↓	3.80	2.105		264	118		382	470
7690	48" x 48"	Q-9	6	2.667		895	134		1,029	1,200
7694	6' x 14' w/2 motors	"	3	5.333		3,225	269		3,494	3,950
7700	For thermostat, add	1 Shee	8	1		31	56		87	120
7800	For transformer 40 VA capacity, add	"	16	.500	↓	19.45	28		47.45	64.50
8000	Multi-blade dampers, parallel blade									
8100	8" x 8"	1 Shee	24	.333	Ea.	82	18.65		100.65	119
8120	12" x 8"		22	.364		82	20.50		102.50	121
8140	16" x 10"		20	.400		103	22.50		125.50	148
8160	18" x 12"		18	.444		113	25		138	162
8180	22" x 12"		15	.533		122	30		152	180
8200	24" x 16"		11	.727		134	40.50		174.50	210
8220	28" x 16"		10	.800		150	45		195	234
8240	30" x 16"		8	1		152	56		208	254
8260	30" x 18"	↓	7	1.143	↓	186	64		250	300
8400	Round damper, butterfly, vol. control w/lever lock reg.									
8410	6" diam.	1 Shee	22	.364	Ea.	25	20.50		45.50	58.50
8412	7" diam.		21	.381		25.50	21.50		47	61
8414	8" diam.		20	.400		27.50	22.50		50	64
8416	9" diam.		19	.421		28	23.50		51.50	67
8418	10" diam.		18	.444		30	25		55	71.50
8420	12" diam.		16	.500		31	28		59	77
8422	14" diam.		14	.571		41	32		73	94
8424	16" diam.		13	.615		44	34.50		78.50	101
8426	18" diam.		12	.667		54	37.50		91.50	117
8428	20" diam.		11	.727		65.50	40.50		106	134
8430	24" diam.		10	.800		81.50	45		126.50	158
8432	30" diam.		9	.889		114	49.50		163.50	202
8434	36" diam.	↓	8	1	↓	152	56		208	253
8500	Round motor operated damper									
8510	6" diam.	1 Shee	22	.364	Ea.	61	20.50		81.50	98.50

23 33 13 – Dampers

23 33 13.13 Volume-Control Dampers

		Crew	Daily Output	Labor-Hours	Unit	Material	2015 Bare Costs Labor	Equipment	Total	Total Incl O&P
8512	7" diam.	1 Shee	21	.381	Ea.	65	21.50		86.50	104
8514	8" diam.		20	.400		67	22.50		89.50	108
8516	10" diam.		18	.444		74	25		99	120
8518	12" diam.		16	.500		79	28		107	130
8520	14" diam.		14	.571		90	32		122	148

23 33 13.16 Fire Dampers

		Crew	Daily Output	Labor-Hours	Unit	Material	2015 Bare Costs Labor	Equipment	Total	Total Incl O&P
0010	**FIRE DAMPERS**									
3000	Fire damper, curtain type, 1-1/2 hr. rated, vertical, 6" x 6"	1 Shee	24	.333	Ea.	24	18.65		42.65	55
3020	8" x 6"		22	.364		24	20.50		44.50	57.50
3021	8" x 8"		22	.364		24	20.50		44.50	57.50
3030	10" x 10"		22	.364		24	20.50		44.50	57.50
3040	12" x 6"		22	.364		24	20.50		44.50	57.50
3044	12" x 12"		20	.400		28.50	22.50		51	65.50
3060	20" x 6"		18	.444		32	25		57	73
3080	12" x 8"		22	.364		24	20.50		44.50	57.50
3100	24" x 8"		16	.500		36.50	28		64.50	83
3120	12" x 10"		21	.381		24	21.50		45.50	59
3140	24" x 10"		15	.533		39	30		69	88.50
3160	36" x 10"		12	.667		50.50	37.50		88	113
3180	16" x 12"		20	.400		33	22.50		55.50	70.50
3200	24" x 12"		13	.615		41.50	34.50		76	98
3220	48" x 12"		10	.800		64	45		109	139
3238	14" x 14"		19	.421		36	23.50		59.50	75.50
3240	16" x 14"		18	.444		44	25		69	86.50
3260	24" x 14"		12	.667		53	37.50		90.50	115
3280	30" x 14"		11	.727		50.50	40.50		91	118
3298	16" x 16"		18	.444		40	25		65	82
3300	18" x 16"		17	.471		43	26.50		69.50	87
3320	24" x 16"		11	.727		46	40.50		86.50	113
3340	36" x 16"		10	.800		58.50	45		103.50	133
3356	18" x 18"		16	.500		44	28		72	91.50
3360	24" x 18"		10	.800		54	45		99	128
3380	48" x 18"		8	1		77.50	56		133.50	171
3398	20" x 20"		10	.800		47.50	45		92.50	121
3400	24" x 20"		8	1		54	56		110	145
3420	36" x 20"		7	1.143		65	64		129	169
3440	24" x 22"		7	1.143		55.50	64		119.50	159
3460	30" x 22"		6	1.333		60.50	74.50		135	181
3478	24" x 24"		8	1		56.50	56		112.50	148
3480	26" x 24"		7	1.143		56.50	64		120.50	160
3484	32" x 24"	Q-9	13	1.231		67	62		129	169
3500	48" x 24"		12	1.333		87	67		154	199
3520	28" x 26"		13	1.231		63.50	62		125.50	164
3540	30" x 28"		12	1.333		70	67		137	180
3560	48" x 30"		11	1.455		102	73		175	224
3580	48" x 36"		10	1.600		136	80.50		216.50	272
3590	32" x 40"		10	1.600		105	80.50		185.50	239
3600	44" x 44"		10	1.600		144	80.50		224.50	282
3620	48" x 48"		8	2		161	101		262	330
3700	U.L. label included in above									
3800	For horizontal operation, add				Ea.	20%				
3900	For cap for blades out of air stream, add					20%				

23 33 13.16 Fire Dampers		Crew	Daily Output	Labor-Hours	Unit	Material	2015 Bare Costs		Total	Total Incl O&P
							Labor	Equipment		
4000	For 10" 22 ga., U.L. approved sleeve, add				Ea.	35%				
4100	For 10" 22 ga., U.L. sleeve, 100% free area, add				↓	65%				
4200	For oversize openings group dampers									
4210	Fire damper, round, Type A, 1-1/2 hour									
4214	4" diam.	1 Shee	20	.400	Ea.	61.50	22.50		84	102
4216	5" diam.		16	.500		50	28		78	98
4218	6" diam.		14.30	.559		50	31.50		81.50	103
4220	7" diam.		13.30	.602		50	33.50		83.50	107
4222	8" diam.		11.40	.702		50	39.50		89.50	115
4224	9" diam.		10.80	.741		55.50	41.50		97	125
4226	10" diam.		10	.800		55.50	45		100.50	130
4228	12" diam.		9.50	.842		62.50	47		109.50	141
4230	14" diam.		8.20	.976		77.50	54.50		132	169
4232	16" diam.		7.60	1.053		77.50	59		136.50	176
4234	18" diam.		7.20	1.111		94	62		156	198
4236	20" diam.		6.40	1.250		97	70		167	213
4238	22" diam.		6.10	1.311		106	73.50		179.50	229
4240	24" diam.		5.70	1.404		111	78.50		189.50	242
4242	26" diam.		5.50	1.455		125	81.50		206.50	262
4244	28" diam.		5.40	1.481		131	83		214	271
4246	30" diam.	↓	5.30	1.509	↓	138	84.50		222.50	280
4502	Power open-spring close, sleeve and actuator motor mounted, 1-1/2 hour									
4506	6" x 6"	1 Shee	24	.333	Ea.	60.50	18.65		79.15	95
4510	8" x 8"		22	.364		214	20.50		234.50	266
4520	16" x 8"		20	.400		340	22.50		362.50	405
4540	18" x 8"		18	.444		345	25		370	420
4560	20" x 8"		16	.500		345	28		373	425
4580	10" x 10"		21	.381		310	21.50		331.50	375
4600	24" x 10"		15	.533		360	30		390	440
4620	30" x 10"		12	.667		370	37.50		407.50	465
4640	12" x 12"		20	.400		340	22.50		362.50	410
4660	18" x 12"		18	.444		355	25		380	430
4680	24" x 12"		13	.615		365	34.50		399.50	455
4700	30" x 12"		11	.727		375	40.50		415.50	475
4720	14" x 14"		17	.471		365	26.50		391.50	440
4740	16" x 14"		18	.444		365	25		390	445
4760	20" x 14"		14	.571		350	32		382	435
4780	24" x 14"		12	.667		375	37.50		412.50	470
4800	30" x 14"		10	.800		400	45		445	510
4820	16" x 16"		16	.500		375	28		403	460
4840	20" x 16"		14	.571		380	32		412	465
4860	24" x 16"		11	.727		385	40.50		425.50	485
4880	30" x 16"		8	1		415	56		471	540
4900	18" x 18"		15	.533		400	30		430	480
5000	24" x 18"		10	.800		405	45		450	515
5020	36" x 18"		7	1.143		525	64		589	675
5040	20" x 20"		13	.615		405	34.50		439.50	500
5060	24" x 20"		8	1		410	56		466	535
5080	30" x 20"		8	1		430	56		486	555
5100	36" x 20"		7	1.143		445	64		509	590
5120	24" x 24"		8	1		425	56		481	555
5130	30" x 24"	↓	7.60	1.053	↓	455	59		514	590
5140	36" x 24"	Q-9	12	1.333		475	67		542	625

23 33 Air Duct Accessories

23 33 13 – Dampers

23 33 13.16 Fire Dampers

		Crew	Daily Output	Labor-Hours	Unit	Material	2015 Bare Costs Labor	2015 Bare Costs Equipment	Total	Total Incl O&P
5141	36" x 30"	Q-9	10	1.600	Ea.	540	80.50		620.50	720
5142	40" x 36"		8	2		890	101		991	1,125
5143	48" x 48"		4	4		1,700	201		1,901	2,175
5150	Damper operator motor, 24 or 120 volt	1 Shee	16	.500		540	28		568	635

23 33 13.25 Exhaust Vent Damper

		Crew	Daily Output	Labor-Hours	Unit	Material	Labor	Equipment	Total	Total Incl O&P
0010	**EXHAUST VENT DAMPER**, Auto, OB with elect. actuator.									
1110	6" x 6"	1 Shee	24	.333	Ea.	64	18.65		82.65	99
1114	8" x 6"		23	.348		65.50	19.45		84.95	102
1116	8" x 8"		22	.364		67.50	20.50		88	106
1118	10" x 6"		23	.348		67.50	19.45		86.95	104
1120	10" x 10"		22	.364		72.50	20.50		93	111
1122	12" x 6"		22	.364		68.50	20.50		89	106
1124	12" x 10"		21	.381		76.50	21.50		98	117
1126	12" x 12"		20	.400		78.50	22.50		101	121
1128	14" x 6"		21	.381		70.50	21.50		92	110
1130	14" x 10"		19	.421		80	23.50		103.50	124
1132	16" x 6"		21	.381		71.50	21.50		93	111
1134	16" x 8"		20	.400		79	22.50		101.50	121
1136	18" x 6"		20	.400		75.50	22.50		98	117
1138	18" x 8"		19	.421		80	23.50		103.50	124

23 33 13.28 Splitter Damper Assembly

		Crew	Daily Output	Labor-Hours	Unit	Material	Labor	Equipment	Total	Total Incl O&P
0010	**SPLITTER DAMPER ASSEMBLY**									
7000	Self locking, 1' rod	1 Shee	24	.333	Ea.	23.50	18.65		42.15	54
7020	3' rod		22	.364		30	20.50		50.50	64
7040	4' rod		20	.400		33.50	22.50		56	71
7060	6' rod		18	.444		41	25		66	83

23 33 13.32 Relief Damper

		Crew	Daily Output	Labor-Hours	Unit	Material	Labor	Equipment	Total	Total Incl O&P
0010	**RELIEF DAMPER**, Electronic bypass with tight seal									
8310	8" x 6"	1 Shee	22	.364	Ea.	176	20.50		196.50	225
8314	10" x 6"		22	.364		176	20.50		196.50	225
8318	10" x 10"		21	.381		183	21.50		204.50	234
8322	12" x 12"		20	.400		187	22.50		209.50	240
8326	12" x 16"		19	.421		194	23.50		217.50	250
8330	16" x 10"		20	.400		204	22.50		226.50	259
8334	16" x 14"		18	.444		209	25		234	268
8338	16" x 18"		17	.471		228	26.50		254.50	290
8342	18" x 12"		18	.444		228	25		253	288
8346	18" x 18"		15	.533		209	30		239	276
8350	20" x 14"		14	.571		212	32		244	283
8354	20" x 18"		13	.615		220	34.50		254.50	295
8358	24" x 12"		13	.615		217	34.50		251.50	291
8362	24" x 18"		10	.800		226	45		271	320
8363	24" x 24"		10	.800		235	45		280	330
8364	24" x 36"		9	.889		280	49.50		329.50	385
8365	24" x 48"		6	1.333		315	74.50		389.50	460
8366	28" x 10"		13	.615		155	34.50		189.50	223
8370	28" x 16"		10	.800		239	45		284	330
8374	30" x 14"		11	.727		163	40.50		203.50	241
8378	30" x 18"		8	1		246	56		302	355
8382	30" x 24"		6	1.333		261	74.50		335.50	400
8390	46" x 36"		4	2		345	112		457	550
8394	48" x 48"		3	2.667		380	149		529	645

For customer support on your Mechanical Cost Data, call 877.795.7394.

327

23 33 Air Duct Accessories

23 33 13 – Dampers

23 33 13.32 Relief Damper		Crew	Daily Output	Labor-Hours	Unit	Material	2015 Bare Costs Labor	Equipment	Total	Total Incl O&P
8396	54" x 36"	1 Shee	2	4	Ea.	400	224		624	780

23 33 19 – Duct Silencers

23 33 19.10 Duct Silencers

		Crew	Daily Output	Labor-Hours	Unit	Material	2015 Bare Costs Labor	Equipment	Total	Total Incl O&P
0010	**DUCT SILENCERS**									
9000	Silencers, noise control for air flow, duct				MCFM	57			57	63
9004	Duct sound trap, packaged									
9010	12" x 18" x 36", 2000 CFM	1 Shee	12	.667	Ea.	470	37.50		507.50	570
9011	24" x 18" x 36", 9000 CFM	Q-9	80	.200		540	10.05		550.05	605
9012	24" x 24" x 36", 9000 CFM		60	.267		805	13.45		818.45	910
9013	24" x 30" x 36", 9000 CFM		50	.320		860	16.10		876.10	970
9014	24" x 36" x 36", 9000 CFM		40	.400		1,075	20		1,095	1,200
9015	24" x 48" x 36", 9000 CFM		28	.571		1,400	29		1,429	1,600
9018	36" x 18" x 36", 6300 CFM		42	.381		1,075	19.20		1,094.20	1,200
9019	36" x 36" x 36", 6300 CFM		21	.762		2,150	38.50		2,188.50	2,425
9020	36" x 48" x 36", 6300 CFM		14	1.143		2,800	57.50		2,857.50	3,175
9021	36" x 60" x 36", 6300 CFM		13	1.231		3,550	62		3,612	4,000
9022	48" x 48" x 36", 6300 CFM		12	1.333		3,775	67		3,842	4,250
9023	24" x 24" x 60", 6000 CFM		3.60	4.444		1,250	224		1,474	1,725
9026	24" x 30" x 60", 7000 CFM		3.50	4.571		1,400	230		1,630	1,875
9030	24" x 36" x 60", 8000 CFM		3.40	4.706		1,525	237		1,762	2,025
9034	24" x 48" x 60", 11,000 CFM		3.30	4.848		1,300	244		1,544	1,825
9038	36" x 18" x 60", 6300 CFM		3.50	4.571		1,025	230		1,255	1,475
9042	36" x 36" x 60", 12,000 CFM		3.30	4.848		1,650	244		1,894	2,200
9046	36" x 48" x 60", 17,000 CFM		3.10	5.161		2,175	260		2,435	2,800
9050	36" x 60" x 60", 20,000 CFM		2.80	5.714		2,650	288		2,938	3,375
9054	48" x 48" x 60", 23,000 CFM		2.60	6.154		3,050	310		3,360	3,850
9056	48" x 60" x 60", 23,000 CFM		2.10	7.619		3,825	385		4,210	4,775

23 33 23 – Turning Vanes

23 33 23.13 Air Turning Vanes

		Crew	Daily Output	Labor-Hours	Unit	Material	2015 Bare Costs Labor	Equipment	Total	Total Incl O&P
0010	**AIR TURNING VANES**									
9400	Turning vane components									
9410	Turning vane rail	1 Shee	160	.050	L.F.	.75	2.80		3.55	5.10
9420	Double thick, factory fab. vane		300	.027		1.17	1.49		2.66	3.57
9428	12" high set		170	.047		1.68	2.63		4.31	5.85
9432	14" high set		160	.050		1.96	2.80		4.76	6.45
9434	16" high set		150	.053		2.24	2.98		5.22	7
9436	18" high set		144	.056		2.52	3.11		5.63	7.50
9438	20" high set		138	.058		2.80	3.24		6.04	8.05
9440	22" high set		130	.062		3.08	3.44		6.52	8.65
9442	24" high set		124	.065		3.36	3.61		6.97	9.20
9444	26" high set		116	.069		3.64	3.86		7.50	9.90
9446	30" high set		112	.071		4.20	4		8.20	10.70

23 33 33 – Duct-Mounting Access Doors

23 33 33.13 Duct Access Doors

		Crew	Daily Output	Labor-Hours	Unit	Material	2015 Bare Costs Labor	Equipment	Total	Total Incl O&P
0010	**DUCT ACCESS DOORS**									
1000	Duct access door, insulated, 6" x 6"	1 Shee	14	.571	Ea.	16.05	32		48.05	66.50
1020	10" x 10"		11	.727		18.45	40.50		58.95	82.50
1040	12" x 12"		10	.800		20	45		65	90.50
1050	12" x 18"		9	.889		37	49.50		86.50	117
1060	16" x 12"		9	.889		29.50	49.50		79	109
1070	18" x 18"		8	1		32.50	56		88.50	121

23 33 33 – Duct-Mounting Access Doors

23 33 33.13 Duct Access Doors		Crew	Daily Output	Labor-Hours	Unit	Material	2015 Bare Costs Labor	Equipment	Total	Total Incl O&P
1074	24" x 18"	1 Shee	8	1	Ea.	44	56		100	134
1080	24" x 24"	↓	8	1	↓	47	56		103	138

23 33 46 – Flexible Ducts

23 33 46.10 Flexible Air Ducts

			Crew	Daily Output	Labor-Hours	Unit	Material	2015 Bare Costs Labor	Equipment	Total	Total Incl O&P
0010	**FLEXIBLE AIR DUCTS**										
1280	Add to labor for elevated installation										
1282	of prefabricated (purchased) ductwork										
1283	10' to 15' high							10%			
1284	15' to 20' high							20%			
1285	20' to 25' high							25%			
1286	25' to 30' high							35%			
1287	30' to 35' high							40%			
1288	35' to 40' high							50%			
1289	Over 40' high							55%			
1300	Flexible, coated fiberglass fabric on corr. resist. metal helix										
1400	pressure to 12" (WG) UL-181										
1500	Noninsulated, 3" diameter		Q-9	400	.040	L.F.	1.12	2.01		3.13	4.31
1520	4" diameter			360	.044		1.16	2.24		3.40	4.70
1540	5" diameter			320	.050		1.30	2.52		3.82	5.30
1560	6" diameter			280	.057		1.50	2.88		4.38	6.05
1580	7" diameter			240	.067		1.53	3.36		4.89	6.85
1600	8" diameter			200	.080		1.91	4.03		5.94	8.25
1620	9" diameter			180	.089		2.05	4.48		6.53	9.10
1640	10" diameter			160	.100		2.46	5.05		7.51	10.40
1660	12" diameter			120	.133		2.94	6.70		9.64	13.50
1680	14" diameter			80	.200		3.56	10.05		13.61	19.30
1700	16" diameter		↓	60	.267	↓	5.05	13.45		18.50	26
1800	For plastic cable tie, add					Ea.	.77			.77	.85
1900	Insulated, 1" thick, PE jacket, 3" diameter	G	Q-9	380	.042	L.F.	2.60	2.12		4.72	6.10
1910	4" diameter	G		340	.047		2.60	2.37		4.97	6.50
1920	5" diameter	G		300	.053		2.60	2.69		5.29	6.95
1940	6" diameter	G		260	.062		2.94	3.10		6.04	7.95
1960	7" diameter	G		220	.073		3.20	3.66		6.86	9.10
1980	8" diameter	G		180	.089		3.49	4.48		7.97	10.70
2000	9" diameter	G		160	.100		3.81	5.05		8.86	11.90
2020	10" diameter	G		140	.114		4.25	5.75		10	13.50
2040	12" diameter	G		100	.160		4.90	8.05		12.95	17.70
2060	14" diameter	G		80	.200		6.05	10.05		16.10	22
2080	16" diameter	G		60	.267		8.35	13.45		21.80	29.50
2100	18" diameter	G	↓	45	.356		9.85	17.90		27.75	38.50
2120	20" diameter	G	Q-10	65	.369	↓	10.85	19.30		30.15	41.50
2500	Insulated, heavy duty, coated fiberglass fabric										
2520	4" diameter	G	Q-9	340	.047	L.F.	3.36	2.37		5.73	7.30
2540	5" diameter	G		300	.053		3.54	2.69		6.23	8
2560	6" diameter	G		260	.062		4.20	3.10		7.30	9.35
2580	7" diameter	G		220	.073		4.58	3.66		8.24	10.65
2600	8" diameter	G		180	.089		5.10	4.48		9.58	12.50
2620	9" diameter	G		160	.100		5.55	5.05		10.60	13.85
2640	10" diameter	G		140	.114		6.40	5.75		12.15	15.85
2660	12" diameter	G		100	.160		7.50	8.05		15.55	20.50
2680	14" diameter	G		80	.200		8.90	10.05		18.95	25
2700	16" diameter	G	↓	60	.267	↓	12.20	13.45		25.65	34

For customer support on your Mechanical Cost Data, call 877.795.7394.

329

23 33 46 – Flexible Ducts

23 33 46.10 Flexible Air Ducts		Crew	Daily Output	Labor-Hours	Unit	Material	2015 Bare Costs Labor	Equipment	Total	Total Incl O&P
2720	18" diameter [G]	Q-9	45	.356	L.F.	14.45	17.90		32.35	43.50
2800	Flexible, aluminum, pressure to 12" (WG) UL-181									
2820	Non-insulated									
2830	3" diameter	Q-9	400	.040	L.F.	1.16	2.01		3.17	4.36
2831	4" diameter		360	.044		1.27	2.24		3.51	4.82
2832	5" diameter		320	.050		1.60	2.52		4.12	5.60
2833	6" diameter		280	.057		2.04	2.88		4.92	6.65
2834	7" diameter		240	.067		2.42	3.36		5.78	7.80
2835	8" diameter		200	.080		2.75	4.03		6.78	9.20
2836	9" diameter		180	.089		3.25	4.48		7.73	10.45
2837	10" diameter		160	.100		3.85	5.05		8.90	11.95
2838	12" diameter		120	.133		5.05	6.70		11.75	15.80
2839	14" diameter		80	.200		5.70	10.05		15.75	21.50
2840	15" diameter		70	.229		7.70	11.50		19.20	26
2841	16" diameter		60	.267		8.90	13.45		22.35	30.50
2842	18" diameter	▼	45	.356		10.10	17.90		28	38.50
2843	20" diameter	Q-10	65	.369	▼	11.65	19.30		30.95	42.50
2880	Insulated, 1" thick with 3/4 lb., PE jacket									
2890	3" diameter [G]	Q-9	380	.042	L.F.	3.14	2.12		5.26	6.70
2891	4" diameter [G]		340	.047		3.14	2.37		5.51	7.05
2892	5" diameter [G]		300	.053		3.58	2.69		6.27	8.05
2893	6" diameter [G]		260	.062		4.24	3.10		7.34	9.40
2894	7" diameter [G]		220	.073		5.10	3.66		8.76	11.25
2895	8" diameter [G]		180	.089		5.25	4.48		9.73	12.60
2896	9" diameter [G]		160	.100		6.05	5.05		11.10	14.35
2897	10" diameter [G]		140	.114		7.10	5.75		12.85	16.60
2898	12" diameter [G]		100	.160		8.35	8.05		16.40	21.50
2899	14" diameter [G]		80	.200		9.45	10.05		19.50	26
2900	15" diameter [G]		70	.229		11.65	11.50		23.15	30.50
2901	16" diameter [G]		60	.267		13.20	13.45		26.65	35
2902	18" diameter [G]	▼	45	.356	▼	15.50	17.90		33.40	44.50
3000	Transitions, 3" deep, square to round									
3010	6" x 6" to round	Q-9	44	.364	Ea.	9.90	18.30		28.20	39
3014	6" x 12" to round		42	.381		17.60	19.20		36.80	49
3018	8" x 8" to round		40	.400		11.55	20		31.55	43.50
3022	9" x 9" to round		38	.421		11.55	21		32.55	45
3026	9" x 12" to round		37	.432		22.50	22		44.50	58.50
3030	9" x 15" to round		36	.444		26.50	22.50		49	63
3034	10" x 10" to round		34	.471		13.75	23.50		37.25	51
3038	10" x 22" to round		32	.500		32.50	25		57.50	74
3042	12" x 12" to round		30	.533		13.75	27		40.75	56
3046	12" x 18" to round		28	.571		32.50	29		61.50	79.50
3050	12" x 24" to round		26	.615		32.50	31		63.50	83
3054	15" x 15" to round		25	.640		18.70	32		50.70	70
3058	15" x 18" to round		24	.667		35	33.50		68.50	90
3062	18" x 18" to round		22	.727		26.50	36.50		63	85
3066	21" x 21" to round		21	.762		29	38.50		67.50	90.50
3070	22" x 22" to round		20	.800		29	40.50		69.50	93.50
3074	24" x 24" to round	▼	18	.889	▼	33	45		78	105
5000	Flexible, aluminum, acoustical, pressure to 2" (wg), NFPA-90A									
5010	Fiberglass insulation 1-1/2" thick, 1/2 lb. density									
5020	Polyethylene jacket, UL approved									
5026	5" diameter [G]	Q-9	300	.053	L.F.	4.29	2.69		6.98	8.80

23 33 46 – Flexible Ducts

23 33 46.10 Flexible Air Ducts		Crew	Daily Output	Labor-Hours	Unit	Material	2015 Bare Costs Labor	Equipment	Total	Total Incl O&P
5030	6" diameter	G Q-9	260	.062	L.F.	5.05	3.10		8.15	10.30
5034	7" diameter	G	220	.073		6.05	3.66		9.71	12.25
5038	8" diameter	G	180	.089		6.15	4.48		10.63	13.65
5042	9" diameter	G	160	.100		7.25	5.05		12.30	15.70
5046	10" diameter	G	140	.114		9.45	5.75		15.20	19.20
5050	12" diameter	G	100	.160		10.45	8.05		18.50	24
5054	14" diameter	G	80	.200		11.20	10.05		21.25	28
5058	16" diameter	G	60	.267		15.85	13.45		29.30	38
5072	Hospital grade, PE jacket, UL approved									
5076	5" diameter	G Q-9	300	.053	L.F.	4.62	2.69		7.31	9.20
5080	6" diameter	G	260	.062		5.50	3.10		8.60	10.80
5084	7" diameter	G	220	.073		6.60	3.66		10.26	12.85
5088	8" diameter	G	180	.089		6.80	4.48		11.28	14.35
5092	9" diameter	G	160	.100		7.90	5.05		12.95	16.40
5096	10" diameter	G	140	.114		10.35	5.75		16.10	20
5100	12" diameter	G	100	.160		11.45	8.05		19.50	25
5104	14" diameter	G	80	.200		12.30	10.05		22.35	29
5108	16" diameter	G	60	.267		17.25	13.45		30.70	39.50
5140	Flexible, aluminum, silencer, pressure to 12" (wg), NFPA-90A									
5144	Fiberglass insulation 1-1/2" thick, 1/2 lb. density									
5148	Aluminum outer shell, UL approved									
5152	5" diameter	G Q-9	290	.055	L.F.	11.85	2.78		14.63	17.25
5156	6" diameter	G	250	.064		13	3.22		16.22	19.20
5160	7" diameter	G	210	.076		13.75	3.84		17.59	21
5164	8" diameter	G	170	.094		14.70	4.74		19.44	23.50
5168	9" diameter	G	150	.107		17.80	5.35		23.15	28
5172	10" diameter	G	130	.123		20	6.20		26.20	31.50
5176	12" diameter	G	90	.178		22.50	8.95		31.45	38
5180	14" diameter	G	70	.229		26.50	11.50		38	46.50
5184	16" diameter	G	50	.320		34	16.10		50.10	62
5200	Hospital grade, aluminum shell, UL approved									
5204	5" diameter	G Q-9	280	.057	L.F.	12.25	2.88		15.13	17.90
5208	6" diameter	G	240	.067		13.50	3.36		16.86	20
5212	7" diameter	G	200	.080		14.30	4.03		18.33	22
5216	8" diameter	G	160	.100		15.30	5.05		20.35	24.50
5220	9" diameter	G	140	.114		18.55	5.75		24.30	29.50
5224	10" diameter	G	130	.123		21	6.20		27.20	32.50
5228	12" diameter	G	80	.200		24.50	10.05		34.55	42
5232	14" diameter	G	60	.267		27.50	13.45		40.95	51
5236	16" diameter	G	40	.400		35.50	20		55.50	70

23 33 53 – Duct Liners

23 33 53.10 Duct Liner Board

		Crew	Daily Output	Labor-Hours	Unit	Material	2015 Bare Costs Labor	Equipment	Total	Total Incl O&P
0010	**DUCT LINER BOARD**									
3340	Board type fiberglass liner, FSK, 1-1/2 lb. density									
3344	1" thick	G Q-14	150	.107	S.F.	.62	5.05		5.67	8.55
3345	1-1/2" thick	G	130	.123		.68	5.80		6.48	9.80
3346	2" thick	G	120	.133		.79	6.30		7.09	10.65
3348	3" thick	G	110	.145		1.02	6.85		7.87	11.75
3350	4" thick	G	100	.160		1.25	7.55		8.80	13.15
3356	3 lb. density, 1" thick	G	150	.107		.79	5.05		5.84	8.70
3358	1-1/2" thick	G	130	.123		1	5.80		6.80	10.15
3360	2" thick	G	120	.133		1.22	6.30		7.52	11.15

For customer support on your Mechanical Cost Data, call 877.795.7394.

331

23 33 53.10 Duct Liner Board		Crew	Daily Output	Labor-Hours	Unit	Material	2015 Bare Costs Labor	Equipment	Total	Total Incl O&P
3362	2-1/2" thick	G Q-14	110	.145	S.F.	1.43	6.85		8.28	12.20
3364	3" thick	G	100	.160		1.64	7.55		9.19	13.55
3366	4" thick	G	90	.178		2.06	8.40		10.46	15.30
3370	6 lb. density, 1" thick	G	140	.114		1.12	5.40		6.52	9.65
3374	1-1/2" thick	G	120	.133		1.50	6.30		7.80	11.45
3378	2" thick	G	100	.160		1.88	7.55		9.43	13.80
3490	Board type, fiberglass liner, 3 lb. density									
3680	No finish									
3700	1" thick	G Q-14	170	.094	S.F.	.44	4.44		4.88	7.40
3710	1-1/2" thick	G	140	.114		.66	5.40		6.06	9.15
3720	2" thick	G	130	.123		.88	5.80		6.68	10
3940	Board type, non-fibrous foam									
3950	Temperature, bacteria and fungi resistant									
3960	1" thick	G Q-14	150	.107	S.F.	2.38	5.05		7.43	10.45
3970	1-1/2" thick	G	130	.123		3.28	5.80		9.08	12.65
3980	2" thick	G	120	.133		3.98	6.30		10.28	14.20

23 34 13.10 Axial Flow HVAC Fans

		Crew	Daily Output	Labor-Hours	Unit	Material	2015 Bare Costs Labor	Equipment	Total	Total Incl O&P
0010	**AXIAL FLOW HVAC FANS**									
0020	Air conditioning and process air handling									
0500	Axial flow, constant speed									
0505	Direct drive, 1/8" S.P.									
0510	12", 1060 CFM, 1/6 HP	Q-20	3	6.667	Ea.	630	340		970	1,225
0514	12", 2095 CFM, 1/2 HP		3	6.667		745	340		1,085	1,325
0518	16", 2490 CFM, 1/3 HP		2.80	7.143		745	365		1,110	1,375
0522	20", 4130 CFM, 3/4 HP		2.60	7.692		885	395		1,280	1,575
0526	22", 4700 CFM, 3/4 HP		2.60	7.692		1,300	395		1,695	2,050
0530	24", 5850 CFM, 1 HP		2.50	8		1,350	410		1,760	2,125
0534	24", 7925 CFM, 1-1/2 HP		2.40	8.333		1,675	425		2,100	2,475
0538	30", 10640 CFM, 2 HP		2.20	9.091		2,150	465		2,615	3,075
0542	30", 14765 CFM, 2-1/2 HP		2.10	9.524		2,500	490		2,990	3,500
0546	36", 16780 CFM, 2 HP		2.10	9.524		2,575	490		3,065	3,575
0550	36", 22920 CFM, 5 HP		1.80	11.111		2,825	570		3,395	4,000
0560	Belt drive, 1/8" S.P.									
0562	15", 2800 CFM, 1/3 HP	Q-20	3.20	6.250	Ea.	1,150	320		1,470	1,750
0564	15", 3400 CFM, 1/2 HP		3	6.667		1,175	340		1,515	1,825
0568	18", 3280 CFM, 1/3 HP		2.80	7.143		1,225	365		1,590	1,900
0572	18", 3900 CFM, 1/2 HP		2.80	7.143		1,250	365		1,615	1,925
0576	18", 5250 CFM, 1 HP		2.70	7.407		1,375	380		1,755	2,075
0584	24", 6430 CFM, 1 HP		2.50	8		1,800	410		2,210	2,600
0588	24", 8860 CFM, 2 HP		2.40	8.333		1,875	425		2,300	2,700
0592	30", 9250 CFM, 1 HP		2.20	9.091		2,300	465		2,765	3,250
0596	30", 16900 CFM, 5 HP		2	10		2,475	510		2,985	3,475
0604	36", 14475 CFM, 2 HP		2.30	8.696		2,825	445		3,270	3,800
0608	36", 20080 CFM, 5 HP		1.80	11.111		2,925	570		3,495	4,075
0616	42", 29000 CFM, 7-1/2 HP		1.40	14.286		3,925	730		4,655	5,450
1500	Vaneaxial, low pressure, 2000 CFM, 1/2 HP		3.60	5.556		2,225	285		2,510	2,875
1520	4,000 CFM, 1 HP		3.20	6.250		2,600	320		2,920	3,325
1540	8,000 CFM, 2 HP		2.80	7.143		3,275	365		3,640	4,175

23 34 HVAC Fans

23 34 13 – Axial HVAC Fans

23 34 13.10 Axial Flow HVAC Fans	Crew	Daily Output	Labor-Hours	Unit	Material	2015 Bare Costs Labor	2015 Bare Costs Equipment	Total	Total Incl O&P
1560 16,000 CFM, 5 HP	Q-20	2.40	8.333	Ea.	4,600	425		5,025	5,725

23 34 14 – Blower HVAC Fans

23 34 14.10 Blower Type HVAC Fans	Crew	Daily Output	Labor-Hours	Unit	Material	Labor	Equipment	Total	Total Incl O&P
0010 **BLOWER TYPE HVAC FANS**									
2000 Blowers, direct drive with motor, complete									
2020 1045 CFM @ .5" S.P., 1/5 HP	Q-20	18	1.111	Ea.	340	57		397	460
2040 1385 CFM @ .5" S.P., 1/4 HP		18	1.111		335	57		392	455
2060 1640 CFM @ .5" S.P., 1/3 HP		18	1.111		305	57		362	420
2080 1760 CFM @ .5" S.P., 1/2 HP		18	1.111		315	57		372	435
2090 4 speed									
2100 1164 to 1739 CFM @ .5" S.P., 1/3 HP	Q-20	16	1.250	Ea.	320	64		384	455
2120 1467 to 2218 CFM @ 1.0" S.P., 3/4 HP	"	14	1.429	"	370	73		443	515
2500 Ceiling fan, right angle, extra quiet, 0.10" S.P.									
2520 95 CFM	Q-20	20	1	Ea.	300	51		351	410
2540 210 CFM		19	1.053		355	54		409	470
2560 385 CFM		18	1.111		450	57		507	580
2580 885 CFM		16	1.250		890	64		954	1,075
2600 1,650 CFM		13	1.538		1,225	79		1,304	1,475
2620 2,960 CFM		11	1.818		1,650	93		1,743	1,950
2640 For wall or roof cap, add	1 Shee	16	.500		300	28		328	375
2660 For straight thru fan, add					10%				
2680 For speed control switch, add	1 Elec	16	.500		164	27.50		191.50	222
7500 Utility set, steel construction, pedestal, 1/4" S.P.									
7520 Direct drive, 150 CFM, 1/8 HP	Q-20	6.40	3.125	Ea.	870	160		1,030	1,200
7540 485 CFM, 1/6 HP		5.80	3.448		1,100	177		1,277	1,475
7560 1950 CFM, 1/2 HP		4.80	4.167		1,275	213		1,488	1,725
7580 2410 CFM, 3/4 HP		4.40	4.545		2,375	233		2,608	2,950
7600 3328 CFM, 1-1/2 HP		3	6.667		2,625	340		2,965	3,425
7680 V-belt drive, drive cover, 3 phase									
7700 800 CFM, 1/4 HP	Q-20	6	3.333	Ea.	980	171		1,151	1,325
7720 1,300 CFM, 1/3 HP		5	4		1,025	205		1,230	1,425
7740 2,000 CFM, 1 HP		4.60	4.348		1,225	223		1,448	1,700
7760 2,900 CFM, 3/4 HP		4.20	4.762		1,650	244		1,894	2,175
7780 3,600 CFM, 3/4 HP		4	5		2,025	256		2,281	2,625
7800 4,800 CFM, 1 HP		3.50	5.714		2,375	293		2,668	3,075
7820 6,700 CFM, 1-1/2 HP		3	6.667		2,950	340		3,290	3,750
7830 7,500 CFM, 2 HP		2.50	8		4,000	410		4,410	5,025
7840 11,000 CFM, 3 HP		2	10		5,325	510		5,835	6,650
7860 13,000 CFM, 3 HP		1.60	12.500		5,425	640		6,065	6,925
7880 15,000 CFM, 5 HP		1	20		5,600	1,025		6,625	7,725
7900 17,000 CFM, 7-1/2 HP		.80	25		6,000	1,275		7,275	8,550
7920 20,000 CFM, 7-1/2 HP		.80	25		7,150	1,275		8,425	9,825

23 34 16 – Centrifugal HVAC Fans

23 34 16.10 Centrifugal Type HVAC Fans	Crew	Daily Output	Labor-Hours	Unit	Material	Labor	Equipment	Total	Total Incl O&P
0010 **CENTRIFUGAL TYPE HVAC FANS**									
0200 In-line centrifugal, supply/exhaust booster									
0220 aluminum wheel/hub, disconnect switch, 1/4" S.P.									
0240 500 CFM, 10" diameter connection	Q-20	3	6.667	Ea.	1,300	340		1,640	1,950
0260 1,380 CFM, 12" diameter connection		2	10		1,375	510		1,885	2,300
0280 1,520 CFM, 16" diameter connection		2	10		1,500	510		2,010	2,425
0300 2,560 CFM, 18" diameter connection		1	20		1,625	1,025		2,650	3,350
0320 3,480 CFM, 20" diameter connection		.80	25		1,925	1,275		3,200	4,075

For customer support on your Mechanical Cost Data, call 877.795.7394.

333

23 34 16 – Centrifugal HVAC Fans

23 34 16.10 Centrifugal Type HVAC Fans		Crew	Daily Output	Labor-Hours	Unit	Material	2015 Bare Costs Labor	Equipment	Total	Total Incl O&P
0326	5,080 CFM, 20" diameter connection	Q-20	.75	26.667	Ea.	2,100	1,375		3,475	4,400
0340	7,500 CFM, 22" diameter connection		.70	28.571		2,500	1,475		3,975	4,975
0350	10,000 CFM, 27" diameter connection		.65	30.769		3,025	1,575		4,600	5,750
3500	Centrifugal, airfoil, motor and drive, complete									
3520	1000 CFM, 1/2 HP	Q-20	2.50	8	Ea.	1,900	410		2,310	2,700
3540	2,000 CFM, 1 HP		2	10		2,150	510		2,660	3,125
3560	4,000 CFM, 3 HP		1.80	11.111		2,725	570		3,295	3,875
3580	8,000 CFM, 7-1/2 HP		1.40	14.286		4,100	730		4,830	5,625
3600	12,000 CFM, 10 HP		1	20		5,450	1,025		6,475	7,550
4000	Single width, belt drive, not incl. motor, capacities									
4020	at 2000 fpm, 2.5" S.P. for indicated motor									
4040	6900 CFM, 5 HP	Q-9	2.40	6.667	Ea.	3,775	335		4,110	4,700
4060	10,340 CFM, 7-1/2 HP		2.20	7.273		5,275	365		5,640	6,350
4080	15,320 CFM, 10 HP		2	8		6,200	405		6,605	7,450
4100	22,780 CFM, 15 HP		1.80	8.889		9,225	450		9,675	10,900
4120	33,840 CFM, 20 HP		1.60	10		12,300	505		12,805	14,300
4140	41,400 CFM, 25 HP		1.40	11.429		16,200	575		16,775	18,800
4160	50,100 CFM, 30 HP		.80	20		20,100	1,000		21,100	23,700
4200	Double width wheel, 12,420 CFM, 7.5 HP		2.20	7.273		5,400	365		5,765	6,500
4220	18,620 CFM, 15 HP		2	8		7,750	405		8,155	9,150
4240	27,580 CFM, 20 HP		1.80	8.889		9,475	450		9,925	11,100
4260	40,980 CFM, 25 HP		1.50	10.667		14,800	535		15,335	17,100
4280	60,920 CFM, 40 HP		1	16		20,000	805		20,805	23,200
4300	74,520 CFM, 50 HP		.80	20		23,700	1,000		24,700	27,600
4320	90,160 CFM, 50 HP		.70	22.857		32,400	1,150		33,550	37,400
4340	110,300 CFM, 60 HP		.50	32		43,800	1,600		45,400	50,500
4360	134,960 CFM, 75 HP		.40	40		58,000	2,025		60,025	67,000
5000	Utility set, centrifugal, V belt drive, motor									
5020	1/4" S.P., 1200 CFM, 1/4 HP	Q-20	6	3.333	Ea.	1,750	171		1,921	2,175
5040	1520 CFM, 1/3 HP		5	4		2,225	205		2,430	2,750
5060	1850 CFM, 1/2 HP		4	5		2,200	256		2,456	2,825
5080	2180 CFM, 3/4 HP		3	6.667		2,600	340		2,940	3,400
5100	1/2" S.P., 3600 CFM, 1 HP		2	10		2,700	510		3,210	3,750
5120	4250 CFM, 1-1/2 HP		1.60	12.500		3,300	640		3,940	4,600
5140	4800 CFM, 2 HP		1.40	14.286		4,000	730		4,730	5,500
5160	6920 CFM, 5 HP		1.30	15.385		4,875	790		5,665	6,575
5180	7700 CFM, 7-1/2 HP		1.20	16.667		5,875	855		6,730	7,775
5200	For explosion proof motor, add					15%				
5300	Fume exhauster without hose and intake nozzle									
5310	630 CFM, 1-1/2 HP	Q-20	2	10	Ea.	885	510		1,395	1,750
5320	Hose extension kit-5'	"	20	1	"	69.50	51		120.50	155
5500	Fans, industrial exhauster, for air which may contain granular matl.									
5520	1000 CFM, 1-1/2 HP	Q-20	2.50	8	Ea.	2,875	410		3,285	3,800
5540	2000 CFM, 3 HP		2	10		3,500	510		4,010	4,625
5560	4000 CFM, 7-1/2 HP		1.80	11.111		4,875	570		5,445	6,225
5580	8000 CFM, 15 HP		1.40	14.286		6,275	730		7,005	8,025
5600	12,000 CFM, 30 HP		1	20		10,200	1,025		11,225	12,800
7000	Roof exhauster, centrifugal, aluminum housing, 12" galvanized									
7020	curb, bird screen, back draft damper, 1/4" S.P.									
7100	Direct drive, 320 CFM, 11" sq. damper	Q-20	7	2.857	Ea.	705	146		851	1,000
7120	600 CFM, 11" sq. damper		6	3.333		900	171		1,071	1,250
7140	815 CFM, 13" sq. damper		5	4		900	205		1,105	1,300
7160	1450 CFM, 13" sq. damper		4.20	4.762		1,450	244		1,694	1,975

23 34 16 – Centrifugal HVAC Fans

23 34 16.10 Centrifugal Type HVAC Fans		Crew	Daily Output	Labor-Hours	Unit	Material	2015 Bare Costs Labor	Equipment	Total	Total Incl O&P
7180	2050 CFM, 16" sq. damper	Q-20	4	5	Ea.	1,725	256		1,981	2,300
7200	V-belt drive, 1650 CFM, 12" sq. damper		6	3.333		1,300	171		1,471	1,675
7220	2750 CFM, 21" sq. damper		5	4		1,550	205		1,755	2,000
7230	3500 CFM, 21" sq. damper		4.50	4.444		1,725	228		1,953	2,250
7240	4910 CFM, 23" sq. damper		4	5		2,125	256		2,381	2,725
7260	8525 CFM, 28" sq. damper		3	6.667		2,800	340		3,140	3,600
7280	13,760 CFM, 35" sq. damper		2	10		3,925	510		4,435	5,100
7300	20,558 CFM, 43" sq. damper	▼	1	20		7,875	1,025		8,900	10,200
7320	For 2 speed winding, add					15%				
7340	For explosionproof motor, add					600			600	660
7360	For belt driven, top discharge, add				▼	15%				
7400	Roof mounted kitchen exhaust, aluminum, centrifugal									
7410	Direct drive, 1/4" S.P., 2 speed, temp to 200°F									
7414	1/3 HP, 2038 CFM	Q-20	7	2.857	Ea.	550	146		696	825
7416	1/2 HP, 2393 CFM		6	3.333		525	171		696	835
7418	3/4 HP, 4329 CFM	▼	5	4	▼	820	205		1,025	1,200
7424	Belt drive, 1/4" S.P., temp to 250°F									
7426	3/4 HP, 2757 CFM	Q-20	5	4	Ea.	790	205		995	1,175
7428	1-1/2 HP, 4913 CFM		4	5		1,125	256		1,381	1,625
7430	1-1/2 HP, 5798 CFM	▼	4	5	▼	1,150	256		1,406	1,675
7450	Upblast, propeller, w/backdraft damper, bird screen & curb									
7454	30,300 CFM @ 3/8" S.P., 5 HP	Q-20	1.60	12.500	Ea.	4,525	640		5,165	5,950
7458	36,000 CFM @ 1/2" S.P., 15 HP	"	1.60	12.500	"	6,425	640		7,065	8,050
8500	Wall exhausters, centrifugal, auto damper, 1/8" S.P.									
8520	Direct drive, 610 CFM, 1/20 HP	Q-20	14	1.429	Ea.	425	73		498	575
8540	796 CFM, 1/12 HP		13	1.538		880	79		959	1,075
8560	822 CFM, 1/6 HP		12	1.667		1,075	85.50		1,160.50	1,300
8580	1,320 CFM, 1/4 HP		12	1.667		1,250	85.50		1,335.50	1,500
8600	1756 CFM, 1/4 HP		11	1.818		1,250	93		1,343	1,525
8620	1983 CFM, 1/4 HP		10	2		1,300	102		1,402	1,575
8640	2900 CFM, 1/2 HP		9	2.222		1,375	114		1,489	1,675
8660	3307 CFM, 3/4 HP		8	2.500		1,475	128		1,603	1,825
8670	5940 CFM, 1 HP	▼	7	2.857	▼	1,550	146		1,696	1,925
9500	V-belt drive, 3 phase									
9520	2,800 CFM, 1/4 HP	Q-20	9	2.222	Ea.	1,925	114		2,039	2,300
9540	3,740 CFM, 1/2 HP		8	2.500		2,000	128		2,128	2,400
9560	4400 CFM, 3/4 HP		7	2.857		2,025	146		2,171	2,450
9580	5700 CFM, 1-1/2 HP	▼	6	3.333	▼	2,100	171		2,271	2,550

23 34 23 – HVAC Power Ventilators

23 34 23.10 HVAC Power Circulators and Ventilators

		Crew	Daily Output	Labor-Hours	Unit	Material	2015 Bare Costs Labor	Equipment	Total	Total Incl O&P
0010	**HVAC POWER CIRCULATORS AND VENTILATORS**									
3000	Paddle blade air circulator, 3 speed switch									
3020	42", 5,000 CFM high, 3000 CFM low [G]	1 Elec	2.40	3.333	Ea.	163	182		345	450
3040	52", 6,500 CFM high, 4000 CFM low [G]	"	2.20	3.636	"	170	199		369	485
3100	For antique white motor, same cost									
3200	For brass plated motor, same cost									
3300	For light adaptor kit, add [G]				Ea.	41			41	45
3310	Industrial grade, reversible, 4 blade									
3312	5500 CFM	Q-20	5	4	Ea.	206	205		411	535
3314	7600 CFM		5	4		213	205		418	545
3316	21,015 CFM	▼	4	5	▼	254	256		510	670
4000	High volume, low speed (HVLS), paddle blade air circulator									

23 34 23.10 HVAC Power Circulators and Ventilators	Crew	Daily Output	Labor-Hours	Unit	Material	2015 Bare Costs Labor	Equipment	Total	Total Incl O&P
4010	Variable speed, reversible, 1 HP motor, motor control panel,								
4020	motor drive cable, control cable with remote, and safety cable.								
4140	8' diameter	Q-5	1.50	10.667	Ea.	4,725	575		5,300	6,075
4150	10' diameter		1.45	11.034		4,750	595		5,345	6,125
4160	12' diameter		1.40	11.429		4,975	615		5,590	6,400
4170	14' diameter		1.35	11.852		5,050	635		5,685	6,525
4180	16' diameter		1.30	12.308		5,100	660		5,760	6,600
4190	18' diameter		1.25	12.800		5,150	690		5,840	6,700
4200	20' diameter		1.20	13.333		5,475	715		6,190	7,100
4210	24' diameter		1	16		5,600	860		6,460	7,475
6000	Propeller exhaust, wall shutter								
6020	Direct drive, one speed, .075" S.P.								
6100	653 CFM, 1/30 HP	Q-20	10	2	Ea.	192	102		294	365
6120	1033 CFM, 1/20 HP		9	2.222		287	114		401	490
6140	1323 CFM, 1/15 HP		8	2.500		320	128		448	545
6160	2444 CFM, 1/4 HP		7	2.857		400	146		546	665
6300	V-belt drive, 3 phase								
6320	6175 CFM, 3/4 HP	Q-20	5	4	Ea.	2,950	205		3,155	3,550
6340	7500 CFM, 3/4 HP		5	4		3,025	205		3,230	3,625
6360	10,100 CFM, 1 HP		4.50	4.444		3,175	228		3,403	3,850
6380	14,300 CFM, 1-1/2 HP		4	5		3,450	256		3,706	4,175
6400	19,800 CFM, 2 HP		3	6.667		1,950	340		2,290	2,650
6420	26,250 CFM, 3 HP		2.60	7.692		2,150	395		2,545	2,975
6440	38,500 CFM, 5 HP		2.20	9.091		2,550	465		3,015	3,500
6460	46,000 CFM, 7-1/2 HP		2	10		2,725	510		3,235	3,775
6480	51,500 CFM, 10 HP		1.80	11.111		2,825	570		3,395	4,000
6490	V-belt drive, 115 V, residential, whole house								
6500	Ceiling-wall, 5200 CFM, 1/4 HP, 30" x 30"	1 Shee	6	1.333	Ea.	545	74.50		619.50	710
6510	7500 CFM, 1/3 HP, 36" x 36"		5	1.600		570	89.50		659.50	760
6520	10,500 CFM, 1/3 HP, 42" x 42"		5	1.600		595	89.50		684.50	785
6530	13,200 CFM, 1/3 HP, 48" x 48"		4	2		655	112		767	890
6540	15,445 CFM, 1/2 HP, 48" x 48"		4	2		680	112		792	920
6550	17,025 CFM, 1/2 HP, 54" x 54"		4	2		1,325	112		1,437	1,625
6560	For two speed motor, add					20%				
6570	Shutter, automatic, ceiling/wall								
6580	30" x 30"	1 Shee	8	1	Ea.	177	56		233	280
6590	36" x 36"		8	1		202	56		258	310
6600	42" x 42"		8	1		252	56		308	365
6610	48" x 48"		7	1.143		274	64		338	400
6620	54" x 54"		6	1.333		360	74.50		434.50	510
6630	Timer, shut off, to 12 hour		20	.400		58	22.50		80.50	98
6650	Residential, bath exhaust, grille, back draft damper								
6660	50 CFM	Q-20	24	.833	Ea.	63.50	42.50		106	135
6670	110 CFM		22	.909		98	46.50		144.50	179
6672	180 CFM		22	.909		108	46.50		154.50	190
6673	210 CFM		22	.909		141	46.50		187.50	226
6674	260 CFM		22	.909		171	46.50		217.50	259
6675	300 CFM		22	.909		129	46.50		175.50	213
6680	Light combination, squirrel cage, 100 watt, 70 CFM		24	.833		112	42.50		154.50	188
6700	Light/heater combination, ceiling mounted								
6710	70 CFM, 1450 watt	Q-20	24	.833	Ea.	162	42.50		204.50	243
6800	Heater combination, recessed, 70 CFM		24	.833		67.50	42.50		110	139
6820	With 2 infrared bulbs		23	.870		105	44.50		149.50	184

23 34 23 – HVAC Power Ventilators

23 34 23.10 HVAC Power Circulators and Ventilators		Crew	Daily Output	Labor-Hours	Unit	Material	2015 Bare Costs		Total	Total Incl O&P
							Labor	Equipment		
6840	Wall mount, 170 CFM	Q-20	22	.909	Ea.	260	46.50		306.50	355
6846	Ceiling mount, 180 CFM		22	.909		263	46.50		309.50	360
6900	Kitchen exhaust, grille, complete, 160 CFM		22	.909		108	46.50		154.50	190
6910	180 CFM		20	1		90.50	51		141.50	178
6920	270 CFM		18	1.111		171	57		228	275
6930	350 CFM		16	1.250		129	64		193	240
6940	Residential roof jacks and wall caps									
6944	Wall cap with back draft damper									
6946	3" & 4" diam. round duct	1 Shee	11	.727	Ea.	26	40.50		66.50	90.50
6948	6" diam. round duct	"	11	.727	"	64	40.50		104.50	133
6958	Roof jack with bird screen and back draft damper									
6960	3" & 4" diam. round duct	1 Shee	11	.727	Ea.	26	40.50		66.50	90.50
6962	3-1/4" x 10" rectangular duct	"	10	.800	"	48.50	45		93.50	122
6980	Transition									
6982	3-1/4" x 10" to 6" diam. round	1 Shee	20	.400	Ea.	32	22.50		54.50	69.50
8020	Attic, roof type									
8030	Aluminum dome, damper & curb									
8080	12" diameter, 1000 CFM (gravity)	1 Elec	10	.800	Ea.	575	44		619	695
8090	16" diameter, 1500 CFM (gravity)		9	.889		690	48.50		738.50	835
8100	20" diameter, 2500 CFM (gravity)		8	1		850	54.50		904.50	1,025
8110	26" diameter, 4000 CFM (gravity)		7	1.143		1,025	62.50		1,087.50	1,225
8120	32" diameter, 6500 CFM (gravity)		6	1.333		1,400	73		1,473	1,650
8130	38" diameter, 8000 CFM (gravity)		5	1.600		2,100	87.50		2,187.50	2,425
8140	50" diameter, 13,000 CFM (gravity)		4	2		3,025	109		3,134	3,500
8160	Plastic, ABS dome									
8180	1050 CFM	1 Elec	14	.571	Ea.	168	31.50		199.50	232
8200	1600 CFM	"	12	.667	"	252	36.50		288.50	335
8240	Attic, wall type, with shutter, one speed									
8250	12" diameter, 1000 CFM	1 Elec	14	.571	Ea.	395	31.50		426.50	480
8260	14" diameter, 1500 CFM		12	.667		430	36.50		466.50	530
8270	16" diameter, 2000 CFM		9	.889		485	48.50		533.50	610
8290	Whole house, wall type, with shutter, one speed									
8300	30" diameter, 4800 CFM	1 Elec	7	1.143	Ea.	1,050	62.50		1,112.50	1,250
8310	36" diameter, 7000 CFM		6	1.333		1,125	73		1,198	1,350
8320	42" diameter, 10,000 CFM		5	1.600		1,275	87.50		1,362.50	1,525
8330	48" diameter, 16,000 CFM		4	2		1,575	109		1,684	1,925
8340	For two speed, add					95			95	105
8350	Whole house, lay-down type, with shutter, one speed									
8360	30" diameter, 4500 CFM	1 Elec	8	1	Ea.	1,100	54.50		1,154.50	1,300
8370	36" diameter, 6500 CFM		7	1.143		1,200	62.50		1,262.50	1,400
8380	42" diameter, 9000 CFM		6	1.333		1,300	73		1,373	1,550
8390	48" diameter, 12,000 CFM		5	1.600		1,500	87.50		1,587.50	1,750
8440	For two speed, add					71.50			71.50	78.50
8450	For 12 hour timer switch, add	1 Elec	32	.250		71.50	13.70		85.20	99

23 34 33 – Air Curtains

23 34 33.10 Air Barrier Curtains

		Crew	Daily Output	Labor-Hours	Unit	Material	Labor	Equipment	Total	Total Incl O&P
0010	**AIR BARRIER CURTAINS**, Incl. motor starters, transformers,									
0050	and door switches									
2450	Conveyor openings or service windows									
3000	Service window, 5' high x 25" wide	2 Shee	5	3.200	Ea.	305	179		484	610
3100	Environmental separation									
3110	Door heights up to 8', low profile, super quiet									

For customer support on your Mechanical Cost Data, call 877.795.7394.

337

23 34 HVAC Fans

23 34 33 – Air Curtains

23 34 33.10 Air Barrier Curtains	Crew	Daily Output	Labor-Hours	Unit	Material	2015 Bare Costs Labor	Equipment	Total	Total Incl O&P
3120	Unheated, variable speed									
3130	36" wide	2 Shee	4	4	Ea.	610	224		834	1,000
3134	42" wide		3.80	4.211		635	236		871	1,050
3138	48" wide		3.60	4.444		655	249		904	1,100
3142	60" wide		3.40	4.706		685	263		948	1,150
3146	72" wide	Q-3	4.60	6.957		850	390		1,240	1,525
3150	96" wide		4.40	7.273		1,300	405		1,705	2,050
3154	120" wide		4.20	7.619		1,400	425		1,825	2,175
3158	144" wide		4	8		1,700	445		2,145	2,550
3200	Door heights up to 10'									
3210	Unheated									
3230	36" wide	2 Shee	3.80	4.211	Ea.	645	236		881	1,075
3234	42" wide		3.60	4.444		665	249		914	1,100
3238	48" wide		3.40	4.706		685	263		948	1,150
3242	60" wide		3.20	5		1,025	280		1,305	1,550
3246	72" wide	Q-3	4.40	7.273		1,075	405		1,480	1,825
3250	96" wide		4.20	7.619		1,275	425		1,700	2,050
3254	120" wide		4	8		1,775	445		2,220	2,625
3258	144" wide		3.80	8.421		1,925	470		2,395	2,800
3300	Door heights up to 12'									
3310	Unheated									
3334	42" wide	2 Shee	3.40	4.706	Ea.	920	263		1,183	1,400
3338	48" wide		3.20	5		925	280		1,205	1,450
3342	60" wide		3	5.333		945	298		1,243	1,500
3346	72" wide	Q-3	4.20	7.619		1,675	425		2,100	2,475
3350	96" wide		4	8		1,825	445		2,270	2,675
3354	120" wide		3.80	8.421		2,225	470		2,695	3,150
3358	144" wide		3.60	8.889		2,375	495		2,870	3,375
3400	Door heights up to 16'									
3410	Unheated									
3438	48" wide	2 Shee	3	5.333	Ea.	1,100	298		1,398	1,675
3442	60" wide	"	2.80	5.714		1,175	320		1,495	1,775
3446	72" wide	Q-3	3.80	8.421		2,000	470		2,470	2,900
3450	96" wide		3.60	8.889		2,075	495		2,570	3,050
3454	120" wide		3.40	9.412		2,825	525		3,350	3,900
3458	144" wide		3.20	10		2,925	560		3,485	4,050
3470	Heated, electric									
3474	48" wide	2 Shee	2.90	5.517	Ea.	1,850	310		2,160	2,500
3478	60" wide	"	2.70	5.926		1,875	330		2,205	2,575
3482	72" wide	Q-3	3.70	8.649		3,175	485		3,660	4,225
3486	96" wide		3.50	9.143		3,275	510		3,785	4,375
3490	120" wide		3.30	9.697		3,325	540		3,865	4,500
3494	144" wide		3.10	10.323		4,475	575		5,050	5,800

338

For customer support on your Mechanical Cost Data, call 877.795.7394.

23 35 16 – Engine Exhaust Systems

23 35 16.10 Engine Exhaust Removal Systems	Crew	Daily Output	Labor-Hours	Unit	Material	2015 Bare Costs Labor	Equipment	Total	Total Incl O&P
0010 **ENGINE EXHAUST REMOVAL SYSTEMS** D3090–320									
0500 Engine exhaust, garage, in-floor system									
0510 Single tube outlet assemblies									
0520 For transite pipe ducting, self-storing tube									
0530 3" tubing adapter plate	1 Shee	16	.500	Ea.	240	28		268	305
0540 4" tubing adapter plate		16	.500		245	28		273	310
0550 5" tubing adapter plate		16	.500		245	28		273	310
0600 For vitrified tile ducting									
0610 3" tubing adapter plate, self-storing tube	1 Shee	16	.500	Ea.	241	28		269	310
0620 4" tubing adapter plate, self-storing tube		16	.500		245	28		273	310
0660 5" tubing adapter plate, self-storing tube		16	.500		245	28		273	310
0800 Two tube outlet assemblies									
0810 For transite pipe ducting, self-storing tube									
0820 3" tubing, dual exhaust adapter plate	1 Shee	16	.500	Ea.	240	28		268	305
0850 For vitrified tile ducting									
0860 3" tubing, dual exhaust, self-storing tube	1 Shee	16	.500	Ea.	298	28		326	375
0870 3" tubing, double outlet, non-storing tubes	"	16	.500	"	298	28		326	375
0900 Accessories for metal tubing, (overhead systems also)									
0910 Adapters, for metal tubing end									
0920 3" tail pipe type				Ea.	49.50			49.50	54.50
0930 4" tail pipe type					53			53	58.50
0940 5" tail pipe type					54			54	59.50
0990 5" diesel stack type					315			315	350
1000 6" diesel stack type					330			330	360
1100 Bullnose (guide) required for in-floor assemblies									
1110 3" tubing size				Ea.	28			28	30.50
1120 4" tubing size					29.50			29.50	32.50
1130 5" tubing size					31.50			31.50	34.50
1150 Plain rings, for tubing end									
1160 3" tubing size				Ea.	24.50			24.50	26.50
1170 4" tubing size				"	41.50			41.50	45.50
1200 Tubing, galvanized, flexible, (for overhead systems also)									
1210 3" ID				L.F.	11.60			11.60	12.75
1220 4" ID					14.20			14.20	15.65
1230 5" ID					16.75			16.75	18.40
1240 6" ID					19.35			19.35	21.50
1250 Stainless steel, flexible, (for overhead system, also)									
1260 3" ID				L.F.	25.50			25.50	28
1270 4" ID					35			35	38.50
1280 5" ID					39.50			39.50	43.50
1290 6" ID					46			46	50.50
1500 Engine exhaust, garage, overhead components, for neoprene tubing									
1510 Alternate metal tubing & accessories see above									
1550 Adapters, for neoprene tubing end									
1560 3" tail pipe, adjustable, neoprene				Ea.	56			56	61.50
1570 3" tail pipe, heavy wall neoprene					61			61	67.50
1580 4" tail pipe, heavy wall neoprene					100			100	110
1590 5" tail pipe, heavy wall neoprene					105			105	116
1650 Connectors, tubing									
1660 3" interior, aluminum				Ea.	24.50			24.50	26.50
1670 4" interior, aluminum					41.50			41.50	45.50
1710 5" interior, neoprene					61			61	67.50
1750 3" spiralock, neoprene					24.50			24.50	26.50

For customer support on your Mechanical Cost Data, call 877.795.7394.

339

23 35 16.10 Engine Exhaust Removal Systems		Crew	Daily Output	Labor-Hours	Unit	Material	2015 Bare Costs Labor	Equipment	Total	Total Incl O&P
1760	4" spiralock, neoprene				Ea.	41.50			41.50	45.50
1780	Y for 3" ID tubing, neoprene, dual exhaust					195			195	215
1790	Y for 4" ID tubing, aluminum, dual exhaust				↓	195			195	215
1850	Elbows, aluminum, splice into tubing for strap									
1860	3" neoprene tubing size				Ea.	48.50			48.50	53.50
1870	4" neoprene tubing size					40.50			40.50	44.50
1900	Flange assemblies, connect tubing to overhead duct				↓	73			73	80
2000	Hardware and accessories									
2020	Cable, galvanized, 1/8" diameter				L.F.	.51			.51	.56
2040	Cleat, tie down cable or rope				Ea.	5.40			5.40	5.90
2060	Pulley					7.40			7.40	8.15
2080	Pulley hook, universal				↓	5.40			5.40	5.90
2100	Rope, nylon, 1/4" diameter				L.F.	.41			.41	.45
2120	Winch, 1" diameter				Ea.	114			114	126
2150	Lifting strap, mounts on neoprene									
2160	3" tubing size				Ea.	27			27	29.50
2170	4" tubing size					27			27	29.50
2180	5" tubing size					27			27	29.50
2190	6" tubing size				↓	27			27	29.50
2200	Tubing, neoprene, 11' lengths									
2210	3" ID				L.F.	10.90			10.90	11.95
2220	4" ID					18.15			18.15	20
2230	5" ID				↓	27			27	29.50
2500	Engine exhaust, thru-door outlet									
2510	3" tube size	1 Carp	16	.500	Ea.	59.50	23.50		83	102
2530	4" tube size	"	16	.500	"	74	23.50		97.50	117
3000	Tubing, exhaust, flex hose, with									
3010	coupler, damper and tail pipe adapter									
3020	Neoprene									
3040	3" x 20'	1 Shee	6	1.333	Ea.	330	74.50		404.50	475
3050	4" x 15'		5.40	1.481		440	83		523	610
3060	4" x 20'		5	1.600		530	89.50		619.50	720
3070	5" x 15'	↓	4.40	1.818	↓	600	102		702	815
3100	Galvanized									
3110	3" x 20'	1 Shee	6	1.333	Ea.	330	74.50		404.50	480
3120	4" x 17'		5.60	1.429	↓	360	80		440	520
3130	4" x 20'		5	1.600		405	89.50		494.50	580
3140	5" x 17'	↓	4.60	1.739	↓	430	97.50		527.50	625
8000	Blower, for tailpipe exhaust system									
8010	Direct drive									
8012	495 CFM, 1/3 HP	Q-9	5	3.200	Ea.	2,150	161		2,311	2,625
8014	1445 CFM, 3/4 HP		4	4		2,675	201		2,876	3,250
8016	1840 CFM, 1-1/2 HP	↓	3	5.333	↓	3,200	269		3,469	3,925
8030	Beltdrive									
8032	1400 CFM, 1 HP	Q-9	4	4	Ea.	2,675	201		2,876	3,250
8034	2023 CFM, 1-1/2 HP		3	5.333		3,700	269		3,969	4,475
8036	2750 CFM, 2 HP		2	8		4,375	405		4,780	5,450
8038	4400 CFM, 3 HP		1.80	8.889		6,450	450		6,900	7,775
8040	7060 CFM, 5 HP	↓	1.40	11.429	↓	9,150	575		9,725	11,000

23 35 Special Exhaust Systems

23 35 43 – Welding Fume Elimination Systems

23 35 43.10 Welding Fume Elimination System Components	Crew	Daily Output	Labor-Hours	Unit	Material	2015 Bare Costs Labor	2015 Bare Costs Equipment	Total	Total Incl O&P
0010 **WELDING FUME ELIMINATION SYSTEM COMPONENTS**									
7500 Welding fume elimination accessories for garage exhaust systems									
7600 Cut off (blast gate)									
7610 3" tubing size, 3" x 6" opening	1 Shee	24	.333	Ea.	26	18.65		44.65	57
7620 4" tubing size, 4" x 8" opening		24	.333		26	18.65		44.65	57
7630 5" tubing size, 5" x 10" opening		24	.333		31.50	18.65		50.15	63
7640 6" tubing size		24	.333		32.50	18.65		51.15	64
7650 8" tubing size		24	.333		45	18.65		63.65	78
7700 Hoods, magnetic, with handle & screen									
7710 3" tubing size, 3" x 6" opening	1 Shee	24	.333	Ea.	107	18.65		125.65	147
7720 4" tubing size, 4" x 8" opening		24	.333		97	18.65		115.65	136
7730 5" tubing size, 5" x 10" opening		24	.333		97	18.65		115.65	136

23 36 Air Terminal Units

23 36 13 – Constant-Air-Volume Units

23 36 13.10 Constant Volume Mixing Boxes

	Crew	Daily Output	Labor-Hours	Unit	Material	Labor	Equipment	Total	Total Incl O&P
0010 **CONSTANT VOLUME MIXING BOXES**									
5180 Mixing box, includes electric or pneumatic motor									
5192 Recommend use with silencer, see Line 23 33 19.10 0010									
5200 Constant volume, 150 to 270 CFM	Q-9	12	1.333	Ea.	700	67		767	875
5210 270 to 600 CFM		11	1.455		720	73		793	905
5230 550 to 1000 CFM		9	1.778		720	89.50		809.50	930
5240 1000 to 1600 CFM		8	2		740	101		841	970
5250 1300 to 1900 CFM		6	2.667		755	134		889	1,025
5260 550 to 2640 CFM		5.60	2.857		835	144		979	1,125
5270 650 to 3120 CFM		5.20	3.077		885	155		1,040	1,200

23 36 16 – Variable-Air-Volume Units

23 36 16.10 Variable Volume Mixing Boxes

	Crew	Daily Output	Labor-Hours	Unit	Material	Labor	Equipment	Total	Total Incl O&P
0010 **VARIABLE VOLUME MIXING BOXES**									
5180 Mixing box, includes electric or pneumatic motor									
5192 Recommend use with attenuator, see Line 23 33 19.10 0010									
5500 VAV Cool only, pneumatic, pressure independent 300 to 600 CFM	Q-9	11	1.455	Ea.	730	73		803	915
5510 500 to 1000 CFM		9	1.778		750	89.50		839.50	960
5520 800 to 1600 CFM		9	1.778		775	89.50		864.50	990
5530 1100 to 2000 CFM		8	2		795	101		896	1,025
5540 1500 to 3000 CFM		7	2.286		845	115		960	1,100
5550 2000 to 4000 CFM		6	2.667		855	134		989	1,150
5560 For electric, w/thermostat, pressure dependent, add					23			23	25.50
5600 VAV, HW coils, damper, actuator and t'stat									
5610 200 CFM	Q-9	11	1.455	Ea.	790	73		863	980
5620 400 CFM		10	1.600		800	80.50		880.50	1,000
5630 600 CFM		10	1.600		800	80.50		880.50	1,000
5640 800 CFM		8	2		820	101		921	1,050
5650 1000 CFM		8	2		820	101		921	1,050
5660 1250 CFM		6	2.667		900	134		1,034	1,200
5670 1500 CFM		6	2.667		900	134		1,034	1,200
5680 2000 CFM		4	4		975	201		1,176	1,375
5684 3000 CFM		3.80	4.211		1,075	212		1,287	1,500
5700 VAV Cool only, fan powered, damper, actuator, thermostat									
5710 200 CFM	Q-9	10	1.600	Ea.	1,075	80.50		1,155.50	1,325

For customer support on your Mechanical Cost Data, call 877.795.7394.

341

23 36 16 – Variable-Air-Volume Units

23 36 16.10 Variable Volume Mixing Boxes		Crew	Daily Output	Labor-Hours	Unit	Material	2015 Bare Costs Labor	Equipment	Total	Total Incl O&P
5720	400 CFM	Q-9	9	1.778	Ea.	1,150	89.50		1,239.50	1,400
5730	600 CFM		9	1.778		1,150	89.50		1,239.50	1,400
5740	800 CFM		7	2.286		1,225	115		1,340	1,525
5750	1000 CFM		7	2.286		1,225	115		1,340	1,525
5760	1250 CFM		5	3.200		1,325	161		1,486	1,725
5770	1500 CFM		5	3.200		1,325	161		1,486	1,725
5780	2000 CFM	▼	4	4	▼	1,475	201		1,676	1,925
5800	VAV Fan powr'd, with HW coils, dampers, actuators, t'stat									
5810	200 CFM	Q-9	10	1.600	Ea.	1,350	80.50		1,430.50	1,600
5820	400 CFM		9	1.778		1,375	89.50		1,464.50	1,650
5830	600 CFM		9	1.778		1,375	89.50		1,464.50	1,650
5840	800 CFM		7	2.286		1,475	115		1,590	1,775
5850	1000 CFM		7	2.286		1,475	115		1,590	1,775
5860	1250 CFM		5	3.200		1,675	161		1,836	2,075
5870	1500 CFM		5	3.200		1,675	161		* 1,836	2,075
5880	2000 CFM	▼	3.50	4.571	▼	1,800	230		2,030	2,325

23 37 13 – Diffusers, Registers, and Grilles

23 37 13.10 Diffusers

			Crew	Daily Output	Labor-Hours	Unit	Material	2015 Bare Costs Labor	Equipment	Total	Total Incl O&P
0010	**DIFFUSERS**, Aluminum, opposed blade damper unless noted										
0100	Ceiling, linear, also for sidewall										
0120	2" wide	R233100-30	1 Shee	32	.250	L.F.	16.20	14		30.20	39.50
0140	3" wide			30	.267		18.45	14.90		33.35	43.50
0160	4" wide	R233700-60		26	.308		21.50	17.20		38.70	50
0180	6" wide			24	.333		26.50	18.65		45.15	58
0200	8" wide			22	.364		31	20.50		51.50	65.50
0220	10" wide			20	.400		35.50	22.50		58	73
0240	12" wide		▼	18	.444	▼	40	25		65	82
0260	For floor or sill application, add						15%				
0500	Perforated, 24" x 24" lay-in panel size, 6" x 6"		1 Shee	16	.500	Ea.	151	28		179	209
0520	8" x 8"			15	.533		159	30		189	221
0530	9" x 9"			14	.571		161	32		193	226
0540	10" x 10"			14	.571		162	32		194	227
0560	12" x 12"			12	.667		168	37.50		205.50	242
0580	15" x 15"			11	.727		171	40.50		211.50	250
0590	16" x 16"			11	.727		189	40.50		229.50	269
0600	18" x 18"			10	.800		202	45		247	291
0610	20" x 20"			10	.800		218	45		263	310
0620	24" x 24"			9	.889		239	49.50		288.50	340
1000	Rectangular, 1 to 4 way blow, 6" x 6"			16	.500		50	28		78	98
1010	8" x 8"			15	.533		58	30		88	109
1014	9" x 9"			15	.533		66	30		96	118
1016	10" x 10"			15	.533		80	30		110	133
1020	12" x 6"			15	.533		71.50	30		101.50	124
1040	12" x 9"			14	.571		75.50	32		107.50	133
1060	12" x 12"			12	.667		84.50	37.50		122	150
1070	14" x 6"			13	.615		77.50	34.50		112	138
1074	14" x 14"			12	.667		128	37.50		165.50	197
1080	18" x 12"			11	.727		137	40.50		177.50	213
1120	15" x 15"		▼	10	.800	▼	110	45		155	190

23 37 13.10 Diffusers		Crew	Daily Output	Labor-Hours	Unit	Material	2015 Bare Costs Labor	Equipment	Total	Total Incl O&P
1140	18" x 15"	1 Shee	9	.889	Ea.	155	49.50		204.50	247
1150	18" x 18"		9	.889		138	49.50		187.50	227
1160	21" x 21"		8	1		215	56		271	320
1170	24" x 12"		10	.800		163	45		208	249
1180	24" x 24"		7	1.143		270	64		334	395
1500	Round, butterfly damper, steel, diffuser size, 6" diameter		18	.444		9	25		34	48
1520	8" diameter		16	.500		9.55	28		37.55	53.50
1540	10" diameter		14	.571		11.85	32		43.85	62
1560	12" diameter		12	.667		15.65	37.50		53.15	74
1580	14" diameter		10	.800		19.55	45		64.55	90
1600	18" diameter	▼	9	.889	▼	47.50	49.50		97	129
1700	Round, steel, adjustable core, annular segmented damper, diffuser size									
1704	6" diameter	1 Shee	18	.444	Ea.	104	25		129	152
1708	8" diameter		16	.500		121	28		149	176
1712	10" diameter		14	.571		142	32		174	205
1716	12" diameter		12	.667		171	37.50		208.50	245
1720	14" diameter		10	.800		157	45		202	241
1724	18" diameter		9	.889		218	49.50		267.50	315
1728	20" diameter		9	.889		251	49.50		300.50	350
1732	24" diameter		8	1		300	56		356	420
1736	30" diameter		7	1.143		525	64		589	680
1740	36" diameter		6	1.333		640	74.50		714.50	820
2000	T bar mounting, 24" x 24" lay-in frame, 6" x 6"		16	.500		61	28		89	110
2020	8" x 8"		14	.571		62	32		94	117
2040	12" x 12"		12	.667		74	37.50		111.50	138
2060	16" x 16"		11	.727		94.50	40.50		135	166
2080	18" x 18"	▼	10	.800	▼	106	45		151	186
2500	Combination supply and return									
2520	21" x 21" supply, 15" x 15" return	Q-9	10	1.600	Ea.	111	80.50		191.50	245
2540	24" x 24" supply, 18" x 18" return		9.50	1.684		140	85		225	283
2560	27" x 27" supply, 18" x 18" return		9	1.778		170	89.50		259.50	325
2580	30" x 30" supply, 21" x 21" return		8.50	1.882		215	95		310	380
2600	33" x 33" supply, 21" x 21" return		8	2		252	101		353	430
2620	36" x 36" supply, 24" x 24" return	▼	7.50	2.133	▼	296	107		403	490
3000	Baseboard, white enameled steel									
3100	18" long	1 Shee	20	.400	Ea.	4.43	22.50		26.93	39
3120	24" long		18	.444		8.65	25		33.65	47.50
3140	48" long	▼	16	.500		15.80	28		43.80	60.50
3400	For matching return, deduct				▼	10%				
4000	Floor, steel, adjustable pattern									
4100	2" x 10"	1 Shee	34	.235	Ea.	5.95	13.15		19.10	26.50
4120	2" x 12"		32	.250		6.55	14		20.55	28.50
4140	2" x 14"		30	.267		7.40	14.90		22.30	31
4200	4" x 10"		28	.286		6.90	16		22.90	32
4220	4" x 12"		26	.308		7.60	17.20		24.80	35
4240	4" x 14"		25	.320		8.60	17.90		26.50	37
4260	6" x 10"		26	.308		10.35	17.20		27.55	38
4280	6" x 12"		24	.333		10.80	18.65		29.45	40.50
4300	6" x 14"	▼	22	.364	▼	11.30	20.50		31.80	43.50
4500	Linear bar diffuser in wall, sill or ceiling									
4510	2" wide	1 Shee	32	.250	L.F.	43	14		57	68.50
4512	3" wide		30	.267		46.50	14.90		61.40	74
4514	4" wide	▼	26	.308		57	17.20		74.20	89.50

For customer support on your Mechanical Cost Data, call 877.795.7394.

343

23 37 13 – Diffusers, Registers, and Grilles

23 37 13.10 Diffusers

		Crew	Daily Output	Labor-Hours	Unit	Material	2015 Bare Costs Labor	Equipment	Total	Total Incl O&P
4516	6" wide	1 Shee	24	.333	L.F.	71	18.65		89.65	107
4520	10" wide		20	.400		94.50	22.50		117	138
4522	12" wide		18	.444		106	25		131	154
5000	Sidewall, aluminum, 3 way dispersion									
5100	8" x 4"	1 Shee	24	.333	Ea.	3.25	18.65		21.90	32
5110	8" x 6"		24	.333		3.48	18.65		22.13	32.50
5120	10" x 4"		22	.364		3.42	20.50		23.92	35
5130	10" x 6"		22	.364		3.66	20.50		24.16	35
5140	10" x 8"		20	.400		5	22.50		27.50	39.50
5160	12" x 4"		18	.444		3.69	25		28.69	42
5170	12" x 6"		17	.471		4.07	26.50		30.57	44.50
5180	12" x 8"		16	.500		5.30	28		33.30	49
5200	14" x 4"		15	.533		4.11	30		34.11	50
5220	14" x 6"		14	.571		4.53	32		36.53	54
5240	14" x 8"		13	.615		5.90	34.50		40.40	59
5260	16" x 6"		12.50	.640		6.45	36		42.45	61.50
6000	For steel diffusers instead of aluminum, deduct					10%				

23 37 13.30 Grilles

		Crew	Daily Output	Labor-Hours	Unit	Material	2015 Bare Costs Labor	Equipment	Total	Total Incl O&P
0010	**GRILLES**									
0020	Aluminum, unless noted otherwise									
0100	Air supply, single deflection, adjustable									
0120	8" x 4"	1 Shee	30	.267	Ea.	6.70	14.90		21.60	30.50
0140	8" x 8"		28	.286		8.15	16		24.15	33.50
0160	10" x 4"		24	.333		7.40	18.65		26.05	36.50
0180	10" x 10"		23	.348		9.40	19.45		28.85	40
0200	12" x 6"		23	.348		8.75	19.45		28.20	39
0220	12" x 12"		22	.364		11.50	20.50		32	43.50
0230	14" x 6"		23	.348		9.75	19.45		29.20	40.50
0240	14" x 8"		23	.348		10.45	19.45		29.90	41
0260	14" x 14"		22	.364		14.10	20.50		34.60	46.50
0270	18" x 8"		23	.348		12.05	19.45		31.50	43
0280	18" x 10"		23	.348		13.80	19.45		33.25	44.50
0300	18" x 18"		21	.381		21.50	21.50		43	56
0320	20" x 12"		22	.364		15.65	20.50		36.15	48.50
0340	20" x 20"		21	.381		26	21.50		47.50	61
0350	24" x 8"		20	.400		15.05	22.50		37.55	50.50
0360	24" x 14"		18	.444		21	25		46	61.50
0380	24" x 24"		15	.533		37	30		67	86
0400	30" x 8"		20	.400		19.35	22.50		41.85	55.50
0420	30" x 10"		19	.421		21.50	23.50		45	59.50
0440	30" x 12"		18	.444		23.50	25		48.50	64
0460	30" x 16"		17	.471		30.50	26.50		57	73.50
0480	30" x 18"		17	.471		34	26.50		60.50	77.50
0500	30" x 30"		14	.571		61	32		93	116
0520	36" x 12"		17	.471		29	26.50		55.50	72
0540	36" x 16"		16	.500		37	28		65	84
0560	36" x 18"		15	.533		41	30		71	91
0570	36" x 20"		15	.533		45	30		75	95
0580	36" x 24"		14	.571		58.50	32		90.50	114
0600	36" x 28"		13	.615		72	34.50		106.50	132
0620	36" x 30"		12	.667		79	37.50		116.50	144
0640	36" x 32"		12	.667		86.50	37.50		124	152

23 37 13 – Diffusers, Registers, and Grilles

23 37 13.30 Grilles		Crew	Daily Output	Labor-Hours	Unit	Material	2015 Bare Costs Labor	Equipment	Total	Total Incl O&P
0660	36" x 34"	1 Shee	11	.727	Ea.	94	40.50		134.50	166
0680	36" x 36"	▼	11	.727		102	40.50		142.50	174
0700	For double deflecting, add					70%				
1000	Air return, steel, 6" x 6"	1 Shee	26	.308		18.70	17.20		35.90	47
1020	10" x 6"		24	.333		18.70	18.65		37.35	49
1040	14" x 6"		23	.348		22.50	19.45		41.95	54.50
1060	10" x 8"		23	.348		22.50	19.45		41.95	54.50
1080	16" x 8"		22	.364		26.50	20.50		47	60
1100	12" x 12"		22	.364		26.50	20.50		47	60
1120	24" x 12"		18	.444		36	25		61	77.50
1140	30" x 12"		16	.500		46	28		74	94
1160	14" x 14"		22	.364		32	20.50		52.50	66
1180	16" x 16"		22	.364		34	20.50		54.50	68.50
1200	18" x 18"		21	.381		38	21.50		59.50	74.50
1220	24" x 18"		16	.500		43.50	28		71.50	91
1240	36" x 18"		15	.533		68	30		98	121
1260	24" x 24"		15	.533		53	30		83	104
1280	36" x 24"		14	.571		75	32		107	132
1300	48" x 24"		12	.667		128	37.50		165.50	197
1320	48" x 30"		11	.727		164	40.50		204.50	243
1340	36" x 36"		13	.615		100	34.50		134.50	163
1360	48" x 36"		11	.727		205	40.50		245.50	288
1380	48" x 48"		8	1		250	56		306	360
2000	Door grilles, 12" x 12"		22	.364		68	20.50		88.50	106
2020	18" x 12"		22	.364		78.50	20.50		99	118
2040	24" x 12"		18	.444		88.50	25		113.50	136
2060	18" x 18"		18	.444		109	25		134	158
2080	24" x 18"		16	.500		124	28		152	179
2100	24" x 24"		15	.533		175	30		205	239
3000	Filter grille with filter, 12" x 12"		24	.333		53	18.65		71.65	86.50
3020	18" x 12"		20	.400		71	22.50		93.50	112
3040	24" x 18"		18	.444		83.50	25		108.50	130
3060	24" x 24"		16	.500		98	28		126	151
3080	30" x 24"		14	.571		119	32		151	180
3100	30" x 30"		13	.615		141	34.50		175.50	208
3950	Eggcrate, framed, 6" x 6" opening		26	.308		15.40	17.20		32.60	43.50
3954	8" x 8" opening		24	.333		18.15	18.65		36.80	48.50
3960	10" x 10" opening		23	.348		23	19.45		42.45	55
3970	12" x 12" opening		22	.364		26	20.50		46.50	59.50
3980	14" x 14" opening		22	.364		33.50	20.50		54	68
3984	16" x 16" opening		21	.381		37.50	21.50		59	73.50
3990	18" x 18" opening		21	.381		45	21.50		66.50	82
4020	22" x 22" opening		17	.471		53	26.50		79.50	98
4040	24" x 24" opening		15	.533		68	30		98	121
4044	28" x 28" opening		14	.571		102	32		134	162
4048	36" x 36" opening		12	.667		117	37.50		154.50	186
4050	48" x 24" opening	▼	12	.667	▼	149	37.50		186.50	220
4060	Eggcrate, lay-in, T-bar system									
4070	48" x 24" sheet	1 Shee	40	.200	Ea.	71	11.20		82.20	95
5000	Transfer grille, vision proof, 8" x 4"		30	.267		12.35	14.90		27.25	36.50
5020	8" x 6"		28	.286		12.35	16		28.35	38
5040	8" x 8"		26	.308		13.85	17.20		31.05	41.50
5060	10" x 6"	▼	24	.333	▼	12.85	18.65		31.50	42.50

23 37 13 – Diffusers, Registers, and Grilles

23 37 13.30 Grilles

		Crew	Daily Output	Labor-Hours	Unit	Material	2015 Bare Costs Labor	Equipment	Total	Total Incl O&P
5080	10" x 10"	1 Shee	23	.348	Ea.	16.30	19.45		35.75	47.50
5090	12" x 6"		23	.348		13.85	19.45		33.30	44.50
5100	12" x 10"		22	.364		17.80	20.50		38.30	50.50
5110	12" x 12"		22	.364		19.25	20.50		39.75	52
5120	14" x 10"		22	.364		19.25	20.50		39.75	52
5140	16" x 8"		21	.381		19.25	21.50		40.75	53.50
5160	18" x 12"		20	.400		25	22.50		47.50	61.50
5170	18" x 18"		21	.381		35.50	21.50		57	71.50
5180	20" x 12"		20	.400		27	22.50		49.50	64
5200	20" x 20"		19	.421		43	23.50		66.50	83
5220	24" x 12"		17	.471		30.50	26.50		57	73.50
5240	24" x 24"		16	.500		58.50	28		86.50	108
5260	30" x 6"		15	.533		27	30		57	75.50
5280	30" x 8"		15	.533		31.50	30		61.50	80
5300	30" x 12"		14	.571		39.50	32		71.50	92.50
5320	30" x 16"		13	.615		47	34.50		81.50	104
5340	30" x 20"		12	.667		59	37.50		96.50	122
5360	30" x 24"		11	.727		73	40.50		113.50	143
5380	30" x 30"		10	.800		96	45		141	175
6000	For steel grilles instead of aluminum in above, deduct					10%				
6200	Plastic, eggcrate, lay-in, T-bar system									
6210	48" x 24" sheet	1 Shee	50	.160	Ea.	19.25	8.95		28.20	34.50
6250	Steel door louver									
6270	With fire link, steel only									
6350	12" x 18"	1 Shee	22	.364	Ea.	208	20.50		228.50	260
6410	12" x 24"		21	.381		232	21.50		253.50	288
6470	18" x 18"		20	.400		231	22.50		253.50	288
6530	18" x 24"		19	.421		266	23.50		289.50	330
6660	24" x 24"		15	.533		269	30		299	340

23 37 13.60 Registers

		Crew	Daily Output	Labor-Hours	Unit	Material	2015 Bare Costs Labor	Equipment	Total	Total Incl O&P
0010	**REGISTERS**									
0980	Air supply									
1000	Ceiling/wall, O.B. damper, anodized aluminum									
1010	One or two way deflection, adj. curved face bars									
1014	6" x 6"	1 Shee	24	.333	Ea.	11.55	18.65		30.20	41.50
1020	8" x 4"		26	.308		11.55	17.20		28.75	39.50
1040	8" x 8"		24	.333		14.45	18.65		33.10	44.50
1060	10" x 6"		20	.400		13.30	22.50		35.80	48.50
1080	10" x 10"		19	.421		17.90	23.50		41.40	55.50
1100	12" x 6"		19	.421		13.45	23.50		36.95	51
1120	12" x 12"		18	.444		21	25		46	61.50
1140	14" x 8"		17	.471		18.40	26.50		44.90	60.50
1160	14" x 14"		18	.444		24.50	25		49.50	65
1170	16" x 16"		17	.471		27.50	26.50		54	70
1180	18" x 8"		18	.444		20.50	25		45.50	60.50
1200	18" x 18"		17	.471		37.50	26.50		64	81.50
1220	20" x 4"		19	.421		18.90	23.50		42.40	57
1240	20" x 6"		18	.444		18.90	25		43.90	59
1260	20" x 8"		18	.444		20.50	25		45.50	61
1280	20" x 20"		17	.471		45.50	26.50		72	90
1290	22" x 22"		15	.533		52.50	30		82.50	103
1300	24" x 4"		17	.471		23.50	26.50		50	65.50

23 37 13 – Diffusers, Registers, and Grilles

23 37 13.60 Registers		Crew	Daily Output	Labor-Hours	Unit	Material	2015 Bare Costs Labor	Equipment	Total	Total Incl O&P
1320	24" x 6"	1 Shee	16	.500	Ea.	23.50	28		51.50	68.50
1340	24" x 8"		13	.615		25.50	34.50		60	80.50
1350	24" x 18"		12	.667		47.50	37.50		85	109
1360	24" x 24"		11	.727		64.50	40.50		105	133
1380	30" x 4"		16	.500		29	28		57	75
1400	30" x 6"		15	.533		29	30		59	77.50
1420	30" x 8"		14	.571		33.50	32		65.50	86
1440	30" x 24"		12	.667		83	37.50		120.50	149
1460	30" x 30"		10	.800		107	45		152	186
1504	4 way deflection, adjustable curved face bars									
1510	6" x 6"	1 Shee	26	.308	Ea.	13.90	17.20		31.10	42
1514	8" x 8"		24	.333		17.35	18.65		36	47.50
1518	10" x 10"		19	.421		21.50	23.50		45	59.50
1522	12" x 6"		19	.421		16.10	23.50		39.60	54
1526	12" x 12"		18	.444		25.50	25		50.50	66
1530	14" x 14"		18	.444		29.50	25		54.50	70.50
1534	16" x 16"		17	.471		33	26.50		59.50	76
1538	18" x 18"		17	.471		45	26.50		71.50	89.50
1542	22" x 22"		16	.500		63	28		91	113
1980	One way deflection, adj. vert. or horiz. face bars									
1990	6" x 6"	1 Shee	26	.308	Ea.	12.65	17.20		29.85	40.50
2000	8" x 4"		26	.308		11.85	17.20		29.05	39.50
2020	8" x 8"		24	.333		14.10	18.65		32.75	44
2040	10" x 6"		20	.400		14.20	22.50		36.70	49.50
2060	10" x 10"		19	.421		16.60	23.50		40.10	54.50
2080	12" x 6"		19	.421		15.35	23.50		38.85	53
2100	12" x 12"		18	.444		20	25		45	60
2120	14" x 8"		17	.471		18.35	26.50		44.85	60
2140	14" x 12"		18	.444		22	25		47	62
2160	14" x 14"		18	.444		25	25		50	65
2180	16" x 6"		18	.444		18.35	25		43.35	58
2200	16" x 12"		18	.444		24	25		49	64.50
2220	16" x 16"		17	.471		32	26.50		58.50	75
2240	18" x 8"		18	.444		21.50	25		46.50	61.50
2260	18" x 12"		17	.471		26	26.50		52.50	68.50
2280	18" x 18"		16	.500		37.50	28		65.50	84.50
2300	20" x 10"		18	.444		25.50	25		50.50	66
2320	20" x 16"		16	.500		35	28		63	81.50
2340	20" x 20"		16	.500		46	28		74	93.50
2360	24" x 12"		15	.533		32	30		62	80.50
2380	24" x 16"		14	.571		42.50	32		74.50	96
2400	24" x 20"		12	.667		53	37.50		90.50	116
2420	24" x 24"		10	.800		65	45		110	140
2440	30" x 12"		13	.615		41.50	34.50		76	98
2460	30" x 16"		13	.615		53.50	34.50		88	112
2480	30" x 24"		11	.727		84	40.50		124.50	154
2500	30" x 30"		9	.889		107	49.50		156.50	194
2520	36" x 12"		11	.727		51.50	40.50		92	119
2540	36" x 24"		10	.800		103	45		148	183
2560	36" x 36"		8	1		173	56		229	276
2600	For 2 way deflect., adj. vert. or horiz. face bars, add					40%				
2700	Above registers in steel instead of aluminum, deduct					10%				
3000	Baseboard, hand adj. damper, enameled steel									

For customer support on your Mechanical Cost Data, call 877.795.7394.

347

23 37 13 - Diffusers, Registers, and Grilles

23 37 13.60 Registers		Crew	Daily Output	Labor-Hours	Unit	Material	2015 Bare Costs Labor	Equipment	Total	Total Incl O&P
3012	8" x 6"	1 Shee	26	.308	Ea.	4.85	17.20		22.05	32
3020	10" x 6"		24	.333		5.30	18.65		23.95	34.50
3040	12" x 5"		23	.348		5.60	19.45		25.05	35.50
3060	12" x 6"		23	.348		5.75	19.45		25.20	36
3080	12" x 8"		22	.364		8.30	20.50		28.80	40
3100	14" x 6"	▼	20	.400	▼	6.20	22.50		28.70	41
4000	Floor, toe operated damper, enameled steel									
4020	4" x 8"	1 Shee	32	.250	Ea.	9	14		23	31.50
4040	4" x 12"		26	.308		10.55	17.20		27.75	38
4060	6" x 8"		28	.286		9.65	16		25.65	35
4080	6" x 14"		22	.364		12.20	20.50		32.70	44.50
4100	8" x 10"		22	.364		11	20.50		31.50	43
4120	8" x 16"		20	.400		14.75	22.50		37.25	50
4140	10" x 10"		20	.400		13.15	22.50		35.65	48.50
4160	10" x 16"		18	.444		19.90	25		44.90	60
4180	12" x 12"		18	.444		16.15	25		41.15	56
4200	12" x 24"		16	.500		27	28		55	72.50
4220	14" x 14"		16	.500		42	28		70	89.50
4240	14" x 20"	▼	15	.533	▼	49.50	30		79.50	99.50
4300	Spiral pipe supply register									
4310	Steel, with air scoop									
4320	4" x 12", for 8" thru 13" diameter duct	1 Shee	25	.320	Ea.	74.50	17.90		92.40	109
4330	4" x 18", for 8" thru 13" diameter duct		18	.444		87	25		112	134
4340	6" x 12", for 14" thru 21" diameter duct		19	.421		78.50	23.50		102	123
4350	6" x 16", for 14" thru 21" diameter duct		18	.444		89	25		114	136
4360	6" x 20", for 14" thru 21" diameter duct		17	.471		98.50	26.50		125	148
4370	6" x 24", for 14" thru 21" diameter duct		16	.500		116	28		144	170
4380	8" x 16", for 22" thru 31" diameter duct		19	.421		94	23.50		117.50	139
4390	8" x 18", for 22" thru 31" diameter duct		18	.444		98.50	25		123.50	146
4400	8" x 24", for 22" thru 31" diameter duct	▼	15	.533	▼	122	30		152	180
4980	Air return									
5000	Ceiling or wall, fixed 45° face blades									
5010	Adjustable O.B. damper, anodized aluminum									
5020	4" x 8"	1 Shee	26	.308	Ea.	11.85	17.20		29.05	39.50
5040	6" x 8"		24	.333		13	18.65		31.65	43
5060	6" x 10"		19	.421		14.20	23.50		37.70	51.50
5070	6" x 12"		19	.421		15.35	23.50		38.85	53
5080	6" x 16"		18	.444		18.35	25		43.35	58
5100	8" x 10"		19	.421		15.45	23.50		38.95	53
5120	8" x 12"		16	.500		17.25	28		45.25	62
5140	10" x 10"		18	.444		16.60	25		41.60	56.50
5160	10" x 16"		17	.471		22	26.50		48.50	64
5170	12" x 12"		18	.444		20	25		45	60
5180	12" x 18"		18	.444		26	25		51	66.50
5184	12" x 24"		14	.571		32	32		64	84
5200	12" x 30"		12	.667		41.50	37.50		79	103
5210	12" x 36"		10	.800		51.50	45		96.50	125
5220	16" x 16"		17	.471		32	26.50		58.50	75
5240	18" x 18"		16	.500		37.50	28		65.50	84.50
5250	18" x 24"		13	.615		48	34.50		82.50	105
5254	18" x 30"		11	.727		60	40.50		100.50	129
5260	18" x 36"		10	.800		72.50	45		117.50	149
5270	20" x 20"	▼	10	.800		46	45		91	119

23 37 13 – Diffusers, Registers, and Grilles

23 37 13.60 Registers		Crew	Daily Output	Labor-Hours	Unit	Material	2015 Bare Costs Labor	Equipment	Total	Total Incl O&P
5274	20" x 24"	1 Shee	12	.667	Ea.	53	37.50		90.50	116
5280	24" x 24"		11	.727		65	40.50		105.50	134
5290	24" x 30"		10	.800		84	45		129	161
5300	24" x 36"		8	1		110	56		166	207
5320	24" x 48"		6	1.333		139	74.50		213.50	267
5340	30" x 30"	↓	6.50	1.231		107	69		176	223
5344	30" x 120"	Q-9	6	2.667*		965	134		1,099	1,250
5360	36" x 36"	1 Shee	6	1.333		173	74.50		247.50	305
5370	40" x 32"	"	6	1.333	↓	160	74.50		234.50	290
5400	Ceiling or wall, removable-reversible core, single deflection									
5402	Adjustable O.B. damper, radiused frame with border, aluminum									
5410	8" x 4"	1 Shee	26	.308	Ea.	36	17.20		53.20	66.50
5414	8" x 6"		24	.333		36	18.65		54.65	68.50
5418	10" x 6"		19	.421		46	23.50		69.50	86.50
5422	10" x 10"		18	.444		52.50	25		77.50	96
5426	12" x 6"		19	.421		46	23.50		69.50	86.50
5430	12" x 12"		18	.444		59	25		84	103
5434	16" x 16"		16	.500		89	28		117	141
5438	18" x 18"		17	.471		112	26.50		138.50	163
5442	20" x 20"		16	.500		135	28		163	191
5446	24" x 12"		15	.533		89	30		119	143
5450	24" x 18"		13	.615		135	34.50		169.50	201
5454	24" x 24"		11	.727		187	40.50		227.50	268
5458	30" x 12"		13	.615		118	34.50		152.50	183
5462	30" x 18"		12.50	.640		171	36		207	243
5466	30" x 24"		11	.727		240	40.50		280.50	325
5470	30" x 30"		9	.889		299	49.50		348.50	405
5474	36" x 12"		11	.727		148	40.50		188.50	225
5478	36" x 18"		10.50	.762		207	42.50		249.50	293
5482	36" x 24"		10	.800		289	45		334	390
5486	36" x 30"		9	.889		390	49.50		439.50	500
5490	36" x 36"	↓	8	1	↓	500	56		556	635
6000	For steel construction instead of aluminum, deduct					10%				

23 37 15 – Louvers

23 37 15.40 HVAC Louvers

23 37 15.40 HVAC Louvers		Crew	Daily Output	Labor-Hours	Unit	Material	2015 Bare Costs Labor	Equipment	Total	Total Incl O&P
0010	**HVAC LOUVERS**									
0100	Aluminum, extruded, with screen, mill finish									
1002	Brick vent, see also Section 04 05 23.19									
1100	Standard, 4" deep, 8" wide, 5" high	1 Shee	24	.333	Ea.	34.50	18.65		53.15	66
1200	Modular, 4" deep, 7-3/4" wide, 5" high		24	.333		36	18.65		54.65	68
1300	Speed brick, 4" deep, 11-5/8" wide, 3-7/8" high		24	.333		36	18.65		54.65	68
1400	Fuel oil brick, 4" deep, 8" wide, 5" high		24	.333	↓	61.50	18.65		80.15	96.50
2000	Cooling tower and mechanical equip., screens, light weight		40	.200	S.F.	15.40	11.20		26.60	34
2020	Standard weight		35	.229		41	12.80		53.80	64.50
2500	Dual combination, automatic, intake or exhaust		20	.400		56	22.50		78.50	95.50
2520	Manual operation		20	.400		41.50	22.50		64	80
2540	Electric or pneumatic operation		20	.400	↓	41.50	22.50		64	80
2560	Motor, for electric or pneumatic	↓	14	.571	Ea.	480	32		512	580
3000	Fixed blade, continuous line									
3100	Mullion type, stormproof	1 Shee	28	.286	S.F.	41.50	16		57.50	70.50
3200	Stormproof		28	.286		41.50	16		57.50	70.50
3300	Vertical line	↓	28	.286	↓	49.50	16		65.50	78.50

23 37 15 – Louvers

23 37 15.40 HVAC Louvers		Crew	Daily Output	Labor- Hours	Unit	Material	2015 Bare Costs Labor	Equipment	Total	Total Incl O&P
3500	For damper to use with above, add					50%	30%			
3520	Motor, for damper, electric or pneumatic	1 Shee	14	.571	Ea.	480	32		512	580
4000	Operating, 45°, manual, electric or pneumatic		24	.333	S.F.	50	18.65		68.65	83.50
4100	Motor, for electric or pneumatic		14	.571	Ea.	480	32		512	580
4200	Penthouse, roof		56	.143	S.F.	24.50	8		32.50	39
4300	Walls		40	.200		58	11.20		69.20	81
5000	Thinline, under 4" thick, fixed blade		40	.200		24	11.20		35.20	43.50
5010	Finishes, applied by mfr. at additional cost, available in colors									
5020	Prime coat only, add				S.F.	3.30			3.30	3.63
5040	Baked enamel finish coating, add					6.10			6.10	6.70
5060	Anodized finish, add					6.60			6.60	7.25
5080	Duranodic finish, add					12			12	13.20
5100	Fluoropolymer finish coating, add					18.90			18.90	21
9000	Stainless steel, fixed blade, continuous line									
9010	Hospital grade									
9110	20 Ga.	1 Shee	22	.364	S.F.	40	20.50		60.50	75
9980	For small orders (under 10 pieces), add				"	25%				

23 37 23 – HVAC Gravity Ventilators

23 37 23.10 HVAC Gravity Air Ventilators

		Crew	Daily Output	Labor- Hours	Unit	Material	2015 Bare Costs Labor	Equipment	Total	Total Incl O&P
0010	**HVAC GRAVITY AIR VENTILATORS**, Includes base									
1280	Rotary ventilators, wind driven, galvanized									
1300	4" neck diameter	Q-9	20	.800	Ea.	64.50	40.50		105	133
1320	5" neck diameter		18	.889		64.50	45		109.50	140
1340	6" neck diameter		16	1		64.50	50.50		115	148
1360	8" neck diameter		14	1.143		74	57.50		131.50	169
1380	10" neck diameter		12	1.333		83.50	67		150.50	195
1400	12" neck diameter		10	1.600		91	80.50		171.50	223
1420	14" neck diameter		10	1.600		153	80.50		233.50	291
1440	16" neck diameter		9	1.778		179	89.50		268.50	335
1460	18" neck diameter		9	1.778		215	89.50		304.50	375
1480	20" neck diameter		8	2		256	101		357	435
1500	24" neck diameter		8	2		390	101		491	580
1520	30" neck diameter		7	2.286		445	115		560	665
1540	36" neck diameter		6	2.667		655	134		789	925
1600	For aluminum, add					300%				
2000	Stationary, gravity, syphon, galvanized									
2100	3" neck diameter, 40 CFM	Q-9	24	.667	Ea.	38	33.50		71.50	93
2120	4" neck diameter, 50 CFM		20	.800		40.50	40.50		81	106
2140	5" neck diameter, 58 CFM		18	.889		43.50	45		88.50	117
2160	6" neck diameter, 66 CFM		16	1		45	50.50		95.50	127
2180	7" neck diameter, 86 CFM		15	1.067		59.50	53.50		113	148
2200	8" neck diameter, 110 CFM		14	1.143		61	57.50		118.50	155
2220	10" neck diameter, 140 CFM		12	1.333		85.50	67		152.50	197
2240	12" neck diameter, 160 CFM		10	1.600		106	80.50		186.50	240
2260	14" neck diameter, 250 CFM		10	1.600		168	80.50		248.50	305
2280	16" neck diameter, 380 CFM		9	1.778		216	89.50		305.50	375
2300	18" neck diameter, 500 CFM		9	1.778		244	89.50		333.50	405
2320	20" neck diameter, 625 CFM		8	2		290	101		391	475
2340	24" neck diameter, 900 CFM		8	2		360	101		461	550
2360	30" neck diameter, 1375 CFM		7	2.286		400	115		515	615
2380	36" neck diameter, 2,000 CFM		6	2.667		440	134		574	690
2400	42" neck diameter, 3000 CFM		4	4		660	201		861	1,025

23 37 23 – HVAC Gravity Ventilators

23 37 23.10 HVAC Gravity Air Ventilators		Crew	Daily Output	Labor-Hours	Unit	Material	2015 Bare Costs		Total	Total Incl O&P
							Labor	Equipment		
2410	48" neck diameter, 4000 CFM	Q-9	3	5.333	Ea.	750	269		1,019	1,225
2500	For aluminum, add					300%				
2520	For stainless steel, add					600%				
2530	For copper, add					600%				
3000	Rotating chimney cap, galvanized, 4" neck diameter	Q-9	20	.800		24.50	40.50		65	88.50
3020	5" neck diameter		18	.889		24.50	45		69.50	95.50
3040	6" neck diameter		16	1		24.50	50.50		75	104
3060	7" neck diameter		15	1.067		31.50	53.50		85	117
3080	8" neck diameter		14	1.143		31.50	57.50		89	123
3100	10" neck diameter		12	1.333		38	67		105	145
3600	Stationary chimney rain cap, galvanized, 3" neck diameter		24	.667		4.32	33.50		37.82	56.50
3620	4" neck diameter		20	.800		4.72	40.50		45.22	66.50
3640	6" neck diameter		16	1		5.60	50.50		56.10	83
3680	8" neck diameter		14	1.143		7.05	57.50		64.55	96
3700	10" neck diameter		12	1.333		8.50	67		75.50	112
3720	12" neck diameter		10	1.600		12.55	80.50		93.05	137
3740	14" neck diameter		10	1.600		19.90	80.50		100.40	145
3760	16" neck diameter		9	1.778		21	89.50		110.50	160
3770	18" neck diameter		9	1.778		33	89.50		122.50	174
3780	20" neck diameter		8	2		44.50	101		145.50	203
3790	24" neck diameter		8	2		56.50	101		157.50	216
4200	Stationary mushroom, aluminum, 16" orifice diameter		10	1.600		620	80.50		700.50	810
4220	26" orifice diameter		6.15	2.602		915	131		1,046	1,200
4230	30" orifice diameter		5.71	2.802		1,350	141		1,491	1,700
4240	38" orifice diameter		5	3.200		1,925	161		2,086	2,350
4250	42" orifice diameter		4.70	3.404		2,550	171		2,721	3,050
4260	50" orifice diameter		4.44	3.604		3,025	181		3,206	3,600
5000	Relief vent									
5500	Rectangular, aluminum, galvanized curb									
5510	intake/exhaust, 0.033" SP									
5580	500 CFM, 12" x 12"	Q-9	8.60	1.860	Ea.	700	93.50		793.50	915
5600	600 CFM, 12" x 16"		8	2		785	101		886	1,025
5620	750 CFM, 12" x 20"		7.20	2.222		840	112		952	1,100
5640	1000 CFM, 12" x 24"		6.60	2.424		880	122		1,002	1,150
5660	1500 CFM, 12" x 36"		5.80	2.759		1,175	139		1,314	1,500
5680	3000 CFM, 20" x 42"		4	4		1,550	201		1,751	2,000
5700	5000 CFM, 20" x 72"		3	5.333		2,225	269		2,494	2,850
5720	6000 CFM, 24" x 72"		2.30	6.957		2,400	350		2,750	3,175
5740	10,000 CFM, 48" x 60"		1.80	8.889		3,050	450		3,500	4,025
5760	12,000 CFM, 48" x 72"		1.60	10		3,600	505		4,105	4,725
5780	13,750 CFM, 60" x 66"		1.40	11.429		3,975	575		4,550	5,250
5800	15,000 CFM, 60" x 72"		1.30	12.308		4,250	620		4,870	5,625
5820	18,000 CFM, 72" x 72"		1.20	13.333		4,825	670		5,495	6,325
5880	Size is throat area, volume is at 500 fpm									
7000	Note: sizes based on exhaust. Intake, with 0.125" SP									
7100	loss, approximately twice listed capacity.									

For customer support on your Mechanical Cost Data, call 877.795.7394.

351

23 38 13.10 Hood and Ventilation Equipment	Crew	Daily Output	Labor-Hours	Unit	Material	2015 Bare Costs Labor	Equipment	Total	Total Incl O&P
0010 **HOOD AND VENTILATION EQUIPMENT**									
2970 Exhaust hood, sst, gutter on all sides, 4' x 4' x 2'	1 Carp	1.80	4.444	Ea.	4,725	209		4,934	5,525
2980 4' x 4' x 7'	"	1.60	5		7,525	235		7,760	8,625
7800 Vent hood, wall canopy with fire protection, 30"	L-3A	9	1.333		425	68.50		493.50	570
7810 Without fire protection, 36"		10	1.200		460	61.50		521.50	600
7820 Island canopy with fire protection, 30"		7	1.714		790	88		878	1,000
7830 Without fire protection, 36"		8	1.500		800	77		877	1,000
7840 Back shelf with fire protection, 30"		11	1.091		400	56		456	525
7850 Without fire protection, black, 36"		12	1		400	51.50		451.50	520
7852 Without fire protection, stainless steel, 36"		12	1		500	51.50		551.50	630
7860 Range hood & CO_2 system, 30"	1 Carp	2.50	3.200		3,975	150		4,125	4,600
7950 Hood fire protection system, electric stove	Q-1	3	5.333		1,875	282		2,157	2,500
7952 Hood fire protection system, gas stove	"	3	5.333		2,200	282		2,482	2,850

23 38 13.20 Kitchen Ventilation

	Crew	Daily Output	Labor-Hours	Unit	Material	2015 Bare Costs Labor	Equipment	Total	Total Incl O&P
0010 **KITCHEN VENTILATION**, Commercial									
6000 Exhaust hoods									
6100 Centrifugal grease extraction									
6110 Water wash type									
6120 Per foot of length	Q-10	7.10	3.380	L.F.	1,775	177		1,952	2,225
6200 Non water wash type									
6210 Per foot of length	Q-10	13.40	1.791	L.F.	1,300	93.50		1,393.50	1,575
6400 Island style ventilator									
6410 Water wash type									
6420 Per foot of length	Q-10	7.10	3.380	L.F.	2,375	177		2,552	2,900
6460 Non water wash type									
6470 Per foot of length	Q-10	13.40	1.791	L.F.	1,025	93.50		1,118.50	1,275

23 41 Particulate Air Filtration
23 41 13 – Panel Air Filters

23 41 13.10 Panel Type Air Filters

	Crew	Daily Output	Labor-Hours	Unit	Material	2015 Bare Costs Labor	Equipment	Total	Total Incl O&P
0010 **PANEL TYPE AIR FILTERS**									
2950 Mechanical media filtration units									
3000 High efficiency type, with frame, non-supported [G]				MCFM	35			35	38.50
3100 Supported type [G]				"	45			45	49.50
5500 Throwaway glass or paper media type				Ea.	3.42			3.42	3.76

23 41 16 – Renewable-Media Air Filters

23 41 16.10 Disposable Media Air Filters

	Crew	Daily Output	Labor-Hours	Unit	Material	2015 Bare Costs Labor	Equipment	Total	Total Incl O&P
0010 **DISPOSABLE MEDIA AIR FILTERS**									
5000 Renewable disposable roll				C.S.F.	1.54			1.54	1.70
5800 Filter, bag type									
5810 90-95% efficiency									
5820 24" x 12" x 29", .75-1.25 MCFM	1 Shee	1.90	4.211	Ea.	56.50	236		292.50	420
5830 24" x 24" x 29", 1.5-2.5 MCFM	"	1.90	4.211	"	99	236		335	470
5850 80-85% efficiency									
5860 24" x 12" x 29", .75-1.25 MCFM	1 Shee	1.90	4.211	Ea.	50.50	236		286.50	415
5870 24" x 24" x 29", 1.5-2.5 MCFM	"	1.90	4.211	"	92	236		328	460

23 41 19 – Washable Air Filters

23 41 19.10 Permanent Air Filters	Crew	Daily Output	Labor-Hours	Unit	Material	2015 Bare Costs Labor	Equipment	Total	Total Incl O&P
0010 **PERMANENT AIR FILTERS**									
4500 Permanent washable	G			MCFM	20			20	22

23 41 23 – Extended Surface Filters

23 41 23.10 Expanded Surface Filters

	Crew	Daily Output	Labor-Hours	Unit	Material	Labor	Equipment	Total	Total Incl O&P
0010 **EXPANDED SURFACE FILTERS**									
4000 Medium efficiency, extended surface	G			MCFM	5.50			5.50	6.05

23 41 33 – High-Efficiency Particulate Filtration

23 41 33.10 HEPA Filters

	Crew	Daily Output	Labor-Hours	Unit	Material	Labor	Equipment	Total	Total Incl O&P
0010 **HEPA FILTERS**									
6000 HEPA filter complete w/particle board,									
6010 kraft paper frame, separator material									
6020 95% DOP efficiency									
6030 12" x 12" x 6", 150 CFM	1 Shee	3.70	2.162	Ea.	44	121		165	234
6034 24" x 12" x 6", 375 CFM		1.90	4.211		73.50	236		309.50	440
6038 24" x 18" x 6", 450 CFM		1.90	4.211		90	236		326	460
6042 24" x 24" x 6", 700 CFM		1.90	4.211		101	236		337	470
6046 12" x 12" x 12", 250 CFM		3.70	2.162		74.50	121		195.50	267
6050 24" x 12" x 12", 500 CFM		1.90	4.211		105	236		341	475
6054 24" x 18" x 12", 875 CFM		1.90	4.211		146	236		382	520
6058 24" x 24" x 12", 1000 CFM		1.90	4.211		158	236		394	535
6100 99% DOP efficiency									
6110 12" x 12" x 6", 150 CFM	1 Shee	3.70	2.162	Ea.	46	121		167	236
6114 24" x 12" x 6", 325 CFM		1.90	4.211		76	236		312	445
6118 24" x 18" x 6", 550 CFM		1.90	4.211		93	236		329	460
6122 24" x 24" x 6", 775 CFM		1.90	4.211		105	236		341	475
6124 24" x 48" x 6", 775 CFM		1.30	6.154		201	345		546	745
6126 12" x 12" x 12", 250 CFM		3.60	2.222		77	124		201	275
6130 24" x 12" x 12", 500 CFM		1.90	4.211		109	236		345	480
6134 24" x 18" x 12", 775 CFM		1.90	4.211		152	236		388	525
6138 24" x 24" x 12", 1100 CFM		1.90	4.211		164	236		400	540
6500 HEPA filter housing, 14 ga. galv. sheet metal									
6510 12" x 12" x 6"	1 Shee	2.50	3.200	Ea.	660	179		839	1,000
6514 24" x 12" x 6"		2	4		740	224		964	1,150
6518 12" x 12" x 12"		2.40	3.333		660	187		847	1,000
6522 14" x 12" x 12"		2.30	3.478		660	195		855	1,025
6526 24" x 18" x 6"		1.90	4.211		940	236		1,176	1,375
6530 24" x 24" x 6"		1.80	4.444		855	249		1,104	1,325
6534 24" x 48" x 6"		1.70	4.706		1,150	263		1,413	1,650
6538 24" x 72" x 6"		1.60	5		1,475	280		1,755	2,050
6542 24" x 18" x 12"		1.80	4.444		940	249		1,189	1,400
6546 24" x 24" x 12"		1.70	4.706		1,050	263		1,313	1,550
6550 24" x 48" x 12"		1.60	5		1,225	280		1,505	1,775
6554 24" x 72" x 12"		1.50	5.333		1,525	298		1,823	2,150
6558 48" x 48" x 6"	Q-9	2.80	5.714		1,700	288		1,988	2,300
6562 48" x 72" x 6"		2.60	6.154		2,025	310		2,335	2,700
6566 48" x 96" x 6"		2.40	6.667		2,575	335		2,910	3,350
6570 48" x 48" x 12"		2.70	5.926		1,700	298		1,998	2,300
6574 48" x 72" x 12"		2.50	6.400		2,025	320		2,345	2,725
6578 48" x 96" x 12"		2.30	6.957		2,575	350		2,925	3,350
6582 114" x 72" x 12"		2	8		3,550	405		3,955	4,525

23 42 Gas-Phase Air Filtration

23 42 13 – Activated-Carbon Air Filtration

23 42 13.10 Charcoal Type Air Filtration	Crew	Daily Output	Labor-Hours	Unit	Material	2015 Bare Costs Labor	Equipment	Total	Total Incl O&P
0010 **CHARCOAL TYPE AIR FILTRATION**									
0050 Activated charcoal type, full flow				MCFM	600			600	660
0060 Full flow, impregnated media 12" deep					225			225	248
0070 HEPA filter & frame for field erection					350			350	385
0080 HEPA filter-diffuser, ceiling install.				↓	300			300	330

23 43 Electronic Air Cleaners

23 43 13 – Washable Electronic Air Cleaners

23 43 13.10 Electronic Air Cleaners

	Crew	Daily Output	Labor-Hours	Unit	Material	2015 Bare Costs Labor	Equipment	Total	Total Incl O&P
0010 **ELECTRONIC AIR CLEANERS**									
2000 Electronic air cleaner, duct mounted									
2150 1000 CFM	1 Shee	4	2	Ea.	420	112		532	635
2200 1200 CFM		3.80	2.105		505	118		623	735
2250 1400 CFM		3.60	2.222		520	124		644	765
2260 2000 CFM	↓	3.20	2.500	↓	585	140		725	860

23 51 Breechings, Chimneys, and Stacks

23 51 13 – Draft Control Devices

23 51 13.13 Draft-Induction Fans

	Crew	Daily Output	Labor-Hours	Unit	Material	2015 Bare Costs Labor	Equipment	Total	Total Incl O&P
0010 **DRAFT-INDUCTION FANS**									
1000 Breeching installation									
1800 Hot gas, 600°F, variable pitch pulley and motor									
1840 6" diam. inlet, 1/4 H.P., 1 phase, 400 CFM	Q-9	6	2.667	Ea.	1,500	134		1,634	1,850
1860 8" diam. inlet, 1/4 H.P., 1phase, 1120 CFM		4	4		2,150	201		2,351	2,650
1870 9" diam. inlet, 3/4 H.P., 1 phase, 1440 CFM		3.60	4.444		2,650	224		2,874	3,250
1880 10" diam. inlet, 3/4 H.P., 1 phase, 2000 CFM		3.30	4.848		2,650	244		2,894	3,275
1900 12" diam. inlet, 3/4 H.P., 3 phase, 2960 CFM		3	5.333		2,925	269		3,194	3,600
1910 14" diam. inlet, 1 H.P., 3 phase, 4160 CFM		2.60	6.154		2,950	310		3,260	3,700
1920 16" diam. inlet, 2 H.P., 3 phase, 5500 CFM		2.30	6.957		3,250	350		3,600	4,100
1950 20" diam. inlet, 3 H.P., 3 phase, 9760 CFM		1.50	10.667		4,675	535		5,210	5,975
1960 22" diam. inlet, 5 H.P., 3 phase, 13,360 CFM		1	16		7,625	805		8,430	9,600
1980 24" diam. inlet, 7-1/2 H.P., 3 phase, 17,760 CFM	↓	.80	20		8,375	1,000		9,375	10,800
2300 For multi-blade damper at fan inlet, add				↓	20%				
3600 Chimney-top installation									
3700 6" size	1 Shee	8	1	Ea.	1,375	56		1,431	1,600
3740 8" size		7	1.143		1,400	64		1,464	1,625
3750 10" size		6.50	1.231		1,825	69		1,894	2,100
3780 13" size	↓	6	1.333		1,875	74.50		1,949.50	2,175
3880 For speed control switch, add					106			106	116
3920 For thermal fan control, add				↓	106			106	116
5500 Flue blade style damper for draft control,									
5510 locking quadrant blade									
5550 8" size	Q-9	8	2	Ea.	405	101		506	600
5560 9" size		7.50	2.133		405	107		512	610
5570 10" size		7	2.286		415	115		530	630
5580 12" size		6.50	2.462		420	124		544	655
5590 14" size		6	2.667		430	134		564	680
5600 16" size		5.50	2.909		445	146		591	715
5610 18" size	↓	5	3.200		465	161		626	755

23 51 Breechings, Chimneys, and Stacks

23 51 13 – Draft Control Devices

23 51 13.13 Draft-Induction Fans

		Crew	Daily Output	Labor-Hours	Unit	Material	2015 Bare Costs Labor	Equipment	Total	Total Incl O&P
5620	20" size	Q-9	4.50	3.556	Ea.	490	179		669	815
5630	22" size		4	4		515	201		716	875
5640	24" size		3.50	4.571		545	230		775	945
5650	27" size		3	5.333		570	269		839	1,050
5660	30" size		2.50	6.400		625	320		945	1,175
5670	32" size		2	8		655	405		1,060	1,325
5680	36" size		1.50	10.667		725	535		1,260	1,625

23 51 13.16 Vent Dampers

		Crew	Daily Output	Labor-Hours	Unit	Material	2015 Bare Costs Labor	Equipment	Total	Total Incl O&P
0010	**VENT DAMPERS**									
5000	Vent damper, bi-metal, gas, 3" diameter	Q-9	24	.667	Ea.	44	33.50		77.50	100
5010	4" diameter		24	.667		44	33.50		77.50	100
5020	5" diameter		23	.696		44	35		79	102
5030	6" diameter		22	.727		44	36.50		80.50	105
5040	7" diameter		21	.762		45.50	38.50		84	109
5050	8" diameter		20	.800		49	40.50		89.50	116

23 51 13.19 Barometric Dampers

		Crew	Daily Output	Labor-Hours	Unit	Material	2015 Bare Costs Labor	Equipment	Total	Total Incl O&P
0010	**BAROMETRIC DAMPERS**									
1000	Barometric, gas fired system only, 6" size for 5" and 6" pipes	1 Shee	20	.400	Ea.	105	22.50		127.50	149
1020	7" size, for 6" and 7" pipes		19	.421		112	23.50		135.50	159
1040	8" size, for 7" and 8" pipes		18	.444		144	25		169	197
1060	9" size, for 8" and 9" pipes		16	.500		162	28		190	221
2000	All fuel, oil, oil/gas, coal									
2020	10" for 9" and 10" pipes	1 Shee	15	.533	Ea.	242	30		272	310
2040	12" for 11" and 12" pipes		15	.533		315	30		345	390
2060	14" for 13" and 14" pipes		14	.571		410	32		442	500
2080	16" for 15" and 16" pipes		13	.615		575	34.50		609.50	685
2100	18" for 17" and 18" pipes		12	.667		765	37.50		802.50	895
2120	20" for 19" and 21" pipes		10	.800		915	45		960	1,075
2140	24" for 22" and 25" pipes	Q-9	12	1.333		1,125	67		1,192	1,325
2160	28" for 26" and 30" pipes		10	1.600		1,375	80.50		1,455.50	1,650
2180	32" for 31" and 34" pipes		8	2		1,775	101		1,876	2,100
3260	For thermal switch for above, add	1 Shee	24	.333		100	18.65		118.65	139

23 51 23 – Gas Vents

23 51 23.10 Gas Chimney Vents

		Crew	Daily Output	Labor-Hours	Unit	Material	2015 Bare Costs Labor	Equipment	Total	Total Incl O&P
0010	**GAS CHIMNEY VENTS**, Prefab metal, U.L. listed									
0020	Gas, double wall, galvanized steel									
0080	3" diameter	Q-9	72	.222	V.L.F.	5.55	11.20		16.75	23
0100	4" diameter		68	.235		7.20	11.85		19.05	26
0120	5" diameter		64	.250		7.80	12.60		20.40	28
0140	6" diameter		60	.267		9.45	13.45		22.90	31
0160	7" diameter		56	.286		15.10	14.40		29.50	38.50
0180	8" diameter		52	.308		17.40	15.50		32.90	42.50
0200	10" diameter		48	.333		36	16.80		52.80	65.50
0220	12" diameter		44	.364		43	18.30		61.30	75.50
0240	14" diameter		42	.381		72	19.20		91.20	109
0260	16" diameter		40	.400		103	20		123	145
0280	18" diameter		38	.421		128	21		149	173
0300	20" diameter	Q-10	36	.667		150	35		185	218
0320	22" diameter		34	.706		187	37		224	262
0340	24" diameter		32	.750		236	39		275	320
0600	For 4", 5" and 6" oval, add					50%				
0650	Gas, double wall, galvanized steel, fittings									

For customer support on your Mechanical Cost Data, call 877.795.7394.

355

23 51 23.10 Gas Chimney Vents		Crew	Daily Output	Labor-Hours	Unit	Material	2015 Bare Costs Labor	Equipment	Total	Total Incl O&P
0660	Elbow 45°, 3" diameter	Q-9	36	.444	Ea.	13.50	22.50		36	49
0670	4" diameter		34	.471		16.35	23.50		39.85	54
0680	5" diameter		32	.500		19.10	25		44.10	59.50
0690	6" diameter		30	.533		23.50	27		50.50	67
0700	7" diameter		28	.571		37.50	29		66.50	85.50
0710	8" diameter		26	.615		50	31		81	103
0720	10" diameter		24	.667		107	33.50		140.50	170
0730	12" diameter		22	.727		114	36.50		150.50	181
0740	14" diameter		21	.762		179	38.50		217.50	256
0750	16" diameter		20	.800		232	40.50		272.50	315
0760	18" diameter		19	.842		305	42.50		347.50	400
0770	20" diameter	Q-10	18	1.333		340	69.50		409.50	480
0780	22" diameter		17	1.412		545	73.50		618.50	715
0790	24" diameter		16	1.500		695	78.50		773.50	885
0916	Adjustable length									
0918	3" diameter, to 12"	Q-9	36	.444	Ea.	12.45	22.50		34.95	47.50
0920	4" diameter, to 12"		34	.471		14.50	23.50		38	52
0924	6" diameter, to 12"		30	.533		18.70	27		45.70	61.50
0928	8" diameter, to 12"		26	.615		35	31		66	86
0930	10" diameter, to 18"		24	.667		107	33.50		140.50	170
0932	12" diameter, to 18"		22	.727		110	36.50		146.50	177
0936	16" diameter, to 18"		20	.800		222	40.50		262.50	305
0938	18" diameter, to 18"		19	.842		254	42.50		296.50	345
0944	24" diameter, to 18"	Q-10	16	1.500		485	78.50		563.50	655
0950	Elbow 90°, adjustable, 3" diameter	Q-9	36	.444		23	22.50		45.50	59.50
0960	4" diameter		34	.471		27	23.50		50.50	65.50
0970	5" diameter		32	.500		33.50	25		58.50	75.50
0980	6" diameter		30	.533		39	27		66	84
0990	7" diameter		28	.571		67	29		96	118
1010	8" diameter		26	.615		68	31		99	123
1020	Wall thimble, 4 to 7" adjustable, 3" diameter		36	.444		14.45	22.50		36.95	50
1022	4" diameter		34	.471		16.25	23.50		39.75	54
1024	5" diameter		32	.500		17.10	25		42.10	57.50
1026	6" diameter		30	.533		17.55	27		44.55	60.50
1028	7" diameter		28	.571		36.50	29		65.50	84
1030	8" diameter		26	.615		51.50	31		82.50	105
1040	Roof flashing, 3" diameter		36	.444		8.50	22.50		31	43.50
1050	4" diameter		34	.471		9.90	23.50		33.40	47
1060	5" diameter		32	.500		29.50	25		54.50	71
1070	6" diameter		30	.533		23.50	27		50.50	67
1080	7" diameter		28	.571		31	29		60	78.50
1090	8" diameter		26	.615		33.50	31		64.50	84
1100	10" diameter		24	.667		45	33.50		78.50	101
1110	12" diameter		22	.727		63	36.50		99.50	125
1120	14" diameter		20	.800		143	40.50		183.50	219
1130	16" diameter		18	.889		201	45		246	290
1140	18" diameter		16	1		229	50.50		279.50	330
1150	20" diameter	Q-10	18	1.333		305	69.50		374.50	440
1160	22" diameter		14	1.714		375	89.50		464.50	550
1170	24" diameter		12	2		440	104		544	640
1200	Tee, 3" diameter	Q-9	27	.593		35	30		65	84.50
1210	4" diameter		26	.615		38.50	31		69.50	90
1220	5" diameter		25	.640		40	32		72	93.50

23 51 Breechings, Chimneys, and Stacks

23 51 23 – Gas Vents

23 51 23.10 Gas Chimney Vents		Crew	Daily Output	Labor-Hours	Unit	Material	2015 Bare Costs Labor	Equipment	Total	Total Incl O&P
1230	6" diameter	Q-9	24	.667	Ea.	45.50	33.50		79	102
1240	7" diameter		23	.696		65.50	35		100.50	126
1250	8" diameter		22	.727		71.50	36.50		108	135
1260	10" diameter		21	.762		191	38.50		229.50	269
1270	12" diameter		20	.800		196	40.50		236.50	277
1280	14" diameter		18	.889		350	45		395	455
1290	16" diameter		16	1		515	50.50		565.50	645
1300	18" diameter		14	1.143		625	57.50		682.50	775
1310	20" diameter	Q-10	17	1.412		855	73.50		928.50	1,050
1320	22" diameter		13	1.846		1,100	96.50		1,196.50	1,350
1330	24" diameter		12	2		1,250	104		1,354	1,525
1460	Tee cap, 3" diameter	Q-9	45	.356		2.36	17.90		20.26	30
1470	4" diameter		42	.381		2.54	19.20		21.74	32.50
1480	5" diameter		40	.400		3.19	20		23.19	34.50
1490	6" diameter		37	.432		4.73	22		26.73	38.50
1500	7" diameter		35	.457		7.95	23		30.95	43.50
1510	8" diameter		34	.471		8.70	23.50		32.20	45.50
1520	10" diameter		32	.500		70	25		95	116
1530	12" diameter		30	.533		70	27		97	118
1540	14" diameter		28	.571		73	29		102	124
1550	16" diameter		25	.640		74.50	32		106.50	132
1560	18" diameter		24	.667		88	33.50		121.50	149
1570	20" diameter	Q-10	27	.889		99.50	46.50		146	180
1580	22" diameter		22	1.091		300	57		357	415
1590	24" diameter		21	1.143		340	59.50		399.50	465
1750	Top, 3" diameter	Q-9	46	.348		13.05	17.50		30.55	41.50
1760	4" diameter		44	.364		13.75	18.30		32.05	43
1770	5" diameter		42	.381		17.50	19.20		36.70	49
1780	6" diameter		40	.400		23.50	20		43.50	57
1790	7" diameter		38	.421		37.50	21		58.50	74
1800	8" diameter		36	.444		46.50	22.50		69	85
1810	10" diameter		34	.471		105	23.50		128.50	151
1820	12" diameter		32	.500		154	25		179	208
1830	14" diameter		30	.533		205	27		232	266
1840	16" diameter		28	.571		263	29		292	335
1850	18" diameter		26	.615		360	31		391	450
1860	20" diameter	Q-10	28	.857		550	45		595	675
1870	22" diameter		22	1.091		840	57		897	1,000
1880	24" diameter		20	1.200		1,050	62.50		1,112.50	1,250
1900	Gas, double wall, galvanized steel, oval									
1904	4" x 1'	Q-9	68	.235	V.L.F.	16.65	11.85		28.50	36.50
1906	5"		64	.250		39.50	12.60		52.10	63
1908	5"/6" x 1'		60	.267		36.50	13.45		49.95	61
1910	Oval fittings									
1912	Adjustable length									
1914	4" diameter to 12" lg.	Q-9	34	.471	Ea.	16.10	23.50		39.60	53.50
1916	5" diameter to 12" lg.		32	.500		44.50	25		69.50	87.50
1918	5"/6" diameter to 12"		30	.533		40.50	27		67.50	85.50
1920	Elbow 45°									
1922	4"	Q-9	34	.471	Ea.	29.50	23.50		53	68.50
1924	5"		32	.500		56	25		81	100
1926	5"/6"		30	.533		53	27		80	99
1930	Elbow 45°, flat									

For customer support on your Mechanical Cost Data, call 877.795.7394.

357

23 51 23 – Gas Vents

23 51 23.10 Gas Chimney Vents		Crew	Daily Output	Labor-Hours	Unit	Material	2015 Bare Costs Labor	Equipment	Total	Total Incl O&P
1932	4"	Q-9	34	.471	Ea.	29.50	23.50		53	68.50
1934	5"		32	.500		55.50	25		80.50	100
1936	5"/6"		30	.533		53	27		80	99
1940	Top									
1942	4"	Q-9	44	.364	Ea.	35	18.30		53.30	66.50
1944	5"		42	.381		24.50	19.20		43.70	56.50
1946	5"/6"		40	.400		59.50	20		79.50	96
1950	Adjustable flashing									
1952	4"	Q-9	34	.471	Ea.	9.15	23.50		32.65	46
1954	5"		32	.500		29	25		54	70.50
1956	5"/6"		30	.533		30.50	27		57.50	74.50
1960	Tee									
1962	4"	Q-9	26	.615	Ea.	43.50	31		74.50	95.50
1964	5"		25	.640		91.50	32		123.50	151
1966	5"/6"		24	.667		84.50	33.50		118	145
1970	Tee with short snout									
1972	4"	Q-9	26	.615	Ea.	43.50	31		74.50	95.50

23 51 26 – All-Fuel Vent Chimneys

23 51 26.10 All-Fuel Vent Chimneys, Press. Tight, Dbl. Wall		Crew	Daily Output	Labor-Hours	Unit	Material	2015 Bare Costs Labor	Equipment	Total	Total Incl O&P
0010	**ALL-FUEL VENT CHIMNEYS, PRESSURE TIGHT, DOUBLE WALL**									
3200	All fuel, pressure tight, double wall, 1" insulation, U.L. listed, 1400°F.									
3210	304 stainless steel liner, aluminized steel outer jacket									
3220	6" diameter	Q-9	60	.267	L.F.	54	13.45		67.45	80
3221	8" diameter		52	.308		61	15.50		76.50	90.50
3222	10" diameter		48	.333		68	16.80		84.80	101
3223	12" diameter		44	.364		78	18.30		96.30	114
3224	14" diameter		42	.381		87.50	19.20		106.70	126
3225	16" diameter		40	.400		99	20		119	140
3226	18" diameter		38	.421		112	21		133	157
3227	20" diameter	Q-10	36	.667		127	35		162	193
3228	24" diameter		32	.750		163	39		202	239
3229	28" diameter		30	.800		187	42		229	270
3230	32" diameter		27	.889		211	46.50		257.50	305
3231	36" diameter		25	.960		236	50		286	335
3232	42" diameter		22	1.091		276	57		333	390
3233	48" diameter		19	1.263		315	66		381	445
3260	For 316 stainless steel liner add					30%				
3280	All fuel, pressure tight, double wall fittings									
3284	304 stainless steel inner, aluminized steel jacket									
3288	Adjustable 20"/29" section									
3292	6" diameter	Q-9	30	.533	Ea.	193	27		220	253
3293	8" diameter		26	.615		201	31		232	269
3294	10" diameter		24	.667		227	33.50		260.50	300
3295	12" diameter		22	.727		257	36.50		293.50	340
3296	14" diameter		21	.762		290	38.50		328.50	380
3297	16" diameter		20	.800		325	40.50		365.50	420
3298	18" diameter		19	.842		370	42.50		412.50	470
3299	20" diameter	Q-10	18	1.333		415	69.50		484.50	565
3300	24" diameter		16	1.500		535	78.50		613.50	710
3301	28" diameter		15	1.600		575	83.50		658.50	765
3302	32" diameter		14	1.714		695	89.50		784.50	900
3303	36" diameter		12	2		775	104		879	1,000

23 51 26 – All-Fuel Vent Chimneys

23 51 26.10 All-Fuel Vent Chimneys, Press. Tight, Dbl. Wall		Crew	Daily Output	Labor-Hours	Unit	Material	2015 Bare Costs Labor	Equipment	Total	Total Incl O&P
3304	42" diameter	Q-10	11	2.182	Ea.	910	114		1,024	1,175
3305	48" diameter	↓	10	2.400	↓	1,025	125		1,150	1,325
3350	Elbow 90° fixed									
3354	6" diameter	Q-9	30	.533	Ea.	405	27		432	485
3355	8" diameter		26	.615		455	31		486	550
3356	10" diameter		24	.667		515	33.50		548.50	615
3357	12" diameter		22	.727		585	36.50		621.50	700
3358	14" diameter		21	.762		660	38.50		698.50	785
3359	16" diameter		20	.800		745	40.50		785.50	880
3360	18" diameter	↓	19	.842		840	42.50		882.50	990
3361	20" diameter	Q-10	18	1.333		955	69.50		1,024.50	1,150
3362	24" diameter		16	1.500		1,225	78.50		1,303.50	1,475
3363	28" diameter		15	1.600		1,375	83.50		1,458.50	1,625
3364	32" diameter		14	1.714		1,850	89.50		1,939.50	2,150
3365	36" diameter		12	2		2,475	104		2,579	2,875
3366	42" diameter		11	2.182		2,950	114		3,064	3,425
3367	48" diameter	↓	10	2.400		3,300	125		3,425	3,825
3380	For 316 stainless steel liner, add				↓	30%				
3400	Elbow 45°									
3404	6" diameter	Q-9	30	.533	Ea.	201	27		228	262
3405	8" diameter		26	.615		227	31		258	297
3406	10" diameter		24	.667		258	33.50		291.50	335
3407	12" diameter		22	.727		294	36.50		330.50	380
3408	14" diameter		21	.762		330	38.50		368.50	425
3409	16" diameter		20	.800		370	40.50		410.50	470
3410	18" diameter	↓	19	.842		420	42.50		462.50	530
3411	20" diameter	Q-10	18	1.333		480	69.50		549.50	630
3412	24" diameter		16	1.500		610	78.50		688.50	795
3413	28" diameter		15	1.600		685	83.50		768.50	885
3414	32" diameter		14	1.714		925	89.50		1,014.50	1,150
3415	36" diameter		12	2		1,250	104		1,354	1,525
3416	42" diameter		11	2.182		1,450	114		1,564	1,775
3417	48" diameter	↓	10	2.400		1,650	125		1,775	2,025
3430	For 316 stainless steel liner, add				↓	30%				
3450	Tee 90°									
3454	6" diameter	Q-9	24	.667	Ea.	237	33.50		270.50	315
3455	8" diameter		22	.727		259	36.50		295.50	340
3456	10" diameter		21	.762		292	38.50		330.50	380
3457	12" diameter		20	.800		340	40.50		380.50	430
3458	14" diameter		18	.889		385	45		430	495
3459	16" diameter		16	1		425	50.50		475.50	540
3460	18" diameter	↓	14	1.143		500	57.50		557.50	640
3461	20" diameter	Q-10	17	1.412		580	73.50		653.50	750
3462	24" diameter		12	2		720	104		824	955
3463	28" diameter		11	2.182		855	114		969	1,125
3464	32" diameter		10	2.400		1,200	125		1,325	1,500
3465	36" diameter		9	2.667		1,350	139		1,489	1,700
3466	42" diameter		6	4		1,575	209		1,784	2,050
3467	48" diameter	↓	5	4.800		1,800	251		2,051	2,350
3480	For Tee Cap, add					35%	20%			
3500	For 316 stainless steel liner, add				↓	30%				
3520	Plate support, galvanized									
3524	6" diameter	Q-9	26	.615	Ea.	118	31		149	178

23 51 26 – All-Fuel Vent Chimneys

23 51 26.10 All-Fuel Vent Chimneys, Press. Tight, Dbl. Wall		Crew	Daily Output	Labor-Hours	Unit	Material	2015 Bare Costs Labor	2015 Bare Costs Equipment	Total	Total Incl O&P
3525	8" diameter	Q-9	22	.727	Ea.	138	36.50		174.50	208
3526	10" diameter		20	.800		150	40.50		190.50	227
3527	12" diameter		18	.889		158	45		203	243
3528	14" diameter		17	.941		187	47.50		234.50	279
3529	16" diameter		16	1		197	50.50		247.50	294
3530	18" diameter		15	1.067		208	53.50		261.50	310
3531	20" diameter	Q-10	16	1.500		218	78.50		296.50	360
3532	24" diameter		14	1.714		227	89.50		316.50	385
3533	28" diameter		13	1.846		278	96.50		374.50	450
3534	32" diameter		12	2		340	104		444	535
3535	36" diameter		10	2.400		410	125		535	640
3536	42" diameter		9	2.667		480	139		619	745
3537	48" diameter		8	3		545	157		702	840
3570	Bellows, lined									
3574	6" diameter	Q-9	30	.533	Ea.	1,300	27		1,327	1,475
3575	8" diameter		26	.615		1,350	31		1,381	1,550
3576	10" diameter		24	.667		1,375	33.50		1,408.50	1,575
3577	12" diameter		22	.727		1,425	36.50		1,461.50	1,600
3578	14" diameter		21	.762		1,450	38.50		1,488.50	1,650
3579	16" diameter		20	.800		1,525	40.50		1,565.50	1,725
3580	18" diameter		19	.842		1,575	42.50		1,617.50	1,800
3581	20" diameter	Q-10	18	1.333		1,625	69.50		1,694.50	1,875
3582	24" diameter		16	1.500		1,875	78.50		1,953.50	2,200
3583	28" diameter		15	1.600		2,000	83.50		2,083.50	2,325
3584	32" diameter		14	1.714		2,250	89.50		2,339.50	2,600
3585	36" diameter		12	2		2,450	104		2,554	2,850
3586	42" diameter		11	2.182		2,750	114		2,864	3,200
3587	48" diameter		10	2.400		3,175	125		3,300	3,700
3590	For all 316 stainless steel construction, add					55%				
3600	Ventilated roof thimble, 304 stainless steel									
3620	6" diameter	Q-9	26	.615	Ea.	247	31		278	320
3624	8" diameter		22	.727		257	36.50		293.50	340
3625	10" diameter		20	.800		264	40.50		304.50	355
3626	12" diameter		18	.889		272	45		317	370
3627	14" diameter		17	.941		282	47.50		329.50	385
3628	16" diameter		16	1		305	50.50		355.50	410
3629	18" diameter		15	1.067		330	53.50		383.50	440
3630	20" diameter	Q-10	16	1.500		345	78.50		423.50	500
3631	24" diameter		14	1.714		380	89.50		469.50	550
3632	28" diameter		13	1.846		410	96.50		506.50	595
3633	32" diameter		12	2		445	104		549	645
3634	36" diameter		10	2.400		475	125		600	715
3635	42" diameter		9	2.667		525	139		664	790
3636	48" diameter		8	3		575	157		732	875
3650	For 316 stainless steel, add					30%				
3670	Exit cone, 316 stainless steel only									
3674	6" diameter	Q-9	46	.348	Ea.	187	17.50		204.50	233
3675	8" diameter		42	.381		193	19.20		212.20	242
3676	10" diameter		40	.400		204	20		224	255
3677	12" diameter		38	.421		217	21		238	272
3678	14" diameter		37	.432		232	22		254	289
3679	16" diameter		36	.444		291	22.50		313.50	355
3680	18" diameter		35	.457		315	23		338	380

For customer support on your Mechanical Cost Data, call 877.795.7394.

23 51 Breechings, Chimneys, and Stacks

23 51 26 – All-Fuel Vent Chimneys

23 51 26.10 All-Fuel Vent Chimneys, Press. Tight, Dbl. Wall		Crew	Daily Output	Labor-Hours	Unit	Material	2015 Bare Costs		Total	Total Incl O&P
							Labor	Equipment		
3681	20" diameter	Q-10	28	.857	Ea.	375	45		420	485
3682	24" diameter		26	.923		495	48		543	620
3683	28" diameter		25	.960		560	50		610	690
3684	32" diameter		24	1		695	52		747	845
3685	36" diameter		22	1.091		845	57		902	1,025
3686	42" diameter		21	1.143		985	59.50		1,044.50	1,175
3687	48" diameter	▼	20	1.200	▼	1,125	62.50		1,187.50	1,325
3720	Roof guide, 304 stainless steel									
3724	6" diameter	Q-9	25	.640	Ea.	85.50	32		117.50	144
3725	8" diameter		21	.762		99.50	38.50		138	169
3726	10" diameter		19	.842		109	42.50		151.50	185
3727	12" diameter		17	.941		113	47.50		160.50	197
3728	14" diameter		16	1		132	50.50		182.50	222
3729	16" diameter		15	1.067		139	53.50		192.50	235
3730	18" diameter	▼	14	1.143		148	57.50		205.50	250
3731	20" diameter	Q-10	15	1.600		155	83.50		238.50	299
3732	24" diameter		13	1.846		162	96.50		258.50	325
3733	28" diameter		12	2	.	197	104		301	375
3734	32" diameter		11	2.182		243	114		357	440
3735	36" diameter		9	2.667		290	139		429	535
3736	42" diameter		8	3		335	157		492	610
3737	48" diameter	▼	7	3.429		385	179		564	700
3750	For 316 stainless steel, add				▼	30%				
3770	Rain cap with bird screen									
3774	6" diameter	Q-9	46	.348	Ea.	290	17.50		307.50	345
3775	8" diameter		42	.381		335	19.20		354.20	400
3776	10" diameter		40	.400		390	20		410	460
3777	12" diameter		38	.421		450	21		471	530
3778	14" diameter		37	.432		510	22		532	600
3779	16" diameter		36	.444		580	22.50		602.50	675
3780	18" diameter	▼	35	.457		675	23		698	775
3781	20" diameter	Q-10	28	.857		765	45		810	910
3782	24" diameter		26	.923		920	48		968	1,075
3783	28" diameter		25	.960		1,100	50		1,150	1,275
3784	32" diameter		24	1		1,250	52		1,302	1,450
3785	36" diameter		22	1.091		1,475	57		1,532	1,700
3786	42" diameter		21	1.143		1,725	59.50		1,784.50	2,000
3787	48" diameter	▼	20	1.200	▼	1,950	62.50		2,012.50	2,250

23 51 26.30 All-Fuel Vent Chimneys, Double Wall, St. Stl.

		Crew	Daily Output	Labor-Hours	Unit	Material	Labor	Equipment	Total	Total Incl O&P
0010	**ALL-FUEL VENT CHIMNEYS, DOUBLE WALL, STAINLESS STEEL**									
7780	All fuel, pressure tight, double wall, 4" insulation, U.L. listed, 1400°F.									
7790	304 stainless steel liner, aluminized steel outer jacket									
7800	6" diameter	Q-9	60	.267	V.L.F.	64	13.45		77.45	90.50
7804	8" diameter	.	52	.308		73	15.50		88.50	104
7806	10" diameter		48	.333		81.50	16.80		98.30	116
7808	12" diameter		44	.364		93.50	18.30		111.80	131
7810	14" diameter	▼	42	.381	▼	105	19.20		124.20	145
7880	For 316 stainless steel liner add				L.F.	30%				
8000	All fuel, double wall, stainless steel fittings									
8010	Roof support 6" diameter	Q-9	30	.533	Ea.	113	27		140	165
8030	8" diameter		26	.615		132	31		163	193
8040	10" diameter	▼	24	.667		139	33.50		172.50	205

For customer support on your Mechanical Cost Data, call 877.795.7394.

23 51 Breechings, Chimneys, and Stacks

23 51 26 – All-Fuel Vent Chimneys

23 51 26.30 All-Fuel Vent Chimneys, Double Wall, St. Stl.	Crew	Daily Output	Labor-Hours	Unit	Material	2015 Bare Costs Labor	Equipment	Total	Total Incl O&P
8050	12" diameter	Q-9	22	.727	Ea.	148	36.50		184.50	218
8060	14" diameter		21	.762		155	38.50		193.50	230
8100	Elbow 45°, 6" diameter		30	.533		241	27		268	305
8140	8" diameter		26	.615		271	31		302	345
8160	10" diameter		24	.667		310	33.50		343.50	390
8180	12" diameter		22	.727		350	36.50		386.50	440
8200	14" diameter		21	.762		395	38.50		433.50	495
8300	Insulated tee, 6" diameter		30	.533		283	27		310	350
8360	8" diameter		26	.615		310	31		341	390
8380	10" diameter		24	.667		350	33.50		383.50	435
8400	12" diameter		22	.727		395	36.50		431.50	490
8420	14" diameter		20	.800		460	40.50		500.50	570
8500	Boot tee, 6" diameter		28	.571		540	29		569	640
8520	8" diameter		24	.667		590	33.50		623.50	695
8530	10" diameter		22	.727		685	36.50		721.50	810
8540	12" diameter		20	.800		810	40.50		850.50	950
8550	14" diameter		18	.889		925	45		970	1,100
8600	Rain cap with bird screen, 6" diameter		30	.533		290	27		317	360
8640	8" diameter		26	.615		335	31		366	420
8660	10" diameter		24	.667		390	33.50		423.50	480
8680	12" diameter		22	.727		450	36.50		486.50	550
8700	14" diameter		21	.762		510	38.50		548.50	625
8800	Flat roof flashing, 6" diameter		30	.533		106	27		133	158
8840	8" diameter		26	.615		116	31		147	175
8860	10" diameter		24	.667		124	33.50		157.50	188
8880	12" diameter		22	.727		136	36.50		172.50	206
8900	14" diameter		21	.762		139	38.50		177.50	212

23 52 Heating Boilers

23 52 13 – Electric Boilers

23 52 13.10 Electric Boilers, ASME

		Crew	Daily Output	Labor-Hours	Unit	Material	2015 Bare Costs Labor	Equipment	Total	Total Incl O&P
0010	**ELECTRIC BOILERS, ASME**, Standard controls and trim R235000-30									
1000	Steam, 6 KW, 20.5 MBH	Q-19	1.20	20	Ea.	3,950	1,075		5,025	5,975
1040	9 KW, 30.7 MBH D3020-102		1.20	20		4,025	1,075		5,100	6,050
1060	18 KW, 61.4 MBH		1.20	20		4,150	1,075		5,225	6,200
1080	24 KW, 81.8 MBH D3020-104		1.10	21.818		4,800	1,175		5,975	7,075
1120	36 KW, 123 MBH		1.10	21.818		5,325	1,175		6,500	7,625
1160	60 KW, 205 MBH		1	24		6,650	1,300		7,950	9,275
1220	112 KW, 382 MBH		.75	32		9,375	1,725		11,100	12,900
1240	148 KW, 505 MBH		.65	36.923		9,600	2,000		11,600	13,600
1260	168 KW, 573 MBH		.60	40		20,400	2,175		22,575	25,800
1280	222 KW, 758 MBH		.55	43.636		23,800	2,350		26,150	29,800
1300	296 KW, 1010 MBH		.45	53.333		25,000	2,875		27,875	32,000
1320	300 KW, 1023 MBH		.40	60		26,400	3,250		29,650	33,900
1340	370 KW, 1263 MBH		.35	68.571		29,200	3,700		32,900	37,700
1360	444 KW, 1515 MBH		.30	80		31,400	4,325		35,725	41,000
1380	518 KW, 1768 MBH	Q-21	.36	88.889		32,600	4,925		37,525	43,300
1400	592 KW, 2020 MBH		.34	94.118		35,400	5,225		40,625	46,800
1420	666 KW, 2273 MBH		.32	100		36,800	5,550		42,350	48,900
1460	740 KW, 2526 MBH		.28	114		38,300	6,350		44,650	51,500
1480	814 KW, 2778 MBH		.25	128		40,600	7,100		47,700	55,500

23 52 13 – Electric Boilers

23 52 13.10 Electric Boilers, ASME	Crew	Daily Output	Labor-Hours	Unit	Material	2015 Bare Costs Labor	Equipment	Total	Total Incl O&P
1500	962 KW, 3283 MBH	Q-21	.22	145	Ea.	44,900	8,075		52,975	61,500
1520	1036 KW, 3536 MBH		.20	160		46,700	8,875		55,575	65,000
1540	1110 KW, 3788 MBH		.19	168		49,600	9,350		58,950	68,500
1560	2070 KW, 7063 MBH		.18	177		68,000	9,875		77,875	90,000
1580	2250 KW, 7677 MBH		.17	188		80,000	10,400		90,400	103,500
1600	2,340 KW, 7984 MBH		.16	200		86,500	11,100		97,600	111,500
2000	Hot water, 7.5 KW, 25.6 MBH	Q-19	1.30	18.462		4,975	1,000		5,975	6,975
2020	15 KW, 51.2 MBH		1.30	18.462		5,000	1,000		6,000	7,000
2040	30 KW, 102 MBH		1.20	20		5,325	1,075		6,400	7,475
2060	45 KW, 164 MBH		1.20	20		5,425	1,075		6,500	7,575
2070	60 KW, 205 MBH		1.20	20		5,525	1,075		6,600	7,700
2080	75 KW, 256 MBH		1.10	21.818		5,900	1,175		7,075	8,275
2100	90 KW, 307 MBH		1.10	21.818		6,000	1,175		7,175	8,375
2120	105 KW, 358 MBH		1	24		6,275	1,300		7,575	8,850
2140	120 KW, 410 MBH		.90	26.667		6,375	1,450		7,825	9,200
2160	135 KW, 461 MBH		.75	32		7,400	1,725		9,125	10,800
2180	150 KW, 512 MBH		.65	36.923		7,500	2,000		9,500	11,200
2200	165 KW, 563 MBH		.60	40		7,600	2,175		9,775	11,600
2220	296 KW, 1010 MBH		.55	43.636		16,100	2,350		18,450	21,300
2280	370 KW, 1263 MBH		.40	60		17,900	3,250		21,150	24,600
2300	444 KW, 1515 MBH		.35	68.571		20,800	3,700		24,500	28,500
2340	518 KW, 1768 MBH	Q-21	.44	72.727		23,100	4,025		27,125	31,500
2360	592 KW, 2020 MBH		.43	74.419		26,100	4,125		30,225	35,000
2400	666 KW, 2273 MBH		.40	80		27,300	4,450		31,750	36,700
2420	740 KW, 2526 MBH		.39	82.051		28,500	4,550		33,050	38,200
2440	814 KW, 2778 MBH		.38	84.211		30,500	4,675		35,175	40,600
2460	888 KW, 3031 MBH		.37	86.486		31,100	4,800		35,900	41,400
2480	962 KW, 3283 MBH		.36	88.889		34,000	4,925		38,925	44,800
2500	1036 KW, 3536 MBH		.34	94.118		35,500	5,225		40,725	47,000
2520	1110 KW, 3788 MBH		.33	96.970		38,800	5,375		44,175	50,500
2540	1440 KW, 4915 MBH		.32	100		43,800	5,550		49,350	56,500
2560	1560 KW, 5323 MBH		.31	103		46,800	5,725		52,525	60,000
2580	1680 KW, 5733 MBH		.30	106		50,000	5,925		55,925	64,000
2600	1800 KW, 6143 MBH		.29	110		53,500	6,125		59,625	67,500
2620	1980 KW, 6757 MBH		.28	114		58,000	6,350		64,350	73,500
2640	2100 KW, 7167 MBH		.27	118		59,500	6,575		66,075	75,500
2660	2220 KW, 7576 MBH		.26	123		65,000	6,825		71,825	82,000
2680	2400 KW, 8191 MBH		.25	128		68,000	7,100		75,100	85,000
2700	2610 KW, 8905 MBH		.24	133		71,000	7,400		78,400	89,000
2720	2790 KW, 9519 MBH		.23	139		75,500	7,725		83,225	94,500
2740	2970 KW, 10133 MBH		.21	152		77,000	8,450		85,450	97,500
2760	3150 KW, 10748 MBH		.19	168		82,500	9,350		91,850	105,000
2780	3240 KW, 11055 MBH		.18	177		84,500	9,875		94,375	108,000
2800	3420 KW, 11669 MBH		.17	188		90,000	10,400		100,400	114,500
2820	3600 KW, 12,283 MBH		.16	200		94,000	11,100		105,100	120,000

For customer support on your Mechanical Cost Data, call 877.795.7394.

363

23 52 16 – Condensing Boilers

23 52 16.24 Condensing Boilers		Crew	Daily Output	Labor-Hours	Unit	Material	2015 Bare Costs Labor	Equipment	Total	Total Incl O&P
0010	**CONDENSING BOILERS**, Cast iron, high efficiency									
0020	Packaged with standard controls, circulator and trim									
0030	Intermittent (spark) pilot, natural or LP gas									
0040	Hot water, DOE MBH output, (AFUE)									
0100	42 MBH, (84.0%) G	Q-5	1.80	8.889	Ea.	1,575	480		2,055	2,475
0120	57 MBH, (84.3%) G		1.60	10		1,775	540		2,315	2,750
0140	85 MBH, (84.0%) G		1.40	11.429		1,950	615		2,565	3,050
0160	112 MBH, (83.7%) G		1.20	13.333		2,175	715		2,890	3,475
0180	140 MBH, (83.3%) G	Q-6	1.60	15		2,450	835		3,285	3,975
0200	167 MBH, (83.0%) G		1.40	17.143		2,775	955		3,730	4,500
0220	194 MBH, (82.7%) G		1.20	20		3,025	1,125		4,150	5,000

23 52 19 – Pulse Combustion Boilers

23 52 19.20 Pulse Type Combustion Boilers

23 52 19.20 Pulse Type Combustion Boilers		Crew	Daily Output	Labor-Hours	Unit	Material	2015 Bare Costs Labor	Equipment	Total	Total Incl O&P
0010	**PULSE TYPE COMBUSTION BOILERS**, High efficiency									
7990	Special feature gas fired boilers									
8000	Pulse combustion, standard controls/trim									
8010	Hot water, DOE MBH output, (AFUE %)									
8030	71 MBH, 95.2%	Q-5	1.60	10	Ea.	3,450	540		3,990	4,600
8050	94 MBH, 95.3% G		1.40	11.429		3,825	615		4,440	5,125
8080	139 MBH, 95.6% G		1.20	13.333		4,300	715		5,015	5,800
8090	207 MBH, 95.4%		1.16	13.793		4,850	740		5,590	6,475
8120	270 MBH, 96.4%		1.12	14.286		6,725	770		7,495	8,550
8130	365 MBH, 91.7%		1.07	14.953		7,600	805		8,405	9,575

23 52 23 – Cast-Iron Boilers

23 52 23.20 Gas-Fired Boilers

23 52 23.20 Gas-Fired Boilers		Crew	Daily Output	Labor-Hours	Unit	Material	2015 Bare Costs Labor	Equipment	Total	Total Incl O&P
0010	**GAS-FIRED BOILERS**, Natural or propane, standard controls, packaged R235000-10									
1000	Cast iron, with insulated jacket									
2000	Steam, gross output, 81 MBH R235000-20	Q-7	1.40	22.857	Ea.	2,450	1,300		3,750	4,675
2020	102 MBH		1.30	24.615		2,675	1,400		4,075	5,075
2040	122 MBH R235000-30		1	32		2,925	1,825		4,750	5,975
2060	163 MBH		.90	35.556		3,275	2,025		5,300	6,650
2080	203 MBH		.90	35.556		3,825	2,025		5,850	7,250
2100	240 MBH		.85	37.647		3,925	2,150		6,075	7,525
2120	280 MBH		.80	40		4,700	2,275		6,975	8,600
2140	320 MBH		.70	45.714		5,000	2,600		7,600	9,425
2160	360 MBH		.63	51.200		5,575	2,925		8,500	10,600
2180	400 MBH		.56	56.838		5,900	3,225		9,125	11,400
2200	440 MBH		.51	62.500		6,350	3,550		9,900	12,400
2220	544 MBH		.45	71.588		10,100	4,075		14,175	17,300
2240	765 MBH		.43	74.419		12,200	4,225		16,425	19,800
2260	892 MBH		.38	84.211		14,100	4,800		18,900	22,700
2280	1275 MBH		.34	94.118		17,300	5,350		22,650	27,100
2300	1530 MBH		.32	100		18,500	5,700		24,200	29,000
2320	1,875 MBH		.30	106		25,200	6,075		31,275	36,900
2340	2170 MBH		.26	122		27,000	6,950		33,950	40,100
2360	2675 MBH		.20	163		28,300	9,300		37,600	45,100
2380	3060 MBH		.19	172		29,300	9,800		39,100	47,100
2400	3570 MBH		.18	181		35,800	10,300		46,100	55,000
2420	4207 MBH		.16	205		39,200	11,700		50,900	60,500
2440	4,720 MBH		.15	207		68,000	11,800		79,800	93,000
2460	5660 MBH		.15	220		77,000	12,600		89,600	103,500

23 52 Heating Boilers

23 52 23 – Cast-Iron Boilers

23 52 23.20 Gas-Fired Boilers

		Crew	Daily Output	Labor-Hours	Unit	Material	2015 Bare Costs Labor	2015 Bare Costs Equipment	Total	Total Incl O&P
2480	6,100 MBH	Q-7	.13	246	Ea.	88,000	14,000		102,000	118,000
2500	6390 MBH		.12	266		93,000	15,200		108,200	125,000
2520	6680 MBH		.11	290		94,000	16,500		110,500	128,500
2540	6,970 MBH		.10	320		98,500	18,200		116,700	136,000
3000	Hot water, gross output, 80 MBH		1.46	21.918		1,975	1,250		3,225	4,050
3020	100 MBH		1.35	23.704		2,550	1,350		3,900	4,850
3040	122 MBH		1.10	29.091		2,675	1,650		4,325	5,425
3060	163 MBH		1	32		3,150	1,825		4,975	6,225
3080	203 MBH		1	32		3,625	1,825		5,450	6,750
3100	240 MBH		.95	33.684		3,700	1,925		5,625	6,950
3120	280 MBH		.90	35.556		4,325	2,025		6,350	7,800
3140	320 MBH		.80	40		4,675	2,275		6,950	8,575
3160	360 MBH		.71	45.070		5,400	2,575		7,975	9,825
3180	400 MBH		.64	50		5,575	2,850		8,425	10,500
3200	440 MBH		.58	54.983		5,950	3,125		9,075	11,300
3220	544 MBH		.51	62.992		9,800	3,575		13,375	16,200
3240	765 MBH		.46	70.022		11,900	3,975		15,875	19,000
3260	1,088 MBH		.40	80		13,500	4,550		18,050	21,800
3280	1,275 MBH		.36	89.888		17,300	5,125		22,425	26,700
3300	1,530 MBH		.31	104		18,200	5,975		24,175	29,000
3320	2,000 MBH		.26	125		21,400	7,100		28,500	34,200
3340	2,312 MBH		.22	148		24,400	8,425		32,825	39,600
3360	2,856 MBH		.20	160		30,800	9,100		39,900	47,600
3380	3,264 MBH		.18	179		32,700	10,200		42,900	51,500
3400	3,996 MBH		.16	195		35,800	11,100		46,900	56,000
3420	4,488 MBH		.15	210		39,400	12,000		51,400	61,500
3440	4,720 MBH		.15	220		74,000	12,600		86,600	100,500
3460	5,520 MBH		.14	228		95,000	13,000		108,000	124,000
3480	6,100 MBH		.13	250		115,500	14,200		129,700	148,500
3500	6,390 MBH		.11	285		121,500	16,300		137,800	158,500
3520	6,680 MBH		.10	310		126,500	17,700		144,200	166,000
3540	6,970 MBH		.09	359		118,500	20,500		139,000	161,500
7000	For tankless water heater, add					10%				
7050	For additional zone valves up to 312 MBH add					194			194	214

23 52 23.30 Gas/Oil Fired Boilers

		Crew	Daily Output	Labor-Hours	Unit	Material	2015 Bare Costs Labor	2015 Bare Costs Equipment	Total	Total Incl O&P
0010	**GAS/OIL FIRED BOILERS**, Combination with burners and controls, packaged									
1000	Cast iron with insulated jacket									
2000	Steam, gross output, 720 MBH	Q-7	.43	74.074	Ea.	14,700	4,225		18,925	22,600
2020	810 MBH		.38	83.990		14,700	4,775		19,475	23,400
2040	1,084 MBH		.34	93.023		17,000	5,300		22,300	26,700
2060	1,360 MBH		.33	98.160		19,500	5,575		25,075	29,800
2080	1,600 MBH		.30	107		20,900	6,100		27,000	32,200
2100	2,040 MBH		.25	130		24,900	7,400		32,300	38,600
2120	2,450 MBH		.21	156		26,600	8,875		35,475	42,600
2140	2,700 MBH		.19	165		28,100	9,425		37,525	45,100
2160	3,000 MBH		.18	175		30,100	10,000		40,100	48,200
2180	3,270 MBH		.17	183		32,900	10,500		43,400	52,000
2200	3,770 MBH		.17	191		67,500	10,900		78,400	90,500
2220	4,070 MBH		.16	200		71,000	11,400		82,400	95,000
2240	4,650 MBH		.15	210		75,000	12,000		87,000	100,500
2260	5,230 MBH		.14	223		83,000	12,700		95,700	110,000
2280	5,520 MBH		.14	235		90,500	13,400		103,900	119,500

23 52 23.30 Gas/Oil Fired Boilers		Crew	Daily Output	Labor-Hours	Unit	Material	2015 Bare Costs Labor	Equipment	Total	Total Incl O&P
2300	5,810 MBH	Q-7	.13	248	Ea.	92,000	14,100		106,100	122,500
2320	6,100 MBH		.12	260		94,000	14,800		108,800	126,000
2340	6,390 MBH		.11	296		97,000	16,900		113,900	132,000
2360	6,680 MBH		.10	320		100,000	18,200		118,200	137,500
2380	6,970 MBH		.09	372		102,500	21,200		123,700	145,000
2900	Hot water, gross output									
2910	200 MBH	Q-6	.62	39.024	Ea.	10,500	2,175		12,675	14,900
2920	300 MBH		.49	49.080		10,500	2,725		13,225	15,700
2930	400 MBH		.41	57.971		12,300	3,225		15,525	18,400
2940	500 MBH		.36	67.039		13,300	3,750		17,050	20,300
3000	584 MBH	Q-7	.44	72.072		14,700	4,100		18,800	22,300
3020	876 MBH		.41	79.012		19,400	4,500		23,900	28,100
3040	1,168 MBH		.31	103		28,500	5,900		34,400	40,200
3060	1,460 MBH		.28	113		36,400	6,425		42,825	49,700
3080	2,044 MBH		.26	122		41,300	6,950		48,250	56,000
3100	2,628 MBH		.21	150		45,300	8,550		53,850	62,500
3120	3,210 MBH		.18	174		48,400	9,950		58,350	68,000
3140	3,796 MBH		.17	186		54,000	10,600		64,600	75,500
3160	4,088 MBH		.16	195		61,500	11,100		72,600	84,500
3180	4,672 MBH		.16	203		67,000	11,600		78,600	91,500
3200	5,256 MBH		.15	217		75,000	12,400		87,400	101,000
3220	6,000 MBH, 179 BHP		.13	256		108,000	14,600		122,600	140,500
3240	7,130 MBH, 213 BHP		.08	385		113,000	21,900		134,900	157,000
3260	9,800 MBH, 286 BHP		.06	533		130,000	30,300		160,300	189,000
3280	10,900 MBH, 325.6 BHP		.05	592		157,500	33,700		191,200	224,500
3290	12,200 MBH, 364.5 BHP		.05	666		172,000	37,900		209,900	246,500
3300	13,500 MBH, 403.3 BHP		.04	727		190,500	41,400		231,900	272,000

23 52 23.40 Oil-Fired Boilers

23 52 23.40 Oil-Fired Boilers		Crew	Daily Output	Labor-Hours	Unit	Material	2015 Bare Costs Labor	Equipment	Total	Total Incl O&P
0010	**OIL-FIRED BOILERS**, Standard controls, flame retention burner, packaged									
1000	Cast iron, with insulated flush jacket									
2000	Steam, gross output, 109 MBH	Q-7	1.20	26.667	Ea.	2,325	1,525		3,850	4,850
2020	144 MBH		1.10	29.091		2,625	1,650		4,275	5,375
2040	173 MBH		1	32		2,925	1,825		4,750	5,975
2060	207 MBH		.90	35.556		3,175	2,025		5,200	6,550
2080	236 MBH		.85	37.647		3,725	2,150		5,875	7,325
2100	300 MBH		.70	45.714		4,700	2,600		7,300	9,100
2120	480 MBH		.50	64		6,075	3,650		9,725	12,200
2140	665 MBH		.45	71.111		8,625	4,050		12,675	15,600
2160	794 MBH		.41	78.049		9,575	4,450		14,025	17,200
2180	1,084 MBH		.38	85.106		10,900	4,850		15,750	19,300
2200	1,360 MBH		.33	98.160		12,500	5,575		18,075	22,100
2220	1,600 MBH		.26	122		14,000	6,950		20,950	25,900
2240	2,175 MBH		.24	133		18,100	7,625		25,725	31,500
2260	2,480 MBH		.21	156		20,600	8,875		29,475	36,000
2280	3,000 MBH		.19	170		23,700	9,675		33,375	40,700
2300	3,550 MBH		.17	187		27,400	10,600		38,000	46,200
2320	3,820 MBH		.16	200		39,000	11,400		50,400	60,000
2340	4,360 MBH		.15	214		42,800	12,200		55,000	65,500
2360	4,940 MBH		.14	225		67,000	12,800		79,800	93,000
2380	5,520 MBH		.14	235		78,500	13,400		91,900	106,500
2400	6,100 MBH		.13	256		90,000	14,600		104,600	121,000
2420	6,390 MBH		.11	290		95,000	16,500		111,500	129,500

23 52 Heating Boilers

23 52 23 – Cast-Iron Boilers

23 52 23.40 Oil-Fired Boilers

		Crew	Daily Output	Labor-Hours	Unit	Material	2015 Bare Costs Labor	Equipment	Total	Total Incl O&P
2440	6,680 MBH	Q-7	.10	313	Ea.	97,500	17,800		115,300	134,500
2460	6,970 MBH	↓	.09	363	↓	101,000	20,700		121,700	142,000
3000	Hot water, same price as steam									
4000	For tankless coil in smaller sizes, add				Ea.	15%				

23 52 23.60 Solid-Fuel Boilers

		Crew	Daily Output	Labor-Hours	Unit	Material	2015 Bare Costs Labor	Equipment	Total	Total Incl O&P
0010	**SOLID-FUEL BOILERS**									
3000	Stoker fired (coal) cast iron with flush jacket and									
3400	insulation, steam or water, gross output, 1280 MBH	Q-6	.36	66.667	Ea.	43,100	3,725		46,825	53,000
3420	1460 MBH		.30	80		47,300	4,450		51,750	58,500
3440	1640 MBH		.28	85.714		50,500	4,775		55,275	62,500
3460	1820 MBH		.26	92.308		54,000	5,150		59,150	67,500
3480	2000 MBH		.25	96		57,500	5,350		62,850	71,000
3500	2360 MBH		.23	104		65,000	5,825		70,825	80,500
3540	2725 MBH	↓	.20	120		72,500	6,700		79,200	89,500
3800	2950 MBH	Q-7	.16	200		121,000	11,400		132,400	150,000
3820	3210 MBH		.15	213		127,000	12,100		139,100	158,000
3840	3480 MBH		.14	228		132,500	13,000		145,500	165,500
3860	3745 MBH		.14	228		139,000	13,000		152,000	172,500
3880	4000 MBH		.13	246		144,500	14,000		158,500	180,000
3900	4200 MBH		.13	246		151,000	14,000		165,000	187,000
3920	4400 MBH		.12	266		157,000	15,200		172,200	195,500
3940	4600 MBH		.12	266	↓	163,000	15,200		178,200	202,000

23 52 26 – Steel Boilers

23 52 26.40 Oil-Fired Boilers

		Crew	Daily Output	Labor-Hours	Unit	Material	2015 Bare Costs Labor	Equipment	Total	Total Incl O&P
0010	**OIL-FIRED BOILERS**, Standard controls, flame retention burner									
5000	Steel, with insulated flush jacket									
7000	Hot water, gross output, 103 MBH	Q-6	1.60	15	Ea.	1,775	835		2,610	3,225
7020	122 MBH		1.45	16.506		1,875	920		2,795	3,475
7040	137 MBH		1.36	17.595		2,000	980		2,980	3,675
7060	168 MBH		1.30	18.405		2,100	1,025		3,125	3,875
7080	225 MBH		1.22	19.704		2,575	1,100		3,675	4,475
7100	315 MBH		.96	25.105		6,550	1,400		7,950	9,350
7120	420 MBH		.70	34.483		6,900	1,925		8,825	10,500
7140	525 MBH		.57	42.403		8,150	2,375		10,525	12,600
7180	735 MBH		.48	50.104		9,575	2,800		12,375	14,700
7220	1,050 MBH		.37	65.753		19,000	3,675		22,675	26,400
7280	2,310 MBH		.21	114		25,800	6,400		32,200	38,100
7320	3,150 MBH	↓	.13	184		30,000	10,300		40,300	48,500
7340	For tankless coil in steam or hot water, add				↓	7%				

23 52 26.70 Packaged Water Tube Boilers

		Crew	Daily Output	Labor-Hours	Unit	Material	2015 Bare Costs Labor	Equipment	Total	Total Incl O&P
0010	**PACKAGED WATER TUBE BOILERS**									
2000	Packaged water tube, #2 oil, steam or hot water, gross output									
2010	200 MBH	Q-6	1.50	16	Ea.	4,025	890		4,915	5,775
2014	275 MBH		1.40	17.143		4,625	955		5,580	6,525
2018	360 MBH		1.10	21.818		6,400	1,225		7,625	8,875
2022	520 MBH		.65	36.923		7,700	2,050		9,750	11,600
2026	600 MBH		.60	40		7,875	2,225		10,100	12,000
2030	720 MBH		.55	43.636		9,350	2,425		11,775	14,000
2034	960 MBH		.48	50		11,600	2,800		14,400	16,900
2040	1200 MBH	Q-7	.50	64		18,500	3,650		22,150	25,800
2044	1440 MBH		.45	71.111		24,800	4,050		28,850	33,400
2060	1600 MBH		.40	80		29,200	4,550		33,750	39,000

For customer support on your Mechanical Cost Data, call 877.795.7394.

367

23 52 Heating Boilers

23 52 26 - Steel Boilers

23 52 26.70 Packaged Water Tube Boilers		Crew	Daily Output	Labor-Hours	Unit	Material	2015 Bare Costs Labor	Equipment	Total	Total Incl O&P
2068	1920 MBH	Q-7	.35	91.429	Ea.	31,700	5,200		36,900	42,700
2072	2160 MBH		.33	96.970		35,100	5,525		40,625	46,900
2080	2400 MBH		.30	106		39,300	6,075		45,375	52,500
2100	3200 MBH		.25	128		44,900	7,275		52,175	60,500
2120	4800 MBH	▼	.20	160	▼	58,500	9,100		67,600	78,000
2200	Gas fired									
2204	200 MBH	Q-6	1.50	16	Ea.	5,650	890		6,540	7,550
2208	275 MBH		1.40	17.143		5,975	955		6,930	8,000
2212	360 MBH		1.10	21.818		6,400	1,225		7,625	8,850
2216	520 MBH		.65	36.923		7,700	2,050		9,750	11,600
2220	600 MBH		.60	40		7,875	2,225		10,100	12,000
2224	720 MBH		.55	43.636		9,350	2,425		11,775	14,000
2228	960 MBH	▼	.48	50		11,600	2,800		14,400	16,900
2232	1220 MBH	Q-7	.50	64		17,300	3,650		20,950	24,500
2236	1440 MBH		.45	71.111		17,800	4,050		21,850	25,700
2240	1680 MBH		.40	80		26,500	4,550		31,050	36,100
2244	1920 MBH		.35	91.429		26,600	5,200		31,800	37,200
2248	2160 MBH		.33	96.970		27,000	5,525		32,525	38,000
2252	2400 MBH	▼	.30	106	▼	30,400	6,075		36,475	42,600

23 52 28 - Swimming Pool Boilers

23 52 28.10 Swimming Pool Heaters

		Crew	Daily Output	Labor-Hours	Unit	Material	2015 Bare Costs Labor	Equipment	Total	Total Incl O&P
0010	**SWIMMING POOL HEATERS**, Not including wiring, external									
0020	piping, base or pad,									
0160	Gas fired, input, 155 MBH	Q-6	1.50	16	Ea.	2,000	890		2,890	3,550
0200	199 MBH		1	24		2,125	1,350		3,475	4,375
0220	250 MBH		.70	34.286		2,325	1,900		4,225	5,425
0240	300 MBH		.60	40		2,425	2,225		4,650	6,050
0260	399 MBH		.50	48		2,725	2,675		5,400	7,050
0280	500 MBH		.40	60		8,900	3,350		12,250	14,800
0300	650 MBH		.35	68.571		9,450	3,825		13,275	16,200
0320	750 MBH		.33	72.727		10,300	4,050		14,350	17,500
0360	990 MBH		.22	109		13,900	6,075		19,975	24,400
0370	1,260 MBH		.21	114		16,300	6,375		22,675	27,500
0380	1,440 MBH		.19	126		17,400	7,050		24,450	29,700
0400	1,800 MBH		.14	171		19,300	9,550		28,850	35,700
0410	2,070 MBH	▼	.13	184		22,800	10,300		33,100	40,600
2000	Electric, 12 KW, 4,800 gallon pool	Q-19	3	8		2,075	435		2,510	2,925
2020	15 KW, 7,200 gallon pool		2.80	8.571		2,100	465		2,565	3,025
2040	24 KW, 9,600 gallon pool		2.40	10		2,425	540		2,965	3,500
2060	30 KW, 12,000 gallon pool		2	12		2,475	650		3,125	3,700
2080	36 KW, 14,400 gallon pool		1.60	15		2,850	810		3,660	4,350
2100	57 KW, 24,000 gallon pool	▼	1.20	20	▼	3,575	1,075		4,650	5,550
9000	To select pool heater: 12 BTUH x S.F. pool area									
9010	X temperature differential = required output									
9050	For electric, KW = gallons x 2.5 divided by 1000									
9100	For family home type pool, double the									
9110	Rated gallon capacity = 1/2°F rise per hour									

23 52 39.13 Scotch Marine Boilers	Crew	Daily Output	Labor-Hours	Unit	Material	2015 Bare Costs Labor	Equipment	Total	Total Incl O&P
0010 SCOTCH MARINE BOILERS									
1000 Packaged fire tube, #2 oil, gross output									
1006 15 PSI steam									
1010 1005 MBH, 30 HP	Q-6	.33	72.072	Ea.	41,100	4,025		45,125	51,500
1014 1675 MBH, 50 HP	"	.26	93.023		51,500	5,200		56,700	64,500
1020 3348 MBH, 100 HP	Q-7	.21	152		69,500	8,675		78,175	89,000
1030 5025 MBH, 150 HP		.17	192		87,000	11,000		98,000	112,500
1040 6696 MBH, 200 HP		.14	223		104,000	12,700		116,700	133,000
1050 8375 MBH, 250 HP		.14	231		122,500	13,200		135,700	155,000
1060 10,044 MBH, 300 HP		.13	251		132,000	14,300		146,300	167,000
1080 16,740 MBH, 500 HP		.08	380		179,500	21,700		201,200	230,000
1100 23,435 MBH, 700 HP		.07	484		220,000	27,600		247,600	283,500
1102 26,780 MBH, 800 HP		.07	484		220,000	27,600		247,600	283,500
1104 30,125 MBH, 900 HP		.06	500		401,500	28,400		429,900	484,500
1106 33,475 MBH, 1,000 HP		.06	524		425,000	29,800		454,800	512,500
1107 36,825 MBH, 1,100 HP		.06	551		456,000	31,400		487,400	549,500
1108 40,170 MBH, 1,200 HP		.06	581		478,500	33,100		511,600	576,000
1110 46,865 MBH, 1,400 HP		.05	603		504,000	34,300		538,300	606,500
1112 53,560 MBH, 1,600 HP		.05	627		547,500	35,700		583,200	656,500
1114 60,250 MBH, 1,800 HP		.05	666		565,500	37,900		603,400	679,500
1116 66,950 MBH, 2,000 HP		.05	695		599,000	39,600		638,600	719,000
1118 73,650 MBH, 2,200 HP		.04	744		647,000	42,300		689,300	776,000
1120 To fire #6, add		.83	38.554		13,300	2,200		15,500	17,900
1140 To fire #6, and gas, add		.42	76.190		19,600	4,325		23,925	28,100
1160 For high pressure, add					21,700			21,700	23,900
1180 For duplex package feed system									
1200 To 3348 MBH boiler, add	Q-7	.54	59.259	Ea.	10,500	3,375		13,875	16,700
1220 To 6696 MBH boiler, add		.41	78.049		12,500	4,450		16,950	20,400
1240 To 10,044 MBH boiler, add		.38	84.211		14,300	4,800		19,100	23,000
1260 To 16,740 MBH boiler, add		.28	114		16,300	6,500		22,800	27,800
1280 To 23,435 MBH boiler, add		.25	128		19,300	7,275		26,575	32,200
1300 150 PSI steam									
1310 1005 MBH, 30 HP	Q-6	.33	72.072	Ea.	48,100	4,025		52,125	59,000
1320 1675 MBH, 50 HP	"	.26	93.023		58,000	5,200		63,200	72,000
1330 3350 MBH, 100 HP	Q-7	.21	152		83,000	8,675		91,675	104,500
1340 5025 MBH, 150 HP		.17	192		106,000	11,000		117,000	133,000
1350 6700 MBH, 200 HP		.14	223		117,500	12,700		130,200	148,500
1360 10,050 MBH, 300 HP		.13	251		165,500	14,300		179,800	203,500
1400 Packaged scotch marine, #6 oil									
1410 15 PSI steam									
1420 1675 MBH, 50 HP	Q-6	.25	97.166	Ea.	63,000	5,425		68,425	77,500
1430 3350 MBH, 100 HP	Q-7	.20	159		82,500	9,050		91,550	104,000
1440 5025 MBH, 150 HP		.16	202		101,000	11,500		112,500	128,500
1450 6700 MBH, 200 HP		.14	233		115,000	13,300		128,300	147,000
1460 8375 MBH, 250 HP		.13	244		132,000	13,900		145,900	166,000
1470 10,050 MBH, 300 HP		.12	262		148,500	14,900		163,400	186,000
1480 13,400 MBH, 400 HP		.09	351		182,500	20,000		202,500	230,500
1500 150 PSI steam									
1510 1675 MBH, 50 HP	Q-6	.25	97.166	Ea.	75,000	5,425		80,425	90,500
1520 3350 MBH, 100 HP	Q-7	.20	159		94,500	9,050		103,550	117,500
1530 5025 MBH, 150 HP		.16	202		117,500	11,500		129,000	146,500
1540 6700 MBH, 200 HP		.14	233		129,000	13,300		142,300	162,000

23 52 Heating Boilers

23 52 39 – Fire-Tube Boilers

23 52 39.13 Scotch Marine Boilers		Crew	Daily Output	Labor-Hours	Unit	Material	2015 Bare Costs Labor	Equipment	Total	Total Incl O&P
1550	8375 MBH, 250 HP	Q-7	.13	244	Ea.	156,500	13,900		170,400	193,000
1560	10,050 MBH, 300 HP		.12	262		174,000	14,900		188,900	214,000
1570	13,400 MBH, 400 HP	↓	.09	351	↓	202,000	20,000		222,000	252,000

23 52 84 – Boiler Blowdown

23 52 84.10 Boiler Blowdown Systems

		Crew	Daily Output	Labor-Hours	Unit	Material	Labor	Equipment	Total	Total Incl O&P
0010	**BOILER BLOWDOWN SYSTEMS**									
1010	Boiler blowdown, auto/manual to 2000 MBH	Q-5	3.75	4.267	Ea.	4,850	229		5,079	5,675
1020	7300 MBH	"	3	5.333	"	5,550	287		5,837	6,525

23 52 88 – Burners

23 52 88.10 Replacement Type Burners

		Crew	Daily Output	Labor-Hours	Unit	Material	Labor	Equipment	Total	Total Incl O&P
0010	**REPLACEMENT TYPE BURNERS**									
0990	Residential, conversion, gas fired, LP or natural									
1000	Gun type, atmospheric input 50 to 225 MBH	Q-1	2.50	6.400	Ea.	720	340		1,060	1,300
1020	100 to 400 MBH		2	8		1,200	425		1,625	1,975
1040	300 to 1000 MBH	↓	1.70	9.412	↓	3,625	495		4,120	4,725
2000	Commercial and industrial, gas/oil, input									
2050	400 MBH	Q-1	1.50	10.667	Ea.	4,175	565		4,740	5,450
2090	670 MBH		1.40	11.429		4,175	605		4,780	5,500
2140	1155 MBH		1.30	12.308		4,650	650		5,300	6,100
2200	1800 MBH		1.20	13.333		4,675	705		5,380	6,200
2260	3000 MBH		1.10	14.545		4,675	770		5,445	6,300
2320	4100 MBH	↓	1	16	↓	4,675	845		5,520	6,425
3000	Flame retention oil fired assembly, input									
3020	.50 to 2.25 GPH	Q-1	2.40	6.667	Ea.	284	350		634	840
3040	2.0 to 5.0 GPH		2	8		315	425		740	985
3060	3.0 to 7.0 GPH		1.80	8.889		570	470		1,040	1,350
3080	6.0 to 12.0 GPH	↓	1.60	10		915	530		1,445	1,800
4600	Gas safety, shut off valve, 3/4" threaded	1 Stpi	20	.400		178	24		202	232
4610	1" threaded		19	.421		172	25		197	228
4620	1-1/4" threaded		15	.533		194	32		226	261
4630	1-1/2" threaded		13	.615		210	37		247	287
4640	2" threaded	↓	11	.727		235	43.50		278.50	325
4650	2-1/2" threaded	Q-1	15	1.067		269	56.50		325.50	380
4660	3" threaded		13	1.231		370	65		435	505
4670	4" flanged	↓	3	5.333		2,625	282		2,907	3,325
4680	6" flanged	Q-2	3	8	↓	5,875	440		6,315	7,125

23 53 Heating Boiler Feedwater Equipment

23 53 24 – Shot Chemical Feeder

23 53 24.10 Chemical Feeder

		Crew	Daily Output	Labor-Hours	Unit	Material	Labor	Equipment	Total	Total Incl O&P
0010	**CHEMICAL FEEDER**									
0200	Shot chem feeder, by pass, in line mount, 300 PSIG, 1.7 gallon	2 Stpi	1.80	8.889	Ea.	252	530		782	1,075
0230	12 gallon		.93	17.204		465	1,025		1,490	2,050
0400	300 lb., ASME, 5 gallon		1.24	12.903		287	770		1,057	1,500
0420	10 gallon	↓	.93	17.204	↓	430	1,025		1,455	2,025

23 54 Furnaces

23 54 13 – Electric-Resistance Furnaces

23 54 13.10 Electric Furnaces

		Crew	Daily Output	Labor-Hours	Unit	Material	2015 Bare Costs Labor	2015 Bare Costs Equipment	Total	Total Incl O&P
0010	**ELECTRIC FURNACES,** Hot air, blowers, std. controls									
0011	not including gas, oil or flue piping									
1000	Electric, UL listed									
1100	34.1 MBH	Q-20	4.40	4.545	Ea.	455	233		688	855
1120	51.6 MBH		4.20	4.762		495	244		739	915
1140	68.3 MBH		4	5		515	256		771	955
1160	85.3 MBH		3.80	5.263		520	270		790	985

23 54 16 – Fuel-Fired Furnaces

23 54 16.13 Gas-Fired Furnaces

		Crew	Daily Output	Labor-Hours	Unit	Material	2015 Bare Costs Labor	2015 Bare Costs Equipment	Total	Total Incl O&P
0010	**GAS-FIRED FURNACES**									
3000	Gas, AGA certified, upflow, direct drive models									
3020	45 MBH input	Q-9	4	4	Ea.	535	201		736	900
3040	60 MBH input		3.80	4.211		535	212		747	915
3060	75 MBH input		3.60	4.444		575	224		799	975
3100	100 MBH input		3.20	5		625	252		877	1,075
3120	125 MBH input		3	5.333		655	269		924	1,125
3130	150 MBH input		2.80	5.714		670	288		958	1,175
3140	200 MBH input		2.60	6.154		2,800	310		3,110	3,550
3160	300 MBH input		2.30	6.957		3,100	350		3,450	3,950
3180	400 MBH input		2	8		3,450	405		3,855	4,425
3200	Gas fired wall furnace									
3204	Horizontal flow									
3208	7.7 MBH	Q-9	7	2.286	Ea.	485	115		600	710
3212	14 MBH		6.50	2.462		485	124		609	725
3216	24 MBH		5	3.200		510	161		671	810
3220	49 MBH		4	4		700	201		901	1,075
3224	65 MBH		3.60	4.444		1,150	224		1,374	1,625
3240	Up-flow									
3244	7.7 MBH	Q-9	7	2.286	Ea.	485	115		600	710
3248	14 MBH		6.50	2.462		485	124		609	725
3252	24 MBH		5	3.200		500	161		661	795
3254	35 MBH		4.50	3.556		780	179		959	1,125
3256	49 MBH		4	4		1,175	201		1,376	1,600
3260	65 MBH		3.60	4.444		1,300	224		1,524	1,775
3500	Gas furnace									
3510	Up-flow									
3514	1200 CFM, 45 MBH	Q-9	4	4	Ea.	1,125	201		1,326	1,550
3518	75 MBH		3.60	4.444		1,250	224		1,474	1,725
3522	100 MBH		3.20	5		1,375	252		1,627	1,875
3526	125 MBH		3	5.333		1,400	269		1,669	1,925
3530	150 MBH		2.80	5.714		1,550	288		1,838	2,175
3540	2000 CFM, 75 MBH		3.60	4.444		1,375	224		1,599	1,850
3544	100 MBH		3.20	5		1,550	252		1,802	2,075
3548	120 MBH		3	5.333		1,625	269		1,894	2,200
3560	Horizontal flow									
3564	1200 CFM, 45 MBH	Q-9	4	4	Ea.	1,300	201		1,501	1,750
3568	60 MBH		3.80	4.211		1,375	212		1,587	1,825
3572	70 MBH		3.60	4.444		1,400	224		1,624	1,900
3576	2000 CFM, 90 MBH		3.30	4.848		1,550	244		1,794	2,075
3580	110 MBH		3.10	5.161		1,800	260		2,060	2,375

For customer support on your Mechanical Cost Data, call 877.795.7394.

371

23 54 16 – Fuel-Fired Furnaces

23 54 16.14 Condensing Furnaces		Crew	Daily Output	Labor-Hours	Unit	Material	2015 Bare Costs Labor	Equipment	Total	Total Incl O&P
0010	**CONDENSING FURNACES**, High efficiency									
0020	Oil fired, packaged, complete									
0030	Upflow									
0040	Output @ 95% A.F.U.E.									
0100	49 MBH @ 1000 CFM	Q-9	3.70	4.324	Ea.	4,450	218		4,668	5,225
0110	73.5 MBH @ 2000 CFM		3.60	4.444		4,650	224		4,874	5,475
0120	96 MBH @ 2000 CFM		3.40	4.706		4,650	237		4,887	5,475
0130	115.6 MBH @ 2000 CFM		3.40	4.706		4,650	237		4,887	5,475
0140	147 MBH @ 2000 CFM		3.30	4.848		10,400	244		10,644	11,900
0150	192 MBH @ 4000 CFM		2.60	6.154		10,800	310		11,110	12,400
0170	231.5 MBH @ 4000 CFM		2.30	6.957		10,800	350		11,150	12,400
0260	For variable speed motor, add					565			565	625
0270	Note: Also available in horizontal, counterflow and lowboy configurations.									

23 54 16.16 Oil-Fired Furnaces

23 54 16.16 Oil-Fired Furnaces		Crew	Daily Output	Labor-Hours	Unit	Material	Labor	Equipment	Total	Total Incl O&P
0010	**OIL-FIRED FURNACES**									
6000	Oil, UL listed, atomizing gun type burner									
6020	56 MBH output	Q-9	3.60	4.444	Ea.	1,725	224		1,949	2,250
6030	84 MBH output		3.50	4.571		1,850	230		2,080	2,400
6040	95 MBH output		3.40	4.706		1,875	237		2,112	2,425
6060	134 MBH output		3.20	5		2,175	252		2,427	2,775
6080	151 MBH output		3	5.333		2,250	269		2,519	2,875
6100	200 MBH input		2.60	6.154		2,450	310		2,760	3,175
6120	300 MBH input		2.30	6.957		3,600	350		3,950	4,500
6140	400 MBH input		2	8		3,900	405		4,305	4,925
6200	Up flow									
6210	105 MBH	Q-9	3.40	4.706	Ea.	1,575	237		1,812	2,100
6220	140 MBH		3.30	4.848		1,725	244		1,969	2,250
6230	168 MBH		2.90	5.517		3,075	278		3,353	3,800
6260	Lo-boy style									
6270	105 MBH	Q-9	3.40	4.706	Ea.	1,575	237		1,812	2,100
6280	140 MBH		3.30	4.848		1,725	244		1,969	2,250
6290	189 MBH		2.80	5.714		3,075	288		3,363	3,825
6320	Down flow									
6330	105 MBH	Q-9	3.40	4.706	Ea.	1,575	237		1,812	2,100
6340	140 MBH		3.30	4.848		1,725	244		1,969	2,250
6350	168 MBH		2.90	5.517		3,275	278		3,553	4,050
6380	Horizontal									
6390	105 MBH	Q-9	3.40	4.706	Ea.	1,575	237		1,812	2,100
6400	140 MBH		3.30	4.848		1,725	244		1,969	2,250
6410	189 MBH		2.80	5.714		2,975	288		3,263	3,700

23 54 16.21 Solid Fuel-Fired Furnaces

23 54 16.21 Solid Fuel-Fired Furnaces			Crew	Daily Output	Labor-Hours	Unit	Material	Labor	Equipment	Total	Total Incl O&P
0010	**SOLID FUEL-FIRED FURNACES**										
6020	Wood fired furnaces										
6030	Includes hot water coil, thermostat, and auto draft control										
6040	24" long firebox	G	Q-9	4	4	Ea.	4,300	201		4,501	5,025
6050	30" long firebox	G		3.60	4.444		5,025	224		5,249	5,875
6060	With fireplace glass doors	G		3.20	5		6,350	252		6,602	7,375
6200	Wood/oil fired furnaces, includes two thermostats										
6210	Includes hot water coil and auto draft control										
6240	24" long firebox	G	Q-9	3.40	4.706	Ea.	5,150	237		5,387	6,025
6250	30" long firebox	G		3	5.333		5,775	269		6,044	6,750
6260	With fireplace glass doors	G		2.80	5.714		7,075	288		7,363	8,225

23 54 Furnaces

23 54 16 – Fuel-Fired Furnaces

23 54 16.21 Solid Fuel-Fired Furnaces		Crew	Daily Output	Labor-Hours	Unit	Material	2015 Bare Costs Labor	2015 Bare Costs Equipment	Total	Total Incl O&P
6400	Wood/gas fired furnaces, includes two thermostats									
6410	Includes hot water coil and auto draft control									
6440	24" long firebox	G Q-9	2.80	5.714	Ea.	5,925	288		6,213	6,975
6450	30" long firebox	G	2.40	6.667		6,625	335		6,960	7,825
6460	With fireplace glass doors	G	2	8		7,900	405		8,305	9,325
6600	Wood/oil/gas fired furnaces, optional accessories									
6610	Hot air plenum	Q-9	16	1	Ea.	108	50.50		158.50	196
6620	Safety heat dump		24	.667		94	33.50		127.50	155
6630	Auto air intake		18	.889		148	45		193	232
6640	Cold air return package		14	1.143		159	57.50		216.50	263
6650	Wood fork					51			51	56
6700	Wood fired outdoor furnace									
6740	24" long firebox	G Q-9	3.80	4.211	Ea.	4,500	212		4,712	5,275
6760	Wood fired outdoor furnace, optional accessories									
6770	Chimney section, stainless steel, 6" ID x 3' lg.	G Q-9	36	.444	Ea.	165	22.50		187.50	216
6780	Chimney cap, stainless steel	G "	40	.400	"	95	20		115	136
6800	Wood fired hot water furnace									
6820	Includes 200 gal. hot water storage, thermostat, and auto draft control									
6840	30" long firebox	G Q-9	2.10	7.619	Ea.	8,275	385		8,660	9,675
6850	Water to air heat exchanger									
6870	Includes mounting kit and blower relay									
6880	140 MBH, 18.75" W x 18.75" L	Q-9	7.50	2.133	Ea.	370	107		477	570
6890	200 MBH, 24" W x 24" L	"	7	2.286	"	525	115		640	755
6900	Water to water heat exchanger									
6940	100 MBH	Q-9	6.50	2.462	Ea.	370	124		494	600
6960	290 MBH	"	6	2.667	"	535	134		669	795
7000	Optional accessories									
7010	Large volume circulation pump, (2 Included)	Q-9	14	1.143	Ea.	242	57.50		299.50	355
7020	Air bleed fittings, (package)		24	.667		50.50	33.50		84	107
7030	Domestic water preheater		6	2.667		219	134		353	445
7040	Smoke pipe kit		4	4		88	201		289	405

23 54 24 – Furnace Components for Cooling

23 54 24.10 Furnace Components and Combinations		Crew	Daily Output	Labor-Hours	Unit	Material	2015 Bare Costs Labor	2015 Bare Costs Equipment	Total	Total Incl O&P
0010	**FURNACE COMPONENTS AND COMBINATIONS**									
0080	Coils, A.C. evaporator, for gas or oil furnaces									
0090	Add-on, with holding charge									
0100	Upflow									
0120	1-1/2 ton cooling	Q-5	4	4	Ea.	220	215		435	565
0130	2 ton cooling		3.70	4.324		241	233		474	615
0140	3 ton cooling		3.30	4.848		375	261		636	810
0150	4 ton cooling		3	5.333		465	287		752	950
0160	5 ton cooling		2.70	5.926		505	320		825	1,025
0300	Downflow									
0330	2-1/2 ton cooling	Q-5	3	5.333	Ea.	295	287		582	760
0340	3-1/2 ton cooling		2.60	6.154		435	330		765	975
0350	5 ton cooling		2.20	7.273		505	390		895	1,150
0600	Horizontal									
0630	2 ton cooling	Q-5	3.90	4.103	Ea.	320	221		541	690
0640	3 ton cooling		3.50	4.571		370	246		616	775
0650	4 ton cooling		3.20	5		460	269		729	910
0660	5 ton cooling		2.90	5.517		460	297		757	955
2000	Cased evaporator coils for air handlers									

For customer support on your Mechanical Cost Data, call 877.795.7394.

373

23 54 24 – Furnace Components for Cooling

23 54 24.10 Furnace Components and Combinations	Crew	Daily Output	Labor-Hours	Unit	Material	2015 Bare Costs Labor	Equipment	Total	Total Incl O&P
2100 1-1/2 ton cooling	Q-5	4.40	3.636	Ea.	267	196		463	59C
2110 2 ton cooling		4.10	3.902		300	210		510	645
2120 2-1/2 ton cooling		3.90	4.103		320	221		541	685
2130 3 ton cooling		3.70	4.324		360	233		593	745
2140 3-1/2 ton cooling		3.50	4.571		485	246		731	905
2150 4 ton cooling		3.20	5		515	269		784	975
2160 5 ton cooling		2.90	5.517		525	297		822	1,025
3010 Air handler, modular									
3100 With cased evaporator cooling coil									
3120 1-1/2 ton cooling	Q-5	3.80	4.211	Ea.	850	226		1,076	1,275
3130 2 ton cooling		3.50	4.571		940	246		1,186	1,400
3140 2-1/2 ton cooling		3.30	4.848		995	261		1,256	1,500
3150 3 ton cooling		3.10	5.161		1,150	278		1,428	1,675
3160 3-1/2 ton cooling		2.90	5.517		1,325	297		1,622	1,900
3170 4 ton cooling		2.50	6.400		1,425	345		1,770	2,075
3180 5 ton cooling		2.10	7.619		1,700	410		2,110	2,475
3500 With no cooling coil									
3520 1-1/2 ton coil size	Q-5	12	1.333	Ea.	620	71.50		691.50	790
3530 2 ton coil size		10	1.600		620	86		706	815
3540 2-1/2 ton coil size		10	1.600		725	86		811	925
3554 3 ton coil size		9	1.778		825	95.50		920.50	1,050
3560 3-1/2 ton coil size		9	1.778		845	95.50		940.50	1,075
3570 4 ton coil size		8.50	1.882		1,050	101		1,151	1,325
3580 5 ton coil size		8	2		1,075	108		1,183	1,325
4000 With heater									
4120 5 kW, 17.1 MBH	Q-5	16	1	Ea.	670	54		724	815
4130 7.5 kW, 25.6 MBH		15.60	1.026		680	55		735	830
4140 10 kW, 34.2 MBH		15.20	1.053		845	56.50		901.50	1,025
4150 12.5 KW, 42.7 MBH		14.80	1.081		950	58		1,008	1,150
4160 15 KW, 51.2 MBH		14.40	1.111		1,025	60		1,085	1,225
4170 25 KW, 85.4 MBH		14	1.143		1,375	61.50		1,436.50	1,600
4180 30 KW, 102 MBH		13	1.231		1,600	66		1,666	1,850

23 55 Fuel-Fired Heaters

23 55 13 – Fuel-Fired Duct Heaters

23 55 13.16 Gas-Fired Duct Heaters

	Crew	Daily Output	Labor-Hours	Unit	Material	2015 Bare Costs Labor	Equipment	Total	Total Incl O&P
0010 **GAS-FIRED DUCT HEATERS**, Includes burner, controls, stainless steel									
0020 heat exchanger. Gas fired, electric ignition									
0030 Indoor installation									
0080 100 MBH output	Q-5	5	3.200	Ea.	2,900	172		3,072	3,425
0100 120 MBH output		4	4		3,175	215		3,390	3,825
0130 200 MBH output		2.70	5.926		4,025	320		4,345	4,900
0140 240 MBH output		2.30	6.957		4,225	375		4,600	5,225
0160 280 MBH output		2	8		4,600	430		5,030	5,700
0180 320 MBH output		1.60	10		5,150	540		5,690	6,475
0300 For powered venter and adapter, add					525			525	575
0502 For required flue pipe, see Section 23 51 23.10									
1000 Outdoor installation, with power venter									
1020 75 MBH output	Q-5	4	4	Ea.	3,525	215		3,740	4,200
1040 94 MBH output		4	4		3,625	215		3,840	4,300
1060 120 MBH output		4	4		3,875	215		4,090	4,575

23 55 Fuel-Fired Heaters

23 55 13 – Fuel-Fired Duct Heaters

23 55 13.16 Gas-Fired Duct Heaters

		Crew	Daily Output	Labor-Hours	Unit	Material	2015 Bare Costs Labor	Equipment	Total	Total Incl O&P
1080	157 MBH output	Q-5	3.50	4.571	Ea.	4,200	246		4,446	4,975
1100	187 MBH output		3	5.333		4,800	287		5,087	5,700
1120	225 MBH output		2.50	6.400		5,000	345		5,345	6,025
1140	300 MBH output		1.80	8.889		7,750	480		8,230	9,250
1160	375 MBH output		1.60	10		8,150	540		8,690	9,775
1180	450 MBH output		1.40	11.429		9,350	615		9,965	11,200
1200	600 MBH output		1	16		10,600	860		11,460	12,900
1300	Aluminized exchanger, subtract					15%				
1500	For two stage gas valve, add					755			755	830

23 55 33 – Fuel-Fired Unit Heaters

23 55 33.13 Oil-Fired Unit Heaters

		Crew	Daily Output	Labor-Hours	Unit	Material	2015 Bare Costs Labor	Equipment	Total	Total Incl O&P
0010	**OIL-FIRED UNIT HEATERS**, Cabinet, grilles, fan, ctrl., burner, no piping									
6000	Oil fired, suspension mounted, 94 MBH output	Q-5	4	4	Ea.	4,825	215		5,040	5,625
6040	140 MBH output		3	5.333		5,075	287		5,362	6,000
6060	184 MBH output		3	5.333		5,375	287		5,662	6,350

23 55 33.16 Gas-Fired Unit Heaters

		Crew	Daily Output	Labor-Hours	Unit	Material	2015 Bare Costs Labor	Equipment	Total	Total Incl O&P
0010	**GAS-FIRED UNIT HEATERS**, Cabinet, grilles, fan, ctrls., burner, no piping									
0022	thermostat, no piping. For flue see Section 23 51 23.10									
1000	Gas fired, floor mounted									
1100	60 MBH output	Q-5	10	1.600	Ea.	870	86		956	1,075
1120	80 MBH output		9	1.778		880	95.50		975.50	1,100
1140	100 MBH output		8	2		960	108		1,068	1,200
1160	120 MBH output		7	2.286		1,150	123		1,273	1,425
1180	180 MBH output		6	2.667		1,375	143		1,518	1,725
1500	Rooftop mounted, power vent, stainless steel exchanger									
1520	75 MBH input	Q-6	4	6	Ea.	5,625	335		5,960	6,700
1540	100 MBH input		3.60	6.667		5,875	370		6,245	7,025
1560	125 MBH input		3.30	7.273		6,125	405		6,530	7,350
1580	150 MBH input		3	8		6,600	445		7,045	7,950
1600	175 MBH input		2.60	9.231		6,725	515		7,240	8,175
1620	225 MBH input		2.30	10.435		7,725	580		8,305	9,375
1640	300 MBH input		1.90	12.632		8,950	705		9,655	10,900
1660	350 MBH input		1.40	17.143		9,750	955		10,705	12,200
1680	450 MBH input		1.20	20		10,200	1,125		11,325	13,000
1720	700 MBH input		.80	30		10,600	1,675		12,275	14,200
1760	1200 MBH input		.30	80		11,000	4,450		15,450	18,800
1900	For aluminized steel exchanger, subtract					10%				
2000	Suspension mounted, propeller fan, 20 MBH output	Q-5	8.50	1.882		1,125	101		1,226	1,400
2020	40 MBH output		7.50	2.133		1,550	115		1,665	1,875
2040	60 MBH output		7	2.286		1,700	123		1,823	2,050
2060	80 MBH output		6	2.667		1,875	143		2,018	2,275
2080	100 MBH output		5.50	2.909		2,050	156		2,206	2,475
2100	130 MBH output		5	3.200		2,225	172		2,397	2,700
2120	140 MBH output		4.50	3.556		2,350	191		2,541	2,875
2140	160 MBH output		4	4		2,450	215		2,665	3,000
2160	180 MBH output		3.50	4.571		2,625	246		2,871	3,275
2180	200 MBH output		3	5.333		2,900	287		3,187	3,600
2200	240 MBH output		2.70	5.926		3,175	320		3,495	3,975
2220	280 MBH output		2.30	6.957		3,500	375		3,875	4,425
2240	320 MBH output		2	8		4,100	430		4,530	5,175
2500	For powered venter and adapter, add					490			490	540
3000	Suspension mounted, blower type, 40 MBH output	Q-5	6.80	2.353		935	127		1,062	1,225

For customer support on your Mechanical Cost Data, call 877.795.7394.

375

23 55 33 – Fuel-Fired Unit Heaters

23 55 33.16 Gas-Fired Unit Heaters	Crew	Daily Output	Labor-Hours	Unit	Material	2015 Bare Costs Labor	2015 Bare Costs Equipment	Total	Total Incl O&P
3020	60 MBH output	Q-5	6.60	2.424	Ea.	960	130		1,090	1,250
3040	84 MBH output		5.80	2.759		995	148		1,143	1,325
3060	104 MBH output		5.20	3.077		1,025	165		1,190	1,375
3080	140 MBH output		4.30	3.721		1,100	200		1,300	1,500
3100	180 MBH output		3.30	4.848		1,300	261		1,561	1,825
3120	240 MBH output		2.50	6.400		1,500	345		1,845	2,175
3140	280 MBH output		2	8		1,675	430		2,105	2,500
4000	Suspension mounted, sealed combustion system,									
4020	Aluminized steel exchanger, powered vent									
4040	100 MBH output	Q-5	5	3.200	Ea.	3,025	172		3,197	3,575
4060	120 MBH output		4.70	3.404		3,600	183		3,783	4,225
4080	160 MBH output		3.70	4.324		4,150	233		4,383	4,925
4100	200 MBH output		2.90	5.517		5,100	297		5,397	6,050
4120	240 MBH output		2.50	6.400		5,450	345		5,795	6,525
4140	320 MBH output		1.70	9.412		6,525	505		7,030	7,950
5000	Wall furnace, 17.5 MBH output		6	2.667		745	143		888	1,025
5020	24 MBH output		5	3.200		745	172		917	1,075
5040	35 MBH output		4	4		790	215		1,005	1,200

23 56 Solar Energy Heating Equipment

23 56 16 – Packaged Solar Heating Equipment

23 56 16.40 Solar Heating Systems

0010	**SOLAR HEATING SYSTEMS**	R235616-60								
0020	System/Package prices, not including connecting									
0030	pipe, insulation, or special heating/plumbing fixtures									
0152	For solar ultraviolet pipe insulation see Section 22 07 19.10									
0500	Hot water, standard package, low temperature									
0540	1 collector, circulator, fittings, 65 gal. tank	G	Q-1	.50	32	Ea.	3,675	1,700		5,375	6,600
0580	2 collectors, circulator, fittings, 120 gal. tank	G		.40	40		5,050	2,125		7,175	8,750
0620	3 collectors, circulator, fittings, 120 gal. tank	G		.34	47.059		6,875	2,475		9,350	11,300
0700	Medium temperature package									
0720	1 collector, circulator, fittings, 80 gal. tank	G	Q-1	.50	32	Ea.	5,025	1,700		6,725	8,075
0740	2 collectors, circulator, fittings, 120 gal. tank	G		.40	40		6,450	2,125		8,575	10,300
0780	3 collectors, circulator, fittings, 120 gal. tank	G		.30	53.333		7,325	2,825		10,150	12,300
0980	For each additional 120 gal. tank, add	G					1,750			1,750	1,925

23 56 19 – Solar Heating Components

23 56 19.50 Solar Heating Ancillary

0010	**SOLAR HEATING ANCILLARY**									
2300	Circulators, air									
2310	Blowers									
2330	100-300 S.F. system, 1/10 HP	G	Q-9	16	1	Ea.	249	50.50		299.50	350
2340	300-500 S.F. system, 1/5 HP	G		15	1.067		330	53.50		383.50	440
2350	Two speed, 100-300 S.F., 1/10 HP	G		14	1.143		142	57.50		199.50	245
2400	Reversible fan, 20" diameter, 2 speed	G		18	.889		113	45		158	193
2550	Booster fan 6" diameter, 120 CFM	G		16	1		36.50	50.50		87	118
2570	6" diameter, 225 CFM	G		16	1		45	50.50		95.50	127
2580	8" diameter, 150 CFM	G		16	1		41	50.50		91.50	123
2590	8" diameter, 310 CFM	G		14	1.143		63.50	57.50		121	158
2600	8" diameter, 425 CFM	G		14	1.143		71	57.50		128.50	166
2650	Rheostat	G		32	.500		15.45	25		40.45	55.50

376

For customer support on your Mechanical Cost Data, call 877.795.7394.

23 56 19.50 Solar Heating Ancillary		Crew	Daily Output	Labor-Hours	Unit	Material	2015 Bare Costs Labor	Equipment	Total	Total Incl O&P
2660	Shutter/damper	G Q-9	12	1.333	Ea.	56	67		123	165
2670	Shutter motor	G	16	1		139	50.50		189.50	230
2800	Circulators, liquid, 1/25 HP, 5.3 GPM	G Q-1	14	1.143		114	60.50		174.50	217
2820	1/20 HP, 17 GPM	G	12	1.333		161	70.50		231.50	283
2850	1/20 HP, 17 GPM, stainless steel	G	12	1.333		254	70.50		324.50	385
2870	1/12 HP, 30 GPM	G	10	1.600		345	84.50		429.50	510
3000	Collector panels, air with aluminum absorber plate									
3010	Wall or roof mount									
3040	Flat black, plastic glazing									
3080	4' x 8'	G Q-9	6	2.667	Ea.	660	134		794	935
3100	4' x 10'	G	5	3.200	"	815	161		976	1,150
3200	Flush roof mount, 10' to 16' x 22" wide	G	96	.167	L.F.	132	8.40		140.40	158
3210	Manifold, by L.F. width of collectors	G	160	.100	"	144	5.05		149.05	167
3300	Collector panels, liquid with copper absorber plate									
3320	Black chrome, tempered glass glazing									
3330	Alum. frame, 4' x 8', 5/32" single glazing	G Q-1	9.50	1.684	Ea.	995	89		1,084	1,225
3390	Alum. frame, 4' x 10', 5/32" single glazing	G	6	2.667		1,150	141		1,291	1,475
3450	Flat black, alum. frame, 3.5' x 7.5'	G	9	1.778		880	94		974	1,100
3500	4' x 8'	G	5.50	2.909		1,050	154		1,204	1,375
3520	4' x 10'	G	10	1.600		1,250	84.50		1,334.50	1,500
3540	4' x 12.5'	G	5	3.200		1,250	169		1,419	1,625
3550	Liquid with fin tube absorber plate									
3560	Alum. frame 4' x 8' tempered glass	G Q-1	10	1.600	Ea.	580	84.50		664.50	770
3580	Liquid with vacuum tubes, 4' x 6'-10"	G	9	1.778		895	94		989	1,125
3600	Liquid, full wetted, plastic, alum. frame, 4' x 10'	G	5	3.200		320	169		489	610
3650	Collector panel mounting, flat roof or ground rack	G	7	2.286		244	121		365	450
3670	Roof clamps	G	70	.229	Set	2.80	12.10		14.90	21.50
3700	Roof strap, teflon	G 1 Plum	205	.039	L.F.	23.50	2.29		25.79	29.50
3900	Differential controller with two sensors									
3930	Thermostat, hard wired	G 1 Plum	8	1	Ea.	101	58.50		159.50	200
3950	Line cord and receptacle	G	12	.667		130	39		169	202
4050	Pool valve system	G	2.50	3.200		161	188		349	460
4070	With 12 VAC actuator	G	2	4		325	235		560	710
4080	Pool pump system, 2" pipe size	G	6	1.333		195	78.50		273.50	330
4100	Five station with digital read-out	G	3	2.667		263	157		420	525
4150	Sensors									
4200	Brass plug, 1/2" MPT	G 1 Plum	32	.250	Ea.	16	14.70		30.70	39.50
4210	Brass plug, reversed	G	32	.250		26.50	14.70		41.20	51
4220	Freeze prevention	G	32	.250		22.50	14.70		37.20	47
4240	Screw attached	G	32	.250		9.85	14.70		24.55	33
4250	Brass, immersion	G	32	.250		13.55	14.70		28.25	37
4300	Heat exchanger									
4315	includes coil, blower, circulator									
4316	and controller for DHW and space hot air									
4330	Fluid to air coil, up flow, 45 MBH	G Q-1	4	4	Ea.	315	211		526	670
4380	70 MBH	G	3.50	4.571		355	242		597	760
4400	80 MBH	G	3	5.333		475	282		757	950
4580	Fluid to fluid package includes two circulating pumps									
4590	expansion tank, check valve, relief valve									
4600	controller, high temperature cutoff and sensors	G Q-1	2.50	6.400	Ea.	800	340		1,140	1,400
4650	Heat transfer fluid									
4700	Propylene glycol, inhibited anti-freeze	G 1 Plum	28	.286	Gal.	15.55	16.75		32.30	42.50
4800	Solar storage tanks, knocked down									

For customer support on your Mechanical Cost Data, call 877.795.7394.

377

23 56 Solar Energy Heating Equipment

23 56 19 – Solar Heating Components

23 56 19.50 Solar Heating Ancillary		Crew	Daily Output	Labor-Hours	Unit	Material	2015 Bare Costs Labor	Equipment	Total	Total Incl O&P
4810	Air, galvanized steel clad, double wall, 4" fiberglass insulation						.			
5120	45 Mil reinforced polypropylene lining,									
5140	4' high, 4' x 4' = 64 C.F./450 gallons	G	Q-9	2	8	Ea.	3,650	405		4,055	4,625
5150	4' x 8' = 128 C.F./900 gallons	G		1.50	10.667		5,450	535		5,985	6,825
5160	4' x 12' = 190 C.F./1300 gallons	G		1.30	12.308		7,275	620		7,895	8,950
5170	8' x 8' = 250 C.F./1700 gallons	G		1	16		7,275	805		8,080	9,225
5190	6'-3" high, 7' x 7' = 306 C.F./2000 gallons	G	Q-10	1.20	20		14,200	1,050		15,250	17,300
5200	7' x 10'-6" = 459 C.F./3000 gallons	G		.80	30		17,800	1,575		19,375	22,000
5210	7' x 14' = 613 C.F./4000 gallons	G		.60	40		21,400	2,100		23,500	26,700
5220	10'-6" x 10'-6" = 689 C.F./4500 gallons	G		.50	48		21,400	2,500		23,900	27,300
5230	10'-6" x 14' = 919 C.F./6000 gallons	G		.40	60		24,900	3,125		28,025	32,200
5240	14' x 14' = 1225 C.F./8000 gallons	G	Q-11	.40	80		28,500	4,250		32,750	37,800
5250	14' x 17'-6" = 1531 C.F./10,000 gallons	G		.30	106		32,100	5,675		37,775	44,000
5260	17'-6" x 17'-6" = 1914 C.F./12,500 gallons	G		.25	128		35,600	6,825		42,425	49,600
5270	17'-6" x 21' = 2297 C.F./15,000 gallons	G		.20	160		39,100	8,525		47,625	56,000
5280	21' x 21' = 2756 C.F./18,000 gallons	G		.18	177		42,700	9,475		52,175	61,500
5290	30 Mil reinforced Hypalon lining, add						.02%			
7000	Solar control valves and vents									
7050	Air purger, 1" pipe size	G	1 Plum	12	.667	Ea.	46.50	39		85.50	111
7070	Air eliminator, automatic 3/4" size	G		32	.250		30	14.70		44.70	55
7090	Air vent, automatic, 1/8" fitting	G		32	.250		17.25	14.70		31.95	41
7100	Manual, 1/8" NPT	G		32	.250		2.44	14.70		17.14	24.50
7120	Backflow preventer, 1/2" pipe size	G		16	.500		75.50	29.50		105	128
7130	3/4" pipe size	G		16	.500		77	29.50		106.50	129
7150	Balancing valve, 3/4" pipe size	G		20	.400		63	23.50		86.50	105
7180	Draindown valve, 1/2" copper tube	G		9	.889		212	52		264	310
7200	Flow control valve, 1/2" pipe size	G		22	.364		120	21.50		141.50	164
7220	Expansion tank, up to 5 gal.	G		32	.250		68	14.70		82.70	97
7250	Hydronic controller (aquastat)	G		8	1		228	58.50		286.50	340
7400	Pressure gauge, 2" dial	G		32	.250		24	14.70		38.70	48.50
7450	Relief valve, temp. and pressure 3/4" pipe size	G		30	.267		21	15.65		36.65	46.50
7500	Solenoid valve, normally closed									
7520	Brass, 3/4" NPT, 24V	G	1 Plum	9	.889	Ea.	123	52		175	214
7530	1" NPT, 24V	G		9	.889		193	52		245	291
7750	Vacuum relief valve, 3/4" pipe size	G		32	.250		29.50	14.70		44.20	54.50
7800	Thermometers									
7820	Digital temperature monitoring, 4 locations	G	1 Plum	2.50	3.200	Ea.	137	188		325	435
7900	Upright, 1/2" NPT	G		8	1		39.50	58.50		98	132
7970	Remote probe, 2" dial	G		8	1		33	58.50		91.50	125
7990	Stem, 2" dial, 9" stem	G		16	.500		21.50	29.50		51	68.50
8250	Water storage tank with heat exchanger and electric element									
8270	66 gal. with 2" x 2 lb. density insulation	G	1 Plum	1.60	5	Ea.	1,600	294		1,894	2,225
8300	80 gal. with 2" x 2 lb. density insulation	G		1.60	5		1,600	294		1,894	2,225
8380	120 gal. with 2" x 2 lb. density insulation	G		1.40	5.714		1,825	335		2,160	2,525
8400	120 gal. with 2" x 2 lb. density insul., 40 S.F. heat coil	G		1.40	5.714		2,325	335		2,660	3,050
8500	Water storage module, plastic									
8600	Tubular, 12" diameter, 4' high	G	1 Carp	48	.167	Ea.	114	7.85		121.85	137
8610	12" diameter, 8' high	G		40	.200		175	9.40		184.40	207
8620	18" diameter, 5' high	G		38	.211		193	9.90		202.90	227
8630	18" diameter, 10' high	G		32	.250		255	11.75		266.75	299
8640	58" diameter, 5' high	G	2 Carp	32	.500		495	23.50		518.50	580
8650	Cap, 12" diameter	G					22			22	24
8660	18" diameter	G					28			28	31

23 57 Heat Exchangers for HVAC

23 57 16 – Steam-to-Water Heat Exchangers

23 57 16.10 Shell/Tube Type Steam-to-Water Heat Exch.	Crew	Daily Output	Labor-Hours	Unit	Material	2015 Bare Costs Labor	Equipment	Total	Total Incl O&P
0010 **SHELL AND TUBE TYPE STEAM-TO-WATER HEAT EXCHANGERS**									
0016 Shell & tube type, 2 or 4 pass, 3/4" O.D. copper tubes,									
0020 C.I. heads, C.I. tube sheet, steel shell									
0100 Hot water 40°F to 180°F, by steam at 10 PSI									
0120 8 GPM	Q-5	6	2.667	Ea.	2,150	143		2,293	2,600
0140 10 GPM		5	3.200		3,250	172		3,422	3,825
0160 40 GPM		4	4		5,025	215		5,240	5,850
0180 64 GPM		2	8		7,700	430		8,130	9,100
0200 96 GPM		1	16		10,300	860		11,160	12,600
0220 120 GPM	Q-6	1.50	16		13,500	890		14,390	16,300
0240 168 GPM		1	24		16,600	1,350		17,950	20,200
0260 240 GPM		.80	30		25,900	1,675		27,575	31,000
0300 600 GPM		.70	34.286		55,500	1,900		57,400	64,000
0500 For bronze head and tube sheet, add					50%				

23 57 19 – Liquid-to-Liquid Heat Exchangers

23 57 19.13 Plate-Type, Liquid-to-Liquid Heat Exchangers

	Crew	Daily Output	Labor-Hours	Unit	Material	2015 Bare Costs Labor	Equipment	Total	Total Incl O&P
0010 **PLATE-TYPE, LIQUID-TO-LIQUID HEAT EXCHANGERS**									
3000 Plate type,									
3100 400 GPM	Q-6	.80	30	Ea.	36,700	1,675		38,375	42,900
3120 800 GPM	"	.50	48		63,500	2,675		66,175	73,500
3140 1200 GPM	Q-7	.34	94.118		94,000	5,350		99,350	111,500
3160 1800 GPM	"	.24	133		125,000	7,575		132,575	149,000

23 57 19.16 Shell-Type, Liquid-to-Liquid Heat Exchangers

	Crew	Daily Output	Labor-Hours	Unit	Material	2015 Bare Costs Labor	Equipment	Total	Total Incl O&P
0010 **SHELL-TYPE, LIQUID-TO-LIQUID HEAT EXCHANGERS**									
1000 Hot water 40°F to 140°F, by water at 200°F									
1020 7 GPM	Q-5	6	2.667	Ea.	2,650	143		2,793	3,125
1040 16 GPM		5	3.200		3,750	172		3,922	4,375
1060 34 GPM		4	4		5,675	215		5,890	6,550
1080 55 GPM		3	5.333		8,225	287		8,512	9,475
1100 74 GPM		1.50	10.667		10,300	575		10,875	12,200
1120 86 GPM		1.40	11.429		13,700	615		14,315	16,000
1140 112 GPM	Q-6	2	12		17,100	670		17,770	19,800
1160 126 GPM		1.80	13.333		21,400	745		22,145	24,600
1180 152 GPM		1	24		27,300	1,350		28,650	32,000

23 61 Refrigerant Compressors

23 61 13 – Centrifugal Refrigerant Compressors

23 61 13.10 Centrifugal Compressors

	Crew	Daily Output	Labor-Hours	Unit	Material	2015 Bare Costs Labor	Equipment	Total	Total Incl O&P
0010 **CENTRIFUGAL COMPRESSORS**									
0100 Refrigeration, centrifugal including switches & protective devices									
0210 200 ton	Q-7	.23	137	Ea.	24,700	7,825		32,525	38,900
0220 300 ton		.17	187		25,900	10,600		36,500	44,600
0230 400 ton		.14	226		29,100	12,900		42,000	51,500
0240 500 ton		.14	237		33,600	13,500		47,100	57,500
0250 600 ton		.13	248		38,500	14,100		52,600	63,500
0260 700 ton		.12	260		40,100	14,800		54,900	66,500
0270 800 ton		.12	271		45,000	15,400		60,400	73,000
0280 900 ton		.11	288		57,000	16,400		73,400	88,000
0290 1000 ton		.11	296		63,500	16,900		80,400	95,500
0300 1100 ton		.10	307		65,000	17,500		82,500	98,000

379

For customer support on your Mechanical Cost Data, call 877.795.7394.

23 61 Refrigerant Compressors

23 61 13 – Centrifugal Refrigerant Compressors

23 61 13.10 Centrifugal Compressors

	23 61 13.10 Centrifugal Compressors	Crew	Daily Output	Labor-Hours	Unit	Material	2015 Bare Costs Labor	Equipment	Total	Total Incl O&P
0310	1200 ton	Q-7	.10	316	Ea.	70,500	18,000		88,500	104,500
0320	1300 ton		.10	326		73,000	18,600		91,600	108,500
0330	1400 ton		.09	340		76,000	19,400		95,400	113,000
0340	1500 ton		.09	351		80,000	20,000		100,000	118,000

23 61 15 – Rotary Refrigerant Compressors

23 61 15.10 Rotary Compressors

		Crew	Daily Output	Labor-Hours	Unit	Material	2015 Bare Costs Labor	Equipment	Total	Total Incl O&P
0010	**ROTARY COMPRESSORS**									
0100	Refrigeration, hermetic, switches and protective devices									
0210	1.25 ton	1 Stpi	3	2.667	Ea.	246	159		405	510
0220	1.42 ton		3	2.667		248	159		407	515
0230	1.68 ton		2.80	2.857		270	171		441	555
0240	2.00 ton		2.60	3.077		278	184		462	585
0250	2.37 ton		2.50	3.200		315	191		506	635
0260	2.67 ton		2.40	3.333		330	199		529	660
0270	3.53 ton		2.30	3.478		360	208		568	710
0280	4.43 ton		2.20	3.636		460	217		677	835
0290	5.08 ton		2.10	3.810		505	228		733	900
0300	5.22 ton		2	4		715	239		954	1,150
0310	7.31 ton	Q-5	3	5.333		1,150	287		1,437	1,675
0320	9.95 ton		2.60	6.154		1,350	330		1,680	1,975
0330	11.6 ton		2.50	6.400		1,725	345		2,070	2,425
0340	15.25 ton		2.30	6.957		2,350	375		2,725	3,175
0350	17.7 ton		2.10	7.619		2,525	410		2,935	3,400

23 61 16 – Reciprocating Refrigerant Compressors

23 61 16.10 Reciprocating Compressors

		Crew	Daily Output	Labor-Hours	Unit	Material	2015 Bare Costs Labor	Equipment	Total	Total Incl O&P
0010	**RECIPROCATING COMPRESSORS**									
0990	Refrigeration, recip. hermetic, switches & protective devices									
1000	10 ton	Q-5	1	16	Ea.	12,800	860		13,660	15,400
1100	20 ton	Q-6	.72	33.333		27,200	1,850		29,050	32,700
1200	30 ton		.64	37.500		28,300	2,100		30,400	34,400
1300	40 ton		.44	54.545		30,500	3,050		33,550	38,100
1400	50 ton		.20	120		32,100	6,700		38,800	45,400
1500	75 ton	Q-7	.27	118		35,500	6,750		42,250	49,200
1600	130 ton	"	.21	152		40,500	8,675		49,175	57,500

23 61 19 – Scroll Refrigerant Compressors

23 61 19.10 Scroll Compressors

		Crew	Daily Output	Labor-Hours	Unit	Material	2015 Bare Costs Labor	Equipment	Total	Total Incl O&P
0010	**SCROLL COMPRESSORS**									
1800	Refrigeration, scroll type									
1810	1.5 ton	1 Stpi	2.70	2.963	Ea.	885	177		1,062	1,225
1820	2.5 ton		2.50	3.200		885	191		1,076	1,275
1830	2.75 ton		2.35	3.404		935	203		1,138	1,325
1840	3 ton		2.30	3.478		940	208		1,148	1,350
1850	3.5 ton		2.25	3.556		965	212		1,177	1,375
1860	4 ton		2.20	3.636		1,075	217		1,292	1,500
1870	5 ton		2.10	3.810		1,175	228		1,403	1,625
1880	6 ton	Q-5	3.70	4.324		1,550	233		1,783	2,050
1900	7.5 ton		3.20	5		2,600	269		2,869	3,250
1910	10 ton		3.10	5.161		2,750	278		3,028	3,450
1920	13 ton		2.90	5.517		3,050	297		3,347	3,800
1960	25 ton		2.20	7.273		5,075	390		5,465	6,200

380

For customer support on your Mechanical Cost Data, call 877.795.7394.

23 62 Packaged Compressor and Condenser Units

23 62 13 – Packaged Air-Cooled Refrigerant Compressor and Condenser Units

23 62 13.10 Packaged Air-Cooled Refrig. Condensing Units

23 62 13.10 Packaged Air-Cooled Refrig. Condensing Units	Crew	Daily Output	Labor-Hours	Unit	Material	2015 Bare Costs Labor	Equipment	Total	Total Incl O&P
0010 **PACKAGED AIR-COOLED REFRIGERANT CONDENSING UNITS**									
0020 Condensing unit									
0030 Air cooled, compressor, standard controls									
0050 1.5 ton	Q-5	2.50	6.400	Ea.	1,250	345		1,595	1,900
0100 2 ton		2.10	7.619		1,350	410		1,760	2,100
0200 2.5 ton		1.70	9.412		1,425	505		1,930	2,350
0300 3 ton		1.30	12.308		1,500	660		2,160	2,650
0350 3.5 ton		1.10	14.545		1,725	780		2,505	3,075
0400 4 ton		.90	17.778		1,975	955		2,930	3,625
0500 5 ton		.60	26.667		2,375	1,425		3,800	4,775
0550 7.5 ton		.55	29.091		3,775	1,575		5,350	6,500
0560 8.5 ton		.53	30.189		4,675	1,625		6,300	7,600
0600 10 ton		.50	32		5,000	1,725		6,725	8,100
0620 12.5 ton		.48	33.333		6,025	1,800		7,825	9,325
0650 15 ton		.40	40		7,950	2,150		10,100	12,000
0700 20 ton	Q-6	.40	60		11,600	3,350		14,950	17,800
0720 25 ton		.35	68.571		17,000	3,825		20,825	24,500
0750 30 ton		.30	80		19,600	4,450		24,050	28,300
0800 40 ton		.20	120		26,900	6,700		33,600	39,700
0840 50 ton		.18	133		29,200	7,425		36,625	43,300
0860 60 ton		.16	150		33,800	8,375		42,175	49,800
0900 70 ton		.14	171		38,700	9,550		48,250	57,000
1000 80 ton		.12	200		43,000	11,200		54,200	64,000
1010 90 ton		.09	266		47,700	14,900		62,600	75,000
1100 100 ton		.09	266		52,500	14,900		67,400	80,000

23 62 23 – Packaged Water-Cooled Refrigerant Compressor and Condenser Units

23 62 23.10 Packaged Water-Cooled Refrigerant Condensing Units

	Crew	Daily Output	Labor-Hours	Unit	Material	Labor	Equipment	Total	Total Incl O&P
0010 **PACKAGED WATER-COOLED REFRIGERANT CONDENSING UNITS**									
2000 Water cooled, compressor, heat exchanger, controls									
2100 5 ton	Q-5	.70	22.857	Ea.	13,600	1,225		14,825	16,900
2110 10 ton		.60	26.667		15,900	1,425		17,325	19,700
2200 15 ton		.50	32		20,300	1,725		22,025	25,000
2300 20 ton	Q-6	.40	60		25,000	3,350		28,350	32,600
2310 30 ton		.30	80		28,700	4,450		33,150	38,300
2400 40 ton		.20	120		31,200	6,700		37,900	44,400

23 63 Refrigerant Condensers

23 63 13 – Air-Cooled Refrigerant Condensers

23 63 13.10 Air-Cooled Refrig. Condensers

	Crew	Daily Output	Labor-Hours	Unit	Material	Labor	Equipment	Total	Total Incl O&P
0010 **AIR-COOLED REFRIG. CONDENSERS**									
0080 Air cooled, belt drive, propeller fan									
0220 45 ton	Q-6	.70	34.286	Ea.	9,800	1,900		11,700	13,700
0240 50 ton		.69	34.985		10,600	1,950		12,550	14,600
0260 54 ton		.64	37.795		11,500	2,100		13,600	15,900
0280 59 ton		.58	41.308		12,700	2,300		15,000	17,500
0300 65 ton		.53	45.541		15,200	2,550		17,750	20,500
0320 73 ton		.47	51.173		16,600	2,850		19,450	22,600
0340 81 ton		.42	56.738		18,300	3,175		21,475	24,900
0360 86 ton		.40	60.302		19,200	3,375		22,575	26,200
0380 88 ton		.39	61.697		20,600	3,450		24,050	27,900

For customer support on your Mechanical Cost Data, call 877.795.7394.

381

23 63 Refrigerant Condensers

23 63 13 – Air-Cooled Refrigerant Condensers

23 63 13.10 Air-Cooled Refrig. Condensers

	23 63 13.10 Air-Cooled Refrig. Condensers	Crew	Daily Output	Labor-Hours	Unit	Material	2015 Bare Costs Labor	2015 Bare Costs Equipment	Total	Total Incl O&P
0400	101 ton	Q-7	.45	70.640	Ea.	22,700	4,025		26,725	31,000
0500	159 ton		.31	102		24,000	5,850		29,850	35,300
0600	228 ton		.22	148		51,000	8,425		59,425	68,500
1500	May be specified single or multi-circuit									
1550	Air cooled, direct drive, propeller fan									
1590	1 ton	Q-5	3.80	4.211	Ea.	1,650	226		1,876	2,175
1600	1-1/2 ton		3.60	4.444		1,975	239		2,214	2,525
1620	2 ton		3.20	5		2,175	269		2,444	2,800
1630	3 ton		2.40	6.667		2,450	360		2,810	3,225
1640	5 ton		2	8		5,400	430		5,830	6,575
1650	8 ton		1.80	8.889		5,625	480		6,105	6,900
1660	10 ton		1.40	11.429		6,325	615		6,940	7,875
1670	12 ton		1.30	12.308		7,700	660		8,360	9,475
1680	14 ton		1.20	13.333		8,800	715		9,515	10,800
1690	16 ton		1.10	14.545		10,200	780		10,980	12,400
1700	21 ton		1	16		11,100	860		11,960	13,500
1720	26 ton		.84	19.002		12,500	1,025		13,525	15,400
1740	30 ton		.70	22.792		16,000	1,225		17,225	19,500
1760	41 ton	Q-6	.77	31.008		17,600	1,725		19,325	22,000
1780	52 ton		.66	36.419		25,100	2,025		27,125	30,700
1800	63 ton		.55	44.037		27,700	2,450		30,150	34,200
1820	76 ton		.45	52.980		31,400	2,950		34,350	39,100
1840	86 ton		.40	60		38,900	3,350		42,250	47,800
1860	97 ton		.35	67.989		44,200	3,800		48,000	54,500
1880	105 ton	Q-7	.44	73.563		51,000	4,175		55,175	62,500
1890	118 ton		.39	82.687		55,000	4,700		59,700	67,500
1900	126 ton		.36	88.154		64,000	5,025		69,025	77,500
1910	136 ton		.34	95.238		71,500	5,425		76,925	86,500
1920	142 ton		.32	99.379		73,500	5,650		79,150	89,000

23 63 30 – Lubricants

23 63 30.10 Lubricant Oils

		Crew	Daily Output	Labor-Hours	Unit	Material	2015 Bare Costs Labor	2015 Bare Costs Equipment	Total	Total Incl O&P
0010	**LUBRICANT OILS**									
8000	Oils									
8100	Lubricating									
8120	Oil, lubricating				Oz.	.39			.39	.43
8500	Refrigeration									
8520	Oil, refrigeration				Gal.	41			41	45
8525	Oil, refrigeration				Qt.	10.20			10.20	11.20

23 63 33 – Evaporative Refrigerant Condensers

23 63 33.10 Evaporative Condensers

		Crew	Daily Output	Labor-Hours	Unit	Material	2015 Bare Costs Labor	2015 Bare Costs Equipment	Total	Total Incl O&P
0010	**EVAPORATIVE CONDENSERS**									
3400	Evaporative, copper coil, pump, fan motor									
3440	10 ton	Q-5	.54	29.630	Ea.	8,350	1,600		9,950	11,600
3460	15 ton		.50	32		8,550	1,725		10,275	12,000
3480	20 ton		.47	34.043		8,800	1,825		10,625	12,500
3500	25 ton		.45	35.556		9,000	1,900		10,900	12,800
3520	30 ton		.42	38.095		9,450	2,050		11,500	13,500
3540	40 ton	Q-6	.49	48.980		10,300	2,725		13,025	15,400
3560	50 ton		.39	61.538		11,700	3,425		15,125	18,100
3580	65 ton		.35	68.571		12,500	3,825		16,325	19,500
3600	80 ton		.33	72.727		13,800	4,050		17,850	21,300
3620	90 ton		.29	82.759		15,100	4,625		19,725	23,600

23 63 Refrigerant Condensers

23 63 33 – Evaporative Refrigerant Condensers

23 63 33.10 Evaporative Condensers		Crew	Daily Output	Labor-Hours	Unit	Material	2015 Bare Costs Labor	Equipment	Total	Total Incl O&P
3640	100 ton	Q-7	.36	88.889	Ea.	16,600	5,050		21,650	25,900
3660	110 ton		.33	96.970		17,500	5,525		23,025	27,500
3680	125 ton		.30	106		18,900	6,075		24,975	30,000
3700	135 ton		.28	114		19,900	6,500		26,400	31,700
3720	150 ton		.25	128		21,600	7,275		28,875	34,800
3740	165 ton		.23	139		22,900	7,925		30,825	37,200
3760	185 ton		.22	145		25,900	8,275		34,175	41,000
3860	For fan damper control, add	Q-5	2	8		785	430		1,215	1,525

23 64 Packaged Water Chillers

23 64 13 – Absorption Water Chillers

23 64 13.13 Direct-Fired Absorption Water Chillers

		Crew	Daily Output	Labor-Hours	Unit	Material	Labor	Equipment	Total	Total Incl O&P
0010	**DIRECT-FIRED ABSORPTION WATER CHILLERS**									
3000	Gas fired, air cooled									
3220	5 ton	Q-5	.60	26.667	Ea.	8,850	1,425		10,275	11,900
3270	10 ton	"	.40	40	"	22,700	2,150		24,850	28,300
4000	Water cooled, duplex									
4130	100 ton	Q-7	.13	246	Ea.	136,000	14,000		150,000	170,500
4140	200 ton		.11	283		183,000	16,100		199,100	226,000
4150	300 ton		.11	299		219,500	17,000		236,500	267,000
4160	400 ton		.10	316		285,500	18,000		303,500	341,000
4170	500 ton		.10	329		359,500	18,800		378,300	423,500
4180	600 ton		.09	340		424,500	19,400		443,900	495,500
4190	700 ton		.09	359		470,500	20,500		491,000	548,500
4200	800 ton		.08	380		568,000	21,700		589,700	657,500
4210	900 ton		.08	400		658,000	22,800		680,800	758,500
4220	1000 ton		.08	421		731,500	24,000		755,500	840,500

23 64 13.16 Indirect-Fired Absorption Water Chillers

		Crew	Daily Output	Labor-Hours	Unit	Material	Labor	Equipment	Total	Total Incl O&P
0010	**INDIRECT-FIRED ABSORPTION WATER CHILLERS**									
0020	Steam or hot water, water cooled									
0050	100 ton	Q-7	.13	240	Ea.	125,000	13,700		138,700	158,000
0100	148 ton		.12	258		219,000	14,700		233,700	263,000
0200	200 ton		.12	275		244,000	15,700		259,700	291,500
0240	250 ton		.11	290		268,500	16,500		285,000	320,000
0300	354 ton		.11	304		335,000	17,300		352,300	394,500
0400	420 ton		.10	323		366,500	18,400		384,900	431,000
0500	665 ton		.09	340		514,500	19,400		533,900	594,500
0600	750 ton		.09	363		601,000	20,700		621,700	692,000
0700	850 ton		.08	385		641,500	21,900		663,400	738,500
0800	955 ton		.08	410		686,000	23,300		709,300	789,500
0900	1125 ton		.08	421		791,500	24,000		815,500	907,000
1000	1250 ton		.07	444		851,500	25,300		876,800	974,500
1100	1465 ton		.07	463		1,009,500	26,400		1,035,900	1,150,500
1200	1660 ton		.07	477		1,187,500	27,200		1,214,700	1,347,500
2000	For two stage unit, add					80%	25%			

For customer support on your Mechanical Cost Data, call 877.795.7394.

383

23 64 Packaged Water Chillers

23 64 16 – Centrifugal Water Chillers

23 64 16.10 Centrifugal Type Water Chillers	Crew	Daily Output	Labor-Hours	Unit	Material	2015 Bare Costs Labor	Equipment	Total	Total Incl O&P
0010 **CENTRIFUGAL TYPE WATER CHILLERS**, With standard controls									
0020 Centrifugal liquid chiller, water cooled									
0030 not including water tower									
0100 2000 ton (twin 1000 ton units)	Q-7	.07	477	Ea.	699,000	27,200		726,200	810,000
0274 Centrifugal, packaged unit, water cooled, not incl. tower									
0278 200 ton	Q-7	.19	172	Ea.	90,000	9,800		99,800	114,000
0279 300 ton		.14	233		94,500	13,300		107,800	124,000
0280 400 ton		.11	283		123,000	16,100		139,100	160,000
0282 450 ton		.11	290		157,500	16,500		174,000	198,000
0290 500 ton		.11	296		165,500	16,900		182,400	208,000
0292 550 ton		.11	304		203,500	17,300		220,800	250,000
0300 600 ton		.10	310		217,000	17,700		234,700	265,500
0302 650 ton		.10	320		228,500	18,200		246,700	279,000
0304 700 ton		.10	326		244,500	18,600		263,100	297,000
0306 750 ton		.10	333		262,000	19,000		281,000	317,000
0310 800 ton		.09	340		279,500	19,400		298,900	336,500
0312 850 ton		.09	351		297,000	20,000		317,000	357,000
0316 900 ton		.09	359		314,500	20,500		335,000	377,000
0318 950 ton		.09	363		332,000	20,700		352,700	396,500
0320 1000 ton		.09	372		349,500	21,200		370,700	416,500
0324 1100 ton		.08	385		360,000	21,900		381,900	429,000
0330 1200 ton		.08	395		382,000	22,500		404,500	454,000
0340 1300 ton		.08	410		413,000	23,300		436,300	489,500
0350 1400 ton		.08	421		445,000	24,000		469,000	525,500
0360 1500 ton		.08	426		469,000	24,300		493,300	552,500
0370 1600 ton		.07	438		499,500	24,900		524,400	587,000
0380 1700 ton		.07	450		531,500	25,600		557,100	623,000
0390 1800 ton		.07	457		563,000	26,000		589,000	658,500
0400 1900 ton		.07	470		593,500	26,800		620,300	693,000
0430 2000 ton		.07	492		626,500	28,000		654,500	731,500
0440 2500 ton		.06	524		781,000	29,800		810,800	904,000

23 64 19 – Reciprocating Water Chillers

23 64 19.10 Reciprocating Type Water Chillers

23 64 19.10 Reciprocating Type Water Chillers	Crew	Daily Output	Labor-Hours	Unit	Material	2015 Bare Costs Labor	Equipment	Total	Total Incl O&P
0010 **RECIPROCATING TYPE WATER CHILLERS**, With standard controls									
0494 Water chillers, integral air cooled condenser									
0538 60 ton cooling	Q-7	.28	115	Ea.	42,000	6,550		48,550	56,000
0546 70 ton cooling		.27	119		52,500	6,825		59,325	68,000
0552 80 ton cooling		.26	123		56,000	7,000		63,000	72,000
0554 90 ton cooling		.25	125		62,500	7,175		69,675	80,000
0600 100 ton cooling		.25	129		73,500	7,350		80,850	92,000
0620 110 ton cooling		.24	132		79,000	7,525		86,525	98,500
0630 130 ton cooling		.24	135		90,000	7,675		97,675	110,500
0640 150 ton cooling		.23	137		100,500	7,850		108,350	122,500
0650 175 ton cooling		.23	140		113,500	8,025		121,525	136,500
0654 190 ton cooling		.22	144		125,000	8,200		133,200	150,000
0660 210 ton cooling		.22	148		132,500	8,425		140,925	158,000
0662 250 ton cooling		.21	151		153,500	8,625		162,125	182,000
0664 275 ton cooling		.21	156		166,500	8,875		175,375	196,500
0666 300 ton cooling		.20	160		179,500	9,100		188,600	211,000
0668 330 ton cooling		.20	164		202,500	9,325		211,825	236,500
0670 360 ton cooling		.19	168		224,000	9,575		233,575	260,500
0672 390 ton cooling		.19	172		248,500	9,850		258,350	288,500

For customer support on your Mechanical Cost Data, call 877.795.7394.

23 64 Packaged Water Chillers

23 64 19 – Reciprocating Water Chillers

23 64 19.10 Reciprocating Type Water Chillers		Crew	Daily Output	Labor-Hours	Unit	Material	2015 Bare Costs Labor	Equipment	Total	Total Incl O&P
0674	420 ton cooling	Q-7	.18	177	Ea.	261,000	10,100		271,100	303,000
0980	Water cooled, multiple compressor, semi-hermetic, tower not incl.									
1000	15 ton cooling	Q-6	.36	65.934	Ea.	18,900	3,675		22,575	26,400
1020	25 ton cooling	Q-7	.41	78.049		19,500	4,450		23,950	28,200
1040	30 ton cooling		.36	89.888		22,700	5,125		27,825	32,700
1060	35 ton cooling		.31	101		24,300	5,800		30,100	35,600
1080	40 ton cooling		.30	108		26,000	6,150		32,150	37,900
1090	45 ton cooling		.29	111		28,200	6,325		34,525	40,600
1100	50 ton cooling		.28	113		35,800	6,475		42,275	49,200
1120	60 ton cooling		.25	125		37,000	7,175		44,175	51,500
1130	75 ton cooling		.23	139		52,500	7,925		60,425	69,500
1140	85 ton cooling		.21	151		57,000	8,625		65,625	76,000
1150	95 ton cooling		.19	164		59,000	9,375		68,375	79,000
1160	100 ton cooling		.18	179		60,500	10,200		70,700	82,000
1170	115 ton cooling		.17	190		61,000	10,800		71,800	83,500
1180	125 ton cooling		.16	196		64,000	11,200		75,200	87,000
1200	145 ton cooling		.16	202		71,500	11,500		83,000	96,000
1210	155 ton cooling	↓	.15	210	↓	75,000	12,000		87,000	100,500
1300	Water cooled, single compressor, semi-hermetic, tower not incl.									
1320	40 ton cooling,	Q-7	.30	108	Ea.	17,300	6,150		23,450	28,300
1340	50 ton cooling		.28	113		21,500	6,475		27,975	33,400
1360	60 ton cooling	↓	.25	125	↓	23,800	7,175		30,975	37,000
1451	Water cooled, dual compressors, semi-hermetic, tower not incl.									
1500	80 ton cooling	Q-7	.14	222	Ea.	29,600	12,600		42,200	51,500
1520	100 ton cooling		.14	228		40,000	13,000		53,000	63,500
1540	120 ton cooling	↓	.14	231	↓	48,500	13,200		61,700	73,500
4000	Packaged chiller, remote air cooled condensers not incl.									
4020	15 ton cooling	Q-7	.30	108	Ea.	15,800	6,150		21,950	26,700
4030	20 ton cooling		.28	115		17,200	6,600		23,800	28,900
4040	25 ton cooling		.25	125		18,200	7,175		25,375	30,800
4050	35 ton cooling		.24	133		19,100	7,625		26,725	32,500
4060	40 ton cooling		.22	144		20,600	8,200		28,800	35,000
4066	45 ton cooling		.21	149		21,700	8,500		30,200	36,600
4070	50 ton cooling		.21	153		30,800	8,750		39,550	47,100
4080	60 ton cooling		.20	164		33,300	9,325		42,625	51,000
4090	75 ton cooling		.18	173		42,700	9,900		52,600	62,000
4100	85 ton cooling		.17	183		44,900	10,500		55,400	65,000
4110	95 ton cooling		.17	193		45,700	11,000		56,700	66,500
4120	105 ton cooling		.16	203		50,000	11,600		61,600	72,500
4130	115 ton cooling		.15	213		51,500	12,100		63,600	75,000
4140	125 ton cooling		.14	223		52,000	12,700		64,700	76,000
4150	145 ton cooling	↓	.14	233	↓	58,000	13,300		71,300	83,500

23 64 23 – Scroll Water Chillers

23 64 23.10 Scroll Water Chillers		Crew	Daily Output	Labor-Hours	Unit	Material	2015 Bare Costs Labor	Equipment	Total	Total Incl O&P
0010	**SCROLL WATER CHILLERS**, With standard controls									
0480	Packaged w/integral air cooled condenser									
0482	10 ton cooling	Q-7	.34	94.118	Ea.	20,300	5,350		25,650	30,500
0490	15 ton cooling		.37	86.486		20,800	4,925		25,725	30,300
0500	20 ton cooling		.34	94.118		21,500	5,350		26,850	31,700
0510	25 ton cooling		.34	94.118		25,300	5,350		30,650	35,900
0515	30 ton cooling		.31	101		27,100	5,800		32,900	38,600
0517	35 ton cooling		.31	104		30,200	5,975		36,175	42,200

23 64 Packaged Water Chillers

23 64 23 – Scroll Water Chillers

23 64 23.10 Scroll Water Chillers

23 64 23.10 Scroll Water Chillers		Crew	Daily Output	Labor-Hours	Unit	Material	2015 Bare Costs Labor	Equipment	Total	Total Incl O&P
0520	40 ton cooling	Q-7	.30	108	Ea.	32,900	6,150		39,050	45,500
0528	45 ton cooling		.29	109		36,500	6,250		42,750	49,600
0536	50 ton cooling		.28	113		38,200	6,425		44,625	51,500
0680	Scroll water cooled, single compressor, hermetic, tower not incl.									
0700	2 ton cooling	Q-5	.57	28.070	Ea.	3,500	1,500		5,000	6,125
0710	5 ton cooling		.57	28.070		4,175	1,500		5,675	6,875
0720	6 ton cooling		.42	38.005		5,250	2,050		7,300	8,850
0740	8 ton cooling		.31	52.117		5,750	2,800		8,550	10,600
0760	10 ton cooling	Q-6	.36	67.039		6,625	3,750		10,375	12,900
0780	15 ton cooling	"	.33	72.727		10,400	4,050		14,450	17,600
0800	20 ton cooling	Q-7	.38	83.990		11,800	4,775		16,575	20,200
0820	30 ton cooling	"	.33	96.096		13,200	5,475		18,675	22,900

23 64 26 – Rotary-Screw Water Chillers

23 64 26.10 Rotary-Screw Type Water Chillers

23 64 26.10 Rotary-Screw Type Water Chillers		Crew	Daily Output	Labor-Hours	Unit	Material	2015 Bare Costs Labor	Equipment	Total	Total Incl O&P
0010	**ROTARY-SCREW TYPE WATER CHILLERS**, With standard controls									
0110	Screw, liquid chiller, air cooled, insulated evaporator									
0120	130 ton	Q-7	.14	228	Ea.	91,500	13,000		104,500	120,000
0124	160 ton		.13	246		112,500	14,000		126,500	144,500
0128	180 ton		.13	250		126,500	14,200		140,700	160,500
0132	210 ton		.12	258		139,000	14,700		153,700	175,000
0136	270 ton		.12	266		159,000	15,200		174,200	198,000
0140	320 ton		.12	275		199,500	15,700		215,200	243,000
0200	Packaged unit, water cooled, not incl. tower									
0210	80 ton	Q-7	.14	223	Ea.	46,000	12,700		58,700	69,500
0220	100 ton		.14	230		57,000	13,100		70,100	82,500
0230	150 ton		.13	240		74,000	13,700		87,700	102,000
0240	200 ton		.13	251		82,000	14,300		96,300	111,500
0250	250 ton		.12	260		100,500	14,800		115,300	133,500
0260	300 ton		.12	266		119,000	15,200		134,200	154,000
0270	350 ton		.12	275		133,500	15,700		149,200	170,000
1450	Water cooled, tower not included									
1560	135 ton cooling, screw compressors	Q-7	.14	235	Ea.	59,500	13,400		72,900	85,500
1580	150 ton cooling, screw compressors		.13	240		69,000	13,700		82,700	96,000
1620	200 ton cooling, screw compressors		.13	250		95,000	14,200		109,200	126,000
1660	291 ton cooling, screw compressors		.12	260		99,000	14,800		113,800	131,500

23 64 33 – Direct Expansion Water Chillers

23 64 33.10 Direct Expansion Type Water Chillers

23 64 33.10 Direct Expansion Type Water Chillers		Crew	Daily Output	Labor-Hours	Unit	Material	2015 Bare Costs Labor	Equipment	Total	Total Incl O&P
0010	**DIRECT EXPANSION TYPE WATER CHILLERS**, With standard controls									
8000	Direct expansion, shell and tube type, for built up systems									
8020	1 ton	Q-5	2	8	Ea.	7,300	430		7,730	8,675
8030	5 ton		1.90	8.421		12,100	455		12,555	14,000
8040	10 ton		1.70	9.412		15,000	505		15,505	17,300
8050	20 ton		1.50	10.667		17,200	575		17,775	19,800
8060	30 ton		1	16		21,800	860		22,660	25,300
8070	50 ton		.90	17.778		35,100	955		36,055	40,100
8080	100 ton	Q-6	.90	26.667		62,500	1,475		63,975	71,000

23 65 13 − Forced-Draft Cooling Towers

23 65 13.10 Forced-Draft Type Cooling Towers	Crew	Daily Output	Labor-Hours	Unit	Material	2015 Bare Costs Labor	Equipment	Total	Total Incl O&P
0010 **FORCED-DRAFT TYPE COOLING TOWERS**, Packaged units D3030−310									
0070 Galvanized steel									
0080 Induced draft, crossflow									
0100 Vertical, belt drive, 61 tons	Q-6	90	.267	TonAC	215	14.85		229.85	260
0150 100 ton		100	.240		208	13.40		221.40	248
0200 115 ton		109	.220		180	12.30		192.30	218
0250 131 ton		120	.200		217	11.15		228.15	255
0260 162 ton		132	.182		175	10.15		185.15	208
1000 For higher capacities, use multiples									
1500 Induced air, double flow									
1900 Vertical, gear drive, 167 ton	Q-6	126	.190	TonAC	170	10.60		180.60	203
2000 297 ton		129	.186		105	10.40		115.40	131
2100 582 ton		132	.182		58.50	10.15		68.65	79.50
2150 849 ton		142	.169		78.50	9.45		87.95	101
2200 1016 ton		150	.160		79	8.90		87.90	100
2500 Blow through, centrifugal type									
2510 50 ton	Q-6	2.28	10.526	Ea.	12,000	585		12,585	14,100
2520 75 ton		1.52	15.789		15,500	880		16,380	18,300
2524 100 ton		1.22	19.672		18,300	1,100		19,400	21,800
2528 125 ton	Q-7	1.31	24.427		21,400	1,400		22,800	25,600
2532 200 ton		.82	39.216		27,900	2,225		30,125	34,100
2536 250 ton		.65	49.231		33,300	2,800		36,100	40,900
2540 300 ton		.54	59.259		38,300	3,375		41,675	47,300
2544 350 ton		.47	68.085		43,300	3,875		47,175	53,500
2548 400 ton		.41	78.049		45,100	4,450		49,550	56,500
2552 450 ton		.36	88.154		54,500	5,025		59,525	67,000
2556 500 ton		.32	100		58,000	5,700		63,700	72,000
2560 550 ton		.30	107		62,500	6,125		68,625	78,000
2564 600 ton		.27	117		65,500	6,700		72,200	82,000
2568 650 ton		.25	127		72,500	7,250		79,750	91,000
2572 700 ton		.23	137		85,000	7,825		92,825	106,000
2576 750 ton		.22	146		91,500	8,350		99,850	113,000
2580 800 ton		.21	152		99,000	8,675		107,675	122,000
2584 850 ton		.20	156		105,000	8,925		113,925	129,000
2588 900 ton		.20	161		111,500	9,200		120,700	136,500
2592 950 ton		.19	170		117,500	9,675		127,175	144,000
2596 1000 ton		.18	180		121,500	10,300		131,800	149,000
2700 Axial fan, induced draft									
2710 50 ton	Q-6	2.28	10.526	Ea.	10,400	585		10,985	12,400
2720 75 ton		1.52	15.789		12,300	880		13,180	14,800
2724 100 ton		1.22	19.672		13,000	1,100		14,100	16,000
2728 125 ton	Q-7	1.31	24.427		16,000	1,400		17,400	19,700
2732 200 ton		.82	39.216		27,100	2,225		29,325	33,200
2736 250 ton		.65	49.231		30,600	2,800		33,400	37,800
2740 300 ton		.54	59.259		35,700	3,375		39,075	44,400
2744 350 ton		.47	68.085		38,300	3,875		42,175	48,000
2748 400 ton		.41	78.049		43,900	4,450		48,350	55,000
2752 450 ton		.36	88.154		48,700	5,025		53,725	61,000
2756 500 ton		.32	100		51,500	5,700		57,200	65,500
2760 550 ton		.30	107		58,000	6,125		64,125	73,000
2764 600 ton		.27	117		61,000	6,700		67,700	77,000
2768 650 ton		.25	127		66,500	7,250		73,750	84,000
2772 700 ton		.23	137		75,500	7,825		83,325	95,000

23 65 Cooling Towers

23 65 13.10 Forced-Draft Type Cooling Towers		Crew	Daily Output	Labor-Hours	Unit	Material	2015 Bare Costs Labor	Equipment	Total	Total Incl O&P
2776	750 ton	Q-7	.22	146	Ea.	78,500	8,350		86,850	99,000
2780	800 ton		.21	152		82,000	8,675		90,675	103,000
2784	850 ton		.20	156		82,500	8,925		91,425	104,500
2788	900 ton		.20	161		85,500	9,200		94,700	108,000
2792	950 ton		.19	170		90,500	9,675		100,175	114,000
2796	1000 ton		.18	180		95,500	10,300		105,800	120,500
3000	For higher capacities, use multiples									
3500	For pumps and piping, add	Q-6	38	.632	TonAC	108	35		143	171
4000	For absorption systems, add				"	75%	75%			
4100	Cooling water chemical feeder	Q-5	3	5.333	Ea.	365	287		652	835
5000	Fiberglass tower on galvanized steel support structure									
5010	Draw thru									
5100	100 ton	Q-6	1.40	17.143	Ea.	13,900	955		14,855	16,800
5120	120 ton		1.20	20		16,200	1,125		17,325	19,500
5140	140 ton		1	24		17,500	1,350		18,850	21,200
5160	160 ton		.80	30		19,400	1,675		21,075	23,900
5180	180 ton		.65	36.923		22,100	2,050		24,150	27,400
5200	200 ton		.48	50		25,000	2,800		27,800	31,700
5300	For stainless steel support structure, add					30%				
5360	For higher capacities, use multiples of each size									
6000	Stainless steel									
6010	Induced draft, crossflow, horizontal, belt drive									
6100	57 ton	Q-6	1.50	16	Ea.	27,100	890		27,990	31,200
6120	91 ton		.99	24.242		32,700	1,350		34,050	38,100
6140	111 ton		.43	55.814		41,600	3,125		44,725	50,500
6160	126 ton		.22	109		41,600	6,075		47,675	55,000
6170	Induced draft, crossflow, vertical, gear drive									
6172	167 ton	Q-6	.75	32	Ea.	53,000	1,775		54,775	61,000
6174	297 ton		.43	55.814		59,000	3,125		62,125	69,500
6176	582 ton		.23	104		93,500	5,825		99,325	112,000
6178	849 ton		.17	141		132,000	7,875		139,875	157,500
6180	1016 ton		.15	160		159,000	8,925		167,925	188,000

23 65 33 – Liquid Coolers

23 65 33.10 Liquid Cooler, Closed Circuit Type Cooling Towers

		Crew	Daily Output	Labor-Hours	Unit	Material	Labor	Equipment	Total	Total Incl O&P
0010	**LIQUID COOLER, CLOSED CIRCUIT TYPE COOLING TOWERS**									
0070	Packaged options include: vibration switch, sump sweeper piping,									
0076	grooved coil connections, fan motor inverter capable,									
0080	304 St. St. cold water basin, external platform w/ladder.									
0100	50 ton	Q-6	2	12	Ea.	32,200	670		32,870	36,400
0150	100 ton	"	1.13	21.239		64,500	1,175		65,675	73,000
0200	150 ton	Q-7	.98	32.653		96,500	1,850		98,350	109,000
0250	200 ton		.78	41.026		128,500	2,325		130,825	145,000
0300	250 ton		.63	50.794		161,000	2,900		163,900	181,500

388

For customer support on your Mechanical Cost Data, call 877.795.7394.

23 72 Air-to-Air Energy Recovery Equipment

23 72 13 – Heat-Wheel Air-to-Air Energy-Recovery Equipment

23 72 13.10 Heat-Wheel Air-to-Air Energy Recovery Equip.

23 72 13.10 Heat-Wheel Air-to-Air Energy Recovery Equip.		Crew	Daily Output	Labor-Hours	Unit	Material	2015 Bare Costs Labor	2015 Bare Costs Equipment	Total	Total Incl O&P
0010	HEAT-WHEEL AIR-TO-AIR ENERGY RECOVERY EQUIPMENT									
0100	Air to air									
4000	Enthalpy recovery wheel									
4010	1000 max CFM	G Q-9	1.20	13.333	Ea.	6,700	670		7,370	8,400
4020	2000 max CFM	G	1	16		7,850	805		8,655	9,850
4030	4000 max CFM	G	.80	20		9,075	1,000		10,075	11,600
4040	6000 max CFM	G	.70	22.857		10,600	1,150		11,750	13,500
4050	8000 max CFM	G Q-10	1	24		11,700	1,250		12,950	14,800
4060	10,000 max CFM	G	.90	26.667		14,000	1,400		15,400	17,500
4070	20,000 max CFM	G	.80	30		25,400	1,575		26,975	30,400
4080	25,000 max CFM	G	.70	34.286		31,100	1,800		32,900	36,900
4090	30,000 max CFM	G	.50	48		34,500	2,500		37,000	41,700
4100	40,000 max CFM	G	.45	53.333		47,600	2,775		50,375	57,000
4110	50,000 max CFM	G	.40	60		55,500	3,125		58,625	66,000

23 72 16 – Heat-Pipe Air-To-Air Energy-Recovery Equipment

23 72 16.10 Heat Pipes

23 72 16.10 Heat Pipes		Crew	Daily Output	Labor-Hours	Unit	Material	2015 Bare Costs Labor	2015 Bare Costs Equipment	Total	Total Incl O&P
0010	HEAT PIPES									
8000	Heat pipe type, glycol, 50% efficient									
8010	100 MBH, 1700 CFM	1 Stpi	.80	10	Ea.	4,300	600		4,900	5,650
8020	160 MBH, 2700 CFM		.60	13.333		5,775	795		6,570	7,550
8030	620 MBH, 4000 CFM		.40	20		8,725	1,200		9,925	11,400

23 73 Indoor Central-Station Air-Handling Units

23 73 13 – Modular Indoor Central-Station Air-Handling Units

23 73 13.10 Air-Handling Units

23 73 13.10 Air-Handling Units		Crew	Daily Output	Labor-Hours	Unit	Material	2015 Bare Costs Labor	2015 Bare Costs Equipment	Total	Total Incl O&P
0010	AIR-HANDLING UNITS, Built-Up									
0100	With cooling/heating coil section, filters, mixing box									
0880	Single zone, horizontal/vertical									
0890	Constant volume									
0900	1600 CFM	Q-5	1.20	13.333	Ea.	5,000	715		5,715	6,575
0906	2000 CFM		1.10	14.545		6,000	780		6,780	7,775
0910	3000 CFM		1	16		8,800	860		9,660	11,000
0916	4000 CFM		.95	16.842		11,500	905		12,405	14,100
0920	5000 CFM	Q-6	1.40	17.143		14,000	955		14,955	16,900
0926	6500 CFM		1.30	18.462		17,500	1,025		18,525	20,900
0930	7500 CFM		1.20	20		19,000	1,125		20,125	22,600
0936	9200 CFM		1.10	21.818		22,000	1,225		23,225	26,000
0940	11,500 CFM		1	24		25,000	1,350		26,350	29,500
0946	13,200 CFM		.90	26.667		30,000	1,475		31,475	35,300
0950	16,500 CFM		.80	30		37,000	1,675		38,675	43,200
0960	19,500 CFM		.70	34.286		43,500	1,900		45,400	51,000
0970	22,000 CFM		.60	40		49,000	2,225		51,225	57,500
0980	27,000 CFM		.50	48		59,000	2,675		61,675	69,000
0990	34,000 CFM		.40	60		75,000	3,350		78,350	87,500
1000	40,000 CFM		.30	80		90,000	4,450		94,450	105,500
1010	47,000 CFM		.20	120		98,500	6,700		105,200	118,500
2300	Multi-zone, horizontal or vertical, blow-thru fan									
2310	3000 CFM	Q-5	1	16	Ea.	6,625	860		7,485	8,600
2314	4000 CFM	"	.95	16.842		8,750	905		9,655	11,000
2318	5000 CFM	Q-6	1.40	17.143		10,900	955		11,855	13,500

For customer support on your Mechanical Cost Data, call 877.795.7394.

389

23 73 13 – Modular Indoor Central-Station Air-Handling Units

23 73 13.10 Air-Handling Units		Crew	Daily Output	Labor-Hours	Unit	Material	2015 Bare Costs Labor	Equipment	Total	Total Incl O&P
2322	6500 CFM	Q-6	1.30	18.462	Ea.	14,100	1,025		15,125	17,200
2326	7500 CFM		1.20	20		16,300	1,125		17,425	19,700
2330	9200 CFM		1.10	21.818		20,100	1,225		21,325	23,900
2334	11,500 CFM		1	24		22,200	1,350		23,550	26,400
2336	13,200 CFM		.90	26.667		25,400	1,475		26,875	30,200
2340	16,500 CFM		.80	30		31,600	1,675		33,275	37,300
2344	19,500 CFM		.70	34.286		37,300	1,900		39,200	44,000
2348	22,000 CFM		.60	40		42,100	2,225		44,325	49,800
2352	27,000 CFM		.50	48		47,100	2,675		49,775	56,000
2356	34,000 CFM		.40	60		59,500	3,350		62,850	70,500
2360	40,000 CFM		.30	80		64,500	4,450		68,950	77,500
2364	47,000 CFM	▼	.20	120	▼	82,000	6,700		88,700	100,000

23 73 13.20 Air Handling Units, Packaged Indoor Type

		Crew	Daily Output	Labor-Hours	Unit	Material	Labor	Equipment	Total	Total Incl O&P
0010	**AIR HANDLING UNITS, PACKAGED INDOOR TYPE**									
0040	Cooling coils may be chilled water or DX									
0050	Heating coils may be hot water, steam or electric									
1090	Constant volume									
1200	2000 CFM	Q-5	1.10	14.545	Ea.	11,000	780		11,780	13,300
1400	5000 CFM	Q-6	.80	30		17,600	1,675		19,275	21,800
1550	10,000 CFM		.54	44.444		25,900	2,475		28,375	32,300
1670	15,000 CFM		.38	63.158		53,000	3,525		56,525	64,000
1700	20,000 CFM		.31	77.419		67,500	4,325		71,825	80,500
1710	30,000 CFM	▼	.27	88.889	▼	91,000	4,950		95,950	107,500
2300	Variable air volume									
2330	2000 CFM	Q-5	1	16	Ea.	15,400	860		16,260	18,300
2340	5000 CFM	Q-6	.70	34.286		21,500	1,900		23,400	26,600
2350	10,000 CFM		.50	48		42,000	2,675		44,675	50,500
2360	15,000 CFM		.40	60		61,000	3,350		64,350	72,000
2370	20,000 CFM		.29	82.759		78,000	4,625		82,625	92,500
2380	30,000 CFM	▼	.20	120	▼	94,000	6,700		100,700	113,500

23 73 39 – Indoor, Direct Gas-Fired Heating and Ventilating Units

23 73 39.10 Make-Up Air Unit

		Crew	Daily Output	Labor-Hours	Unit	Material	Labor	Equipment	Total	Total Incl O&P
0010	**MAKE-UP AIR UNIT**									
0020	Indoor suspension, natural/LP gas, direct fired,									
0032	standard control. For flue see Section 23 51 23.10									
0040	70°F temperature rise, MBH is input									
0100	75 MBH input	Q-6	3.60	6.667	Ea.	4,925	370		5,295	5,975
0120	100 MBH input		3.40	7.059		5,125	395		5,520	6,225
0140	125 MBH input		3.20	7.500		5,425	420		5,845	6,600
0160	150 MBH input		3	8		5,975	445		6,420	7,250
0180	175 MBH input		2.80	8.571		6,125	480		6,605	7,475
0200	200 MBH input		2.60	9.231		6,475	515		6,990	7,900
0220	225 MBH input		2.40	10		6,750	560		7,310	8,250
0240	250 MBH input		2	12		7,725	670		8,395	9,500
0260	300 MBH input		1.90	12.632		7,925	705		8,630	9,800
0280	350 MBH input		1.75	13.714		8,875	765		9,640	10,900
0300	400 MBH input		1.60	15		15,400	835		16,235	18,200
0320	840 MBH input		1.20	20		29,100	1,125		30,225	33,700
0340	1200 MBH input	▼	.88	27.273		31,700	1,525		33,225	37,200
0600	For discharge louver assembly, add					5%				
0700	For filters, add					10%				
0800	For air shut-off damper section, add					30%				

23 74 Packaged Outdoor HVAC Equipment

23 74 13 – Packaged, Outdoor, Central-Station Air-Handling Units

23 74 13.10 Packaged, Outdoor Type, Central-Station AHU	Crew	Daily Output	Labor-Hours	Unit	Material	2015 Bare Costs Labor	Equipment	Total	Total Incl O&P
0010 **PACKAGED, OUTDOOR TYPE, CENTRAL-STATION AIR-HANDLING UNITS**									
2900 Cooling coils may be chilled water or DX									
2910 Heating coils may be hot water, steam or electric									
3000 Weathertight, ground or rooftop									
3010 Constant volume									
3100 2000 CFM	Q-6	1	24	Ea.	23,700	1,350		25,050	28,000
3140 5000 CFM		.79	30.380		28,100	1,700		29,800	33,500
3150 10,000 CFM		.52	46.154		57,500	2,575		60,075	67,500
3160 15,000 CFM		.41	58.537		84,500	3,275		87,775	97,500
3170 20,000 CFM		.30	80		115,500	4,450		119,950	133,500
3180 30,000 CFM		.24	100		147,000	5,575		152,575	170,500
3200 Variable air volume									
3210 2000 CFM	Q-6	.96	25	Ea.	15,200	1,400		16,600	18,800
3240 5000 CFM		.73	32.877		30,800	1,825		32,625	36,600
3250 10,000 CFM		.50	48		66,000	2,675		68,675	77,000
3260 15,000 CFM		.40	60		100,500	3,350		103,850	115,500
3270 20,000 CFM		.29	82.759		130,500	4,625		135,125	150,500
3280 30,000 CFM		.20	120		154,500	6,700		161,200	180,000

23 74 23 – Packaged, Outdoor, Heating-Only Makeup-Air Units

23 74 23.16 Packaged, Ind.-Fired, In/Outdoor

23 74 23.16 Packaged, Ind.-Fired, In/Outdoor	Crew	Daily Output	Labor-Hours	Unit	Material	2015 Bare Costs Labor	Equipment	Total	Total Incl O&P
0010 **PACKAGED, IND.-FIRED, IN/OUTDOOR**, Heat-Only Makeup-Air Units									
1000 Rooftop unit, natural gas, gravity vent, S.S. exchanger									
1010 70°F temperature rise, MBH is input									
1020 250 MBH	Q-6	4	6	Ea.	15,400	335		15,735	17,400
1040 400 MBH		3.60	6.667		17,300	370		17,670	19,600
1060 550 MBH		3.30	7.273		17,800	405		18,205	20,200
1080 750 MBH		3	8		21,700	445		22,145	24,500
1100 1000 MBH		2.60	9.231		24,300	515		24,815	27,500
1120 1750 MBH		2.30	10.435		30,400	580		30,980	34,300
1140 2500 MBH		1.90	12.632		41,600	705		42,305	46,900
1160 3250 MBH		1.40	17.143		50,000	955		50,955	56,500
1180 4000 MBH		1.20	20		65,000	1,125		66,125	73,000
1200 6000 MBH		1	24		82,000	1,350		83,350	92,000
1600 For cleanable filters, add					5%				
1700 For electric modulating gas control, add					10%				

23 74 33 – Dedicated Outdoor-Air Units

23 74 33.10 Rooftop Air Conditioners

23 74 33.10 Rooftop Air Conditioners		Crew	Daily Output	Labor-Hours	Unit	Material	2015 Bare Costs Labor	Equipment	Total	Total Incl O&P
0010 **ROOFTOP AIR CONDITIONERS**, Standard controls, curb, economizer										
1000 Single zone, electric cool, gas heat										
1100 3 ton cooling, 60 MBH heating	D3050-150	Q-5	.70	22.857	Ea.	3,925	1,225		5,150	6,175
1120 4 ton cooling, 95 MBH heating	D3050-155		.61	26.403		4,575	1,425		6,000	7,200
1140 5 ton cooling, 112 MBH heating	R236000-10		.56	28.521		5,125	1,525		6,650	7,950
1145 6 ton cooling, 140 MBH heating	R236000-20		.52	30.769		5,875	1,650		7,525	8,975
1150 7.5 ton cooling, 170 MBH heating			.50	32.258		6,825	1,725		8,550	10,200
1156 8.5 ton cooling, 170 MBH heating			.46	34.783		8,125	1,875		10,000	11,800
1160 10 ton cooling, 200 MBH heating		Q-6	.67	35.982		9,425	2,000		11,425	13,400
1170 12.5 ton cooling, 230 MBH heating			.63	37.975		11,900	2,125		14,025	16,300
1180 15 ton cooling, 270 MBH heating			.57	42.032		13,600	2,350		15,950	18,500
1190 17.5 ton cooling, 330 MBH heating			.52	45.889		16,100	2,550		18,650	21,700
1200 20 ton cooling, 360 MBH heating		Q-7	.67	47.976		23,800	2,725		26,525	30,300
1210 25 ton cooling, 450 MBH heating			.56	57.554		27,100	3,275		30,375	34,800
1220 30 ton cooling, 540 MBH heating			.47	68.376		30,800	3,900		34,700	39,800

23 74 33 – Dedicated Outdoor-Air Units

23 74 33.10 Rooftop Air Conditioners	Crew	Daily Output	Labor-Hours	Unit	Material	2015 Bare Costs Labor	Equipment	Total	Total Incl O&P
1240	40 ton cooling, 675 MBH heating	Q-7	.35	91.168	Ea.	40,200	5,175		45,375	52,000
1260	50 ton cooling, 810 MBH heating		.28	113		47,100	6,475		53,575	62,000
1265	60 ton cooling, 900 MBH heating		.23	136		57,000	7,775		64,775	74,000
1270	80 ton cooling, 1000 MBH heating		.20	160		87,500	9,100		96,600	110,000
1275	90 ton cooling, 1200 MBH heating		.17	188		96,500	10,700		107,200	122,500
1280	100 ton cooling, 1350 MBH heating	▼	.14	228	▼	115,000	13,000		128,000	146,000
1300	Electric cool, electric heat									
1310	5 ton	Q-5	.63	25.600	Ea.	4,325	1,375		5,700	6,850
1314	10 ton	Q-6	.74	32.432		7,875	1,800		9,675	11,400
1318	15 ton	"	.64	37.795		11,800	2,100		13,900	16,200
1322	20 ton	Q-7	.74	43.185		15,800	2,450		18,250	21,000
1326	25 ton		.62	51.696		19,700	2,950		22,650	26,200
1330	30 ton		.52	61.657		23,700	3,500		27,200	31,300
1334	40 ton		.39	82.051		31,500	4,675		36,175	41,800
1338	50 ton		.31	102		37,400	5,825		43,225	50,000
1342	60 ton		.26	123		44,900	7,000		51,900	60,000
1346	75 ton	▼	.21	153	▼	56,000	8,750		64,750	75,000
2000	Multizone, electric cool, gas heat, economizer									
2100	15 ton cooling, 360 MBH heating	Q-7	.61	52.545	Ea.	65,000	3,000		68,000	76,000
2120	20 ton cooling, 360 MBH heating		.53	60.038		70,000	3,425		73,425	82,000
2140	25 ton cooling, 450 MBH heating		.45	71.910		83,500	4,100		87,600	98,000
2160	28 ton cooling, 450 MBH heating		.41	79.012		95,500	4,500		100,000	112,000
2180	30 ton cooling, 540 MBH heating		.37	85.562		107,000	4,875		111,875	125,000
2200	40 ton cooling, 540 MBH heating		.28	113		122,500	6,475		128,975	144,500
2210	50 ton cooling, 540 MBH heating		.23	142		152,000	8,100		160,100	179,500
2220	70 ton cooling, 1500 MBH heating		.16	198		164,500	11,300		175,800	198,000
2240	80 ton cooling, 1500 MBH heating		.14	228		188,000	13,000		201,000	226,500
2260	90 ton cooling, 1500 MBH heating		.13	256		197,500	14,600		212,100	239,000
2280	105 ton cooling, 1500 MBH heating	▼	.11	290		217,500	16,500		234,000	264,000
2400	For hot water heat coil, deduct					5%				
2500	For steam heat coil, deduct					2%				
2600	For electric heat, deduct				▼	3%	5%			
4000	Single zone, electric cool, gas heat, variable air volume									
4100	12.5 ton cooling, 230 MBH heating	Q-6	.55	43.716	Ea.	14,500	2,450		16,950	19,700
4120	18 ton cooling, 330 MBH heating	"	.46	52.747		19,400	2,950		22,350	25,800
4140	25 ton cooling, 450 MBH heating	Q-7	.48	66.116		30,900	3,750		34,650	39,700
4160	40 ton cooling, 675 MBH heating		.31	104		47,200	5,975		53,175	61,000
4180	60 ton cooling, 900 MBH heating		.20	157		69,000	8,975		77,975	89,500
4200	80 ton cooling, 1000 MBH heating	▼	.15	209	▼	101,500	11,900		113,400	130,000
5000	Single zone, electric cool only									
5050	3 ton cooling	Q-5	.88	18.203	Ea.	2,700	980		3,680	4,450
5060	4 ton cooling		.76	21.108		3,225	1,125		4,350	5,275
5070	5 ton cooling		.70	22.792		3,575	1,225		4,800	5,800
5080	6 ton cooling		.65	24.502		4,375	1,325		5,700	6,825
5090	7.5 ton cooling		.62	25.806		4,725	1,400		6,125	7,300
5100	8.5 ton cooling	▼	.59	26.981		5,450	1,450		6,900	8,200
5110	10 ton cooling	Q-6	.83	28.812		6,375	1,600		7,975	9,450
5120	12.5 ton cooling		.80	30		8,775	1,675		10,450	12,200
5130	15 ton cooling		.71	33.613		11,800	1,875		13,675	15,800
5140	17.5 ton cooling		.65	36.697		12,600	2,050		14,650	16,900
5150	20 ton cooling	Q-7	.83	38.415		23,000	2,175		25,175	28,600
5160	25 ton cooling		.70	45.977		24,400	2,625		27,025	30,900
5170	30 ton cooling		.59	54.701		28,200	3,100		31,300	35,700

23 74 33 - Dedicated Outdoor-Air Units

23 74 33.10 Rooftop Air Conditioners	Crew	Daily Output	Labor-Hours	Unit	Material	2015 Bare Costs Labor	Equipment	Total	Total Incl O&P
5180	40 ton cooling	Q-7	.44	73.059	Ea.	37,200	4,150		41,350	47,200
5190	50 ton cooling		.38	84.211		43,500	4,800		48,300	55,000
5200	60 ton cooling		.28	114		53,500	6,500		60,000	69,000
5400	For low heat, add					7%				
5410	For high heat, add					10%				
6000	Single zone electric cooling with variable volume distribution									
6020	20 ton cooling	Q-7	.72	44.199	Ea.	26,100	2,525		28,625	32,500
6030	25 ton cooling		.61	52.893		29,000	3,000		32,000	36,500
6040	30 ton cooling		.51	62.868		33,100	3,575		36,675	41,900
6050	40 ton cooling		.38	83.990		38,100	4,775		42,875	49,100
6060	50 ton cooling		.31	104		44,300	5,975		50,275	58,000
6070	60 ton cooling		.25	125		53,500	7,175		60,675	69,500
6200	For low heat, add					7%				
6210	For high heat, add					10%				
7000	Multizone, cool/heat, variable volume distribution									
7100	50 ton cooling	Q-7	.20	164	Ea.	113,000	9,325		122,325	138,000
7110	70 ton cooling		.14	228		158,000	13,000		171,000	193,500
7120	90 ton cooling		.11	296		181,500	16,900		198,400	225,500
7130	105 ton cooling		.10	333		203,000	19,000		222,000	252,000
7140	120 ton cooling		.08	380		232,500	21,700		254,200	288,000
7150	140 ton cooling		.07	444		248,500	25,300		273,800	311,500
7300	Penthouse unit, cool/heat, variable volume distribution									
7400	150 ton cooling	Q-7	.10	310	Ea.	406,500	17,700		424,200	473,500
7410	170 ton cooling		.10	320		434,500	18,200		452,700	505,500
7420	200 ton cooling		.10	326		511,000	18,600		529,600	590,000
7430	225 ton cooling		.09	340		559,000	19,400		578,400	644,000
7440	250 ton cooling		.09	359		577,000	20,500		597,500	665,500
7450	270 ton cooling		.09	363		623,000	20,700		643,700	716,500
7460	300 ton cooling		.09	372		692,500	21,200		713,700	793,500
9400	Sub assemblies for assembly systems									
9410	Ductwork, per ton, rooftop 1 zone units	Q-9	.96	16.667	Ton	177	840		1,017	1,475
9420	Ductwork, per ton, rooftop multizone units		.50	32		268	1,600		1,868	2,750
9430	Ductwork, VAV cooling/ton, rooftop multizone, 1 zone/ton		.32	50		425	2,525		2,950	4,325
9440	Ductwork per ton, packaged water & air cooled units		1.26	12.698		69	640		709	1,050
9450	Ductwork per ton, split system remote condensing units		1.32	12.121		65	610		675	1,000
9500	Ductwork package for residential units, 1 ton		.80	20	Ea.	280	1,000		1,280	1,850
9520	2 ton		.40	40		560	2,025		2,585	3,700
9530	3 ton		.27	59.259		840	2,975		3,815	5,475
9540	4 ton	Q-10	.30	80		1,125	4,175		5,300	7,600
9550	5 ton		.23	104		1,400	5,450		6,850	9,875
9560	6 ton		.20	120		1,675	6,275		7,950	11,400
9570	7 ton		.17	141		1,950	7,375		9,325	13,500
9580	8 ton		.15	160		2,250	8,350		10,600	15,300

For customer support on your Mechanical Cost Data, call 877.795.7394.

393

23 76 Evaporative Air-Cooling Equipment

23 76 13 – Direct Evaporative Air Coolers

23 76 13.10 Evaporative Coolers		Crew	Daily Output	Labor-Hours	Unit	Material	2015 Bare Costs Labor	Equipment	Total	Total Incl O&P
0010	**EVAPORATIVE COOLERS**, (Swamp Coolers) ducted, not incl. duct.									
0100	Side discharge style, capacities at .25" S.P.									
0120	1785 CFM, 1/3 HP, 115 V [G]	Q-9	5	3.200	Ea.	460	161		621	750
0140	2740 CFM, 1/3 HP, 115 V [G]		4.50	3.556		545	179		724	875
0160	3235 CFM, 1/2 HP, 115 V [G]		4	4		550	201		751	915
0180	3615 CFM, 1/2 HP, 230 V [G]		3.60	4.444		680	224		904	1,075
0200	4215 CFM, 3/4 HP, 230 V [G]		3.20	5		720	252		972	1,175
0220	5255 CFM, 1 HP, 115/230 V [G]		3	5.333		1,275	269		1,544	1,800
0240	6090 CFM, 1 HP, 230/460 V [G]		2.80	5.714		1,900	288		2,188	2,550
0260	8300 CFM, 1-1/2 HP, 230/460 V [G]		2.60	6.154		1,900	310		2,210	2,575
0280	8360 CFM, 1-1/2 HP, 230/460 V [G]		2.20	7.273		1,900	365		2,265	2,650
0300	9725 CFM, 2 HP, 230/460 V [G]		1.80	8.889		2,150	450		2,600	3,050
0320	11,715 CFM, 3 HP, 230/460 V [G]		1.40	11.429		2,400	575		2,975	3,525
0340	14,410 CFM, 5 HP, 230/460 V [G]		1	16		2,600	805		3,405	4,075
0400	For two-speed motor, add					5%				
0500	For down discharge style, add					10%				

23 76 26 – Fan Coil Evaporators for Coolers/Freezers

23 76 26.10 Evaporators

		Crew	Daily Output	Labor-Hours	Unit	Material	Labor	Equipment	Total	Total Incl O&P
0010	**EVAPORATORS**, DX coils, remote compressors not included.									
1000	Coolers, reach-in type, above freezing temperatures									
1300	Shallow depth, wall mount, 7 fins per inch, air defrost									
1310	600 BTUH, 8" fan, rust-proof core	Q-5	3.80	4.211	Ea.	680	226		906	1,100
1320	900 BTUH, 8" fan, rust-proof core		3.30	4.848		790	261		1,051	1,250
1330	1200 BTUH, 8" fan, rust-proof core		3	5.333		890	287		1,177	1,400
1340	1800 BTUH, 8" fan, rust-proof core		2.40	6.667		995	360		1,355	1,650
1350	2500 BTUH, 10" fan, rust-proof core		2.20	7.273		1,050	390		1,440	1,775
1360	3500 BTUH, 10" fan		2	8		1,200	430		1,630	1,975
1370	4500 BTUH, 10" fan		1.90	8.421		1,250	455		1,705	2,050
1390	For pan drain, add					59			59	64.50
1600	Undercounter refrigerators, ceiling or wall mount,									
1610	8 fins per inch, air defrost, rust-proof core									
1630	800 BTUH, one 6" fan	Q-5	5	3.200	Ea.	305	172		477	595
1640	1300 BTUH, two 6" fans		4	4		370	215		585	730
1650	1700 BTUH, two 6" fans		3.60	4.444		415	239		654	815
2000	Coolers, reach-in and walk-in types, above freezing temperatures									
2600	Two-way discharge, ceiling mount, 150-4100 CFM,									
2610	above 34°F applications, air defrost									
2630	900 BTUH, 7 fins per inch, 8" fan	Q-5	4	4	Ea.	550	215		765	930
2660	2500 BTUH, 7 fins per inch, 10" fan		2.60	6.154		865	330		1,195	1,450
2690	5500 BTUH, 7 fins per inch, 12" fan		1.70	9.412		1,100	505		1,605	1,975
2720	8500 BTUH, 8 fins per inch, 16" fan		1.20	13.333		1,575	715		2,290	2,800
2750	15,000 BTUH, 7 fins per inch, 18" fan		1.10	14.545		2,500	780		3,280	3,925
2770	24,000 BTUH, 7 fins per inch, two 16" fans		1	16		3,750	860		4,610	5,400
2790	30,000 BTUH, 7 fins per inch, two 18" fans		.90	17.778		4,650	955		5,605	6,550
2850	Two-way discharge, low profile, ceiling mount,									
2860	8 fins per inch, 200-570 CFM, air defrost									
2880	800 BTUH, one 6" fan	Q-5	4	4	Ea.	370	215		585	730
2890	1300 BTUH, two 6" fans		3.50	4.571		440	246		686	850
2900	1800 BTUH, two 6" fans		3.30	4.848		460	261		721	900
2910	2700 BTUH, three 6" fans		2.70	5.926		540	320		860	1,075
3000	Coolers, walk-in type, above freezing temperatures									
3300	General use, ceiling mount, 108-2080 CFM									

23 76 26.10 Evaporators		Crew	Daily Output	Labor-Hours	Unit	Material	2015 Bare Costs Labor	Equipment	Total	Total Incl O&P
3320	600 BTUH, 7 fins per inch, 6" fan	Q-5	5.20	3.077	Ea.	325	165		490	610
3340	1200 BTUH, 7 fins per inch, 8" fan		3.70	4.324		450	233		683	845
3360	1800 BTUH, 7 fins per inch, 10" fan		2.70	5.926		630	320		950	1,175
3380	3500 BTUH, 7 fins per inch, 12" fan		2.40	6.667		930	360		1,290	1,575
3400	5500 BTUH, 8 fins per inch, 12" fan		2	8		1,025	430		1,455	1,800
3430	8500 BTUH, 8 fins per inch, 16" fan		1.40	11.429		1,550	615		2,165	2,625
3460	15,000 BTUH, 7 fins per inch, two 16" fans	▼	1.10	14.545	▼	2,475	780		3,255	3,900
3640	Low velocity, high latent load, ceiling mount,									
3650	1050-2420 CFM, air defrost, 6 fins per inch									
3670	6700 BTUH, two 10" fans	Q-5	1.30	12.308	Ea.	2,025	660		2,685	3,250
3680	10,000 BTUH, three 10" fans		1	16		2,575	860		3,435	4,150
3690	13,500 BTUH, three 10" fans		1	16		2,950	860		3,810	4,550
3700	18,000 BTUH, four 10" fans		1	16		3,975	860		4,835	5,675
3710	26,500 BTUH, four 10" fans	▼	.80	20		4,650	1,075		5,725	6,750
3730	For electric defrost, add				▼	17%				
5000	Freezers and coolers, reach-in type, above 34°F									
5030	to sub-freezing temperature range, low latent load,									
5050	air defrost, 7 fins per inch									
5070	1200 BTUH, 8" fan, rustproof core	Q-5	4.40	3.636	Ea.	605	196		801	960
5080	1500 BTUH, 8" fan, rustproof core		4.20	3.810		615	205		820	990
5090	1800 BTUH, 8" fan, rustproof core		3.70	4.324		635	233		868	1,050
5100	2500 BTUH, 10" fan, rustproof core		2.90	5.517		690	297		987	1,200
5110	3500 BTUH, 10" fan, rustproof core		2.50	6.400		865	345		1,210	1,475
5120	4500 BTUH, 12" fan, rustproof core	▼	2.10	7.619	▼	1,075	410		1,485	1,800
6000	Freezers and coolers, walk-in type									
6050	1960-18,000 CFM, medium profile									
6060	Standard motor, 6 fins per inch, to -30°F, all aluminum									
6080	10,500 BTUH, air defrost, one 18" fan	Q-5	1.40	11.429	Ea.	1,650	615		2,265	2,725
6090	12,500 BTUH, air defrost, one 18" fan		1.40	11.429		1,775	615		2,390	2,875
6100	16,400 BTUH, air defrost, one 18" fan		1.30	12.308		1,800	660		2,460	2,975
6110	20,900 BTUH, air defrost, one 18" fans		1	16		2,100	860		2,960	3,600
6120	27,000 BTUH, air defrost, two 18" fans	▼	1	16		3,100	860		3,960	4,700
6130	32,900 BTUH, air defrost, two 20" fans	Q-6	1.30	18.462		3,450	1,025		4,475	5,350
6140	39,000 BTUH, air defrost, two 20" fans		1.25	19.200		4,025	1,075		5,100	6,050
6150	44,100 BTUH, air defrost, three 20" fans		1	24		4,550	1,350		5,900	7,050
6155	53,000 BTUH, air defrost, three 20" fans		1	24		4,550	1,350		5,900	7,050
6160	66,200 BTUH, air defrost, four 20" fans		1	24		5,175	1,350		6,525	7,725
6170	78,000 BTUH, air defrost, four 20" fans		.80	30		6,450	1,675		8,125	9,625
6180	88,200 BTUH, air defrost, four 24" fans		.60	40		6,875	2,225		9,100	10,900
6190	110,000 BTUH, air defrost, four 24" fans	▼	.50	48		7,775	2,675		10,450	12,600
6330	Hot gas defrost, standard motor units, add					35%				
6370	Electric defrost, 230V, standard motor, add					16%				
6410	460V, standard or high capacity motor, add				▼	25%				
6800	Eight fins per inch, increases BTUH 75%									
6810	Standard capacity, add				Ea.	5%				
6830	Hot gas & electric defrost not recommended									
7000	800-4150 CFM, 12" fans									
7010	Four fins per inch									
7030	3400 BTUH, 1 fan	Q-5	2.80	5.714	Ea.	555	305		860	1,075
7040	4200 BTUH, 1 fan		2.40	6.667		635	360		995	1,250
7050	5300 BTUH, 1 fan		1.90	8.421		825	455		1,280	1,600
7060	6800 BTUH, 1 fan		1.70	9.412		990	505		1,495	1,875
7070	8400 BTUH, 2 fans	▼	1.60	10	▼	1,125	540		1,665	2,050

23 76 Evaporative Air-Cooling Equipment

23 76 26 – Fan Coil Evaporators for Coolers/Freezers

23 76 26.10 Evaporators		Crew	Daily Output	Labor-Hours	Unit	Material	2015 Bare Costs Labor	Equipment	Total	Total Incl O&P
7080	10,500 BTUH, 2 fans	Q-5	1.40	11.429	Ea.	1,200	615		1,815	2,250
7090	13,000 BTUH, 3 fans		1.30	12.308		1,600	660		2,260	2,750
7100	17,000 BTUH, 4 fans		1.20	13.333		2,025	715		2,740	3,300
7110	21,500 BTUH, 5 fans		1	16		2,525	860		3,385	4,075
7160	Six fins per inch increases BTUH 35%, add					5%				
7180	Eight fins per inch increases BTUH 53%,									
7190	not recommended for below 35°F, add				Ea.	8%				
7210	For 12°F temperature differential, add					20%				
7230	For adjustable thermostat control, add					141			141	155
7240	For hot gas defrost, except 8 fin models									
7250	on applications below 35°F, add				Ea.	58%				
7260	For electric defrost, except 8 fin models									
7270	on applications below 35°F, add				Ea.	36%				
8000	Freezers, pass-thru door uprights, walk-in storage									
8300	Low temperature, thin profile, electric defrost									
8310	138-1800 CFM, 5 fins per inch									
8320	900 BTUH, one 6" fan	Q-5	3.30	4.848	Ea.	430	261		691	865
8340	1500 BTUH, two 6" fans		2.20	7.273		640	390		1,030	1,300
8360	2600 BTUH, three 6" fans		1.90	8.421		920	455		1,375	1,675
8370	3300 BTUH, four 6" fans		1.70	9.412		1,025	505		1,530	1,900
8380	4400 BTUH, five 6" fans		1.40	11.429		1,425	615		2,040	2,500
8400	7000 BTUH, three 10" fans		1.30	12.308		1,725	660		2,385	2,900
8410	8700 BTUH, four 10" fans		1.10	14.545		2,050	780		2,830	3,450
8450	For air defrost, deduct					255			255	281

23 81 Decentralized Unitary HVAC Equipment

23 81 13 – Packaged Terminal Air-Conditioners

23 81 13.10 Packaged Cabinet Type Air-Conditioners

		Crew	Daily Output	Labor-Hours	Unit	Material	2015 Bare Costs Labor	Equipment	Total	Total Incl O&P
0010	**PACKAGED CABINET TYPE AIR-CONDITIONERS**, Cabinet, wall sleeve,									
0100	louver, electric heat, thermostat, manual changeover, 208 V									
0200	6,000 BTUH cooling, 8800 BTU heat	Q-5	6	2.667	Ea.	775	143		918	1,075
0220	9,000 BTUH cooling, 13,900 BTU heat		5	3.200		1,200	172		1,372	1,575
0240	12,000 BTUH cooling, 13,900 BTU heat		4	4		1,350	215		1,565	1,825
0260	15,000 BTUH cooling, 13,900 BTU heat		3	5.333		1,450	287		1,737	2,025
0280	18,000 BTUH cooling, 10 KW heat		2.40	6.667		1,900	360		2,260	2,650
0300	24,000 BTUH cooling, 10 KW heat		1.90	8.421		1,925	455		2,380	2,800
0320	30,000 BTUH cooling, 10 KW heat		1.40	11.429		2,125	615		2,740	3,275
0340	36,000 BTUH cooling, 10 KW heat		1.25	12.800		2,250	690		2,940	3,525
0360	42,000 BTUH cooling, 10 KW heat		1	16		2,750	860		3,610	4,325
0380	48,000 BTUH cooling, 10 KW heat		.90	17.778		2,900	955		3,855	4,650
0500	For hot water coil, increase heat by 10%, add					5%	10%			
1000	For steam, increase heat output by 30%, add					8%	10%			

23 81 19 – Self-Contained Air-Conditioners

23 81 19.10 Window Unit Air Conditioners

		Crew	Daily Output	Labor-Hours	Unit	Material	2015 Bare Costs Labor	Equipment	Total	Total Incl O&P
0010	**WINDOW UNIT AIR CONDITIONERS**									
4000	Portable/window, 15 amp 125 V grounded receptacle required									
4060	5000 BTUH	1 Carp	8	1	Ea.	295	47		342	400
4340	6000 BTUH		8	1		340	47		387	450
4480	8000 BTUH		6	1.333		460	62.50		522.50	605
4500	10,000 BTUH		6	1.333		590	62.50		652.50	745

23 81 Decentralized Unitary HVAC Equipment

23 81 19 – Self-Contained Air-Conditioners

23 81 19.10 Window Unit Air Conditioners	Crew	Daily Output	Labor-Hours	Unit	Material	2015 Bare Costs Labor	Equipment	Total	Total Incl O&P
4520	12,000 BTUH	L-2	8	2	Ea.	725	82.50		807.50	920
4600	Window/thru-the-wall, 15 amp 230 V grounded receptacle required									
4780	18,000 BTUH	L-2	6	2.667	Ea.	970	110		1,080	1,250
4940	25,000 BTUH		4	4		1,175	165		1,340	1,550
4960	29,000 BTUH		4	4		1,325	165		1,490	1,725

23 81 19.20 Self-Contained Single Package

		Crew	Daily Output	Labor-Hours	Unit	Material	Labor	Equipment	Total	Total Incl O&P
0010	**SELF-CONTAINED SINGLE PACKAGE**									
0100	Air cooled, for free blow or duct, not incl. remote condenser									
0110	Constant volume									
0200	3 ton cooling	Q-5	1	16	Ea.	3,750	860		4,610	5,400
0210	4 ton cooling	"	.80	20		4,075	1,075		5,150	6,100
0220	5 ton cooling	Q-6	1.20	20		4,450	1,125		5,575	6,575
0230	7.5 ton cooling	"	.90	26.667		6,475	1,475		7,950	9,375
0240	10 ton cooling	Q-7	1	32		7,125	1,825		8,950	10,600
0250	15 ton cooling		.95	33.684		9,050	1,925		10,975	12,900
0260	20 ton cooling		.90	35.556		12,800	2,025		14,825	17,200
0270	25 ton cooling		.85	37.647		19,000	2,150		21,150	24,100
0280	30 ton cooling		.80	40		25,300	2,275		27,575	31,200
0300	40 ton cooling		.60	53.333		36,100	3,025		39,125	44,300
0320	50 ton cooling		.50	64		47,200	3,650		50,850	57,500
0340	60 ton cooling	Q-8	.40	80		56,000	4,550	144	60,694	68,500
0380	With hot water heat, variable air volume									
0390	10 ton cooling	Q-7	.96	33.333	Ea.	10,100	1,900		12,000	14,000
0400	20 ton cooling		.86	37.209		17,900	2,125		20,025	22,900
0410	30 ton cooling		.76	42.105		33,700	2,400		36,100	40,600
0420	40 ton cooling		.58	55.172		46,700	3,150		49,850	56,500
0430	50 ton cooling		.48	66.667		60,500	3,800		64,300	72,000
0440	60 ton cooling	Q-8	.38	84.211		73,500	4,800	152	78,452	88,500
0490	For duct mounting no price change									
0500	For steam heating coils, add				Ea.	10%	10%			
0550	Packaged, with electric heat									
0552	Constant volume									
0560	5 ton cooling	Q-6	1.20	20	Ea.	7,950	1,125		9,075	10,400
0570	10 ton cooling	Q-7	1	32		12,500	1,825		14,325	16,500
0580	20 ton cooling		.90	35.556		20,500	2,025		22,525	25,700
0590	30 ton cooling		.80	40		38,100	2,275		40,375	45,300
0600	40 ton cooling		.60	53.333		50,000	3,025		53,025	59,500
0610	50 ton cooling		.50	64		66,500	3,650		70,150	78,500
0700	Variable air volume									
0710	10 ton cooling	Q-7	1	32	Ea.	14,800	1,825		16,625	19,100
0720	20 ton cooling		.90	35.556		24,300	2,025		26,325	29,800
0730	30 ton cooling		.80	40		44,000	2,275		46,275	52,000
0740	40 ton cooling		.60	53.333		57,000	3,025		60,025	67,000
0750	50 ton cooling		.50	64		75,000	3,650		78,650	88,000
0760	60 ton cooling	Q-8	.40	80		85,500	4,550	144	90,194	101,000
1000	Water cooled for free blow or duct, not including tower									
1010	Constant volume									
1100	3 ton cooling	Q-6	1	24	Ea.	3,725	1,350		5,075	6,125
1120	5 ton cooling		1	24		4,850	1,350		6,200	7,350
1130	7.5 ton cooling		.80	30		6,850	1,675		8,525	10,100
1140	10 ton cooling	Q-7	.90	35.556		9,450	2,025		11,475	13,500
1150	15 ton cooling		.85	37.647		13,700	2,150		15,850	18,300

For customer support on your Mechanical Cost Data, call 877.795.7394.

397

23 81 Decentralized Unitary HVAC Equipment

23 81 19 – Self-Contained Air-Conditioners

23 81 19.20 Self-Contained Single Package		Crew	Daily Output	Labor-Hours	Unit	Material	2015 Bare Costs Labor	Equipment	Total	Total Incl O&P
1160	20 ton cooling	Q-7	.80	40	Ea.	28,600	2,275		30,875	34,800
1170	25 ton cooling		.75	42.667		33,700	2,425		36,125	40,800
1180	30 ton cooling		.70	45.714		37,900	2,600		40,500	45,600
1200	40 ton cooling		.40	80		48,000	4,550		52,550	60,000
1220	50 ton cooling		.30	106		59,500	6,075		65,575	74,500
1240	60 ton cooling	Q-8	.30	106		69,500	6,075	192	75,767	86,000
1300	For hot water or steam heat coils, add					12%	10%			
2000	Water cooled with electric heat, not including tower									
2010	Constant volume									
2100	5 ton cooling	Q-6	1.20	20	Ea.	8,350	1,125		9,475	10,900
2120	10 ton cooling	Q-7	1	32		14,800	1,825		16,625	19,100
2130	15 ton cooling	"	.90	35.556		21,300	2,025		23,325	26,600
2400	Variable air volume									
2440	10 ton cooling	Q-7	1	32	Ea.	17,100	1,825		18,925	21,600
2450	15 ton cooling	"	.90	35.556	"	25,200	2,025		27,225	30,800
2600	Water cooled, hot water coils, not including tower									
2610	Variable air volume									
2640	10 ton cooling	Q-7	1	32	Ea.	12,700	1,825		14,525	16,800
2650	15 ton cooling	"	.80	40	"	18,900	2,275		21,175	24,100

23 81 23 – Computer-Room Air-Conditioners

23 81 23.10 Computer Room Units

		Crew	Daily Output	Labor-Hours	Unit	Material	2015 Bare Costs Labor	Equipment	Total	Total Incl O&P
0010	**COMPUTER ROOM UNITS** D3050-185									
1000	Air cooled, includes remote condenser but not									
1020	interconnecting tubing or refrigerant									
1080	3 ton	Q-5	.50	32	Ea.	18,800	1,725		20,525	23,300
1120	5 ton		.45	35.556		20,100	1,900		22,000	25,000
1160	6 ton		.30	53.333		37,000	2,875		39,875	45,000
1200	8 ton		.27	59.259		37,400	3,175		40,575	46,000
1240	10 ton		.25	64		39,100	3,450		42,550	48,300
1260	12 ton		.24	66.667		40,500	3,575		44,075	50,000
1280	15 ton		.22	72.727		43,100	3,900		47,000	53,500
1290	18 ton		.20	80		49,300	4,300		53,600	61,000
1300	20 ton	Q-6	.26	92.308		51,500	5,150		56,650	64,500
1320	22 ton		.24	100		52,000	5,575		57,575	66,000
1360	30 ton		.21	114		64,500	6,375		70,875	80,000
2200	Chilled water, for connection to									
2220	existing chiller system of adequate capacity									
2260	5 ton	Q-5	.74	21.622	Ea.	14,300	1,175		15,475	17,600
2280	6 ton		.52	30.769		14,400	1,650		16,050	18,400
2300	8 ton		.50	32		14,600	1,725		16,325	18,700
2320	10 ton		.49	32.653		14,700	1,750		16,450	18,900
2330	12 ton		.49	32.990		15,100	1,775		16,875	19,300
2360	15 ton		.48	33.333		15,400	1,800		17,200	19,600
2400	20 ton		.46	34.783		16,300	1,875		18,175	20,800
2440	25 ton	Q-6	.63	38.095		17,400	2,125		19,525	22,300
2460	30 ton		.57	42.105		18,900	2,350		21,250	24,400
2480	35 ton		.52	46.154		19,100	2,575		21,675	24,900
2500	45 ton		.46	52.174		20,200	2,900		23,100	26,600
2520	50 ton		.42	57.143		23,100	3,175		26,275	30,200
2540	60 ton		.38	63.158		24,600	3,525		28,125	32,400
2560	77 ton		.35	68.571		41,600	3,825		45,425	51,500
2580	95 ton		.32	75		48,500	4,175		52,675	60,000

23 81 Decentralized Unitary HVAC Equipment

23 81 23 – Computer-Room Air-Conditioners

23 81 23.10 Computer Room Units		Crew	Daily Output	Labor-Hours	Unit	Material	2015 Bare Costs Labor	Equipment	Total	Total Incl O&P
4000	Glycol system, complete except for interconnecting tubing									
4060	3 ton	Q-5	.40	40	Ea.	23,600	2,150		25,750	29,200
4100	5 ton		.38	42.105		25,800	2,275		28,075	31,800
4120	6 ton		.25	64		38,300	3,450		41,750	47,300
4140	8 ton		.23	69.565		41,400	3,750		45,150	51,000
4160	10 ton		.21	76.190		44,000	4,100		48,100	54,500
4180	12 ton		.19	84.211		47,700	4,525		52,225	59,500
4200	15 ton	Q-6	.26	92.308		54,500	5,150		59,650	68,000
4240	20 ton		.24	100		60,000	5,575		65,575	74,500
4280	22 ton		.23	104		63,000	5,825		68,825	78,500
4430	30 ton		.22	109		78,000	6,075		84,075	95,000
8000	Water cooled system, not including condenser,									
8020	water supply or cooling tower									
8060	3 ton	Q-5	.62	25.806	Ea.	19,100	1,400		20,500	23,100
8100	5 ton		.54	29.630		20,800	1,600		22,400	25,300
8120	6 ton		.35	45.714		29,700	2,450		32,150	36,400
8140	8 ton		.33	48.485		34,100	2,600		36,700	41,400
8160	10 ton		.31	51.613		35,200	2,775		37,975	42,900
8180	12 ton		.29	55.172		38,200	2,975		41,175	46,500
8200	15 ton		.27	59.259		41,300	3,175		44,475	50,000
8240	20 ton	Q-6	.38	63.158		44,300	3,525		47,825	54,000
8280	22 ton		.35	68.571		46,600	3,825		50,425	57,000
8300	30 ton		.30	80		58,000	4,450		62,450	70,000

23 81 26 – Split-System Air-Conditioners

23 81 26.10 Split Ductless Systems		Crew	Daily Output	Labor-Hours	Unit	Material	2015 Bare Costs Labor	Equipment	Total	Total Incl O&P
0010	**SPLIT DUCTLESS SYSTEMS**									
0100	Cooling only, single zone									
0110	Wall mount									
0120	3/4 ton cooling	Q-5	2	8	Ea.	1,100	430		1,530	1,875
0130	1 ton cooling		1.80	8.889		1,225	480		1,705	2,075
0140	1-1/2 ton cooling		1.60	10		1,975	540		2,515	2,975
0150	2 ton cooling		1.40	11.429		2,250	615		2,865	3,400
1000	Ceiling mount									
1020	2 ton cooling	Q-5	1.40	11.429	Ea.	2,000	615		2,615	3,125
1030	3 ton cooling	"	1.20	13.333	"	2,550	715		3,265	3,875
3000	Multizone									
3010	Wall mount									
3020	2 @ 3/4 ton cooling	Q-5	1.80	8.889	Ea.	3,375	480		3,855	4,425
5000	Cooling/Heating									
5010	Wall mount									
5110	1 ton cooling	Q-5	1.70	9.412	Ea.	1,300	505		1,805	2,200
5120	1-1/2 ton cooling	"	1.50	10.667	"	1,975	575		2,550	3,050
7000	Accessories for all split ductless systems									
7010	Add for ambient frost control	Q-5	8	2	Ea.	126	108		234	300
7020	Add for tube/wiring kit, (Line sets)									
7030	15' kit	Q-5	32	.500	Ea.	103	27		130	154
7036	25' kit		28	.571		166	30.50		196.50	229
7040	35' kit		24	.667		174	36		210	245
7050	50' kit		20	.800		261	43		304	350

For customer support on your Mechanical Cost Data, call 877.795.7394.

399

23 81 29.10 Heat Pump, Gas Driven	Crew	Daily Output	Labor-Hours	Unit	Material	2015 Bare Costs Labor	2015 Bare Costs Equipment	Total	Total Incl O&P
0010 **HEAT PUMP, GAS DRIVEN,** Variable refrigerant volume (VRV) type									
0020 Not including interconnecting tubing or multi-zone controls									
1000 For indoor fan VRV type AHU see 23 82 19.40									
1010 Multi-zone split									
1020 Outdoor unit									
1100 8 tons cooling, for up to 17 zones	Q-5	1.30	12.308	Ea.	29,100	660		29,760	33,000
1110 Isolation rails		2.60	6.154	Pair	1,025	330		1,355	1,625
1160 15 tons cooling, for up to 33 zones		1	16	Ea.	38,600	860		39,460	43,700
1170 Isolation rails		2.60	6.154	Pair	1,200	330		1,530	1,825
2000 Packaged unit									
2020 Outdoor unit									
2200 11 tons cooling	Q-5	1.30	12.308	Ea.	37,200	660		37,860	41,900
2210 Roof curb adapter	"	2.60	6.154		645	330		975	1,200
2220 Thermostat	1 Stpi	1.30	6.154		385	370		755	980

23 81 43 – Air-Source Unitary Heat Pumps

23 81 43.10 Air-Source Heat Pumps

	Crew	Daily Output	Labor-Hours	Unit	Material	2015 Bare Costs Labor	2015 Bare Costs Equipment	Total	Total Incl O&P
0010 **AIR-SOURCE HEAT PUMPS,** Not including interconnecting tubing									
1000 Air to air, split system, not including curbs, pads, fan coil and ductwork									
1010 For curbs/pads see Section 23 91 00									
1012 Outside condensing unit only, for fan coil see Section 23 82 19.10									
1015 1.5 ton cooling, 7 MBH heat @ 0°F	Q-5	2.40	6.667	Ea.	2,175	360		2,535	2,950
1020 2 ton cooling, 8.5 MBH heat @ 0°F		2	8		2,400	430		2,830	3,300
1030 2.5 ton cooling, 10 MBH heat @ 0°F		1.60	10		2,525	540		3,065	3,575
1040 3 ton cooling, 13 MBH heat @ 0°F		1.20	13.333		2,775	715		3,490	4,125
1050 3.5 ton cooling, 18 MBH heat @ 0°F		1	16		3,000	860		3,860	4,600
1054 4 ton cooling, 24 MBH heat @ 0°F		.80	20		3,250	1,075		4,325	5,200
1060 5 ton cooling, 27 MBH heat @ 0°F		.50	32		3,600	1,725		5,325	6,550
1080 7.5 ton cooling, 33 MBH heat @ 0°F		.45	35.556		6,425	1,900		8,325	9,950
1100 10 ton cooling, 50 MBH heat @ 0°F	Q-6	.64	37.500		8,500	2,100		10,600	12,500
1120 15 ton cooling, 64 MBH heat @ 0°F		.50	48		11,800	2,675		14,475	17,100
1130 20 ton cooling, 85 MBH heat @ 0°F		.35	68.571		17,000	3,825		20,825	24,500
1140 25 ton cooling, 119 MBH heat @ 0°F		.25	96		20,400	5,350		25,750	30,600
1500 Single package, not including curbs, pads, or plenums									
1502 1/2 ton cooling, supplementary heat included	Q-5	8	2	Ea.	2,450	108		2,558	2,850
1504 3/4 ton cooling, supplementary heat supplementary heat included		6	2.667		2,575	143		2,718	3,050
1506 1 ton cooling, supplementary heat included		4	4		2,975	215		3,190	3,600
1510 1.5 ton cooling, 5 MBH heat @ 0°F		1.55	10.323		3,050	555		3,605	4,200
1520 2 ton cooling, 6.5 MBH heat @ 0°F		1.50	10.667		3,100	575		3,675	4,300
1540 2.5 ton cooling, 8 MBH heat @ 0°F		1.40	11.429		3,175	615		3,790	4,425
1560 3 ton cooling, 10 MBH heat @ 0°F		1.20	13.333		3,575	715		4,290	5,000
1570 3.5 ton cooling, 11 MBH heat @ 0°F		1	16		3,900	860		4,760	5,575
1580 4 ton cooling, 13 MBH heat @ 0°F		.96	16.667		4,200	895		5,095	5,975
1620 5 ton cooling, 27 MBH heat @ 0°F		.65	24.615		4,800	1,325		6,125	7,275
1640 7.5 ton cooling, 35 MBH heat @ 0°F		.40	40		7,325	2,150		9,475	11,300
1648 10 ton cooling, 45 MBH heat @ 0°F	Q-6	.40	60		9,925	3,350		13,275	16,000
1652 12 ton cooling, 50 MBH heat @ 0°F	"	.36	66.667		12,200	3,725		15,925	19,000
1696 Supplementary electric heat coil incl., unless noted otherwise									
6000 Air to water, single package, excluding. storage tank and ductwork									
6010 Includes circulating water pump, air duct connections, digital temperature									
6020 controller with remote tank temp. probe and sensor for storage tank.									
6040 Water heating - air cooling capacity									
6110 35.5 MBH heat water, 2.3 Ton cool air	Q-5	1.60	10	Ea.	11,400	540		11,940	13,400

For customer support on your Mechanical Cost Data, call 877.795.7394.

23 81 Decentralized Unitary HVAC Equipment

23 81 43 – Air-Source Unitary Heat Pumps

23 81 43.10 Air-Source Heat Pumps		Crew	Daily Output	Labor-Hours	Unit	Material	2015 Bare Costs Labor	Equipment	Total	Total Incl O&P
6120	58 MBH heat water, 3.8 Ton cool air	Q-5	1.10	14.545	Ea.	13,000	780		13,780	15,500
6130	76 MBH heat water, 4.9 Ton cool air		.87	18.391		15,500	990		16,490	18,600
6140	98 MBH heat water, 6.5 Ton cool air		.62	25.806		20,100	1,400		21,500	24,200
6150	113 MBH heat water, 7.4 Ton cool air		.59	27.119		22,000	1,450		23,450	26,400
6160	142 MBH heat water, 9.2 Ton cool air		.52	30.769		26,700	1,650		28,350	31,900
6170	171 MBH heat water, 11.1 Ton cool air	▼	.49	32.653	▼	31,400	1,750		33,150	37,200

23 81 46 – Water-Source Unitary Heat Pumps

23 81 46.10 Water Source Heat Pumps

		Crew	Daily Output	Labor-Hours	Unit	Material	2015 Bare Costs Labor	Equipment	Total	Total Incl O&P
0010	**WATER SOURCE HEAT PUMPS**, Not incl. connecting tubing or water source									
2000	Water source to air, single package									
2100	1 ton cooling, 13 MBH heat @ 75°F	Q-5	2	8	Ea.	2,075	430		2,505	2,950
2120	1.5 ton cooling, 17 MBH heat @ 75°F		1.80	8.889		2,400	480		2,880	3,350
2140	2 ton cooling, 19 MBH heat @ 75°F		1.70	9.412		2,575	505		3,080	3,600
2160	2.5 ton cooling, 25 MBH heat @ 75°F		1.60	10		2,800	540		3,340	3,900
2180	3 ton cooling, 27 MBH heat @ 75°F		1.40	11.429		2,875	615		3,490	4,075
2190	3.5 ton cooling, 29 MBH heat @ 75°F		1.30	12.308		3,100	660		3,760	4,425
2200	4 ton cooling, 31 MBH heat @ 75°F		1.20	13.333		3,475	715		4,190	4,900
2220	5 ton cooling, 29 MBH heat @ 75°F		.90	17.778		3,825	955		4,780	5,650
2240	7.5 ton cooling, 35 MBH heat @ 75°F		.60	26.667		7,275	1,425		8,700	10,200
2250	8.5 ton cooling, 40 MBH heat @ 75°F		.58	27.586		8,800	1,475		10,275	12,000
2260	10 ton cooling, 50 MBH heat @ 75°F	▼	.53	30.189		9,225	1,625		10,850	12,600
2280	15 ton cooling, 64 MBH heat @ 75°F	Q-6	.47	51.064		14,000	2,850		16,850	19,700
2300	20 ton cooling, 100 MBH heat @ 75°F		.41	58.537		15,300	3,275		18,575	21,700
2310	25 ton cooling, 100 MBH heat @ 75°F		.32	75		20,800	4,175		24,975	29,200
2320	30 ton cooling, 128 MBH heat @ 75°F		.24	102		22,800	5,700		28,500	33,700
2340	40 ton cooling, 200 MBH heat @ 75°F		.21	117		32,200	6,525		38,725	45,300
2360	50 ton cooling, 200 MBH heat @ 75°F	▼	.15	160	▼	36,400	8,925		45,325	53,500
3960	For supplementary heat coil, add					10%				
4000	For increase in capacity thru use									
4020	of solar collector, size boiler at 60%									

23 82 Convection Heating and Cooling Units

23 82 13 – Valance Heating and Cooling Units

23 82 13.16 Valance Units

		Crew	Daily Output	Labor-Hours	Unit	Material	2015 Bare Costs Labor	Equipment	Total	Total Incl O&P
0010	**VALANCE UNITS**									
6000	Valance units, complete with 1/2" cooling coil, enclosure									
6020	2 tube	Q-5	18	.889	L.F.	33.50	48		81.50	109
6040	3 tube		16	1		40	54		94	125
6060	4 tube		16	1		48.50	54		102.50	135
6080	5 tube		15	1.067		52	57.50		109.50	144
6100	6 tube		15	1.067		57.50	57.50		115	150
6120	8 tube	▼	14	1.143		83.50	61.50		145	185
6200	For 3/4" cooling coil, add					10%				

For customer support on your Mechanical Cost Data, call 877.795.7394.

401

23 82 16.10 Flanged Coils	Crew	Daily Output	Labor-Hours	Unit	Material	2015 Bare Costs Labor	Equipment	Total	Total Incl O&P
0010 **FLANGED COILS**									
0100 Basic water, DX, or condenser coils									
0110 Copper tubes, alum. fins, galv. end sheets									
0112 H is finned height, L is finned length									
0120 3/8" x .016 tube, .0065 aluminum fins									
0130 2 row, 8 fins per inch									
0140 4" H x 12" L	Q-5	48	.333	Ea.	620	17.95		637.95	710
0150 4" H x 24" L		38.38	.417		670	22.50		692.50	775
0160 4" H x 48" L		19.25	.831		730	44.50		774.50	870
0170 4" H x 72" L		12.80	1.250		850	67		917	1,025
0180 6" H x 12" L		48	.333		655	17.95		672.95	745
0190 6" H x 24" L		25.60	.625		710	33.50		743.50	830
0200 6" H x 48" L		12.80	1.250		780	67		847	960
0210 6" H x 72" L		8.53	1.876		920	101		1,021	1,150
0220 10" H x 12" L		30.73	.521		720	28		748	835
0230 10" H x 24" L		15.33	1.044		795	56		851	955
0240 10" H x 48" L		7.69	2.081		890	112		1,002	1,150
0250 10" H x 72" L		5.12	3.125		1,050	168		1,218	1,400
0260 12" H x 12" L		25.60	.625		750	33.50		783.50	875
0270 12" H x 24" L		12.80	1.250		835	67		902	1,025
0280 12" H x 48" L		6.40	2.500		945	134		1,079	1,250
0290 12" H x 72" L		4.27	3.747		1,125	202		1,327	1,550
0300 16" H x 16" L		14.38	1.113		835	60		895	1,000
0310 16" H x 24" L		9.59	1.668		915	89.50		1,004.50	1,125
0320 16" H x 48" L		4.80	3.333		1,050	179		1,229	1,425
0330 16" H x 72" L		3.20	5		1,275	269		1,544	1,800
0340 20" H x 24" L		7.69	2.081		995	112		1,107	1,275
0350 20" H x 48" L		3.84	4.167		1,175	224		1,399	1,625
0360 20" H x 72" L		2.56	6.250		1,425	335		1,760	2,075
0370 24" H x 24" L		6.40	2.500		1,075	134		1,209	1,375
0380 24" H x 48" L		3.20	5		1,275	269		1,544	1,800
0390 24" H x 72" L		2.13	7.512		1,575	405		1,980	2,325
0400 30" H x 28" L		4.39	3.645		1,250	196		1,446	1,675
0410 30" H x 48" L		2.56	6.250		1,450	335		1,785	2,100
0420 30" H x 72" L		1.71	9.357		1,775	505		2,280	2,700
0430 30" H x 84" L		1.46	10.959		1,950	590		2,540	3,050
0440 34" H x 32" L		3.39	4.720		1,375	254		1,629	1,900
0450 34" H x 48" L		2.26	7.080		1,575	380		1,955	2,300
0460 34" H x 72" L		1.51	10.596		1,925	570		2,495	2,950
0470 34" H x 84" L		1.29	12.403		2,100	665		2,765	3,325
0600 For 10 fins per inch, add					1%				
0610 For 12 fins per inch, add					3%				
0620 For 4 row, add				Ea.	30%	50%			
0630 For 6 row, add					60%	100%			
0640 For 8 row, add					80%	150%			
1100 1/2" x .017 tube, .0065 aluminum fins									
1110 2 row, 8 fins per inch									
1120 5" H x 6" L	Q-5	40	.400	Ea.	620	21.50		641.50	715
1130 5" H x 20" L		37.10	.431		675	23		698	775
1140 5" H x 45" L		16.41	.975		745	52.50		797.50	900
1150 5" H x 90" L		8.21	1.949		1,000	105		1,105	1,250
1160 7.5" H x 5" L		38.50	.416		620	22.50		642.50	720

23 82 Convection Heating and Cooling Units

23 82 16 – Air Coils

23 82 16.10 Flanged Coils		Crew	Daily Output	Labor-Hours	Unit	Material	2015 Bare Costs Labor	Equipment	Total	Total Incl O&P
1170	7.5" H x 20" L	Q-5	24.62	.650	Ea.	720	35		755	845
1180	7.5" H x 45" L		10.94	1.463		810	78.50		888.50	1,000
1190	7.5" H x 90" L		5.46	2.930		1,100	158		1,258	1,450
1200	12.5" H x 10" L		29.49	.543		735	29		764	855
1210	12.5" H x 20" L		14.75	1.085		820	58.50		878.50	995
1220	12.5" H x 45" L		6.55	2.443		940	131		1,071	1,225
1230	12.5" H x 90" L		3.28	4.878		1,325	262		1,587	1,850
1240	15" H x 15" L		16.41	.975		795	52.50		847.50	955
1250	15" H x 30" L		8.21	1.949		920	105		1,025	1,150
1260	15" H x 45" L		5.46	2.930		1,000	158		1,158	1,350
1270	15" H x 90" L		2.73	5.861		1,425	315		1,740	2,050
1280	20" H x 20" L		9.22	1.735		950	93.50		1,043.50	1,200
1290	20" H x 45" L		4.10	3.902		1,150	210		1,360	1,575
1300	20" H x 90" L		2.05	7.805		1,650	420		2,070	2,425
1310	25" H x 25" L		5.90	2.712		1,100	146		1,246	1,425
1320	25" H x 45" L		3.28	4.878		1,300	262		1,562	1,825
1330	25" H x 90" L		1.64	9.756		1,850	525		2,375	2,825
1340	30" H x 30" L		4.10	3.902		1,250	210		1,460	1,700
1350	30" H x 45" L		2.73	5.861		1,425	315		1,740	2,050
1360	30" H x 90" L		1.37	11.679		2,050	630		2,680	3,225
1370	35" H x 35" L		3.01	5.316		1,800	286		2,086	2,400
1380	35" H x 45" L		2.34	6.838		1,575	370		1,945	2,275
1390	35" H x 90" L		1.17	13.675		2,275	735		3,010	3,600
1400	42.5" H x 45" L		1.93	8.290		1,775	445		2,220	2,625
1410	42.5" H x 75" L		1.16	13.793		2,300	740		3,040	3,650
1420	42.5" H x 105" L		.83	19.277		2,800	1,025		3,825	4,650
1600	For 10 fins per inch, add					2%				
1610	For 12 fins per inch, add					5%				
1620	For 4 row, add					35%	50%			
1630	For 6 row, add					65%	100%			
1640	For 8 row, add					90%	150%			
2010	5/8" x .020" tube, .0065 aluminum fins									
2020	2 row, 8 fins per inch									
2030	6" H x 6" L	Q-5	38	.421	Ea.	555	22.50		577.50	645
2040	6" H x 24" L		24.96	.641		620	34.50		654.50	735
2050	6" H x 48" L		12.48	1.282		695	69		764	865
2060	6" H x 96" L		6.24	2.564		950	138		1,088	1,250
2070	12" H x 12" L		24.96	.641		645	34.50		679.50	760
2080	12" H x 24" L		12.48	1.282		730	69		799	905
2090	12" H x 48" L		6.24	2.564		840	138		978	1,125
2100	12" H x 96" L		3.12	5.128		1,600	276		1,876	2,200
2110	18" H x 24" L		8.32	1.923		835	103		938	1,075
2120	18" H x 48" L		4.16	3.846		995	207		1,202	1,400
2130	18" H x 96" L		2.08	7.692		1,450	415		1,865	2,225
2140	24" H x 24" L		6.24	2.564		940	138		1,078	1,225
2150	24" H x 48" L		3.12	5.128		1,150	276		1,426	1,700
2160	24" H x 96" L		1.56	10.256		1,700	550		2,250	2,675
2170	30" H x 30" L		3.99	4.010		1,100	216		1,316	1,550
2180	30" H x 48" L		2.50	6.400		1,300	345		1,645	1,950
2190	30" H x 96" L		1.25	12.800		1,925	690		2,615	3,175
2200	30" H x 120" L		1	16		2,225	860		3,085	3,750
2210	36" H x 36" L		2.77	5.776		1,300	310		1,610	1,925
2220	36" H x 48" L		2.08	7.692		1,450	415		1,865	2,225

For customer support on your Mechanical Cost Data, call 877.795.7394.

403

23 82 16 – Air Coils

23 82 16.10 Flanged Coils		Crew	Daily Output	Labor-Hours	Unit	Material	2015 Bare Costs Labor	Equipment	Total	Total Incl O&P
2230	36" H x 90" L	Q-5	1.11	14.414	Ea.	2,100	775		2,875	3,475
2240	36" H x 120" L		.83	19.277		2,525	1,025		3,550	4,350
2250	42" H x 48" L		1.78	8.989		1,600	485		2,085	2,500
2260	42" H x 90" L		.95	16.842		2,325	905		3,230	3,925
2270	42" H x 120" L		.71	22.535		2,800	1,200		4,000	4,925
2280	51" H x 48" L		1.47	10.884		1,875	585		2,460	2,950
2290	51" H x 96" L		.73	21.918		2,825	1,175		4,000	4,875
2300	51" H x 120" L		.59	27.119		3,275	1,450		4,725	5,800
2500	For 10 fins per inch, add					3%				
2510	For 12 fins per inch, add					6%				
2520	For 4 row, add				Ea.	40%	50%			
2530	For 6 row, add					80%	100%			
2540	For 8 row, add					110%	150%			
3000	Hot water booster coils									
3010	Copper tubes, alum. fins, galv. end sheets									
3012	H is finned height, L is finned length									
3020	1/2" x .017" tube, .0065 alum. fins									
3030	1 row, 10 fins per inch									
3040	5"H x 10"L	Q-5	40	.400	Ea.	335	21.50		356.50	405
3050	5"H x 25"L		33.18	.482		380	26		406	460
3060	10"H x 10"L		39	.410		365	22		387	435
3070	10"H x 20"L		20.73	.772		405	41.50		446.50	515
3080	10"H x 30"L		13.83	1.157		440	62		502	580
3090	15"H x 15"L		18.44	.868		410	46.50		456.50	520
3110	15"H x 20"L		13.83	1.157		445	62		507	585
3120	15"H x 30"L		9.22	1.735		490	93.50		583.50	680
3130	20"H x 20"L		10.37	1.543		485	83		568	660
3140	20"H x 30"L		6.91	2.315		540	125		665	785
3150	20"H x 40"L		5.18	3.089		580	166		746	890
3160	25"H x 25"L		6.64	2.410		560	130		690	815
3170	25"H x 35"L		4.74	3.376		615	182		797	950
3180	25"H x 45"L		3.69	4.336		665	233		898	1,075
3190	30"H x 30"L		4.61	3.471		640	187		827	985
3200	30"H x 40"L		3.46	4.624		700	249		949	1,150
3210	30"H x 50"L		2.76	5.797		770	310		1,080	1,325
3300	For 2 row, add					20%	50%			
3400	1/2" x .020" tube, .008 alum. fins									
3410	2 row, 10 fins per inch									
3420	20" H x 27" L	Q-5	7.27	2.201	Ea.	695	118		813	945
3430	20" H x 36" L		5.44	2.941		760	158		918	1,075
3440	20" H x 45" L		4.35	3.678		830	198		1,028	1,200
3450	25" H x 45" L		3.49	4.585		945	247		1,192	1,425
3460	25" H x 54" L		2.90	5.517		1,125	297		1,422	1,700
3470	30" H x 54" L		2.42	6.612		1,275	355		1,630	1,925
3480	35" H x 54" L		2.07	7.729		1,400	415		1,815	2,175
3490	35" H x 63" L		1.78	8.989		1,525	485		2,010	2,400
3500	40" H x 72" L		1.36	11.765		1,800	635		2,435	2,925
3510	55" H x 63" L		1.13	14.159		2,200	760		2,960	3,550
3520	55" H x 72" L		.99	16.162		2,350	870		3,220	3,900
3530	55" H x 81" L		.88	18.182		2,600	980		3,580	4,325
3540	75" H x 81" L		.64	25		3,375	1,350		4,725	5,750
3550	75" H x 90" L		.58	27.586		3,575	1,475		5,050	6,200
3560	75" H x 108" L		.48	33.333		4,150	1,800		5,950	7,250

For customer support on your Mechanical Cost Data, call 877.795.7394.

23 82 16 – Air Coils

23 82 16.10 Flanged Coils	Crew	Daily Output	Labor-Hours	Unit	Material	2015 Bare Costs Labor	Equipment	Total	Total Incl O&P
3800	5/8" x .020 tubing, .0065 aluminum fins								
3810	1 row, 10 fins per inch								
3820	6" H x 12" L	Q-5	38	.421	Ea.	305	22.50		327.50	370
3830	6" H x 24" L		28.80	.556		340	30		370	420
3840	6" H x 30" L		23.04	.694		355	37.50		392.50	445
3850	12" H x 12" L		28.80	.556		340	30		370	420
3860	12" H x 24" L		14.40	1.111		390	60		450	515
3870	12" H x 30" L		11.52	1.389		410	74.50		484.50	565
3880	18" H x 18" L		12.80	1.250		390	67		457	525
3890	18" H x 24" L		9.60	1.667		430	89.50		519.50	610
3900	18" H x 36" L		6.40	2.500		480	134		614	735
3910	24" H x 24" L		7.20	2.222		480	120		600	705
3920	24" H x 30" L		4.80	3.333		515	179		694	840
3930	24" H x 36" L		5.05	3.168		540	170		710	850
3940	30" H x 30" L		4.61	3.471		565	187		752	905
3950	30" H x 36" L		3.85	4.159		600	224		824	1,000
3960	30" H x 42" L		3.29	4.863		635	262		897	1,100
3970	36" H x 36" L		3.20	5		660	269		929	1,125
3980	36" H x 42" L		2.74	5.839		700	315		1,015	1,250
4100	For 2 row, add					30%	50%			
4400	Steam and hot water heating coils								
4410	Copper tubes, alum. fins. galv. end sheets								
4412	H is finned height, L is finned length								
4420	5/8" tubing, to 175 PSIG working pressure								
4430	1 row, 8 fins per inch								
4440	6"H x 6"L	Q-5	38	.421	Ea.	610	22.50		632.50	705
4450	6"H x 12"L		37	.432		625	23.50		648.50	720
4460	6"H x 24"L		24.48	.654		685	35		720	810
4470	6"H x 36"L		16.32	.980		725	52.50		777.50	880
4480	6"H x 48"L		12.24	1.307		770	70.50		840.50	950
4490	6"H x 84"L		6.99	2.289		1,175	123		1,298	1,450
4500	6" H x 108" L		5.44	2.941		1,275	158		1,433	1,650
4510	12" H x 12" L		24.48	.654		740	35		775	870
4520	12" H x 24" L		12.24	1.307		835	70.50		905.50	1,025
4530	12" H x 36" L		8.16	1.961		860	105		965	1,100
4540	12" H x 48" L		6.12	2.614		975	141		1,116	1,275
4550	12" H x 84" L		3.50	4.571		1,525	246		1,771	2,050
4560	12" H x 108" L		2.72	5.882		1,775	315		2,090	2,425
4570	18" H x 18" L		10.88	1.471		920	79		999	1,125
4580	18" H x 24" L		8.16	1.961		1,000	105		1,105	1,250
4590	18" H x 48" L		4.08	3.922		1,200	211		1,411	1,650
4600	18" H x 84" L		2.33	6.867		1,875	370		2,245	2,625
4610	18" H x 108" L		1.81	8.840		2,250	475		2,725	3,200
4620	24" H x 24" L		6.12	2.614		1,225	141		1,366	1,525
4630	24" H x 48" L		3.06	5.229		1,450	281		1,731	2,025
4640	24" H x 84" L		1.75	9.143		2,350	490		2,840	3,325
4650	24" H x 108" L		1.36	11.765		2,775	635		3,410	4,000
4660	36" H x 36" L		2.72	5.882		1,675	315		1,990	2,300
4670	36" H x 48" L		2.04	7.843		1,850	420		2,270	2,675
4680	36" H x 84" L		1.17	13.675		2,950	735		3,685	4,350
4690	36" H x 108" L		.91	17.582		3,550	945		4,495	5,325
4700	48"H x 48"L		1.53	10.458		2,425	560		2,985	3,525
4710	48"H x 84"L		.87	18.391		3,850	990		4,840	5,725

For customer support on your Mechanical Cost Data, call 877.795.7394.

405

23 82 Convection Heating and Cooling Units

23 82 16 – Air Coils

23 82 16.10 Flanged Coils		Crew	Daily Output	Labor-Hours	Unit	Material	2015 Bare Costs Labor	Equipment	Total	Total Incl O&P
4720	48"H x 108"L	Q-5	.68	23.529	Ea.	4,600	1,275		5,875	6,950
4730	48"H x 126"L	↓	.58	27.586	↓	4,950	1,475		6,425	7,700
4910	For 10 fins per inch, add					2%				
4920	For 12 fins per inch, add					5%				
4930	For 2 row, add					30%	50%			
4940	For 3 row, add					50%	100%			
4950	For 4 row, add					80%	123%			
4960	For 230 PSIG heavy duty, add					20%	30%			
6000	Steam, heavy duty, 200 PSIG									
6010	Copper tubes, alum. fins, galv. end sheets									
6012	H is finned height, L is finned length									
6020	1" x .035" tubing, .010 aluminum fins									
6030	1 row, 8 fins per inch									
6040	12"H x 12"L	Q-5	24.48	.654	Ea.	790	35		825	920
6050	12"H x 24"L		12.24	1.307		945	70.50		1,015.50	1,150
6060	12"H x 36"L		8.16	1.961		1,075	105		1,180	1,325
6070	12"H x 48"L		6.12	2.614		1,225	141		1,366	1,525
6080	12"H x 84"L		3.50	4.571		1,750	246		1,996	2,300
6090	12"H x 108"L		2.72	5.882		2,100	315		2,415	2,775
6100	18"H x 24"L		8.16	1.961		1,200	105		1,305	1,475
6110	18"H x 36"L		5.44	2.941		1,400	158		1,558	1,775
6120	18"H x 48"L		4.08	3.922		1,600	211		1,811	2,100
6130	18"H x 84"L		2.33	6.867		2,325	370		2,695	3,100
6140	18"H x 108"L		1.81	8.840		2,775	475		3,250	3,775
6150	24"H x 24"L		6.12	2.614		1,400	141		1,541	1,725
6160	24"H x 36"L		4.08	3.922		1,650	211		1,861	2,150
6170	24"H x 48"L		3.06	5.229		1,925	281		2,206	2,525
6180	24"H x 84"L		1.75	9.143		2,800	490		3,290	3,850
6190	24"H x 108"L		1.36	11.765		3,400	635		4,035	4,700
6200	30"H x 36"L		3.26	4.908		2,050	264		2,314	2,650
6210	30"H x 48"L		2.45	6.531		2,375	350		2,725	3,150
6220	30"H x 84"L		1.40	11.429		3,450	615		4,065	4,725
6230	30"H x 108"L		1.09	14.679		4,200	790		4,990	5,825
6240	36"H x 36"L		2.72	5.882		2,325	315		2,640	3,025
6250	36"H x 48"L		2.04	7.843		2,700	420		3,120	3,600
6260	36"H x 84"L		1.17	13.675		3,975	735		4,710	5,475
6270	36"H x 108"L		.91	17.582		4,850	945		5,795	6,750
6280	36"H x 120"L		.82	19.512		5,225	1,050		6,275	7,325
6290	48"H x 48"L		1.53	10.458		3,575	560		4,135	4,775
6300	48"H x 84"L		.87	18.391		4,750	990		5,740	6,725
6310	48"H x 108"L		.68	23.529		6,350	1,275		7,625	8,875
6320	48"H x 120"L	↓	.61	26.230		6,850	1,400		8,250	9,675
6400	For 10 fins per inch, add					4%				
6410	For 12 fins per inch, add				↓	10%				

23 82 16.20 Duct Heaters

		Crew	Daily Output	Labor-Hours	Unit	Material	2015 Bare Costs Labor	Equipment	Total	Total Incl O&P
0010	**DUCT HEATERS**, Electric, 480 V, 3 Ph.									
0020	Finned tubular insert, 500°F									
0100	8" wide x 6" high, 4.0 kW	Q-20	16	1.250	Ea.	765	64		829	945
0120	12" high, 8.0 kW		15	1.333		1,275	68.50		1,343.50	1,500
0140	18" high, 12.0 kW		14	1.429		1,775	73		1,848	2,050
0160	24" high, 16.0 kW		13	1.538		2,300	79		2,379	2,650
0180	30" high, 20.0 kW	↓	12	1.667	↓	2,800	85.50		2,885.50	3,200

23 82 Convection Heating and Cooling Units

23 82 16 – Air Coils

23 82 16.20 Duct Heaters		Crew	Daily Output	Labor-Hours	Unit	Material	2015 Bare Costs Labor	Equipment	Total	Total Incl O&P
0300	12" wide x 6" high, 6.7 kW	Q-20	15	1.333	Ea.	815	68.50		883.50	1,000
0320	12" high, 13.3 kW		14	1.429		1,325	73		1,398	1,550
0340	18" high, 20.0 kW		13	1.538		1,850	79		1,929	2,150
0360	24" high, 26.7 kW		12	1.667		2,375	85.50		2,460.50	2,725
0380	30" high, 33.3 kW		11	1.818		2,900	93		2,993	3,325
0500	18" wide x 6" high, 13.3 kW		14	1.429		880	73		953	1,075
0520	12" high, 26.7 kW		13	1.538		1,500	79		1,579	1,775
0540	18" high, 40.0 kW		12	1.667		2,000	85.50		2,085.50	2,325
0560	24" high, 53.3 kW		11	1.818		2,675	93		2,768	3,075
0580	30" high, 66.7 kW		10	2		3,325	102		3,427	3,825
0700	24" wide x 6" high, 17.8 kW		13	1.538		965	79		1,044	1,200
0720	12" high, 35.6 kW		12	1.667		1,625	85.50		1,710.50	1,925
0740	18" high, 53.3 kW		11	1.818		2,300	93		2,393	2,675
0760	24" high, 71.1 kW		10	2		2,950	102		3,052	3,400
0780	30" high, 88.9 kW		9	2.222		3,625	114		3,739	4,175
0900	30" wide x 6" high, 22.2 kW		12	1.667		1,025	85.50		1,110.50	1,250
0920	12" high, 44.4 kW		11	1.818		1,725	93		1,818	2,050
0940	18" high, 66.7 kW		10	2		2,425	102		2,527	2,825
0960	24" high, 88.9 kW		9	2.222		3,125	114		3,239	3,625
0980	30" high, 111.0 kW		8	2.500		3,825	128		3,953	4,425
1400	Note decreased kW available for									
1410	each duct size at same cost									
1420	See line 5000 for modifications and accessories									
2000	Finned tubular flange with insulated									
2020	terminal box, 500°F									
2100	12" wide x 36" high, 54 kW	Q-20	10	2	Ea.	3,775	102		3,877	4,300
2120	40" high, 60 kW		9	2.222		4,350	114		4,464	4,950
2200	24" wide x 36" high, 118.8 kW		9	2.222		4,825	114		4,939	5,500
2220	40" high, 132 kW		8	2.500		4,825	128		4,953	5,500
2400	36" wide x 8" high, 40 kW		11	1.818		1,925	93		2,018	2,275
2420	16" high, 80 kW		10	2		2,575	102		2,677	2,975
2440	24" high, 120 kW		9	2.222		3,350	114		3,464	3,875
2460	32" high, 160 kW		8	2.500		4,300	128		4,428	4,925
2480	36" high, 180 kW		7	2.857		5,300	146		5,446	6,050
2500	40" high, 200 kW		6	3.333		5,875	171		6,046	6,700
2600	40" wide x 8" high, 45 kW		11	1.818		2,000	93		2,093	2,350
2620	16" high, 90 kW		10	2		2,775	102		2,877	3,200
2640	24" high, 135 kW		9	2.222		3,500	114		3,614	4,025
2660	32" high, 180 kW		8	2.500		4,675	128		4,803	5,350
2680	36" high, 202.5 kW		7	2.857		5,375	146		5,521	6,150
2700	40" high, 225 kW		6	3.333		6,125	171		6,296	7,000
2800	48" wide x 8" high, 54.8 kW		10	2		2,100	102		2,202	2,475
2820	16" high, 109.8 kW		9	2.222		2,925	114		3,039	3,375
2840	24" high, 164.4 kW		8	2.500		3,725	128		3,853	4,275
2860	32" high, 219.2 kW		7	2.857		4,950	146		5,096	5,675
2880	36" high, 246.6 kW		6	3.333		5,700	171		5,871	6,525
2900	40" high, 274 kW		5	4		6,450	205		6,655	7,375
3000	56" wide x 8" high, 64 kW		9	2.222		2,400	114		2,514	2,800
3020	16" high, 128 kW		8	2.500		3,325	128		3,453	3,850
3040	24" high, 192 kW		7	2.857		4,025	146		4,171	4,650
3060	32" high, 256 kW		6	3.333		5,625	171		5,796	6,425
3080	36" high, 288kW		5	4		6,450	205		6,655	7,400
3100	40" high, 320kW		4	5		7,125	256		7,381	8,250

For customer support on your Mechanical Cost Data, call 877.795.7394.

407

23 82 Convection Heating and Cooling Units

23 82 16 – Air Coils

23 82 16.20 Duct Heaters		Crew	Daily Output	Labor-Hours	Unit	Material	2015 Bare Costs Labor	Equipment	Total	Total Incl O&P
3200	64" wide x 8" high, 74kW	Q-20	8	2.500	Ea.	2,475	128		2,603	2,900
3220	16" high, 148kW		7	2.857		3,425	146		3,571	3,975
3240	24" high, 222kW		6	3.333		4,350	171		4,521	5,025
3260	32" high, 296kW		5	4		5,825	205		6,030	6,700
3280	36" high, 333kW		4	5		6,950	256		7,206	8,050
3300	40" high, 370kW		3	6.667		7,675	340		8,015	8,950
3800	Note decreased kW available for									
3820	each duct size at same cost									
5000	Duct heater modifications and accessories									
5120	T.C.O. limit auto or manual reset	Q-20	42	.476	Ea.	126	24.50		150.50	175
5140	Thermostat		28	.714		535	36.50		571.50	645
5160	Overheat thermocouple (removable)		7	2.857		760	146		906	1,050
5180	Fan interlock relay		18	1.111		183	57		240	289
5200	Air flow switch		20	1		158	51		209	252
5220	Split terminal box cover		100	.200		52.50	10.25		62.75	73
8000	To obtain BTU multiply kW by 3413									

23 82 19 – Fan Coil Units

23 82 19.10 Fan Coil Air Conditioning

		Crew	Daily Output	Labor-Hours	Unit	Material	2015 Bare Costs Labor	Equipment	Total	Total Incl O&P
0010	**FAN COIL AIR CONDITIONING**									
0030	Fan coil AC, cabinet mounted, filters and controls									
0100	Chilled water, 1/2 ton cooling	Q-5	8	2	Ea.	815	108		923	1,050
0110	3/4 ton cooling		7	2.286		850	123		973	1,125
0120	1 ton cooling		6	2.667		950	143		1,093	1,275
0140	1.5 ton cooling		5.50	2.909		1,050	156		1,206	1,375
0150	2 ton cooling		5.25	3.048		1,350	164		1,514	1,725
0160	2.5 ton cooling		5	3.200		1,975	172		2,147	2,425
0180	3 ton cooling		4	4		2,175	215		2,390	2,725
0262	For hot water coil, add					40%	10%			
0300	Console, 2 pipe with electric heat									
0310	1/2 ton cooling	Q-5	8	2	Ea.	1,300	108		1,408	1,575
0315	3/4 ton cooling		7	2.286		1,425	123		1,548	1,750
0320	1 ton cooling		6	2.667		1,525	143		1,668	1,925
0330	1.5 ton cooling		5.50	2.909		1,900	156		2,056	2,300
0340	2 ton cooling		4.70	3.404		2,500	183		2,683	3,025
0345	2-1/2 ton cooling		4.70	3.404		3,625	183		3,808	4,275
0350	3 ton cooling		4	4		4,925	215		5,140	5,750
0360	4 ton cooling		4	4		5,125	215		5,340	5,975
0940	Direct expansion, for use w/air cooled condensing unit, 1.5 ton cooling		5	3.200		680	172		852	1,000
0950	2 ton cooling		4.80	3.333		730	179		909	1,075
0960	2.5 ton cooling		4.40	3.636		775	196		971	1,150
0970	3 ton cooling		3.80	4.211		960	226		1,186	1,400
0980	3.5 ton cooling		3.60	4.444		970	239		1,209	1,425
0990	4 ton cooling		3.40	4.706		1,100	253		1,353	1,575
1000	5 ton cooling		3	5.333		1,300	287		1,587	1,875
1020	7.5 ton cooling		3	5.333		2,450	287		2,737	3,125
1040	10 ton cooling	Q-6	2.60	9.231		2,800	515		3,315	3,850
1042	12.5 ton cooling		2.30	10.435		3,550	580		4,130	4,775
1050	15 ton cooling		1.60	15		3,950	835		4,785	5,600
1060	20 ton cooling		.70	34.286		5,400	1,900		7,300	8,825
1070	25 ton cooling		.65	36.923		6,475	2,050		8,525	10,200
1080	30 ton cooling		.60	40		7,050	2,225		9,275	11,100
1500	For hot water coil, add					40%	10%			

23 82 Convection Heating and Cooling Units

23 82 19 — Fan Coil Units

23 82 19.10 Fan Coil Air Conditioning

		Crew	Daily Output	Labor-Hours	Unit	Material	2015 Bare Costs Labor	Equipment	Total	Total Incl O&P
1512	For condensing unit add see Section 23 62									
3000	Chilled water, horizontal unit, housing, 2 pipe, fan control, no valves									
3100	1/2 ton cooling	Q-5	8	2	Ea.	1,275	108		1,383	1,550
3110	1 ton cooling		6	2.667		1,475	143		1,618	1,850
3120	1.5 ton cooling		5.50	2.909		1,700	156		1,856	2,100
3130	2 ton cooling		5.25	3.048		2,050	164		2,214	2,500
3140	3 ton cooling		4	4		2,425	215		2,640	3,000
3150	3.5 ton cooling		3.80	4.211		2,675	226		2,901	3,300
3160	4 ton cooling		3.80	4.211		2,675	226		2,901	3,300
3170	5 ton cooling		3.40	4.706		3,125	253		3,378	3,800
3180	6 ton cooling		3.40	4.706		3,125	253		3,378	3,800
3190	7 ton cooling		2.90	5.517		3,400	297		3,697	4,175
3200	8 ton cooling		2.90	5.517		3,400	297		3,697	4,175
3210	10 ton cooling	Q-6	2.80	8.571		3,550	480		4,030	4,625
3212	12.5 ton cooling		2.50	9.600		4,050	535		4,585	5,250
3214	15 ton cooling		2.30	10.435		4,400	580		4,980	5,700
3216	20 ton cooling		2.10	11.429		5,050	635		5,685	6,525
3218	25 ton cooling		1.80	13.333		8,100	745		8,845	10,000
3220	30 ton cooling		1.60	15		8,700	835		9,535	10,900
3240	For hot water coil, add					40%	10%			
4000	With electric heat, 2 pipe									
4100	1/2 ton cooling	Q-5	8	2	Ea.	1,475	108		1,583	1,775
4105	3/4 ton cooling		7	2.286		1,625	123		1,748	1,975
4110	1 ton cooling		6	2.667		1,800	143		1,943	2,200
4120	1.5 ton cooling		5.50	2.909		2,000	156		2,156	2,425
4130	2 ton cooling		5.25	3.048		2,525	164		2,689	3,025
4135	2.5 ton cooling		4.80	3.333		3,475	179		3,654	4,100
4140	3 ton cooling		4	4		4,725	215		4,940	5,525
4150	3.5 ton cooling		3.80	4.211		5,650	226		5,876	6,550
4160	4 ton cooling		3.60	4.444		6,525	239		6,764	7,525
4170	5 ton cooling		3.40	4.706		7,950	253		8,203	9,125
4180	6 ton cooling		3.20	5		7,950	269		8,219	9,150
4190	7 ton cooling		2.90	5.517		8,125	297		8,422	9,375
4200	8 ton cooling		2.20	7.273		8,700	390		9,090	10,200
4210	10 ton cooling	Q-6	2.80	8.571		9,425	480		9,905	11,100

23 82 19.20 Heating and Ventilating Units

		Crew	Daily Output	Labor-Hours	Unit	Material	2015 Bare Costs Labor	Equipment	Total	Total Incl O&P
0010	**HEATING AND VENTILATING UNITS**, Classroom units									
0020	Includes filter, heating/cooling coils, standard controls									
0080	750 CFM, 2 tons cooling	Q-6	2	12	Ea.	4,050	670		4,720	5,475
0100	1000 CFM, 2-1/2 tons cooling		1.60	15		4,800	835		5,635	6,550
0120	1250 CFM, 3 tons cooling		1.40	17.143		4,950	955		5,905	6,900
0140	1500 CFM, 4 tons cooling		.80	30		5,300	1,675		6,975	8,350
0160	2000 CFM 5 tons cooling		.50	48		6,225	2,675		8,900	10,900
0500	For electric heat, add					35%				
1000	For no cooling, deduct					25%	10%			

23 82 19.40 Fan Coil Air Conditioning

			Crew	Daily Output	Labor-Hours	Unit	Material	2015 Bare Costs Labor	Equipment	Total	Total Incl O&P
0010	**FAN COIL AIR CONDITIONING**, Variable refrigerant volume (VRV) type										
0020	Not including interconnecting tubing or multi-zone controls										
0030	For VRV condensing unit see section 23 81 29.10										
0050	Indoor type, ducted										
0100	Vertical concealed										
0130	1 ton cooling	G	Q-5	2.97	5.387	Ea.	2,475	290		2,765	3,150

For customer support on your Mechanical Cost Data, call 877.795.7394.

409

23 82 19 – Fan Coil Units

23 82 19.40 Fan Coil Air Conditioning		Crew	Daily Output	Labor-Hours	Unit	Material	2015 Bare Costs Labor	Equipment	Total	Total Incl O&P
0140	1.5 ton cooling	G	Q-5	2.77	5.776	Ea.	2,525	310		2,835	3,250
0150	2 ton cooling	G		2.70	5.926		2,725	320		3,045	3,475
0160	2.5 ton cooling	G		2.60	6.154		2,875	330		3,205	3,650
0170	3 ton cooling	G		2.50	6.400		2,925	345		3,270	3,725
0180	3.5 ton cooling	G		2.30	6.957		3,075	375		3,450	3,950
0190	4 ton cooling	G		2.19	7.306		3,075	395		3,470	4,000
0200	4.5 ton cooling	G	▼	2.08	7.692	▼	3,400	415		3,815	4,350
0230	Outside air connection possible										
1100	Ceiling concealed										
1130	.6 ton cooling	G	Q-5	2.60	6.154	Ea.	1,625	330		1,955	2,275
1140	.75 ton cooling	G		2.60	6.154		1,675	330		2,005	2,350
1150	1 ton cooling	G	▼	2.60	6.154		1,775	330		2,105	2,450
1152	Screening door	G	1 Stpi	5.20	1.538		98	92		190	247
1160	1.5 ton cooling	G	Q-5	2.60	6.154		1,825	330		2,155	2,525
1162	Screening door	G	1 Stpi	5.20	1.538		113	92		205	264
1170	2 ton cooling	G	Q-5	2.60	6.154		2,150	330		2,480	2,850
1172	Screening door	G	1 Stpi	5.20	1.538		98	92		190	247
1180	2.5 ton cooling	G	Q-5	2.60	6.154		2,500	330		2,830	3,250
1182	Screening door	G	1 Stpi	5.20	1.538		170	92		262	325
1190	3 ton cooling	G	Q-5	2.60	6.154		2,650	330		2,980	3,425
1192	Screening door	G	1 Stpi	5.20	1.538		170	92		262	325
1200	4 ton cooling	G	Q-5	2.30	6.957		2,725	375		3,100	3,550
1202	Screening door	G	1 Stpi	5.20	1.538		170	92		262	325
1220	6 ton cooling	G	Q-5	2.08	7.692		4,625	415		5,040	5,700
1230	8 ton cooling	G	"	1.89	8.466	▼	5,200	455		5,655	6,400
1260	Outside air connection possible										
4050	Indoor type, duct-free										
4100	Ceiling mounted cassette										
4130	.75 ton cooling	G	Q-5	3.46	4.624	Ea.	1,875	249		2,124	2,450
4140	1 ton cooling	G		2.97	5.387		2,000	290		2,290	2,625
4150	1.5 ton cooling	G		2.77	5.776		2,050	310		2,360	2,725
4160	2 ton cooling	G		2.60	6.154		2,200	330		2,530	2,925
4170	2.5 ton cooling	G		2.50	6.400		2,250	345		2,595	3,000
4180	3 ton cooling	G		2.30	6.957		2,425	375		2,800	3,225
4190	4 ton cooling	G	▼	2.08	7.692		2,700	415		3,115	3,600
4198	For ceiling cassette decoration panel, add	G	1 Stpi	5.20	1.538	▼	294	92		386	465
4230	Outside air connection possible										
4500	Wall mounted										
4530	.6 ton cooling	G	Q-5	5.20	3.077	Ea.	995	165		1,160	1,350
4540	.75 ton cooling	G		5.20	3.077		1,050	165		1,215	1,400
4550	1 ton cooling	G		4.16	3.846		1,200	207		1,407	1,600
4560	1.5 ton cooling	G		3.77	4.244		1,300	228		1,528	1,800
4570	2 ton cooling	G	▼	3.46	4.624		1,400	249		1,649	1,925
4590	Condensate pump for the above air handlers	G	1 Stpi	6.46	1.238	▼	211	74		285	345
4700	Floor standing unit										
4730	1 ton cooling	G	Q-5	2.60	6.154	Ea.	1,675	330		2,005	2,350
4740	1.5 ton cooling	G		2.60	6.154		1,900	330		2,230	2,600
4750	2 ton cooling	G	▼	2.60	6.154	▼	2,050	330		2,380	2,750
4800	Floor standing unit, concealed										
4830	1 ton cooling	G	Q-5	2.60	6.154	Ea.	1,600	330		1,930	2,275
4840	1.5 ton cooling	G		2.60	6.154		1,825	330		2,155	2,500
4850	2 ton cooling	G	▼	2.60	6.154	▼	1,950	330		2,280	2,650
4880	Outside air connection possible										

For customer support on your Mechanical Cost Data, call 877.795.7394.

23 82 19 – Fan Coil Units

23 82 19.40 Fan Coil Air Conditioning	Crew	Daily Output	Labor-Hours	Unit	Material	2015 Bare Costs Labor	Equipment	Total	Total Incl O&P
8000	Accessories									
8100	Branch divergence pipe fitting								
8110	Capacity under 76 MBH	1 Stpi	2.50	3.200	Ea.	151	191		342	455
8120	76 MBH to 112 MBH		2.50	3.200		178	191		369	485
8130	112 MBH to 234 MBH		2.50	3.200		505	191		696	845
8200	Header pipe fitting								
8210	Max 4 branches								
8220	Capacity under 76 MBH	1 Stpi	1.70	4.706	Ea.	221	281		502	670
8260	Max 8 branches								
8270	76 MBH to 112 MBH	1 Stpi	1.70	4.706	Ea.	450	281		731	920
8280	112 MBH to 234 MBH	"	1.30	6.154	"	720	370		1,090	1,350

23 82 27 – Infrared Units

23 82 27.10 Infrared Type Heating Units

		Crew	Daily Output	Labor-Hours	Unit	Material	2015 Bare Costs Labor	Equipment	Total	Total Incl O&P
0010	**INFRARED TYPE HEATING UNITS**									
0020	Gas fired, unvented, electric ignition, 100% shutoff.									
0030	Piping and wiring not included									
0100	Input, 30 MBH	Q-5	6	2.667	Ea.	945	143		1,088	1,250
0120	45 MBH		5	3.200		950	172		1,122	1,300
0140	50 MBH		4.50	3.556		950	191		1,141	1,350
0160	60 MBH		4	4		950	215		1,165	1,375
0180	75 MBH		3	5.333		950	287		1,237	1,475
0200	90 MBH		2.50	6.400		950	345		1,295	1,575
0220	105 MBH		2	8		950	430		1,380	1,700
0240	120 MBH		2	8		1,575	430		2,005	2,375
1000	Gas fired, vented, electric ignition, tubular									
1020	Piping and wiring not included, 20' to 80' lengths									
1030	Single stage, input, 60 MBH	Q-6	4.50	5.333	Ea.	1,425	297		1,722	2,025
1040	80 MBH		3.90	6.154		1,425	345		1,770	2,100
1050	100 MBH		3.40	7.059		1,425	395		1,820	2,175
1060	125 MBH		2.90	8.276		1,425	460		1,885	2,275
1070	150 MBH		2.70	8.889		1,425	495		1,920	2,325
1080	170 MBH		2.50	9.600		1,425	535		1,960	2,375
1090	200 MBH		2.20	10.909		1,625	610		2,235	2,725
1100	Note: Final pricing may vary due to									
1110	tube length and configuration package selected									
1130	Two stage, input, 60 MBH high, 45 MBH low	Q-6	4.50	5.333	Ea.	1,750	297		2,047	2,375
1140	80 MBH high, 60 MBH low		3.90	6.154		1,750	345		2,095	2,450
1150	100 MBH high, 65 MBH low		3.40	7.059		1,750	395		2,145	2,525
1160	125 MBH high, 95 MBH low		2.90	8.276		1,750	460		2,210	2,625
1170	150 MBH high, 100 MBH low		2.70	8.889		1,750	495		2,245	2,675
1180	170 MBH high, 125 MBH low		2.50	9.600		1,950	535		2,485	2,950
1190	200 MBH high, 150 MBH low		2.20	10.909		1,950	610		2,560	3,075
1220	Note: Final pricing may vary due to									
1230	tube length and configuration package selected									
2000	Electric, single or three phase									
2050	6 kW, 20,478 BTU	1 Elec	2.30	3.478	Ea.	705	190		895	1,050
2100	13.5 KW, 40,956 BTU		2.20	3.636		870	199		1,069	1,250
2150	24 KW, 81,912 BTU		2	4		1,750	219		1,969	2,250
3000	Oil fired, pump, controls, fusible valve, oil supply tank									
3050	91,000 BTU	Q-5	2.50	6.400	Ea.	6,425	345		6,770	7,600
3080	105,000 BTU		2.25	7.111		6,800	380		7,180	8,075
3110	119,000 BTU		2	8		7,450	430		7,880	8,850

For customer support on your Mechanical Cost Data, call 877.795.7394.

411

23 82 29 – Radiators

23 82 29.10 Hydronic Heating		Crew	Daily Output	Labor-Hours	Unit	Material	2015 Bare Costs Labor	2015 Bare Costs Equipment	Total	Total Incl O&P
0010	**HYDRONIC HEATING**, Terminal units, not incl. main supply pipe									
1000	Radiation									
1100	Panel, baseboard, C.I., including supports, no covers	Q-5	46	.348	L.F.	38	18.70		56.70	69.50
3000	Radiators, cast iron									
3100	Free standing or wall hung, 6 tube, 25" high	Q-5	96	.167	Section	44	8.95		52.95	62
3150	4 tube 25" high		96	.167		34	8.95		42.95	51
3200	4 tube, 19" high		96	.167		31.50	8.95		40.45	48
3250	Adj. brackets, 2 per wall radiator up to 30 sections	1 Stpi	32	.250	Ea.	53	14.95		67.95	81
3500	Recessed, 20" high x 5" deep, without grille	Q-5	60	.267	Section	37	14.35		51.35	62
3600	For inlet grille, add				"	2.58			2.58	2.84
9500	To convert SFR to BTU rating: Hot water, 150 x SFR									
9510	Forced hot water, 180 x SFR; steam, 240 x SFR									

23 82 33 – Convectors

23 82 33.10 Convector Units		Crew	Daily Output	Labor-Hours	Unit	Material	2015 Bare Costs Labor	2015 Bare Costs Equipment	Total	Total Incl O&P
0010	**CONVECTOR UNITS**, Terminal units, not incl. main supply pipe									
2204	Convector, multifin, 2 pipe w/cabinet									
2210	17" H x 24" L	Q-5	10	1.600	Ea.	99	86		185	239
2214	17" H x 36" L		8.60	1.860		148	100		248	315
2218	17" H x 48" L		7.40	2.162		198	116		314	395
2222	21" H x 24" L		9	1.778		101	95.50		196.50	255
2226	21" H x 36" L		8.20	1.951		151	105		256	325
2228	21" H x 48" L		6.80	2.353		201	127		328	410
2240	For knob operated damper, add					140%				
2241	For metal trim strips, add	Q-5	64	.250	Ea.	12.90	13.45		26.35	34.50
2243	For snap-on inlet grille, add					10%	10%			
2245	For hinged access door, add	Q-5	64	.250	Ea.	35.50	13.45		48.95	59.50
2246	For air chamber, auto-venting, add	"	58	.276	"	7.75	14.85		22.60	31

23 82 36 – Finned-Tube Radiation Heaters

23 82 36.10 Finned Tube Radiation		Crew	Daily Output	Labor-Hours	Unit	Material	2015 Bare Costs Labor	2015 Bare Costs Equipment	Total	Total Incl O&P
0010	**FINNED TUBE RADIATION**, Terminal units, not incl. main supply pipe									
1150	Fin tube, wall hung, 14" slope top cover, with damper									
1200	1-1/4" copper tube, 4-1/4" alum. fin	Q-5	38	.421	L.F.	43	22.50		65.50	81
1250	1-1/4" steel tube, 4-1/4" steel fin		36	.444		39	24		63	79
1255	2" steel tube, 4-1/4" steel fin		32	.500		43	27		70	87.50
1260	21", two tier, slope top, w/damper									
1262	1-1/4" copper tube, 4-1/4" alum. fin	Q-5	30	.533	L.F.	50.50	28.50		79	99
1264	1-1/4" steel tube, 4-1/4" steel fin		28	.571		47	30.50		77.50	98.50
1266	2" steel tube, 4-1/4" steel fin		25	.640		50.50	34.50		85	108
1270	10", flat top, with damper									
1272	1-1/4" copper tube, 4-1/4" alum. fin	Q-5	38	.421	L.F.	21	22.50		43.50	57
1274	1-1/4" steel tube, 4-1/4" steel fin		36	.444		40	24		64	80
1276	2" steel tube, 4-1/4" steel fin		32	.500		43	27		70	87.50
1280	17", two tier, flat top, w/damper									
1282	1-1/4" copper tube, 4-1/4" alum. fin	Q-5	30	.533	L.F.	49.50	28.50		78	98
1284	1-1/4" steel tube, 4-1/4" steel fin		28	.571		46.50	30.50		77	97.50
1286	2" steel tube, 4-1/4" steel fin		25	.640		50.50	34.50		85	108
1301	Rough in wall hung, steel fin w/supply & balance valves	1 Stpi	.91	8.791	Ea.	305	525		830	1,125
1310	Baseboard, pkgd, 1/2" copper tube, alum. fin, 7" high	Q-5	60	.267	L.F.	11.20	14.35		25.55	34
1320	3/4" copper tube, alum. fin, 7" high		58	.276		7.35	14.85		22.20	30.50
1340	1" copper tube, alum. fin, 8-7/8" high		56	.286		18.85	15.35		34.20	43.50
1360	1-1/4" copper tube, alum. fin, 8-7/8" high		54	.296		28	15.95		43.95	54.50

412

For customer support on your Mechanical Cost Data, call 877.795.7394.

23 82 Convection Heating and Cooling Units

23 82 36 – Finned-Tube Radiation Heaters

23 82 36.10 Finned Tube Radiation		Crew	Daily Output	Labor-Hours	Unit	Material	2015 Bare Costs Labor	Equipment	Total	Total Incl O&P
1380	1-1/4" IPS steel tube with steel fins	Q-5	52	.308	L.F.	27.50	16.55		44.05	55.50
1381	Rough in baseboard panel & fin tube, supply & balance valves	1 Stpi	1.06	7.547	Ea.	239	450		689	945
1500	Note: fin tube may also require corners, caps, etc.									

23 82 39 – Unit Heaters

23 82 39.13 Cabinet Unit Heaters

		Crew	Daily Output	Labor-Hours	Unit	Material	Labor	Equipment	Total	Total Incl O&P
0010	**CABINET UNIT HEATERS**									
5400	Cabinet, horizontal, hot water, blower type									
5420	20 MBH	Q-5	10	1.600	Ea.	895	86		981	1,100
5430	60 MBH		7	2.286		1,550	123		1,673	1,875
5440	100 MBH		5	3.200		1,825	172		1,997	2,250
5450	120 MBH		4	4		1,900	215		2,115	2,425

23 82 39.16 Propeller Unit Heaters

		Crew	Daily Output	Labor-Hours	Unit	Material	Labor	Equipment	Total	Total Incl O&P
0010	**PROPELLER UNIT HEATERS**									
3950	Unit heaters, propeller, 115 V 2 psi steam, 60°F entering air									
4000	Horizontal, 12 MBH	Q-5	12	1.333	Ea.	360	71.50		431.50	505
4020	28.2 MBH		10	1.600		470	86		556	645
4040	36.5 MBH		8	2		545	108		653	760
4060	43.9 MBH		8	2		545	108		653	760
4080	56.5 MBH		7.50	2.133		585	115		700	815
4100	65.6 MBH		7	2.286		600	123		723	845
4120	87.6 MBH		6.50	2.462		650	132		782	915
4140	96.8 MBH		6	2.667		795	143		938	1,100
4160	133.3 MBH		5	3.200		865	172		1,037	1,200
4180	157.6 MBH		4	4		1,050	215		1,265	1,500
4200	197.7 MBH		3	5.333		1,200	287		1,487	1,750
4220	257.2 MBH		2.50	6.400		1,475	345		1,820	2,150
4240	286.9 MBH		2	8		1,650	430		2,080	2,475
4260	364 MBH		1.80	8.889		2,075	480		2,555	3,000
4270	404 MBH		1.60	10		2,150	540		2,690	3,175
4300	Vertical diffuser same price									
4310	Vertical flow, 40 MBH	Q-5	11	1.455	Ea.	540	78		618	715
4314	58.5 MBH		8	2		560	108		668	775
4318	92 MBH		7	2.286		715	123		838	970
4322	109.7 MBH		6	2.667		845	143		988	1,150
4326	131.0 MBH		4	4		845	215		1,060	1,250
4330	160.0 MBH		3	5.333		895	287		1,182	1,425
4334	194.0 MBH		2.20	7.273		1,025	390		1,415	1,750
4338	212.0 MBH		2.10	7.619		1,275	410		1,685	2,025
4342	247.0 MBH		1.96	8.163		1,525	440		1,965	2,350
4346	297.0 MBH		1.80	8.889		1,600	480		2,080	2,500
4350	333.0 MBH	Q-6	1.90	12.632		1,750	705		2,455	3,000
4354	420 MBH, (460 V)		1.80	13.333		2,150	745		2,895	3,500
4358	500 MBH, (460 V)		1.71	14.035		2,875	785		3,660	4,325
4362	570 MBH, (460 V)		1.40	17.143		3,925	955		4,880	5,775
4366	620 MBH, (460 V)		1.30	18.462		3,925	1,025		4,950	5,875
4370	960 MBH, (460 V)		1.10	21.818		7,450	1,225		8,675	10,000
4400	Unit heaters, propeller, hot water, 115 V, 60 Deg. F entering air									
4404	Horizontal									
4408	6 MBH	Q-5	11	1.455	Ea.	360	78		438	515
4412	12 MBH		11	1.455		360	78		438	515
4416	16 MBH		10	1.600		405	86		491	575
4420	24 MBH		9	1.778		470	95.50		565.50	660

For customer support on your Mechanical Cost Data, call 877.795.7394.

413

23 82 Convection Heating and Cooling Units

23 82 39 – Unit Heaters

23 82 39.16 Propeller Unit Heaters		Crew	Daily Output	Labor-Hours	Unit	Material	2015 Bare Costs Labor	Equipment	Total	Total Incl O&P
4424	29 MBH	Q-5	8	2	Ea.	545	108		653	760
4428	47 MBH		7	2.286		600	123		723	845
4432	63 MBH		6	2.667		650	143		793	930
4436	81 MBH		5.50	2.909		795	156		951	1,100
4440	90 MBH		5	3.200		865	172		1,037	1,200
4460	133 MBH		4	4		1,050	215		1,265	1,500
4464	139 MBH		3.50	4.571		1,200	246		1,446	1,700
4468	198 MBH		2.50	6.400		1,475	345		1,820	2,150
4472	224 MBH		2	8		1,650	430		2,080	2,475
4476	273 MBH		1.50	10.667		2,075	575		2,650	3,150
4500	Vertical									
4504	8 MBH	Q-5	13	1.231	Ea.	775	66		841	950
4508	12 MBH		13	1.231		775	66		841	950
4512	16 MBH		13	1.231		775	66		841	950
4516	30 MBH		12	1.333		775	71.50		846.50	960
4520	43 MBH		11	1.455		800	78		878	1,000
4524	57 MBH		8	2		985	108		1,093	1,225
4528	68 MBH		7	2.286		1,025	123		1,148	1,300
4540	105 MBH		6	2.667		1,200	143		1,343	1,550
4544	123 MBH		5	3.200		1,275	172		1,447	1,650
4550	140 MBH		5	3.200		1,475	172		1,647	1,875
4554	156 MBH		4	4		1,825	215		2,040	2,325
4560	210 MBH		4	4		2,300	215		2,515	2,850
4564	223 MBH		3	5.333		2,500	287		2,787	3,175
4568	257 MBH		2	8		2,500	430		2,930	3,400

23 83 Radiant Heating Units

23 83 16 – Radiant-Heating Hydronic Piping

23 83 16.10 Radiant Floor Heating

		Crew	Daily Output	Labor-Hours	Unit	Material	2015 Bare Costs Labor	Equipment	Total	Total Incl O&P
0010	**RADIANT FLOOR HEATING**									
0100	Tubing, PEX (cross-linked polyethylene)									
0110	Oxygen barrier type for systems with ferrous materials									
0120	1/2"	Q-5	800	.020	L.F.	.96	1.08		2.04	2.68
0130	3/4"		535	.030		1.35	1.61		2.96	3.92
0140	1"		400	.040		2.11	2.15		4.26	5.55
0200	Non barrier type for ferrous free systems									
0210	1/2"	Q-5	800	.020	L.F.	.61	1.08		1.69	2.29
0220	3/4"		535	.030		1.11	1.61		2.72	3.65
0230	1"		400	.040		1.90	2.15		4.05	5.35
1000	Manifolds									
1110	Brass									
1120	With supply and return valves, flow meter, thermometer,									
1122	auto air vent and drain/fill valve.									
1130	1", 2 circuit	Q-5	14	1.143	Ea.	257	61.50		318.50	375
1140	1", 3 circuit		13.50	1.185		294	63.50		357.50	420
1150	1", 4 circuit		13	1.231		320	66		386	450
1154	1", 5 circuit		12.50	1.280		365	69		434	510
1158	1", 6 circuit		12	1.333		415	71.50		486.50	565
1162	1", 7 circuit		11.50	1.391		455	75		530	615
1166	1", 8 circuit		11	1.455		500	78		578	670
1172	1", 9 circuit		10.50	1.524		540	82		622	720

23 83 16 – Radiant-Heating Hydronic Piping

23 83 16.10 Radiant Floor Heating		Crew	Daily Output	Labor-Hours	Unit	Material	2015 Bare Costs Labor	2015 Bare Costs Equipment	Total	Total Incl O&P
1174	1", 10 circuit	Q-5	10	1.600	Ea.	580	86		666	770
1178	1", 11 circuit		9.50	1.684		605	90.50		695.50	800
1182	1", 12 circuit	↓	9	1.778	↓	670	95.50		765.50	880
1610	Copper manifold header, (cut to size)									
1620	1" header, 12 – 1/2" sweat outlets	Q-5	3.33	4.805	Ea.	91	258		349	490
1630	1-1/4" header, 12 – 1/2" sweat outlets		3.20	5		106	269		375	520
1640	1-1/4" header, 12 – 3/4" sweat outlets		3	5.333		114	287		401	560
1650	1-1/2" header, 12 – 3/4" sweat outlets		3.10	5.161		137	278		415	570
1660	2" header, 12 – 3/4" sweat outlets	↓	2.90	5.517	↓	201	297		498	670
3000	Valves									
3110	Thermostatic zone valve actuator with end switch	Q-5	40	.400	Ea.	38.50	21.50		60	75
3114	Thermostatic zone valve actuator	"	36	.444	"	81	24		105	125
3120	Motorized straight zone valve with operator complete									
3130	3/4"	Q-5	35	.457	Ea.	130	24.50		154.50	180
3140	1"		32	.500		141	27		168	196
3150	1-1/4"	↓	29.60	.541	↓	179	29		208	241
3500	4 Way mixing valve, manual, brass									
3530	1"	Q-5	13.30	1.203	Ea.	180	64.50		244.50	296
3540	1-1/4"		11.40	1.404		195	75.50		270.50	330
3550	1-1/2"		11	1.455		249	78		327	390
3560	2"		10.60	1.509		350	81		431	515
3800	Mixing valve motor, 4 way for valves, 1" and 1-1/4"		34	.471		310	25.50		335.50	380
3810	Mixing valve motor, 4 way for valves, 1-1/2" and 2"	↓	30	.533	↓	355	28.50		383.50	435
5000	Radiant floor heating, zone control panel									
5120	4 Zone actuator valve control, expandable	Q-5	20	.800	Ea.	163	43		206	244
5130	6 Zone actuator valve control, expandable		18	.889		226	48		274	320
6070	Thermal track, straight panel for long continuous runs, 5.333 S.F.		40	.400		26.50	21.50		48	62
6080	Thermal track, utility panel, for direction reverse at run end, 5.333 S.F.		40	.400		26.50	21.50		48	62
6090	Combination panel, for direction reverse plus straight run, 5.333 S.F.	↓	40	.400		26.50	21.50		48	62
7000	PEX tubing fittings									
7100	Compression type									
7116	Coupling									
7120	1/2" x 1/2"	1 Stpi	27	.296	Ea.	6.70	17.70		24.40	34
7124	3/4" x 3/4"	"	23	.348	"	10.60	21		31.60	43
7130	Adapter									
7132	1/2" x female sweat 1/2"	1 Stpi	27	.296	Ea.	4.36	17.70		22.06	31.50
7134	1/2" x female sweat 3/4"		26	.308		4.88	18.40		23.28	33.50
7136	5/8" x female sweat 3/4"	↓	24	.333		7	19.90		26.90	37.50
7140	Elbow									
7142	1/2" x female sweat 1/2"	1 Stpi	27	.296	Ea.	6.65	17.70		24.35	34
7144	1/2" x female sweat 3/4"		26	.308		7.80	18.40		26.20	36.50
7146	5/8" x female sweat 3/4"	↓	24	.333	↓	8.75	19.90		28.65	39.50
7200	Insert type									
7206	PEX x male NPT									
7210	1/2" x 1/2"	1 Stpi	29	.276	Ea.	2.59	16.50		19.09	28
7220	3/4" x 3/4"		27	.296		3.81	17.70		21.51	30.50
7230	1" x 1"	↓	26	.308	↓	6.45	18.40		24.85	35
7300	PEX coupling									
7310	1/2" x 1/2"	1 Stpi	30	.267	Ea.	1.76	15.95		17.71	26
7320	3/4" x 3/4"		29	.276		2.15	16.50		18.65	27.50
7330	1" x 1"	↓	28	.286	↓	5.85	17.05		22.90	32.50
7400	PEX stainless crimp ring									
7410	1/2" x 1/2"	1 Stpi	86	.093	Ea.	.37	5.55		5.92	8.80

23 83 Radiant Heating Units

23 83 16 – Radiant-Heating Hydronic Piping

23 83 16.10 Radiant Floor Heating	Crew	Daily Output	Labor-Hours	Unit	Material	2015 Bare Costs Labor	Equipment	Total	Total Incl O&P
7420	3/4" x 3/4"	1 Stpi	84	.095	Ea.	.51	5.70		6.21	9.15
7430	1" x 1"	↓	82	.098	↓	.73	5.85		6.58	9.60

23 83 33 – Electric Radiant Heaters

23 83 33.10 Electric Heating

	23 83 33.10 Electric Heating	Crew	Daily Output	Labor-Hours	Unit	Material	Labor	Equipment	Total	Total Incl O&P
0010	**ELECTRIC HEATING**, not incl. conduit or feed wiring									
1100	Rule of thumb: Baseboard units, including control	1 Elec	4.40	1.818	kW	104	99.50		203.50	263
1300	Baseboard heaters, 2' long, 350 watt		8	1	Ea.	28.50	54.50		.83	114
1400	3' long, 750 watt		8	1		33	54.50		87.50	119
1600	4' long, 1000 watt		6.70	1.194		39	65.50		104.50	141
1800	5' long, 935 watt		5.70	1.404		48	77		125	168
2000	6' long, 1500 watt		5	1.600		54.50	87.50		142	191
2200	7' long, 1310 watt		4.40	1.818		63	99.50		162.50	219
2400	8' long, 2000 watt		4	2		66	109		175	237
2600	9' long, 1680 watt		3.60	2.222		102	122		224	294
2800	10' long, 1875 watt	↓	3.30	2.424	↓	190	133		323	410
2950	Wall heaters with fan, 120 to 277 volt									
3160	Recessed, residential, 750 watt	1 Elec	6	1.333	Ea.	134	73		207	256
3170	1000 watt		6	1.333		134	73		207	256
3180	1250 watt		5	1.600		134	87.50		221.50	278
3190	1500 watt		4	2		134	109		243	310
3210	2000 watt		4	2		134	109		243	310
3230	2500 watt		3.50	2.286		335	125		460	555
3240	3000 watt		3	2.667		495	146		641	760
3250	4000 watt		2.70	2.963		495	162		657	790
3260	Commercial, 750 watt		6	1.333		192	73		265	320
3270	1000 watt		6	1.333		192	73		265	320
3280	1250 watt		5	1.600		192	87.50		279.50	345
3290	1500 watt		4	2		192	109		301	375
3300	2000 watt		4	2		192	109		301	375
3310	2500 watt		3.50	2.286		192	125		317	400
3320	3000 watt		3	2.667		345	146		491	600
3330	4000 watt		2.70	2.963		345	162		507	625
3600	Thermostats, integral		16	.500		30	27.50		57.50	74
3800	Line voltage, 1 pole		8	1		19.20	54.50		73.70	103
3810	2 pole	↓	8	1	↓	29.50	54.50		84	115
4000	Heat trace system, 400 degree									
4020	115 V, 2.5 watts per L.F.	1 Elec	530	.015	L.F.	7.90	.83		8.73	9.90
4030	5 watts per L.F.		530	.015		7.90	.83		8.73	9.90
4050	10 watts per L.F.		530	.015		7.90	.83		8.73	9.90
4060	208 V, 5 watts per L.F.		530	.015		7.90	.83		8.73	9.90
4080	480 V, 8 watts per L.F.	↓	530	.015	↓	7.90	.83		8.73	9.90
4200	Heater raceway									
4260	Heat transfer cement									
4280	1 gallon				Ea.	60			60	66
4300	5 gallon				"	234			234	257
4320	Snap band, clamp									
4340	3/4" pipe size	1 Elec	470	.017	Ea.		.93		.93	1.39
4360	1" pipe size		444	.018			.99		.99	1.48
4380	1-1/4" pipe size		400	.020			1.09		1.09	1.64
4400	1-1/2" pipe size		355	.023			1.23		1.23	1.85
4420	2" pipe size		320	.025			1.37		1.37	2.05
4440	3" pipe size	↓	160	.050	↓		2.74		2.74	4.10

23 83 33 – Electric Radiant Heaters

23 83 33.10 Electric Heating	Crew	Daily Output	Labor-Hours	Unit	Material	2015 Bare Costs Labor	Equipment	Total	Total Incl O&P
4460	4" pipe size	1 Elec	100	.080	Ea.		4.38		4.38	6.55
4480	Thermostat NEMA 3R, 22 amp, 0-150 Deg, 10' cap.		8	1		241	54.50		295.50	345
4500	Thermostat NEMA 4X, 25 amp, 40 Deg, 5-1/2' cap.		7	1.143		241	62.50		303.50	360
4520	Thermostat NEMA 4X, 22 amp, 25-325 Deg, 10' cap.		7	1.143		650	62.50		712.50	810
4540	Thermostat NEMA 4X, 22 amp, 15-140 Deg,		6	1.333		545	73		618	710
4580	Thermostat NEMA 4,7,9, 22 amp, 25-325 Deg, 10' cap.		3.60	2.222		800	122		922	1,050
4600	Thermostat NEMA 4,7,9, 22 amp, 15-140 Deg,		3	2.667		795	146		941	1,100
4720	Fiberglass application tape, 36 yard roll		11	.727		77.50	40		117.50	145
5000	Radiant heating ceiling panels, 2' x 4', 500 watt		16	.500		305	27.50		332.50	375
5050	750 watt		16	.500		310	27.50		337.50	385
5200	For recessed plaster frame, add		32	.250		111	13.70		124.70	143
5300	Infrared quartz heaters, 120 volts, 1000 watts		6.70	1.194		284	65.50		349.50	410
5350	1500 watt		5	1.600		284	87.50		371.50	440
5400	240 volts, 1500 watt		5	1.600		284	87.50		371.50	440
5450	2000 watt		4	2		284	109		393	475
5500	3000 watt		3	2.667		325	146		471	575
5550	4000 watt		2.60	3.077		325	168		493	605
5570	Modulating control		.80	10		146	545		691	980
5600	Unit heaters, heavy duty, with fan & mounting bracket									
5650	Single phase, 208-240-277 volt, 3 kW	1 Elec	6	1.333	Ea.	445	73		518	600
5750	5 kW		5.50	1.455		460	79.50		539.50	625
5800	7 kW		5	1.600		710	87.50		797.50	910
5850	10 kW		4	2		810	109		919	1,050
5950	15 kW		3.80	2.105		1,300	115		1,415	1,600
6000	480 volt, 3 kW		6	1.333		425	73		498	575
6020	4 kW		5.80	1.379		445	75.50		520.50	605
6040	5 kW		5.50	1.455		460	79.50		539.50	625
6060	7 kW		5	1.600		710	87.50		797.50	910
6080	10 kW		4	2		780	109		889	1,025
6100	13 kW		3.80	2.105		1,300	115		1,415	1,600
6120	15 kW		3.70	2.162		1,300	118		1,418	1,600
6140	20 kW		3.50	2.286		1,700	125		1,825	2,025
6300	3 phase, 208-240 volt, 5 kW		5.50	1.455		435	79.50		514.50	600
6320	7 kW		5	1.600		670	87.50		757.50	865
6340	10 kW		4	2		720	109		829	960
6360	15 kW		3.70	2.162		1,200	118		1,318	1,500
6380	20 kW		3.50	2.286		1,725	125		1,850	2,075
6400	25 kW		3.30	2.424		2,025	133		2,158	2,425
6500	480 volt, 5 kW		5.50	1.455		605	79.50		684.50	790
6520	7 kW		5	1.600		765	87.50		852.50	970
6540	10 kW		4	2		810	109		919	1,050
6560	13 kW		3.80	2.105		1,300	115		1,415	1,600
6580	15 kW		3.70	2.162		1,700	118		1,818	2,025
6600	20 kW		3.50	2.286		1,700	125		1,825	2,025
6620	25 kW		3.30	2.424		2,025	133		2,158	2,425
6630	30 kW		3	2.667		2,350	146		2,496	2,825
6640	40 kW		2	4		3,000	219		3,219	3,625
6650	50 kW		1.60	5		3,625	274		3,899	4,375
6800	Vertical discharge heaters, with fan									
6820	Single phase, 208-240-277 volt, 10 kW	1 Elec	4	2	Ea.	750	109		859	990
6840	15 kW		3.70	2.162		1,250	118		1,368	1,550
6900	3 phase, 208-240 volt, 10 kW		4	2		720	109		829	960
6920	15 kW		3.70	2.162		1,200	118		1,318	1,500

For customer support on your Mechanical Cost Data, call 877.795.7394.

417

23 83 33 – Electric Radiant Heaters

23 83 33.10 Electric Heating		Crew	Daily Output	Labor-Hours	Unit	Material	2015 Bare Costs Labor	Equipment	Total	Total Incl O&P
6940	20 kW	1 Elec	3.50	2.286	Ea.	1,725	125		1,850	2,075
6960	25 kW		3.30	2.424		2,025	133		2,158	2,425
6980	30 kW		3	2.667		2,350	146		2,496	2,825
7000	40 kW		2	4		3,000	219		3,219	3,625
7020	50 kW		1.60	5		3,625	274		3,899	4,375
7100	480 volt, 10 kW		4	2		810	109		919	1,050
7120	15 kW		3.70	2.162		1,300	118		1,418	1,600
7140	20 kW		3.50	2.286		1,700	125		1,825	2,025
7160	25 kW		3.30	2.424		2,025	133		2,158	2,425
7180	30 kW		3	2.667		2,350	146		2,496	2,825
7200	40 kW		2	4		3,000	219		3,219	3,625
7220	50 kW	▼	1.60	5	▼	3,625	274		3,899	4,375
7900	Cabinet convector heaters, 240 volt, three phase,									
7920	3' long, 2000 watt	1 Elec	5.30	1.509	Ea.	1,900	82.50		1,982.50	2,225
7940	3000 watt		5.30	1.509		2,000	82.50		2,082.50	2,300
7960	4000 watt		5.30	1.509		2,050	82.50		2,132.50	2,375
7980	6000 watt		4.60	1.739		2,125	95		2,220	2,475
8000	8000 watt		4.60	1.739		2,200	95		2,295	2,575
8020	4' long, 4000 watt		4.60	1.739		2,075	95		2,170	2,425
8040	6000 watt		4	2		2,150	109		2,259	2,550
8060	8000 watt		4	2		2,225	109		2,334	2,625
8080	10,000 watt	▼	4	2	▼	2,250	109		2,359	2,650
8100	Available also in 208 or 277 volt									
8200	Cabinet unit heaters, 120 to 277 volt, 1 pole,									
8220	wall mounted, 2 kW	1 Elec	4.60	1.739	Ea.	1,875	95		1,970	2,200
8230	3 kW		4.60	1.739		1,950	95		2,045	2,300
8240	4 kW		4.40	1.818		2,000	99.50		2,099.50	2,350
8250	5 kW		4.40	1.818		2,075	99.50		2,174.50	2,450
8260	6 kW		4.20	1.905		2,125	104		2,229	2,475
8270	8 kW		4	2		2,150	109		2,259	2,550
8280	10 kW		3.80	2.105		2,225	115		2,340	2,600
8290	12 kW		3.50	2.286		2,250	125		2,375	2,650
8300	13.5 kW		2.90	2.759		2,700	151		2,851	3,200
8310	16 kW		2.70	2.963		2,725	162		2,887	3,250
8320	20 kW		2.30	3.478		3,475	190		3,665	4,100
8330	24 kW		1.90	4.211		3,525	230		3,755	4,225
8350	Recessed, 2 kW		4.40	1.818		1,875	99.50		1,974.50	2,200
8370	3 kW		4.40	1.818		1,950	99.50		2,049.50	2,300
8380	4 kW		4.20	1.905		2,000	104		2,104	2,350
8390	5 kW		4.20	1.905		2,075	104		2,179	2,425
8400	6 kW		4	2		2,100	109		2,209	2,500
8410	8 kW		3.80	2.105		2,200	115		2,315	2,600
8420	10 kW		3.50	2.286		2,675	125		2,800	3,125
8430	12 kW		2.90	2.759		2,675	151		2,826	3,175
8440	13.5 kW		2.70	2.963		2,700	162		2,862	3,225
8450	16 kW		2.30	3.478		2,875	190		3,065	3,450
8460	20 kW		1.90	4.211		3,275	230		3,505	3,950
8470	24 kW		1.60	5		3,325	274		3,599	4,050
8490	Ceiling mounted, 2 kW		3.20	2.500		1,875	137		2,012	2,250
8510	3 kW		3.20	2.500		1,950	137		2,087	2,350
8520	4 kW		3	2.667		2,000	146		2,146	2,425
8530	5 kW		3	2.667		2,075	146		2,221	2,500
8540	6 kW		2.80	2.857		2,075	156		2,231	2,525

23 83 Radiant Heating Units

23 83 33 – Electric Radiant Heaters

23 83 33.10 Electric Heating	Crew	Daily Output	Labor-Hours	Unit	Material	2015 Bare Costs Labor	Equipment	Total	Total Incl O&P
8550	8 kW	1 Elec	2.40	3.333	Ea.	2,150	182		2,332	2,650
8560	10 kW		2.20	3.636		2,200	199		2,399	2,725
8570	12 kW		2	4		2,250	219		2,469	2,800
8580	13.5 kW		1.50	5.333		2,275	292		2,567	2,950
8590	16 kW		1.30	6.154		2,350	335		2,685	3,075
8600	20 kW		.90	8.889		3,275	485		3,760	4,325
8610	24 kW	▼	.60	13.333	▼	3,325	730		4,055	4,750
8630	208 to 480 V, 3 pole									
8650	Wall mounted, 2 kW	1 Elec	4.60	1.739	Ea.	1,975	95		2,070	2,325
8670	3 kW		4.60	1.739		2,075	95		2,170	2,425
8680	4 kW		4.40	1.818		2,125	99.50		2,224.50	2,475
8690	5 kW		4.40	1.818		2,175	99.50		2,274.50	2,550
8700	6 kW		4.20	1.905		2,200	104		2,304	2,575
8710	8 kW		4	2		2,275	109		2,384	2,675
8720	10 kW		3.80	2.105		2,325	115		2,440	2,725
8730	12 kW		3.50	2.286		2,375	125		2,500	2,775
8740	13.5 kW		2.90	2.759		2,400	151		2,551	2,875
8750	16 kW		2.70	2.963		2,450	162		2,612	2,950
8760	20 kW		2.30	3.478		3,400	190		3,590	4,000
8770	24 kW		1.90	4.211		3,600	230		3,830	4,300
8790	Recessed, 2 kW		4.40	1.818		1,975	99.50		2,074.50	2,325
8810	3 kW		4.40	1.818		2,000	99.50		2,099.50	2,350
8820	4 kW		4.20	1.905		2,075	104		2,179	2,450
8830	5 kW		4.20	1.905		2,100	104		2,204	2,475
8840	6 kW		4	2		2,125	109		2,234	2,500
8850	8 kW		3.80	2.105		2,200	115		2,315	2,600
8860	10 kW		3.50	2.286		2,350	125		2,475	2,750
8870	12 kW		2.90	2.759		2,450	151		2,601	2,925
8880	13.5 kW		2.70	2.963		2,500	162		2,662	3,000
8890	16 kW		2.30	3.478		3,625	190		3,815	4,250
8900	20 kW		1.90	4.211		3,675	230		3,905	4,400
8920	24 kW		1.60	5		3,725	274		3,999	4,500
8940	Ceiling mount, 2 kW		3.20	2.500		2,075	137		2,212	2,475
8950	3 kW		3.20	2.500		2,200	137		2,337	2,625
8960	4 kW		3	2.667		2,200	146		2,346	2,650
8970	5 kW		3	2.667		2,275	146		2,421	2,725
8980	6 kW		2.80	2.857		2,300	156		2,456	2,750
8990	8 kW		2.40	3.333		2,375	182		2,557	2,875
9000	10 kW		2.20	3.636		2,425	199		2,624	2,950
9020	13.5 kW		1.50	5.333		2,500	292		2,792	3,175
9030	16 kW		1.30	6.154		3,625	335		3,960	4,475
9040	20 kW		.90	8.889		3,675	485		4,160	4,775
9060	24 kW	▼	.60	13.333	▼	4,075	730		4,805	5,600

For customer support on your Mechanical Cost Data, call 877.795.7394.

419

23 84 13 – Humidifiers

23 84 13.10 Humidifier Units	Crew	Daily Output	Labor-Hours	Unit	Material	2015 Bare Costs Labor	Equipment	Total	Total Incl O&P
0010 **HUMIDIFIER UNITS**									
0520 Steam, room or duct, filter, regulators, auto. controls, 220 V									
0540 11 lb. per hour	Q-5	6	2.667	Ea.	2,700	143		2,843	3,200
0560 22 lb. per hour		5	3.200		2,975	172		3,147	3,525
0580 33 lb. per hour		4	4		3,050	215		3,265	3,700
0600 50 lb. per hour		4	4		3,550	215		3,765	4,225
0620 100 lb. per hour		3	5.333		4,475	287		4,762	5,350
0640 150 lb. per hour		2.50	6.400		6,400	345		6,745	7,575
0660 200 lb. per hour	▼	2	8	▼	8,025	430		8,455	9,475
0700 With blower									
0720 11 lb. per hour	Q-5	5.50	2.909	Ea.	4,300	156		4,456	4,950
0740 22 lb. per hour		4.75	3.368		4,525	181		4,706	5,250
0760 33 lb. per hour		3.75	4.267		4,650	229		4,879	5,475
0780 50 lb. per hour		3.50	4.571		5,325	246		5,571	6,225
0800 100 lb. per hour		2.75	5.818		5,950	315		6,265	7,025
0820 150 lb. per hour		2	8		8,675	430		9,105	10,200
0840 200 lb. per hour	▼	1.50	10.667	▼	9,700	575		10,275	11,600
1000 Steam for duct installation, manifold with pneum. controls									
1010 3 – 191 lb./hr., 24" x 24"	Q-5	5	3.200	Ea.	2,075	172		2,247	2,550
1020 3 – 191 lb./hr., 48" x 48"		4.60	3.478		4,000	187		4,187	4,675
1030 65 – 334 lb./hr., 36" x 36"		4.80	3.333		3,900	179		4,079	4,575
1040 65 – 334 lb./hr., 72" x 72"		4	4		4,225	215		4,440	4,975
1050 100 – 690 lb./hr., 48" x 48"		4	4		5,475	215		5,690	6,350
1060 100 – 690 lb./hr., 84" x 84"		3.60	4.444		5,650	239		5,889	6,575
1070 220 – 2000 lb./hr., 72" x 72"		3.80	4.211		5,725	226		5,951	6,650
1080 220 – 2000 lb./hr., 144" x 144"	▼	3.40	4.706		6,700	253		6,953	7,750
1090 For electrically controlled unit, add				▼	805			805	885
1092 For additional manifold, add					25%	15%			
1200 Electric steam, self generating, with manifold									
1210 1 – 96 lb./hr., 12" x 72"	Q-5	3.60	4.444	Ea.	10,300	239		10,539	11,700
1220 20 – 168 lb./hr., 12" x 72"		3.60	4.444		15,800	239		16,039	17,700
1290 For room mounting, add	▼	16	1	▼	257	54		311	365
5000 Furnace type, wheel bypass									
5020 10 GPD	1 Stpi	4	2	Ea.	260	120		380	465
5040 14 GPD		3.80	2.105		293	126		419	510
5060 19 GPD	▼	3.60	2.222		405	133		538	645

23 84 16 – Mechanical Dehumidification Units

23 84 16.10 Dehumidifier Units

23 84 16.10 Dehumidifier Units	Crew	Daily Output	Labor-Hours	Unit	Material	2015 Bare Costs Labor	Equipment	Total	Total Incl O&P
0010 **DEHUMIDIFIER UNITS**									
6000 Self contained with filters and standard controls									
6040 1.5 lb./hr., 50 CFM	1 Plum	8	1	Ea.	7,950	58.50		8,008.50	8,825
6060 3 lb./hr., 150 CFM	Q-1	12	1.333		8,300	70.50		8,370.50	9,225
6065 6 lb./hr., 150 CFM		9	1.778		13,500	94		13,594	15,000
6070 16 to 20 lb./hr., 600 CFM		5	3.200		25,700	169		25,869	28,600
6080 30 to 40 lb./hr., 1125 CFM		4	4		40,000	211		40,211	44,300
6090 60 to 75 lb./hr., 2250 CFM		3	5.333		52,000	282		52,282	58,000
6100 120 to 155 lb./hr., 4500 CFM		2	8		92,500	425		92,925	102,000
6110 240 to 310 lb./hr., 9000 CFM	▼	1.50	10.667		132,500	565		133,065	146,500
6120 400 to 515 lb./hr., 15,000 CFM	Q-2	1.60	15		166,500	820		167,320	184,500
6130 530 to 690 lb./hr., 20,000 CFM		1.40	17.143		180,500	940		181,440	200,000
6140 800 to 1030 lb./hr., 30,000 CFM		1.20	20		205,500	1,100		206,600	227,500
6150 1060 to 1375 lb./hr., 40,000 CFM	▼	1	24	▼	286,000	1,325		287,325	316,500

420

For customer support on your Mechanical Cost Data, call 877.795.7394.

23 91 10 – Prefabricated Curbs, Pads and Stands

23 91 10.10 Prefabricated Pads and Stands	Crew	Daily Output	Labor-Hours	Unit	Material	2015 Bare Costs Labor	Equipment	Total	Total Incl O&P
0010 **PREFABRICATED PADS AND STANDS**									
6000 Pad, fiberglass reinforced concrete with polystyrene foam core									
6050 Condenser, 2" thick, 20" x 38"	1 Shee	8	1	Ea.	27.50	56		83.50	116
6070 24" x 24"		16	.500		23	28		51	68.50
6090 24" x 36"		12	.667		35	37.50		72.50	95.50
6110 24" x 42"		8	1		41	56		97	131
6150 26" x 36"	Q-9	8	2		38.50	101		139.50	197
6170 28" x 38"	1 Shee	8	1		43.50	56		99.50	134
6190 30" x 30"		12	.667		36.50	37.50		74	97.50
6220 30" x 36"		8	1		43.50	56		99.50	134
6240 30" x 40"	Q-9	8	2		47.50	101		148.50	206
6260 32" x 32"		7	2.286		41	115		156	221
6280 36" x 36"		8	2		50.50	101		151.50	210
6300 36" x 40"		7	2.286		57.50	115		172.50	240
6320 36" x 48"		7	2.286		67.50	115		182.50	250
6340 36" x 54"		6	2.667		81.50	134		215.50	295
6360 36" x 60" x 3"		7	2.286		96.50	115		211.50	282
6400 41" x 47" x 3"		6	2.667		90	134		224	305
6490 26" round		10	1.600		21.50	80.50		102	147
6550 30" round		9	1.778		28.50	89.50		118	169
6600 36" round		8	2		39	101		140	197
6700 For 3" thick pad, add					20%	5%			
8800 Stand, corrosion-resistant plastic									
8850 Heat pump, 6" high	1 Shee	32	.250	Ea.	5.50	14		19.50	27.50
8870 12" high	"	32	.250	"	12.50	14		26.50	35.50

For customer support on your Mechanical Cost Data, call 877.795.7394.

421

Division Notes

	CREW	DAILY OUTPUT	LABOR-HOURS	UNIT	BARE COSTS				TOTAL INCL O&P
					MAT.	LABOR	EQUIP.	TOTAL	

Estimating Tips

26 05 00 Common Work Results for Electrical

- Conduit should be taken off in three main categories—power distribution, branch power, and branch lighting—so the estimator can concentrate on systems and components, therefore making it easier to ensure all items have been accounted for.

- For cost modifications for elevated conduit installation, add the percentages to labor according to the height of installation, and only to the quantities exceeding the different height levels, not to the total conduit quantities.

- Remember that aluminum wiring of equal ampacity is larger in diameter than copper and may require larger conduit.

- If more than three wires at a time are being pulled, deduct percentages from the labor hours of that grouping of wires.

- When taking off grounding systems, identify separately the type and size of wire, and list each unique type of ground connection.

- The estimator should take the weights of materials into consideration when completing a takeoff. Topics to consider include: How will the materials be supported? What methods of support are available? How high will the support structure have to reach? Will the final support structure be able to withstand the total burden? Is the support material included or separate from the fixture, equipment, and material specified?

- Do not overlook the costs for equipment used in the installation. If scaffolding or highlifts are available in the field, contractors may use them in lieu of the proposed ladders and rolling staging.

26 20 00 Low-Voltage Electrical Transmission

- Supports and concrete pads may be shown on drawings for the larger equipment, or the support system may be only a piece of plywood for the back of a panelboard. In either case, it must be included in the costs.

26 40 00 Electrical and Cathodic Protection

- When taking off cathodic protections systems, identify the type and size of cable, and list each unique type of anode connection.

26 50 00 Lighting

- Fixtures should be taken off room by room, using the fixture schedule, specifications, and the ceiling plan. For large concentrations of lighting fixtures in the same area, deduct the percentages from labor hours.

Reference Numbers

Reference numbers are shown in shaded boxes at the beginning of some major classifications. These numbers refer to related items in the Reference Section. The reference information may be an estimating procedure, an alternate pricing method, or technical information.

Note: Not all subdivisions listed here necessarily appear in this publication. ■

---

*Note: **Trade Service**, in part, has been used as a reference source for some of the material prices used in Division 26.*

26 05 33.95 Cutting and Drilling	Crew	Daily Output	Labor-Hours	Unit	Material	2015 Bare Costs Labor	Equipment	Total	Total Incl O&P
0010 **CUTTING AND DRILLING**									
0100 Hole drilling to 10' high, concrete wall									
0110 8" thick, 1/2" pipe size	R-31	12	.667	Ea.	.24	36.50	4.59	41.33	60
0120 3/4" pipe size		12	.667		.24	36.50	4.59	41.33	60
0130 1" pipe size		9.50	.842		.38	46	5.80	52.18	76
0140 1-1/4" pipe size		9.50	.842		.38	46	5.80	52.18	76
0150 1-1/2" pipe size		9.50	.842		.38	46	5.80	52.18	76
0160 2" pipe size		4.40	1.818		.51	99.50	12.55	112.56	163
0170 2-1/2" pipe size		4.40	1.818		.51	99.50	12.55	112.56	163
0180 3" pipe size		4.40	1.818		.51	99.50	12.55	112.56	163
0190 3-1/2" pipe size		3.30	2.424		.66	133	16.70	150.36	218
0200 4" pipe size		3.30	2.424		.66	133	16.70	150.36	218
0500 12" thick, 1/2" pipe size		9.40	.851		.35	46.50	5.85	52.70	76.50
0520 3/4" pipe size		9.40	.851		.35	46.50	5.85	52.70	76.50
0540 1" pipe size		7.30	1.096		.56	60	7.55	68.11	99
0560 1-1/4" pipe size		7.30	1.096		.56	60	7.55	68.11	99
0570 1-1/2" pipe size		7.30	1.096		.56	60	7.55	68.11	99
0580 2" pipe size		3.60	2.222		.76	122	15.30	138.06	200
0590 2-1/2" pipe size		3.60	2.222		.76	122	15.30	138.06	200
0600 3" pipe size		3.60	2.222		.76	122	15.30	138.06	200
0610 3-1/2" pipe size		2.80	2.857		.99	156	19.70	176.69	257
0630 4" pipe size		2.50	3.200		.99	175	22	197.99	288
0650 16" thick, 1/2" pipe size		7.60	1.053		.47	57.50	7.25	65.22	95
0670 3/4" pipe size		7	1.143		.47	62.50	7.85	70.82	103
0690 1" pipe size		6	1.333		.75	73	9.20	82.95	120
0710 1-1/4" pipe size		5.50	1.455		.75	79.50	10	90.25	131
0730 1-1/2" pipe size		5.50	1.455		.75	79.50	10	90.25	131
0750 2" pipe size		3	2.667		1.02	146	18.35	165.37	240
0770 2-1/2" pipe size		2.70	2.963		1.02	162	20.50	183.52	267
0790 3" pipe size		2.50	3.200		1.02	175	22	198.02	288
0810 3-1/2" pipe size		2.30	3.478		1.32	190	24	215.32	315
0830 4" pipe size		2	4		1.32	219	27.50	247.82	360
0850 20" thick, 1/2" pipe size		6.40	1.250		.59	68.50	8.60	77.69	112
0870 3/4" pipe size		6	1.333		.59	73	9.20	82.79	120
0890 1" pipe size		5	1.600		.94	87.50	11	99.44	144
0910 1-1/4" pipe size		4.80	1.667		.94	91	11.50	103.44	151
0930 1-1/2" pipe size		4.60	1.739		.94	95	12	107.94	157
0950 2" pipe size		2.70	2.963		1.27	162	20.50	183.77	267
0970 2-1/2" pipe size		2.40	3.333		1.27	182	23	206.27	300
0990 3" pipe size		2.20	3.636		1.27	199	25	225.27	325
1010 3-1/2" pipe size		2	4		1.64	219	27.50	248.14	360
1030 4" pipe size		1.70	4.706		1.64	257	32.50	291.14	420
1050 24" thick, 1/2" pipe size		5.50	1.455		.71	79.50	10	90.21	131
1070 3/4" pipe size		5.10	1.569		.71	86	10.80	97.51	142
1090 1" pipe size		4.30	1.860		1.13	102	12.80	115.93	167
1110 1-1/4" pipe size		4	2		1.13	109	13.80	123.93	180
1130 1-1/2" pipe size		4	2		1.13	109	13.80	123.93	180
1150 2" pipe size		2.40	3.333		1.52	182	23	206.52	300
1170 2-1/2" pipe size		2.20	3.636		1.52	199	25	225.52	325
1190 3" pipe size		2	4		1.52	219	27.50	248.02	360
1210 3-1/2" pipe size		1.80	4.444		1.97	243	30.50	275.47	400
1230 4" pipe size		1.50	5.333		1.97	292	37	330.97	480
1500 Brick wall, 8" thick, 1/2" pipe size		18	.444		.24	24.50	3.06	27.80	40

26 05 33 – Raceway and Boxes for Electrical Systems

26 05 33.95 Cutting and Drilling	Crew	Daily Output	Labor-Hours	Unit	Material	2015 Bare Costs Labor	Equipment	Total	Total Incl O&P
1520	3/4" pipe size	R-31	18	.444	Ea.	.24	24.50	3.06	27.80	40
1540	1" pipe size		13.30	.602		.38	33	4.14	37.52	54.50
1560	1-1/4" pipe size		13.30	.602		.38	33	4.14	37.52	54.50
1580	1-1/2" pipe size		13.30	.602		.38	33	4.14	37.52	54.50
1600	2" pipe size		5.70	1.404		.51	77	9.65	87.16	126
1620	2-1/2" pipe size		5.70	1.404		.51	77	9.65	87.16	126
1640	3" pipe size		5.70	1.404		.51	77	9.65	87.16	126
1660	3-1/2" pipe size		4.40	1.818		.66	99.50	12.55	112.71	164
1680	4" pipe size		4	2		.66	109	13.80	123.46	180
1700	12" thick, 1/2" pipe size		14.50	.552		.35	30	3.80	34.15	49.50
1720	3/4" pipe size		14.50	.552		.35	30	3.80	34.15	49.50
1740	1" pipe size		11	.727		.56	40	5	45.56	65.50
1760	1-1/4" pipe size		11	.727		.56	40	5	45.56	65.50
1780	1-1/2" pipe size		11	.727		.56	40	5	45.56	65.50
1800	2" pipe size		5	1.600		.76	87.50	11	99.26	144
1820	2-1/2" pipe size		5	1.600		.76	87.50	11	99.26	144
1840	3" pipe size		5	1.600		.76	87.50	11	99.26	144
1860	3-1/2" pipe size		3.80	2.105		.99	115	14.50	130.49	190
1880	4" pipe size		3.30	2.424		.99	133	16.70	150.69	218
1900	16" thick, 1/2" pipe size		12.30	.650		.47	35.50	4.48	40.45	59
1920	3/4" pipe size		12.30	.650		.47	35.50	4.48	40.45	59
1940	1" pipe size		9.30	.860		.75	47	5.95	53.70	78
1960	1-1/4" pipe size		9.30	.860		.75	47	5.95	53.70	78
1980	1-1/2" pipe size		9.30	.860		.75	47	5.95	53.70	78
2000	2" pipe size		4.40	1.818		1.02	99.50	12.55	113.07	164
2010	2-1/2" pipe size		4.40	1.818		1.02	99.50	12.55	113.07	164
2030	3" pipe size		4.40	1.818		1.02	99.50	12.55	113.07	164
2050	3-1/2" pipe size		3.30	2.424		1.32	133	16.70	151.02	219
2070	4" pipe size		3	2.667		1.32	146	18.35	165.67	240
2090	20" thick, 1/2" pipe size		10.70	.748		.59	41	5.15	46.74	68
2110	3/4" pipe size		10.70	.748		.59	41	5.15	46.74	68
2130	1" pipe size		8	1		.94	54.50	6.90	62.34	90.50
2150	1-1/4" pipe size		8	1		.94	54.50	6.90	62.34	90.50
2170	1-1/2" pipe size		8	1		.94	54.50	6.90	62.34	90.50
2190	2" pipe size		4	2		1.27	109	13.80	124.07	181
2210	2-1/2" pipe size		4	2		1.27	109	13.80	124.07	181
2230	3" pipe size		4	2		1.27	109	13.80	124.07	181
2250	3-1/2" pipe size		3	2.667		1.64	146	18.35	165.99	241
2270	4" pipe size		2.70	2.963		1.64	162	20.50	184.14	267
2290	24" thick, 1/2" pipe size		9.40	.851		.71	46.50	5.85	53.06	76.50
2310	3/4" pipe size		9.40	.851		.71	46.50	5.85	53.06	76.50
2330	1" pipe size		7.10	1.127		1.13	61.50	7.75	70.38	102
2350	1-1/4" pipe size		7.10	1.127		1.13	61.50	7.75	70.38	102
2370	1-1/2" pipe size		7.10	1.127		1.13	61.50	7.75	70.38	102
2390	2" pipe size		3.60	2.222		1.52	122	15.30	138.82	201
2410	2-1/2" pipe size		3.60	2.222		1.52	122	15.30	138.82	201
2430	3" pipe size		3.60	2.222		1.52	122	15.30	138.82	201
2450	3-1/2" pipe size		2.80	2.857		1.97	156	19.70	177.67	258
2470	4" pipe size		2.50	3.200		1.97	175	22	198.97	289
3000	Knockouts to 8' high, metal boxes & enclosures									
3020	With hole saw, 1/2" pipe size	1 Elec	53	.151	Ea.		8.25		8.25	12.35
3040	3/4" pipe size		47	.170			9.30		9.30	13.95
3050	1" pipe size		40	.200			10.95		10.95	16.40

For customer support on your Mechanical Cost Data, call 877.795.7394.

425

26 05 33 – Raceway and Boxes for Electrical Systems

26 05 33.95 Cutting and Drilling

		Crew	Daily Output	Labor-Hours	Unit	Material	2015 Bare Costs Labor	Equipment	Total	Total Incl O&P
3060	1-1/4" pipe size	1 Elec	36	.222	Ea.		12.15		12.15	18.20
3070	1-1/2" pipe size		32	.250			13.70		13.70	20.50
3080	2" pipe size		27	.296			16.20		16.20	24.50
3090	2-1/2" pipe size		20	.400			22		22	33
4010	3" pipe size		16	.500			27.50		27.50	41
4030	3-1/2" pipe size		13	.615			33.50		33.50	50.50
4050	4" pipe size	▼	11	.727	▼		40		40	59.50

26 05 80 – Wiring Connections

26 05 80.10 Motor Connections

		Crew	Daily Output	Labor-Hours	Unit	Material	2015 Bare Costs Labor	Equipment	Total	Total Incl O&P
0010	**MOTOR CONNECTIONS**									
0020	Flexible conduit and fittings, 115 volt, 1 phase, up to 1 HP motor	1 Elec	8	1	Ea.	6.35	54.50		60.85	89
0050	2 HP motor		6.50	1.231		12.55	67.50		80.05	115
0100	3 HP motor		5.50	1.455		11.05	79.50		90.55	131
0110	230 volt, 3 phase, 3 HP motor		6.78	1.180		7.65	64.50		72.15	105
0112	5 HP motor		5.47	1.463		6.50	80		86.50	127
0114	7-1/2 HP motor		4.61	1.735		9.55	95		104.55	153
0120	10 HP motor		4.20	1.905		21	104		125	180
0150	15 HP motor		3.30	2.424		21	133		154	223
0200	25 HP motor		2.70	2.963		29	162		191	275
0400	50 HP motor		2.20	3.636		59	199		258	365
0600	100 HP motor		1.50	5.333		140	292		432	590
1500	460 volt, 5 HP motor, 3 phase		8	1		6.85	54.50		61.35	89.50
1520	10 HP motor		8	1		6.85	54.50		61.35	89.50
1530	25 HP motor		6	1.333		13.45	73		86.45	124
1540	30 HP motor		6	1.333		13.45	73		86.45	124
1550	40 HP motor		5	1.600		20	87.50		107.50	153
1560	50 HP motor		5	1.600		23	87.50		110.50	156
1570	60 HP motor		3.80	2.105		25	115		140	201
1580	75 HP motor		3.50	2.286		34.50	125		159.50	225
1590	100 HP motor		2.50	3.200		53	175		228	320
1600	125 HP motor		2	4		68	219		287	405
1610	150 HP motor		1.80	4.444		72.50	243		315.50	445
1620	200 HP motor		1.50	5.333		116	292		408	565
2005	460 Volt, 5 HP motor, 3 Phase, w/sealtite		8	1		10.80	54.50		65.30	94
2010	10 HP motor		8	1		10.80	54.50		65.30	94
2015	25 HP motor		6	1.333		21	73		94	132
2020	30 HP motor		6	1.333		21	73		94	132
2025	40 HP motor		5	1.600		37.50	87.50		125	173
2030	50 HP motor		5	1.600		38.50	87.50		126	173
2035	60 HP motor		3.80	2.105		59.50	115		174.50	238
2040	75 HP motor		3.50	2.286		62.50	125		187.50	256
2045	100 HP motor		2.50	3.200		85.50	175		260.50	355
2055	150 HP motor		1.80	4.444		91	243		334	465
2060	200 HP motor	▼	1.50	5.333	▼	605	292		897	1,100

26 05 90 – Residential Applications

26 05 90.10 Residential Wiring	Crew	Daily Output	Labor-Hours	Unit	Material	2015 Bare Costs Labor	2015 Bare Costs Equipment	Total	Total Incl O&P
0010 **RESIDENTIAL WIRING**									
0020 20' avg. runs and #14/2 wiring incl. unless otherwise noted									
1000 Service & panel, includes 24' SE-AL cable, service eye, meter,									
1010 Socket, panel board, main bkr., ground rod, 15 or 20 amp									
1020 1-pole circuit breakers, and misc. hardware									
1100 100 amp, with 10 branch breakers	1 Elec	1.19	6.723	Ea.	560	370		930	1,175
1110 With PVC conduit and wire		.92	8.696		605	475		1,080	1,375
1120 With RGS conduit and wire		.73	10.959		765	600		1,365	1,750
1150 150 amp, with 14 branch breakers		1.03	7.767		860	425		1,285	1,575
1170 With PVC conduit and wire		.82	9.756		955	535		1,490	1,850
1180 With RGS conduit and wire		.67	11.940		1,275	655		1,930	2,375
1200 200 amp, with 18 branch breakers	2 Elec	1.80	8.889		1,150	485		1,635	2,000
1220 With PVC conduit and wire		1.46	10.959		1,250	600		1,850	2,275
1230 With RGS conduit and wire		1.24	12.903		1,650	705		2,355	2,850
1800 Lightning surge suppressor	1 Elec	32	.250		50.50	13.70		64.20	76.50
2000 Switch devices									
2100 Single pole, 15 amp, Ivory, with a 1-gang box, cover plate,									
2110 Type NM (Romex) cable	1 Elec	17.10	.468	Ea.	12.35	25.50		37.85	52
2120 Type MC (BX) cable		14.30	.559		21.50	30.50		52	69.50
2130 EMT & wire		5.71	1.401		30	76.50		106.50	148
2150 3-way, #14/3, type NM cable		14.55	.550		14.85	30		44.85	61.50
2170 Type MC cable		12.31	.650		27	35.50		62.50	83
2180 EMT & wire		5	1.600		32	87.50		119.50	166
2200 4-way, #14/3, type NM cable		14.55	.550		21	30		51	68
2220 Type MC cable		12.31	.650		33	35.50		68.50	90
2230 EMT & wire		5	1.600		38.50	87.50		126	173
2250 S.P., 20 amp, #12/2, type NM cable		13.33	.600		22	33		55	73
2270 Type MC cable		11.43	.700		28.50	38.50		67	89
2280 EMT & wire		4.85	1.649		41	90		131	180
2290 S.P. rotary dimmer, 600W, no wiring		17	.471		29.50	25.50		55	71
2300 S.P. rotary dimmer, 600W, type NM cable		14.55	.550		34.50	30		64.50	83
2320 Type MC cable		12.31	.650		43.50	35.50		79	102
2330 EMT & wire		5	1.600		53.50	87.50		141	190
2350 3-way rotary dimmer, type NM cable		13.33	.600		28.50	33		61.50	80.50
2370 Type MC cable		11.43	.700		37.50	38.50		76	98.50
2380 EMT & wire		4.85	1.649		47.50	90		137.50	188
2400 Interval timer wall switch, 20 amp, 1-30 min., #12/2									
2410 Type NM cable	1 Elec	14.55	.550	Ea.	56	30		86	107
2420 Type MC cable		12.31	.650		60	35.50		95.50	120
2430 EMT & wire		5	1.600		75	87.50		162.50	214
2500 Decorator style									
2510 S.P., 15 amp, type NM cable	1 Elec	17.10	.468	Ea.	16.25	25.50		41.75	56.50
2520 Type MC cable		14.30	.559		25	30.50		55.50	73.50
2530 EMT & wire		5.71	1.401		34	76.50		110.50	152
2550 3-way, #14/3, type NM cable		14.55	.550		18.75	30		48.75	65.50
2570 Type MC cable		12.31	.650		30.50	35.50		66	87
2580 EMT & wire		5	1.600		36	87.50		123.50	171
2600 4-way, #14/3, type NM cable		14.55	.550		25	30		55	72.50
2620 Type MC cable		12.31	.650		37	35.50		72.50	94
2630 EMT & wire		5	1.600		42	87.50		129.50	178
2650 S.P., 20 amp, #12/2, type NM cable		13.33	.600		26	33		59	77.50
2670 Type MC cable		11.43	.700		32.50	38.50		71	93

For customer support on your Mechanical Cost Data, call 877.795.7394.

427

26 05 90.10 Residential Wiring		Crew	Daily Output	Labor-Hours	Unit	Material	2015 Bare Costs Labor	Equipment	Total	Total Incl O&P
2680	EMT & wire	1 Elec	4.85	1.649	Ea.	45	90		135	185
2700	S.P., slide dimmer, type NM cable		17.10	.468		28.50	25.50		54	70
2720	Type MC cable		14.30	.559		37.50	30.50		68	87.50
2730	EMT & wire		5.71	1.401		47.50	76.50		124	168
2750	S.P., touch dimmer, type NM cable		17.10	.468		32	25.50		57.50	74
2770	Type MC cable		14.30	.559		41	30.50		71.50	91.50
2780	EMT & wire		5.71	1.401		51.50	76.50		128	172
2800	3-way touch dimmer, type NM cable		13.33	.600		52	33		85	106
2820	Type MC cable		11.43	.700		60.50	38.50		99	125
2830	EMT & wire	↓	4.85	1.649	↓	71	90		161	213
3000	Combination devices									
3100	S.P. switch/15 amp recpt., Ivory, 1-gang box, plate									
3110	Type NM cable	1 Elec	11.43	.700	Ea.	24.50	38.50		63	84.50
3120	Type MC cable		10	.800		33.50	44		77.50	102
3130	EMT & wire		4.40	1.818		43.50	99.50		143	197
3150	S.P. switch/pilot light, type NM cable		11.43	.700		24.50	38.50		63	84.50
3170	Type MC cable		10	.800		33.50	44		77.50	103
3180	EMT & wire		4.43	1.806		43.50	99		142.50	196
3190	2-S.P. switches, 2-#14/2, no wiring		14	.571		6.90	31.50		38.40	54.50
3200	2-S.P. switches, 2-#14/2, type NM cables		10	.800		27.50	44		71.50	95.50
3220	Type MC cable		8.89	.900		41	49		90	119
3230	EMT & wire		4.10	1.951		46.50	107		153.50	211
3250	3-way switch/15 amp recpt., #14/3, type NM cable		10	.800		32	44		76	101
3270	Type MC cable		8.89	.900		44	49		93	122
3280	EMT & wire		4.10	1.951		49	107		156	214
3300	2-3 way switches, 2-#14/3, type NM cables		8.89	.900		41.50	49		90.50	120
3320	Type MC cable		8	1		60.50	54.50		115	149
3330	EMT & wire		4	2		56.50	109		165.50	226
3350	S.P. switch/20 amp recpt., #12/2, type NM cable		10	.800		32	44		76	101
3370	Type MC cable		8.89	.900		36	49		85	114
3380	EMT & wire	↓	4.10	1.951	↓	51	107		158	216
3400	Decorator style									
3410	S.P. switch/15 amp recpt., type NM cable	1 Elec	11.43	.700	Ea.	28.50	38.50		67	88.50
3420	Type MC cable		10	.800		37	44		81	107
3430	EMT & wire		4.40	1.818		47.50	99.50		147	201
3450	S.P. switch/pilot light, type NM cable		11.43	.700		28.50	38.50		67	88.50
3470	Type MC cable		10	.800		37.50	44		81.50	107
3480	EMT & wire		4.40	1.818		47.50	99.50		147	202
3500	2-S.P. switches, 2-#14/2, type NM cables		10	.800		31.50	44		75.50	100
3520	Type MC cable		8.89	.900		44.50	49		93.50	123
3530	EMT & wire		4.10	1.951		50.50	107		157.50	216
3550	3-way/15 amp recpt., #14/3, type NM cable		10	.800		36	44		80	105
3570	Type MC cable		8.89	.900		47.50	49		96.50	127
3580	EMT & wire		4.10	1.951		53	107		160	218
3650	2-3 way switches, 2-#14/3, type NM cables		8.89	.900		45.50	49		94.50	124
3670	Type MC cable		8	1		64.50	54.50		119	153
3680	EMT & wire		4	2		60.50	109		169.50	231
3700	S.P. switch/20 amp recpt., #12/2, type NM cable		10	.800		35.50	44		79.50	105
3720	Type MC cable		8.89	.900		39.50	49		88.50	118
3730	EMT & wire	↓	4.10	1.951	↓	55	107		162	221
4000	Receptacle devices									
4010	Duplex outlet, 15 amp recpt., Ivory, 1-gang box, plate									
4015	Type NM cable	1 Elec	14.55	.550	Ea.	10.80	30		40.80	57

428

For customer support on your Mechanical Cost Data, call 877.795.7394.

26 05 90.10 Residential Wiring	Crew	Daily Output	Labor-Hours	Unit	Material	2015 Bare Costs Labor	Equipment	Total	Total Incl O&P
4020 Type MC cable	1 Elec	12.31	.650	Ea.	19.75	35.50		55.25	75
4030 EMT & wire		5.33	1.501		28.50	82		110.50	155
4050 With #12/2, type NM cable		12.31	.650		13.30	35.50		48.80	68
4070 Type MC cable		10.67	.750		19.85	41		60.85	83.50
4080 EMT & wire		4.71	1.699		32.50	93		125.50	175
4100 20 amp recpt., #12/2, type NM cable		12.31	.650		19.95	35.50		55.45	75.50
4120 Type MC cable		10.67	.750		26.50	41		67.50	90.50
4130 EMT & wire	▼	4.71	1.699	▼	39	93		132	182
4140 For GFI see Section 26 05 90.10 line 4300 below									
4150 Decorator style, 15 amp recpt., type NM cable	1 Elec	14.55	.550	Ea.	14.70	30		44.70	61
4170 Type MC cable		12.31	.650		23.50	35.50		59	79.50
4180 EMT & wire		5.33	1.501		32.50	82		114.50	159
4200 With #12/2, type NM cable		12.31	.650		17.20	35.50		52.70	72.50
4220 Type MC cable		10.67	.750		23.50	41		64.50	87.50
4230 EMT & wire		4.71	1.699		36.50	93		129.50	179
4250 20 amp recpt. #12/2, type NM cable		12.31	.650		24	35.50		59.50	79.50
4270 Type MC cable		10.67	.750		30.50	41		71.50	95
4280 EMT & wire		4.71	1.699		43	93		136	187
4300 GFI, 15 amp recpt., type NM cable		12.31	.650		41.50	35.50		77	99
4320 Type MC cable		10.67	.750		50.50	41		91.50	117
4330 EMT & wire		4.71	1.699		59	93		152	204
4350 GFI with #12/2, type NM cable		10.67	.750		44	41		85	110
4370 Type MC cable		9.20	.870		50.50	47.50		98	127
4380 EMT & wire		4.21	1.900		63	104		167	226
4400 20 amp recpt., #12/2 type NM cable		10.67	.750		52	41		93	119
4420 Type MC cable		9.20	.870		58.50	47.50		106	136
4430 EMT & wire		4.21	1.900		71.50	104		175.50	235
4500 Weather-proof cover for above receptacles, add	▼	32	.250	▼	3.46	13.70		17.16	24.50
4550 Air conditioner outlet, 20 amp-240 volt recpt.									
4560 30' of #12/2, 2 pole circuit breaker									
4570 Type NM cable	1 Elec	10	.800	Ea.	63	44		107	135
4580 Type MC cable		9	.889		70.50	48.50		119	151
4590 EMT & wire		4	2		82.50	109		191.50	255
4600 Decorator style, type NM cable		10	.800		68	44		112	141
4620 Type MC cable		9	.889		75.50	48.50		124	156
4630 EMT & wire	▼	4	2	▼	87	109		196	260
4650 Dryer outlet, 30 amp-240 volt recpt., 20' of #10/3									
4660 2 pole circuit breaker									
4670 Type NM cable	1 Elec	6.41	1.248	Ea.	61.50	68.50		130	170
4680 Type MC cable		5.71	1.401		64.50	76.50		141	186
4690 EMT & wire	▼	3.48	2.299	▼	76	126		202	272
4700 Range outlet, 50 amp-240 volt recpt., 30' of #8/3									
4710 Type NM cable	1 Elec	4.21	1.900	Ea.	90.50	104		194.50	256
4720 Type MC cable		4	2		124	109		233	300
4730 EMT & wire		2.96	2.703		108	148		256	340
4750 Central vacuum outlet, Type NM cable		6.40	1.250		58.50	68.50		127	167
4770 Type MC cable		5.71	1.401		69.50	76.50		146	192
4780 EMT & wire	▼	3.48	2.299	▼	83.50	126		209.50	280
4800 30 amp-110 volt locking recpt., #10/2 circ. bkr.									
4810 Type NM cable	1 Elec	6.20	1.290	Ea.	67	70.50		137.50	180
4830 EMT & wire	"	3.20	2.500	"	94	137		231	310
4900 Low voltage outlets									
4910 Telephone recpt., 20' of 4/C phone wire	1 Elec	26	.308	Ea.	9.30	16.85		26.15	35.50

26 05 90 – Residential Applications

26 05 90.10 Residential Wiring	Crew	Daily Output	Labor-Hours	Unit	Material	2015 Bare Costs Labor	Equipment	Total	Total Incl O&P
4920 TV recpt., 20' of RG59U coax wire, F type connector	1 Elec	16	.500	Ea.	18.75	27.50		46.25	61.50
4950 Door bell chime, transformer, 2 buttons, 60' of bellwire									
4970 Economy model	1 Elec	11.50	.696	Ea.	51.50	38		89.50	114
4980 Custom model		11.50	.696		99	38		137	166
4990 Luxury model, 3 buttons		9.50	.842		281	46		327	380
6000 Lighting outlets									
6050 Wire only (for fixture), type NM cable	1 Elec	32	.250	Ea.	6.90	13.70		20.60	28
6070 Type MC cable		24	.333		12.30	18.25		30.55	41
6080 EMT & wire		10	.800		20	44		64	87.50
6100 Box (4"), and wire (for fixture), type NM cable		25	.320		14.45	17.50		31.95	42
6120 Type MC cable		20	.400		19.90	22		41.90	55
6130 EMT & wire		11	.727		27.50	40		67.50	90
6200 Fixtures (use with lines 6050 or 6100 above)									
6210 Canopy style, economy grade	1 Elec	40	.200	Ea.	32	10.95		42.95	51.50
6220 Custom grade		40	.200		53.50	10.95		64.45	75.50
6250 Dining room chandelier, economy grade		19	.421		79.50	23		102.50	122
6260 Custom grade		19	.421		315	23		338	380
6270 Luxury grade		15	.533		715	29		744	830
6310 Kitchen fixture (fluorescent), economy grade		30	.267		71.50	14.60		86.10	101
6320 Custom grade		25	.320		219	17.50		236.50	267
6350 Outdoor, wall mounted, economy grade		30	.267		30	14.60		44.60	55
6360 Custom grade		30	.267		120	14.60		134.60	154
6370 Luxury grade		25	.320		248	17.50		265.50	299
6410 Outdoor PAR floodlights, 1 lamp, 150 watt		20	.400		32.50	22		54.50	68.50
6420 2 lamp, 150 watt each		20	.400		53.50	22		75.50	92
6430 For infrared security sensor, add		32	.250		132	13.70		145.70	166
6450 Outdoor, quartz-halogen, 300 watt flood		20	.400		39	22		61	75.50
6600 Recessed downlight, round, pre-wired, 50 or 75 watt trim		30	.267		80	14.60		94.60	110
6610 With shower light trim		30	.267		89	14.60		103.60	120
6620 With wall washer trim		28	.286		99.50	15.65		115.15	133
6630 With eye-ball trim		28	.286		99.50	15.65		115.15	133
6700 Porcelain lamp holder		40	.200		2.96	10.95		13.91	19.65
6710 With pull switch		40	.200		6.55	10.95		17.50	23.50
6750 Fluorescent strip, 2-20 watt tube, wrap around diffuser, 24"		24	.333		49.50	18.25		67.75	82
6770 2-34 watt tubes, 48"		20	.400		103	22		125	146
6800 Bathroom heat lamp, 1-250 watt		28	.286		44	15.65		59.65	71.50
6810 2-250 watt lamps		28	.286		70	15.65		85.65	101
6820 For timer switch, see Section 26 05 90.10 line 2400									
6900 Outdoor post lamp, incl. post, fixture, 35' of #14/2									
6910 Type NMC cable	1 Elec	3.50	2.286	Ea.	258	125		383	470
6920 Photo-eye, add		27	.296		32	16.20		48.20	60
6950 Clock dial time switch, 24 hr., w/enclosure, type NM cable		11.43	.700		73.50	38.50		112	139
6970 Type MC cable		11	.727		82.50	40		122.50	150
6980 EMT & wire		4.85	1.649		91	90		181	235
7000 Alarm systems									
7050 Smoke detectors, box, #14/3, type NM cable	1 Elec	14.55	.550	Ea.	34.50	30		64.50	83
7070 Type MC cable		12.31	.650		44	35.50		79.50	102
7080 EMT & wire		5	1.600		49.50	87.50		137	186
7090 For relay output to security system, add					12			12	13.20
8000 Residential equipment									
8050 Disposal hook-up, incl. switch, outlet box, 3' of flex									
8060 20 amp-1 pole circ. bkr., and 25' of #12/2									
8070 Type NM cable	1 Elec	10	.800	Ea.	32.50	44		76.50	101

26 05 90 – Residential Applications

26 05 90.10 Residential Wiring	Crew	Daily Output	Labor-Hours	Unit	Material	2015 Bare Costs Labor	Equipment	Total	Total Incl O&P
8080	Type MC cable	1 Elec	8	1	Ea.	39.50	54.50		94	126
8090	EMT & wire	↓	5	1.600	↓	54.50	87.50		142	191
8100	Trash compactor or dishwasher hook-up, incl. outlet box,									
8110	3' of flex, 15 amp-1 pole circ. bkr., and 25' of #14/2									
8120	Type NM cable	1 Elec	10	.800	Ea.	24	44		68	91.50
8130	Type MC cable		8	1		34	54.50		88.50	120
8140	EMT & wire	↓	5	1.600	↓	46	87.50		133.50	182
8150	Hot water sink dispensor hook-up, use line 8100									
8200	Vent/exhaust fan hook-up, type NM cable	1 Elec	32	.250	Ea.	6.90	13.70		20.60	28
8220	Type MC cable		24	.333		12.30	18.25		30.55	41
8230	EMT & wire	↓	10	.800	↓	20	44		64	87.50
8250	Bathroom vent fan, 50 CFM (use with above hook-up)									
8260	Economy model	1 Elec	15	.533	Ea.	23	29		52	69
8270	Low noise model		15	.533		41	29		70	88.50
8280	Custom model	↓	12	.667	↓	127	36.50		163.50	195
8300	Bathroom or kitchen vent fan, 110 CFM									
8310	Economy model	1 Elec	15	.533	Ea.	67.50	29		96.50	118
8320	Low noise model	"	15	.533	"	92	29		121	145
8350	Paddle fan, variable speed (w/o lights)									
8360	Economy model (AC motor)	1 Elec	10	.800	Ea.	109	44		153	186
8362	With light kit		10	.800		150	44		194	231
8370	Custom model (AC motor)		10	.800		227	44		271	315
8372	With light kit		10	.800		268	44		312	360
8380	Luxury model (DC motor)		8	1		330	54.50		384.50	445
8382	With light kit		8	1		375	54.50		429.50	490
8390	Remote speed switch for above, add	↓	12	.667	↓	37.50	36.50		74	95.50
8500	Whole house exhaust fan, ceiling mount, 36", variable speed									
8510	Remote switch, incl. shutters, 20 amp-1 pole circ. bkr.									
8520	30' of #12/2, type NM cable	1 Elec	4	2	Ea.	1,300	109		1,409	1,600
8530	Type MC cable		3.50	2.286		1,300	125		1,425	1,600
8540	EMT & wire	↓	3	2.667	↓	1,325	146		1,471	1,675
8600	Whirlpool tub hook-up, incl. timer switch, outlet box									
8610	3' of flex, 20 amp-1 pole GFI circ. bkr.									
8620	30' of #12/2, type NM cable	1 Elec	5	1.600	Ea.	142	87.50		229.50	287
8630	Type MC cable		4.20	1.905		145	104		249	315
8640	EMT & wire	↓	3.40	2.353	↓	158	129		287	365
8650	Hot water heater hook-up, incl. 1-2 pole circ. bkr., box;									
8660	3' of flex, 20' of #10/2, type NM cable	1 Elec	5	1.600	Ea.	33	87.50		120.50	168
8670	Type MC cable		4.20	1.905		43	104		147	204
8680	EMT & wire	↓	3.40	2.353	↓	48.50	129		177.50	246
9000	Heating/air conditioning									
9050	Furnace/boiler hook-up, incl. firestat, local on-off switch									
9060	Emergency switch, and 40' of type NM cable	1 Elec	4	2	Ea.	52.50	109		161.50	222
9070	Type MC cable		3.50	2.286		65.50	125		190.50	260
9080	EMT & wire	↓	1.50	5.333	↓	83	292		375	525
9100	Air conditioner hook-up, incl. local 60 amp disc. switch									
9110	3' sealtite, 40 amp, 2 pole circuit breaker									
9130	40' of #8/2, type NM cable	1 Elec	3.50	2.286	Ea.	163	125		288	365
9140	Type MC cable		3	2.667		216	146		362	455
9150	EMT & wire	↓	1.30	6.154	↓	204	335		539	730
9200	Heat pump hook-up, 1-40 & 1-100 amp 2 pole circ. bkr.									
9210	Local disconnect switch, 3' sealtite									
9220	40' of #8/2 & 30' of #3/2									

26 05 Common Work Results for Electrical

26 05 90 – Residential Applications

26 05 90.10 Residential Wiring		Crew	Daily Output	Labor-Hours	Unit	Material	2015 Bare Costs Labor	Equipment	Total	Total Incl O&P
9230	Type NM cable	1 Elec	1.30	6.154	Ea.	520	335		855	1,075
9240	Type MC cable		1.08	7.407		535	405		940	1,200
9250	EMT & wire	↓	.94	8.511	↓	585	465		1,050	1,350
9500	Thermostat hook-up, using low voltage wire									
9520	Heating only, 25' of #18-3	1 Elec	24	.333	Ea.	8.80	18.25		27.05	37
9530	Heating/cooling, 25' of #18-4	"	20	.400	"	11.45	22		33.45	45.50

26 24 Switchboards and Panelboards

26 24 19 – Motor-Control Centers

26 24 19.40 Motor Starters and Controls

		Crew	Daily Output	Labor-Hours	Unit	Material	2015 Bare Costs Labor	Equipment	Total	Total Incl O&P
0010	**MOTOR STARTERS AND CONTROLS**									
0050	Magnetic, FVNR, with enclosure and heaters, 480 volt									
0080	2 HP, size 00	1 Elec	3.50	2.286	Ea.	203	125		328	410
0100	5 HP, size 0		2.30	3.478		272	190		462	585
0200	10 HP, size 1	↓	1.60	5		275	274		549	715
0300	25 HP, size 2	2 Elec	2.20	7.273		520	400		920	1,175
0400	50 HP, size 3		1.80	8.889		845	485		1,330	1,650
0500	100 HP, size 4		1.20	13.333		1,875	730		2,605	3,150
0600	200 HP, size 5		.90	17.778		4,375	970		5,345	6,275
0610	400 HP, size 6	↓	.80	20		19,000	1,100		20,100	22,600
0620	NEMA 7, 5 HP, size 0	1 Elec	1.60	5		1,400	274		1,674	1,950
0630	10 HP, size 1	"	1.10	7.273		1,475	400		1,875	2,225
0640	25 HP, size 2	2 Elec	1.80	8.889		2,375	485		2,860	3,350
0650	50 HP, size 3		1.20	13.333		3,575	730		4,305	5,025
0660	100 HP, size 4		.90	17.778		5,775	970		6,745	7,800
0670	200 HP, size 5	↓	.50	32		13,800	1,750		15,550	17,800
0700	Combination, with motor circuit protectors, 5 HP, size 0	1 Elec	1.80	4.444		880	243		1,123	1,325
0800	10 HP, size 1	"	1.30	6.154		915	335		1,250	1,500
0900	25 HP, size 2	2 Elec	2	8		1,275	440		1,715	2,050
1000	50 HP, size 3		1.32	12.121		1,850	665		2,515	3,025
1200	100 HP, size 4	↓	.80	20		4,000	1,100		5,100	6,050
1220	NEMA 7, 5 HP, size 0	1 Elec	1.30	6.154		2,825	335		3,160	3,600
1230	10 HP, size 1	"	1	8		2,900	440		3,340	3,825
1240	25 HP, size 2	2 Elec	1.32	12.121		3,850	665		4,515	5,250
1250	50 HP, size 3		.80	20		6,375	1,100		7,475	8,675
1260	100 HP, size 4		.60	26.667		9,925	1,450		11,375	13,100
1270	200 HP, size 5	↓	.40	40		21,600	2,200		23,800	27,100
1400	Combination, with fused switch, 5 HP, size 0	1 Elec	1.80	4.444		610	243		853	1,025
1600	10 HP, size 1	"	1.30	6.154		650	335		985	1,225
1800	25 HP, size 2	2 Elec	2	8		1,050	440		1,490	1,800
2000	50 HP, size 3		1.32	12.121		1,775	665		2,440	2,975
2200	100 HP, size 4	↓	.80	20	↓	3,125	1,100		4,225	5,075
3500	Magnetic FVNR with NEMA 12, enclosure & heaters, 480 volt									
3600	5 HP, size 0	1 Elec	2.20	3.636	Ea.	231	199		430	550
3700	10 HP, size 1	"	1.50	5.333		350	292		642	820
3800	25 HP, size 2	2 Elec	2	8		650	440		1,090	1,375
3900	50 HP, size 3		1.60	10		1,000	545		1,545	1,925
4000	100 HP, size 4		1	16		2,400	875		3,275	3,925
4100	200 HP, size 5	↓	.80	20		5,750	1,100		6,850	7,975
4200	Combination, with motor circuit protectors, 5 HP, size 0	1 Elec	1.70	4.706		765	257		1,022	1,225
4300	10 HP, size 1	"	1.20	6.667		795	365		1,160	1,425

26 24 Switchboards and Panelboards

26 24 19 – Motor-Control Centers

26 24 19.40 Motor Starters and Controls		Crew	Daily Output	Labor-Hours	Unit	Material	2015 Bare Costs Labor	Equipment	Total	Total Incl O&P
4400	25 HP, size 2	2 Elec	1.80	8.889	Ea.	1,200	485		1,685	2,025
4500	50 HP, size 3		1.20	13.333		1,950	730		2,680	3,225
4600	100 HP, size 4	↓	.74	21.622		4,375	1,175		5,550	6,600
4700	Combination, with fused switch, 5 HP, size 0	1 Elec	1.70	4.706		740	257		997	1,200
4800	10 HP, size 1	"	1.20	6.667		770	365		1,135	1,400
4900	25 HP, size 2	2 Elec	1.80	8.889		1,175	485		1,660	2,025
5000	50 HP, size 3		1.20	13.333		1,875	730		2,605	3,175
5100	100 HP, size 4	↓	.74	21.622	↓	3,825	1,175		5,000	5,975
5200	Factory installed controls, adders to size 0 thru 5									
5300	Start-stop push button	1 Elec	32	.250	Ea.	48.50	13.70		62.20	74
5400	Hand-off-auto-selector switch		32	.250		48.50	13.70		62.20	74
5500	Pilot light		32	.250		91	13.70		104.70	121
5600	Start-stop-pilot		32	.250		139	13.70		152.70	174
5700	Auxiliary contact, NO or NC		32	.250		66.50	13.70		80.20	93.50
5800	NO-NC	↓	32	.250	↓	133	13.70		146.70	167

26 29 Low-Voltage Controllers

26 29 13 – Enclosed Controllers

26 29 13.20 Control Stations

26 29 13.20 Control Stations		Crew	Daily Output	Labor-Hours	Unit	Material	2015 Bare Costs Labor	Equipment	Total	Total Incl O&P
0010	**CONTROL STATIONS**									
0050	NEMA 1, heavy duty, stop/start	1 Elec	8	1	Ea.	135	54.50		189.50	231
0100	Stop/start, pilot light		6.20	1.290		184	70.50		254.50	310
0200	Hand/off/automatic		6.20	1.290		100	70.50		170.50	216
0400	Stop/start/reverse		5.30	1.509		182	82.50		264.50	325
0500	NEMA 7, heavy duty, stop/start		6	1.333		425	73		498	575
0600	Stop/start, pilot light	↓	4	2	↓	515	109		624	735

26 29 23 – Variable-Frequency Motor Controllers

26 29 23.10 Variable Frequency Drives/Adj. Frequency Drives

26 29 23.10 Variable Frequency Drives/Adj. Frequency Drives			Crew	Daily Output	Labor-Hours	Unit	Material	2015 Bare Costs Labor	Equipment	Total	Total Incl O&P
0010	**VARIABLE FREQUENCY DRIVES/ADJ. FREQUENCY DRIVES**										
0100	Enclosed (NEMA 1), 460 volt, for 3 HP motor size	G	1 Elec	.80	10	Ea.	1,700	545		2,245	2,700
0110	5 HP motor size	G		.80	10		1,925	545		2,470	2,925
0120	7.5 HP motor size	G		.67	11.940		2,300	655		2,955	3,500
0130	10 HP motor size	G	↓	.67	11.940		2,675	655		3,330	3,900
0140	15 HP motor size	G	2 Elec	.89	17.978		3,250	985		4,235	5,050
0150	20 HP motor size	G		.89	17.978		3,800	985		4,785	5,675
0160	25 HP motor size	G		.67	23.881		4,700	1,300		6,000	7,100
0170	30 HP motor size	G		.67	23.881		5,900	1,300		7,200	8,425
0180	40 HP motor size	G		.67	23.881		7,025	1,300		8,325	9,700
0190	50 HP motor size	G		.53	30.189		8,675	1,650		10,325	12,000
0200	60 HP motor size	G	R-3	.56	35.714		10,500	1,950	246	12,696	14,700
0210	75 HP motor size	G		.56	35.714		12,100	1,950	246	14,296	16,500
0220	100 HP motor size	G		.50	40		14,100	2,175	276	16,551	19,100
0230	125 HP motor size	G		.50	40		15,700	2,175	276	18,151	20,900
0240	150 HP motor size	G		.50	40		17,800	2,175	276	20,251	23,200
0250	200 HP motor size	G	↓	.42	47.619		23,400	2,575	330	26,305	30,000
1100	Custom-engineered, 460 volt, for 3 HP motor size	G	1 Elec	.56	14.286		2,725	780		3,505	4,175
1110	5 HP motor size	G		.56	14.286		2,725	780		3,505	4,175
1120	7.5 HP motor size	G		.47	17.021		2,875	930		3,805	4,550
1130	10 HP motor size	G	↓	.47	17.021		3,000	930		3,930	4,700
1140	15 HP motor size	G	2 Elec	.62	25.806	↓	3,750	1,400		5,150	6,250

For customer support on your Mechanical Cost Data, call 877.795.7394.

433

26 29 23 – Variable-Frequency Motor Controllers

26 29 23.10 Variable Frequency Drives/Adj. Frequency Drives		Crew	Daily Output	Labor-Hours	Unit	Material	2015 Bare Costs Labor	Equipment	Total	Total Incl O&P
1150	20 HP motor size	G	2 Elec	.62	25.806	Ea.	4,225	1,400		5,625	6,775
1160	25 HP motor size	G		.47	34.043		4,950	1,850		6,800	8,225
1170	30 HP motor size	G		.47	34.043		6,175	1,850		8,025	9,600
1180	40 HP motor size	G		.47	34.043		7,300	1,850		9,150	10,800
1190	50 HP motor size	G		.37	43.243		8,375	2,375		10,750	12,800
1200	60 HP motor size	G	R-3	.39	51.282		12,600	2,775	355	15,730	18,500
1210	75 HP motor size	G		.39	51.282		13,400	2,775	355	16,530	19,400
1220	100 HP motor size	G		.35	57.143		14,700	3,100	395	18,195	21,300
1230	125 HP motor size	G		.35	57.143		15,700	3,100	395	19,195	22,400
1240	150 HP motor size	G		.35	57.143		18,000	3,100	395	21,495	24,900
1250	200 HP motor size	G		.29	68.966		23,500	3,750	475	27,725	32,000
2000	For complex & special design systems to meet specific										
2010	requirements, obtain quote from vendor.										

Estimating Tips

- When estimating material costs for electronic safety and security systems, it is always prudent to obtain manufacturers' quotations for equipment prices and special installation requirements that affect the total cost.

- Fire alarm systems consist of control panels, annunciator panels, battery with rack, charger, and fire alarm actuating and indicating devices. Some fire alarm systems include speakers, telephone lines, door closer controls, and other components. Be careful not to overlook the costs related to installation for these items. Also be aware of costs for integrated automation instrumentation and terminal devices, control equipment, control wiring, and programming.

- Security equipment includes items such as CCTV, access control, and other detection and identification systems to perform alert and alarm functions. Be sure to consider the costs related to installation for this security equipment, such as for integrated automation instrumentation and terminal devices, control equipment, control wiring, and programming.

Reference Numbers

Reference numbers are shown in shaded boxes at the beginning of some major classifications. These numbers refer to related items in the Reference Section. The reference information may be an estimating procedure, an alternate pricing method, or technical information.

Note: Not all subdivisions listed here necessarily appear in this publication. ■

28 13 Access Control

28 13 53 – Security Access Detection

28 13 53.13 Security Access Metal Detectors	Crew	Daily Output	Labor-Hours	Unit	Material	2015 Bare Costs Labor	Equipment	Total	Total Incl O&P
0010 **SECURITY ACCESS METAL DETECTORS**									
0240 Metal detector, hand-held, wand type, unit only				Ea.	89			89	98
0250 Metal detector, walk through portal type, single zone	1 Elec	2	4		3,700	219		3,919	4,400
0260 Multi-zone	"	2	4	↓	4,700	219		4,919	5,500

28 13 53.16 Security Access X-Ray Equipment

0010 **SECURITY ACCESS X-RAY EQUIPMENT**									
0290 X-ray machine, desk top, for mail/small packages/letters	1 Elec	4	2	Ea.	3,425	109		3,534	3,950
0300 Conveyor type, incl monitor		2	4		16,000	219		16,219	17,900
0310 Includes additional features		2	4	↓	28,600	219		28,819	31,700
0320 X-ray machine, large unit, for airports, incl monitor	2 Elec	1	16		40,000	875		40,875	45,300
0330 Full console	"	.50	32	↓	68,500	1,750		70,250	78,000

28 13 53.23 Security Access Explosive Detection Equipment

0010 **SECURITY ACCESS EXPLOSIVE DETECTION EQUIPMENT**									
0270 Explosives detector, walk through portal type	1 Elec	2	4	Ea.	44,200	219		44,419	48,900
0280 Hand-held, battery operated				"				25,500	28,100

28 16 Intrusion Detection

28 16 16 – Intrusion Detection Systems Infrastructure

28 16 16.50 Intrusion Detection

	Crew	Daily Output	Labor-Hours	Unit	Material	2015 Bare Costs Labor	Equipment	Total	Total Incl O&P
0010 **INTRUSION DETECTION**, not including wires & conduits									
0100 Burglar alarm, battery operated, mechanical trigger	1 Elec	4	2	Ea.	278	109		387	470
0200 Electrical trigger		4	2		330	109		439	530
0400 For outside key control, add		8	1		84	54.50		138.50	175
0600 For remote signaling circuitry, add		8	1		125	54.50		179.50	219
0800 Card reader, flush type, standard		2.70	2.963		930	162		1,092	1,275
1000 Multi-code	↓	2.70	2.963	↓	1,200	162		1,362	1,575

28 23 Video Surveillance

28 23 23 – Video Surveillance Systems Infrastructure

28 23 23.50 Video Surveillance Equipments

	Crew	Daily Output	Labor-Hours	Unit	Material	2015 Bare Costs Labor	Equipment	Total	Total Incl O&P
0010 **VIDEO SURVEILLANCE EQUIPMENTS**									
0200 Video cameras, wireless, hidden in exit signs, clocks, etc., incl. receiver	1 Elec	3	2.667	Ea.	108	146		254	340
0210 Accessories for video recorder, single camera		3	2.667		183	146		329	420
0220 For multiple cameras		3	2.667		1,750	146		1,896	2,150
0230 Video cameras, wireless, for under vehicle searching, complete	↓	2	4	↓	10,200	219		10,419	11,500

28 31 Fire Detection and Alarm

28 31 23 – Fire Detection and Alarm Annunciation Panels and Fire Stations

28 31 23.50 Alarm Panels and Devices

	Crew	Daily Output	Labor-Hours	Unit	Material	2015 Bare Costs Labor	Equipment	Total	Total Incl O&P
0010 **ALARM PANELS AND DEVICES**, not including wires & conduits									
3594 Fire, alarm control panel									
3600 4 zone	2 Elec	2	8	Ea.	400	440		840	1,100
3800 8 zone		1	16		780	875		1,655	2,150
4000 12 zone	↓	.67	23.988	↓	2,400	1,300		3,700	4,600

28 31 Fire Detection and Alarm

28 31 23 – Fire Detection and Alarm Annunciation Panels and Fire Stations

28 31 23.50 Alarm Panels and Devices	Crew	Daily Output	Labor-Hours	Unit	Material	2015 Bare Costs Labor	Equipment	Total	Total Incl O&P
4020	Alarm device	1 Elec	8	1	Ea.	238	54.50		292.50	345
4050	Actuating device		8	1		335	54.50		389.50	450
4200	Battery and rack		4	2		410	109		519	620
4400	Automatic charger		8	1		585	54.50		639.50	720
4600	Signal bell		8	1		78.50	54.50		133	169
4800	Trouble buzzer or manual station		8	1		83.50	54.50		138	174
5600	Strobe and horn		5.30	1.509		152	82.50		234.50	291
5800	Fire alarm horn		6.70	1.194		61	65.50		126.50	165
6000	Door holder, electro-magnetic		4	2		103	109		212	277
6200	Combination holder and closer		3.20	2.500		123	137		260	340
6600	Drill switch		8	1		370	54.50		424.50	490
6800	Master box		2.70	2.963		6,400	162		6,562	7,300
7000	Break glass station		8	1		55.50	54.50		110	143
7800	Remote annunciator, 8 zone lamp		1.80	4.444		209	243		452	595
8000	12 zone lamp	2 Elec	2.60	6.154		335	335		670	875
8200	16 zone lamp	"	2.20	7.273		420	400		820	1,050

28 31 43 – Fire Detection Sensors

28 31 43.50 Fire and Heat Detectors

		Crew	Daily Output	Labor-Hours	Unit	Material	Labor	Equipment	Total	Total Incl O&P
0010	**FIRE & HEAT DETECTORS**									
5000	Detector, rate of rise	1 Elec	8	1	Ea.	51	54.50		105.50	138
5100	Fixed temperature	"	8	1	"	51	54.50		105.50	138

28 31 46 – Smoke Detection Sensors

28 31 46.50 Smoke Detectors

		Crew	Daily Output	Labor-Hours	Unit	Material	Labor	Equipment	Total	Total Incl O&P
0010	**SMOKE DETECTORS**									
5200	Smoke detector, ceiling type	1 Elec	6.20	1.290	Ea.	110	70.50		180.50	227
5400	Duct type	"	3.20	2.500	"	325	137		462	565

28 33 Gas Detection and Alarm

28 33 33 – Gas Detection Sensors

28 33 33.50 Tank Leak Detection Systems

		Crew	Daily Output	Labor-Hours	Unit	Material	Labor	Equipment	Total	Total Incl O&P
0010	**TANK LEAK DETECTION SYSTEMS** Liquid and vapor									
0100	For hydrocarbons and hazardous liquids/vapors									
0120	Controller, data acquisition, incl. printer, modem, RS232 port									
0140	24 channel, for use with all probes				Ea.	4,175			4,175	4,575
0160	9 channel, for external monitoring				"	915			915	1,000
0200	Probes									
0210	Well monitoring									
0220	Liquid phase detection				Ea.	475			475	525
0230	Hydrocarbon vapor, fixed position					480			480	530
0240	Hydrocarbon vapor, float mounted					480			480	530
0250	Both liquid and vapor hydrocarbon					480			480	530
0300	Secondary containment, liquid phase									
0310	Pipe trench/manway sump				Ea.	820			820	905
0320	Double wall pipe and manual sump					825			825	905
0330	Double wall fiberglass annular space					325			325	355
0340	Double wall steel tank annular space					335			335	370

For customer support on your Mechanical Cost Data, call 877.795.7394.

437

28 33 33.50 Tank Leak Detection Systems	Crew	Daily Output	Labor-Hours	Unit	Material	2015 Bare Costs Labor	Equipment	Total	Total Incl O&P
0500	Accessories									
0510	Modem, non-dedicated phone line				Ea.	305			305	335
0600	Monitoring, internal									
0610	Automatic tank gauge, incl. overfill				Ea.	1,150			1,150	1,275
0620	Product line				"	1,150			1,150	1,275
0700	Monitoring, special									
0710	Cathodic protection				Ea.	725			725	795
0720	Annular space chemical monitor				"	985			985	1,075

438

For customer support on your Mechanical Cost Data, call 877.795.7394.

Estimating Tips

31 05 00 Common Work Results for Earthwork

■ Estimating the actual cost of performing earthwork requires careful consideration of the variables involved. This includes items such as type of soil, whether water will be encountered, dewatering, whether banks need bracing, disposal of excavated earth, and length of haul to fill or spoil sites, etc. If the project has large quantities of cut or fill, consider raising or lowering the site to reduce costs, while paying close attention to the effect on site drainage and utilities.

■ If the project has large quantities of fill, creating a borrow pit on the site can significantly lower the costs.

■ It is very important to consider what time of year the project is scheduled for completion. Bad weather can create large cost overruns from dewatering, site repair, and lost productivity from cold weather.

Reference Numbers

Reference numbers are shown in shaded boxes at the beginning of some major classifications. These numbers refer to related items in the Reference Section. The reference information may be an estimating procedure, an alternate pricing method, or technical information.

Note: Not all subdivisions listed here necessarily appear in this publication. ■

31 05 Common Work Results for Earthwork

31 05 23 – Cement and Concrete for Earthwork

31 05 23.30 Plant Mixed Bituminous Concrete		Crew	Daily Output	Labor-Hours	Unit	Material	2015 Bare Costs Labor	Equipment	Total	Total Incl O&P
0010	**PLANT MIXED BITUMINOUS CONCRETE**									
0020	Asphaltic concrete plant mix (145 lb. per C.F.)				Ton	70			70	77
0040	Asphaltic concrete less than 300 tons add trucking costs									
0050	See Section 31 23 23.20 for hauling costs									
0200	All weather patching mix, hot				Ton	69			69	76
0250	Cold patch					79.50			79.50	87.50
0300	Berm mix				↓	69			69	76

31 23 Excavation and Fill

31 23 16 – Excavation

31 23 16.13 Excavating, Trench

31 23 16.13 Excavating, Trench		Crew	Daily Output	Labor-Hours	Unit	Material	2015 Bare Costs Labor	Equipment	Total	Total Incl O&P
0010	**EXCAVATING, TRENCH** G1030–805									
0011	Or continuous footing									
0020	Common earth with no sheeting or dewatering included									
0050	1' to 4' deep, 3/8 C.Y. excavator	B-11C	150	.107	B.C.Y.		4.70	2.43	7.13	9.85
0060	1/2 C.Y. excavator	B-11M	200	.080			3.53	1.96	5.49	7.55
0090	4' to 6' deep, 1/2 C.Y. excavator	"	200	.080			3.53	1.96	5.49	7.55
0100	5/8 C.Y. excavator	B-12Q	250	.064			2.86	2.36	5.22	6.95
0110	3/4 C.Y. excavator	B-12F	300	.053			2.38	2.18	4.56	6.05
0300	1/2 C.Y. excavator, truck mounted	B-12J	200	.080			3.57	4.41	7.98	10.30
0500	6' to 10' deep, 3/4 C.Y. excavator	B-12F	225	.071			3.18	2.91	6.09	8.05
0600	1 C.Y. excavator, truck mounted	B-12K	400	.040			1.79	2.51	4.30	5.50
0900	10' to 14' deep, 3/4 C.Y. excavator	B-12F	200	.080			3.57	3.27	6.84	9.05
1000	1-1/2 C.Y. excavator	B-12B	540	.030			1.32	1.91	3.23	4.12
1300	14' to 20' deep, 1 C.Y. excavator	B-12A	320	.050			2.23	2.54	4.77	6.20
1340	20' to 24' deep, 1 C.Y. excavator	"	288	.056			2.48	2.82	5.30	6.90
1352	4' to 6' deep, 1/2 C.Y. excavator w/trench box	B-13H	188	.085			3.80	5.10	8.90	11.45
1354	5/8 C.Y. excavator	"	235	.068			3.04	4.10	7.14	9.15
1356	3/4 C.Y. excavator	B-13G	282	.057			2.53	2.61	5.14	6.75
1362	6' to 10' deep, 3/4 C.Y. excavator w/trench box		212	.075			3.37	3.47	6.84	8.95
1374	10' to 14' deep, 3/4 C.Y. excavator w/trench box	↓	188	.085			3.80	3.91	7.71	10.10
1376	1-1/2 C.Y. excavator	B-13E	508	.032			1.41	2.19	3.60	4.56
1381	14' to 20' deep, 1 C.Y. excavator w/trench box	B-13D	301	.053			2.37	2.97	5.34	6.90
1386	20' to 24' deep, 1 C.Y. excavator w/trench box	"	271	.059			2.64	3.30	5.94	7.65
1400	By hand with pick and shovel 2' to 6' deep, light soil	1 Clab	8	1			37.50		37.50	58
1500	Heavy soil	"	4	2	↓		75		75	116
1700	For tamping backfilled trenches, air tamp, add	A-1G	100	.080	E.C.Y.		3.01	.56	3.57	5.25
1900	Vibrating plate, add	B-18	180	.133	"		5.10	.26	5.36	8.15
2100	Trim sides and bottom for concrete pours, common earth		1500	.016	S.F.		.61	.03	.64	.97
2300	Hardpan	↓	600	.040	"		1.53	.08	1.61	2.45
5020	Loam & Sandy clay with no sheeting or dewatering included									
5050	1' to 4' deep, 3/8 C.Y. tractor loader/backhoe	B-11C	162	.099	B.C.Y.		4.36	2.25	6.61	9.10
5060	1/2 C.Y. excavator	B-11M	216	.074			3.27	1.81	5.08	7
5080	4' to 6' deep, 1/2 C.Y. excavator	"	216	.074			3.27	1.81	5.08	7
5090	5/8 C.Y. excavator	B-12Q	276	.058			2.59	2.14	4.73	6.30
5100	3/4 C.Y. excavator	B-12F	324	.049			2.20	2.02	4.22	5.60
5130	1/2 C.Y. excavator, truck mounted	B-12J	216	.074			3.31	4.08	7.39	9.55
5140	6' to 10' deep, 3/4 C.Y. excavator	B-12F	243	.066			2.94	2.69	5.63	7.45
5160	1 C.Y. excavator, truck mounted	B-12K	432	.037			1.65	2.32	3.97	5.05
5190	10' to 14' deep, 3/4 C.Y. excavator	B-12F	216	.074			3.31	3.03	6.34	8.40
5210	1-1/2 C.Y. excavator	B-12B	583	.027	↓		1.23	1.77	3	3.81

31 23 16 – Excavation

31 23 16.13 Excavating, Trench

	Crew	Daily Output	Labor-Hours	Unit	Material	2015 Bare Costs Labor	Equipment	Total	Total Incl O&P
5250	14' to 20' deep, 1 C.Y. excavator	B-12A	346	.046	B.C.Y.		2.06	2.35	4.41	5.75
5300	20' to 24' deep, 1 C.Y. excavator	"	311	.051			2.30	2.61	4.91	6.40
5352	4' to 6' deep, 1/2 C.Y. excavator w/trench box	B-13H	205	.078			3.48	4.70	8.18	10.45
5354	5/8 C.Y. excavator	"	257	.062			2.78	3.75	6.53	8.35
5356	3/4 C.Y. excavator	B-13G	308	.052			2.32	2.39	4.71	6.15
5362	6' to 10' deep, 3/4 C.Y. excavator w/trench box		231	.069			3.09	3.19	6.28	8.20
5370	10' to 14' deep, 3/4 C.Y. excavator w/trench box		205	.078			3.48	3.59	7.07	9.25
5374	1-1/2 C.Y. excavator	B-13E	554	.029			1.29	2.01	3.30	4.18
5382	14' to 20' deep, 1 C.Y. excavator w/trench box	B-13D	329	.049			2.17	2.72	4.89	6.30
5392	20' to 24' deep, 1 C.Y. excavator w/trench box	"	295	.054			2.42	3.03	5.45	7.05
6020	Sand & gravel with no sheeting or dewatering included									
6050	1' to 4' deep, 3/8 C.Y. excavator	B-11C	165	.097	B.C.Y.		4.28	2.21	6.49	9
6060	1/2 C.Y. excavator	B-11M	220	.073			3.21	1.78	4.99	6.85
6080	4' to 6' deep, 1/2 C.Y. excavator	"	220	.073			3.21	1.78	4.99	6.85
6090	5/8 C.Y. excavator	B-12Q	275	.058			2.60	2.14	4.74	6.35
6100	3/4 C.Y. excavator	B-12F	330	.048			2.16	1.98	4.14	5.50
6130	1/2 C.Y. excavator, truck mounted	B-12J	220	.073			3.25	4.01	7.26	9.35
6140	6' to 10' deep, 3/4 C.Y. excavator	B-12F	248	.065			2.88	2.64	5.52	7.30
6160	1 C.Y. excavator, truck mounted	B-12K	440	.036			1.62	2.28	3.90	4.98
6190	10' to 14' deep, 3/4 C.Y. excavator	B-12F	220	.073			3.25	2.98	6.23	8.25
6210	1-1/2 C.Y. excavator	B-12B	594	.027			1.20	1.73	2.93	3.75
6250	14' to 20' deep, 1 C.Y. excavator	B-12A	352	.045			2.03	2.31	4.34	5.65
6300	20' to 24' deep, 1 C.Y. excavator	"	317	.050			2.25	2.56	4.81	6.25
6352	4' to 6' deep, 1/2 C.Y. excavator w/trench box	B-13H	209	.077			3.42	4.61	8.03	10.25
6354	5/8 C.Y. excavator	"	261	.061			2.74	3.69	6.43	8.25
6356	3/4 C.Y. excavator	B-13G	314	.051			2.28	2.34	4.62	6.05
6362	6' to 10' deep, 3/4 C.Y. excavator w/trench box		236	.068			3.03	3.12	6.15	8.05
6370	10' to 14' deep, 3/4 C.Y. excavator w/trench box		209	.077			3.42	3.52	6.94	9.05
6374	1-1/2 C.Y. excavator	B-13E	564	.028			1.27	1.97	3.24	4.10
6382	14' to 20' deep, 1 C.Y. excavator w/trench box	B-13D	334	.048			2.14	2.68	4.82	6.20
6392	20' to 24' deep, 1 C.Y. excavator w/trench box	"	301	.053			2.37	2.97	5.34	6.90
7020	Dense hard clay with no sheeting or dewatering included									
7050	1' to 4' deep, 3/8 C.Y. excavator	B-11C	132	.121	B.C.Y.		5.35	2.76	8.11	11.20
7060	1/2 C.Y. excavator	B-11M	176	.091			4.01	2.23	6.24	8.55
7080	4' to 6' deep, 1/2 C.Y. excavator	"	176	.091			4.01	2.23	6.24	8.55
7090	5/8 C.Y. excavator	B-12Q	220	.073			3.25	2.68	5.93	7.90
7100	3/4 C.Y. excavator	B-12F	264	.061			2.71	2.48	5.19	6.85
7130	1/2 C.Y. excavator, truck mounted	B-12J	176	.091			4.06	5	9.06	11.70
7140	6' to 10' deep, 3/4 C.Y. excavator	B-12F	198	.081			3.61	3.31	6.92	9.15
7160	1 C.Y. excavator, truck mounted	B-12K	352	.045			2.03	2.85	4.88	6.25
7190	10' to 14' deep, 3/4 C.Y. excavator	B-12F	176	.091			4.06	3.72	7.78	10.30
7210	1-1/2 C.Y. excavator	B-12B	475	.034			1.50	2.17	3.67	4.68
7250	14' to 20' deep, 1 C.Y. excavator	B-12A	282	.057			2.53	2.88	5.41	7.05
7300	20' to 24' deep, 1 C.Y. excavator	"	254	.063			2.81	3.20	6.01	7.80

31 23 16.14 Excavating, Utility Trench

		Crew	Daily Output	Labor-Hours	Unit	Material	2015 Bare Costs Labor	Equipment	Total	Total Incl O&P
0010	**EXCAVATING, UTILITY TRENCH**									
0011	Common earth									
0050	Trenching with chain trencher, 12 H.P., operator walking									
0100	4" wide trench, 12" deep	B-53	800	.010	L.F.		.49	.09	.58	.83
0150	18" deep		750	.011			.52	.09	.61	.89
0200	24" deep		700	.011			.56	.10	.66	.95
0300	6" wide trench, 12" deep		650	.012			.60	.10	.70	1.03

For customer support on your Mechanical Cost Data, call 877.795.7394.

441

31 23 16 – Excavation

31 23 16.14 Excavating, Utility Trench

		Crew	Daily Output	Labor-Hours	Unit	Material	2015 Bare Costs Labor	Equipment	Total	Total Incl O&P
0350	18" deep	B-53	600	.013	L.F.		.65	.11	.76	1.10
0400	24" deep		550	.015			.71	.12	.83	1.21
0450	36" deep		450	.018			.86	.15	1.01	1.48
0600	8" wide trench, 12" deep		475	.017			.82	.14	.96	1.40
0650	18" deep		400	.020			.97	.17	1.14	1.67
0700	24" deep		350	.023			1.11	.19	1.30	1.90
0750	36" deep	↓	300	.027	↓		1.30	.23	1.53	2.22
1000	Backfill by hand including compaction, add									
1050	4" wide trench, 12" deep	A-1G	800	.010	L.F.		.38	.07	.45	.66
1100	18" deep		530	.015			.57	.11	.68	.99
1150	24" deep		400	.020			.75	.14	.89	1.32
1300	6" wide trench, 12" deep		540	.015			.56	.10	.66	.97
1350	18" deep		405	.020			.74	.14	.88	1.29
1400	24" deep		270	.030			1.11	.21	1.32	1.94
1450	36" deep		180	.044			1.67	.31	1.98	2.91
1600	8" wide trench, 12" deep		400	.020			.75	.14	.89	1.32
1650	18" deep		265	.030			1.14	.21	1.35	1.98
1700	24" deep		200	.040			1.50	.28	1.78	2.62
1750	36" deep	↓	135	.059	↓		2.23	.42	2.65	3.89
2000	Chain trencher, 40 H.P. operator riding									
2050	6" wide trench and backfill, 12" deep	B-54	1200	.007	L.F.		.32	.28	.60	.80
2100	18" deep		1000	.008			.39	.34	.73	.96
2150	24" deep		975	.008			.40	.34	.74	.99
2200	36" deep		900	.009			.43	.37	.80	1.07
2250	48" deep		750	.011			.52	.45	.97	1.28
2300	60" deep		650	.012			.60	.52	1.12	1.48
2400	8" wide trench and backfill, 12" deep		1000	.008			.39	.34	.73	.96
2450	18" deep		950	.008			.41	.35	.76	1.01
2500	24" deep		900	.009			.43	.37	.80	1.07
2550	36" deep		800	.010			.49	.42	.91	1.20
2600	48" deep		650	.012			.60	.52	1.12	1.48
2700	12" wide trench and backfill, 12" deep		975	.008			.40	.34	.74	.99
2750	18" deep		860	.009			.45	.39	.84	1.12
2800	24" deep		800	.010			.49	.42	.91	1.20
2850	36" deep		725	.011			.54	.46	1	1.32
3000	16" wide trench and backfill, 12" deep		835	.010			.47	.40	.87	1.15
3050	18" deep		750	.011			.52	.45	.97	1.28
3100	24" deep	↓	700	.011	↓		.56	.48	1.04	1.37
3200	Compaction with vibratory plate, add								35%	35%
5100	Hand excavate and trim for pipe bells after trench excavation									
5200	8" pipe	1 Clab	155	.052	L.F.		1.94		1.94	2.99
5300	18" pipe	"	130	.062	"		2.31		2.31	3.56

31 23 16.16 Structural Excavation for Minor Structures

		Crew	Daily Output	Labor-Hours	Unit	Material	2015 Bare Costs Labor	Equipment	Total	Total Incl O&P
0010	**STRUCTURAL EXCAVATION FOR MINOR STRUCTURES**									
0015	Hand, pits to 6' deep, sandy soil	1 Clab	8	1	B.C.Y.		37.50		37.50	58
0100	Heavy soil or clay		4	2			75		75	116
0300	Pits 6' to 12' deep, sandy soil		5	1.600			60		60	92.50
0500	Heavy soil or clay		3	2.667			100		100	154
0700	Pits 12' to 18' deep, sandy soil		4	2			75		75	116
0900	Heavy soil or clay	↓	2	4	↓		150		150	231
1500	For wet or muck hand excavation, add to above								50%	50%

31 23 Excavation and Fill

31 23 16 – Excavation

31 23 16.30 Drilling and Blasting Rock

		Crew	Daily Output	Labor-Hours	Unit	Material	2015 Bare Costs Labor	Equipment	Total	Total Incl O&P
0010	**DRILLING AND BLASTING ROCK**									
0020	Rock, open face, under 1500 C.Y.	B-47	225	.107	B.C.Y.	3.20	4.47	7.15	14.82	18.20
0100	Over 1500 C.Y.		300	.080		3.20	3.35	5.35	11.90	14.55
0200	Areas where blasting mats are required, under 1500 C.Y.		175	.137		3.20	5.75	9.20	18.15	22.50
0250	Over 1500 C.Y.		250	.096		3.20	4.03	6.45	13.68	16.75
2200	Trenches, up to 1500 C.Y.		22	1.091		9.30	45.50	73	127.80	161
2300	Over 1500 C.Y.		26	.923		9.30	38.50	62	109.80	137

31 23 19 – Dewatering

31 23 19.20 Dewatering Systems

		Crew	Daily Output	Labor-Hours	Unit	Material	2015 Bare Costs Labor	Equipment	Total	Total Incl O&P
0010	**DEWATERING SYSTEMS**									
0020	Excavate drainage trench, 2' wide, 2' deep	B-11C	90	.178	C.Y.		7.85	4.05	11.90	16.40
0100	2' wide, 3' deep, with backhoe loader	"	135	.119			5.25	2.70	7.95	10.95
0200	Excavate sump pits by hand, light soil	1 Clab	7.10	1.127			42.50		42.50	65
0300	Heavy soil	"	3.50	2.286			86		86	132
0500	Pumping 8 hr., attended 2 hrs. per day, including 20 L.F.									
0550	of suction hose & 100 L.F. discharge hose									
0600	2" diaphragm pump used for 8 hours	B-10H	4	3	Day		139	18.70	157.70	232
0650	4" diaphragm pump used for 8 hours	B-10I	4	3			139	30.50	169.50	245
0800	8 hrs. attended, 2" diaphragm pump	B-10H	1	12			555	75	630	925
0900	3" centrifugal pump	B-10J	1	12			555	84	639	940
1000	4" diaphragm pump	B-10I	1	12			555	122	677	980
1100	6" centrifugal pump	B-10K	1	12			555	370	925	1,250
1300	CMP, incl. excavation 3' deep, 12" diameter	B-6	115	.209	L.F.	11.05	8.60	3.17	22.82	29
1400	18" diameter		100	.240	"	16.70	9.90	3.64	30.24	37.50
1600	Sump hole construction, incl. excavation and gravel, pit		1250	.019	C.F.	1.09	.79	.29	2.17	2.73
1700	With 12" gravel collar, 12" pipe, corrugated, 16 ga.		70	.343	L.F.	21.50	14.15	5.20	40.85	50.50
1800	15" pipe, corrugated, 16 ga.		55	.436		28	18	6.60	52.60	65.50
1900	18" pipe, corrugated, 16 ga.		50	.480		32	19.80	7.30	59.10	74
2000	24" pipe, corrugated, 14 ga.		40	.600		38.50	25	9.10	72.60	90.50
2200	Wood lining, up to 4' x 4', add		300	.080	SFCA	16.25	3.30	1.21	20.76	24.50
9950	See Section 31 23 19.40 for wellpoints									
9960	See Section 31 23 19.30 for deep well systems									

31 23 19.30 Wells

		Crew	Daily Output	Labor-Hours	Unit	Material	2015 Bare Costs Labor	Equipment	Total	Total Incl O&P
0010	**WELLS**									
0011	For dewatering 10' to 20' deep, 2' diameter									
0020	with steel casing, minimum	B-6	165	.145	V.L.F.	38	6	2.21	46.21	53
0050	Average		98	.245		43	10.10	3.72	56.82	66.50
0100	Maximum		49	.490		47.50	20	7.45	74.95	91.50
0300	For dewatering pumps see 01 54 33 in Reference Section									
0500	For domestic water wells, see Section 33 21 13.10									

31 23 19.40 Wellpoints

		Crew	Daily Output	Labor-Hours	Unit	Material	2015 Bare Costs Labor	Equipment	Total	Total Incl O&P
0010	**WELLPOINTS** R312319-90									
0011	For equipment rental, see 01 54 33 in Reference Section									
0100	Installation and removal of single stage system									
0110	Labor only, .75 labor-hours per L.F.	1 Clab	10.70	.748	LF Hdr		28		28	43.50
0200	2.0 labor-hours per L.F.	"	4	2	"		75		75	116
0400	Pump operation, 4 @ 6 hr. shifts									
0410	Per 24 hour day	4 Eqlt	1.27	25.197	Day		1,225		1,225	1,850
0500	Per 168 hour week, 160 hr. straight, 8 hr. double time		.18	177	Week		8,650		8,650	13,100
0550	Per 4.3 week month		.04	800	Month		38,900		38,900	59,000
0600	Complete installation, operation, equipment rental, fuel &									

31 23 Excavation and Fill

31 23 19 – Dewatering

31 23 19.40 Wellpoints

		Crew	Daily Output	Labor-Hours	Unit	Material	2015 Bare Costs Labor	2015 Bare Costs Equipment	Total	Total Incl O&P
0610	removal of system with 2" wellpoints 5' O.C.									
0700	100' long header, 6" diameter, first month	4 Eqlt	3.23	9.907	LF Hdr	159	480		639	905
0800	Thereafter, per month		4.13	7.748		127	375		502	710
1000	200' long header, 8" diameter, first month		6	5.333		145	259		404	555
1100	Thereafter, per month		8.39	3.814		71.50	185		256.50	360
1300	500' long header, 8" diameter, first month		10.63	3.010		55.50	146		201.50	283
1400	Thereafter, per month		20.91	1.530		39.50	74.50		114	157
1600	1,000' long header, 10" diameter, first month		11.62	2.754		47.50	134		181.50	256
1700	Thereafter, per month		41.81	.765		24	37		61	82.50
1900	Note: above figures include pumping 168 hrs. per week									
1910	and include the pump operator and one stand-by pump.									

31 23 23 – Fill

31 23 23.13 Backfill

		Crew	Daily Output	Labor-Hours	Unit	Material	2015 Bare Costs Labor	2015 Bare Costs Equipment	Total	Total Incl O&P
0010	**BACKFILL** G1030–805									
0015	By hand, no compaction, light soil	1 Clab	14	.571	L.C.Y.		21.50		21.50	33
0100	Heavy soil		11	.727	"		27.50		27.50	42
0300	Compaction in 6" layers, hand tamp, add to above		20.60	.388	E.C.Y.		14.60		14.60	22.50
0400	Roller compaction operator walking, add	B-10A	100	.120			5.55	1.80	7.35	10.45
0500	Air tamp, add	B-9D	190	.211			8	1.40	9.40	13.85
0600	Vibrating plate, add	A-1D	60	.133			5	.60	5.60	8.35
0800	Compaction in 12" layers, hand tamp, add to above	1 Clab	34	.235			8.85		8.85	13.60
0900	Roller compaction operator walking, add	B-10A	150	.080			3.70	1.20	4.90	6.95
1000	Air tamp, add	B-9	285	.140			5.35	.82	6.17	9.10
1100	Vibrating plate, add	A-1E	90	.089			3.34	.52	3.86	5.70
1300	Dozer backfilling, bulk, up to 300' haul, no compaction	B-10B	1200	.010	L.C.Y.		.46	1.16	1.62	1.97
1400	Air tamped, add	B-11B	80	.200	E.C.Y.		8.60	3.77	12.37	17.30
1600	Compacting backfill, 6" to 12" lifts, vibrating roller	B-10C	800	.015			.69	2.23	2.92	3.51
1700	Sheepsfoot roller	B-10D	750	.016			.74	2.42	3.16	3.79
1900	Dozer backfilling, trench, up to 300' haul, no compaction	B-10B	900	.013	L.C.Y.		.62	1.54	2.16	2.63
2000	Air tamped, add	B-11B	80	.200	E.C.Y.		8.60	3.77	12.37	17.30
2200	Compacting backfill, 6" to 12" lifts, vibrating roller	B-10C	700	.017			.79	2.55	3.34	4.01
2300	Sheepsfoot roller	B-10D	650	.018			.85	2.79	3.64	4.37

31 23 23.16 Fill By Borrow and Utility Bedding

		Crew	Daily Output	Labor-Hours	Unit	Material	2015 Bare Costs Labor	2015 Bare Costs Equipment	Total	Total Incl O&P
0010	**FILL BY BORROW AND UTILITY BEDDING**									
0049	Utility bedding, for pipe & conduit, not incl. compaction									
0050	Crushed or screened bank run gravel	B-6	150	.160	L.C.Y.	25.50	6.60	2.43	34.53	41
0100	Crushed stone 3/4" to 1/2"		150	.160		23.50	6.60	2.43	32.53	39
0200	Sand, dead or bank		150	.160		17.85	6.60	2.43	26.88	32.50
0500	Compacting bedding in trench	A-1D	90	.089	E.C.Y.		3.34	.40	3.74	5.60
0600	If material source exceeds 2 miles, add for extra mileage.									
0610	See Section 31 23 23.20 for hauling mileage add.									

31 23 23.17 General Fill

		Crew	Daily Output	Labor-Hours	Unit	Material	2015 Bare Costs Labor	2015 Bare Costs Equipment	Total	Total Incl O&P
0010	**GENERAL FILL**									
0011	Spread dumped material, no compaction									
0020	By dozer, no compaction	B-10B	1000	.012	L.C.Y.		.56	1.39	1.95	2.38
0100	By hand	1 Clab	12	.667	"		25		25	38.50
0500	Gravel fill, compacted, under floor slabs, 4" deep	B-37	10000	.005	S.F.	.42	.19	.02	.63	.77
0600	6" deep		8600	.006		.63	.22	.02	.87	1.05
0700	9" deep		7200	.007		1.05	.27	.02	1.34	1.58
0800	12" deep		6000	.008		1.47	.32	.03	1.82	2.14
1000	Alternate pricing method, 4" deep		120	.400	E.C.Y.	31.50	15.90	1.31	48.71	60.50
1100	6" deep		160	.300		31.50	11.95	.98	44.43	54

31 23 23 – Fill

31 23 23.17 General Fill	Crew	Daily Output	Labor-Hours	Unit	Material	2015 Bare Costs Labor	2015 Bare Costs Equipment	Total	Total Incl O&P
1200	9" deep	B-37	200	.240	E.C.Y.	31.50	9.55	.79	41.84	50
1300	12" deep	↓	220	.218	↓	31.50	8.70	.72	40.92	48.50
1400	Granular fill				L.C.Y.	21			21	23

31 23 23.20 Hauling

		Crew	Daily Output	Labor-Hours	Unit	Material	Labor	Equipment	Total	Total Incl O&P
0010	**HAULING**									
0011	Excavated or borrow, loose cubic yards									
0012	no loading equipment, including hauling, waiting, loading/dumping									
0013	time per cycle (wait, load, travel, unload or dump & return)									
0014	8 C.Y. truck, 15 MPH ave, cycle 0.5 miles, 10 min. wait/Ld./Uld.	B-34A	320	.025	L.C.Y.		1	1.29	2.29	2.92
0016	cycle 1 mile		272	.029			1.18	1.51	2.69	3.44
0018	cycle 2 miles		208	.038			1.54	1.98	3.52	4.50
0020	cycle 4 miles		144	.056			2.23	2.86	5.09	6.50
0022	cycle 6 miles		112	.071			2.86	3.67	6.53	8.35
0024	cycle 8 miles		88	.091			3.64	4.67	8.31	10.65
0026	20 MPH ave, cycle 0.5 mile		336	.024			.95	1.22	2.17	2.79
0028	cycle 1 mile		296	.027			1.08	1.39	2.47	3.17
0030	cycle 2 miles		240	.033			1.33	1.71	3.04	3.90
0032	cycle 4 miles		176	.045			1.82	2.34	4.16	5.30
0034	cycle 6 miles		136	.059			2.36	3.02	5.38	6.90
0036	cycle 8 miles		112	.071			2.86	3.67	6.53	8.35
0044	25 MPH ave, cycle 4 miles		192	.042			1.67	2.14	3.81	4.88
0046	cycle 6 miles		160	.050			2	2.57	4.57	5.85
0048	cycle 8 miles		128	.063			2.50	3.21	5.71	7.30
0050	30 MPH ave, cycle 4 miles		216	.037			1.48	1.90	3.38	4.33
0052	cycle 6 miles		176	.045			1.82	2.34	4.16	5.30
0054	cycle 8 miles		144	.056			2.23	2.86	5.09	6.50
0114	15 MPH ave, cycle 0.5 mile, 15 min. wait/Ld./Uld.		224	.036			1.43	1.84	3.27	4.18
0116	cycle 1 mile		200	.040			1.60	2.06	3.66	4.68
0118	cycle 2 miles		168	.048			1.91	2.45	4.36	5.55
0120	cycle 4 miles		120	.067			2.67	3.43	6.10	7.80
0122	cycle 6 miles		96	.083			3.34	4.28	7.62	9.75
0124	cycle 8 miles		80	.100			4.01	5.15	9.16	11.70
0126	20 MPH ave, cycle 0.5 mile		232	.034			1.38	1.77	3.15	4.04
0128	cycle 1 mile		208	.038			1.54	1.98	3.52	4.50
0130	cycle 2 miles		184	.043			1.74	2.23	3.97	5.10
0132	cycle 4 miles		144	.056			2.23	2.86	5.09	6.50
0134	cycle 6 miles		112	.071			2.86	3.67	6.53	8.35
0136	cycle 8 miles		96	.083			3.34	4.28	7.62	9.75
0144	25 MPH ave, cycle 4 miles		152	.053			2.11	2.71	4.82	6.15
0146	cycle 6 miles		128	.063			2.50	3.21	5.71	7.30
0148	cycle 8 miles		112	.071			2.86	3.67	6.53	8.35
0150	30 MPH ave, cycle 4 miles		168	.048			1.91	2.45	4.36	5.55
0152	cycle 6 miles		144	.056			2.23	2.86	5.09	6.50
0154	cycle 8 miles		120	.067			2.67	3.43	6.10	7.80
0214	15 MPH ave, cycle 0.5 mile, 20 min wait/Ld./Uld.		176	.045			1.82	2.34	4.16	5.30
0216	cycle 1 mile		160	.050			2	2.57	4.57	5.85
0218	cycle 2 miles		136	.059			2.36	3.02	5.38	6.90
0220	cycle 4 miles		104	.077			3.08	3.95	7.03	9
0222	cycle 6 miles		88	.091			3.64	4.67	8.31	10.65
0224	cycle 8 miles		72	.111			4.45	5.70	10.15	13.05
0226	20 MPH ave, cycle 0.5 mile		176	.045			1.82	2.34	4.16	5.30
0228	cycle 1 mile	↓	168	.048	↓		1.91	2.45	4.36	5.55

For customer support on your Mechanical Cost Data, call 877.795.7394.

445

31 23 23.20 Hauling		Crew	Daily Output	Labor-Hours	Unit	Material	2015 Bare Costs Labor	2015 Bare Costs Equipment	Total	Total Incl O&P
0230	cycle 2 miles	B-34A	144	.056	L.C.Y.		2.23	2.86	5.09	6.50
0232	cycle 4 miles		120	.067			2.67	3.43	6.10	7.80
0234	cycle 6 miles		96	.083			3.34	4.28	7.62	9.75
0236	cycle 8 miles		88	.091			3.64	4.67	8.31	10.65
0244	25 MPH ave, cycle 4 miles		128	.063			2.50	3.21	5.71	7.30
0246	cycle 6 miles		112	.071			2.86	3.67	6.53	8.35
0248	cycle 8 miles		96	.083			3.34	4.28	7.62	9.75
0250	30 MPH ave, cycle 4 miles		136	.059			2.36	3.02	5.38	6.90
0252	cycle 6 miles		120	.067			2.67	3.43	6.10	7.80
0254	cycle 8 miles		104	.077			3.08	3.95	7.03	9
0314	15 MPH ave, cycle 0.5 mile, 25 min wait/Ld./Uld.		144	.056			2.23	2.86	5.09	6.50
0316	cycle 1 mile		128	.063			2.50	3.21	5.71	7.30
0318	cycle 2 miles		112	.071			2.86	3.67	6.53	8.35
0320	cycle 4 miles		96	.083			3.34	4.28	7.62	9.75
0322	cycle 6 miles		80	.100			4.01	5.15	9.16	11.70
0324	cycle 8 miles		64	.125			5	6.45	11.45	14.60
0326	20 MPH ave, cycle 0.5 mile		144	.056			2.23	2.86	5.09	6.50
0328	cycle 1 mile		136	.059			2.36	3.02	5.38	6.90
0330	cycle 2 miles		120	.067			2.67	3.43	6.10	7.80
0332	cycle 4 miles		104	.077			3.08	3.95	7.03	9
0334	cycle 6 miles		88	.091			3.64	4.67	8.31	10.65
0336	cycle 8 miles		80	.100			4.01	5.15	9.16	11.70
0344	25 MPH ave, cycle 4 miles		112	.071			2.86	3.67	6.53	8.35
0346	cycle 6 miles		96	.083			3.34	4.28	7.62	9.75
0348	cycle 8 miles		88	.091			3.64	4.67	8.31	10.65
0350	30 MPH ave, cycle 4 miles		112	.071			2.86	3.67	6.53	8.35
0352	cycle 6 miles		104	.077			3.08	3.95	7.03	9
0354	cycle 8 miles		96	.083			3.34	4.28	7.62	9.75
0414	15 MPH ave, cycle 0.5 mile, 30 min wait/Ld./Uld.		120	.067			2.67	3.43	6.10	7.80
0416	cycle 1 mile		112	.071			2.86	3.67	6.53	8.35
0418	cycle 2 miles		96	.083			3.34	4.28	7.62	9.75
0420	cycle 4 miles		80	.100			4.01	5.15	9.16	11.70
0422	cycle 6 miles		72	.111			4.45	5.70	10.15	13.05
0424	cycle 8 miles		64	.125			5	6.45	11.45	14.60
0426	20 MPH ave, cycle 0.5 mile		120	.067			2.67	3.43	6.10	7.80
0428	cycle 1 mile		112	.071			2.86	3.67	6.53	8.35
0430	cycle 2 miles		104	.077			3.08	3.95	7.03	9
0432	cycle 4 miles		88	.091			3.64	4.67	8.31	10.65
0434	cycle 6 miles		80	.100			4.01	5.15	9.16	11.70
0436	cycle 8 miles		72	.111			4.45	5.70	10.15	13.05
0444	25 MPH ave, cycle 4 miles		96	.083			3.34	4.28	7.62	9.75
0446	cycle 6 miles		88	.091			3.64	4.67	8.31	10.65
0448	cycle 8 miles		80	.100			4.01	5.15	9.16	11.70
0450	30 MPH ave, cycle 4 miles		96	.083			3.34	4.28	7.62	9.75
0452	cycle 6 miles		88	.091			3.64	4.67	8.31	10.65
0454	cycle 8 miles		80	.100			4.01	5.15	9.16	11.70
0514	15 MPH ave, cycle 0.5 mile, 35 min wait/Ld./Uld.		104	.077			3.08	3.95	7.03	9
0516	cycle 1 mile		96	.083			3.34	4.28	7.62	9.75
0518	cycle 2 miles		88	.091			3.64	4.67	8.31	10.65
0520	cycle 4 miles		72	.111			4.45	5.70	10.15	13.05
0522	cycle 6 miles		64	.125			5	6.45	11.45	14.60
0524	cycle 8 miles		56	.143			5.70	7.35	13.05	16.75
0526	20 MPH ave, cycle 0.5 mile		104	.077			3.08	3.95	7.03	9

31 23 23 – Fill

31 23 23.20 Hauling		Crew	Daily Output	Labor-Hours	Unit	Material	2015 Bare Costs Labor	Equipment	Total	Total Incl O&P
0528	cycle 1 mile	B-34A	96	.083	L.C.Y.		3.34	4.28	7.62	9.75
0530	cycle 2 miles		96	.083			3.34	4.28	7.62	9.75
0532	cycle 4 miles		80	.100			4.01	5.15	9.16	11.70
0534	cycle 6 miles		72	.111			4.45	5.70	10.15	13.05
0536	cycle 8 miles		64	.125			5	6.45	11.45	14.60
0544	25 MPH ave, cycle 4 miles		88	.091			3.64	4.67	8.31	10.65
0546	cycle 6 miles		80	.100			4.01	5.15	9.16	11.70
0548	cycle 8 miles		72	.111			4.45	5.70	10.15	13.05
0550	30 MPH ave, cycle 4 miles		88	.091			3.64	4.67	8.31	10.65
0552	cycle 6 miles		80	.100			4.01	5.15	9.16	11.70
0554	cycle 8 miles		72	.111			4.45	5.70	10.15	13.05
1014	12 C.Y. truck, cycle 0.5 mile, 15 MPH ave, 15 min. wait/Ld./Uld.	B-34B	336	.024			.95	2.06	3.01	3.70
1016	cycle 1 mile		300	.027			1.07	2.30	3.37	4.14
1018	cycle 2 miles		252	.032			1.27	2.74	4.01	4.94
1020	cycle 4 miles		180	.044			1.78	3.84	5.62	6.90
1022	cycle 6 miles		144	.056			2.23	4.80	7.03	8.65
1024	cycle 8 miles		120	.067			2.67	5.75	8.42	10.40
1025	cycle 10 miles		96	.083			3.34	7.20	10.54	12.95
1026	20 MPH ave, cycle 0.5 mile		348	.023			.92	1.99	2.91	3.57
1028	cycle 1 mile		312	.026			1.03	2.21	3.24	3.99
1030	cycle 2 miles		276	.029			1.16	2.50	3.66	4.51
1032	cycle 4 miles		216	.037			1.48	3.20	4.68	5.75
1034	cycle 6 miles		168	.048			1.91	4.11	6.02	7.40
1036	cycle 8 miles		144	.056			2.23	4.80	7.03	8.65
1038	cycle 10 miles		120	.067			2.67	5.75	8.42	10.40
1040	25 MPH ave, cycle 4 miles		228	.035			1.41	3.03	4.44	5.45
1042	cycle 6 miles		192	.042			1.67	3.60	5.27	6.50
1044	cycle 8 miles		168	.048			1.91	4.11	6.02	7.40
1046	cycle 10 miles		144	.056			2.23	4.80	7.03	8.65
1050	30 MPH ave, cycle 4 miles		252	.032			1.27	2.74	4.01	4.94
1052	cycle 6 miles		216	.037			1.48	3.20	4.68	5.75
1054	cycle 8 miles		180	.044			1.78	3.84	5.62	6.90
1056	cycle 10 miles		156	.051			2.05	4.43	6.48	8
1060	35 MPH ave, cycle 4 miles		264	.030			1.21	2.62	3.83	4.71
1062	cycle 6 miles		228	.035			1.41	3.03	4.44	5.45
1064	cycle 8 miles		204	.039			1.57	3.39	4.96	6.10
1066	cycle 10 miles		180	.044			1.78	3.84	5.62	6.90
1068	cycle 20 miles		120	.067			2.67	5.75	8.42	10.40
1069	cycle 30 miles		84	.095			3.81	8.25	12.06	14.80
1070	cycle 40 miles		72	.111			4.45	9.60	14.05	17.30
1072	40 MPH ave, cycle 6 miles		240	.033			1.33	2.88	4.21	5.20
1074	cycle 8 miles		216	.037			1.48	3.20	4.68	5.75
1076	cycle 10 miles		192	.042			1.67	3.60	5.27	6.50
1078	cycle 20 miles		120	.067			2.67	5.75	8.42	10.40
1080	cycle 30 miles		96	.083			3.34	7.20	10.54	12.95
1082	cycle 40 miles		72	.111			4.45	9.60	14.05	17.30
1084	cycle 50 miles		60	.133			5.35	11.50	16.85	20.50
1094	45 MPH ave, cycle 8 miles		216	.037			1.48	3.20	4.68	5.75
1096	cycle 10 miles		204	.039			1.57	3.39	4.96	6.10
1098	cycle 20 miles		132	.061			2.43	5.25	7.68	9.40
1100	cycle 30 miles		108	.074			2.97	6.40	9.37	11.55
1102	cycle 40 miles		84	.095			3.81	8.25	12.06	14.80
1104	cycle 50 miles		72	.111			4.45	9.60	14.05	17.30

For customer support on your Mechanical Cost Data, call 877.795.7394.

447

31 23 23 – Fill

31 23 23.20 Hauling		Crew	Daily Output	Labor-Hours	Unit	Material	2015 Bare Costs		Total	Total Incl O&P
							Labor	Equipment		
1106	50 MPH ave, cycle 10 miles	B-34B	216	.037	L.C.Y.		1.48	3.20	4.68	5.75
1108	cycle 20 miles		144	.056			2.23	4.80	7.03	8.65
1110	cycle 30 miles		108	.074			2.97	6.40	9.37	11.55
1112	cycle 40 miles		84	.095			3.81	8.25	12.06	14.80
1114	cycle 50 miles		72	.111			4.45	9.60	14.05	17.30
1214	15 MPH ave, cycle 0.5 mile, 20 min. wait/Ld./Uld.		264	.030			1.21	2.62	3.83	4.71
1216	cycle 1 mile		240	.033			1.33	2.88	4.21	5.20
1218	cycle 2 miles		204	.039			1.57	3.39	4.96	6.10
1220	cycle 4 miles		156	.051			2.05	4.43	6.48	8
1222	cycle 6 miles		132	.061			2.43	5.25	7.68	9.40
1224	cycle 8 miles		108	.074			2.97	6.40	9.37	11.55
1225	cycle 10 miles		96	.083			3.34	7.20	10.54	12.95
1226	20 MPH ave, cycle 0.5 mile		264	.030			1.21	2.62	3.83	4.71
1228	cycle 1 mile		252	.032			1.27	2.74	4.01	4.94
1230	cycle 2 miles		216	.037			1.48	3.20	4.68	5.75
1232	cycle 4 miles		180	.044			1.78	3.84	5.62	6.90
1234	cycle 6 miles		144	.056			2.23	4.80	7.03	8.65
1236	cycle 8 miles		132	.061			2.43	5.25	7.68	9.40
1238	cycle 10 miles		108	.074			2.97	6.40	9.37	11.55
1240	25 MPH ave, cycle 4 miles		192	.042			1.67	3.60	5.27	6.50
1242	cycle 6 miles		168	.048			1.91	4.11	6.02	7.40
1244	cycle 8 miles		144	.056			2.23	4.80	7.03	8.65
1246	cycle 10 miles		132	.061			2.43	5.25	7.68	9.40
1250	30 MPH ave, cycle 4 miles		204	.039			1.57	3.39	4.96	6.10
1252	cycle 6 miles		180	.044			1.78	3.84	5.62	6.90
1254	cycle 8 miles		156	.051			2.05	4.43	6.48	8
1256	cycle 10 miles		144	.056			2.23	4.80	7.03	8.65
1260	35 MPH ave, cycle 4 miles		216	.037			1.48	3.20	4.68	5.75
1262	cycle 6 miles		192	.042			1.67	3.60	5.27	6.50
1264	cycle 8 miles		168	.048			1.91	4.11	6.02	7.40
1266	cycle 10 miles		156	.051			2.05	4.43	6.48	8
1268	cycle 20 miles		108	.074			2.97	6.40	9.37	11.55
1269	cycle 30 miles		72	.111			4.45	9.60	14.05	17.30
1270	cycle 40 miles		60	.133			5.35	11.50	16.85	20.50
1272	40 MPH ave, cycle 6 miles		192	.042			1.67	3.60	5.27	6.50
1274	cycle 8 miles		180	.044			1.78	3.84	5.62	6.90
1276	cycle 10 miles		156	.051			2.05	4.43	6.48	8
1278	cycle 20 miles		108	.074			2.97	6.40	9.37	11.55
1280	cycle 30 miles		84	.095			3.81	8.25	12.06	14.80
1282	cycle 40 miles		72	.111			4.45	9.60	14.05	17.30
1284	cycle 50 miles		60	.133			5.35	11.50	16.85	20.50
1294	45 MPH ave, cycle 8 miles		180	.044			1.78	3.84	5.62	6.90
1296	cycle 10 miles		168	.048			1.91	4.11	6.02	7.40
1298	cycle 20 miles		120	.067			2.67	5.75	8.42	10.40
1300	cycle 30 miles		96	.083			3.34	7.20	10.54	12.95
1302	cycle 40 miles		72	.111			4.45	9.60	14.05	17.30
1304	cycle 50 miles		60	.133			5.35	11.50	16.85	20.50
1306	50 MPH ave, cycle 10 miles		180	.044			1.78	3.84	5.62	6.90
1308	cycle 20 miles		132	.061			2.43	5.25	7.68	9.40
1310	cycle 30 miles		96	.083			3.34	7.20	10.54	12.95
1312	cycle 40 miles		84	.095			3.81	8.25	12.06	14.80
1314	cycle 50 miles		72	.111			4.45	9.60	14.05	17.30
1414	15 MPH ave, cycle 0.5 mile, 25 min. wait/Ld./Uld.		204	.039			1.57	3.39	4.96	6.10

448

31 23 23 – Fill

31 23 23.20 Hauling	Crew	Daily Output	Labor-Hours	Unit	Material	2015 Bare Costs Labor	Equipment	Total	Total Incl O&P
1416	cycle 1 mile	B-34B	192	.042	L.C.Y.		1.67	3.60	5.27	6.50
1418	cycle 2 miles		168	.048			1.91	4.11	6.02	7.40
1420	cycle 4 miles		132	.061			2.43	5.25	7.68	9.40
1422	cycle 6 miles		120	.067			2.67	5.75	8.42	10.40
1424	cycle 8 miles		96	.083			3.34	7.20	10.54	12.95
1425	cycle 10 miles		84	.095			3.81	8.25	12.06	14.80
1426	20 MPH ave, cycle 0.5 mile		216	.037			1.48	3.20	4.68	5.75
1428	cycle 1 mile		204	.039			1.57	3.39	4.96	6.10
1430	cycle 2 miles		180	.044			1.78	3.84	5.62	6.90
1432	cycle 4 miles		156	.051			2.05	4.43	6.48	8
1434	cycle 6 miles		132	.061			2.43	5.25	7.68	9.40
1436	cycle 8 miles		120	.067			2.67	5.75	8.42	10.40
1438	cycle 10 miles		96	.083			3.34	7.20	10.54	12.95
1440	25 MPH ave, cycle 4 miles		168	.048			1.91	4.11	6.02	7.40
1442	cycle 6 miles		144	.056			2.23	4.80	7.03	8.65
1444	cycle 8 miles		132	.061			2.43	5.25	7.68	9.40
1446	cycle 10 miles		108	.074			2.97	6.40	9.37	11.55
1450	30 MPH ave, cycle 4 miles		168	.048			1.91	4.11	6.02	7.40
1452	cycle 6 miles		156	.051			2.05	4.43	6.48	8
1454	cycle 8 miles		132	.061			2.43	5.25	7.68	9.40
1456	cycle 10 miles		120	.067			2.67	5.75	8.42	10.40
1460	35 MPH ave, cycle 4 miles		180	.044			1.78	3.84	5.62	6.90
1462	cycle 6 miles		156	.051			2.05	4.43	6.48	8
1464	cycle 8 miles		144	.056			2.23	4.80	7.03	8.65
1466	cycle 10 miles		132	.061			2.43	5.25	7.68	9.40
1468	cycle 20 miles		96	.083			3.34	7.20	10.54	12.95
1469	cycle 30 miles		72	.111			4.45	9.60	14.05	17.30
1470	cycle 40 miles		60	.133			5.35	11.50	16.85	20.50
1472	40 MPH ave, cycle 6 miles		168	.048			1.91	4.11	6.02	7.40
1474	cycle 8 miles		156	.051			2.05	4.43	6.48	8
1476	cycle 10 miles		144	.056			2.23	4.80	7.03	8.65
1478	cycle 20 miles		96	.083			3.34	7.20	10.54	12.95
1480	cycle 30 miles		84	.095			3.81	8.25	12.06	14.80
1482	cycle 40 miles		60	.133			5.35	11.50	16.85	20.50
1484	cycle 50 miles		60	.133			5.35	11.50	16.85	20.50
1494	45 MPH ave, cycle 8 miles		156	.051			2.05	4.43	6.48	8
1496	cycle 10 miles		144	.056			2.23	4.80	7.03	8.65
1498	cycle 20 miles		108	.074			2.97	6.40	9.37	11.55
1500	cycle 30 miles		84	.095			3.81	8.25	12.06	14.80
1502	cycle 40 miles		72	.111			4.45	9.60	14.05	17.30
1504	cycle 50 miles		60	.133			5.35	11.50	16.85	20.50
1506	50 MPH ave, cycle 10 miles		156	.051			2.05	4.43	6.48	8
1508	cycle 20 miles		120	.067			2.67	5.75	8.42	10.40
1510	cycle 30 miles		96	.083			3.34	7.20	10.54	12.95
1512	cycle 40 miles		72	.111			4.45	9.60	14.05	17.30
1514	cycle 50 miles		60	.133			5.35	11.50	16.85	20.50
1614	15 MPH, cycle 0.5 mile, 30 min. wait/Ld./Uld.		180	.044			1.78	3.84	5.62	6.90
1616	cycle 1 mile		168	.048			1.91	4.11	6.02	7.40
1618	cycle 2 miles		144	.056			2.23	4.80	7.03	8.65
1620	cycle 4 miles		120	.067			2.67	5.75	8.42	10.40
1622	cycle 6 miles		108	.074			2.97	6.40	9.37	11.55
1624	cycle 8 miles		84	.095			3.81	8.25	12.06	14.80
1625	cycle 10 miles		84	.095			3.81	8.25	12.06	14.80

For customer support on your Mechanical Cost Data, call 877.795.7394.

449

31 23 23.20 Hauling		Crew	Daily Output	Labor-Hours	Unit	Material	2015 Bare Costs Labor	Equipment	Total	Total Incl O&P
1626	20 MPH ave, cycle 0.5 mile	B-34B	180	.044	L.C.Y.		1.78	3.84	5.62	6.90
1628	cycle 1 mile		168	.048			1.91	4.11	6.02	7.40
1630	cycle 2 miles		156	.051			2.05	4.43	6.48	8
1632	cycle 4 miles		132	.061			2.43	5.25	7.68	9.40
1634	cycle 6 miles		120	.067			2.67	5.75	8.42	10.40
1636	cycle 8 miles		108	.074			2.97	6.40	9.37	11.55
1638	cycle 10 miles		96	.083			3.34	7.20	10.54	12.95
1640	25 MPH ave, cycle 4 miles		144	.056			2.23	4.80	7.03	8.65
1642	cycle 6 miles		132	.061			2.43	5.25	7.68	9.40
1644	cycle 8 miles		108	.074			2.97	6.40	9.37	11.55
1646	cycle 10 miles		108	.074			2.97	6.40	9.37	11.55
1650	30 MPH ave, cycle 4 miles		144	.056			2.23	4.80	7.03	8.65
1652	cycle 6 miles		132	.061			2.43	5.25	7.68	9.40
1654	cycle 8 miles		120	.067			2.67	5.75	8.42	10.40
1656	cycle 10 miles		108	.074			2.97	6.40	9.37	11.55
1660	35 MPH ave, cycle 4 miles		156	.051			2.05	4.43	6.48	8
1662	cycle 6 miles		144	.056			2.23	4.80	7.03	8.65
1664	cycle 8 miles		132	.061			2.43	5.25	7.68	9.40
1666	cycle 10 miles		120	.067			2.67	5.75	8.42	10.40
1668	cycle 20 miles		84	.095			3.81	8.25	12.06	14.80
1669	cycle 30 miles		72	.111			4.45	9.60	14.05	17.30
1670	cycle 40 miles		60	.133			5.35	11.50	16.85	20.50
1672	40 MPH, cycle 6 miles		144	.056			2.23	4.80	7.03	8.65
1674	cycle 8 miles		132	.061			2.43	5.25	7.68	9.40
1676	cycle 10 miles		120	.067			2.67	5.75	8.42	10.40
1678	cycle 20 miles		96	.083			3.34	7.20	10.54	12.95
1680	cycle 30 miles		72	.111			4.45	9.60	14.05	17.30
1682	cycle 40 miles		60	.133			5.35	11.50	16.85	20.50
1684	cycle 50 miles		48	.167			6.70	14.40	21.10	26
1694	45 MPH ave, cycle 8 miles		144	.056			2.23	4.80	7.03	8.65
1696	cycle 10 miles		132	.061			2.43	5.25	7.68	9.40
1698	cycle 20 miles		96	.083			3.34	7.20	10.54	12.95
1700	cycle 30 miles		84	.095			3.81	8.25	12.06	14.80
1702	cycle 40 miles		60	.133			5.35	11.50	16.85	20.50
1704	cycle 50 miles		60	.133			5.35	11.50	16.85	20.50
1706	50 MPH ave, cycle 10 miles		132	.061			2.43	5.25	7.68	9.40
1708	cycle 20 miles		108	.074			2.97	6.40	9.37	11.55
1710	cycle 30 miles		84	.095			3.81	8.25	12.06	14.80
1712	cycle 40 miles		72	.111			4.45	9.60	14.05	17.30
1714	cycle 50 miles		60	.133			5.35	11.50	16.85	20.50
2000	Hauling, 8 C.Y. truck, small project cost per hour	B-34A	8	1	Hr.		40	51.50	91.50	117
2100	12 C.Y. Truck	B-34B	8	1			40	86.50	126.50	156
2150	16.5 C.Y. Truck	B-34C	8	1			40	91	131	161
2175	18 C.Y. 8 wheel Truck	B-34I	8	1			40	109	149	180
2200	20 C.Y. Truck	B-34D	8	1			40	92.50	132.50	163
2300	Grading at dump, or embankment if required, by dozer	B-10B	1000	.012	L.C.Y.		.56	1.39	1.95	2.38
2310	Spotter at fill or cut, if required	1 Clab	8	1	Hr.		37.50		37.50	58
9014	18 C.Y. truck, 8 wheels,15 min. wait/Ld./Uld.,15 MPH, cycle 0.5 mi.	B-34I	504	.016	L.C.Y.		.64	1.72	2.36	2.86
9016	cycle 1 mile		450	.018			.71	1.93	2.64	3.20
9018	cycle 2 miles		378	.021			.85	2.30	3.15	3.81
9020	cycle 4 miles		270	.030			1.19	3.22	4.41	5.35
9022	cycle 6 miles		216	.037			1.48	4.02	5.50	6.65
9024	cycle 8 miles		180	.044			1.78	4.83	6.61	8

For customer support on your Mechanical Cost Data, call 877.795.7394.

31 23 23.20 Hauling	Crew	Daily Output	Labor-Hours	Unit	Material	2015 Bare Costs Labor	Equipment	Total	Total Incl O&P
9025	cycle 10 miles	B-34I	144	.056	L.C.Y.		2.23	6.05	8.28	10
9026	20 MPH ave, cycle 0.5 mile		522	.015			.61	1.67	2.28	2.76
9028	cycle 1 mile		468	.017			.68	1.86	2.54	3.07
9030	cycle 2 miles		414	.019			.77	2.10	2.87	3.48
9032	cycle 4 miles		324	.025			.99	2.68	3.67	4.45
9034	cycle 6 miles		252	.032			1.27	3.45	4.72	5.70
9036	cycle 8 miles		216	.037			1.48	4.02	5.50	6.65
9038	cycle 10 miles		180	.044			1.78	4.83	6.61	8
9040	25 MPH ave, cycle 4 miles		342	.023			.94	2.54	3.48	4.21
9042	cycle 6 miles		288	.028			1.11	3.02	4.13	5
9044	cycle 8 miles		252	.032			1.27	3.45	4.72	5.70
9046	cycle 10 miles		216	.037			1.48	4.02	5.50	6.65
9050	30 MPH ave, cycle 4 miles		378	.021			.85	2.30	3.15	3.81
9052	cycle 6 miles		324	.025			.99	2.68	3.67	4.45
9054	cycle 8 miles		270	.030			1.19	3.22	4.41	5.35
9056	cycle 10 miles		234	.034			1.37	3.71	5.08	6.15
9060	35 MPH ave, cycle 4 miles		396	.020			.81	2.19	3	3.63
9062	cycle 6 miles		342	.023			.94	2.54	3.48	4.21
9064	cycle 8 miles		288	.028			1.11	3.02	4.13	5
9066	cycle 10 miles		270	.030			1.19	3.22	4.41	5.35
9068	cycle 20 miles		162	.049			1.98	5.35	7.33	8.90
9070	cycle 30 miles		126	.063			2.54	6.90	9.44	11.45
9072	cycle 40 miles		90	.089			3.56	9.65	13.21	16
9074	40 MPH ave, cycle 6 miles		360	.022			.89	2.41	3.30	4.01
9076	cycle 8 miles		324	.025			.99	2.68	3.67	4.45
9078	cycle 10 miles		288	.028			1.11	3.02	4.13	5
9080	cycle 20 miles		180	.044			1.78	4.83	6.61	8
9082	cycle 30 miles		144	.056			2.23	6.05	8.28	10
9084	cycle 40 miles		108	.074			2.97	8.05	11.02	13.35
9086	cycle 50 miles		90	.089			3.56	9.65	13.21	16
9094	45 MPH ave, cycle 8 miles		324	.025			.99	2.68	3.67	4.45
9096	cycle 10 miles		306	.026			1.05	2.84	3.89	4.70
9098	cycle 20 miles		198	.040			1.62	4.39	6.01	7.30
9100	cycle 30 miles		144	.056			2.23	6.05	8.28	10
9102	cycle 40 miles		126	.063			2.54	6.90	9.44	11.45
9104	cycle 50 miles		108	.074			2.97	8.05	11.02	13.35
9106	50 MPH ave, cycle 10 miles		324	.025			.99	2.68	3.67	4.45
9108	cycle 20 miles		216	.037			1.48	4.02	5.50	6.65
9110	cycle 30 miles		162	.049			1.98	5.35	7.33	8.90
9112	cycle 40 miles		126	.063			2.54	6.90	9.44	11.45
9114	cycle 50 miles		108	.074			2.97	8.05	11.02	13.35
9214	20 min. wait/Ld./Uld.,15 MPH, cycle 0.5 mi.		396	.020			.81	2.19	3	3.63
9216	cycle 1 mile		360	.022			.89	2.41	3.30	4.01
9218	cycle 2 miles		306	.026			1.05	2.84	3.89	4.70
9220	cycle 4 miles		234	.034			1.37	3.71	5.08	6.15
9222	cycle 6 miles		198	.040			1.62	4.39	6.01	7.30
9224	cycle 8 miles		162	.049			1.98	5.35	7.33	8.90
9225	cycle 10 miles		144	.056			2.23	6.05	8.28	10
9226	20 MPH ave, cycle 0.5 mile		396	.020			.81	2.19	3	3.63
9228	cycle 1 mile		378	.021			.85	2.30	3.15	3.81
9230	cycle 2 miles		324	.025			.99	2.68	3.67	4.45
9232	cycle 4 miles		270	.030			1.19	3.22	4.41	5.35
9234	cycle 6 miles		216	.037			1.48	4.02	5.50	6.65

For customer support on your Mechanical Cost Data, call 877.795.7394.

451

31 23 23 – Fill

31 23 23.20 Hauling		Crew	Daily Output	Labor-Hours	Unit	Material	2015 Bare Costs		Total	Total Incl O&P
							Labor	Equipment		
9236	cycle 8 miles	B-34I	198	.040	L.C.Y.		1.62	4.39	6.01	7.30
9238	cycle 10 miles		162	.049			1.98	5.35	7.33	8.90
9240	25 MPH ave, cycle 4 miles		288	.028			1.11	3.02	4.13	5
9242	cycle 6 miles		252	.032			1.27	3.45	4.72	5.70
9244	cycle 8 miles		216	.037			1.48	4.02	5.50	6.65
9246	cycle 10 miles		198	.040			1.62	4.39	6.01	7.30
9250	30 MPH ave, cycle 4 miles		306	.026			1.05	2.84	3.89	4.70
9252	cycle 6 miles		270	.030			1.19	3.22	4.41	5.35
9254	cycle 8 miles		234	.034			1.37	3.71	5.08	6.15
9256	cycle 10 miles		216	.037			1.48	4.02	5.50	6.65
9260	35 MPH ave, cycle 4 miles		324	.025			.99	2.68	3.67	4.45
9262	cycle 6 miles		288	.028			1.11	3.02	4.13	5
9264	cycle 8 miles		252	.032			1.27	3.45	4.72	5.70
9266	cycle 10 miles		234	.034			1.37	3.71	5.08	6.15
9268	cycle 20 miles		162	.049			1.98	5.35	7.33	8.90
9270	cycle 30 miles		108	.074			2.97	8.05	11.02	13.35
9272	cycle 40 miles		90	.089			3.56	9.65	13.21	16
9274	40 MPH ave, cycle 6 miles		288	.028			1.11	3.02	4.13	5
9276	cycle 8 miles		270	.030			1.19	3.22	4.41	5.35
9278	cycle 10 miles		234	.034			1.37	3.71	5.08	6.15
9280	cycle 20 miles		162	.049			1.98	5.35	7.33	8.90
9282	cycle 30 miles		126	.063			2.54	6.90	9.44	11.45
9284	cycle 40 miles		108	.074			2.97	8.05	11.02	13.35
9286	cycle 50 miles		90	.089			3.56	9.65	13.21	16
9294	45 MPH ave, cycle 8 miles		270	.030			1.19	3.22	4.41	5.35
9296	cycle 10 miles		252	.032			1.27	3.45	4.72	5.70
9298	cycle 20 miles		180	.044			1.78	4.83	6.61	8
9300	cycle 30 miles		144	.056			2.23	6.05	8.28	10
9302	cycle 40 miles		108	.074			2.97	8.05	11.02	13.35
9304	cycle 50 miles		90	.089			3.56	9.65	13.21	16
9306	50 MPH ave, cycle 10 miles		270	.030			1.19	3.22	4.41	5.35
9308	cycle 20 miles		198	.040			1.62	4.39	6.01	7.30
9310	cycle 30 miles		144	.056			2.23	6.05	8.28	10
9312	cycle 40 miles		126	.063			2.54	6.90	9.44	11.45
9314	cycle 50 miles		108	.074			2.97	8.05	11.02	13.35
9414	25 min. wait/Ld./Uld.,15 MPH, cycle 0.5 mi.		306	.026			1.05	2.84	3.89	4.70
9416	cycle 1 mile		288	.028			1.11	3.02	4.13	5
9418	cycle 2 miles		252	.032			1.27	3.45	4.72	5.70
9420	cycle 4 miles		198	.040			1.62	4.39	6.01	7.30
9422	cycle 6 miles		180	.044			1.78	4.83	6.61	8
9424	cycle 8 miles		144	.056			2.23	6.05	8.28	10
9425	cycle 10 miles		126	.063			2.54	6.90	9.44	11.45
9426	20 MPH ave, cycle 0.5 mile		324	.025			.99	2.68	3.67	4.45
9428	cycle 1 mile		306	.026			1.05	2.84	3.89	4.70
9430	cycle 2 miles		270	.030			1.19	3.22	4.41	5.35
9432	cycle 4 miles		234	.034			1.37	3.71	5.08	6.15
9434	cycle 6 miles		198	.040			1.62	4.39	6.01	7.30
9436	cycle 8 miles		180	.044			1.78	4.83	6.61	8
9438	cycle 10 miles		144	.056			2.23	6.05	8.28	10
9440	25 MPH ave, cycle 4 miles		252	.032			1.27	3.45	4.72	5.70
9442	cycle 6 miles		216	.037			1.48	4.02	5.50	6.65
9444	cycle 8 miles		198	.040			1.62	4.39	6.01	7.30
9446	cycle 10 miles		180	.044			1.78	4.83	6.61	8

31 23 23 – Fill

31 23 23.20 Hauling		Crew	Daily Output	Labor-Hours	Unit	Material	2015 Bare Costs Labor	Equipment	Total	Total Incl O&P
9450	30 MPH ave, cycle 4 miles	B-34I	252	.032	L.C.Y.		1.27	3.45	4.72	5.70
9452	cycle 6 miles		234	.034			1.37	3.71	5.08	6.15
9454	cycle 8 miles		198	.040			1.62	4.39	6.01	7.30
9456	cycle 10 miles		180	.044			1.78	4.83	6.61	8
9460	35 MPH ave, cycle 4 miles		270	.030			1.19	3.22	4.41	5.35
9462	cycle 6 miles		234	.034			1.37	3.71	5.08	6.15
9464	cycle 8 miles		216	.037			1.48	4.02	5.50	6.65
9466	cycle 10 miles		198	.040			1.62	4.39	6.01	7.30
9468	cycle 20 miles		144	.056			2.23	6.05	8.28	10
9470	cycle 30 miles		108	.074			2.97	8.05	11.02	13.35
9472	cycle 40 miles		90	.089			3.56	9.65	13.21	16
9474	40 MPH ave, cycle 6 miles		252	.032			1.27	3.45	4.72	5.70
9476	cycle 8 miles		234	.034			1.37	3.71	5.08	6.15
9478	cycle 10 miles		216	.037			1.48	4.02	5.50	6.65
9480	cycle 20 miles		144	.056			2.23	6.05	8.28	10
9482	cycle 30 miles		126	.063			2.54	6.90	9.44	11.45
9484	cycle 40 miles		90	.089			3.56	9.65	13.21	16
9486	cycle 50 miles		90	.089			3.56	9.65	13.21	16
9494	45 MPH ave, cycle 8 miles		234	.034			1.37	3.71	5.08	6.15
9496	cycle 10 miles		216	.037			1.48	4.02	5.50	6.65
9498	cycle 20 miles		162	.049			1.98	5.35	7.33	8.90
9500	cycle 30 miles		126	.063			2.54	6.90	9.44	11.45
9502	cycle 40 miles		108	.074			2.97	8.05	11.02	13.35
9504	cycle 50 miles		90	.089			3.56	9.65	13.21	16
9506	50 MPH ave, cycle 10 miles		234	.034			1.37	3.71	5.08	6.15
9508	cycle 20 miles		180	.044			1.78	4.83	6.61	8
9510	cycle 30 miles		144	.056			2.23	6.05	8.28	10
9512	cycle 40 miles		108	.074			2.97	8.05	11.02	13.35
9514	cycle 50 miles		90	.089			3.56	9.65	13.21	16
9614	30 min. wait/Ld./Uld.,15 MPH, cycle 0.5 mi.		270	.030			1.19	3.22	4.41	5.35
9616	cycle 1 mile		252	.032			1.27	3.45	4.72	5.70
9618	cycle 2 miles		216	.037			1.48	4.02	5.50	6.65
9620	cycle 4 miles		180	.044			1.78	4.83	6.61	8
9622	cycle 6 miles		162	.049			1.98	5.35	7.33	8.90
9624	cycle 8 miles		126	.063			2.54	6.90	9.44	11.45
9625	cycle 10 miles		126	.063			2.54	6.90	9.44	11.45
9626	20 MPH ave, cycle 0.5 mile		270	.030			1.19	3.22	4.41	5.35
9628	cycle 1 mile		252	.032			1.27	3.45	4.72	5.70
9630	cycle 2 miles		234	.034			1.37	3.71	5.08	6.15
9632	cycle 4 miles		198	.040			1.62	4.39	6.01	7.30
9634	cycle 6 miles		180	.044			1.78	4.83	6.61	8
9636	cycle 8 miles		162	.049			1.98	5.35	7.33	8.90
9638	cycle 10 miles		144	.056			2.23	6.05	8.28	10
9640	25 MPH ave, cycle 4 miles		216	.037			1.48	4.02	5.50	6.65
9642	cycle 6 miles		198	.040			1.62	4.39	6.01	7.30
9644	cycle 8 miles		180	.044			1.78	4.83	6.61	8
9646	cycle 10 miles		162	.049			1.98	5.35	7.33	8.90
9650	30 MPH ave, cycle 4 miles		216	.037			1.48	4.02	5.50	6.65
9652	cycle 6 miles		198	.040			1.62	4.39	6.01	7.30
9654	cycle 8 miles		180	.044			1.78	4.83	6.61	8
9656	cycle 10 miles		162	.049			1.98	5.35	7.33	8.90
9660	35 MPH ave, cycle 4 miles		234	.034			1.37	3.71	5.08	6.15
9662	cycle 6 miles		216	.037			1.48	4.02	5.50	6.65

For customer support on your Mechanical Cost Data, call 877.795.7394.

453

31 23 23 – Fill

31 23 23.20 Hauling		Crew	Daily Output	Labor-Hours	Unit	Material	2015 Bare Costs Labor	Equipment	Total	Total Incl O&P
9664	cycle 8 miles	B-34I	198	.040	L.C.Y.		1.62	4.39	6.01	7.30
9666	cycle 10 miles		180	.044			1.78	4.83	6.61	8
9668	cycle 20 miles		126	.063			2.54	6.90	9.44	11.45
9670	cycle 30 miles		108	.074			2.97	8.05	11.02	13.35
9672	cycle 40 miles		90	.089			3.56	9.65	13.21	16
9674	40 MPH ave, cycle 6 miles		216	.037			1.48	4.02	5.50	6.65
9676	cycle 8 miles		198	.040			1.62	4.39	6.01	7.30
9678	cycle 10 miles		180	.044			1.78	4.83	6.61	8
9680	cycle 20 miles		144	.056			2.23	6.05	8.28	10
9682	cycle 30 miles		108	.074			2.97	8.05	11.02	13.35
9684	cycle 40 miles		90	.089			3.56	9.65	13.21	16
9686	cycle 50 miles		72	.111			4.45	12.05	16.50	20
9694	45 MPH ave, cycle 8 miles		216	.037			1.48	4.02	5.50	6.65
9696	cycle 10 miles		198	.040			1.62	4.39	6.01	7.30
9698	cycle 20 miles		144	.056			2.23	6.05	8.28	10
9700	cycle 30 miles		126	.063			2.54	6.90	9.44	11.45
9702	cycle 40 miles		108	.074			2.97	8.05	11.02	13.35
9704	cycle 50 miles		90	.089			3.56	9.65	13.21	16
9706	50 MPH ave, cycle 10 miles		198	.040			1.62	4.39	6.01	7.30
9708	cycle 20 miles		162	.049			1.98	5.35	7.33	8.90
9710	cycle 30 miles		126	.063			2.54	6.90	9.44	11.45
9712	cycle 40 miles		108	.074			2.97	8.05	11.02	13.35
9714	cycle 50 miles		90	.089			3.56	9.65	13.21	16

31 41 Shoring

31 41 16 – Sheet Piling

31 41 16.10 Sheet Piling Systems

		Crew	Daily Output	Labor-Hours	Unit	Material	2015 Bare Costs Labor	Equipment	Total	Total Incl O&P
0010	**SHEET PILING SYSTEMS**									
0020	Sheet piling steel, not incl. wales, 22 psf, 15' excav., left in place	B-40	10.81	5.920	Ton	1,600	282	345	2,227	2,575
0100	Drive, extract & salvage R314116-45		6	10.667		510	510	625	1,645	2,025
0300	20' deep excavation, 27 psf, left in place		12.95	4.942		1,600	235	289	2,124	2,425
0400	Drive, extract & salvage		6.55	9.771		510	465	570	1,545	1,925
1200	15' deep excavation, 22 psf, left in place		983	.065	S.F.	18.60	3.10	3.81	25.51	29.50
1300	Drive, extract & salvage		545	.117		5.70	5.60	6.85	18.15	22.50
1500	20' deep excavation, 27 psf, left in place		960	.067		23.50	3.18	3.90	30.58	34.50
1600	Drive, extract & salvage		485	.132		7.45	6.30	7.70	21.45	26.50
2100	Rent steel sheet piling and wales, first month				Ton	310			310	340
2200	Per added month					31			31	34
2300	Rental piling left in place, add to rental					1,150			1,150	1,275
2500	Wales, connections & struts, 2/3 salvage					480			480	530
3900	Wood, solid sheeting, incl. wales, braces and spacers, R314116-40									
3910	drive, extract & salvage, 8' deep excavation	B-31	330	.121	S.F.	1.89	4.83	.66	7.38	10.25
4000	10' deep, 50 S.F./hr. in & 150 S.F./hr. out		300	.133		1.95	5.30	.73	7.98	11.15
4100	12' deep, 45 S.F./hr. in & 135 S.F./hr. out		270	.148		2	5.90	.81	8.71	12.20
4200	14' deep, 42 S.F./hr. in & 126 S.F./hr. out		250	.160		2.06	6.40	.88	9.34	13.05
4300	16' deep, 40 S.F./hr. in & 120 S.F./hr. out		240	.167		2.13	6.65	.91	9.69	13.60
4400	18' deep, 38 S.F./hr. in & 114 S.F./hr. out		230	.174		2.20	6.95	.95	10.10	14.10
4520	Left in place, 8' deep, 55 S.F./hr.		440	.091		3.41	3.62	.50	7.53	9.90
4540	10' deep, 50 S.F./hr.		400	.100		3.58	3.99	.55	8.12	10.70
4560	12' deep, 45 S.F./hr.		360	.111		3.78	4.43	.61	8.82	11.65
4565	14' deep, 42 S.F./hr.		335	.119		4.01	4.76	.65	9.42	12.50

For customer support on your Mechanical Cost Data, call 877.795.7394.

31 41 Shoring

31 41 16 – Sheet Piling

31 41 16.10 Sheet Piling Systems	Crew	Daily Output	Labor-Hours	Unit	Material	2015 Bare Costs Labor	Equipment	Total	Total Incl O&P
4570	16' deep, 40 S.F./hr.	B-31	320	.125	S.F.	4.26	4.98	.68	9.92	13.10
4700	Alternate pricing, left in place, 8' deep		1.76	22.727	M.B.F.	765	905	124	1,794	2,375
4800	Drive, extract and salvage, 8' deep		1.32	30.303	"	680	1,200	166	2,046	2,775

31 52 Cofferdams

31 52 16 – Timber Cofferdams

31 52 16.10 Cofferdams

		Crew	Daily Output	Labor-Hours	Unit	Material	2015 Bare Costs Labor	Equipment	Total	Total Incl O&P
0010	**COFFERDAMS**									
0011	Incl. mobilization and temporary sheeting									
0080	Soldier beams & lagging H piles with 3" wood sheeting									
0090	horizontal between piles, including removal of wales & braces									
0100	No hydrostatic head, 15' deep, 1 line of braces, minimum	B-50	545	.206	S.F.	8.50	9.30	4.48	22.28	29
0200	Maximum		495	.226		9.45	10.25	4.93	24.63	32
1300	No hydrostatic head, left in place, 15' dp., 1 line of braces, min.		635	.176		11.35	8	3.85	23.20	29
1400	Maximum		575	.195		12.15	8.80	4.25	25.20	32
2350	Lagging only, 3" thick wood between piles 8' O.C., minimum	B-46	400	.120		1.89	5.05	.11	7.05	10.10
2370	Maximum		250	.192		2.84	8.10	.18	11.12	15.95
2400	Open sheeting no bracing, for trenches to 10' deep, min.		1736	.028		.85	1.17	.03	2.05	2.79
2450	Maximum		1510	.032		.95	1.34	.03	2.32	3.17
2500	Tie-back method, add to open sheeting, add, minimum								20%	20%
2550	Maximum								60%	60%
2700	Tie-backs only, based on tie-backs total length, minimum	B-46	86.80	.553	L.F.	14.60	23.50	.52	38.62	53
2750	Maximum		38.50	1.247	"	25.50	52.50	1.17	79.17	112
3500	Tie-backs only, typical average, 25' long		2	24	Ea.	645	1,000	22.50	1,667.50	2,300
3600	35' long		1.58	30.380	"	855	1,275	28.50	2,158.50	2,975

Division Notes

	CREW	DAILY OUTPUT	LABOR-HOURS	UNIT	BARE COSTS				TOTAL INCL O&P
					MAT.	LABOR	EQUIP.	TOTAL	

Estimating Tips

32 01 00 Operations and Maintenance of Exterior Improvements

- Recycling of asphalt pavement is becoming very popular and is an alternative to removal and replacement. It can be a good value engineering proposal if removed pavement can be recycled, either at the project site or at another site that is reasonably close to the project site. Sections on repair of flexible and rigid pavement are included.

32 10 00 Bases, Ballasts, and Paving

- When estimating paving, keep in mind the project schedule. Also note that prices for asphalt and concrete are generally higher in the cold seasons. Lines for pavement markings, including tactile warning systems and fence lines, are included.

32 90 00 Planting

- The timing of planting and guarantee specifications often dictate the costs for establishing tree and shrub growth and a stand of grass or ground cover. Establish the work performance schedule to coincide with the local planting season. Maintenance and growth guarantees can add from 20%–100% to the total landscaping cost and can be contractually cumbersome. The cost to replace trees and shrubs can be as high as 5% of the total cost, depending on the planting zone, soil conditions, and time of year.

Reference Numbers

Reference numbers are shown in shaded boxes at the beginning of some major classifications. These numbers refer to related items in the Reference Section. The reference information may be an estimating procedure, an alternate pricing method, or technical information.

Note: Not all subdivisions listed here necessarily appear in this publication. ■

32 12 Flexible Paving

32 12 16 – Asphalt Paving

32 12 16.13 Plant-Mix Asphalt Paving

		Crew	Daily Output	Labor-Hours	Unit	Material	2015 Bare Costs Labor	2015 Bare Costs Equipment	Total	Total Incl O&P
0010	**PLANT-MIX ASPHALT PAVING**									
0020	And large paved areas with no hauling included									
0025	See Section 31 23 23.20 for hauling costs									
0080	Binder course, 1-1/2" thick	B-25	7725	.011	S.Y.	5.70	.47	.35	6.52	7.40
0120	2" thick		6345	.014		7.60	.57	.43	8.60	9.70
0130	2-1/2" thick		5620	.016		9.50	.65	.48	10.63	11.95
0160	3" thick		4905	.018		11.40	.74	.55	12.69	14.30
0170	3-1/2" thick		4520	.019		13.30	.80	.60	14.70	16.55
0200	4" thick		4140	.021		15.25	.88	.66	16.79	18.80
0300	Wearing course, 1" thick	B-25B	10575	.009		3.78	.38	.28	4.44	5.05
0340	1-1/2" thick		7725	.012		6.35	.52	.38	7.25	8.20
0380	2" thick		6345	.015		8.50	.64	.46	9.60	10.90
0420	2-1/2" thick		5480	.018		10.50	.74	.54	11.78	13.25
0460	3" thick		4900	.020		12.55	.82	.60	13.97	15.70
0470	3-1/2" thick		4520	.021		14.70	.89	.65	16.24	18.30
0480	4" thick		4140	.023		16.80	.98	.71	18.49	21
0500	Open graded friction course	B-25C	5000	.010		2.13	.41	.47	3.01	3.48
0800	Alternate method of figuring paving costs									
0810	Binder course, 1-1/2" thick	B-25	630	.140	Ton	70	5.75	4.31	80.06	90.50
0811	2" thick		690	.128		70	5.25	3.93	79.18	89.50
0812	3" thick		800	.110		70	4.55	3.39	77.94	87.50
0813	4" thick		900	.098		70	4.04	3.01	77.05	86.50
0850	Wearing course, 1" thick	B-25B	575	.167		69	7.05	5.15	81.20	92.50
0851	1-1/2" thick		630	.152		69	6.40	4.68	80.08	91
0852	2" thick		690	.139		69	5.85	4.28	79.13	89.50
0853	2-1/2" thick		765	.125		69	5.30	3.86	78.16	88.50
0854	3" thick		800	.120		69	5.05	3.69	77.74	88
1000	Pavement replacement over trench, 2" thick	B-37	90	.533	S.Y.	7.85	21	1.75	30.60	43
1050	4" thick		70	.686		15.55	27.50	2.25	45.30	61.50
1080	6" thick		55	.873		25	34.50	2.86	62.36	83.50

32 12 16.14 Asphaltic Concrete Paving

		Crew	Daily Output	Labor-Hours	Unit	Material	2015 Bare Costs Labor	2015 Bare Costs Equipment	Total	Total Incl O&P
0011	**ASPHALTIC CONCRETE PAVING**, parking lots & driveways									
0015	No asphalt hauling included									
0018	Use 6.05 C.Y. per inch per M.S.F. for hauling									
0020	6" stone base, 2" binder course, 1" topping	B-25C	9000	.005	S.F.	1.86	.23	.26	2.35	2.68
0025	2" binder course, 2" topping		9000	.005		2.28	.23	.26	2.77	3.14
0030	3" binder course, 2" topping		9000	.005		2.71	.23	.26	3.20	3.61
0035	4" binder course, 2" topping		9000	.005		3.13	.23	.26	3.62	4.07
0040	1.5" binder course, 1" topping		9000	.005		1.65	.23	.26	2.14	2.45
0042	3" binder course, 1" topping		9000	.005		2.29	.23	.26	2.78	3.15
0045	3" binder course, 3" topping		9000	.005		3.13	.23	.26	3.62	4.07
0050	4" binder course, 3" topping		9000	.005		3.55	.23	.26	4.04	4.53
0055	4" binder course, 4" topping		9000	.005		3.96	.23	.26	4.45	4.99
0300	Binder course, 1-1/2" thick		35000	.001		.64	.06	.07	.77	.86
0400	2" thick		25000	.002		.82	.08	.09	.99	1.13
0500	3" thick		15000	.003		1.27	.14	.16	1.57	1.78
0600	4" thick		10800	.004		1.67	.19	.22	2.08	2.36
0800	Sand finish course, 3/4" thick		41000	.001		.32	.05	.06	.43	.49
0900	1" thick		34000	.001		.40	.06	.07	.53	.61
1000	Fill pot holes, hot mix, 2" thick	B-16	4200	.008		.85	.30	.16	1.31	1.57
1100	4" thick		3500	.009		1.25	.35	.20	1.80	2.13
1120	6" thick		3100	.010		1.67	.40	.22	2.29	2.70

32 12 Flexible Paving

32 12 16 – Asphalt Paving

32 12 16.14 Asphaltic Concrete Paving	Crew	Daily Output	Labor-Hours	Unit	Material	2015 Bare Costs Labor	Equipment	Total	Total Incl O&P
1140	Cold patch, 2" thick	B-51	3000	.016	S.F.	.99	.61	.08	1.68	2.12
1160	4" thick		2700	.018		1.89	.68	.09	2.66	3.22
1180	6" thick		1900	.025		2.94	.96	.13	4.03	4.86

32 84 Planting Irrigation

32 84 23 – Underground Sprinklers

32 84 23.10 Sprinkler Irrigation System

		Crew	Daily Output	Labor-Hours	Unit	Material	2015 Bare Costs Labor	Equipment	Total	Total Incl O&P
0010	**SPRINKLER IRRIGATION SYSTEM**									
0011	For lawns									
0100	Golf course with fully automatic system	C-17	.05	1600	9 holes	100,000	78,500		178,500	231,000
0200	24' diam. head at 15' O.C. incl. piping, auto oper., minimum	B-20	70	.343	Head	27	14.40		41.40	51.50
0300	Maximum		40	.600		45	25		70	88.50
0600	Sprinkler irrigation sys, golf course, auto sys, 60' dia HD		23	1.043		150	44		194	233
0800	Residential system, custom, 1" supply		2000	.012	S.F.	.26	.50		.76	1.07
0900	1-1/2" supply		1800	.013	"	.49	.56		1.05	1.40

For customer support on your Mechanical Cost Data, call 877.795.7394.

459

Division Notes

	CREW	DAILY OUTPUT	LABOR-HOURS	UNIT	BARE COSTS				TOTAL INCL O&P
					MAT.	LABOR	EQUIP.	TOTAL	

Estimating Tips

33 10 00 Water Utilities
33 30 00 Sanitary Sewerage Utilities
33 40 00 Storm Drainage Utilities

- Never assume that the water, sewer, and drainage lines will go in at the early stages of the project. Consider the site access needs before dividing the site in half with open trenches, loose pipe, and machinery obstructions. Always inspect the site to establish that the site drawings are complete. Check off all existing utilities on your drawings as you locate them. Be especially careful with underground utilities because appurtenances are sometimes buried during regrading or repaving operations. If you find any discrepancies, mark up the site plan for further research. Differing site conditions can be very costly if discovered later in the project.

- See also Section 33 01 00 for restoration of pipe where removal/replacement may be undesirable. Use of new types of piping materials can reduce the overall project cost. Owners/design engineers should consider the installing contractor as a valuable source of current information on utility products and local conditions that could lead to significant cost savings.

Reference Numbers

Reference numbers are shown in shaded boxes at the beginning of some major classifications. These numbers refer to related items in the Reference Section. The reference information may be an estimating procedure, an alternate pricing method, or technical information.

Note: Not all subdivisions listed here necessarily appear in this publication. ■

*Note: **Trade Service**, in part, has been used as a reference source for some of the material prices used in Division 33.*

33 01 10 – Operation and Maintenance of Water Utilities

33 01 10.10 Corrosion Resistance	Crew	Daily Output	Labor-Hours	Unit	Material	2015 Bare Costs Labor	Equipment	Total	Total Incl O&P
0010 **CORROSION RESISTANCE**									
0012 Wrap & coat, add to pipe, 4" diameter				L.F.	2.21			2.21	2.4
0020 5" diameter					2.66			2.66	2.93
0040 6" diameter					3.28			3.28	3.6
0060 8" diameter					4.04			4.04	4.4
0080 10" diameter					4.91			4.91	5.4
0100 12" diameter					6.20			6.20	6.8
0120 14" diameter					7.30			7.30	8
0140 16" diameter					9.75			9.75	10.7
0160 18" diameter					10			10	11
0180 20" diameter					10.50			10.50	11.5
0200 24" diameter					12.95			12.95	14.2
0220 Small diameter pipe, 1" diameter, add					1.02			1.02	1.1
0240 2" diameter					1.38			1.38	1.5
0260 2-1/2" diameter					1.62			1.62	1.7
0280 3" diameter					2.01			2.01	2.2
0300 Fittings, field covered, add				S.F.	10.40			10.40	11.4
0500 Coating, bituminous, per diameter inch, 1 coat, add				L.F.	.63			.63	.6
0540 3 coat					1.92			1.92	2.1
0560 Coal tar epoxy, per diameter inch, 1 coat, add					.22			.22	.2
0600 3 coat					.68			.68	.7
1000 Polyethylene H.D. extruded, .025" thk., 1/2" diameter add					.10			.10	.1
1020 3/4" diameter					.15			.15	.1
1040 1" diameter					.20			.20	.2
1060 1-1/4" diameter					.25			.25	.2
1080 1-1/2" diameter					.30			.30	.3
1100 .030" thk., 2" diameter					.40			.40	.4
1120 2-1/2" diameter					.50			.50	.5
1140 .035" thk., 3" diameter					.60			.60	.6
1160 3-1/2" diameter					.70			.70	.7
1180 4" diameter					.80			.80	.8
1200 .040" thk., 5" diameter					1			1	1.1
1220 6" diameter					1.20			1.20	1.3
1240 8" diameter					1.60			1.60	1.7
1260 10" diameter					2			2	2.2
1280 12" diameter					2.40			2.40	2.6
1300 .060" thk., 14" diameter					2.80			2.80	3.0
1320 16" diameter					3.20			3.20	3.5
1340 18" diameter					3.60			3.60	3.9
1360 20" diameter					4			4	4.4
1380 Fittings, field wrapped, add				S.F.	4.79			4.79	5.2

33 01 10.20 Pipe Repair

	Crew	Daily Output	Labor-Hours	Unit	Material	2015 Bare Costs Labor	Equipment	Total	Total Incl O&P
0010 **PIPE REPAIR**									
0020 Not including excavation or backfill									
0100 Clamp, stainless steel, lightweight, for steel pipe									
0110 3" long, 1/2" diameter pipe	1 Plum	34	.235	Ea.	12.80	13.80		26.60	35
0120 3/4" diameter pipe		32	.250		13.05	14.70		27.75	36.5
0130 1" diameter pipe		30	.267		14.25	15.65		29.90	39
0140 1-1/4" diameter pipe		28	.286		14.45	16.75		31.20	41.5
0150 1-1/2" diameter pipe		26	.308		15.35	18.05		33.40	44.5
0160 2" diameter pipe		24	.333		16.50	19.55		36.05	47.5
0170 2-1/2" diameter pipe		23	.348		17.50	20.50		38	50.5

33 01 10.20 Pipe Repair	Crew	Daily Output	Labor-Hours	Unit	Material	2015 Bare Costs Labor	Equipment	Total	Total Incl O&P
0180	3" diameter pipe	1 Plum	22	.364	Ea.	21	21.50		42.50	55
0190	3-1/2" diameter pipe	↓	21	.381		25	22.50		47.50	61.50
0200	4" diameter pipe	B-20	44	.545		25	23		48	63
0210	5" diameter pipe		42	.571		29.50	24		53.50	69.50
0220	6" diameter pipe		38	.632		34	26.50		60.50	78.50
0230	8" diameter pipe		30	.800		40.50	33.50		74	96
0240	10" diameter pipe		28	.857		115	36		151	183
0250	12" diameter pipe		24	1		120	42		162	197
0260	14" diameter pipe		22	1.091		125	46		171	209
0270	16" diameter pipe		20	1.200		135	50.50		185.50	227
0280	18" diameter pipe		18	1.333		145	56		201	246
0290	20" diameter pipe		16	1.500		150	63		213	262
0300	24" diameter pipe	↓	14	1.714		170	72		242	298
0360	For 6" long, add					100%	40%			
0370	For 9" long, add					200%	100%			
0380	For 12" long, add					300%	150%			
0390	For 18" long, add				↓	500%	200%			
0400	Pipe Freezing for live repairs of systems 3/8 inch to 6 inch									
0410	Note: Pipe Freezing can also be used to install a valve into a live system									
0420	Pipe Freezing each side 3/8 inch	2 Skwk	8	2	Ea.	540	97.50		637.50	745
0425	Pipe Freezing each side 3/8 inch, second location same kit		8	2		22.50	97.50		120	175
0430	Pipe Freezing each side 3/4 inch		8	2		495	97.50		592.50	695
0435	Pipe Freezing each side 3/4 inch, second location same kit		8	2		22.50	97.50		120	175
0440	Pipe Freezing each side 1-1/2 inch		6	2.667		495	130		625	745
0445	Pipe Freezing each side 1-1/2 inch , second location same kit		6	2.667		22.50	130		152.50	225
0450	Pipe Freezing each side 2 inch		6	2.667		890	130		1,020	1,175
0455	Pipe Freezing each side 2 inch, second location same kit		6	2.667		22.50	130		152.50	225
0460	Pipe Freezing each side 2-1/2 inch-3 inch		6	2.667		925	130		1,055	1,225
0465	Pipe Freeze each side 2-1/2 -3 inch, second location same kit		6	2.667		45.50	130		175.50	250
0470	Pipe Freezing each side 4 inch		4	4		1,475	195		1,670	1,925
0475	Pipe Freezing each side 4 inch, second location same kit		4	4		72	195		267	380
0480	Pipe Freezing each side 5-6 inch		4	4		4,000	195		4,195	4,700
0485	Pipe Freezing each side 5-6 inch, second location same kit	↓	4	4		216	195		411	540
0490	Pipe Freezing extra 20 lb. CO$_2$ cylinders (3/8" to 2" - 1 ea, 3" -2 ea)					222			222	244
0500	Pipe Freezing extra 50 lb. CO$_2$ cylinders (4" - 2 ea, 5"-6" -6 ea)				↓	440			440	485
1000	Clamp, stainless steel, with threaded service tap									
1040	Full seal for iron, steel, PVC pipe									
1100	6" long, 2" diameter pipe	1 Plum	17	.471	Ea.	87	27.50		114.50	137
1110	2-1/2" diameter pipe		16	.500		90	29.50		119.50	144
1120	3" diameter pipe		15.60	.513		104	30		134	161
1130	3-1/2" diameter pipe	↓	15	.533		110	31.50		141.50	169
1140	4" diameter pipe	B-20	32	.750		117	31.50		148.50	177
1150	6" diameter pipe		28	.857		145	36		181	216
1160	8" diameter pipe		21	1.143		171	48		219	262
1170	10" diameter pipe		20	1.200		224	50.50		274.50	325
1180	12" diameter pipe	↓	17	1.412		261	59		320	380
1200	8" long, 2" diameter pipe	1 Plum	11.72	.683		178	40		218	257
1210	2-1/2" diameter pipe		11	.727		183	42.50		225.50	266
1220	3" diameter pipe		10.75	.744		189	43.50		232.50	274
1230	3-1/2" diameter pipe	↓	10.34	.774		199	45.50		244.50	288
1240	4" diameter pipe	B-20	22	1.091		206	46		252	297
1250	6" diameter pipe		19.31	1.243		234	52		286	340
1260	8" diameter pipe	↓	14.48	1.657		264	69.50		333.50	400

33 01 10.20 Pipe Repair		Crew	Daily Output	Labor-Hours	Unit	Material	2015 Bare Costs Labor	Equipment	Total	Total Incl O&P
1270	10" diameter pipe	B-20	13.80	1.739	Ea.	330	73		403	475
1280	12" diameter pipe	▼	11.72	2.048		370	86		456	535
1300	12" long, 2" diameter pipe	1 Plum	9.44	.847		257	50		307	360
1310	2-1/2" diameter pipe		8.89	.900		262	53		315	370
1320	3" diameter pipe		8.67	.923		269	54		323	380
1330	3-1/2" diameter pipe	▼	8.33	.960		280	56.50		336.50	395
1340	4" diameter pipe	B-20	17.78	1.350		450	56.50		506.50	585
1350	6" diameter pipe		15.56	1.542		340	64.50		404.50	470
1360	8" diameter pipe		11.67	2.057		390	86.50		476.50	560
1370	10" diameter pipe		11.11	2.160		485	90.50		575.50	675
1380	12" diameter pipe	▼	9.44	2.542		555	107		662	775
1400	20" long, 2" diameter pipe	1 Plum	8.10	.988		299	58		357	420
1410	2-1/2" diameter pipe		7.62	1.050		325	61.50		386.50	450
1420	3" diameter pipe		7.43	1.077		365	63		428	495
1430	3-1/2" diameter pipe	▼	7.14	1.120		380	66		446	520
1440	4" diameter pipe	B-20	15.24	1.575		450	66		516	595
1450	6" diameter pipe		13.33	1.800		525	75.50		600.50	690
1460	8" diameter pipe		10	2.400		590	101		691	805
1470	10" diameter pipe		9.52	2.521		715	106		821	950
1480	12" diameter pipe	▼	8.10	2.963	▼	835	124		959	1,100
1600	Clamp, stainless steel, single section									
1640	Full seal for iron, steel, PVC pipe									
1700	6" long, 2" diameter pipe	1 Plum	17	.471	Ea.	87	27.50		114.50	137
1710	2-1/2" diameter pipe		16	.500		90	29.50		119.50	144
1720	3" diameter pipe		15.60	.513		104	30		134	161
1730	3-1/2" diameter pipe	▼	15	.533		110	31.50		141.50	169
1740	4" diameter pipe	B-20	32	.750		117	31.50		148.50	177
1750	6" diameter pipe		27	.889		145	37.50		182.50	218
1760	8" diameter pipe		21	1.143		171	48		219	262
1770	10" diameter pipe		20	1.200		224	50.50		274.50	325
1780	12" diameter pipe	▼	17	1.412		261	59		320	380
1800	8" long, 2" diameter pipe	1 Plum	11.72	.683		118	40		158	191
1805	2-1/2" diameter pipe		11.03	.725		123	42.50		165.50	200
1810	3" diameter pipe		10.76	.743		129	43.50		172.50	208
1815	3-1/2" diameter pipe	▼	10.34	.774		139	45.50		184.50	222
1820	4" diameter pipe	B-20	22.07	1.087		146	45.50		191.50	231
1825	6" diameter pipe		19.31	1.243		174	52		226	273
1830	8" diameter pipe		14.48	1.657		204	69.50		273.50	330
1835	10" diameter pipe		13.79	1.740		272	73		345	410
1840	12" diameter pipe	▼	11.72	2.048		310	86		396	470
1850	12" long, 2" diameter pipe	1 Plum	9.44	.847		197	50		247	291
1855	2-1/2" diameter pipe		8.89	.900		201	53		254	300
1860	3" diameter pipe		8.67	.923		209	54		263	310
1865	3-1/2" diameter pipe	▼	8.33	.960		220	56.50		276.50	325
1870	4" diameter pipe	B-20	17.78	1.350		231	56.50		287.50	340
1875	6" diameter pipe		15.56	1.542		278	64.50		342.50	405
1880	8" diameter pipe		11.67	2.057		330	86.50		416.50	495
1885	10" diameter pipe		11.11	2.160		425	90.50		515.50	610
1890	12" diameter pipe	▼	9.44	2.542		495	107		602	710
1900	20" long, 2" diameter pipe	1 Plum	8.10	.988		239	58		297	350
1905	2-1/2" diameter pipe		7.62	1.050		264	61.50		325.50	385
1910	3" diameter pipe		7.43	1.077		305	63		368	430
1915	3-1/2" diameter pipe	▼	7.14	1.120		320	66		386	450

33 01 Operation and Maintenance of Utilities

33 01 10 – Operation and Maintenance of Water Utilities

33 01 10.20 Pipe Repair

		Crew	Daily Output	Labor-Hours	Unit	Material	2015 Bare Costs Labor	Equipment	Total	Total Incl O&P
1920	4" diameter pipe	B-20	15.24	1.575	Ea.	390	66		456	530
1925	6" diameter pipe		13.33	1.800		465	75.50		540.50	625
1930	8" diameter pipe		10	2.400		530	101		631	740
1935	10" diameter pipe		9.52	2.521		655	106		761	885
1940	12" diameter pipe		8.10	2.963		775	124		899	1,050
2000	Clamp, stainless steel, two section									
2040	Full seal, for iron, steel, PVC pipe									
2100	6" long, 4" diameter pipe	B-20	24	1	Ea.	230	42		272	320
2110	6" diameter pipe		20	1.200		264	50.50		314.50	370
2120	8" diameter pipe		13	1.846		300	77.50		377.50	450
2130	10" diameter pipe		12	2		305	84		389	465
2140	12" diameter pipe		10	2.400		390	101		491	585
2200	9" long, 4" diameter pipe		16	1.500		300	63		363	425
2210	6" diameter pipe		13	1.846		340	77.50		417.50	495
2220	8" diameter pipe		9	2.667		375	112		487	585
2230	10" diameter pipe		8	3		495	126		621	740
2240	12" diameter pipe		7	3.429		565	144		709	845
2250	14" diameter pipe		6.40	3.750		760	157		917	1,075
2260	16" diameter pipe		6	4		705	168		873	1,050
2270	18" diameter pipe		5	4.800		800	201		1,001	1,200
2280	20" diameter pipe		4.60	5.217		885	219		1,104	1,300
2290	24" diameter pipe		4	6		1,175	252		1,427	1,700
2320	For 12" long, add to 9"					15%	25%			
2330	For 18" long, add to 9"					70%	55%			
8000	For internal cleaning and inspection, see Section 33 01 30.16									
8100	For pipe testing, see Section 23 05 93.50									

33 01 30 – Operation and Maintenance of Sewer Utilities

33 01 30.16 TV Inspection of Sewer Pipelines

		Crew	Daily Output	Labor-Hours	Unit	Material	2015 Bare Costs Labor	Equipment	Total	Total Incl O&P
0010	**TV INSPECTION OF SEWER PIPELINES**									
0100	Pipe internal cleaning & inspection, cleaning, pressure pipe systems									
0120	Pig method, lengths 1000' to 10,000'									
0140	4" diameter thru 24" diameter, minimum				L.F.				3.60	4.14
0160	Maximum				"				18	21
6000	Sewage/sanitary systems									
6100	Power rodder with header & cutters									
6110	Mobilization charge, minimum				Total				695	800
6120	Mobilization charge, maximum				"				9,125	10,600
6140	Cleaning 4"-12" diameter				L.F.				3.39	3.90
6190	14"-24" diameter								3.98	4.59
6240	30" diameter								5.80	6.65
6250	36" diameter								6.75	7.75
6260	48" diameter								7.70	8.85
6270	60" diameter								8.70	9.95
6280	72" diameter								9.65	11.10
9000	Inspection, television camera with video									
9060	up to 500 linear feet				Total				715	820

33 01 30.71 Rehabilitation of Sewer Utilities

		Crew	Daily Output	Labor-Hours	Unit	Material	2015 Bare Costs Labor	Equipment	Total	Total Incl O&P
0010	**REHABILITATION OF SEWER UTILITIES**									
0011	300' runs, replace with HDPE pipe									
0020	Not including excavation, backfill, shoring, or dewatering									
0100	6" to 15" diameter, minimum				L.F.				104	114
0200	Maximum								214	236

For customer support on your Mechanical Cost Data, call 877.795.7394.

465

33 01 Operation and Maintenance of Utilities

33 01 30 – Operation and Maintenance of Sewer Utilities

33 01 30.71 Rehabilitation of Sewer Utilities	Crew	Daily Output	Labor-Hours	Unit	Material	2015 Bare Costs Labor	Equipment	Total	Total Incl O&P
0300	18" to 36" diameter, minimum				L.F.				208	229
0400	Maximum				↓				435	510
0500	Mobilize and demobilize, minimum				Job				3,000	3,300
0600	Maximum				"				32,100	35,400

33 01 30.74 HDPE Pipe Lining	Crew	Daily Output	Labor-Hours	Unit	Material	2015 Bare Costs Labor	Equipment	Total	Total Incl O&P
0010	**HDPE PIPE LINING**, excludes cleaning and video inspection									
0020	Pipe relined with one pipe size smaller than original (4" for 6")									
0100	6" diameter, original size	B-6B	600	.080	L.F.	2.47	3.06	1.50	7.03	9.10
0150	8" diameter, original size		600	.080		5.85	3.06	1.50	10.41	12.80
0200	10" diameter, original size		600	.080		8.55	3.06	1.50	13.11	15.75
0250	12" diameter, original size		400	.120		14.65	4.59	2.26	21.50	25.50
0300	14" diameter, original size	↓	400	.120		17.75	4.59	2.26	24.60	29
0350	16" diameter, original size	B-6C	300	.160		27.50	6.10	5.35	38.95	45.50
0400	18" diameter, original size	"	300	.160	↓	32.50	6.10	5.35	43.95	51
1000	Pipe HDPE lining, make service line taps	B-6	4	6	Ea.	97	248	91	436	585

33 05 Common Work Results for Utilities

33 05 23 – Trenchless Utility Installation

33 05 23.19 Microtunneling

		Crew	Daily Output	Labor-Hours	Unit	Material	2015 Bare Costs Labor	Equipment	Total	Total Incl O&P
0010	**MICROTUNNELING**									
0011	Not including excavation, backfill, shoring,									
0020	or dewatering, average 50'/day, slurry method									
0100	24" to 48" outside diameter, minimum				L.F.				875	965
0110	Adverse conditions, add				%				50%	50%
1000	Rent microtunneling machine, average monthly lease				Month				97,500	107,000
1010	Operating technician				Day				630	705
1100	Mobilization and demobilization, minimum				Job				41,200	45,900
1110	Maximum				"				445,500	490,500

33 05 23.20 Horizontal Boring

		Crew	Daily Output	Labor-Hours	Unit	Material	2015 Bare Costs Labor	Equipment	Total	Total Incl O&P
0010	**HORIZONTAL BORING**									
0011	Casing only, 100' minimum,									
0020	not incl. jacking pits or dewatering									
0100	Roadwork, 1/2" thick wall, 24" diameter casing	B-42	20	3.200	L.F.	121	136	68	325	420
0200	36" diameter		16	4		223	170	85	478	605
0300	48" diameter		15	4.267		310	181	90.50	581.50	725
0500	Railroad work, 24" diameter		15	4.267		121	181	90.50	392.50	515
0600	36" diameter		14	4.571		223	194	97	514	655
0700	48" diameter	↓	12	5.333		310	226	113	649	820
0900	For ledge, add								20%	20%
1000	Small diameter boring, 3", sandy soil	B-82	900	.018	↓	22	.77	.09	22.86	25.50
1040	Rocky soil	"	500	.032		22	1.38	.17	23.55	26.50
1100	Prepare jacking pits, incl. mobilization & demobilization, minimum				Ea.				3,225	3,700
1101	Maximum				"				22,000	25,500

33 05 26 – Utility Identification

33 05 26.05 Utility Connection

		Crew	Daily Output	Labor-Hours	Unit	Material	2015 Bare Costs Labor	Equipment	Total	Total Incl O&P
0010	**UTILITY CONNECTION**									
0020	Water, sanitary, stormwater, gas, single connection	B-14	1	48	Ea.	3,225	1,900	365	5,490	6,875
0030	Telecommunication	"	1	48	"	395	1,900	365	2,660	3,750

466

For customer support on your Mechanical Cost Data, call 877.795.7394.

33 05 Common Work Results for Utilities

33 05 26 – Utility Identification

33 05 26.10 Utility Accessories		Crew	Daily Output	Labor-Hours	Unit	Material	2015 Bare Costs Labor	2015 Bare Costs Equipment	Total	Total Incl O&P
0010	**UTILITY ACCESSORIES**	G1030–805								
0400	Underground tape, detectable, reinforced, alum. foil core, 2"	1 Clab	150	.053	C.L.F.	6.50	2.01		8.51	10.25
0500	6"	"	140	.057	"	27.50	2.15		29.65	33.50

33 11 Water Utility Distribution Piping

33 11 13 – Public Water Utility Distribution Piping

33 11 13.15 Water Supply, Ductile Iron Pipe

		Crew	Daily Output	Labor-Hours	Unit	Material	2015 Bare Costs Labor	2015 Bare Costs Equipment	Total	Total Incl O&P
0010	**WATER SUPPLY, DUCTILE IRON PIPE**	R331113-80								
0020	Not including excavation or backfill									
2000	Pipe, class 50 water piping, 18' lengths									
2020	Mechanical joint, 4" diameter	B-21A	200	.200	L.F.	30.50	9.40	2.37	42.27	50.50
2040	6" diameter		160	.250		32	11.75	2.96	46.71	56
2060	8" diameter		133.33	.300		44.50	14.05	3.55	62.10	74.50
2080	10" diameter		114.29	.350		58.50	16.40	4.14	79.04	93.50
2100	12" diameter		105.26	.380		79	17.85	4.50	101.35	119
2120	14" diameter		100	.400		93	18.75	4.74	116.49	136
2140	16" diameter		72.73	.550		94.50	26	6.50	127	151
2160	18" diameter		68.97	.580		126	27	6.85	159.85	188
2170	20" diameter		57.14	.700		127	33	8.30	168.30	199
2180	24" diameter		47.06	.850		141	40	10.05	191.05	227
3000	Push-on joint, 4" diameter		400	.100		21	4.69	1.18	26.87	31.50
3020	6" diameter		333.33	.120		17	5.65	1.42	24.07	29
3040	8" diameter		200	.200		23	9.40	2.37	34.77	42.50
3060	10" diameter		181.82	.220		36.50	10.30	2.60	49.40	59
3080	12" diameter		160	.250		38.50	11.75	2.96	53.21	63.50
3100	14" diameter		133.33	.300		42.50	14.05	3.55	60.10	72.50
3120	16" diameter		114.29	.350		51.50	16.40	4.14	72.04	86
3140	18" diameter		100	.400		57	18.75	4.74	80.49	96
3160	20" diameter		88.89	.450		59.50	21	5.35	85.85	103
3180	24" diameter	↓	76.92	.520	↓	59	24.50	6.15	89.65	109
8000	Piping, fittings, mechanical joint, AWWA C110									
8006	90° bend, 4" diameter	B-20A	16	2	Ea.	155	91.50		246.50	310
8020	6" diameter		12.80	2.500		229	114		343	425
8040	8" diameter	↓	10.67	2.999		450	137		587	705
8060	10" diameter	B-21A	11.43	3.500		620	164	41.50	825.50	975
8080	12" diameter		10.53	3.799		880	178	45	1,103	1,300
8100	14" diameter		10	4		1,200	188	47.50	1,435.50	1,650
8120	16" diameter		7.27	5.502		1,525	258	65	1,848	2,150
8140	18" diameter		6.90	5.797		2,125	272	68.50	2,465.50	2,850
8160	20" diameter		5.71	7.005		2,650	330	83	3,063	3,525
8180	24" diameter		4.70	8.511		4,200	400	101	4,701	5,325
8200	Wye or tee, 4" diameter	B-20A	10.67	2.999		355	137		492	600
8220	6" diameter		8.53	3.751		535	171		706	850
8240	8" diameter	↓	7.11	4.501		850	206		1,056	1,250
8260	10" diameter	B-21A	7.62	5.249		1,225	246	62	1,533	1,800
8280	12" diameter		7.02	5.698		1,625	267	67.50	1,959.50	2,250
8300	14" diameter		6.67	5.997		2,600	281	71	2,952	3,375
8320	16" diameter		4.85	8.247		2,900	385	97.50	3,382.50	3,875
8340	18" diameter		4.60	8.696		3,900	410	103	4,413	5,025
8360	20" diameter	↓	3.81	10.499		5,475	490	124	6,089	6,900

For customer support on your Mechanical Cost Data, call 877.795.7394.

467

33 11 13.15 Water Supply, Ductile Iron Pipe

		Crew	Daily Output	Labor-Hours	Unit	Material	2015 Bare Costs Labor	Equipment	Total	Total Incl O&P
8380	24" diameter	B-21A	3.14	12.739	Ea.	9,275	600	151	10,026	11,300
8398	45° bends, 4" diameter	B-20A	16	2		255	91.50		346.50	420
8400	6" diameter		12.80	2.500		199	114		313	395
8405	8" diameter		10.67	2.999		291	137		428	530
8410	12" diameter	B-21A	10.53	3.799		730	178	45	953	1,125
8420	16" diameter		7.27	5.502		1,275	258	65	1,598	1,875
8430	20" diameter		5.71	7.005		2,075	330	83	2,488	2,875
8440	24" diameter		4.70	8.511		2,900	400	101	3,401	3,900
8450	Decreaser, 6" x 4" diameter	B-20A	14.22	2.250		207	103		310	385
8460	8" x 6" diameter	"	11.64	2.749		310	126		436	530
8470	10" x 6" diameter	B-21A	13.33	3.001		390	141	35.50	566.50	685
8480	12" x 6" diameter		12.70	3.150		550	148	37.50	735.50	870
8490	16" x 6" diameter		10	4		890	188	47.50	1,125.50	1,325
8500	20" x 6" diameter		8.42	4.751		1,625	223	56.50	1,904.50	2,175
8552	For water utility valves see Section 33 12 16									
8700	Joint restraint, ductile iron mechanical joints									
8710	4" diameter	B-20A	32	1	Ea.	29	45.50		74.50	102
8720	6" diameter		25.60	1.250		31.50	57		88.50	122
8730	8" diameter		21.33	1.500		30	68.50		98.50	137
8740	10" diameter		18.28	1.751		38.50	80		118.50	165
8750	12" diameter		16.84	1.900		50	87		137	187
8760	14" diameter		16	2		53	91.50		144.50	198
8770	16" diameter		11.64	2.749		58	126		184	255
8780	18" diameter		11.03	2.901		73	133		206	282
8785	20" diameter		9.14	3.501		78	160		238	330
8790	24" diameter		7.53	4.250		104	194		298	410
9600	Steel sleeve with tap, 4" diameter	B-20	3	8		435	335		770	995
9620	6" diameter		2	12		475	505		980	1,300
9630	8" diameter		2	12		515	505		1,020	1,350

33 11 13.25 Water Supply, Polyvinyl Chloride Pipe

		Crew	Daily Output	Labor-Hours	Unit	Material	2015 Bare Costs Labor	Equipment	Total	Total Incl O&P
0010	**WATER SUPPLY, POLYVINYL CHLORIDE PIPE** R331113-80									
0020	Not including excavation or backfill, unless specified									
2100	PVC pipe, Class 150, 1-1/2" diameter	Q-1A	750	.013	L.F.	.45	.79		1.24	1.69
2120	2" diameter		686	.015		.68	.86		1.54	2.05
2140	2-1/2" diameter		500	.020		1.30	1.18		2.48	3.22
2160	3" diameter	B-20	430	.056		1.40	2.34		3.74	5.15
3010	AWWA C905, PR 100, DR 25									
3030	14" diameter	B-20A	213	.150	L.F.	13.65	6.85		20.50	25.50
3040	16" diameter		200	.160		19.25	7.30		26.55	32
3050	18" diameter		160	.200		24.50	9.15		33.65	40.50
3060	20" diameter		133	.241		30	11		41	50
3070	24" diameter		107	.299		43	13.65		56.65	68.50
3080	30" diameter		80	.400		81	18.30		99.30	117
3090	36" diameter		80	.400		126	18.30		144.30	166
3100	42" diameter		60	.533		168	24.50		192.50	222
3200	48" diameter		60	.533		220	24.50		244.50	279
4520	Pressure pipe Class 150, SDR 18, AWWA C900, 4" diameter		380	.084		2.73	3.85		6.58	8.85
4530	6" diameter		316	.101		5.35	4.63		9.98	12.95
4540	8" diameter		264	.121		8.80	5.55		14.35	18.10
4550	10" diameter		220	.145		14.05	6.65		20.70	25.50
4560	12" diameter		186	.172		19.90	7.85		27.75	34
8000	Fittings with rubber gasket									

468

For customer support on your Mechanical Cost Data, call 877.795.7394.

33 11 13 − Public Water Utility Distribution Piping

33 11 13.25 Water Supply, Polyvinyl Chloride Pipe	Crew	Daily Output	Labor-Hours	Unit	Material	2015 Bare Costs Labor	Equipment	Total	Total Incl O&P
8003 Class 150, DR 18									
8006 90° Bend , 4" diameter	B-20	100	.240	Ea.	42.50	10.05		52.55	62
8020 6" diameter		90	.267		75	11.20		86.20	100
8040 8" diameter		80	.300		145	12.60		157.60	178
8060 10" diameter		50	.480		330	20		350	395
8080 12" diameter		30	.800		425	33.50		458.50	515
8100 Tee, 4" diameter		90	.267		58.50	11.20		69.70	82
8120 6" diameter		80	.300		144	12.60		156.60	177
8140 8" diameter		70	.343		185	14.40		199.40	225
8160 10" diameter		40	.600		590	25		615	690
8180 12" diameter		20	1.200		765	50.50		815.50	920
8200 45° Bend, 4" diameter		100	.240		42	10.05		52.05	62
8220 6" diameter		90	.267		73	11.20		84.20	97.50
8240 8" diameter		50	.480		139	20		159	183
8260 10" diameter		50	.480		278	20		298	335
8280 12" diameter		30	.800		360	33.50		393.50	445
8300 Reducing tee 6" x 4"		100	.240		104	10.05		114.05	131
8320 8" x 6"		90	.267		166	11.20		177.20	199
8330 10" x 6"		90	.267		196	11.20		207.20	232
8340 10" x 8"		90	.267		216	11.20		227.20	255
8350 12" x 6"		90	.267		245	11.20		256.20	286
8360 12" x 8"		90	.267		265	11.20		276.20	310
8400 Tapped service tee (threaded type) 6" x 6" x 3/4"		100	.240		95.50	10.05		105.55	121
8430 6" x 6" x 1"		90	.267		95.50	11.20		106.70	122
8440 6" x 6" x 1-1/2"		90	.267		95.50	11.20		106.70	122
8450 6" x 6" x 2"		90	.267		95.50	11.20		106.70	122
8460 8" x 8" x 3/4"		90	.267		140	11.20		151.20	171
8470 8" x 8" x 1"		90	.267		140	11.20		151.20	171
8480 8" x 8" x 1-1/2"		90	.267		140	11.20		151.20	171
8490 8" x 8" x 2"		90	.267		140	11.20		151.20	171
8500 Repair coupling 4"		100	.240		26	10.05		36.05	44
8520 6" diameter		90	.267		40.50	11.20		51.70	62
8540 8" diameter		50	.480		96.50	20		116.50	137
8560 10" diameter		50	.480		203	20		223	254
8580 12" diameter		50	.480		296	20		316	355
8600 Plug end 4"		100	.240		23	10.05		33.05	41
8620 6" diameter		90	.267		41	11.20		52.20	62.50
8640 8" diameter		50	.480		69	20		89	107
8660 10" diameter		50	.480		97	20		117	137
8680 12" diameter	▼	50	.480	▼	119	20		139	162
8700 PVC pipe, joint restraint									
8710 4" diameter	B-20A	32	1	Ea.	45.50	45.50		91	120
8720 6" diameter		25.60	1.250		56	57		113	149
8730 8" diameter		21.33	1.500		74.50	68.50		143	186
8740 10" diameter		18.28	1.751		130	80		210	265
8750 12" diameter		16.84	1.900		145	87		232	291
8760 14" diameter		16	2		204	91.50		295.50	365
8770 16" diameter		11.64	2.749		265	126		391	485
8780 18" diameter		11.03	2.901		340	133		473	575
8785 20" diameter		9.14	3.501		420	160		580	705
8790 24" diameter	▼	7.53	4.250	▼	490	194		684	835

For customer support on your Mechanical Cost Data, call 877.795.7394.

469

33 11 13.35 Water Supply, HDPE

	Crew	Daily Output	Labor-Hours	Unit	Material	2015 Bare Costs Labor	2015 Bare Costs Equipment	Total	Total Incl O&P
0010 **WATER SUPPLY, HDPE**									
0011 Butt fusion joints, SDR 21 40' lengths not including excavation or backfill									
0100 4" diameter	B-22A	400	.100	L.F.	2.47	4.30	1.64	8.41	11.15
0200 6" diameter		380	.105		5.85	4.53	1.73	12.11	15.30
0300 8" diameter		320	.125		8.55	5.40	2.06	16.01	19.90
0400 10" diameter		300	.133		14.65	5.75	2.19	22.59	27.50
0500 12" diameter		260	.154		17.75	6.60	2.53	26.88	32.50
0600 14" diameter	B-22B	220	.182		27.50	7.80	5	40.30	47.50
0700 16" diameter		180	.222		32.50	9.55	6.15	48.20	57
0800 18" diameter		140	.286		38	12.30	7.90	58.20	69
0900 24" diameter		100	.400		59	17.20	11.05	87.25	104
1000 Fittings									
1100 Elbows, 90 degrees									
1200 4" diameter	B-22A	32	1.250	Ea.	18.45	54	20.50	92.95	126
1300 6" diameter		28	1.429		59	61.50	23.50	144	186
1400 8" diameter		24	1.667		121	71.50	27.50	220	274
1500 10" diameter		18	2.222		385	95.50	36.50	517	610
1600 12" diameter		12	3.333		400	143	55	598	720
1700 14" diameter	B-22B	9	4.444		490	191	123	804	965
1800 16" diameter		6	6.667		570	287	184	1,041	1,275
1900 18" diameter		4	10		660	430	276	1,366	1,700
2000 24" diameter		3	13.333		985	575	370	1,930	2,350
2100 Tees									
2200 4" diameter	B-22A	30	1.333	Ea.	23.50	57.50	22	103	138
2300 6" diameter		26	1.538		58	66	25.50	149.50	194
2400 8" diameter		22	1.818		146	78	30	254	315
2500 10" diameter		15	2.667		193	115	44	352	435
2600 12" diameter		10	4		545	172	66	783	935
2700 14" diameter	B-22B	8	5		640	215	138	993	1,175
2800 16" diameter		6	6.667		755	287	184	1,226	1,475
2900 18" diameter		4	10		985	430	276	1,691	2,050
3000 24" diameter		2	20		1,500	860	555	2,915	3,575
4100 Caps									
4110 4" diameter	B-22A	34	1.176	Ea.	13.60	50.50	19.35	83.45	114
4120 6" diameter		30	1.333		35.50	57.50	22	115	151
4130 8" diameter		26	1.538		52	66	25.50	143.50	187
4150 10" diameter		20	2		111	86	33	230	290
4160 12" diameter		14	2.857		124	123	47	294	380

33 12 Water Utility Distribution Equipment

33 12 13 – Water Service Connections

33 12 13.15 Tapping, Crosses and Sleeves

	Crew	Daily Output	Labor-Hours	Unit	Material	2015 Bare Costs Labor	2015 Bare Costs Equipment	Total	Total Incl O&P
0010 **TAPPING, CROSSES AND SLEEVES**									
4000 Drill and tap pressurized main (labor only)									
4100 6" main, 1" to 2" service	Q-1	3	5.333	Ea.		282		282	425
4150 8" main, 1" to 2" service	"	2.75	5.818	"		305		305	465
4500 Tap and insert gate valve									
4600 8" main, 4" branch	B-21	3.20	8.750	Ea.		380	43	423	635
4650 6" branch		2.70	10.370			450	51	501	745
4700 10" Main, 4" branch		2.70	10.370			450	51	501	745

470

For customer support on your Mechanical Cost Data, call 877.795.7394.

33 12 Water Utility Distribution Equipment

33 12 13 – Water Service Connections

33 12 13.15 Tapping, Crosses and Sleeves

		Crew	Daily Output	Labor-Hours	Unit	Material	2015 Bare Costs Labor	2015 Bare Costs Equipment	Total	Total Incl O&P
4750	6" branch	B-21	2.35	11.915	Ea.		515	58.50	573.50	860
4800	12" main, 6" branch		2.35	11.915			515	58.50	573.50	860
4850	8" branch		2.35	11.915			515	58.50	573.50	860
7020	Crosses, 4" x 4"		37	.757		1,375	33	3.72	1,411.72	1,550
7030	6" x 4"		25	1.120		1,150	48.50	5.50	1,204	1,350
7040	6" x 6"		25	1.120		1,175	48.50	5.50	1,229	1,375
7060	8" x 6"		21	1.333		1,500	58	6.55	1,564.55	1,750
7080	8" x 8"		21	1.333		1,800	58	6.55	1,864.55	2,075
7100	10" x 6"		21	1.333		2,600	58	6.55	2,664.55	2,950
7120	10" x 10"		21	1.333		2,650	58	6.55	2,714.55	3,025
7140	12" x 6"		18	1.556		2,600	67.50	7.65	2,675.15	2,950
7160	12" x 12"		18	1.556		3,575	67.50	7.65	3,650.15	4,050
7180	14" x 6"		16	1.750		6,475	76	8.60	6,559.60	7,250
7200	14" x 14"		16	1.750		6,900	76	8.60	6,984.60	7,700
7220	16" x 6"		14	2		6,900	86.50	9.85	6,996.35	7,750
7240	16" x 10"		14	2		7,000	86.50	9.85	7,096.35	7,850
7260	16" x 16"		14	2		7,425	86.50	9.85	7,521.35	8,325
7280	18" x 6"		10	2.800		9,875	121	13.80	10,009.80	11,100
7300	18" x 12"		10	2.800		10,100	121	13.80	10,234.80	11,300
7320	18" x 18"		10	2.800		9,650	121	13.80	9,784.80	10,800
7340	20" x 6"		8	3.500		8,475	152	17.20	8,644.20	9,575
7360	20" x 12"		8	3.500		8,800	152	17.20	8,969.20	9,925
7380	20" x 20"		8	3.500		13,400	152	17.20	13,569.20	15,000
7400	24" x 6"		6	4.667		12,400	202	23	12,625	13,900
7420	24" x 12"		6	4.667		11,600	202	23	11,825	13,100
7440	24" x 18"		6	4.667		17,400	202	23	17,625	19,500
7460	24" x 24"		6	4.667		18,000	202	23	18,225	20,100
7600	Cut-in sleeves with rubber gaskets, 4"		18	1.556		430	67.50	7.65	505.15	585
7620	6"		12	2.333		505	101	11.50	617.50	725
7640	8"		10	2.800		650	121	13.80	784.80	915
7660	10"		10	2.800		960	121	13.80	1,094.80	1,250
7680	12"		9	3.111		1,225	135	15.30	1,375.30	1,575
7800	Cut-in valves with rubber gaskets, 4"		18	1.556		505	67.50	7.65	580.15	665
7820	6"		12	2.333		675	101	11.50	787.50	910
7840	8"		10	2.800		1,050	121	13.80	1,184.80	1,375
7860	10"		10	2.800		1,150	121	13.80	1,284.80	1,450
7880	12"		9	3.111		1,300	135	15.30	1,450.30	1,650
7900	Tapping Valve 4 inch, MJ, ductile iron		18	1.556		555	67.50	7.65	630.15	720
7920	6 inch, MJ, ductile iron		12	2.333		780	101	11.50	892.50	1,025
8000	Sleeves with rubber gaskets, 4" x 4"		37	.757		845	33	3.72	881.72	985
8010	6" x 4"		25	1.120		900	48.50	5.50	954	1,075
8020	6" x 6"		25	1.120		1,125	48.50	5.50	1,179	1,325
8030	8" x 4"		21	1.333		960	58	6.55	1,024.55	1,150
8040	8" x 6"		21	1.333		1,075	58	6.55	1,139.55	1,275
8060	8" x 8"		21	1.333		1,250	58	6.55	1,314.55	1,475
8070	10" x 4"		21	1.333		1,075	58	6.55	1,139.55	1,300
8080	10" x 6"		21	1.333		1,200	58	6.55	1,264.55	1,425
8090	10" x 8"		21	1.333		1,325	58	6.55	1,389.55	1,550
8100	10" x 10"		21	1.333		1,950	58	6.55	2,014.55	2,250
8110	12" x 4"		18	1.556		1,150	67.50	7.65	1,225.15	1,375
8120	12" x 6"		18	1.556		1,175	67.50	7.65	1,250.15	1,400
8130	12" x 8"		18	1.556		1,475	67.50	7.65	1,550.15	1,725
8135	12" x 10"		18	1.556		2,325	67.50	7.65	2,400.15	2,650

For customer support on your Mechanical Cost Data, call 877.795.7394.

33 12 13 – Water Service Connections

33 12 13.15 Tapping, Crosses and Sleeves

		Crew	Daily Output	Labor-Hours	Unit	Material	2015 Bare Costs Labor	Equipment	Total	Total Incl O&P
8140	12" x 12"	B-21	18	1.556	Ea.	2,350	67.50	7.65	2,425.15	2,700
8160	14" x 6"		16	1.750		1,900	76	8.60	1,984.60	2,225
8180	14" x 14"		16	1.750		2,500	76	8.60	2,584.60	2,875
8200	16" x 6"		14	2		2,100	86.50	9.85	2,196.35	2,450
8220	16" x 10"		14	2		3,525	86.50	9.85	3,621.35	4,025
8240	16" x 16"		14	2		8,650	86.50	9.85	8,746.35	9,675
8260	18" x 6"		10	2.800		2,250	121	13.80	2,384.80	2,675
8280	18" x 12"		10	2.800		3,625	121	13.80	3,759.80	4,175
8300	18" x 18"		10	2.800		11,600	121	13.80	11,734.80	13,000
8320	20" x 6"		8	3.500		2,550	152	17.20	2,719.20	3,075
8340	20" x 12"		8	3.500		4,400	152	17.20	4,569.20	5,100
8360	20" x 20"		8	3.500		11,000	152	17.20	11,169.20	12,400
8380	24" x 6"		6	4.667		2,700	202	23	2,925	3,300
8400	24" x 12"		6	4.667		4,400	202	23	4,625	5,175
8420	24" x 18"		6	4.667		11,700	202	23	11,925	13,200
8440	24" x 24"	▼	6	4.667		14,800	202	23	15,025	16,600
8800	Hydrant valve box, 6' long	B-20	20	1.200		280	50.50		330.50	390
8820	8' long		18	1.333		355	56		411	475
8830	Valve box w/lid 4' deep		14	1.714		107	72		179	228
8840	Valve box and large base w/lid	▼	14	1.714	▼	325	72		397	470

33 12 16 – Water Utility Distribution Valves

33 12 16.10 Valves

		Crew	Daily Output	Labor-Hours	Unit	Material	2015 Bare Costs Labor	Equipment	Total	Total Incl O&P
0010	**VALVES**, water distribution									
0011	See Sections 22 05 23.20 and 22 05 23.60									
3000	Butterfly valves with boxes, cast iron, mech. jt.									
3100	4" diameter	B-6	6	4	Ea.	575	165	60.50	800.50	955
3140	6" diameter		6	4		750	165	60.50	975.50	1,150
3180	8" diameter		6	4		940	165	60.50	1,165.50	1,350
3300	10" diameter		6	4		1,275	165	60.50	1,500.50	1,725
3340	12" diameter		6	4		1,700	165	60.50	1,925.50	2,200
3400	14" diameter		4	6		3,400	248	91	3,739	4,200
3440	16" diameter		4	6		5,025	248	91	5,364	6,000
3460	18" diameter		4	6		6,225	248	91	6,564	7,325
3480	20" diameter		4	6		8,200	248	91	8,539	9,475
3500	24" diameter	▼	4	6	▼	13,700	248	91	14,039	15,600
3600	With lever operator									
3610	4" diameter	B-6	6	4	Ea.	470	165	60.50	695.50	835
3614	6" diameter		6	4		645	165	60.50	870.50	1,025
3616	8" diameter		6	4		835	165	60.50	1,060.50	1,250
3618	10" diameter		6	4		1,175	165	60.50	1,400.50	1,600
3620	12" diameter		6	4		1,600	165	60.50	1,825.50	2,100
3622	14" diameter		4	6		3,075	248	91	3,414	3,850
3624	16" diameter		4	6		4,700	248	91	5,039	5,625
3626	18" diameter		4	6		5,900	248	91	6,239	6,950
3628	20" diameter		4	6		7,875	248	91	8,214	9,125
3630	24" diameter	▼	4	6	▼	13,400	248	91	13,739	15,200
3700	Check valves, flanged									
3710	4" diameter	B-6	6	4	Ea.	605	165	60.50	830.50	985
3714	6" diameter		6	4		1,025	165	60.50	1,250.50	1,475
3716	8" diameter		6	4		1,950	165	60.50	2,175.50	2,475
3718	10" diameter		6	4		3,325	165	60.50	3,550.50	3,975
3720	12" diameter		6	4		5,175	165	60.50	5,400.50	6,025

33 12 16 – Water Utility Distribution Valves

33 12 16.10 Valves		Crew	Daily Output	Labor-Hours	Unit	Material	2015 Bare Costs Labor	Equipment	Total	Total Incl O&P
3726	18" diameter	B-6	4	6	Ea.	23,700	248	91	24,039	26,600
3730	24" diameter	↓	4	6	↓	52,000	248	91	52,339	58,000
3800	Gate valves, C.I., 250 PSI, mechanical joint, w/boxes									
3810	4" diameter	B-6	6	4	Ea.	880	165	60.50	1,105.50	1,275
3814	6" diameter		6	4		1,350	165	60.50	1,575.50	1,800
3816	8" diameter		6	4		2,300	165	60.50	2,525.50	2,850
3818	10" diameter		6	4		4,150	165	60.50	4,375.50	4,875
3820	12" diameter		6	4		5,625	165	60.50	5,850.50	6,500
3822	14" diameter		4	6		12,300	248	91	12,639	14,100
3824	16" diameter		4	6		17,300	248	91	17,639	19,500
3826	18" diameter		4	6		21,100	248	91	21,439	23,700
3828	20" diameter		4	6		23,200	248	91	23,539	26,000
3830	24" diameter		4	6		25,200	248	91	25,539	28,200
3831	30" diameter		4	6		39,200	248	91	39,539	43,600
3832	36" diameter	↓	4	6	↓	57,000	248	91	57,339	63,000

33 12 16.20 Valves		Crew	Daily Output	Labor-Hours	Unit	Material	2015 Bare Costs Labor	Equipment	Total	Total Incl O&P
0010	**VALVES**									
0011	Special trim or use									
9000	Valves, gate valve, N.R.S. PIV with post, 4" diameter	B-6	6	4	Ea.	1,875	165	60.50	2,100.50	2,375
9020	6" diameter		6	4		2,350	165	60.50	2,575.50	2,900
9040	8" diameter		6	4		3,275	165	60.50	3,500.50	3,950
9060	10" diameter		6	4		5,150	165	60.50	5,375.50	5,975
9080	12" diameter		6	4		6,600	165	60.50	6,825.50	7,600
9100	14" diameter		6	4		13,100	165	60.50	13,325.50	14,700
9120	OS&Y, 4" diameter		6	4		730	165	60.50	955.50	1,125
9140	6" diameter		6	4		990	165	60.50	1,215.50	1,425
9160	8" diameter		6	4		1,550	165	60.50	1,775.50	2,025
9180	10" diameter		6	4		2,475	165	60.50	2,700.50	3,025
9200	12" diameter		6	4		3,725	165	60.50	3,950.50	4,425
9220	14" diameter		4	6		5,075	248	91	5,414	6,075
9400	Check valves, rubber disc, 2-1/2" diameter		6	4		710	165	60.50	935.50	1,100
9420	3" diameter		6	4		725	165	60.50	950.50	1,125
9440	4" diameter		6	4		605	165	60.50	830.50	985
9480	6" diameter		6	4		1,025	165	60.50	1,250.50	1,475
9500	8" diameter		6	4		1,950	165	60.50	2,175.50	2,475
9520	10" diameter		6	4		3,325	165	60.50	3,550.50	3,975
9540	12" diameter		6	4		5,175	165	60.50	5,400.50	6,025
9542	14" diameter		4	6		5,250	248	91	5,589	6,250
9700	Detector check valves, reducing, 4" diameter		6	4		1,350	165	60.50	1,575.50	1,800
9720	6" diameter		6	4		2,075	165	60.50	2,300.50	2,625
9740	8" diameter		6	4		3,050	165	60.50	3,275.50	3,675
9760	10" diameter		6	4		5,400	165	60.50	5,625.50	6,275
9800	Galvanized, 4" diameter		6	4		1,575	165	60.50	1,800.50	2,075
9820	6" diameter		6	4		2,125	165	60.50	2,350.50	2,675
9840	8" diameter		6	4		3,350	165	60.50	3,575.50	4,000
9860	10" diameter	↓	6	4	↓	5,125	165	60.50	5,350.50	5,975

For customer support on your Mechanical Cost Data, call 877.795.7394.

473

33 16 13 – Aboveground Water Utility Storage Tanks

33 16 13.13 Steel Water Storage Tanks	Crew	Daily Output	Labor-Hours	Unit	Material	2015 Bare Costs Labor	Equipment	Total	Total Incl O&P
0010 **STEEL WATER STORAGE TANKS**									
0910 Steel, ground level, ht./diam. less than 1, not incl. fdn., 100,000 gallons				Ea.				202,000	244,500
1000 250,000 gallons								295,500	324,000
1200 500,000 gallons								417,000	458,500
1250 750,000 gallons								538,000	591,500
1300 1,000,000 gallons								558,000	725,500
1500 2,000,000 gallons								1,043,000	1,148,000
1600 4,000,000 gallons								2,121,000	2,333,000
1800 6,000,000 gallons								3,095,000	3,405,000
1850 8,000,000 gallons								4,068,000	4,475,000
1910 10,000,000 gallons								5,050,000	5,554,500
2100 Steel standpipes, ht./diam. more than 1, 100' to overflow, no fdn.									
2200 500,000 gallons				Ea.				546,500	600,500
2400 750,000 gallons								722,500	794,500
2500 1,000,000 gallons								1,060,500	1,167,000
2700 1,500,000 gallons								1,749,000	1,923,000
2800 2,000,000 gallons								2,327,000	2,559,000

33 16 13.16 Prestressed Conc. Water Storage Tanks

	Crew	Daily Output	Labor-Hours	Unit	Material	Labor	Equipment	Total	Total Incl O&P
0010 **PRESTRESSED CONC. WATER STORAGE TANKS**									
0020 Not including fdn., pipe or pumps, 250,000 gallons				Ea.				299,000	329,500
0100 500,000 gallons								487,000	536,000
0300 1,000,000 gallons								707,000	807,500
0400 2,000,000 gallons								1,072,000	1,179,000
0600 4,000,000 gallons								1,706,000	1,877,000
0700 6,000,000 gallons								2,266,000	2,493,000
0750 8,000,000 gallons								2,924,000	3,216,000
0800 10,000,000 gallons								3,533,000	3,886,000

33 16 13.23 Plastic-Coated Fabric Pillow Water Tanks

	Crew	Daily Output	Labor-Hours	Unit	Material	Labor	Equipment	Total	Total Incl O&P
0010 **PLASTIC-COATED FABRIC PILLOW WATER TANKS**									
7000 Water tanks, vinyl coated fabric pillow tanks, freestanding, 5,000 gallons	4 Clab	4	8	Ea.	3,675	300		3,975	4,525
7100 Supporting embankment not included, 25,000 gallons	6 Clab	2	24		13,200	900		14,100	15,900
7200 50,000 gallons	8 Clab	1.50	42.667		18,500	1,600		20,100	22,800
7300 100,000 gallons	9 Clab	.90	80		42,300	3,000		45,300	51,000
7400 150,000 gallons		.50	144		60,500	5,425		65,925	75,500
7500 200,000 gallons		.40	180		75,000	6,775		81,775	93,000
7600 250,000 gallons		.30	240		105,500	9,025		114,525	130,000

33 16 19 – Elevated Water Utility Storage Tanks

33 16 19.50 Elevated Water Storage Tanks

	Crew	Daily Output	Labor-Hours	Unit	Material	Labor	Equipment	Total	Total Incl O&P
0010 **ELEVATED WATER STORAGE TANKS**									
0011 Not incl. pipe, pumps or foundation									
3000 Elevated water tanks, 100' to bottom capacity line, incl. painting									
3010 50,000 gallons				Ea.				185,000	204,000
3300 100,000 gallons								280,000	307,500
3400 250,000 gallons								751,500	826,500
3600 500,000 gallons								1,336,000	1,470,000
3700 750,000 gallons								1,622,000	1,783,500
3900 1,000,000 gallons								2,322,000	2,556,000

33 21 13.10 Wells and Accessories	Crew	Daily Output	Labor-Hours	Unit	Material	2015 Bare Costs Labor	2015 Bare Costs Equipment	Total	Total Incl O&P
0010	**WELLS & ACCESSORIES**								
0011	Domestic								
0100	Drilled, 4" to 6" diameter	B-23	120	.333	L.F.		12.65	23.50	36.15	45.50
0200	8" diameter	"	95.20	.420	"		15.95	30	45.95	57.50
0400	Gravel pack well, 40' deep, incl. gravel & casing, complete									
0500	24" diameter casing x 18" diameter screen	B-23	.13	307	Total	38,800	11,700	21,900	72,400	85,000
0600	36" diameter casing x 18" diameter screen		.12	333	"	40,000	12,700	23,700	76,400	89,500
0800	Observation wells, 1-1/4" riser pipe	↓	163	.245	V.L.F.	20.50	9.35	17.45	47.30	56
0900	For flush Buffalo roadway box, add	1 Skwk	16.60	.482	Ea.	51	23.50		74.50	92
1200	Test well, 2-1/2" diameter, up to 50' deep (15 to 50 GPM)	B-23	1.51	26.490	"	805	1,000	1,875	3,680	4,525
1300	Over 50' deep, add	"	121.80	.328	L.F.	21.50	12.50	23.50	57.50	68
1500	Pumps, installed in wells to 100' deep, 4" submersible									
1510	1/2 H.P.	Q-1	3.22	4.969	Ea.	350	263		613	780
1520	3/4 H.P.		2.66	6.015		400	320		720	920
1600	1 H.P.	↓	2.29	6.987		425	370		795	1,025
1700	1-1/2 H.P.	Q-22	1.60	10		800	530	410	1,740	2,125
1800	2 H.P.		1.33	12.030		830	635	490	1,955	2,425
1900	3 H.P.		1.14	14.035		1,150	740	575	2,465	3,025
2000	5 H.P.		1.14	14.035		1,600	740	575	2,915	3,500
2050	Remove and install motor only, 4 H.P.		1.14	14.035		1,100	740	575	2,415	2,950
3000	Pump, 6" submersible, 25' to 150' deep, 25 H.P., 249 to 297 GPM	↓	.89	17.978	↓	3,350	950	735	5,035	5,900
3100	25' to 500' deep, 30 H.P., 100 to 300 GPM		.73	21.918		4,100	1,150	895	6,145	7,225
8110	Well screen assembly, stainless steel, 2" diameter	B-23A	273	.088	L.F.	81	3.75	9.85	94.60	106
8120	3" diameter		253	.095		142	4.04	10.60	156.64	174
8130	4" diameter		200	.120		177	5.10	13.45	195.55	217
8140	5" diameter		168	.143		187	6.10	16	209.10	232
8150	6" diameter		126	.190		212	8.10	21.50	241.60	269
8160	8" diameter		98.50	.244		288	10.40	27.50	325.90	360
8170	10" diameter		73	.329		360	14	37	411	455
8180	12" diameter		62.50	.384		410	16.35	43	469.35	525
8190	14" diameter		54.30	.442		460	18.85	49.50	528.35	590
8200	16" diameter		48.30	.497		510	21	55.50	586.50	655
8210	18" diameter		39.20	.612		635	26	68.50	729.50	815
8220	20" diameter		31.20	.769		710	33	86	829	930
8230	24" diameter		23.80	1.008		855	43	113	1,011	1,125
8240	26" diameter		21	1.143		930	48.50	128	1,106.50	1,250
8244	Well casing or drop pipe, PVC, 0.5" dia		550	.044		1.55	1.86	4.88	8.29	9.90
8245	0.75" diameter		550	.044		1.13	1.86	4.88	7.87	9.45
8246	1" diameter		550	.044		1.17	1.86	4.88	7.91	9.50
8247	1.25" diameter		520	.046		1.35	1.97	5.15	8.47	10.20
8248	1.5" diameter		490	.049		1.58	2.09	5.50	9.17	11
8249	1.75" diameter		380	.063		1.48	2.69	7.05	11.22	13.55
8250	2" diameter		280	.086		1.65	3.65	9.60	14.90	17.95
8252	3" diameter		260	.092		3.39	3.93	10.35	17.67	21
8254	4" diameter		205	.117		4.90	4.99	13.10	22.99	27.50
8255	5" diameter		170	.141		3.20	6	15.80	25	30
8256	6" diameter		130	.185		8.50	7.85	20.50	36.85	44
8258	8" diameter		100	.240		12.95	10.20	27	50.15	59.50
8260	10" diameter		73	.329		9	14	37	60	72
8262	12" diameter		62	.387		12	16.50	43.50	72	85.50
8300	Slotted PVC, 1-1/4" diameter		521	.046		2.40	1.96	5.15	9.51	11.30
8310	1-1/2" diameter		488	.049		2.85	2.10	5.50	10.45	12.40
8320	2" diameter	↓	273	.088	↓	3.67	3.75	9.85	17.27	20.50

For customer support on your Mechanical Cost Data, call 877.795.7394.

475

33 21 13 – Public Water Supply Wells

33 21 13.10 Wells and Accessories	Crew	Daily Output	Labor-Hours	Unit	Material	2015 Bare Costs Labor	Equipment	Total	Total Incl O&P
8330	3" diameter	B-23A	253	.095	L.F.	5.85	4.04	10.60	20.49	24.50
8340	4" diameter		200	.120		6.75	5.10	13.45	25.30	30
8350	5" diameter		168	.143		14.05	6.10	16	36.15	42.50
8360	6" diameter		126	.190		16.65	8.10	21.50	46.25	54.50
8370	8" diameter		98.50	.244		26	10.40	27.50	63.90	74.50
8400	Artificial gravel pack, 2" screen, 6" casing	B-23B	174	.138		3.60	5.90	17.45	26.95	32
8405	8" casing		111	.216		4.63	9.20	27.50	41.33	49
8410	10" casing		74.50	.322		6.60	13.70	40.50	60.80	73.50
8415	12" casing		60	.400		9.25	17.05	50.50	76.80	91.50
8420	14" casing		50.20	.478		10.75	20.50	60.50	91.75	109
8425	16" casing		40.70	.590		13.65	25	74.50	113.15	136
8430	18" casing		36	.667		16.40	28.50	84.50	129.40	154
8435	20" casing		29.50	.814		19.25	34.50	103	156.75	187
8440	24" casing		25.70	.934		22	40	118	180	215
8445	26" casing		24.60	.976		24.50	41.50	123	189	227
8450	30" casing		20	1.200		28	51	152	231	276
8455	36" casing		16.40	1.463		31.50	62.50	185	279	335
8500	Develop well		8	3	Hr.	262	128	380	770	900
8550	Pump test well		8	3		82.50	128	380	590.50	700
8560	Standby well	B-23A	8	3		78	128	335	541	650
8570	Standby, drill rig		8	3			128	335	463	565
8580	Surface seal well, concrete filled		1	24	Ea.	850	1,025	2,675	4,550	5,450
8590	Well test pump, install & remove	B-23	1	40			1,525	2,850	4,375	5,475
8600	Well sterilization, chlorine	2 Clab	1	16		122	600		722	1,050
8610	Well water pressure switch	1 Clab	12	.667		50.50	25		75.50	94
8630	Well, water pressure switch with manual reset	"	12	.667		35.50	25		60.50	77.50
9950	See Section 31 23 19.40 for wellpoints									
9960	See Section 31 23 19.30 for drainage wells									

33 21 13.20 Water Supply Wells, Pumps

		Crew	Daily Output	Labor-Hours	Unit	Material	Labor	Equipment	Total	Total Incl O&P
0010	**WATER SUPPLY WELLS, PUMPS**									
0011	With pressure control									
1000	Deep well, jet, 42 gal. galvanized tank									
1040	3/4 HP	1 Plum	.80	10	Ea.	1,100	585		1,685	2,100
3000	Shallow well, jet, 30 gal. galvanized tank									
3040	1/2 HP	1 Plum	2	4	Ea.	895	235		1,130	1,350

33 31 Sanitary Utility Sewerage Piping

33 31 13 – Public Sanitary Utility Sewerage Piping

33 31 13.15 Sewage Collection, Concrete Pipe

		Crew	Daily Output	Labor-Hours	Unit	Material	Labor	Equipment	Total	Total Incl O&P
0010	**SEWAGE COLLECTION, CONCRETE PIPE**									
0020	See Section 33 41 13.60 for sewage/drainage collection, concrete pipe									

33 31 13.25 Sewage Collection, Polyvinyl Chloride Pipe

		Crew	Daily Output	Labor-Hours	Unit	Material	Labor	Equipment	Total	Total Incl O&P
0010	**SEWAGE COLLECTION, POLYVINYL CHLORIDE PIPE**									
0020	Not including excavation or backfill									
2000	20' lengths, SDR 35, B&S, 4" diameter	B-20	375	.064	L.F.	1.45	2.68		4.13	5.75
2040	6" diameter		350	.069		3.28	2.88		6.16	8.05
2080	13' lengths, SDR 35, B&S, 8" diameter		335	.072		6.90	3.01		9.91	12.25
2120	10" diameter	B-21	330	.085		11.40	3.68	.42	15.50	18.60
2160	12" diameter		320	.088		12.75	3.79	.43	16.97	20.50
2200	15" diameter		240	.117		12.95	5.05	.57	18.57	22.50

33 31 Sanitary Utility Sewerage Piping

33 31 13 – Public Sanitary Utility Sewerage Piping

33 31 13.25 Sewage Collection, Polyvinyl Chloride Pipe	Crew	Daily Output	Labor-Hours	Unit	Material	2015 Bare Costs Labor	Equipment	Total	Total Incl O&P
4000	Piping, DWV PVC, no exc./bkfill., 10' L, Sch 40, 4" diameter	B-20	375	.064	L.F.	3.84	2.68		6.52	8.35
4010	6" diameter		350	.069		7.65	2.88		10.53	12.85
4020	8" diameter		335	.072		12	3.01		15.01	17.85

33 41 Storm Utility Drainage Piping

33 41 13 – Public Storm Utility Drainage Piping

33 41 13.40 Piping, Storm Drainage, Corrugated Metal

		Crew	Daily Output	Labor-Hours	Unit	Material	2015 Bare Costs Labor	Equipment	Total	Total Incl O&P
0010	**PIPING, STORM DRAINAGE, CORRUGATED METAL**									
0020	Not including excavation or backfill									
2000	Corrugated metal pipe, galvanized									
2020	Bituminous coated with paved invert, 20' lengths									
2040	8" diameter, 16 ga.	B-14	330	.145	L.F.	8.65	5.80	1.10	15.55	19.60
2060	10" diameter, 16 ga.		260	.185		9	7.35	1.40	17.75	22.50
2080	12" diameter, 16 ga.		210	.229		11.05	9.10	1.73	21.88	28
2100	15" diameter, 16 ga.		200	.240		15.15	9.55	1.82	26.52	33.50
2120	18" diameter, 16 ga.		190	.253		16.70	10.05	1.92	28.67	36
2140	24" diameter, 14 ga.		160	.300		21.50	11.95	2.28	35.73	44.50
2160	30" diameter, 14 ga.	B-13	120	.467		27.50	19.15	6.15	52.80	67
2180	36" diameter, 12 ga.		120	.467		35.50	19.15	6.15	60.80	75.50
2200	48" diameter, 12 ga.		100	.560		52.50	23	7.35	82.85	101
2220	60" diameter, 10 ga.	B-13B	75	.747		80	30.50	15.05	125.55	151
2240	72" diameter, 8 ga.	"	45	1.244		95.50	51	25	171.50	211
2500	Galvanized, uncoated, 20' lengths									
2520	8" diameter, 16 ga.	B-14	355	.135	L.F.	7.80	5.40	1.03	14.23	18
2540	10" diameter, 16 ga.		280	.171		8.95	6.80	1.30	17.05	21.50
2560	12" diameter, 16 ga.		220	.218		9.95	8.70	1.66	20.31	26
2580	15" diameter, 16 ga.		220	.218		12.45	8.70	1.66	22.81	29
2600	18" diameter, 16 ga.		205	.234		15.05	9.30	1.78	26.13	33
2620	24" diameter, 14 ga.		175	.274		18.95	10.90	2.08	31.93	40
2640	30" diameter, 14 ga.	B-13	130	.431		25	17.65	5.65	48.30	61
2660	36" diameter, 12 ga.		130	.431		32	17.65	5.65	55.30	68.50
2680	48" diameter, 12 ga.		110	.509		47.50	21	6.70	75.20	91.50
2690	60" diameter, 10 ga.	B-13B	78	.718		72	29.50	14.45	115.95	140
2780	End sections, 8" diameter	B-14	35	1.371	Ea.	73	54.50	10.40	137.90	175
2785	10" diameter		35	1.371		77	54.50	10.40	141.90	179
2790	12" diameter		35	1.371		114	54.50	10.40	178.90	220
2800	18" diameter		30	1.600		115	63.50	12.15	190.65	237
2810	24" diameter	B-13	25	2.240		215	92	29.50	336.50	410
2820	30" diameter		25	2.240		330	92	29.50	451.50	535
2825	36" diameter		20	2.800		480	115	37	632	745
2830	48" diameter		10	5.600		950	230	73.50	1,253.50	1,475
2835	60" diameter	B-13B	5	11.200		1,650	460	226	2,336	2,750
2840	72" diameter	"	4	14		1,975	575	282	2,832	3,350

33 41 13.60 Sewage/Drainage Collection, Concrete Pipe

		Crew	Daily Output	Labor-Hours	Unit	Material	2015 Bare Costs Labor	Equipment	Total	Total Incl O&P
0010	**SEWAGE/DRAINAGE COLLECTION, CONCRETE PIPE**									
0020	Not including excavation or backfill									
1000	Non-reinforced pipe, extra strength, B&S or T&G joints									
1010	6" diameter	B-14	265.04	.181	L.F.	6	7.20	1.37	14.57	19.15
1020	8" diameter		224	.214		6.60	8.50	1.63	16.73	22
1030	10" diameter		216	.222		7.30	8.85	1.69	17.84	23.50
1040	12" diameter		200	.240		8.20	9.55	1.82	19.57	25.50

For customer support on your Mechanical Cost Data, call 877.795.7394.

477

33 41 13 – Public Storm Utility Drainage Piping

33 41 13.60 Sewage/Drainage Collection, Concrete Pipe	Crew	Daily Output	Labor-Hours	Unit	Material	2015 Bare Costs Labor	Equipment	Total	Total Incl O&P
1050	15" diameter	B-14	180	.267	L.F.	12	10.60	2.02	24.62	31.50
1060	18" diameter		144	.333		15	13.25	2.53	30.78	40
1070	21" diameter		112	.429		17	17.05	3.25	37.30	48.50
1080	24" diameter		100	.480		22	19.10	3.64	44.74	57.50
2000	Reinforced culvert, class 3, no gaskets									
2010	12" diameter	B-14	150	.320	L.F.	11	12.75	2.43	26.18	34.50
2020	15" diameter		150	.320		14	12.75	2.43	29.18	37.50
2030	18" diameter		132	.364		18	14.45	2.76	35.21	45
2035	21" diameter		120	.400		22	15.90	3.04	40.94	52
2040	24" diameter		100	.480		26	19.10	3.64	48.74	62
2045	27" diameter	B-13	92	.609		37	25	8	70	87.50
2050	30" diameter		88	.636		42	26	8.35	76.35	95
2060	36" diameter		72	.778		56	32	10.25	98.25	122
2070	42" diameter	B-13B	72	.778		75	32	15.65	122.65	149
2080	48" diameter		64	.875		89	36	17.60	142.60	172
2090	60" diameter		48	1.167		136	48	23.50	207.50	249
2100	72" diameter		40	1.400		206	57.50	28	291.50	345
2120	84" diameter		32	1.750		275	71.50	35.50	382	455
2140	96" diameter		24	2.333		330	95.50	47	472.50	565
2200	With gaskets, class 3, 12" diameter	B-21	168	.167		12.10	7.20	.82	20.12	25.50
2220	15" diameter		160	.175		15.40	7.60	.86	23.86	29.50
2230	18" diameter		152	.184		19.80	8	.91	28.71	35.50
2240	24" diameter		136	.206		33	8.90	1.01	42.91	51.50
2260	30" diameter	B-13	88	.636		50	26	8.35	84.35	104
2270	36" diameter	"	72	.778		65.50	32	10.25	107.75	132
2290	48" diameter	B-13B	64	.875		102	36	17.60	155.60	186
2310	72" diameter	"	40	1.400		225	57.50	28	310.50	365
2330	Flared ends, 12" diameter	B-21	31	.903	Ea.	226	39	4.44	269.44	315
2340	15" diameter		25	1.120		267	48.50	5.50	321	375
2400	18" diameter		20	1.400		305	60.50	6.90	372.40	440
2420	24" diameter		14	2		370	86.50	9.85	466.35	555
2440	36" diameter	B-13	10	5.600		790	230	73.50	1,093.50	1,300
3080	Radius pipe, add to pipe prices, 12" to 60" diameter				L.F.	50%				
3090	Over 60" diameter, add				"	20%				
3500	Reinforced elliptical, 8' lengths, C507 class 3									
3520	14" x 23" inside, round equivalent 18" diameter	B-21	82	.341	L.F.	41	14.80	1.68	57.48	70
3530	24" x 38" inside, round equivalent 30" diameter	B-13	58	.966		62	39.50	12.70	114.20	142
3540	29" x 45" inside, round equivalent 36" diameter		52	1.077		78	44	14.15	136.15	169
3550	38" x 60" inside, round equivalent 48" diameter		38	1.474		134	60.50	19.40	213.90	261
3560	48" x 76" inside, round equivalent 60" diameter		26	2.154		186	88.50	28.50	303	370
3570	58" x 91" inside, round equivalent 72" diameter		22	2.545		272	104	33.50	409.50	495
3780	Concrete slotted pipe, class 4 mortar joint									
3800	12" diameter	B-21	168	.167	L.F.	28	7.20	.82	36.02	43
3840	18" diameter	"	152	.184	"	32	8	.91	40.91	48.50
3900	Concrete slotted pipe, Class 4 O-ring joint									
3940	12" diameter	B-21	168	.167	L.F.	28	7.20	.82	36.02	43
3960	18" diameter	"	152	.184	"	32	8	.91	40.91	48.50

33 42 Culverts

33 42 16 – Concrete Culverts

33 42 16.15 Oval Arch Culverts

		Crew	Daily Output	Labor-Hours	Unit	Material	2015 Bare Costs Labor	2015 Bare Costs Equipment	Total	Total Incl O&P
0010	**OVAL ARCH CULVERTS**									
3000	Corrugated galvanized or aluminum, coated & paved									
3020	17" x 13", 16 ga., 15" equivalent	B-14	200	.240	L.F.	13.20	9.55	1.82	24.57	31
3040	21" x 15", 16 ga., 18" equivalent		150	.320		16.05	12.75	2.43	31.23	40
3060	28" x 20", 14 ga., 24" equivalent		125	.384		24	15.25	2.91	42.16	53
3080	35" x 24", 14 ga., 30" equivalent		100	.480		29.50	19.10	3.64	52.24	66
3100	42" x 29", 12 ga., 36" equivalent	B-13	100	.560		35.50	23	7.35	65.85	82
3120	49" x 33", 12 ga., 42" equivalent		90	.622		41	25.50	8.20	74.70	93.50
3140	57" x 38", 12 ga., 48" equivalent		75	.747		57	30.50	9.80	97.30	120
3160	Steel, plain oval arch culverts, plain									
3180	17" x 13", 16 ga., 15" equivalent	B-14	225	.213	L.F.	11.95	8.50	1.62	22.07	28
3200	21" x 15", 16 ga., 18" equivalent		175	.274		14.45	10.90	2.08	27.43	35
3220	28" x 20", 14 ga., 24" equivalent		150	.320		21.50	12.75	2.43	36.68	45.50
3240	35" x 24", 14 ga., 30" equivalent	B-13	108	.519		26.50	21.50	6.80	54.80	69
3260	42" x 29", 12 ga., 36" equivalent		108	.519		32	21.50	6.80	60.30	75.50
3280	49" x 33", 12 ga., 42" equivalent		92	.609		37	25	8	70	88
3300	57" x 38", 12 ga., 48" equivalent		75	.747		51.50	30.50	9.80	91.80	114
3320	End sections, 17" x 13"		22	2.545	Ea.	144	104	33.50	281.50	355
3340	42" x 29"		17	3.294	"	395	135	43.50	573.50	690
3360	Multi-plate arch, steel	B-20	1690	.014	Lb.	1.25	.60		1.85	2.30

33 44 Storm Utility Water Drains

33 44 13 – Utility Area Drains

33 44 13.13 Catchbasins

		Crew	Daily Output	Labor-Hours	Unit	Material	2015 Bare Costs Labor	2015 Bare Costs Equipment	Total	Total Incl O&P
0010	**CATCHBASINS**									
0011	Not including footing & excavation									
1600	Frames & grates, C.I., 24" square, 500 lb.	B-6	7.80	3.077	Ea.	340	127	46.50	513.50	620
1700	26" D shape, 600 lb.		7	3.429		505	142	52	699	830
1800	Light traffic, 18" diameter, 100 lb.		10	2.400		123	99	36.50	258.50	325
1900	24" diameter, 300 lb.		8.70	2.759		196	114	42	352	435
2000	36" diameter, 900 lb.		5.80	4.138		570	171	63	804	955
2100	Heavy traffic, 24" diameter, 400 lb.		7.80	3.077		244	127	46.50	417.50	515
2200	36" diameter, 1150 lb.		3	8		795	330	121	1,246	1,525
2300	Mass. State standard, 26" diameter, 475 lb.		7	3.429		266	142	52	460	565
2400	30" diameter, 620 lb.		7	3.429		345	142	52	539	655
2500	Watertight, 24" diameter, 350 lb.		7.80	3.077		320	127	46.50	493.50	595
2600	26" diameter, 500 lb.		7	3.429		425	142	52	619	745
2700	32" diameter, 575 lb.		6	4		850	165	60.50	1,075.50	1,250
2800	3 piece cover & frame, 10" deep,									
2900	1200 lb., for heavy equipment	B-6	3	8	Ea.	1,050	330	121	1,501	1,825
3000	Raised for paving 1-1/4" to 2" high									
3100	4 piece expansion ring									
3200	20" to 26" diameter	1 Clab	3	2.667	Ea.	158	100		258	330
3300	30" to 36" diameter	"	3	2.667	"	217	100		317	395
3320	Frames and covers, existing, raised for paving, 2", including									
3340	row of brick, concrete collar, up to 12" wide frame	B-6	18	1.333	Ea.	45	55	20	120	156
3360	20" to 26" wide frame		11	2.182		67.50	90	33	190.50	249
3380	30" to 36" wide frame		9	2.667		83.50	110	40.50	234	305
3400	Inverts, single channel brick	D-1	3	5.333		97	225		322	450
3500	Concrete		5	3.200		104	135		239	320
3600	Triple channel, brick		2	8		148	335		483	680

For customer support on your Mechanical Cost Data, call 877.795.7394.

479

33 44 13 – Utility Area Drains

33 44 13.13 Catchbasins	Crew	Daily Output	Labor-Hours	Unit	Material	2015 Bare Costs Labor	2015 Bare Costs Equipment	Total	Total Incl O&P
3700 Concrete	D-1	3	5.333	Ea.	139	225		364	500

33 46 Subdrainage

33 46 16 – Subdrainage Piping

33 46 16.25 Piping, Subdrainage, Corrugated Metal

	Crew	Daily Output	Labor-Hours	Unit	Material	Labor	Equipment	Total	Total Incl O&P
0010 **PIPING, SUBDRAINAGE, CORRUGATED METAL**									
0021 Not including excavation and backfill									
2010 Aluminum, perforated									
2020 6" diameter, 18 ga.	B-20	380	.063	L.F.	6.50	2.65		9.15	11.25
2200 8" diameter, 16 ga.	"	370	.065		8.55	2.72		11.27	13.60
2220 10" diameter, 16 ga.	B-21	360	.078		10.70	3.37	.38	14.45	17.35
2240 12" diameter, 16 ga.		285	.098		11.95	4.26	.48	16.69	20
2260 18" diameter, 16 ga.	↓	205	.137	↓	17.95	5.90	.67	24.52	29.50
3000 Uncoated galvanized, perforated									
3020 6" diameter, 18 ga.	B-20	380	.063	L.F.	6.05	2.65		8.70	10.75
3200 8" diameter, 16 ga.	"	370	.065		8.30	2.72		11.02	13.35
3220 10" diameter, 16 ga.	B-21	360	.078		8.80	3.37	.38	12.55	15.30
3240 12" diameter, 16 ga.		285	.098		9.80	4.26	.48	14.54	17.85
3260 18" diameter, 16 ga.	↓	205	.137	↓	15	5.90	.67	21.57	26.50
4000 Steel, perforated, asphalt coated									
4020 6" diameter 18 ga.	B-20	380	.063	L.F.	6.50	2.65		9.15	11.25
4030 8" diameter 18 ga.	"	370	.065		8.55	2.72		11.27	13.60
4040 10" diameter 16 ga.	B-21	360	.078		10.05	3.37	.38	13.80	16.65
4050 12" diameter 16 ga.		285	.098		11.05	4.26	.48	15.79	19.35
4060 18" diameter 16 ga.	↓	205	.137	↓	17.10	5.90	.67	23.67	28.50

33 46 16.30 Piping, Subdrainage, Plastic

	Crew	Daily Output	Labor-Hours	Unit	Material	Labor	Equipment	Total	Total Incl O&P
0010 **PIPING, SUBDRAINAGE, PLASTIC**									
0020 Not including excavation and backfill									
2100 Perforated PVC, 4" diameter	B-14	314	.153	L.F.	1.45	6.10	1.16	8.71	12.25
2110 6" diameter		300	.160		3.28	6.35	1.21	10.84	14.70
2120 8" diameter		290	.166		6.30	6.60	1.26	14.16	18.40
2130 10" diameter		280	.171		9.10	6.80	1.30	17.20	22
2140 12" diameter	↓	270	.178	↓	12.75	7.05	1.35	21.15	26.50

33 49 Storm Drainage Structures

33 49 13 – Storm Drainage Manholes, Frames, and Covers

33 49 13.10 Storm Drainage Manholes, Frames and Covers

	Crew	Daily Output	Labor-Hours	Unit	Material	Labor	Equipment	Total	Total Incl O&P
0010 **STORM DRAINAGE MANHOLES, FRAMES & COVERS**									
0020 Excludes footing, excavation, backfill (See line items for frame & cover)									
0050 Brick, 4' inside diameter, 4' deep	D-1	1	16	Ea.	520	675		1,195	1,600
0100 6' deep		.70	22.857		740	960		1,700	2,300
0150 8' deep		.50	32	↓	955	1,350		2,305	3,100
0200 For depths over 8', add		4	4	V.L.F.	83.50	168		251.50	350
0400 Concrete blocks (radial), 4' I.D., 4' deep		1.50	10.667	Ea.	415	450		865	1,150
0500 6' deep		1	16		565	675		1,240	1,650
0600 8' deep		.70	22.857	↓	710	960		1,670	2,250
0700 For depths over 8', add	↓	5.50	2.909	V.L.F.	77	122		199	272
0800 Concrete, cast in place, 4' x 4', 8" thick, 4' deep	C-14H	2	24	Ea.	505	1,100	15.60	1,620.60	2,275
0900 6' deep	↓	1.50	32	↓	730	1,475	21	2,226	3,100

480

33 49 13 – Storm Drainage Manholes, Frames, and Covers

33 49 13.10 Storm Drainage Manholes, Frames and Covers	Crew	Daily Output	Labor-Hours	Unit	Material	2015 Bare Costs Labor	Equipment	Total	Total Incl O&P
1000	8' deep	C-14H	1	48	Ea.	1,050	2,225	31	3,306	4,575
1100	For depths over 8', add	↓	8	6	V.L.F.	119	278	3.90	400.90	560
1110	Precast, 4' I.D., 4' deep	B-22	4.10	7.317	Ea.	725	320	50.50	1,095.50	1,350
1120	6' deep		3	10		925	440	69	1,434	1,775
1130	8' deep		2	15	↓	1,075	660	103	1,838	2,300
1140	For depths over 8', add	↓	16	1.875	V.L.F.	127	82.50	12.90	222.40	280
1150	5' I.D., 4' deep	B-6	3	8	Ea.	1,650	330	121	2,101	2,450
1160	6' deep		2	12		1,925	495	182	2,602	3,075
1170	8' deep		1.50	16	↓	2,400	660	243	3,303	3,925
1180	For depths over 8', add		12	2	V.L.F.	280	82.50	30.50	393	470
1190	6' I.D., 4' deep		2	12	Ea.	2,150	495	182	2,827	3,325
1200	6' deep		1.50	16		2,600	660	243	3,503	4,125
1210	8' deep		1	24		3,200	990	365	4,555	5,425
1220	For depths over 8', add	↓	8	3	V.L.F.	380	124	45.50	549.50	655
1250	Slab tops, precast, 8" thick									
1300	4' diameter manhole	B-6	8	3	Ea.	252	124	45.50	421.50	515
1400	5' diameter manhole		7.50	3.200		410	132	48.50	590.50	710
1500	6' diameter manhole	↓	7	3.429		635	142	52	829	975
3800	Steps, heavyweight cast iron, 7" x 9"	1 Bric	40	.200		19.25	9.25		28.50	35
3900	8" x 9"		40	.200		23	9.25		32.25	39.50
3928	12" x 10-1/2"		40	.200		27	9.25		36.25	43.50
4000	Standard sizes, galvanized steel		40	.200		22	9.25		31.25	38
4100	Aluminum		40	.200		24	9.25		33.25	40.50
4150	Polyethylene	↓	40	.200	↓	26	9.25		35.25	42.50

33 51 Natural-Gas Distribution

33 51 13 – Natural-Gas Piping

33 51 13.10 Piping, Gas Service and Distribution, P.E.	Crew	Daily Output	Labor-Hours	Unit	Material	2015 Bare Costs Labor	Equipment	Total	Total Incl O&P
0010	**PIPING, GAS SERVICE AND DISTRIBUTION, POLYETHYLENE**									
0020	Not including excavation or backfill									
1000	60 psi coils, compression coupling @ 100', 1/2" diameter, SDR 11	B-20A	608	.053	L.F.	.45	2.41		2.86	4.16
1010	1" diameter, SDR 11		544	.059		1.05	2.69		3.74	5.25
1040	1-1/4" diameter, SDR 11		544	.059		1.62	2.69		4.31	5.85
1100	2" diameter, SDR 11		488	.066		2.82	3		5.82	7.65
1160	3" diameter, SDR 11	↓	408	.078		6.45	3.59		10.04	12.55
1500	60 PSI 40' joints with coupling, 3" diameter, SDR 11	B-21A	408	.098		10.05	4.60	1.16	15.81	19.35
1540	4" diameter, SDR 11		352	.114		14.25	5.35	1.35	20.95	25
1600	6" diameter, SDR 11		328	.122		33.50	5.70	1.44	40.64	47.50
1640	8" diameter, SDR 11	↓	272	.147	↓	51.50	6.90	1.74	60.14	69

For customer support on your Mechanical Cost Data, call 877.795.7394.

481

33 52 16.13 Gasoline Piping	Crew	Daily Output	Labor-Hours	Unit	Material	2015 Bare Costs Labor	2015 Bare Costs Equipment	Total	Total Incl O&P
0010 GASOLINE PIPING									
0020 Primary containment pipe, fiberglass-reinforced									
0030 Plastic pipe 15' & 30' lengths									
0040 2" diameter	Q-6	425	.056	L.F.	5.95	3.15		9.10	11.25
0050 3" diameter		400	.060		10.35	3.35		13.70	16.45
0060 4" diameter		375	.064		13.65	3.57		17.22	20.50
0100 Fittings									
0110 Elbows, 90° & 45°, bell-ends, 2"	Q-6	24	1	Ea.	43.50	56		99.50	132
0120 3" diameter		22	1.091		55	61		116	153
0130 4" diameter		20	1.200		70	67		137	178
0200 Tees, bell ends, 2"		21	1.143		60.50	63.50		124	163
0210 3" diameter		18	1.333		64	74.50		138.50	183
0220 4" diameter		15	1.600		84	89		173	228
0230 Flanges bell ends, 2"		24	1		33.50	56		89.50	121
0240 3" diameter		22	1.091		39	61		100	135
0250 4" diameter		20	1.200		45	67		112	151
0260 Sleeve couplings, 2"		21	1.143		12.40	63.50		75.90	110
0270 3" diameter		18	1.333		17.80	74.50		92.30	132
0280 4" diameter		15	1.600		23	89		112	160
0290 Threaded adapters 2"		21	1.143		17.95	63.50		81.45	116
0300 3" diameter		18	1.333		34	74.50		108.50	150
0310 4" diameter		15	1.600		38	89		127	177
0320 Reducers, 2"		27	.889		27.50	49.50		77	105
0330 3" diameter		22	1.091		27.50	61		88.50	122
0340 4" diameter		20	1.200		37	67		104	142
1010 Gas station product line for secondary containment (double wall)									
1100 Fiberglass reinforced plastic pipe 25' lengths									
1120 Pipe, plain end, 3" diameter	Q-6	375	.064	L.F.	26.50	3.57		30.07	34.50
1130 4" diameter		350	.069		32	3.82		35.82	41.50
1140 5" diameter		325	.074		35.50	4.12		39.62	45
1150 6" diameter		300	.080		39	4.46		43.46	50
1200 Fittings									
1230 Elbows, 90° & 45°, 3" diameter	Q-6	18	1.333	Ea.	136	74.50		210.50	262
1240 4" diameter		16	1.500		167	83.50		250.50	310
1250 5" diameter		14	1.714		184	95.50		279.50	345
1260 6" diameter		12	2		204	112		316	390
1270 Tees, 3" diameter		15	1.600		166	89		255	315
1280 4" diameter		12	2		202	112		314	390
1290 5" diameter		9	2.667		315	149		464	575
1300 6" diameter		6	4		375	223		598	750
1310 Couplings, 3" diameter		18	1.333		55	74.50		129.50	173
1320 4" diameter		16	1.500		119	83.50		202.50	257
1330 5" diameter		14	1.714		214	95.50		309.50	380
1340 6" diameter		12	2		315	112		427	520
1350 Cross-over nipples, 3" diameter		18	1.333		10.60	74.50		85.10	124
1360 4" diameter		16	1.500		12.70	83.50		96.20	140
1370 5" diameter		14	1.714		15.90	95.50		111.40	162
1380 6" diameter		12	2		19.05	112		131.05	189
1400 Telescoping, reducers, concentric 4" x 3"		18	1.333		48	74.50		122.50	165
1410 5" x 4"		17	1.412		95.50	78.50		174	224
1420 6" x 5"		16	1.500		234	83.50		317.50	385

33 61 13 – Underground Hydronic Energy Distribution

33 61 13.20 Pipe Conduit, Prefabricated/Preinsulated	Crew	Daily Output	Labor-Hours	Unit	Material	2015 Bare Costs Labor	Equipment	Total	Total Incl O&P
0010 **PIPE CONDUIT, PREFABRICATED/PREINSULATED**									
0020 Does not include trenching, fittings or crane.									
0300 For cathodic protection, add 12 to 14%									
0310 of total built-up price (casing plus service pipe)									
0580 Polyurethane insulated system, 250°F. max. temp.									
0620 Black steel service pipe, standard wt., 1/2" insulation									
0660 3/4" diam. pipe size	Q-17	54	.296	L.F.	57	15.95	1.07	74.02	88
0670 1" diam. pipe size		50	.320		63	17.20	1.16	81.36	96.50
0680 1-1/4" diam. pipe size		47	.340		70	18.30	1.23	89.53	106
0690 1-1/2" diam. pipe size		45	.356		76	19.10	1.28	96.38	114
0700 2" diam. pipe size		42	.381		79	20.50	1.38	100.88	119
0710 2-1/2" diam. pipe size		34	.471		80	25.50	1.70	107.20	128
0720 3" diam. pipe size		28	.571		93	30.50	2.06	125.56	151
0730 4" diam. pipe size		22	.727		117	39	2.63	158.63	191
0740 5" diam. pipe size		18	.889		149	48	3.21	200.21	240
0750 6" diam. pipe size	Q-18	23	1.043		173	58	2.50	233.50	282
0760 8" diam. pipe size		19	1.263		254	70.50	3.03	327.53	390
0770 10" diam. pipe size		16	1.500		320	83.50	3.60	407.10	485
0780 12" diam. pipe size		13	1.846		400	103	4.43	507.43	600
0790 14" diam. pipe size		11	2.182		445	122	5.25	572.25	680
0800 16" diam. pipe size		10	2.400		510	134	5.75	649.75	775
0810 18" diam. pipe size		8	3		590	167	7.20	764.20	910
0820 20" diam. pipe size		7	3.429		655	191	8.25	854.25	1,025
0830 24" diam. pipe size		6	4		805	223	9.60	1,037.60	1,225
0900 For 1" thick insulation, add					10%				
0940 For 1-1/2" thick insulation, add					13%				
0980 For 2" thick insulation, add					20%				
1500 Gland seal for system, 3/4" diam. pipe size	Q-17	32	.500	Ea.	775	27	1.81	803.81	890
1510 1" diam. pipe size		32	.500		775	27	1.81	803.81	890
1540 1-1/4" diam. pipe size		30	.533		825	28.50	1.93	855.43	955
1550 1-1/2" diam. pipe size		30	.533		825	28.50	1.93	855.43	955
1560 2" diam. pipe size		28	.571		980	30.50	2.06	1,012.56	1,125
1570 2-1/2" diam. pipe size		26	.615		1,050	33	2.22	1,085.22	1,225
1580 3" diam. pipe size		24	.667		1,125	36	2.41	1,163.41	1,300
1590 4" diam. pipe size		22	.727		1,350	39	2.63	1,391.63	1,525
1600 5" diam. pipe size		19	.842		1,650	45.50	3.04	1,698.54	1,900
1610 6" diam. pipe size	Q-18	26	.923		1,750	51.50	2.22	1,803.72	2,000
1620 8" diam. pipe size		25	.960		2,050	53.50	2.30	2,105.80	2,325
1630 10" diam. pipe size		23	1.043		2,375	58	2.50	2,435.50	2,725
1640 12" diam. pipe size		21	1.143		2,625	63.50	2.74	2,691.24	3,000
1650 14" diam. pipe size		19	1.263		2,950	70.50	3.03	3,023.53	3,350
1660 16" diam. pipe size		18	1.333		3,500	74.50	3.20	3,577.70	3,975
1670 18" diam. pipe size		16	1.500		3,700	83.50	3.60	3,787.10	4,200
1680 20" diam. pipe size		14	1.714		4,200	95.50	4.11	4,299.61	4,775
1690 24" diam. pipe size		12	2		4,675	112	4.80	4,791.80	5,325
2000 Elbow, 45° for system									
2020 3/4" diam. pipe size	Q-17	14	1.143	Ea.	485	61.50	4.13	550.63	635
2040 1" diam. pipe size		13	1.231		500	66	4.44	570.44	655
2050 1-1/4" diam. pipe size		11	1.455		570	78	5.25	653.25	750
2060 1-1/2" diam. pipe size		9	1.778		595	95.50	6.40	696.90	805
2070 2" diam. pipe size		6	2.667		620	143	9.65	772.65	915
2080 2-1/2" diam. pipe size		4	4		670	215	14.45	899.45	1,075
2090 3" diam. pipe size		3.50	4.571		775	246	16.50	1,037.50	1,250

33 61 13 – Underground Hydronic Energy Distribution

33 61 13.20 Pipe Conduit, Prefabricated/Preinsulated

		Crew	Daily Output	Labor-Hours	Unit	Material	2015 Bare Costs Labor	Equipment	Total	Total Incl O&P
2100	4" diam. pipe size	Q-17	3	5.333	Ea.	905	287	19.25	1,211.25	1,450
2110	5" diam. pipe size	↓	2.80	5.714		1,175	305	20.50	1,500.50	1,775
2120	6" diam. pipe size	Q-18	4	6		1,325	335	14.40	1,674.40	1,975
2130	8" diam. pipe size		3	8		1,925	445	19.20	2,389.20	2,800
2140	10" diam. pipe size		2.40	10		2,350	560	24	2,934	3,450
2150	12" diam. pipe size		2	12		3,075	670	29	3,774	4,425
2160	14" diam. pipe size		1.80	13.333		3,825	745	32	4,602	5,375
2170	16" diam. pipe size		1.60	15		4,550	835	36	5,421	6,350
2180	18" diam. pipe size		1.30	18.462		5,725	1,025	44.50	6,794.50	7,900
2190	20" diam. pipe size		1	24		7,200	1,350	57.50	8,607.50	10,000
2200	24" diam. pipe size	↓	.70	34.286		9,050	1,900	82.50	11,032.50	12,900
2260	For elbow, 90°, add					25%				
2300	For tee, straight, add					85%	30%			
2340	For tee, reducing, add					170%	30%			
2380	For weldolet, straight, add					50%				
2400	Polyurethane insulation, 1"									
2410	FRP carrier and casing									
2420	4"	Q-5	18	.889	L.F.	54.50	48		102.50	132
2422	6"	Q-6	23	1.043		94.50	58		152.50	192
2424	8"		19	1.263		154	70.50		224.50	275
2426	10"		16	1.500		207	83.50		290.50	355
2428	12"	↓	13	1.846	↓	270	103		373	450
2430	FRP carrier and PVC casing									
2440	4"	Q-5	18	.889	L.F.	21.50	48		69.50	95.50
2444	8"	Q-6	19	1.263		45	70.50		115.50	155
2446	10"		16	1.500		64.50	83.50		148	197
2448	12"	↓	13	1.846	↓	82.50	103		185.50	246
2450	PVC carrier and casing									
2460	4"	Q-1	36	.444	L.F.	10.45	23.50		33.95	47
2462	6"	"	29	.552		14.55	29		43.55	60
2464	8"	Q-2	36	.667		20.50	36.50		57	77.50
2466	10"		32	.750		28	41		69	92.50
2468	12"	↓	31	.774	↓	32.50	42.50		75	99.50

33 63 13 – Underground Steam and Condensate Distribution Piping

33 63 13.10 Calcium Silicate Insulated System

		Crew	Daily Output	Labor-Hours	Unit	Material	2015 Bare Costs Labor	Equipment	Total	Total Incl O&P
0010	**CALCIUM SILICATE INSULATED SYSTEM**									
0011	High temp. (1200 degrees F)									
2840	Steel casing with protective exterior coating									
2850	6-5/8" diameter	Q-18	52	.462	L.F.	106	25.50	1.11	132.61	157
2860	8-5/8" diameter		50	.480		115	27	1.15	143.15	169
2870	10-3/4" diameter		47	.511		135	28.50	1.23	164.73	193
2880	12-3/4" diameter		44	.545		147	30.50	1.31	178.81	209
2890	14" diameter		41	.585		166	32.50	1.40	199.90	234
2900	16" diameter		39	.615		178	34.50	1.48	213.98	250
2910	18" diameter		36	.667		196	37	1.60	234.60	274
2920	20" diameter		34	.706		221	39.50	1.69	262.19	305
2930	22" diameter		32	.750		305	42	1.80	348.80	400
2940	24" diameter		29	.828		350	46	1.99	397.99	455
2950	26" diameter	↓	26	.923	↓	400	51.50	2.22	453.72	520

33 63 13 – Underground Steam and Condensate Distribution Piping

33 63 13.10 Calcium Silicate Insulated System	Crew	Daily Output	Labor-Hours	Unit	Material	2015 Bare Costs Labor	2015 Bare Costs Equipment	Total	Total Incl O&P
2960	28" diameter	Q-18	23	1.043	L.F.	495	58	2.50	555.50	635
2970	30" diameter		21	1.143		530	63.50	2.74	596.24	685
2980	32" diameter		19	1.263		595	70.50	3.03	668.53	765
2990	34" diameter		18	1.333		600	74.50	3.20	677.70	775
3000	36" diameter		16	1.500		645	83.50	3.60	732.10	840
3040	For multi-pipe casings, add					10%				
3060	For oversize casings, add					2%				
3400	Steel casing gland seal, single pipe									
3420	6-5/8" diameter	Q-18	25	.960	Ea.	1,475	53.50	2.30	1,530.80	1,700
3440	8-5/8" diameter		23	1.043		1,725	58	2.50	1,785.50	2,000
3450	10-3/4" diameter		21	1.143		1,925	63.50	2.74	1,991.24	2,225
3460	12-3/4" diameter		19	1.263		2,275	70.50	3.03	2,348.53	2,600
3470	14" diameter		17	1.412		2,525	78.50	3.39	2,606.89	2,900
3480	16" diameter		16	1.500		2,950	83.50	3.60	3,037.10	3,375
3490	18" diameter		15	1.600		3,275	89	3.84	3,367.84	3,750
3500	20" diameter		13	1.846		3,625	103	4.43	3,732.43	4,150
3510	22" diameter		12	2		4,075	112	4.80	4,191.80	4,650
3520	24" diameter		11	2.182		4,575	122	5.25	4,702.25	5,225
3530	26" diameter		10	2.400		5,225	134	5.75	5,364.75	5,950
3540	28" diameter		9.50	2.526		5,900	141	6.05	6,047.05	6,725
3550	30" diameter		9	2.667		6,000	149	6.40	6,155.40	6,825
3560	32" diameter		8.50	2.824		6,750	157	6.80	6,913.80	7,675
3570	34" diameter		8	3		7,350	167	7.20	7,524.20	8,325
3580	36" diameter		7	3.429		7,825	191	8.25	8,024.25	8,900
3620	For multi-pipe casings, add					5%				
4000	Steel casing anchors, single pipe									
4020	6-5/8" diameter	Q-18	8	3	Ea.	1,325	167	7.20	1,499.20	1,700
4040	8-5/8" diameter		7.50	3.200		1,375	178	7.70	1,560.70	1,800
4050	10-3/4" diameter		7	3.429		1,800	191	8.25	1,999.25	2,275
4060	12-3/4" diameter		6.50	3.692		1,925	206	8.85	2,139.85	2,450
4070	14" diameter		6	4		2,275	223	9.60	2,507.60	2,850
4080	16" diameter		5.50	4.364		2,625	243	10.45	2,878.45	3,275
4090	18" diameter		5	4.800		2,975	268	11.50	3,254.50	3,700
4100	20" diameter		4.50	5.333		3,275	297	12.80	3,584.80	4,075
4110	22" diameter		4	6		3,650	335	14.40	3,999.40	4,550
4120	24" diameter		3.50	6.857		3,975	380	16.45	4,371.45	4,975
4130	26" diameter		3	8		4,500	445	19.20	4,964.20	5,650
4140	28" diameter		2.50	9.600		4,950	535	23	5,508	6,275
4150	30" diameter		2	12		5,250	670	29	5,949	6,800
4160	32" diameter		1.50	16		6,275	890	38.50	7,203.50	8,300
4170	34" diameter		1	24		7,050	1,350	57.50	8,457.50	9,850
4180	36" diameter		1	24		7,675	1,350	57.50	9,082.50	10,500
4220	For multi-pipe, add					5%	20%			
4800	Steel casing elbow									
4820	6-5/8" diameter	Q-18	15	1.600	Ea.	1,800	89	3.84	1,892.84	2,125
4830	8-5/8" diameter		15	1.600		1,925	89	3.84	2,017.84	2,275
4850	10-3/4" diameter		14	1.714		2,325	95.50	4.11	2,424.61	2,700
4860	12-3/4" diameter		13	1.846		2,750	103	4.43	2,857.43	3,175
4870	14" diameter		12	2		2,950	112	4.80	3,066.80	3,425
4880	16" diameter		11	2.182		3,225	122	5.25	3,352.25	3,750
4890	18" diameter		10	2.400		3,675	134	5.75	3,814.75	4,250
4900	20" diameter		9	2.667		3,925	149	6.40	4,080.40	4,550
4910	22" diameter		8	3		4,225	167	7.20	4,399.20	4,900

For customer support on your Mechanical Cost Data, call 877.795.7394.

485

33 63 13 – Underground Steam and Condensate Distribution Piping

33 63 13.10 Calcium Silicate Insulated System	Crew	Daily Output	Labor-Hours	Unit	Material	2015 Bare Costs Labor	Equipment	Total	Total Incl O&P
4920	24" diameter	Q-18	7	3.429	Ea.	4,650	191	8.25	4,849.25	5,425
4930	26" diameter		6	4		5,100	223	9.60	5,332.60	5,950
4940	28" diameter		5	4.800		5,500	268	11.50	5,779.50	6,475
4950	30" diameter		4	6		5,525	335	14.40	5,874.40	6,600
4960	32" diameter		3	8		6,275	445	19.20	6,739.20	7,600
4970	34" diameter		2	12		6,900	670	29	7,599	8,625
4980	36" diameter		2	12		7,350	670	29	8,049	9,100
5500	Black steel service pipe, std. wt., 1" thick insulation									
5510	3/4" diameter pipe size	Q-17	54	.296	L.F.	62	15.95	1.07	79.02	93
5540	1" diameter pipe size		50	.320		64	17.20	1.16	82.36	98
5550	1-1/4" diameter pipe size		47	.340		70	18.30	1.23	89.53	106
5560	1-1/2" diameter pipe size		45	.356		74	19.10	1.28	94.38	112
5570	2" diameter pipe size		42	.381		81	20.50	1.38	102.88	122
5580	2-1/2" diameter pipe size		34	.471		85	25.50	1.70	112.20	133
5590	3" diameter pipe size		28	.571		93	30.50	2.06	125.56	151
5600	4" diameter pipe size		22	.727		98	39	2.63	139.63	170
5610	5" diameter pipe size		18	.889		107	48	3.21	158.21	194
5620	6" diameter pipe size	Q-18	23	1.043		115	58	2.50	175.50	218
6000	Black steel service pipe, std. wt., 1-1/2" thick insul.									
6010	3/4" diameter pipe size	Q-17	54	.296	L.F.	62	15.95	1.07	79.02	93
6040	1" diameter pipe size		50	.320		67	17.20	1.16	85.36	101
6050	1-1/4" diameter pipe size		47	.340		73	18.30	1.23	92.53	109
6060	1-1/2" diameter pipe size		45	.356		79	19.10	1.28	99.38	117
6070	2" diameter pipe size		42	.381		85	20.50	1.38	106.88	126
6080	2-1/2" diameter pipe size		34	.471		89	25.50	1.70	116.20	138
6090	3" diameter pipe size		28	.571		99	30.50	2.06	131.56	158
6100	4" diameter pipe size		22	.727		107	39	2.63	148.63	180
6110	5" diameter pipe size		18	.889		115	48	3.21	166.21	203
6120	6" diameter pipe size	Q-18	23	1.043		130	58	2.50	190.50	234
6130	8" diameter pipe size		19	1.263		180	70.50	3.03	253.53	305
6140	10" diameter pipe size		16	1.500		240	83.50	3.60	327.10	395
6150	12" diameter pipe size		13	1.846		300	103	4.43	407.43	490
6190	For 2" thick insulation, add					15%				
6220	For 2-1/2" thick insulation, add					25%				
6260	For 3" thick insulation, add					30%				
6800	Black steel service pipe, ex. hvy. wt., 1" thick insul.									
6820	3/4" diameter pipe size	Q-17	50	.320	L.F.	64	17.20	1.16	82.36	98
6840	1" diameter pipe size		47	.340		68	18.30	1.23	87.53	104
6850	1-1/4" diameter pipe size		44	.364		76	19.55	1.31	96.86	114
6860	1-1/2" diameter pipe size		42	.381		78	20.50	1.38	99.88	119
6870	2" diameter pipe size		40	.400		85	21.50	1.44	107.94	128
6880	2-1/2" diameter pipe size		31	.516		89	28	1.86	118.86	142
6890	3" diameter pipe size		27	.593		97	32	2.14	131.14	157
6900	4" diameter pipe size		21	.762		102	41	2.75	145.75	177
6910	5" diameter pipe size		17	.941		110	50.50	3.40	163.90	201
6920	6" diameter pipe size	Q-18	22	1.091		120	61	2.62	183.62	227
7400	Black steel service pipe, ex. hvy. wt., 1-1/2" thick insul.									
7420	3/4" diameter pipe size	Q-17	50	.320	L.F.	55	17.20	1.16	73.36	88
7440	1" diameter pipe size		47	.340		60	18.30	1.23	79.53	95
7450	1-1/4" diameter pipe size		44	.364		67	19.55	1.31	87.86	104
7460	1-1/2" diameter pipe size		42	.381		72	20.50	1.38	93.88	112
7470	2" diameter pipe size		40	.400		77	21.50	1.44	99.94	119
7480	2-1/2" diameter pipe size		31	.516		82	28	1.86	111.86	134

33 63 Steam Energy Distribution

33 63 13 – Underground Steam and Condensate Distribution Piping

33 63 13.10 Calcium Silicate Insulated System		Crew	Daily Output	Labor-Hours	Unit	Material	2015 Bare Costs Labor	2015 Bare Costs Equipment	Total	Total Incl O&P
7490	3" diameter pipe size	Q-17	27	.593	L.F.	91	32	2.14	125.14	150
7500	4" diameter pipe size		21	.762		110	41	2.75	153.75	186
7510	5" diameter pipe size		17	.941		152	50.50	3.40	205.90	247
7520	6" diameter pipe size	Q-18	22	1.091		160	61	2.62	223.62	271
7530	8" diameter pipe size		18	1.333		215	74.50	3.20	292.70	355
7540	10" diameter pipe size		15	1.600		275	89	3.84	367.84	445
7550	12" diameter pipe size		13	1.846		330	103	4.43	437.43	525
7590	For 2" thick insulation, add					13%				
7640	For 2-1/2" thick insulation, add					18%				
7680	For 3" thick insulation, add					24%				

33 63 13.20 Combined Steam Pipe With Condensate Return

		Crew	Daily Output	Labor-Hours	Unit	Material	Labor	Equipment	Total	Total Incl O&P
0010	**COMBINED STEAM PIPE WITH CONDENSATE RETURN**									
9010	8" and 4" in 24" case	Q-18	100	.240	L.F.	395	13.40	.58	408.98	450
9020	6" and 3" in 20" case		104	.231		365	12.85	.55	378.40	425
9030	3" and 1-1/2" in 16" case		110	.218		275	12.15	.52	287.67	325
9040	2" and 1-1/4" in 12-3/4" case		114	.211		235	11.75	.51	247.26	277
9050	1-1/2" and 1-1/4" in 10-3/4" case		116	.207		235	11.55	.50	247.05	277
9100	Steam pipe only (no return)									
9110	6" in 18" case	Q-18	108	.222	L.F.	300	12.40	.53	312.93	350
9120	4" in 14" case		112	.214		235	11.95	.51	247.46	278
9130	3" in 14" case		114	.211		220	11.75	.51	232.26	260
9140	2-1/2" in 12-3/4" case		118	.203		205	11.35	.49	216.84	244
9150	2" in 12-3/4" case		122	.197		195	10.95	.47	206.42	232

For customer support on your Mechanical Cost Data, call 877.795.7394.

487

Division Notes

	CREW	DAILY OUTPUT	LABOR-HOURS	UNIT	BARE COSTS				TOTAL INCL O&P
					MAT.	LABOR	EQUIP.	TOTAL	

Estimating Tips

This section involves equipment and construction costs for air noise and odor pollution control systems. These systems may be interrelated and care must be taken that the complete systems are estimated. For example, air pollution equipment may include dust and air-entrained particles that have to be collected. The vacuum systems could be noisy, requiring silencers to reduce noise pollution, and the collected solids have to be disposed of to prevent solid pollution.

Reference Numbers

Reference numbers are shown in shaded boxes at the beginning of some major classifications. These numbers refer to related items in the Reference Section. The reference information may be an estimating procedure, an alternate pricing method, or technical information.

Note: Not all subdivisions listed here necessarily appear in this publication. ■

44 11 Particulate Control Equipment

44 11 16 – Fugitive Dust Barrier Systems

44 11 16.10 Dust Collection Systems	Crew	Daily Output	Labor-Hours	Unit	Material	2015 Bare Costs Labor	Equipment	Total	Total Incl O&P
0010 **DUST COLLECTION SYSTEMS** Commercial/Industrial									
0120 Central vacuum units									
0130 Includes stand, filters and motorized shaker									
0200 500 CFM, 10" inlet, 2 HP	Q-20	2.40	8.333	Ea.	4,175	425		4,600	5,250
0220 1000 CFM, 10" inlet, 3 HP		2.20	9.091		4,400	465		4,865	5,525
0240 1500 CFM, 10" inlet, 5 HP		2	10		4,625	510		5,135	5,875
0260 3000 CFM, 13" inlet, 10 HP		1.50	13.333		12,900	685		13,585	15,300
0280 5000 CFM, 16" inlet, 2 @ 10 HP		1	20		13,700	1,025		14,725	16,700
1000 Vacuum tubing, galvanized									
1100 2-1/8" OD, 16 ga.	Q-9	440	.036	L.F.	2.95	1.83		4.78	6.05
1110 2-1/2" OD, 16 ga.		420	.038		3.24	1.92		5.16	6.50
1120 3" OD, 16 ga.		400	.040		4.13	2.01		6.14	7.60
1130 3-1/2" OD, 16 ga.		380	.042		5.85	2.12		7.97	9.70
1140 4" OD, 16 ga.		360	.044		6	2.24		8.24	10
1150 5" OD, 14 ga.		320	.050		10.80	2.52		13.32	15.70
1160 6" OD, 14 ga.		280	.057		12.10	2.88		14.98	17.70
1170 8" OD, 14 ga.		200	.080		18.30	4.03		22.33	26
1180 10" OD, 12 ga.		160	.100		35.50	5.05		40.55	46.50
1190 12" OD, 12 ga.		120	.133		47	6.70		53.70	62
1200 14" OD, 12 ga.		80	.200		53.50	10.05		63.55	74
1940 Hose, flexible wire reinforced rubber									
1956 3" diam.	Q-9	400	.040	L.F.	9.15	2.01		11.16	13.15
1960 4" diam.		360	.044		10.60	2.24		12.84	15.05
1970 5" diam.		320	.050		13.75	2.52		16.27	19
1980 6" diam.		280	.057		14.40	2.88		17.28	20
2000 90° Elbow, slip fit									
2110 2-1/8" diam.	Q-9	70	.229	Ea.	10.85	11.50		22.35	29.50
2120 2-1/2" diam.		65	.246		15.20	12.40		27.60	35.50
2130 3" diam.		60	.267		20.50	13.45		33.95	43
2140 3-1/2" diam.		55	.291		25.50	14.65		40.15	50.50
2150 4" diam.		50	.320		31.50	16.10		47.60	59
2160 5" diam.		45	.356		60	17.90		77.90	93.50
2170 6" diam.		40	.400		82.50	20		102.50	122
2180 8" diam.		30	.533		157	27		184	213
2400 45° Elbow, slip fit									
2410 2-1/8" diam.	Q-9	70	.229	Ea.	9.50	11.50		21	28
2420 2-1/2" diam.		65	.246		14	12.40		26.40	34.50
2430 3" diam.		60	.267		16.90	13.45		30.35	39
2440 3-1/2" diam.		55	.291		21.50	14.65		36.15	46
2450 4" diam.		50	.320		27.50	16.10		43.60	55
2460 5" diam.		45	.356		46.50	17.90		64.40	78.50
2470 6" diam.		40	.400		63	20		83	100
2480 8" diam.		35	.457		123	23		146	171
2800 90° TY, slip fit thru 6" diam.									
2810 2-1/8" diam.	Q-9	42	.381	Ea.	17.75	19.20		36.95	49
2820 2-1/2" diam.		39	.410		26.50	20.50		47	60.50
2830 3" diam.		36	.444		37	22.50		59.50	74.50
2840 3-1/2" diam.		33	.485		39	24.50		63.50	80
2850 4" diam.		30	.533		41.50	27		68.50	87
2860 5" diam.		27	.593		70.50	30		100.50	123
2870 6" diam.		24	.667		101	33.50		134.50	163
2880 8" diam., butt end					155			155	170
2890 10" diam., butt end					475			475	525

490

For customer support on your Mechanical Cost Data, call 877.795.7394.

44 11 16 – Fugitive Dust Barrier Systems

44 11 16.10 Dust Collection Systems		Crew	Daily Output	Labor-Hours	Unit	Material	2015 Bare Costs Labor	Equipment	Total	Total Incl O&P
2900	12" diam., butt end				Ea.	475			475	525
2910	14" diam., butt end					855			855	940
2920	6" x 4" diam., butt end					106			106	116
2930	8" x 4" diam., butt end					200			200	220
2940	10" x 4" diam., butt end					263			263	289
2950	12" x 4" diam., butt end					300			300	330
3100	90° Elbow, butt end segmented									
3110	8" diam., butt end, segmented				Ea.	157			157	172
3120	10" diam., butt end, segmented					390			390	430
3130	12" diam., butt end, segmented					495			495	545
3140	14" diam., butt end, segmented					620			620	680
3200	45° Elbow, butt end segmented									
3210	8" diam., butt end, segmented				Ea.	123			123	136
3220	10" diam., butt end, segmented					282			282	310
3230	12" diam., butt end, segmented					288			288	315
3240	14" diam., butt end, segmented					585			585	645
3400	All butt end fittings require one coupling per joint.									
3410	Labor for fitting included with couplings.									
3460	Compression coupling, galvanized, neoprene gasket									
3470	2-1/8" diam.	Q-9	44	.364	Ea.	10.85	18.30		29.15	40
3480	2-1/2" diam.		44	.364		10.85	18.30		29.15	40
3490	3" diam.		38	.421		15.60	21		36.60	49.50
3500	3-1/2" diam.		35	.457		17.45	23		40.45	54
3510	4" diam.		33	.485		18.95	24.50		43.45	58.50
3520	5" diam.		29	.552		21.50	28		49.50	66.50
3530	6" diam.		26	.615		25.50	31		56.50	75.50
3540	8" diam.		22	.727		46	36.50		82.50	107
3550	10" diam.		20	.800		67.50	40.50		108	136
3560	12" diam.		18	.889		84	45		129	161
3570	14" diam.		16	1		119	50.50		169.50	208
3800	Air gate valves, galvanized									
3810	2-1/8" diam.	Q-9	30	.533	Ea.	111	27		138	163
3820	2-1/2" diam.		28	.571		119	29		148	175
3830	3" diam.		26	.615		127	31		158	188
3840	4" diam.		23	.696		147	35		182	216
3850	6" diam.		18	.889		205	45		250	294

44 11 96 – Dry Plate Electrostatic Precipitator Equipment

44 11 96.10 Electrostatic Precipitators		Crew	Daily Output	Labor-Hours	Unit	Material	2015 Bare Costs Labor	Equipment	Total	Total Incl O&P
0010	**ELECTROSTATIC PRECIPITATORS**									
4100	Smoke pollution control									
4110	Exhaust air volume									
4120	1 to 2.4 MCFM	Q-10	2.15	11.163	MCFM	8,950	585		9,535	10,700
4130	2.4 to 4.8 MCFM		2.70	8.889		10,200	465		10,665	11,900
4140	4.8 to 7.2 MCFM		3.60	6.667		10,400	350		10,750	11,900
4150	7.2 to 12 MCFM		5.40	4.444		10,800	232		11,032	12,300

For customer support on your Mechanical Cost Data, call 877.795.7394.

491

Division Notes

	CREW	DAILY OUTPUT	LABOR-HOURS	UNIT	BARE COSTS				TOTAL INCL O&P
					MAT.	LABOR	EQUIP.	TOTAL	

Estimating Tips

This division contains information about water and wastewater equipment and systems, which was formerly located in Division 44. The main areas of focus are total wastewater treatment plants and components of wastewater treatment plants. In addition, there are assemblies such as sewage treatment lagoons that can be found in some publications under G30 Site Mechanical Utilities. Also included in this section are oil/water separators for wastewater treatment.

Reference Numbers

Reference numbers are shown in shaded boxes at the beginning of some major classifications. These numbers refer to related items in the Reference Section. The reference information may be an estimating procedure, an alternate pricing method, or technical information.

Note: Not all subdivisions listed here necessarily appear in this publication. ■

46 23 Grit Removal And Handling Equipment

46 23 23 – Vortex Grit Removal Equipment

46 23 23.10 Rainwater Filters	Crew	Daily Output	Labor-Hours	Unit	Material	2015 Bare Costs Labor	2015 Bare Costs Equipment	Total	Total Incl O&P
0010 **RAINWATER FILTERS**									
0100 42 gal./min	B-21	3.50	8	Ea.	45,100	345	39.50	45,484.50	50,000
0200 65 gal./min		3.50	8		45,200	345	39.50	45,584.50	50,500
0300 208 gal./min		3.50	8		45,200	345	39.50	45,584.50	50,500

46 25 Oil and Grease Separation and Removal Equipment

46 25 13 – Coalescing Oil-Water Separators

46 25 13.20 Oil/Water Separators

	Crew	Daily Output	Labor-Hours	Unit	Material	2015 Bare Costs Labor	2015 Bare Costs Equipment	Total	Total Incl O&P
0010 **OIL/WATER SEPARATORS**									
0020 Underground, tank only									
0030 Excludes excavation, backfill, & piping									
0100 200 GPM	B-21	3.50	8	Ea.	44,900	345	39.50	45,284.50	50,000
0110 400 GPM		3.25	8.615		62,500	375	42.50	62,917.50	69,500
0120 600 GPM		2.75	10.182		70,000	440	50	70,490	77,500
0130 800 GPM		2.50	11.200		79,000	485	55	79,540	88,000
0140 1000 GPM		2	14		88,500	605	69	89,174	98,500
0150 1200 GPM		1.50	18.667		101,000	810	92	101,902	112,500
0160 1500 GPM		1	28		119,500	1,225	138	120,863	133,000

46 25 16 – API Oil-Water Separators

46 25 16.10 Oil/Water Separators

	Crew	Daily Output	Labor-Hours	Unit	Material	2015 Bare Costs Labor	2015 Bare Costs Equipment	Total	Total Incl O&P
0010 **OIL/WATER SEPARATORS**									
0020 Complete system, not including excavation and backfill									
0100 Treated capacity 0.2 cubic feet per second	B-22	1	30	Ea.	5,200	1,325	207	6,732	7,950
0200 0.5 cubic foot per second	B-13	.75	74.667		8,300	3,050	980	12,330	14,900
0300 1.0 cubic foot per second		.60	93.333		15,400	3,825	1,225	20,450	24,100
0400 2.4 – 3 cubic feet per second		.30	186		23,500	7,650	2,450	33,600	40,200
0500 11 cubic feet per second		.17	329		84,000	13,500	4,325	101,825	118,000
0600 22 cubic feet per second		.10	560		126,500	23,000	7,375	156,875	182,000

Assemblies Section

Table of Contents

How RSMeans Assemblies Data Works

Assemblies estimating provides a fast and reasonably accurate way to develop construction costs. An assembly is the grouping of individual work items, with appropriate quantities, to provide a cost for a major construction component in a convenient unit of measure.

An assemblies estimate is often used during early stages of design development to compare the cost impact of various design alternatives on total building cost.

Assemblies estimates are also used as an efficient tool to verify construction estimates.

Assemblies estimates do not require a completed design or detailed drawings. Instead, they are based on the general size of the structure and other known parameters of the project. The degree of accuracy of an assemblies estimate is generally within +/- 15%.

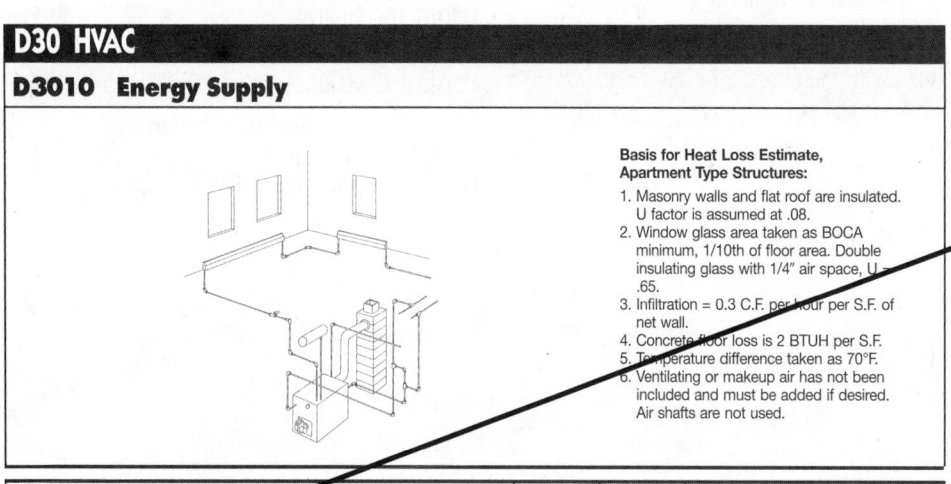

D30 HVAC

D3010 Energy Supply

Basis for Heat Loss Estimate, Apartment Type Structures:

1. Masonry walls and flat roof are insulated. U factor is assumed at .08.
2. Window glass area taken as BOCA minimum, 1/10th of floor area. Double insulating glass with 1/4" air space, U = .65.
3. Infiltration = 0.3 C.F. per hour per S.F. of net wall.
4. Concrete floor loss is 2 BTUH per S.F.
5. Temperature difference taken as 70°F.
6. Ventilating or makeup air has not been included and must be added if desired. Air shafts are not used.

RSMeans assemblies are identified by a **unique 12-character identifier**. The assemblies are numbered using UNIFORMAT II, ASTM Standard E1557. The first 6 characters represent this system to Level 4. The last 6 characters represent further breakdown by RSMeans in order to arrange items in understandable groups of similar tasks. Line numbers are consistent across all RSMeans publications, so a line number in any RSMeans assemblies data set will always refer to the same work.

System Components	QUANTITY	UNIT	COST EACH		
			MAT.	INST.	TOTAL
SYSTEM D3010 510 1760					
HEATING SYSTEM, FIN TUBE RADIATION, FORCED HOT WATER					
1,000 S.F. AREA, 10,000 C.F. VOLUME					
Boiler, oil fired, CI, burner, ctrls/insul/breech/pipe/ftng/valves, 109 MBH	1.000	Ea.	4,462.50	4,025	8,487.50
Circulating pump, CI flange connection, 1/12 HP	1.000	Ea.	400	213	613
Expansion tank, painted steel, ASME 18 Gal capacity	1.000	Ea.	785	93	878
Storage tank, steel, above ground, 275 Gal capacity w/supports	1.000	Ea.	540	260	800
Copper tubing type L, solder joint, hanger 10' OC, 3/4" diam	100.000	L.F.	570	935	1,505
Radiation, 3/4" copper tube w/alum fin baseboard pkg, 7" high	30.000	L.F.	243	675	918
Pipe covering, calcium silicate w/cover, 1' wall, 3/4" diam	100.000	L.F.	448	690	1,138
TOTAL			7,448.50	6,891	14,339.50
COST PER S.F.			7.45	6.89	14.34

Information is available in the Reference Section to assist the estimator with estimating procedures, alternate pricing methods, and additional technical information.

The **Reference Box** indicates the exact location of the information in the Reference Section at the back of the book. The "R" stands for "reference," and the remaining characters are the RSMeans line numbers.

D3010 510	Apartment Building Heating - Fin Tube Radiation		COST PER S.F.		
			MAT.	INST.	TOTAL
1740	Heating systems, fin tube radiation, forced hot water				
1760	1,000 S.F. area, 10,000 C.F. volume		7.45	6.90	14.35
1800	10,000 S.F. area, 100,000 C.F. volume	R235000 -10	3.13	4.13	7.26
1840	20,000 S.F. area, 200,000 C.F. volume		3.45	4.64	8.09
1880	30,000 S.F. area, 300,000 C.F. volume	R235000 -20	3.34	4.49	7.83
1890					

RSMeans assemblies descriptions appear in two formats: narrative and table. **Narrative descriptions** are shown in a hierarchical structure to make them readable. In order to read a complete description, read up through the indents to the top of the section. Include everything that is above and to the left that is not contradicted by information below.

For supplemental customizable square foot estimating forms, visit: **http://www.reedconstructiondata.com/rsmeans/assemblies**

st assemblies consist of three major elements: a graphic, the stem components, and the cost data itself. The **Graphic** is a visual presentation showing the typical appearance of the assembly in estion, frequently accompanied by additional explanatory technical formation describing the class of items. The **System Components** is isting of the individual tasks that make up the assembly, including e quantity and unit of measure for each item, along with the cost material and installation. The **Assemblies Data** below lists prices for her similar systems with dimensional and/or size variations.

All RSMeans assemblies costs represent the cost for the installing contractor. An allowance for profit has been added to all material, labor, and equipment rental costs.

A markup for labor burdens, including workers' compensation, fixed overhead, and business overhead, is included with installation costs.

The data published in RSMeans print books represents a "national average" cost. This data should be modified to the project location using the **City Cost Indexes** or **Location Factors** tables found in the Reference Section in the back of the book.

All RSMeans assemblies data includes a typical **Unit of Measure** used for estimating that item. For instance, for continuous footings or foundation walls the unit is linear feet (L.F.). For spread footings the unit is each (Ea.). The estimator needs to take special care that the unit in the data matches the unit in the takeoff. Abbreviations and unit conversions can be found in the Reference Section.

System components are listed separately to detail what is included in the development of the total system price.

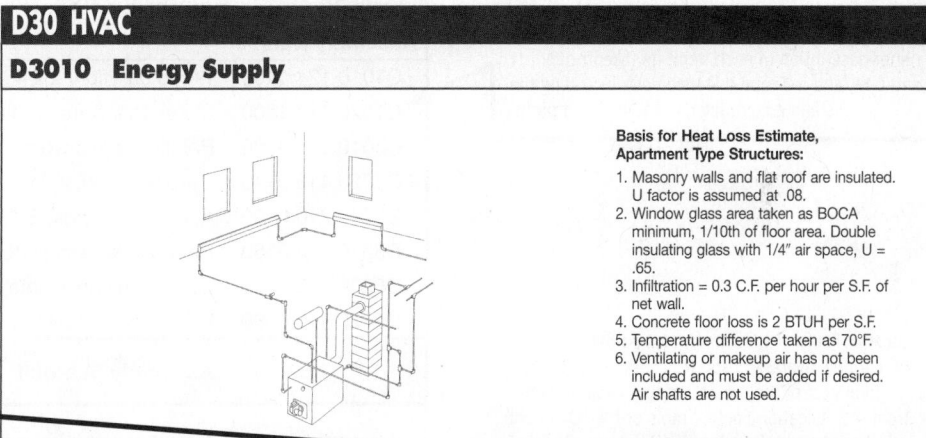

D30 HVAC

D3010 Energy Supply

Basis for Heat Loss Estimate, Apartment Type Structures:

1. Masonry walls and flat roof are insulated. U factor is assumed at .08.
2. Window glass area taken as BOCA minimum, 1/10th of floor area. Double insulating glass with 1/4" air space, U = .65.
3. Infiltration = 0.3 C.F. per hour per S.F. of net wall.
4. Concrete floor loss is 2 BTUH per S.F.
5. Temperature difference taken as 70°F.
6. Ventilating or makeup air has not been included and must be added if desired. Air shafts are not used.

System Components	QUANTITY	UNIT	COST EACH		
			MAT.	INST.	TOTAL
SYSTEM D3010 510 1760					
HEATING SYSTEM, FIN TUBE RADIATION, FORCED HOT WATER					
1,000 S.F. AREA, 10,000 C.F. VOLUME					
Boiler, oil fired, CI, burner, ctrls/insul/breech/pipe/ftng/valves, 109 MBH	1.000	Ea.	4,462.50	4,025	8,487.50
Circulating pump, CI flange connection, 1/12 HP	1.000	Ea.	400	213	613
Expansion tank, painted steel, ASME 18 Gal capacity	1.000	Ea.	785	93	878
Storage tank, steel, above ground, 275 Gal capacity w/supports	1.000	Ea.	540	260	800
Copper tubing type L, solder joint, hanger 10' OC, 3/4" diam	100.000	L.F.	570	935	1,505
Radiation, 3/4" copper tube w/alum fin baseboard pkg, 7" high	30.000	L.F.	243	675	918
Pipe covering, calcium silicate w/cover, 1' wall, 3/4" diam	100.000	L.F.	448	690	1,138
TOTAL			7,448.50	6,891	14,339.50
COST PER S.F.			7.45	6.89	14.34

D3010 510	Apartment Building Heating - Fin Tube Radiation		COST PER S.F.		
			MAT.	INST.	TOTAL
1740	Heating systems, fin tube radiation, forced hot water				
1760	1,000 S.F. area, 10,000 C.F. volume		7.45	6.90	14.35
1800	10,000 S.F. area, 100,000 C.F. volume	R235000 -10	3.13	4.13	7.26
1840	20,000 S.F. area, 200,000 C.F. volume		3.45	4.64	8.09
1880	30,000 S.F. area, 300,000 C.F. volume	R235000 -20	3.34	4.49	7.83
1890					

How RSMeans Assemblies Data Works (Continued)

Sample Estimate

This sample demonstrates the elements of an estimate, including a tally of the RSMeans data lines. Published assemblies costs include all markups for labor burden and profit for the installing contractor. This estimate adds a summary of the markups applied by a general contractor on the installing contractors' work. These figures represent the total cost to the owner. The RSMeans location factor is added at the bottom of the estimate to adjust the cost of the work to a specific location

Work performed: The body of the estimate shows the RSMeans data selected, including line numbers, a brief description of each item, its takeoff quantity and unit, and the total installed cost, including the installing contractor's overhead and profit.

Location Factor: RSMeans published data is based on national average costs. If necessary, adjust the total cost of the project using a location factor from the "Location Factor" table or the "City Cost Indexes" table, found in the Reference Section. Use location factors if the work is general, covering the work of multiple trades. If the work is by a single trade (e.g. masonry) use the more specific data found in the City Cost Indexes.

To adjust costs by location factors, multiply the base cost by the factor and divide by 100.

Contingency: A factor for contingency may be added to any estimate to represent the cost of unknowns that may occur between the time that the estimate is performed and the time the project is constructed. The amount of the allowance will depend on the stage of design at which the estimate is done, and the contractor's assessment of the risk involved.

Project Name:	Interior Fit-out, ABC Office
Location:	**Anywhere, USA**
Assembly Number	**Description**
C1010 124 1200	Wood partition, 2 x 4 @ 16" OC w/5/8" FR gypsum board
C1020 114 1800	Metal door & frame, flush hollow core, 3'-0" x 7'-0"
C3010 230 0080	Painting, brushwork, primer & 2 coats
C3020 410 0140	Carpet, tufted, nylon, roll goods, 12' wide, 26 oz
C3030 210 6000	Acoustic ceilings, 24" x 48" tile, tee grid suspension
D5020 125 0560	Receptacles incl plate, box, conduit, wire, 20 A duplex
D5020 125 0720	Light switch incl plate, box, conduit, wire, 20 A single pole
D5020 210 0560	Fluorescent fixtures, recess mounted, 20 per 1000 SF
Assembly Subtotal	
	Sales Tax @
	General Requirements @
Subtotal A	
	GC Overhead @
Subtotal B	
	GC Profit @
Subtotal C	
Adjusted by Location Factor	
	Architects Fee @
	Contingency @
Project Total Cost	

This estimate is based on an interactive spreadsheet. A copy of this spreadsheet is located on the RSMeans website at **http://www.reedconstructiondata.com/rsmeans/extras/546011.** You are free to download it and adjust it to your methodology.

Date: 1/1/2015		STD
Qty.	Unit	Subtotal
560.000	S.F.	$2,693.60
2.000	Ea.	$2,450.00
1,120.000	S.F.	$1,344.00
240.000	S.F.	$1,003.20
200.000	S.F.	$1,140.00
8.000	Ea.	$2,164.00
2.000	Ea.	$545.00
200.000	S.F.	$2,140.00
		$13,479.80
5 %		$ 337.00
7 %		$ 943.59
		$14,760.38
5 %		$ 738.02
		$15,498.40
5 %		$ 774.92
		$16,273.32
118.1		$ 19,218.79
8 %		$ 1,537.50
15 %		$ 2,882.82
		$ 23,639.11

Sales Tax: If the work is subject to state or local sales taxes, the amount must be added to the estimate. In a conceptual estimate it can be assumed that one half of the total represents material costs. Therefore, apply the sales tax rate to 50% of the assembly subtotal.

General Requirements: This item covers project-wide needs provided by the General Contractor. These items vary by project, but may include temporary facilities and utilities, security, testing, project cleanup, etc. In assemblies estimates a percentage is used, typically between 5% and 15% of project cost.

General Contractor Overhead: This entry represents the General Contractor's markup on all work to cover project administration costs.

General Contractor Profit: This entry represents the GC's profit on all work performed. The value included here can vary widely by project, and is influenced by the GC's perception of the project's financial risk and market conditions.

Architects Fee: If appropriate, add the design cost to the project estimate. These fees vary based on project complexity and size. Typical design and engineering fees can be found in the reference section.

A2020 Basement Walls

A2020 220	Subdrainage Piping	COST PER L.F.		
		MAT.	INST.	TOTAL
2000	Piping, excavation & backfill excluded, PVC, perforated			
2110	3" diameter	1.60	4.14	5.74
2130	4" diameter	1.60	4.14	5.74
2140	5" diameter	3.61	4.43	8.04
2150	6" diameter	3.61	4.43	8.04
3000	Metal alum. or steel, perforated asphalt coated			
3150	6" diameter	7.15	4.08	11.23
3160	8" diameter	9.40	4.19	13.59
3170	10" diameter	11.75	5.60	17.35
3180	12" diameter	13.15	7.10	20.25
3220	18" diameter	19.75	9.85	29.60
4000	Corrugated HDPE tubing			
4130	4" diameter	.55	.77	1.32
4150	6" diameter	1.57	1.03	2.60
4160	8" diameter	2.97	1.32	4.29
4180	12" diameter	7.05	4.56	11.61
4200	15" diameter	8.35	5.15	13.50
4220	18" diameter	13.30	7.35	20.65

In this closed-loop indirect collection system, fluid with a low freezing temperature, such as propylene glycol, transports heat from the collectors to water storage. The transfer fluid is contained in a closed-loop consisting of collectors, supply and return piping, and a remote heat exchanger. The heat exchanger transfers heat energy from the fluid in the collector loop to potable water circulated in a storage loop. A typical two-or-three panel system contains 5 to 6 gallons of heat transfer fluid.

When the collectors become approximately 20°F warmer than the storage temperature, a controller activates the circulator on the collector and storage loops. The circulators will move the fluid and potable water through the heat exchanger until heat collection no longer occurs. At that point, the system shuts down. Since the heat transfer medium is a fluid with a very low freezing temperature, there is no need for it to be drained from the system between periods of collection.

504

D2020 Domestic Water Distribution

System Components	QUANTITY	UNIT	COST EACH		
			MAT.	INST.	TOTAL
SYSTEM D2020 265 2760					
SOLAR, CLOSED LOOP, ADD-ON HOT WATER SYS., EXTERNAL HEAT EXCHANGER					
3/4" TUBING, TWO 3'X7' BLACK CHROME COLLECTORS					
A,B,G,L,K,M Heat exchanger fluid-fluid pkg incl 2 circulators, expansion tank,					
Check valve, relief valve, controller, hi temp cutoff, & 2 sensors	1.000	Ea.	880	510	1,390
C Thermometer, 2" dial	3.000	Ea.	72	133.50	205.50
D, T Fill & drain valve, brass, 3/4" connection	1.000	Ea.	10.70	29.50	40.20
E Air vent, manual, 1/8" fitting	2.000	Ea.	5.36	44	49.36
F Air purger	1.000	Ea.	51.50	59	110.50
H Strainer, Y type, bronze body, 3/4" IPS	1.000	Ea.	37.50	38	75.50
I Valve, gate, bronze, NRS, soldered 3/4" diam	6.000	Ea.	420	213	633
J Neoprene vent flashing	2.000	Ea.	19	71	90
N-1, N Relief valve temp & press, 150 psi 210°F self-closing 3/4" IPS	1.000	Ea.	23	23.50	46.50
O Pipe covering, urethane, ultraviolet cover, 1" wall 3/4" diam	20.000	L.F.	62.20	128	190.20
P Pipe covering, fiberglass, all service jacket, 1" wall, 3/4" diam	50.000	L.F.	49.50	255	304.50
Q Collector panel solar energy blk chrome on copper, 1/8" temp glass 3'x7'	2.000	Ea.	2,200	268	2,468
Roof clamps for solar energy collector panels	2.000	Set	6.16	36.50	42.66
R Valve, swing check, bronze, regrinding disc, 3/4" diam	2.000	Ea.	220	71	291
S Pressure gauge, 60 psi, 2" dial	1.000	Ea.	26.50	22	48.50
U Valve, water tempering, bronze, sweat connections, 3/4" diam	1.000	Ea.	139	35.50	174.50
W-2, V Tank water storage w/heating element, drain, relief valve, existing	1.000	Ea.			
Copper tubing type L, solder joint, hanger 10' OC 3/4" diam	20.000	L.F.	114	187	301
Copper tubing, type M, solder joint, hanger 10' OC 3/4" diam	70.000	L.F.	317.80	637	954.80
Sensor wire, #22-2 conductor multistranded	.500	C.L.F.	7.83	32.75	40.58
Solar energy heat transfer fluid, propylene glycol anti-freeze	6.000	Gal.	102.60	153	255.60
Wrought copper fittings & solder, 3/4" diam	76.000	Ea.	205.96	2,850	3,055.96
TOTAL			4,970.61	5,797.25	10,767.86

D2020 265	Solar, Closed Loop, Add-On Hot Water Systems	COST EACH		
		MAT.	INST.	TOTAL
2550	Solar, closed loop, add-on hot water system, external heat exchanger			
2570	3/8" tubing, 3 ea. 4' x 4'-4" vacuum tube collectors	5,525	5,400	10,925
2580	1/2" tubing, 4 ea 4 x 4'-4" vacuum tube collectors R235616 -60	6,325	5,850	12,175
2600	2 ea. 3'x7' black chrome collectors	4,575	5,525	10,100
2620	3 ea. 3'x7' black chrome collectors	5,675	5,675	11,350
2640	2 ea. 3'x7' flat black collectors	4,300	5,525	9,825
2660	3 ea. 3'x7' flat black collectors	5,275	5,700	10,975
2700	3/4" tubing, 3 ea. 3'x7' black chrome collectors	6,075	5,950	12,025
2720	3 ea. 3'x7' flat black absorber plate collectors	5,675	5,975	11,650
2740	2 ea. 4'x9' flat black w/plastic glazing collectors	5,075	6,000	11,075
2760	2 ea. 3'x7' black chrome collectors	4,975	5,800	10,775
2780	1" tubing, 4 ea 2'x9' plastic absorber & glazing collectors	6,100	6,750	12,850
2800	4 ea. 3'x7' black chrome absorber collectors	7,925	6,775	14,700
2820	4 ea. 3'x7' flat black absorber collectors	7,400	6,800	14,200

In the drainback indirect-collection system, the heat transfer fluid is distilled water contained in a loop consisting of collectors, supply and return piping, and an unpressurized holding tank. A large heat exchanger containing incoming potable water is immersed in the holding tank. When a controller activates solar collection, the distilled water is pumped through the collectors and heated and pumped back down to the holding tank. When the temperature differential between the water in the collectors and water in storage is such that collection no longer occurs, the pump turns off and gravity causes the distilled water in the collector loop to drain back to the holding tank. All the loop piping is pitched so that the water can drain out of the collectors and piping and not freeze there. As hot water is needed in the home, incoming water first flows through the holding tank with the immersed heat exchanger and is warmed and then flows through a conventional heater for any supplemental heating that is necessary.

D2020 Domestic Water Distribution

System Components	QUANTITY	UNIT	COST EACH		
			MAT.	INST.	TOTAL
SYSTEM D2020 270 2760					
SOLAR, DRAINBACK, ADD ON, HOT WATER, IMMERSED HEAT EXCHANGER					
3/4" TUBING, THREE EA 3'X7' BLACK CHROME COLLECTOR					
A, B Differential controller 2 sensors, thermostat, solar energy system	1.000	Ea.	143	59	202
C Thermometer 2" dial	3.000	Ea.	72	133.50	205.50
D, T Fill & drain valve, brass, 3/4" connection	1.000	Ea.	10.70	29.50	40.20
E-1 Automatic air vent 1/8" fitting	1.000	Ea.	18.95	22	40.95
H Strainer, Y type, bronze body, 3/4" IPS	1.000	Ea.	37.50	38	75.50
I Valve, gate, bronze, NRS, soldered 3/4" diam	2.000	Ea.	140	71	211
J Neoprene vent flashing	2.000	Ea.	19	71	90
L Circulator, solar heated liquid, 1/20 HP	1.000	Ea.	177	106	283
N Relief valve temp. & press. 150 psi 210°F self-closing 3/4" IPS	1.000	Ea.	23	23.50	46.50
O Pipe covering, urethane, ultraviolet cover, 1" wall, 3/4" diam	20.000	L.F.	62.20	128	190.20
P Pipe covering, fiberglass, all service jacket, 1" wall, 3/4" diam	50.000	L.F.	49.50	255	304.50
Q Collector panel solar energy blk chrome on copper, 1/8" temp glas 3'x7'	3.000	Ea.	3,300	402	3,702
Roof clamps for solar energy collector panels	3.000	Set	9.24	54.75	63.99
R Valve, swing check, bronze, regrinding disc, 3/4" diam	1.000	Ea.	110	35.50	145.50
U Valve, water tempering, bronze sweat connections, 3/4" diam	1.000	Ea.	139	35.50	174.50
V Tank, water storage w/heating element, drain, relief valve, existing	1.000	Ea.			
W Tank, water storage immersed heat exchr elec 2"x1/2# insul 120 gal	1.000	Ea.	2,025	505	2,530
X Valve, globe, bronze, rising stem, 3/4" diam, soldered	3.000	Ea.	327	106.50	433.50
Y Flow control valve	1.000	Ea.	132	32	164
Z Valve, ball, bronze, solder 3/4" diam, solar loop flow control	1.000	Ea.	25.50	35.50	61
Copper tubing, type L, solder joint, hanger 10' OC 3/4" diam	20.000	L.F.	114	187	301
Copper tubing, type M, solder joint, hanger 10' OC 3/4" diam	70.000	L.F.	317.80	637	954.80
Sensor wire, #22-2 conductor, multistranded	.500	C.L.F.	7.83	32.75	40.58
Wrought copper fittings & solder, 3/4" diam	76.000	Ea.	205.96	2,850	3,055.96
TOTAL			7,466.18	5,850	13,316.18

D2020 270	Solar, Drainback, Hot Water Systems		COST EACH		
			MAT.	INST.	TOTAL
2550	Solar, drainback, hot water, immersed heat exchanger	R235616 -60			
2560	3/8" tubing, 3 ea. 4' x 4'-4" vacuum tube collectors		6,700	5,250	11,950
2580	1/2" tubing, 4 ea 4' x 4'-4" vacuum tube collectors, 80 gal tank		7,500	5,700	13,200
2600	120 gal tank		7,750	5,775	13,525
2640	2 ea. 3'x7' blk chrome collectors, 80 gal tank		5,750	5,375	11,125
2660	3 ea. 3'x7' blk chrome collectors, 120 gal tank		7,100	5,575	12,675
2700	2 ea. 3'x7' flat blk collectors, 120 gal tank		5,725	5,450	11,175
2720	3 ea. 3'x7' flat blk collectors, 120 gal tank		6,700	5,600	12,300
2760	3/4" tubing, 3 ea 3'x7' black chrome collectors, 120 gal tank		7,475	5,850	13,325
2780	3 ea. 3'x7' flat black absorber collectors, 120 gal tank		7,050	5,875	12,925
2800	2 ea. 4'x9' flat blk w/plastic glazing collectors 120 gal tank		6,475	5,900	12,375
2840	1" tubing, 4 ea. 2'x9' plastic absorber & glazing collectors, 120 gal tank		7,575	6,650	14,225
2860	4 ea. 3'x7' black chrome absorber collectors, 120 gal tank		9,425	6,675	16,100
2880	4 ea. 3'x7' flat black absorber collectors, 120 gal tank		8,875	6,700	15,575

For customer support on your Mechanical Cost Data, call 877.795.7394.

507

In the draindown direct-collection system, incoming domestic water is heated in the collectors. When the controller activates solar collection, domestic water is first heated as it flows through the collectors and is then pumped to storage. When conditions are no longer suitable for heat collection, the pump shuts off and the water in the loop drains down and out of the system by means of solenoid valves and properly pitched piping.

D2020 Domestic Water Distribution

System Components	QUANTITY	UNIT	COST EACH		
			MAT.	INST.	TOTAL
SYSTEM D2020 275 2760					
SOLAR, DRAINDOWN, HOT WATER, DIRECT COLLECTION					
3/4″ TUBING, THREE 3′X7′ BLACK CHROME COLLECTORS					
A, B Differential controller, 2 sensors, thermostat, solar energy system	1.000	Ea.	143	59	202
A-1 Solenoid valve, solar heating loop, brass, 3/4″ diam, 24 volts	3.000	Ea.	405	237	642
B-1 Solar energy sensor, freeze prevention	1.000	Ea.	25	22	47
C Thermometer, 2″ dial	3.000	Ea.	72	133.50	205.50
E-1 Vacuum relief valve, 3/4″ diam	1.000	Ea.	32.50	22	54.50
F-1 Air vent, automatic, 1/8″ fitting	1.000	Ea.	18.95	22	40.95
H Strainer, Y type, bronze body, 3/4″ IPS	1.000	Ea.	37.50	38	75.50
I Valve, gate, bronze, NRS, soldered, 3/4″ diam	2.000	Ea.	140	71	211
J Vent flashing neoprene	2.000	Ea.	19	71	90
K Circulator, solar heated liquid, 1/25 HP	1.000	Ea.	126	91	217
N Relief valve temp & press 150 psi 210°F self-closing 3/4″ IPS	1.000	Ea.	23	23.50	46.50
O Pipe covering, urethane, ultraviolet cover, 1″ wall, 3/4″ diam	20.000	L.F.	62.20	128	190.20
P Pipe covering, fiberglass, all service jacket, 1″ wall, 3/4″ diam	50.000	L.F.	49.50	255	304.50
Roof clamps for solar energy collector panels	3.000	Set	9.24	54.75	63.99
Q Collector panel solar energy blk chrome on copper, 1/8″ temp glass 3′x7′	3.000	Ea.	3,300	402	3,702
R Valve, swing check, bronze, regrinding disc, 3/4″ diam, soldered	2.000	Ea.	220	71	291
T Drain valve, brass, 3/4″ connection	2.000	Ea.	21.40	59	80.40
U Valve, water tempering, bronze, sweat connections, 3/4″ diam	1.000	Ea.	139	35.50	174.50
W-2, W Tank, water storage elec elem 2″x1/2# insul 120 gal	1.000	Ea.	2,025	505	2,530
X Valve, globe, bronze, rising stem, 3/4″ diam, soldered	1.000	Ea.	109	35.50	144.50
Copper tubing, type L, solder joints, hangers 10′ OC 3/4″ diam	20.000	L.F.	114	187	301
Copper tubing, type M, solder joints, hangers 10′ OC 3/4″ diam	70.000	L.F.	317.80	637	954.80
Sensor wire, #22-2 conductor, multistranded	.500	C.L.F.	7.83	32.75	40.58
Wrought copper fittings & solder, 3/4″ diam	76.000	Ea.	205.96	2,850	3,055.96
TOTAL			7,622.88	6,042.50	13,665.38

D2020 275	Solar, Draindown, Hot Water Systems	COST EACH		
		MAT.	INST.	TOTAL
2550	Solar, draindown, hot water			
2560	3/8″ tubing, 3 ea. 4′ x 4′-4″ vacuum tube collectors, 80 gal tank	6,675	5,750	12,425
2580	1/2″ tubing, 4 ea. 4′ x 4′-4″ vacuum tube collectors, 80 gal tank	7,650	5,900	13,550
2600	120 gal tank	7,900	5,975	13,875
2640	2 ea. 3′x7′ black chrome collectors, 80 gal tank	5,900	5,575	11,475
2660	3 ea. 3′x7′ black chrome collectors, 120 gal tank	7,250	5,775	13,025
2700	2 ea. 3′x7′ flat black collectors, 120 gal tank	5,875	5,650	11,525
2720	3 ea. 3′x7′ flat black collectors, 120 gal tank	6,775	5,775	12,550
2760	3/4″ tubing, 3 ea. 3′x7′ black chrome collectors, 120 gal tank	7,625	6,050	13,675
2780	3 ea. 3′x7′ flat collectors, 120 gal tank	7,225	6,075	13,300
2800	2 ea. 4′x9′ flat black & plastic glazing collectors, 120 gal tank	6,625	6,075	12,700
2840	1″ tubing, 4 ea. 2′x9′ plastic absorber & glazing collectors, 120 gal tank	11,300	8,000	19,300
2860	4 ea. 3′x7′ black chrome absorber collectors, 120 gal tank	10,800	7,575	18,375
2880	4 ea. 3′x7′ flat black absorber collectors, 120 gal tank	9,175	6,875	16,050

(Note next to row 2580/2600: R235616-60)

In the recirculation system, (a direct-collection system), incoming domestic water is heated in the collectors. When the controller activates solar collection, domestic water is heated as it flows through the collectors and then it flows back to storage. When conditions are not suitable for heat collection, the pump shuts off and the flow of the water stops. In this type of system, water remains in the collector loop at all times. A "frost sensor" at the collector activates the circulation of warm water from storage through the collectors when protection from freezing is required.

D2020 Domestic Water Distribution

System Components	QUANTITY	UNIT	COST EACH		
			MAT.	INST.	TOTAL
SYSTEM D2020 280 2820					
SOLAR, RECIRCULATION, HOT WATER					
3/4″ TUBING, TWO 3′X7′ BLACK CHROME COLLECTORS					
A, B Differential controller 2 sensors, thermostat, for solar energy system	1.000	Ea.	143	59	202
A-1 Solenoid valve, solar heating loop, brass, 3/4″ IPS, 24 volts	2.000	Ea.	270	158	428
B-1 Solar energy sensor freeze prevention	1.000	Ea.	25	22	47
C Thermometer, 2″ dial	3.000	Ea.	72	133.50	205.50
D Drain valve, brass, 3/4″ connection	1.000	Ea.	10.70	29.50	40.20
F-1 Air vent, automatic, 1/8″ fitting	1.000	Ea.	18.95	22	40.95
H Strainer, Y type, bronze body, 3/4″ IPS	1.000	Ea.	37.50	38	75.50
I Valve, gate, bronze, 125 lb, soldered 3/4″ diam	3.000	Ea.	210	106.50	316.50
J Vent flashing, neoprene	2.000	Ea.	19	71	90
L Circulator, solar heated liquid, 1/20 HP	1.000	Ea.	177	106	283
N Relief valve, temp & press 150 psi 210° F self-closing 3/4″ IPS	2.000	Ea.	46	47	93
O Pipe covering, urethane, ultraviolet cover, 1″ wall, 3/4″ diam	20.000	L.F.	62.20	128	190.20
P Pipe covering, fiberglass, all service jacket, 1″ wall 3/4″ diam	50.000	L.F.	49.50	255	304.50
Q Collector panel solar energy blk chrome on copper, 1/8″ temp glass 3′x7′	2.000	Ea.	2,200	268	2,468
Roof clamps for solar energy collector panels	2.000	Set	6.16	36.50	42.66
R Valve, swing check, bronze, 125 lb, regrinding disc, soldered 3/4″ diam	2.000	Ea.	220	71	291
U Valve, water tempering, bronze, sweat connections, 3/4″ diam	1.000	Ea.	139	35.50	174.50
V Tank, water storage, w/heating element, drain, relief valve, existing	1.000	Ea.			
X Valve, globe, bronze, 125 lb, 3/4″ diam	2.000	Ea.	218	71	289
Copper tubing, type L, solder joints, hangers 10′ OC 3/4″ diam	20.000	L.F.	114	187	301
Copper tubing, type M, solder joints, hangers 10′ OC 3/4″ diam	70.000	L.F.	317.80	637	954.80
Wrought copper fittings & solder, 3/4″ diam	76.000	Ea.	205.96	2,850	3,055.96
Sensor wire, #22-2 conductor, multistranded	.500	C.L.F.	7.83	32.75	40.58
TOTAL			4,569.60	5,364.25	9,933.85

D2020 280	Solar Recirculation, Domestic Hot Water Systems		COST EACH		
			MAT.	INST.	TOTAL
2550	Solar recirculation, hot water				
2560	3/8″ tubing, 3 ea. 4′ x 4′-4″ vacuum tube collectors		5,125	4,975	10,100
2580	1/2″ tubing, 4 ea. 4′ x 4′-4″ vacuum tube collectors	R235616	5,925	5,425	11,350
2640	2 ea. 3′x7′ black chrome collectors	-60	4,200	5,100	9,300
2660	3 ea. 3′x7′ black chrome collectors		5,300	5,250	10,550
2700	2 ea. 3′x7′ flat black collectors		3,925	5,100	9,025
2720	3 ea. 3′x7′ flat black collectors		4,900	5,275	10,175
2760	3/4″ tubing, 3 ea. 3′x7′ black chrome collectors		5,675	5,525	11,200
2780	3 ea. 3′x7′ flat black absorber plate collectors		5,275	5,550	10,825
2800	2 ea. 4′x9′ flat black w/plastic glazing collectors		4,675	5,550	10,225
2820	2 ea. 3′x7′ black chrome collectors		4,575	5,375	9,950
2840	1″ tubing, 4 ea. 2′x9′ black plastic absorber & glazing collectors		5,925	6,300	12,225
2860	4 ea. 3′x7′ black chrome absorber collectors		7,775	6,325	14,100
2880	4 ea. 3′x7′ flat black absorber collectors		7,225	6,350	13,575

511

For customer support on your Mechanical Cost Data, call 877.795.7394.

The thermosyphon domestic hot water system, a direct collection system, operates under city water pressure and does not require pumps for system operation. An insulated water storage tank is located above the collectors. As the sun heats the collectors, warm water in them rises by means of natural convection; the colder water in the storage tank flows into the collectors by means of gravity. As long as the sun is shining the water continues to flow through the collectors and to become warmer.

To prevent freezing, the system must be drained or the collectors covered with an insulated lid when the temperature drops below 32°F.

D2020 Domestic Water Distribution

System Components	QUANTITY	UNIT	COST EACH		
			MAT.	INST.	TOTAL
SYSTEM D2020 285 0960					
SOLAR, THERMOSYPHON, WATER HEATER					
3/4" TUBING, TWO 3'X7' BLACK CHROME COLLECTORS					
D-1 Framing lumber, fir, 2" x 6" x 8', tank cradle	.008	M.B.F.	5.88	7.40	13.28
F-1 Framing lumber, fir, 2" x 4" x 24', sleepers	.016	M.B.F.	11.20	22.40	33.60
I Valve, gate, bronze, 125 lb, soldered 1/2" diam	2.000	Ea.	119	59	178
J Vent flashing, neoprene	4.000	Ea.	38	142	180
O Pipe covering, urethane, ultraviolet cover, 1" wall, 1/2"diam	40.000	L.F.	95.20	250	345.20
P Pipe covering fiberglass all service jacket 1" wall 1/2" diam	160.000	L.F.	145.60	782.40	928
Q Collector panel solar, blk chrome on copper, 3/16" temp glass 3'-6" x 7.5'	2.000	Ea.	1,930	284	2,214
Y Flow control valve, globe, bronze, 125#, soldered, 1/2" diam	1.000	Ea.	105	29.50	134.50
U Valve, water tempering, bronze,sweat connections, 1/2" diam	1.000	Ea.	108	29.50	137.50
W Tank, water storage, solar energy system, 80 Gal, 2" x 1/2 lb insul	1.000	Ea.	1,775	445	2,220
Copper tubing type L, solder joints, hangers 10' OC 1/2" diam	150.000	L.F.	610.50	1,312.50	1,923
Copper tubing type M, solder joints, hangers 10' OC 1/2" diam	50.000	L.F.	168	422.50	590.50
Sensor wire, #22-2 gauge multistranded	.500	C.L.F.	7.83	32.75	40.58
Wrought copper fittings & solder, 1/2" diam	75.000	Ea.	90.75	2,662.50	2,753.25
TOTAL			5,209.96	6,481.45	11,691.41

D2020 285	Thermosyphon, Hot Water		COST EACH		
			MAT.	INST.	TOTAL
0960 / 0970	Solar, thermosyphon, hot water, two collector system	R235616 -60	5,200	6,475	11,675

This domestic hot water pre-heat system includes heat exchanger with a circulating pump, blower, air-to-water, coil and controls, mounted in the upper collector manifold. Heat from the hot air coming out of the collectors is transferred through the heat exchanger. For each degree of DHW preheating gained, one degree less heating is needed from the fuel fired water heater. The system is simple, inexpensive to operate and can provide a substantial portion of DHW requirements for modest additional cost.

D2020 Domestic Water Distribution

System Components	QUANTITY	UNIT	COST EACH		
			MAT.	INST.	TOTAL
SYSTEM D2020 290 2560					
SOLAR, HOT WATER, AIR TO WATER HEAT EXCHANGE					
THREE COLLECTORS, OPTICAL BLACK ON ALUM., 7.5' x 3.5', 80 GAL TANK					
A, B Differential controller, 2 sensors, thermostat, solar energy system	1.000	Ea.	143	59	202
C Thermometer, 2" dial	2.000	Ea.	48	89	137
C-1 Heat exchanger, air to fluid, up flow 70 MBH	1.000	Ea.	395	365	760
E Air vent, manual, for solar energy system 1/8" fitting	1.000	Ea.	2.68	22	24.68
F Air purger	1.000	Ea.	51.50	59	110.50
G Expansion tank	1.000	Ea.	75	22	97
H Strainer, Y type, bronze body, 1/2" IPS	1.000	Ea.	30.50	36	66.50
I Valve, gate, bronze, 125 lb, NRS, soldered 1/2" diam	5.000	Ea.	297.50	147.50	445
N Relief valve, temp & pressure solar 150 psi 210°F self-closing	1.000	Ea.	23	23.50	46.50
P Pipe covering, fiberglass, all service jacket, 1" wall, 1/2" diam	60.000	L.F.	54.60	293.40	348
Q Collector panel solar energy, air, black on alum plate, 7.5' x 3.5'	3.000	Ea.	2,895	426	3,321
B-1, R-1 Shutter damper for solar heater circulator	1.000	Ea.	61.50	103	164.50
B-1, R-1 Shutter motor for solar heater circulator	1.000	Ea.	153	77	230
R Backflow preventer, 1/2" pipe size	2.000	Ea.	166	89	255
T Drain valve, brass, 3/4" connection	1.000	Ea.	10.70	29.50	40.20
U Valve, water tempering, bronze, sweat connections, 1/2" diam	1.000	Ea.	108	29.50	137.50
W Tank, water storage, solar energy system, 80 Gal, 2" x 2 lb insul	1.000	Ea.	1,775	445	2,220
V Tank, water storage, w/heating element, drain, relief valve, existing	1.000	System			
X Valve, globe, bronze, 125 lb, rising stem, 1/2" diam	1.000	Ea.	105	29.50	134.50
Copper tubing type M, solder joints, hangers 10' OC, 1/2" diam	50.000	L.F.	168	422.50	590.50
Copper tubing type L, solder joints, hangers 10' OC 1/2" diam	10.000	L.F.	40.70	87.50	128.20
Wrought copper fittings & solder, 1/2" diam	10.000	Ea.	12.10	355	367.10
Sensor wire, #22-2 conductor multistranded	.500	C.L.F.	7.83	32.75	40.58
Q-1, Q-2 Ductwork, fiberglass, aluminized jacket, 1-1/2" thick, 8" diam	32.000	S.F.	180.80	219.20	400
Q-3 Manifold for flush mount solar energy collector panels, air	6.000	L.F.	954	46.20	1,000.20
TOTAL			7,758.41	3,508.05	11,266.46

D2020 290	Solar Hot Water, Air To Water Heat Exchange		COST EACH		
			MAT.	INST.	TOTAL
2550	Solar hot water, air to water heat exchange				
2560	Three collectors, optical black on aluminum, 7.5' x 3.5', 80 Gal tank		7,750	3,500	11,250
2580	Four collectors, optical black on aluminum, 7.5' x 3.5', 80 Gal tank	R235616 -60	9,100	3,725	12,825
2600	Four collectors, optical black on aluminum, 7.5' x 3.5', 120 Gal tank		9,350	3,775	13,125

515

In this closed-loop indirect collection system, fluid with a low freezing temperature, such as propylene glycol, transports heat from the collectors to water storage. The transfer fluid is contained in a closed-loop consisting of collectors, supply and return piping, and a heat exchanger immersed in the storage tank. A typical two-or-three panel system contains 5 to 6 gallons of heat transfer fluid.

When the collectors become approximately 20°F warmer than the storage temperature, a controller activates the circulator. The circulator moves the fluid continuously through the collectors until the temperature difference between the collectors and storage is such that heat collection no longer occurs; at that point, the circulator shuts off. Since the heat transfer fluid has a very low freezing temperature, there is no need for it to be drained from the collectors between periods of collection.

D2020 Domestic Water Distribution

System Components	QUANTITY	UNIT	COST EACH		
			MAT.	INST.	TOTAL
SYSTEM D2020 295 2760					
SOLAR, CLOSED LOOP, HOT WATER SYSTEM, IMMERSED HEAT EXCHANGER					
3/4" TUBING, THREE 3' X 7' BLACK CHROME COLLECTORS					
A, B Differential controller, 2 sensors, thermostat, solar energy system	1.000	Ea.	143	59	202
C Thermometer 2" dial	3.000	Ea.	72	133.50	205.50
D, T Fill & drain valves, brass, 3/4" connection	3.000	Ea.	32.10	88.50	120.60
E Air vent, manual, 1/8" fitting	1.000	Ea.	2.68	22	24.68
F Air purger	1.000	Ea.	51.50	59	110.50
G Expansion tank	1.000	Ea.	75	22	97
I Valve, gate, bronze, NRS, soldered 3/4" diam	3.000	Ea.	210	106.50	316.50
J Neoprene vent flashing	2.000	Ea.	19	71	90
K Circulator, solar heated liquid, 1/25 HP	1.000	Ea.	126	91	217
N-1, N Relief valve, temp & press 150 psi 210°F self-closing 3/4" IPS	2.000	Ea.	46	47	93
O Pipe covering, urethane ultraviolet cover, 1" wall, 3/4" diam	20.000	L.F.	62.20	128	190.20
P Pipe covering, fiberglass, all service jacket, 1" wall, 3/4" diam	50.000	L.F.	49.50	255	304.50
Roof clamps for solar energy collector panel	3.000	Set	9.24	54.75	63.99
Q Collector panel solar blk chrome on copper, 1/8" temp glass, 3'x7'	3.000	Ea.	3,300	402	3,702
R-1 Valve, swing check, bronze, regrinding disc, 3/4" diam, soldered	1.000	Ea.	110	35.50	145.50
S Pressure gauge, 60 psi, 2-1/2" dial	1.000	Ea.	26.50	22	48.50
U Valve, water tempering, bronze, sweat connections, 3/4" diam	1.000	Ea.	139	35.50	174.50
W-2, W Tank, water storage immersed heat exchr elec elem 2"x2# insul 120 Gal	1.000	Ea.	2,025	505	2,530
X Valve, globe, bronze, rising stem, 3/4" diam, soldered	1.000	Ea.	109	35.50	144.50
Copper tubing type L, solder joint, hanger 10' OC 3/4" diam	20.000	L.F.	114	187	301
Copper tubing, type M, solder joint, hanger 10' OC 3/4" diam	70.000	L.F.	317.80	637	954.80
Sensor wire, #22-2 conductor multistranded	.500	C.L.F.	7.83	32.75	40.58
Solar energy heat transfer fluid, propylene glycol, anti-freeze	6.000	Gal.	102.60	153	255.60
Wrought copper fittings & solder, 3/4" diam	76.000	Ea.	205.96	2,850	3,055.96
TOTAL			7,355.91	6,032.50	13,388.41

D2020 295	Solar, Closed Loop, Hot Water Systems	COST EACH		
		MAT.	INST.	TOTAL
2550	Solar, closed loop, hot water system, immersed heat exchanger			
2560	3/8" tubing, 3 ea. 4' x 4'-4" vacuum tube collectors, 80 gal. tank	6,600	5,450	12,050
2580	1/2" tubing, 4 ea. 4' x 4'-4" vacuum tube collectors, 80 gal. tank	7,400	5,900	13,300
2600	120 gal. tank	7,650	5,950	13,600
2640	2 ea. 3'x7' black chrome collectors, 80 gal. tank R235616-60	5,725	5,625	11,350
2660	120 gal. tank	5,900	5,625	11,525
2700	2 ea. 3'x7' flat black collectors, 120 gal. tank	5,650	5,625	11,275
2720	3 ea. 3'x7' flat black collectors, 120 gal. tank	6,600	5,800	12,400
2760	3/4" tubing, 3 ea. 3'x7' black chrome collectors, 120 gal. tank	7,350	6,025	13,375
2780	3 ea. 3'x7' flat black collectors, 120 gal. tank	6,950	6,050	13,000
2800	2 ea. 4'x9' flat black w/plastic glazing collectors 120 gal. tank	6,350	6,075	12,425
2840	1" tubing, 4 ea. 2'x9' plastic absorber & glazing collectors 120 gal. tank	7,275	6,800	14,075
2860	4 ea. 3'x7' black chrome collectors, 120 gal. tank	9,100	6,825	15,925
2880	4 ea. 3'x7' flat black absorber collectors, 120 gal. tank	8,575	6,850	15,425

Basis for Heat Loss Estimate, Apartment Type Structures:

1. Masonry walls and flat roof are insulated. U factor is assumed at .08.
2. Window glass area taken as BOCA minimum, 1/10th of floor area. Double insulating glass with 1/4" air space, U = .65.
3. Infiltration = 0.3 C.F. per hour per S.F. of net wall.
4. Concrete floor loss is 2 BTUH per S.F.
5. Temperature difference taken as 70°F.
6. Ventilating or makeup air has not been included and must be added if desired. Air shafts are not used.

System Components	QUANTITY	UNIT	COST EACH		
			MAT.	INST.	TOTAL
SYSTEM D3010 510 1760					
HEATING SYSTEM, FIN TUBE RADIATION, FORCED HOT WATER					
1,000 S.F. AREA, 10,000 C.F. VOLUME					
Boiler, oil fired, CI, burner, ctrls/insul/breech/pipe/ftng/valves, 109 MBH	1.000	Ea.	4,462.50	4,025	8,487.50
Circulating pump, CI flange connection, 1/12 HP	1.000	Ea.	400	213	613
Expansion tank, painted steel, ASME 18 Gal capacity	1.000	Ea.	785	93	878
Storage tank, steel, above ground, 275 Gal capacity w/supports	1.000	Ea.	540	260	800
Copper tubing type L, solder joint, hanger 10' OC, 3/4" diam	100.000	L.F.	570	935	1,505
Radiation, 3/4" copper tube w/alum fin baseboard pkg, 7" high	30.000	L.F.	243	675	918
Pipe covering, calcium silicate w/cover, 1' wall, 3/4' diam	100.000	L.F.	448	690	1,138
TOTAL			7,448.50	6,891	14,339.50
COST PER S.F.			7.45	6.89	14.34

D3010 510	Apartment Building Heating - Fin Tube Radiation		COST PER S.F.		
			MAT.	INST.	TOTAL
1740	Heating systems, fin tube radiation, forced hot water				
1760	1,000 S.F. area, 10,000 C.F. volume		7.45	6.90	14.35
1800	10,000 S.F. area, 100,000 C.F. volume	R235000 -10	3.13	4.13	7.26
1840	20,000 S.F. area, 200,000 C.F. volume		3.45	4.64	8.09
1880	30,000 S.F. area, 300,000 C.F. volume	R235000 -20	3.34	4.49	7.83
1890					

D3010 Energy Supply

Fin Tube Radiator

Basis for Heat Loss Estimate, Factory or Commercial Type Structures:

1. Walls and flat roof are of lightly insulated concrete block or metal. U factor is assumed at .17.
2. Windows and doors are figured at 1-1/2 S.F. per linear foot of building perimeter. The U factor for flat single glass is 1.13.
3. Infiltration is approximately 5 C.F. per hour per S.F. of wall.
4. Concrete floor loss is 2 BTUH per S.F. of floor.
5. Temperature difference is assumed at 70°F.
6. Ventilation or makeup air has not been included and must be added if desired.

System Components	QUANTITY	UNIT	COST EACH MAT.	COST EACH INST.	COST EACH TOTAL
SYSTEM D3010 520 1960					
HEATING SYSTEM, FIN TUBE RADIATION, FORCED HOT WATER					
1,000 S.F. BLDG., ONE FLOOR					
Boiler, oil fired, CI, burner/ctrls/insul/breech/pipe/ftngs/valves, 109 MBH	1.000	Ea.	4,462.50	4,025	8,487.50
Expansion tank, painted steel, ASME 18 Gal capacity	1.000	Ea.	2,675	108	2,783
Storage tank, steel, above ground, 550 Gal capacity w/supports	1.000	Ea.	4,125	480	4,605
Circulating pump, CI flanged, 1/8 HP	1.000	Ea.	670	213	883
Pipe, steel, black, sch. 40, threaded, cplg & hngr 10'OC, 1-1/2" diam.	260.000	L.F.	2,028	4,147	6,175
Pipe covering, calcium silicate w/cover , 1" wall, 1-1/2" diam	260.000	L.F.	1,180.40	1,846	3,026.40
Radiation, steel 1-1/4" tube & 4-1/4" fin w/cover & damper, wall hung	42.000	L.F.	1,806	1,512	3,318
Rough in, steel fin tube radiation w/supply & balance valves	4.000	Set	1,340	3,180	4,520
TOTAL			18,286.90	15,511	33,797.90
COST PER S.F.			18.29	15.51	33.80

D3010 520	Commercial Building Heating - Fin Tube Radiation		COST PER S.F. MAT.	COST PER S.F. INST.	COST PER S.F. TOTAL
1940	Heating systems, fin tube radiation, forced hot water				
1960	1,000 S.F. bldg, one floor		18.30	15.50	33.80
2000	10,000 S.F., 100,000 C.F., total two floors	R235000 -10	4.89	5.85	10.74
2040	100,000 S.F., 1,000,000 C.F., total three floors		1.97	2.62	4.59
2080	1,000,000 S.F., 10,000,000 C.F., total five floors	R235000 -20	1.07	1.37	2.44
2090					

Unit Heater

Basis for Heat Loss Estimate, Factory or Commercial Type Structures:

1. Walls and flat roof are of lightly insulated concrete block or metal. U factor is assumed at .17.
2. Windows and doors are figured at 1-1/2 S.F. per linear foot of building perimeter. The U factor for flat single glass is 1.13.
3. Infiltration is approximately 5 C.F. per hour per S.F. of wall.
4. Concrete floor loss is 2 BTUH per S.F. of floor.
5. Temperature difference is assumed at 70°F.
6. Ventilation or makeup air has not been included and must be added if desired.

System Components	QUANTITY	UNIT	COST EACH		
			MAT.	INST.	TOTAL
SYSTEM D3010 530 1880					
HEATING SYSTEM, TERMINAL UNIT HEATERS, FORCED HOT WATER					
1,000 S.F. BLDG., ONE FLOOR					
Boiler oil fired, CI, burner/ctrls/insul/breech/pipe/ftngs/valves, 109 MBH	1.000	Ea.	4,462.50	4,025	8,487.50
Expansion tank, painted steel, ASME 18 Gal capacity	1.000	Ea.	2,675	108	2,783
Storage tank, steel, above ground, 550 Gal capacity w/supports	1.000	Ea.	4,125	480	4,605
Circulating pump, CI, flanged, 1/8 HP	1.000	Ea.	670	213	883
Pipe, steel, black, schedule 40, threaded, cplg & hngr 10' OC 1-1/2" diam	260.000	L.F.	2,028	4,147	6,175
Pipe covering, calcium silicate w/cover, 1" wall, 1-1/2" diam	260.000	L.F.	1,180.40	1,846	3,026.40
Unit heater, 1 speed propeller, horizontal, 200° EWT, 26.9 MBH	2.000	Ea.	1,200	324	1,524
Unit heater piping hookup with controls	2.000	Set	1,580	3,050	4,630
TOTAL			17,920.90	14,193	32,113.90
COST PER S.F.			17.92	14.19	32.11

D3010 530	Commercial Bldg. Heating - Terminal Unit Heaters		COST PER S.F.		
			MAT.	INST.	TOTAL
1860	Heating systems, terminal unit heaters, forced hot water				
1880	1,000 S.F. bldg., one floor		17.90	14.20	32.10
1920	10,000 S.F. bldg., 100,000 C.F. total two floors		4.41	4.85	9.26
1960	100,000 S.F. bldg., 1,000,000 C.F. total three floors	R235000 -10	2.01	2.39	4.40
2000	1,000,000 S.F. bldg., 10,000,000 C.F. total five floors		1.36	1.54	2.90
2010		R235000 -20			

Many styles of active solar energy systems exist. Those shown on the following page represent the majority of systems now being installed in different regions of the country.

The five active domestic hot water (DHW) systems which follow are typically specified as two or three panel systems with additional variations being the type of glazing and size of the storage tanks. Various combinations have been costed for the user's evaluation and comparison. The basic specifications from which the following systems were developed satisfy the construction detail requirements specified in the HUD Intermediate Minimum Property Standards (IMPS). If these standards are not complied with, the renewable energy system's costs could be significantly lower than shown.

To develop the system's specifications and costs it was necessary to make a number of assumptions about the systems. Certain systems are more appropriate to one climatic region than another or the systems may require modifications to be usable in particular locations. Specific instances in which the systems are not appropriate throughout the country as specified, or in which modification will be needed include the following:

• The freeze protection mechanisms provided in the DHW systems vary greatly. In harsh climates, where freeze is a major concern, a closed-loop indirect collection system may be more appropriate than a direct collection system.

• The thermosyphon water heater system described cannot be used when temperatures drop below 32°F.

• In warm climates it may be necessary to modify the systems installed to prevent overheating.

For each renewable resource (solar) system a schematic diagram and descriptive summary of the system is provided along with a list of all the components priced as part of the system. The costs were developed based on these specifications.

Considerations affecting costs which may increase or decrease beyond the estimates presented here include the following:

• Special structural qualities (allowance for earthquake, future expansion, high winds, and unusual spans or shapes);

• Isolated building site or rough terrain that would affect the transportation of personnel, material, or equipment;

• Unusual climatic conditions during the construction process;

• Substitution of other materials or system components for those used in the system specifications.

521

In this closed-loop indirect collection system, fluid with a low freezing temperature, propylene glycol, transports heat from the collectors to water storage. The transfer fluid is contained in a closed loop consisting of collectors, supply and return piping, and a heat exchanger immersed in the storage tank.

When the collectors become approximately 20°F warmer than the storage temperature, the controller activates the circulator. The circulator moves the fluid continuously until the temperature difference between fluid in the collectors and storage is such that the collection will no longer occur and then the circulator turns off. Since the heat transfer fluid has a very low freezing temperature, there is no need for it to be drained from the collectors between periods of collection.

D3010 Energy Supply

System Components	QUANTITY	UNIT	COST EACH		
			MAT.	INST.	TOTAL
SYSTEM D3010 650 2750					
SOLAR, CLOSED LOOP, SPACE/HOT WATER					
1″ TUBING, TEN 3′X7′ BLK CHROME ON COPPER ABSORBER COLLECTORS					
A, B Differential controller 2 sensors, thermostat, solar energy system	2.000	Ea.	286	118	404
C Thermometer, 2″ dial	10.000	Ea.	240	445	685
C-1 Heat exchanger, solar energy system, fluid to air, up flow, 80 MBH	1.000	Ea.	525	425	950
D, T Fill & drain valves, brass, 3/4″ connection	5.000	Ea.	53.50	147.50	201
D-1 Fan center	1.000	Ea.	130	146	276
E Air vent, manual, 1/8″ fitting	1.000	Ea.	2.68	22	24.68
E-2 Thermostat, 2 stage for sensing room temperature	1.000	Ea.	205	90	295
F Air purger	2.000	Ea.	103	118	221
F-1 Controller, liquid temperature, solar energy system	1.000	Ea.	118	142	260
G Expansion tank	2.000	Ea.	150	44	194
I Valve, gate, bronze, 125 lb, soldered, 1″ diam	5.000	Ea.	420	187.50	607.50
J Vent flashing, neoprene	2.000	Ea.	19	71	90
K Circulator, solar heated liquid, 1/25 HP	2.000	Ea.	252	182	434
L Circulator, solar heated liquid, 1/20 HP	1.000	Ea.	177	106	283
N Relief valve, temp & pressure 150 psi 210°F self-closing	4.000	Ea.	92	94	186
N-1 Relief valve, pressure poppet, bronze, 30 psi, 3/4″ IPS	3.000	Ea.	241.50	76.50	318
O Pipe covering, urethane, ultraviolet cover, 1″ wall, 1″ diam	50.000	L.F.	155.50	320	475.50
P Pipe covering, fiberglass, all service jacket, 1″ wall, 1″ diam	60.000	L.F.	59.40	306	365.40
Q Collector panel solar energy blk chrome on copper 1/8″ temp glass 3′x7′	10.000	Ea.	11,000	1,340	12,340
Roof clamps for solar energy collector panel	10.000	Set	30.80	182.50	213.30
R Valve, swing check, bronze, 125 lb, regrinding disc, 3/4″ & 1″ diam	4.000	Ea.	632	150	782
S Pressure gage, 0-60 psi, for solar energy system	2.000	Ea.	53	44	97
U Valve, water tempering, bronze, sweat connections, 3/4″ diam	1.000	Ea.	139	35.50	174.50
W-2, W-1, W Tank, water storage immersed heat xchr elec elem 2″x1/2# ins 120 gal	4.000	Ea.	8,100	2,020	10,120
X Valve, globe, bronze, 125 lb, rising stem, 1″ diam	3.000	Ea.	507	112.50	619.50
Y Valve, flow control	1.000	Ea.	132	32	164
Copper tubing, type M, solder joint, hanger 10′ OC 1″ diam	110.000	L.F.	797.50	1,116.50	1,914
Copper tubing, type L, solder joint, hanger 10′ OC 3/4″ diam	20.000	L.F.	114	187	301
Wrought copper fittings & solder, 3/4″ & 1″ diam	121.000	Ea.	804.65	5,384.50	6,189.15
Sensor, wire, #22-2 conductor, multistranded	.700	C.L.F.	10.96	45.85	56.81
Ductwork, galvanized steel, for heat exchanger	8.000	Lb.	5.76	65.20	70.96
Solar energy heat transfer fluid propylene glycol, anti-freeze	25.000	Gal.	427.50	637.50	1,065
TOTAL			25,983.75	14,393.55	40,377.30

D3010 650	Solar, Closed Loop, Space/Hot Water Systems		COST EACH		
			MAT.	INST.	TOTAL
2540	Solar, closed loop, space/hot water				
2550	1/2″ tubing, 12 ea. 4′x4′4″ vacuum tube collectors		25,100	13,400	38,500
2600	3/4″ tubing, 12 ea. 4′x4′4″ vacuum tube collectors		26,300	13,800	40,100
2650	10 ea. 3′ x 7′ black chrome absorber collectors	R235616 -60	25,400	13,400	38,800
2700	10 ea. 3′ x 7′ flat black absorber collectors		24,100	13,500	37,600
2750	1″ tubing, 10 ea. 3′ x 7′ black chrome absorber collectors		26,000	14,400	40,400
2800	10 ea. 3′ x 7′ flat black absorber collectors		24,800	14,500	39,300
2850	6 ea. 4′ x 9′ flat black w/plastic glazing collectors		22,000	14,400	36,400
2900	12 ea. 2′ x 9′ plastic absorber and glazing collectors		22,800	14,600	37,400

523

For customer support on your Mechanical Cost Data, call 877.795.7394.

This draindown pool system uses a differential thermostat similar to those used in solar domestic hot water and space heating applications. To heat the pool, the pool water passes through the conventional pump-filter loop and then flows through the collectors. When collection is not possible, or when the pool temperature is reached, all water drains from the solar loop back to the pool through the existing piping. The modes are controlled by solenoid valves or other automatic valves in conjunction with a vacuum breaker relief valve, which facilitates draindown.

D3010 Energy Supply

System Components	QUANTITY	UNIT	COST EACH		
			MAT.	INST.	TOTAL
SYSTEM D3010 660 2640					
SOLAR SWIMMING POOL HEATER, ROOF MOUNTED COLLECTORS					
TEN 4' X 10' FULLY WETTED UNGLAZED PLASTIC ABSORBERS					
A Differential thermostat/controller, 110V, adj pool pump system	1.000	Ea.	355	355	710
A-1 Solenoid valve, PVC, normally 1 open 1 closed (included)	2.000	Ea.			
B Sensor, thermistor type (included)	2.000	Ea.			
E-1 Valve, vacuum relief	1.000	Ea.	32.50	22	54.50
Q Collector panel, solar energy, plastic, liquid full wetted, 4' x 10'	10.000	Ea.	3,550	2,550	6,100
R Valve, ball check, PVC, socket, 1-1/2" diam	1.000	Ea.	118	35.50	153.50
Z Valve, ball, PVC, socket, 1-1/2" diam	3.000	Ea.	276	106.50	382.50
Pipe, PVC, sch 40, 1-1/2" diam	80.000	L.F.	604	1,576	2,180
Pipe fittings, PVC sch 40, socket joint, 1-1/2" diam	10.000	Ea.	20.40	355	375.40
Sensor wire, #22-2 conductor, multistranded	.500	C.L.F.	7.83	32.75	40.58
Roof clamps for solar energy collector panels	10.000	Set	30.80	182.50	213.30
Roof strap, teflon for solar energy collector panels	26.000	L.F.	676	89.96	765.96
TOTAL			5,670.53	5,305.21	10,975.74

D3010 660	Solar Swimming Pool Heater Systems		COST EACH		
			MAT.	INST.	TOTAL
2530	Solar swimming pool heater systems, roof mounted collectors				
2540	10 ea. 3'x7' black chrome absorber, 1/8" temp. glass		13,100	4,100	17,200
2560	10 ea. 4'x8' black chrome absorber, 3/16" temp. glass	R235616 -60	14,600	4,875	19,475
2580	10 ea. 3'8"x6' flat black absorber, 3/16" temp. glass		11,800	4,175	15,975
2600	10 ea. 4'x9' flat black absorber, plastic glazing		13,600	5,075	18,675
2620	10 ea. 2'x9' rubber absorber, plastic glazing		15,000	5,500	20,500
2640	10 ea. 4'x10' fully wetted unglazed plastic absorber		5,675	5,300	10,975
2660	Ground mounted collectors				
2680	10 ea. 3'x7' black chrome absorber, 1/8" temp. glass		13,100	4,575	17,675
2700	10 ea. 4'x8' black chrome absorber, 3/16" temp. glass		14,600	5,350	19,950
2720	10 ea. 3'8"x6' flat blk absorber, 3/16" temp. glass		11,800	4,650	16,450
2740	10 ea. 4'x9' flat blk absorber, plastic glazing		13,600	5,550	19,150
2760	10 ea. 2'x9' rubber absorber, plastic glazing		14,900	5,775	20,675
2780	10 ea. 4'x10' fully wetted unglazed plastic absorber		5,700	5,775	11,475

The complete Solar Air Heating System provides maximum savings of conventional fuel with both space heating and year-round domestic hot water heating. It allows for the home air conditioning to operate simultaneously and independently from the solar domestic water heating in summer. The system's modes of operation are:

Mode 1: The building is heated directly from the collectors with air circulated by the Solar Air Mover.

Mode 2: When heat is not needed in the building, the dampers change within the air mover to circulate the air from the collectors to the rock storage bin.

Mode 3: When heat is not available from the collector array and is available in rock storage, the air mover draws heated air from rock storage and directs it into the building. When heat is not available from the collectors or the rock storage bin, the auxiliary heating unit will provide heat for the building. The size of the collector array is typically 25% the size of the main floor area.

D3010 Energy Supply

System Components	QUANTITY	UNIT	COST EACH		
			MAT.	INST.	TOTAL
SYSTEM D3010 675 1210					
SOLAR, SPACE/HOT WATER, AIR TO WATER HEAT EXCHANGE					
A, B Differential controller 2 sensors thermos., solar energy sys liquid loop	1.000	Ea.	143	59	202
A-1, B Differential controller 2 sensors 6 station solar energy sys air loop	1.000	Ea.	289	236	525
B-1 Solar energy sensor, freeze prevention	1.000	Ea.	25	22	47
C Thermometer for solar energy system, 2" dial	2.000	Ea.	48	89	137
C-1 Heat exchanger, solar energy system, air to fluid, up flow, 70 MBH	1.000	Ea.	395	365	760
D Drain valve, brass, 3/4" connection	2.000	Ea.	21.40	59	80.40
E Air vent, manual, for solar energy system 1/8" fitting	1.000	Ea.	2.68	22	24.68
E-1 Thermostat, 2 stage for sensing room temperature	1.000	Ea.	205	90	295
F Air purger	1.000	Ea.	51.50	59	110.50
G Expansion tank, for solar energy system	1.000	Ea.	75	22	97
I Valve, gate, bronze, 125 lb, NRS, soldered 3/4" diam	2.000	Ea.	140	71	211
K Circulator, solar heated liquid, 1/25 HP	1.000	Ea.	126	91	217
N Relief valve temp & press 150 psi 210°F self-closing, 3/4" IPS	1.000	Ea.	23	23.50	46.50
N-1 Relief valve, pressure, poppet, bronze, 30 psi, 3/4" IPS	1.000	Ea.	80.50	25.50	106
P Pipe covering, fiberglass, all service jacket, 1" wall, 3/4" diam	60.000	L.F.	59.40	306	365.40
Q-3 Manifold for flush mount solar energy collector panels	20.000	L.F.	3,180	154	3,334
Q Collector panel solar energy, air, black on alum. plate, 7.5' x 3.5'	10.000	Ea.	9,650	1,420	11,070
R Valve, swing check, bronze, 125 lb, regrinding disc, 3/4" diam	2.000	Ea.	220	71	291
S Pressure gage, 2" dial, for solar energy system	1.000	Ea.	26.50	22	48.50
U Valve, water tempering, bronze, sweat connections, 3/4" diam	1.000	Ea.	139	35.50	174.50
W-2, W Tank, water storage, solar, elec element 2"x1/2# insul, 80 Gal	1.000	Ea.	1,775	445	2,220
X Valve, globe, bronze, 125 lb, soldered, 3/4" diam	1.000	Ea.	109	35.50	144.50
Copper tubing type L, solder joints, hangers 10' OC 3/4" diam	10.000	L.F.	57	93.50	150.50
Copper tubing type M, solder joints, hangers 10' OC 3/4" diam	60.000	L.F.	272.40	546	818.40
Wrought copper fittings & solder, 3/4" diam	26.000	Ea.	70.46	975	1,045.46
Sensor wire, #22-2 conductor multistranded	1.200	C.L.F.	18.78	78.60	97.38
Q-1 Duct work, rigid fiberglass, rectangular	400.000	S.F.	348	2,180	2,528
Duct work, spiral preformed, steel, PVC coated both sides, 12" x 10"	8.000	Ea.	208	412	620
R-2 Shutter/damper for solar heater circulator	9.000	Ea.	553.50	927	1,480.50
K-1 Shutter motor for solar heater circulator blower	9.000	Ea.	1,377	693	2,070
K-1 Fan, solar energy heated air circulator, space & DHW system	1.000	Ea.	1,700	2,450	4,150
Z-1 Tank, solar energy air storage, 6'-3"H 7'x7' = 306 CF/2000 Gal	1.000	Ea.	15,700	1,600	17,300
Z-2 Crushed stone 1-1/2"	11.000	C.Y.	286	89.32	375.32
C-2 Thermometer, remote probe, 2" dial	1.000	Ea.	36.50	88.50	125
R-1 Solenoid valve	1.000	Ea.	135	79	214
V, R-4, R-3 Furnace, supply diffusers, return grilles, existing	1.000	Ea.	15,700	1,600	17,300
TOTAL			53,246.62	15,534.92	68,781.54

D3010 675	Air To Water Heat Exchange		COST EACH		
			MAT.	INST.	TOTAL
1210	Solar, air to water heat exchange, for space/hot water heating	R235616 -60	53,000	15,600	68,600
1220					

527

Small Electric Boiler
System Considerations:
1. Terminal units are fin tube baseboard radiation rated at 720 BTU/hr with 200° water temperature or 820 BTU/hr steam.
2. Primary use being for residential or smaller supplementary areas, the floor levels are based on 7-1/2' ceiling heights.
3. All distribution piping is copper for boilers through 205 MBH. All piping for larger systems is steel pipe.

Boiler   **Baseboard Radiation**

System Components	QUANTITY	UNIT	COST EACH		
			MAT.	INST.	TOTAL
SYSTEM D3020 102 1120					
SMALL HEATING SYSTEM, HYDRONIC, ELECTRIC BOILER					
1,480 S.F., 61 MBH, STEAM, 1 FLOOR					
Boiler, electric steam, std cntrls, trim, ftngs and valves, 18 KW, 61.4 MBH	1.000	Ea.	5,032.50	1,787.50	6,820
Copper tubing type L, solder joint, hanger 10'OC, 1-1/4" diam	160.000	L.F.	1,816	1,960	3,776
Radiation, 3/4" copper tube w/alum fin baseboard pkg 7" high	60.000	L.F.	486	1,350	1,836
Rough in baseboard panel or fin tube with valves & traps	10.000	Set	2,630	6,800	9,430
Pipe covering, calcium silicate w/cover, 1" wall 1-1/4" diam	160.000	L.F.	716.80	1,136	1,852.80
Low water cut-off, quick hookup, in gage glass tappings	1.000	Ea.	281	45	326
TOTAL			10,962.30	13,078.50	24,040.80
COST PER S.F.			7.41	8.84	16.25

D3020 102	Small Heating Systems, Hydronic, Electric Boilers		COST PER S.F.		
			MAT.	INST.	TOTAL
1100	Small heating systems, hydronic, electric boilers				
1120	Steam, 1 floor, 1480 S.F., 61 M.B.H.		7.43	8.85	16.28
1160	3,000 S.F., 123 M.B.H.	R235000 -10	5.65	7.75	13.40
1200	5,000 S.F., 205 M.B.H.		4.93	7.15	12.08
1240	2 floors, 12,400 S.F., 512 M.B.H.	R235000 -20	3.92	7.10	11.02
1280	3 floors, 24,800 S.F., 1023 M.B.H.		4.41	7.05	11.46
1320	34,750 S.F., 1,433 M.B.H.	R235000 -30	4.04	6.85	10.89
1360	Hot water, 1 floor, 1,000 S.F., 41 M.B.H.		12.55	4.91	17.46
1400	2,500 S.F., 103 M.B.H.		8.80	8.85	17.65
1440	2 floors, 4,850 S.F., 205 M.B.H.		8.50	10.60	19.10
1480	3 floors, 9,700 S.F., 410 M.B.H.		8.80	10.95	19.75

D3020 Heat Generating Systems

Boiler

Unit Heater

Large Electric Boiler System Considerations:

1. Terminal units are all unit heaters of the same size. Quantities are varied to accommodate total requirements.
2. All air is circulated through the heaters a minimum of three times per hour.
3. As the capacities are adequate for commercial use, floor levels are based on 10′ ceiling heights.
4. All distribution piping is black steel pipe.

System Components	QUANTITY	UNIT	COST EACH		
			MAT.	INST.	TOTAL
SYSTEM D3020 104 1240					
LARGE HEATING SYSTEM, HYDRONIC, ELECTRIC BOILER					
9,280 S.F., 135 KW, 461 MBH, 1 FLOOR					
Boiler, electric hot water, std ctrls, trim, ftngs, valves, 135 KW, 461 MBH	1.000	Ea.	12,225	3,900	16,125
Expansion tank, painted steel, 60 Gal capacity ASME	1.000	Ea.	4,175	217	4,392
Circulating pump, CI, close cpld, 50 GPM, 2 HP, 2″ pipe conn	1.000	Ea.	2,925	425	3,350
Unit heater, 1 speed propeller, horizontal, 200° EWT, 72.7 MBH	7.000	Ea.	6,125	1,652	7,777
Unit heater piping hookup with controls	7.000	Set	5,530	10,675	16,205
Pipe, steel, black, schedule 40, welded, 2-1/2″ diam	380.000	L.F.	5,339	10,773	16,112
Pipe covering, calcium silicate w/cover, 1″ wall, 2-1/2″ diam	380.000	L.F.	2,109	2,793	4,902
TOTAL			38,428	30,435	68,863
COST PER S.F.			4.14	3.28	7.42

D3020 104	Large Heating Systems, Hydronic, Electric Boilers		COST PER S.F.		
			MAT.	INST.	TOTAL
1230	Large heating systems, hydronic, electric boilers				
1240	9,280 S.F., 135 K.W., 461 M.B.H., 1 floor		4.14	3.28	7.42
1280	14,900 S.F., 240 K.W., 820 M.B.H., 2 floors	R235000 -10	5.55	5.35	10.90
1320	18,600 S.F., 296 K.W., 1,010 M.B.H., 3 floors		5.45	5.75	11.20
1360	26,100 S.F., 420 K.W., 1,432 M.B.H., 4 floors	R235000 -20	5.35	5.65	11
1400	39,100 S.F., 666 K.W., 2,273 M.B.H., 4 floors		4.59	4.73	9.32
1440	57,700 S.F., 900 K.W., 3,071 M.B.H., 5 floors	R235000 -30	4.39	4.67	9.06
1480	111,700 S.F., 1,800 K.W., 6,148 M.B.H., 6 floors		3.89	3.99	7.88
1520	149,000 S.F., 2,400 K.W., 8,191 M.B.H., 8 floors		3.84	4	7.84
1560	223,300 S.F., 3,600 K.W., 12,283 M.B.H., 14 floors		4.10	4.53	8.63

529

For customer support on your Mechanical Cost Data, call 877.795.7394.

Unit Heater

Fossil Fuel Boiler System Considerations:

1. Terminal units are horizontal unit heaters. Quantities are varied to accommodate total heat loss per building.
2. Unit heater selection was determined by their capacity to circulate the building volume a minimum of three times per hour in addition to the BTU output.
3. Systems shown are forced hot water. Steam boilers cost slightly more than hot water boilers. However, this is compensated for by the smaller size or fewer terminal units required with steam.
4. Floor levels are based on 10' story heights.
5. MBH requirements are gross boiler output.

System Components	QUANTITY	UNIT	COST EACH		
			MAT.	INST.	TOTAL
SYSTEM D3020 108 1280					
HEATING SYSTEM, HYDRONIC, FOSSIL FUEL, TERMINAL UNIT HEATERS					
CAST IRON BOILER, GAS, 80 MBH, 1,070 S.F. BUILDING					
Boiler, gas, hot water, CI, burner, controls, insulation, breeching, 80 MBH	1.000	Ea.	2,283.75	1,968.75	4,252.50
Pipe, steel, black, schedule 40, threaded, cplg & hngr 10'OC, 2" diam	200.000	L.F.	1,980	3,990	5,970
Unit heater, 1 speed propeller, horizontal, 200° EWT, 72.7 MBH	2.000	Ea.	1,750	472	2,222
Unit heater piping hookup with controls	2.000	Set	1,580	3,050	4,630
Expansion tank, painted steel, ASME, 18 Gal capacity	1.000	Ea.	2,675	108	2,783
Circulating pump, CI, flange connection, 1/12 HP	1.000	Ea.	400	213	613
Pipe covering, calcium silicate w/cover, 1" wall, 2" diam	200.000	L.F.	1,040	1,470	2,510
TOTAL			11,708.75	11,271.75	22,980.50
COST PER S.F.			10.94	10.53	21.47

D3020 108	Heating Systems, Unit Heaters		COST PER S.F.		
			MAT.	INST.	TOTAL
1260	Heating systems, hydronic, fossil fuel, terminal unit heaters,				
1280	Cast iron boiler, gas, 80 M.B.H., 1,070 S.F. bldg.		10.93	10.56	21.49
1320	163 M.B.H., 2,140 S.F. bldg.		7.40	7.15	14.55
1360	544 M.B.H., 7,250 S.F. bldg.	R235000 -10	5.50	5.05	10.55
1400	1,088 M.B.H., 14,500 S.F. bldg.		4.57	4.76	9.33
1440	3,264 M.B.H., 43,500 S.F. bldg.	R235000 -20	3.74	3.60	7.34
1480	5,032 M.B.H., 67,100 S.F. bldg.		4.32	3.71	8.03
1520	Oil, 109 M.B.H., 1,420 S.F. bldg.	R235000 -30	12.05	9.40	21.45
1560	235 M.B.H., 3,150 S.F. bldg.		7.60	6.65	14.25
1600	940 M.B.H., 12,500 S.F. bldg.		5.45	4.37	9.82
1640	1,600 M.B.H., 21,300 S.F. bldg.		5.05	4.18	9.23
1680	2,480 M.B.H., 33,100 S.F. bldg.		5.05	3.78	8.83
1720	3,350 M.B.H., 44,500 S.F. bldg.		4.46	3.85	8.31
1760	Coal, 148 M.B.H., 1,975 S.F. bldg.		10.35	6.10	16.45
1800	300 M.B.H., 4,000 S.F. bldg.		8.20	4.80	13
1840	2,360 M.B.H., 31,500 S.F. bldg.		5.50	3.82	9.32

Fin Tube Radiator

Fossil Fuel Boiler System Considerations:

1. Terminal units are commercial steel fin tube radiation. Quantities are varied to accommodate total heat loss per building.
2. Systems shown are forced hot water. Steam boilers cost slightly more than hot water boilers. However, this is compensated for by the smaller size or fewer terminal units required with steam.
3. Floor levels are based on 10′ story heights.
4. MBH requirements are gross boiler output.

System Components	QUANTITY	UNIT	COST EACH		
			MAT.	INST.	TOTAL
SYSTEM D3020 110 3240					
HEATING SYSTEM, HYDRONIC, FOSSIL FUEL, FIN TUBE RADIATION					
CAST IRON BOILER, GAS, 80 MBH, 1,070 S.F. BLDG.					
Boiler, gas, hot water, CI, burner, controls, insulation, breeching, 80 MBH	1.000	Ea.	2,283.75	1,968.75	4,252.50
Pipe, steel, black, schedule 40, threaded, 2″ diam	200.000	L.F.	1,980	3,990	5,970
Radiation, steel 1-1/4″ tube & 4-1/4″ fin w/cover & damper, wall hung	80.000	L.F.	3,440	2,880	6,320
Rough-in, wall hung steel fin radiation with supply & balance valves	8.000	Set	2,680	6,360	9,040
Expansion tank, painted steel, ASME, 18 Gal capacity	1.000	Ea.	2,675	108	2,783
Circulating pump, CI flanged, 1/12 HP	1.000	Ea.	400	213	613
Pipe covering, calcium silicate w/cover, 1″ wall, 2″ diam	200.000	L.F.	1,040	1,470	2,510
TOTAL			14,498.75	16,989.75	31,488.50
COST PER S.F.			13.55	15.88	29.43

D3020 110	Heating System, Fin Tube Radiation		COST PER S.F.		
			MAT.	INST.	TOTAL
3230	Heating systems, hydronic, fossil fuel, fin tube radiation				
3240	Cast iron boiler, gas, 80 MBH, 1,070 S.F. bldg.		13.55	15.89	29.44
3280	169 M.B.H., 2,140 S.F. bldg.		8.60	10.05	18.65
3320	544 M.B.H., 7,250 S.F. bldg.	R235000 -10	7.35	8.60	15.95
3360	1,088 M.B.H., 14,500 S.F. bldg.		6.55	8.45	15
3400	3,264 M.B.H., 43,500 S.F. bldg.	R235000 -20	5.90	7.45	13.35
3440	5,032 M.B.H., 67,100 S.F. bldg.		6.50	7.55	14.05
3480	Oil, 109 M.B.H., 1,420 S.F. bldg.	R235000 -30	16.55	17.35	33.90
3520	235 M.B.H., 3,150 S.F. bldg.		9.45	10.15	19.60
3560	940 M.B.H., 12,500 S.F. bldg.		7.50	8.05	15.55
3600	1,600 M.B.H., 21,300 S.F. bldg.		7.20	8	15.20
3640	2,480 M.B.H., 33,100 S.F. bldg.		7.20	7.60	14.80
3680	3,350 M.B.H., 44,500 S.F. bldg.		6.60	7.70	14.30
3720	Coal, 148 M.B.H., 1,975 S.F. bldg.		12.15	9.60	21.75
3760	300 M.B.H., 4,000 S.F. bldg.		10	8.25	18.25
3800	2,360 M.B.H., 31,500 S.F. bldg.		7.60	7.60	15.20
4080	Steel boiler, oil, 97 M.B.H., 1,300 S.F. bldg.		12.30	14.25	26.55
4120	315 M.B.H., 4,550 S.F. bldg.		6.45	7.25	13.70
4160	525 M.B.H., 7,000 S.F. bldg.		9.10	8.30	17.40
4200	1,050 M.B.H., 14,000 S.F. bldg.		8.20	8.30	16.50
4240	2,310 M.B.H., 30,800 S.F. bldg.		7.20	7.65	14.85
4280	3,150 M.B.H., 42,000 S.F. bldg.		6.70	7.75	14.45

Electric Boiler, Hot Water

System Components	QUANTITY	UNIT	COST EACH		
			MAT.	INST.	TOTAL
SYSTEM D3020 126 1010					
BOILER, ELECTRIC, HOT WATER, 15 kW, 52 MBH					
Boilers, electric, ASME, hot water, 15 KW, 52 MBH	1.000	Ea.	5,500	1,500	7,000
Pipe, black steel, Sch 40, threaded, W/coupling & hangers, 10' OC, 3/4" dia	10.000	L.F.	49.45	133.98	183.43
Pipe, black steel, Sch 40, threaded, W/coupling & hangers, 10' OC, 1" dia	20.000	L.F.	119.33	288.10	407.43
Elbow, 90°, black, straight, 3/4" dia.	6.000	Ea.	26.34	303	329.34
Elbow, 90°, black, straight, 1" dia.	6.000	Ea.	45.60	327	372.60
Tee, black, straight, 3/4" dia.	1.000	Ea.	6.95	79	85.95
Tee, black, straight, 1" dia.	2.000	Ea.	23.90	177	200.90
Tee, black, reducing, 1" dia.	2.000	Ea.	38.40	177	215.40
Union, black with brass seat, 3/4" dia.	1.000	Ea.	18.50	54.50	73
Union, black with brass seat, 1" dia.	2.000	Ea.	48	118	166
Pipe nipples, 3/4" diam.	3.000	Ea.	6.45	17.48	23.93
Pipe nipples, 1" diam.	3.000	Ea.	8.33	20.10	28.43
Valves, bronze, gate, N.R.S., threaded, class 150, 3/4" size	2.000	Ea.	179	71	250
Valves, bronze, gate, N.R.S., threaded, class 150, 1" size	2.000	Ea.	224	75	299
Thermometer, stem type, 9" case, 8" stem, 3/4" NPT	1.000	Ea.	478	52	530
Tank, steel, liquid expansion, ASME, painted, 15 gallon capacity	1.000	Ea.	700	76.50	776.50
Pump, circulating, bronze, flange connection, 3/4" to 1-1/2" size, 1/8 HP	1.000	Ea.	1,075	213	1,288
Insulation, fiberglass pipe covering, 1" wall, 1" IPS	20.000	L.F.	21.40	107	128.40
TOTAL			8,568.65	3,789.66	12,358.31

D3020 126	Electric Boiler, Hot Water	COST EACH		
		MAT.	INST.	TOTAL
1010	Boiler, electric, hot water, 15 kW, 52 MBH	8,575	3,800	12,375
1020	30 kW, 103 MBH	9,225	4,000	13,225
1030	60 kW, 205 MBH	9,450	4,125	13,575
1040	120 kW, 410 MBH	11,200	4,950	16,150
1050	150 kW, 510 MBH	18,200	9,500	27,700
1060	210 kW, 716 MBH	20,700	10,400	31,100
1070	296 kW, 1010 MBH	30,200	10,800	41,000
1080	444 kW, 1515 MBH	36,100	14,500	50,600
1090	666 kW, 2273 MBH	43,600	15,800	59,400
1100	900 kW, 3071 MBH	55,000	18,300	73,300
1110	1320 kW, 4505 MBH	63,500	19,200	82,700
1120	2100 kW, 7167 MBH	88,000	23,400	111,400

D3020 Heat Generating Systems

D3020 126	Electric Boiler, Hot Water	COST EACH		
		MAT.	INST.	TOTAL
1130	2610 kW, 8905 MBH	111,000	28,100	139,100
1140	3600 kW, 12,283 MBH	136,500	33,600	170,100

D3020 128	Electric Boiler, Steam	COST EACH		
		MAT.	INST.	TOTAL
1010	Boiler, electric, steam, 18 kW, 61.4 MBH	5,900	3,625	9,525
1020	36 kW, 123 MBH	7,400	3,850	11,250
1030	60 kW, 205 MBH	15,400	7,825	23,225
1040	120 kW, 409 MBH	18,400	8,625	27,025
1050	150 kW, 512 MBH	22,000	13,000	35,000
1060	210 kW, 716 MBH	38,400	13,900	52,300
1070	300 kW, 1023 MBH	41,200	15,300	56,500
1080	510 kW, 1740 MBH	50,500	19,900	70,400
1090	720 kW, 2456 MBH	55,000	20,900	75,900
1100	1080 kW, 3685 MBH	62,000	24,800	86,800
1110	1260 kW, 4300 MBH	69,000	26,400	95,400
1120	1620 kW, 5527 MBH	72,000	29,900	101,900
1130	2070 kW, 7063 MBH	95,500	31,400	126,900
1140	2340 kW, 7984 MBH	123,000	35,600	158,600

D3020 Heat Generating Systems

Cast Iron Boiler, Hot Water, Gas Fired

System Components	QUANTITY	UNIT	COST EACH		
			MAT.	INST.	TOTAL
SYSTEM D3020 130 1010					
BOILER, CAST IRON, GAS, HOT WATER, 100 MBH					
Boilers, gas fired, std controls, CI, insulated, HW, gross output 100 MBH	1.000	Ea.	2,825	2,025	4,850
Pipe, black steel, Sch 40, threaded, W/coupling & hangers, 10' OC, 3/4" dia.	20.000	L.F.	96.75	262.13	358.88
Pipe, black steel, Sch 40, threaded, W/coupling & hangers, 10' OC, 1" dia.	20.000	L.F.	119.33	288.10	407.43
Elbow, 90°, black, straight, 3/4" dia.	9.000	Ea.	39.51	454.50	494.01
Elbow, 90°, black, straight, 1" dia.	6.000	Ea.	45.60	327	372.60
Tee, black, straight, 3/4" dia.	2.000	Ea.	13.90	158	171.90
Tee, black, straight, 1" dia.	2.000	Ea.	23.90	177	200.90
Tee, black, reducing, 1" dia.	2.000	Ea.	38.40	177	215.40
Pipe cap, black, 3/4" dia.	1.000	Ea.	4.98	22	26.98
Union, black with brass seat, 3/4" dia.	2.000	Ea.	37	109	146
Union, black with brass seat, 1" dia.	2.000	Ea.	48	118	166
Pipe nipples, black, 3/4" dia.	5.000	Ea.	10.75	29.13	39.88
Pipe nipples, black, 1" dia.	3.000	Ea.	8.33	20.10	28.43
Valves, bronze, gate, N.R.S., threaded, class 150, 3/4" size	2.000	Ea.	179	71	250
Valves, bronze, gate, N.R.S., threaded, class 150, 1" size	2.000	Ea.	224	75	299
Gas cock, brass, 3/4" size	1.000	Ea.	17.05	32	49.05
Thermometer, stem type, 9" case, 8" stem, 3/4" NPT	2.000	Ea.	478	52	530
Tank, steel, liquid expansion, ASME, painted, 15 gallon capacity	1.000	Ea.	700	76.50	776.50
Pump, circulating, bronze, flange connection, 3/4" to 1-1/2" size, 1/8 HP	1.000	Ea.	1,075	213	1,288
Vent chimney, all fuel, pressure tight, double wall, SS, 6" dia.	20.000	L.F.	1,190	410	1,600
Vent chimney, elbow, 90° fixed, 6" dia.	2.000	Ea.	890	82	972
Vent chimney, Tee, 6" dia.	2.000	Ea.	522	103	625
Vent chimney, ventilated roof thimble, 6" dia.	1.000	Ea.	272	47.50	319.50
Vent chimney, flat roof flashing, 6" dia.	1.000	Ea.	117	41	158
Vent chimney, stack cap, 6" diameter	1.000	Ea.	320	27	347
Insulation, fiberglass pipe covering, 1" wall, 1" IPS	20.000	L.F.	21.40	107	128.40
TOTAL			9,316.90	5,503.96	14,820.86

D3020 130	Boiler, Cast Iron, Hot Water, Gas	COST EACH		
		MAT.	INST.	TOTAL
1010	Boiler, cast iron, gas, hot water, 100 MBH	9,325	5,500	14,825
1020	200 MBH	10,700	6,325	17,025
1030	320 MBH	10,800	6,850	17,650
1040	440 MBH	14,600	8,875	23,475
1050	544 MBH	24,800	13,800	38,600
1060	765 MBH	30,000	14,900	44,900

D3020 Heat Generating Systems

D3020 130	Boiler, Cast Iron, Hot Water, Gas	COST EACH		
		MAT.	INST.	TOTAL
1070	1088 MBH	32,000	15,900	47,900
1080	1530 MBH	39,000	21,000	60,000
1090	2312 MBH	45,900	24,700	70,600
1100	2856 MBH	60,000	26,700	86,700
1110	3808 MBH	66,500	31,300	97,800
1120	4720 MBH	110,500	35,100	145,600
1130	6100 MBH	156,500	39,100	195,600
1140	6970 MBH	160,000	48,300	208,300

D3020 134	Boiler, Cast Iron, Steam, Gas	COST EACH		
		MAT.	INST.	TOTAL
1010	Boiler, cast iron, gas, steam, 100 MBH	7,800	5,350	13,150
1020	200 MBH	15,600	10,000	25,600
1030	320 MBH	17,400	11,300	28,700
1040	544 MBH	26,400	17,600	44,000
1050	765 MBH	30,000	18,400	48,400
1060	1275 MBH	35,700	20,200	55,900
1070	2675 MBH	51,000	29,400	80,400
1080	3570 MBH	63,000	33,000	96,000
1090	4720 MBH	98,500	35,200	133,700
1100	6100 MBH	124,000	41,000	165,000
1110	6970 MBH	136,000	48,100	184,100

D3020 136	Boiler, Cast Iron, Hot Water, Gas/Oil	COST EACH		
		MAT.	INST.	TOTAL
1010	Boiler, cast iron, gas & oil, hot water, 584 MBH	30,300	15,400	45,700
1020	876 MBH	38,600	16,500	55,100
1030	1168 MBH	48,600	18,600	67,200
1040	1460 MBH	59,000	22,400	81,400
1050	2044 MBH	64,500	23,200	87,700
1060	2628 MBH	77,500	28,100	105,600
1070	3210 MBH	80,500	30,200	110,700
1080	3796 MBH	87,000	31,200	118,200
1090	4672 MBH	101,500	32,700	134,200
1100	5256 MBH	123,500	39,600	163,100
1110	6000 MBH	148,000	40,300	188,300
1120	9800 MBH	185,500	69,000	254,500
1130	12,200 MBH	231,500	80,500	312,000
1140	13,500 MBH	252,000	85,500	337,500

D3020 138	Boiler, Cast Iron, Steam, Gas/Oil	COST EACH		
		MAT.	INST.	TOTAL
1010	Boiler, cast iron, gas & oil, steam, 810 MBH	33,100	20,000	53,100
1020	1360 MBH	38,400	21,300	59,700
1030	2040 MBH	47,600	27,400	75,000
1040	2700 MBH	51,000	30,400	81,400
1050	3270 MBH	60,000	34,000	94,000
1060	3770 MBH	98,000	34,700	132,700
1070	4650 MBH	110,000	39,400	149,400
1080	5520 MBH	127,000	41,500	168,500
1090	6100 MBH	131,000	43,600	174,600
1100	6970 MBH	140,500	53,500	194,000

For customer support ion your Mechanical Cost Data, call 877.795.7394.

535

D3020 Heat Generating Systems

Base Mounted End—Suction Pump

System Components	QUANTITY	UNIT	COST EACH		
			MAT.	INST.	TOTAL
SYSTEM D3020 330 1010					
PUMP, BASE MTD. WITH MOTOR, END-SUCTION, 2-1/2" SIZE, 3 HP, TO 150 GPM					
Pump, circulating, CI, base mounted, 2-1/2" size, 3 HP, to 150 GPM	1.000	Ea.	8,675	710	9,385
Pipe, black steel, Sch. 40, on yoke & roll hangers, 10' O.C., 2-1/2" dia	12.000	L.F.	168.60	340.20	508.80
Elbow, 90°, weld joint, steel, 2-1/2" pipe size	1.000	Ea.	30.50	167.95	198.45
Flange, weld neck, 150 LB, 2-1/2" pipe size	8.000	Ea.	280	671.76	951.76
Valve, iron body, gate, 125 lb., N.R.S., flanged, 2-1/2" size	1.000	Ea.	810	255	1,065
Strainer, Y type, iron body, flanged, 125 lb., 2-1/2" pipe size	1.000	Ea.	127	260	387
Multipurpose valve, CI body, 2" size	1.000	Ea.	505	90	595
Expansion joint, flanged spool, 6" F to F, 2-1/2" dia	2.000	Ea.	588	210	798
T-O-L, weld joint, socket, 1/4" pipe size, nozzle	2.000	Ea.	15.60	116.52	132.12
T-O-L, weld joint, socket, 1/2" pipe size, nozzle	2.000	Ea.	15.60	121.78	137.38
Control gauges, pressure or vacuum, 3-1/2" diameter dial	2.000	Ea.	22.80	45	67.80
Pressure/temperature relief plug, 316 SS, 3/4" OD, 7-1/2" insertion	2.000	Ea.	153	45	198
Insulation, fiberglass pipe covering, 1-1/2" wall, 2-1/2" IPS	12.000	L.F.	30	78	108
Water flow	1.000	Ea.	1,325	1,075	2,400
Pump balancing	1.000	Ea.		315	315
TOTAL			12,746.10	4,501.21	17,247.31

D3020 330	Circulating Pump Systems, End Suction	COST EACH		
		MAT.	INST.	TOTAL
1010	Pump, base mtd with motor, end-suction, 2-1/2" size, 3 HP, to 150 GPM	12,700	4,500	17,200
1020	3" size, 5 HP, to 225 GPM	15,300	5,100	20,400
1030	4" size, 7-1/2 HP, to 350 GPM	18,100	6,050	24,150
1040	5" size, 15 HP, to 1000 GPM	26,500	8,950	35,450
1050	6" size, 25 HP, to 1550 GPM	35,100	10,800	45,900

D3020 340	Circulating Pump Systems, Double Suction	COST EACH		
		MAT.	INST.	TOTAL
1010	Pump, base mtd w/motor, double suction, 6" size, 50 HP, to 1200 GPM	23,500	15,400	38,900
1020	8" size, 75 HP, to 2500 GPM	42,600	20,600	63,200
1030	8" size, 100 HP, to 3000 GPM	57,000	23,000	80,000
1040	10" size, 150 HP, to 4000 GPM	82,500	26,100	108,600

D3030 Cooling Generating Systems

Chilled Water Supply & Return Piping

Air Cooled Water Chiller Unit

Design Assumptions: The chilled water, air cooled systems priced, utilize reciprocating hermetic compressors and propeller-type condenser fans. Piping with pumps and expansion tanks is included based on a two pipe system. No ducting is included and the fan-coil units are cooling only. Water treatment and balancing are not included. Chilled water piping is insulated. Area distribution is through the use of multiple fan coil units. Fewer but larger fan coil units with duct distribution would be approximately the same S.F. cost.

System Components	QUANTITY	UNIT	COST EACH		
			MAT.	INST.	TOTAL
SYSTEM D3030 110 1200					
PACKAGED CHILLER, AIR COOLED, WITH FAN COIL UNIT					
APARTMENT CORRIDORS, 3,000 S.F., 5.50 TON					
Fan coil air conditioning unit, cabinet mounted & filters chilled water	1.000	Ea.	4,399.20	595.73	4,994.93
Water chiller, air conditioning unit, air cooled	1.000	Ea.	6,490	2,220.63	8,710.63
Chilled water unit coil connections	1.000	Ea.	1,300	1,625	2,925
Chilled water distribution piping	440.000	L.F.	10,120	22,000	32,120
TOTAL			22,309.20	26,441.36	48,750.56
COST PER S.F.			7.44	8.81	16.25

*Cooling requirements would lead to choosing a water cooled unit.

D3030 110	Chilled Water, Air Cooled Condenser Systems	COST PER S.F.		
		MAT.	INST.	TOTAL
1180	Packaged chiller, air cooled, with fan coil unit			
1200	Apartment corridors, 3,000 S.F., 5.50 ton	7.43	8.80	16.23
1240	6,000 S.F., 11.00 ton	5.95	7	12.95
1280	10,000 S.F., 18.33 ton	5.05	5.30	10.35
1320	20,000 S.F., 36.66 ton	3.97	3.83	7.80
1360	40,000 S.F., 73.33 ton	5.05	3.83	8.88
1440	Banks and libraries, 3,000 S.F., 12.50 ton	10.75	10.20	20.95
1480	6,000 S.F., 25.00 ton	9.80	8.30	18.10
1520	10,000 S.F., 41.66 ton	7.70	5.75	13.45
1560	20,000 S.F., 83.33 ton	8.55	5.15	13.70
1600	40,000 S.F., 167 ton*			
1680	Bars and taverns, 3,000 S.F., 33.25 ton	18.80	11.95	30.75
1720	6,000 S.F., 66.50 ton	16.90	9.10	26
1760	10,000 S.F., 110.83 ton	13.80	2.72	16.52
1800	20,000 S.F., 220 ton*			
1840	40,000 S.F., 440 ton*			
1920	Bowling alleys, 3,000 S.F., 17.00 ton	13.45	11.45	24.90
1960	6,000 S.F., 34.00 ton	10.50	8	18.50
2000	10,000 S.F., 56.66 ton	9.10	5.70	14.80
2040	20,000 S.F., 113.33 ton	9.80	5.20	15
2080	40,000 S.F., 227 ton*			
2160	Department stores, 3,000 S.F., 8.75 ton	10.30	9.80	20.10
2200	6,000 S.F., 17.50 ton	7.55	7.50	15.05
2240	10,000 S.F., 29.17 ton	6.20	5.40	11.60

537

For customer support on your Mechanical Cost Data, call 877.795.7394.

D3030 Cooling Generating Systems

D3030 110	Chilled Water, Air Cooled Condenser Systems	COST PER S.F.		
		MAT.	INST.	TOTAL
2280	20,000 S.F., 58.33 ton	5.20	3.94	9.14
2320	40,000 S.F., 116.66 ton	6.40	3.94	10.34
2400	Drug stores, 3,000 S.F., 20.00 ton	15.20	11.80	27
2440	6,000 S.F., 40.00 ton	12.40	8.75	21.15
2480	10,000 S.F., 66.66 ton	13.05	7.40	20.45
2520	20,000 S.F., 133.33 ton	11.65	5.45	17.10
2560	40,000 S.F., 267 ton*			
2640	Factories, 2,000 S.F., 10.00 ton	9.50	9.85	19.35
2680	6,000 S.F., 20.00 ton	8.55	7.90	16.45
2720	10,000 S.F., 33.33 ton	6.70	5.50	12.20
2760	20,000 S.F., 66.66 ton	7.50	4.93	12.43
2800	40,000 S.F., 133.33 ton	6.90	4.04	10.94
2880	Food supermarkets, 3,000 S.F., 8.50 ton	10.10	9.75	19.85
2920	6,000 S.F., 17.00 ton	7.45	7.45	14.90
2960	10,000 S.F., 28.33 ton	5.85	5.25	11.10
3000	20,000 S.F., 56.66 ton	5	3.87	8.87
3040	40,000 S.F., 113.33 ton	6.20	3.95	10.15
3120	Medical centers, 3,000 S.F., 7.00 ton	9.10	9.50	18.60
3160	6,000 S.F., 14.00 ton	6.70	7.25	13.95
3200	10,000 S.F., 23.33 ton	5.90	5.50	11.40
3240	20,000 S.F., 46.66 ton	4.73	4.02	8.75
3280	40,000 S.F., 93.33 ton	5.65	3.89	9.54
3360	Offices, 3,000 S.F., 9.50 ton	9	9.70	18.70
3400	6,000 S.F., 19.00 ton	8.30	7.85	16.15
3440	10,000 S.F., 31.66 ton	6.50	5.45	11.95
3480	20,000 S.F., 63.33 ton	7.55	5.05	12.60
3520	40,000 S.F., 126.66 ton	6.80	4.08	10.88
3600	Restaurants, 3,000 S.F., 15.00 ton	12	10.60	22.60
3640	6,000 S.F., 30.00 ton	10	8.05	18.05
3680	10,000 S.F., 50.00 ton	8.75	5.85	14.60
3720	20,000 S.F., 100.00 ton	10	5.50	15.50
3760	40,000 S.F., 200 ton*			
3840	Schools and colleges, 3,000 S.F., 11.50 ton	10.25	10.05	20.30
3880	6,000 S.F., 23.00 ton	9.30	8.15	17.45
3920	10,000 S.F., 38.33 ton	7.30	5.65	12.95
3960	20,000 S.F., 76.66 ton	8.40	5.20	13.60
4000	40,000 S.F., 153 ton*			

For customer support on your Mechanical Cost Data, call 877.795.7394.

D3030 Cooling Generating Systems

General: Water cooled chillers are available in the same sizes as air cooled units. They are also available in larger capacities.

Design Assumptions: The chilled water systems with water cooled condenser include reciprocating hermetic compressors, water cooling tower, pumps, piping and expansion tanks and are based on a two pipe system. Chilled water piping is insulated. No ducts are included and fan-coil units are cooling only. Area distribution is through use of multiple fan coil units. Fewer but larger fan coil units with duct distribution would be approximately the same S.F. cost. Water treatment and balancing are not included.

System Components	QUANTITY	UNIT	COST EACH MAT.	COST EACH INST.	COST EACH TOTAL
SYSTEM D3030 115 1320					
PACKAGED CHILLER, WATER COOLED, WITH FAN COIL UNIT					
APARTMENT CORRIDORS, 4,000 S.F., 7.33 TON					
Fan coil air conditioner unit, cabinet mounted & filters, chilled water	2.000	Ea.	5,863.20	793.98	6,657.18
Water chiller, water cooled, 1 compressor, hermetic scroll,	1.000	Ea.	5,793.70	3,870.10	9,663.80
Cooling tower, draw thru single flow, belt drive	1.000	Ea.	1,737.21	164.93	1,902.14
Cooling tower pumps & piping	1.000	System	864.94	388.49	1,253.43
Chilled water unit coil connections	2.000	Ea.	2,600	3,250	5,850
Chilled water distribution piping	520.000	L.F.	11,960	26,000	37,960
TOTAL			28,819.05	34,467.50	63,286.55
COST PER S.F.			7.20	8.62	15.82

*Cooling requirements would lead to choosing a water cooled unit.

D3030 115	Chilled Water, Cooling Tower Systems		COST PER S.F. MAT.	COST PER S.F. INST.	COST PER S.F. TOTAL
1300	Packaged chiller, water cooled, with fan coil unit				
1320	Apartment corridors, 4,000 S.F., 7.33 ton		7.20	8.63	15.83
1360	6,000 S.F., 11.00 ton	R236000	5.90	7.40	13.30
1400	10,000 S.F., 18.33 ton	-20	5.75	5.45	11.20
1440	20,000 S.F., 26.66 ton		4.65	4.06	8.71
1480	40,000 S.F., 73.33 ton		4.96	4.14	9.10
1520	60,000 S.F., 110.00 ton		5.25	4.25	9.50
1600	Banks and libraries, 4,000 S.F., 16.66 ton		12.45	9.35	21.80
1640	6,000 S.F., 25.00 ton		10.50	8.20	18.70
1680	10,000 S.F., 41.66 ton		8.40	6.05	14.45
1720	20,000 S.F., 83.33 ton		8.20	5.80	14
1760	40,000 S.F., 166.66 ton		9.05	7	16.05
1800	60,000 S.F., 250.00 ton		9.25	7.45	16.70
1880	Bars and taverns, 4,000 S.F., 44.33 ton		19.80	11.60	31.40
1920	6,000 S.F., 66.50 ton		22.50	12	34.50
1960	10,000 S.F., 110.83 ton		18.20	9.45	27.65
2000	20,000 S.F., 221.66 ton		21.50	9.55	31.05
2040	40,000 S.F., 440 ton*				
2080	60,000 S.F., 660 ton*				
2160	Bowling alleys, 4,000 S.F., 22.66 ton		14.30	10.20	24.50
2200	6,000 S.F., 34.00 ton		12.75	9	21.75
2240	10,000 S.F., 56.66 ton		11.10	6.55	17.65
2280	20,000 S.F., 113.33 ton		10.25	6.05	16.30
2320	40,000 S.F., 226.66 ton		12.20	6.95	19.15

539

D3030 Cooling Generating Systems

D3030 115	Chilled Water, Cooling Tower Systems	COST PER S.F.		
		MAT.	INST.	TOTAL
2360	60,000 S.F., 340 ton			
2440	Department stores, 4,000 S.F., 11.66 ton	7.95	9.30	17.25
2480	6,000 S.F., 17.50 ton	9.20	7.65	16.85
2520	10,000 S.F., 29.17 ton	6.95	5.65	12.60
2560	20,000 S.F., 58.33 ton	5.70	4.19	9.89
2600	40,000 S.F., 116.66 ton	6.25	4.32	10.57
2640	60,000 S.F., 175.00 ton	8.20	6.80	15
2720	Drug stores, 4,000 S.F., 26.66 ton	15.55	10.50	26.05
2760	6,000 S.F., 40.00 ton	13.30	9	22.30
2800	10,000 S.F., 66.66 ton	14.50	8.10	22.60
2840	20,000 S.F., 133.33 ton	11.75	6.35	18.10
2880	40,000 S.F., 266.67 ton	12.35	7.85	20.20
2920	60,000 S.F., 400 ton*			
3000	Factories, 4,000 S.F., 13.33 ton	10.70	8.90	19.60
3040	6,000 S.F., 20.00 ton	9.15	7.70	16.85
3080	10,000 S.F., 33.33 ton	7.60	5.85	13.45
3120	20,000 S.F., 66.66 ton	8.10	5.25	13.35
3160	40,000 S.F., 133.33 ton	6.95	4.47	11.42
3200	60,000 S.F., 200.00 ton	8.35	7.15	15.50
3280	Food supermarkets, 4,000 S.F., 11.33 ton	7.80	9.20	17
3320	6,000 S.F., 17.00 ton	8.15	7.50	15.65
3360	10,000 S.F., 28.33 ton	6.85	5.65	12.50
3400	20,000 S.F., 56.66 ton	5.90	4.21	10.11
3440	40,000 S.F., 113.33 ton	6.20	4.34	10.54
3480	60,000 S.F., 170.00 ton	8.10	6.85	14.95
3560	Medical centers, 4,000 S.F., 9.33 ton	6.80	8.45	15.25
3600	6,000 S.F., 14.00 ton	8	7.35	15.35
3640	10,000 S.F., 23.33 ton	6.30	5.45	11.75
3680	20,000 S.F., 46.66 ton	5.30	4.11	9.41
3720	40,000 S.F., 93.33 ton	5.65	4.26	9.91
3760	60,000 S.F., 140.00 ton	6.90	6.90	13.80
3840	Offices, 4,000 S.F., 12.66 ton	10.40	8.85	19.25
3880	6,000 S.F., 19.00 ton	9.05	7.85	16.90
3920	10,000 S.F., 31.66 ton	7.55	5.90	13.45
3960	20,000 S.F., 63.33 ton	8	5.25	13.25
4000	40,000 S.F., 126.66 ton	7.50	6.70	14.20
4040	60,000 S.F., 190.00 ton	8.10	7.10	15.20
4120	Restaurants, 4,000 S.F., 20.00 ton	12.80	9.50	22.30
4160	6,000 S.F., 30.00 ton	11.50	8.45	19.95
4200	10,000 S.F., 50.00 ton	10.20	6.30	16.50
4240	20,000 S.F., 100.00 ton	9.65	6.05	15.70
4280	40,000 S.F., 200.00 ton	9.60	6.75	16.35
4320	60,000 S.F., 300.00 ton	10.30	7.70	18
4400	Schools and colleges, 4,000 S.F., 15.33 ton	11.75	9.15	20.90
4440	6,000 S.F., 23.00 ton	9.95	8.05	18
4480	10,000 S.F., 38.33 ton	7.95	5.95	13.90
4520	20,000 S.F., 76.66 ton	7.95	5.75	13.70
4560	40,000 S.F., 153.33 ton	8.50	6.85	15.35
4600	60,000 S.F., 230.00 ton	8.55	7.15	15.70

D3030 Cooling Generating Systems

Chilled Water Piping

Reciprocating Chiller, Water Cooled

Condenser Water Piping

System Components	QUANTITY	UNIT	COST EACH		
			MAT.	INST.	TOTAL
SYSTEM D3030 130 1010					
CHILLER, RECIPROCATING, WATER COOLED, STD. CONTROLS, 60 TON					
Water chiller, reciprocating, water cooled, 60 ton cooling	1.000	Ea.	40,700	10,800	51,500
Pipe, black steel, welded, Sch. 40, on yoke & roll hgrs, 10' O.C., 4" dia	40.000	L.F.	960	1,448.80	2,408.80
Elbow, 90°, weld joint, butt, 4" pipe size	8.000	Ea.	424	2,141.60	2,565.60
Flange, welding neck, 150 lb., 4" pipe size	16.000	Ea.	744	2,149.60	2,893.60
Valves, iron body, butterfly, lug type, lever actuator, 4" size	4.000	Ea.	912	1,020	1,932
Expansion joints, bellows type, flange spool, 6" F to F, 4" dia	4.000	Ea.	1,420	620	2,040
T-O-L socket, weld joint, 1/4" pipe size, nozzle	4.000	Ea.	31.20	233.04	264.24
T-O-L socket, weld joint, 3/4" pipe size, nozzle	4.000	Ea.	36	256.08	292.08
Thermometers, stem type, 9" case, 8" stem, 3/4" NPT	4.000	Ea.	956	104	1,060
Gauges, pressure or vacuum, 3-1/2" diameter dial	4.000	Ea.	45.60	90	135.60
Insulation, fiberglass pipe covering, 1-1/2" wall, 4" IPS	20.000	L.F.	59.40	168	227.40
Chiller balancing	1.000	Ea.		615	615
TOTAL			46,288.20	19,646.12	65,934.32

D3030 130	Reciprocating Chiller, Water Cooled	COST EACH		
		MAT.	INST.	TOTAL
1010	Chiller, reciprocating, water cooled, std. controls, 60 Ton	46,300	19,700	66,000
1020	100 Ton	51,000	29,600	80,600
1030	150 Ton	83,500	32,600	116,100
1040	200 Ton	113,000	34,400	147,400

D3030 135	Reciprocating Chiller, Air Cooled	COST EACH		
		MAT.	INST.	TOTAL
1010	Chiller, reciprocating, air cooled, std. controls, 20 Ton	25,800	11,600	37,400
1020	30 Ton	32,000	12,200	44,200
1030	40 Ton	38,600	13,100	51,700
1040	60 Ton	49,000	14,700	63,700
1050	70 Ton	60,500	15,100	75,600
1060	80 Ton	64,500	15,400	79,900
1070	90 Ton	72,000	15,600	87,600
1080	100 Ton	84,000	15,900	99,900
1090	110 Ton	91,000	17,200	108,200
1100	130 Ton	103,000	17,400	120,400

D3030 Cooling Generating Systems

D3030 135	Reciprocating Chiller, Air Cooled	COST EACH		
		MAT.	INST.	TOTAL
1110	150 Ton	114,500	17,600	132,100
1120	175 Ton	128,500	17,900	146,400
1130	210 Ton	149,500	19,400	168,900

Chilled Water Piping

Centrifugal Chiller, Water Cooled, Hermetic

Condenser Water Piping

System Components	QUANTITY	UNIT	COST EACH		
			MAT.	INST.	TOTAL
SYSTEM D3030 140 1010					
CHILLER, CENTRIF., WATER COOLED, PKGD. HERMETIC, STD. CONTROLS, 200 TON					
Water chiller, cntfgl, water cooled, not incl. tower, 200 ton	1.000	Ea.	99,000	14,800	113,800
Pipe, black steel, welded, Sch. 40, on yoke & roll hgrs, 10' OC, 6" dia	40.000	L.F.	1,900	2,270.40	4,170.40
Elbow, 90°, weld joint, 6" pipe size	8.000	Ea.	928	3,261.20	4,189.20
Flange, welding neck, 150 lb., 6" pipe size	16.000	Ea.	1,120	3,285.60	4,405.60
Valves, iron body, butterfly, lug type, lever actuator, 6" size	4.000	Ea.	1,540	1,580	3,120
Expansion joints, bellows type, flange spool, 6" F to F, 6" dia	4.000	Ea.	1,800	764	2,564
T-O-L socket, weld jiont, 1/4" pipe size, nozzle	4.000	Ea.	31.20	233.04	264.24
T-O-L socket, weld jiont, 3/4" pipe size, nozzle	4.000	Ea.	36	256.08	292.08
Thermometers, stem type, 9" case, 8" stem, 3/4" NPT	4.000	Ea.	956	104	1,060
Gauges, pressure or vacuum, 3-1/2" diameter dial	4.000	Ea.	45.60	90	135.60
Insulation, fiberglass pipe covering, 1-1/2" wall, 6" IPS	40.000	L.F.	140.40	426	566.40
Chiller balancing	1.000	Ea.		615	615
TOTAL			107,497.20	27,685.32	135,182.52

D3030 140	Centrifugal Chiller, Water Cooled	COST EACH		
		MAT.	INST.	TOTAL
1010	Chiller, centrif., water cooled, pkgd. hermetic, std controls, 200 Ton	107,500	27,700	135,200
1020	400 Ton	152,000	41,800	193,800
1030	1,000 Ton	426,000	61,500	487,500
1040	1,500 Ton	580,500	76,500	657,000

543

For customer support on your Mechanical Cost Data, call 877.795.7394.

D3030 Cooling Generating Systems

Chilled Water Piping
(Condenser Water Piping Similar)

Gas Absorption Chiller, Water Cooled

Gas Piping

System Components	QUANTITY	UNIT	COST EACH MAT.	COST EACH INST.	COST EACH TOTAL
SYSTEM D3030 145 1010					
CHILLER, GAS FIRED ABSORPTION, WATER COOLED, STD. CONTROLS, 800 TON					
Absorption water chiller, gas fired, water cooled, 800 ton	1.000	Ea.	625,000	32,700	657,700
Pipe, black steel, weld, Sch. 40, on yoke & roll hangers, 10' O.C., 10" dia	20.000	L.F.	1,960	1,702.80	3,662.80
Pipe, black steel, welded, Sch. 40, 14" dia., (two rod roll type hanger)	44.000	L.F.	4,752	5,993.68	10,745.68
Pipe, black steel, weld, Sch. 40, on yoke & roll hangers, 10' O.C., 4" dia	20.000	L.F.	480	724.40	1,204.40
Elbow, 90°, weld joint, 10" pipe size	4.000	Ea.	1,740	2,724	4,464
Elbow, 90°, weld joint, 14" pipe size	6.000	Ea.	6,450	6,159	12,609
Elbow, 90°, weld joint, 4" pipe size	3.000	Ea.	159	803.10	962.10
Flange, welding neck, 150 lb., 10" pipe size	8.000	Ea.	1,592	2,724.40	4,316.40
Flange, welding neck, 150 lb., 14" pipe size	14.000	Ea.	5,810	6,357.40	12,167.40
Flange, welding neck, 150 lb., 4" pipe size	3.000	Ea.	139.50	403.05	542.55
Valves, iron body, butterfly, lug type, gear operated, 10" size	2.000	Ea.	2,550	990	3,540
Valves, iron body, butterfly, lug type, gear operated, 14" size	5.000	Ea.	16,875	4,325	21,200
Valves, semi-steel, lubricated plug valve, flanged, 4" pipe size	1.000	Ea.	520	425	945
Expansion joints, bellows type, flange spool, 6" F to F, 10" dia	2.000	Ea.	1,430	520	1,950
Expansion joints, bellows type, flange spool, 10" F to F, 14" dia	2.000	Ea.	2,500	680	3,180
T-O-L socket, weld joint, 1/4" pipe size, nozzle	4.000	Ea.	31.20	233.04	264.24
T-O-L socket, weld joint, 3/4" pipe size, nozzle	4.000	Ea.	36	256.08	292.08
Thermometers, stem type, 9" case, 8" stem, 3/4" NPT	4.000	Ea.	956	104	1,060
Gauges, pressure or vacuum, 3-1/2" diameter dial	4.000	Ea.	45.60	90	135.60
Insulation, fiberglass pipe covering, 2" wall, 10" IPS	20.000	L.F.	156	335	491
Vent chimney, all fuel, pressure tight, double wall, SS, 12" dia	20.000	L.F.	1,720	560	2,280
Chiller balancing	1.000	Ea.		615	615
TOTAL			**674,902.30**	**69,424.95**	**744,327.25**

D3030 145	Gas Absorption Chiller, Water Cooled		COST EACH MAT.	COST EACH INST.	COST EACH TOTAL
1010	Chiller, gas fired absorption, water cooled, std. ctrls, 800 Ton		675,000	69,500	744,500
1020	1000 Ton		887,500	89,500	977,000

D3030 150	Steam Absorption Chiller, Water Cooled		COST EACH MAT.	COST EACH INST.	COST EACH TOTAL
1010	Chiller, steam absorption, water cooled, std. ctrls, 750 Ton		729,500	82,000	811,500
1020	955 Ton		853,000	98,000	951,000
1030	1465 Ton		1,235,500	112,500	1,348,000
1040	1660 Ton		1,492,000	128,000	1,620,000

D3030 Cooling Generating Systems

Cold Water Fill

Drain To Waste Piping

Cooling Tower

Condenser Water Piping

System Components	QUANTITY	UNIT	COST EACH		
			MAT.	INST.	TOTAL
SYSTEM D3030 310 1010					
COOLING TOWER, GALVANIZED STEEL, PACKAGED UNIT, DRAW THRU, 60 TON					
Cooling towers, draw thru, single flow, belt drive, 60 tons	1.000	Ea.	14,220	1,350	15,570
Pipe, black steel, welded, Sch. 40, on yoke & roll hangers, 10' OC, 4" dia.	40.000	L.F.	960	1,448.80	2,408.80
Tubing, copper, Type L, couplings & hangers 10' OC, 1-1/2" dia.	20.000	L.F.	285	273	558
Elbow, 90°, copper, cu x cu, 1-1/2" dia.	3.000	Ea.	46.50	163.50	210
Elbow, 90°, weld joint, 4" pipe size	6.000	Ea.	318	1,606.20	1,924.20
Flange, welding neck, 150 lb., 4" pipe size	6.000	Ea.	279	806.10	1,085.10
Valves, bronze, gate, N.R.S., soldered, 125 psi, 1-1/2" size	1.000	Ea.	157	54.50	211.50
Valves, iron body, butterfly, lug type, lever actuator, 4" size	2.000	Ea.	456	510	966
Electric heat trace system, 400 degree, 115v, 5 watts per L.F.	160.000	L.F.	1,384	198.40	1,582.40
Insulation, fiberglass pipe covering, , 1-1/2" wall, 4" IPS	40.000	L.F.	118.80	336	454.80
Insulation, fiberglass pipe covering, 1" wall, 1-1/2" IPS	20.000	L.F.	24.80	112	136.80
Cooling tower condenser control system	1.000	Ea.	6,075	2,925	9,000
Cooling tower balancing	1.000	Ea.		470	470
TOTAL			24,324.10	10,253.50	34,577.60

D3030 310	Galvanized Draw Through Cooling Tower		COST EACH		
			MAT.	INST.	TOTAL
1010	Cooling tower, galvanized steel, packaged unit, draw thru, 60 Ton		24,300	10,300	34,600
1020	110 Ton		33,400	12,100	45,500
1030	300 Ton		49,000	18,200	67,200
1040	600 Ton		60,500	27,300	87,800
1050	1000 Ton		125,500	40,100	165,600

D3030 320	Fiberglass Draw Through Cooling Tower		COST EACH		
			MAT.	INST.	TOTAL
1010	Cooling tower, fiberglass, packaged unit, draw thru, 100 Ton		25,400	10,400	35,800
1020	120 Ton		29,300	11,700	41,000
1030	140 Ton		33,700	15,500	49,200
1040	160 Ton		43,700	20,600	64,300
1050	180 Ton		63,000	29,800	92,800

D3030 330	Stainless Steel Draw Through Cooling Tower		COST EACH		
			MAT.	INST.	TOTAL
1010	Cooling tower, stainless steel, packaged unit, draw thru, 60 Ton		39,900	10,300	50,200
1020	110 Ton		47,500	12,100	59,600
1030	300 Ton		79,500	18,200	97,700
1040	600 Ton		125,500	26,900	152,400
1050	1000 Ton		213,000	40,200	253,200

For customer support on your Mechanical Cost Data, call 877.795.7394.

545

D3040 Distribution Systems

Cooling Coil

Two-Way Valve Piping

Three-Way Valve Piping

Heating Coil

Service Lights (typ.)

Filters

Heating Coils

Cooling Coil

Fan

Fan

18 Ga. Exterior with 7.5# per S.F. Insulation

Double Walled Access Doors

Sound Absorptive 24 Ga. Perforated Inner Panels

Built-Up Air Handling Unit

System Components	QUANTITY	UNIT	COST EACH		
			MAT.	INST.	TOTAL
SYSTEM D3040 106 1010					
AHU, FIELD FAB., BLT. UP, COOL/HEAT COILS, FLTRS., CONST. VOL., 40,000 CFM					
AHU, Built-up, cool/heat coils, filters, mix box, constant volume	1.000	Ea.	99,000	6,725	105,725
Pipe, black steel, welded, Sch. 40, on yoke & roll hgrs, 10' OC, 3" dia.	26.000	L.F.	442	805.48	1,247.48
Pipe, black steel, welded, Sch. 40, on yoke & roll hgrs, 10' OC, 4" dia	20.000	L.F.	480	724.40	1,204.40
Elbow, 90°, weld joint, 3" pipe size	5.000	Ea.	160	955.25	1,115.25
Elbow, 90°, weld joint, 4" pipe size	4.000	Ea.	212	1,070.80	1,282.80
Tee, welded, reducing on outlet, 3" by 2-1/2" pipe	3.000	Ea.	390	1,007.70	1,397.70
Tee, welded, reducing on outlet, 4" by 3" pipe	2.000	Ea.	274	892	1,166
Reducer, eccentric, weld joint, 3" pipe size	4.000	Ea.	260	671.80	931.80
Reducer, eccentric, weld joint, 4" pipe size	4.000	Ea.	332	894.40	1,226.40
Flange, weld neck, 150 LB, 2-1/2" pipe size	3.000	Ea.	105	251.91	356.91
Flange, weld neck, 150 lb., 3" pipe size	14.000	Ea.	539	1,337.56	1,876.56
Flange, weld neck, 150 lb., 4" pipe size	10.000	Ea.	465	1,343.50	1,808.50
Valves, iron body, butterfly, lug type, lever actuator, 3" size	2.000	Ea.	418	320	738
Valves, iron body, butterfly, lug type, lever actuator, 4" size	2.000	Ea.	456	510	966
Strainer, Y type, iron body, flanged, 125 lb., 3" pipe size	1.000	Ea.	204	289	493
Strainer, Y type, iron body, flanged, 125 lb., 4" pipe size	1.000	Ea.	335	435	770
Valve, electric motor actuated, iron body, 3 way, flanged, 2-1/2" pipe	1.000	Ea.	1,550	345	1,895
Valve, electric motor actuated, iron body, two way, flanged, 3" pipe size	1.000	Ea.	1,625	345	1,970
Expansion joints, bellows type, flange spool, 6" F to F, 3" dia	2.000	Ea.	660	228	888
Expansion joints, bellows type, flange spool, 6" F to F, 4" dia	2.000	Ea.	710	310	1,020
T-O-L socket, weld joint, 1/4" pipe size, nozzle	4.000	Ea.	31.20	233.04	264.24
T-O-L socket, weld joint, 1/2" pipe size, nozzle	12.000	Ea.	93.60	730.68	824.28
Thermometers, stem type, 9" case, 8" stem, 3/4" NPT	4.000	Ea.	956	104	1,060
Pressure/temperature relief plug, 316 SS, 3/4"OD, 7-1/2" insertion	4.000	Ea.	306	90	396
Insulation, fiberglass pipe covering, 1-1/2" wall, 3" IPS	26.000	L.F.	67.86	179.40	247.26
Insulation, fiberglass pipe covering, 1-1/2" wall, 4" IPS	20.000	L.F.	59.40	168	227.40
Fan coil control system	1.000	Ea.	1,375	650	2,025
Re-heat coil balancing	2.000	Ea.		268	268
Air conditioner balancing	1.000	Ea.		820	820
TOTAL			111,506.06	22,704.92	134,210.98

D3040 106	Field Fabricated Air Handling Unit	COST EACH		
		MAT.	INST.	TOTAL
1010	AHU, Field fab, blt up, cool/heat coils, fltrs, const vol, 40,000 CFM	111,500	22,700	134,200
1020	60,000 CFM	107,500	35,300	142,800
1030	75,000 CFM	130,500	37,200	167,700

546

D3040 Distribution Systems

D3040 108	Field Fabricated VAV Air Handling Unit	COST EACH		
		MAT.	INST.	TOTAL
1010	AHU, Field fab, blt up, cool/heat coils, fltrs, VAV, 75,000 CFM	132,500	42,600	175,100
1020	100,000 CFM	204,000	57,000	261,000
1030	150,000 CFM	336,000	90,500	426,500

547

For customer support on your Mechanical Cost Data, call 877.795.7394.

D3040 Distribution Systems

| Chilled Water Piping | Central Station Air Handling Unit | Hot Water Piping |

System Components			COST EACH		
	QUANTITY	UNIT	MAT.	INST.	TOTAL
SYSTEM D3040 110 1010					
AHU, CENTRAL STATION, COOL/HEAT COILS, CONST. VOL., FLTRS., 2,000 CFM					
Central station air handling unit, chilled water, 2000 CFM	1.000	Ea.	12,100	1,175	13,275
Pipe, black steel, Sch 40, threaded, W/cplgs & hangers, 10' OC, 1" dia.	26.000	L.F.	180.38	435.50	615.88
Pipe, black steel, Sch 40, threaded, W/cplg & hangers, 10' OC, 1-1/4" dia.	20.000	L.F.	176.80	373.10	549.90
Elbow, 90°, black, straight, 1" dia.	4.000	Ea.	30.40	218	248.40
Elbow, 90°, black, straight, 1-1/4" dia.	4.000	Ea.	50.20	232	282.20
Tee, black, reducing, 1" dia.	8.000	Ea.	153.60	708	861.60
Tee, black, reducing, 1-1/4" dia.	8.000	Ea.	272	728	1,000
Reducer, concentric, black, 1" dia.	4.000	Ea.	55.60	190	245.60
Reducer, concentric, black, 1-1/4" dia.	4.000	Ea.	62.20	196	258.20
Union, black with brass seat, 1" dia.	4.000	Ea.	96	236	332
Union, black with brass seat, 1-1/4" dia.	2.000	Ea.	69	122	191
Pipe nipple, black, 1" dia	13.000	Ea.	36.08	87.10	123.18
Pipe nipple, black, 1-1/4" dia	12.000	Ea.	40.80	86.10	126.90
Valves, bronze, gate, N.R.S., threaded, class 150, 1" size	2.000	Ea.	224	75	299
Valves, bronze, gate, N.R.S., threaded, class 150, 1-1/4" size	2.000	Ea.	304	95	399
Strainer, Y type, bronze body, screwed, 150 lb., 1" pipe size	1.000	Ea.	55	42.50	97.50
Strainer, Y type, bronze body, screwed, 150 lb., 1-1/4" pipe size	1.000	Ea.	89	48	137
Valve, electric motor actuated, brass, 3 way, screwed, 3/4" pipe	1.000	Ea.	410	38.50	448.50
Valve, electric motor actuated, brass, 2 way, screwed, 1" pipe size	1.000	Ea.	665	37	702
Thermometers, stem type, 9" case, 8" stem, 3/4" NPT	4.000	Ea.	956	104	1,060
Pressure/temperature relief plug, 316 SS, 3/4"OD, 7-1/2" insertion	4.000	Ea.	306	90	396
Insulation, fiberglass pipe covering, 1" wall, 1" IPS	26.000	L.F.	27.82	139.10	166.92
Insulation, fiberglass pipe covering, 1" wall, 1-1/4" IPS	20.000	L.F.	23.20	112	135.20
Fan coil control system	1.000	Ea.	1,375	650	2,025
Re-heat coil balancing	2.000	Ea.		268	268
Multizone A/C balancing	1.000	Ea.		615	615
TOTAL			17,758.08	7,100.90	24,858.98

D3040 110	Central Station Air Handling Unit	COST EACH		
		MAT.	INST.	TOTAL
1010	AHU, Central station, cool/heat coils, constant vol, fltrs, 2,000 CFM	17,800	7,100	24,900
1020	5,000 CFM	26,500	9,150	35,650
1030	10,000 CFM	38,900	12,800	51,700
1040	15,000 CFM	72,000	17,300	89,300
1050	20,000 CFM	87,500	20,200	107,700

D3040 Distribution Systems

D3040 112	Central Station Air Handling Unit, VAV	COST EACH		
		MAT.	INST.	TOTAL
1010	AHU, Central station, cool/heat coils, VAV, fltrs, 5,000 CFM	30,900	9,500	40,400
1020	10,000 CFM	56,500	13,100	69,600
1030	15,000 CFM	80,500	17,000	97,500
1040	20,000 CFM	99,000	20,600	119,600
1050	30,000 CFM	117,000	23,700	140,700

D3040 Distribution Systems

Two-Way Valve Piping

Three-Way Valve Piping

Return Air Fan Section

Exhaust Section

Mixing Box

Filter Section

Coil Section

Roof Mounting Curb

Supply Air Fan Section

Rooftop Air Handling Unit

System Components	QUANTITY	UNIT	COST EACH		
			MAT.	INST.	TOTAL
SYSTEM D3040 114 1010					
AHU, ROOFTOP, COOL/HEAT COILS, FLTRS., CONST. VOL., 2,000 CFM					
AHU, Built-up, cool/heat coils, filter, mix box, rooftop, constant volume	1.000	Ea.	26,000	2,025	28,025
Pipe, black steel, Sch 40, threaded, W/cplgs & hangers, 10' OC, 1" dia.	26.000	L.F.	180.38	435.50	615.88
Pipe, black steel, Sch 40, threaded, W/cplg & hangers, 10' OC, 1-1/4" dia.	20.000	L.F.	176.80	373.10	549.90
Elbow, 90°, black, straight, 1" dia.	4.000	Ea.	30.40	218	248.40
Elbow, 90°, black, straight, 1-1/4" dia.	4.000	Ea.	50.20	232	282.20
Tee, black, reducing, 1" dia.	8.000	Ea.	153.60	708	861.60
Tee, black, reducing, 1-1/4" dia.	8.000	Ea.	272	728	1,000
Reducer, black, concentric, 1" dia.	4.000	Ea.	55.60	190	245.60
Reducer, black, concentric, 1-1/4" dia.	4.000	Ea.	62.20	196	258.20
Union, black with brass seat, 1" dia.	4.000	Ea.	96	236	332
Union, black with brass seat, 1-1/4" dia.	2.000	Ea.	69	122	191
Pipe nipples, 1" dia	13.000	Ea.	36.08	87.10	123.18
Pipe nipples, 1-1/4" dia	12.000	Ea.	40.80	86.10	126.90
Valves, bronze, gate, N.R.S., threaded, class 150, 1" size	2.000	Ea.	224	75	299
Valves, bronze, gate, N.R.S., threaded, class 150, 1-1/4" size	2.000	Ea.	304	95	399
Strainer, Y type, bronze body, screwed, 150 lb., 1" pipe size	1.000	Ea.	55	42.50	97.50
Strainer, Y type, bronze body, screwed, 150 lb., 1-1/4" pipe size	1.000	Ea.	89	48	137
Valve, electric motor actuated, brass, 3 way, screwed, 3/4" pipe	1.000	Ea.	410	38.50	448.50
Valve, electric motor actuated, brass, 2 way, screwed, 1" pipe size	1.000	Ea.	665	37	702
Thermometers, stem type, 9" case, 8" stem, 3/4" NPT	4.000	Ea.	956	104	1,060
Pressure/temperature relief plug, 316 SS, 3/4"OD, 7-1/2" insertion	4.000	Ea.	306	90	396
Insulation, fiberglass pipe covering, 1" wall, 1" IPS	26.000	L.F.	27.82	139.10	166.92
Insulation, fiberglass pipe covering, 1" wall, 1-1/4" IPS	20.000	L.F.	23.20	112	135.20
Insulation pipe covering .016" aluminum jacket finish, add	45.000	S.F.	45.45	263.25	308.70
Fan coil control system	1.000	Ea.	1,375	650	2,025
Re-heat coil balancing	2.000	Ea.		268	268
Rooftop unit heat/cool balancing	1.000	Ea.		480	480
TOTAL			31,703.53	8,079.15	39,782.68

D3040 114	Rooftop Air Handling Unit	COST EACH		
		MAT.	INST.	TOTAL
1010	AHU, Rooftop, cool/heat coils, constant vol., fltrs, 2,000 CFM	31,700	8,075	39,775
1020	5,000 CFM	38,200	9,375	47,575

D3040 Distribution Systems

D3040 114	Rooftop Air Handling Unit	COST EACH		
		MAT.	INST.	TOTAL
1030	10,000 CFM	74,000	13,100	87,100
1040	15,000 CFM	106,000	17,200	123,200
1050	20,000 CFM	140,500	20,700	161,200

D3040 116	Rooftop Air Handling Unit, VAV	COST EACH		
		MAT.	INST.	TOTAL
1010	AHU, Rooftop, cool/heat coils, VAV, fltrs, 5,000 CFM	41,100	9,600	50,700
1020	10,000 CFM	83,500	13,300	96,800
1030	15,000 CFM	124,000	17,300	141,300
1040	20,000 CFM	157,000	21,000	178,000
1050	30,000 CFM	183,500	24,100	207,600

D3040 Distribution Systems

Galvanized Steel Duct
Fiberglass Duct Insulation
Flexible Fiberglass Duct

Horizontal Fan Coil Air Conditioning System

System Components	QUANTITY	UNIT	COST EACH		
			MAT.	INST.	TOTAL
SYSTEM D3040 124 1010					
FAN COIL A/C SYSTEM, HORIZONTAL, W/HOUSING, CONTROLS, 2 PIPE, 1/2 TON					
Fan coil A/C, horizontal housing, filters, chilled water, 1/2 ton cooling	1.000	Ea.	1,400	162	1,562
Pipe, black steel, Sch 40, threaded, W/cplgs & hangers, 10' OC, 3/4" dia.	20.000	L.F.	92.45	250.48	342.93
Elbow, 90°, black, straight, 3/4" dia.	6.000	Ea.	26.34	303	329.34
Tee, black, straight, 3/4" dia.	2.000	Ea.	13.90	158	171.90
Union, black with brass seat, 3/4" dia.	2.000	Ea.	37	109	146
Pipe nipples, 3/4" diam	3.000	Ea.	6.45	17.48	23.93
Valves, bronze, gate, N.R.S., threaded, class 150, 3/4" size	2.000	Ea.	179	71	250
Circuit setter, balance valve, bronze body, threaded, 3/4" pipe size	1.000	Ea.	87	36	123
Insulation, fiberglass pipe covering, 1" wall, 3/4" IPS	20.000	L.F.	19.80	102	121.80
Ductwork, 12" x 8" fabricated, galvanized steel, 12 LF	55.000	Lb.	39.60	448.25	487.85
Insulation, ductwork, blanket type, fiberglass, 1" thk, 1-1/2 LB density	40.000	S.F.	199.60	156.40	356
Diffusers, aluminum, OB damper, ceiling, perf, 24"x24" panel size, 6"x6"	2.000	Ea.	332	86	418
Ductwork, flexible, fiberglass fabric, insulated, 1"thk, PE jacket, 6" dia	16.000	L.F.	51.68	75.84	127.52
Round volume control damper 6" dia.	2.000	Ea.	55	62	117
Fan coil unit balancing	1.000	Ea.		89	89
Re-heat coil balancing	1.000	Ea.		134	134
Diffuser/register, high, balancing	2.000	Ea.		246	246
TOTAL			2,539.82	2,506.45	5,046.27

D3040 118	Fan Coil A/C Unit, Two Pipe	COST EACH		
		MAT.	INST.	TOTAL
1010	Fan coil A/C system, cabinet mounted, controls, 2 pipe, 1/2 Ton	1,350	1,350	2,700
1020	1 Ton	1,650	1,500	3,150
1030	1-1/2 Ton	1,750	1,525	3,275
1040	2 Ton	2,850	1,850	4,700
1050	3 Ton	3,775	1,925	5,700

D3040 120	Fan Coil A/C Unit, Two Pipe, Electric Heat	COST EACH		
		MAT.	INST.	TOTAL
1010	Fan coil A/C system, cabinet mntd, elect. ht, controls, 2 pipe, 1/2 Ton	1,875	1,350	3,225
1020	1 Ton	2,300	1,500	3,800
1030	1-1/2 Ton	2,675	1,525	4,200
1040	2 Ton	4,125	1,875	6,000
1050	3 Ton	6,800	1,925	8,725

D3040 Distribution Systems

D3040 122	Fan Coil A/C Unit, Four Pipe	COST EACH		
		MAT.	INST.	TOTAL
1010	Fan coil A/C system, cabinet mounted, controls, 4 pipe, 1/2 Ton	2,200	2,600	4,800
1020	1 Ton	2,525	2,775	5,300
1030	1-1/2 Ton	2,675	2,800	5,475
1040	2 Ton	3,350	2,875	6,225
1050	3 Ton	4,800	3,100	7,900

D3040 124	Fan Coil A/C, Horizontal, Duct Mount, 2 Pipe	COST EACH		
		MAT.	INST.	TOTAL
1010	Fan coil A/C system, horizontal w/housing, controls, 2 pipe, 1/2 Ton	2,550	2,500	5,050
1020	1 Ton	3,425	3,650	7,075
1030	1-1/2 Ton	4,275	4,750	9,025
1040	2 Ton	5,725	5,725	11,450
1050	3 Ton	7,275	7,600	14,875
1060	3-1/2 Ton	7,550	7,850	15,400
1070	4 Ton	7,975	8,925	16,900
1080	5 Ton	9,650	11,300	20,950
1090	6 Ton	9,875	12,200	22,075
1100	7 Ton	9,850	12,600	22,450
1110	8 Ton	9,850	12,900	22,750
1120	10 Ton	10,400	13,400	23,800

D3040 126	Fan Coil A/C, Horiz., Duct Mount, 2 Pipe, Elec. Ht.	COST EACH		
		MAT.	INST.	TOTAL
1010	Fan coil A/C system, horiz. hsng, elect. ht, ctrls, 2 pipe, 1/2 Ton	2,775	2,500	5,275
1020	1 Ton	3,775	3,650	7,425
1030	1-1/2 Ton	4,600	4,750	9,350
1040	2 Ton	6,250	5,725	11,975
1050	3 Ton	9,800	7,850	17,650
1060	3-1/2 Ton	10,800	7,850	18,650
1070	4 Ton	12,200	8,950	21,150
1080	5 Ton	15,000	11,300	26,300
1090	6 Ton	15,200	12,200	27,400
1100	7 Ton	15,600	12,800	28,400
1110	8 Ton	15,800	13,400	29,200
1120	10 Ton	16,900	12,600	29,500

D3040 128	Fan Coil A/C, Horizontal, Duct Mount, 4 Pipe	COST EACH		
		MAT.	INST.	TOTAL
1010	Fan coil A/C system, horiz. w/cabinet, controls, 4 pipe, 1/2 Ton	3,550	3,750	7,300
1020	1 Ton	4,550	5,050	9,600
1030	1-1/2 Ton	5,500	6,025	11,525
1040	2 Ton	6,550	6,775	13,325
1050	3 Ton	8,400	9,025	17,425
1060	3.5 Ton	9,925	11,000	20,925
1070	4 Ton	9,900	11,200	21,100
1080	5 Ton	12,300	15,300	27,600
1090	6 Ton	13,900	18,200	32,100
1100	7 Ton	15,800	20,500	36,300
1110	8 Ton	16,400	21,600	38,000
1120	10 Ton	19,400	26,500	45,900

553

For customer support on your Mechanical Cost Data, call 877.795.7394.

D3040 Distribution Systems

| VAV Terminal | Fan Powered VAV Terminal |

System Components			COST EACH		
	QUANTITY	UNIT	MAT.	INST.	TOTAL
SYSTEM D3040 132 1010					
VAV TERMINAL, COOLING ONLY, WITH ACTUATOR/CONTROLS, 200 CFM					
Mixing box variable volume 300 to 600 CFM, cool only	1.000	Ea.	885.50	123.20	1,008.70
Ductwork, 12" x 8" fabricated, galvanized steel, 12 LF	55.000	Lb.	39.60	448.25	487.85
Insulation, ductwork, blanket type, fiberglass, 1" thk, 1-1/2 LB density	40.000	S.F.	199.60	156.40	356
Diffusers, aluminum, OB damper, ceiling, perf, 24"x24" panel size, 6"x6"	2.000	Ea.	332	86	418
Ductwork, flexible, fiberglass fabric, insulated, 1"thk, PE jacket, 6" dia	16.000	L.F.	51.68	75.84	127.52
Round volume control damper 6" dia.	2.000	Ea.	55	62	117
VAV box balancing	1.000	Ea.		82	82
Diffuser/register, high, balancing	2.000	Ea.		246	246
TOTAL			1,563.38	1,279.69	2,843.07

D3040 132	VAV Terminal, Cooling Only	COST EACH		
		MAT.	INST.	TOTAL
1010	VAV Terminal, cooling only, with actuator/controls, 200 CFM	1,575	1,275	2,850
1020	400 CFM	2,075	2,275	4,350
1030	600 CFM	2,700	3,375	6,075
1040	800 CFM	3,000	4,100	7,100
1050	1000 CFM	3,450	4,750	8,200
1060	1250 CFM	4,150	6,050	10,200
1070	1500 CFM	4,675	7,150	11,825
1080	2000 CFM	6,100	10,100	16,200

D3040 134	VAV Terminal, Hot Water Reheat	COST EACH		
		MAT.	INST.	TOTAL
1010	VAV Terminal, cooling, HW reheat, with actuator/controls, 200 CFM	2,625	2,375	5,000
1020	400 CFM	3,150	3,275	6,425
1030	600 CFM	3,750	4,475	8,225
1040	800 CFM	4,100	5,225	9,325
1050	1000 CFM	4,500	5,875	10,375
1060	1250 CFM	5,300	7,225	12,525
1070	1500 CFM	6,050	8,575	14,625
1080	2000 CFM	7,525	11,600	19,125

D3040 136	Fan Powered VAV Terminal, Cooling Only	COST EACH		
		MAT.	INST.	TOTAL
1010	VAV Terminal, cooling, fan powrd, with actuator/controls, 200 CFM	1,875	1,275	3,150
1020	400 CFM	2,475	2,125	4,600

554

For customer support on your Mechanical Cost Data, call 877.795.7394.

D3040 Distribution Systems

D3040 136	Fan Powered VAV Terminal, Cooling Only	COST EACH		
		MAT.	INST.	TOTAL
1030	600 CFM	3,075	3,350	6,425
1040	800 CFM	3,450	4,050	7,500
1050	1000 CFM	3,850	4,675	8,525
1060	1250 CFM	4,700	6,150	10,850
1070	1500 CFM	5,125	7,200	12,325
1080	2000 CFM	6,675	10,500	17,175

D3040 138	Fan Powered VAV Terminal, Hot Water Reheat	COST EACH		
		MAT.	INST.	TOTAL
1010	VAV Terminal, cool, HW reht, fan powrd, with actuator/ctrls, 200 CFM	3,225	2,375	5,600
1020	400 CFM	3,800	3,275	7,075
1030	600 CFM	4,400	4,475	8,875
1040	800 CFM	4,775	5,150	9,925
1050	1000 CFM	5,175	5,800	10,975
1060	1250 CFM	6,125	7,275	13,400
1070	1500 CFM	6,875	8,650	15,525
1080	2000 CFM	8,450	11,900	20,350

D3040 Distribution Systems

Axial Flow Centrifugal Fan
Belt Drive

Belt Drive Utility Set

Centrifugal Roof Exhaust Fan
Direct Drive

System Components	QUANTITY	UNIT	COST EACH		
			MAT.	INST.	TOTAL
SYSTEM D3040 220 1010					
FAN SYSTEM, IN-LINE CENTRIFUGAL, 500 CFM					
Fans, in-line centrifugal, supply/exhaust, 500 CFM, 10" dia conn	1.000	Ea.	1,425	520	1,945
Louver, alum., W/screen, damper, mill fin, fxd blade, cont line, stormproof	4.050	S.F.	186.30	99.23	285.53
Motor for louver damper, electric or pneumatic	1.000	Ea.	530	49	579
Ductwork, 12" x 8" fabricated, galvanized steel, 60 LF	200.000	Lb.	182	2,028	2,210
Grilles, aluminum, air supply, single deflection, adjustable, 12" x 6"	1.000	Ea.	28.80	88.50	117.30
Fire damper, curtain type, vertical, 12" x 8"	1.000	Ea.	26.50	31	57.50
Duct access door, insulated, 10" x 10"	1.000	Ea.	20.50	62	82.50
Duct access door, insulated, 16" x 12"	1.000	Ea.	32.50	76	108.50
Pressure controller/switch, air flow controller	1.000	Ea.	228	60	288
Fan control switch and mounting base	1.000	Ea.	140	45	185
Fan balancing	1.000	Ea.		410	410
Diffuser/register balancing	3.000	Ea.		246	246
TOTAL			2,799.60	3,714.73	6,514.33

D3040 220	Centrifugal In Line Fan Systems	COST EACH		
		MAT.	INST.	TOTAL
1010	Fan system, in-line centrifugal, 500 CFM	2,800	3,725	6,525
1020	1300 CFM	3,300	6,425	9,725
1030	1500 CFM	3,675	7,425	11,100
1040	2500 CFM	5,600	22,700	28,300
1050	3500 CFM	7,075	30,600	37,675
1060	5000 CFM	9,250	45,300	54,550
1070	7500 CFM	10,700	46,300	57,000
1080	10,000 CFM	12,900	54,500	67,400

D3040 230	Utility Set Fan Systems	COST EACH		
		MAT.	INST.	TOTAL
1010	Utility fan set system, belt drive, 2000 CFM	2,775	12,100	14,875
1020	3500 CFM	4,675	21,200	25,875
1030	5000 CFM	5,900	29,500	35,400
1040	7500 CFM	8,275	34,200	42,475
1050	10,000 CFM	11,100	44,900	56,000
1060	15,000 CFM	11,200	45,800	57,000
1070	20,000 CFM	13,700	52,000	65,700

D3040 Distribution Systems

D3040 240	Roof Exhaust Fan Systems	COST EACH		
		MAT.	INST.	TOTAL
1010	Roof vent. system, centrifugal, alum., galv curb, BDD, 500 CFM	1,400	2,425	3,825
1020	800 CFM	1,675	4,700	6,375
1030	1500 CFM	2,500	6,975	9,475
1040	2750 CFM	3,525	15,800	19,325
1050	3500 CFM	4,300	20,300	24,600
1060	5000 CFM	6,025	33,500	39,525
1070	8500 CFM	7,950	42,900	50,850
1080	13,800 CFM	11,300	62,500	73,800

D3040 Distribution Systems

Commercial/Industrial Dust Collection System

System Components	QUANTITY	UNIT	COST EACH		
			MAT.	INST.	TOTAL
SYSTEM D3040 250 1010					
COMMERCIAL/INDUSTRIAL VACUUM DUST COLLECTION SYSTEM, 500 CFM					
Dust collection central vac unit inc. stand, filter & shaker, 500 CFM 2 HP	1.000	Ea.	4,600	650	5,250
Galvanized tubing, 16 ga., 4" OD	80.000	L.F.	528	273.60	801.60
Galvanized tubing, 90° ell, 4" dia.	3.000	Ea.	103.50	73.50	177
Galvanized tubing, 45° ell, 4" dia.	4.000	Ea.	122	98	220
Galvanized tubing, TY slip fit; 4" dia.	4.000	Ea.	184	164	348
Flexible rubber hose, 4" dia.	32.000	L.F.	372.80	109.44	482.24
Galvanized air gate valve, 4" dia.	4.000	Ea.	648	214	862
Galvanized compression coupling, 4" dia.	36.000	Ea.	756	1,350	2,106
Pipe hangers, steel, 4" pipe size	16.000	Ea.	400	185.60	585.60
TOTAL			7,714.30	3,118.14	10,832.44

D3040 250	Commercial/Industrial Vacuum Dust Collection	COST EACH		
		MAT.	INST.	TOTAL
1010	Commercial/industrial, vacuum dust collection system, 500 CFM	7,725	3,125	10,850
1020	1000 CFM	10,700	3,550	14,250
1030	1500 CFM	13,300	3,925	17,225
1040	3000 CFM	29,100	5,500	34,600
1050	5000 CFM	38,900	8,250	47,150

D3040 Distribution Systems

Commercial Kitchen Exhaust/Make-up Air Rooftop System

System Components	QUANTITY	UNIT	COST EACH		
			MAT.	INST.	TOTAL
SYSTEM D3040 260 1010					
COMMERCIAL KITCHEN EXHAUST/MAKE-UP AIR SYSTEM, ROOFTOP, GAS, 2000 CFM					
Make-up air rooftop unit, gas, 750 MBH	1.000	Ea.	5,450.20	154.58	5,604.78
Pipe, black steel, Sch 40, thrded, nipples, W/cplngs/hngrs, 10' OC, 1" dia.	20.000	L.F.	122.10	294.80	416.90
Elbow, 90°, black steel, straight, 3/4" dia.	2.000	Ea.	8.78	101	109.78
Elbow, 90°, black steel, straight, 1" dia.	3.000	Ea.	22.80	163.50	186.30
Tee, black steel, reducing, 1" dia.	1.000	Ea.	19.20	88.50	107.70
Union, black with brass seat, 1" dia.	1.000	Ea.	24	59	83
Pipe nipples, black, 3/4" dia	2.000	Ea.	8.60	23.30	31.90
Pipe cap, malleable iron, black, 1" dia.	1.000	Ea.	6.05	23.50	29.55
Gas cock, brass, 1" size	1.000	Ea.	34	37.50	71.50
Ductwork, 20" x 14" fabricated, galv. steel, 30 LF	220.000	Lb.	149.60	1,650	1,799.60
Ductwork, 18" x 10" fabricated, galv. steel, 12 LF	72.000	Lb.	48.96	540	588.96
Ductwork, 30" x 10" fabricated, galv. steel, 30 LF	260.000	Lb.	176.80	1,950	2,126.80
Ductwork, 16" x 10" fabricated, galv. steel, 12 LF	70.000	Lb.	47.60	525	572.60
Kitchen ventilation, island style, water wash, Stainless Steel	5.000	L.F.	13,125	1,350	14,475
Control system, Make-up Air Unit/Exhauster	1.000	Ea.	3,050	2,550	5,600
Rooftop unit heat/cool balancing	1.000	Ea.		480	480
Roof fan balancing	1.000	Ea.		274	274
TOTAL			22,293.69	10,264.68	32,558.37

D3040 260	Kitchen Exhaust/Make-Up Air	COST EACH		
		MAT.	INST.	TOTAL
1010	Commercial kitchen exhaust/make-up air system, rooftop, gas, 2000 CFM	22,300	10,300	32,600
1020	3000 CFM	33,900	14,200	48,100
1030	5000 CFM	39,100	16,900	56,000
1040	8000 CFM	57,000	24,900	81,900
1050	12,000 CFM	74,000	29,400	103,400
1060	16,000 CFM	91,500	44,800	136,300

559

D3040 Distribution Systems

Plate Heat Exchanger   Shell and Tube Heat Exchanger

System Components	QUANTITY	UNIT	COST EACH		
			MAT.	INST.	TOTAL
SYSTEM D3040 620 1010					
SHELL & TUBE HEAT EXCHANGER, 40 GPM					
Heat exchanger, 4 pass, 3/4" O.D. copper tubes, by steam at 10 psi, 40 GPM	1.000	Ea.	5,525	325	5,850
Pipe, black steel, Sch 40, threaded, W/couplings & hangers, 10' OC, 2" dia.	40.000	L.F.	499.95	1,007.48	1,507.43
Elbow, 90°, straight, 2" dia.	16.000	Ea.	456	1,136	1,592
Reducer, black steel, concentric, 2" dia.	2.000	Ea.	64	122	186
Valves, bronze, gate, rising stem, threaded, class 150, 2" size	6.000	Ea.	1,500	387	1,887
Valves, bronze, globe, class 150, rising stem, threaded, 2" size	2.000	Ea.	1,030	129	1,159
Strainers, Y type, bronze body, screwed, 125 lb, 2" pipe size	2.000	Ea.	316	111	427
Valve, electric motor actuated, brass, 2 way, screwed, 1-1/2" pipe size	1.000	Ea.	730	54.50	784.50
Union, black with brass seat, 2" dia.	8.000	Ea.	400	600	1,000
Tee, black, straight, 2" dia.	9.000	Ea.	324	1,044	1,368
Tee, black, reducing run and outlet, 2" dia.	4.000	Ea.	264	464	728
Pipe nipple, black, 2" dia.	21.000	Ea.	4.95	9.98	14.93
Thermometers, stem type, 9" case, 8" stem, 3/4" NPT	2.000	Ea.	478	52	530
Gauges, pressure or vacuum, 3-1/2" diameter dial	2.000	Ea.	22.80	45	67.80
Insulation, fiberglass pipe covering, 1" wall, 2" IPS	40.000	L.F.	53.60	234	287.60
Heat exchanger control system	1.000	Ea.	2,950	2,275	5,225
Coil balancing	1.000	Ea.		133	133
TOTAL			14,618.30	8,128.96	22,747.26

D3040 610	Heat Exchanger, Plate Type	COST EACH		
		MAT.	INST.	TOTAL
1010	Plate heat exchanger, 400 GPM	57,000	16,400	73,400
1020	800 GPM	91,000	20,900	111,900
1030	1200 GPM	137,500	28,300	165,800
1040	1800 GPM	186,000	35,300	221,300

D3040 620	Heat Exchanger, Shell & Tube	COST EACH		
		MAT.	INST.	TOTAL
1010	Shell & tube heat exchanger, 40 GPM	14,600	8,125	22,725
1020	96 GPM	26,800	14,400	41,200
1030	240 GPM	48,800	19,600	68,400
1040	600 GPM	94,500	27,500	122,000

D3050 Terminal & Package Units

| Cabinet Unit Heater | Gas Unit Heater | Hydronic Unit Heater |

System Components	QUANTITY	UNIT	COST EACH		
			MAT.	INST.	TOTAL
SYSTEM D3050 120 1010					
SPACE HEATER, SUSPENDED, GAS FIRED, PROPELLER FAN, 20 MBH					
Space heater, propeller fan, 20 MBH output	1.000	Ea.	1,250	153	1,403
Pipe, black steel, Sch 40, threaded, W/coupling & hangers, 10' OC, 3/4" dia	20.000	L.F.	90.30	244.65	334.95
Elbow, 90°, black steel, straight, 1/2" dia.	3.000	Ea.	10.89	142.50	153.39
Elbow, 90°, black steel, straight, 3/4" dia.	3.000	Ea.	13.17	151.50	164.67
Tee, black steel, reducing, 3/4" dia.	1.000	Ea.	12	79	91
Union, black with brass seat, 3/4" dia.	1.000	Ea.	18.50	54.50	73
Pipe nipples, black, 1/2" dia	2.000	Ea.	7.32	22.50	29.82
Pipe nipples, black, 3/4" dia	2.000	Ea.	4.30	11.65	15.95
Pipe cap, black, 3/4" dia.	1.000	Ea.	4.98	22	26.98
Gas cock, brass, 3/4" size	1.000	Ea.	17.05	32	49.05
Thermostat, 1 set back, electric, timed	1.000	Ea.	41	90	131
Wiring, thermostat hook-up,25' of #18-3	1.000	Ea.	9.70	27.50	37.20
Vent chimney, prefab metal, U.L. listed, gas, double wall, galv. st, 4" dia	12.000	L.F.	95.40	217.20	312.60
Vent chimney, gas, double wall, galv. steel, elbow 90°, 4" dia.	2.000	Ea.	59	72	131
Vent chimney, gas, double wall, galv. steel, Tee, 4" dia.	1.000	Ea.	42.50	47.50	90
Vent chimney, gas, double wall, galv. steel, T cap, 4" dia.	1.000	Ea.	2.79	29.50	32.29
Vent chimney, gas, double wall, galv. steel, roof flashing, 4" dia.	1.000	Ea.	10.90	36	46.90
Vent chimney, gas, double wall, galv. steel, top, 4" dia.	1.000	Ea.	15.10	28	43.10
TOTAL			1,704.90	1,461	3,165.90

D3050 120	Unit Heaters, Gas	COST EACH		
		MAT.	INST.	TOTAL
1010	Space heater, suspended, gas fired, propeller fan, 20 MBH	1,700	1,450	3,150
1020	60 MBH	2,375	1,525	3,900
1030	100 MBH	2,800	1,600	4,400
1040	160 MBH	3,300	1,775	5,075
1050	200 MBH	4,025	1,950	5,975
1060	280 MBH	4,700	2,100	6,800
1070	320 MBH	6,325	2,375	8,700

D3050 130	Unit Heaters, Hydronic	COST EACH		
		MAT.	INST.	TOTAL
1010	Space heater, suspended, horiz. mount, HW, prop. fan, 20 MBH	2,125	1,575	3,700
1020	60 MBH	2,875	1,775	4,650

561

D30 HVAC

D3050 Terminal & Package Units

D3050 130	Unit Heaters, Hydronic	COST EACH		
		MAT.	INST.	TOTAL
1030	100 MBH	3,450	1,925	5,375
1040	150 MBH	3,900	2,125	6,025
1050	200 MBH	4,425	2,400	6,825
1060	300 MBH	5,225	2,800	8,025

D3050 140	Cabinet Unit Heaters, Hydronic	COST EACH		
		MAT.	INST.	TOTAL
1010	Unit heater, cabinet type, horizontal blower, hot water, 20 MBH	2,500	1,525	4,025
1020	60 MBH	3,625	1,700	5,325
1030	100 MBH	4,275	1,875	6,150
1040	120 MBH	4,375	1,925	6,300

D3050 Terminal & Package Units

Roof — **Roof Top Unit**

Return Duct

Insulated Supply Duct
Finish Ceiling

Return Grille

Supply Diff.

System Description: Rooftop single zone units are electric cooling and gas heat. Duct systems are low velocity, galvanized steel supply and return. Price variations between sizes are due to several factors. Jumps in the cost of the rooftop unit occur when the manufacturer shifts from the largest capacity unit on a small frame to the smallest capacity on the next larger frame, or changes from one compressor to two. As the unit capacity increases for larger areas the duct distribution grows in proportion. For most applications there is a tradeoff point where it is less expensive and more efficient to utilize smaller units with short simple distribution systems. Larger units also require larger initial supply and return ducts which can create a space problem. Supplemental heat may be desired in colder locations. The table below is based on one unit supplying the area listed. The 10,000 S.F. unit for bars and taverns is not listed because a nominal 110 ton unit would be required and this is above the normal single zone rooftop capacity.

System Components	QUANTITY	UNIT	COST EACH		
			MAT.	INST.	TOTAL
SYSTEM D3050 150 1280					
ROOFTOP, SINGLE ZONE, AIR CONDITIONER					
APARTMENT CORRIDORS, 500 S.F., .92 TON					
Rooftop air conditioner, 1 zone, electric cool, standard controls, curb	1.000	Ea.	1,989.50	851	2,840.50
Ductwork package for rooftop single zone units	1.000	System	179.40	1,173	1,352.40
TOTAL			2,168.90	2,024	4,192.90
COST PER S.F.			4.34	4.05	8.39

*Size would suggest multiple units.

D3050 150	Rooftop Single Zone Unit Systems	COST PER S.F.		
		MAT.	INST.	TOTAL
1260	Rooftop, single zone, air conditioner			
1280	Apartment corridors, 500 S.F., .92 ton	4.35	4.05	8.40
1320	1,000 S.F., 1.83 ton	4.32	4.02	8.34
1360	1500 S.F., 2.75 ton	2.67	3.32	5.99
1400	3,000 S.F., 5.50 ton	2.42	3.19	5.61
1440	5,000 S.F., 9.17 ton	2.27	2.89	5.16
1480	10,000 S.F., 18.33 ton	2.76	2.72	5.48
1560	Banks or libraries, 500 S.F., 2.08 ton	9.80	9.15	18.95
1600	1,000 S.F., 4.17 ton	6.05	7.55	13.60
1640	1,500 S.F., 6.25 ton	5.50	7.25	12.75
1680	3,000 S.F., 12.50 ton	5.15	6.55	11.70
1720	5,000 S.F., 20.80 ton	6.25	6.15	12.40
1760	10,000 S.F., 41.67 ton	5.40	6.10	11.50
1840	Bars and taverns, 500 S.F. 5.54 ton	13.55	12.20	25.75
1880	1,000 S.F., 11.08 ton	12.60	10.40	23
1920	1,500 S.F., 16.62 ton	12.10	9.70	21.80
1960	3,000 S.F., 33.25 ton	13.60	9.25	22.85
2000	5,000 S.F., 55.42 ton	12.60	9.25	21.85
2040	10,000 S.F., 110.83 ton*			
2080	Bowling alleys, 500 S.F., 2.83 ton	8.25	10.25	18.50
2120	1,000 S.F., 5.67 ton	7.50	9.85	17.35
2160	1,500 S.F., 8.50 ton	7	8.95	15.95
2200	3,000 S.F., 17.00 ton	6.75	8.55	15.30
2240	5,000 S.F., 28.33 ton	7.50	8.35	15.85
2280	10,000 S.F., 56.67 ton	7	8.35	15.35
2360	Department stores, 500 S.F., 1.46 ton	6.90	6.40	13.30

R236000 -20 (between rows 1320 and 1360)

D3050 Terminal & Package Units

D3050 150	Rooftop Single Zone Unit Systems	COST PER S.F.		
		MAT.	INST.	TOTAL
2400	1,000 S.F., 2.92 ton	4.25	5.30	9.55
2440	1,500 S.F., 4.37 ton	3.85	5.05	8.90
2480	3,000 S.F., 8.75 ton	3.60	4.60	8.20
2520	5,000 S.F., 14.58 ton	3.47	4.41	7.88
2560	10,000 S.F., 29.17 ton	3.87	4.29	8.16
2640	Drug stores, 500 S.F., 3.33 ton	9.70	12.05	21.75
2680	1,000 S.F., 6.67 ton	8.80	11.60	20.40
2720	1,500 S.F., 10.00 ton	8.25	10.50	18.75
2760	3,000 S.F., 20.00 ton	10.05	9.90	19.95
2800	5,000 S.F., 33.33 ton	8.85	9.80	18.65
2840	10,000 S.F., 66.67 ton	8.25	9.80	18.05
2920	Factories, 500 S.F., 1.67 ton	7.85	7.35	15.20
2960	1,000 S.F., 3.33 ton	4.86	6.05	10.91
3000	1,500 S.F., 5.00 ton	4.40	5.80	10.20
3040	3,000 S.F., 10.00 ton	4.12	5.25	9.37
3080	5,000 S.F., 16.67 ton	3.96	5.05	9.01
3120	10,000 S.F., 33.33 ton	4.42	4.90	9.32
3200	Food supermarkets, 500 S.F., 1.42 ton	6.65	6.25	12.90
3240	1,000 S.F., 2.83 ton	4.09	5.10	9.19
3280	1,500 S.F., 4.25 ton	3.74	4.93	8.67
3320	3,000 S.F., 8.50 ton	3.50	4.47	7.97
3360	5,000 S.F., 14.17 ton	3.36	4.28	7.64
3400	10,000 S.F., 28.33 ton	3.75	4.16	7.91
3480	Medical centers, 500 S.F., 1.17 ton	5.50	5.15	10.65
3520	1,000 S.F., 2.33 ton	5.50	5.15	10.65
3560	1,500 S.F., 3.50 ton	3.41	4.23	7.64
3600	3,000 S.F., 7.00 ton	3.09	4.07	7.16
3640	5,000 S.F., 11.67 ton	2.89	3.69	6.58
3680	10,000 S.F., 23.33 ton	3.51	3.45	6.96
3760	Offices, 500 S.F., 1.58 ton	7.45	6.95	14.40
3800	1,000 S.F., 3.17 ton	4.62	5.75	10.37
3840	1,500 S.F., 4.75 ton	4.18	5.50	9.68
3880	3,000 S.F., 9.50 ton	3.91	5	8.91
3920	5,000 S.F., 15.83 ton	3.77	4.79	8.56
3960	10,000 S.F., 31.67 ton	4.20	4.66	8.86
4000	Restaurants, 500 S.F., 2.50 ton	11.80	11	22.80
4040	1,000 S.F., 5.00 ton	6.60	8.70	15.30
4080	1,500 S.F., 7.50 ton	6.20	7.90	14.10
4120	3,000 S.F., 15.00 ton	5.95	7.55	13.50
4160	5,000 S.F., 25.00 ton	7.55	7.40	14.95
4200	10,000 S.F., 50.00 ton	6.20	7.35	13.55
4240	Schools and colleges, 500 S.F., 1.92 ton	9.05	8.45	17.50
4280	1,000 S.F., 3.83 ton	5.60	6.95	12.55
4320	1,500 S.F., 5.75 ton	5.05	6.65	11.70
4360	3,000 S.F., 11.50 ton	4.74	6.05	10.79
4400	5,000 S.F., 19.17 ton	5.75	5.70	11.45
4440	10,000 S.F., 38.33 ton	4.98	5.65	10.63
5000				
5100				
5200				
9000	Components of ductwork packages for above systems, per ton of cooling:			
9010	Ductwork; galvanized steel, 120 pounds			
9020	Insulation; fiberglass 2" thick, FRK faced, 52 S.F.			
9030	Diffusers; aluminum, 24" x 12", one			
9040	Registers; aluminum, return with dampers, one			

564

For customer support on your Mechanical Cost Data, call 877.795.7394.

Roof — **Roof Top Unit**

Return Ducts

Insulated Supply Ducts

Finish Ceiling

Return Grille (Typ.)

Supply Diff. (Typ.)

System Description: Rooftop units are multizone with up to 12 zones, and include electric cooling, gas heat, thermostats, filters, supply and return fans complete. Duct systems are low velocity, galvanized steel supply and return with insulated supplies.

Multizone units cost more per ton of cooling than single zone. However, they offer flexibility where load conditions are varied due to heat generating areas or exposure to radiational heating. For example, perimeter offices on the "sunny side" may require cooling at the same

time "shady side" or central offices may require heating. It is possible to accomplish similar results using duct heaters in branches of the single zone unit. However, heater location could be a problem and total system operating energy efficiency could be lower.

System Components	QUANTITY	UNIT	COST EACH		
			MAT.	INST.	TOTAL
SYSTEM D3050 155 1280					
ROOFTOP, MULTIZONE, AIR CONDITIONER					
APARTMENT CORRIDORS, 3,000 S.F., 5.50 TON					
Rooftop multizone unit, standard controls, curb	1.000	Ea.	31,460	1,991	33,451
Ductwork package for rooftop multizone units	1.000	System	1,617	13,475	15,092
TOTAL			33,077	15,466	48,543
COST PER S.F.			11.03	5.16	16.19

Note A: Small single zone unit recommended.
Note B: A combination of multizone units recommended.

D3050 155	Rooftop Multizone Unit Systems		COST PER S.F.		
			MAT.	INST.	TOTAL
1240	Rooftop, multizone, air conditioner				
1260	Apartment corridors, 1,500 S.F., 2.75 ton. See Note A.				
1280	3,000 S.F., 5.50 ton	R236000 -20	11.03	5.17	16.20
1320	10,000 S.F., 18.30 ton		7.60	4.95	12.55
1360	15,000 S.F., 27.50 ton		7.40	4.93	12.33
1400	20,000 S.F., 36.70 ton		7.70	4.95	12.65
1440	25,000 S.F., 45.80 ton		7.20	4.97	12.17
1520	Banks or libraries, 1,500 S.F., 6.25 ton		25	11.70	36.70
1560	3,000 S.F., 12.50 ton		21	11.45	32.45
1600	10,000 S.F., 41.67 ton		16.35	11.30	27.65
1640	15,000 S.F., 62.50 ton		12	11.25	23.25
1680	20,000 S.F., 83.33 ton		12	11.25	23.25
1720	25,000 S.F., 104.00 ton		10.70	11.20	21.90
1800	Bars and taverns, 1,500 S.F., 16.62 ton		54.50	16.90	71.40
1840	3,000 S.F., 33.24 ton		45	16.30	61.30
1880	10,000 S.F., 110.83 ton		27	16.20	43.20
1920	15,000 S.F., 165 ton, See Note B.				
1960	20,000 S.F., 220 ton, See Note B.				
2000	25,000 S.F., 275 ton, See Note B.				
2080	Bowling alleys, 1,500 S.F., 8.50 ton		34	15.95	49.95
2120	3,000 S.F., 17.00 ton		28.50	15.60	44.10
2160	10,000 S.F., 56.70 ton		22.50	15.40	37.90
2200	15,000 S.F., 85.00 ton		16.35	15.25	31.60
2240	20,000 S.F., 113.00 ton		14.50	15.20	29.70
2280	25,000 S.F., 140.00 ton see Note B.				
2360	Department stores, 1,500 S.F., 4.37 ton, See Note A.				

565

D3050 Terminal & Package Units

D3050 155	Rooftop Multizone Unit Systems	COST PER S.F.		
		MAT.	INST.	TOTAL
2400	3,000 S.F., 8.75 ton	17.55	8.20	25.75
2440	10,000 S.F., 29.17 ton	12.30	7.85	20.15
2480	15,000 S.F., 43.75 ton	11.45	7.90	19.35
2520	20,000 S.F., 58.33 ton	8.40	7.85	16.25
2560	25,000 S.F., 72.92 ton	8.40	7.85	16.25
2640	Drug stores, 1,500 S.F., 10.00 ton	40	18.75	58.75
2680	3,000 S.F., 20.00 ton	27.50	18.05	45.55
2720	10,000 S.F., 66.66 ton	19.20	17.95	37.15
2760	15,000 S.F., 100.00 ton	17.15	17.90	35.05
2800	20,000 S.F., 135 ton, See Note B.			
2840	25,000 S.F., 165 ton, See Note B.			
2920	Factories, 1,500 S.F., 5 ton, See Note A.			
2960	3,000 S.F., 10.00 ton	20	9.40	29.40
3000	10,000 S.F., 33.33 ton	14.05	9	23.05
3040	15,000 S.F., 50.00 ton	13.10	9.05	22.15
3080	20,000 S.F., 66.66 ton	9.60	9	18.60
3120	25,000 S.F., 83.33 ton	9.60	9	18.60
3200	Food supermarkets, 1,500 S.F., 4.25 ton, See Note A.			
3240	3,000 S.F., 8.50 ton	17.05	7.95	25
3280	10,000 S.F., 28.33 ton	11.45	7.65	19.10
3320	15,000 S.F., 42.50 ton	11.15	7.70	18.85
3360	20,000 S.F., 56.67 ton	8.15	7.65	15.80
3400	25,000 S.F., 70.83 ton	8.15	7.65	15.80
3480	Medical centers, 1,500 S.F., 3.5 ton, See Note A.			
3520	3,000 S.F., 7.00 ton	14.05	6.55	20.60
3560	10,000 S.F., 23.33 ton	9.25	6.30	15.55
3600	15,000 S.F., 35.00 ton	9.80	6.30	16.10
3640	20,000 S.F., 46.66 ton	9.15	6.35	15.50
3680	25,000 S.F., 58.33 ton	6.70	6.30	13
3760	Offices, 1,500 S.F., 4.75 ton, See Note A.			
3800	3,000 S.F., 9.50 ton	19.05	8.90	27.95
3840	10,000 S.F., 31.66 ton	13.35	8.55	21.90
3880	15,000 S.F., 47.50 ton	12.45	8.60	21.05
3920	20,000 S.F., 63.33 ton	9.10	8.55	17.65
3960	25,000 S.F., 79.16 ton	9.10	8.55	17.65
4000	Restaurants, 1,500 S.F., 7.50 ton	30	14.05	44.05
4040	3,000 S.F., 15.00 ton	25.50	13.75	39.25
4080	10,000 S.F., 50.00 ton	19.65	13.55	33.20
4120	15,000 S.F., 75.00 ton	14.40	13.45	27.85
4160	20,000 S.F., 100.00 ton	12.85	13.45	26.30
4200	25,000 S.F., 125 ton, See Note B.			
4240	Schools and colleges, 1,500 S.F., 5.75 ton	23	10.80	33.80
4280	3,000 S.F., 11.50 ton	19.40	10.55	29.95
4320	10,000 S.F., 38.33 ton	15.05	10.40	25.45
4360	15,000 S.F., 57.50 ton	11.05	10.35	21.40
4400	20,000 S.F.,76.66 ton	11.05	10.35	21.40
4440	25,000 S.F., 95.83 ton	10.35	10.35	20.70
4450				
9000	Components of ductwork packages for above systems, per ton of cooling:			
9010	Ductwork; galvanized steel, 240 pounds			
9020	Insulation; fiberglass, 2" thick, FRK faced, 104 S.F.			
9030	Diffusers; aluminum, 24" x 12", two			
9040	Registers; aluminum, return with dampers, one			

D3050 Terminal & Package Units

System Description: Self-contained, single package water cooled units include cooling tower, pump, piping allowance. Systems for 1000 S.F. and up include duct and diffusers to provide for even distribution of air. Smaller units distribute air through a supply air plenum, which is integral with the unit.

Returns are not ducted and supplies are not insulated.

Hot water or steam heating coils are included but piping to boiler and the boiler itself is not included.

Where local codes or conditions permit single pass cooling for the smaller units, deduct 10%.

System Components	QUANTITY	UNIT	COST EACH		
			MAT.	INST.	TOTAL
SYSTEM D3050 160 1300					
SELF-CONTAINED, WATER COOLED UNIT					
APARTMENT CORRIDORS, 500 S.F., .92 TON					
Self contained, water cooled, single package air conditioner unit	1.000	Ea.	1,377.60	680.40	2,058
Ductwork package for water or air cooled packaged units	1.000	System	69.92	897	966.92
Cooling tower, draw thru single flow, belt drive	1.000	Ea.	218.04	20.70	238.74
Cooling tower pumps & piping	1.000	System	108.56	48.76	157.32
TOTAL			1,774.12	1,646.86	3,420.98
COST PER S.F.			3.55	3.29	6.84

D3050 160	Self-contained, Water Cooled Unit Systems	COST PER S.F.		
		MAT.	INST.	TOTAL
1280	Self-contained, water cooled unit	3.56	3.29	6.85
1300	Apartment corridors, 500 S.F., .92 ton	3.55	3.30	6.85
1320	1,000 S.F., 1.83 ton	3.54	3.28	6.82
1360	3,000 S.F., 5.50 ton	2.94	2.75	5.69
1400	5,000 S.F., 9.17 ton	2.89	2.55	5.44
1440	10,000 S.F., 18.33 ton	3.96	2.28	6.24
1520	Banks or libraries, 500 S.F., 2.08 ton	7.75	3.40	11.15
1560	1,000 S.F., 4.17 ton	6.70	6.20	12.90
1600	3,000 S.F., 12.50 ton	6.55	5.75	12.30
1640	5,000 S.F., 20.80 ton	9	5.15	14.15
1680	10,000 S.F., 41.66 ton	7.85	5.15	13
1760	Bars and taverns, 500 S.F., 5.54 ton	16.85	5.75	22.60
1800	1,000 S.F., 11.08 ton	16.95	9.95	26.90
1840	3,000 S.F., 33.25 ton	21	7.80	28.80
1880	5,000 S.F., 55.42 ton	19.90	8.15	28.05
1920	10,000 S.F., 110.00 ton	19.35	8.05	27.40
2000	Bowling alleys, 500 S.F., 2.83 ton	10.50	4.63	15.13
2040	1,000 S.F., 5.66 ton	9.05	8.45	17.50
2080	3,000 S.F., 17.00 ton	12.25	7.05	19.30
2120	5,000 S.F., 28.33 ton	11.10	6.75	17.85
2160	10,000 S.F., 56.66 ton	10.40	6.90	17.30
2200	Department stores, 500 S.F., 1.46 ton	5.40	2.39	7.79
2240	1,000 S.F., 2.92 ton	4.67	4.36	9.03
2280	3,000 S.F., 8.75 ton	4.59	4.04	8.63
2320	5,000 S.F., 14.58 ton	6.30	3.61	9.91
2360	10,000 S.F., 29.17 ton	5.70	3.48	9.18

R236000 -20

D3050 Terminal & Package Units

D3050 160	Self-contained, Water Cooled Unit Systems	COST PER S.F.		
		MAT.	INST.	TOTAL
2440	Drug-stores, 500 S.F., 3.33 ton	12.40	5.45	17.85
2480	1,000 S.F., 6.66 ton	10.70	9.95	20.65
2520	3,000 S.F., 20.00 ton	14.40	8.25	22.65
2560	5,000 S.F., 33.33 ton	15.10	8.15	23.25
2600	10,000 S.F., 66.66 ton	12.25	8.15	20.40
2680	Factories, 500 S.F., 1.66 ton	6.20	2.72	8.92
2720	1,000 S.F. 3.37 ton	5.40	5.05	10.45
2760	3,000 S.F., 10.00 ton	5.25	4.63	9.88
2800	5,000 S.F., 16.66 ton	7.20	4.13	11.33
2840	10,000 S.F., 33.33 ton	6.55	3.98	10.53
2920	Food supermarkets, 500 S.F., 1.42 ton	5.25	2.31	7.56
2960	1,000 S.F., 2.83 ton	5.50	5.05	10.55
3000	3,000 S.F., 8.50 ton	4.54	4.23	8.77
3040	5,000 S.F., 14.17 ton	4.46	3.92	8.38
3080	10,000 S.F., 28.33 ton	5.55	3.38	8.93
3160	Medical centers, 500 S.F., 1.17 ton	4.33	1.90	6.23
3200	1,000 S.F., 2.33 ton	4.52	4.17	8.69
3240	3,000 S.F., 7.00 ton	3.74	3.49	7.23
3280	5,000 S.F., 11.66 ton	3.68	3.22	6.90
3320	10,000 S.F., 23.33 ton	5.05	2.88	7.93
3400	Offices, 500 S.F., 1.58 ton	5.90	2.59	8.49
3440	1,000 S.F., 3.17 ton	6.10	5.70	11.80
3480	3,000 S.F., 9.50 ton	4.98	4.39	9.37
3520	5,000 S.F., 15.83 ton	6.80	3.93	10.73
3560	10,000 S.F., 31.67 ton	6.20	3.79	9.99
3640	Restaurants, 500 S.F., 2.50 ton	9.30	4.09	13.39
3680	1,000 S.F., 5.00 ton	8	7.50	15.50
3720	3,000 S.F., 15.00 ton	7.90	6.95	14.85
3760	5,000 S.F., 25.00 ton	10.80	6.20	17
3800	10,000 S.F., 50.00 ton	6.75	5.70	12.45
3880	Schools and colleges, 500 S.F., 1.92 ton	7.10	3.13	10.23
3920	1,000 S.F., 3.83 ton	6.15	5.75	11.90
3960	3,000 S.F., 11.50 ton	6.05	5.30	11.35
4000	5,000 S.F., 19.17 ton	8.25	4.75	13
4040	10,000 S.F., 38.33 ton	7.25	4.75	12
4050				
9000	Components of ductwork packages for above systems, per ton of cooling:			
9010	Ductwork; galvanized steel, 108 pounds			
9020	Diffusers, aluminum, 24" x 12", two			

D3050 Terminal & Package Units

System Description: Self-contained air cooled units with remote air cooled condenser and interconnecting tubing. Systems for 1000 S.F. and up include duct and diffusers. Smaller units distribute air directly.

Returns are not ducted and supplies are not insulated.

Potential savings may be realized by using a single zone rooftop system or through-the-wall unit, especially in the smaller capacities, if the application permits.

Hot water or steam heating coils are included but piping to boiler and the boiler itself is not included.

Condenserless models are available for 15% less where remote refrigerant source is available.

System Components	QUANTITY	UNIT	COST EACH MAT.	COST EACH INST.	COST EACH TOTAL
SYSTEM D3050 165 1320					
SELF-CONTAINED, AIR COOLED UNIT					
APARTMENT CORRIDORS, 500 S.F., .92 TON					
Air cooled, package unit	1.000	Ea.	1,394	442	1,836
Ductwork package for water or air cooled packaged units	1.000	System	69.92	897	966.92
Refrigerant piping	1.000	System	381.80	598	979.80
Air cooled condenser, direct drive, propeller fan	1.000	Ea.	829.25	167.40	996.65
TOTAL			2,674.97	2,104.40	4,779.37
COST PER S.F.			5.35	4.21	9.56

D3050 165	Self-contained, Air Cooled Unit Systems		COST PER S.F. MAT.	COST PER S.F. INST.	COST PER S.F. TOTAL
1300	Self-contained, air cooled unit				
1320	Apartment corridors, 500 S.F., .92 ton		5.35	4.20	9.55
1360	1,000 S.F., 1.83 ton		5.30	4.17	9.47
1400	3,000 S.F., 5.50 ton	R236000 -20	5.05	3.90	8.95
1440	5,000 S.F., 9.17 ton		3.75	3.71	7.46
1480	10,000 S.F., 18.33 ton		3.38	3.40	6.78
1560	Banks or libraries, 500 S.F., 2.08 ton		11.60	5.40	17
1600	1,000 S.F., 4.17 ton		11.45	8.85	20.30
1640	3,000 S.F., 12.50 ton		8.55	8.45	17
1680	5,000 S.F., 20.80 ton		7.80	7.75	15.55
1720	10,000 S.F., 41.66 ton		8.55	7.55	16.10
1800	Bars and taverns, 500 S.F., 5.54 ton		27.50	11.80	39.30
1840	1,000 S.F., 11.08 ton		22.50	17	39.50
1880	3,000 S.F., 33.25 ton		23	14.65	37.65
1920	5,000 S.F., 55.42 ton		23	14.70	37.70
1960	10,000 S.F., 110.00 ton		23.50	14.60	38.10
2040	Bowling alleys, 500 S.F., 2.83 ton		15.90	7.40	23.30
2080	1,000 S.F., 5.66 ton		15.60	12.05	27.65
2120	3,000 S.F., 17.00 ton		10.50	10.50	21
2160	5,000 S.F., 28.33 ton		11.90	10.25	22.15
2200	10,000 S.F., 56.66 ton		11.90	10.25	22.15
2240	Department stores, 500 S.F., 1.46 ton		8.20	3.81	12.01
2280	1,000 S.F., 2.92 ton		8.05	6.20	14.25
2320	3,000 S.F., 8.75 ton		5.95	5.90	11.85

569

D30 HVAC

D3050 Terminal & Package Units

D3050 165	Self-contained, Air Cooled Unit Systems	COST PER S.F.		
		MAT.	INST.	TOTAL
2360	5,000 S.F., 14.58 ton	5.95	5.90	11.85
2400	10,000 S.F., 29.17 ton	6.10	5.30	11.40
2480	Drug stores, 500 S.F., 3.33 ton	18.70	8.70	27.40
2520	1,000 S.F., 6.66 ton	18.35	14.15	32.50
2560	3,000 S.F., 20.00 ton	12.50	12.40	24.90
2600	5,000 S.F., 33.33 ton	13.95	12.10	26.05
2640	10,000 S.F., 66.66 ton	14.15	12.10	26.25
2720	Factories, 500 S.F., 1.66 ton	9.50	4.39	13.89
2760	1,000 S.F., 3.33 ton	9.25	7.10	16.35
2800	3,000 S.F., 10.00 ton	6.80	6.75	13.55
2840	5,000 S.F., 16.66 ton	6.15	6.20	12.35
2880	10,000 S.F., 33.33 ton	6.95	6.05	13
2960	Food supermarkets, 500 S.F., 1.42 ton	7.95	3.70	11.65
3000	1,000 S.F., 2.83 ton	8.15	6.45	14.60
3040	3,000 S.F., 8.50 ton	7.80	6	13.80
3080	5,000 S.F., 14.17 ton	5.80	5.70	11.50
3120	10,000 S.F., 28.33 ton	5.95	5.15	11.10
3200	Medical centers, 500 S.F., 1.17 ton	6.60	3.06	9.66
3240	1,000 S.F., 2.33 ton	6.75	5.35	12.10
3280	3,000 S.F., 7.00 ton	6.45	4.96	11.41
3320	5,000 S.F., 16.66 ton	4.80	4.72	9.52
3360	10,000 S.F., 23.33 ton	4.32	4.32	8.64
3440	Offices, 500 S.F., 1.58 ton	8.90	4.14	13.04
3480	1,000 S.F., 3.16 ton	9.20	7.25	16.45
3520	3,000 S.F., 9.50 ton	6.50	6.40	12.90
3560	5,000 S.F., 15.83 ton	5.95	5.90	11.85
3600	10,000 S.F., 31.66 ton	6.65	5.75	12.40
3680	Restaurants, 500 S.F., 2.50 ton	14	6.50	20.50
3720	1,000 S.F., 5.00 ton	13.80	10.60	24.40
3760	3,000 S.F., 15.00 ton	10.25	10.10	20.35
3800	5,000 S.F., 25.00 ton	10.45	9.05	19.50
3840	10,000 S.F., 50.00 ton	10.85	9.05	19.90
3920	Schools and colleges, 500 S.F., 1.92 ton	10.75	5	15.75
3960	1,000 S.F., 3.83 ton	10.60	8.15	18.75
4000	3,000 S.F., 11.50 ton	7.85	7.75	15.60
4040	5,000 S.F., 19.17 ton	7.10	7.10	14.20
4080	10,000 S.F., 38.33 ton	7.90	6.95	14.85
4090				
9000	Components of ductwork packages for above systems, per ton of cooling:			
9010	Ductwork; galvanized steel, 102 pounds			
9020	Diffusers; aluminum, 24" x 12", two			

570

General: Split systems offer several important advantages which should be evaluated when a selection is to be made. They provide a greater degree of flexibility in component selection which permits an accurate match-up of the proper equipment size and type with the particular needs of the building. This allows for maximum use of modern energy saving concepts in heating and cooling. Outdoor installation of the air cooled condensing unit allows space savings in the building and also isolates the equipment operating sounds from building occupants.

Design Assumptions: The systems below are comprised of a direct expansion air handling unit and air cooled condensing unit with interconnecting copper tubing. Ducts and diffusers are also included for distribution of air. Systems are priced for cooling only. Heat can be added as desired either by putting hot water/steam coils into the air unit or into the duct supplying the particular area of need. Gas fired duct furnaces are also available. Refrigerant liquid line is insulated.

System Components	QUANTITY	UNIT	COST EACH		
			MAT.	INST.	TOTAL
SYSTEM D3050 170 1280					
SPLIT SYSTEM, AIR COOLED CONDENSING UNIT					
APARTMENT CORRIDORS, 1,000 S.F., 1.80 TON					
Fan coil AC unit, cabinet mntd & filters direct expansion air cool	1.000	Ea.	530.70	159.21	689.91
Ductwork package, for split system, remote condensing unit	1.000	System	65.42	855.53	920.95
Refrigeration piping	1.000	System	353.19	988.20	1,341.39
Condensing unit, air cooled, incls compressor & standard controls	1.000	Ea.	1,677.50	634.40	2,311.90
TOTAL			2,626.81	2,637.34	5,264.15
COST PER S.F.			2.63	2.64	5.27

*Cooling requirements would lead to choosing a water cooled unit.

D3050 170	Split Systems With Air Cooled Condensing Units		COST PER S.F.		
			MAT.	INST.	TOTAL
1260	Split system, air cooled condensing unit				
1280	Apartment corridors, 1,000 S.F., 1.83 ton		2.63	2.63	5.26
1320	2,000 S.F., 3.66 ton	R236000 -20	1.95	2.67	4.62
1360	5,000 S.F., 9.17 ton		2.05	3.32	5.37
1400	10,000 S.F., 18.33 ton		2.36	3.62	5.98
1440	20,000 S.F., 36.66 ton		2.70	3.69	6.39
1520	Banks and libraries, 1,000 S.F., 4.17 ton		4.43	6.05	10.48
1560	2,000 S.F., 8.33 ton		4.67	7.55	12.22
1600	5,000 S.F., 20.80 ton		5.35	8.25	13.60
1640	10,000 S.F., 41.66 ton		6.10	8.40	14.50
1680	20,000 S.F., 83.32 ton		6.75	8.75	15.50
1760	Bars and taverns, 1,000 S.F., 11.08 ton		10.95	11.90	22.85
1800	2,000 S.F., 22.16 ton		14.90	14.25	29.15
1840	5,000 S.F., 55.42 ton		13.90	13.35	27.25
1880	10,000 S.F., 110.84 ton		14.85	13.95	28.80
1920	20,000 S.F., 220 ton*				
2000	Bowling alleys, 1,000 S.F., 5.66 ton		6.10	11.30	17.40
2040	2,000 S.F., 11.33 ton		6.35	10.25	16.60
2080	5,000 S.F., 28.33 ton		7.30	11.20	18.50
2120	10,000 S.F., 56.66 ton		8.35	11.40	19.75
2160	20,000 S.F., 113.32 ton		9.70	12.40	22.10
2320	Department stores, 1,000 S.F., 2.92 ton		3.11	4.16	7.27
2360	2,000 S.F., 5.83 ton		3.14	5.85	8.99
2400	5,000 S.F., 14.58 ton		3.27	5.30	8.57

D3050 Terminal & Package Units

D3050 170	Split Systems With Air Cooled Condensing Units	COST PER S.F.		
		MAT.	INST.	TOTAL
2440	10,000 S.F., 29.17 ton	3.76	5.80	9.56
2480	20,000 S.F., 58.33 ton	4.29	5.90	10.19
2560	Drug stores, 1,000 S.F., 6.66 ton	7.15	13.30	20.45
2600	2,000 S.F., 13.32 ton	7.50	12.10	19.60
2640	5,000 S.F., 33.33 ton	9.80	13.45	23.25
2680	10,000 S.F., 66.66 ton	9.85	14	23.85
2720	20,000 S.F., 133.32 ton*			
2800	Factories, 1,000 S.F., 3.33 ton	3.56	4.75	8.31
2840	2,000 S.F., 6.66 ton	3.58	6.65	10.23
2880	5,000 S.F., 16.66 ton	4.30	6.60	10.90
2920	10,000 S.F., 33.33 ton	4.90	6.70	11.60
2960	20,000 S.F., 66.66 ton	4.90	7	11.90
3040	Food supermarkets, 1,000 S.F., 2.83 ton	3.03	4.04	7.07
3080	2,000 S.F., 5.66 ton	3.04	5.65	8.69
3120	5,000 S.F., 14.66 ton	3.18	5.15	8.33
3160	10,000 S.F., 28.33 ton	3.64	5.60	9.24
3200	20,000 S.F., 56.66 ton	4.16	5.70	9.86
3280	Medical centers, 1,000 S.F., 2.33 ton	2.68	3.26	5.94
3320	2,000 S.F., 4.66 ton	2.51	4.65	7.16
3360	5,000 S.F., 11.66 ton	2.62	4.23	6.85
3400	10,000 S.F., 23.33 ton	3	4.63	7.63
3440	20,000 S.F., 46.66 ton	3.43	4.71	8.14
3520	Offices, 1,000 S.F., 3.17 ton	3.38	4.53	7.91
3560	2,000 S.F., 6.33 ton	3.41	6.35	9.76
3600	5,000 S.F., 15.83 ton	3.55	5.75	9.30
3640	10,000 S.F., 31.66 ton	4.08	6.30	10.38
3680	20,000 S.F., 63.32 ton	4.66	6.65	11.31
3760	Restaurants, 1,000 S.F., 5.00 ton	5.40	10	15.40
3800	2,000 S.F., 10.00 ton	5.60	9.05	14.65
3840	5,000 S.F., 25.00 ton	6.45	9.90	16.35
3880	10,000 S.F., 50.00 ton	7.35	10.10	17.45
3920	20,000 S.F., 100.00 ton	8.55	10.95	19.50
4000	Schools and colleges, 1,000 S.F., 3.83 ton	4.07	5.60	9.67
4040	2,000 S.F., 7.66 ton	4.30	6.95	11.25
4080	5,000 S.F., 19.17 ton	4.93	7.60	12.53
4120	10,000 S.F., 38.33 ton	5.65	7.75	13.40
4160	20,000 S.F., 76.66 ton	5.65	8.05	13.70
5000				
5100				
5200				
9000	Components of ductwork packages for above systems, per ton of cooling:			
9010	Ductwork; galvanized steel, 102 pounds			
9020	Diffusers; aluminum, 24" x 12", two			

D3050 Terminal & Package Units

Rooftop Air Conditioner Unit

System Components	QUANTITY	UNIT	COST EACH		
			MAT.	INST.	TOTAL
SYSTEM D3050 175 1010					
A/C, ROOFTOP, DX COOL, GAS HEAT, CURB, ECONOMIZER, FILTERS., 5 TON					
Roof top A/C, curb, economizer, sgl zone, elec cool, gas ht, 5 ton, 112 MBH	1.000	Ea.	5,625	2,325	7,950
Pipe, black steel, Sch 40, threaded, W/cplgs & hangers, 10' OC, 1" dia	20.000	L.F.	116.55	281.40	397.95
Elbow, 90°, black, straight, 3/4" dia.	3.000	Ea.	13.17	151.50	164.67
Elbow, 90°, black, straight, 1" dia.	3.000	Ea.	22.80	163.50	186.30
Tee, black, reducing, 1" dia.	1.000	Ea.	19.20	88.50	107.70
Union, black with brass seat, 1" dia.	1.000	Ea.	24	59	83
Pipe nipple, black, 3/4" dia.	2.000	Ea.	8.60	23.30	31.90
Pipe nipple, black, 1" dia	2.000	Ea.	5.55	13.40	18.95
Cap, black, 1" dia.	1.000	Ea.	6.05	23.50	29.55
Gas cock, brass, 1" size	1.000	Ea.	34	37.50	71.50
Control system, pneumatic, Rooftop A/C unit	1.000	Ea.	3,050	2,550	5,600
Rooftop unit heat/cool balancing	1.000	Ea.		480	480
TOTAL			8,924.92	6,196.60	15,121.52

D3050 175	Rooftop Air Conditioner, Const. Volume	COST EACH		
		MAT.	INST.	TOTAL
1010	A/C, Rooftop, DX cool, gas heat, curb, economizer, fltrs, 5 Ton	8,925	6,200	15,125
1020	7-1/2 Ton	10,800	6,450	17,250
1030	12-1/2 Ton	16,500	7,075	23,575
1040	18 Ton	21,200	7,750	28,950
1050	25 Ton	33,200	8,825	42,025
1060	40 Ton	47,700	11,800	59,500

D3050 180	Rooftop Air Conditioner, Variable Air Volume	COST EACH		
		MAT.	INST.	TOTAL
1010	A/C, Rooftop, DX cool, gas heat, curb, ecmizr, fltrs, VAV, 12-1/2 Ton	19,400	7,550	26,950
1020	18 Ton	24,700	8,325	33,025
1030	25 Ton	37,400	9,550	46,950
1040	40 Ton	55,500	13,000	68,500
1050	60 Ton	79,500	17,600	97,100
1060	80 Ton	115,500	22,100	137,600

D3050 Terminal & Package Units

Computer rooms impose special requirements on air conditioning systems. A prime requirement is reliability, due to the potential monetary loss that could be incurred by a system failure. A second basic requirement is the tolerance of control with which temperature and humidity are regulated, and dust eliminated. As the air conditioning system reliability is so vital, the additional cost of reserve capacity and redundant components is often justified.

System Descriptions: Computer areas may be environmentally controlled by one of three methods as follows:

1. Self-contained Units
These are units built to higher standards of performance and reliability.

They usuallycontain alarms and controls to indicate component operation failure, filter change, etc. It should be remembered that these units in the room will occupy space that is relatively expensive to build and that all alterations and service of the equipment will also have to be accomplished within the computer area.

2. Decentralized Air Handling Units In operation these are similar to the self-contained units except that their cooling capability comes from remotely located refrigeration equipment as refrigerant or chilled water. As no compressors or refrigerating equipment are required in the air units, they are smaller and require less service than

self-contained units. An added plus for this type of system occurs if some of the computer components themselves also require chilled water for cooling.

3. Central System Supply Cooling is obtained from a central source which, since it is not located within the computer room, may have excess capacity and permit greater flexibility without interfering with the computer components. System performance criteria must still be met.

Note: The costs shown below do not include an allowance for ductwork or piping.

D3050 185	Computer Room Cooling Units	COST EACH		
		MAT.	INST.	TOTAL
0560	Computer room unit, air cooled, includes remote condenser			
0580	3 ton	20,700	2,600	23,300
0600	5 ton	22,100	2,875	24,975
0620	8 ton	41,200	4,800	46,000
0640	10 ton	43,100	5,200	48,300
0660	15 ton	47,400	5,900	53,300
0680	20 ton	57,500	8,425	65,925
0700	23 ton	70,500	9,625	80,125
0800	Chilled water, for connection to existing chiller system			
0820	5 ton	15,800	1,750	17,550
0840	8 ton	16,100	2,600	18,700
0860	10 ton	16,200	2,650	18,850
0880	15 ton	16,900	2,700	19,600
0900	20 ton	18,000	2,825	20,825
0920	23 ton	19,100	3,200	22,300
1000	Glycol system, complete except for interconnecting tubing			
1020	3 ton	25,900	3,250	29,150
1040	5 ton	28,400	3,425	31,825
1060	8 ton	45,500	5,650	51,150
1080	10 ton	48,400	6,175	54,575
1100	15 ton	60,000	7,775	67,775
1120	20 ton	66,000	8,425	74,425
1140	23 ton	69,500	8,775	78,275
1240	Water cooled, not including condenser water supply or cooling tower			
1260	3 ton	21,000	2,100	23,100
1280	5 ton	22,900	2,400	25,300
1300	8 ton	37,500	3,925	41,425
1320	15 ton	45,400	4,800	50,200
1340	20 ton	48,800	5,325	54,125
1360	23 ton	51,000	5,775	56,775

D3050 201	Packaged A/C, Elec. Ht., Const. Volume	COST EACH		
		MAT.	INST.	TOTAL
1010	A/C Packaged, DX, air cooled, electric heat, constant vol., 5 Ton	9,650	4,450	14,100
1020	10 Ton	14,800	6,150	20,950
1030	20 Ton	24,200	8,050	32,250
1040	30 Ton	43,700	8,925	52,625

D3050 Terminal & Package Units

D3050 201	Packaged A/C, Elec. Ht., Const. Volume	COST EACH		
		MAT.	INST.	TOTAL
1050	40 Ton	57,000	10,300	67,300
1060	50 Ton	75,000	11,500	86,500

D3050 202	Packaged Air Conditioner, Elect. Heat, VAV	COST EACH		
		MAT.	INST.	TOTAL
1010	A/C Packaged, DX, air cooled, electric heat, VAV, 10 Ton	16,800	7,525	24,325
1020	20 Ton	27,400	9,900	37,300
1030	30 Ton	49,200	10,800	60,000
1040	40 Ton	63,500	12,400	75,900
1050	50 Ton	83,500	13,800	97,300
1060	60 Ton	95,000	16,200	111,200

D3050 203	Packaged A/C, Hot Wtr. Heat, Const. Volume	COST EACH		
		MAT.	INST.	TOTAL
1010	A/C Packaged, DX, air cooled, H/W heat, constant vol., 5 Ton	8,525	6,550	15,075
1020	10 Ton	11,900	8,600	20,500
1030	20 Ton	20,200	11,500	31,700
1040	30 Ton	36,900	14,300	51,200
1050	40 Ton	50,500	16,100	66,600
1060	50 Ton	65,000	17,500	82,500

D3050 204	Packaged A/C, Hot Wtr. Heat, VAV	COST EACH		
		MAT.	INST.	TOTAL
1010	A/C Packaged, DX, air cooled, H/W heat, VAV, 10 Ton	13,800	9,825	23,625
1020	20 Ton	23,800	12,900	36,700
1030	30 Ton	42,300	16,000	58,300
1040	40 Ton	57,500	17,300	74,800
1050	50 Ton	73,000	20,000	93,000
1060	60 Ton	87,500	22,500	110,000

D3050 Terminal & Package Units

Condenser Supply Water Piping

Self—Contained Air Conditioner Unit, Water Cooled

Condenser Return Water Piping

System Components	QUANTITY	UNIT	COST EACH		
			MAT.	INST.	TOTAL
SYSTEM D3050 210 1010					
A/C, SELF CONTAINED, SINGLE PKG., WATER COOLED, ELECT. HEAT, 5 TON					
Self-contained, water cooled, elect. heat, not inc tower, 5 ton, const vol	1.000	Ea.	9,175	1,675	10,850
Pipe, black steel, Sch 40, threaded, W/cplg & hangers, 10' OC, 1-1/4" dia.	20.000	L.F.	146.20	308.53	454.73
Elbow, 90°, black, straight, 1-1/4" dia.	6.000	Ea.	75.30	348	423.30
Tee, black, straight, 1-1/4" dia.	2.000	Ea.	38.70	182	220.70
Tee, black, reducing, 1-1/4" dia.	2.000	Ea.	68	182	250
Thermometers, stem type, 9" case, 8" stem, 3/4" NPT	2.000	Ea.	478	52	530
Union, black with brass seat, 1-1/4" dia.	2.000	Ea.	69	122	191
Pipe nipple, black, 1-1/4" dia	3.000	Ea.	10.20	21.53	31.73
Valves, bronze, gate, N.R.S., threaded, class 150, 1-1/4" size	2.000	Ea.	304	95	399
Circuit setter, bal valve, bronze body, threaded, 1-1/4" pipe size	1.000	Ea.	165	48	213
Insulation, fiberglass pipe covering, 1" wall, 1-1/4" IPS	20.000	L.F.	23.20	112	135.20
Control system, pneumatic, A/C Unit with heat	1.000	Ea.	3,050	2,550	5,600
Re-heat coil balancing	1.000	Ea.		134	134
Rooftop unit heat/cool balancing	1.000	Ea.		480	480
TOTAL			13,602.60	6,310.06	19,912.66

D3050 210	AC Unit, Package, Elec. Ht., Water Cooled	COST EACH		
		MAT.	INST.	TOTAL
1010	A/C, Self contained, single pkg., water cooled, elect. heat, 5 Ton	13,600	6,300	19,900
1020	10 Ton	21,300	7,775	29,075
1030	20 Ton	28,900	9,850	38,750
1040	30 Ton	47,500	10,800	58,300
1050	40 Ton	60,500	11,900	72,400
1060	50 Ton	79,000	14,200	93,200

D3050 215	AC Unit, Package, Elec. Ht., Water Cooled, VAV	COST EACH		
		MAT.	INST.	TOTAL
1010	A/C, Self contained, single pkg., water cooled, elect. ht, VAV, 10 Ton	23,800	7,775	31,575
1020	20 Ton	33,100	9,850	42,950
1030	30 Ton	54,000	10,800	64,800
1040	40 Ton	68,000	11,900	79,900
1050	50 Ton	88,500	14,200	102,700
1060	60 Ton	101,500	17,200	118,700

D3050 Terminal & Package Units

D3050 220	AC Unit, Package, Hot Water Coil, Water Cooled	COST EACH		
		MAT.	INST.	TOTAL
1010	A/C, Self contn'd, single pkg., water cool, H/W ht, const. vol, 5 Ton	12,200	8,775	20,975
1020	10 Ton	18,100	10,400	28,500
1030	20 Ton	43,300	13,000	56,300
1040	30 Ton	56,000	16,500	72,500
1050	40 Ton	69,000	20,000	89,000
1060	50 Ton	84,500	24,700	109,200

D3050 225	AC Unit, Package, HW Coil, Water Cooled, VAV	COST EACH		
		MAT.	INST.	TOTAL
1010	A/C, Self contn'd, single pkg., water cool, H/W ht, VAV, 10 Ton	20,600	9,825	30,425
1020	20 Ton	29,400	12,600	42,000
1030	30 Ton	47,300	15,800	63,100
1040	40 Ton	62,500	17,200	79,700
1050	50 Ton	79,000	20,300	99,300
1060	60 Ton	93,500	22,000	115,500

577

For customer support on your Mechanical Cost Data, call 877.795.7394.

D3050 Terminal & Package Units

Note: Water source units do not include an allowance for any type of water source or piping to it.

Heat Pump, Air/Air, Rooftop w/ Economizer

System Components	QUANTITY	UNIT	COST EACH		
			MAT.	INST.	TOTAL
SYSTEM D3050 245 1010					
HEAT PUMP, ROOFTOP, AIR/AIR, CURB., ECONOMIZER., SUP. ELECT. HEAT, 10 TON					
Heat pump, air to air, rooftop, curb, economizer, elect ht, 10 ton cooling	1.000	Ea.	66,000	5,050	71,050
Ductwork, 32" x 12" fabricated, galvanized steel, 16 LF	176.000	Lb.	126.72	1,434.40	1,561.12
Insulation, ductwork, blanket type, fiberglass, 1" thk, 1-1/2 LB density	117.000	S.F.	583.83	457.47	1,041.30
Rooftop unit heat/cool balancing	1.000	Ea.		480	480
TOTAL			66,710.55	7,421.87	74,132.42

D3050 230	Heat Pump, Water Source, Central Station	COST EACH		
		MAT.	INST.	TOTAL
1010	Heat pump, central station, water source, constant vol., 5 Ton	5,575	3,525	9,100
1020	10 Ton	12,100	4,925	17,025
1030	20 Ton	19,100	9,175	28,275
1040	30 Ton	27,700	13,400	41,100
1050	40 Ton	38,000	14,600	52,600
1060	50 Ton	43,100	19,700	62,800

D3050 235	Heat Pump, Water Source, Console	COST EACH		
		MAT.	INST.	TOTAL
1020	Heat pump, console, water source, 1 Ton	2,900	1,925	4,825
1030	1-1/2 Ton	3,225	2,000	5,225
1040	2 Ton	4,200	2,375	6,575
1050	3 Ton	4,525	2,525	7,050
1060	3-1/2 Ton	4,800	2,600	7,400

D3050 240	Heat Pump, Horizontal, Duct Mounted	COST EACH		
		MAT.	INST.	TOTAL
1020	Heat pump, horizontal, ducted, water source, 1 Ton	3,325	4,100	7,425
1030	1-1/2 Ton	3,850	5,125	8,975
1040	2 Ton	4,875	6,150	11,025
1050	3 Ton	5,525	8,050	13,575
1060	3-1/2 Ton	5,825	8,300	14,125

D3050 Terminal & Package Units

D3050 245	Heat Pump, Rooftop, Air/Air	COST EACH		
		MAT.	INST.	TOTAL
1010	Heat pump, rooftop, air/air, curb, economizer, sup. elect heat, 10 Ton	66,500	7,425	73,925
1020	20 Ton	95,500	12,700	108,200
1030	30 Ton	128,500	17,700	146,200
1040	40 Ton	143,000	20,300	163,300
1050	50 Ton	167,000	22,800	189,800
1060	60 Ton	176,000	25,900	201,900

579

For customer support on your Mechanical Cost Data, call 877.795.7394.

D3050   Terminal & Package Units

Thru—Wall Heat Pump   Thru—Wall Air Conditioner

System Components	QUANTITY	UNIT	COST EACH		
			MAT.	INST.	TOTAL
SYSTEM D3050 255 1010 A/C UNIT, THRU-THE-WALL, SUPPLEMENTAL ELECT. HEAT, CABINET, LOUVER, 1/2 TON					
A/C unit, thru-wall, electric heat., cabinet, louver, 1/2 ton	1.000	Ea.	850	217	1,067
TOTAL			850	217	1,067

D3050 255	Thru-Wall A/C Unit	COST EACH		
		MAT.	INST.	TOTAL
1010	A/C Unit, thru-the-wall, supplemental elect. heat, cabinet, louver, 1/2 Ton	850	217	1,067
1020	3/4 Ton	1,325	260	1,585
1030	1 Ton	1,500	325	1,825
1040	1-1/2 Ton	2,100	540	2,640
1050	2 Ton	2,125	685	2,810

D3050 260	Thru-Wall Heat Pump	COST EACH		
		MAT.	INST.	TOTAL
1010	Heat pump, thru-the-wall, cabinet, louver, 1/2 Ton	2,700	162	2,862
1020	3/4 Ton	2,825	217	3,042
1030	1 Ton	3,275	325	3,600
1040	Supplemental. elect. heat, 1-1/2 Ton	3,350	840	4,190
1050	2 Ton	3,425	865	4,290

D3090 Other HVAC Systems/Equip

Roof Exhaust Fan

10″ Dia. Fiberglass 90° Elbows

10″ Dia. Fiberglass Reinforced Duct

Fume Hood

Fume Hood Exhaust

System Components	QUANTITY	UNIT	COST EACH		
			MAT.	INST.	TOTAL
SYSTEM D3090 310 1010					
FUME HOOD EXHAUST SYSTEM, 3′ LONG, 1000 CFM					
Fume hood	3.000	L.F.	2,655	1,440	4,095
Ductwork, plastic, FM approved for acid fume & smoke, straight, 10″ dia	60.000	L.F.	1,380	870	2,250
Ductwork, plastic, FM approved for acid fume & smoke, elbow 90°, 10″ dia	3.000	Ea.	351	474	825
Fan, corrosive fume resistant plastic, 1630 CFM, 1/2 HP	1.000	Ea.	6,500	390	6,890
Fan control switch & mounting base	1.000	Ea.	140	45	185
Roof fan balancing	1.000	Ea.		274	274
TOTAL			11,026	3,493	14,519

D3090 310	Fume Hood Exhaust Systems	COST EACH		
		MAT.	INST.	TOTAL
1010	Fume hood exhaust system, 3′ long, 1000 CFM	11,000	3,500	14,500
1020	4′ long, 2000 CFM	12,500	4,400	16,900
1030	6′ long, 3500 CFM	16,300	6,600	22,900
1040	6′ long, 5000 CFM	25,600	7,750	33,350
1050	10′ long, 8000 CFM	31,900	10,600	42,500

D3090 Other HVAC Systems/Equip

Cast Iron Garage Exhaust System

Dual Exhaust System

System Components	QUANTITY	UNIT	COST EACH		
			MAT.	INST.	TOTAL
SYSTEM D3090 320 1040					
GARAGE, EXHAUST, SINGLE 3″ EXHAUST OUTLET, CARS & LIGHT TRUCKS					
A Outlet top assy, for engine exhaust system with adapters and ftngs, 3″ diam	1.000	Ea.	265	43	308
F Bullnose (guide) for engine exhaust system, 3″ diam	1.000	Ea.	30.50		30.50
G Galvanized flexible tubing for engine exhaust system, 3″ diam	8.000	L.F.	102		102
H Adapter for metal tubing end of engine exhaust system, 3″ tail pipe	1.000	Ea.	54.50		54.50
J Pipe, sewer, cast iron, push-on joint, 8″ diam.	18.000	L.F.	1,044	783	1,827
Excavating utility trench, chain trencher, 8″ wide, 24″ deep	18.000	L.F.		34.20	34.20
Backfill utility trench by hand, incl. compaction, 8″ wide 24″ deep	18.000	L.F.		47.16	47.16
Stand for blower, concrete over polystyrene core, 6″ high	1.000	Ea.	6.05	21.50	27.55
AC&V duct spiral reducer 10″x8″	1.000	Ea.	22.50	38.50	61
AC&V duct spiral reducer 12″x10″	1.000	Ea.	26	51.50	77.50
AC&V duct spiral preformed 45° elbow, 8″ diam	1.000	Ea.	8.90	44	52.90
AC&V utility fan, belt drive, 3 phase, 2000 CFM, 1 HP	2.000	Ea.	2,700	680	3,380
Safety switch, heavy duty fused, 240V, 3 pole, 30 amp	1.000	Ea.	130	205	335
TOTAL			4,389.45	1,947.86	6,337.31

D3090 320	Garage Exhaust Systems	COST PER BAY		
		MAT.	INST.	TOTAL
1040	Garage, single 3″ exhaust outlet, cars & light trucks, one bay	4,400	1,950	6,350
1060	Additional bays up to seven bays	1,025	525	1,550
1500	4″ outlet, trucks, one bay	4,425	1,950	6,375
1520	Additional bays up to six bays	1,075	525	1,600
1600	5″ outlet, diesel trucks, one bay	4,750	1,950	6,700
1650	Additional single bays up to six	1,500	620	2,120
1700	Two adjoining bays	4,750	1,950	6,700
2000	Dual exhaust, 3″ outlets, pair of adjoining bays	5,500	2,450	7,950
2100	Additional pairs of adjoining bays	1,675	620	2,295

System Components	QUANTITY	UNIT	COST EACH		
			MAT.	INST.	TOTAL
SYSTEM D3090 750 1000					
SEISMIC PAD FOR 20-38 SELF-CONTAINED UNIT					
Form equipment pad	28.000	L.F.	32.20	120.68	152.88
Reinforcing in hold down pad, #3 to #7	60.000	Lb.	31.80	37.80	69.60
Concrete 3000 PSI	.866	C.Y.	96.99		96.99
Place concrete pad	.866	C.Y.		32.13	32.13
Finish concrete equipment pad	47.000	S.F.		39.01	39.01
Drill for 3/8 inch slab anchors	10.000	Ea.	.60	91.50	92.10
Exisiting slab anchors 3/8 inch	10.000	Ea.	9.70	41.30	51
Equipment anchors 3/8 inch for plate/foot	8.000	Ea.	13.04	35.28	48.32
TOTAL			184.33	397.70	582.03

D3090 750	Seismic Concrete Equipment Pads	COST PER EACH		
		MAT.	INST.	TOTAL
0990	Seismic pad for self-contained AC units for plate/foot attachment			
1000	Seismic pad for 20-38 tons Self-contained AC unit without economizer .25g	184	395	579
1010	0.5 g seismic acceleration	184	395	579
1020	0.75 g seismic acceleration	184	395	579
1030	1.0 g seismic acceleration	184	395	579
1040	1.5 g seismic acceleration	188	450	638
1050	2.0 g seismic acceleration	204	620	824
1060	2.5 g seismic acceleration	216	780	996
1100	Seismic pad for 40-80 tons Self contained AC unit without economizer .25g	287	575	862
1110	0.5 g seismic acceleration	287	575	862
1120	0.75 g seismic acceleration	287	575	862
1130	1.0 g seismic acceleration	287	575	862
1140	1.5 g seismic acceleration	315	820	1,135
1150	2.0 g seismic acceleration	375	1,275	1,650

D3090 Other HVAC Systems/Equip

D3090 750	Seismic Concrete Equipment Pads	COST PER EACH		
		MAT.	INST.	TOTAL
1160	2.5 g seismic acceleration	440	1,800	2,240
1200	Seismic pad for 80-110 tons Self contained AC unit without economizer .25g	315	620	935
1210	0.5 g seismic acceleration	315	620	935
1220	0.75 g seismic acceleration	315	620	935
1230	1.0 g seismic acceleration	315	620	935
1240	1.5 g seismic acceleration	345	880	1,225
1250	2.0 g seismic acceleration	410	1,375	1,785
1260	2.5 g seismic acceleration	480	1,950	2,430
1990	Seismic pad for self-contained AC units for isolator/snubber bolts			
2000	Seismic pad for 20-38 tons Self contained AC unit without economizer .25g	197	430	627
2010	0.5 g seismic acceleration	197	430	627
2020	0.75 g seismic acceleration	197	430	627
2030	1.0 g seismic acceleration	197	430	627
2040	1.5 g seismic acceleration	201	485	686
2050	2.0 g seismic acceleration	220	660	880
2060	2.5 g seismic acceleration	233	820	1,053
2100	Seismic pad for 40-80 tons Self contained AC unit without economizer .25g	305	620	925
2110	0.5 g seismic acceleration	305	620	925
2120	0.75 g seismic acceleration	305	620	925
2130	1.0 g seismic acceleration	305	620	925
2140	1.5 g seismic acceleration	330	865	1,195
2150	2.0 g seismic acceleration	400	1,325	1,725
2160	2.5 g seismic acceleration	470	1,875	2,345
2200	Seismic pad for 80-110 tons Self contained AC unit without economizer .25g	330	665	995
2210	0.5 g seismic acceleration	330	665	995
2220	0.75 g seismic acceleration	330	665	995
2230	1.0 g seismic acceleration	330	665	995
2240	1.5 g seismic acceleration	365	925	1,290
2250	2.0 g seismic acceleration	435	1,425	1,860
2260	2.5 g seismic acceleration	510	2,025	2,535
2990	Seismic pad for self-contained AC economizer units for plate/foot attachment			
3000	Seismic pad for 20-38 tons Self contained AC unit economizer .25 g-2.0 g	123	232	355
3010	2.5 g seismic acceleration	131	310	441
3100	Seismic pad for 40-80 tons Self contained AC unit economizer .25-1.5 g	164	310	474
3110	2.0 g seismic acceleration	170	355	525
3120	2.5 g seismic acceleration	177	420	597
3190	Seismic pad for self-contained AC unit economizer for isolator/snubber bolts			
3200	Seismic pad for 20-38 tons Self contained AC unit economizer .25 g-2.0 g	127	245	372
3210	2.5 g seismic acceleration	139	330	469
3300	Seismic pad for 40-80 tons Self contained AC unit economizer .25-1.5 g	173	335	508
3310	2.0 g seismic acceleration	179	385	564
3320	2.5 g seismic acceleration	187	450	637

G1030 Site Earthwork

Trenching Systems are shown on a cost per linear foot basis. The systems include: excavation; backfill and removal of spoil; and compaction for various depths and trench bottom widths. The backfill has been reduced to accommodate a pipe of suitable diameter and bedding.

The slope for trench sides varies from none to 1:1.

The Expanded System Listing shows Trenching Systems that range from 2' to 12' in width. Depths range from 2' to 25'.

System Components	QUANTITY	UNIT	COST PER L.F.		
			EQUIP.	LABOR	TOTAL
SYSTEM G1030 805 1310					
TRENCHING COMMON EARTH, NO SLOPE, 2' WIDE, 2' DP, 3/8 C.Y. BUCKET					
Excavation, trench, hyd. backhoe, track mtd., 3/8 C.Y. bucket	.148	B.C.Y.	.40	1.07	1.47
Backfill and load spoil, from stockpile	.153	L.C.Y.	.13	.32	.45
Compaction by vibrating plate, 6" lifts, 4 passes	.118	E.C.Y.	.03	.39	.42
Remove excess spoil, 8 C.Y. dump truck, 2 mile roundtrip	.040	L.C.Y.	.15	.16	.31
TOTAL			.71	1.94	2.65

G1030 805	Trenching Common Earth	COST PER L.F.		
		EQUIP.	LABOR	TOTAL
1310	Trenching, common earth, no slope, 2' wide, 2' deep, 3/8 C.Y. bucket	.71	1.94	2.65
1320	3' deep, 3/8 C.Y. bucket	1	2.93	3.93
1330	4' deep, 3/8 C.Y. bucket	1.29	3.90	5.19
1340	6' deep, 3/8 C.Y. bucket	1.67	5.05	6.72
1350	8' deep, 1/2 C.Y. bucket	2.21	6.70	8.91
1360	10' deep, 1 C.Y. bucket	3.46	7.95	11.41
1400	4' wide, 2' deep, 3/8 C.Y. bucket	1.60	3.83	5.43
1410	3' deep, 3/8 C.Y. bucket	2.20	5.80	8
1420	4' deep, 1/2 C.Y. bucket	2.56	6.50	9.06
1430	6' deep, 1/2 C.Y. bucket	4.12	10.45	14.57
1440	8' deep, 1/2 C.Y. bucket	6.60	13.45	20.05
1450	10' deep, 1 C.Y. bucket	7.95	16.70	24.65
1460	12' deep, 1 C.Y. bucket	10.25	21.50	31.75
1470	15' deep, 1-1/2 C.Y. bucket	9.40	19.05	28.45
1480	18' deep, 2-1/2 C.Y. bucket	13	26.50	39.50
1520	6' wide, 6' deep, 5/8 C.Y. bucket w/trench box	8.50	15.45	23.95
1530	8' deep, 3/4 C.Y. bucket	11.30	20.50	31.80
1540	10' deep, 1 C.Y. bucket	11.50	21	32.50
1550	12' deep, 1-1/2 C.Y. bucket	12.35	22.50	34.85
1560	16' deep, 2-1/2 C.Y. bucket	16.90	28.50	45.40
1570	20' deep, 3-1/2 C.Y. bucket	22	34	56
1580	24' deep, 3-1/2 C.Y. bucket	26	40.50	66.50
1640	8' wide, 12' deep, 1-1/2 C.Y. bucket w/trench box	17.30	28.50	45.80
1650	15' deep, 1-1/2 C.Y. bucket	22.50	37.50	60
1660	18' deep, 2-1/2 C.Y. bucket	24.50	38	62.50
1680	24' deep, 3-1/2 C.Y. bucket	35.50	52.50	88
1730	10' wide, 20' deep, 3-1/2 C.Y. bucket w/trench box	28.50	50	78.50
1740	24' deep, 3-1/2 C.Y. bucket	42.50	60	102.50
1780	12' wide, 20' deep, 3-1/2 C.Y. bucket w/trench box	45	63.50	108.50
1790	25' deep, bucket	55.50	81	136.50
1800	1/2 to 1 slope, 2' wide, 2' deep, 3/8 C.Y. bucket	1	2.93	3.93
1810	3' deep, 3/8 C.Y. bucket	1.67	5.15	6.82
1820	4' deep, 3/8 C.Y. bucket	2.49	7.85	10.34
1840	6' deep, 3/8 C.Y. bucket	4	12.70	16.70

G1030 Site Earthwork

G1030 805	Trenching Common Earth	COST PER L.F.		
		EQUIP.	LABOR	TOTAL
1860	8' deep, 1/2 C.Y. bucket	6.35	20.50	26.85
1880	10' deep, 1 C.Y. bucket	11.90	28	39.90
2300	4' wide, 2' deep, 3/8 C.Y. bucket	1.90	4.81	6.71
2310	3' deep, 3/8 C.Y. bucket	2.87	8	10.87
2320	4' deep, 1/2 C.Y. bucket	3.61	9.90	13.51
2340	6' deep, 1/2 C.Y. bucket	6.90	18.55	25.45
2360	8' deep, 1/2 C.Y. bucket	12.75	27	39.75
2380	10' deep, 1 C.Y. bucket	17.60	38.50	56.10
2400	12' deep, 1 C.Y. bucket	24	52	76
2430	15' deep, 1-1/2 C.Y. bucket	26.50	55.50	82
2460	18' deep, 2-1/2 C.Y. bucket	46	87	133
2840	6' wide, 6' deep, 5/8 C.Y. bucket w/trench box	12.55	23	35.55
2860	8' deep, 3/4 C.Y. bucket	18.20	34.50	52.70
2880	10' deep, 1 C.Y. bucket	18.55	35	53.55
2900	12' deep, 1-1/2 C.Y. bucket	23.50	46	69.50
2940	16' deep, 2-1/2 C.Y. bucket	38.50	67.50	106
2980	20' deep, 3-1/2 C.Y. bucket	55	91	146
3020	24' deep, 3-1/2 C.Y. bucket	77.50	125	202.50
3100	8' wide, 12' deep, 1-1/2 C.Y. bucket w/trench box	29	52.50	81.50
3120	15' deep, 1-1/2 C.Y. bucket	42	76	118
3140	18' deep, 2-1/2 C.Y. bucket	53	90.50	143.50
3180	24' deep, 3-1/2 C.Y. bucket	86.50	137	223.50
3270	10' wide, 20' deep, 3-1/2 C.Y. bucket w/trench box	55.50	105	160.50
3280	24' deep, 3-1/2 C.Y. bucket	96	149	245
3370	12' wide, 20' deep, 3-1/2 C.Y. bucket w/trench box	80.50	122	202.50
3380	25' deep, 3-1/2 C.Y. bucket	112	174	286
3500	1 to 1 slope, 2' wide, 2' deep, 3/8 C.Y. bucket	1.29	3.90	5.19
3520	3' deep, 3/8 C.Y. bucket	3.70	8.80	12.50
3540	4' deep, 3/8 C.Y. bucket	3.68	11.75	15.43
3560	6' deep, 1/2 C.Y. bucket	4	12.70	16.70
3580	8' deep, 1/2 C.Y. bucket	7.90	25.50	33.40
3600	10' deep, 1 C.Y. bucket	20.50	48.50	69
3800	4' wide, 2' deep, 3/8 C.Y. bucket	2.20	5.80	8
3820	3' deep, 3/8 C.Y. bucket	3.54	10.20	13.74
3840	4' deep, 1/2 C.Y. bucket	4.64	13.30	17.94
3860	6' deep, 1/2 C.Y. bucket	9.65	26.50	36.15
3880	8' deep, 1/2 C.Y. bucket	18.95	41	59.95
3900	10' deep, 1 C.Y. bucket	27.50	60	87.50
3920	12' deep, 1 C.Y. bucket	40	86.50	126.50
3940	15' deep, 1-1/2 C.Y. bucket	43.50	92.50	136
3960	18' deep, 2-1/2 C.Y. bucket	63.50	120	183.50
4030	6' wide, 6' deep, 5/8 C.Y. bucket w/trench box	16.50	31	47.50
4040	8' deep, 3/4 C.Y. bucket	23.50	42.50	66
4050	10' deep, 1 C.Y. bucket	27	52	79
4060	12' deep, 1-1/2 C.Y. bucket	36	70.50	106.50
4070	16' deep, 2-1/2 C.Y. bucket	60.50	107	167.50
4080	20' deep, 3-1/2 C.Y. bucket	88.50	149	237.50
4090	24' deep, 3-1/2 C.Y. bucket	129	209	338
4500	8' wide, 12' deep, 1-1/2 C.Y. bucket w/trench box	41	76	117
4550	15' deep, 1-1/2 C.Y. bucket	62	114	176
4600	18' deep, 2-1/2 C.Y. bucket	80.50	140	220.50
4650	24' deep, 3-1/2 C.Y. bucket	138	221	359
4800	10' wide, 20' deep, 3-1/2 C.Y. bucket w/trench box	82.50	160	242.50
4850	24' deep, 3-1/2 C.Y. bucket	147	233	380
4950	12' wide, 20' deep, 3-1/2 C.Y. bucket w/trench box	116	180	296
4980	25' deep, 3-1/2 C.Y. bucket	167	263	430

G1030   Site Earthwork

Trenching Systems are shown on a cost per linear foot basis. The systems include: excavation; backfill and removal of spoil; and compaction for various depths and trench bottom widths. The backfill has been reduced to accommodate a pipe of suitable diameter and bedding.

The slope for trench sides varies from none to 1:1.

The Expanded System Listing shows Trenching Systems that range from 2' to 12' in width. Depths range from 2' to 25'.

System Components	QUANTITY	UNIT	COST PER L.F. EQUIP.	COST PER L.F. LABOR	COST PER L.F. TOTAL
SYSTEM G1030 806 1310					
TRENCHING LOAM & SANDY CLAY, NO SLOPE, 2' WIDE, 2' DP, 3/8 C.Y. BUCKET					
Excavation, trench, hyd. backhoe, track mtd., 3/8 C.Y. bucket	.148	B.C.Y.	.37	.98	1.35
Backfill and load spoil, from stockpile	.165	L.C.Y.	.14	.35	.49
Compaction by vibrating plate 18" wide, 6" lifts, 4 passes	.118	E.C.Y.	.03	.39	.42
Remove excess spoil, 8 C.Y. dump truck, 2 mile roundtrip	.042	L.C.Y.	.16	.17	.33
TOTAL			.71	1.94	2.65

G1030 806	Trenching Loam & Sandy Clay	COST PER L.F. EQUIP.	COST PER L.F. LABOR	COST PER L.F. TOTAL
1310	Trenching, loam & sandy clay, no slope, 2' wide, 2' deep, 3/8 C.Y. bucket	.70	1.89	2.59
1320	3' deep, 3/8 C.Y. bucket	1.08	3.13	4.21
1330	4' deep, 3/8 C.Y. bucket	1.27	3.81	5.10
1340	6' deep, 3/8 C.Y. bucket	1.80	4.51	6.30
1350	8' deep, 1/2 C.Y. bucket	2.38	5.95	8.35
1360	10' deep, 1 C.Y. bucket	2.69	6.40	9.10
1400	4' wide, 2' deep, 3/8 C.Y. bucket	1.60	3.75	5.35
1410	3' deep, 3/8 C.Y. bucket	2.18	5.65	7.85
1420	4' deep, 1/2 C.Y. bucket	2.55	6.40	8.95
1430	6' deep, 1/2 C.Y. bucket	4.39	9.35	13.75
1440	8' deep, 1/2 C.Y. bucket	6.45	13.25	19.70
1450	10' deep, 1 C.Y. bucket	6.45	13.60	20
1460	12' deep, 1 C.Y. bucket	8.10	16.90	25
1470	15' deep, 1-1/2 C.Y. bucket	9.80	19.70	29.50
1480	18' deep, 2-1/2 C.Y. bucket	11.50	21.50	33
1520	6' wide, 6' deep, 5/8 C.Y. bucket w/trench box	8.25	15.20	23.50
1530	8' deep, 3/4 C.Y. bucket	10.90	19.95	31
1540	10' deep, 1 C.Y. bucket	10.55	20.50	31
1550	12' deep, 1-1/2 C.Y. bucket	12.05	23	35
1560	16' deep, 2-1/2 C.Y. bucket	16.50	28.50	45
1570	20' deep, 3-1/2 C.Y. bucket	20	34	54
1580	24' deep, 3-1/2 C.Y. bucket	25.50	41.50	67
1640	8' wide, 12' deep, 1-1/4 C.Y. bucket w/trench box	17.05	29	46
1650	15' deep, 1-1/2 C.Y. bucket	21	36.50	57.50
1660	18' deep, 2-1/2 C.Y. bucket	25	41.50	66.50
1680	24' deep, 3-1/2 C.Y. bucket	34.50	53.50	88
1730	10' wide, 20' deep, 3-1/2 C.Y. bucket w/trench box	34.50	53.50	88
1740	24' deep, 3-1/2 C.Y. bucket	43.50	66	110
1780	12' wide, 20' deep, 3-1/2 C.Y. bucket w/trench box	42	63.50	106
1790	25' deep, 3-1/2 C.Y. bucket	54.50	82	137
1800	1/2:1 slope, 2' wide, 2' deep, 3/8 C.Y. bucket	.98	2.86	3.84
1810	3' deep, 3/8 C.Y. bucket	1.63	5	6.65
1820	4' deep, 3/8 C.Y. bucket	2.43	7.65	10.10
1840	6' deep, 3/8 C.Y. bucket	4.31	11.30	15.60

G1030 Site Earthwork

G1030 806	Trenching Loam & Sandy Clay	COST PER L.F.		
		EQUIP.	LABOR	TOTAL
1860	8' deep, 1/2 C.Y. bucket	6.85	18.10	25
1880	10' deep, 1 C.Y. bucket	9.15	22.50	31.50
2300	4' wide, 2' deep, 3/8 C.Y. bucket	1.88	4.70	6.60
2310	3' deep, 3/8 C.Y. bucket	2.83	7.80	10.65
2320	4' deep, 1/2 C.Y. bucket	3.57	9.70	13.25
2340	6' deep, 1/2 C.Y. bucket	7.35	16.65	24
2360	8' deep, 1/2 C.Y. bucket	12.40	27	39.50
2380	10' deep, 1 C.Y. bucket	14.10	31	45
2400	12' deep, 1 C.Y. bucket	23.50	51.50	75
2430	15' deep, 1-1/2 C.Y. bucket	27.50	57.50	85
2460	18' deep, 2-1/2 C.Y. bucket	45.50	88	134
2840	6' wide, 6' deep, 5/8 C.Y. bucket w/trench box	11.95	23	35
2860	8' deep, 3/4 C.Y. bucket	17.50	34	51.50
2880	10' deep, 1 C.Y. bucket	18.90	38	57
2900	12' deep, 1-1/2 C.Y. bucket	23.50	46.50	70
2940	16' deep, 2-1/2 C.Y. bucket	37.50	68.50	106
2980	20' deep, 3-1/2 C.Y. bucket	53	92.50	146
3020	24' deep, 3-1/2 C.Y. bucket	75	127	202
3100	8' wide, 12' deep, 1-1/2 C.Y. bucket w/trench box	28.50	52.50	81
3120	15' deep, 1-1/2 C.Y. bucket	39	74	113
3140	18' deep, 2-1/2 C.Y. bucket	52	91.50	144
3180	24' deep, 3-1/2 C.Y. bucket	84	139	223
3270	10' wide, 20' deep, 3-1/2 C.Y. bucket w/trench box	67.50	112	180
3280	24' deep, 3-1/2 C.Y. bucket	93	151	244
3320	12' wide, 20' deep, 3-1/2 C.Y. bucket w/trench box	75	122	197
3380	25' deep, 3-1/2 C.Y. bucket w/trench box	102	164	266
3500	1:1 slope, 2' wide, 2' deep, 3/8 C.Y. bucket	1.27	3.81	5.10
3520	3' deep, 3/8 C.Y. bucket	2.28	7.15	9.45
3540	4' deep, 3/8 C.Y. bucket	3.59	11.45	15.05
3560	6' deep, 1/2 C.Y. bucket	4.31	11.30	15.60
3580	8' deep, 1/2 C.Y. bucket	11.35	30	41.50
3600	10' deep, 1 C.Y. bucket	15.65	39	54.50
3800	4' wide, 2' deep, 3/8 C.Y. bucket	2.18	5.65	7.85
3820	3' deep, 1/2 C.Y. bucket	3.48	9.95	13.45
3840	4' deep, 1/2 C.Y. bucket	4.58	13.05	17.65
3860	6' deep, 1/2 C.Y. bucket	10.30	24	34.50
3880	8' deep, 1/2 C.Y. bucket	18.40	40.50	59
3900	10' deep, 1 C.Y. bucket	22	49	71
3920	12' deep, 1 C.Y. bucket	31.50	69	101
3940	15' deep, 1-1/2 C.Y. bucket	45.50	95.50	141
3960	18' deep, 2-1/2 C.Y. bucket	62	121	183
4030	6' wide, 6' deep, 5/8 C.Y. bucket w/trench box	15.70	31	46.50
4040	8' deep, 3/4 C.Y. bucket	24	47.50	71.50
4050	10' deep, 1 C.Y. bucket	27	56	83
4060	12' deep, 1-1/2 C.Y. bucket	35	70.50	106
4070	16' deep, 2-1/2 C.Y. bucket	59	108	167
4080	20' deep, 3-1/2 C.Y. bucket	86	151	237
4090	24' deep, 3-1/2 C.Y. bucket	125	212	335
4500	8' wide, 12' deep, 1-1/4 C.Y. bucket w/trench box	40	76.50	117
4550	15' deep, 1-1/2 C.Y. bucket	56.50	111	168
4600	18' deep, 2-1/2 C.Y. bucket	78.50	142	221
4650	24' deep, 3-1/2 C.Y. bucket	134	224	360
4800	10' wide, 20' deep, 3-1/2 C.Y. bucket w/trench box	100	171	271
4850	24' deep, 3-1/2 C.Y. bucket	143	237	380
4950	12' wide, 20' deep, 3-1/2 C.Y. bucket w/trench box	108	181	289
4980	25' deep, 3-1/2 C.Y. bucket w/trench box	162	267	430

589

For customer support on your Mechanical Cost Data, call 877.795.7394.

G1030 Site Earthwork

Trenching Systems are shown on a cost per linear foot basis. The systems include: excavation; backfill and removal of spoil; and compaction for various depths and trench bottom widths. The backfill has been reduced to accommodate a pipe of suitable diameter and bedding.

The slope for trench sides varies from none to 1:1.

The Expanded System Listing shows Trenching Systems that range from 2' to 12' in width. Depths range from 2' to 25'.

System Components			COST PER L.F.		
	QUANTITY	UNIT	EQUIP.	LABOR	TOTAL
SYSTEM G1030 807 1310					
TRENCHING SAND & GRAVEL, NO SLOPE, 2' WIDE, 2' DEEP, 3/8 C.Y. BUCKET					
Excavation, trench, hyd. backhoe, track mtd., 3/8 C.Y. bucket	.148	B.C.Y.	.36	.97	1.33
Backfill and load spoil, from stockpile	.140	L.C.Y.	.11	.30	.41
Compaction by vibrating plate 18" wide, 6" lifts, 4 passes	.118	E.C.Y.	.03	.39	.42
Remove excess spoil, 8 C.Y. dump truck, 2 mile roundtrip	.035	L.C.Y.	.13	.14	.27
TOTAL			.63	1.80	2.43

G1030 807	Trenching Sand & Gravel	COST PER L.F.		
		EQUIP.	LABOR	TOTAL
1310	Trenching, sand & gravel, no slope, 2' wide, 2' deep, 3/8 C.Y. bucket	.63	1.80	2.43
1320	3' deep, 3/8 C.Y. bucket	1.01	3.01	4.02
1330	4' deep, 3/8 C.Y. bucket	1.18	3.62	4.80
1340	6' deep, 3/8 C.Y. bucket	1.70	4.30	6
1350	8' deep, 1/2 C.Y. bucket	2.25	5.70	7.95
1360	10' deep, 1 C.Y. bucket	2.50	6.05	8.55
1400	4' wide, 2' deep, 3/8 C.Y. bucket	1.46	3.54	5
1410	3' deep, 3/8 C.Y. bucket	2.01	5.35	7.35
1420	4' deep, 1/2 C.Y. bucket	2.33	6.05	8.40
1430	6' deep, 1/2 C.Y. bucket	4.16	8.95	13.10
1440	8' deep, 1/2 C.Y. bucket	6.05	12.60	18.65
1450	10' deep, 1 C.Y. bucket	6.05	12.85	18.90
1460	12' deep, 1 C.Y. bucket	7.65	16	23.50
1470	15' deep, 1-1/2 C.Y. bucket	9.25	18.55	28
1480	18' deep, 2-1/2 C.Y. bucket	10.85	20.50	31.50
1520	6' wide, 6' deep, 5/8 C.Y. bucket w/trench box	7.75	14.35	22
1530	8' deep, 3/4 C.Y. bucket	10.30	18.95	29.50
1540	10' deep, 1 C.Y. bucket	9.95	19.15	29
1550	12' deep, 1-1/2 C.Y. bucket	11.40	21.50	33
1560	16' deep, 2 C.Y. bucket	15.75	27	43
1570	20' deep, 3-1/2 C.Y. bucket	19.10	32	51
1580	24' deep, 3-1/2 C.Y. bucket	24	39	63
1640	8' wide, 12' deep, 1-1/2 C.Y. bucket w/trench box	16.05	27	43
1650	15' deep, 1-1/2 C.Y. bucket	19.50	34.50	54
1660	18' deep, 2-1/2 C.Y. bucket	23.50	39	62.50
1680	24' deep, 3-1/2 C.Y. bucket	32.50	50	82.50
1730	10' wide, 20' deep, 3-1/2 C.Y. bucket w/trench box	32.50	50	82.50
1740	24' deep, 3-1/2 C.Y. bucket	41	62	103
1780	12' wide, 20' deep, 3-1/2 C.Y. bucket w/trench box	39.50	59.50	99
1790	25' deep, 3-1/2 C.Y. bucket	51	77	128
1800	1/2:1 slope, 2' wide, 2' deep, 3/8 C.Y. bucket	.91	2.72	3.63
1810	3' deep, 3/8 C.Y. bucket	1.53	4.77	6.30
1820	4' deep, 3/8 C.Y. bucket	2.27	7.30	9.55
1840	6' deep, 3/8 C.Y. bucket	4.10	10.80	14.90

G1030 Site Earthwork

G1030 807	Trenching Sand & Gravel	COST PER L.F.		
		EQUIP.	LABOR	TOTAL
1860	8' deep, 1/2 C.Y. bucket	6.55	17.30	24
1880	10' deep, 1 C.Y. bucket	8.55	21.50	30
2300	4' wide, 2' deep, 3/8 C.Y. bucket	1.74	4.45	6.20
2310	3' deep, 3/8 C.Y. bucket	2.62	7.40	10
2320	4' deep, 1/2 C.Y. bucket	3.29	9.20	12.50
2340	6' deep, 1/2 C.Y. bucket	7	15.95	23
2360	8' deep, 1/2 C.Y. bucket	11.75	25.50	37.50
2380	10' deep, 1 C.Y. bucket	13.35	29.50	43
2400	12' deep, 1 C.Y. bucket	22.50	49	71.50
2430	15' deep, 1-1/2 C.Y. bucket	26	54.50	80.50
2460	18' deep, 2-1/2 C.Y. bucket	43	83	126
2840	6' wide, 6' deep, 5/8 C.Y. bucket w/trench box	11.35	22	33.50
2860	8' deep, 3/4 C.Y. bucket	16.60	32	48.50
2880	10' deep, 1 C.Y. bucket	17.85	36	54
2900	12' deep, 1-1/2 C.Y. bucket	22.50	44	66.50
2940	16' deep, 2 C.Y. bucket	36	64.50	101
2980	20' deep, 3-1/2 C.Y. bucket	50.50	87	138
3020	24' deep, 3-1/2 C.Y. bucket	71.50	119	191
3100	8' wide, 12' deep, 1-1/4 C.Y. bucket w/trench box	27	50	77
3120	15' deep, 1-1/2 C.Y. bucket	36.50	69.50	106
3140	18' deep, 2-1/2 C.Y. bucket	49	86	135
3180	24' deep, 3-1/2 C.Y. bucket	80	130	210
3270	10' wide, 20' deep, 3-1/2 C.Y. bucket w/trench box	64	105	169
3280	24' deep, 3-1/2 C.Y. bucket	88	142	230
3370	12' wide, 20' deep, 3-1/2 C.Y. bucket w/trench box	71.50	117	189
3380	25' deep, 3-1/2 C.Y. bucket	103	164	267
3500	1:1 slope, 2' wide, 2' deep, 3/8 C.Y. bucket	2.05	4.46	6.50
3520	3' deep, 3/8 C.Y. bucket	2.14	6.85	9
3540	4' deep, 3/8 C.Y. bucket	3.36	10.95	14.30
3560	6' deep, 3/8 C.Y. bucket	4.10	10.80	14.90
3580	8' deep, 1/2 C.Y. bucket	10.80	29	40
3600	10' deep, 1 C.Y. bucket	14.65	36.50	51
3800	4' wide, 2' deep, 3/8 C.Y. bucket	2.01	5.35	7.35
3820	3' deep, 3/8 C.Y. bucket	3.24	9.50	12.75
3840	4' deep, 1/2 C.Y. bucket	4.23	12.35	16.60
3860	6' deep, 1/2 C.Y. bucket	9.85	23	33
3880	8' deep, 1/2 C.Y. bucket	17.45	38.50	56
3900	10' deep, 1 C.Y. bucket	20.50	46	66.50
3920	12' deep, 1 C.Y. bucket	30	65.50	95.50
3940	15' deep, 1-1/2 C.Y. bucket	43	90	133
3960	18' deep, 2-1/2 C.Y. bucket	59	114	173
4030	6' wide, 6' deep, 5/8 C.Y. bucket w/trench box	14.90	29.50	44.50
4040	8' deep, 3/4 C.Y. bucket	23	45.50	68.50
4050	10' deep, 1 C.Y. bucket	25.50	53	78.50
4060	12' deep, 1-1/2 C.Y. bucket	33	66.50	99.50
4070	16' deep, 2 C.Y. bucket	56.50	102	159
4080	20' deep, 3-1/2 C.Y. bucket	81.50	142	224
4090	24' deep, 3-1/2 C.Y. bucket	119	199	320
4500	8' wide, 12' deep, 1-1/2 C.Y. bucket w/trench box	38	72.50	111
4550	15' deep, 1-1/2 C.Y. bucket	53.50	105	159
4600	18' deep, 2-1/2 C.Y. bucket	74.50	134	209
4650	24' deep, 3-1/2 C.Y. bucket	127	210	335
4800	10' wide, 20' deep, 3-1/2 C.Y. bucket w/trench box	95	160	255
4850	24' deep, 3-1/2 C.Y. bucket	135	222	355
4950	12' wide, 20' deep, 3-1/2 C.Y. bucket w/trench box	102	170	272
4980	25' deep, 3-1/2 C.Y. bucket	154	251	405

591

For customer support on your Mechanical Cost Data, call 877.795.7394.

The Pipe Bedding System is shown for various pipe diameters. Compacted bank sand is used for pipe bedding and to fill 12″ over the pipe. No backfill is included. Various side slopes are shown to accommodate different soil conditions. Pipe sizes vary from 6″ to 84″ diameter.

System Components	QUANTITY	UNIT	COST PER L.F.		
			MAT.	INST.	TOTAL
SYSTEM G1030 815 1440					
PIPE BEDDING, SIDE SLOPE 0 TO 1, 1′ WIDE, PIPE SIZE 6″ DIAMETER					
Borrow, bank sand, 2 mile haul, machine spread	.086	C.Y.	1.68	.70	2.38
Compaction, vibrating plate	.086	C.Y.		.22	.22
TOTAL			1.68	.92	2.60

G1030 815	Pipe Bedding	COST PER L.F.		
		MAT.	INST.	TOTAL
1440	Pipe bedding, side slope 0 to 1, 1′ wide, pipe size 6″ diameter	1.68	.92	2.60
1460	2′ wide, pipe size 8″ diameter	3.63	1.97	5.60
1480	Pipe size 10″ diameter	3.70	2.01	5.71
1500	Pipe size 12″ diameter	3.78	2.05	5.83
1520	3′ wide, pipe size 14″ diameter	6.15	3.33	9.48
1540	Pipe size 15″ diameter	6.20	3.35	9.55
1560	Pipe size 16″ diameter	6.25	3.39	9.64
1580	Pipe size 18″ diameter	6.35	3.46	9.81
1600	4′ wide, pipe size 20″ diameter	9.05	4.92	13.97
1620	Pipe size 21″ diameter	9.15	4.96	14.11
1640	Pipe size 24″ diameter	9.35	5.05	14.40
1660	Pipe size 30″ diameter	9.50	5.15	14.65
1680	6′ wide, pipe size 32″ diameter	16.25	8.85	25.10
1700	Pipe size 36″ diameter	16.65	9	25.65
1720	7′ wide, pipe size 48″ diameter	26.50	14.35	40.85
1740	8′ wide, pipe size 60″ diameter	32	17.45	49.45
1760	10′ wide, pipe size 72″ diameter	45	24.50	69.50
1780	12′ wide, pipe size 84″ diameter	59	32	91
2140	Side slope 1/2 to 1, 1′ wide, pipe size 6″ diameter	3.13	1.69	4.82
2160	2′ wide, pipe size 8″ diameter	5.30	2.88	8.18
2180	Pipe size 10″ diameter	5.70	3.09	8.79
2200	Pipe size 12″ diameter	6.05	3.27	9.32
2220	3′ wide, pipe size 14″ diameter	8.70	4.73	13.43
2240	Pipe size 15″ diameter	8.90	4.83	13.73
2260	Pipe size 16″ diameter	9.15	4.96	14.11
2280	Pipe size 18″ diameter	9.60	5.20	14.80
2300	4′ wide, pipe size 20″ diameter	12.70	6.90	19.60
2320	Pipe size 21″ diameter	13	7.05	20.05
2340	Pipe size 24″ diameter	13.80	7.50	21.30
2360	Pipe size 30″ diameter	15.30	8.30	23.60
2380	6′ wide, pipe size 32″ diameter	22.50	12.25	34.75
2400	Pipe size 36″ diameter	24	13.05	37.05
2420	7′ wide, pipe size 48″ diameter	37.50	20.50	58
2440	8′ wide, pipe size 60″ diameter	47.50	26	73.50
2460	10′ wide, pipe size 72″ diameter	65	35.50	100.50
2480	12′ wide, pipe size 84″ diameter	85.50	46	131.50
2620	Side slope 1 to 1, 1′ wide, pipe size 6″ diameter	4.58	2.49	7.07
2640	2′ wide, pipe size 8″ diameter	7.05	3.83	10.88

G10 Site Preparation

G1030 Site Earthwork

G1030 815	Pipe Bedding	MAT.	INST.	TOTAL
2660	Pipe size 10" diameter	7.65	4.15	11.80
2680	Pipe size 12" diameter	8.35	4.53	12.88
2700	3' wide, pipe size 14" diameter	11.30	6.15	17.45
2720	Pipe size 15" diameter	11.65	6.30	17.95
2740	Pipe size 16" diameter	12.05	6.55	18.60
2760	Pipe size 18" diameter	12.90	7	19.90
2780	4' wide, pipe size 20" diameter	16.35	8.85	25.20
2800	Pipe size 21" diameter	16.80	9.10	25.90
2820	Pipe size 24" diameter	18.25	9.90	28.15
2840	Pipe size 30" diameter	21	11.40	32.40
2860	6' wide, pipe size 32" diameter	29	15.65	44.65
2880	Pipe size 36" diameter	31.50	17	48.50
2900	7' wide, pipe size 48" diameter	48.50	26.50	75
2920	8' wide, pipe size 60" diameter	63	34	97
2940	10' wide, pipe size 72" diameter	85.50	46.50	132
2960	12' wide, pipe size 84" diameter	112	60.50	172.50

Cost header: COST PER L.F.

G3010 Water Supply

G3010 110	Water Distribution Piping	COST PER L.F.		
		MAT.	INST.	TOTAL
2000	Piping, excav. & backfill excl., ductile iron class 250, mech. joint			
2130	4" diameter	33.50	16.85	50.35
2150	6" diameter	35	21	56
2160	8" diameter	49	25.50	74.50
2170	10" diameter	64	29.50	93.50
2180	12" diameter	87	32	119
2210	16" diameter	104	46.50	150.50
2220	18" diameter	139	49	188
3000	Tyton joint			
3130	4" diameter	23	8.45	31.45
3150	6" diameter	18.70	10.10	28.80
3160	8" diameter	25.50	16.85	42.35
3170	10" diameter	40.50	18.55	59.05
3180	12" diameter	42.50	21	63.50
3210	16" diameter	56.50	29.50	86
3220	18" diameter	62.50	33.50	96
3230	20" diameter	65	38	103
3250	24" diameter	65	44	109
4000	Copper tubing, type K			
4050	3/4" diameter	6.95	3.19	10.14
4060	1" diameter	9.25	3.99	13.24
4080	1-1/2" diameter	14.70	4.82	19.52
4090	2" diameter	22.50	5.55	28.05
4110	3" diameter	46.50	9.55	56.05
4130	4" diameter	76.50	13.45	89.95
4150	6" diameter	165	25	190
5000	Polyvinyl chloride class 160, S.D.R. 26			
5130	1-1/2" diameter	.50	1.19	1.69
5150	2" diameter	.75	1.30	2.05
5160	4" diameter	3	5.85	8.85
5170	6" diameter	5.90	7.05	12.95
5180	8" diameter	9.65	8.45	18.10
6000	Polyethylene 160 psi, S.D.R. 7			
6050	3/4" diameter	.51	1.70	2.21
6060	1" diameter	.66	1.84	2.50
6080	1-1/2" diameter	1.49	1.98	3.47
6090	2" diameter	2.42	2.45	4.87

G3020 Sanitary Sewer

G3020 110	Drainage & Sewage Piping	COST PER L.F.		
		MAT.	INST.	TOTAL
2000	Piping, excavation & backfill excluded, PVC, plain			
2130	4" diameter	1.60	4.14	5.74
2150	6" diameter	3.61	4.43	8.04
2160	8" diameter	7.60	4.63	12.23
2900	Box culvert, precast, 8' long			
3000	6' x 3'	279	35	314
3020	6' x 7'	355	39	394
3040	8' x 3'	350	43	393
3060	8' x 8'	425	48.50	473.50
3080	10' x 3'	485	44	529
3100	10' x 8'	560	60.50	620.50
3120	12' x 3'	725	48.50	773.50
3140	12' x 8'	930	72	1,002
4000	Concrete, nonreinforced			
4150	6" diameter	6.60	12.55	19.15
4160	8" diameter	7.25	14.90	22.15
4170	10" diameter	8.05	15.40	23.45
4180	12" diameter	9	16.65	25.65
4200	15" diameter	13.20	18.50	31.70
4220	18" diameter	16.50	23.50	40
4250	24" diameter	24	33.50	57.50
4400	Reinforced, no gasket			
4580	12" diameter	12.10	22	34.10
4600	15" diameter	15.40	22	37.40
4620	18" diameter	19.80	25	44.80
4650	24" diameter	28.50	33.50	62
4670	30" diameter	46	49	95
4680	36" diameter	61.50	60.50	122
4690	42" diameter	82.50	66.50	149
4700	48" diameter	98	74.50	172.50
4720	60" diameter	150	99	249
4730	72" diameter	227	119	346
4740	84" diameter	305	149	454
4800	With gasket			
4980	12" diameter	13.30	12	25.30
5000	15" diameter	16.95	12.60	29.55
5020	18" diameter	22	13.25	35.25
5050	24" diameter	36.50	14.80	51.30
5070	30" diameter	55	49	104
5080	36" diameter	72	60.50	132.50
5090	42" diameter	89.50	59.50	149
5100	48" diameter	112	74.50	186.50
5120	60" diameter	182	95.50	277.50
5130	72" diameter	248	119	367
5140	84" diameter	365	198	563
5700	Corrugated metal, alum. or galv. bit. coated			
5760	8" diameter	9.50	10.10	19.60
5770	10" diameter	9.90	12.80	22.70
5780	12" diameter	12.15	15.85	28
5800	15" diameter	16.65	16.65	33.30
5820	18" diameter	18.35	17.50	35.85
5850	24" diameter	23.50	21	44.50
5870	30" diameter	30.50	36.50	67
5880	36" diameter	39	36.50	75.50
5900	48" diameter	58	43	101
5920	60" diameter	87.50	63.50	151
5930	72" diameter	105	106	211
6000	Plain			

For customer support on your Mechanical Cost Data, call 877.795.7394.

595

G3020 Sanitary Sewer

G3020 110	Drainage & Sewage Piping	COST PER L.F.		
		MAT.	INST.	TOTAL
6060	8" diameter	8.60	9.40	18
6070	10" diameter	9.85	11.90	21.75
6080	12" diameter	10.95	15.10	26.05
6100	15" diameter	13.70	15.10	28.80
6120	18" diameter	16.55	16.25	32.80
6140	24" diameter	21	19.05	40.05
6170	30" diameter	27.50	33.50	61
6180	36" diameter	35	33.50	68.50
6200	48" diameter	52	39.50	91.50
6220	60" diameter	79	61	140
6230	72" diameter	176	156	332
6300	Steel or alum. oval arch, coated & paved invert			
6400	15" equivalent diameter	14.50	16.65	31.15
6420	18" equivalent diameter	17.65	22	39.65
6450	24" equivalent diameter	26.50	26.50	53
6470	30" equivalent diameter	32.50	33.50	66
6480	36" equivalent diameter	39	43	82
6490	42" equivalent diameter	45.50	48	93.50
6500	48" equivalent diameter	62.50	58	120.50
6600	Plain			
6700	15" equivalent diameter	13.15	14.80	27.95
6720	18" equivalent diameter	15.90	19.05	34.95
6750	24" equivalent diameter	23.50	22	45.50
6770	30" equivalent diameter	29	40	69
6780	36" equivalent diameter	35.50	40	75.50
6790	42" equivalent diameter	41	47	88
6800	48" equivalent diameter	56.50	58	114.50
8000	Polyvinyl chloride SDR 35			
8130	4" diameter	1.60	4.14	5.74
8150	6" diameter	3.61	4.43	8.04
8160	8" diameter	7.60	4.63	12.23
8170	10" diameter	12.50	6.10	18.60
8180	12" diameter	14	6.30	20.30
8200	15" diameter	14.25	8.40	22.65

G3030 Storm Sewer

Manhole   Catch Basin

The Manhole and Catch Basin System includes: excavation with a backhoe; a formed concrete footing; frame and cover; cast iron steps and compacted backfill.

The Expanded System Listing shows manholes that have a 4', 5' and 6' inside diameter riser. Depths range from 4' to 14'. Construction material shown is either concrete, concrete block, precast concrete, or brick.

System Components	QUANTITY	UNIT	COST PER EACH		
			MAT.	INST.	TOTAL
SYSTEM G3030 210 1920					
MANHOLE/CATCH BASIN, BRICK, 4' I.D. RISER, 4' DEEP					
Excavation, hydraulic backhoe, 3/8 C.Y. bucket	14.815	B.C.Y.		119.26	119.26
Trim sides and bottom of excavation	64.000	S.F.		62.08	62.08
Forms in place, manhole base, 4 uses	20.000	SFCA	16.40	106	122.40
Reinforcing in place footings, #4 to #7	.019	Ton	20.43	23.75	44.18
Concrete, 3000 psi	.925	C.Y.	103.60		103.60
Place and vibrate concrete, footing, direct chute	.925	C.Y.		49.26	49.26
Catch basin or MH, brick, 4' ID, 4' deep	1.000	Ea.	575	1,025	1,600
Catch basin or MH steps; heavy galvanized cast iron	1.000	Ea.	21	14.10	35.10
Catch basin or MH frame and cover	1.000	Ea.	268	245.50	513.50
Fill, granular	12.954	L.C.Y.	297.94		297.94
Backfill, spread with wheeled front end loader	12.954	L.C.Y.		34.07	34.07
Backfill compaction, 12" lifts, air tamp	12.954	E.C.Y.		117.88	117.88
TOTAL			1,302.37	1,796.90	3,099.27

G3030 210	Manholes & Catch Basins	COST PER EACH		
		MAT.	INST.	TOTAL
1920	Manhole/catch basin, brick, 4' I.D. riser, 4' deep	1,300	1,775	3,075
1940	6' deep	1,850	2,500	4,350
1960	8' deep	2,450	3,425	5,875
1980	10' deep	2,900	4,200	7,100
3000	12' deep	3,625	4,575	8,200
3020	14' deep	4,450	6,425	10,875
3200	Block, 4' I.D. riser, 4' deep	1,175	1,450	2,625
3220	6' deep	1,650	2,050	3,700
3240	8' deep	2,200	2,850	5,050
3260	10' deep	2,625	3,500	6,125
3280	12' deep	3,325	4,450	7,775
3300	14' deep	4,150	5,425	9,575
4620	Concrete, cast-in-place, 4' I.D. riser, 4' deep	1,275	2,475	3,750
4640	6' deep	1,825	3,325	5,150
4660	8' deep	2,550	4,800	7,350
4680	10' deep	3,075	5,950	9,025
4700	12' deep	3,875	7,375	11,250
4720	14' deep	4,800	8,825	13,625
5820	Concrete, precast, 4' I.D. riser, 4' deep	1,525	1,325	2,850
5840	6' deep	2,050	1,775	3,825

G3030 Storm Sewer

G3030 210	Manholes & Catch Basins	COST PER EACH		
		MAT.	INST.	TOTAL
5860	8' deep	2,575	2,475	5,050
5880	10' deep	3,125	3,050	6,175
5900	12' deep	3,925	3,775	7,700
5920	14' deep	4,875	4,775	9,650
6000	5' I.D. riser, 4' deep	2,600	1,500	4,100
6020	6' deep	3,225	2,100	5,325
6040	8' deep	4,175	2,775	6,950
6060	10' deep	5,300	3,525	8,825
6080	12' deep	6,525	4,525	11,050
6100	14' deep	7,900	5,500	13,400
6200	6' I.D. riser, 4' deep	3,325	1,975	5,300
6220	6' deep	4,150	2,600	6,750
6240	8' deep	5,275	3,675	8,950
6260	10' deep	6,675	4,650	11,325
6280	12' deep	8,200	5,850	14,050
6300	14' deep	9,850	7,100	16,950

G3060 Fuel Distribution

G3060 110	Gas Service Piping	COST PER L.F.		
		MAT.	INST.	TOTAL
2000	Piping, excavation & backfill excluded, polyethylene			
2070	1-1/4" diam, SDR 10	1.78	4.09	5.87
2090	2" diam, SDR 11	3.10	4.56	7.66
2110	3" diam, SDR 11.5	7.10	5.45	12.55
2130	4" diam,SDR 11	15.65	9.60	25.25
2150	6" diam, SDR 21	37	10.30	47.30
2160	8" diam, SDR 21	56.50	12.40	68.90
3000	Steel, schedule 40, plain end, tarred & wrapped			
3060	1" diameter	5.25	9.20	14.45
3090	2" diameter	8.20	9.90	18.10
3110	3" diameter	13.60	10.65	24.25
3130	4" diameter	17.65	16.55	34.20
3140	5" diameter	25.50	19.15	44.65
3150	6" diameter	31.50	23.50	55
3160	8" diameter	49.50	30	79.50

599

For customer support on your Mechanical Cost Data, call 877.795.7394.

G30 Site Mechanical Utilities

G3060 Fuel Distribution

Fiberglass Underground Fuel Storage Tank, Double Wall

System Components	QUANTITY	UNIT	COST EACH		
			MAT.	INST.	TOTAL
SYSTEM G3060 310 1010					
STORAGE TANK, FUEL, UNDERGROUND, DOUBLE WALL FIBERGLASS, 600 GAL.					
Excavating trench, no sheeting or dewatering, 6'-10 ' D, 3/4 CY hyd backhoe	18.000	Ea.		144.90	144.90
Corrosion resistance wrap & coat, small diam. pipe, 1" diam., add	120.000	Ea.	134.40		134.40
Fiberglass-reinforced elbows 90 & 45 degree, 2 inch	7.000	Ea.	332.50	588	920.50
Fiberglass-reinforced threaded adapters, 2 inch	2.000	Ea.	39.50	192	231.50
Remote tank gauging system, 30', 5" pointer travel	1.000	Ea.	5,625	360	5,985
Corrosion resistance wrap & coat, small diam. pipe. 4 " diam. add	30.000	Ea.	72.90		72.90
Pea gravel	9.000	Ea.	1,062	454.50	1,516.50
Reinforcing in hold down pad, #3 to #7	120.000	Ea.	63.60	75.60	139.20
Tank leak detector system, 8 channel, external monitoring	1.000	Ea.	1,000		1,000
Tubbing copper, type L, 3/4" diam.	120.000	Ea.	684	1,122	1,806
Elbow 90 degree ,wrought cu x cu 3/4 " diam.	4.000	Ea.	10.84	150	160.84
Pipe, black steel welded, sch 40, 3" diam.	30.000	Ea.	510	929.40	1,439.40
Elbow 90 degree, steel welded butt joint, 3 " diam.	3.000	Ea.	96	573.15	669.15
Fiberglass-reinforced plastic pipe, 2 inch	156.000	Ea.	1,014	741	1,755
Fiberglass-reinforced plastic pipe, 3 inch	36.000	Ea.	1,044	194.40	1,238.40
Fiberglass-reinforced elbows 90 & 45 degree, 3 inch	3.000	Ea.	450	336	786
Tank leak detection system, probes, well monitoring, liquid phase detection	2.000	Ea.	1,050		1,050
Double wall fiberglass annular space	1.000	Ea.	355		355
Flange, weld neck, 150 lb, 3 " pipe size	1.000	Ea.	38.50	95.54	134.04
Vent protector/breather, 2" diam.	1.000	Ea.	33.50	22.50	56
Corrosion resistance wrap & coat, small diam. pipe, 2" diam., add	36.000	Ea.	54.72		54.72
Concrete hold down pad, 6' x 5' x 8" thick	30.000	Ea.	88.80	46.20	135
600 gallon capacity, underground storage tank, double wall fiberglass	1.000	Ea.	7,925	535	8,460
Tank, for hold-downs 500-4000 gal., add	1.000	Ea.	490	172	662
Foot valve, single poppet, 3/4" diam.	1.000	Ea.	78.50	40	118.50
Fuel fill box, locking inner cover, 3" diam.	1.000	Ea.	133	144	277
Fuel oil specialties, valve,ball chk, globe type fusible, 3/4" diam.	2.000	Ea.	154	72	226
TOTAL			22,539.76	6,988.19	29,527.95

G3060 310	Fiberglass Fuel Tank, Double Wall	COST EACH		
		MAT.	INST.	TOTAL
1010	Storage tank, fuel, underground, double wall fiberglass, 600 Gal.	22,500	6,975	29,475
1020	2500 Gal.	35,700	9,600	45,300
1030	4000 Gal.	41,800	11,900	53,700
1040	6000 Gal.	48,800	13,400	62,200

600

For customer support on your Mechanical Cost Data, call 877.795.7394.

G3060 Fuel Distribution

G3060 310	Fiberglass Fuel Tank, Double Wall	COST EACH		
		MAT.	INST.	TOTAL
1050	8000 Gal.	56,500	16,200	72,700
1060	10,000 Gal.	64,000	18,600	82,600
1070	15,000 Gal.	87,500	21,400	108,900
1080	20,000 Gal.	110,500	25,900	136,400
1090	25,000 Gal.	140,500	30,000	170,500
1100	30,000 Gal.	160,500	31,800	192,300

G3060 Fuel Distribution

Steel Underground Fuel Storage Tank, Double Wall

System Components	QUANTITY	UNIT	COST EACH		
			MAT.	INST.	TOTAL
SYSTEM G3060 320 1010					
STORAGE TANK, FUEL, UNDERGROUND, DOUBLE WALL STEEL, 500 GAL.					
Fiberglass-reinforced plastic pipe, 2 inch	156.000	Ea.	1,014	741	1,755
Fiberglass-reinforced plastic pipe, 3 inch	36.000	Ea.	1,044	194.40	1,238.40
Fiberglass-reinforced elbows 90 & 45 degree, 3 inch	3.000	Ea.	450	336	786
Tank leak detection system, probes, well monitoring, liquid phase detection	2.000	Ea.	1,050		1,050
Flange, weld neck, 150 lb, 3" pipe size	1.000	Ea.	38.50	95.54	134.04
Vent protector/breather, 2" diam.	1.000	Ea.	33.50	22.50	56
Concrete hold down pad, 6' x 5' x 8" thick	30.000	Ea.	88.80	46.20	135
Tanks, for hold-downs, 500-2000 gal, add	1.000	Ea.	201	172	373
Double wall steel tank annular space	1.000	Ea.	370		370
Foot valve, single poppet, 3/4" diam.	1.000	Ea.	78.50	40	118.50
Fuel fill box, locking inner cover, 3" diam.	1.000	Ea.	133	144	277
Fuel oil specialties, valve, ball chk, globe type fusible, 3/4" diam.	2.000	Ea.	154	72	226
Pea gravel	9.000	Ea.	1,062	454.50	1,516.50
Reinforcing in hold down pad, #3 to #7	120.000	Ea.	63.60	75.60	139.20
Tank, for manways, add	1.000	Ea.	2,000		2,000
Tank leak detector system, 8 channel, external monitoring	1.000	Ea.	1,000		1,000
Tubbing copper, type L, 3/4" diam.	120.000	Ea.	684	1,122	1,806
Elbow 90 degree, wrought cu x cu 3/4" diam.	8.000	Ea.	21.68	300	321.68
Pipe, black steel welded, sch 40, 3" diam.	30.000	Ea.	510	929.40	1,439.40
Elbow 90 degree, steel welded butt joint, 3" diam.	3.000	Ea.	96	573.15	669.15
Excavating trench, no sheeting or dewatering, 6'-10' D, 3/4 CY hyd backhoe	18.000	Ea.		144.90	144.90
Fiberglass-reinforced elbows 90 & 45 degree, 2 inch	7.000	Ea.	332.50	588	920.50
Fiberglass-reinforced threaded adapters, 2 inch	2.000	Ea.	39.50	192	231.50
500 gallon capacity, underground storage tank, double wall steel	1.000	Ea.	6,150	540	6,690
Remote tank gauging system, 30', 5" pointer travel	1.000	Ea.	5,625	360	5,985
TOTAL			22,239.58	7,143.19	29,382.77

G3060 320	Steel Fuel Tank, Double Wall	COST EACH		
		MAT.	INST.	TOTAL
1010	Storage tank, fuel, underground, double wall steel, 500 Gal.	22,200	7,125	29,325
1020	2000 Gal.	29,900	9,325	39,225
1030	4000 Gal.	40,700	12,000	52,700
1040	6000 Gal.	55,000	13,300	68,300
1050	8000 Gal.	59,000	16,400	75,400
1060	10,000 Gal.	66,000	18,800	84,800

G3060 Fuel Distribution

G3060 320	Steel Fuel Tank, Double Wall	COST EACH		
		MAT.	INST.	TOTAL
1070	15,000 Gal.	75,000	21,900	96,900
1080	20,000 Gal.	87,000	26,200	113,200
1090	25,000 Gal.	135,500	30,000	165,500
1100	30,000 Gal.	154,000	31,900	185,900
1110	40,000 Gal.	194,000	40,000	234,000
1120	50,000 Gal.	234,500	48,000	282,500

G3060 325	Steel Tank, Above Ground, Single Wall	COST EACH		
		MAT.	INST.	TOTAL
1010	Storage tank, fuel, above ground, single wall steel, 550 Gal.	14,200	5,000	19,200
1020	2000 Gal.	21,800	5,125	26,925
1030	5000 Gal.	32,900	7,825	40,725
1040	10,000 Gal.	50,500	8,350	58,850
1050	15,000 Gal.	61,500	8,600	70,100
1060	20,000 Gal.	76,000	8,875	84,875
1070	25,000 Gal.	87,000	9,100	96,100
1080	30,000 Gal.	102,000	9,475	111,475

G3060 330	Steel Tank, Above Ground, Double Wall	COST EACH		
		MAT.	INST.	TOTAL
1010	Storage tank, fuel, above ground, double wall steel, 500 Gal.	14,300	5,050	19,350
1020	2000 Gal.	22,500	5,175	27,675
1030	4000 Gal.	31,400	5,275	36,675
1040	6000 Gal.	36,700	8,125	44,825
1050	8000 Gal.	43,700	8,350	52,050
1060	10,000 Gal.	47,200	8,500	55,700
1070	15,000 Gal.	64,500	8,800	73,300
1080	20,000 Gal.	71,500	9,100	80,600
1090	25,000 Gal.	84,000	9,375	93,375
1100	30,000 Gal.	90,500	9,725	100,225

603

For customer support on your Mechanical Cost Data, call 877.795.7394.

Reference Section

All the reference information is in one section, making it easy to find what you need to know ... and easy to use the book on a daily basis. This section is visually identified by a vertical black bar on the page edges.

In this Reference Section, we've included Equipment Rental Costs, a listing of rental and operating costs; Crew Listings, a full listing of all crews and equipment, and their costs; Historical Cost Indexes for cost comparisons over time; City Cost Indexes and Location Factors for adjusting costs to the region you are in; Reference Tables, where you will find explanations, estimating information and procedures, or technical data; Change Orders, information on pricing changes to contract documents; Square Foot Costs that allow you to make a rough estimate for the overall cost of a project; and an explanation of all the Abbreviations in the book.

Table of Contents

Estimating Tips

- This section contains the average costs to rent and operate hundreds of pieces of construction equipment. This is useful information when estimating the time and material requirements of any particular operation in order to establish a unit or total cost. Equipment costs include not only rental, but also operating costs for equipment under normal use.

Rental Costs

- Equipment rental rates are obtained from the following industry sources throughout North America: contractors, suppliers, dealers, manufacturers, and distributors.

- Rental rates vary throughout the country, with larger cities generally having lower rates. Lease plans for new equipment are available for periods in excess of six months, with a percentage of payments applying toward purchase.

- Monthly rental rates vary from 2% to 5% of the purchase price of the equipment depending on the anticipated life of the equipment and its wearing parts.

- Weekly rental rates are about 1/3 the monthly rates, and daily rental rates are about 1/3 the weekly rate.

- Rental rates can also be treated as reimbursement costs for contractor-owned equipment. Owned equipment costs include depreciation, loan payments, interest, taxes, insurance, storage, and major repairs.

Operating Costs

- The operating costs include parts and labor for routine servicing, such as repair and replacement of pumps, filters and worn lines. Normal operating expendables, such as fuel, lubricants, tires and electricity (where applicable), are also included.

- Extraordinary operating expendables with highly variable wear patterns, such as diamond bits and blades, are excluded. These costs can be found as material costs in the Unit Price section.

- The hourly operating costs listed do not include the operator's wages.

Equipment Cost/Day

- Any power equipment required by a crew is shown in the Crew Listings with a daily cost.

- The daily cost of equipment needed by a crew is based on dividing the weekly rental rate by 5 (number of working days in the week), and then adding the hourly operating cost times 8 (the number of hours in a day). This "Equipment Cost/ Day" is shown in the far right column of the Equipment Rental pages.

- If equipment is needed for only one or two days, it is best to develop your own cost by including components for daily rent and hourly operating cost. This is important when the listed Crew for a task does not contain the equipment needed, such as a crane for lifting mechanical heating/ cooling equipment up onto a roof.

- If the quantity of work is less than the crew's Daily Output shown for a Unit Price line item that includes a bare unit equipment cost, it is recommended to estimate one day's rental cost and operating cost for equipment shown in the Crew Listing for that line item.

Mobilization/ Demobilization

- The cost to move construction equipment from an equipment yard or rental company to the job site and back again is not included in equipment rental costs listed in the Reference Section, nor in the bare equipment cost of any unit price line item, nor in any equipment costs shown in the Crew listings.

- Mobilization (to the site) and demobilization (from the site) costs can be found in the Unit Price Section.

- If a piece of equipment is already at the job site, it is not appropriate to utilize mobilization/demobilization. costs again in an estimate. ■

01 54 33 | Equipment Rental

		UNIT	HOURLY OPER. COST	RENT PER DAY	RENT PER WEEK	RENT PER MONTH	EQUIPMENT COST/DAY
10	**0010**	**CONCRETE EQUIPMENT RENTAL** without operators R015433 -10						**10**
	0200	Bucket, concrete lightweight, 1/2 C.Y.	Ea.	.80	23.50	70	210	20.40
	0300	1 C.Y.		.90	27.50	83	249	23.80
	0400	1-1/2 C.Y.		1.10	36.50	110	330	30.80
	0500	2 C.Y.		1.25	45	135	405	37
	0580	8 C.Y.		6.10	257	770	2,300	202.80
	0600	Cart, concrete, self-propelled, operator walking, 10 C.F.		3.25	56.50	170	510	60
	0700	Operator riding, 18 C.F.		5.35	95	285	855	99.80
	0800	Conveyer for concrete, portable, gas, 16" wide, 26' long		12.60	123	370	1,100	174.80
	0900	46' long		13.00	148	445	1,325	193
	1000	56' long		13.10	157	470	1,400	198.80
	1100	Core drill, electric, 2-1/2 H.P., 1" to 8" bit diameter		1.77	68.50	205	615	55.15
	1150	11 H.P., 8" to 18" cores		5.95	113	340	1,025	115.60
	1200	Finisher, concrete floor, gas, riding trowel, 96" wide		12.75	145	435	1,300	189
	1300	Gas, walk-behind, 3 blade, 36" trowel		2.15	20	60	180	29.20
	1400	4 blade, 48" trowel		4.20	27.50	83	249	50.20
	1500	Float, hand-operated (Bull float) 48" wide		.08	13.35	40	120	8.65
	1570	Curb builder, 14 H.P., gas, single screw		14.65	248	745	2,225	266.20
	1590	Double screw		15.35	290	870	2,600	296.80
	1600	Floor grinder, concrete and terrazzo, electric, 22" path		2.57	158	475	1,425	115.55
	1700	Edger, concrete, electric, 7" path		1.04	51.50	155	465	39.30
	1750	Vacuum pick-up system for floor grinders, wet/dry		1.49	81.50	245	735	60.90
	1800	Mixer, powered, mortar and concrete, gas, 6 C.F., 18 H.P.		8.65	118	355	1,075	140.20
	1900	10 C.F., 25 H.P.		10.80	143	430	1,300	172.40
	2000	16 C.F.		11.15	165	495	1,475	188.20
	2100	Concrete, stationary, tilt drum, 2 C.Y.		6.90	232	695	2,075	194.20
	2120	Pump, concrete, truck mounted 4" line 80' boom		25.00	865	2,600	7,800	720
	2140	5" line, 110' boom		32.40	1,150	3,435	10,300	946.20
	2160	Mud jack, 50 C.F. per hr.		7.30	125	375	1,125	133.40
	2180	225 C.F. per hr.		9.50	145	435	1,300	163
	2190	Shotcrete pump rig, 12 C.Y./hr.		14.85	220	660	1,975	250.80
	2200	35 C.Y./hr.		17.50	235	705	2,125	281
	2600	Saw, concrete, manual, gas, 18 H.P.		6.75	45	135	405	81
	2650	Self-propelled, gas, 30 H.P.		13.30	102	305	915	167.40
	2675	V-groove crack chaser, manual, gas, 6 H.P.		2.35	17.35	52	156	29.20
	2700	Vibrators, concrete, electric, 60 cycle, 2 H.P.		.46	8.65	26	78	8.90
	2800	3 H.P.		.60	11.65	35	105	11.80
	2900	Gas engine, 5 H.P.		2.00	16	48	144	25.60
	3000	8 H.P.		2.75	15.35	46	138	31.20
	3050	Vibrating screed, gas engine, 8 H.P.		2.91	71.50	215	645	66.30
	3120	Concrete transit mixer, 6 x 4, 250 H.P., 8 C.Y., rear discharge		61.55	570	1,715	5,150	835.40
	3200	Front discharge		72.45	700	2,095	6,275	998.60
	3300	6 x 6, 285 H.P., 12 C.Y., rear discharge		71.50	660	1,980	5,950	968
	3400	Front discharge		74.00	705	2,120	6,350	1,016
20	**0010**	**EARTHWORK EQUIPMENT RENTAL** without operators R015433 -10						**20**
	0040	Aggregate spreader, push type 8' to 12' wide	Ea.	3.05	25.50	76	228	39.60
	0045	Tailgate type, 8' wide		2.90	32.50	98	294	42.80
	0055	Earth auger, truck-mounted, for fence & sign posts, utility poles		19.15	440	1,320	3,950	417.20
	0060	For borings and monitoring wells		46.65	675	2,030	6,100	779.20
	0070	Portable, trailer mounted		3.30	32.50	98	294	46
	0075	Truck-mounted, for caissons, water wells		100.15	2,900	8,705	26,100	2,542
	0080	Horizontal boring machine, 12" to 36" diameter, 45 H.P.		24.20	192	575	1,725	308.60
	0090	12" to 48" diameter, 65 H.P.		34.35	335	1,000	3,000	474.80
	0095	Auger, for fence posts, gas engine, hand held		.60	6	18	54	8.40
	0100	Excavator, diesel hydraulic, crawler mounted, 1/2 C.Y. cap.		24.65	420	1,255	3,775	448.20
	0120	5/8 C.Y. capacity		32.55	550	1,645	4,925	589.40
	0140	3/4 C.Y. capacity		35.60	615	1,850	5,550	654.80
	0150	1 C.Y. capacity		49.70	690	2,075	6,225	812.60

01 54 33	Equipment Rental	UNIT	HOURLY OPER. COST	RENT PER DAY	RENT PER WEEK	RENT PER MONTH	EQUIPMENT COST/DAY	
0200	1-1/2 C.Y. capacity	Ea.	59.80	920	2,760	8,275	1,030	20
0300	2 C.Y. capacity		67.35	1,050	3,180	9,550	1,175	
0320	2-1/2 C.Y. capacity		102.55	1,300	3,925	11,800	1,605	
0325	3-1/2 C.Y. capacity		144.45	2,150	6,420	19,300	2,440	
0330	4-1/2 C.Y. capacity		173.40	2,625	7,880	23,600	2,963	
0335	6 C.Y. capacity		219.80	2,900	8,680	26,000	3,494	
0340	7 C.Y. capacity		222.05	3,025	9,070	27,200	3,590	
0342	Excavator attachments, bucket thumbs		3.20	245	735	2,200	172.60	
0345	Grapples		2.75	193	580	1,750	138	
0346	Hydraulic hammer for boom mounting, 4000 ft lb.		12.40	350	1,045	3,125	308.20	
0347	5000 ft lb.		14.40	425	1,280	3,850	371.20	
0348	8000 ft lb.		21.30	625	1,875	5,625	545.40	
0349	12,000 ft lb.		23.35	750	2,245	6,725	635.80	
0350	Gradall type, truck mounted, 3 ton @ 15' radius, 5/8 C.Y.		43.65	890	2,665	8,000	882.20	
0370	1 C.Y. capacity		47.85	1,025	3,095	9,275	1,002	
0400	Backhoe-loader, 40 to 45 H.P., 5/8 C.Y. capacity		14.90	240	720	2,150	263.20	
0450	45 H.P. to 60 H.P., 3/4 C.Y. capacity		23.40	295	885	2,650	364.20	
0460	80 H.P., 1-1/4 C.Y. capacity		25.60	310	935	2,800	391.80	
0470	112 H.P., 1-1/2 C.Y. capacity		40.95	645	1,930	5,800	713.60	
0482	Backhoe-loader attachment, compactor, 20,000 lb.		5.80	142	425	1,275	131.40	
0485	Hydraulic hammer, 750 ft lb.		3.30	96.50	290	870	84.40	
0486	Hydraulic hammer, 1200 ft lb.		6.30	217	650	1,950	180.40	
0500	Brush chipper, gas engine, 6" cutter head, 35 H.P.		11.10	105	315	945	151.80	
0550	Diesel engine, 12" cutter head, 130 H.P.		27.55	285	855	2,575	391.40	
0600	15" cutter head, 165 H.P.		33.10	335	1,005	3,025	465.80	
0750	Bucket, clamshell, general purpose, 3/8 C.Y.		1.30	38.50	115	345	33.40	
0800	1/2 C.Y.		1.40	45	135	405	38.20	
0850	3/4 C.Y.		1.55	55	165	495	45.40	
0900	1 C.Y.		1.60	58.50	175	525	47.80	
0950	1-1/2 C.Y.		2.55	81.50	245	735	69.40	
1000	2 C.Y.		2.70	90	270	810	75.60	
1010	Bucket, dragline, medium duty, 1/2 C.Y.		.75	23.50	70	210	20	
1020	3/4 C.Y.		.75	24.50	73	219	20.60	
1030	1 C.Y.		.80	26	78	234	22	
1040	1-1/2 C.Y.		1.20	40	120	360	33.60	
1050	2 C.Y.		1.30	45	135	405	37.40	
1070	3 C.Y.		1.95	61.50	185	555	52.60	
1200	Compactor, manually guided 2-drum vibratory smooth roller, 7.5 H.P.		7.25	203	610	1,825	180	
1250	Rammer/tamper, gas, 8"		2.75	46.50	140	420	50	
1260	15"		3.05	53.50	160	480	56.40	
1300	Vibratory plate, gas, 18" plate, 3000 lb. blow		2.70	24.50	73	219	36.20	
1350	21" plate, 5000 lb. blow		3.40	32.50	98	294	46.80	
1370	Curb builder/extruder, 14 H.P., gas, single screw		14.65	248	745	2,225	266.20	
1390	Double screw		15.35	290	870	2,600	296.80	
1500	Disc harrow attachment, for tractor		.44	73.50	221	665	47.70	
1810	Feller buncher, shearing & accumulating trees, 100 H.P.		48.30	755	2,260	6,775	838.40	
1860	Grader, self-propelled, 25,000 lb.		39.50	640	1,925	5,775	701	
1910	30,000 lb.		43.15	650	1,955	5,875	736.20	
1920	40,000 lb.		64.90	1,100	3,275	9,825	1,174	
1930	55,000 lb.		84.15	1,700	5,130	15,400	1,699	
1950	Hammer, pavement breaker, self-propelled, diesel, 1000 to 1250 lb.		29.90	350	1,055	3,175	450.20	
2000	1300 to 1500 lb.		44.85	705	2,110	6,325	780.80	
2050	Pile driving hammer, steam or air, 4150 ft lb. @ 225 bpm		10.40	475	1,420	4,250	367.20	
2100	8750 ft lb. @ 145 bpm		12.40	660	1,975	5,925	494.20	
2150	15,000 ft lb. @ 60 bpm		14.00	790	2,370	7,100	586	
2200	24,450 ft lb. @ 111 bpm		15.00	875	2,630	7,900	646	
2250	Leads, 60' high for pile driving hammers up to 20,000 ft lb.		3.30	80.50	242	725	74.80	
2300	90' high for hammers over 20,000 ft lb.		4.95	141	424	1,275	124.40	

01 54 33 | Equipment Rental

		UNIT	HOURLY OPER. COST	RENT PER DAY	RENT PER WEEK	RENT PER MONTH	EQUIPMENT COST/DAY	
20	2350	Diesel type hammer, 22,400 ft lb.	Ea.	22.55	460	1,380	4,150	456.40	**2**
	2400	41,300 ft lb.		32.00	540	1,625	4,875	581	
	2450	141,000 ft lb.		51.65	960	2,875	8,625	988.20	
	2500	Vib. elec. hammer/extractor, 200 kW diesel generator, 34 H.P.		54.90	635	1,900	5,700	819.20	
	2550	80 H.P.		100.65	925	2,775	8,325	1,360	
	2600	150 H.P.		190.90	1,775	5,300	15,900	2,587	
	2800	Log chipper, up to 22" diameter, 600 H.P.		63.30	640	1,915	5,750	889.40	
	2850	Logger, for skidding & stacking logs, 150 H.P.		52.75	815	2,445	7,325	911	
	2860	Mulcher, diesel powered, trailer mounted		24.05	212	635	1,900	319.40	
	2900	Rake, spring tooth, with tractor		18.23	345	1,042	3,125	354.25	
	3000	Roller, vibratory, tandem, smooth drum, 20 H.P.		8.80	145	435	1,300	157.40	
	3050	35 H.P.		11.10	247	740	2,225	236.80	
	3100	Towed type vibratory compactor, smooth drum, 50 H.P.		26.00	315	940	2,825	396	
	3150	Sheepsfoot, 50 H.P.		27.35	345	1,035	3,100	425.80	
	3170	Landfill compactor, 220 H.P.		91.00	1,475	4,440	13,300	1,616	
	3200	Pneumatic tire roller, 80 H.P.		16.80	350	1,050	3,150	344.40	
	3250	120 H.P.		25.20	600	1,800	5,400	561.60	
	3300	Sheepsfoot vibratory roller, 240 H.P.		68.95	1,100	3,270	9,800	1,206	
	3320	340 H.P.		100.85	1,575	4,730	14,200	1,753	
	3350	Smooth drum vibratory roller, 75 H.P.		24.85	585	1,755	5,275	549.80	
	3400	125 H.P.		32.55	710	2,135	6,400	687.40	
	3410	Rotary mower, brush, 60", with tractor		22.60	315	950	2,850	370.80	
	3420	Rototiller, walk-behind, gas, 5 H.P.		1.96	51.50	155	465	46.70	
	3422	8 H.P.		3.08	80	240	720	72.65	
	3440	Scrapers, towed type, 7 C.Y. capacity		5.80	113	340	1,025	114.40	
	3450	10 C.Y. capacity		6.60	157	470	1,400	146.80	
	3500	15 C.Y. capacity		7.10	180	540	1,625	164.80	
	3525	Self-propelled, single engine, 14 C.Y. capacity		114.76	1,600	4,830	14,500	1,884	
	3550	Dual engine, 21 C.Y. capacity		173.80	2,175	6,490	19,500	2,688	
	3600	31 C.Y. capacity		231.70	3,075	9,195	27,600	3,693	
	3640	44 C.Y. capacity		287.80	3,950	11,885	35,700	4,679	
	3650	Elevating type, single engine, 11 C.Y. capacity		71.45	985	2,955	8,875	1,163	
	3700	22 C.Y. capacity		139.55	2,350	7,025	21,100	2,521	
	3710	Screening plant 110 H.P. w/5' x 10' screen		25.35	400	1,200	3,600	442.80	
	3720	5' x 16' screen		29.65	505	1,515	4,550	540.20	
	3850	Shovel, crawler-mounted, front-loading, 7 C.Y. capacity		258.85	3,200	9,610	28,800	3,993	
	3855	12 C.Y. capacity		386.45	4,400	13,220	39,700	5,736	
	3860	Shovel/backhoe bucket, 1/2 C.Y.		2.50	66.50	200	600	60	
	3870	3/4 C.Y.		2.55	75	225	675	65.40	
	3880	1 C.Y.		2.65	83.50	250	750	71.20	
	3890	1-1/2 C.Y.		2.75	96.50	290	870	80	
	3910	3 C.Y.		3.15	130	390	1,175	103.20	
	3950	Stump chipper, 18" deep, 30 H.P.		7.82	205	615	1,850	185.55	
	4110	Dozer, crawler, torque converter, diesel 80 H.P.		29.05	400	1,200	3,600	472.40	
	4150	105 H.P.		35.60	530	1,590	4,775	602.80	
	4200	140 H.P.		51.25	795	2,380	7,150	886	
	4260	200 H.P.		77.25	1,275	3,845	11,500	1,387	
	4310	300 H.P.		100.65	1,825	5,460	16,400	1,897	
	4360	410 H.P.		132.60	2,250	6,720	20,200	2,405	
	4370	500 H.P.		171.15	3,225	9,650	29,000	3,299	
	4380	700 H.P.		280.90	4,550	13,660	41,000	4,979	
	4400	Loader, crawler, torque conv., diesel, 1-1/2 C.Y., 80 H.P.		31.25	455	1,360	4,075	522	
	4450	1-1/2 to 1-3/4 C.Y., 95 H.P.		34.60	590	1,765	5,300	629.80	
	4510	1-3/4 to 2-1/4 C.Y., 130 H.P.		53.80	875	2,630	7,900	956.40	
	4530	2-1/2 to 3-1/4 C.Y., 190 H.P.		65.95	1,100	3,300	9,900	1,188	
	4560	3-1/2 to 5 C.Y., 275 H.P.		87.80	1,425	4,295	12,900	1,561	
	4610	Front end loader, 4WD, articulated frame, diesel, 1 to 1-1/4 C.Y., 70 H.P.		19.75	235	705	2,125	299	
	4620	1-1/2 to 1-3/4 C.Y., 95 H.P.		24.80	300	900	2,700	378.40	

01 54 33 | Equipment Rental

		UNIT	HOURLY OPER. COST	RENT PER DAY	RENT PER WEEK	RENT PER MONTH	EQUIPMENT COST/DAY
4650	1-3/4 to 2 C.Y., 130 H.P.	Ea.	29.15	380	1,140	3,425	461.20
4710	2-1/2 to 3-1/2 C.Y., 145 H.P.		33.10	425	1,275	3,825	519.80
4730	3 to 4-1/2 C.Y., 185 H.P.		42.40	550	1,650	4,950	669.20
4760	5-1/4 to 5-3/4 C.Y., 270 H.P.		67.50	905	2,710	8,125	1,082
4810	7 to 9 C.Y., 475 H.P.		114.05	1,700	5,125	15,400	1,937
4870	9 - 11 C.Y., 620 H.P.		160.55	2,775	8,290	24,900	2,942
4880	Skid steer loader, wheeled, 10 C.F., 30 H.P. gas		10.45	150	450	1,350	173.60
4890	1 C.Y., 78 H.P., diesel		20.20	262	785	2,350	318.60
4892	Skid-steer attachment, auger		.48	80.50	242	725	52.25
4893	Backhoe		.66	111	332	995	71.70
4894	Broom		.74	124	372	1,125	80.30
4895	Forks		.22	36	108	325	23.35
4896	Grapple		.56	92.50	278	835	60.10
4897	Concrete hammer		1.06	177	531	1,600	114.70
4898	Tree spade		1.03	172	515	1,550	111.25
4899	Trencher		.54	90	270	810	58.30
4900	Trencher, chain, boom type, gas, operator walking, 12 H.P.		5.00	46.50	140	420	68
4910	Operator riding, 40 H.P.		20.15	290	870	2,600	335.20
5000	Wheel type, diesel, 4' deep, 12" wide		85.95	845	2,535	7,600	1,195
5100	6' deep, 20" wide		92.75	1,925	5,740	17,200	1,890
5150	Chain type, diesel, 5' deep, 8" wide		38.35	560	1,675	5,025	641.80
5200	Diesel, 8' deep, 16" wide		155.95	3,600	10,770	32,300	3,402
5202	Rock trencher, wheel type, 6" wide x 18" deep		21.50	335	1,000	3,000	372
5206	Chain type, 18" wide x 7' deep		112.00	2,800	8,395	25,200	2,575
5210	Tree spade, self-propelled		14.38	267	800	2,400	275.05
5250	Truck, dump, 2-axle, 12 ton, 8 C.Y. payload, 220 H.P.		34.40	227	680	2,050	411.20
5300	Three axle dump, 16 ton, 12 C.Y. payload, 400 H.P.		61.00	340	1,015	3,050	691
5310	Four axle dump, 25 ton, 18 C.Y. payload, 450 H.P.		71.75	490	1,475	4,425	869
5350	Dump trailer only, rear dump, 16-1/2 C.Y.		5.45	138	415	1,250	126.60
5400	20 C.Y.		5.90	157	470	1,400	141.20
5450	Flatbed, single axle, 1-1/2 ton rating		25.55	68.50	205	615	245.40
5500	3 ton rating		30.65	96.50	290	870	303.20
5550	Off highway rear dump, 25 ton capacity		73.90	1,275	3,800	11,400	1,351
5600	35 ton capacity		82.50	1,425	4,260	12,800	1,512
5610	50 ton capacity		102.65	1,700	5,080	15,200	1,837
5620	65 ton capacity		105.65	1,700	5,090	15,300	1,863
5630	100 ton capacity		151.40	2,850	8,570	25,700	2,925
6000	Vibratory plow, 25 H.P., walking		8.45	60	180	540	103.60
0010	**GENERAL EQUIPMENT RENTAL** without operators R015433 -10	Ea.					
0150	Aerial lift, scissor type, to 15' high, 1000 lb. cap., electric		3.05	51.50	155	465	55.40
0160	To 25' high, 2000 lb. capacity		3.45	66.50	200	600	67.60
0170	Telescoping boom to 40' high, 500 lb. capacity, diesel		13.55	320	965	2,900	301.40
0180	To 45' high, 500 lb. capacity		14.75	350	1,055	3,175	329
0190	To 60' high, 600 lb. capacity		17.20	495	1,490	4,475	435.60
0195	Air compressor, portable, 6.5 CFM, electric		.67	12.65	38	114	12.95
0196	Gasoline		.81	19	57	171	17.90
0200	Towed type, gas engine, 60 CFM		13.55	48.50	145	435	137.40
0300	160 CFM		15.80	50	150	450	156.40
0400	Diesel engine, rotary screw, 250 CFM		17.05	108	325	975	201.40
0500	365 CFM		23.10	132	395	1,175	263.80
0550	450 CFM		29.35	165	495	1,475	333.80
0600	600 CFM		51.70	228	685	2,050	550.60
0700	750 CFM		51.90	237	710	2,125	557.20
0800	For silenced models, small sizes, add to rent		3%	5%	5%	5%	
0900	Large sizes, add to rent		5%	7%	7%	7%	
0930	Air tools, breaker, pavement, 60 lb.	Ea.	.50	9.65	29	87	9.80
0940	80 lb.		.50	10.35	31	93	10.20
0950	Drills, hand (jackhammer) 65 lb.		.60	17.35	52	156	15.20

For customer support on your Mechanical Cost Data, call 877.795.7394.

611

		UNIT	HOURLY OPER. COST	RENT PER DAY	RENT PER WEEK	RENT PER MONTH	EQUIPMENT COST/DAY
01 54 33	**Equipment Rental**						
40	0960	Track or wagon, swing boom, 4" drifter	Ea.	62.50	880	2,640	7,925	1,028
	0970	5" drifter		74.45	1,075	3,235	9,700	1,243
	0975	Track mounted quarry drill, 6" diameter drill		128.10	1,600	4,765	14,300	1,978
	0980	Dust control per drill		1.02	23.50	71	213	22.35
	0990	Hammer, chipping, 12 lb.		.55	26	78	234	20
	1000	Hose, air with couplings, 50' long, 3/4" diameter		.03	5	15	45	3.25
	1100	1" diameter		.04	6.35	19	57	4.10
	1200	1-1/2" diameter		.05	9	27	81	5.80
	1300	2" diameter		.07	12	36	108	7.75
	1400	2-1/2" diameter		.11	19	57	171	12.30
	1410	3" diameter		.14	23	69	207	14.90
	1450	Drill, steel, 7/8" x 2'		.05	8.65	26	78	5.60
	1460	7/8" x 6'		.05	9	27	81	5.80
	1520	Moil points		.02	3.33	10	30	2.15
	1525	Pneumatic nailer w/accessories		.58	38.50	115	345	27.65
	1530	Sheeting driver for 60 lb. breaker		.04	6	18	54	3.90
	1540	For 90 lb. breaker		.12	8	24	72	5.75
	1550	Spade, 25 lb.		.45	6.65	20	60	7.60
	1560	Tamper, single, 35 lb.		.55	36.50	109	325	26.20
	1570	Triple, 140 lb.		.82	54.50	164	490	39.35
	1580	Wrenches, impact, air powered, up to 3/4" bolt		.40	12.65	38	114	10.80
	1590	Up to 1-1/4" bolt		.50	23.50	70	210	18
	1600	Barricades, barrels, reflectorized, 1 to 99 barrels		.03	4.60	13.80	41.50	3
	1610	100 to 200 barrels		.02	3.53	10.60	32	2.30
	1620	Barrels with flashers, 1 to 99 barrels		.03	5.25	15.80	47.50	3.40
	1630	100 to 200 barrels		.03	4.20	12.60	38	2.75
	1640	Barrels with steady burn type C lights		.04	7	21	63	4.50
	1650	Illuminated board, trailer mounted, with generator		3.50	130	390	1,175	106
	1670	Portable barricade, stock, with flashers, 1 to 6 units		.03	5.25	15.80	47.50	3.40
	1680	25 to 50 units		.03	4.90	14.70	44	3.20
	1685	Butt fusion machine, wheeled, 1.5 HP electric, 2" - 8" diameter pipe		2.63	167	500	1,500	121.05
	1690	Tracked, 20 HP diesel, 4"-12" diameter pipe		11.14	490	1,465	4,400	382.10
	1695	83 HP diesel, 8" - 24" diameter pipe		30.46	975	2,930	8,800	829.70
	1700	Carts, brick, hand powered, 1000 lb. capacity		.50	83.50	251	755	54.20
	1800	Gas engine, 1500 lb., 7-1/2' lift		4.17	115	345	1,025	102.35
	1822	Dehumidifier, medium, 6 lb./hr., 150 CFM		1.00	62	186	560	45.20
	1824	Large, 18 lb./hr., 600 CFM		2.02	126	378	1,125	91.75
	1830	Distributor, asphalt, trailer mounted, 2000 gal., 38 H.P. diesel		9.95	325	980	2,950	275.60
	1840	3000 gal., 38 H.P. diesel		11.45	355	1,070	3,200	305.60
	1850	Drill, rotary hammer, electric		1.02	26	78	234	23.75
	1860	Carbide bit, 1-1/2" diameter, add to electric rotary hammer		.02	3.61	10.84	32.50	2.35
	1865	Rotary, crawler, 250 H.P.		148.50	2,050	6,185	18,600	2,425
	1870	Emulsion sprayer, 65 gal., 5 H.P. gas engine		2.91	97	291	875	81.50
	1880	200 gal., 5 H.P. engine		7.85	162	485	1,450	159.80
	1900	Floor auto-scrubbing machine, walk-behind, 28" path		4.98	325	970	2,900	233.85
	1930	Floodlight, mercury vapor, or quartz, on tripod, 1000 watt		.44	20.50	62	186	15.90
	1940	2000 watt		.83	41	123	370	31.25
	1950	Floodlights, trailer mounted with generator, 1 - 300 watt light		3.65	71.50	215	645	72.20
	1960	2 - 1000 watt lights		4.85	96.50	290	870	96.80
	2000	4 - 300 watt lights		4.55	93.50	280	840	92.40
	2005	Foam spray rig, incl. box trailer, compressor, generator, proportioner		35.03	490	1,465	4,400	573.25
	2020	Forklift, straight mast, 12' lift, 5000 lb., 2 wheel drive, gas		26.35	202	605	1,825	331.80
	2040	21' lift, 5000 lb., 4 wheel drive, diesel		20.90	240	720	2,150	311.20
	2050	For rough terrain, 42' lift, 35' reach, 9000 lb., 110 H.P.		29.55	485	1,450	4,350	526.40
	2060	For plant, 4 ton capacity, 80 H.P., 2 wheel drive, gas		16.15	93.50	280	840	185.20
	2080	10 ton capacity, 120 H.P., 2 wheel drive, diesel		24.10	162	485	1,450	289.80
	2100	Generator, electric, gas engine, 1.5 kW to 3 kW		3.75	11.35	34	102	36.80
	2200	5 kW		4.80	15.35	46	138	47.60

01 54 33 | Equipment Rental

		UNIT	HOURLY OPER. COST	RENT PER DAY	RENT PER WEEK	RENT PER MONTH	EQUIPMENT COST/DAY	
2300	10 kW	Ea.	9.10	36.50	110	330	94.80	40
2400	25 kW		10.60	85	255	765	135.80	
2500	Diesel engine, 20 kW		12.10	66.50	200	600	136.80	
2600	50 kW		23.35	103	310	930	248.80	
2700	100 kW		43.30	128	385	1,150	423.40	
2800	250 kW		85.15	235	705	2,125	822.20	
2850	Hammer, hydraulic, for mounting on boom, to 500 ft lb.		2.55	75	225	675	65.40	
2860	1000 ft lb.		4.35	127	380	1,150	110.80	
2900	Heaters, space, oil or electric, 50 MBH		2.01	7.65	23	69	20.70	
3000	100 MBH		3.76	10.65	32	96	36.50	
3100	300 MBH		11.03	38.50	115	345	111.25	
3150	500 MBH		18.15	45	135	405	172.20	
3200	Hose, water, suction with coupling, 20' long, 2" diameter		.02	3	9	27	1.95	
3210	3" diameter		.03	4.33	13	39	2.85	
3220	4" diameter		.03	5	15	45	3.25	
3230	6" diameter		.11	17.65	53	159	11.50	
3240	8" diameter		.27	44.50	133	400	28.75	
3250	Discharge hose with coupling, 50' long, 2" diameter		.01	1.33	4	12	.90	
3260	3" diameter		.01	2.33	7	21	1.50	
3270	4" diameter		.02	3.67	11	33	2.35	
3280	6" diameter		.06	9.35	28	84	6.10	
3290	8" diameter		.18	30	90	270	19.45	
3295	Insulation blower		.78	6	18	54	9.85	
3300	Ladders, extension type, 16' to 36' long		.14	22.50	68	204	14.70	
3400	40' to 60' long		.19	31	93	279	20.10	
3405	Lance for cutting concrete		2.35	71.50	215	645	61.80	
3407	Lawn mower, rotary, 22", 5 H.P.		1.80	49	147	440	43.80	
3408	48" self propelled		2.98	75	225	675	68.85	
3410	Level, electronic, automatic, with tripod and leveling rod		1.20	75.50	226	680	54.80	
3430	Laser type, for pipe and sewer line and grade		.73	48.50	145	435	34.85	
3440	Rotating beam for interior control		.78	52	156	470	37.45	
3460	Builder's optical transit, with tripod and rod		.09	14.65	44	132	9.50	
3500	Light towers, towable, with diesel generator, 2000 watt		4.55	93.50	280	840	92.40	
3600	4000 watt		4.85	96.50	290	870	96.80	
3700	Mixer, powered, plaster and mortar, 6 C.F., 7 H.P.		2.70	20	60	180	33.60	
3800	10 C.F., 9 H.P.		2.80	31.50	94	282	41.20	
3850	Nailer, pneumatic		.58	38.50	115	345	27.65	
3900	Paint sprayers complete, 8 CFM		1.04	69.50	208	625	49.90	
4000	17 CFM		1.90	127	380	1,150	91.20	
4020	Pavers, bituminous, rubber tires, 8' wide, 50 H.P., diesel		31.15	490	1,465	4,400	542.20	
4030	10' wide, 150 H.P.		106.10	1,800	5,370	16,100	1,923	
4050	Crawler, 8' wide, 100 H.P., diesel		88.65	1,800	5,385	16,200	1,786	
4060	10' wide, 150 H.P.		113.15	2,200	6,610	19,800	2,227	
4070	Concrete paver, 12' to 24' wide, 250 H.P.		101.40	1,550	4,675	14,000	1,746	
4080	Placer-spreader-trimmer, 24' wide, 300 H.P.		150.25	2,650	7,960	23,900	2,794	
4100	Pump, centrifugal gas pump, 1-1/2" diam., 65 GPM		3.90	48.50	145	435	60.20	
4200	2" diameter, 130 GPM		5.35	53.50	160	480	74.80	
4300	3" diameter, 250 GPM		5.65	55	165	495	78.20	
4400	6" diameter, 1500 GPM		30.45	172	515	1,550	346.60	
4500	Submersible electric pump, 1-1/4" diameter, 55 GPM		.39	16.35	49	147	12.90	
4600	1-1/2" diameter, 83 GPM		.43	18.65	56	168	14.65	
4700	2" diameter, 120 GPM		1.50	23.50	70	210	26	
4800	3" diameter, 300 GPM		2.69	41.50	125	375	46.50	
4900	4" diameter, 560 GPM		12.34	158	475	1,425	193.70	
5000	6" diameter, 1590 GPM		18.40	212	635	1,900	274.20	
5100	Diaphragm pump, gas, single, 1-1/2" diameter		1.18	49.50	148	445	39.05	
5200	2" diameter		4.25	61.50	185	555	71	
5300	3" diameter		4.25	61.50	185	555	71	

01 54 33 | Equipment Rental

		UNIT	HOURLY OPER. COST	RENT PER DAY	RENT PER WEEK	RENT PER MONTH	EQUIPMENT COST/DAY
40 5400	Double, 4" diameter	Ea.	6.40	105	315	945	114.20
5450	Pressure washer 5 GPM, 3000 psi		4.90	51.50	155	465	70.20
5460	7 GPM, 3000 psi		6.45	60	180	540	87.60
5500	Trash pump, self-priming, gas, 2" diameter		4.60	21	63	189	49.40
5600	Diesel, 4" diameter		9.15	88.50	265	795	126.20
5650	Diesel, 6" diameter		25.25	147	440	1,325	290
5655	Grout Pump		26.35	268	805	2,425	371.80
5700	Salamanders, L.P. gas fired, 100,000 Btu		3.96	13.65	41	123	39.90
5705	50,000 Btu		2.21	10.35	31	93	23.90
5720	Sandblaster, portable, open top, 3 C.F. capacity		.55	26.50	80	240	20.40
5730	6 C.F. capacity		.95	40	120	360	31.60
5740	Accessories for above		.13	21.50	65	195	14.05
5750	Sander, floor		.70	14.35	43	129	14.20
5760	Edger		.50	14.35	43	129	12.60
5800	Saw, chain, gas engine, 18" long		2.30	21.50	64	192	31.20
5900	Hydraulic powered, 36" long		.75	65	195	585	45
5950	60" long		.75	66.50	200	600	46
6000	Masonry, table mounted, 14" diameter, 5 H.P.		1.32	56.50	170	510	44.55
6050	Portable cut-off, 8 H.P.		2.50	33.50	100	300	40
6100	Circular, hand held, electric, 7-1/4" diameter		.23	4.67	14	42	4.65
6200	12" diameter		.23	8	24	72	6.65
6250	Wall saw, w/hydraulic power, 10 H.P.		9.70	61.50	185	555	114.60
6275	Shot blaster, walk-behind, 20" wide		4.75	293	880	2,650	214
6280	Sidewalk broom, walk-behind		2.52	78.50	235	705	67.15
6300	Steam cleaner, 100 gallons per hour		3.70	76.50	230	690	75.60
6310	200 gallons per hour		5.35	95	285	855	99.80
6340	Tar Kettle/Pot, 400 gallons		15.60	75	225	675	169.80
6350	Torch, cutting, acetylene-oxygen, 150' hose, excludes gases		.30	15	45	135	11.40
6360	Hourly operating cost includes tips and gas		19.00				152
6410	Toilet, portable chemical		.13	21	63	189	13.65
6420	Recycle flush type		.15	25	75	225	16.20
6430	Toilet, fresh water flush, garden hose,		.18	30.50	91	273	19.65
6440	Hoisted, non-flush, for high rise		.15	24.50	74	222	16
6465	Tractor, farm with attachment		21.50	297	890	2,675	350
6480	Trailers, platform, flush deck, 2 axle, 3 ton capacity		1.45	20	60	180	23.60
6500	25 ton capacity		5.45	117	350	1,050	113.60
6600	40 ton capacity		7.00	163	490	1,475	154
6700	3 axle, 50 ton capacity		7.55	180	540	1,625	168.40
6800	75 ton capacity		9.40	235	705	2,125	216.20
6810	Trailer mounted cable reel for high voltage line work		5.45	260	779	2,325	199.40
6820	Trailer mounted cable tensioning rig		10.85	515	1,550	4,650	396.80
6830	Cable pulling rig		73.05	2,900	8,680	26,000	2,320
6900	Water tank trailer, engine driven discharge, 5000 gallons		7.00	142	425	1,275	141
6925	10,000 gallons		9.50	197	590	1,775	194
6950	Water truck, off highway, 6000 gallons		88.15	770	2,315	6,950	1,168
7010	Tram car for high voltage line work, powered, 2 conductor		6.60	141	423	1,275	137.40
7020	Transit (builder's level) with tripod		.09	14.65	44	132	9.50
7030	Trench box, 3000 lb., 6' x 8'		.56	93.50	280	840	60.50
7040	7200 lb., 6' x 20'		.75	125	375	1,125	81
7050	8000 lb., 8' x 16'		1.08	180	540	1,625	116.65
7060	9500 lb., 8' x 20'		1.21	201	603	1,800	130.30
7065	11,000 lb., 8' x 24'		1.27	211	633	1,900	136.75
7070	12,000 lb., 10' x 20'		1.50	249	748	2,250	161.60
7100	Truck, pickup, 3/4 ton, 2 wheel drive		13.65	58.50	175	525	144.20
7200	4 wheel drive		13.95	73.50	220	660	155.60
7250	Crew carrier, 9 passenger		19.30	86.50	260	780	206.40
7290	Flat bed truck, 20,000 lb. GVW		21.70	125	375	1,125	248.60
7300	Tractor, 4 x 2, 220 H.P.		30.20	197	590	1,775	359.60

01 54 33 | Equipment Rental

		UNIT	HOURLY OPER. COST	RENT PER DAY	RENT PER WEEK	RENT PER MONTH	EQUIPMENT COST/DAY	
7410	330 H.P.	Ea.	44.75	270	810	2,425	520	**40**
7500	6 x 4, 380 H.P.		51.35	315	945	2,825	599.80	
7600	450 H.P.		62.30	380	1,145	3,425	727.40	
7610	Tractor, with A frame, boom and winch, 225 H.P.		33.00	272	815	2,450	427	
7620	Vacuum truck, hazardous material, 2500 gallons		12.05	290	870	2,600	270.40	
7625	5,000 gallons		14.26	405	1,220	3,650	358.10	
7650	Vacuum, HEPA, 16 gallon, wet/dry		.85	18	54	162	17.60	
7655	55 gallon, wet/dry		.80	27	81	243	22.60	
7660	Water tank, portable		.16	26.50	80	240	17.30	
7690	Sewer/catch basin vacuum, 14 C.Y., 1500 gallons		18.53	610	1,830	5,500	514.25	
7700	Welder, electric, 200 amp		3.88	16.35	49	147	40.85	
7800	300 amp		5.71	20	60	180	57.70	
7900	Gas engine, 200 amp		14.65	23.50	70	210	131.20	
8000	300 amp		16.38	24.50	74	222	145.85	
8100	Wheelbarrow, any size		.08	13	39	117	8.45	
8200	Wrecking ball, 4000 lb.		2.35	68.50	205	615	59.80	
0010	**HIGHWAY EQUIPMENT RENTAL** without operators							**50**
0050	Asphalt batch plant, portable drum mixer, 100 ton/hr.	Ea.	78.06	1,425	4,270	12,800	1,478	
0060	200 ton/hr.		89.18	1,500	4,530	13,600	1,619	
0070	300 ton/hr.		106.34	1,775	5,325	16,000	1,916	
0100	Backhoe attachment, long stick, up to 185 H.P., 10.5' long		.35	23.50	70	210	16.80	
0140	Up to 250 H.P., 12' long		.38	25	75	225	18.05	
0180	Over 250 H.P., 15' long		.53	35	105	315	25.25	
0200	Special dipper arm, up to 100 H.P., 32' long		1.08	71.50	215	645	51.65	
0240	Over 100 H.P., 33' long		1.34	89.50	268	805	64.30	
0280	Catch basin/sewer cleaning truck, 3 ton, 9 C.Y., 1000 gal.		42.70	385	1,160	3,475	573.60	
0300	Concrete batch plant, portable, electric, 200 C.Y./hr.		22.87	515	1,550	4,650	492.95	
0520	Grader/dozer attachment, ripper/scarifier, rear mounted, up to 135 H.P.		2.90	56.50	170	510	57.20	
0540	Up to 180 H.P.		3.90	88.50	265	795	84.20	
0580	Up to 250 H.P.		5.00	117	350	1,050	110	
0700	Pvmt. removal bucket, for hyd. excavator, up to 90 H.P.		1.90	53.50	160	480	47.20	
0740	Up to 200 H.P.		2.15	73.50	220	660	61.20	
0780	Over 200 H.P.		2.25	85	255	765	69	
0900	Aggregate spreader, self-propelled, 187 H.P.		53.00	685	2,050	6,150	834	
1000	Chemical spreader, 3 C.Y.		3.30	43.50	130	390	52.40	
1900	Hammermill, traveling, 250 H.P.		78.83	2,075	6,240	18,700	1,879	
2000	Horizontal borer, 3" diameter, 13 H.P. gas driven		6.35	55	165	495	83.80	
2150	Horizontal directional drill, 20,000 lb. thrust, 78 H.P. diesel		31.15	675	2,025	6,075	654.20	
2160	30,000 lb. thrust, 115 H.P.		38.90	1,025	3,100	9,300	931.20	
2170	50,000 lb. thrust, 170 H.P.		55.65	1,325	3,960	11,900	1,237	
2190	Mud trailer for HDD, 1500 gallons, 175 H.P., gas		31.90	152	455	1,375	346.20	
2200	Hydromulcher, diesel, 3000 gallon, for truck mounting		23.40	245	735	2,200	334.20	
2300	Gas, 600 gallon		8.60	98.50	295	885	127.80	
2400	Joint & crack cleaner, walk behind, 25 H.P.		3.95	50	150	450	61.60	
2500	Filler, trailer mounted, 400 gallons, 20 H.P.		9.50	210	630	1,900	202	
3000	Paint striper, self-propelled, 40 gallon, 22 H.P.		7.30	155	465	1,400	151.40	
3100	120 gallon, 120 H.P.		23.00	395	1,185	3,550	421	
3200	Post drivers, 6" I-Beam frame, for truck mounting		19.05	380	1,135	3,400	379.40	
3400	Road sweeper, self-propelled, 8' wide, 90 H.P.		39.80	575	1,720	5,150	662.40	
3450	Road sweeper, vacuum assisted, 4 C.Y., 220 gallons		75.75	635	1,900	5,700	986	
4000	Road mixer, self-propelled, 130 H.P.		46.85	765	2,295	6,875	833.80	
4100	310 H.P.		83.25	2,100	6,315	18,900	1,929	
4220	Cold mix paver, incl. pug mill and bitumen tank, 165 H.P.		97.00	2,275	6,855	20,600	2,147	
4240	Pavement brush, towed		3.20	93.50	280	840	81.60	
4250	Paver, asphalt, wheel or crawler, 130 H.P., diesel		96.65	2,275	6,795	20,400	2,132	
4300	Paver, road widener, gas 1' to 6', 67 H.P.		48.50	895	2,680	8,050	924	
4400	Diesel, 2' to 14', 88 H.P.		62.70	1,050	3,135	9,400	1,129	
4600	Slipform pavers, curb and gutter, 2 track, 75 H.P.		57.20	900	2,700	8,100	997.60	

R015433 -10

For customer support on your Mechanical Cost Data, call 877.795.7394.

615

01 54 33 | Equipment Rental

		UNIT	HOURLY OPER. COST	RENT PER DAY	RENT PER WEEK	RENT PER MONTH	EQUIPMENT COST/DAY
50	4700	4 track, 165 H.P.	Ea.	41.50	740	2,220	6,650	776
	4800	Median barrier, 215 H.P.		60.50	1,050	3,150	9,450	1,114
	4901	Trailer, low bed, 75 ton capacity		10.10	235	705	2,125	221.80
	5000	Road planer, walk behind, 10" cutting width, 10 H.P.		3.70	33.50	100	300	49.60
	5100	Self-propelled, 12" cutting width, 64 H.P.		10.40	113	340	1,025	151.20
	5120	Traffic line remover, metal ball blaster, truck mounted, 115 H.P.		47.95	730	2,190	6,575	821.60
	5140	Grinder, truck mounted, 115 H.P.		54.60	790	2,375	7,125	911.80
	5160	Walk-behind, 11 H.P.		4.20	53.50	160	480	65.60
	5200	Pavement profiler, 4' to 6' wide, 450 H.P.		254.50	3,325	9,960	29,900	4,028
	5300	8' to 10' wide, 750 H.P.		399.20	4,350	13,055	39,200	5,805
	5400	Roadway plate, steel, 1" x 8' x 20'		.08	13.35	40	120	8.65
	5600	Stabilizer, self-propelled, 150 H.P.		49.30	600	1,795	5,375	753.40
	5700	310 H.P.		99.05	1,675	5,030	15,100	1,798
	5800	Striper, truck mounted, 120 gallon paint, 460 H.P.		64.20	485	1,455	4,375	804.60
	5900	Thermal paint heating kettle, 115 gallons		8.20	25.50	77	231	81
	6000	Tar kettle, 330 gallon, trailer mounted		12.25	56.50	170	510	132
	7000	Tunnel locomotive, diesel, 8 to 12 ton		31.85	585	1,750	5,250	604.80
	7005	Electric, 10 ton		26.00	665	1,995	5,975	607
	7010	Muck cars, 1/2 C.Y. capacity		2.00	24.50	73	219	30.60
	7020	1 C.Y. capacity		2.25	32.50	97	291	37.40
	7030	2 C.Y. capacity		2.35	36.50	110	330	40.80
	7040	Side dump, 2 C.Y. capacity		2.60	45	135	405	47.80
	7050	3 C.Y. capacity		3.45	50	150	450	57.60
	7060	5 C.Y. capacity		4.95	63.50	190	570	77.60
	7100	Ventilating blower for tunnel, 7-1/2 H.P.		2.05	51.50	155	465	47.40
	7110	10 H.P.		2.28	51.50	155	465	49.25
	7120	20 H.P.		3.48	67.50	202	605	68.25
	7140	40 H.P.		5.76	96.50	290	870	104.10
	7160	60 H.P.		8.77	152	455	1,375	161.15
	7175	75 H.P.		11.66	207	620	1,850	217.30
	7180	200 H.P.		23.65	305	910	2,725	371.20
	7800	Windrow loader, elevating		55.00	1,300	3,935	11,800	1,227
60	0010	**LIFTING AND HOISTING EQUIPMENT RENTAL** without operators						
	0120	Aerial lift truck, 2 person, to 80'	Ea.	26.30	705	2,120	6,350	634.40
	0140	Boom work platform, 40' snorkel		12.00	275	825	2,475	261
	0150	Crane, flatbed mounted, 3 ton capacity		16.35	188	565	1,700	243.80
	0200	Crane, climbing, 106' jib, 6000 lb. capacity, 410 fpm		39.35	1,650	4,960	14,900	1,307
	0300	101' jib, 10,250 lb. capacity, 270 fpm		45.95	2,100	6,280	18,800	1,624
	0500	Tower, static, 130' high, 106' jib, 6200 lb. capacity at 400 fpm		43.20	1,900	5,730	17,200	1,492
	0600	Crawler mounted, lattice boom, 1/2 C.Y., 15 tons at 12' radius		36.98	625	1,870	5,600	669.85
	0700	3/4 C.Y., 20 tons at 12' radius		49.31	780	2,340	7,025	862.50
	0800	1 C.Y., 25 tons at 12' radius		65.75	1,050	3,120	9,350	1,150
	0900	1-1/2 C.Y., 40 tons at 12' radius		65.75	1,050	3,150	9,450	1,156
	1000	2 C.Y., 50 tons at 12' radius		69.70	1,225	3,680	11,000	1,294
	1100	3 C.Y., 75 tons at 12' radius		74.70	1,425	4,305	12,900	1,459
	1200	100 ton capacity, 60' boom		84.60	1,650	4,950	14,900	1,667
	1300	165 ton capacity, 60' boom		107.75	1,925	5,785	17,400	2,019
	1400	200 ton capacity, 70' boom		130.55	2,400	7,210	21,600	2,486
	1500	350 ton capacity, 80' boom		182.90	3,625	10,845	32,500	3,632
	1600	Truck mounted, lattice boom, 6 x 4, 20 tons at 10' radius		37.11	1,075	3,260	9,775	948.90
	1700	25 tons at 10' radius		40.09	1,175	3,550	10,700	1,031
	1800	8 x 4, 30 tons at 10' radius		43.61	1,250	3,780	11,300	1,105
	1900	40 tons at 12' radius		46.70	1,325	3,950	11,900	1,164
	2000	60 tons at 15' radius		53.07	1,400	4,180	12,500	1,261
	2050	82 tons at 15' radius		59.80	1,475	4,460	13,400	1,370
	2100	90 tons at 15' radius		67.36	1,625	4,860	14,600	1,511
	2200	115 tons at 15' radius		76.11	1,800	5,430	16,300	1,695
	2300	150 tons at 18' radius		83.65	1,900	5,720	17,200	1,813

R015433-10, R015433-15, R312316-45

01 54 33 | Equipment Rental

		UNIT	HOURLY OPER. COST	RENT PER DAY	RENT PER WEEK	RENT PER MONTH	EQUIPMENT COST/DAY	
2350	165 tons at 18' radius	Ea.	90.05	2,025	6,060	18,200	1,932	60
2400	Truck mounted, hydraulic, 12 ton capacity		42.60	520	1,565	4,700	653.80	
2500	25 ton capacity		44.95	630	1,885	5,650	736.60	
2550	33 ton capacity		45.50	645	1,930	5,800	750	
2560	40 ton capacity		58.85	750	2,250	6,750	920.80	
2600	55 ton capacity		76.80	855	2,570	7,700	1,128	
2700	80 ton capacity		100.45	1,375	4,105	12,300	1,625	
2720	100 ton capacity		94.20	1,425	4,250	12,800	1,604	
2740	120 ton capacity		100.05	1,525	4,575	13,700	1,715	
2760	150 ton capacity		127.60	2,000	5,995	18,000	2,220	
2800	Self-propelled, 4 x 4, with telescoping boom, 5 ton		17.55	225	675	2,025	275.40	
2900	12-1/2 ton capacity		32.30	360	1,075	3,225	473.40	
3000	15 ton capacity		33.00	375	1,130	3,400	490	
3050	20 ton capacity		35.95	445	1,335	4,000	554.60	
3100	25 ton capacity		37.55	500	1,495	4,475	599.40	
3150	40 ton capacity		46.00	560	1,675	5,025	703	
3200	Derricks, guy, 20 ton capacity, 60' boom, 75' mast		27.55	405	1,214	3,650	463.20	
3300	100' boom, 115' mast		43.21	695	2,090	6,275	763.70	
3400	Stiffleg, 20 ton capacity, 70' boom, 37' mast		30.11	525	1,580	4,750	556.90	
3500	100' boom, 47' mast		46.29	845	2,530	7,600	876.30	
3550	Helicopter, small, lift to 1250 lb. maximum, w/pilot		103.70	3,275	9,800	29,400	2,790	
3600	Hoists, chain type, overhead, manual, 3/4 ton		.10	.33	1	3	1	
3900	10 ton		.70	6	18	54	9.20	
4000	Hoist and tower, 5000 lb. cap., portable electric, 40' high		4.95	233	699	2,100	179.40	
4100	For each added 10' section, add		.11	18.35	55	165	11.90	
4200	Hoist and single tubular tower, 5000 lb. electric, 100' high		6.70	325	976	2,925	248.80	
4300	For each added 6'-6" section, add		.19	31.50	95	285	20.50	
4400	Hoist and double tubular tower, 5000 lb., 100' high		7.19	360	1,075	3,225	272.50	
4500	For each added 6'-6" section, add		.21	35	105	315	22.70	
4550	Hoist and tower, mast type, 6000 lb., 100' high		7.76	370	1,115	3,350	285.10	
4570	For each added 10' section, add		.13	21.50	65	195	14.05	
4600	Hoist and tower, personnel, electric, 2000 lb., 100' @ 125 fpm		16.31	990	2,970	8,900	724.50	
4700	3000 lb., 100' @ 200 fpm		18.62	1,125	3,360	10,100	820.95	
4800	3000 lb., 150' @ 300 fpm		20.62	1,250	3,760	11,300	916.95	
4900	4000 lb., 100' @ 300 fpm		21.38	1,275	3,840	11,500	939.05	
5000	6000 lb., 100' @ 275 fpm	▼	23.06	1,350	4,030	12,100	990.50	
5100	For added heights up to 500', add	L.F.	.01	1.67	5	15	1.10	
5200	Jacks, hydraulic, 20 ton	Ea.	.05	2	6	18	1.60	
5500	100 ton		.40	11.65	35	105	10.20	
6100	Jacks, hydraulic, climbing w/50' jackrods, control console, 30 ton cap.		2.01	134	402	1,200	96.50	
6150	For each added 10' jackrod section, add		.05	3.33	10	30	2.40	
6300	50 ton capacity		3.23	215	646	1,950	155.05	
6350	For each added 10' jackrod section, add		.06	4	12	36	2.90	
6500	125 ton capacity		8.45	565	1,690	5,075	405.60	
6550	For each added 10' jackrod section, add		.58	38.50	115	345	27.65	
6600	Cable jack, 10 ton capacity with 200' cable		1.68	112	336	1,000	80.65	
6650	For each added 50' of cable, add	▼	.20	13.35	40	120	9.60	
0010	**WELLPOINT EQUIPMENT RENTAL** without operators R015433 -10							70
0020	Based on 2 months rental							
0100	Combination jetting & wellpoint pump, 60 H.P. diesel	Ea.	18.40	330	996	3,000	346.40	
0200	High pressure gas jet pump, 200 H.P., 300 psi	"	44.06	284	851	2,550	522.70	
0300	Discharge pipe, 8" diameter	L.F.	.01	.54	1.62	4.86	.40	
0350	12" diameter		.01	.79	2.38	7.15	.55	
0400	Header pipe, flows up to 150 GPM, 4" diameter		.01	.49	1.47	4.41	.35	
0500	400 GPM, 6" diameter		.01	.57	1.72	5.15	.40	
0600	800 GPM, 8" diameter		.01	.79	2.38	7.15	.55	
0700	1500 GPM, 10" diameter	▼	.01	.84	2.51	7.55	.60	

01 54 33 | Equipment Rental

			UNIT	HOURLY OPER. COST	RENT PER DAY	RENT PER WEEK	RENT PER MONTH	EQUIPMENT COST/DAY
70	0800	2500 GPM, 12" diameter	L.F.	.02	1.58	4.74	14.20	1.10
	0900	4500 GPM, 16" diameter		.03	2.02	6.07	18.20	1.45
	0950	For quick coupling aluminum and plastic pipe, add	↓	.03	2.09	6.28	18.85	1.50
	1100	Wellpoint, 25' long, with fittings & riser pipe, 1-1/2" or 2" diameter	Ea.	.06	4.18	12.54	37.50	3
	1200	Wellpoint pump, diesel powered, 4" suction, 20 H.P.		7.83	191	574	1,725	177.45
	1300	6" suction, 30 H.P.		10.70	237	712	2,125	228
	1400	8" suction, 40 H.P.		14.45	325	976	2,925	310.80
	1500	10" suction, 75 H.P.		22.27	380	1,141	3,425	406.35
	1600	12" suction, 100 H.P.		31.72	605	1,810	5,425	615.75
	1700	12" suction, 175 H.P.	↓	47.41	670	2,010	6,025	781.30
80	0010	**MARINE EQUIPMENT RENTAL** without operators						
	0200	Barge, 400 Ton, 30' wide x 90' long	Ea.	17.50	1,050	3,180	9,550	776
	0240	800 Ton, 45' wide x 90' long		21.25	1,300	3,870	11,600	944
	2000	Tugboat, diesel, 100 H.P.		38.20	217	650	1,950	435.60
	2040	250 H.P.		79.25	395	1,180	3,550	870
	2080	380 H.P.		156.80	1,175	3,535	10,600	1,961
	3000	Small work boat, gas, 16-foot, 50 H.P.		17.15	60	180	540	173.20
	4000	Large, diesel, 48-foot, 200 H.P.	↓	89.40	1,250	3,740	11,200	1,463

(0010) R015433 -10

Crew No.	Bare Costs		Incl. Subs O&P		Cost Per Labor-Hour	
Crew A-1	Hr.	Daily	Hr.	Daily	Bare Costs	Incl. O&P
1 Building Laborer	$37.60	$300.80	$57.85	$462.80	$37.60	$57.85
1 Concrete Saw, Gas Manual		81.00		89.10	10.13	11.14
8 L.H., Daily Totals		$381.80		$551.90	$47.73	$68.99
Crew A-1A	Hr.	Daily	Hr.	Daily	Bare Costs	Incl. O&P
1 Skilled Worker	$48.65	$389.20	$75.15	$601.20	$48.65	$75.15
1 Shot Blaster, 20"		214.00		235.40	26.75	29.43
8 L.H., Daily Totals		$603.20		$836.60	$75.40	$104.58
Crew A-1B	Hr.	Daily	Hr.	Daily	Bare Costs	Incl. O&P
1 Building Laborer	$37.60	$300.80	$57.85	$462.80	$37.60	$57.85
1 Concrete Saw		167.40		184.14	20.93	23.02
8 L.H., Daily Totals		$468.20		$646.94	$58.52	$80.87
Crew A-1C	Hr.	Daily	Hr.	Daily	Bare Costs	Incl. O&P
1 Building Laborer	$37.60	$300.80	$57.85	$462.80	$37.60	$57.85
1 Chain Saw, Gas, 18"		31.20		34.32	3.90	4.29
8 L.H., Daily Totals		$332.00		$497.12	$41.50	$62.14
Crew A-1D	Hr.	Daily	Hr.	Daily	Bare Costs	Incl. O&P
1 Building Laborer	$37.60	$300.80	$57.85	$462.80	$37.60	$57.85
1 Vibrating Plate, Gas, 18"		36.20		39.82	4.53	4.98
8 L.H., Daily Totals		$337.00		$502.62	$42.13	$62.83
Crew A-1E	Hr.	Daily	Hr.	Daily	Bare Costs	Incl. O&P
1 Building Laborer	$37.60	$300.80	$57.85	$462.80	$37.60	$57.85
1 Vibrating Plate, Gas, 21"		46.80		51.48	5.85	6.43
8 L.H., Daily Totals		$347.60		$514.28	$43.45	$64.28
Crew A-1F	Hr.	Daily	Hr.	Daily	Bare Costs	Incl. O&P
1 Building Laborer	$37.60	$300.80	$57.85	$462.80	$37.60	$57.85
1 Rammer/Tamper, Gas, 8"		50.00		55.00	6.25	6.88
8 L.H., Daily Totals		$350.80		$517.80	$43.85	$64.72
Crew A-1G	Hr.	Daily	Hr.	Daily	Bare Costs	Incl. O&P
1 Building Laborer	$37.60	$300.80	$57.85	$462.80	$37.60	$57.85
1 Rammer/Tamper, Gas, 15"		56.40		62.04	7.05	7.75
8 L.H., Daily Totals		$357.20		$524.84	$44.65	$65.61
Crew A-1H	Hr.	Daily	Hr.	Daily	Bare Costs	Incl. O&P
1 Building Laborer	$37.60	$300.80	$57.85	$462.80	$37.60	$57.85
1 Exterior Steam Cleaner		75.60		83.16	9.45	10.40
8 L.H., Daily Totals		$376.40		$545.96	$47.05	$68.25
Crew A-1J	Hr.	Daily	Hr.	Daily	Bare Costs	Incl. O&P
1 Building Laborer	$37.60	$300.80	$57.85	$462.80	$37.60	$57.85
1 Cultivator, Walk-Behind, 5 H.P.		46.70		51.37	5.84	6.42
8 L.H., Daily Totals		$347.50		$514.17	$43.44	$64.27
Crew A-1K	Hr.	Daily	Hr.	Daily	Bare Costs	Incl. O&P
1 Building Laborer	$37.60	$300.80	$57.85	$462.80	$37.60	$57.85
1 Cultivator, Walk-Behind, 8 H.P.		72.65		79.92	9.08	9.99
8 L.H., Daily Totals		$373.45		$542.72	$46.68	$67.84
Crew A-1M	Hr.	Daily	Hr.	Daily	Bare Costs	Incl. O&P
1 Building Laborer	$37.60	$300.80	$57.85	$462.80	$37.60	$57.85
1 Snow Blower, Walk-Behind		67.15		73.86	8.39	9.23
8 L.H., Daily Totals		$367.95		$536.66	$45.99	$67.08

Crew No.	Bare Costs		Incl. Subs O&P		Cost Per Labor-Hour	
Crew A-2	Hr.	Daily	Hr.	Daily	Bare Costs	Incl. O&P
2 Laborers	$37.60	$601.60	$57.85	$925.60	$38.10	$58.27
1 Truck Driver (light)	39.10	312.80	59.10	472.80		
1 Flatbed Truck, Gas, 1.5 Ton		245.40		269.94	10.23	11.25
24 L.H., Daily Totals		$1159.80		$1668.34	$48.33	$69.51
Crew A-2A	Hr.	Daily	Hr.	Daily	Bare Costs	Incl. O&P
2 Laborers	$37.60	$601.60	$57.85	$925.60	$38.10	$58.27
1 Truck Driver (light)	39.10	312.80	59.10	472.80		
1 Flatbed Truck, Gas, 1.5 Ton		245.40		269.94		
1 Concrete Saw		167.40		184.14	17.20	18.92
24 L.H., Daily Totals		$1327.20		$1852.48	$55.30	$77.19
Crew A-2B	Hr.	Daily	Hr.	Daily	Bare Costs	Incl. O&P
1 Truck Driver (light)	$39.10	$312.80	$59.10	$472.80	$39.10	$59.10
1 Flatbed Truck, Gas, 1.5 Ton		245.40		269.94	30.68	33.74
8 L.H., Daily Totals		$558.20		$742.74	$69.78	$92.84
Crew A-3A	Hr.	Daily	Hr.	Daily	Bare Costs	Incl. O&P
1 Equip. Oper. (light)	$48.60	$388.80	$73.75	$590.00	$48.60	$73.75
1 Pickup Truck, 4 x 4, 3/4 Ton		155.60		171.16	19.45	21.40
8 L.H., Daily Totals		$544.40		$761.16	$68.05	$95.14
Crew A-3B	Hr.	Daily	Hr.	Daily	Bare Costs	Incl. O&P
1 Equip. Oper. (medium)	$50.60	$404.80	$76.75	$614.00	$45.33	$68.65
1 Truck Driver (heavy)	40.05	320.40	60.55	484.40		
1 Dump Truck, 12 C.Y., 400 H.P.		691.00		760.10		
1 F.E. Loader, W.M., 2.5 C.Y.		519.80		571.78	75.67	83.24
16 L.H., Daily Totals		$1936.00		$2430.28	$121.00	$151.89
Crew A-3C	Hr.	Daily	Hr.	Daily	Bare Costs	Incl. O&P
1 Equip. Oper. (light)	$48.60	$388.80	$73.75	$590.00	$48.60	$73.75
1 Loader, Skid Steer, 78 H.P.		318.60		350.46	39.83	43.81
8 L.H., Daily Totals		$707.40		$940.46	$88.42	$117.56
Crew A-3D	Hr.	Daily	Hr.	Daily	Bare Costs	Incl. O&P
1 Truck Driver (light)	$39.10	$312.80	$59.10	$472.80	$39.10	$59.10
1 Pickup Truck, 4 x 4, 3/4 Ton		155.60		171.16		
1 Flatbed Trailer, 25 Ton		113.60		124.96	33.65	37.02
8 L.H., Daily Totals		$582.00		$768.92	$72.75	$96.11
Crew A-3E	Hr.	Daily	Hr.	Daily	Bare Costs	Incl. O&P
1 Equip. Oper. (crane)	$51.70	$413.60	$78.45	$627.60	$45.88	$69.50
1 Truck Driver (heavy)	40.05	320.40	60.55	484.40		
1 Pickup Truck, 4 x 4, 3/4 Ton		155.60		171.16	9.72	10.70
16 L.H., Daily Totals		$889.60		$1283.16	$55.60	$80.20
Crew A-3F	Hr.	Daily	Hr.	Daily	Bare Costs	Incl. O&P
1 Equip. Oper. (crane)	$51.70	$413.60	$78.45	$627.60	$45.88	$69.50
1 Truck Driver (heavy)	40.05	320.40	60.55	484.40		
1 Pickup Truck, 4 x 4, 3/4 Ton		155.60		171.16		
1 Truck Tractor, 6x4, 380 H.P.		599.80		659.78		
1 Lowbed Trailer, 75 Ton		221.80		243.98	61.08	67.18
16 L.H., Daily Totals		$1711.20		$2186.92	$106.95	$136.68

Crew No.	Bare Costs Hr.	Daily	Incl. Subs O&P Hr.	Daily	Cost Per Labor-Hour Bare Costs	Incl. O&P
Crew A-3G	Hr.	Daily	Hr.	Daily	Bare Costs	Incl. O&P
1 Equip. Oper. (crane)	$51.70	$413.60	$78.45	$627.60	$45.88	$69.50
1 Truck Driver (heavy)	40.05	320.40	60.55	484.40		
1 Pickup Truck, 4 x 4, 3/4 Ton		155.60		171.16		
1 Truck Tractor, 6x4, 450 H.P.		727.40		800.14		
1 Lowbed Trailer, 75 Ton		221.80		243.98	69.05	75.95
16 L.H., Daily Totals		$1838.80		$2327.28	$114.93	$145.46
Crew A-3H	Hr.	Daily	Hr.	Daily	Bare Costs	Incl. O&P
1 Equip. Oper. (crane)	$51.70	$413.60	$78.45	$627.60	$51.70	$78.45
1 Hyd. Crane, 12 Ton (Daily)		860.80		946.88	107.60	118.36
8 L.H., Daily Totals		$1274.40		$1574.48	$159.30	$196.81
Crew A-3I	Hr.	Daily	Hr.	Daily	Bare Costs	Incl. O&P
1 Equip. Oper. (crane)	$51.70	$413.60	$78.45	$627.60	$51.70	$78.45
1 Hyd. Crane, 25 Ton (Daily)		989.60		1088.56	123.70	136.07
8 L.H., Daily Totals		$1403.20		$1716.16	$175.40	$214.52
Crew A-3J	Hr.	Daily	Hr.	Daily	Bare Costs	Incl. O&P
1 Equip. Oper. (crane)	$51.70	$413.60	$78.45	$627.60	$51.70	$78.45
1 Hyd. Crane, 40 Ton (Daily)		1221.00		1343.10	152.63	167.89
8 L.H., Daily Totals		$1634.60		$1970.70	$204.32	$246.34
Crew A-3K	Hr.	Daily	Hr.	Daily	Bare Costs	Incl. O&P
1 Equip. Oper. (crane)	$51.70	$413.60	$78.45	$627.60	$48.45	$73.50
1 Equip. Oper. (oiler)	45.20	361.60	68.55	548.40		
1 Hyd. Crane, 55 Ton (Daily)		1469.00		1615.90		
1 P/U Truck, 3/4 Ton (Daily)		167.20		183.92	102.26	112.49
16 L.H., Daily Totals		$2411.40		$2975.82	$150.71	$185.99
Crew A-3L	Hr.	Daily	Hr.	Daily	Bare Costs	Incl. O&P
1 Equip. Oper. (crane)	$51.70	$413.60	$78.45	$627.60	$48.45	$73.50
1 Equip. Oper. (oiler)	45.20	361.60	68.55	548.40		
1 Hyd. Crane, 80 Ton (Daily)		2174.00		2391.40		
1 P/U Truck, 3/4 Ton (Daily)		167.20		183.92	146.32	160.96
16 L.H., Daily Totals		$3116.40		$3751.32	$194.78	$234.46
Crew A-3M	Hr.	Daily	Hr.	Daily	Bare Costs	Incl. O&P
1 Equip. Oper. (crane)	$51.70	$413.60	$78.45	$627.60	$48.45	$73.50
1 Equip. Oper. (oiler)	45.20	361.60	68.55	548.40		
1 Hyd. Crane, 100 Ton (Daily)		2169.00		2385.90		
1 P/U Truck, 3/4 Ton (Daily)		167.20		183.92	146.01	160.61
16 L.H., Daily Totals		$3111.40		$3745.82	$194.46	$234.11
Crew A-3N	Hr.	Daily	Hr.	Daily	Bare Costs	Incl. O&P
1 Equip. Oper. (crane)	$51.70	$413.60	$78.45	$627.60	$51.70	$78.45
1 Tower Cane (monthly)		1128.00		1240.80	141.00	155.10
8 L.H., Daily Totals		$1541.60		$1868.40	$192.70	$233.55
Crew A-3P	Hr.	Daily	Hr.	Daily	Bare Costs	Incl. O&P
1 Equip. Oper. (light)	$48.60	$388.80	$73.75	$590.00	$48.60	$73.75
1 A.T. Forklift, 42' lift		526.40		579.04	65.80	72.38
8 L.H., Daily Totals		$915.20		$1169.04	$114.40	$146.13
Crew A-3Q	Hr.	Daily	Hr.	Daily	Bare Costs	Incl. O&P
1 Equip. Oper. (light)	$48.60	$388.80	$73.75	$590.00	$48.60	$73.75
1 Pickup Truck, 4 x 4, 3/4 Ton		155.60		171.16		
1 Flatbed trailer, 3 Ton		23.60		25.96	22.40	24.64
8 L.H., Daily Totals		$568.00		$787.12	$71.00	$98.39

Crew No.	Bare Costs Hr.	Daily	Incl. Subs O&P Hr.	Daily	Cost Per Labor-Hour Bare Costs	Incl. O&P
Crew A-4	Hr.	Daily	Hr.	Daily	Bare Costs	Incl. O&P
2 Carpenters	$46.95	$751.20	$72.25	$1156.00	$44.75	$68.43
1 Painter, Ordinary	40.35	322.80	60.80	486.40		
24 L.H., Daily Totals		$1074.00		$1642.40	$44.75	$68.43
Crew A-5	Hr.	Daily	Hr.	Daily	Bare Costs	Incl. O&P
2 Laborers	$37.60	$601.60	$57.85	$925.60	$37.77	$57.99
.25 Truck Driver (light)	39.10	78.20	59.10	118.20		
.25 Flatbed Truck, Gas, 1.5 Ton		61.35		67.48	3.41	3.75
18 L.H., Daily Totals		$741.15		$1111.29	$41.17	$61.74
Crew A-6	Hr.	Daily	Hr.	Daily	Bare Costs	Incl. O&P
1 Instrument Man	$48.65	$389.20	$75.15	$601.20	$46.88	$71.92
1 Rodman/Chainman	45.10	360.80	68.70	549.60		
1 Level, Electronic		54.80		60.28	3.42	3.77
16 L.H., Daily Totals		$804.80		$1211.08	$50.30	$75.69
Crew A-7	Hr.	Daily	Hr.	Daily	Bare Costs	Incl. O&P
1 Chief of Party	$60.90	$487.20	$93.20	$745.60	$51.55	$79.02
1 Instrument Man	48.65	389.20	75.15	601.20		
1 Rodman/Chainman	45.10	360.80	68.70	549.60		
1 Level, Electronic		54.80		60.28	2.28	2.51
24 L.H., Daily Totals		$1292.00		$1956.68	$53.83	$81.53
Crew A-8	Hr.	Daily	Hr.	Daily	Bare Costs	Incl. O&P
1 Chief of Party	$60.90	$487.20	$93.20	$745.60	$49.94	$76.44
1 Instrument Man	48.65	389.20	75.15	601.20		
2 Rodmen/Chainmen	45.10	721.60	68.70	1099.20		
1 Level, Electronic		54.80		60.28	1.71	1.88
32 L.H., Daily Totals		$1652.80		$2506.28	$51.65	$78.32
Crew A-9	Hr.	Daily	Hr.	Daily	Bare Costs	Incl. O&P
1 Asbestos Foreman	$52.85	$422.80	$82.30	$658.40	$52.41	$81.60
7 Asbestos Workers	52.35	2931.60	81.50	4564.00		
64 L.H., Daily Totals		$3354.40		$5222.40	$52.41	$81.60
Crew A-10A	Hr.	Daily	Hr.	Daily	Bare Costs	Incl. O&P
1 Asbestos Foreman	$52.85	$422.80	$82.30	$658.40	$52.52	$81.77
2 Asbestos Workers	52.35	837.60	81.50	1304.00		
24 L.H., Daily Totals		$1260.40		$1962.40	$52.52	$81.77
Crew A-10B	Hr.	Daily	Hr.	Daily	Bare Costs	Incl. O&P
1 Asbestos Foreman	$52.85	$422.80	$82.30	$658.40	$52.48	$81.70
3 Asbestos Workers	52.35	1256.40	81.50	1956.00		
32 L.H., Daily Totals		$1679.20		$2614.40	$52.48	$81.70
Crew A-10C	Hr.	Daily	Hr.	Daily	Bare Costs	Incl. O&P
3 Asbestos Workers	$52.35	$1256.40	$81.50	$1956.00	$52.35	$81.50
1 Flatbed Truck, Gas, 1.5 Ton		245.40		269.94	10.23	11.25
24 L.H., Daily Totals		$1501.80		$2225.94	$62.58	$92.75
Crew A-10D	Hr.	Daily	Hr.	Daily	Bare Costs	Incl. O&P
2 Asbestos Workers	$52.35	$837.60	$81.50	$1304.00	$50.40	$77.50
1 Equip. Oper. (crane)	51.70	413.60	78.45	627.60		
1 Equip. Oper. (oiler)	45.20	361.60	68.55	548.40		
1 Hydraulic Crane, 33 Ton		750.00		825.00	23.44	25.78
32 L.H., Daily Totals		$2362.80		$3305.00	$73.84	$103.28

Crew A-11

Crew No.	Bare Costs Hr.	Daily	Incl. Subs O&P Hr.	Daily	Cost Per Labor-Hour Bare Costs	Incl. O&P
1 Asbestos Foreman	$52.85	$422.80	$82.30	$658.40	$52.41	$81.60
7 Asbestos Workers	52.35	2931.60	81.50	4564.00		
2 Chip. Hammers, 12 Lb., Elec.		40.00		44.00	.63	.69
64 L.H., Daily Totals		$3394.40		$5266.40	$53.04	$82.29

Crew A-12

Crew No.	Bare Costs Hr.	Daily	Incl. Subs O&P Hr.	Daily	Cost Per Labor-Hour Bare Costs	Incl. O&P
1 Asbestos Foreman	$52.85	$422.80	$82.30	$658.40	$52.41	$81.60
7 Asbestos Workers	52.35	2931.60	81.50	4564.00		
1 Trk-Mtd Vac, 14 CY, 1500 Gal.		514.25		565.67		
1 Flatbed Truck, 20,000 GVW		248.60		273.46	11.92	13.11
64 L.H., Daily Totals		$4117.25		$6061.53	$64.33	$94.71

Crew A-13

Crew No.	Bare Costs Hr.	Daily	Incl. Subs O&P Hr.	Daily	Cost Per Labor-Hour Bare Costs	Incl. O&P
1 Equip. Oper. (light)	$48.60	$388.80	$73.75	$590.00	$48.60	$73.75
1 Trk-Mtd Vac, 14 CY, 1500 Gal.		514.25		565.67		
1 Flatbed Truck, 20,000 GVW		248.60		273.46	95.36	104.89
8 L.H., Daily Totals		$1151.65		$1429.14	$143.96	$178.64

Crew B-1

Crew No.	Bare Costs Hr.	Daily	Incl. Subs O&P Hr.	Daily	Cost Per Labor-Hour Bare Costs	Incl. O&P
1 Labor Foreman (outside)	$39.60	$316.80	$60.95	$487.60	$38.27	$58.88
2 Laborers	37.60	601.60	57.85	925.60		
24 L.H., Daily Totals		$918.40		$1413.20	$38.27	$58.88

Crew B-1A

Crew No.	Bare Costs Hr.	Daily	Incl. Subs O&P Hr.	Daily	Cost Per Labor-Hour Bare Costs	Incl. O&P
1 Labor Foreman (outside)	$39.60	$316.80	$60.95	$487.60	$38.27	$58.88
2 Laborers	37.60	601.60	57.85	925.60		
2 Cutting Torches		22.80		25.08		
2 Sets of Gases		304.00		334.40	13.62	14.98
24 L.H., Daily Totals		$1245.20		$1772.68	$51.88	$73.86

Crew B-1B

Crew No.	Bare Costs Hr.	Daily	Incl. Subs O&P Hr.	Daily	Cost Per Labor-Hour Bare Costs	Incl. O&P
1 Labor Foreman (outside)	$39.60	$316.80	$60.95	$487.60	$41.63	$63.77
2 Laborers	37.60	601.60	57.85	925.60		
1 Equip. Oper. (crane)	51.70	413.60	78.45	627.60		
2 Cutting Torches		22.80		25.08		
2 Sets of Gases		304.00		334.40		
1 Hyd. Crane, 12 Ton		653.80		719.18	30.64	33.71
32 L.H., Daily Totals		$2312.60		$3119.46	$72.27	$97.48

Crew B-1C

Crew No.	Bare Costs Hr.	Daily	Incl. Subs O&P Hr.	Daily	Cost Per Labor-Hour Bare Costs	Incl. O&P
1 Labor Foreman (outside)	$39.60	$316.80	$60.95	$487.60	$38.27	$58.88
2 Laborers	37.60	601.60	57.85	925.60		
1 Aerial Lift Truck, 60' Boom		435.60		479.16	18.15	19.97
24 L.H., Daily Totals		$1354.00		$1892.36	$56.42	$78.85

Crew B-1D

Crew No.	Bare Costs Hr.	Daily	Incl. Subs O&P Hr.	Daily	Cost Per Labor-Hour Bare Costs	Incl. O&P
2 Laborers	$37.60	$601.60	$57.85	$925.60	$37.60	$57.85
1 Small Work Boat, Gas, 50 H.P.		173.20		190.52		
1 Pressure Washer, 7 GPM		87.60		96.36	16.30	17.93
16 L.H., Daily Totals		$862.40		$1212.48	$53.90	$75.78

Crew B-1E

Crew No.	Bare Costs Hr.	Daily	Incl. Subs O&P Hr.	Daily	Cost Per Labor-Hour Bare Costs	Incl. O&P
1 Labor Foreman (outside)	$39.60	$316.80	$60.95	$487.60	$38.10	$58.63
3 Laborers	37.60	902.40	57.85	1388.40		
1 Work Boat, Diesel, 200 H.P.		1463.00		1609.30		
2 Pressure Washer, 7 GPM		175.20		192.72	51.19	56.31
32 L.H., Daily Totals		$2857.40		$3678.02	$89.29	$114.94

Crew B-1F

Crew No.	Bare Costs Hr.	Daily	Incl. Subs O&P Hr.	Daily	Cost Per Labor-Hour Bare Costs	Incl. O&P
2 Skilled Workers	$48.65	$778.40	$75.15	$1202.40	$44.97	$69.38
1 Laborer	37.60	300.80	57.85	462.80		
1 Small Work Boat, Gas, 50 H.P.		173.20		190.52		
1 Pressure Washer, 7 GPM		87.60		96.36	10.87	11.95
24 L.H., Daily Totals		$1340.00		$1952.08	$55.83	$81.34

Crew B-1G

Crew No.	Bare Costs Hr.	Daily	Incl. Subs O&P Hr.	Daily	Cost Per Labor-Hour Bare Costs	Incl. O&P
2 Laborers	$37.60	$601.60	$57.85	$925.60	$37.60	$57.85
1 Small Work Boat, Gas, 50 H.P.		173.20		190.52	10.82	11.91
16 L.H., Daily Totals		$774.80		$1116.12	$48.42	$69.76

Crew B-1H

Crew No.	Bare Costs Hr.	Daily	Incl. Subs O&P Hr.	Daily	Cost Per Labor-Hour Bare Costs	Incl. O&P
2 Skilled Workers	$48.65	$778.40	$75.15	$1202.40	$44.97	$69.38
1 Laborer	37.60	300.80	57.85	462.80		
1 Small Work Boat, Gas, 50 H.P.		173.20		190.52	7.22	7.94
24 L.H., Daily Totals		$1252.40		$1855.72	$52.18	$77.32

Crew B-1J

Crew No.	Bare Costs Hr.	Daily	Incl. Subs O&P Hr.	Daily	Cost Per Labor-Hour Bare Costs	Incl. O&P
1 Labor Foreman (inside)	$38.10	$304.80	$58.65	$469.20	$37.85	$58.25
1 Laborer	37.60	300.80	57.85	462.80		
16 L.H., Daily Totals		$605.60		$932.00	$37.85	$58.25

Crew B-1K

Crew No.	Bare Costs Hr.	Daily	Incl. Subs O&P Hr.	Daily	Cost Per Labor-Hour Bare Costs	Incl. O&P
1 Carpenter Foreman (inside)	$47.45	$379.60	$73.05	$584.40	$47.20	$72.65
1 Carpenter	46.95	375.60	72.25	578.00		
16 L.H., Daily Totals		$755.20		$1162.40	$47.20	$72.65

Crew B-2

Crew No.	Bare Costs Hr.	Daily	Incl. Subs O&P Hr.	Daily	Cost Per Labor-Hour Bare Costs	Incl. O&P
1 Labor Foreman (outside)	$39.60	$316.80	$60.95	$487.60	$38.00	$58.47
4 Laborers	37.60	1203.20	57.85	1851.20		
40 L.H., Daily Totals		$1520.00		$2338.80	$38.00	$58.47

Crew B-2A

Crew No.	Bare Costs Hr.	Daily	Incl. Subs O&P Hr.	Daily	Cost Per Labor-Hour Bare Costs	Incl. O&P
1 Labor Foreman (outside)	$39.60	$316.80	$60.95	$487.60	$38.27	$58.88
2 Laborers	37.60	601.60	57.85	925.60		
1 Aerial Lift Truck, 60' Boom		435.60		479.16	18.15	19.97
24 L.H., Daily Totals		$1354.00		$1892.36	$56.42	$78.85

Crew B-3

Crew No.	Bare Costs Hr.	Daily	Incl. Subs O&P Hr.	Daily	Cost Per Labor-Hour Bare Costs	Incl. O&P
1 Labor Foreman (outside)	$39.60	$316.80	$60.95	$487.60	$40.92	$62.42
2 Laborers	37.60	601.60	57.85	925.60		
1 Equip. Oper. (medium)	50.60	404.80	76.75	614.00		
2 Truck Drivers (heavy)	40.05	640.80	60.55	968.80		
1 Crawler Loader, 3 C.Y.		1188.00		1306.80		
2 Dump Trucks, 12 C.Y., 400 H.P.		1382.00		1520.20	53.54	58.90
48 L.H., Daily Totals		$4534.00		$5823.00	$94.46	$121.31

Crew B-3A

Crew No.	Bare Costs Hr.	Daily	Incl. Subs O&P Hr.	Daily	Cost Per Labor-Hour Bare Costs	Incl. O&P
4 Laborers	$37.60	$1203.20	$57.85	$1851.20	$40.20	$61.63
1 Equip. Oper. (medium)	50.60	404.80	76.75	614.00		
1 Hyd. Excavator, 1.5 C.Y.		1030.00		1133.00	25.75	28.32
40 L.H., Daily Totals		$2638.00		$3598.20	$65.95	$89.95

Crew B-3B

Crew No.	Bare Costs Hr.	Daily	Incl. Subs O&P Hr.	Daily	Cost Per Labor-Hour Bare Costs	Incl. O&P
2 Laborers	$37.60	$601.60	$57.85	$925.60	$41.46	$63.25
1 Equip. Oper. (medium)	50.60	404.80	76.75	614.00		
1 Truck Driver (heavy)	40.05	320.40	60.55	484.40		
1 Backhoe Loader, 80 H.P.		391.80		430.98		
1 Dump Truck, 12 C.Y., 400 H.P.		691.00		760.10	33.84	37.22
32 L.H., Daily Totals		$2409.60		$3215.08	$75.30	$100.47

Crew No.	Bare Costs		Incl. Subs O&P		Cost Per Labor-Hour	

Crew B-3C	Hr.	Daily	Hr.	Daily	Bare Costs	Incl. O&P
3 Laborers	$37.60	$902.40	$57.85	$1388.40	$40.85	$62.58
1 Equip. Oper. (medium)	50.60	404.80	76.75	614.00		
1 Crawler Loader, 4 C.Y.		1561.00		1717.10	48.78	53.66
32 L.H., Daily Totals		$2868.20		$3719.50	$89.63	$116.23

Crew B-4	Hr.	Daily	Hr.	Daily	Bare Costs	Incl. O&P
1 Labor Foreman (outside)	$39.60	$316.80	$60.95	$487.60	$38.34	$58.82
4 Laborers	37.60	1203.20	57.85	1851.20		
1 Truck Driver (heavy)	40.05	320.40	60.55	484.40		
1 Truck Tractor, 220 H.P.		359.60		395.56		
1 Flatbed Trailer, 40 Ton		154.00		169.40	10.70	11.77
48 L.H., Daily Totals		$2354.00		$3388.16	$49.04	$70.59

Crew B-5	Hr.	Daily	Hr.	Daily	Bare Costs	Incl. O&P
1 Labor Foreman (outside)	$39.60	$316.80	$60.95	$487.60	$41.60	$63.69
4 Laborers	37.60	1203.20	57.85	1851.20		
2 Equip. Oper. (medium)	50.60	809.60	76.75	1228.00		
1 Air Compressor, 250 cfm		201.40		221.54		
2 Breakers, Pavement, 60 lb.		19.60		21.56		
2 -50' Air Hoses, 1.5"		11.60		12.76		
1 Crawler Loader, 3 C.Y.		1188.00		1306.80	25.37	27.90
56 L.H., Daily Totals		$3750.20		$5129.46	$66.97	$91.60

Crew B-5A	Hr.	Daily	Hr.	Daily	Bare Costs	Incl. O&P
1 Labor Foreman (outside)	$39.60	$316.80	$60.95	$487.60	$41.26	$63.03
6 Laborers	37.60	1804.80	57.85	2776.80		
2 Equip. Oper. (medium)	50.60	809.60	76.75	1228.00		
1 Equip. Oper. (light)	48.60	388.80	73.75	590.00		
2 Truck Drivers (heavy)	40.05	640.80	60.55	968.80		
1 Air Compressor, 365 cfm		263.80		290.18		
2 Breakers, Pavement, 60 lb.		19.60		21.56		
8 -50' Air Hoses, 1"		32.80		36.08		
2 Dump Trucks, 8 C.Y., 220 H.P.		822.40		904.64	11.86	13.05
96 L.H., Daily Totals		$5099.40		$7303.66	$53.12	$76.08

Crew B-5B	Hr.	Daily	Hr.	Daily	Bare Costs	Incl. O&P
1 Powderman	$48.65	$389.20	$75.15	$601.20	$45.00	$68.38
2 Equip. Oper. (medium)	50.60	809.60	76.75	1228.00		
3 Truck Drivers (heavy)	40.05	961.20	60.55	1453.20		
1 F.E. Loader, W.M.,2.5 C.Y.		519.80		571.78		
3 Dump Trucks, 12 C.Y., 400 H.P.		2073.00		2280.30		
1 Air Compressor, 365 CFM		263.80		290.18	59.51	65.46
48 L.H., Daily Totals		$5016.60		$6424.66	$104.51	$133.85

Crew B-5C	Hr.	Daily	Hr.	Daily	Bare Costs	Incl. O&P
3 Laborers	$37.60	$902.40	$57.85	$1388.40	$42.55	$64.80
1 Equip. Oper. (medium)	50.60	404.80	76.75	614.00		
2 Truck Drivers (heavy)	40.05	640.80	60.55	968.80		
1 Equip. Oper. (crane)	51.70	413.60	78.45	627.60		
1 Equip. Oper. (oiler)	45.20	361.60	68.55	548.40		
2 Dump Trucks, 12 C.Y., 400 H.P.		1382.00		1520.20		
1 Crawler Loader, 4 C.Y.		1561.00		1717.10		
1 S.P. Crane, 4x4, 25 Ton		599.40		659.34	55.35	60.88
64 L.H., Daily Totals		$6265.60		$8043.84	$97.90	$125.69

Crew B-5D	Hr.	Daily	Hr.	Daily	Bare Costs	Incl. O&P
1 Labor Foreman (outside)	$39.60	$316.80	$60.95	$487.60	$41.41	$63.30
4 Laborers	37.60	1203.20	57.85	1851.20		
2 Equip. Oper. (medium)	50.60	809.60	76.75	1228.00		
1 Truck Driver (heavy)	40.05	320.40	60.55	484.40		
1 Air Compressor, 250 cfm		201.40		221.54		
2 Breakers, Pavement, 60 lb.		19.60		21.56		
2 -50' Air Hoses, 1.5"		11.60		12.76		
1 Crawler Loader, 3 C.Y.		1188.00		1306.80		
1 Dump Truck, 12 C.Y., 400 H.P.		691.00		760.10	32.99	36.29
64 L.H., Daily Totals		$4761.60		$6373.96	$74.40	$99.59

Crew B-6	Hr.	Daily	Hr.	Daily	Bare Costs	Incl. O&P
2 Laborers	$37.60	$601.60	$57.85	$925.60	$41.27	$63.15
1 Equip. Oper. (light)	48.60	388.80	73.75	590.00		
1 Backhoe Loader, 48 H.P.		364.20		400.62	15.18	16.69
24 L.H., Daily Totals		$1354.60		$1916.22	$56.44	$79.84

Crew B-6A	Hr.	Daily	Hr.	Daily	Bare Costs	Incl. O&P
.5 Labor Foreman (outside)	$39.60	$158.40	$60.95	$243.80	$43.20	$66.03
1 Laborer	37.60	300.80	57.85	462.80		
1 Equip. Oper. (medium)	50.60	404.80	76.75	614.00		
1 Vacuum Truck, 5000 Gal.		358.10		393.91	17.91	19.70
20 L.H., Daily Totals		$1222.10		$1714.51	$61.10	$85.73

Crew B-6B	Hr.	Daily	Hr.	Daily	Bare Costs	Incl. O&P
2 Labor Foremen (outside)	$39.60	$633.60	$60.95	$975.20	$38.27	$58.88
4 Laborers	37.60	1203.20	57.85	1851.20		
1 S.P. Crane, 4x4, 5 Ton		275.40		302.94		
1 Flatbed Truck, Gas, 1.5 Ton		245.40		269.94		
1 Butt Fusion Mach., 4"-12" diam.		382.10		420.31	18.81	20.69
48 L.H., Daily Totals		$2739.70		$3819.59	$57.08	$79.57

Crew B-6C	Hr.	Daily	Hr.	Daily	Bare Costs	Incl. O&P
2 Labor Foremen (outside)	$39.60	$633.60	$60.95	$975.20	$38.27	$58.88
4 Laborers	37.60	1203.20	57.85	1851.20		
1 S.P. Crane, 4x4, 12 Ton		473.40		520.74		
1 Flatbed Truck, Gas, 3 Ton		303.20		333.52		
1 Butt Fusion Mach., 8"-24" diam.		829.70		912.67	33.46	36.81
48 L.H., Daily Totals		$3443.10		$4593.33	$71.73	$95.69

Crew B-7	Hr.	Daily	Hr.	Daily	Bare Costs	Incl. O&P
1 Labor Foreman (outside)	$39.60	$316.80	$60.95	$487.60	$40.10	$61.52
4 Laborers	37.60	1203.20	57.85	1851.20		
1 Equip. Oper. (medium)	50.60	404.80	76.75	614.00		
1 Brush Chipper, 12", 130 H.P.		391.40		430.54		
1 Crawler Loader, 3 C.Y.		1188.00		1306.80		
2 Chain Saws, Gas, 36" Long		90.00		99.00	34.78	38.26
48 L.H., Daily Totals		$3594.20		$4789.14	$74.88	$99.77

Crew B-7A	Hr.	Daily	Hr.	Daily	Bare Costs	Incl. O&P
2 Laborers	$37.60	$601.60	$57.85	$925.60	$41.27	$63.15
1 Equip. Oper. (light)	48.60	388.80	73.75	590.00		
1 Rake w/Tractor		354.25		389.68		
2 Chain Saw, Gas, 18"		62.40		68.64	17.36	19.10
24 L.H., Daily Totals		$1407.05		$1973.92	$58.63	$82.27

Crew No.		Bare Costs		Incl. Subs O&P	Cost Per Labor-Hour	
Crew B-7B	Hr.	Daily	Hr.	Daily	Bare Costs	Incl. O&P
1 Labor Foreman (outside)	$39.60	$316.80	$60.95	$487.60	$40.09	$61.38
4 Laborers	37.60	1203.20	57.85	1851.20		
1 Equip. Oper. (medium)	50.60	404.80	76.75	614.00		
1 Truck Driver (heavy)	40.05	320.40	60.55	484.40		
1 Brush Chipper, 12", 130 H.P.		391.40		430.54		
1 Crawler Loader, 3 C.Y.		1188.00		1306.80		
2 Chain Saws, Gas, 36" Long		90.00		99.00		
1 Dump Truck, 8 C.Y., 220 H.P.		411.20		452.32	37.15	40.87
56 L.H., Daily Totals		$4325.80		$5725.86	$77.25	$102.25
Crew B-7C	Hr.	Daily	Hr.	Daily	Bare Costs	Incl. O&P
1 Labor Foreman (outside)	$39.60	$316.80	$60.95	$487.60	$40.09	$61.38
4 Laborers	37.60	1203.20	57.85	1851.20		
1 Equip. Oper. (medium)	50.60	404.80	76.75	614.00		
1 Truck Driver (heavy)	40.05	320.40	60.55	484.40		
1 Brush Chipper, 12", 130 H.P.		391.40		430.54		
1 Crawler Loader, 3 C.Y.		1188.00		1306.80		
2 Chain Saws, Gas, 36" Long		90.00		99.00		
1 Dump Truck, 12 C.Y., 400 H.P.		691.00		760.10	42.15	46.37
56 L.H., Daily Totals		$4605.60		$6033.64	$82.24	$107.74
Crew B-8	Hr.	Daily	Hr.	Daily	Bare Costs	Incl. O&P
1 Labor Foreman (outside)	$39.60	$316.80	$60.95	$487.60	$42.66	$64.97
2 Laborers	37.60	601.60	57.85	925.60		
2 Equip. Oper. (medium)	50.60	809.60	76.75	1228.00		
1 Equip. Oper. (oiler)	45.20	361.60	68.55	548.40		
2 Truck Drivers (heavy)	40.05	640.80	60.55	968.80		
1 Hyd. Crane, 25 Ton		736.60		810.26		
1 Crawler Loader, 3 C.Y.		1188.00		1306.80		
2 Dump Trucks, 12 C.Y., 400 H.P.		1382.00		1520.20	51.67	56.83
64 L.H., Daily Totals		$6037.00		$7795.66	$94.33	$121.81
Crew B-9	Hr.	Daily	Hr.	Daily	Bare Costs	Incl. O&P
1 Labor Foreman (outside)	$39.60	$316.80	$60.95	$487.60	$38.00	$58.47
4 Laborers	37.60	1203.20	57.85	1851.20		
1 Air Compressor, 250 cfm		201.40		221.54		
2 Breakers, Pavement, 60 lb.		19.60		21.56		
2 -50' Air Hoses, 1.5"		11.60		12.76	5.82	6.40
40 L.H., Daily Totals		$1752.60		$2594.66	$43.81	$64.87
Crew B-9A	Hr.	Daily	Hr.	Daily	Bare Costs	Incl. O&P
2 Laborers	$37.60	$601.60	$57.85	$925.60	$38.42	$58.75
1 Truck Driver (heavy)	40.05	320.40	60.55	484.40		
1 Water Tank Trailer, 5000 Gal.		141.00		155.10		
1 Truck Tractor, 220 H.P.		359.60		395.56		
2 -50' Discharge Hoses, 3"		3.00		3.30	20.98	23.08
24 L.H., Daily Totals		$1425.60		$1963.96	$59.40	$81.83
Crew B-9B	Hr.	Daily	Hr.	Daily	Bare Costs	Incl. O&P
2 Laborers	$37.60	$601.60	$57.85	$925.60	$38.42	$58.75
1 Truck Driver (heavy)	40.05	320.40	60.55	484.40		
2 -50' Discharge Hoses, 3"		3.00		3.30		
1 Water Tank Trailer, 5000 Gal.		141.00		155.10		
1 Truck Tractor, 220 H.P.		359.60		395.56		
1 Pressure Washer		70.20		77.22	23.91	26.30
24 L.H., Daily Totals		$1495.80		$2041.18	$62.33	$85.05

Crew No.		Bare Costs		Incl. Subs O&P	Cost Per Labor-Hour	
Crew B-9D	Hr.	Daily	Hr.	Daily	Bare Costs	Incl. O&P
1 Labor Foreman (outside)	$39.60	$316.80	$60.95	$487.60	$38.00	$58.47
4 Common Laborers	37.60	1203.20	57.85	1851.20		
1 Air Compressor, 250 cfm		201.40		221.54		
2 -50' Air Hoses, 1.5"		11.60		12.76		
2 Air Powered Tampers		52.40		57.64	6.63	7.30
40 L.H., Daily Totals		$1785.40		$2630.74	$44.63	$65.77
Crew B-10	Hr.	Daily	Hr.	Daily	Bare Costs	Incl. O&P
1 Equip. Oper. (medium)	$50.60	$404.80	$76.75	$614.00	$46.27	$70.45
.5 Laborer	37.60	150.40	57.85	231.40		
12 L.H., Daily Totals		$555.20		$845.40	$46.27	$70.45
Crew B-10A	Hr.	Daily	Hr.	Daily	Bare Costs	Incl. O&P
1 Equip. Oper. (medium)	$50.60	$404.80	$76.75	$614.00	$46.27	$70.45
.5 Laborer	37.60	150.40	57.85	231.40		
1 Roller, 2-Drum, W.B., 7.5 H.P.		180.00		198.00	15.00	16.50
12 L.H., Daily Totals		$735.20		$1043.40	$61.27	$86.95
Crew B-10B	Hr.	Daily	Hr.	Daily	Bare Costs	Incl. O&P
1 Equip. Oper. (medium)	$50.60	$404.80	$76.75	$614.00	$46.27	$70.45
.5 Laborer	37.60	150.40	57.85	231.40		
1 Dozer, 200 H.P.		1387.00		1525.70	115.58	127.14
12 L.H., Daily Totals		$1942.20		$2371.10	$161.85	$197.59
Crew B-10C	Hr.	Daily	Hr.	Daily	Bare Costs	Incl. O&P
1 Equip. Oper. (medium)	$50.60	$404.80	$76.75	$614.00	$46.27	$70.45
.5 Laborer	37.60	150.40	57.85	231.40		
1 Dozer, 200 H.P.		1387.00		1525.70		
1 Vibratory Roller, Towed, 23 Ton		396.00		435.60	148.58	163.44
12 L.H., Daily Totals		$2338.20		$2806.70	$194.85	$233.89
Crew B-10D	Hr.	Daily	Hr.	Daily	Bare Costs	Incl. O&P
1 Equip. Oper. (medium)	$50.60	$404.80	$76.75	$614.00	$46.27	$70.45
.5 Laborer	37.60	150.40	57.85	231.40		
1 Dozer, 200 H.P.		1387.00		1525.70		
1 Sheepsft. Roller, Towed		425.80		468.38	151.07	166.17
12 L.H., Daily Totals		$2368.00		$2839.48	$197.33	$236.62
Crew B-10E	Hr.	Daily	Hr.	Daily	Bare Costs	Incl. O&P
1 Equip. Oper. (medium)	$50.60	$404.80	$76.75	$614.00	$46.27	$70.45
.5 Laborer	37.60	150.40	57.85	231.40		
1 Tandem Roller, 5 Ton		157.40		173.14	13.12	14.43
12 L.H., Daily Totals		$712.60		$1018.54	$59.38	$84.88
Crew B-10F	Hr.	Daily	Hr.	Daily	Bare Costs	Incl. O&P
1 Equip. Oper. (medium)	$50.60	$404.80	$76.75	$614.00	$46.27	$70.45
.5 Laborer	37.60	150.40	57.85	231.40		
1 Tandem Roller, 10 Ton		236.80		260.48	19.73	21.71
12 L.H., Daily Totals		$792.00		$1105.88	$66.00	$92.16
Crew B-10G	Hr.	Daily	Hr.	Daily	Bare Costs	Incl. O&P
1 Equip. Oper. (medium)	$50.60	$404.80	$76.75	$614.00	$46.27	$70.45
.5 Laborer	37.60	150.40	57.85	231.40		
1 Sheepsfoot Roller, 240 H.P.		1206.00		1326.60	100.50	110.55
12 L.H., Daily Totals		$1761.20		$2172.00	$146.77	$181.00

Crew B-10H

Crew B-10H	Hr.	Daily	Hr.	Daily	Bare Costs	Incl. O&P
1 Equip. Oper. (medium)	$50.60	$404.80	$76.75	$614.00	$46.27	$70.45
.5 Laborer	37.60	150.40	57.85	231.40		
1 Diaphragm Water Pump, 2"		71.00		78.10		
1 -20' Suction Hose, 2"		1.95		2.15		
2 -50' Discharge Hoses, 2"		1.80		1.98	6.23	6.85
12 L.H., Daily Totals		$629.95		$927.63	$52.50	$77.30

Crew B-10I

Crew B-10I	Hr.	Daily	Hr.	Daily	Bare Costs	Incl. O&P
1 Equip. Oper. (medium)	$50.60	$404.80	$76.75	$614.00	$46.27	$70.45
.5 Laborer	37.60	150.40	57.85	231.40		
1 Diaphragm Water Pump, 4"		114.20		125.62		
1 -20' Suction Hose, 4"		3.25		3.58		
2 -50' Discharge Hoses, 4"		4.70		5.17	10.18	11.20
12 L.H., Daily Totals		$677.35		$979.76	$56.45	$81.65

Crew B-10J

Crew B-10J	Hr.	Daily	Hr.	Daily	Bare Costs	Incl. O&P
1 Equip. Oper. (medium)	$50.60	$404.80	$76.75	$614.00	$46.27	$70.45
.5 Laborer	37.60	150.40	57.85	231.40		
1 Centrifugal Water Pump, 3"		78.20		86.02		
1 -20' Suction Hose, 3"		2.85		3.13		
2 -50' Discharge Hoses, 3"		3.00		3.30	7.00	7.70
12 L.H., Daily Totals		$639.25		$937.86	$53.27	$78.15

Crew B-10K

Crew B-10K	Hr.	Daily	Hr.	Daily	Bare Costs	Incl. O&P
1 Equip. Oper. (medium)	$50.60	$404.80	$76.75	$614.00	$46.27	$70.45
.5 Laborer	37.60	150.40	57.85	231.40		
1 Centr. Water Pump, 6"		346.60		381.26		
1 -20' Suction Hose, 6"		11.50		12.65		
2 -50' Discharge Hoses, 6"		12.20		13.42	30.86	33.94
12 L.H., Daily Totals		$925.50		$1252.73	$77.13	$104.39

Crew B-10L

Crew B-10L	Hr.	Daily	Hr.	Daily	Bare Costs	Incl. O&P
1 Equip. Oper. (medium)	$50.60	$404.80	$76.75	$614.00	$46.27	$70.45
.5 Laborer	37.60	150.40	57.85	231.40		
1 Dozer, 80 H.P.		472.40		519.64	39.37	43.30
12 L.H., Daily Totals		$1027.60		$1365.04	$85.63	$113.75

Crew B-10M

Crew B-10M	Hr.	Daily	Hr.	Daily	Bare Costs	Incl. O&P
1 Equip. Oper. (medium)	$50.60	$404.80	$76.75	$614.00	$46.27	$70.45
.5 Laborer	37.60	150.40	57.85	231.40		
1 Dozer, 300 H.P.		1897.00		2086.70	158.08	173.89
12 L.H., Daily Totals		$2452.20		$2932.10	$204.35	$244.34

Crew B-10N

Crew B-10N	Hr.	Daily	Hr.	Daily	Bare Costs	Incl. O&P
1 Equip. Oper. (medium)	$50.60	$404.80	$76.75	$614.00	$46.27	$70.45
.5 Laborer	37.60	150.40	57.85	231.40		
1 F.E. Loader, T.M., 1.5 C.Y		522.00		574.20	43.50	47.85
12 L.H., Daily Totals		$1077.20		$1419.60	$89.77	$118.30

Crew B-10O

Crew B-10O	Hr.	Daily	Hr.	Daily	Bare Costs	Incl. O&P
1 Equip. Oper. (medium)	$50.60	$404.80	$76.75	$614.00	$46.27	$70.45
.5 Laborer	37.60	150.40	57.85	231.40		
1 F.E. Loader, T.M., 2.25 C.Y.		956.40		1052.04	79.70	87.67
12 L.H., Daily Totals		$1511.60		$1897.44	$125.97	$158.12

Crew B-10P

Crew B-10P	Hr.	Daily	Hr.	Daily	Bare Costs	Incl. O&P
1 Equip. Oper. (medium)	$50.60	$404.80	$76.75	$614.00	$46.27	$70.45
.5 Laborer	37.60	150.40	57.85	231.40		
1 Crawler Loader, 3 C.Y.		1188.00		1306.80	99.00	108.90
12 L.H., Daily Totals		$1743.20		$2152.20	$145.27	$179.35

Crew B-10Q

Crew B-10Q	Hr.	Daily	Hr.	Daily	Bare Costs	Incl. O&P
1 Equip. Oper. (medium)	$50.60	$404.80	$76.75	$614.00	$46.27	$70.45
.5 Laborer	37.60	150.40	57.85	231.40		
1 Crawler Loader, 4 C.Y.		1561.00		1717.10	130.08	143.09
12 L.H., Daily Totals		$2116.20		$2562.50	$176.35	$213.54

Crew B-10R

Crew B-10R	Hr.	Daily	Hr.	Daily	Bare Costs	Incl. O&P
1 Equip. Oper. (medium)	$50.60	$404.80	$76.75	$614.00	$46.27	$70.45
.5 Laborer	37.60	150.40	57.85	231.40		
1 F.E. Loader, W.M., 1 C.Y.		299.00		328.90	24.92	27.41
12 L.H., Daily Totals		$854.20		$1174.30	$71.18	$97.86

Crew B-10S

Crew B-10S	Hr.	Daily	Hr.	Daily	Bare Costs	Incl. O&P
1 Equip. Oper. (medium)	$50.60	$404.80	$76.75	$614.00	$46.27	$70.45
.5 Laborer	37.60	150.40	57.85	231.40		
1 F.E. Loader, W.M., 1.5 C.Y.		378.40		416.24	31.53	34.69
12 L.H., Daily Totals		$933.60		$1261.64	$77.80	$105.14

Crew B-10T

Crew B-10T	Hr.	Daily	Hr.	Daily	Bare Costs	Incl. O&P
1 Equip. Oper. (medium)	$50.60	$404.80	$76.75	$614.00	$46.27	$70.45
.5 Laborer	37.60	150.40	57.85	231.40		
1 F.E. Loader, W.M., 2.5 C.Y.		519.80		571.78	43.32	47.65
12 L.H., Daily Totals		$1075.00		$1417.18	$89.58	$118.10

Crew B-10U

Crew B-10U	Hr.	Daily	Hr.	Daily	Bare Costs	Incl. O&P
1 Equip. Oper. (medium)	$50.60	$404.80	$76.75	$614.00	$46.27	$70.45
.5 Laborer	37.60	150.40	57.85	231.40		
1 F.E. Loader, W.M., 5.5 C.Y.		1082.00		1190.20	90.17	99.18
12 L.H., Daily Totals		$1637.20		$2035.60	$136.43	$169.63

Crew B-10V

Crew B-10V	Hr.	Daily	Hr.	Daily	Bare Costs	Incl. O&P
1 Equip. Oper. (medium)	$50.60	$404.80	$76.75	$614.00	$46.27	$70.45
.5 Laborer	37.60	150.40	57.85	231.40		
1 Dozer, 700 H.P.		4979.00		5476.90	414.92	456.41
12 L.H., Daily Totals		$5534.20		$6322.30	$461.18	$526.86

Crew B-10W

Crew B-10W	Hr.	Daily	Hr.	Daily	Bare Costs	Incl. O&P
1 Equip. Oper. (medium)	$50.60	$404.80	$76.75	$614.00	$46.27	$70.45
.5 Laborer	37.60	150.40	57.85	231.40		
1 Dozer, 105 H.P.		602.80		663.08	50.23	55.26
12 L.H., Daily Totals		$1158.00		$1508.48	$96.50	$125.71

Crew B-10X

Crew B-10X	Hr.	Daily	Hr.	Daily	Bare Costs	Incl. O&P
1 Equip. Oper. (medium)	$50.60	$404.80	$76.75	$614.00	$46.27	$70.45
.5 Laborer	37.60	150.40	57.85	231.40		
1 Dozer, 410 H.P.		2405.00		2645.50	200.42	220.46
12 L.H., Daily Totals		$2960.20		$3490.90	$246.68	$290.91

Crew B-10Y

Crew B-10Y	Hr.	Daily	Hr.	Daily	Bare Costs	Incl. O&P
1 Equip. Oper. (medium)	$50.60	$404.80	$76.75	$614.00	$46.27	$70.45
.5 Laborer	37.60	150.40	57.85	231.40		
1 Vibr. Roller, Towed, 12 Ton		549.80		604.78	45.82	50.40
12 L.H., Daily Totals		$1105.00		$1450.18	$92.08	$120.85

Crew B-11A

Crew B-11A	Hr.	Daily	Hr.	Daily	Bare Costs	Incl. O&P
1 Equipment Oper. (med.)	$50.60	$404.80	$76.75	$614.00	$44.10	$67.30
1 Laborer	37.60	300.80	57.85	462.80		
1 Dozer, 200 H.P.		1387.00		1525.70	86.69	95.36
16 L.H., Daily Totals		$2092.60		$2602.50	$130.79	$162.66

Crew B-11B

Crew No.	Bare Costs Hr.	Daily	Incl. Subs O&P Hr.	Daily	Cost Per Labor-Hour Bare Costs	Incl. O&P
1 Equipment Oper. (light)	$48.60	$388.80	$73.75	$590.00	$43.10	$65.80
1 Laborer	37.60	300.80	57.85	462.80		
1 Air Powered Tamper		26.20		28.82		
1 Air Compressor, 365 cfm		263.80		290.18		
2 -50' Air Hoses, 1.5"		11.60		12.76	18.85	20.73
16 L.H., Daily Totals		$991.20		$1384.56	$61.95	$86.53

Crew B-11C

Crew No.	Bare Costs Hr.	Daily	Incl. Subs O&P Hr.	Daily	Cost Per Labor-Hour Bare Costs	Incl. O&P
1 Equipment Oper. (med.)	$50.60	$404.80	$76.75	$614.00	$44.10	$67.30
1 Laborer	37.60	300.80	57.85	462.80		
1 Backhoe Loader, 48 H.P.		364.20		400.62	22.76	25.04
16 L.H., Daily Totals		$1069.80		$1477.42	$66.86	$92.34

Crew B-11J

Crew No.	Bare Costs Hr.	Daily	Incl. Subs O&P Hr.	Daily	Cost Per Labor-Hour Bare Costs	Incl. O&P
1 Equipment Oper. (med.)	$50.60	$404.80	$76.75	$614.00	$44.10	$67.30
1 Laborer	37.60	300.80	57.85	462.80		
1 Grader, 30,000 Lbs.		736.20		809.82		
1 Ripper, Beam & 1 Shank		84.20		92.62	51.27	56.40
16 L.H., Daily Totals		$1526.00		$1979.24	$95.38	$123.70

Crew B-11K

Crew No.	Bare Costs Hr.	Daily	Incl. Subs O&P Hr.	Daily	Cost Per Labor-Hour Bare Costs	Incl. O&P
1 Equipment Oper. (med.)	$50.60	$404.80	$76.75	$614.00	$44.10	$67.30
1 Laborer	37.60	300.80	57.85	462.80		
1 Trencher, Chain Type, 8' D		3402.00		3742.20	212.63	233.89
16 L.H., Daily Totals		$4107.60		$4819.00	$256.73	$301.19

Crew B-11L

Crew No.	Bare Costs Hr.	Daily	Incl. Subs O&P Hr.	Daily	Cost Per Labor-Hour Bare Costs	Incl. O&P
1 Equipment Oper. (med.)	$50.60	$404.80	$76.75	$614.00	$44.10	$67.30
1 Laborer	37.60	300.80	57.85	462.80		
1 Grader, 30,000 Lbs.		736.20		809.82	46.01	50.61
16 L.H., Daily Totals		$1441.80		$1886.62	$90.11	$117.91

Crew B-11M

Crew No.	Bare Costs Hr.	Daily	Incl. Subs O&P Hr.	Daily	Cost Per Labor-Hour Bare Costs	Incl. O&P
1 Equipment Oper. (med.)	$50.60	$404.80	$76.75	$614.00	$44.10	$67.30
1 Laborer	37.60	300.80	57.85	462.80		
1 Backhoe Loader, 80 H.P.		391.80		430.98	24.49	26.94
16 L.H., Daily Totals		$1097.40		$1507.78	$68.59	$94.24

Crew B-11N

Crew No.	Bare Costs Hr.	Daily	Incl. Subs O&P Hr.	Daily	Cost Per Labor-Hour Bare Costs	Incl. O&P
1 Labor Foreman (outside)	$39.60	$316.80	$60.95	$487.60	$42.34	$64.19
2 Equipment Operators (med.)	50.60	809.60	76.75	1228.00		
6 Truck Drivers (heavy)	40.05	1922.40	60.55	2906.40		
1 F.E. Loader, W.M., 5.5 C.Y.		1082.00		1190.20		
1 Dozer, 410 H.P.		2405.00		2645.50		
6 Dump Trucks, Off Hwy., 50 Ton		11022.00		12124.20	201.51	221.67
72 L.H., Daily Totals		$17557.80		$20581.90	$243.86	$285.86

Crew B-11Q

Crew No.	Bare Costs Hr.	Daily	Incl. Subs O&P Hr.	Daily	Cost Per Labor-Hour Bare Costs	Incl. O&P
1 Equipment Operator (med.)	$50.60	$404.80	$76.75	$614.00	$46.27	$70.45
.5 Laborer	37.60	150.40	57.85	231.40		
1 Dozer, 140 H.P.		886.00		974.60	73.83	81.22
12 L.H., Daily Totals		$1441.20		$1820.00	$120.10	$151.67

Crew B-11R

Crew No.	Bare Costs Hr.	Daily	Incl. Subs O&P Hr.	Daily	Cost Per Labor-Hour Bare Costs	Incl. O&P
1 Equipment Operator (med.)	$50.60	$404.80	$76.75	$614.00	$46.27	$70.45
.5 Laborer	37.60	150.40	57.85	231.40		
1 Dozer, 200 H.P.		1387.00		1525.70	115.58	127.14
12 L.H., Daily Totals		$1942.20		$2371.10	$161.85	$197.59

Crew B-11S

Crew No.	Bare Costs Hr.	Daily	Incl. Subs O&P Hr.	Daily	Cost Per Labor-Hour Bare Costs	Incl. O&P
1 Equipment Operator (med.)	$50.60	$404.80	$76.75	$614.00	$46.27	$70.45
.5 Laborer	37.60	150.40	57.85	231.40		
1 Dozer, 300 H.P.		1897.00		2086.70		
1 Ripper, Beam & 1 Shank		84.20		92.62	165.10	181.61
12 L.H., Daily Totals		$2536.40		$3024.72	$211.37	$252.06

Crew B-11T

Crew No.	Bare Costs Hr.	Daily	Incl. Subs O&P Hr.	Daily	Cost Per Labor-Hour Bare Costs	Incl. O&P
1 Equipment Operator (med.)	$50.60	$404.80	$76.75	$614.00	$46.27	$70.45
.5 Laborer	37.60	150.40	57.85	231.40		
1 Dozer, 410 H.P.		2405.00		2645.50		
1 Ripper, Beam & 2 Shanks		110.00		121.00	209.58	230.54
12 L.H., Daily Totals		$3070.20		$3611.90	$255.85	$300.99

Crew B-11U

Crew No.	Bare Costs Hr.	Daily	Incl. Subs O&P Hr.	Daily	Cost Per Labor-Hour Bare Costs	Incl. O&P
1 Equipment Operator (med.)	$50.60	$404.80	$76.75	$614.00	$46.27	$70.45
.5 Laborer	37.60	150.40	57.85	231.40		
1 Dozer, 520 H.P.		3299.00		3628.90	274.92	302.41
12 L.H., Daily Totals		$3854.20		$4474.30	$321.18	$372.86

Crew B-11V

Crew No.	Bare Costs Hr.	Daily	Incl. Subs O&P Hr.	Daily	Cost Per Labor-Hour Bare Costs	Incl. O&P
3 Laborers	$37.60	$902.40	$57.85	$1388.40	$37.60	$57.85
1 Roller, 2-Drum, W.B., 7.5 H.P.		180.00		198.00	7.50	8.25
24 L.H., Daily Totals		$1082.40		$1586.40	$45.10	$66.10

Crew B-11W

Crew No.	Bare Costs Hr.	Daily	Incl. Subs O&P Hr.	Daily	Cost Per Labor-Hour Bare Costs	Incl. O&P
1 Equipment Operator (med.)	$50.60	$404.80	$76.75	$614.00	$40.73	$61.67
1 Common Laborer	37.60	300.80	57.85	462.80		
10 Truck Drivers (heavy)	40.05	3204.00	60.55	4844.00		
1 Dozer, 200 H.P.		1387.00		1525.70		
1 Vibratory Roller, Towed, 23 Ton		396.00		435.60		
10 Dump Trucks, 8 C.Y., 220 H.P.		4112.00		4523.20	61.41	67.55
96 L.H., Daily Totals		$9804.60		$12405.30	$102.13	$129.22

Crew B-11Y

Crew No.	Bare Costs Hr.	Daily	Incl. Subs O&P Hr.	Daily	Cost Per Labor-Hour Bare Costs	Incl. O&P
1 Labor Foreman (outside)	$39.60	$316.80	$60.95	$487.60	$42.16	$64.49
5 Common Laborers	37.60	1504.00	57.85	2314.00		
3 Equipment Operators (med.)	50.60	1214.40	76.75	1842.00		
1 Dozer, 80 H.P.		472.40		519.64		
2 Roller, 2-Drum, W.B., 7.5 H.P.		360.00		396.00		
4 Vibrating Plate, Gas, 21"		187.20		205.92	14.16	15.58
72 L.H., Daily Totals		$4054.80		$5765.16	$56.32	$80.07

Crew B-12A

Crew No.	Bare Costs Hr.	Daily	Incl. Subs O&P Hr.	Daily	Cost Per Labor-Hour Bare Costs	Incl. O&P
1 Equip. Oper. (crane)	$51.70	$413.60	$78.45	$627.60	$44.65	$68.15
1 Laborer	37.60	300.80	57.85	462.80		
1 Hyd. Excavator, 1 C.Y.		812.60		893.86	50.79	55.87
16 L.H., Daily Totals		$1527.00		$1984.26	$95.44	$124.02

Crew B-12B

Crew No.	Bare Costs Hr.	Daily	Incl. Subs O&P Hr.	Daily	Cost Per Labor-Hour Bare Costs	Incl. O&P
1 Equip. Oper. (crane)	$51.70	$413.60	$78.45	$627.60	$44.65	$68.15
1 Laborer	37.60	300.80	57.85	462.80		
1 Hyd. Excavator, 1.5 C.Y.		1030.00		1133.00	64.38	70.81
16 L.H., Daily Totals		$1744.40		$2223.40	$109.03	$138.96

Crew B-12C

Crew No.	Bare Costs Hr.	Daily	Incl. Subs O&P Hr.	Daily	Cost Per Labor-Hour Bare Costs	Incl. O&P
1 Equip. Oper. (crane)	$51.70	$413.60	$78.45	$627.60	$44.65	$68.15
1 Laborer	37.60	300.80	57.85	462.80		
1 Hyd. Excavator, 2 C.Y.		1175.00		1292.50	73.44	80.78
16 L.H., Daily Totals		$1889.40		$2382.90	$118.09	$148.93

Crew No.	Bare Costs Hr.	Daily	Incl. Subs O&P Hr.	Daily	Bare Costs	Incl. O&P
Crew B-12D						
1 Equip. Oper. (crane)	$51.70	$413.60	$78.45	$627.60	$44.65	$68.15
1 Laborer	37.60	300.80	57.85	462.80		
1 Hyd. Excavator, 3.5 C.Y.		2440.00		2684.00	152.50	167.75
16 L.H., Daily Totals		$3154.40		$3774.40	$197.15	$235.90
Crew B-12E						
1 Equip. Oper. (crane)	$51.70	$413.60	$78.45	$627.60	$44.65	$68.15
1 Laborer	37.60	300.80	57.85	462.80		
1 Hyd. Excavator, .5 C.Y.		448.20		493.02	28.01	30.81
16 L.H., Daily Totals		$1162.60		$1583.42	$72.66	$98.96
Crew B-12F						
1 Equip. Oper. (crane)	$51.70	$413.60	$78.45	$627.60	$44.65	$68.15
1 Laborer	37.60	300.80	57.85	462.80		
1 Hyd. Excavator, .75 C.Y.		654.80		720.28	40.92	45.02
16 L.H., Daily Totals		$1369.20		$1810.68	$85.58	$113.17
Crew B-12G						
1 Equip. Oper. (crane)	$51.70	$413.60	$78.45	$627.60	$44.65	$68.15
1 Laborer	37.60	300.80	57.85	462.80		
1 Crawler Crane, 15 Ton		669.85		736.84		
1 Clamshell Bucket, .5 C.Y.		38.20		42.02	44.25	48.68
16 L.H., Daily Totals		$1422.45		$1869.26	$88.90	$116.83
Crew B-12H						
1 Equip. Oper. (crane)	$51.70	$413.60	$78.45	$627.60	$44.65	$68.15
1 Laborer	37.60	300.80	57.85	462.80		
1 Crawler Crane, 25 Ton		1150.00		1265.00		
1 Clamshell Bucket, 1 C.Y.		47.80		52.58	74.86	82.35
16 L.H., Daily Totals		$1912.20		$2407.98	$119.51	$150.50
Crew B-12I						
1 Equip. Oper. (crane)	$51.70	$413.60	$78.45	$627.60	$44.65	$68.15
1 Laborer	37.60	300.80	57.85	462.80		
1 Crawler Crane, 20 Ton		862.50		948.75		
1 Dragline Bucket, .75 C.Y.		20.60		22.66	55.19	60.71
16 L.H., Daily Totals		$1597.50		$2061.81	$99.84	$128.86
Crew B-12J						
1 Equip. Oper. (crane)	$51.70	$413.60	$78.45	$627.60	$44.65	$68.15
1 Laborer	37.60	300.80	57.85	462.80		
1 Gradall, 5/8 C.Y.		882.20		970.42	55.14	60.65
16 L.H., Daily Totals		$1596.60		$2060.82	$99.79	$128.80
Crew B-12K						
1 Equip. Oper. (crane)	$51.70	$413.60	$78.45	$627.60	$44.65	$68.15
1 Laborer	37.60	300.80	57.85	462.80		
1 Gradall, 3 Ton, 1 C.Y.		1002.00		1102.20	62.63	68.89
16 L.H., Daily Totals		$1716.40		$2192.60	$107.28	$137.04
Crew B-12L						
1 Equip. Oper. (crane)	$51.70	$413.60	$78.45	$627.60	$44.65	$68.15
1 Laborer	37.60	300.80	57.85	462.80		
1 Crawler Crane, 15 Ton		669.85		736.84		
1 F.E. Attachment, .5 C.Y.		60.00		66.00	45.62	50.18
16 L.H., Daily Totals		$1444.25		$1893.23	$90.27	$118.33

Crew No.	Bare Costs Hr.	Daily	Incl. Subs O&P Hr.	Daily	Bare Costs	Incl. O&P
Crew B-12M						
1 Equip. Oper. (crane)	$51.70	$413.60	$78.45	$627.60	$44.65	$68.15
1 Laborer	37.60	300.80	57.85	462.80		
1 Crawler Crane, 20 Ton		862.50		948.75		
1 F.E. Attachment, .75 C.Y.		65.40		71.94	57.99	63.79
16 L.H., Daily Totals		$1642.30		$2111.09	$102.64	$131.94
Crew B-12N						
1 Equip. Oper. (crane)	$51.70	$413.60	$78.45	$627.60	$44.65	$68.15
1 Laborer	37.60	300.80	57.85	462.80		
1 Crawler Crane, 25 Ton		1150.00		1265.00		
1 F.E. Attachment, 1 C.Y.		71.20		78.32	76.33	83.96
16 L.H., Daily Totals		$1935.60		$2433.72	$120.97	$152.11
Crew B-12O						
1 Equip. Oper. (crane)	$51.70	$413.60	$78.45	$627.60	$44.65	$68.15
1 Laborer	37.60	300.80	57.85	462.80		
1 Crawler Crane, 40 Ton		1156.00		1271.60		
1 F.E. Attachment, 1.5 C.Y.		80.00		88.00	77.25	84.97
16 L.H., Daily Totals		$1950.40		$2450.00	$121.90	$153.13
Crew B-12P						
1 Equip. Oper. (crane)	$51.70	$413.60	$78.45	$627.60	$44.65	$68.15
1 Laborer	37.60	300.80	57.85	462.80		
1 Crawler Crane, 40 Ton		1156.00		1271.60		
1 Dragline Bucket, 1.5 C.Y.		33.60		36.96	74.35	81.78
16 L.H., Daily Totals		$1904.00		$2398.96	$119.00	$149.94
Crew B-12Q						
1 Equip. Oper. (crane)	$51.70	$413.60	$78.45	$627.60	$44.65	$68.15
1 Laborer	37.60	300.80	57.85	462.80		
1 Hyd. Excavator, 5/8 C.Y.		589.40		648.34	36.84	40.52
16 L.H., Daily Totals		$1303.80		$1738.74	$81.49	$108.67
Crew B-12S						
1 Equip. Oper. (crane)	$51.70	$413.60	$78.45	$627.60	$44.65	$68.15
1 Laborer	37.60	300.80	57.85	462.80		
1 Hyd. Excavator, 2.5 C.Y.		1605.00		1765.50	100.31	110.34
16 L.H., Daily Totals		$2319.40		$2855.90	$144.96	$178.49
Crew B-12T						
1 Equip. Oper. (crane)	$51.70	$413.60	$78.45	$627.60	$44.65	$68.15
1 Laborer	37.60	300.80	57.85	462.80		
1 Crawler Crane, 75 Ton		1459.00		1604.90		
1 F.E. Attachment, 3 C.Y.		103.20		113.52	97.64	107.40
16 L.H., Daily Totals		$2276.60		$2808.82	$142.29	$175.55
Crew B-12V						
1 Equip. Oper. (crane)	$51.70	$413.60	$78.45	$627.60	$44.65	$68.15
1 Laborer	37.60	300.80	57.85	462.80		
1 Crawler Crane, 75 Ton		1459.00		1604.90		
1 Dragline Bucket, 3 C.Y.		52.60		57.86	94.47	103.92
16 L.H., Daily Totals		$2226.00		$2753.16	$139.13	$172.07
Crew B-12Y						
1 Equip. Oper. (crane)	$51.70	$413.60	$78.45	$627.60	$42.30	$64.72
2 Laborers	37.60	601.60	57.85	925.60		
1 Hyd. Excavator, 3.5 C.Y.		2440.00		2684.00	101.67	111.83
24 L.H., Daily Totals		$3455.20		$4237.20	$143.97	$176.55

Crew B-12Z

	Bare Costs Hr.	Daily	Incl. Subs O&P Hr.	Daily	Cost Per LH Bare Costs	Incl. O&P
1 Equip. Oper. (crane)	$51.70	$413.60	$78.45	$627.60	$42.30	$64.72
2 Laborers	37.60	601.60	57.85	925.60		
1 Hyd. Excavator, 2.5 C.Y.		1605.00		1765.50	66.88	73.56
24 L.H., Daily Totals		$2620.20		$3318.70	$109.18	$138.28

Crew B-13

	Bare Costs Hr.	Daily	Incl. Subs O&P Hr.	Daily	Cost Per LH Bare Costs	Incl. O&P
1 Labor Foreman (outside)	$39.60	$316.80	$60.95	$487.60	$40.99	$62.76
4 Laborers	37.60	1203.20	57.85	1851.20		
1 Equip. Oper. (crane)	51.70	413.60	78.45	627.60		
1 Equip. Oper. (oiler)	45.20	361.60	68.55	548.40		
1 Hyd. Crane, 25 Ton		736.60		810.26	13.15	14.47
56 L.H., Daily Totals		$3031.80		$4325.06	$54.14	$77.23

Crew B-13A

	Bare Costs Hr.	Daily	Incl. Subs O&P Hr.	Daily	Cost Per LH Bare Costs	Incl. O&P
1 Labor Foreman (outside)	$39.60	$316.80	$60.95	$487.60	$42.30	$64.46
2 Laborers	37.60	601.60	57.85	925.60		
2 Equipment Operators (med.)	50.60	809.60	76.75	1228.00		
2 Truck Drivers (heavy)	40.05	640.80	60.55	968.80		
1 Crawler Crane, 75 Ton		1459.00		1604.90		
1 Crawler Loader, 4 C.Y.		1561.00		1717.10		
2 Dump Trucks, 8 C.Y., 220 H.P.		822.40		904.64	68.61	75.48
56 L.H., Daily Totals		$6211.20		$7836.64	$110.91	$139.94

Crew B-13B

	Bare Costs Hr.	Daily	Incl. Subs O&P Hr.	Daily	Cost Per LH Bare Costs	Incl. O&P
1 Labor Foreman (outside)	$39.60	$316.80	$60.95	$487.60	$40.99	$62.76
4 Laborers	37.60	1203.20	57.85	1851.20		
1 Equip. Oper. (crane)	51.70	413.60	78.45	627.60		
1 Equip. Oper. (oiler)	45.20	361.60	68.55	548.40		
1 Hyd. Crane, 55 Ton		1128.00		1240.80	20.14	22.16
56 L.H., Daily Totals		$3423.20		$4755.60	$61.13	$84.92

Crew B-13C

	Bare Costs Hr.	Daily	Incl. Subs O&P Hr.	Daily	Cost Per LH Bare Costs	Incl. O&P
1 Labor Foreman (outside)	$39.60	$316.80	$60.95	$487.60	$40.99	$62.76
4 Laborers	37.60	1203.20	57.85	1851.20		
1 Equip. Oper. (crane)	51.70	413.60	78.45	627.60		
1 Equip. Oper. (oiler)	45.20	361.60	68.55	548.40		
1 Crawler Crane, 100 Ton		1667.00		1833.70	29.77	32.74
56 L.H., Daily Totals		$3962.20		$5348.50	$70.75	$95.51

Crew B-13D

	Bare Costs Hr.	Daily	Incl. Subs O&P Hr.	Daily	Cost Per LH Bare Costs	Incl. O&P
1 Laborer	$37.60	$300.80	$57.85	$462.80	$44.65	$68.15
1 Equip. Oper. (crane)	51.70	413.60	78.45	627.60		
1 Hyd. Excavator, 1 C.Y.		812.60		893.86		
1 Trench Box		81.00		89.10	55.85	61.44
16 L.H., Daily Totals		$1608.00		$2073.36	$100.50	$129.59

Crew B-13E

	Bare Costs Hr.	Daily	Incl. Subs O&P Hr.	Daily	Cost Per LH Bare Costs	Incl. O&P
1 Laborer	$37.60	$300.80	$57.85	$462.80	$44.65	$68.15
1 Equip. Oper. (crane)	51.70	413.60	78.45	627.60		
1 Hyd. Excavator, 1.5 C.Y.		1030.00		1133.00		
1 Trench Box		81.00		89.10	69.44	76.38
16 L.H., Daily Totals		$1825.40		$2312.50	$114.09	$144.53

Crew B-13F

	Bare Costs Hr.	Daily	Incl. Subs O&P Hr.	Daily	Cost Per LH Bare Costs	Incl. O&P
1 Laborer	$37.60	$300.80	$57.85	$462.80	$44.65	$68.15
1 Equip. Oper. (crane)	51.70	413.60	78.45	627.60		
1 Hyd. Excavator, 3.5 C.Y.		2440.00		2684.00		
1 Trench Box		81.00		89.10	157.56	173.32
16 L.H., Daily Totals		$3235.40		$3863.50	$202.21	$241.47

Crew B-13G

	Bare Costs Hr.	Daily	Incl. Subs O&P Hr.	Daily	Cost Per LH Bare Costs	Incl. O&P
1 Laborer	$37.60	$300.80	$57.85	$462.80	$44.65	$68.15
1 Equip. Oper. (crane)	51.70	413.60	78.45	627.60		
1 Hyd. Excavator, .75 C.Y.		654.80		720.28		
1 Trench Box		81.00		89.10	45.99	50.59
16 L.H., Daily Totals		$1450.20		$1899.78	$90.64	$118.74

Crew B-13H

	Bare Costs Hr.	Daily	Incl. Subs O&P Hr.	Daily	Cost Per LH Bare Costs	Incl. O&P
1 Laborer	$37.60	$300.80	$57.85	$462.80	$44.65	$68.15
1 Equip. Oper. (crane)	51.70	413.60	78.45	627.60		
1 Gradall, 5/8 C.Y.		882.20		970.42		
1 Trench Box		81.00		89.10	60.20	66.22
16 L.H., Daily Totals		$1677.60		$2149.92	$104.85	$134.37

Crew B-13I

	Bare Costs Hr.	Daily	Incl. Subs O&P Hr.	Daily	Cost Per LH Bare Costs	Incl. O&P
1 Laborer	$37.60	$300.80	$57.85	$462.80	$44.65	$68.15
1 Equip. Oper. (crane)	51.70	413.60	78.45	627.60		
1 Gradall, 3 Ton, 1 C.Y.		1002.00		1102.20		
1 Trench Box		81.00		89.10	67.69	74.46
16 L.H., Daily Totals		$1797.40		$2281.70	$112.34	$142.61

Crew B-13J

	Bare Costs Hr.	Daily	Incl. Subs O&P Hr.	Daily	Cost Per LH Bare Costs	Incl. O&P
1 Laborer	$37.60	$300.80	$57.85	$462.80	$44.65	$68.15
1 Equip. Oper. (crane)	51.70	413.60	78.45	627.60		
1 Hyd. Excavator, 2.5 C.Y.		1605.00		1765.50		
1 Trench Box		81.00		89.10	105.38	115.91
16 L.H., Daily Totals		$2400.40		$2945.00	$150.03	$184.06

Crew B-13K

	Bare Costs Hr.	Daily	Incl. Subs O&P Hr.	Daily	Cost Per LH Bare Costs	Incl. O&P
2 Equip. Opers. (crane)	$51.70	$827.20	$78.45	$1255.20	$51.70	$78.45
1 Hyd. Excavator, .75 C.Y.		654.80		720.28		
1 Hyd. Hammer, 4000 ft-lb		308.20		339.02		
1 Hyd. Excavator, .75 C.Y.		654.80		720.28	101.11	111.22
16 L.H., Daily Totals		$2445.00		$3034.78	$152.81	$189.67

Crew B-13L

	Bare Costs Hr.	Daily	Incl. Subs O&P Hr.	Daily	Cost Per LH Bare Costs	Incl. O&P
2 Equip. Opers. (crane)	$51.70	$827.20	$78.45	$1255.20	$51.70	$78.45
1 Hyd. Excavator, 1.5 C.Y.		1030.00		1133.00		
1 Hyd. Hammer, 5000 ft-lb		371.20		408.32		
1 Hyd. Excavator, .75 C.Y.		654.80		720.28	128.50	141.35
16 L.H., Daily Totals		$2883.20		$3516.80	$180.20	$219.80

Crew B-13M

	Bare Costs Hr.	Daily	Incl. Subs O&P Hr.	Daily	Cost Per LH Bare Costs	Incl. O&P
2 Equip. Opers. (crane)	$51.70	$827.20	$78.45	$1255.20	$51.70	$78.45
1 Hyd. Excavator, 2.5 C.Y.		1605.00		1765.50		
1 Hyd. Hammer, 8000 ft-lb		545.40		599.94		
1 Hyd. Excavator, 1.5 C.Y.		1030.00		1133.00	198.78	218.65
16 L.H., Daily Totals		$4007.60		$4753.64	$250.47	$297.10

Crew B-13N

	Bare Costs Hr.	Daily	Incl. Subs O&P Hr.	Daily	Cost Per LH Bare Costs	Incl. O&P
2 Equip. Opers. (crane)	$51.70	$827.20	$78.45	$1255.20	$51.70	$78.45
1 Hyd. Excavator, 3.5 C.Y.		2440.00		2684.00		
1 Hyd. Hammer, 12,000 ft-lb		635.80		699.38		
1 Hyd. Excavator, 1.5 C.Y.		1030.00		1133.00	256.61	282.27
16 L.H., Daily Totals		$4933.00		$5771.58	$308.31	$360.72

For customer support on your Mechanical Cost Data, call 877.795.7394.

627

Crew B-14

	Bare Costs		Incl. Subs O&P		Cost Per Labor-Hour	
Crew B-14	Hr.	Daily	Hr.	Daily	Bare Costs	Incl. O&P
1 Labor Foreman (outside)	$39.60	$316.80	$60.95	$487.60	$39.77	$61.02
4 Laborers	37.60	1203.20	57.85	1851.20		
1 Equip. Oper. (light)	48.60	388.80	73.75	590.00		
1 Backhoe Loader, 48 H.P.		364.20		400.62	7.59	8.35
48 L.H., Daily Totals		$2273.00		$3329.42	$47.35	$69.36

Crew B-14A

	Bare Costs		Incl. Subs O&P		Cost Per Labor-Hour	
Crew B-14A	Hr.	Daily	Hr.	Daily	Bare Costs	Incl. O&P
1 Equip. Oper. (crane)	$51.70	$413.60	$78.45	$627.60	$47.00	$71.58
.5 Laborer	37.60	150.40	57.85	231.40		
1 Hyd. Excavator, 4.5 C.Y.		2963.00		3259.30	246.92	271.61
12 L.H., Daily Totals		$3527.00		$4118.30	$293.92	$343.19

Crew B-14B

	Bare Costs		Incl. Subs O&P		Cost Per Labor-Hour	
Crew B-14B	Hr.	Daily	Hr.	Daily	Bare Costs	Incl. O&P
1 Equip. Oper. (crane)	$51.70	$413.60	$78.45	$627.60	$47.00	$71.58
.5 Laborer	37.60	150.40	57.85	231.40		
1 Hyd. Excavator, 6 C.Y.		3494.00		3843.40	291.17	320.28
12 L.H., Daily Totals		$4058.00		$4702.40	$338.17	$391.87

Crew B-14C

	Bare Costs		Incl. Subs O&P		Cost Per Labor-Hour	
Crew B-14C	Hr.	Daily	Hr.	Daily	Bare Costs	Incl. O&P
1 Equip. Oper. (crane)	$51.70	$413.60	$78.45	$627.60	$47.00	$71.58
.5 Laborer	37.60	150.40	57.85	231.40		
1 Hyd. Excavator, 7 C.Y.		3590.00		3949.00	299.17	329.08
12 L.H., Daily Totals		$4154.00		$4808.00	$346.17	$400.67

Crew B-14F

	Bare Costs		Incl. Subs O&P		Cost Per Labor-Hour	
Crew B-14F	Hr.	Daily	Hr.	Daily	Bare Costs	Incl. O&P
1 Equip. Oper. (crane)	$51.70	$413.60	$78.45	$627.60	$47.00	$71.58
.5 Laborer	37.60	150.40	57.85	231.40		
1 Hyd. Shovel, 7 C.Y.		3993.00		4392.30	332.75	366.02
12 L.H., Daily Totals		$4557.00		$5251.30	$379.75	$437.61

Crew B-14G

	Bare Costs		Incl. Subs O&P		Cost Per Labor-Hour	
Crew B-14G	Hr.	Daily	Hr.	Daily	Bare Costs	Incl. O&P
1 Equip. Oper. (crane)	$51.70	$413.60	$78.45	$627.60	$47.00	$71.58
.5 Laborer	37.60	150.40	57.85	231.40		
1 Hyd. Shovel, 12 C.Y.		5736.00		6309.60	478.00	525.80
12 L.H., Daily Totals		$6300.00		$7168.60	$525.00	$597.38

Crew B-14J

	Bare Costs		Incl. Subs O&P		Cost Per Labor-Hour	
Crew B-14J	Hr.	Daily	Hr.	Daily	Bare Costs	Incl. O&P
1 Equip. Oper. (medium)	$50.60	$404.80	$76.75	$614.00	$46.27	$70.45
.5 Laborer	37.60	150.40	57.85	231.40		
1 F.E. Loader, 8 C.Y.		1937.00		2130.70	161.42	177.56
12 L.H., Daily Totals		$2492.20		$2976.10	$207.68	$248.01

Crew B-14K

	Bare Costs		Incl. Subs O&P		Cost Per Labor-Hour	
Crew B-14K	Hr.	Daily	Hr.	Daily	Bare Costs	Incl. O&P
1 Equip. Oper. (medium)	$50.60	$404.80	$76.75	$614.00	$46.27	$70.45
.5 Laborer	37.60	150.40	57.85	231.40		
1 F.E. Loader, 10 C.Y.		2942.00		3236.20	245.17	269.68
12 L.H., Daily Totals		$3497.20		$4081.60	$291.43	$340.13

Crew B-15

	Bare Costs		Incl. Subs O&P		Cost Per Labor-Hour	
Crew B-15	Hr.	Daily	Hr.	Daily	Bare Costs	Incl. O&P
1 Equipment Oper. (med.)	$50.60	$404.80	$76.75	$614.00	$42.71	$64.79
.5 Laborer	37.60	150.40	57.85	231.40		
2 Truck Drivers (heavy)	40.05	640.80	60.55	968.80		
2 Dump Trucks, 12 C.Y., 400 H.P.		1382.00		1520.20		
1 Dozer, 200 H.P.		1387.00		1525.70	98.89	108.78
28 L.H., Daily Totals		$3965.00		$4860.10	$141.61	$173.57

Crew B-16

	Bare Costs		Incl. Subs O&P		Cost Per Labor-Hour	
Crew B-16	Hr.	Daily	Hr.	Daily	Bare Costs	Incl. O&P
1 Labor Foreman (outside)	$39.60	$316.80	$60.95	$487.60	$38.71	$59.30
2 Laborers	37.60	601.60	57.85	925.60		
1 Truck Driver (heavy)	40.05	320.40	60.55	484.40		
1 Dump Truck, 12 C.Y., 400 H.P.		691.00		760.10	21.59	23.75
32 L.H., Daily Totals		$1929.80		$2657.70	$60.31	$83.05

Crew B-17

	Bare Costs		Incl. Subs O&P		Cost Per Labor-Hour	
Crew B-17	Hr.	Daily	Hr.	Daily	Bare Costs	Incl. O&P
2 Laborers	$37.60	$601.60	$57.85	$925.60	$40.96	$62.50
1 Equip. Oper. (light)	48.60	388.80	73.75	590.00		
1 Truck Driver (heavy)	40.05	320.40	60.55	484.40		
1 Backhoe Loader, 48 H.P.		364.20		400.62		
1 Dump Truck, 8 C.Y., 220 H.P.		411.20		452.32	24.23	26.65
32 L.H., Daily Totals		$2086.20		$2852.94	$65.19	$89.15

Crew B-17A

	Bare Costs		Incl. Subs O&P		Cost Per Labor-Hour	
Crew B-17A	Hr.	Daily	Hr.	Daily	Bare Costs	Incl. O&P
2 Labor Foremen (outside)	$39.60	$633.60	$60.95	$975.20	$40.41	$62.24
6 Laborers	37.60	1804.80	57.85	2776.80		
1 Skilled Worker Foreman (out)	50.65	405.20	78.25	626.00		
1 Skilled Worker	48.65	389.20	75.15	601.20		
80 L.H., Daily Totals		$3232.80		$4979.20	$40.41	$62.24

Crew B-17B

	Bare Costs		Incl. Subs O&P		Cost Per Labor-Hour	
Crew B-17B	Hr.	Daily	Hr.	Daily	Bare Costs	Incl. O&P
2 Laborers	$37.60	$601.60	$57.85	$925.60	$40.96	$62.50
1 Equip. Oper. (light)	48.60	388.80	73.75	590.00		
1 Truck Driver (heavy)	40.05	320.40	60.55	484.40		
1 Backhoe Loader, 48 H.P.		364.20		400.62		
1 Dump Truck, 12 C.Y., 400 H.P.		691.00		760.10	32.98	36.27
32 L.H., Daily Totals		$2366.00		$3160.72	$73.94	$98.77

Crew B-18

	Bare Costs		Incl. Subs O&P		Cost Per Labor-Hour	
Crew B-18	Hr.	Daily	Hr.	Daily	Bare Costs	Incl. O&P
1 Labor Foreman (outside)	$39.60	$316.80	$60.95	$487.60	$38.27	$58.88
2 Laborers	37.60	601.60	57.85	925.60		
1 Vibrating Plate, Gas, 21"		46.80		51.48	1.95	2.15
24 L.H., Daily Totals		$965.20		$1464.68	$40.22	$61.03

Crew B-19

	Bare Costs		Incl. Subs O&P		Cost Per Labor-Hour	
Crew B-19	Hr.	Daily	Hr.	Daily	Bare Costs	Incl. O&P
1 Pile Driver Foreman (outside)	$48.10	$384.80	$76.15	$609.20	$47.64	$74.20
4 Pile Drivers	46.10	1475.20	73.00	2336.00		
2 Equip. Oper. (crane)	51.70	827.20	78.45	1255.20		
1 Equip. Oper. (oiler)	45.20	361.60	68.55	548.40		
1 Crawler Crane, 40 Ton		1156.00		1271.60		
1 Lead, 90' High		124.40		136.84		
1 Hammer, Diesel, 22k ft-lb		456.40		502.04	27.14	29.85
64 L.H., Daily Totals		$4785.60		$6659.28	$74.78	$104.05

Crew B-19A

	Bare Costs		Incl. Subs O&P		Cost Per Labor-Hour	
Crew B-19A	Hr.	Daily	Hr.	Daily	Bare Costs	Incl. O&P
1 Pile Driver Foreman (outside)	$48.10	$384.80	$76.15	$609.20	$47.64	$74.20
4 Pile Drivers	46.10	1475.20	73.00	2336.00		
2 Equip. Oper. (crane)	51.70	827.20	78.45	1255.20		
1 Equip. Oper. (oiler)	45.20	361.60	68.55	548.40		
1 Crawler Crane, 75 Ton		1459.00		1604.90		
1 Lead, 90' high		124.40		136.84		
1 Hammer, Diesel, 41k ft-lb		581.00		639.10	33.82	37.20
64 L.H., Daily Totals		$5213.20		$7129.64	$81.46	$111.40

Crew No.	Bare Costs		Incl. Subs O&P		Cost Per Labor-Hour	
Crew B-19B	Hr.	Daily	Hr.	Daily	Bare Costs	Incl. O&P
1 Pile Driver Foreman (outside)	$48.10	$384.80	$76.15	$609.20	$47.64	$74.20
4 Pile Drivers	46.10	1475.20	73.00	2336.00		
2 Equip. Oper. (crane)	51.70	827.20	78.45	1255.20		
1 Equip. Oper. (oiler)	45.20	361.60	68.55	548.40		
1 Crawler Crane, 40 Ton		1156.00		1271.60		
1 Lead, 90' High		124.40		136.84		
1 Hammer, Diesel, 22k ft-lb		456.40		502.04		
1 Barge, 400 Ton		776.00		853.60	39.26	43.19
64 L.H., Daily Totals		$5561.60		$7512.88	$86.90	$117.39
Crew B-19C	Hr.	Daily	Hr.	Daily	Bare Costs	Incl. O&P
1 Pile Driver Foreman (outside)	$48.10	$384.80	$76.15	$609.20	$47.64	$74.20
4 Pile Drivers	46.10	1475.20	73.00	2336.00		
2 Equip. Oper. (crane)	51.70	827.20	78.45	1255.20		
1 Equip. Oper. (oiler)	45.20	361.60	68.55	548.40		
1 Crawler Crane, 75 Ton		1459.00		1604.90		
1 Lead, 90' High		124.40		136.84		
1 Hammer, Diesel, 41k ft-lb		581.00		639.10		
1 Barge, 400 Ton		776.00		853.60	45.94	50.54
64 L.H., Daily Totals		$5989.20		$7983.24	$93.58	$124.74
Crew B-20	Hr.	Daily	Hr.	Daily	Bare Costs	Incl. O&P
1 Labor Foreman (outside)	$39.60	$316.80	$60.95	$487.60	$41.95	$64.65
1 Skilled Worker	48.65	389.20	75.15	601.20		
1 Laborer	37.60	300.80	57.85	462.80		
24 L.H., Daily Totals		$1006.80		$1551.60	$41.95	$64.65
Crew B-20A	Hr.	Daily	Hr.	Daily	Bare Costs	Incl. O&P
1 Labor Foreman (outside)	$39.60	$316.80	$60.95	$487.60	$45.71	$69.59
1 Laborer	37.60	300.80	57.85	462.80		
1 Plumber	58.70	469.60	88.65	709.20		
1 Plumber Apprentice	46.95	375.60	70.90	567.20		
32 L.H., Daily Totals		$1462.80		$2226.80	$45.71	$69.59
Crew B-21	Hr.	Daily	Hr.	Daily	Bare Costs	Incl. O&P
1 Labor Foreman (outside)	$39.60	$316.80	$60.95	$487.60	$43.34	$66.62
1 Skilled Worker	48.65	389.20	75.15	601.20		
1 Laborer	37.60	300.80	57.85	462.80		
.5 Equip. Oper. (crane)	51.70	206.80	78.45	313.80		
.5 S.P. Crane, 4x4, 5 Ton		137.70		151.47	4.92	5.41
28 L.H., Daily Totals		$1351.30		$2016.87	$48.26	$72.03
Crew B-21A	Hr.	Daily	Hr.	Daily	Bare Costs	Incl. O&P
1 Labor Foreman (outside)	$39.60	$316.80	$60.95	$487.60	$46.91	$71.36
1 Laborer	37.60	300.80	57.85	462.80		
1 Plumber	58.70	469.60	88.65	709.20		
1 Plumber Apprentice	46.95	375.60	70.90	567.20		
1 Equip. Oper. (crane)	51.70	413.60	78.45	627.60		
1 S.P. Crane, 4x4, 12 Ton		473.40		520.74	11.84	13.02
40 L.H., Daily Totals		$2349.80		$3375.14	$58.74	$84.38
Crew B-21B	Hr.	Daily	Hr.	Daily	Bare Costs	Incl. O&P
1 Labor Foreman (outside)	$39.60	$316.80	$60.95	$487.60	$40.82	$62.59
3 Laborers	37.60	902.40	57.85	1388.40		
1 Equip. Oper. (crane)	51.70	413.60	78.45	627.60		
1 Hyd. Crane, 12 Ton		653.80		719.18	16.34	17.98
40 L.H., Daily Totals		$2286.60		$3222.78	$57.16	$80.57

Crew No.	Bare Costs		Incl. Subs O&P		Cost Per Labor-Hour	
Crew B-21C	Hr.	Daily	Hr.	Daily	Bare Costs	Incl. O&P
1 Labor Foreman (outside)	$39.60	$316.80	$60.95	$487.60	$40.99	$62.76
4 Laborers	37.60	1203.20	57.85	1851.20		
1 Equip. Oper. (crane)	51.70	413.60	78.45	627.60		
1 Equip. Oper. (oiler)	45.20	361.60	68.55	548.40		
2 Cutting Torches		22.80		25.08		
2 Sets of Gases		304.00		334.40		
1 Lattice Boom Crane, 90 Ton		1511.00		1662.10	32.82	36.10
56 L.H., Daily Totals		$4133.00		$5536.38	$73.80	$98.86
Crew B-22	Hr.	Daily	Hr.	Daily	Bare Costs	Incl. O&P
1 Labor Foreman (outside)	$39.60	$316.80	$60.95	$487.60	$43.90	$67.41
1 Skilled Worker	48.65	389.20	75.15	601.20		
1 Laborer	37.60	300.80	57.85	462.80		
.75 Equip. Oper. (crane)	51.70	310.20	78.45	470.70		
.75 S.P. Crane, 4x4, 5 Ton		206.55		227.21	6.88	7.57
30 L.H., Daily Totals		$1523.55		$2249.51	$50.78	$74.98
Crew B-22A	Hr.	Daily	Hr.	Daily	Bare Costs	Incl. O&P
1 Labor Foreman (outside)	$39.60	$316.80	$60.95	$487.60	$43.03	$66.05
1 Skilled Worker	48.65	389.20	75.15	601.20		
2 Laborers	37.60	601.60	57.85	925.60		
1 Equipment Operator, Crane	51.70	413.60	78.45	627.60		
1 S.P. Crane, 4x4, 5 Ton		275.40		302.94		
1 Butt Fusion Mach., 4"-12" diam.		382.10		420.31	16.44	18.08
40 L.H., Daily Totals		$2378.70		$3365.25	$59.47	$84.13
Crew B-22B	Hr.	Daily	Hr.	Daily	Bare Costs	Incl. O&P
1 Labor Foreman (outside)	$39.60	$316.80	$60.95	$487.60	$43.03	$66.05
1 Skilled Worker	48.65	389.20	75.15	601.20		
2 Laborers	37.60	601.60	57.85	925.60		
1 Equip. Oper. (crane)	51.70	413.60	78.45	627.60		
1 S.P. Crane, 4x4, 5 Ton		275.40		302.94		
1 Butt Fusion Mach., 8"-24" diam.		829.70		912.67	27.63	30.39
40 L.H., Daily Totals		$2826.30		$3857.61	$70.66	$96.44
Crew B-22C	Hr.	Daily	Hr.	Daily	Bare Costs	Incl. O&P
1 Skilled Worker	$48.65	$389.20	$75.15	$601.20	$43.13	$66.50
1 Laborer	37.60	300.80	57.85	462.80		
1 Butt Fusion Mach., 2"-8" diam.		121.05		133.16	7.57	8.32
16 L.H., Daily Totals		$811.05		$1197.16	$50.69	$74.82
Crew B-23	Hr.	Daily	Hr.	Daily	Bare Costs	Incl. O&P
1 Labor Foreman (outside)	$39.60	$316.80	$60.95	$487.60	$38.00	$58.47
4 Laborers	37.60	1203.20	57.85	1851.20		
1 Drill Rig, Truck-Mounted		2542.00		2796.20		
1 Flatbed Truck, Gas, 3 Ton		303.20		333.52	71.13	78.24
40 L.H., Daily Totals		$4365.20		$5468.52	$109.13	$136.71
Crew B-23A	Hr.	Daily	Hr.	Daily	Bare Costs	Incl. O&P
1 Labor Foreman (outside)	$39.60	$316.80	$60.95	$487.60	$42.60	$65.18
1 Laborer	37.60	300.80	57.85	462.80		
1 Equip. Oper. (medium)	50.60	404.80	76.75	614.00		
1 Drill Rig, Truck-Mounted		2542.00		2796.20		
1 Pickup Truck, 3/4 Ton		144.20		158.62	111.93	123.12
24 L.H., Daily Totals		$3708.60		$4519.22	$154.53	$188.30

For customer support on your Mechanical Cost Data, call 877.795.7394.

629

| Crew No. | Bare Costs | | Incl. Subs O&P | | Cost Per Labor-Hour | |

Crew B-23B

Crew B-23B	Hr.	Daily	Hr.	Daily	Bare Costs	Incl. O&P
1 Labor Foreman (outside)	$39.60	$316.80	$60.95	$487.60	$42.60	$65.18
1 Laborer	37.60	300.80	57.85	462.80		
1 Equip. Oper. (medium)	50.60	404.80	76.75	614.00		
1 Drill Rig, Truck-Mounted		2542.00		2796.20		
1 Pickup Truck, 3/4 Ton		144.20		158.62		
1 Centr. Water Pump, 6"		346.60		381.26	126.37	139.00
24 L.H., Daily Totals		$4055.20		$4900.48	$168.97	$204.19

Crew B-24

Crew B-24	Hr.	Daily	Hr.	Daily	Bare Costs	Incl. O&P
1 Cement Finisher	$45.00	$360.00	$66.65	$533.20	$43.18	$65.58
1 Laborer	37.60	300.80	57.85	462.80		
1 Carpenter	46.95	375.60	72.25	578.00		
24 L.H., Daily Totals		$1036.40		$1574.00	$43.18	$65.58

Crew B-25

Crew B-25	Hr.	Daily	Hr.	Daily	Bare Costs	Incl. O&P
1 Labor Foreman (outside)	$39.60	$316.80	$60.95	$487.60	$41.33	$63.29
7 Laborers	37.60	2105.60	57.85	3239.60		
3 Equip. Oper. (medium)	50.60	1214.40	76.75	1842.00		
1 Asphalt Paver, 130 H.P.		2132.00		2345.20		
1 Tandem Roller, 10 Ton		236.80		260.48		
1 Roller, Pneum. Whl., 12 Ton		344.40		378.84	30.83	33.91
88 L.H., Daily Totals		$6350.00		$8553.72	$72.16	$97.20

Crew B-25B

Crew B-25B	Hr.	Daily	Hr.	Daily	Bare Costs	Incl. O&P
1 Labor Foreman (outside)	$39.60	$316.80	$60.95	$487.60	$42.10	$64.41
7 Laborers	37.60	2105.60	57.85	3239.60		
4 Equip. Oper. (medium)	50.60	1619.20	76.75	2456.00		
1 Asphalt Paver, 130 H.P.		2132.00		2345.20		
2 Tandem Rollers, 10 Ton		473.60		520.96		
1 Roller, Pneum. Whl., 12 Ton		344.40		378.84	30.73	33.80
96 L.H., Daily Totals		$6991.60		$9428.20	$72.83	$98.21

Crew B-25C

Crew B-25C	Hr.	Daily	Hr.	Daily	Bare Costs	Incl. O&P
1 Labor Foreman (outside)	$39.60	$316.80	$60.95	$487.60	$42.27	$64.67
3 Laborers	37.60	902.40	57.85	1388.40		
2 Equip. Oper. (medium)	50.60	809.60	76.75	1228.00		
1 Asphalt Paver, 130 H.P.		2132.00		2345.20		
1 Tandem Roller, 10 Ton		236.80		260.48	49.35	54.28
48 L.H., Daily Totals		$4397.60		$5709.68	$91.62	$118.95

Crew B-25D

Crew B-25D	Hr.	Daily	Hr.	Daily	Bare Costs	Incl. O&P
1 Labor Foreman (outside)	$39.60	$316.80	$60.95	$487.60	$42.39	$64.83
3 Laborers	37.60	902.40	57.85	1388.40		
2.125 Equip. Oper. (medium)	50.60	860.20	76.75	1304.75		
.125 Truck Driver (heavy)	40.05	40.05	60.55	60.55		
.125 Truck Tractor, 6x4, 380 H.P.		74.97		82.47		
.125 Dist. Tanker, 3000 Gallon		38.20		42.02		
1 Asphalt Paver, 130 H.P.		2132.00		2345.20		
1 Tandem Roller, 10 Ton		236.80		260.48	49.64	54.60
50 L.H., Daily Totals		$4601.43		$5971.47	$92.03	$119.43

Crew B-25E

Crew B-25E	Hr.	Daily	Hr.	Daily	Bare Costs	Incl. O&P
1 Labor Foreman (outside)	$39.60	$316.80	$60.95	$487.60	$42.50	$64.97
3 Laborers	37.60	902.40	57.85	1388.40		
2.250 Equip. Oper. (medium)	50.60	910.80	76.75	1381.50		
.25 Truck Driver (heavy)	40.05	80.10	60.55	121.10		
.25 Truck Tractor, 6x4, 380 H.P.		149.95		164.94		
.25 Dist. Tanker, 3000 Gallon		76.40		84.04		
1 Asphalt Paver, 130 H.P.		2132.00		2345.20		
1 Tandem Roller, 10 Ton		236.80		260.48	49.91	54.90
52 L.H., Daily Totals		$4805.25		$6233.27	$92.41	$119.87

Crew B-26

Crew B-26	Hr.	Daily	Hr.	Daily	Bare Costs	Incl. O&P
1 Labor Foreman (outside)	$39.60	$316.80	$60.95	$487.60	$42.18	$64.57
6 Laborers	37.60	1804.80	57.85	2776.80		
2 Equip. Oper. (medium)	50.60	809.60	76.75	1228.00		
1 Rodman (reinf.)	52.55	420.40	82.10	656.80		
1 Cement Finisher	45.00	360.00	66.65	533.20		
1 Grader, 30,000 Lbs.		736.20		809.82		
1 Paving Mach. & Equip.		2794.00		3073.40	40.12	44.13
88 L.H., Daily Totals		$7241.80		$9565.62	$82.29	$108.70

Crew B-26A

Crew B-26A	Hr.	Daily	Hr.	Daily	Bare Costs	Incl. O&P
1 Labor Foreman (outside)	$39.60	$316.80	$60.95	$487.60	$42.18	$64.57
6 Laborers	37.60	1804.80	57.85	2776.80		
2 Equip. Oper. (medium)	50.60	809.60	76.75	1228.00		
1 Rodman (reinf.)	52.55	420.40	82.10	656.80		
1 Cement Finisher	45.00	360.00	66.65	533.20		
1 Grader, 30,000 Lbs.		736.20		809.82		
1 Paving Mach. & Equip.		2794.00		3073.40		
1 Concrete Saw		167.40		184.14	42.02	46.22
88 L.H., Daily Totals		$7409.20		$9749.76	$84.20	$110.79

Crew B-26B

Crew B-26B	Hr.	Daily	Hr.	Daily	Bare Costs	Incl. O&P
1 Labor Foreman (outside)	$39.60	$316.80	$60.95	$487.60	$42.88	$65.59
6 Laborers	37.60	1804.80	57.85	2776.80		
3 Equip. Oper. (medium)	50.60	1214.40	76.75	1842.00		
1 Rodman (reinf.)	52.55	420.40	82.10	656.80		
1 Cement Finisher	45.00	360.00	66.65	533.20		
1 Grader, 30,000 Lbs.		736.20		809.82		
1 Paving Mach. & Equip.		2794.00		3073.40		
1 Concrete Pump, 110' Boom		946.20		1040.82	46.63	51.29
96 L.H., Daily Totals		$8592.80		$11220.44	$89.51	$116.88

Crew B-26C

Crew B-26C	Hr.	Daily	Hr.	Daily	Bare Costs	Incl. O&P
1 Labor Foreman (outside)	$39.60	$316.80	$60.95	$487.60	$41.34	$63.35
6 Laborers	37.60	1804.80	57.85	2776.80		
1 Equip. Oper. (medium)	50.60	404.80	76.75	614.00		
1 Rodman (reinf.)	52.55	420.40	82.10	656.80		
1 Cement Finisher	45.00	360.00	66.65	533.20		
1 Paving Mach. & Equip.		2794.00		3073.40		
1 Concrete Saw		167.40		184.14	37.02	40.72
80 L.H., Daily Totals		$6268.20		$8325.94	$78.35	$104.07

Crew B-27

Crew B-27	Hr.	Daily	Hr.	Daily	Bare Costs	Incl. O&P
1 Labor Foreman (outside)	$39.60	$316.80	$60.95	$487.60	$38.10	$58.63
3 Laborers	37.60	902.40	57.85	1388.40		
1 Berm Machine		296.80		326.48	9.28	10.20
32 L.H., Daily Totals		$1516.00		$2202.48	$47.38	$68.83

Crew B-28

Crew B-28	Hr.	Daily	Hr.	Daily	Bare Costs	Incl. O&P
2 Carpenters	$46.95	$751.20	$72.25	$1156.00	$43.83	$67.45
1 Laborer	37.60	300.80	57.85	462.80		
24 L.H., Daily Totals		$1052.00		$1618.80	$43.83	$67.45

Crew B-29

Crew B-29	Hr.	Daily	Hr.	Daily	Bare Costs	Incl. O&P
1 Labor Foreman (outside)	$39.60	$316.80	$60.95	$487.60	$40.99	$62.76
4 Laborers	37.60	1203.20	57.85	1851.20		
1 Equip. Oper. (crane)	51.70	413.60	78.45	627.60		
1 Equip. Oper. (oiler)	45.20	361.60	68.55	548.40		
1 Gradall, 5/8 C.Y.		882.20		970.42	15.75	17.33
56 L.H., Daily Totals		$3177.40		$4485.22	$56.74	$80.09

For customer support on your **Mechanical Cost Data**, call 877.795.7394.

Crew B-30

Crew No.	Bare Costs Hr.	Bare Costs Daily	Incl. Subs O&P Hr.	Incl. Subs O&P Daily	Cost Per Labor-Hour Bare Costs	Cost Per Labor-Hour Incl. O&P
1 Equip. Oper. (medium)	$50.60	$404.80	$76.75	$614.00	$43.57	$65.95
2 Truck Drivers (heavy)	40.05	640.80	60.55	968.80		
1 Hyd. Excavator, 1.5 C.Y.		1030.00		1133.00		
2 Dump Trucks, 12 C.Y., 400 H.P.		1382.00		1520.20	100.50	110.55
24 L.H., Daily Totals		$3457.60		$4236.00	$144.07	$176.50

Crew B-31

Crew No.	Bare Costs Hr.	Bare Costs Daily	Incl. Subs O&P Hr.	Incl. Subs O&P Daily	Cost Per Labor-Hour Bare Costs	Cost Per Labor-Hour Incl. O&P
1 Labor Foreman (outside)	$39.60	$316.80	$60.95	$487.60	$39.87	$61.35
3 Laborers	37.60	902.40	57.85	1388.40		
1 Carpenter	46.95	375.60	72.25	578.00		
1 Air Compressor, 250 cfm		201.40		221.54		
1 Sheeting Driver		5.75		6.33		
2 -50' Air Hoses, 1.5"		11.60		12.76	5.47	6.02
40 L.H., Daily Totals		$1813.55		$2694.63	$45.34	$67.37

Crew B-32

Crew No.	Bare Costs Hr.	Bare Costs Daily	Incl. Subs O&P Hr.	Incl. Subs O&P Daily	Cost Per Labor-Hour Bare Costs	Cost Per Labor-Hour Incl. O&P
1 Laborer	$37.60	$300.80	$57.85	$462.80	$47.35	$72.03
3 Equip. Oper. (medium)	50.60	1214.40	76.75	1842.00		
1 Grader, 30,000 Lbs.		736.20		809.82		
1 Tandem Roller, 10 Ton		236.80		260.48		
1 Dozer, 200 H.P.		1387.00		1525.70	73.75	81.13
32 L.H., Daily Totals		$3875.20		$4900.80	$121.10	$153.15

Crew B-32A

Crew No.	Bare Costs Hr.	Bare Costs Daily	Incl. Subs O&P Hr.	Incl. Subs O&P Daily	Cost Per Labor-Hour Bare Costs	Cost Per Labor-Hour Incl. O&P
1 Laborer	$37.60	$300.80	$57.85	$462.80	$46.27	$70.45
2 Equip. Oper. (medium)	50.60	809.60	76.75	1228.00		
1 Grader, 30,000 Lbs.		736.20		809.82		
1 Roller, Vibratory, 25 Ton		687.40		756.14	59.32	65.25
24 L.H., Daily Totals		$2534.00		$3256.76	$105.58	$135.70

Crew B-32B

Crew No.	Bare Costs Hr.	Bare Costs Daily	Incl. Subs O&P Hr.	Incl. Subs O&P Daily	Cost Per Labor-Hour Bare Costs	Cost Per Labor-Hour Incl. O&P
1 Laborer	$37.60	$300.80	$57.85	$462.80	$46.27	$70.45
2 Equip. Oper. (medium)	50.60	809.60	76.75	1228.00		
1 Dozer, 200 H.P.		1387.00		1525.70		
1 Roller, Vibratory, 25 Ton		687.40		756.14	86.43	95.08
24 L.H., Daily Totals		$3184.80		$3972.64	$132.70	$165.53

Crew B-32C

Crew No.	Bare Costs Hr.	Bare Costs Daily	Incl. Subs O&P Hr.	Incl. Subs O&P Daily	Cost Per Labor-Hour Bare Costs	Cost Per Labor-Hour Incl. O&P
1 Labor Foreman (outside)	$39.60	$316.80	$60.95	$487.60	$44.43	$67.82
2 Laborers	37.60	601.60	57.85	925.60		
3 Equip. Oper. (medium)	50.60	1214.40	76.75	1842.00		
1 Grader, 30,000 Lbs.		736.20		809.82		
1 Tandem Roller, 10 Ton		236.80		260.48		
1 Dozer, 200 H.P.		1387.00		1525.70	49.17	54.08
48 L.H., Daily Totals		$4492.80		$5851.20	$93.60	$121.90

Crew B-33A

Crew No.	Bare Costs Hr.	Bare Costs Daily	Incl. Subs O&P Hr.	Incl. Subs O&P Daily	Cost Per Labor-Hour Bare Costs	Cost Per Labor-Hour Incl. O&P
1 Equip. Oper. (medium)	$50.60	$404.80	$76.75	$614.00	$46.89	$71.35
.5 Laborer	37.60	150.40	57.85	231.40		
.25 Equip. Oper. (medium)	50.60	101.20	76.75	153.50		
1 Scraper, Towed, 7 C.Y.		114.40		125.84		
1.25 Dozers, 300 H.P.		2371.25		2608.38	177.55	195.30
14 L.H., Daily Totals		$3142.05		$3733.11	$224.43	$266.65

Crew B-33B

Crew No.	Bare Costs Hr.	Bare Costs Daily	Incl. Subs O&P Hr.	Incl. Subs O&P Daily	Cost Per Labor-Hour Bare Costs	Cost Per Labor-Hour Incl. O&P
1 Equip. Oper. (medium)	$50.60	$404.80	$76.75	$614.00	$46.89	$71.35
.5 Laborer	37.60	150.40	57.85	231.40		
.25 Equip. Oper. (medium)	50.60	101.20	76.75	153.50		
1 Scraper, Towed, 10 C.Y.		146.80		161.48		
1.25 Dozers, 300 H.P.		2371.25		2608.38	179.86	197.85
14 L.H., Daily Totals		$3174.45		$3768.76	$226.75	$269.20

Crew B-33C

Crew No.	Bare Costs Hr.	Bare Costs Daily	Incl. Subs O&P Hr.	Incl. Subs O&P Daily	Cost Per Labor-Hour Bare Costs	Cost Per Labor-Hour Incl. O&P
1 Equip. Oper. (medium)	$50.60	$404.80	$76.75	$614.00	$46.89	$71.35
.5 Laborer	37.60	150.40	57.85	231.40		
.25 Equip. Oper. (medium)	50.60	101.20	76.75	153.50		
1 Scraper, Towed, 15 C.Y.		164.80		181.28		
1.25 Dozers, 300 H.P.		2371.25		2608.38	181.15	199.26
14 L.H., Daily Totals		$3192.45		$3788.55	$228.03	$270.61

Crew B-33D

Crew No.	Bare Costs Hr.	Bare Costs Daily	Incl. Subs O&P Hr.	Incl. Subs O&P Daily	Cost Per Labor-Hour Bare Costs	Cost Per Labor-Hour Incl. O&P
1 Equip. Oper. (medium)	$50.60	$404.80	$76.75	$614.00	$46.89	$71.35
.5 Laborer	37.60	150.40	57.85	231.40		
.25 Equip. Oper. (medium)	50.60	101.20	76.75	153.50		
1 S.P. Scraper, 14 C.Y.		1884.00		2072.40		
.25 Dozer, 300 H.P.		474.25		521.67	168.45	185.29
14 L.H., Daily Totals		$3014.65		$3592.97	$215.33	$256.64

Crew B-33E

Crew No.	Bare Costs Hr.	Bare Costs Daily	Incl. Subs O&P Hr.	Incl. Subs O&P Daily	Cost Per Labor-Hour Bare Costs	Cost Per Labor-Hour Incl. O&P
1 Equip. Oper. (medium)	$50.60	$404.80	$76.75	$614.00	$46.89	$71.35
.5 Laborer	37.60	150.40	57.85	231.40		
.25 Equip. Oper. (medium)	50.60	101.20	76.75	153.50		
1 S.P. Scraper, 21 C.Y.		2688.00		2956.80		
.25 Dozer, 300 H.P.		474.25		521.67	225.88	248.46
14 L.H., Daily Totals		$3818.65		$4477.38	$272.76	$319.81

Crew B-33F

Crew No.	Bare Costs Hr.	Bare Costs Daily	Incl. Subs O&P Hr.	Incl. Subs O&P Daily	Cost Per Labor-Hour Bare Costs	Cost Per Labor-Hour Incl. O&P
1 Equip. Oper. (medium)	$50.60	$404.80	$76.75	$614.00	$46.89	$71.35
.5 Laborer	37.60	150.40	57.85	231.40		
.25 Equip. Oper. (medium)	50.60	101.20	76.75	153.50		
1 Elev. Scraper, 11 C.Y.		1163.00		1279.30		
.25 Dozer, 300 H.P.		474.25		521.67	116.95	128.64
14 L.H., Daily Totals		$2293.65		$2799.88	$163.83	$199.99

Crew B-33G

Crew No.	Bare Costs Hr.	Bare Costs Daily	Incl. Subs O&P Hr.	Incl. Subs O&P Daily	Cost Per Labor-Hour Bare Costs	Cost Per Labor-Hour Incl. O&P
1 Equip. Oper. (medium)	$50.60	$404.80	$76.75	$614.00	$46.89	$71.35
.5 Laborer	37.60	150.40	57.85	231.40		
.25 Equip. Oper. (medium)	50.60	101.20	76.75	153.50		
1 Elev. Scraper, 22 C.Y.		2521.00		2773.10		
.25 Dozer, 300 H.P.		474.25		521.67	213.95	235.34
14 L.H., Daily Totals		$3651.65		$4293.68	$260.83	$306.69

Crew B-33H

Crew No.	Bare Costs Hr.	Bare Costs Daily	Incl. Subs O&P Hr.	Incl. Subs O&P Daily	Cost Per Labor-Hour Bare Costs	Cost Per Labor-Hour Incl. O&P
.5 Laborer	$37.60	$150.40	$57.85	$231.40	$46.89	$71.35
1 Equipment Operator (med.)	50.60	404.80	76.75	614.00		
.25 Equipment Operator (med.)	50.60	101.20	76.75	153.50		
1 S.P. Scraper, 44 C.Y.		4679.00		5146.90		
.25 Dozer, 410 H.P.		601.25		661.38	377.16	414.88
14 L.H., Daily Totals		$5936.65		$6807.18	$424.05	$486.23

Crew B-33J

Crew No.	Bare Costs Hr.	Bare Costs Daily	Incl. Subs O&P Hr.	Incl. Subs O&P Daily	Cost Per Labor-Hour Bare Costs	Cost Per Labor-Hour Incl. O&P
1 Equipment Operator (med.)	$50.60	$404.80	$76.75	$614.00	$50.60	$76.75
1 S.P. Scraper, 14 C.Y.		1884.00		2072.40	235.50	259.05
8 L.H., Daily Totals		$2288.80		$2686.40	$286.10	$335.80

Crew B-33K

Crew No.	Bare Costs Hr.	Bare Costs Daily	Incl. Subs O&P Hr.	Incl. Subs O&P Daily	Cost Per Labor-Hour Bare Costs	Cost Per Labor-Hour Incl. O&P
1 Equipment Operator (med.)	$50.60	$404.80	$76.75	$614.00	$46.89	$71.35
.25 Equipment Operator (med.)	50.60	101.20	76.75	153.50		
.5 Laborer	37.60	150.40	57.85	231.40		
1 S.P. Scraper, 31 C.Y.		3693.00		4062.30		
.25 Dozer, 410 H.P.		601.25		661.38	306.73	337.41
14 L.H., Daily Totals		$4950.65		$5722.57	$353.62	$408.76

Crews

Crew No.	Bare Costs Hr.	Daily	Incl. Subs O&P Hr.	Daily	Cost Per Labor-Hour Bare Costs	Incl. O&P
Crew B-34A						
1 Truck Driver (heavy)	$40.05	$320.40	$60.55	$484.40	$40.05	$60.55
1 Dump Truck, 8 C.Y., 220 H.P.		411.20		452.32	51.40	56.54
8 L.H., Daily Totals		$731.60		$936.72	$91.45	$117.09
Crew B-34B						
1 Truck Driver (heavy)	$40.05	$320.40	$60.55	$484.40	$40.05	$60.55
1 Dump Truck, 12 C.Y., 400 H.P.		691.00		760.10	86.38	95.01
8 L.H., Daily Totals		$1011.40		$1244.50	$126.43	$155.56
Crew B-34C						
1 Truck Driver (heavy)	$40.05	$320.40	$60.55	$484.40	$40.05	$60.55
1 Truck Tractor, 6x4, 380 H.P.		599.80		659.78		
1 Dump Trailer, 16.5 C.Y.		126.60		139.26	90.80	99.88
8 L.H., Daily Totals		$1046.80		$1283.44	$130.85	$160.43
Crew B-34D						
1 Truck Driver (heavy)	$40.05	$320.40	$60.55	$484.40	$40.05	$60.55
1 Truck Tractor, 6x4, 380 H.P.		599.80		659.78		
1 Dump Trailer, 20 C.Y.		141.20		155.32	92.63	101.89
8 L.H., Daily Totals		$1061.40		$1299.50	$132.68	$162.44
Crew B-34E						
1 Truck Driver (heavy)	$40.05	$320.40	$60.55	$484.40	$40.05	$60.55
1 Dump Truck, Off Hwy., 25 Ton		1351.00		1486.10	168.88	185.76
8 L.H., Daily Totals		$1671.40		$1970.50	$208.93	$246.31
Crew B-34F						
1 Truck Driver (heavy)	$40.05	$320.40	$60.55	$484.40	$40.05	$60.55
1 Dump Truck, Off Hwy., 35 Ton		1512.00		1663.20	189.00	207.90
8 L.H., Daily Totals		$1832.40		$2147.60	$229.05	$268.45
Crew B-34G						
1 Truck Driver (heavy)	$40.05	$320.40	$60.55	$484.40	$40.05	$60.55
1 Dump Truck, Off Hwy., 50 Ton		1837.00		2020.70	229.63	252.59
8 L.H., Daily Totals		$2157.40		$2505.10	$269.68	$313.14
Crew B-34H						
1 Truck Driver (heavy)	$40.05	$320.40	$60.55	$484.40	$40.05	$60.55
1 Dump Truck, Off Hwy., 65 Ton		1863.00		2049.30	232.88	256.16
8 L.H., Daily Totals		$2183.40		$2533.70	$272.93	$316.71
Crew B-34I						
1 Truck Driver (heavy)	$40.05	$320.40	$60.55	$484.40	$40.05	$60.55
1 Dump Truck, 18 C.Y., 450 H.P.		869.00		955.90	108.63	119.49
8 L.H., Daily Totals		$1189.40		$1440.30	$148.68	$180.04
Crew B-34J						
1 Truck Driver (heavy)	$40.05	$320.40	$60.55	$484.40	$40.05	$60.55
1 Dump Truck, Off Hwy., 100 Ton		2925.00		3217.50	365.63	402.19
8 L.H., Daily Totals		$3245.40		$3701.90	$405.68	$462.74
Crew B-34K						
1 Truck Driver (heavy)	$40.05	$320.40	$60.55	$484.40	$40.05	$60.55
1 Truck Tractor, 6x4, 450 H.P.		727.40		800.14		
1 Lowbed Trailer, 75 Ton		221.80		243.98	118.65	130.51
8 L.H., Daily Totals		$1269.60		$1528.52	$158.70	$191.07

Crew No.	Bare Costs Hr.	Daily	Incl. Subs O&P Hr.	Daily	Cost Per Labor-Hour Bare Costs	Incl. O&P
Crew B-34L						
1 Equip. Oper. (light)	$48.60	$388.80	$73.75	$590.00	$48.60	$73.75
1 Flatbed Truck, Gas, 1.5 Ton		245.40		269.94	30.68	33.74
8 L.H., Daily Totals		$634.20		$859.94	$79.28	$107.49
Crew B-34M						
1 Equip. Oper. (light)	$48.60	$388.80	$73.75	$590.00	$48.60	$73.75
1 Flatbed Truck, Gas, 3 Ton		303.20		333.52	37.90	41.69
8 L.H., Daily Totals		$692.00		$923.52	$86.50	$115.44
Crew B-34N						
1 Truck Driver (heavy)	$40.05	$320.40	$60.55	$484.40	$45.33	$68.65
1 Equip. Oper. (medium)	50.60	404.80	76.75	614.00		
1 Truck Tractor, 6x4, 380 H.P.		599.80		659.78		
1 Flatbed Trailer, 40 Ton		154.00		169.40	47.11	51.82
16 L.H., Daily Totals		$1479.00		$1927.58	$92.44	$120.47
Crew B-34P						
1 Pipe Fitter	$59.75	$478.00	$90.20	$721.60	$49.82	$75.35
1 Truck Driver (light)	39.10	312.80	59.10	472.80		
1 Equip. Oper. (medium)	50.60	404.80	76.75	614.00		
1 Flatbed Truck, Gas, 3 Ton		303.20		333.52		
1 Backhoe Loader, 48 H.P.		364.20		400.62	27.81	30.59
24 L.H., Daily Totals		$1863.00		$2542.54	$77.63	$105.94
Crew B-34Q						
1 Pipe Fitter	$59.75	$478.00	$90.20	$721.60	$50.18	$75.92
1 Truck Driver (light)	39.10	312.80	59.10	472.80		
1 Equip. Oper. (crane)	51.70	413.60	78.45	627.60		
1 Flatbed Trailer, 25 Ton		113.60		124.96		
1 Dump Truck, 8 C.Y., 220 H.P.		411.20		452.32		
1 Hyd. Crane, 25 Ton		736.60		810.26	52.56	57.81
24 L.H., Daily Totals		$2465.80		$3209.54	$102.74	$133.73
Crew B-34R						
1 Pipe Fitter	$59.75	$478.00	$90.20	$721.60	$50.18	$75.92
1 Truck Driver (light)	39.10	312.80	59.10	472.80		
1 Equip. Oper. (crane)	51.70	413.60	78.45	627.60		
1 Flatbed Trailer, 25 Ton		113.60		124.96		
1 Dump Truck, 8 C.Y., 220 H.P.		411.20		452.32		
1 Hyd. Crane, 25 Ton		736.60		810.26		
1 Hyd. Excavator, 1 C.Y.		812.60		893.86	86.42	95.06
24 L.H., Daily Totals		$3278.40		$4103.40	$136.60	$170.97
Crew B-34S						
2 Pipe Fitters	$59.75	$956.00	$90.20	$1443.20	$52.81	$79.85
1 Truck Driver (heavy)	40.05	320.40	60.55	484.40		
1 Equip. Oper. (crane)	51.70	413.60	78.45	627.60		
1 Flatbed Trailer, 40 Ton		154.00		169.40		
1 Truck Tractor, 6x4, 380 H.P.		599.80		659.78		
1 Hyd. Crane, 80 Ton		1625.00		1787.50		
1 Hyd. Excavator, 2 C.Y.		1175.00		1292.50	111.06	122.16
32 L.H., Daily Totals		$5243.80		$6464.38	$163.87	$202.01
Crew B-34T						
2 Pipe Fitters	$59.75	$956.00	$90.20	$1443.20	$52.81	$79.85
1 Truck Driver (heavy)	40.05	320.40	60.55	484.40		
1 Equip. Oper. (crane)	51.70	413.60	78.45	627.60		
1 Flatbed Trailer, 40 Ton		154.00		169.40		
1 Truck Tractor, 6x4, 380 H.P.		599.80		659.78		
1 Hyd. Crane, 80 Ton		1625.00		1787.50	74.34	81.77
32 L.H., Daily Totals		$4068.80		$5171.88	$127.15	$161.62

Crew No.	Bare Costs		Incl. Subs O&P		Cost Per Labor-Hour	
Crew B-34U	Hr.	Daily	Hr.	Daily	Bare Costs	Incl. O&P
1 Truck Driver (heavy)	$40.05	$320.40	$60.55	$484.40	$44.33	$67.15
1 Equip. Oper. (light)	48.60	388.80	73.75	590.00		
1 Truck Tractor, 220 H.P.		359.60		395.56		
1 Flatbed Trailer, 25 Ton		113.60		124.96	29.57	32.53
16 L.H., Daily Totals		$1182.40		$1594.92	$73.90	$99.68
Crew B-34V	Hr.	Daily	Hr.	Daily	Bare Costs	Incl. O&P
1 Truck Driver (heavy)	$40.05	$320.40	$60.55	$484.40	$46.78	$70.92
1 Equip. Oper. (crane)	51.70	413.60	78.45	627.60		
1 Equip. Oper. (light)	48.60	388.80	73.75	590.00		
1 Truck Tractor, 6x4, 450 H.P.		727.40		800.14		
1 Equipment Trailer, 50 Ton		168.40		185.24		
1 Pickup Truck, 4 x 4, 3/4 Ton		155.60		171.16	43.81	48.19
24 L.H., Daily Totals		$2174.20		$2858.54	$90.59	$119.11
Crew B-34W	Hr.	Daily	Hr.	Daily	Bare Costs	Incl. O&P
5 Truck Drivers (heavy)	$40.05	$1602.00	$60.55	$2422.00	$43.66	$66.21
2 Equip. Opers. (crane)	51.70	827.20	78.45	1255.20		
1 Equip. Oper. (mechanic)	51.65	413.20	78.35	626.80		
1 Laborer	37.60	300.80	57.85	462.80		
4 Truck Tractor, 6x4, 380 H.P.		2399.20		2639.12		
2 Equipment Trailer, 50 Ton		336.80		370.48		
2 Flatbed Trailer, 40 Ton		308.00		338.80		
1 Pickup Truck, 4 x 4, 3/4 Ton		155.60		171.16		
1 S.P. Crane, 4x4, 20 Ton		554.60		610.06	52.14	57.36
72 L.H., Daily Totals		$6897.40		$8896.42	$95.80	$123.56
Crew B-35	Hr.	Daily	Hr.	Daily	Bare Costs	Incl. O&P
1 Labor Foreman (outside)	$39.60	$316.80	$60.95	$487.60	$46.91	$71.60
1 Skilled Worker	48.65	389.20	75.15	601.20		
1 Welder (plumber)	58.70	469.60	88.65	709.20		
1 Laborer	37.60	300.80	57.85	462.80		
1 Equip. Oper. (crane)	51.70	413.60	78.45	627.60		
1 Equip. Oper. (oiler)	45.20	361.60	68.55	548.40		
1 Welder, Electric, 300 amp		57.70		63.47		
1 Hyd. Excavator, .75 C.Y.		654.80		720.28	14.84	16.33
48 L.H., Daily Totals		$2964.10		$4220.55	$61.75	$87.93
Crew B-35A	Hr.	Daily	Hr.	Daily	Bare Costs	Incl. O&P
1 Labor Foreman (outside)	$39.60	$316.80	$60.95	$487.60	$45.58	$69.64
2 Laborers	37.60	601.60	57.85	925.60		
1 Skilled Worker	48.65	389.20	75.15	601.20		
1 Welder (plumber)	58.70	469.60	88.65	709.20		
1 Equip. Oper. (crane)	51.70	413.60	78.45	627.60		
1 Equip. Oper. (oiler)	45.20	361.60	68.55	548.40		
1 Welder, Gas Engine, 300 amp		145.85		160.44		
1 Crawler Crane, 75 Ton		1459.00		1604.90	28.66	31.52
56 L.H., Daily Totals		$4157.25		$5664.94	$74.24	$101.16
Crew B-36	Hr.	Daily	Hr.	Daily	Bare Costs	Incl. O&P
1 Labor Foreman (outside)	$39.60	$316.80	$60.95	$487.60	$43.20	$66.03
2 Laborers	37.60	601.60	57.85	925.60		
2 Equip. Oper. (medium)	50.60	809.60	76.75	1228.00		
1 Dozer, 200 H.P.		1387.00		1525.70		
1 Aggregate Spreader		39.60		43.56		
1 Tandem Roller, 10 Ton		236.80		260.48	41.59	45.74
40 L.H., Daily Totals		$3391.40		$4470.94	$84.78	$111.77

Crew No.	Bare Costs		Incl. Subs O&P		Cost Per Labor-Hour	
Crew B-36A	Hr.	Daily	Hr.	Daily	Bare Costs	Incl. O&P
1 Labor Foreman (outside)	$39.60	$316.80	$60.95	$487.60	$45.31	$69.09
2 Laborers	37.60	601.60	57.85	925.60		
4 Equip. Oper. (medium)	50.60	1619.20	76.75	2456.00		
1 Dozer, 200 H.P.		1387.00		1525.70		
1 Aggregate Spreader		39.60		43.56		
1 Tandem Roller, 10 Ton		236.80		260.48		
1 Roller, Pneum. Whl., 12 Ton		344.40		378.84	35.85	39.44
56 L.H., Daily Totals		$4545.40		$6077.78	$81.17	$108.53
Crew B-36B	Hr.	Daily	Hr.	Daily	Bare Costs	Incl. O&P
1 Labor Foreman (outside)	$39.60	$316.80	$60.95	$487.60	$44.66	$68.03
2 Laborers	37.60	601.60	57.85	925.60		
4 Equip. Oper. (medium)	50.60	1619.20	76.75	2456.00		
1 Truck Driver (heavy)	40.05	320.40	60.55	484.40		
1 Grader, 30,000 Lbs.		736.20		809.82		
1 F.E. Loader, Crl. 1.5 C.Y.		629.80		692.78		
1 Dozer, 300 H.P.		1897.00		2086.70		
1 Roller, Vibratory, 25 Ton		687.40		756.14		
1 Truck Tractor, 6x4, 450 H.P.		727.40		800.14		
1 Water Tank Trailer, 5000 Gal.		141.00		155.10	75.29	82.82
64 L.H., Daily Totals		$7676.80		$9654.28	$119.95	$150.85
Crew B-36C	Hr.	Daily	Hr.	Daily	Bare Costs	Incl. O&P
1 Labor Foreman (outside)	$39.60	$316.80	$60.95	$487.60	$46.29	$70.35
3 Equip. Oper. (medium)	50.60	1214.40	76.75	1842.00		
1 Truck Driver (heavy)	40.05	320.40	60.55	484.40		
1 Grader, 30,000 Lbs.		736.20		809.82		
1 Dozer, 300 H.P.		1897.00		2086.70		
1 Roller, Vibratory, 25 Ton		687.40		756.14		
1 Truck Tractor, 6x4, 450 H.P.		727.40		800.14		
1 Water Tank Trailer, 5000 Gal.		141.00		155.10	104.72	115.20
40 L.H., Daily Totals		$6040.60		$7421.90	$151.01	$185.55
Crew B-36D	Hr.	Daily	Hr.	Daily	Bare Costs	Incl. O&P
1 Labor Foreman (outside)	$39.60	$316.80	$60.95	$487.60	$47.85	$72.80
3 Equip. Oper. (medium)	50.60	1214.40	76.75	1842.00		
1 Grader, 30,000 Lbs.		736.20		809.82		
1 Dozer, 300 H.P.		1897.00		2086.70		
1 Roller, Vibratory, 25 Ton		687.40		756.14	103.77	114.15
32 L.H., Daily Totals		$4851.80		$5982.26	$151.62	$186.95
Crew B-37	Hr.	Daily	Hr.	Daily	Bare Costs	Incl. O&P
1 Labor Foreman (outside)	$39.60	$316.80	$60.95	$487.60	$39.77	$61.02
4 Laborers	37.60	1203.20	57.85	1851.20		
1 Equip. Oper. (light)	48.60	388.80	73.75	590.00		
1 Tandem Roller, 5 Ton		157.40		173.14	3.28	3.61
48 L.H., Daily Totals		$2066.20		$3101.94	$43.05	$64.62
Crew B-37A	Hr.	Daily	Hr.	Daily	Bare Costs	Incl. O&P
2 Laborers	$37.60	$601.60	$57.85	$925.60	$38.10	$58.27
1 Truck Driver (light)	39.10	312.80	59.10	472.80		
1 Flatbed Truck, Gas, 1.5 Ton		245.40		269.94		
1 Tar Kettle, T.M.		132.00		145.20	15.73	17.30
24 L.H., Daily Totals		$1291.80		$1813.54	$53.83	$75.56
Crew B-37B	Hr.	Daily	Hr.	Daily	Bare Costs	Incl. O&P
3 Laborers	$37.60	$902.40	$57.85	$1388.40	$37.98	$58.16
1 Truck Driver (light)	39.10	312.80	59.10	472.80		
1 Flatbed Truck, Gas, 1.5 Ton		245.40		269.94		
1 Tar Kettle, T.M.		132.00		145.20	11.79	12.97
32 L.H., Daily Totals		$1592.60		$2276.34	$49.77	$71.14

Crew No.	Bare Costs		Incl. Subs O&P		Cost Per Labor-Hour	

Crew B-37C

Crew B-37C	Hr.	Daily	Hr.	Daily	Bare Costs	Incl. O&P
2 Laborers	$37.60	$601.60	$57.85	$925.60	$38.35	$58.48
2 Truck Drivers (light)	39.10	625.60	59.10	945.60		
2 Flatbed Trucks, Gas, 1.5 Ton		490.80		539.88		
1 Tar Kettle, T.M.		132.00		145.20	19.46	21.41
32 L.H., Daily Totals		$1850.00		$2556.28	$57.81	$79.88

Crew B-37D

Crew B-37D	Hr.	Daily	Hr.	Daily	Bare Costs	Incl. O&P
1 Laborer	$37.60	$300.80	$57.85	$462.80	$38.35	$58.48
1 Truck Driver (light)	39.10	312.80	59.10	472.80		
1 Pickup Truck, 3/4 Ton		144.20		158.62	9.01	9.91
16 L.H., Daily Totals		$757.80		$1094.22	$47.36	$68.39

Crew B-37E

Crew B-37E	Hr.	Daily	Hr.	Daily	Bare Costs	Incl. O&P
3 Laborers	$37.60	$902.40	$57.85	$1388.40	$41.46	$63.18
1 Equip. Oper. (light)	48.60	388.80	73.75	590.00		
1 Equip. Oper. (medium)	50.60	404.80	76.75	614.00		
2 Truck Drivers (light)	39.10	625.60	59.10	945.60		
4 Barrels w/ Flasher		13.60		14.96		
1 Concrete Saw		167.40		184.14		
1 Rotary Hammer Drill		23.75		26.13		
1 Hammer Drill Bit		2.35		2.59		
1 Loader, Skid Steer, 30 H.P.		173.60		190.96		
1 Conc. Hammer Attach.		114.70		126.17		
1 Vibrating Plate, Gas, 18"		36.20		39.82		
2 Flatbed Trucks, Gas, 1.5 Ton		490.80		539.88	18.26	20.08
56 L.H., Daily Totals		$3344.00		$4662.64	$59.71	$83.26

Crew B-37F

Crew B-37F	Hr.	Daily	Hr.	Daily	Bare Costs	Incl. O&P
3 Laborers	$37.60	$902.40	$57.85	$1388.40	$37.98	$58.16
1 Truck Driver (light)	39.10	312.80	59.10	472.80		
4 Barrels w/ Flasher		13.60		14.96		
1 Concrete Mixer, 10 C.F.		172.40		189.64		
1 Air Compressor, 60 cfm		137.40		151.14		
1 -50' Air Hose, 3/4"		3.25		3.58		
1 Spade (Chipper)		7.60		8.36		
1 Flatbed Truck, Gas, 1.5 Ton		245.40		269.94	18.11	19.93
32 L.H., Daily Totals		$1794.85		$2498.82	$56.09	$78.09

Crew B-37G

Crew B-37G	Hr.	Daily	Hr.	Daily	Bare Costs	Incl. O&P
1 Labor Foreman (outside)	$39.60	$316.80	$60.95	$487.60	$39.77	$61.02
4 Laborers	37.60	1203.20	57.85	1851.20		
1 Equip. Oper. (light)	48.60	388.80	73.75	590.00		
1 Berm Machine		296.80		326.48		
1 Tandem Roller, 5 Ton		157.40		173.14	9.46	10.41
48 L.H., Daily Totals		$2363.00		$3428.42	$49.23	$71.43

Crew B-37H

Crew B-37H	Hr.	Daily	Hr.	Daily	Bare Costs	Incl. O&P
1 Labor Foreman (outside)	$39.60	$316.80	$60.95	$487.60	$39.77	$61.02
4 Laborers	37.60	1203.20	57.85	1851.20		
1 Equip. Oper. (light)	48.60	388.80	73.75	590.00		
1 Tandem Roller, 5 Ton		157.40		173.14		
1 Flatbed Trucks, Gas, 1.5 Ton		245.40		269.94		
1 Tar Kettle, T.M.		132.00		145.20	11.14	12.26
48 L.H., Daily Totals		$2443.60		$3517.08	$50.91	$73.27

Crew B-37I

Crew B-37I	Hr.	Daily	Hr.	Daily	Bare Costs	Incl. O&P
3 Laborers	$37.60	$902.40	$57.85	$1388.40	$41.46	$63.18
1 Equip. Oper. (light)	48.60	388.80	73.75	590.00		
1 Equip. Oper. (medium)	50.60	404.80	76.75	614.00		
2 Truck Drivers (light)	39.10	625.60	59.10	945.60		
4 Barrels w/ Flasher		13.60		14.96		
1 Concrete Saw		167.40		184.14		
1 Rotary Hammer Drill		23.75		26.13		
1 Hammer Drill Bit		2.35		2.59		
1 Air Compressor, 60 cfm		137.40		151.14		
1 -50' Air Hose, 3/4"		3.25		3.58		
1 Spade (Chipper)		7.60		8.36		
1 Loader, Skid Steer, 30 H.P.		173.60		190.96		
1 Conc. Hammer Attach.		114.70		126.17		
1 Concrete Mixer, 10 C.F.		172.40		189.64		
1 Vibrating Plate, Gas, 18"		36.20		39.82		
2 Flatbed Trucks, Gas, 1.5 Ton		490.80		539.88	23.98	26.38
56 L.H., Daily Totals		$3664.65		$5015.35	$65.44	$89.56

Crew B-37J

Crew B-37J	Hr.	Daily	Hr.	Daily	Bare Costs	Incl. O&P
1 Labor Foreman (outside)	$39.60	$316.80	$60.95	$487.60	$39.77	$61.02
4 Laborers	37.60	1203.20	57.85	1851.20		
1 Equip. Oper. (light)	48.60	388.80	73.75	590.00		
1 Air Compressor, 60 cfm		137.40		151.14		
1 -50' Air Hose, 3/4"		3.25		3.58		
2 Concrete Mixer, 10 C.F.		344.80		379.28		
2 Flatbed Trucks, Gas, 1.5 Ton		490.80		539.88		
1 Shot Blaster, 20"		214.00		235.40	24.80	27.28
48 L.H., Daily Totals		$3099.05		$4238.07	$64.56	$88.29

Crew B-37K

Crew B-37K	Hr.	Daily	Hr.	Daily	Bare Costs	Incl. O&P
1 Labor Foreman (outside)	$39.60	$316.80	$60.95	$487.60	$39.77	$61.02
4 Laborers	37.60	1203.20	57.85	1851.20		
1 Equip. Oper. (light)	48.60	388.80	73.75	590.00		
1 Air Compressor, 60 cfm		137.40		151.14		
1 -50' Air Hose, 3/4"		3.25		3.58		
2 Flatbed Trucks, Gas, 1.5 Ton		490.80		539.88		
1 Shot Blaster, 20"		214.00		235.40	17.61	19.37
48 L.H., Daily Totals		$2754.25		$3858.80	$57.38	$80.39

Crew B-38

Crew B-38	Hr.	Daily	Hr.	Daily	Bare Costs	Incl. O&P
1 Labor Foreman (outside)	$39.60	$316.80	$60.95	$487.60	$42.80	$65.43
2 Laborers	37.60	601.60	57.85	925.60		
1 Equip. Oper. (light)	48.60	388.80	73.75	590.00		
1 Equip. Oper. (medium)	50.60	404.80	76.75	614.00		
1 Backhoe Loader, 48 H.P.		364.20		400.62		
1 Hyd. Hammer, (1200 lb.)		180.40		198.44		
1 F.E. Loader, W.M., 4 C.Y.		669.20		736.12		
1 Pvmt. Rem. Bucket		61.20		67.32	31.88	35.06
40 L.H., Daily Totals		$2987.00		$4019.70	$74.67	$100.49

Crew B-39

Crew B-39	Hr.	Daily	Hr.	Daily	Bare Costs	Incl. O&P
1 Labor Foreman (outside)	$39.60	$316.80	$60.95	$487.60	$39.77	$61.02
4 Laborers	37.60	1203.20	57.85	1851.20		
1 Equip. Oper. (light)	48.60	388.80	73.75	590.00		
1 Air Compressor, 250 cfm		201.40		221.54		
2 Breakers, Pavement, 60 lb.		19.60		21.56		
2 -50' Air Hoses, 1.5"		11.60		12.76	4.85	5.33
48 L.H., Daily Totals		$2141.40		$3184.66	$44.61	$66.35

Crew No.	Bare Costs Hr.	Daily	Incl. Subs O&P Hr.	Daily	Cost Per Labor-Hour Bare Costs	Incl. O&P
Crew B-40	Hr.	Daily	Hr.	Daily	Bare Costs	Incl. O&P
1 Pile Driver Foreman (outside)	$48.10	$384.80	$76.15	$609.20	$47.64	$74.20
4 Pile Drivers	46.10	1475.20	73.00	2336.00		
2 Equip. Oper. (crane)	51.70	827.20	78.45	1255.20		
1 Equip. Oper. (oiler)	45.20	361.60	68.55	548.40		
1 Crawler Crane, 40 Ton		1156.00		1271.60		
1 Vibratory Hammer & Gen.		2587.00		2845.70	58.48	64.33
64 L.H., Daily Totals		$6791.80		$8866.10	$106.12	$138.53
Crew B-40B	Hr.	Daily	Hr.	Daily	Bare Costs	Incl. O&P
1 Labor Foreman (outside)	$39.60	$316.80	$60.95	$487.60	$41.55	$63.58
3 Laborers	37.60	902.40	57.85	1388.40		
1 Equip. Oper. (crane)	51.70	413.60	78.45	627.60		
1 Equip. Oper. (oiler)	45.20	361.60	68.55	548.40		
1 Lattice Boom Crane, 40 Ton		1164.00		1280.40	24.25	26.68
48 L.H., Daily Totals		$3158.40		$4332.40	$65.80	$90.26
Crew B-41	Hr.	Daily	Hr.	Daily	Bare Costs	Incl. O&P
1 Labor Foreman (outside)	$39.60	$316.80	$60.95	$487.60	$38.95	$59.84
4 Laborers	37.60	1203.20	57.85	1851.20		
.25 Equip. Oper. (crane)	51.70	103.40	78.45	156.90		
.25 Equip. Oper. (oiler)	45.20	90.40	68.55	137.10		
.25 Crawler Crane, 40 Ton		289.00		317.90	6.57	7.22
44 L.H., Daily Totals		$2002.80		$2950.70	$45.52	$67.06
Crew B-42	Hr.	Daily	Hr.	Daily	Bare Costs	Incl. O&P
1 Labor Foreman (outside)	$39.60	$316.80	$60.95	$487.60	$42.44	$66.35
4 Laborers	37.60	1203.20	57.85	1851.20		
1 Equip. Oper. (crane)	51.70	413.60	78.45	627.60		
1 Equip. Oper. (oiler)	45.20	361.60	68.55	548.40		
1 Welder	52.65	421.20	91.45	731.60		
1 Hyd. Crane, 25 Ton		736.60		810.26		
1 Welder, Gas Engine, 300 amp		145.85		160.44		
1 Horz. Boring Csg. Mch.		474.80		522.28	21.21	23.33
64 L.H., Daily Totals		$4073.65		$5739.38	$63.65	$89.68
Crew B-43	Hr.	Daily	Hr.	Daily	Bare Costs	Incl. O&P
1 Labor Foreman (outside)	$39.60	$316.80	$60.95	$487.60	$41.55	$63.58
3 Laborers	37.60	902.40	57.85	1388.40		
1 Equip. Oper. (crane)	51.70	413.60	78.45	627.60		
1 Equip. Oper. (oiler)	45.20	361.60	68.55	548.40		
1 Drill Rig, Truck-Mounted		2542.00		2796.20	52.96	58.25
48 L.H., Daily Totals		$4536.40		$5848.20	$94.51	$121.84
Crew B-44	Hr.	Daily	Hr.	Daily	Bare Costs	Incl. O&P
1 Pile Driver Foreman (outside)	$48.10	$384.80	$76.15	$609.20	$46.69	$72.86
4 Pile Drivers	46.10	1475.20	73.00	2336.00		
2 Equip. Oper. (crane)	51.70	827.20	78.45	1255.20		
1 Laborer	37.60	300.80	57.85	462.80		
1 Crawler Crane, 40 Ton		1156.00		1271.60		
1 Lead, 60' High		74.80		82.28		
1 Hammer, Diesel, 15K ft.-lbs.		586.00		644.60	28.39	31.23
64 L.H., Daily Totals		$4804.80		$6661.68	$75.08	$104.09
Crew B-45	Hr.	Daily	Hr.	Daily	Bare Costs	Incl. O&P
1 Equip. Oper. (medium)	$50.60	$404.80	$76.75	$614.00	$45.33	$68.65
1 Truck Driver (heavy)	40.05	320.40	60.55	484.40		
1 Dist. Tanker, 3000 Gallon		305.60		336.16		
1 Truck Tractor, 6x4, 380 H.P.		599.80		659.78	56.59	62.25
16 L.H., Daily Totals		$1630.60		$2094.34	$101.91	$130.90
Crew B-46	Hr.	Daily	Hr.	Daily	Bare Costs	Incl. O&P
1 Pile Driver Foreman (outside)	$48.10	$384.80	$76.15	$609.20	$42.18	$65.95
2 Pile Drivers	46.10	737.60	73.00	1168.00		
3 Laborers	37.60	902.40	57.85	1388.40		
1 Chain Saw, Gas, 36" Long		45.00		49.50	.94	1.03
48 L.H., Daily Totals		$2069.80		$3215.10	$43.12	$66.98
Crew B-47	Hr.	Daily	Hr.	Daily	Bare Costs	Incl. O&P
1 Blast Foreman (outside)	$39.60	$316.80	$60.95	$487.60	$41.93	$64.18
1 Driller	37.60	300.80	57.85	462.80		
1 Equip. Oper. (light)	48.60	388.80	73.75	590.00		
1 Air Track Drill, 4"		1028.00		1130.80		
1 Air Compressor, 600 cfm		550.60		605.66		
2 -50' Air Hoses, 3"		29.80		32.78	67.02	73.72
24 L.H., Daily Totals		$2614.80		$3309.64	$108.95	$137.90
Crew B-47A	Hr.	Daily	Hr.	Daily	Bare Costs	Incl. O&P
1 Drilling Foreman (outside)	$39.60	$316.80	$60.95	$487.60	$45.50	$69.32
1 Equip. Oper. (heavy)	51.70	413.60	78.45	627.60		
1 Equip. Oper. (oiler)	45.20	361.60	68.55	548.40		
1 Air Track Drill, 5"		1243.00		1367.30	51.79	56.97
24 L.H., Daily Totals		$2335.00		$3030.90	$97.29	$126.29
Crew B-47C	Hr.	Daily	Hr.	Daily	Bare Costs	Incl. O&P
1 Laborer	$37.60	$300.80	$57.85	$462.80	$43.10	$65.80
1 Equip. Oper. (light)	48.60	388.80	73.75	590.00		
1 Air Compressor, 750 cfm		557.20		612.92		
2 -50' Air Hoses, 3"		29.80		32.78		
1 Air Track Drill, 4"		1028.00		1130.80	100.94	111.03
16 L.H., Daily Totals		$2304.60		$2829.30	$144.04	$176.83
Crew B-47E	Hr.	Daily	Hr.	Daily	Bare Costs	Incl. O&P
1 Labor Foreman (outside)	$39.60	$316.80	$60.95	$487.60	$38.10	$58.63
3 Laborers	37.60	902.40	57.85	1388.40		
1 Flatbed Truck, Gas, 3 Ton		303.20		333.52	9.47	10.42
32 L.H., Daily Totals		$1522.40		$2209.52	$47.58	$69.05
Crew B-47G	Hr.	Daily	Hr.	Daily	Bare Costs	Incl. O&P
1 Labor Foreman (outside)	$39.60	$316.80	$60.95	$487.60	$40.85	$62.60
2 Laborers	37.60	601.60	57.85	925.60		
1 Equip. Oper. (light)	48.60	388.80	73.75	590.00		
1 Air Track Drill, 4"		1028.00		1130.80		
1 Air Compressor, 600 cfm		550.60		605.66		
2 -50' Air Hoses, 3"		29.80		32.78		
1 Gunite Pump Rig		371.80		408.98	61.88	68.07
32 L.H., Daily Totals		$3287.40		$4181.42	$102.73	$130.67
Crew B-47H	Hr.	Daily	Hr.	Daily	Bare Costs	Incl. O&P
1 Skilled Worker Foreman (out)	$50.65	$405.20	$78.25	$626.00	$49.15	$75.92
3 Skilled Workers	48.65	1167.60	75.15	1803.60		
1 Flatbed Truck, Gas, 3 Ton		303.20		333.52	9.47	10.42
32 L.H., Daily Totals		$1876.00		$2763.12	$58.63	$86.35

Crew No.	Bare Costs		Incl. Subs O&P		Cost Per Labor-Hour	

Crew B-48	Hr.	Daily	Hr.	Daily	Bare Costs	Incl. O&P
1 Labor Foreman (outside)	$39.60	$316.80	$60.95	$487.60	$42.56	$65.04
3 Laborers	37.60	902.40	57.85	1388.40		
1 Equip. Oper. (crane)	51.70	413.60	78.45	627.60		
1 Equip. Oper. (oiler)	45.20	361.60	68.55	548.40		
1 Equip. Oper. (light)	48.60	388.80	73.75	590.00		
1 Centr. Water Pump, 6"		346.60		381.26		
1 -20' Suction Hose, 6"		11.50		12.65		
1 -50' Discharge Hose, 6"		6.10		6.71		
1 Drill Rig, Truck-Mounted		2542.00		2796.20	51.90	57.09
56 L.H., Daily Totals		$5289.40		$6838.82	$94.45	$122.12

Crew B-49	Hr.	Daily	Hr.	Daily	Bare Costs	Incl. O&P
1 Labor Foreman (outside)	$39.60	$316.80	$60.95	$487.60	$44.27	$68.02
3 Laborers	37.60	902.40	57.85	1388.40		
2 Equip. Oper. (crane)	51.70	827.20	78.45	1255.20		
2 Equip. Oper. (oilers)	45.20	723.20	68.55	1096.80		
1 Equip. Oper. (light)	48.60	388.80	73.75	590.00		
2 Pile Drivers	46.10	737.60	73.00	1168.00		
1 Hyd. Crane, 25 Ton		736.60		810.26		
1 Centr. Water Pump, 6"		346.60		381.26		
1 -20' Suction Hose, 6"		11.50		12.65		
1 -50' Discharge Hose, 6"		6.10		6.71		
1 Drill Rig, Truck-Mounted		2542.00		2796.20	41.40	45.53
88 L.H., Daily Totals		$7538.80		$9993.08	$85.67	$113.56

Crew B-50	Hr.	Daily	Hr.	Daily	Bare Costs	Incl. O&P
2 Pile Driver Foremen (outside)	$48.10	$769.60	$76.15	$1218.40	$45.30	$70.66
6 Pile Drivers	46.10	2212.80	73.00	3504.00		
2 Equip. Oper. (crane)	51.70	827.20	78.45	1255.20		
1 Equip. Oper. (oiler)	45.20	361.60	68.55	548.40		
3 Laborers	37.60	902.40	57.85	1388.40		
1 Crawler Crane, 40 Ton		1156.00		1271.60		
1 Lead, 60' High		74.80		82.28		
1 Hammer, Diesel, 15K ft.-lbs.		586.00		644.60		
1 Air Compressor, 600 cfm		550.60		605.66		
2 -50' Air Hoses, 3"		29.80		32.78		
1 Chain Saw, Gas, 36" Long		45.00		49.50	21.81	23.99
112 L.H., Daily Totals		$7515.80		$10600.82	$67.11	$94.65

Crew B-51	Hr.	Daily	Hr.	Daily	Bare Costs	Incl. O&P
1 Labor Foreman (outside)	$39.60	$316.80	$60.95	$487.60	$38.18	$58.58
4 Laborers	37.60	1203.20	57.85	1851.20		
1 Truck Driver (light)	39.10	312.80	59.10	472.80		
1 Flatbed Truck, Gas, 1.5 Ton		245.40		269.94	5.11	5.62
48 L.H., Daily Totals		$2078.20		$3081.54	$43.30	$64.20

Crew B-52	Hr.	Daily	Hr.	Daily	Bare Costs	Incl. O&P
1 Carpenter Foreman (outside)	$48.95	$391.60	$75.35	$602.80	$43.61	$66.75
1 Carpenter	46.95	375.60	72.25	578.00		
3 Laborers	37.60	902.40	57.85	1388.40		
1 Cement Finisher	45.00	360.00	66.65	533.20		
.5 Rodman (reinf.)	52.55	210.20	82.10	328.40		
.5 Equip. Oper. (medium)	50.60	202.40	76.75	307.00		
.5 Crawler Loader, 3 C.Y.		594.00		653.40	10.61	11.67
56 L.H., Daily Totals		$3036.20		$4391.20	$54.22	$78.41

Crew B-53	Hr.	Daily	Hr.	Daily	Bare Costs	Incl. O&P
1 Equip. Oper. (light)	$48.60	$388.80	$73.75	$590.00	$48.60	$73.75
1 Trencher, Chain, 12 H.P.		68.00		74.80	8.50	9.35
8 L.H., Daily Totals		$456.80		$664.80	$57.10	$83.10

Crew B-54	Hr.	Daily	Hr.	Daily	Bare Costs	Incl. O&P
1 Equip. Oper. (light)	$48.60	$388.80	$73.75	$590.00	$48.60	$73.75
1 Trencher, Chain, 40 H.P.		335.20		368.72	41.90	46.09
8 L.H., Daily Totals		$724.00		$958.72	$90.50	$119.84

Crew B-54A	Hr.	Daily	Hr.	Daily	Bare Costs	Incl. O&P
.17 Labor Foreman (outside)	$39.60	$53.86	$60.95	$82.89	$49.00	$74.45
1 Equipment Operator (med.)	50.60	404.80	76.75	614.00		
1 Wheel Trencher, 67 H.P.		1195.00		1314.50	127.67	140.44
9.36 L.H., Daily Totals		$1653.66		$2011.39	$176.67	$214.89

Crew B-54B	Hr.	Daily	Hr.	Daily	Bare Costs	Incl. O&P
.25 Labor Foreman (outside)	$39.60	$79.20	$60.95	$121.90	$48.40	$73.59
1 Equipment Operator (med.)	50.60	404.80	76.75	614.00		
1 Wheel Trencher, 150 H.P.		1890.00		2079.00	189.00	207.90
10 L.H., Daily Totals		$2374.00		$2814.90	$237.40	$281.49

Crew B-54C	Hr.	Daily	Hr.	Daily	Bare Costs	Incl. O&P
1 Laborer	$37.60	$300.80	$57.85	$462.80	$44.10	$67.30
1 Equipment Operator (med.)	50.60	404.80	76.75	614.00		
1 Wheel Trencher, 67 H.P.		1195.00		1314.50	74.69	82.16
16 L.H., Daily Totals		$1900.60		$2391.30	$118.79	$149.46

Crew B-54D	Hr.	Daily	Hr.	Daily	Bare Costs	Incl. O&P
1 Laborer	$37.60	$300.80	$57.85	$462.80	$44.10	$67.30
1 Equipment Operator (med.)	50.60	404.80	76.75	614.00		
1 Rock Trencher, 6" Width		372.00		409.20	23.25	25.57
16 L.H., Daily Totals		$1077.60		$1486.00	$67.35	$92.88

Crew B-54E	Hr.	Daily	Hr.	Daily	Bare Costs	Incl. O&P
1 Laborer	$37.60	$300.80	$57.85	$462.80	$44.10	$67.30
1 Equipment Operator (med.)	50.60	404.80	76.75	614.00		
1 Rock Trencher, 18" Width		2575.00		2832.50	160.94	177.03
16 L.H., Daily Totals		$3280.60		$3909.30	$205.04	$244.33

Crew B-55	Hr.	Daily	Hr.	Daily	Bare Costs	Incl. O&P
2 Laborers	$37.60	$601.60	$57.85	$925.60	$38.10	$58.27
1 Truck Driver (light)	39.10	312.80	59.10	472.80		
1 Truck-Mounted Earth Auger		779.20		857.12		
1 Flatbed Truck, Gas, 3 Ton		303.20		333.52	45.10	49.61
24 L.H., Daily Totals		$1996.80		$2589.04	$83.20	$107.88

Crew B-56	Hr.	Daily	Hr.	Daily	Bare Costs	Incl. O&P
1 Laborer	$37.60	$300.80	$57.85	$462.80	$43.10	$65.65
1 Equip. Oper. (light)	48.60	388.80	73.75	590.00		
1 Air Track Drill, 4"		1028.00		1130.80		
1 Air Compressor, 600 cfm		550.60		605.66		
1 -50' Air Hose, 3"		14.90		16.39	99.59	109.55
16 L.H., Daily Totals		$2283.10		$2805.65	$142.69	$175.35

Crews

Crew No.	Bare Costs Hr.	Daily	Incl. Subs O&P Hr.	Daily	Cost Per Labor-Hour Bare Costs	Incl. O&P
Crew B-57	Hr.	Daily	Hr.	Daily	Bare Costs	Incl. O&P
1 Labor Foreman (outside)	$39.60	$316.80	$60.95	$487.60	$43.38	$66.23
2 Laborers	37.60	601.60	57.85	925.60		
1 Equip. Oper. (crane)	51.70	413.60	78.45	627.60		
1 Equip. Oper. (light)	48.60	388.80	73.75	590.00		
1 Equip. Oper. (oiler)	45.20	361.60	68.55	548.40		
1 Crawler Crane, 25 Ton		1150.00		1265.00		
1 Clamshell Bucket, 1 C.Y.		47.80		52.58		
1 Centr. Water Pump, 6"		346.60		381.26		
1 -20' Suction Hose, 6"		11.50		12.65		
20 -50' Discharge Hoses, 6"		122.00		134.20	34.96	38.45
48 L.H., Daily Totals		$3760.30		$5024.89	$78.34	$104.69
Crew B-58	Hr.	Daily	Hr.	Daily	Bare Costs	Incl. O&P
2 Laborers	$37.60	$601.60	$57.85	$925.60	$41.27	$63.15
1 Equip. Oper. (light)	48.60	388.80	73.75	590.00		
1 Backhoe Loader, 48 H.P.		364.20		400.62		
1 Small Helicopter, w/ Pilot		2790.00		3069.00	131.43	144.57
24 L.H., Daily Totals		$4144.60		$4985.22	$172.69	$207.72
Crew B-59	Hr.	Daily	Hr.	Daily	Bare Costs	Incl. O&P
1 Truck Driver (heavy)	$40.05	$320.40	$60.55	$484.40	$40.05	$60.55
1 Truck Tractor, 220 H.P.		359.60		395.56		
1 Water Tank Trailer, 5000 Gal.		141.00		155.10	62.58	68.83
8 L.H., Daily Totals		$821.00		$1035.06	$102.63	$129.38
Crew B-59A	Hr.	Daily	Hr.	Daily	Bare Costs	Incl. O&P
2 Laborers	$37.60	$601.60	$57.85	$925.60	$38.42	$58.75
1 Truck Driver (heavy)	40.05	320.40	60.55	484.40		
1 Water Tank Trailer, 5000 Gal.		141.00		155.10		
1 Truck Tractor, 220 H.P.		359.60		395.56	20.86	22.94
24 L.H., Daily Totals		$1422.60		$1960.66	$59.27	$81.69
Crew B-60	Hr.	Daily	Hr.	Daily	Bare Costs	Incl. O&P
1 Labor Foreman (outside)	$39.60	$316.80	$60.95	$487.60	$44.13	$67.31
2 Laborers	37.60	601.60	57.85	925.60		
1 Equip. Oper. (crane)	51.70	413.60	78.45	627.60		
2 Equip. Oper. (light)	48.60	777.60	73.75	1180.00		
1 Equip. Oper. (oiler)	45.20	361.60	68.55	548.40		
1 Crawler Crane, 40 Ton		1156.00		1271.60		
1 Lead, 60' High		74.80		82.28		
1 Hammer, Diesel, 15K ft.-lbs.		586.00		644.60		
1 Backhoe Loader, 48 H.P.		364.20		400.62	38.95	42.84
56 L.H., Daily Totals		$4652.20		$6168.30	$83.08	$110.15
Crew B-61	Hr.	Daily	Hr.	Daily	Bare Costs	Incl. O&P
1 Labor Foreman (outside)	$39.60	$316.80	$60.95	$487.60	$40.20	$61.65
3 Laborers	37.60	902.40	57.85	1388.40		
1 Equip. Oper. (light)	48.60	388.80	73.75	590.00		
1 Cement Mixer, 2 C.Y.		194.20		213.62		
1 Air Compressor, 160 cfm		156.40		172.04	8.77	9.64
40 L.H., Daily Totals		$1958.60		$2851.66	$48.97	$71.29
Crew B-62	Hr.	Daily	Hr.	Daily	Bare Costs	Incl. O&P
2 Laborers	$37.60	$601.60	$57.85	$925.60	$41.27	$63.15
1 Equip. Oper. (light)	48.60	388.80	73.75	590.00		
1 Loader, Skid Steer, 30 H.P.		173.60		190.96	7.23	7.96
24 L.H., Daily Totals		$1164.00		$1706.56	$48.50	$71.11

Crew No.	Bare Costs Hr.	Daily	Incl. Subs O&P Hr.	Daily	Cost Per Labor-Hour Bare Costs	Incl. O&P
Crew B-62A	Hr.	Daily	Hr.	Daily	Bare Costs	Incl. O&P
2 Laborers	$37.60	$601.60	$57.85	$925.60	$41.27	$63.15
1 Equip. Oper. (light)	48.60	388.80	73.75	590.00		
1 Loader, Skid Steer, 30 H.P.		173.60		190.96		
1 Trencher Attachment		58.30		64.13	9.66	10.63
24 L.H., Daily Totals		$1222.30		$1770.69	$50.93	$73.78
Crew B-63	Hr.	Daily	Hr.	Daily	Bare Costs	Incl. O&P
4 Laborers	$37.60	$1203.20	$57.85	$1851.20	$39.80	$61.03
1 Equip. Oper. (light)	48.60	388.80	73.75	590.00		
1 Loader, Skid Steer, 30 H.P.		173.60		190.96	4.34	4.77
40 L.H., Daily Totals		$1765.60		$2632.16	$44.14	$65.80
Crew B-63B	Hr.	Daily	Hr.	Daily	Bare Costs	Incl. O&P
1 Labor Foreman (inside)	$38.10	$304.80	$58.65	$469.20	$40.48	$62.02
2 Laborers	37.60	601.60	57.85	925.60		
1 Equip. Oper. (light)	48.60	388.80	73.75	590.00		
1 Loader, Skid Steer, 78 H.P.		318.60		350.46	9.96	10.95
32 L.H., Daily Totals		$1613.80		$2335.26	$50.43	$72.98
Crew B-64	Hr.	Daily	Hr.	Daily	Bare Costs	Incl. O&P
1 Laborer	$37.60	$300.80	$57.85	$462.80	$38.35	$58.48
1 Truck Driver (light)	39.10	312.80	59.10	472.80		
1 Power Mulcher (small)		151.80		166.98		
1 Flatbed Truck, Gas, 1.5 Ton		245.40		269.94	24.82	27.31
16 L.H., Daily Totals		$1010.80		$1372.52	$63.17	$85.78
Crew B-65	Hr.	Daily	Hr.	Daily	Bare Costs	Incl. O&P
1 Laborer	$37.60	$300.80	$57.85	$462.80	$38.35	$58.48
1 Truck Driver (light)	39.10	312.80	59.10	472.80		
1 Power Mulcher (Large)		319.40		351.34		
1 Flatbed Truck, Gas, 1.5 Ton		245.40		269.94	35.30	38.83
16 L.H., Daily Totals		$1178.40		$1556.88	$73.65	$97.31
Crew B-66	Hr.	Daily	Hr.	Daily	Bare Costs	Incl. O&P
1 Equip. Oper. (light)	$48.60	$388.80	$73.75	$590.00	$48.60	$73.75
1 Loader-Backhoe, 40 H.P.		263.20		289.52	32.90	36.19
8 L.H., Daily Totals		$652.00		$879.52	$81.50	$109.94
Crew B-67	Hr.	Daily	Hr.	Daily	Bare Costs	Incl. O&P
1 Millwright	$49.15	$393.20	$72.35	$578.80	$48.88	$73.05
1 Equip. Oper. (light)	48.60	388.80	73.75	590.00		
1 Forklift, R/T, 4,000 Lb.		311.20		342.32	19.45	21.40
16 L.H., Daily Totals		$1093.20		$1511.12	$68.33	$94.44
Crew B-67B	Hr.	Daily	Hr.	Daily	Bare Costs	Incl. O&P
1 Millwright Foreman (inside)	$49.65	$397.20	$73.10	$584.80	$49.40	$72.72
1 Millwright	49.15	393.20	72.35	578.80		
16 L.H., Daily Totals		$790.40		$1163.60	$49.40	$72.72
Crew B-68	Hr.	Daily	Hr.	Daily	Bare Costs	Incl. O&P
2 Millwrights	$49.15	$786.40	$72.35	$1157.60	$48.97	$72.82
1 Equip. Oper. (light)	48.60	388.80	73.75	590.00		
1 Forklift, R/T, 4,000 Lb.		311.20		342.32	12.97	14.26
24 L.H., Daily Totals		$1486.40		$2089.92	$61.93	$87.08
Crew B-68A	Hr.	Daily	Hr.	Daily	Bare Costs	Incl. O&P
1 Millwright Foreman (inside)	$49.65	$397.20	$73.10	$584.80	$49.32	$72.60
2 Millwrights	49.15	786.40	72.35	1157.60		
1 Forklift, 8,000 Lb.		185.20		203.72	7.72	8.49
24 L.H., Daily Totals		$1368.80		$1946.12	$57.03	$81.09

Crews

Crew No.	Bare Costs		Incl. Subs O&P		Cost Per Labor-Hour	
Crew B-68B	Hr.	Daily	Hr.	Daily	Bare Costs	Incl. O&P
1 Millwright Foreman (inside)	$49.65	$397.20	$73.10	$584.80	$53.54	$79.86
2 Millwrights	49.15	786.40	72.35	1157.60		
2 Electricians	54.70	875.20	81.95	1311.20		
2 Plumbers	58.70	939.20	88.65	1418.40		
1 Forklift, 5,000 Lb.		331.80		364.98	5.92	6.52
56 L.H., Daily Totals		$3329.80		$4836.98	$59.46	$86.37

Crew No.	Bare Costs		Incl. Subs O&P		Cost Per Labor-Hour	
Crew B-68C	Hr.	Daily	Hr.	Daily	Bare Costs	Incl. O&P
1 Millwright Foreman (inside)	$49.65	$397.20	$73.10	$584.80	$53.05	$79.01
1 Millwright	49.15	393.20	72.35	578.80		
1 Electrician	54.70	437.60	81.95	655.60		
1 Plumber	58.70	469.60	88.65	709.20		
1 Forklift, 5,000 Lb.		331.80		364.98	10.37	11.41
32 L.H., Daily Totals		$2029.40		$2893.38	$63.42	$90.42

Crew No.	Bare Costs		Incl. Subs O&P		Cost Per Labor-Hour	
Crew B-68D	Hr.	Daily	Hr.	Daily	Bare Costs	Incl. O&P
1 Labor Foreman (inside)	$38.10	$304.80	$58.65	$469.20	$41.43	$63.42
1 Laborer	37.60	300.80	57.85	462.80		
1 Equip. Oper. (light)	48.60	388.80	73.75	590.00		
1 Forklift, 5,000 Lb.		331.80		364.98	13.82	15.21
24 L.H., Daily Totals		$1326.20		$1886.98	$55.26	$78.62

Crew No.	Bare Costs		Incl. Subs O&P		Cost Per Labor-Hour	
Crew B-68E	Hr.	Daily	Hr.	Daily	Bare Costs	Incl. O&P
1 Struc. Steel Foreman (inside)	$53.15	$425.20	$92.30	$738.40	$52.75	$91.62
3 Struc. Steel Workers	52.65	1263.60	91.45	2194.80		
1 Welder	52.65	421.20	91.45	731.60		
1 Forklift, 8,000 Lb.		185.20		203.72	4.63	5.09
40 L.H., Daily Totals		$2295.20		$3868.52	$57.38	$96.71

Crew No.	Bare Costs		Incl. Subs O&P		Cost Per Labor-Hour	
Crew B-68F	Hr.	Daily	Hr.	Daily	Bare Costs	Incl. O&P
1 Skilled Worker Foreman (out)	$50.65	$405.20	$78.25	$626.00	$49.32	$76.18
2 Skilled Workers	48.65	778.40	75.15	1202.40		
1 Forklift, 5,000 Lb.		331.80		364.98	13.82	15.21
24 L.H., Daily Totals		$1515.40		$2193.38	$63.14	$91.39

Crew No.	Bare Costs		Incl. Subs O&P		Cost Per Labor-Hour	
Crew B-68G	Hr.	Daily	Hr.	Daily	Bare Costs	Incl. O&P
2 Structural Steel Workers	$52.65	$842.40	$91.45	$1463.20	$52.65	$91.45
1 Forklift, 5,000 Lb.		331.80		364.98	20.74	22.81
16 L.H., Daily Totals		$1174.20		$1828.18	$73.39	$114.26

Crew No.	Bare Costs		Incl. Subs O&P		Cost Per Labor-Hour	
Crew B-69	Hr.	Daily	Hr.	Daily	Bare Costs	Incl. O&P
1 Labor Foreman (outside)	$39.60	$316.80	$60.95	$487.60	$41.55	$63.58
3 Laborers	37.60	902.40	57.85	1388.40		
1 Equip. Oper. (crane)	51.70	413.60	78.45	627.60		
1 Equip. Oper. (oiler)	45.20	361.60	68.55	548.40		
1 Hyd. Crane, 80 Ton		1625.00		1787.50	33.85	37.24
48 L.H., Daily Totals		$3619.40		$4839.50	$75.40	$100.82

Crew No.	Bare Costs		Incl. Subs O&P		Cost Per Labor-Hour	
Crew B-69A	Hr.	Daily	Hr.	Daily	Bare Costs	Incl. O&P
1 Labor Foreman (outside)	$39.60	$316.80	$60.95	$487.60	$41.33	$62.98
3 Laborers	37.60	902.40	57.85	1388.40		
1 Equip. Oper. (medium)	50.60	404.80	76.75	614.00		
1 Concrete Finisher	45.00	360.00	66.65	533.20		
1 Curb/Gutter Paver, 2-Track		997.60		1097.36	20.78	22.86
48 L.H., Daily Totals		$2981.60		$4120.56	$62.12	$85.84

Crew No.	Bare Costs		Incl. Subs O&P		Cost Per Labor-Hour	
Crew B-69B	Hr.	Daily	Hr.	Daily	Bare Costs	Incl. O&P
1 Labor Foreman (outside)	$39.60	$316.80	$60.95	$487.60	$41.33	$62.98
3 Laborers	37.60	902.40	57.85	1388.40		
1 Equip. Oper. (medium)	50.60	404.80	76.75	614.00		
1 Cement Finisher	45.00	360.00	66.65	533.20		
1 Curb/Gutter Paver, 4-Track		776.00		853.60	16.17	17.78
48 L.H., Daily Totals		$2760.00		$3876.80	$57.50	$80.77

Crew No.	Bare Costs		Incl. Subs O&P		Cost Per Labor-Hour	
Crew B-70	Hr.	Daily	Hr.	Daily	Bare Costs	Incl. O&P
1 Labor Foreman (outside)	$39.60	$316.80	$60.95	$487.60	$43.46	$66.39
3 Laborers	37.60	902.40	57.85	1388.40		
3 Equip. Oper. (medium)	50.60	1214.40	76.75	1842.00		
1 Grader, 30,000 Lbs.		736.20		809.82		
1 Ripper, Beam & 1 Shank		84.20		92.62		
1 Road Sweeper, S.P., 8' wide		662.40		728.64		
1 F.E. Loader, W.M., 1.5 C.Y.		378.40		416.24	33.24	36.56
56 L.H., Daily Totals		$4294.80		$5765.32	$76.69	$102.95

Crew No.	Bare Costs		Incl. Subs O&P		Cost Per Labor-Hour	
Crew B-70A	Hr.	Daily	Hr.	Daily	Bare Costs	Incl. O&P
1 Laborer	$37.60	$300.80	$57.85	$462.80	$48.00	$72.97
4 Equip. Oper. (medium)	50.60	1619.20	76.75	2456.00		
1 Grader, 40,000 Lbs.		1174.00		1291.40		
1 F.E. Loader, W.M., 2.5 C.Y.		519.80		571.78		
1 Dozer, 80 H.P.		472.40		519.64		
1 Roller, Pneum. Whl., 12 Ton		344.40		378.84	62.77	69.04
40 L.H., Daily Totals		$4430.60		$5680.46	$110.77	$142.01

Crew No.	Bare Costs		Incl. Subs O&P		Cost Per Labor-Hour	
Crew B-71	Hr.	Daily	Hr.	Daily	Bare Costs	Incl. O&P
1 Labor Foreman (outside)	$39.60	$316.80	$60.95	$487.60	$43.46	$66.39
3 Laborers	37.60	902.40	57.85	1388.40		
3 Equip. Oper. (medium)	50.60	1214.40	76.75	1842.00		
1 Pvmt. Profiler, 750 H.P.		5805.00		6385.50		
1 Road Sweeper, S.P., 8' wide		662.40		728.64		
1 F.E. Loader, W.M., 1.5 C.Y.		378.40		416.24	122.25	134.47
56 L.H., Daily Totals		$9279.40		$11248.38	$165.70	$200.86

Crew No.	Bare Costs		Incl. Subs O&P		Cost Per Labor-Hour	
Crew B-72	Hr.	Daily	Hr.	Daily	Bare Costs	Incl. O&P
1 Labor Foreman (outside)	$39.60	$316.80	$60.95	$487.60	$44.35	$67.69
3 Laborers	37.60	902.40	57.85	1388.40		
4 Equip. Oper. (medium)	50.60	1619.20	76.75	2456.00		
1 Pvmt. Profiler, 750 H.P.		5805.00		6385.50		
1 Hammermill, 250 H.P.		1879.00		2066.90		
1 Windrow Loader		1227.00		1349.70		
1 Mix Paver 165 H.P.		2147.00		2361.70		
1 Roller, Pneum. Whl., 12 Ton		344.40		378.84	178.16	195.98
64 L.H., Daily Totals		$14240.80		$16874.64	$222.51	$263.67

Crew No.	Bare Costs		Incl. Subs O&P		Cost Per Labor-Hour	
Crew B-73	Hr.	Daily	Hr.	Daily	Bare Costs	Incl. O&P
1 Labor Foreman (outside)	$39.60	$316.80	$60.95	$487.60	$45.98	$70.05
2 Laborers	37.60	601.60	57.85	925.60		
5 Equip. Oper. (medium)	50.60	2024.00	76.75	3070.00		
1 Road Mixer, 310 H.P.		1929.00		2121.90		
1 Tandem Roller, 10 Ton		236.80		260.48		
1 Hammermill, 250 H.P.		1879.00		2066.90		
1 Grader, 30,000 Lbs.		736.20		809.82		
.5 F.E. Loader, W.M., 1.5 C.Y.		189.20		208.12		
.5 Truck Tractor, 220 H.P.		179.80		197.78		
.5 Water Tank Trailer, 5000 Gal.		70.50		77.55	81.57	89.73
64 L.H., Daily Totals		$8162.90		$10225.75	$127.55	$159.78

Crew No.	Bare Costs		Incl. Subs O&P		Cost Per Labor-Hour	

Crew B-74

	Hr.	Daily	Hr.	Daily	Bare Costs	Incl. O&P
1 Labor Foreman (outside)	$39.60	$316.80	$60.95	$487.60	$44.96	$68.36
1 Laborer	37.60	300.80	57.85	462.80		
4 Equip. Oper. (medium)	50.60	1619.20	76.75	2456.00		
2 Truck Drivers (heavy)	40.05	640.80	60.55	968.80		
1 Grader, 30,000 Lbs.		736.20		809.82		
1 Ripper, Beam & 1 Shank		84.20		92.62		
2 Stabilizers, 310 H.P.		3596.00		3955.60		
1 Flatbed Truck, Gas, 3 Ton		303.20		333.52		
1 Chem. Spreader, Towed		52.40		57.64		
1 Roller, Vibratory, 25 Ton		687.40		756.14		
1 Water Tank Trailer, 5000 Gal.		141.00		155.10		
1 Truck Tractor, 220 H.P.		359.60		395.56	93.13	102.44
64 L.H., Daily Totals		$8837.60		$10931.20	$138.09	$170.80

Crew B-75

	Hr.	Daily	Hr.	Daily	Bare Costs	Incl. O&P
1 Labor Foreman (outside)	$39.60	$316.80	$60.95	$487.60	$45.66	$69.48
1 Laborer	37.60	300.80	57.85	462.80		
4 Equip. Oper. (medium)	50.60	1619.20	76.75	2456.00		
1 Truck Driver (heavy)	40.05	320.40	60.55	484.40		
1 Grader, 30,000 Lbs.		736.20		809.82		
1 Ripper, Beam & 1 Shank		84.20		92.62		
2 Stabilizers, 310 H.P.		3596.00		3955.60		
1 Dist. Tanker, 3000 Gallon		305.60		336.16		
1 Truck Tractor, 6x4, 380 H.P.		599.80		659.78		
1 Roller, Vibratory, 25 Ton		687.40		756.14	107.31	118.04
56 L.H., Daily Totals		$8566.40		$10500.92	$152.97	$187.52

Crew B-76

	Hr.	Daily	Hr.	Daily	Bare Costs	Incl. O&P
1 Dock Builder Foreman (outside)	$48.10	$384.80	$76.15	$609.20	$47.47	$74.07
5 Dock Builders	46.10	1844.00	73.00	2920.00		
2 Equip. Oper. (crane)	51.70	827.20	78.45	1255.20		
1 Equip. Oper. (oiler)	45.20	361.60	68.55	548.40		
1 Crawler Crane, 50 Ton		1294.00		1423.40		
1 Barge, 400 Ton		776.00		853.60		
1 Hammer, Diesel, 15K ft.-lbs.		586.00		644.60		
1 Lead, 60' High		74.80		82.28		
1 Air Compressor, 600 cfm		550.60		605.66		
2 -50' Air Hoses, 3"		29.80		32.78	45.99	50.59
72 L.H., Daily Totals		$6728.80		$8975.12	$93.46	$124.65

Crew B-76A

	Hr.	Daily	Hr.	Daily	Bare Costs	Incl. O&P
1 Labor Foreman (outside)	$39.60	$316.80	$60.95	$487.60	$40.56	$62.15
5 Laborers	37.60	1504.00	57.85	2314.00		
1 Equip. Oper. (crane)	51.70	413.60	78.45	627.60		
1 Equip. Oper. (oiler)	45.20	361.60	68.55	548.40		
1 Crawler Crane, 50 Ton		1294.00		1423.40		
1 Barge, 400 Ton		776.00		853.60	32.34	35.58
64 L.H., Daily Totals		$4666.00		$6254.60	$72.91	$97.73

Crew B-77

	Hr.	Daily	Hr.	Daily	Bare Costs	Incl. O&P
1 Labor Foreman (outside)	$39.60	$316.80	$60.95	$487.60	$38.30	$58.72
3 Laborers	37.60	902.40	57.85	1388.40		
1 Truck Driver (light)	39.10	312.80	59.10	472.80		
1 Crack Cleaner, 25 H.P.		61.60		67.76		
1 Crack Filler, Trailer Mtd.		202.00		222.20		
1 Flatbed Truck, Gas, 3 Ton		303.20		333.52	14.17	15.59
40 L.H., Daily Totals		$2098.80		$2972.28	$52.47	$74.31

Crew B-78

	Hr.	Daily	Hr.	Daily	Bare Costs	Incl. O&P
1 Labor Foreman (outside)	$39.60	$316.80	$60.95	$487.60	$38.18	$58.58
4 Laborers	37.60	1203.20	57.85	1851.20		
1 Truck Driver (light)	39.10	312.80	59.10	472.80		
1 Paint Striper, S.P., 40 Gallon		151.40		166.54		
1 Flatbed Truck, Gas, 3 Ton		303.20		333.52		
1 Pickup Truck, 3/4 Ton		144.20		158.62	12.48	13.72
48 L.H., Daily Totals		$2431.60		$3470.28	$50.66	$72.30

Crew B-78A

	Hr.	Daily	Hr.	Daily	Bare Costs	Incl. O&P
1 Equip. Oper. (light)	$48.60	$388.80	$73.75	$590.00	$48.60	$73.75
1 Line Rem. (Metal Balls) 115 H.P.		821.60		903.76	102.70	112.97
8 L.H., Daily Totals		$1210.40		$1493.76	$151.30	$186.72

Crew B-78B

	Hr.	Daily	Hr.	Daily	Bare Costs	Incl. O&P
2 Laborers	$37.60	$601.60	$57.85	$925.60	$38.82	$59.62
.25 Equip. Oper. (light)	48.60	97.20	73.75	147.50		
1 Pickup Truck, 3/4 Ton		144.20		158.62		
1 Line Rem.,11 H.P.,Walk Behind		65.60		72.16		
.25 Road Sweeper, S.P., 8' wide		165.60		182.16	20.86	22.94
18 L.H., Daily Totals		$1074.20		$1486.04	$59.68	$82.56

Crew B-78C

	Hr.	Daily	Hr.	Daily	Bare Costs	Incl. O&P
1 Labor Foreman (outside)	$39.60	$316.80	$60.95	$487.60	$38.18	$58.58
4 Laborers	37.60	1203.20	57.85	1851.20		
1 Truck Driver (light)	39.10	312.80	59.10	472.80		
1 Paint Striper, T.M., 120 Gal.		804.60		885.06		
1 Flatbed Truck, Gas, 3 Ton		303.20		333.52		
1 Pickup Truck, 3/4 Ton		144.20		158.62	26.08	28.69
48 L.H., Daily Totals		$3084.80		$4188.80	$64.27	$87.27

Crew B-78D

	Hr.	Daily	Hr.	Daily	Bare Costs	Incl. O&P
2 Labor Foremen (outside)	$39.60	$633.60	$60.95	$975.20	$38.15	$58.59
7 Laborers	37.60	2105.60	57.85	3239.60		
1 Truck Driver (light)	39.10	312.80	59.10	472.80		
1 Paint Striper, T.M., 120 Gal.		804.60		885.06		
1 Flatbed Truck, Gas, 3 Ton		303.20		333.52		
3 Pickup Trucks, 3/4 Ton		432.60		475.86		
1 Air Compressor, 60 cfm		137.40		151.14		
1 -50' Air Hose, 3/4"		3.25		3.58		
1 Breakers, Pavement, 60 lb.		9.80		10.78	21.14	23.25
80 L.H., Daily Totals		$4742.85		$6547.53	$59.29	$81.84

Crew B-78E

	Hr.	Daily	Hr.	Daily	Bare Costs	Incl. O&P
2 Labor Foremen (outside)	$39.60	$633.60	$60.95	$975.20	$38.06	$58.47
9 Laborers	37.60	2707.20	57.85	4165.20		
1 Truck Driver (light)	39.10	312.80	59.10	472.80		
1 Paint Striper, T.M., 120 Gal.		804.60		885.06		
1 Flatbed Truck, Gas, 3 Ton		303.20		333.52		
4 Pickup Trucks, 3/4 Ton		576.80		634.48		
2 Air Compressor, 60 cfm		274.80		302.28		
2 -50' Air Hose, 3/4"		6.50		7.15		
2 Breakers, Pavement, 60 lb.		19.60		21.56	20.68	22.75
96 L.H., Daily Totals		$5639.10		$7797.25	$58.74	$81.22

For customer support on your **Mechanical Cost Data**, call 877.795.7394.

639

Crews

Crew B-78F	Bare Costs Hr.	Daily	Incl. Subs O&P Hr.	Daily	Cost Per Labor-Hour Bare Costs	Incl. O&P
2 Labor Foremen (outside)	$39.60	$633.60	$60.95	$975.20	$37.99	$58.38
11 Laborers	37.60	3308.80	57.85	5090.80		
1 Truck Driver (light)	39.10	312.80	59.10	472.80		
1 Paint Striper, T.M., 120 Gal.		804.60		885.06		
1 Flatbed Truck, Gas, 3 Ton		303.20		333.52		
7 Pickup Trucks, 3/4 Ton		1009.40		1110.34		
3 Air Compressor, 60 cfm		412.20		453.42		
3 -50' Air Hose, 3/4"		9.75		10.73		
3 Breakers, Pavement, 60 lb.		29.40		32.34	22.93	25.23
112 L.H., Daily Totals		$6823.75		$9364.20	$60.93	$83.61

Crew B-79	Bare Costs Hr.	Daily	Incl. Subs O&P Hr.	Daily	Cost Per Labor-Hour Bare Costs	Incl. O&P
1 Labor Foreman (outside)	$39.60	$316.80	$60.95	$487.60	$38.30	$58.72
3 Laborers	37.60	902.40	57.85	1388.40		
1 Truck Driver (light)	39.10	312.80	59.10	472.80		
1 Paint Striper, T.M., 120 Gal.		804.60		885.06		
1 Heating Kettle, 115 Gallon		81.00		89.10		
1 Flatbed Truck, Gas, 3 Ton		303.20		333.52		
2 Pickup Trucks, 3/4 Ton		288.40		317.24	36.93	40.62
40 L.H., Daily Totals		$3009.20		$3973.72	$75.23	$99.34

Crew B-79A	Bare Costs Hr.	Daily	Incl. Subs O&P Hr.	Daily	Cost Per Labor-Hour Bare Costs	Incl. O&P
1.5 Equip. Oper. (light)	$48.60	$583.20	$73.75	$885.00	$48.60	$73.75
.5 Line Remov. (Grinder) 115 H.P.		455.90		501.49		
1 Line Rem. (Metal Balls) 115 H.P.		821.60		903.76	106.46	117.10
12 L.H., Daily Totals		$1860.70		$2290.25	$155.06	$190.85

Crew B-79B	Bare Costs Hr.	Daily	Incl. Subs O&P Hr.	Daily	Cost Per Labor-Hour Bare Costs	Incl. O&P
1 Laborer	$37.60	$300.80	$57.85	$462.80	$37.60	$57.85
1 Set of Gases		152.00		167.20	19.00	20.90
8 L.H., Daily Totals		$452.80		$630.00	$56.60	$78.75

Crew B-79C	Bare Costs Hr.	Daily	Incl. Subs O&P Hr.	Daily	Cost Per Labor-Hour Bare Costs	Incl. O&P
1 Labor Foreman (outside)	$39.60	$316.80	$60.95	$487.60	$38.10	$58.47
5 Laborers	37.60	1504.00	57.85	2314.00		
1 Truck Driver (light)	39.10	312.80	59.10	472.80		
1 Paint Striper, T.M., 120 Gal.		804.60		885.06		
1 Heating Kettle, 115 Gallon		81.00		89.10		
1 Flatbed Truck, Gas, 3 Ton		303.20		333.52		
3 Pickup Trucks, 3/4 Ton		432.60		475.86		
1 Air Compressor, 60 cfm		137.40		151.14		
1 -50' Air Hose, 3/4"		3.25		3.58		
1 Breakers, Pavement, 60 lb.		9.80		10.78	31.64	34.80
56 L.H., Daily Totals		$3905.45		$5223.44	$69.74	$93.28

Crew B-79D	Bare Costs Hr.	Daily	Incl. Subs O&P Hr.	Daily	Cost Per Labor-Hour Bare Costs	Incl. O&P
2 Labor Foremen (outside)	$39.60	$633.60	$60.95	$975.20	$38.29	$58.78
5 Laborers	37.60	1504.00	57.85	2314.00		
1 Truck Driver (light)	39.10	312.80	59.10	472.80		
1 Paint Striper, T.M., 120 Gal.		804.60		885.06		
1 Heating Kettle, 115 Gallon		81.00		89.10		
1 Flatbed Truck, Gas, 3 Ton		303.20		333.52		
4 Pickup Trucks, 3/4 Ton		576.80		634.48		
1 Air Compressor, 60 cfm		137.40		151.14		
1 -50' Air Hose, 3/4"		3.25		3.58		
1 Breakers, Pavement, 60 lb.		9.80		10.78	29.94	32.93
64 L.H., Daily Totals		$4366.45		$5869.65	$68.23	$91.71

Crew B-79E	Bare Costs Hr.	Daily	Incl. Subs O&P Hr.	Daily	Cost Per Labor-Hour Bare Costs	Incl. O&P
2 Labor Foremen (outside)	$39.60	$633.60	$60.95	$975.20	$38.15	$58.59
7 Laborers	37.60	2105.60	57.85	3239.60		
1 Truck Driver (light)	39.10	312.80	59.10	472.80		
1 Paint Striper, T.M., 120 Gal.		804.60		885.06		
1 Heating Kettle, 115 Gallon		81.00		89.10		
1 Flatbed Truck, Gas, 3 Ton		303.20		333.52		
5 Pickup Trucks, 3/4 Ton		721.00		793.10		
2 Air Compressors, 60 cfm		274.80		302.28		
2 -50' Air Hoses, 3/4"		6.50		7.15		
2 Breakers, Pavement, 60 lb.		19.60		21.56	27.63	30.40
80 L.H., Daily Totals		$5262.70		$7119.37	$65.78	$88.99

Crew B-80	Bare Costs Hr.	Daily	Incl. Subs O&P Hr.	Daily	Cost Per Labor-Hour Bare Costs	Incl. O&P
1 Labor Foreman (outside)	$39.60	$316.80	$60.95	$487.60	$41.23	$62.91
1 Laborer	37.60	300.80	57.85	462.80		
1 Truck Driver (light)	39.10	312.80	59.10	472.80		
1 Equip. Oper. (light)	48.60	388.80	73.75	590.00		
1 Flatbed Truck, Gas, 3 Ton		303.20		333.52		
1 Earth Auger, Truck-Mtd.		417.20		458.92	22.51	24.76
32 L.H., Daily Totals		$2039.60		$2805.64	$63.74	$87.68

Crew B-80A	Bare Costs Hr.	Daily	Incl. Subs O&P Hr.	Daily	Cost Per Labor-Hour Bare Costs	Incl. O&P
3 Laborers	$37.60	$902.40	$57.85	$1388.40	$37.60	$57.85
1 Flatbed Truck, Gas, 3 Ton		303.20		333.52	12.63	13.90
24 L.H., Daily Totals		$1205.60		$1721.92	$50.23	$71.75

Crew B-80B	Bare Costs Hr.	Daily	Incl. Subs O&P Hr.	Daily	Cost Per Labor-Hour Bare Costs	Incl. O&P
3 Laborers	$37.60	$902.40	$57.85	$1388.40	$40.35	$61.83
1 Equip. Oper. (light)	48.60	388.80	73.75	590.00		
1 Crane, Flatbed Mounted, 3 Ton		243.80		268.18	7.62	8.38
32 L.H., Daily Totals		$1535.00		$2246.58	$47.97	$70.21

Crew B-80C	Bare Costs Hr.	Daily	Incl. Subs O&P Hr.	Daily	Cost Per Labor-Hour Bare Costs	Incl. O&P
2 Laborers	$37.60	$601.60	$57.85	$925.60	$38.10	$58.27
1 Truck Driver (light)	39.10	312.80	59.10	472.80		
1 Flatbed Truck, Gas, 1.5 Ton		245.40		269.94		
1 Manual Fence Post Auger, Gas		8.40		9.24	10.57	11.63
24 L.H., Daily Totals		$1168.20		$1677.58	$48.67	$69.90

Crew B-81	Bare Costs Hr.	Daily	Incl. Subs O&P Hr.	Daily	Cost Per Labor-Hour Bare Costs	Incl. O&P
1 Laborer	$37.60	$300.80	$57.85	$462.80	$42.75	$65.05
1 Equip. Oper. (medium)	50.60	404.80	76.75	614.00		
1 Truck Driver (heavy)	40.05	320.40	60.55	484.40		
1 Hydromulcher, T.M., 3000 Gal.		334.20		367.62		
1 Truck Tractor, 220 H.P.		359.60		395.56	28.91	31.80
24 L.H., Daily Totals		$1719.80		$2324.38	$71.66	$96.85

Crew B-81A	Bare Costs Hr.	Daily	Incl. Subs O&P Hr.	Daily	Cost Per Labor-Hour Bare Costs	Incl. O&P
1 Laborer	$37.60	$300.80	$57.85	$462.80	$38.35	$58.48
1 Truck Driver (light)	39.10	312.80	59.10	472.80		
1 Hydromulcher, T.M., 600 Gal.		127.80		140.58		
1 Flatbed Truck, Gas, 3 Ton		303.20		333.52	26.94	29.63
16 L.H., Daily Totals		$1044.60		$1409.70	$65.29	$88.11

Crew B-82	Bare Costs Hr.	Daily	Incl. Subs O&P Hr.	Daily	Cost Per Labor-Hour Bare Costs	Incl. O&P
1 Laborer	$37.60	$300.80	$57.85	$462.80	$43.10	$65.80
1 Equip. Oper. (light)	48.60	388.80	73.75	590.00		
1 Horiz. Borer, 6 H.P.		83.80		92.18	5.24	5.76
16 L.H., Daily Totals		$773.40		$1144.98	$48.34	$71.56

For customer support on your Mechanical Cost Data, call 877.795.7394.

Crew No.	Bare Costs		Incl. Subs O&P		Cost Per Labor-Hour	

Crew B-82A

Crew B-82A	Hr.	Daily	Hr.	Daily	Bare Costs	Incl. O&P
2 Laborers	$37.60	$601.60	$57.85	$925.60	$43.10	$65.80
2 Equip. Opers. (light)	48.60	777.60	73.75	1180.00		
2 Dump Truck, 8 C.Y., 220 H.P.		822.40		904.64		
1 Flatbed Trailer, 25 Ton		113.60		124.96		
1 Horiz. Dir. Drill, 20k lb. Thrust		654.20		719.62		
1 Mud Trailer for HDD, 1500 Gal.		346.20		380.82		
1 Pickup Truck, 4 x 4, 3/4 Ton		155.60		171.16		
1 Flatbed trailer, 3 Ton		23.60		25.96		
1 Loader, Skid Steer, 78 H.P.		318.60		350.46	76.07	83.68
32 L.H., Daily Totals		$3813.40		$4783.22	$119.17	$149.48

Crew B-82B	Hr.	Daily	Hr.	Daily	Bare Costs	Incl. O&P
2 Laborers	$37.60	$601.60	$57.85	$925.60	$43.10	$65.80
2 Equip. Opers. (light)	48.60	777.60	73.75	1180.00		
2 Dump Truck, 8 C.Y., 220 H.P.		822.40		904.64		
1 Flatbed Trailer, 25 Ton		113.60		124.96		
1 Horiz. Dir. Drill, 30k lb. Thrust		931.20		1024.32		
1 Mud Trailer for HDD, 1500 Gal.		346.20		380.82		
1 Pickup Truck, 4 x 4, 3/4 Ton		155.60		171.16		
1 Flatbed trailer, 3 Ton		23.60		25.96		
1 Loader, Skid Steer, 78 H.P.		318.60		350.46	84.72	93.20
32 L.H., Daily Totals		$4090.40		$5087.92	$127.83	$159.00

Crew B-82C	Hr.	Daily	Hr.	Daily	Bare Costs	Incl. O&P
2 Laborers	$37.60	$601.60	$57.85	$925.60	$43.10	$65.80
2 Equip. Opers. (light)	48.60	777.60	73.75	1180.00		
2 Dump Truck, 8 C.Y., 220 H.P.		822.40		904.64		
1 Flatbed Trailer, 25 Ton		113.60		124.96		
1 Horiz. Dir. Drill, 50k lb. Thrust		1237.00		1360.70		
1 Mud Trailer for HDD, 1500 Gal.		346.20		380.82		
1 Pickup Truck, 4 x 4, 3/4 Ton		155.60		171.16		
1 Flatbed trailer, 3 Ton		23.60		25.96		
1 Loader, Skid Steer, 78 H.P.		318.60		350.46	94.28	103.71
32 L.H., Daily Totals		$4396.20		$5424.30	$137.38	$169.51

Crew B-82D	Hr.	Daily	Hr.	Daily	Bare Costs	Incl. O&P
1 Equip. Oper. (light)	$48.60	$388.80	$73.75	$590.00	$48.60	$73.75
1 Mud Trailer for HDD, 1500 Gal.		346.20		380.82	43.27	47.60
8 L.H., Daily Totals		$735.00		$970.82	$91.88	$121.35

Crew B-83	Hr.	Daily	Hr.	Daily	Bare Costs	Incl. O&P
1 Tugboat Captain	$50.60	$404.80	$76.75	$614.00	$44.10	$67.30
1 Tugboat Hand	37.60	300.80	57.85	462.80		
1 Tugboat, 250 H.P.		870.00		957.00	54.38	59.81
16 L.H., Daily Totals		$1575.60		$2033.80	$98.47	$127.11

Crew B-84	Hr.	Daily	Hr.	Daily	Bare Costs	Incl. O&P
1 Equip. Oper. (medium)	$50.60	$404.80	$76.75	$614.00	$50.60	$76.75
1 Rotary Mower/Tractor		370.80		407.88	46.35	50.98
8 L.H., Daily Totals		$775.60		$1021.88	$96.95	$127.74

Crew B-85	Hr.	Daily	Hr.	Daily	Bare Costs	Incl. O&P
3 Laborers	$37.60	$902.40	$57.85	$1388.40	$40.69	$62.17
1 Equip. Oper. (medium)	50.60	404.80	76.75	614.00		
1 Truck Driver (heavy)	40.05	320.40	60.55	484.40		
1 Aerial Lift Truck, 80'		634.40		697.84		
1 Brush Chipper, 12", 130 H.P.		391.40		430.54		
1 Pruning Saw, Rotary		6.65		7.32	25.81	28.39
40 L.H., Daily Totals		$2660.05		$3622.49	$66.50	$90.56

Crew B-86	Hr.	Daily	Hr.	Daily	Bare Costs	Incl. O&P
1 Equip. Oper. (medium)	$50.60	$404.80	$76.75	$614.00	$50.60	$76.75
1 Stump Chipper, S.P.		185.55		204.10	23.19	25.51
8 L.H., Daily Totals		$590.35		$818.11	$73.79	$102.26

Crew B-86A	Hr.	Daily	Hr.	Daily	Bare Costs	Incl. O&P
1 Equip. Oper. (medium)	$50.60	$404.80	$76.75	$614.00	$50.60	$76.75
1 Grader, 30,000 Lbs.		736.20		809.82	92.03	101.23
8 L.H., Daily Totals		$1141.00		$1423.82	$142.63	$177.98

Crew B-86B	Hr.	Daily	Hr.	Daily	Bare Costs	Incl. O&P
1 Equip. Oper. (medium)	$50.60	$404.80	$76.75	$614.00	$50.60	$76.75
1 Dozer, 200 H.P.		1387.00		1525.70	173.38	190.71
8 L.H., Daily Totals		$1791.80		$2139.70	$223.97	$267.46

Crew B-87	Hr.	Daily	Hr.	Daily	Bare Costs	Incl. O&P
1 Laborer	$37.60	$300.80	$57.85	$462.80	$48.00	$72.97
4 Equip. Oper. (medium)	50.60	1619.20	76.75	2456.00		
2 Feller Bunchers, 100 H.P.		1676.80		1844.48		
1 Log Chipper, 22" Tree		889.40		978.34		
1 Dozer, 105 H.P.		602.80		663.08		
1 Chain Saw, Gas, 36" Long		45.00		49.50	80.35	88.39
40 L.H., Daily Totals		$5134.00		$6454.20	$128.35	$161.35

Crew B-88	Hr.	Daily	Hr.	Daily	Bare Costs	Incl. O&P
1 Laborer	$37.60	$300.80	$57.85	$462.80	$48.74	$74.05
6 Equip. Oper. (medium)	50.60	2428.80	76.75	3684.00		
2 Feller Bunchers, 100 H.P.		1676.80		1844.48		
1 Log Chipper, 22" Tree		889.40		978.34		
2 Log Skidders, 50 H.P.		1822.00		2004.20		
1 Dozer, 105 H.P.		602.80		663.08		
1 Chain Saw, Gas, 36" Long		45.00		49.50	89.93	98.92
56 L.H., Daily Totals		$7765.60		$9686.40	$138.67	$172.97

Crew B-89	Hr.	Daily	Hr.	Daily	Bare Costs	Incl. O&P
1 Equip. Oper. (light)	$48.60	$388.80	$73.75	$590.00	$43.85	$66.42
1 Truck Driver (light)	39.10	312.80	59.10	472.80		
1 Flatbed Truck, Gas, 3 Ton		303.20		333.52		
1 Concrete Saw		167.40		184.14		
1 Water Tank, 65 Gal.		17.30		19.03	30.49	33.54
16 L.H., Daily Totals		$1189.50		$1599.49	$74.34	$99.97

Crew B-89A	Hr.	Daily	Hr.	Daily	Bare Costs	Incl. O&P
1 Skilled Worker	$48.65	$389.20	$75.15	$601.20	$43.13	$66.50
1 Laborer	37.60	300.80	57.85	462.80		
1 Core Drill (Large)		115.60		127.16	7.22	7.95
16 L.H., Daily Totals		$805.60		$1191.16	$50.35	$74.45

Crew B-89B	Hr.	Daily	Hr.	Daily	Bare Costs	Incl. O&P
1 Equip. Oper. (light)	$48.60	$388.80	$73.75	$590.00	$43.85	$66.42
1 Truck Driver (light)	39.10	312.80	59.10	472.80		
1 Wall Saw, Hydraulic, 10 H.P.		114.60		126.06		
1 Generator, Diesel, 100 kW		423.40		465.74		
1 Water Tank, 65 Gal.		17.30		19.03		
1 Flatbed Truck, Gas, 3 Ton		303.20		333.52	53.66	59.02
16 L.H., Daily Totals		$1560.10		$2007.15	$97.51	$125.45

641

For customer support on your Mechanical Cost Data, call 877.795.7394.

Crew No.	Bare Costs		Incl. Subs O&P		Cost Per Labor-Hour	

Crew B-90

	Hr.	Daily	Hr.	Daily	Bare Costs	Incl. O&P
1 Labor Foreman (outside)	$39.60	$316.80	$60.95	$487.60	$41.21	$62.89
3 Laborers	37.60	902.40	57.85	1388.40		
2 Equip. Oper. (light)	48.60	777.60	73.75	1180.00		
2 Truck Drivers (heavy)	40.05	640.80	60.55	968.80		
1 Road Mixer, 310 H.P.		1929.00		2121.90		
1 Dist. Truck, 2000 Gal.		275.60		303.16	34.45	37.89
64 L.H., Daily Totals		$4842.20		$6449.86	$75.66	$100.78

Crew B-90A

	Hr.	Daily	Hr.	Daily	Bare Costs	Incl. O&P
1 Labor Foreman (outside)	$39.60	$316.80	$60.95	$487.60	$45.31	$69.09
2 Laborers	37.60	601.60	57.85	925.60		
4 Equip. Oper. (medium)	50.60	1619.20	76.75	2456.00		
2 Graders, 30,000 Lbs.		1472.40		1619.64		
1 Tandem Roller, 10 Ton		236.80		260.48		
1 Roller, Pneum. Whl., 12 Ton		344.40		378.84	36.67	40.34
56 L.H., Daily Totals		$4591.20		$6128.16	$81.99	$109.43

Crew B-90B

	Hr.	Daily	Hr.	Daily	Bare Costs	Incl. O&P
1 Labor Foreman (outside)	$39.60	$316.80	$60.95	$487.60	$44.43	$67.82
2 Laborers	37.60	601.60	57.85	925.60		
3 Equip. Oper. (medium)	50.60	1214.40	76.75	1842.00		
1 Roller, Pneum. Whl., 12 Ton		344.40		378.84		
1 Road Mixer, 310 H.P.		1929.00		2121.90	47.36	52.10
48 L.H., Daily Totals		$4406.20		$5755.94	$91.80	$119.92

Crew B-90C

	Hr.	Daily	Hr.	Daily	Bare Costs	Incl. O&P
1 Labor Foreman (outside)	$39.60	$316.80	$60.95	$487.60	$42.00	$64.02
4 Laborers	37.60	1203.20	57.85	1851.20		
3 Equip. Oper. (medium)	50.60	1214.40	76.75	1842.00		
3 Truck Drivers (heavy)	40.05	961.20	60.55	1453.20		
3 Road Mixers, 310 H.P.		5787.00		6365.70	65.76	72.34
88 L.H., Daily Totals		$9482.60		$11999.70	$107.76	$136.36

Crew B-90D

	Hr.	Daily	Hr.	Daily	Bare Costs	Incl. O&P
1 Labor Foreman (outside)	$39.60	$316.80	$60.95	$487.60	$41.32	$63.07
6 Laborers	37.60	1804.80	57.85	2776.80		
3 Equip. Oper. (medium)	50.60	1214.40	76.75	1842.00		
3 Truck Drivers (heavy)	40.05	961.20	60.55	1453.20		
3 Road Mixers, 310 H.P.		5787.00		6365.70	55.64	61.21
104 L.H., Daily Totals		$10084.20		$12925.30	$96.96	$124.28

Crew B-90E

	Hr.	Daily	Hr.	Daily	Bare Costs	Incl. O&P
1 Labor Foreman (outside)	$39.60	$316.80	$60.95	$487.60	$42.43	$64.79
4 Laborers	37.60	1203.20	57.85	1851.20		
3 Equip. Oper. (medium)	50.60	1214.40	76.75	1842.00		
1 Truck Driver (heavy)	40.05	320.40	60.55	484.40		
1 Road Mixers, 310 H.P.		1929.00		2121.90	26.79	29.47
72 L.H., Daily Totals		$4983.80		$6787.10	$69.22	$94.27

Crew B-91

	Hr.	Daily	Hr.	Daily	Bare Costs	Incl. O&P
1 Labor Foreman (outside)	$39.60	$316.80	$60.95	$487.60	$44.66	$68.03
2 Laborers	37.60	601.60	57.85	925.60		
4 Equip. Oper. (medium)	50.60	1619.20	76.75	2456.00		
1 Truck Driver (heavy)	40.05	320.40	60.55	484.40		
1 Dist. Tanker, 3000 Gallon		305.60		336.16		
1 Truck Tractor, 6x4, 380 H.P.		599.80		659.78		
1 Aggreg. Spreader, S.P.		834.00		917.40		
1 Roller, Pneum. Whl., 12 Ton		344.40		378.84		
1 Tandem Roller, 10 Ton		236.80		260.48	36.26	39.89
64 L.H., Daily Totals		$5178.60		$6906.26	$80.92	$107.91

Crew B-91B

	Hr.	Daily	Hr.	Daily	Bare Costs	Incl. O&P
1 Laborer	$37.60	$300.80	$57.85	$462.80	$44.10	$67.30
1 Equipment Oper. (med.)	50.60	404.80	76.75	614.00		
1 Road Sweeper, Vac. Assist.		986.00		1084.60	61.63	67.79
16 L.H., Daily Totals		$1691.60		$2161.40	$105.72	$135.09

Crew B-91C

	Hr.	Daily	Hr.	Daily	Bare Costs	Incl. O&P
1 Laborer	$37.60	$300.80	$57.85	$462.80	$38.35	$58.48
1 Truck Driver (light)	39.10	312.80	59.10	472.80		
1 Catch Basin Cleaning Truck		573.60		630.96	35.85	39.44
16 L.H., Daily Totals		$1187.20		$1566.56	$74.20	$97.91

Crew B-91D

	Hr.	Daily	Hr.	Daily	Bare Costs	Incl. O&P
1 Labor Foreman (outside)	$39.60	$316.80	$60.95	$487.60	$43.13	$65.77
5 Laborers	37.60	1504.00	57.85	2314.00		
5 Equip. Oper. (medium)	50.60	2024.00	76.75	3070.00		
2 Truck Drivers (heavy)	40.05	640.80	60.55	968.80		
1 Aggreg. Spreader, S.P.		834.00		917.40		
2 Truck Tractor, 6x4, 380 H.P.		1199.60		1319.56		
2 Dist. Tanker, 3000 Gallon		611.20		672.32		
2 Pavement Brush, Towed		163.20		179.52		
2 Roller, Pneum. Whl., 12 Ton		688.80		757.68	33.62	36.99
104 L.H., Daily Totals		$7982.40		$10686.88	$76.75	$102.76

Crew B-92

	Hr.	Daily	Hr.	Daily	Bare Costs	Incl. O&P
1 Labor Foreman (outside)	$39.60	$316.80	$60.95	$487.60	$38.10	$58.63
3 Laborers	37.60	902.40	57.85	1388.40		
1 Crack Cleaner, 25 H.P.		61.60		67.76		
1 Air Compressor, 60 cfm		137.40		151.14		
1 Tar Kettle, T.M.		132.00		145.20		
1 Flatbed Truck, Gas, 3 Ton		303.20		333.52	19.82	21.80
32 L.H., Daily Totals		$1853.40		$2573.62	$57.92	$80.43

Crew B-93

	Hr.	Daily	Hr.	Daily	Bare Costs	Incl. O&P
1 Equip. Oper. (medium)	$50.60	$404.80	$76.75	$614.00	$50.60	$76.75
1 Feller Buncher, 100 H.P.		838.40		922.24	104.80	115.28
8 L.H., Daily Totals		$1243.20		$1536.24	$155.40	$192.03

Crew B-94A

	Hr.	Daily	Hr.	Daily	Bare Costs	Incl. O&P
1 Laborer	$37.60	$300.80	$57.85	$462.80	$37.60	$57.85
1 Diaphragm Water Pump, 2"		71.00		78.10		
1 -20' Suction Hose, 2"		1.95		2.15		
2 -50' Discharge Hoses, 2"		1.80		1.98	9.34	10.28
8 L.H., Daily Totals		$375.55		$545.02	$46.94	$68.13

Crew B-94B

	Hr.	Daily	Hr.	Daily	Bare Costs	Incl. O&P
1 Laborer	$37.60	$300.80	$57.85	$462.80	$37.60	$57.85
1 Diaphragm Water Pump, 4"		114.20		125.62		
1 -20' Suction Hose, 4"		3.25		3.58		
2 -50' Discharge Hoses, 4"		4.70		5.17	15.27	16.80
8 L.H., Daily Totals		$422.95		$597.16	$52.87	$74.65

Crew B-94C

	Hr.	Daily	Hr.	Daily	Bare Costs	Incl. O&P
1 Laborer	$37.60	$300.80	$57.85	$462.80	$37.60	$57.85
1 Centrifugal Water Pump, 3"		78.20		86.02		
1 -20' Suction Hose, 3"		2.85		3.13		
2 -50' Discharge Hoses, 3"		3.00		3.30	10.51	11.56
8 L.H., Daily Totals		$384.85		$555.26	$48.11	$69.41

Crew No.	Bare Costs Hr.	Bare Costs Daily	Incl. Subs O&P Hr.	Incl. Subs O&P Daily	Cost Per LH Bare Costs	Cost Per LH Incl. O&P
Crew B-94D	Hr.	Daily	Hr.	Daily	Bare Costs	Incl. O&P
1 Laborer	$37.60	$300.80	$57.85	$462.80	$37.60	$57.85
1 Centr. Water Pump, 6"		346.60		381.26		
1 -20' Suction Hose, 6"		11.50		12.65		
2 -50' Discharge Hoses, 6"		12.20		13.42	46.29	50.92
8 L.H., Daily Totals		$671.10		$870.13	$83.89	$108.77

Crew C-1	Hr.	Daily	Hr.	Daily	Bare Costs	Incl. O&P
3 Carpenters	$46.95	$1126.80	$72.25	$1734.00	$44.61	$68.65
1 Laborer	37.60	300.80	57.85	462.80		
32 L.H., Daily Totals		$1427.60		$2196.80	$44.61	$68.65

Crew C-2	Hr.	Daily	Hr.	Daily	Bare Costs	Incl. O&P
1 Carpenter Foreman (outside)	$48.95	$391.60	$75.35	$602.80	$45.73	$70.37
4 Carpenters	46.95	1502.40	72.25	2312.00		
1 Laborer	37.60	300.80	57.85	462.80		
48 L.H., Daily Totals		$2194.80		$3377.60	$45.73	$70.37

Crew C-2A	Hr.	Daily	Hr.	Daily	Bare Costs	Incl. O&P
1 Carpenter Foreman (outside)	$48.95	$391.60	$75.35	$602.80	$45.40	$69.43
3 Carpenters	46.95	1126.80	72.25	1734.00		
1 Cement Finisher	45.00	360.00	66.65	533.20		
1 Laborer	37.60	300.80	57.85	462.80		
48 L.H., Daily Totals		$2179.20		$3332.80	$45.40	$69.43

Crew C-3	Hr.	Daily	Hr.	Daily	Bare Costs	Incl. O&P
1 Rodman Foreman (outside)	$54.55	$436.40	$85.20	$681.60	$48.57	$75.38
4 Rodmen (reinf.)	52.55	1681.60	82.10	2627.20		
1 Equip. Oper. (light)	48.60	388.80	73.75	590.00		
2 Laborers	37.60	601.60	57.85	925.60		
3 Stressing Equipment		30.60		33.66		
.5 Grouting Equipment		81.50		89.65	1.75	1.93
64 L.H., Daily Totals		$3220.50		$4947.71	$50.32	$77.31

Crew C-4	Hr.	Daily	Hr.	Daily	Bare Costs	Incl. O&P
1 Rodman Foreman (outside)	$54.55	$436.40	$85.20	$681.60	$53.05	$82.88
3 Rodmen (reinf.)	52.55	1261.20	82.10	1970.40		
3 Stressing Equipment		30.60		33.66	.96	1.05
32 L.H., Daily Totals		$1728.20		$2685.66	$54.01	$83.93

Crew C-4A	Hr.	Daily	Hr.	Daily	Bare Costs	Incl. O&P
2 Rodmen (reinf.)	$52.55	$840.80	$82.10	$1313.60	$52.55	$82.10
4 Stressing Equipment		40.80		44.88	2.55	2.81
16 L.H., Daily Totals		$881.60		$1358.48	$55.10	$84.91

Crew C-5	Hr.	Daily	Hr.	Daily	Bare Costs	Incl. O&P
1 Rodman Foreman (outside)	$54.55	$436.40	$85.20	$681.60	$51.66	$80.09
4 Rodmen (reinf.)	52.55	1681.60	82.10	2627.20		
1 Equip. Oper. (crane)	51.70	413.60	78.45	627.60		
1 Equip. Oper. (oiler)	45.20	361.60	68.55	548.40		
1 Hyd. Crane, 25 Ton		736.60		810.26	13.15	14.47
56 L.H., Daily Totals		$3629.80		$5295.06	$64.82	$94.55

Crew C-6	Hr.	Daily	Hr.	Daily	Bare Costs	Incl. O&P
1 Labor Foreman (outside)	$39.60	$316.80	$60.95	$487.60	$39.17	$59.83
4 Laborers	37.60	1203.20	57.85	1851.20		
1 Cement Finisher	45.00	360.00	66.65	533.20		
2 Gas Engine Vibrators		62.40		68.64	1.30	1.43
48 L.H., Daily Totals		$1942.40		$2940.64	$40.47	$61.26

Crew C-7	Hr.	Daily	Hr.	Daily	Bare Costs	Incl. O&P
1 Labor Foreman (outside)	$39.60	$316.80	$60.95	$487.60	$40.93	$62.46
5 Laborers	37.60	1504.00	57.85	2314.00		
1 Cement Finisher	45.00	360.00	66.65	533.20		
1 Equip. Oper. (medium)	50.60	404.80	76.75	614.00		
1 Equip. Oper. (oiler)	45.20	361.60	68.55	548.40		
2 Gas Engine Vibrators		62.40		68.64		
1 Concrete Bucket, 1 C.Y.		23.80		26.18		
1 Hyd. Crane, 55 Ton		1128.00		1240.80	16.86	18.55
72 L.H., Daily Totals		$4161.40		$5832.82	$57.80	$81.01

Crew C-7A	Hr.	Daily	Hr.	Daily	Bare Costs	Incl. O&P
1 Labor Foreman (outside)	$39.60	$316.80	$60.95	$487.60	$38.46	$58.91
5 Laborers	37.60	1504.00	57.85	2314.00		
2 Truck Drivers (heavy)	40.05	640.80	60.55	968.80		
2 Conc. Transit Mixers		2032.00		2235.20	31.75	34.92
64 L.H., Daily Totals		$4493.60		$6005.60	$70.21	$93.84

Crew C-7B	Hr.	Daily	Hr.	Daily	Bare Costs	Incl. O&P
1 Labor Foreman (outside)	$39.60	$316.80	$60.95	$487.60	$40.56	$62.15
5 Laborers	37.60	1504.00	57.85	2314.00		
1 Equipment Operator, Crane	51.70	413.60	78.45	627.60		
1 Equipment Oiler	45.20	361.60	68.55	548.40		
1 Conc. Bucket, 2 C.Y.		37.00		40.70		
1 Lattice Boom Crane, 165 Ton		1932.00		2125.20	30.77	33.84
64 L.H., Daily Totals		$4565.00		$6143.50	$71.33	$95.99

Crew C-7C	Hr.	Daily	Hr.	Daily	Bare Costs	Incl. O&P
1 Labor Foreman (outside)	$39.60	$316.80	$60.95	$487.60	$41.10	$62.96
5 Laborers	37.60	1504.00	57.85	2314.00		
2 Equipment Operators (med.)	50.60	809.60	76.75	1228.00		
2 F.E. Loaders, W.M., 4 C.Y.		1338.40		1472.24	20.91	23.00
64 L.H., Daily Totals		$3968.80		$5501.84	$62.01	$85.97

Crew C-7D	Hr.	Daily	Hr.	Daily	Bare Costs	Incl. O&P
1 Labor Foreman (outside)	$39.60	$316.80	$60.95	$487.60	$39.74	$60.99
5 Laborers	37.60	1504.00	57.85	2314.00		
1 Equip. Oper. (medium)	50.60	404.80	76.75	614.00		
1 Concrete Conveyer		198.80		218.68	3.55	3.90
56 L.H., Daily Totals		$2424.40		$3634.28	$43.29	$64.90

Crew C-8	Hr.	Daily	Hr.	Daily	Bare Costs	Incl. O&P
1 Labor Foreman (outside)	$39.60	$316.80	$60.95	$487.60	$41.86	$63.51
3 Laborers	37.60	902.40	57.85	1388.40		
2 Cement Finishers	45.00	720.00	66.65	1066.40		
1 Equip. Oper. (medium)	50.60	404.80	76.75	614.00		
1 Concrete Pump (Small)		720.00		792.00	12.86	14.14
56 L.H., Daily Totals		$3064.00		$4348.40	$54.71	$77.65

Crew C-8A	Hr.	Daily	Hr.	Daily	Bare Costs	Incl. O&P
1 Labor Foreman (outside)	$39.60	$316.80	$60.95	$487.60	$40.40	$61.30
3 Laborers	37.60	902.40	57.85	1388.40		
2 Cement Finishers	45.00	720.00	66.65	1066.40		
48 L.H., Daily Totals		$1939.20		$2942.40	$40.40	$61.30

For customer support on your Mechanical Cost Data, call 877.795.7394.

643

Crew No.	Bare Costs Hr.	Daily	Incl. Subs O&P Hr.	Daily	Cost Per Labor-Hour Bare Costs	Incl. O&P
Crew C-8B	Hr.	Daily	Hr.	Daily	Bare Costs	Incl. O&P
1 Labor Foreman (outside)	$39.60	$316.80	$60.95	$487.60	$40.60	$62.25
3 Laborers	37.60	902.40	57.85	1388.40		
1 Equip. Oper. (medium)	50.60	404.80	76.75	614.00		
1 Vibrating Power Screed		66.30		72.93		
1 Roller, Vibratory, 25 Ton		687.40		756.14		
1 Dozer, 200 H.P.		1387.00		1525.70	53.52	58.87
40 L.H., Daily Totals		$3764.70		$4844.77	$94.12	$121.12
Crew C-8C	Hr.	Daily	Hr.	Daily	Bare Costs	Incl. O&P
1 Labor Foreman (outside)	$39.60	$316.80	$60.95	$487.60	$41.33	$62.98
3 Laborers	37.60	902.40	57.85	1388.40		
1 Cement Finisher	45.00	360.00	66.65	533.20		
1 Equip. Oper. (medium)	50.60	404.80	76.75	614.00		
1 Shotcrete Rig, 12 C.Y./hr		250.80		275.88		
1 Air Compressor, 160 cfm		156.40		172.04		
4 -50' Air Hoses, 1"		16.40		18.04		
4 -50' Air Hoses, 2"		31.00		34.10	9.47	10.42
48 L.H., Daily Totals		$2438.60		$3523.26	$50.80	$73.40
Crew C-8D	Hr.	Daily	Hr.	Daily	Bare Costs	Incl. O&P
1 Labor Foreman (outside)	$39.60	$316.80	$60.95	$487.60	$42.70	$64.80
1 Laborer	37.60	300.80	57.85	462.80		
1 Cement Finisher	45.00	360.00	66.65	533.20		
1 Equipment Oper. (light)	48.60	388.80	73.75	590.00		
1 Air Compressor, 250 cfm		201.40		221.54		
2 -50' Air Hoses, 1"		8.20		9.02	6.55	7.21
32 L.H., Daily Totals		$1576.00		$2304.16	$49.25	$72.00
Crew C-8E	Hr.	Daily	Hr.	Daily	Bare Costs	Incl. O&P
1 Labor Foreman (outside)	$39.60	$316.80	$60.95	$487.60	$41.00	$62.48
3 Laborers	37.60	902.40	57.85	1388.40		
1 Cement Finisher	45.00	360.00	66.65	533.20		
1 Equipment Oper. (light)	48.60	388.80	73.75	590.00		
1 Shotcrete Rig, 35 C.Y./hr		281.00		309.10		
1 Air Compressor, 250 cfm		201.40		221.54		
4 -50' Air Hoses, 1"		16.40		18.04		
4 -50' Air Hoses, 2"		31.00		34.10	11.04	12.14
48 L.H., Daily Totals		$2497.80		$3581.98	$52.04	$74.62
Crew C-10	Hr.	Daily	Hr.	Daily	Bare Costs	Incl. O&P
1 Laborer	$37.60	$300.80	$57.85	$462.80	$42.53	$63.72
2 Cement Finishers	45.00	720.00	66.65	1066.40		
24 L.H., Daily Totals		$1020.80		$1529.20	$42.53	$63.72
Crew C-10B	Hr.	Daily	Hr.	Daily	Bare Costs	Incl. O&P
3 Laborers	$37.60	$902.40	$57.85	$1388.40	$40.56	$61.37
2 Cement Finishers	45.00	720.00	66.65	1066.40		
1 Concrete Mixer, 10 C.F.		172.40		189.64		
2 Trowels, 48" Walk-Behind		100.40		110.44	6.82	7.50
40 L.H., Daily Totals		$1895.20		$2754.88	$47.38	$68.87
Crew C-10C	Hr.	Daily	Hr.	Daily	Bare Costs	Incl. O&P
1 Laborer	$37.60	$300.80	$57.85	$462.80	$42.53	$63.72
2 Cement Finishers	45.00	720.00	66.65	1066.40		
1 Trowel, 48" Walk-Behind		50.20		55.22	2.09	2.30
24 L.H., Daily Totals		$1071.00		$1584.42	$44.63	$66.02

Crew No.	Bare Costs Hr.	Daily	Incl. Subs O&P Hr.	Daily	Cost Per Labor-Hour Bare Costs	Incl. O&P
Crew C-10D	Hr.	Daily	Hr.	Daily	Bare Costs	Incl. O&P
1 Laborer	$37.60	$300.80	$57.85	$462.80	$42.53	$63.72
2 Cement Finishers	45.00	720.00	66.65	1066.40		
1 Vibrating Power Screed		66.30		72.93		
1 Trowel, 48" Walk-Behind		50.20		55.22	4.85	5.34
24 L.H., Daily Totals		$1137.30		$1657.35	$47.39	$69.06
Crew C-10E	Hr.	Daily	Hr.	Daily	Bare Costs	Incl. O&P
1 Laborer	$37.60	$300.80	$57.85	$462.80	$42.53	$63.72
2 Cement Finishers	45.00	720.00	66.65	1066.40		
1 Vibrating Power Screed		66.30		72.93		
1 Cement Trowel, 96" Ride-On		189.00		207.90	10.64	11.70
24 L.H., Daily Totals		$1276.10		$1810.03	$53.17	$75.42
Crew C-10F	Hr.	Daily	Hr.	Daily	Bare Costs	Incl. O&P
1 Laborer	$37.60	$300.80	$57.85	$462.80	$42.53	$63.72
2 Cement Finishers	45.00	720.00	66.65	1066.40		
1 Aerial Lift Truck, 60' Boom		435.60		479.16	18.15	19.97
24 L.H., Daily Totals		$1456.40		$2008.36	$60.68	$83.68
Crew C-11	Hr.	Daily	Hr.	Daily	Bare Costs	Incl. O&P
1 Struc. Steel Foreman (outside)	$54.65	$437.20	$94.95	$759.60	$51.94	$87.85
6 Struc. Steel Workers	52.65	2527.20	91.45	4389.60		
1 Equip. Oper. (crane)	51.70	413.60	78.45	627.60		
1 Equip. Oper. (oiler)	45.20	361.60	68.55	548.40		
1 Lattice Boom Crane, 150 Ton		1813.00		1994.30	25.18	27.70
72 L.H., Daily Totals		$5552.60		$8319.50	$77.12	$115.55
Crew C-12	Hr.	Daily	Hr.	Daily	Bare Costs	Incl. O&P
1 Carpenter Foreman (outside)	$48.95	$391.60	$75.35	$602.80	$46.52	$71.40
3 Carpenters	46.95	1126.80	72.25	1734.00		
1 Laborer	37.60	300.80	57.85	462.80		
1 Equip. Oper. (crane)	51.70	413.60	78.45	627.60		
1 Hyd. Crane, 12 Ton		653.80		719.18	13.62	14.98
48 L.H., Daily Totals		$2886.60		$4146.38	$60.14	$86.38
Crew C-13	Hr.	Daily	Hr.	Daily	Bare Costs	Incl. O&P
1 Struc. Steel Worker	$52.65	$421.20	$91.45	$731.60	$50.75	$85.05
1 Welder	52.65	421.20	91.45	731.60		
1 Carpenter	46.95	375.60	72.25	578.00		
1 Welder, Gas Engine, 300 amp		145.85		160.44	6.08	6.68
24 L.H., Daily Totals		$1363.85		$2201.64	$56.83	$91.73
Crew C-14	Hr.	Daily	Hr.	Daily	Bare Costs	Incl. O&P
1 Carpenter Foreman (outside)	$48.95	$391.60	$75.35	$602.80	$46.18	$70.93
5 Carpenters	46.95	1878.00	72.25	2890.00		
4 Laborers	37.60	1203.20	57.85	1851.20		
4 Rodmen (reinf.)	52.55	1681.60	82.10	2627.20		
2 Cement Finishers	45.00	720.00	66.65	1066.40		
1 Equip. Oper. (crane)	51.70	413.60	78.45	627.60		
1 Equip. Oper. (oiler)	45.20	361.60	68.55	548.40		
1 Hyd. Crane, 80 Ton		1625.00		1787.50	11.28	12.41
144 L.H., Daily Totals		$8274.60		$12001.10	$57.46	$83.34

For customer support on your Mechanical Cost Data, call 877.795.7394.

Crew No.	Bare Costs		Incl. Subs O&P		Cost Per Labor-Hour	
Crew C-14A	Hr.	Daily	Hr.	Daily	Bare Costs	Incl. O&P
1 Carpenter Foreman (outside)	$48.95	$391.60	$75.35	$602.80	$47.25	$72.75
16 Carpenters	46.95	6009.60	72.25	9248.00		
4 Rodmen (reinf.)	52.55	1681.60	82.10	2627.20		
2 Laborers	37.60	601.60	57.85	925.60		
1 Cement Finisher	45.00	360.00	66.65	533.20		
1 Equip. Oper. (medium)	50.60	404.80	76.75	614.00		
1 Gas Engine Vibrator		31.20		34.32		
1 Concrete Pump (Small)		720.00		792.00	3.76	4.13
200 L.H., Daily Totals		$10200.40		$15377.12	$51.00	$76.89
Crew C-14B	Hr.	Daily	Hr.	Daily	Bare Costs	Incl. O&P
1 Carpenter Foreman (outside)	$48.95	$391.60	$75.35	$602.80	$47.16	$72.52
16 Carpenters	46.95	6009.60	72.25	9248.00		
4 Rodmen (reinf.)	52.55	1681.60	82.10	2627.20		
2 Laborers	37.60	601.60	57.85	925.60		
2 Cement Finishers	45.00	720.00	66.65	1066.40		
1 Equip. Oper. (medium)	50.60	404.80	76.75	614.00		
1 Gas Engine Vibrator		31.20		34.32		
1 Concrete Pump (Small)		720.00		792.00	3.61	3.97
208 L.H., Daily Totals		$10560.40		$15910.32	$50.77	$76.49
Crew C-14C	Hr.	Daily	Hr.	Daily	Bare Costs	Incl. O&P
1 Carpenter Foreman (outside)	$48.95	$391.60	$75.35	$602.80	$45.08	$69.36
6 Carpenters	46.95	2253.60	72.25	3468.00		
2 Rodmen (reinf.)	52.55	840.80	82.10	1313.60		
4 Laborers	37.60	1203.20	57.85	1851.20		
1 Cement Finisher	45.00	360.00	66.65	533.20		
1 Gas Engine Vibrator		31.20		34.32	.28	.31
112 L.H., Daily Totals		$5080.40		$7803.12	$45.36	$69.67
Crew C-14D	Hr.	Daily	Hr.	Daily	Bare Costs	Incl. O&P
1 Carpenter Foreman (outside)	$48.95	$391.60	$75.35	$602.80	$46.80	$71.97
18 Carpenters	46.95	6760.80	72.25	10404.00		
2 Rodmen (reinf.)	52.55	840.80	82.10	1313.60		
2 Laborers	37.60	601.60	57.85	925.60		
1 Cement Finisher	45.00	360.00	66.65	533.20		
1 Equip. Oper. (medium)	50.60	404.80	76.75	614.00		
1 Gas Engine Vibrator		31.20		34.32		
1 Concrete Pump (Small)		720.00		792.00	3.76	4.13
200 L.H., Daily Totals		$10110.80		$15219.52	$50.55	$76.10
Crew C-14E	Hr.	Daily	Hr.	Daily	Bare Costs	Incl. O&P
1 Carpenter Foreman (outside)	$48.95	$391.60	$75.35	$602.80	$46.44	$71.68
2 Carpenters	46.95	751.20	72.25	1156.00		
4 Rodmen (reinf.)	52.55	1681.60	82.10	2627.20		
3 Laborers	37.60	902.40	57.85	1388.40		
1 Cement Finisher	45.00	360.00	66.65	533.20		
1 Gas Engine Vibrator		31.20		34.32	.35	.39
88 L.H., Daily Totals		$4118.00		$6341.92	$46.80	$72.07
Crew C-14F	Hr.	Daily	Hr.	Daily	Bare Costs	Incl. O&P
1 Labor Foreman (outside)	$39.60	$316.80	$60.95	$487.60	$42.76	$64.06
2 Laborers	37.60	601.60	57.85	925.60		
6 Cement Finishers	45.00	2160.00	66.65	3199.20		
1 Gas Engine Vibrator		31.20		34.32	.43	.48
72 L.H., Daily Totals		$3109.60		$4646.72	$43.19	$64.54

Crew No.	Bare Costs		Incl. Subs O&P		Cost Per Labor-Hour	
Crew C-14G	Hr.	Daily	Hr.	Daily	Bare Costs	Incl. O&P
1 Labor Foreman (outside)	$39.60	$316.80	$60.95	$487.60	$42.11	$63.32
2 Laborers	37.60	601.60	57.85	925.60		
4 Cement Finishers	45.00	1440.00	66.65	2132.80		
1 Gas Engine Vibrator		31.20		34.32	.56	.61
56 L.H., Daily Totals		$2389.60		$3580.32	$42.67	$63.93
Crew C-14H	Hr.	Daily	Hr.	Daily	Bare Costs	Incl. O&P
1 Carpenter Foreman (outside)	$48.95	$391.60	$75.35	$602.80	$46.33	$71.08
2 Carpenters	46.95	751.20	72.25	1156.00		
1 Rodman (reinf.)	52.55	420.40	82.10	656.80		
1 Laborer	37.60	300.80	57.85	462.80		
1 Cement Finisher	45.00	360.00	66.65	533.20		
1 Gas Engine Vibrator		31.20		34.32	.65	.71
48 L.H., Daily Totals		$2255.20		$3445.92	$46.98	$71.79
Crew C-14L	Hr.	Daily	Hr.	Daily	Bare Costs	Incl. O&P
1 Carpenter Foreman (outside)	$48.95	$391.60	$75.35	$602.80	$43.84	$67.24
6 Carpenters	46.95	2253.60	72.25	3468.00		
4 Laborers	37.60	1203.20	57.85	1851.20		
1 Cement Finisher	45.00	360.00	66.65	533.20		
1 Gas Engine Vibrator		31.20		34.32	.33	.36
96 L.H., Daily Totals		$4239.60		$6489.52	$44.16	$67.60
Crew C-14M	Hr.	Daily	Hr.	Daily	Bare Costs	Incl. O&P
1 Carpenter Foreman (outside)	$48.95	$391.60	$75.35	$602.80	$45.77	$70.13
2 Carpenters	46.95	751.20	72.25	1156.00		
1 Rodman (reinf.)	52.55	420.40	82.10	656.80		
2 Laborers	37.60	601.60	57.85	925.60		
1 Cement Finisher	45.00	360.00	66.65	533.20		
1 Equip. Oper. (medium)	50.60	404.80	76.75	614.00		
1 Gas Engine Vibrator		31.20		34.32		
1 Concrete Pump (Small)		720.00		792.00	11.74	12.91
64 L.H., Daily Totals		$3680.80		$5314.72	$57.51	$83.04
Crew C-15	Hr.	Daily	Hr.	Daily	Bare Costs	Incl. O&P
1 Carpenter Foreman (outside)	$48.95	$391.60	$75.35	$602.80	$44.24	$67.64
2 Carpenters	46.95	751.20	72.25	1156.00		
3 Laborers	37.60	902.40	57.85	1388.40		
2 Cement Finishers	45.00	720.00	66.65	1066.40		
1 Rodman (reinf.)	52.55	420.40	82.10	656.80		
72 L.H., Daily Totals		$3185.60		$4870.40	$44.24	$67.64
Crew C-16	Hr.	Daily	Hr.	Daily	Bare Costs	Incl. O&P
1 Labor Foreman (outside)	$39.60	$316.80	$60.95	$487.60	$41.86	$63.51
3 Laborers	37.60	902.40	57.85	1388.40		
2 Cement Finishers	45.00	720.00	66.65	1066.40		
1 Equip. Oper. (medium)	50.60	404.80	76.75	614.00		
1 Gunite Pump Rig		371.80		408.98		
2 -50' Air Hoses, 3/4"		6.50		7.15		
2 -50' Air Hoses, 2"		15.50		17.05	7.03	7.74
56 L.H., Daily Totals		$2737.80		$3989.58	$48.89	$71.24
Crew C-16A	Hr.	Daily	Hr.	Daily	Bare Costs	Incl. O&P
1 Laborer	$37.60	$300.80	$57.85	$462.80	$44.55	$66.97
2 Cement Finishers	45.00	720.00	66.65	1066.40		
1 Equip. Oper. (medium)	50.60	404.80	76.75	614.00		
1 Gunite Pump Rig		371.80		408.98		
2 -50' Air Hoses, 3/4"		6.50		7.15		
2 -50' Air Hoses, 2"		15.50		17.05		
1 Aerial Lift Truck, 60' Boom		435.60		479.16	25.92	28.51
32 L.H., Daily Totals		$2255.00		$3055.54	$70.47	$95.49

Crew C-17

Crew No.	Bare Costs Hr.	Daily	Incl. Subs O&P Hr.	Daily	Cost Per Labor-Hour Bare Costs	Incl. O&P
2 Skilled Worker Foremen (out)	$50.65	$810.40	$78.25	$1252.00	$49.05	$75.77
8 Skilled Workers	48.65	3113.60	75.15	4809.60		
80 L.H., Daily Totals		$3924.00		$6061.60	$49.05	$75.77

Crew C-17A

Crew No.	Bare Costs Hr.	Daily	Incl. Subs O&P Hr.	Daily	Cost Per Labor-Hour Bare Costs	Incl. O&P
2 Skilled Worker Foremen (out)	$50.65	$810.40	$78.25	$1252.00	$49.08	$75.80
8 Skilled Workers	48.65	3113.60	75.15	4809.60		
.125 Equip. Oper. (crane)	51.70	51.70	78.45	78.45		
.125 Hyd. Crane, 80 Ton		203.13		223.44	2.51	2.76
81 L.H., Daily Totals		$4178.82		$6363.49	$51.59	$78.56

Crew C-17B

Crew No.	Bare Costs Hr.	Daily	Incl. Subs O&P Hr.	Daily	Cost Per Labor-Hour Bare Costs	Incl. O&P
2 Skilled Worker Foremen (out)	$50.65	$810.40	$78.25	$1252.00	$49.11	$75.84
8 Skilled Workers	48.65	3113.60	75.15	4809.60		
.25 Equip. Oper. (crane)	51.70	103.40	78.45	156.90		
.25 Hyd. Crane, 80 Ton		406.25		446.88		
.25 Trowel, 48" Walk-Behind		12.55		13.81	5.11	5.62
82 L.H., Daily Totals		$4446.20		$6679.18	$54.22	$81.45

Crew C-17C

Crew No.	Bare Costs Hr.	Daily	Incl. Subs O&P Hr.	Daily	Cost Per Labor-Hour Bare Costs	Incl. O&P
2 Skilled Worker Foremen (out)	$50.65	$810.40	$78.25	$1252.00	$49.15	$75.87
8 Skilled Workers	48.65	3113.60	75.15	4809.60		
.375 Equip. Oper. (crane)	51.70	155.10	78.45	235.35		
.375 Hyd. Crane, 80 Ton		609.38		670.31	7.34	8.08
83 L.H., Daily Totals		$4688.48		$6967.26	$56.49	$83.94

Crew C-17D

Crew No.	Bare Costs Hr.	Daily	Incl. Subs O&P Hr.	Daily	Cost Per Labor-Hour Bare Costs	Incl. O&P
2 Skilled Worker Foremen (out)	$50.65	$810.40	$78.25	$1252.00	$49.18	$75.90
8 Skilled Workers	48.65	3113.60	75.15	4809.60		
.5 Equip. Oper. (crane)	51.70	206.80	78.45	313.80		
.5 Hyd. Crane, 80 Ton		812.50		893.75	9.67	10.64
84 L.H., Daily Totals		$4943.30		$7269.15	$58.85	$86.54

Crew C-17E

Crew No.	Bare Costs Hr.	Daily	Incl. Subs O&P Hr.	Daily	Cost Per Labor-Hour Bare Costs	Incl. O&P
2 Skilled Worker Foremen (out)	$50.65	$810.40	$78.25	$1252.00	$49.05	$75.77
8 Skilled Workers	48.65	3113.60	75.15	4809.60		
1 Hyd. Jack with Rods		96.50		106.15	1.21	1.33
80 L.H., Daily Totals		$4020.50		$6167.75	$50.26	$77.10

Crew C-18

Crew No.	Bare Costs Hr.	Daily	Incl. Subs O&P Hr.	Daily	Cost Per Labor-Hour Bare Costs	Incl. O&P
.125 Labor Foreman (outside)	$39.60	$39.60	$60.95	$60.95	$37.82	$58.19
1 Laborer	37.60	300.80	57.85	462.80		
1 Concrete Cart, 10 C.F.		60.00		66.00	6.67	7.33
9 L.H., Daily Totals		$400.40		$589.75	$44.49	$65.53

Crew C-19

Crew No.	Bare Costs Hr.	Daily	Incl. Subs O&P Hr.	Daily	Cost Per Labor-Hour Bare Costs	Incl. O&P
.125 Labor Foreman (outside)	$39.60	$39.60	$60.95	$60.95	$37.82	$58.19
1 Laborer	37.60	300.80	57.85	462.80		
1 Concrete Cart, 18 C.F.		99.80		109.78	11.09	12.20
9 L.H., Daily Totals		$440.20		$633.53	$48.91	$70.39

Crew C-20

Crew No.	Bare Costs Hr.	Daily	Incl. Subs O&P Hr.	Daily	Cost Per Labor-Hour Bare Costs	Incl. O&P
1 Labor Foreman (outside)	$39.60	$316.80	$60.95	$487.60	$40.40	$61.70
5 Laborers	37.60	1504.00	57.85	2314.00		
1 Cement Finisher	45.00	360.00	66.65	533.20		
1 Equip. Oper. (medium)	50.60	404.80	76.75	614.00		
2 Gas Engine Vibrators		62.40		68.64		
1 Concrete Pump (Small)		720.00		792.00	12.23	13.45
64 L.H., Daily Totals		$3368.00		$4809.44	$52.63	$75.15

Crew C-21

Crew No.	Bare Costs Hr.	Daily	Incl. Subs O&P Hr.	Daily	Cost Per Labor-Hour Bare Costs	Incl. O&P
1 Labor Foreman (outside)	$39.60	$316.80	$60.95	$487.60	$40.40	$61.70
5 Laborers	37.60	1504.00	57.85	2314.00		
1 Cement Finisher	45.00	360.00	66.65	533.20		
1 Equip. Oper. (medium)	50.60	404.80	76.75	614.00		
2 Gas Engine Vibrators		62.40		68.64		
1 Concrete Conveyer		198.80		218.68	4.08	4.49
64 L.H., Daily Totals		$2846.80		$4236.12	$44.48	$66.19

Crew C-22

Crew No.	Bare Costs Hr.	Daily	Incl. Subs O&P Hr.	Daily	Cost Per Labor-Hour Bare Costs	Incl. O&P
1 Rodman Foreman (outside)	$54.55	$436.40	$85.20	$681.60	$52.74	$82.28
4 Rodmen (reinf.)	52.55	1681.60	82.10	2627.20		
.125 Equip. Oper. (crane)	51.70	51.70	78.45	78.45		
.125 Equip. Oper. (oiler)	45.20	45.20	68.55	68.55		
.125 Hyd. Crane, 25 Ton		92.08		101.28	2.19	2.41
42 L.H., Daily Totals		$2306.97		$3557.08	$54.93	$84.69

Crew C-23

Crew No.	Bare Costs Hr.	Daily	Incl. Subs O&P Hr.	Daily	Cost Per Labor-Hour Bare Costs	Incl. O&P
2 Skilled Worker Foremen (out)	$50.65	$810.40	$78.25	$1252.00	$49.01	$75.44
6 Skilled Workers	48.65	2335.20	75.15	3607.20		
1 Equip. Oper. (crane)	51.70	413.60	78.45	627.60		
1 Equip. Oper. (oiler)	45.20	361.60	68.55	548.40		
1 Lattice Boom Crane, 90 Ton		1511.00		1662.10	18.89	20.78
80 L.H., Daily Totals		$5431.80		$7697.30	$67.90	$96.22

Crew C-23A

Crew No.	Bare Costs Hr.	Daily	Incl. Subs O&P Hr.	Daily	Cost Per Labor-Hour Bare Costs	Incl. O&P
1 Labor Foreman (outside)	$39.60	$316.80	$60.95	$487.60	$42.34	$64.73
2 Laborers	37.60	601.60	57.85	925.60		
1 Equip. Oper. (crane)	51.70	413.60	78.45	627.60		
1 Equip. Oper. (oiler)	45.20	361.60	68.55	548.40		
1 Crawler Crane, 100 Ton		1667.00		1833.70		
3 Conc. Buckets, 8 C.Y.		608.40		669.24	56.88	62.57
40 L.H., Daily Totals		$3969.00		$5092.14	$99.22	$127.30

Crew C-24

Crew No.	Bare Costs Hr.	Daily	Incl. Subs O&P Hr.	Daily	Cost Per Labor-Hour Bare Costs	Incl. O&P
2 Skilled Worker Foremen (out)	$50.65	$810.40	$78.25	$1252.00	$49.01	$75.44
6 Skilled Workers	48.65	2335.20	75.15	3607.20		
1 Equip. Oper. (crane)	51.70	413.60	78.45	627.60		
1 Equip. Oper. (oiler)	45.20	361.60	68.55	548.40		
1 Lattice Boom Crane, 150 Ton		1813.00		1994.30	22.66	24.93
80 L.H., Daily Totals		$5733.80		$8029.50	$71.67	$100.37

Crew C-25

Crew No.	Bare Costs Hr.	Daily	Incl. Subs O&P Hr.	Daily	Cost Per Labor-Hour Bare Costs	Incl. O&P
2 Rodmen (reinf.)	$52.55	$840.80	$82.10	$1313.60	$41.30	$66.70
2 Rodmen Helpers	30.05	480.80	51.30	820.80		
32 L.H., Daily Totals		$1321.60		$2134.40	$41.30	$66.70

Crew C-27

Crew No.	Bare Costs Hr.	Daily	Incl. Subs O&P Hr.	Daily	Cost Per Labor-Hour Bare Costs	Incl. O&P
2 Cement Finishers	$45.00	$720.00	$66.65	$1066.40	$45.00	$66.65
1 Concrete Saw		167.40		184.14	10.46	11.51
16 L.H., Daily Totals		$887.40		$1250.54	$55.46	$78.16

Crew C-28

Crew No.	Bare Costs Hr.	Daily	Incl. Subs O&P Hr.	Daily	Cost Per Labor-Hour Bare Costs	Incl. O&P
1 Cement Finisher	$45.00	$360.00	$66.65	$533.20	$45.00	$66.65
1 Portable Air Compressor, Gas		17.90		19.69	2.24	2.46
8 L.H., Daily Totals		$377.90		$552.89	$47.24	$69.11

Crew C-29

Crew No.	Bare Costs Hr.	Daily	Incl. Subs O&P Hr.	Daily	Cost Per Labor-Hour Bare Costs	Incl. O&P
1 Laborer	$37.60	$300.80	$57.85	$462.80	$37.60	$57.85
1 Pressure Washer		70.20		77.22	8.78	9.65
8 L.H., Daily Totals		$371.00		$540.02	$46.38	$67.50

Crews

Crew No.	Bare Costs Hr.	Bare Costs Daily	Incl. Subs O&P Hr.	Incl. Subs O&P Daily	Cost Per Labor-Hour Bare Costs	Cost Per Labor-Hour Incl. O&P
Crew C-30	Hr.	Daily	Hr.	Daily	Bare Costs	Incl. O&P
1 Laborer	$37.60	$300.80	$57.85	$462.80	$37.60	$57.85
1 Concrete Mixer, 10 C.F.		172.40		189.64	21.55	23.70
8 L.H., Daily Totals		$473.20		$652.44	$59.15	$81.56
Crew C-31	Hr.	Daily	Hr.	Daily	Bare Costs	Incl. O&P
1 Cement Finisher	$45.00	$360.00	$66.65	$533.20	$45.00	$66.65
1 Grout Pump		371.80		408.98	46.48	51.12
8 L.H., Daily Totals		$731.80		$942.18	$91.47	$117.77
Crew C-32	Hr.	Daily	Hr.	Daily	Bare Costs	Incl. O&P
1 Cement Finisher	$45.00	$360.00	$66.65	$533.20	$41.30	$62.25
1 Laborer	37.60	300.80	57.85	462.80		
1 Crack Chaser Saw, Gas, 6 H.P.		29.20		32.12		
1 Vacuum Pick-Up System		60.90		66.99	5.63	6.19
16 L.H., Daily Totals		$750.90		$1095.11	$46.93	$68.44
Crew D-1	Hr.	Daily	Hr.	Daily	Bare Costs	Incl. O&P
1 Bricklayer	$46.15	$369.20	$70.45	$563.60	$42.10	$64.28
1 Bricklayer Helper	38.05	304.40	58.10	464.80		
16 L.H., Daily Totals		$673.60		$1028.40	$42.10	$64.28
Crew D-2	Hr.	Daily	Hr.	Daily	Bare Costs	Incl. O&P
3 Bricklayers	$46.15	$1107.60	$70.45	$1690.80	$43.28	$66.12
2 Bricklayer Helpers	38.05	608.80	58.10	929.60		
.5 Carpenter	46.95	187.80	72.25	289.00		
44 L.H., Daily Totals		$1904.20		$2909.40	$43.28	$66.12
Crew D-3	Hr.	Daily	Hr.	Daily	Bare Costs	Incl. O&P
3 Bricklayers	$46.15	$1107.60	$70.45	$1690.80	$43.10	$65.83
2 Bricklayer Helpers	38.05	608.80	58.10	929.60		
.25 Carpenter	46.95	93.90	72.25	144.50		
42 L.H., Daily Totals		$1810.30		$2764.90	$43.10	$65.83
Crew D-4	Hr.	Daily	Hr.	Daily	Bare Costs	Incl. O&P
1 Bricklayer	$46.15	$369.20	$70.45	$563.60	$42.71	$65.10
2 Bricklayer Helpers	38.05	608.80	58.10	929.60		
1 Equip. Oper. (light)	48.60	388.80	73.75	590.00		
1 Grout Pump, 50 C.F./hr.		133.40		146.74	4.17	4.59
32 L.H., Daily Totals		$1500.20		$2229.94	$46.88	$69.69
Crew D-5	Hr.	Daily	Hr.	Daily	Bare Costs	Incl. O&P
1 Bricklayer	46.15	369.20	70.45	563.60	46.15	70.45
8 L.H., Daily Totals		$369.20		$563.60	$46.15	$70.45
Crew D-6	Hr.	Daily	Hr.	Daily	Bare Costs	Incl. O&P
3 Bricklayers	$46.15	$1107.60	$70.45	$1690.80	$42.29	$64.59
3 Bricklayer Helpers	38.05	913.20	58.10	1394.40		
.25 Carpenter	46.95	93.90	72.25	144.50		
50 L.H., Daily Totals		$2114.70		$3229.70	$42.29	$64.59
Crew D-7	Hr.	Daily	Hr.	Daily	Bare Costs	Incl. O&P
1 Tile Layer	$42.80	$342.40	$63.30	$506.40	$38.17	$56.45
1 Tile Layer Helper	33.55	268.40	49.60	396.80		
16 L.H., Daily Totals		$610.80		$903.20	$38.17	$56.45
Crew D-8	Hr.	Daily	Hr.	Daily	Bare Costs	Incl. O&P
3 Bricklayers	$46.15	$1107.60	$70.45	$1690.80	$42.91	$65.51
2 Bricklayer Helpers	38.05	608.80	58.10	929.60		
40 L.H., Daily Totals		$1716.40		$2620.40	$42.91	$65.51
Crew D-9	Hr.	Daily	Hr.	Daily	Bare Costs	Incl. O&P
3 Bricklayers	$46.15	$1107.60	$70.45	$1690.80	$42.10	$64.28
3 Bricklayer Helpers	38.05	913.20	58.10	1394.40		
48 L.H., Daily Totals		$2020.80		$3085.20	$42.10	$64.28
Crew D-10	Hr.	Daily	Hr.	Daily	Bare Costs	Incl. O&P
1 Bricklayer Foreman (outside)	$48.15	$385.20	$73.55	$588.40	$46.01	$70.14
1 Bricklayer	46.15	369.20	70.45	563.60		
1 Bricklayer Helper	38.05	304.40	58.10	464.80		
1 Equip. Oper. (crane)	51.70	413.60	78.45	627.60		
1 S.P. Crane, 4x4, 12 Ton		473.40		520.74	14.79	16.27
32 L.H., Daily Totals		$1945.80		$2765.14	$60.81	$86.41
Crew D-11	Hr.	Daily	Hr.	Daily	Bare Costs	Incl. O&P
1 Bricklayer Foreman (outside)	$48.15	$385.20	$73.55	$588.40	$44.12	$67.37
1 Bricklayer	46.15	369.20	70.45	563.60		
1 Bricklayer Helper	38.05	304.40	58.10	464.80		
24 L.H., Daily Totals		$1058.80		$1616.80	$44.12	$67.37
Crew D-12	Hr.	Daily	Hr.	Daily	Bare Costs	Incl. O&P
1 Bricklayer Foreman (outside)	$48.15	$385.20	$73.55	$588.40	$42.60	$65.05
1 Bricklayer	46.15	369.20	70.45	563.60		
2 Bricklayer Helpers	38.05	608.80	58.10	929.60		
32 L.H., Daily Totals		$1363.20		$2081.60	$42.60	$65.05
Crew D-13	Hr.	Daily	Hr.	Daily	Bare Costs	Incl. O&P
1 Bricklayer Foreman (outside)	$48.15	$385.20	$73.55	$588.40	$44.84	$68.48
1 Bricklayer	46.15	369.20	70.45	563.60		
2 Bricklayer Helpers	38.05	608.80	58.10	929.60		
1 Carpenter	46.95	375.60	72.25	578.00		
1 Equip. Oper. (crane)	51.70	413.60	78.45	627.60		
1 S.P. Crane, 4x4, 12 Ton		473.40		520.74	9.86	10.85
48 L.H., Daily Totals		$2625.80		$3807.94	$54.70	$79.33
Crew E-1	Hr.	Daily	Hr.	Daily	Bare Costs	Incl. O&P
1 Welder Foreman (outside)	$54.65	$437.20	$94.95	$759.60	$51.97	$86.72
1 Welder	52.65	421.20	91.45	731.60		
1 Equip. Oper. (light)	48.60	388.80	73.75	590.00		
1 Welder, Gas Engine, 300 amp		145.85		160.44	6.08	6.68
24 L.H., Daily Totals		$1393.05		$2241.64	$58.04	$93.40
Crew E-2	Hr.	Daily	Hr.	Daily	Bare Costs	Incl. O&P
1 Struc. Steel Foreman (outside)	$54.65	$437.20	$94.95	$759.60	$51.74	$86.82
4 Struc. Steel Workers	52.65	1684.80	91.45	2926.40		
1 Equip. Oper. (crane)	51.70	413.60	78.45	627.60		
1 Equip. Oper. (oiler)	45.20	361.60	68.55	548.40		
1 Lattice Boom Crane, 90 Ton		1511.00		1662.10	26.98	29.68
56 L.H., Daily Totals		$4408.20		$6524.10	$78.72	$116.50
Crew E-3	Hr.	Daily	Hr.	Daily	Bare Costs	Incl. O&P
1 Struc. Steel Foreman (outside)	$54.65	$437.20	$94.95	$759.60	$53.32	$92.62
1 Struc. Steel Worker	52.65	421.20	91.45	731.60		
1 Welder	52.65	421.20	91.45	731.60		
1 Welder, Gas Engine, 300 amp		145.85		160.44	6.08	6.68
24 L.H., Daily Totals		$1425.45		$2383.24	$59.39	$99.30

For customer support on your Mechanical Cost Data, call 877.795.7394.

647

Crew No.	Bare Costs		Incl. Subs O&P		Cost Per Labor-Hour	

Crew E-3A	Hr.	Daily	Hr.	Daily	Bare Costs	Incl. O&P
1 Struc. Steel Foreman (outside)	$54.65	$437.20	$94.95	$759.60	$53.32	$92.62
1 Struc. Steel Worker	52.65	421.20	91.45	731.60		
1 Welder	52.65	421.20	91.45	731.60		
1 Welder, Gas Engine, 300 amp		145.85		160.44		
1 Aerial Lift Truck, 40' Boom		301.40		331.54	18.64	20.50
24 L.H., Daily Totals		$1726.85		$2714.78	$71.95	$113.12

Crew E-4	Hr.	Daily	Hr.	Daily	Bare Costs	Incl. O&P
1 Struc. Steel Foreman (outside)	$54.65	$437.20	$94.95	$759.60	$53.15	$92.33
3 Struc. Steel Workers	52.65	1263.60	91.45	2194.80		
1 Welder, Gas Engine, 300 amp		145.85		160.44	4.56	5.01
32 L.H., Daily Totals		$1846.65		$3114.84	$57.71	$97.34

Crew E-5	Hr.	Daily	Hr.	Daily	Bare Costs	Incl. O&P
2 Struc. Steel Foremen (outside)	$54.65	$874.40	$94.95	$1519.20	$52.21	$88.56
5 Struc. Steel Workers	52.65	2106.00	91.45	3658.00		
1 Equip. Oper. (crane)	51.70	413.60	78.45	627.60		
1 Welder	52.65	421.20	91.45	731.60		
1 Equip. Oper. (oiler)	45.20	361.60	68.55	548.40		
1 Lattice Boom Crane, 90 Ton		1511.00		1662.10		
1 Welder, Gas Engine, 300 amp		145.85		160.44	20.71	22.78
80 L.H., Daily Totals		$5833.65		$8907.33	$72.92	$111.34

Crew E-6	Hr.	Daily	Hr.	Daily	Bare Costs	Incl. O&P
3 Struc. Steel Foremen (outside)	$54.65	$1311.60	$94.95	$2278.80	$52.25	$88.76
9 Struc. Steel Workers	52.65	3790.80	91.45	6584.40		
1 Equip. Oper. (crane)	51.70	413.60	78.45	627.60		
1 Welder	52.65	421.20	91.45	731.60		
1 Equip. Oper. (oiler)	45.20	361.60	68.55	548.40		
1 Equip. Oper. (light)	48.60	388.80	73.75	590.00		
1 Lattice Boom Crane, 90 Ton		1511.00		1662.10		
1 Welder, Gas Engine, 300 amp		145.85		160.44		
1 Air Compressor, 160 cfm		156.40		172.04		
2 Impact Wrenches		36.00		39.60	14.45	15.89
128 L.H., Daily Totals		$8536.85		$13394.98	$66.69	$104.65

Crew E-7	Hr.	Daily	Hr.	Daily	Bare Costs	Incl. O&P
1 Struc. Steel Foreman (outside)	$54.65	$437.20	$94.95	$759.60	$52.21	$88.56
4 Struc. Steel Workers	52.65	1684.80	91.45	2926.40		
1 Equip. Oper. (crane)	51.70	413.60	78.45	627.60		
1 Equip. Oper. (oiler)	45.20	361.60	68.55	548.40		
1 Welder Foreman (outside)	54.65	437.20	94.95	759.60		
2 Welders	52.65	842.40	91.45	1463.20		
1 Lattice Boom Crane, 90 Ton		1511.00		1662.10		
2 Welder, Gas Engine, 300 amp		291.70		320.87	22.53	24.79
80 L.H., Daily Totals		$5979.50		$9067.77	$74.74	$113.35

Crew E-8	Hr.	Daily	Hr.	Daily	Bare Costs	Incl. O&P
1 Struc. Steel Foreman (outside)	$54.65	$437.20	$94.95	$759.60	$52.00	$87.87
4 Struc. Steel Workers	52.65	1684.80	91.45	2926.40		
1 Welder Foreman (outside)	54.65	437.20	94.95	759.60		
4 Welders	52.65	1684.80	91.45	2926.40		
1 Equip. Oper. (crane)	51.70	413.60	78.45	627.60		
1 Equip. Oper. (oiler)	45.20	361.60	68.55	548.40		
1 Equip. Oper. (light)	48.60	388.80	73.75	590.00		
1 Lattice Boom Crane, 90 Ton		1511.00		1662.10		
4 Welder, Gas Engine, 300 amp		583.40		641.74	20.14	22.15
104 L.H., Daily Totals		$7502.40		$11441.84	$72.14	$110.02

Crew E-9	Hr.	Daily	Hr.	Daily	Bare Costs	Incl. O&P
2 Struc. Steel Foremen (outside)	$54.65	$874.40	$94.95	$1519.20	$52.25	$88.76
5 Struc. Steel Workers	52.65	2106.00	91.45	3658.00		
1 Welder Foreman (outside)	54.65	437.20	94.95	759.60		
5 Welders	52.65	2106.00	91.45	3658.00		
1 Equip. Oper. (crane)	51.70	413.60	78.45	627.60		
1 Equip. Oper. (oiler)	45.20	361.60	68.55	548.40		
1 Equip. Oper. (light)	48.60	388.80	73.75	590.00		
1 Lattice Boom Crane, 90 Ton		1511.00		1662.10		
5 Welder, Gas Engine, 300 amp		729.25		802.17	17.50	19.25
128 L.H., Daily Totals		$8927.85		$13825.08	$69.75	$108.01

Crew E-10	Hr.	Daily	Hr.	Daily	Bare Costs	Incl. O&P
1 Welder Foreman (outside)	$54.65	$437.20	$94.95	$759.60	$53.65	$93.20
1 Welder	52.65	421.20	91.45	731.60		
1 Welder, Gas Engine, 300 amp		145.85		160.44		
1 Flatbed Truck, Gas, 3 Ton		303.20		333.52	28.07	30.87
16 L.H., Daily Totals		$1307.45		$1985.16	$81.72	$124.07

Crew E-11	Hr.	Daily	Hr.	Daily	Bare Costs	Incl. O&P
2 Painters, Struc. Steel	$41.45	$663.20	$72.95	$1167.20	$42.27	$69.38
1 Building Laborer	37.60	300.80	57.85	462.80		
1 Equip. Oper. (light)	48.60	388.80	73.75	590.00		
1 Air Compressor, 250 cfm		201.40		221.54		
1 Sandblaster, Portable, 3 C.F.		20.40		22.44		
1 Set Sand Blasting Accessories		14.05		15.46	7.37	8.11
32 L.H., Daily Totals		$1588.65		$2479.43	$49.65	$77.48

Crew E-11A	Hr.	Daily	Hr.	Daily	Bare Costs	Incl. O&P
2 Painters, Struc. Steel	$41.45	$663.20	$72.95	$1167.20	$42.27	$69.38
1 Building Laborer	37.60	300.80	57.85	462.80		
1 Equip. Oper. (light)	48.60	388.80	73.75	590.00		
1 Air Compressor, 250 cfm		201.40		221.54		
1 Sandblaster, Portable, 3 C.F.		20.40		22.44		
1 Set Sand Blasting Accessories		14.05		15.46		
1 Aerial Lift Truck, 60' Boom		435.60		479.16	20.98	23.08
32 L.H., Daily Totals		$2024.25		$2958.59	$63.26	$92.46

Crew E-11B	Hr.	Daily	Hr.	Daily	Bare Costs	Incl. O&P
2 Painters, Struc. Steel	$41.45	$663.20	$72.95	$1167.20	$40.17	$67.92
1 Building Laborer	37.60	300.80	57.85	462.80		
2 Paint Sprayer, 8 C.F.M.		99.80		109.78		
1 Aerial Lift Truck, 60' Boom		435.60		479.16	22.31	24.54
24 L.H., Daily Totals		$1499.40		$2218.94	$62.48	$92.46

Crew E-12	Hr.	Daily	Hr.	Daily	Bare Costs	Incl. O&P
1 Welder Foreman (outside)	$54.65	$437.20	$94.95	$759.60	$51.63	$84.35
1 Equip. Oper. (light)	48.60	388.80	73.75	590.00		
1 Welder, Gas Engine, 300 amp		145.85		160.44	9.12	10.03
16 L.H., Daily Totals		$971.85		$1510.04	$60.74	$94.38

Crew E-13	Hr.	Daily	Hr.	Daily	Bare Costs	Incl. O&P
1 Welder Foreman (outside)	$54.65	$437.20	$94.95	$759.60	$52.63	$87.88
.5 Equip. Oper. (light)	48.60	194.40	73.75	295.00		
1 Welder, Gas Engine, 300 amp		145.85		160.44	12.15	13.37
12 L.H., Daily Totals		$777.45		$1215.04	$64.79	$101.25

Crew E-14	Hr.	Daily	Hr.	Daily	Bare Costs	Incl. O&P
1 Welder Foreman (outside)	$54.65	$437.20	$94.95	$759.60	$54.65	$94.95
1 Welder, Gas Engine, 300 amp		145.85		160.44	18.23	20.05
8 L.H., Daily Totals		$583.05		$920.03	$72.88	$115.00

Crew E-16

Crew No.	Bare Costs Hr.	Daily	Incl. Subs O&P Hr.	Daily	Cost Per Labor-Hour Bare Costs	Incl. O&P
1 Welder Foreman (outside)	$54.65	$437.20	$94.95	$759.60	$53.65	$93.20
1 Welder	52.65	421.20	91.45	731.60		
1 Welder, Gas Engine, 300 amp		145.85		160.44	9.12	10.03
16 L.H., Daily Totals		$1004.25		$1651.64	$62.77	$103.23

Crew E-17

Crew No.	Bare Costs Hr.	Daily	Incl. Subs O&P Hr.	Daily	Cost Per Labor-Hour Bare Costs	Incl. O&P
1 Struc. Steel Foreman (outside)	$54.65	$437.20	$94.95	$759.60	$53.65	$93.20
1 Structural Steel Worker	52.65	421.20	91.45	731.60		
16 L.H., Daily Totals		$858.40		$1491.20	$53.65	$93.20

Crew E-18

Crew No.	Bare Costs Hr.	Daily	Incl. Subs O&P Hr.	Daily	Cost Per Labor-Hour Bare Costs	Incl. O&P
1 Struc. Steel Foreman (outside)	$54.65	$437.20	$94.95	$759.60	$52.64	$89.21
3 Structural Steel Workers	52.65	1263.60	91.45	2194.80		
1 Equipment Operator (med.)	50.60	404.80	76.75	614.00		
1 Lattice Boom Crane, 20 Ton		948.90		1043.79	23.72	26.09
40 L.H., Daily Totals		$3054.50		$4612.19	$76.36	$115.30

Crew E-19

Crew No.	Bare Costs Hr.	Daily	Incl. Subs O&P Hr.	Daily	Cost Per Labor-Hour Bare Costs	Incl. O&P
1 Struc. Steel Foreman (outside)	$54.65	$437.20	$94.95	$759.60	$51.97	$86.72
1 Structural Steel Worker	52.65	421.20	91.45	731.60		
1 Equip. Oper. (light)	48.60	388.80	73.75	590.00		
1 Lattice Boom Crane, 20 Ton		948.90		1043.79	39.54	43.49
24 L.H., Daily Totals		$2196.10		$3124.99	$91.50	$130.21

Crew E-20

Crew No.	Bare Costs Hr.	Daily	Incl. Subs O&P Hr.	Daily	Cost Per Labor-Hour Bare Costs	Incl. O&P
1 Struc. Steel Foreman (outside)	$54.65	$437.20	$94.95	$759.60	$51.85	$87.40
5 Structural Steel Workers	52.65	2106.00	91.45	3658.00		
1 Equip. Oper. (crane)	51.70	413.60	78.45	627.60		
1 Equip. Oper. (oiler)	45.20	361.60	68.55	548.40		
1 Lattice Boom Crane, 40 Ton		1164.00		1280.40	18.19	20.01
64 L.H., Daily Totals		$4482.40		$6874.00	$70.04	$107.41

Crew E-22

Crew No.	Bare Costs Hr.	Daily	Incl. Subs O&P Hr.	Daily	Cost Per Labor-Hour Bare Costs	Incl. O&P
1 Skilled Worker Foreman (out)	$50.65	$405.20	$78.25	$626.00	$49.32	$76.18
2 Skilled Workers	48.65	778.40	75.15	1202.40		
24 L.H., Daily Totals		$1183.60		$1828.40	$49.32	$76.18

Crew E-24

Crew No.	Bare Costs Hr.	Daily	Incl. Subs O&P Hr.	Daily	Cost Per Labor-Hour Bare Costs	Incl. O&P
3 Structural Steel Workers	$52.65	$1263.60	$91.45	$2194.80	$52.14	$87.78
1 Equipment Operator (med.)	50.60	404.80	76.75	614.00		
1 Hyd. Crane, 25 Ton		736.60		810.26	23.02	25.32
32 L.H., Daily Totals		$2405.00		$3619.06	$75.16	$113.10

Crew E-25

Crew No.	Bare Costs Hr.	Daily	Incl. Subs O&P Hr.	Daily	Cost Per Labor-Hour Bare Costs	Incl. O&P
1 Welder Foreman (outside)	$54.65	$437.20	$94.95	$759.60	$54.65	$94.95
1 Cutting Torch		11.40		12.54	1.43	1.57
8 L.H., Daily Totals		$448.60		$772.14	$56.08	$96.52

Crew F-3

Crew No.	Bare Costs Hr.	Daily	Incl. Subs O&P Hr.	Daily	Cost Per Labor-Hour Bare Costs	Incl. O&P
4 Carpenters	$46.95	$1502.40	$72.25	$2312.00	$47.90	$73.49
1 Equip. Oper. (crane)	51.70	413.60	78.45	627.60		
1 Hyd. Crane, 12 Ton		653.80		719.18	16.34	17.98
40 L.H., Daily Totals		$2569.80		$3658.78	$64.25	$91.47

Crew F-4

Crew No.	Bare Costs Hr.	Daily	Incl. Subs O&P Hr.	Daily	Cost Per Labor-Hour Bare Costs	Incl. O&P
4 Carpenters	$46.95	$1502.40	$72.25	$2312.00	$47.45	$72.67
1 Equip. Oper. (crane)	51.70	413.60	78.45	627.60		
1 Equip. Oper. (oiler)	45.20	361.60	68.55	548.40		
1 Hyd. Crane, 55 Ton		1128.00		1240.80	23.50	25.85
48 L.H., Daily Totals		$3405.60		$4728.80	$70.95	$98.52

Crew F-5

Crew No.	Bare Costs Hr.	Daily	Incl. Subs O&P Hr.	Daily	Cost Per Labor-Hour Bare Costs	Incl. O&P
1 Carpenter Foreman (outside)	$48.95	$391.60	$75.35	$602.80	$47.45	$73.03
3 Carpenters	46.95	1126.80	72.25	1734.00		
32 L.H., Daily Totals		$1518.40		$2336.80	$47.45	$73.03

Crew F-6

Crew No.	Bare Costs Hr.	Daily	Incl. Subs O&P Hr.	Daily	Cost Per Labor-Hour Bare Costs	Incl. O&P
2 Carpenters	$46.95	$751.20	$72.25	$1156.00	$44.16	$67.73
2 Building Laborers	37.60	601.60	57.85	925.60		
1 Equip. Oper. (crane)	51.70	413.60	78.45	627.60		
1 Hyd. Crane, 12 Ton		653.80		719.18	16.34	17.98
40 L.H., Daily Totals		$2420.20		$3428.38	$60.51	$85.71

Crew F-7

Crew No.	Bare Costs Hr.	Daily	Incl. Subs O&P Hr.	Daily	Cost Per Labor-Hour Bare Costs	Incl. O&P
2 Carpenters	$46.95	$751.20	$72.25	$1156.00	$42.27	$65.05
2 Building Laborers	37.60	601.60	57.85	925.60		
32 L.H., Daily Totals		$1352.80		$2081.60	$42.27	$65.05

Crew G-1

Crew No.	Bare Costs Hr.	Daily	Incl. Subs O&P Hr.	Daily	Cost Per Labor-Hour Bare Costs	Incl. O&P
1 Roofer Foreman (outside)	$42.10	$336.80	$71.85	$574.80	$37.51	$64.04
4 Roofers Composition	40.10	1283.20	68.45	2190.40		
2 Roofer Helpers	30.05	480.80	51.30	820.80		
1 Application Equipment		193.00		212.30		
1 Tar Kettle/Pot		169.80		186.78		
1 Crew Truck		206.40		227.04	10.16	11.18
56 L.H., Daily Totals		$2670.00		$4212.12	$47.68	$75.22

Crew G-2

Crew No.	Bare Costs Hr.	Daily	Incl. Subs O&P Hr.	Daily	Cost Per Labor-Hour Bare Costs	Incl. O&P
1 Plasterer	$42.95	$343.60	$64.40	$515.20	$39.55	$59.78
1 Plasterer Helper	38.10	304.80	57.10	456.80		
1 Building Laborer	37.60	300.80	57.85	462.80		
1 Grout Pump, 50 C.F./hr.		133.40		146.74	5.56	6.11
24 L.H., Daily Totals		$1082.60		$1581.54	$45.11	$65.90

Crew G-2A

Crew No.	Bare Costs Hr.	Daily	Incl. Subs O&P Hr.	Daily	Cost Per Labor-Hour Bare Costs	Incl. O&P
1 Roofer Composition	$40.10	$320.80	$68.45	$547.60	$35.92	$59.20
1 Roofer Helper	30.05	240.40	51.30	410.40		
1 Building Laborer	37.60	300.80	57.85	462.80		
1 Foam Spray Rig, Trailer-Mtd.		573.25		630.58		
1 Pickup Truck, 3/4 Ton		144.20		158.62	29.89	32.88
24 L.H., Daily Totals		$1579.45		$2209.99	$65.81	$92.08

Crew G-3

Crew No.	Bare Costs Hr.	Daily	Incl. Subs O&P Hr.	Daily	Cost Per Labor-Hour Bare Costs	Incl. O&P
2 Sheet Metal Workers	$55.95	$895.20	$85.50	$1368.00	$46.77	$71.67
2 Building Laborers	37.60	601.60	57.85	925.60		
32 L.H., Daily Totals		$1496.80		$2293.60	$46.77	$71.67

Crew G-4

Crew No.	Bare Costs Hr.	Daily	Incl. Subs O&P Hr.	Daily	Cost Per Labor-Hour Bare Costs	Incl. O&P
1 Labor Foreman (outside)	$39.60	$316.80	$60.95	$487.60	$38.27	$58.88
2 Building Laborers	37.60	601.60	57.85	925.60		
1 Flatbed Truck, Gas, 1.5 Ton		245.40		269.94		
1 Air Compressor, 160 cfm		156.40		172.04	16.74	18.42
24 L.H., Daily Totals		$1320.20		$1855.18	$55.01	$77.30

Crew G-5

Crew No.	Bare Costs Hr.	Daily	Incl. Subs O&P Hr.	Daily	Cost Per Labor-Hour Bare Costs	Incl. O&P
1 Roofer Foreman (outside)	$42.10	$336.80	$71.85	$574.80	$36.48	$62.27
2 Roofers Composition	40.10	641.60	68.45	1095.20		
2 Roofer Helpers	30.05	480.80	51.30	820.80		
1 Application Equipment		193.00		212.30	4.83	5.31
40 L.H., Daily Totals		$1652.20		$2703.10	$41.31	$67.58

Crew No.	Bare Costs Hr.	Bare Costs Daily	Incl. Subs O&P Hr.	Incl. Subs O&P Daily	Cost Per Labor-Hour Bare Costs	Cost Per Labor-Hour Incl. O&P
Crew G-6A						
2 Roofers Composition	$40.10	$641.60	$68.45	$1095.20	$40.10	$68.45
1 Small Compressor, Electric		12.95		14.24		
2 Pneumatic Nailers		55.30		60.83	4.27	4.69
16 L.H., Daily Totals		$709.85		$1170.28	$44.37	$73.14
Crew G-7						
1 Carpenter	$46.95	$375.60	$72.25	$578.00	$46.95	$72.25
1 Small Compressor, Electric		12.95		14.24		
1 Pneumatic Nailer		27.65		30.41	5.08	5.58
8 L.H., Daily Totals		$416.20		$622.66	$52.02	$77.83
Crew H-1						
2 Glaziers	$45.10	$721.60	$68.70	$1099.20	$48.88	$80.08
2 Struc. Steel Workers	52.65	842.40	91.45	1463.20		
32 L.H., Daily Totals		$1564.00		$2562.40	$48.88	$80.08
Crew H-2						
2 Glaziers	$45.10	$721.60	$68.70	$1099.20	$42.60	$65.08
1 Building Laborer	37.60	300.80	57.85	462.80		
24 L.H., Daily Totals		$1022.40		$1562.00	$42.60	$65.08
Crew H-3						
1 Glazier	$45.10	$360.80	$68.70	$549.60	$40.27	$61.83
1 Helper	35.45	283.60	54.95	439.60		
16 L.H., Daily Totals		$644.40		$989.20	$40.27	$61.83
Crew H-4						
1 Carpenter	$46.95	$375.60	$72.25	$578.00	$43.90	$67.27
1 Carpenter Helper	35.45	283.60	54.95	439.60		
.5 Electrician	54.70	218.80	81.95	327.80		
20 L.H., Daily Totals		$878.00		$1345.40	$43.90	$67.27
Crew J-1						
3 Plasterers	$42.95	$1030.80	$64.40	$1545.60	$41.01	$61.48
2 Plasterer Helpers	38.10	609.60	57.10	913.60		
1 Mixing Machine, 6 C.F.		140.20		154.22	3.50	3.86
40 L.H., Daily Totals		$1780.60		$2613.42	$44.52	$65.34
Crew J-2						
3 Plasterers	$42.95	$1030.80	$64.40	$1545.60	$41.34	$61.80
2 Plasterer Helpers	38.10	609.60	57.10	913.60		
1 Lather	43.00	344.00	63.40	507.20		
1 Mixing Machine, 6 C.F.		140.20		154.22	2.92	3.21
48 L.H., Daily Totals		$2124.60		$3120.62	$44.26	$65.01
Crew J-3						
1 Terrazzo Worker	$42.95	$343.60	$63.50	$508.00	$39.23	$58.00
1 Terrazzo Helper	35.50	284.00	52.50	420.00		
1 Floor Grinder, 22" Path		115.55		127.11		
1 Terrazzo Mixer		188.20		207.02	18.98	20.88
16 L.H., Daily Totals		$931.35		$1262.13	$58.21	$78.88
Crew J-4						
2 Cement Finishers	$45.00	$720.00	$66.65	$1066.40	$42.53	$63.72
1 Laborer	37.60	300.80	57.85	462.80		
1 Floor Grinder, 22" Path		115.55		127.11		
1 Floor Edger, 7" Path		39.30		43.23		
1 Vacuum Pick-Up System		60.90		66.99	8.99	9.89
24 L.H., Daily Totals		$1236.55		$1766.53	$51.52	$73.61

Crew No.	Bare Costs Hr.	Bare Costs Daily	Incl. Subs O&P Hr.	Incl. Subs O&P Daily	Cost Per Labor-Hour Bare Costs	Cost Per Labor-Hour Incl. O&P
Crew J-4A						
2 Cement Finishers	$45.00	$720.00	$66.65	$1066.40	$41.30	$62.25
2 Laborers	37.60	601.60	57.85	925.60		
1 Floor Grinder, 22" Path		115.55		127.11		
1 Floor Edger, 7" Path		39.30		43.23		
1 Vacuum Pick-Up System		60.90		66.99		
1 Floor Auto Scrubber		233.85		257.24	14.05	15.46
32 L.H., Daily Totals		$1771.20		$2486.56	$55.35	$77.70
Crew J-4B						
1 Laborer	$37.60	$300.80	$57.85	$462.80	$37.60	$57.85
1 Floor Auto Scrubber		233.85		257.24	29.23	32.15
8 L.H., Daily Totals		$534.65		$720.03	$66.83	$90.00
Crew J-6						
2 Painters	$40.35	$645.60	$60.80	$972.80	$41.73	$63.30
1 Building Laborer	37.60	300.80	57.85	462.80		
1 Equip. Oper. (light)	48.60	388.80	73.75	590.00		
1 Air Compressor, 250 cfm		201.40		221.54		
1 Sandblaster, Portable, 3 C.F.		20.40		22.44		
1 Set Sand Blasting Accessories		14.05		15.46	7.37	8.11
32 L.H., Daily Totals		$1571.05		$2285.03	$49.10	$71.41
Crew J-7						
2 Painters	$40.35	$645.60	$60.80	$972.80	$40.35	$60.80
1 Floor Belt Sander		14.20		15.62		
1 Floor Sanding Edger		12.60		13.86	1.68	1.84
16 L.H., Daily Totals		$672.40		$1002.28	$42.02	$62.64
Crew K-1						
1 Carpenter	$46.95	$375.60	$72.25	$578.00	$43.02	$65.67
1 Truck Driver (light)	39.10	312.80	59.10	472.80		
1 Flatbed Truck, Gas, 3 Ton		303.20		333.52	18.95	20.84
16 L.H., Daily Totals		$991.60		$1384.32	$61.98	$86.52
Crew K-2						
1 Struc. Steel Foreman (outside)	$54.65	$437.20	$94.95	$759.60	$48.80	$81.83
1 Struc. Steel Worker	52.65	421.20	91.45	731.60		
1 Truck Driver (light)	39.10	312.80	59.10	472.80		
1 Flatbed Truck, Gas, 3 Ton		303.20		333.52	12.63	13.90
24 L.H., Daily Totals		$1474.40		$2297.52	$61.43	$95.73
Crew L-1						
1 Electrician	$54.70	$437.60	$81.95	$655.60	$56.70	$85.30
1 Plumber	58.70	469.60	88.65	709.20		
16 L.H., Daily Totals		$907.20		$1364.80	$56.70	$85.30
Crew L-2						
1 Carpenter	$46.95	$375.60	$72.25	$578.00	$41.20	$63.60
1 Carpenter Helper	35.45	283.60	54.95	439.60		
16 L.H., Daily Totals		$659.20		$1017.60	$41.20	$63.60
Crew L-3						
1 Carpenter	$46.95	$375.60	$72.25	$578.00	$51.14	$77.99
.5 Electrician	54.70	218.80	81.95	327.80		
.5 Sheet Metal Worker	55.95	223.80	85.50	342.00		
16 L.H., Daily Totals		$818.20		$1247.80	$51.14	$77.99

Crew L-3A

Crew No.	Bare Costs		Incl. Subs O&P		Cost Per Labor-Hour	
	Hr.	Daily	Hr.	Daily	Bare Costs	Incl. O&P
1 Carpenter Foreman (outside)	$48.95	$391.60	$75.35	$602.80	$51.28	$78.73
.5 Sheet Metal Worker	55.95	223.80	85.50	342.00		
12 L.H., Daily Totals		$615.40		$944.80	$51.28	$78.73

Crew L-4

Crew No.	Hr.	Daily	Hr.	Daily	Bare Costs	Incl. O&P
2 Skilled Workers	$48.65	$778.40	$75.15	$1202.40	$44.25	$68.42
1 Helper	35.45	283.60	54.95	439.60		
24 L.H., Daily Totals		$1062.00		$1642.00	$44.25	$68.42

Crew L-5

Crew No.	Hr.	Daily	Hr.	Daily	Bare Costs	Incl. O&P
1 Struc. Steel Foreman (outside)	$54.65	$437.20	$94.95	$759.60	$52.80	$90.09
5 Struc. Steel Workers	52.65	2106.00	91.45	3658.00		
1 Equip. Oper. (crane)	51.70	413.60	78.45	627.60		
1 Hyd. Crane, 25 Ton		736.60		810.26	13.15	14.47
56 L.H., Daily Totals		$3693.40		$5855.46	$65.95	$104.56

Crew L-5A

Crew No.	Hr.	Daily	Hr.	Daily	Bare Costs	Incl. O&P
1 Struc. Steel Foreman (outside)	$54.65	$437.20	$94.95	$759.60	$52.91	$89.08
2 Structural Steel Workers	52.65	842.40	91.45	1463.20		
1 Equip. Oper. (crane)	51.70	413.60	78.45	627.60		
1 S.P. Crane, 4x4, 25 Ton		599.40		659.34	18.73	20.60
32 L.H., Daily Totals		$2292.60		$3509.74	$71.64	$109.68

Crew L-5B

Crew No.	Hr.	Daily	Hr.	Daily	Bare Costs	Incl. O&P
1 Struc. Steel Foreman (outside)	$54.65	$437.20	$94.95	$759.60	$53.97	$85.46
2 Structural Steel Workers	52.65	842.40	91.45	1463.20		
2 Electricians	54.70	875.20	81.95	1311.20		
2 Steamfitters/Pipefitters	59.75	956.00	90.20	1443.20		
1 Equip. Oper. (crane)	51.70	413.60	78.45	627.60		
1 Equip. Oper. (oiler)	45.20	361.60	68.55	548.40		
1 Hyd. Crane, 80 Ton		1625.00		1787.50	22.57	24.83
72 L.H., Daily Totals		$5511.00		$7940.70	$76.54	$110.29

Crew L-6

Crew No.	Hr.	Daily	Hr.	Daily	Bare Costs	Incl. O&P
1 Plumber	$58.70	$469.60	$88.65	$709.20	$57.37	$86.42
.5 Electrician	54.70	218.80	81.95	327.80		
12 L.H., Daily Totals		$688.40		$1037.00	$57.37	$86.42

Crew L-7

Crew No.	Hr.	Daily	Hr.	Daily	Bare Costs	Incl. O&P
2 Carpenters	$46.95	$751.20	$72.25	$1156.00	$45.39	$69.52
1 Building Laborer	37.60	300.80	57.85	462.80		
.5 Electrician	54.70	218.80	81.95	327.80		
28 L.H., Daily Totals		$1270.80		$1946.60	$45.39	$69.52

Crew L-8

Crew No.	Hr.	Daily	Hr.	Daily	Bare Costs	Incl. O&P
2 Carpenters	$46.95	$751.20	$72.25	$1156.00	$49.30	$75.53
.5 Plumber	58.70	234.80	88.65	354.60		
20 L.H., Daily Totals		$986.00		$1510.60	$49.30	$75.53

Crew L-9

Crew No.	Hr.	Daily	Hr.	Daily	Bare Costs	Incl. O&P
1 Labor Foreman (inside)	$38.10	$304.80	$58.65	$469.20	$42.96	$68.17
2 Building Laborers	37.60	601.60	57.85	925.60		
1 Struc. Steel Worker	52.65	421.20	91.45	731.60		
.5 Electrician	54.70	218.80	81.95	327.80		
36 L.H., Daily Totals		$1546.40		$2454.20	$42.96	$68.17

Crew L-10

Crew No.	Hr.	Daily	Hr.	Daily	Bare Costs	Incl. O&P
1 Struc. Steel Foreman (outside)	$54.65	$437.20	$94.95	$759.60	$53.00	$88.28
1 Structural Steel Worker	52.65	421.20	91.45	731.60		
1 Equip. Oper. (crane)	51.70	413.60	78.45	627.60		
1 Hyd. Crane, 12 Ton		653.80		719.18	27.24	29.97
24 L.H., Daily Totals		$1925.80		$2837.98	$80.24	$118.25

Crew L-11

Crew No.	Hr.	Daily	Hr.	Daily	Bare Costs	Incl. O&P
2 Wreckers	$37.60	$601.60	$62.10	$993.60	$43.88	$69.10
1 Equip. Oper. (crane)	51.70	413.60	78.45	627.60		
1 Equip. Oper. (light)	48.60	388.80	73.75	590.00		
1 Hyd. Excavator, 2.5 C.Y.		1605.00		1765.50		
1 Loader, Skid Steer, 78 H.P.		318.60		350.46	60.11	66.12
32 L.H., Daily Totals		$3327.60		$4327.16	$103.99	$135.22

Crew M-1

Crew No.	Hr.	Daily	Hr.	Daily	Bare Costs	Incl. O&P
3 Elevator Constructors	$76.50	$1836.00	$114.20	$2740.80	$72.67	$108.49
1 Elevator Apprentice	61.20	489.60	91.35	730.80		
5 Hand Tools		46.00		50.60	1.44	1.58
32 L.H., Daily Totals		$2371.60		$3522.20	$74.11	$110.07

Crew M-3

Crew No.	Hr.	Daily	Hr.	Daily	Bare Costs	Incl. O&P
1 Electrician Foreman (outside)	$56.70	$453.60	$84.95	$679.60	$57.56	$86.48
1 Common Laborer	37.60	300.80	57.85	462.80		
.25 Equipment Operator (med.)	50.60	101.20	76.75	153.50		
1 Elevator Constructor	76.50	612.00	114.20	913.60		
1 Elevator Apprentice	61.20	489.60	91.35	730.80		
.25 S.P. Crane, 4x4, 20 Ton		138.65		152.51	4.08	4.49
34 L.H., Daily Totals		$2095.85		$3092.82	$61.64	$90.97

Crew M-4

Crew No.	Hr.	Daily	Hr.	Daily	Bare Costs	Incl. O&P
1 Electrician Foreman (outside)	$56.70	$453.60	$84.95	$679.60	$56.94	$85.58
1 Common Laborer	37.60	300.80	57.85	462.80		
.25 Equipment Operator, Crane	51.70	103.40	78.45	156.90		
.25 Equip. Oper. (oiler)	45.20	90.40	68.55	137.10		
1 Elevator Constructor	76.50	612.00	114.20	913.60		
1 Elevator Apprentice	61.20	489.60	91.35	730.80		
.25 S.P. Crane, 4x4, 40 Ton		175.75		193.32	4.88	5.37
36 L.H., Daily Totals		$2225.55		$3274.13	$61.82	$90.95

Crew Q-1

Crew No.	Hr.	Daily	Hr.	Daily	Bare Costs	Incl. O&P
1 Plumber	$58.70	$469.60	$88.65	$709.20	$52.83	$79.78
1 Plumber Apprentice	46.95	375.60	70.90	567.20		
16 L.H., Daily Totals		$845.20		$1276.40	$52.83	$79.78

Crew Q-1A

Crew No.	Hr.	Daily	Hr.	Daily	Bare Costs	Incl. O&P
.25 Plumber Foreman (outside)	$60.70	$121.40	$91.65	$183.30	$59.10	$89.25
1 Plumber	58.70	469.60	88.65	709.20		
10 L.H., Daily Totals		$591.00		$892.50	$59.10	$89.25

Crew Q-1C

Crew No.	Hr.	Daily	Hr.	Daily	Bare Costs	Incl. O&P
1 Plumber	$58.70	$469.60	$88.65	$709.20	$52.08	$78.77
1 Plumber Apprentice	46.95	375.60	70.90	567.20		
1 Equip. Oper. (medium)	50.60	404.80	76.75	614.00		
1 Trencher, Chain Type, 8' D		3402.00		3742.20	141.75	155.93
24 L.H., Daily Totals		$4652.00		$5632.60	$193.83	$234.69

Crew Q-2

Crew No.	Hr.	Daily	Hr.	Daily	Bare Costs	Incl. O&P
2 Plumbers	$58.70	$939.20	$88.65	$1418.40	$54.78	$82.73
1 Plumber Apprentice	46.95	375.60	70.90	567.20		
24 L.H., Daily Totals		$1314.80		$1985.60	$54.78	$82.73

Crew No.	Bare Costs		Incl. Subs O&P		Cost Per Labor-Hour	

Crew Q-3

	Hr.	Daily	Hr.	Daily	Bare Costs	Incl. O&P
1 Plumber Foreman (inside)	$59.20	$473.60	$89.40	$715.20	$55.89	$84.40
2 Plumbers	58.70	939.20	88.65	1418.40		
1 Plumber Apprentice	46.95	375.60	70.90	567.20		
32 L.H., Daily Totals		$1788.40		$2700.80	$55.89	$84.40

Crew Q-4

	Hr.	Daily	Hr.	Daily	Bare Costs	Incl. O&P
1 Plumber Foreman (inside)	$59.20	$473.60	$89.40	$715.20	$55.89	$84.40
1 Plumber	58.70	469.60	88.65	709.20		
1 Welder (plumber)	58.70	469.60	88.65	709.20		
1 Plumber Apprentice	46.95	375.60	70.90	567.20		
1 Welder, Electric, 300 amp		57.70		63.47	1.80	1.98
32 L.H., Daily Totals		$1846.10		$2764.27	$57.69	$86.38

Crew Q-5

	Hr.	Daily	Hr.	Daily	Bare Costs	Incl. O&P
1 Steamfitter	$59.75	$478.00	$90.20	$721.60	$53.77	$81.20
1 Steamfitter Apprentice	47.80	382.40	72.20	577.60		
16 L.H., Daily Totals		$860.40		$1299.20	$53.77	$81.20

Crew Q-6

	Hr.	Daily	Hr.	Daily	Bare Costs	Incl. O&P
2 Steamfitters	$59.75	$956.00	$90.20	$1443.20	$55.77	$84.20
1 Steamfitter Apprentice	47.80	382.40	72.20	577.60		
24 L.H., Daily Totals		$1338.40		$2020.80	$55.77	$84.20

Crew Q-7

	Hr.	Daily	Hr.	Daily	Bare Costs	Incl. O&P
1 Steamfitter Foreman (inside)	$60.25	$482.00	$91.00	$728.00	$56.89	$85.90
2 Steamfitters	59.75	956.00	90.20	1443.20		
1 Steamfitter Apprentice	47.80	382.40	72.20	577.60		
32 L.H., Daily Totals		$1820.40		$2748.80	$56.89	$85.90

Crew Q-8

	Hr.	Daily	Hr.	Daily	Bare Costs	Incl. O&P
1 Steamfitter Foreman (inside)	$60.25	$482.00	$91.00	$728.00	$56.89	$85.90
1 Steamfitter	59.75	478.00	90.20	721.60		
1 Welder (steamfitter)	59.75	478.00	90.20	721.60		
1 Steamfitter Apprentice	47.80	382.40	72.20	577.60		
1 Welder, Electric, 300 amp		57.70		63.47	1.80	1.98
32 L.H., Daily Totals		$1878.10		$2812.27	$58.69	$87.88

Crew Q-9

	Hr.	Daily	Hr.	Daily	Bare Costs	Incl. O&P
1 Sheet Metal Worker	$55.95	$447.60	$85.50	$684.00	$50.35	$76.95
1 Sheet Metal Apprentice	44.75	358.00	68.40	547.20		
16 L.H., Daily Totals		$805.60		$1231.20	$50.35	$76.95

Crew Q-10

	Hr.	Daily	Hr.	Daily	Bare Costs	Incl. O&P
2 Sheet Metal Workers	$55.95	$895.20	$85.50	$1368.00	$52.22	$79.80
1 Sheet Metal Apprentice	44.75	358.00	68.40	547.20		
24 L.H., Daily Totals		$1253.20		$1915.20	$52.22	$79.80

Crew Q-11

	Hr.	Daily	Hr.	Daily	Bare Costs	Incl. O&P
1 Sheet Metal Foreman (inside)	$56.45	$451.60	$86.25	$690.00	$53.27	$81.41
2 Sheet Metal Workers	55.95	895.20	85.50	1368.00		
1 Sheet Metal Apprentice	44.75	358.00	68.40	547.20		
32 L.H., Daily Totals		$1704.80		$2605.20	$53.27	$81.41

Crew Q-12

	Hr.	Daily	Hr.	Daily	Bare Costs	Incl. O&P
1 Sprinkler Installer	$56.15	$449.20	$84.95	$679.60	$50.52	$76.45
1 Sprinkler Apprentice	44.90	359.20	67.95	543.60		
16 L.H., Daily Totals		$808.40		$1223.20	$50.52	$76.45

Crew Q-13

	Hr.	Daily	Hr.	Daily	Bare Costs	Incl. O&P
1 Sprinkler Foreman (inside)	$56.65	$453.20	$85.70	$685.60	$53.46	$80.89
2 Sprinkler Installers	56.15	898.40	84.95	1359.20		
1 Sprinkler Apprentice	44.90	359.20	67.95	543.60		
32 L.H., Daily Totals		$1710.80		$2588.40	$53.46	$80.89

Crew Q-14

	Hr.	Daily	Hr.	Daily	Bare Costs	Incl. O&P
1 Asbestos Worker	$52.35	$418.80	$81.50	$652.00	$47.13	$73.38
1 Asbestos Apprentice	41.90	335.20	65.25	522.00		
16 L.H., Daily Totals		$754.00		$1174.00	$47.13	$73.38

Crew Q-15

	Hr.	Daily	Hr.	Daily	Bare Costs	Incl. O&P
1 Plumber	$58.70	$469.60	$88.65	$709.20	$52.83	$79.78
1 Plumber Apprentice	46.95	375.60	70.90	567.20		
1 Welder, Electric, 300 amp		57.70		63.47	3.61	3.97
16 L.H., Daily Totals		$902.90		$1339.87	$56.43	$83.74

Crew Q-16

	Hr.	Daily	Hr.	Daily	Bare Costs	Incl. O&P
2 Plumbers	$58.70	$939.20	$88.65	$1418.40	$54.78	$82.73
1 Plumber Apprentice	46.95	375.60	70.90	567.20		
1 Welder, Electric, 300 amp		57.70		63.47	2.40	2.64
24 L.H., Daily Totals		$1372.50		$2049.07	$57.19	$85.38

Crew Q-17

	Hr.	Daily	Hr.	Daily	Bare Costs	Incl. O&P
1 Steamfitter	$59.75	$478.00	$90.20	$721.60	$53.77	$81.20
1 Steamfitter Apprentice	47.80	382.40	72.20	577.60		
1 Welder, Electric, 300 amp		57.70		63.47	3.61	3.97
16 L.H., Daily Totals		$918.10		$1362.67	$57.38	$85.17

Crew Q-17A

	Hr.	Daily	Hr.	Daily	Bare Costs	Incl. O&P
1 Steamfitter	$59.75	$478.00	$90.20	$721.60	$53.08	$80.28
1 Steamfitter Apprentice	47.80	382.40	72.20	577.60		
1 Equip. Oper. (crane)	51.70	413.60	78.45	627.60		
1 Hyd. Crane, 12 Ton		653.80		719.18		
1 Welder, Electric, 300 amp		57.70		63.47	29.65	32.61
24 L.H., Daily Totals		$1985.50		$2709.45	$82.73	$112.89

Crew Q-18

	Hr.	Daily	Hr.	Daily	Bare Costs	Incl. O&P
2 Steamfitters	$59.75	$956.00	$90.20	$1443.20	$55.77	$84.20
1 Steamfitter Apprentice	47.80	382.40	72.20	577.60		
1 Welder, Electric, 300 amp		57.70		63.47	2.40	2.64
24 L.H., Daily Totals		$1396.10		$2084.27	$58.17	$86.84

Crew Q-19

	Hr.	Daily	Hr.	Daily	Bare Costs	Incl. O&P
1 Steamfitter	$59.75	$478.00	$90.20	$721.60	$54.08	$81.45
1 Steamfitter Apprentice	47.80	382.40	72.20	577.60		
1 Electrician	54.70	437.60	81.95	655.60		
24 L.H., Daily Totals		$1298.00		$1954.80	$54.08	$81.45

Crew Q-20

	Hr.	Daily	Hr.	Daily	Bare Costs	Incl. O&P
1 Sheet Metal Worker	$55.95	$447.60	$85.50	$684.00	$51.22	$77.95
1 Sheet Metal Apprentice	44.75	358.00	68.40	547.20		
.5 Electrician	54.70	218.80	81.95	327.80		
20 L.H., Daily Totals		$1024.40		$1559.00	$51.22	$77.95

Crew Q-21

	Hr.	Daily	Hr.	Daily	Bare Costs	Incl. O&P
2 Steamfitters	$59.75	$956.00	$90.20	$1443.20	$55.50	$83.64
1 Steamfitter Apprentice	47.80	382.40	72.20	577.60		
1 Electrician	54.70	437.60	81.95	655.60		
32 L.H., Daily Totals		$1776.00		$2676.40	$55.50	$83.64

652

For customer support on your Mechanical Cost Data, call 877.795.7394.

Crew Q-22

Crew No.	Bare Costs Hr.	Daily	Incl. Subs O&P Hr.	Daily	Cost Per L.H. Bare Costs	Incl. O&P
1 Plumber	$58.70	$469.60	$88.65	$709.20	$52.83	$79.78
1 Plumber Apprentice	46.95	375.60	70.90	567.20		
1 Hyd. Crane, 12 Ton		653.80		719.18	40.86	44.95
16 L.H., Daily Totals		$1499.00		$1995.58	$93.69	$124.72

Crew Q-22A

Crew No.	Bare Costs Hr.	Daily	Incl. Subs O&P Hr.	Daily	Cost Per L.H. Bare Costs	Incl. O&P
1 Plumber	$58.70	$469.60	$88.65	$709.20	$48.74	$73.96
1 Plumber Apprentice	46.95	375.60	70.90	567.20		
1 Laborer	37.60	300.80	57.85	462.80		
1 Equip. Oper. (crane)	51.70	413.60	78.45	627.60		
1 Hyd. Crane, 12 Ton		653.80		719.18	20.43	22.47
32 L.H., Daily Totals		$2213.40		$3085.98	$69.17	$96.44

Crew Q-23

Crew No.	Bare Costs Hr.	Daily	Incl. Subs O&P Hr.	Daily	Cost Per L.H. Bare Costs	Incl. O&P
1 Plumber Foreman (outside)	$60.70	$485.60	$91.65	$733.20	$56.67	$85.68
1 Plumber	58.70	469.60	88.65	709.20		
1 Equip. Oper. (medium)	50.60	404.80	76.75	614.00		
1 Lattice Boom Crane, 20 Ton		948.90		1043.79	39.54	43.49
24 L.H., Daily Totals		$2308.90		$3100.19	$96.20	$129.17

Crew R-1

Crew No.	Bare Costs Hr.	Daily	Incl. Subs O&P Hr.	Daily	Cost Per L.H. Bare Costs	Incl. O&P
1 Electrician Foreman	$55.20	$441.60	$82.70	$661.60	$48.37	$73.08
3 Electricians	54.70	1312.80	81.95	1966.80		
2 Helpers	35.45	567.20	54.95	879.20		
48 L.H., Daily Totals		$2321.60		$3507.60	$48.37	$73.08

Crew R-1A

Crew No.	Bare Costs Hr.	Daily	Incl. Subs O&P Hr.	Daily	Cost Per L.H. Bare Costs	Incl. O&P
1 Electrician	$54.70	$437.60	$81.95	$655.60	$45.08	$68.45
1 Helper	35.45	283.60	54.95	439.60		
16 L.H., Daily Totals		$721.20		$1095.20	$45.08	$68.45

Crew R-2

Crew No.	Bare Costs Hr.	Daily	Incl. Subs O&P Hr.	Daily	Cost Per L.H. Bare Costs	Incl. O&P
1 Electrician Foreman	$55.20	$441.60	$82.70	$661.60	$48.84	$73.84
3 Electricians	54.70	1312.80	81.95	1966.80		
2 Helpers	35.45	567.20	54.95	879.20		
1 Equip. Oper. (crane)	51.70	413.60	78.45	627.60		
1 S.P. Crane, 4x4, 5 Ton		275.40		302.94	4.92	5.41
56 L.H., Daily Totals		$3010.60		$4438.14	$53.76	$79.25

Crew R-3

Crew No.	Bare Costs Hr.	Daily	Incl. Subs O&P Hr.	Daily	Cost Per L.H. Bare Costs	Incl. O&P
1 Electrician Foreman	$55.20	$441.60	$82.70	$661.60	$54.30	$81.55
1 Electrician	54.70	437.60	81.95	655.60		
.5 Equip. Oper. (crane)	51.70	206.80	78.45	313.80		
.5 S.P. Crane, 4x4, 5 Ton		137.70		151.47	6.88	7.57
20 L.H., Daily Totals		$1223.70		$1782.47	$61.19	$89.12

Crew R-4

Crew No.	Bare Costs Hr.	Daily	Incl. Subs O&P Hr.	Daily	Cost Per L.H. Bare Costs	Incl. O&P
1 Struc. Steel Foreman (outside)	$54.65	$437.20	$94.95	$759.60	$53.46	$90.25
3 Struc. Steel Workers	52.65	1263.60	91.45	2194.80		
1 Electrician	54.70	437.60	81.95	655.60		
1 Welder, Gas Engine, 300 amp		145.85		160.44	3.65	4.01
40 L.H., Daily Totals		$2284.25		$3770.43	$57.11	$94.26

Crew R-5

Crew No.	Bare Costs Hr.	Daily	Incl. Subs O&P Hr.	Daily	Cost Per L.H. Bare Costs	Incl. O&P
1 Electrician Foreman	$55.20	$441.60	$82.70	$661.60	$47.75	$72.20
4 Electrician Linemen	54.70	1750.40	81.95	2622.40		
2 Electrician Operators	54.70	875.20	81.95	1311.20		
4 Electrician Groundmen	35.45	1134.40	54.95	1758.40		
1 Crew Truck		206.40		227.04		
1 Flatbed Truck, 20,000 GVW		248.60		273.46		
1 Pickup Truck, 3/4 Ton		144.20		158.62		
.2 Hyd. Crane, 55 Ton		225.60		248.16		
.2 Hyd. Crane, 12 Ton		130.76		143.84		
.2 Earth Auger, Truck-Mtd.		83.44		91.78		
1 Tractor w/Winch		427.00		469.70	16.66	18.32
88 L.H., Daily Totals		$5667.60		$7966.20	$64.40	$90.53

Crew R-6

Crew No.	Bare Costs Hr.	Daily	Incl. Subs O&P Hr.	Daily	Cost Per L.H. Bare Costs	Incl. O&P
1 Electrician Foreman	$55.20	$441.60	$82.70	$661.60	$47.75	$72.20
4 Electrician Linemen	54.70	1750.40	81.95	2622.40		
2 Electrician Operators	54.70	875.20	81.95	1311.20		
4 Electrician Groundmen	35.45	1134.40	54.95	1758.40		
1 Crew Truck		206.40		227.04		
1 Flatbed Truck, 20,000 GVW		248.60		273.46		
1 Pickup Truck, 3/4 Ton		144.20		158.62		
.2 Hyd. Crane, 55 Ton		225.60		248.16		
.2 Hyd. Crane, 12 Ton		130.76		143.84		
.2 Earth Auger, Truck-Mtd.		83.44		91.78		
1 Tractor w/Winch		427.00		469.70		
3 Cable Trailers		598.20		658.02		
.5 Tensioning Rig		198.40		218.24		
.5 Cable Pulling Rig		1160.00		1276.00	38.89	42.78
88 L.H., Daily Totals		$7624.20		$10118.46	$86.64	$114.98

Crew R-7

Crew No.	Bare Costs Hr.	Daily	Incl. Subs O&P Hr.	Daily	Cost Per L.H. Bare Costs	Incl. O&P
1 Electrician Foreman	$55.20	$441.60	$82.70	$661.60	$38.74	$59.58
5 Electrician Groundmen	35.45	1418.00	54.95	2198.00		
1 Crew Truck		206.40		227.04	4.30	4.73
48 L.H., Daily Totals		$2066.00		$3086.64	$43.04	$64.31

Crew R-8

Crew No.	Bare Costs Hr.	Daily	Incl. Subs O&P Hr.	Daily	Cost Per L.H. Bare Costs	Incl. O&P
1 Electrician Foreman	$55.20	$441.60	$82.70	$661.60	$48.37	$73.08
3 Electrician Linemen	54.70	1312.80	81.95	1966.80		
2 Electrician Groundmen	35.45	567.20	54.95	879.20		
1 Pickup Truck, 3/4 Ton		144.20		158.62		
1 Crew Truck		206.40		227.04	7.30	8.03
48 L.H., Daily Totals		$2672.20		$3893.26	$55.67	$81.11

Crew R-9

Crew No.	Bare Costs Hr.	Daily	Incl. Subs O&P Hr.	Daily	Cost Per L.H. Bare Costs	Incl. O&P
1 Electrician Foreman	$55.20	$441.60	$82.70	$661.60	$45.14	$68.54
1 Electrician Lineman	54.70	437.60	81.95	655.60		
2 Electrician Operators	54.70	875.20	81.95	1311.20		
4 Electrician Groundmen	35.45	1134.40	54.95	1758.40		
1 Pickup Truck, 3/4 Ton		144.20		158.62		
1 Crew Truck		206.40		227.04	5.48	6.03
64 L.H., Daily Totals		$3239.40		$4772.46	$50.62	$74.57

Crew R-10

Crew No.	Bare Costs Hr.	Daily	Incl. Subs O&P Hr.	Daily	Cost Per L.H. Bare Costs	Incl. O&P
1 Electrician Foreman	$55.20	$441.60	$82.70	$661.60	$51.58	$77.58
4 Electrician Linemen	54.70	1750.40	81.95	2622.40		
1 Electrician Groundman	35.45	283.60	54.95	439.60		
1 Crew Truck		206.40		227.04		
3 Tram Cars		412.20		453.42	12.89	14.18
48 L.H., Daily Totals		$3094.20		$4404.06	$64.46	$91.75

Crew R-11

Crew No.	Bare Costs		Incl. Subs O&P		Cost Per Labor-Hour	
Crew R-11	Hr.	Daily	Hr.	Daily	Bare Costs	Incl. O&P
1 Electrician Foreman	$55.20	$441.60	$82.70	$661.60	$51.90	$78.11
4 Electricians	54.70	1750.40	81.95	2622.40		
1 Equip. Oper. (crane)	51.70	413.60	78.45	627.60		
1 Common Laborer	37.60	300.80	57.85	462.80		
1 Crew Truck		206.40		227.04		
1 Hyd. Crane, 12 Ton		653.80		719.18	15.36	16.90
56 L.H., Daily Totals		$3766.60		$5320.62	$67.26	$95.01

Crew R-12

Crew No.	Hr.	Daily	Hr.	Daily	Bare Costs	Incl. O&P
1 Carpenter Foreman (inside)	$47.45	$379.60	$73.05	$584.40	$44.45	$69.24
4 Carpenters	46.95	1502.40	72.25	2312.00		
4 Common Laborers	37.60	1203.20	57.85	1851.20		
1 Equip. Oper. (medium)	50.60	404.80	76.75	614.00		
1 Steel Worker	52.65	421.20	91.45	731.60		
1 Dozer, 200 H.P.		1387.00		1525.70		
1 Pickup Truck, 3/4 Ton		144.20		158.62	17.40	19.14
88 L.H., Daily Totals		$5442.40		$7777.52	$61.85	$88.38

Crew R-13

Crew No.	Hr.	Daily	Hr.	Daily	Bare Costs	Incl. O&P
1 Electrician Foreman	$55.20	$441.60	$82.70	$661.60	$52.84	$79.37
3 Electricians	54.70	1312.80	81.95	1966.80		
.25 Equip. Oper. (crane)	51.70	103.40	78.45	156.90		
1 Equipment Oiler	45.20	361.60	68.55	548.40		
.25 Hydraulic Crane, 33 Ton		187.50		206.25	4.46	4.91
42 L.H., Daily Totals		$2406.90		$3539.95	$57.31	$84.28

Crew R-15

Crew No.	Hr.	Daily	Hr.	Daily	Bare Costs	Incl. O&P
1 Electrician Foreman	$55.20	$441.60	$82.70	$661.60	$53.77	$80.71
4 Electricians	54.70	1750.40	81.95	2622.40		
1 Equipment Oper. (light)	48.60	388.80	73.75	590.00		
1 Aerial Lift Truck, 40' Boom		301.40		331.54	6.28	6.91
48 L.H., Daily Totals		$2882.20		$4205.54	$60.05	$87.62

Crew R-18

Crew No.	Hr.	Daily	Hr.	Daily	Bare Costs	Incl. O&P
.25 Electrician Foreman	$55.20	$110.40	$82.70	$165.40	$42.89	$65.39
1 Electrician	54.70	437.60	81.95	655.60		
2 Helpers	35.45	567.20	54.95	879.20		
26 L.H., Daily Totals		$1115.20		$1700.20	$42.89	$65.39

Crew R-19

Crew No.	Hr.	Daily	Hr.	Daily	Bare Costs	Incl. O&P
.5 Electrician Foreman	$55.20	$220.80	$82.70	$330.80	$54.80	$82.10
2 Electricians	54.70	875.20	81.95	1311.20		
20 L.H., Daily Totals		$1096.00		$1642.00	$54.80	$82.10

Crew R-21

Crew No.	Hr.	Daily	Hr.	Daily	Bare Costs	Incl. O&P
1 Electrician Foreman	$55.20	$441.60	$82.70	$661.60	$54.72	$82.01
3 Electricians	54.70	1312.80	81.95	1966.80		
.1 Equip. Oper. (medium)	50.60	40.48	76.75	61.40		
.1 S.P. Crane, 4x4, 25 Ton		59.94		65.93	1.83	2.01
32.8 L.H., Daily Totals		$1854.82		$2755.73	$56.55	$84.02

Crew R-22

Crew No.	Hr.	Daily	Hr.	Daily	Bare Costs	Incl. O&P
.66 Electrician Foreman	$55.20	$291.46	$82.70	$436.66	$46.51	$70.47
2 Electricians	54.70	875.20	81.95	1311.20		
2 Helpers	35.45	567.20	54.95	879.20		
37.28 L.H., Daily Totals		$1733.86		$2627.06	$46.51	$70.47

Crew R-30

Crew No.	Bare Costs		Incl. Subs O&P		Cost Per Labor-Hour	
Crew R-30	Hr.	Daily	Hr.	Daily	Bare Costs	Incl. O&P
.25 Electrician Foreman (outside)	$56.70	$113.40	$84.95	$169.90	$44.33	$67.35
1 Electrician	54.70	437.60	81.95	655.60		
2 Laborers, (Semi-Skilled)	37.60	601.60	57.85	925.60		
26 L.H., Daily Totals		$1152.60		$1751.10	$44.33	$67.35

Crew R-31

Crew No.	Hr.	Daily	Hr.	Daily	Bare Costs	Incl. O&P
1 Electrician	$54.70	$437.60	$81.95	$655.60	$54.70	$81.95
1 Core Drill, Electric, 2.5 H.P.		55.15		60.66	6.89	7.58
8 L.H., Daily Totals		$492.75		$716.26	$61.59	$89.53

Crew W-41E

Crew No.	Hr.	Daily	Hr.	Daily	Bare Costs	Incl. O&P
.5 Plumber Foreman (outside)	$60.70	$242.80	$91.65	$366.60	$50.66	$76.93
1 Plumber	58.70	469.60	88.65	709.20		
1 Laborer	37.60	300.80	57.85	462.80		
20 L.H., Daily Totals		$1013.20		$1538.60	$50.66	$76.93

Historical Cost Indexes

The table below lists both the RSMeans® historical cost index based on Jan. 1, 1993 = 100 as well as the computed value of an index based on Jan. 1, 2015 costs. Since the Jan. 1, 2015 figure is estimated, space is left to write in the actual index figures as they become available through either the quarterly *RSMeans Construction Cost Indexes* or as printed in the *Engineering News-Record.* To compute the actual index based on Jan. 1, 2015 = 100, divide the historical cost index for a particular year by the actual Jan. 1, 2015 construction cost index. Space has been left to advance the index figures as the year progresses.

Year	Historical Cost Index Jan. 1, 1993 = 100		Current Index Based on Jan. 1, 2015 = 100		Year	Historical Cost Index Jan. 1, 1993 = 100	Current Index Based on Jan. 1, 2015 = 100		Year	Historical Cost Index Jan. 1, 1993 = 100	Current Index Based on Jan. 1, 2015 = 100	
	Est.	Actual	Est.	Actual		Actual	Est.	Actual		Actual	Est.	Actual
Oct 2015*					July 2000	120.9	58.5		July 1982	76.1	36.8	
July 2015*					1999	117.6	56.9		1981	70.0	33.9	
April 2015*					1998	115.1	55.7		1980	62.9	30.4	
Jan 2015*	206.7		100.0	100.0	1997	112.8	54.6		1979	57.8	28.0	
July 2014		204.9	99.1		1996	110.2	53.3		1978	53.5	25.9	
2013		201.2	97.3		1995	107.6	52.1		1977	49.5	23.9	
2012		194.6	94.1		1994	104.4	50.5		1976	46.9	22.7	
2011		191.2	92.5		1993	101.7	49.2		1975	44.8	21.7	
2010		183.5	88.8		1992	99.4	48.1		1974	41.4	20.0	
2009		180.1	87.1		1991	96.8	46.8		1973	37.7	18.2	
2008		180.4	87.3		1990	94.3	45.6		1972	34.8	16.8	
2007		169.4	82.0		1989	92.1	44.6		1971	32.1	15.5	
2006		162.0	78.4		1988	89.9	43.5		1970	28.7	13.9	
2005		151.6	73.3		1987	87.7	42.4		1969	26.9	13.0	
2004		143.7	69.5		1986	84.2	40.7		1968	24.9	12.0	
2003		132.0	63.9		1985	82.6	40.0		1967	23.5	11.4	
2002		128.7	62.3		1984	82.0	39.7		1966	22.7	11.0	
2001		125.1	60.5		1983	80.2	38.8		1965	21.7	10.5	

Adjustments to Costs

The "Historical Cost Index" can be used to convert national average building costs at a particular time to the approximate building costs for some other time.

Example:

Estimate and compare construction costs for different years in the same city.

To estimate the national average construction cost of a building in 1970, knowing that it cost $900,000 in 2015:

INDEX in 1970 = 28.7

INDEX in 2015 = 206.7

Note: The city cost indexes for Canada can be used to convert U.S. national averages to local costs in Canadian dollars.

Example:

To estimate and compare the cost of a building in Toronto, ON in 2015 with the known cost of $600,000 (US$) in New York, NY in 2015:

INDEX Toronto = 110.9

INDEX New York = 131.8

$$\frac{\text{INDEX Toronto}}{\text{INDEX New York}} \times \text{Cost New York} = \text{Cost Toronto}$$

$$\frac{110.9}{131.8} \times \$600,000 = .841 \times \$600,000 = \$504,600$$

The construction cost of the building in Toronto is $504,600 (CN$).

Time Adjustment Using the Historical Cost Indexes:

$$\frac{\text{Index for Year A}}{\text{Index for Year B}} \times \text{Cost in Year B} = \text{Cost in Year A}$$

$$\frac{\text{INDEX 1970}}{\text{INDEX 2015}} \times \text{Cost 2015} = \text{Cost 1970}$$

$$\frac{28.7}{206.7} \times \$900,000 = .139 \times \$900,000 = \$125,100$$

The construction cost of the building in 1970 is $125,100.

*Historical Cost Index updates and other resources are provided on the following website.
http://info.thegordiangroup.com/RSMeans.html

How to Use the City Cost Indexes

What you should know before you begin

RSMeans City Cost Indexes (CCI) are an extremely useful tool to use when you want to compare costs from city to city and region to region.

This publication contains average construction cost indexes for 731 U.S. and Canadian cities covering over 930 three-digit zip code locations, as listed directly under each city.

Keep in mind that a City Cost Index number is a percentage ratio of a specific city's cost to the national average cost of the same item at a stated time period.

In other words, these index figures represent relative construction factors (or, if you prefer, multipliers) for Material and Installation costs, as well as the weighted average for Total In Place costs for each CSI MasterFormat division. Installation costs include both labor and equipment rental costs. When estimating equipment rental rates only, for a specific location, use 01 54 33 EQUIPMENT RENTAL COSTS in the Reference Section at the back of the book.

The 30 City Average Index is the average of 30 major U.S. cities and serves as a National Average.

Index figures for both material and installation are based on the 30 major city average of 100 and represent the cost relationship as of July 1, 2014. The index for each division is computed from representative material and labor quantities for that division. The weighted average for each city is a weighted total of the components listed above it, but does not include relative productivity between trades or cities.

As changes occur in local material prices, labor rates, and equipment rental rates (including fuel costs), the impact of these changes should be accurately measured by the change in the City Cost Index for each particular city (as compared to the 30 City Average).

Therefore, if you know (or have estimated) building costs in one city today, you can easily convert those costs to expected building costs in another city.

In addition, by using the Historical Cost Index, you can easily convert National Average building costs at a particular time to the approximate building costs for some other time. The City Cost Indexes can then be applied to calculate the costs for a particular city.

Quick Calculations

Location Adjustment Using the City Cost Indexes:

$$\frac{\text{Index for City A}}{\text{Index for City B}} \times \text{Cost in City B} = \text{Cost in City A}$$

Time Adjustment for the National Average Using the Historical Cost Index:

$$\frac{\text{Index for Year A}}{\text{Index for Year B}} \times \text{Cost in Year B} = \text{Cost in Year A}$$

Adjustment from the National Average:

$$\frac{\text{Index for City A}}{100} \times \text{National Average Cost} = \text{Cost in City A}$$

Since each of the other RSMeans publications contains many different items, any *one* item multiplied by the particular city index may give incorrect results. However, the larger the number of items compiled, the closer the results should be to actual costs for that particular city.

The City Cost Indexes for Canadian cities are calculated using Canadian material and equipment prices and labor rates, in Canadian dollars. Therefore, indexes for Canadian cities can be used to convert U.S. National Average prices to local costs in Canadian dollars.

How to use this section

1. Compare costs from city to city.

In using the RSMeans Indexes, remember that an index number is not a fixed number but a ratio: It's a percentage ratio of a building component's cost at any stated time to the National Average cost of that same component at the same time period. Put in the form of an equation:

$$\frac{\text{Specific City Cost}}{\text{National Average Cost}} \times 100 = \text{City Index Number}$$

Therefore, when making cost comparisons between cities, do not subtract one city's index number from the index number of another city and read the result as a percentage difference. Instead, divide one city's index number by that of the other city. The resulting number may then be used as a multiplier to calculate cost differences from city to city.

The formula used to find cost differences between cities for the purpose of comparison is as follows:

$$\frac{\text{City A Index}}{\text{City B Index}} \times \text{City B Cost (Known)} = \text{City A Cost (Unknown)}$$

In addition, you can use RSMeans CCI to calculate and compare costs division by division between cities using the same basic formula. (Just be sure that you're comparing similar divisions.)

2. Compare a specific city's construction costs with the National Average.

When you're studying construction location feasibility, it's advisable to compare a prospective project's cost index with an index of the National Average cost.

For example, divide the weighted average index of construction costs of a specific city by that of the 30 City Average, which = 100.

$$\frac{\text{City Index}}{100} = \% \text{ of National Average}$$

As a result, you get a ratio that indicates the relative cost of construction in that city in comparison with the National Average.

3. Convert U.S. National Average to actual costs in Canadian City.

$$\frac{\text{Index for Canadian City}}{100} \times \text{National Average Cost} = \text{Cost in Canadian City in \$ CAN}$$

4. Adjust construction cost data based on a National Average.

When you use a source of construction cost data which is based on a National Average (such as RSMeans cost data publications), it is necessary to adjust those costs to a specific location.

$$\frac{\text{City Index}}{100} \times \frac{\text{"Book" Cost Based on}}{\text{National Average Costs}} = \frac{\text{City Cost}}{\text{(Unknown)}}$$

5. When applying the City Cost Indexes to demolition projects, use the appropriate division installation index.
For example, for removal of existing doors and windows, use Division 8 (Openings) index.

What you might like to know about how we developed the Indexes

The information presented in the CCI is organized according to the Construction Specifications Institute (CSI) MasterFormat 2014 classification system.

To create a reliable index, RSMeans researched the building type most often constructed in the United States and Canada. Because it was concluded that no one type of building completely represented the building construction industry, nine different types of buildings were combined to create a composite model.

The exact material, labor, and equipment quantities are based on detailed analyses of these nine building types, and then each quantity is weighted in proportion to expected usage. These various material items, labor hours, and equipment rental rates are thus combined to form a composite building representing as closely as possible the actual usage of materials, labor, and equipment used in the North American building construction industry.

The following structures were chosen to make up that composite model:

1. Factory, 1 story
2. Office, 2–4 story
3. Store, Retail
4. Town Hall, 2–3 story
5. High School, 2–3 story
6. Hospital, 4–8 story
7. Garage, Parking
8. Apartment, 1–3 story
9. Hotel/Motel, 2–3 story

For the purposes of ensuring the timeliness of the data, the components of the index for the composite model have been streamlined. They currently consist of:

- specific quantities of 66 commonly used construction materials;
- specific labor-hours for 21 building construction trades; and
- specific days of equipment rental for 6 types of construction equipment (normally used to install the 66 material items by the 21 trades.) Fuel costs and routine maintenance costs are included in the equipment cost.

A sophisticated computer program handles the updating of all costs for each city on a quarterly basis. Material and equipment price quotations are gathered quarterly from cities in the United States and Canada. These prices and the latest negotiated labor wage rates for 21 different building trades are used to compile the quarterly update of the City Cost Index.

The 30 major U.S. cities used to calculate the National Average are:

Atlanta, GA	Memphis, TN
Baltimore, MD	Milwaukee, WI
Boston, MA	Minneapolis, MN
Buffalo, NY	Nashville, TN
Chicago, IL	New Orleans, LA
Cincinnati, OH	New York, NY
Cleveland, OH	Philadelphia, PA
Columbus, OH	Phoenix, AZ
Dallas, TX	Pittsburgh, PA
Denver, CO	St. Louis, MO
Detroit, MI	San Antonio, TX
Houston, TX	San Diego, CA
Indianapolis, IN	San Francisco, CA
Kansas City, MO	Seattle, WA
Los Angeles, CA	Washington, DC

What the CCI does not indicate

The weighted average for each city is a total of the divisional components weighted to reflect typical usage, but it does not include the productivity variations between trades or cities.

In addition, the CCI does not take into consideration factors such as the following:

- managerial efficiency
- competitive conditions
- automation
- restrictive union practices
- unique local requirements
- regional variations due to specific building codes

For customer support on your Mechanical Cost Data, call 877.795.7394.

657

United States / Alabama

DIVISION		UNITED STATES 30 CITY AVERAGE			ANNISTON 362			BIRMINGHAM 350-352			BUTLER 369			DECATUR 356			DOTHAN 363		
		MAT.	INST.	TOTAL	MAT.	INST.	TOTAL	MAT.	INST.	TOTAL	MAT.	INST.	TOTAL	MAT.	INST.	TOTAL	MAT.	INST.	TOTAL
015433	CONTRACTOR EQUIPMENT		100.0	100.0		101.0	101.0		101.1	101.1		98.2	98.2		101.0	101.0		98.2	98.2
0241, 31 - 34	SITE & INFRASTRUCTURE, DEMOLITION	100.0	100.0	100.0	90.2	93.1	92.3	97.4	93.8	94.8	103.5	87.1	91.8	88.8	91.5	90.7	101.2	87.1	91.1
0310	Concrete Forming & Accessories	100.0	100.0	100.0	90.1	47.1	53.0	93.1	75.9	78.2	86.4	42.6	48.6	94.6	45.4	52.1	95.5	42.0	49.4
0320	Concrete Reinforcing	100.0	100.0	100.0	87.8	86.3	87.1	94.7	86.9	90.7	92.8	45.9	68.9	88.6	77.0	82.7	92.8	45.4	68.7
0330	Cast-in-Place Concrete	100.0	100.0	100.0	101.8	49.7	80.4	110.0	74.6	95.4	99.3	54.8	81.0	103.4	64.1	87.2	99.3	46.3	77.5
03	CONCRETE	100.0	100.0	100.0	102.1	57.0	79.9	102.2	78.4	90.5	103.0	49.3	76.6	98.3	59.3	79.1	102.3	46.0	74.7
04	MASONRY	100.0	100.0	100.0	100.1	66.8	79.3	98.8	76.0	84.6	105.9	48.6	70.2	97.3	51.3	68.6	107.1	39.7	65.1
05	METALS	100.0	100.0	100.0	104.0	91.5	100.1	104.0	94.1	101.0	102.8	76.6	94.8	106.2	86.1	100.0	102.9	75.3	94.4
06	WOOD, PLASTICS & COMPOSITES	100.0	100.0	100.0	90.5	42.7	63.7	97.4	76.1	85.4	85.0	42.4	61.1	98.9	42.9	67.6	97.4	42.4	66.6
07	THERMAL & MOISTURE PROTECTION	100.0	100.0	100.0	97.9	52.9	79.5	99.7	81.5	92.2	98.0	56.7	81.0	96.7	58.3	81.0	98.0	49.1	77.9
08	OPENINGS	100.0	100.0	100.0	98.6	50.1	87.3	103.2	76.0	96.8	98.6	44.2	85.9	106.5	50.4	93.5	98.6	44.2	86.0
0920	Plaster & Gypsum Board	100.0	100.0	100.0	90.8	41.4	57.4	94.8	75.8	81.9	87.9	41.0	56.2	95.8	41.6	59.2	98.6	41.0	59.7
0950, 0980	Ceilings & Acoustic Treatment	100.0	100.0	100.0	80.4	41.4	54.7	89.8	75.8	80.6	80.4	41.0	54.5	87.2	41.6	57.2	80.4	41.0	54.5
0960	Flooring	100.0	100.0	100.0	91.5	40.2	76.8	102.5	76.7	95.1	95.8	56.0	84.4	99.7	50.6	85.6	100.6	27.5	79.7
0970, 0990	Wall Finishes & Painting/Coating	100.0	100.0	100.0	103.7	31.5	60.1	103.9	66.5	81.3	103.7	53.3	73.3	99.6	66.8	79.8	103.7	53.3	73.3
09	FINISHES	100.0	100.0	100.0	87.0	43.0	62.8	95.4	74.8	84.1	89.9	45.7	65.5	91.9	47.7	67.6	92.5	40.1	63.6
COVERS	DIVS. 10 - 14, 25, 28, 41, 43, 44, 46	100.0	100.0	100.0	100.0	68.3	93.6	100.0	88.4	97.7	100.0	41.4	88.2	100.0	43.2	88.5	100.0	41.5	88.2
21, 22, 23	FIRE SUPPRESSION, PLUMBING & HVAC	100.0	100.0	100.0	100.0	55.4	82.0	100.0	70.4	88.1	97.3	33.3	71.5	100.0	40.8	76.1	97.3	33.2	71.5
26, 27, 3370	ELECTRICAL, COMMUNICATIONS & UTIL.	100.0	100.0	100.0	93.1	57.3	74.2	99.0	61.6	79.3	95.0	40.2	66.0	94.2	64.9	78.7	93.8	57.4	74.6
MF2014	WEIGHTED AVERAGE	100.0	100.0	100.0	98.6	61.8	82.6	100.6	76.7	90.2	98.9	49.4	77.3	99.8	57.9	81.5	99.0	49.4	77.4

Alabama

DIVISION		EVERGREEN 364			GADSDEN 359			HUNTSVILLE 357-358			JASPER 355			MOBILE 365-366			MONTGOMERY 360-361		
		MAT.	INST.	TOTAL	MAT.	INST.	TOTAL	MAT.	INST.	TOTAL	MAT.	INST.	TOTAL	MAT.	INST.	TOTAL	MAT.	INST.	TOTAL
015433	CONTRACTOR EQUIPMENT		98.2	98.2		101.0	101.0		101.0	101.0		101.0	101.0		98.2	98.2		98.2	98.2
0241, 31 - 34	SITE & INFRASTRUCTURE, DEMOLITION	104.0	87.4	92.2	94.9	92.7	93.3	88.6	92.7	91.5	94.6	92.4	93.1	96.4	88.1	90.5	96.5	88.1	90.5
0310	Concrete Forming & Accessories	83.0	44.2	49.5	86.9	42.5	48.6	94.6	69.9	73.3	92.1	34.6	42.5	94.5	54.0	59.6	94.1	44.1	51.0
0320	Concrete Reinforcing	92.9	45.9	69.0	94.0	86.8	90.3	88.6	81.6	85.1	88.6	85.6	87.1	90.6	83.6	87.0	95.8	86.1	90.9
0330	Cast-in-Place Concrete	99.3	57.3	82.1	103.4	66.7	88.3	100.7	69.5	87.9	114.5	44.1	85.6	104.1	66.4	88.6	105.2	56.5	85.2
03	CONCRETE	103.3	50.9	77.5	102.9	60.8	82.2	97.0	72.9	85.2	106.6	49.4	78.5	98.9	65.3	82.4	100.2	57.9	79.4
04	MASONRY	105.9	53.2	73.0	95.6	62.4	74.9	98.8	67.7	79.4	92.9	41.2	60.7	103.7	54.2	72.9	100.0	49.7	68.6
05	METALS	102.9	76.5	94.7	104.0	93.2	100.7	106.2	90.6	101.4	104.0	90.2	99.7	105.0	91.6	100.9	104.0	91.9	100.3
06	WOOD, PLASTICS & COMPOSITES	81.3	42.4	59.5	89.3	38.7	61.0	98.9	72.6	84.2	95.9	31.6	59.9	95.9	52.8	71.8	95.3	42.4	65.6
07	THERMAL & MOISTURE PROTECTION	97.9	56.9	81.1	96.8	71.7	86.5	96.6	77.6	88.8	96.8	45.4	75.7	97.6	70.1	86.3	97.9	66.6	85.0
08	OPENINGS	98.6	44.2	85.9	102.8	49.0	90.3	106.5	67.4	97.4	102.8	48.0	90.0	102.2	58.8	92.1	103.3	53.1	91.7
0920	Plaster & Gypsum Board	87.2	41.0	56.0	88.2	37.3	53.8	95.8	72.2	79.9	92.1	30.0	50.1	95.2	51.8	65.9	94.6	41.0	58.4
0950, 0980	Ceilings & Acoustic Treatment	80.4	41.0	54.5	83.8	37.3	53.2	89.0	72.2	78.0	83.8	30.0	48.4	85.6	51.8	63.4	88.0	41.0	57.1
0960	Flooring	93.9	56.0	83.1	95.5	76.7	90.1	99.7	76.7	93.1	97.8	40.2	81.3	100.5	58.5	88.5	97.3	58.5	86.2
0970, 0990	Wall Finishes & Painting/Coating	103.7	53.3	73.3	99.6	60.7	76.1	99.6	65.9	79.3	99.6	38.6	62.8	107.3	54.3	75.3	103.3	53.3	73.1
09	FINISHES	89.2	46.8	65.9	89.4	48.7	67.0	92.3	70.7	80.4	90.5	34.4	59.6	92.7	53.5	71.1	92.8	46.1	67.0
COVERS	DIVS. 10 - 14, 25, 28, 41, 43, 44, 46	100.0	42.9	88.5	100.0	79.8	95.9	100.0	85.3	97.0	100.0	39.8	87.8	100.0	83.1	96.6	100.0	79.6	95.9
21, 22, 23	FIRE SUPPRESSION, PLUMBING & HVAC	97.3	35.6	72.4	102.0	36.3	75.5	100.0	61.6	84.5	102.0	61.4	85.6	99.9	60.7	84.1	100.0	34.2	73.4
26, 27, 3370	ELECTRICAL, COMMUNICATIONS & UTIL.	92.5	40.2	64.8	94.3	61.6	77.0	95.3	64.9	79.3	93.8	61.3	76.6	95.7	59.1	76.3	96.3	60.9	77.6
MF2014	WEIGHTED AVERAGE	98.6	50.7	77.7	99.9	60.0	82.5	99.8	72.1	87.8	100.2	57.5	81.6	99.9	64.4	84.9	99.9	57.1	81.2

Alabama / Alaska

DIVISION		PHENIX CITY 368			SELMA 367			TUSCALOOSA 354			ANCHORAGE 995-996			FAIRBANKS 997			JUNEAU 998		
		MAT.	INST.	TOTAL	MAT.	INST.	TOTAL	MAT.	INST.	TOTAL	MAT.	INST.	TOTAL	MAT.	INST.	TOTAL	MAT.	INST.	TOTAL
015433	CONTRACTOR EQUIPMENT		98.2	98.2		98.2	98.2		101.0	101.0		114.7	114.7		114.7	114.7		114.7	114.7
0241, 31 - 34	SITE & INFRASTRUCTURE, DEMOLITION	107.8	88.2	93.9	101.0	88.1	91.8	89.1	92.9	91.8	127.2	129.6	128.9	119.2	129.6	126.6	135.6	129.6	131.3
0310	Concrete Forming & Accessories	90.1	39.1	46.1	87.6	42.6	48.8	94.5	47.9	54.3	123.7	119.8	120.3	131.6	119.9	121.5	130.1	119.8	121.2
0320	Concrete Reinforcing	92.8	61.4	76.8	92.8	85.2	89.0	88.6	86.8	87.7	149.7	110.7	129.9	150.9	110.7	130.5	135.7	110.7	123.0
0330	Cast-in-Place Concrete	99.3	57.0	81.9	99.3	46.5	77.6	104.9	68.4	89.9	132.0	117.8	126.2	128.3	118.2	124.1	133.2	117.8	126.9
03	CONCRETE	106.2	51.3	79.2	101.7	53.7	78.1	99.0	63.8	81.7	138.6	116.7	127.9	123.8	116.9	120.4	136.0	116.7	126.5
04	MASONRY	105.9	41.1	65.5	109.6	38.7	65.4	97.6	65.5	77.6	184.6	125.3	147.6	189.5	125.3	149.5	176.6	125.3	144.6
05	METALS	102.8	81.9	96.4	102.8	89.3	98.7	105.3	93.2	101.6	113.6	103.4	110.5	117.6	103.4	113.3	116.3	103.4	112.3
06	WOOD, PLASTICS & COMPOSITES	90.2	35.4	59.5	87.0	42.4	62.0	98.9	44.4	68.4	129.2	118.7	123.4	136.1	118.7	126.4	129.4	118.7	123.4
07	THERMAL & MOISTURE PROTECTION	98.3	63.8	84.1	97.8	52.1	79.1	96.7	73.7	87.3	159.5	118.0	142.5	166.5	119.1	147.1	168.0	118.0	147.5
08	OPENINGS	98.6	43.5	85.8	98.6	53.1	88.0	106.5	58.8	95.4	132.7	115.6	128.7	128.9	115.7	125.8	129.8	115.6	126.5
0920	Plaster & Gypsum Board	92.2	33.9	52.8	90.1	41.0	56.9	95.8	43.1	60.2	140.9	119.1	126.2	164.8	119.1	133.9	148.1	119.1	128.5
0950, 0980	Ceilings & Acoustic Treatment	80.4	33.9	49.8	80.4	41.0	54.5	89.0	43.1	58.9	119.6	119.1	119.3	119.2	119.1	119.2	125.9	119.1	121.4
0960	Flooring	97.6	58.5	86.4	96.2	28.5	76.8	99.7	76.7	93.1	134.6	133.0	134.1	127.2	133.0	128.9	131.0	133.0	131.6
0970, 0990	Wall Finishes & Painting/Coating	103.7	53.3	73.3	103.7	53.3	73.3	99.6	45.9	67.2	136.8	116.3	124.4	136.1	116.3	124.1	132.1	116.3	122.6
09	FINISHES	91.4	42.2	64.3	90.0	40.2	62.6	92.3	51.2	69.7	134.3	122.2	127.7	132.4	122.8	127.1	134.4	122.2	127.7
COVERS	DIVS. 10 - 14, 25, 28, 41, 43, 44, 46	100.0	79.3	95.8	100.0	41.4	88.2	100.0	81.4	96.2	100.0	112.8	102.6	100.0	112.8	102.6	100.0	112.8	102.6
21, 22, 23	FIRE SUPPRESSION, PLUMBING & HVAC	97.3	34.5	72.0	97.3	33.6	71.6	100.0	34.0	73.4	100.3	105.0	102.2	100.2	108.0	103.4	100.3	105.0	102.2
26, 27, 3370	ELECTRICAL, COMMUNICATIONS & UTIL.	94.4	69.5	81.3	93.6	40.2	65.4	94.8	61.6	77.3	117.7	117.8	117.7	130.0	117.8	123.5	119.9	117.8	118.8
MF2014	WEIGHTED AVERAGE	99.5	54.6	79.9	98.7	49.9	77.4	99.8	61.2	83.0	121.1	115.6	118.7	121.0	116.4	119.0	121.3	115.6	118.8

DIVISION		ALASKA KETCHIKAN 999			ARIZONA CHAMBERS 865			FLAGSTAFF 860			GLOBE 855			KINGMAN 864			MESA/TEMPE 852		
		MAT.	INST.	TOTAL	MAT.	INST.	TOTAL	MAT.	INST.	TOTAL	MAT.	INST.	TOTAL	MAT.	INST.	TOTAL	MAT.	INST.	TOTAL
015433	CONTRACTOR EQUIPMENT		114.7	114.7		93.0	93.0		93.0	93.0		92.2	92.2		93.0	93.0		92.2	92.2
0241, 31 - 34	SITE & INFRASTRUCTURE, DEMOLITION	173.8	129.6	142.4	69.6	96.1	88.4	87.3	96.3	93.7	97.3	95.6	96.1	69.6	96.3	88.6	89.0	95.8	93.8
0310	Concrete Forming & Accessories	122.9	119.8	120.2	98.8	58.1	63.7	104.4	65.1	70.5	99.0	58.1	63.7	97.0	65.1	69.5	102.2	69.3	73.9
0320	Concrete Reinforcing	117.4	110.7	114.0	97.1	85.2	91.1	97.0	85.3	91.0	108.0	85.2	96.4	97.2	85.3	91.1	108.7	85.3	96.8
0330	Cast-in-Place Concrete	259.2	117.8	201.2	90.7	73.2	83.5	90.8	73.4	83.7	94.3	72.0	85.1	90.4	73.4	83.4	95.0	72.3	85.7
03	CONCRETE	208.7	116.7	163.5	95.0	68.6	82.0	114.1	71.8	93.3	110.0	68.2	89.5	94.6	71.8	83.4	101.8	73.3	87.8
04	MASONRY	198.4	125.3	152.8	92.5	62.3	73.7	92.6	62.4	73.8	109.9	62.2	80.1	92.5	62.4	73.7	110.1	62.2	80.3
05	METALS	117.8	103.4	113.4	96.2	75.6	89.8	96.7	76.3	90.4	93.3	76.0	88.0	96.8	76.3	90.5	93.6	76.8	88.4
06	WOOD, PLASTICS & COMPOSITES	126.3	118.7	122.1	101.0	55.0	75.2	107.3	64.2	83.1	97.6	55.0	73.8	96.1	64.2	78.2	101.1	69.8	83.6
07	THERMAL & MOISTURE PROTECTION	169.3	118.0	148.3	94.6	65.0	82.5	96.2	68.1	84.7	101.9	63.0	85.9	94.6	65.6	82.7	101.2	65.1	86.4
08	OPENINGS	130.8	115.6	127.3	108.1	65.3	98.2	108.3	70.3	99.4	100.0	65.3	91.9	108.3	70.3	99.5	100.1	73.3	93.8
0920	Plaster & Gypsum Board	152.7	119.1	130.0	90.2	53.7	65.5	93.6	63.2	73.0	96.6	53.7	67.6	82.9	63.2	69.6	98.7	68.9	78.6
0950, 0980	Ceilings & Acoustic Treatment	112.7	119.1	116.9	99.7	53.7	69.5	100.5	63.2	76.0	88.9	53.7	65.8	100.5	63.2	76.0	88.9	68.9	75.8
0960	Flooring	127.2	133.0	128.9	93.3	39.5	77.9	95.4	39.7	79.5	103.5	39.5	85.2	92.1	53.7	81.1	104.7	49.5	88.9
0970, 0990	Wall Finishes & Painting/Coating	133.5	116.3	123.1	98.3	55.1	72.2	98.3	55.1	72.2	103.4	55.1	74.3	98.3	55.1	72.2	103.4	55.1	74.3
09	FINISHES	133.4	122.2	127.2	93.7	52.9	71.2	96.6	58.4	75.6	97.1	53.0	72.8	92.5	60.7	75.0	96.8	63.6	78.5
COVERS	DIVS. 10 - 14, 25, 28, 41, 43, 44, 46	100.0	112.8	102.6	100.0	82.3	96.4	100.0	83.3	96.6	100.0	82.4	96.4	100.0	83.3	96.6	100.0	84.0	96.8
21, 22, 23	FIRE SUPPRESSION, PLUMBING & HVAC	98.4	105.0	101.1	97.0	78.9	89.7	100.2	79.0	91.6	95.2	78.9	88.7	97.0	79.0	89.7	100.0	79.0	91.5
26, 27, 3370	ELECTRICAL, COMMUNICATIONS & UTIL.	129.9	117.8	123.5	104.5	70.9	86.7	103.4	61.3	81.1	97.5	61.2	78.3	104.5	61.3	81.7	94.2	61.3	76.8
MF2014	WEIGHTED AVERAGE	132.0	115.6	124.8	97.6	71.5	86.2	101.3	71.8	88.4	98.8	70.0	86.3	97.6	72.0	86.4	98.6	72.8	87.4

DIVISION		ARIZONA PHOENIX 850,853			PRESCOTT 863			SHOW LOW 859			TUCSON 856 - 857			ARKANSAS BATESVILLE 725			CAMDEN 717		
		MAT.	INST.	TOTAL	MAT.	INST.	TOTAL	MAT.	INST.	TOTAL	MAT.	INST.	TOTAL	MAT.	INST.	TOTAL	MAT.	INST.	TOTAL
015433	CONTRACTOR EQUIPMENT		92.7	92.7		93.0	93.0		92.2	92.2		92.2	92.2		89.1	89.1		89.1	89.1
0241, 31 - 34	SITE & INFRASTRUCTURE, DEMOLITION	89.4	96.0	94.1	76.0	96.0	90.2	99.2	95.6	96.6	85.2	95.8	92.7	77.1	84.9	82.6	78.3	84.4	82.7
0310	Concrete Forming & Accessories	103.1	68.3	73.1	100.4	52.5	59.1	106.3	69.1	74.2	102.5	68.1	72.9	85.9	44.9	50.5	83.7	30.7	38.0
0320	Concrete Reinforcing	107.0	85.4	96.0	97.0	85.2	91.0	108.7	85.2	96.8	89.9	85.3	87.5	87.4	67.5	77.3	93.6	67.4	80.3
0330	Cast-in-Place Concrete	95.1	72.4	85.8	90.7	73.1	83.5	94.3	72.1	85.2	97.8	72.3	87.3	77.4	45.3	64.2	82.6	38.3	64.4
03	CONCRETE	101.4	73.0	87.4	100.1	66.1	83.4	112.4	73.1	93.1	100.0	72.8	86.6	83.5	50.1	67.1	87.5	41.3	64.8
04	MASONRY	97.5	64.1	76.7	92.6	62.3	73.7	109.9	62.2	80.1	95.5	62.2	74.8	99.8	40.7	63.0	116.6	31.4	63.5
05	METALS	95.1	77.6	89.7	96.7	75.3	90.1	93.1	76.1	87.8	94.3	76.7	88.9	99.5	66.9	89.4	103.4	66.5	92.4
06	WOOD, PLASTICS & COMPOSITES	102.1	68.3	83.2	102.4	47.4	71.6	105.4	69.8	85.6	101.4	68.3	82.8	89.9	45.5	65.0	90.3	29.7	56.4
07	THERMAL & MOISTURE PROTECTION	100.9	67.0	87.0	95.1	64.2	82.4	102.1	65.0	86.9	102.4	64.2	86.7	98.2	43.9	76.0	94.2	35.7	70.2
08	OPENINGS	102.0	72.5	95.2	108.3	61.1	97.3	99.2	73.3	93.2	96.3	72.5	90.7	97.7	45.8	85.6	102.1	40.6	87.8
0920	Plaster & Gypsum Board	101.0	67.4	78.2	90.3	45.9	60.3	100.9	68.9	79.3	104.0	67.4	79.2	83.3	44.3	56.9	85.0	28.0	46.5
0950, 0980	Ceilings & Acoustic Treatment	96.5	67.4	77.3	98.8	45.9	64.0	88.9	68.9	75.8	89.8	67.4	75.0	83.2	44.3	57.6	83.3	28.0	47.0
0960	Flooring	104.9	51.7	89.7	94.1	39.5	78.5	106.2	45.0	88.7	95.2	44.2	80.6	96.7	59.1	85.9	99.6	39.6	82.4
0970, 0990	Wall Finishes & Painting/Coating	103.4	61.7	78.3	98.3	55.1	72.2	103.4	55.1	74.3	104.5	55.1	74.7	107.4	41.2	67.4	104.4	50.4	71.8
09	FINISHES	98.9	63.8	79.6	94.2	48.5	69.0	98.7	62.7	78.8	94.9	61.6	76.5	86.6	46.7	64.6	88.5	33.5	58.2
COVERS	DIVS. 10 - 14, 25, 28, 41, 43, 44, 46	100.0	83.9	96.7	100.0	81.4	96.3	100.0	84.0	96.8	100.0	83.9	96.7	100.0	39.5	87.8	100.0	35.6	87.0
21, 22, 23	FIRE SUPPRESSION, PLUMBING & HVAC	99.9	79.0	91.5	100.2	78.9	91.6	95.2	79.0	88.7	100.0	79.0	91.5	95.3	51.6	77.7	95.3	53.1	78.3
26, 27, 3370	ELECTRICAL, COMMUNICATIONS & UTIL.	101.0	66.7	82.9	103.0	61.2	80.9	94.6	61.2	77.0	96.5	61.3	77.9	96.5	63.5	79.0	96.8	62.1	78.5
MF2014	WEIGHTED AVERAGE	99.2	73.8	88.1	99.2	68.9	86.0	98.9	72.5	87.4	97.4	72.4	86.5	94.6	54.6	77.1	97.1	50.1	76.6

DIVISION		ARKANSAS FAYETTEVILLE 727			FORT SMITH 729			HARRISON 726			HOT SPRINGS 719			JONESBORO 724			LITTLE ROCK 720 - 722		
		MAT.	INST.	TOTAL	MAT.	INST.	TOTAL	MAT.	INST.	TOTAL	MAT.	INST.	TOTAL	MAT.	INST.	TOTAL	MAT.	INST.	TOTAL
015433	CONTRACTOR EQUIPMENT		89.1	89.1		89.1	89.1		89.1	89.1		89.1	89.1		108.3	108.3		89.1	89.1
0241, 31 - 34	SITE & INFRASTRUCTURE, DEMOLITION	76.4	86.3	83.5	81.8	86.7	85.3	82.1	84.9	84.1	81.4	85.8	84.5	103.2	101.1	101.7	93.1	86.8	88.6
0310	Concrete Forming & Accessories	81.2	38.5	44.4	101.3	57.3	63.4	90.6	44.8	51.1	81.1	35.7	41.9	89.6	48.2	53.9	96.4	60.0	65.0
0320	Concrete Reinforcing	87.4	71.3	79.2	88.4	72.0	80.1	87.0	63.7	75.1	91.8	67.5	79.4	84.5	70.1	77.2	93.4	68.3	80.6
0330	Cast-in-Place Concrete	77.4	49.4	65.5	88.5	70.3	81.0	85.8	43.4	68.4	84.5	38.8	65.7	84.3	55.0	72.2	87.3	70.3	80.3
03	CONCRETE	83.2	49.3	66.5	91.3	65.0	78.4	90.8	48.7	70.1	91.2	43.7	67.9	88.1	56.0	72.3	92.8	65.5	79.4
04	MASONRY	90.5	41.4	59.9	96.9	50.9	68.3	100.1	39.0	62.0	87.6	33.5	53.9	91.9	41.9	60.7	96.8	50.9	68.2
05	METALS	99.5	67.8	89.7	101.8	70.6	92.2	100.6	66.5	90.1	103.8	66.9	92.5	95.9	80.0	91.0	102.0	69.4	92.0
06	WOOD, PLASTICS & COMPOSITES	85.8	35.5	57.6	107.6	59.0	80.4	96.0	45.5	67.7	87.5	35.7	58.5	93.9	48.3	68.4	99.9	62.5	78.9
07	THERMAL & MOISTURE PROTECTION	99.0	45.3	77.0	99.4	57.5	82.2	98.5	42.6	75.6	94.4	37.9	71.2	103.3	49.1	81.1	95.7	57.8	80.2
08	OPENINGS	97.7	44.7	85.4	98.5	56.2	88.7	98.5	46.5	86.4	102.1	42.3	88.2	100.6	54.0	89.7	99.5	58.4	89.9
0920	Plaster & Gypsum Board	81.8	34.0	49.5	89.4	58.3	68.4	88.3	44.3	58.6	84.0	34.2	50.4	97.7	46.9	63.4	94.6	61.8	72.5
0950, 0980	Ceilings & Acoustic Treatment	83.2	34.0	50.9	84.8	58.3	67.4	84.8	44.3	58.2	83.3	34.2	51.0	86.2	46.9	60.8	88.0	61.8	70.8
0960	Flooring	94.0	59.1	84.0	102.9	60.8	90.9	98.8	59.1	87.5	98.5	59.1	87.2	68.9	53.4	64.5	101.0	60.8	89.5
0970, 0990	Wall Finishes & Painting/Coating	107.4	29.4	60.3	107.4	53.8	75.1	107.4	41.2	67.4	104.4	55.2	74.7	94.9	48.4	66.8	108.8	55.2	76.5
09	FINISHES	85.7	40.0	60.5	89.9	57.6	72.1	88.6	46.8	65.6	88.4	41.0	62.3	83.7	48.5	64.3	92.7	59.8	74.6
COVERS	DIVS. 10 - 14, 25, 28, 41, 43, 44, 46	100.0	48.7	89.6	100.0	78.8	95.7	100.0	49.1	89.7	100.0	36.5	87.2	100.0	45.2	88.9	100.0	79.2	95.8
21, 22, 23	FIRE SUPPRESSION, PLUMBING & HVAC	95.3	50.2	77.1	100.1	50.6	80.1	95.3	49.4	76.8	95.3	48.7	76.5	100.3	52.5	81.0	99.9	56.2	82.3
26, 27, 3370	ELECTRICAL, COMMUNICATIONS & UTIL.	90.1	51.2	69.5	93.7	64.5	78.3	95.0	37.1	64.4	99.0	67.8	82.5	100.2	63.5	80.8	100.7	70.2	84.6
MF2014	WEIGHTED AVERAGE	93.4	52.0	75.4	97.2	61.6	81.7	95.9	50.4	76.0	96.4	51.8	76.9	96.7	59.1	80.3	98.5	64.0	83.4

For customer support on your Mechanical Cost Data, call 877.795.7394.

659

City Cost Indexes

ARKANSAS / CALIFORNIA

DIVISION		PINE BLUFF 716 MAT.	INST.	TOTAL	RUSSELLVILLE 728 MAT.	INST.	TOTAL	TEXARKANA 718 MAT.	INST.	TOTAL	WEST MEMPHIS 723 MAT.	INST.	TOTAL	ALHAMBRA 917-918 MAT.	INST.	TOTAL	ANAHEIM 928 MAT.	INST.	TOTAL
015433	CONTRACTOR EQUIPMENT		89.1	89.1		89.1	89.1		89.9	89.9		108.3	108.3		100.1	100.1		100.8	100.8
0241, 31 - 34	SITE & INFRASTRUCTURE, DEMOLITION	83.7	86.8	85.9	78.3	84.9	83.0	94.9	87.1	89.3	110.5	101.1	103.8	97.9	110.9	107.1	99.4	108.3	105.7
0310	Concrete Forming & Accessories	80.7	60.0	62.8	86.6	54.7	59.1	87.3	40.1	46.6	95.5	48.5	54.9	117.1	116.4	116.5	105.4	124.2	121.6
0320	Concrete Reinforcing	93.5	68.2	80.7	88.0	67.3	77.5	93.1	67.5	80.1	84.5	70.1	77.2	106.6	113.3	110.0	93.8	113.2	103.7
0330	Cast-in-Place Concrete	84.5	70.3	78.7	81.1	45.3	66.4	92.1	43.3	72.0	88.4	55.1	74.7	95.1	120.8	105.6	92.9	123.3	105.4
03	CONCRETE	92.1	65.5	79.0	86.7	54.4	70.8	89.8	47.2	68.9	95.5	56.2	76.2	101.1	116.5	108.7	101.8	120.8	111.2
04	MASONRY	125.0	50.9	78.8	96.1	37.3	59.5	102.4	33.1	59.2	79.4	41.9	56.0	122.7	121.1	121.7	80.0	118.8	104.2
05	METALS	104.6	69.4	93.8	99.5	66.5	89.3	96.8	66.9	87.6	95.0	80.3	90.5	85.3	100.9	90.1	102.7	101.1	102.2
06	WOOD, PLASTICS & COMPOSITES	87.0	62.5	73.3	91.3	59.0	73.2	95.5	42.0	65.5	100.1	48.3	71.1	100.0	112.1	106.7	101.9	122.6	113.5
07	THERMAL & MOISTURE PROTECTION	94.5	57.8	79.5	99.2	43.9	76.5	95.2	44.5	74.4	103.7	49.1	81.3	95.1	118.1	104.5	99.4	122.0	108.7
08	OPENINGS	102.1	58.4	91.9	97.7	55.7	87.9	107.3	46.5	93.2	100.6	54.0	89.7	91.8	114.8	97.2	104.2	120.6	108.0
0920	Plaster & Gypsum Board	83.6	61.8	68.9	83.3	58.3	66.4	86.6	40.7	55.6	100.2	46.9	64.2	97.4	112.5	107.6	107.1	123.2	118.0
0950, 0980	Ceilings & Acoustic Treatment	83.3	61.8	69.2	83.2	58.3	66.8	86.6	40.7	56.4	84.3	46.9	59.7	100.2	112.5	108.3	105.0	123.2	117.0
0960	Flooring	98.2	60.8	87.5	96.2	59.1	85.6	100.4	51.0	86.3	71.0	53.4	66.0	95.4	110.2	99.6	101.9	112.0	104.8
0970, 0990	Wall Finishes & Painting/Coating	104.4	55.2	74.7	107.4	35.0	63.7	104.4	30.3	59.7	94.9	50.4	68.0	100.5	111.5	107.1	98.9	106.8	103.7
09	FINISHES	88.3	59.8	72.6	86.7	54.1	68.7	90.7	40.4	62.9	85.0	48.8	65.0	99.8	114.0	107.6	100.5	120.1	111.3
COVERS	DIVS. 10 - 14, 25, 28, 41, 43, 44, 46	100.0	79.2	95.8	100.0	41.0	88.1	100.0	32.6	86.4	100.0	45.2	88.9	100.0	115.0	102.1	100.0	112.0	102.4
21, 22, 23	FIRE SUPPRESSION, PLUMBING & HVAC	100.1	52.9	81.1	95.3	51.5	77.6	100.1	53.5	81.3	95.6	65.8	83.6	95.2	115.6	103.4	100.0	118.6	107.5
26, 27, 3370	ELECTRICAL, COMMUNICATIONS & UTIL.	97.0	70.2	82.8	93.7	44.7	67.8	98.9	36.0	65.7	101.9	65.7	82.8	118.4	120.5	119.5	92.2	105.7	99.3
MF2014	WEIGHTED AVERAGE	99.4	63.3	83.6	94.6	53.7	76.8	98.2	49.2	76.8	96.1	62.4	81.4	98.3	114.8	105.5	99.4	114.9	106.2

CALIFORNIA

DIVISION		BAKERSFIELD 932-933 MAT.	INST.	TOTAL	BERKELEY 947 MAT.	INST.	TOTAL	EUREKA 955 MAT.	INST.	TOTAL	FRESNO 936-938 MAT.	INST.	TOTAL	INGLEWOOD 903-905 MAT.	INST.	TOTAL	LONG BEACH 906-908 MAT.	INST.	TOTAL
015433	CONTRACTOR EQUIPMENT		98.8	98.8		100.3	100.3		98.5	98.5		98.8	98.8		96.5	96.5		96.5	96.5
0241, 31 - 34	SITE & INFRASTRUCTURE, DEMOLITION	100.4	106.0	104.4	118.4	107.6	110.7	110.8	103.9	105.9	103.1	105.4	104.7	89.3	103.5	99.4	95.9	103.5	101.3
0310	Concrete Forming & Accessories	104.0	124.3	121.5	115.9	149.9	145.2	114.8	133.6	131.0	104.3	134.2	130.1	111.8	115.4	114.9	106.5	115.4	114.1
0320	Concrete Reinforcing	103.4	113.2	108.4	94.5	114.4	104.6	102.5	114.1	108.4	80.6	113.7	97.5	102.7	113.3	108.1	101.9	113.3	107.7
0330	Cast-in-Place Concrete	94.0	122.6	105.7	123.4	124.8	124.0	100.8	118.0	107.9	99.4	118.9	107.4	84.5	121.4	99.7	96.2	121.4	106.5
03	CONCRETE	97.9	120.6	109.1	107.3	132.7	119.8	112.8	123.2	117.9	98.6	123.8	111.0	90.9	116.3	103.4	99.8	116.3	107.9
04	MASONRY	100.2	118.3	111.5	113.4	134.3	126.5	104.0	132.6	121.8	103.6	122.0	115.1	76.3	121.2	104.3	85.3	121.2	107.7
05	METALS	105.3	100.9	104.0	106.9	100.5	104.9	102.6	98.9	101.4	105.0	100.5	104.1	94.4	101.5	96.6	94.3	101.5	96.5
06	WOOD, PLASTICS & COMPOSITES	96.0	122.7	111.0	111.4	154.9	135.8	116.6	138.5	128.9	108.1	138.5	125.1	103.8	110.7	107.6	96.9	110.7	104.6
07	THERMAL & MOISTURE PROTECTION	98.1	116.1	105.5	105.2	137.2	118.3	103.4	119.8	110.1	88.7	116.0	99.9	98.3	118.0	106.4	98.5	118.0	106.5
08	OPENINGS	97.8	116.1	102.0	93.0	140.6	104.1	103.5	113.5	105.8	98.1	124.7	104.3	88.8	114.1	94.7	88.8	114.1	94.7
0920	Plaster & Gypsum Board	101.9	123.2	116.3	111.6	156.0	141.6	113.4	139.5	131.0	100.6	139.5	126.9	103.8	111.0	108.6	99.9	111.0	107.4
0950, 0980	Ceilings & Acoustic Treatment	96.3	123.2	114.0	111.6	156.0	139.0	110.1	139.5	129.4	92.8	139.5	123.5	100.3	111.0	107.4	100.3	111.0	107.4
0960	Flooring	106.8	110.2	107.7	108.2	127.6	113.8	106.0	116.5	109.0	113.3	133.5	119.1	104.4	110.2	106.0	101.8	110.2	104.2
0970, 0990	Wall Finishes & Painting/Coating	111.4	102.0	105.7	102.5	143.9	127.5	100.5	46.2	67.7	129.0	107.2	115.8	100.9	111.5	107.3	100.9	111.5	107.3
09	FINISHES	100.0	120.5	111.3	105.4	147.2	128.4	106.1	123.6	115.8	102.1	133.2	119.2	102.7	113.2	108.5	101.8	113.2	108.1
COVERS	DIVS. 10 - 14, 25, 28, 41, 43, 44, 46	100.0	111.9	102.4	100.0	126.4	105.3	100.0	122.2	104.5	100.0	122.2	104.5	100.0	110.4	102.1	100.0	110.4	102.1
21, 22, 23	FIRE SUPPRESSION, PLUMBING & HVAC	100.1	117.1	107.0	95.3	147.8	116.5	95.2	110.6	101.4	100.2	111.3	104.7	94.8	115.6	103.2	94.8	115.6	103.2
26, 27, 3370	ELECTRICAL, COMMUNICATIONS & UTIL.	103.6	102.5	103.1	109.4	145.0	128.2	99.1	116.7	108.4	93.8	100.8	97.5	103.8	120.5	112.6	103.5	120.5	112.5
MF2014	WEIGHTED AVERAGE	100.7	113.6	106.3	102.7	135.1	116.8	102.1	116.4	108.4	100.1	115.3	106.7	94.8	114.1	103.2	96.2	114.1	104.0

CALIFORNIA

DIVISION		LOS ANGELES 900-902 MAT.	INST.	TOTAL	MARYSVILLE 959 MAT.	INST.	TOTAL	MODESTO 953 MAT.	INST.	TOTAL	MOJAVE 935 MAT.	INST.	TOTAL	OAKLAND 946 MAT.	INST.	TOTAL	OXNARD 930 MAT.	INST.	TOTAL
015433	CONTRACTOR EQUIPMENT		100.1	100.1		98.5	98.5		98.5	98.5		98.8	98.8		100.3	100.3		97.7	97.7
0241, 31 - 34	SITE & INFRASTRUCTURE, DEMOLITION	96.6	107.1	104.1	107.2	105.0	105.7	102.2	105.1	104.3	96.4	106.0	103.2	124.7	107.6	112.5	103.3	104.0	103.8
0310	Concrete Forming & Accessories	108.6	124.3	122.2	104.6	134.4	130.3	100.7	134.3	129.7	116.1	112.1	112.6	104.9	149.9	143.7	107.5	124.3	122.0
0320	Concrete Reinforcing	103.6	113.4	108.6	102.5	113.8	108.2	106.2	113.8	110.0	97.3	113.1	105.3	96.7	114.4	105.7	95.5	113.1	104.4
0330	Cast-in-Place Concrete	91.2	122.0	103.9	112.6	119.2	115.3	100.8	119.2	108.4	86.0	122.4	101.0	117.0	124.8	120.2	100.1	122.7	109.4
03	CONCRETE	95.9	120.6	108.0	113.9	124.0	118.9	104.7	124.0	114.2	91.0	115.1	102.8	106.8	132.7	119.5	99.2	120.6	109.7
04	MASONRY	90.5	122.6	110.5	104.9	123.1	116.2	102.9	123.1	115.5	101.9	118.3	112.1	120.6	134.3	129.2	104.8	116.5	112.3
05	METALS	101.1	102.9	101.7	102.0	101.8	102.0	99.1	101.7	99.9	102.8	100.3	102.0	101.4	100.4	101.1	100.7	100.7	100.7
06	WOOD, PLASTICS & COMPOSITES	103.0	122.6	113.9	103.0	138.5	122.9	98.5	138.5	120.9	107.4	106.8	107.1	98.8	154.9	130.2	102.1	122.7	113.6
07	THERMAL & MOISTURE PROTECTION	97.6	121.2	107.3	102.9	121.5	110.5	102.5	118.7	109.1	94.4	110.9	101.2	103.1	137.2	117.1	97.4	119.8	106.6
08	OPENINGS	95.4	120.7	101.3	102.8	125.3	108.0	101.6	125.4	107.1	93.2	107.5	96.5	93.1	140.6	104.1	95.7	120.6	101.5
0920	Plaster & Gypsum Board	103.6	123.2	116.9	105.0	139.5	128.3	107.5	139.5	129.1	110.8	106.8	108.1	105.7	156.0	139.7	104.7	123.2	117.2
0950, 0980	Ceilings & Acoustic Treatment	110.5	123.2	118.8	109.2	139.5	129.1	105.0	139.5	127.7	94.2	106.8	102.5	108.9	156.0	139.9	96.6	123.2	114.1
0960	Flooring	102.0	112.0	104.9	102.0	117.6	106.4	102.4	112.8	105.3	112.6	110.2	111.9	104.3	127.6	110.3	104.8	112.0	106.9
0970, 0990	Wall Finishes & Painting/Coating	100.0	111.5	106.9	100.5	115.8	109.7	100.5	117.0	110.5	111.1	102.0	105.6	102.5	143.9	127.5	111.1	101.1	105.1
09	FINISHES	104.2	120.6	113.3	103.1	131.5	118.8	102.2	130.9	118.0	101.8	109.9	106.2	104.3	147.2	127.9	99.4	119.6	110.5
COVERS	DIVS. 10 - 14, 25, 28, 41, 43, 44, 46	100.0	111.9	102.4	100.0	122.2	104.5	100.0	122.2	104.5	100.0	107.8	101.6	100.0	126.3	105.3	100.0	112.0	102.5
21, 22, 23	FIRE SUPPRESSION, PLUMBING & HVAC	99.9	118.6	107.5	95.2	109.0	100.8	100.0	111.3	104.6	95.3	115.1	103.3	100.1	147.8	119.4	100.1	118.6	107.6
26, 27, 3370	ELECTRICAL, COMMUNICATIONS & UTIL.	102.0	121.3	112.2	95.7	107.4	101.9	98.2	104.0	101.3	92.0	99.9	96.2	108.5	139.3	124.8	97.7	107.3	102.8
MF2014	WEIGHTED AVERAGE	99.2	117.5	107.2	101.3	115.9	107.7	100.8	115.7	107.3	96.8	109.9	102.5	103.1	134.3	116.7	99.6	114.4	106.0

CALIFORNIA

DIVISION		PALM SPRINGS 922			PALO ALTO 943			PASADENA 910 - 912			REDDING 960			RICHMOND 948			RIVERSIDE 925		
		MAT.	INST.	TOTAL	MAT.	INST.	TOTAL	MAT.	INST.	TOTAL	MAT.	INST.	TOTAL	MAT.	INST.	TOTAL	MAT.	INST.	TOTAL
015433	CONTRACTOR EQUIPMENT		99.6	99.6		100.3	100.3		100.1	100.1		98.5	98.5		100.3	100.3		99.6	99.6
0241, 31 - 34	SITE & INFRASTRUCTURE, DEMOLITION	90.9	106.3	101.8	114.4	107.6	109.6	94.9	110.9	106.3	125.1	105.0	110.8	123.8	107.6	112.3	97.9	106.3	103.9
0310	Concrete Forming & Accessories	101.9	115.3	113.4	103.0	146.3	140.3	106.2	115.3	114.1	107.8	142.7	137.9	118.7	149.7	145.5	105.8	124.2	121.7
0320	Concrete Reinforcing	107.8	113.1	110.5	94.5	114.4	104.6	107.5	113.3	110.5	118.7	113.8	116.2	94.5	114.3	104.6	104.7	113.1	109.0
0330	Cast-in-Place Concrete	88.6	123.2	102.8	104.4	124.8	112.8	90.2	120.8	102.8	118.0	119.2	118.5	120.0	124.8	122.0	96.3	123.3	107.4
03	CONCRETE	96.6	116.8	106.5	96.5	131.1	113.5	96.6	116.0	106.2	122.7	127.7	125.2	108.8	132.6	120.5	102.6	120.8	111.6
04	MASONRY	77.6	118.5	103.1	97.4	134.3	120.4	106.5	121.1	115.6	130.0	123.1	125.7	113.2	132.8	125.4	78.6	118.1	103.2
05	METALS	103.3	100.8	102.5	98.9	100.3	99.3	85.3	100.8	90.1	101.6	101.8	101.7	98.9	100.2	99.3	102.8	101.0	102.3
06	WOOD, PLASTICS & COMPOSITES	96.6	110.8	104.5	96.1	150.2	126.4	86.5	110.6	100.0	110.6	149.8	132.6	115.3	154.9	137.5	101.9	122.6	113.5
07	THERMAL & MOISTURE PROTECTION	99.1	118.7	107.2	102.7	137.5	117.0	94.9	117.4	104.1	121.6	122.1	121.8	103.3	135.9	116.7	99.6	121.1	108.4
08	OPENINGS	100.1	114.1	103.4	93.1	136.9	103.3	91.8	114.0	97.0	114.1	131.5	118.1	93.1	139.4	103.9	102.8	120.6	107.0
0920	Plaster & Gypsum Board	102.0	111.0	108.1	103.9	151.1	135.9	90.2	111.0	104.3	106.2	151.1	136.6	113.6	156.0	142.3	106.3	123.2	117.7
0950, 0980	Ceilings & Acoustic Treatment	101.7	111.0	107.8	107.1	151.1	136.0	100.2	111.0	107.3	131.5	151.1	144.4	107.1	156.0	139.3	109.2	123.2	118.4
0960	Flooring	104.6	107.2	105.3	102.4	127.6	109.6	90.8	110.2	96.3	98.0	117.6	103.6	109.9	127.6	115.0	106.0	112.0	107.7
0970, 0990	Wall Finishes & Painting/Coating	97.2	111.3	105.7	102.5	143.9	127.5	100.5	111.5	107.1	115.9	115.8	115.8	102.5	143.9	127.5	97.2	106.8	103.0
09	FINISHES	99.1	112.8	106.6	102.7	144.4	125.6	97.3	113.2	106.0	109.3	138.2	125.3	106.9	147.2	129.1	102.1	120.1	112.0
COVERS	DIVS. 10 - 14, 25, 28, 41, 43, 44, 46	100.0	110.7	102.2	100.0	125.9	105.2	100.0	110.3	102.1	100.0	123.4	104.7	100.0	126.3	105.3	100.0	112.0	102.4
21, 22, 23	FIRE SUPPRESSION, PLUMBING & HVAC	95.2	115.5	103.4	95.3	145.5	115.6	95.2	115.6	103.4	100.2	109.0	103.7	95.3	147.8	116.5	100.0	118.6	107.5
26, 27, 3370	ELECTRICAL, COMMUNICATIONS & UTIL.	95.3	105.4	100.7	108.4	150.2	130.5	115.1	120.5	118.0	99.2	107.4	103.5	109.0	132.7	121.5	91.9	105.4	99.1
MF2014	WEIGHTED AVERAGE	97.2	112.0	103.6	98.9	134.5	114.4	96.4	114.6	104.3	107.8	117.7	112.1	101.8	133.1	115.4	99.4	114.6	106.0

CALIFORNIA

DIVISION		SACRAMENTO 942,956 - 958			SALINAS 939			SAN BERNARDINO 923 - 924			SAN DIEGO 919 - 921			SAN FRANCISCO 940 - 941			SAN JOSE 951		
		MAT.	INST.	TOTAL	MAT.	INST.	TOTAL	MAT.	INST.	TOTAL	MAT.	INST.	TOTAL	MAT.	INST.	TOTAL	MAT.	INST.	TOTAL
015433	CONTRACTOR EQUIPMENT		99.9	99.9		98.8	98.8		99.6	99.6		100.1	100.1		110.7	110.7		99.3	99.3
0241, 31 - 34	SITE & INFRASTRUCTURE, DEMOLITION	100.6	113.3	109.6	116.3	105.7	108.8	77.7	106.3	98.0	101.7	104.0	103.4	127.0	113.6	117.5	133.5	100.4	109.9
0310	Concrete Forming & Accessories	103.3	137.0	132.4	111.0	137.4	133.7	109.6	115.2	114.4	105.6	113.3	112.2	104.6	150.9	144.6	107.0	149.8	143.9
0320	Concrete Reinforcing	89.4	113.8	101.8	96.0	114.2	105.2	104.7	113.0	108.9	104.2	113.1	108.7	110.1	115.0	112.6	93.5	114.5	104.1
0330	Cast-in-Place Concrete	99.0	120.3	107.7	98.9	119.6	107.4	66.6	123.2	89.8	98.9	107.8	102.6	120.0	126.4	122.6	117.1	124.2	120.0
03	CONCRETE	97.0	125.2	110.9	108.5	125.6	116.9	76.8	116.8	96.4	101.8	110.7	106.2	110.3	134.4	122.1	110.8	132.9	121.7
04	MASONRY	98.4	123.1	113.8	101.6	128.1	118.1	84.9	115.9	104.2	97.4	115.6	108.7	121.1	141.5	133.8	131.4	134.4	133.3
05	METALS	97.0	96.1	96.7	105.5	102.9	104.7	102.8	100.5	102.1	101.1	101.4	101.2	107.6	110.8	108.6	97.5	107.5	100.6
06	WOOD, PLASTICS & COMPOSITES	92.9	141.7	120.2	107.3	141.5	126.4	105.8	110.8	108.6	100.6	110.2	106.0	98.8	155.1	130.3	111.3	154.7	135.6
07	THERMAL & MOISTURE PROTECTION	110.8	121.3	115.1	95.0	126.1	107.7	98.3	117.7	106.3	99.3	107.1	102.5	104.9	141.9	120.1	98.9	139.3	115.4
08	OPENINGS	106.3	127.2	111.1	96.9	133.3	105.4	100.2	114.1	103.4	99.1	111.8	102.1	97.3	140.7	107.4	92.8	140.5	103.9
0920	Plaster & Gypsum Board	101.0	142.6	129.1	105.8	142.6	130.6	108.2	111.0	110.1	97.8	110.3	106.3	108.3	156.0	140.6	102.7	156.0	138.7
0950, 0980	Ceilings & Acoustic Treatment	107.1	142.6	130.4	94.2	142.6	126.0	105.0	111.0	109.0	107.7	110.3	109.4	117.1	156.0	142.7	103.3	156.0	138.0
0960	Flooring	102.6	117.6	106.9	107.1	121.1	111.1	107.8	110.2	108.5	98.5	112.0	102.4	103.4	127.6	110.3	95.4	127.6	104.6
0970, 0990	Wall Finishes & Painting/Coating	100.2	116.7	110.1	112.3	143.9	131.4	97.2	104.0	101.3	97.1	111.5	105.8	102.5	153.2	133.1	100.8	143.9	126.8
09	FINISHES	101.4	133.6	119.1	101.4	137.0	121.0	100.5	112.5	107.1	102.9	112.9	108.4	106.5	148.3	129.6	100.7	147.0	126.2
COVERS	DIVS. 10 - 14, 25, 28, 41, 43, 44, 46	100.0	123.0	104.6	100.0	122.6	104.6	100.0	108.1	101.6	100.0	109.7	102.0	100.0	126.8	105.4	100.0	125.9	105.2
21, 22, 23	FIRE SUPPRESSION, PLUMBING & HVAC	100.0	120.8	108.4	95.3	117.7	104.4	95.2	115.6	103.4	99.9	116.9	106.8	100.1	175.0	130.3	100.0	147.3	119.1
26, 27, 3370	ELECTRICAL, COMMUNICATIONS & UTIL.	103.5	108.7	106.2	93.2	120.9	107.8	95.3	103.4	99.6	101.2	98.5	99.8	108.6	157.8	134.6	101.1	155.9	130.1
MF2014	WEIGHTED AVERAGE	100.5	119.3	108.7	100.2	121.6	109.5	95.1	111.3	102.1	100.5	109.6	104.5	105.2	145.4	122.7	102.5	136.6	117.4

CALIFORNIA

DIVISION		SAN LUIS OBISPO 934			SAN MATEO 944			SAN RAFAEL 949			SANTA ANA 926 - 927			SANTA BARBARA 931			SANTA CRUZ 950		
		MAT.	INST.	TOTAL	MAT.	INST.	TOTAL	MAT.	INST.	TOTAL	MAT.	INST.	TOTAL	MAT.	INST.	TOTAL	MAT.	INST.	TOTAL
015433	CONTRACTOR EQUIPMENT		98.8	98.8		100.3	100.3		100.3	100.3		99.6	99.6		98.8	98.8		99.3	99.3
0241, 31 - 34	SITE & INFRASTRUCTURE, DEMOLITION	108.4	106.0	106.7	121.3	107.6	111.6	113.2	113.4	113.4	89.4	106.3	101.4	103.3	106.0	105.2	133.1	100.0	109.6
0310	Concrete Forming & Accessories	117.9	115.4	115.8	109.0	149.8	144.2	112.9	149.9	144.9	109.9	115.3	114.5	108.3	124.2	122.1	107.0	137.6	133.4
0320	Concrete Reinforcing	97.3	113.1	105.4	94.5	114.6	104.7	95.1	114.6	105.1	108.3	113.1	110.8	95.5	113.2	104.5	115.7	114.2	114.9
0330	Cast-in-Place Concrete	106.3	122.5	112.9	116.2	124.8	119.7	135.2	124.1	130.7	85.0	123.2	100.7	99.8	122.6	109.1	116.3	121.4	118.4
03	CONCRETE	106.9	116.6	111.7	105.4	132.7	118.8	126.0	132.4	129.1	94.2	116.8	105.3	99.0	120.6	109.6	113.9	126.5	120.1
04	MASONRY	103.5	118.3	112.7	112.9	137.3	128.1	93.2	137.4	120.7	74.6	118.8	102.1	102.2	119.1	112.7	135.3	128.2	130.9
05	METALS	103.5	100.6	102.6	98.7	100.6	99.3	100.1	98.8	99.7	102.9	100.8	102.2	101.2	100.9	101.1	104.5	106.3	105.0
06	WOOD, PLASTICS & COMPOSITES	109.9	110.9	110.5	104.0	154.9	132.5	101.8	154.7	131.4	107.8	110.8	109.5	102.1	122.7	113.6	111.3	141.6	128.3
07	THERMAL & MOISTURE PROTECTION	95.1	116.9	104.0	103.1	138.8	117.7	107.2	138.9	120.2	99.4	118.3	107.2	94.6	118.9	104.5	98.8	129.2	111.3
08	OPENINGS	95.1	109.7	98.5	93.1	139.4	103.8	103.8	139.3	112.1	99.5	114.1	102.9	96.6	120.6	102.2	94.0	133.4	103.2
0920	Plaster & Gypsum Board	112.2	111.0	111.4	108.9	156.0	140.8	111.5	156.0	141.6	109.6	111.0	110.6	104.7	123.2	117.2	110.1	142.6	132.0
0950, 0980	Ceilings & Acoustic Treatment	94.2	111.0	105.3	107.1	156.0	139.3	115.4	156.0	142.1	105.0	111.0	109.0	96.6	123.2	114.1	106.7	142.6	130.3
0960	Flooring	113.5	110.2	112.6	105.1	127.6	111.5	115.0	121.1	116.8	108.3	110.2	108.8	106.2	108.6	106.9	99.6	121.1	105.7
0970, 0990	Wall Finishes & Painting/Coating	111.1	101.7	105.5	102.5	143.9	127.5	98.9	142.6	125.3	97.2	106.8	103.0	111.1	101.1	105.1	101.0	143.9	126.9
09	FINISHES	103.2	112.3	108.2	104.7	147.2	128.1	107.6	145.8	128.6	102.0	112.8	107.9	100.0	119.0	110.5	103.7	137.1	122.1
COVERS	DIVS. 10 - 14, 25, 28, 41, 43, 44, 46	100.0	120.6	104.2	100.0	126.4	105.3	100.0	125.7	105.2	100.0	110.7	102.2	100.0	112.2	102.5	100.0	122.9	104.6
21, 22, 23	FIRE SUPPRESSION, PLUMBING & HVAC	95.3	115.6	103.5	95.3	142.2	114.2	95.3	169.5	125.2	95.2	114.3	102.9	100.1	118.6	107.6	100.0	117.8	107.2
26, 27, 3370	ELECTRICAL, COMMUNICATIONS & UTIL.	92.0	106.4	99.6	108.4	148.0	129.3	105.4	120.0	113.1	95.4	105.7	100.8	91.0	110.5	101.3	100.3	120.9	111.2
MF2014	WEIGHTED AVERAGE	99.4	112.0	104.9	100.9	134.6	115.6	103.5	136.7	118.0	96.9	111.8	103.4	98.9	115.1	106.0	104.5	121.8	112.0

For customer support on your Mechanical Cost Data, call 877.795.7394.

661

CALIFORNIA / COLORADO

DIVISION		SANTA ROSA 954 MAT.	INST.	TOTAL	STOCKTON 952 MAT.	INST.	TOTAL	SUSANVILLE 961 MAT.	INST.	TOTAL	VALLEJO 945 MAT.	INST.	TOTAL	VAN NUYS 913-916 MAT.	INST.	TOTAL	ALAMOSA 811 MAT.	INST.	TOTAL
015433	CONTRACTOR EQUIPMENT		99.0	99.0		98.5	98.5		98.5	98.5		100.3	100.3		100.1	100.1		93.8	93.8
0241, 31 - 34	SITE & INFRASTRUCTURE, DEMOLITION	103.0	105.1	104.5	102.0	105.1	104.2	131.8	105.0	112.8	100.5	113.2	109.5	111.4	110.9	111.0	134.7	88.0	101.5
0310	Concrete Forming & Accessories	102.9	148.1	141.9	104.8	136.6	132.2	109.1	142.8	138.1	103.4	147.7	141.6	112.7	115.3	115.0	105.0	67.5	72.7
0320	Concrete Reinforcing	103.4	114.8	109.2	106.2	113.8	110.1	118.7	113.2	115.9	96.3	114.5	105.6	107.5	113.3	110.5	105.1	76.4	90.5
0330	Cast-in-Place Concrete	110.6	120.8	114.8	98.2	119.2	106.9	107.4	119.3	112.3	107.7	121.5	113.4	95.1	120.8	105.7	101.1	78.4	91.8
03	CONCRETE	113.7	130.9	122.1	103.7	125.0	114.2	125.2	127.7	126.4	102.2	130.5	116.1	110.2	116.0	113.0	113.5	73.3	93.8
04	MASONRY	103.1	137.2	124.3	102.8	123.1	115.5	127.9	112.7	118.4	71.6	133.1	109.9	122.7	121.1	121.7	126.0	71.9	92.3
05	METALS	103.2	105.0	103.8	99.3	101.8	100.0	100.6	101.1	100.7	100.1	98.1	99.5	84.4	100.8	89.5	98.2	80.2	92.7
06	WOOD, PLASTICS & COMPOSITES	98.0	154.5	129.6	104.3	141.5	125.1	112.5	149.8	133.4	90.6	154.7	126.5	95.0	110.6	103.7	98.0	67.5	80.9
07	THERMAL & MOISTURE PROTECTION	99.8	136.2	114.7	102.6	119.9	109.7	122.2	120.4	121.4	105.0	135.3	117.4	95.7	117.4	104.6	104.9	78.1	93.9
08	OPENINGS	101.0	140.4	110.2	101.6	127.1	107.5	113.9	131.5	118.0	105.6	140.5	113.7	91.6	114.0	96.8	98.3	74.1	92.7
0920	Plaster & Gypsum Board	103.9	156.0	139.2	107.5	142.6	131.2	107.1	151.1	136.9	105.8	156.0	139.8	95.3	111.0	105.9	79.0	66.3	70.4
0950, 0980	Ceilings & Acoustic Treatment	105.0	156.0	138.6	112.5	142.6	132.3	124.1	151.1	141.9	117.3	156.0	142.8	97.7	111.0	106.5	96.3	66.3	76.6
0960	Flooring	105.0	115.2	107.9	102.4	112.8	105.3	98.4	122.1	105.2	110.6	127.6	115.4	93.2	110.2	98.0	113.5	54.8	96.7
0970, 0990	Wall Finishes & Painting/Coating	97.2	142.6	124.6	100.5	117.0	110.5	115.9	115.8	115.8	99.8	142.6	125.6	100.5	111.5	107.1	114.1	39.8	69.3
09	FINISHES	101.1	143.6	124.5	103.8	132.7	119.7	108.9	139.0	125.5	104.6	145.8	127.3	99.3	113.2	107.0	102.5	61.7	80.0
COVERS	DIVS. 10 - 14, 25, 28, 41, 43, 44, 46	100.0	123.8	104.8	100.0	122.5	104.6	100.0	123.5	104.7	100.0	124.3	104.9	100.0	110.3	102.1	100.0	88.5	97.7
21, 22, 23	FIRE SUPPRESSION, PLUMBING & HVAC	95.2	167.3	124.3	100.0	111.3	104.6	95.4	109.0	100.9	100.1	127.5	111.1	95.2	115.6	103.4	95.2	72.0	85.9
26, 27, 3370	ELECTRICAL, COMMUNICATIONS & UTIL.	95.7	114.3	105.5	98.2	110.6	104.8	99.6	118.4	109.5	100.9	122.4	112.3	115.1	120.5	118.0	99.1	72.5	85.0
MF2014	WEIGHTED AVERAGE	100.8	134.7	115.6	100.9	117.2	108.0	106.8	118.1	111.7	100.1	127.2	111.9	99.1	114.6	105.9	102.2	73.7	89.8

COLORADO

DIVISION		BOULDER 803 MAT.	INST.	TOTAL	COLORADO SPRINGS 808-809 MAT.	INST.	TOTAL	DENVER 800-802 MAT.	INST.	TOTAL	DURANGO 813 MAT.	INST.	TOTAL	FORT COLLINS 805 MAT.	INST.	TOTAL	FORT MORGAN 807 MAT.	INST.	TOTAL
015433	CONTRACTOR EQUIPMENT		97.5	97.5		95.8	95.8		100.4	100.4		93.8	93.8		97.5	97.5		97.5	97.5
0241, 31 - 34	SITE & INFRASTRUCTURE, DEMOLITION	94.8	96.4	95.9	96.9	94.7	95.3	96.1	102.6	100.7	128.5	88.0	99.7	107.1	96.0	99.2	97.7	95.9	96.4
0310	Concrete Forming & Accessories	103.1	79.9	83.1	93.6	79.4	81.3	100.0	76.0	79.3	111.4	67.7	73.7	100.7	74.7	78.3	103.6	74.9	78.9
0320	Concrete Reinforcing	98.9	76.6	87.5	98.1	80.2	89.0	98.1	80.2	89.0	105.1	76.5	90.5	99.0	76.6	87.6	99.1	76.5	87.6
0330	Cast-in-Place Concrete	101.8	80.0	92.9	104.6	88.3	97.9	99.0	82.8	92.4	116.3	78.5	100.8	114.9	78.9	100.1	99.9	78.9	91.3
03	CONCRETE	103.1	79.6	91.5	106.3	82.9	94.8	100.7	79.5	90.3	116.1	73.4	95.1	113.5	76.7	95.4	101.5	76.8	89.4
04	MASONRY	91.0	72.1	79.2	91.3	81.9	85.4	93.3	72.2	80.1	113.6	71.9	87.6	108.5	75.7	88.1	104.9	72.2	84.5
05	METALS	96.0	83.5	92.2	99.1	85.7	95.0	101.6	85.6	96.7	98.2	80.4	92.7	97.2	80.4	92.0	95.7	80.3	91.0
06	WOOD, PLASTICS & COMPOSITES	102.2	83.0	91.4	91.7	77.3	83.6	99.8	77.2	87.2	107.6	67.5	85.1	99.6	77.2	87.0	102.2	77.2	88.2
07	THERMAL & MOISTURE PROTECTION	103.1	81.7	94.3	103.9	85.6	96.4	102.4	75.8	91.5	104.9	78.1	93.9	103.5	73.5	91.2	103.0	81.2	94.1
08	OPENINGS	98.2	82.5	94.5	102.4	80.5	97.3	103.0	80.4	97.8	105.3	74.1	98.0	98.2	79.4	93.8	98.1	79.4	93.8
0920	Plaster & Gypsum Board	105.1	82.7	89.9	87.9	76.7	80.3	100.7	76.8	84.5	92.5	66.3	74.8	99.1	76.7	84.0	105.1	76.7	85.9
0950, 0980	Ceilings & Acoustic Treatment	96.9	82.7	87.5	104.4	76.7	86.2	107.3	76.8	87.2	96.3	66.3	76.6	96.9	76.7	83.6	96.9	76.7	83.6
0960	Flooring	99.7	83.8	95.2	92.0	68.2	85.2	96.7	84.8	93.3	118.2	54.8	100.0	96.6	54.8	84.6	100.1	54.8	87.1
0970, 0990	Wall Finishes & Painting/Coating	101.7	66.6	80.5	101.4	40.5	64.7	101.7	75.7	86.0	114.1	39.8	69.3	101.7	40.2	64.6	101.7	53.6	72.6
09	FINISHES	101.7	79.4	89.4	98.8	72.8	84.5	102.4	77.3	88.6	105.0	61.7	81.1	100.5	67.4	82.3	101.8	68.9	83.7
COVERS	DIVS. 10 - 14, 25, 28, 41, 43, 44, 46	100.0	89.4	97.9	100.0	92.2	98.4	100.0	88.8	97.7	100.0	88.4	97.7	100.0	88.8	97.7	100.0	88.8	97.7
21, 22, 23	FIRE SUPPRESSION, PLUMBING & HVAC	95.3	77.3	88.1	100.2	88.2	95.4	100.0	79.9	91.9	95.2	83.1	90.3	100.1	77.2	90.9	95.3	77.2	88.0
26, 27, 3370	ELECTRICAL, COMMUNICATIONS & UTIL.	97.4	84.5	90.6	100.8	82.1	90.9	102.6	83.2	92.4	98.5	69.4	83.1	97.4	84.5	90.6	97.8	84.5	90.7
MF2014	WEIGHTED AVERAGE	97.8	81.3	90.6	100.4	84.0	93.2	100.8	81.7	92.5	102.7	75.7	90.9	101.3	79.0	91.6	98.3	79.0	89.9

COLORADO

DIVISION		GLENWOOD SPRINGS 816 MAT.	INST.	TOTAL	GOLDEN 804 MAT.	INST.	TOTAL	GRAND JUNCTION 815 MAT.	INST.	TOTAL	GREELEY 806 MAT.	INST.	TOTAL	MONTROSE 814 MAT.	INST.	TOTAL	PUEBLO 810 MAT.	INST.	TOTAL
015433	CONTRACTOR EQUIPMENT		96.7	96.7		97.5	97.5		96.7	96.7		97.5	97.5		95.2	95.2		93.8	93.8
0241, 31 - 34	SITE & INFRASTRUCTURE, DEMOLITION	143.2	95.1	109.0	107.3	96.2	99.4	128.4	94.7	104.5	94.0	95.4	95.0	137.1	91.2	104.5	120.6	91.0	99.6
0310	Concrete Forming & Accessories	102.0	75.0	78.7	96.1	74.8	77.7	110.2	74.4	79.3	98.6	78.4	81.2	101.4	74.8	78.4	107.3	79.4	83.3
0320	Concrete Reinforcing	104.0	76.5	90.0	99.1	76.4	87.6	104.3	76.3	90.1	98.9	76.1	86.8	103.9	76.4	89.9	100.6	80.2	90.2
0330	Cast-in-Place Concrete	101.1	77.9	91.6	100.0	78.9	91.3	111.9	77.1	97.6	96.1	59.1	80.9	101.1	77.8	91.5	100.4	87.8	95.2
03	CONCRETE	118.5	76.5	97.9	111.9	76.7	94.6	112.5	75.9	94.5	98.3	71.3	85.0	109.7	76.3	93.3	102.7	82.7	92.9
04	MASONRY	98.9	72.1	82.2	107.7	71.8	85.3	132.6	71.5	94.5	102.6	49.0	69.2	106.2	71.9	84.9	95.4	81.7	86.9
05	METALS	97.9	80.9	92.6	95.9	80.1	91.0	99.5	79.2	93.2	97.2	77.5	91.1	97.1	80.0	91.8	101.1	86.2	96.5
06	WOOD, PLASTICS & COMPOSITES	93.2	77.3	84.3	94.3	77.2	84.7	105.3	77.3	89.6	96.8	83.0	89.1	94.3	77.4	84.8	100.9	77.6	87.8
07	THERMAL & MOISTURE PROTECTION	104.8	79.2	94.3	103.9	75.4	92.2	104.0	68.5	89.5	102.9	66.4	87.9	105.0	79.2	94.4	103.5	83.6	95.3
08	OPENINGS	104.2	79.4	98.5	98.2	78.8	93.7	105.0	78.8	98.9	98.1	82.4	94.5	105.5	79.5	99.4	100.2	80.6	95.6
0920	Plaster & Gypsum Board	117.8	76.7	90.0	96.9	76.7	83.3	131.0	76.7	94.3	97.6	82.7	87.5	78.3	76.7	77.2	83.9	76.7	79.1
0950, 0980	Ceilings & Acoustic Treatment	95.5	76.7	83.2	96.9	76.7	83.6	95.5	76.7	83.2	96.9	82.7	87.5	96.3	76.7	83.4	96.9	76.7	83.6
0960	Flooring	112.8	50.4	95.0	94.7	54.8	83.3	117.5	54.8	99.6	95.7	54.8	84.0	115.8	45.4	95.7	114.6	84.8	106.1
0970, 0990	Wall Finishes & Painting/Coating	114.1	66.5	85.4	101.7	66.6	80.5	114.1	66.6	85.4	101.7	25.0	55.4	114.1	39.8	69.3	114.1	37.5	67.9
09	FINISHES	107.9	69.6	86.8	100.4	71.0	84.2	109.3	70.4	87.9	99.3	69.2	82.7	103.2	65.7	82.5	103.7	75.6	88.2
COVERS	DIVS. 10 - 14, 25, 28, 41, 43, 44, 46	100.0	88.8	97.7	100.0	88.8	97.7	100.0	88.8	97.7	100.0	89.4	97.9	100.0	89.1	97.8	100.0	92.7	98.5
21, 22, 23	FIRE SUPPRESSION, PLUMBING & HVAC	95.2	83.0	90.3	95.3	76.7	87.8	99.9	82.5	92.9	100.1	77.2	90.8	95.2	83.0	90.3	99.9	76.8	90.6
26, 27, 3370	ELECTRICAL, COMMUNICATIONS & UTIL.	95.7	69.4	81.8	97.8	84.5	90.7	98.2	53.8	74.7	97.4	84.5	90.6	98.2	56.4	76.1	99.1	73.0	85.3
MF2014	WEIGHTED AVERAGE	102.3	78.1	91.8	99.8	78.9	90.7	104.8	75.3	91.9	98.9	75.5	88.7	101.4	75.4	90.1	101.1	80.3	92.0

For customer support on your Mechanical Cost Data, call 877.795.7394.

City Cost Indexes

COLORADO / CONNECTICUT

DIVISION		SALIDA 812 MAT.	INST.	TOTAL	BRIDGEPORT 066 MAT.	INST.	TOTAL	BRISTOL 060 MAT.	INST.	TOTAL	HARTFORD 061 MAT.	INST.	TOTAL	MERIDEN 064 MAT.	INST.	TOTAL	NEW BRITAIN 060 MAT.	INST.	TOTAL
015433	CONTRACTOR EQUIPMENT		95.2	95.2		100.9	100.9		100.9	100.9		100.9	100.9		101.3	101.3		100.9	100.9
0241, 31 - 34	SITE & INFRASTRUCTURE, DEMOLITION	128.2	91.5	102.1	109.8	105.1	106.4	108.9	105.0	106.2	104.4	105.0	104.9	106.7	105.8	106.0	109.1	105.0	106.2
0310	Concrete Forming & Accessories	110.1	74.7	79.6	98.9	124.5	121.0	98.9	124.3	120.8	95.7	124.3	120.4	98.6	124.3	120.8	99.2	124.3	120.9
0320	Concrete Reinforcing	103.7	76.4	89.8	105.1	127.3	116.4	105.1	127.3	116.4	104.6	127.3	116.2	105.1	127.3	116.4	105.1	127.3	116.4
0330	Cast-in-Place Concrete	115.8	77.8	100.2	107.9	127.1	115.8	101.1	127.1	111.8	103.0	127.1	112.9	97.3	127.1	109.5	102.8	127.1	112.0
03	CONCRETE	111.2	76.3	94.1	106.6	125.7	116.0	103.4	125.6	114.3	104.0	125.6	114.6	101.6	125.6	113.4	104.2	125.6	114.7
04	MASONRY	133.9	71.9	95.3	111.8	134.0	125.6	102.9	134.0	122.3	107.7	134.0	124.1	102.5	134.0	122.1	105.2	134.0	123.1
05	METALS	96.7	79.9	91.6	100.5	125.5	108.2	100.5	125.4	108.1	105.6	125.4	111.7	97.4	125.4	106.0	96.5	125.4	105.4
06	WOOD, PLASTICS & COMPOSITES	101.9	77.4	88.2	98.8	123.4	112.6	98.8	123.4	112.6	92.3	123.4	109.7	98.8	123.4	112.6	98.8	123.4	112.6
07	THERMAL & MOISTURE PROTECTION	104.0	79.2	93.8	98.6	129.1	111.1	98.7	126.0	109.9	102.8	126.0	112.3	98.7	126.0	109.9	98.7	126.0	109.9
08	OPENINGS	98.4	79.5	94.0	101.2	132.1	108.4	101.2	132.1	108.4	100.3	132.1	107.7	103.6	132.1	110.2	101.2	132.1	108.4
0920	Plaster & Gypsum Board	78.6	76.7	77.3	101.7	123.5	116.4	101.7	123.5	116.4	97.4	123.5	115.0	103.2	123.5	116.9	101.7	123.5	116.4
0950, 0980	Ceilings & Acoustic Treatment	96.3	76.7	83.4	86.4	123.5	110.8	86.4	123.5	110.8	89.7	123.5	111.9	90.4	123.5	112.2	86.4	123.5	110.8
0960	Flooring	120.6	45.4	99.1	101.9	131.2	110.3	101.9	131.2	110.3	100.5	131.2	109.2	101.9	131.2	110.3	101.9	131.2	110.3
0970, 0990	Wall Finishes & Painting/Coating	114.1	41.7	70.4	104.9	122.9	115.8	104.9	122.9	115.8	105.9	122.9	116.2	104.9	122.9	115.8	104.9	122.9	115.8
09	FINISHES	103.6	65.9	82.8	97.2	125.0	112.6	97.3	125.0	112.6	96.3	125.0	112.1	98.4	125.0	113.1	97.3	125.0	112.6
COVERS	DIVS. 10 - 14, 25, 28, 41, 43, 44, 46	100.0	89.2	97.8	100.0	112.4	102.5	100.0	112.4	102.5	100.0	112.4	102.5	100.0	112.4	102.5	100.0	112.4	102.5
21, 22, 23	FIRE SUPPRESSION, PLUMBING & HVAC	95.2	72.0	85.8	100.0	116.4	106.6	100.0	116.4	106.6	100.0	116.4	106.6	95.2	116.4	103.8	100.0	116.4	106.6
26, 27, 3370	ELECTRICAL, COMMUNICATIONS & UTIL.	98.4	72.5	84.7	98.5	112.2	105.7	98.5	111.8	105.6	97.9	112.8	105.8	98.4	111.8	105.5	98.6	111.8	105.6
MF2014	WEIGHTED AVERAGE	101.9	75.3	90.3	101.3	120.8	109.8	100.5	120.6	109.3	101.4	120.8	109.8	98.9	120.7	108.4	100.1	120.6	109.1

CONNECTICUT

DIVISION		NEW HAVEN 065 MAT.	INST.	TOTAL	NEW LONDON 063 MAT.	INST.	TOTAL	NORWALK 068 MAT.	INST.	TOTAL	STAMFORD 069 MAT.	INST.	TOTAL	WATERBURY 067 MAT.	INST.	TOTAL	WILLIMANTIC 062 MAT.	INST.	TOTAL
015433	CONTRACTOR EQUIPMENT		101.3	101.3		101.3	101.3		100.9	100.9		100.9	100.9		100.9	100.9		100.9	100.9
0241, 31 - 34	SITE & INFRASTRUCTURE, DEMOLITION	108.9	105.8	106.7	101.1	105.8	104.4	109.5	105.1	106.4	110.2	105.1	106.6	109.5	105.0	106.3	109.5	105.0	106.3
0310	Concrete Forming & Accessories	98.6	124.3	120.8	98.6	124.3	120.8	98.9	124.9	121.3	98.9	124.9	121.3	98.9	124.3	120.8	98.9	124.1	120.6
0320	Concrete Reinforcing	105.1	127.3	116.4	82.4	127.3	105.2	105.1	127.4	116.5	105.1	127.4	116.5	105.1	127.3	116.4	105.1	127.2	116.4
0330	Cast-in-Place Concrete	104.5	127.1	113.8	89.1	127.1	104.7	106.2	128.6	115.4	107.9	128.6	116.4	107.9	127.1	115.8	100.8	125.8	111.0
03	CONCRETE	118.7	125.6	122.1	91.4	125.6	108.2	105.8	126.4	115.9	106.6	126.4	116.3	106.6	125.6	115.9	103.3	125.0	114.0
04	MASONRY	103.2	134.0	122.4	101.5	134.0	121.7	102.7	135.4	123.1	103.4	135.4	123.3	103.4	134.0	122.5	102.7	134.0	122.2
05	METALS	96.8	125.4	105.6	96.5	125.4	105.4	100.5	126.1	108.3	100.5	126.1	108.1	100.5	125.4	108.1	100.2	125.2	107.9
06	WOOD, PLASTICS & COMPOSITES	98.8	123.4	112.6	98.8	123.4	112.6	98.8	123.4	112.6	98.8	123.4	112.6	98.8	123.4	112.6	98.8	123.4	112.6
07	THERMAL & MOISTURE PROTECTION	98.8	126.1	110.0	98.7	126.0	109.9	98.8	129.6	111.4	98.7	129.6	111.4	98.7	126.1	110.0	98.9	124.6	109.5
08	OPENINGS	101.2	132.1	108.4	104.0	132.1	110.6	101.2	132.1	108.4	101.2	132.1	108.4	101.2	132.1	108.4	104.0	132.1	110.6
0920	Plaster & Gypsum Board	101.7	123.5	116.4	101.7	123.5	116.4	101.7	123.5	116.4	101.7	123.5	116.4	101.7	123.5	116.4	101.7	123.5	116.4
0950, 0980	Ceilings & Acoustic Treatment	86.4	123.5	110.8	84.5	123.5	110.2	86.4	123.5	110.8	86.4	123.5	110.8	86.4	123.5	110.8	84.5	123.5	110.2
0960	Flooring	101.9	131.2	110.3	101.9	131.2	110.3	101.9	131.2	110.3	101.9	131.2	110.3	101.9	131.2	110.3	101.9	133.2	110.9
0970, 0990	Wall Finishes & Painting/Coating	104.9	122.9	115.8	104.9	122.9	115.8	104.9	122.9	115.8	104.9	122.9	115.8	104.9	122.9	115.8	104.9	122.9	115.8
09	FINISHES	97.3	125.0	112.6	96.4	125.0	112.2	97.3	125.0	112.6	97.4	125.0	112.6	97.2	125.0	112.5	97.2	125.4	112.6
COVERS	DIVS. 10 - 14, 25, 28, 41, 43, 44, 46	100.0	112.4	102.5	100.0	112.4	102.5	100.0	112.6	102.6	100.0	112.6	102.6	100.0	112.4	102.5	100.0	112.4	102.5
21, 22, 23	FIRE SUPPRESSION, PLUMBING & HVAC	100.0	116.4	106.6	95.2	116.4	103.8	100.0	116.4	106.6	100.0	116.4	106.6	100.0	116.4	106.6	100.0	116.1	106.5
26, 27, 3370	ELECTRICAL, COMMUNICATIONS & UTIL.	98.4	111.8	105.5	95.3	111.8	104.0	98.5	167.2	134.8	98.5	167.2	134.8	98.0	112.2	105.5	98.5	109.6	104.4
MF2014	WEIGHTED AVERAGE	101.6	120.7	109.9	97.1	120.7	107.4	100.8	128.7	113.0	100.9	128.7	113.0	100.9	120.7	109.5	100.8	120.2	109.2

D.C. / DELAWARE / FLORIDA

DIVISION		WASHINGTON 200 - 205 MAT.	INST.	TOTAL	DOVER 199 MAT.	INST.	TOTAL	NEWARK 197 MAT.	INST.	TOTAL	WILMINGTON 198 MAT.	INST.	TOTAL	DAYTONA BEACH 321 MAT.	INST.	TOTAL	FORT LAUDERDALE 333 MAT.	INST.	TOTAL
015433	CONTRACTOR EQUIPMENT		105.5	105.5		118.0	118.0		118.0	118.0		118.2	118.2		98.2	98.2		91.0	91.0
0241, 31 - 34	SITE & INFRASTRUCTURE, DEMOLITION	103.8	94.5	97.2	102.3	113.3	110.1	102.6	113.3	110.2	99.5	113.6	109.5	105.9	89.2	94.0	96.4	77.1	82.7
0310	Concrete Forming & Accessories	97.4	78.3	80.9	96.3	102.2	101.4	91.5	102.2	101.5	98.1	102.2	101.6	97.6	68.1	72.1	95.7	68.4	72.1
0320	Concrete Reinforcing	105.4	93.8	99.5	94.0	102.9	98.5	91.5	102.9	97.3	93.4	102.9	98.2	91.2	76.6	83.8	88.2	72.1	80.0
0330	Cast-in-Place Concrete	115.4	88.0	104.2	104.6	104.6	104.6	90.0	104.6	96.0	99.2	104.6	101.4	90.2	70.8	82.2	94.6	77.1	87.4
03	CONCRETE	107.7	85.9	97.0	102.2	104.0	103.1	94.6	104.0	99.2	99.6	104.0	101.8	91.6	71.8	81.9	94.5	73.3	84.0
04	MASONRY	98.2	80.0	86.9	107.5	98.1	101.6	103.2	98.1	100.0	108.5	98.1	102.0	99.1	65.7	78.3	102.8	68.3	81.3
05	METALS	99.9	108.6	102.6	102.0	116.7	106.5	103.5	116.7	107.5	102.0	116.7	106.5	104.7	91.1	100.5	102.0	89.6	98.2
06	WOOD, PLASTICS & COMPOSITES	95.3	76.2	84.6	94.0	101.9	98.4	95.8	101.9	99.2	92.0	101.9	97.6	96.3	69.2	81.1	84.3	67.2	74.7
07	THERMAL & MOISTURE PROTECTION	101.7	85.8	95.2	98.8	110.0	103.4	101.7	110.0	105.1	97.9	110.0	102.9	96.2	73.6	86.9	100.3	77.2	90.8
08	OPENINGS	99.9	87.9	97.1	91.3	110.9	95.9	91.6	110.9	96.1	89.9	110.9	94.8	98.6	67.4	91.3	97.5	65.5	90.1
0920	Plaster & Gypsum Board	108.0	75.5	86.0	97.1	101.8	100.3	99.5	101.8	101.1	101.7	101.8	101.8	95.6	68.7	77.4	105.4	66.6	79.2
0950, 0980	Ceilings & Acoustic Treatment	112.2	75.5	88.1	91.3	101.8	98.2	88.7	101.8	97.3	92.4	101.8	98.6	84.9	68.7	74.2	87.9	66.6	73.9
0960	Flooring	104.7	93.6	101.5	99.8	109.7	102.7	96.8	109.7	100.5	104.3	109.7	105.9	110.5	73.6	100.0	105.1	69.3	94.9
0970, 0990	Wall Finishes & Painting/Coating	108.0	82.2	92.4	97.2	106.4	102.8	97.3	106.4	102.8	95.0	106.4	101.9	104.5	74.1	86.2	98.9	70.2	81.6
09	FINISHES	100.8	80.5	89.6	96.2	103.6	100.3	95.0	103.6	99.7	98.7	103.6	101.4	96.9	69.3	81.7	94.9	68.1	80.1
COVERS	DIVS. 10 - 14, 25, 28, 41, 43, 44, 46	100.0	98.6	99.7	100.0	88.0	97.6	100.0	88.0	97.6	100.0	88.0	97.6	100.0	84.4	96.9	100.0	87.4	97.5
21, 22, 23	FIRE SUPPRESSION, PLUMBING & HVAC	100.1	92.1	96.9	100.0	117.4	107.0	100.1	117.4	107.1	100.1	117.4	107.1	99.9	76.1	90.3	100.0	66.5	86.5
26, 27, 3370	ELECTRICAL, COMMUNICATIONS & UTIL.	101.5	105.8	103.8	96.6	111.9	104.7	98.7	111.9	105.7	99.8	111.9	106.2	95.8	54.9	74.2	96.0	72.7	83.7
MF2014	WEIGHTED AVERAGE	101.1	91.9	97.1	99.3	109.2	103.6	98.8	109.2	103.3	99.3	109.2	103.6	99.0	72.9	87.6	98.6	72.6	87.2

For customer support on your Mechanical Cost Data, call 877.795.7394.

663

City Cost Indexes

FLORIDA

DIVISION		FORT MYERS 339,341			GAINESVILLE 326,344			JACKSONVILLE 320,322			LAKELAND 338			MELBOURNE 329			MIAMI 330 - 332,340		
		MAT.	INST.	TOTAL	MAT.	INST.	TOTAL	MAT.	INST.	TOTAL	MAT.	INST.	TOTAL	MAT.	INST.	TOTAL	MAT.	INST.	TOTAL
015433	CONTRACTOR EQUIPMENT		98.2	98.2		98.2	98.2		98.2	98.2		98.2	98.2		98.2	98.2		91.0	91.0
0241, 31 - 34	SITE & INFRASTRUCTURE, DEMOLITION	107.3	88.3	93.8	113.9	88.3	95.7	106.0	88.6	93.6	109.2	88.7	94.6	112.7	88.5	95.5	99.5	76.9	83.4
0310	Concrete Forming & Accessories	91.5	74.8	77.1	92.7	54.3	59.6	97.4	54.7	60.5	88.0	75.3	77.1	93.8	69.8	73.1	100.7	68.7	73.1
0320	Concrete Reinforcing	89.3	92.7	91.0	96.8	64.8	80.5	91.2	64.9	77.8	91.5	93.6	92.5	92.3	76.6	84.3	94.3	72.2	83.0
0330	Cast-in-Place Concrete	98.7	68.2	86.2	103.6	62.6	86.7	91.1	68.3	81.8	100.9	69.6	88.1	108.7	73.1	94.1	95.5	78.1	88.3
03	CONCRETE	95.1	76.9	86.2	102.6	60.7	82.0	92.1	62.8	77.7	96.8	77.8	87.5	103.0	73.3	88.4	96.1	73.7	85.1
04	MASONRY	95.8	61.9	74.7	114.0	61.2	81.1	99.1	61.2	75.5	114.1	75.8	90.2	97.5	69.8	80.2	103.1	71.9	83.7
05	METALS	104.2	96.8	102.0	103.6	85.0	97.9	103.2	85.4	97.7	104.1	97.8	102.2	113.4	91.3	106.6	102.4	88.5	98.1
06	WOOD, PLASTICS & COMPOSITES	81.1	77.4	79.0	89.9	51.8	68.5	96.3	51.8	71.4	76.5	77.4	77.0	91.5	69.2	79.0	90.6	67.2	77.5
07	THERMAL & MOISTURE PROTECTION	100.1	80.0	91.9	96.5	61.7	82.2	96.4	62.3	82.4	100.1	84.6	93.7	96.6	76.0	88.2	101.6	71.6	89.3
08	OPENINGS	98.4	74.4	92.8	96.9	52.7	86.6	98.6	55.8	88.6	98.4	75.1	93.0	97.8	70.7	91.5	99.7	65.5	91.7
0920	Plaster & Gypsum Board	101.0	77.1	84.9	91.3	50.7	63.9	95.6	50.7	65.3	97.4	77.1	83.7	91.3	68.7	76.0	103.2	66.6	78.4
0950, 0980	Ceilings & Acoustic Treatment	82.7	77.1	79.0	79.5	50.7	60.6	84.9	50.7	62.5	82.7	77.1	79.0	83.3	68.7	73.7	92.2	66.6	75.4
0960	Flooring	102.3	55.1	88.8	108.1	43.1	89.5	110.5	64.5	97.4	100.3	56.5	87.8	108.3	73.6	98.4	107.3	73.0	97.5
0970, 0990	Wall Finishes & Painting/Coating	103.7	67.6	81.9	104.5	67.6	82.3	104.5	67.6	82.3	103.7	67.6	81.9	104.5	91.9	96.9	96.0	70.2	80.5
09	FINISHES	94.4	70.7	81.3	95.3	52.5	71.7	97.0	56.7	74.8	93.5	71.0	81.1	95.9	72.3	82.9	97.1	69.2	81.7
COVERS	DIVS. 10 - 14, 25, 28, 41, 43, 44, 46	100.0	72.6	94.5	100.0	82.8	96.5	100.0	81.0	96.2	100.0	72.6	94.5	100.0	85.8	97.1	100.0	88.0	97.6
21, 22, 23	FIRE SUPPRESSION, PLUMBING & HVAC	97.4	64.3	84.1	98.8	64.1	84.8	99.9	64.2	85.5	97.4	80.8	90.7	99.9	78.2	91.2	100.0	66.4	86.4
26, 27, 3370	ELECTRICAL, COMMUNICATIONS & UTIL.	98.1	62.8	79.5	96.1	71.8	83.3	95.5	62.2	77.9	96.3	61.6	77.9	97.0	67.8	81.6	100.1	74.9	86.8
MF2014	WEIGHTED AVERAGE	98.6	72.6	87.3	100.3	66.7	85.7	98.8	66.4	84.7	99.4	77.8	90.0	101.7	76.3	90.6	99.8	73.1	88.2

FLORIDA

DIVISION		ORLANDO 327 - 328,347			PANAMA CITY 324			PENSACOLA 325			SARASOTA 342			ST. PETERSBURG 337			TALLAHASSEE 323		
		MAT.	INST.	TOTAL	MAT.	INST.	TOTAL	MAT.	INST.	TOTAL	MAT.	INST.	TOTAL	MAT.	INST.	TOTAL	MAT.	INST.	TOTAL
015433	CONTRACTOR EQUIPMENT		98.2	98.2		98.2	98.2		98.2	98.2		98.2	98.2		98.2	98.2		98.2	98.2
0241, 31 - 34	SITE & INFRASTRUCTURE, DEMOLITION	107.6	88.4	93.9	117.6	87.4	96.1	117.6	87.9	96.5	114.0	88.4	95.8	110.9	88.1	94.7	105.1	87.7	92.7
0310	Concrete Forming & Accessories	101.4	72.0	76.1	96.7	43.9	51.1	94.6	51.8	57.7	96.0	75.0	77.9	94.9	51.0	57.1	99.9	44.1	51.8
0320	Concrete Reinforcing	96.6	73.9	85.0	95.3	72.5	83.7	97.7	73.0	85.1	92.5	93.5	93.0	91.5	86.1	88.7	98.3	64.7	81.2
0330	Cast-in-Place Concrete	112.1	70.6	95.1	95.7	56.7	79.7	118.2	65.9	96.8	106.6	69.5	91.4	102.0	64.5	86.6	97.2	56.4	80.5
03	CONCRETE	103.4	73.0	88.4	100.8	55.4	78.5	110.7	62.2	86.9	100.2	77.5	89.1	98.5	63.8	81.4	97.3	54.0	76.0
04	MASONRY	100.6	65.7	78.8	103.9	47.3	68.6	124.7	54.5	80.9	99.8	75.8	84.8	157.9	49.0	90.0	103.6	53.8	72.6
05	METALS	102.4	89.6	98.5	104.4	86.5	98.9	105.6	87.9	100.2	105.2	97.5	102.8	105.0	92.9	101.3	102.2	84.6	96.8
06	WOOD, PLASTICS & COMPOSITES	95.8	75.2	84.2	95.1	41.8	65.2	92.7	51.5	69.7	95.7	77.4	85.4	85.5	50.1	65.7	95.0	41.3	64.9
07	THERMAL & MOISTURE PROTECTION	94.9	74.3	86.4	96.7	56.6	80.2	96.6	62.3	82.5	98.1	84.6	92.5	100.3	57.8	82.8	102.6	71.5	89.9
08	OPENINGS	101.4	69.2	93.9	96.5	46.5	84.9	96.5	56.9	87.3	99.7	74.2	93.8	98.4	60.9	89.6	100.2	47.3	87.9
0920	Plaster & Gypsum Board	99.7	74.9	82.9	94.5	40.4	57.9	97.4	50.5	65.7	97.8	77.1	83.8	103.5	49.0	66.7	108.1	40.0	62.0
0950, 0980	Ceilings & Acoustic Treatment	91.4	74.9	80.5	83.3	40.4	55.1	83.3	50.5	61.7	86.2	77.1	80.3	84.5	49.0	61.2	94.1	40.0	58.5
0960	Flooring	104.1	73.6	95.4	110.1	43.0	90.9	106.1	63.0	93.7	111.1	57.9	95.9	104.1	55.2	90.1	111.1	62.0	97.0
0970, 0990	Wall Finishes & Painting/Coating	103.4	70.2	83.4	104.5	64.9	80.6	104.5	67.6	82.3	109.5	67.6	84.2	103.7	60.8	77.8	99.5	67.6	80.3
09	FINISHES	98.2	72.4	84.0	97.5	44.8	68.5	96.4	54.8	73.5	99.9	71.2	84.1	95.9	51.7	71.6	100.7	48.5	72.0
COVERS	DIVS. 10 - 14, 25, 28, 41, 43, 44, 46	100.0	85.1	97.0	100.0	46.2	89.1	100.0	46.3	89.1	100.0	72.6	94.5	100.0	55.8	91.1	100.0	65.8	93.1
21, 22, 23	FIRE SUPPRESSION, PLUMBING & HVAC	99.9	56.6	82.4	99.9	52.3	80.7	99.9	52.5	80.8	99.9	64.8	85.7	100.0	58.4	83.2	100.0	38.9	75.4
26, 27, 3370	ELECTRICAL, COMMUNICATIONS & UTIL.	97.6	60.0	77.7	94.5	59.3	75.9	98.5	55.8	75.9	97.2	61.6	78.4	96.3	61.6	77.9	103.3	60.0	80.4
MF2014	WEIGHTED AVERAGE	100.5	70.0	87.2	100.2	57.8	81.7	102.7	61.2	84.6	100.8	74.3	89.2	102.6	63.3	85.5	100.8	56.9	81.7

FLORIDA / GEORGIA

DIVISION		TAMPA 335 - 336,346			WEST PALM BEACH 334,349			ALBANY 317,398			ATHENS 306			ATLANTA 300 - 303,399			AUGUSTA 308 - 309		
		MAT.	INST.	TOTAL	MAT.	INST.	TOTAL	MAT.	INST.	TOTAL	MAT.	INST.	TOTAL	MAT.	INST.	TOTAL	MAT.	INST.	TOTAL
015433	CONTRACTOR EQUIPMENT		98.2	98.2		91.0	91.0		91.9	91.9		94.2	94.2		94.7	94.7		94.2	94.2
0241, 31 - 34	SITE & INFRASTRUCTURE, DEMOLITION	111.4	88.6	95.2	93.2	77.1	81.8	98.8	78.7	84.5	100.9	93.4	95.6	97.6	94.8	95.7	94.5	93.2	93.6
0310	Concrete Forming & Accessories	97.7	75.6	78.6	99.2	68.1	72.4	90.4	43.6	50.0	92.9	45.9	52.4	96.7	73.5	76.7	94.2	65.5	69.4
0320	Concrete Reinforcing	88.2	93.6	91.0	90.7	71.8	81.1	90.6	80.2	85.3	95.3	77.9	86.5	94.6	81.2	87.8	95.7	71.8	83.6
0330	Cast-in-Place Concrete	99.7	69.7	87.4	90.0	74.3	83.5	92.4	54.5	76.8	107.8	55.6	86.4	107.8	70.8	92.6	101.9	49.9	80.5
03	CONCRETE	97.0	77.9	87.6	91.3	72.1	81.9	93.1	55.8	74.8	104.8	56.0	80.8	102.2	74.1	88.4	97.5	61.7	79.9
04	MASONRY	103.9	75.8	86.4	102.3	66.6	80.0	102.5	49.1	69.2	81.5	53.8	64.2	95.2	66.2	77.1	95.5	43.3	63.0
05	METALS	104.0	98.1	102.2	101.1	89.3	97.4	105.0	85.5	99.0	91.9	73.3	86.2	92.8	77.0	88.0	91.6	71.6	85.5
06	WOOD, PLASTICS & COMPOSITES	89.3	77.4	82.6	89.2	67.2	76.9	85.0	38.0	58.7	91.8	40.9	63.3	96.1	75.8	84.7	93.3	70.9	80.8
07	THERMAL & MOISTURE PROTECTION	100.5	84.6	94.0	100.1	70.2	87.8	95.9	60.3	81.3	94.1	52.8	77.2	94.0	72.0	85.0	93.7	57.0	78.7
08	OPENINGS	99.7	79.2	94.9	96.8	65.5	89.5	91.1	44.2	80.2	89.5	45.7	79.3	94.7	71.9	89.4	89.5	62.2	83.1
0920	Plaster & Gypsum Board	106.1	77.1	86.5	110.2	66.6	80.7	97.5	36.5	56.3	98.7	39.4	58.7	100.9	75.4	83.6	99.8	70.4	79.9
0950, 0980	Ceilings & Acoustic Treatment	87.9	77.1	80.8	82.7	66.6	72.1	84.0	36.5	52.8	97.2	39.4	59.2	97.2	75.4	82.9	98.1	70.4	79.9
0960	Flooring	105.1	56.5	91.2	106.8	66.4	95.3	110.9	47.8	92.9	96.4	53.9	84.2	97.7	65.2	88.4	96.6	46.6	82.3
0970, 0990	Wall Finishes & Painting/Coating	103.7	67.6	81.9	98.9	70.2	81.6	105.3	52.8	73.6	106.0	46.0	69.8	106.0	85.1	93.4	106.0	46.0	69.8
09	FINISHES	97.4	71.0	82.8	94.7	67.5	79.7	98.1	43.2	67.9	95.4	45.7	68.0	95.7	73.4	83.4	95.2	60.5	76.1
COVERS	DIVS. 10 - 14, 25, 28, 41, 43, 44, 46	100.0	85.1	97.0	100.0	87.4	97.5	100.0	80.2	96.0	100.0	77.3	95.4	100.0	85.6	97.1	100.0	78.5	95.7
21, 22, 23	FIRE SUPPRESSION, PLUMBING & HVAC	100.0	80.8	92.2	97.4	62.7	83.4	99.9	68.0	87.0	95.2	69.6	84.9	99.9	70.6	88.1	100.0	61.2	84.4
26, 27, 3370	ELECTRICAL, COMMUNICATIONS & UTIL.	96.0	61.6	77.8	97.2	72.7	84.3	96.9	58.7	76.7	99.7	69.5	83.8	99.0	71.9	84.7	100.4	61.8	80.0
MF2014	WEIGHTED AVERAGE	100.1	78.4	90.6	97.4	71.1	85.9	98.5	61.3	82.3	95.5	63.9	81.7	97.6	74.4	87.5	96.3	63.6	82.1

City Cost Indexes

GEORGIA

DIVISION		COLUMBUS 318-319			DALTON 307			GAINESVILLE 305			MACON 310-312			SAVANNAH 313-314			STATESBORO 304		
		MAT.	INST.	TOTAL	MAT.	INST.	TOTAL	MAT.	INST.	TOTAL	MAT.	INST.	TOTAL	MAT.	INST.	TOTAL	MAT.	INST.	TOTAL
015433	CONTRACTOR EQUIPMENT		91.9	91.9		106.1	106.1		94.2	94.2		102.3	102.3		92.8	92.8		93.4	93.4
0241, 31 - 34	SITE & INFRASTRUCTURE, DEMOLITION	98.7	78.7	84.5	101.1	97.8	98.8	100.8	93.3	95.5	99.8	93.5	95.3	100.6	80.1	86.1	102.0	77.5	84.6
0310	Concrete Forming & Accessories	90.3	54.3	59.3	85.7	46.7	52.1	96.4	42.9	50.2	89.9	51.9	57.1	92.1	50.0	55.8	80.3	51.4	55.4
0320	Concrete Reinforcing	90.9	80.6	85.7	94.8	73.9	84.2	95.1	77.8	86.3	92.1	80.3	86.1	98.1	71.9	84.7	94.4	41.8	67.6
0330	Cast-in-Place Concrete	92.1	53.9	76.4	104.7	50.2	82.3	113.3	52.9	88.5	90.9	65.6	80.5	100.2	55.2	81.7	107.6	59.7	87.9
03	CONCRETE	93.0	60.4	77.0	103.9	54.6	79.7	106.7	53.7	80.6	92.6	63.3	78.2	98.0	57.3	78.0	104.2	54.0	79.6
04	MASONRY	102.6	55.2	73.1	83.8	36.7	54.4	89.5	54.2	67.5	115.9	44.5	71.4	98.8	50.9	69.0	84.7	40.6	57.2
05	METALS	104.5	86.3	98.9	92.8	82.8	89.7	91.2	72.5	85.4	100.1	85.9	95.7	101.3	82.4	95.5	96.3	72.6	89.0
06	WOOD, PLASTICS & COMPOSITES	85.0	52.1	66.6	75.2	47.5	59.7	95.9	37.9	63.4	91.8	51.2	69.0	96.0	45.9	68.0	68.5	53.9	60.3
07	THERMAL & MOISTURE PROTECTION	95.9	63.5	82.6	96.1	51.5	77.8	94.1	54.8	78.0	94.4	62.2	81.2	95.0	56.6	79.2	94.7	51.1	76.8
08	OPENINGS	91.1	59.0	83.7	90.1	49.5	80.6	89.5	40.0	77.9	90.0	52.8	81.3	94.9	48.4	84.1	91.1	40.6	79.4
0920	Plaster & Gypsum Board	97.5	51.1	66.1	86.2	46.2	59.2	100.9	36.3	57.3	103.0	50.1	67.2	95.6	44.7	61.2	86.9	52.8	63.9
0950, 0980	Ceilings & Acoustic Treatment	84.0	51.1	62.4	109.3	46.2	67.8	97.2	36.3	57.2	79.2	50.1	60.1	91.5	44.7	60.8	105.7	52.8	71.0
0960	Flooring	110.9	50.4	93.6	96.8	46.9	82.5	97.6	46.6	83.0	87.9	47.8	76.5	108.6	46.4	90.8	115.6	44.9	95.4
0970, 0990	Wall Finishes & Painting/Coating	105.3	66.5	81.9	96.3	59.8	74.2	106.0	46.0	69.8	107.7	52.8	74.6	103.8	58.3	76.3	103.5	39.2	64.7
09	FINISHES	98.0	53.5	73.5	103.0	47.1	72.2	95.9	41.6	66.0	86.7	49.9	66.4	98.2	48.9	71.0	106.9	49.0	75.0
COVERS	DIVS. 10 - 14, 25, 28, 41, 43, 44, 46	100.0	81.8	96.3	100.0	22.8	84.4	100.0	39.4	87.8	100.0	80.2	96.0	100.0	78.0	95.6	100.0	43.5	88.6
21, 22, 23	FIRE SUPPRESSION, PLUMBING & HVAC	99.9	63.4	85.2	95.2	57.0	79.8	95.2	69.5	84.8	99.9	66.7	86.6	100.0	61.8	84.6	95.7	56.7	80.0
26, 27, 3370	ELECTRICAL, COMMUNICATIONS & UTIL.	97.1	69.6	82.5	109.7	67.7	87.5	99.7	69.5	83.8	96.1	61.2	77.7	101.7	57.4	78.3	99.9	57.4	77.4
MF2014	WEIGHTED AVERAGE	98.4	65.4	84.0	97.2	59.0	80.5	96.0	61.6	81.0	97.2	64.7	83.0	99.2	60.9	82.5	97.3	55.8	79.2

GEORGIA / HAWAII / IDAHO

DIVISION		VALDOSTA 316			WAYCROSS 315			HILO 967			HONOLULU 968			STATES & POSS., GUAM 969			BOISE 836 - 837		
		MAT.	INST.	TOTAL	MAT.	INST.	TOTAL	MAT.	INST.	TOTAL	MAT.	INST.	TOTAL	MAT.	INST.	TOTAL	MAT.	INST.	TOTAL
015433	CONTRACTOR EQUIPMENT		91.9	91.9		91.9	91.9		99.5	99.5		99.5	99.5		164.5	164.5		98.2	98.2
0241, 31 - 34	SITE & INFRASTRUCTURE, DEMOLITION	107.9	78.8	87.2	104.7	77.4	85.3	144.8	106.5	117.6	155.4	106.5	120.6	184.9	103.7	127.2	85.5	96.3	93.2
0310	Concrete Forming & Accessories	81.0	44.1	49.2	82.8	64.7	67.2	111.2	135.6	132.2	123.6	135.6	134.0	113.7	63.6	70.5	100.8	77.4	80.7
0320	Concrete Reinforcing	92.8	76.4	84.4	92.8	72.7	82.5	123.5	117.9	120.7	132.5	117.9	125.1	214.4	30.2	120.7	99.2	80.4	89.6
0330	Cast-in-Place Concrete	90.5	55.9	76.3	102.2	48.2	80.0	198.3	125.7	168.5	163.7	125.7	148.1	171.9	105.5	144.7	91.6	88.4	90.3
03	CONCRETE	98.1	55.8	77.3	101.2	61.4	81.6	158.1	127.7	143.2	150.8	127.7	139.5	159.9	72.3	116.9	99.1	81.9	90.7
04	MASONRY	108.9	50.5	72.5	109.7	39.3	65.8	105.5	128.3	136.7	150.6	128.3	136.7	213.9	43.5	107.7	121.4	84.2	98.2
05	METALS	104.1	84.0	97.9	103.1	77.7	95.3	108.1	107.3	107.8	120.7	107.3	116.5	139.2	76.2	119.8	102.2	80.9	95.6
06	WOOD, PLASTICS & COMPOSITES	73.4	37.6	53.4	75.0	72.4	73.5	114.5	139.2	128.3	134.4	139.2	137.1	127.1	67.4	93.7	93.1	75.9	83.4
07	THERMAL & MOISTURE PROTECTION	96.1	62.4	82.3	95.9	50.3	77.2	120.8	124.6	122.4	138.4	124.6	132.7	140.4	65.2	109.6	94.6	81.4	89.2
08	OPENINGS	87.3	42.5	76.9	87.4	57.8	80.5	111.3	132.8	116.3	121.8	132.8	124.4	116.6	54.1	102.1	99.1	71.2	92.6
0920	Plaster & Gypsum Board	90.4	36.2	53.8	90.4	72.0	78.0	109.7	140.1	130.3	151.3	140.1	143.7	219.9	55.5	108.8	92.4	75.1	80.7
0950, 0980	Ceilings & Acoustic Treatment	81.5	36.2	51.7	79.6	72.0	74.6	121.5	140.1	133.8	132.0	140.1	137.3	237.3	55.5	117.8	99.1	75.1	83.3
0960	Flooring	104.7	47.8	88.4	105.9	29.9	84.1	116.0	139.9	122.8	132.2	139.9	134.4	135.5	46.4	110.0	96.3	83.7	92.7
0970, 0990	Wall Finishes & Painting/Coating	105.3	52.2	73.2	105.3	46.0	69.5	110.4	143.8	130.3	119.4	143.8	134.2	115.9	35.9	67.5	103.2	39.5	64.7
09	FINISHES	95.6	43.6	67.0	95.2	57.4	74.4	113.7	138.6	127.4	128.4	138.6	134.0	188.8	58.9	117.2	96.0	74.8	84.3
COVERS	DIVS. 10 - 14, 25, 28, 41, 43, 44, 46	100.0	77.1	95.4	100.0	51.5	90.2	100.0	116.0	103.2	100.0	116.0	103.2	100.0	75.2	95.0	100.0	87.4	97.5
21, 22, 23	FIRE SUPPRESSION, PLUMBING & HVAC	99.9	70.0	87.9	96.9	57.1	80.8	100.2	107.7	103.2	100.3	107.7	103.3	102.6	37.4	76.3	100.0	71.8	88.6
26, 27, 3370	ELECTRICAL, COMMUNICATIONS & UTIL.	95.0	56.2	74.5	99.6	57.4	77.3	106.5	121.1	114.2	107.9	121.1	114.9	153.5	41.2	94.1	98.5	72.8	84.9
MF2014	WEIGHTED AVERAGE	98.5	61.4	82.3	98.3	59.4	81.4	114.8	120.2	117.2	119.5	120.2	119.8	136.2	58.0	102.1	100.1	78.6	90.7

IDAHO / ILLINOIS

DIVISION		COEUR D'ALENE 838			IDAHO FALLS 834			LEWISTON 835			POCATELLO 832			TWIN FALLS 833			BLOOMINGTON 617		
		MAT.	INST.	TOTAL	MAT.	INST.	TOTAL	MAT.	INST.	TOTAL	MAT.	INST.	TOTAL	MAT.	INST.	TOTAL	MAT.	INST.	TOTAL
015433	CONTRACTOR EQUIPMENT		92.8	92.8		98.2	98.2		92.8	92.8		98.2	98.2		98.2	98.2		101.6	101.6
0241, 31 - 34	SITE & INFRASTRUCTURE, DEMOLITION	83.1	91.7	89.2	83.1	96.1	92.4	89.9	92.5	91.7	85.8	96.3	93.3	92.1	97.2	95.8	97.4	98.4	98.1
0310	Concrete Forming & Accessories	112.6	81.1	85.5	94.7	78.1	80.4	117.9	82.2	87.1	101.0	77.1	80.4	102.0	54.7	61.2	84.3	117.2	112.7
0320	Concrete Reinforcing	106.2	96.5	101.3	101.0	80.0	90.3	106.2	96.8	101.4	99.6	80.2	89.7	101.3	79.9	90.4	94.6	110.7	102.8
0330	Cast-in-Place Concrete	99.0	87.2	94.1	87.2	74.3	81.9	102.9	86.0	95.9	94.1	88.3	91.7	96.6	63.9	83.1	102.8	114.8	107.7
03	CONCRETE	105.7	86.1	96.1	91.3	77.2	84.4	109.3	86.2	98.0	98.3	81.7	90.1	105.8	63.3	84.9	100.1	115.2	107.5
04	MASONRY	122.9	83.9	98.6	116.5	81.0	94.4	123.4	85.9	100.0	118.8	81.1	95.3	121.6	81.1	96.4	120.3	118.3	119.1
05	METALS	96.3	86.9	93.4	110.0	79.1	100.5	95.7	88.0	93.4	110.1	80.3	100.9	110.1	78.9	100.5	97.6	113.6	102.5
06	WOOD, PLASTICS & COMPOSITES	96.8	81.0	88.0	86.8	78.8	82.3	102.4	81.0	90.4	93.1	75.9	83.4	94.2	47.2	67.9	86.4	115.4	102.6
07	THERMAL & MOISTURE PROTECTION	148.3	80.8	120.6	94.3	71.7	85.0	148.5	81.4	121.0	94.7	73.7	86.1	95.4	74.7	86.9	98.4	112.8	104.3
08	OPENINGS	117.8	73.0	107.4	102.8	68.0	94.7	117.8	75.6	108.0	99.8	66.3	92.0	102.8	48.8	90.2	94.9	104.9	97.2
0920	Plaster & Gypsum Board	161.4	80.5	106.7	78.5	78.1	78.2	162.5	80.5	107.0	80.7	75.1	76.9	82.3	45.6	57.5	91.3	115.8	107.9
0950, 0980	Ceilings & Acoustic Treatment	129.0	80.5	97.1	97.2	78.1	84.6	129.0	80.5	97.1	104.7	75.1	85.2	99.7	45.6	64.1	88.0	115.8	106.3
0960	Flooring	138.3	45.9	111.9	96.4	43.1	81.2	141.2	93.8	127.7	99.6	83.7	95.1	100.7	43.1	84.3	93.8	118.7	100.9
0970, 0990	Wall Finishes & Painting/Coating	123.8	67.8	90.0	103.3	40.3	65.3	123.8	67.8	90.0	103.1	41.1	65.7	103.3	37.3	63.4	93.8	126.4	113.5
09	FINISHES	161.8	73.3	113.0	93.3	68.7	79.7	162.9	82.8	118.8	96.6	75.0	84.7	96.7	49.7	70.8	93.3	118.9	107.4
COVERS	DIVS. 10 - 14, 25, 28, 41, 43, 44, 46	100.0	87.4	97.5	100.0	47.7	89.4	100.0	87.6	97.5	100.0	87.4	97.5	100.0	44.2	88.7	100.0	105.2	101.0
21, 22, 23	FIRE SUPPRESSION, PLUMBING & HVAC	99.5	82.6	92.7	100.9	71.7	89.1	100.7	85.6	94.6	99.9	71.8	88.6	99.9	69.3	87.6	95.1	108.8	100.7
26, 27, 3370	ELECTRICAL, COMMUNICATIONS & UTIL.	91.0	77.6	83.9	90.8	70.4	80.0	88.9	80.7	84.5	96.2	70.4	82.6	92.3	60.1	75.3	94.2	94.6	94.4
MF2014	WEIGHTED AVERAGE	107.9	82.2	96.7	99.8	74.7	88.9	108.7	84.9	98.3	101.1	77.4	90.8	102.2	67.3	87.0	97.5	109.4	102.7

For customer support on your Mechanical Cost Data, call 877.795.7394.

665

ILLINOIS

DIVISION		CARBONDALE 629			CENTRALIA 628			CHAMPAIGN 618 - 619			CHICAGO 606 - 608			DECATUR 625			EAST ST. LOUIS 620 - 622		
		MAT.	INST.	TOTAL	MAT.	INST.	TOTAL	MAT.	INST.	TOTAL	MAT.	INST.	TOTAL	MAT.	INST.	TOTAL	MAT.	INST.	TOTAL
015433	CONTRACTOR EQUIPMENT		108.0	108.0		108.0	108.0		102.4	102.4		94.3	94.3		102.4	102.4		108.0	108.0
0241, 31 - 34	SITE & INFRASTRUCTURE, DEMOLITION	98.7	99.5	99.2	99.0	100.1	99.8	106.4	99.0	101.2	107.0	95.7	99.0	93.2	99.1	97.4	101.2	99.8	100.2
0310	Concrete Forming & Accessories	90.0	109.0	106.4	91.6	112.8	109.9	90.7	115.2	111.8	96.6	156.7	148.4	91.8	114.9	111.8	87.6	114.8	111.1
0320	Concrete Reinforcing	91.2	111.6	101.6	91.2	111.8	101.7	94.6	105.1	99.9	94.3	158.1	129.1	89.3	102.7	96.1	91.1	109.9	100.6
0330	Cast-in-Place Concrete	97.2	102.5	99.4	97.7	117.3	105.7	119.2	109.5	115.2	107.5	149.4	124.7	106.1	110.5	107.9	99.3	117.0	106.6
03	CONCRETE	90.1	107.9	98.9	90.6	114.8	102.5	113.1	111.3	112.2	102.6	153.2	127.5	101.3	111.1	106.1	91.7	115.2	103.2
04	MASONRY	83.6	108.6	99.2	83.6	116.7	104.2	145.3	117.7	128.1	101.1	156.6	135.7	79.6	114.7	101.5	83.9	116.6	104.3
05	METALS	96.2	119.6	103.4	96.2	120.8	103.8	97.6	108.7	101.0	94.3	134.3	106.6	99.9	108.1	102.4	97.3	119.4	104.1
06	WOOD, PLASTICS & COMPOSITES	91.7	106.1	99.8	94.1	109.7	102.8	93.5	113.9	104.9	104.0	155.9	133.1	91.9	113.9	104.2	89.2	112.1	102.0
07	THERMAL & MOISTURE PROTECTION	97.1	101.5	98.9	97.2	111.7	103.1	99.0	113.2	104.9	98.0	145.1	117.3	103.2	109.3	105.7	97.2	110.2	102.5
08	OPENINGS	89.4	114.2	95.2	89.4	116.2	95.7	95.5	111.1	99.2	105.3	158.7	117.8	100.9	110.6	103.2	89.5	116.9	95.9
0920	Plaster & Gypsum Board	97.2	106.2	103.3	98.3	110.0	106.2	93.7	114.2	107.6	92.7	157.6	136.6	100.0	114.2	109.6	96.1	112.4	107.2
0950, 0980	Ceilings & Acoustic Treatment	91.4	106.2	101.2	91.4	110.0	103.6	88.0	114.2	105.2	99.2	157.6	137.6	98.1	114.2	108.7	91.4	112.4	105.2
0960	Flooring	122.2	120.2	121.6	123.2	115.7	121.0	96.7	120.2	103.4	96.9	148.0	111.5	109.1	116.7	111.3	121.2	115.7	119.6
0970, 0990	Wall Finishes & Painting/Coating	112.6	97.4	103.5	112.6	106.6	109.0	93.8	110.0	103.6	91.9	153.0	128.8	102.5	106.5	104.9	112.6	103.5	107.1
09	FINISHES	101.8	108.6	105.5	102.2	112.6	107.9	95.2	115.9	106.6	98.0	155.6	129.7	101.8	115.0	109.1	101.4	113.8	108.2
COVERS	DIVS. 10 - 14, 25, 28, 41, 43, 44, 46	100.0	102.6	100.5	100.0	104.1	100.8	100.0	104.4	100.9	100.0	124.5	105.0	100.0	104.3	100.9	100.0	104.4	100.9
21, 22, 23	FIRE SUPPRESSION, PLUMBING & HVAC	95.1	106.1	99.5	95.1	95.3	95.2	95.1	105.5	99.3	99.8	133.9	113.6	99.9	98.5	99.4	99.9	98.8	99.4
26, 27, 3370	ELECTRICAL, COMMUNICATIONS & UTIL.	95.4	107.7	101.9	96.8	107.7	102.6	97.4	94.9	96.1	98.3	133.9	117.1	99.5	90.4	94.7	96.4	103.5	100.1
MF2014	WEIGHTED AVERAGE	94.7	107.9	100.5	95.0	108.5	100.9	101.0	107.6	103.9	99.8	139.7	117.2	99.2	104.9	101.7	96.4	108.7	101.8

ILLINOIS

DIVISION		EFFINGHAM 624			GALESBURG 614			JOLIET 604			KANKAKEE 609			LA SALLE 613			NORTH SUBURBAN 600 - 603		
		MAT.	INST.	TOTAL	MAT.	INST.	TOTAL	MAT.	INST.	TOTAL	MAT.	INST.	TOTAL	MAT.	INST.	TOTAL	MAT.	INST.	TOTAL
015433	CONTRACTOR EQUIPMENT		102.4	102.4		101.6	101.6		92.5	92.5		92.5	92.5		101.6	101.6		92.5	92.5
0241, 31 - 34	SITE & INFRASTRUCTURE, DEMOLITION	97.6	98.9	98.5	99.9	98.3	98.8	106.9	94.9	98.4	100.5	94.6	96.3	99.2	99.2	99.2	106.1	94.9	98.1
0310	Concrete Forming & Accessories	96.3	114.0	111.6	90.6	116.6	113.1	98.4	159.6	151.2	91.8	143.2	136.2	104.6	125.3	122.4	97.8	154.0	146.3
0320	Concrete Reinforcing	92.2	99.9	96.1	94.1	110.6	102.5	99.0	150.8	125.4	99.8	147.4	124.0	94.3	144.7	119.9	99.0	156.5	128.3
0330	Cast-in-Place Concrete	105.7	108.4	106.8	105.9	106.2	106.0	107.4	146.5	123.5	100.1	132.5	113.4	105.7	122.5	112.6	107.5	141.2	121.3
03	CONCRETE	102.1	109.4	105.7	103.2	112.0	107.5	102.7	152.1	126.9	96.7	139.3	117.7	104.1	127.9	115.8	102.7	148.8	125.4
04	MASONRY	88.2	108.5	100.8	120.5	118.0	118.9	104.3	148.5	131.8	100.6	141.2	125.9	120.5	125.1	123.4	101.1	144.1	127.9
05	METALS	97.0	105.6	99.7	97.6	113.0	102.4	92.2	129.1	103.6	92.2	126.6	102.8	97.7	131.6	108.1	93.4	131.4	105.1
06	WOOD, PLASTICS & COMPOSITES	94.2	113.9	105.2	93.3	115.5	105.7	105.7	161.0	136.7	97.8	142.2	122.7	108.8	123.7	117.1	104.0	155.7	132.9
07	THERMAL & MOISTURE PROTECTION	102.7	106.6	104.3	98.5	107.6	102.2	97.8	141.6	115.8	97.0	135.7	112.9	98.7	117.8	106.5	98.2	139.3	115.1
08	OPENINGS	94.9	109.9	98.4	94.9	111.0	98.6	102.8	159.6	116.0	95.4	148.4	107.7	94.9	129.5	102.9	102.9	158.2	115.8
0920	Plaster & Gypsum Board	99.7	114.2	109.5	93.7	115.9	108.7	89.9	162.9	139.2	86.7	143.4	125.0	100.1	124.3	116.5	92.7	157.3	136.4
0950, 0980	Ceilings & Acoustic Treatment	91.4	114.2	106.4	88.0	115.9	106.3	99.2	162.9	141.0	99.2	143.4	128.3	88.0	124.3	111.9	99.2	157.3	137.4
0960	Flooring	110.1	120.2	113.0	96.6	118.7	102.9	96.5	140.3	109.1	93.7	131.9	104.6	102.8	122.6	108.5	96.9	140.3	109.3
0970, 0990	Wall Finishes & Painting/Coating	102.5	104.2	103.5	93.8	95.1	94.5	90.1	152.5	127.8	90.1	126.4	112.0	93.8	126.4	113.5	91.9	152.5	128.4
09	FINISHES	100.6	115.2	108.7	94.6	115.4	106.1	97.4	156.9	130.2	95.7	137.9	119.0	97.3	124.3	112.2	97.9	152.7	128.1
COVERS	DIVS. 10 - 14, 25, 28, 41, 43, 44, 46	100.0	70.8	94.1	100.0	105.1	101.0	100.0	124.5	105.0	100.0	120.8	104.2	100.0	102.5	100.5	100.0	122.3	104.5
21, 22, 23	FIRE SUPPRESSION, PLUMBING & HVAC	95.2	102.8	98.2	95.1	105.3	99.2	99.9	131.2	112.5	95.1	129.0	108.7	95.1	123.4	106.6	99.8	128.6	111.4
26, 27, 3370	ELECTRICAL, COMMUNICATIONS & UTIL.	97.3	107.6	102.8	95.1	86.6	90.6	97.5	137.9	118.9	92.3	137.1	116.0	92.2	137.1	116.0	97.3	128.5	113.8
MF2014	WEIGHTED AVERAGE	97.3	106.0	101.1	98.2	106.7	101.9	99.2	138.4	116.3	95.6	131.7	111.3	98.4	124.4	109.7	99.3	135.1	114.9

ILLINOIS

DIVISION		PEORIA 615 - 616			QUINCY 623			ROCK ISLAND 612			ROCKFORD 610 - 611			SOUTH SUBURBAN 605			SPRINGFIELD 626 - 627		
		MAT.	INST.	TOTAL	MAT.	INST.	TOTAL	MAT.	INST.	TOTAL	MAT.	INST.	TOTAL	MAT.	INST.	TOTAL	MAT.	INST.	TOTAL
015433	CONTRACTOR EQUIPMENT		101.6	101.6		102.4	102.4		101.6	101.6		101.6	101.6		92.5	92.5		102.4	102.4
0241, 31 - 34	SITE & INFRASTRUCTURE, DEMOLITION	100.4	98.3	98.9	96.5	98.7	98.0	98.0	97.4	97.5	99.8	99.9	99.7	106.1	94.9	98.1	99.0	99.1	99.1
0310	Concrete Forming & Accessories	93.7	117.4	114.2	94.2	111.8	109.4	92.3	104.2	102.5	98.1	131.3	126.7	97.8	154.0	146.3	92.6	115.3	112.1
0320	Concrete Reinforcing	91.7	110.8	101.4	91.8	105.3	98.7	94.1	103.7	99.0	86.7	136.9	112.2	99.0	156.5	128.3	94.2	105.1	99.7
0330	Cast-in-Place Concrete	102.7	114.1	107.3	105.9	102.9	104.7	103.6	97.3	101.0	105.1	126.4	113.8	107.5	141.2	121.3	100.8	109.2	104.2
03	CONCRETE	100.2	115.0	107.5	101.7	107.6	104.6	101.0	102.0	101.5	100.9	130.4	115.4	102.7	148.8	125.4	99.1	111.3	105.1
04	MASONRY	120.1	118.1	118.8	111.7	104.1	106.9	120.3	97.0	105.8	94.1	135.8	120.0	101.1	144.1	127.9	91.0	117.8	107.7
05	METALS	100.4	113.8	104.5	97.1	108.0	100.5	97.7	108.4	101.0	100.4	128.2	108.9	93.4	131.4	105.1	97.5	109.3	101.1
06	WOOD, PLASTICS & COMPOSITES	101.2	115.5	109.2	91.8	113.9	104.2	95.0	104.4	100.3	101.1	128.3	116.4	104.0	155.7	132.9	89.2	113.9	103.0
07	THERMAL & MOISTURE PROTECTION	99.3	112.4	104.7	102.7	103.5	103.0	98.5	98.3	98.4	101.8	120.9	113.0	98.2	139.3	115.1	104.4	112.2	107.6
08	OPENINGS	101.3	116.0	104.7	95.8	111.4	99.4	94.9	103.2	96.8	101.3	132.1	108.5	102.9	158.2	115.8	103.0	111.2	104.9
0920	Plaster & Gypsum Board	97.1	115.9	109.8	98.3	114.2	109.0	93.7	104.5	101.0	97.1	129.1	118.8	92.7	157.3	136.4	98.7	114.2	109.2
0950, 0980	Ceilings & Acoustic Treatment	93.0	115.9	108.1	91.4	114.2	106.4	88.0	104.5	98.8	93.0	129.1	116.7	99.2	157.3	137.4	102.1	114.2	110.1
0960	Flooring	99.9	118.7	105.3	109.1	108.1	108.8	97.7	105.6	100.0	99.9	122.6	106.4	96.9	140.3	109.3	113.5	107.8	111.9
0970, 0990	Wall Finishes & Painting/Coating	93.8	126.4	113.5	102.5	106.5	104.9	93.8	95.1	94.5	93.8	133.1	117.5	91.9	152.5	128.4	101.1	106.5	104.3
09	FINISHES	97.2	118.9	109.2	100.1	111.8	106.6	94.9	103.7	99.7	97.2	130.0	115.3	97.9	152.7	128.1	104.9	113.5	109.6
COVERS	DIVS. 10 - 14, 25, 28, 41, 43, 44, 46	100.0	105.2	101.0	100.0	71.2	94.2	100.0	99.0	99.8	100.0	114.3	102.9	100.0	122.3	104.5	100.0	104.5	100.9
21, 22, 23	FIRE SUPPRESSION, PLUMBING & HVAC	99.9	104.9	101.9	95.2	101.0	97.5	95.1	99.9	97.1	100.0	116.5	106.7	99.8	128.6	111.4	99.9	103.3	101.3
26, 27, 3370	ELECTRICAL, COMMUNICATIONS & UTIL.	96.1	96.7	96.4	94.7	80.8	87.3	87.3	95.5	91.6	96.4	131.1	114.7	97.3	128.5	113.8	102.4	92.0	96.9
MF2014	WEIGHTED AVERAGE	100.5	109.3	104.4	98.1	101.0	99.3	97.2	100.5	98.6	99.5	124.8	110.5	99.3	135.1	114.9	100.0	106.5	102.8

INDIANA

DIVISION		ANDERSON 460 MAT.	INST.	TOTAL	BLOOMINGTON 474 MAT.	INST.	TOTAL	COLUMBUS 472 MAT.	INST.	TOTAL	EVANSVILLE 476-477 MAT.	INST.	TOTAL	FORT WAYNE 467-468 MAT.	INST.	TOTAL	GARY 463-464 MAT.	INST.	TOTAL
015433	CONTRACTOR EQUIPMENT		97.0	97.0		86.5	86.5		86.5	86.5		116.0	116.0		97.0	97.0		97.0	97.0
0241, 31 - 34	SITE & INFRASTRUCTURE, DEMOLITION	93.9	96.1	95.5	85.8	94.4	91.9	82.2	94.3	90.8	91.0	123.9	114.4	94.8	96.0	95.7	94.5	99.6	98.1
0310	Concrete Forming & Accessories	97.8	81.5	83.7	101.0	80.7	83.5	95.0	78.6	80.8	94.4	82.7	84.3	96.1	75.2	78.1	97.9	116.3	113.8
0320	Concrete Reinforcing	95.8	82.8	89.2	86.8	80.9	83.8	87.2	80.9	84.0	95.1	78.1	86.4	95.8	75.5	85.5	95.8	110.8	103.5
0330	Cast-in-Place Concrete	109.4	80.2	97.4	103.6	78.2	93.1	103.1	71.2	90.0	99.0	87.9	94.4	116.3	82.8	102.5	114.3	113.9	114.1
03	CONCRETE	100.8	81.7	91.5	105.0	79.7	92.6	104.2	76.3	90.5	105.3	83.8	94.7	104.0	78.4	91.4	103.2	114.2	108.6
04	MASONRY	94.1	80.1	85.4	95.6	76.5	83.7	95.4	76.5	83.6	91.2	82.5	85.8	98.5	77.9	85.6	95.6	114.6	107.5
05	METALS	92.8	89.3	91.7	95.8	77.6	90.2	95.9	77.0	90.0	89.3	85.0	87.9	92.8	85.9	90.7	92.8	108.7	97.7
06	WOOD, PLASTICS & COMPOSITES	100.2	81.9	89.9	112.9	80.9	95.0	107.7	78.2	91.2	93.5	82.0	87.0	99.9	74.7	85.8	97.7	115.7	107.8
07	THERMAL & MOISTURE PROTECTION	107.7	75.9	94.7	95.1	78.3	88.2	94.7	78.1	87.9	99.4	83.7	93.0	107.5	77.7	95.3	106.3	108.6	107.2
08	OPENINGS	98.5	82.1	94.7	105.8	81.1	100.1	101.5	79.6	96.4	99.1	80.8	94.9	98.5	74.2	92.9	98.5	120.3	103.6
0920	Plaster & Gypsum Board	102.5	81.6	88.4	97.5	81.0	86.4	95.0	78.2	83.6	93.4	80.8	84.9	101.8	74.2	83.1	96.1	116.4	109.8
0950, 0980	Ceilings & Acoustic Treatment	87.9	81.6	83.7	80.9	81.0	81.0	80.9	78.2	79.1	85.5	80.8	82.4	87.9	74.2	78.9	87.9	116.4	106.6
0960	Flooring	100.6	85.0	96.2	105.5	75.0	96.8	100.8	75.0	93.4	100.3	80.7	94.7	100.6	78.8	94.4	100.6	123.4	107.1
0970, 0990	Wall Finishes & Painting/Coating	105.6	69.4	83.7	95.3	82.4	87.5	95.3	82.4	87.5	101.2	84.9	91.4	105.6	73.9	86.5	105.6	123.9	116.7
09	FINISHES	95.6	81.0	87.6	94.8	79.8	86.5	93.0	78.2	84.9	93.8	82.6	87.6	95.4	75.7	84.5	94.7	118.7	107.9
COVERS	DIVS. 10 - 14, 25, 28, 41, 43, 44, 46	100.0	91.8	98.3	100.0	90.8	98.1	100.0	90.5	98.1	100.0	95.9	99.2	100.0	91.7	98.3	100.0	106.1	101.2
21, 22, 23	FIRE SUPPRESSION, PLUMBING & HVAC	100.0	78.8	91.5	99.7	78.9	91.3	94.9	78.8	88.4	99.9	79.6	91.7	100.0	72.1	88.7	100.0	106.6	102.7
26, 27, 3370	ELECTRICAL, COMMUNICATIONS & UTIL.	87.0	88.4	87.7	98.9	88.0	93.2	98.1	87.9	92.7	95.1	88.3	91.5	87.7	79.2	83.2	98.7	107.6	103.4
MF2014	WEIGHTED AVERAGE	97.0	83.8	91.3	99.3	81.8	91.6	97.2	80.9	90.1	97.1	86.9	92.6	97.6	79.1	89.5	98.3	110.4	103.6

INDIANA

DIVISION		INDIANAPOLIS 461-462 MAT.	INST.	TOTAL	KOKOMO 469 MAT.	INST.	TOTAL	LAFAYETTE 479 MAT.	INST.	TOTAL	LAWRENCEBURG 470 MAT.	INST.	TOTAL	MUNCIE 473 MAT.	INST.	TOTAL	NEW ALBANY 471 MAT.	INST.	TOTAL
015433	CONTRACTOR EQUIPMENT		93.7	93.7		97.0	97.0		86.5	86.5		104.4	104.4		95.0	95.0		94.5	94.5
0241, 31 - 34	SITE & INFRASTRUCTURE, DEMOLITION	93.5	99.3	97.7	90.3	96.0	94.4	83.1	94.3	91.0	81.2	110.4	102.0	85.6	94.6	92.0	78.2	96.8	91.4
0310	Concrete Forming & Accessories	98.6	85.6	87.4	101.2	76.6	80.0	92.6	82.6	84.0	91.5	75.4	77.6	92.6	81.0	82.6	89.5	73.3	75.5
0320	Concrete Reinforcing	99.4	83.0	91.1	86.7	81.1	83.9	86.8	82.7	84.7	86.1	74.2	80.1	96.0	82.7	89.3	87.4	77.1	82.2
0330	Cast-in-Place Concrete	101.9	86.4	95.5	108.3	82.7	97.8	103.7	81.8	94.7	97.0	74.8	87.9	108.8	78.6	96.4	100.1	74.0	89.4
03	CONCRETE	100.9	85.1	93.1	97.5	80.1	89.0	104.5	82.1	93.5	97.4	74.5	86.6	103.5	81.0	92.4	102.8	74.6	89.0
04	MASONRY	95.8	80.3	86.2	93.7	79.1	84.6	101.2	80.3	88.2	79.9	74.7	76.7	97.7	80.1	86.7	86.9	68.3	75.3
05	METALS	93.5	80.5	89.5	89.3	88.3	89.0	94.3	78.4	89.4	91.0	83.5	88.7	97.6	89.2	95.0	92.9	81.0	89.3
06	WOOD, PLASTICS & COMPOSITES	99.7	86.2	92.2	103.4	75.0	87.5	104.7	83.1	92.6	91.9	75.2	82.6	106.3	81.5	92.4	94.1	74.3	83.0
07	THERMAL & MOISTURE PROTECTION	100.7	80.8	92.5	107.3	75.7	94.3	94.7	80.6	88.9	100.1	76.7	90.5	97.7	76.8	89.2	86.8	69.6	79.7
08	OPENINGS	105.7	84.5	100.8	93.3	77.9	89.8	99.9	82.8	95.9	101.4	75.3	95.3	99.0	81.9	95.0	98.7	76.3	93.5
0920	Plaster & Gypsum Board	97.0	85.8	89.5	107.5	74.5	85.2	92.2	83.3	86.1	71.5	75.0	73.9	93.4	81.6	85.4	91.6	73.8	79.6
0950, 0980	Ceilings & Acoustic Treatment	93.8	85.8	88.6	87.9	74.5	79.1	76.7	83.3	81.0	89.7	75.0	80.0	81.7	81.6	81.6	85.5	73.8	77.8
0960	Flooring	101.4	85.0	96.7	104.6	93.3	101.3	99.7	88.8	96.6	74.1	85.0	77.2	99.9	85.0	95.6	97.9	62.8	87.9
0970, 0990	Wall Finishes & Painting/Coating	103.6	82.4	90.8	105.6	71.7	85.2	95.3	93.1	94.0	96.2	71.6	81.4	95.3	69.4	79.7	101.2	82.7	90.0
09	FINISHES	96.1	85.4	90.2	97.3	78.8	87.1	91.4	85.0	87.9	83.7	77.2	80.1	92.4	80.6	85.9	93.1	72.7	81.8
COVERS	DIVS. 10 - 14, 25, 28, 41, 43, 44, 46	100.0	93.3	98.7	100.0	91.1	98.2	100.0	91.1	98.2	100.0	43.6	88.6	100.0	90.8	98.1	100.0	43.1	88.5
21, 22, 23	FIRE SUPPRESSION, PLUMBING & HVAC	99.9	79.4	91.6	95.2	79.0	88.7	94.9	79.4	88.6	95.8	74.1	87.1	99.7	78.7	91.2	95.2	75.2	87.1
26, 27, 3370	ELECTRICAL, COMMUNICATIONS & UTIL.	101.0	88.4	94.4	91.4	78.7	84.7	97.6	82.5	89.6	93.0	73.5	82.7	90.9	79.6	84.9	93.6	75.8	84.2
MF2014	WEIGHTED AVERAGE	99.1	84.8	92.9	94.8	81.5	89.0	96.9	82.7	90.7	94.0	77.7	86.9	97.9	82.3	91.1	95.1	75.4	86.5

INDIANA / IOWA

DIVISION		SOUTH BEND 465-466 MAT.	INST.	TOTAL	TERRE HAUTE 478 MAT.	INST.	TOTAL	WASHINGTON 475 MAT.	INST.	TOTAL	BURLINGTON 526 MAT.	INST.	TOTAL	CARROLL 514 MAT.	INST.	TOTAL	CEDAR RAPIDS 522-524 MAT.	INST.	TOTAL
015433	CONTRACTOR EQUIPMENT		105.3	105.3		116.0	116.0		116.0	116.0		99.4	99.4		99.4	99.4		96.1	96.1
0241, 31 - 34	SITE & INFRASTRUCTURE, DEMOLITION	96.7	96.2	96.3	92.9	124.1	115.1	92.4	121.3	113.0	98.3	96.5	97.0	87.5	96.5	93.9	100.0	96.5	96.5
0310	Concrete Forming & Accessories	99.5	81.4	83.9	95.4	82.2	84.0	96.2	78.8	81.2	94.8	75.5	78.2	82.6	50.4	54.8	100.5	82.3	84.8
0320	Concrete Reinforcing	97.3	79.9	88.4	95.1	82.9	88.9	87.9	48.8	68.0	91.0	82.7	86.8	91.8	78.7	85.1	91.7	84.2	87.9
0330	Cast-in-Place Concrete	106.6	79.8	95.6	95.9	85.6	91.6	104.3	84.8	96.3	111.7	55.1	88.4	111.7	59.7	90.3	111.9	79.7	98.7
03	CONCRETE	98.3	81.8	90.2	108.4	83.7	96.2	114.3	75.3	95.1	102.6	70.8	86.9	101.3	60.2	81.1	102.7	82.3	92.7
04	MASONRY	105.7	77.8	88.3	98.9	79.6	86.9	91.3	80.8	84.8	102.3	64.9	79.0	104.2	74.1	85.4	108.2	78.0	89.4
05	METALS	92.8	99.9	95.0	90.0	87.4	89.2	84.6	68.2	79.5	88.8	93.0	90.1	88.9	88.7	88.8	91.3	92.2	91.5
06	WOOD, PLASTICS & COMPOSITES	94.8	81.6	87.4	95.7	82.1	88.1	96.1	78.9	86.5	95.1	76.6	84.8	81.6	44.9	61.1	101.9	82.6	91.1
07	THERMAL & MOISTURE PROTECTION	101.9	80.6	93.2	99.5	80.8	91.8	99.6	82.2	91.8	102.9	73.9	91.0	103.2	69.4	89.3	100.3	80.9	94.5
08	OPENINGS	96.0	80.5	92.4	99.7	82.2	95.6	96.4	67.6	89.7	94.8	70.4	89.1	99.3	53.4	88.6	100.3	81.9	96.0
0920	Plaster & Gypsum Board	90.8	81.3	84.4	93.4	81.0	85.0	93.4	77.7	82.8	102.3	75.9	84.5	97.7	43.2	60.9	106.9	82.3	90.3
0950, 0980	Ceilings & Acoustic Treatment	88.9	81.3	83.9	85.5	81.0	82.6	79.6	77.7	78.3	98.7	75.9	83.7	98.7	43.2	62.3	101.2	82.3	88.8
0960	Flooring	99.3	91.1	96.9	100.3	85.0	95.9	101.2	76.4	94.1	107.9	37.8	87.8	102.2	33.2	82.5	124.0	78.3	110.9
0970, 0990	Wall Finishes & Painting/Coating	99.2	87.1	91.9	101.2	83.1	90.3	101.2	83.4	90.5	106.3	78.5	89.5	106.3	78.7	89.6	107.6	72.2	86.3
09	FINISHES	95.4	83.9	89.1	93.8	82.8	87.7	93.1	79.6	85.6	102.5	68.9	84.0	98.7	48.3	71.0	108.4	80.6	93.1
COVERS	DIVS. 10 - 14, 25, 28, 41, 43, 44, 46	100.0	92.8	98.5	100.0	93.7	98.7	100.0	95.3	99.0	100.0	86.8	97.3	100.0	65.1	92.9	100.0	91.8	98.3
21, 22, 23	FIRE SUPPRESSION, PLUMBING & HVAC	99.9	77.1	90.7	99.9	79.6	91.7	95.2	77.6	88.1	95.4	72.6	86.2	95.4	72.6	86.2	100.2	80.5	92.2
26, 27, 3370	ELECTRICAL, COMMUNICATIONS & UTIL.	99.6	86.6	92.7	93.3	86.7	89.8	93.8	84.2	88.8	100.6	67.7	83.2	101.3	81.3	90.7	98.1	80.6	88.8
MF2014	WEIGHTED AVERAGE	98.0	84.4	92.1	97.9	86.5	92.9	95.8	81.8	89.7	97.2	75.3	87.6	97.0	71.2	85.8	100.0	83.2	92.7

IOWA

DIVISION		COUNCIL BLUFFS 515			CRESTON 508			DAVENPORT 527 - 528			DECORAH 521			DES MOINES 500 - 503,509			DUBUQUE 520		
		MAT.	INST.	TOTAL	MAT.	INST.	TOTAL	MAT.	INST.	TOTAL	MAT.	INST.	TOTAL	MAT.	INST.	TOTAL	MAT.	INST.	TOTAL
015433	CONTRACTOR EQUIPMENT		95.5	95.5		99.4	99.4		99.4	99.4		99.4	99.4		101.1	101.1		94.9	94.9
0241, 31 - 34	SITE & INFRASTRUCTURE, DEMOLITION	103.8	91.5	95.0	93.9	95.5	95.1	98.7	98.6	98.6	96.9	95.5	95.9	102.9	99.8	100.7	97.9	92.4	94.0
0310	Concrete Forming & Accessories	82.0	71.2	72.7	78.4	65.8	67.5	100.0	91.3	92.5	92.3	45.4	51.8	95.4	81.8	83.6	83.3	74.9	76.1
0320	Concrete Reinforcing	93.6	79.0	86.1	89.3	82.4	85.8	91.7	97.0	94.4	91.0	76.5	83.6	98.2	86.7	92.3	90.4	84.1	87.2
0330	Cast-in-Place Concrete	116.4	73.6	98.8	115.2	64.1	94.2	107.8	90.1	100.5	108.6	58.0	87.8	101.4	87.5	95.7	109.6	98.6	105.1
03	CONCRETE	105.0	74.3	89.9	102.9	69.3	86.4	100.7	92.5	96.7	100.5	57.0	79.1	100.0	85.4	92.8	99.4	85.5	92.6
04	MASONRY	109.8	75.0	88.1	108.1	81.9	91.7	105.3	86.8	93.7	124.0	71.4	91.2	101.0	81.2	88.7	109.2	73.9	87.2
05	METALS	96.3	89.3	94.1	93.8	90.5	92.8	91.3	102.7	94.8	89.0	85.8	88.0	99.9	96.5	98.8	89.8	92.0	90.5
06	WOOD, PLASTICS & COMPOSITES	80.4	70.9	75.1	74.1	61.6	67.1	101.9	91.0	95.8	92.2	37.1	61.3	91.8	81.1	85.8	82.1	73.5	77.3
07	THERMAL & MOISTURE PROTECTION	103.3	67.1	88.5	104.2	77.6	93.3	103.4	87.5	96.8	103.1	53.4	82.7	98.5	79.4	90.6	103.6	73.9	91.4
08	OPENINGS	99.3	75.9	93.9	108.4	63.9	98.1	100.3	90.8	98.1	97.8	48.6	86.4	102.7	86.5	98.9	99.3	79.3	94.7
0920	Plaster & Gypsum Board	97.7	70.3	79.2	93.6	60.5	71.2	106.9	90.8	96.0	101.2	35.2	56.6	90.4	80.6	83.8	97.7	72.9	81.0
0950, 0980	Ceilings & Acoustic Treatment	98.7	70.3	80.0	90.8	60.5	70.9	101.2	90.8	94.4	98.7	35.2	57.0	94.2	80.6	85.2	98.7	72.9	81.8
0960	Flooring	100.8	80.7	95.0	97.4	33.2	79.1	110.3	97.0	106.5	107.6	47.2	90.3	102.5	89.6	98.8	114.9	70.1	104.1
0970, 0990	Wall Finishes & Painting/Coating	102.2	65.8	80.2	101.0	78.7	87.5	106.3	95.5	99.8	106.3	32.9	62.0	96.2	88.6	91.6	106.7	63.8	80.8
09	FINISHES	99.4	72.1	84.4	94.2	60.2	75.4	104.3	92.7	97.9	102.2	42.6	69.4	97.1	83.7	89.7	103.7	73.5	87.1
COVERS	DIVS. 10 - 14, 25, 28, 41, 43, 44, 46	100.0	88.6	97.7	100.0	69.5	93.8	100.0	94.5	98.9	100.0	82.9	96.5	100.0	92.1	98.4	100.0	90.2	98.0
21, 22, 23	FIRE SUPPRESSION, PLUMBING & HVAC	100.2	74.2	89.7	95.2	79.7	89.0	100.2	92.8	97.2	95.4	72.9	86.3	99.9	79.7	91.7	100.2	75.3	90.1
26, 27, 3370	ELECTRICAL, COMMUNICATIONS & UTIL.	103.6	81.5	91.9	93.8	81.3	87.2	95.6	91.0	93.2	98.1	44.3	69.7	105.4	81.3	92.7	102.1	78.2	89.4
MF2014	WEIGHTED AVERAGE	100.7	78.1	90.9	98.2	77.3	89.1	99.0	93.1	96.4	98.0	64.2	83.3	100.5	85.2	93.8	99.1	80.2	90.9

IOWA

DIVISION		FORT DODGE 505			MASON CITY 504			OTTUMWA 525			SHENANDOAH 516			SIBLEY 512			SIOUX CITY 510 - 511		
		MAT.	INST.	TOTAL	MAT.	INST.	TOTAL	MAT.	INST.	TOTAL	MAT.	INST.	TOTAL	MAT.	INST.	TOTAL	MAT.	INST.	TOTAL
015433	CONTRACTOR EQUIPMENT		99.4	99.4		99.4	99.4		94.9	94.9		95.5	95.5		99.4	99.4		99.4	99.4
0241, 31 - 34	SITE & INFRASTRUCTURE, DEMOLITION	102.2	94.4	96.6	102.3	95.4	97.4	98.1	90.5	92.7	102.2	90.5	93.9	108.1	94.2	98.2	109.9	95.0	99.3
0310	Concrete Forming & Accessories	78.9	45.1	49.7	83.1	45.5	50.6	90.3	73.7	75.9	83.6	56.5	60.2	84.0	37.9	44.2	100.5	66.4	71.1
0320	Concrete Reinforcing	89.3	67.2	78.1	89.2	81.8	85.4	91.0	86.1	88.5	93.6	68.4	80.7	93.6	65.3	79.2	91.7	77.9	84.7
0330	Cast-in-Place Concrete	108.2	44.2	81.9	108.2	57.5	87.4	112.4	53.7	88.3	112.5	59.8	90.9	110.3	44.7	83.4	111.0	55.5	88.2
03	CONCRETE	98.2	50.4	74.7	98.5	57.9	78.5	102.1	69.9	86.3	102.4	61.0	82.1	101.4	47.0	74.6	102.0	65.8	84.2
04	MASONRY	106.9	37.9	63.9	120.5	69.8	88.9	105.8	57.8	75.9	109.4	75.0	88.0	128.6	38.1	72.2	102.2	54.6	72.6
05	METALS	93.9	81.3	90.1	94.0	88.7	92.4	88.8	91.7	89.7	95.3	83.3	91.6	89.1	79.7	86.2	91.3	87.4	90.1
06	WOOD, PLASTICS & COMPOSITES	74.5	45.2	58.1	78.4	37.1	55.3	89.3	80.9	84.6	82.1	52.8	65.7	82.9	35.9	56.6	101.9	66.6	82.1
07	THERMAL & MOISTURE PROTECTION	103.5	57.1	84.5	103.0	63.5	86.8	103.7	64.6	87.7	102.6	65.2	87.2	102.9	48.3	80.5	103.4	64.3	87.4
08	OPENINGS	101.9	49.5	89.7	93.5	50.1	83.4	99.3	77.7	94.3	90.3	54.7	82.0	95.9	43.8	83.8	100.3	65.9	92.3
0920	Plaster & Gypsum Board	93.6	43.5	59.8	93.6	35.2	54.1	98.8	80.6	86.5	97.7	51.6	66.5	97.7	33.9	54.6	106.9	65.7	79.0
0950, 0980	Ceilings & Acoustic Treatment	90.8	43.5	59.7	90.8	35.2	54.2	98.7	80.6	86.8	98.7	51.6	67.7	98.7	33.9	56.1	101.2	65.7	77.9
0960	Flooring	98.9	47.2	84.1	100.9	47.2	85.6	117.9	50.2	98.5	101.5	34.3	82.3	103.1	34.0	83.3	110.3	53.7	94.1
0970, 0990	Wall Finishes & Painting/Coating	101.0	61.7	77.3	101.0	30.1	58.2	106.7	78.7	89.8	102.2	65.8	80.2	106.3	61.7	79.4	106.3	63.5	80.5
09	FINISHES	96.1	46.1	68.5	96.7	42.0	66.5	104.9	70.4	85.9	99.5	52.3	73.4	101.9	38.0	66.7	105.8	63.6	82.5
COVERS	DIVS. 10 - 14, 25, 28, 41, 43, 44, 46	100.0	81.7	96.3	100.0	86.5	97.3	100.0	82.4	96.4	100.0	66.1	93.2	100.0	80.8	96.1	100.0	88.2	97.6
21, 22, 23	FIRE SUPPRESSION, PLUMBING & HVAC	95.2	65.5	83.3	95.2	70.5	85.2	95.4	70.9	85.5	95.4	73.5	86.6	95.4	68.0	84.3	100.2	75.8	90.4
26, 27, 3370	ELECTRICAL, COMMUNICATIONS & UTIL.	100.3	41.9	69.4	99.4	55.6	76.3	100.4	72.0	85.4	98.1	81.5	89.3	98.1	41.4	68.1	98.1	73.0	84.8
MF2014	WEIGHTED AVERAGE	97.9	58.2	80.6	97.6	65.8	83.8	97.9	73.7	87.3	97.6	71.2	86.1	98.3	56.4	80.0	99.7	72.6	87.9

IOWA / KANSAS

DIVISION		SPENCER 513 (IOWA)			WATERLOO 506 - 507 (IOWA)			BELLEVILLE 669 (KANSAS)			COLBY 677			DODGE CITY 678			EMPORIA 668		
		MAT.	INST.	TOTAL	MAT.	INST.	TOTAL	MAT.	INST.	TOTAL	MAT.	INST.	TOTAL	MAT.	INST.	TOTAL	MAT.	INST.	TOTAL
015433	CONTRACTOR EQUIPMENT		99.4	99.4		99.4	99.4		103.4	103.4		103.4	103.4		103.4	103.4		101.6	101.6
0241, 31 - 34	SITE & INFRASTRUCTURE, DEMOLITION	108.1	94.2	98.2	107.6	95.3	98.8	111.0	95.0	99.7	112.3	95.2	100.1	114.9	94.9	100.7	103.0	92.5	95.5
0310	Concrete Forming & Accessories	90.2	37.9	45.1	94.1	55.1	60.5	96.2	54.1	59.8	99.5	58.3	64.0	93.1	58.2	63.0	87.2	65.3	68.3
0320	Concrete Reinforcing	93.6	67.1	80.1	89.9	83.9	86.8	96.6	55.1	75.5	98.9	55.2	76.7	96.5	55.1	75.4	95.3	55.7	75.2
0330	Cast-in-Place Concrete	110.3	44.7	83.4	115.8	59.2	92.6	124.2	55.7	96.1	127.1	55.8	97.8	129.3	55.5	99.0	120.1	53.6	92.8
03	CONCRETE	101.8	47.3	75.0	104.4	63.1	84.1	119.3	56.2	88.3	119.9	58.1	89.5	121.3	57.9	90.2	111.5	60.5	86.5
04	MASONRY	128.6	38.1	72.2	107.7	73.1	86.1	99.2	58.7	73.9	108.1	59.9	78.1	118.9	58.9	81.5	105.5	68.5	82.4
05	METALS	89.0	80.5	86.4	96.4	91.1	94.7	95.6	77.3	90.0	96.0	77.7	90.4	97.4	76.7	91.1	95.3	78.9	90.3
06	WOOD, PLASTICS & COMPOSITES	89.2	35.9	59.3	91.6	48.0	67.2	97.9	52.1	72.3	103.6	57.9	78.0	95.8	57.9	74.5	88.7	66.6	76.3
07	THERMAL & MOISTURE PROTECTION	103.8	48.2	81.0	103.3	71.3	90.2	94.8	60.7	80.8	99.1	61.6	83.7	99.1	60.7	83.4	93.1	75.9	86.0
08	OPENINGS	107.5	44.4	92.9	94.4	58.2	86.0	98.6	48.4	86.9	103.2	51.5	91.2	103.1	51.5	91.3	96.3	56.3	87.0
0920	Plaster & Gypsum Board	98.8	33.9	54.9	101.9	46.5	64.4	97.2	50.6	65.7	99.7	56.5	70.5	94.0	56.5	68.7	94.4	65.5	74.9
0950, 0980	Ceilings & Acoustic Treatment	98.7	33.9	56.1	94.2	46.5	62.8	86.3	50.6	62.8	84.0	56.5	65.9	84.0	56.5	65.9	86.3	65.5	72.7
0960	Flooring	105.7	34.0	85.2	106.2	64.9	94.4	103.0	39.2	84.8	103.1	39.2	84.9	99.6	39.2	82.3	98.0	36.4	80.4
0970, 0990	Wall Finishes & Painting/Coating	106.3	61.7	79.4	101.0	77.8	87.0	100.0	38.9	63.1	104.7	38.9	65.0	104.7	38.9	65.0	100.0	38.9	63.1
09	FINISHES	102.8	38.0	67.1	100.3	57.6	76.8	96.9	49.3	70.6	96.6	52.7	72.4	94.9	52.7	71.6	94.0	57.3	73.8
COVERS	DIVS. 10 - 14, 25, 28, 41, 43, 44, 46	100.0	80.8	96.1	100.0	87.8	97.5	100.0	40.8	88.0	100.0	41.4	88.2	100.0	41.4	88.2	100.0	40.5	88.0
21, 22, 23	FIRE SUPPRESSION, PLUMBING & HVAC	95.4	68.0	84.3	100.0	79.8	91.9	95.2	71.6	85.7	95.2	69.0	84.6	100.0	69.0	87.5	95.2	73.4	86.4
26, 27, 3370	ELECTRICAL, COMMUNICATIONS & UTIL.	99.9	41.4	69.0	95.8	55.7	74.6	96.5	67.7	81.3	100.0	64.5	81.2	96.8	73.8	84.6	93.9	71.4	82.0
MF2014	WEIGHTED AVERAGE	99.9	56.6	81.0	99.5	71.8	87.4	99.6	64.9	84.4	101.1	65.0	85.4	102.7	66.0	86.7	97.9	69.2	85.4

For customer support on your Mechanical Cost Data, call 877.795.7394.

City Cost Indexes

KANSAS

DIVISION		FORT SCOTT 667			HAYS 676			HUTCHINSON 675			INDEPENDENCE 673			KANSAS CITY 660 - 662			LIBERAL 679		
		MAT.	INST.	TOTAL	MAT.	INST.	TOTAL	MAT.	INST.	TOTAL	MAT.	INST.	TOTAL	MAT.	INST.	TOTAL	MAT.	INST.	TOTAL
015433	CONTRACTOR EQUIPMENT		102.5	102.5		103.4	103.4		103.4	103.4		103.4	103.4		99.9	99.9		103.4	103.4
0241, 31 - 34	SITE & INFRASTRUCTURE, DEMOLITION	99.8	93.4	95.2	117.7	95.2	101.7	96.2	95.0	95.4	116.0	95.1	101.1	94.4	93.1	93.5	117.3	95.0	101.5
0310	Concrete Forming & Accessories	104.2	76.8	80.6	97.2	58.3	63.7	87.9	58.2	62.2	107.9	66.3	72.0	100.9	94.2	95.1	93.5	58.2	63.0
0320	Concrete Reinforcing	94.6	93.3	94.0	96.5	55.2	75.5	96.5	55.1	75.4	95.9	62.7	79.0	91.8	97.2	94.6	97.8	55.1	76.1
0330	Cast-in-Place Concrete	111.4	54.5	88.0	100.3	55.8	82.0	92.9	53.1	76.6	129.9	53.4	98.5	95.4	97.4	96.3	100.3	53.1	80.9
03	CONCRETE	106.6	72.9	90.1	110.4	58.1	84.7	92.4	57.1	75.0	122.6	62.2	92.9	98.4	96.2	97.3	112.5	57.1	85.2
04	MASONRY	106.7	60.3	77.8	118.0	59.9	81.8	107.9	58.6	77.2	105.3	64.5	79.9	107.8	99.4	102.6	116.6	58.6	80.4
05	METALS	95.3	92.8	94.5	95.6	77.7	90.1	95.4	76.7	89.6	95.3	80.5	90.8	102.9	100.1	102.1	95.9	76.7	90.0
06	WOOD, PLASTICS & COMPOSITES	108.7	83.5	94.6	100.6	57.9	76.7	90.9	57.9	72.4	113.7	68.0	88.1	104.4	94.5	98.8	96.4	57.9	74.8
07	THERMAL & MOISTURE PROTECTION	94.0	80.8	88.6	99.4	61.6	83.9	98.0	60.6	82.7	99.1	80.5	91.5	93.7	98.0	95.5	99.5	60.6	83.6
08	OPENINGS	96.3	78.2	92.1	103.1	51.5	91.1	103.0	51.5	91.0	103.0	51.5	91.0	97.7	89.0	95.7	103.1	51.5	91.1
0920	Plaster & Gypsum Board	99.7	82.9	88.4	97.3	56.5	69.7	93.0	56.5	68.3	106.9	67.0	79.9	92.9	94.2	93.8	94.8	56.5	68.9
0950, 0980	Ceilings & Acoustic Treatment	86.3	82.9	84.1	84.0	56.5	65.9	84.0	56.5	65.9	84.0	67.0	72.8	86.3	94.2	91.5	84.0	56.5	65.9
0960	Flooring	113.7	39.4	92.5	102.0	39.2	84.0	96.8	39.2	80.3	107.3	39.2	87.8	91.4	98.7	93.5	99.8	39.2	82.5
0970, 0990	Wall Finishes & Painting/Coating	101.8	38.9	63.8	104.7	38.9	65.0	104.7	38.9	65.0	104.7	38.9	65.0	108.5	67.7	83.9	104.7	38.9	65.0
09	FINISHES	99.6	66.9	81.6	96.4	52.7	72.3	92.4	52.7	70.5	98.9	58.7	76.8	94.3	91.0	92.5	95.7	52.7	72.0
COVERS	DIVS. 10 - 14, 25, 28, 41, 43, 44, 46	100.0	46.2	89.1	100.0	41.4	88.2	100.0	41.4	88.2	100.0	42.5	88.4	100.0	61.9	92.3	100.0	41.4	88.2
21, 22, 23	FIRE SUPPRESSION, PLUMBING & HVAC	95.2	68.3	84.3	95.2	69.0	84.6	95.2	69.0	84.6	95.2	70.1	85.1	99.9	96.4	98.5	95.2	67.2	83.9
26, 27, 3370	ELECTRICAL, COMMUNICATIONS & UTIL.	93.2	71.4	81.6	98.9	67.7	82.4	93.9	67.7	80.0	96.1	71.4	83.0	98.7	97.2	97.9	96.8	73.8	84.6
MF2014	WEIGHTED AVERAGE	97.9	73.0	87.0	100.5	65.4	85.2	96.5	65.0	82.8	100.9	69.2	87.1	99.5	94.8	97.5	100.4	65.5	85.2

KANSAS / KENTUCKY

DIVISION		SALINA 674			TOPEKA 664 - 666			WICHITA 670 - 672			ASHLAND 411 - 412			BOWLING GREEN 421 - 422			CAMPTON 413 - 414		
		MAT.	INST.	TOTAL	MAT.	INST.	TOTAL	MAT.	INST.	TOTAL	MAT.	INST.	TOTAL	MAT.	INST.	TOTAL	MAT.	INST.	TOTAL
015433	CONTRACTOR EQUIPMENT		103.4	103.4		101.6	101.6		103.4	103.4		97.3	97.3		94.5	94.5		101.1	101.1
0241, 31 - 34	SITE & INFRASTRUCTURE, DEMOLITION	105.2	95.2	98.1	98.2	91.5	93.4	101.8	93.8	96.1	112.3	86.3	93.8	78.6	97.1	91.8	87.0	98.3	95.0
0310	Concrete Forming & Accessories	89.7	54.9	59.7	98.4	40.9	48.8	95.3	51.0	57.1	87.9	105.2	102.8	85.7	83.0	83.4	89.4	82.6	83.6
0320	Concrete Reinforcing	95.9	59.4	77.3	94.9	99.5	97.2	95.3	77.3	86.1	88.8	108.7	98.9	86.2	91.3	88.8	87.0	101.8	94.5
0330	Cast-in-Place Concrete	112.3	53.7	88.2	100.4	45.5	77.9	105.3	51.7	83.3	91.1	101.1	95.2	90.4	95.5	92.5	100.8	73.6	89.6
03	CONCRETE	107.3	56.7	82.5	99.4	55.1	77.7	101.6	57.4	79.9	98.6	104.8	101.6	96.8	88.9	92.9	100.7	83.2	92.1
04	MASONRY	134.8	59.9	88.2	102.5	55.9	73.4	106.3	50.2	71.4	97.8	103.7	101.5	99.7	80.2	87.5	96.9	69.0	79.5
05	METALS	97.3	80.6	92.1	99.4	96.9	98.6	99.4	84.2	94.7	92.1	110.3	97.7	93.6	87.3	91.7	92.9	92.0	92.6
06	WOOD, PLASTICS & COMPOSITES	92.4	52.1	69.8	97.1	37.3	63.6	95.1	51.3	70.6	76.0	105.2	92.3	88.6	83.1	85.5	87.1	89.0	88.2
07	THERMAL & MOISTURE PROTECTION	98.6	61.4	83.3	96.7	67.4	84.7	97.2	56.1	80.4	91.0	97.5	93.7	86.6	80.2	84.0	99.7	67.5	86.5
08	OPENINGS	103.1	48.5	90.4	100.8	55.4	90.2	105.3	56.2	93.9	97.9	99.6	98.3	98.7	79.3	94.2	100.2	87.6	97.3
0920	Plaster & Gypsum Board	93.0	50.6	64.3	96.8	35.3	55.2	91.8	49.8	63.4	60.9	105.4	91.0	87.4	82.9	84.3	87.4	88.1	87.9
0950, 0980	Ceilings & Acoustic Treatment	84.0	50.6	62.0	92.0	35.3	54.7	88.8	49.8	63.2	80.8	105.4	97.0	85.5	82.9	83.8	85.5	88.1	87.2
0960	Flooring	98.2	61.9	87.8	103.4	73.7	94.9	105.4	63.0	93.3	79.4	102.4	86.0	95.8	86.0	93.0	98.0	41.2	81.7
0970, 0990	Wall Finishes & Painting/Coating	104.7	38.9	65.0	101.8	52.5	72.1	103.6	48.0	70.0	103.1	101.2	102.0	101.2	72.7	84.0	101.2	65.4	79.6
09	FINISHES	93.6	53.6	71.6	98.5	45.9	69.5	97.7	52.3	72.7	80.9	104.8	94.1	91.8	82.8	86.8	92.6	74.0	82.4
COVERS	DIVS. 10 - 14, 25, 28, 41, 43, 44, 46	100.0	86.0	97.2	100.0	47.2	89.3	100.0	83.0	96.6	100.0	94.1	98.8	100.0	62.3	92.4	100.0	55.8	91.1
21, 22, 23	FIRE SUPPRESSION, PLUMBING & HVAC	100.0	69.1	87.5	100.0	69.1	87.5	99.8	64.8	85.7	95.0	92.5	94.0	99.9	82.7	93.0	95.2	79.8	89.0
26, 27, 3370	ELECTRICAL, COMMUNICATIONS & UTIL.	96.6	73.8	84.5	97.5	71.8	83.9	100.9	73.8	86.5	91.4	97.3	94.5	94.0	81.5	87.4	91.4	66.1	78.0
MF2014	WEIGHTED AVERAGE	101.5	67.6	86.7	99.5	66.0	84.9	100.7	66.0	85.6	94.6	99.3	96.6	96.2	83.9	90.8	95.8	78.6	88.3

KENTUCKY

DIVISION		CORBIN 407 - 409			COVINGTON 410			ELIZABETHTOWN 427			FRANKFORT 406			HAZARD 417 - 418			HENDERSON 424		
		MAT.	INST.	TOTAL	MAT.	INST.	TOTAL	MAT.	INST.	TOTAL	MAT.	INST.	TOTAL	MAT.	INST.	TOTAL	MAT.	INST.	TOTAL
015433	CONTRACTOR EQUIPMENT		101.1	101.1		104.4	104.4		94.5	94.5		101.1	101.1		101.1	101.1		116.0	116.0
0241, 31 - 34	SITE & INFRASTRUCTURE, DEMOLITION	87.9	98.1	95.1	82.6	112.3	103.7	73.1	97.2	90.2	88.9	99.2	96.2	84.8	99.6	95.3	81.1	124.0	111.6
0310	Concrete Forming & Accessories	84.5	70.6	72.5	85.0	82.0	82.4	80.7	80.5	80.5	94.2	71.8	74.8	86.1	83.5	83.8	92.5	81.5	83.0
0320	Concrete Reinforcing	86.0	69.8	77.7	85.7	91.1	88.5	86.6	90.8	88.8	97.7	90.8	94.2	87.4	107.6	97.7	86.3	84.8	85.6
0330	Cast-in-Place Concrete	93.6	58.6	79.2	96.5	97.6	96.9	81.8	73.0	78.2	89.4	69.1	81.1	97.0	82.5	91.0	79.9	90.6	84.3
03	CONCRETE	93.3	66.8	80.3	98.9	89.7	94.4	88.6	80.0	84.4	93.8	74.8	84.5	97.3	87.7	92.6	93.9	85.4	89.7
04	MASONRY	94.4	61.9	74.1	111.8	101.5	105.4	83.5	75.4	78.5	94.6	79.4	85.1	94.8	71.5	80.3	103.6	92.9	96.9
05	METALS	90.4	78.2	86.6	91.0	96.9	92.8	92.8	87.2	91.1	93.0	87.8	91.4	92.9	94.2	93.3	84.4	86.9	85.1
06	WOOD, PLASTICS & COMPOSITES	76.9	75.9	76.4	85.0	71.7	77.6	83.7	82.0	82.7	91.4	68.7	78.7	84.1	89.0	86.8	91.3	79.0	84.4
07	THERMAL & MOISTURE PROTECTION	99.3	63.8	84.7	100.3	92.3	97.0	86.3	76.9	82.5	101.0	75.1	90.4	99.7	69.5	87.3	99.0	91.3	95.9
08	OPENINGS	97.9	65.6	90.4	102.4	81.7	97.6	98.7	83.2	95.1	104.5	76.7	98.0	100.6	79.9	95.8	96.8	79.9	92.9
0920	Plaster & Gypsum Board	92.3	74.6	80.3	68.8	71.4	70.6	86.7	81.7	83.3	96.5	67.2	76.7	86.7	88.1	87.6	90.2	77.8	81.8
0950, 0980	Ceilings & Acoustic Treatment	80.4	74.6	76.6	88.8	71.4	77.4	85.5	81.7	83.0	89.6	67.2	74.8	85.5	88.1	87.2	79.6	77.8	78.4
0960	Flooring	97.2	41.2	81.2	71.8	88.6	76.6	93.4	86.0	91.3	106.0	54.4	91.2	96.4	42.0	80.9	99.5	86.0	95.6
0970, 0990	Wall Finishes & Painting/Coating	107.2	49.4	72.3	96.2	83.2	88.3	101.2	76.2	86.1	109.1	79.8	91.4	101.2	65.4	79.6	101.2	96.0	98.1
09	FINISHES	90.7	63.6	75.8	82.7	82.0	82.4	90.6	80.7	85.1	97.6	68.8	81.7	91.9	74.8	82.5	91.4	83.8	87.2
COVERS	DIVS. 10 - 14, 25, 28, 41, 43, 44, 46	100.0	48.4	89.6	100.0	98.7	99.7	100.0	84.2	96.8	100.0	65.5	93.0	100.0	56.6	91.2	100.0	65.4	93.0
21, 22, 23	FIRE SUPPRESSION, PLUMBING & HVAC	95.2	71.4	85.6	95.9	92.4	94.5	95.5	80.1	89.3	100.1	82.3	92.9	95.2	80.7	89.4	95.5	77.6	88.3
26, 27, 3370	ELECTRICAL, COMMUNICATIONS & UTIL.	91.4	80.1	85.4	95.2	79.3	86.8	91.2	80.2	85.4	101.3	87.1	93.8	91.4	58.4	73.9	93.4	80.3	86.5
MF2014	WEIGHTED AVERAGE	94.0	71.8	84.3	95.9	91.4	93.9	92.8	82.0	88.1	98.1	80.7	90.5	95.2	78.7	88.0	93.7	86.1	90.4

For customer support on your Mechanical Cost Data, call 877.795.7394.

669

KENTUCKY

DIVISION		LEXINGTON 403 - 405			LOUISVILLE 400 - 402			OWENSBORO 423			PADUCAH 420			PIKEVILLE 415 - 416			SOMERSET 425 - 426		
		MAT.	INST.	TOTAL	MAT.	INST.	TOTAL	MAT.	INST.	TOTAL	MAT.	INST.	TOTAL	MAT.	INST.	TOTAL	MAT.	INST.	TOTAL
015433	CONTRACTOR EQUIPMENT		101.1	101.1		94.5	94.5		116.0	116.0		116.0	116.0		97.3	97.3		101.1	101.1
0241, 31 - 34	SITE & INFRASTRUCTURE, DEMOLITION	90.1	100.6	97.5	84.8	97.3	93.7	91.0	124.5	114.8	83.7	123.4	111.9	123.2	85.3	96.3	77.9	98.8	92.8
0310	Concrete Forming & Accessories	96.4	73.0	76.2	95.8	82.0	83.9	90.9	85.1	85.9	88.9	79.9	81.1	96.8	88.9	90.0	87.0	75.9	77.4
0320	Concrete Reinforcing	94.4	92.2	93.3	96.8	92.6	94.7	86.3	91.8	89.1	86.8	86.2	86.5	89.3	108.4	99.0	86.6	90.8	88.7
0330	Cast-in-Place Concrete	95.8	92.7	94.5	96.4	73.9	87.2	93.2	91.1	92.4	85.2	86.5	85.7	100.1	96.7	98.7	79.9	98.3	87.5
03	CONCRETE	96.5	83.7	90.2	97.1	81.3	89.4	106.3	88.5	97.5	98.6	83.5	91.2	112.7	96.0	104.5	83.6	86.6	85.1
04	MASONRY	93.7	73.3	81.0	91.6	78.8	83.6	95.9	91.5	93.2	98.8	88.4	92.3	95.3	96.8	96.2	89.8	76.7	81.6
05	METALS	92.8	89.0	91.6	93.9	88.2	92.1	85.8	90.6	87.3	82.9	86.7	84.1	92.0	108.8	97.2	92.8	87.6	91.2
06	WOOD, PLASTICS & COMPOSITES	92.0	68.7	79.0	90.9	83.1	86.5	89.2	83.3	85.9	86.8	79.1	82.5	85.5	88.3	87.1	84.6	75.9	79.7
07	THERMAL & MOISTURE PROTECTION	99.5	84.1	93.2	96.3	78.2	88.9	99.4	92.8	96.7	99.1	78.8	90.8	91.7	79.5	86.7	99.0	71.2	87.6
08	OPENINGS	98.1	73.9	92.5	93.5	84.3	91.4	96.8	85.7	94.2	96.0	77.5	91.7	98.6	87.8	96.1	99.5	77.7	94.4
0920	Plaster & Gypsum Board	101.9	67.2	78.4	98.0	82.9	87.8	88.7	82.2	84.3	87.7	77.9	81.1	63.8	88.1	80.2	86.7	74.6	78.5
0950, 0980	Ceilings & Acoustic Treatment	84.5	67.2	73.1	88.9	82.9	84.9	79.6	82.2	81.3	79.6	77.9	78.5	80.8	88.1	85.6	85.5	74.6	78.3
0960	Flooring	101.9	70.3	92.9	98.6	86.0	95.0	98.9	86.0	95.2	97.8	59.4	86.8	83.4	102.4	88.8	96.7	41.2	80.8
0970, 0990	Wall Finishes & Painting/Coating	107.2	81.0	91.4	104.7	76.2	87.5	101.2	96.0	98.1	101.2	79.7	88.2	103.1	81.2	89.9	101.2	76.2	86.1
09	FINISHES	94.4	72.3	82.2	94.9	82.3	88.0	91.5	86.0	88.5	90.7	76.1	82.7	83.3	90.9	87.5	91.2	69.4	79.2
COVERS	DIVS. 10 - 14, 25, 28, 41, 43, 44, 46	100.0	86.5	97.3	100.0	84.9	96.9	100.0	102.9	100.6	100.0	61.8	92.3	100.0	56.3	91.2	100.0	58.5	91.6
21, 22, 23	FIRE SUPPRESSION, PLUMBING & HVAC	100.0	79.5	91.7	100.0	80.9	92.3	99.9	78.6	91.4	95.5	81.1	89.7	95.0	87.2	91.8	95.5	77.4	88.2
26, 27, 3370	ELECTRICAL, COMMUNICATIONS & UTIL.	94.2	81.5	87.5	99.4	81.5	89.9	93.5	81.4	87.1	95.8	79.9	87.4	94.4	73.4	83.3	91.8	80.2	85.6
MF2014	WEIGHTED AVERAGE	96.6	81.4	90.0	96.6	83.3	90.8	96.3	88.9	93.1	93.9	84.5	89.8	96.9	88.6	93.3	93.2	79.9	87.4

LOUISIANA

DIVISION		ALEXANDRIA 713 - 714			BATON ROUGE 707 - 708			HAMMOND 704			LAFAYETTE 705			LAKE CHARLES 706			MONROE 712		
		MAT.	INST.	TOTAL	MAT.	INST.	TOTAL	MAT.	INST.	TOTAL	MAT.	INST.	TOTAL	MAT.	INST.	TOTAL	MAT.	INST.	TOTAL
015433	CONTRACTOR EQUIPMENT		89.9	89.9		89.5	89.5		90.1	90.1		90.1	90.1		89.5	89.5		89.9	89.9
0241, 31 - 34	SITE & INFRASTRUCTURE, DEMOLITION	101.2	87.1	91.1	104.0	87.2	92.0	102.1	87.8	91.9	103.2	88.1	92.5	103.9	87.0	91.9	101.2	87.0	91.1
0310	Concrete Forming & Accessories	82.9	43.4	48.9	97.6	60.8	65.8	78.4	46.2	50.7	96.0	54.1	59.9	96.8	56.7	62.2	82.4	43.3	48.7
0320	Concrete Reinforcing	94.9	54.3	74.2	101.1	58.9	79.6	97.9	59.3	78.3	99.3	58.9	78.7	99.3	59.1	78.9	93.8	54.0	73.6
0330	Cast-in-Place Concrete	96.2	49.9	77.2	97.2	58.8	81.4	95.8	44.0	74.5	95.3	47.9	75.8	100.3	67.3	86.8	96.2	56.8	80.0
03	CONCRETE	95.5	48.8	72.6	98.9	60.4	80.0	95.2	48.9	72.5	96.4	53.7	75.4	98.8	61.5	80.5	95.3	51.1	73.6
04	MASONRY	121.3	53.4	79.0	94.9	54.1	69.5	95.7	51.2	68.0	95.7	52.4	68.7	95.0	58.8	72.5	115.6	47.9	73.4
05	METALS	95.4	70.4	87.7	100.6	72.1	91.8	95.7	71.2	88.1	94.8	72.1	87.8	94.8	72.6	87.9	95.4	69.9	87.6
06	WOOD, PLASTICS & COMPOSITES	90.5	41.7	63.2	99.0	63.8	79.3	84.2	47.0	63.3	105.7	55.7	77.7	103.8	57.7	78.0	89.8	42.2	63.1
07	THERMAL & MOISTURE PROTECTION	95.6	62.9	82.2	96.2	65.4	83.5	96.5	61.4	82.1	97.1	63.8	83.4	96.1	66.8	84.1	95.6	60.8	81.3
08	OPENINGS	109.1	44.7	94.1	97.9	59.1	88.9	93.0	51.7	83.4	96.7	53.2	86.6	96.7	54.9	87.0	109.1	47.6	94.8
0920	Plaster & Gypsum Board	84.2	40.4	54.6	96.7	63.0	74.0	97.5	45.7	62.5	105.7	54.7	71.2	105.7	56.8	72.6	83.8	40.9	54.8
0950, 0980	Ceilings & Acoustic Treatment	84.1	40.4	55.4	91.3	63.0	72.7	94.1	45.7	62.3	92.4	54.7	67.6	93.3	56.8	69.3	84.1	40.9	55.7
0960	Flooring	98.5	65.7	89.1	101.8	65.7	91.5	96.3	65.7	87.5	104.9	65.7	93.7	104.9	72.9	95.7	98.1	57.1	86.4
0970, 0990	Wall Finishes & Painting/Coating	104.4	64.4	80.3	94.7	46.3	65.5	99.4	48.3	68.6	99.4	57.9	74.4	99.4	50.0	69.6	104.4	48.1	70.4
09	FINISHES	89.8	48.9	67.2	94.7	60.2	75.7	94.8	49.7	70.0	98.1	56.3	75.0	98.3	58.6	76.4	89.6	45.5	65.3
COVERS	DIVS. 10 - 14, 25, 28, 41, 43, 44, 46	100.0	50.0	89.9	100.0	80.8	96.1	100.0	45.4	89.0	100.0	79.4	95.8	100.0	80.3	96.0	100.0	45.7	89.0
21, 22, 23	FIRE SUPPRESSION, PLUMBING & HVAC	100.1	55.3	82.1	100.0	59.5	83.7	95.3	42.8	74.1	100.1	61.1	84.4	100.1	62.0	84.7	100.1	54.3	81.6
26, 27, 3370	ELECTRICAL, COMMUNICATIONS & UTIL.	95.7	57.9	75.7	100.8	61.6	80.1	96.7	56.0	75.2	97.9	67.4	81.8	97.4	65.0	80.3	97.7	59.6	77.6
MF2014	WEIGHTED AVERAGE	99.4	57.2	81.0	99.1	63.7	83.7	95.8	54.5	77.8	97.9	63.0	82.7	98.1	65.0	83.7	99.3	56.5	80.6

LOUISIANA / MAINE

DIVISION		NEW ORLEANS 700 - 701			SHREVEPORT 710 - 711			THIBODAUX 703			AUGUSTA 043			BANGOR 044			BATH 045		
		MAT.	INST.	TOTAL	MAT.	INST.	TOTAL	MAT.	INST.	TOTAL	MAT.	INST.	TOTAL	MAT.	INST.	TOTAL	MAT.	INST.	TOTAL
015433	CONTRACTOR EQUIPMENT		90.4	90.4		89.9	89.9		90.1	90.1		100.9	100.9		100.9	100.9		100.9	100.9
0241, 31 - 34	SITE & INFRASTRUCTURE, DEMOLITION	103.4	91.1	94.7	105.5	87.1	92.4	104.4	88.1	92.8	91.6	101.1	98.4	93.3	101.3	99.0	91.0	101.1	98.2
0310	Concrete Forming & Accessories	95.1	65.5	69.5	98.5	44.2	51.7	90.4	64.6	68.1	95.9	94.8	94.9	93.1	96.6	96.1	88.9	94.9	94.1
0320	Concrete Reinforcing	100.2	60.1	79.8	98.3	54.4	76.0	97.9	59.1	78.2	96.1	106.7	101.5	87.7	108.1	98.1	86.8	107.9	97.5
0330	Cast-in-Place Concrete	98.0	69.4	86.3	101.1	53.6	81.6	102.9	51.6	81.8	101.5	58.8	84.0	82.2	110.6	93.9	82.2	58.9	72.7
03	CONCRETE	99.0	66.3	82.9	100.4	50.5	75.9	100.5	59.7	80.4	103.1	84.1	93.8	94.1	102.9	98.4	94.1	84.4	89.4
04	MASONRY	97.7	59.8	74.1	107.1	50.5	71.8	120.5	48.8	75.8	107.4	63.6	80.1	118.0	103.9	109.2	125.6	86.1	101.0
05	METALS	110.3	73.3	98.9	99.5	70.6	90.6	95.6	71.6	88.2	106.0	86.6	100.1	96.1	89.3	94.0	94.6	88.4	92.7
06	WOOD, PLASTICS & COMPOSITES	97.9	66.3	80.2	101.7	42.7	68.7	92.3	69.7	79.6	91.1	105.3	99.0	90.1	96.4	93.6	84.6	105.3	96.2
07	THERMAL & MOISTURE PROTECTION	94.9	68.8	84.2	95.3	59.8	80.8	96.4	61.8	82.2	100.1	65.0	85.7	97.8	84.4	92.3	97.8	71.8	87.1
08	OPENINGS	97.8	65.4	90.2	108.0	45.3	93.4	97.6	64.1	89.8	106.3	87.9	102.0	102.8	83.0	98.2	102.8	87.8	99.3
0920	Plaster & Gypsum Board	101.4	65.7	77.2	94.0	41.5	58.5	99.7	69.2	79.0	100.6	104.8	103.5	99.4	95.7	96.9	95.5	104.8	101.8
0950, 0980	Ceilings & Acoustic Treatment	98.2	65.7	76.8	88.1	41.5	57.4	94.1	69.2	77.7	97.4	104.8	102.3	86.2	95.7	92.5	84.3	104.8	97.8
0960	Flooring	105.7	65.7	94.2	101.9	61.3	90.3	102.3	44.1	85.7	103.9	55.3	90.0	97.1	114.3	102.0	95.3	55.3	83.9
0970, 0990	Wall Finishes & Painting/Coating	101.9	64.1	79.1	100.4	44.6	66.7	100.7	49.6	69.9	112.7	63.3	82.9	105.0	44.1	68.3	105.0	44.1	68.3
09	FINISHES	101.2	65.1	81.3	94.5	46.3	67.9	97.2	59.8	76.6	100.2	85.8	92.3	95.7	94.8	95.2	94.2	83.7	88.4
COVERS	DIVS. 10 - 14, 25, 28, 41, 43, 44, 46	100.0	83.1	96.6	100.0	78.1	95.6	100.0	81.4	96.2	100.0	100.5	100.1	100.0	109.8	102.0	100.0	100.5	100.1
21, 22, 23	FIRE SUPPRESSION, PLUMBING & HVAC	100.0	65.1	85.9	99.9	58.6	83.3	95.3	61.7	81.8	99.9	64.2	85.5	100.1	74.3	89.7	95.4	64.3*	82.9
26, 27, 3370	ELECTRICAL, COMMUNICATIONS & UTIL.	102.1	73.3	86.9	103.5	66.6	84.0	95.3	73.3	83.7	100.9	80.1	89.9	97.2	75.8	85.9	95.2	80.1	87.2
MF2014	WEIGHTED AVERAGE	101.4	69.5	87.5	101.0	59.6	83.0	98.2	65.4	83.9	102.1	79.5	92.3	99.0	89.5	94.9	97.5	81.9	90.7

MAINE

DIVISION		HOULTON 047 MAT.	INST.	TOTAL	KITTERY 039 MAT.	INST.	TOTAL	LEWISTON 042 MAT.	INST.	TOTAL	MACHIAS 046 MAT.	INST.	TOTAL	PORTLAND 040-041 MAT.	INST.	TOTAL	ROCKLAND 048 MAT.	INST.	TOTAL
015433	CONTRACTOR EQUIPMENT		100.9	100.9		100.9	100.9		100.9	100.9		100.9	100.9		100.9	100.9		100.9	100.9
0241, 31-34	SITE & INFRASTRUCTURE, DEMOLITION	93.0	102.2	99.5	84.8	101.2	96.5	90.9	101.3	98.3	92.3	102.2	99.4	90.1	101.3	98.1	88.7	102.2	98.3
0310	Concrete Forming & Accessories	96.7	101.6	101.0	89.2	95.5	94.6	98.5	96.7	96.9	93.9	101.6	100.5	98.5	96.7	96.9	94.9	101.6	100.7
0320	Concrete Reinforcing	87.7	106.5	97.2	84.1	107.9	96.2	107.8	108.1	107.9	87.7	106.5	97.2	102.2	108.1	105.2	87.7	106.5	97.2
0330	Cast-in-Place Concrete	82.3	71.7	77.9	81.2	59.8	72.4	83.9	110.7	94.9	82.2	71.1	77.7	96.7	110.7	102.4	83.9	71.1	78.7
03	CONCRETE	95.1	91.5	93.4	89.7	85.0	87.4	94.7	102.9	98.7	94.6	91.3	93.0	101.9	102.9	102.4	92.0	91.3	91.6
04	MASONRY	101.3	76.3	85.7	113.3	87.1	97.0	101.9	103.9	103.2	101.3	76.3	85.7	112.8	103.9	107.2	95.2	76.3	83.4
05	METALS	94.8	86.6	92.3	85.5	88.6	86.4	99.5	89.3	96.4	94.8	86.5	92.3	107.9	89.3	102.2	94.7	86.5	92.2
06	WOOD, PLASTICS & COMPOSITES	94.0	105.3	100.3	89.2	105.3	98.2	96.0	96.4	96.2	91.0	105.3	99.0	93.7	96.4	95.2	91.9	105.3	99.4
07	THERMAL & MOISTURE PROTECTION	97.9	72.6	87.5	99.2	71.1	87.7	97.6	84.4	92.2	97.8	71.6	87.1	100.1	84.4	93.7	97.5	71.7	86.9
08	OPENINGS	102.9	86.1	99.0	102.0	87.8	98.7	106.2	83.0	100.8	102.9	86.1	99.0	104.5	83.0	99.5	102.8	86.1	98.9
0920	Plaster & Gypsum Board	101.6	104.8	103.8	100.0	104.8	103.3	104.9	95.7	98.7	100.1	104.8	103.3	102.5	95.7	97.9	100.1	104.8	103.3
0950, 0980	Ceilings & Acoustic Treatment	84.3	104.8	97.8	94.2	104.8	101.2	95.4	95.7	95.6	84.3	104.8	97.8	99.1	95.7	96.9	84.3	104.8	97.8
0960	Flooring	98.3	51.7	85.0	99.8	57.7	87.7	100.0	114.3	104.1	97.5	51.7	84.4	98.6	114.3	103.1	97.9	51.7	84.7
0970, 0990	Wall Finishes & Painting/Coating	105.0	137.1	124.4	96.4	38.4	61.4	105.0	44.1	68.3	105.0	137.1	124.4	102.0	44.1	67.1	105.0	137.1	124.4
09	FINISHES	95.9	97.8	97.0	96.5	83.7	89.5	98.9	94.8	96.6	95.4	97.8	96.7	98.8	94.8	96.6	95.2	97.8	96.6
COVERS	DIVS. 10-14, 25, 28, 41, 43, 44, 46	100.0	106.4	101.3	100.0	100.8	100.2	100.0	109.8	102.0	100.0	106.4	101.3	100.0	109.8	102.0	100.0	106.4	101.3
21, 22, 23	FIRE SUPPRESSION, PLUMBING & HVAC	95.4	73.2	86.4	95.3	76.6	87.8	100.1	74.3	89.7	95.4	73.2	86.4	100.0	74.3	89.6	95.4	73.2	86.4
26, 27, 3370	ELECTRICAL, COMMUNICATIONS & UTIL.	99.2	80.1	89.1	97.0	80.1	88.1	99.3	80.1	89.1	99.2	80.1	89.1	98.9	80.1	89.0	99.1	80.1	89.1
MF2014	WEIGHTED AVERAGE	97.2	85.6	92.1	95.2	84.7	90.6	99.6	90.1	95.5	97.1	85.5	92.0	102.0	90.1	96.8	96.3	85.5	91.6

MAINE / MARYLAND

DIVISION		WATERVILLE 049 MAT.	INST.	TOTAL	ANNAPOLIS 214 MAT.	INST.	TOTAL	BALTIMORE 210-212 MAT.	INST.	TOTAL	COLLEGE PARK 207-208 MAT.	INST.	TOTAL	CUMBERLAND 215 MAT.	INST.	TOTAL	EASTON 216 MAT.	INST.	TOTAL
015433	CONTRACTOR EQUIPMENT		100.9	100.9		99.8	99.8		103.4	103.4		105.5	105.5		99.8	99.8		99.8	99.8
0241, 31-34	SITE & INFRASTRUCTURE, DEMOLITION	92.8	101.1	98.7	103.0	91.5	94.9	101.5	95.6	97.3	100.5	94.5	96.2	94.3	91.8	92.5	101.4	88.6	92.3
0310	Concrete Forming & Accessories	88.4	94.8	93.9	97.2	73.6	76.8	101.4	71.1	75.3	84.3	70.2	72.2	91.4	81.0	82.4	89.2	70.2	72.8
0320	Concrete Reinforcing	87.7	106.7	97.3	100.3	80.7	90.3	105.6	80.7	92.9	105.4	78.1	91.5	86.6	72.3	79.3	86.0	79.0	82.4
0330	Cast-in-Place Concrete	82.3	58.8	72.6	110.0	75.7	95.9	107.5	76.4	94.7	116.5	77.5	100.4	91.7	85.6	89.2	101.8	48.1	79.8
03	CONCRETE	95.6	84.1	90.0	104.8	76.6	90.9	104.8	75.8	90.6	107.5	75.5	91.8	90.2	81.8	86.1	98.0	65.2	81.9
04	MASONRY	112.2	63.6	81.9	103.5	71.6	83.6	100.1	71.7	82.4	111.8	69.0	85.1	99.1	83.8	89.5	113.4	42.9	69.4
05	METALS	94.8	86.6	92.3	104.0	93.4	100.8	100.6	93.9	98.5	86.8	97.4	90.0	98.5	90.5	96.1	98.8	87.2	95.2
06	WOOD, PLASTICS & COMPOSITES	84.0	105.3	95.9	92.9	75.4	83.1	99.6	72.1	84.2	80.0	70.0	74.4	84.9	80.1	82.2	82.7	77.4	79.7
07	THERMAL & MOISTURE PROTECTION	97.9	65.0	84.4	100.2	78.8	91.4	101.9	78.6	92.3	101.9	78.7	92.4	100.2	78.9	91.5	100.3	59.0	83.4
08	OPENINGS	102.9	87.9	99.4	105.8	80.3	99.9	99.0	78.5	94.2	94.3	74.8	89.7	99.7	78.0	94.7	98.0	71.4	91.8
0920	Plaster & Gypsum Board	95.5	104.8	101.8	100.5	74.9	83.2	102.3	71.3	81.4	99.0	69.0	78.8	102.3	79.8	87.1	102.3	77.0	85.2
0950, 0980	Ceilings & Acoustic Treatment	84.3	104.8	97.8	91.3	74.9	80.6	94.0	71.3	79.1	102.2	69.0	80.4	93.3	79.8	84.4	93.3	77.0	82.6
0960	Flooring	95.0	55.3	83.6	99.8	77.2	93.3	100.7	77.2	94.0	97.8	79.7	92.6	95.5	90.8	94.2	94.7	51.7	82.4
0970, 0990	Wall Finishes & Painting/Coating	105.0	63.3	79.8	92.1	79.2	84.3	97.9	79.2	86.6	108.0	77.6	89.7	96.5	73.4	82.5	96.5	77.6	85.1
09	FINISHES	94.3	85.8	89.6	94.8	74.3	83.5	100.1	72.3	84.8	95.3	71.9	82.4	96.5	81.9	88.5	96.7	69.1	81.5
COVERS	DIVS. 10-14, 25, 28, 41, 43, 44, 46	100.0	100.5	100.1	100.0	86.2	97.2	100.0	86.2	97.2	100.0	85.5	97.1	100.0	90.5	98.1	100.0	75.3	95.0
21, 22, 23	FIRE SUPPRESSION, PLUMBING & HVAC	95.4	64.2	82.8	100.1	80.5	92.2	100.0	80.6	92.2	95.4	83.6	90.7	95.2	73.0	86.3	95.2	68.3	84.4
26, 27, 3370	ELECTRICAL, COMMUNICATIONS & UTIL.	99.2	80.1	89.1	99.5	91.9	95.5	102.0	91.9	96.6	101.1	99.7	100.4	98.3	81.8	89.6	97.8	65.2	80.5
MF2014	WEIGHTED AVERAGE	97.5	79.5	89.7	101.5	82.0	93.0	100.8	81.9	92.6	97.2	83.4	91.2	96.8	81.8	90.2	98.3	68.6	85.3

MARYLAND / MASSACHUSETTS

DIVISION		ELKTON 219 MAT.	INST.	TOTAL	HAGERSTOWN 217 MAT.	INST.	TOTAL	SALISBURY 218 MAT.	INST.	TOTAL	SILVER SPRING 209 MAT.	INST.	TOTAL	WALDORF 206 MAT.	INST.	TOTAL	BOSTON 020-022, 024 MAT.	INST.	TOTAL
015433	CONTRACTOR EQUIPMENT		99.8	99.8		99.8	99.8		99.8	99.8		98.1	98.1		98.1	98.1		106.5	106.5
0241, 31-34	SITE & INFRASTRUCTURE, DEMOLITION	88.4	89.5	89.2	92.6	92.3	92.4	101.3	88.6	92.3	89.0	87.6	88.0	95.4	87.3	89.7	98.9	109.1	106.2
0310	Concrete Forming & Accessories	95.4	81.0	82.9	90.3	75.1	77.2	103.9	51.7	58.9	92.8	70.4	73.5	100.2	68.7	73.0	102.7	143.8	138.1
0320	Concrete Reinforcing	86.0	105.3	95.8	86.6	72.3	79.3	86.0	63.5	74.6	104.1	77.9	90.8	104.8	78.0	91.1	108.9	155.8	132.8
0330	Cast-in-Place Concrete	82.5	75.2	79.5	87.4	85.7	86.7	101.8	46.7	79.2	119.3	79.3	102.8	133.6	76.5	110.2	104.0	151.2	123.4
03	CONCRETE	83.3	84.3	83.8	86.8	79.2	83.0	99.0	53.7	76.8	105.9	76.1	91.2	116.5	74.4	95.8	105.7	147.5	126.2
04	MASONRY	98.3	58.9	73.8	105.3	83.8	91.9	113.1	47.3	72.1	110.9	72.3	86.8	95.4	67.2	77.9	106.6	164.2	142.5
05	METALS	98.8	101.3	99.6	98.7	90.7	96.2	98.8	81.2	93.4	91.0	93.8	91.9	91.0	93.9	91.9	100.2	132.4	110.1
06	WOOD, PLASTICS & COMPOSITES	90.0	86.8	88.3	84.0	71.9	77.2	100.6	55.1	75.1	86.5	69.2	76.8	94.0	69.2	80.1	98.6	143.3	123.6
07	THERMAL & MOISTURE PROTECTION	99.9	73.5	89.1	100.0	76.8	90.5	100.6	63.7	85.4	105.4	84.9	97.0	105.8	82.7	96.4	103.4	151.9	123.3
08	OPENINGS	98.0	81.3	94.1	98.0	73.0	92.2	98.2	61.7	89.7	85.7	74.3	83.0	86.2	74.3	83.5	101.4	146.2	111.8
0920	Plaster & Gypsum Board	105.2	86.7	92.7	102.3	71.3	81.4	111.6	54.0	72.6	105.1	69.0	80.7	108.3	69.0	81.8	107.5	144.1	132.2
0950, 0980	Ceilings & Acoustic Treatment	93.3	86.7	89.0	94.3	71.3	79.2	93.3	54.0	67.5	112.2	69.0	83.8	112.2	69.0	83.8	106.1	144.1	131.1
0960	Flooring	97.0	58.5	86.0	95.2	90.8	93.9	100.7	64.7	90.4	104.1	79.7	97.2	107.7	79.0	99.5	97.9	182.7	122.2
0970, 0990	Wall Finishes & Painting/Coating	96.5	77.6	85.1	96.5	77.6	85.1	96.5	77.6	85.1	115.8	77.6	92.8	115.8	77.6	92.8	104.4	155.9	135.4
09	FINISHES	96.9	76.9	85.9	96.4	77.5	86.0	99.8	56.3	75.8	96.9	71.4	82.8	98.6	70.7	83.2	104.2	152.7	130.9
COVERS	DIVS. 10-14, 25, 28, 41, 43, 44, 46	100.0	59.9	91.9	100.0	89.6	97.9	100.0	38.7	87.6	100.0	84.7	96.9	100.0	83.0	96.6	100.0	119.4	103.9
21, 22, 23	FIRE SUPPRESSION, PLUMBING & HVAC	95.2	80.5	89.3	100.0	86.2	94.4	95.2	59.9	81.0	95.4	84.8	91.2	95.4	82.3	90.1	100.1	132.4	113.1
26, 27, 3370	ELECTRICAL, COMMUNICATIONS & UTIL.	99.6	91.9	95.5	98.1	81.8	89.5	96.5	67.1	81.0	98.4	99.7	99.1	95.7	99.7	97.8	100.8	135.4	119.1
MF2014	WEIGHTED AVERAGE	95.9	81.9	89.8	97.6	83.3	91.4	98.7	62.1	82.7	96.5	83.2	90.7	97.1	81.7	90.4	101.6	139.5	118.1

For customer support on your Mechanical Cost Data, call 877.795.7394.

671

City Cost Indexes

MASSACHUSETTS

DIVISION		BROCKTON 023			BUZZARDS BAY 025			FALL RIVER 027			FITCHBURG 014			FRAMINGHAM 017			GREENFIELD 013		
		MAT.	INST.	TOTAL	MAT.	INST.	TOTAL	MAT.	INST.	TOTAL	MAT.	INST.	TOTAL	MAT.	INST.	TOTAL	MAT.	INST.	TOTAL
015433	CONTRACTOR EQUIPMENT		102.7	102.7		102.7	102.7		103.7	103.7		100.9	100.9		101.9	101.9		100.9	100.9
0241, 31 - 34	SITE & INFRASTRUCTURE, DEMOLITION	94.5	105.7	102.4	84.8	105.6	99.6	93.5	105.8	102.2	86.3	105.6	100.0	83.0	105.3	98.9	90.0	104.3	100.1
0310	Concrete Forming & Accessories	101.6	138.7	133.6	99.2	138.5	133.1	101.6	138.8	133.6	93.8	130.8	125.7	101.2	138.8	133.6	92.1	113.2	110.3
0320	Concrete Reinforcing	106.4	155.6	131.4	85.3	160.1	123.4	106.4	160.2	133.7	83.9	151.4	118.3	83.9	155.6	124.9	87.3	124.4	106.2
0330	Cast-in-Place Concrete	97.7	151.5	119.8	81.2	151.5	110.1	94.5	152.0	118.1	85.9	150.5	112.4	85.9	148.3	111.5	88.3	131.2	105.9
03	CONCRETE	101.8	145.2	123.1	86.2	145.9	115.5	100.3	146.2	122.9	85.9	140.4	112.7	88.7	144.1	116.0	89.6	120.9	104.9
04	MASONRY	101.7	160.0	138.0	94.4	160.0	135.3	102.6	159.9	138.3	101.0	159.0	137.2	107.2	160.1	140.1	105.4	136.0	124.4
05	METALS	97.4	130.0	107.5	92.3	131.7	104.4	97.4	132.1	108.1	95.5	125.5	104.7	95.5	130.0	106.2	97.8	110.4	101.7
06	WOOD, PLASTICS & COMPOSITES	99.1	138.4	121.1	96.1	138.4	119.8	99.1	138.7	121.3	93.7	128.3	113.1	100.1	138.2	121.4	91.5	110.9	102.4
07	THERMAL & MOISTURE PROTECTION	100.9	148.7	120.5	100.1	147.0	119.4	100.8	146.3	119.5	99.2	141.5	116.5	99.3	149.1	119.7	99.2	121.9	108.5
08	OPENINGS	102.2	143.5	111.9	97.9	139.2	107.5	102.2	139.3	110.9	104.6	137.0	112.2	94.9	143.4	106.2	104.7	115.0	107.1
0920	Plaster & Gypsum Board	93.7	139.0	124.3	89.5	139.0	123.0	93.7	139.0	124.3	100.3	128.6	119.4	102.8	139.0	127.3	101.0	110.7	107.5
0950, 0980	Ceilings & Acoustic Treatment	102.2	139.0	126.4	88.2	139.0	121.6	102.2	139.0	126.4	89.0	128.6	115.0	89.0	139.0	121.9	97.4	110.7	106.1
0960	Flooring	98.6	182.7	122.7	96.7	182.7	121.3	97.7	182.7	122.0	98.1	182.7	122.3	99.7	182.7	123.5	97.3	154.0	113.5
0970, 0990	Wall Finishes & Painting/Coating	99.9	155.1	133.2	99.9	155.1	133.2	99.9	155.1	133.2	99.5	155.1	133.1	100.5	155.1	133.5	99.5	115.8	109.4
09	FINISHES	98.9	149.1	126.6	94.2	149.1	124.4	98.7	149.3	126.6	94.6	143.1	121.3	95.3	148.9	124.8	96.7	121.0	110.0
COVERS	DIVS. 10 - 14, 25, 28, 41, 43, 44, 46	100.0	117.9	103.6	100.0	117.9	103.6	100.0	118.5	103.7	100.0	109.0	101.8	100.0	117.5	103.5	100.0	104.6	100.9
21, 22, 23	FIRE SUPPRESSION, PLUMBING & HVAC	100.1	111.1	104.6	95.4	111.1	101.7	100.1	111.2	104.6	96.0	113.6	103.1	96.0	126.1	108.1	96.0	102.8	98.7
26, 27, 3370	ELECTRICAL, COMMUNICATIONS & UTIL.	99.3	100.1	99.7	96.3	100.1	98.3	99.2	100.1	99.7	100.1	105.1	102.8	96.6	128.9	113.7	100.1	98.9	99.5
MF2014	WEIGHTED AVERAGE	99.9	128.1	112.2	94.4	128.1	109.1	99.7	128.3	112.2	96.5	126.5	109.6	95.7	135.1	112.9	97.7	112.3	104.1

MASSACHUSETTS

DIVISION		HYANNIS 026			LAWRENCE 019			LOWELL 018			NEW BEDFORD 027			PITTSFIELD 012			SPRINGFIELD 010 - 011		
		MAT.	INST.	TOTAL	MAT.	INST.	TOTAL	MAT.	INST.	TOTAL	MAT.	INST.	TOTAL	MAT.	INST.	TOTAL	MAT.	INST.	TOTAL
015433	CONTRACTOR EQUIPMENT		102.7	102.7		102.7	102.7		100.9	100.9		103.7	103.7		100.9	100.9		100.9	100.9
0241, 31 - 34	SITE & INFRASTRUCTURE, DEMOLITION	90.9	105.6	101.3	95.2	105.7	102.7	94.2	105.6	102.3	92.1	105.8	101.8	95.2	104.2	101.6	94.6	104.4	101.6
0310	Concrete Forming & Accessories	93.5	138.5	132.3	102.4	139.0	134.0	99.0	139.0	133.5	101.6	138.8	133.6	99.0	112.2	110.4	99.3	113.6	111.6
0320	Concrete Reinforcing	85.3	160.1	123.4	103.8	150.4	127.5	104.6	150.4	127.9	106.4	160.2	133.7	86.6	122.3	104.8	104.6	124.4	114.7
0330	Cast-in-Place Concrete	89.1	151.5	114.7	99.4	148.9	119.7	90.3	148.9	114.4	83.3	152.0	111.5	98.5	129.7	111.3	94.0	131.7	109.5
03	CONCRETE	92.4	145.9	118.7	103.7	143.5	123.2	95.1	143.3	118.8	95.0	146.2	120.2	96.2	119.5	107.6	96.9	121.2	108.8
04	MASONRY	100.6	160.0	137.6	112.3	160.7	142.5	100.2	160.0	137.5	100.6	159.9	137.6	100.9	133.2	121.0	100.5	136.9	123.2
05	METALS	93.7	131.7	105.4	98.2	128.3	107.5	98.2	126.0	106.7	97.4	132.1	108.1	98.0	109.5	101.5	100.9	110.4	103.8
06	WOOD, PLASTICS & COMPOSITES	89.3	138.4	116.8	100.6	138.4	121.8	99.8	138.4	121.4	99.1	138.7	121.3	99.8	110.9	106.0	99.8	110.9	106.0
07	THERMAL & MOISTURE PROTECTION	100.4	147.0	119.5	99.7	148.5	119.7	99.5	148.6	119.7	100.7	146.3	119.4	99.6	120.7	108.2	99.5	122.2	108.8
08	OPENINGS	98.5	139.2	108.0	99.0	142.1	109.0	105.9	142.1	114.4	102.2	139.3	110.9	105.9	114.4	107.9	105.9	115.0	108.0
0920	Plaster & Gypsum Board	85.2	139.0	121.6	105.6	139.0	128.2	105.6	139.0	128.2	93.7	139.0	124.3	105.6	110.7	109.0	105.6	110.7	109.0
0950, 0980	Ceilings & Acoustic Treatment	93.8	139.0	123.5	99.3	139.0	125.4	99.3	139.0	125.4	102.2	139.0	126.4	99.3	110.7	106.8	99.3	110.7	106.8
0960	Flooring	94.4	182.7	119.6	100.2	182.7	123.8	100.2	182.7	123.8	97.7	182.7	122.0	100.6	154.0	115.9	99.7	154.0	115.3
0970, 0990	Wall Finishes & Painting/Coating	99.9	155.1	133.2	99.6	155.1	133.1	99.5	155.1	133.1	99.9	155.1	133.2	99.5	115.8	109.4	100.9	115.8	109.9
09	FINISHES	94.4	149.1	124.6	98.8	149.1	126.5	98.7	149.1	126.5	98.6	149.3	126.5	98.8	120.3	116.9	98.7	121.2	111.1
COVERS	DIVS. 10 - 14, 25, 28, 41, 43, 44, 46	100.0	117.9	103.6	100.0	118.1	103.6	100.0	118.1	103.6	100.0	118.5	103.7	100.0	103.7	100.7	100.0	104.9	101.0
21, 22, 23	FIRE SUPPRESSION, PLUMBING & HVAC	100.1	111.1	104.6	100.1	124.8	110.0	100.1	125.7	110.4	100.1	111.2	104.6	100.1	101.4	100.6	100.1	103.3	101.4
26, 27, 3370	ELECTRICAL, COMMUNICATIONS & UTIL.	96.8	100.1	98.5	99.1	128.9	114.9	99.6	128.9	115.1	100.0	100.1	100.1	99.6	98.9	99.2	99.7	98.9	99.2
MF2014	WEIGHTED AVERAGE	97.0	128.1	110.6	100.3	134.6	115.3	99.5	134.5	114.8	99.1	128.3	111.8	99.7	111.3	104.8	100.2	112.6	105.6

DIVISION		MASSACHUSETTS WORCESTER 015 - 016			MICHIGAN ANN ARBOR 481			BATTLE CREEK 490			BAY CITY 487			DEARBORN 481			DETROIT 482		
		MAT.	INST.	TOTAL	MAT.	INST.	TOTAL	MAT.	INST.	TOTAL	MAT.	INST.	TOTAL	MAT.	INST.	TOTAL	MAT.	INST.	TOTAL
015433	CONTRACTOR EQUIPMENT		100.9	100.9		110.7	110.7		102.8	102.8		110.7	110.7		110.7	110.7		98.8	98.8
0241, 31 - 34	SITE & INFRASTRUCTURE, DEMOLITION	94.6	105.6	102.4	82.3	98.8	94.0	93.3	87.5	89.2	73.9	97.8	90.9	82.1	98.9	94.1	94.7	100.6	98.9
0310	Concrete Forming & Accessories	99.6	130.7	126.5	98.3	111.6	109.8	97.6	87.6	89.0	98.4	87.1	88.6	98.2	115.8	113.3	100.8	115.8	113.7
0320	Concrete Reinforcing	104.6	150.6	128.0	92.3	120.1	106.5	91.4	93.3	92.4	92.3	119.2	106.0	92.3	120.2	106.5	94.9	120.1	107.8
0330	Cast-in-Place Concrete	93.5	150.5	116.9	88.8	107.4	96.4	99.0	100.4	99.6	85.0	90.6	87.3	86.8	109.9	96.3	94.1	109.9	100.6
03	CONCRETE	96.6	140.2	118.0	93.5	112.0	102.6	98.1	92.5	95.4	91.7	95.2	93.4	92.6	114.7	103.5	96.6	113.6	105.0
04	MASONRY	100.0	159.0	136.8	104.4	108.4	106.9	105.2	87.2	94.0	104.0	87.0	93.4	104.3	111.6	108.9	97.9	111.6	106.5
05	METALS	101.0	125.2	108.4	94.2	118.8	101.8	98.3	87.6	95.0	94.8	115.4	101.2	94.3	119.1	101.9	94.9	102.1	97.1
06	WOOD, PLASTICS & COMPOSITES	100.3	128.3	116.0	97.5	112.8	106.0	96.5	87.2	91.3	97.5	86.5	91.3	97.5	116.8	108.3	103.0	116.8	110.7
07	THERMAL & MOISTURE PROTECTION	99.5	141.5	116.7	100.4	108.2	103.6	94.6	84.8	90.6	98.2	92.4	95.8	98.9	115.7	105.8	97.7	115.7	105.1
08	OPENINGS	105.9	136.7	113.1	99.4	110.4	101.9	96.6	82.3	93.3	99.4	93.9	98.1	99.4	112.6	102.5	99.2	112.9	102.4
0920	Plaster & Gypsum Board	105.6	128.6	121.2	103.8	112.4	109.6	93.7	82.9	86.4	103.8	85.3	91.3	103.8	116.6	112.4	100.5	116.6	111.3
0950, 0980	Ceilings & Acoustic Treatment	99.3	128.6	118.5	88.7	112.4	104.3	96.3	82.9	87.5	89.7	85.3	86.8	88.7	116.6	107.0	91.2	116.6	107.9
0960	Flooring	100.2	179.8	123.0	97.5	117.6	103.2	105.7	91.9	101.7	97.5	81.5	92.9	96.9	113.9	101.8	95.9	113.9	101.1
0970, 0990	Wall Finishes & Painting/Coating	99.5	155.1	133.1	96.2	98.7	97.7	107.9	81.9	92.2	96.2	82.0	87.6	96.2	100.8	99.0	97.3	100.8	99.4
09	FINISHES	98.7	142.5	122.9	92.8	111.6	103.1	98.0	87.6	92.3	92.6	84.8	88.3	92.6	114.3	104.5	93.9	114.3	105.1
COVERS	DIVS. 10 - 14, 25, 28, 41, 43, 44, 46	100.0	109.0	101.8	100.0	105.8	101.2	100.0	96.4	99.3	100.0	94.1	98.8	100.0	107.1	101.4	100.0	107.1	101.4
21, 22, 23	FIRE SUPPRESSION, PLUMBING & HVAC	100.1	113.6	105.5	100.0	98.8	99.5	100.0	87.4	94.9	100.0	83.6	93.4	100.0	109.2	103.7	100.0	110.2	104.1
26, 27, 3370	ELECTRICAL, COMMUNICATIONS & UTIL.	99.7	105.1	102.5	96.3	108.8	102.9	94.4	83.7	88.8	95.2	90.3	92.6	96.3	107.0	101.9	98.2	107.0	102.9
MF2014	WEIGHTED AVERAGE	100.2	126.4	111.6	97.1	107.5	101.7	98.3	87.6	93.7	96.6	91.7	94.5	97.0	111.0	103.1	97.8	109.6	102.9

MICHIGAN

DIVISION		FLINT 484 - 485			GAYLORD 497			GRAND RAPIDS 493,495			IRON MOUNTAIN 498 - 499			JACKSON 492			KALAMAZOO 491		
		MAT.	INST.	TOTAL	MAT.	INST.	TOTAL	MAT.	INST.	TOTAL	MAT.	INST.	TOTAL	MAT.	INST.	TOTAL	MAT.	INST.	TOTAL
015433	CONTRACTOR EQUIPMENT		110.7	110.7		104.7	104.7		102.8	102.8		93.7	93.7		104.7	104.7		102.8	102.8
0241, 31 - 34	SITE & INFRASTRUCTURE, DEMOLITION	71.7	98.0	90.4	88.5	85.0	86.0	93.2	87.6	89.2	96.1	93.7	94.4	109.4	87.3	93.7	93.6	87.5	89.3
0310	Concrete Forming & Accessories	101.3	89.9	91.5	95.5	77.3	79.8	97.1	84.6	86.3	87.8	85.0	85.4	92.5	86.7	87.5	97.6	87.3	88.7
0320	Concrete Reinforcing	92.3	119.6	106.2	85.2	118.6	102.2	97.8	93.2	95.5	85.0	99.8	92.5	82.8	119.4	101.4	91.4	93.4	92.4
0330	Cast-in-Place Concrete	89.4	92.4	90.6	98.7	86.1	93.5	104.1	97.6	101.4	116.8	85.4	103.9	98.6	92.2	96.0	101.0	98.8	100.1
03	CONCRETE	94.0	97.1	95.5	94.6	88.9	91.8	101.6	90.2	96.0	104.3	87.9	96.2	89.3	95.4	92.3	101.6	91.8	96.8
04	MASONRY	104.5	96.5	99.5	116.2	79.1	93.1	102.0	87.4	92.9	101.2	86.1	91.8	95.1	91.6	92.9	103.8	89.1	94.6
05	METALS	94.3	116.3	101.0	99.7	109.9	102.8	95.3	87.1	92.8	99.0	89.6	96.1	99.9	114.1	104.2	98.3	86.5	94.7
06	WOOD, PLASTICS & COMPOSITES	100.9	88.2	93.8	89.4	76.8	82.4	97.7	83.2	89.6	85.0	85.1	85.0	88.1	84.7	86.2	96.5	87.2	91.3
07	THERMAL & MOISTURE PROTECTION	98.2	96.3	97.4	93.2	74.5	85.5	97.4	76.6	88.9	96.7	81.4	90.4	92.5	92.6	92.5	94.6	85.1	90.7
08	OPENINGS	99.4	95.6	98.5	97.5	75.1	92.3	103.4	84.7	99.0	104.6	76.0	98.0	96.5	93.1	95.7	96.6	81.7	93.1
0920	Plaster & Gypsum Board	105.2	87.1	93.0	93.4	74.9	80.9	95.4	78.8	84.2	51.7	85.2	74.3	91.7	82.9	85.8	93.7	82.9	86.4
0950, 0980	Ceilings & Acoustic Treatment	88.7	87.1	87.6	95.4	74.9	81.9	104.3	78.8	87.5	94.4	85.2	88.3	95.4	82.9	87.2	96.3	82.9	87.5
0960	Flooring	97.5	95.8	97.0	97.8	92.4	96.2	105.5	85.7	99.8	122.4	93.8	114.2	96.5	82.7	92.6	105.7	82.7	99.1
0970, 0990	Wall Finishes & Painting/Coating	96.2	85.4	89.7	103.8	82.0	90.6	109.7	80.8	92.3	126.1	72.1	93.5	103.8	96.8	99.6	107.9	81.9	92.2
09	FINISHES	92.1	89.8	90.9	97.4	79.1	87.3	101.6	84.2	92.0	100.0	85.3	91.9	98.3	86.2	91.6	98.0	85.6	91.2
COVERS	DIVS. 10 - 14, 25, 28, 41, 43, 44, 46	100.0	95.5	99.1	100.0	88.6	97.7	100.0	96.4	99.3	100.0	87.2	97.4	100.0	100.9	100.2	100.0	96.4	99.3
21, 22, 23	FIRE SUPPRESSION, PLUMBING & HVAC	100.0	89.6	95.8	95.6	78.4	88.6	100.0	83.0	93.1	95.5	85.3	91.4	95.6	88.4	92.7	100.0	80.8	92.3
26, 27, 3370	ELECTRICAL, COMMUNICATIONS & UTIL.	96.3	96.0	96.1	92.4	79.1	85.4	99.7	79.1	88.8	98.7	84.3	91.1	96.3	108.1	102.9	94.3	81.8	87.7
MF2014	WEIGHTED AVERAGE	96.8	96.0	96.4	97.2	83.6	91.3	99.7	85.1	93.3	99.3	86.2	93.6	96.5	95.1	95.9	98.7	85.7	93.0

MICHIGAN / MINNESOTA

DIVISION		LANSING 488 - 489			MUSKEGON 494			ROYAL OAK 480,483			SAGINAW 486			TRAVERSE CITY 496			BEMIDJI 566		
		MAT.	INST.	TOTAL	MAT.	INST.	TOTAL	MAT.	INST.	TOTAL	MAT.	INST.	TOTAL	MAT.	INST.	TOTAL	MAT.	INST.	TOTAL
015433	CONTRACTOR EQUIPMENT		110.7	110.7		102.8	102.8		96.2	96.2		110.7	110.7		93.7	93.7		98.2	98.2
0241, 31 - 34	SITE & INFRASTRUCTURE, DEMOLITION	95.6	98.0	97.3	91.2	87.5	88.5	86.5	98.4	95.0	74.9	97.8	91.2	82.4	93.2	90.0	95.4	97.5	96.9
0310	Concrete Forming & Accessories	98.2	89.7	90.9	98.0	84.1	86.0	94.1	112.6	110.1	98.3	87.6	89.1	87.8	75.2	76.9	86.1	89.6	89.1
0320	Concrete Reinforcing	94.3	119.4	107.1	92.1	93.3	92.7	83.7	119.9	102.1	92.3	119.2	106.0	86.3	92.5	89.5	93.7	106.6	100.2
0330	Cast-in-Place Concrete	103.5	92.1	98.8	98.6	94.8	97.1	77.8	106.3	89.5	87.7	90.5	88.9	91.2	78.2	85.9	106.0	103.0	104.8
03	CONCRETE	100.8	96.9	98.9	96.4	89.1	92.8	80.5	110.8	95.4	93.0	95.4	94.2	85.8	79.8	82.9	97.6	98.5	98.1
04	MASONRY	106.6	94.2	98.9	102.2	88.5	93.7	98.2	112.0	106.8	106.0	87.0	94.1	99.2	82.4	88.7	102.5	103.7	103.3
05	METALS	93.9	115.6	100.6	96.0	87.2	93.3	97.3	98.0	97.5	94.3	115.2	100.7	99.0	88.2	95.7	92.1	120.0	100.7
06	WOOD, PLASTICS & COMPOSITES	96.3	88.8	92.1	93.2	83.4	87.7	93.0	113.1	104.3	93.8	87.6	90.3	85.0	75.1	79.4	71.3	85.5	79.3
07	THERMAL & MOISTURE PROTECTION	98.6	89.6	94.9	93.6	78.9	87.6	96.8	111.2	102.7	99.0	92.6	96.4	95.8	71.7	85.9	104.7	95.4	100.9
08	OPENINGS	103.2	95.1	101.3	95.8	84.5	93.2	99.3	109.4	101.7	97.4	94.5	96.7	104.6	70.6	96.7	100.2	107.2	101.8
0920	Plaster & Gypsum Board	96.9	87.7	90.7	75.7	79.0	77.9	101.3	112.7	109.0	103.8	86.4	92.1	51.7	74.9	67.3	102.2	85.3	90.8
0950, 0980	Ceilings & Acoustic Treatment	94.0	87.7	89.8	97.1	79.0	85.2	88.0	112.7	104.3	88.7	86.4	87.2	94.4	74.9	81.6	122.8	85.3	98.2
0960	Flooring	106.6	85.7	100.6	104.5	85.7	99.1	94.8	113.9	100.3	97.5	81.5	92.9	122.4	94.9	114.5	104.4	120.8	109.1
0970, 0990	Wall Finishes & Painting/Coating	108.0	80.8	91.6	106.3	80.8	90.9	97.6	96.8	97.1	96.2	82.0	87.5	126.1	44.7	76.9	102.2	95.9	98.4
09	FINISHES	98.3	87.6	92.4	94.7	83.2	88.4	91.7	111.5	102.6	92.4	85.5	88.6	99.0	74.1	86.3	106.2	95.2	100.1
COVERS	DIVS. 10 - 14, 25, 28, 41, 43, 44, 46	100.0	100.6	100.1	100.0	96.0	99.2	100.0	103.0	100.8	100.0	94.3	98.8	100.0	84.6	96.9	100.0	96.9	99.4
21, 22, 23	FIRE SUPPRESSION, PLUMBING & HVAC	99.9	88.6	95.3	99.9	82.3	92.8	95.6	107.2	100.3	100.0	83.1	93.2	95.5	77.7	88.3	95.6	83.4	90.6
26, 27, 3370	ELECTRICAL, COMMUNICATIONS & UTIL.	98.6	93.3	95.8	94.8	79.1	86.5	98.4	104.4	101.6	93.9	90.5	92.1	94.0	79.1	86.1	104.3	104.6	104.5
MF2014	WEIGHTED AVERAGE	99.3	94.7	97.3	97.2	84.8	91.8	94.9	106.9	100.1	96.4	91.8	94.4	96.3	80.1	89.3	98.2	98.2	98.2

MINNESOTA

DIVISION		BRAINERD 564			DETROIT LAKES 565			DULUTH 556 - 558			MANKATO 560			MINNEAPOLIS 553 - 555			ROCHESTER 559		
		MAT.	INST.	TOTAL	MAT.	INST.	TOTAL	MAT.	INST.	TOTAL	MAT.	INST.	TOTAL	MAT.	INST.	TOTAL	MAT.	INST.	TOTAL
015433	CONTRACTOR EQUIPMENT		100.8	100.8		98.2	98.2		99.1	99.1		100.8	100.8		102.6	102.6		99.1	99.1
0241, 31 - 34	SITE & INFRASTRUCTURE, DEMOLITION	96.9	102.2	100.7	93.6	97.9	96.6	100.6	99.4	99.7	93.8	102.9	100.3	97.9	104.7	102.8	97.4	99.0	98.5
0310	Concrete Forming & Accessories	87.7	91.6	91.1	83.1	91.6	90.4	98.7	109.9	108.4	95.6	99.1	98.6	100.7	128.6	124.8	99.3	106.5	105.5
0320	Concrete Reinforcing	92.5	106.7	99.7	93.7	106.6	100.2	97.4	107.1	102.3	92.4	116.2	104.5	92.2	116.6	104.6	91.7	116.4	104.3
0330	Cast-in-Place Concrete	115.2	106.7	111.7	102.9	106.1	104.2	114.8	106.0	111.2	106.2	107.3	106.6	109.6	117.4	112.8	112.3	99.4	107.0
03	CONCRETE	101.7	100.7	101.2	95.2	100.4	97.8	106.5	108.7	107.6	97.2	106.1	101.5	103.4	122.8	112.9	102.5	106.8	104.6
04	MASONRY	126.4	112.0	117.4	126.3	110.1	116.2	107.6	117.6	113.9	114.5	110.4	112.0	109.0	126.7	120.1	104.5	114.1	110.5
05	METALS	93.2	120.0	101.5	92.1	120.0	100.6	101.7	122.4	108.0	93.1	125.5	103.0	99.1	128.9	108.3	100.4	127.3	108.7
06	WOOD, PLASTICS & COMPOSITES	89.4	85.3	87.1	68.4	85.5	78.0	98.2	109.3	104.4	98.8	96.4	97.5	106.7	127.4	118.3	103.2	105.3	104.4
07	THERMAL & MOISTURE PROTECTION	103.4	107.1	105.2	104.6	108.3	106.1	101.6	114.9	107.1	103.8	99.0	101.8	102.8	127.4	112.9	104.9	100.9	103.2
08	OPENINGS	86.9	107.1	91.6	100.2	107.2	101.8	102.5	116.5	105.8	91.5	115.5	97.1	97.1	133.6	105.6	96.4	121.5	102.2
0920	Plaster & Gypsum Board	86.9	85.3	85.8	101.1	85.3	90.5	90.2	110.0	103.6	91.5	96.8	95.1	95.8	128.6	117.9	97.6	105.8	103.2
0950, 0980	Ceilings & Acoustic Treatment	59.3	85.3	76.4	122.8	85.3	98.2	96.1	110.0	105.3	59.3	96.8	84.0	98.2	128.6	118.2	93.2	105.8	101.5
0960	Flooring	103.3	120.8	108.3	103.2	120.8	108.2	101.9	122.1	107.6	105.0	120.8	109.5	97.9	120.8	104.4	102.9	120.8	108.0
0970, 0990	Wall Finishes & Painting/Coating	96.0	95.9	95.9	102.2	95.9	98.4	101.9	110.3	107.0	108.3	104.9	106.3	98.2	126.1	115.1	97.2	104.9	101.9
09	FINISHES	89.4	96.4	93.2	105.5	96.5	100.6	98.3	112.5	106.1	91.1	103.4	97.9	98.6	127.6	114.6	97.4	109.1	103.8
COVERS	DIVS. 10 - 14, 25, 28, 41, 43, 44, 46	100.0	98.3	99.7	100.0	98.6	99.7	100.0	97.6	99.5	100.0	98.6	99.7	100.0	109.7	102.0	100.0	102.8	100.6
21, 22, 23	FIRE SUPPRESSION, PLUMBING & HVAC	94.8	86.1	91.3	95.6	86.0	91.7	99.8	96.5	98.5	94.8	87.7	91.9	99.9	115.6	106.2	99.9	96.2	98.4
26, 27, 3370	ELECTRICAL, COMMUNICATIONS & UTIL.	101.8	102.4	102.1	104.0	67.5	84.7	98.4	102.4	100.6	108.6	89.2	98.3	101.2	111.4	106.6	98.6	89.2	93.6
MF2014	WEIGHTED AVERAGE	96.9	100.5	98.5	98.9	95.1	97.3	101.3	107.3	103.9	97.2	101.3	99.0	100.4	120.0	108.9	99.9	104.7	102.0

For customer support on your Mechanical Cost Data, call 877.795.7394.

673

MINNESOTA / MISSISSIPPI

DIVISION		SAINT PAUL 550-551			ST. CLOUD 563			THIEF RIVER FALLS 567			WILLMAR 562			WINDOM 561			BILOXI 395		
		MAT.	INST.	TOTAL	MAT.	INST.	TOTAL	MAT.	INST.	TOTAL	MAT.	INST.	TOTAL	MAT.	INST.	TOTAL	MAT.	INST.	TOTAL
015433	CONTRACTOR EQUIPMENT		99.1	99.1		100.8	100.8		98.2	98.2		100.8	100.8		100.8	100.8		98.8	98.8
0241, 31 - 34	SITE & INFRASTRUCTURE, DEMOLITION	95.2	100.1	98.7	92.5	104.3	100.9	94.4	97.5	96.6	91.8	102.0	99.1	86.0	101.0	96.6	100.6	88.4	91.9
0310	Concrete Forming & Accessories	93.9	123.0	119.0	84.9	117.7	113.2	86.8	89.0	88.7	84.7	91.1	90.2	89.1	84.8	85.4	94.7	52.9	58.7
0320	Concrete Reinforcing	98.9	116.6	107.9	92.6	116.4	104.7	94.0	106.5	100.4	92.2	116.0	104.3	92.2	115.3	104.0	95.0	58.3	76.4
0330	Cast-in-Place Concrete	124.0	116.7	121.0	101.7	114.8	107.1	105.1	88.1	98.1	103.3	85.3	95.9	89.4	66.3	79.9	108.4	54.0	86.0
03	CONCRETE	111.2	120.1	115.5	92.9	116.9	104.7	96.3	93.1	94.7	92.9	94.9	93.9	83.6	85.5	84.5	101.6	55.9	79.2
04	MASONRY	111.7	126.7	121.1	111.1	119.9	116.6	102.4	104.5	103.7	115.4	112.0	113.3	125.8	92.1	104.8	102.7	44.3	66.3
05	METALS	101.1	128.6	109.6	93.9	126.8	104.0	92.2	119.4	100.6	93.0	124.5	102.7	92.9	122.4	102.0	94.9	81.7	90.8
06	WOOD, PLASTICS & COMPOSITES	93.4	119.9	108.2	86.7	114.9	102.5	72.3	85.5	79.7	86.4	85.5	85.9	90.7	82.8	86.3	99.1	56.1	75.0
07	THERMAL & MOISTURE PROTECTION	104.1	126.4	113.3	103.6	117.5	109.3	105.6	98.4	102.7	103.4	108.6	105.5	103.4	89.6	97.7	97.4	56.1	80.4
08	OPENINGS	99.8	129.5	106.7	91.5	126.7	99.7	100.2	107.2	101.8	89.0	91.9	89.7	92.7	90.4	92.2	99.6	57.7	89.8
0920	Plaster & Gypsum Board	88.9	120.9	110.6	86.9	115.8	106.4	101.8	85.3	90.7	86.9	85.6	86.0	86.9	82.8	84.1	101.7	55.2	70.3
0950, 0980	Ceilings & Acoustic Treatment	95.6	120.9	112.3	59.3	115.8	96.4	122.8	85.3	98.2	59.3	85.6	76.6	59.3	82.8	74.7	87.3	55.2	66.2
0960	Flooring	103.7	120.8	108.6	100.0	120.8	105.9	104.1	120.8	108.8	101.6	120.8	107.1	103.3	120.8	108.8	101.4	55.1	88.1
0970, 0990	Wall Finishes & Painting/Coating	105.3	126.1	117.8	108.3	126.1	119.1	102.2	95.9	98.4	102.2	95.9	98.4	102.2	104.9	103.8	99.5	43.4	65.7
09	FINISHES	97.9	123.1	111.8	88.9	119.2	105.6	106.0	95.2	100.0	88.9	94.2	91.8	89.1	91.3	90.3	93.3	52.4	70.8
COVERS	DIVS. 10 - 14, 25, 28, 41, 43, 44, 46	100.0	108.6	101.7	100.0	104.3	100.9	100.0	96.8	99.3	100.0	98.2	99.6	100.0	93.7	98.7	100.0	55.4	91.0
21, 22, 23	FIRE SUPPRESSION, PLUMBING & HVAC	99.9	113.8	105.5	99.6	113.8	105.3	95.6	83.1	90.5	94.8	106.8	99.6	94.8	84.1	90.5	100.0	49.1	79.5
26, 27, 3370	ELECTRICAL, COMMUNICATIONS & UTIL.	97.1	111.4	104.7	101.8	111.4	106.9	101.3	67.5	83.5	101.8	85.2	93.0	108.6	89.2	98.3	103.8	56.9	79.0
MF2014	WEIGHTED AVERAGE	101.4	117.9	108.6	96.9	116.0	105.2	97.8	92.4	95.4	95.5	101.3	98.0	95.9	92.3	94.3	99.2	58.1	81.3

MISSISSIPPI

DIVISION		CLARKSDALE 386			COLUMBUS 397			GREENVILLE 387			GREENWOOD 389			JACKSON 390 - 392			LAUREL 394		
		MAT.	INST.	TOTAL	MAT.	INST.	TOTAL	MAT.	INST.	TOTAL	MAT.	INST.	TOTAL	MAT.	INST.	TOTAL	MAT.	INST.	TOTAL
015433	CONTRACTOR EQUIPMENT		98.8	98.8		98.8	98.8		98.8	98.8		98.8	98.8		98.8	98.8		98.8	98.8
0241, 31 - 34	SITE & INFRASTRUCTURE, DEMOLITION	98.3	88.2	91.1	98.9	88.4	91.5	103.9	88.4	92.9	101.0	87.9	91.7	97.6	88.4	91.1	104.4	88.3	92.9
0310	Concrete Forming & Accessories	85.0	49.2	54.2	83.3	51.2	55.6	81.7	63.7	66.1	94.2	49.2	55.4	93.6	57.5	62.5	83.3	52.9	57.1
0320	Concrete Reinforcing	99.8	41.6	70.1	101.8	42.5	71.6	100.3	60.3	79.9	99.8	52.5	75.7	100.2	56.1	77.7	102.4	39.1	70.2
0330	Cast-in-Place Concrete	105.6	53.5	84.2	110.5	58.6	89.2	108.7	55.6	86.9	113.5	53.2	88.7	98.7	56.8	81.5	108.1	55.2	86.4
03	CONCRETE	98.7	50.9	75.2	103.2	53.7	78.9	104.2	61.6	83.3	105.2	52.7	79.4	97.7	58.5	78.5	106.0	52.7	79.8
04	MASONRY	100.6	45.0	66.0	131.6	51.6	81.7	149.8	49.5	87.3	101.3	44.9	66.1	109.2	49.5	72.0	127.1	48.1	77.9
05	METALS	95.4	73.4	88.6	91.9	74.2	86.5	96.4	82.5	92.1	95.4	74.8	89.0	101.8	80.8	95.4	92.0	73.3	86.3
06	WOOD, PLASTICS & COMPOSITES	82.1	51.2	64.8	83.9	52.2	66.1	78.7	68.1	72.8	95.0	51.2	70.5	96.3	59.6	75.8	85.1	54.6	68.0
07	THERMAL & MOISTURE PROTECTION	95.7	47.2	75.8	97.3	51.5	78.5	96.0	60.0	81.2	96.1	47.1	76.0	97.5	59.2	81.8	97.4	53.1	79.3
08	OPENINGS	97.0	49.4	85.9	98.9	49.5	87.4	97.0	61.4	88.7	97.0	50.7	86.2	103.7	55.6	92.5	95.6	50.3	85.0
0920	Plaster & Gypsum Board	90.2	50.2	63.1	91.3	51.2	64.2	89.8	67.6	74.8	100.9	50.2	66.6	95.0	58.8	70.5	91.3	53.7	65.8
0950, 0980	Ceilings & Acoustic Treatment	82.9	50.2	61.4	82.0	51.2	61.8	85.7	67.6	73.8	82.9	50.2	61.4	89.7	58.8	69.4	82.0	53.7	63.4
0960	Flooring	105.7	52.8	90.6	95.5	59.3	85.1	104.1	52.8	89.4	111.1	52.8	94.4	99.3	55.1	86.7	94.2	52.8	82.4
0970, 0990	Wall Finishes & Painting/Coating	104.4	46.1	69.2	99.5	53.0	71.4	104.4	65.9	81.2	104.4	46.1	69.2	96.5	65.9	78.0	99.5	60.9	76.2
09	FINISHES	94.0	49.9	69.7	89.1	53.2	69.3	94.6	62.9	77.1	97.4	49.9	71.2	93.2	58.3	74.0	89.2	54.3	70.0
COVERS	DIVS. 10 - 14, 25, 28, 41, 43, 44, 46	100.0	55.7	91.1	100.0	56.8	91.3	100.0	58.5	91.6	100.0	55.7	91.1	100.0	57.5	91.4	100.0	37.5	87.4
21, 22, 23	FIRE SUPPRESSION, PLUMBING & HVAC	97.9	41.3	75.1	97.4	37.5	73.3	100.0	47.6	78.8	97.9	39.3	74.3	100.1	57.0	82.7	97.5	45.8	76.6
26, 27, 3370	ELECTRICAL, COMMUNICATIONS & UTIL.	97.1	46.4	70.3	100.8	49.1	73.5	97.1	62.7	78.9	97.1	43.4	68.8	105.4	62.7	82.8	102.6	56.9	78.5
MF2014	WEIGHTED AVERAGE	97.3	52.6	77.8	98.8	53.9	79.2	101.0	61.7	83.9	98.4	52.2	78.3	100.7	62.2	83.9	98.9	55.8	80.1

MISSISSIPPI / MISSOURI

DIVISION		MCCOMB 396			MERIDIAN 393			TUPELO 388			BOWLING GREEN 633			CAPE GIRARDEAU 637			CHILLICOTHE 646		
		MAT.	INST.	TOTAL	MAT.	INST.	TOTAL	MAT.	INST.	TOTAL	MAT.	INST.	TOTAL	MAT.	INST.	TOTAL	MAT.	INST.	TOTAL
015433	CONTRACTOR EQUIPMENT		98.8	98.8		98.8	98.8		98.8	98.8		106.9	106.9		106.9	106.9		101.9	101.9
0241, 31 - 34	SITE & INFRASTRUCTURE, DEMOLITION	92.2	88.0	89.2	96.2	88.7	90.8	95.6	88.1	90.3	89.6	95.6	93.9	91.3	95.4	94.2	105.2	94.6	97.7
0310	Concrete Forming & Accessories	83.3	50.9	55.4	80.4	63.3	65.7	82.2	51.2	55.5	92.3	87.3	88.0	85.2	85.1	85.1	89.0	93.0	92.5
0320	Concrete Reinforcing	103.0	41.0	71.4	101.8	56.1	78.5	97.6	52.5	74.7	101.4	102.8	102.1	102.7	88.7	95.5	96.8	106.5	101.8
0330	Cast-in-Place Concrete	96.0	53.0	78.3	102.7	59.2	84.8	105.6	54.9	84.8	91.2	84.7	88.5	90.2	86.0	88.5	103.0	85.8	95.9
03	CONCRETE	92.9	51.3	72.5	97.2	61.9	79.8	98.2	54.2	76.6	95.8	90.5	93.2	94.9	87.3	91.2	105.6	93.6	99.7
04	MASONRY	133.0	44.4	77.7	102.4	53.5	71.9	138.4	47.9	82.0	117.6	98.7	105.8	113.9	80.0	92.8	102.9	98.8	100.3
05	METALS	92.2	71.6	85.8	93.0	81.0	89.3	95.3	75.1	89.1	94.6	114.4	100.7	95.4	107.0	99.1	95.1	108.5	99.2
06	WOOD, PLASTICS & COMPOSITES	83.9	54.0	67.2	81.2	65.1	72.2	79.3	52.2	64.1	90.8	86.0	88.1	83.5	85.5	84.6	93.6	93.3	93.4
07	THERMAL & MOISTURE PROTECTION	96.9	46.4	76.2	97.0	61.6	82.5	95.7	52.9	78.2	100.3	98.9	99.7	100.2	84.2	93.6	98.3	94.5	96.8
08	OPENINGS	99.0	51.0	87.8	98.9	64.1	90.8	97.0	50.6	86.2	97.7	97.5	97.6	97.7	82.9	94.2	95.0	95.7	95.2
0920	Plaster & Gypsum Board	91.3	53.0	65.4	91.3	64.5	73.2	89.8	51.2	63.7	95.2	85.6	88.7	94.2	85.1	88.0	102.0	92.9	95.8
0950, 0980	Ceilings & Acoustic Treatment	82.0	53.0	63.0	83.9	64.5	71.2	82.9	51.2	62.0	92.6	85.6	88.0	92.6	85.1	87.6	89.1	92.9	91.6
0960	Flooring	95.5	52.8	83.3	94.2	55.1	83.0	104.4	52.8	89.6	96.4	96.8	96.5	93.3	84.0	90.6	97.3	102.5	98.8
0970, 0990	Wall Finishes & Painting/Coating	99.5	42.3	65.0	99.5	76.1	85.4	104.4	58.7	76.8	97.1	102.8	100.5	97.1	73.2	82.7	87.9	106.3	99.0
09	FINISHES	88.5	51.0	67.8	88.8	63.7	75.0	93.5	52.6	71.0	97.8	89.7	93.4	96.7	83.0	89.2	100.2	96.1	97.9
COVERS	DIVS. 10 - 14, 25, 28, 41, 43, 44, 46	100.0	59.1	91.7	100.0	59.5	91.8	100.0	56.6	91.3	100.0	88.2	97.6	100.0	93.5	98.7	100.0	88.5	97.7
21, 22, 23	FIRE SUPPRESSION, PLUMBING & HVAC	97.4	34.6	72.1	100.0	59.1	83.5	98.1	43.2	76.0	95.2	96.2	95.6	99.9	96.8	98.7	95.3	97.0	96.0
26, 27, 3370	ELECTRICAL, COMMUNICATIONS & UTIL.	99.2	49.4	72.9	102.6	57.8	78.9	96.9	49.3	71.7	96.9	82.6	89.3	96.9	101.5	99.3	92.4	78.6	85.1
MF2014	WEIGHTED AVERAGE	97.4	51.7	77.5	97.6	64.1	83.0	98.8	54.9	79.7	97.3	94.4	96.0	98.2	92.4	95.7	97.6	94.5	96.3

674

For customer support on your Mechanical Cost Data, call 877.795.7394.

City Cost Indexes

MISSOURI

DIVISION		COLUMBIA 652			FLAT RIVER 636			HANNIBAL 634			HARRISONVILLE 647			JEFFERSON CITY 650 - 651			JOPLIN 648		
		MAT.	INST.	TOTAL	MAT.	INST.	TOTAL	MAT.	INST.	TOTAL	MAT.	INST.	TOTAL	MAT.	INST.	TOTAL	MAT.	INST.	TOTAL
015433	CONTRACTOR EQUIPMENT		108.0	108.0		106.9	106.9		106.9	106.9		101.9	101.9		108.0	108.0		105.7	105.7
0241, 31 - 34	SITE & INFRASTRUCTURE, DEMOLITION	102.8	97.2	98.8	92.2	95.6	94.6	87.6	95.7	93.4	96.6	95.7	95.9	102.3	97.2	98.6	105.6	99.7	101.4
0310	Concrete Forming & Accessories	86.9	81.8	82.5	98.3	84.3	86.2	90.5	79.0	80.6	86.2	98.7	97.0	98.3	81.8	84.1	101.3	75.3	78.9
0320	Concrete Reinforcing	96.3	119.0	107.9	102.7	107.7	105.3	100.9	97.7	99.3	96.4	115.3	106.0	98.5	111.9	105.3	99.9	98.5	99.2
0330	Cast-in-Place Concrete	96.1	87.8	92.7	94.2	93.8	94.0	86.3	86.0	86.2	105.5	103.9	104.8	102.0	87.0	95.9	111.5	77.9	97.4
03	CONCRETE	93.2	92.2	92.7	98.8	93.2	96.0	92.1	86.3	89.2	101.8	104.0	102.9	98.9	90.6	94.8	105.2	81.4	93.5
04	MASONRY	154.1	90.0	114.2	114.5	81.4	93.9	109.0	101.3	104.2	97.4	103.4	101.2	112.0	90.0	98.3	96.5	85.9	89.9
05	METALS	98.4	121.2	105.4	94.5	115.6	101.0	94.6	111.9	99.9	95.6	113.2	101.0	98.0	118.3	104.2	98.2	100.5	98.9
06	WOOD, PLASTICS & COMPOSITES	91.7	78.7	84.4	99.1	82.6	89.9	88.9	73.8	80.4	89.9	97.5	94.2	100.0	78.7	88.1	106.0	74.2	88.2
07	THERMAL & MOISTURE PROTECTION	96.2	87.5	92.6	100.5	93.8	97.7	100.1	97.0	98.8	97.6	103.2	99.9	101.8	87.6	96.0	97.7	78.9	90.0
08	OPENINGS	99.5	95.0	98.4	97.7	97.2	97.6	97.7	82.3	94.1	95.1	102.8	96.9	99.1	93.1	97.7	96.2	78.5	92.1
0920	Plaster & Gypsum Board	87.2	78.0	81.0	101.3	82.1	88.3	94.5	73.0	79.9	97.3	97.3	97.3	91.3	78.0	82.3	108.2	73.2	84.5
0950, 0980	Ceilings & Acoustic Treatment	93.2	78.0	83.2	92.6	82.1	85.7	92.6	73.0	79.7	89.1	97.3	94.5	97.2	78.0	84.6	90.0	73.2	78.9
0960	Flooring	97.6	98.6	97.9	99.4	84.0	95.0	95.8	96.8	96.1	92.6	102.5	95.4	104.4	98.6	102.7	123.8	74.9	109.8
0970, 0990	Wall Finishes & Painting/Coating	105.2	77.1	88.2	97.1	78.4	85.8	97.1	102.8	100.5	92.5	103.8	99.3	102.3	77.1	87.1	87.5	74.9	79.9
09	FINISHES	92.8	83.0	87.4	99.7	82.2	90.1	97.4	83.1	89.5	97.8	99.5	98.7	97.2	83.0	89.4	107.0	75.3	89.6
COVERS	DIVS. 10 - 14, 25, 28, 41, 43, 44, 46	100.0	96.2	99.2	100.0	91.8	98.3	100.0	87.8	97.5	100.0	90.6	98.1	100.0	96.2	99.2	100.0	87.9	97.6
21, 22, 23	FIRE SUPPRESSION, PLUMBING & HVAC	99.8	96.4	98.4	95.2	97.8	96.2	95.2	97.5	96.1	95.2	100.0	97.2	99.8	98.1	99.1	100.1	72.1	88.8
26, 27, 3370	ELECTRICAL, COMMUNICATIONS & UTIL.	94.8	86.5	90.4	101.1	101.5	101.3	95.7	82.6	88.8	98.8	99.6	99.2	100.4	86.5	93.0	90.4	77.5	83.6
MF2014	WEIGHTED AVERAGE	100.2	93.9	97.5	98.1	95.0	96.8	96.2	92.4	94.6	97.1	101.5	99.1	99.9	93.7	97.2	99.5	81.7	91.8

MISSOURI

DIVISION		KANSAS CITY 640 - 641			KIRKSVILLE 635			POPLAR BLUFF 639			ROLLA 654 - 655			SEDALIA 653			SIKESTON 638		
		MAT.	INST.	TOTAL	MAT.	INST.	TOTAL	MAT.	INST.	TOTAL	MAT.	INST.	TOTAL	MAT.	INST.	TOTAL	MAT.	INST.	TOTAL
015433	CONTRACTOR EQUIPMENT		103.3	103.3		99.1	99.1		101.5	101.5		108.0	108.0		99.9	99.9		101.5	101.5
0241, 31 - 34	SITE & INFRASTRUCTURE, DEMOLITION	98.9	98.2	98.4	91.8	91.4	91.5	79.1	95.4	90.7	101.5	97.1	98.3	99.0	92.8	94.6	82.4	95.4	91.7
0310	Concrete Forming & Accessories	100.1	105.4	104.6	83.2	80.6	81.0	83.3	85.0	84.7	95.0	91.9	92.3	92.4	81.4	82.9	84.3	84.9	84.9
0320	Concrete Reinforcing	94.9	119.8	107.6	101.6	98.8	100.2	104.7	94.7	99.6	96.7	94.7	95.7	95.3	106.2	100.8	104.0	98.4	101.2
0330	Cast-in-Place Concrete	103.7	106.6	104.9	94.1	83.8	89.9	72.4	86.5	78.2	98.3	96.2	97.5	102.7	83.8	94.9	77.4	86.5	81.1
03	CONCRETE	101.1	108.7	104.8	110.8	86.0	98.6	85.9	88.1	86.9	95.2	95.0	95.1	109.8	87.7	99.0	89.5	88.7	89.1
04	MASONRY	99.5	107.5	104.5	121.0	88.9	101.0	112.5	80.1	92.3	126.3	87.9	102.4	132.9	88.3	105.1	112.2	80.1	92.2
05	METALS	105.9	115.8	108.9	94.2	103.8	97.2	94.8	101.7	96.9	97.8	110.5	101.7	96.5	108.0	100.1	95.1	103.2	97.6
06	WOOD, PLASTICS & COMPOSITES	105.3	105.0	105.1	77.0	78.7	77.9	76.1	85.5	81.4	100.3	92.8	96.1	93.1	78.7	85.1	77.7	85.5	82.1
07	THERMAL & MOISTURE PROTECTION	97.8	106.8	101.5	106.8	88.2	99.2	105.2	85.9	97.3	96.4	93.4	95.1	102.5	93.1	98.7	105.3	84.5	96.8
08	OPENINGS	102.5	109.2	104.0	102.7	85.8	98.8	103.7	84.5	99.2	99.5	88.3	96.9	104.5	87.4	100.5	103.7	85.5	99.5
0920	Plaster & Gypsum Board	105.2	105.0	105.0	89.8	78.0	81.8	90.2	85.1	86.7	89.3	92.5	91.5	82.5	78.0	79.5	92.0	85.1	87.3
0950, 0980	Ceilings & Acoustic Treatment	96.7	105.0	102.1	90.9	78.0	82.4	92.6	85.1	87.6	93.2	92.5	92.7	93.2	78.0	83.2	92.6	85.1	87.6
0960	Flooring	98.0	109.0	101.2	74.4	95.8	80.5	88.3	82.0	86.5	101.2	95.8	99.6	78.8	95.8	83.7	88.7	82.0	86.8
0970, 0990	Wall Finishes & Painting/Coating	92.5	112.2	104.4	92.5	82.1	86.2	92.0	73.2	80.7	105.2	100.6	102.5	105.2	106.3	105.9	92.0	73.2	80.7
09	FINISHES	102.0	106.6	104.5	96.5	82.7	88.9	96.5	82.7	88.9	94.3	93.4	93.8	91.9	85.4	88.3	97.1	82.7	89.2
COVERS	DIVS. 10 - 14, 25, 28, 41, 43, 44, 46	100.0	100.6	100.1	100.0	87.1	97.4	100.0	93.6	98.7	100.0	89.0	97.8	100.0	90.2	98.0	100.0	94.2	98.8
21, 22, 23	FIRE SUPPRESSION, PLUMBING & HVAC	100.0	104.3	101.7	95.2	97.3	96.1	95.2	96.3	95.6	95.0	98.2	96.3	95.0	95.9	95.4	95.2	96.3	95.6
26, 27, 3370	ELECTRICAL, COMMUNICATIONS & UTIL.	100.3	100.1	100.2	95.8	82.5	88.8	96.1	101.5	99.0	93.2	86.5	89.6	94.7	128.8	112.7	95.3	101.5	98.6
MF2014	WEIGHTED AVERAGE	101.4	105.7	103.3	99.5	89.9	95.3	96.2	92.0	94.4	97.9	94.6	96.5	100.2	97.3	98.9	96.7	92.2	94.8

DIVISION		MISSOURI									MONTANA								
		SPRINGFIELD 656 - 658			ST. JOSEPH 644 - 645			ST. LOUIS 630 - 631			BILLINGS 590 - 591			BUTTE 597			GREAT FALLS 594		
		MAT.	INST.	TOTAL	MAT.	INST.	TOTAL	MAT.	INST.	TOTAL	MAT.	INST.	TOTAL	MAT.	INST.	TOTAL	MAT.	INST.	TOTAL
015433	CONTRACTOR EQUIPMENT		102.5	102.5		101.9	101.9		107.9	107.9		98.5	98.5		98.2	98.2		98.2	98.2
0241, 31 - 34	SITE & INFRASTRUCTURE, DEMOLITION	101.6	95.1	97.0	100.5	93.9	95.8	91.7	98.9	96.8	95.0	96.3	95.9	102.0	96.1	97.8	105.7	96.2	98.9
0310	Concrete Forming & Accessories	101.4	79.1	82.2	100.0	89.6	91.0	98.4	104.8	103.9	98.4	71.5	75.2	85.5	71.8	73.7	98.4	71.5	75.2
0320	Concrete Reinforcing	92.7	118.7	105.9	93.8	115.0	104.6	93.9	112.7	103.4	90.2	81.7	85.9	97.8	81.9	89.7	90.2	81.8	85.9
0330	Cast-in-Place Concrete	104.3	79.6	94.2	103.6	103.3	103.5	90.2	105.0	96.3	129.3	67.6	103.9	142.3	69.0	112.2	150.4	63.7	114.8
03	CONCRETE	105.8	87.4	96.8	100.9	99.7	100.3	94.4	107.1	100.7	109.8	72.9	91.7	113.8	73.5	94.0	119.5	71.5	96.0
04	MASONRY	102.9	89.8	94.8	99.3	96.1	97.3	95.4	111.9	105.7	128.0	73.7	94.2	123.5	76.6	94.3	128.0	75.4	95.2
05	METALS	102.8	108.8	104.7	101.9	112.4	105.2	100.3	120.2	106.4	105.8	90.0	100.9	100.1	90.1	97.1	103.2	90.0	99.2
06	WOOD, PLASTICS & COMPOSITES	101.8	77.1	88.0	106.0	87.6	95.7	99.0	103.0	101.2	98.6	70.5	82.8	85.7	70.8	77.4	99.9	70.5	83.4
07	THERMAL & MOISTURE PROTECTION	100.7	82.2	93.2	98.2	95.6	97.1	100.2	106.9	103.0	106.5	69.9	91.5	106.3	71.0	91.8	106.9	68.5	91.2
08	OPENINGS	107.2	84.3	101.9	100.6	98.6	100.1	97.6	110.9	100.7	97.3	68.7	90.6	95.6	70.0	89.6	98.5	68.0	91.4
0920	Plaster & Gypsum Board	90.4	76.3	80.9	109.7	87.0	94.4	102.2	103.1	102.8	98.0	69.8	78.9	98.7	70.2	79.4	107.3	69.8	82.0
0950, 0980	Ceilings & Acoustic Treatment	93.2	76.3	82.1	95.8	87.0	90.0	97.6	103.1	101.2	81.5	69.8	73.8	88.3	70.2	76.4	90.0	69.8	76.7
0960	Flooring	99.9	74.9	92.8	101.5	107.3	103.2	99.2	97.6	98.7	106.6	70.2	96.2	104.4	74.2	95.7	111.5	74.2	100.8
0970, 0990	Wall Finishes & Painting/Coating	99.0	74.9	84.4	87.9	86.7	87.2	97.1	106.0	102.5	105.2	86.6	94.0	103.4	51.2	71.9	103.4	86.6	93.3
09	FINISHES	96.7	78.0	86.4	103.2	91.7	96.9	100.8	103.1	102.1	96.6	72.6	83.4	97.2	69.7	82.0	101.1	73.3	85.8
COVERS	DIVS. 10 - 14, 25, 28, 41, 43, 44, 46	100.0	94.8	99.0	100.0	96.9	99.4	100.0	102.4	100.5	100.0	94.7	98.9	100.0	94.8	98.9	100.0	94.7	98.9
21, 22, 23	FIRE SUPPRESSION, PLUMBING & HVAC	99.8	74.6	89.6	100.1	89.0	95.6	100.0	107.4	103.0	100.1	74.7	89.9	100.2	75.6	90.3	100.7	70.7	88.3
26, 27, 3370	ELECTRICAL, COMMUNICATIONS & UTIL.	99.2	69.7	83.6	98.8	78.6	88.2	99.2	108.3	104.0	98.5	72.2	84.6	105.5	71.7	87.7	97.7	71.9	84.1
MF2014	WEIGHTED AVERAGE	101.7	83.7	93.8	100.6	93.5	97.5	98.7	107.8	102.7	102.7	77.1	91.5	102.6	77.3	91.6	104.1	76.2	91.9

For customer support on your Mechanical Cost Data, call 877.795.7394.

675

MONTANA

DIVISION		HAVRE 595 MAT.	INST.	TOTAL	HELENA 596 MAT.	INST.	TOTAL	KALISPELL 599 MAT.	INST.	TOTAL	MILES CITY 593 MAT.	INST.	TOTAL	MISSOULA 598 MAT.	INST.	TOTAL	WOLF POINT 592 MAT.	INST.	TOTAL
015433	CONTRACTOR EQUIPMENT		98.2	98.2		98.2	98.2		98.2	98.2		98.2	98.2		98.2	98.2		98.2	98.2
0241, 31 - 34	SITE & INFRASTRUCTURE, DEMOLITION	109.2	95.8	99.7	98.5	95.8	96.6	92.2	96.0	94.9	98.5	95.4	96.3	85.1	95.8	92.7	115.5	95.6	101.3
0310	Concrete Forming & Accessories	78.4	69.0	70.3	101.5	65.2	70.2	88.7	71.4	73.8	96.8	67.6	71.6	88.7	71.4	73.8	89.2	67.9	70.8
0320	Concrete Reinforcing	98.6	81.7	90.0	102.5	81.7	91.9	100.4	83.4	91.8	98.2	81.8	89.8	99.5	83.4	91.3	99.6	81.8	90.5
0330	Cast-in-Place Concrete	153.3	60.4	115.1	115.0	62.9	93.6	123.6	62.7	98.6	135.3	60.3	104.5	104.9	64.9	88.4	151.5	60.6	114.2
03	CONCRETE	122.7	69.3	96.5	106.7	68.4	87.9	102.6	71.5	87.3	110.4	68.6	89.9	90.0	72.2	81.3	126.5	68.9	98.2
04	MASONRY	124.5	69.8	90.4	120.1	75.7	92.4	122.2	77.0	94.0	129.7	65.3	89.6	147.9	76.6	103.4	131.0	66.0	90.5
05	METALS	96.3	89.6	94.2	102.1	88.8	98.0	96.1	90.6	94.4	95.4	89.9	93.7	96.6	90.7	94.8	95.5	89.2	93.6
06	WOOD, PLASTICS & COMPOSITES	77.4	70.5	73.5	102.9	62.3	80.2	89.1	70.5	78.7	96.9	70.5	82.1	89.1	70.5	78.7	88.5	70.5	78.4
07	THERMAL & MOISTURE PROTECTION	106.6	68.0	90.8	103.9	70.7	90.3	105.9	72.8	92.3	106.2	65.7	89.6	105.5	79.1	94.7	107.1	66.0	90.3
08	OPENINGS	95.6	68.0	89.2	99.8	65.2	91.7	95.6	68.5	89.3	95.1	68.0	88.8	95.6	68.3	89.2	95.1	67.4	88.7
0920	Plaster & Gypsum Board	94.8	69.8	77.9	96.0	61.4	72.6	98.7	69.8	79.2	106.7	69.8	81.8	98.7	69.8	79.2	102.1	69.8	80.3
0950, 0980	Ceilings & Acoustic Treatment	88.3	69.8	76.2	90.8	61.4	71.5	88.3	69.8	76.2	87.5	69.8	75.9	88.3	69.8	76.2	87.5	69.8	75.9
0960	Flooring	101.8	74.2	93.9	106.7	59.1	93.1	106.2	74.2	97.1	111.3	70.2	99.5	106.2	74.2	97.1	107.6	70.2	96.9
0970, 0990	Wall Finishes & Painting/Coating	103.4	48.6	70.4	99.7	47.2	68.0	103.4	47.2	69.5	103.4	47.2	69.5	103.4	62.0	78.4	103.4	48.6	70.4
09	FINISHES	96.5	67.7	80.6	98.6	61.6	78.2	97.2	69.1	81.7	100.0	65.7	81.1	96.7	70.7	82.3	99.6	66.0	81.1
COVERS	DIVS. 10 - 14, 25, 28, 41, 43, 44, 46	100.0	70.8	94.1	100.0	94.0	98.8	100.0	93.9	98.8	100.0	90.4	98.1	100.0	93.7	98.7	100.0	90.6	98.1
21, 22, 23	FIRE SUPPRESSION, PLUMBING & HVAC	95.4	67.8	84.3	100.1	70.1	88.0	95.4	69.2	84.8	95.4	69.6	85.0	100.2	69.2	87.7	95.4	70.0	85.1
26, 27, 3370	ELECTRICAL, COMMUNICATIONS & UTIL.	97.7	71.2	83.7	104.6	71.3	87.0	102.2	69.3	84.8	97.7	76.9	86.7	103.2	67.7	84.5	97.7	76.9	86.7
MF2014	WEIGHTED AVERAGE	101.2	73.1	89.0	102.4	73.8	89.9	99.0	75.3	88.7	100.1	74.0	88.7	99.9	75.5	89.3	102.3	74.2	90.0

NEBRASKA

DIVISION		ALLIANCE 693 MAT.	INST.	TOTAL	COLUMBUS 686 MAT.	INST.	TOTAL	GRAND ISLAND 688 MAT.	INST.	TOTAL	HASTINGS 689 MAT.	INST.	TOTAL	LINCOLN 683 - 685 MAT.	INST.	TOTAL	MCCOOK 690 MAT.	INST.	TOTAL
015433	CONTRACTOR EQUIPMENT		97.8	97.8		101.6	101.6		101.6	101.6		101.6	101.6		101.6	101.6		101.6	101.6
0241, 31 - 34	SITE & INFRASTRUCTURE, DEMOLITION	100.5	98.8	99.3	100.2	92.3	94.6	105.0	93.1	96.5	103.7	92.3	95.6	92.7	93.0	92.9	103.8	92.2	95.6
0310	Concrete Forming & Accessories	87.3	56.4	60.7	95.5	67.4	71.2	95.1	72.4	75.5	98.2	71.9	75.5	95.7	68.2	71.9	93.0	56.1	61.2
0320	Concrete Reinforcing	108.8	85.1	96.7	98.3	76.5	87.2	97.7	76.4	86.9	97.7	77.1	87.2	96.7	76.0	86.1	101.8	76.2	88.8
0330	Cast-in-Place Concrete	113.7	60.0	91.6	115.7	60.0	92.8	122.5	67.2	99.8	122.5	63.2	98.2	92.4	69.6	83.1	122.6	59.3	96.6
03	CONCRETE	123.9	63.5	94.2	108.8	67.3	88.4	113.7	72.1	93.3	113.9	70.6	92.6	95.6	71.0	83.5	112.9	62.0	87.9
04	MASONRY	115.6	74.1	89.7	122.7	77.8	94.7	115.4	76.1	90.9	124.9	88.0	101.9	104.8	67.2	81.4	110.7	73.9	87.8
05	METALS	101.5	78.7	94.5	96.1	84.6	92.6	97.8	86.3	94.3	98.7	85.8	94.7	99.9	85.5	95.5	96.3	83.6	92.4
06	WOOD, PLASTICS & COMPOSITES	87.0	52.2	67.5	101.2	66.7	81.9	100.4	72.2	84.6	104.1	72.2	86.3	98.4	66.7	80.7	95.7	52.2	71.4
07	THERMAL & MOISTURE PROTECTION	104.3	67.8	89.4	100.8	69.8	88.1	100.9	72.8	89.4	100.9	83.4	93.8	98.1	69.9	86.5	98.9	66.6	85.6
08	OPENINGS	94.5	58.6	86.1	95.4	66.1	88.6	95.5	69.7	89.5	95.5	69.1	89.3	107.8	63.0	97.4	95.0	56.6	86.1
0920	Plaster & Gypsum Board	82.7	50.7	61.1	90.9	65.6	73.8	90.1	71.3	77.4	91.9	71.3	78.0	96.2	65.6	75.5	93.7	50.7	64.7
0950, 0980	Ceilings & Acoustic Treatment	91.9	50.7	64.8	80.2	65.6	70.6	80.2	71.3	74.3	80.2	71.3	74.3	90.4	65.6	74.1	88.0	50.7	63.5
0960	Flooring	99.9	82.1	94.8	92.9	100.4	95.0	92.7	107.2	96.8	93.9	100.4	95.7	101.0	83.6	96.0	97.6	82.1	93.1
0970, 0990	Wall Finishes & Painting/Coating	168.5	54.8	99.9	85.5	63.7	72.4	85.5	67.6	74.7	85.5	63.7	72.4	103.8	77.9	88.1	93.8	47.7	66.0
09	FINISHES	98.4	59.5	77.0	89.9	71.9	80.0	90.0	77.1	82.9	90.6	75.2	82.1	96.9	71.7	83.0	95.6	58.6	75.2
COVERS	DIVS. 10 - 14, 25, 28, 41, 43, 44, 46	100.0	65.3	93.0	100.0	86.1	97.2	100.0	88.7	97.7	100.0	86.7	97.3	100.0	88.1	97.6	100.0	64.5	92.8
21, 22, 23	FIRE SUPPRESSION, PLUMBING & HVAC	95.3	70.5	85.3	95.2	76.0	87.4	100.0	78.0	91.1	95.2	76.0	87.5	99.9	78.0	91.0	95.1	75.9	87.4
26, 27, 3370	ELECTRICAL, COMMUNICATIONS & UTIL.	92.3	69.4	80.2	93.6	82.3	87.6	92.2	67.0	78.9	91.6	83.6	87.4	105.4	67.0	85.1	94.4	69.4	81.2
MF2014	WEIGHTED AVERAGE	101.0	70.5	87.7	98.4	77.1	89.1	100.0	77.1	90.0	99.4	79.8	90.9	100.5	74.9	89.4	98.7	71.1	86.7

NEBRASKA / NEVADA

DIVISION		NORFOLK 687 MAT.	INST.	TOTAL	NORTH PLATTE 691 MAT.	INST.	TOTAL	OMAHA 680 - 681 MAT.	INST.	TOTAL	VALENTINE 692 MAT.	INST.	TOTAL	CARSON CITY 897 MAT.	INST.	TOTAL	ELKO 898 MAT.	INST.	TOTAL
015433	CONTRACTOR EQUIPMENT		92.8	92.8		101.6	101.6		92.8	92.8		95.9	95.9		98.2	98.2		98.2	98.2
0241, 31 - 34	SITE & INFRASTRUCTURE, DEMOLITION	83.6	91.5	89.2	105.2	92.3	96.0	90.0	92.1	91.5	88.6	96.5	94.2	84.5	99.0	94.8	66.9	97.6	88.8
0310	Concrete Forming & Accessories	81.7	71.1	72.5	95.4	71.5	74.8	94.3	72.4	75.4	83.7	55.8	59.6	103.3	92.8	94.2	109.2	81.3	85.2
0320	Concrete Reinforcing	98.4	66.5	82.2	101.2	77.1	88.9	99.7	76.6	87.9	101.8	66.1	83.7	103.2	118.4	111.0	102.7	113.2	108.1
0330	Cast-in-Place Concrete	116.7	62.4	94.4	122.6	64.5	98.7	101.1	75.5	90.6	108.4	57.2	87.4	100.4	85.7	94.4	95.1	80.2	89.0
03	CONCRETE	107.3	67.5	87.8	113.0	70.8	92.3	100.1	74.5	87.5	110.2	58.9	85.0	100.9	95.2	98.1	95.4	87.2	91.4
04	MASONRY	129.4	79.1	98.0	98.6	87.9	92.0	106.8	79.0	89.4	110.9	73.9	87.8	117.3	94.6	103.2	115.0	71.9	88.1
05	METALS	99.6	73.5	91.6	95.5	85.4	92.4	99.9	78.9	93.4	107.6	73.4	97.1	94.0	104.7	97.3	96.9	99.4	97.7
06	WOOD, PLASTICS & COMPOSITES	84.0	71.7	77.1	97.8	72.2	83.5	94.1	72.0	81.7	81.7	51.7	64.9	92.1	94.4	93.4	100.6	81.8	90.1
07	THERMAL & MOISTURE PROTECTION	100.8	73.6	89.6	98.9	82.8	92.3	95.5	78.7	88.6	99.6	67.6	86.5	103.5	92.3	99.0	99.9	76.1	90.1
08	OPENINGS	97.2	66.5	90.0	94.3	69.1	88.4	100.0	71.5	93.3	96.9	55.6	87.3	99.5	107.9	101.4	100.9	85.9	97.4
0920	Plaster & Gypsum Board	90.8	71.3	77.6	93.7	71.3	78.6	100.5	71.6	81.0	95.6	50.7	65.3	100.9	94.1	96.3	105.3	81.2	89.0
0950, 0980	Ceilings & Acoustic Treatment	92.0	71.3	78.4	88.0	71.3	77.0	93.0	71.6	78.9	104.9	50.7	69.3	94.0	94.1	94.1	94.4	81.2	85.7
0960	Flooring	117.4	100.4	112.5	98.5	100.4	99.1	113.4	81.9	104.4	127.4	79.8	113.8	101.9	107.9	103.6	105.4	51.1	89.9
0970, 0990	Wall Finishes & Painting/Coating	151.7	63.7	98.6	93.8	63.7	75.6	126.2	67.2	90.6	167.2	66.1	106.1	101.2	80.8	88.9	102.6	95.9	98.5
09	FINISHES	108.8	75.2	90.3	95.9	75.2	84.5	106.0	73.3	88.0	118.2	59.7	86.0	97.9	94.3	95.9	97.6	77.2	86.3
COVERS	DIVS. 10 - 14, 25, 28, 41, 43, 44, 46	100.0	85.5	97.1	100.0	66.8	93.3	100.0	87.6	97.5	100.0	62.5	92.4	100.0	89.8	97.9	100.0	62.8	92.5
21, 22, 23	FIRE SUPPRESSION, PLUMBING & HVAC	95.0	75.8	87.2	99.9	76.0	90.2	99.9	76.8	90.6	94.8	75.7	87.1	100.0	80.1	92.0	97.7	79.6	90.4
26, 27, 3370	ELECTRICAL, COMMUNICATIONS & UTIL.	92.4	82.3	87.1	92.6	77.1	84.4	99.1	82.3	90.2	89.7	86.6	88.0	103.2	96.7	99.7	100.7	96.6	98.6
MF2014	WEIGHTED AVERAGE	100.0	76.7	89.8	99.0	78.3	90.0	100.2	78.6	90.8	101.2	72.5	88.7	99.7	93.5	97.0	98.3	84.9	92.4

For customer support on your Mechanical Cost Data, call 877.795.7394.

DIVISION		NEVADA								NEW HAMPSHIRE								
		ELY 893			LAS VEGAS 889 - 891			RENO 894 - 895			CHARLESTON 036			CLAREMONT 037			CONCORD 032 - 033		
		MAT.	INST.	TOTAL	MAT.	INST.	TOTAL	MAT.	INST.	TOTAL	MAT.	INST.	TOTAL	MAT.	INST.	TOTAL	MAT.	INST.	TOTAL
015433	CONTRACTOR EQUIPMENT		98.2	98.2		98.2	98.2		98.2	98.2		100.9	100.9		100.9	100.9		100.9	100.9
0241, 31 - 34	SITE & INFRASTRUCTURE, DEMOLITION	72.2	99.1	91.3	75.4	101.0	93.6	72.2	99.0	91.3	86.9	99.6	96.0	80.9	99.6	94.2	93.7	103.9	100.9
0310	Concrete Forming & Accessories	102.3	105.7	105.2	103.5	109.6	108.8	98.7	92.7	93.5	86.8	81.1	81.9	92.6	81.1	82.7	95.5	93.6	93.8
0320	Concrete Reinforcing	101.5	111.6	106.7	93.7	119.9	107.1	96.0	119.9	108.1	84.1	93.5	88.9	84.1	93.5	88.9	95.3	94.1	94.7
0330	Cast-in-Place Concrete	102.1	105.4	103.4	98.9	107.8	102.5	108.0	85.7	98.8	97.1	70.5	86.2	89.2	70.5	81.6	112.2	90.5	103.3
03	CONCRETE	103.4	106.4	104.9	98.7	110.6	104.6	103.1	95.4	99.3	99.1	80.0	89.7	91.1	80.0	85.6	106.6	92.5	99.6
04	MASONRY	119.7	102.5	109.0	108.3	104.0	105.6	114.3	94.6	102.0	97.5	84.7	89.5	96.9	84.7	89.3	110.9	104.2	106.7
05	METALS	96.9	103.3	98.9	104.5	108.6	105.8	98.4	105.1	100.5	92.8	90.1	92.0	92.8	90.1	92.0	98.8	93.8	97.3
06	WOOD, PLASTICS & COMPOSITES	91.3	106.8	100.0	89.8	107.9	99.9	85.9	94.4	90.6	87.4	90.2	89.0	93.6	90.2	91.7	95.4	92.4	93.7
07	THERMAL & MOISTURE PROTECTION	100.3	95.7	98.4	113.8	102.6	109.2	99.9	92.3	96.8	98.8	72.3	88.0	98.7	72.3	87.9	101.9	91.5	97.6
08	OPENINGS	100.8	98.8	100.3	99.9	115.6	103.5	98.7	102.3	99.6	102.0	76.4	96.0	103.3	76.4	97.0	104.9	87.2	100.8
0920	Plaster & Gypsum Board	101.0	107.0	105.1	97.5	108.1	104.7	92.0	94.1	93.4	100.0	89.3	92.8	101.0	89.3	93.1	103.9	91.6	95.6
0950, 0980	Ceilings & Acoustic Treatment	94.4	107.0	102.7	102.7	108.1	106.3	99.4	94.1	95.9	94.2	89.3	91.0	94.2	89.3	91.0	99.9	91.6	94.4
0960	Flooring	103.0	57.2	89.9	94.6	107.9	98.4	99.9	107.9	102.2	98.4	32.1	79.4	100.6	32.1	81.0	102.7	116.9	106.8
0970, 0990	Wall Finishes & Painting/Coating	102.6	121.0	113.7	105.4	121.0	114.8	102.6	80.8	89.4	96.4	45.7	65.8	96.4	46.0	65.9	95.8	95.4	95.6
09	FINISHES	96.8	98.9	98.0	95.8	110.4	103.9	95.6	94.3	94.9	95.3	69.7	81.2	95.6	69.8	81.4	97.4	97.9	97.7
COVERS	DIVS. 10 - 14, 25, 28, 41, 43, 44, 46	100.0	63.6	92.6	100.0	104.4	100.9	100.0	89.8	97.9	100.0	92.2	98.4	100.0	92.2	98.4	100.0	104.3	100.9
21, 22, 23	FIRE SUPPRESSION, PLUMBING & HVAC	97.7	104.5	100.5	100.1	104.4	101.8	100.0	80.1	92.0	95.3	39.6	72.8	95.3	39.7	72.9	99.9	84.4	93.7
26, 27, 3370	ELECTRICAL, COMMUNICATIONS & UTIL.	101.0	109.1	105.3	105.4	119.9	113.1	101.4	96.7	98.9	98.5	52.9	74.4	98.5	52.9	74.4	99.5	83.6	91.1
MF2014	WEIGHTED AVERAGE	99.4	102.2	100.6	100.9	108.7	104.3	99.6	93.3	96.8	96.7	69.5	84.8	95.9	69.5	84.4	101.1	92.5	97.4

DIVISION		NEW HAMPSHIRE															NEW JERSEY		
		KEENE 034			LITTLETON 035			MANCHESTER 031			NASHUA 030			PORTSMOUTH 038			ATLANTIC CITY 082,084		
		MAT.	INST.	TOTAL	MAT.	INST.	TOTAL	MAT.	INST.	TOTAL	MAT.	INST.	TOTAL	MAT.	INST.	TOTAL	MAT.	INST.	TOTAL
015433	CONTRACTOR EQUIPMENT		100.9	100.9		100.9	100.9		100.9	100.9		100.9	100.9		100.9	100.9		99.1	99.1
0241, 31 - 34	SITE & INFRASTRUCTURE, DEMOLITION	94.0	99.9	98.2	81.0	99.9	94.4	93.6	103.9	100.9	95.7	103.9	101.5	89.2	104.0	99.7	96.4	105.2	102.7
0310	Concrete Forming & Accessories	91.3	83.0	84.2	102.9	83.0	85.8	96.9	94.0	94.4	99.3	94.0	94.8	88.2	94.3	93.4	109.0	127.5	125.0
0320	Concrete Reinforcing	84.1	93.6	88.9	84.8	93.6	89.3	103.7	94.1	98.8	105.1	94.1	99.5	84.1	94.2	89.2	76.3	118.1	97.6
0330	Cast-in-Place Concrete	97.6	72.8	87.4	87.6	72.7	81.5	110.6	113.1	111.7	92.4	113.1	100.9	87.6	113.2	98.1	86.1	133.0	105.3
03	CONCRETE	98.7	81.6	90.3	90.5	81.6	86.1	107.2	100.4	103.9	98.9	100.4	99.7	90.6	100.6	95.5	93.1	126.5	109.5
04	MASONRY	101.3	88.3	93.2	107.9	88.3	95.7	102.9	104.2	103.7	101.4	104.2	103.1	97.2	104.2	101.5	111.8	132.5	124.4
05	METALS	93.5	90.7	92.6	93.5	90.7	92.6	101.3	94.3	99.2	98.8	94.3	97.5	95.0	94.9	94.9	95.6	106.1	98.8
06	WOOD, PLASTICS & COMPOSITES	92.0	90.2	91.0	103.5	90.2	96.0	95.3	92.4	93.7	101.9	92.4	96.6	88.7	92.4	90.8	109.6	127.2	119.4
07	THERMAL & MOISTURE PROTECTION	99.2	75.3	89.4	98.8	73.8	88.5	101.5	95.0	98.8	99.6	95.0	97.7	99.3	115.0	105.7	105.9	124.2	113.4
08	OPENINGS	100.4	81.2	96.0	104.4	77.1	98.0	105.2	87.2	101.0	105.4	87.2	101.2	106.2	84.0	101.0	101.7	124.4	107.0
0920	Plaster & Gypsum Board	100.3	89.3	92.9	114.6	89.3	97.5	104.7	91.6	95.8	110.0	91.6	97.5	100.0	91.6	94.3	105.4	127.5	120.3
0950, 0980	Ceilings & Acoustic Treatment	94.2	89.3	91.0	94.2	89.3	91.0	101.7	91.6	95.0	105.1	91.6	96.2	95.1	91.6	92.8	81.3	127.5	111.6
0960	Flooring	100.2	52.6	86.6	109.9	32.1	87.6	98.6	116.9	103.8	103.7	116.9	107.4	98.5	116.9	103.8	105.4	159.9	121.0
0970, 0990	Wall Finishes & Painting/Coating	96.4	45.7	65.8	96.4	59.9	74.4	100.7	95.4	97.5	100.7	95.4	95.8	96.4	95.4	95.8	93.9	130.2	115.8
09	FINISHES	97.0	74.5	84.6	100.2	72.2	84.8	98.6	97.9	98.2	101.7	97.9	99.6	96.2	97.9	97.1	96.0	134.5	117.2
COVERS	DIVS. 10 - 14, 25, 28, 41, 43, 44, 46	100.0	93.4	98.7	100.0	98.6	99.7	100.0	104.3	100.9	100.0	104.3	100.9	100.0	104.3	100.9	100.0	111.5	102.3
21, 22, 23	FIRE SUPPRESSION, PLUMBING & HVAC	95.3	43.3	74.3	95.3	63.8	82.6	99.9	84.4	93.7	100.1	84.4	93.8	100.1	84.4	93.8	99.7	123.6	109.3
26, 27, 3370	ELECTRICAL, COMMUNICATIONS & UTIL.	98.5	63.2	79.9	99.8	55.9	76.6	99.8	83.6	91.2	101.2	83.6	91.9	99.1	83.6	90.9	91.1	138.1	116.0
MF2014	WEIGHTED AVERAGE	97.2	73.3	86.7	97.1	76.3	88.0	101.4	93.8	98.1	100.5	93.8	97.6	97.9	94.3	96.4	98.2	124.8	109.8

DIVISION		NEW JERSEY																	
		CAMDEN 081			DOVER 078			ELIZABETH 072			HACKENSACK 076			JERSEY CITY 073			LONG BRANCH 077		
		MAT.	INST.	TOTAL	MAT.	INST.	TOTAL	MAT.	INST.	TOTAL	MAT.	INST.	TOTAL	MAT.	INST.	TOTAL	MAT.	INST.	TOTAL
015433	CONTRACTOR EQUIPMENT		99.1	99.1		100.9	100.9		100.9	100.9		100.9	100.9		99.1	99.1		98.7	98.7
0241, 31 - 34	SITE & INFRASTRUCTURE, DEMOLITION	97.8	105.5	103.3	102.0	106.3	105.1	106.1	106.3	106.3	103.0	106.3	105.3	93.4	106.3	102.5	97.5	106.0	103.6
0310	Concrete Forming & Accessories	100.2	127.5	123.8	97.8	128.3	124.1	110.2	128.4	125.9	97.8	128.3	124.1	101.8	128.3	124.7	102.4	128.0	124.5
0320	Concrete Reinforcing	100.1	118.3	109.4	77.2	137.2	107.7	77.2	137.2	107.7	77.2	137.2	107.7	100.1	137.2	119.0	77.2	137.1	107.7
0330	Cast-in-Place Concrete	83.5	132.9	103.8	101.5	127.7	112.2	87.1	131.4	105.3	99.2	131.4	112.4	79.2	127.7	99.1	88.0	132.9	106.4
03	CONCRETE	94.0	126.5	110.0	98.8	128.7	113.5	94.5	130.1	112.0	97.0	130.0	113.2	92.1	128.6	110.0	96.2	130.2	112.9
04	MASONRY	101.3	132.0	120.4	93.0	132.5	117.6	108.7	132.5	123.5	96.9	132.5	119.1	87.0	132.5	115.4	101.3	132.0	120.5
05	METALS	101.1	106.1	102.6	93.3	116.2	100.4	94.8	116.2	101.4	93.4	116.1	100.4	98.9	113.9	103.5	93.4	113.7	99.7
06	WOOD, PLASTICS & COMPOSITES	98.4	127.2	114.5	96.2	127.2	113.5	111.7	127.2	120.4	96.2	127.2	113.5	97.1	127.2	114.0	98.2	127.1	114.4
07	THERMAL & MOISTURE PROTECTION	105.8	123.3	113.0	100.6	132.8	113.8	100.8	133.4	114.1	100.4	125.5	110.7	101.6	127.7	107.7	100.3	124.6	104.7
08	OPENINGS	104.1	124.5	108.8	105.5	127.7	110.6	103.6	127.7	109.2	102.9	127.7	108.7	101.6	127.7	107.7	97.8	127.6	104.7
0920	Plaster & Gypsum Board	101.5	127.5	119.0	100.9	127.5	118.9	107.7	127.5	121.0	100.9	127.5	118.9	104.3	127.5	120.0	102.7	127.5	119.4
0950, 0980	Ceilings & Acoustic Treatment	91.5	127.5	115.1	83.8	127.5	112.5	85.7	127.5	113.1	83.8	127.5	112.5	94.9	127.5	116.3	83.8	127.5	112.5
0960	Flooring	101.5	159.9	118.2	93.6	178.1	117.8	98.5	178.1	121.3	93.6	178.1	117.8	94.6	178.1	118.5	94.8	178.1	118.6
0970, 0990	Wall Finishes & Painting/Coating	93.9	130.2	115.8	98.4	132.5	119.0	98.4	132.5	119.0	98.4	132.5	119.0	98.5	132.5	119.0	98.5	130.2	117.6
09	FINISHES	96.4	134.5	117.4	93.7	136.9	117.5	96.9	136.9	118.9	93.5	136.9	117.4	96.6	136.9	118.8	94.5	137.5	118.2
COVERS	DIVS. 10 - 14, 25, 28, 41, 43, 44, 46	100.0	111.5	102.3	100.0	118.1	103.7	100.0	118.1	103.7	100.0	118.1	103.7	100.0	118.1	103.7	100.0	111.3	102.3
21, 22, 23	FIRE SUPPRESSION, PLUMBING & HVAC	100.0	123.6	109.5	99.7	127.4	110.3	100.1	125.5	110.3	99.7	127.4	110.9	100.1	127.4	111.1	99.7	127.1	110.8
26, 27, 3370	ELECTRICAL, COMMUNICATIONS & UTIL.	96.0	138.1	118.2	92.8	137.7	116.5	93.4	137.7	116.8	92.8	140.1	117.8	97.6	140.1	120.1	92.5	130.8	112.7
MF2014	WEIGHTED AVERAGE	99.5	124.8	110.5	97.9	127.8	110.9	98.8	127.6	111.3	97.6	128.1	110.7	97.9	127.9	111.0	97.1	126.3	109.8

For customer support on your **Mechanical Cost Data**, call 877.795.7394.

677

NEW JERSEY

DIVISION		NEW BRUNSWICK 088-089			NEWARK 070-071			PATERSON 074-075			POINT PLEASANT 087			SUMMIT 079			TRENTON 085-086		
		MAT.	INST.	TOTAL	MAT.	INST.	TOTAL	MAT.	INST.	TOTAL	MAT.	INST.	TOTAL	MAT.	INST.	TOTAL	MAT.	INST.	TOTAL
015433	CONTRACTOR EQUIPMENT		98.7	98.7		100.9	100.9		100.9	100.9		98.7	98.7		100.9	100.9		98.7	98.7
0241, 31 - 34	SITE & INFRASTRUCTURE, DEMOLITION	109.2	106.0	107.0	107.8	106.3	106.8	105.0	106.3	105.9	110.9	106.0	107.4	103.7	106.3	105.6	95.8	106.0	103.0
0310	Concrete Forming & Accessories	103.3	128.3	124.8	97.6	128.4	124.2	99.9	128.2	124.3	97.7	127.9	123.7	100.6	128.4	124.5	98.6	127.7	123.7
0320	Concrete Reinforcing	77.2	137.2	107.7	99.8	137.2	118.8	100.1	137.2	119.0	77.2	137.1	107.7	77.2	137.2	107.7	99.8	112.4	106.2
0330	Cast-in-Place Concrete	106.3	133.2	117.4	108.8	131.5	118.1	100.8	131.4	113.4	106.3	132.8	117.2	84.3	131.4	103.6	101.8	132.8	114.5
03	CONCRETE	110.4	130.5	120.3	105.8	130.1	117.7	102.2	130.0	115.9	110.1	130.1	119.9	91.5	130.1	110.5	102.5	125.5	113.8
04	MASONRY	109.3	132.5	123.7	99.6	132.5	120.1	93.6	132.5	117.9	97.0	132.0	118.8	95.5	132.5	118.6	103.9	132.0	121.4
05	METALS	95.6	113.8	101.2	100.8	116.3	105.5	94.2	116.1	100.9	95.6	113.5	101.1	93.3	116.2	100.4	100.8	105.2	102.1
06	WOOD, PLASTICS & COMPOSITES	103.0	127.1	116.5	93.9	127.2	112.5	98.8	127.2	114.7	95.9	127.1	113.4	100.1	127.2	115.3	95.6	127.1	113.3
07	THERMAL & MOISTURE PROTECTION	106.1	132.0	116.7	101.1	133.3	114.3	100.7	125.5	110.9	106.2	124.4	113.7	101.0	133.3	114.2	105.5	124.3	113.2
08	OPENINGS	96.1	127.6	103.5	104.4	127.7	109.8	98.3	127.7	112.8	98.2	129.2	105.4	110.2	127.7	114.2	106.1	122.2	109.8
0920	Plaster & Gypsum Board	102.6	127.5	119.4	101.1	127.5	118.9	104.3	127.5	119.9	98.3	127.5	118.0	102.7	127.5	119.4	98.7	127.5	118.1
0950, 0980	Ceilings & Acoustic Treatment	81.3	127.5	111.6	97.6	127.5	117.2	94.9	127.5	116.3	81.3	127.5	111.6	83.8	127.5	112.5	93.1	127.5	115.7
0960	Flooring	103.0	178.1	124.5	95.7	178.1	119.2	94.6	178.1	118.5	100.5	159.9	117.5	94.9	178.1	118.7	102.1	172.2	122.1
0970, 0990	Wall Finishes & Painting/Coating	93.9	132.5	117.2	99.8	132.5	119.5	98.4	132.5	119.0	93.9	130.2	115.8	98.4	132.5	119.0	99.1	130.2	117.8
09	FINISHES	96.0	136.8	118.5	95.6	136.9	118.3	96.7	136.9	118.9	94.7	134.5	116.6	94.6	136.9	117.9	96.6	136.5	118.6
COVERS	DIVS. 10 - 14, 25, 28, 41, 43, 44, 46	100.0	118.0	103.6	100.0	118.1	103.7	100.0	118.1	103.7	100.0	108.8	101.8	100.0	118.1	103.7	100.0	111.3	102.3
21, 22, 23	FIRE SUPPRESSION, PLUMBING & HVAC	99.7	127.4	110.8	100.1	127.4	111.1	100.1	127.4	111.1	99.7	127.0	110.8	99.7	125.5	110.1	100.1	126.9	110.9
26, 27, 3370	ELECTRICAL, COMMUNICATIONS & UTIL.	91.8	136.8	115.6	101.2	140.1	121.8	97.6	137.7	118.8	91.1	130.8	112.1	93.4	137.7	116.8	99.3	135.4	118.4
MF2014	WEIGHTED AVERAGE	99.7	127.6	111.9	101.2	128.4	113.0	99.6	127.8	111.9	99.2	125.9	110.8	97.9	127.6	110.8	100.9	125.1	111.5

NEW JERSEY / NEW MEXICO

DIVISION		NEW JERSEY VINELAND 080,083			ALBUQUERQUE 870-872			CARRIZOZO 883			CLOVIS 881			FARMINGTON 874			GALLUP 873		
		MAT.	INST.	TOTAL	MAT.	INST.	TOTAL	MAT.	INST.	TOTAL	MAT.	INST.	TOTAL	MAT.	INST.	TOTAL	MAT.	INST.	TOTAL
015433	CONTRACTOR EQUIPMENT		99.1	99.1		110.1	110.1		110.1	110.1		110.1	110.1		110.1	110.1		110.1	110.1
0241, 31 - 34	SITE & INFRASTRUCTURE, DEMOLITION	100.7	105.5	104.1	87.6	103.3	98.7	105.8	103.3	104.0	94.1	103.3	100.6	94.1	103.3	100.6	102.1	103.3	102.9
0310	Concrete Forming & Accessories	95.2	127.6	123.1	101.7	64.6	69.7	99.2	64.6	69.3	99.2	64.5	69.2	101.7	64.6	69.7	101.7	64.6	69.7
0320	Concrete Reinforcing	76.3	116.1	96.6	98.4	68.9	83.4	107.0	68.9	87.6	108.2	68.9	88.2	107.6	68.9	87.9	103.0	68.9	85.6
0330	Cast-in-Place Concrete	92.8	133.0	109.3	98.9	70.4	87.2	95.8	70.4	85.4	95.7	70.4	85.3	99.8	70.4	87.7	93.9	70.4	84.3
03	CONCRETE	97.9	126.2	111.8	100.2	68.5	84.6	117.0	68.5	93.2	105.8	68.4	87.4	103.9	68.5	86.5	110.0	68.5	89.6
04	MASONRY	99.4	132.0	119.7	105.0	58.7	76.1	103.4	58.7	75.5	103.4	58.7	75.5	113.7	58.7	79.4	99.4	58.7	74.0
05	METALS	95.5	105.5	98.6	105.2	86.8	99.5	99.5	86.8	95.6	99.2	86.7	95.3	102.9	86.8	97.9	102.0	86.8	97.3
06	WOOD, PLASTICS & COMPOSITES	92.8	127.2	112.1	97.5	65.5	79.6	93.1	65.5	77.6	93.1	65.5	77.6	97.7	65.5	79.7	97.7	65.5	79.7
07	THERMAL & MOISTURE PROTECTION	105.7	124.2	113.3	94.8	73.5	86.0	100.3	73.5	89.3	99.2	73.5	88.7	94.9	73.5	86.2	95.9	73.5	86.7
08	OPENINGS	97.7	124.1	103.8	101.2	67.7	93.4	98.6	67.7	91.4	98.7	67.7	91.5	103.6	67.7	95.3	103.6	67.7	95.3
0920	Plaster & Gypsum Board	96.9	127.5	117.5	95.8	64.2	74.4	78.6	64.2	68.8	78.6	64.2	68.8	89.4	64.2	72.3	89.4	64.2	72.3
0950, 0980	Ceilings & Acoustic Treatment	81.3	127.5	111.6	95.5	64.2	74.9	96.3	64.2	75.2	96.3	64.2	75.2	92.6	64.2	73.9	92.6	64.2	73.9
0960	Flooring	99.6	159.9	116.9	101.9	66.0	91.6	101.2	66.0	91.2	101.2	66.0	91.2	103.7	66.0	92.9	103.7	66.0	92.9
0970, 0990	Wall Finishes & Painting/Coating	93.9	130.2	115.8	113.2	66.6	85.1	103.3	66.6	81.1	103.3	66.6	81.1	107.1	66.6	82.7	107.1	66.6	82.7
09	FINISHES	93.5	134.5	116.1	95.9	64.8	78.8	96.6	64.8	79.1	95.4	64.8	78.5	94.8	64.8	78.3	96.0	64.8	78.8
COVERS	DIVS. 10 - 14, 25, 28, 41, 43, 44, 46	100.0	111.5	102.3	100.0	82.9	96.5	100.0	82.9	96.5	100.0	82.9	96.5	100.0	82.9	96.5	100.0	82.9	96.5
21, 22, 23	FIRE SUPPRESSION, PLUMBING & HVAC	99.7	123.6	109.3	100.1	69.0	87.6	97.2	69.0	85.8	97.2	68.7	85.7	100.0	69.0	87.5	97.1	69.0	85.8
26, 27, 3370	ELECTRICAL, COMMUNICATIONS & UTIL.	91.1	138.1	116.0	91.2	71.2	80.6	92.8	71.2	81.4	90.2	71.2	80.2	88.8	71.2	79.5	88.0	71.2	79.1
MF2014	WEIGHTED AVERAGE	97.5	124.7	109.4	99.6	72.6	87.8	100.3	72.6	88.2	98.3	72.5	87.1	100.1	72.6	88.1	99.5	72.6	87.8

NEW MEXICO

DIVISION		LAS CRUCES 880			LAS VEGAS 877			ROSWELL 882			SANTA FE 875			SOCORRO 878			TRUTH/CONSEQUENCES 879		
		MAT.	INST.	TOTAL	MAT.	INST.	TOTAL	MAT.	INST.	TOTAL	MAT.	INST.	TOTAL	MAT.	INST.	TOTAL	MAT.	INST.	TOTAL
015433	CONTRACTOR EQUIPMENT		86.0	86.0		110.1	110.1		110.1	110.1		110.1	110.1		110.1	110.1		86.0	86.0
0241, 31 - 34	SITE & INFRASTRUCTURE, DEMOLITION	94.0	82.9	86.1	93.6	103.3	100.5	96.3	103.3	101.3	98.8	103.3	102.0	89.8	103.3	99.4	108.3	82.9	90.3
0310	Concrete Forming & Accessories	95.7	63.4	67.8	101.7	64.6	69.7	99.2	64.6	69.3	100.4	64.6	69.5	101.7	64.6	69.7	99.3	63.4	68.3
0320	Concrete Reinforcing	104.6	68.7	86.4	104.7	68.9	86.5	108.2	68.9	88.2	103.8	68.9	86.1	106.8	68.9	87.5	100.8	68.7	84.5
0330	Cast-in-Place Concrete	90.4	62.7	79.0	97.1	70.4	86.1	95.8	70.4	85.3	105.3	70.4	91.0	95.1	70.4	85.0	104.2	62.7	87.1
03	CONCRETE	84.9	65.0	75.1	101.2	68.5	85.1	106.5	68.5	87.8	104.0	68.5	86.5	100.2	68.5	84.6	94.1	65.0	79.8
04	MASONRY	99.2	58.3	73.7	99.7	58.7	74.1	114.3	58.7	79.6	104.1	58.7	75.8	99.6	58.7	74.1	97.2	58.3	73.0
05	METALS	98.1	80.3	92.6	101.7	86.8	97.1	100.4	86.8	96.2	99.0	86.8	95.2	102.0	86.8	97.3	101.6	80.3	95.0
06	WOOD, PLASTICS & COMPOSITES	82.5	64.4	72.4	97.7	65.5	79.7	93.1	65.5	77.6	99.3	65.5	80.4	97.7	65.5	79.7	89.1	64.4	75.3
07	THERMAL & MOISTURE PROTECTION	86.0	68.6	78.8	94.5	73.5	85.9	99.3	73.5	88.7	96.9	73.5	87.3	94.5	73.5	85.9	83.3	68.6	77.2
08	OPENINGS	91.5	67.1	85.8	99.9	67.7	92.4	98.5	67.7	91.4	102.0	67.7	94.0	99.8	67.7	92.3	93.1	67.1	87.0
0920	Plaster & Gypsum Board	77.5	64.2	68.5	89.4	64.2	72.3	78.6	64.2	68.8	98.9	64.2	75.4	89.4	64.2	72.3	91.1	64.2	72.9
0950, 0980	Ceilings & Acoustic Treatment	84.2	64.2	71.0	92.6	64.2	73.9	96.3	64.2	75.2	93.0	64.2	74.0	92.6	64.2	73.9	82.4	64.2	70.4
0960	Flooring	131.2	66.0	112.5	103.7	66.0	92.9	101.2	66.0	91.2	110.8	66.0	98.0	103.7	66.0	92.9	135.3	66.0	115.4
0970, 0990	Wall Finishes & Painting/Coating	90.7	66.6	76.2	107.1	66.6	82.7	103.3	66.6	81.1	111.8	66.6	84.5	107.1	66.6	82.7	97.7	66.6	79.0
09	FINISHES	105.2	63.9	82.5	94.7	64.8	78.2	95.5	64.8	78.6	99.7	64.8	80.5	94.6	64.8	78.2	107.5	63.9	83.5
COVERS	DIVS. 10 - 14, 25, 28, 41, 43, 44, 46	100.0	80.2	96.0	100.0	82.9	96.5	100.0	82.9	96.5	100.0	82.9	96.5	100.0	82.9	96.5	100.0	80.2	96.0
21, 22, 23	FIRE SUPPRESSION, PLUMBING & HVAC	100.3	68.7	87.6	97.1	69.0	85.8	99.9	69.0	87.5	100.0	69.0	87.5	97.1	69.0	85.8	97.1	68.7	85.6
26, 27, 3370	ELECTRICAL, COMMUNICATIONS & UTIL.	92.1	71.2	81.0	90.6	71.2	80.4	91.7	71.2	80.9	102.9	71.2	86.2	88.5	71.2	79.4	92.3	71.2	81.2
MF2014	WEIGHTED AVERAGE	96.1	69.4	84.4	98.0	72.6	86.9	100.0	72.6	88.0	100.8	72.6	88.5	97.6	72.6	86.7	97.5	69.4	85.2

City Cost Indexes

DIVISION		NEW MEXICO TUCUMCARI 884			NEW YORK ALBANY 120 - 122			NEW YORK BINGHAMTON 137 - 139			NEW YORK BRONX 104			NEW YORK BROOKLYN 112			NEW YORK BUFFALO 140 - 142		
		MAT.	INST.	TOTAL	MAT.	INST.	TOTAL	MAT.	INST.	TOTAL	MAT.	INST.	TOTAL	MAT.	INST.	TOTAL	MAT.	INST.	TOTAL
015433	CONTRACTOR EQUIPMENT		110.1	110.1		112.9	112.9		114.1	114.1		110.6	110.6		113.3	113.3		97.1	97.1
0241, 31 - 34	SITE & INFRASTRUCTURE, DEMOLITION	93.8	103.3	100.5	83.8	106.3	99.8	95.9	94.1	94.6	108.8	120.8	117.3	120.3	127.6	125.5	98.3	98.2	98.2
0310	Concrete Forming & Accessories	99.2	64.5	69.2	100.1	100.6	100.5	100.8	88.5	90.2	98.4	175.3	164.8	107.2	182.9	172.5	97.3	116.5	113.9
0320	Concrete Reinforcing	106.0	68.9	87.1	104.0	103.8	103.9	93.7	97.8	95.8	103.9	184.4	144.9	95.1	205.9	151.5	97.8	102.1	100.0
0330	Cast-in-Place Concrete	95.7	70.3	85.3	91.9	111.3	99.9	102.7	127.9	113.1	95.9	173.4	127.7	104.8	172.1	132.4	106.7	119.7	112.0
03	CONCRETE	105.0	68.4	87.0	99.0	105.4	102.1	95.8	105.2	100.4	96.1	174.8	134.8	107.5	181.3	143.7	103.0	114.1	108.4
04	MASONRY	114.6	58.7	79.8	101.0	112.6	108.2	109.7	127.8	121.0	92.7	177.5	145.5	120.2	177.4	155.9	105.1	121.0	115.0
05	METALS	99.2	86.7	95.3	104.1	110.3	106.0	96.3	118.9	103.3	99.9	151.6	115.8	104.4	150.1	118.4	99.6	95.4	98.3
06	WOOD, PLASTICS & COMPOSITES	93.1	65.5	77.6	98.7	97.9	98.3	104.9	84.7	93.6	94.4	174.9	139.5	106.9	185.1	150.6	99.4	116.4	108.9
07	THERMAL & MOISTURE PROTECTION	99.1	73.5	88.6	105.9	105.8	105.9	107.3	103.6	105.8	108.5	163.2	131.0	108.4	164.1	131.2	101.8	110.6	105.4
08	OPENINGS	98.5	67.7	91.3	102.6	94.0	100.6	93.0	85.0	91.1	87.9	175.5	108.3	90.4	180.0	111.2	97.7	103.9	99.1
0920	Plaster & Gypsum Board	78.6	64.2	68.8	97.2	97.6	97.5	107.4	83.8	91.4	99.2	176.9	151.7	102.8	187.6	160.2	98.3	116.7	110.7
0950, 0980	Ceilings & Acoustic Treatment	96.3	64.2	75.2	92.4	97.6	95.8	91.1	83.8	86.3	83.8	176.9	145.0	87.1	187.6	153.2	102.0	116.7	111.6
0960	Flooring	101.2	66.0	91.2	97.1	113.7	101.9	106.8	103.6	105.9	98.3	186.8	123.6	113.9	186.8	134.8	101.3	121.2	107.0
0970, 0990	Wall Finishes & Painting/Coating	103.3	66.6	81.1	103.8	94.3	98.0	93.3	98.8	96.6	102.9	157.4	135.8	123.4	157.4	143.9	98.8	112.4	107.0
09	FINISHES	95.4	64.8	78.5	94.7	101.9	98.7	94.9	91.1	92.8	94.0	175.9	139.1	108.4	181.9	148.9	100.6	117.9	110.2
COVERS	DIVS. 10 - 14, 25, 28, 41, 43, 44, 46	100.0	82.9	96.5	100.0	99.0	99.8	100.0	96.2	99.2	100.0	135.0	107.1	100.0	135.4	107.1	100.0	105.6	101.1
21, 22, 23	FIRE SUPPRESSION, PLUMBING & HVAC	97.2	68.7	85.7	100.0	102.8	101.1	100.5	90.2	96.3	100.2	165.5	126.5	99.7	165.4	126.2	100.0	96.7	98.7
26, 27, 3370	ELECTRICAL, COMMUNICATIONS & UTIL.	92.8	71.2	81.4	98.7	104.1	101.6	99.9	105.4	102.8	97.0	181.9	141.9	99.7	181.9	143.1	100.2	102.7	101.5
MF2014	WEIGHTED AVERAGE	99.0	72.5	87.4	100.1	104.7	102.1	98.5	101.4	99.8	97.7	166.1	127.5	102.8	168.5	131.4	100.3	106.2	102.9

DIVISION		NEW YORK ELMIRA 148 - 149			NEW YORK FAR ROCKAWAY 116			NEW YORK FLUSHING 113			NEW YORK GLENS FALLS 128			NEW YORK HICKSVILLE 115,117,118			NEW YORK JAMAICA 114		
		MAT.	INST.	TOTAL	MAT.	INST.	TOTAL	MAT.	INST.	TOTAL	MAT.	INST.	TOTAL	MAT.	INST.	TOTAL	MAT.	INST.	TOTAL
015433	CONTRACTOR EQUIPMENT		116.0	116.0		113.3	113.3		113.3	113.3		112.9	112.9		113.3	113.3		113.3	113.3
0241, 31 - 34	SITE & INFRASTRUCTURE, DEMOLITION	97.1	94.1	95.0	123.4	127.6	126.4	123.4	127.6	126.4	73.6	105.8	96.5	113.3	126.3	122.6	117.7	127.6	124.7
0310	Concrete Forming & Accessories	81.3	93.2	91.6	93.5	175.1	163.9	97.3	175.1	164.4	85.7	90.3	89.7	90.0	154.5	145.6	97.3	175.1	164.4
0320	Concrete Reinforcing	97.4	95.7	96.5	95.1	205.9	151.5	96.7	205.9	152.3	95.4	94.4	94.9	95.1	210.2	153.7	95.1	205.9	151.5
0330	Cast-in-Place Concrete	93.7	103.2	97.6	113.3	172.1	137.5	113.3	172.1	137.5	85.3	106.7	94.1	96.3	165.5	124.7	104.8	172.1	132.4
03	CONCRETE	90.8	98.6	94.6	113.6	177.8	145.1	114.1	177.8	145.4	88.3	97.5	92.9	99.4	167.1	132.6	106.8	177.8	141.7
04	MASONRY	103.5	102.1	102.6	124.1	177.4	157.4	118.1	177.4	155.1	100.7	105.6	103.7	114.6	166.8	147.1	122.2	177.4	156.6
05	METALS	96.7	117.3	103.0	104.4	150.1	118.4	104.4	150.1	118.4	97.7	106.1	100.3	105.8	149.1	119.2	104.4	150.1	118.4
06	WOOD, PLASTICS & COMPOSITES	85.0	92.2	89.0	90.1	174.6	137.4	94.8	174.6	139.5	87.1	87.6	87.3	86.6	152.6	123.6	94.8	174.6	139.5
07	THERMAL & MOISTURE PROTECTION	103.9	94.1	99.9	108.3	163.0	130.7	108.3	163.0	130.7	98.8	97.2	98.2	107.9	156.3	127.8	108.1	163.0	130.6
08	OPENINGS	99.4	88.9	97.0	89.0	174.4	108.8	89.0	174.4	108.8	93.3	86.2	91.7	89.0	163.4	106.3	89.0	174.4	108.8
0920	Plaster & Gypsum Board	97.8	91.7	93.7	91.6	176.9	149.2	94.1	176.9	150.0	89.1	87.0	87.7	91.2	154.2	133.8	94.1	176.9	150.0
0950, 0980	Ceilings & Acoustic Treatment	95.4	91.7	93.0	76.2	176.9	142.4	76.2	176.9	142.4	82.1	87.0	85.3	75.2	154.2	127.2	76.2	176.9	142.4
0960	Flooring	94.6	103.6	97.1	109.2	186.8	131.4	110.6	186.8	132.4	86.2	111.3	93.3	108.3	185.1	130.2	110.6	186.8	132.4
0970, 0990	Wall Finishes & Painting/Coating	98.9	90.0	93.5	123.4	157.4	143.9	123.4	157.4	143.9	101.1	87.4	92.8	123.4	157.4	143.9	123.4	157.4	143.9
09	FINISHES	95.5	94.7	95.0	103.5	175.7	143.3	104.3	175.7	143.6	86.1	93.1	90.0	102.1	159.7	133.8	103.8	175.7	143.4
COVERS	DIVS. 10 - 14, 25, 28, 41, 43, 44, 46	100.0	99.4	99.9	100.0	134.2	106.9	100.0	134.2	106.9	100.0	96.2	99.2	100.0	128.2	105.7	100.0	134.2	106.9
21, 22, 23	FIRE SUPPRESSION, PLUMBING & HVAC	95.5	91.5	93.9	95.0	165.3	123.4	95.0	165.3	123.4	95.5	95.7	95.6	99.7	154.4	121.8	95.0	165.3	123.4
26, 27, 3370	ELECTRICAL, COMMUNICATIONS & UTIL.	96.5	96.2	96.3	107.0	181.9	146.6	107.0	181.9	146.6	93.6	100.5	97.3	99.0	143.4	122.5	97.9	181.9	142.3
MF2014	WEIGHTED AVERAGE	96.7	97.4	97.0	102.6	166.8	130.6	102.5	166.8	130.5	94.0	98.6	96.0	100.9	153.3	123.7	100.8	166.8	129.6

DIVISION		NEW YORK JAMESTOWN 147			NEW YORK KINGSTON 124			NEW YORK LONG ISLAND CITY 111			NEW YORK MONTICELLO 127			NEW YORK MOUNT VERNON 105			NEW YORK NEW ROCHELLE 108		
		MAT.	INST.	TOTAL	MAT.	INST.	TOTAL	MAT.	INST.	TOTAL	MAT.	INST.	TOTAL	MAT.	INST.	TOTAL	MAT.	INST.	TOTAL
015433	CONTRACTOR EQUIPMENT		93.9	93.9		113.3	113.3		113.3	113.3		113.3	113.3		110.6	110.6		110.6	110.6
0241, 31 - 34	SITE & INFRASTRUCTURE, DEMOLITION	98.4	94.2	95.4	140.6	122.9	128.0	121.3	127.6	125.8	135.5	122.7	126.4	115.2	118.5	117.6	114.6	118.5	117.4
0310	Concrete Forming & Accessories	81.3	86.9	86.1	87.1	104.4	102.0	101.6	175.1	165.0	94.7	102.9	101.8	88.8	139.1	132.1	103.6	139.0	134.2
0320	Concrete Reinforcing	97.6	98.9	98.2	95.8	140.6	118.6	95.1	205.9	151.5	95.1	140.1	118.0	102.9	183.3	143.8	103.0	183.3	143.9
0330	Cast-in-Place Concrete	97.2	102.1	99.2	115.2	135.1	123.4	108.1	172.1	134.4	107.8	124.8	114.8	107.1	141.3	121.1	107.0	141.3	121.1
03	CONCRETE	93.8	94.3	94.0	110.8	121.5	116.0	109.9	177.8	143.3	105.5	117.2	111.2	105.4	146.9	125.8	105.0	146.9	125.6
04	MASONRY	111.8	99.9	104.4	116.0	141.3	131.8	116.9	177.4	154.6	108.6	128.7	121.1	98.1	147.3	128.8	98.1	147.3	128.8
05	METALS	94.0	92.1	93.5	106.1	117.2	109.5	104.4	150.1	118.4	106.0	116.2	109.2	99.6	135.6	110.7	99.9	135.6	110.9
06	WOOD, PLASTICS & COMPOSITES	83.5	84.0	83.8	88.4	96.4	92.9	100.9	174.6	142.2	96.1	96.4	96.3	84.6	136.2	113.5	101.4	136.2	120.9
07	THERMAL & MOISTURE PROTECTION	103.4	93.5	99.3	122.3	137.2	128.4	108.3	163.0	130.7	122.0	132.1	126.1	109.4	144.1	123.6	109.5	144.1	123.7
08	OPENINGS	99.2	85.6	96.1	95.9	117.3	100.9	89.0	174.4	108.8	91.1	117.3	97.2	87.9	148.6	102.0	88.0	148.6	102.0
0920	Plaster & Gypsum Board	88.5	83.2	84.9	91.4	96.3	94.7	98.7	176.9	151.5	92.1	96.3	94.9	94.9	137.0	123.3	106.3	137.0	127.0
0950, 0980	Ceilings & Acoustic Treatment	92.0	83.2	86.2	72.9	96.3	88.3	76.2	176.9	142.4	72.9	96.3	88.3	82.1	137.0	118.2	82.1	137.0	118.2
0960	Flooring	97.7	103.6	99.4	103.8	72.5	94.9	112.1	186.8	133.5	106.1	72.5	96.5	90.2	186.8	117.8	97.1	186.8	122.8
0970, 0990	Wall Finishes & Painting/Coating	100.2	94.2	96.6	134.4	116.8	123.8	123.4	157.4	143.9	134.4	116.8	123.8	101.1	157.4	135.1	101.1	157.4	135.1
09	FINISHES	94.6	89.9	92.0	101.0	97.1	98.8	105.1	175.7	144.0	101.4	96.3	98.6	91.2	149.2	123.2	94.6	149.2	124.7
COVERS	DIVS. 10 - 14, 25, 28, 41, 43, 44, 46	100.0	98.6	99.7	100.0	111.9	102.4	100.0	134.2	106.9	100.0	110.8	102.2	100.0	125.2	105.1	100.0	121.9	104.4
21, 22, 23	FIRE SUPPRESSION, PLUMBING & HVAC	95.4	87.2	92.1	95.4	122.3	106.3	99.7	165.3	126.2	95.4	116.7	104.0	95.5	134.1	111.1	95.5	134.1	111.1
26, 27, 3370	ELECTRICAL, COMMUNICATIONS & UTIL.	95.4	96.3	95.9	95.1	111.0	103.5	98.5	181.9	142.6	95.1	111.0	103.5	95.2	155.2	126.9	95.2	155.2	126.9
MF2014	WEIGHTED AVERAGE	96.8	92.5	94.9	102.6	118.5	109.6	102.4	166.8	130.5	101.1	115.1	107.2	97.5	141.4	116.6	97.9	141.3	116.8

For customer support on your Mechanical Cost Data, call 877.795.7394.

679

City Cost Indexes

NEW YORK

DIVISION		NEW YORK 100 - 102			NIAGARA FALLS 143			PLATTSBURGH 129			POUGHKEEPSIE 125 - 126			QUEENS 110			RIVERHEAD 119		
		MAT.	INST.	TOTAL	MAT.	INST.	TOTAL	MAT.	INST.	TOTAL	MAT.	INST.	TOTAL	MAT.	INST.	TOTAL	MAT.	INST.	TOTAL
015433	CONTRACTOR EQUIPMENT		111.1	111.1		93.9	93.9		98.9	98.9		113.3	113.3		113.3	113.3		113.3	113.3
0241, 31 - 34	SITE & INFRASTRUCTURE, DEMOLITION	117.3	121.5	120.2	100.5	95.5	97.0	106.1	103.0	103.9	136.5	123.5	127.2	116.6	127.6	124.4	114.5	125.9	122.6
0310	Concrete Forming & Accessories	102.6	183.3	172.2	81.3	114.4	109.8	91.1	95.4	94.9	87.1	165.0	154.3	90.1	175.1	163.4	94.6	153.4	145.3
0320	Concrete Reinforcing	109.9	210.4	161.0	96.3	102.5	99.5	99.9	102.7	101.3	95.8	141.0	118.8	96.7	205.9	152.3	96.9	184.2	141.3
0330	Cast-in-Place Concrete	107.8	177.7	136.5	100.6	120.3	108.7	104.4	104.4	104.4	111.5	137.9	122.4	99.6	172.1	129.4	97.9	164.5	125.3
03	CONCRETE	106.2	184.2	144.5	95.9	113.3	104.4	102.5	99.3	101.0	107.9	149.5	128.3	102.6	177.8	139.6	100.1	161.4	130.2
04	MASONRY	102.5	177.5	149.2	119.9	127.5	124.6	95.0	101.6	99.1	108.6	144.5	131.0	111.1	177.4	152.4	119.9	166.3	148.8
05	METALS	113.1	151.9	125.0	96.7	93.8	95.8	102.0	91.2	98.7	106.1	119.8	110.3	104.4	150.1	118.4	106.3	138.4	116.2
06	WOOD, PLASTICS & COMPOSITES	98.5	185.4	147.2	83.4	110.3	98.5	93.8	93.5	93.6	88.4	174.6	136.7	86.7	174.6	135.9	91.8	152.6	125.9
07	THERMAL & MOISTURE PROTECTION	108.6	164.5	131.5	103.5	112.9	107.4	116.3	100.5	109.8	122.3	146.8	132.3	107.9	163.0	130.5	108.8	156.1	128.2
08	OPENINGS	93.7	180.9	114.0	99.2	100.7	99.6	101.6	90.8	99.1	95.9	159.5	110.7	89.0	174.4	108.8	89.0	157.5	104.9
0920	Plaster & Gypsum Board	106.0	187.6	161.2	88.5	110.3	103.3	109.4	92.6	98.0	91.4	176.9	149.2	91.2	176.9	149.1	92.5	154.2	134.2
0950, 0980	Ceilings & Acoustic Treatment	102.2	187.6	158.4	92.0	110.3	104.1	98.1	92.6	94.5	72.9	176.9	141.2	76.2	176.9	142.4	76.1	154.2	127.5
0960	Flooring	99.5	186.8	124.5	97.7	121.2	104.4	108.9	113.7	110.3	103.8	167.2	122.0	108.3	186.8	130.7	109.2	142.7	118.8
0970, 0990	Wall Finishes & Painting/Coating	102.9	157.4	135.8	100.2	112.4	107.6	130.0	91.4	106.7	134.4	117.2	124.0	123.4	157.4	143.9	123.4	157.4	143.9
09	FINISHES	99.9	182.1	145.2	94.7	115.9	106.4	97.9	98.0	98.0	100.8	163.0	135.1	102.5	175.7	142.8	102.7	151.3	129.5
COVERS	DIVS. 10 - 14, 25, 28, 41, 43, 44, 46	100.0	136.1	107.3	100.0	107.1	101.4	100.0	98.0	99.6	100.0	121.6	104.4	100.0	134.2	106.9	100.0	127.6	105.6
21, 22, 23	FIRE SUPPRESSION, PLUMBING & HVAC	100.1	165.5	126.5	95.4	100.0	97.2	95.4	96.6	95.9	95.4	127.0	108.1	99.7	165.3	126.2	99.9	151.4	120.7
26, 27, 3370	ELECTRICAL, COMMUNICATIONS & UTIL.	104.7	181.9	145.5	94.0	100.3	97.3	91.6	90.3	90.9	95.1	119.9	108.2	99.0	181.9	142.8	100.6	133.6	118.1
MF2014	WEIGHTED AVERAGE	103.4	168.7	131.8	97.8	106.4	101.5	99.1	96.7	98.0	101.8	136.7	117.0	100.9	166.8	129.6	101.6	148.2	121.9

NEW YORK

DIVISION		ROCHESTER 144 - 146			SCHENECTADY 123			STATEN ISLAND 103			SUFFERN 109			SYRACUSE 130 - 132			UTICA 133 - 135		
		MAT.	INST.	TOTAL	MAT.	INST.	TOTAL	MAT.	INST.	TOTAL	MAT.	INST.	TOTAL	MAT.	INST.	TOTAL	MAT.	INST.	TOTAL
015433	CONTRACTOR EQUIPMENT		116.7	116.7		112.9	112.9		110.6	110.6		110.6	110.6		112.9	112.9		112.9	112.9
0241, 31 - 34	SITE & INFRASTRUCTURE, DEMOLITION	85.9	109.9	102.9	84.2	106.3	99.9	119.9	120.8	120.6	111.4	116.3	114.9	94.8	105.6	102.4	73.5	104.3	95.4
0310	Concrete Forming & Accessories	98.7	98.2	98.3	103.0	100.6	100.9	88.3	183.3	170.2	96.9	135.2	129.9	99.8	91.0	92.2	101.0	87.9	89.7
0320	Concrete Reinforcing	100.1	95.7	97.9	94.4	103.8	99.2	103.9	210.4	158.1	103.0	140.9	122.3	94.7	96.4	95.6	94.7	95.7	95.2
0330	Cast-in-Place Concrete	94.2	104.4	98.4	102.9	111.3	106.4	107.1	173.5	134.4	103.8	138.2	117.9	95.4	105.9	99.7	87.3	104.7	94.4
03	CONCRETE	99.3	100.6	99.9	102.9	105.4	104.1	107.1	182.7	144.2	102.0	136.3	118.9	98.9	97.8	98.4	96.9	95.9	96.4
04	MASONRY	107.0	104.9	105.7	97.7	112.6	107.0	104.7	177.5	150.1	97.6	142.6	125.7	101.5	105.2	103.8	93.1	103.9	99.8
05	METALS	104.1	107.3	105.1	101.9	110.3	104.5	97.8	151.8	114.4	97.8	119.4	104.5	99.8	105.6	101.6	97.7	105.1	100.0
06	WOOD, PLASTICS & COMPOSITES	97.6	97.6	97.6	107.4	97.9	102.1	83.3	185.4	140.4	94.0	136.2	117.6	101.3	88.5	94.2	101.3	84.7	92.0
07	THERMAL & MOISTURE PROTECTION	103.5	101.7	102.7	100.3	105.8	102.6	108.9	164.3	131.6	109.3	142.1	122.8	102.3	97.6	100.4	90.6	97.6	93.5
08	OPENINGS	105.3	92.2	102.2	99.4	94.0	98.1	87.9	181.2	109.6	88.0	139.0	99.8	95.0	85.6	92.8	98.0	83.4	94.6
0920	Plaster & Gypsum Board	107.2	97.4	100.6	98.8	97.6	98.0	95.0	187.6	157.6	98.5	137.0	124.5	98.0	88.0	91.2	98.0	84.0	88.5
0950, 0980	Ceilings & Acoustic Treatment	100.4	97.4	98.5	88.7	97.6	94.6	83.8	187.6	152.0	82.1	137.0	118.2	91.1	88.0	89.0	91.1	84.0	86.4
0960	Flooring	93.8	113.3	99.4	92.9	113.7	98.9	94.2	186.8	120.7	93.5	185.1	119.7	94.6	102.1	96.8	92.1	102.2	95.0
0970, 0990	Wall Finishes & Painting/Coating	100.3	99.1	99.6	101.1	94.3	97.0	102.9	157.4	135.8	101.1	124.9	115.5	98.5	99.8	99.3	91.3	99.8	96.4
09	FINISHES	100.4	101.3	100.9	91.6	101.9	97.3	93.2	182.1	142.2	92.3	140.0	118.6	94.0	93.5	93.7	92.1	91.1	91.6
COVERS	DIVS. 10 - 14, 25, 28, 41, 43, 44, 46	100.0	99.8	100.0	100.0	99.0	99.8	100.0	136.1	107.3	100.0	123.6	104.8	100.0	96.5	99.3	100.0	96.2	99.2
21, 22, 23	FIRE SUPPRESSION, PLUMBING & HVAC	100.1	90.3	96.1	100.2	102.8	101.2	100.2	165.5	126.5	95.5	123.3	106.7	100.2	91.9	96.9	100.2	92.2	97.0
26, 27, 3370	ELECTRICAL, COMMUNICATIONS & UTIL.	98.7	94.6	96.6	98.1	104.1	101.3	97.0	181.9	141.9	102.8	119.9	111.8	100.0	102.2	101.1	98.1	102.2	100.3
MF2014	WEIGHTED AVERAGE	101.1	99.1	100.2	99.3	104.7	101.7	99.2	168.4	129.4	97.6	129.0	111.3	98.9	98.0	98.5	97.1	97.1	97.1

NEW YORK / NORTH CAROLINA

DIVISION		WATERTOWN 136			WHITE PLAINS 106			YONKERS 107			ASHEVILLE 287 - 288			CHARLOTTE 281 - 282			DURHAM 277		
		MAT.	INST.	TOTAL	MAT.	INST.	TOTAL	MAT.	INST.	TOTAL	MAT.	INST.	TOTAL	MAT.	INST.	TOTAL	MAT.	INST.	TOTAL
015433	CONTRACTOR EQUIPMENT		112.9	112.9		110.6	110.6		110.6	110.6		96.3	96.3		96.3	96.3		101.7	101.7
0241, 31 - 34	SITE & INFRASTRUCTURE, DEMOLITION	80.9	105.7	98.5	108.5	118.5	115.6	116.7	118.5	118.0	102.4	76.9	84.3	105.6	76.9	85.2	100.4	85.7	89.9
0310	Concrete Forming & Accessories	85.9	94.3	93.1	101.9	139.1	134.0	102.2	143.9	138.2	96.0	41.3	48.8	99.7	42.8	50.6	100.2	44.3	52.0
0320	Concrete Reinforcing	95.3	96.5	95.9	103.0	183.3	143.9	107.1	183.3	145.9	93.0	63.0	77.7	98.8	58.5	78.3	91.2	57.9	74.2
0330	Cast-in-Place Concrete	101.5	107.6	104.1	95.1	141.3	114.1	106.3	141.4	120.7	115.4	51.3	89.1	120.2	49.8	91.3	101.6	47.1	79.2
03	CONCRETE	109.3	99.9	104.7	95.7	146.9	120.9	105.2	149.1	126.7	106.0	50.6	78.8	108.6	49.9	79.8	98.7	49.5	74.5
04	MASONRY	94.2	107.9	102.8	97.1	147.3	128.4	102.1	147.3	130.3	93.6	43.7	62.5	101.9	51.2	70.3	86.4	37.9	56.2
05	METALS	97.8	105.6	100.2	99.3	135.6	110.5	108.9	135.7	117.1	103.1	82.3	96.7	104.0	80.9	96.9	121.4	79.9	108.6
06	WOOD, PLASTICS & COMPOSITES	83.0	92.2	88.1	99.5	136.2	120.0	99.3	142.7	123.6	97.9	40.3	65.6	103.0	41.9	68.8	96.7	44.7	67.6
07	THERMAL & MOISTURE PROTECTION	90.9	100.1	94.7	109.2	144.1	123.5	109.5	144.8	124.0	106.7	43.7	80.8	100.8	46.6	78.6	106.9	45.7	81.8
08	OPENINGS	98.0	89.4	96.0	88.0	148.6	102.1	91.1	151.7	105.2	97.1	44.1	84.8	102.5	45.0	89.1	105.8	47.1	92.2
0920	Plaster & Gypsum Board	88.7	91.7	90.8	101.3	137.0	125.4	105.6	143.6	131.3	100.1	38.3	58.3	100.0	39.9	59.4	105.9	42.8	63.3
0950, 0980	Ceilings & Acoustic Treatment	91.1	91.7	91.5	82.1	137.0	118.2	100.5	143.6	128.9	85.7	38.3	54.5	88.8	39.9	56.7	88.1	42.8	58.3
0960	Flooring	86.0	102.2	90.6	95.6	186.8	121.7	95.2	186.8	121.4	102.2	42.7	85.2	101.6	43.4	84.9	103.9	42.7	86.4
0970, 0990	Wall Finishes & Painting/Coating	91.3	96.2	94.3	101.1	157.4	135.1	101.1	157.4	135.1	113.6	40.3	69.3	113.8	49.4	74.9	105.4	37.4	64.3
09	FINISHES	89.7	95.7	93.0	92.9	149.2	123.9	98.0	153.0	128.3	95.6	40.7	65.3	95.5	42.9	66.5	96.0	42.9	66.7
COVERS	DIVS. 10 - 14, 25, 28, 41, 43, 44, 46	100.0	97.5	99.5	100.0	125.2	105.1	100.0	126.3	105.3	100.0	77.7	95.5	100.0	78.1	95.6	100.0	72.1	94.4
21, 22, 23	FIRE SUPPRESSION, PLUMBING & HVAC	100.2	85.9	94.4	100.3	134.1	114.0	100.3	134.1	114.0	100.4	53.4	81.4	100.0	54.1	81.5	100.5	52.9	81.3
26, 27, 3370	ELECTRICAL, COMMUNICATIONS & UTIL.	100.0	90.4	94.9	95.2	155.2	126.9	102.9	163.8	135.1	102.2	55.7	77.6	101.2	58.5	78.6	96.1	55.6	74.7
MF2014	WEIGHTED AVERAGE	98.5	96.3	97.5	97.6	141.4	116.7	102.1	143.7	120.2	100.8	55.3	80.9	101.9	56.8	82.2	102.9	55.3	82.1

NORTH CAROLINA

DIVISION		ELIZABETH CITY 279			FAYETTEVILLE 283			GASTONIA 280			GREENSBORO 270,272-274			HICKORY 286			KINSTON 285		
		MAT.	INST.	TOTAL	MAT.	INST.	TOTAL	MAT.	INST.	TOTAL	MAT.	INST.	TOTAL	MAT.	INST.	TOTAL	MAT.	INST.	TOTAL
015433	CONTRACTOR EQUIPMENT		106.0	106.0		101.7	101.7		96.3	96.3		101.7	101.7		101.7	101.7		101.7	101.7
0241, 31 - 34	SITE & INFRASTRUCTURE, DEMOLITION	104.6	87.4	92.4	101.7	85.6	90.3	102.3	76.9	84.2	100.3	85.7	89.9	101.2	85.5	90.1	100.3	85.6	89.8
0310	Concrete Forming & Accessories	85.3	42.5	48.4	95.6	60.3	65.2	103.1	38.7	47.5	99.9	44.4	52.1	92.0	37.1	44.6	88.2	42.1	48.4
0320	Concrete Reinforcing	89.2	45.9	67.2	96.8	58.0	77.0	93.5	56.5	74.7	90.1	58.0	73.8	93.0	56.1	74.2	92.5	45.8	68.7
0330	Cast-in-Place Concrete	101.8	47.0	79.3	121.0	48.3	91.2	112.8	52.7	88.1	100.8	47.6	79.0	115.4	48.1	87.7	111.4	44.8	84.0
03	CONCRETE	98.7	46.4	73.0	108.1	57.1	83.1	104.5	48.7	77.1	98.1	49.8	74.4	105.7	46.3	76.5	102.4	45.5	74.4
04	MASONRY	98.6	48.0	67.1	97.3	38.9	60.9	98.3	50.9	68.7	82.5	41.4	56.9	82.3	43.7	58.2	89.1	48.1	63.6
05	METALS	107.1	75.8	97.5	124.2	80.0	110.6	103.9	80.0	96.5	113.7	80.0	103.4	103.2	78.7	95.7	102.0	74.8	93.6
06	WOOD, PLASTICS & COMPOSITES	80.1	43.6	59.7	97.1	66.8	80.1	107.0	36.6	67.6	96.4	44.9	67.6	92.1	35.2	60.2	88.6	42.9	63.0
07	THERMAL & MOISTURE PROTECTION	106.2	43.3	80.4	106.3	45.9	81.5	106.9	45.2	81.6	106.7	42.8	80.5	107.0	41.7	80.2	106.8	42.8	80.6
08	OPENINGS	102.6	38.4	87.7	97.2	59.1	88.4	109.0	41.0	87.0	105.8	47.2	92.2	97.2	36.8	83.1	97.3	43.1	84.7
0920	Plaster & Gypsum Board	98.6	41.0	59.7	104.7	65.6	78.3	107.0	34.5	58.0	107.5	43.0	63.9	100.1	33.1	54.8	99.8	41.0	60.1
0950, 0980	Ceilings & Acoustic Treatment	88.1	41.0	57.1	86.5	65.6	72.7	89.0	34.5	53.2	88.1	43.0	58.4	85.7	33.1	51.1	89.0	41.0	57.4
0960	Flooring	95.7	23.2	75.0	102.4	42.7	85.3	105.4	42.7	87.4	103.9	39.7	85.6	102.1	32.8	82.3	99.6	22.6	77.6
0970, 0990	Wall Finishes & Painting/Coating	105.4	42.0	67.1	113.6	33.2	65.0	113.6	40.3	69.3	105.4	31.6	60.9	113.6	40.3	69.3	113.6	39.2	68.7
09	FINISHES	93.0	38.7	63.0	96.5	55.3	73.8	98.1	38.6	65.3	96.3	41.9	66.3	95.8	35.4	62.5	95.5	38.0	63.8
COVERS	DIVS. 10 - 14, 25, 28, 41, 43, 44, 46	100.0	79.4	95.8	100.0	74.3	94.8	100.0	77.3	95.4	100.0	75.9	95.1	100.0	77.1	95.4	100.0	71.6	94.3
21, 22, 23	FIRE SUPPRESSION, PLUMBING & HVAC	95.6	51.1	77.7	100.2	52.7	81.0	100.4	52.4	81.0	100.4	53.1	81.3	95.6	52.2	78.1	95.6	51.2	77.7
26, 27, 3370	ELECTRICAL, COMMUNICATIONS & UTIL.	95.9	34.5	63.4	101.4	50.7	74.6	101.6	57.4	78.2	95.2	55.7	74.3	99.6	57.4	77.3	99.4	46.1	71.2
MF2014	WEIGHTED AVERAGE	99.3	51.6	78.5	104.5	58.1	84.2	101.6	55.1	81.3	101.3	55.7	81.4	98.8	53.9	79.2	98.4	52.7	78.5

NORTH CAROLINA / NORTH DAKOTA

DIVISION		MURPHY 289			RALEIGH 275 - 276			ROCKY MOUNT 278			WILMINGTON 284			WINSTON-SALEM 271			BISMARCK 585		
		MAT.	INST.	TOTAL	MAT.	INST.	TOTAL	MAT.	INST.	TOTAL	MAT.	INST.	TOTAL	MAT.	INST.	TOTAL	MAT.	INST.	TOTAL
015433	CONTRACTOR EQUIPMENT		96.3	96.3		101.7	101.7		101.7	101.7		96.3	96.3		101.7	101.7		98.2	98.2
0241, 31 - 34	SITE & INFRASTRUCTURE, DEMOLITION	103.5	76.7	84.5	101.3	85.7	90.2	102.6	85.7	90.6	103.7	77.1	84.8	100.6	85.7	90.0	101.4	96.9	98.2
0310	Concrete Forming & Accessories	103.7	39.3	48.2	98.9	46.6	53.8	91.9	44.2	50.8	97.5	48.8	55.5	101.9	48.0	55.4	104.9	40.8	49.6
0320	Concrete Reinforcing	92.6	45.4	68.6	96.2	56.0	75.7	89.2	55.8	72.2	93.7	57.9	75.5	90.1	57.9	73.7	99.9	92.7	96.2
0330	Cast-in-Place Concrete	119.4	44.4	88.6	106.7	51.4	84.0	99.6	47.7	78.3	115.0	49.5	88.1	103.4	50.0	81.4	102.6	47.5	79.9
03	CONCRETE	109.2	44.1	77.2	101.7	51.7	77.1	99.5	49.3	74.9	105.9	52.4	79.6	99.4	52.2	76.2	101.5	54.2	78.3
04	MASONRY	85.3	41.2	57.8	86.6	42.2	58.9	76.7	39.5	53.5	82.8	41.7	57.2	82.7	38.6	55.2	114.6	56.4	78.3
05	METALS	100.9	74.2	92.7	105.7	79.2	97.5	106.3	78.4	97.7	102.7	79.9	95.7	110.8	79.9	101.3	103.4	88.4	98.8
06	WOOD, PLASTICS & COMPOSITES	107.8	39.7	69.7	95.1	47.5	68.5	87.4	44.7	63.5	100.1	49.0	71.5	96.4	49.4	70.1	95.3	34.7	61.4
07	THERMAL & MOISTURE PROTECTION	106.9	40.8	79.8	100.5	47.0	78.5	106.6	40.8	79.7	106.7	46.9	82.2	106.7	43.5	80.8	112.4	51.3	87.4
08	OPENINGS	97.1	38.9	83.5	104.8	47.4	91.4	101.8	41.9	87.9	97.3	51.0	86.5	105.8	49.7	92.8	108.0	50.1	94.5
0920	Plaster & Gypsum Board	106.1	37.6	59.8	99.8	45.7	63.3	100.5	42.8	61.5	102.4	47.2	65.1	107.5	47.6	67.0	98.8	33.0	54.3
0950, 0980	Ceilings & Acoustic Treatment	85.7	37.6	54.1	88.8	45.7	60.5	85.6	42.8	57.5	86.5	47.2	60.7	88.1	47.6	61.5	112.9	33.0	60.4
0960	Flooring	105.7	24.1	82.4	101.2	42.7	84.5	99.4	21.8	77.2	102.9	44.4	86.2	103.9	42.7	86.4	94.9	72.1	91.6
0970, 0990	Wall Finishes & Painting/Coating	113.6	39.4	68.8	104.2	36.9	63.5	105.4	38.5	65.0	113.6	38.7	68.4	105.4	36.7	64.0	101.9	29.4	58.1
09	FINISHES	97.7	36.1	63.7	95.5	44.7	67.5	93.9	39.3	63.8	96.3	46.8	69.0	96.3	45.7	68.4	102.7	42.8	69.7
COVERS	DIVS. 10 - 14, 25, 28, 41, 43, 44, 46	100.0	76.9	95.3	100.0	72.6	94.5	100.0	72.5	94.5	100.0	73.8	94.7	100.0	78.7	95.7	100.0	82.5	96.5
21, 22, 23	FIRE SUPPRESSION, PLUMBING & HVAC	95.6	50.9	77.6	100.0	52.5	80.8	95.6	52.7	78.3	100.4	54.9	82.0	100.4	53.3	81.4	100.1	71.9	88.7
26, 27, 3370	ELECTRICAL, COMMUNICATIONS & UTIL.	103.2	29.3	64.1	98.3	40.4	67.7	97.9	39.7	67.1	102.4	50.7	75.1	95.2	55.7	74.3	102.5	71.6	86.2
MF2014	WEIGHTED AVERAGE	99.6	48.3	77.2	100.5	54.1	80.3	98.5	52.3	78.3	100.3	55.9	81.0	101.0	56.5	81.6	103.1	66.2	87.0

NORTH DAKOTA

DIVISION		DEVILS LAKE 583			DICKINSON 586			FARGO 580 - 581			GRAND FORKS 582			JAMESTOWN 584			MINOT 587		
		MAT.	INST.	TOTAL	MAT.	INST.	TOTAL	MAT.	INST.	TOTAL	MAT.	INST.	TOTAL	MAT.	INST.	TOTAL	MAT.	INST.	TOTAL
015433	CONTRACTOR EQUIPMENT		98.2	98.2		98.2	98.2		98.2	98.2		98.2	98.2		98.2	98.2		98.2	98.2
0241, 31 - 34	SITE & INFRASTRUCTURE, DEMOLITION	105.7	94.4	97.7	113.6	92.9	98.9	102.6	96.9	98.5	109.6	92.9	97.7	104.7	92.9	96.3	107.1	96.9	99.8
0310	Concrete Forming & Accessories	101.2	35.4	44.4	90.8	34.9	42.5	99.3	41.5	49.4	94.5	34.2	42.5	92.3	34.1	42.1	90.5	67.0	70.2
0320	Concrete Reinforcing	100.2	93.1	96.6	101.1	92.8	96.9	96.9	92.8	94.8	98.7	92.9	95.7	100.8	91.3	95.9	102.1	93.2	97.6
0330	Cast-in-Place Concrete	126.3	46.0	93.3	114.3	44.4	85.6	113.6	49.2	87.2	114.3	44.1	85.5	124.7	44.1	91.6	114.3	46.5	86.4
03	CONCRETE	110.6	51.3	81.5	109.6	50.2	80.4	105.9	55.1	80.9	106.8	49.8	78.8	109.1	48.7	79.5	105.5	65.6	85.9
04	MASONRY	121.7	64.9	86.3	124.3	60.4	84.5	114.2	56.4	78.2	115.9	64.4	83.8	135.1	32.8	71.3	114.2	65.4	83.8
05	METALS	99.9	87.4	96.0	99.8	82.5	94.5	103.0	88.7	98.6	99.9	82.1	94.4	99.8	63.7	88.7	100.1	89.3	96.8
06	WOOD, PLASTICS & COMPOSITES	95.2	32.0	59.8	83.1	32.0	54.5	92.9	35.2	60.6	87.4	32.0	56.4	85.0	32.0	55.3	82.8	70.1	75.7
07	THERMAL & MOISTURE PROTECTION	107.6	49.8	83.9	108.1	48.5	83.7	107.4	51.6	84.5	107.4	49.8	84.0	107.4	41.5	80.4	107.5	57.4	87.0
08	OPENINGS	100.7	42.7	87.2	100.7	42.7	87.2	100.5	50.3	88.9	100.7	42.7	87.2	100.7	32.2	84.8	100.9	69.3	93.5
0920	Plaster & Gypsum Board	118.4	30.2	58.8	109.1	30.2	55.8	96.6	33.5	54.0	110.5	30.2	56.2	110.2	30.2	56.1	109.1	69.4	82.3
0950, 0980	Ceilings & Acoustic Treatment	114.3	30.2	59.0	114.3	30.2	59.0	112.1	33.5	60.5	114.3	30.2	59.0	114.3	30.2	59.0	114.3	69.4	84.8
0960	Flooring	107.9	35.9	87.3	101.4	35.9	82.7	102.8	72.1	94.0	103.3	35.9	84.0	102.1	35.9	83.2	101.1	89.3	97.8
0970, 0990	Wall Finishes & Painting/Coating	104.4	22.5	54.9	104.4	31.5	60.4	101.2	70.7	82.8	104.4	28.2	58.4	104.4	22.5	54.9	104.4	27.9	58.2
09	FINISHES	108.2	32.2	66.3	105.9	33.2	65.8	103.8	47.7	72.9	106.1	32.8	65.7	105.3	32.2	65.0	105.0	67.0	84.1
COVERS	DIVS. 10 - 14, 25, 28, 41, 43, 44, 46	100.0	32.4	86.3	100.0	32.6	86.4	100.0	82.6	96.5	100.0	32.5	86.4	100.0	80.6	96.1	100.0	86.4	97.3
21, 22, 23	FIRE SUPPRESSION, PLUMBING & HVAC	95.6	74.0	86.9	95.6	66.2	83.7	100.1	77.0	90.8	100.4	33.3	73.3	95.6	35.0	71.2	100.4	60.2	84.2
26, 27, 3370	ELECTRICAL, COMMUNICATIONS & UTIL.	98.3	35.8	65.3	102.7	72.5	89.1	102.4	68.4	84.4	102.1	53.5	76.4	98.3	35.7	65.2	105.5	75.7	89.8
MF2014	WEIGHTED AVERAGE	102.0	58.6	83.1	102.8	60.9	84.6	102.7	67.6	87.4	102.7	51.6	80.4	102.1	45.2	77.3	102.7	71.3	89.0

For customer support on your Mechanical Cost Data, call 877.795.7394.

681

NORTH DAKOTA / OHIO

DIVISION		WILLISTON 588 MAT.	INST.	TOTAL	AKRON 442 - 443 MAT.	INST.	TOTAL	ATHENS 457 MAT.	INST.	TOTAL	CANTON 446 - 447 MAT.	INST.	TOTAL	CHILLICOTHE 456 MAT.	INST.	TOTAL	CINCINNATI 451 - 452 MAT.	INST.	TOTAL
015433	CONTRACTOR EQUIPMENT		98.2	98.2		94.4	94.4		90.4	90.4		94.4	94.4		99.7	99.7		99.5	99.5
0241, 31 - 34	SITE & INFRASTRUCTURE, DEMOLITION	107.5	92.9	97.1	98.1	101.6	100.6	111.0	91.6	97.2	98.2	101.0	100.2	97.0	102.6	101.0	94.1	102.4	100.0
0310	Concrete Forming & Accessories	96.3	34.9	43.3	99.0	94.3	95.0	95.0	84.3	85.8	99.0	83.7	85.8	97.4	92.3	93.0	99.3	79.2	82.0
0320	Concrete Reinforcing	103.1	92.8	97.9	99.5	92.1	95.7	93.1	87.0	90.0	99.5	75.8	87.5	90.1	80.4	85.1	95.5	79.2	87.2
0330	Cast-in-Place Concrete	114.3	44.4	85.6	94.0	98.0	95.7	111.5	96.5	105.4	94.9	95.0	95.0	101.2	99.3	100.4	93.2	92.4	92.9
03	CONCRETE	106.9	50.2	79.1	97.3	94.4	95.9	107.6	88.6	98.3	97.7	85.7	91.8	101.9	92.4	97.2	96.3	84.0	90.3
04	MASONRY	108.8	60.4	78.7	93.3	96.2	95.1	86.4	90.4	88.9	94.0	84.3	87.9	94.0	100.5	98.1	93.7	85.0	88.2
05	METALS	100.0	82.5	94.6	95.4	81.6	91.1	103.0	80.0	95.9	95.4	74.5	88.9	94.9	85.7	92.1	97.2	84.6	93.3
06	WOOD, PLASTICS & COMPOSITES	88.8	32.0	57.0	99.1	93.6	96.0	85.7	84.9	85.3	99.5	82.8	90.2	98.4	89.6	93.5	100.9	76.8	87.4
07	THERMAL & MOISTURE PROTECTION	107.7	48.5	83.5	110.8	96.5	104.9	99.4	94.8	97.5	112.0	91.4	103.5	101.2	97.4	99.7	99.2	88.7	94.9
08	OPENINGS	100.8	42.7	87.3	110.7	93.3	106.7	101.7	82.3	97.2	104.3	77.9	98.2	93.8	83.5	91.4	102.0	77.4	96.3
0920	Plaster & Gypsum Board	110.5	30.2	56.2	96.9	93.2	94.4	91.4	84.1	86.5	98.0	82.1	87.2	93.3	89.5	90.7	94.9	76.3	82.3
0950, 0980	Ceilings & Acoustic Treatment	114.3	30.2	59.0	96.9	93.2	93.1	103.3	84.1	90.7	93.0	82.1	85.8	97.9	89.5	92.4	98.7	76.3	84.0
0960	Flooring	104.1	35.9	84.6	96.9	93.4	95.9	121.3	99.0	114.9	97.1	81.7	92.7	98.3	97.7	98.1	99.3	90.7	96.8
0970, 0990	Wall Finishes & Painting/Coating	104.4	31.5	60.4	96.3	106.3	102.3	100.5	99.1	99.6	96.3	83.2	88.4	97.8	92.9	94.9	97.8	83.9	89.4
09	FINISHES	106.2	33.2	66.0	98.0	95.3	96.5	101.8	89.0	94.8	98.2	82.9	89.8	98.8	93.3	95.8	99.2	81.1	89.3
COVERS	DIVS. 10 - 14, 25, 28, 41, 43, 44, 46	100.0	32.6	86.4	100.0	98.2	99.6	100.0	52.0	90.3	100.0	95.4	99.1	100.0	92.6	98.5	100.0	89.3	97.8
21, 22, 23	FIRE SUPPRESSION, PLUMBING & HVAC	95.6	66.2	83.7	100.1	94.6	97.9	95.3	51.3	77.6	100.1	81.4	92.5	95.8	95.2	95.6	100.0	83.3	93.3
26, 27, 3370	ELECTRICAL, COMMUNICATIONS & UTIL.	102.5	72.5	86.6	99.3	93.9	96.5	95.5	98.1	96.9	98.5	92.7	95.5	95.0	85.6	90.0	93.9	79.4	86.3
MF2014	WEIGHTED AVERAGE	101.2	60.9	83.7	99.9	94.2	97.4	99.6	80.4	91.2	99.3	85.6	93.3	96.8	93.0	95.1	98.3	84.5	92.2

OHIO

DIVISION		CLEVELAND 441 MAT.	INST.	TOTAL	COLUMBUS 430 - 432 MAT.	INST.	TOTAL	DAYTON 453 - 454 MAT.	INST.	TOTAL	HAMILTON 450 MAT.	INST.	TOTAL	LIMA 458 MAT.	INST.	TOTAL	LORAIN 440 MAT.	INST.	TOTAL
015433	CONTRACTOR EQUIPMENT		94.7	94.7		93.6	93.6		94.7	94.7		99.7	99.7		92.9	92.9		94.4	94.4
0241, 31 - 34	SITE & INFRASTRUCTURE, DEMOLITION	98.0	102.2	101.0	95.5	98.3	97.5	92.9	101.8	99.2	92.8	102.1	99.4	104.6	91.6	95.4	97.5	102.9	101.4
0310	Concrete Forming & Accessories	99.1	99.6	99.5	100.3	84.5	86.7	99.3	79.0	81.8	99.4	79.4	82.1	95.0	86.9	88.1	99.1	86.8	88.5
0320	Concrete Reinforcing	100.0	92.5	96.2	102.7	82.6	92.5	95.5	81.3	88.3	95.5	79.2	87.2	93.1	81.5	87.2	99.5	92.4	95.9
0330	Cast-in-Place Concrete	92.2	106.5	98.1	94.1	93.7	93.9	86.7	86.4	86.6	92.9	92.7	92.8	102.4	97.5	100.4	89.5	104.1	95.5
03	CONCRETE	96.5	99.8	98.2	97.9	87.1	92.6	93.3	81.8	87.6	96.2	84.2	90.3	100.4	89.3	94.9	95.2	93.2	94.2
04	MASONRY	97.8	105.6	102.7	96.6	94.0	95.0	93.2	83.5	87.1	93.5	85.5	88.5	117.7	86.4	98.2	89.9	104.6	99.1
05	METALS	96.9	84.4	93.1	96.9	80.7	91.9	96.4	78.6	91.0	96.5	84.5	92.8	103.0	81.5	96.4	96.0	82.8	91.9
06	WOOD, PLASTICS & COMPOSITES	98.2	97.1	97.6	97.1	82.3	88.8	102.2	77.1	88.1	100.9	76.8	87.4	85.6	86.3	86.0	99.1	80.9	88.9
07	THERMAL & MOISTURE PROTECTION	109.5	108.6	109.1	100.9	94.9	98.4	104.7	87.6	97.7	101.4	88.8	96.2	99.0	95.0	97.3	111.9	103.0	108.2
08	OPENINGS	100.6	95.2	99.3	102.4	80.0	97.2	101.3	78.1	95.9	98.9	77.4	93.9	101.7	79.8	96.6	104.3	86.4	100.2
0920	Plaster & Gypsum Board	96.2	96.8	96.6	93.5	81.8	85.6	94.9	76.6	82.6	94.9	76.3	82.3	91.4	85.5	87.4	96.9	80.1	85.5
0950, 0980	Ceilings & Acoustic Treatment	91.3	96.8	94.9	97.7	81.8	87.3	99.7	76.6	84.5	98.7	76.3	84.0	102.4	85.5	91.3	93.0	80.1	84.5
0960	Flooring	96.7	105.0	99.1	92.0	90.3	91.5	102.0	80.7	95.9	99.3	90.7	96.8	120.3	92.2	112.3	97.1	105.0	99.3
0970, 0990	Wall Finishes & Painting/Coating	96.3	105.1	101.6	95.5	92.9	94.0	97.8	85.0	90.1	97.8	84.7	89.9	100.5	82.7	89.8	96.3	105.1	101.6
09	FINISHES	97.5	100.9	99.4	94.1	85.9	89.6	100.2	79.3	88.7	99.1	81.4	89.3	100.8	87.2	93.3	98.0	91.1	94.2
COVERS	DIVS. 10 - 14, 25, 28, 41, 43, 44, 46	100.0	102.5	100.5	100.0	93.3	98.6	100.0	89.1	97.8	100.0	89.5	97.9	100.0	95.2	99.0	100.0	100.3	100.1
21, 22, 23	FIRE SUPPRESSION, PLUMBING & HVAC	100.0	100.5	100.2	100.1	92.0	96.8	100.9	86.6	95.2	100.6	83.6	93.7	95.3	89.0	92.8	100.1	92.1	96.9
26, 27, 3370	ELECTRICAL, COMMUNICATIONS & UTIL.	98.8	105.0	102.1	97.6	87.8	92.4	92.6	83.0	87.5	93.0	83.3	87.9	95.8	80.3	87.6	98.7	88.0	93.0
MF2014	WEIGHTED AVERAGE	99.0	100.3	99.6	98.6	89.2	94.5	98.0	84.4	92.1	97.9	85.1	92.3	100.0	86.8	94.3	98.9	93.0	96.3

OHIO

DIVISION		MANSFIELD 448 - 449 MAT.	INST.	TOTAL	MARION 433 MAT.	INST.	TOTAL	SPRINGFIELD 455 MAT.	INST.	TOTAL	STEUBENVILLE 439 MAT.	INST.	TOTAL	TOLEDO 434 - 436 MAT.	INST.	TOTAL	YOUNGSTOWN 444 - 445 MAT.	INST.	TOTAL
015433	CONTRACTOR EQUIPMENT		94.4	94.4		93.2	93.2		94.7	94.7		98.1	98.1		96.0	96.0		94.4	94.4
0241, 31 - 34	SITE & INFRASTRUCTURE, DEMOLITION	93.9	101.5	99.3	91.8	97.1	95.6	93.2	100.5	98.4	133.1	106.9	114.5	94.8	99.0	97.8	98.0	102.0	100.8
0310	Concrete Forming & Accessories	89.1	83.7	84.5	96.7	81.2	83.3	99.3	83.5	85.7	98.3	89.4	90.7	100.3	97.1	97.6	99.0	87.5	89.1
0320	Concrete Reinforcing	90.6	76.4	83.4	94.7	82.5	88.5	95.5	81.3	88.3	92.3	85.5	88.8	102.7	85.4	93.9	99.5	85.6	92.4
0330	Cast-in-Place Concrete	87.1	94.7	90.2	85.9	92.7	88.7	89.1	86.3	87.9	93.3	94.1	93.6	94.1	100.0	96.5	93.1	96.9	94.7
03	CONCRETE	89.7	85.7	87.7	89.9	85.1	87.6	94.4	83.8	89.2	93.8	89.7	91.8	97.9	95.5	96.7	96.9	89.8	93.4
04	MASONRY	92.4	97.0	95.3	98.8	95.5	96.7	93.4	83.5	87.2	86.2	94.2	91.2	104.8	99.9	101.7	93.6	93.6	93.6
05	METALS	96.2	75.5	89.8	96.0	78.7	90.6	96.4	78.4	90.9	92.5	79.5	88.5	96.7	85.5	93.3	95.4	78.7	90.3
06	WOOD, PLASTICS & COMPOSITES	86.7	80.9	83.4	92.8	79.8	85.5	103.6	83.6	92.4	88.4	88.3	88.3	97.1	96.9	97.0	99.1	86.1	91.8
07	THERMAL & MOISTURE PROTECTION	111.3	94.4	104.4	100.5	85.2	94.2	104.6	88.2	97.9	112.8	96.4	106.1	102.5	102.7	102.6	112.1	94.3	104.8
08	OPENINGS	104.9	76.0	98.2	96.6	76.5	91.9	99.3	79.0	94.6	97.0	83.7	94.0	99.7	91.3	97.7	104.3	85.1	99.8
0920	Plaster & Gypsum Board	90.7	80.1	83.5	91.4	79.3	83.2	94.9	83.3	87.1	90.7	87.5	88.6	93.5	96.9	95.8	96.9	85.5	89.2
0950, 0980	Ceilings & Acoustic Treatment	93.9	80.1	84.8	97.7	79.3	85.6	99.7	83.3	88.9	94.8	87.5	90.0	97.7	96.9	97.2	93.0	85.5	88.1
0960	Flooring	92.5	107.5	96.8	90.9	107.5	95.6	102.0	80.7	95.9	118.6	100.4	113.4	91.2	99.7	93.6	97.1	93.9	96.2
0970, 0990	Wall Finishes & Painting/Coating	96.3	88.5	91.6	95.5	47.5	66.5	97.8	85.0	90.1	107.8	103.1	104.9	95.6	102.0	99.4	96.3	92.6	94.1
09	FINISHES	95.8	88.2	91.6	93.1	82.6	87.3	100.2	83.2	90.8	110.0	92.6	100.4	93.8	98.1	96.2	98.1	88.8	93.0
COVERS	DIVS. 10 - 14, 25, 28, 41, 43, 44, 46	100.0	96.2	99.2	100.0	54.4	90.8	100.0	89.8	97.9	100.0	95.0	99.0	100.0	98.7	99.7	100.0	96.4	99.3
21, 22, 23	FIRE SUPPRESSION, PLUMBING & HVAC	95.3	88.9	92.7	95.3	91.0	93.6	100.9	81.3	93.0	95.7	92.0	94.2	100.1	100.4	100.2	100.1	87.1	94.8
26, 27, 3370	ELECTRICAL, COMMUNICATIONS & UTIL.	96.2	79.0	87.1	91.9	79.0	85.1	92.6	87.8	90.1	86.9	114.1	101.3	97.6	104.7	101.4	98.7	86.0	92.0
MF2014	WEIGHTED AVERAGE	96.7	87.3	92.6	95.1	85.3	90.8	98.0	84.7	92.2	96.8	94.9	96.0	98.7	98.1	98.4	99.2	89.0	94.6

DIVISION		OHIO ZANESVILLE 437-438			OKLAHOMA ARDMORE 734			CLINTON 736			DURANT 747			ENID 737			GUYMON 739		
		MAT.	INST.	TOTAL	MAT.	INST.	TOTAL	MAT.	INST.	TOTAL	MAT.	INST.	TOTAL	MAT.	INST.	TOTAL	MAT.	INST.	TOTAL
015433	CONTRACTOR EQUIPMENT		93.2	93.2		82.7	82.7		81.8	81.8		81.8	81.8		81.8	81.8		81.8	81.8
0241, 31 - 34	SITE & INFRASTRUCTURE, DEMOLITION	94.5	98.7	97.5	96.2	91.6	92.9	97.6	90.2	92.3	95.5	89.9	91.5	99.3	90.2	92.8	101.6	90.0	93.3
0310	Concrete Forming & Accessories	93.7	81.3	83.0	94.0	40.4	47.8	92.5	46.6	53.0	85.5	44.1	49.8	96.2	34.1	42.7	99.8	44.9	52.4
0320	Concrete Reinforcing	94.1	85.1	89.5	90.8	79.8	85.2	91.3	79.8	85.5	95.8	80.2	87.9	90.7	79.8	85.2	91.3	79.8	85.5
0330	Cast-in-Place Concrete	90.5	91.7	91.0	97.5	43.4	75.3	94.3	45.1	74.1	91.7	42.8	71.6	94.3	46.4	74.6	94.3	43.1	73.3
03	CONCRETE	93.5	85.3	89.5	95.2	49.3	72.7	94.7	52.7	74.1	92.3	50.8	71.9	95.3	47.6	71.8	98.2	51.2	75.1
04	MASONRY	95.5	85.9	89.5	103.4	57.1	74.5	129.7	57.1	84.4	95.6	62.3	74.9	110.1	57.1	77.0	105.9	53.7	73.3
05	METALS	97.4	80.4	92.1	102.5	70.1	92.5	102.6	70.1	92.6	93.5	70.7	86.5	104.0	70.0	93.6	103.2	69.8	92.9
06	WOOD, PLASTICS & COMPOSITES	88.5	79.8	83.6	97.4	37.9	64.0	96.4	46.2	68.3	88.1	43.0	62.9	100.1	29.3	60.5	104.0	46.0	71.5
07	THERMAL & MOISTURE PROTECTION	100.6	92.6	97.3	108.9	60.6	89.1	109.1	61.5	89.6	99.7	61.8	84.2	109.2	59.9	89.0	109.5	57.3	88.1
08	OPENINGS	96.6	79.3	92.5	102.9	49.0	90.3	102.9	53.3	91.4	96.3	52.0	86.0	102.9	43.9	89.1	103.0	49.2	90.5
0920	Plaster & Gypsum Board	88.2	79.3	82.2	89.8	36.6	53.8	89.4	45.2	59.6	79.4	41.9	54.1	90.5	27.8	48.1	90.7	44.9	59.8
0950, 0980	Ceilings & Acoustic Treatment	97.7	79.3	85.6	85.7	36.6	53.4	85.7	45.2	59.1	82.4	41.9	55.8	85.7	27.8	47.6	86.5	44.9	59.2
0960	Flooring	89.3	90.3	89.6	107.0	43.2	88.7	105.8	41.1	87.3	102.8	61.9	91.1	107.7	41.1	88.6	109.3	24.4	85.0
0970, 0990	Wall Finishes & Painting/Coating	95.5	92.9	94.0	104.7	51.3	72.5	104.7	51.3	72.5	103.9	51.3	72.1	104.7	51.3	72.5	104.7	33.0	61.5
09	FINISHES	92.4	83.6	87.5	92.9	40.1	63.8	92.7	44.7	66.3	90.2	46.6	66.2	93.5	34.6	61.1	94.5	38.9	63.9
COVERS	DIVS. 10 - 14, 25, 28, 41, 43, 44, 46	100.0	88.5	97.7	100.0	77.6	95.5	100.0	78.6	95.7	100.0	77.9	95.5	100.0	76.7	95.3	100.0	77.4	95.4
21, 22, 23	FIRE SUPPRESSION, PLUMBING & HVAC	95.3	89.9	93.1	95.6	65.8	83.6	95.6	65.9	83.6	95.5	65.4	83.4	100.3	65.8	86.4	95.6	63.8	82.7
26, 27, 3370	ELECTRICAL, COMMUNICATIONS & UTIL.	92.3	83.4	87.5	93.6	70.9	81.6	94.6	70.9	82.1	96.5	70.9	83.0	94.6	70.9	82.1	96.2	61.5	77.9
MF2014	WEIGHTED AVERAGE	95.6	86.5	91.6	98.3	61.9	82.4	99.6	63.1	83.7	95.1	63.5	81.3	100.2	60.5	82.9	99.4	59.7	82.1

DIVISION		OKLAHOMA LAWTON 735			MCALESTER 745			MIAMI 743			MUSKOGEE 744			OKLAHOMA CITY 730 - 731			PONCA CITY 746		
		MAT.	INST.	TOTAL	MAT.	INST.	TOTAL	MAT.	INST.	TOTAL	MAT.	INST.	TOTAL	MAT.	INST.	TOTAL	MAT.	INST.	TOTAL
015433	CONTRACTOR EQUIPMENT		82.7	82.7		81.8	81.8		89.9	89.9		89.9	89.9		83.0	83.0		81.8	81.8
0241, 31 - 34	SITE & INFRASTRUCTURE, DEMOLITION	95.7	91.6	92.8	89.1	90.1	89.9	90.5	87.3	88.2	90.8	87.1	88.2	95.1	92.0	92.9	95.9	90.1	91.8
0310	Concrete Forming & Accessories	99.8	44.3	52.0	83.5	43.6	49.1	96.9	66.2	70.4	101.4	33.4	42.7	97.7	57.9	63.4	92.2	46.3	52.6
0320	Concrete Reinforcing	91.0	79.8	85.3	95.5	79.8	87.5	94.0	79.9	86.8	94.9	79.3	87.0	96.2	79.9	87.9	94.9	79.9	87.2
0330	Cast-in-Place Concrete	91.2	46.4	72.8	80.4	45.2	65.9	84.3	47.5	69.2	85.3	45.8	69.1	92.9	46.9	74.0	94.2	44.7	73.8
03	CONCRETE	91.5	52.1	72.1	82.9	51.3	67.4	87.6	63.0	75.5	89.4	47.7	68.9	94.9	58.3	76.9	94.4	52.4	73.8
04	MASONRY	105.8	57.1	75.4	114.3	56.1	78.1	98.4	56.3	72.1	116.9	45.2	72.2	109.8	56.4	76.5	90.9	56.1	69.2
05	METALS	108.0	70.1	96.4	93.4	70.0	86.2	93.4	82.1	89.9	94.8	80.4	90.4	100.7	70.2	91.3	93.4	70.3	86.3
06	WOOD, PLASTICS & COMPOSITES	103.0	43.1	69.5	85.6	43.0	61.7	101.0	72.9	85.3	105.6	30.7	63.6	97.0	61.8	77.3	96.3	46.0	68.1
07	THERMAL & MOISTURE PROTECTION	108.9	61.3	89.4	99.4	60.6	83.5	99.8	64.3	85.2	99.9	48.6	78.8	101.1	62.9	85.4	99.9	66.2	86.1
08	OPENINGS	104.6	51.4	92.2	96.2	51.8	85.9	96.2	68.3	89.7	96.2	42.4	83.7	104.0	61.6	94.2	96.2	53.1	86.2
0920	Plaster & Gypsum Board	92.4	42.0	58.3	78.4	41.9	53.7	84.8	72.6	76.5	86.9	29.0	47.8	98.0	61.2	73.1	83.7	44.9	57.5
0950, 0980	Ceilings & Acoustic Treatment	93.2	42.0	59.6	82.4	41.9	55.8	82.4	72.6	75.9	90.8	29.0	50.2	96.4	61.2	73.3	82.4	44.9	57.8
0960	Flooring	109.7	41.1	90.1	101.8	41.1	84.4	108.8	61.9	95.4	111.3	39.8	90.8	106.2	41.1	87.6	105.9	41.1	87.4
0970, 0990	Wall Finishes & Painting/Coating	104.7	51.3	72.5	103.9	37.8	64.0	103.9	77.5	87.9	103.9	34.2	61.8	106.4	51.3	73.2	103.9	51.3	72.1
09	FINISHES	95.5	42.8	66.5	89.2	41.0	62.6	92.0	67.3	78.4	94.9	33.0	60.8	97.6	53.7	73.4	91.8	44.3	65.6
COVERS	DIVS. 10 - 14, 25, 28, 41, 43, 44, 46	100.0	78.2	95.6	100.0	77.9	95.5	100.0	81.6	96.3	100.0	76.1	95.2	100.0	80.0	96.0	100.0	78.2	95.6
21, 22, 23	FIRE SUPPRESSION, PLUMBING & HVAC	100.3	65.9	86.4	95.5	61.6	81.8	95.5	61.9	81.9	100.3	60.6	84.3	100.1	65.5	86.2	95.5	61.8	81.9
26, 27, 3370	ELECTRICAL, COMMUNICATIONS & UTIL.	96.2	70.9	82.8	94.9	64.5	78.8	96.3	64.6	79.5	94.5	52.4	72.2	101.7	70.9	85.4	94.5	70.4	81.7
MF2014	WEIGHTED AVERAGE	100.7	62.8	84.2	94.5	60.4	79.7	94.8	67.5	82.9	97.4	55.7	79.2	100.3	65.7	85.2	95.1	62.1	80.7

DIVISION		OKLAHOMA POTEAU 749			SHAWNEE 748			TULSA 740 - 741			WOODWARD 738			OREGON BEND 977			EUGENE 974		
		MAT.	INST.	TOTAL	MAT.	INST.	TOTAL	MAT.	INST.	TOTAL	MAT.	INST.	TOTAL	MAT.	INST.	TOTAL	MAT.	INST.	TOTAL
015433	CONTRACTOR EQUIPMENT		89.1	89.1		81.8	81.8		89.9	89.9		81.8	81.8		98.8	98.8		98.8	98.8
0241, 31 - 34	SITE & INFRASTRUCTURE, DEMOLITION	77.2	85.8	83.3	99.0	90.1	92.7	97.1	87.5	90.3	97.9	90.2	92.4	105.4	102.9	103.6	96.1	102.9	100.9
0310	Concrete Forming & Accessories	90.1	41.0	47.8	85.4	43.7	49.5	101.5	40.3	48.7	92.6	46.8	53.1	110.1	98.8	100.4	106.5	98.7	99.8
0320	Concrete Reinforcing	95.9	79.9	87.8	94.9	79.8	87.2	95.1	79.8	87.3	90.7	79.9	85.2	93.3	99.6	96.5	97.4	99.6	98.5
0330	Cast-in-Place Concrete	84.3	44.9	68.1	97.2	42.9	74.9	92.9	46.2	73.7	94.3	45.3	74.2	104.4	102.7	103.7	101.1	102.6	101.7
03	CONCRETE	89.9	50.9	70.7	96.0	50.6	73.7	94.7	51.0	73.2	95.0	52.8	74.3	107.1	100.0	103.6	98.8	100.0	99.4
04	MASONRY	98.7	56.2	72.2	115.6	56.1	78.5	99.3	60.0	74.8	98.6	57.1	72.7	104.1	103.3	103.6	101.1	103.3	102.5
05	METALS	93.4	81.7	89.8	93.3	69.9	86.1	98.1	81.4	93.0	102.7	70.5	92.8	91.9	96.4	93.3	92.6	96.2	93.7
06	WOOD, PLASTICS & COMPOSITES	93.0	39.1	62.8	88.0	43.0	62.8	104.8	36.6	66.6	96.5	46.0	68.2	101.7	98.4	99.9	97.6	98.4	98.1
07	THERMAL & MOISTURE PROTECTION	99.9	60.5	83.7	99.9	59.3	83.2	99.8	62.2	84.4	109.1	66.6	91.7	107.9	94.9	102.6	107.2	91.6	100.8
08	OPENINGS	96.2	49.9	86.5	96.2	51.8	85.9	97.8	47.5	86.1	102.9	53.1	91.3	96.9	102.3	98.1	97.1	102.3	98.4
0920	Plaster & Gypsum Board	82.3	37.8	52.2	79.4	41.9	54.1	86.9	35.2	51.9	89.6	44.9	59.4	107.4	98.2	101.2	106.0	98.2	100.7
0950, 0980	Ceilings & Acoustic Treatment	82.4	37.8	53.1	82.4	41.9	55.8	90.8	35.2	54.3	86.5	44.9	59.2	91.3	98.2	95.8	92.3	98.2	96.1
0960	Flooring	105.2	61.9	92.8	102.8	32.7	82.8	110.0	42.8	90.8	105.8	43.2	87.9	110.5	103.4	108.5	108.9	103.4	107.3
0970, 0990	Wall Finishes & Painting/Coating	103.9	51.3	72.1	103.9	34.1	61.8	103.9	40.5	65.6	104.7	51.3	72.5	106.4	74.8	87.3	106.4	74.8	87.3
09	FINISHES	89.9	44.3	64.8	90.4	39.2	62.2	94.7	38.9	63.9	93.0	44.9	66.5	102.7	96.8	99.5	101.2	96.8	98.8
COVERS	DIVS. 10 - 14, 25, 28, 41, 43, 44, 46	100.0	77.7	95.5	100.0	77.9	95.5	100.0	78.8	95.7	100.0	78.5	95.7	100.0	99.6	99.9	100.0	99.6	99.9
21, 22, 23	FIRE SUPPRESSION, PLUMBING & HVAC	95.5	61.8	81.9	95.5	65.4	83.4	100.3	63.5	85.4	95.6	65.9	83.6	95.1	100.8	97.4	99.9	100.8	100.2
26, 27, 3370	ELECTRICAL, COMMUNICATIONS & UTIL.	94.6	64.6	78.7	96.6	70.9	83.0	96.5	64.6	79.6	96.1	70.9	82.8	101.6	96.1	98.7	100.1	96.1	98.0
MF2014	WEIGHTED AVERAGE	94.3	61.4	80.0	96.5	61.8	81.4	98.1	61.6	82.2	98.3	63.4	83.1	98.9	99.4	99.1	98.6	99.3	98.9

For customer support on your Mechanical Cost Data, call 877.795.7394.

683

City Cost Indexes

OREGON

DIVISION		KLAMATH FALLS 976			MEDFORD 975			PENDLETON 978			PORTLAND 970 - 972			SALEM 973			VALE 979		
		MAT.	INST.	TOTAL	MAT.	INST.	TOTAL	MAT.	INST.	TOTAL	MAT.	INST.	TOTAL	MAT.	INST.	TOTAL	MAT.	INST.	TOTAL
015433	CONTRACTOR EQUIPMENT		98.8	98.8		98.8	98.8		96.2	96.2		98.8	98.8		98.8	98.8		96.2	96.2
0241, 31 - 34	SITE & INFRASTRUCTURE, DEMOLITION	109.2	102.9	104.7	103.4	102.9	103.0	102.5	96.4	98.2	98.5	102.9	101.6	91.9	102.9	99.7	90.5	96.4	94.7
0310	Concrete Forming & Accessories	102.9	98.6	99.2	102.0	98.6	99.1	103.3	99.0	99.6	107.7	98.9	100.1	106.0	98.8	99.8	109.9	98.7	100.2
0320	Concrete Reinforcing	93.3	99.6	96.5	94.9	99.6	97.3	92.6	99.7	96.2	98.1	99.7	98.9	103.6	99.6	101.6	90.4	99.6	95.1
0330	Cast-in-Place Concrete	104.4	102.6	103.7	104.4	102.6	103.7	105.2	103.9	104.7	103.9	102.7	103.4	97.9	102.7	99.9	82.9	103.8	91.5
03	CONCRETE	109.7	99.9	104.9	104.6	99.9	102.3	91.4	100.6	95.9	100.3	100.1	100.2	98.2	100.0	99.1	76.9	100.4	88.5
04	MASONRY	117.5	103.3	108.6	98.2	103.3	101.4	107.8	103.4	105.0	102.5	103.3	103.0	109.3	103.3	105.6	106.1	103.4	104.4
05	METALS	91.9	96.2	93.2	92.2	96.2	93.4	98.1	96.9	97.7	93.5	96.5	94.4	98.7	96.4	98.0	98.0	96.6	97.5
06	WOOD, PLASTICS & COMPOSITES	92.6	98.4	95.9	91.5	98.4	95.3	94.7	98.5	96.8	98.6	98.4	98.5	95.5	98.4	97.1	103.2	98.5	100.6
07	THERMAL & MOISTURE PROTECTION	108.2	92.9	101.9	107.8	92.9	101.7	101.1	93.5	98.0	107.1	96.7	102.9	104.9	94.9	100.8	100.5	94.4	98.0
08	OPENINGS	96.9	102.3	98.1	99.7	102.3	100.3	93.2	102.4	95.4	95.0	102.3	96.7	98.7	102.3	99.5	93.2	94.1	93.4
0920	Plaster & Gypsum Board	102.9	98.2	99.7	102.2	98.2	99.5	90.2	98.2	95.6	105.5	98.2	100.5	102.1	98.2	99.4	95.9	98.2	97.4
0950, 0980	Ceilings & Acoustic Treatment	98.9	98.2	98.4	105.4	98.2	100.7	64.4	98.2	86.6	94.2	98.2	96.9	99.3	98.2	98.5	64.4	98.2	86.6
0960	Flooring	107.6	103.4	106.4	107.1	103.4	106.1	74.7	103.4	82.9	106.4	103.4	105.5	107.2	103.4	106.1	76.7	103.4	84.4
0970, 0990	Wall Finishes & Painting/Coating	106.4	70.4	84.6	106.4	70.4	84.6	96.6	72.8	82.2	106.2	72.8	86.0	104.6	74.8	86.6	96.6	74.8	83.4
09	FINISHES	103.4	96.3	99.5	103.8	96.3	99.6	72.3	96.7	85.7	100.8	96.6	98.5	100.0	96.8	98.2	72.7	96.9	86.0
COVERS	DIVS. 10 - 14, 25, 28, 41, 43, 44, 46	100.0	99.5	99.9	100.0	99.5	99.9	100.0	90.1	98.0	100.0	99.6	99.9	100.0	99.6	99.9	100.0	99.9	100.0
21, 22, 23	FIRE SUPPRESSION, PLUMBING & HVAC	95.1	100.7	97.4	99.9	100.7	100.2	97.0	113.2	103.5	99.9	100.8	100.2	99.9	100.8	100.2	97.0	98.0	97.4
26, 27, 3370	ELECTRICAL, COMMUNICATIONS & UTIL.	100.2	80.8	89.9	103.9	80.8	91.6	92.3	97.1	94.8	100.4	103.2	101.9	108.0	96.1	101.7	92.3	97.0	94.8
MF2014	WEIGHTED AVERAGE	99.8	97.1	98.6	100.0	97.1	98.8	94.8	101.4	97.7	98.8	100.4	99.5	100.5	99.4	100.0	92.9	98.1	95.2

PENNSYLVANIA

DIVISION		ALLENTOWN 181			ALTOONA 166			BEDFORD 155			BRADFORD 167			BUTLER 160			CHAMBERSBURG 172		
		MAT.	INST.	TOTAL	MAT.	INST.	TOTAL	MAT.	INST.	TOTAL	MAT.	INST.	TOTAL	MAT.	INST.	TOTAL	MAT.	INST.	TOTAL
015433	CONTRACTOR EQUIPMENT		112.9	112.9		112.9	112.9		109.6	109.6		112.9	112.9		112.9	112.9		112.1	112.1
0241, 31 - 34	SITE & INFRASTRUCTURE, DEMOLITION	93.6	104.5	101.3	96.8	104.4	102.2	101.4	100.1	100.5	92.4	103.5	100.3	88.1	105.8	100.7	89.2	101.2	97.8
0310	Concrete Forming & Accessories	99.2	113.0	111.1	84.2	80.4	80.9	84.1	80.8	81.3	86.4	81.9	82.5	85.7	96.0	94.6	90.2	80.0	81.4
0320	Concrete Reinforcing	94.7	108.1	101.5	91.8	102.7	97.3	93.1	80.9	86.9	93.7	103.2	98.5	92.4	109.1	100.9	91.7	103.0	97.4
0330	Cast-in-Place Concrete	86.5	104.9	94.0	96.3	86.1	92.1	106.7	70.1	91.6	92.0	92.6	92.3	85.1	96.7	89.9	91.1	69.8	82.3
03	CONCRETE	93.6	110.1	101.7	89.1	88.1	88.6	102.1	78.5	90.5	95.4	91.0	93.2	81.3	99.9	90.4	101.9	82.2	92.3
04	MASONRY	97.4	101.0	99.7	100.5	64.5	78.0	114.1	87.0	97.2	97.6	88.4	91.9	102.6	98.9	100.3	102.6	87.4	93.1
05	METALS	100.0	123.0	107.1	93.9	116.8	101.0	97.6	104.0	99.6	97.9	116.5	103.6	93.6	121.9	102.3	97.5	114.6	102.7
06	WOOD, PLASTICS & COMPOSITES	100.7	115.8	109.1	79.2	83.2	81.4	84.0	80.7	82.2	85.5	80.1	82.5	80.6	95.5	89.0	88.0	80.8	84.0
07	THERMAL & MOISTURE PROTECTION	102.3	119.1	109.2	101.4	91.0	97.1	103.0	87.7	96.7	102.3	91.7	97.9	101.1	100.7	100.9	99.3	72.5	88.3
08	OPENINGS	95.0	115.0	99.6	88.7	89.3	88.8	98.0	81.6	94.2	95.2	91.0	94.2	88.7	104.8	92.4	94.1	82.8	91.4
0920	Plaster & Gypsum Board	95.9	116.0	109.5	87.7	82.5	84.2	92.3	80.0	84.0	88.5	79.2	82.2	87.7	95.2	92.7	96.7	80.0	85.4
0950, 0980	Ceilings & Acoustic Treatment	82.8	116.0	104.6	87.0	82.5	84.0	95.0	80.0	85.1	85.4	79.2	81.3	87.9	95.2	92.7	84.0	80.0	81.4
0960	Flooring	94.6	95.9	95.0	88.2	97.5	90.9	93.1	88.2	91.7	89.4	105.3	94.0	89.0	81.0	86.7	97.9	46.2	83.1
0970, 0990	Wall Finishes & Painting/Coating	98.5	70.5	81.6	94.0	109.2	103.2	100.2	81.6	89.0	98.5	94.6	96.1	94.0	109.2	103.2	105.9	81.6	91.2
09	FINISHES	91.8	105.5	99.4	90.0	85.8	87.7	95.0	80.8	87.2	89.9	85.5	87.5	89.9	93.6	91.9	92.1	73.7	82.0
COVERS	DIVS. 10 - 14, 25, 28, 41, 43, 44, 46	100.0	105.6	101.1	100.0	95.4	99.1	100.0	98.3	99.7	100.0	99.2	99.8	100.0	102.4	100.5	100.0	96.1	99.2
21, 22, 23	FIRE SUPPRESSION, PLUMBING & HVAC	100.2	112.1	105.0	99.7	82.5	92.8	95.2	87.3	92.0	95.5	90.7	93.6	95.0	93.9	94.6	95.5	87.1	92.1
26, 27, 3370	ELECTRICAL, COMMUNICATIONS & UTIL.	99.1	98.4	98.8	89.3	112.2	101.4	94.3	112.2	103.8	92.7	112.2	103.0	89.9	111.2	101.1	89.6	86.3	87.8
MF2014	WEIGHTED AVERAGE	97.9	108.6	102.5	94.6	91.9	93.4	98.2	91.3	95.2	95.7	96.4	96.0	92.5	102.0	96.6	96.2	88.0	92.6

PENNSYLVANIA

DIVISION		DOYLESTOWN 189			DUBOIS 158			ERIE 164 - 165			GREENSBURG 156			HARRISBURG 170 - 171			HAZLETON 182		
		MAT.	INST.	TOTAL	MAT.	INST.	TOTAL	MAT.	INST.	TOTAL	MAT.	INST.	TOTAL	MAT.	INST.	TOTAL	MAT.	INST.	TOTAL
015433	CONTRACTOR EQUIPMENT		93.9	93.9		109.6	109.6		112.9	112.9		109.6	109.6		112.1	112.1		112.9	112.9
0241, 31 - 34	SITE & INFRASTRUCTURE, DEMOLITION	106.9	90.2	96.5	106.1	100.5	102.1	93.7	105.2	101.9	97.8	102.9	101.4	89.6	103.5	99.5	86.9	104.7	99.6
0310	Concrete Forming & Accessories	83.2	128.2	122.0	83.5	84.6	84.4	98.5	88.1	89.5	90.7	95.9	95.2	100.4	87.9	89.6	80.8	88.7	87.6
0320	Concrete Reinforcing	91.6	130.9	111.6	92.6	103.5	98.1	93.7	103.3	98.6	92.6	109.1	101.0	101.4	106.2	103.9	91.9	107.1	99.6
0330	Cast-in-Place Concrete	81.7	86.7	83.8	102.9	92.9	98.8	94.7	81.8	89.4	99.0	96.3	97.9	93.2	95.8	94.3	81.7	93.1	86.4
03	CONCRETE	89.2	114.2	101.5	103.6	92.2	98.0	88.2	90.0	89.1	97.2	99.6	98.4	97.7	95.4	96.6	86.3	95.0	90.6
04	MASONRY	100.7	128.1	117.7	114.4	91.1	99.9	89.6	92.5	91.4	124.8	98.9	108.7	102.5	89.6	94.5	110.2	95.7	101.2
05	METALS	97.5	123.8	105.6	97.6	116.1	103.3	94.1	117.0	101.2	97.5	120.6	104.6	104.0	121.4	109.4	99.7	120.6	106.1
06	WOOD, PLASTICS & COMPOSITES	81.0	130.7	108.8	82.9	83.1	83.0	96.9	86.6	91.1	90.9	95.4	93.4	95.8	86.7	90.7	79.6	86.6	83.5
07	THERMAL & MOISTURE PROTECTION	100.1	131.4	112.9	103.2	96.5	100.5	101.9	92.6	98.1	102.8	100.7	102.0	102.5	110.2	105.7	101.8	106.6	103.8
08	OPENINGS	97.2	137.9	106.7	98.0	92.7	96.7	88.8	92.2	89.6	97.9	104.7	99.5	100.6	94.0	99.0	95.6	92.3	94.8
0920	Plaster & Gypsum Board	86.2	131.4	116.7	91.2	82.5	85.3	95.9	86.0	89.2	93.5	95.2	94.6	100.6	86.0	90.7	86.6	86.0	86.2
0950, 0980	Ceilings & Acoustic Treatment	82.0	131.4	114.5	95.0	82.5	86.7	82.8	86.0	84.9	94.2	95.2	94.8	92.6	86.0	88.3	83.7	86.0	85.2
0960	Flooring	79.4	134.3	95.1	92.9	105.3	96.4	93.2	91.8	92.8	96.2	68.0	88.2	103.2	91.7	99.9	87.0	94.6	89.2
0970, 0990	Wall Finishes & Painting/Coating	98.0	69.0	80.5	100.2	106.6	104.1	103.5	94.6	98.1	100.0	106.6	104.1	106.4	89.6	96.3	98.5	106.4	103.2
09	FINISHES	83.2	122.4	104.8	95.2	89.5	92.1	92.2	88.8	90.4	95.6	91.7	93.4	96.9	88.0	92.0	88.1	89.5	88.9
COVERS	DIVS. 10 - 14, 25, 28, 41, 43, 44, 46	100.0	70.6	94.1	100.0	99.1	99.8	100.0	100.8	100.2	100.0	102.2	100.4	100.0	97.4	99.5	100.0	101.0	100.2
21, 22, 23	FIRE SUPPRESSION, PLUMBING & HVAC	95.0	127.4	108.1	95.2	88.4	92.4	99.7	92.9	97.0	95.2	91.1	93.5	100.1	92.1	96.9	95.5	99.2	97.0
26, 27, 3370	ELECTRICAL, COMMUNICATIONS & UTIL.	92.1	128.1	111.1	94.9	112.2	104.0	91.1	96.7	94.0	94.9	112.2	104.1	96.8	88.3	92.3	93.6	91.7	92.6
MF2014	WEIGHTED AVERAGE	94.9	120.6	106.1	98.5	96.8	97.8	94.5	95.8	95.1	98.2	100.9	99.4	99.8	95.6	98.0	95.4	98.3	96.6

PENNSYLVANIA

DIVISION		INDIANA 157			JOHNSTOWN 159			KITTANNING 162			LANCASTER 175 - 176			LEHIGH VALLEY 180			MONTROSE 188		
		MAT.	INST.	TOTAL	MAT.	INST.	TOTAL	MAT.	INST.	TOTAL	MAT.	INST.	TOTAL	MAT.	INST.	TOTAL	MAT.	INST.	TOTAL
015433	CONTRACTOR EQUIPMENT		109.6	109.6		109.6	109.6		112.9	112.9		112.1	112.1		112.9	112.9		112.9	112.9
0241, 31 - 34	SITE & INFRASTRUCTURE, DEMOLITION	95.9	101.2	99.7	101.9	102.2	102.1	90.7	105.7	101.4	81.6	103.5	97.2	90.7	104.3	100.4	89.4	102.1	98.4
0310	Concrete Forming & Accessories	84.7	86.3	86.1	83.5	84.8	84.6	85.7	95.9	94.5	92.3	87.5	88.2	92.8	112.7	110.0	81.8	88.5	87.6
0320	Concrete Reinforcing	91.8	109.2	100.6	93.1	108.9	101.2	92.4	109.2	101.0	91.4	106.1	98.9	91.9	108.1	100.1	96.3	106.3	101.4
0330	Cast-in-Place Concrete	97.1	95.9	96.6	107.6	92.4	101.4	88.4	96.5	91.7	77.5	97.9	85.9	88.4	104.7	95.1	86.7	90.9	88.4
03	CONCRETE	94.7	95.2	94.9	102.9	93.2	98.2	83.8	99.7	91.6	89.8	96.0	92.8	92.6	109.8	101.1	91.3	93.7	92.5
04	MASONRY	110.1	100.1	103.9	110.9	90.6	98.3	105.4	100.1	102.1	108.2	90.1	97.0	97.3	101.0	99.6	97.3	95.1	95.9
05	METALS	97.7	120.1	104.6	97.6	118.9	104.2	93.7	121.9	102.4	97.5	120.8	104.7	99.7	122.0	106.5	98.0	114.1	102.9
06	WOOD, PLASTICS & COMPOSITES	84.8	83.1	83.8	82.9	83.1	83.0	80.6	95.5	89.0	90.8	86.7	88.5	92.2	115.8	105.4	80.4	88.4	84.9
07	THERMAL & MOISTURE PROTECTION	102.7	98.2	100.9	103.0	95.1	99.7	101.1	101.0	101.1	98.8	97.1	98.1	102.2	102.8	102.4	101.8	91.7	97.7
08	OPENINGS	98.0	94.4	97.1	98.0	90.9	96.3	88.7	101.2	91.6	94.1	98.6	95.1	95.6	115.0	100.1	92.2	93.1	92.4
0920	Plaster & Gypsum Board	92.7	82.5	85.8	91.0	82.5	85.2	87.7	95.2	92.7	98.5	86.0	90.1	89.5	116.0	107.4	87.1	87.8	87.6
0950, 0980	Ceilings & Acoustic Treatment	95.0	82.5	86.7	94.2	82.5	86.5	87.9	95.2	92.7	84.0	86.0	85.3	83.7	116.0	105.0	85.4	87.8	87.0
0960	Flooring	93.7	105.3	97.0	92.9	98.2	94.4	89.0	105.3	93.6	98.8	91.7	96.8	91.9	95.9	93.0	87.6	57.9	79.1
0970, 0990	Wall Finishes & Painting/Coating	100.2	106.6	104.1	100.2	109.2	105.6	94.0	106.6	101.6	105.9	57.1	76.5	98.5	66.4	79.1	98.5	106.4	103.2
09	FINISHES	94.8	90.0	92.1	94.6	88.8	91.4	90.1	97.4	94.1	92.0	84.5	87.9	90.2	104.1	97.9	88.9	84.7	86.6
COVERS	DIVS. 10 - 14, 25, 28, 41, 43, 44, 46	100.0	100.6	100.1	100.0	99.3	99.9	100.0	102.2	100.5	100.0	97.4	99.5	100.0	106.0	101.2	100.0	101.2	100.2
21, 22, 23	FIRE SUPPRESSION, PLUMBING & HVAC	95.2	90.8	93.4	95.2	88.7	92.5	95.0	96.9	95.8	95.5	92.3	94.2	95.5	112.0	102.2	95.5	98.4	96.7
26, 27, 3370	ELECTRICAL, COMMUNICATIONS & UTIL.	94.9	112.2	104.1	94.9	112.2	104.0	89.3	112.2	101.4	91.0	44.0	66.2	93.6	143.4	119.9	92.7	97.6	95.3
MF2014	WEIGHTED AVERAGE	97.1	99.3	98.0	98.1	97.2	97.7	92.9	103.3	98.1	94.1	89.0	92.4	95.8	114.0	103.7	94.7	96.9	95.7

PENNSYLVANIA

DIVISION		NEW CASTLE 161			NORRISTOWN 194			OIL CITY 163			PHILADELPHIA 190 - 191			PITTSBURGH 150 - 152			POTTSVILLE 179		
		MAT.	INST.	TOTAL	MAT.	INST.	TOTAL	MAT.	INST.	TOTAL	MAT.	INST.	TOTAL	MAT.	INST.	TOTAL	MAT.	INST.	TOTAL
015433	CONTRACTOR EQUIPMENT		112.9	112.9		98.8	98.8		112.9	112.9		98.2	98.2		110.8	110.8		112.1	112.1
0241, 31 - 34	SITE & INFRASTRUCTURE, DEMOLITION	88.5	105.8	100.8	95.8	100.6	99.2	87.1	103.7	98.9	101.6	100.0	100.5	101.3	104.9	103.8	84.4	103.4	97.9
0310	Concrete Forming & Accessories	85.7	95.6	94.3	83.2	129.7	123.3	85.7	83.4	83.7	98.9	140.1	134.4	98.5	96.7	97.0	83.3	89.7	88.8
0320	Concrete Reinforcing	91.3	92.5	91.9	90.2	137.9	114.5	92.4	92.3	92.4	100.6	137.9	119.6	93.6	109.4	101.6	90.7	102.7	96.8
0330	Cast-in-Place Concrete	85.9	96.5	90.2	85.0	127.1	102.3	83.4	95.5	88.4	98.0	132.2	112.1	102.9	96.5	100.3	82.6	100.9	90.1
03	CONCRETE	81.6	96.4	88.9	89.2	130.1	109.3	80.1	90.5	85.2	99.4	136.4	117.6	100.5	100.1	100.3	93.3	97.3	95.3
04	MASONRY	101.9	97.7	99.3	110.4	124.9	119.4	101.8	95.8	98.0	96.7	130.7	117.9	103.0	102.1	102.4	102.1	92.2	95.9
05	METALS	93.7	113.6	99.8	99.2	130.0	108.7	93.7	111.5	99.2	102.0	130.2	110.7	99.0	121.3	105.9	97.7	118.8	104.2
06	WOOD, PLASTICS & COMPOSITES	80.6	95.9	89.2	79.9	130.6	108.3	80.6	80.1	80.3	98.6	141.9	122.9	100.6	95.8	97.9	80.2	88.1	84.6
07	THERMAL & MOISTURE PROTECTION	101.1	98.0	99.8	101.4	131.1	113.5	101.0	94.4	98.3	102.8	135.2	116.1	103.1	101.6	102.4	98.9	106.7	102.1
08	OPENINGS	88.7	96.6	90.5	86.9	140.1	99.2	88.7	80.3	86.7	98.6	146.2	109.7	101.8	105.0	102.5	94.1	91.8	93.6
0920	Plaster & Gypsum Board	87.7	95.6	93.0	85.6	131.4	116.5	87.7	79.2	82.0	96.6	143.1	128.0	99.8	95.6	96.9	93.5	87.5	89.5
0950, 0980	Ceilings & Acoustic Treatment	87.9	95.6	92.9	84.5	131.4	115.3	87.9	79.2	82.2	94.0	143.1	126.2	95.0	95.6	95.4	84.0	87.5	86.3
0960	Flooring	89.0	57.5	80.0	92.9	139.2	106.1	89.0	105.3	93.6	98.4	139.2	110.1	99.5	106.6	101.5	95.2	91.7	94.2
0970, 0990	Wall Finishes & Painting/Coating	94.0	109.2	103.2	98.6	147.8	128.3	94.0	106.6	101.6	99.8	154.7	133.0	100.2	119.2	111.7	105.9	106.4	106.2
09	FINISHES	90.0	89.6	89.8	90.7	133.0	114.0	89.8	87.8	88.7	98.9	141.8	122.5	97.7	100.2	99.1	90.5	91.3	90.9
COVERS	DIVS. 10 - 14, 25, 28, 41, 43, 44, 46	100.0	102.4	100.5	100.0	117.5	103.5	100.0	100.6	100.1	100.0	120.8	104.2	100.0	102.2	100.4	100.0	98.7	99.7
21, 22, 23	FIRE SUPPRESSION, PLUMBING & HVAC	95.0	93.6	94.4	95.2	127.3	108.2	95.0	92.9	94.2	100.0	131.8	112.9	99.9	100.4	100.1	95.5	98.6	96.8
26, 27, 3370	ELECTRICAL, COMMUNICATIONS & UTIL.	89.9	98.1	94.2	93.0	148.4	122.3	91.8	111.1	102.0	97.1	148.4	124.2	97.2	112.2	105.1	89.2	95.5	92.5
MF2014	WEIGHTED AVERAGE	92.5	97.9	94.8	95.0	129.5	110.0	92.4	96.9	94.4	99.7	133.6	114.5	99.9	104.6	102.0	94.9	98.5	96.5

PENNSYLVANIA

DIVISION		READING 195 - 196			SCRANTON 184 - 185			STATE COLLEGE 168			STROUDSBURG 183			SUNBURY 178			UNIONTOWN 154		
		MAT.	INST.	TOTAL	MAT.	INST.	TOTAL	MAT.	INST.	TOTAL	MAT.	INST.	TOTAL	MAT.	INST.	TOTAL	MAT.	INST.	TOTAL
015433	CONTRACTOR EQUIPMENT		118.0	118.0		112.9	112.9		112.1	112.1		112.9	112.9		112.9	112.9		109.6	109.6
0241, 31 - 34	SITE & INFRASTRUCTURE, DEMOLITION	100.3	112.7	109.1	94.1	104.8	101.7	84.4	103.1	97.7	88.5	102.2	98.2	96.6	104.5	102.2	96.5	102.9	101.0
0310	Concrete Forming & Accessories	98.8	90.2	91.4	99.3	88.6	90.0	84.4	80.3	80.8	87.2	89.4	89.1	96.1	88.5	89.6	77.6	96.0	93.5
0320	Concrete Reinforcing	91.5	104.7	98.2	94.7	106.6	100.8	93.0	103.4	98.3	95.0	111.6	103.4	93.2	106.2	99.8	92.6	109.1	101.0
0330	Cast-in-Place Concrete	76.2	97.3	84.9	90.3	93.1	91.5	87.1	65.6	78.2	85.1	72.6	80.0	90.2	95.5	92.4	97.1	96.4	96.8
03	CONCRETE	88.2	96.7	92.4	95.4	94.9	95.1	95.5	81.0	88.4	90.1	88.9	89.5	96.7	95.5	96.1	94.4	99.7	97.0
04	MASONRY	99.5	93.0	95.4	97.7	95.7	96.4	103.1	77.9	87.4	95.1	100.4	98.4	102.3	89.7	94.5	127.0	98.9	109.5
05	METALS	99.5	121.2	106.2	102.1	120.5	107.8	97.7	117.6	103.8	99.7	115.9	104.7	97.4	120.0	104.4	97.4	120.6	104.5
06	WOOD, PLASTICS & COMPOSITES	98.5	88.1	92.6	100.7	86.2	92.6	87.6	83.2	85.1	86.4	88.4	87.5	88.9	88.1	88.5	76.7	95.4	87.2
07	THERMAL & MOISTURE PROTECTION	101.8	112.1	106.0	102.2	95.6	99.5	101.5	90.6	97.0	102.0	85.3	95.2	100.0	104.9	102.0	102.6	100.7	101.8
08	OPENINGS	91.3	98.9	93.1	95.0	91.9	94.3	92.0	89.3	91.4	95.6	88.2	93.9	94.2	92.8	93.9	97.9	104.7	99.5
0920	Plaster & Gypsum Board	96.8	87.5	90.5	98.0	85.6	89.6	89.6	82.5	84.8	88.0	87.8	87.9	92.7	87.5	89.2	89.2	95.2	93.2
0950, 0980	Ceilings & Acoustic Treatment	76.9	87.5	83.9	91.1	85.6	87.5	82.9	82.5	82.6	82.0	87.8	85.8	80.7	87.5	85.2	94.2	95.2	94.8
0960	Flooring	96.8	91.7	95.4	94.6	107.0	98.2	92.3	97.5	93.8	89.9	51.3	78.9	95.9	87.7	93.5	90.4	105.3	94.7
0970, 0990	Wall Finishes & Painting/Coating	97.3	106.4	102.8	98.5	106.4	103.2	98.5	109.2	105.0	98.5	64.6	78.0	105.9	94.3	98.9	100.2	109.2	105.6
09	FINISHES	92.0	90.8	91.3	93.9	93.1	93.4	89.4	85.8	87.4	88.9	80.0	84.0	91.3	88.7	89.9	93.2	97.5	95.6
COVERS	DIVS. 10 - 14, 25, 28, 41, 43, 44, 46	100.0	100.3	100.1	100.0	101.0	100.2	100.0	93.1	98.6	100.0	62.4	92.4	100.0	97.9	99.6	100.0	102.2	100.4
21, 22, 23	FIRE SUPPRESSION, PLUMBING & HVAC	100.1	108.8	103.7	100.2	99.2	99.8	95.0	85.9	91.6	95.5	100.0	97.3	95.5	92.0	94.1	95.2	91.1	93.5
26, 27, 3370	ELECTRICAL, COMMUNICATIONS & UTIL.	99.6	95.5	97.5	99.2	97.7	98.4	91.9	112.2	102.6	93.6	143.3	119.9	89.5	90.8	90.2	92.0	112.2	102.7
MF2014	WEIGHTED AVERAGE	97.1	102.1	99.3	98.6	99.2	98.9	95.3	92.8	94.2	95.3	101.4	97.9	95.8	95.8	95.8	97.3	101.7	99.2

For customer support on your Mechanical Cost Data, call 877.795.7394.

685

PENNSYLVANIA

DIVISION		WASHINGTON 153			WELLSBORO 169			WESTCHESTER 193			WILKES-BARRE 186 - 187			WILLIAMSPORT 177			YORK 173 - 174		
		MAT.	INST.	TOTAL	MAT.	INST.	TOTAL	MAT.	INST.	TOTAL	MAT.	INST.	TOTAL	MAT.	INST.	TOTAL	MAT.	INST.	TOTAL
015433	CONTRACTOR EQUIPMENT		109.6	109.6		112.9	112.9		98.8	98.8		112.9	112.9		112.9	112.9		112.1	112.1
0241, 31 - 34	SITE & INFRASTRUCTURE, DEMOLITION	96.6	102.9	101.1	95.9	102.0	100.3	101.6	97.8	98.9	86.5	104.7	99.5	87.9	103.4	98.9	85.2	103.5	98.2
0310	Concrete Forming & Accessories	84.8	96.2	94.6	85.8	86.7	86.6	89.8	128.2	122.9	89.9	88.9	89.1	92.5	61.0	65.3	86.9	88.1	87.9
0320	Concrete Reinforcing	92.6	109.3	101.1	93.0	106.2	99.7	89.3	113.6	101.7	93.7	107.1	100.6	92.5	64.6	78.3	93.2	106.2	99.8
0330	Cast-in-Place Concrete	97.1	96.4	96.8	91.2	88.9	90.3	94.1	126.4	107.4	81.7	93.0	86.4	76.4	74.2	75.5	83.0	98.1	89.2
03	CONCRETE	94.9	99.8	97.3	97.8	92.2	95.1	96.9	124.5	110.5	87.2	95.1	91.1	84.7	68.2	76.6	94.5	96.3	95.4
04	MASONRY	109.4	100.5	103.9	103.6	89.7	94.9	104.7	124.9	117.3	110.6	95.3	101.0	93.9	93.7	93.8	103.6	90.1	95.2
05	METALS	97.3	120.9	104.6	97.8	114.2	102.9	99.2	117.2	104.8	97.9	121.0	105.0	97.5	100.4	98.4	99.1	121.4	105.9
06	WOOD, PLASTICS & COMPOSITES	84.9	95.4	90.8	84.9	88.1	86.7	87.1	130.6	111.5	88.9	86.6	87.6	85.2	51.0	66.1	84.1	86.7	85.5
07	THERMAL & MOISTURE PROTECTION	102.7	101.1	102.1	102.5	89.6	97.2	101.7	129.9	113.3	101.8	106.4	103.7	99.4	102.4	100.6	99.0	110.4	103.7
08	OPENINGS	97.9	104.7	99.5	95.1	93.0	94.6	86.9	126.1	96.0	92.2	98.8	93.8	94.2	53.5	84.7	94.1	94.0	94.1
0920	Plaster & Gypsum Board	92.4	95.2	94.2	87.9	87.5	87.6	87.0	131.4	117.0	88.8	86.0	86.9	93.5	49.3	63.6	94.3	86.0	88.7
0950, 0980	Ceilings & Acoustic Treatment	94.2	95.2	94.8	82.9	87.5	85.9	84.5	131.4	115.3	85.4	86.0	85.8	84.0	49.3	61.2	83.1	86.0	85.0
0960	Flooring	93.8	105.3	97.1	89.1	50.8	78.2	95.7	139.2	108.1	90.7	94.6	91.8	94.8	49.7	81.9	96.4	91.7	95.1
0970, 0990	Wall Finishes & Painting/Coating	100.2	109.2	105.6	98.5	106.4	103.2	98.6	147.8	128.3	98.5	106.4	103.2	105.9	106.4	106.2	105.9	89.6	96.1
09	FINISHES	94.6	97.8	96.4	89.5	82.2	85.5	92.1	131.0	113.6	89.9	90.8	90.4	91.1	62.3	75.2	90.8	88.1	89.3
COVERS	DIVS. 10 - 14, 25, 28, 41, 43, 44, 46	100.0	102.2	100.4	100.0	97.9	99.6	100.0	117.3	103.5	100.0	100.8	100.2	100.0	94.6	98.9	100.0	97.6	99.5
21, 22, 23	FIRE SUPPRESSION, PLUMBING & HVAC	95.2	97.2	96.0	95.5	91.1	93.7	95.2	125.8	107.6	95.5	99.0	96.9	95.5	93.1	94.5	100.2	92.4	97.1
26, 27, 3370	ELECTRICAL, COMMUNICATIONS & UTIL.	94.3	112.2	103.8	92.7	86.5	89.5	92.9	114.9	104.5	93.6	91.7	92.6	90.0	74.6	81.8	91.0	88.3	89.5
MF2014	WEIGHTED AVERAGE	97.0	103.2	99.7	96.3	92.6	94.7	95.9	121.5	107.1	95.1	98.7	96.6	93.8	83.0	89.1	96.7	95.9	96.3

DIVISION		PUERTO RICO SAN JUAN 009			RHODE ISLAND NEWPORT 028			RHODE ISLAND PROVIDENCE 029			SOUTH CAROLINA AIKEN 298			SOUTH CAROLINA BEAUFORT 299			SOUTH CAROLINA CHARLESTON 294		
		MAT.	INST.	TOTAL	MAT.	INST.	TOTAL	MAT.	INST.	TOTAL	MAT.	INST.	TOTAL	MAT.	INST.	TOTAL	MAT.	INST.	TOTAL
015433	CONTRACTOR EQUIPMENT		90.5	90.5		102.4	102.4		102.4	102.4		101.3	101.3		101.3	101.3		101.3	101.3
0241, 31 - 34	SITE & INFRASTRUCTURE, DEMOLITION	133.8	91.2	103.5	89.2	105.0	100.4	91.4	105.0	101.1	118.9	87.4	96.5	114.3	85.6	93.9	99.7	86.3	90.2
0310	Concrete Forming & Accessories	92.4	17.8	28.1	101.5	122.4	119.5	99.8	122.4	119.3	97.5	67.4	71.5	96.4	37.9	46.0	95.4	63.0	67.5
0320	Concrete Reinforcing	188.3	12.7	98.9	106.4	149.2	128.2	102.3	149.2	126.2	93.7	65.2	79.2	92.8	25.8	58.7	92.7	59.3	75.7
0330	Cast-in-Place Concrete	103.8	30.4	73.7	79.5	124.1	97.8	95.5	124.1	107.2	79.2	69.8	75.4	79.2	47.2	66.1	92.9	49.5	75.1
03	CONCRETE	108.7	22.2	66.2	93.2	127.5	110.0	100.4	127.5	113.7	102.7	68.9	86.1	99.9	40.9	70.9	94.4	58.9	77.0
04	MASONRY	90.1	16.9	44.5	95.4	132.5	118.5	101.3	132.5	120.7	79.6	61.4	68.3	93.9	33.3	56.1	95.1	40.9	64.4
05	METALS	118.4	34.6	92.6	97.4	125.6	106.1	103.1	125.6	110.0	101.8	83.8	96.3	101.8	68.2	91.5	103.8	80.4	96.6
06	WOOD, PLASTICS & COMPOSITES	94.4	17.5	51.3	99.0	120.9	111.3	100.1	120.9	111.7	97.0	68.6	81.1	95.3	37.8	63.1	94.0	68.4	79.7
07	THERMAL & MOISTURE PROTECTION	129.1	21.6	85.0	100.6	121.5	109.2	100.5	121.5	109.1	102.7	66.4	87.8	102.4	38.8	76.3	101.6	47.3	79.3
08	OPENINGS	152.0	15.4	120.2	102.2	128.2	108.3	107.3	128.2	112.2	99.0	64.8	91.1	99.0	36.7	84.5	102.9	63.3	93.7
0920	Plaster & Gypsum Board	159.8	14.9	61.9	92.9	120.9	111.9	94.5	120.9	112.4	104.0	67.4	79.3	107.4	35.7	59.0	109.0	67.3	80.8
0950, 0980	Ceilings & Acoustic Treatment	225.7	14.9	87.2	93.2	120.9	111.4	90.1	120.9	110.4	86.4	67.4	73.9	89.8	35.7	54.3	89.8	67.3	75.0
0960	Flooring	224.3	16.6	164.9	97.7	141.5	110.2	98.0	141.5	110.4	105.4	68.4	94.8	106.8	51.2	90.9	106.5	58.2	92.7
0970, 0990	Wall Finishes & Painting/Coating	213.5	16.8	94.8	99.9	129.2	117.6	97.7	129.2	116.7	109.5	70.3	85.8	109.5	34.7	64.3	109.5	66.9	83.8
09	FINISHES	210.6	17.6	104.2	96.6	126.9	113.3	93.7	126.9	112.0	98.3	68.0	81.6	99.5	39.8	66.6	97.8	63.5	78.9
COVERS	DIVS. 10 - 14, 25, 28, 41, 43, 44, 46	100.0	17.5	83.3	100.0	108.6	101.7	100.0	108.6	101.7	100.0	71.8	94.3	100.0	70.4	94.0	100.0	69.0	93.7
21, 22, 23	FIRE SUPPRESSION, PLUMBING & HVAC	103.3	14.0	67.3	100.1	110.6	104.3	99.9	110.6	104.2	95.7	63.1	82.5	95.7	36.5	71.8	100.5	53.7	81.6
26, 27, 3370	ELECTRICAL, COMMUNICATIONS & UTIL.	125.9	13.2	66.3	100.0	99.0	99.5	99.6	99.0	99.2	96.9	65.9	80.5	100.7	33.6	65.2	99.0	88.8	93.6
MF2014	WEIGHTED AVERAGE	122.7	24.4	79.8	98.4	117.5	106.7	100.6	117.5	108.0	98.6	69.1	85.7	99.3	44.9	75.6	99.9	65.2	84.8

DIVISION		SOUTH CAROLINA COLUMBIA 290 - 292			SOUTH CAROLINA FLORENCE 295			SOUTH CAROLINA GREENVILLE 296			SOUTH CAROLINA ROCK HILL 297			SOUTH CAROLINA SPARTANBURG 293			SOUTH DAKOTA ABERDEEN 574		
		MAT.	INST.	TOTAL	MAT.	INST.	TOTAL	MAT.	INST.	TOTAL	MAT.	INST.	TOTAL	MAT.	INST.	TOTAL	MAT.	INST.	TOTAL
015433	CONTRACTOR EQUIPMENT		101.3	101.3		101.3	101.3		101.3	101.3		101.3	101.3		101.3	101.3		98.2	98.2
0241, 31 - 34	SITE & INFRASTRUCTURE, DEMOLITION	99.8	86.3	90.2	108.8	86.3	92.8	104.1	85.9	91.2	101.7	85.0	89.8	103.9	86.0	91.1	99.2	93.7	95.3
0310	Concrete Forming & Accessories	94.4	45.1	51.9	83.2	45.3	50.5	95.0	45.1	52.0	93.2	38.7	46.2	98.0	45.3	52.5	94.8	36.6	44.6
0320	Concrete Reinforcing	95.8	59.1	77.1	92.3	59.3	75.5	92.2	44.5	67.9	93.0	44.1	68.1	92.2	57.0	74.3	96.1	38.1	66.6
0330	Cast-in-Place Concrete	95.8	50.1	77.0	79.2	49.4	66.9	79.2	49.3	66.9	79.2	43.9	64.7	79.2	49.4	66.9	105.6	43.5	80.1
03	CONCRETE	96.2	51.1	74.1	94.2	51.0	73.0	93.0	48.2	71.0	90.9	43.3	67.5	93.2	50.6	72.3	102.8	40.7	72.3
04	MASONRY	91.5	37.6	57.9	79.8	40.9	55.6	77.4	40.9	54.7	101.6	34.6	59.8	79.8	40.9	55.6	115.9	55.4	78.2
05	METALS	100.9	79.6	94.3	102.6	80.0	95.6	102.6	74.5	93.9	101.8	71.2	92.4	102.6	79.1	95.4	95.9	63.6	86.0
06	WOOD, PLASTICS & COMPOSITES	98.0	44.8	68.2	79.7	44.8	60.2	93.7	44.8	66.4	92.0	38.9	62.3	98.0	44.8	68.2	99.6	36.0	64.0
07	THERMAL & MOISTURE PROTECTION	97.2	43.3	75.1	101.9	44.8	78.5	101.8	44.8	78.4	101.6	39.7	76.3	101.8	44.8	78.5	99.2	47.7	78.1
08	OPENINGS	104.6	50.5	92.0	99.1	50.5	87.8	99.0	47.0	86.9	99.0	39.4	85.2	99.0	50.0	87.6	99.1	36.7	84.6
0920	Plaster & Gypsum Board	102.3	42.9	62.2	97.1	42.9	60.5	102.6	42.9	62.3	101.9	36.8	57.9	105.5	42.9	63.2	103.7	34.3	56.8
0950, 0980	Ceilings & Acoustic Treatment	91.3	42.9	59.5	87.3	42.9	58.1	86.4	42.9	57.8	86.4	36.8	53.8	86.4	42.9	57.8	93.2	34.3	54.5
0960	Flooring	100.5	43.6	84.2	98.4	43.6	82.7	104.3	57.2	90.8	103.3	44.4	86.5	105.6	57.2	91.8	108.4	50.3	91.8
0970, 0990	Wall Finishes & Painting/Coating	106.2	66.9	82.5	109.5	66.9	83.8	109.5	66.9	83.8	109.5	39.2	67.0	109.5	66.9	83.8	102.6	37.1	63.1
09	FINISHES	95.9	46.6	68.7	94.4	47.1	68.3	96.3	49.3	70.4	95.7	39.4	64.6	97.1	49.3	70.8	100.3	38.8	66.4
COVERS	DIVS. 10 - 14, 25, 28, 41, 43, 44, 46	100.0	66.3	93.2	100.0	66.3	93.2	100.0	66.3	93.2	100.0	64.6	92.9	100.0	66.3	93.2	100.0	40.5	88.0
21, 22, 23	FIRE SUPPRESSION, PLUMBING & HVAC	100.0	53.0	81.1	100.5	53.1	81.4	100.5	53.0	81.3	95.7	44.2	74.9	100.5	53.1	81.3	100.1	39.8	75.8
26, 27, 3370	ELECTRICAL, COMMUNICATIONS & UTIL.	99.0	59.7	78.3	96.8	59.7	77.2	99.1	57.4	77.0	99.1	56.6	76.6	99.1	57.4	77.1	101.7	51.1	75.0
MF2014	WEIGHTED AVERAGE	99.3	56.5	80.6	98.2	56.9	80.2	98.3	55.8	79.8	97.8	50.3	77.1	98.6	56.7	80.3	100.5	49.6	78.3

For customer support on your **Mechanical Cost Data**, call 877.795.7394.

SOUTH DAKOTA

DIVISION		MITCHELL 573			MOBRIDGE 576			PIERRE 575			RAPID CITY 577			SIOUX FALLS 570 - 571			WATERTOWN 572		
		MAT.	INST.	TOTAL	MAT.	INST.	TOTAL	MAT.	INST.	TOTAL	MAT.	INST.	TOTAL	MAT.	INST.	TOTAL	MAT.	INST.	TOTAL
015433	CONTRACTOR EQUIPMENT		98.2	98.2		98.2	98.2		98.2	98.2		98.2	98.2		99.2	99.2		98.2	98.2
0241, 31 - 34	SITE & INFRASTRUCTURE, DEMOLITION	96.0	93.7	94.3	95.9	93.7	94.3	100.5	93.7	95.7	97.7	93.8	94.9	94.3	95.4	95.0	95.8	93.7	94.3
0310	Concrete Forming & Accessories	94.0	37.0	44.8	84.9	36.8	43.4	97.7	38.4	46.6	102.2	36.7	45.7	98.8	40.5	48.5	81.6	36.6	42.8
0320	Concrete Reinforcing	95.5	46.3	70.5	98.0	38.1	67.5	99.0	71.1	84.8	89.9	71.3	80.4	98.0	71.2	84.3	92.9	38.2	65.1
0330	Cast-in-Place Concrete	102.6	42.3	77.8	102.6	43.5	78.3	98.5	42.2	75.4	101.8	42.8	77.6	92.3	42.9	72.0	102.6	45.9	79.3
03	CONCRETE	100.5	41.8	71.7	100.3	40.7	71.0	99.0	47.3	73.6	99.8	46.7	73.7	95.9	48.4	72.6	99.2	41.5	70.9
04	MASONRY	103.9	53.3	72.4	113.4	55.4	77.2	117.4	53.5	77.6	113.5	56.4	77.9	105.1	53.6	73.0	141.1	57.4	88.9
05	METALS	95.0	63.8	85.4	95.0	63.6	85.3	98.4	79.1	92.4	97.8	79.7	92.2	98.2	79.3	92.4	95.0	64.1	85.5
06	WOOD, PLASTICS & COMPOSITES	98.5	36.5	63.8	87.6	36.2	58.8	104.5	36.8	66.6	103.5	33.3	64.2	96.8	39.3	64.6	83.8	36.0	57.0
07	THERMAL & MOISTURE PROTECTION	98.9	45.6	77.1	99.0	47.8	78.0	101.8	45.9	78.9	99.6	47.8	78.3	101.9	48.2	79.9	98.8	48.3	78.1
08	OPENINGS	97.7	37.3	83.6	100.4	36.3	85.5	105.7	46.7	92.0	103.1	44.8	89.6	106.5	48.1	92.9	97.7	36.5	83.4
0920	Plaster & Gypsum Board	102.1	34.8	56.7	96.2	34.6	54.5	100.1	35.1	56.2	103.2	31.6	54.8	90.7	37.7	54.9	93.9	34.3	53.6
0950, 0980	Ceilings & Acoustic Treatment	89.9	34.8	53.7	93.2	34.6	54.7	93.3	35.1	55.0	95.7	31.6	53.6	90.6	37.7	55.8	89.9	34.3	53.3
0960	Flooring	107.9	50.3	91.4	103.5	50.3	88.3	107.4	35.8	86.9	107.6	78.6	99.3	103.8	74.9	95.5	102.1	50.3	87.3
0970, 0990	Wall Finishes & Painting/Coating	102.6	40.6	65.2	102.6	41.7	65.8	105.7	44.6	68.8	102.6	44.6	67.6	101.7	44.6	67.2	102.6	37.1	63.1
09	FINISHES	99.1	39.5	66.2	97.7	39.4	65.6	100.8	37.3	65.8	100.5	43.9	69.3	97.1	46.5	69.2	96.2	38.8	64.6
COVERS	DIVS. 10 - 14, 25, 28, 41, 43, 44, 46	100.0	37.6	87.4	100.0	40.5	88.0	100.0	75.8	95.1	100.0	75.7	95.1	100.0	76.2	95.2	100.0	40.5	88.0
21, 22, 23	FIRE SUPPRESSION, PLUMBING & HVAC	95.3	39.4	72.8	95.3	39.9	73.0	100.0	64.5	85.7	100.1	65.0	85.9	100.0	38.6	75.2	95.3	39.9	73.0
26, 27, 3370	ELECTRICAL, COMMUNICATIONS & UTIL.	99.9	44.0	70.3	101.7	44.8	71.7	104.8	51.8	76.8	98.0	51.8	73.6	101.3	75.1	87.5	99.0	44.8	70.4
MF2014	WEIGHTED AVERAGE	97.8	48.4	76.3	98.5	48.8	76.8	101.7	58.4	82.8	100.4	59.5	82.6	100.0	57.7	81.5	99.0	49.1	77.2

TENNESSEE

DIVISION		CHATTANOOGA 373 - 374			COLUMBIA 384			COOKEVILLE 385			JACKSON 383			JOHNSON CITY 376			KNOXVILLE 377 - 379		
		MAT.	INST.	TOTAL	MAT.	INST.	TOTAL	MAT.	INST.	TOTAL	MAT.	INST.	TOTAL	MAT.	INST.	TOTAL	MAT.	INST.	TOTAL
015433	CONTRACTOR EQUIPMENT		103.8	103.8		98.2	98.2		98.2	98.2		104.5	104.5		97.7	97.7		97.7	97.7
0241, 31 - 34	SITE & INFRASTRUCTURE, DEMOLITION	104.5	97.9	99.8	89.1	86.9	87.5	94.5	86.4	88.8	97.7	96.8	97.1	111.1	86.6	93.7	90.7	87.3	88.3
0310	Concrete Forming & Accessories	97.1	56.4	62.0	82.1	62.0	64.8	82.2	35.4	41.8	89.0	44.9	51.0	83.7	38.9	45.1	95.6	61.0	65.7
0320	Concrete Reinforcing	91.2	67.4	79.1	87.4	61.2	74.1	87.4	61.2	74.1	87.4	62.0	74.4	91.7	60.3	75.7	91.2	62.3	76.5
0330	Cast-in-Place Concrete	101.3	62.7	85.5	94.2	51.6	76.7	106.7	43.5	80.7	104.2	49.3	81.6	81.6	60.2	72.8	95.2	65.7	83.1
03	CONCRETE	96.0	62.3	79.4	96.7	59.8	78.6	106.9	45.2	76.6	97.7	51.6	75.1	103.3	52.3	78.3	93.4	64.4	79.1
04	MASONRY	109.7	51.7	73.5	127.7	54.6	82.1	121.9	43.8	73.2	127.9	47.2	77.6	125.7	44.8	75.3	86.6	55.3	67.1
05	METALS	98.7	89.2	95.8	96.4	86.0	93.2	96.5	85.9	93.2	98.8	85.9	94.8	95.9	85.2	92.6	99.3	87.0	95.5
06	WOOD, PLASTICS & COMPOSITES	104.0	57.3	77.8	70.1	64.4	66.9	70.3	33.7	49.8	85.4	45.4	63.0	76.9	36.9	54.5	91.0	60.6	73.9
07	THERMAL & MOISTURE PROTECTION	99.4	59.9	83.2	93.6	59.2	79.5	94.0	45.7	74.2	96.0	54.1	78.8	94.5	52.6	77.3	92.5	61.6	79.8
08	OPENINGS	99.7	57.6	89.9	92.6	57.9	84.6	92.7	42.2	80.9	100.3	51.6	89.0	96.2	45.1	84.3	93.3	58.6	85.2
0920	Plaster & Gypsum Board	82.8	56.5	65.0	87.7	63.7	71.5	87.7	32.1	50.1	89.3	44.1	58.8	102.4	35.5	57.2	109.0	59.8	75.8
0950, 0980	Ceilings & Acoustic Treatment	96.6	56.5	70.2	77.0	63.7	68.3	77.0	32.1	47.5	86.2	44.1	58.5	93.0	35.5	55.2	93.8	59.8	71.5
0960	Flooring	99.1	58.9	87.7	89.3	20.5	69.7	89.4	58.6	80.6	87.1	40.8	73.9	96.7	40.6	80.7	101.3	55.8	88.3
0970, 0990	Wall Finishes & Painting/Coating	102.5	70.7	83.3	94.0	33.4	57.4	94.0	36.5	59.3	95.8	49.6	67.9	99.6	42.1	64.9	99.6	81.2	88.5
09	FINISHES	97.2	58.2	75.7	89.8	52.0	69.0	90.3	38.4	61.7	89.7	43.7	64.4	101.9	38.1	66.7	94.5	62.0	76.6
COVERS	DIVS. 10 - 14, 25, 28, 41, 43, 44, 46	100.0	41.8	88.3	100.0	47.9	89.5	100.0	41.0	88.1	100.0	43.0	88.5	100.0	75.5	95.1	100.0	81.7	96.3
21, 22, 23	FIRE SUPPRESSION, PLUMBING & HVAC	100.2	62.6	85.0	97.4	78.1	89.6	97.4	72.6	87.4	100.1	68.0	87.1	99.9	59.1	83.4	99.9	67.9	87.0
26, 27, 3370	ELECTRICAL, COMMUNICATIONS & UTIL.	102.3	69.2	84.8	94.2	57.4	74.7	95.9	61.6	77.8	101.0	58.5	78.5	92.5	46.3	68.1	98.1	56.0	75.8
MF2014	WEIGHTED AVERAGE	100.0	66.2	85.2	96.9	66.1	83.5	98.1	59.1	81.1	99.9	61.2	83.0	99.8	56.5	80.9	96.6	66.9	83.7

TENNESSEE / TEXAS

DIVISION		MCKENZIE 382			MEMPHIS 375,380 - 381			NASHVILLE 370 - 372			ABILENE 795 - 796			AMARILLO 790 - 791			AUSTIN 786 - 787		
		MAT.	INST.	TOTAL	MAT.	INST.	TOTAL	MAT.	INST.	TOTAL	MAT.	INST.	TOTAL	MAT.	INST.	TOTAL	MAT.	INST.	TOTAL
015433	CONTRACTOR EQUIPMENT		98.2	98.2		102.8	102.8		103.7	103.7		89.9	89.9		89.9	89.9		89.4	89.4
0241, 31 - 34	SITE & INFRASTRUCTURE, DEMOLITION	94.3	86.5	88.7	94.1	93.7	93.8	98.9	97.5	97.9	99.4	88.7	91.8	97.3	88.6	91.1	102.4	88.4	92.4
0310	Concrete Forming & Accessories	90.0	38.1	45.2	94.3	64.9	69.0	98.9	65.8	70.3	99.0	65.5	70.1	97.9	57.1	62.7	98.6	59.1	64.5
0320	Concrete Reinforcing	87.5	61.9	74.5	99.7	68.7	83.9	96.0	67.7	81.6	92.9	52.5	72.3	99.0	51.3	74.7	96.5	48.5	72.1
0330	Cast-in-Place Concrete	104.4	57.3	85.1	94.1	61.3	80.6	90.3	65.9	80.3	96.6	63.7	83.1	94.1	63.7	81.6	93.4	66.3	82.3
03	CONCRETE	105.5	51.3	78.9	94.1	65.8	80.2	92.4	67.6	80.2	95.0	63.0	79.3	96.7	59.0	78.2	95.7	60.2	78.3
04	MASONRY	126.2	46.2	76.3	99.8	62.7	76.7	93.8	59.0	72.1	104.2	63.6	78.9	109.7	63.6	81.0	107.6	56.5	75.7
05	METALS	96.5	86.1	93.3	100.1	90.6	97.2	100.4	89.9	97.2	105.3	69.8	94.4	101.6	69.2	91.7	100.5	66.3	90.0
06	WOOD, PLASTICS & COMPOSITES	79.1	36.3	55.1	95.4	67.0	79.5	101.6	66.6	82.0	101.1	68.5	82.8	100.8	57.1	76.4	94.4	59.9	75.0
07	THERMAL & MOISTURE PROTECTION	94.0	48.7	75.4	95.5	63.4	82.4	95.7	61.5	81.7	100.0	67.7	86.7	97.9	65.3	84.5	104.2	64.8	88.1
08	OPENINGS	92.7	44.1	81.4	100.1	66.0	92.2	98.5	65.9	90.9	97.1	63.3	89.2	101.1	56.8	90.8	104.3	55.6	92.9
0920	Plaster & Gypsum Board	90.5	34.8	52.8	95.6	66.3	75.8	95.8	66.0	75.6	84.0	68.0	73.2	92.1	56.3	67.9	86.7	59.2	68.1
0950, 0980	Ceilings & Acoustic Treatment	77.0	34.8	49.2	93.9	66.3	75.8	95.5	66.0	76.1	88.9	68.0	75.2	98.0	56.3	70.6	87.9	59.2	69.0
0960	Flooring	92.1	39.3	77.0	99.4	36.1	81.3	97.7	64.2	88.2	110.9	76.5	101.1	106.5	76.5	97.9	105.7	64.1	93.8
0970, 0990	Wall Finishes & Painting/Coating	94.0	49.6	67.2	97.0	55.7	72.1	100.5	72.0	83.3	106.6	56.1	76.1	100.8	56.1	73.8	104.3	48.8	70.8
09	FINISHES	91.3	38.2	62.1	97.3	58.5	75.9	100.2	66.0	81.4	93.7	67.1	79.0	96.8	60.3	76.7	95.6	58.7	75.2
COVERS	DIVS. 10 - 14, 25, 28, 41, 43, 44, 46	100.0	27.4	85.3	100.0	82.4	96.4	100.0	82.9	96.5	100.0	82.2	96.4	100.0	68.6	93.7	100.0	80.5	96.1
21, 22, 23	FIRE SUPPRESSION, PLUMBING & HVAC	97.4	68.3	85.6	100.0	73.9	89.5	100.0	84.0	93.5	100.3	49.0	79.6	100.0	54.9	81.8	100.1	60.1	84.0
26, 27, 3370	ELECTRICAL, COMMUNICATIONS & UTIL.	95.6	62.9	78.3	101.7	65.3	82.4	97.6	63.1	79.4	99.3	48.7	72.5	101.1	64.1	81.6	96.8	63.9	79.4
MF2014	WEIGHTED AVERAGE	98.2	59.2	81.2	99.0	71.3	86.9	98.4	74.1	87.8	99.7	62.2	83.3	100.2	63.2	84.1	100.0	63.6	84.1

687

TEXAS

DIVISION		BEAUMONT 776 - 777			BROWNWOOD 768			BRYAN 778			CHILDRESS 792			CORPUS CHRISTI 783 - 784			DALLAS 752 - 753		
		MAT.	INST.	TOTAL	MAT.	INST.	TOTAL	MAT.	INST.	TOTAL	MAT.	INST.	TOTAL	MAT.	INST.	TOTAL	MAT.	INST.	TOTAL
015433	CONTRACTOR EQUIPMENT		91.5	91.5		89.9	89.9		91.5	91.5		89.9	89.9		96.6	96.6		99.1	99.1
0241, 31 - 34	SITE & INFRASTRUCTURE, DEMOLITION	90.0	89.9	89.9	105.2	89.6	94.1	81.8	90.5	88.0	109.7	88.0	94.3	140.1	84.4	100.5	104.3	89.8	94.0
0310	Concrete Forming & Accessories	101.8	61.9	67.4	99.0	60.5	65.8	81.0	67.1	69.0	97.3	65.5	69.8	100.2	58.1	63.9	101.2	65.8	70.7
0320	Concrete Reinforcing	95.9	65.2	80.3	92.9	52.1	72.1	98.3	48.6	73.0	93.0	52.1	72.2	88.0	48.5	67.9	100.2	52.5	75.9
0330	Cast-in-Place Concrete	89.5	64.7	79.3	101.4	63.6	85.9	71.7	65.4	69.1	99.1	63.7	84.5	109.8	64.4	91.2	98.0	64.6	84.3
03	CONCRETE	96.3	64.2	80.5	101.2	60.7	81.3	80.9	63.7	72.5	104.0	62.9	83.8	100.1	59.9	80.4	99.0	64.1	81.9
04	MASONRY	106.8	63.9	80.0	141.8	56.6	88.7	146.3	60.4	92.7	108.5	56.6	76.2	91.0	56.6	69.6	102.4	56.7	73.9
05	METALS	95.2	75.5	89.1	99.9	69.1	90.5	94.8	70.1	87.2	102.7	69.2	92.4	94.7	77.9	89.5	100.4	80.0	94.1
06	WOOD, PLASTICS & COMPOSITES	106.1	61.8	81.3	99.5	61.9	78.4	73.5	68.5	70.7	100.4	68.5	82.6	116.0	58.4	83.7	104.1	68.6	84.2
07	THERMAL & MOISTURE PROTECTION	101.6	66.7	87.5	98.2	65.2	84.7	93.9	66.3	82.5	100.5	64.8	85.9	104.5	66.2	88.8	93.0	66.4	82.1
08	OPENINGS	94.0	62.3	86.7	94.7	59.7	86.6	97.2	61.6	88.9	94.2	63.3	87.1	109.1	54.7	96.5	99.5	63.5	91.1
0920	Plaster & Gypsum Board	102.3	61.2	74.5	87.8	61.2	69.8	90.3	68.0	75.2	83.6	68.0	73.1	98.4	57.5	70.7	96.7	68.0	77.3
0950, 0980	Ceilings & Acoustic Treatment	98.2	61.2	73.9	80.6	61.2	67.8	91.2	68.0	76.0	87.3	68.0	74.6	92.7	57.5	69.6	97.8	68.0	78.2
0960	Flooring	112.4	73.0	101.1	95.3	64.1	86.3	85.8	64.1	79.6	109.1	64.1	96.2	118.7	64.1	103.1	102.8	64.1	91.7
0970, 0990	Wall Finishes & Painting/Coating	95.1	59.4	73.6	104.2	56.1	75.1	92.5	62.2	74.2	106.6	56.1	76.1	120.1	56.5	81.7	107.0	56.1	76.3
09	FINISHES	93.7	63.4	77.0	87.7	60.7	72.8	82.3	66.2	73.4	94.0	64.6	77.8	105.9	58.7	79.9	100.0	64.7	80.5
COVERS	DIVS. 10 - 14, 25, 28, 41, 43, 44, 46	100.0	83.4	96.7	100.0	81.4	96.3	100.0	82.7	96.5	100.0	82.2	96.4	100.0	82.6	96.5	100.0	82.5	96.5
21, 22, 23	FIRE SUPPRESSION, PLUMBING & HVAC	100.2	64.1	85.6	95.4	50.0	77.1	95.4	66.5	83.8	95.5	54.9	79.1	100.2	58.9	83.5	100.0	62.4	84.9
26, 27, 3370	ELECTRICAL, COMMUNICATIONS & UTIL.	95.2	70.6	82.2	93.4	42.8	66.7	93.4	67.4	79.7	99.2	64.1	80.7	92.8	60.0	75.5	95.2	64.6	79.0
MF2014	WEIGHTED AVERAGE	97.4	68.6	84.9	98.9	59.4	81.7	94.9	68.2	83.3	99.3	64.3	84.0	100.9	63.5	84.6	99.5	67.4	85.5

TEXAS

DIVISION		DEL RIO 788			DENTON 762			EASTLAND 764			EL PASO 798 - 799,885			FORT WORTH 760 - 761			GALVESTON 775		
		MAT.	INST.	TOTAL	MAT.	INST.	TOTAL	MAT.	INST.	TOTAL	MAT.	INST.	TOTAL	MAT.	INST.	TOTAL	MAT.	INST.	TOTAL
015433	CONTRACTOR EQUIPMENT		89.4	89.4		95.8	95.8		89.9	89.9		89.9	89.9		89.9	89.9		100.5	100.5
0241, 31 - 34	SITE & INFRASTRUCTURE, DEMOLITION	121.8	88.4	98.0	105.5	81.3	88.3	108.0	87.6	93.5	101.7	87.6	91.7	102.1	88.7	92.6	107.2	88.4	93.8
0310	Concrete Forming & Accessories	96.5	57.8	63.1	106.2	65.3	70.9	99.8	65.3	70.0	97.8	63.4	68.1	98.0	65.6	70.1	89.8	64.6	68.1
0320	Concrete Reinforcing	88.7	48.4	68.2	94.3	52.1	72.8	93.1	50.9	71.6	99.2	52.0	75.2	98.4	52.4	75.0	97.8	62.4	79.8
0330	Cast-in-Place Concrete	118.2	63.2	95.6	78.5	64.6	72.8	107.2	63.6	89.3	88.4	63.7	78.3	95.0	63.7	82.1	95.2	66.4	83.4
03	CONCRETE	121.3	58.6	90.5	80.0	63.9	72.1	105.9	62.6	84.6	94.0	61.9	78.3	97.0	63.1	80.4	97.5	66.4	82.2
04	MASONRY	106.9	56.5	75.5	152.1	56.7	92.6	106.8	56.6	75.5	99.0	57.7	73.3	104.7	56.6	74.7	102.5	60.4	76.3
05	METALS	94.4	66.2	85.7	99.5	80.6	93.7	99.7	68.5	90.1	98.5	67.4	89.0	101.5	69.5	91.7	96.3	88.3	93.9
06	WOOD, PLASTICS & COMPOSITES	96.5	58.3	75.1	111.3	68.6	87.4	106.1	68.5	85.1	91.3	65.7	76.9	97.3	68.5	81.2	89.4	64.7	75.6
07	THERMAL & MOISTURE PROTECTION	101.3	65.3	86.5	96.0	66.7	84.0	98.6	65.9	85.1	99.8	64.4	85.3	96.0	65.9	83.6	93.0	67.1	82.4
08	OPENINGS	101.8	54.7	90.8	112.2	63.4	100.8	70.4	63.0	68.7	93.3	58.1	85.1	99.4	63.4	91.0	101.6	63.2	92.7
0920	Plaster & Gypsum Board	94.7	57.5	69.6	92.1	68.0	75.8	87.8	68.0	74.4	95.9	65.1	75.1	91.8	68.0	75.7	96.8	64.0	74.7
0950, 0980	Ceilings & Acoustic Treatment	89.1	57.5	68.3	84.7	68.0	73.8	80.6	68.0	72.3	91.4	65.1	74.1	90.5	68.0	75.7	94.6	64.0	74.5
0960	Flooring	100.7	64.1	90.2	90.4	64.1	82.9	121.4	64.1	105.0	110.4	64.1	97.1	117.9	64.1	102.5	100.1	64.1	89.8
0970, 0990	Wall Finishes & Painting/Coating	106.4	48.8	71.6	115.6	56.1	79.7	105.8	56.1	75.8	106.2	53.7	74.5	103.2	56.1	74.7	103.4	60.0	77.2
09	FINISHES	98.2	57.7	75.9	85.8	64.7	74.2	95.4	64.6	78.6	97.3	63.0	78.4	98.4	64.6	79.8	91.4	63.7	76.2
COVERS	DIVS. 10 - 14, 25, 28, 41, 43, 44, 46	100.0	80.5	96.1	100.0	82.5	96.5	100.0	82.2	96.4	100.0	81.0	96.2	100.0	82.3	96.4	100.0	84.5	96.9
21, 22, 23	FIRE SUPPRESSION, PLUMBING & HVAC	95.4	57.3	80.0	95.4	42.6	74.1	95.4	49.9	77.1	100.0	50.3	80.0	100.0	58.4	83.2	95.4	66.7	83.8
26, 27, 3370	ELECTRICAL, COMMUNICATIONS & UTIL.	94.9	68.8	81.1	95.9	64.6	79.3	93.3	64.5	78.1	99.1	57.4	77.0	94.7	64.5	78.8	95.0	67.9	80.7
MF2014	WEIGHTED AVERAGE	100.8	63.2	84.4	98.9	62.5	83.0	95.9	63.2	81.6	98.0	61.7	82.2	99.3	65.3	84.5	97.0	70.0	85.2

TEXAS

DIVISION		GIDDINGS 789			GREENVILLE 754			HOUSTON 770 - 772			HUNTSVILLE 773			LAREDO 780			LONGVIEW 756		
		MAT.	INST.	TOTAL	MAT.	INST.	TOTAL	MAT.	INST.	TOTAL	MAT.	INST.	TOTAL	MAT.	INST.	TOTAL	MAT.	INST.	TOTAL
015433	CONTRACTOR EQUIPMENT		89.4	89.4		96.7	96.7		100.4	100.4		91.5	91.5		89.4	89.4		91.4	91.4
0241, 31 - 34	SITE & INFRASTRUCTURE, DEMOLITION	107.8	88.4	94.0	97.9	85.4	89.0	106.0	88.2	93.4	96.6	89.9	91.8	102.1	88.4	92.3	95.8	92.3	93.3
0310	Concrete Forming & Accessories	94.1	57.9	62.9	92.3	63.5	67.5	91.8	64.6	68.4	87.8	61.6	65.2	96.6	58.0	63.3	88.4	64.9	68.1
0320	Concrete Reinforcing	89.2	45.9	67.2	100.7	52.1	76.0	97.5	61.4	79.1	98.5	50.6	74.1	88.7	48.8	68.4	99.6	52.0	75.4
0330	Cast-in-Place Concrete	100.2	63.2	85.0	89.3	63.9	78.9	92.3	66.4	81.7	98.6	64.6	84.6	84.6	63.3	75.8	103.8	62.8	86.9
03	CONCRETE	97.6	58.2	78.2	91.6	63.6	77.8	95.2	66.2	80.9	103.9	61.3	83.0	92.5	58.8	75.9	107.5	62.3	85.3
04	MASONRY	115.7	56.5	78.8	160.0	56.6	95.6	102.3	66.5	80.0	145.1	58.9	91.4	100.2	63.5	77.3	155.7	56.6	93.9
05	METALS	93.9	65.3	85.1	97.7	79.3	92.1	99.2	88.2	95.8	94.7	69.2	86.8	97.0	66.9	87.8	91.1	67.8	83.9
06	WOOD, PLASTICS & COMPOSITES	95.6	58.3	74.7	93.5	68.6	79.5	91.7	64.7	76.6	82.0	61.8	70.7	96.5	58.3	75.1	87.2	68.4	76.7
07	THERMAL & MOISTURE PROTECTION	102.0	64.5	86.6	93.0	66.0	81.9	92.6	67.8	82.4	94.9	65.9	83.0	100.2	66.5	86.4	94.5	65.2	82.5
08	OPENINGS	100.8	54.0	89.9	97.7	63.4	89.7	104.3	62.8	94.6	97.2	58.5	88.2	101.6	54.7	90.7	87.8	63.3	82.1
0920	Plaster & Gypsum Board	93.7	57.5	69.2	90.3	68.0	75.3	99.1	64.0	75.4	94.2	61.2	71.9	95.8	57.5	69.9	89.1	68.0	74.9
0950, 0980	Ceilings & Acoustic Treatment	89.1	57.5	68.3	93.6	68.0	76.8	99.6	64.0	76.2	91.2	61.2	71.5	93.3	57.5	69.8	90.3	68.0	75.7
0960	Flooring	101.1	64.1	90.5	98.6	64.1	88.7	101.2	64.1	90.6	89.4	64.1	82.1	100.5	64.1	90.1	103.5	64.1	92.2
0970, 0990	Wall Finishes & Painting/Coating	106.4	48.8	71.6	107.0	56.1	76.3	103.4	62.2	78.5	92.5	59.4	72.5	106.4	53.2	74.3	96.8	56.1	72.2
09	FINISHES	97.1	57.7	75.4	96.6	64.6	79.0	98.9	64.0	79.7	84.7	61.6	72.0	97.7	58.2	75.9	101.7	64.5	81.2
COVERS	DIVS. 10 - 14, 25, 28, 41, 43, 44, 46	100.0	80.8	96.1	100.0	82.3	96.4	100.0	84.5	96.9	100.0	81.0	96.2	100.0	80.5	96.1	100.0	82.0	96.4
21, 22, 23	FIRE SUPPRESSION, PLUMBING & HVAC	95.4	64.6	83.0	95.3	49.0	76.6	100.1	66.7	86.6	95.4	65.9	83.5	100.2	64.1	85.6	95.3	33.6	70.4
26, 27, 3370	ELECTRICAL, COMMUNICATIONS & UTIL.	91.8	63.9	77.1	92.0	64.6	77.5	96.9	67.5	81.4	93.4	68.0	80.0	95.1	60.9	77.0	92.2	55.6	72.9
MF2014	WEIGHTED AVERAGE	97.7	64.0	83.0	98.8	64.0	83.6	99.4	70.5	86.8	98.0	66.7	84.4	98.3	64.5	83.6	98.5	58.7	81.2

City Cost Indexes

SOUTH DAKOTA

DIVISION		MITCHELL 573			MOBRIDGE 576			PIERRE 575			RAPID CITY 577			SIOUX FALLS 570 - 571			WATERTOWN 572		
		MAT.	INST.	TOTAL	MAT.	INST.	TOTAL	MAT.	INST.	TOTAL	MAT.	INST.	TOTAL	MAT.	INST.	TOTAL	MAT.	INST.	TOTAL
015433	CONTRACTOR EQUIPMENT		98.2	98.2		98.2	98.2		98.2	98.2		98.2	98.2		99.2	99.2		98.2	98.2
0241, 31 - 34	SITE & INFRASTRUCTURE, DEMOLITION	96.0	93.7	94.3	95.9	93.7	94.3	100.5	93.7	95.7	97.7	93.8	94.9	94.3	95.4	95.0	95.8	93.7	94.3
0310	Concrete Forming & Accessories	94.0	37.0	44.8	84.9	36.8	43.4	97.7	38.4	46.6	102.2	36.7	45.7	98.8	40.5	48.5	81.6	36.6	42.8
0320	Concrete Reinforcing	95.5	46.3	70.5	98.0	38.1	67.5	99.0	71.1	84.8	89.9	71.3	80.4	98.0	71.2	84.3	92.9	38.2	65.1
0330	Cast-in-Place Concrete	102.6	42.3	77.8	102.6	43.5	78.3	98.5	42.2	75.4	101.8	42.8	77.6	92.3	42.9	72.0	102.6	45.9	79.3
03	CONCRETE	100.5	41.8	71.7	100.3	40.7	71.0	99.0	47.3	73.6	99.8	46.7	73.7	95.9	48.4	72.6	99.2	41.5	70.9
04	MASONRY	103.9	53.3	72.4	113.4	55.4	77.2	117.4	53.5	77.6	113.5	56.4	77.9	105.1	53.6	73.0	141.1	57.4	88.9
05	METALS	95.0	63.8	85.4	95.0	63.6	85.3	98.4	79.1	92.4	97.8	79.7	92.2	98.2	79.3	92.4	95.0	64.1	85.5
06	WOOD, PLASTICS & COMPOSITES	98.5	36.5	63.8	87.6	36.2	58.8	104.5	36.8	66.6	103.5	33.3	64.2	96.8	39.3	64.6	83.8	36.0	57.0
07	THERMAL & MOISTURE PROTECTION	98.9	45.6	77.1	99.0	47.8	78.0	101.8	45.9	78.9	99.6	47.8	78.3	101.9	48.2	79.9	98.8	48.3	78.1
08	OPENINGS	97.7	37.3	83.6	100.4	36.3	85.5	105.7	46.7	92.0	103.1	44.8	89.6	106.5	48.1	92.9	97.7	36.5	83.4
0920	Plaster & Gypsum Board	102.1	34.8	56.7	96.2	34.6	54.5	100.1	35.1	56.2	103.2	31.6	54.8	90.7	37.7	54.9	93.9	34.3	53.6
0950, 0980	Ceilings & Acoustic Treatment	89.9	34.8	53.7	93.2	34.6	54.7	93.3	35.1	55.0	95.7	31.6	53.6	90.6	37.7	55.8	89.9	34.3	53.3
0960	Flooring	107.9	50.3	91.4	103.5	50.3	88.3	107.4	35.8	86.9	107.6	78.6	99.3	103.8	74.9	95.5	102.1	50.3	87.3
0970, 0990	Wall Finishes & Painting/Coating	102.6	40.6	65.2	102.6	41.7	65.8	105.7	44.6	68.8	102.6	44.6	67.6	101.7	44.6	67.2	102.6	37.1	63.1
09	FINISHES	99.1	39.5	66.2	97.7	39.4	65.6	100.8	37.3	65.8	100.5	43.9	69.3	97.1	46.5	69.2	96.2	38.8	64.6
COVERS	DIVS. 10 - 14, 25, 28, 41, 43, 44, 46	100.0	37.6	87.4	100.0	40.5	88.0	100.0	75.8	95.1	100.0	75.7	95.1	100.0	76.2	95.2	100.0	40.5	88.0
21, 22, 23	FIRE SUPPRESSION, PLUMBING & HVAC	95.3	39.4	72.8	95.3	39.9	73.0	100.0	64.5	85.7	100.1	65.0	85.9	100.0	38.6	75.2	95.3	39.9	73.0
26, 27, 3370	ELECTRICAL, COMMUNICATIONS & UTIL.	99.9	44.0	70.3	101.7	44.8	71.7	104.8	51.8	76.8	98.0	51.8	73.6	101.3	75.1	87.5	99.0	44.8	70.4
MF2014	WEIGHTED AVERAGE	97.8	48.4	76.3	98.5	48.8	76.8	101.7	58.4	82.8	99.5	59.5	82.6	100.0	57.7	81.5	99.0	49.1	77.2

TENNESSEE

DIVISION		CHATTANOOGA 373 - 374			COLUMBIA 384			COOKEVILLE 385			JACKSON 383			JOHNSON CITY 376			KNOXVILLE 377 - 379		
		MAT.	INST.	TOTAL	MAT.	INST.	TOTAL	MAT.	INST.	TOTAL	MAT.	INST.	TOTAL	MAT.	INST.	TOTAL	MAT.	INST.	TOTAL
015433	CONTRACTOR EQUIPMENT		103.8	103.8		98.2	98.2		98.2	98.2		104.5	104.5		97.7	97.7		97.7	97.7
0241, 31 - 34	SITE & INFRASTRUCTURE, DEMOLITION	104.5	97.9	99.8	89.1	86.9	87.5	94.5	86.4	88.8	97.7	96.8	97.1	111.1	86.6	93.7	90.7	87.3	88.3
0310	Concrete Forming & Accessories	97.1	56.4	62.0	82.1	62.0	64.8	82.2	35.4	41.8	89.0	44.9	51.0	83.7	38.9	45.1	95.6	61.0	65.7
0320	Concrete Reinforcing	91.2	67.4	79.1	87.4	61.2	74.1	87.4	61.2	74.1	87.4	62.0	74.4	91.7	60.3	75.7	91.2	62.3	76.5
0330	Cast-in-Place Concrete	101.3	62.7	85.5	94.2	51.6	76.7	106.7	43.5	80.7	104.2	49.3	81.6	103.3	52.3	78.3	93.4	64.4	83.1
03	CONCRETE	96.0	62.3	79.4	96.7	59.8	78.6	106.9	45.2	76.6	97.7	51.6	75.1	103.3	52.3	78.3	93.4	64.4	83.1
04	MASONRY	109.7	51.7	73.5	127.7	54.6	82.1	121.9	43.8	73.2	127.9	47.2	77.6	125.7	44.8	75.3	86.6	55.3	67.1
05	METALS	98.7	89.2	95.8	96.4	86.0	93.2	96.5	85.9	93.2	98.8	85.9	94.8	95.9	85.2	92.6	99.3	87.0	95.5
06	WOOD, PLASTICS & COMPOSITES	104.0	57.3	77.8	70.1	64.4	66.9	70.3	33.7	49.8	85.4	45.4	63.0	76.9	36.9	54.5	91.0	60.6	73.9
07	THERMAL & MOISTURE PROTECTION	99.4	59.9	83.2	93.6	59.2	79.5	94.0	45.7	74.2	96.0	54.1	78.8	94.5	52.6	77.3	92.5	61.6	79.8
08	OPENINGS	99.7	57.6	89.9	92.6	57.9	84.6	92.7	42.2	80.9	100.3	51.6	89.0	96.2	45.1	84.3	93.3	58.6	85.2
0920	Plaster & Gypsum Board	82.8	56.5	65.0	87.7	63.7	71.5	87.7	32.1	50.1	89.3	44.1	58.8	102.4	35.5	57.2	109.0	59.8	75.8
0950, 0980	Ceilings & Acoustic Treatment	96.6	56.5	70.2	77.0	63.7	68.3	77.0	32.1	47.5	86.2	44.1	58.5	93.0	35.5	55.2	93.8	59.8	71.5
0960	Flooring	99.1	58.9	87.7	89.3	20.5	69.7	89.4	58.6	80.6	87.1	40.8	73.9	96.7	40.6	80.7	101.3	55.8	88.3
0970, 0990	Wall Finishes & Painting/Coating	102.5	70.7	83.3	94.0	33.4	57.4	94.0	36.5	59.3	95.8	49.6	67.9	99.6	42.1	64.9	99.6	81.2	88.5
09	FINISHES	97.2	58.2	75.7	89.8	52.0	69.0	90.3	38.4	61.7	89.7	43.7	64.4	101.9	38.1	66.7	94.5	62.0	76.6
COVERS	DIVS. 10 - 14, 25, 28, 41, 43, 44, 46	100.0	41.8	88.3	100.0	47.9	89.5	100.0	41.0	88.1	100.0	43.0	88.5	100.0	75.5	95.1	100.0	81.7	96.3
21, 22, 23	FIRE SUPPRESSION, PLUMBING & HVAC	100.2	62.6	85.0	97.4	78.1	89.6	97.4	72.6	87.4	100.1	68.0	87.1	99.9	59.1	83.4	99.9	67.9	87.0
26, 27, 3370	ELECTRICAL, COMMUNICATIONS & UTIL.	102.3	69.2	84.8	94.2	57.4	74.7	95.9	61.6	77.8	101.0	58.5	78.5	92.5	46.3	68.1	98.1	56.0	75.8
MF2014	WEIGHTED AVERAGE	100.0	66.2	85.2	96.9	66.1	83.5	98.1	59.1	81.1	99.7	61.2	83.0	99.8	56.5	80.9	96.6	66.9	83.7

TENNESSEE / TEXAS

DIVISION		MCKENZIE 382			MEMPHIS 375,380 - 381			NASHVILLE 370 - 372			ABILENE 795 - 796			AMARILLO 790 - 791			AUSTIN 786 - 787		
		MAT.	INST.	TOTAL	MAT.	INST.	TOTAL	MAT.	INST.	TOTAL	MAT.	INST.	TOTAL	MAT.	INST.	TOTAL	MAT.	INST.	TOTAL
015433	CONTRACTOR EQUIPMENT		98.2	98.2		102.8	102.8		103.7	103.7		89.9	89.9		89.9	89.9		89.4	89.4
0241, 31 - 34	SITE & INFRASTRUCTURE, DEMOLITION	94.3	86.5	88.7	94.1	93.7	93.8	98.9	97.5	97.9	99.4	88.7	91.8	97.3	88.6	91.1	102.4	88.4	92.4
0310	Concrete Forming & Accessories	90.0	38.1	45.2	94.3	64.9	69.0	98.9	65.8	70.3	99.0	65.5	70.1	97.9	57.1	62.7	98.6	59.1	64.5
0320	Concrete Reinforcing	87.5	61.9	74.5	99.7	68.7	83.9	96.0	67.7	81.6	92.9	52.5	72.3	99.0	51.3	74.7	96.5	48.5	72.1
0330	Cast-in-Place Concrete	104.4	57.3	85.1	94.1	61.3	80.6	90.3	65.9	80.3	96.6	63.7	83.1	94.1	63.7	81.6	93.4	66.3	82.3
03	CONCRETE	105.5	51.3	78.9	94.1	65.8	80.2	92.4	67.6	80.2	95.0	63.0	79.3	96.7	59.0	78.2	95.7	60.2	78.3
04	MASONRY	126.2	46.2	76.3	99.8	62.7	76.7	93.8	59.0	72.1	104.2	63.6	78.9	109.7	63.6	81.0	107.6	56.5	75.7
05	METALS	96.5	86.1	93.3	100.1	90.6	97.2	100.4	89.9	97.2	105.3	69.8	94.4	101.6	69.1	91.7	100.5	66.3	90.0
06	WOOD, PLASTICS & COMPOSITES	79.1	36.3	55.1	95.4	67.0	79.5	101.6	66.6	82.0	101.1	68.5	82.8	100.8	57.1	76.4	94.4	59.9	75.0
07	THERMAL & MOISTURE PROTECTION	94.0	48.7	75.4	95.5	63.4	82.4	95.7	61.5	81.7	100.0	67.7	86.7	97.9	65.3	84.5	104.2	64.8	88.1
08	OPENINGS	92.7	44.1	81.4	100.1	66.0	92.2	98.5	65.9	90.9	97.1	63.3	89.2	101.1	56.8	90.8	104.3	55.6	92.9
0920	Plaster & Gypsum Board	90.5	34.8	52.8	95.6	66.3	75.8	95.5	66.0	75.6	84.0	68.0	73.2	92.1	56.3	67.9	86.7	59.2	68.1
0950, 0980	Ceilings & Acoustic Treatment	77.0	34.8	49.2	93.9	66.3	75.8	95.5	66.0	76.1	88.9	68.0	75.2	98.0	56.3	70.6	87.9	59.2	69.0
0960	Flooring	92.1	39.3	77.0	99.4	36.1	81.3	97.7	64.2	88.2	110.9	76.5	101.1	106.5	76.5	97.9	105.7	64.1	93.8
0970, 0990	Wall Finishes & Painting/Coating	94.0	49.6	67.2	97.0	55.7	72.1	100.5	72.0	83.3	106.6	56.1	76.1	100.8	56.1	73.8	104.3	48.8	70.8
09	FINISHES	91.3	38.2	62.1	97.3	58.5	75.9	100.2	66.0	81.4	93.7	67.1	79.0	96.8	60.3	76.7	95.6	58.7	75.2
COVERS	DIVS. 10 - 14, 25, 28, 41, 43, 44, 46	100.0	27.4	85.3	100.0	82.4	96.4	100.0	82.9	96.5	100.0	82.2	96.4	100.0	68.6	93.7	100.0	80.5	96.1
21, 22, 23	FIRE SUPPRESSION, PLUMBING & HVAC	97.4	68.3	85.6	100.0	73.9	89.5	100.0	84.0	93.5	100.3	49.0	79.6	100.0	54.9	81.8	100.1	60.1	84.0
26, 27, 3370	ELECTRICAL, COMMUNICATIONS & UTIL.	95.6	62.9	78.3	101.7	65.3	82.4	97.6	63.1	79.4	99.3	48.7	72.5	101.1	64.1	81.6	96.8	63.9	79.4
MF2014	WEIGHTED AVERAGE	98.2	59.2	81.2	99.0	71.3	86.9	98.4	74.1	87.8	99.7	62.2	83.3	100.2	63.2	84.1	100.0	63.6	84.1

For customer support on your Mechanical Cost Data, call 877.795.7394.

TEXAS

DIVISION		LUBBOCK 793-794			LUFKIN 759			MCALLEN 785			MCKINNEY 750			MIDLAND 797			ODESSA 797		
		MAT.	INST.	TOTAL	MAT.	INST.	TOTAL	MAT.	INST.	TOTAL	MAT.	INST.	TOTAL	MAT.	INST.	TOTAL	MAT.	INST.	TOTAL
015433	CONTRACTOR EQUIPMENT		98.5	98.5		91.4	91.4		96.8	96.8		96.7	96.7		98.5	98.5		89.9	89.9
0241, 31 - 34	SITE & INFRASTRUCTURE, DEMOLITION	123.8	86.1	97.0	91.2	94.0	93.1	144.1	84.4	101.7	94.6	85.3	88.0	126.7	86.1	97.8	99.7	88.0	91.4
0310	Concrete Forming & Accessories	97.9	57.2	62.8	91.7	61.7	65.9	101.2	57.5	63.5	91.3	65.3	68.8	101.9	65.4	70.5	98.9	65.4	70.0
0320	Concrete Reinforcing	94.1	52.5	72.9	101.3	64.9	82.8	88.2	48.4	67.9	100.7	50.9	75.3	95.1	52.1	73.2	92.9	52.1	72.1
0330	Cast-in-Place Concrete	96.8	64.7	83.6	92.9	64.0	81.0	119.3	64.1	96.6	83.8	63.8	75.6	102.9	64.6	87.2	96.6	63.6	83.1
03	CONCRETE	93.7	60.4	77.4	99.8	63.6	82.0	107.9	59.5	84.1	87.0	63.3	75.3	98.2	64.0	81.4	95.0	62.8	79.2
04	MASONRY	103.6	64.0	78.9	119.4	58.8	81.6	106.8	56.4	75.4	172.2	56.6	100.2	121.5	56.7	81.1	104.2	56.6	74.6
05	METALS	109.0	81.3	100.5	97.9	72.4	90.0	94.3	77.5	89.2	97.6	78.8	91.8	107.2	80.7	99.0	104.6	69.1	93.7
06	WOOD, PLASTICS & COMPOSITES	101.1	57.3	76.5	95.3	61.8	76.5	114.6	58.4	83.1	92.3	68.6	79.0	106.0	68.6	85.1	101.1	68.5	82.8
07	THERMAL & MOISTURE PROTECTION	89.3	66.2	79.8	94.3	65.6	82.5	104.8	59.8	86.3	92.8	66.0	81.8	89.5	65.4	79.7	100.0	65.9	86.0
08	OPENINGS	108.5	57.2	96.6	66.7	63.0	65.9	105.7	54.7	93.9	97.7	63.0	89.6	107.5	63.4	97.2	97.1	63.3	89.2
0920	Plaster & Gypsum Board	84.4	56.3	65.4	88.0	61.2	69.9	99.4	57.5	71.1	90.0	68.0	75.1	85.9	68.0	73.8	84.0	68.0	73.2
0950, 0980	Ceilings & Acoustic Treatment	90.6	56.3	68.1	84.4	61.2	69.1	93.3	57.5	69.8	93.6	68.0	76.8	88.1	68.0	74.9	88.9	68.0	75.2
0960	Flooring	104.4	64.1	92.8	138.2	64.1	117.0	118.1	63.8	102.6	98.2	64.1	88.4	105.7	64.1	93.8	110.9	64.1	97.5
0970, 0990	Wall Finishes & Painting/Coating	119.0	56.1	81.0	96.8	56.1	72.2	120.1	48.8	77.1	107.0	56.1	76.3	119.0	56.1	81.0	106.6	56.1	76.1
09	FINISHES	96.4	58.0	75.2	110.2	61.2	83.2	106.4	57.8	79.6	96.2	64.6	78.8	96.8	64.7	79.1	93.7	64.6	77.7
COVERS	DIVS. 10 - 14, 25, 28, 41, 43, 44, 46	100.0	81.3	96.2	100.0	80.9	96.1	100.0	80.7	96.1	100.0	82.3	96.4	100.0	81.2	96.2	100.0	80.9	96.1
21, 22, 23	FIRE SUPPRESSION, PLUMBING & HVAC	99.8	50.7	80.0	95.3	64.4	82.8	95.4	58.7	80.6	95.3	45.4	75.2	95.0	49.1	76.5	100.3	49.1	79.6
26, 27, 3370	ELECTRICAL, COMMUNICATIONS & UTIL.	97.9	64.2	80.1	93.5	70.6	81.4	92.6	36.3	62.8	92.1	64.6	77.5	97.9	64.2	80.1	99.4	64.1	80.7
MF2014	WEIGHTED AVERAGE	101.6	63.6	85.0	95.6	67.8	83.5	101.0	59.8	83.0	98.7	63.1	83.2	101.5	64.2	85.2	99.6	63.1	83.7

TEXAS

DIVISION		PALESTINE 758			SAN ANGELO 769			SAN ANTONIO 781 - 782			TEMPLE 765			TEXARKANA 755			TYLER 757		
		MAT.	INST.	TOTAL	MAT.	INST.	TOTAL	MAT.	INST.	TOTAL	MAT.	INST.	TOTAL	MAT.	INST.	TOTAL	MAT.	INST.	TOTAL
015433	CONTRACTOR EQUIPMENT		91.4	91.4		89.9	89.9		91.9	91.9		89.9	89.9		91.4	91.4		91.4	91.4
0241, 31 - 34	SITE & INFRASTRUCTURE, DEMOLITION	96.6	92.6	93.8	101.5	89.7	93.1	100.9	92.0	94.5	89.8	89.1	89.3	85.9	92.0	90.2	95.1	92.4	93.2
0310	Concrete Forming & Accessories	82.7	65.4	67.8	99.3	58.7	64.3	95.0	58.1	63.2	102.8	57.8	64.0	99.0	57.4	63.1	93.6	65.4	69.3
0320	Concrete Reinforcing	98.8	52.1	75.1	92.7	52.1	72.1	93.6	49.8	71.3	92.9	49.1	70.6	98.7	52.3	75.1	99.6	52.4	75.6
0330	Cast-in-Place Concrete	84.9	63.0	75.9	95.7	64.8	83.0	85.6	65.2	77.2	78.5	63.1	72.2	85.5	62.7	76.2	101.9	63.0	85.9
03	CONCRETE	101.6	62.6	82.4	96.7	60.3	78.8	91.7	59.7	76.0	82.9	58.7	71.0	92.8	59.0	76.2	107.1	62.7	85.3
04	MASONRY	113.8	56.6	78.2	138.0	58.7	88.5	96.8	65.3	77.2	151.0	56.4	92.1	176.5	56.6	101.8	166.0	56.6	97.8
05	METALS	97.6	68.3	88.6	100.1	69.1	90.6	96.9	68.6	88.2	99.8	66.5	89.6	91.0	67.8	83.9	97.5	68.4	88.5
06	WOOD, PLASTICS & COMPOSITES	85.3	68.4	75.8	99.8	58.3	76.5	94.7	57.4	73.8	108.9	58.3	80.6	99.7	58.2	76.5	97.0	68.4	81.0
07	THERMAL & MOISTURE PROTECTION	94.8	65.3	82.7	98.0	66.0	84.9	96.4	68.2	84.8	97.7	64.8	84.2	94.1	64.2	81.9	94.6	65.3	82.6
08	OPENINGS	66.7	63.3	65.9	94.7	57.8	86.1	102.7	54.4	91.5	67.1	57.0	64.7	87.7	57.8	80.8	66.6	63.4	65.9
0920	Plaster & Gypsum Board	85.2	68.0	73.6	87.8	57.5	67.3	98.0	56.5	70.0	87.8	57.5	67.3	93.4	57.5	69.1	88.0	68.0	74.5
0950, 0980	Ceilings & Acoustic Treatment	84.4	68.0	73.6	80.6	57.5	65.4	102.7	56.5	72.3	80.6	57.5	65.4	90.3	57.5	68.7	84.4	68.0	73.6
0960	Flooring	129.9	64.1	111.1	95.3	64.1	86.4	104.1	64.1	92.6	123.2	64.1	106.3	111.1	64.1	97.6	140.1	64.1	118.4
0970, 0990	Wall Finishes & Painting/Coating	96.8	56.1	72.2	104.2	56.1	75.1	106.3	53.2	74.2	105.8	48.8	71.4	96.8	56.1	72.2	111.1	64.5	85.4
09	FINISHES	108.0	64.5	84.0	87.5	59.1	71.8	103.4	58.1	78.4	95.1	57.7	74.5	104.0	58.5	78.9	111.1	64.5	85.4
COVERS	DIVS. 10 - 14, 25, 28, 41, 43, 44, 46	100.0	82.0	96.4	100.0	81.0	96.2	100.0	81.2	96.2	100.0	81.1	96.2	100.0	80.9	96.1	100.0	82.0	96.4
21, 22, 23	FIRE SUPPRESSION, PLUMBING & HVAC	95.3	59.6	80.9	95.4	50.1	77.1	100.0	63.2	85.1	95.4	54.7	79.0	95.3	37.0	71.8	95.3	62.4	82.0
26, 27, 3370	ELECTRICAL, COMMUNICATIONS & UTIL.	89.8	55.6	71.7	97.4	44.8	69.6	96.4	60.9	77.6	94.5	59.8	76.2	93.4	60.1	75.8	92.2	55.5	72.8
MF2014	WEIGHTED AVERAGE	95.0	64.4	81.7	98.5	59.6	81.5	98.5	65.1	84.0	94.7	61.7	80.3	98.0	58.4	80.7	98.5	65.0	83.9

TEXAS / UTAH

DIVISION		VICTORIA 779			WACO 766 - 767			WAXAHACHIE 751			WHARTON 774			WICHITA FALLS 763			LOGAN 843		
		MAT.	INST.	TOTAL	MAT.	INST.	TOTAL	MAT.	INST.	TOTAL	MAT.	INST.	TOTAL	MAT.	INST.	TOTAL	MAT.	INST.	TOTAL
015433	CONTRACTOR EQUIPMENT		99.3	99.3		89.9	89.9		96.7	96.7		100.5	100.5		89.9	89.9		97.5	97.5
0241, 31 - 34	SITE & INFRASTRUCTURE, DEMOLITION	111.8	85.7	93.2	98.6	88.7	91.6	96.0	85.4	88.4	117.1	87.8	96.2	99.3	88.7	91.8	96.0	95.7	95.8
0310	Concrete Forming & Accessories	90.0	59.0	63.3	101.2	65.5	70.4	91.3	65.5	69.0	84.9	61.7	64.9	101.2	65.5	70.4	104.7	58.1	64.5
0320	Concrete Reinforcing	93.9	48.5	70.8	92.5	48.5	70.1	100.7	50.9	75.4	97.7	50.6	73.7	92.5	51.3	71.5	99.6	81.3	90.3
0330	Cast-in-Place Concrete	106.9	65.6	89.9	84.9	66.8	77.4	88.4	63.9	78.3	109.8	65.5	91.6	90.7	63.7	79.6	87.3	73.4	81.6
03	CONCRETE	104.9	60.9	83.3	89.7	63.3	76.7	90.6	63.4	77.2	109.1	62.5	86.2	92.4	62.8	77.9	106.0	68.3	87.5
04	MASONRY	120.0	58.9	81.9	105.2	56.6	74.9	160.7	56.6	95.8	103.6	58.9	75.8	105.7	63.6	79.5	105.8	63.2	79.3
05	METALS	94.9	80.7	90.5	102.2	67.8	91.6	97.7	79.0	92.0	96.3	81.6	91.8	102.2	69.8	92.2	102.2	79.0	95.0
06	WOOD, PLASTICS & COMPOSITES	92.5	58.4	73.4	107.1	68.5	85.5	92.3	68.6	79.0	83.1	62.0	71.3	107.1	68.5	85.5	85.3	54.9	68.3
07	THERMAL & MOISTURE PROTECTION	96.8	66.9	84.6	98.5	65.8	85.1	92.9	66.0	81.9	93.4	66.6	82.4	98.5	67.7	85.9	97.4	67.9	85.3
08	OPENINGS	101.4	56.1	90.8	77.7	62.4	74.1	97.7	63.0	89.6	101.6	58.6	91.6	77.7	63.3	74.4	94.3	56.5	85.5
0920	Plaster & Gypsum Board	93.8	57.5	69.3	88.2	68.0	74.6	90.4	68.0	75.3	92.5	61.2	71.3	88.2	68.0	74.6	78.9	53.5	61.7
0950, 0980	Ceilings & Acoustic Treatment	95.4	57.5	70.5	82.2	68.0	72.9	95.3	68.0	77.4	94.6	61.2	72.6	82.2	68.0	72.9	97.2	53.5	68.5
0960	Flooring	99.5	64.1	89.4	122.4	64.1	105.8	98.2	64.1	88.4	97.7	64.1	88.1	123.4	76.5	110.0	101.2	47.7	85.9
0970, 0990	Wall Finishes & Painting/Coating	103.6	59.4	76.9	105.8	48.8	71.4	107.0	56.1	76.3	103.4	59.4	76.8	108.5	56.1	76.8	103.3	60.7	77.5
09	FINISHES	89.4	59.6	73.0	95.8	63.8	78.1	96.8	64.6	79.1	90.8	61.7	74.8	96.3	67.1	80.2	96.2	55.3	73.6
COVERS	DIVS. 10 - 14, 25, 28, 41, 43, 44, 46	100.0	80.8	96.1	100.0	82.3	96.4	100.0	82.3	96.4	100.0	81.3	96.2	100.0	69.8	93.9	100.0	85.1	97.0
21, 22, 23	FIRE SUPPRESSION, PLUMBING & HVAC	95.4	65.9	83.5	100.2	54.7	81.8	95.3	58.5	80.5	95.4	60.5	83.1	100.2	54.2	81.6	99.9	69.5	87.6
26, 27, 3370	ELECTRICAL, COMMUNICATIONS & UTIL.	99.8	57.6	77.5	97.6	63.6	79.6	92.1	64.6	77.5	98.6	68.0	82.4	99.3	60.1	78.6	97.5	73.0	84.6
MF2014	WEIGHTED AVERAGE	99.0	65.6	84.4	96.5	64.1	82.4	98.6	65.9	84.4	98.8	67.7	85.3	97.1	64.4	82.8	99.8	70.1	86.9

For customer support on your Mechanical Cost Data, call 877.795.7394.

689

DIVISION		UTAH											VERMONT					
		OGDEN			PRICE			PROVO			SALT LAKE CITY			BELLOWS FALLS			BENNINGTON		
		842,844			845			846 - 847			840 - 841			051			052		
		MAT.	INST.	TOTAL	MAT.	INST.	TOTAL	MAT.	INST.	TOTAL	MAT.	INST.	TOTAL	MAT.	INST.	TOTAL	MAT.	INST.	TOTAL
015433	CONTRACTOR EQUIPMENT		97.5	97.5		96.6	96.6		96.6	96.6		97.5	97.5		100.9	100.9		100.9	100.9
0241, 31 - 34	SITE & INFRASTRUCTURE, DEMOLITION	84.9	95.7	92.6	93.3	93.8	93.7	92.3	94.1	93.6	84.5	95.6	92.4	88.2	100.5	96.9	87.5	100.5	96.7
0310	Concrete Forming & Accessories	104.7	58.1	64.5	107.1	48.3	56.4	106.3	58.0	64.7	107.1	58.0	64.8	98.3	85.5	87.3	95.9	106.6	105.2
0320	Concrete Reinforcing	99.2	81.3	90.1	106.8	81.1	93.7	107.7	81.3	94.3	101.5	81.3	91.2	83.2	85.7	84.5	83.2	85.7	84.5
0330	Cast-in-Place Concrete	88.7	73.4	82.4	87.4	59.5	75.9	87.4	73.3	81.6	96.9	73.3	87.2	87.0	110.7	96.7	87.0	110.7	96.7
03	CONCRETE	95.9	68.3	82.3	107.3	59.0	83.6	105.9	68.2	87.4	114.8	68.2	91.9	91.0	94.2	92.6	90.8	103.6	97.1
04	MASONRY	99.7	63.2	77.0	110.9	61.3	80.0	111.1	63.2	81.2	113.1	63.2	82.0	103.6	95.2	98.4	112.4	95.2	101.6
05	METALS	102.7	79.0	95.4	99.4	76.6	92.4	100.3	78.9	93.7	106.2	78.9	97.8	94.8	88.3	92.8	94.7	88.2	92.7
06	WOOD, PLASTICS & COMPOSITES	85.3	54.9	68.3	88.6	44.3	63.8	86.8	54.9	68.9	87.2	54.9	69.1	103.7	83.5	92.4	100.6	112.4	107.2
07	THERMAL & MOISTURE PROTECTION	96.4	67.9	84.7	99.0	59.9	82.9	99.0	67.9	86.2	103.4	67.9	88.9	97.0	82.7	91.1	97.0	81.1	90.4
08	OPENINGS	94.3	56.5	85.5	98.2	48.9	86.7	98.2	56.5	88.5	96.2	56.5	86.9	105.0	86.5	100.7	105.0	102.3	104.3
0920	Plaster & Gypsum Board	78.9	53.5	61.7	81.7	42.5	55.2	79.2	53.5	61.8	90.5	53.5	65.5	106.0	82.4	90.1	104.2	112.2	109.6
0950, 0980	Ceilings & Acoustic Treatment	97.2	53.5	68.5	97.2	42.5	61.3	97.2	53.5	68.5	91.5	53.5	66.5	91.5	82.4	85.6	91.5	112.2	105.1
0960	Flooring	99.0	47.7	84.3	102.3	35.5	83.2	102.0	47.7	86.5	103.2	47.7	87.3	100.6	104.6	101.8	99.7	104.6	101.1
0970, 0990	Wall Finishes & Painting/Coating	103.3	60.7	77.5	103.3	37.5	63.6	103.3	62.6	78.7	106.7	62.6	80.1	97.4	105.8	102.5	97.4	105.8	102.5
09	FINISHES	94.3	55.3	72.8	97.3	43.5	67.7	96.8	55.5	74.0	96.6	55.5	74.0	95.9	90.7	93.0	95.4	107.8	102.2
COVERS	DIVS. 10 - 14, 25, 28, 41, 43, 44, 46	100.0	85.1	97.0	100.0	45.8	89.0	100.0	85.0	97.0	100.0	85.0	97.0	100.0	97.6	99.5	100.0	100.7	100.1
21, 22, 23	FIRE SUPPRESSION, PLUMBING & HVAC	99.9	69.5	87.6	97.5	67.7	85.5	99.9	69.4	87.6	100.1	69.4	87.7	95.3	96.4	95.7	95.3	96.4	95.7
26, 27, 3370	ELECTRICAL, COMMUNICATIONS & UTIL.	97.9	73.0	84.7	103.5	73.0	87.4	98.4	59.8	78.0	100.5	59.8	79.0	101.5	89.3	95.1	101.5	61.2	80.2
MF2014	WEIGHTED AVERAGE	98.1	70.1	85.9	100.2	64.6	84.7	100.3	68.2	86.3	102.2	68.3	87.4	97.2	93.0	95.3	97.5	93.5	95.8

DIVISION		VERMONT																
		BRATTLEBORO			BURLINGTON			GUILDHALL			MONTPELIER			RUTLAND			ST. JOHNSBURY		
		053			054			059			056			057			058		
		MAT.	INST.	TOTAL	MAT.	INST.	TOTAL	MAT.	INST.	TOTAL	MAT.	INST.	TOTAL	MAT.	INST.	TOTAL	MAT.	INST.	TOTAL
015433	CONTRACTOR EQUIPMENT		100.9	100.9		100.9	100.9		100.9	100.9		100.9	100.9		100.9	100.9		100.9	100.9
0241, 31 - 34	SITE & INFRASTRUCTURE, DEMOLITION	88.9	100.5	97.1	93.0	100.4	98.2	87.2	99.0	95.6	91.5	100.1	97.6	91.7	100.1	97.7	87.3	99.0	95.6
0310	Concrete Forming & Accessories	98.6	85.4	87.2	98.0	87.6	89.0	96.1	78.3	80.7	97.9	84.1	86.0	98.9	87.6	89.1	94.4	78.3	80.5
0320	Concrete Reinforcing	82.3	85.7	84.0	101.1	85.6	93.2	83.9	85.6	84.8	94.1	85.6	89.8	103.9	85.6	94.6	82.3	85.6	84.0
0330	Cast-in-Place Concrete	89.7	110.7	98.3	100.9	109.6	104.5	84.3	101.0	91.2	100.9	109.6	104.5	85.2	109.6	95.2	84.3	101.0	91.2
03	CONCRETE	93.0	94.1	93.6	101.2	94.7	98.0	88.6	87.6	88.1	100.1	93.1	96.7	94.3	94.7	94.5	88.2	87.6	87.9
04	MASONRY	111.5	95.2	101.3	115.7	93.7	102.0	111.8	78.2	90.8	109.6	93.7	99.7	93.9	93.7	93.8	138.8	78.2	101.0
05	METALS	94.7	88.2	92.7	103.5	87.5	98.6	94.8	87.3	92.5	99.6	87.4	95.9	100.6	87.4	96.5	94.8	87.3	92.5
06	WOOD, PLASTICS & COMPOSITES	104.0	83.5	92.5	99.9	88.2	93.3	100.0	83.5	90.8	97.7	83.5	89.7	104.2	88.2	95.2	94.8	83.5	88.5
07	THERMAL & MOISTURE PROTECTION	97.1	78.0	89.3	103.8	80.4	94.2	96.8	70.7	86.1	103.3	79.9	93.7	97.2	80.4	90.3	96.7	70.7	86.1
08	OPENINGS	105.0	86.5	100.7	108.4	85.0	103.0	105.0	82.4	99.7	105.8	82.4	100.3	108.4	85.0	102.9	105.0	82.4	99.7
0920	Plaster & Gypsum Board	106.0	82.4	90.1	109.7	87.2	94.5	112.4	82.4	92.2	108.9	82.4	91.0	106.6	87.2	93.5	113.8	82.4	92.6
0950, 0980	Ceilings & Acoustic Treatment	91.5	82.4	85.6	101.0	87.2	91.9	91.5	82.4	85.6	95.1	82.4	86.8	95.9	87.2	90.2	91.5	82.4	85.6
0960	Flooring	100.7	104.6	101.8	102.4	104.6	103.1	103.6	104.6	103.9	103.2	104.6	103.6	100.6	104.6	101.8	106.9	104.6	106.3
0970, 0990	Wall Finishes & Painting/Coating	97.4	105.8	102.5	102.9	85.2	92.2	97.4	85.2	90.0	101.7	85.2	91.7	97.4	85.2	90.0	97.4	85.2	90.0
09	FINISHES	96.0	90.7	93.1	99.7	90.8	94.8	97.5	84.2	90.2	97.9	88.0	92.5	97.1	90.8	93.6	98.7	84.2	90.7
COVERS	DIVS. 10 - 14, 25, 28, 41, 43, 44, 46	100.0	97.6	99.5	100.0	97.5	99.5	100.0	91.8	98.3	100.0	97.0	99.4	100.0	97.5	99.5	100.0	91.8	98.3
21, 22, 23	FIRE SUPPRESSION, PLUMBING & HVAC	95.3	96.4	95.7	99.9	71.8	86.6	95.3	64.0	82.7	95.1	71.8	85.7	100.1	71.8	88.7	95.3	64.0	82.7
26, 27, 3370	ELECTRICAL, COMMUNICATIONS & UTIL.	101.5	89.3	95.1	102.2	61.2	80.5	101.5	61.2	80.2	99.7	61.2	79.3	101.6	61.2	80.2	101.5	61.2	80.2
MF2014	WEIGHTED AVERAGE	97.8	92.8	95.6	102.4	83.6	94.2	97.4	77.8	88.8	99.5	82.8	92.2	99.7	83.6	92.7	98.6	77.8	89.6

DIVISION		VERMONT			VIRGINIA													
		WHITE RIVER JCT.			ALEXANDRIA			ARLINGTON			BRISTOL			CHARLOTTESVILLE			CULPEPER		
		050			223			222			242			229			227		
		MAT.	INST.	TOTAL	MAT.	INST.	TOTAL	MAT.	INST.	TOTAL	MAT.	INST.	TOTAL	MAT.	INST.	TOTAL	MAT.	INST.	TOTAL
015433	CONTRACTOR EQUIPMENT		100.9	100.9		103.0	103.0		101.7	101.7		101.7	101.7		106.1	106.1		101.7	101.7
0241, 31 - 34	SITE & INFRASTRUCTURE, DEMOLITION	91.5	99.2	97.0	114.3	89.8	96.9	124.5	87.7	98.3	108.8	86.1	92.6	113.5	88.1	95.4	111.8	87.6	94.6
0310	Concrete Forming & Accessories	93.1	78.9	80.8	92.2	72.8	75.5	91.2	72.0	74.7	87.1	42.5	48.6	85.4	48.7	53.7	82.5	71.4	72.9
0320	Concrete Reinforcing	83.2	85.6	84.4	83.0	87.0	85.1	93.7	84.4	88.9	93.7	68.4	80.8	93.1	72.8	82.7	93.7	84.2	88.9
0330	Cast-in-Place Concrete	89.7	101.9	94.7	105.1	81.8	95.6	102.3	80.3	93.3	101.8	48.5	79.9	105.9	55.5	85.2	104.8	69.2	90.2
03	CONCRETE	94.7	88.2	91.5	102.0	79.6	91.0	106.7	78.3	92.8	102.9	51.2	77.5	103.5	57.2	80.8	100.6	74.1	87.6
04	MASONRY	124.5	79.7	96.6	91.3	72.8	79.8	106.3	70.7	84.1	95.4	51.4	68.0	120.3	53.2	78.5	107.5	70.7	84.6
05	METALS	94.8	87.3	92.5	105.7	97.0	103.0	104.3	97.2	102.1	103.1	85.7	97.8	103.4	90.2	99.3	103.5	95.3	101.0
06	WOOD, PLASTICS & COMPOSITES	97.4	83.5	89.6	95.0	71.0	81.6	91.5	71.0	80.0	83.7	40.0	59.2	82.2	45.6	61.7	80.8	71.0	75.3
07	THERMAL & MOISTURE PROTECTION	97.1	71.4	86.6	102.7	81.5	94.0	104.7	80.6	94.8	104.2	60.1	86.1	103.8	67.7	89.0	104.0	78.1	93.4
08	OPENINGS	105.0	82.4	99.7	100.9	75.2	94.9	99.0	75.2	93.5	102.0	46.5	89.1	100.2	52.0	89.0	100.5	75.2	94.6
0920	Plaster & Gypsum Board	102.8	82.4	89.0	107.4	69.9	82.1	104.2	69.9	81.0	99.7	38.0	58.0	99.7	43.0	61.4	99.9	69.9	79.6
0950, 0980	Ceilings & Acoustic Treatment	91.5	82.4	85.6	91.4	69.9	77.3	89.8	69.9	76.7	88.9	38.0	55.4	88.9	43.0	58.8	89.8	69.9	76.7
0960	Flooring	98.6	104.6	100.4	106.0	82.9	99.4	104.5	81.9	98.0	101.0	66.5	91.1	99.7	66.5	90.2	99.7	81.9	94.6
0970, 0990	Wall Finishes & Painting/Coating	97.4	85.2	90.0	124.0	80.0	97.5	124.0	80.0	97.5	109.6	54.3	76.2	109.6	76.1	89.3	124.0	76.1	95.1
09	FINISHES	95.2	84.6	89.3	100.1	74.6	86.0	100.0	73.8	85.6	96.5	47.2	69.3	96.2	53.5	72.6	96.9	73.2	83.9
COVERS	DIVS. 10 - 14, 25, 28, 41, 43, 44, 46	100.0	92.3	98.4	100.0	88.6	97.7	100.0	87.6	97.5	100.0	67.8	93.5	100.0	79.3	95.8	100.0	87.6	97.5
21, 22, 23	FIRE SUPPRESSION, PLUMBING & HVAC	95.3	64.8	83.0	100.3	86.9	94.9	100.3	85.4	94.3	95.6	50.5	77.4	95.6	69.3	85.0	95.6	72.9	86.4
26, 27, 3370	ELECTRICAL, COMMUNICATIONS & UTIL.	101.5	61.2	80.2	97.1	102.4	99.9	94.7	99.2	97.1	96.8	35.4	64.3	96.7	71.9	83.6	99.3	99.2	99.2
MF2014	WEIGHTED AVERAGE	98.5	78.3	89.7	101.1	85.5	94.3	101.9	84.0	94.1	99.4	54.9	80.0	100.5	67.1	86.0	99.9	80.5	91.4

VIRGINIA

DIVISION		FAIRFAX 220 - 221			FARMVILLE 239			FREDERICKSBURG 224 - 225			GRUNDY 246			HARRISONBURG 228			LYNCHBURG 245		
		MAT.	INST.	TOTAL	MAT.	INST.	TOTAL	MAT.	INST.	TOTAL	MAT.	INST.	TOTAL	MAT.	INST.	TOTAL	MAT.	INST.	TOTAL
015433	CONTRACTOR EQUIPMENT		101.7	101.7		106.1	106.1		101.7	101.7		101.7	101.7		101.7	101.7		101.7	101.7
0241, 31 - 34	SITE & INFRASTRUCTURE, DEMOLITION	123.1	87.8	98.0	109.2	87.6	93.8	111.4	87.7	94.5	106.6	84.8	91.1	120.1	86.3	96.1	107.5	86.3	92.4
0310	Concrete Forming & Accessories	85.4	72.0	73.9	98.0	43.7	51.2	85.4	69.4	71.6	90.3	36.4	43.8	81.4	42.6	47.9	87.1	60.3	64.0
0320	Concrete Reinforcing	93.7	87.0	90.3	90.9	49.9	70.0	94.4	87.0	90.6	92.4	50.4	71.0	93.7	64.9	79.0	93.1	68.5	80.5
0330	Cast-in-Place Concrete	102.3	80.6	93.4	103.1	53.5	82.7	103.9	69.3	89.7	101.8	47.5	79.5	102.3	58.0	84.1	101.8	57.7	83.7
03	CONCRETE	106.3	78.8	92.8	101.2	49.9	76.0	100.3	73.8	87.3	101.6	44.4	73.5	104.1	53.8	79.4	101.5	62.3	82.3
04	MASONRY	106.2	71.4	84.5	103.5	44.9	66.9	106.6	70.7	84.2	96.5	47.3	65.8	104.2	50.3	70.6	112.0	53.2	75.4
05	METALS	103.6	97.0	101.5	101.0	77.1	93.6	103.5	96.2	101.3	103.1	69.9	92.9	103.4	83.9	97.4	103.3	86.3	98.1
06	WOOD, PLASTICS & COMPOSITES	83.7	71.0	76.6	96.4	44.1	67.1	83.7	68.1	75.0	86.7	35.1	57.8	79.8	39.0	56.9	83.7	62.6	71.9
07	THERMAL & MOISTURE PROTECTION	104.6	73.7	91.9	104.3	49.3	81.8	104.0	78.5	93.6	104.2	45.3	80.0	104.4	63.9	87.8	104.0	63.9	87.6
08	OPENINGS	99.0	75.2	93.5	100.7	43.0	87.3	100.2	73.6	94.0	102.0	34.4	86.3	100.5	49.4	88.6	100.5	58.8	90.8
0920	Plaster & Gypsum Board	99.9	69.9	79.6	108.6	41.5	63.2	99.9	66.9	77.6	99.7	32.9	54.5	99.7	36.9	57.3	99.7	61.3	73.7
0950, 0980	Ceilings & Acoustic Treatment	89.8	69.9	76.7	84.8	41.5	56.3	89.8	66.9	74.8	88.9	32.9	52.1	88.9	36.9	54.8	88.9	61.3	70.8
0960	Flooring	101.4	81.9	95.8	106.5	60.3	93.3	101.4	81.9	95.8	102.3	33.7	82.6	99.4	74.0	92.1	101.0	66.5	91.1
0970, 0990	Wall Finishes & Painting/Coating	124.0	80.0	97.5	112.8	38.7	68.1	124.0	76.1	95.1	109.6	35.1	64.6	124.0	54.3	81.9	109.6	54.3	76.2
09	FINISHES	98.5	74.0	85.0	98.0	46.0	69.4	97.4	71.5	83.1	96.7	35.1	62.8	97.3	47.8	70.1	96.3	60.2	76.4
COVERS	DIVS. 10 - 14, 25, 28, 41, 43, 44, 46	100.0	87.8	97.5	100.0	45.7	89.0	100.0	83.4	96.6	100.0	43.7	88.6	100.0	67.6	93.5	100.0	70.9	94.1
21, 22, 23	FIRE SUPPRESSION, PLUMBING & HVAC	95.6	85.7	91.6	95.5	43.4	74.5	95.6	85.1	91.3	95.6	62.1	82.1	95.6	66.9	84.0	95.6	69.0	84.8
26, 27, 3370	ELECTRICAL, COMMUNICATIONS & UTIL.	97.9	99.2	98.6	90.1	48.2	68.0	94.9	99.2	97.2	96.8	44.6	69.2	96.9	96.4	96.7	97.9	52.7	74.0
MF2014	WEIGHTED AVERAGE	100.7	84.0	93.4	98.7	52.4	78.5	99.5	82.7	92.1	99.3	52.5	78.9	100.2	67.2	85.8	99.9	65.6	85.0

VIRGINIA

DIVISION		NEWPORT NEWS 236			NORFOLK 233 - 235			PETERSBURG 238			PORTSMOUTH 237			PULASKI 243			RICHMOND 230 - 232		
		MAT.	INST.	TOTAL	MAT.	INST.	TOTAL	MAT.	INST.	TOTAL	MAT.	INST.	TOTAL	MAT.	INST.	TOTAL	MAT.	INST.	TOTAL
015433	CONTRACTOR EQUIPMENT		106.1	106.1		106.7	106.7		106.1	106.1		106.0	106.0		101.7	101.7		106.1	106.1
0241, 31 - 34	SITE & INFRASTRUCTURE, DEMOLITION	108.2	88.8	94.4	108.9	89.8	95.3	111.7	89.0	95.6	106.7	88.6	93.8	105.9	85.5	91.4	103.6	89.0	93.2
0310	Concrete Forming & Accessories	97.3	64.1	68.7	97.9	64.3	68.9	90.6	57.2	61.8	86.4	53.1	57.7	90.3	38.7	45.8	97.6	57.2	62.8
0320	Concrete Reinforcing	90.6	72.0	81.2	94.7	72.1	83.2	90.3	73.0	81.5	90.3	72.0	81.0	92.4	64.6	78.3	97.5	73.0	85.0
0330	Cast-in-Place Concrete	100.2	66.1	86.2	105.2	66.1	89.1	106.3	53.7	84.7	99.2	66.0	85.6	101.8	48.2	79.8	98.1	53.7	79.8
03	CONCRETE	98.4	67.6	83.3	101.5	67.7	84.9	103.6	60.4	82.4	97.1	62.6	80.2	101.6	48.7	75.6	98.5	60.4	79.8
04	MASONRY	97.8	53.9	70.5	103.8	53.9	72.7	111.9	52.6	75.0	103.8	53.9	72.7	91.5	48.2	64.5	102.5	52.6	71.4
05	METALS	103.3	89.8	99.2	103.1	89.9	99.1	101.1	90.7	97.9	102.2	89.1	98.2	103.2	82.5	96.8	107.1	90.7	102.0
06	WOOD, PLASTICS & COMPOSITES	95.3	66.2	79.0	93.2	66.2	78.1	86.4	58.6	70.8	82.9	51.5	65.3	86.7	36.8	58.7	95.0	58.6	74.6
07	THERMAL & MOISTURE PROTECTION	104.3	68.4	89.6	100.9	68.4	87.5	104.3	68.3	89.5	104.3	66.8	88.9	104.2	51.1	82.4	102.2	68.3	88.3
08	OPENINGS	101.1	62.2	92.1	100.4	62.2	91.5	100.4	59.0	90.8	101.2	54.1	90.2	102.0	42.2	88.1	104.8	59.0	94.1
0920	Plaster & Gypsum Board	109.4	64.3	78.9	105.5	64.3	77.6	101.3	56.4	71.0	101.6	49.1	66.1	99.7	34.7	55.7	104.2	56.4	71.9
0950, 0980	Ceilings & Acoustic Treatment	88.2	64.3	72.5	88.9	64.3	72.7	85.7	56.4	66.4	88.2	49.1	62.5	88.9	34.7	53.3	90.6	56.4	68.1
0960	Flooring	106.5	66.5	95.1	105.6	66.5	94.4	102.5	71.2	93.6	99.5	66.5	90.0	102.3	66.5	92.0	103.4	71.2	94.2
0970, 0990	Wall Finishes & Painting/Coating	112.8	42.0	70.1	113.0	76.1	90.7	112.8	76.1	90.6	112.8	76.1	90.6	109.6	54.3	76.2	109.6	76.1	89.5
09	FINISHES	98.7	62.2	78.6	98.2	66.0	80.5	96.3	61.2	76.9	95.6	56.7	74.2	96.7	43.7	67.5	98.5	61.2	77.9
COVERS	DIVS. 10 - 14, 25, 28, 41, 43, 44, 46	100.0	82.2	96.4	100.0	82.2	96.4	100.0	79.6	95.9	100.0	72.0	94.3	100.0	44.0	88.7	100.0	79.6	95.9
21, 22, 23	FIRE SUPPRESSION, PLUMBING & HVAC	100.3	63.6	85.5	100.0	64.3	85.6	95.5	67.5	84.2	100.3	64.3	85.8	95.6	62.2	82.1	99.9	67.5	86.8
26, 27, 3370	ELECTRICAL, COMMUNICATIONS & UTIL.	92.8	63.5	77.3	95.5	58.3	75.8	93.0	71.9	81.8	91.2	58.3	73.8	96.8	54.4	74.4	96.8	71.9	83.6
MF2014	WEIGHTED AVERAGE	100.0	68.2	86.1	100.5	68.2	86.4	99.4	68.7	86.0	99.4	65.3	84.5	99.0	57.3	80.9	101.3	68.7	87.1

DIVISION		VIRGINIA									WASHINGTON								
		ROANOKE 240 - 241			STAUNTON 244			WINCHESTER 226			CLARKSTON 994			EVERETT 982			OLYMPIA 985		
		MAT.	INST.	TOTAL	MAT.	INST.	TOTAL	MAT.	INST.	TOTAL	MAT.	INST.	TOTAL	MAT.	INST.	TOTAL	MAT.	INST.	TOTAL
015433	CONTRACTOR EQUIPMENT		101.7	101.7		106.1	106.1		101.7	101.7		91.2	91.2		101.6	101.6		101.6	101.6
0241, 31 - 34	SITE & INFRASTRUCTURE, DEMOLITION	106.4	86.3	92.1	109.9	87.7	94.1	118.7	87.6	96.6	99.2	90.1	92.7	91.5	108.4	103.5	93.4	108.4	104.1
0310	Concrete Forming & Accessories	96.8	60.5	65.5	89.9	50.0	55.5	83.8	69.5	71.4	112.7	67.4	73.6	113.4	100.4	102.2	99.0	100.3	100.1
0320	Concrete Reinforcing	93.4	68.5	80.7	93.1	48.9	70.6	93.1	86.4	89.7	112.0	87.2	99.4	104.2	98.8	101.5	110.8	98.7	104.6
0330	Cast-in-Place Concrete	115.7	57.7	91.9	105.9	51.5	83.6	102.3	69.3	88.7	96.3	81.9	90.4	103.5	106.2	104.6	98.9	106.2	101.9
03	CONCRETE	105.9	62.4	84.5	102.9	52.0	77.9	103.5	73.7	88.9	107.2	76.3	92.0	98.7	101.6	100.1	98.9	101.6	100.2
04	MASONRY	97.7	53.2	70.0	106.9	51.4	72.3	101.7	59.7	75.5	102.3	81.5	89.3	110.6	100.8	104.5	103.0	99.1	100.6
05	METALS	105.5	86.5	99.7	103.4	80.5	96.3	103.5	94.7	100.8	88.3	81.6	86.2	101.8	92.4	98.9	101.2	92.1	98.4
06	WOOD, PLASTICS & COMPOSITES	95.9	62.6	77.3	86.7	49.9	66.1	82.2	68.1	74.3	113.2	63.8	85.5	111.7	100.0	105.2	91.1	100.0	96.1
07	THERMAL & MOISTURE PROTECTION	103.8	66.8	88.7	103.8	52.0	82.6	104.5	75.6	92.7	143.8	77.2	116.5	103.1	101.7	102.9	103.5	96.8	100.7
08	OPENINGS	100.9	58.8	91.1	100.5	48.1	88.3	102.1	73.7	95.5	119.0	68.7	107.3	101.3	99.8	101.0	106.9	99.9	105.3
0920	Plaster & Gypsum Board	107.4	61.3	76.2	99.7	47.5	64.4	99.9	66.9	77.6	145.7	62.5	89.5	109.4	99.9	103.0	101.3	99.9	100.4
0950, 0980	Ceilings & Acoustic Treatment	91.4	61.3	71.6	88.9	47.5	61.7	89.8	66.9	74.8	97.6	62.5	74.5	99.8	99.9	99.9	98.1	99.9	99.3
0960	Flooring	106.0	66.5	94.7	101.9	39.5	84.0	100.7	81.9	95.4	91.6	47.1	78.9	113.8	91.6	107.4	103.8	91.7	100.3
0970, 0990	Wall Finishes & Painting/Coating	109.6	54.3	76.2	109.6	34.5	64.3	124.0	87.2	101.8	100.0	58.5	75.0	95.4	91.5	93.0	99.0	91.5	94.5
09	FINISHES	99.0	61.1	78.1	96.6	46.6	69.0	97.9	72.7	84.0	110.3	61.3	83.3	104.7	97.6	100.8	98.9	97.6	98.2
COVERS	DIVS. 10 - 14, 25, 28, 41, 43, 44, 46	100.0	70.9	94.1	100.0	70.9	94.1	100.0	87.3	97.4	100.0	75.4	95.0	100.0	99.0	99.8	100.0	100.5	100.1
21, 22, 23	FIRE SUPPRESSION, PLUMBING & HVAC	100.3	64.8	86.0	95.6	58.9	80.8	95.6	87.7	92.4	95.6	83.4	90.7	100.1	99.1	99.7	100.0	99.1	99.6
26, 27, 3370	ELECTRICAL, COMMUNICATIONS & UTIL.	96.8	54.4	74.4	95.6	74.9	84.7	95.3	99.2	97.4	92.4	94.8	93.7	104.9	98.5	101.5	104.4	99.1	101.6
MF2014	WEIGHTED AVERAGE	101.5	65.1	85.6	99.7	61.8	83.2	100.0	82.2	92.2	101.5	80.2	92.2	101.6	99.6	100.7	101.2	99.4	100.4

For customer support on your Mechanical Cost Data, call 877.795.7394.

691

WASHINGTON

DIVISION		RICHLAND 993			SEATTLE 980 - 981,987			SPOKANE 990 - 992			TACOMA 983 - 984			VANCOUVER 986			WENATCHEE 988		
		MAT.	INST.	TOTAL	MAT.	INST.	TOTAL	MAT.	INST.	TOTAL	MAT.	INST.	TOTAL	MAT.	INST.	TOTAL	MAT.	INST.	TOTAL
015433	CONTRACTOR EQUIPMENT		91.2	91.2		101.6	101.6		91.2	91.2		101.6	101.6		96.7	96.7		101.6	101.6
0241, 31 - 34	SITE & INFRASTRUCTURE, DEMOLITION	101.8	91.0	94.1	95.1	106.7	103.4	101.1	91.0	93.9	94.5	108.4	104.4	104.2	96.1	98.4	103.2	107.1	106.0
0310	Concrete Forming & Accessories	112.8	78.5	83.2	108.3	100.9	102.0	118.2	78.1	83.6	104.2	100.4	100.9	104.9	90.5	92.5	105.8	77.6	81.5
0320	Concrete Reinforcing	107.4	87.4	97.2	105.1	98.9	101.9	108.2	87.3	97.6	103.1	98.8	100.9	103.9	98.4	101.1	103.9	87.7	95.6
0330	Cast-in-Place Concrete	96.5	84.4	91.5	100.9	106.4	103.2	100.3	84.2	93.7	106.4	106.2	106.3	110.8	98.4	110.4	108.6	77.3	95.7
03	CONCRETE	106.8	82.2	94.7	99.6	102.1	100.8	109.1	82.0	95.8	100.4	101.6	101.0	110.7	94.6	102.8	108.4	79.6	94.2
04	MASONRY	104.3	84.3	91.8	108.2	100.9	103.6	104.9	84.3	92.0	106.2	100.9	102.9	106.7	95.6	99.8	108.6	91.4	97.9
05	METALS	88.7	84.9	87.5	102.5	94.2	99.9	90.9	84.5	89.0	103.6	92.3	100.1	101.0	91.8	98.2	101.0	86.2	96.4
06	WOOD, PLASTICS & COMPOSITES	113.4	76.1	92.5	104.9	100.0	102.2	122.4	76.1	96.5	101.0	100.0	100.4	92.9	90.2	91.4	102.5	75.8	87.5
07	THERMAL & MOISTURE PROTECTION	145.0	78.8	117.8	102.1	101.8	102.0	141.4	79.5	116.0	103.5	98.5	101.4	104.0	91.6	98.9	104.1	81.5	94.8
08	OPENINGS	121.4	71.3	109.7	100.1	99.9	100.0	121.9	71.8	110.3	102.0	99.9	101.5	98.3	92.8	97.0	101.5	71.0	94.4
0920	Plaster & Gypsum Board	145.7	75.2	98.1	103.6	99.9	101.1	137.7	75.2	95.5	107.8	99.9	102.5	105.7	90.1	95.1	110.2	74.9	86.4
0950, 0980	Ceilings & Acoustic Treatment	104.2	75.2	85.1	110.2	99.9	103.5	99.5	75.2	83.5	103.1	99.9	101.0	100.2	90.1	93.5	95.4	74.9	82.0
0960	Flooring	92.0	80.1	88.6	108.3	106.8	107.8	91.2	85.3	89.5	107.2	91.7	102.8	113.5	102.2	110.2	109.9	62.6	96.4
0970, 0990	Wall Finishes & Painting/Coating	100.0	67.6	80.4	96.3	91.5	93.4	100.2	70.9	82.5	95.4	91.5	93.0	97.6	70.8	81.4	95.4	72.6	81.6
09	FINISHES	112.0	76.7	92.6	105.0	100.6	102.6	109.8	78.1	92.3	103.4	97.6	100.2	101.5	90.6	95.5	103.8	73.5	87.1
COVERS	DIVS. 10 - 14, 25, 28, 41, 43, 44, 46	100.0	85.4	97.0	100.0	100.5	100.1	100.0	85.3	97.0	100.0	100.5	100.1	100.0	60.7	92.1	100.0	77.2	95.4
21, 22, 23	FIRE SUPPRESSION, PLUMBING & HVAC	100.5	107.6	103.3	100.0	115.1	106.1	100.4	84.3	93.9	100.2	99.2	99.8	100.3	87.0	94.9	95.4	86.7	91.9
26, 27, 3370	ELECTRICAL, COMMUNICATIONS & UTIL.	89.7	94.8	92.4	102.6	107.0	104.9	88.0	79.0	83.3	104.7	99.1	101.7	109.7	101.0	105.1	105.5	98.4	101.8
MF2014	WEIGHTED AVERAGE	103.1	89.3	97.1	101.4	104.7	102.8	103.4	82.3	94.2	101.8	99.7	100.9	102.7	92.1	98.1	101.6	86.6	95.0

DIVISION		WASHINGTON			WEST VIRGINIA														
		YAKIMA 989			BECKLEY 258 - 259			BLUEFIELD 247 - 248			BUCKHANNON 262			CHARLESTON 250 - 253			CLARKSBURG 263 - 264		
		MAT.	INST.	TOTAL	MAT.	INST.	TOTAL	MAT.	INST.	TOTAL	MAT.	INST.	TOTAL	MAT.	INST.	TOTAL	MAT.	INST.	TOTAL
015433	CONTRACTOR EQUIPMENT		101.6	101.6		101.7	101.7		101.7	101.7		101.7	101.7		101.7	101.7		101.7	101.7
0241, 31 - 34	SITE & INFRASTRUCTURE, DEMOLITION	96.8	107.8	104.7	100.5	89.9	92.9	100.6	89.9	92.9	106.8	89.9	94.8	101.5	90.7	93.8	107.4	89.9	95.0
0310	Concrete Forming & Accessories	104.6	95.8	97.0	83.5	93.6	92.2	87.0	93.6	92.7	86.4	94.3	93.2	96.7	94.1	94.4	83.8	94.2	92.8
0320	Concrete Reinforcing	103.5	87.4	95.3	92.7	89.0	90.8	92.0	84.1	88.0	92.6	84.2	88.3	101.0	89.1	94.9	92.6	84.1	88.3
0330	Cast-in-Place Concrete	113.6	83.9	101.4	98.1	103.0	100.1	99.6	102.9	101.0	99.3	101.8	100.3	96.8	103.0	99.3	108.8	100.0	105.2
03	CONCRETE	105.3	89.9	97.7	97.6	96.4	97.0	97.4	95.5	96.5	100.4	95.4	98.0	99.1	96.7	97.9	104.4	94.8	99.7
04	MASONRY	99.8	83.1	89.4	98.9	98.9	98.9	93.3	98.9	96.8	104.5	98.9	101.0	95.8	97.9	97.1	108.4	98.9	102.4
05	METALS	101.6	86.3	96.9	100.9	98.1	100.0	103.4	96.3	101.2	103.6	97.1	101.6	99.6	98.9	99.4	103.6	96.9	101.5
06	WOOD, PLASTICS & COMPOSITES	101.4	100.0	100.6	83.5	93.4	89.1	85.6	93.4	90.0	84.9	93.4	89.7	98.4	93.4	95.6	81.5	93.4	88.2
07	THERMAL & MOISTURE PROTECTION	103.6	82.7	95.0	102.9	90.6	97.9	104.0	90.6	98.5	104.2	91.7	99.1	99.4	90.4	95.7	104.1	90.6	98.6
08	OPENINGS	101.5	81.0	96.7	101.3	85.8	97.7	102.5	84.6	98.3	102.5	84.6	98.4	103.6	85.8	99.4	102.5	85.1	98.5
0920	Plaster & Gypsum Board	107.4	99.9	102.4	97.0	93.0	94.3	98.9	93.0	94.9	99.3	93.0	95.1	100.3	93.0	95.4	97.2	93.0	94.4
0950, 0980	Ceilings & Acoustic Treatment	97.3	99.9	99.0	83.9	93.0	89.9	87.3	93.0	91.1	88.9	93.0	91.6	92.2	93.0	92.7	88.9	93.0	91.6
0960	Flooring	108.1	58.0	93.8	102.4	118.0	106.8	98.9	118.0	104.3	98.6	118.0	104.1	106.4	118.0	109.7	97.6	118.0	103.4
0970, 0990	Wall Finishes & Painting/Coating	95.4	70.9	80.6	107.4	91.5	97.8	109.6	94.6	100.6	109.6	94.3	100.3	106.9	95.9	100.2	109.6	94.3	100.3
09	FINISHES	102.5	86.6	93.7	94.7	98.6	96.9	94.8	98.9	97.1	95.7	98.9	97.5	99.1	98.9	99.0	95.0	98.9	97.2
COVERS	DIVS. 10 - 14, 25, 28, 41, 43, 44, 46	100.0	97.6	99.5	100.0	56.9	91.3	100.0	56.9	91.3	100.0	101.5	100.3	100.0	101.2	100.2	100.0	101.5	100.3
21, 22, 23	FIRE SUPPRESSION, PLUMBING & HVAC	100.2	107.2	103.0	95.5	93.5	94.7	95.6	92.8	94.4	95.6	95.1	95.4	100.0	93.4	97.4	95.6	95.0	95.3
26, 27, 3370	ELECTRICAL, COMMUNICATIONS & UTIL.	107.8	94.9	101.0	93.9	91.1	92.4	95.7	91.1	93.3	97.1	96.1	96.6	99.1	91.1	94.8	97.1	96.1	96.6
MF2014	WEIGHTED AVERAGE	102.0	94.1	98.6	97.9	93.3	95.9	98.3	92.9	96.0	99.6	95.5	97.8	99.9	94.8	97.6	100.1	95.4	98.1

WEST VIRGINIA

DIVISION		GASSAWAY 266			HUNTINGTON 255 - 257			LEWISBURG 249			MARTINSBURG 254			MORGANTOWN 265			PARKERSBURG 261		
		MAT.	INST.	TOTAL	MAT.	INST.	TOTAL	MAT.	INST.	TOTAL	MAT.	INST.	TOTAL	MAT.	INST.	TOTAL	MAT.	INST.	TOTAL
015433	CONTRACTOR EQUIPMENT		101.7	101.7		101.7	101.7		101.7	101.7		101.7	101.7		101.7	101.7		101.7	101.7
0241, 31 - 34	SITE & INFRASTRUCTURE, DEMOLITION	104.2	89.9	94.0	105.1	90.9	95.0	116.3	89.8	97.5	104.2	90.2	94.3	101.6	90.5	93.7	109.8	90.6	96.2
0310	Concrete Forming & Accessories	85.8	92.2	91.4	95.3	95.9	95.8	84.2	93.5	92.2	83.6	82.8	82.9	84.2	93.9	92.6	88.6	90.3	90.1
0320	Concrete Reinforcing	92.6	84.0	88.3	94.1	92.9	93.5	92.6	83.9	88.2	92.7	80.6	86.6	92.6	84.1	88.3	92.0	83.9	87.9
0330	Cast-in-Place Concrete	104.0	103.5	103.7	106.9	102.5	105.1	99.7	102.9	101.0	102.8	96.7	100.3	99.3	101.6	100.2	101.5	96.4	99.4
03	CONCRETE	100.8	95.1	98.0	102.8	98.0	100.4	107.1	95.4	101.4	101.2	87.9	94.7	97.2	95.2	96.2	102.5	91.8	97.2
04	MASONRY	109.2	95.9	100.9	96.4	100.8	99.1	95.9	95.9	95.9	99.7	89.5	93.3	126.5	98.9	109.3	82.9	91.7	88.4
05	METALS	103.5	96.8	101.4	103.3	100.4	102.4	103.5	96.0	101.2	101.3	94.2	99.1	103.6	96.6	101.5	104.2	96.6	101.9
06	WOOD, PLASTICS & COMPOSITES	84.0	90.7	87.8	94.7	95.3	95.1	81.8	93.4	88.3	83.5	81.6	82.5	81.8	93.4	88.3	85.0	89.1	87.3
07	THERMAL & MOISTURE PROTECTION	104.0	90.6	98.5	103.0	90.8	98.0	104.8	89.8	98.7	103.1	79.3	93.4	104.0	90.6	98.5	103.9	89.0	97.8
08	OPENINGS	100.7	83.2	96.6	100.5	87.6	97.5	102.5	84.6	98.4	103.3	73.4	96.3	103.8	85.1	99.4	101.4	82.3	96.9
0920	Plaster & Gypsum Board	98.6	90.2	92.9	104.8	95.0	98.1	97.2	93.0	94.4	97.6	80.9	86.3	97.2	93.0	94.4	99.7	88.5	92.1
0950, 0980	Ceilings & Acoustic Treatment	88.9	90.2	89.8	86.4	95.0	92.0	88.9	93.0	91.6	86.4	80.9	82.8	88.9	93.0	91.6	88.9	88.5	88.7
0960	Flooring	98.4	118.0	104.0	109.8	121.7	113.2	97.7	118.0	103.5	102.4	118.0	106.8	97.7	118.0	103.5	101.5	118.0	106.2
0970, 0990	Wall Finishes & Painting/Coating	109.6	95.9	101.3	107.4	94.6	99.7	109.6	75.4	88.9	107.4	41.4	67.5	109.6	94.3	100.3	109.6	94.6	100.6
09	FINISHES	95.2	97.5	96.5	98.5	100.8	99.8	96.1	96.8	96.5	95.6	83.1	88.7	94.6	98.9	97.0	96.6	95.9	96.2
COVERS	DIVS. 10 - 14, 25, 28, 41, 43, 44, 46	100.0	101.2	100.2	100.0	101.8	100.4	100.0	56.9	91.3	100.0	97.7	99.5	100.0	65.8	93.1	100.0	100.5	100.1
21, 22, 23	FIRE SUPPRESSION, PLUMBING & HVAC	95.6	93.9	94.9	100.3	90.5	96.4	95.6	93.5	94.7	95.5	86.0	91.7	95.6	95.0	95.3	100.3	88.3	95.4
26, 27, 3370	ELECTRICAL, COMMUNICATIONS & UTIL.	97.1	91.1	93.9	97.8	95.3	96.5	93.0	91.1	92.0	100.1	81.8	90.4	97.3	96.1	96.7	97.2	92.7	94.8
MF2014	WEIGHTED AVERAGE	99.5	93.9	97.1	100.7	95.7	98.5	99.8	92.4	96.6	99.3	86.4	93.7	100.2	94.4	97.6	100.1	91.8	96.5

City Cost Indexes

WEST VIRGINIA / WISCONSIN

DIVISION		PETERSBURG 268 MAT.	INST.	TOTAL	ROMNEY 267 MAT.	INST.	TOTAL	WHEELING 260 MAT.	INST.	TOTAL	BELOIT 535 MAT.	INST.	TOTAL	EAU CLAIRE 547 MAT.	INST.	TOTAL	GREEN BAY 541 - 543 MAT.	INST.	TOTAL
015433	CONTRACTOR EQUIPMENT		101.7	101.7		101.7	101.7		101.7	101.7		102.3	102.3		100.4	100.4		98.2	98.2
0241, 31 - 34	SITE & INFRASTRUCTURE, DEMOLITION	100.9	90.5	93.5	103.6	90.5	94.3	110.6	90.4	96.2	94.7	108.3	104.3	96.1	102.9	101.0	99.7	98.9	99.2
0310	Concrete Forming & Accessories	87.4	91.6	91.0	83.4	91.8	90.6	90.3	93.2	92.8	100.2	100.3	100.3	97.1	99.4	99.1	105.5	98.7	99.7
0320	Concrete Reinforcing	92.0	82.5	87.2	92.6	81.0	86.7	91.4	84.0	87.7	92.3	116.0	104.3	90.4	99.2	94.9	88.7	92.4	90.6
0330	Cast-in-Place Concrete	99.3	99.1	99.2	104.0	99.1	102.0	101.5	100.1	100.9	100.6	100.9	100.7	104.0	99.2	102.0	107.6	99.7	104.3
03	CONCRETE	97.3	93.0	95.2	100.6	92.8	96.8	102.5	94.4	98.5	98.8	103.5	101.1	99.7	99.4	99.6	102.8	98.0	100.5
04	MASONRY	99.5	98.9	99.1	96.2	98.9	97.9	107.6	93.9	99.1	100.6	104.3	102.9	94.7	102.1	99.3	126.7	101.4	110.9
05	METALS	103.7	95.8	101.2	103.7	95.5	101.2	104.4	96.8	102.1	93.1	107.7	97.6	95.0	101.4	97.0	97.4	98.2	97.6
06	WOOD, PLASTICS & COMPOSITES	85.9	90.7	88.6	80.9	90.7	86.4	86.7	93.4	90.5	99.9	99.2	99.5	105.1	99.3	101.9	110.0	99.3	104.0
07	THERMAL & MOISTURE PROTECTION	104.0	86.0	96.6	104.1	83.4	95.6	104.3	90.0	98.4	102.1	96.8	99.9	102.4	86.7	95.9	104.3	86.0	96.8
08	OPENINGS	103.8	82.8	98.9	103.7	82.4	98.8	102.2	85.1	98.2	102.5	108.8	103.9	103.7	96.2	102.0	100.1	96.0	99.2
0920	Plaster & Gypsum Board	99.3	90.2	93.2	96.8	90.2	92.4	99.7	93.0	95.2	94.5	99.6	97.9	102.9	99.6	100.6	98.7	99.6	99.3
0950, 0980	Ceilings & Acoustic Treatment	88.9	90.2	89.8	88.9	90.2	89.8	88.9	93.0	91.6	88.1	99.6	95.6	90.4	99.6	96.4	81.9	99.6	93.5
0960	Flooring	99.4	118.0	104.7	97.5	118.0	103.4	102.3	118.0	106.8	99.8	118.1	105.1	91.4	114.1	97.9	110.2	114.1	111.3
0970, 0990	Wall Finishes & Painting/Coating	109.6	94.3	100.3	109.6	94.3	100.3	109.6	94.3	100.3	98.3	97.2	97.7	94.5	86.4	89.6	105.4	83.2	92.0
09	FINISHES	95.4	97.3	96.4	94.7	97.3	96.1	96.9	98.2	97.6	97.4	103.1	100.5	94.6	101.2	98.3	99.0	100.7	99.9
COVERS	DIVS. 10 - 14, 25, 28, 41, 43, 44, 46	100.0	65.5	93.0	100.0	101.2	100.2	100.0	95.4	99.1	100.0	96.6	99.3	100.0	96.8	99.3	100.0	96.5	99.3
21, 22, 23	FIRE SUPPRESSION, PLUMBING & HVAC	95.6	93.7	94.8	95.6	93.4	94.7	100.3	93.6	97.6	100.1	98.6	99.5	100.1	87.6	95.0	100.3	85.1	94.2
26, 27, 3370	ELECTRICAL, COMMUNICATIONS & UTIL.	100.5	81.8	90.6	99.8	81.8	90.3	94.4	96.1	95.3	101.0	86.7	93.4	103.8	84.9	93.8	98.2	82.0	89.6
MF2014	WEIGHTED AVERAGE	99.3	91.3	95.8	99.4	92.2	96.3	101.2	94.3	98.3	98.9	100.7	99.7	99.3	95.3	97.5	101.2	93.4	97.8

WISCONSIN

DIVISION		KENOSHA 531 MAT.	INST.	TOTAL	LA CROSSE 546 MAT.	INST.	TOTAL	LANCASTER 538 MAT.	INST.	TOTAL	MADISON 537 MAT.	INST.	TOTAL	MILWAUKEE 530,532 MAT.	INST.	TOTAL	NEW RICHMOND 540 MAT.	INST.	TOTAL
015433	CONTRACTOR EQUIPMENT		100.2	100.2		100.4	100.4		102.3	102.3		102.3	102.3		90.5	90.5		100.8	100.8
0241, 31 - 34	SITE & INFRASTRUCTURE, DEMOLITION	100.3	105.2	103.8	90.1	102.8	99.1	93.9	108.2	104.1	92.4	108.3	103.7	94.6	98.5	97.4	94.4	103.3	100.8
0310	Concrete Forming & Accessories	108.0	100.5	101.6	84.1	99.1	97.1	99.5	97.9	98.2	102.0	100.1	100.4	104.1	118.0	116.1	92.4	100.9	99.7
0320	Concrete Reinforcing	92.1	97.9	95.0	90.1	91.5	90.9	93.5	91.3	92.3	97.9	91.6	94.7	95.7	98.2	97.0	87.8	99.0	93.5
0330	Cast-in-Place Concrete	109.4	104.3	107.3	93.5	95.9	94.5	100.0	100.9	100.4	91.1	100.8	95.1	99.8	109.6	103.8	108.3	82.5	97.7
03	CONCRETE	103.5	102.4	103.0	90.8	96.8	93.7	98.5	97.8	98.1	95.7	98.8	97.2	99.7	110.5	105.0	97.5	94.3	95.9
04	MASONRY	97.9	114.1	108.0	93.8	105.9	101.4	100.7	104.3	102.9	101.5	104.3	103.2	104.9	120.6	114.7	121.6	102.3	109.6
05	METALS	93.9	101.2	96.2	94.9	98.2	95.9	90.6	95.0	92.0	91.9	97.2	93.6	97.5	94.8	96.7	95.1	99.7	96.5
06	WOOD, PLASTICS & COMPOSITES	105.0	96.4	100.2	90.1	99.3	95.3	99.1	99.2	99.2	96.7	99.2	98.1	104.7	118.5	112.5	94.0	102.5	98.7
07	THERMAL & MOISTURE PROTECTION	102.2	108.7	104.9	101.8	90.2	97.1	101.9	82.5	93.9	102.8	105.1	103.8	100.8	114.0	106.2	103.6	92.2	98.9
08	OPENINGS	96.9	101.1	97.9	103.7	85.0	99.4	97.9	85.8	95.1	107.9	100.5	106.2	101.0	113.2	103.8	89.1	97.5	91.1
0920	Plaster & Gypsum Board	85.3	96.7	93.0	97.7	99.6	99.0	93.3	99.6	97.5	99.6	99.6	99.6	101.0	119.3	113.4	87.7	103.0	98.1
0950, 0980	Ceilings & Acoustic Treatment	88.1	96.7	93.7	89.5	99.6	96.1	83.1	99.6	93.9	94.0	99.6	97.6	95.9	119.3	111.3	58.5	103.0	87.8
0960	Flooring	118.8	110.4	116.4	85.6	114.1	93.7	99.5	110.5	102.7	103.2	110.5	105.3	98.2	114.0	102.7	103.2	114.1	106.3
0970, 0990	Wall Finishes & Painting/Coating	108.3	115.5	112.6	94.5	77.9	84.5	98.3	66.4	79.1	102.4	97.2	99.3	94.4	122.9	111.6	108.3	86.4	95.1
09	FINISHES	102.5	103.6	103.1	91.6	100.3	96.4	96.0	96.5	96.3	98.6	101.9	100.4	100.9	118.6	110.7	90.0	101.3	96.2
COVERS	DIVS. 10 - 14, 25, 28, 41, 43, 44, 46	100.0	101.1	100.2	100.0	96.7	99.3	100.0	49.7	89.8	100.0	99.7	99.9	100.0	104.6	100.9	100.0	96.1	99.2
21, 22, 23	FIRE SUPPRESSION, PLUMBING & HVAC	100.3	97.0	98.9	100.1	87.5	95.0	95.3	87.3	92.1	100.0	95.6	98.2	100.0	103.5	101.4	94.8	86.9	91.6
26, 27, 3370	ELECTRICAL, COMMUNICATIONS & UTIL.	101.5	98.7	100.1	104.2	85.1	94.1	100.8	85.1	92.5	102.1	92.1	96.8	100.8	99.5	100.1	101.9	85.1	93.0
MF2014	WEIGHTED AVERAGE	99.5	102.2	100.7	97.8	94.5	96.3	96.7	92.5	94.8	99.1	99.0	99.1	100.0	107.2	103.1	96.8	94.6	95.9

WISCONSIN

DIVISION		OSHKOSH 549 MAT.	INST.	TOTAL	PORTAGE 539 MAT.	INST.	TOTAL	RACINE 534 MAT.	INST.	TOTAL	RHINELANDER 545 MAT.	INST.	TOTAL	SUPERIOR 548 MAT.	INST.	TOTAL	WAUSAU 544 MAT.	INST.	TOTAL
015433	CONTRACTOR EQUIPMENT		98.2	98.2		102.3	102.3		102.3	102.3		98.2	98.2		100.8	100.8		98.2	98.2
0241, 31 - 34	SITE & INFRASTRUCTURE, DEMOLITION	91.5	98.9	96.8	85.0	108.0	101.4	94.5	109.0	104.8	103.3	98.8	100.1	91.3	103.3	99.9	87.4	98.9	95.6
0310	Concrete Forming & Accessories	88.6	98.1	96.8	91.7	98.9	97.9	100.5	100.5	100.5	86.2	97.8	96.2	90.5	91.9	91.7	88.1	98.7	97.2
0320	Concrete Reinforcing	88.8	92.3	90.6	93.5	91.4	92.5	92.3	97.9	95.1	89.0	90.9	90.0	87.8	91.4	89.6	89.0	91.3	90.2
0330	Cast-in-Place Concrete	99.7	99.4	99.6	85.9	101.8	92.4	98.7	107.0	102.1	113.1	99.3	107.4	102.0	97.9	100.3	93.0	99.6	95.7
03	CONCRETE	92.4	97.6	94.9	86.4	98.5	92.4	97.9	102.3	100.1	104.7	97.2	101.0	91.8	94.2	93.0	87.3	97.8	92.5
04	MASONRY	107.9	101.4	103.8	99.6	104.3	102.5	100.6	114.0	109.0	125.3	101.4	110.4	120.8	104.5	110.6	107.4	101.4	103.6
05	METALS	95.4	96.7	95.8	91.3	96.0	92.7	94.8	101.2	96.8	95.3	96.7	95.7	96.1	98.2	96.8	95.1	97.7	95.9
06	WOOD, PLASTICS & COMPOSITES	89.8	99.3	95.1	89.2	99.2	94.8	96.0	96.4	98.1	87.3	99.3	94.0	92.3	90.1	91.0	89.2	99.3	94.9
07	THERMAL & MOISTURE PROTECTION	103.4	85.7	96.1	101.4	92.6	97.7	102.1	108.3	104.7	104.2	83.5	95.7	103.3	92.2	98.7	103.3	84.4	95.5
08	OPENINGS	95.9	95.5	95.8	98.1	92.3	96.8	102.5	101.1	102.2	96.0	87.1	93.9	88.6	91.1	89.2	96.1	95.7	96.0
0920	Plaster & Gypsum Board	86.2	99.6	95.2	86.4	99.6	95.3	94.5	96.7	96.0	86.2	99.6	95.2	87.6	90.2	89.4	86.2	99.6	95.2
0950, 0980	Ceilings & Acoustic Treatment	81.9	99.6	93.5	85.6	99.6	94.8	88.1	96.7	93.7	81.9	99.6	93.5	59.3	90.2	79.6	81.9	99.6	93.5
0960	Flooring	101.6	114.1	105.2	96.2	118.1	102.5	99.8	110.4	102.8	100.9	114.1	104.7	104.4	122.4	109.6	101.5	114.1	105.1
0970, 0990	Wall Finishes & Painting/Coating	102.2	101.4	101.7	98.3	66.4	79.1	98.3	115.5	108.7	102.2	60.4	76.9	96.0	102.8	100.1	102.2	83.2	90.7
09	FINISHES	94.0	100.9	97.8	94.1	98.6	96.6	97.4	103.6	100.8	94.8	97.0	96.0	89.3	98.6	94.4	93.7	100.7	97.5
COVERS	DIVS. 10 - 14, 25, 28, 41, 43, 44, 46	100.0	64.6	92.8	100.0	63.4	92.6	100.0	101.1	100.2	100.0	55.0	90.9	100.0	94.1	98.8	100.0	96.5	99.3
21, 22, 23	FIRE SUPPRESSION, PLUMBING & HVAC	95.6	88.2	91.3	95.3	95.3	95.3	100.1	97.0	98.8	95.6	86.7	92.0	94.8	89.4	92.6	95.6	86.4	91.9
26, 27, 3370	ELECTRICAL, COMMUNICATIONS & UTIL.	102.5	80.4	90.9	104.8	92.1	98.1	100.8	98.6	99.6	101.8	81.8	91.3	107.1	99.4	102.8	103.7	81.8	92.1
MF2014	WEIGHTED AVERAGE	96.9	91.9	94.7	95.3	96.6	95.9	99.1	102.4	100.5	99.3	91.2	95.8	96.6	96.3	96.5	96.2	93.5	95.0

For customer support on your Mechanical Cost Data, call 877.795.7394.

693

WYOMING

DIVISION		CASPER 826			CHEYENNE 820			NEWCASTLE 827			RAWLINS 823			RIVERTON 825			ROCK SPRINGS 829 - 831		
		MAT.	INST.	TOTAL	MAT.	INST.	TOTAL	MAT.	INST.	TOTAL	MAT.	INST.	TOTAL	MAT.	INST.	TOTAL	MAT.	INST.	TOTAL
015433	CONTRACTOR EQUIPMENT		98.2	98.2		98.2	98.2		98.2	98.2		98.2	98.2		98.2	98.2		98.2	98.2
0241, 31 - 34	SITE & INFRASTRUCTURE, DEMOLITION	98.6	94.4	95.6	92.5	94.4	93.8	85.1	94.0	91.5	98.4	94.0	95.3	92.1	94.0	93.5	89.2	94.0	92.6
0310	Concrete Forming & Accessories	102.5	50.1	57.3	104.4	53.5	60.5	94.8	62.2	66.7	99.2	62.2	67.3	93.7	62.1	66.5	101.5	62.1	67.5
0320	Concrete Reinforcing	105.7	70.6	87.8	98.0	70.6	84.1	105.5	70.8	87.9	105.2	70.8	87.7	106.1	70.6	88.0	106.1	70.6	88.0
0330	Cast-in-Place Concrete	102.8	72.0	90.2	95.0	72.1	85.6	95.9	61.6	81.8	96.0	61.6	81.9	96.0	60.0	81.2	95.9	59.9	81.1
03	CONCRETE	101.7	62.1	82.3	98.6	63.7	81.4	99.0	63.9	81.8	111.8	63.9	88.3	106.7	63.3	85.4	99.5	63.2	81.7
04	MASONRY	108.2	51.9	73.1	103.7	51.9	71.4	100.3	54.8	71.9	100.3	54.8	71.9	100.3	50.5	69.3	160.4	50.4	91.8
05	METALS	99.3	72.1	90.9	101.7	72.1	92.6	97.8	72.0	89.9	97.9	72.0	89.9	98.0	71.6	89.8	98.7	71.1	90.2
06	WOOD, PLASTICS & COMPOSITES	97.2	49.7	70.6	96.5	54.3	72.9	86.4	66.9	75.5	90.4	66.9	77.2	85.3	66.9	75.0	95.4	66.9	79.4
07	THERMAL & MOISTURE PROTECTION	105.9	55.1	85.1	100.1	55.6	81.8	101.8	55.8	82.9	103.2	55.8	83.7	102.7	59.9	85.1	101.9	59.9	84.7
08	OPENINGS	108.0	53.3	95.3	108.3	55.8	96.1	112.6	61.7	100.8	112.3	61.7	100.5	112.5	61.7	100.7	113.0	61.3	101.0
0920	Plaster & Gypsum Board	97.4	48.1	64.1	89.3	52.9	64.7	85.2	65.8	72.1	85.5	65.8	72.2	85.2	65.8	72.1	97.3	65.8	76.0
0950, 0980	Ceilings & Acoustic Treatment	99.2	48.1	65.6	91.0	52.9	66.0	92.9	65.8	75.1	92.9	65.8	75.1	92.9	65.8	75.1	92.9	65.8	75.1
0960	Flooring	112.0	56.4	96.1	109.9	56.4	94.6	103.5	56.4	90.0	106.2	56.4	92.0	103.0	56.4	89.7	108.4	56.4	93.5
0970, 0990	Wall Finishes & Painting/Coating	104.7	43.1	67.5	111.2	43.1	70.0	107.1	55.4	75.9	107.1	55.4	75.9	107.1	55.4	75.9	107.1	55.4	75.9
09	FINISHES	103.2	49.6	73.6	100.0	52.3	73.7	95.0	60.8	76.1	97.2	60.8	77.1	95.7	60.8	76.5	98.1	60.8	77.5
COVERS	DIVS. 10 - 14, 25, 28, 41, 43, 44, 46	100.0	87.6	97.5	100.0	88.2	97.6	100.0	89.1	97.8	100.0	89.1	97.8	100.0	81.4	96.2	100.0	80.8	96.1
21, 22, 23	FIRE SUPPRESSION, PLUMBING & HVAC	99.9	64.1	85.4	99.9	64.1	85.4	97.4	63.4	83.7	97.4	63.4	83.7	97.4	63.4	83.7	99.8	63.3	85.1
26, 27, 3370	ELECTRICAL, COMMUNICATIONS & UTIL.	103.3	64.1	82.6	105.8	68.6	86.1	104.5	69.0	85.7	104.5	69.0	85.7	104.5	62.9	82.5	102.4	62.9	81.5
MF2014	WEIGHTED AVERAGE	102.0	63.9	85.4	101.5	65.2	85.7	99.9	66.9	85.5	101.8	66.9	86.6	101.0	65.4	85.5	103.7	65.3	87.0

WYOMING / CANADA

DIVISION		SHERIDAN 828			WHEATLAND 822			WORLAND 824			YELLOWSTONE NAT'L PA 821			BARRIE, ONTARIO			BATHURST, NEW BRUNSWICK		
		MAT.	INST.	TOTAL	MAT.	INST.	TOTAL	MAT.	INST.	TOTAL	MAT.	INST.	TOTAL	MAT.	INST.	TOTAL	MAT.	INST.	TOTAL
015433	CONTRACTOR EQUIPMENT		98.2	98.2		98.2	98.2		98.2	98.2		98.2	98.2		102.1	102.1		102.2	102.2
0241, 31 - 34	SITE & INFRASTRUCTURE, DEMOLITION	92.4	94.4	93.8	89.6	94.0	92.6	87.1	94.0	92.0	87.2	94.2	92.2	118.5	100.4	105.6	103.4	96.6	98.6
0310	Concrete Forming & Accessories	102.2	50.6	57.7	96.8	52.7	58.8	96.9	62.0	66.8	96.9	63.1	67.8	124.4	89.4	94.2	104.4	63.7	69.3
0320	Concrete Reinforcing	106.1	70.9	88.2	105.5	70.8	87.8	106.1	70.6	88.0	107.9	70.6	88.9	173.2	83.8	127.7	137.0	56.5	96.0
0330	Cast-in-Place Concrete	99.2	72.0	88.1	100.2	61.5	84.3	95.9	59.9	81.1	95.9	61.6	81.8	165.6	89.0	134.2	131.6	61.9	103.0
03	CONCRETE	106.9	62.4	85.0	103.7	59.7	82.1	99.2	63.2	81.6	99.5	64.3	82.2	150.3	88.4	119.9	124.6	62.4	94.1
04	MASONRY	100.6	56.3	73.0	100.7	54.7	72.0	100.3	50.4	69.2	100.4	53.4	71.1	171.5	98.0	125.7	160.8	64.9	101.1
05	METALS	101.6	72.4	92.6	97.8	71.7	89.8	98.0	71.5	89.8	98.6	71.5	90.3	110.2	90.2	104.0	109.0	71.5	97.5
06	WOOD, PLASTICS & COMPOSITES	97.9	50.4	71.3	88.3	54.3	69.2	88.3	66.9	76.3	88.3	66.9	76.3	119.8	88.2	102.1	101.8	64.0	80.7
07	THERMAL & MOISTURE PROTECTION	102.9	56.3	83.8	102.1	55.8	83.1	101.9	57.5	83.7	101.4	58.8	83.9	111.8	89.1	102.5	106.0	62.2	88.1
08	OPENINGS	113.2	52.7	99.1	111.1	55.8	98.3	112.8	61.7	100.9	105.6	61.7	95.4	93.9	86.5	92.2	87.9	55.9	80.5
0920	Plaster & Gypsum Board	109.4	48.8	68.5	85.2	52.8	63.3	85.2	65.8	72.1	85.4	65.8	72.1	151.1	87.7	108.2	137.6	62.7	87.0
0950, 0980	Ceilings & Acoustic Treatment	95.7	48.8	64.9	92.9	52.8	66.6	92.9	65.8	75.1	93.7	65.8	75.4	91.5	87.7	89.0	106.4	62.7	77.7
0960	Flooring	107.5	56.4	92.9	105.0	56.4	91.1	105.0	56.4	91.1	105.0	56.4	91.1	127.2	98.9	119.1	107.9	47.2	90.6
0970, 0990	Wall Finishes & Painting/Coating	109.4	43.1	69.4	107.1	34.4	63.3	107.1	55.4	75.9	107.1	55.4	75.9	116.2	89.9	100.3	117.6	51.6	77.8
09	FINISHES	102.8	50.0	73.7	95.8	51.0	71.1	95.5	60.8	76.3	95.7	61.5	76.8	115.0	91.5	101.1	111.3	59.7	82.8
COVERS	DIVS. 10 - 14, 25, 28, 41, 43, 44, 46	100.0	87.7	97.5	100.0	80.3	96.0	100.0	81.4	96.2	100.0	82.4	96.4	139.2	74.4	126.1	131.1	65.3	117.8
21, 22, 23	FIRE SUPPRESSION, PLUMBING & HVAC	97.4	64.1	83.9	97.4	63.3	83.6	97.4	63.3	83.6	97.4	64.9	84.3	101.8	99.3	100.8	102.0	69.8	89.0
26, 27, 3370	ELECTRICAL, COMMUNICATIONS & UTIL.	107.4	63.0	83.9	104.5	68.6	85.5	104.5	62.9	82.5	103.2	62.9	81.9	116.8	90.5	102.9	112.7	61.8	85.8
MF2014	WEIGHTED AVERAGE	102.6	64.3	85.9	100.4	64.3	84.7	100.0	65.3	84.9	99.3	66.3	84.9	117.7	93.0	106.9	111.5	67.3	92.2

CANADA

DIVISION		BRANDON, MANITOBA			BRANTFORD, ONTARIO			BRIDGEWATER, NOVA SCOTIA			CALGARY, ALBERTA			CAP-DE-LA-MADELEINE, QUEBEC			CHARLESBOURG, QUEBEC		
		MAT.	INST.	TOTAL	MAT.	INST.	TOTAL	MAT.	INST.	TOTAL	MAT.	INST.	TOTAL	MAT.	INST.	TOTAL	MAT.	INST.	TOTAL
015433	CONTRACTOR EQUIPMENT		104.5	104.5		102.1	102.1		101.9	101.9		107.1	107.1		102.8	102.8		102.8	102.8
0241, 31 - 34	SITE & INFRASTRUCTURE, DEMOLITION	133.0	99.3	109.1	118.0	100.1	105.7	102.6	98.2	99.5	124.0	105.4	110.7	98.4	99.9	99.4	98.4	99.9	99.4
0310	Concrete Forming & Accessories	146.5	72.5	82.7	124.9	96.9	100.7	97.1	71.6	75.1	124.4	97.4	101.1	130.7	88.0	93.8	130.7	88.0	93.8
0320	Concrete Reinforcing	185.9	53.2	118.4	165.4	82.5	123.2	140.6	46.7	92.8	135.6	71.9	103.2	140.6	78.8	109.2	140.6	78.8	109.2
0330	Cast-in-Place Concrete	133.4	76.7	110.1	153.6	110.3	135.8	157.8	72.2	122.7	189.0	107.3	155.5	124.0	97.8	113.2	124.0	97.8	113.2
03	CONCRETE	145.6	71.0	109.0	141.8	98.8	120.7	139.4	67.9	104.3	159.1	96.1	128.2	124.3	89.8	107.4	124.3	89.8	107.4
04	MASONRY	233.5	67.6	130.1	168.0	102.6	127.2	163.4	71.9	106.4	196.3	91.4	130.9	163.9	87.1	116.0	163.9	87.1	116.0
05	METALS	125.6	76.8	110.6	109.1	90.9	103.5	108.2	74.2	97.7	137.0	89.3	122.3	107.4	87.6	101.3	107.4	87.6	101.3
06	WOOD, PLASTICS & COMPOSITES	154.2	73.5	109.0	123.7	95.7	108.1	92.9	71.1	80.7	102.3	97.2	99.4	135.3	88.0	108.8	135.3	88.0	108.8
07	THERMAL & MOISTURE PROTECTION	135.0	71.8	109.1	113.8	94.7	106.0	108.6	69.9	92.8	117.8	94.6	108.3	107.4	89.9	100.3	107.4	89.9	100.3
08	OPENINGS	105.5	64.8	96.1	92.4	92.7	92.5	86.1	64.9	81.1	89.9	86.2	89.1	93.6	81.1	90.7	93.6	81.1	90.7
0920	Plaster & Gypsum Board	119.3	72.2	87.4	130.1	95.5	106.7	130.7	70.0	89.7	139.4	96.6	110.5	157.0	87.4	109.9	157.0	87.4	109.9
0950, 0980	Ceilings & Acoustic Treatment	109.6	72.2	85.0	96.5	95.5	95.8	96.5	70.0	79.1	138.8	96.6	111.0	96.5	87.4	90.5	96.5	87.4	90.5
0960	Flooring	138.6	70.2	119.1	120.5	98.9	114.3	102.7	67.2	92.5	122.3	92.6	113.8	120.5	99.5	114.5	120.5	99.5	114.5
0970, 0990	Wall Finishes & Painting/Coating	126.5	58.7	85.6	118.3	98.6	106.4	118.3	64.1	85.6	127.4	112.4	118.3	118.3	91.7	102.2	118.3	91.7	102.2
09	FINISHES	122.5	71.0	94.1	112.4	97.9	104.4	106.6	70.3	86.6	125.5	98.7	110.7	115.3	90.8	101.8	115.3	90.8	101.8
COVERS	DIVS. 10 - 14, 25, 28, 41, 43, 44, 46	131.1	67.6	118.3	131.1	76.6	120.1	131.1	66.6	118.1	131.1	96.9	124.2	131.1	85.2	121.9	131.1	85.2	121.9
21, 22, 23	FIRE SUPPRESSION, PLUMBING & HVAC	102.1	84.1	94.9	102.0	102.3	102.1	102.0	84.4	94.9	101.2	92.9	97.8	102.3	91.0	97.8	102.3	91.0	97.8
26, 27, 3370	ELECTRICAL, COMMUNICATIONS & UTIL.	118.6	69.6	92.7	111.5	89.9	100.1	117.1	64.4	89.2	113.5	101.3	107.1	111.2	72.6	90.8	111.2	72.6	90.8
MF2014	WEIGHTED AVERAGE	125.2	75.8	103.7	114.9	96.9	107.1	112.9	74.6	96.2	123.5	95.7	111.4	112.4	87.6	101.6	112.4	87.6	101.6

CANADA

DIVISION		CHARLOTTETOWN, PRINCE EDWARD ISLAND			CHICOUTIMI, QUEBEC			CORNER BROOK, NEWFOUNDLAND			CORNWALL, ONTARIO			DALHOUSIE, NEW BRUNSWICK			DARTMOUTH, NOVA SCOTIA		
		MAT.	INST.	TOTAL	MAT.	INST.	TOTAL	MAT.	INST.	TOTAL	MAT.	INST.	TOTAL	MAT.	INST.	TOTAL	MAT.	INST.	TOTAL
015433	CONTRACTOR EQUIPMENT		101.8	101.8		102.8	102.8		103.3	103.3		102.1	102.1		102.2	102.2		101.9	101.9
0241, 31 - 34	SITE & INFRASTRUCTURE, DEMOLITION	121.3	95.7	103.1	100.7	99.1	99.6	137.7	97.1	108.9	116.1	100.2	104.8	100.9	96.6	97.8	124.8	98.2	105.9
0310	Concrete Forming & Accessories	110.4	58.2	65.4	133.2	93.1	98.6	121.8	61.7	69.9	122.6	89.6	94.1	103.5	63.9	69.3	111.8	71.6	77.2
0320	Concrete Reinforcing	141.5	46.8	93.3	103.3	95.9	99.5	170.3	48.6	108.4	165.4	82.2	123.1	142.7	56.6	98.9	178.2	46.7	111.3
0330	Cast-in-Place Concrete	166.8	61.2	123.4	124.0	99.8	114.0	155.3	70.6	120.5	138.2	100.6	122.7	131.8	62.0	103.2	149.8	72.2	118.0
03	CONCRETE	147.1	57.9	103.3	114.6	95.9	105.4	182.2	63.2	123.7	134.3	92.1	113.6	132.5	62.5	98.1	162.1	67.9	115.8
04	MASONRY	178.4	62.6	106.3	162.7	95.0	120.5	229.3	64.1	126.4	166.8	94.0	121.4	160.2	64.9	100.8	244.8	71.9	137.1
05	METALS	133.9	67.6	113.5	108.8	92.2	103.7	126.0	72.9	109.7	109.1	89.6	103.1	104.4	71.8	94.3	125.9	74.2	110.0
06	WOOD, PLASTICS & COMPOSITES	97.8	57.8	75.4	137.4	93.6	112.9	128.3	60.8	90.5	122.1	89.0	103.6	100.2	64.0	80.0	116.1	71.1	90.9
07	THERMAL & MOISTURE PROTECTION	121.3	60.4	96.3	106.5	96.4	102.3	139.8	61.8	107.8	113.6	89.3	103.6	110.8	62.2	90.9	136.8	69.9	109.4
08	OPENINGS	88.2	50.0	79.3	92.2	80.6	89.5	112.4	57.1	99.5	93.6	86.2	91.9	89.1	55.9	81.4	95.3	64.9	88.2
0920	Plaster & Gypsum Board	131.6	56.3	80.7	159.4	93.2	114.6	149.0	59.3	88.4	189.4	88.6	121.3	141.2	62.7	88.2	144.6	70.0	94.2
0950, 0980	Ceilings & Acoustic Treatment	120.2	56.3	78.2	105.6	93.2	97.4	110.5	59.3	76.9	99.0	88.6	92.2	98.9	62.7	75.1	117.2	70.0	86.2
0960	Flooring	111.7	62.9	97.8	123.6	99.5	116.8	121.0	57.2	102.8	120.5	97.5	113.9	106.5	72.1	96.7	116.1	67.2	102.1
0970, 0990	Wall Finishes & Painting/Coating	125.6	42.9	75.7	117.6	106.4	110.8	126.4	61.8	87.4	118.3	91.9	102.4	120.7	51.6	79.0	126.4	64.1	88.8
09	FINISHES	117.1	57.8	84.4	118.4	96.5	106.4	122.0	60.9	88.4	121.0	91.6	104.8	110.6	64.5	85.2	120.1	70.3	92.6
COVERS	DIVS. 10 - 14, 25, 28, 41, 43, 44, 46	131.1	64.8	117.7	131.1	86.5	122.1	131.1	65.6	117.9	131.1	74.0	119.6	131.1	65.3	117.8	131.1	66.6	118.1
21, 22, 23	FIRE SUPPRESSION, PLUMBING & HVAC	102.3	63.8	86.8	102.0	92.3	98.1	102.1	71.8	89.9	102.3	100.1	101.4	102.0	69.8	89.0	102.1	84.4	95.0
26, 27, 3370	ELECTRICAL, COMMUNICATIONS & UTIL.	110.6	52.3	79.8	109.7	86.8	97.6	116.0	58.3	85.5	112.4	90.9	101.0	114.5	58.4	84.8	120.7	64.4	90.9
MF2014	WEIGHTED AVERAGE	120.0	62.8	95.1	111.4	92.9	103.3	129.6	67.6	102.6	115.0	93.3	105.2	111.9	67.5	92.5	126.1	74.6	103.7

CANADA

DIVISION		EDMONTON, ALBERTA			FORT MCMURRAY, ALBERTA			FREDERICTON, NEW BRUNSWICK			GATINEAU, QUEBEC			GRANBY, QUEBEC			HALIFAX, NOVA SCOTIA		
		MAT.	INST.	TOTAL	MAT.	INST.	TOTAL	MAT.	INST.	TOTAL	MAT.	INST.	TOTAL	MAT.	INST.	TOTAL	MAT.	INST.	TOTAL
015433	CONTRACTOR EQUIPMENT		107.1	107.1		104.4	104.4		102.2	102.2		102.8	102.8		102.8	102.8		101.9	101.9
0241, 31 - 34	SITE & INFRASTRUCTURE, DEMOLITION	144.1	105.4	116.6	123.1	101.3	107.6	105.0	96.6	99.1	98.2	99.8	99.4	98.7	99.8	99.5	109.2	98.3	101.5
0310	Concrete Forming & Accessories	123.2	97.4	100.9	122.8	92.3	96.5	120.3	64.2	71.9	130.7	87.8	93.7	130.7	87.8	93.7	108.4	78.8	82.9
0320	Concrete Reinforcing	134.2	71.9	102.5	152.9	71.8	111.7	132.8	56.7	94.1	148.7	78.8	113.2	148.7	78.8	113.2	144.3	62.7	102.7
0330	Cast-in-Place Concrete	183.9	107.3	152.4	203.9	105.1	163.3	128.3	62.0	101.1	122.3	97.8	112.2	126.4	97.7	114.6	149.4	81.9	121.7
03	CONCRETE	156.4	96.1	126.8	163.6	93.1	129.0	126.6	62.7	95.2	124.8	89.8	107.6	126.7	89.7	108.5	136.4	77.3	107.4
04	MASONRY	187.2	91.4	127.5	207.9	89.1	133.8	182.6	66.5	110.2	163.7	87.1	116.0	164.1	87.1	116.1	185.9	84.7	122.8
05	METALS	136.9	89.3	122.3	135.4	88.9	121.1	129.1	72.3	111.6	107.4	87.4	101.2	107.4	87.4	101.2	134.9	81.5	118.4
06	WOOD, PLASTICS & COMPOSITES	105.2	97.2	100.7	117.6	91.8	103.2	116.7	64.0	87.2	135.3	88.0	108.8	135.3	88.0	108.8	97.1	77.8	86.3
07	THERMAL & MOISTURE PROTECTION	122.7	94.6	111.2	121.9	91.9	109.6	116.0	63.2	94.4	107.4	89.9	100.3	107.4	88.3	99.6	119.0	79.4	102.8
08	OPENINGS	89.2	86.2	88.5	93.6	83.3	91.2	89.3	54.8	81.2	93.6	76.3	89.6	93.6	76.3	89.6	93.5	72.0	88.5
0920	Plaster & Gypsum Board	134.3	96.6	108.8	128.2	91.1	103.1	137.4	62.7	86.9	126.6	87.4	100.1	129.0	87.4	100.8	127.0	77.0	93.2
0950, 0980	Ceilings & Acoustic Treatment	139.5	96.6	111.3	104.9	91.1	95.8	116.1	62.7	81.0	96.5	87.4	90.5	96.5	87.4	90.5	118.7	77.0	91.3
0960	Flooring	122.5	92.6	114.0	120.5	92.6	112.5	116.8	75.3	104.9	120.5	99.5	114.5	120.5	99.5	114.5	109.2	88.3	103.3
0970, 0990	Wall Finishes & Painting/Coating	122.4	112.4	116.3	118.4	95.5	104.5	123.1	65.8	88.5	118.3	91.7	102.2	118.3	91.7	102.2	124.9	87.4	102.2
09	FINISHES	126.6	98.7	111.2	115.3	93.0	103.0	117.7	66.7	89.6	111.2	90.8	100.0	111.5	90.8	100.1	115.6	81.8	97.0
COVERS	DIVS. 10 - 14, 25, 28, 41, 43, 44, 46	131.1	96.9	124.2	131.1	95.2	123.9	131.1	65.3	117.8	131.1	85.2	121.9	131.1	85.2	121.9	131.1	68.5	118.5
21, 22, 23	FIRE SUPPRESSION, PLUMBING & HVAC	101.1	92.9	97.8	102.4	99.0	101.0	102.3	79.3	93.0	102.3	91.0	97.8	102.0	91.0	97.6	100.6	77.6	91.3
26, 27, 3370	ELECTRICAL, COMMUNICATIONS & UTIL.	110.6	101.3	105.7	105.8	85.4	95.1	116.1	77.1	95.5	111.2	72.6	90.8	111.8	72.6	91.1	117.0	81.2	98.1
MF2014	WEIGHTED AVERAGE	123.1	95.7	111.2	123.6	92.8	110.2	117.4	72.5	97.9	112.1	87.4	101.3	112.3	87.4	101.4	119.6	80.9	102.7

CANADA

DIVISION		HAMILTON, ONTARIO			HULL, QUEBEC			JOLIETTE, QUEBEC			KAMLOOPS, BRITISH COLUMBIA			KINGSTON, ONTARIO			KITCHENER, ONTARIO		
		MAT.	INST.	TOTAL	MAT.	INST.	TOTAL	MAT.	INST.	TOTAL	MAT.	INST.	TOTAL	MAT.	INST.	TOTAL	MAT.	INST.	TOTAL
015433	CONTRACTOR EQUIPMENT		108.9	108.9		102.8	102.8		102.8	102.8		106.1	106.1		104.4	104.4		104.3	104.3
0241, 31 - 34	SITE & INFRASTRUCTURE, DEMOLITION	115.5	112.4	113.3	98.2	99.8	99.4	98.8	99.9	99.6	120.5	103.3	108.3	116.1	104.0	107.5	99.6	104.8	103.3
0310	Concrete Forming & Accessories	120.9	94.7	98.3	130.7	87.8	93.7	130.7	88.0	93.8	123.4	90.8	95.3	122.8	89.6	94.2	113.0	87.2	90.8
0320	Concrete Reinforcing	146.4	92.0	118.7	148.7	78.8	113.2	140.6	78.8	109.2	110.3	76.8	93.2	165.4	82.2	123.1	100.8	91.9	96.3
0330	Cast-in-Place Concrete	143.5	101.0	126.1	122.3	97.8	112.2	127.4	97.8	115.2	110.3	101.6	106.7	138.2	100.6	122.7	129.0	93.6	114.4
03	CONCRETE	132.8	96.3	114.8	124.8	89.8	107.6	125.9	89.8	108.2	132.9	92.0	112.8	136.2	92.2	114.6	130.8	90.4	100.8
04	MASONRY	183.0	101.2	132.0	163.7	87.1	116.0	164.1	87.1	116.1	170.7	95.0	123.5	173.5	94.1	124.0	150.8	97.3	117.4
05	METALS	123.4	92.5	113.9	107.4	87.4	101.2	107.4	87.6	101.3	109.8	87.5	102.9	110.8	89.5	104.3	117.4	92.2	109.6
06	WOOD, PLASTICS & COMPOSITES	100.4	93.9	96.8	135.3	88.0	108.8	135.3	88.0	108.8	105.2	89.4	96.4	122.1	89.2	103.7	109.7	85.5	96.1
07	THERMAL & MOISTURE PROTECTION	118.2	95.7	109.0	107.4	89.9	100.2	107.4	89.9	100.2	123.3	86.8	108.3	113.6	90.4	104.1	108.6	92.8	102.1
08	OPENINGS	91.3	91.6	91.4	93.6	76.3	89.6	93.6	81.1	90.7	90.2	85.7	89.2	93.6	85.9	91.8	84.7	85.3	84.8
0920	Plaster & Gypsum Board	149.5	93.6	111.7	126.6	87.4	100.1	157.0	87.4	109.9	113.6	88.5	96.6	192.8	88.7	122.4	120.2	84.9	96.3
0950, 0980	Ceilings & Acoustic Treatment	122.1	93.6	103.4	96.5	87.4	90.5	96.5	87.4	90.5	96.5	88.5	91.2	112.4	88.7	96.8	98.6	84.9	89.6
0960	Flooring	123.1	102.3	117.2	120.5	99.5	114.5	120.5	99.5	114.5	119.6	55.6	101.3	120.5	97.5	113.9	110.5	102.3	108.1
0970, 0990	Wall Finishes & Painting/Coating	121.6	102.0	110.2	118.3	91.7	102.2	118.3	91.7	102.2	118.3	84.0	98.1	118.3	84.9	98.1	115.3	92.0	101.8
09	FINISHES	122.0	97.0	108.2	111.2	90.8	100.0	115.3	90.8	101.8	111.7	84.2	96.6	124.4	90.9	105.9	106.4	90.2	97.5
COVERS	DIVS. 10 - 14, 25, 28, 41, 43, 44, 46	131.1	98.8	124.6	131.1	85.2	121.9	131.1	85.2	121.9	131.1	95.6	124.0	131.1	74.0	119.6	131.1	97.0	124.2
21, 22, 23	FIRE SUPPRESSION, PLUMBING & HVAC	101.3	89.2	96.4	102.0	91.0	97.6	102.0	91.0	97.6	102.0	94.4	98.9	102.3	100.3	101.5	100.4	87.7	95.3
26, 27, 3370	ELECTRICAL, COMMUNICATIONS & UTIL.	108.3	100.5	104.2	112.9	72.6	91.6	111.8	72.6	91.1	115.1	81.9	97.6	112.4	89.6	100.3	111.5	98.0	104.4
MF2014	WEIGHTED AVERAGE	117.0	96.7	108.2	112.2	87.4	101.4	112.5	87.6	101.7	114.5	90.6	104.1	116.0	93.4	106.2	109.6	92.9	102.3

For customer support on your Mechanical Cost Data, call 877.795.7394.

695

CANADA

DIVISION		LAVAL, QUEBEC			LETHBRIDGE, ALBERTA			LLOYDMINSTER, ALBERTA			LONDON, ONTARIO			MEDICINE HAT, ALBERTA			MONCTON, NEW BRUNSWICK		
		MAT.	INST.	TOTAL	MAT.	INST.	TOTAL	MAT.	INST.	TOTAL	MAT.	INST.	TOTAL	MAT.	INST.	TOTAL	MAT.	INST.	TOTAL
015433	CONTRACTOR EQUIPMENT		102.8	102.8		104.4	104.4		104.4	104.4		104.4	104.4		104.4	104.4		102.2	102.2
0241, 31 - 34	SITE & INFRASTRUCTURE, DEMOLITION	98.7	99.8	99.5	115.6	101.9	105.9	115.4	101.3	105.4	113.1	104.9	107.3	114.2	101.4	105.1	102.8	96.8	98.5
0310	Concrete Forming & Accessories	130.8	87.8	93.8	124.1	92.4	96.8	122.3	82.6	88.1	119.1	88.5	92.7	124.1	82.6	88.3	104.4	64.5	69.9
0320	Concrete Reinforcing	148.7	78.8	113.2	152.9	71.8	111.7	152.9	71.8	111.7	132.8	91.9	112.0	152.9	71.8	111.7	137.0	61.5	98.6
0330	Cast-in-Place Concrete	126.4	97.8	114.6	153.0	105.1	133.3	141.9	101.4	125.3	147.4	98.6	127.4	141.9	101.4	125.3	126.8	78.7	107.0
03	CONCRETE	126.7	89.8	108.6	139.5	93.2	116.7	134.1	87.5	111.3	132.4	92.7	112.9	134.3	87.5	111.3	122.3	69.7	96.5
04	MASONRY	164.0	87.1	116.1	181.7	89.1	124.0	163.4	82.5	113.0	182.6	98.7	130.3	163.4	82.5	113.0	160.5	65.0	101.0
05	METALS	107.4	87.4	101.3	128.8	88.9	116.5	110.0	88.8	103.4	124.9	91.8	114.7	110.0	88.7	103.4	109.0	80.4	100.2
06	WOOD, PLASTICS & COMPOSITES	135.5	88.0	108.9	121.3	91.8	104.8	117.6	82.0	97.7	108.2	86.6	96.1	121.3	82.0	99.3	101.8	64.0	80.7
07	THERMAL & MOISTURE PROTECTION	107.9	89.9	100.5	119.1	91.9	108.0	116.0	87.5	104.3	119.1	93.5	108.6	122.3	87.5	108.0	110.2	65.3	91.8
08	OPENINGS	93.6	76.3	89.6	93.6	83.3	91.2	93.6	77.9	89.9	86.4	86.6	86.4	93.6	77.9	89.9	87.9	61.0	81.6
0920	Plaster & Gypsum Board	129.3	87.4	100.9	119.1	91.1	100.2	114.4	81.0	91.8	152.4	86.1	107.6	117.2	81.0	92.8	137.6	62.7	87.0
0950, 0980	Ceilings & Acoustic Treatment	96.5	87.4	90.5	104.1	91.1	95.5	96.5	81.0	86.3	121.1	86.1	98.1	96.5	81.0	86.3	106.4	62.7	77.7
0960	Flooring	120.5	99.5	114.5	120.5	92.6	112.5	120.5	92.6	112.5	117.2	102.3	113.0	120.5	92.6	112.5	107.9	72.1	97.7
0970, 0990	Wall Finishes & Painting/Coating	118.3	91.7	102.2	118.2	104.2	109.8	118.4	81.2	96.0	122.0	99.4	108.3	118.2	81.2	95.9	117.6	51.6	77.8
09	FINISHES	111.5	90.8	100.1	112.9	94.0	102.5	110.7	84.2	96.0	120.5	91.9	104.7	110.9	84.0	96.1	111.3	64.5	85.5
COVERS	DIVS. 10 - 14, 25, 28, 41, 43, 44, 46	131.1	85.2	121.9	131.1	95.2	123.9	131.1	91.9	123.2	131.1	97.5	124.3	131.1	91.9	123.2	131.1	65.3	117.8
21, 22, 23	FIRE SUPPRESSION, PLUMBING & HVAC	100.4	91.0	96.6	102.2	95.6	99.6	102.3	95.6	99.6	101.4	86.6	95.4	102.0	92.3	98.1	102.0	70.1	89.1
26, 27, 3370	ELECTRICAL, COMMUNICATIONS & UTIL.	112.9	72.6	91.6	107.1	85.4	95.7	104.9	85.4	94.6	107.0	97.5	102.0	104.9	85.4	94.6	117.2	61.8	88.0
MF2014	WEIGHTED AVERAGE	112.1	87.4	101.3	118.4	92.2	107.0	113.4	88.9	102.8	116.4	93.3	106.3	113.6	88.2	102.5	111.8	70.1	93.6

CANADA

DIVISION		MONTREAL, QUEBEC			MOOSE JAW, SASKATCHEWAN			NEW GLASGOW, NOVA SCOTIA			NEWCASTLE, NEW BRUNSWICK			NORTH BAY, ONTARIO			OSHAWA, ONTARIO		
		MAT.	INST.	TOTAL	MAT.	INST.	TOTAL	MAT.	INST.	TOTAL	MAT.	INST.	TOTAL	MAT.	INST.	TOTAL	MAT.	INST.	TOTAL
015433	CONTRACTOR EQUIPMENT		104.5	104.5		100.6	100.6		101.9	101.9		102.2	102.2		102.1	102.1		104.3	104.3
0241, 31 - 34	SITE & INFRASTRUCTURE, DEMOLITION	112.1	99.4	103.1	115.5	95.9	101.6	118.5	98.2	104.1	103.4	96.6	98.6	133.9	99.8	109.7	110.7	103.9	105.9
0310	Concrete Forming & Accessories	124.3	93.4	97.7	107.1	60.5	66.9	111.7	71.6	77.1	104.4	63.9	69.5	147.4	87.0	95.3	118.4	89.1	93.1
0320	Concrete Reinforcing	134.0	96.0	114.7	107.9	61.9	84.5	170.3	46.7	107.4	137.0	56.6	96.1	201.7	81.8	140.7	159.6	84.4	121.3
0330	Cast-in-Place Concrete	161.7	101.2	136.8	137.7	71.1	110.4	149.8	72.2	118.0	131.6	62.0	103.0	144.3	87.1	120.8	149.1	87.9	124.0
03	CONCRETE	139.7	96.5	118.4	117.9	65.1	92.0	160.9	67.9	115.2	124.6	62.5	94.1	163.4	86.3	125.5	135.6	88.0	112.2
04	MASONRY	167.9	95.1	122.5	162.0	64.4	101.1	228.9	71.9	131.1	160.8	64.9	101.1	235.3	89.9	144.7	153.8	94.4	116.8
05	METALS	128.0	92.7	117.1	106.3	73.9	96.3	123.6	74.2	108.4	109.0	71.8	97.5	124.6	89.2	113.7	108.6	91.4	103.3
06	WOOD, PLASTICS & COMPOSITES	114.4	93.9	102.9	102.7	59.1	78.3	116.1	71.1	90.9	101.8	64.0	80.7	157.4	87.6	118.3	116.7	88.2	100.7
07	THERMAL & MOISTURE PROTECTION	116.5	96.9	108.5	106.8	64.1	89.3	136.8	69.9	109.4	110.2	62.2	90.5	143.5	85.6	119.8	109.4	87.4	100.4
08	OPENINGS	90.1	82.4	88.3	89.3	56.8	81.7	95.3	64.9	88.2	87.9	55.9	80.5	104.2	83.7	99.4	90.4	87.9	89.8
0920	Plaster & Gypsum Board	135.1	93.2	106.8	110.7	57.8	74.9	142.7	70.0	93.6	137.6	62.7	87.0	140.3	87.1	104.4	124.7	87.7	99.7
0950, 0980	Ceilings & Acoustic Treatment	126.1	93.2	104.4	96.5	57.8	71.0	109.6	70.0	83.6	106.4	62.7	77.7	109.6	87.1	94.8	95.3	87.7	90.3
0960	Flooring	122.0	99.5	115.6	110.6	63.3	97.1	116.1	67.2	102.1	109.7	72.1	97.7	138.6	97.5	126.9	113.7	104.9	111.2
0970, 0990	Wall Finishes & Painting/Coating	120.8	106.4	112.1	118.3	66.8	87.2	126.4	64.1	88.8	117.6	51.6	77.8	126.4	91.2	105.1	115.3	106.7	110.1
09	FINISHES	119.9	96.7	107.1	107.1	61.4	81.9	118.2	70.3	91.8	111.3	64.5	85.5	125.1	89.8	105.7	107.9	93.9	100.2
COVERS	DIVS. 10 - 14, 25, 28, 41, 43, 44, 46	131.1	87.1	122.3	131.1	64.6	117.7	131.1	66.6	118.1	131.1	65.3	117.8	131.1	72.7	119.3	131.1	97.2	124.3
21, 22, 23	FIRE SUPPRESSION, PLUMBING & HVAC	101.2	92.4	97.7	102.3	76.1	91.8	102.1	84.4	95.0	102.0	69.8	89.0	102.1	98.1	100.5	100.4	100.7	100.5
26, 27, 3370	ELECTRICAL, COMMUNICATIONS & UTIL.	112.6	86.8	99.0	113.9	62.8	86.9	116.3	64.4	88.9	112.0	61.8	85.5	116.4	90.8	102.9	112.6	90.5	100.9
MF2014	WEIGHTED AVERAGE	117.8	93.2	107.1	110.7	69.5	92.7	124.1	74.6	102.5	111.6	67.9	92.5	127.3	91.1	111.5	112.3	94.3	104.4

CANADA

DIVISION		OTTAWA, ONTARIO			OWEN SOUND, ONTARIO			PETERBOROUGH, ONTARIO			PORTAGE LA PRAIRIE, MANITOBA			PRINCE ALBERT, SASKATCHEWAN			PRINCE GEORGE, BRITISH COLUMBIA		
		MAT.	INST.	TOTAL	MAT.	INST.	TOTAL	MAT.	INST.	TOTAL	MAT.	INST.	TOTAL	MAT.	INST.	TOTAL	MAT.	INST.	TOTAL
015433	CONTRACTOR EQUIPMENT		104.3	104.3		102.1	102.1		102.1	102.1		104.5	104.5		100.6	100.6		106.1	106.1
0241, 31 - 34	SITE & INFRASTRUCTURE, DEMOLITION	108.2	104.6	105.7	118.5	100.3	105.5	118.0	100.2	105.3	116.6	99.3	104.3	110.9	96.1	100.3	124.0	103.3	109.3
0310	Concrete Forming & Accessories	121.2	91.4	95.5	124.4	85.5	90.9	124.9	88.1	93.1	124.3	72.0	79.2	107.1	60.3	66.8	112.7	85.5	89.3
0320	Concrete Reinforcing	138.9	91.9	115.0	173.2	83.8	127.7	165.4	82.3	123.1	152.9	53.2	102.2	112.7	61.8	86.8	110.3	76.8	93.2
0330	Cast-in-Place Concrete	145.6	99.9	126.8	165.6	83.1	131.7	153.6	88.7	127.0	141.9	76.1	114.9	124.8	71.0	102.7	138.2	101.6	123.2
03	CONCRETE	132.6	94.4	113.8	150.3	84.6	118.1	141.8	87.4	115.1	127.8	70.6	99.7	112.6	65.0	89.2	145.4	89.7	118.0
04	MASONRY	168.0	98.8	124.9	171.5	95.6	124.2	168.0	96.7	123.5	166.5	66.6	104.2	161.1	64.4	100.8	172.9	95.0	124.3
05	METALS	127.9	91.7	116.8	110.2	90.0	104.0	109.1	89.7	103.1	110.0	76.7	99.7	106.3	73.7	96.3	109.8	87.6	102.9
06	WOOD, PLASTICS & COMPOSITES	109.4	90.6	98.9	119.8	84.3	99.9	123.7	86.3	102.8	121.1	73.5	94.4	102.7	59.1	78.3	105.2	82.1	92.3
07	THERMAL & MOISTURE PROTECTION	124.4	93.8	111.8	111.8	86.4	101.4	113.8	91.2	104.5	107.3	71.3	92.5	106.7	63.0	88.8	117.2	86.1	104.4
08	OPENINGS	93.9	88.7	92.7	93.9	83.1	91.4	92.4	85.5	90.8	93.6	64.8	86.9	88.2	56.8	80.9	90.2	81.8	88.2
0920	Plaster & Gypsum Board	168.3	90.2	115.5	151.1	83.7	105.6	130.1	85.8	100.1	114.2	72.2	85.8	110.7	57.8	74.9	113.6	81.0	91.6
0950, 0980	Ceilings & Acoustic Treatment	125.6	90.2	102.4	91.5	83.7	86.4	96.5	85.8	89.4	96.5	72.2	80.5	96.5	57.8	71.0	96.5	81.0	86.3
0960	Flooring	112.3	97.5	108.1	127.2	98.9	119.1	120.5	97.5	113.9	120.5	70.2	106.1	110.6	63.3	97.1	116.0	76.2	104.6
0970, 0990	Wall Finishes & Painting/Coating	123.2	94.2	105.7	116.2	89.9	100.3	118.3	93.5	103.3	118.4	58.7	82.4	118.3	57.0	81.3	118.3	84.0	97.6
09	FINISHES	122.6	92.9	106.2	115.0	88.6	100.5	112.4	90.5	100.3	110.5	70.8	88.6	107.1	60.3	81.3	110.6	83.4	95.6
COVERS	DIVS. 10 - 14, 25, 28, 41, 43, 44, 46	131.1	95.7	124.0	139.2	73.2	125.9	131.1	74.2	119.6	131.1	67.2	118.2	131.1	64.6	117.7	131.1	94.8	123.8
21, 22, 23	FIRE SUPPRESSION, PLUMBING & HVAC	101.4	87.2	95.7	101.8	98.1	100.3	102.0	101.7	101.9	102.0	83.6	94.5	102.3	68.6	88.7	102.0	94.4	98.9
26, 27, 3370	ELECTRICAL, COMMUNICATIONS & UTIL.	107.3	98.7	102.8	118.2	89.6	103.1	111.5	90.4	100.4	113.6	60.1	85.3	113.9	62.8	86.9	111.7	81.9	96.0
MF2014	WEIGHTED AVERAGE	117.3	94.0	107.2	117.8	91.2	106.2	114.9	93.1	105.4	113.4	74.2	96.3	109.9	67.7	91.5	115.4	89.9	104.3

CANADA

DIVISION		QUEBEC, QUEBEC			RED DEER, ALBERTA			REGINA, SASKATCHEWAN			RIMOUSKI, QUEBEC			ROUYN-NORANDA, QUEBEC			SAINT HYACINTHE, QUEBEC		
		MAT.	INST.	TOTAL	MAT.	INST.	TOTAL	MAT.	INST.	TOTAL	MAT.	INST.	TOTAL	MAT.	INST.	TOTAL	MAT.	INST.	TOTAL
015433	CONTRACTOR EQUIPMENT		104.9	104.9		104.4	104.4		100.6	100.6		102.8	102.8		102.8	102.8		102.8	102.8
0241, 31 - 34	SITE & INFRASTRUCTURE, DEMOLITION	110.8	99.5	102.8	114.2	101.4	105.1	124.4	98.0	105.6	98.7	99.1	99.0	98.2	99.8	99.4	98.7	99.8	99.5
0310	Concrete Forming & Accessories	125.6	93.7	98.1	139.6	82.6	90.4	117.6	90.9	94.6	130.7	93.1	98.3	130.7	87.8	93.7	130.7	87.8	93.7
0320	Concrete Reinforcing	138.3	96.0	116.8	152.9	71.8	111.7	138.0	84.8	111.0	106.0	95.9	100.9	148.7	78.8	113.2	148.7	78.8	113.2
0330	Cast-in-Place Concrete	141.8	101.6	125.2	141.9	101.4	125.3	171.5	95.7	140.4	128.6	99.8	116.8	122.3	97.8	112.2	126.4	97.8	114.6
03	CONCRETE	131.0	96.7	114.1	135.3	87.5	111.8	148.9	91.4	120.6	121.1	95.9	108.7	124.8	89.8	107.6	126.7	89.8	108.5
04	MASONRY	163.0	95.1	120.7	163.4	82.5	113.0	183.5	96.7	129.4	163.6	95.0	120.9	163.7	87.1	116.0	164.0	87.1	116.1
05	METALS	128.0	93.0	117.2	110.0	88.7	103.4	129.3	86.4	116.1	106.9	92.2	102.4	107.4	87.4	101.2	107.4	87.4	101.2
06	WOOD, PLASTICS & COMPOSITES	115.8	93.9	103.6	121.3	82.0	99.3	101.9	91.9	96.3	135.3	93.6	112.0	135.3	88.0	108.8	135.3	88.0	108.8
07	THERMAL & MOISTURE PROTECTION	118.4	97.0	109.6	132.2	87.5	113.9	123.2	86.0	107.9	107.4	96.4	102.9	107.4	89.9	100.2	107.7	89.9	100.4
08	OPENINGS	93.0	90.1	92.3	93.6	77.9	89.9	91.9	80.7	89.3	93.2	80.6	90.2	93.6	76.3	89.6	93.6	76.3	89.6
0920	Plaster & Gypsum Board	142.1	93.2	109.0	117.2	81.0	92.8	130.5	91.6	104.2	156.8	93.2	113.8	126.4	87.4	100.0	128.7	87.4	100.8
0950, 0980	Ceilings & Acoustic Treatment	126.1	93.2	104.4	96.5	81.0	86.3	125.4	91.6	103.2	95.7	93.2	94.0	95.7	87.4	90.2	95.7	87.4	90.2
0960	Flooring	125.8	96.5	118.3	123.0	92.6	114.3	120.0	65.5	104.4	120.5	99.5	114.5	120.5	99.5	114.5	120.5	99.5	114.5
0970, 0990	Wall Finishes & Painting/Coating	127.8	106.4	114.9	118.2	81.2	95.9	123.9	92.9	105.2	118.3	106.4	111.1	118.3	91.7	102.2	118.3	91.7	102.2
09	FINISHES	122.2	96.8	108.2	111.7	84.0	96.4	122.7	87.2	103.1	115.1	96.5	104.9	111.0	90.8	99.9	111.3	90.8	100.0
COVERS	DIVS. 10 - 14, 25, 28, 41, 43, 44, 46	131.1	87.3	122.3	131.1	91.9	123.2	131.1	72.4	119.3	131.1	86.5	122.1	131.1	85.2	121.9	131.1	85.2	121.9
21, 22, 23	FIRE SUPPRESSION, PLUMBING & HVAC	101.2	92.4	97.7	102.0	92.3	98.1	100.7	84.8	94.3	102.0	92.3	98.1	102.0	91.0	97.6	98.8	91.0	95.7
26, 27, 3370	ELECTRICAL, COMMUNICATIONS & UTIL.	111.4	86.8	98.4	104.9	85.4	94.6	114.4	92.4	102.8	111.8	86.8	98.6	111.8	72.6	91.1	112.5	72.6	91.4
MF2014	WEIGHTED AVERAGE	117.1	93.6	106.8	114.1	88.2	102.8	120.7	89.0	106.9	111.8	92.9	103.6	112.0	87.4	101.3	111.6	87.4	101.1

CANADA

DIVISION		SAINT JOHN, NEW BRUNSWICK			SARNIA, ONTARIO			SASKATOON, SASKATCHEWAN			SAULT STE MARIE, ONTARIO			SHERBROOKE, QUEBEC			SOREL, QUEBEC		
		MAT.	INST.	TOTAL	MAT.	INST.	TOTAL	MAT.	INST.	TOTAL	MAT.	INST.	TOTAL	MAT.	INST.	TOTAL	MAT.	INST.	TOTAL
015433	CONTRACTOR EQUIPMENT		102.2	102.2		102.1	102.1		100.6	100.6		102.1	102.1		102.8	102.8		102.8	102.8
0241, 31 - 34	SITE & INFRASTRUCTURE, DEMOLITION	103.5	98.0	99.6	116.5	100.3	105.0	111.7	98.0	102.0	106.7	99.8	101.8	98.7	99.8	99.5	98.8	99.9	99.6
0310	Concrete Forming & Accessories	124.6	67.0	74.9	123.6	95.3	99.1	107.4	90.9	93.2	112.8	87.9	91.3	130.7	87.8	93.7	130.7	88.0	93.8
0320	Concrete Reinforcing	137.0	62.0	98.8	117.4	83.6	100.2	114.5	84.8	99.4	106.1	82.1	93.9	148.7	78.8	113.2	140.6	78.8	109.2
0330	Cast-in-Place Concrete	129.7	80.1	109.3	141.7	102.1	125.5	135.4	95.7	119.1	127.4	87.2	110.9	126.4	97.8	114.6	127.4	97.8	115.2
03	CONCRETE	125.2	71.4	98.8	128.6	95.4	112.3	119.6	91.4	105.7	113.1	86.8	100.2	126.7	89.8	108.5	125.9	89.8	108.2
04	MASONRY	180.8	73.7	114.0	179.1	96.9	129.5	168.8	96.7	123.8	164.5	94.4	120.8	164.1	87.1	116.1	164.1	87.1	116.1
05	METALS	108.9	81.7	100.5	109.1	90.1	103.2	104.5	86.4	99.0	108.3	90.2	102.7	107.4	87.4	101.2	107.4	87.6	101.3
06	WOOD, PLASTICS & COMPOSITES	125.7	65.4	91.9	122.6	94.7	107.0	100.5	91.9	95.7	109.6	88.3	97.6	135.3	88.0	108.8	135.3	88.0	108.8
07	THERMAL & MOISTURE PROTECTION	110.5	70.9	94.3	113.9	94.7	106.0	107.1	86.0	98.4	112.7	88.4	102.7	107.4	89.9	100.3	107.4	89.9	100.2
08	OPENINGS	87.8	60.6	81.5	94.7	89.4	93.4	88.9	80.7	87.0	86.5	85.6	86.3	93.6	76.3	89.6	93.6	81.1	90.7
0920	Plaster & Gypsum Board	152.7	64.2	92.9	150.6	94.4	112.6	123.5	91.6	101.9	119.6	87.8	98.1	128.7	87.4	100.8	156.8	87.4	109.9
0950, 0980	Ceilings & Acoustic Treatment	111.5	64.2	80.4	100.7	94.4	96.6	117.4	91.6	100.4	96.5	87.8	90.8	95.7	87.4	90.2	95.7	87.4	90.2
0960	Flooring	119.5	72.1	105.9	120.5	106.6	116.5	120.0	65.5	98.0	113.4	100.8	109.8	120.5	99.5	114.5	120.5	99.5	114.5
0970, 0990	Wall Finishes & Painting/Coating	117.6	80.2	95.0	118.3	105.7	110.7	120.7	92.9	103.9	118.3	97.9	106.0	118.3	91.7	102.2	118.3	91.7	102.2
09	FINISHES	117.9	69.0	91.0	116.1	98.9	106.6	114.3	87.2	99.3	108.1	91.5	99.0	111.3	90.8	100.0	115.1	90.8	101.7
COVERS	DIVS. 10 - 14, 25, 28, 41, 43, 44, 46	131.1	66.3	118.0	131.1	75.7	119.9	131.1	74.1	119.6	131.1	96.4	124.1	131.1	85.2	121.9	131.1	85.2	121.9
21, 22, 23	FIRE SUPPRESSION, PLUMBING & HVAC	102.0	79.8	93.0	102.0	107.8	104.3	100.5	84.8	94.2	102.0	94.8	99.1	102.3	91.0	97.8	102.0	91.0	97.6
26, 27, 3370	ELECTRICAL, COMMUNICATIONS & UTIL.	120.3	88.2	103.3	114.2	92.7	102.8	116.2	92.4	103.6	113.0	90.8	101.3	111.8	72.6	91.1	111.8	72.6	91.1
MF2014	WEIGHTED AVERAGE	114.0	77.9	98.3	114.8	97.5	107.3	111.1	89.1	101.5	110.3	92.1	102.3	112.4	87.4	101.5	112.5	87.6	101.7

CANADA

DIVISION		ST CATHARINES, ONTARIO			ST JEROME, QUEBEC			ST JOHNS, NEWFOUNDLAND			SUDBURY, ONTARIO			SUMMERSIDE, PRINCE EDWARD ISLAND			SYDNEY, NOVA SCOTIA		
		MAT.	INST.	TOTAL	MAT.	INST.	TOTAL	MAT.	INST.	TOTAL	MAT.	INST.	TOTAL	MAT.	INST.	TOTAL	MAT.	INST.	TOTAL
015433	CONTRACTOR EQUIPMENT		102.1	102.1		102.8	102.8		103.3	103.3		102.1	102.1		101.8	101.8		101.9	101.9
0241, 31 - 34	SITE & INFRASTRUCTURE, DEMOLITION	100.0	101.4	101.0	98.2	99.8	99.4	123.4	99.9	106.7	100.1	100.9	100.7	128.4	95.7	105.2	114.3	98.2	102.8
0310	Concrete Forming & Accessories	110.9	94.2	96.5	130.7	87.8	93.7	123.9	84.8	90.1	107.0	89.0	91.5	112.0	58.3	65.7	111.7	71.6	77.1
0320	Concrete Reinforcing	101.7	92.0	96.7	148.7	78.8	113.2	158.9	75.5	116.4	102.5	91.3	96.8	168.1	46.8	106.4	170.3	46.7	107.4
0330	Cast-in-Place Concrete	123.2	100.6	113.9	122.3	97.8	112.2	167.5	96.4	138.3	124.2	97.0	113.0	142.3	61.2	109.0	115.6	72.2	97.8
03	CONCRETE	108.0	95.9	102.1	124.8	89.8	107.6	152.6	87.2	120.5	118.4	92.2	100.4	170.1	57.9	115.0	144.7	67.9	106.9
04	MASONRY	150.3	100.7	119.4	163.7	87.1	116.0	187.7	92.4	128.3	150.4	96.6	116.8	228.1	62.6	125.0	226.2	71.9	130.1
05	METALS	107.7	92.2	102.9	107.4	87.4	101.2	136.3	84.3	120.3	107.7	91.5	102.7	123.6	67.6	106.4	123.6	74.2	108.4
06	WOOD, PLASTICS & COMPOSITES	107.4	93.9	99.8	135.3	88.0	108.8	114.7	82.8	96.8	103.6	88.3	95.0	116.4	57.8	83.6	116.1	71.1	90.9
07	THERMAL & MOISTURE PROTECTION	108.6	96.5	103.6	107.4	89.9	100.2	127.1	91.4	112.5	108.0	91.7	101.3	136.1	61.1	105.4	136.8	69.9	109.4
08	OPENINGS	84.1	90.2	85.5	93.6	76.3	89.6	93.7	74.5	89.2	84.8	85.6	85.0	108.0	50.0	94.5	95.3	64.9	88.2
0920	Plaster & Gypsum Board	108.5	93.6	98.4	126.4	87.4	100.0	143.9	82.1	102.1	112.9	87.8	95.9	143.1	56.3	84.4	142.7	70.0	93.6
0950, 0980	Ceilings & Acoustic Treatment	95.3	93.6	94.1	95.7	87.4	90.2	120.6	82.1	95.3	90.3	87.8	88.6	109.6	56.3	74.6	109.6	70.0	83.6
0960	Flooring	109.3	98.9	106.3	120.5	99.5	114.5	116.0	59.3	99.8	107.3	100.8	105.5	116.2	62.9	100.9	116.1	67.2	102.1
0970, 0990	Wall Finishes & Painting/Coating	115.3	116.0	115.7	118.3	91.7	102.2	120.4	95.9	105.6	115.3	93.6	102.2	126.4	42.9	76.0	126.4	64.1	88.8
09	FINISHES	103.8	97.8	100.4	111.0	90.8	99.9	122.0	81.4	99.6	102.7	91.6	96.5	119.2	57.8	85.4	118.2	70.3	91.8
COVERS	DIVS. 10 - 14, 25, 28, 41, 43, 44, 46	131.1	75.2	119.8	131.1	85.2	121.9	131.1	72.6	119.3	131.1	97.1	124.3	131.1	64.8	117.7	131.1	66.6	118.1
21, 22, 23	FIRE SUPPRESSION, PLUMBING & HVAC	100.4	87.8	95.3	102.0	91.0	97.6	100.8	82.1	93.3	101.9	87.8	96.2	102.1	63.8	86.7	102.1	84.4	95.0
26, 27, 3370	ELECTRICAL, COMMUNICATIONS & UTIL.	112.9	98.8	105.5	112.4	72.6	91.3	114.4	83.2	97.9	110.3	99.2	104.4	115.4	52.3	82.0	116.3	64.4	88.9
MF2014	WEIGHTED AVERAGE	107.6	94.5	101.9	112.1	87.4	101.3	122.7	85.2	106.4	107.7	93.0	101.3	126.7	62.8	98.8	122.1	74.6	101.4

City Cost Indexes

		CANADA																
	DIVISION	THUNDER BAY, ONTARIO			TIMMINS, ONTARIO			TORONTO, ONTARIO			TROIS RIVIERES, QUEBEC			TRURO, NOVA SCOTIA			VANCOUVER, BRITISH COLUMBIA		
		MAT.	INST.	TOTAL	MAT.	INST.	TOTAL	MAT.	INST.	TOTAL	MAT.	INST.	TOTAL	MAT.	INST.	TOTAL	MAT.	INST.	TOTAL
015433	CONTRACTOR EQUIPMENT		102.1	102.1		102.1	102.1		104.4	104.4		102.8	102.8		101.9	101.9		112.7	112.7
0241, 31 - 34	SITE & INFRASTRUCTURE, DEMOLITION	104.8	101.3	102.3	118.0	99.8	105.1	133.8	105.6	113.7	114.8	99.9	104.2	102.8	98.2	99.5	118.2	107.6	110.6
0310	Concrete Forming & Accessories	118.3	92.2	95.8	124.9	87.0	92.2	124.3	100.8	104.0	155.5	88.0	97.3	97.1	71.6	75.1	121.1	91.0	95.1
0320	Concrete Reinforcing	90.9	90.8	90.9	165.4	81.8	122.9	146.0	94.4	119.7	170.3	78.8	123.8	140.6	46.7	92.8	134.1	79.2	106.2
0330	Cast-in-Place Concrete	135.6	99.7	120.8	153.6	87.1	126.3	136.8	111.8	126.5	119.7	97.8	110.7	159.5	72.2	123.7	143.5	97.9	124.8
03	CONCRETE	115.8	94.5	105.3	141.8	86.3	114.5	129.7	103.1	116.7	147.4	89.8	119.1	140.2	67.9	104.7	137.1	91.4	114.6
04	MASONRY	151.0	100.6	119.6	168.0	89.9	119.3	182.0	110.9	137.7	230.7	87.1	141.2	163.6	71.9	106.5	166.7	91.4	119.7
05	METALS	107.5	91.1	102.5	109.1	89.2	103.0	132.5	93.9	120.7	122.7	87.6	111.9	108.2	74.2	97.7	139.3	90.3	124.2
06	WOOD, PLASTICS & COMPOSITES	116.7	90.9	102.2	123.7	87.6	103.5	115.1	98.9	106.0	173.0	88.0	125.4	92.9	71.1	80.7	104.8	91.5	97.4
07	THERMAL & MOISTURE PROTECTION	108.8	93.8	102.7	113.8	85.6	102.2	123.9	102.8	115.2	135.2	89.9	116.6	108.6	69.9	92.8	125.2	86.9	109.5
08	OPENINGS	83.3	88.0	84.4	92.4	83.7	90.4	90.5	96.8	92.0	105.5	81.1	99.8	86.1	64.9	81.1	88.9	86.9	88.4
0920	Plaster & Gypsum Board	139.8	90.5	106.5	130.1	87.1	101.1	133.0	98.8	109.9	173.4	87.4	115.3	130.7	70.0	89.7	129.7	90.6	103.3
0950, 0980	Ceilings & Acoustic Treatment	90.3	90.5	90.4	96.5	87.1	90.3	114.9	98.8	104.3	108.8	87.4	94.7	96.5	70.0	79.1	132.2	90.6	104.8
0960	Flooring	113.7	105.9	111.4	120.5	97.5	113.9	117.0	108.4	114.6	119.7	99.5	127.5	102.7	67.2	92.5	124.6	96.0	116.4
0970, 0990	Wall Finishes & Painting/Coating	115.3	94.6	102.8	118.3	91.2	101.9	120.4	106.7	112.1	126.4	91.7	105.4	118.3	64.1	85.6	121.7	95.6	106.0
09	FINISHES	108.5	95.1	101.1	112.4	89.8	99.9	116.9	103.0	109.2	128.7	90.8	107.8	106.6	70.3	86.6	123.8	92.7	106.6
COVERS	DIVS. 10 - 14, 25, 28, 41, 43, 44, 46	131.1	75.2	119.8	131.1	72.7	119.3	131.1	101.2	125.1	131.1	85.2	121.9	131.1	66.6	118.1	131.1	94.9	123.8
21, 22, 23	FIRE SUPPRESSION, PLUMBING & HVAC	100.4	88.0	95.4	102.0	98.1	100.4	101.0	97.3	99.5	102.1	91.0	97.7	102.0	84.4	94.9	101.2	80.5	92.8
26, 27, 3370	ELECTRICAL, COMMUNICATIONS & UTIL.	111.5	97.0	103.8	113.0	90.8	101.3	108.8	101.0	104.7	116.6	72.6	93.4	111.5	64.4	86.6	109.0	83.3	95.4
MF2014	WEIGHTED AVERAGE	108.8	93.4	102.1	115.1	91.1	104.6	118.3	101.3	110.9	124.8	87.6	108.6	112.5	74.6	96.0	119.5	89.1	106.2

		CANADA																
	DIVISION	VICTORIA, BRITISH COLUMBIA			WHITEHORSE, YUKON			WINDSOR, ONTARIO			WINNIPEG, MANITOBA			YARMOUTH, NOVA SCOTIA			YELLOWKNIFE, NWT		
		MAT.	INST.	TOTAL	MAT.	INST.	TOTAL	MAT.	INST.	TOTAL	MAT.	INST.	TOTAL	MAT.	INST.	TOTAL	MAT.	INST.	TOTAL
015433	CONTRACTOR EQUIPMENT		109.3	109.3		102.1	102.1		102.1	102.1		106.5	106.5		101.9	101.9		101.9	101.9
0241, 31 - 34	SITE & INFRASTRUCTURE, DEMOLITION	123.0	107.1	111.7	142.3	96.7	109.9	95.9	101.3	99.7	118.2	100.9	105.9	118.3	98.2	104.0	154.1	101.0	116.4
0310	Concrete Forming & Accessories	112.7	90.4	93.5	129.4	59.6	69.2	118.3	90.5	94.3	128.5	69.3	77.4	111.7	71.6	77.1	132.7	80.2	87.4
0320	Concrete Reinforcing	112.0	79.1	95.2	160.8	60.0	109.5	99.6	91.9	95.7	132.9	57.9	94.7	170.3	46.7	107.4	151.4	62.7	106.2
0330	Cast-in-Place Concrete	140.3	97.2	122.6	209.3	70.3	152.2	126.2	101.9	116.2	163.9	75.0	127.4	148.3	72.2	117.0	211.2	88.1	160.6
03	CONCRETE	150.1	90.7	120.9	182.4	64.1	124.3	109.6	94.7	102.3	144.4	69.7	107.7	160.1	67.9	114.8	182.0	80.0	131.9
04	MASONRY	171.7	91.3	121.6	260.3	62.8	137.2	150.5	99.9	119.0	176.6	70.8	110.7	228.8	71.9	131.0	260.3	74.8	144.7
05	METALS	107.0	86.3	100.6	135.3	73.6	116.3	107.6	91.7	102.7	141.7	75.0	121.2	123.6	74.2	108.4	137.9	78.7	119.7
06	WOOD, PLASTICS & COMPOSITES	104.4	91.4	97.1	117.6	58.3	84.4	116.7	89.0	101.2	113.6	70.2	89.3	116.1	71.1	90.9	124.0	81.9	100.4
07	THERMAL & MOISTURE PROTECTION	116.9	86.7	104.5	139.1	62.0	107.5	108.6	93.9	102.6	120.8	71.8	100.7	136.8	69.9	109.4	151.5	78.4	121.5
08	OPENINGS	90.5	83.2	88.8	107.6	55.5	95.4	83.1	87.5	84.1	85.5	62.6	80.2	95.3	64.9	88.2	102.2	68.9	94.4
0920	Plaster & Gypsum Board	115.8	90.6	98.8	150.5	56.8	87.2	125.7	88.6	100.6	129.5	68.6	88.3	142.7	70.0	93.6	163.7	81.1	107.9
0950, 0980	Ceilings & Acoustic Treatment	98.1	90.6	93.2	144.6	56.8	86.9	90.3	88.6	89.2	136.2	68.6	91.8	109.6	70.0	83.6	138.9	81.1	100.9
0960	Flooring	116.4	76.2	104.9	141.0	61.3	118.2	113.7	103.1	110.6	117.0	75.6	105.2	116.1	67.2	102.1	135.5	92.5	123.2
0970, 0990	Wall Finishes & Painting/Coating	120.7	95.6	105.5	143.1	56.0	90.5	115.3	95.5	103.3	124.5	56.5	83.4	126.4	64.1	88.8	135.3	80.3	102.1
09	FINISHES	112.1	89.3	99.5	140.9	59.3	95.9	106.3	93.4	99.2	122.3	69.5	93.2	118.2	70.3	91.8	137.5	82.0	106.9
COVERS	DIVS. 10 - 14, 25, 28, 41, 43, 44, 46	131.1	71.5	119.1	131.1	63.9	117.6	131.1	74.7	119.7	131.1	68.7	118.5	131.1	66.6	118.1	131.1	67.1	118.2
21, 22, 23	FIRE SUPPRESSION, PLUMBING & HVAC	102.0	80.5	93.3	102.5	74.8	91.3	100.4	88.0	95.4	101.1	66.9	87.3	102.1	84.4	95.0	102.7	94.0	99.2
26, 27, 3370	ELECTRICAL, COMMUNICATIONS & UTIL.	113.1	82.7	97.0	133.5	62.2	95.8	116.0	99.0	107.0	113.3	68.0	89.4	116.3	64.4	88.9	130.5	85.2	106.5
MF2014	WEIGHTED AVERAGE	115.7	87.2	103.3	135.3	68.5	106.2	108.2	93.4	101.7	121.0	71.8	99.5	124.0	74.6	102.5	135.3	84.2	113.0

Costs shown in RSMeans cost data publications are based on national averages for materials and installation. To adjust these costs to a specific location, simply multiply the base cost by the factor and divide by 100 for that city. The data is arranged alphabetically by state and postal zip code numbers. For a city not listed, use the factor for a nearby city with similar economic characteristics.

STATE/ZIP	CITY	MAT.	INST.	TOTAL
ALABAMA				
350-352	Birmingham	100.6	76.7	90.2
354	Tuscaloosa	99.8	61.2	83.0
355	Jasper	100.2	57.5	81.6
356	Decatur	99.8	57.9	81.5
357-358	Huntsville	99.8	72.1	87.8
359	Gadsden	99.9	60.0	82.5
360-361	Montgomery	99.9	57.1	81.2
362	Anniston	98.6	61.8	82.6
363	Dothan	99.0	49.4	77.4
364	Evergreen	98.6	50.7	77.7
365-366	Mobile	99.9	65.4	84.9
367	Selma	98.7	49.9	77.4
368	Phenix City	99.5	54.6	79.9
369	Butler	98.9	49.4	77.3
ALASKA				
995-996	Anchorage	121.1	115.6	118.7
997	Fairbanks	121.0	116.4	119.0
998	Juneau	121.3	115.6	118.8
999	Ketchikan	132.0	115.6	124.8
ARIZONA				
850,853	Phoenix	99.2	73.8	88.1
851,852	Mesa/Tempe	98.6	72.8	87.4
855	Globe	98.8	70.0	86.3
856-857	Tucson	97.4	72.4	86.5
859	Show Low	98.9	72.5	87.4
860	Flagstaff	101.3	71.8	88.4
863	Prescott	99.2	68.9	86.0
864	Kingman	97.6	72.0	86.4
865	Chambers	97.6	71.5	86.2
ARKANSAS				
716	Pine Bluff	99.4	63.3	83.6
717	Camden	97.1	50.1	76.6
718	Texarkana	98.2	49.2	76.8
719	Hot Springs	96.4	51.8	76.9
720-722	Little Rock	98.5	64.0	83.4
723	West Memphis	96.1	62.4	81.4
724	Jonesboro	96.7	59.1	80.3
725	Batesville	94.6	54.6	77.1
726	Harrison	95.9	50.4	76.0
727	Fayetteville	93.4	52.0	75.4
728	Russellville	94.6	53.7	76.8
729	Fort Smith	97.2	61.6	81.7
CALIFORNIA				
900-902	Los Angeles	99.2	117.5	107.2
903-905	Inglewood	94.8	114.1	103.2
906-908	Long Beach	96.2	114.1	104.0
910-912	Pasadena	96.4	114.6	104.3
913-916	Van Nuys	99.1	114.6	105.9
917-918	Alhambra	98.3	114.8	105.5
919-921	San Diego	100.5	109.6	104.5
922	Palm Springs	97.2	112.0	103.6
923-924	San Bernardino	95.1	111.3	102.1
925	Riverside	99.4	114.6	106.0
926-927	Santa Ana	96.9	111.8	103.4
928	Anaheim	99.4	114.9	106.2
930	Oxnard	99.6	114.4	106.0
931	Santa Barbara	98.9	115.1	106.0
932-933	Bakersfield	100.7	113.6	106.3
934	San Luis Obispo	99.4	112.0	104.9
935	Mojave	96.8	109.8	102.5
936-938	Fresno	100.1	115.3	106.7
939	Salinas	100.2	121.6	109.5
940-941	San Francisco	105.2	145.4	122.7
942,956-958	Sacramento	100.5	119.3	108.7
943	Palo Alto	98.9	134.5	114.4
944	San Mateo	100.9	134.6	115.6
945	Vallejo	100.1	127.2	111.9
946	Oakland	103.1	134.3	116.7
947	Berkeley	102.7	135.1	116.8
948	Richmond	101.8	133.1	115.4
949	San Rafael	103.5	136.7	118.0
950	Santa Cruz	104.5	121.8	112.0

STATE/ZIP	CITY	MAT.	INST.	TOTAL
CALIFORNIA (CONT'D)				
951	San Jose	102.5	136.6	117.4
952	Stockton	100.9	117.2	108.0
953	Modesto	100.8	115.7	107.3
954	Santa Rosa	100.8	134.7	115.6
955	Eureka	102.1	116.4	108.4
959	Marysville	101.3	115.9	107.7
960	Redding	107.8	117.7	112.1
961	Susanville	106.8	118.1	111.7
COLORADO				
800-802	Denver	100.8	81.7	92.5
803	Boulder	97.8	81.3	90.6
804	Golden	99.8	78.9	90.7
805	Fort Collins	101.3	79.0	91.6
806	Greeley	98.9	75.5	88.7
807	Fort Morgan	98.3	79.0	89.9
808-809	Colorado Springs	100.4	84.0	93.2
810	Pueblo	101.1	80.3	92.0
811	Alamosa	102.2	73.7	89.8
812	Salida	101.9	75.3	90.3
813	Durango	102.7	75.7	90.9
814	Montrose	101.4	75.4	90.1
815	Grand Junction	104.8	75.3	91.9
816	Glenwood Springs	102.3	78.1	91.8
CONNECTICUT				
060	New Britain	100.1	120.6	109.1
061	Hartford	101.4	120.8	109.8
062	Willimantic	100.8	120.2	109.2
063	New London	97.1	120.7	107.4
064	Meriden	98.9	120.7	108.4
065	New Haven	101.6	120.7	109.9
066	Bridgeport	101.3	120.8	109.8
067	Waterbury	100.9	120.7	109.5
068	Norwalk	100.8	128.7	113.0
069	Stamford	100.9	128.7	113.0
D.C.				
200-205	Washington	101.1	91.9	97.1
DELAWARE				
197	Newark	98.8	109.2	103.3
198	Wilmington	99.3	109.2	103.6
199	Dover	99.3	109.2	103.6
FLORIDA				
320,322	Jacksonville	98.8	66.4	84.7
321	Daytona Beach	99.0	72.9	87.6
323	Tallahassee	100.8	56.9	81.7
324	Panama City	100.2	57.8	81.7
325	Pensacola	102.7	61.2	84.6
326,344	Gainesville	100.3	66.7	85.7
327-328,347	Orlando	100.5	70.0	87.2
329	Melbourne	101.7	76.3	90.6
330-332,340	Miami	99.8	73.1	88.2
333	Fort Lauderdale	98.6	72.6	87.2
334,349	West Palm Beach	97.4	71.1	85.9
335-336,346	Tampa	100.1	78.4	90.6
337	St. Petersburg	102.6	63.3	85.5
338	Lakeland	99.4	77.8	90.0
339,341	Fort Myers	98.6	72.6	87.3
342	Sarasota	100.8	74.3	89.2
GEORGIA				
300-303,399	Atlanta	97.6	74.4	87.5
304	Statesboro	97.3	55.8	79.2
305	Gainesville	96.0	61.6	81.0
306	Athens	95.5	63.9	81.7
307	Dalton	97.2	59.0	80.5
308-309	Augusta	96.3	63.6	82.1
310-312	Macon	97.2	64.7	83.0
313-314	Savannah	99.2	60.9	82.5
315	Waycross	98.3	59.4	81.4
316	Valdosta	98.5	61.4	82.3
317,398	Albany	98.5	61.3	82.3
318-319	Columbus	98.4	65.4	84.0

Location Factors

STATE/ZIP	CITY	MAT.	INST.	TOTAL
HAWAII				
967	Hilo	114.8	120.2	117.2
968	Honolulu	119.5	120.2	119.8
STATES & POSS.				
969	Guam	136.2	58.0	102.1
IDAHO				
832	Pocatello	101.1	77.4	90.8
833	Twin Falls	102.2	67.3	87.0
834	Idaho Falls	99.8	74.7	88.9
835	Lewiston	108.7	84.9	98.3
836-837	Boise	100.1	78.6	90.7
838	Coeur d'Alene	107.9	82.2	96.7
ILLINOIS				
600-603	North Suburban	99.3	135.1	114.9
604	Joliet	99.2	138.4	116.3
605	South Suburban	99.3	135.1	114.9
606-608	Chicago	99.8	139.7	117.2
609	Kankakee	95.6	131.7	111.3
610-611	Rockford	99.5	124.8	110.5
612	Rock Island	97.2	100.5	98.6
613	La Salle	98.4	124.4	109.7
614	Galesburg	98.2	106.7	101.9
615-616	Peoria	100.5	109.3	104.4
617	Bloomington	97.5	109.4	102.7
618-619	Champaign	101.0	107.6	103.9
620-622	East St. Louis	96.4	108.7	101.8
623	Quincy	98.1	101.0	99.3
624	Effingham	97.3	106.0	101.1
625	Decatur	99.2	104.9	101.7
626-627	Springfield	100.0	106.5	102.8
628	Centralia	95.0	108.5	100.9
629	Carbondale	94.7	107.9	100.5
INDIANA				
460	Anderson	97.0	83.8	91.3
461-462	Indianapolis	99.1	84.8	92.9
463-464	Gary	98.3	110.4	103.6
465-466	South Bend	98.0	84.4	92.1
467-468	Fort Wayne	97.6	79.1	89.5
469	Kokomo	94.8	81.5	89.0
470	Lawrenceburg	94.0	77.7	86.9
471	New Albany	95.1	75.4	86.5
472	Columbus	97.2	80.9	90.1
473	Muncie	97.9	82.3	91.1
474	Bloomington	99.3	81.8	91.6
475	Washington	95.8	81.8	89.7
476-477	Evansville	97.1	86.9	92.6
478	Terre Haute	97.9	86.5	92.9
479	Lafayette	96.9	82.7	90.7
IOWA				
500-503,509	Des Moines	100.5	85.2	93.8
504	Mason City	97.6	65.8	83.8
505	Fort Dodge	97.9	58.2	80.6
506-507	Waterloo	99.5	71.8	87.4
508	Creston	98.2	77.3	89.1
510-511	Sioux City	99.7	72.6	87.9
512	Sibley	98.3	56.4	80.0
513	Spencer	99.9	56.6	81.0
514	Carroll	97.0	71.2	85.8
515	Council Bluffs	100.7	78.1	90.9
516	Shenandoah	97.6	71.2	86.1
520	Dubuque	99.1	80.2	90.9
521	Decorah	98.0	64.2	83.3
522-524	Cedar Rapids	100.0	83.2	92.7
525	Ottumwa	97.9	73.7	87.3
526	Burlington	97.2	75.3	87.6
527-528	Davenport	99.0	93.1	96.4
KANSAS				
660-662	Kansas City	99.5	94.8	97.5
664-666	Topeka	99.5	66.0	84.9
667	Fort Scott	97.9	73.0	87.0
668	Emporia	97.9	69.2	85.4
669	Belleville	99.6	64.9	84.4
670-672	Wichita	100.7	66.0	85.6
673	Independence	100.9	69.2	87.1
674	Salina	101.5	67.6	86.7
675	Hutchinson	96.5	65.0	82.8
676	Hays	100.5	65.4	85.2
677	Colby	101.1	65.0	85.4

STATE/ZIP	CITY	MAT.	INST.	TOTAL
KANSAS (CONT'D)				
678	Dodge City	102.7	66.0	86.7
679	Liberal	100.4	65.5	85.2
KENTUCKY				
400-402	Louisville	96.6	83.3	90.8
403-405	Lexington	96.6	81.4	90.0
406	Frankfort	98.1	80.7	90.5
407-409	Corbin	94.0	71.8	84.3
410	Covington	95.9	91.4	93.9
411-412	Ashland	94.6	99.3	96.6
413-414	Campton	95.8	78.6	88.3
415-416	Pikeville	96.9	88.6	93.3
417-418	Hazard	95.2	78.7	88.0
420	Paducah	93.9	84.5	89.8
421-422	Bowling Green	96.2	83.9	90.8
423	Owensboro	96.3	88.9	93.1
424	Henderson	93.7	86.1	90.4
425-426	Somerset	93.2	79.9	87.4
427	Elizabethtown	92.8	82.0	88.1
LOUISIANA				
700-701	New Orleans	101.4	69.5	87.5
703	Thibodaux	98.2	65.4	83.9
704	Hammond	95.8	54.5	77.8
705	Lafayette	97.9	63.0	82.7
706	Lake Charles	98.1	65.0	83.7
707-708	Baton Rouge	99.1	63.7	83.7
710-711	Shreveport	101.0	59.6	83.0
712	Monroe	99.3	56.5	80.6
713-714	Alexandria	99.4	57.2	81.0
MAINE				
039	Kittery	95.2	84.7	90.6
040-041	Portland	102.0	90.1	96.8
042	Lewiston	99.6	90.1	95.5
043	Augusta	102.1	79.5	92.3
044	Bangor	99.0	89.5	94.9
045	Bath	97.5	81.9	90.7
046	Machias	97.1	85.5	92.0
047	Houlton	97.2	85.6	92.1
048	Rockland	96.3	85.5	91.6
049	Waterville	97.5	79.5	89.7
MARYLAND				
206	Waldorf	97.1	81.7	90.4
207-208	College Park	97.2	83.4	91.2
209	Silver Spring	96.5	83.2	90.7
210-212	Baltimore	100.8	81.9	92.6
214	Annapolis	101.5	82.0	93.0
215	Cumberland	96.8	81.8	90.2
216	Easton	98.3	68.6	85.3
217	Hagerstown	97.6	83.3	91.4
218	Salisbury	98.7	62.1	82.7
219	Elkton	95.9	81.9	89.8
MASSACHUSETTS				
010-011	Springfield	100.2	112.6	105.6
012	Pittsfield	99.7	111.3	104.8
013	Greenfield	97.7	112.3	104.1
014	Fitchburg	96.5	126.5	109.6
015-016	Worcester	100.2	126.4	111.6
017	Framingham	95.7	135.1	112.9
018	Lowell	99.5	134.5	114.8
019	Lawrence	100.3	134.6	115.3
020-022, 024	Boston	101.6	139.5	118.1
023	Brockton	99.9	128.1	112.2
025	Buzzards Bay	94.4	128.1	109.1
026	Hyannis	97.0	128.1	110.6
027	New Bedford	99.1	128.3	111.8
MICHIGAN				
480,483	Royal Oak	94.9	106.9	100.1
481	Ann Arbor	97.1	107.5	101.7
482	Detroit	97.8	109.6	102.9
484-485	Flint	96.8	96.0	96.4
486	Saginaw	96.4	91.8	94.4
487	Bay City	96.6	91.7	94.5
488-489	Lansing	99.3	94.7	97.3
490	Battle Creek	98.3	87.6	93.7
491	Kalamazoo	98.7	85.7	93.0
492	Jackson	96.5	95.1	95.9
493,495	Grand Rapids	99.7	85.1	93.3
494	Muskegon	97.2	84.8	91.8

Location Factors

STATE/ZIP	CITY	MAT.	INST.	TOTAL	STATE/ZIP	CITY	MAT.	INST.	TOTAL
MICHIGAN (CONT'D)					**NEW HAMPSHIRE (CONT'D)**				
496	Traverse City	96.3	80.1	89.3	032-033	Concord	101.1	92.5	97.4
497	Gaylord	97.2	83.6	91.3	034	Keene	97.2	73.3	86.7
498-499	Iron mountain	99.3	86.2	93.6	035	Littleton	97.1	76.3	88.0
					036	Charleston	96.7	69.5	84.8
MINNESOTA					037	Claremont	95.9	69.5	84.4
550-551	Saint Paul	101.4	117.9	108.6	038	Portsmouth	97.9	94.3	96.4
553-555	Minneapolis	100.4	120.0	108.9					
556-558	Duluth	101.3	107.3	103.9	**NEW JERSEY**				
559	Rochester	99.9	104.7	102.0	070-071	Newark	101.2	128.4	113.0
560	Mankato	97.2	101.3	99.0	072	Elizabeth	98.8	127.6	111.3
561	Windom	95.9	92.3	94.3	073	Jersey City	97.9	127.9	111.0
562	Willmar	95.5	101.3	98.0	074-075	Paterson	99.6	127.8	111.9
563	St. Cloud	96.9	116.0	105.2	076	Hackensack	97.6	128.1	110.9
564	Brainerd	96.9	100.5	98.5	077	Long Branch	97.1	126.3	109.8
565	Detroit Lakes	98.9	95.1	97.3	078	Dover	97.9	127.8	110.9
566	Bemidji	98.2	98.2	98.2	079	Summit	97.9	127.6	110.8
567	Thief River Falls	97.8	92.4	95.4	080,083	Vineland	97.5	124.7	109.4
					081	Camden	99.5	124.8	110.5
MISSISSIPPI					082,084	Atlantic City	98.2	124.8	109.8
386	Clarksdale	97.3	52.6	77.8	085-086	Trenton	100.9	125.1	111.5
387	Greenville	101.0	61.7	83.9	087	Point Pleasant	99.2	125.9	110.8
388	Tupelo	98.8	54.9	79.7	088-089	New Brunswick	99.7	127.6	111.9
389	Greenwood	98.4	52.2	78.3					
390-392	Jackson	100.7	62.2	83.9	**NEW MEXICO**				
393	Meridian	97.6	64.1	83.0	870-872	Albuquerque	99.6	72.6	87.8
394	Laurel	98.9	55.8	80.1	873	Gallup	99.5	72.6	87.8
395	Biloxi	99.2	58.1	81.3	874	Farmington	100.1	72.6	88.1
396	Mccomb	97.4	51.7	77.5	875	Santa Fe	100.8	72.6	88.5
397	Columbus	98.8	53.9	79.2	877	Las Vegas	98.0	72.6	86.9
					878	Socorro	97.6	72.6	86.7
MISSOURI					879	Truth/Consequences	97.5	69.4	85.2
630-631	St. Louis	98.7	107.8	102.7	880	Las Cruces	96.1	69.4	84.4
633	Bowling Green	97.3	94.4	96.0	881	Clovis	98.3	72.5	87.1
634	Hannibal	96.2	92.4	94.6	882	Roswell	100.0	72.6	88.0
635	Kirksville	99.5	89.9	95.3	883	Carrizozo	100.3	72.6	88.2
636	Flat River	98.1	95.0	96.8	884	Tucumcari	99.0	72.5	87.4
637	Cape Girardeau	98.2	92.4	95.7					
638	Sikeston	96.7	92.2	94.8	**NEW YORK**				
639	Poplar Bluff	96.2	92.0	94.4	100-102	New York	103.4	168.7	131.8
640-641	Kansas City	101.4	105.7	103.3	103	Staten Island	99.2	168.4	129.4
644-645	St. Joseph	100.6	93.5	97.5	104	Bronx	97.7	166.1	127.5
646	Chillicothe	97.6	94.5	96.3	105	Mount Vernon	97.5	141.4	116.6
647	Harrisonville	97.1	101.5	99.1	106	White Plains	97.6	141.4	116.7
648	Joplin	99.5	81.7	91.8	107	Yonkers	102.1	143.7	120.2
650-651	Jefferson City	99.9	93.7	97.2	108	New Rochelle	97.9	141.3	116.8
652	Columbia	100.2	93.9	97.5	109	Suffern	97.6	129.0	111.3
653	Sedalia	100.2	97.3	98.9	110	Queens	100.9	166.8	129.6
654-655	Rolla	97.9	94.6	96.5	111	Long Island City	102.4	166.8	130.5
656-658	Springfield	101.7	83.7	93.8	112	Brooklyn	102.8	168.5	131.4
					113	Flushing	102.5	166.8	130.5
MONTANA					114	Jamaica	100.8	166.8	129.6
590-591	Billings	102.7	77.1	91.5	115,117,118	Hicksville	100.9	153.3	123.7
592	Wolf Point	102.3	74.2	90.0	116	Far Rockaway	102.6	166.8	130.6
593	Miles City	100.1	74.0	88.7	119	Riverhead	101.6	148.2	121.9
594	Great Falls	104.1	76.2	91.9	120-122	Albany	100.1	104.7	102.1
595	Havre	101.2	73.1	89.0	123	Schenectady	99.3	104.7	101.7
596	Helena	102.4	73.8	89.9	124	Kingston	102.6	118.5	109.6
597	Butte	102.6	77.3	91.6	125-126	Poughkeepsie	101.8	136.7	117.0
598	Missoula	99.9	75.5	89.3	127	Monticello	101.1	115.1	107.2
599	Kalispell	99.0	75.3	88.7	128	Glens Falls	94.0	98.6	96.0
					129	Plattsburgh	99.1	96.7	98.0
NEBRASKA					130-132	Syracuse	98.9	98.0	98.5
680-681	Omaha	100.2	78.6	90.8	133-135	Utica	97.1	97.1	97.1
683-685	Lincoln	100.5	74.9	89.4	136	Watertown	98.5	96.3	97.5
686	Columbus	98.4	77.1	89.1	137-139	Binghamton	98.5	101.4	99.8
687	Norfolk	100.0	76.7	89.8	140-142	Buffalo	100.3	106.2	102.9
688	Grand Island	100.0	77.1	90.0	143	Niagara Falls	97.8	106.4	101.5
689	Hastings	99.4	79.8	90.9	144-146	Rochester	101.1	99.1	100.2
690	Mccook	98.7	71.1	86.7	147	Jamestown	96.8	92.5	94.9
691	North Platte	99.0	78.3	90.0	148-149	Elmira	96.7	97.4	97.0
692	Valentine	101.2	72.5	88.7					
693	Alliance	101.0	70.5	87.7	**NORTH CAROLINA**				
					270,272-274	Greensboro	101.3	55.7	81.4
NEVADA					271	Winston-Salem	101.0	56.5	81.6
889-891	Las Vegas	100.9	108.7	104.3	275-276	Raleigh	100.5	54.1	80.3
893	Ely	99.4	102.2	100.6	277	Durham	102.9	55.3	82.1
894-895	Reno	99.6	93.3	96.8	278	Rocky Mount	98.5	52.3	78.3
897	Carson City	99.7	93.5	97.0	279	Elizabeth City	99.3	51.6	78.5
898	Elko	98.3	84.9	92.4	280	Gastonia	101.6	55.1	81.3
					281-282	Charlotte	101.9	56.8	82.2
NEW HAMPSHIRE					283	Fayetteville	104.5	58.1	84.2
030	Nashua	100.5	93.8	97.6	284	Wilmington	100.3	55.9	81.0
031	Manchester	101.4	93.8	98.1	285	Kinston	98.4	52.7	78.5

Location Factors

STATE/ZIP	CITY	MAT.	INST.	TOTAL		STATE/ZIP	CITY	MAT.	INST.	TOTAL
NORTH CAROLINA (CONT'D)						**PENNSYLVANIA (CONT'D)**				
286	Hickory	98.8	53.9	79.2		177	Williamsport	93.8	83.0	89.1
287-288	Asheville	100.8	55.3	80.9		178	Sunbury	95.8	95.8	95.8
289	Murphy	99.6	48.3	77.2		179	Pottsville	94.9	98.5	96.5
						180	Lehigh Valley	95.8	114.0	103.7
NORTH DAKOTA						181	Allentown	97.9	108.6	102.5
580-581	Fargo	102.7	67.6	87.4		182	Hazleton	95.4	98.3	96.6
582	Grand Forks	102.7	51.6	80.4		183	Stroudsburg	95.3	101.4	97.9
583	Devils Lake	102.0	58.6	83.1		184-185	Scranton	98.6	99.2	98.9
584	Jamestown	102.1	45.2	77.3		186-187	Wilkes-Barre	95.1	98.7	96.6
585	Bismarck	103.1	66.2	87.0		188	Montrose	94.7	96.9	95.7
586	Dickinson	102.8	60.9	84.6		189	Doylestown	94.9	120.6	106.1
587	Minot	102.7	71.3	89.0		190-191	Philadelphia	99.7	133.6	114.5
588	Williston	101.2	60.9	83.7		193	Westchester	95.9	121.5	107.1
						194	Norristown	95.0	129.5	110.0
OHIO						195-196	Reading	97.1	102.1	99.3
430-432	Columbus	98.6	89.2	94.5						
433	Marion	95.1	85.3	90.8		**PUERTO RICO**				
434-436	Toledo	98.7	98.1	98.4		009	San Juan	122.7	24.4	79.8
437-438	Zanesville	95.6	86.5	91.6						
439	Steubenville	96.8	94.9	96.0		**RHODE ISLAND**				
440	Lorain	98.9	93.0	96.3		028	Newport	98.4	117.5	106.7
441	Cleveland	99.0	100.3	99.6		029	Providence	100.6	117.5	108.0
442-443	Akron	99.9	94.2	97.4						
444-445	Youngstown	99.2	89.0	94.7		**SOUTH CAROLINA**				
446-447	Canton	99.3	85.6	93.3		290-292	Columbia	99.3	56.5	80.6
448-449	Mansfield	96.7	87.3	92.6		293	Spartanburg	98.6	56.7	80.3
450	Hamilton	97.9	85.1	92.3		294	Charleston	99.9	65.2	84.8
451-452	Cincinnati	98.3	84.5	92.2		295	Florence	98.2	56.9	80.2
453-454	Dayton	98.0	84.4	92.1		296	Greenville	98.3	55.8	79.8
455	Springfield	98.0	84.7	92.2		297	Rock Hill	97.8	50.3	77.1
456	Chillicothe	96.8	93.0	95.1		298	Aiken	98.6	69.1	85.7
457	Athens	99.6	80.4	91.2		299	Beaufort	99.3	44.9	75.6
458	Lima	100.0	86.8	94.3						
						SOUTH DAKOTA				
OKLAHOMA						570-571	Sioux Falls	100.0	57.7	81.5
730-731	Oklahoma City	100.3	65.7	85.2		572	Watertown	99.0	49.1	77.2
734	Ardmore	98.3	61.9	82.4		573	Mitchell	97.8	48.4	76.3
735	Lawton	100.7	62.8	84.2		574	Aberdeen	100.5	49.6	78.3
736	Clinton	99.6	63.1	83.7		575	Pierre	101.7	58.4	82.8
737	Enid	100.2	60.5	82.9		576	Mobridge	98.5	48.8	76.8
738	Woodward	98.3	63.4	83.1		577	Rapid City	100.4	59.5	82.6
739	Guymon	99.4	59.7	82.1						
740-741	Tulsa	98.1	61.6	82.2		**TENNESSEE**				
743	Miami	94.8	67.5	82.9		370-372	Nashville	98.4	74.1	87.8
744	Muskogee	97.4	55.7	79.2		373-374	Chattanooga	100.0	66.2	85.2
745	Mcalester	94.5	60.4	79.7		375,380-381	Memphis	99.0	71.3	86.9
746	Ponca City	95.1	62.1	80.7		376	Johnson City	99.8	56.5	80.9
747	Durant	95.1	63.5	81.3		377-379	Knoxville	96.6	66.9	83.7
748	Shawnee	96.5	61.8	81.4		382	Mckenzie	98.2	59.2	81.2
749	Poteau	94.3	61.4	80.0		383	Jackson	99.9	61.2	83.0
						384	Columbia	96.9	66.1	83.5
OREGON						385	Cookeville	98.1	59.1	81.1
970-972	Portland	98.8	100.4	99.5						
973	Salem	100.5	99.4	100.0		**TEXAS**				
974	Eugene	98.6	99.3	98.9		750	Mckinney	98.7	63.1	83.2
975	Medford	100.0	97.1	98.8		751	Waxahackie	98.6	65.9	84.4
976	Klamath Falls	99.8	97.1	98.6		752-753	Dallas	99.5	67.4	85.5
977	Bend	98.9	99.4	99.1		754	Greenville	98.8	64.0	83.6
978	Pendleton	94.8	101.4	97.7		755	Texarkana	98.0	58.4	80.7
979	Vale	92.9	98.1	95.2		756	Longview	98.5	58.7	81.2
						757	Tyler	98.5	65.0	83.9
PENNSYLVANIA						758	Palestine	95.0	64.4	81.7
150-152	Pittsburgh	99.9	104.6	102.0		759	Lufkin	95.6	67.8	83.5
153	Washington	97.0	103.2	99.7		760-761	Fort Worth	99.3	65.3	84.5
154	Uniontown	97.3	101.7	99.2		762	Denton	98.9	62.5	83.0
155	Bedford	98.2	91.3	95.2		763	Wichita Falls	97.1	64.4	82.8
156	Greensburg	98.2	100.9	99.4		764	Eastland	95.9	63.2	81.6
157	Indiana	97.1	99.3	98.0		765	Temple	94.7	61.7	80.3
158	Dubois	98.5	96.8	97.8		766-767	Waco	96.5	64.1	82.4
159	Johnstown	98.1	97.2	97.7		768	Brownwood	98.9	59.4	81.7
160	Butler	92.5	102.0	96.6		769	San Angelo	98.5	59.6	81.5
161	New Castle	92.5	97.9	94.8		770-772	Houston	99.4	70.5	86.8
162	Kittanning	92.9	103.3	97.4		773	Huntsville	98.0	66.7	84.4
163	Oil City	92.4	96.9	94.4		774	Wharton	98.9	67.7	85.3
164-165	Erie	94.5	95.8	95.1		775	Galveston	97.0	70.0	85.2
166	Altoona	94.6	91.9	93.4		776-777	Beaumont	97.4	68.6	84.9
167	Bradford	95.7	96.4	96.0		778	Bryan	94.9	68.2	83.3
168	State College	95.3	92.8	94.2		779	Victoria	99.0	65.6	84.4
169	Wellsboro	96.3	92.6	94.7		780	Laredo	98.3	64.5	83.6
170-171	Harrisburg	99.8	95.6	98.0		781-782	San Antonio	98.5	65.1	84.0
172	Chambersburg	96.2	88.0	92.6		783-784	Corpus Christi	100.9	63.5	84.6
173-174	York	96.7	95.9	96.3		785	Mc Allen	101.0	59.8	83.0
175-176	Lancaster	95.1	89.0	92.4		786-787	Austin	100.0	63.6	84.1

702

Location Factors

STATE/ZIP	CITY	MAT.	INST.	TOTAL
TEXAS (CONT'D)				
788	Del Rio	100.8	63.2	84.4
789	Giddings	97.7	64.0	83.0
790-791	Amarillo	100.2	63.2	84.1
792	Childress	99.3	64.3	84.0
793-794	Lubbock	101.6	63.6	85.0
795-796	Abilene	99.7	62.2	83.3
797	Midland	101.5	64.2	85.2
798-799,885	El Paso	98.0	61.7	82.2
UTAH				
840-841	Salt Lake City	102.2	68.3	87.4
842,844	Ogden	98.1	70.1	85.9
843	Logan	99.8	70.1	86.9
845	Price	100.2	64.6	84.7
846-847	Provo	100.3	68.2	86.3
VERMONT				
050	White River Jct.	98.5	78.3	89.7
051	Bellows Falls	97.2	93.0	95.3
052	Bennington	97.5	93.5	95.8
053	Brattleboro	97.8	92.8	95.6
054	Burlington	102.4	83.6	94.2
056	Montpelier	99.5	82.8	92.2
057	Rutland	99.7	83.6	92.7
058	St. Johnsbury	98.6	77.8	89.6
059	Guildhall	97.4	77.8	88.8
VIRGINIA				
220-221	Fairfax	100.7	84.0	93.4
222	Arlington	101.9	84.0	94.1
223	Alexandria	101.1	85.5	94.3
224-225	Fredericksburg	99.5	82.7	92.1
226	Winchester	100.0	82.2	92.2
227	Culpeper	99.9	80.5	91.4
228	Harrisonburg	100.2	67.2	85.8
229	Charlottesville	100.5	67.1	86.0
230-232	Richmond	101.3	68.7	87.1
233-235	Norfolk	100.5	68.2	86.4
236	Newport News	100.0	68.2	86.1
237	Portsmouth	99.4	65.3	84.5
238	Petersburg	99.4	68.7	86.0
239	Farmville	98.7	52.4	78.5
240-241	Roanoke	101.5	65.1	85.6
242	Bristol	99.4	54.9	80.0
243	Pulaski	99.0	57.3	80.9
244	Staunton	99.7	61.8	83.2
245	Lynchburg	99.9	65.6	85.0
246	Grundy	99.3	52.5	78.9
WASHINGTON				
980-981,987	Seattle	101.4	104.7	102.8
982	Everett	101.6	99.6	100.7
983-984	Tacoma	101.8	99.7	100.9
985	Olympia	101.2	99.4	100.4
986	Vancouver	102.7	92.1	98.1
988	Wenatchee	101.6	86.6	95.0
989	Yakima	102.0	94.1	98.6
990-992	Spokane	103.4	82.3	94.2
993	Richland	103.1	89.3	97.1
994	Clarkston	101.5	80.2	92.2
WEST VIRGINIA				
247-248	Bluefield	98.3	92.9	96.0
249	Lewisburg	99.8	92.4	96.6
250-253	Charleston	99.9	94.8	97.6
254	Martinsburg	99.3	86.4	93.7
255-257	Huntington	100.7	95.7	98.5
258-259	Beckley	97.9	93.3	95.9
260	Wheeling	101.2	94.3	98.2
261	Parkersburg	100.1	91.8	96.5
262	Buckhannon	99.6	95.5	97.8
263-264	Clarksburg	100.1	95.4	98.1
265	Morgantown	100.2	94.4	97.6
266	Gassaway	99.5	93.9	97.1
267	Romney	99.4	92.2	96.3
268	Petersburg	99.3	91.3	95.8
WISCONSIN				
530,532	Milwaukee	100.0	107.2	103.1
531	Kenosha	99.5	102.2	100.7
534	Racine	99.1	102.4	100.5
535	Beloit	98.9	100.7	99.7
537	Madison	99.1	99.0	99.1

STATE/ZIP	CITY	MAT.	INST.	TOTAL
WISCONSIN (CONT'D)				
538	Lancaster	96.7	92.5	94.8
539	Portage	95.3	96.6	95.9
540	New Richmond	96.8	94.6	95.9
541-543	Green Bay	101.2	93.4	97.8
544	Wausau	96.2	93.5	95.0
545	Rhinelander	99.3	91.2	95.8
546	La Crosse	97.8	94.5	96.3
547	Eau Claire	99.3	95.3	97.5
548	Superior	96.6	96.3	96.5
549	Oshkosh	96.9	91.9	94.7
WYOMING				
820	Cheyenne	101.5	65.2	85.7
821	Yellowstone Nat'l Park	99.3	66.3	84.9
822	Wheatland	100.4	64.3	84.7
823	Rawlins	101.8	66.9	86.6
824	Worland	100.0	65.3	84.9
825	Riverton	101.0	65.4	85.5
826	Casper	102.0	63.9	85.4
827	Newcastle	99.9	66.9	85.5
828	Sheridan	102.6	64.3	85.9
829-831	Rock Springs	103.7	65.3	87.0
CANADIAN FACTORS (reflect Canadian currency)				
ALBERTA				
	Calgary	123.5	95.7	111.4
	Edmonton	123.1	95.7	111.2
	Fort McMurray	123.6	92.8	110.2
	Lethbridge	118.4	92.2	107.0
	Lloydminster	113.4	88.9	102.8
	Medicine Hat	113.6	88.2	102.5
	Red Deer	114.1	88.2	102.8
BRITISH COLUMBIA				
	Kamloops	114.5	90.6	104.1
	Prince George	115.4	89.9	104.3
	Vancouver	119.5	89.1	106.2
	Victoria	115.7	87.2	103.3
MANITOBA				
	Brandon	125.2	75.8	103.7
	Portage la Prairie	113.4	74.2	96.3
	Winnipeg	121.0	71.8	99.5
NEW BRUNSWICK				
	Bathurst	111.5	67.3	92.2
	Dalhousie	111.9	67.5	92.5
	Fredericton	117.4	72.5	97.9
	Moncton	111.8	70.1	93.6
	Newcastle	111.6	67.9	92.5
	St. John	114.0	77.9	98.3
NEWFOUNDLAND				
	Corner Brook	129.6	67.6	102.6
	St Johns	122.7	85.2	106.4
NORTHWEST TERRITORIES				
	Yellowknife	135.3	84.2	113.0
NOVA SCOTIA				
	Bridgewater	112.9	74.6	96.2
	Dartmouth	126.1	74.6	103.7
	Halifax	119.6	80.9	102.7
	New Glasgow	124.1	74.6	102.5
	Sydney	122.1	74.6	101.4
	Truro	112.5	74.6	96.0
	Yarmouth	124.0	74.6	102.5
ONTARIO				
	Barrie	117.7	93.0	106.9
	Brantford	114.9	96.9	107.1
	Cornwall	115.0	93.3	105.5
	Hamilton	117.0	96.7	108.2
	Kingston	116.0	93.4	106.2
	Kitchener	109.6	92.9	102.3
	London	116.4	93.3	106.3
	North Bay	127.3	91.1	111.5
	Oshawa	112.3	94.3	104.4
	Ottawa	117.3	94.0	107.2
	Owen Sound	117.8	91.2	106.2
	Peterborough	114.9	93.1	105.4
	Sarnia	114.8	97.5	107.3
	Sault Ste Marie	110.3	92.1	102.3

Location Factors

STATE/ZIP	CITY	MAT.	INST.	TOTAL
ONTARIO (CONT'D)				
	St. Catharines	107.6	94.5	101.9
	Sudbury	107.7	93.0	101.3
	Thunder Bay	108.8	93.4	102.1
	Timmins	115.1	91.1	104.6
	Toronto	118.3	101.3	110.9
	Windsor	108.2	93.4	101.7
PRINCE EDWARD ISLAND				
	Charlottetown	120.0	62.8	95.1
	Summerside	126.7	62.8	98.8
QUEBEC				
	Cap-de-la-Madeleine	112.4	87.6	101.6
	Charlesbourg	112.4	87.6	101.6
	Chicoutimi	111.4	92.9	103.3
	Gatineau	112.1	87.4	101.3
	Granby	112.3	87.4	101.4
	Hull	112.2	87.4	101.4
	Joliette	112.5	87.6	101.7
	Laval	112.1	87.4	101.3
	Montreal	117.8	93.2	107.1
	Quebec	117.1	93.6	106.8
	Rimouski	111.8	92.9	103.6
	Rouyn-Noranda	112.0	87.4	101.3
	Saint Hyacinthe	111.6	87.4	101.1
	Sherbrooke	112.4	87.4	101.5
	Sorel	112.5	87.6	101.7
	St Jerome	112.1	87.4	101.3
	Trois Rivieres	124.8	87.6	108.6
SASKATCHEWAN				
	Moose Jaw	110.7	69.5	92.7
	Prince Albert	109.9	67.7	91.5
	Regina	120.7	89.0	106.9
	Saskatoon	111.1	89.1	101.5
YUKON				
	Whitehorse	135.3	68.5	106.2

704

R011105-05   Tips for Accurate Estimating

1. Use pre-printed or columnar forms for orderly sequence of dimensions and locations and for recording telephone quotations.

2. Use only the front side of each paper or form except for certain pre-printed summary forms.

3. Be consistent in listing dimensions: For example, length x width x height. This helps in rechecking to ensure that, the total length of partitions is appropriate for the building area.

4. Use printed (rather than measured) dimensions where given.

5. Add up multiple printed dimensions for a single entry where possible.

6. Measure all other dimensions carefully.

7. Use each set of dimensions to calculate multiple related quantities.

8. Convert foot and inch measurements to decimal feet when listing. Memorize decimal equivalents to .01 parts of a foot (1/8″ equals approximately .01′).

9. Do not "round off" quantities until the final summary.

10. Mark drawings with different colors as items are taken off.

11. Keep similar items together, different items separate.

12. Identify location and drawing numbers to aid in future checking for completeness.

13. Measure or list everything on the drawings or mentioned in the specifications.

14. It may be necessary to list items not called for to make the job complete.

15. Be alert for: Notes on plans such as N.T.S. (not to scale); changes in scale throughout the drawings; reduced size drawings; discrepancies between the specifications and the drawings.

16. Develop a consistent pattern of performing an estimate. For example:
    a. Start the quantity takeoff at the lower floor and move to the next higher floor.
    b. Proceed from the main section of the building to the wings.
    c. Proceed from south to north or vice versa, clockwise or counterclockwise.
    d. Take off floor plan quantities first, elevations next, then detail drawings.

17. List all gross dimensions that can be either used again for different quantities, or used as a rough check of other quantities for verification (exterior perimeter, gross floor area, individual floor areas, etc.).

18. Utilize design symmetry or repetition (repetitive floors, repetitive wings, symmetrical design around a center line, similar room layouts, etc.). Note: Extreme caution is needed here so as not to omit or duplicate an area.

19. Do not convert units until the final total is obtained. For instance, when estimating concrete work, keep all units to the nearest cubic foot, then summarize and convert to cubic yards.

20. When figuring alternatives, it is best to total all items involved in the basic system, then total all items involved in the alternates. Therefore you work with positive numbers in all cases. When adds and deducts are used, it is often confusing whether to add or subtract a portion of an item; especially on a complicated or involved alternate.

R011105-10   Unit Gross Area Requirements

The figures in the table below indicate typical ranges in square feet as a function of the "occupant" unit. This table is best used in the preliminary design stages to help determine the probable size requirement for the total project.

Building Type	Unit	Gross Area in S.F.		
		1/4	Median	3/4
Apartments	Unit	660	860	1,100
Auditorium & Play Theaters	Seat	18	25	38
Bowling Alleys	Lane		940	
Churches & Synagogues	Seat	20	28	39
Dormitories	Bed	200	230	275
Fraternity & Sorority Houses	Bed	220	315	370
Garages, Parking	Car	325	355	385
Hospitals	Bed	685	850	1,075
Hotels	Rental Unit	475	600	710
Housing for the elderly	Unit	515	635	755
Housing, Public	Unit	700	875	1,030
Ice Skating Rinks	Total	27,000	30,000	36,000
Motels	Rental Unit	360	465	620
Nursing Homes	Bed	290	350	450
Restaurants	Seat	23	29	39
Schools, Elementary	Pupil	65	77	90
Junior High & Middle		85	110	129
Senior High		102	130	145
Vocational		110	135	195
Shooting Ranges	Point		450	
Theaters & Movies	Seat		15	

R011105-20   Floor Area Ratios

Table below lists commonly used gross to net area and net to gross area ratios expressed in % for various building types.

Building Type	Gross to Net Ratio	Net to Gross Ratio	Building Type	Gross to Net Ratio	Net to Gross Ratio
Apartment	156	64	School Buildings (campus type)		
Bank	140	72	Administrative	150	67
Church	142	70	Auditorium	142	70
Courthouse	162	61	Biology	161	62
Department Store	123	81	Chemistry	170	59
Garage	118	85	Classroom	152	66
Hospital	183	55	Dining Hall	138	72
Hotel	158	63	Dormitory	154	65
Laboratory	171	58	Engineering	164	61
Library	132	76	Fraternity	160	63
Office	135	75	Gymnasium	142	70
Restaurant	141	70	Science	167	60
Warehouse	108	93	Service	120	83
			Student Union	172	59

The gross area of a building is the total floor area based on outside dimensions.

The net area of a building is the usable floor area for the function intended and excludes such items as stairways, corridors and mechanical rooms. In the case of a commercial building, it might be considered as the "leasable area."

R011105-30 Occupancy Determinations

Function of Space	SF/Person Required
Accessory storage areas, mechanical equipment rooms	300
Agriculture Building	300
Aircraft Hangars	500
Airport Terminal	
Baggage claim	20
Baggage handling	300
Concourse	100
Waiting areas	15
Assembly	
Gaming floors (keno, slots, etc.)	11
Exhibit Gallery and Museum	30
Assembly w/ fixed seats	load determined by seat number
Assembly w/o fixed seats	
Concentrated (chairs only-not fixed)	7
Standing space	5
Unconcentrated (tables and chairs)	15
Bowling centers, allow 5 persons for each lane including 15 feet of runway, and for additional areas	7
Business areas	100
Courtrooms-other than fixed seating areas	40
Day care	35
Dormitories	50
Educational	
Classroom areas	20
Shops and other vocational room areas	50
Exercise rooms	50
Fabrication and Manufacturing areas where hazardous materials are used	200
Industrial areas	100
Institutional areas	
Inpatient treatment areas	240
Outpatient areas	100
Sleeping areas	120
Kitchens commercial	200
Library	
Reading rooms	50
Stack area	100
Mercantile	
Areas on other floors	60
Basement and grade floor areas	30
Storage, stock, shipping areas	300
Parking garages	200
Residential	200
Skating rinks, swimming pools	
Rink and pool	50
Decks	15
Stages and platforms	15
Warehouses	500

R011105-50   Metric Conversion Factors

Description: This table is primarily for converting customary U.S. units in the left hand column to SI metric units in the right hand column. In addition, conversion factors for some commonly encountered Canadian and non-SI metric units are included.

If You Know		Multiply By		To Find
Length	Inches	x	25.4[a]	=	Millimeters
	Feet	x	0.3048[a]	=	Meters
	Yards	x	0.9144[a]	=	Meters
	Miles (statute)	x	1.609	=	Kilometers
Area	Square inches	x	645.2	=	Square millimeters
	Square feet	x	0.0929	=	Square meters
	Square yards	x	0.8361	=	Square meters
Volume (Capacity)	Cubic inches	x	16,387	=	Cubic millimeters
	Cubic feet	x	0.02832	=	Cubic meters
	Cubic yards	x	0.7646	=	Cubic meters
	Gallons (U.S. liquids)[b]	x	0.003785	=	Cubic meters[c]
	Gallons (Canadian liquid)[b]	x	0.004546	=	Cubic meters[c]
	Ounces (U.S. liquid)[b]	x	29.57	=	Milliliters[c, d]
	Quarts (U.S. liquid)[b]	x	0.9464	=	Liters[c, d]
	Gallons (U.S. liquid)[b]	x	3.785	=	Liters[c, d]
Force	Kilograms force[d]	x	9.807	=	Newtons
	Pounds force	x	4.448	=	Newtons
	Pounds force	x	0.4536	=	Kilograms force[d]
	Kips	x	4448	=	Newtons
	Kips	x	453.6	=	Kilograms force[d]
Pressure, Stress, Strength (Force per unit area)	Kilograms force per square centimeter[d]	x	0.09807	=	Megapascals
	Pounds force per square inch (psi)	x	0.006895	=	Megapascals
	Kips per square inch	x	6.895	=	Megapascals
	Pounds force per square inch (psi)	x	0.07031	=	Kilograms force per square centimeter[d]
	Pounds force per square foot	x	47.88	=	Pascals
	Pounds force per square foot	x	4.882	=	Kilograms force per square meter[d]
Bending Moment Or Torque	Inch-pounds force	x	0.01152	=	Meter-kilograms force[d]
	Inch-pounds force	x	0.1130	=	Newton-meters
	Foot-pounds force	x	0.1383	=	Meter-kilograms force[d]
	Foot-pounds force	x	1.356	=	Newton-meters
	Meter-kilograms force[d]	x	9.807	=	Newton-meters
Mass	Ounces (avoirdupois)	x	28.35	=	Grams
	Pounds (avoirdupois)	x	0.4536	=	Kilograms
	Tons (metric)	x	1000	=	Kilograms
	Tons, short (2000 pounds)	x	907.2	=	Kilograms
	Tons, short (2000 pounds)	x	0.9072	=	Megagrams[e]
Mass per Unit Volume	Pounds mass per cubic foot	x	16.02	=	Kilograms per cubic meter
	Pounds mass per cubic yard	x	0.5933	=	Kilograms per cubic meter
	Pounds mass per gallon (U.S. liquid)[b]	x	119.8	=	Kilograms per cubic meter
	Pounds mass per gallon (Canadian liquid)[b]	x	99.78	=	Kilograms per cubic meter
Temperature	Degrees Fahrenheit	(F-32)/1.8		=	Degrees Celsius
	Degrees Fahrenheit	(F+459.67)/1.8		=	Degrees Kelvin
	Degrees Celsius	C+273.15		=	Degrees Kelvin

[a]The factor given is exact
[b]One U.S. gallon = 0.8327 Canadian gallon
[c]1 liter = 1000 milliliters = 1000 cubic centimeters
  1 cubic decimeter = 0.001 cubic meter
[d]Metric but not SI unit
[e]Called "tonne" in England and
  "metric ton" in other metric countries

708

R011105-60   Weights and Measures

Measures of Length
1 Mile = 1760 Yards = 5280 Feet
1 Yard = 3 Feet = 36 inches
1 Foot = 12 Inches
1 Mil = 0.001 Inch
1 Fathom = 2 Yards = 6 Feet
1 Rod = 5.5 Yards = 16.5 Feet
1 Hand = 4 Inches
1 Span = 9 Inches
1 Micro-inch = One Millionth Inch or 0.000001 Inch
1 Micron = One Millionth Meter + 0.00003937 Inch

Surveyor's Measure
1 Mile = 8 Furlongs = 80 Chains
1 Furlong = 10 Chains = 220 Yards
1 Chain = 4 Rods = 22 Yards = 66 Feet = 100 Links
1 Link = 7.92 Inches

Square Measure
1 Square Mile = 640 Acres = 6400 Square Chains
1 Acre = 10 Square Chains = 4840 Square Yards = 43,560 Sq. Ft.
1 Square Chain = 16 Square Rods = 484 Square Yards = 4356 Sq. Ft.
1 Square Rod = 30.25 Square Yards = 272.25 Square Feet = 625 Square Lines
1 Square Yard = 9 Square Feet
1 Square Foot = 144 Square Inches
An Acre equals a Square 208.7 Feet per Side

Cubic Measure
1 Cubic Yard = 27 Cubic Feet
1 Cubic Foot = 1728 Cubic Inches
1 Cord of Wood = 4 x 4 x 8 Feet = 128 Cubic Feet
1 Perch of Masonry = 16½ x 1½ x 1 Foot = 24.75 Cubic Feet

Avoirdupois or Commercial Weight
1 Gross or Long Ton = 2240 Pounds
1 Net or Short ton = 2000 Pounds
1 Pound = 16 Ounces = 7000 Grains
1 Ounce = 16 Drachms = 437.5 Grains
1 Stone = 14 Pounds

Shipping Measure
For Measuring Internal Capacity of a Vessel:
    1 Register Ton = 100 Cubic Feet

For Measurement of Cargo:
    Approximately 40 Cubic Feet of Merchandise is considered a Shipping Ton, unless that bulk would weigh more than 2000 Pounds, in which case Freight Charge may be based upon weight.

40 Cubic Feet = 32.143 U.S. Bushels = 31.16 Imp. Bushels

Liquid Measure
1 Imperial Gallon = 1.2009 U.S. Gallon = 277.42 Cu. In.
1 Cubic Foot = 7.48 U.S. Gallons

R011110-10   Architectural Fees

Tabulated below are typical percentage fees by project size, for good professional architectural service. Fees may vary from those listed depending upon degree of design difficulty and economic conditions in any particular area.

Rates can be interpolated horizontally and vertically. Various portions of the same project requiring different rates should be adjusted proportionally. For alterations, add 50% to the fee for the first $500,000 of project cost and add 25% to the fee for project cost over $500,000.

Architectural fees tabulated below include Structural, Mechanical and Electrical Engineering Fees. They do not include the fees for special consultants such as kitchen planning, security, acoustical, interior design, etc.

Civil Engineering fees are included in the Architectural fee for project sites requiring minimal design such as city sites. However, separate Civil Engineering fees must be added when utility connections require design, drainage calculations are needed, stepped foundations are required, or provisions are required to protect adjacent wetlands.

Building Types	Total Project Size in Thousands of Dollars						
	100	250	500	1,000	5,000	10,000	50,000
Factories, garages, warehouses, repetitive housing	9.0%	8.0%	7.0%	6.2%	5.3%	4.9%	4.5%
Apartments, banks, schools, libraries, offices, municipal buildings	12.2	12.3	9.2	8.0	7.0	6.6	6.2
Churches, hospitals, homes, laboratories, museums, research	15.0	13.6	12.7	11.9	9.5	8.8	8.0
Memorials, monumental work, decorative furnishings	—	16.0	14.5	13.1	10.0	9.0	8.3

R011110-30   Engineering Fees

Typical **Structural Engineering Fees** based on type of construction and total project size. These fees are included in Architectural Fees.

Type of Construction	Total Project Size (in thousands of dollars)			
	$500	$500-$1,000	$1,000-$5,000	Over $5000
Industrial buildings, factories & warehouses	Technical payroll times 2.0 to 2.5	1.60%	1.25%	1.00%
Hotels, apartments, offices, dormitories, hospitals, public buildings, food stores		2.00%	1.70%	1.20%
Museums, banks, churches and cathedrals		2.00%	1.75%	1.25%
Thin shells, prestressed concrete, earthquake resistive		2.00%	1.75%	1.50%
Parking ramps, auditoriums, stadiums, convention halls, hangars & boiler houses		2.50%	2.00%	1.75%
Special buildings, major alterations, underpinning & future expansion		Add to above 0.5%	Add to above 0.5%	Add to above 0.5%

For complex reinforced concrete or unusually complicated structures, add 20% to 50%.

Typical **Mechanical and Electrical Engineering Fees** are based on the size of the subcontract. The fee structure for both are shown below. These fees are included in Architectural Fees.

Type of Construction	Subcontract Size							
	$25,000	$50,000	$100,000	$225,000	$350,000	$500,000	$750,000	$1,000,000
Simple structures	6.4%	5.7%	4.8%	4.5%	4.4%	4.3%	4.2%	4.1%
Intermediate structures	8.0	7.3	6.5	5.6	5.1	5.0	4.9	4.8
Complex structures	10.1	9.0	9.0	8.0	7.5	7.5	7.0	7.0

For renovations, add 15% to 25% to applicable fee.

R012153-10   Repair and Remodeling

Cost figures are based on new construction utilizing the most cost-effective combination of labor, equipment and material with the work scheduled in proper sequence to allow the various trades to accomplish their work in an efficient manner.

The costs for repair and remodeling work must be modified due to the following factors that may be present in any given repair and remodeling project.

1. Equipment usage curtailment due to the physical limitations of the project, with only hand-operated equipment being used.

2. Increased requirement for shoring and bracing to hold up the building while structural changes are being made and to allow for temporary storage of construction materials on above-grade floors.

3. Material handling becomes more costly due to having to move within the confines of an enclosed building. For multi-story construction, low capacity elevators and stairwells may be the only access to the upper floors.

4. Large amount of cutting and patching and attempting to match the existing construction is required. It is often more economical to remove entire walls rather than create many new door and window openings. This sort of trade-off has to be carefully analyzed.

5. Cost of protection of completed work is increased since the usual sequence of construction usually cannot be accomplished.

6. Economies of scale usually associated with new construction may not be present. If small quantities of components must be custom fabricated due to job requirements, unit costs will naturally increase. Also, if only small work areas are available at a given time, job scheduling between trades becomes difficult and subcontractor quotations may reflect the excessive start-up and shut-down phases of the job.

7. Work may have to be done on other than normal shifts and may have to be done around an existing production facility which has to stay in production during the course of the repair and remodeling.

8. Dust and noise protection of adjoining non-construction areas can involve substantial special protection and alter usual construction methods.

9. Job may be delayed due to unexpected conditions discovered during demolition or removal. These delays ultimately increase construction costs.

10. Piping and ductwork runs may not be as simple as for new construction. Wiring may have to be snaked through walls and floors.

11. Matching "existing construction" may be impossible because materials may no longer be manufactured. Substitutions may be expensive.

12. Weather protection of existing structure requires additional temporary structures to protect building at openings.

13. On small projects, because of local conditions, it may be necessary to pay a tradesman for a minimum of four hours for a task that is completed in one hour.

All of the above areas can contribute to increased costs for a repair and remodeling project. Each of the above factors should be considered in the planning, bidding and construction stage in order to minimize the increased costs associated with repair and remodeling jobs.

711

R012909-80   Sales Tax by State

State sales tax on materials is tabulated below (5 states have no sales tax). Many states allow local jurisdictions, such as a county or city, to levy additional sales tax.

Some projects may be sales tax exempt, particularly those constructed with public funds.

State	Tax (%)	State	Tax (%)	State	Tax (%)	State	Tax (%)
Alabama	4	Illinois	6.25	Montana	0	Rhode Island	7
Alaska	0	Indiana	7	Nebraska	5.5	South Carolina	6
Arizona	5.6	Iowa	6	Nevada	6.85	South Dakota	4
Arkansas	6	Kansas	6.3	New Hampshire	0	Tennessee	7
California	7.5	Kentucky	6	New Jersey	7	Texas	6.25
Colorado	2.9	Louisiana	4	New Mexico	5.13	Utah	4.7
Connecticut	6.35	Maine	5	New York	4	Vermont	6
Delaware	0	Maryland	6	North Carolina	4.75	Virginia	5
District of Columbia	6	Massachusetts	6.25	North Dakota	5	Washington	6.5
Florida	6	Michigan	6	Ohio	5.5	West Virginia	6
Georgia	4	Minnesota	6.88	Oklahoma	4.5	Wisconsin	5
Hawaii	4	Mississippi	7	Oregon	0	Wyoming	4
Idaho	6	Missouri	4.23	Pennsylvania	6	Average	5.04 %

Sales Tax by Province (Canada)

GST - a value-added tax, which the government imposes on most goods and services provided in or imported into Canada. PST - a retail sales tax, which three of the provinces impose on the price of most goods and some

services. QST - a value-added tax, similar to the federal GST, which Quebec imposes. HST - Five provinces have combined their retail sales tax with the federal GST into one harmonized tax.

Province	PST (%)	QST (%)	GST(%)	HST(%)
Alberta	0	0	5	0
British Columbia	7	0	5	0
Manitoba	8	0	5	0
New Brunswick	0	0	0	13
Newfoundland	0	0	0	13
Northwest Territories	0	0	5	0
Nova Scotia	0	0	0	15
Ontario	0	0	0	13
Prince Edward Island	0	0	0	14
Quebec	0	9.975	5	0
Saskatchewan	5	0	5	0
Yukon	0	0	5	0

For customer support on your Mechanical Cost Data, call 877.795.7394.

R012909-85 Unemployment Taxes and Social Security Taxes

State unemployment tax rates vary not only from state to state, but also with the experience rating of the contractor. The federal unemployment tax rate is 6.0% of the first $7,000 of wages. This is reduced by a credit of up to 5.4% for timely payment to the state. The minimum federal unemployment tax is 0.6% after all credits.

Social security (FICA) for 2015 is estimated at time of publication to be 7.65% of wages up to $117,000.

R012909-90 Overtime

One way to improve the completion date of a project or eliminate negative float from a schedule is to compress activity duration times. This can be achieved by increasing the crew size or working overtime with the proposed crew.

To determine the costs of working overtime to compress activity duration times, consider the following examples. Below is an overtime efficiency and cost chart based on a five, six, or seven day week with an eight through twelve hour day. Payroll percentage increases for time and one half and double time are shown for the various working days.

Days per Week	Hours per Day	Production Efficiency					Payroll Cost Factors	
		1st Week	2nd Week	3rd Week	4th Week	Average 4 Weeks	@ 1-1/2 Times	@ 2 Times
5	8	100%	100%	100%	100%	100 %	1.000	1.000
	9	100	100	95	90	96	1.056	1.111
	10	100	95	90	85	93	1.100	1.200
	11	95	90	75	65	81	1.136	1.273
	12	90	85	70	60	76	1.167	1.333
6	8	100	100	95	90	96	1.083	1.167
	9	100	95	90	85	93	1.130	1.259
	10	95	90	85	80	88	1.167	1.333
	11	95	85	70	65	79	1.197	1.394
	12	90	80	65	60	74	1.222	1.444
7	8	100	95	85	75	89	1.143	1.286
	9	95	90	80	70	84	1.183	1.365
	10	90	85	75	65	79	1.214	1.429
	11	85	80	65	60	73	1.240	1.481
	12	85	75	60	55	69	1.262	1.524

R013113-40   Builder's Risk Insurance

Builder's Risk Insurance is insurance on a building during construction. Premiums are paid by the owner or the contractor. Blasting, collapse and underground insurance would raise total insurance costs above those listed. Floater policy for materials delivered to the job runs $.75 to $1.25 per $100 value. Contractor equipment insurance runs $.50 to $1.50 per $100 value. Insurance for miscellaneous tools to $1,500 value runs from $3.00 to $7.50 per $100 value.

Tabulated below are New England Builder's Risk insurance rates in dollars per $100 value for $1,000 deductible. For $25,000 deductible, rates can be reduced 13% to 34%. On contracts over $1,000,000, rates may be lower than those tabulated. Policies are written annually for the total completed value in place. For "all risk" insurance (excluding flood, earthquake and certain other perils) add $.025 to total rates below.

Coverage	Frame Construction (Class 1)			Brick Construction (Class 4)			Fire Resistive (Class 6)		
	Range		Average	Range		Average	Range		Average
Fire Insurance	$.350 to $.850		$.600	$.158 to $.189		$.174	$.052 to $.080		$.070
Extended Coverage	.115 to .200		.158	.080 to .105		.101	.081 to .105		.100
Vandalism	.012 to .016		.014	.008 to .011		.011	.008 to .011		.010
Total Annual Rate	$.477 to $1.066		$.772	$.246 to $.305		$.286	$.141 to $.196		$.180

R013113-50   General Contractor's Overhead

There are two distinct types of overhead on a construction project: Project Overhead and Main Office Overhead. Project Overhead includes those costs at a construction site not directly associated with the installation of construction materials. Examples of Project Overhead costs include the following:

1. Superintendent
2. Construction office and storage trailers
3. Temporary sanitary facilities
4. Temporary utilities
5. Security fencing
6. Photographs
7. Clean up
8. Performance and payment bonds

The above Project Overhead items are also referred to as General Requirements and therefore are estimated in Division 1. Division 1 is the first division listed in the CSI MasterFormat but it is usually the last division estimated. The sum of the costs in Divisions 1 through 49 is referred to as the sum of the direct costs.

All construction projects also include indirect costs. The primary components of indirect costs are the contractor's Main Office Overhead and profit. The amount of the Main Office Overhead expense varies depending on the following:

1. Owner's compensation
2. Project managers and estimator's wages
3. Clerical support wages
4. Office rent and utilities
5. Corporate legal and accounting costs
6. Advertising
7. Automobile expenses
8. Association dues
9. Travel and entertainment expenses

These costs are usually calculated as a percentage of annual sales volume. This percentage can range from 35% for a small contractor doing less than $500,000 to 5% for a large contractor with sales in excess of $100 million.

R013113-60   Workers' Compensation Insurance Rates by Trade

The table below tabulates the national averages for workers' compensation insurance rates by trade and type of building. The average "Insurance Rate" is multiplied by the "% of Building Cost" for each trade. This produces the "Workers' Compensation" cost by % of total labor cost, to be added for each trade by building type to determine the weighted average workers' compensation rate for the building types analyzed.

Trade	Insurance Rate (% Labor Cost) Range		Average	% of Building Cost Office Bldgs.	Schools & Apts.	Mfg.	Workers' Compensation Office Bldgs.	Schools & Apts.	Mfg.
Excavation, Grading, etc.	3.4 % to	21.2%	9.7%	4.8%	4.9%	4.5%	0.47%	0.48%	0.44%
Piles & Foundations	6.1 to	28.3	14.3	7.1	5.2	8.7	1.02	0.74	1.24
Concrete	4.2 to	35.8	12.9	5.0	14.8	3.7	0.65	1.91	0.48
Masonry	4.9 to	36.4	13.7	6.9	7.5	1.9	0.95	1.03	0.26
Structural Steel	6.1 to	101.1	31.7	10.7	3.9	17.6	3.39	1.24	5.58
Miscellaneous & Ornamental Metals	4.3 to	31.5	12.2	2.8	4.0	3.6	0.34	0.49	0.44
Carpentry & Millwork	5.0 to	36.8	14.9	3.7	4.0	0.5	0.55	0.60	0.07
Metal or Composition Siding	6.1 to	69.2	18.4	2.3	0.3	4.3	0.42	0.06	0.79
Roofing	6.1 to	100.3	31.7	2.3	2.6	3.1	0.73	0.82	0.98
Doors & Hardware	4.1 to	36.8	11.3	0.9	1.4	0.4	0.10	0.16	0.05
Sash & Glazing	4.8 to	30.7	13.3	3.5	4.0	1.0	0.47	0.53	0.13
Lath & Plaster	3.1 to	36.2	10.9	3.3	6.9	0.8	0.36	0.75	0.09
Tile, Marble & Floors	3 to	24.8	8.9	2.6	3.0	0.5	0.23	0.27	0.04
Acoustical Ceilings	3.8 to	33.8	8.4	2.4	0.2	0.3	0.20	0.02	0.03
Painting	4.5 to	36.1	11.7	1.5	1.6	1.6	0.18	0.19	0.19
Interior Partitions	5.0 to	36.8	14.9	3.9	4.3	4.4	0.58	0.64	0.66
Miscellaneous Items	2.5 to	132.3	13.4	5.2	3.7	9.7	0.70	0.50	1.30
Elevators	1.3 to	12.8	5.3	2.1	1.1	2.2	0.11	0.06	0.12
Sprinklers	2.9 to	18.5	7.3	0.5	—	2.0	0.04	—	0.15
Plumbing	2.2 to	19.3	7.0	4.9	7.2	5.2	0.34	0.50	0.36
Heat., Vent., Air Conditioning	3.8 to	18.1	8.8	13.5	11.0	12.9	1.19	0.97	1.14
Electrical	2.4 to	13.5	5.8	10.1	8.4	11.1	0.59	0.49	0.64
Total	1.3 % to	132.3%	—	100.0%	100.0%	100.0%	13.61%	12.45%	15.18%
			Overall Weighted Average	13.75%					

Workers' Compensation Insurance Rates by States

The table below lists the weighted average workers' compensation base rate for each state with a factor comparing this with the national average of 13.5%.

State	Weighted Average	Factor	State	Weighted Average	Factor	State	Weighted Average	Factor
Alabama	20.6%	153	Kentucky	12.5%	93	North Dakota	8.1%	60
Alaska	12.5	93	Louisiana	19.6	145	Ohio	8.1	60
Arizona	13.8	102	Maine	11.4	84	Oklahoma	10.3	76
Arkansas	7.7	57	Maryland	13.5	100	Oregon	11.3	84
California	27.1	201	Massachusetts	11.5	85	Pennsylvania	18.3	136
Colorado	7.3	54	Michigan	13.5	100	Rhode Island	14.7	109
Connecticut	25.9	192	Minnesota	22.8	169	South Carolina	16.9	125
Delaware	14.0	104	Mississippi	14.0	104	South Dakota	14.1	104
District of Columbia	10.4	77	Missouri	13.9	103	Tennessee	12.2	90
Florida	10.8	80	Montana	8.3	61	Texas	9.3	69
Georgia	29.5	219	Nebraska	17.0	126	Utah	8.8	65
Hawaii	8.7	64	Nevada	9.2	68	Vermont	13.3	99
Idaho	10.5	78	New Hampshire	18.8	139	Virginia	8.7	64
Illinois	26.9	199	New Jersey	15.7	116	Washington	9.5	70
Indiana	5.0	37	New Mexico	15.3	113	West Virginia	7.4	55
Iowa	15.5	115	New York	19.0	141	Wisconsin	13.2	98
Kansas	9.2	68	North Carolina	17.3	128	Wyoming	6.5	48
			Weighted Average for U.S. is	13.7% of payroll = 100%				

Rates in the following table are the base or manual costs per $100 of payroll for workers' compensation in each state. Rates are usually applied to straight time wages only and not to premium time wages and bonuses.

The weighted average skilled worker rate for 35 trades is 13.5%. For bidding purposes, apply the full value of workers' compensation directly to total labor costs, or if labor is 38%, materials 42% and overhead and profit 20% of total cost, carry 38/80 × 13.5% =6.4% of cost (before overhead and profit) into overhead. Rates vary not only from state to state but also with the experience rating of the contractor.

Rates are the most current available at the time of publication.

715

R013113-80 Performance Bond

This table shows the cost of a Performance Bond for a construction job scheduled to be completed in 12 months. Add 1% of the premium cost per month for jobs requiring more than 12 months to complete. The rates are "standard" rates offered to contractors that the bonding company considers financially sound and capable of doing the work. Preferred rates are offered by some bonding companies based upon financial strength of the contractor. Actual rates vary from contractor to contractor and from bonding company to bonding company. Contractors should prequalify through a bonding agency before submitting a bid on a contract that requires a bond.

Contract Amount			Building Construction Class B Projects			Highways & Bridges					
						Class A New Construction			Class A-1 Highway Resurfacing		
First	$ 100,000	bid	$25.00 per M			$15.00 per M			$9.40 per M		
Next	400,000	bid	$ 2,500	plus	$15.00 per M	$ 1,500	plus	$10.00 per M	$ 940	plus	$7.20 per M
Next	2,000,000	bid	8,500	plus	10.00 per M	5,500	plus	7.00 per M	3,820	plus	5.00 per M
Next	2,500,000	bid	28,500	plus	7.50 per M	19,500	plus	5.50 per M	15,820	plus	4.50 per M
Next	2,500,000	bid	47,250	plus	7.00 per M	33,250	plus	5.00 per M	28,320	plus	4.50 per M
Over	7,500,000	bid	64,750	plus	6.00 per M	45,750	plus	4.50 per M	39,570	plus	4.00 per M

R015423-10 Steel Tubular Scaffolding

On new construction, tubular scaffolding is efficient up to 60' high or five stories. Above this it is usually better to use a hung scaffolding if construction permits. Swing scaffolding operations may interfere with tenants. In this case, the tubular is more practical at all heights.

In repairing or cleaning the front of an existing building the cost of tubular scaffolding per S.F. of building front increases as the height increases above the first tier. The first tier cost is relatively high due to leveling and alignment.

The minimum efficient crew for erecting and dismantling is three workers. They can set up and remove 18 frame sections per day up to 5 stories high. For 6 to 12 stories high, a crew of four is most efficient. Use two or more on top and two on the bottom for handing up or hoisting. They can also set up and remove 18 frame sections per day. At 7' horizontal spacing, this will run about 800 S.F. per day of erecting and dismantling. Time for placing and removing planks must be added to the above. A crew of three can place and remove 72 planks per day up to 5 stories. For over 5 stories, a crew of four can place and remove 80 planks per day.

The table below shows the number of pieces required to erect tubular steel scaffolding for 1000 S.F. of building frontage. This area is made up of a scaffolding system that is 12 frames (11 bays) long by 2 frames high.

For jobs under twenty-five frames, add 50% to rental cost. Rental rates will be lower for jobs over three months duration. Large quantities for long periods can reduce rental rates by 20%.

Description of Component	Number of Pieces for 1000 S.F. of Building Front	Unit
5' Wide Standard Frame, 6'-4" High	24	Ea.
Leveling Jack & Plate	24	
Cross Brace	44	
Side Arm Bracket, 21"	12	
Guardrail Post	12	
Guardrail, 7' section	22	
Stairway Section	2	
Stairway Starter Bar	1	
Stairway Inside Handrail	2	
Stairway Outside Handrail	2	
Walk-Thru Frame Guardrail	2	

Scaffolding is often used as falsework over 15' high during construction of cast-in-place concrete beams and slabs. Two foot wide scaffolding is generally used for heavy beam construction. The span between frames depends upon the load to be carried with a maximum span of 5'.

Heavy duty shoring frames with a capacity of 10,000#/leg can be spaced up to 10' O.C. depending upon form support design and loading.

Scaffolding used as horizontal shoring requires less than half the material required with conventional shoring.

On new construction, erection is done by carpenters.

Rolling towers supporting horizontal shores can reduce labor and speed the job. For maintenance work, catwalks with spans up to 70' can be supported by the rolling towers.

R015433-10   Contractor Equipment

Rental Rates shown elsewhere in the book pertain to late model high quality machines in excellent working condition, rented from equipment dealers. Rental rates from contractors may be substantially lower than the rental rates from equipment dealers depending upon economic conditions; for older, less productive machines, reduce rates by a maximum of 15%. Any overtime must be added to the base rates. For shift work, rates are lower. Usual rule of thumb is 150% of one shift rate for two shifts; 200% for three shifts.

For periods of less than one week, operated equipment is usually more economical to rent than renting bare equipment and hiring an operator.

Costs to move equipment to a job site (mobilization) or from a job site (demobilization) are not included in rental rates, nor in any Equipment costs on any Unit Price line items or crew listings. These costs can be found elsewhere. If a piece of equipment is already at a job site, it is not appropriate to utilize mob/demob costs in an estimate again.

Rental rates vary throughout the country with larger cities generally having lower rates. Lease plans for new equipment are available for periods in excess of six months with a percentage of payments applying toward purchase.

Rental rates can also be treated as reimbursement costs for contractor-owned equipment. Owned equipment costs include depreciation, loan payments, interest, taxes, insurance, storage, and major repairs.

Monthly rental rates vary from 2% to 5% of the cost of the equipment depending on the anticipated life of the equipment and its wearing parts. Weekly rates are about 1/3 the monthly rates and daily rental rates about 1/3 the weekly rate.

The hourly operating costs for each piece of equipment include costs to the user such as fuel, oil, lubrication, normal expendables for the equipment, and a percentage of mechanic's wages chargeable to maintenance. The hourly operating costs listed do not include the operator's wages.

The daily cost for equipment used in the standard crews is figured by dividing the weekly rate by five, then adding eight times the hourly operating cost to give the total daily equipment cost, not including the operator. This figure is in the right hand column of the Equipment listings under Equipment Cost/Day.

Pile Driving rates shown for pile hammer and extractor do not include leads, crane, boiler or compressor. Vibratory pile driving requires an added field specialist during set-up and pile driving operation for the electric model. The hydraulic model requires a field specialist for set-up only. Up to 125 reuses of sheet piling are possible using vibratory drivers. For normal conditions, crane capacity for hammer type and size are as follows.

Crane Capacity	Hammer Type and Size		
	Air or Steam	Diesel	Vibratory
25 ton	to 8,750 ft.-lb.		70 H.P.
40 ton	15,000 ft.-lb.	to 32,000 ft.-lb.	170 H.P.
60 ton	25,000 ft.-lb.		300 H.P.
100 ton		112,000 ft.-lb.	

Cranes should be specified for the job by size, building and site characteristics, availability, performance characteristics, and duration of time required.

Backhoes & Shovels rent for about the same as equivalent size cranes but maintenance and operating expense is higher. Crane operators rate must be adjusted for high boom heights. Average adjustments: for 150' boom add 2% per hour; over 185', add 4% per hour; over 210', add 6% per hour; over 250', add 8% per hour and over 295', add 12% per hour.

Tower Cranes of the climbing or static type have jibs from 50' to 200' and capacities at maximum reach range from 4,000 to 14,000 pounds. Lifting capacities increase up to maximum load as the hook radius decreases.

Typical rental rates, based on purchase price are about 2% to 3% per month.

Erection and dismantling runs between 500 and 2000 labor hours. Climbing operation takes 10 labor hours per 20' climb. Crane dead time is about 5 hours per 40' climb. If crane is bolted to side of the building add cost of ties and extra mast sections. Climbing cranes have from 80' to 180' of mast while static cranes have 80' to 800' of mast.

Truck Cranes can be converted to tower cranes by using tower attachments. Mast heights over 400' have been used.

A single 100' high material **Hoist and Tower** can be erected and dismantled in about 400 labor hours; a double 100' high hoist and tower in about 600 labor hours. Erection times for additional heights are 3 and 4 labor hours per vertical foot respectively up to 150', and 4 to 5 labor hours per vertical foot over 150' high. A 40' high portable Buck hoist takes about 160 labor hours to erect and dismantle. Additional heights take 2 labor hours per vertical foot to 80' and 3 labor hours per vertical foot for the next 100'. Most material hoists do not meet local code requirements for carrying personnel.

A 150' high **Personnel Hoist** requires about 500 to 800 labor hours to erect and dismantle. Budget erection time at 5 labor hours per vertical foot for all trades. Local code requirements or labor scarcity requiring overtime can add up to 50% to any of the above erection costs.

Earthmoving Equipment: The selection of earthmoving equipment depends upon the type and quantity of material, moisture content, haul distance, haul road, time available, and equipment available. Short haul cut and fill operations may require dozers only, while another operation may require excavators, a fleet of trucks, and spreading and compaction equipment. Stockpiled material and granular material are easily excavated with front end loaders. Scrapers are most economically used with hauls between 300' and 1-1/2 miles if adequate haul roads can be maintained. Shovels are often used for blasted rock and any material where a vertical face of 8' or more can be excavated. Special conditions may dictate the use of draglines, clamshells, or backhoes. Spreading and compaction equipment must be matched to the soil characteristics, the compaction required and the rate the fill is being supplied.

717

R015436-50   Mobilization

Costs to move rented construction equipment to a job site from an equipment dealer's or contractor's yard (mobilization), or to move the equipment off the job site (demobilization), are not included in the rental or operating rates, nor in the equipment cost on a unit price line or in a crew listing. These costs can be found consolidated in the Mobilization section of the data and elsewhere in particular site work sections. If a piece of equipment is already on the job site, it is not appropriate to include mob/demob costs in a new estimate that requires use of that equipment. The following table identifies approximate sizes of rented construction equipment that would be hauled on a towed trailer. Because this listing is not all-encompassing, the user can infer as to what size trailer might be required for a piece of equipment not listed.

3-ton Trailer	20-ton Trailer	40-ton Trailer	50-ton Trailer
20 H.P. Excavator	110 H.P. Excavator	200 H.P. Excavator	270 H.P. Excavator
50 H.P. Skid Steer	165 H.P. Dozer	300 H.P. Dozer	Small Crawler Crane
35 H.P. Roller	150 H.P. Roller	400 H.P. Scraper	500 H.P. Scraper
40 H.P. Trencher	Backhoe	450 H.P. Art. Dump Truck	500 H.P. Art. Dump Truck

R024119-10   Demolition Defined

Whole Building Demolition - Demolition of the whole building with no concern for any particular building element, component, or material type being demolished. This type of demolition is accomplished with large pieces of construction equipment that break up the structure, load it into trucks and haul it to a disposal site, but disposal or dump fees are not included. Demolition of below-grade foundation elements, such as footings, foundation walls, grade beams, slabs on grade, etc., is not included. Certain mechanical equipment containing flammable liquids or ozone-depleting refrigerants, electric lighting elements, communication equipment components, and other building elements may contain hazardous waste, and must be removed, either selectively or carefully, as hazardous waste before the building can be demolished.

Foundation Demolition - Demolition of below-grade foundation footings, foundation walls, grade beams, and slabs on grade. This type of demolition is accomplished by hand or pneumatic hand tools, and does not include saw cutting, or handling, loading, hauling, or disposal of the debris.

Gutting - Removal of building interior finishes and electrical/mechanical systems down to the load-bearing and sub-floor elements of the rough building frame, with no concern for any particular building element, component, or material type being demolished. This type of demolition is accomplished by hand or pneumatic hand tools, and includes loading into trucks, but not hauling, disposal or dump fees, scaffolding, or shoring. Certain mechanical equipment containing flammable liquids or ozone-depleting refrigerants, electric lighting elements, communication equipment components, and other building elements may contain hazardous waste, and must be removed, either selectively or carefully, as hazardous waste, before the building is gutted.

Selective Demolition - Demolition of a selected building element, component, or finish, with some concern for surrounding or adjacent elements, components, or finishes (see the first Subdivision (s) at the beginning of appropriate Divisions). This type of demolition is accomplished by hand or pneumatic hand tools, and does not include handling, loading, storing, hauling, or disposal of the debris, scaffolding, or shoring. "Gutting" methods may be used in order to save time, but damage that is caused to surrounding or adjacent elements, components, or finishes may have to be repaired at a later time.

Careful Removal - Removal of a piece of service equipment, building element or component, or material type, with great concern for both the removed item and surrounding or adjacent elements, components or finishes. The purpose of careful removal may be to protect the removed item for later re-use, preserve a higher salvage value of the removed item, or replace an item while taking care to protect surrounding or adjacent elements, components, connections, or finishes from cosmetic and/or structural damage. An approximation of the time required to perform this type of removal is 1/3 to 1/2 the time it would take to install a new item of like kind (see Reference Number R220105-10). This type of removal is accomplished by hand or pneumatic hand tools, and does not include loading, hauling, or storing the removed item, scaffolding, shoring, or lifting equipment.

Cutout Demolition - Demolition of a small quantity of floor, wall, roof, or other assembly, with concern for the appearance and structural integrity of the surrounding materials. This type of demolition is accomplished by hand or pneumatic hand tools, and does not include saw cutting, handling, loading, hauling, or disposal of debris, scaffolding, or shoring.

Rubbish Handling - Work activities that involve handling, loading or hauling of debris. Generally, the cost of rubbish handling must be added to the cost of all types of demolition, with the exception of whole building demolition.

Minor Site Demolition - Demolition of site elements outside the footprint of a building. This type of demolition is accomplished by hand or pneumatic hand tools, or with larger pieces of construction equipment, and may include loading a removed item onto a truck (check the Crew for equipment used). It does not include saw cutting, hauling or disposal of debris, and, sometimes, handling or loading.

R024119-20   Dumpsters

Dumpster rental costs on construction sites are presented in two ways.

The cost per week rental includes the delivery of the dumpster; its pulling or emptying once per week, and its final removal. The assumption is made that the dumpster contractor could choose to empty a dumpster by simply bringing in an empty unit and removing the full one. These costs also include the disposal of the materials in the dumpster.

The Alternate Pricing can be used when actual planned conditions are not approximated by the weekly numbers. For example, these lines can be used when a dumpster is needed for 4 weeks and will need to be emptied 2 or 3 times per week. Conversely the Alternate Pricing lines can be used when a dumpster will be rented for several weeks or months but needs to be emptied only a few times over this period.

R026510-20   Underground Storage Tank Removal

Underground Storage Tank Removal can be divided into two categories: Non-Leaking and Leaking. Prior to removing an underground storage tank, tests should be made, with the proper authorities present, to determine whether a tank has been leaking or the surrounding soil has been contaminated.

To safely remove Liquid Underground Storage Tanks:
1. Excavate to the top of the tank.
2. Disconnect all piping.
3. Open all tank vents and access ports.
4. Remove all liquids and/or sludge.
5. Purge the tank with an inert gas.
6. Provide access to the inside of the tank and clean out the interior using proper personal protective equipment (PPE).
7. Excavate soil surrounding the tank using proper PPE for on-site personnel.
8. Pull and properly dispose of the tank.
9. Clean up the site of all contaminated material.
10. Install new tanks or close the excavation.

719

R028213-20   Asbestos Removal Process

Asbestos removal is accomplished by a specialty contractor who understands the federal and state regulations regarding the handling and disposal of the material. The process of asbestos removal is divided into many individual steps. An accurate estimate can be calculated only after all the steps have been priced.

The steps are generally as follows:

1. Obtain an asbestos abatement plan from an industrial hygienist.
2. Monitor the air quality in and around the removal area and along the path of travel between the removal area and transport area. This establishes the background contamination.
3. Construct a two part decontamination chamber at entrance to removal area.
4. Install a HEPA filter to create a negative pressure in the removal area.
5. Install wall, floor and ceiling protection as required by the plan, usually 2 layers of fireproof 6 mil polyethylene.

6. Industrial hygienist visually inspects work area to verify compliance with plan.
7. Provide temporary supports for conduit and piping affected by the removal process.
8. Proceed with asbestos removal and bagging process. Monitor air quality as described in Step #2. Discontinue operations when contaminate levels exceed applicable standards.
9. Document the legal disposal of materials in accordance with EPA standards.
10. Thoroughly clean removal area including all ledges, crevices and surfaces.
11. Post abatement inspection by industrial hygienist to verify plan compliance.
12. Provide a certificate from a licensed industrial hygienist attesting that contaminate levels are within acceptable standards before returning area to regular use.

R078413-30   Firestopping

Firestopping is the sealing of structural, mechanical, electrical and other penetrations through fire-rated assemblies. The basic components of firestop systems are safing insulation and firestop sealant on both sides of wall penetrations and the top side of floor penetrations.

Pipe penetrations are assumed to be through concrete, grout, or joint compound and can be sleeved or unsleeved. Costs for the penetrations and sleeves are not included. An annular space of 1″ is assumed. Escutcheons are not included.

Metallic pipe is assumed to be copper, aluminum, cast iron or similar metallic material. Insulated metallic pipe is assumed to be covered with a thermal insulating jacket of varying thickness and materials.

Non-metallic pipe is assumed to be PVC, CPVC, FR Polypropylene or similar plastic piping material. Intumescent firestop sealant or wrap strips are included. Collars on both sides of wall penetrations and a sheet metal plate on the underside of floor penetrations are included.

Ductwork is assumed to be sheet metal, stainless steel or similar metallic material. Duct penetrations are assumed to be through concrete, grout or joint compound. Costs for penetrations and sleeves are not included. An annular space of 1/2″ is assumed.

Multi-trade openings include costs for sheet metal forms, firestop mortar, wrap strips, collars and sealants as necessary.

Structural penetrations joints are assumed to be 1/2″ or less. CMU walls are assumed to be within 1-1/2″ of metal deck. Drywall walls are assumed to be tight to the underside of metal decking.

Metal panel, glass or curtain wall systems include a spandrel area of 5′ filled with mineral wool foil-faced insulation. Fasteners and stiffeners are included.

Finishes — R0991 Painting

R099100-10 Painting Estimating Techniques

Proper estimating methodology is needed to obtain an accurate painting estimate. There is no known reliable shortcut or square foot method. The following steps should be followed:

- List all surfaces to be painted, with an accurate quantity (area) of each. Items having similar surface condition, finish, application method and accessibility may be grouped together.
- List all the tasks required for each surface to be painted, including surface preparation, masking, and protection of adjacent surfaces. Surface preparation may include minor repairs, washing, sanding and puttying.
- Select the proper Means line for each task. Review and consider all adjustments to labor and materials for type of paint and location of work. Apply the height adjustment carefully. For instance, when applying the adjustment for work over 8' high to a wall that is 12' high, apply the adjustment only to the area between 8' and 12' high, and not to the entire wall.

When applying more than one percent (%) adjustment, apply each to the base cost of the data, rather than applying one percentage adjustment on top of the other.

When estimating the cost of painting walls and ceilings remember to add the brushwork for all cut-ins at inside corners and around windows and doors as a LF measure. One linear foot of cut-in with brush equals one square foot of painting.

All items for spray painting include the labor for roll-back.

Deduct for openings greater than 100 SF, or openings that extend from floor to ceiling and are greater than 5' wide. Do not deduct small openings.

The cost of brushes, rollers, ladders and spray equipment are considered to be part of a painting contractor's overhead, and should not be added to the estimate. The cost of rented equipment such as scaffolding and swing staging should be added to the estimate.

Plumbing — R2201 Operation & Maintenance of Plumbing

R220102-20 Labor Adjustment Factors

Labor Adjustment Factors are provided for Divisions 21, 22, and 23 to assist the mechanical estimator account for the various complexities and special conditions of any particular project. While a single percentage has been entered on each line of Division 22 01 02.20, it should be understood that these are just suggested midpoints of ranges of values commonly used by mechanical estimators. They may be increased or decreased depending on the severity of the special conditions.

The group for "existing occupied buildings", has been the subject of requests for explanation. Actually there are two stages to this group: buildings that are existing and "finished" but unoccupied, and those that also are occupied. Buildings that are "finished" may result in higher labor costs due to the workers having to be more careful not to damage finished walls, ceilings, floors, etc. and may necessitate special protective coverings and barriers. Also corridor bends and doorways may not accommodate long pieces of pipe or larger pieces of equipment. Work above an already hung ceiling can be very time consuming. The addition of occupants may force the work to be done on premium time (nights and/or weekends), eliminate the possible use of some preferred tools such as pneumatic drivers, powder charged drivers etc. The estimator should evaluate the access to the work area and just how the work is going to be accomplished to arrive at an increase in labor costs over "normal" new construction productivity.

R220105-10 Demolition (Selective vs. Removal for Replacement)

Demolition can be divided into two basic categories.

One type of demolition involves the removal of material with no concern for its replacement. The labor-hours to estimate this work are found under "Selective Demolition" in the Fire Protection, Plumbing and HVAC Divisions. It is selective in that individual items or all the material installed as a system or trade grouping such as plumbing or heating systems are removed. This may be accomplished by the easiest way possible, such as sawing, torch cutting, or sledge hammer as well as simple unbolting.

The second type of demolition is the removal of some item for repair or replacement. This removal may involve careful draining, opening of unions,

disconnecting and tagging of electrical connections, capping of pipes/ducts to prevent entry of debris or leakage of the material contained as well as transport of the item away from its in-place location to a truck/dumpster. An approximation of the time required to accomplish this type of demolition is to use half of the time indicated as necessary to install a new unit. For example; installation of a new pump might be listed as requiring 6 labor-hours so if we had to estimate the removal of the old pump we would allow an additional 3 hours for a total of 9 hours. That is, the complete replacement of a defective pump with a new pump would be estimated to take 9 labor-hours.

R220523-80 Valve Materials

VALVE MATERIALS

Bronze:
Bronze is one of the oldest materials used to make valves. It is most commonly used in hot and cold water systems and other non-corrosive services. It is often used as a seating surface in larger iron body valves to ensure tight closure.

Carbon Steel:
Carbon steel is a high strength material. Therefore, valves made from this metal are used in higher pressure services, such as steam lines up to 600 psi at 850° F. Many steel valves are available with butt-weld ends for economy and are generally used in high pressure steam service as well as other higher pressure non-corrosive services.

Forged Steel:
Valves from tough carbon steel are used in service up to 2000 psi and temperatures up to 1000° F in Gate, Globe and Check valves.

Iron:
Valves are normally used in medium to large pipe lines to control non-corrosive fluid and gases, where pressures do not exceed 250 psi at 450° F or 500 psi cold water, oil or gas.

Stainless Steel:
Developed steel alloys can be used in over 90% corrosive services.

Plastic PVC:
This is used in a great variety of valves generally in high corrosive service with lower temperatures and pressures.

VALVE SERVICE PRESSURES

Pressure ratings on valves provide an indication of the safe operating pressure for a valve at some elevated temperature. This temperature is dependent upon the materials used and the fabrication of the valve. When specific data is not available, a good "rule-of-thumb" to follow is the temperature of saturated steam on the primary rating indicated on the valve body. Example: The valve has the number 150S printed on the side indicating 150 psi and hence, a maximum operating temperature of 367° F (temperature of saturated steam and 150 psi).

DEFINITIONS
1. "WOG" – Water, oil, gas (cold working pressures).
2. "SWP" – Steam working pressure.
3. 100% area (full port) – means the area through the valve is equal to or greater than the area of standard pipe.
4. "Standard Opening" – means that the area through the valve is less than the area of standard pipe and therefore these valves should be used only where restriction of flow is unimportant.
5. "Round Port" – means the valve has a full round opening through the plug and body, of the same size and area as standard pipe.
6. "Rectangular Port" – valves have rectangular shaped ports through the plug body. The area of the port is either equal to 100% of the area of standard pipe, or restricted (standard opening). In either case it is clearly marked.
7. "ANSI" – American National Standards Institute.

R220523-90 Valve Selection Considerations

INTRODUCTION: In any piping application, valve performance is critical. Valves should be selected to give the best performance at the lowest cost.

The following is a list of performance characteristics generally expected of valves.
1. Stopping flow or starting it.
2. Throttling flow (Modulation).
3. Flow direction changing.
4. Checking backflow (Permitting flow in only one direction).
5. Relieving or regulating pressure.

In order to properly select the right valve, some facts must be determined.

A. What liquid or gas will flow through the valve?
B. Does the fluid contain suspended particles?
C. Does the fluid remain in liquid form at all times?

D. Which metals does fluid corrode?
E. What are the pressure and temperature limits? (As temperature and pressure rise, so will the price of the valve.)
F. Is there constant line pressure?
G. Is the valve merely an on-off valve?
H. Will checking of backflow be required?
I. Will the valve operate frequently or infrequently?

Valves are classified by design type into such classifications as Gate, Globe, Angle, Check, Ball, Butterfly and Plug. They are also classified by end connection, stem, pressure restrictions and material such as bronze, cast iron, etc. Each valve has a specific use. A quality valve used correctly will provide a lifetime of trouble- free service, but a high quality valve installed in the wrong service may require frequent attention.

STEM TYPES
(OS & Y)—Rising Stem-Outside Screw and Yoke

Offers a visual indication of whether the valve is open or closed. Recommended where high temperatures, corrosives, and solids in the line might cause damage to inside-valve stem threads. The stem threads are engaged by the yoke bushing so the stem rises through the hand wheel as it is turned.

(R.S.)—Rising Stem-Inside Screw

Adequate clearance for operation must be provided because both the hand wheel and the stem rise.
The valve wedge position is indicated by the position of the stem and hand wheel.

(N.R.S.)—Non-Rising Stem-Inside Screw

A minimum clearance is required for operating this type of valve. Excessive wear or damage to stem threads inside the valve may be caused by heat, corrosion, and solids. Because the hand wheel and stem do not rise, wedge position cannot be visually determined.

VALVE TYPES
Gate Valves

Provide full flow, minute pressure drop, minimum turbulence and minimum fluid trapped in the line.
They are normally used where operation is infrequent.

Globe Valves

Globe valves are designed for throttling and/or frequent operation with positive shut-off. Particular attention must be paid to the several types of seating materials available to avoid unnecessary wear. The seats must be compatible with the fluid in service and may be composition or metal. The configuration of the Globe valve opening causes turbulence which results in increased resistance. Most bronze Globe valves are rising stem-inside screw, but they are also available on O.S. & Y.

Angle Valves

The fundamental difference between the Angle valve and the Globe valve is the fluid flow through the Angle valve. It makes a 90° turn and offers less resistance to flow than the Globe valve while replacing an elbow. An Angle valve thus reduces the number of joints and installation time.

723

Check Valves

Check valves are designed to prevent backflow by automatically seating when the direction of fluid is reversed.

Swing Check valves are generally installed with Gate-valves, as they provide comparable full flow. Usually recommended for lines where flow velocities are low and should not be used on lines with pulsating flow. Recommended for horizontal installation, or in vertical lines only where flow is upward.

Lift Check Valves

These are commonly used with Globe and Angle valves since they have similar diaphragm seating arrangements and are recommended for preventing backflow of steam, air, gas and water, and on vapor lines with high flow velocities. For horizontal lines, horizontal lift checks should be used and vertical lift checks for vertical lines.

Ball Valves

Ball valves are light and easily installed, yet because of modern elastomeric seats, provide tight closure. Flow is controlled by rotating up to 90° a drilled ball which fits tightly against resilient seals. This ball seats with flow in either direction, and valve handle indicates the degree of opening. Recommended for frequent operation readily adaptable to automation, ideal for installation where space is limited.

Butterfly Valves

Butterfly valves provide bubble-tight closure with excellent throttling characteristics. They can be used for full-open, closed and for throttling applications.

The Butterfly valve consists of a disc within the valve body which is controlled by a shaft. In its closed position, the valve disc seals against a resilient seat. The disc position throughout the full 90° rotation is visually indicated by the position of the operator.

A Butterfly valve is only a fraction of the weight of a Gate valve and requires no gaskets between flanges in most cases. Recommended for frequent operation and adaptable to automation where space is limited.

Wafer and Lug type bodies when installed between two pipe flanges, can be easily removed from the line. The pressure of the bolted flanges holds the valve in place.

Locating lugs makes installation easier.

Plug Valves

Lubricated plug valves, because of the wide range of service to which they are adapted, may be classified as all purpose valves. They can be safely used at all pressure and vacuums, and at all temperatures up to the limits of available lubricants. They are the most satisfactory valves for the handling of gritty suspensions and many other destructive, erosive, corrosive and chemical solutions.

R221113-50 Pipe Material Considerations

1. Malleable fittings should be used for gas service.
2. Malleable fittings are used where there are stresses/strains due to expansion and vibration.
3. Cast fittings may be broken as an aid to disassembling of heating lines frozen by long use, temperature and minerals.
4. Cast iron pipe is extensively used for underground and submerged service.
5. Type M (light wall) copper tubing is available in hard temper only and is used for nonpressure and less severe applications than K and L.

6. Type L (medium wall) copper tubing, available hard or soft for interior service.
7. Type K (heavy wall) copper tubing, available in hard or soft temper for use where conditions are severe. For underground and interior service.
8. Hard drawn tubing requires fewer hangers or supports but should not be bent. Silver brazed fittings are recommended, however soft solder is normally used.
9. Type DMV (very light wall) copper tubing designed for drainage, waste and vent plus other non-critical pressure services.

Domestic/Imported Pipe and Fittings Cost

The prices shown in this publication for steel/cast iron pipe and steel, cast iron, malleable iron fittings are based on domestic production sold at the normal trade discounts. The above listed items of foreign manufacture may be available at prices of 1/3 to 1/2 those shown. Some imported items after minor machining or finishing operations are being sold as domestic to further complicate the system.

Caution: Most pipe prices in this book also include a coupling and pipe hangers which for the larger sizes can add significantly to the per foot cost and should be taken into account when comparing "book cost" with quoted supplier's cost.

725

R221113-70 Piping to 10' High

When taking off pipe, it is important to identify the different material types and joining procedures, as well as distances between supports and components required for proper support.

During the takeoff, measure through all fittings. Do not subtract the lengths of the fittings, valves, or strainers, etc. This added length plus the final rounding of the totals will compensate for nipples and waste.

When rounding off totals always increase the actual amount to correspond with manufacturer's shipping lengths.

A. Both red brass and yellow brass pipe are normally furnished in 12' lengths, plain end. The Unit Price section includes in the linear foot costs two field threads and one coupling per 10' length. A carbon steel clevis type hanger assembly every 10' is also prorated into the linear foot costs, including both material and labor.

B. Cast iron soil pipe is furnished in either 5' or 10' lengths. For pricing purposes, the Unit Price section features 10' lengths with a joint and a carbon steel clevis hanger assembly every 5' prorated into the per foot costs of both material and labor.

Three methods of joining are considered: lead and oakum poured joints or push-on gasket type joints for the bell and spigot pipe and a joint clamp for the no-hub soil pipe. The labor and material costs for each of these individual joining procedures are also prorated into the linear costs per foot.

C. Copper tubing covers types K, L, M, and DWV which are furnished in 20' lengths. Means pricing data is based on a tubing cut each length and a coupling and two soft soldered joints every 10'. A carbon steel, clevis type hanger assembly every 10' is also prorated into the per foot costs. The prices for refrigeration tubing are for materials only. Labor for full lengths may be based on the type L labor but short cut measures in tight areas can increase the installation labor-hours from 20 to 40%.

D. Corrosion-resistant piping does not lend itself to one particular standard of hanging or support assembly due to its diversity of application and placement. The several varieties of corrosion-resistant piping do not include any material or labor costs for hanger assemblies (See the Unit Price section for appropriate selection).

E. Glass pipe is furnished in standard lengths either 5' or 10' long, beaded on one end. Special orders for diverse lengths beaded on both ends are also available. For pricing purposes, R.S. Means features 10' lengths with a coupling and a carbon steel band hanger assembly every 10' prorated into the per foot linear costs.

Glass pipe is also available with conical ends and standard lengths ranging from 6" through 3' in 6" increments, then up to 10' in 12" increments. Special lengths can be customized for particular installation requirements.

For pricing purposes, Means has based the labor and material pricing on 10' lengths. Included in these costs per linear foot are the prorated costs for a flanged assembly every 10' consisting of two flanges, a gasket, two insertable seals, and the required number of bolts and nuts. A carbon steel band hanger assembly based on 10' center lines has also been prorated into the costs per foot for labor and materials.

F. Plastic pipe of several compositions and joining methods are considered. Fiberglass reinforced pipe (FRP) is priced, based on 10' lengths (20' lengths are also available), with coupling and epoxy joints every 10'. FRP is furnished in both "General Service" and "High Strength." A carbon steel clevis hanger assembly, 3 for every 10', is built into the prorated labor and material costs on a per foot basis.

The PVC and CPVC pipe schedules 40, 80 and 120, plus SDR ratings are all based on 20' lengths with a coupling installed every 10', as well as a carbon steel clevis hanger assembly every 3'. The PVC and

ABS type DWV piping is based on 10' lengths with solvent weld couplings every 10', and with carbon steel clevis hanger assemblies, 3 for every 10'. The rest of the plastic piping in this section is based on flexible 100' coils and does not include any coupling or supports.

This section ends with PVC drain and sewer piping based on 10' lengths with bell and spigot ends and 0-ring type, push-on joints.

G. Stainless steel piping includes both weld end and threaded piping, both in the type 304 and 316 specification and in the following schedules, 5, 10, 40, 80, and 160. Although this piping is usually furnished in 20' lengths, this cost grouping has a joint (either heli-arc butt-welded or threads and coupling) every 10'. A carbon steel clevis type hanger assembly is also included at 10' intervals and prorated into the linear foot costs.

H. Carbon steel pipe includes both black and galvanized. This section encompasses schedules 40 (standard) and 80 (extra heavy).

Several common methods of joining steel pipe — such as thread and coupled, butt welded, and flanged (150 lb. weld neck flanges) are also included.

For estimating purposes, it is assumed that the piping is purchased in 20' lengths and that a compatible joint is made up every 10'. These joints are prorated into the labor and material costs per linear foot. The following hanger and support assemblies every 10' are also included: carbon steel clevis for the T & C pipe, and single rod roll type for both the welded and flanged piping. All of these hangers are oversized to accommodate pipe insulation 3/4" thick through 5" pipe size and 1-1/2" thick from 6" through 12" pipe size.

I. Grooved joint steel pipe is priced both black and galvanized, in schedules 10, 40, and 80, furnished in 20' lengths. This section describes two joining methods: cut groove and roll groove. The schedule 10 piping is roll-grooved, while the heavier schedules are cut-grooved. The labor and material costs are prorated into per linear foot prices, including a coupled joint every 10', as well as a carbon steel clevis hanger assembly.

Notes:

The pipe hanger assemblies mentioned in the preceding paragraphs include the described hanger; appropriately sized steel, box-type insert and nut; plus 18" of threaded hanger rod.

C clamps are used when the pipe is to be supported from steel shapes rather than anchored in the slab. C clamps are slightly less costly than inserts. However, to save time in estimating, it is advisable to use the given line number cost, rather than substituting a C clamp for the insert.

Add to piping labor for elevated installation:

10' to 14.5' high	10%	30' to 34.5' high	40%
15' to 19.5' high	20%	35' to 39.5' high	50%
20' to 24.5' high	25%	Over 40' and higher	55%
25' to 29.5' high	35%		

When using the percentage adds for elevated piping installations as shown above, bear in mind that the given heights are for the pipe supports, even though the insert, anchor, or clamp may be several feet higher than the pipe itself.

An allowance has been included in the piping installation time for testing and minor tightening of leaking joints, fittings, stuffing boxes, packing glands, etc. For extraordinary test requirements such as x-rays, prolonged pressure or demonstration tests, a percentage of the piping labor, based on the estimator's experience, must be added to the labor total. A testing service specializing in weld x-rays should be consulted for pricing if it is an estimate requirement. Equipment installation time includes start-up with associated adjustments.

726

For customer support on your Mechanical Cost Data, call 877.795.7394.

R230500-10 Subcontractors

On the unit cost pages of the R.S. Means Cost Data books, the last column is entitled "Total Incl. O&P". This is normally the cost of the installing contractor. In the HVAC Division, this is the cost of the mechanical contractor. If the particular work being estimated is to be performed by a sub to the mechanical contractor, the mechanical's profit and handling charge (usually 10%) is added to the total of the last column.

Heating, Ventilating & A.C. | R2331 HVAC Ducts & Casings

R233100-30 Duct Fabrication/Installation

The labor cost for fabricated sheet metal duct includes both the cost of fabrication and installation of the duct. The split is approximately 40% for fabrication, 60% for installation. It is for this reason that the percentage add for elevated installation of fabricated duct is less than the percentage add for prefabricated/preformed duct.

Example: assume a piece of duct cost $100 installed (labor only)

Sheet Metal Fabrication = 40% = $40

Installation = 60% = $60

The add for elevated installation is:

$$\frac{\text{Based on total labor}}{\text{(fabrication \& installation)}} = \$100 \times 6\% = \$6.00$$

$$\frac{\text{Based on installation cost only}}{\text{(Material purchased prefabricated)}} = \$60 \times 10\% = \$6.00$$

The $6.00 markup (10' to 15' high) is the same.

727

Heating, Ventilating & A.C. R2331 HVAC Ducts & Casings

R233100-40 Steel Sheet Metal Calculator (Weight in Lb./Ft. of Length)

Gauge	26	24	22	20	18	16	Gauge	26	24	22	20	18	16
Wt.-Lb./S.F.	.906	1.156	1.406	1.656	2.156	2.656	Wt.-Lb./S.F.	.906	1.156	1.406	1.656	2.156	2.656
SMACNA Max. Dimension – Long Side	*	30″	54″	84″	85″ Up	*	SMACNA Max. Dimension – Long Side	*	30″	54″	84″	85″ Up	*
Sum-2 sides							**Sum-2 Sides**						
2	.3	.40	.50	.60	.80	.90	56	9.3	12.0	14.0	16.2	21.3	25.2
3	.5	.65	.80	.90	1.1	1.4	57	9.5	12.3	14.3	16.5	21.7	25.7
4	.7	.85	1.0	1.2	1.5	1.8	58	9.7	12.5	14.5	16.8	22.0	26.1
5	.8	1.1	1.3	1.5	1.9	2.3	59	9.8	12.7	14.8	17.1	22.4	26.6
6	1.0	1.3	1.5	1.7	2.3	2.7	60	10.0	12.9	15.0	17.4	22.8	27.0
7	1.2	1.5	1.8	2.0	2.7	3.2	61	10.2	13.1	15.3	17.7	23.2	27.5
8	1.3	1.7	2.0	2.3	3.0	3.6	62	10.3	13.3	15.5	18.0	23.6	27.9
9	1.5	1.9	2.3	2.6	3.4	4.1	63	10.5	13.5	15.8	18.3	24.0	28.4
10	1.7	2.2	2.5	2.9	3.8	4.5	64	10.7	13.7	16.0	18.6	24.3	28.8
11	1.8	2.4	2.8	3.2	4.2	5.0	65	10.8	13.9	16.3	18.9	24.7	29.3
12	2.0	2.6	3.0	3.5	4.6	5.4	66	11.0	14.1	16.5	19.1	25.1	29.7
13	2.2	2.8	3.3	3.8	4.9	5.9	67	11.2	14.3	16.8	19.4	25.5	30.2
14	2.3	3.0	3.5	4.1	5.3	6.3	68	11.3	14.6	17.0	19.7	25.8	30.6
15	2.5	3.2	3.8	4.4	5.7	6.8	69	11.5	14.8	17.3	20.0	26.2	31.1
16	2.7	3.4	4.0	4.6	6.1	7.2	70	11.7	15.0	17.5	20.3	26.6	31.5
17	2.8	3.7	4.3	4.9	6.5	7.7	71	11.8	15.2	17.8	20.6	27.0	32.0
18	3.0	3.9	4.5	5.2	6.8	8.1	72	12.0	15.4	18.0	20.9	27.4	32.4
19	3.2	4.1	4.8	5.5	7.2	8.6	73	12.2	15.6	18.3	21.2	27.7	32.9
20	3.3	4.3	5.0	5.8	7.6	9.0	74	12.3	15.8	18.5	21.5	28.1	33.3
21	3.5	4.5	5.3	6.1	8.0	9.5	75	12.5	16.1	18.8	21.8	28.5	33.8
22	3.7	4.7	5.5	6.4	8.4	9.9	76	12.7	16.3	19.0	22.0	28.9	34.2
23	3.8	5.0	5.8	6.7	8.7	10.4	77	12.8	16.5	19.3	22.3	29.3	34.7
24	4.0	5.2	6.0	7.0	9.1	10.8	78	13.0	16.7	19.5	22.6	29.6	35.1
25	4.2	5.4	6.3	7.3	9.5	11.3	79	13.2	16.9	19.8	22.9	30.0	35.6
26	4.3	5.6	6.5	7.5	9.9	11.7	80	13.3	17.1	20.0	23.2	30.4	36.0
27	4.5	5.8	6.8	7.8	10.3	12.2	81	13.5	17.3	20.3	23.5	30.8	36.5
28	4.7	6.0	7.0	8.1	10.6	12.6	82	13.7	17.5	20.5	23.8	31.2	36.9
29	4.8	6.2	7.3	8.4	11.0	13.1	83	13.8	17.8	20.8	24.1	31.5	37.4
30	5.0	6.5	7.5	8.7	11.4	13.5	84	14.0	18.0	21.0	24.4	31.9	37.8
31	5.2	6.7	7.8	9.0	11.8	14.0	85	14.2	18.2	21.3	24.7	32.3	38.3
32	5.3	6.9	8.0	9.3	12.2	14.4	86	14.3	18.4	21.5	24.9	32.7	38.7
33	5.5	7.1	8.3	9.6	12.5	14.9	87	14.5	18.6	21.8	25.2	33.1	39.2
34	5.7	7.3	8.5	9.9	12.9	15.3	88	14.7	18.8	22.0	25.5	33.4	39.6
35	5.8	7.5	8.8	10.2	13.3	15.8	89	14.8	19.0	22.3	25.8	33.8	40.1
36	6.0	7.8	9.0	10.4	13.7	16.2	90	15.0	19.3	22.5	26.1	34.2	40.5
37	6.2	8.0	9.3	10.7	14.1	16.7	91	15.2	19.5	22.8	26.4	34.6	41.0
38	6.3	8.2	9.5	11.0	14.4	17.1	92	15.3	19.7	23.0	26.7	35.0	41.4
39	6.5	8.4	9.8	11.3	14.8	17.6	93	15.5	19.9	23.3	27.0	35.3	41.9
40	6.7	8.6	10.0	11.6	15.2	18.0	94	15.7	20.1	23.5	27.3	35.7	42.3
41	6.8	8.8	10.3	11.9	15.6	18.5	95	15.8	20.3	23.8	27.6	36.1	42.8
42	7.0	9.0	10.5	12.2	16.0	18.9	96	16.0	20.5	24.0	27.8	36.5	43.2
43	7.2	9.2	10.8	12.5	16.3	19.4	97	16.2	20.8	24.3	28.1	36.9	43.7
44	7.3	9.5	11.0	12.8	16.7	19.8	98	16.3	21.0	24.5	28.4	37.2	44.1
45	7.5	9.7	11.3	13.1	17.1	20.3	99	16.5	21.2	24.8	28.7	37.6	44.6
46	7.7	9.9	11.5	13.3	17.5	20.7	100	16.7	21.4	25.0	29.0	38.0	45.0
47	7.8	10.1	11.8	13.6	17.9	21.2	101	16.8	21.6	25.3	29.3	38.4	45.5
48	8.0	10.3	12.0	13.9	18.2	21.6	102	17.0	21.8	25.5	29.6	38.8	45.9
49	8.2	10.5	12.3	14.2	18.6	22.1	103	17.2	22.0	25.8	29.9	39.1	46.4
50	8.3	10.7	12.5	14.5	19.0	22.5	104	17.3	22.3	26.0	30.2	39.5	46.8
51	8.5	11.0	12.8	14.8	19.4	23.0	105	17.5	22.5	26.3	30.5	39.9	47.3
52	8.7	11.2	13.0	15.1	19.8	23.4	106	17.7	22.7	26.5	30.7	40.3	47.7
53	8.8	11.4	13.3	15.4	20.1	23.9	107	17.8	22.9	26.8	31.0	40.7	48.2
54	9.0	11.6	13.5	15.7	20.5	24.3	108	18.0	23.1	27.0	31.3	41.0	48.6
55	9.2	11.8	13.8	16.0	20.9	24.8	109	18.2	23.3	27.3	31.6	41.4	49.1
							110	18.3	23.5	27.5	31.9	41.8	49.5

Example: If duct is 34″ x 20″ x 15′ long, 34″ is greater than 30″ maximum, for 24 ga. so must be 22 ga. 34″ + 20″ = 54″ going across from 54″ find 13.5 lb. per foot. 13.5 x 15′ = 202.5 lbs. For S.F. of surface area 202.5 ÷ 1.406 = 144 S.F.

Note: Figures include an allowance for scrap.
*Do Not use unless engineer specified. 26GA is very light and sometimes used for toilet exhaust. 16 GA is heavy plate and mostly specified for plenums and hoods.

R233100-50   Ductwork Packages (per Ton of Cooling)

System	Sheet Metal	Insulation	Diffusers	Return Register
Roof Top Unit Single Zone	120 Lbs.	52 S.F.	1	1
Roof Top Unit Multizone	240 Lbs.	104 S.F.	2	1
Self-contained Air or Water Cooled	108 Lbs.	—	2	—
Split System Air Cooled	102 Lbs.	—	2	—

Systems reflect most common usage.
Refer to system graphics for duct layout.

Heating, Ventilating & A.C. R2337 Air Outlets & Inlets

R233700-60   Diffuser Evaluation

$CFM = V \times A_n \times K$ where V = Outlet velocity in feet per minute. A_n = Neck area in square feet and K = Diffuser delivery factor. An undersized diffuser for a desired CFM will produce a high velocity and noise level. When air moves past people at a velocity in excess of 25 FPM, an annoying draft is felt. An oversized diffuser will result in low velocity with poor mixing. Consideration must be given to avoid vertical stratification or horizontal areas of stagnation.

729

R235000-10 Heating Systems

Heating Systems

The basic function of a heating system is to bring an enclosed volume up to a desired temperature and then maintain that temperature within a reasonable range. To accomplish this, the selected system must have sufficient capacity to offset transmission losses resulting from the temperature difference on the interior and exterior of the enclosing walls in addition to losses due to cold air infiltration through cracks, crevices and around doors and windows. The amount of heat to be furnished is dependent upon the building size, construction, temperature difference, air leakage, use, shape, orientation and exposure. Air circulation is also an important consideration. Circulation will prevent stratification which could result in heat losses through uneven temperatures at various levels. For example, the most

efficient use of unit heaters can usually be achieved by circulating the space volume through the total number of units once every 20 minutes or 3 times an hour. This general rule must, of course, be adapted for special cases such as large buildings with low ratios of heat transmitting surface to cubical volume. The type of occupancy of a building will have considerable bearing on the number of heat transmitting units and the location selected. It is axiomatic, however, that the basis of any successful heating system is to provide the maximum amount of heat at the points of maximum heat loss such as exposed walls, windows, and doors. Large roof areas, wind direction, and wide doorways create problems of excessive heat loss and require special consideration and treatment.

Heat Transmission

Heat transfer is an important parameter to be considered during selection of the exterior wall style, material and window area. A high rate of transfer will permit greater heat loss during the wintertime with the resultant increase in heating energy costs and a greater rate of heat gain in the summer with proportionally greater cooling cost. Several terms are used to describe various aspects of heat transfer. However, for general estimating purposes this book lists U values for systems of construction materials. U is the "overall heat transfer coefficient." It is defined as the heat flow per hour through one square foot when the temperature difference in the air on either side of the structure wall, roof, ceiling or floor is one degree Fahrenheit. The structural segment may be a single homogeneous material or a composite.

Total heat transfer is found using the following equation:

$Q = AU(T_2 - T_1)$ where

Q = Heat flow, BTU per hour
A = Area, square feet
U = Overall heat transfer coefficient
$(T_2 - T_1)$ = Difference in temperature of air on each side of the construction component. (Also abbreviated TD)

Note that heat can flow through all surfaces of any building and this flow is in addition to heat gain or loss due to ventilation, infiltration and generation (appliances, machinery, people).

R235000-20 Heating Approximations for Quick Estimating

Oil Piping & Boiler Room Piping

Small System . . . . 20 to 30% of Boiler

Complex System

with Pumps, Headers, Etc. . . . . 80 to 110% of Boiler

Breeching With Insulation:

Small . . . . 10 to 15% of Boiler

Large . . . . 15 to 25% of Boiler

Coils: . . . . 15 to 30% of Containing Unit

Balancing (Independent) . . . . 1/2% of H.V.A.C. Estimate

Quality/Complexity Adjustment: For all heating installations add these adjustments to the estimate to more closely allow for the equipment and conditions of the particular job under consideration.

Economy installation, add . . . . 0 to 5% of System

Good quality, medium complexity, add . . . . 5 to 15% of System

Above average quality and complexity, add . . . . 15 to 25% of System

730

For customer support on your Mechanical Cost Data, call 877.795.7394.

R235000-30 The Basics of a Heating System

The function of a heating system is to achieve and maintain a desired temperature in a room or building by replacing the amount of heat being dissipated. There are four kinds of heating systems: hot-water, steam, warm-air and electric resistance. Each has certain essential and similar elements with the exception of electric resistance heating.

The basic elements of a heating system are:

A. A **combustion chamber** in which fuel is burned and heat transferred to a conveying medium.

B. The **"fluid"** used for conveying the heat (water, steam or air).

C. **Conductors** or pipes for transporting the fluid to specific desired locations.

D. A means of disseminating the heat, sometimes called **terminal units.**

A. The **combustion chamber** in a furnace heats air which is then distributed. This is called a warm-air system.

The combustion chamber in a boiler heats water which is either distributed as hot water or steam and this is termed a hydronic system.

The maximum allowable working pressures are limited by ASME "Code for Heating Boilers" to 15 PSI for steam and 160 PSI for hot water heating boilers, with a maximum temperature limitation of 250° F. Hot water boilers are generally rated for a working pressure of 30 PSI. High pressure boilers are governed by the ASME "Code for Power Boilers" which is used almost universally for boilers operating over 15 PSIG. High pressure boilers used for a combination of heating/process loads are usually designed for 150 PSIG.

Boiler ratings are usually indicated as either Gross or Net Output. The Gross Load is equal to the Net Load plus a piping and pickup allowance. When this allowance cannot be determined, divide the gross output rating by 1.25 for a value equal to or greater than the net heat loss requirement of the building.

B. Of the three **fluids** used, steam carries the greatest amount of heat per unit volume. This is due to the fact that it gives up its latent heat of vaporization at a temperature considerably above room temperature. Another advantage is that the pressure to produce a positive circulation is readily available. Piping conducts the steam to terminal units and returns condensate to the boiler.

The **steam system** is well adapted to large buildings because of its positive circulation, its comparatively economical installation and its ability to deliver large quantities of heat. Nearly all large office buildings, stores, hotels, and industrial buildings are so heated, in addition to many residences.

Hot water, when used as the heat carrying fluid, gives up a portion of its sensible heat and then returns to the boiler or heating apparatus for reheating. As the heat conveyed by each pound of water is about one-fiftieth of the heat conveyed by a pound of steam, it is necessary to circulate about fifty times as much water as steam by weight (although only one-thirtieth as much by volume). The hot water system is usually, although not necessarily, designed to operate at temperatures below that of the ordinary steam system and so the amount of heat transfer surface must be correspondingly greater. A temperature of 190° F to 200° F is normally the maximum. Circulation in small buildings may depend on the difference in density between hot water and the cool water returning to the boiler; circulating pumps are normally used to maintain a desired rate of flow. Pumps permit a greater degree of flexibility and better control.

In **warm-air** furnace systems, cool air is taken from one or more points in the building, passed over the combustion chamber and flue gas passages and then distributed through a duct system. A disadvantage of this system is that the ducts take up much more building volume than steam or hot water pipes. Advantages of this system are the relative ease with which humidification can be accomplished by the evaporation of water as the air circulates through the heater, and the lack of need for expensive disseminating units as the warm air simply becomes part of the interior atmosphere of the building.

C. **Conductors** (pipes and ducts) have been lightly treated in the discussion of conveying fluids. For more detailed information such as sizing and distribution methods, the reader is referred to technical publications such as the American Society of Heating, Refrigerating and Air-Conditioning Engineers "Handbook of Fundamentals."

D. **Terminal units** come in an almost infinite variety of sizes and styles, but the basic principles of operation are very limited. As previously mentioned, warm-air systems require only a simple register or diffuser to mix heated air with that present in the room. Special application items such as radiant coils and infrared heaters are available to meet particular conditions but are not usually considered for general heating needs. Most heating is accomplished by having air flow over coils or pipes containing the heat transporting medium (steam, hot-water, electricity). These units, while varied, may be separated into two general types, (1) radiator/convectors and (2) unit heaters.

Radiator/convectors may be cast, fin-tube or pipe assemblies. They may be direct, indirect, exposed, concealed or mounted within a cabinet enclosure, upright or baseboard style. These units are often collectively referred to as "radiatiors" or "radiation" although none gives off heat either entirely by radiation or by convection but rather a combination of both. The air flows over the units as a gravity "current." It is necessary to have one or more heat-emitting units in each room. The most efficient placement is low along an outside wall or under a window to counteract the cold coming into the room and achieve an even distribution.

In contrast to radiator/convectors which operate most effectively against the walls of smaller rooms, **unit heaters** utilize a fan to move air over heating coils and are very effective in locations of relatively large volume. Unit heaters, while usually suspended overhead, may be floor mounted. They also may take in fresh outside air for ventilation. The heat distributed by unit heaters may be from a remote source and conveyed by a fluid or it may be from the combustion of fuel in each individual heater. In the latter case the only piping required would be for fuel, however, a vent for the product of combustion would be necessary.

The following list gives may of the advantages of unit heaters for applications other than office or residential:

a. Large capacity so smaller number of units are required,
b. Piping system simplified, **c.** Space saved where they are located overhead out of the way, **d.** Rapid heating directed where needed with effective wide distribution, **e.** Difference between floor and ceiling temperature reduced, **f.** Circulation of air obtained, and ventilation with introduction of fresh air possible, **g.** Heat output flexible and easily controlled.

731

R235616-60   Solar Heating (Space and Hot Water)

Collectors should face as close to due South as possible, however, variations of up to 20 degrees on either side of true South are acceptable. Local climate and collector type may influence the choice between east or west deviations. Obviously they should be located so they are not shaded from the sun's rays. Incline collectors at a slope of latitude minus 5 degrees for domestic hot water and latitude plus 15 degrees for space heating.

Flat plate collectors consist of a number of components as follows: Insulation to reduce heat loss through the bottom and sides of the collector. The enclosure which contains all the components in this assembly is usually weatherproof and prevents dust, wind and water from coming in contact with the absorber plate. The cover plate usually consists of one or more layers of a variety of glass or plastic and reduces the reradiation by creating an air space which traps the heat between the cover and the absorber plates.

The absorber plate must have a good thermal bond with the fluid passages. The absorber plate is usually metallic and treated with a surface coating which improves absorptivity. Black or dark paints or selective coatings are used for this purpose, and the design of this passage and plate combination helps determine a solar system's effectiveness.

Heat transfer fluid passage tubes are attached above and below or integral with an absorber plate for the purpose of transferring thermal energy from the absorber plate to a heat transfer medium. The heat exchanger is a device for transferring thermal energy from one fluid to another.

Piping and storage tanks should be well insulated to minimize heat losses.

Size domestic water heating storage tanks to hold 20 gallons of water per user, minimum, plus 10 gallons per dishwasher or washing machine. For domestic water heating an optimum collector size is approximately 3/4 square foot of area per gallon of water storage. For space heating of residences and small commercial applications the collector is commonly sized between 30% and 50% of the internal floor area. For space heating of large commercial applications, collector areas less than 30% of the internal floor area can still provide significant heat reductions.

A supplementary heat source is recommended for Northern states for December through February.

The solar energy transmission per square foot of collector surface varies greatly with the material used. Initial cost, heat transmittance and useful life are obviously interrelated.

R236000-10   Air Conditioning

General: The purpose of air conditioning is to control the environment of a space so that comfort is provided for the occupants and/or conditions are suitable for the processes or equipment contained therein. The several items which should be evaluated to define system objectives are:

   Temperature Control
   Humidity Control
   Cleanliness
   Odor, smoke and fumes
   Ventilation

Efforts to control the above parameters must also include consideration of the degree or tolerance of variation, the noise level introduced, the velocity of air motion and the energy requirements to accomplish the desired results.

The variation in **temperature** and **humidity** is a function of the sensor and the controller. The controller reacts to a signal from the sensor and produces the appropriate suitable response in either the terminal unit, the conductor of the transporting medium (air, steam, chilled water, etc.), or the source (boiler, evaporating coils, etc.).

The **noise level** is a by-product of the energy supplied to moving components of the system. Those items which usually contribute the most noise are pumps, blowers, fans, compressors and diffusers. The level of noise can be partially controlled through use of vibration pads, isolators, proper sizing, shields, baffles and sound absorbing liners.

Some **air motion** is necessary to prevent stagnation and stratification. The maximum acceptable velocity varies with the degree of heating or cooling which is taking place. Most people feel air moving past them at velocities in excess of 25 FPM as an annoying draft. However, velocities up to 45 FPM may be acceptable in certain cases. Ventilation, expressed as air changes per hour and percentage of fresh air, is usually an item regulated by local codes.

Selection of the system to be used for a particular application is usually a trade-off. In some cases the building size, style, or room available for mechanical use limits the range of possibilities. Prime factors influencing the decision are first cost and total life (operating, maintenance and replacement costs). The accuracy with which each parameter is determined will be an important measure of the reliability of the decision and subsequent satisfactory operation of the installed system.

Heat delivery may be desired from an air conditioning system. Heating capability usually is added as follows: A gas fired burner or hot water/steam/electric coils may be added to the air handling unit directly and heat all air equally. For limited or localized heat requirements the water/steam/electric coils may be inserted into the duct branch supplying the cold areas. Gas fired duct furnaces are also available.

Note: When water or steam coils are used the cost of the piping and boiler must also be added. For a rough estimate use the cost per square foot of the appropriate sized hydronic system with unit heaters. This will provide a cost for the boiler and piping, and the unit heaters of the system would equate to the approximate cost of the heating coils.

R236000-20 Air Conditioning Requirements

BTUs per hour per S.F. of floor area and S.F. per ton of air conditioning.

Type of Building	BTU/Hr per S.F.	S.F. per Ton	Type of Building	BTU/Hr per S.F.	S.F. per Ton	Type of Building	BTU/Hr per S.F.	S.F. per Ton
Apartments, Individual	26	450	Dormitory, Rooms	40	300	Libraries	50	240
Corridors	22	550	Corridors	30	400	Low Rise Office, Exterior	38	320
Auditoriums & Theaters	40	300/18*	Dress Shops	43	280	Interior	33	360
Banks	50	240	Drug Stores	80	150	Medical Centers	28	425
Barber Shops	48	250	Factories	40	300	Motels	28	425
Bars & Taverns	133	90	High Rise Office—Ext. Rms.	46	263	Office (small suite)	43	280
Beauty Parlors	66	180	Interior Rooms	37	325	Post Office, Individual Office	42	285
Bowling Alleys	68	175	Hospitals, Core	43	280	Central Area	46	260
Churches	36	330/20*	Perimeter	46	260	Residences	20	600
Cocktail Lounges	68	175	Hotel, Guest Rooms	44	275	Restaurants	60	200
Computer Rooms	141	85	Corridors	30	400	Schools & Colleges	46	260
Dental Offices	52	230	Public Spaces	55	220	Shoe Stores	55	220
Dept. Stores, Basement	34	350	Industrial Plants, Offices	38	320	Shop'g. Ctrs., Supermarkets	34	350
Main Floor	40	300	General Offices	34	350	Retail Stores	48	250
Upper Floor	30	400	Plant Areas	40	300	Specialty	60	200

*Persons per ton

12,000 BTU = 1 ton of air conditioning

733

R236000-30 Psychrometric Table

Dewpoint or Saturation Temperature (F)

Relative humidity (%)	32	35	40	45	50	55	60	65	70	75	80	85	90	95	100
100	32	35	40	45	50	55	60	65	70	75	80	85	90	95	100
90	30	33	37	42	47	52	57	62	67	72	77	82	87	92	97
80	27	30	34	39	44	49	54	58	64	68	73	78	83	88	93
70	24	27	31	36	40	45	50	55	60	64	69	74	79	84	88
60	20	24	28	32	36	41	46	51	55	60	65	69	74	79	83
50	16	20	24	28	33	36	41	46	50	55	60	64	69	73	78
40	12	15	18	23	27	31	35	40	45	49	53	58	62	67	71
30	8	10	14	18	21	25	29	33	37	42	46	50	54	59	62
20	6	7	8	9	13	16	20	24	28	31	35	40	43	48	52
10	4	4	5	5	6	8	9	10	13	17	20	24	27	30	34
	32	35	40	45	50	55	60	65	70	75	80	85	90	95	100

Dry bulb temperature (F)

This table shows the relationship between RELATIVE HUMIDITY, DRY BULB TEMPERATURE AND DEWPOINT.

As an example, assume that the thermometer in a room reads 75° F, and we know that the relative humidity is 50%. The chart shows the dewpoint temperature to be 55° F. That is, any surface colder than 55° F will "sweat" or collect condensing moisture. This surface could be the outside of an uninsulated chilled water pipe in the summertime, or the inside surface of a wall or deck in the wintertime. After determining the extreme ambient parameters, the table at the left is useful in determining which surfaces need insulation or vapor barrier protection.

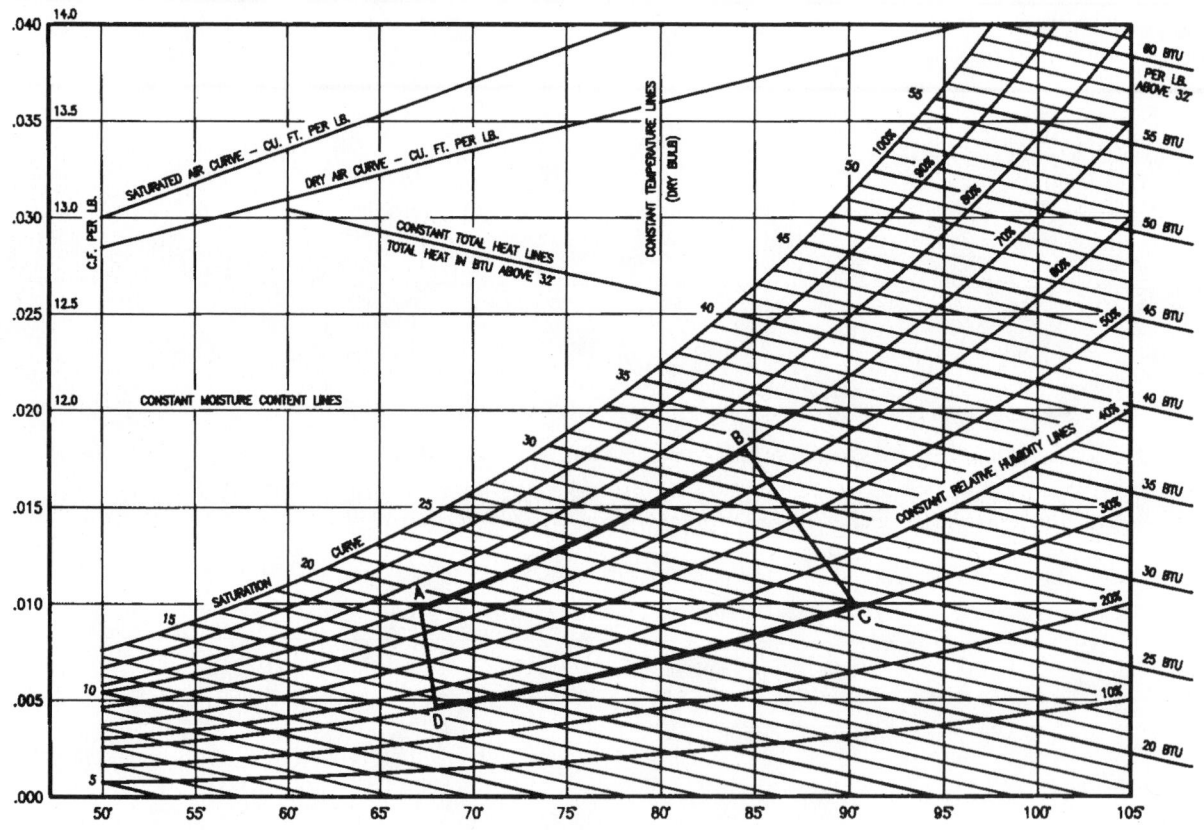

TEMPERATURE DEGREES FAHRENHEIT
TOTAL PRESSURE = 14.696 LB. PER SQ. IN. ABS.

Psychrometric chart showing different variables based on one pound of dry air. Space marked A B C D is temperature-humidity range which is most comfortable for majority of people.

734

For customer support on your Mechanical Cost Data, call 877.795.7394.

R236000-90   Quality/Complexity Adjustment for Air Conditioning Systems

Economy installation, add . . . . 0 to 5%

Good quality, medium complexity, add . . . . 5 to 15%

Above average quality and complexity, add . . . . 15 to 25%

Add the above adjustments to the estimate to more closely allow for the equipment and conditions of the particular job under consideration.

735

Fig. R238313-11

R238313-10   Heat Trace Systems

Before you can determine the cost of a HEAT TRACE installation, the method of attachment must be established. There are (4) common methods:

1. Cable is simply attached to the pipe with polyester tape every 12'.
2. Cable is attached with a continuous cover of 2" wide aluminum tape.
3. Cable is attached with factory extruded heat transfer cement and covered with metallic raceway with clips every 10'.
4. Cable is attached between layers of pipe insulation using either clips or polyester tape.

Example: Components for method 3 must include:
- A. Heat trace cable by voltage and watts per linear foot.
- B. Heat transfer cement, 1 gallon per 60 linear feet of cover.
- C. Metallic raceway by size and type.
- D. Raceway clips by size of pipe.

When taking off linear foot lengths of cable add the following for each valve in the system. (E)

In all of the above methods each component of the system must be priced individually.

SCREWED OR WELDED VALVE:			FLANGED VALVE:			BUTTERFLY VALVES:		
1/2"	=	6"	1/2"	=	1' -0"	1/2"	=	0'
3/4"	=	9"	3/4"	=	1' -6"	3/4"	=	0'
1"	=	1' -0"	1"	=	2' -0"	1"	=	1' -0"
1-1/2"	=	1' -6"	1-1/2"	=	2' -6"	1-1/2"	=	1' -6"
2"	=	2'	2"	=	2' -6"	2"	=	2' -0"
2-1/2"	=	2' -6"	2-1/2"	=	3' -0"	2-1/2"	=	2' -6"
3"	=	2' -6"	3"	=	3' -6"	3"	=	2' -6"
4"	=	4' -0"	4"	=	4' -0"	4"	=	3' -0"
6"	=	7' -0"	6"	=	8' -0"	6"	=	3' -6"
8"	=	9' -6"	8"	=	11' -0"	8"	=	4' -0"
10"	=	12' -6"	10"	=	14' -0"	10"	=	4' -0"
12"	=	15' -0"	12"	=	16' -6"	12"	=	5' -0"
14"	=	18' -0"	14"	=	19' -6"	14"	=	5' -6"
16"	=	21' -6"	16"	=	23' -0"	16"	=	6' -0"
18"	=	25' -6"	18"	=	27' -0"	18"	=	6' -6"
20"	=	28' -6"	20"	=	30' -0"	20"	=	7' -0"
24"	=	34' -0"	24"	=	36' -0"	24"	=	8' -0"
30"	=	40' -0"	30"	=	42' -0"	30"	=	10' -0"

736

For customer support on your Mechanical Cost Data, call 877.795.7394.

R238313-10   Heat Trace Systems (cont.)

Add the following quantities of heat transfer cement to linear foot totals for each valve:

Nominal Valve Size	Gallons of Cement per Valve
1/2"	0.14
3/4"	0.21
1"	0.29
1-1/2"	0.36
2"	0.43
2-1/2"	0.70
3"	0.71
4"	1.00
6"	1.43
8"	1.48
10"	1.50
12"	1.60
14"	1.75
16"	2.00
18"	2.25
20"	2.50
24"	3.00
30"	3.75

The following must be added to the list of components to accurately price HEAT TRACE systems:

1. Expediter fitting and clamp fasteners (F)
2. Junction box and nipple connected to expediter fitting (G)
3. Field installed terminal blocks within junction box
4. Ground lugs
5. Piping from power source to expediter fitting
6. Controls
7. Thermostats
8. Branch wiring
9. Cable splices
10. End of cable terminations
11. Branch piping fittings and boxes

Deduct the following percentages from labor if cable lengths in the same area exceed:

150' to 250'	10%	351' to 500'	20%
251' to 350'	15%	Over 500'	25%

Add the following percentages to labor for elevated installations:

15' to 20' high	10%	31' to 35' high	40%
21' to 25' high	20%	36' to 40' high	50%
26' to 30' high	30%	Over 40' high	60%

R238313-20   Spiral-Wrapped Heat Trace Cable (Pitch Table)

In order to increase the amount of heat, occasionally heat trace cable is wrapped in a spiral fashion around a pipe; increasing the number of feet of heater cable per linear foot of pipe.

Engineers first determine the heat loss per foot of pipe (based on the insulating material, its thickness, and the temperature differential across it). A ratio is then calculated by the formula:

$$\text{Feet of Heat Trace per Foot of Pipe} = \frac{\text{Watts/Foot of Heat Loss}}{\text{Watts/Foot of the Cable}}$$

The linear distance between wraps (pitch) is then taken from a chart or table. Generally, the pitch is listed on a drawing leaving the estimator to calculate the total length of heat tape required. An approximation may be taken from this table.

Feet of Heat Trace Per Foot of Pipe

	Nominal Pipe Size in Inches															
Pitch In Inches	1	1¼	1½	2	2½	3	4	6	8	10	12	14	16	18	20	24
3.5	1.80															
4	1.65															
5	1.46	1.60	1.80													
6	1.34	1.45	1.55	1.75												
7	1.25	1.35	1.43	1.57	1.75											
8	1.20	1.28	1.34	1.45	1.60	1.80										
9	1.16	1.23	1.28	1.37	1.51	1.68										
10	1.13	1.19	1.24	1.32	1.44	1.57	1.82									
15	1.06	1.08	1.10	1.15	1.21	1.29	1.42	1.78								
20	1.04	1.05	1.06	1.08	1.13	1.17	1.25	1.49	1.73							
25		1.04	1.04	1.06	1.08	1.11	1.17	1.33	1.51	1.72						
30				1.04	1.05	1.07	1.12	1.24	1.37	1.54	1.70	1.80				
35					1.06	1.06	1.09	1.17	1.28	1.42	1.54	1.64	1.78			
40						1.05	1.07	1.14	1.22	1.33	1.44	1.52	1.64	1.75		
50							1.05	1.09	1.15	1.22	1.29	1.35	1.44	1.53	1.64	1.83
60								1.06	1.11	1.16	1.21	1.25	1.31	1.39	1.46	1.62
70								1.05	1.08	1.12	1.17	1.19	1.24	1.30	1.35	1.47
80									1.06	1.09	1.13	1.15	1.19	1.24	1.28	1.38
90									1.04	1.06	1.10	1.13	1.16	1.19	1.23	1.32
100										1.05	1.08	1.10	1.13	1.15	1.19	1.23

Note: Common practice would normally limit the lower end of the table to 5% of additional heat and above 80% an engineer would likely opt for two (2) parallel cables.

R312319-90 Wellpoints

A single stage wellpoint system is usually limited to dewatering an average 15′ depth below normal ground water level. Multi-stage systems are employed for greater depth with the pumping equipment installed only at the lowest header level. Ejectors with unlimited lift capacity can be economical when two or more stages of wellpoints can be replaced or when horizontal clearance is restricted, such as in deep trenches or tunneling projects, and where low water flows are expected. Wellpoints are usually spaced on 2-1/2′ to 10′ centers along a header pipe. Wellpoint spacing, header size, and pump size are all determined by the expected flow as dictated by soil conditions.

In almost all soils encountered in wellpoint dewatering, the wellpoints may be jetted into place. Cemented soils and stiff clays may require sand wicks about 12″ in diameter around each wellpoint to increase efficiency and eliminate weeping into the excavation. These sand wicks require 1/2 to 3 C.Y. of washed filter sand and are installed by using a 12″ diameter steel casing and hole puncher jetted into the ground 2′ deeper than the wellpoint. Rock may require predrilled holes.

Labor required for the complete installation and removal of a single stage wellpoint system is in the range of 3/4 to 2 labor-hours per linear foot of header, depending upon jetting conditions, wellpoint spacing, etc.

Continuous pumping is necessary except in some free draining soil where temporary flooding is permissible (as in trenches which are backfilled after each day's work). Good practice requires provision of a stand-by pump during the continuous pumping operation.

Systems for continuous trenching below the water table should be installed three to four times the length of expected daily progress to ensure uninterrupted digging, and header pipe size should not be changed during the job.

For pervious free draining soils, deep wells in place of wellpoints may be economical because of lower installation and maintenance costs. Daily production ranges between two to three wells per day, for 25′ to 40′ depths, to one well per day for depths over 50′.

Detailed analysis and estimating for any dewatering problem is available at no cost from wellpoint manufacturers. Major firms will quote "sufficient equipment" quotes or their affiliates offer lump sum proposals to cover complete dewatering responsibility.

Description for 200′ System with 8″ Header		Quantities
	Wellpoints 25′ long, 2″ diameter @ 5′ O.C.	40 Each
	Header pipe, 8″ diameter	200 L.F.
	Discharge pipe, 8″ diameter	100 L.F.
	8″ valves	3 Each
Equipment & Material	Combination jetting & wellpoint pump (standby)	1 Each
	Wellpoint pump, 8″ diameter	1 Each
	Transportation to and from site	1 Day
	Fuel for 30 days x 60 gal./day	1800 Gallons
	Lubricants for 30 days x 16 lbs./day	480 Lbs.
	Sand for points	40 C.Y.
	Technician to supervise installation	1 Week
Labor	Labor for installation and removal of system	300 Labor-hours
	4 Operators straight time 40 hrs./wk. for 4.33 wks.	693 Hrs.
	4 Operators overtime 2 hrs./wk. for 4.33 wks.	35 Hrs.

R314116-40   Wood Sheet Piling

Wood sheet piling may be used for depths to 20' where there is no ground water. If moderate ground water is encountered Tongue & Groove sheeting will help to keep it out. When considerable ground water is present, steel sheeting must be used.

For estimating purposes on trench excavation, sizes are as follows:

Depth	Sheeting	Wales	Braces	B.F. per S.F.
To 8'	3 x 12's	6 x 8's, 2 line	6 x 8's, @ 10'	4.0 @ 8'
8' x 12'	3 x 12's	10 x 10's, 2 line	10 x 10's, @ 9'	5.0 average
12' to 20'	3 x 12's	12 x 12's, 3 line	12 x 12's, @ 8'	7.0 average

Sheeting to be toed in at least 2' depending upon soil conditions. A five person crew with an air compressor and sheeting driver can drive and brace 440 SF/day at 8' deep, 360 SF/day at 12' deep, and 320 SF/day at 16' deep.

For normal soils, piling can be pulled in 1/3 the time to install. Pulling difficulty increases with the time in the ground. Production can be increased by high pressure jetting.

R314116-45   Steel Sheet Piling

Limiting weights are 22 to 38#/S.F. of wall surface with 27#/S.F. average for usual types and sizes. (Weights of piles themselves are from 30.7#/L.F. to 57#/L.F. but they are 15" to 21" wide.) Lightweight sections 12" to 28" wide from 3 ga. to 12 ga. thick are also available for shallow excavations. Piles may be driven two at a time with an impact or vibratory hammer (use vibratory to pull) hung from a crane without leads. A reasonable estimate of the life of steel sheet piling is 10 uses with up to 125 uses possible if a vibratory hammer is used. Used piling costs from 50% to 80% of new piling depending on location and market conditions. Sheet piling and H piles can be rented for about 30% of the delivered mill price for the first month and 5% per month thereafter. Allow 1 labor-hour per pile for cleaning and trimming after driving. These costs increase with depth and hydrostatic head. Vibratory drivers are faster in wet granular soils and are excellent for pile extraction. Pulling difficulty increases with the time in the ground and may cost more than driving. It is often economical to abandon the sheet piling, especially if it can be used as the outer wall form. Allow about 1/3 additional length or more for toeing into ground. Add bracing, waler and strut costs. Waler costs can equal the cost per ton of sheeting.

R331113-80   Piping Designations

There are several systems currently in use to describe pipe and fittings. The following paragraphs will help to identify and clarify classifications of piping systems used for water distribution.

Piping may be classified by schedule. Piping schedules include 5S, 10S, 10, 20, 30, Standard, 40, 60, Extra Strong, 80, 100, 120, 140, 160 and Double Extra Strong. These schedules are dependent upon the pipe wall thickness. The wall thickness of a particular schedule may vary with pipe size.

Ductile iron pipe for water distribution is classified by Pressure Classes such as Class 150, 200, 250, 300 and 350. These classes are actually the rated water working pressure of the pipe in pounds per square inch (psi). The pipe in these pressure classes is designed to withstand the rated water working pressure plus a surge allowance of 100 psi.

The American Water Works Association (AWWA) provides standards for various types of **plastic pipe.** C-900 is the specification for polyvinyl chloride (PVC) piping used for water distribution in sizes ranging from 4" through 12". C-901 is the specification for polyethylene (PE) pressure pipe, tubing and fittings used for water distribution in sizes ranging from 1/2" through 3". C-905 is the specification for PVC piping sizes 14" and greater.

PVC pressure-rated pipe is identified using the standard dimensional ratio (SDR) method. This method is defined by the American Society for Testing and Materials (ASTM) Standard D 2241. This pipe is available in SDR numbers 64, 41, 32.5, 26, 21, 17, and 13.5. Pipe with an SDR of 64 will have the thinnest wall while pipe with an SDR of 13.5 will have the thickest wall. When the pressure rating (PR) of a pipe is given in psi, it is based on a line supplying water at 73 degrees F.

The National Sanitation Foundation (NSF) seal of approval is applied to products that can be used with potable water. These products have been tested to ANSI/NSF Standard 14.

Valves and strainers are classified by American National Standards Institute (ANSI) Classes. These Classes are 125, 150, 200, 250, 300, 400, 600, 900, 1500 and 2500. Within each class there is an operating pressure range dependent upon temperature. Design parameters should be compared to the appropriate material dependent, pressure-temperature rating chart for accurate valve selection.

Change Orders

Change Order Considerations

A change order is a written document, usually prepared by the design professional, and signed by the owner, the architect/engineer, and the contractor. A change order states the agreement of the parties to: an addition, deletion, or revision in the work; an adjustment in the contract sum, if any; or an adjustment in the contract time, if any. Change orders, or "extras" in the construction process occur after execution of the construction contract and impact architects/ engineers, contractors, and owners.

Change orders that are properly recognized and managed can ensure orderly, professional, and profitable progress for all who are involved in the project. There are many causes for change orders and change order requests. In all cases, change orders or change order requests should be addressed promptly and in a precise and prescribed manner. The following paragraphs include information regarding change order pricing and procedures.

The Causes of Change Orders

Reasons for issuing change orders include:

- Unforeseen field conditions that require a change in the work
- Correction of design discrepancies, errors, or omissions in the contract documents
- Owner-requested changes, either by design criteria, scope of work, or project objectives
- Completion date changes for reasons unrelated to the construction process
- Changes in building code interpretations, or other public authority requirements that require a change in the work
- Changes in availability of existing or new materials and products

Procedures

Properly written contract documents must include the correct change order procedures for all parties—owners, design professionals and contractors—to follow in order to avoid costly delays and litigation.

Being "in the right" is not always a sufficient or acceptable defense. The contract provisions requiring notification and documentation must be adhered to within a defined or reasonable time frame.

The appropriate method of handling change orders is by a written proposal and acceptance by all parties involved. Prior to starting work on a project, all parties should identify their

authorized agents who may sign and accept change orders, as well as any limits placed on their authority.

Time may be a critical factor when the need for a change arises. For such cases, the contractor might be directed to proceed on a "time and materials" basis, rather than wait for all paperwork to be processed—a delay that could impede progress. In this situation, the contractor must still follow the prescribed change order procedures including, but not limited to, notification and documentation.

Lack of documentation can be very costly, especially if legal judgments are to be made, and if certain field personnel are no longer available. For time and material change orders, the contractor should keep accurate daily records of all labor and material allocated to the change.

Owners or awarding authorities who do considerable and continual building construction (such as the federal government) realize the inevitability of change orders for numerous reasons, both predictable and unpredictable. As a result, the federal government, the American Institute of Architects (AIA), the Engineers Joint Contract Documents Committee (EJCDC) and other contractor, legal, and technical organizations have developed standards and procedures to be followed by all parties to achieve contract continuance and timely completion, while being financially fair to all concerned.

Pricing Change Orders

When pricing change orders, regardless of their cause, the most significant factor is when the change occurs. The need for a change may be perceived in the field or requested by the architect/engineer *before* any of the actual installation has begun, or may evolve or appear *during* construction when the item of work in question is partially installed. In the latter cases, the original sequence of construction is disrupted, along with all contiguous and supporting systems. Change orders cause the greatest impact when they occur *after* the installation has been completed and must be uncovered, or even replaced. Post-completion changes may be caused by necessary design changes, product failure, or changes in the owner's requirements that are not discovered until the building or the systems begin to function.

Specified procedures of notification and record keeping must be adhered to and enforced regardless of the stage of construction: *before, during,* or *after* installation. Some bidding documents anticipate change orders by requiring that unit prices including overhead and profit percentages—for additional as well as deductible changes—be listed. Generally these unit prices do not fully take into account the ripple effect, or impact on other trades, and should be used for general guidance only.

When pricing change orders, it is important to classify the time frame in which the change occurs. There are two basic time frames for change orders: *pre-installation change orders*, which occur before the start of construction, and *post-installation change orders*, which involve reworking after the original installation. Change orders that occur between these stages may be priced according to the extent of work completed using a combination of techniques developed for pricing *pre-* and *post-installation* changes.

Factors To Consider When Pricing Change Orders

As an estimator begins to prepare a change order, the following questions should be reviewed to determine their impact on the final price.

General

- *Is the change order work* pre-installation *or* post-installation?

  Change order work costs vary according to how much of the installation has been completed. Once workers have the project scoped in their minds, even though they have not started, it can be difficult to refocus. Consequently they may spend more than the normal amount of time understanding the change. Also, modifications to work in place, such as trimming or refitting, usually take more time than was initially estimated. The greater the amount of work in place, the more reluctant workers are to change it. Psychologically they may resent the change and as a result the rework takes longer than normal. Post-installation change order estimates must include demolition of existing work as required to accomplish the change. If the work is performed at a later time, additional obstacles, such as building finishes, may be present which must be protected. Regardless of whether the change occurs pre-installation or post-installation, attempt to isolate the identifiable factors and price them separately. For example, add shipping costs that may be required pre-installation or any demolition required post-installation. Then analyze the potential impact on productivity of psychological and/or learning curve factors and adjust the output rates accordingly. One approach is to break down the typical workday into segments and quantify the impact on each segment.

Change Order Installation Efficiency

The labor-hours expressed (for new construction) are based on average installation time, using an efficiency level. For change order situations, adjustments to this efficiency level should reflect the daily labor-hour allocation for that particular occurrence.

• *Will the change substantially delay the original completion date?*

A significant change in the project may cause the original completion date to be extended. The extended schedule may subject the contractor to new wage rates dictated by relevant labor contracts. Project supervision and other project overhead must also be extended beyond the original completion date. The schedule extension may also put installation into a new weather season. For example, underground piping scheduled for October installation was delayed until January. As a result, frost penetrated the trench area, thereby changing the degree of difficulty of the task. Changes and delays may have a ripple effect throughout the project. This effect must be analyzed and negotiated with the owner.

• *What is the net effect of a deduct change order?*

In most cases, change orders resulting in a deduction or credit reflect only bare costs. The contractor may retain the overhead and profit based on the original bid.

Materials

• *Will you have to pay more or less for the new material, required by the change order, than you paid for the original purchase?*

The same material prices or discounts will usually apply to materials purchased for change orders as new construction. In some instances, however, the contractor may forfeit the advantages of competitive pricing for change orders. Consider the following example:

A contractor purchased over $20,000 worth of fan coil units for an installation, and obtained the maximum discount. Some time later it was determined the project required an additional matching unit. The contractor has to purchase this unit from the original supplier to ensure a match. The supplier at this time may not discount the unit because of the small quantity, and the fact that he is no longer in a competitive situation. The impact of quantity on purchase can add between 0% and 25% to material prices and/or subcontractor quotes.

• *If materials have been ordered or delivered to the job site, will they be subject to a cancellation charge or restocking fee?*

Check with the supplier to determine if ordered materials are subject to a cancellation charge. Delivered materials not used as result of a change order may be subject to a restocking fee if returned to the supplier. Common restocking charges run between 20% and 40%. Also, delivery charges to return the goods to the supplier must be added.

Labor

• *How efficient is the existing crew at the actual installation?*

Is the same crew that performed the initial work going to do the change order? Possibly the change consists of the installation of a unit identical to one already installed; therefore, the change should take less time. Be sure to consider this potential productivity increase and modify the productivity rates accordingly.

• *If the crew size is increased, what impact will that have on supervision requirements?*

Under most bargaining agreements or management practices, there is a point at which a working foreman is replaced by a nonworking foreman. This replacement increases project overhead by adding a nonproductive worker. If additional workers are added to accelerate the project or to perform changes while maintaining the schedule, be sure to add additional supervision time if warranted. Calculate the hours involved and the additional cost directly if possible.

• *What are the other impacts of increased crew size?*

The larger the crew, the greater the potential for productivity to decrease. Some of the factors that cause this productivity loss are: overcrowding (producing restrictive conditions in the working space), and possibly a shortage of any special tools and equipment required. Such factors affect not only the crew working on the elements directly involved in the change order, but other crews whose movement may also be hampered. As the crew increases, check its basic composition for changes by the addition or deletion of apprentices or nonworking foreman, and quantify the potential effects of equipment shortages or other logistical factors.

• *As new crews, unfamiliar with the project, are brought onto the site, how long will it take them to become oriented to the project requirements?*

The orientation time for a new crew to become 100% effective varies with the site and type of project. Orientation is easiest at a new construction site, and most difficult at existing, very restrictive renovation sites. The type of work also affects orientation time. When all elements of the work are exposed, such as concrete or masonry work, orientation is decreased. When the work is concealed or less visible, such as existing electrical systems, orientation takes longer. Usually orientation can be accomplished in one day or less. Costs for added orientation should be itemized and added to the total estimated cost.

• *How much actual production can be gained by working overtime?*

Short term overtime can be used effectively to accomplish more work in a day. However, as overtime is scheduled to run beyond several weeks, studies have shown marked decreases in output. The following chart shows the effect of long term overtime on worker efficiency. If the anticipated change requires extended overtime to keep the job on schedule, these factors can be used as a guide to predict the impact on time and cost. Add project overhead, particularly supervision, that may also be incurred.

Days per Week	Hours per Day	Production Efficiency					Payroll Cost Factors	
		1 Week	2 Weeks	3 Weeks	4 Weeks	Average 4 Weeks	@ 1-1/2 Times	@ 2 Times
5	8	100%	100%	100%	100%	100%	100%	100%
	9	100	100	95	90	96.25	105.6	111.1
	10	100	95	90	85	91.25	110.0	120.0
	11	95	90	75	65	81.25	113.6	127.3
	12	90	85	70	60	76.25	116.7	133.3
6	8	100	100	95	90	96.25	108.3	116.7
	9	100	95	90	85	92.50	113.0	125.9
	10	95	90	85	80	87.50	116.7	133.3
	11	95	85	70	65	78.75	119.7	139.4
	12	90	80	65	60	73.75	122.2	144.4
7	8	100	95	85	75	88.75	114.3	128.6
	9	95	90	80	70	83.75	118.3	136.5
	10	90	85	75	65	78.75	121.4	142.9
	11	85	80	65	60	72.50	124.0	148.1
	12	85	75	60	55	68.75	126.2	152.4

Effects of Overtime

Caution: Under many labor agreements, Sundays and holidays are paid at a higher premium than the normal overtime rate.

The use of long-term overtime is counterproductive on almost any construction job; that is, the longer the period of overtime, the lower the actual production rate. Numerous studies have been conducted, and while they have resulted in slightly different numbers, all reach the same conclusion. The figure above tabulates the effects of overtime work on efficiency.

As illustrated, there can be a difference between the *actual* payroll cost per hour and the *effective* cost per hour for overtime work. This is due to the reduced production efficiency with the increase in weekly hours beyond 40. This difference between actual and effective cost results from overtime work over a prolonged period. Short-term overtime work does not result in as great a reduction in efficiency and, in such cases, effective cost may not vary significantly from the actual payroll cost. As the total hours per week are increased on a regular basis, more time is lost due to fatigue, lowered morale, and an increased accident rate.

As an example, assume a project where workers are working 6 days a week, 10 hours per day. From the figure above (based on productivity studies), the average effective productive hours over a 4-week period are:

$$0.875 \times 60 = 52.5$$

Depending upon the locale and day of week, overtime hours may be paid at time and a half or double time. For time and a half, the overall (average) *actual* payroll cost (including regular and overtime hours) is determined as follows:

$$\frac{40 \text{ reg. hrs.} + (20 \text{ overtime hrs. x } 1.5)}{60 \text{ hrs.}} = 1.167$$

Based on 60 hours, the payroll cost per hour will be 116.7% of the normal rate at 40 hours per week. However, because the effective production (efficiency) for 60 hours is reduced to the equivalent of 52.5 hours, the effective cost of overtime is calculated as follows:

For time and a half:

$$\frac{40 \text{ reg. hrs.} + (20 \text{ overtime hrs. x } 1.5)}{52.5 \text{ hrs.}} = 1.33$$

Installed cost will be 133% of the normal rate (for labor).

Thus, when figuring overtime, the actual cost per unit of work will be higher than the apparent overtime payroll dollar increase, due to the reduced productivity of the longer workweek. These efficiency calculations are true only for those cost factors determined by hours worked. Costs that are applied weekly or monthly, such as equipment rentals, will not be similarly affected.

Equipment

- *What equipment is required to complete the change order?*

Change orders may require extending the rental period of equipment already on the job site, or the addition of special equipment brought in to accomplish the change work. In either case, the additional rental charges and operator labor charges must be added.

Summary

The preceding considerations and others you deem appropriate should be analyzed and applied to a change order estimate. The impact of each should be quantified and listed on the estimate to form an audit trail.

Change orders that are properly identified, documented, and managed help to ensure the orderly, professional and profitable progress of the work. They also minimize potential claims or disputes at the end of the project.

Estimating Tips

- The cost figures in this section were derived from approximately 11,000 projects contained in the RSMeans database of completed construction projects. They include the contractor's overhead and profit, but do not generally include architectural fees or land costs. The figures have been adjusted to January of the current year. New projects are added to our files each year, and outdated projects are discarded. For this reason, certain costs may not show a uniform annual progression. In no case are all subdivisions of a project listed.

- These projects were located throughout the U.S. and reflect a tremendous variation in square foot (S.F.) and cubic foot (C.F.) costs. This is due to differences, not only in labor and material costs, but also in individual owners' requirements. For instance, a bank in a large city would have different features than one in a rural area. This is true of all the different types of buildings analyzed. Therefore, caution should be exercised when using these square foot costs. For example, for courthouses, costs in the database are local courthouse costs and will not apply to the larger, more elaborate federal courthouses. As a general rule, the projects in the 1/4 column do not include any site work or equipment, while the projects in the 3/4 column may include both equipment and site work.

The median figures do not generally include site work.

- None of the figures "go with" any others. All individual cost items were computed and tabulated separately. Thus, the sum of the median figures for plumbing, HVAC, and electrical will not normally total up to the total mechanical and electrical costs arrived at by separate analysis and tabulation of the projects.

- Each building was analyzed as to total and component costs and percentages. The figures were arranged in ascending order with the results tabulated as shown. The 1/4 column shows that 25% of the projects had lower costs and 75% had higher. The 3/4 column shows that 75% of the projects had lower costs and 25% had higher. The median column shows that 50% of the projects had lower costs and 50% had higher.

- There are two times when square foot costs are useful. The first is in the conceptual stage when no details are available. Then, square foot costs make a useful starting point. The second is after the bids are in and the costs can be worked back into their appropriate categories for information purposes. As soon as details become available in the project design, the square foot approach should be discontinued and the project priced as to its particular components. When more precision is required, or

for estimating the replacement cost of specific buildings, the current edition of *RSMeans Square Foot Costs* should be used.

- In using the figures in this section, it is recommended that the median column be used for preliminary figures if no additional information is available. The median figures, when multiplied by the total city construction cost index figures (see City Cost Indexes) and then multiplied by the project size modifier at the end of this section, should present a fairly accurate base figure, which would then have to be adjusted in view of the estimator's experience, local economic conditions, code requirements, and the owner's particular requirements. There is no need to factor the percentage figures, as these should remain constant from city to city. All tabulations mentioning air conditioning had at least partial air conditioning.

- The editors of this book would greatly appreciate receiving cost figures on one or more of your recent projects, which would then be included in the averages for next year. All cost figures received will be kept confidential, except that they will be averaged with other similar projects to arrive at square foot cost figures for next year's book. See the last page of the book for details and the discount available for submitting one or more of your projects. ∎

50 17 00 | S.F. Costs

			UNIT	UNIT COSTS			% OF TOTAL			
				1/4	MEDIAN	3/4	1/4	MEDIAN	3/4	
01	0010	**APARTMENTS Low Rise (1 to 3 story)**	S.F.	74.50	94	125				01
	0020	Total project cost	C.F.	6.70	8.85	10.95				
	0100	Site work	S.F.	5.45	8.70	15.30	6.05%	10.55%	13.95%	
	0500	Masonry		1.47	3.62	5.95	1.54%	3.92%	6.50%	
	1500	Finishes		7.90	10.85	13.45	9.05%	10.75%	12.85%	
	1800	Equipment		2.44	3.70	5.50	2.71%	3.99%	5.95%	
	2720	Plumbing		5.80	7.45	9.50	6.65%	8.95%	10.05%	
	2770	Heating, ventilating, air conditioning		3.70	4.56	6.70	4.20%	5.60%	7.60%	
	2900	Electrical		4.33	5.75	7.80	5.20%	6.65%	8.35%	
	3100	Total: Mechanical & Electrical		15.40	19.95	24.50	16.05%	18.20%	23%	
	9000	Per apartment unit, total cost	Apt.	69,500	106,000	156,500				
	9500	Total: Mechanical & Electrical	"	13,100	20,700	27,000				
02	0010	**APARTMENTS Mid Rise (4 to 7 story)**	S.F.	99	119	147				02
	0020	Total project costs	C.F.	7.70	10.65	14.55				
	0100	Site work	S.F.	3.95	8	15.45	5.25%	6.70%	9.20%	
	0500	Masonry		4.13	9.05	12.40	5.10%	7.25%	10.50%	
	1500	Finishes		12.95	17.35	20.50	10.70%	13.50%	17.70%	
	1800	Equipment		2.77	4.30	5.65	2.54%	3.47%	4.31%	
	2500	Conveying equipment		2.24	2.76	3.33	2.05%	2.27%	2.69%	
	2720	Plumbing		5.80	9.30	9.85	5.70%	7.20%	8.95%	
	2900	Electrical		6.50	8.85	10.75	6.35%	7.20%	8.95%	
	3100	Total: Mechanical & Electrical		21	26	31.50	18.25%	21%	23%	
	9000	Per apartment unit, total cost	Apt.	112,000	132,000	218,500				
	9500	Total: Mechanical & Electrical	"	21,100	24,400	25,600				
03	0010	**APARTMENTS High Rise (8 to 24 story)**	S.F.	112	129	155				03
	0020	Total project costs	C.F.	10.90	12.65	14.95				
	0100	Site work	S.F.	4.07	6.60	9.20	2.58%	4.84%	6.15%	
	0500	Masonry		6.50	11.80	14.65	4.74%	9.65%	11.05%	
	1500	Finishes		12.45	15.55	18.35	9.75%	11.80%	13.70%	
	1800	Equipment		3.61	4.44	5.85	2.78%	3.49%	4.35%	
	2500	Conveying equipment		2.55	3.87	5.25	2.23%	2.78%	3.37%	
	2720	Plumbing		7.15	9.75	11.95	6.80%	7.20%	10.45%	
	2900	Electrical		7.70	9.75	13.15	6.45%	7.65%	8.80%	
	3100	Total: Mechanical & Electrical		23	29.50	35.50	17.95%	22.50%	24.50%	
	9000	Per apartment unit, total cost	Apt.	116,500	128,500	178,000				
	9500	Total: Mechanical & Electrical	"	23,200	28,800	30,500				
04	0010	**AUDITORIUMS**	S.F.	117	161	234				04
	0020	Total project costs	C.F.	7.30	10.15	14.55				
	2720	Plumbing	S.F.	6.90	10.20	12.20	5.85%	7.20%	8.70%	
	2900	Electrical		9.40	13.55	22.50	6.85%	9.50%	11.40%	
	3100	Total: Mechanical & Electrical		62.50	83	101	24.50%	27.50%	31%	
05	0010	**AUTOMOTIVE SALES**	S.F.	86.50	119	146				05
	0020	Total project costs	C.F.	5.70	6.80	8.85				
	2720	Plumbing	S.F.	3.93	6.85	7.45	2.89%	6.05%	6.50%	
	2770	Heating, ventilating, air conditioning		6.05	9.25	10	4.61%	10%	10.35%	
	2900	Electrical		6.95	10.90	16	7.25%	8.80%	12.15%	
	3100	Total: Mechanical & Electrical		19.35	31	37.50	17.30%	20.50%	22%	
06	0010	**BANKS**	S.F.	170	211	268				06
	0020	Total project costs	C.F.	12.10	16.45	21.50				
	0100	Site work	S.F.	19.40	30.50	43	7.90%	12.95%	17%	
	0500	Masonry		7.70	16.50	29.50	3.36%	6.90%	10.05%	
	1500	Finishes		15.15	23	28.50	5.85%	8.65%	11.70%	
	1800	Equipment		6.35	14	28.50	1.34%	5.55%	10.50%	
	2720	Plumbing		5.30	7.55	11.05	2.82%	3.90%	4.93%	
	2770	Heating, ventilating, air conditioning		10.05	13.45	17.90	4.86%	7.15%	8.50%	
	2900	Electrical		16.05	21.50	28	8.25%	10.20%	12.20%	
	3100	Total: Mechanical & Electrical		38.50	51.50	62	16.55%	19.45%	24%	
	3500	See also division 11 22 00								

For customer support on your Mechanical Cost Data, call 877.795.7394.

50 17 00 | S.F. Costs

		UNIT	UNIT COSTS			% OF TOTAL		
			1/4	MEDIAN	3/4	1/4	MEDIAN	3/4
3	0010	**CHURCHES**	S.F.	114	145	190				13
	0020	Total project costs	C.F.	7.05	8.90	11.75			
	1800	Equipment	S.F.	1.22	3.19	6.80	.83%	2.04%	4.30%
	2720	Plumbing		4.43	6.20	9.15	3.51%	4.96%	6.25%
	2770	Heating, ventilating, air conditioning		10.35	13.50	19.15	7.50%	10%	12%
	2900	Electrical		9.60	13.20	18	7.30%	8.80%	10.95%
	3100	Total: Mechanical & Electrical	▼	29.50	39.50	53	18.30%	22%	25%
	3500	See also division 11 91 00							
5	0010	**CLUBS, COUNTRY**	S.F.	122	147	185				15
	0020	Total project costs	C.F.	9.85	12	16.55			
	2720	Plumbing	S.F.	7.40	10.95	25	5.60%	7.90%	10%
	2900	Electrical		9.60	13.15	17.15	7%	8.95%	11%
	3100	Total: Mechanical & Electrical	▼	51	64	67.50	19%	26.50%	29.50%
7	0010	**CLUBS, SOCIAL Fraternal**	S.F.	103	141	188				17
	0020	Total project costs	C.F.	6.10	9.25	11			
	2720	Plumbing	S.F.	6.15	7.65	11.55	5.60%	6.90%	8.55%
	2770	Heating, ventilating, air conditioning		7.15	10.70	13.75	8.20%	9.25%	14.40%
	2900	Electrical		7.15	12.05	13.75	5.95%	9.30%	10.55%
	3100	Total: Mechanical & Electrical	▼	38	41	52	21%	23%	23.50%
8	0010	**CLUBS, Y.M.C.A.**	S.F.	134	167	252				18
	0020	Total project costs	C.F.	5.65	9.45	14.05			
	2720	Plumbing	S.F.	7.75	15.40	17.25	5.65%	7.60%	10.85%
	2900	Electrical		10.05	13	22.50	6.45%	7.95%	8.90%
	3100	Total: Mechanical & Electrical	▼	41	46.50	76	20.50%	21.50%	28.50%
9	0010	**COLLEGES Classrooms & Administration**	S.F.	118	167	216				19
	0020	Total project costs	C.F.	7.10	12.50	20			
	0500	Masonry	S.F.	9.30	17.15	21	5.10%	8.05%	10.50%
	2720	Plumbing		5.30	12.90	24	5.10%	6.60%	8.95%
	2900	Electrical		10.05	15.85	22	7.70%	9.85%	12%
	3100	Total: Mechanical & Electrical	▼	42.50	58.50	77.50	23%	28%	31.50%
1	0010	**COLLEGES Science, Engineering, Laboratories**	S.F.	230	269	310				21
	0020	Total project costs	C.F.	13.20	19.25	22			
	1800	Equipment	S.F.	6.25	29	31.50	2%	6.45%	12.65%
	2900	Electrical		18.95	27	41.50	7.10%	9.40%	12.10%
	3100	Total: Mechanical & Electrical	▼	70.50	83.50	129	28.50%	31.50%	41%
	3500	See also division 11 53 00							
3	0010	**COLLEGES Student Unions**	S.F.	147	199	241				23
	0020	Total project costs	C.F.	8.20	10.70	13.25			
	3100	Total: Mechanical & Electrical	S.F.	55	59.50	70.50	23.50%	26%	29%
5	0010	**COMMUNITY CENTERS**	S.F.	122	150	203				25
	0020	Total project costs	C.F.	7.90	11.30	14.65			
	1800	Equipment	S.F.	2.45	4.80	7.90	1.47%	3.01%	5.30%
	2720	Plumbing		5.75	10.05	13.70	4.85%	7%	8.95%
	2770	Heating, ventilating, air conditioning		8.65	13.35	19.15	6.80%	10.35%	12.90%
	2900	Electrical		10.25	13.55	19.90	7.15%	8.90%	10.40%
	3100	Total: Mechanical & Electrical	▼	34	43	61.50	18.80%	23%	30%
8	0010	**COURT HOUSES**	S.F.	175	206	284				28
	0020	Total project costs	C.F.	13.40	16	20			
	2720	Plumbing	S.F.	8.30	11.60	13.15	5.95%	7.45%	8.20%
	2900	Electrical		18.55	21	30.50	9.05%	10.65%	12.15%
	3100	Total: Mechanical & Electrical	▼	52	68	74.50	22.50%	26.50%	30%
0	0010	**DEPARTMENT STORES**	S.F.	64.50	87.50	110				30
	0020	Total project costs	C.F.	3.46	4.48	6.10			
	2720	Plumbing	S.F.	2.01	2.54	3.85	1.82%	4.21%	5.90%
	2770	Heating, ventilating, air conditioning	▼	4.88	9.05	13.65	8.20%	9.10%	14.80%

50 17 00 | S.F. Costs

			UNIT	UNIT COSTS			% OF TOTAL		
				1/4	MEDIAN	3/4	1/4	MEDIAN	3/4
30	2900	Electrical	S.F.	7.40	10.15	12	9.05%	12.15%	14.95%	3
	3100	Total: Mechanical & Electrical	↓	13.05	16.65	29	13.20%	21.50%	50%	
31	0010	**DORMITORIES Low Rise (1 to 3 story)**	S.F.	122	170	212				3
	0020	Total project costs	C.F.	6.90	11.20	16.75				
	2720	Plumbing	S.F.	7.35	9.85	12.45	8.05%	9%	9.65%	
	2770	Heating, ventilating, air conditioning		7.80	9.35	12.45	4.61%	8.05%	10%	
	2900	Electrical		8.10	12.35	16.90	6.40%	8.65%	9.50%	
	3100	Total: Mechanical & Electrical	↓	42	45	70.50	22%	25%	27%	
	9000	Per bed, total cost	Bed	52,000	57,500	123,500				
32	0010	**DORMITORIES Mid Rise (4 to 8 story)**	S.F.	150	196	242				3
	0020	Total project costs	C.F.	16.55	18.20	22				
	2900	Electrical	S.F.	15.95	18.15	24.50	8.20%	10.20%	11.95%	
	3100	Total: Mechanical & Electrical	"	44.50	89	90.50	25%	30.50%	35.50%	
	9000	Per bed, total cost	Bed	21,400	48,800	282,500				
34	0010	**FACTORIES**	S.F.	57	85	131				3
	0020	Total project costs	C.F.	3.65	5.45	9.05				
	0100	Site work	S.F.	6.50	11.85	18.75	6.95%	11.45%	17.95%	
	2720	Plumbing		3.07	5.70	9.45	3.73%	6.05%	8.10%	
	2770	Heating, ventilating, air conditioning		5.95	8.55	11.55	5.25%	8.45%	11.35%	
	2900	Electrical		7.05	11.20	17.05	8.10%	10.50%	14.20%	
	3100	Total: Mechanical & Electrical	↓	20	32.50	41	21%	28.50%	35.50%	
36	0010	**FIRE STATIONS**	S.F.	113	156	213				3
	0020	Total project costs	C.F.	6.60	9.05	12.05				
	0500	Masonry	S.F.	15.40	30	38	7.50%	10.95%	15.55%	
	1140	Roofing		3.68	9.95	11.35	1.90%	4.94%	5.05%	
	1580	Painting		2.86	4.27	4.37	1.37%	1.57%	2.07%	
	1800	Equipment		1.40	2.69	4.97	.62%	1.63%	3.42%	
	2720	Plumbing		6.30	10.05	14.25	5.85%	7.35%	9.45%	
	2770	Heating, ventilating, air conditioning		6.25	10.15	15.65	5.15%	7.40%	9.40%	
	2900	Electrical		8.15	14.35	19.35	6.90%	8.60%	10.60%	
	3100	Total: Mechanical & Electrical	↓	41.50	53	60	18.40%	23%	27%	
37	0010	**FRATERNITY HOUSES & Sorority Houses**	S.F.	113	145	199				3
	0020	Total project costs	C.F.	11.25	11.70	14.10				
	2720	Plumbing	S.F.	8.55	9.75	17.90	6.80%	8%	10.85%	
	2900	Electrical	↓	7.45	16.10	19.70	6.60%	9.90%	10.65%	
38	0010	**FUNERAL HOMES**	S.F.	119	162	295				3
	0020	Total project costs	C.F.	12.15	13.55	26				
	2900	Electrical	S.F.	5.25	9.65	10.55	3.58%	4.44%	5.95%	
39	0010	**GARAGES, COMMERCIAL (Service)**	S.F.	67.50	104	144				3
	0020	Total project costs	C.F.	4.43	6.55	9.50				
	1800	Equipment	S.F.	3.80	8.55	13.30	2.21%	4.62%	6.80%	
	2720	Plumbing		4.67	7.20	13.10	5.45%	7.85%	10.65%	
	2730	Heating & ventilating		4.60	7.95	10.15	5.25%	6.85%	8.20%	
	2900	Electrical		6.40	9.75	14.10	7.15%	9.25%	10.85%	
	3100	Total: Mechanical & Electrical	↓	12.10	27	40	12.35%	17.40%	26%	
40	0010	**GARAGES, MUNICIPAL (Repair)**	S.F.	105	135	192				4
	0020	Total project costs	C.F.	6.20	7.85	13.50				
	0500	Masonry	S.F.	5.20	18.20	28	4.03%	9.15%	12.50%	
	2720	Plumbing		4.44	8.55	16.05	3.59%	6.70%	7.95%	
	2730	Heating & ventilating		7.60	11	21	6.15%	7.45%	13.50%	
	2900	Electrical		7.65	12.35	22.50	6.90%	9.40%	13%	
	3100	Total: Mechanical & Electrical	↓	34.50	56.50	68	21.50%	25.50%	35.50%	
41	0010	**GARAGES, PARKING**	S.F.	39.50	56	96.50				4
	0020	Total project costs	C.F.	3.62	4.92	7.15				

50 17 00	S.F. Costs		UNIT COSTS			% OF TOTAL			
		UNIT	1/4	MEDIAN	3/4	1/4	MEDIAN	3/4	
41	2720 Plumbing	S.F.	.66	1.69	2.61	1.72%	2.70%	3.85%	41
	2900 Electrical		2.12	2.80	4.08	4.52%	5.40%	6.35%	
	3100 Total: Mechanical & Electrical	↓	4.11	6.05	7.50	7%	8.90%	11.05%	
	3200								
	9000 Per car, total cost	Car	16,300	20,400	26,000				
43	0010 **GYMNASIUMS**	S.F.	108	143	196				43
	0020 Total project costs	C.F.	5.35	7.25	8.90				
	1800 Equipment	S.F.	2.55	4.78	8.60	1.76%	3.26%	6.70%	
	2720 Plumbing		5.90	7.90	10.40	4.65%	6.40%	7.75%	
	2770 Heating, ventilating, air conditioning		6.40	11.10	22.50	5.15%	9.05%	11.10%	
	2900 Electrical		8.50	11.25	14.95	6.75%	8.50%	10.30%	
	3100 Total: Mechanical & Electrical	↓	29	40.50	49.50	19.75%	23.50%	29%	
	3500 See also division 11 66 00								
46	0010 **HOSPITALS**	S.F.	206	258	355				46
	0020 Total project costs	C.F.	15.60	19.40	28				
	1800 Equipment	S.F.	5.20	10.05	17.30	.80%	2.53%	4.80%	
	2720 Plumbing		17.70	25	32	7.60%	9.10%	10.85%	
	2770 Heating, ventilating, air conditioning		26	33.50	46.50	7.80%	12.95%	16.65%	
	2900 Electrical		22.50	30.50	45.50	10%	11.75%	14.10%	
	3100 Total: Mechanical & Electrical	↓	66.50	92.50	137	28%	33.50%	37%	
	9000 Per bed or person, total cost	Bed	238,000	328,500	378,500				
	9900 See also division 11 71 00								
48	0010 **HOUSING For the Elderly**	S.F.	101	128	157				48
	0020 Total project costs	C.F.	7.20	10	12.80				
	0100 Site work	S.F.	7.05	10.95	16.05	5.05%	7.90%	12.10%	
	0500 Masonry		1.91	11.55	16.85	1.30%	6.05%	11%	
	1800 Equipment		2.45	3.37	5.35	1.88%	3.23%	4.43%	
	2510 Conveying systems		2.29	3.31	4.49	1.78%	2.20%	2.81%	
	2720 Plumbing		7.50	9.60	12.10	8.15%	9.55%	10.50%	
	2730 Heating, ventilating, air conditioning		3.86	5.45	8.15	3.30%	5.60%	7.25%	
	2900 Electrical		7.55	10.25	13.10	7.30%	8.50%	10.25%	
	3100 Total: Mechanical & Electrical	↓	26	32	41	18.10%	22.50%	29%	
	9000 Per rental unit, total cost	Unit	94,000	110,000	122,500				
	9500 Total: Mechanical & Electrical	"	21,000	24,100	28,100				
50	0010 **HOUSING Public (Low Rise)**	S.F.	85	118	154				50
	0020 Total project costs	C.F.	6.75	9.45	11.75				
	0100 Site work	S.F.	10.85	15.60	25.50	8.35%	11.75%	16.50%	
	1800 Equipment		2.31	3.77	5.75	2.26%	3.03%	4.24%	
	2720 Plumbing		6.15	8.10	10.25	7.15%	9.05%	11.60%	
	2730 Heating, ventilating, air conditioning		3.08	6	6.55	4.26%	6.05%	6.45%	
	2900 Electrical		5.15	7.65	10.65	5.10%	6.55%	8.25%	
	3100 Total: Mechanical & Electrical	↓	24.50	31.50	35	14.50%	17.55%	26.50%	
	9000 Per apartment, total cost	Apt.	93,500	106,500	133,500				
	9500 Total: Mechanical & Electrical	"	19,900	24,600	27,200				
51	0010 **ICE SKATING RINKS**	S.F.	72.50	170	187				51
	0020 Total project costs	C.F.	5.35	5.45	6.30				
	2720 Plumbing	S.F.	2.72	5.10	5.20	3.12%	3.23%	5.65%	
	2900 Electrical	↓	7.80	11.95	12.65	6.30%	10.15%	15.05%	
52	0010 **JAILS**	S.F.	189	286	370				52
	0020 Total project costs	C.F.	19.95	28	32.50				
	1800 Equipment	S.F.	8.65	25.50	43.50	2.80%	6.95%	9.80%	
	2720 Plumbing		21.50	28.50	37.50	7%	8.90%	13.35%	
	2770 Heating, ventilating, air conditioning		19.95	26.50	51.50	7.50%	9.45%	17.75%	
	2900 Electrical		23.50	31.50	40	9.40%	11.70%	15.25%	
	3100 Total: Mechanical & Electrical	↓	62	110	131	28%	31%	36%	
53	0010 **LIBRARIES**	S.F.	143	190	248				53
	0020 Total project costs	C.F.	9.55	12	15.30				

For customer support on your Mechanical Cost Data, call 877.795.7394.

747

50 17 00 | S.F. Costs

			UNIT	UNIT COSTS			% OF TOTAL			
				1/4	MEDIAN	3/4	1/4	MEDIAN	3/4	
53	0500	Masonry	S.F.	9.50	19.35	32.50	5.60%	7.15%	10.95%	53
	1800	Equipment		1.91	5.15	7.75	.28%	1.39%	4.07%	
	2720	Plumbing		5.15	7.30	10.20	3.38%	4.60%	5.70%	
	2770	Heating, ventilating, air conditioning		11.45	19.35	23.50	7.80%	10.95%	12.80%	
	2900	Electrical		14.40	19	25.50	8.40%	10.55%	12%	
	3100	Total: Mechanical & Electrical		47	57	67	21%	24%	28%	
54	0010	**LIVING, ASSISTED**	S.F.	129	154	180				54
	0020	Total project costs	C.F.	10.90	12.70	14.45				
	0500	Masonry	S.F.	3.75	4.53	5.55	2.36%	3.16%	3.86%	
	1800	Equipment		2.87	3.42	4.38	2.12%	2.45%	2.87%	
	2720	Plumbing		10.85	14.50	15.05	6.05%	8.15%	10.60%	
	2770	Heating, ventilating, air conditioning		12.85	13.45	14.75	7.95%	9.35%	9.70%	
	2900	Electrical		12.70	14.25	16.50	9%	9.95%	10.65%	
	3100	Total: Mechanical & Electrical		35	41	48	24%	28.50%	31.50%	
55	0010	**MEDICAL CLINICS**	S.F.	132	162	207				55
	0020	Total project costs	C.F.	9.65	12.50	16.60				
	1800	Equipment	S.F.	1.68	7.50	11.65	1.05%	2.94%	6.35%	
	2720	Plumbing		8.75	12.30	16.50	6.15%	8.40%	10.10%	
	2770	Heating, ventilating, air conditioning		10.45	13.70	20	6.65%	8.85%	11.35%	
	2900	Electrical		11.30	16.10	21	8.10%	10%	12.20%	
	3100	Total: Mechanical & Electrical		36	49	67.50	22%	27%	33.50%	
	3500	See also division 11 71 00								
57	0010	**MEDICAL OFFICES**	S.F.	124	154	190				57
	0020	Total project costs	C.F.	9.25	12.50	16.95				
	1800	Equipment	S.F.	4.09	8.10	11.35	.66%	4.87%	6.50%	
	2720	Plumbing		6.85	10.55	14.25	5.60%	6.80%	8.50%	
	2770	Heating, ventilating, air conditioning		8.25	11.95	15.75	6.10%	8%	9.70%	
	2900	Electrical		9.90	14.50	20.50	7.50%	9.80%	11.70%	
	3100	Total: Mechanical & Electrical		27	39.50	56	19.30%	22.50%	27.50%	
59	0010	**MOTELS**	S.F.	78	113	148				59
	0020	Total project costs	C.F.	6.95	9.30	15.25				
	2720	Plumbing	S.F.	7.90	10.10	12.05	9.45%	10.60%	12.55%	
	2770	Heating, ventilating, air conditioning		4.83	7.20	12.90	5.60%	5.60%	10%	
	2900	Electrical		7.40	9.35	11.65	7.45%	9.05%	10.45%	
	3100	Total: Mechanical & Electrical		22.50	31.50	54	18.50%	24%	25.50%	
	5000									
	9000	Per rental unit, total cost	Unit	39,800	75,500	82,000				
	9500	Total: Mechanical & Electrical	"	7,750	11,700	13,600				
60	0010	**NURSING HOMES**	S.F.	123	158	197				60
	0020	Total project costs	C.F.	9.65	12.05	16.45				
	1800	Equipment	S.F.	3.34	5.10	8.55	2%	3.62%	4.99%	
	2720	Plumbing		10.50	15.90	19.20	8.75%	10.10%	12.70%	
	2770	Heating, ventilating, air conditioning		11.05	16.80	22.50	9.70%	11.45%	11.80%	
	2900	Electrical		12.15	15.20	21	9.40%	10.60%	12.50%	
	3100	Total: Mechanical & Electrical		29	40.50	68	26%	28%	30%	
	9000	Per bed or person, total cost	Bed	54,500	68,000	88,000				
61	0010	**OFFICES Low Rise (1 to 4 story)**	S.F.	103	135	175				61
	0020	Total project costs	C.F.	7.35	10.15	13.40				
	0100	Site work	S.F.	8.30	14.45	21.50	5.95%	9.70%	13.55%	
	0500	Masonry		4	8.05	14.70	2.61%	5.40%	8.45%	
	1800	Equipment		.93	2.15	5.85	.57%	1.50%	3.42%	
	2720	Plumbing		3.69	5.70	8.35	3.66%	4.50%	6.10%	
	2770	Heating, ventilating, air conditioning		8.15	11.40	16.65	7.20%	10.30%	11.70%	
	2900	Electrical		8.45	12.10	17.20	7.45%	9.65%	11.40%	
	3100	Total: Mechanical & Electrical		23.50	32.50	48.50	18.20%	22%	27%	
62	0010	**OFFICES Mid Rise (5 to 10 story)**	S.F.	109	132	180				62
	0020	Total project costs	C.F.	7.75	9.90	14				

For customer support on your Mechanical Cost Data, call 877.795.7394.

50 17 00 | S.F. Costs		UNIT	UNIT COSTS			% OF TOTAL			
			1/4	MEDIAN	3/4	1/4	MEDIAN	3/4	
2720	Plumbing	S.F.	3.30	5.10	7.35	2.83%	3.74%	4.50%	62
2770	Heating, ventilating, air conditioning		8.30	11.85	18.95	7.65%	9.40%	11%	
2900	Electrical		8.10	10.40	15.70	6.35%	7.80%	10%	
3100	Total: Mechanical & Electrical		21	27	51.50	18.95%	21%	27.50%	
0010	**OFFICES High Rise (11 to 20 story)**	S.F.	134	169	208				63
0020	Total project costs	C.F.	9.40	11.75	16.85				
2900	Electrical	S.F.	8.15	9.95	14.80	5.80%	7.85%	10.50%	
3100	Total: Mechanical & Electrical		26.50	35.50	59.50	16.90%	23.50%	34%	
0010	**POLICE STATIONS**	S.F.	161	212	270				64
0020	Total project costs	C.F.	12.85	15.70	21.50				
0500	Masonry	S.F.	16.10	26.50	33.50	7.80%	9.10%	11.35%	
1800	Equipment		2.31	11.25	17.80	.98%	3.35%	6.70%	
2720	Plumbing		9	17.95	22.50	5.65%	6.90%	10.75%	
2770	Heating, ventilating, air conditioning		13.20	18.70	26.50	5.85%	10.55%	11.70%	
2900	Electrical		17.60	25.50	33	9.80%	11.70%	14.50%	
3100	Total: Mechanical & Electrical		67	70.50	94	28.50%	31.50%	32.50%	
0010	**POST OFFICES**	S.F.	127	157	200				65
0020	Total project costs	C.F.	7.65	9.70	11				
2720	Plumbing	S.F.	5.15	7.10	8.95	4.24%	5.30%	5.60%	
2770	Heating, ventilating, air conditioning		8	11.05	12.30	6.65%	7.15%	9.35%	
2900	Electrical		10.50	14.75	17.50	7.25%	9%	11%	
3100	Total: Mechanical & Electrical		30.50	39.50	45	16.25%	18.80%	22%	
0010	**POWER PLANTS**	S.F.	880	1,175	2,150				66
0020	Total project costs	C.F.	24.50	53	113				
2900	Electrical	S.F.	62.50	132	197	9.30%	12.75%	21.50%	
8100	Total: Mechanical & Electrical		155	505	1,125	32.50%	32.50%	52.50%	
0010	**RELIGIOUS EDUCATION**	S.F.	103	137	170				67
0020	Total project costs	C.F.	5.70	8.15	10.20				
2720	Plumbing	S.F.	4.28	6.05	8.60	4.40%	5.30%	7.10%	
2770	Heating, ventilating, air conditioning		10.80	12.25	17.30	10.05%	11.45%	12.35%	
2900	Electrical		8.15	11.65	17.10	7.70%	9.05%	10.35%	
3100	Total: Mechanical & Electrical		35.50	45	53	22%	23.50%	26%	
0010	**RESEARCH Laboratories & Facilities**	S.F.	157	222	325				69
0020	Total project costs	C.F.	11.40	22.50	27				
1800	Equipment	S.F.	6.80	13.40	32.50	1.22%	4.80%	9.35%	
2720	Plumbing		11.20	19.70	31.50	6.15%	8.30%	10.80%	
2770	Heating, ventilating, air conditioning		13.80	46.50	55	7.25%	16.50%	17.50%	
2900	Electrical		18.45	29.50	48.50	9.45%	11.15%	14.60%	
3100	Total: Mechanical & Electrical		59	104	147	29.50%	36%	41%	
0010	**RESTAURANTS**	S.F.	149	192	249				70
0020	Total project costs	C.F.	12.50	16.40	21.50				
1800	Equipment	S.F.	8.80	23.50	35.50	6.10%	13%	15.65%	
2720	Plumbing		11.80	14.30	18.75	6.10%	8.15%	9%	
2770	Heating, ventilating, air conditioning		14.95	21	25	9.20%	12%	12.40%	
2900	Electrical		15.70	19.40	25	8.35%	10.55%	11.55%	
3100	Total: Mechanical & Electrical		48.50	52	67.50	21%	25%	29.50%	
9000	Per seat unit, total cost	Seat	5,450	7,275	8,600				
9500	Total: Mechanical & Electrical	"	1,375	1,825	2,150				
0010	**RETAIL STORES**	S.F.	69.50	93.50	124				72
0020	Total project costs	C.F.	4.71	6.70	9.35				
2720	Plumbing	S.F.	2.52	4.20	7.15	3.26%	4.60%	6.80%	
2770	Heating, ventilating, air conditioning		5.45	7.45	11.20	6.75%	8.75%	10.15%	
2900	Electrical		6.25	8.55	12.35	7.25%	9.90%	11.60%	
3100	Total: Mechanical & Electrical		16.65	21.50	28.50	17.05%	21%	23.50%	
0010	**SCHOOLS Elementary**	S.F.	114	141	172				74
0020	Total project costs	C.F.	7.40	9.50	12.25				
0500	Masonry	S.F.	8	17.30	25.50	4.89%	10.50%	14%	
1800	Equipment		2.63	4.96	9.40	1.83%	3.13%	4.61%	

50 17 00 | S.F. Costs		UNIT	UNIT COSTS			% OF TOTAL		
			1/4	MEDIAN	3/4	1/4	MEDIAN	3/4
74 2720	Plumbing	S.F.	6.50	9.20	12.30	5.70%	7.15%	9.35%	74
2730	Heating, ventilating, air conditioning		9.80	15.55	22.50	8.15%	10.80%	14.90%
2900	Electrical		10.70	14.25	18.20	8.45%	10.05%	11.85%
3100	Total: Mechanical & Electrical	↓	39	47.50	60	25%	27.50%	30%
9000	Per pupil, total cost	Ea.	13,000	19,300	43,300			
9500	Total: Mechanical & Electrical	"	3,675	4,650	11,700			
76 0010	**SCHOOLS Junior High & Middle**	S.F.	117	145	177				76
0020	Total project costs	C.F.	7.40	9.60	10.75			
0500	Masonry	S.F.	12.55	18.80	23	7.55%	11.10%	14.30%
1800	Equipment		3.20	6	8.80	1.80%	3.03%	4.80%
2720	Plumbing		6.80	8.40	10.40	5.30%	6.80%	7.25%
2770	Heating, ventilating, air conditioning		13.60	16.55	29	8.90%	11.55%	14.20%
2900	Electrical		11.90	14.70	18.60	8.05%	9.55%	10.60%
3100	Total: Mechanical & Electrical	↓	37.50	48	59.50	23%	25.50%	29.50%
9000	Per pupil, total cost	Ea.	14,800	19,400	26,100			
78 0010	**SCHOOLS Senior High**	S.F.	123	150	188				78
0020	Total project costs	C.F.	7.35	10.30	17.40			
1800	Equipment	S.F.	3.24	7.55	10.45	1.88%	2.67%	4.30%
2720	Plumbing		6.80	10.25	18.70	5.60%	6.90%	8.30%
2770	Heating, ventilating, air conditioning		11.75	15.95	24.50	8.95%	11.60%	15%
2900	Electrical		12.30	16.40	24	8.70%	10.35%	12.50%
3100	Total: Mechanical & Electrical	↓	40.50	48	77.50	24%	26.50%	28.50%
9000	Per pupil, total cost	Ea.	11,500	23,300	29,200			
80 0010	**SCHOOLS Vocational**	S.F.	98.50	143	177				80
0020	Total project costs	C.F.	6.10	8.80	12.15			
0500	Masonry	S.F.	4.33	14.25	22	3.20%	4.61%	10.95%
1800	Equipment		3.08	7.65	10.60	1.24%	3.10%	4.26%
2720	Plumbing		6.30	9.35	13.80	5.40%	6.90%	8.55%
2770	Heating, ventilating, air conditioning		8.80	16.40	27.50	8.60%	11.90%	14.65%
2900	Electrical		10.25	14	19.15	8.45%	11%	13.20%
3100	Total: Mechanical & Electrical	↓	38.50	67.50	86.50	27.50%	29.50%	31%
9000	Per pupil, total cost	Ea.	13,700	36,700	54,500			
83 0010	**SPORTS ARENAS**	S.F.	86	115	177				83
0020	Total project costs	C.F.	4.67	8.35	10.80			
2720	Plumbing	S.F.	4.31	7.55	15.95	4.35%	6.35%	9.40%
2770	Heating, ventilating, air conditioning		10.75	12.70	17.60	8.80%	10.20%	13.55%
2900	Electrical	↓	7.95	12.15	15.70	7.65%	9.75%	11.90%
85 0010	**SUPERMARKETS**	S.F.	79.50	92	108				85
0020	Total project costs	C.F.	4.42	5.35	8.10			
2720	Plumbing	S.F.	4.44	5.60	6.50	5.40%	6%	7.45%
2770	Heating, ventilating, air conditioning		6.55	8.65	10.55	8.60%	8.65%	9.60%
2900	Electrical		9.30	11.35	13.50	10.40%	12.45%	13.60%
3100	Total: Mechanical & Electrical	↓	25.50	27	36.50	23.50%	27.50%	28.50%
86 0010	**SWIMMING POOLS**	S.F.	129	300	460				86
0020	Total project costs	C.F.	10.30	12.85	14			
2720	Plumbing	S.F.	11.90	13.60	18.25	4.80%	9.70%	20.50%
2900	Electrical		9.75	15.65	34.50	6.05%	6.95%	7.60%
3100	Total: Mechanical & Electrical	↓	61	80.50	106	11.15%	14.10%	23.50%
87 0010	**TELEPHONE EXCHANGES**	S.F.	160	243	315				87
0020	Total project costs	C.F.	10.65	17.05	23.50			
2720	Plumbing	S.F.	7.20	11.15	16.30	4.52%	5.80%	6.90%
2770	Heating, ventilating, air conditioning		16.75	33.50	42	11.80%	16.05%	18.40%
2900	Electrical		17.40	27.50	49	10.90%	14%	17.85%
3100	Total: Mechanical & Electrical	↓	51.50	97.50	138	29.50%	33.50%	44.50%
91 0010	**THEATERS**	S.F.	102	138	203				91
0020	Total project costs	C.F.	4.96	7.35	10.80			

50 17 00 | S.F. Costs

		UNIT	UNIT COSTS			% OF TOTAL			
			1/4	MEDIAN	3/4	1/4	MEDIAN	3/4	
•1	2720	Plumbing	S.F.	3.34	3.88	15.85	2.92%	4.70%	6.80%	91
	2770	Heating, ventilating, air conditioning		10.45	12.65	15.65	8%	12.25%	13.40%	
	2900	Electrical		9.40	12.70	26	8.05%	9.35%	12.25%	
	3100	Total: Mechanical & Electrical	↓	24	32.50	38.50	21.50%	25.50%	27.50%	
4	0010	**TOWN HALLS City Halls & Municipal Buildings**	S.F.	112	152	199				94
	0020	Total project costs	C.F.	8.10	12.60	18.45				
	2720	Plumbing	S.F.	5	9.35	16.45	4.31%	5.95%	7.95%	
	2770	Heating, ventilating, air conditioning		9.05	17.95	26.50	7.05%	9.05%	13.45%	
	2900	Electrical		11.40	15.70	21.50	8.05%	9.45%	11.65%	
	3100	Total: Mechanical & Electrical	↓	39.50	50	76.50	22%	26.50%	31%	
7	0010	**WAREHOUSES & Storage Buildings**	S.F.	44.50	63.50	95				97
	0020	Total project costs	C.F.	2.34	3.67	6.05				
	0100	Site work	S.F.	4.62	9.15	13.80	6.05%	12.95%	19.55%	
	0500	Masonry		2.54	6.35	13.70	3.60%	7.35%	12%	
	1800	Equipment		.68	1.55	8.70	.72%	1.69%	5.55%	
	2720	Plumbing		1.49	2.68	5	2.90%	4.80%	6.55%	
	2730	Heating, ventilating, air conditioning		1.70	4.80	6.45	2.41%	5%	8.90%	
	2900	Electrical		2.65	4.98	8.20	5.15%	7.20%	10.05%	
	3100	Total: Mechanical & Electrical	↓	7.40	11.35	22.50	13.30%	18.90%	26%	
9	0010	**WAREHOUSE & OFFICES Combination**	S.F.	55	73.50	101				99
	0020	Total project costs	C.F.	2.82	4.09	6.05				
	1800	Equipment	S.F.	.95	1.84	2.74	.42%	1.19%	2.40%	
	2720	Plumbing		2.13	3.70	5.50	3.74%	4.76%	6.30%	
	2770	Heating, ventilating, air conditioning		3.36	5.25	7.35	5%	5.65%	10.05%	
	2900	Electrical		3.75	5.55	8.65	5.85%	8%	10%	
	3100	Total: Mechanical & Electrical	↓	10.45	15.85	25	14.55%	19.95%	24.50%	

Square Foot Project Size Modifier

One factor that affects the S.F. cost of a particular building is the size. In general, for buildings built to the same specifications in the same locality, the larger building will have the lower S.F. cost. This is due mainly to the decreasing contribution of the exterior walls, plus the economy of scale usually achievable in larger buildings. The area conversion scale shown below will give a factor to convert costs for the typical size building to an adjusted cost for the particular project.

The square foot base size table lists the median costs, most typical project size in our accumulated data, and the range in size of the projects.

The size factor for your project is determined by dividing your project area in S.F. by the typical project size for the particular building type. With this factor, enter the area conversion scale at the appropriate size factor and determine the appropriate cost multiplier for your building size.

Example: Determine the cost per S.F. for a 100,000 S.F. Mid-rise apartment building.

$$\frac{\text{Proposed building area} = 100,000 \text{ S.F.}}{\text{Typical size from below} = 50,000 \text{ S.F.}} = 2.00$$

Enter area conversion scale at 2.0, intersect curve, read horizontally the appropriate cost multiplier of .94. Size adjusted cost becomes .94 × $119.00 = $112.00 based on national average costs.

Note: For size factors less than .50, the cost multiplier is 1.1
For size factors greater than 3.5, the cost multiplier is .90

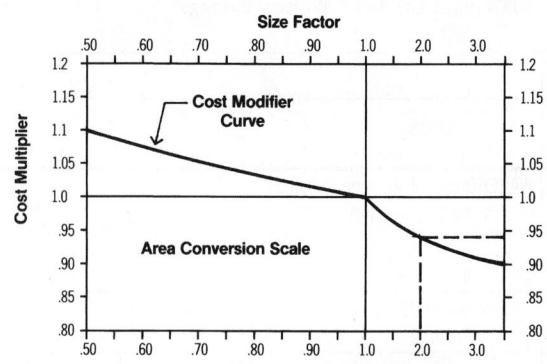

Square Foot Base Size							
Building Type	Median Cost per S.F.	Typical Size Gross S.F.	Typical Range Gross S.F.	Building Type	Median Cost per S.F.	Typical Size Gross S.F.	Typical Range Gross S.F.
Apartments, Low Rise	$ 94.00	21,000	9,700 - 37,200	Jails	$ 286.00	40,000	5,500 - 145,000
Apartments, Mid Rise	119.00	50,000	32,000 - 100,000	Libraries	190.00	12,000	7,000 - 31,000
Apartments, High Rise	129.00	145,000	95,000 - 600,000	Living, Assisted	154.00	32,300	23,500 - 50,300
Auditoriums	161.00	25,000	7,600 - 39,000	Medical Clinics	162.00	7,200	4,200 - 15,700
Auto Sales	119.00	20,000	10,800 - 28,600	Medical Offices	154.00	6,000	4,000 - 15,000
Banks	211.00	4,200	2,500 - 7,500	Motels	113.00	40,000	15,800 - 120,000
Churches	145.00	17,000	2,000 - 42,000	Nursing Homes	158.00	23,000	15,000 - 37,000
Clubs, Country	147.00	6,500	4,500 - 15,000	Offices, Low Rise	135.00	20,000	5,000 - 80,000
Clubs, Social	141.00	10,000	6,000 - 13,500	Offices, Mid Rise	132.00	120,000	20,000 - 300,000
Clubs, YMCA	167.00	28,300	12,800 - 39,400	Offices, High Rise	169.00	260,000	120,000 - 800,000
Colleges (Class)	167.00	50,000	15,000 - 150,000	Police Stations	212.00	10,500	4,000 - 19,000
Colleges (Science Lab)	269.00	45,600	16,600 - 80,000	Post Offices	157.00	12,400	6,800 - 30,000
College (Student Union)	199.00	33,400	16,000 - 85,000	Power Plants	1175.00	7,500	1,000 - 20,000
Community Center	150.00	9,400	5,300 - 16,700	Religious Education	137.00	9,000	6,000 - 12,000
Court Houses	206.00	32,400	17,800 - 106,000	Research	222.00	19,000	6,300 - 45,000
Dept. Stores	87.50	90,000	44,000 - 122,000	Restaurants	192.00	4,400	2,800 - 6,000
Dormitories, Low Rise	170.00	25,000	10,000 - 95,000	Retail Stores	93.50	7,200	4,000 - 17,600
Dormitories, Mid Rise	196.00	85,000	20,000 - 200,000	Schools, Elementary	141.00	41,000	24,500 - 55,000
Factories	85.00	26,400	12,900 - 50,000	Schools, Jr. High	145.00	92,000	52,000 - 119,000
Fire Stations	156.00	5,800	4,000 - 8,700	Schools, Sr. High	150.00	101,000	50,500 - 175,000
Fraternity Houses	145.00	12,500	8,200 - 14,800	Schools, Vocational	143.00	37,000	20,500 - 82,000
Funeral Homes	162.00	10,000	4,000 - 20,000	Sports Arenas	115.00	15,000	5,000 - 40,000
Garages, Commercial	104.00	9,300	5,000 - 13,600	Supermarkets	92.00	44,000	12,000 - 60,000
Garages, Municipal	135.00	8,300	4,500 - 12,600	Swimming Pools	300.00	20,000	10,000 - 32,000
Garages, Parking	56.00	163,000	76,400 - 225,300	Telephone Exchange	243.00	4,500	1,200 - 10,600
Gymnasiums	143.00	19,200	11,600 - 41,000	Theaters	138.00	10,500	8,800 - 17,500
Hospitals	258.00	55,000	27,200 - 125,000	Town Halls	152.00	10,800	4,800 - 23,400
House (Elderly)	128.00	37,000	21,000 - 66,000	Warehouses	63.50	25,000	8,000 - 72,000
Housing (Public)	118.00	36,000	14,400 - 74,400	Warehouse & Office	73.50	25,000	8,000 - 72,000
Ice Rinks	170.00	29,000	27,200 - 33,600				

Abbreviations

A	Area Square Feet; Ampere	Brk., brk	Brick	Csc	Cosecant
AAFES	Army and Air Force Exchange Service	brkt	Bracket	C.S.F.	Hundred Square Feet
		Brng.	Bearing	CSI	Construction Specifications Institute
ABS	Acrylonitrile Butadiene Stryrene; Asbestos Bonded Steel	Brs.	Brass		
		Brz.	Bronze	CT	Current Transformer
A.C., AC	Alternating Current; Air-Conditioning; Asbestos Cement; Plywood Grade A & C	Bsn.	Basin	CTS	Copper Tube Size
		Btr.	Better	Cu	Copper, Cubic
		Btu	British Thermal Unit	Cu. Ft.	Cubic Foot
		BTUH	BTU per Hour	cw	Continuous Wave
ACI	American Concrete Institute	Bu.	bushels	C.W.	Cool White; Cold Water
ACR	Air Conditioning Refrigeration	BUR	Built-up Roofing	Cwt.	100 Pounds
ADA	Americans with Disabilities Act	BX	Interlocked Armored Cable	C.W.X.	Cool White Deluxe
AD	Plywood, Grade A & D	°C	degree centegrade	C.Y.	Cubic Yard (27 cubic feet)
Addit.	Additional	c	Conductivity, Copper Sweat	C.Y./Hr.	Cubic Yard per Hour
Adj.	Adjustable	C	Hundred; Centigrade	Cyl.	Cylinder
af	Audio-frequency	C/C	Center to Center, Cedar on Cedar	d	Penny (nail size)
AFUE	Annual Fuel Utilization Efficiency	C-C	Center to Center	D	Deep; Depth; Discharge
AGA	American Gas Association	Cab	Cabinet	Dis., Disch.	Discharge
Agg.	Aggregate	Cair.	Air Tool Laborer	Db	Decibel
A.H., Ah	Ampere Hours	Cal.	caliper	Dbl.	Double
A hr.	Ampere-hour	Calc	Calculated	DC	Direct Current
A.H.U., AHU	Air Handling Unit	Cap.	Capacity	DDC	Direct Digital Control
A.I.A.	American Institute of Architects	Carp.	Carpenter	Demob.	Demobilization
AIC	Ampere Interrupting Capacity	C.B.	Circuit Breaker	d.f.t.	Dry Film Thickness
Allow.	Allowance	C.C.A.	Chromate Copper Arsenate	d.f.u.	Drainage Fixture Units
alt., alt	Alternate	C.C.F.	Hundred Cubic Feet	D.H.	Double Hung
Alum.	Aluminum	cd	Candela	DHW	Domestic Hot Water
a.m.	Ante Meridiem	cd/sf	Candela per Square Foot	DI	Ductile Iron
Amp.	Ampere	CD	Grade of Plywood Face & Back	Diag.	Diagonal
Anod.	Anodized	CDX	Plywood, Grade C & D, exterior glue	Diam., Dia	Diameter
ANSI	American National Standards Institute			Distrib.	Distribution
		Cefi.	Cement Finisher	Div.	Division
APA	American Plywood Association	Cem.	Cement	Dk.	Deck
Approx.	Approximate	CF	Hundred Feet	D.L.	Dead Load; Diesel
Apt.	Apartment	C.F.	Cubic Feet	DLH	Deep Long Span Bar Joist
Asb.	Asbestos	CFM	Cubic Feet per Minute	dlx	Deluxe
A.S.B.C.	American Standard Building Code	CFRP	Carbon Fiber Reinforced Plastic	Do.	Ditto
Asbe.	Asbestos Worker	c.g.	Center of Gravity	DOP	Dioctyl Phthalate Penetration Test (Air Filters)
ASCE.	American Society of Civil Engineers	CHW	Chilled Water; Commercial Hot Water		
A.S.H.R.A.E.	American Society of Heating, Refrig. & AC Engineers			Dp., dp	Depth
		C.I., CI	Cast Iron	D.P.S.T.	Double Pole, Single Throw
ASME	American Society of Mechanical Engineers	C.I.P., CIP	Cast in Place	Dr.	Drive
		Circ.	Circuit	DR	Dimension Ratio
ASTM	American Society for Testing and Materials	C.L.	Carload Lot	Drink.	Drinking
		CL	Chain Link	D.S.	Double Strength
Attchmt.	Attachment	Clab.	Common Laborer	D.S.A.	Double Strength A Grade
Avg., Ave.	Average	Clam	Common maintenance laborer	D.S.B.	Double Strength B Grade
AWG	American Wire Gauge	C.L.F.	Hundred Linear Feet	Dty.	Duty
AWWA	American Water Works Assoc.	CLF	Current Limiting Fuse	DWV	Drain Waste Vent
Bbl.	Barrel	CLP	Cross Linked Polyethylene	DX	Deluxe White, Direct Expansion
B&B, BB	Grade B and Better; Balled & Burlapped	cm	Centimeter	dyn	Dyne
		CMP	Corr. Metal Pipe	e	Eccentricity
B&S	Bell and Spigot	CMU	Concrete Masonry Unit	E	Equipment Only; East; emissivity
B.&W.	Black and White	CN	Change Notice	Ea.	Each
b.c.c.	Body-centered Cubic	Col.	Column	EB	Encased Burial
B.C.Y.	Bank Cubic Yards	CO₂	Carbon Dioxide	Econ.	Economy
BE	Bevel End	Comb.	Combination	E.C.Y	Embankment Cubic Yards
B.F.	Board Feet	comm.	Commercial, Communication	EDP	Electronic Data Processing
Bg. cem.	Bag of Cement	Compr.	Compressor	EIFS	Exterior Insulation Finish System
BHP	Boiler Horsepower; Brake Horsepower	Conc.	Concrete	E.D.R.	Equiv. Direct Radiation
		Cont., cont	Continuous; Continued, Container	Eq.	Equation
B.I.	Black Iron	Corr.	Corrugated	EL	elevation
bidir.	bidirectional	Cos	Cosine	Elec.	Electrician; Electrical
Bit., Bitum.	Bituminous	Cot	Cotangent	Elev.	Elevator; Elevating
Bit., Conc.	Bituminous Concrete	Cov.	Cover	EMT	Electrical Metallic Conduit; Thin Wall Conduit
Bk.	Backed	C/P	Cedar on Paneling		
Bkrs.	Breakers	CPA	Control Point Adjustment	Eng.	Engine, Engineered
Bldg., bldg	Building	Cplg.	Coupling	EPDM	Ethylene Propylene Diene Monomer
Blk.	Block	CPM	Critical Path Method		
Bm.	Beam	CPVC	Chlorinated Polyvinyl Chloride	EPS	Expanded Polystyrene
Boil.	Boilermaker	C.Pr.	Hundred Pair	Eqhv.	Equip. Oper., Heavy
bpm	Blows per Minute	CRC	Cold Rolled Channel	Eqlt.	Equip. Oper., Light
BR	Bedroom	Creos.	Creosote	Eqmd.	Equip. Oper., Medium
Brg.	Bearing	Crpt.	Carpet & Linoleum Layer	Eqmm.	Equip. Oper., Master Mechanic
Brhe.	Bricklayer Helper	CRT	Cathode-ray Tube	Eqol.	Equip. Oper., Oilers
Bric.	Bricklayer	CS	Carbon Steel, Constant Shear Bar Joist	Equip.	Equipment
				ERW	Electric Resistance Welded

For customer support on your Mechanical Cost Data, call 877.795.7394.

753

E.S.	Energy Saver	H	High Henry	Lath.	Lather
Est.	Estimated	HC	High Capacity	Lav.	Lavatory
esu	Electrostatic Units	H.D., HD	Heavy Duty; High Density	lb.; #	Pound
E.W.	Each Way	H.D.O.	High Density Overlaid	L.B., LB	Load Bearing; L Conduit Body
EWT	Entering Water Temperature	HDPE	High density polyethelene plastic	L. & E.	Labor & Equipment
Excav.	Excavation	Hdr.	Header	lb./hr.	Pounds per Hour
excl	Excluding	Hdwe.	Hardware	lb./L.F.	Pounds per Linear Foot
Exp., exp	Expansion, Exposure	H.I.D., HID	High Intensity Discharge	lbf/sq.in.	Pound-force per Square Inch
Ext., ext	Exterior; Extension	Help.	Helper Average	L.C.L.	Less than Carload Lot
Extru.	Extrusion	HEPA	High Efficiency Particulate Air	L.C.Y.	Loose Cubic Yard
f.	Fiber stress		Filter	Ld.	Load
F	Fahrenheit; Female; Fill	Hg	Mercury	LE	Lead Equivalent
Fab., fab	Fabricated; fabric	HIC	High Interrupting Capacity	LED	Light Emitting Diode
FBGS	Fiberglass	HM	Hollow Metal	L.F.	Linear Foot
F.C.	Footcandles	HMWPE	high molecular weight	L.F. Nose	Linear Foot of Stair Nosing
f.c.c.	Face-centered Cubic		polyethylene	L.F. Rsr	Linear Foot of Stair Riser
f'c.	Compressive Stress in Concrete;	HO	High Output	Lg.	Long; Length; Large
	Extreme Compressive Stress	Horiz.	Horizontal	L & H	Light and Heat
F.E.	Front End	H.P., HP	Horsepower; High Pressure	LH	Long Span Bar Joist
FEP	Fluorinated Ethylene Propylene	H.P.F.	High Power Factor	L.H.	Labor Hours
	(Teflon)	Hr.	Hour	L.L., LL	Live Load
F.G.	Flat Grain	Hrs./Day	Hours per Day	L.L.D.	Lamp Lumen Depreciation
F.H.A.	Federal Housing Administration	HSC	High Short Circuit	lm	Lumen
Fig.	Figure	Ht.	Height	lm/sf	Lumen per Square Foot
Fin.	Finished	Htg.	Heating	lm/W	Lumen per Watt
FIPS	Female Iron Pipe Size	Htrs.	Heaters	LOA	Length Over All
Fixt.	Fixture	HVAC	Heating, Ventilation & Air-	log	Logarithm
FJP	Finger jointed and primed		Conditioning	L-O-L	Lateralolet
Fl. Oz.	Fluid Ounces	Hvy.	Heavy	long.	longitude
Flr.	Floor	HW	Hot Water	L.P., LP	Liquefied Petroleum; Low Pressure
FM	Frequency Modulation;	Hyd.; Hydr.	Hydraulic	L.P.F.	Low Power Factor
	Factory Mutual	Hz	Hertz (cycles)	LR	Long Radius
Fmg.	Framing	I.	Moment of Inertia	L.S.	Lump Sum
FM/UL	Factory Mutual/Underwriters Labs	IBC	International Building Code	Lt.	Light
Fdn.	Foundation	I.C.	Interrupting Capacity	Lt. Ga.	Light Gauge
FNPT	Female National Pipe Thread	ID	Inside Diameter	L.T.L.	Less than Truckload Lot
Fori.	Foreman, Inside	I.D.	Inside Dimension; Identification	Lt. Wt.	Lightweight
Foro.	Foreman, Outside	I.F.	Inside Frosted	L.V.	Low Voltage
Fount.	Fountain	I.M.C.	Intermediate Metal Conduit	M	Thousand; Material; Male;
fpm	Feet per Minute	In.	Inch		Light Wall Copper Tubing
FPT	Female Pipe Thread	Incan.	Incandescent	M²CA	Meters Squared Contact Area
Fr	Frame	Incl.	Included; Including	m/hr.; M.H.	Man-hour
F.R.	Fire Rating	Int.	Interior	mA	Milliampere
FRK	Foil Reinforced Kraft	Inst.	Installation	Mach.	Machine
FSK	Foil/scrim/kraft	Insul., insul	Insulation/Insulated	Mag. Str.	Magnetic Starter
FRP	Fiberglass Reinforced Plastic	I.P.	Iron Pipe	Maint.	Maintenance
FS	Forged Steel	I.P.S., IPS	Iron Pipe Size	Marb.	Marble Setter
FSC	Cast Body; Cast Switch Box	IPT	Iron Pipe Threaded	Mat; Mat'l.	Material
Ft., ft	Foot; Feet	I.W.	Indirect Waste	Max.	Maximum
Ftng.	Fitting	J	Joule	MBF	Thousand Board Feet
Ftg.	Footing	J.I.C.	Joint Industrial Council	MBH	Thousand BTU's per hr.
Ft lb.	Foot Pound	K	Thousand; Thousand Pounds;	MC	Metal Clad Cable
Furn.	Furniture		Heavy Wall Copper Tubing, Kelvin	MCC	Motor Control Center
FVNR	Full Voltage Non-Reversing	K.A.H.	Thousand Amp. Hours	M.C.F.	Thousand Cubic Feet
FVR	Full Voltage Reversing	kcmil	Thousand Circular Mils	MCFM	Thousand Cubic Feet per Minute
FXM	Female by Male	KD	Knock Down	M.C.M.	Thousand Circular Mils
Fy.	Minimum Yield Stress of Steel	K.D.A.T.	Kiln Dried After Treatment	MCP	Motor Circuit Protector
g	Gram	kg	Kilogram	MD	Medium Duty
G	Gauss	kG	Kilogauss	MDF	Medium-density fibreboard
Ga.	Gauge	kgf	Kilogram Force	M.D.O.	Medium Density Overlaid
Gal., gal.	Gallon	kHz	Kilohertz	Med.	Medium
gpm, GPM	Gallon per Minute	Kip	1000 Pounds	MF	Thousand Feet
Galv., galv	Galvanized	KJ	Kiljoule	M.F.B.M.	Thousand Feet Board Measure
GC/MS	Gas Chromatograph/Mass	K.L.	Effective Length Factor	Mfg.	Manufacturing
	Spectrometer	K.L.F.	Kips per Linear Foot	Mfrs.	Manufacturers
Gen.	General	Km	Kilometer	mg	Milligram
GFI	Ground Fault Interrupter	KO	Knock Out	MGD	Million Gallons per Day
GFRC	Glass Fiber Reinforced Concrete	K.S.F.	Kips per Square Foot	MGPH	Thousand Gallons per Hour
Glaz.	Glazier	K.S.I.	Kips per Square Inch	MH, M.H.	Manhole; Metal Halide; Man-Hour
GPD	Gallons per Day	kV	Kilovolt	MHz	Megahertz
gpf	Gallon per flush	kVA	Kilovolt Ampere	Mi.	Mile
GPH	Gallons per Hour	kVAR	Kilovar (Reactance)	MI	Malleable Iron; Mineral Insulated
GPM	Gallons per Minute	KW	Kilowatt	MIPS	Male Iron Pipe Size
GR	Grade	KWh	Kilowatt-hour	mj	Mechanical Joint
Gran.	Granular	L	Labor Only; Length; Long;	m	Meter
Grnd.	Ground		Medium Wall Copper Tubing	mm	Millimeter
GVW	Gross Vehicle Weight	Lab.	Labor	Mill.	Millwright
GWB	Gypsum wall board	lat	Latitude	Min., min.	Minimum, minute

Misc.	Miscellaneous	PDCA	Painting and Decorating Contractors of America	SC	Screw Cover
ml	Milliliter, Mainline	P.E., PE	Professional Engineer;	SCFM	Standard Cubic Feet per Minute
M.L.F.	Thousand Linear Feet		Porcelain Enamel;	Scaf.	Scaffold
Mo.	Month		Polyethylene; Plain End	Sch., Sched.	Schedule
Mobil.	Mobilization	P.E.C.I.	Porcelain Enamel on Cast Iron	S.C.R.	Modular Brick
Mog.	Mogul Base	Perf.	Perforated	S.D.	Sound Deadening
MPH	Miles per Hour	PEX	Cross linked polyethylene	SDR	Standard Dimension Ratio
MPT	Male Pipe Thread	Ph.	Phase	S.E.	Surfaced Edge
MRGWB	Moisture Resistant Gypsum Wallboard	P.I.	Pressure Injected	Sel.	Select
MRT	Mile Round Trip	Pile.	Pile Driver	SER, SEU	Service Entrance Cable
ms	Millisecond	Pkg.	Package	S.F.	Square Foot
M.S.F.	Thousand Square Feet	Pl.	Plate	S.F.C.A.	Square Foot Contact Area
Mstz.	Mosaic & Terrazzo Worker	Plah.	Plasterer Helper	S.F. Flr.	Square Foot of Floor
M.S.Y.	Thousand Square Yards	Plas.	Plasterer	S.F.G.	Square Foot of Ground
Mtd., mtd., mtd	Mounted	plf	Pounds Per Linear Foot	S.F. Hor.	Square Foot Horizontal
Mthe.	Mosaic & Terrazzo Helper	Pluh.	Plumbers Helper	SFR	Square Feet of Radiation
Mtng.	Mounting	Plum.	Plumber	S.F. Shlf.	Square Foot of Shelf
Mult.	Multi; Multiply	Ply.	Plywood	S4S	Surface 4 Sides
M.V.A.	Million Volt Amperes	p.m.	Post Meridiem	Shee.	Sheet Metal Worker
M.V.A.R.	Million Volt Amperes Reactance	Pntd.	Painted	Sin.	Sine
MV	Megavolt	Pord.	Painter, Ordinary	Skwk.	Skilled Worker
MW	Megawatt	pp	Pages	SL	Saran Lined
MXM	Male by Male	PP, PPL	Polypropylene	S.L.	Slimline
MYD	Thousand Yards	P.P.M.	Parts per Million	Sldr.	Solder
N	Natural; North	Pr.	Pair	SLH	Super Long Span Bar Joist
nA	Nanoampere	P.E.S.B.	Pre-engineered Steel Building	S.N.	Solid Neutral
NA	Not Available; Not Applicable	Prefab.	Prefabricated	SO	Stranded with oil resistant inside insulation
N.B.C.	National Building Code	Prefin.	Prefinished	S-O-L	Socketolet
NC	Normally Closed	Prop.	Propelled	sp	Standpipe
NEMA	National Electrical Manufacturers Assoc.	PSF, psf	Pounds per Square Foot	S.P.	Static Pressure; Single Pole; Self-Propelled
NEHB	Bolted Circuit Breaker to 600V.	PSI, psi	Pounds per Square Inch	Spri.	Sprinkler Installer
NFPA	National Fire Protection Association	PSIG	Pounds per Square Inch Gauge	spwg	Static Pressure Water Gauge
NLB	Non-Load-Bearing	PSP	Plastic Sewer Pipe	S.P.D.T.	Single Pole, Double Throw
NM	Non-Metallic Cable	Pspr.	Painter, Spray	SPF	Spruce Pine Fir; Sprayed Polyurethane Foam
nm	Nanometer	Psst.	Painter, Structural Steel		
No.	Number	P.T.	Potential Transformer	S.P.S.T.	Single Pole, Single Throw
NO	Normally Open	P. & T.	Pressure & Temperature	SPT	Standard Pipe Thread
N.O.C.	Not Otherwise Classified	Ptd.	Painted	Sq.	Square; 100 Square Feet
Nose.	Nosing	Ptns.	Partitions	Sq. Hd.	Square Head
NPT	National Pipe Thread	Pu	Ultimate Load	Sq. In.	Square Inch
NQOD	Combination Plug-on/Bolt on Circuit Breaker to 240V.	PVC	Polyvinyl Chloride	S.S.	Single Strength; Stainless Steel
N.R.C., NRC	Noise Reduction Coefficient/ Nuclear Regulator Commission	Pvmt.	Pavement	S.S.B.	Single Strength B Grade
		PRV	Pressure Relief Valve	sst, ss	Stainless Steel
N.R.S.	Non Rising Stem	Pwr.	Power	Sswk.	Structural Steel Worker
ns	Nanosecond	Q	Quantity Heat Flow	Sswl.	Structural Steel Welder
nW	Nanowatt	Qt.	Quart	St.; Stl.	Steel
OB	Opposing Blade	Quan., Qty.	Quantity	STC	Sound Transmission Coefficient
OC	On Center	Q.C.	Quick Coupling	Std.	Standard
OD	Outside Diameter	r	Radius of Gyration	Stg.	Staging
O.D.	Outside Dimension	R	Resistance	STK	Select Tight Knot
ODS	Overhead Distribution System	R.C.P.	Reinforced Concrete Pipe	STP	Standard Temperature & Pressure
O.G.	Ogee	Rect.	Rectangle	Stpi.	Steamfitter, Pipefitter
O.H.	Overhead	recpt.	receptacle	Str.	Strength; Starter; Straight
O&P	Overhead and Profit	Reg.	Regular	Strd.	Stranded
Oper.	Operator	Reinf.	Reinforced	Struct.	Structural
Opng.	Opening	Req'd.	Required	Sty.	Story
Orna.	Ornamental	Res.	Resistant	Subj.	Subject
OSB	Oriented Strand Board	Resi.	Residential	Subs.	Subcontractors
OS&Y	Outside Screw and Yoke	RF	Radio Frequency	Surf.	Surface
OSHA	Occupational Safety and Health Act	RFID	Radio-frequency identification	Sw.	Switch
		Rgh.	Rough	Swbd.	Switchboard
Ovhd.	Overhead	RGS	Rigid Galvanized Steel	S.Y.	Square Yard
OWG	Oil, Water or Gas	RHW	Rubber, Heat & Water Resistant; Residential Hot Water	Syn.	Synthetic
Oz.	Ounce			S.Y.P.	Southern Yellow Pine
P	Pole; Applied Load; Projection	rms	Root Mean Square	Sys.	System
p.	Page	Rnd.	Round	t.	Thickness
Pape.	Paperhanger	Rodm.	Rodman	T	Temperature; Ton
P.A.P.R.	Powered Air Purifying Respirator	Rofc.	Roofer, Composition	Tan	Tangent
PAR	Parabolic Reflector	Rofp.	Roofer, Precast	T.C.	Terra Cotta
PB., PB	Push Button	Rohe.	Roofer Helpers (Composition)	T & C	Threaded and Coupled
Pc., Pcs.	Piece, Pieces	Rots.	Roofer, Tile & Slate	T.D.	Temperature Difference
P.C.	Portland Cement; Power Connector	R.O.W.	Right of Way	Tdd	Telecommunications Device for the Deaf
P.C.F.	Pounds per Cubic Foot	RPM	Revolutions per Minute		
PCM	Phase Contrast Microscopy	R.S.	Rapid Start	T.E.M.	Transmission Electron Microscopy
		Rsr	Riser	temp	Temperature, Tempered, Temporary
		RT	Round Trip	TFFN	Nylon Jacketed Wire
		S.	Suction; Single Entrance; South		
		SBS	Styrene Butadiere Styrene		

755

TFE	Tetrafluoroethylene (Teflon)	U.L., UL	Underwriters Laboratory	w/	With
T. & G.	Tongue & Groove;	Uld.	unloading	W.C., WC	Water Column; Water Closet
	Tar & Gravel	Unfin.	Unfinished	W.F.	Wide Flange
Th., Thk.	Thick	UPS	Uninterruptible Power Supply	W.G.	Water Gauge
Thn.	Thin	URD	Underground Residential	Wldg.	Welding
Thrded	Threaded		Distribution	W. Mile	Wire Mile
Tilf.	Tile Layer, Floor	US	United States	W-O-L	Weldolet
Tilh.	Tile Layer, Helper	USGBC	U.S. Green Building Council	W.R.	Water Resistant
THHN	Nylon Jacketed Wire	USP	United States Primed	Wrck.	Wrecker
THW.	Insulated Strand Wire	UTMCD	Uniform Traffic Manual For Control	W.S.P.	Water, Steam, Petroleum
THWN	Nylon Jacketed Wire		Devices	WT., Wt.	Weight
T.L., TL	Truckload	UTP	Unshielded Twisted Pair	WWF	Welded Wire Fabric
T.M.	Track Mounted	V	Volt	XFER	Transfer
Tot.	Total	VA	Volt Amperes	XFMR	Transformer
T-O-L	Threadolet	VAT	Vinyl Asbestos Tile	XHD	Extra Heavy Duty
tmpd	Tempered	V.C.T.	Vinyl Composition Tile	XHHW,	Cross-Linked Polyethylene Wire
TPO	Thermoplastic Polyolefin	VAV	Variable Air Volume	XLPE	Insulation
T.S.	Trigger Start	VC	Veneer Core	XLP	Cross-linked Polyethylene
Tr.	Trade	VDC	Volts Direct Current	Xport	Transport
Transf.	Transformer	Vent.	Ventilation	Y	Wye
Trhv.	Truck Driver, Heavy	Vert.	Vertical	yd	Yard
Trlr	Trailer	V.F.	Vinyl Faced	yr	Year
Trlt.	Truck Driver, Light	V.G.	Vertical Grain	Δ	Delta
TTY	Teletypewriter	VHF	Very High Frequency	%	Percent
TV	Television	VHO	Very High Output	~	Approximately
T.W.	Thermoplastic Water Resistant	Vib.	Vibrating	Ø	Phase; diameter
	Wire	VLF	Vertical Linear Foot	@	At
UCI	Uniform Construction Index	VOC	Volitile Organic Compound	#	Pound; Number
UF	Underground Feeder	Vol.	Volume	<	Less Than
UGND	Underground Feeder	VRP	Vinyl Reinforced Polyester	>	Greater Than
UHF	Ultra High Frequency	W	Wire; Watt; Wide; West	Z	zone
U.I.	United Inch				

For customer support on your Mechanical Cost Data, call 877.795.7394.

For customer support on your Mechanical Cost Data, call 877.795.7394.

For customer support on your Mechanical Cost Data, call 877.795.7394.

For customer support on your Mechanical Cost Data, call 877.795.7394.

For customer support on your Mechanical Cost Data, call 877.795.7394.

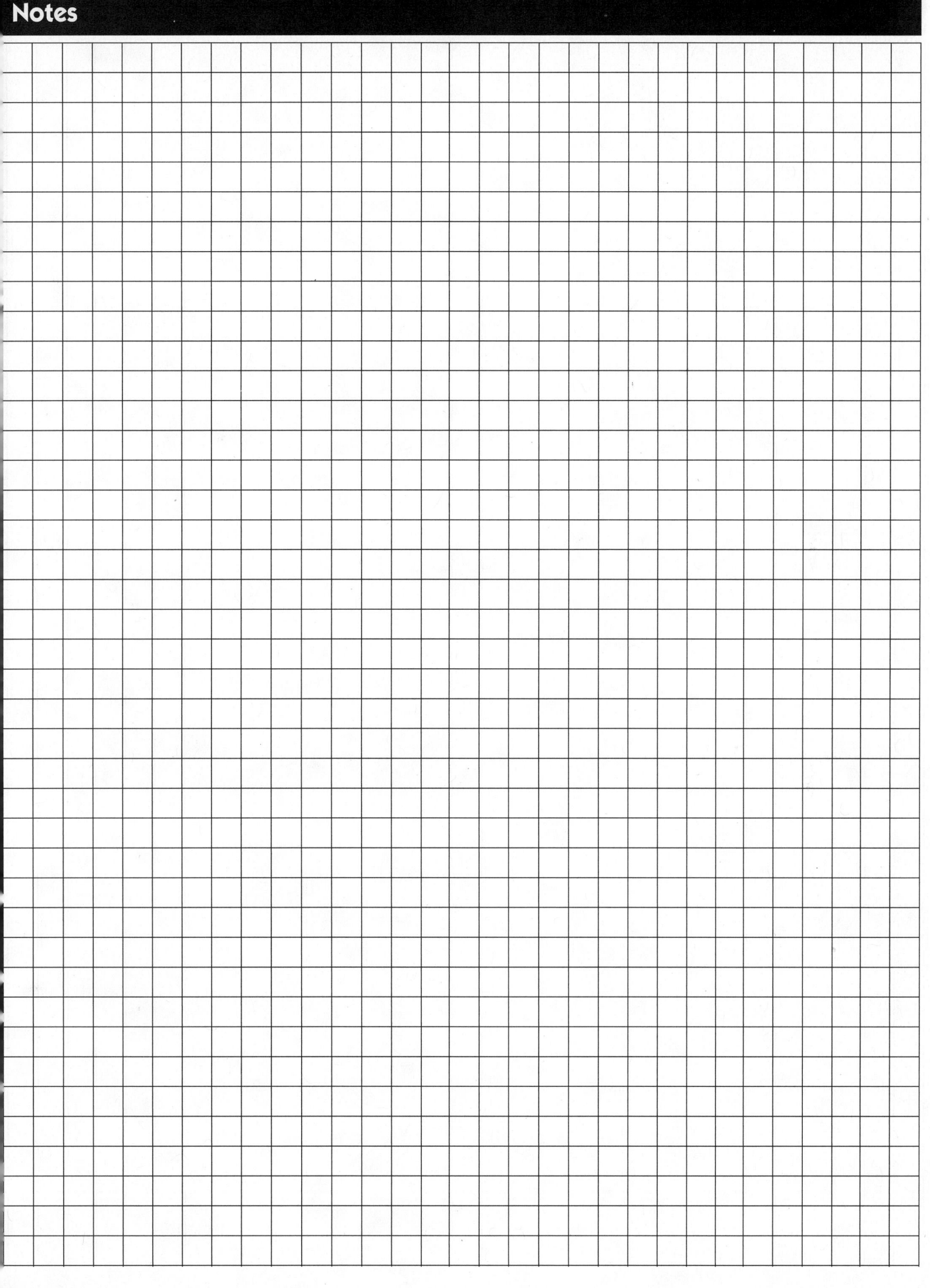

Other RSMeans Products & Services

RSMeans—
a tradition of excellence in construction cost information and services since 1942

For more information visit the RSMeans website at **www.rsmeans.com** Unit prices according to the latest MasterForma

Book Selection Guide   The following table provides definitive information on the content of each cost data publication. The number of lines of data provided in each unit price or assemblies division, as well as the number of crews, is listed for each book. The presence of other elements such as reference tables, square foot models, equipment rental costs, historical cost indexes, and city cost indexes, is also indicated. You can use the table to help select the RSMeans book that has the quantity and type of information you most need in your work.

Unit Cost Divisions	Building Construction	Mechanical	Electrical	Commercial Renovation	Square Foot	Site Work Landsc.	Green Building	Interior	Concrete Masonry	Open Shop	Heavy Construction	Light Commercial	Facilities Construction	Plumbing	Residential
1	573	393	410	506		509	191	312	456	572	501	246	1042	403	178
2	777	279	85	733		993	178	399	213	776	734	481	1221	286	257
3	1690	340	230	1085		1471	987	355	2032	1689	1685	481	1788	316	388
4	957	21	0	922		720	179	612	1155	925	612	529	1172	0	441
5	1893	158	155	1090		841	1799	1095	720	1893	1037	979	1909	204	721
6	2452	18	18	2110		110	589	1528	281	2448	123	2140	2124	22	2660
7	1584	215	128	1623		581	757	532	524	1581	26	1317	1685	227	1037
8	2082	81	45	2609		265	1123	1753	105	2077	0	2193	2879	0	1526
9	1971	72	26	1767		313	449	2056	390	1911	15	1659	2220	54	1435
10	1026	17	10	642		215	27	842	158	1026	29	511	1118	235	224
11	1087	208	165	546		124	54	932	28	1071	0	230	1107	169	110
12	544	0	2	319		217	137	1654	.14	532	0	352	1704	23	310
13	719	140	137	242		343	111	248	66	695	244	83	746	94	79
14	273	36	0	221		0	0	257	0	273	0	12	293	16	6
21	90	0	16	37		0	0	250	0	87	0	68	426	431	220
22	1157	7544	150	1190		1568	1067	838	20	1121	1677	857	7440	9347	712
23	1177	6971	580	928		158	902	776	38	1101	111	878	5187	1918	469
26	1354	454	10081	1021		793	611	1136	55	1349	562	1298	10030	399	617
27	72	0	297	34		13	0	71	0	72	39	52	279	0	22
28	96	58	136	71		0	21	78	0	99	0	40	151	44	25
31	1498	735	610	803		3217	289	7	1208	1443	3256	600	1560	660	611
32	822	53	8	886		4426	356	405	291	793	1841	420	1701	165	468
33	530	1079	538	252		2178	41	0	237	522	2159	128	1699	1286	154
34	107	0	47	4		190	0	0	31	62	193	0	128	0	0
35	18	0	0	0		327	0	0	0	18	442	0	84	0	0
41	60	0	0	32		7	0	22	0	60	30	0	67	14	0
44	75	79	0	0		0	0	0	0	0	0	0	75	75	0
46	23	16	0	0		274	261	0	0	23	264	0	33	33	0
48	12	0	25	0		0	25	0	0	12	17	12	12	0	12
Totals	24719	18967	13899	19673		19853	10154	16158	8022	24231	15597	15566	49880	16421	12682

Assem Div	Building Construction	Mechanical	Electrical	Commercial Renovation	Square Foot	Site Work Landscape	Assemblies	Green Building	Interior	Concrete Masonry	Heavy Construction	Light Commercial	Facilities Construction	Plumbing	Asm Div	Residential
A		15	0	188	150	577	598	0	0	536	571	154	24	0	1	376
B		0	0	848	2498	0	5658	56	329	1975	368	2089	174	0	2	211
C		0	0	647	926	0	1304	0	1629	146	0	816	250	0	3	588
D		1067	941	712	1859	72	2538	330	825	0	0	1345	1105	1088	4	851
E		0	0	86	260	0	300	0	5	0	0	257	5	0	5	392
F		0	0	0	114	0	114	0	0	0	0	114	3	0	6	357
G		527	447	318	249	3365	729	0	0	535	1350	205	293	677	7	307
															8	760
															9	80
															10	0
															11	0
															12	0
Totals		1609	1388	2799	6056	4014	11241	386	2788	3192	2289	4980	1854	1765		3922

Reference Section	Building Construction Costs	Mechanical	Electrical	Commercial Renovation	Square Foot	Site Work Landscape	Assem.	Green Building	Interior	Concrete Masonry	Open Shop	Heavy Construction	Light Commercial	Facilities Construction	Plumbing	Resi.
Reference Tables	yes	yes	yes	yes	no	yes	yes	yes	yes	yes	yes	yes	yes	yes	yes	yes
Models					111			25					50			28
Crews	571	571	571	550		571		571	571	571	548	571	548	550	571	548
Equipment Rental Costs	yes	yes	yes	yes		yes		yes	yes	yes	yes	yes	yes	yes	yes	yes
Historical Cost Indexes	yes	yes	yes	yes		yes		yes	yes	yes	yes	yes	yes	yes	yes	no
City Cost Indexes	yes	yes	yes	yes		yes		yes	yes	yes	yes	yes	yes	yes	yes	yes

For more information visit the RSMeans website at **www.rsmeans.com**

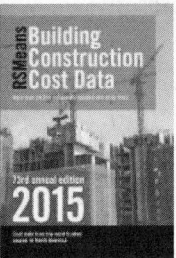

RSMeans Building Construction Cost Data 2015

Offers you unchallenged unit price reliability in an easy-to-use format. Whether used for verifying complete, finished estimates or for periodic checks, it supplies more cost facts better and faster than any comparable source. More than 24,700 unit prices have been updated for 2015. The City Cost Indexes and Location Factors cover more than 930 areas, for indexing to any project location in North America. Order and get *RSMeans Quarterly Update Service* FREE.

.06.95 | Available Sept. 2014 | Catalog no. 60015

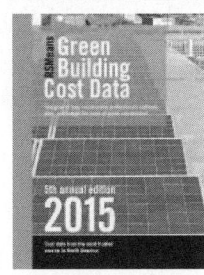

RSMeans Green Building Cost Data 2015

Estimate, plan, and budget the costs of green building for both new commercial construction and renovation work with this fifth edition of *RSMeans Green Building Cost Data*. More than 10,000 unit costs for a wide array of green building products plus assemblies costs. Easily identified cross references to LEED and Green Globes building rating systems criteria.

$170.95 | Available Nov. 2014 | Catalog no. 60555

RSMeans Mechanical Cost Data 2015

Total unit and systems price guidance for mechanical construction . . . materials, parts, fittings, and complete labor cost information. Includes prices for piping, heating, air conditioning, ventilation, and all related construction.

Plus new 2015 unit costs for:

- Thousands of installed HVAC/controls, sub-assemblies and assemblies
- "On-site" Location Factors for more than 930 cities and towns in the U.S. and Canada
- Crews, labor, and equipment

.03.95 | Available Oct. 2014 | Catalog no. 60025

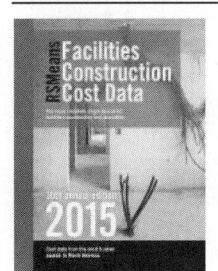

RSMeans Facilities Construction Cost Data 2015

For the maintenance and construction of commercial, industrial, municipal, and institutional properties. Costs are shown for new and remodeling construction and are broken down into materials, labor, equipment, and overhead and profit. Special emphasis is given to sections on mechanical, electrical, furnishings, site work, building maintenance, finish work, and demolition.

More than 49,800 unit costs, plus assemblies costs and a comprehensive Reference Section are included.

$524.95 | Available Nov. 2014 | Catalog no. 60205

RSMeans Square Foot Costs 2015
Accurate and Easy-to-Use

- **Updated price information** based on nationwide figures from suppliers, estimators, labor experts, and contractors.
- Green building models
- Realistic graphics, offering true-to-life illustrations of building projects
- Extensive information on using square foot cost data, including sample estimates and alternate pricing methods

18.95 | Available Oct. 2014 | Catalog no. 60055

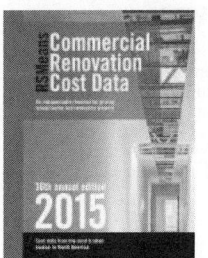

RSMeans Commercial Renovation Cost Data 2015
Commercial/Multi-family Residential

Use this valuable tool to estimate commercial and multi-family residential renovation and remodeling.

Includes: Updated costs for hundreds of unique methods, materials, and conditions that only come up in repair and remodeling, PLUS:

- Unit costs for more than 19,600 construction components
- Installed costs for more than 2,700 assemblies
- More than 930 "on-site" localization factors for the U.S. and Canada

$166.95 | Available Oct. 2014 | Catalog no. 60045

RSMeans Electrical Cost Data 2015

Pricing information for every part of electrical cost planning. More than 13,800 unit and systems costs with design tables; clear specifications and drawings; engineering guides; illustrated estimating procedures; complete labor-hour and materials costs for better scheduling and procurement; and the latest electrical products and construction methods.

- A variety of special electrical systems, including cathodic protection
- Costs for maintenance, demolition, HVAC/mechanical, specialties, equipment, and more

208.95 | Available Oct. 2014 | Catalog no. 60035

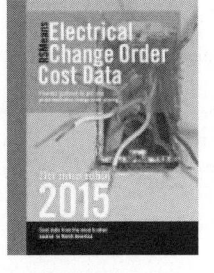

RSMeans Electrical Change Order Cost Data 2015

RSMeans Electrical Change Order Cost Data provides you with electrical unit prices exclusively for pricing change orders—based on the recent, direct experience of contractors and suppliers. Analyze and check your own change order estimates against the experience others have had doing the same work. It also covers productivity analysis, and change order cost justifications. With useful information for calculating the effects of change orders and dealing with their administration.

$199.95 | Available Dec. 2014 | Catalog no. 60235

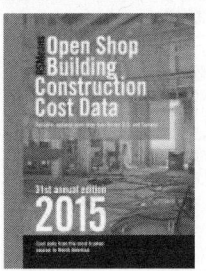

RSMeans Assemblies Cost Data 2015

RSMeans Assemblies Cost Data takes the guesswork out of preliminary or conceptual estimates. Now you don't have to try to calculate the assembled cost by working up individual component costs. We've done all the work for you.

Presents detailed illustrations, descriptions, specifications, and costs for every conceivable building assembly—over 350 types in all—arranged in the easy-to-use UNIFORMAT II system. Each illustrated "assembled" cost includes a complete grouping of materials and associated installation costs, including the installing contractor's overhead and profit.

334.95 | Available Sept. 2014 | Catalog no. 60065

RSMeans Open Shop Building Construction Cost Data 2015

The latest costs for accurate budgeting and estimating of new commercial and residential construction . . . renovation work . . . change orders . . . cost engineering.

RSMeans Open Shop "BCCD" will assist you to:

- Develop benchmark prices for change orders.
- Plug gaps in preliminary estimates and budgets.
- Estimate complex projects.
- Substantiate invoices on contracts.
- Price ADA-related renovations.

$175.95 | Available Dec. 2014 | Catalog no. 60155

Unit prices according to the latest MasterFormat!

Annual Cost Guides

For more information visit the RSMeans website at **www.rsmeans.com**

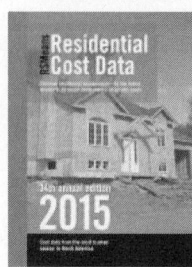

RSMeans Residential
Cost Data 2015

Contains square foot costs for 28 basic home models with the look of today, plus hundreds of custom additions and modifications you can quote right off the page. Includes more than 3,900 costs for 89 residential systems. Complete with blank estimating forms, sample estimates, and step-by-step instructions.

Contains line items for cultured stone and brick, PVC trim, lumber, and TPO roofing.

$148.95 | Available Oct. 2014 | Catalog no. 60175

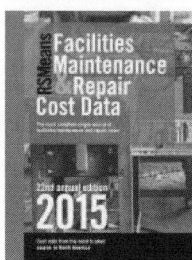

RSMeans Facilities Maintenance & Repair
Cost Data 2015

RSMeans Facilities Maintenance & Repair Cost Data gives you a complete system to manage and plan your facility repair and maintenance costs and budget efficiently. Guidelines for auditing a facility and developing an annual maintenance plan. Budgeting is included, along with reference tables on cost and management, and information on frequency and productivity of maintenance operations.

The only nationally recognized source of maintenance and repair costs. Developed in cooperation with the Civil Engineering Research Laboratory (CERL) of the Army Corps of Engineers.

$461.95 | Available Nov. 2014 | Catalog no. 60305

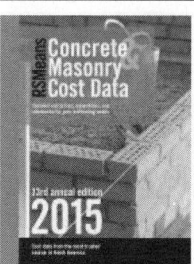

RSMeans Concrete & Masonry
Cost Data 2015

Provides you with cost facts for virtually all concrete/masonry estimating needs, from complicated formwork to various sizes and face finishes of brick and block—all in great detail. The comprehensive Unit Price Section contains more than 8,000 selected entries. Also contains an Assemblies [Cost] Section, and a detailed Reference Section that supplements the cost data.

$188.95 | Available Dec. 2014 | Catalog no. 60115

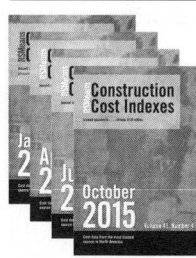

RSMeans Construction
Cost Indexes 2015

What materials and labor costs will change unexpectedly this year? By how much?

- Breakdowns for 318 major cities
- National averages for 30 key cities
- Expanded five major city indexes
- Historical construction cost indexes

$359.95 per year (subscription) | Catalog no. 50145
$94.95 individual quarters | Catalog no. 60145 A,B,C,D

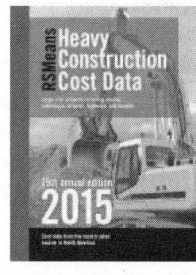

RSMeans Heavy Construction
Cost Data 2015

A comprehensive guide to heavy construction costs. Includes costs for highly specialized projects such as tunnels, dams, highways, airports, and waterways. Information on labor rates, equipment, and materials costs is included. Features unit price costs, systems costs, and numerous reference tables for costs and design.

$204.95 | Available Dec. 2014 | Catalog no. 60165

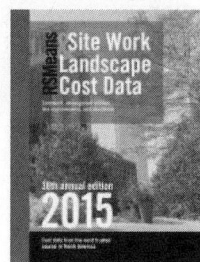

RSMeans Site Work & Landscape
Cost Data 2015

Includes unit and assemblies costs for earthwork, sewerage, piped utilities, site improvements, drainage, paving, trees and shrubs, street openings/repairs, underground tanks, and more. Contains more than 60 types of assemblies costs for accurate conceptual estimates.

Includes:

- Estimating for infrastructure improvements
- Environmentally-oriented construction
- ADA-mandated handicapped access
- Hazardous waste line items

$199.95 | Available Dec. 2014 | Catalog no. 60285

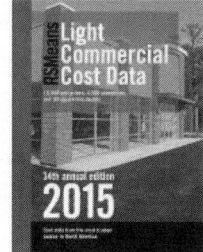

RSMeans Light Commercial
Cost Data 2015

Specifically addresses the light commercial market, which is a specialized niche in the construction industry. Aids you, the owner/designer/contractor, in preparing all types of estimates—from budgets to detailed bids. Includes new advances in methods and materials.

Assemblies Section allows you to evaluate alternatives in the early stages of design/planning.

More than 15,500 unit costs ensure that you have the prices you need, when you need them.

$153.95 | Available Nov. 2014 | Catalog no. 60185

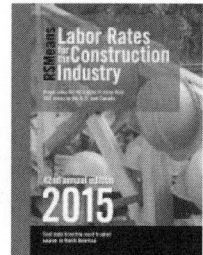

RSMeans Labor Rates for the
Construction Industry 2015

Complete information for estimating labor costs, making comparisons, and negotiating wage rates by trade for more than 300 U.S. and Canadian cities. With 46 construction trades in each city, and historical wage rates included for comparison. Each city chart lists the county and is alphabetically arranged with handy visual flip tabs for quick reference.

$450.95 | Available Dec. 2014 | Catalog no. 60125

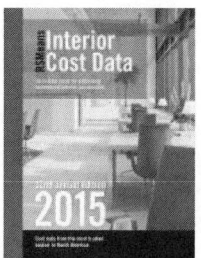

RSMeans Interior
Cost Data 2015

Provides you with prices and guidance needed to make accurate interior work estimates. Contains costs on materials, equipment, hardware, custom installations, furnishings, and labor for new and remodel commercial and industrial interior construction, including updated information on office furnishings, and reference information.

$208.95 | Available Nov. 2014 | Catalog no. 60095

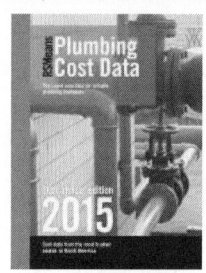

RSMeans Plumbing
Cost Data 2015

Comprehensive unit prices and assemblies for plumbing, irrigation systems, commercial and residential fire protection, point-of-use water heaters, and the latest approved materials. This publication and its companion, *RSMeans Mechanical Cost Data*, provide full-range cost estimating coverage for all the mechanical trades.

Contains updated costs for potable water, radiant heat systems, and high efficiency fixtures.

$206.95 | Available Oct. 2014 | Catalog no. 60215

SMeans data titles are also available in online format! Go to RSMeansOnline.com for more details.

For more information visit the RSMeans website at **www.rsmeans.com**

RSMeans Online and RSMeans CostWorks CD Comparison Matrix

Both the RSMeans CostWorks CD package and the RSMeans Online estimating tool provide the same comprehensive, up-to-date, and localized cost information to improve planning, estimating, and budgeting with RSMeans—the industry's most-trusted source of construction cost data.

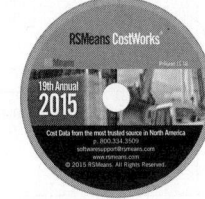

In addition to key differences noted in the comparison matrix below, both products let you do the following:

• Quickly locate costs in the searchable RSMeans database.

• Instantly adjust the data to your specific location for accurate costs anywhere in the U.S. and Canada.

• Create reliable, updated cost estimates in minutes.

• Manage, share, and export your estimates to MS Excel® with ease.

• Access a wide range of reference materials, including crew rates and national average labor rates.

Both the CD and Online options are available in a convenient one-year subscription!

Use the handy comparison matrix below to help you decide the best product for your cost estimating needs!

Why us?

For more than 70 years, RSMeans has provided the quality cost data construction professionals everywhere depend on to estimate with confidence, save time, improve decision-making, and increase profits.

Improve planning and decision-making

Back your estimates with complete, accurate, and up-to-date cost data for informed business decisions.
- Verify construction costs from third parties.
- Check validity of subcontractor proposals.
- Evaluate material and assembly alternatives.

Increase profits

Use RSMeans to estimate projects quickly and accurately, so you can gain an edge over your competition.
- Create accurate and competitive bids.
- Minimize the risk of cost overruns.
- Reduce variability.
- Gain control over costs.

The following matrix summarizes the key comparison points between RSMeans' two electronic cost data products.

	Question	CostWorks CD	RSMeans Online.com
1	How is the data presented?	• The data is organized by book title and sold at comparable book cost. • Several packaged bundles of five titles are available at a cost savings. • Available in a convenient one-year subscription	• The data is organized by book title and sold at comparable book cost. • Several packaged bundles of five titles are available at a cost savings. • RSMeans Online also offers a Complete Library title with all online data at a cost savings. • Available in a convenient one-year subscription
2	Where is the data stored?	With a typical install, the data is read off of your hard drive or off of the CD.	All data is stored on our secure server.
3	How are updates delivered?	Updates are available on a quarterly basis for download from the Costworks HomePort website.	Data is updated automatically on a quarterly basis, and users are notified when updates occur.
4	Where are my projects and estimates stored?	Upon installation, a "My CostWorks Projects" folder is created within the "My Documents" directory. All projects are defaulted to save here unless otherwise specified.	Estimates are stored on our secure server.
5	Does my IT team need to assist me in installing and running the program?	Yes, depending on the level of your permission. See #11 for system requirements.	While the product is entirely Web-based, you may need IT support with permissions for certain tasks. See #11 for system requirements.
6	Can users collaborate on projects/estimates?	Yes, you can copy project folders between computers.	You can create shared folders that can be viewed, copied, and edited by anyone within your account workgroup.

For more information visit the RSMeans website at **www.rsmeans.com**

Comparison matrix continued

7	Can I create and use my own data with the RSMeans database?	Custom lines cannot be added to the CD database. With the Estimator tool, however, you can create and add custom cost lines to any estimate. You can also save custom lines as an Estimator template for use in creating new estimates.	You can create • a custom line within an estimate • a custom cost book consisting of custom unit (not assembly) cost lines, no more than once per quarter **Note** that custom data is searchable using the Include My Custom Data feature.
8	Can I enter my own trade/labor rates and markups?	Although you cannot adjust the trade labor rates, you can add custom markups for each report template using the Estimator tool.	For each quarterly update, you can create one custom unit cost book. You can edit the labor rates by individual trade and the crew equipment costs, and also edit the associated material and equipment markups.
9	What are my reporting options?	The Estimator tool lets you create editable custom reports and report templates. Reports can also be exported to MS Excel®.	You can generate pre-formatted estimate summaries and cost lists in both PDF and Excel formats.
10	How do I find specific cost data?	You can search by CSI line number and by keyword within a division.	You can search by CSI line number and by keyword within a division. There are several additional search options, including keyword and advanced search using special characters.
11	What are the system requirements?	• Operating system – Windows XP or later • PC requirements - 256 MB RAM (recommended), 1024 x 768 High Color Monitor, 800MHz Computer (recommended), 24X CD-ROM, 250 MB Hard Disk Space	• Resolution – Recommended resolution is 1024 x 768 or higher • Compatibility - Internet Explorer version 8.x, 9.x and Mozilla Firefox version 10 and higher • iPad - iOS 5 and higher

2015 RSMeans Seminar Schedule
Note: call for exact dates and details.

Location	Dates	Location	Dates
Seattle, WA	January and August	Jacksonville FL	September
Dallas/Ft. Worth, TX	January	Dallas, TX	September
Austin, TX	February	Charleston, SC	October
Anchorage, AK	March and September	Houston, TX	October
Las Vegas, NV	March and October	Atlanta, GA	November
Washington, DC	April and September	Baltimore, MD	November
Phoenix, AZ	April	Orlando, FL	November
Kansas City, MO	April	San Diego, CA	December
Toronto	May	San Antonio, TX	December
Denver, CO	May	Raleigh, NC	December
San Francisco, CA	June		
Bethesda, MD	June		
Columbus, GA	June		
El Segundo, CA	August		

1-800-334-3509, Press 1

Professional Development

For more information visit the RSMeans website at **www.rsmeans.com**

eLearning Training Sessions

Learn how to use *RSMeans Online®* or *RSMeans CostWorks®* CD from the convenience of your home or office. Our eLearning training sessions let you join a training conference call and share the instructors' desktops, so you can view the presentation and step-by-step instructions on your own computer screen. The live webinars are held from 9 a.m. to 4 p.m. eastern standard time, with a one-hour break for lunch.

For these sessions, you must have a computer with high speed Internet access and a compatible Web browser. Learn more at www.rsmeansonline.com or call for a schedule: 1-800-334-3509 and press 1. Webinars are generally held on selected Wednesdays each month.

RSMeans Online®
$299 per person

RSMeans CostWorks® CD
$299 per person

RSMeans Online™ Training

Construction estimating is vital to the decision-making process in each state of every project. RSMeansOnline works the way you do. It's systematic, flexible and intuitive. In this one day class you will see how you can estimate any phase of any project faster and better.

Some of what you'll learn:

- Customizing RSMeansOnline
- Making the most of RSMeans "Circle Reference" numbers
- How to integrate your cost data
- Generate reports, exporting estimates to MS Excel, sharing, collaborating and more

Also available: RSMeans Online® training webinar

Facilities Construction Estimating

In this *two-day* course, professionals working in facilities management can get help with their daily challenges to establish budgets for all phases of a project.

Some of what you'll learn:

- Determining the full scope of a project
- Identifying the scope of risks and opportunities
- Creative solutions to estimating issues
- Organizing estimates for presentation and discussion
- Special techniques for repair/remodel and maintenance projects
- Negotiating project change orders

Who should attend: facility managers, engineers, contractors, facility tradespeople, planners, and project managers.

Construction Cost Estimating: Concepts and Practice

This *one-day* introductory course to improve estimating skills and effectiveness starts with the details of interpreting bid documents, ending with the summary of the estimate and bid submission.

Some of what you'll learn:

- Using the plans and specifications for creating estimates
- The takeoff process—deriving all tasks with correct quantities
- Developing pricing, using various sources; how subcontractor pricing fits in
- Summarizing the estimate to arrive at the final number
- Formulas for area and cubic measure, adding waste and adjusting productivity to specific projects
- Evaluating subcontractors' proposals and prices
- Adding Insurance and bonds
- Understanding how labor costs are calculated
- Submitting bids and proposals

Who should attend: project managers, architects, engineers, owner's representatives, contractors, and anyone who's responsible for budgeting or estimating construction projects.

Maintenance & Repair Estimating for Facilities

This *two-day* course teaches attendees how to plan, budget, and estimate the cost of ongoing and preventive maintenance and repair for existing buildings and grounds.

Some of what you'll learn:

- The most financially favorable maintenance, repair, and replacement scheduling and estimating
- Auditing and value engineering facilities
- Preventive planning and facilities upgrading
- Determining both in-house and contract-out service costs
- Annual, asset-protecting M&R plan

Who should attend: facility managers, maintenance supervisors, buildings and grounds superintendents, plant managers, planners, estimators, and others involved in facilities planning and budgeting.

Practical Project Management for Construction Professionals

In this *two-day* course, acquire the essential knowledge and develop the skills to effectively and efficiently execute the day-to-day responsibilities of the construction project manager.

Covers:

- General conditions of the construction contract
- Contract modifications: change orders and construction change directives
- Negotiations with subcontractors and vendors
- Effective writing: notification and communications
- Dispute resolution: claims and liens

Who should attend: architects, engineers, owner's representatives, project managers.

Mechanical & Electrical Estimating

This *two-day* course teaches attendees how to prepare more accurate and complete mechanical/electrical estimates, avoiding the pitfalls of omission and double-counting, while understanding the composition and rationale within the RSMeans mechanical/electrical database.

Some of what you'll learn:

- The unique way mechanical and electrical systems are interrelated
- M&E estimates—conceptual, planning, budgeting, and bidding stages
- Order of magnitude, square foot, assemblies, and unit price estimating
- Comparative cost analysis of equipment and design alternatives

Who should attend: architects, engineers, facilities managers, mechanical and electrical contractors, and others who need a highly reliable method for developing, understanding, and evaluating mechanical and electrical contracts.

Professional Development

For more information visit the RSMeans website at **www.rsmeans.com**

Unit Price Estimating

This interactive *two-day* seminar teaches attendees how to interpret project information and process it into final, detailed estimates with the greatest accuracy level.

The most important credential an estimator can take to the job is the ability to visualize construction and estimate accurately.

Some of what you'll learn:
- Interpreting the design in terms of cost
- The most detailed, time-tested methodology for accurate pricing
- Key cost drivers—material, labor, equipment, staging, and subcontracts
- Understanding direct and indirect costs for accurate job cost accounting and change order management

Who should attend: corporate and government estimators and purchasers, architects, engineers, and others who need to produce accurate project estimates.

RSMeans CostWorks® CD Training

This one-day course helps users become more familiar with the functionality of *RSMeans CostWork*s program. Each menu, icon, screen, and function found in the program is explained in depth. Time is devoted to hands-on estimating exercises.

Some of what you'll learn:
- Searching the database using all navigation methods
- Exporting RSMeans data to your preferred spreadsheet format
- Viewing crews, assembly components, and much more
- Automatically regionalizing the database

This training session requires you to bring a laptop computer to class.

When you register for this course you will receive an outline for your laptop requirements.

Also offering web training for RSMeans CostWorks CD!

Facilities Estimating Using RSMeans CostWorks® CD

Combines hands-on skill building with best estimating practices and real-life problems. Brings you up-to-date with key concepts, and provides tips, pointers, and guidelines to save time and avoid cost oversights and errors.

Some of what you'll learn:
- Estimating process concepts
- Customizing and adapting RSMeans cost data
- Establishing scope of work to account for all known variables
- Budget estimating: when, why, and how
- Site visits: what to look for—what you can't afford to overlook
- How to estimate repair and remodeling variables

This training session requires you to bring a laptop computer to class.

Who should attend: facility managers, architects, engineers, contractors, facility tradespeople, planners, project managers and anyone involved with JOC, SABRE, or IDIQ.

Conceptual Estimating Using RSMeans CostWorks® CD

This *two-day* class uses the leading industry data and a powerful software package to develop highly accurate conceptual estimates for your construction projects. All attendees must bring a laptop computer loaded with the current year *Square Foot Costs* and the *Assemblies Cost Data* CostWorks titles.

Some of what you'll learn:
- Introduction to conceptual estimating
- Types of conceptual estimates
- Helpful hints
- Order of magnitude estimating
- Square foot estimating
- Assemblies estimating

Who should attend: architects, engineers, contractors, construction estimators, owner's representatives, and anyone looking for an electronic method for performing square foot estimating.

Assessing Scope of Work for Facility Construction Estimating

This *two-day* practical training program addresses the vital importance of understanding the SCOPE of projects in order to produce accurate cost estimates for facility repair and remodeling.

Some of what you'll learn:
- Discussions of site visits, plans/specs, record drawings of facilities, and site-specific lists
- Review of CSI divisions, including means, methods, materials, and the challenges of scoping each topic
- Exercises in SCOPE identification and SCOPE writing for accurate estimating of projects
- Hands-on exercises that require SCOPE, take-off, and pricing

Who should attend: corporate and government estimators, planners, facility managers, and others who need to produce accurate project estimates.

Unit Price Estimating Using RSMeans CostWorks® CD

Step-by-step instruction and practice problems to identify and track key cost drivers—material, labor, equipment, staging, and subcontractors—for each specific task. Learn the most detailed, time-tested methodology for accurately "pricing" these variables, their impact on each other and on total cost.

Some of what you'll learn:
- Unit price cost estimating
- Order of magnitude, square foot, and assemblies estimating
- Quantity takeoff
- Direct and indirect construction costs
- Development of contractor's bill rates
- How to use *RSMeans Building Construction Cost Data*

This training session requires you to bring a laptop computer to class.

Who should attend: architects, engineers, corporate and government estimators, facility managers, and government procurement staff.

Registration Information

Register early . . . Save up to $100! Register 30 days before the start date of a seminar and save $100 off your total fee. *Note: This discount can be applied only once per order. It cannot be applied to team discount registrations or any other special offer.*

How to register Register by phone today! The RSMeans toll-free number for making reservations is **1-800-334-3509, Press 1.**

Two-day seminar registration fee - $935. One-day *RSMeans CostWorks*® training registration fee - $375. To register by mail, complete the registration form and return, with your full fee, to: RSMeans Seminars, 700 Longwater Drive, Norwell, MA 02061.

Government pricing All federal government employees save off the regular seminar price. Other promotional discounts cannot be combined with the government discount.

Team discount program for two to four seminar registrations. Call for pricing: 1-800-334-3509, Press 1

Multiple course discounts When signing up for two or more courses, call for pricing.

Refund policy Cancellations will be accepted up to ten business days prior to the seminar start. There are no refunds for cancellations received later than ten working days prior to the first day of the seminar. A $150 processing fee will be applied for all cancellations. Written notice of cancellation is required. Substitutions can be made at any time before the session starts. **No-shows are subject to the full seminar fee.**

AACE approved courses Many seminars described and offered here have been approved for 14 hours (1.4 recertification credits) of credit by the AACE International Certification Board toward meeting the continuing education requirements for recertification as a Certified Cost Engineer/Certified Cost Consultant.

AIA Continuing Education We are registered with the AIA Continuing Education System (AIA/CES) and are committed to developing quality learning activities in accordance with the CES criteria. Many seminars meet the AIA/CES criteria for Quality Level 2. AIA members may receive (14) learning units (LUs) for each two-day RSMeans course.

Daily course schedule The first day of each seminar session begins at 8:30 a.m. and ends at 4:30 p.m. The second day begins at 8:00 a.m. and ends at 4:00 p.m. Participants are urged to bring a hand-held calculator, since many actual problems will be worked out in each session.

Continental breakfast Your registration includes the cost of a continental breakfast, and a morning and afternoon refreshment break. These informal segments allow you to discuss topics of mutual interest with other seminar attendees. (You are free to make your own lunch and dinner arrangements.)

Hotel/transportation arrangements RSMeans arranges to hold a block of rooms at most host hotels. To take advantage of special group rates when making your reservation, be sure to mention that you are attending the RSMeans seminar. You are, of course, free to stay at the lodging place of your choice. (**Hotel reservations and transportation arrangements should be made directly by seminar attendees.**)

Important Class sizes are limited, so please register as soon as possible.

Note: Pricing subject to change.

Registration Form

ADDS-1000

Call 1-800-334-3509, Press 1 to register or FAX this form 1-800-632-6732. Visit our website: www.rsmeans.com

Please register the following people for the RSMeans construction seminars as shown here. We understand that we must make our own hotel reservations if overnight stays are necessary.

☐ Full payment of $_____ enclosed.

☐ Bill me.

Please print name of registrant(s).

To appear on certificate of completion)

_____

_____

_____

P.O. #: _____
    GOVERNMENT AGENCIES MUST SUPPLY PURCHASE ORDER NUMBER OR
    TRAINING FORM.

Firm name_____

Address_____

City/State/Zip_____

Telephone no._____ Fax no._____

E-mail address_____

Charge registration(s) to: ☐ MasterCard ☐ VISA ☐ American Express

Account no._____ Exp. date_____

Cardholder's signature_____

Seminar name_____

Seminar City _____

Please mail check to: RSMeans Seminars, 700 Longwater Drive, Norwell, MA 02061 USA

Professional Development

For more information visit the RSMeans website at **www.rsmeans.com**

Managing the Profitable Construction Business: The Contractor's Guide to Success and Survival Strategies, 2nd Ed.

by Thomas C. Scheifler, Ph.D.

Learn from a team of construction business veterans led by Thomas C. Schleifer, who is commonly referred to as a construction business "turnaround" expert due to the number of construction companies he has rescued from financial distress. His financial acumen, combined with his practical, hands-on experience, has made him a sought-after private consultant.

$60.00 | 288 pages, hardcover | Catalog no. 67370

Construction Project Safety

by John Schaufelberger, Ken-Yu Lin

This essential introduction to construction safety for construction management personnel takes a project-based approach to present potential hazards in construction and their mitigation or prevention.

Beginning with an introduction to Accident Prevention Programs and OSHA compliance requirements, the book integrates safety instruction into the building process by following a building project from site construction through interior finish. Reinforcing this applied approach are photographs, drawings, contract documentation, and an online 3D BIM model to help visualize the on-site scenarios.

$85.00 | 320 pages, hardcover | Catalog no. 67368

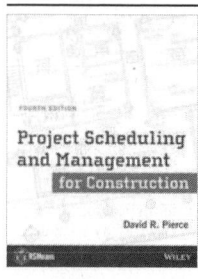

Project Control: Integrating Cost & Schedule in Construction

by Wayne J. Del Pico

Written by a seasoned professional in the field, *Project Control: Integrating Cost and Schedule in Construction* fills a void in the area of project control as applied in the construction industry today. It demonstrates how productivity models for an individual project are created, monitored, and controlled, and how corrective actions are implemented as deviations from the baseline occur.

$65.00 | 240 pages, softcover | Catalog no. 67366

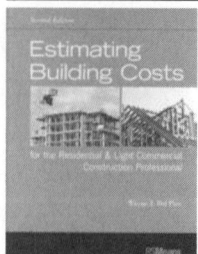

Project Scheduling and Management for Construction, 4th Edition

by David R. Pierce

First published in 1988 by RSMeans, the new edition of *Project Scheduling and Management for Construction* has been substantially revised for both professionals and students enrolled in construction management and civil engineering programs. While retaining its emphasis on developing practical, professional-level scheduling skills, the new edition is a relatable, real world case study.

$95.00 | 272 pages, softcover | Catalog no. 67367

Estimating Building Costs, 2nd Edition
For the Residential & Light Commercial Construction Professional

by Wayne J. Del Pico

Explains, in detail, how to put together a reliable estimate that can be used not only for budgeting, but also for developing a schedule, managing a project, dealing with contingencies, and, ultimately, making a profit.

Completely revised and updated to reflect the CSI MasterFormat 2010 system.

$75.00 | 528 pages, softcover | Catalog no. 67343A

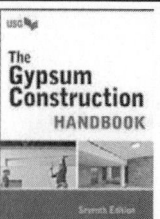

The Gypsum Construction Handbook, 7th Edition

by USG

The tried-and-true Gypsum Construction Handbook is a systematic guide to selecting and using gypsum drywall, veneer plaster, tile backers, ceilings, and conventional plaster building materials. A widely respected training text for aspiring architects and engineers, the book provides detailed product information and efficient installation methodology.

The 7th edition features updates in gypsum products, including ultralight panels, glass-mat panels, paper-faced plastic bead, and ultralightweight joint compound, and modern specialty acoustical and ceiling product guidelines. This comprehensive reference also incorporates the latest in sustainable products.

$40.00 | 576 pages, softcover | Catalog no. 67357B

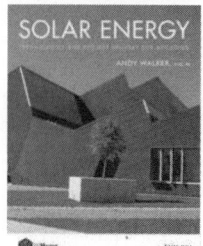

Solar Energy: Technologies & Project Delivery for Buildings

by Andy Walker

An authoritative reference on the design of solar energy systems in building projects, with applications, operating principles, and simple tools for the construction, engineering, and design professionals, the book simplifies the solar design and engineering process and provides sample documentation for the complete design of a solar energy system for buildings.

$85.00 | 320 pages, hardcover | Catalog no. 67365

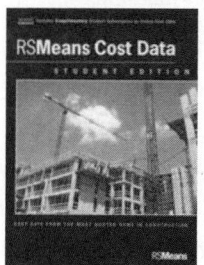

RSMeans Cost Data, Student Edition

Provides a thorough introduction to cost estimating in a print and online package. Clear explanations and example-driven. The ideal reference for students and new professionals. Features include:

- Commercial and residential construction cost data in print and online formats
- Complete how-to guidance on the essentials of cost estimating
- A supplemental website with plans, problem sets, and a full sample estimate

$110.00 | 512 pages, softcover | Catalog no. 67363

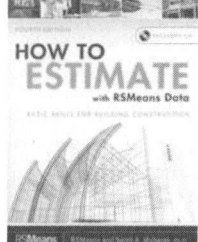

How to Estimate with Means Data & CostWorks, 4th Edition

by RSMeans and Saleh A. Mubarak, Ph.D.

This step-by-step guide takes you through all the major construction items with extensive coverage of site work, concrete and masonry, wood and metal framing, doors and windows, and other divisions. The only construction cost estimating handbook that uses the most popular source of construction data, RSMeans, this indispensible guide features access to the instructional version of *CostWorks* in electronic form, enabling you to practice techniques to solve real-world estimating problems.

$75.00 | 320 pages, softcover | Includes CostWorks CD | Catalog no. 67324C

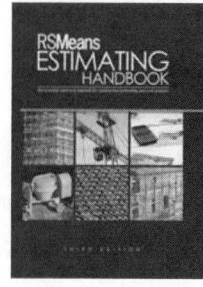

RSMeans Estimating Handbook, 3rd Edition

Widely used in the industry for tasks ranging from routine estimates to special cost analysis projects. This handbook will help construction professionals:

- Evaluate architectural plans and specifications
- Prepare accurate quantity takeoffs
- Compare design alternatives and costs
- Perform value engineering
- Double-check estimates and quotes
- Estimate change orders

$110.00 | Over 976 pages, hardcover | Catalog No. 67276B

Reference Books

RSMeans data titles are also available in Online format. Go to RSMeansOnline.com for more details.

For more information visit the RSMeans website at **www.rsmeans.com**

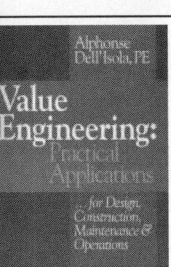

Value Engineering: Practical Applications

For Design, Construction, Maintenance & Operations

by Alphonse Dell'Isola, PE

A tool for immediate application—for engineers, architects, facility managers, owners, and contractors. Includes making the case for VE—the management briefing; integrating VE into planning, budgeting, and design; conducting life cycle costing; using VE methodology in design review and consultant selection; and case studies.

$90.00 | Over 427 pages, illustrated, softcover | Catalog no. 67319A

Risk Management for Design and Construction

Introduces risk as a central pillar of project management and shows how a project manager can prepare to deal with uncertainty. Includes:

* Integrated cost and schedule risk analysis
* An introduction to a ready-to-use system of analyzing a project's risks and tools to proactively manage risks
* A methodology that was developed and used by the Washington State Department of Transportation
* Case studies and examples on the proper application of principles
* Combining value analysis with risk analysis

$125.00 | Over 288 pages, softcover | Catalog no. 67359

How Your House Works, 2nd Edition

by Charlie Wing

Knowledge of your home's systems helps you control repair and construction costs and makes sure the correct elements are being installed or replaced. This book uncovers the mysteries behind just about every major appliance and building element in your house. See-through, cross-section drawings in full color show you exactly how these things should be put together and how they function, including what to check if they don't work. It just might save you having to call in a professional.

$22.95 | 208 pages, softcover | Catalog no. 67351A

Construction Business Management

by Nick Ganaway

Only 43% of construction firms stay in business after four years. Make sure your company thrives with valuable guidance from a pro with 25 years of success as a commercial contractor. Find out what it takes to build all aspects of a business that is profitable, enjoyable, and enduring. With a bonus chapter on retail construction.

$60.00 | 201 pages, softcover | Catalog no. 67352

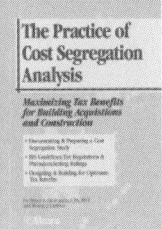

The Practice of Cost Segregation Analysis

by Bruce A. Desrosiers and Wayne J. Del Pico

This expert guide walks you through the practice of cost segregation analysis, which enables property owners to defer taxes and benefit from "accelerated cost recovery" through depreciation deductions on assets that are properly identified and classified.

With a glossary of terms, sample cost segregation estimates for various building types, key information resources, and updates via a dedicated website, this book is a critical resource for anyone involved in cost segregation analysis.

$105.00 | Over 240 pages | Catalog no. 67345

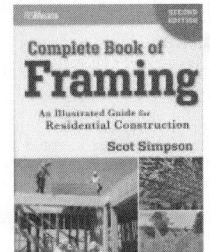

Complete Book of Framing, 2nd Edition

by Scot Simpson

This updated, easy-to-learn guide to rough carpentry and framing is written by an expert with more than thirty years of framing experience. Starting with the basics, this book begins with types of lumber, nails, and what tools are needed, followed by detailed, fully illustrated steps for framing each building element. Framer-Friendly Tips throughout the book show how to get a task done right—and more easily.

$29.95 | 368 pages, softcover | Catalog no. 67353A

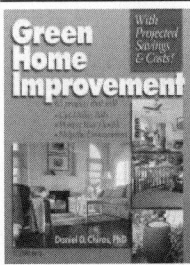

Green Home Improvement

by Daniel D. Chiras, PhD

With energy costs rising and environmental awareness increasing, people are looking to make their homes greener. This book, with 65 projects and actual costs and projected savings, helps homeowners prioritize their green improvements.

Projects range from simple water savers that cost only a few dollars, to bigger-ticket items such as HVAC systems. With color photos and cost estimates, each project compares options and describes the work involved, the benefits, and the savings.

$34.95 | 320 pages, illustrated, softcover | Catalog no. 67355

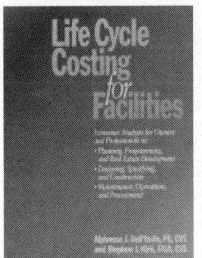

Life Cycle Costing for Facilities

by Alphonse Dell'Isola and Dr. Stephen Kirk

Guidance for achieving higher quality design and construction projects at lower costs! Cost-cutting efforts often sacrifice quality to yield the cheapest product. Life cycle costing enables building designers and owners to achieve both. The authors of this book show how LCC can work for a variety of projects—from roads to HVAC upgrades to different types of buildings.

$110.00 | 396 pages, hardcover | Catalog no. 67341

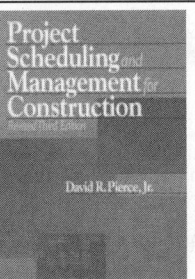

Project Scheduling and Management for Construction, 3rd Edition

by David R. Pierce, Jr.

A comprehensive, yet easy-to-follow guide to construction project scheduling and control—from vital project management principles through the latest scheduling, tracking, and controlling techniques. The author is a leading authority on scheduling, with years of field and teaching experience at leading academic institutions. Spend a few hours with this book and come away with a solid understanding of this essential management topic.

$64.95 | Over 286 pages, illustrated, hardcover | Catalog no. 67247B

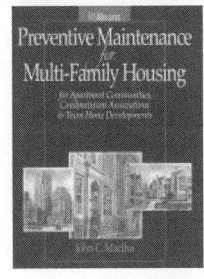

Preventive Maintenance for Multi-Family Housing

by John C. Maciha

Prepared by one of the nation's leading experts on multi-family housing.

This complete PM system for apartment and condominium communities features expert guidance, checklists for buildings and grounds maintenance tasks and their frequencies, and a dedicated website featuring customizable electronic forms. A must-have for anyone involved with multi-family housing maintenance and upkeep.

$95.00 | 290 pages | Catalog no. 67346

Reference Books

For more information visit the RSMeans website at **www.rsmeans.com**

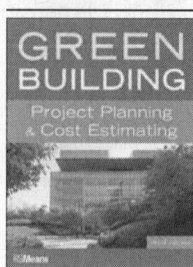

Green Building: Project Planning & Cost Estimating, 3rd Edition

Since the widely read first edition of this book, green building has gone from a growing trend to a major force in design and construction.

This new edition has been updated with the latest in green building technologies, design concepts, standards, and costs. Full-color with all new case studies—plus a new chapter on commercial real estate.

$110.00 | 480 pages, softcover | Catalog no. 67338B

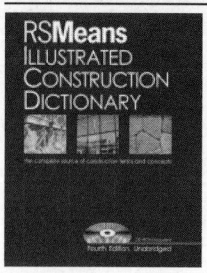

RSMeans Illustrated Construction Dictionary Unabridged 4th Edition, with CD-ROM

Long regarded as the industry's finest, *Means Illustrated Construction Dictionary* is now even better. With nearly 20,000 terms and more than 1,400 illustrations and photos, it is the clear choice for the most comprehensive and current information. The companion CD-ROM that comes with this new edition adds extra features, such as larger graphics and expanded definitions.

$110.00 | Over 880 pages, illust., hardcover | Catalog no. 67292B

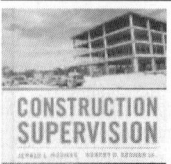

Construction Supervision

Inspires excellence with proven tactics and techniques applied by thousands of construction supervisors. Leadership guidelines carve out a practical blueprint for motivating work performance and increasing productivity through effective communication. Features:

- A unique focus on field supervision and crew management
- Coverage of supervision from the foreman to the superintendent level
- An overview of technical skills whose mastery will build confidence and success for the supervisor
- A detailed view of "soft" management and communication skills

$95.00 | 464 pages, softcover | Catalog no. 67358

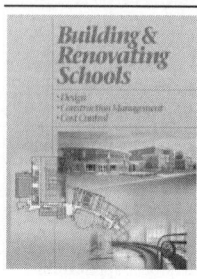

Building & Renovating Schools

This all-inclusive guide covers every step of the school construction process—from initial planning, needs assessment, and design, right through moving into the new facility. A must-have resource for anyone concerned with new school construction or renovation. With square foot cost models for elementary, middle, and high school facilities, and real-life case studies of recently completed school projects.

The contributors to this book—architects, construction project managers, contractors, and estimators who specialize in school construction—provide start-to-finish expert guidance on the process.

$110.00 | Over 412 pages, hardcover | Catalog no. 67342

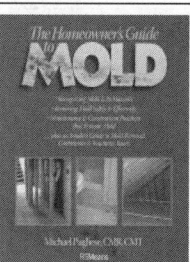

The Homeowner's Guide to Mold

By Michael Pugliese

Mold, whether caused by leaks, humidity or flooding, is a real health and financial issue—for homeowners and contractors. This full-color book explains:

- Construction and maintenance practices to prevent mold
- How to inspect for and remove mold
- Mold remediation procedures and costs
- What to do after a flood
- How to deal with insurance companies

$21.95 | 144 pages, softcover | Catalog no. 67344

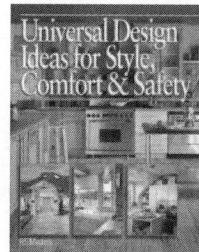

Universal Design Ideas for Style, Comfort & Safety

by RSMeans and Lexicon Consulting, Inc.

Incorporating universal design when building or remodeling helps people of any age and physical ability more fully and safely enjoy their living spaces. This book shows how universal design can be artfully blended into the most attractive homes. It discusses specialized products like adjustable countertops and chair lifts, as well as simple ways to enhance a home's safety and comfort. With color photos and expert guidance, every area of the home is covered. Includes budget estimates that give an idea how much projects will cost.

$21.95 | 160 pages, illustrated, softcover | Catalog no. 67354

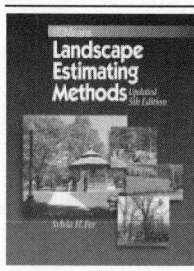

Landscape Estimating Methods, 5th Edition

Answers questions about preparing competitive landscape construction estimates, with up-to-date cost estimates and the new MasterFormat classification system. Expanded and revised to address the latest materials and methods, including new coverage on approaches to green building. Includes:

- Step-by-step explanation of the estimating process
- Sample forms and worksheets that save time and prevent errors

$75.00 | 336 pages, softcover | Catalog no. 67295C

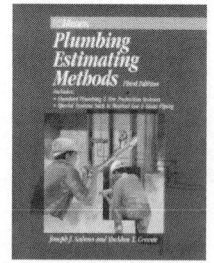

Plumbing Estimating Methods, 3rd Edition

by Joseph J. Galeno and Sheldon T. Greene

Updated and revised! This practical guide walks you through a plumbing estimate, from basic materials and installation methods through change order analysis. *Plumbing Estimating Methods* covers residential, commercial, industrial, and medical systems, and features sample takeoff and estimate forms and detailed illustrations of systems and components.

$65.00 | 381+ pages, softcover | Catalog no. 67283B

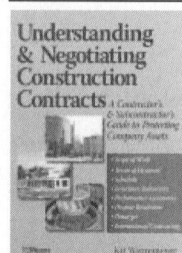

Understanding & Negotiating Construction Contracts

by Kit Werremeyer

Take advantage of the author's 30 years' experience in small-to-large (including international) construction projects. Learn how to identify, understand, and evaluate high risk terms and conditions typically found in all construction contracts—then, negotiate to lower or eliminate the risk, improve terms of payment, and reduce exposure to claims and disputes. The author avoids "legalese" and gives real-life examples from actual projects.

$80.00 | 320 pages, softcover | Catalog no. 67350

Reference Books